COUNTY FINDER MAP OF MISSOURI

County	No.	County	No.	County	No.
Adair	17	Greene	83	Ozark	105
Andrew	11	Grundy	15	Pemiscot	114
Atchison	1	Harrison	4	Perry	75
Audrain	42	Henry	60	Pettis	51
Barry	102	Hickory	67	Phelps	70
Barton	76	Holt	10	Pike	36
Bates	59	Howard	40	Platte	28
Benton	61	Howell	106	Polk	78
Bollinger	93	Iron	90	Pulaski	69
Boone	41	Jackson	37	Putnam	6
Buchanan	20	Jasper	81	Ralls	35
Butler	110	Jefferson	58	Randolph	33
Caldwell	22	Johnson	50	Ray	30
Callaway	43	Knox	18	Reynolds	89
Camden	68	Laclede	80	Ripley	109
Cape Girardeau	94	Lafayette	38	St. Charles	47
Carroll	31	Lawrence	98	St. Clair	66
Carter	108	Lewis	19	St. Francois	73
Cass	49	Lincoln	46	St. Louis	48
Cedar	77	Linn	24	Ste. Genevieve	74
Chariton	32	Livingston	23	Saline	39
Christian	99	Macon	25	Schuyler	7
Clark	9	Madison	92	Scotland	8
Clay	29	Maries	64	Scott	95
Clinton	21	Marion	27	Shannon	88
Cole	54	McDonald	101	Shelby	26
Cooper	52	Mercer	5	Stoddard	111
Crawford	71	Miller	63	Stone	103
Dade	82	Mississippi	96	Sullivan	16
Dallas	79	Moniteau	53	Taney	104
Daviess	14	Monroe	34	Texas	86
DeKalb	13	Montgomery	44	Vernon	65
Dent	87	Morgan	62	Warren	45
Douglas	100	New Madrid	112	Washington	72
Dunklin	113	Newton	97	Wayne	91
Franklin	57	Nodaway	2	Webster	84
Gasconade	56	Oregon	107	Worth	3
Gentry	12	Osage	55	Wright	85

STEYERMARK'S
Flora of Missouri

STEYERMARK'S
Flora of Missouri
Volume 2

Revised edition

George Yatskievych

The Missouri Botanical Garden Press

St. Louis, Missouri

in cooperation with

The Missouri Department of Conservation

Jefferson City, Missouri

PRINTING HISTORY

Steyermark's Flora of Missouri, revised edition

Volume 1, published by the Missouri Department of Conservation, Jefferson City, Missouri, in cooperation with the Missouri Botanical Garden Press, St. Louis, Missouri.

Published February 1999

Volume 2, published by the Missouri Botanical Garden Press in cooperation with the Missouri Department of Conservation.

Published June 2006

Copyright © 2006 jointly by the Missouri Botanical Garden and the Conservation Commission of the State of Missouri

ISBN: 1-930723-49-0

Technical editing by Julianna M. Schroeder
Page composition by Kay Yatskievych

Missouri Botanical Garden Press
P.O. Box 299
St. Louis, MO 63166-0299

Missouri Department of Conservation
P.O. Box 180
Jefferson City, MO 65102-0108

CONTENTS

Preface vii

Taxonomic Summary of Volume 2 x

Acknowledgments xii

Contributors of Treatments xiv

Illustration Acknowledgments xv

Dicots (part 1): Acanthaceae
   through Fabaceae (first part) 1

Glossary 1085

Literature Cited 1103

Index 1139

PREFACE

The parties that met in 1987 to sign a cooperative agreement for a joint Flora of Missouri Project doubtless never imagined that the project would take on a life of its own and still be going strong 18 years later. Originally, the Missouri Department of Conservation and Missouri Botanical Garden conceived a much more limited attempt to correct and update the late Julian Steyermark's (1963) wonderful floristic manual. What evolved, however, has become an ongoing effort to collect data on the spontaneous flora of the state and to provide this information to those with an interest or need for it. Of course, the revision of *Steyermark's Flora of Missouri* is still central to the goals of the project, but even this turned out to be a lengthier and far more complex process than was predicted at the outset.

After the project formally began in August 1987, it required three years to review the literature pertinent to Missouri plants and to create a new *Catalogue of the Flora of Missouri* with updated nomenclature and inclusion of the nearly 300 species discovered growing in the state since 1963 (Yatskievych and Turner, 1990). By then it had been discovered that the original illustrations from Steyermark's *Flora of Missouri* had degraded over time and were no longer usable in the new edition. It took a dozen freelance artists about three years to completely reillustrate the Flora in nearly 600 full-page plates.

Thus, about six years had passed before writing of the text was ever begun. Once it became apparent that a really encyclopedic treatment of the flora would not fit into a single volume, the scope of the revision grew into two volumes and later into three. The writing, editing, and production of the first volume (Yatskievych, 1999) consumed another six years, and the present second volume required a further six years of effort. Everyone involved with the project is hopeful that the third and final volume can be produced in a shorter time than the first two, but the inherent complexity of these large undertakings makes it impossible to predict an endpoint with confidence.

The staffing of the project also has changed as time has gone by. Over the years, a number of talented volunteers have contributed to all aspects of the work. Beginning in 1990, funding was received from the Missouri Department of Conservation for the hiring of a project assistant, a situation that thankfully has persisted at some level through the completion of Volume 2. The series of assistants, along with the volunteers, has been instrumental in entering the more than 120,000 specimen records presently in the Flora of Missouri database and in managing the flow of herbarium specimens through the project's offices. In 1999, further funding was provided that permitted the creation of a postdoctoral position to help with the writing of treatments, and these funds were renewed through 2002. Finally, in 2003, funding through a grant to the Missouri Department of Conservation (to develop a comprehensive state wildlife conservation plan) from the U.S. Fish and Wildlife Foundation allowed the creation of a one-year position for a database manager to upgrade and thoroughly redesign the Flora of Missouri database.

From its beginning, the Flora of Missouri Project was an active collaboration between a state agency and a private not-for-profit organization. To date, no other state has quite succeeded in emulating this arrangement for a state-sponsored program of floristic and taxonomic research, although individuals in a number of states have inquired about how Missouri was able to accomplish this feat. A major change in the Flora of Missouri Project's administration occurred at the end of August 2005, following a reorganization within the Missouri Department of Conservation that led to a number of programmatic and staffing changes within the agency. The result has been a matriculation of the Flora of Missouri Project director position from the ranks of state employment to the payroll of the Missouri Botanical Garden. The Department is taking a less active role in the management of the project in the future, but is continuing its financial support for the present in the form of grants to the Garden.

In the past few years, various researchers have published new information on Missouri's vegetation and flora that has changed our thinking on various topics. Given the length of time involved in completing the revised edition, this is not surprising. In order to avoid major changes in terminology in mid-project, some of these changes have not been adopted in the present work, but readers should be aware that they exist.

The most profound changes that have occurred concern the way that landscapes are viewed ecologically in the state. One marvelous recent resource has been the development of an ecoregional classification system by the Missouri Resource Assessment Program that treats the vegetational associations of the Missouri landscape in a much more precise fashion than was possible before the advent of satellite imagery and computerized Geographic Information Systems. The introductory sections of the *Atlas of Missouri Ecoregions* (Nigh and Schroeder, 2002) present many of the topics included in the introductory chapters in Volume 1 of the present work in a more detailed and updated fashion. However, the actual classification of land-type associations is too detailed to be adapted for use in describing habitats of species in a state-level floristic manual. At its coarsest level of focus, that of Ecological Sections, the classification of Nigh and Schroeder is roughly congruent with the system of Natural Divisions (Thom and Wilson, 1980, 1983) that is discussed and used in *Steyermark's Flora of Missouri*.

A second recent landmark publication on Missouri vegetation is the revised *Terrestrial Natural Communities of Missouri* classification system produced by the Missouri Natural Areas Committee (Nelson, 2005). This work also includes a wonderful series of introductory essays that will be of use to everyone interested in the Missouri flora. The major vegetational categories in the original edition of that work (Nelson, 1985) were adapted for use in the habitat summaries in *Steyermark's Flora of Missouri* and are summarized in Volume 1. The revised classification contains profound changes in its details that have resulted from the numerous ecological studies completed around the state during the intervening 15 years. However, users of *Steyermark's Flora* will note relatively few differences in the major vegetational categories. The biggest changes at the higher levels pertain to the understanding of savannas. Previously, the term savanna was used to encompass all woodlands with relatively incomplete tree canopies and representing a transition zone between prairies and forests. In the revised classification, the term savanna has become restricted to the most prairie-like extreme with widely scattered trees in a matrix of prairie flora. A new category designated as woodlands encompasses most of the plant communities formerly treated as savannas and also includes those communities previously classified as dry upland forests. To avoid confusion for readers in interpreting which portions of *Steyermark's Flora* follow the old system as opposed to the revised classification of terrestrial natural communities, the decision was made to retain the initial classification consistently through the three volumes. Readers should be aware that most of the species listed as growing in upland prairies also occur in savannas (in the new restricted sense), that the term savanna in the Flora refers mostly to habitats now classified as woodlands, and that plants ascribed to dry upland forests occur in woodlands in the new classification.

As with our understanding of ecological processes and classifications, there also have been numerous taxonomic studies in recent years that have generated new hypotheses on the circumscriptions of plant families. In Volume 1 of *Steyermark's Flora of Missouri*, the system of plant families followed was essentially that of the late Arthur Cronquist (1981, 1991), and this remains mostly true in the present volume. An exponential increase in the volume of molecular phylogenetic studies has resulted in a number of realignments of genera with different families than those in which they had been included previously. Of greater concern has been the new system of ordinal and familial classification proposed by the Angiosperm Phylogeny Group (1998, 2003), which itself is an evolving work in progress. In Volume 2, deviations from the Cronquistian system to conform to newer, primarily molecular-based theories have been made relatively conservatively. In some cases, these changes have resulted from overwhelming evidence that certain families as traditionally circumscribed are unnatural; in other cases the changes were accepted as a matter of expediency in constructing a more easily used floristic manual.

The alterations in familial classification from the system of Cronquist that have affected the contents of the present volume include: 1) a broad concept of the legumes as a single family Fabaceae with three subfamilies, instead of the recognition of three related families (Caesalpiniaceae, Fabaceae, and Mimosaceae); 2) a broad concept of the Ericaceae to include the achlorophyllous herbs previously segregated into the Monotropaceae; 3) the breakup of the traditional Boraginaceae into three families in Missouri: Boraginaceae, Ehretiaceae, and Heliotropiaceae; 4) a broad concept of the Convolvulaceae to include the parasitic Cuscutaceae; and 5) the submersion of the families Buddlejaceae and Callitrichaceae into the morphologically diverse Scrophulariaceae. Discussions justifying the first four of these changes are included in the treatments of the appropriate families. The fifth exception will be discussed in the treatment of the Scrophulariaceae in Volume 3. Table 1 provides a taxonomic summary of the families, genera, species, and infraspecific taxa included in Volume 2.

Table 1. Summary of the Missouri flora treated in Volume 2. The order of 45 dicot families in the following table is the same as in the text. Native species are those considered to be native in all or part of their ranges in Missouri (they may be native in one part of the state and introduced in another). Those species discussed in the text for which the distributional status (native vs. non-native) in the state is uncertain are treated as native for purposes of this summary. Introduced species include all of the non-native species treated in the text, both those that are naturalized widely and those that are known from one or few collections or from populations that may not have persisted to the present. Infraspecific taxa are those varieties and subspecies (excluding forms) in addition to the one that is understood to exist for each species accepted as part of the flora.

Family	Genera	Species			Infraspecific Taxa		Hybrids	Total Taxa
		Native	Introduced	Total	Native	Introduced		
Acanthaceae	3	6	0	6	0	0	0	6
Aceraceae	1	4	1	5	5	0	0	10
Aizoaceae	1	0	1	1	0	0	0	1
Amaranthaceae	5	8	12	20	0	1	2	23
Anacardiaceae	3	6	1	7	3	0	0	10
Annonaceae	1	1	0	1	0	0	0	1
Apiaceae	41	36	21	57	3	0	0	60
Apocynaceae	4	6	2	8	1	0	1	10
Aquifoliaceae	1	3	1	4	0	0	0	4
Araliaceae	3	4	1	5	0	0	0	5
Aristolochiaceae	2	3	0	3	0	0	0	3
Asclepiadaceae	4	20	2	22	0	0	0	22
Asteraceae	104	235	94	329	34	1	40	404
tribe Anthemideae	8	5	17	22	1	0	0	23
tribe Astereae	18	74	4	78	20	0	11	109
tribe Cardueae	6	7	17	24	0	0	1	25
tribe Cichorieae	17	20	23	43	0	1	0	44
tribe Eupatorieae	8	26	1	27	5	0	6	38
tribe Gnaphalieae	5	5	5	10	1	0	0	11
tribe Heliantheae	33	78	24	102	6	0	9	117
tribe Inuleae	1	0	1	1	0	0	0	1
tribe Plucheeae	1	2	0	2	0	0	0	2
tribe Senecioneae	5	12	2	14	0	0	4	18
tribe Vernonieae	2	6	0	6	1	0	9	16
Balsaminaceae	1	2	1	3	0	0	0	3
Berberidaceae	3	3	2	5	0	0	0	5
Betulaceae	5	5	1	6	1	0	0	7
Bignoniaceae	3	3	2	5	0	0	2	7
Boraginaceae	13	10	14	24	2	0	0	26
Brassicaceae	41	34	54	88	2	0	2	92

Family	Genera	Species			Infraspecific Taxa		Hybrids	Total Taxa
		Native	Introduced	Total	Native	Introduced		
Buxaceae	1	0	1	1	0	0	0	1
Cabombaceae	2	2	0	2	0	0	0	2
Cactaceae	1	3	0	3	0	0	0	3
Calycanthaceae	1	0	1	1	0	0	0	1
Campanulaceae	3	13	1	14	1	0	3	18
Cannabaceae	2	1	3	4	1	0	0	5
Caprifoliaceae	5	18	6	24	0	0	2	26
Caryophyllaceae	21	21	34	55	0	1	2	58
Celastraceae	2	4	3	7	0	0	0	7
Ceratophyllaceae	1	2	0	2	0	0	0	2
Chenopodiaceae	11	12	30	42	1	2	1	46
Cistaceae	2	5	0	5	0	0	0	5
Cleomaceae	2	2	1	3	1	0	0	4
Clusiaceae	2	16	0	16	1	0	1	18
Convolvulaceae	8	19	10	29	0	0	2	31
Cornaceae	2	8	0	8	1	0	4	13
Crassulaceae	1	3	5	8	0	0	0	8
Cucurbitaceae	8	5	5	10	0	1	0	11
Dipsacaceae	1	0	2	2	0	0	0	2
Ebenaceae	1	1	0	1	0	0	0	1
Ehretiaceae	1	0	1	1	0	0	0	1
Elaeagnaceae	1	0	2	2	0	0	0	2
Elatinaceae	2	2	0	2	1	0	0	3
Ericaceae	5	8	1	9	0	0	0	9
Euphorbiaceae	7	32	8	40	1	0	0	41
Fabaceae (excluding Faboideae)	11	11	5	16	0	0	0	16
subfamily Caesalpinioideae	6	8	2	10	0	0	0	10
subfamily Mimosoideae	5	3	3	6	0	0	0	6
Total Vol. 2	343	577	329	906	59	6	62	1033

ACKNOWLEDGMENTS

During the course of my years with the Flora of Missouri Project, so many kind individuals and groups have touched my life and made the work what it is today. Without the legion of botanists, conservationists, and plant enthusiasts who have provided most of the new information that has gone into the books and databases, this project would not have been possible.

First and foremost, gratitude must be showered upon the Missouri Botanical Garden, where I have been headquartered since 1987. Peter Raven's continuing vision of a centralized botanical survey for the state has been the pillar around which the Flora of Missouri Project's activities have been built. Others at the Missouri Botanical Garden directly involved in the administration of the project have included Bob Magill, Mike Olson, and Jim Solomon. I am also indebted to several of the curators in the Garden's Research Division who have lent support and expertise to the research, including Ihsan Al-Shehbaz, Fred Barrie, Tom Croat, Gerrit Davidse, Roy Gereau, Shirley Graham, Peter Hoch, Peter Jorgensen, Richard Keating, Gordon McPherson, Jim Miller, John Pruski, Doug Stevens, Peter Stevens, Charlotte Taylor, Nick Turland, Carmen Ulloa, and Jim Zarucchi. Trisha Consiglio developed the custom scripts that allow the generation of county dot maps for Volume 2 electronically directly from the Flora of Missouri database. A number of other present and former students and staff at the Garden also helped with the research in important ways, including James Beck, the late Rick Clinebell, Joe Ditto, Eric Feltz, Ron Liesner, Sandy Lopez, Lúcia Lohmann, Kim McCue, Patrick Sweeney, and Andrea Voyeur. As always, the Missouri Botanical Garden Library's staff and volunteers have been a joy to work with, especially Rosemary Armbruster, Andrew Colligan, Julie Crawford, Doug Holland, Victoria McMichael, Linda Oestry, Mary Stiffler, and Zoltan Tomory.

The Missouri Conservation Commission and its Department of Conservation are to be commended for continuing to support floristic research in the state. I am grateful for the guidance from and patience of each of my successive supervisors at the Department: Jim Wilson, Rick Thom, Don Kurz, David Urich, Eric Kurzejeski, and Vicki Heidy. Publication of the volumes would not be possible without the unstinting support of the Department's Outreach and Education Division, where past and present staff, including Kathy Love, Bernadette Dryden, Libby Block, Tracy Ritter, Dan Witter, and Eric Kurzejeski, have promoted the Flora of Missouri Project over the years. A number of present and former biologists with the Missouri Department of Conservation have been particularly helpful with plant records and other data, including Mike Arduser, Cindy Becker, Jeff Briggler, David Bruns, Dan Drees, John George, Bob Gillespie, Jenny Grabner, Greg Gremaud, Brad Jacobs, Randy Jensen, Emily Kathol, Karen Kramer, Sherry Leis, Norman Murray, Tom Nagel, Tim Nigh, Rhonda Rimer, Larry Rizzo, C. D. Scott, Mike Skinner, and especially Tim Smith.

Similarly, a number of past and present naturalists and biologists at the Missouri Department of Natural Resources have provided new records, including Michael Currier, Wanda Doolen, Ken McCarty, Ron Mullikin, Bruce Schuette, and Tim Vogt. Paul McKenzie of the U.S. Fish and Wildlife Service's Columbia Field Office continues to contribute important information and specimens to the Project and is also a wonderful source of information on plant rarity. David Moore and Paul Nelson of the Mark Twain National Forest also have provided data and support. Doug Ladd and his associates at the Missouri Chapter of The Nature Conservancy have been very helpful both in providing new records and for insightful discussions on the vegetation and flora.

My project assistants, interns, and volunteers have worked thousands of hours at jobs ranging from routine to very unusual and challenging, and all have performed above and beyond the call of duty. Since the publication of Volume 1, John Archer, Alan Brant, Ann Larson, Nancy Parker, Timothy Rye, Bill Summers, Pat Walker, Michelle Williams, and Kathleen Wood have assisted with all aspects of the work. Special thanks go to Rex Hill, who has served as the

Flora of Missouri Database Manager since 2003. During the period when funds for postdoctoral support were available, Alan Whittemore and David Bogler assisted ably with the research and writing of selected treatments.

Information on the artists and funding sources for the plates of species illustrations are listed separately, as are the names and addresses of those specialists who contributed treatments to the volumes. A number of other plant taxonomists have helped to provide data or to review drafts of various plant groups for Volume 2. It is a pleasure to acknowledge their assistance. They include: Luc Brouillet (University of Montreal), Craig Freeman (University of Kansas), Steven Hill (Illinois Natural History Survey), John Knox (Washington and Lee University), Eric Lamont (Riverhead, New York), Geoff Levin (Illinois Natural History Survey), Guy Nesom (Botanical Research Institute of Texas), Bruce Parfitt (University of Michigan–Flint), Donald Pinkava (Arizona State University), John Semple (University of Waterloo), and Michael Windham (Utah Museum of Natural History).

The raw data upon which this revision is based originated from the hundreds of thousands of specimens of Missouri plants in various herbaria around the state and elsewhere. In addition to my co-workers at the Missouri Botanical Garden, I am grateful to the curators and staff members of the herbaria consulted both within Missouri and elsewhere in the United States for their courtesy in allowing me to examine and borrow specimens, especially: Allan Bornstein (Southeast Missouri State University); Michelle Bowe and Paul Redfearn (Missouri State University); Norlan Henderson (University of Missouri–Kansas City, the collection now in the care of Powell Gardens); Lisa Hooper (Truman State University); Robin Kennedy and Bāādi Tadych (University of Missouri–Columbia); Craig Freeman, Ron McGregor, and Caleb Morse (University of Kansas); Pat Holmgren, Jackie Kallunki, Robbin Moran, and Thomas Zanoni (New York Botanical Garden); Chris Niezgoda and Fred Barrie (Field Museum); Emily Wood, Walter Kittredge, and David Boufford (Harvard University); Barney Lipscomb and Guy Nesom (Botanical Research Institute of Texas); and Rusty Russell (Smithsonian Institution).

Since the start of the project in 1987, collectors have added about 5,000 new Missouri specimens to the Missouri Botanical Garden herbarium annually. When one adds to this the new accessions at other herbaria around the state, it becomes apparent that a large number of amateur and professional botanists have generated valuable new information on the state's flora, including numerous distributional records. This lively interest, along with my interactions with these botanists, provided a strong impetus to see the work through to completion. In particular, the questions, discussions, and continual prodding by members of the Missouri Native Plant Society and the Botany Group of the Webster Groves Nature Study Society have provided the incentive that has kept me going at times when the project seemed a bit overwhelming. The list of these supporters is too long to attempt a full register here, but I would like to acknowledge some amateurs and professionals who have been especially productive in their collecting activities during the past six years: Alan Brant, Carl Darigo, Jack and Pat Harris, Nels Holmberg, Rachel Katz, Paul McKenzie, Jeannie Moe, Jim Sullivan, Bill Summers, and Dan Tenaglia.

The process of turning a manuscript into a published volume is lengthy and involved. I am indebted to my editor, Julie Schroeder, for her attention to detail, and to my beloved wife, Kay Yatskievych, who formatted the pages and produced the final product in record time. At the Garden, Victoria Hollowell, Amy McPherson, Diana Gunter, and Beth Parada provided invaluable technical assistance on matters of production.

No list of acknowledgments could be complete without an acknowledgment of the late Julian Steyermark's role in the project. It was with his blessing that this revision became possible, and his support at the beginning was invaluable in getting me started. I doubt that Julian Steyermark would have imagined how the Flora of Missouri Project would blossom over time, but I suspect that he would have been proud to see that his labor of love continues to this day.

CONTRIBUTORS OF TREATMENTS

Ihsan Al-Shehbaz
Missouri Botanical Garden
P.O. Box 299
St. Louis, Missouri 63166

David J. Bogler
Missouri Botanical Garden
P.O. Box 299
St. Louis, Missouri 63166

Alan E. Brant
Missouri Botanical Garden
P.O. Box 299
St. Louis, Missouri 63166

Kuo-fang Chung
Department of Biology
Campus Box 1137
Washington University
1 Brookings Drive
St. Louis, Missouri 63130

Ronald L. Hartman
Department of Botany 3165
University of Wyoming
1000 East University Avenue
Laramie, Wyoming 82071

Mark H. Mayfield
Division of Biology
Kansas State University
Manhattan, Kansas 66506

John Pruski
Missouri Botanical Garden
P.O. Box 299
St. Louis, Missouri 63166

Richard K. Rabeler
University of Michigan Herbarium
3600 Varsity Drive
Ann Arbor, Michigan 48108

Sarah M. Tofari
Indianapolis Parks and Recreation
Holliday Park
6363 Spring Mill Road
Indianapolis, Indiana 46260

Alan Whittemore
U.S. National Arboretum
3501 New York Ave NE
Washington, DC 20002

ILLUSTRATION ACKNOWLEDGMENTS

The revised edition features entirely new plates, and includes considerably more species illustrations than were in Steyermark's first edition. These drawings are a major contribution to the manual, and are an invaluable resource for both learning about various species and verifying specimen determinations. The 586 new plates of line drawings to appear in *Steyermark's Flora of Missouri* would not have been possible without generous contributions from the following funding sources:

    The Academy of Science of St. Louis
    Timon Primm Memorial Fund (at the Missouri Botanical Garden)
    Mrs. John S. Lehmann
    The John Allan Love Foundation
    The Missouri Department of Conservation
    The Wednesday Club of St. Louis

The following artists applied their considerable talents to the completion of the 193 plates that appear in Volume 2. Their dedication and patience with the author's often detailed instructions are greatly appreciated.

Phyllis Bick	Plates 202–223, 301–351, 359–387
Linda Ellis	Plates 224–300, 374
Sheila Flinchpaugh	Plates 200, 201
Kate Johnson	Plates 197–199, 352–358
Ellen Lissant	Plates 195, 196

CLASS MAGNOLIOPSIDA (dicots)

Herbaceous or woody plants. Embryos mostly with 2 cotyledons. Stems usually with the vascular tissues in a ring of small bundles or a cylinder in cross-section, this usually enclosing a central pith (except in larger stems of woody species). Leaves with pinnately or palmately branching, non-parallel veins (except in Plantaginaceae and a few genera scattered in other families, like *Eryngium* [Apiaceae]), these sometimes reduced to a single midvein. Sepals, petals, stamens, and pistils, when present, in multiples of 1–many, but usually 4–5 (rarely 3). About 325 families.

For decades, most botanists have acknowledged that the monocots represent a specialized group that evolved from dicotyledonous ancestors. However, it has been convenient to maintain the two major groups of angiosperms until recently. Molecular and morphological studies (summarized in Judd et al., 2002) have helped to determine the origins and relationships of the monocots within the flowering plants. Those dicots that presumably arose before the dicot/monocot split are now sometimes treated informally as separate from the remaining dicots and assigned names like paleoherbs and the magnoliid complex. In Missouri, the paleoherbs are represented only by the Nymphaeaceae, but the magnoliid complex is represented by several families, including the Annonaceae, Aristolochiaceae, Calycanthaceae, Lauraceae, Magnoliaceae, and Saururaceae. Because the formal classification of these primitive angiosperm families remains somewhat ambiguous, the dicots are treated in the traditional broad sense in the present work.

In the dicots, morphologically similar families sometimes are separable by examination of the type of placentation in the ovary; that is, the pattern of attachment of the ovule(s) to the carpel walls. This is accomplished by slicing carefully across the ovary or fruit with a sharp knife, scalpel, or single-edged razor blade. Species with axile placentation have the ovules attached in the central portion of the crosswall(s) separating the locules of a compound ovary. Taxa with parietal placentation have the ovules attached in one or more clumps along the outer wall. Less common patterns involving 1-locular ovaries include basal placentation, in which the ovules are all attached at the base of the ovary, and free-central placentation, in which the ovules are attached along a longitudinal column in the center of the ovary.

A KEY TO FAMILIES
WILL APPEAR IN VOLUME 3
OF THE PRESENT WORK

ACANTHACEAE (Acanthus Family)
(Wasshausen, 1998)

Plants annual or more commonly perennial herbs (shrubs or small trees elsewhere). Stems often somewhat angled, usually bluntly 4-angled in Missouri species. Leaves opposite, lacking stipules, simple. Inflorescences axillary and sometimes also terminal, often reduced to clusters of flowers in the leaf axils, but in some genera in elongate spikes or small panicles. Flowers zygomorphic, perfect, hypogynous, subtended by 1 or more often conspicuous bracts. Calyces actinomorphic, (4)5-lobed, sometimes lobed nearly to the base. Corollas zygomorphic to nearly actinomorphic, 5-lobed or sometimes 2-lipped. Stamens 2 or 4, the filaments fused to the corolla tube, the anther sacs often appearing relatively distinct from each other and sometimes attached asymmetrically with one sac appearing terminal and the other sometimes appearing lateral. Staminodes sometimes also present, (1)2, particularly in some genera with only 2 functional stamens. Pistil 1 per flower, of 2 fused carpels. Ovary superior, with 2 locules, the placentation axile. Style 1 per flower, the stigmas 2. Ovules 1–8 per locule. Fruits capsules, 2-valved, explosively dehiscent longitudinally. Seeds variously shaped, often flattened, the outer coat usually becoming somewhat sticky or gelatinous when moistened, sometimes with appressed hairs that become erect when moistened. About 250 genera, about 2,600 species, cosmopolitan, but most diverse in tropical and subtropical regions.

The Acanthaceae can be difficult to separate from the Scrophulariaceae. In addition to their unusual, elastically dehiscent capsules, members of the Acanthaceae usually have capsules with fewer seeds and flowers subtended by often conspicuous bracts. At maturity, the fruits of most species dehisce suddenly by snapping open and discharging the seeds through the action of a springlike structure associated with each seed. Many species produce both chasmogamous (open-flowering) and cleistogamous (producing fruit without opening) flowers, the latter often produced toward the end of the growing season. Corolla measurements in the keys and descriptions below refer to chasmogamous flowers. Cleistogamous flowers have the corolla reduced in size, somewhat thickened, more or less club-shaped, and white to light yellow or pale green. Also, in a number of species, chasmogamous flowers are heterostylous; that is, some plants have flowers with longer styles and the stigmas positioned above the stamens, whereas others have flowers with shorter styles and the anthers positioned above the stigmas. This mixture of morphologies promotes cross-pollination by the insects that pollinate the flowers.

1. Calyx lobes 10–30 mm long; corollas nearly actinomorphic, with 5 lobes; stamens 4 . . . . 3. RUELLIA
1. Calyx lobes 2–9 mm long; corollas strongly 2-lipped; stamens 2
   2. Leaves with the petioles 10–70 mm long; stalk of each inflorescence shorter than the petiole of the subtending leaf; flowers subtended by oblong to obovate bracts longer than the calyces . . . . 1. DICLIPTERA
   2. Leaves sessile or with petioles 1–10 mm long; stalk of each inflorescence much longer than the petiole of the subtending leaf; flowers subtended by triangular bracts shorter than the calyces . . . . 2. JUSTICIA

1. Dicliptera Juss.

About 150 species, nearly worldwide, mostly in tropical and warm-temperate regions.

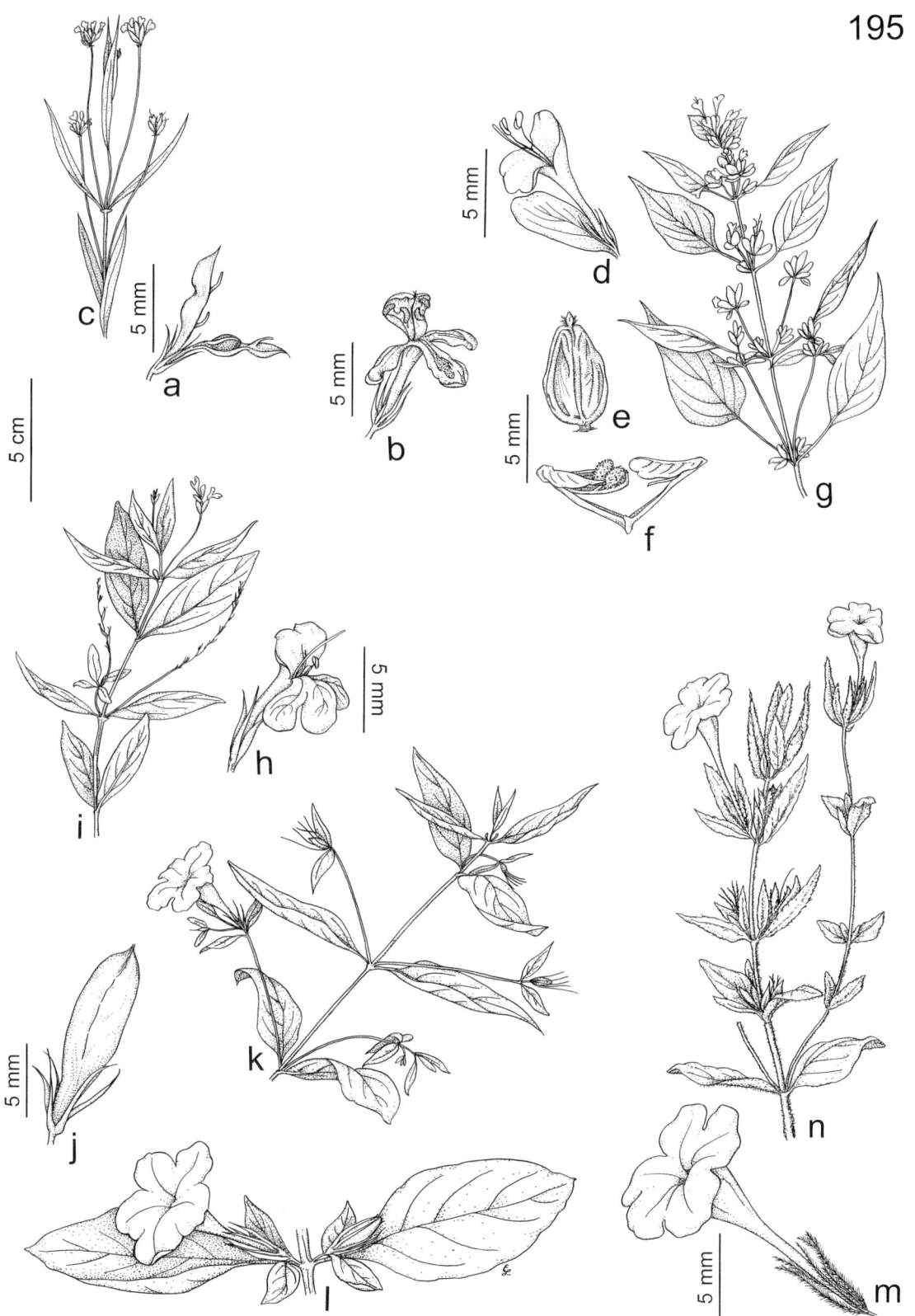

Plate 195. Acanthaceae. *Justicia americana*, **a)** fruit, **b)** flower, **c)** habit. *Dicliptera brachiata*, **d)** flower, **e)** fruit, **f)** dehiscing fruit with seeds, **g)** habit. *Justicia ovata*, **h)** flower, **i)** habit. *Ruellia pedunculata,* **j)** fruit, **k)** habit. *Ruellia strepens,* **l)** flower and leaves. *Ruellia humilis,* **m)** flower, **n)** habit.

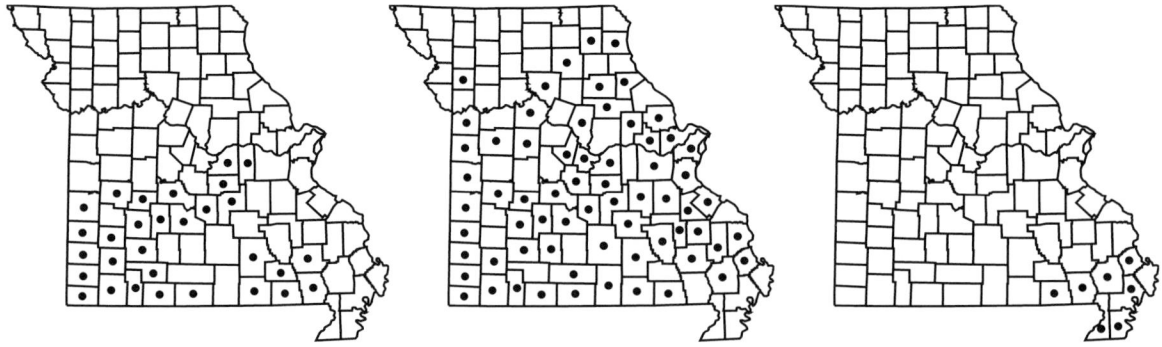

798. Dicliptera brachiata   799. Justicia americana   800. Justicia ovata

1. Dicliptera brachiata (Pursh) Spreng.

Pl. 195 d–g; Map 798

Plants annual. Stems 30–80 cm long, erect or ascending, with usually many branches, glabrous or hairy. Petioles 1–7 cm long. Leaf blades (2–)5–12 cm long, elliptic to ovate, rarely lanceolate, usually tapered at the tip and base, the margins entire or nearly so, hairy, the surfaces sparsely hairy. Inflorescences few-flowered axillary clusters toward the branch tips, subtended by reduced leaves, sometimes appearing as short spikes, the inflorescence stalk usually absent or nearly so, much shorter than the petiole of the subtending leaf. Flowers subtended by bracts 6–10 mm long, these longer than the calyx, oblong to obovate. Calyces 4–6 mm long, deeply lobed, the lobes 2–4 mm long, narrowly lanceolate, long-tapered at the tip, glabrous or hairy along the margins. Corollas 11–15 mm long, the tube 4–6 mm long, strongly 2-lipped, the lips usually with 1 or 2 shallow notches at the tip, hairy on the outer surface, at least when young, pink (drying purple) or less commonly white, the lower lip often with purple spots. Stamens 2, the anther sacs spreading. Staminodes 2, linear, inconspicuous, and hidden in the corolla tube. Fruits 4–6 mm long, elliptic to broadly ovate, flattened, the valves spreading after dehiscence. Seeds 2 or 4, 2.0–2.5 mm long, broadly ovate to circular in outline, flattened, the surface with numerous small spines, reddish brown to black. $2n=80$. August–October.

Scattered in the southern half of the state, mostly in the Ozark and Unglaciated Plains Divisions (southeastern U.S. west to Kansas and Texas). Bottomland forests and banks of streams and rivers; occasionally moist depressions of roadsides.

2. Justicia L.

Plants perennial, with rhizomes, often forming dense colonies. Aerial stems erect from frequently spreading bases, often rooting at the lower nodes, unbranched or branched, glabrous. Leaves sessile or with the petiole 1–10 mm long, glabrous. Inflorescences spikes or headlike clusters, the inflorescence stalk 3–15 cm long, much longer than the petiole of the subtending leaf. Flowers subtended by 3 bracts 0.5–1.5 mm long, these shorter than the calyx, triangular. Calyces lobed nearly to the base, the lobes sharply pointed at the tip, glabrous or minutely hairy along the margins. Corollas strongly 2-lipped, the lower lip 3-lobed, the upper lip entire or shallowly notched at the tip, glabrous or hairy on the outer surface, at least when young, white or light purple. Stamens 2, the anther sacs noticeably asymmetrical. Staminodes absent. Fruits club-shaped from a narrow, stalklike base, the valves arched outward after dehiscence. Seeds usually 4, more or less circular in outline, somewhat flattened, reddish brown to brown. About 300 species, nearly worldwide, mostly in tropical and warm-temperate regions.

1. Flowers densely clustered into a headlike inflorescence; seeds with the surfaces papillose, lacking a differentiated margin .... 1. J. AMERICANA
1. Flowers loosely spaced along the elongate axis of a spicate inflorescence; seeds with the surfaces smooth or nearly so, ringed with a well-differentiated thinner margin .... 2. J. OVATA

1. Justicia americana (L.) Vahl (water willow)

J. americana var. *subcoriacea* Fernald

Pl. 195 a–c; Map 799

Aerial stems 30–100 cm long. Leaf blades 1–16 cm long, the lowermost leaves often reduced, mostly narrowly elliptic or lanceolate, rarely broader, tapered at the tip and base. Inflorescences axillary headlike clusters, the inflorescence stalk slightly shorter than to longer than the subtending leaf. Calyces 4–8 mm long, the lobes narrowly lanceolate. Corollas 8–15 mm long, the lower lip white or pale purple with purple markings, the upper lip usually light purple, rarely white. Fruits 8–13 mm long. Seeds 2.0–3.5 mm long, reddish brown to brown, lacking a differentiated margin, the surfaces papillose. $2n=26$. May–October.

Common nearly throughout the state, but uncommon or absent from portions of the Glaciated Plains and Mississippi Lowlands Divisions (eastern U.S. and adjacent Canada west to Wisconsin and Texas). Usually emergent aquatics along banks of spring branches, streams, and rivers, less commonly margins of ponds, lakes, and sloughs; also ditches.

Water willow is a characteristic species of gravel bars and other stream banks throughout much of Missouri. The dense colonies of aerial stems frequently are unbranched and often do not flower every year.

2. Justicia ovata (Walter) Lindau **var. lanceolata** (Chapm.) R.W. Long (lance-leaved water willow)

Pl. 195 h, i; Map 800

Aerial stems 10–60 cm long. Leaf blades 1–11 cm long, mostly linear to narrowly elliptic, the lowermost leaves sometimes elliptic to obovate, usually tapered at the tip and base. Inflorescences axillary spikes with the flowers single at each node and all on one side of the axis. Calyces 4–7 mm long, the lobes linear or narrowly lanceolate. Corollas 8–12 mm long, light purple, the lower lip with darker purple markings. Fruits 10–13 mm long. Seeds 2.0–2.6 mm long, brown, with a well-differentiated, thinner, lighter-colored margin, the surfaces smooth. $2n=14$. May–August.

Uncommon in the Mississippi Lowlands Division (Florida to Texas north to Tennessee and Missouri). Swamps, bottomland forests, and sloughs; also ditches and banks of canals; sometimes emergent aquatics.

Yatskievych and Turner (1990) incorrectly attributed Missouri specimens to var. *ovata*, which is restricted to the Atlantic and Gulf Coastal Plains. It differs from var. *lanceolata* in having flowers paired (vs. single) at the nodes of the spike and calyces 7–10 mm (vs. 4–7 mm) long.

3. Ruellia L.

Plants perennial. Aerial stems spreading or ascending, unbranched or branched. Inflorescences few-flowered clusters, these either nearly sessile in the axils of main stem leaves or associated with reduced leaves at the tip of an axillary branch (inflorescence stalk). Flowers subtended by bracts 1–12 mm long, these shorter than the calyx, linear to narrowly elliptic or oblanceolate. Calyces deeply lobed, the lobes 10–30 mm long, sharply pointed at the tip. Corollas nearly actinomorphic, with a slender tube expanded fairly abruptly into 5 lobes, glabrous or hairy, light purple to lavender, rarely white. Stamens 4, usually included in the corolla throat, the anther sacs similar in size and parallel. Staminodes usually absent, rarely 1. Fruits narrowly oblong-elliptic or club-shaped, the valves ascending to arched outward after dehiscence. Seeds (4–)6–16, 2.5–3.5 mm in diameter, more or less circular in outline, somewhat flattened, reddish brown to brown, usually somewhat sticky when fresh and appearing minutely hairy when moistened. About 250 species, nearly worldwide, mostly in tropical and warm-temperate regions.

801. Ruellia humilis   802. Ruellia pedunculata   803. Ruellia strepens

The last monographer of *Ruellia* in eastern North America, M. L. Fernald (1945), divided most of the species into a number of infraspecific taxa, based on differences in flower color, pubescence, presence of cleistogamous flowers, and corolla length. Some recent students of the genus (Long, 1970; Turner, 1991) have suggested that most of these taxa intergrade too much to allow recognition, and although Steyermark (1963) accepted several varieties and forms for the Missouri species, he also noted that widespread intermediates made recognition of infraspecific taxa difficult. Accordingly, the varieties and forms treated by Steyermark are not accepted in the present treatment.

1. Stems glabrous or more commonly minutely hairy in 2 narrow, longitudinal bands on opposite sides; calyx lobes 2–4 mm wide, narrowly lanceolate .... 3. R. STREPENS
1. Stems evenly hairy on all sides; calyx lobes 0.5–1.5 mm wide, linear, often needlelike or bristlelike
    2. Leaves sessile or with petioles 1–3 mm long; fruits glabrous; flower clusters in the axils of main stem leaves .... 1. R. HUMILIS
    2. Leaves with petioles 3–15 mm long; fruits finely hairy; flower clusters in the axils of reduced leaves at the tip of axillary branches (inflorescence stalks) .... 2. R. PEDUNCULATA

1. Ruellia humilis Nutt. (wild petunia, fringeleaf ruellia)
    R. humilis f. *alba* (Steyerm.) Steyerm.
    R. humilis var. *expansa* Fernald
    R. humilis var. *frondosa* Fernald
    R. humilis f. *grisea* Fernald
    R. humilis var. *longiflora* (A. Gray) Fernald
            Pl. 195 m, n; Map 801

Stems 7–55 cm long, unbranched when young but often becoming branched later in the season, evenly hairy on all sides with spreading hairs 1–2 mm long, sometimes appearing nearly glabrous with age. Leaves sessile or with petioles 1–3 mm long, the leaf blades 2–7 cm long, narrowly to broadly ovate, pointed to rounded at the tip, angled to rounded at the base, hairy on both sides and along the margins. Inflorescences consisting of flower clusters in the axils of main stem leaves, the flowers subtended by hairy, narrowly lanceolate to lanceolate bracts 10–18 mm long. Calyx lobes 15–30 mm long, 0.5–1.5 mm wide, linear, often needlelike or bristlelike, hairy. Corollas 3–7 cm long, 2.0–3.5 cm wide. Fruits 10–14 mm long, glabrous. $2n=34$. May–October.

Scattered nearly throughout Missouri, but apparently absent from portions of the Mississippi Lowlands Division (northeastern U.S. west to Nebraska and Texas). Openings of mesic to dry upland forests, savannas, prairies, glades, and ledges of bluffs; also dry pastures and roadsides.

This species is extremely variable with regard to degree of hairiness, corolla length, and leaf shape.

2. Ruellia pedunculata Torr. ex A. Gray
(wild petunia)

R. pedunculata f. *baueri* Steyerm.

Pl. 195 j, k; Map 802

Stems 15–70 cm long, branched by flowering time, evenly hairy on all sides with recurved hairs 0.1–0.3 mm long. Leaves with petioles 3–15 mm long, the blades of main stem leaves 2–11 cm long (those of flowering branches 1–5 cm), lanceolate to ovate, mostly tapered to a sharp point at the tip, angled to rounded at the base, hairy on both sides, sometimes sparsely so. Inflorescences consisting of flower clusters at and near the tip of axillary branches (inflorescence stalks) to 20 cm long, the flowers subtended by hairy, lanceolate to ovate bracts 7–25 mm long. Calyx lobes 10–30 mm long, 0.5–1.5 mm wide, linear, often needlelike or bristlelike, hairy. Corollas 3–6 cm long, 2–3 cm wide. Fruits 13–20 mm long, finely hairy. $2n=34$. May–September.

Scattered in the Ozark and Ozark Border Divisions and uncommon in the Big Rivers and Mississippi Lowlands Divisions (southeastern U.S. west to Missouri and Texas). Mesic to dry upland forests, ledges of bluffs, and glades; less commonly bottomland forests and banks of streams; also roadsides.

The closely related *R. pinetorum* Fernald of the southeastern United States is sometimes treated as a subspecies of *R. pedunculata* (Wasshausen, 1966; Long, 1970; Turner, 1991). It differs in its glabrous fruits and more or less sessile leaves.

3. Ruellia strepens L. (wild petunia, smooth ruellia, limestone ruellia)

R. strepens f. *alba* Steyerm.
R. strepens f. *cleistantha* (A. Gray) S. McCoy

Pl. 195 l; Map 803

Stems 15–100 cm long, unbranched or branched, glabrous or more commonly minutely hairy in 2 narrow, longitudinal bands on opposite sides, the hairs 0.1–0.2 mm long, appearing crinkled. Leaves with petioles 3–20 mm long, the blades of main stem leaves 2–16 cm long, ovate or broadly lanceolate to elliptic or less commonly obovate, mostly tapered to a sharp point at the tip, tapered or less commonly rounded at the base, hairy on both sides and minutely hairy along the margins. Inflorescences consisting of flower clusters in the axils of main stem leaves and usually also at and near the tip of axillary branches (inflorescence stalks) to 8 cm long, the flowers subtended by hairy lanceolate to obovate bracts (3–)10–40 mm long. Calyx lobes 9–20 mm long, 2–4 mm wide, narrowly lanceolate, sparsely to less commonly densely hairy on the back (especially along the midnerve), with a fringe of white hairs 1–2 mm long along the margins. Corollas 3–7 cm long, 2–4 cm wide. Fruits 10–20 mm long, glabrous. $2n=34$. May–October.

Scattered nearly throughout the state (northeastern U.S. west to Nebraska and Texas). Bottomland forests, mesic upland forests, banks of streams and rivers, margins of ponds and lakes, less commonly bottomland prairies and fens; also pastures, moist roadsides, and railroads.

ACERACEAE (Maple Family)

Two genera, about 115 species, widespread in temperate portions of the Northern Hemisphere and in mountains in the tropics.

1. Acer L. (maple)
(Murray, 1975)

Plants mostly monoecious or dioecious, occasionally with perfect flowers mixed with the pistillate ones, shrubs or more commonly small or large trees. Leaves opposite, petiolate, lacking stipules (these fused to the petiole bases and usually not apparent, except sometimes in *A. saccharum*), the leaf blades usually palmately lobed, less commonly pinnately compound. Inflorescences terminal or lateral toward the branch tips, sometimes axillary, ranging from small clusters to racemes or small panicles. Flowers actinomorphic, hypogynous, the staminate ones often perigynous. Calyces of 4 or 5(6) sepals, these sometimes fused, often colored.

Corollas absent or of 4 or 5(6) free petals. Stamens 3–8, usually strongly exserted, the filaments sometimes attached to the margin of a nectar disc. Pistil 1 per flower, superior, of 2 fused carpels, usually with 2 locules, flattened at right angles to the septum. Styles 2 per flower or sometimes 1 and deeply 2-lobed, the stigmas 2. Ovules usually 2 per locule. Fruits consisting of 2 samaras that are initially fused at the base but break apart at maturity and are dispersed independently, each with a single basal seed and a terminal wing. About 115 species, widespread in temperate portions of the Northern Hemisphere and in mountains in the tropics.

Emerging morphological and molecular evidence suggests that the Aceraceae (and Hippocastanaceae) might best be treated within an expanded circumscription of the Sapindaceae (Judd et al., 2002). The traditional classification is followed here in anticipation of further studies to resolve the phylogenetic relationships within this family complex.

Maples are important components of many deciduous forest communities. The wood of various species is of commercial importance for lumber, for boards and slats, for veneers, as pulpwood for paper, and in the construction of furniture and musical instruments. Species valued for timber production usually are divided into two groups: hard maples with harder wood that is better suited for structural uses, which include the *A. saccharum* complex; and soft maples with more brittle wood, which include *A. negundo*, *A. rubrum*, and *A. saccharinum*. Although *A. saccharum* is the species best known for the use of its sap for a sugar and maple syrup, most other species also yield sap with similar properties (but inferior in quality) to that of sugar maple.

Numerous species also are cultivated as ornamentals because of their interesting leaf shapes and colors, especially their bright coloration during the autumn. A large number of cultivars exist for some taxa. In addition to the species treated below, several additional maples are cultivated commonly in Missouri but have not been documented as escapes. These include *A. palmatum* Thunb. (Japanese maple), a shrub or small tree with leaves having strongly tapered tips on the lobes and relatively small, widely spreading samaras; and *A. pseudoplatanus* L. (sycamore maple), which has relatively showy flowers in small panicles that are not produced until the trees are mostly leafed out. Settergren and McDermott (1962) suggested that *A. platanoides* L. (Norway maple), a shade tree distinguished by its milky sap and widely spreading samaras, probably had become naturalized somewhere in Missouri, but although this species is commonly cultivated as a shade tree in the state it has not yet been found established outside cultivation.

1. Leaves pinnately compound, sometimes with only 3 leaflets . . . . 2. A. NEGUNDO
1. Leaves palmately lobed
    2. Sinuses between the main leaf lobes rounded or U-shaped
        3. Leaf blades silvery white on the undersurface, the central lobe noticeably narrowed toward the base; flowers produced before the leaves
        . . . . 4. A. SACCHARINUM
        3. Leaf blades lighter green on the undersurface than on the upper surface, but not silvery white, the central lobe broadest at the base or with nearly parallel sides; flowers produced during expansion of the leaves
        . . . . 5. A. SACCHARUM
    2. Sinuses between the main leaf lobes angled or V-shaped
        4. Leaf blades relatively deeply lobed, the lateral lobes cut $^1/_2$–$^2/_3$ of the way to the base, the central lobe noticeably narrowed toward the base; flowers lacking petals; fruits with the wings 3.0–6.5 cm long, broadly spreading
        . . . . 4. A. SACCHARINUM

4. Leaf blades relatively shallowly lobed, the lateral lobes cut ¼–½ of the way to the base, the central lobe broadest at or just above the base; flowers with small petals; fruits with the wings 1.5–3.0(–4.0) cm long, narrowly spreading, sometimes appearing parallel or nearly so
   5. Flowers produced during or after expansion of the leaves, individually short-stalked in dense, narrow panicles from branch tips, appearing pale yellowish green to yellowish white . . . . 1. A. GINNALA
   5. Flowers produced before the leaves, sessile or nearly so in dense clusters from buds along the branches (the stalks elongating greatly after flowering as the fruits mature), appearing orangish red to purplish red . . . . . . 3. A. RUBRUM

1. Acer ginnala Maxim. (Amur maple, Siberian maple)

A. tataricum L. ssp. ginnala (Maxim.) Wesm.

Map 804

Plants monoecious, shrubs or small trees to 3(8) m tall with spreading branches, the bark of young trees more or less roughened and gray, eventually becoming separated into long thin plates on older trees. Twigs yellowish brown to brown, the winter buds ovate, pointed at the tip, with 2 or 4 overlapping outer scales usually obscuring 2 or more inner scales. Leaf blades 2–8 cm long, broadly triangular-ovate in outline, the undersurface lighter green than the green to bluish green upper surface, glabrous or hairy along the main veins, usually with 3 main lobes (rarely with 2 additional short lobes), these tapered to sharply pointed tips and with the sinuses angled or V-shaped, the lateral lobes cut ⅓–½ of the way to the base, the central lobe often longer than the lateral ones and broadest at or just above the base, the margins irregularly toothed (sometimes appearing doubly toothed). Inflorescences produced during or after expansion of the leaves; dense, narrow panicles from branch tips; the flowers individually short-stalked. Calyces 1.8–2.5 mm long, the sepals fused only at the very base, the 4 lobes oblong-elliptic, rounded at the tips, whitish green, the margins scarious and usually with a fringe of short, curly hairs. Petals 4, 1.5–2.0 mm long, narrowly spatulate to oblanceolate, yellowish white. Staminate flowers with (7)8 stamens inserted on the margin of a nectar disc. Pistillate flowers with the ovary densely hairy. Fruits dispersing mostly long after the leaves are mature, the samaras 2.0–3.5 cm long, glabrous or sparsely hairy, the wings 1.5–3.0 cm long, narrowly spreading, sometimes appearing parallel or nearly so. $2n=26$. April–June.

Introduced, known thus far from only from Crawford, Franklin, Howell, and Lincoln Counties (native of eastern Asia; naturalized sporadically in the northeastern U.S. west to Minnesota). Disturbed portions of mesic upland forests and old fields.

Vegetatively, A. ginnala can be difficult to distinguish from glabrescent individuals of A. rubrum. It may be distinguished by its more compact and usually shrubby growth form, as well as its winter buds with usually fewer observable bud scales. Amur maple was first reported for Missouri by Ebinger and McClain (1991), who documented an extensive population near Elsberry in fields and forests surrounding the Plant Materials Center of the Natural Resources Conservation Service. It is fairly widely cultivated in northeastern North America and Europe, but it appears to escape relatively uncommonly. In the autumn, the foliage turns a beautiful bright crimson color. A number of cultivars (mostly differing in leaf shape and coloration) exist and several poorly defined subspecies also have been described. Naturalized plants in the midwestern United States correspond to ssp. *ginnala*.

2. Acer negundo L. (box elder, ash-leaved maple)

Pl. 196 a, b; Map 805

Plants dioecious, small to medium trees to 25 m tall with ascending to spreading branches, the bark of young trees smooth and gray to light brown, eventually becoming separated into a network of long thin ridges on older trees. Twigs green to olive green, sometimes glaucous and appearing pale purple to purple, glabrous or densely hairy, the winter buds bluntly pointed at the tip, with 2 or 4 overlapping scales. Leaf blades 7–15 cm long, broadly triangular-ovate to oblong-ovate in general outline, pinnately compound, those of fertile branches and seedlings with 3–5 leaflets on a short rachis, those of vigorous vegetative shoots with 5–9 leaflets. Leaflets 5–10 cm long, oblong to ovate (terminal leaflet sometimes obovate), angled or tapered to sharply pointed tips, the upper surface light green, the undersurface pale grayish green, usually somewhat hairy when young, becoming sparsely hairy or glabrous at maturity, the margins coarsely and irregularly toothed and occasion-

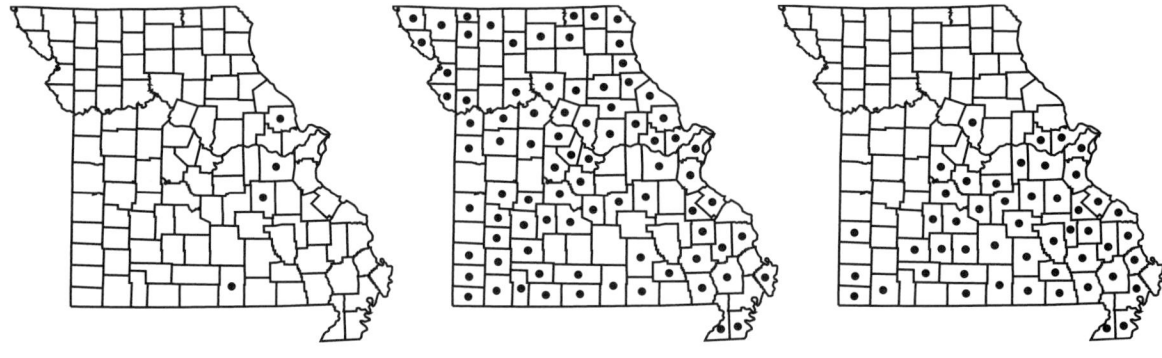

804. Acer ginnala   805. Acer negundo   806. Acer rubrum

ally also with a few shallow lobes. Inflorescences produced before the leaves or during leaf development, the staminate ones umbellate clusters from buds along the branches, the individual flowers with long drooping stalks, the pistillate ones narrow drooping racemes from at or near branch tips. Calyces 1–3 mm long, the sepals fused only at the very base, the 5 lobes obovate to narrowly oblong-elliptic, rounded at the tips, yellowish green, hairy. Petals absent. Staminate flowers with 3–6 stamens and lacking a nectar disc. Pistillate flowers with the ovary glabrous or less commonly hairy. Fruits dispersing long after the leaves are mature, often not until the following autumn, the samaras 2.5–4.5 cm long, glabrous or less commonly hairy, the wings 2–4 cm long, spreading at less than a 90° angle. $2n=26$. April–May.

Common nearly throughout Missouri (U.S., Canada, Mexico, Guatemala). Bottomland forests, banks of streams, mesic upland forests in bottoms of ravines, and bases of bluffs; also shaded ditches and moist roadsides.

Box elder has fallen into disfavor horticulturally, for although it grows easily under a variety of site conditions, the branches of older trees break easily during storms, and the trees are susceptible to a number of fungal diseases and insect pests, notably the box elder bug, *Leptocoris trivittatus* (Say) (Wagner, 1975). The leaves turn yellow in the autumn and are shed earlier than those of most other trees. Unlike other Missouri species of *Acer*, which are pollinated by both insects and wind, the flowers of *A. negundo* are only wind-pollinated and are a cause of hay fever during the spring.

Acer negundo has been treated as consisting of a number of varieties by many botanists (Murray, 1975), each occupying a portion of the species' overall distribution. Four of these taxa were accepted as occurring in Missouri by Steyermark (1963), who nevertheless had difficulties with infraspecific determinations of some specimens. For convenience, the species is here treated as a complex of only two varieties, which allows nearly all specimens to be classified relatively easily.

1. Twigs glabrous or glaucous
   .... 2A. VAR. NEGUNDO
1. Twigs densely short-hairy
   .... 2B. VAR. TEXANUM

2a. var. negundo

A. negundo var. *violaceum* (G. Kirchn.) Jaeger

Twigs glabrous or glaucous, green to olive green or appearing pale purple to purple. Ovaries and fruits glabrous. $2n=26$. April–May.

Common nearly throughout Missouri (mostly eastern U.S.; Canada). Bottomland forests, banks of streams, mesic upland forests in bottoms of ravines, and bases of bluffs; also shaded ditches and moist roadsides.

Although the separation of plants with glabrous vs. hairy twigs is relatively easy in Missouri material, there exists a nearly continuous range of variation from trees with nonglaucous twigs to those with grayish purple glaucousness on the twigs (sometimes separated as var. *violaceum*). Glaucous specimens are scattered throughout the range of the species in the state.

2b. var. texanum Pax

A. negundo var. *interius* (Britton) Sarg.

Twigs densely short-hairy, not glaucous, green to olive green. Ovaries and fruits glabrous or short-hairy. April–May.

Scattered in the western half of the state and disjunctly in Lincoln County (mostly western and southern U.S.; Canada, Mexico). Bottomland forests, banks of streams, mesic upland forests in bottoms of ravines, and bases of bluffs; also shaded ditches and moist roadsides.

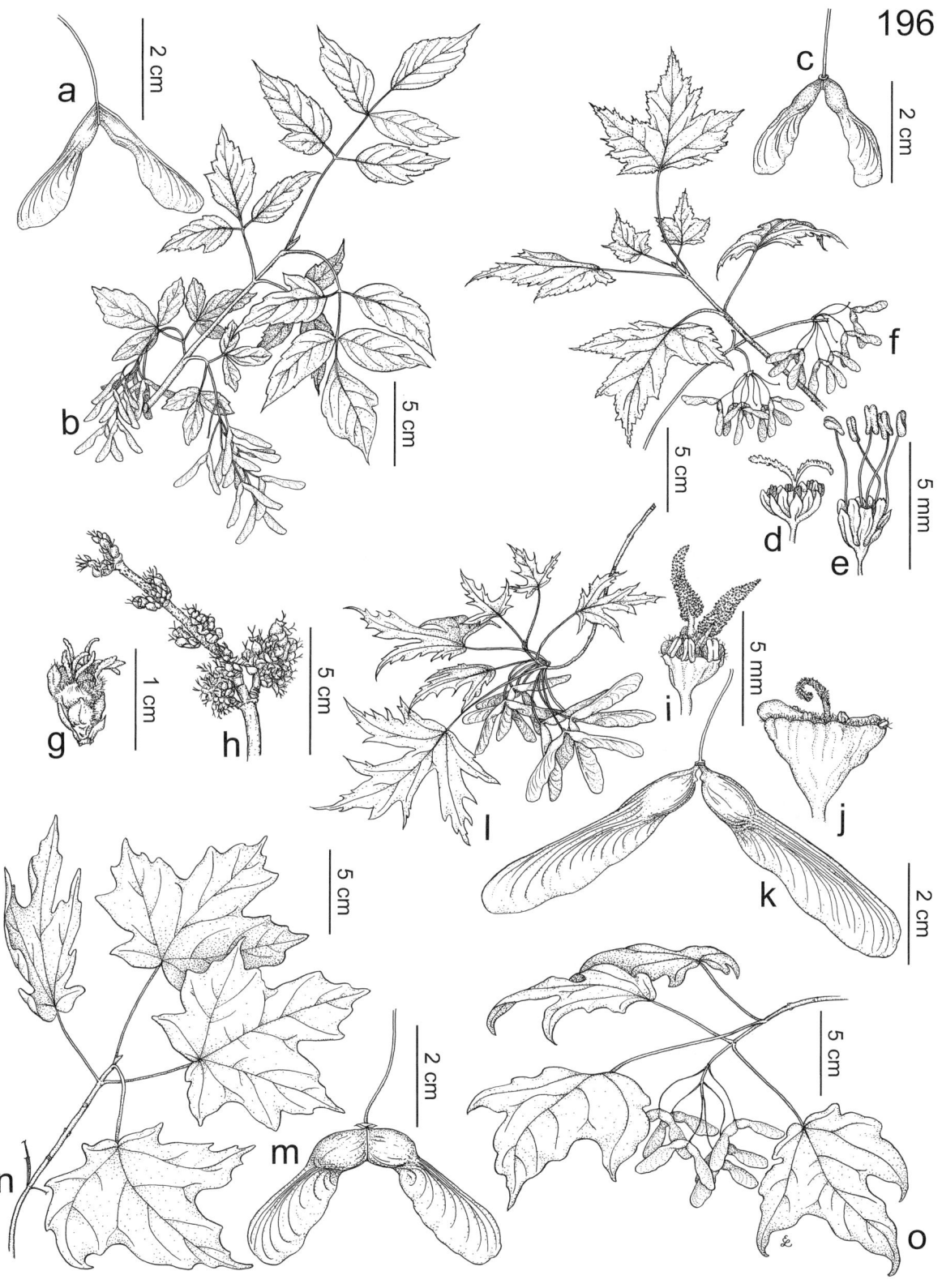

Plate 196. Aceraceae. *Acer negundo*, **a)** fruit, **b)** fruiting branch. *Acer rubrum*, **c)** fruit, **d)** pistillate flower, **e)** staminate flower, **f)** fruiting branch. *Acer saccharinum*, **g)** flower, **h)** flowering branch, **i)** flower, **j)** young fruit, **k)** mature fruit, **l)** fruiting branch. *Acer saccharum* ssp. *saccharum*, **m)** fruit, **n)** branch. *Acer saccharum* ssp. *nigrum*, **o)** fruiting branch.

The var. *interius* is said to differ from var. *texanum* in its glabrous or nearly glabrous (vs. minutely hairy) ovaries and fruits, as well as its hairy (vs. glabrous) petioles. In Missouri material, there is no correlation between these characters, and a range of pubescence densities exists on the petioles. McGregor (1986b) suggested that individuals from Kansas and Missouri with hairy twigs and glabrous or nearly glabrous fruits might represent hybrids between var. *negundo* and var. *texanum,* but this requires further study.

3. Acer rubrum L. (red maple)

Pl. 196 c–f; Map 806

Plants monoecious or dioecious, small to medium trees to 15 m tall with usually spreading branches, the bark of young trees smooth and gray, eventually becoming dark gray to brown and separated into long thin plates or ridges on older trees. Twigs red and shiny, the winter buds ovate, bluntly pointed at the tip, with 4–10 overlapping scales. Leaf blades 5–12 cm long, broadly triangular-ovate to nearly semicircular in outline, the undersurface lighter than the green to dark green upper surface and often strongly white-glaucous, glabrous or hairy, with 3 or 5 main lobes, these tapered to sharply pointed tips and with the sinuses angled or V-shaped, the lateral lobes cut $1/4$–$1/2$ of the way to the base, the central lobe shorter than to slightly longer than the lateral ones and broadest at or just above the base, the margins irregularly toothed (sometimes appearing doubly toothed). Inflorescences produced before the leaves, dense clusters from lateral buds along the branches, the flowers sessile or nearly so (the stalks elongating greatly after flowering as the fruits mature). Calyces 1.4–2.2 mm long, the sepals fused only at the very base, the 4 or 5 lobes oblong-elliptic, rounded at the tips, red to purplish red, glabrous, the margins not scarious. Petals 4 or 5, 1.6–2.4 mm long, narrowly oblong to linear, orangish red to purplish red. Staminate flowers with 5–8 stamens inserted on the margin of a nectar disc. Pistillate flowers with the ovary glabrous. Fruits dispersing mostly after the leaves are mature, the samaras 2–4 cm long, glabrous, the wings 1.5–3.0 (–4.0) cm long, narrowly spreading, sometimes appearing parallel or nearly so. $2n$=78, 91, 104. March–April.

Scattered to common in the Ozark, Ozark Border, and Mississippi Lowlands Divisions (eastern U.S. west to Illinois, Missouri, and Texas; Canada). Swamps, bottomland forests, mesic to dry upland forests, sinkhole ponds, banks of streams, and ledges of bluffs.

Acer rubrum has been considered a minor component of most forest ecosystems in which it occurs. However, Abrams (1998) has documented an explosive increase in the abundance of this species throughout much of its range during the past several decades that is analogous to the situation apparent in Missouri for *A. saccharum.* Abrams attributed this to an opportunistic increase of the species resulting from the cumulative effects of land management patterns in the region, including logging practices, land clearing for agriculture, diseases of other forest trees, and fire suppression.

In addition to the use of its wood for veneers, implements, furniture construction, and pulp for papermaking, as well as the use of its sap as a low-grade substitute for that of sugar maple in syrup production, red maple has also been used historically as a source of tannins for ink production, and an extract of the bark was used for preparing reddish brown and black dyes (Steyermark, 1963). The leaves turn a bright crimson to orangish red color in the autumn and the species is cultivated as an ornamental for its foliage. The species is highly variable in leaf size, shape, and coloration. Some forms with 3-lobed leaves have been segregated as var. *trilobum,* but as noted by Steyermark, there is no correlation between this character and others involving fruit size and leaf pubescence. In Missouri, two ecologically distinctive but morphologically overlapping varieties may be recognized. Plants in flower (prior to the development of leaves) cannot be separated into varieties based on morphology, although the habitat in which a tree is growing may provide clues, as var. *drummondii* is unknown from upland sites in Missouri.

1. Leaf blades with the undersurface densely hairy at maturity, at least along the main veins; fruits with the wings 2–3(–4) cm long .... 3A. VAR. DRUMMONDII
1. Leaf blades with the undersurface glabrous to sparsely hairy when young, usually glabrous at maturity; fruits with the wings 1.5–2.5 cm long .... 3B. VAR. RUBRUM

3a. var. drummondii (Hook. & Arn. ex Nutt.) Sarg.

A. rubrum f. *rotundata* Sarg.

Leaf blades with the undersurfaces usually strongly whitened and also densely and persistently hairy at maturity, at least along the main veins. Fruits with the wings 2–3(–4) cm long. $2n$=78. March–April.

Uncommon in the Mississippi Lowlands Division west locally to Ripley and Wayne Counties (southeastern U.S. west to Missouri and Texas). Swamps, bottomland forests, and sinkhole ponds.

3b. var. rubrum

A. rubrum var. *trilobum* Torr. & A. Gray ex K. Koch

A. rubrum var. *tridens* A.W. Wood

A. rubrum f. *tomentosum* (Desf.) Dans.

Leaf blades with the undersurfaces lighter than the upper surface and often strongly white-glaucous, glabrous to sparsely hairy when young, glabrous or rarely sparsely hairy along the main veins at maturity. Fruits with the wings 1.5–2.5 cm long. $2n=78, 91, 104$. March–April.

Scattered to common in the Ozark, Ozark Border, and Mississippi Lowlands Divisions (eastern U.S. west to Illinois, Missouri, and Texas; Canada). Mesic to dry upland forests and ledges of bluffs, rarely in bottomland forests, sinkhole ponds, and banks of streams.

The leaf blades in var. *rubrum* generally are glabrous or nearly so at maturity, the few hairs that were present initially having disappeared during development. Steyermark (1963) noted the existence of a few specimens from Camden, Douglas, and Shannon Counties that retained moderate leaf pubescence but had the smaller fruits typical of var. *rubrum*. He assigned these to f. *tomentosa,* which occurs sporadically throughout the overall range of the species. Vegetative samples of such plants might easily be mistaken for var. *drummondii*.

4. Acer saccharinum L. (silver maple, soft maple)

Pl. 196 g–l; Map 807

Plants monoecious, medium to large trees to 30 m tall with ascending to spreading branches, the bark of young trees smooth and light gray, becoming separated into long thin plates and ridges on older trees. Twigs red to yellowish brown, the winter buds elliptic-ovate, blunt at the tip, with 6–10 overlapping scales. Leaf blades 8–15 cm long, broadly triangular-ovate in outline, the undersurface silvery white and glabrous or sparsely hairy when young, with 5 deep main lobes, these tapered to sharply pointed tips and with the sinuses usually angled or V-shaped (occasionally those of individual leaves appearing bluntly angled to nearly rounded), the lateral lobes cut $1/2$–$2/3$ of the way to the base, the central lobe noticeably narrowed toward the base, the margins toothed and with smaller lobes. Inflorescences produced before the leaves; small, dense, staminate and pistillate clusters from lateral buds along the branches; the flowers sessile or very short-stalked (the stalks elongating greatly after flowering as the fruits mature). Calyces 4–6 mm long, the sepals fused with only 5 shallow lobes apparent, yellowish green, sometimes grayish pink when young. Corollas absent. Staminate flowers with (3–)5(–7) stamens and lacking a nectar disc. Pistillate flowers with the ovary densely hairy. Fruits dispersing before the leaves are mature, the samaras 3.5–7.0 cm long, glabrous or sparsely hairy, the wings 3.5–6.5 cm long, spreading at about 90–120° and slightly incurved. $2n=52$. January–April.

Scattered to common nearly throughout the state (eastern U.S. west to South Dakota and Oklahoma; Canada). Bottomland forests, mesic upland forests in ravine bottoms; banks of streams and riverbanks, and margins of ponds and lakes; also roadsides, ditches, vacant lots, and moist, open, disturbed areas.

The leaves of *A. saccharinum* turn yellow in the autumn. The twigs produce a disagreeable odor when bruised or broken. The species forms huge often nearly monocultural stands in disturbed bottomland forests and on the river sides of levees in the Big Rivers Division, and it often colonizes gravel bars along larger streams and rivers elsewhere in the state.

The wood of silver maple is of minor commercial importance regionally for use in making furniture and as pulpwood for paper. The sap has also been used to a small extent for maple syrup production. However, the species is extremely important horticulturally as a shade tree, both because of its beauty and its ability to grow quickly into a large tree. Drawbacks to its cultivation are the tendency of older, brittle branches to break during storms and the large crop of seeds that germinate to become weeds in gardens, alleys, and vacant lots. It is also somewhat prone to attack by insect pests and is considered by foresters to have a relatively short average life span for a maple species.

5. Acer saccharum Marshall (sugar maple, hard maple)

Pl. 196 m–o; Map 808

Plants monoecious or sometimes dioecious, medium to large trees to 30 m tall with ascending to spreading branches, the bark of young trees smooth and gray to dark brown, becoming furrowed and/or separated into narrow, thick plates on older trees. Twigs gray to reddish to orangish brown, the winter buds elliptic-ovate, sharply

807. Acer saccharinum   808. Acer saccharum   809. Trianthema portulacastrum

pointed at the tip, with 6–12 overlapping scales. Leaf blades 7–15 cm long, broadly triangular-ovate in outline, sometimes wider than long, the undersurface light green and sometimes glaucous, glabrous, or often with minute tufts of hairs in the axils of the main veins, less commonly hairy on the surface or along the veins, with (3)5 main lobes, these tapered to sharply pointed tips and with the sinuses rounded or U-shaped, the margins undulate or more commonly toothed and with smaller lobes, the central lobe cut $^1/_3$–$^1/_2$ of the way to the blade base, broadest at the base or with nearly parallel sides. Inflorescences produced during expansion of the leaves, umbellate staminate and pistillate clusters from buds at or near the branch tips, the individual flowers with long, drooping, hairy stalks. Calyces 2.5–6.0 mm long, the sepals fused more than $^1/_2$ of the way to the tip, with 5 shallow lobes, greenish yellow, usually hairy toward the tips. Corollas absent. Staminate flowers with 5–8 stamens inserted on the margin of a nectar disc. Pistillate flowers with the ovary glabrous or nearly so. Fruits dispersing after the leaves are mature, the samaras 2.5–4.0 cm long, glabrous, the wings 2.0–3.5 cm long, spreading at about 90–120° (rarely much narrower and nearly parallel). $2n=26$. April–May.

Common throughout Missouri (eastern U.S. west locally to Idaho and Arizona; Canada, Mexico). Mesic to dry upland forests, margins of glades, ledges and bases of bluffs, and banks of streams, rarely bottomland forests; also moist to dry, shaded, disturbed areas.

This species is a major timber tree in the eastern United States. However, as noted by Settergren and McDermott (1962), most of the hard maple that is sold in Missouri (principally as veneers and boards for flooring) is imported from elsewhere, because Missouri sugar maples tend to have unsightly discolorations in their wood. The species is the principal source of maple syrup and also is cultivated as a shade tree. The leaves turn various shades of yellow, orange, and red in the autumn.

The *A. saccharum* complex consists of a bewildering assortment of names and taxa, many of which have been treated as separate species at different times in the past. It is possible that this is an example of a species whose distribution became fragmented at some point in the past (perhaps in response to glaciation), with subsequent morphological divergence of geographically isolated groups of populations over time. These segregates then presumably came into contact again and hybridized as secondary hardwood forests developed following the rapid destruction of primary forests in the eastern United States for agriculture and logging during the past two centuries. The present treatment follows that of Desmarais (1952), whose detailed study of morphological variation provides a good starting point for further investigations into interrelationships among taxa in the group. Readers should note that there is marked intergradation between the four subspecies recognized below.

The analysis of Desmarais (1952) is notable for documenting the large range of specimens that are intermediate between the black and sugar maple morphotypes, which many botanists continue to treat as distinct species. Regional floras of midwestern states, including Voss (1985) and the present volume, also have noted the existence of large numbers of sugar maple trees with at least some black maple characteristics. In addition to morphological features, such trees seem to combine the faster growth rate of sugar maple with the greater drought resistance of black maples. The decline of oak regeneration and concomitant invasion of the oak-hickory-dominated forests by sugar maples in Missouri has been of grave concern to foresters and ecologists (Wuenscher and Valiunas, 1967; Nigh et al., 1985; Pallardy et al., 1988, 1991). It has been attributed to fire suppression following European settlement of the region that promoted succession to a more mesic forest type. It also may have been aided by the ecological adaptations of sugar maple stocks potentially containing genetic materials of black maples conferring greater drought resistance on such trees.

1. Leaf blades 3–10 cm long, the lobes blunt to rounded at the tips; ovary and young fruit hairy (becoming nearly glabrous at maturity); bark light gray, smooth, becoming shallowly furrowed and sometimes with scaly ridges on older trees . . . . 5A. SSP. FLORIDANUM
1. Leaf blades 6–15 cm long, the lobes tapered to sharply pointed tips (sometimes bluntly pointed in ssp. *schneckii*); ovary and fruit glabrous; bark gray to dark gray or brown, somewhat roughened, becoming deeply furrowed and sometimes with peeling ridges on older trees
    2. Leaf blades with the undersurface yellowish green to green; sinuses between the main leaf lobes mostly forming angles of greater than 90° . . . . 5B. SSP. NIGRUM
    2. Leaf blades with the undersurface pale green, bluish green, grayish green, or whitish, sometimes glaucous; sinuses between the main leaf lobes mostly forming angles of less than 90°
        3. Lower surface of the leaf blades glabrous or hairy, the lobes tapered to sharply pointed tips and often with secondary lobes or teeth . . . . 5C. SSP. SACCHARUM
        3. Lower surface of the leaf blades hairy along the veins, the lobes tapered but bluntly pointed or rounded at the tips, usually lacking or with very few secondary lobes or teeth . . . . 5D. SSP. SCHNECKII

5a. ssp. floridanum (Chapm.) Desmarais (southern sugar maple; Florida maple)
A. saccharum var. *floridanum* (Chapm.) Small & A. Heller
A. nigrum F. Michx. var. *floridanum* (Chapm.) Fosberg
A. floridanum (Chapm.) Pax
A. barbatum Michx.

Bark light gray, smooth, becoming shallowly furrowed and sometimes with scaly ridges on older trees. Leaf blades 3–10 cm long, the undersurface pale green to whitish-glaucous and usually also hairy, the lobes blunt to rounded at the tips, usually with few secondary lobes or teeth, the sinuses between the main leaf lobes forming angles of less than 90°, the margins often slightly curled under. Calyx with dense white hairs on the inner surface that often extend past the lobes. Flower stalks elongating to 3–6 cm, usually hairy. Ovary and young fruit hairy (becoming nearly glabrous at maturity). April–May.

Uncommon in southern and central Missouri (southeastern U.S. west to Oklahoma and Texas). Mesic upland forests, ledges and bases of bluffs, and banks of streams, rarely bottomland forests.

This uncommon taxon is perhaps the most distinctive element within the sugar maple complex in Missouri, by virtue of its beechlike bark and relatively small leaves. It is most common along the Atlantic and Gulf Coastal Plains and is at the edge of its range in Missouri. See the treatment of ssp. *schneckii* for discussion of specimens that have intermediate leaf characters between ssp. *floridanum* and ssp. *saccharum*.

5b. ssp. nigrum (F. Michx.) Desmarais (black maple)
A. saccharum var. *nigrum* (F. Michx.) Britton
A. saccharum var. *viride* (Schmidt) Voss
A. nigrum F. Michx.
A. nigrum f. *pubescens* Deam
A. nigrum var. *palmeri* Sarg.

Pl. 196 o; Map 808

Bark gray to dark gray or brown, somewhat roughened, becoming deeply furrowed and sometimes with peeling ridges on older trees. Leaf blades 6–15 cm long, the undersurface pale yellowish green to green and usually also hairy, the lobes tapered to sharply pointed tips, usually lacking or with very few secondary lobes or teeth, the sinuses between the main leaf lobes mostly forming angles of greater than 90°, the margins often slightly curled under. Calyx frequently hairy, but without dense white hairs on the inner surface that extend past the lobes. Flower stalks elongating to 5–10 cm, usually hairy. Ovary and young fruit glabrous. $2n=26$. April–May.

Scattered in northern Missouri, uncommon south of the Missouri River (northeastern U.S. west to Minnesota and Arkansas; Canada). Mesic to dry upland forests, margins of glades, ledges and bases of bluffs, and banks of streams.

Trees with the typical black maple morphology are most common in northern Missouri, but they may also be found occasionally elsewhere in the state. In addition to the characters in the key to subspecies above, such trees also have the petioles abruptly enlarged at the base and often with small, leaflike, stipular outgrowths. The leaves are duller on the upper surface than those of ssp. *saccharum* and tend to be both less deeply divided and less toothed along the margins (although these differences are difficult to quantify). For further discussion, see the treatment of ssp. *schneckii* and ssp. *saccharum*.

5c. ssp. saccharum (sugar maple)
  A. *saccharum* f. *glaucum* (Schmidt) Pax
  A. *nigrum* F. Michx. var. *glaucum* (Schmidt) Fosberg

Pl. 196 m, n; Map 808

Bark gray to dark gray or brown, somewhat roughened, becoming deeply furrowed and sometimes with peeling ridges on older trees. Leaf blades 6–15 cm long, the undersurface pale green, bluish green, grayish green, or whitish, sometimes glaucous, glabrous or hairy, the lobes tapered to sharply pointed tips, often with secondary lobes or teeth, the sinuses between the main leaf lobes mostly forming angles of less than 90°, the margins sometimes slightly curled under. Calyx frequently hairy, but without dense white hairs on the inner surface that extend past the lobes. Flower stalks elongating to 5–10 cm, usually hairy. Ovary and young fruit glabrous. $2n=26$. April–May.

Common throughout Missouri (eastern U.S. west locally to Idaho and Arizona; Canada). Mesic to dry upland forests, margins of glades, ledges and bases of bluffs, and banks of streams, rarely bottomland forests; also moist to dry, shaded, disturbed areas.

As noted above, many of the trees classified as ssp. *saccharum* in Missouri (particularly those colonizing drier upland forests) possess at least some characteristics of black maples, including leaf blades less strongly divided than the typical phase and with hairy undersurfaces. Such trees still have leaves with pale leaf undersurfaces and somewhat shiny upper surfaces, and they also lack the strongly expanded petiole bases of ssp. *nigrum*. This morphological race is the one most commonly interfering with oak regeneration in drier upland forest sites.

5d. ssp. schneckii (Rehder) Desmarais
  A. *saccharum* var. *schneckii* Rehder
  A. *saccharum* f. *schneckii* (Rehder) Deam
  A. *saccharum* f. *rugelii* (Pax) E.J. Palmer & Steyerm.
  A. *nigrum* F. Michx. var. *schneckii* (Rehder) Fosberg

Bark gray to dark gray or brown, somewhat roughened, becoming deeply furrowed and sometimes with peeling ridges on older trees. Leaf blades 6–15 cm long, the undersurface pale green, bluish green, grayish green, or whitish, sometimes glaucous and hairy along the veins, the lobes tapered but bluntly pointed or rounded at the very tip, usually lacking or with very few secondary lobes or teeth, the sinuses between the main leaf lobes mostly forming angles of less than 90°, the margins sometimes slightly curled under. Calyx frequently hairy, but without dense white hairs on the inner surface that extend past the lobes. Flower stalks elongating to 5–10 cm, usually hairy. Ovary and young fruit glabrous. April–May.

Uncommon in eastern and southern Missouri (Indiana to Wisconsin south to Tennessee and Missouri). Mesic to dry upland forests, margins of glades, ledges and bases of bluffs, and banks of streams.

Trees of ssp. *schneckii* tend to combine features of ssp. *floridanum,* ssp. *nigrum,* and ssp. *saccharum,* although visually they resemble ssp. *nigrum* most closely. They have the relatively shallowly lobed, dull leaves typical of ssp. *nigrum* but lack the yellowish green undersurfaces. The lobes are tapered but are blunt or even rounded at the very tips. In Missouri, trees are usually found as scattered individuals in areas where other subspecies are more common. Thus, there are some questions as to the status of this subspecies. Some authors, such as Gleason and Cronquist (1991) have expanded the definition of ssp. *schneckii* to include plants otherwise assignable to ssp. *saccharum* but with hairs on the leaf undersurfaces. However, this definition seems too broad and deviates from the original concept of ssp. *schneckii* (Desmarais, 1952; Steyermark, 1963); if it were adopted, nearly all Missouri sugar maples would have to be included in it.

AIZOACEAE (Fig-Marigold Family)

About 60 genera, about 2,500 species, nearly worldwide, but most diverse in the Old World, particularly South Africa and Australia.

The genera *Glinus* and *Mollugo* were treated in the family Aizoaceae by Steyermark (1963) and many earlier authors, but they are now considered by nearly all botanists to belong to a separate family, Molluginaceae (Bogle, 1970; Cronquist, 1981, 1991). Also, the monotypic genus *Geocarpon* was treated in the Aizoaceae by some earlier authors, but it is now generally accepted as belonging in the Caryophyllaceae (Palmer and Steyermark, 1950; Steyermark, 1963).

1. Trianthema L. (sea purslane)

Twenty to 25 species, nearly worldwide, mostly in tropical and warm-temperate regions.

Barker (1986) reported *Sesuvium verrucosum* Raf. from Missouri without citation of a locality or voucher specimen. This species is native to Africa and Australia and occurs as a weed in disturbed areas with saline soils from the southwestern United States eastward to Kansas, Oklahoma, and Texas. The genus *Sesuvium* differs from *Trianthema* in its ovaries with 3–5 locules, 3–5 styles, and numerous ovules, as well as its leaves lacking expanded stipular bases.

1. Trianthema portulacastrum L. (sea purslane, horse purslane)

Pl. 202 e, f; Map 809

Plants perennial herbs (sometimes woody at the base farther south), succulent, with a short taproot. Stems 15–100 cm long, highly branched from the base, ascending to spreading, glabrous or sparsely pubescent in longitudinal lines. Leaves opposite, with the two leaves at each node unequal in size, petiolate. Stipules fused to the petioles and appearing as narrow wings of tissue at the petiole base, the petioles thus appearing to wrap around the nodes. Leaf blades 1–4 cm long, obovate to broadly elliptic or nearly circular, flattened but thick, rounded or abruptly narrowed at each end, usually with a short point at the tip, glabrous or sparsely hairy. Flowers solitary or in small clusters in the leaf axils, deeply perigynous, each with a pair of minute bracts that are fused to the calyx, often hidden by and sometimes fused with the expanded petiole bases. Calyx 5-lobed, the lobes 2.0–2.5 mm long, arched inward over the flower top, green on the outer surface, tinged pink or purple on the inner surface. Petals absent. Stamens 5–10, the filaments fused to the calyx tube. Pistil 1 per flower, the ovary superior but sometimes appearing nearly inferior in fruit, consisting of 1 carpel, with 1 locule (sometimes incompletely 2-locular), the placentation parietal. Style 1(2), the stigma 1(2). Ovules 3–9. Fruits capsules, 3.5–4.5 mm long, cylindrical, with the margin extended into a pair of irregular wings at the tip, circumscissilely dehiscent by a ring at the middle, the uppermost 1 or 2 seeds usually dispersed with the cap, the other 2 or 3 seeds dispersing from the capsule base. Seeds 1.6–2.0 mm wide, somewhat flattened, broadly kidney-shaped (the embryo appearing curved or coiled), with a small aril, the surface wrinkled, reddish brown to black. $2n=26, 36$. May–October.

Introduced, uncommon in western Missouri (apparently native to Africa and Australia; introduced and weedy in the New World tropics; in the U.S. from Florida to California north to New Jersey, Oklahoma, and Missouri). Railroads and open, disturbed areas.

Although the leaves of this species are prepared as a vegetable in some parts of Africa and Asia, they are also widely used medicinally (ranging from cathartic to abortifacient) and are reputedly poisonous to livestock (Bogle, 1970).

AMARANTHACEAE (Amaranth Family)

Plants annual or perennial herbs (sometimes woody elsewhere), often monoecious or dioecious, not or only slightly succulent, often with a taproot, glabrous or hairy, often tinged with pink to purple pigmentation. Stems spreading to erect. Leaves alternate or opposite, simple, the margins entire or sometimes somewhat wavy (occasionally minutely sharply toothed in *Iresine*). Stipules absent (paired stipulelike axillary spines present in *Amaranthus spinosus*). Inflorescences axillary and/or terminal; dense spikes, spikelike racemes, or panicles, sometimes reduced to small, axillary clusters (globose heads or solitary flowers elsewhere), the main axis occasionally broadened and flattened (fasciated) with flowers across the surface. Flowers sessile or very short-stalked, with 1–3 small, papery to scalelike or hardened (sometimes appearing spine-tipped) bracts (1 bract and usually 2 additional bracteoles), imperfect or perfect, hypogynous. Calyx absent or more commonly of (2–)4 or 5 sepals, these free or fused either most of their length or only at the base, green and often somewhat hardened (sometimes appearing spine-tipped) or white, yellow, pink, red, or purple and papery, persistent at fruiting. Petals absent. Stamens (1–)4 or 5, absent or reduced to minute staminodes in pistillate flowers, the filaments sometimes fused, at least toward the base, the anthers attached toward their midpoints, usually yellow. Pistil 1 per flower (absent in staminate flowers), the ovary superior, consisting of 2 or 3 fused carpels, with 1 locule, the placentation usually basal. Style absent or 1, often very short, the stigmas 1–3, slender or capitate, occasionally lobed. Ovules 1 (several in *Celosia*). Fruits mostly capsules (occasionally indehiscent and achenelike), not winged at the tip, sometimes beaked, indehiscent or more commonly with irregular or circumscissile dehiscence. Seeds 1 (2–6 in *Celosia*), minute, often somewhat flattened, circular in outline or nearly so (the embryo appearing curved or coiled but not always easily observed). Sixty-five to 69 genera, 900–1,000 species, nearly worldwide, but most diverse in tropical and subtropical regions.

The Amaranthaceae are here treated in the traditional sense as a family separate from the Chenopodiaceae. However, a number of morphological and molecular phylogenetic studies (Rodman, 1990, 1993; Manhart and Rettig, 1994; Downie et al., 1997; see also Judd et al., 2002) have presented evidence to suggest that the Chenopodiaceae as thus circumscribed are paraphyletic; that is, that the genera of Amaranthaceae represent a specialized subgroup within the lineage of Chenopodiaceae rather than a separate sister clade. Because some of the conclusions of these papers are contradictory and a few relationships among genera are yet controversial (such as the placement of *Spinacia* L.), it seems premature to combine these families in a floristic treatment until more detailed studies can be completed. The morphological features that generally separate the Amaranthaceae from Chenopodiaceae include their stamens with the filaments fused basally (vs. free) and papery (vs. herbaceous) perianth and bracts, but numerous exceptions exist.

Members of the Amaranthaceae are nearly all wind-pollinated. Pollen grains of most Amaranthaceae and Chenopodiaceae are virtually indistinguishable morphologically, and the two families are usually lumped into a single pollen class in projects that monitor airborne spores and pollen for air quality and hay fever reports. In addition to members of the genera included below, some species of *Gomphrena* L., globe amaranth, are cultivated in gardens for their ornamental foliage and inflorescences.

In the keys and descriptions below, measurements of bract length refer to the bract, which often is larger and more conspicuous than the two bracteoles that usually are also present, although sometimes the three structures are essentially indistinguishable.

1. Leaves alternate
    2. Flowers imperfect (rarely a few perfect flowers present in monoecious species); stamens with the filaments free to the base; ovule 1 per ovary; style absent or very short, the stigmas tapered and slender; fruits 1-seeded
    .... 1. AMARANTHUS
    2. Flowers perfect; stamens with the filaments fused toward the base; ovules 2 to several per ovary; style well developed, the stigmas capitate; fruits 2–6-seeded .... 2. CELOSIA
1. Leaves opposite
    3. Inflorescences small, axillary clusters; stems spreading to very loosely ascending, pubescent with stellate hairs; leaf blades 0.5–3.0 cm long
    .... 5. TIDESTROMIA
    3. Inflorescences spikes or panicles; stems erect or strongly ascending, glabrous or pubescent with simple hairs; leaf blades (2–)3–14 cm long
        4. Leaves sessile or nearly so, the blades somewhat thickened or leathery, linear to narrowly oblong-lanceolate or narrowly oblanceolate, the undersurface densely pubescent with woolly hairs; sepals fused most of their length into a persistent tube, this becoming somewhat hardened after flowering .... 3. FROELICHIA
        4. Leaves short- to long-petiolate, the blades thin and herbaceous, lanceolate to more commonly ovate, the undersurface glabrous or sparsely and inconspicuously hairy; sepals free or rarely fused only at the very base, not becoming hardened after flowering .... 4. IRESINE

1. Amaranthus L. (pigweed, amaranth)

Plants annual, monoecious or dioecious (rarely a few perfect flowers present in monoecious species). Stems erect to spreading, glabrous or variously pubescent, green to yellowish white or reddish purple, often with pink to purple longitudinal lines or ridges. Leaves alternate, short- to long-petiolate. Leaf blades 0.8–20.0 cm long, variously shaped, usually pubescent with short, curved hairs when young, usually becoming nearly glabrous (except often along the margins) at maturity. Inflorescences axillary and/or terminal, dense spikes, spikelike racemes, panicles (often appearing as spicate branches arranged racemosely along a central spicate axis), or axillary clusters. Flowers imperfect. Sepals 1–5 or sometimes absent in pistillate flowers, free, more or less similar or differentiated into inner and outer series, often with an awnlike extension, this sometimes becoming somewhat hardened and spinelike at maturity, herbaceous to hardened or leathery, green, at least centrally, but often with thin, papery margins. Staminate flowers with the stamens 3–5, the filaments free. Pistillate flowers with the ovary ovoid. Ovule 1. Style absent or very short, the 2 or 3 stigmas tapered and slender, persistent. Fruits mostly with papery walls, ovoid or ellipsoid, with 2 or 3 small beaks, indehiscent or more commonly with irregular or circumscissile dehiscence, 1-seeded. Seeds often somewhat flattened, circular or nearly so in outline, rounded or angled along the rim, the surface smooth (sometimes somewhat roughened in *A. blitum* and *A. viridis*), shiny. Sixty to 70 species, nearly worldwide.

The genus *Amaranthus* has great economic importance. Several of the Missouri species (the weedy amaranths) are serious weeds in crop fields. All of the pigweeds produce copious quantities of potentially allergenic, wind-dispersed pollen that are major causes of hay fever during some times of the year. On the plus side, however, certain species have a long history of cultivation for food. Three species, *A. caudatus, A. cruentus,* and *A. hypochondriacus,* have histories of cultivation in Latin America as pseudo–grain crops that stretch back to long before European colonization of the Americas, and these cereal amaranths also subsequently became cultivated in parts of Europe and Asia (Sauer, 1967; Robertson, 1981; Costea et al., 2001 a, b). Red-pigmented forms of these species and others, principally *A. hybridus,* also have been cultivated as garden ornamentals, and in Latin America they are used to produce a red dye for religious ceremonies (Robertson, 1981; Heiser, 1985). Yet other species, such as *A. tricolor,* have been used widely for food, fresh as greens and boiled as potherbs, and also as a green supplement to some livestock feeds. That these uses persist today is evidenced by the fact that all of the species mentioned have been recorded outside of cultivation in Missouri. In fact, grain amaranths are available to consumers in many grocery and health food stores, the seeds dried and ground into flour, popped for use in baked goods and cereals, and powdered as a component of health food beverages (Robertson, 1981). They are quite nutritious, a good source of protein, and contain high levels of lysine. Wild food enthusiasts are cautioned, however, that amaranths growing in the wild occasionally can absorb and accumulate in their foliage toxic substances present in the soil.

The species of *Amaranthus* considered native in Missouri and the central portion of the United States occur in a variety of habitats with moist, bare soil. The natural limits of the distributional ranges of these species are not well understood, as they have been spread in recent times by agriculture and other human activities as well as by natural means. These plants are primary colonizers of both habitats prone to frequent natural disturbance, especially in the floodplains of major rivers and streams, and also anthropogenically altered sites, like crop fields and railroad embankments. Distributional limits of these disturbophiles had already expanded both westward across the Great Plains and eastward to the Atlantic seaboard as a result of European colonization by the time that botanists began paying attention to the biogeography of such species (Sauer, 1957, 1967, 1972).

The weedy amaranths frequently grow in mixed populations, and a number of hybrids have been reported. These hybrids are sterile, producing inflorescences with many bracts but with the flowers lacking or vestigial, or in some cases with reduced flowers having shrunken, misshapen, or absent fruits, so detecting such plants is not difficult. However, evaluating the presumed parentage of such sterile individuals with confidence is often not possible. Carl Sauer, a specialist on the genus, determined one or a few specimens as representing each of the following hybrid combinations in Missouri: *A. hybridus* × *A. palmeri, A. hybridus* × *A. retroflexus, A. hybridus* × *A. rudis, A. hybridus* × *A. tuberculatus,* and *A. retroflexus* × *A. rudis* (note, however that *A. rudis* is reduced to synonymy under *A. tuberculatus* in the present treatment). Other combinations are to be expected where two or more species grow together, but apparently they are rare, even in large mixed populations.

Identification of *Amaranthus* species is difficult, especially when only staminate flowers are present, not only because of the presence of occasional hybrids but also because determination depends on observation of details of the small, morphologically similar flowers and fruits. Within an inflorescence, the numerous flowers are at various stages of development, and fruits often continue to mature in the plant press during drying. Taxonomists traditionally have divided the genus into two main subgroups, the dioecious subgenus *Acnida* (L.) Aellen ex K.R. Robertson and the monoecious subgenus *Amaranthus.* Recently, a few authors (Costea et al., 2001a) have segregated subgenus *Albersia* (Kunth) Gren. & Godr. from subgenus *Amaranthus,* based on its mostly axillary clusters of flowers (vs. mostly terminal spikes or panicles), indehiscent (vs. circumscissile) fruits, and minute differences in leaf and seed anatomy, but most botanists still treat this group as section *Blitopsis* Dumort. of subgenus *Amaranthus.* In spite of the instinctive appeal of these groupings, the validity of maintaining such infrageneric taxa is drawn into question by the existence of various hybrids between them.

1. Stems with conspicuous paired spines at most nodes . . . . 12. A. SPINOSUS
1. Stems unarmed (but bracts and sepals sometimes becoming spinelike at maturity)
    2. Plants dioecious
        3. Flowers pistillate
            4. Sepals absent or irregularly 1 or 2, linear to lanceolate; stigmas 3 or 4 . . . . 14. A. TUBERCULATUS
            4. Sepals 5, at least the inner ones oblanceolate to spatulate; stigmas 2(3)
                5. Sepals rounded or bluntly pointed at the tip, the midrib not or only slightly extending beyond the main body as a minute, sharp point; outer sepals 2.0–2.5 mm long; inner sepals only slightly shorter than the outer sepals and bracts . . . . 2. A. ARENICOLA
                5. At least the outer sepals sharply pointed at the tip, the midrib extending beyond the main body as a short awn, usually spinelike at maturity; outer sepals 3–4 mm long; inner sepals conspicuously shorter than the outer sepals and bracts . . . . . 9. A. PALMERI
        3. Flowers staminate
            6. Bracts 4–6 mm long, slightly to more commonly conspicuously longer than the sepals . . . . 9. A. PALMERI
            6. Bracts 1.5–2.5 mm long, shorter than to about as long as the sepals
                7. Bracts with the midrib not or only slightly extending beyond the main body as a minute, sharp point . . . . 2. A. ARENICOLA
                7. Bracts with the midrib extending beyond the main body as a short awn, usually somewhat spinelike at maturity
                . . . . 14. A. TUBERCULATUS
    2. Plants monoecious (the staminate and pistillate flowers either on different parts of the plant or intermingled in each inflorescence)
        8. Main inflorescences mostly small, axillary clusters (short, terminal spikes occasionally also present)
            9. Stems spreading to less commonly loosely ascending, often forming dense mats; bracts of the pistillate flowers about as long as the 4 or 5 sepals . . . . 3. A. BLITOIDES
            9. Stems loosely to strongly ascending or erect, often forming dense, irregularly globose masses; bracts of the pistillate flowers longer than the 3 sepals
                10. Stems loosely to strongly ascending, often forming dense, irregularly globose masses; leaf blades 0.5–4.0(–8.0) cm long, elliptic to obovate, rounded or shallowly and minutely notched at the tip; bracts 2.0–2.6 mm long; sepals 0.8–2.3 mm long . . . . 1. A. ALBUS
                10. Stems erect or strongly ascending, not forming globose masses; leaf blades 3–15 cm long, ovate to broadly triangular-ovate, narrowed or tapered to a bluntly or sharply pointed tip (often minutely notched at the very tip); bracts 3–6 mm long; sepals 2–4 mm long . . . . 13. A. TRICOLOR
        8. Main inflorescences elongate terminal spikes, these often grouped into panicles (additional axillary clusters, spikes, and/or panicles also often present)

11. Sepals 2 or 3
     12. Leaf blades truncate to noticeably notched at the tip (the midvein sometimes extending as a short, sharp point) . . . . 4. A. BLITUM
     12. Leaf blades narrowed to tapered to a bluntly or sharply pointed tip (sometimes minutely notched at the very tip)
          13. Fruits 1.4–1.7 mm long, indehiscent; seeds angled along the rim; sepals 0.9–1.2 mm long, shorter than the fruit, oblanceolate to oblong-oblanceolate; bracts 0.6–0.9 mm long, shorter than the sepals and fruit; axillary inflorescences elongate spikes . . . . 15. A. VIRIDIS
          13. Fruits 1.8–2.5 mm long, with circumscissile dehiscence; seeds rounded along the rim; sepals 2–4 mm long, as long as or longer than the fruit, lanceolate to linear-lanceolate or ovate to elliptic-ovate; bracts 3.0–7.5 mm long, longer than the sepals and fruit; axillary inflorescences mostly more or less globose clusters
               14. Bracts and sepals lanceolate to narrowly oblong-lanceolate or linear-lanceolate, the thin, papery margins only slightly broader than the strongly thickened green midrib . . . . 10. A. POWELLII
               14. Bracts and sepals ovate to elliptic-ovate, with thin, papery margins much broader than the slightly thickened green midrib
                    . . . . 13. A. TRICOLOR
11. Sepals (4)5
     15. Pistillate flowers with the main body of the bracts (excluding the awnlike or spinelike extension of the midrib) conspicuously longer than the fruits; inflorescence usually stiffly erect at the tip
          16. Pistillate flowers with the sepals narrowed or tapered to a sharply pointed, stiff, erect tip, often tapered to a short, sharp extension of the midrib; staminate flowers usually with 3 stamens; terminal inflorescence a solitary spike or panicle of few long spikes (these branching from near the panicle base) . . . . 10. A. POWELLII
          16. Pistillate flowers with the sepals rounded, truncate or shallowly notched at the soft, outward-curved tip, sometimes with an abrupt, minute, sharp extension of the midrib; staminate flowers with (4)5 stamens; terminal inflorescence usually a panicle with numerous clusters of short, dense spikes (these branching along most of the panicle axis)
               . . . . 11. A. RETROFLEXUS
     15. Pistillate flowers with the main body of the bracts (excluding the awnlike or spinelike extension of the midrib) shorter than to slightly longer than the fruits; inflorescences stiffly erect or nodding or drooping at the tip
          17. Pistillate flowers with the bracts (including the awnlike or spinelike extension of the midrib) conspicuously longer than the fruits; sepals sharply pointed at the tip; inflorescences dull green, occasionally dull reddish-tinged . . . . 7. A. HYBRIDUS
          17. Pistillate flowers with the bracts (including the short, awnlike extension of the midrib, if present) shorter than to about as long as the fruits; sepals of at least the pistillate flowers rounded to bluntly pointed at the tip; inflorescences bright green, yellow, or red
               18. Main inflorescence with the tip stiffly straight, erect or nearly so
                    . . . . 8. A. HYPOCHONDRIACUS
               18. Main inflorescence with the tip nodding or drooping

19. Sepals of the pistillate flowers 1.5–2.0 mm long, obovate to spatulate, noticeably overlapping, at least the outer ones curved outward at the tip; stigmas spreading . . . . 5. A. CAUDATUS

19. Sepals of the pistillate flowers 1.0–1.6 mm long, narrowly oblong to narrowly oblong-elliptic, slightly and inconspicuously overlapping, erect at the tip; stigmas erect or nearly so . . . . 6. A. CRUENTUS

1. Amaranthus albus L. (tumbleweed)
A. albus var. *pubescens* (Uline & W.L. Bray) Fernald

Pl. 198 a, b; Map 810

Plants monoecious. Stems 30–100 cm long, loosely to strongly ascending, often forming dense, irregularly globose masses, glabrous or sparsely to moderately pubescent with inconspicuous, mostly multicellular hairs, unarmed. Leaves short- to long-petiolate. Leaf blades 0.5–4.0(–8.0) cm long, elliptic to obovate, rounded or shallowly and minutely notched at the tip (the midvein sometimes extending as a minute, sharp point), tapered at the base, glabrous. Inflorescences grayish green to green, all or nearly all axillary; mostly dense, small, globose clusters, with a short, terminal spike occasionally also present. Bracts 2.0–2.6 mm long, at least those of the pistillate flowers conspicuously longer than the sepals, lanceolate to oblong-lanceolate, narrowed or tapered to a sharply pointed tip, with a somewhat thickened midrib, green or sometimes with narrow, thin, papery margins, the midrib extending beyond the main body as a short awn, somewhat spinelike at maturity. Staminate flowers with 3 more or less similar sepals, these 1.2–2.3 mm long, erect or very slightly outward-curved, lanceolate to narrowly oblong-elliptic, narrowed or tapered to a minute, awnlike extension of the midrib at the tip. Stamens 3. Pistillate flowers with 3 more or less similar sepals (the third sepal occasionally slightly longer than the other 2), these 0.8–1.2 mm long, erect or very slightly outward-curved, lanceolate to narrowly oblong-elliptic, narrowed or tapered to a blunt or sharp point. Stigmas 3, more or less erect from a thickened base. Fruits 0.9–1.5 mm long, with circumscissile dehiscence, the surface often finely wrinkled when dry. Seeds 0.7–1.0 mm in diameter, rounded along the rim, the surface reddish brown to black. $2n=32$. July–October.

Scattered and sporadic, mostly in the western and southern halves of the state (native of North America; introduced widely in Central America, South America, Europe, Asia, and Africa). Banks of streams and rivers, exposed bases of bluffs, and open-soil areas of upland prairies; also roadsides, railroads, margins of crop fields, fallow fields, and open, disturbed areas.

The original distributional limits of this species are poorly understood, but it is generally considered native in Missouri. Hairy plants are encountered occasionally nearly throughout the North American range of the species. These have been called var. *pubescens* by some authors (Steyermark, 1963) but are unworthy of formal taxonomic recognition.

2. Amaranthus arenicola I.M. Johnst.
(sandhills pigweed, sandhills amaranth)

Pl. 197 e, f; Map 811

Plants dioecious. Stems 50–200 cm long, erect or ascending, glabrous or nearly so, unarmed. Leaves mostly long-petiolate. Leaf blades 1–8 cm long, narrowly oblong-lanceolate to oblong-elliptic, narrowed or tapered to a bluntly or sharply pointed tip, narrowed or tapered at the base, glabrous. Inflorescences dull or grayish green, occasionally dull reddish-tinged, mostly terminal, the axillary inflorescences mostly elongate spikes, the terminal inflorescence a panicle with few to numerous ascending branches, the flowers often grouped into discontinuous clusters or regions along the basal portions of the spikes, the tip usually straight or nearly so at maturity, the main axis and branches glabrous or nearly so. Bracts 1.5–2.5 mm long, shorter than (in staminate plants) to about as long as (in pistillate plants) the sepals, lanceolate to narrowly ovate, narrowed or tapered to a sharply pointed tip, with a somewhat thickened green midrib and relatively broad, thin, papery margins, the midrib not or only slightly extending beyond the main body as a minute, sharp point. Staminate flowers with 5 more or less similar sepals, these 3–5 mm long, erect or very slightly outward-curved, lanceolate to oblong-elliptic, narrowed to a minute, sharp point at the tip, the inner ones otherwise bluntly pointed to shallowly and minutely notched, the outer ones bluntly to sharply pointed. Stamens 5. Pistillate flowers with 5 sepals, these outward-curved at the tip, oblanceolate to spatulate, rounded or bluntly pointed at the tip, the midrib not or only slightly extending beyond the main body as a minute, sharp point, the inner ones 1.5–2.0 mm long, the outer ones 2.0–2.5 mm long, bluntly pointed at the tip. Stigmas 2(3), spread-

810. Amaranthus albus 811. Amaranthus arenicola 812. Amaranthus blitoides

ing. Fruits 1.4–1.7 mm long, circumscissilely dehiscent at about the midpoint, the surface smooth when dry. Seeds 1.0–1.3 mm in diameter, rounded along the rim, the surface reddish brown to black. $2n=32$. June–October.

Introduced, known only from the St. Louis metropolitan areas (native of Iowa to Wyoming south to Texas and New Mexico; introduced eastward to Virginia and New Jersey and westward to California). Railroads and open, sandy, disturbed areas.

The name *A. torreyi* (A. Gray) Benth. ex S. Watson, used for this species by most earlier authors (Steyermark, 1963), was originally described based on a mixture of specimens now referred to several western *Amaranthus* species and long misapplied to *A. arenicola* (Sauer, 1955).

3. Amaranthus blitoides S. Watson (tumbleweed, prostrate amaranth, spreading pigweed)

Pl. 198 f; Map 812

Plants monoecious. Stems 10–70 cm long, spreading to less commonly loosely ascending, often forming dense mats, glabrous or sparsely pubescent with inconspicuous, mostly multicellular hairs, unarmed. Leaves short- to long-petiolate. Leaf blades 0.5–2.0(4.0) cm long, oblong-elliptic to obovate, rounded or shallowly notched at the tip (the midvein sometimes extending as a minute, sharp point), tapered at the base, glabrous. Inflorescences grayish green to green, all or nearly all axillary; mostly dense, small, globose clusters, with a short, terminal spike rarely also present. Bracts 1.0–2.5 mm long, those of the pistillate flowers about as long as the sepals, lanceolate to oblong-lanceolate, narrowed or tapered to a sharply pointed tip, with a somewhat thickened midrib, green or sometimes with narrow, thin, papery margins, the midrib sometimes extending beyond the main body as a minute awn, somewhat spine-like at maturity. Staminate flowers with 4 or 5 more or less similar sepals, these 1.3–2.0 mm long, erect or very slightly outward-curved, lanceolate to narrowly oblong-elliptic, narrowed or tapered to a minute, awnlike extension of the midrib at the tip. Stamens 3. Pistillate flowers with 4 or 5 sepals (the inner sepals somewhat shorter and narrower than the outer ones), these 1.2–2.7 mm long, erect or somewhat outward-curved, lanceolate to ovate or narrowly oblong-elliptic, narrowed or tapered to a blunt or sharp point. Stigmas 3, spreading. Fruits 1.5–2.1 mm long, with circumscissile dehiscence, the surface smooth or nearly so when dry. Seeds 1.3–1.7 mm in diameter, rounded along the rim, the surface black. $2n=32$. July–October.

Scattered nearly throughout Missouri, but uncommon to absent in the southeastern quarter of the state (western U.S. [including Alaska] east to North Dakota and Texas; Canada; introduced eastward to Maine and Florida, as well as to Mexico, Caribbean Islands, Europe, and Asia). Banks of streams and rivers, bases of bluffs, and talus slopes; also roadsides, railroads, pastures, and open, disturbed areas.

Steyermark (1963) and many earlier authors called this species *A. graecizans* L., but that name has been shown to apply to a species native to the Mediterranean region with only two North American records (from New Jersey), apparently from plants introduced in ship's ballast during the nineteenth century (Sauer and Davidson, 1961; Costea et al., 2001b).

4. Amaranthus blitum L. ssp. **emarginatus** (Moq. ex Uline & W.L. Bray) Carretero, Muñoz Garm. & Pedrol

Map 813

Plants monoecious. Stems 20–70 cm long, spreading or ascending, glabrous or nearly so, unarmed. Leaves mostly long-petiolate. Leaf

813. Amaranthus blitum   814. Amaranthus caudatus   815. Amaranthus cruentus

blades 1–4(–6) cm long, oblong-obovate to spatulate, truncate to noticeably notched at the tip (the midvein sometimes extending as a short, sharp point), tapered at the base, glabrous. Inflorescences yellowish green, occasionally reddish-tinged, axillary and terminal; the axillary inflorescences mostly dense, small, globose clusters, the terminal inflorescence a spike or panicle with few to rarely numerous ascending branches; the flowers often grouped into discontinuous clusters or regions along the basal portions of the spikes; the tip often somewhat curved or nodding; the main axis and branches glabrous or nearly so. Bracts 1.5–1.8 mm long, shorter than to about as long as the sepals, narrowly ovate to ovate, narrowed or tapered to a sharply pointed tip, with a somewhat thickened green midrib and relatively narrow, thin, papery margins, the midrib not or only slightly extending beyond the main body as a minute, sharp point. Staminate flowers with 2(3) more or less similar sepals, these 1.0–1.8 mm long, erect or very slightly outward-curved, lanceolate to oblong-ovate, narrowed or tapered to a blunt or sharp point at the tip. Stamens 2 or 3. Pistillate flowers with 2 or 3 more or less similar sepals (the third sepal occasionally reduced), these 1.0–1.8 mm long, erect or very slightly outward-curved, lanceolate to oblong-ovate, narrowed or tapered to a blunt or sharp point. Stigmas (2)3, spreading. Fruits 1.4–2.0 mm long, indehiscent, the surface smooth when dry. Seeds 0.8–1.2 mm in diameter, rounded along the rim, the surface reddish brown to black. $2n=34$. July–October.

Introduced, known thus far from a single specimen from Cape Girardeau County (native possibly of Europe, Asia, Africa; introduced widely in tropical regions and sporadically in the eastern U.S.). Weed in a greenhouse.

This taxon was first reported for Missouri by Costea et al. (2001b). These authors treated *A. blitoides* as consisting of three subspecies, with ssp. *oleraceus* (L.) Costea (previously known as *A. lividus* L.) representing a robust cultigen derived from ssp. *blitum* and grown as a vegetable crop in the Old World. The two wild subspecies are also widespread weeds. The ssp. *blitum* has larger seeds (1.1–1.8 mm) and rounded to truncate cotyledons, whereas ssp. *emarginatus* has smaller seeds (0.8–1.1 mm) and bluntly to sharply pointed cotyledons. Within ssp. *emarginatus*, these authors further distinguished two varieties: var. *emarginatus*, with spreading stems, smaller leaves, and mostly axillary inflorescences; and var. *pseudogracilis*, with ascending stems, larger leaves, and usually mostly terminal inflorescences. The latter variety apparently is found in the United States mostly as a weed in crop fields and greenhouses, whereas var. *emarginatus* was said to be more widely distributed in moist, sandy, disturbed areas, often along bodies of water. Whether these varieties are merely ecological forms or represent true genetic variants is still not fully understood.

For a discussion of differences between *A. blitum* and the closely related *A. viridis*, see the treatment of that species.

5. Amaranthus caudatus L. (purple amaranth, love-lies-bleeding, tassel flower)

Pl. 198 e; Map 814

Plants monoecious. Stems 30–200 cm long, erect or ascending, glabrous or nearly so, unarmed. Leaves long-petiolate. Leaf blades 2–20 cm long, lanceolate to ovate or elliptic, narrowed or tapered to a bluntly or sharply pointed tip, narrowed or tapered at the base, the surfaces glabrous or the undersurface sparsely pubescent mostly along the veins with inconspicuous, mostly crinkled, multicellular hairs. Inflorescences usually red, less commonly mostly bright green, axillary and terminal, the axillary inflorescences short to long spikes, the terminal inflorescence usually a panicle with numerous clusters of short to long, dense spikes (these branching along mostly the lower half of the panicle axis), the flowers mostly continuous along the spikes, the tip curved or nodding, the main axis and branches sparsely to moderately pubescent with mostly crinkled, multicellular hairs. Bracts 1.7–2.5 mm long, shorter than to about as long as the fruits, lanceolate to ovate, narrowed or tapered to a sharply pointed tip, with a slightly thickened green midrib and broad, thin, papery margins, the midrib extending beyond the main body as a short awn, often somewhat spinelike at maturity. Staminate flowers with 5 more or less similar sepals, these 1.7–2.5 mm long, straight, oblong-elliptic to ovate, narrowed or tapered to a bluntly or more commonly sharply pointed tip, usually with a minute, awnlike extension of the midrib. Stamens 5. Pistillate flowers with 5 more or less similar sepals, these 1.5–2.0 mm long, noticeably overlapping, at least the outer ones outward-curved at the tip, oblong-obovate to spatulate, rounded to bluntly pointed, often with a minute, awnlike extension of the midrib. Stigmas 3, spreading. Fruits 1.6–2.5 mm long, with circumscissile dehiscence, the surface smooth or finely wrinkled above the midpoint when dry. Seeds 0.9–1.2 mm in diameter, angled along the rim, the surface pale whitish yellow, often reddish-tinged, especially along the rim, occasionally reddish brown. $2n=32, 34$. July–October.

Introduced, uncommon, known thus far only from the city of St. Louis (originated in South America, widely cultivated in tropical and warm-temperate regions, escaping sporadically in the U.S. and Canada). Gardens and open, disturbed areas.

Amaranthus caudatus is one of the cultigens derived long ago in the Andean region from selected strains of *A. hybridus*. It is one of the principal crop amaranths, but in the United States it is cultivated more commonly as an ornamental for its long, drooping, red inflorescences.

6. Amaranthus cruentus L.

Map 815

Plants monoecious. Stems 40–200 cm long, erect or ascending, sparsely to moderately pubescent toward the tip with inconspicuous, mostly crinkled, multicellular hairs, unarmed. Leaves long-petiolate. Leaf blades 2–25 cm long, lanceolate to ovate or elliptic, narrowed or tapered to a usually sharply pointed tip, narrowed or tapered at the base, the surfaces glabrous or the undersurface sparsely pubescent mostly along the veins with inconspicuous, mostly crinkled, multicellular hairs. Inflorescences usually red, less commonly bright green or yellow, axillary and terminal, the axillary inflorescences short to long spikes, the terminal inflorescence usually a panicle with numerous clusters of short to long, dense spikes (these branching along most of the panicle axis), the flowers mostly continuous along the spikes, the tip curved or nodding, the main axis and branches moderately to densely pubescent with mostly crinkled, multicellular hairs. Bracts 1.2–2.0 mm long, shorter than to about as long as the fruits, lanceolate to ovate, narrowed or tapered to a sharply pointed tip, with a slightly thickened green midrib and broad, thin, papery margins, the midrib extending beyond the main body as a short awn, often somewhat spinelike at maturity. Staminate flowers with 5 more or less similar sepals, these 1.0–1.7 mm long, straight, oblong-elliptic to oblong-ovate, narrowed or tapered to a bluntly or more commonly sharply pointed tip, usually with a minute, awnlike extension of the midrib. Stamens 5. Pistillate flowers with 5 more or less similar sepals, these 1.0–1.6 mm long, slightly and inconspicuously overlapping, erect at the tip, narrowly oblong to narrowly oblong-elliptic, rounded to bluntly pointed, often with a minute, awnlike extension of the midrib. Stigmas (2)3, erect or nearly so. Fruits 1.4–1.8 mm long, with circumscissile dehiscence, the surface smooth or finely wrinkled above the midpoint when dry. Seeds 1.0–1.3 mm in diameter, angled along the rim, the surface dark brown. $2n=32, 34$. July–October.

Introduced, uncommon and sporadic in Missouri (originated in South America, widely cultivated in tropical and warm-temperate regions, escaping sporadically in the U.S.). Gardens and open, disturbed areas.

Amaranthus cruentus is one of the cultigens derived long ago in the Andean region from selected strains of *A. hybridus*. Like *A. caudatus,* it is one of the principal crop amaranths, but in the United States it is cultivated more commonly as an ornamental for its long, drooping, red inflorescences.

Steyermark (1963) reported two specimens from Boone County as representing escapes of *A. hypochondriacus* L. (as *A. hybridus* var. *hypochondriacus*), but these specimens subsequently were redetermined as *A. cruentus* (Costea et al., 2001a).

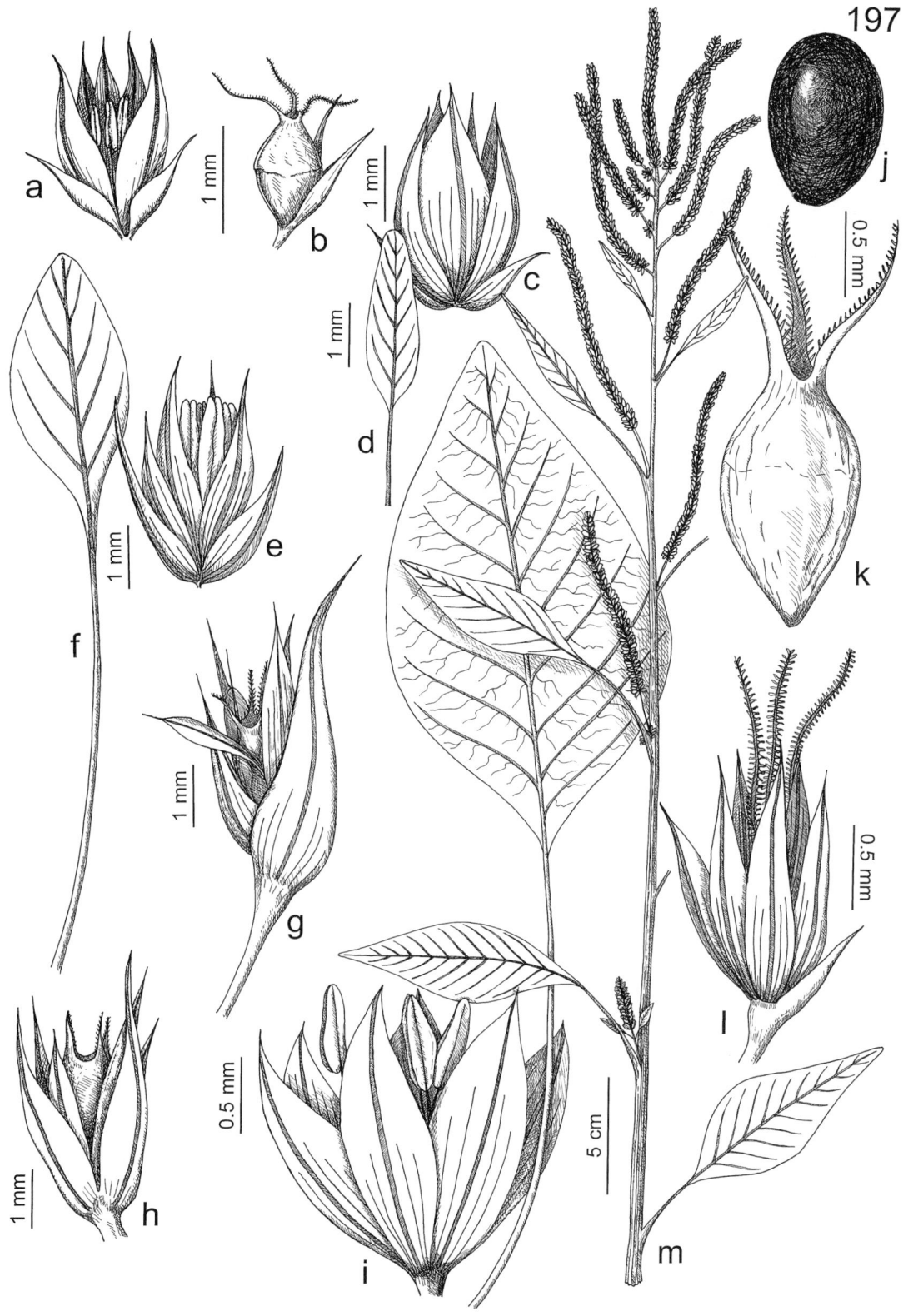

Plate 197. Amaranthaceae. *Amaranthus tuberculatus*, **a)** flower, **b)** fruit. *Amaranthus palmeri*, **c)** flower, **d)** leaf. *Amaranthus arenicola*, **e)** staminate flower, **f)** leaf. *Amaranthus retroflexus*, **g)** fruit. *Amaranthus powellii*, **h)** fruit. *Amaranthus tuberculatus* (*rudis* phase), **i)** staminate flower, **j)** seed, **k)** fruit, **l)** pistillate flower, **m)** inflorescence and lower leaf.

816. Amaranthus hybridus
817. Amaranthus hypochondriacus
818. Amaranthus palmeri

7. Amaranthus hybridus L. (green amaranth, slender pigweed)

Pl. 198 g, h; Map 816

Plants monoecious. Stems 30–200 cm long, erect or ascending, sparsely (toward the base) to densely (toward the tip) pubescent with mostly crinkled, multicellular hairs, unarmed. Leaves short- to long-petiolate. Leaf blades 2–15 cm long, lanceolate to ovate or elliptic, narrowed or tapered to a bluntly or sharply pointed tip (often minutely notched at the very tip), narrowed or tapered at the base, the surfaces sparsely to moderately pubescent mostly along the veins with inconspicuous, mostly crinkled, multicellular hairs, the upper surface sometimes glabrous or nearly so. Inflorescences dull or grayish green, occasionally dull reddish-tinged, axillary and terminal, the axillary inflorescences short spikes or less commonly dense globose clusters, the terminal inflorescence usually a panicle with numerous clusters of short, dense spikes (these branching along most of the panicle axis), the flowers mostly continuous along the spikes, the tip often curved or nodding, the main axis and branches moderately to densely pubescent with mostly crinkled, multicellular hairs. Bracts 2–4 mm long, the main body (excluding the awn) shorter than to slightly longer than the fruits, but the entire structure (including the awnlike or spinelike extension of the midrib) conspicuously longer than the fruits, lanceolate to ovate, narrowed or tapered to a sharply pointed tip, with a strongly thickened green midrib and narrow or broad, thin, papery margins, the midrib usually extending beyond the main body as a short awn, spinelike at maturity. Staminate flowers with (4)5 more or less similar sepals, these 1.5–2.3 mm long, straight, elliptic to oblong-ovate, narrowed or tapered to a sharply pointed tip, usually tapered to a short, awnlike extension of the midrib. Stamens (4)5. Pistillate flowers with (4)5 more or less similar sepals, these 1.5–2.3 mm long, straight or nearly so, lanceolate to oblong-elliptic or oblong-ovate, narrowed or tapered to a relatively stiff, sharply pointed tip, often tapered to a short, awnlike extension of the midrib. Stigmas 3, erect or less commonly somewhat spreading. Fruits 1.2–2.0 mm long, with circumscissile dehiscence, the surface finely wrinkled above the midpoint when dry. Seeds 0.8–1.2 mm in diameter, rounded along the rim, the surface reddish brown to more commonly black. $2n=32$. July–October.

Scattered nearly throughout the state (nearly throughout North America, Central America, South America; introduced in Europe, Asia, Africa, Australia). Banks of streams and rivers, bottomland forests, margins of sloughs, and bases of bluffs; also fallow fields, crop fields, gardens, pastures, levees, farmyards, roadsides, railroads, and open, disturbed areas.

8. Amaranthus hypochondriacus L. (Prince's feather)

A. hybridus L. var. *hypochondriacus* (L.) B.L. Rob.

Map 817

Plants monoecious. Stems 30–250 cm long, erect or ascending, glabrous or sparsely to moderately pubescent toward the tip with inconspicuous, mostly crinkled, multicellular hairs, unarmed. Leaves long-petiolate. Leaf blades 2–15 cm long, lanceolate to ovate or elliptic, narrowed or tapered to a sharply or rarely bluntly pointed tip, narrowed or tapered at the base, the surfaces glabrous or the undersurface sparsely pubescent mostly along the veins with inconspicuous, mostly crinkled, multicellular hairs. Inflorescences usually red or yellow, less commonly mostly bright green, axillary and terminal, the axillary inflorescences short to long spikes, the terminal inflorescence usually a panicle with few to many clusters of short to long, dense spikes (these branching along mostly the lower half of the panicle axis), the flowers mostly continuous along the spikes, the tip stiffly straight or nearly so, the main axis and branches moderately pubescent with mostly crinkled, multicellular hairs. Bracts 2.6–3.4 mm long, about as long as to slightly longer than the fruits, lanceolate to ovate, narrowed or tapered to a sharply pointed tip, with a moderately thickened green midrib and broad, thin, papery margins, the midrib extending beyond the main body as a short awn, often somewhat spinelike at maturity. Staminate flowers with (4)5 more or less similar sepals, these 3.0–3.3 mm long, straight, narrowly oblong to ovate, narrowed or tapered to a bluntly or more commonly sharply pointed tip, usually with a minute, awnlike extension of the midrib. Stamens 5. Pistillate flowers with (4)5 more or less similar sepals, these 1.7–2.3 mm long, slightly and inconspicuously overlapping, more or less straight at the tip, oblong-elliptic to ovate, bluntly to more commonly pointed, usually with a short, awnlike extension of the midrib. Stigmas 3, spreading from a thickened, erect base. Fruits 2.0–2.4 mm long, with circumscissile dehiscence, the surface smooth or finely wrinkled when dry. Seeds 1.0–1.3 mm in diameter, angled along the rim, the surface pale dark brown to black or less commonly whitish yellow. $2n=32$. July–October.

Introduced, known thus far only from a single historical collection from Jackson County (originated in Latin America, widely cultivated in tropical and warm-temperate regions, escaping sporadically in the U.S.). Gardens and open, disturbed areas.

As noted in the treatment of *A. cruentus*, Steyermark's (1963) original report of this taxon was based on two Boone County specimens that subsequently were redetermined as *A. cruentus*. However, Costea et al. (2001a) reported a new record from Jackson County, which Steyermark had misdetermined earlier as *A. caudatus*. *Amaranthus hypochondriacus* is a principal crop amaranth, one of the cultigens derived long ago in Latin America, probably from selected strains of *A. powellii* or, more likely, from hybrids between *A. cruentus* and *A. powellii*.

9. Amaranthus palmeri S. Watson

Pl. 197 c, d; Map 818

Plants dioecious. Stems 30–100(–150) cm long, erect or ascending, glabrous (rarely hairy elsewhere), unarmed. Leaves mostly long-petiolate. Leaf blades 1–10 cm long, lanceolate to elliptic or ovate, narrowed or short-tapered to a bluntly or sharply pointed tip, rounded or narrowed at the base, glabrous. Inflorescences dull or dark green, axillary and terminal, the axillary inflorescences mostly elongate spikes, the terminal inflorescence a panicle with few to numerous ascending branches, the tip often somewhat curved or nodding at maturity, the main axis and branches glabrous or nearly so. Bracts 4–6 mm long, slightly to more commonly conspicuously longer than the sepals, narrowly ovate to ovate, tapered to a sharply pointed tip, with a somewhat thickened green midrib and relatively broad, thin, papery margins, the midrib extending beyond the main body as a conspicuous, spinelike awn. Staminate flowers with 5 sepals, these lanceolate to oblong-elliptic, erect or very slightly outward-curved, the inner ones 2.5–3.0 mm long, bluntly pointed or often minutely notched at the tip and with the midrib extending only slightly as a minute, sharp point, the outer ones 3.5–4.0 mm long, narrowed or tapered to a sharply pointed tip, the midrib extending beyond the main body as a short, spinelike awn. Stamens 5. Pistillate flowers with 5 sepals, these outward-curved at the tip, the inner ones 2–3 mm long, oblanceolate to spatulate, rounded or bluntly pointed at the tip, the midrib not or only slightly extending beyond the main body as a minute, sharp point, the outer ones 3–4 mm long, sharply pointed at the tip, the midrib extending beyond the main body as a short, spinelike awn. Stigmas 2(3), spreading. Fruits 1.5–2.2 mm long, circumscissilely dehiscent at about the midpoint, the surface somewhat wrinkled when dry. Seeds 1.0–1.3 mm in diameter, rounded along the rim, the surface reddish brown to black. $2n=32$, 34. July–October.

Introduced, scattered sporadically mostly in the southern half of the state (native of southwestern U.S. east to Oklahoma and Louisiana, Mexico; introduced eastward and northward to Massachusetts and Florida, Canada). Fallow fields, margins of crop fields, railroads, roadsides, and open, disturbed areas, frequently in sandy soils.

10. Amaranthus powellii S. Watson ssp. powellii

Pl. 197 h; Map 819

Plants monoecious. Stems 50–200 cm long, erect or ascending, sparsely to moderately pubescent toward the tip with mostly crinkled, multicellular hairs, unarmed. Leaves mostly long-petiolate. Leaf blades 2–12 cm long, lanceolate to ovate or elliptic, narrowed or tapered to a usually bluntly pointed tip (often minutely notched at the very tip), narrowed or tapered at the base, the undersurface

819. Amaranthus powellii   820. Amaranthus retroflexus   821. Amaranthus spinosus

sparsely pubescent along the main veins with inconspicuous, mostly crinkled, multicellular hairs. Inflorescences dull or dark green, axillary and terminal, the axillary inflorescences mostly dense globose clusters, the terminal inflorescence a spike or panicle with relatively few, long, ascending branches from near the base, the flowers mostly continuous along the spikes, the tip straight and usually stiffly erect, the main axis and branches often densely pubescent with mostly crinkled, multicellular hairs. Bracts 4–7 mm long, the main body (excluding the awn) conspicuously longer than the sepals and fruits, lanceolate to narrowly oblong-lanceolate or linear-lanceolate, narrowed or tapered to a sharply pointed tip, with a strongly thickened green midrib and relatively narrow, thin, papery margins, the midrib extending beyond the main body as a short awn, spinelike at maturity. Staminate flowers with 3–5 more or less similar sepals, these 1.5–3.0 mm long, erect or ascending, lanceolate to narrowly oblong-lanceolate, narrowed or tapered to a sharply pointed tip, often tapered to a short, awnlike extension of the midrib. Stamens 3(–5). Pistillate flowers with 3–5 more or less similar sepals, these 2–4 mm long, erect, lanceolate to narrowly oblong-lanceolate, narrowed or tapered to a stiff, sharply pointed tip, often tapered to a short, awnlike extension of the midrib. Stigmas 3, spreading from a short, thickened base. Fruits 1.8–2.2 mm long, with circumscissile dehiscence, the surface somewhat roughened or finely wrinkled above the midpoint when dry. Seeds 1.0–1.3 mm in diameter, rounded along the rim, the surface reddish brown to black. $2n=34$. June–October.

Introduced, uncommon in eastern and southwestern Missouri (native of western U.S. east to South Dakota and Texas; Canada, Mexico, South America; introduced widely eastward in North America to the East Coast, also Europe). Roadsides, railroads, and open, disturbed areas.

11. **Amaranthus retroflexus** L. (rough green amaranth, rough pigweed)

Pl. 197 g; Map 820

Plants monoecious. Stems 30–200 cm long, erect or ascending, sparsely to densely pubescent toward the tip with mostly crinkled, multicellular hairs, unarmed. Leaves mostly long-petiolate. Leaf blades 2–12 cm long, lanceolate to ovate or elliptic, narrowed or tapered to a usually bluntly pointed tip (often minutely notched at the very tip), narrowed or tapered at the base, the surfaces sparsely to moderately pubescent mostly along the veins with inconspicuous, mostly crinkled, multicellular hairs, the upper surface sometimes glabrous or nearly so. Inflorescences dull or grayish green occasionally dull reddish-tinged; axillary and terminal; the axillary inflorescences short spikes or less commonly dense, globose clusters; the terminal inflorescence usually a panicle with numerous clusters of short, dense spikes (these branching along most of the panicle axis); the flowers mostly continuous along the spikes; the tip straight and usually stiffly erect; the main axis and branches moderately to densely pubescent with mostly crinkled, multicellular hairs. Bracts 4–8 mm long, the main body (excluding the awn) conspicuously longer than the sepals and fruits, lanceolate to ovate, narrowed or tapered to a sharply pointed tip, with a strongly thickened green midrib and relatively narrow, thin, papery margins, the midrib extending beyond the main body as a short awn, spinelike at maturity. Staminate flowers with (4)5 more or less similar sepals, these 2.0–2.6 mm long, slightly outward-curved, lanceolate to oblong-ovate, narrowed or tapered to a sharply pointed tip, usually tapered to a short, awnlike extension of the midrib at the tip. Stamens (4)5. Pistillate flowers with (4)5 more or less similar sepals, these 2.2–3.0 mm long, outward-curved, narrowly oblong to narrowly oblong-elliptic, trun-

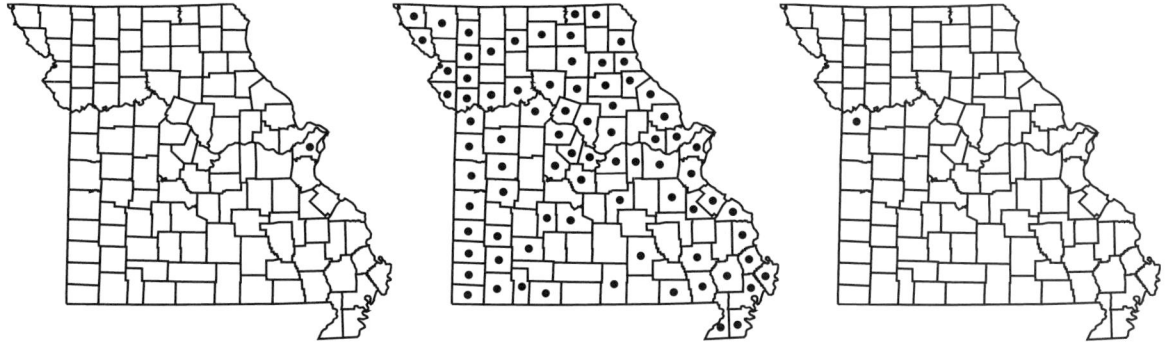

822. Amaranthus tricolor   823. Amaranthus tuberculatus   824. Amaranthus viridis

cate, rounded or shallowly notched at the soft tip, sometimes with an abrupt, minute, sharp extension of the midrib. Stigmas 3, erect or ascending. Fruits 1.8–2.2 mm long, with circumscissile dehiscence, the surface finely wrinkled above the midpoint when dry. Seeds 0.9–1.2 mm in diameter, rounded along the rim, the surface reddish brown to more commonly black. 2n=34. June–October.

Scattered, mostly south of the Missouri River (native of most of North America and portions of South America; introduced in Europe, Asia, Africa). Tops of bluffs and upland prairies; crop fields, fallow fields, gardens, levees, pastures, farmyards, feedlots, roadsides, railroads, and open, disturbed areas.

12. Amaranthus spinosus L. (spiny pigweed, thorny amaranth, careless weed)

Pl. 198 c, d; Map 821

Plants monoecious. Stems 30–150 cm long, erect or ascending, glabrous or nearly so, with conspicuous pairs of slender, straight, spreading spines at most nodes. Leaves mostly long-petiolate. Leaf blades 2–10 cm long, narrowly ovate to ovate, narrowed or tapered to a bluntly or sharply pointed tip, narrowed or tapered at the base, glabrous or nearly so. Inflorescences dull or grayish green, occasionally dull reddish-tinged; axillary and terminal; the axillary inflorescences dense, small, globose clusters or elongate spikes; the terminal inflorescence a spike or panicle with few to rarely numerous ascending branches; the flowers often grouped into discontinuous clusters or regions along the basal portions of the spikes; the tip straight or somewhat curved at maturity; the main axis and branches glabrous or nearly so. Bracts 0.5–1.0 mm long, shorter than the sepals, lanceolate to narrowly lanceolate or linear, narrowed or tapered to a sharply pointed tip, with a somewhat thickened green midrib and relatively narrow, thin, papery margins, the midrib not or only slightly extending beyond the main body as a minute, sharp point. Staminate flowers with 5 more or less similar sepals, these 1.0–1.6 mm long, somewhat outward-curved, oblong-lanceolate, narrowed or tapered to a sharp point at the tip. Stamens 5. Pistillate flowers with 5 more or less similar sepals, these 1.0–1.6 mm long, somewhat outward-curved, oblong to oblong-elliptic or oblong-lanceolate, narrowed or tapered to a blunt or sharp point. Stigmas 3, ascending to erect. Fruits 1.5–2.0 mm long, dehiscing irregularly or less commonly indehiscent or with more or less circumscissile dehiscence, the surface somewhat roughened or wrinkled above the midpoint when dry. Seeds 0.7–1.0 mm in diameter, rounded along the rim, the surface black. 2n=32, 34. June–October.

Introduced, scattered, mostly south of the Missouri River (probably originally a native of New World tropics, now distributed nearly throughout tropical and warm-temperate regions, in North America introduced north to California, Nebraska, Minnesota, and Maine; also Canada). Bottomland forests, banks of streams and rivers, and margins of sloughs; also fallow fields, crop fields, barnyards, feedlots, pastures, roadsides, railroads, and open, disturbed areas.

13. Amaranthus tricolor L. (Chinese spinach, Malabar spinach)

Map 822

Plants monoecious. Stems 40–150(–200) cm long, erect or ascending, glabrous, unarmed. Leaves long-petiolate. Leaf blades 3–15 cm long, ovate to broadly triangular-ovate, narrowed or tapered to a bluntly or sharply pointed tip (often minutely notched at the very tip), tapered at the base, glabrous. Inflorescences green, often reddish-tinged; axillary and terminal; the axillary inflorescences dense, globose clusters, present at nearly every node; the terminal inflorescence sometimes reduced or absent, otherwise a spike or panicle with relatively few, long, ascending branches from near the base; the flowers mostly continuous along the spikes; the tip straight or often curved to nodding; the main axis and branches glabrous. Bracts 3–6 mm long, longer than the sepals and fruits, ovate to elliptic-ovate, narrowed or tapered to a sharply pointed tip, with a narrow, slightly thickened, green midrib and relatively broad, thin, papery margins, the midrib extending beyond the main body as a short awn, spinelike at maturity. Staminate flowers with 3 more or less similar sepals, these 2.5–4.0 mm long, ascending or somewhat outward-curved at the tip, lanceolate to ovate to elliptic-ovate, narrowed or tapered to a sharply pointed tip, often tapered to a short, awnlike extension of the midrib. Stamens 3. Pistillate flowers with 3 more or less similar sepals, these 2–4 mm long, somewhat outward-curved at the tip, ovate to elliptic-ovate, narrowed or tapered to a stiff, sharply pointed tip, often tapered to a short, awnlike extension of the midrib. Stigmas 2 or 3, spreading from a short, thickened base. Fruits 1.8–2.5 mm long, with circumscissile dehiscence, the surface finely wrinkled when dry. Seeds 1.0–1.3 mm in diameter, rounded along the rim, the surface reddish brown to black. $2n=34$. August–October.

Introduced, known only from the city of St. Louis (native of Asia and Malesia; introduced widely in tropical and warm-temperate regions nearly worldwide, in the U.S. sporadically from Michigan to Louisiana). Railroads.

This species, which is cultivated as a salad green and potherb in Asia, was first reported for Missouri by Mühlenbach (1979). Aellen (1959) proposed a complicated infraspecific classification for this species. The single specimen from Missouri is apparently ssp. *tricolor*, but it does not key well to these subspecies.

14. Amaranthus tuberculatus (Moq.) J.D. Sauer (water hemp)

A. rudis J.D. Sauer
A. tuberculatus var. *rudis* (J.D. Sauer) Costea & Tardif
A. tamariscinus Nutt.
Acnida tamariscina (Nutt.) A.W. Wood var. *tuberculata* (Moq.) Uline & W.L. Bray
Acnida tamariscina var. *prostrata* Uline & W.L. Bray

Pl. 197 a, b, i–m; Map 823

Plants dioecious. Stems (20–)50–200(–350) cm long, usually ascending or erect, glabrous or nearly so, unarmed. Leaves long-petiolate basally, with progressively shorter petioles toward the stem tip. Leaf blades 2–20 cm long, lanceolate to elliptic or ovate, narrowed or tapered to a sharply pointed, bluntly pointed or minutely notched tip, narrowed or tapered at the base, glabrous. Inflorescences dull or grayish green, occasionally dull reddish-tinged, axillary and terminal, the axillary inflorescences mostly elongate spikes or small panicles of elongate spikes at maturity, the terminal inflorescence a panicle with usually numerous ascending branches, the flowers often grouped into discontinuous clusters or regions along the spikes, the tip straight or somewhat curved at maturity, the main axis and branches glabrous to densely but inconspicuously pubescent with mostly crinkled multicellular hairs. Bracts 1.0–2.5 mm long, shorter than to about as long as the sepals, lanceolate to narrowly ovate, with a somewhat thickened green midrib and relatively broad, thin, papery margins, the midrib usually extending beyond the main body as a short awn, usually somewhat spinelike at maturity. Staminate flowers with 5 unequal (the outer 2 more strongly pointed, slightly longer than and strongly overlapping the inner 3) to more or less similar sepals, these 2.2–3.0 mm long, erect or very slightly outward-curved, lanceolate to oblong-elliptic, all tapered and with a short, awnlike extension of the midrib at the tip, or the inner ones sometimes bluntly pointed to shallowly and minutely notched. Stamens 5. Pistillate flowers with the sepals absent or irregularly 1 or 2 and 0.5–2.0 mm long, linear to lanceolate, tapered to a short, awnlike extension of the midrib, this often somewhat spiny at maturity. Stigmas 3 or 4, spreading. Fruits 1–2 mm long, indehiscent or circumscissilely dehiscent at about the midpoint, the surface smooth or more commonly noticeably wrinkled when dry. Seeds 0.8–1.2 mm in diameter, rounded along the rim, the surface reddish brown to black. $2n=32$. July–October.

Scattered nearly throughout the state (Wisconsin to Indiana and Alabama west to North Dakota and Texas; Canada; introduced farther east and west to the Atlantic and Pacific seaboards). Banks of streams and rivers, sloughs, and margins of ponds and lakes; also crop fields, fallow fields, banks of ditches, roadsides, railroads, and moist, open, disturbed areas.

Traditionally, *A. rudis* has been treated as a separate species, based on a number of subtle flower and fruit differences, with intermediate plants written off as hybrids between *A. rudis* and *A. tuberculatus*. Pratt and Clark (2001) studied

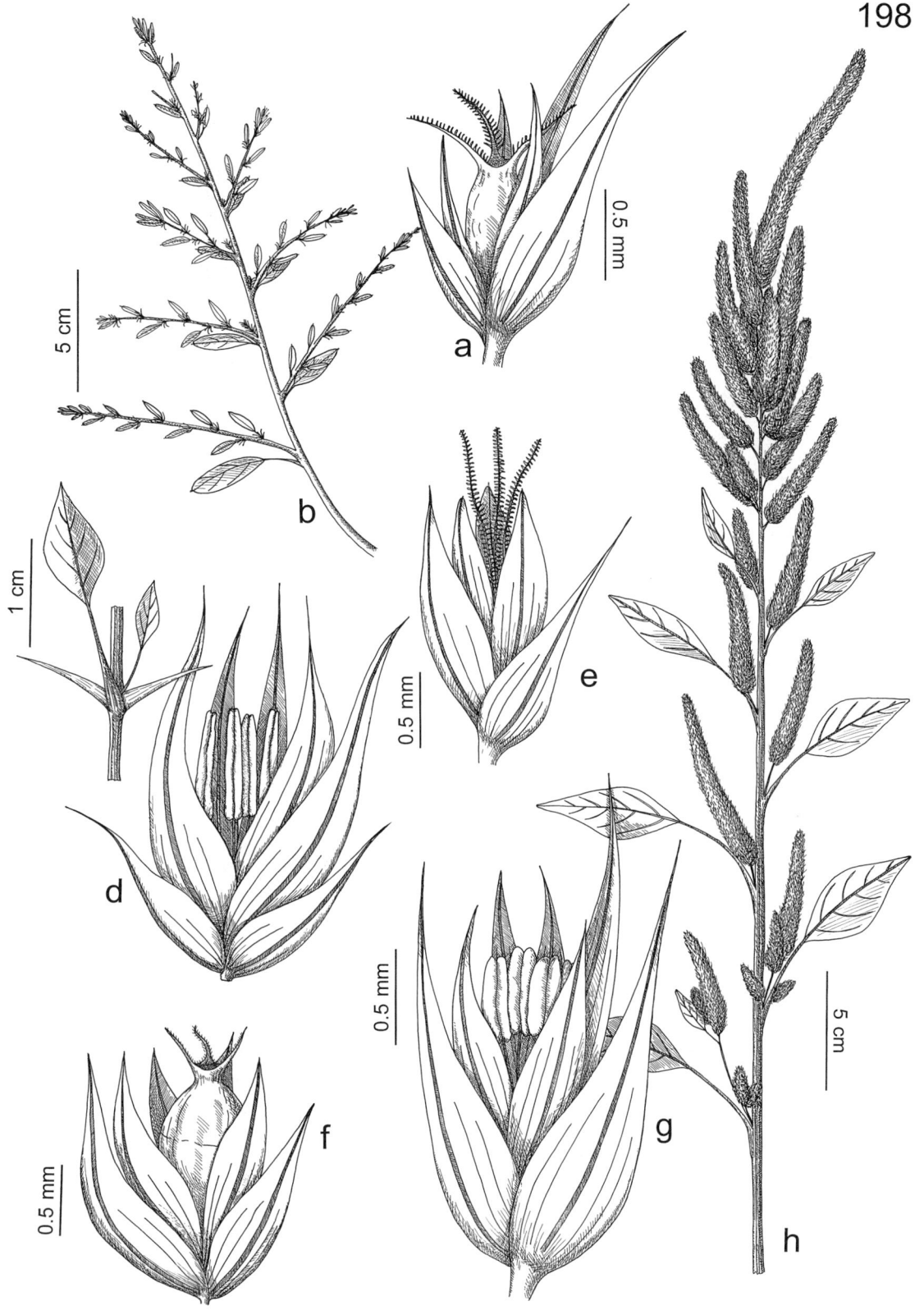

Plate 198. Amaranthaceae. *Amaranthus albus*, **a)** pistillate flower, **b)** habit. *Amaranthus spinosus*, **c)** node, **d)** staminate flower. *Amaranthus caudatus*, **e)** pistillate flower. *Amaranthus blitoides*, **f)** pistillate flower. *Amaranthus hybridus*, **g)** staminate flower, **h)** habit.

both morphological and allozymic variation in the complex and concluded that there was no morphological or genetic basis for maintaining more than a single biological species of water hemp. Plants traditionally referred to as *A. tuberculatus* apparently are merely a form in which the sepals of pistillate flowers are reduced or absent and the fruits tend not to dehisce. Costea and Tardif (2003) treated the two entities as varieties of *A. tuberculatus,* noting that var. *tuberculatus* and var. *rudis* are more or less distinguishable in the eastern and western portions respectively of the overall range. However, in the midwestern states, the two taxa are neither morphologically nor ecologically distinguishable. Steyermark (1963) and most earlier authors called plants having pistillate flowers with 1 or 2 well-developed sepals *A. tamariscinus* Nutt., but Sauer (1972) showed that this name actually refers to a sterile hybrid between what he named *A. rudis* and one of the members of the monoecious *A. hybridus* complex.

Amaranthus tuberculatus is extremely variable morphologically. Rare plants with spreading to loosely ascending stems and poorly developed terminal spikes were once called var. *prostrata*. Sauer (1955) concluded that these represented stunted ecological variants that grew from seeds germinating too late in the season for proper development to have occurred.

15. Amaranthus viridis L.

A. gracilis Poir.

Map 824

Plants monoecious. Stems 30–120 cm long, ascending or more or less spreading with ascending branches, glabrous or sparsely pubescent with inconspicuous mostly multicellular hairs toward the tip, unarmed. Leaves mostly long-petiolate. Leaf blades 2–8 cm long, elliptic to ovate or broadly ovate, narrowed or tapered to a usually bluntly pointed tip (often minutely notched at the very tip), narrowed or tapered at the base, the undersurface glabrous or sparsely pubescent along the main veins with inconspicuous mostly multicellular hairs. Inflorescences dull or dark green, axillary and terminal, the axillary inflorescences of elongate spikes, the terminal inflorescence a spike or panicle with few to several, long, ascending branches from near the base, the flowers often grouped into discontinuous clusters or regions along the basal portions of the spikes, the tip somewhat curved or nodding, the main axis and branches sparsely to moderately pubescent with inconspicuous, mostly multicellular hairs. Bracts 0.6–0.9 mm long, shorter than the sepals and fruits, ovate to oblong-ovate, narrowed or tapered to a sharply pointed tip, with a somewhat thickened green midrib and relatively broad, thin, papery margins, the midrib sometimes extending beyond the main body as a minute, short point, not spinelike. Staminate flowers with 3 more or less similar sepals, these 0.9–1.2 mm long, erect or ascending, oblanceolate to oblong-oblanceolate, abruptly tapered to a sharply pointed tip, the midrib sometimes extending beyond the main body as a minute, short point, not spinelike. Stamens 3. Pistillate flowers with 3 more or less similar sepals, these 0.9–1.2 mm long, erect or ascending, oblanceolate to oblong-oblanceolate, abruptly tapered to a sharply pointed tip, the midrib sometimes extending beyond the main body as a minute, short point, not spinelike. Stigmas 3, erect. Fruits 1.4–1.7 mm long, indehiscent, the surface usually strongly wrinkled when dry. Seeds 1.0–1.3 mm in diameter, angled along the rim, the surface reddish brown to black. $2n=34$. July–October.

Introduced, known thus far from a single historical collection from Jackson County (originally probably native to South America; now widely introduced in tropical and warm-temperate regions nearly worldwide; in the U.S. introduced in states along the Atlantic coast and west along the southern tier of states to Arizona). Open, disturbed areas.

Amaranthus viridis somewhat resembles *A. blitum* in its slender, somewhat flexuous spikes, small flowers, indehiscent fruits, relatively weak stems, and broad leaves. Aside from quantitative differences noted in the two descriptions, it differs most notably in having leaf blades at most minutely notched at the tip, the axillary inflorescences elongate of spikes usually nearly as long as the terminal ones, and in its 3 sepals that are broadest above the midpoint.

2. Celosia L.

About 65 species, southern U.S. to South America, Caribbean Islands, Asia, Africa.

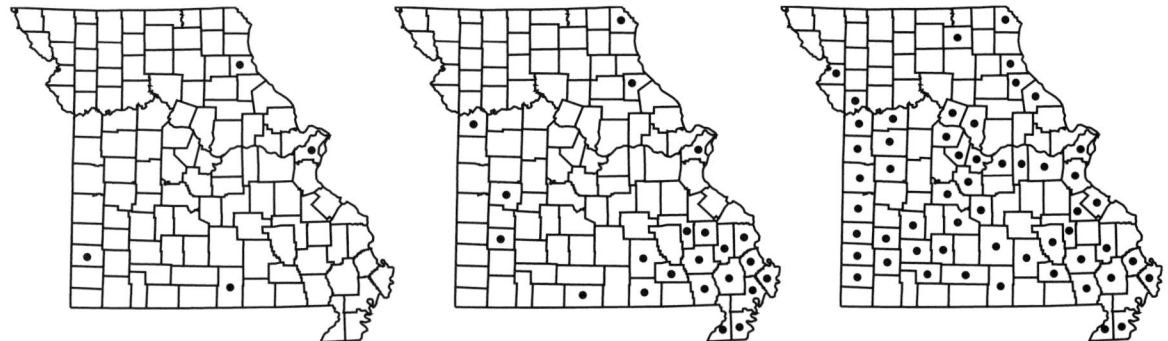

825. Celosia argentea   826. Froelichia floridana   827. Froelichia gracilis

1. Celosia argentea L.

Pl. 199 e; Map 825

Plants annual (perennial herbs or shrubs elsewhere), glabrous. Stems 30–120 cm long, erect or ascending, green to yellowish white or reddish purple, often with pink to purple longitudinal lines or ridges. Leaves alternate, short- to long-petiolate. Leaf blades 3–15 cm long, herbaceous, narrowly lanceolate to elliptic-ovate, narrowed to less commonly rounded at the base, narrowed or tapered to a sharply pointed tip, the margins entire or nearly so. Inflorescences terminal and sometimes also axillary, dense spikes, these sometimes grouped into panicles, the main axis sometimes broadened and flattened (fasciated) with flowers across the surface. Bracts similar to the sepals in size and shape. Flowers perfect. Sepals 5, free, all similar in size and shape, 6–8 mm long, oblong-lanceolate to lanceolate, tapered to a sharply pointed but unawned tip, papery or scalelike, silvery white, yellow, pink, red, or purple. Stamens 5, the filaments fused toward the base. Ovary narrowly ovoid to nearly globose. Ovules 2 to several. Style well developed, persistent, the 2 or 3 stigmas capitate. Fruits mostly with papery walls, the main body 2.5–4.0 mm long, ovoid to globose, tapered to a single beak, the dehiscence usually circumscissile at about the midpoint, 2–6-seeded. Seeds 1.0–1.5 mm long, somewhat flattened, circular or nearly so in outline, rounded to bluntly angled along the rim, the surface reddish brown to black, shiny. $2n=36, 72$. July–October.

Introduced, uncommon and sporadic in Missouri (native distribution not known, widely but sporadically escaping from cultivation in the U.S.). Banks of streams and rivers; also railroads and open, disturbed areas.

Celosia argentea has an extremely long history of cultivation as an ornamental, and its natural range had already become obscured by the presence of widespread weedy escapes by the time that botanists became interested in the species' wild origin. Today, plants are widely distributed in tropical and subtropical regions of both the Old World and the New World. Numerous cultivars have been developed for horticultural uses, and plants are quite variable in size, color, and inflorescence patterns. In Missouri, it is doubtful whether populations persist for very long after becoming established outside cultivation.

1. Inflorescences spikes, these sometimes grouped into small panicles, the flowers attached to a more or less tubular, stemlike axis . . . . 1A. VAR. ARGENTEA
1. Inflorescences fasciated, the main axis flattened and broaded into an irregular fan-shaped structure with flowers along the surface . . . . 1B. VAR. CRISTATA

1a. var. argentea

Stems relatively slender, inconspicuously ridged. Inflorescences spikes, these sometimes grouped into small panicles, the flowers attached to a more or less tubular, stemlike axis. $2n=36, 72$ (mostly $2n=72$). July–October.

Introduced, known thus far only from St. Louis County (native distribution not known, widely but escaping sporadically in the U.S.). Banks of streams; also open, disturbed areas.

Plants of var. *argentea* tend to be taller and with more slender stems than those of var. *cristata*. Sometimes, partial fusion or expansion of the inflorescences can be observed, and although such plants are probably best referred to var. *argentea*, their inflorescence structure is somewhat intermediate between the two varieties.

1b. var. cristata (L.) Kuntze (cockscomb)
C. *cristata* L.
C. *argentea* f. *cristata* (L.) Schinz

Pl. 199 e

Stems relatively stout, often strongly ridged or fluted. Inflorescences fasciated, the main axis flattened and broadened into an irregular fan-shaped structure with flowers along the surface. $2n=36$. July–October.

Introduced, uncommon and sporadic in Missouri (cultivated widely in tropical to temperate regions; escaping sporadically in the eastern U.S. west to Kansas and Louisiana). Banks of streams and rivers; also railroads, gardens, and open, disturbed areas.

The var. *cristata* is a cultigen that apparently was developed long ago through selection and breeding of various forms of the tetraploid ($2n=36$) cytotype of var. *argentea* (Robertson, 1981).

3. Froelichia Moench (cottonweed, snake-cotton)

Plants annual (perennial herbs elsewhere), but with the taproot sometimes appearing somewhat woody. Aerial stems erect or strongly ascending, somewhat angled, densely pubescent with unbranched, usually woolly hairs, especially the nodes often appearing cobwebby, the hairs becoming shorter near the tip, occasionally also somewhat sticky. Leaves opposite, positioned mostly below the stem midpoint, sessile or nearly so. Leaf blades 5–14 cm long, somewhat thickened or leathery, linear to narrowly oblong-lanceolate or narrowly oblanceolate, narrowed or tapered at the base, narrowed or tapered to a sharply or bluntly pointed tip, the margins entire, the upper surface densely pubescent with silky or woolly hairs but frequently becoming nearly glabrous at maturity, the undersurface densely persistently pubescent with woolly hairs. Inflorescences terminal, narrow panicles with pairs of dense spikes ascending from the nodes of the main axis, these sometimes reduced to short, dense clusters and the inflorescence then appearing as an interrupted spike, the lateral branches often terminating in simple spikes. Bracts similar in texture but much shorter than the sepals, broadly ovate, papery or scalelike, glabrous. Flowers perfect. Sepals 5, fused most of their length into a persistent, conical to flask-shaped, papery tube, this becoming somewhat hardened and developing winglike longitudinal ridges or rows of spines and also basal tubercles after flowering, white to yellowish white, densely pubescent with woolly hairs, the lobes 1–2 mm long, oblong to lanceolate, bluntly to sharply pointed, greenish white, sometimes pinkish-tinged, more or less glabrous. Stamens 5, the filaments fused nearly their entire length, persistent, the anthers appearing sessile in the sinuses between 5 short, strap-shaped lobes. Ovary ovoid. Ovule 1. Style well developed, persistent, the stigma 1, capitate. Fruits shorter than and hidden within the persistent calyx and anther tubes, with membranous walls, ovoid, glabrous, indehiscent, 1-seeded. Seeds somewhat flattened, circular or nearly so in outline, the surface shiny. Twelve to 20 species, U.S. and adjacent Canada to South America, Caribbean Islands.

To observe the crests and tubercles on the fruiting calyx, it is necessary to first remove the thick woolly covering of hair.

1. Calyx tube at fruiting (4.5–)5.0–6.0 mm long, flask-shaped, symmetric at the tip, with deeply toothed longitudinal wings, the basal tubercles blunt; blades of largest leaves (5–)10–30 mm wide . . . . 1. F. FLORIDANA
1. Calyx tube at fruiting 3.5–4.5 mm long, flask-shaped to more commonly conical, asymmetric at the tip, with longitudinal rows of individual short spines, the basal tubercles often also spiny; blades of largest leaves 5–10(–15 mm) wide
 . . . . 2. F. GRACILIS

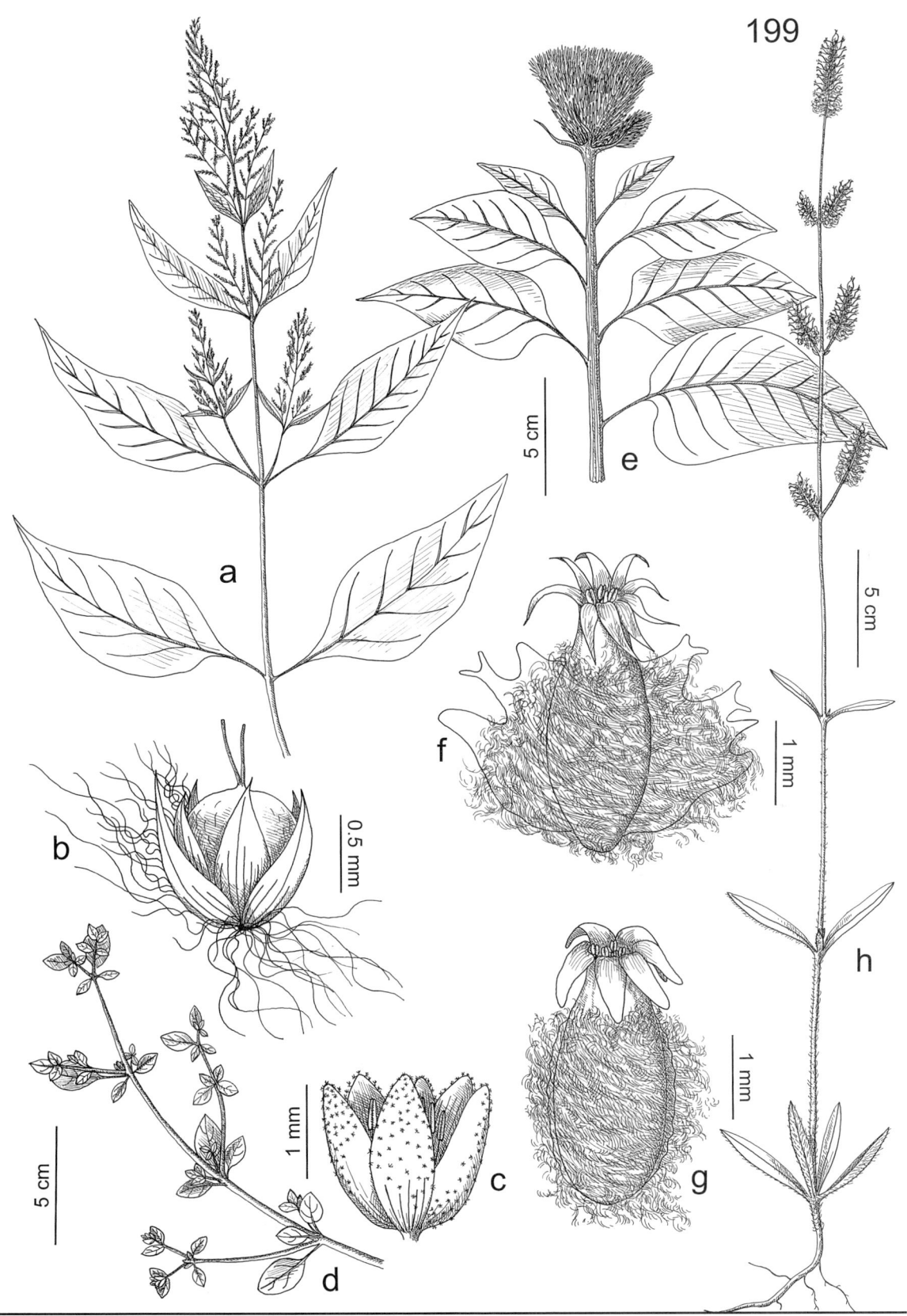

Plate 199. Amaranthaceae. *Iresine rhizomatosa*, **a)** habit, **b)** fruit. *Tidestromia lanuginosa*, **c)** flower, **d)** habit. *Celosia argentea* var. *cristata*, **e)** habit. *Froelichia floridana*, **f)** fruit. *Froelichia gracilis*, **g)** fruit, **h)** habit.

1. Froelichia floridana (Nutt.) Moq. **var. campestris** (Small) Fernald (field cottonweed, field snake-cotton)

Pl. 199 f; Map 826

Stems (30–)50–140(–200) cm long, usually relatively stout (to 7 mm in diameter), unbranched or few-branched at the base (but often with more branches above the midpoint). Leaf blades 2–14 cm long, those of the largest leaves (5–)10–30 mm wide, elliptic-lanceolate to oblanceolate, less commonly narrowly oblong-lanceolate. Inflorescences with the spikes 1–8 cm long, 10–14 mm in diameter, often short-stalked, the flowers in a dense, 5-ranked spiral. Calyx tube at fruiting (4.5–)5.0–6.0 mm long, flask-shaped, symmetric at the tip, with deeply toothed longitudinal wings, the basal tubercles blunt. Seeds 1.5–2.0 mm long, reddish brown. $2n$=about 78. May–September.

Scattered in the southeastern portion of the state, uncommon farther north and west (Ohio and Kentucky west to South Dakota, Colorado, and New Mexico). Sand prairies and banks of streams and rivers; also fallow fields, pastures, railroads, roadsides, and open, sandy, disturbed areas.

Froelichia floridana is usually treated as comprising two or more varieties, but the genus is in need of more detailed taxonomic study. The var. *floridana* is widespread along the Atlantic and Gulf Coastal Plains and occurs inland to Kentucky and Tennessee. It is generally more robust (often to 2 m tall) than var. *campestris*, with leaves that are mostly widest below the midpoint and more gradually tapered to the sharply pointed tip. For a discussion of possible hybridization with *F. gracilis*, see the treatment of that species.

2. Froelichia gracilis (Hook.) Moq. (slender cottonweed, slender snake-cotton)

Pl. 199 g, h; Map 827

Calyx tube at fruiting 3.5–4.5 mm long, flask-shaped to more commonly conical, asymmetric at the tip, with longitudinal rows of individual short spines, the basal tubercles often also spiny. Stems 15–45(–70) cm long, usually relatively slender (to 3 mm in diameter), usually several- to many-branched at the base. Leaf blades 2–8(–12) cm long, those of the largest leaves 5–10(–15) mm wide, linear to narrowly lanceolate or narrowly oblanceolate. Inflorescences with the spikes and/or clusters 0.7–3.0 cm long, 5–10 mm in diameter, sessile, the flowers in a dense, 3-ranked spiral. Calyx tube at fruiting 3.5–4.5 mm long, flask-shaped to more commonly conical, asymmetric at the tip, with longitudinal rows of individual short spines, the basal tubercles often also spiny. Seeds 1.2–1.5 mm long, tan to yellowish brown or reddish brown. $2n$=54. May–September.

Scattered nearly throughout the state, but apparently absent from the northwestern portion of the Glaciated Plains Division (Indiana to Arkansas west to Wyoming and Arizona; Mexico; introduced eastward to the Atlantic seaboard, also Canada). Sand prairies, upland prairies, tops of bluffs, openings of mesic upland forests, and banks of streams and rivers; also fallow fields, pastures, roadsides, railroads, and open, sandy or rocky, disturbed areas.

Froelichia gracilis is often found growing in mixed populations with *F. floridana* but is more widely distributed in Missouri. Steyermark (1963) cited a single specimen from Jasper County as somewhat morphologically intermediate between *F. gracilis* and *F. floridana* and suggested that it might represent an interspecific hybrid. Such plants probably are more common but presumably have been overlooked by collectors.

4. Iresine P. Browne

Sixty-five to 80 species, North America to South America, Caribbean Islands, Australia.

1. Iresine rhizomatosa Standl. (bloodleaf)

Pl. 199 a, b; Map 828

Plants perennial herbs (annual or woody elsewhere), dioecious, with slender rhizomes. Aerial stems 40–150 cm long, erect or strongly ascending, somewhat angled or slightly ridged longitudinally, inconspicuously pubescent with short, unbranched hairs, often only at the slightly swollen nodes. Leaves opposite, short- to more commonly long-petiolate. Leaf blades 5–14 cm long, thin and herbaceous, lanceolate to more commonly ovate, tapered abruptly at the base, gradually narrowed or tapered to a sharply pointed tip, the margins entire or occasionally minutely and sharply toothed, the surfaces glabrous or sparsely and inconspicuously pubescent with unbranched hairs, the undersurface sometimes appearing pebbled when dry. Inflorescences terminal and sometimes

828. Iresine rhizomatosa   829. Tidestromia lanuginosa   830. Cotinus obovatus

also axillary, the pyramidal panicles with numerous short spikes along the ultimate branches, those of staminate plants usually with more spreading branches than those of pistillate plants. Bracts similar in texture but somewhat shorter than the sepals, papery or scalelike, glabrous. Flowers imperfect, the pistillate ones with dense, long (to 5 mm at fruiting), woolly to cobwebby hairs at the base. Sepals 5, free or rarely fused only at the very base, all similar in size and shape, 1.2–1.5 mm long, narrowly ovate, narrowed to a bluntly or sharply pointed but unawned tip, papery or scalelike, not becoming hardened after flowering, silvery white, glabrous. Staminate flowers with 5 stamens, the filaments fused toward the base, usually alternating with 5 minute, triangular teeth, sometimes a highly reduced rudimentary pistil also present. Pistillate flowers with the ovary oblong to nearly circular in outline, flattened (elliptic in cross-section), sometimes 5 highly reduced staminodes also present. Ovule 1. Style absent or very short, persistent, the stigmas 2, slender. Fruits with papery walls, 2.0–2.5 mm long, more or less circular in outline, flattened (elliptic in cross-section), more or less rounded at the tip, minutely beaked, glabrous, indehiscent, 1-seeded. Seeds 0.4–0.6 mm long, more or less globose, the surface reddish brown to black, shiny. August–October.

Scattered in the Ozark, Ozark Border, and Unglaciated Plains Divisions (eastern [mostly southeastern] U.S. west to Kansas and Texas). Bottomland forests, banks of streams and rivers, and bases of bluffs; rarely also roadsides and disturbed shaded areas.

5. Tidestromia Standl. (honeysweet)

About 6 species, southern U.S., Mexico, Caribbean Islands.
The common name honeysweet refers to the sweet fragrance of the tiny flowers.

1. Tidestromia lanuginosa (Nutt.) Standl.
ssp. lanuginosa

Pl. 199 c, d; Map 829

Plants annual (perennial herbs elsewhere). Stems 10–60 cm long, spreading to very loosely ascending, usually reddish purple, without longitudinal lines or ridges, densely pubescent with stellate hairs, sometimes the lower portion becoming nearly glabrous with age, the nodes somewhat swollen. Leaves opposite (rarely whorled toward the stem tip, rarely alternate toward the base), short- to long-petiolate, those of a pair usually joined by an inconspicuous ridge around the node. Leaf blades 0.5–3.0 cm long, somewhat thickened and leathery, obovate to spatulate or nearly circular, narrowed or tapered at the base, rounded or bluntly pointed at the tip, the margins entire, the surfaces densely pubescent with stellate hairs, appearing grayish green, sometimes becoming nearly glabrous with age. Inflorescences in small, sessile, axillary clusters of 2–5 flowers. Bracts similar in texture but about 1/3 as long as the sepals, papery or scalelike, hairy. Flowers perfect. Sepals 5, free, 1–3 mm long, the outer 3 larger than the inner 2, lanceolate to narrowly ovate, tapered to a sharply pointed but unawned tip, papery or scalelike, translucent, straw-colored to yellow to yellowish brown, hairy, sometimes becoming glabrous and shiny at maturity. Stamens 5, the filaments fused toward the base, usually alternating with 5 minute, triangular teeth. Ovary more or less globose. Ovule 1. Style short, persistent, the stigma 1, capitate, sometimes somewhat 2-lobed. Fruits with rigid walls, 1.5–2.0 mm long, more or less globose, tapered abruptly to the minute beak, glabrous, indehiscent, 1-seeded. Seeds 1.2–1.6 mm long, more or less globose, the surface yellowish brown, shiny. July–October.

Introduced, known only from historical collections from Jackson County (southwestern U.S. east to South Dakota and Louisiana; Mexico, Caribbean Islands; introduced in Illinois, Missouri, and Pennsylvania). Railroads and open, disturbed areas.

The recently segregated ssp. *eliassoniana* Sánchez-del Pino & Flores Olvera includes plants of the southwestern United States and adjacent Mexico that differ from ssp. *lanuginosa* in having microscopically spiny (vs. smooth) pollen grains and in subtle details of trichome morphology.

ANACARDIACEAE (Cashew Family)
Contributed by David J. Bogler and George Yatskievych

Plants trees, shrubs, or lianas, usually dioecious, with well-developed resin canals in the bark and leaves and acrid or milky sap. Leaves alternate, pinnately compound, trifoliate, or less commonly appearing simple. Stipules absent or obscure. Inflorescences terminal or axillary panicles or clusters. Flowers small, actinomorphic, almost always imperfect, usually hypogynous. Sepals usually 5, fused at the base. Petals usually 5, distinct. Stamens 5–10, the filaments usually distinct, reduced or absent in pistillate flowers, the anthers attached above the base. Cuplike nectar disk present between stamens and pistil. Pistil of usually 3 fused carpels, but almost always with only 1 carpel fertile and fully developed; reduced or absent in staminate flowers. Ovary with a single fertile ovule (or rarely 1 ovule per carpel), often appearing 1-locular, the placentation axile. Styles 3, distinct or united. Stigmas 3, capitate. Fruits drupes, often somewhat flattened, often waxy or hairy, the stone 1-seeded, bony. Sixty to 80 genera, about 600 species, chiefly tropical, but extending into temperate areas of North and South America, Asia, and Europe.

The family includes a number of important fruits and nuts, including pistachio (*Pistacia vera* L.), mango (*Mangifera indica* L.), and cashew (*Anacardium occidentale* L.). A small number of species are cultivated as ornamentals. However, many members are poisonous or cause severe dermatitis.

1. Leaves simple; fruiting panicles with elongate, plumose-hairy, sterile stalks; styles sublateral, unequal; fruit asymmetrical (swollen toward the tip on the side opposite the minute beak) . . . . 1. COTINUS
1. Leaves trifoliate or pinnately compound; fruiting panicles not plumose-hairy; styles terminal, more or less equal; fruit symmetrical
    2. Leaves trifoliate or pinnately compound; inflorescence dense, terminal, or lateral on previous year's wood; fruits red, noticeably pubescent with red glandular hairs . . . . 2. RHUS
    2. Leaves trifoliate; inflorescence loose, axillary; fruits whitish or yellowish, glabrous or inconspicuously hairy . . . . 3. TOXICODENDRON

1. Cotinus Adans. (smoke tree)

Three or 4 species in eastern North America, southern Europe, and eastern subtropical Asia.

1. **Cotinus obovatus** Raf. (American smoke tree, chittam-wood)
C. americanus Nutt.
C. cotinoides (Nutt.) Britton
Rhus americanus (Nutt.) Sudw.
R. cotinoides Nutt.

Pl. 200 e, f; Map 830

Plants shrubs or trees 2–5(–10) m tall, dioecious or nearly so, with yellow wood and aromatic, resinous sap. Twigs at first glaucous, reddish, becoming brown at maturity. Leaves alternate, simple, the petioles 0.5–6.0 cm long. Leaf blades 10–17 cm long, 4–9 cm wide, obovate to elliptic-obovate, entire, with a rounded to bluntly pointed tip, the upper surface glabrous, somewhat glaucous, the lower surface sparsely hairy to nearly glabrous. Inflorescences terminal panicles, to 30 cm long and 15 cm broad, loosely flowered. Flowers mostly sterile and shed early in development, leaving behind elongate stalks 1.0–3.5 cm long, these with spreading purple plumose hairs 1.0–1.2 mm long. Sepals 5, 0.9–1.2 mm long, lanceolate, persistent at fruiting. Petals 5, 1.4–1.6 mm long, oblanceolate, rounded at the tip, yellowish to greenish white. Staminate flowers with the stamens 5, the filaments short, the anthers broadly ovoid. Pistillate flowers with the styles 3, appearing sublateral, unequal, the style of the fertile carpel longer than those of the 2 inconspicuous sterile carpels. Ovary with 1 locule. Fruits 3.5–4.2 mm long, 2.5–3.2 mm wide, asymmetrically ovoid (appearing swollen toward the tip on the side opposite the minute beak), flattened, the outer layer with a network of veins, the fleshy middle layer very thin. Stone more or less kidney-shaped, smooth. $2n=15$. May.

Uncommon in the western portion of the Ozark Division (Georgia to Kentucky west to Oklahoma and Texas). Glades and tops of bluffs, usually on calcareous substrates.

American smoke tree is cultivated in the eastern U.S., where it is valued for its interesting smokelike texture of the fruiting panicles and intense yellow to red fall foliage. The long hairs of the sterile pedicels are jointed, spotted with purple, and translucent, which diffuses the light and generates the purple puff-of-smoke effect. The wood yields a clear yellowish orange dye that once was used to color silk and wool, and many of the larger trees were cut down for this purpose.

The Eurasian smoke tree, *C. coggygria* Scop., also is cultivated in the United States and occasionally has escaped in the northeastern states and Utah. It differs from *C. obovatus* most visibly in its more oval leaves and even showier inflorescences.

2. **Rhus** L. (sumac)

Plants shrubs or small trees, almost always dioecious, usually with long-creeping branched rhizomes and forming dense colonies, often pubescent with glandular and nonglandular hairs. Leaves trifoliate or pinnately compound, the rachis sometimes winged. Leaflets with the margins entire or toothed. Inflorescences dense terminal panicles, occasionally relatively small and appearing as dense clusters of flowers, developing before or after the leaves expand. Flowers all fertile, the stalks not plumose-hairy. Sepals 5, united at the base, usually persistent at fruiting. Petals 5, often hairy on the inner surface, greenish white to yellow. Staminate flowers with the stamens 5, the anthers ovoid, usually shorter than the filaments. Pistillate flowers with the styles 3, appearing terminal, equal in length or nearly so, short, sometimes fused toward the base. Ovary with 1 locule. Fruits globose or nearly so, often flattened, red or reddish, noticeably pubescent with dense red glandular hairs, sometimes also with nonglandular hairs, the outer layer and resinous fleshy to waxy middle layer readily detachable from the smooth stone. About 100 species, widespread in the Northern Hemisphere.

Native Americans apparently used the species of *Rhus* that grow in Missouri more or less interchangeably for various medicinal purposes, including to control vomiting and as a poultice for skin ailments (Moerman, 1998). The fruits also were chewed as a breath freshener. The fruits of sumacs are relished by a variety of wildlife. The leaves of the Missouri species turn bright red or less commonly reddish orange in autumn.

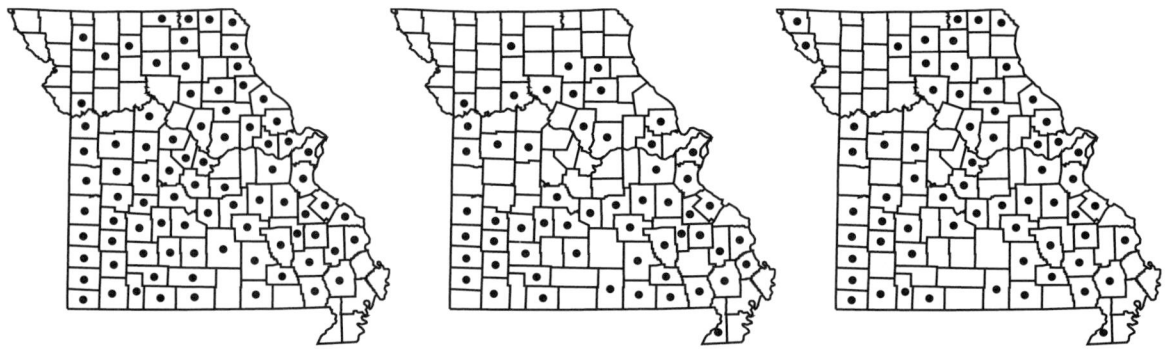

831. Rhus aromatica   832. Rhus copallinum   833. Rhus glabra

1. Leaves trifoliate .... 1. R. AROMATICA
1. Leaves pinnately compound, with 7–21 leaflets
   2. Leaf rachis winged between leaflets .... 2. R. COPALLINUM
   2. Leaf rachis without wings
      3. Branches glabrous; fruits with dense, short, red papillae .... 3. R. GLABRA
      3. Branches densely hairy; fruits with dense red hairs .... 4. R. TYPHINA

1. Rhus aromatica Aiton (fragrant sumac, aromatic sumac)

Pl. 200 a, b; Map 831

Plants shrubs. Stems 0.5–1.5 m long, erect or ascending. Branches nearly glabrous to densely hairy, aromatic when bruised. Leaves trifoliate, the petiole 1.0–2.5 cm long. Leaflets variable in shape and lobing, nearly glabrous to densely pubescent, the terminal leaflet sessile, broadly ovate to rhombic, 4–9 cm long, 2–8 cm wide, scalloped or toothed near the tip, entire and angled at the base. Inflorescences terminal, small panicles with spicate branches, 2–6 cm long, 1–3 cm wide, the branches occasionally relatively small and appearing as dense clusters of flowers. Flower stalks 1–3 mm long. Sepals lanceolate, 1.0–1.4 mm long, 0.3–0.4 mm wide, broadly rounded at the tip, the surfaces glabrous, reddish brown, the margins with nonglandular hairs. Petals oblong-obovate, 1.6–2.5 mm long, rounded at the tip, glabrous or hairy on the inner surface, yellow. Fruits 5–7 mm long, 4–6 mm wide, red, slightly flattened, pubescent with dense, minute, stout, red glandular hairs and sparse to dense, white to colorless nonglandular hairs. 2n=30. March–May.

Scattered to common throughout the state (eastern U.S. west to South Dakota and Texas; Canada). Glades, tops of bluffs, savannas, and openings of mesic to dry upland forests; also old fields and roadsides.

The species of *Rhus* with flowers in spikes sometimes have been treated separately from those with flowers in a terminal panicle, as the genus or subgenus *Schmaltzia*. The trifoliate-leaved species at times have been placed in *Lobadium*, which is now considered a section of *Rhus*. Barkley (1937) recognized four species of *Rhus* with trifoliate leaves and reddish fruits, which he separated on the basis of overall size, length of the bracts, shape and size of the leaflets, relative pubescence, length of the flower stalks, and time of flowering relative to leaf emergence. Two of these are found in the United States: *R. aromatica* in the eastern portion of the country, and *R. trilobata* in the Great Plains region and farther west. The transition from one species to the other occurs in a broad region that includes Missouri, Arkansas, Oklahoma, and Kansas. Numerous varieties and combinations in both *R. aromatica* and *R. trilobata* have been proposed, but there is general agreement that a careful biosystematic study of the entire complex is needed before any of these names can be applied with confidence.

The fruits of both *R. aromatica* and *R. trilobata* sometimes are steeped in hot water to make a pleasant beverage with a somewhat lemony flavor. However, because these species contain trace amounts of the same chemical substances that are produced more abundantly in *Toxicodendron*, a very small percentage of individuals who are hypersensitive to urushiols develop a strong allergic reaction to drinking the tea.

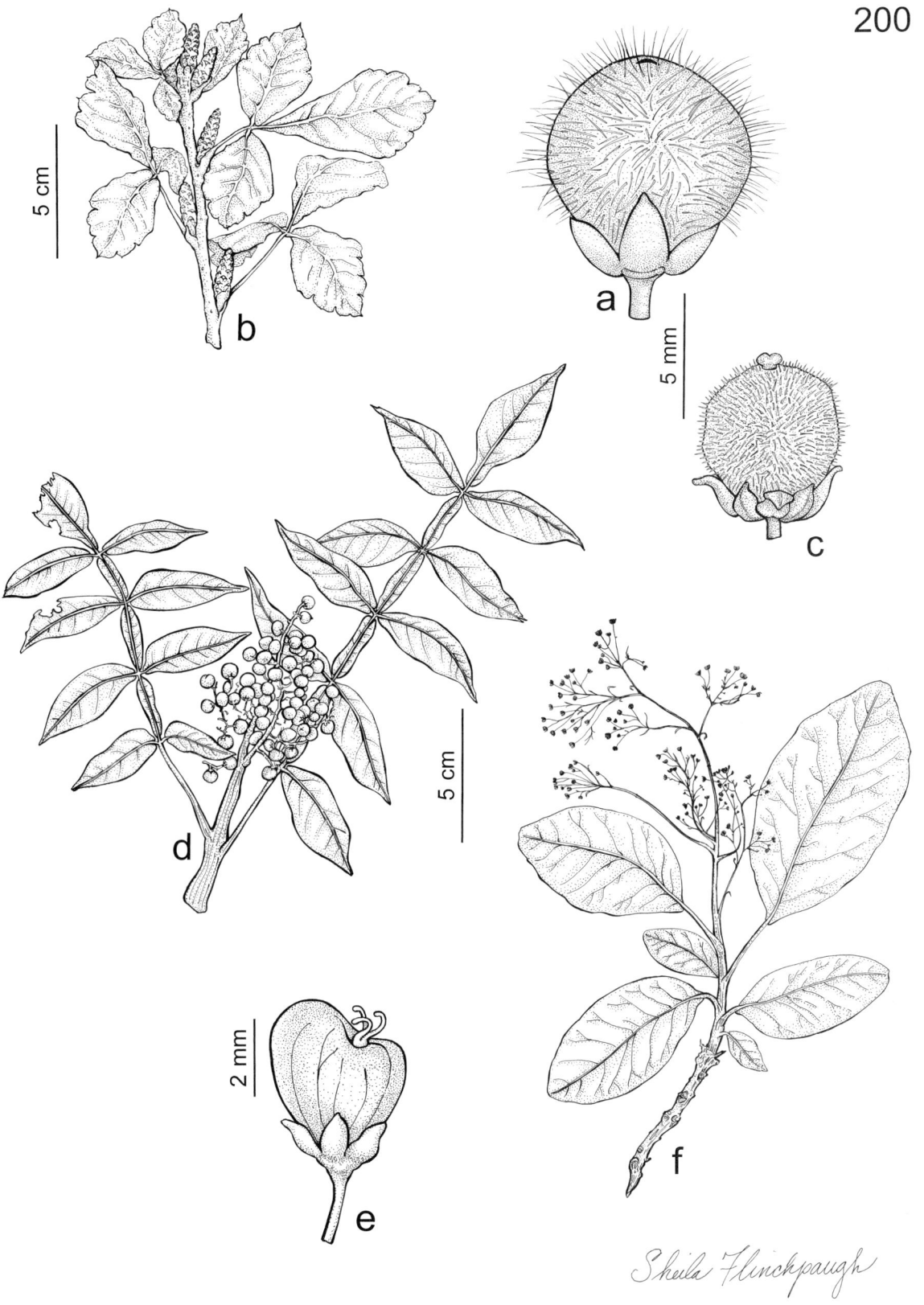

Plate 200. Anacardiaceae. *Rhus aromatica*, **a)** fruit, **b)** inflorescence. *Rhus copallinum*, **c)** fruit, **d)** fruiting branch. *Cotinus obovatus*, **e)** fruit, f) fruiting branch.

1. Terminal leaflet rhombic to ovate, not 3-lobed, the upper margins evenly and bluntly to sharply toothed, pointed at the tip; flowers produced before expansion of the leaves, the stalks 1–2 mm long; petals glabrous on the inner surface; fruits 4–5 mm diameter
   . . . . 1A. VAR. AROMATICA
1. Terminal leaflet broadly obovate, more or less 3-lobed, the upper margins scalloped or with broad to rounded teeth, angled to rounded at the tip; flowering at or after expansion of the leaves, the stalks 2–3 mm long; petals usually hairy on the inner surface; fruits 5–6 mm diameter
   . . . . 1B. VAR. SEROTINA

1a. var. aromatica
   R. aromatica Marshall var. *illinoensis* (Greene) Rehder
   R. canadensis Marshall
   Schmaltzia crenata (Mill.) Greene

Mature leaflets 2–7 cm long, 1.5–5.0 cm wide, sessile, both surfaces sparsely to densely pubescent with silky or velvety hairs, the margins also hairy. Terminal leaflet the largest, rhombic to ovate, the lower leaf margins entire, the upper leaf margins more or less evenly and bluntly to sharply toothed, pointed at the tip. Flowering before expansion of the leaves. Outer floral bracts stiff, reddish brown at the base, the central portion glabrous, the margins thin and hairy. Flower stalks 1–2 mm long, glabrous. Sepals 1.2–1.4 mm long, about 1.2 mm wide. Petals 1.6–2.5 mm long, glabrous on the inner surface. Fruits 4–5 mm wide, usually not shiny. Late March–mid-April.

Scattered to common throughout the state (eastern U.S. west to South Dakota and Texas; Canada). Glades, tops of bluffs, and openings of mesic to dry upland forests; also old fields and roadsides.

Rhus aromatica var. *aromatica* is the common taxon throughout eastern North America. It can usually be identified by the leaves alone, when fully mature leaves are available. The terminal leaflet is roughly diamond-shaped. The lower leaf margins are entire, the upper leaf margins with sharp to blunt, evenly spaced teeth. The tip is usually angled or tapered to a relatively sharp point. The flowering stems generally are straighter and less branched, and the leaves are somewhat thinner and less leathery than in *R. aromatica* var. *serotina*. In Missouri, var. *aromatica* consistently flowers about 2–3 weeks earlier than does var. *serotina*, before the leaves have emerged (Spellman and Dunn, 1965). Many herbarium specimens were collected after the flowering period, so it is difficult to determine how constant this character is throughout the state. Scattered populations of var. *aromatica* in eastern and northern states apparently flower with or after the emergence of the leaves.

Midwestern plants with densely hairy leaves have been called var. *illinoensis*. This variant, which is rare in Missouri, grades into the less-hairy forms considered more typical of var. *aromatica*.

1b. var. serotina (Greene) Rehder
   Schmaltzia serotina Greene
   R. canadensis var. *serotina* (Greene) E.J. Palmer & Steyerm.
   R. trilobata var. *serotina* (Greene) F.A. Barkley

Mature leaflets 2.0–7.5 cm long, 3–7 cm wide, sessile, sparsely pubescent with appressed, silky hairs above and below. Terminal leaflet largest, broadly obovate, variously lobed above, rounded, blunt to occasionally acute. Flowering at or after expansion of the leaves. Bracts reddish brown, densely woolly at the base and along the margins. Flower stalks 2–3 mm long, densely woolly to nearly glabrous. Sepals 1.0–1.2 mm long, about 1 mm wide. Petals 2.2–2.4 mm long, usually hairy on the inner surface. Fruits 5–6 mm wide, often somewhat shiny. Late April–early May.

Scattered south of the Missouri River, uncommon northward to Monroe and Pike Counties, apparently absent from the Mississippi Lowlands Division (Illinois to South Dakota and Texas). Glades, tops of bluffs, and openings of mesic to dry upland forests.

The native range of this variety in Missouri has become obscured as a result of its use for wildlife plantings in woodland restorations and other places managed for natural features.

The var. *serotina* is a variable taxon but is distinguished from var. *aromatica* by several characters. In Missouri, var. *serotina* flowers later than var. *aromatica*, after the leaves have made their appearance. The name *serotina* refers to this later flowering. Spellman and Dunn (1965) recorded phenological data on populations for four years and found that the peak of flowering was consistently three weeks apart. The stems in var. *serotina* tend to be more branched and twisted, as in *R. trilobata*. The mature leaves are about the same size as in var. *aromatica*, but the terminal leaflet has a tendency to be 3-lobed, as in *R. trilobata*, and the teeth are more irregularly spaced and rounded. In addition, the flower stalks of var. *serotina* are usually somewhat longer than in var. *aromatica,* and the petals are usually hairy on the inner surface. Although var. *serotina* is distinct in Missouri and Arkansas, populations in Oklahoma and Kansas grade into *R. trilobata,* which is the common species in the Great Plains and western North America. Indeed, Spellman and Dunn (1965) agreed with Barkley (1937) that var. *serotina* had more in common with *R. trilobata* than with *R. aromatica.* They pointed out that var. *serotina* often occurs in more xeric situations, such as rocky outcrops and steep bluffs, much like *R. trilobata* does farther west.

This taxon originally was described as a distinct species, *Schmaltzia serotina,* with the type specimen from Independence (Jackson County). Since then, various botanists have treated it as a variety of the eastern *R. aromatica,* then called *R. canadensis* (now considered a synonym of *R. aromatica*), and the western *R. trilobata.* Some authors have gone further and treated *R. trilobata* as a variety of *R. aromatica* (McGregor, 1986c). However, *R. trilobata* appears to be a distinct and more or less uniform species, at least throughout much of the central Great Plains region, with small- to medium-sized 3-lobed leaflets. Some populations of *R. trilobata* flower before the leaves emerge, and some after the leaves emerge. Further complications arise from the fact that several varieties of *R. trilobata* have been described for the southwestern United States and California, which are in need of further study. Perhaps all these taxa eventually will be treated as varieties of a single widespread, polymorphic species.

2. Rhus copallinum L. (dwarf sumac, winged sumac, shining sumac)
Schmaltzia copallina Small
R. copallinum var. *latifolia* Engl.
Pl. 200 c, d; Map 832

Plants shrubs or small trees. Stems 1.5–6.0 m long, ascending. Young branches densely hairy, becoming glabrous or nearly so with age, the older branches with prominent lenticels. Leaves pinnately compound, with 7–13 leaflets, the petiole 3–6 cm long, densely hairy, the rachis narrowly winged between leaflets (the wing interrupted where the leaflets are attached). Leaflets 3.0–8.5 cm long, 1.5–3.0 cm wide, sessile, broadly lanceolate to oblong, the margins entire or with a few teeth distally, the upper surface dark green, glabrous or nearly so except for the densely hairy main veins, shiny, the undersurface light green to grayish green, sparsely to moderately felty-hairy and with scattered, minute, reddish-brown glandular hairs. Inflorescences terminal, dense, ovoid panicles 12–18 cm long, 6–8 cm wide. Sepals 0.8–1.1 mm long, ovate, bluntly to sharply pointed at the tip, moderately to densely hairy. Petals 1.5–2.5 mm long, oblong, rounded at the tip, yellowish green. Fruits 4–6 mm long, 4–5 mm wide, somewhat flattened, red, with dense, minute, stout, red glandular hairs and sparse to moderate, longer, white to colorless nonglandular hairs. June–July.

Scattered to common nearly throughout the state, but uncommon or absent from northern portions of the Glaciated Plains Division (eastern U.S. west to Minnesota and Texas; Canada). Glades, upland prairies, savannas, and openings of mesic to dry upland forests; also old fields, railroads, roadsides, and open, disturbed areas.

The species epithet has been spelled *copallina* in some of the botanical literature, but the original spelling that Linnaeus used should be retained. Steyermark (1963) noted that Missouri plants fall into var. *latifolia,* which has fewer but broader leaflets, as opposed to the leaves of var. *copallinum,* which have mostly 11–23 leaflets that are only 1–2 cm wide. There is broad intergradation between these varieties and among the other named variants within this species.

3. Rhus glabra L. (smooth sumac)
Schmaltzia glabra Small
Pl. 201 a–c; Map 833

Plants large shrubs or rarely small trees. Stems 2–5 m tall, ascending. Young branches glabrous, glaucous (but note that inflorescence branches usually are sparsely hairy), the older branches usually with prominent lenticels. Leaves pinnately compound with 11–21 leaflets, the petiole 6–11 mm long, glabrous, reddish purple, the rachis not winged. Leaflets 5–13 cm long, 1.5–3.0 cm wide, lanceolate to elliptic-lanceolate, sessile or very short-stalked, the margins toothed, the upper surface dark green, glabrous, shiny, the undersurface light green, glabrous, glaucous. Inflorescences terminal, dense, ovoid panicles, 10–25 cm long, 5–10 cm wide. Sepals 1.6–2.0 mm long, narrowly ovate, sharply pointed at the tip. Petals 2.0–2.5 mm long, oblanceolate, rounded at the tip, greenish yellow, sparsely hairy on the inner surface. Fruits 4–6 mm long, 4–5 mm wide, somewhat flattened, red, with dense, minute, stout, red glandular hairs. May–June.

Scattered nearly throughout the state (U.S., Canada, Mexico). Open woods, brushy areas along roadsides, railroads, and fencerows.

834. Rhus typhina   835. Toxicodendron pubescens   836. Toxicodendron radicans

This species has been investigated as a potential source of tannins and oils (Campbell, 1984). It is too aggressive for most home gardens, but a cut-leaved cultivar, f. *laciniata* (Carr.) B.L. Rob., apparently is less aggressive and sometimes is cultivated. Steyermark (1963) mentioned the northern var. *borealis* Britton with somewhat hairy branches, but this is now considered a putative hybrid between *R. glabra* and *R. typhina* and treated as *R.* ×*borealis* (Britton) Greene. Because *R. typhina* is rare in Missouri, it is unlikely that this hybrid will be recorded from the state in the future.

4. Rhus typhina L. (staghorn sumac)
R. hirta (L.) Sudw.
Schmaltzia hirta Small

Pl. 201 d; Map 834

Plants tall shrubs or small trees 2–6 m tall, with stout branches. Young branches, petioles, and leaf rachis densely pubescent with woolly or felty hairs. Leaves pinnately compound, with 9–25 leaflets, rarely 2 times pinnately compound, 9–40 cm long, the petioles 3–10 cm long, densely woolly or felty, the rachis not winged. Leaflets 7–10 cm long, 1.0–4.5 cm wide, (appearing dissected in bipinnate forms), lanceolate to narrowly oblong, sessile or very short-stalked, the margins toothed, the upper surface dark green, glabrous or nearly so, shiny, the undersurface light green, moderately to densely hairy along the veins, also glaucous. Inflorescences terminal, dense, ovoid panicles, 7–25 cm long, 3–4 cm wide. Sepals 1.2–1.5 mm long, narrowly ovate, sharply pointed at the tip. Petals 1.5–2.5 mm long, oblong-oblanceolate, rounded at the tip, greenish yellow, sparsely hairy on both surfaces. Fruits 4–5 mm long, 4–5 mm wide, somewhat flattened, red, with dense, slender, straight, red hairs 1–2 mm long. $2n=60$. June–July.

Introduced, known thus far from Greene County and the St. Louis metropolitan area (eastern U.S. west to Wisconsin and Mississippi, introduced farther west; Canada). Railroads and open, disturbed areas.

The name *R. typhina* L. is widely used and deeply entrenched in the botanical literature, but for a time it appeared that, as a consequence of a minor change in wording in a past International Code of Botanical Nomenclature, an older name, *R. hirta*, might have to be used for this taxon (Reveal, 1991b; Kartesz and Gandhi, 1991). The basis for this name, *Datisca hirta* L., was based upon a monstrous cultivated form of the species in which the inflorescence had reverted to leaves. Fortunately, that epithet was rejected formally from any further use at the 1999 International Botanical Congress in St. Louis (Greuter et al., 2000), thus stabilizing the use of the name *R. typhina*.

The original reports of this species in Missouri were of escaped plants of *R. typhina* f. *dissecta* Rehder (var. *laciniata* A.W. Wood), a cultivar with dissected, 2 times pinnately compound leaves that is sometimes planted as an ornamental in gardens.

3. **Toxicodendron** Mill. (poison ivy, poison oak)

Plants shrubs or lianas, dioecious, climbing by masses of aerial roots, often with rhizomes. Leaves trifoliate (pinnately compound elsewhere). Leaflets with the margins entire or coarsely toothed to irregularly lobed. Inflorescences axillary panicles, loose and drooping, developing

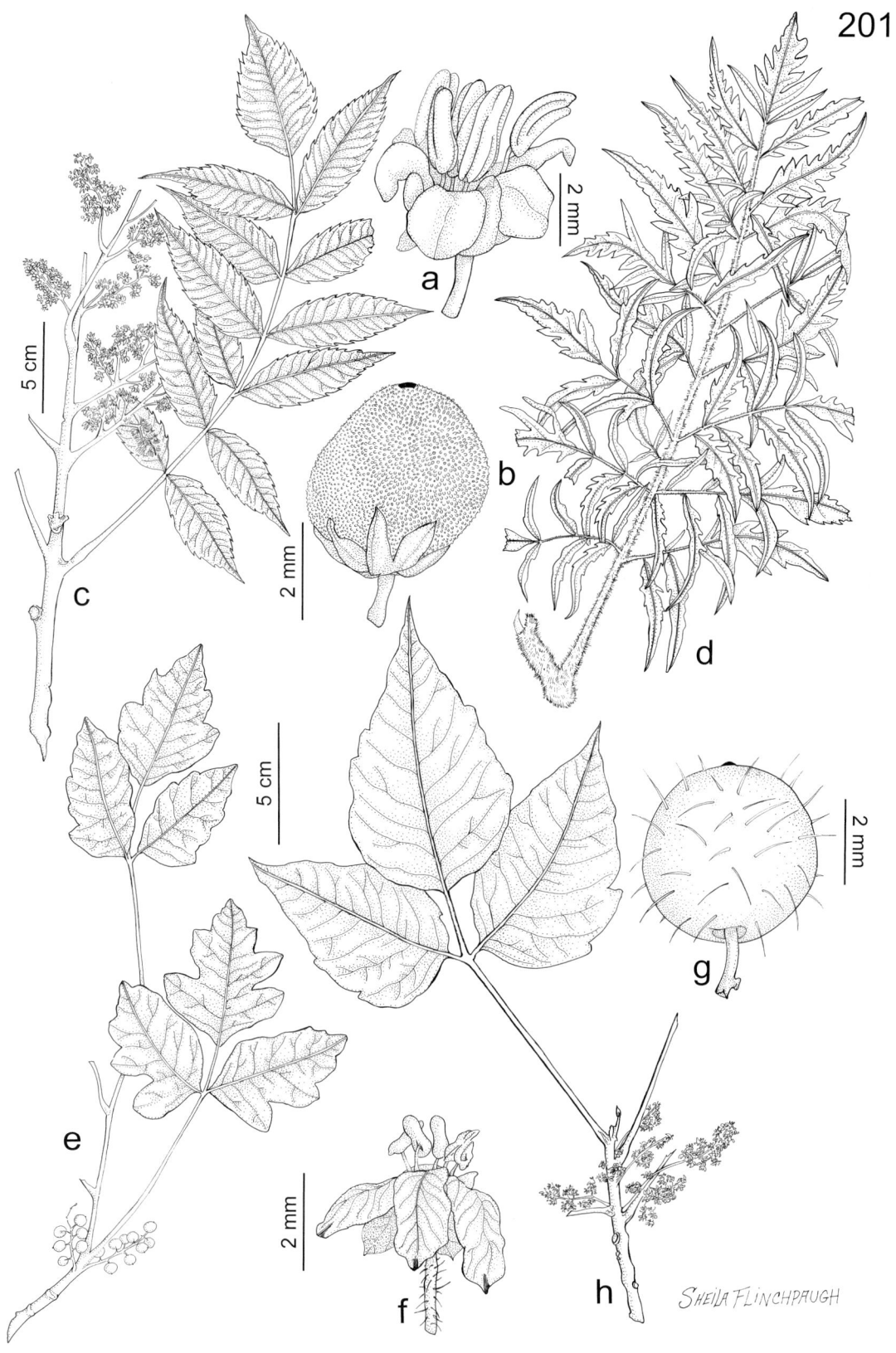

Plate 201. Anacardiaceae. *Rhus glabra*, **a)** flower, **b)** fruit, **c)** flowering branch. *Rhus typhina* (f. *dissecta*), **d)** branch. *Toxicodendron pubescens*, **e)** fruiting branch. *Toxicodendron radicans*, **f)** flower, **g)** fruit, **h)** flowering branch.

after the leaves expand. Flowers all fertile, the stalks not plumose-hairy. Sepals 5, united at the base, greenish to cream-colored. Petals 5, glabrous, cream-colored. Staminate flowers with the stamens 5, the anthers ovoid to more or less oblong in outline. Pistillate flowers with the styles free, appearing terminal, equal in length or nearly so, short, sometimes fused toward the base. Fruits globose to subglobose drupes, glabrous or pubescent with inconspicuous nonglandular hairs, sometimes with minute tubercles, the thin outer layer often splitting open at maturity to expose the waxy middle layer, this appearing white or nearly so, often spotted with black resin ducts, the stone with a few longitudinal ridges. About 15 species, mostly temperate North America and eastern Asia, less commonly South America.

Toxicodendron, which means poison tree, has a well-deserved reputation as a plant to be avoided. All parts of the plant contain a resinous oil, urushiol, that causes an irritating rash in many people. The oil contains a mixture of catechols with long alkyl side-chains that penetrate the skin, interact with proteins, and trigger an immune response. The rash usually appears 24–48 hours after contact with the oil, and until washed off, the oil may spread to other parts of the body. Affected skin reddens, itches, and tiny blisters often appear. The rash usually fades away in 2–3 weeks. Symptoms may develop after merely brushing the plant, or by coming into contact with clothing, tools, or pets that have touched the plant. Some people seem to be more susceptible than others, and sensitivity often increases with repeated exposures. Although people do not develop symptoms without direct contact with the plant, burning poison ivy leaves can release droplets of the oil, which can then be carried by the smoke to the eyes, throat, and lungs. If contact with the plant is suspected, immediate washing with cold water may help remove some of the oils. Cleaning the exposed area with isopropyl alcohol may also be helpful. Washing with soap and hot water may actually spread the oils. Once a rash has started, there is no proven cure, but the discomfort can be soothed with cooling compresses or calamine lotion. Scratching the blisters will not spread the rash, but it can lead to infection and should be resisted. A doctor should be consulted in extreme cases, if the rash spreads over much of the body, causes significant swelling, or affects the eyes, throat, or internal organs. Many folk cures and myths about poison ivy exist. Most are harmless, but so-called preventative measures such as eating small pieces of the leaves to induce immunity are dangerous and not recommended.

The best medicine is prevention, learning to recognize poison ivy, and avoiding contact with it. The old adage "leaflets three, let it be" is useful, although there are a few plants with trifoliate leaves that are harmless. *Toxicodendron radicans* and *T. pubescens* are distinguished by the distinctive petiole-like stalk of the terminal leaflet. The whitish berries are another distinguishing character, although they are green while maturing and are not produced on the staminate plants. *Toxicodendron radicans* often grows straight up a tree trunk, adhering by dense aerial roots (these produced in response to contact of the stem with a substrate), but it takes on a variety of growth forms in different habitats. Among the plants frequently mistaken for poison ivy, fragrant sumac (*Rhus aromatica*) has trifoliate leaves, but the terminal leaflet lacks the distinct stalk found in poison ivy, and the berries are reddish. The leaves of box elder (*Acer negundo*) have 3–7 leaflets, but the leaves are opposite on the stem rather than alternate as in poison ivy. Virginia creeper (*Parthenocissus quinquefolia*) climbs like *T. radicans* but usually has 5 leaflets instead of three (except in seedlings) and has blue berries.

The name poison sumac sometimes has been attributed to Missouri plants, but in fact this species does not occur in the state. True poison sumac, *Toxicodendron vernix* (L.) Kuntze, has pinnately compound leaves with 7–13 entire-margined leaflets and occurs in swamps and bogs to the east and north of the state.

1. Plants low shrubs, not climbing, not producing aerial roots; leaflets usually with 3–7 deep, more or less rounded lobes; fruits inconspicuously hairy and/or with minute papillae .... 1. T. PUBESCENS
1. Plants often climbing (but can appear shrubby in open habitats), usually producing aerial roots; leaflets entire or coarsely and bluntly saw-toothed, sometimes irregularly lobed; fruits glabrous, rarely with minute papillae .... 2. T. RADICANS

1. Toxicodendron pubescens Mill. (eastern poison oak)

Rhus toxicodendron L.
T. toxicarium (Salisb.) Gillis
R. toxicodendron var. *quercifolium* Michx.
T. quercifolium (Michx.) Greene

Pl. 201 e; Map 835

Plants low shrubs 0.3–0.5 m tall, spreading by rhizomes, not climbing, not producing aerial roots. Leaves often clustered near the tip of the stem, the petiole 5–15 cm long. Leaflet stalks 1–3 mm long in lateral leaflets, 0.5–2.0 cm long in the terminal leaflet. Leaflets 3–10 cm long, 3.0–7.5 cm wide, ovate to obovate, angled at the base, rounded to broadly and bluntly pointed at the tip, the margins with 3–7 coarse, deep, rounded lobes, those of the lateral leaflets usually somewhat asymmetrical (deeper on the lower side than on the upper side) and sometimes with a few blunt teeth, the upper surface sparsely hairy, the undersurface sparsely to moderately hairy, especially along the veins. Inflorescences 1–4 cm long. Sepals 1.0–1.2 mm long, narrowly ovate, green. Petals 2.2–2.4 mm long, oblanceolate, cream-colored with dark veins. Ovary densely hairy. Fruits 4.0–4.5 mm long, 4.0–4.5 mm wide, globose, greenish white to tan, the outer layer separating at maturity to reveal a powdery white middle layer, the stone kidney-shaped with a smooth surface. $2n=30$. May–June.

Uncommon in the southern portion of the Ozark Division and in the northern part of the Mississippi Lowlands (southeastern U.S. west to Kansas and Texas). Glades, openings of dry upland forests, sand prairies, and sand savannas; also roadsides.

Toxicodendron pubescens is recognized by a combination of its shrubby, nonclimbing habit, deeply lobed leaves, hairy fruits, and occurrence in low-nutrient, sandy or rocky soils. It produces aerial stems from rhizomes, but it never climbs or produces aerial roots as in *T. radicans*. In its typical form, the leaflets of *T. pubescens* are deeply lobed and reminiscent of the leaves of white oak or blackjack oak, hence the common name poison oak. Unfortunately, this common name is frequently misapplied to lobed forms of *T. radicans* as well. According to Gillis (1971), the lobes are deeper on staminate plants than on pistillate plants. Although the fruits are usually sparsely hairy, occasionally they are nearly glabrous. A few putative hybrid specimens between *T. pubescens* and *T. radicans* have been identified, but the relative rarity of hybrids was a factor in the recognition by Gillis (1971) of *T. pubescens* as a distinct species.

The nomenclatural history of *T. pubescens* is quite complex (Fernald, 1941b; Gillis, 1971; Kartesz and Gandhi, 1991; Reveal, 1991a). The lobate-leaved, pubescent forms of poison ivy originally were described as *Rhus toxicodendron* L. However, when classified into the segregate genus *Toxicodendron*, the name *Toxicodendron toxicodendron* (L.) Britton is not allowed under the International Code of Botanical Nomenclature because it would form a tautonym (a binomial in which the generic and specific epithets are identical). Gillis (1971) recognized poison oak under the name *T. toxicarium*, which was in turn an illegitimate combination, leaving the next available legitimate name within *Toxicodendron* as *T. pubescens*.

2. Toxicodendron radicans (L.) Kuntze (poison ivy)

Rhus radicans L.

Pl. 201 f–h; Map 836

Plants shrubs 0.5–3.0 m tall or more often lianas to 30 m or more, spreading by rhizomes, often with abundant aerial rootlets. Leaves scattered along the branches, the petiole 6–15 cm long. Leaflet stalks 2–5 mm long in lateral leaflets, 2–5 cm long in terminal leaflets. Leaflets 5–18 cm long, 4–16 cm wide, ovate, angled to truncate, often unequally so on lateral leaflets, usually angled or tapered to a sharply pointed tip, the margins entire or coarsely and bluntly saw-toothed, sometimes irregularly lobed, those of the lateral leaflets usually somewhat asymmetrical (deeper on the lower side than on the upper side), the upper surface glabrous to sparsely hairy, the undersurface glabrous or sparsely to moderately hairy, especially along veins. Inflorescences 4–12 cm long. Sepals 1.2–1.4 mm long, narrowly ovate-triangular. Petals 2.4–2.6 mm long, narrowly elliptic, yellowish white with dark veins. Fruits 3–4 mm long, 4–5 mm wide, subglobose, the outer layer shiny, green at first (July–August) and turning creamy yellow or tan at maturity (September), glabrous, rarely with sparse, minute papillae or hairs, the outer layer becoming papery at maturity and separating, revealing a powdery white middle layer spotted with small, black lines (resin ducts). $2n=30$. May–July.

Common throughout the state (eastern U.S. west to South Dakota and Texas; Canada, Mexico, Central America, Asia). Bottomland forests, mesic upland forests, sand savannas, thickets in upland prairies and loess hill prairies, banks of streams, rivers, and spring branches, margins of ponds, lakes, and sinkhole ponds, and ledges of bluffs; also fencerows, railroads, roadsides, and shaded, disturbed areas.

This species is extremely variable in growth form. To the north of Missouri, in sand dunes along the Great Lakes, it takes on a low, nearly herbaceous habit, but in Missouri it produces woody stems. In open areas the plants tend to remain shrubby, but when growing under a tree canopy, the plants usually become stout lianas that climb to the tops of trees and have thick-barked stems that can eventually reach 10 cm in diameter. Steyermark (1963) noted that a plant in Indiana measured more than 50 m in length. Thick masses of adventitious roots are produced relatively evenly (vs. usually in discrete clumps or patches in *Parthenocissus*) wherever a climbing stem makes contact with a substrate and these act to anchor the plant securely to rocks, tree trunks, etc. Although poison ivy is not recommended for cultivation in gardens, it has a long history of being grown as an ornamental in parts of Europe, presumably for its bright fall foliage.

Toxicodendron radicans exhibits a wide range of leaf forms and pubescence types that are difficult to categorize into discrete subspecies. Although Barkley (1937) had lumped all of the temperate North American plants into the single var. *radicans*, Gillis (1971) recognized a number of subspecies that are roughly correlated with broad geographical regions. There exist many intermediate, transitional, or possibly hybrid plants. The three subspecies said to occur in Missouri are separated in the following key. Although these subspecies may be distinct in some parts of their range in other states, they are not clearly distinguishable throughout Missouri. Their inclusion as distinct taxa in Missouri can only be considered tentative at present.

1. Leaflets with the undersurface mostly glabrous or with small tufts of hairs in the vein axils, the margins entire; fruits minutely roughened or with sparse, minute papillae or hairs .... 2C. SSP. RADICANS
1. Leaflets variously pubescent, lacking tufts of hairs in the vein axils, the margins rarely entire or more commonly variously toothed or lobed; fruits completely glabrous
    2. Leaflets sparsely hairy on the undersurface ...... 2A. SSP. NEGUNDO
    2. Leaflets densely hairy (often woolly or felty) on the undersurface .... 2B. SSP. PUBENS

2a. ssp. negundo (Greene) Gillis
  T. negundo Greene
  T. radicans (L.) Kuntze var. *negundo* (Greene) Reveal
  Rhus radicans var. *vulgaris* (Michx.) DC. f. *negundo* (Greene) Fernald

Leaflets ovate to elliptic, the margins usually toothed or with 1 or 2 large lobes, the upper surface glabrous or with a line of short, curly hairs along the midvein, the undersurface glabrous or sparsely hairy, lacking tufts of hairs in the vein axils. Fruits glabrous. $2n=30$. May–July.

Common throughout the state (northeastern U.S. west to South Dakota and Texas; Canada). Bottomland forests, mesic upland forests, thickets in upland prairies and loess hill prairies, banks of streams, rivers, and spring branches, margins of ponds, lakes, and sinkhole ponds, and ledges of bluffs; also fencerows, railroads, roadsides, and shaded, disturbed areas.

In the Midwest, ssp. *negundo* is the prevalent subspecies, and it is by far the most common subspecies in Missouri. It is distinguished by not having the characteristics of other subspecies, rather than by any particular character of its own (Gillis, 1971). The margins of the leaflets are quite variable, toothed and often irregularly lobed, especially on the lateral leaflets, but rarely are they entire, as in ssp. *radicans*. The fruits are usually glabrous but rarely can be sparsely and minutely papillose. The undersurface of the leaflets usually is pubescent along the veins, but whether the hairs are tufted in the vein axils or more generally distributed can be difficult to diagnose. Following Fernald (1941b), Steyermark (1963) recognized two specimens in Missouri with entire leaves as *Rhus radicans* var. *radicans* f. *hypomalaca* Fernald. Although Gillis (1971) suggested that f. *hypomalaca* represents an intermediate between ssp. *radicans* and ssp. *negundo,* it seems more likely that the character of having entire leaves is part of the range of variation exhibited within ssp. *negundo.*

2b. ssp. pubens (Engelm. ex S. Watson) Gillis
Rhus toxicodendron var. *pubens* Engelm. ex S. Watson
T. radicans (L.) Kuntze var. *pubens* (Engelm. ex S. Watson) Reveal

Leaflets ovate, the margins toothed, notched, or irregularly lobed, the upper surface roughened or rarely glabrous, the undersurface densely hairy (often woolly or felty), lacking well-defined tufts of hairs in the vein axils. Fruits glabrous.

Uncommon and widely scattered in Missouri (Kansas, Oklahoma, and Texas east to Virginia, Tennessee, and Mississippi). Bottomland forests, mesic upland forests, margins of ponds and lakes, and ledges of bluffs; also roadsides and shaded, disturbed areas.

This subspecies was not recognized by Barkley (1937) or Steyermark (1963). It is said to differ from ssp. *negundo* in its broad, hairier leaves, and from ssp. *radicans* by its usually toothed, hairy leaves and glabrous fruits (Gillis, 1971). It occurs most commonly and in its purest form in the evergreen forests of the western Gulf Coastal Plain, but it supposedly hybridizes with both of these subspecies to the north and west, including populations in the state of Missouri. In actual practice, it is often impossible to clearly distinguish ssp. *pubens* from ssp. *negundo* in Missouri, as many populations are somewhat intermediate in leaflet morphology.

2c. ssp. radicans

Leaflets narrowly to broadly ovate, the margins entire, the undersurface mostly glabrous or with small tufts of hairs in the vein axils, the undersides of leaflets with tufts of hairs in the vein axils. Fruits minutely roughened or with sparse, minute papillae or hairs.

Uncommon, known thus far only from a single collection from Scott County (eastern U.S. west to Illinois and Texas; Canada). Sand savannas.

Plants attributed to ssp. *radicans* were first collected in Missouri in 1992. At the only site known to date, it occurs in a mixed population with plants of ssp. *negundo* and *T. pubescens* in deep, sandy soil. Plants of all three taxa occur in this degraded sand savanna as low, spreading shrubs.

ANNONACEAE (Custard Apple Family)

About 112 genera, about 2,150 species, tropical and subtropical regions worldwide, with one temperate North American genus.

1. Asimina Adans.
(Kral, 1960; Wilbur, 1970)

About 10 species, eastern U.S. and adjacent Canada.

Asimina is the only temperate genus in a large, otherwise tropical family. The flowers are only successfully fertilized via outcrossing, often resulting in low fruit set. The fruits are dispersed by mammals.

1. Asimina triloba (L.) Dunal (pawpaw)
Pl. 202 g–i; Map 837

Plants small trees, rarely to 10 m tall, often forming colonies from root suckers. Bark thin, smooth, becoming roughened (warty) or rarely scaly on older trees, gray or less commonly dark brown, usually with white to light gray patches. Twigs light to dark reddish brown, glabrous or sparsely hairy, the buds lacking scales, reddish brown, hairy, the terminal bud flattened and narrowly ovate in outline, the lateral buds nearly globose. Leaves alternate, simple, 10–35 cm long, oblanceolate to oblong-obovate, gradually narrowed or tapered to a petiole 5–20 mm long at the base, abruptly tapered to a short point at the tip, glabrous at maturity (hairy when very young), the veins prominent, the margins entire, the upper surface green and usually shiny, the undersurface

52 ANNONACEAE

837. Asimina triloba

838. Ammi majus

839. Ammoselinum butleri

pale green. Stipules absent. Flowers produced before or as the leaves develop, solitary at the nodes of the leafless second year's growth along branches, perfect, hypogynous, with short, hairy stalks. Sepals 3, 8–12 mm long, broadly triangular, sharply pointed at the tip, green, usually hairy, and shed early. Petals in two unequal and strongly overlapping series of 3, the outer series 2.1–2.7 cm long, broadly ovate, the tips spreading and curled outward, the inner series 1.0–1.4 cm long, ovate, erect or with the tips somewhat spreading, the surface of both series with prominent veins, dark purplish brown to maroon at maturity (often green during development), hairy on the outer (under) surface, the inner series with nectar-producing glands at the base of the inner surface. Stamens numerous, small, free, densely packed around the elongated receptacle, not clearly differentiated into an anther and filament. Pistils 3–5(–15), hairy, with 1 carpel, the ovary superior and with 12–16 ovules, the style short, the stigma globose. Fruits single or 2–4 in a spreading or drooping cluster, berries, 4–13 cm long, ellipsoid to more or less cylindrical, rounded at the ends, pale green, turning yellowish and then brownish black with age. Seeds 2–10, 2.0–2.8 cm long, flattened, elliptic-ovate in outline, dark brown, shiny, embedded in a white to light yellow pulp. $2n=18$. March–May.

Scattered to common nearly throughout the state (eastern U.S. and adjacent Canada west to Nebraska and Texas). Bottomland forests, mesic upland forests in ravines, banks of streams and rivers, and bases of bluffs.

Plants of *Asimina* are aromatic. The leaves have an odor similar to that of motor oil when bruised. The flowers have an aroma similar to fermenting grapes. The bruised or opened fruits also have a perfumy aroma. The fruits of pawpaws have been called Indiana bananas and Missouri bananas and are considered a delicacy by wild foods enthusiasts, although others may find the taste nauseating. If not already eaten by birds or mammals, they may be harvested in September and early October, when they are slightly soft to the touch. The pulp has a creamy texture and a strong, sweet, exotic flavor. It may be eaten raw or made into ice cream, sherbet, pudding, pie, jelly, or other confections. Care must be taken, as some individuals develop dermatitis when handling the fruits or have an allergic reaction to the ingested pulp. The seeds are poisonous. Aside from the fruits, most other parts of the plant (especially the bark) contain a variety of poisonous chemicals, including alkaloids (such as asiminine and analobine) and a group of fatty acid lactones called acetogins (notably asimicin). The acetogins of pawpaw are under investigation as potential anticancer and antitumor drugs, and also for possible applications as pesticides. Other uses of pawpaw include cultivation as a small shade tree (especially the cultivars with thicker trunks and reduced suckering). Pioneers and some Native Americans used strips of the tough, thin bark for weaving into cloth and for fish stringers.

The leaves of *Asimina* are a principal larval food source for the zebra swallowtail butterfly (*Euritides marcellus* Cramer). Although most members of the Annonaceae are pollinated by beetles, it is possible that *A. triloba* is pollinated by flies. Robertson (1889) reported visitation of flowers on Illinois plants by several different small flies, and Joe Smentowski (personal communication) of Brentwood, Missouri, reported seeing small, nocturnal flies attracted by the odor of the flowers.

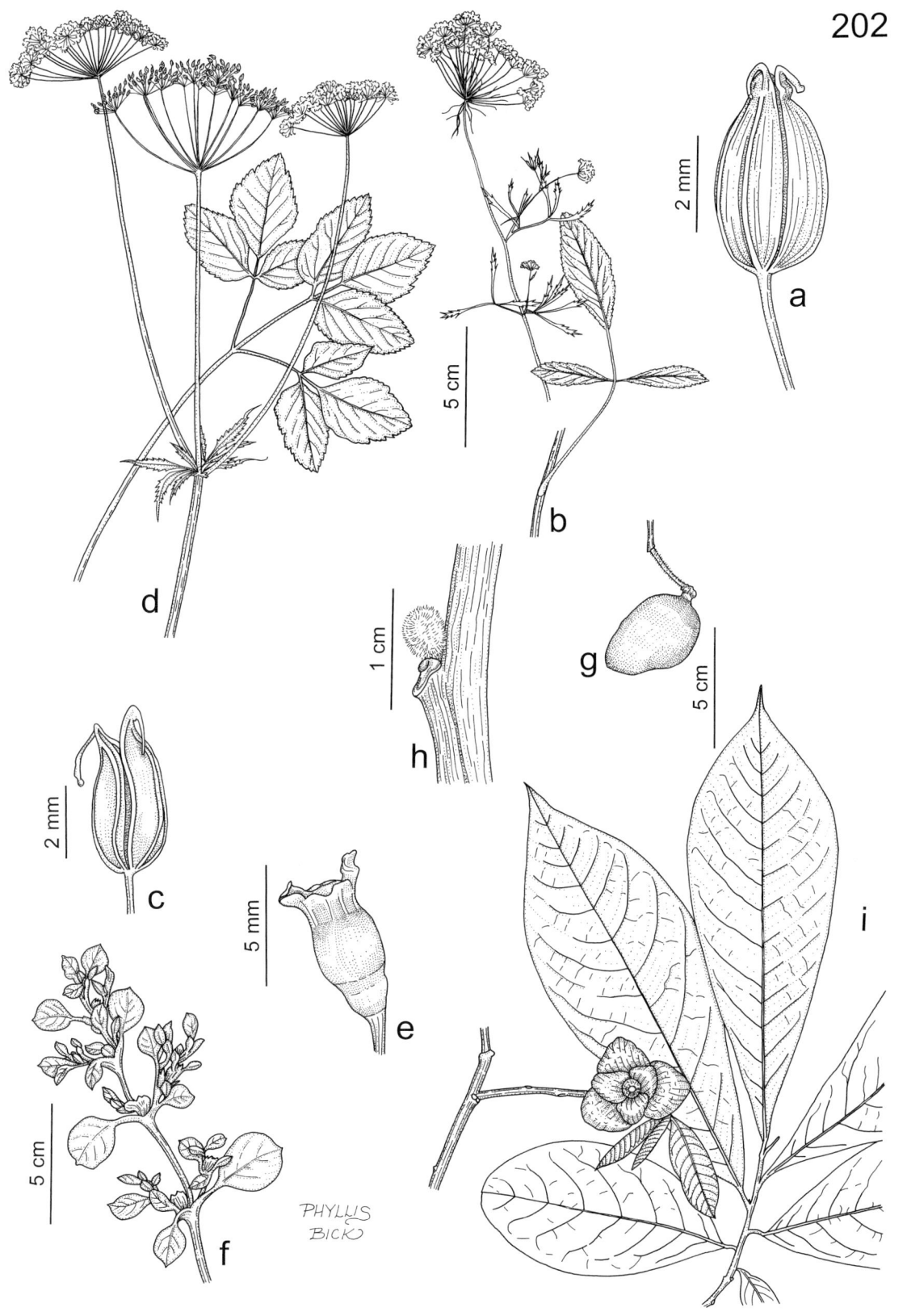

Plate 202. Aizioaceae, Annonaceae, Apiaceae. *Ammi majus*, **a)** fruit, **b)** habit. *Aegopodium podagraria*, **c)** fruit, **d)** habit. *Trianthema portulacastrum*, **e)** fruit, **f)** habit. *Asimina triloba*, **g)** fruit, **h)** winter bud, **i)** flower and leaves.

APIACEAE (UMBELLIFERAE) (Carrot Family)
(Mathias and Constance, 1944–1945)

Plants annual or perennial herbs. Stems usually branched, often hollow between the nodes. Leaves alternate and/or basal, less commonly opposite or whorled, variously compound or simple, petiolate or less commonly sessile, rarely perfoliate or peltate, the petiole base often expanded and somewhat sheathing, the stipules partially fused to the petiole base or lacking. Inflorescences simple or compound umbels of small flowers or sometimes condensed into heads, rarely appearing as small spikes, often with bracts subtending the flowers and at the branch points. Flowers mostly perfect (functionally staminate or pistillate flowers mixed with the perfect ones in some genera, or occasionally the outermost flowers of an umbel functionally sterile and with enlarged corollas), epigynous, actinomorphic or occasionally the outermost flowers of the inflorescence with some of the petals enlarged. Sepals reduced to 5 small teeth (slightly larger in *Eryngium* and some genera with enlarged perianth in the outer flowers), sometimes absent, when present usually persistent in fruit. Petals 5, the tips usually curved or curled inward. Stamens 5, the filaments free. Pistil 1 per flower, composed of 2 fused carpels, the ovary inferior with a swollen nectar disc at the tip, the styles 2, often expanded at the base. Fruits schizocarps consisting of 2 mericarps that separate along the inner side at maturity, the tips of the inner sides often remaining attached for some time to a slender, sometimes Y-shaped stalk. Mericarps indehiscent, 1-seeded, often somewhat flattened, with 5 more or less prominent ribs, some or all of these sometimes winged. Three hundred to 450 genera, 3,000–3,550 species, nearly cosmopolitan, most diverse in temperate portions of Northern Hemisphere and mountainous regions of the tropics.

Genera and species in this family can be difficult to determine. As with many large, morphologically complex families, a set of specialized terms has evolved to describe variation in the Apiaceae and to facilitate discrimination between genera. In species with compound umbels, the branches (stalks) of the inflorescence are called *rays*. When present, the *involucre* subtending an inflorescence consists of *bracts,* whereas the *involucel* subtending an individual *umbellet* of a compound inflorescence is composed of *bractlets* (sometimes referred to as *bracteoles* in the literature). The small petals (0.5–3.0 mm in most Missouri species) are usually incurved or rolled inward, making size measurements difficult, and petal lengths thus are not useful for differentiating most genera. In the fruits, the usually persistent stalk positioned between the mericarps, to which they often remain attached after beginning to break apart, is called a *carpophore*. Details of the fruits often require viewing under magnification. The sides of the mericarps that are joined prior to breaking apart are known as *commissures,* and the two ribs along the edges of each commissure are referred to as *lateral ribs,* to distinguish them from the *dorsal rib* along the opposite margin and the two *intermediate ribs* on the mericarp face between the lateral and dorsal ones. Fruits that are *laterally flattened* have narrow commissures relative to the sides and individual mericarps that appear relatively plump, whereas fruits that are *dorsally flattened* have broad commissures and individual mericarps that appear strongly flattened. In most species, longitudinal secretory canals known as *oil tubes* occur between the ribs; these contain aromatic oils or resins and, although usually visible on the fruit surface, are most easily seen in cross-sectioned samples.

The relationship between the Apiaceae and Araliaceae, as traditionally circumscribed, continues to be imperfectly understood, and some authors have advocated uniting the two into a single family, Apiaceae (Thorne, 1973; Judd et al., 1994, 2002). Molecular studies of a number of independent gene sequences (summarized in Plunkett et al., 1997; Downie et al., 2000) have suggested that although the situation is complicated, most of the genera can be classified into two lineages that correspond roughly to Apiaceae and Araliaceae. However, the genera traditionally included in Apiaceae subfamily Hydrocotyloideae, whose relationships were controversial even before the advent of molecular systematics (Cronquist, 1981), have been shown to represent several groups with different affinities. One cluster, including *Hydrocotyle* and some other non-Missouri genera, appears to be more closely related to the Araliaceae than Apiaceae, and probably is better classified as a morphologically aberrant basal group in that family. Because further studies involving larger samples of additional species will be necessary to resolve the complex relationships among the umbellifer genera, the traditional classification of the two families continues to be followed in the present treatment.

Most Apiaceae contain aromatic oils in secretory canals present throughout the plants (although not evenly distributed in all parts; canals in the mericarps, for instance, are known as oil tubes), and many species emit characteristic strong odors when bruised or crushed. In part because of these terpenoid oils, many species are used as food, garnishes, and/or spices, including dill (*Anethium graveolens* L.), celery (*Apium graveolens* L.), caraway (*Carum carvi* L.), coriander (*Coriandrum sativum* L.), carrot (*Daucus carota* L.), parsley (*Petroselinum crispum* (Mill.) Nyman), and parsnip (*Pastinaca sativa* L.). A number of species also are cultivated as garden ornamentals. However, many members of the family are extremely poisonous when ingested, especially the hemlocks (*Cicuta, Conium*), and inexperienced individuals should resist the urge to harvest plants from the wild for culinary experiments. After contact with the skin, a number of species also cause dermatitis, in particular phototoxic reactions (those brought on and mediated by continued exposure of the skin to sunlight or other strong light sources). The information presented for individual taxa below was taken primarily from Kingsbury (1964) and Roth et al. (1994) but should not be considered an exhaustive account of all toxic species in the family. Unless sure knowledge of the identity and properties of a given species is known, all wild members of the Apiaceae should be considered potentially poisonous or toxic to the skin.

Steyermark (1963) reported *Aegopodium podagraria* L. (goutweed) (Pl. 202 c, d) from Missouri, based on a single specimen from Jefferson County. This specimen, cited as being collected by Bill Bauer in 1940, could not be located during the present study. *Aegopodium podagraria* is commonly grown in gardens as a ground cover and spreads vigorously by branched rhizomes, often forming irregular patches of basal rosettes of large, glabrous leaves, these 1 or 2 times ternately compound, with large, oblong to ovate, coarsely toothed leaflets. The white-petaled flowers are in dense compound umbels usually lacking bracts and bractlets, and the flattened fruits are inconspicuously ribbed. Steyermark did not state whether the specimen he reported had green or variegated leaves (var. *variegatum* L.H. Bailey), but forms with mottled or white-margined leaves are common in gardens. Goutweed can persist at old homesites and eventually may become naturalized in Missouri.

1. Leaves all simple, sometimes deeply lobed
    2. Stems prostrate, rooting at the nodes
        3. Leaves mostly opposite or whorled at the nodes of the creeping stem, the blades lanceolate to ovate, narrowed, rounded, or truncate at the base; inflorescences subtended by 2–5 linear bracts . . . . 18. ERYNGIUM
        3. Leaves mostly solitary at the nodes of the creeping stem, the blades circular to very broadly ovate or broadly kidney-shaped, peltate or deeply cordate at the base; inflorescences without bracts . . . . 22. HYDROCOTYLE
    2. Stems erect or ascending
        4. Leaf margins entire; at least the uppermost leaves perfoliate, the lower ones sessile . . . . 8. BUPLEURUM
        4. Leaf margins toothed or lobed, sometimes with spiny or threadlike teeth; leaves sessile or petiolate, usually sheathing the stem but not perfoliate

5. Leaf margins with threadlike or spiny teeth or lobes; flowers sessile, in dense cylindrical or globose heads, each flower subtended by a small bractlet .... 18. ERYNGIUM
5. Leaf margins toothed and/or lobed but not spiny; flowers some or all noticeably stalked, in compound umbels (the umbellets sometimes appearing nearly headlike), the individual flowers without bractlets
6. Umbellets with all staminate flowers or with a mixture of perfect and staminate flowers; ovaries and fruits densely pubescent with hooked bristles, the fruits with mericarps lacking ribs .... 34. SANICULA
6. Umbellets with all or nearly all perfect flowers; ovaries and fruits glabrous, the fruits with mericarps noticeably ribbed
7. Flowers all with stalks 1–5 mm long; mericarps with the ribs noticeably winged .... 38. THASPIUM
7. Central flower of each umbellet sessile or with the stalk less than 0.5 mm long, the other flowers with stalks 1–4 mm long; mericarps with the ribs angled and often somewhat corky, but lacking wings.................................. 41. ZIZIA
1. At least some of the leaves 1 or more times compound
8. Ovaries and fruits variously pubescent with short to long hairs or bristles, the hairs or bristles sometimes hooked
9. Leaf blades at most pinnately, ternately, or palmately 1 time compound, the leaflets often variously lobed
10. Fruits inconspicuously and sometimes sparsely pubescent with short hairs lacking hooked or barbed tips; stems 60–200(–300) cm long, stout and coarsely ridged; leaves with the sheathing bases, at least those of the median and upper leaves, moderately to strongly inflated
.... 21. HERACLEUM
10. Fruits conspicuously pubescent with hairs or bristles having hooked or barbed tips; stems 15–100 cm long, relatively slender (except rarely in *Torilis*) and finely ridged; leaves with the sheathing bases not or only slightly inflated
11. Leaf blades palmately compound with 3 or 5(7) leaflets, these elliptic-lanceolate to obovate in outline, sometimes irregularly few-lobed; fruits with the hairs or bristles having hooked tips
.... 34. SANICULA
11. Leaf blades pinnately compound, the leaflets lanceolate to oblong-ovate, usually deeply pinnately lobed; fruits with the hairs or bristles having minutely barbed tips .... 39. TORILIS
9. Leaf blades 2 or more times compound or dissected
12. Fruits conspicuously pubescent with hairs or bristles having hooked or barbed tips (the hairs usually also apparent [but shorter] on ovaries at flowering time)
13. Leaves glabrous, the blades finely pinnately 2–4 times dissected into narrowly linear, often threadlike ultimate segments; stems glabrous .... 36. SPERMOLEPIS
13. Leaves sparsely to moderately hairy, the blades 1 or 2 times pinnately compound, the leaflets lanceolate to oblong-ovate or ovate, usually deeply pinnately lobed, the lobes linear to elliptic-lanceolate, not threadlike; stems usually hairy

14. Fruits with secondary ribs between the primary ones, these winged and with a row of flattened bristles having minutely barbed or hooked tips, the primary ribs wingless and finely pubescent with straight hairs; involucre with the bracts leaflike, pinnately 1 or 2 times dissected with linear segments, rarely entire .... 16. DAUCUS
14. Fruits ribless or with slender nervelike ribs, these wingless, the surface covered with hairs or bristles having barbed or hooked tips; involucre absent or with the bracts inconspicuous, entire, narrowly triangular to more commonly linear, often hairlike
   15. Fruits tapered to a short beak at the tip, covered with short, stout hairs having hooked tips; sheathing bases of at least the uppermost leaves hairy along the margins .... 5. ANTHRISCUS
   15. Fruits narrowed to a blunt, beakless tip, covered with long bristles having minutely barbed tips; sheathing bases of the leaves all glabrous along the margins .... 39. TORILIS
12. Fruits pubescent with short hairs lacking hooked or barbed tips
   16. Leaf blades 2 or 3 times compound with distinct leaflets, these all or mostly 10 mm or more wide (except sometimes on the uppermost leaves)
      17. Fruits 4–6 mm long, oblong-elliptic to broadly ovate-elliptic in outline, shallowly cordate at the base, flattened dorsally, the lateral ribs with broad, papery wings; inflorescences with 18 to numerous rays .... 4. ANGELICA
      17. Fruits 10–24 mm long, linear to narrowly oblong-oblanceolate in outline, long-tapered at the base, flattened laterally, the ribs narrow, angled, lacking wings; inflorescences with 3–6 (–8) rays .... 28. OSMORHIZA
   16. Leaf blades 2–4 times compound and/or dissected, the leaflets or ultimate segments 0.5–4.0 mm wide
      18. Leaves all basal; petals yellow .... 26. LOMATIUM
      18. Leaves alternate along the stems and sometimes also basal; petals white, pink, or purple, sometimes drying yellow
         19. Fruits 4–10 mm long, linear to narrowly oblong-elliptic in outline, with 5 narrow to broad, low, blunt ribs, these lacking wings or bristles, the entire mericarp usually finely and uniformly hairy; involucre absent .... 10. CHAEROPHYLLUM
         19. Fruits 3–5 mm long, oblong-elliptic to oblong-ovate in outline, with secondary ribs between the primary ones, these winged and with a row of flattened bristles, the primary ribs wingless and finely pubescent; involucre of 4–15 leaflike bracts .... 16. DAUCUS
8. Ovaries and fruits glabrous, sometimes minutely roughened with minute teeth or tubercles
   20. Leaves at most once palmately, ternately, or pinnately compound, the leaflets usually discrete but sometimes lobed or relatively narrow

21. Leaflets with the margins entire
    22. Fruits 4–7 mm long, strongly flattened dorsally, the lateral ribs with broad, spreading wings; leaves with the rachis and leaflets lacking cross-veins or partitions; leaflets (2–)5–45 mm wide, glaucous on the undersurface; plants perennial, with clusters of tuberous-thickened roots . . . . 29. OXYPOLIS
    22. Fruits 2–4 mm long, slightly flattened laterally, the ribs lacking wings, but the lateral ribs with small, corky extensions over the commissures; leaves with the rachis or leaflets having irregular fine cross-veins or partitions; leaflets 1–6 mm wide, not glaucous; plants annual, with fibrous roots
        23. Leaf blades (except those of simple leaves) palmately compound, the 3–5 leaflets all attached at the tip of the petiole; fruits with a short but noticeable beak . . . . 15. CYNOSCIADIUM
        23. Leaf blades (except those of simple leaves) pinnately compound, the 3–11 leaflets attached along a rachis; fruits lacking a beak . . . . 25. LIMNOSCIADIUM
21. Leaflets lobed or with the margins toothed
    24. Leaves with the sheathing bases, at least those of the median and upper leaves, moderately to strongly inflated; stems noticeably pubescent, the hairs easily visible to the naked eye; fruits 7–12 mm long . . . . 21. HERACLEUM
    24. Leaves with the sheathing bases not or only slightly inflated; stems glabrous or inconspicuously short-hairy, the hairs visible only with magnification; fruits 1.5–7.0(–8.0) mm long
        25. Leaves with no more than 3 leaflets (these sometimes lobed)
            26. Petals white; fruits narrowly oblong or narrowly oblong-elliptic in outline
                27. Leaflets or ultimate segments oblong-lanceolate to elliptic or obovate, the margins sharply and often irregularly toothed (often doubly toothed), the teeth lacking spiny tips; fruits 4–8 mm long, narrowly oblong-elliptic in outline; involucre absent or of 1 spreading to ascending bract . . . . 14. CRYPTOTAENIA
                27. Leaflets or ultimate segments linear to narrowly oblong, the margins finely and regularly toothed, the teeth with minute, thickened, spiny tips; fruits 2–5 mm long, narrowly oblong in outline; involucre of 3–12 spreading to loosely reflexed bracts . . . . 19. FALCARIA
            26. Petals yellow or dark purple; fruits 2–4 mm long, ovate to oblong-ovate in outline
                28. Flowers all with stalks 1 mm or longer; fruits with the ribs winged . . . . 38. THASPIUM
                28. Central flower of each umbellet sessile or with a stalk less than 0.5 mm long (the other flowers with stalks 1–4 mm long); fruits with the ribs lacking wings . . . . 41. ZIZIA
        25. At least some of the leaves with 5–21 leaflets
            29. Petals yellow; involucel and involucre both absent; leaflets or ultimate divisions rounded or narrowed to a blunt point at the tip . . . . 30. PASTINACA
            29. Petals white; involucel present; involucre present (except in *Apium*); leaflets or ultimate divisions narrowed or tapered to a sharp point at the tip (rounded in *Apium*)
                30. Involucel absent; leaflets or ultimate divisions obovate or narrowly to broadly wedge-shaped, rounded at the tip . . . . 6. APIUM

30. Involucel present; leaflets or ultimate divisions linear to narrowly oblong-lanceolate or ovate to broadly triangular, sharply pointed at the tip
    31. Leaflets with relatively few teeth (less than 8 teeth per side) distributed mostly above the middle . . . . 29. OXYPOLIS
    31. Leaflets with numerous teeth distributed more or less equally along the margins
        32. Fruits with narrow, inconspicuous ribs lacking wings; stems spreading, at least toward the base; leaf blades (at least those of the upper leaves) irregularly divided and toothed. . . . . 7. BERULA
        32. Fruits with conspicuous ribs having short, thick, corky wings; stems erect or ascending; leaf blades (except those of submerged leaves) not divided, finely and regularly toothed. . . . . 35. SIUM
20. At least some of the leaves 2 or more times compound or dissected, the leaflets discrete or sometimes not discernable
    33. Leaves all basal or near the stem base
        34. Petals white; fruits with the ribs wingless or narrowly and inconspicuously winged; involucre of 1 leaflike, dissected bract; rays 1–4 . . 17. ERIGENIA
        34. Petals yellow; fruits with the lateral ribs broadly winged; involucre absent or rarely of 1–3 narrowly triangular, entire bracts; rays 10 to numerous . . . . 26. LOMATIUM
    33. Leaves alternate along the stems and sometimes also basal
        35. Leaves dissected, not divided into well-differentiated leaflets, the ultimate segments linear to narrowly elliptic (sometimes threadlike); fruits with the ribs lacking wings
        36. Involucre of 2–15 bracts, these sometimes relatively inconspicuous
            37. Rays 2–4; fruits 8–10 mm long, linear to narrowly oblong in outline . . . . 40. TREPOCARPUS
            37. Rays 4–25; fruits 1–5 mm long, oblong-elliptic to broadly ovate to nearly circular in outline
                38. Plants of aquatic habitats, with only the submerged leaves dissected 2 or more times, those of emergent stems 1 time pinnately compound (the leaflets often lobed); stems spreading with ascending apical portions and branches, sometimes rooting at the lower nodes . . . . 7. BERULA
                38. Plants in wet to dry habitats, but the leaves all similar in division pattern (the uppermost leaves generally smaller than those lower on the plant), all dissected 2 or more times; stems erect or ascending, not rooting at the lower nodes
                    39. Involucel of 8–14 bractlets; plants perennial, with clusters of mostly 2 or 3 tuberous-thickened roots. . . . 31. PERIDERIDIA
                    39. Involucel of 1–5(–7) bractlets; plants annual or biennial, with fibrous roots or a somewhat tuberous-thickened taproot
                        40. Leaf blades with the ultimate divisions linear to narrowly elliptic, but not threadlike; sepals absent; fruits oblong-elliptic in outline; plants biennial, with somewhat tuberous-thickened taproots . . . . 9. CARUM

40. Leaf blades with the ultimate divisions very fine and threadlike; sepals minute, narrowly to broadly triangular scales; fruits broadly ovate to nearly circular in outline; plants annual, with finely fibrous roots . . . . 33. PTILIMNIUM

36. Involucre absent or at most of 1 bract
   41. Involucel absent; petals yellow
      42. Fruits 4–6 mm long, flattened dorsally, at least the lateral ribs narrowly but noticeably winged; petioles with the sheathing portion up to 3 cm long . . . . 3. ANETHUM
      42. Fruits 3.5–4.0 mm long, slightly flattened laterally, the ribs more or less angled but not winged; petioles of the larger leaves with the sheathing portion 3–10 cm long . . . . 20. FOENICULUM
   41. Involucel of 1–14 bractlets, these sometimes inconspicuous; petals white
      43. Involucel of 8–14 inconspicuous bractlets; plants perennial, with clusters of mostly 2 or 3 tuberous-thickened roots . . . . 31. PERIDERIDIA
      43. Involucel of 1–6 bractlets, these conspicuous or inconspicuous; plants annual, with relatively slender taproots
         44. Sepals minute triangular teeth (visible under magnification), those of the outermost flowers of each umbellet usually somewhat enlarged, ovate (note that the sepals persist on the fruits); petals of the outermost flowers of each umbellet enlarged and spreading; fruits at maturity with the mericarps not or tardily separating . . . . 13. CORIANDRUM
         44. Sepals absent; outer flowers of the umbellets without enlarged sepals or petals, the petals all incurved or erect; fruits at maturity with the mericarps separating readily
            45. Fruits 4–10 mm long, linear to narrowly oblong-elliptic in outline, not roughened; involucel of 4–6 bractlets, these mostly longer than the flower stalks but often shorter than the fruit stalks, usually fused together at the very base, elliptic-ovate to oblong-obovate . . . . 10. CHAEROPHYLLUM
            45. Fruits 1.5–3.0 mm long, oblong or ovate in outline, sometimes roughened with minute teeth or tubercles; involucel of 1–3 bractlets, these mostly shorter than the flower stalks, not fused, linear to narrowly lanceolate
               46. Inflorescences axillary, sessile or minutely stalked; leaf blades with the ultimate divisions linear but not threadlike; fruits 2–3 mm long, oblong in outline, with noticeable, bluntly angled ribs, the surface smooth or sometimes roughened with minute teeth along the ribs, lacking tubercles . . . . 2. AMMOSELINUM
               46. Inflorescences terminal and axillary, short- to more commonly long-stalked; leaf blades with the ultimate divisions narrowly linear, usually threadlike; fruits 1.5–2.0 mm long, ovate in outline, with inconspicuous, narrow, rounded ribs, the surface smooth or sometimes roughened with dense, minute tubercles, lacking teeth along the ribs . . . . 36. SPERMOLEPIS

35. Leaves compound, the ultimate segments mostly broader (variously narrowly elliptic to broadly obovate), at least the larger leaves consisting of more or less well-differentiated leaflets, these entire to deeply lobed; fruits with the ribs winged or lacking wings
    47. Leaflets all entire, not toothed along the margins (note that occasional leaflets may have 1 or 2 deep lobes, but no teeth) . . . . 37. TAENIDIA
    47. Some or all of the leaflets incised or toothed along the margins
        48. Involucre of numerous (more than 20) bracts, these sometimes inconspicuous
            49. Involucre with the bracts 1 or 2 times pinnately dissected; rays numerous (more than 20); fruits 1.5–2.5 mm long, the ribs shallow and angled, lacking wings; petals white . . . . 1. AMMI
            49. Involucre with the bracts entire; rays 12–20; fruits 4–7 mm long, the lateral and sometimes also the intermediate and dorsal ribs having narrow, corky wings; petals yellow to greenish yellow . . . 23. LEVISTICUM
        48. Involucre absent or of 1–12 bracts, these sometimes inconspicuous
            50. Involucel absent
                51. Sepals absent; petals yellow, occasionally tinged with red; fruits 5–7 mm long, flattened dorsally, the lateral ribs with thin, broad wings . . . . 30. PASTINACA
                51. Sepals minute triangular scales; petals white; fruits 1.0–4.5 mm long, flattened laterally, the ribs sometimes thick and corky, but lacking wings
                    52. Fruits 1.0–1.5 mm long, the ribs narrow, more or less rounded, not corky; plants with taproots; stems 30–100 cm long, the base not thickened; rays 7–17 . . . . 6. APIUM
                    52. Fruits 2.0–4.5 mm long, the ribs blunt and somewhat corky; plants with at least some of the main roots tuberous-thickened; stems 50–200 cm long, the base usually somewhat thickened; rays numerous . . . . 11. CICUTA
            50. Involucel of 2–14 bractlets, these sometimes inconspicuous
                53. Stems mottled or spotted with purple; involucre with the bracts often fused at the base . . . . 12. CONIUM
                53. Stems green, rarely slightly purplish-tinged, lacking purple mottling or spots; involucre with the bracts free
                    54. Leaflets with the main lateral veins ending mostly in the sinuses between the teeth; stems usually somewhat thickened at the base . . . . 11. CICUTA
                    54. Leaflets with the main lateral veins ending mostly at the tips of the teeth or lobes; stems not noticeably thickened at the base
                        55. Leaf blades short-hairy on the upper and/or lower surface(s), sometimes hairy only along the margins or roughened with minute teeth along the main veins
                            56. Rays 1–6(–8); petals white; fruits linear to narrowly oblong-lanceolate or narrowly oblong-oblanceolate in outline, flattened laterally, the ribs lacking wings; sepals absent
                                57. Rays 1–4; fruits 4–10 mm long; plants annual, with slender taproots . . . . 10. CHAEROPHYLLUM

57. Rays 3–6(–8); fruits 10–24 mm long; plants perennial, with clusters of somewhat tuberous-thickened roots .... 28. OSMORHIZA
56. Rays 8–20; petals cream-colored to yellow; fruits oblong-elliptic to ovate in outline, flattened dorsally, the lateral and sometimes other ribs winged; sepals present or sometimes very small to absent in *Thaspium*
 58. Leaf blades roughened with minute teeth on the undersurface along the veins, otherwise glabrous; fruits 5–11 mm long; plants with somewhat tuberous-thickened taproots .... 32. POLYTAENIA
 58. Leaf blades short-hairy along the margins and usually also the surfaces, not roughened along the veins; fruits 3–6 mm long; plants with clusters of fibrous roots, lacking taproots .... 38. THASPIUM
55. Leaf blades glabrous and smooth on the surfaces and margins
 59. Sepals absent; rays 1–4(–5); fruits linear to narrowly oblong-lanceolate in outline .... 10. CHAEROPHYLLUM
 59. Sepals minute triangular scales; rays (4–)6 to numerous; fruits broadly oblong-elliptic to oblong, ovate or ovate-elliptic in outline (narrowly oblong in *Falcaria,* with mostly 10–25 rays)
  60. Stems ascending from spreading bases, rooting at the lower nodes; petioles and also often stems somewhat inflated .... 27. OENANTHE
  60. Stems erect or ascending, not rooting at the lower nodes; petioles and stems not inflated
   61. Fruits narrowly oblong in outline; larger leaf blades with the main axis (rachis) usually winged below the attachment points of the leaflets .... 19. FALCARIA
   61. Fruits broadly oblong-elliptic to ovate or ovate-elliptic in outline; leaf blades without winged tissue below the leaflets
    62. Petals white; fruits 4–7 mm long; plants with expanded fibrous bases and taproots .... 24. LIGUSTICUM
    62. Petals yellow; fruits 2–4 mm long; plants with clusters of some what tuberous-thickened roots .... 41. ZIZIA

1. Ammi L.

Three to 5 species, Europe, Asia, Africa.

1. Ammi majus L. (bishop's weed)

Pl. 202 a, b; Map 838

Plants annual. Stems 20–80 cm long, erect or ascending, glabrous. Leaves alternate and also basal (basal rosette usually present at flowering), glabrous, short- to long-petiolate, the sheathing bases not inflated. Leaf blades 2–20 cm long, oblong to broadly triangular-ovate in outline, those of the basal and lowermost stem leaves ternately or pinnately 1 time compound, the leaflets 10–20 mm long, elliptic-lanceolate, narrowed at the base, rounded or narrowed to a sharp point at the tip, finely toothed along the margins; the blades of the median and upper stem leaves 2 times pinnately dissected, the ultimate segments 2–20 mm long, narrowly linear, narrowed to sharply pointed tips. Inflorescences terminal and sometimes also axillary, compound umbels, mostly long-stalked, the stalks roughened. Involucre of numerous 1 or 2 times pinnately dissected bracts, these mostly longer than the rays, spreading to reflexed at flowering, with thin, papery margins and sharply pointed tips. Rays numerous, 2–7 cm long, roughened. Involucel of numerous entire bractlets, these mostly slightly shorter than the flower stalks, with thin, papery margins and sharply pointed tips. Flowers mostly numerous in each umbellet, the stalks 3–12 mm long. Sepals minute triangular teeth. Petals ovate to obovate, broadly rounded to more commonly notched or 2-lobed at the tip, white. Ovaries glabrous. Fruits 1.5–2.5 mm long, oblong-elliptic in outline, flattened laterally, glabrous, dark brown, each mericarp with 5 shallow, angled ribs lacking wings. $2n=22$. May–July.

Introduced, known thus far only from the city of St. Louis (native of Europe, Asia; widely cultivated as an ornamental, escaped sporadically in the U.S.). Railroads and open, disturbed areas.

Plants of *A. majus*, particularly the immature fruits, reportedly contain furanocoumarins and related compounds similar to those found in *Heracleum* and may cause phototoxic dermatitis in some individuals. The plants are sometimes grown for the cut-flower trade or for use in dried flower arrangements.

2. Ammoselinum Torr. & A. Gray

Four species, U.S., Mexico, South America.

1. Ammoselinum butleri (S. Watson) J.M. Coult. & Rose (sand parsley)

Map 839

Plants annual, glaucous. Stems 15–40 cm long, erect or more commonly spreading to loosely ascending, glabrous, sometimes slightly roughened toward the tip. Leaves alternate (basal leaves usually absent at flowering), glabrous, petiolate, the sheathing bases not inflated. Leaf blades 4–20 cm long, oblong to ovate-triangular in outline, ternately or ternately then pinnately 2 times compound or dissected, the ultimate segments 1–8 mm long, linear, rounded to bluntly pointed at the tip, sometimes with a minute, sharp point, entire along the margins. Inflorescences axillary, compound umbels, sessile or minutely stalked. Involucre absent. Rays 2–6, noticeably unequal in length, variously absent or up to 2 cm long, glabrous or slightly roughened. Involucel of 1–3 bractlets, these mostly shorter than the flower stalks, linear to narrowly elliptic-lanceolate. Flowers (1–)2–10 in each umbellet, the stalks noticeably unequal in length, 1–6 mm long. Sepals absent. Petals ovate, rounded at the tip, white. Ovaries glabrous. Fruits 2–3 mm long, oblong in outline, somewhat flattened laterally, glabrous or roughened with minute teeth along the ribs, dark brown, each mericarp with 5 bluntly angled ribs lacking wings, the lateral ribs of adjacent mericarps often appearing fused prior to breakup of the fruit. April–May.

Introduced, uncommon in the Mississippi Lowlands Division (Texas, Oklahoma, Arkansas, Mississippi, and Louisiana; introduced in Kansas, Missouri, and North Carolina). Lawns and disturbed, sandy, open places.

Steyermark (1963) excluded *A. butleri* from the flora because he was unable to locate specimens collected in the state. Hudson (1997) confirmed the presence of the species in Missouri. Its finely divided, membranous foliage might be confused with that of *Chaerophyllum procumbens* but differs in being strongly glaucous, which renders plants of sand parsley grayish green when fresh.

Steyermark (1963) also noted an earlier report of the closely related *A. popei* Torr. & A. Gray from Missouri but excluded this species from the flora because no specimens could be located from the state. *Ammoselinum popei* is a superficially similar species of the southern states and adjacent Mexico that differs from *A. butleri* in its stalked umbels and fruits with the lateral ribs having corky appendages.

3. Anethum L.

Two to 4 species, Europe, Asia.

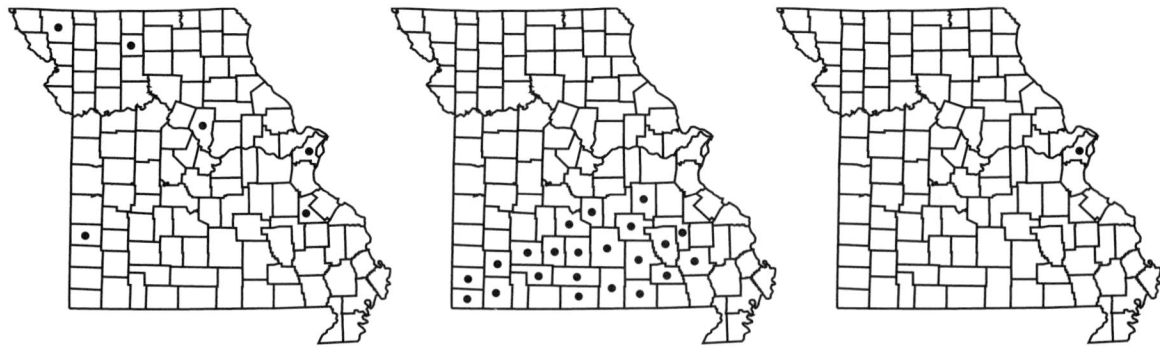

840. Anethum graveolens   841. Angelica venenosa   842. Anthriscus caucalis

1. Anethum graveolens L. (dill)

Pl. 203 h, i; Map 840

Plants annual, glaucous, with somewhat thickened taproots. Stems 40–170 cm long, erect or ascending, glabrous, noticeably longitudinally ridged. Leaves alternate and occasionally also basal (a few basal leaves sometimes persisting at flowering), glabrous, short- to long-petiolate, the upper leaves often nearly sessile, the sheathing bases 1–3 cm long, slightly to moderately inflated, sometimes turning tan and papery with age. Leaf blades 4–35 cm long, ovate in outline, pinnately 3 to several times dissected, the ultimate segments 4–20 mm long, narrowly linear to threadlike, sharply pointed at the tip, entire along the margins. Inflorescences terminal and usually also axillary, compound umbels, mostly long-stalked. Involucre absent. Rays 10 to numerous, 3–10 cm long. Involucel absent. Flowers numerous in each umbellet, the stalks 6–10 mm long. Sepals absent. Petals broadly ovate, rounded to bluntly pointed at the tip, yellow. Ovaries glabrous. Fruits 4–6 mm long, narrowly ovate-elliptic in outline, flattened dorsally, glabrous, brown, each mericarp with 5 thin ribs, the lateral and often also dorsal and intermediate ones narrowly but noticeably winged. $2n=22$. June–August.

Introduced, uncommon and widely scattered in Missouri (native of Europe; widely cultivated nearly worldwide and escaped sporadically everywhere).

Dill has a strong odor somewhat reminiscent of anise (*Pimpinella anisum* L., another Apiaceae). The foliage, inflorescences, and seeds have long been used as a flavoring and garnish in cooked foods and salads, and in the manufacture of dill pickles.

4. Angelica L. (angelica)

About 110 species, U.S., Canada; Europe, Asia, Africa.

1. Angelica venenosa (Greenway) Fernald
(wood angelica, hairy angelica)

Pl. 203 a, b; Map 841

Plants perennial, with tuberous-thickened taproots. Stems 40–150 cm long, erect or ascending, glabrous. Leaves alternate and sometimes also basal (1 or 2 basal leaves sometimes present at flowering), glabrous, short- to long-petiolate, the sheathing bases not or only slightly inflated, the uppermost leaves sometimes reduced to bladeless, somewhat inflated sheaths. Leaf blades (4–)10–25 cm long, triangular-ovate in outline, pinnately or ternately then pinnately 2 or 3 times compound with distinct leaflets, these 15–50 mm long, mostly (except on the uppermost leaves) 10 mm or more wide, ovate or narrowly ovate to elliptic, occasionally with 1 or 2 basal lobes, narrowed or tapered at the base, narrowed to a blunt or sharp point at the tip, finely toothed along the margins. Inflorescences mostly terminal, compound umbels, long-stalked, the stalks moderately to densely short-hairy. Involucre absent. Rays 18 to numerous, 1–8 cm long, moderately to densely short-hairy. Involucel of several entire bractlets, these mostly slightly shorter than the flower stalks, linear, short-hairy. Flowers 8 to numerous in each umbellet, the stalks 2–10 mm long. Sepals absent or minute triangular teeth. Petals obovate, rounded at the tip, white. Ovaries short-hairy. Fruits 4–6 mm long, oblong-elliptic to broadly ovate-elliptic in outline, shallowly cordate

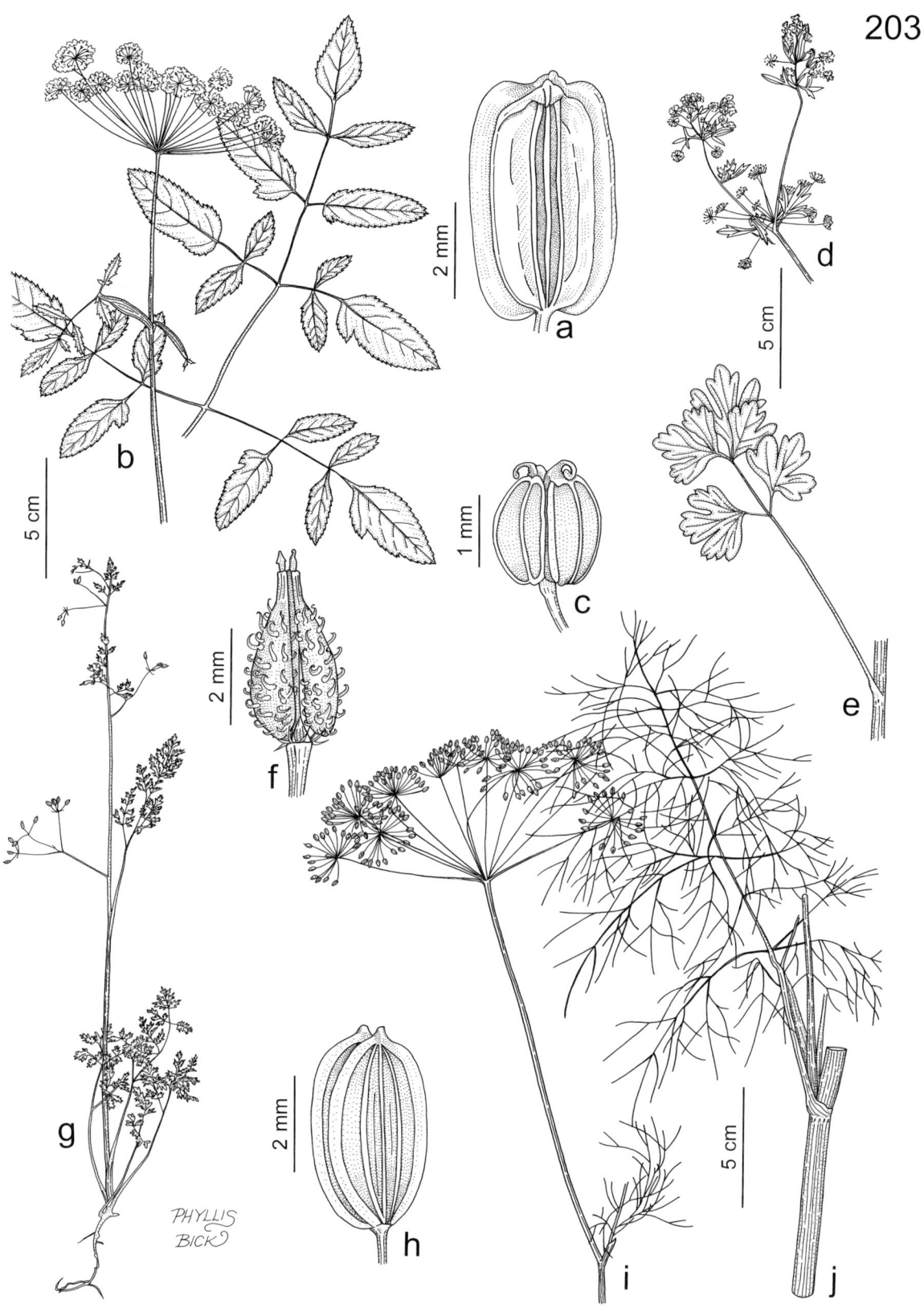

Plate 203. Apiaceae. *Angelica venosa*, **a)** fruit, **b)** habit. *Apium graveolens*, **c)** fruit, **d)** inflorescence, **e)** leaf. *Anthriscis caucalis*, **f)** fruit, **g)** habit. *Anethum graveolens*, **h)** fruit, **i)** inflorescence and leaf.

at the base, flattened dorsally, sparsely to moderately short-hairy, dark brown with usually lighter ribs, each mericarp with the dorsal and intermediate ribs not or narrowly winged, the lateral ribs with broad, papery wings wider than the main body. $2n=22$. May–July.

Scattered, mostly in the Ozark Division (eastern U.S. west to Minnesota and Arkansas). Mesic to dry upland forests, savannas, edges of prairies, and banks of streams and rivers.

Apparently, *A. venenosa* has not been tested phytochemically, although other species in the genus are known to contain furanocoumarins and related compounds similar to those found in *Heracleum* and may cause phototoxic dermatitis in some individuals. Steyermark (1963) also noted an anecdotal report in the historical literature (Greenway, 1793) of a boy in Virginia being poisoned by eating a small quantity of the root of this species. The symptoms described were similar to those reported for *Conium* poisoning, and perhaps the plant was misdetermined by the original author.

5. Anthriscus Pers. (chervil)

Ten to 12 species, Europe, Asia, Africa.

1. Anthriscus caucalis M. Bieb. (bur chervil)
 A. scandicina (Weber) Mansf.
Pl. 203 f, g; Map 842

Plants annual. Stems 40–90 cm long, erect or ascending, glabrous or sparsely short-hairy. Leaves alternate (basal leaves usually absent at flowering), sparsely to moderately pubescent with short, spreading hairs, petiolate, the sheathing bases not or only slightly inflated, at least the uppermost hairy along the margins. Leaf blades 2–15 cm long, triangular-ovate in outline, 2 or 3 times pinnately compound, the leaflets 3–9 mm long, ovate, dissected into several deep lobes, these 1–4 mm long, linear to lanceolate or narrowly oblong, narrowed to a sharp point at the tip, the margins otherwise entire. Inflorescences terminal and lateral, compound umbels, relatively short-stalked or less commonly sessile, the stalks moderately to densely pubescent with short, spreading hairs. Involucre absent. Rays 3–6, 0.8–2.0 cm long, glabrous or sparsely pubescent with short, spreading hairs, especially at the base and tip. Involucel of 2–5 entire bractlets, these shorter than the flower stalks, linear to narrowly lanceolate, short-hairy along the margins, sparsely to moderately pubescent with short, stiff, white hairs on both surfaces. Flowers 3–7 in each umbellet, the stalks 2–9 mm long. Sepals absent. Petals narrowly oblong-obovate, rounded at the tip, white. Ovaries with minute, hooked hairs. Fruits 3–4 mm long, ovate in outline, rounded at the base, tapered to a short beak at the tip, flattened laterally, the body covered with short, stout, hooked hairs, dark brown, the mericarps narrowed along the commissures, lacking ribs. $2n=14$. May–June.

Introduced, uncommon, known thus far only from St. Louis (native of Europe; introduced widely but sporadically in the U.S. and Canada). Railroads.

Species of *Anthriscus* reportedly contain furanocoumarins and related compounds similar to those found in *Heracleum,* and may cause phototoxic dermatitis in some individuals.

6. Apium L.

Twenty to 25 species, nearly worldwide.

1. Apium graveolens L. (celery)
Pl. 203 c–e; Map 843

Plants perennial, with taproots, glabrous. Stems 30–100 cm long, erect or ascending, not thickened at the base. Leaves alternate and often also basal (a few basal leaves often present at flowering), short- to long-petiolate, the sheathing bases not or only slightly inflated. Leaf blades 4–18 cm long, oblong to broadly obovate in outline, 1 time pinnately compound with 3–9 leaflets, the leaflets 8–50 mm long, obovate or narrowly to broadly wedge-shaped, 1 or 2 times ternately lobed and/or toothed, the lobes mostly obovate or wedge-shaped, broadly to narrowly tapered at the base, mostly rounded at the tip. Inflorescences terminal and lateral, compound umbels, short-stalked to sessile or less commonly long-stalked. Involucre absent, but the subtending leaves sometimes appearing as an

843. Apium graveolens 844. Berula erecta 845. Bupleurum rotundifolium

involucre in sessile inflorescences. Rays 7–15, 0.7–2.5 cm long. Involucel absent. Flowers 7–17 in each umbellet, the stalks 1–6 mm long. Sepals minute triangular teeth. Petals broadly ovate, rounded at the tip, white. Ovaries glabrous. Fruits 1.0–1.5 mm long, broadly oblong-elliptic to depressed-circular in outline, rounded at the base, flattened laterally, glabrous, brown, each mericarp narrowed along the commissures, with 5 ribs, these narrow, more or less rounded, lacking wings. 2n=22. June–August.

Introduced, known thus far only from St. Louis (native of Europe, Asia; introduced sporadically in the New World). Railroads.

Apium graveolens is sometimes divided into two or more subspecies. The edible celery, with thick juicy petioles, is var. *dulce* (Mill.) Pers. The var. *graveolens* contains weedy plants with slender petioles, which usually become spread as seed contaminants. The few specimens from Missouri appear to represent the latter phase.

7. Berula (water parsnip)

One species, North America, Europe, Asia, Africa.

1. Berula erecta (Huds.) Coville **var. incisum** (Torr.) Cronquist

Pl. 204 d, e; Map 844

Plants perennial, with fibrous roots, glabrous. Stems 20–80 cm long, spreading with ascending apical portions and branches, sometimes rooting at the lower nodes. Leaves alternate and sometimes also basal (a few basal leaves sometimes present at flowering), short- to long-petiolate, the sheathing bases not or only slightly inflated. Leaf blades 2–25 cm long, narrowly oblong-elliptic in outline, those of submerged leaves pinnately dissected with linear segments, otherwise 1 time pinnately compound with (5–)7–21 leaflets, somewhat dimorphic (the basal and lower stem leaves with broader, less divided leaflets than those of the median and upper leaves), the leaflets 4–40 mm long, ranging from ovate and shallowly toothed and/or lobed to narrowly oblong-lanceolate and deeply toothed and lobed, the teeth sharply pointed. Inflorescences terminal and lateral, compound umbels, mostly relatively long-stalked. Involucre of 6–8 bracts, these shorter than the rays, spreading at flowering, linear, with narrow, white, entire or minutely toothed margins and sharply pointed tips. Rays 6–15, 0.5–3.0 cm long. Involucel of 4–8 bractlets, these shorter than to slightly longer than the flower stalks, linear to lanceolate, with thin, papery margins and sharply pointed tips. Flowers 5–17 in each umbellet, the stalks 2–5 mm long. Sepals absent or minute triangular teeth. Petals obovate, rounded at the tip, white. Ovaries glabrous. Fruits 1.5–2.0 mm long, broadly elliptic to circular in outline, rounded at the base, flattened laterally, glabrous, brown, each mericarp narrowed along the commissures, with 5 narrow ribs, these inconspicuous at maturity, lacking wings. 2n=12. June–October.

Uncommon, known thus far only from Saline and Lafayette Counties (U.S., Mexico). Fens; frequently emergent aquatics.

This species was first reported by Gremaud (1988) from a small, isolated fen complex in an otherwise highly agriculturalized area in the southwestern portion of the Glaciated Plains Division. The Saline County locality is equally isolated. North American plants are all var. *incisum*.

The var. *erecta* occurs in portions of Europe and Asia and differs in details of its leaf dissection.

8. Bupleurum L.

About 180 species, U.S. and Canada (1 species), Europe, Asia, Africa.

1. Bupleurum rotundifolium L.
(thoroughwax, hare's ear)
Pl. 204 i, j; Map 845

Plants annual, often somewhat glaucous. Stems 10–50 cm long, erect or ascending, glabrous. Leaves alternate and usually also basal (a few basal leaves usually present at flowering), glabrous, sessile or nearly so (in basal leaves), the sheathing bases absent or poorly developed, not inflated. Leaf blades 1.5–8.0 cm long, simple, ovate to elliptic-ovate in outline, rounded at the tip, the margins entire, the basal and lowermost stem leaves with the bases clasping the stems and with more or less rounded auricles, those of the median and upper leaves with the bases perfoliate. Inflorescences terminal and axillary, compound umbels, mostly long-stalked. Involucre absent. Rays 4–10, 0.5–1.5 cm long. Involucel of 5 bractlets, these fused at the base, 5–15 mm long, broadly ovate, tapered to sharp points at the tip. Flowers 10–12 in each umbellet, the stalks 1–3 mm long, giving the umbellets a dense, headlike appearance. Sepals absent. Petals broadly oblong to nearly circular, rounded at the tip, yellow to greenish yellow. Ovaries glabrous. Fruits 2.5–3.0 mm long, oblong-elliptic in outline, slightly flattened laterally, glabrous, purplish brown to black, each mericarp with 5 slender, inconspicuous ribs lacking wings. $2n=16$. May–July.

Introduced, uncommon and sporadic (native of Europe, Asia; introduced in the eastern U.S. west to South Dakota and Texas). Glades and dry upland forests; also roadsides, railroads, pastures, and open, disturbed areas.

9. Carum L.

About 30 species, Europe, Asia, Africa.

1. Carum carvi L. (caraway)
Pl. 204 f–h; Map 846

Plants biennial, with somewhat tuberous-thickened taproots, glabrous. Stems 30–100 cm long, erect or ascending. Leaves alternate and sometimes also basal (a few basal leaves sometimes present at flowering), the basal and lower stem leaves mostly long-petiolate, the median and upper leaves short-petiolate to sessile, the sheathing bases not or only slightly inflated. Leaf blades 2–15 cm long, oblong-lanceolate to ovate in outline, 2 or 3 times pinnately compound or dissected, the ultimate segments 3–15 mm long, linear or narrowly elliptic, entire or with few teeth or lobes, narrowed or tapered to a sharp point at the tip. Inflorescences terminal and axillary, compound umbels, mostly long-stalked. Involucre absent or of 1–3 bracts, these shorter than the rays, spreading to ascending at flowering, linear, entire or occasionally with a few linear lobes. Rays 7–15, 0.3–4.0 cm long, often noticeably unequal in length. Involucel absent or of 1–3 bractlets, these mostly shorter than the flower stalks, similar to the bracts but smaller. Flowers 11 to numerous in each umbellet, the stalks 1–12 mm long, unequal in length. Sepals absent. Petals obovate, rounded or shallowly notched at the tip, white, rarely tinged with pink. Ovaries glabrous. Fruits 3–5 mm long, oblong-elliptic in outline, flattened laterally, glabrous, dark brown with pale ribs, each mericarp somewhat narrowed along the commissures, with 5 conspicuous ribs, these lacking wings. $2n=20$. May–July.

Introduced, known thus far only from St. Louis (native of Europe, Asia; introduced sporadically in the northeastern U.S. and Canada). Habitat unknown, but presumably disturbed, open areas.

This species is included in the flora with some reservations. The specimens on which Steyermark (1963) based his reports from Boone and Jackson Counties could not be located during the present study. However, a single vegetative specimen ex-

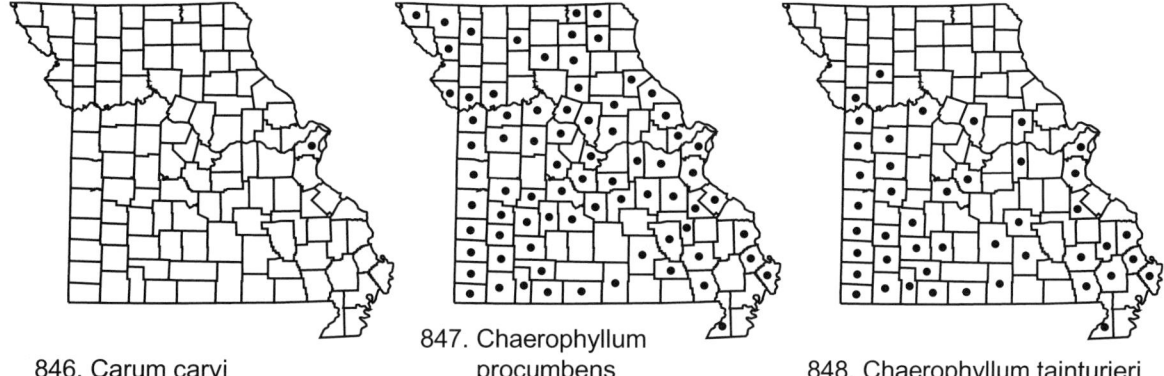

846. Carum carvi
847. Chaerophyllum procumbens
848. Chaerophyllum tainturieri

ists from the city of St. Louis that unfortunately lacks sufficient label data to determine whether the gathering originated from a cultivated or spontaneous occurrence. *Carum carvi* is to be expected to escape sporadically from cultivation in Missouri, particularly in urban areas. Caraway fruits are used as a flavoring in some baked goods and alcoholic beverages. They also are a component of the fragrances in some soaps and perfumes.

10. Chaerophyllum L. (chervil)

Plants annual, glabrous or hairy. Stems spreading to erect. Leaves alternate and usually also basal (1 to several leaves usually present at flowering), short- to long-petiolate, the uppermost leaves sometimes nearly sessile, the sheathing bases not or only slightly inflated. Leaf blades broadly ovate to oblong-ovate in outline, pinnately or ternately then pinnately 3 times compound, the ultimate leaflets entire to few-toothed or deeply pinnately lobed or dissected, the leaflets or lobes linear to narrowly obovate, mostly narrowed at the base, rounded or narrowed to a blunt or sharp point at the tip. Inflorescences terminal and axillary, compound or sometimes simple (1-rayed) umbels, short- to long-stalked or often sessile. Involucre absent (in sessile umbels, the subtending leaf sometimes appearing as a bract). Rays 1–4(–5), usually unequal in length, elongating as the fruits develop, ascending. Involucel of 4–6 bractlets, these mostly longer than the flower stalks but often shorter than the fruit stalks, usually fused together at the very base, elliptic-ovate to oblong-obovate, usually hairy along the margins. Flowers 2–15 in each umbellet, sessile or short-stalked at flowering (the umbellets thus appearing headlike at flowering), often elongating unequally as the fruits develop. Sepals absent. Petals obovate, rounded or shallowly notched at the tip, white. Ovaries glabrous or hairy. Fruits linear to narrowly oblong-elliptic in outline, rounded or narrowed at the base, narrowed or short-tapered at the tip, flattened laterally, glabrous or hairy, brown to dark brown with usually lighter ribs, the mericarps sometimes somewhat arched or curved, sometimes somewhat narrowed along the commissures, with 5 narrow to broad, low, blunt ribs, these lacking wings. About 35 species, U.S., Canada; Europe, Asia, Africa.

1. Stems glabrous or sparsely hairy toward the base, sometimes also at the nodes; fruit stalks not thickened toward the tips .... 1. C. PROCUMBENS
1. Stems moderately to densely hairy toward the base, sparsely to moderately hairy toward the tip (sometimes becoming nearly glabrous with age); fruit stalks somewhat thickened toward the tips .... 2. C. TAINTURIERI

1. Chaerophyllum procumbens (L.) Crantz
(wild chervil)

Pl. 204 a, b; Map 847

Stems 10–60 cm long, erect to more commonly spreading to loosely ascending, glabrous or sparsely hairy toward the base, sometimes also at the nodes. Leaf blades 1–12 cm long, glabrous or the undersurface sparsely hairy along the veins, the ultimate segments 1–8 mm long, 1–4 mm wide. Rays 0.3–1.5 cm long at flowering, elongating to 5.5 cm at fruiting. Flowers 2–6 in each umbellet, sessile or the stalks to 2 mm long at flowering, these elongating unequally to 11 mm as the fruits develop, the fruit stalks uniformly linear, slender. Fruits 5–10 mm long. $2n=22$. March–May.

Scattered to common nearly throughout the state (northeastern U.S. and adjacent Canada west to Iowa and Arkansas). Bottomland forest, mesic upland forests, banks of spring branches, streams, and rivers, bluffs, and occasionally margins of glades; also pastures, margins of crop fields, fencerows, railroads, roadsides, yards, and moist, disturbed areas.

Chaerophyllum procumbens is often classified into two varieties (Mathias and Constance, 1944–1945; Steyermark, 1963). The var. *shortii* Torr. & A. Gray has hairy fruits that tend to taper toward the tip, whereas var. *procumbens* has glabrous fruits that are narrowed more uniformly. These variants occur together throughout the range of the species, and plants with hairy and glabrous fruits are found growing intermingled in many populations. Thus recognition of these minor variants seems unwarranted.

2. Chaerophyllum tainturieri Hook. (wild chervil)

Pl. 204 c; Map 848

Stems 10–70 cm long, erect or ascending, less commonly spreading at the base, moderately to densely hairy toward the base, sparsely to moderately hairy toward the tip, sometimes becoming nearly glabrous with age. Leaf blades 1–12 cm long, glabrous or the undersurface sparsely to moderately hairy, the ultimate segments 1–6 mm long, 1–3 mm wide. Rays 0.3–1.5 cm long at flowering, elongating to 7.0 cm at fruiting. Flowers 2–15 in each umbellet, sessile or the stalks to 2 mm long at flowering, these elongating unequally to 10 mm as the fruits develop, the fruit stalks narrowly club-shaped, somewhat thickened toward the tips. Fruits 4–8 mm long. $2n=22$. March–June.

Scattered, mostly south of the Missouri River (southeastern U.S. west to Kansas and Texas). Glades, upland prairies, and less commonly openings of mesic to dry upland forests; also fallow fields, pastures, railroads, roadsides, and open, disturbed areas.

Morphological variation within *C. tainturieri* is complex, and most earlier authors (Steyermark, 1963) have accepted one or more segregates or infraspecific taxa. Mathias and Constance (1944–1945) segregated populations with narrow leaflets, 6–15 (vs. 1–6) flowers in each umbellet, and involucels with reflexed (vs. spreading to ascending) bractlets as *C. texanum* J.M. Coult. & Rose, but after further study of more specimens, they concluded that such plants merely represented part of the overall spectrum of variation found within *C. tainturieri* (Mathias and Constance, 1951). Mathias and Constance (1944–1945) also accepted var. *dasycarpum* S. Watson for plants with hairy fruits, which occur sporadically throughout the species range. Interestingly, hairy-fruited individuals can be found among plants with both fewer-flowered umbellets and those with more flowers. Steyermark (1963) further resurrected the name var. *floridanum* J.M. Coult. & Rose for plants having fruits with the ribs narrower than the intervening spaces. In light of the apparent lack of correlations between the distributions of these character states among populations and their seemingly random dispersal in the species range, formal taxonomic recognition of these segregates and varieties does not seem prudent.

11. Cicuta L. (water hemlock)
(Mulligan, 1980)

Four species, North America, Asia.

1. Cicuta maculata L. (common water hemlock, spotted cowbane)

Pl. 205 a, b; Map 849

Plants perennial, glabrous, often glaucous, with at least some of the main roots tuberous-thickened. Stems 50–200 cm long, erect or ascending, sometimes purple-spotted or mottled toward the base, the base usually somewhat thickened, hollow but cross-partitioned into chambers. Leaves alternate and sometimes also basal (1 or few basal leaves

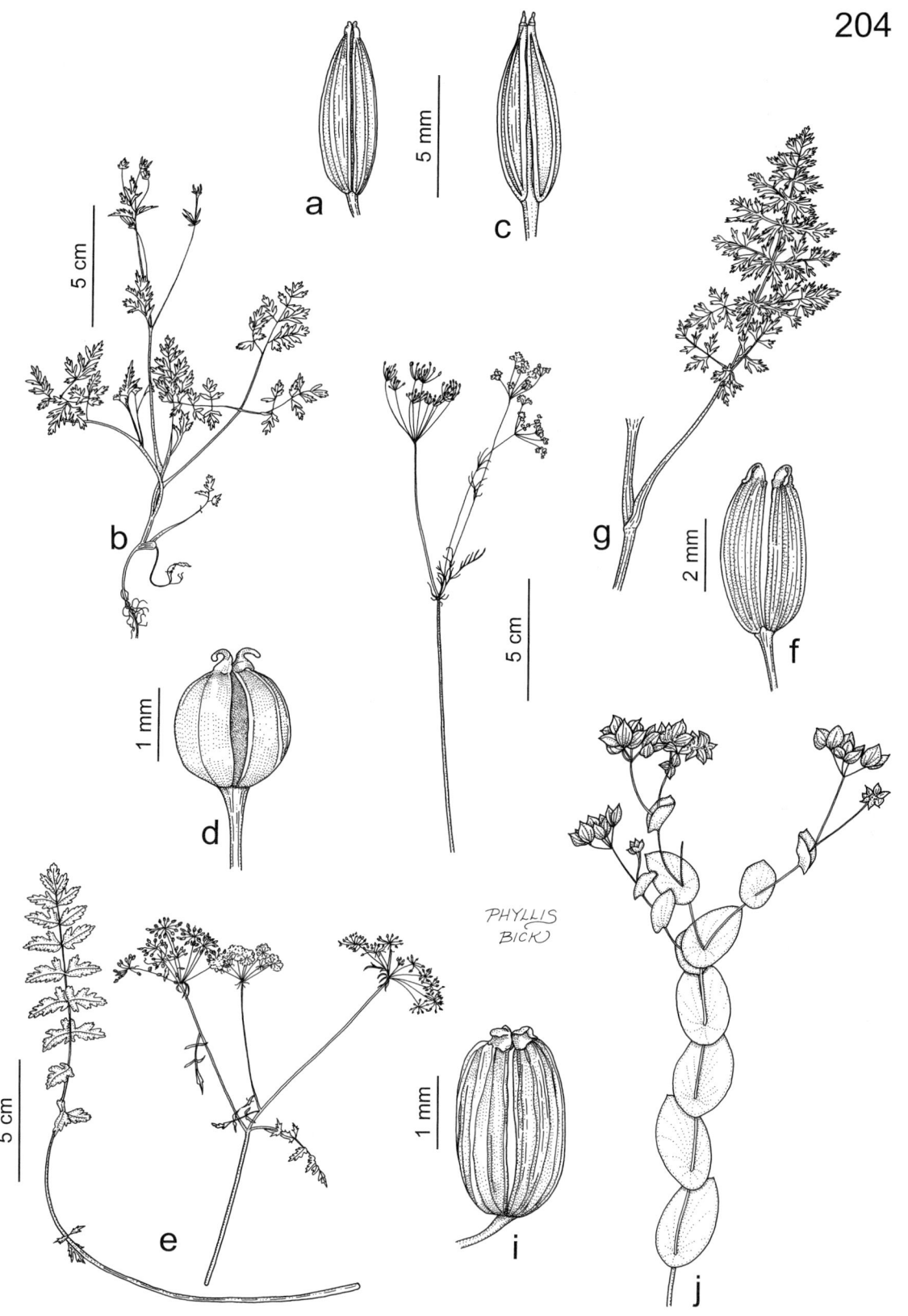

Plate 204. Apiaceae. *Chaerophyllum procumbens*, **a)** fruit, **b)** habit. *Chaerophyllum tainturieri*, **c)** fruit. *Berula erecta*, **d)** fruit, **e)** habit. *Carum carvi*, **f)** fruit, **g)** leaf, **h)** inflorescence. *Bupleurum rotundifolium*, **i)** fruit, **j)** habit.

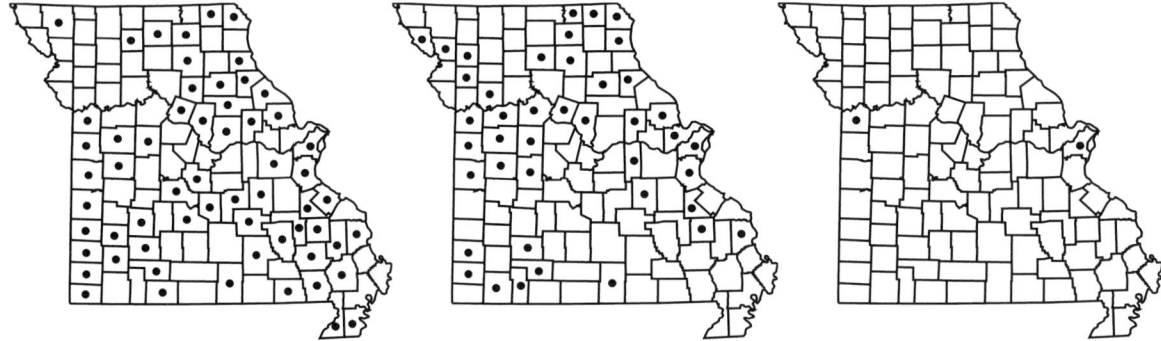

849. Cicuta maculata   850. Conium maculatum   851. Coriandrum sativum

sometimes present at flowering), the basal and lower stem leaves mostly long-petiolate, the median and upper leaves short-petiolate to nearly sessile, at least the lowermost sheathing bases not or only slightly inflated. Leaf blades 2–40 cm long, broadly ovate to triangular-ovate in outline, those of the basal and lowermost stem leaves 2 or 3 times pinnately compound, the ultimate leaflets 20–120 mm long, narrowly lanceolate to oblong-lanceolate, narrowed or tapered at the base, finely to coarsely toothed along the margins, occasionally with 1 or 2 basal lobes, the margins usually also roughened with minute teeth, the main lateral veins mostly ending in the sinuses between the teeth; those of the median and upper leaves progressively reduced, 1 or 2 times pinnately compound, the uppermost occasionally simple, the leaflets similar to those of the lower leaves. Inflorescences terminal and axillary, compound umbels, mostly long-stalked. Involucre absent or less commonly of 1–4 bracts, these shorter than the rays, spreading to ascending at flowering, linear, with sharply pointed tips. Rays usually numerous, 1.5–6.5 cm long, often unequal in length. Involucel absent or more commonly of 3–7 bractlets, these mostly shorter than the flower stalks, linear to broadly lanceolate, with thin, white, papery margins, tapered to sharply pointed tips. Flowers mostly numerous in each umbellet, the stalks 2–10 mm long. Sepals minute triangular teeth. Petals obovate, narrowed abruptly to a slender, pointed extension at the tip, white. Ovaries glabrous. Fruits 2.0–4.5 mm long, broadly oblong-elliptic in outline, flattened laterally, glabrous, dark brown to reddish brown with pale ribs, the mericarps with 5 ribs, these blunt and somewhat corky. $2n=22$. May–September.

Scattered nearly throughout the state (U.S., Canada, Mexico). Sloughs, banks of streams, rivers, and spring branches, margins of ponds and lakes, bottomland prairies, moist depressions of upland prairies, and openings of bottomland forests; also roadsides, railroads, and ditches.

Cicuta maculata is perhaps the most poisonous of all North American plants. Although all parts are toxic, the tuberous roots, swollen lower stems, and new growth are considered the most poisonous parts of the plants. The active compounds, a series of polyacetylenes (including cicutoxin and cicutol), are quite different from the alkaloids found in *Conium* and produce much more violent symptoms. Ingestion of a walnut-sized portion of the root has been suggested as sufficient to kill an animal the size of a cow.

In his revision of *Cicuta* in North America, Mulligan (1980) recognized four varieties of *C. maculata*, two of which were reported for Missouri. These are differentiated mainly based on differences in details of fruit morphology, and flowering specimens from Missouri cannot be determined below the species level. Thus, the distributions listed below for var. *bolanderi* and var. *maculata* are relatively inexact. A third variety, var. *angustifolia* Hook., is unique in its subglobose fruits, shorter styles (mostly less than 1 mm), and median and upper leaves with narrower main leaflets (more than 5 times as long as wide). It is widely distributed to the west of the state but has been collected in southeastern Kansas and eventually may be found in Missouri. The other variety, var. *victorinii* (Fernald) B. Boivin, occurs mostly in the Pacific Northwest and has ribless fruits.

1. Mericarps with the dorsal and intermediate ribs much narrower than the space between the ribs, narrowed along the commissures . . . . 1A. VAR. BOLANDERI
1. Mericarps with the dorsal and intermediate ribs wider than to about as wide as the space between the ribs, not narrowed along the commissures
. . . . 1B. VAR. MACULATA

Plate 205. Apiaceae. *Cicuta maculata*, **a)** fruit, **b)** flowering branch. *Conium maculatum*, **c)** fruit, **d)** leaf. *Coriandrum sativum*, **e)** fruit, **f)** habit.

1a. var. bolanderi (S. Watson) G.A. Mulligan

Median and upper stem leaves with the principal leaflets mostly less than 5 times as long as wide, usually coarsely toothed along the margins. Styles mostly more than 1 mm long. Fruits longer than wide, abruptly and unevenly narrowed along the commissures. Mericarps with the dorsal and intermediate ribs much narrower than the space between the ribs, the lateral ribs larger than the dorsal and intermediate ribs, larger than the oil tubes. $2n=22$. May–September.

Uncommon, mostly in the western and southern halves of the state (U.S. [mostly central states]; Mexico). Banks of streams, margins of ponds and lakes, and moist depressions of upland prairies; also roadsides and ditches.

1b. var. maculata

  C. maculata var. *curtissii* (J.M. Coult. & Rose) Fernald

Median and upper stem leaves with the principal leaflets mostly less than 5 times as long as wide, finely to coarsely toothed along the margins. Styles mostly more than 1 mm long. Fruits longer than wide, not narrowed along the commissures. Mericarps with the dorsal and intermediate ribs wider than to as wide as the space between the ribs, the lateral ribs larger than the dorsal and intermediate ribs, but smaller than the oil tubes. $2n=22$. May–September.

Scattered nearly throughout the state (U.S. [mostly eastern half], Canada, Mexico). Sloughs, banks of streams, rivers, and spring branches, margins of ponds and lakes, bottomland prairies, moist depressions of upland prairies, and openings of bottomland forests; also roadsides, railroads, and ditches.

12. Conium L.

Four to 6 species, Europe, Asia, Africa; introduced nearly worldwide.

1. Conium maculatum L. (poison hemlock)

Pl. 205 c, d; Map 850

Plants biennial, glabrous, often glaucous. Stems 50–300 cm long, erect or ascending, purple-spotted or mottled. Leaves alternate and usually also basal (1 to several basal leaves usually present at flowering), the basal and lower stem leaves long-petiolate, the median and upper leaves short-petiolate to nearly sessile, at least the lowermost sheathing bases somewhat inflated. Leaf blades 3–40 cm long, broadly ovate to ovate-triangular in outline, those of the basal and lowermost stem leaves 3 or 4 times pinnately compound, the ultimate leaflets 5–35 mm long, oblong-lanceolate to ovate, narrowed at the base, pinnately lobed, the lobes narrowed or tapered to a blunt or sharp point at the tip; those of the median and upper leaves progressively reduced, 2 or 3 times pinnately compound, the leaflets similar to those of the lower leaves. Inflorescences mostly terminal, compound umbels or more commonly loose clusters or panicles of compound umbels, mostly long-stalked. Involucre of 4–8 bracts, these 2–6 mm long, much shorter than the rays, spreading to reflexed at flowering, narrowly lanceolate to narrowly ovate, with broad, thin, papery margins and sharply pointed tips, some adjacent bracts sometimes fused toward the base. Rays numerous, 1.5–3.0 cm long. Involucel of 4–9 bractlets, these shorter than the flower stalks, similar to the bracts but smaller. Flowers mostly numerous in each umbellet, the stalks 3–6 mm long. Sepals absent. Petals obovate, rounded or notched at the tip, white. Ovaries glabrous. Fruits 2.5–3.5 mm long, ovate to broadly elliptic-ovate in outline, flattened laterally, glabrous, dark brown with pale ribs, the mericarps often slightly narrowed along the commissures, with 5 ribs, these blunt and somewhat corky. $2n=22$. May–August.

Introduced, scattered to common nearly throughout the state (native of Europe, Asia, Africa; widely introduced in North America). Banks of streams, rivers, and spring branches; also roadsides, railroads, ditches, pastures, fencerows, and open, disturbed areas.

Conium maculatum contains toxic alkaloids, including coniin and conicein, and is extremely poisonous to humans and other animals when ingested. This species has been implicated as the hemlock with which the Greek philosopher Socrates was poisoned. Although plants are not considered particularly palatable to livestock, landowners should take care to remove plants from pastures and fencerows, where they have a tendency to accumulate when other plants are suppressed or eliminated by grazing.

13. Coriandrum L.

Two or 3 species, Europe, Asia, Africa.

1. Coriandrum sativum L. (coriander)

Pl. 205 e, f; Map 851

Plants annual. Stems 20–70 cm long, erect or ascending, glabrous. Leaves alternate and usually also basal (a few basal leaves present at flowering), the basal and lower stem leaves mostly long-petiolate, the median and upper leaves short-petiolate or sessile, the sheathing bases not or only slightly inflated. Leaf blades 3–15 cm long, narrowly oblong to broadly ovate in outline, those of the basal and lowermost stem leaves 1 or 2 times pinnately compound (rarely more divided or simple), the leaflets 10–20 mm long, broadly obovate to fan-shaped, narrowed at the base, palmately toothed or lobed, rounded or bluntly pointed at the tip; those of the median and upper stem leaves progressively more divided, 2–3 times pinnately dissected, the ultimate segments linear, entire or with few teeth or lobes, mostly sharply pointed at the tip. Inflorescences terminal and axillary, compound umbels, mostly long-stalked. Involucre absent or of 1 inconspicuous bract, this shorter than the rays, spreading at flowering, linear, with a sharply pointed tip. Rays 2–8, 1.0–2.5 cm long. Involucel of 3–5 bractlets, these shorter than to longer than the flower stalks and unequal in size, linear, and sharply pointed at the tip. Flowers 11 to numerous in each umbellet, the stalks 2–5 mm long. Sepals mostly minute triangular teeth, but those of the outermost flowers of each umbellet usually somewhat enlarged, to 1 mm long, narrowly ovate. Petals obovate, rounded or shallowly notched at the tip, white or pale pink, some or all of those of the outermost flowers of each umbellet enlarged to 4 mm long, narrowly obovate, rounded at the tip, spreading. Ovaries glabrous. Fruits 1.5–2.5 mm long, broadly oblong-elliptic to nearly circular in outline, not flattened, glabrous, brown with pale ribs, the mericarps not or tardily separating, with 5 low, narrow, blunt ribs lacking wings and sometimes also with faint additional ribs. $2n=22$. May–June.

Introduced, uncommon, known thus far from Jackson County and St. Louis (native of Europe, Asia, Africa; widely but sporadically introduced in North America, Central America, Caribbean Islands). Roadsides, railroads, and open, disturbed areas.

Coriander has a long history as a spice and food additive. In addition to its use as a flavoring in cooked and baked foods, the seeds are used to flavor confections and curry powder, as well as gin and some other alcoholic beverages. An extract also has been used for fragrance in some soaps, bath oils, shampoos, and potpourri.

14. Cryptotaenia DC. (honewort)

Four to 6 species, U.S. (including Hawaii), Canada; Europe, Asia, Africa.

1. Cryptotaenia canadensis (L.) DC. (honewort, wild chervil)

Pl. 206 e, f; Map 852

Plants perennial, with fibrous roots, glabrous. Stems 30–100 cm long, erect or ascending. Leaves alternate and sometimes also basal (1 or a few basal leaves often present at flowering), short- to long-petiolate, the uppermost leaves sometimes nearly sessile, the sheathing bases not or only slightly inflated. Leaf blades 3–13 cm long, broadly ovate to depressed-ovate in outline, 1 time compound with 3 leaflets, the central leaflet sometimes with a pair of deep basal lobes, the lateral leaflets sometimes 1 or both with a single basal lobe, the leaflets or lobes 30–150 mm long, oblong-lanceolate to elliptic or obovate, short- to long-tapered at the base, coarsely to finely, sharply and often irregularly toothed (often doubly toothed) along the margins, tapered to a sharp point at the tip. Inflorescences terminal and axillary, compound umbels, often grouped into small panicles with ascending branches, mostly relatively long-stalked. Involucre absent or of 1 bract, this inconspicuous, shorter than the rays, spreading to ascending at flowering, linear, with a sharply pointed tip. Rays 2–7, 0.5–5.0 cm long, unequal in length, ascending. Involucel absent or of 1 or 2 bractlets, these shorter than the flower stalks, similar to the bracts. Flowers 2–10 in each umbellet, the stalks 2–30 mm long, unequal in length. Sepals absent or consisting of minute teeth. Petals obovate, rounded or with an abrupt, minute point at the tip, white.

852. Cryptotaenia canadensis   853. Cynosciadium digitatum   854. Daucus carota

Ovaries glabrous. Fruits 4–7(–8) mm long, narrowly oblong-elliptic in outline, narrowed at the base, tapered to a short beak at the tip, flattened laterally, glabrous, dark brown with lighter, greenish yellow ribs, the mericarps sometimes somewhat arched or curved, somewhat narrowed along the commissures, with 5 narrow, blunt ribs, these lacking wings. $2n=20$. May–August.

Scattered nearly throughout the state (eastern U.S. west to North Dakota, Colorado, and Texas; Canada). Bottomland forests, mesic upland forests, ledges of sheltered bluffs, fens, margins of sloughs, and banks of streams and rivers; also roadsides and railroads.

15. Cynosciadium DC.

One species, southeastern U.S.

1. Cynosciadium digitatum DC.

Pl. 206 a, b; Map 853

Plants annual, glabrous. Stems 30–80(–120) cm long, erect or ascending. Leaves alternate and usually also basal (1 or a few basal leaves usually present at flowering), short-petiolate or nearly sessile, the sheathing bases not or only slightly inflated. Leaf blades 2–15 cm long, the basal and lower stem leaves (rarely also the uppermost ones) simple, the median and upper leaves palmately compound with 3–5 leaflets, linear or narrowly elliptic (when simple) to broadly fan-shaped in outline, the leaflets 20–150 mm long, 1–6 mm wide, linear to narrowly elliptic, entire, gradually narrowed or tapered at the base and tip, with irregular, fine cross-veins or partitions. Inflorescences terminal and axillary, compound umbels, short- to long-stalked or rarely sessile. Involucre absent or of 3–5 bracts, these 2–10 mm long, spreading to reflexed at flowering, entire, linear. Rays 2–10, usually unequal in length, 1–4 cm long, less commonly 1 or more umbellets sessile. Involucel absent or of 2–5 bractlets, these shorter than the flower stalks, entire, linear. Flowers 2–20 in each umbellet, the stalks 2–12 mm long. Sepals minute teeth, sometimes slightly enlarged and ovate-triangular. Petals broadly ovate, rounded or bluntly pointed at the tip, white. Ovaries glabrous. Fruits 2–3 mm long, broadly ovate-elliptic in outline, rounded at the base, tapered abruptly to a short beak at the tip, slightly flattened laterally, glabrous, dark brown with lighter ribs, each mericarp with 5 prominent, corky ribs, the dorsal and intermediate ribs relatively narrow and lacking wings, the lateral ribs broader than the others, with small, corky extensions over the commissures. May–June.

Uncommon, restricted to the Mississippi Lowlands Division (southeastern U.S. west to Missouri and Texas). Swamps, bottomland forests, sloughs, and banks of streams, also ditches, moist roadsides, and margins of rice fields.

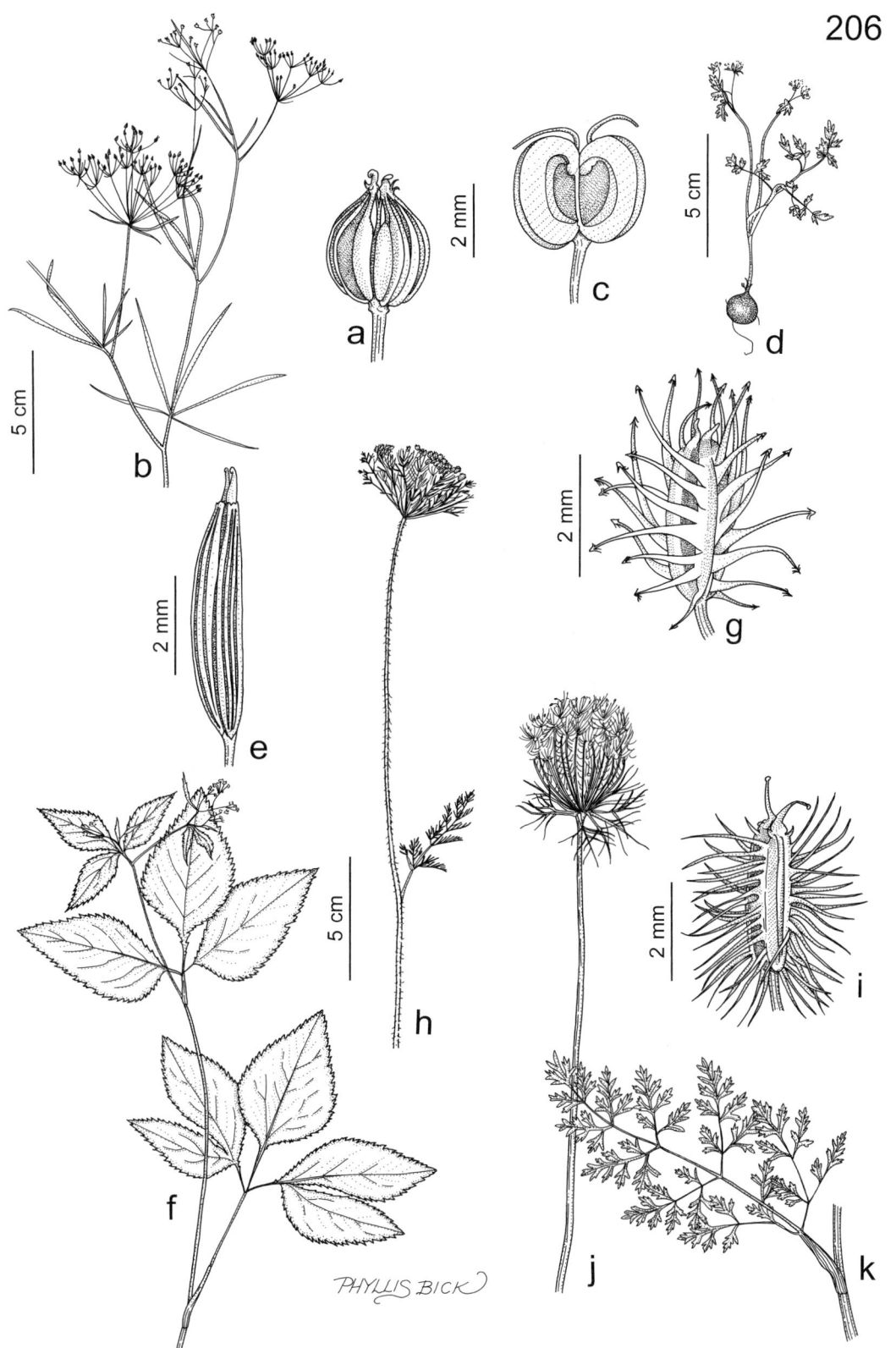

Plate 206. Apiaceae. *Cynosciadum digitatum*, **a)** fruit, **b)** habit. *Erigenia bulbosa*, **c)** fruit cross section, **d)** habit. *Cryptotaenia canadensis*, **e)** fruit, **f)** habit. *Daucus pusillus*, **g)** fruit, **h)** habit. *Daucus carota*, **i)** fruit, **j)** infructescence, **k)** leaf.

16. Daucus L. (carrot)

Plants annual or biennial. Stems erect or ascending, sparsely to densely pubescent with spreading to recurved, broad-based, stiff hairs. Leaves alternate and sometimes also basal (1 or a few basal leaves often present at flowering), short- to long-petiolate (mostly short-petiolate above the base), the sheathing bases not or only slightly inflated. Leaf blades oblong to triangular-ovate in outline, 2–4 times pinnately compound and/or dissected, the ultimate leaflets or segments linear to lanceolate, short-tapered to a sharp point at the tip, the margins entire or few-toothed or -lobed, sparsely to moderately hairy, especially along the margins and veins. Inflorescences terminal and axillary, compound umbels, long-stalked, the stalks moderately to densely pubescent with mostly recurved, broad-based, stiff hairs. Involucre of 4–15 bracts, these shorter than to more commonly longer than the rays, leaflike, pinnately 1–2 times dissected with linear segments, rarely entire, glabrous or sparsely hairy. Rays mostly numerous, roughened or hairy, the innermost ones shorter than the outer ones, the umbels thus appearing more or less flat-topped at flowering. Involucel of 5–13 bracts, these shorter than to more commonly slightly longer than the rays, entire or less commonly pinnately few-lobed toward the tip, glabrous or sparsely hairy. Flowers 5–20 in each umbellet, the stalks roughened, unequal in length. Sepals absent or consisting of minute triangular teeth. Petals obovate, notched into 2 unequal lobes at the tip, those of the outermost flowers in some umbels often somewhat enlarged, white or rarely pink, sometimes drying yellow, the innermost flower of each umbellet usually with a dark purple or rarely pink corolla. Ovaries hairy. Fruits oblong-elliptic to oblong-ovate in outline, flattened dorsally, brown, each mericarp with 5 inconspicuous primary ribs, these slender, nervelike, brown, lacking wings, hairy, and also with 4 secondary ribs between the primary ones, these, slender, straw-colored, prominently winged, the wing margins with a row of flattened bristles having pointed or minutely barbed or hooked tips. About 60 species, Europe, Asia, Africa, Australia; introduced nearly worldwide.

1. Fruiting umbels with the rays curving upward and inward, the umbels thus becoming more or less oblong in outline as the fruits develop; bractlets of the umbellets with broad, thin, white margins, at least toward the base, linear, appressed-ascending at fruiting.................................... 1. D. CAROTA
1. Fruiting umbels with the rays curving upward only slightly (or not), the umbels thus remaining more or less flat-topped as the fruits develop; bractlets of the umbellets without thin white margins (occasionally with very thin white margins toward the base), mostly narrowly elliptic-lanceolate, spreading to reflexed at fruiting .... 2. D. PUSILLUS

1. Daucus carota L. ssp. carota (Queen Anne's lace, wild carrot)

Pl. 206 i–k; Map 854

Stems 40–150 cm long. Leaf blades 5–20 cm long, the ultimate segments 2–12 mm long, 0.5–2.0 mm wide. Involucre with the bracts 4–40 mm long, spreading to loosely ascending at flowering, spreading to loosely reflexed at fruiting. Rays 3.0–7.5 cm long, spreading to loosely ascending at flowering, curving upward and inward at fruiting, the umbels thus becoming more or less oblong in outline as the fruits develop. Involucel with the bractlets (or their lobes) linear, with broad, thin, white margins, at least toward the base, appressed-ascending at fruiting. Flowers sessile (the central flower in each umbellet) or the stalks 1–8 mm long. Fruits 3–4 mm long, the mericarps with the flattened bristles of the winged secondary ribs usually very minutely barbed or hooked at the tip (barely visible at 10× magnification), appearing merely pointed at lower magnification. $2n=18$. May–October.

Introduced, common nearly throughout the state, except apparently uncommon in the Mississippi Lowlands Division (native of Europe, Asia; naturalized widely in North America, Central America, Caribbean Islands). Banks of streams and rivers, tops of bluffs, glades, and occasionally

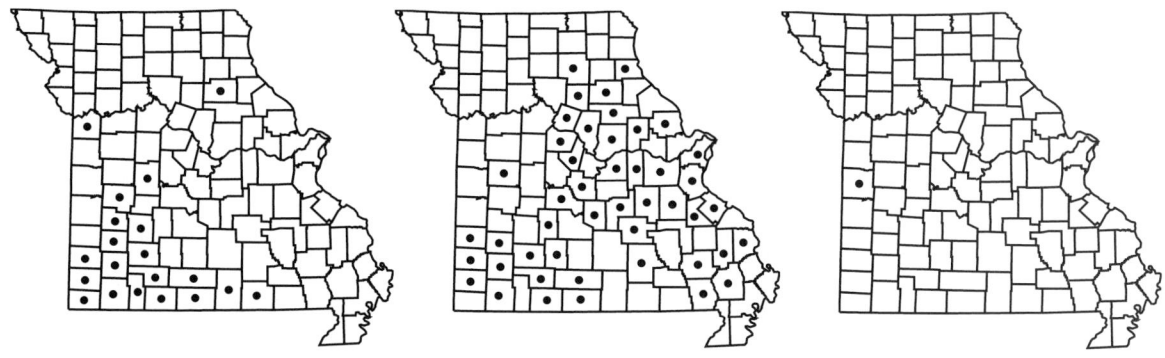

855. Daucus pusillus   856. Erigenia bulbosa   857. Eryngium leavenworthii

mesic to dry upland forests; also margins of crop fields, fallow fields, old fields, pastures, fencerows, roadsides, railroads, and open, disturbed areas.

Daucus carota is a polymorphic species consisting of several closely related subspecies (St. Pierre et al., 1990), notably ssp. *carota* (Queen Anne's lace), and ssp. *sativus* (Hoffm.) Arcang. (cultivated carrots). There are numerous carrot cultivars differing in root size, shape, and color, but all possess expanded taproots rich in starch and sugar, and most have yellowish corollas. Carrots also are high in beta carotene, which is converted easily into vitamin A in the human digestive tract, and in addition to its consumption raw, cooked, or baked, the subspecies is a commercial source of carotene for dietary supplements. The inflorescences of ssp. *carota* are sometimes used fresh as cut flowers or dried in floral arrangements. However, this weedy subspecies can produce furanocoumarins that may cause phototoxic dermatitis in some individuals. There is disagreement on whether ssp. *carota* should be considered mildly poisonous, but it is known that plants are capable of producing a series of polyacetylenes (see the treatment of *Cicuta*), principally falcarinol, that act as natural pesticides and have been documented to cause mild symptoms in livestock in Europe. Steyermark (1963) noted that the milk from cows grazing on *Daucus* has a bitter flavor. Several trivial color mutants of Queen Anne's lace have been named as forms, including those with all of the flowers pink to purple (f. *roseus* Farw.) and those with all of the flowers white, including the central flower of each umbellet (f. *epurpuratus* Millsp.).

2. Daucus pusillus Michx. (small wild carrot)

Pl. 206 g, h; Map 855

Stems 10–60(–80) cm long. Leaf blades 3–14 cm long, the ultimate segments 1–6 mm long, 0.5–1.0 mm wide. Involucre with the bracts 3–40 mm long, spreading to loosely ascending at flowering, spreading to loosely reflexed at fruiting. Rays 0.4–4.0 cm long, spreading to loosely ascending at flowering, curving upward only slightly (or not at all) at fruiting, the umbels thus remaining more or less flat-topped as the fruits develop. Involucel with the bractlets (or their lobes) linear to more commonly narrowly elliptic-lanceolate, lacking papery white margins, occasionally with very narrow, white margins toward the base, spreading to reflexed at fruiting. Flowers sessile (the central flower in each umbellet) or the stalks 1–9 mm long. Fruits 3–5 mm long, the mericarps with the flattened bristles of the winged secondary ribs minutely barbed or hooked at the tip (usually easily visible at 10× magnification). $2n=22$. April–June.

Scattered in the southwestern quarter of the state; sporadic north of the Missouri River (southern and western U.S. north to South Carolina, Kansas, and Washington; Canada, Mexico).

17. Erigenia Nutt. (harbinger of spring)

One species, eastern U.S.

1. Erigenia bulbosa (Michx.) Nutt. (harbinger of spring, pepper and salt)

Pl. 206 c, d; Map 856

Plants perennial, with globose tubers, glabrous. Stems 5–15 cm long at flowering, elongating to 25 cm at fruiting, erect to loosely ascending. Leaves mostly basal, sometimes also 1 or 2 alternate just above the stem base, mostly short-petiolate, the sheathing bases slightly inflated. Leaf blades 3–11 cm long, broadly ovate in outline, 2–4 times ternately compound, the leaflets 3–12 mm long, linear or narrowly spatulate, sometimes with a pair of basal lobes, narrowed at the base, rounded or abruptly narrowed to a sharp point at the tip, entire along the margins. Inflorescences terminal, simple or more commonly compound umbels, sessile. Involucre of 1 leaflike bract. Rays 1–4, 0.6–1.5 cm long at flowering, elongating to 4 cm at fruiting. Involucel of 3–7 bractlets, these longer than the flower stalks, these narrowly oblanceolate to narrowly spatulate, entire or toothed. Flowers 3–5 in each umbellet, the stalks 0.5–2.0 mm long. Sepals absent. Petals oblanceolate to obovate, rounded at the tip, white. Ovaries glabrous. Fruits 2–3 mm long, depressed-circular in outline, sometimes shallowly cordate at the base, shallowly and narrowly notched at the tip, strongly flattened laterally, glabrous, dark brown, each mericarp strongly curved at maturity, with 5 thin ribs sometimes narrowly winged, the commissures corky-thickened. $2n=20$. January–April.

Common in the eastern and southern halves of the state (eastern [mostly northeastern] U.S. west to Minnesota and Oklahoma). Bottomland forests, mesic upland forests, mostly in ravines and valleys, bases of bluffs, and banks of streams and rivers.

Harbinger of spring is perhaps Missouri's earliest-blooming native wildflower and a welcome sign of the impending spring season for outdoor enthusiasts. The common name pepper and salt is derived from the striking contrast between the white petals and dark reddish purple to nearly black anthers. The species is more common in Missouri than the specimen record indicates, for the plants flower early and are relatively inconspicuous in fruit. The small, deeply seated tubers supposedly are edible.

18. Eryngium L. (eryngo)
(Mathias and Constance, 1941b)

Plants annual or perennial. Stems erect, ascending, or prostrate. Leaves alternate (sometimes also basal) or opposite to whorled, the sheathing bases relatively short, not inflated. Leaf blades simple, entire or palmately lobed, the margins entire or toothed, the teeth sometimes spiny. Inflorescences terminal or axillary, dense cylindrical or globose heads of numerous flowers, these solitary or in small panicles, short- to long-stalked. Involucre subtending each head of 4–10 bracts. Flowers each subtended by a bractlet. Sepals ovate or oblong, entire or with deep, narrow, spiny teeth or lobes at the tip. Petals rounded at the tip, white, purple, or blue. Ovaries with small scales or tubercles. Fruits oblong, obovate, or nearly circular in outline, usually slightly flattened laterally, brown, each mericarp with 5 angles, lacking ribs, variously with small scales or tubercles. About 250 species, nearly worldwide.

1. Stems prostrate, creeping, rooting at the nodes; leaves basal and opposite or whorled, entire to lobed or toothed, but not spiny; heads grayish blue
.... 2. E. PROSTRATUM
1. Stems erect or ascending; leaves basal and alternate, the margins with spiny lobes or teeth or threadlike teeth; heads grayish white or purple
    2. Leaves with netted venation, not or slightly glaucous, mostly purple-tinged; heads more or less cylindrical, purple, with conspicuous, spiny-lobed bracts at the base and a terminal crown of spiny-lobed bracts.... 1. E. LEAVENWORTHII
    2. Leaves with parallel venation, strongly glaucous, not purple-tinged; heads more or less globose, grayish white, with inconspicuous, entire bracts at the base, lacking a terminal cluster of bracts .... 3. E. YUCCIFOLIUM

1. Eryngium leavenworthii Torr. & A. Gray
(Leavenworth eryngo)
Map 857

Plants annual, glabrous, mostly purple-tinged, not or only slightly glaucous. Stems 40–100 cm long, erect or ascending. Leaves alternate and rarely basal (basal leaves usually withered or absent at flowering), sessile or the basal leaves sometimes short-petiolate, the venation netted, the margins with a thickened whitened band. Basal

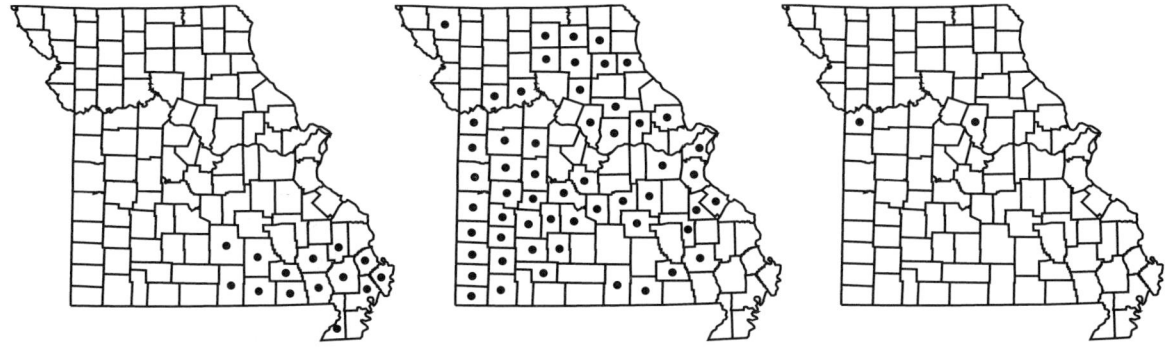

858. Eryngium prostratum   859. Eryngium yuccifolium   860. Falcaria vulgaris

and lower stem leaves with the blades 3–7 cm long, unlobed, broadly oblanceolate, tapered at the base, rounded to bluntly pointed at the tip, the margins spiny-toothed. Median and upper leaves with the blades 2.5–7.5 cm long, broadly obovate to nearly circular in outline, deeply palmately 3–7-lobed, the lobes pinnately lobed and/or toothed, the lobes or teeth with spiny tips. Inflorescences more or less cylindrical heads 2.0–3.5 cm long, terminal, solitary or few in small, loose clusters, short-stalked, purple, the branch points with reduced, leaflike bracts. Bracts subtending each head 4–8, 25–40 mm long, oblong-elliptic, narrowed or short-tapered to a spiny tip, the margins with narrow, spiny teeth or lobes. Terminal bracts (actually enlarged bractlets) 4–8, more or less similar to the bracts, but 10–35 mm long. Flowers numerous, sessile, each subtended by a bractlet similar to the bracts, but 8–10 mm long. Sepals 4–6 mm long, oblong, with 3–5 narrow, spiny teeth or lobes at the tip. Petals oblong, rounded at the tip, purple (rarely white elsewhere). Ovaries with ascending, papery, linear scales. Fruits 2–4 mm long, oblong in outline, with appressed, linear scales. $2n=12$. July–September.

Introduced, known thus far only from a single site in Bates County (Kansas to Texas). Pastures (potentially upland prairies).

This species was excluded from the flora by Steyermark (1963), based on unreadable and questionable label data on the single historical specimen he was able to locate. This specimen probably originated from Anderson County, Kansas. More recently, the species was reported from Bates County by Dierker (1992). At this site, plants apparently were introduced as contaminants in hay brought into a degraded upland prairie serving as a pasture. Dierker noted that the plant grows natively in Kansas in several counties adjacent to Missouri and natural occurrences might eventually be discovered in the state.

2. **Eryngium prostratum** Nutt. (spreading eryngo)

Pl. 207 d, e; Map 858

Plants perennial, with thin, fibrous roots, glabrous, not glaucous. Stems 7–50 cm long, prostrate, creeping, rooting at the nodes. Leaves opposite or more commonly whorled and basal (rooted nodes may produce a new basal rosette with age), the blades entire or with a pair of narrow basal lobes, the venation netted. Basal leaves long-petiolate, the blades 1–4 cm long, lanceolate to broadly ovate, narrowed or rounded at the base, rounded to bluntly pointed at the tip, the margins entire or with few, relatively coarse, blunt teeth. Leaves at the nodes of the stems similar to the basal leaves, but sessile or mostly short-petiolate, the blades 0.4–2.5 cm long. Inflorescences more or less cylindrical heads 0.4–0.9 cm long, axillary, solitary, mostly long-stalked, grayish blue. Bracts subtending each head 5–10, 2–12 mm long, linear to narrowly lanceolate or narrowly oblanceolate, sharply pointed at the tip, the margins entire or with a few minute teeth. Terminal bracts absent. Flowers numerous, sessile, each subtended by a narrowly lanceolate bractlet 0.5–0.8 mm long. Sepals 0.5–0.8 mm long, ovate to broadly ovate, entire, usually with broad, thin, pale margins. Petals obovate, rounded or shallowly notched at the tip, blue or less commonly white. Ovaries with minute, white tubercles. Fruits 0.5–0.8 mm long, broadly oblong-obovate in outline, with minute, white tubercles. $2n=16$. May–November.

Scattered in the southeastern quarter of the state (southeastern U.S. west to Missouri and Texas). Swamps, bottomland forests, margins of ponds, sinkhole ponds, and lakes, banks of streams and rivers, and fens; also pastures.

This diminutive creeping species was recorded by Steyermark (1963) from moist to wet habitats in the Mississippi Lowlands Division and margins of scattered sinkhole ponds in the eastern half of the Ozarks. In more recent years, however, it appears to have spread to a number of additional wetland communities within the original range and is probably more abundant now than it was prior to the 1960s.

3. Eryngium yuccifolium Michx. (rattlesnake master, button snakeroot)

Pl. 207 f, g; Map 859

Plants perennial, with tuberous roots, glabrous, strongly glaucous. Stems 40–150 cm long, erect or ascending. Leaves alternate and usually also basal (several basal leaves usually present at flowering), sessile, grasslike, the venation parallel. Leaf blades 4–100 cm long (those of basal and lower leaves 15–100 cm), linear, sharply pointed at the tip, the margins with widely spaced, spiny or threadlike teeth, less commonly entire. Inflorescences globose heads 1.0–2.5 cm long, terminal, solitary or in small panicles or loose clusters, mostly long-stalked, grayish white, the branch points with reduced, leaflike bracts, these with spiny teeth or lobes. Bracts subtending each head 6–10, 4–16 mm long, narrowly ovate, sharply pointed at the tip, the margins entire or finely toothed. Terminal bracts absent. Flowers numerous, sessile, each subtended by a bractlet similar to the bracts, but 6–10 mm long. Sepals 2.5–3.0 mm long, ovate, entire. Petals oblong, rounded at the tip, white. Ovaries with appressed, lanceolate scales. Fruits 4–8 mm long, oblong in outline, with ascending, papery, lanceolate scales mostly along the angles. $2n=96$. June–August.

Scattered to common nearly throughout the state, but apparently absent from the Mississippi Lowlands Division (southeastern U.S. west to Kansas and Texas). Upland prairies, glades, savannas, and openings of mesic to dry upland forests.

Eryngium yuccifolium was widely used in Native American medicine, principally as a diuretic for kidney ailments, as a sedative and pain reliever, and as a tonic (Moerman, 1998). The common names rattlesnake master and snakeroot refer to the belief in its efficacy as a remedy against snakebites. The species also was an important fiber plant and was used in cordage, bags, cloth, and sandals. Gordon (1999) recently studied the archaeological record of slippers woven from the leaves of *E. yuccifolium*.

The foliage of *E. yuccifolium* can be mistaken for that of a monocot, but among Missouri monocots only *Yucca* has threadlike teeth or processes along the leaf margins, and the leaves in that genus are more leathery, restricted to a basal rosette, and the marginal processes are fibrous and never spinelike.

19. Falcaria Fabr.

One or 5 species, Europe, Asia.

1. Falcaria vulgaris Bernh. (sickleweed)
F. sioides (Wibel) Asch.

Pl. 207 a–c; Map 860

Plants biennial or perennial, glabrous, glaucous, with long taproots. Stems 30–100 cm long, erect or ascending. Leaves alternate and sometimes also basal (1 or more basal leaves sometimes present at flowering), the basal and lowermost stem leaves usually long-petiolate, the median and upper leaves short-petiolate or sessile, the sheathing bases not or only slightly inflated. Leaf blades 1–35 cm long, broadly oblong to triangular-ovate in outline, those of the basal and lowermost stem leaves ternately or ternately then pinnately 1–2 (–3) times compound or lobed, less commonly ternately lobed or compound then 2-lobed, the main axis (rachis) often winged below the attachment points of the leaflets, the ultimate leaflets or segments 20–250 mm long, narrowly oblong to linear, narrowed at the base, tapered to a sharp point at the tip, the margins finely toothed (the teeth with minute, thickened, spiny tips); the leaflets of the median and upper stem leaves progressively reduced, 1 or 2 times ternately compound or lobed, the uppermost leaves often simple. Inflorescences terminal and axillary, compound umbels, mostly long-stalked. Involucre of 3–12 bracts, these shorter than the rays, spreading to loosely reflexed at flowering, linear to narrowly lanceolate, entire, with sharply pointed tips. Rays (4–)10–25, 0.8–4.0 cm long. Involucel of 3–11 bractlets, these shorter than to longer than the flower stalks, similar to the bracts but smaller. Flowers 5 to numerous in each umbellet, the stalks 4–10 mm long. Sepals minute triangular teeth. Petals obovate, rounded or shallowly notched at the tip, white. Ovaries glabrous. Fruits 2–5 mm long, narrowly oblong in outline, flattened laterally, glabrous, brown with pale ribs, the mericarps often somewhat arched or curved, with 5 broad, flattened ribs (wider than the spaces between them), these lacking wings. $2n=22$. July–September.

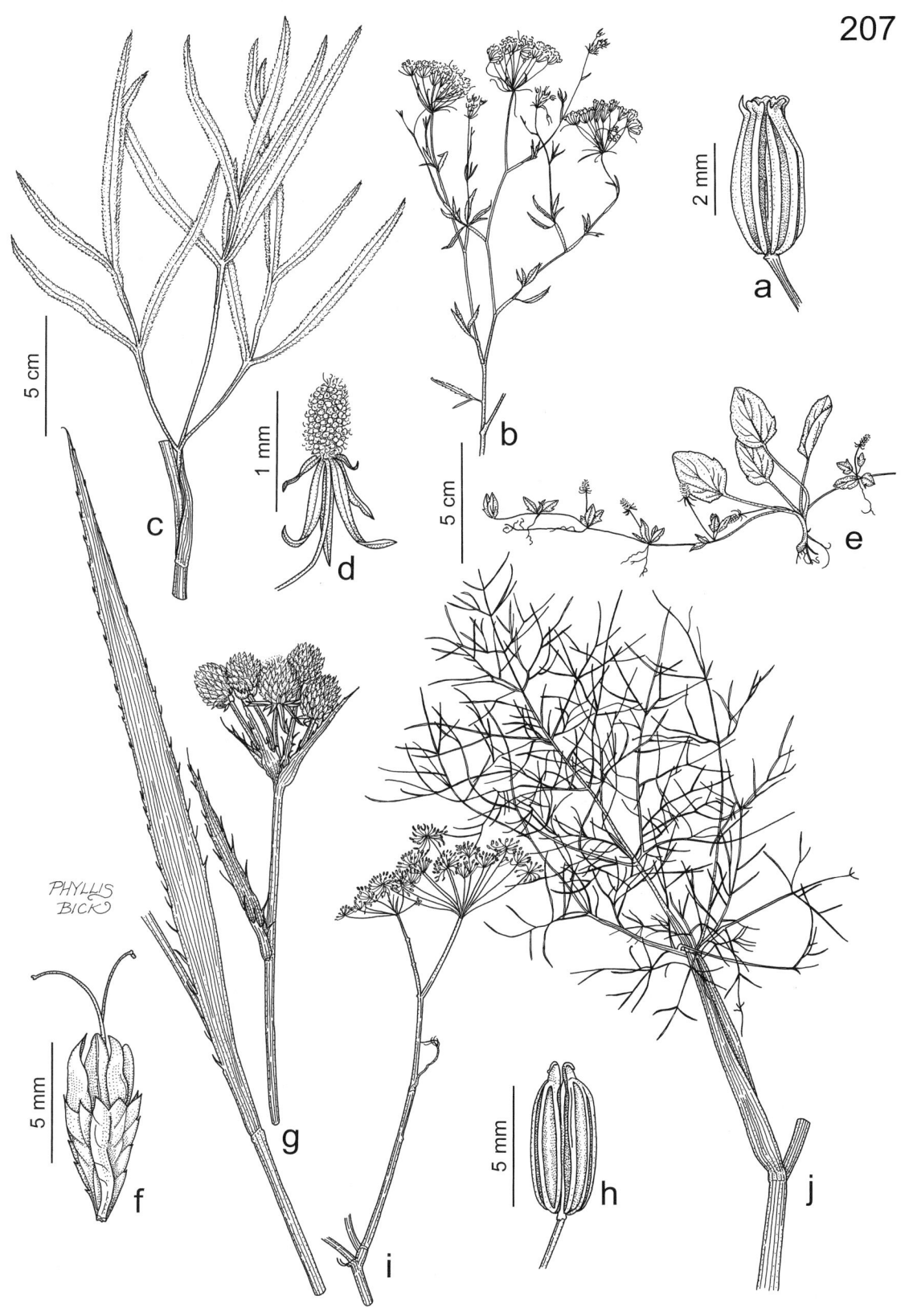

Plate 207. Apiaceae. *Falcaria vulgaris*, **a)** fruit, **b)** habit, **c)** leaf. *Eryngium prostratum*, **d)** inflorescence, **e)** habit. *Eryngium yuccifolium*, **f)** fruit, **g)** leaf and inflorescence. *Foeniculum vulgare*, **h)** fruit, **i)** inflorescence, **j)** leaf.

861. Foeniculum vulgare   862. Heracleum sphondylium   863. Hydrocotyle ranunculoides

Introduced, uncommon, known thus far only from Boone and Jackson Counties (native of Europe, Asia; introduced widely but sporadically in the northeastern U.S.). Roadsides, fencerows, and open, disturbed areas.

20. Foeniculum Mill.

Five species, Europe, Asia, Africa.

1. Foeniculum vulgare Mill. (fennel)

Pl. 207 h–j; Map 861

Plants functionally annual or biennial in Missouri (perennial farther south), glabrous, glaucous, with taproots. Stems 50–200 cm long, erect or ascending. Leaves alternate and sometimes also basal (1 or 2 basal leaves occasionally present at flowering), short-petiolate to nearly sessile (basal leaves sometimes long-petiolate), the sheathing bases of at least the larger leaves 3–10 cm long, slightly to moderately inflated, turning tan and papery with age, the uppermost leaves sometimes reduced to bladeless sheaths. Leaf blades 3–30 cm long, ovate to broadly triangular-ovate in outline, those of the basal to median stem leaves pinnately 3–5 times dissected, the ultimate segments 4–40 mm long, linear, short-tapered to an abrupt, sharp point at the tip; those of the upper stem leaves progressively reduced, 1 or 2 times dissected or sometimes bladeless. Inflorescences terminal and axillary, compound umbels, short- to long-stalked. Involucre absent. Rays mostly numerous, 1.0–6.5 cm long, unequal in length. Involucel absent. Flowers 12 to numerous in each umbellet, the stalks 2–10 mm long, unequal in length. Sepals absent. Petals obovate, rounded or bluntly pointed at the tip, yellow. Ovaries glabrous. Fruits 3.5–4.0 mm long, oblong in outline, slightly flattened laterally, glabrous, dark brown with lighter, yellowish ribs, each mericarp with 5 ribs, these more or less angled but lacking wings. $2n=22$. May–September.

Introduced, uncommon, widely scattered (native of Europe, Asia, Africa; introduced widely in North America, Central America, South America, Caribbean Islands). Railroads, roadsides, and open, disturbed areas.

The fruits and herbage of fennel are used to flavor various foods, and the fruits are also sometimes used to flavor confections and liquors. The young foliage and stems, as well as the roots, are sometimes eaten raw in salads or cooked. Aromatic oils with an odor similar to that of anise (*Pimpinella anisum* L., another Apiaceae) are extracted from the fruits as an ingredient in perfume, soap, and bath oil fragrances. Medicinally, the species has been used in tonics, mostly for various gastrointestinal ailments.

21. Heracleum L.

About 65 species, U.S., Canada; Europe, Asia, Africa.

1. **Heracleum sphondylium** L. **ssp. montanum** (Schleich. ex Gaudin) Briq. (cow parsnip, masterwort)
   H. maximum W. Bartram
   H. lanatum Michx.

Pl. 208 e, f; Map 862

Plants perennial, usually with taproots. Stems 60–200(–300) cm long, erect, stout, strongly ridged, sparsely to densely pubescent with spreading hairs. Leaves alternate and usually also basal (1 or a few basal leaves usually present at flowering) short- to long-petiolate, the uppermost leaves sometimes appearing sessile, the sheathing bases, at least those of the median and upper leaves, moderately to strongly inflated, moderately to densely pubescent with mostly spreading hairs. Leaf blades 10–50 cm long, broadly ovate to circular-cordate in outline, those of the uppermost leaves sometimes only 7 cm long, simple and 3-lobed; those of the main leaves 1 time ternately compound, the leaflets (40–)90–400 mm long, ovate to nearly circular, pubescent with mostly spreading hairs on the undersurface, especially along the main veins and basally, often shallowly to deeply 3-lobed and with shallow lobes and teeth along the margins, the lobes narrowed or tapered to an abrupt, sharp point at the tip. Inflorescences terminal and axillary, compound umbels, mostly long-stalked. Involucre of 5–10 bracts, these 5–20 mm long, lanceolate, long-tapered to sharply pointed tips, hairy, usually shed before the fruits mature. Rays 15 to numerous, 5–10 cm long, unequal in length, hairy. Involucel of 4–10 bractlets, these shorter than to about as long as the flower stalks, similar to the bracts but smaller. Flowers numerous in each umbellet, the stalks 5–12 mm long, unequal in length, elongating to 20 mm at fruiting, hairy. Sepals absent or minute triangular scales. Petals obovate, rounded or shallowly notched at the tip, white, those of the outermost flowers of the umbel sometimes somewhat enlarged and more deeply notched. Ovaries sparsely to densely hairy. Fruits 7–12 mm long, oblong-obovate in outline, narrowed or rounded at the base, broadly notched at the tip, strongly flattened dorsally, usually sparsely hairy, tan to straw-colored with usually prominent, reddish brown oil tubes between the ribs, each mericarp with the intermediate and dorsal ribs slender and nervelike, hardly raised from the surface, the lateral ribs with thin, broad wings. $2n=22$. May–July.

Uncommon to scattered in the Glaciated Plains Division, with isolated occurrences in southern Missouri (U.S. [including Alaska]; Canada, Siberia). Bottomland forests, mesic upland forests in ravines, and banks of streams; also roadsides.

Steyermark (1963) and many other authors have treated American plants as a distinct species, *H. maximum*, but the differences between this and the Eurasian *S. sphondylium* are not very well marked (Brummitt, 1971). The var. *sphondylium* differs in its slightly more divided leaves and umbels often with more rays.

Although plants of this species are considered only mildly poisonous when ingested, compared with some other Apiaceae, the sap of cow parsnip contains furanocoumarins, which can cause severe phototoxic dermatitis in some people, resulting variously in irritation, rash, and potentially serious blisters. The closely related *H. mantegazzianum* Sommier & Levier (giant cow parsnip), a Eurasian species capable of reaching 5 m in height that has escaped from cultivation in a few northeastern states and Canada, causes an even more severe reaction.

22. **Hydrocotyle** L. (water pennywort)

Plants perennial, with thin, fibrous roots, glabrous. Stems 5–30 (or more) cm long, prostrate, creeping, rooting at the nodes. Leaves solitary or occasionally appearing paired at the nodes, long-petiolate, glabrous. Leaf blades peltate or deeply cordate at the base, circular to broadly depressed-oval or broadly kidney-shaped, the margins scalloped and/or lobed. Inflorescences simple umbels or spikes, sometimes appearing headlike on smaller plants, axillary, the stalks shorter than the petioles. Bractlets 0.5–1.2 mm long, lanceolate to oblong-lanceolate. Sepals absent. Petals ovate, rounded or bluntly pointed at the tip, white. Ovaries glabrous. Fruits circular to depressed-circular in outline, rounded or truncate to shallowly cordate at the base and tip, strongly flattened laterally, glabrous, tan to brown, the mericarps with the ribs inconspicuous or the dorsal and lateral ribs somewhat corky-thickened. About 130 species, nearly worldwide.

864. Hydrocotyle verticillata   865. Levisticum officinale   866. Ligusticum canadense

1. Leaf blades deeply cordate at the base, with 2–4 lobes in addition to the scalloped margins; mericarps with the ribs all inconspicuous .... 1. H. RANUNCULOIDES
1. Leaf blades peltate, without prominent lobes in addition to the scalloped margins; mericarps with the dorsal and lateral ribs somewhat corky-thickened
.... 2. H. VERTICILLATA

1. Hydrocotyle ranunculoides L.f. (buttercup pennywort)

Map 863

Leaf blades 1.0–5.5 cm wide, deeply cordate at the base, broadly depressed-oval to more commonly broadly kidney-shaped, the margins finely scalloped and 2–4-lobed up to half the way to the base. Inflorescences of simple umbels of 4–10 flowers, these short-stalked. Fruits 1–3 mm long, circular to depressed-circular in outline, rounded or truncate to rarely very shallowly cordate at the base and tip, each mericarp with the ribs all inconspicuous and not thickened. $2n=24$. May–August.

Possibly introduced, known thus far only from a single specimen from Stoddard County (eastern U.S. west to Kansas and Texas, also Washington, Oregon, California, and Arizona; Canada, Mexico, Central America, South America). Emergent aquatic along the margin of a slough.

This species was first discovered in the state by Stan Hudson in 2000. It is known to occur in adjacent states, including Arkansas, Illinois, Kansas, Oklahoma, and Tennessee, and was thus to be expected in Missouri. However, the population in Stoddard County grows near an office building and former homesite. Thus, it is unclear whether the species is a native component of the aquatic plant community at the site or whether it escaped from historical cultivation in the vicinity.

2. Hydrocotyle verticillata Thunb. (whorled pennywort)

Pl. 208 c, d; Map 864

Leaf blades 0.5–5.5 cm in diameter, peltate, circular or nearly so, the margins finely scalloped or with a few very shallow, blunt lobes. Inflorescences spikes (often appearing headlike on smaller plants) with 1 or 2(–4) whorls of 2–7 flowers, these sessile or nearly so. Fruits 1–3 mm long, depressed-circular in outline, rounded or truncate to rarely very shallowly cordate at the base, shallowly notched at the tip, strongly flattened laterally, glabrous, tan to brown, each mericarp with the lateral ribs somewhat corky-thickened, the intermediate ribs narrow and unwinged, and the dorsal ribs somewhat corky-thickened, tapered to a thin edge. $2n=\pm 88$. May–August.

Uncommon, known thus far only from Oregon and Ozark Counties (eastern and southwestern U.S., Hawaii; Mexico, Central America, South America, Caribbean Islands, Africa). Banks of spring branches and rivers, usually in sandy soil, sometimes emergent aquatics in shallow water.

The single Oregon County specimen consists of robust nonflowering material collected at Morgan Spring, where aquatic plants formerly were raised commercially in concrete troughs, and from which materials washed into the spring branch and adjacent Eleven Point River during one or more floods. The plants have not been relocated since the original discovery by Paul Redfearn in 1970. Because *H. verticillata* and the superficially similar *H. umbellata* L. (water pennywort) cannot be distinguished reliably from vegetative samples, there is a possibility that the Oregon County specimen is actually the latter species. *Hydrocotyle umbellata* is widespread in North America (and elsewhere in the New World) and occurs natively as close as Fulton County, Arkansas, which is just to the south of Oregon County, Missouri. Like *H.*

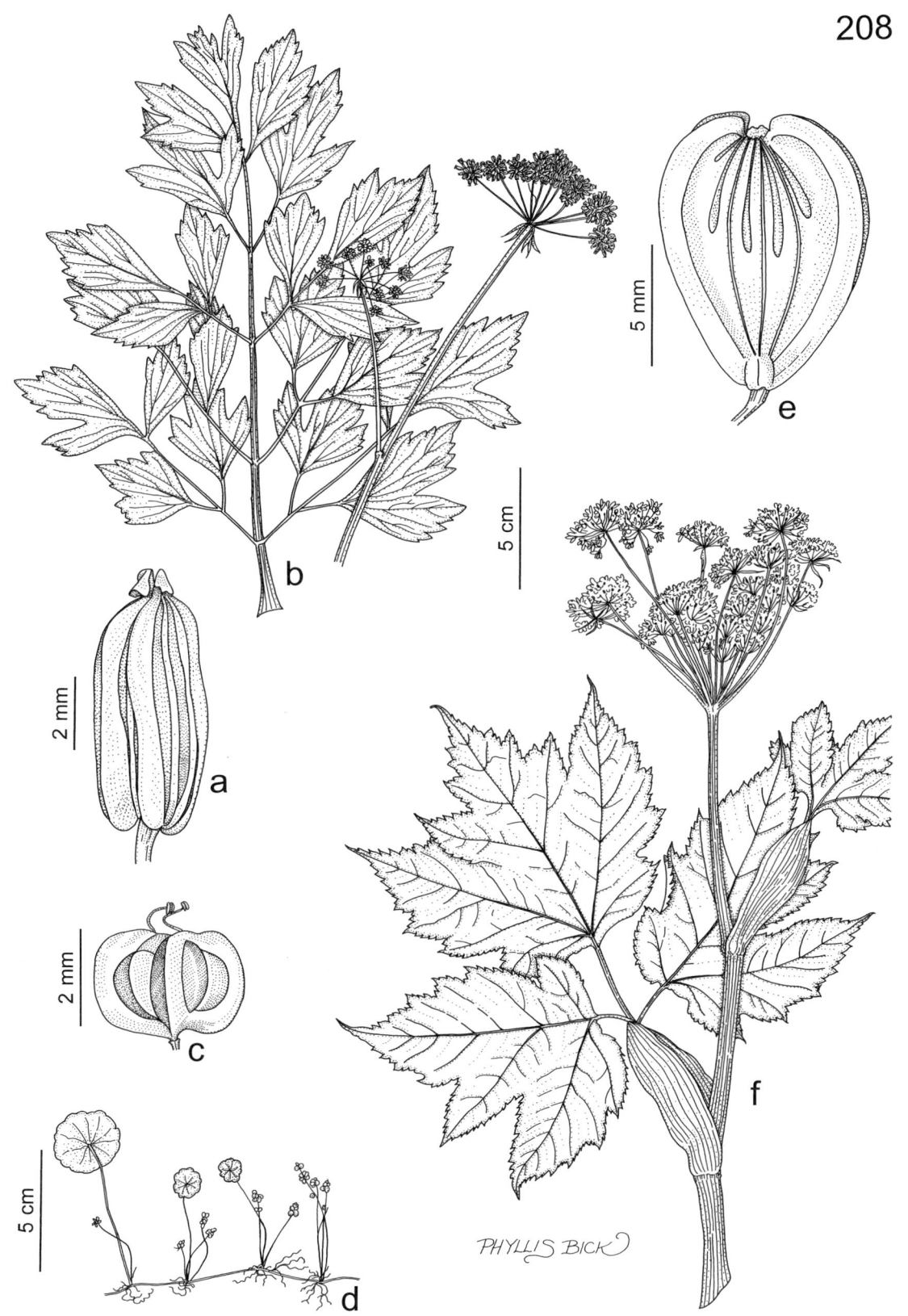

Plate 208. Apiaceae. *Levisticum officinale*, **a)** fruit, **b)** inflorescence and leaf. *Hydrocotyle verticillata*, **c)** fruit, **d)** habit. *Heracleum sphondylium,* **e)** fruit, **f)** habit.

verticillata, it is sold in the horticultural trade for cultivation in ponds and wet areas. The two species differ in inflorescence type (*H. umbellata* has simple umbels with long-stalked flowers) and fruits (*H. umbellata* tends to have slightly smaller fruits with more rounded ribs and somewhat cordate bases). This species should be searched for in southern Missouri.

Mathias and Constance (1944–1945) recognized several varieties in *H. verticillata*, differing in minor characters of the leaves and inflorescences. Missouri plants are all referable to var. *verticillata*. The validity of these taxa requires further study.

23. Levisticum Hill

One or 3 species, Asia.

1. Levisticum officinale W.D.J. Koch (lovage)
L. paludapifolium (Lam.) Asch.

Pl. 208 a, b; Map 865

Plants annual. Stems 80–200 cm long, erect or ascending, glabrous, sometimes slightly purplish-tinged. Leaves alternate and sometimes also basal (a few basal leaves occasionally present at flowering), glabrous, often somewhat glaucous, mostly short-petiolate, the sheathing bases not or only slightly inflated. Leaf blades 4–60 cm long, ovate to broadly triangular-ovate in outline, those of the basal and lowermost stem leaves ternately or more commonly pinnately 2 or 3 times compound, the leaflets 30–110 mm long, narrowly oblong to broadly ovate, narrowed at the base, usually coarsely toothed or lobed, the lobes narrowed or tapered to a sharp point at the tip; the leaflets of the median and upper stem leaves progressively reduced, the uppermost leaflets 1 time pinnately compound or lobed, sometimes simple. Inflorescences terminal and axillary, compound umbels, mostly long-stalked, the stalks glabrous or minutely roughened. Involucre of numerous bracts, these shorter than the rays, spreading to reflexed at flowering, lanceolate to narrowly lanceolate, with thin, papery margins and sharply pointed tips. Rays 12–20, 0.8–3.0 cm long, roughened. Involucel of numerous bractlets, these mostly longer than the flower stalks, similar to the bracts but smaller. Flowers mostly numerous in each umbellet, the stalks 1–5 mm long, roughened. Sepals absent. Petals obovate, rounded or shallowly notched at the tip, yellow to greenish yellow. Ovaries glabrous. Fruits 4–7 mm long, oblong-elliptic in outline, flattened dorsally, glabrous, dark brown with pale ribs, the mericarps with the lateral and sometimes also the intermediate and dorsal ribs with narrow, corky wings. $2n=22$. June–August.

Introduced, uncommon and sporadic (native of Europe; introduced sporadically in the U.S.). Roadsides and open, disturbed areas.

Lovage is cultivated as an herb, for its medicinal, culinary, and other uses. Medicinally it has been used mainly as a sedative and anticonvulsant, and to treat sores. Fruits are sometimes steeped in brandy for use as a digestive aid. The foliage and fruits are used as a flavoring on salads and in baked goods, as well as soups and stews. The aromatic oils also provide fragrance for soaps, bath oils, and potpourri. The common name lovage apparently arose because of the plant's reputation in folklore as an ingredient in love potions.

24. Ligusticum L. (lovage)

Forty to 50 species, North America, Europe, Asia.

1. Ligusticum canadense (L.) Britton
(angelico, nondo)

Pl. 209 i, j; Map 866

Plants perennial, with expanded, fibrous bases and taproots. Stems 40–150 cm long, erect or ascending, glabrous. Leaves alternate and sometimes also basal (1 or a few basal leaves usually present at flowering), glabrous, short- to long-petiolate, the sheathing bases not or only slightly inflated. Leaf blades (4–)10–24 cm long, ovate to triangular-ovate in outline, the basal and lower stem leaves ternately (less commonly ternately then pinnately) 3 or 4 times compound, the upper leaves reduced and occasionally simple, the leaflets 25–120 mm long, lanceolate to ovate or oblong-ovate, occasionally with 1 or 2 basal lobes, narrowed or tapered at the base, narrowed or tapered to a sharp point at the tip, finely to more coarsely toothed along

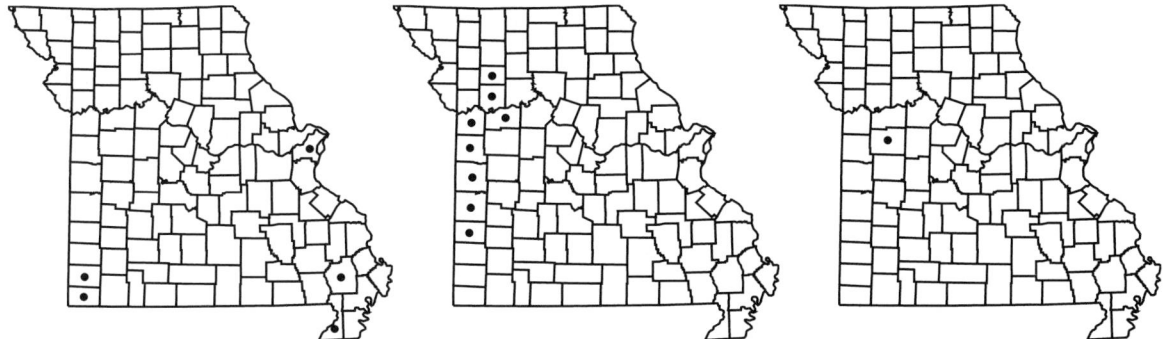

867. Limnosciadium pinnatum   868. Lomatium foeniculaceum   869. Oenanthe javanica

the margins. Inflorescences terminal and often also axillary, the terminal ones sometimes in clusters or alternate on a continuation of the stem, mostly compound umbels, long-stalked, the stalks glabrous or sparsely hairy. Involucre absent. Rays 6–14, unequal in length, 2–5 cm long, glabrous or less commonly sparsely hairy. Involucel of 2–5 entire bractlets, these shorter than the flower stalks, linear or narrowly oblong, glabrous. Flowers 6–14 in each umbellet, the stalks 2–4 mm long. Sepals minute triangular teeth. Petals obovate, rounded or bluntly pointed at the tip, white. Ovaries glabrous. Fruits 4–7 mm long, ovate-elliptic in outline, rounded at the base, slightly flattened laterally, glabrous, dark brown with usually lighter ribs, each mericarp with 5 narrowly winged ribs. $2n=22$. May–July.

Scattered, mostly in the Ozark Division (eastern U.S. west to Missouri and Arkansas). Mesic to dry upland forests, savannas, ledges of bluffs, and banks of streams and rivers.

25. Limnosciadium Mathias and Constance

Two species, central U.S.

Mathias and Constance (1941a) segregated *Limnosciadium* from the superficially similar *Cynosciadium* based on differences in leaf division patterns and details of fruit morphology. They cited these genera as an example of convergent vegetative morphologies in the Apiaceae. Steyermark (1963) disagreed with their conclusions, treating only a single genus, but most botanists since then have accepted the concept that, within the family, subtle differences in the plane of fruit compression and the shape of the carpophore can provide important clues to generic relationships and classification.

1. Limnosciadium pinnatum (DC.) Mathias & Constance

Cynosciadium pinnatum DC.

Pl. 209 g, h; Map 867

Plants annual, glabrous. Stems 6–50(–80) cm long, loosely ascending to erect. Leaves alternate and often also basal (1 or 2 basal leaves often present at flowering), short- to long-petiolate, the sheathing bases not or only slightly inflated. Leaf blades 2–20 cm long, pinnately compound with 3–9 widely spaced leaflets, the basal and uppermost stem leaves (or occasionally nearly all the leaves) sometimes simple, linear (when simple) to lanceolate or ovate-elliptic in outline, the leaflets 6–100 mm long, the terminal leaflet noticeably longer than the others, 1–6 mm wide, linear to narrowly elliptic-lanceolate, entire, gradually narrowed or tapered at the base and tip, with irregular, fine cross-veins or partitions (these mostly on the rachis in compound leaves). Inflorescences terminal and axillary, compound umbels, sessile to long-stalked. Involucre of 2–8 bracts or rarely absent, the bracts 2–6 mm long, reflexed at flowering, entire, linear to narrowly lanceolate or triangular. Rays 3–12, usually unequal in length, 0.5–2.5 cm long. Involucel of 3–7 bractlets, these shorter than the flower stalks, entire, linear or narrowly triangular. Flowers 4–20 in each umbellet, the stalks 2–8 mm long. Sepals minute, ovate or triangular, with thin, white margins. Petals obovate, rounded at the tip, white. Ovaries glabrous. Fruits 2–4 mm long, oblong-elliptic in outline, rounded at the base, rounded and lacking a beak at the tip, slightly flattened laterally, glabrous, dark brown with lighter ribs, each mericarp with 5 blunt, corky ribs, the dorsal and intermediate ribs lacking wings, the lateral ribs broader than the others, with small, corky extensions over the commissures. May–July.

Uncommon in southwestern and southeastern Missouri; also introduced in St. Louis (Iowa to Kansas south to Louisiana and Texas). Moist depressions of glades and upland prairies; also railroads, roadsides, and ditches.

26. Lomatium Raf. (wild parsley)
(Mathias, 1938; Theobald, 1966)

About 74 species, western and central U.S., Canada.

1. Lomatium foeniculaceum (Nutt.) J.M. Coult. & Rose (hairy parsley, wild parsley)
Pl. 209 e, f; Map 868

Plants perennial, with tuberous-thickened taproots. Aerial stems absent (only the inflorescence stalk present). Leaves in a basal rosette, short- to long-petiolate, the sheathing bases not or only slightly inflated, tinged with purple. Leaf blades 4–19 cm long, broadly ovate-triangular to oblong-ovate in outline, pinnately or ternately then pinnately 3 or 4 times compound, the ultimate leaflets or segments 2–9 mm long, 0.5–1.0 mm wide, linear, tapered abruptly to a minute, sharp point at the tip, entire, glabrous or more commonly sparsely to densely hairy. Inflorescences usually solitary, long-stalked, compound umbels arising from the rosette, the stalk 7–40 cm long, glabrous or sparsely to densely hairy. Involucre absent or rarely of 1–3 bracts, these 6–10 mm long, lanceolate-triangular, densely hairy, with broad, thin, white margins. Rays 10 to numerous, unequal in length, 0.4–3.0 cm long (also often 1 umbel sessile) at flowering, elongating to 9 cm at fruiting, glabrous or more commonly sparsely to densely hairy. Involucel an irregular cup of 7–19 elliptic-lanceolate hairy bractlets fused variously less than or more than half of the way to the tip, these shorter than to longer than the flower stalks, the cup tearing between bractlets (usually between the shorter ones) as the flowers mature, leaving an irregular, somewhat spathelike structure at the base of the umbel. Flowers numerous in each umbellet, the stalks 1–15 mm long. Sepals minute triangular teeth or more commonly absent. Petals obovate, rounded or bluntly pointed at the tip, yellow. Ovaries glabrous or hairy. Fruits 5–9 mm long, oblong-ovate in outline, rounded at the base, flattened laterally, glabrous or hairy, light brown to tan or straw-colored with prominent reddish brown oil tubes, the mericarps with the lateral and intermediate ribs relatively flat and inconspicuous, the lateral ribs with broad, corky wings. $2n=22$. April–May.

Uncommon in the Unglaciated Plains Division and the adjacent southwestern portion of the Glaciated Plains (western U.S. east to North Dakota, Missouri, and Texas; Canada). Glades and rocky portions of upland prairies, on calcareous substrates.

Mathias (1938) classified the *L. foeniculaceum* complex as consisting of several closely related species. Theobald (1966) reduced these to subspecies of three western species, of which only *L. foeniculaceum* occurs to the east of California. Three of Theobald's five subspecies of *L. foeniculaceum* occur only to the west of the Great Plains and differ from Missouri plants in several subtle characters, including hairy petals, free bractlets, and/or more tightly sheathing petioles. Missouri is in a zone of geographic overlap between Theobald's other two closely related subspecies, and both he and McGregor (1986d) indicated that populations from this region were not as sharply separable into subspecies as were plants growing outside the zone of overlap. It is notable that Steyermark (1963), who was familiar only with Missouri specimens, chose not to recognize segregates in the complex. Future studies may determine that the plants with glabrous and hairy ovaries are unworthy of taxonomic recognition.

1. Ovaries and young fruits glabrous or rarely with a few hairs; mature fruits glabrous .... 1. SSP. DAUCIFOLIUM
1. Ovaries and young fruits moderately to densely hairy; mature fruits sparsely to moderately hairy
.... 2. SSP. FOENICULACEUM

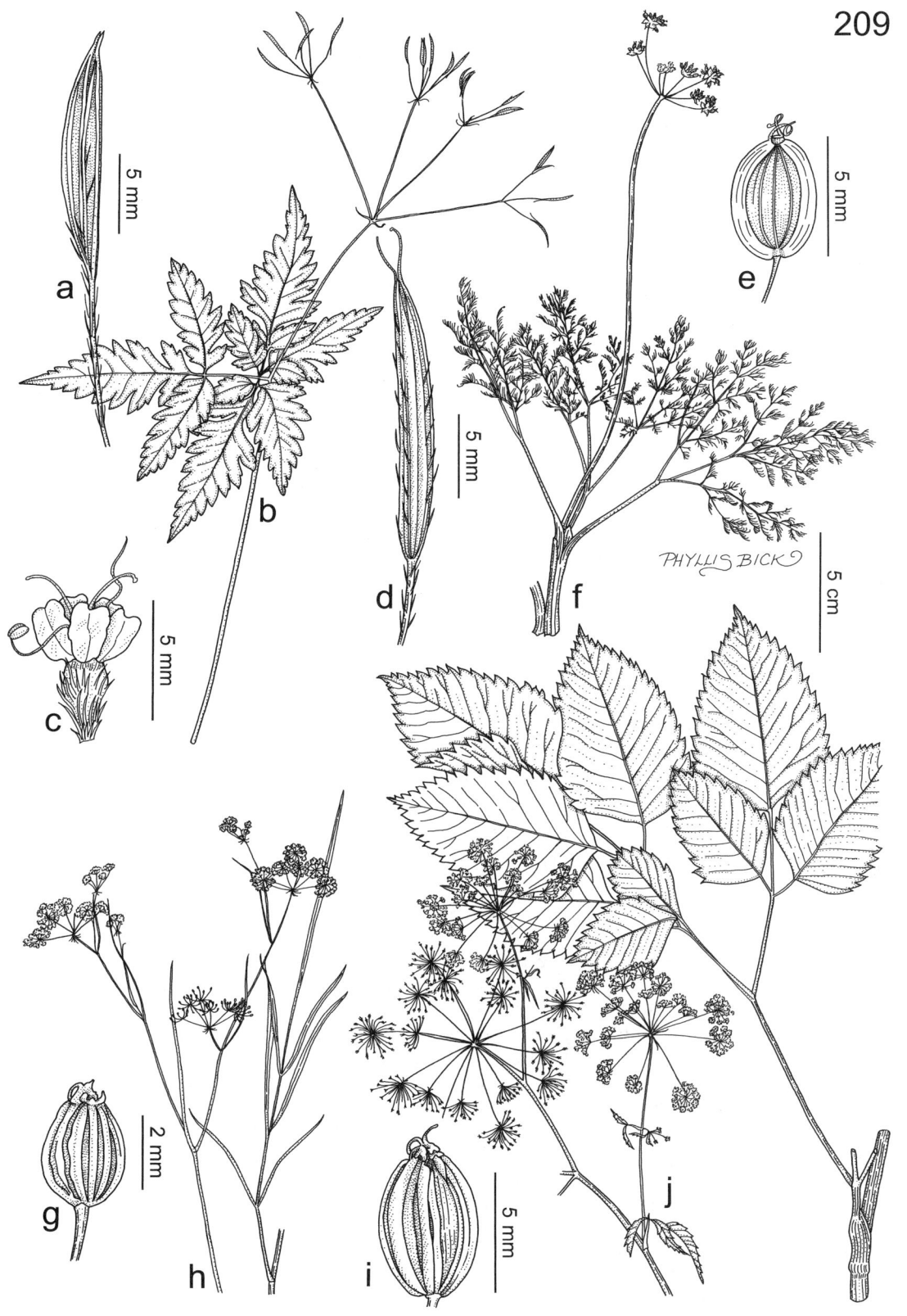

Plate 209. Apiaceae. *Osmorhiza claytonii*, **a)** fruit, **b)** habit. *Osmorhiza longistylis*, **c)** flower, **d)** fruit. *Lomatium foeniculaceum* ssp. *daucifolium*, **e)** fruit, **f)** habit. *Limnosciadium pinnatum*, **g)** fruit, **h)** inflorescence and leaf. *Ligusticum canadense*, **i)** fruit, **j)** inflorescence and leaf.

1. ssp. daucifolium (Nutt. ex Torr. & A. Gray) W.L. Theob.

L. foeniculaceum var. *daucifolium* (Nutt. ex Torr. & A. Gray) Cronquist

Pl. 209 e, f

Involucel with the bractlets mostly fused to above the middle. Ovaries and young fruits glabrous or rarely with a few hairs. Mature fruits glabrous. $2n=22$. April–May.

Uncommon in the Unglaciated Plains Division and the adjacent southwestern portion of the Glaciated Plains (Nebraska, Kansas, Missouri, Oklahoma, and Texas). Glades and rocky portions of upland prairies, on calcareous substrates.

2. ssp. foeniculaceum

Involucel with the bractlets mostly fused to below the middle. Ovaries and young fruits moderately to densely hairy. Mature fruits sparsely to moderately hairy. $2n=22$. April–May.

Uncommon, known thus far from Cass and Jackson Counties (Montana east to North Dakota, south to Texas; Canada). Glades and rocky portions of upland prairies, on calcareous substrates.

The distribution in the United States of ssp. *foeniculaceum* is mainly from Montana to South Dakota, south through the northern Great Plains to central Nebraska, with disjunct populations in eastern Kansas and adjacent Missouri, as well as in a small area of northwestern Texas and adjacent Oklahoma and Colorado. In Missouri, this subspecies appears to be far less abundant than does ssp. *daucifolium*.

27. Oenanthe L. (water dropwort)

Thirty to 40 species, North America, Europe, Asia, Africa, Malesia.

Most species of *Oenanthe* that have been tested produce a series of polyacetylenes similar to those in *Cicuta* and thus are highly poisonous (hence the common name dropwort). The sole exception is *O. javanica,* which is eaten as a vegetable in parts of Asia and Indonesia. Plants introduced for horticulture in the United States may not have originated from edible stocks and thus should be avoided for culinary purposes without certain knowledge of the edibility of the particular plants in question. In addition to *O. javanica,* several species in the genus occasionally are cultivated in bog gardens and other aquatic horticultural applications, including *O. aquatica* (L.) Poir. (water fennel, which apparently has escaped in Ohio and Washington, D.C.) and *O. pimpinelloides* L.

1. Oenanthe javanica (Blume) DC. (water celery)

Map 869

Plants perennial, glabrous, with fibrous roots. Stems 30–150 cm long, spreading with ascending branches and tips, somewhat inflated, rooting at the lower nodes. Leaves alternate and sometimes also basal (1 or 2 basal leaves sometimes present at flowering), short- to long-petiolate, the petioles somewhat inflated. Leaf blades 3–20 cm long, ovate to triangular-ovate in outline, 1–2(–3) times pinnately compound, the leaflets 10–50 mm long, narrowly lanceolate to broadly ovate, rounded, narrowed, or tapered (sometimes unequally) at the base, finely to more commonly coarsely toothed along the margins, occasionally with 1 or 2 basal lobes. Inflorescences opposite the leaves and occasionally also terminal, mostly long-stalked. Involucre absent or less commonly of 1 or 2 bracts, these shorter than the rays, spreading to ascending at flowering, linear, with sharply pointed tips. Rays (4–)6–20, 0.5–3.0 cm long, strongly angled and with entire or minutely toothed, pale angles or narrow wings. Involucel of 7–13 bractlets, these shorter than to more commonly longer than the flower stalks, linear, sometimes with thin, white, papery margins, tapered to sharply pointed tips. Flowers 5 to numerous in each umbellet, the stalks 1–5 mm long. Sepals minute triangular teeth. Petals obovate, appearing shallowly notched but narrowed abruptly to a slender, pointed extension at the tip, white. Ovaries glabrous. Fruits 2–3 mm long, oblong in outline, somewhat flattened laterally, glabrous, the mericarps with 5 ribs, these blunt, broad, and tan or light yellow to straw-colored, all or mostly obscuring the reddish brown surfaces between them. $2n=20, 42$. June–October.

Introduced, known thus far only from a single locality in Johnson County (native of Asia, Malesia; cultivated sporadically in the U.S.). Margins of lakes; emergent aquatics.

The Missouri station apparently represents the first report of this species having escaped from cultivation in the United States. Interestingly, an uncomfortably closely related species, *O. sarmentosa* C. Presl, which is reported as native from Alaska through westernmost Canada to California, may not prove to be distinct from *O. javanica* upon further study, but may rather be shown to represent a North American extension in the range of this widespread Asian species.

28. Osmorhiza Raf. (sweet cicely)
(Lowry and Jones, 1984)

Plants perennial, with clusters of somewhat tuberous-thickened roots. Stems erect or ascending, sometimes from spreading bases, glabrous or sparsely to densely hairy. Leaves alternate and often also basal (1 or a few basal leaves often present at flowering), short- to long-petiolate, the uppermost leaves often sessile or nearly so, the sheathing bases not or only slightly inflated. Leaf blades broadly ovate to triangular-ovate in outline, ternately or ternately then pinnately 2 or 3 times compound with distinct leaflets, these mostly 1 cm or more wide, ovate to lanceolate or oblong-lanceolate, shallowly to moderately several-lobed and sometimes also with a pair of deep, basal lobes, the margins also toothed, shallowly cordate to long-tapered at the base, narrowed or tapered to a sharp point at the tip, the upper and undersurface sparsely to moderately pubescent with stiff, spreading hairs, especially along the main veins and margins. Inflorescences terminal and axillary, compound umbels, mostly long-stalked, the umbellets sometimes all or partially of staminate flowers. Involucre of 1–6 bracts, rarely absent, the bracts 3–20 mm long, usually reflexed, linear to lanceolate, with a sharply pointed tip, densely pubescent with stiff, spreading or ascending hairs along the margins and midvein. Rays 3–6(–8), often somewhat unequal in length, usually ascending. Involucel of 4–7 bractlets, these shorter than the flower stalks, similar to the bracts. Flowers 7–19 in each umbellet (usually 7 or fewer perfect, the rest staminate), the stalks 2–5 mm long at flowering (the central flower usually sessile and staminate), those of the perfect flowers elongating to 12 mm at fruiting. Sepals absent. Petals oblanceolate to obovate, rounded or with an abrupt, minute point at the tip, white. Ovaries glabrous or hairy. Fruits linear to narrowly oblong-oblanceolate in outline, long-tapered at the base, tapered to a beak at the tip, flattened laterally, glabrous or pubescent with stiff, ascending hairs (bristles) along the ribs, greenish brown to dark brown or black with green to dark brown ribs, the mericarps sometimes slightly arched or curved, somewhat narrowed along the commissures, with 5 narrow, angled ribs, these lacking wings. Ten species, North America, South America, Asia.

1. Styles 0.5–1.5 mm long at fruiting, shorter than the petals at flowering
   .... 1. O. CLAYTONII
1. Styles 3–4 mm long at fruiting, longer than the petals at flowering
   .... 2. O. LONGISTYLIS

1. Osmorhiza claytonii (Michx.) C.B. Clarke
(sweet cicely, woolly sweet cicely)
Pl. 209 a, b; Map 870

Plants (especially the roots) not or only slightly anise-scented. Stems 30–90 cm long, sparsely to more commonly densely pubescent with spreading hairs, rarely glabrous. Leaf blades 4–30 cm long. Leaflets 3–8 cm long. Rays 2–8 cm long. Styles shorter than the petals at flowering, 0.5–1.5 mm long at fruiting. Fruits (12–)15–24 mm long. $2n=22$. April–June.

Scattered nearly throughout the state, most commonly north of the Missouri River (northeastern U.S. west to North Dakota and Arkansas; Canada). Bottomland forests, mesic upland forests, often in ravines, and banks of streams.

Of the two species of *Osmorhiza* in Missouri, *O. claytonii* appears to be by far the less common.

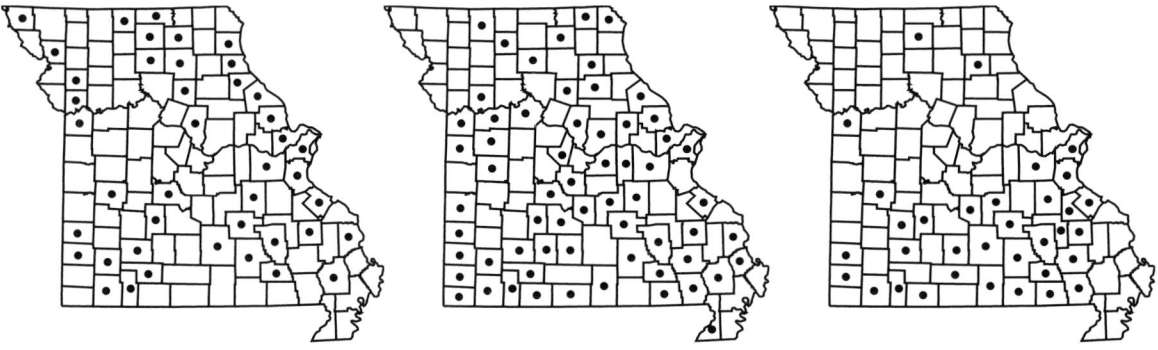

870. Osmorhiza claytonii   871. Osmorhiza longistylis   872. Oxypolis rigidior

Although *O. claytonii* tends to be a somewhat shorter, more slender plant than *O. longistylis*, vegetatively the two species can be difficult to separate. The strong anise-scent of the latter species, particularly of the roots, is a useful character in separating the two, although care must be taken, as the roots of *O. claytonii* occasionally have a faint odor of anise (*Pimpinella anisum* L.) as well.

2. Osmorhiza longistylis (Torr.) DC. (anise root, sweet anise, smooth sweet cicely)

O. longistylis var. *brachycoma* S.F. Blake
O. longistylis var. *villicaulis* Fernald
Pl. 209 c, d; Map 871

Plants (especially the roots) usually strongly anise-scented. Stems 50–100 cm long, sparsely to densely pubescent with spreading hairs, sometimes glabrous. Leaf blades 4–25 cm long. Leaflets 3–10 cm long. Rays 1.5–5.0 cm long. Styles longer than the petals at flowering, 3–4 mm long at fruiting. Fruits 10–22 mm long. $2n=22$. April–June.

Scattered to common nearly throughout the state (eastern U.S. west to Montana, Colorado, and Texas; Canada). Bottomland forests, mesic upland forests, often in ravines, and banks of streams; also shady, moist, disturbed areas.

The strong odor of anise, particularly by the roots in this species, is caused by production of an aromatic oil. A root extract (or the grated roots) of *O. longistylis* occasionally is used as a flavoring in place of anise.

29. Oxypolis Raf. (hog fennel)

Six species, North America.

1. Oxypolis rigidior (L.) Raf. (cowbane, common water dropwort)

O. rigidior var. *ambigua* (Nutt.) B.L. Rob.
Pl. 210 e, f; Map 872

Plants perennial, glabrous, with clusters of tuberous-thickened roots. Stems (30–)60–150 cm long, erect or ascending, often somewhat glaucous. Leaves alternate and rarely also basal (basal leaves almost never present at flowering), long- to mostly short-petiolate, the uppermost leaves often sessile or nearly so, sometimes reduced to nearly bladeless sheaths, the sheathing bases not or only slightly inflated. Blades of the main leaves 4–30 cm long (uppermost leaves sometimes highly reduced and nearly bladeless), ovate to broadly triangular in outline, 1 time pinnately compound with (3–)5–9 leaflets, the leaflets 35–150 mm long, (2–)5–45 mm wide, linear or lanceolate to oblong or narrowly obovate, narrowed at the base, entire or few-toothed (less than 8 teeth per side, mostly above the middle), narrowed to a sharp point at the tip, glaucous on the undersurface. Inflorescences terminal and axillary, compound umbels, short- to long-stalked. Involucre of 1–3 bracts, these 2–20 mm long, linear, sometimes with narrow, white margins, often shed by flowering. Rays 10 to numerous, 2–12 cm long. Involucel of 2–9 bractlets, these mostly shorter than the flower stalks, entire, linear. Flowers 10 to numerous in each umbellet, the stalks 4–20 mm long. Sepals minute triangular scales. Petals ovate, narrowed or tapered to a short, slender tip, white. Ovaries glabrous. Fruits 4–7 mm long, oblong-elliptic in outline, strongly flattened dorsally, glabrous, tan to straw-colored, with prominent, reddish brown oil tubes filling the spaces between the ribs, each mericarp with the intermediate and dorsal ribs low and blunt, the lateral ribs with broad, thin wings. $2n=32$. July–September.

Scattered, mostly south of the Missouri, mainly in the Ozark Division (eastern U.S. west to Minnesota and Texas; Canada). Bottomland prairies, moist depressions of upland prairies, banks of rivers, streams, and spring branches, fens, moist bases of bluffs, and occasionally openings of bottomland forests; also roadsides, ditches, and moist, disturbed areas.

30. Pastinaca L. (parsnip)

Fourteen species, Europe, Asia.

1. Pastinaca sativa L. (parsnip, wild parsnip)
Pl. 210 g, h; Map 873

Plants biennial. Stems 35–150(–200) cm long, erect, relatively stout, usually strongly ridged, glabrous or sparsely to moderately short-hairy. Leaves alternate and often also basal (a few basal leaves usually present at flowering), glabrous or the undersurface short-hairy, short- to long-petiolate, the sheathing bases of lower leaves not or only slightly inflated, those of the middle and upper leaves sometimes somewhat more inflated. Blades of the basal and main leaves 5–55 cm long (uppermost leaves usually highly reduced and nearly bladeless), oblong-elliptic to ovate in outline, pinnately 1 or 2 times compound, the leaflets 10–200 mm long, lanceolate or ovate to nearly circular, narrowed to less commonly shallowly cordate at the base, toothed (often coarsely so) and often shallowly to deeply lobed, rounded or narrowed to a blunt point at the tip. Inflorescences terminal and axillary, compound umbels, these sometimes appearing in loose clusters of 3 or 5, short- to long-stalked, the stalks glabrous or minutely hairy. Involucre absent. Rays 5 to numerous, 2–10 cm long, unequal in length, glabrous or minutely hairy. Involucel absent. Flowers 8 to numerous in each umbellet, the stalks 2–10 mm long (the central flower sometimes sessile), elongated slightly at fruiting, glabrous or minutely hairy. Sepals absent. Petals ovate, narrowed or tapered to a short, slender tip, yellow, occasionally tinged with red. Ovaries glabrous. Fruits 5–7 mm long, oblong-elliptic or broadly elliptic to slightly oblong-obovate in outline, flattened dorsally, glabrous, tan to light greenish brown with usually prominent reddish brown oil tubes between the ribs, each mericarp with the intermediate and dorsal ribs slender and nervelike, hardly raised from the surface, the lateral ribs with thin, broad wings. $2n=22$. May–October.

Introduced, scattered to common nearly throughout the state, but apparently still absent from the Mississippi Lowlands Division and adjacent Ozarks (native of Europe; introduced widely in the U.S. and Canada). Disturbed portions of upland prairies and banks of streams; also pastures, fallow fields, old fields, fencerows, roadsides, railroads, and open, disturbed areas.

Although the tuberous roots of the cultivated form of parsnip are edible, the aboveground portions of both cultivated and weedy races have been reported to contain furanocoumarins and related compounds similar to those found in *Heracleum* and may cause phototoxic dermatitis in some individuals.

31. Perideridia Rchb.
(Chuang and Constance, 1969)

About 14 species, U.S., Asia.

Downie et al. (2004) recently have shown that the single Asian species included in *Perideridia* by some authors, *P. neurophylla* (Maxim.) T.I. Chuang & Constance, is only distantly related to the North American members of the genus and should be reclassified in a different tribe of Apiaceae in the monotypic genus *Pterygopleurum* Kitag.

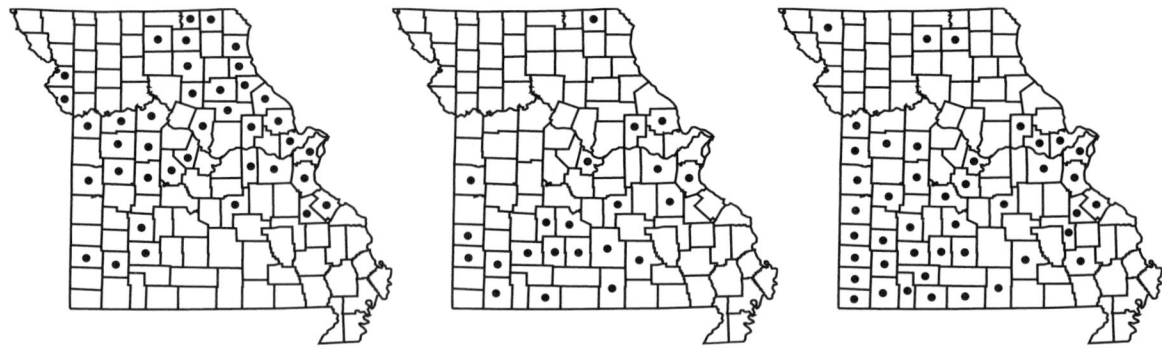

873. Pastinaca sativa   874. Perideridia americana   875. Polytaenia nuttallii

1. Perideridia americana (Nutt. ex DC.) Rchb.
Pl. 210 a, b; Map 874

Plants perennial, glabrous, with clusters of mostly 2 or 3 tuberous-thickened roots. Stems 50–120 cm long, erect or ascending. Leaves alternate and occasionally also basal (1 or 2 basal leaves rarely present at flowering), mostly short-petiolate, the uppermost leaves often sessile or nearly so, the sheathing bases not or only slightly inflated. Leaf blades 3–15 cm long (basal leaves sometimes longer), narrowly oblong-ovate to ovate-triangular in outline, pinnately or ternately then pinnately (1–)2–4 times compound or dissected, the leaflets or segments 5–50 mm long, linear to narrowly oblong, mostly narrowed at the base, the margins entire, narrowed to a sharp point at the tip. Inflorescences terminal and axillary, compound umbels, short- to long-stalked. Involucre absent or more commonly of 1–6 bracts, these 1–5 mm long, narrowly elliptic to narrowly triangular, usually shed by fruiting. Rays 6–20, 2–9 cm long. Involucel of 8–14 bractlets, these mostly shorter than the flower stalks, narrowly triangular to narrowly ovate. Flowers 15–25 in each umbellet, the stalks 3–10 mm long, elongated to 16 mm at fruiting. Sepals minute triangular scales. Petals broadly elliptic-obovate, shallowly notched or narrowed or tapered abruptly to a short, slender tip, white. Ovaries glabrous. Fruits 3–5 mm long, elliptic-ovate in outline, rounded at the base, flattened laterally, glabrous, brown to greenish brown, the mericarps narrowed along the commissures, with 5 slender and nervelike ribs, these hardly raised from the surface. $2n=40$. April–July.

Scattered in the Ozark and Unglaciated Plains Divisions, with a few populations in the eastern portion of the Ozark Border and Glaciated Plains (Ohio to Tennessee west to Missouri, Oklahoma, and Arkansas). Mesic to dry upland forests, tops and exposed ledges of bluffs, glades, and upland prairies, on calcareous substrates; also roadsides.

32. Polytaenia DC.

Two species, U.S.

1. Polytaenia nuttallii DC. (prairie parsley)
Pl. 210 c, d; Map 875

Plants perennial, with somewhat tuberous-thickened taproots. Stems 40–100 cm long, erect or ascending, relatively stout, minutely hairy and/or roughened, especially around the nodes and toward the tip. Leaves alternate and usually also basal (several basal leaves usually present at flowering), mostly short- to long-petiolate, the uppermost leaves often sessile or nearly so, the sheathing bases slightly to moderately inflated. Leaf blades 2–18 cm long, oblong-ovate to broadly ovate-triangular in outline, pinnately or less commonly ternately then pinnately 2 times compound, the leaflets 10–50 mm long, oblong or ovate, deeply pinnately lobed, narrowed or rounded at the base, the segments or lobes narrowly oblong to oblanceolate, less commonly ovate, entire or with few coarse teeth or lobes, narrowed to a blunt or sharp point or rounded at the tip, the undersurface roughened with minute teeth along the veins. Inflorescences terminal and axillary, compound umbels, short- to long-stalked, the stalks minutely hairy and/or roughened. Involucre absent. Rays

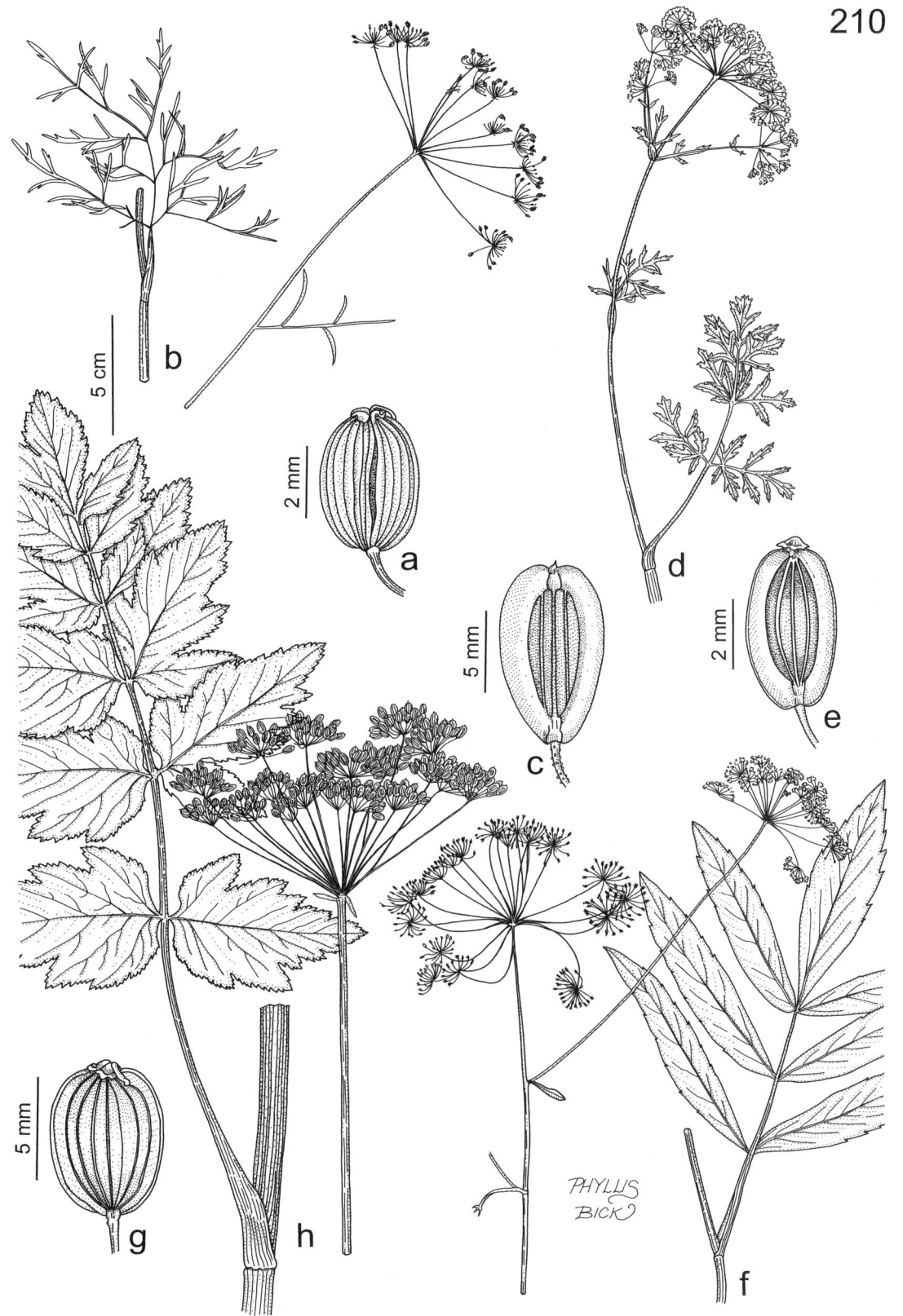

Plate 210. Apiaceae. *Perideridia americana*, **a)** fruit, **b)** inflorescence and leaf. *Polytaenia nuttallii*, **c)** fruit, **d)** habit. *Oxypolis rigidior*, **e)** fruit, **f)** inflorescence and leaf. *Pastinaca sativa*, **g)** fruit, **h)** inflorescence and leaf.

10–20, 1–4 cm long, minutely hairy and/or roughened. Involucel of 2–8 bractlets, these shorter than to longer than the flower stalks, linear, usually hairlike. Flowers mostly numerous in each umbellet, the stalks 3–5 mm long, minutely hairy and/or roughened. Sepals minute, ovate to triangular scales. Petals oblong to narrowly obovate, shallowly notched or narrowed or tapered abruptly to a short, slender tip, yellow. Ovaries glabrous. Fruits 5–11 mm long, oblong-elliptic or oblong-obovate in outline, rounded at the base, strongly flattened dorsally, glabrous, yellowish brown to dark brown, the mericarps with the dorsal and intermediate ribs nervelike and obscure, the lateral ribs with thick, corky, lighter-colored wings. $2n=22$. April–June.

Scattered, mostly south of the Missouri River, but apparently absent from the Mississippi Lowlands Division (Michigan to Louisiana west to North Dakota and Texas). Upland prairies, glades, savannas, and openings of dry upland forests, often (but not always) on calcareous substrates; also roadsides and railroads.

33. Ptilimnium Raf. (mock bishop's weed)
(Easterly, 1957)

Plants annual, glabrous. Stems erect or ascending. Leaves alternate (basal leaves usually absent at flowering), short-petiolate or more commonly sessile, the sheathing bases not or only slightly inflated. Leaf blades narrowly ovate to oblong-ovate in outline, pinnately 2–3 times dissected, the usually numerous ultimate segments narrowly linear, usually threadlike. Inflorescences terminal and axillary, compound umbels, short- to more commonly long-stalked. Involucre of few to several bracts. Rays 4–25. Involucel of 2–5(–7) bractlets, these mostly shorter than the flower stalks, linear. Flowers 5 to numerous in each umbellet. Sepals minute, narrowly to broadly triangular scales. Petals obovate, often appearing shallowly notched, but narrowed or tapered abruptly to a short, slender tip, white. Ovaries glabrous. Fruits broadly ovate to nearly circular in outline, slightly flattened laterally, glabrous, the mericarps with the 5 ribs rounded. Four species, eastern and central U.S.

1. Bracts of the involucre mostly divided into 3 threadlike segments; sepals minute, broadly triangular scales . . . . 1. P. CAPILLACEUM
1. Bracts of the involucre mostly entire; sepals minute, narrowly triangular scales
    2. Main segments of the leaf blade in whorls along the rachis; fruits 2–4 mm long, with relatively narrow, inconspicuous, dark brown surfaces between the lateral ribs and the commissures (also between the ribs) . . . . . 2. P. COSTATUM
    2. Main segments of the leaf blade alternate or opposite along the rachis; fruits 1.0–1.5 mm long, with relatively broad, conspicuous, straw-colored to light yellowish brown surfaces between the lateral ribs and the commissures (areas between the ribs have reddish brown oil tubes) . . . . 3. P. NUTTALLII

1. Ptilimnium capillaceum (Michx.) Raf.
(Atlantic mock bishop's weed)
Pl. 211 h; Map 876

Stems 10–80 cm long. Leaf blades 4–12 cm long, the main segments opposite or in whorls of 3 along the rachis, the ultimate divisions 3–25 mm long. Involucre of 2–9 bracts, these 4–12 mm long, mostly divided from near the base into 3 threadlike segments. Rays 4–20, 1.0–3.5 cm long. Flowers 5–20 in each umbellet, the stalks 3–6(–12) mm long. Sepals minute, broadly triangular scales. Styles at fruiting 0.2–0.5 mm long, loosely ascending. Fruits 1.5–3.0 mm long, with relatively broad, conspicuous, straw-colored to light yellowish brown surfaces between the lateral ribs and the commissures, the ribs similarly colored, the areas between the ribs with reddish brown oil tubes. $2n=14, 28$. June–August.

Uncommon, widely scattered in southeastern Missouri and introduced locally in St. Louis (eastern U.S. west to South Dakota and Texas). Natural habitat unknown, but presumably moist depressions or seeps in openings of mesic to dry upland forests; also fencerows and railroads.

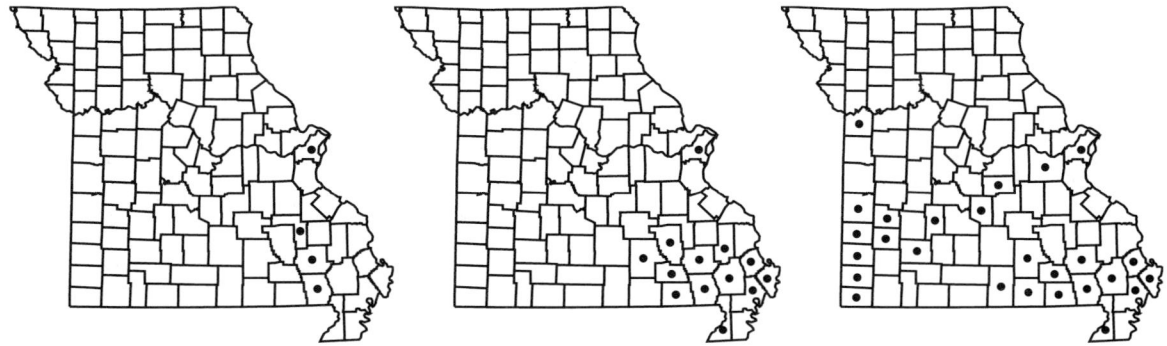

876. Ptilimnium capillaceum   877. Ptilimnium costatum   878. Ptilimnium nuttallii

2. **Ptilimnium costatum** (Elliott) Raf. (big mock bishop's weed)

Pl. 211 a–c; Map 877

Stems 60–150 cm long. Leaf blades 5–15 cm long, the main segments mostly in whorls of 3–5 along the rachis, the ultimate divisions 3–18 mm long. Involucre of 4–15 bracts, these 5–12 mm long, mostly entire. Rays 12–25, 1.5–4.5 cm long. Flowers 15 to numerous in each umbellet, the stalks 2–6(–12) mm long. Sepals minute narrowly triangular scales. Styles at fruiting 1–3 mm long, spreading. Fruits 2–4 mm long, with relatively narrow, inconspicuous, dark brown surfaces between the lateral ribs and the commissures, and also between the ribs, the oil tubes inconspicuous. $2n=22, 32$. July–September.

Uncommon in the southeastern quarter of the state north locally to St. Louis County (southeastern U.S., west to Missouri and Texas). Swamps, sloughs, bottomland forests, banks of streams, acid seeps, and fens; also ditches.

3. **Ptilimnium nuttallii** (DC.) Britton (Ozark mock bishop's weed)

Pl. 211 d, e; Map 878

Stems 30–70 cm long. Leaf blades 3–10 cm long, the main segments alternate or opposite along the rachis, the ultimate divisions 4–60 mm long. Involucre of 6–15 bracts, these 4–15(–30) mm long, mostly entire. Rays 15 to numerous, 1.0–3.5 cm long. Flowers mostly numerous in each umbellet, the stalks 3–8 mm long. Sepals minute, narrowly triangular scales. Styles at fruiting 0.5–1.0 mm long, usually curved downward. Fruits 1.0–1.5 mm long, with relatively broad, conspicuous, straw-colored to light yellowish brown surfaces between the lateral ribs and the commissures, the ribs similarly colored, the areas between the ribs with reddish brown oil tubes. $2n=14$. June–August.

Scattered south of the Missouri River (Kentucky to Alabama west to Kansas and Texas). Bottomland prairies, moist depressions of upland prairies and glades, banks of streams, fens, and swamps; also ditches, roadsides, and railroads.

Occasional robust specimens, particularly some from southeastern Missouri, have the largest inflorescences on the plants (usually those terminal on the stems) with some of the involucral bracts 3-lobed, as in *P. capillaceum*. The remaining umbels of such plants have involucres with more typical, undivided bracts. When these unusual plants are encountered, characters of the sepal width and fruit morphology should be used to verify the determination.

34. Sanicula L. (black snakeroot, sanicle)
(Shan and Constance, 1951; Pryer and Phillippe, 1989)

Plants biennial or perennial, glabrous (variously hairy elsewhere), with fibrous or somewhat tuberous-thickened clusters of roots. Stems loosely ascending to erect. Leaves alternate and usually also basal (1 to several basal leaves usually present at flowering), the uppermost sometimes appearing opposite, the basal and lower ones long-petiolate, the median and upper leaves short-petiolate to sessile, the sheathing bases not or only slightly inflated. Leaf blades broadly triangular to ovate or nearly circular in outline, deeply palmately 3- or 5(7)-lobed and/or compound, the lobes or leaflets elliptic-lanceolate to obovate in outline, narrowed or

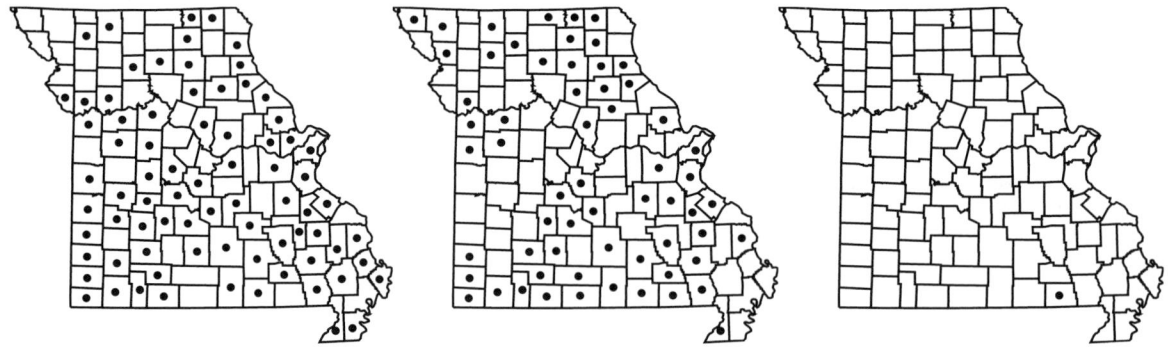

879. Sanicula canadensis   880. Sanicula odorata   881. Sanicula smallii

tapered to bluntly or sharply pointed tips, tapered at the base, those of the lower leaves often irregularly few-lobed, the margins sharply and doubly toothed, the teeth often with light-colored, slender, spiny tips. Inflorescences terminal and usually also axillary, compound umbels, these often grouped into loose clusters or small panicles, short- to more commonly long-stalked, the branch points with pairs of leaflike bracts. Involucre of (1)2(3) bracts, these leaflike, usually 3-lobed. Rays 2 or 3(–7), usually unequal in length, loosely ascending or spreading. Umbellets sometimes of 2 noticeably different types, some with a mixture of longer-stalked staminate flowers and shorter-stalked or sessile perfect flowers (this in all species), others with only staminate flowers (this sometimes in *S. odorata*). Involucel of (2–)3–9 bractlets, these minute, shorter than to longer than the flower stalks, lanceolate to ovate-triangular, the margins entire or less commonly few-toothed. Staminate flowers 1–27 per umbellet; pistillate flowers usually 3 per umbellet, sometimes appearing fewer at fruiting. Sepals minute, narrowly lanceolate to triangular scales, these fused toward the base. Petals oblanceolate to ovate, tapered abruptly to a short, slender tip, white or greenish yellow. Ovaries densely pubescent with hooked hairs (bristles). Fruits oblong-ovate to nearly circular in outline, somewhat flattened laterally, densely pubescent with hooked bristles, these with expanded, somewhat inflated bases, the mericarps lacking ribs. About 40 species, nearly worldwide.

1. Styles noticeably longer than the bristles of the fruit, up to twice as long as the sepals; staminate flowers mostly 12–25 per umbellet . . . . 2. S. ODORATA
1. Styles shorter than the bristles of the fruit, shorter than to slightly longer than the sepals; staminate flowers mostly 1–7 per umbellet
   2. Styles shorter than to about as long as the sepals; fruits 2–5 mm long, the stalks 1.0–1.5 mm long; plants biennial, with thin, fibrous roots
       . . . . 1. S. CANADENSIS
   2. Styles slightly longer than the sepals; fruits 4–6 mm long, sessile; plants perennial, with somewhat tuberous-thickened roots . . . . 3. S. SMALLII

1. Sanicula canadensis L. (Canada snakeroot)
Pl. 211 i, j; Map 879

Plants biennial, with fibrous roots. Stems 15–100 cm long. Leaf blades 1.5–14.0 cm long, deeply palmately 3- or 5(7)-lobed and/or compound. Involucre with the bracts 8–35 mm long. Rays 0.2–3.0 cm long. Umbellets all with a mixture of staminate and pistillate flowers, the staminate flowers usually 1–7 per umbellet. Sepals 0.4–1.1 mm long, fused only at the base, narrowly lanceolate, with the tip tapered to a sharp point. Petals greenish white, usually shorter than the sepals. Stamens with the anthers white. Styles shorter than the bristles of the fruit, shorter than to about as long as the sepals. Fruits 2–5 mm long, the stalks 1.0–1.5 mm long. $2n=16$. May–July.

Scattered to common throughout the state (eastern U.S. west to Wyoming and Texas;

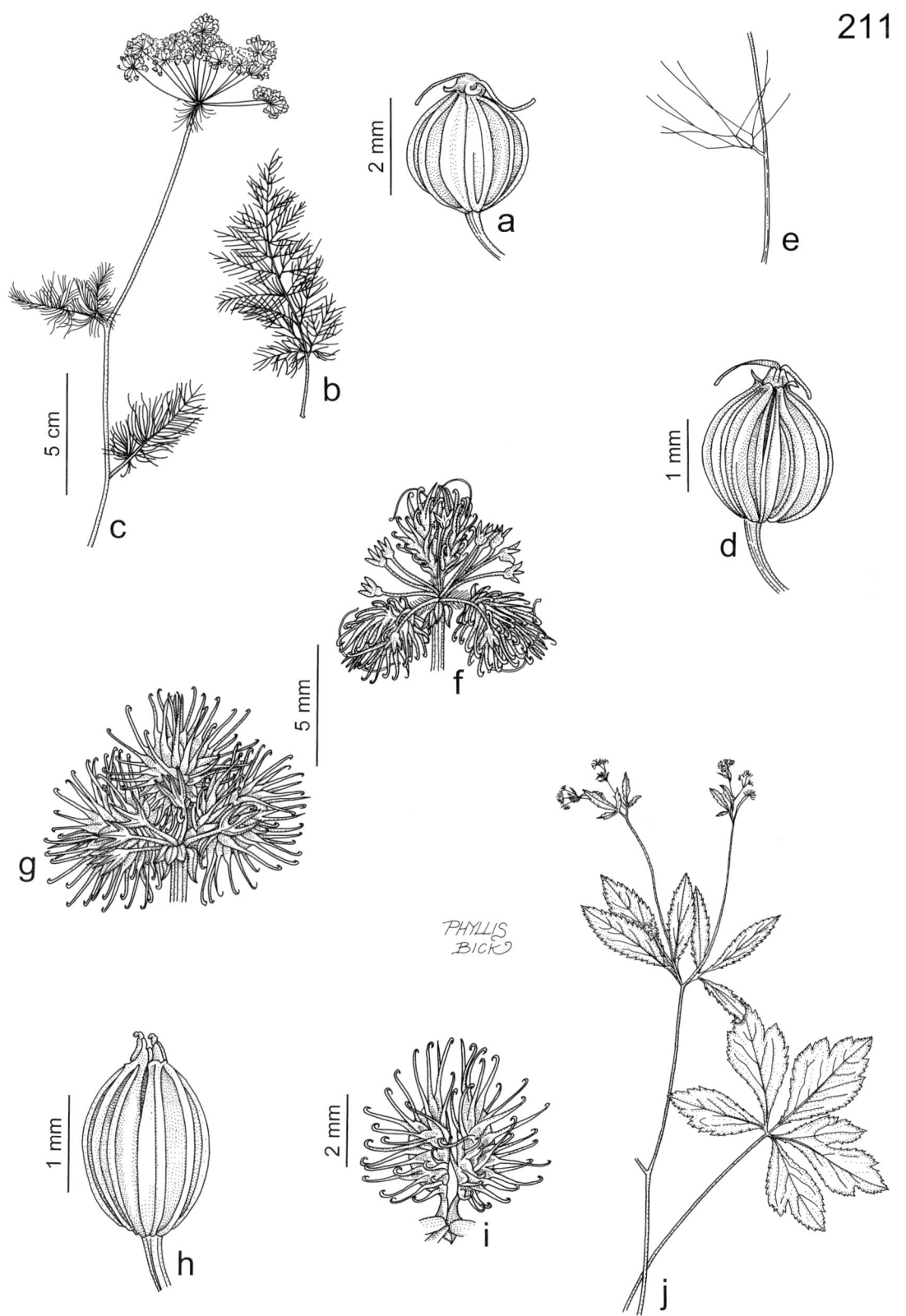

Plate 211. Apiaceae. *Ptilimnium costatum,* **a)** fruit, **b)** leaf, **c)** habit. *Ptilimnium nuttallii,* **d)** fruit, **e)** leaf. *Sanicula odorata,* **f)** fruits. *Sanicula smallii,* **g)** fruits. *Ptilimnium capillaceum,* **h)** fruit. *Sanicula canadensis,* **i)** fruit, **j)** inflorescence and leaf.

Canada). Bottomland forests, mesic upland forests, acid seeps, banks of streams and spring branches, and margins of upland prairies and glades; rarely also shaded, disturbed areas.

Steyermark treated a few specimens scattered through the range of *S. canadensis* in Missouri as var. *grandis* Fernald. This variety originally was defined to include robust plants with larger leaves and clusters of fruits and was thought to be unworthy of formal taxonomic recognition by Shan and Constance (1951) and most later authors. Phillippe (1978) and Pryer and Phillippe (1989) redefined var. *grandis* to include plants generally intermediate between *S. canadensis* and *S. marilandica* in inflorescence, fruit, and stylar morphology, and Phillippe cited a single specimen from St. Louis County as conforming to this concept of the taxon. This specimen was relocated through the kindness of Dr. Bruno Wallnöfer of the Natural History Museum in Vienna, Austria, and shown not to have been collected in Missouri after all. Also, it is of interest that although Phillippe and Pryer and Phillippe suggested that their variety did not represent plants of hybrid origin, many of the specimens from other states annotated by Phillippe as var. *grandis* earlier had been ascribed provisionally to *S. canadensis* × *marilandica* by Shan and Constance. The identity of these morphologically intermediate plants remains in question and requires further study using modern methods of genetic analysis. Thus, it seems premature to recognize varieties within *S. canadensis* at the present time, and even if the var. *grandis* were to be recognized, it would have to be excluded from the Missouri flora.

2. Sanicula odorata (Raf.) Pryer & Phillippe
(cluster-sanicle)
S. gregaria E.P. Bicknell

Pl. 211 f; Map 880

Plants perennial, with relatively slender, fibrous roots. Stems 20–80 cm long. Leaf blades 2–12 cm long, deeply palmately 3- or 5(7)-lobed and/or compound. Involucre with the bracts 8–40 mm long. Rays 1–6 cm long. Umbellets with a mixture of staminate and pistillate flowers or more commonly some of the umbellets all staminate, the staminate flowers mostly 12–25 per umbellet. Sepals 0.4–0.7 mm long, fused in the basal third, triangular, with the tip narrowed to a short, usually blunt point. Petals yellowish green, longer than the sepals. Stamens with the anthers yellow. Styles noticeably longer than the bristles of the fruit, up to twice as long as the sepals. Fruits 3–5 mm long, the stalks 0.5–1.0 mm long. 2n=16. April–June.

Scattered nearly throughout the state (eastern U.S. west to North Dakota and Texas; Canada). Bottomland forests, mesic upland forests, and banks of streams and rivers.

Phillippe (1978) and Pryer and Phillippe (1989) discussed the rationale for adopting the epithet *S. odorata* for this species. Earlier authors (Shan and Constance, 1951) had considered the name, as originally described, to be of ambiguous application, apparently compiled from elements of several other species, and had urged that it be rejected as a source of confusion. Phillippe (1978) and Pryer and Phillippe (1989) combined historical research and a process of elimination to conclude that the name can only apply to the taxon otherwise known as *S. gregaria*, and their arguments are accepted here.

Shan and Constance (1951) mapped an occurrence of *S. marilandica* L. (black snakeroot) from northeastern Missouri but failed to cite a specimen to support their record. Although Gleason and Cronquist (1991) accepted this literature report, Steyermark (1963) did not mention the species as occurring in Missouri. No specimens to support its occurrence were discovered during the present study, nor were any located by Phillippe (1978) in his dissertation research on the genus. Although it has been excluded from the flora for the present, this widespread species eventually may be found in northern Missouri, as it is known to grow in adjacent portions of Illinois, Iowa, and Nebraska. *Sanicula marilandica* is similar to *S. odorata* in having long styles and relatively numerous staminate flowers, but it differs in having sepals 1.0–1.5 long, narrowly lanceolate, with the tip tapered to a sharp point; stamens with the anthers greenish white; and petals white to greenish white.

3. Sanicula smallii E.P. Bicknell (southern sanicle)

Pl. 211 g; Map 881

Plants perennial, with somewhat tuberous-thickened roots. Stems 20–60 cm long. Leaf blades 1.5–12.0 cm long, deeply palmately mostly 3- or 5-lobed and/or compound. Involucre with the bracts 6–40 mm long. Rays 0.2–3.0 cm long. Umbellets all with a mixture of staminate and pistillate flowers, the staminate flowers usually 1–7 per umbellet. Sepals 1.0–1.5 mm long, fused only at the base, narrowly lanceolate, with the tip tapered to a sharp point. Petals greenish white, usually shorter than the sepals. Stamens with the anthers white. Styles shorter than the bristles of the fruit, slightly longer than the sepals. Fruits 4–6 mm long, sessile. 2n=16. May–June.

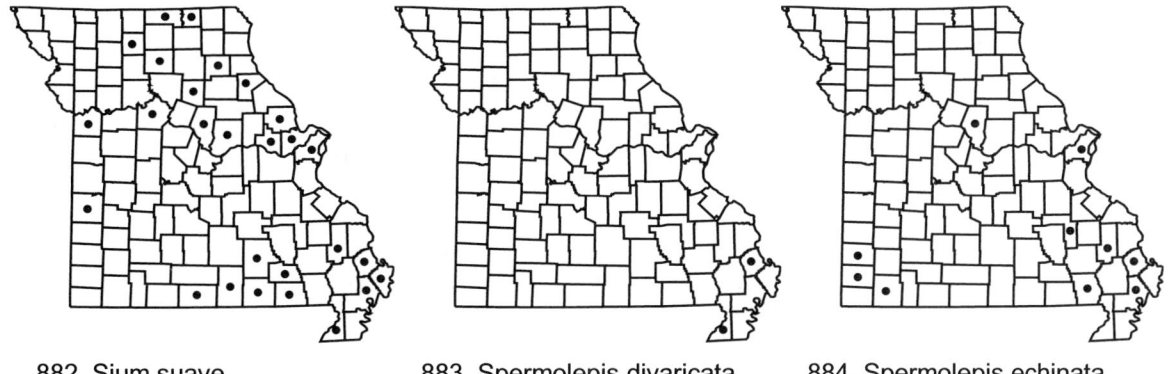

882. Sium suave 883. Spermolepis divaricata 884. Spermolepis echinata

Uncommon, known thus far only from a single historical collection from Ripley County (southeastern U.S. west to Missouri and Texas). Bottomland forests.

35. Sium L.

Eight to 14 species, U.S., Canada; Europe, Asia, Africa.

1. Sium suave Walter (water parsley, water parsnip)

Pl. 212 h–j; Map 882

Plants perennial, glabrous, with fibrous roots. Stems 50–200 cm long, erect or ascending, often relatively stout and strongly ridged. Leaves alternate and sometimes also basal (1 or a few basal leaves sometimes present at flowering), long- to mostly short-petiolate, the uppermost leaves often sessile or nearly so, the sheathing bases not or only slightly inflated. Leaf blades 2–30 cm long, narrowly oblong to broadly ovate in outline, 1 time pinnately compound with (3–)7–17 leaflets (the leaflets several times pinnately or dichotomously dissected in submerged leaves), those of the uppermost leaves sometimes simple, the leaflets 10–100 mm long, those of emergent leaves linear to lanceolate or narrowly ovate, narrowed or rounded at the base, sharply and often finely many-toothed along the margins, narrowed or tapered to a sharp point at the tip, those of submerged leaves in aquatic individuals usually deeply and finely dissected into linear segments. Inflorescences terminal and axillary, compound umbels, short- to long-stalked. Involucre of 6–10 bracts, these 3–15 mm long, linear to lanceolate, the broader ones usually with irregular, white, papery margins. Rays 10–20, 1.0–4.5 cm long. Involucel of 4–8 bractlets, these mostly shorter than the flower stalks, entire, linear to narrowly lanceolate, sometimes with narrow, white margins. Flowers mostly numerous in each umbellet, the stalks 3–5 mm long. Sepals absent or minute triangular scales. Petals elliptic-obovate, shallowly notched or narrowed or tapered abruptly to a short, slender tip, white. Ovaries glabrous. Fruits 2–3 mm long, broadly elliptic to nearly circular in outline, slightly flattened laterally, glabrous, tan to straw-colored, with prominent reddish brown oil tubes filling the spaces between the ribs, the mericarps slightly narrowed along the commissures, with 5 prominent ribs, these with short, thick, corky wings. $2n=12$. July–September.

Scattered nearly throughout the state but absent from many of the western counties (U.S., Canada). Bottomland prairies, moist depressions of upland prairies, marshes, swamps, bottomland forests, margins of ponds and lakes, and banks of rivers, streams, and spring branches; also roadsides and ditches; often emergent aquatics.

Species of *Sium* reportedly contain polyacetylenes similar to those found in *Cicuta* and should be considered poisonous.

36. Spermolepis Raf. (scale-seed)

Plants annual, glabrous. Stems loosely ascending to erect. Leaves alternate and occasionally also basal (1 basal leaf rarely present at flowering), short-petiolate or more commonly sessile, the sheathing bases not or only slightly inflated. Leaf blades ovate to oblong-ovate in outline, pinnately 2–4 times dissected, the usually numerous ultimate segments 1–25 mm long, narrowly linear, usually threadlike. Inflorescences terminal and axillary, compound umbels, short- to more commonly long-stalked. Involucre absent. Rays 3–15, ascending or spreading. Involucel of 1–3 bractlets, these mostly shorter than the flower stalks, linear, the margins sometimes minutely toothed. Flowers 1–6 in each umbellet. Sepals absent. Petals oblong to ovate, rounded at the tip, white. Ovaries glabrous or densely pubescent with hooked hairs (bristles). Fruits 1.5–2.0 mm long, ovate in outline, flattened laterally, slightly narrowed along the commissures, glabrous, roughened with minute tubercles, or pubescent with hooked bristles, the mericarps with 5 inconspicuous, narrow, rounded ribs, these more or less obscured by the tubercles or bristles (when present). Four species, North America, Hawaii, South America.

1. Ovaries and fruits with dense, short, hooked bristles . . . . 2. S. ECHINATA
1. Ovaries and fruits glabrous or roughened with dense, minute tubercles
    2. Rays equal or only slightly unequal in length, spreading to loosely ascending
       . . . . 1. S. DIVARICATA
    2. Rays strongly unequal in length, strongly ascending . . . . 3. S. INERMIS

1. Spermolepis divaricata (Walter) Raf. ex Ser. (forked scale-seed)

Pl. 212 d, e; Map 883

Stems 7–60 cm long, strongly ascending to erect. Leaf blades 0.5–5.0 cm long, oblong to oblong-ovate in outline. Rays 3–7, 5–35 mm long, equal or only slightly unequal in length, spreading to loosely ascending, straight or somewhat curved. Flower stalks 2–15 mm long, the central flower of some umbellets sessile. Ovaries and fruits roughened with dense, minute tubercles. $2n=16$. April–June.

Uncommon in the Mississippi Lowlands Division (southeastern U.S. west to Kansas and New Mexico; Mexico). Sand prairies; also fallow fields, sandy roadsides, and railroads.

2. Spermolepis echinata (Nutt. ex DC.) A. Heller (hooked scale-seed)

Pl. 212 f, g; Map 884

Stems 5–40 cm long, loosely ascending to erect. Leaf blades 0.7–2.5 cm long, mostly ovate in outline. Rays 5–15, 1–12 mm long, unequal in length, strongly ascending, straight, the central umbellet of most umbels sessile and 1- or 2-flowered. Flower stalks 1–7 mm long, the central umbellets often with the flowers sessile or nearly so. Ovaries and fruits with dense, short, hooked bristles. $2n=16$. April–June.

Uncommon in the southern third of the state; introduced locally in Boone County and St. Louis (southern U.S. and adjacent Mexico north to Maryland [and disjunctly New York], Illinois, and California). Sand prairies, rock outcrops in upland prairies, igneous and chert glades, and rarely rocky banks of rivers; also fallow fields, roadsides, railroads, and sandy, open, disturbed areas.

The eastern and northern limits of this species' natural distribution are poorly understood, but apparently most localities in the southeastern United States, as well as the occurrence in New York, represent nonnative plants.

3. Spermolepis inermis (Nutt. ex DC.) Mathias & Constance (western scale-seed)

Pl. 212 a–c; Map 885

Stems 8–60 cm long, strongly ascending to erect. Leaf blades 1–5 cm long, oblong to oblong-ovate in outline. Rays 5–11, 1–13 mm long, unequal in length, strongly ascending, straight, the central umbellet of most umbels sessile or nearly so and 1–3-flowered. Flower stalks 1–6 mm long, the central umbellets often with the flowers sessile or nearly so. Ovaries and fruits glabrous and smooth or roughened with dense, minute tubercles. $2n=16$. May–June.

Scattered in the western half of the state, uncommonly eastward to St. Louis, Scott, and

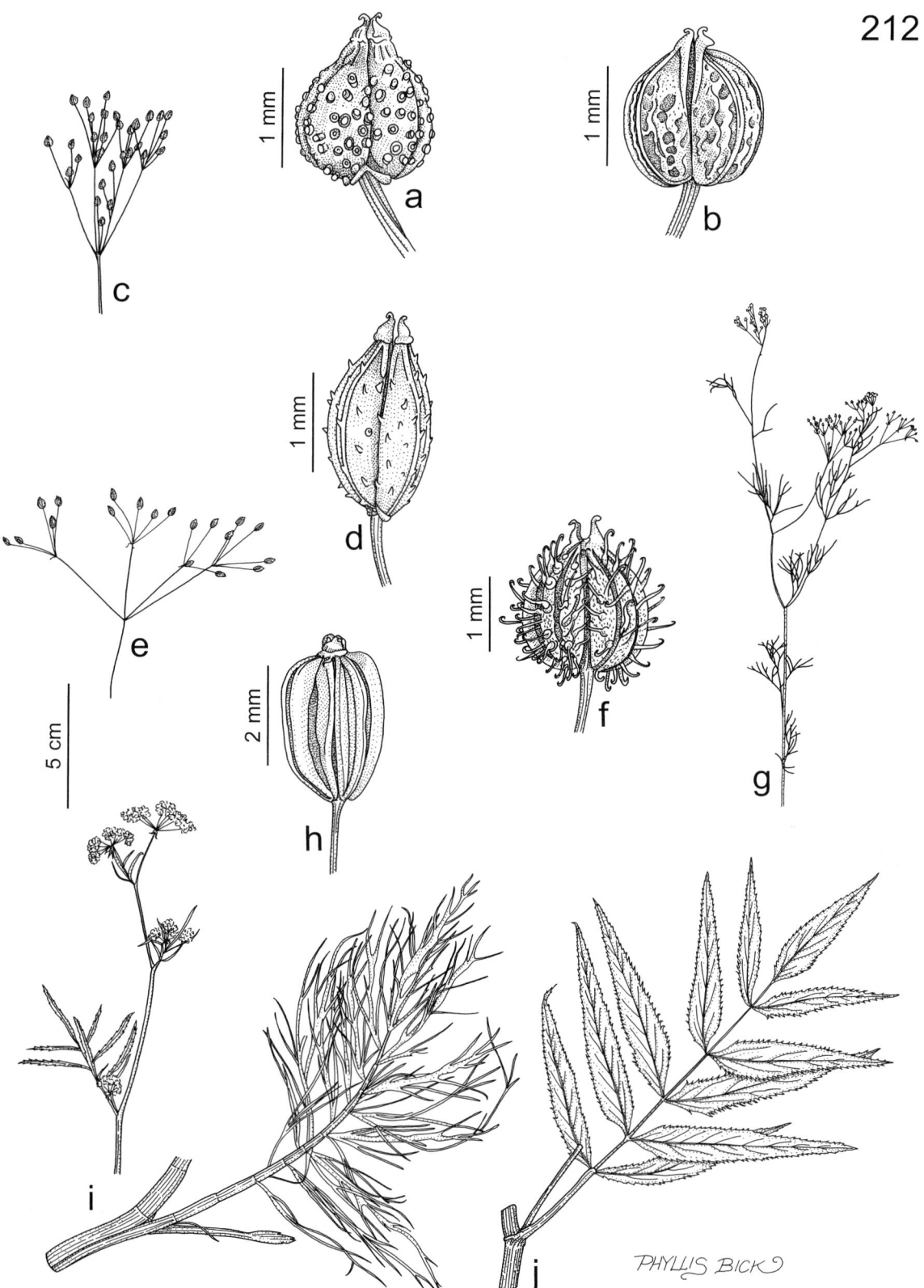

Plate 212. Apiaceae. *Spermolepis inermis*, **a)** fruit (tuberculate form), **b)** fruit (smooth form), **c)** inflorescence. *Spermolepis divaricata*, **d)** fruit, **e)** inflorescence. *Spermolepis echinata*, **f)** fruit, **g)** habit. *Sium suave*, **h)** fruit, **i)** inflorescence and more divided leaf, **j)** less divided leaf.

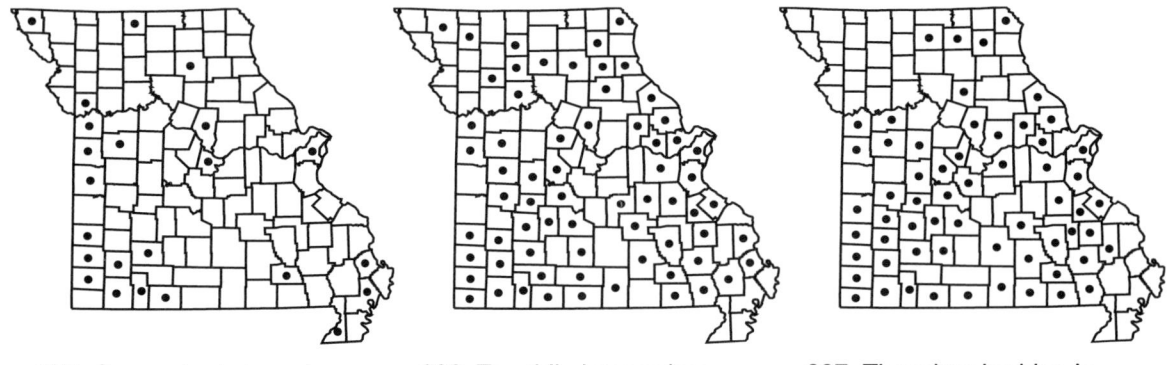

885. Spermolepis inermis   886. Taenidia integerrima   887. Thaspium barbinode

Dunklin Counties (Illinois to Mississippi west to Nebraska and New Mexico, introduced farther east to North Carolina and Florida; Mexico). Sand prairies, upland prairies, and glades; also fallow fields, roadsides, and railroads.

37. Taenidia Drude
(Guthrie, 1969)

Two species, eastern U.S., Canada.

1. Taenidia integerrima (L.) Drude (yellow pimpernel)

Pl. 213 a–c; Map 886

Plants perennial, glabrous, with somewhat tuberous-thickened taproots. Stems 30–85 cm long, erect or ascending, usually purple-tinged toward the base, glaucous. Leaves alternate and usually also basal (1 or 2 basal leaves usually present at flowering), short- to long-petiolate, the uppermost leaves usually sessile or nearly so, the sheathing bases not or only slightly inflated. Leaf blades 3–15 cm long, ovate or triangular-ovate to obovate in outline, ternately or pinnately 2 or 3 times compound, the leaflets 10–45 mm long, lanceolate, ovate, elliptic, or less commonly obovate, the margins entire or with 1 or 2 deep lobes (but not toothed), narrowed or rounded at the base, narrowed to a blunt or sharp point at the tip, glaucous on the undersurface. Inflorescences terminal and axillary, compound umbels, mostly long-stalked. Involucre absent. Rays 9–20, 0.8–8.0 cm long, the middle ones bearing umbels of all or mostly reduced sterile flowers and shorter than those with fertile flowers. Involucel absent. Flowers numerous in each umbellet, the central ones sterile and very short-stalked, the fertile ones with stalks 5–12 mm long. Sepals absent. Petals ovate, narrowed or tapered abruptly to a short, slender tip, yellow. Ovaries glabrous. Fruits 3–5 mm long, ovate to oblong-elliptic in outline, flattened laterally, glabrous, dark brown to nearly black, the mericarps with the 5 ribs narrow and rounded, lighter brown. $2n=22$. May–July.

Scattered to common nearly throughout the state (eastern U.S. west to Minnesota and Texas; Canada). Upland prairies, glades, tops and ledges of bluffs, mesic to dry upland forests, savannas, rocky banks of streams and rivers, and occasionally margins of ponds and lakes; also roadsides.

38. Thaspium Nutt. (meadow parsnip)

Plants perennial, with fibrous roots. Stems erect or ascending. Leaves alternate and usually also basal (2 to few basal leaves usually present at flowering), the basal and lower stem leaves long-petiolate, the median and upper leaves short-petiolate to nearly sessile, the sheathing bases not or only slightly inflated. Leaf blades ovate to more or less circular in outline, simple or variously lobed or compound. Inflorescences terminal and axillary, compound umbels, short- to long-stalked. Involucre absent or rarely of 1–3 minute, triangular bracts. Rays 6–18, unequal or nearly equal in length. Involucel of 4–9 bractlets or rarely absent, the bractlets shorter than the flower stalks. Flowers 9–19 (to numerous) in each umbellet, all with short stalks (1–4 mm long). Sepals absent or more commonly minute, triangular to obovate scales. Petals obovate, narrowed or tapered abruptly to a short, slender tip, cream-colored to yellow or dark purple. Ovaries glabrous. Fruits ovate to oblong-ovate in outline, flattened (sometimes only slightly) dorsally, glabrous, the mericarps with 5 ribs, these all or nearly all (development on the dorsal and intermediate ribs sometimes irregular) with prominent, somewhat corky, light yellow to straw-colored wings, and prominent reddish brown oil tubes filling the spaces between the ribs. Three species, U.S., Canada.

The genera *Thaspium* and *Zizia* are sometimes cited in botany textbooks as an example of genera whose morphologies have converged over time to the point that they are often difficult to distinguish. They may be determined at flowering by the sessile or nearly sessile central flower of each umbellet in *Zizia* vs. the noticeably short-stalked central flowers in *Thaspium*, but this character requires practice and patience to discern with confidence. Plants of *Zizia* also have rootstocks with clusters of slightly to moderately tuberous-thickened roots, whereas the roots of *Thaspium* species are fibrous and not fleshy or thickened.

Swink and Wilhelm (1994) suggested a small suite of vegetative characters for differentiating species of *Thaspium* and *Zizia*: 1) the leaflets in *T. barbinode* lack a white marginal band and are short-hairy (they call them scaberulous); 2) in *Z. aurea*, the teeth along the leaf margins are abruptly tapered to relatively blunt tips, whereas those in *T. trifoliatum* are narrowed or tapered more gradually to the bluntly or sharply pointed tips; 3) the leaf sheaths in *T. trifoliatum* tend to be slightly more inflated than those of *Z. aptera*. Although *T. barbinode* is generally easily separated from the other three Missouri species in these genera, in practice the width of the leaf sheath tends to vary somewhat based on position of the leaf on the stem, whereas the shapes of the marginal teeth vary from the base to the tip of the leaflets, making both of these characters very difficult to interpret in individual plants.

Because classification of the genera of Apiaceae has relied so strongly on characters of fruit morphology, the different fruit types (strongly winged in *Thaspium* vs. unwinged or only slightly winged in *Zizia*) have caused some authors to consider the two genera to be relatively distantly related within the tribe Apieae of the subfamily Apioideae. However, Lindsey (1975) questioned whether the genera are truly distinct, based on her examination of a suite of morphological and anatomical characters. Preliminary molecular analyses (Downie et al., 2000) have not fully resolved the genera of Apioideae but have tended to reinforce the hypothesis that *Thaspium* and *Zizia* are closely related.

Occasional plants will not key well using the characters presented below. See the treatment of *T. trifoliatum* for further discussion.

1. Basal leaves ternately or ternately then pinnately 2 or 3 times compound, the margins hairy, not whitened . . . . 1. T. BARBINODE
1. Basal leaves simple or ternately 1 time lobed or compound, the margins glabrous, with a narrow, white border . . . . 2. T. TRIFOLIATUM

1. Thaspium barbinode (Michx.) Nutt.
Pl. 213 f–h; Map 887

Stems 40–110 cm long, usually pubescent with a band of short, white hairs (visible to the naked eye) at the base of at least the uppermost leaf sheaths, sometimes also sparsely and minutely hairy toward the tip. Basal leaves with the blades 6–25 cm long, ternately or ternately then pinnately 2 or 3 times compound, the leaflets 1–12 cm long, lanceolate to ovate, tapered to rounded (often unequally) at the base, sometimes with 1 or 2 lobes toward the base, the margins otherwise finely to coarsely toothed, short-hairy, not whitened, the surfaces glabrous or more commonly sparsely to moderately pubescent with straight, white, speading hairs. Stem leaves similar to the basal

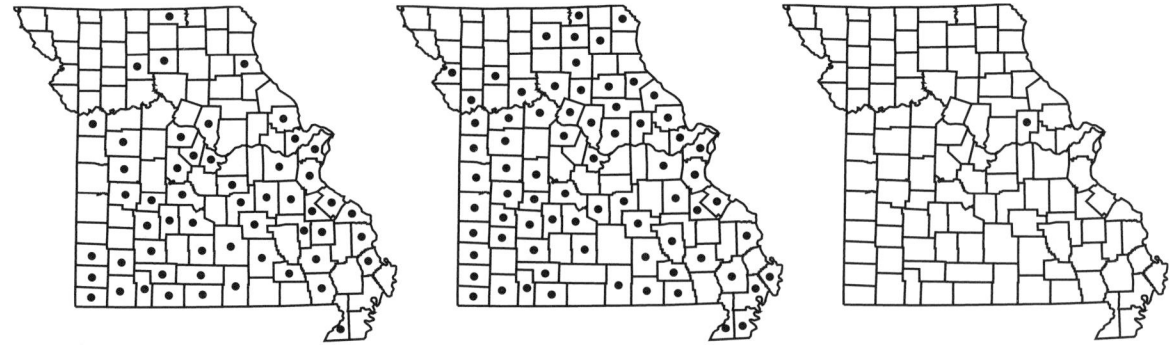

888. Thaspium trifoliatum   889. Torilis arvensis   890. Torilis japonica

leaves, gradually reduced in size and degree of dissection toward the stem tip. Rays 8–16, 1–4 cm long, more or less equal in length. Involucel of linear bractlets. Flower stalks 2–5 mm long. Petals pale yellow or cream-colored. Fruits 3–6 mm long. $2n=22$. April–June.

Scattered throughout most of the state but absent from the western portion of the Glaciated Plains Division and the Mississippi Lowlands (eastern U.S. west to Minnesota and Texas; Canada). Bottomland forests, mesic upland forests, banks of streams, rivers, and spring branches, fens, bases and ledges of bluffs, and margins of glades; also roadsides and railroads.

For a discussion of possible hybridization with *T. trifoliatum*, see the treatment of that species.

2. Thaspium trifoliatum (L.) A. Gray

Pl. 213 i–k; Map 888

Stems 20–80 cm long, usually pubescent with a band of minute hairs (visible only with magnification) at the base of at least the uppermost leaf sheaths, otherwise glabrous. Basal leaves with the blades 3–9 cm long, simple or ternately 1 time lobed or compound, the leaflets (when present) 1–7 cm long, broadly ovate to oblong-obovate or lanceolate, narrowed to rounded (often unequally) or cordate at the base, sometimes with 1 or 2 lobes toward the base, the margins otherwise finely toothed, glabrous, with a narrow, white border, the surfaces glabrous or slightly roughened along the main veins. Stem leaves similar to the basal leaves, mostly ternately 1 time lobed or compound, less commonly simple or ternately 2 times lobed or compound, those just above the stem base usually somewhat larger than the basal leaves, the median and upper leaves somewhat reduced in size. Rays 6–10(–18), 0.5–3.0 cm long, unequal in length. Involucel of linear to ovate-triangular bractlets. Flower stalks 1–4 mm long. Petals bright yellow or dark purple. Fruits 3–4 mm long. $2n=22$. April–June.

Scattered nearly throughout the state, but more common south of the Missouri River (eastern U.S. west to Minnesota and Texas; Canada). Mesic to dry upland forests, upland prairies, savannas, glades, ledges and tops of bluffs, and less commonly banks of streams; also old fields, roadsides, and railroads.

Steyermark (1963) maintained that *T. trifoliatum* is totally glabrous, but examination of specimens reveals that on many plants the bases of the leaf sheaths have a ring of pubescence similar to that described for *T. barbinode*. In *T. barbinode*, the hairs are slightly longer and white, and are thus usually visible to the naked eye, whereas in *T. trifoliatum* the hairs are minute and can only be discerned with magnification. Some authors (McGregor, 1986d) refer to the nodes of *T. barbinode* as having a beard of short hairs and those of the latter species as merely roughened. A few specimens from Adair and Montgomery Counties are anomalous for either of the two species and may represent hybrids between them. These plants have inflorescences with too many rays and leaves that are too divided to be comfortably placed in *T. trifoliatum*, but the rays are somewhat unequal in length and the leaflets have white margins that are glabrous or only slightly hairy, and thus are atypical for *T. barbinode*. Flower color also seems to vary from pale to bright yellow in such plants. Lindsey (1982) documented that species of both *Thaspium* and *Zizia* are highly outcrossing in nature. Further studies are needed to evaluate the potential for hybridization between *T. barbinode* and *T. trifoliatum*.

Two varieties, differing only in the amount of anthocyanin production, are accepted within *T. trifoliatum* by most authors. The two varieties differ somewhat in overall distribution, with var.

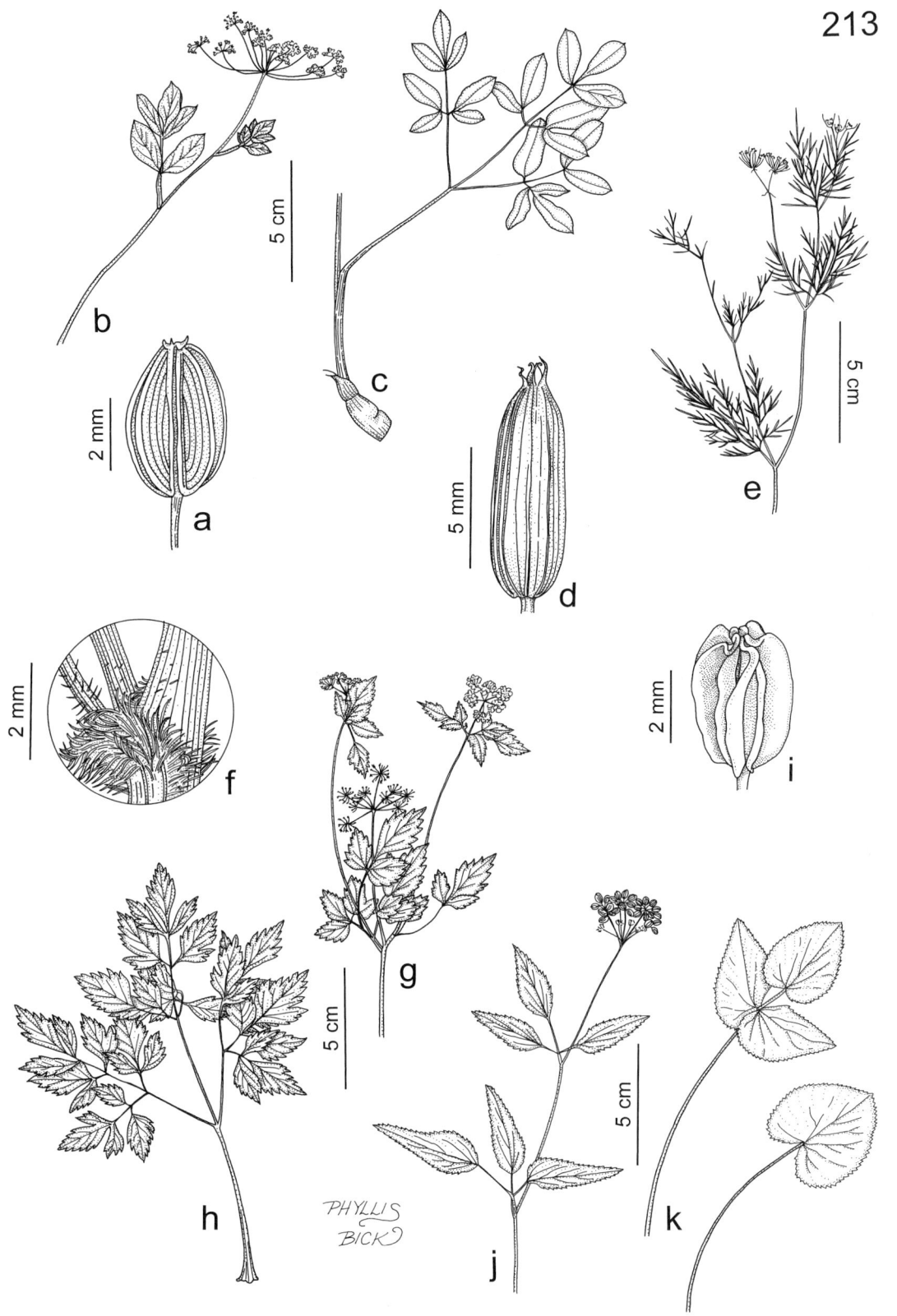

Plate 213. Apiaceae. *Taenidia integerrima*, **a)** fruit, **b)** inflorescence, **c)** leaf. *Trepocarpus aethusae*, **d)** fruit, **e)** habit. *Thaspium barbinode*, **f)** bearded node, **g)** inflorescences, **h)** leaf. *Thaspium trifoliatum*, **i)** fruit, **j)** inflorescence, **k)** leaves.

trifoliatum occupying the southeastern portion of the range of the more widespread var. *flavum*.

1. Petals bright yellow; stems and leaves green, usually not purple-tinged .... 2A. VAR. FLAVUM
1. Petals dark purple; stems and leaves often purple-tinged .... 2B. VAR. TRIFOLIATUM

2a. var. flavum S.F. Blake

Stems and leaves green, usually not purple-tinged. Petals bright yellow. 2n=22. April–June.

Scattered nearly throughout the state, but more common south of the Missouri River (eastern U.S. west to Minnesota and Texas; Canada). Mesic to dry upland forests, upland prairies, savannas, glades, ledges and tops of bluffs, and less commonly banks of streams; also old fields, roadsides, and railroads.

2b. var. trifoliatum

Stems and leaves often purple-tinged. Petals dark purple. 2n=22. April–June.

Uncommon and widely scattered in the Ozark Division (eastern [mostly southeastern] U.S. west to Kansas and Louisiana). Mesic to dry upland forests and banks of streams; also roadsides.

39. Torilis Adans. (hedge-parsley)

Plants annual. Stems loosely ascending to erect, pubescent with stiff, white, usually downwardly appressed hairs. Leaves alternate and usually also basal (several basal leaves usually present at flowering), the basal and lower stem leaves short- to long-petiolate, the median and upper leaves short-petiolate or more commonly sessile or nearly so, the sheathing bases not or only slightly inflated, with glabrous margins. Leaf blades 3–15 cm long (uppermost leaves sometimes only 0.5 cm), narrowly oblong-ovate to ovate-triangular in outline, 1 or 2 times pinnately compound, the leaflets 5–60 mm long, lanceolate to oblong-ovate, usually deeply pinnately dissected, mostly narrowed at the base, the lobes with the margins mostly sharply few-toothed, narrowed to a sharp point at the tip, moderately to densely pubescent with stiff, white hairs on both surfaces. Inflorescences terminal and/or lateral, compound umbels (often appearing simple and headlike in *T. nodosa*), short- to long-stalked (sometimes appearing sessile in *T. nodosa*). Involucre absent or of 1–10 bracts, these hairy. Rays 2–10, pubescent with stiff, white, usually upwardly appressed hairs. Involucel of 3–8 bractlets, these longer than the flower stalks, narrowly elliptic to more commonly linear, hairy. Flowers 5–20 in each umbellet. Sepals absent or minute triangular scales. Petals obovate to fan-shaped, rounded or shallowly notched at the tip, white. Ovaries hairy. Fruits oblong-elliptic to ovate in outline, narrowed or rounded at the base, narrowed to a blunt, beakless tip, flattened laterally, grayish brown to grayish green, the mericarps somewhat narrowed along the commissures, the slender and nervelike ribs usually appressed-hairy, usually obscured by a dense covering of long bristles, these more or less spreading, mostly with minute, terminal barbs. About 15 species, Europe, Asia, Africa.

1. Inflorescences all lateral, consisting of dense, often apparently simple umbels opposite the leaves, the umbels appearing sessile or nearly so, dense and headlike .... 3. T. NODOSA
1. Inflorescences terminal and sometimes also axillary, consisting of noticeably compound umbels, these short- to long-stalked, relatively long-rayed, the individual umbellets often appearing dense and headlike
    2. Involucre absent or of 1(2) bract(s); bristles 0.6–1.1 mm long, about as long as the width of the body of the mericarp, spreading at nearly 90° from the fruit .... 1. T. ARVENSIS

2. Involucre of 3–10 bracts, often 1 per ray; bristles 0.4–0.8 mm long, slightly shorter than the width of the body of the mericarp, loosely ascending
.... 2. T. JAPONICA

1. Torilis arvensis (Huds.) Link (field hedge-parsley, hemlock chervil)

Pl. 214 c–e; Map 889

Stems 20–100 cm long, usually relatively slender. Inflorescences terminal and sometimes also axillary, consisting of noticeably compound umbels, these short- to long-stalked. Involucre absent or of 1(2) bract(s), the bract(s) 1–12 mm long, narrowly triangular to more commonly linear, often hairlike. Rays 3–10, 0.5–2.5 cm long. Flower stalks 1–4 mm long, the umbellets thus often appearing dense and headlike. Fruits 3–5 mm long, the bristles all elongate, 0.6–1.1 mm long, about as long as the width of the body of the mericarp, spreading at nearly 90 from the fruit and straight or slightly arched. $2n=12$. June–September.

Introduced, scattered to common throughout the state (native of Europe, Asia; widely introduced in the eastern U.S. west to Wisconsin and Texas, and disjunctly California and Oregon; also Canada, Caribbean Islands, and South America). Banks of streams and rivers, disturbed portions of glades, upland prairies, and savannas, and regenerating clear-cuts in mesic upland forests; also roadsides, railroads, old fields, and open, disturbed areas.

Steyermark (1963) and many earlier authors mistakenly treated this species under the name *T. japonica,* in spite of the detailed descriptions and key provided by Mathias and Constance (1944–1945), who themselves apparently reversed the distributional range statements of the two taxa. Mühlenbach (1979) was the first to note the problems with older treatments.

Torilis arvensis has become much more abundant in Missouri during recent decades. Steyermark (1963) noted that it was first collected in the state in 1909 (in Jasper County) and suggested that although at the time of his writing the species was reasonably common in southern and central Missouri, it was still uncommon and local farther north. By now, it undoubtedly is present in every county and continues to spread mainly along roadsides and railroads.

2. Torilis japonica (Houtt.) DC. (Japanese hedge-parsley)

Map 890

Stems 20–60 cm long, usually relatively stout toward the base. Inflorescences terminal and sometimes also axillary, consisting of noticeably compound umbels, these short- to long-stalked. Involucre of 3–10 bracts, these 1–8 mm long, narrowly triangular to more commonly linear, often hairlike. Rays 5–10, 0.5–1.5 cm long. Flower stalks 1–3 mm long, the umbellets thus often appearing dense and headlike. Fruits 2–4 mm long, the bristles all elongate, 0.4–0.8 mm long, slightly shorter than the width of the mericarp, spreading at nearly 90° from the fruit and straight or slightly arched. $2n=16$. June–August.

Introduced, known thus far from a single specimen from Montgomery County (native of Europe, Asia, Africa; introduced widely but sporadically in the eastern U.S.). Roadsides.

The first specimen from Missouri was collected in 1988 but was not noted until the present research.

3. Torilis nodosa (L.) Gaertn. (knotted hedge-parsley)

Pl. 214 a, b; Map 891

Stems 8–50 cm long, relatively slender. Inflorescences lateral, consisting of dense, compound umbels situated opposite the leaves, usually appearing as simple, headlike umbels, these short-stalked or appearing sessile. Involucre absent or of 1 bract, this 1–4 mm long, linear, hairlike. Rays 2–4, 0.03–0.20 cm long, inconspicuous, stout. Flower stalks absent or to 0.5 mm long. Fruits 3–5 mm long, all or some of the bristles of some fruits in each inflorescence usually reduced, often to warty papillae, the full-sized bristles mostly 0.5–1.4 mm long, mostly longer than the width of the mericarp, usually loosely ascending and somewhat incurved. $2n=22, 24$. May–June.

Introduced, known thus far only from St. Louis City and County (native of Europe, Asia, Africa; introduced sporadically in the southern U.S. north to Oregon, Iowa, Michigan, and New Jersey; also Canada, Caribbean Islands, South America). Railroads and open, disturbed areas.

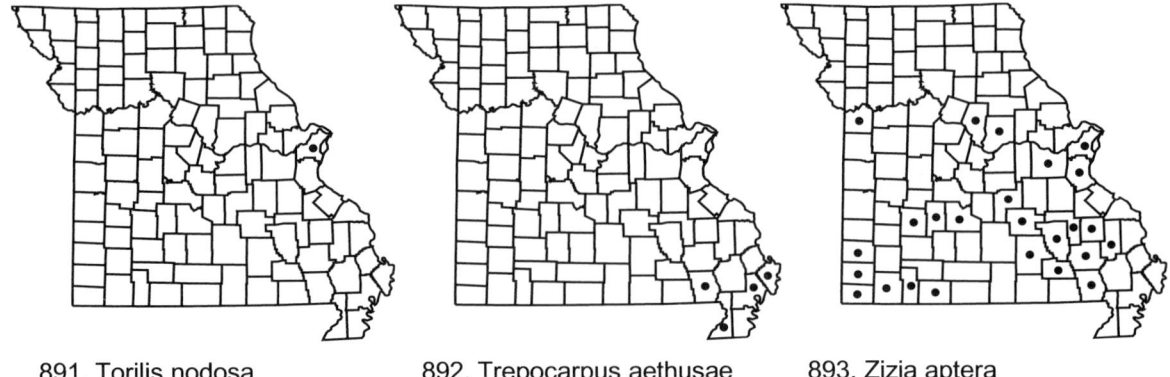

891. Torilis nodosa   892. Trepocarpus aethusae   893. Zizia aptera

40. Trepocarpus Nutt.

One species, southeastern U.S.

1. Trepocarpus aethusae Nutt. ex DC.

Pl. 213 d, e; Map 892

Plants annual, glabrous. Stems 25–60 cm long, erect or ascending, sometimes from a spreading base. Leaves alternate and rarely also basal (basal leaves usually absent at flowering), short-petiolate or sessile, the sheathing bases not or only slightly inflated. Leaf blades 3–12 cm long, ovate to broadly triangular-ovate in outline, 3 or 4 times pinnately compound or dissected, the ultimate segments 2–12 mm long, linear, narrowed or short-tapered to a sharp point at the tip. Inflorescences opposite the leaves or occasionally terminal, compound umbels, short- to more commonly long-stalked. Involucre of 2–4 bracts, these 2–12 mm long, linear, occasionally with a few linear lobes. Rays 2–4, 0.5–1.5 cm long. Involucel of 2–6 bractlets, these shorter than to longer than the flower stalks, linear. Flowers 2–8 in each umbellet, sessile or the stalks to 3 mm long, the umbellets thus sometimes appearing headlike. Sepals minute, narrowly triangular scales, unequal in length. Petals obovate, appearing shallowly notched, but narrowed or tapered abruptly to a short, slender tip, white. Ovaries glabrous. Fruits 8–10 mm long, linear to narrowly oblong in outline, slightly flattened laterally, glabrous, light brown to greenish brown, the mericarps with the 5 primary ribs inconspicuous and nervelike, but with 4 conspicuous, rounded, corky-thickened secondary nerves, these light tan to straw-colored. May–June.

Scattered in the Mississippi Lowlands Division (southeastern U.S. west to Missouri and Texas). Bottomland forests, sloughs, and perhaps swamps; rarely also ditches; sometimes emergent aquatics.

This attractive species with its delicate, thin-textured leaves likely has been overlooked by plant collectors in southeastern Missouri because of its inconspicuous inflorescences. Originally, it was reported for the state based on a single collection from Mississippi County (Oskins, 1981; Doolen, 1984) that was never accessioned in an herbarium. However, searches for plants in recent years have disclosed that it is reasonably common in some of the few remnant bottomland forests of the Mississippi Lowlands Division. The presence of *T. aethusae* in southernmost Illinois also was reported only relatively recently (Wilm and Taft, 1997).

41. Zizia W.D.J. Koch (golden Alexanders)

Plants perennial, glabrous (or minutely hairy or roughened at some of the nodes), with clusters of slightly to moderately tuberous-thickened roots. Stems erect or ascending, sometimes from spreading bases. Leaves alternate and usually also basal (2 to few basal leaves usually present at flowering), the basal and lower stem leaves long-petiolate, the median and upper leaves short-petiolate to nearly sessile, the sheathing bases not or only slightly in-

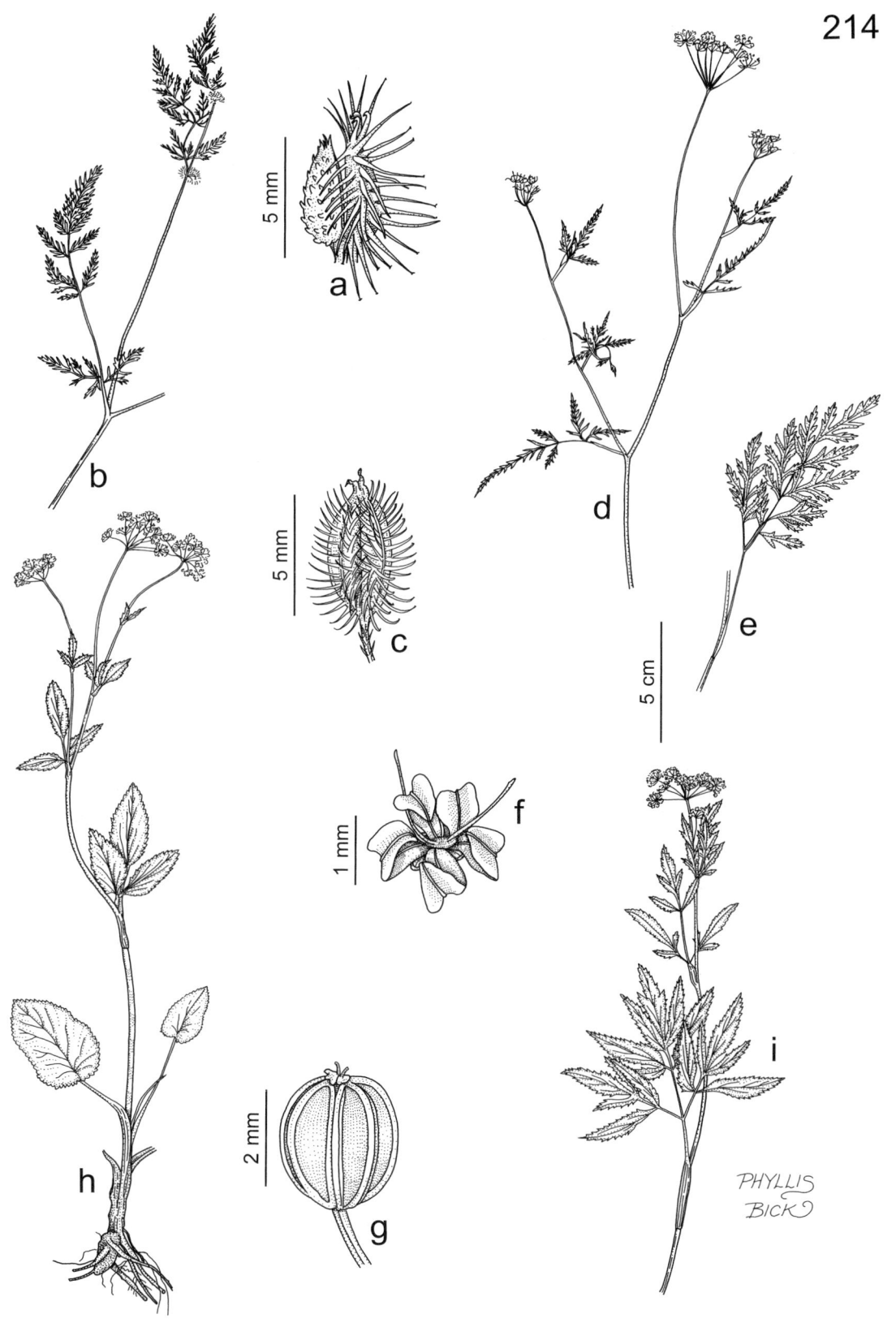

Plate 214. Apiaceae. *Torilis nodosa*, **a)** fruit, **b)** habit. *Torilis arvensis*, **c)** fruit, **d)** inflorescence, **e)** leaf. *Zizia aurea*, **f)** flower, **i)** habit. *Zizia aptera*, **g)** fruit, **h)** habit.

flated. Leaf blades ovate to more or less circular in outline, simple or 1–3 times ternately lobed or compound, the margins glabrous, with a narrow, white border. Inflorescences terminal and axillary, compound umbels, short- to long-stalked. Involucre absent. Rays 10–18, more or less unequal in length. Involucel of 3–9 bractlets, these shorter than the flower stalks, linear, sometimes reduced and fused basally into a low, irregular crown. Flowers 11–19 (to numerous) in each umbellet, the central flower of each umbellet sessile or with a stalk less than 0.5 mm long, the others with stalks 1–4 mm long. Sepals minute triangular scales. Petals obovate, narrowed or tapered abruptly to a short, slender tip, bright yellow. Ovaries glabrous. Fruits 2–4 mm long, ovate to oblong-ovate in outline, flattened laterally, glabrous, the mericarps reddish brown, the 5 ribs tan to yellowish brown, angled and often somewhat corky, but lacking wings. Four species, North America.

Zizia is frequently confused with *Thaspium*. For further discussion, see the treatment of that genus.

1. Basal leaves simple or less commonly ternately 1 time lobed or compound
   .... 1. Z. APTERA
1. Basal leaves 1 or 2 times ternately compound .... 2. Z. AUREA

1. Zizia aptera (A. Gray) Fernald (heart-leaved golden Alexanders, heart-leaved meadow parsnip)

Pl. 214 g, h; Map 893

Stems 20–60 cm long. Basal leaves with the blades 2–15 cm long, simple or 1 time ternately lobed or compound, the leaflets (when present) 1–9 cm long, broadly ovate to oblong-obovate or lanceolate, narrowed to rounded (often unequally) or cordate at the base, sometimes with 1 or 2 lobes toward the base. Stem leaves similar to the basal leaves, mostly 1 time ternately compound, less commonly simple or 1 or 2 times ternately lobed, those just above the stem base usually about as big as or somewhat larger than the basal leaves, the uppermost leaves somewhat reduced in size. Rays 11–16, 0.6–3.2(–5.0) cm long. $2n=22$. April–June.

Scattered, mostly south of the Missouri River (U.S. [except for some southern and northeastern states]; Canada). Mesic to dry upland forests, upland prairies, glades, ledges and tops of bluffs, and occasionally banks of streams; also roadsides.

2. Zizia aurea (L.) W.D.J. Koch (common golden Alexanders, common meadow parsnip)

Z. aurea f. *obtusifolia* (Bissell) Fernald

Pl. 214 f, i; Map 894

Stems 30–110(–150) cm long. Basal leaves with the blades 4–14(–20) cm long, 1 or 2 times ternately compound, the leaflets 1–12 cm long, broadly ovate to oblong-obovate or lanceolate, narrowed to rounded (often unequally) or cordate at the base, sometimes with 1 or 2 lobes toward the base. Stem leaves similar to the basal leaves, gradually reduced toward the stem tip, becoming more finely divided with somewhat narrower leaflets and/or segments, the uppermost often only ternately 1–3 times deeply lobed. Rays 10–21, 0.6–5.0 cm long. $2n=22$. April–June.

Scattered nearly throughout the state, but apparently absent from most of the Mississippi Lowlands Division (eastern U.S. west to Montana and Texas; Canada). Bottomland forests, mesic upland forests, upland prairies, glades, savannas, banks of streams, rivers, and spring branches, bases, ledges, and tops of bluffs, and rarely fens; also roadsides.

Zizia aurea grows in the widest variety of habitats of any of the *Thaspium* and *Zizia* species in Missouri. It can be grown easily in gardens from seed and the foliage and flowers are quite attractive, but if allowed to produce seed it can become overly aggressive.

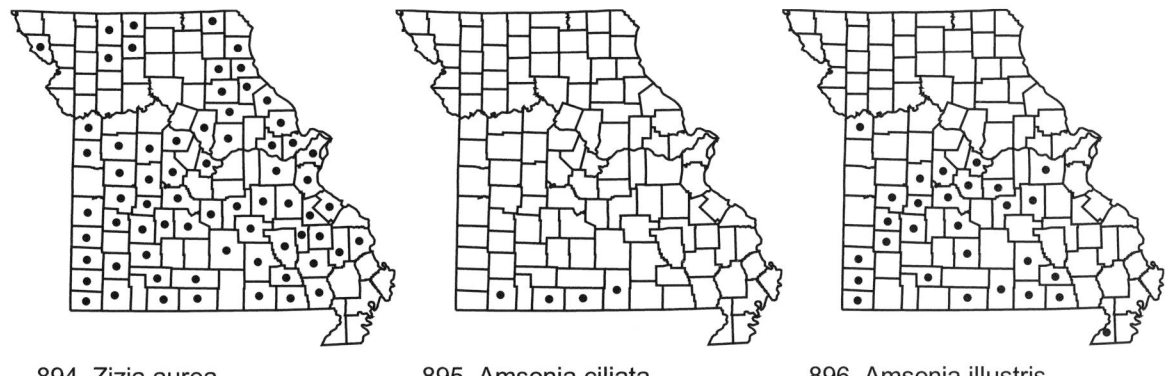

894. Zizia aurea   895. Amsonia ciliata   896. Amsonia illustris

APOCYNACEAE (Dogbane Family)
Contributed by David J. Bogler

Plants herbs (shrubs or trees elsewhere), often with rhizomes or a woody rootstock, or sometimes lianas. Stems sometimes twining, usually with milky sap. Leaves opposite or alternate, rarely appearing whorled, simple, or entire. Stipules absent or inconspicuous. Inflorescences axillary or terminal clusters or panicles, or the flowers solitary. Flowers perfect, hypogynous, actinomorphic. Calyces deeply 5-lobed. Corollas 5-lobed, commonly funnelform to trumpet-shaped, the interior of the tube sometimes with hairs, scalelike appendages, or outgrowths, the lobes overlapping and contorted (twisted) in bud. Stamens 5, alternating with the corolla lobes, the filaments short, fused to the corolla tube, lacking appendages, the anthers forming a close ring around the stigma, free from the stigma but often more or less held in place by sticky secretions. Pistil of 2 carpels, these free below but fused above the ovary. Each ovary superior, 1-locular, with numerous ovules, the placentation parietal. Style 1 per flower, the stigma capitate or somewhat conical, sometimes slightly 2-lobed, often relatively large, sometimes with a small, cuplike wing or other outgrowths. Fruits follicles (berries or capsules elsewhere), potentially 2 per flower. Seeds usually numerous, glabrous or with a tuft of silky hairs at the end opposite the attachment point. About 300 genera, about 2,000 species, nearly worldwide, but most diverse in tropical regions.

The milkweed family, Asclepiadaceae, is included in the Apocynaceae in some of the recent botanical literature, but the two groups are treated in the traditional sense as separate families here. See the treatment of Asclepiadaceae for further discussion.

A number of species of Apocynaceae are popular ornamentals in gardens and greenhouses, including members of some native genera, as well as the exotic genera *Allamanda* L. (allamanda), *Carissa* L. (natal plum), *Catharanthus* G. Don (Madagascar periwinkle, rosy periwinkle), *Nerium* L. (oleander), *Pachypodium* Lindl. (pachypodium, Madagascar palm), *Plumeria* L. (frangipani), *Thevetia* L. (yellow oleander), and *Vinca* L. (periwinkle). In the tropics, a number of genera include timber trees and fiber plants. Members of the Apocynaceae contain a diverse assortment of alkaloids and other compounds. Most of the species are considered poisonous. Because of this complex biochemistry, some species have been used medicinally for a variety of ailments. Some of these medicinal uses have been investigated and have resulted in the development of important pharmaceuticals. For example, *Catharanthus roseus* (L.) G. Don is the source of alkaloids important in the treatment of leukemia and certain cancers; reserpine, which is used to treat schizophrenia and hypertension, comes from *Rauvolfia* L.; and *Strophanthus* DC. is the source of strophanthin, which is used in treating heart disease and as a precursor in the manufacture of cortisone. A number of genera also have been investigated as possible sources of latex for producing rubber.

1. Stems upright; flowers in terminal clusters
    2. Leaves alternate; corolla saucer-shaped; seeds naked .... 1. AMSONIA
    2. Leaves strictly opposite; corolla bell-shaped; seeds with a tuft of hairs at the tip .... 2. APOCYNUM
1. Stems trailing or vinelike; flowers solitary or in axillary clusters
    3. Stems twining; flowers in loose axillary clusters or small panicles, the corolla yellow; seeds with a tuft of hairs at the tip .... 3. TRACHELOSPERMUM
    3. Stems not twining, creeping and sometimes scrambling over other vegetation; flowers solitary in the leaf axils, the corolla blue to bluish lavender (rarely white); seeds naked .... 4. VINCA

1. Amsonia Walter (blue star)

Plants perennial herbs, often somewhat woody at the base. Stems usually several, erect or ascending, unbranched or few-branched toward the tip, often becoming more branched after flowering, usually hollow at maturity. Leaves alternate, sometimes appearing subopposite or nearly whorled at some nodes, sessile or short-petiolate, those of the lowermost nodes usually reduced and scalelike. Leaf blades linear to elliptic or ovate, narrowed or tapered to a sharply pointed tip, the margins sometimes curled under, sometimes minutely hairy along the margins. Stipules absent. Inflorescences terminal or occasionally subterminal, branched loose clusters of few to many flowers. Calyx lobes triangular to linear-triangular, glabrous. Corollas trumpet-shaped, glabrous or hairy on the outer surface, lacking appendages, light blue, the throat slightly swollen and hairy within, the lobes abruptly spreading, overlapping, lanceolate. Stamens attached near the top of the corolla tube, the anthers incurved but free from and positioned above the stigma, arrowhead-shaped, with a pair of prominent triangular basal lobes. Nectar glands absent. Style elongate, the stigma capitate, encircled by a small, cuplike wing above the midpoint. Fruits slender, straight or curved, erect to pendulous, glabrous. Seeds numerous, cylindrical with broadly rounded to more commonly truncate ends, lacking a tuft of hairs. Five to 20 species, southern U.S. and adjacent Mexico, eastern Asia.

Amsonia is one of the few wholly temperate genera in the family. Within the Apocynaceae, the genus is distinguished by the occurrence of free, unappendaged anthers, glabrous seeds, sinistrorse (left-overlapping) corolla lobes, and alternate leaves. It may be related to *Vinca* and *Catharanthus* (Endress and Bruyns, 2000), or perhaps to *Haplophyton,* a small genus of 3 species in the southwestern United States and Mexico that also has alternate leaves (Woodson, 1928; Rosatti, 1989). Within the genus *Amsonia* are two major groups, one in the southeastern United States and eastern Asia (where a single species occurs), and the other in the southwestern United States and northern Mexico. Within these groups the species are highly variable and circumscription of species is difficult. In the southeastern United States, there are 2–8 taxa, but it is not clear whether or at what taxonomic level some of them should be recognized.

Several species of *Amsonia,* including the three species in Missouri, are cultivated as ornamentals in gardens.

1. Leaf blades narrowly linear to linear; corolla glabrous externally ...... 1. A. CILIATA
1. Leaf blades lanceolate to ovate, not linear; corolla hairy externally
    2. Leaves shining on upper surface; calyces sparsely hairy; fruits slightly constricted between the seeds .... 2. A. ILLUSTRIS
    2. Leaves dull on upper surface; calyces glabrous; fruits not constricted between seeds .... 3. A. TABERNAEMONTANA

1. Amsonia ciliata Walter var. **filifolia** A.W. Wood (ciliate blue star)

Tabernaemontana angustifolia Aiton
A. angustifolia (Aiton) Michx.
A. ciliata var. *tenuifolia* (Raf.) Woodson

Pl. 215 f, g; Map 895

Stems 20–50 cm long, 2–3 mm in diameter at the base, glabrous or sparsely hairy, the branches elongating to 10–25 cm long after flowering. Leaves alternate or appearing nearly whorled at some nodes, sessile. Leaf blades mostly 1.5–5.0 cm long, those of the main stem leaves narrowly linear to linear (the lowermost sometimes narrowly oblanceolate), glabrous, the upper surface somewhat shiny, the undersurface pale green and often somewhat glaucous. Inflorescences loose terminal clusters, 4–6 cm long at flowering, positioned above the leaves. Flower stalks 8–10 mm long, often tinged with blue. Calyces glabrous, the lobes 1.0–1.2 mm long, triangular. Corollas densely hairy internally, glabrous externally, the tube 6–7 mm long, about 1 mm wide at base, the throat 2.0–2.4 mm wide, the lobes 5–8 mm long, 1.5–2.5 mm wide. Fruits 7–13 cm long, more or less erect and held above the leaves, not constricted between the seeds, often dehiscing by a slit toward the tip. Seeds 4–7 mm long, 1.2–1.6 mm wide, the surface with wrinkled and pitted or with low, corky ridges, dark brown. April–May.

Uncommon in the southern portion of the Ozark Division, thus far only in counties along the Arkansas border (southeastern U.S. west to Missouri and Texas). Calcareous glades, tops of bluffs, and gravelly banks of streams and rivers.

This species is readily distinguished from other species in Missouri by its very narrow, almost threadlike leaves, inflorescence and follicles produced above the leaves, and externally glabrous corolla. Woodson (1928) treated *A. ciliata* as comprising three varieties, the narrower-leaved (15–30 times longer than broad) var. *tenuifolia* (= var. *filifolia*), the broader-leaved (4–15 times longer than broad) var. *ciliata,* and the larger-flowered corolla tube (9–12 mm vs. 6–8 mm long) var. *texana* (A. Gray) Coulter. The distinctness of at least the first two of these varieties requires further study.

2. Amsonia illustris Woodson (shining blue star)

Pl. 215 c–e; Map 896

Stems 60–100 cm long, 5–10 mm in diameter at the base, glabrous, the branches elongating to 10–35 cm after flowering. Leaves alternate or appearing subopposite at some nodes, short-petiolate. Leaf blades mostly 4–12 cm long, 1.5–3.0 cm wide, those of the main stem leaves narrowly lanceolate to lanceolate, the upper surface shiny, the undersurface pale green, glabrous or sparsely hairy along the main veins, occasionally somewhat glaucous. Inflorescences loose, terminal and sometimes also subterminal clusters, 7–12 cm long at flowering, barely surpassing the branches and foliage. Flower stalks 2–8 mm long. Calyces sparsely hairy, the lobes 1–2 mm long, narrowly to less commonly broadly triangular. Corollas densely hairy internally, sparsely hairy externally, the tube 6–8 mm long, about 1.5 mm wide at base, the throat 3.0–3.5 mm wide, the lobes 5–10 mm long, 1.5–3.0 mm wide. Fruits 7–15 cm long, spreading to pendulous at maturity and positioned among the leaves, slightly constricted between the seeds and sometimes breaking between them. Seeds 6–9 mm long, 2–3 mm wide, the surface usually with low, corky ridges and tubercles, dark brown. $2n=22$. April–June.

Scattered in the southern half of the state, mostly in the Ozark Division (Missouri and Arkansas west to Kansas, Oklahoma, and Texas). Gravelly banks of streams and rivers, ledges and tops of bluffs, and occasionally glades and openings of dry upland forests.

In general, this species is recognized by the lustrous upper leaf surface, sparsely hairy calyx, and constricted follicles that are drooping at maturity. It is closely related to *A. tabernaemontana* and could be viewed as a variety of that species. However, the distinctive leaf and flower characters along with the tendency toward a streamside habitat support recognition of this taxon at the species level (Woodson, 1929).

3. Amsonia tabernaemontana Walter (blue star, willow amsonia)

Pl. 215 h, i; Map 897

Stems to 50–110 cm long, 7–11 mm in diameter at the base, the branches elongating to 10–35 cm after flowering. Leaves alternate or appearing subopposite at some nodes, short-petiolate. Leaf blades 6–15 cm long, 2–5 cm wide, those of the main stem leaves lanceolate to broadly ovate, not linear, the upper surface dull, the undersurface glabrous or finely hairy along the veins, often pale green, sometimes somewhat glaucous. Inflorescences loose, terminal and sometimes also subterminal clusters, 7–12 cm long at flowering, barely or not surpassing the branches and foliage. Flower stalks 2–7 mm long. Calyces glabrous, the lobes narrowly to broadly triangular. Corollas densely hairy internally, sparsely to moderately hairy toward the tip externally, the tube 7–8 mm long, about 1.5 mm

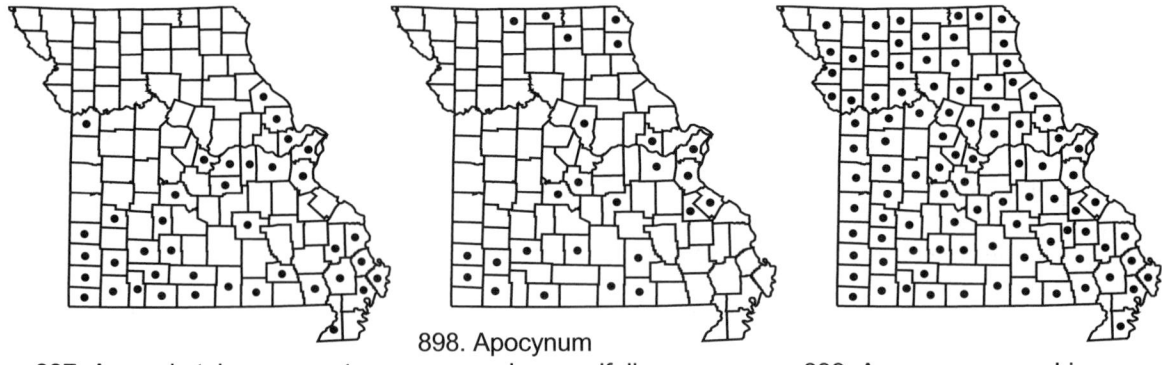

897. Amsonia tabernaemontana
898. Apocynum androsaemifolium
899. Apocynum cannabinum

wide at base, the throat 2.5–3.5 mm wide, the lobes 5–8 mm long, 1.2–2.0 mm wide. Fruits 8–12 cm long, erect or ascending at maturity and positioned among or slightly above the leaves, not constricted between the seeds. Seeds 8–10 mm long, the surface usually with low, interrupted, corky ridges and tubercles. 2n=16, 22, 32. April–May.

Scattered in the southern half of the state and north locally in eastern Missouri to Pike County (eastern U.S. west to Kansas and Texas). Bottomland forests, mesic to dry upland forests, glades, bottomland prairies, banks of streams and rivers; also pastures, ditches, levees, railroads, roadsides, and open, disturbed areas.

Woodson (1928) recognized three varieties within *A. tabernaemontana,* with var. *tabernaemontana* having relatively broad leaves and the other two having narrower leaves (much like those of *A. illustris*). The leaf pubescence and flower density characteristics said to separate his var. *gattingeri* and var. *salicifolia* do not appear to work well for many of the Missouri specimens, and these are combined in the present treatment. Although var. *salicifolia* continues to be by far the more common of the two varieties in Missouri, the range limits of the two taxa have been somewhat blurred by their apparent escape from ornamental plantings and indiscriminate use in habitat restoration projects around the state.

1. Leaf blades 1–3 cm wide, lanceolate to elliptic-lanceolate, narrowly angled or tapered at the base
 . . . . 3A. VAR. SALICIFOLIA
1. Leaf blades 3.0–6.5 cm wide, ovate to broadly ovate, angled to rounded at the base . . . . . 3B. VAR. TABERNAEMONTANA

3a. var. salicifolia (Pursh) Woodson
 A. salicifolia Pursh
 A. tabernaemontana var. *gattingeri* Woodson
 Leaf blades 1–3 cm wide, lanceolate to elliptic-lanceolate, narrowly angled or tapered at the base, angled or tapered to a sharply pointed tip, the undersurface sparsely hairy along the veins to nearly glabrous, often pale green and more or less glaucous. Fruits 6–14 cm long. Seeds mostly 8–9 mm long. April–May.

Scattered in the southern half of the state and north locally in eastern Missouri to Pike County (Ohio, Virginia, and Georgia west to Kansas, Oklahoma, and Texas). Bottomland forests, mesic to dry upland forests, glades, bottomland prairies, banks of streams and rivers; also pastures, ditches, levees, railroads, roadsides, and open, disturbed areas.

3b. var. tabernaemontana (willow amsonia)
 Leaf blades 3.0–6.5 cm wide, ovate to broadly ovate, angled to rounded at the base, tapered to a sharply pointed tip, the undersurface pale green, sparsely hairy along the veins. Fruits 5–10 cm long. Seeds mostly 9–10 mm long. April–May.

Uncommon in the southern portion of the state, mostly in the Ozark Division (eastern U.S. west to Kansas and Texas). Bottomland forests and banks of streams; also rarely ditches.

This variety occurs mostly in southwestern Missouri and tends to grow in richer, more shaded and mesic habitats than does var. *salicifolia*.

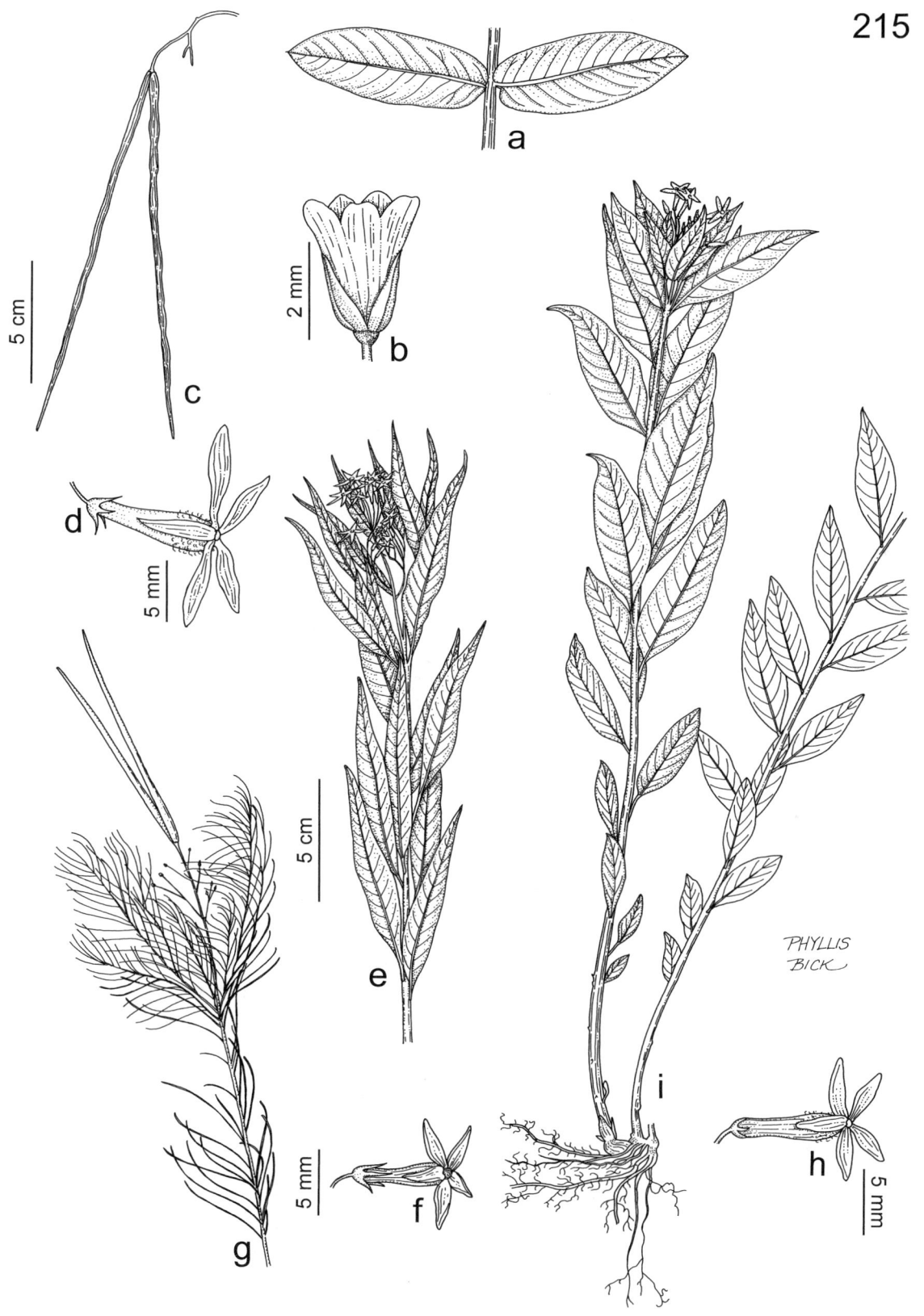

Plate 215. Apocynaceae. *Apocynum cannabinum*, **a)** leaves, **b)** flower. *Amsonia illustris*, **c)** fruit, **d)** flower, **e)** habit. *Amsonia ciliata*, **f)** flower, **g)** fruiting habit. *Amsonia tabernaemontana*, **h)** flower, **i)** habit.

2. Apocynum L. (dogbane, Indian hemp)

Plants perennial herbs, sometimes somewhat woody at the base. Stems solitary or few (but plants sometimes forming large colonies from elongate, branching rhizomes), erect or ascending, often branched dichotomously (usually above the midpoint), glabrous. Leaves opposite or sometimes some of them subopposite or alternate, short-petiolate or sessile. Leaf blades ovate to narrowly lanceolate, glabrous or pubescent with nonglandular hairs, rounded to angled and often with a minute, abrupt, sharp point at the tip, often somewhat asymmetrical at the angled to rounded or less commonly shallowly cordate base. Inflorescences terminal or axillary, branched loose clusters of few to many flowers. Flowers sweetly fragrant. Calyx lobes sharply pointed at the tip, glabrous or with nonglandular hairs. Corollas bell-shaped to urn-shaped, bearing usually 5 small, scalelike appendages internally at the base, white to sometimes tinged or lined with pink, mostly glabrous, the lobes less than half as long as the tube, erect to spreading. Stamens attached near the base of the corolla tube, incurved to form a cone over the stigma, the anthers adhering to the stigma by sticky secretions; anthers arrowhead-shaped with short, triangular, basal lobes. Nectar glands 5, positioned around the ovary bases alternating with the stamens. Styles very short, the stigma more or less ovoid, encircled by a narrow thickening or rim around the midpoint. Fruits slender, elongate. Seeds numerous, narrowly cylindrical, somewhat tapered toward the base, with a tuft of hairs at the truncate tip. About seven species, North America.

Estimates of the number of species of *Apocynum* range from 2 to more than 80, depending on whether Old World taxa are included, how hybrids are treated, and what characters are emphasized. Following Woodson's (1930) treatment, most authors recognize about 7 species that are native to North America, with the Old World species consigned to such segregate genera as *Trachomitum* Woodson and *Poacynum* Baill. In general, characters of size, shape, and pubescence of the leaves are not considered reliable. Woodson emphasized the angle at which the leaves are held (a character that unfortunately is often difficult to see in pressed specimens), the relative length of the calyx lobes and corolla tube, and the size and orientation of the follicles. Variation in flower color from pink to white with various markings also has been noted, but this also is lost frequently when specimens are pressed and dried. In actual practice, it is quite difficult to separate the species, and there are many specimens that are intermediate between the extremes of variation. The most reliable character for separating the species appears to be the length of the corolla tube relative to the calyx lobes.

The pollination biology of *Apocynum* remains something of a mystery. Production of follicles is relatively uncommon despite a high visitation rate by insects. Woodson (1930) suggested the low fruit set indicated that the flowers were self-incompatible. Lipow and Wyatt (1999) found that self-pollinated flowers of *A. cannabinum* produced no fruit, and they concluded that this species was indeed self-incompatible. There appears to be a failure of early embryo development in self-pollinated flowers, and the physical separation of the anthers from the receptive lateral portion of the stigma prevents direct transfer of pollen as the anthers dehisce. The flowers produce nectar and attract a wide assortment of flies, bees, and butterflies, but rarely do the insects affect pollination. Johnson et al. (1998) investigated reproductive biology and hybridization in *A.* ×*floribundum* and its presumed parental species in Colorado. Caged plants from which pollinators were excluded failed to produce any follicles. Fruit set was very low in populations of the uncaged parental species and was almost nonexistent in the uncaged hybrids. Insect visitors were found to be carrying little or no *Apocynum* pollen, although Waddington (1976) had earlier found *A. sibiricum* pollen on the proboscis of butterfly visitors. On the other hand, population genetic studies involving allozyme variation (Johnson et al., 1998) indicated that the hybrid populations are more heterozygous than the parental species, which had low levels of heterozygosity, suggesting a long history of inbreeding. Populations of all of the species often consist of one or few large clones, with few unique genotypes.

Apocynum has had a long history of use by humans. Fibers derived from the plants have great strength and formerly were used as a source of thread and cord for making fishnets, bags, and clothing. The fibers can also be used to make paper. The latex can be used to produce a type of rubber and also contains powerful alkaloids that stimulate the heart much like digitalis. All of the species produce abundant nectar and are considered good bee plants. However, some species, such as *A. cannabinum,* can be troublesome weeds in crop fields, pastures, and gardens, and may have allelopathic effects on cultivated plants. Yield reductions of up to 45 percent have been reported from infested crop fields in Kansas (McGregor, 1984).

1. Leaves spreading to drooping on living plants, all petiolate; corollas 4–7 mm long, the tube usually more than twice as long as the calyx, whitish pink with pink veins or stripes on the inner surface, the lobes spreading to recurved
.... 1. A. ANDROSAEMIFOLIUM
1. Leaves mostly ascending on living plants, the lowermost leaves sessile (the upper ones sessile or short-petiolate); corollas 2–5 mm long, the tube about as long as the calyx, white or pale green, the lobes erect .... 2. A. CANNABINUM

1. Apocynum androsaemifolium L. (spreading dogbane, pink-flowered dogbane)
A. ambigens Greene

Pl. 216 j, k; Map 898

Stems 0.3–1.5 m tall, usually glaucous. Leaves drooping to spreading, all petiolate, the petioles 3–5 mm long. Leaf blades 1–9 cm long, 1–5 cm wide, broadly ovate to elliptic, rounded to broadly angled at the base, never cordate or clasping, usually broadly angled at the tip, the upper surface glabrous, the undersurface often sparsely hairy, whitish green. Flowers erect to drooping. Calyces glabrous, the lobes 1.5–2.5 mm long, ovate-triangular. Corollas bell-shaped, 4–7 mm long, whitish to pink, with pink veins or stripes on the inner surface, the tube 3–4 mm long, usually more than twice as long as the calyx, 4–5 mm wide at the tip, the lobes 2–3 mm long, spreading or recurved. Filaments 0.8–1.1 mm long, hairy. Anthers 2.5–3.2 mm long. Styles 0.7–1.0 mm long, the stigma about 1 mm long. Fruits 6–9 cm long, pendulous, straight. Seeds with the body 2–3 mm long, the apical tuft of hairs 15–20 mm long. $2n=16, 22$. May–September.

Scattered mostly south of the Missouri River, uncommon in the Glaciated Plains Division, apparently absent from the Mississippi Lowlands Division (U.S., including Alaska; Canada). Upland prairies, openings of mesic to dry upland forests, and banks of streams and spring branches; also pastures, fencerows, roadsides, and open, disturbed areas.

Woodson (1930) separated *A. androsaemifolium* into three varieties based upon differences in leaf pubescence and corolla shape. These scarcely seem worthy of recognition. Missouri specimens are referable to var. *androsaemifolium,* based on their glabrous leaves and more broadly bell-shaped flowers.

Woodson (1930) emphasized the importance of hybridization in the taxonomy and reproductive biology of *Apocynum.* One of the species and most of the varieties found in eastern North America were thought to have originated from hybridization events. His *A. medium* Greene (now known as *A. ×floribundum* Greene; Pl. 216 i) was presumed to be the result of hybridization between *A. androsaemifolium* and *A. cannabinum* and was identified as having characters intermediate between the two parental species. It has spreading leaves (as opposed to drooping or ascending), pinkish corollas 4–5 mm long, usually reduced fertility, and seeds 3–4 mm long (when formed). Woodson (1930) described five varieties of *A. medium* based mostly on the presence and location of pubescence. He observed that hybrids frequently were sterile or partially so and often produced abnormal pollen. Anderson (1936) investigated the hybrid origin of *A. medium* by growing seeds from plants identified by Woodson as *A. androsaemifolium, A. cannabinum,* and *A. medium*. Specimens from supposed hybrids did indeed have lower pollen viability. The progeny of the parental species bred true, but the progeny of the putative hybrids were quite variable in appearance. Some of the mature plants grown from seeds of *A. medium* were identified as *A. cannabinum* or *A. androsaemifolium.* Anderson concluded that hybridization was indeed present in *Apocynum,* that *A. medium* is of hybrid origin, and that relatively high levels of hybridization tended to obscure species limits, at least in the Midwest. Some other authors (Gleason and Cronquist, 1963, 1991) have suggested that plants called *A.* ×*medium* or *A.* ×*floribundum* represent the results of rampant hybridization between *A. androsaemifolium* and two of Woodson's other species, *A. cannabinum* and *A. sibiricum*. See the treatment of *A. cannabinum* for further discussions of difficulties in differentiating the putative parental taxa.

2. Apocynum cannabinum L. (Indian hemp, prairie dogbane)
A. cannabinum var. *glaberrimum* A. DC.
A. cannabinum var. *pubescens* (Mitch. ex R. Br.) A. DC.
A. cordigerum Greene
A. hypericifolium Aiton
A. hypericifolium var. *cordigerum* (Greene) Bég. & Belosersky
A. sibiricum Jacq.
A. sibiricum var. *cordigerum* (Greene) Fernald
A. missouriense Greene
A. leuconeuron Greene

Pl. 215 a, b, 216 f–h; Map 899

Stems 0.2–1.0 m tall, often reddish-tinged. Leaves mostly ascending to spreading (excluding drought-stressed individuals), the petioles of the upper and sometimes median leaves 5–7 mm long, the lowermost leaves sessile or subsessile. Leaf blades 1.5–14.0 cm long, 0.5–8.0 cm wide, narrowly lanceolate to broadly ovate or elliptic, angled, rounded or shallowly cordate at the base, rounded to more commonly angled or short-tapered to a sharply pointed tip, the upper surface glabrous or sparsely to moderately hairy, the undersurface glabrous or sparsely to relatively densely hairy, pale green. Flowers erect or less commonly drooping. Calyces glabrous, the lobes 1.5–2.5 mm long. Corollas 2–5 mm long, narrowly bell-shaped to urn-shaped, white to pale greenish, the tube 1.5–2.5 mm long, about as long as the calyx, 1.5–2.0 mm wide at the tip, the lobes 1.5–2.0 mm long, erect. Filaments 0.2–0.4 mm long, hairy. Anthers 1.3–1.6 mm long. Fruits 4–20 cm long, pendulous, straight or nearly so. Seeds with the body 3–5 mm long, with the apical tuft of hairs 8–30 mm long. $2n=16, 22$. June–August.

Common throughout the state (U.S., Canada). Bottomland forests, openings of mesic to dry upland forests, savannas, depressions in and around glades, bottomland prairies, swales in upland prairies, margins of swamps, oxbows, ponds, and lakes, fens, and banks of streams and rivers; also pastures, fencerows, ditches, levees, roadsides, railroads, and open, disturbed areas.

Apocynum cannabinum is distinguished from *A. androsaemifolium* by the relatively small flowers in relatively dense clusters. See the treatment of the latter species for a discussion of problems in identification caused by interspecific hybridization between them.

The leaves are highly variable in shape and pubescence, and these characters were used to describe a number of varieties said to be present in Missouri (Woodson, 1930), including the glabrous var. *glaberrimum* A. DC.; var. *cannabinum*, with a somewhat hairy leaf undersurface; and var. *pubescens* (Mitchell) A. DC., with hairs on the inflorescence and both leaf surfaces. Steyermark (1963) treated these varieties somewhat differently. All specimens that were glabrous or only lightly hairy on the leaf undersurface were placed in var. *cannabinum*, and only plants with denser and more evenly distributed hairs were included in var. *pubescens*. In fact, pubescence is extremely variable in Missouri populations and both glabrous and hairy plants are scattered throughout the state. Thus separation of varieties based on this character does not appear to be warranted.

Apocynum sibiricum Jacq. has been distinguished from *A. cannabinum* based primarily on its sessile or short-petiolate leaves with more or less clasping bases (Woodson, 1930; Steyermark, 1963). It has been distinguished further by having a supposed northern Missouri distribution, foliaceous bracts subtending the inflorescence (or, as sometimes stated, the inflorescence mostly overtopped by the leaves), follicles 4–10 cm (vs. 12–20 cm) long, and seeds with the apical tuft of hairs 8–12 mm (vs. 20–30 mm) long. Many if not all of the specimens attributed to *A. cannabinum* also have sessile leaves on the lower part of the main stem, with the leaves of the upper stems and branches more or less petiolate. Unfortunately, most collectors do not include the lower stems and leaves in their specimens. Thus, many specimens probably are determined as *A. sibiricum* if the lower leaves are present and as *A. cannabinum* if these were not collected. The other characters used to define *A. sibiricum* also can be found in various combinations in specimens with leaves typical of *A. cannabinum* (McGregor, 1984), and plants that can be keyed more or less to either species occur nearly statewide. Following Hartman (1986a), *A. sibiricum* is here included reluctantly in a broadly circumscribed *A. cannabinum*. Further research may reveal that there are two biological entities involved

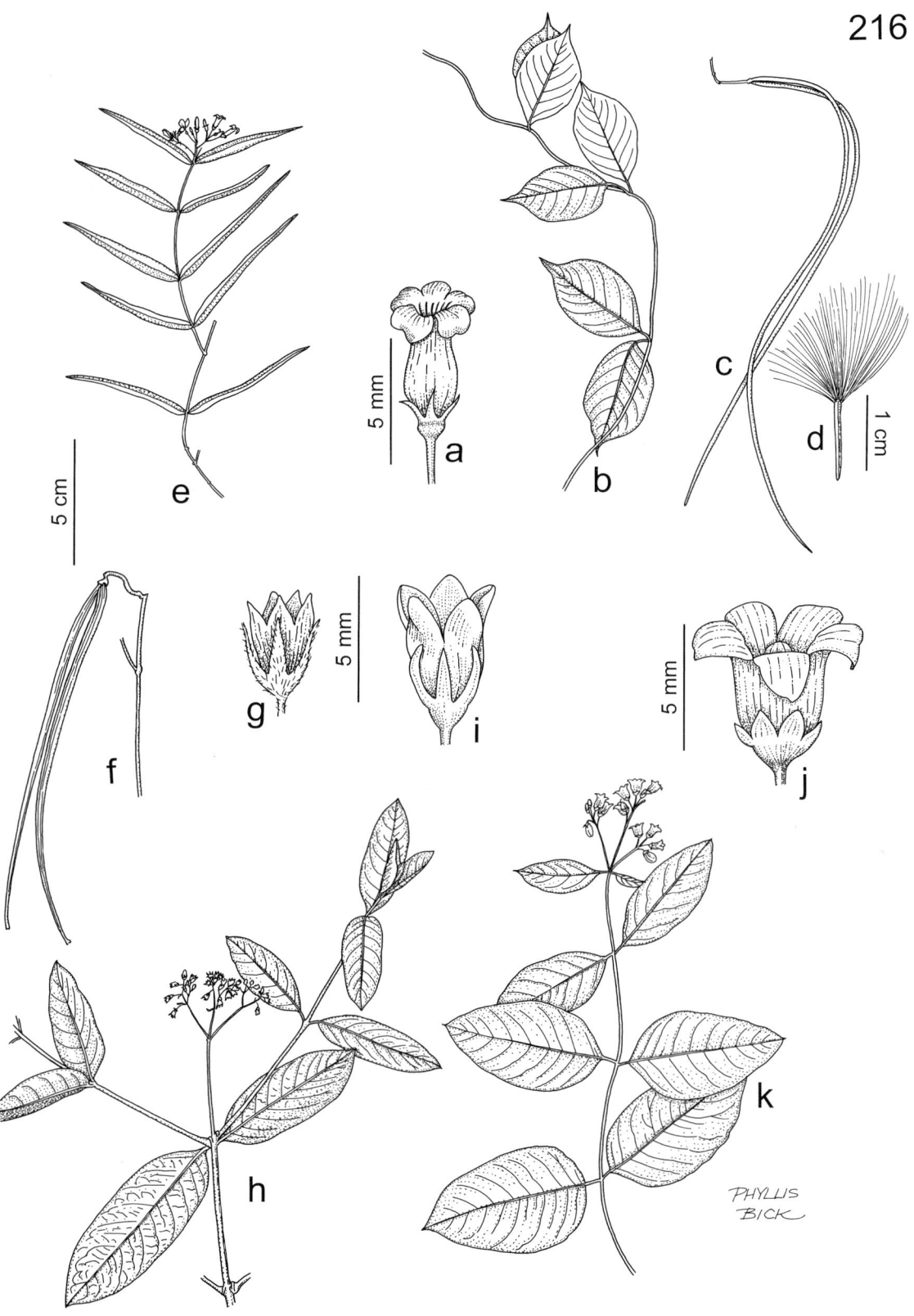

Plate 216. Apocynaceae. *Trachelospermum difforme*, **a**) flower, **b**) stem with broader leaves, **c**) fruit, **d**) seed, **e**) stem tip with narrower leaves and inflorescence. *Apocynum cannabinum*, **f**) fruit, **g**) flower, **h**) habit. *Apocynum ×floribundum*, **i**) flower. *Apocynum androsaemifolium*, **j**) flower, **k**) habit.

900. Trachelospermum difforme   901. Vinca major   902. Vinca minor

in the complex, with hybridization between them obscuring their morphological separation.

The dried rhizomes of *A. cannabinum* were once prescribed as a cardiac stimulant, and they also have been used as a laxative and to induce vomiting. Plants also were used to make a dark tan and black dye. It can be a troublesome weed in cultivated fields, pastures, and gardens.

3. Trachelospermum Lemaire

Ten to 30 species, southeastern U.S., Asia.

Only one species occurs natively in the New World. The remainder are native to eastern and southeastern Asia. The Confederate jasmine, *T. jasminoides* (Lindl.) Lem., is cultivated as an ornamental in the warmer parts of the United States and occasionally escapes along the Gulf Coastal Plain, but it has not been reported from farther north.

1. Trachelospermum difforme (Walter) A. Gray (climbing dogbane)

Echites difformis Walter

Pl. 216 a–e; Map 900

Plants lianas or sometimes appearing herbaceous, sometimes more or less emergent aquatics. Stems 2–5 m or more, woody or mostly herbaceous, twining, often rooting at the nodes, glabrous or with sparse, short, reddish hairs, especially when young. Leaves opposite, the petioles 4–5 mm long. Leaf blades 2.5–9.0 cm long, 0.5–5.5 cm wide, variable in shape from broadly ovate or nearly circular to lanceolate or narrowly oblong-elliptic, narrowed or more commonly tapered to a sharply pointed tip, angled, tapered, or rarely rounded at the base, the upper surface glabrous and sometimes somewhat shiny, the undersurface sparsely to densely short-hairy. Inflorescences axillary, loose clusters or small panicles. Calyces glabrous or sparsely hairy, the lobes 2.5–3.0 mm long, narrowly ovate-triangular. Corollas funnelform to trumpet-shaped, lacking appendages, the tube 5–7 mm long, the lobes 1.5–2.0 mm long, yellow with red coloration on the inner surface toward the top of the tube and the lobes, sparsely hairy inside the tube. Stamens attached near the midpoint of the corolla tube, the anthers incurved and adhering to the stigma, appearing arrowhead-shaped with a pair of slender basal lobes. Nectar glands 5, positioned around the ovary bases alternating with the stamens. Style elongate, the stigma appearing somewhat umbrella-shaped with a broadly club-shaped body that is expanded into a basal wing. Fruits 10–23 cm long, usually pendulous, slender, reddish, dehiscing along the longitudinal suture. Seeds 0.8–11 mm long, truncate to tapered at the base, the tip with a dense tuft of yellowish gray hairs 14–15 mm long. June–July.

Uncommon in the Mississippi Lowlands Division and southern portion of the Ozarks north locally to Iron County (southeastern U.S. west to Missouri, Oklahoma, and Texas). Bottomland forests, swamps, banks of streams and rivers, oxbows, and rarely glades; also ditches and roadsides.

The leaves of *T. difforme* are quite variable, ranging from broadly ovate or nearly circular to narrowly oblong-elliptic or lanceolate. Some specimens have both broad and narrow leaves on the same plant, sometimes on the same branch. The leaves generally are evergreen farther south, but in Missouri are frequently deciduous, and the aboveground portions often die back partially or completely during especially cold winters.

Woodson (1935) and Steyermark (1963) reported a record of *T. difforme* from St. Louis County, but although Woodson himself apparently collected this specimen in 1931 along the margins of Creve Coeur Lake, it could not be located during the present study. The existence of this New World taxon may be due to the ancient biogeographic connection between eastern Asia and eastern North America. Alternatively, *T. difforme* may be more closely related the tropical American genus *Secondatia* than to the Asiatic species. Woodson (1935) suggested that the distribution of *T. difforme* matches the ancient continental shorelines of the Cretaceous Period, much like that of *Taxodium distichum* (bald cypress).

4. Vinca L. (periwinkle)
(Stearn, 1973)

Plants perennial herbs, sometimes somewhat woody at the base, the sap not milky. Stems prostrate, forming dense mats, not twining, sometimes with loosely ascending branches or scrambling on other plants, occasionally climbing up tree trunks, rooting at the nodes. Leaves opposite, petiolate or rarely sessile. Leaf blades ovate-triangular or ovate to narrowly lanceolate, glabrous or pubescent with nonglandular hairs, rounded to angled or slightly tapered to a bluntly or sharply pointed tip, sometimes with a minute, abrupt, sharp point at the tip, angled, rounded, truncate, or shallowly cordate at the base, the main veins and adjacent tissue sometimes appearing whitened or lighter green on the upper surface, glabrous on both surfaces (the margins sometimes hairy). Inflorescences of solitary axillary flowers. Calyx lobes narrowly triangular to linear, sharply pointed at the tip, glabrous. Corollas trumpet-shaped, lacking appendages but usually with a narrow ridge or slightly thickened band around the tip of the tube (along the base of the lobes), the tube funnelform, blue or bluish lavender, rarely white, hairy on the inner surface near the stamens, the lobes slightly shorter than to longer than the tube, spreading, often somewhat asymmetrical toward the tip. Stamens attached toward the midpoint of the corolla tube, the anthers incurved but free from and positioned above the stigma, narrowly obovate, hairy. Nectar glands 2, positioned on opposite sides of the ovary bases. Style gradually broadened toward the tip, the stigma more or less conical, with a narrow, disclike rim or wing of tissue at the base and minute tufts of hairs at the tip and base. Fruits rarely produced, 2–7 cm long, erect or ascending, relatively stout, narrowed or tapered above and below the seed(s), dehiscing longitudinally only with age. Seeds 1(2–5) per fruit, 3–7 mm long, mostly narrowly elliptic-ovate in outline, flattened, with a deep, longitudinal groove (the lateral margins appearing somewhat inrolled), the surface otherwise finely wrinkled, glabrous, dark brown to nearly black. About 6 or 7 species, Europe, adjacent southwestern Asia, northern Africa; widely escaped from cultivation in temperate and subtropical regions elsewhere.

1. Leaf blades ovate to triangular-ovate, broadly rounded to truncate or shallowly cordate at the base (those of the smallest leaves sometimes appearing angled at the base), the margins finely hairy; petioles of the larger leaves 9–18 mm long; corolla tube 12–20 mm long, the lobes 15–25 mm long . . . . 1. V. MAJOR
1. Leaf blades lanceolate to ovate or elliptic, mostly angled at the base, the margins glabrous; petioles 1–2 mm long; corolla tube 8–12 mm long, the lobes 9–14 mm long . . . . 2. V. MINOR

1. Vinca major L. (greater periwinkle)

Map. 901

Stems 10–150 cm or more long. Leaves more or less evergreen, herbaceous to somewhat leathery, the petioles (3–)9–18 mm long. Leaf blades 1.5–7.0 cm long, 0.8–5.5 cm wide, ovate to triangular-ovate, narrowed or slightly tapered to a usually sharply pointed tip, broadly rounded to truncate or shallowly cordate at the base, the smallest leaves with blades sometimes appearing angled at the base, the margins densely but inconspicuously pubescent with short, stiff, somewhat appressed-ascending hairs. Flower stalks (12–)16–50 mm long. Calyx lobes 9–12 mm long. Corollas with the tube 12–20 mm long, the lobes 15–25 mm long. $2n=92$. April–May.

Introduced, known thus far only from St. Charles County (native of Europe, Asia; introduced widely but sporadically in the U.S. and Canada). Bottomland forests and mesic upland forests; also potentially abandoned homesites and cemeteries.

Vinca major is cultivated less commonly in Missouri than is *V. minor*. It differs from that species in its generally more robust habit with more ascending flowering branches, but the leaves are not as leathery and tend to become deciduous or at least somewhat damaged in very cold weather. Its overall distribution in North America tends to be more southern and coastal than that of *V. minor*. Greater periwinkle has long been known to become invasive on a local scale in mesic forested habitats in other parts of the eastern and midwestern United States. It was not recorded outside of cultivation in Missouri until late in 2004, when Lia Bollmann, a biologist with the Missouri Department of Conservation, discovered a few large (more than 5 m in diameter) patches at Weldon Springs Conservation Area. Attempts are under way to control the species at this site.

2. Vinca minor L. (common periwinkle, myrtle)

Pl. 217 a–c; Map 902

Stems 10–150 m or more long. Leaves evergreen, strongly leathery, the petioles 1–2 mm long. Leaf blades 1.8–5.0 cm long, 0.8–2.5 cm wide, lanceolate, elliptic, or ovate, narrowed to a sharply or bluntly pointed tip, angled at the sometimes slightly asymmetrical base, the margins glabrous. Flower stalks 9–12 mm long. Calyx lobes 4–5 mm long. Corollas with the tube 8–12 mm long, the lobes 9–14 mm long. $2n=46$. March–May.

Introduced, scattered, mostly south of the Missouri River (native of Europe, Asia; introduced widely but sporadically in the U.S. and Canada). Bottomland forests and mesic upland forests; also abandoned homesites, cemeteries, ditches, roadsides, and shaded, disturbed areas.

Vinca minor is the hardiest of the periwinkles and has long been cultivated in North America as an evergreen ground cover. The species is relatively aggressive, however, potentially covering large areas and climbing into trees. Although available very commonly in the horticultural trade, it is not recommended here for planting. Because it almost never produces fruits in North America, periwinkle usually escapes from cultivation by pieces of plants that are discarded by gardeners or that are washed into drainages from adjacent plantings. The species might be confused with a superficially similar invasive exotic, *Euonymus hederacea* Champ. ex Benth. (*E. fortunei* (Turcz.) Hand.-Mazz.) (wintercreeper, Celastraceae), but the leaves in that species have blunt teeth along the margins and the flowers have small, yellowish white, 4-parted corollas.

AQUIFOLIACEAE (Holly Family)

Four genera, about 420 species, nearly worldwide.

1. Ilex L. (holly)

Plants dioecious, shrubs or small to medium-sized trees. Leaves alternate (in some species sometimes appearing fascicled on short spur shoots), simple, short-petiolate, the margins of the blades usually toothed. Flowers small, short-stalked, in small, axillary clusters, occasionally

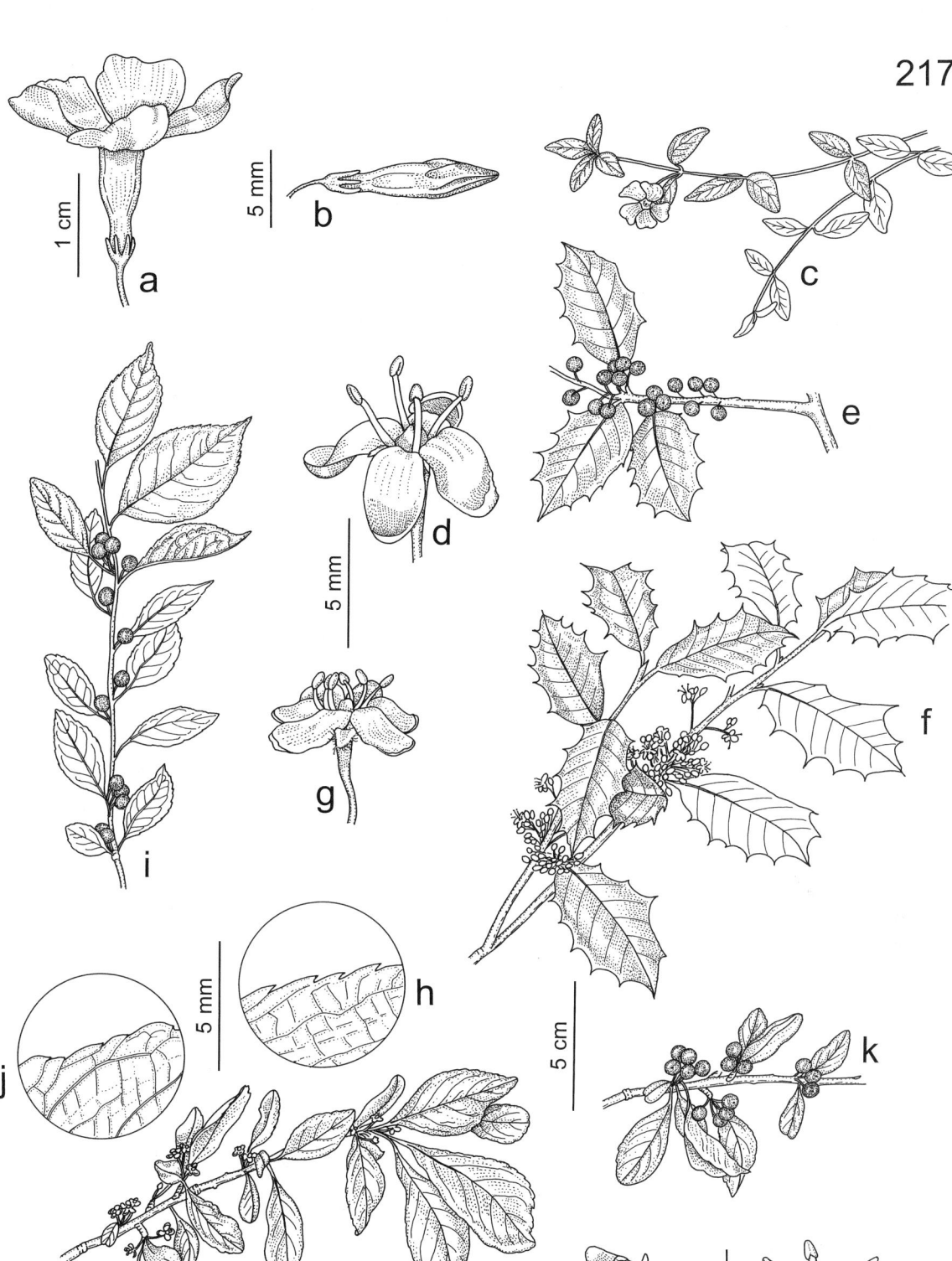

Plate 217. Apocynaceae, Aquifoliaceae. *Vinca minor*, **a)** flower, **b)** flower bud, **c)** habit. *Ilex opaca*, **d)** flower, **e)** fruiting branch, **f)** habit. *Ilex verticillata*, **g)** flower, **h)** leaf margin detail, **i)** habit. *Ilex decidua*, **j)** leaf margin detail, **k)** fruiting branch, **l)** pistillate flower, **m)** staminate flower, **n)** habit.

solitary, hypogynous. Calyx 4–8-lobed, persistent under the fruit. Corolla deeply 4–8-lobed (sometimes appearing as distinct petals), white to greenish white, the lobes rounded, spreading, the tips sometimes reflexed. Stamens 4–8, fused to the base of the corolla, the pistillate flowers with staminodes similar in appearance to stamens, but smaller. Pistil 1 per flower, of 2–5(–9) fused carpels, in staminate flowers reduced to a minute protuberance. Ovary superior, with 2–5(–9) locules. Style absent or minute. Stigma capitate to disc-shaped, often shallowly 2–5(–9)-lobed. Fruits berrylike drupes, globose, usually bright red at maturity (rarely yellow or brown), with 2–10 nutlets, each containing 1 seed in a hard coating, these more or less wedge-shaped, white to light yellow or light tan. About 400 species, nearly worldwide.

Most species of *Ilex* retain their mature fruits through the winter months and many species thus provide important winter food for birds. However, the berries reputedly are poisonous to humans (Steyermark, 1963). The leaves of several species have been used for teas.

1. Tips of the leaf blades with a spine; margins of the leaf blades usually also with few to several spine-tipped teeth, less commonly entire .... 2. I. OPACA
1. Tips of the leaf blades spineless; margins of the leaf blades with several to numerous minute, spineless teeth, these sometimes gland-tipped
    2. Tips of the leaf blades short-tapered to a sharp point; margins of the leaf blades with the teeth sharply pointed and straight to spreading (appearing saw-toothed); calyx lobes noticeably hairy along the margins .... 3. I. VERTICILLATA
    2. Tips of the leaf blades narrowed or short-tapered to a blunt point; margins of the leaf blades with the teeth incurved (appearing more or less scalloped and rounded or bluntly pointed); calyx lobes glabrous or minutely and inconspicuously hairy (the pubescence visible only with strong magnification of more than 10×)
        3. Leaves deciduous, the blades relatively thin, herbaceous, oblanceolate to obovate in outline .... 1. I. DECIDUA
        3. Leaves evergreen, the blades relatively thick, leathery, narrowly to broadly elliptic in outline .... 4. I. VOMITORIA

1. Ilex decidua Walter (possum haw, deciduous holly)

Pl. 217 j–n; Map 903

Plants shrubs or small trees, to 7 m tall. Bark smooth, gray to grayish brown. Leaves deciduous, the blades 2–7 cm long, oblanceolate to narrowly elliptic-obovate, rounded to narrowed or short-tapered to a blunt (spineless) point at the tip, gradually narrowed or tapered at the base, relatively thin, herbaceous, the margins with numerous, fine, incurved teeth (appearing more or less scalloped and rounded or bluntly pointed), these often minutely gland-tipped; the upper surface dull, glabrous, the undersurface sparsely to moderately hairy along the veins. Staminate flowers in clusters of 2–12, usually at the tips of spur shoots, with 4 or 5 perianth lobes and stamens. Pistillate flowers solitary or in clusters of 2–6, with (4)5 perianth lobes and staminodes.

Calyx 0.5–0.8 mm long, the lobes sharply pointed, glabrous. Corollas 1.5–2.2 mm long. Fruits 5–8 mm in diameter, with 4(–8) nutlets, these grooved longitudinally on the back. $2n=40$. April–May.

Scattered to common south of the Missouri River and also to the north in counties along the Mississippi River (southeastern U.S. west to Kansas and Texas). Swamps, bottomland forests, mesic to dry upland forests, ledges and tops of bluffs, margins of glades, sloughs, margins of ponds and lakes, and banks of streams and rivers.

2. Ilex opaca Aiton **var. opaca** (American holly)

Pl. 217 d–f; 904

Plants small trees, to 15 m tall. Bark smooth to roughened or with fine, irregular, warty ridges, light gray to less commonly light brown. Leaves evergreen, the blades 4–12 cm long, elliptic to

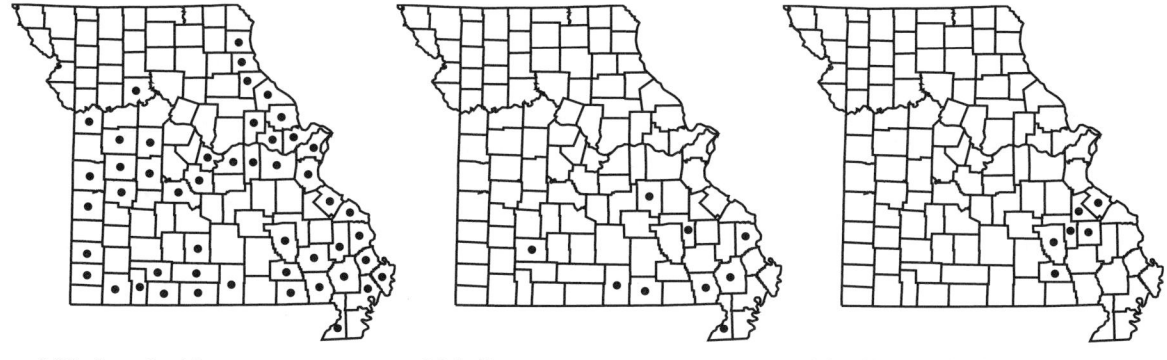

903. Ilex decidua   904. Ilex opaca   905. Ilex verticillata

oblong-obovate, narrowed or tapered to a spiny tip, rounded or narrowed at the base, relatively thick, leathery, the margins entire or more commonly with few to several, spreading, spine-tipped teeth, the upper surface usually shiny, glabrous, the undersurface dull, glabrous or sparsely hairy along the midvein. Staminate flowers in clusters of 3–10, with 4 perianth lobes and stamens. Pistillate flowers solitary or in clusters of 2 or 3, with 4 perianth lobes and staminodes. Calyces 0.8–1.8 mm long, the lobes sharply pointed and sometimes with a few teeth, hairy along the margins. Corollas 2–3 mm long. Fruits 7–12 mm in diameter, with 4 nutlets, these grooved longitudinally on the back. $2n=36$. May–June.

Uncommon, with native populations restricted to the Crowley's Ridge Section of the Mississippi Lowlands Division and occasional introduced occurrences present farther north and west in the state (southeastern U.S. west to Missouri and Texas; also north to Maine along the Atlantic Coastal Plain). Acid seeps and mesic upland forests; also margins of crop fields.

American holly is one of the few evergreen species in the native flora and is commonly cultivated in the eastern half of the country as an ornamental tree and for Christmas decorations. The wood is sometimes used in implements, furniture, inlays, and piano keys. The natural occurrences of *I. opaca* are restricted to acid seeps in drainages of Crowley's Ridge, where they are part of the unusual mesic upland forest assemblage that includes *Fagus* and *Liquidambar*. Following a series of programs in 1957 by Julian Steyermark promoting the preservation of this species in Missouri, a large committee that included members representing the Federated Garden Clubs of Missouri, the Women's League of Missouri, the Missouri Conservation Commission, and The Nature Conservancy raised funds in the early 1960s to purchase and fence two tracts of land in Stoddard County containing the best remaining populations in the state and best acid seep communities in the Crowley's Ridge area (Edgington, 1960; Ashley, 1963). This preserve originally was deeded to The Nature Conservancy, subsequently donated to the Department of Conservation in 1975, and dedicated as a state Natural Area in 1976.

Plants that become established as escapes from cultivation elsewhere in the state via bird-dispersed seeds generally do not reproduce well (probably because of limited seed production), and such occurrences usually contain only one or a few trees. In contrast, cultivated groves in which both staminate and pistillate trees are present generally produce abundant fruits and volunteer seedlings are relatively common under and around the parent trees. Trivial forms of minor horticultural interest include the entire-leaved f. *subintegra* Weath. and the yellow-fruited f. *xanthocarpa* Rehder. The var. *arenicola* (Ashe) Ashe, known as scrub holly and hummock holly, is endemic to central Florida and differs in its narrower leaves and mostly shrubby habit.

3. Ilex verticillata (L.) A. Gray **var. padifolia** (Willd.) Torr. & A. Gray (winterberry, black alder)

Pl. 217 g–i; Map 905

Plants usually shrubs, rarely small trees, to 5 m tall. Bark smooth, gray to grayish brown. Leaves deciduous, the blades 2–9 cm long, narrowly to broadly elliptic, short-tapered to a sharp (but spineless) point at the tip, narrowed or rounded at the base, relatively thin, herbaceous, the margins with numerous, fine, sharply pointed, straight to spreading teeth (appearing saw-toothed, but spineless); the upper surface dull, glabrous, the undersurface sparsely to moderately

906. Ilex vomitoria   907. Aralia nudicaulis   908. Aralia racemosa

hairy along the veins. Staminate flowers in clusters of 2–10, with 4–6 perianth lobes and stamens. Pistillate flowers solitary or in clusters of 2 or 3, with 5–8 perianth lobes and staminodes. Calyx 0.7–1.4 mm long, the lobes rounded to bluntly pointed, hairy, at least along the margins. Corollas 1.5–2.5 mm long. Fruits 5–7 mm in diameter, with 5–10 nutlets, these smooth on the back. $2n=36$. April–June.

Uncommon to scattered, restricted to the eastern portion of the Ozark Division and adjacent counties of the Ozark Border (eastern U.S. and adjacent Canada west to Ohio, Missouri, and Mississippi). Banks of streams, ledges of bluffs, and steep, rocky hillsides in mesic upland forests, on igneous and sandstone substrates.

4. Ilex vomitoria Aiton var. **vomitoria**
(yaupon)

Map 906

Plants shrubs or less commonly small trees, to 6 m tall. Bark roughened, gray to reddish brown, often mottled or developing small plates with age. Leaves evergreen, the blades 1–4 cm long, narrowly to broadly elliptic or oblong-elliptic, rounded (spineless) at the tip, rounded at the base, relatively thick, leathery, the margins with several to numerous fine, incurved teeth (appearing more or less scalloped and rounded or bluntly pointed), glabrous, the upper surface shiny, the undersurface dull. Staminate flowers in clusters of 3–10, with 4 perianth lobes and stamens. Pistillate flowers solitary or in clusters of 2 or 3, with 4 perianth lobes and staminodes. Calyx 0.5–0.8 mm long, the lobes broadly rounded, glabrous or minutely and inconspicuously hairy. Corollas 1.8–2.5 mm long. Fruits 4–7 mm in diameter, with 4 nutlets, these grooved longitudinally on the back. $2n=40$. May.

Introduced, known from a single historical collection from Dunklin County (southeastern U.S. west to Arkansas and Texas, mostly along the Atlantic and Gulf Coastal Plains). Railroad.

The single Missouri specimen was collected in September 1897 and consists only of vegetative material. It is not known whether the plant or plants ever flowered and the description above was taken from non-Missouri specimens for flower and fruit characters.

This species is occasionally cultivated in Missouri as an ornamental shrub or trimmed into a hedge. Alston and Schultes (1951) discussed the historical ceremonial use of *I. vomitoria* by Native Americans in the southeastern United States, who prepared a concentrated infusion known as black drink from the leaves (which contain caffeine) for use as an emetic and stimulant beverage. The other variety, var. *chiapensis* Sharp, occurs disjunctly in southern Mexico, and differs in its hairy foliage.

ARALIACEAE (Ginseng Family)

Plants perennial, trees, shrubs, lianas, or herbs, sometimes with rhizomes. Leaves alternate (less commonly opposite or whorled in *Hedera*) or basal from rhizomes, simple or variously compound, petiolate, the petiole base often expanded and somewhat sheathing, the stipules partially fused to the petiole base or lacking. Inflorescences umbels of small flowers, these solitary or arranged into compound umbels, racemes, or panicles, usually with small, lanceolate bracts subtending the flowers and at the branch points. Flowers mostly perfect (functionally staminate or pistillate flowers sometimes mixed with the perfect ones), epigynous, actinomorphic. Sepals reduced to an inconspicuous crown or 5 small teeth, sometimes absent, when present usually persistent in fruit. Petals 5, often shed quickly after the flower opens. Stamens 5, the filaments free. Pistil 1 per flower, composed of 2–5 fused carpels, the ovary inferior with a nectar disc at the tip, the styles 1–5, sometimes slightly expanded at the base, persistent in fruit. Fruits berrylike drupes, with 1 stone per carpel. Sixty to 70 genera, 700–1,300 species, nearly cosmopolitan, most diverse in tropical portions of South America and Asia and Malesia.

Recent molecular and morphological phylogenetic studies (summarized in Judd et al., 1994, 2002) suggest that Apiaceae and Araliaceae might better be treated as a single family (under the name Apiaceae), but relationships among some groups of genera are still not clearly understood. Thus, the traditional classification as two separate families is followed in the present work. For further discussion, see the introductory portion of the Apiaceae treatment.

A number of species in the family are cultivated as ornamentals and/or used medicinally. Stem pith of the Asian species, *Tetrapanax papyriferus* (Hook.) K. Koch, is the source of rice paper (Graham, 1966).

1. Plants lianas, less commonly bushy shrubs; leaves evergreen, simple, entire or more commonly 3-lobed .... 2. HEDERA
1. Plants perennial herbs or (in *Aralia spinosa*) shrubs or small trees; leaves deciduous, compound
    2. Leaves all basal from rhizomes or alternate along the aerial stems, 2 or more times compound, pinnately so or more commonly with 3 main divisions, these in turn 1 or 2 times pinnately compound; inflorescences consisting of 2 to numerous umbels arranged into compound umbels, racemes, or panicles .... 1. ARALIA
    2. Leaves in a single whorl of (1–)3(–5) at the tip of a short aerial stem, 1 time palmately compound with 3–5 leaflets; inflorescences consisting of a single umbel .... 3. PANAX

1. Aralia L. (spikenard)

Plants perennial herbs, shrubs, or trees, with long-creeping rhizomes. Leaves all basal from rhizomes or alternate along the aerial stems (sometimes appearing fascicled along short branches in *A. spinosa*), deciduous, 2 or more times compound, pinnately compound or with 3 main divisions, these in turn 1 or 2 times pinnately compound. Leaflets tapered or narrowed to a sharp point at the tip, the bases rounded or narrowed, sometimes cordate, sometimes slightly asymmetrical, the margins toothed. Inflorescences consisting of 2 to numerous umbels arranged into compound umbels, racemes, or panicles. Sepals 5 low, broadly triangular teeth. Petals oblong-elliptic. Styles (4)5(6), sometimes fused together toward the base. Fruits globose, somewhat (4)5(6)-lobed, with (4)5(6) stones. Thirty to 38 species, North America, Asia, Malesia.

1. Plants shrubs or trees, the stems spiny .... 3. A. SPINOSA
1. Plants perennial herbs (sometimes slightly woody at the base in *A. racemosa*), without spines
    2. Aerial stems absent; leaf usually solitary from the tip of the rhizome, the 3 primary divisions each pinnately divided into 3–5 leaflets; inflorescence solitary from the rhizome, a compound umbel with (2)3(–7) umbels
    .... 1. A. NUDICAULIS
    2. Aerial stems present; leaves 2 or more along each aerial stem, the 3 primary divisions each pinnately divided into 9–21 leaflets; inflorescences at the tip of the aerial stem and sometimes also axillary, panicles or less commonly racemes of umbels .... 2. A. RACEMOSA

1. Aralia nudicaulis L. (wild sarsaparilla)

Pl. 218 b; Map 907

Plants perennial herbs. Aerial stems absent. Leaves usually solitary from the tip of the rhizome, the petioles 15–35 cm long. Leaf blades 15–40 cm long, the 3 primary divisions each pinnately divided into 3–5 leaflets, the leaflets 3–15 cm long, narrowly elliptic to broadly ovate, the margins singly toothed, the upper surface green, the undersurface slightly lighter green, minutely hairy along the larger veins. Inflorescence solitary from the rhizome, a compound umbel with (2)3(–7) umbels, the main stalk 8–25 cm long, minutely hairy toward the tip. Petals 1–2 mm long, white or green. Fruits 6–10 mm long, purplish black. $2n=24$. May–June.

Uncommon in northeastern Missouri south locally to Jefferson County (northeastern U.S. west to South Dakota and Colorado, south locally to Georgia and Missouri; Canada). Ledges of moist, shaded bluffs, on both calcareous substrates and sandstone.

Aralia nudicaulis has been used medicinally and as a beverage as a substitute for true sarsaparilla (this name actually refers to several neotropical species of *Smilax*). The rootstocks were used by the Iroquois and other northeastern tribes as a remedy for various blood, kidney, stomach, and other ailments and was brewed into a tea or drink (Arnason et al., 1981). The fruits were fermented into a wine. As with *A. racemosa*, *A. nudicaulis* is an ingredient in root beer.

2. Aralia racemosa L. (American spikenard)

Pl. 218 a; Map 908

Plants perennial herbs, sometimes slightly woody at the base. Aerial stems 60–240 cm long, erect to spreading, often branched, without spines. Leaves 2 or more along each aerial stem, alternate. Leaf blades 30–80 cm long, the 3 primary divisions each pinnately divided into 9–21 leaflets, the leaflets 2–18 cm long, ovate, the margins doubly toothed, green, hairy along the veins when young, sometimes becoming glabrous or nearly so at maturity. Inflorescences at the tip of the aerial stem and sometimes also axillary, panicles or less commonly racemes of often numerous umbels, the branches hairy. Petals 0.8–1.0 mm long, white. Fruits 5–6 mm long, reddish purple to blackish purple. $2n=24$. June–August.

Scattered nearly throughout the state but absent from the Unglaciated Plains Division and the southwestern portion of the Glaciated Plains, also absent from most of the Mississippi Lowlands (eastern and southwestern U.S. and adjacent Mexico; Canada). Mesic upland forests, often toward the base of north-facing slopes, moist ledges of bluffs, often along streams.

The aromatic, spicy rootstocks of *A. racemosa* have been used medicinally for respiratory ailments, rheumatic fever, syphilis, and skin problems (apparently without any medical basis) and are an ingredient in root beer. Use by Native Americans was similar to that of *A. nudicaulis* (Arnason et al., 1981). The foliage and inflorescences of the species are attractive, but the plants sometimes are considered too large and bushy for most garden situations.

3. Aralia spinosa L. (Hercules' club, devil's walking stick, tear-blanket)

Pl. 218 c–f; Map 909

Plants shrubs or trees, often forming small colonies. Trunks to 12 m tall, erect or ascending, usually unbranched except for short, spurlike branches, with stout, straight, light brown spines, these often densest in patches or rings below the leaf scars, the bark with shallow furrows and longitudinal plates, dark brown, sometimes with light orangish brown areas along the plates. Leaf scars linear, U-shaped, with several bundle scars in a single row. Leaves alternate, sometimes appearing fascicled along short branches. Leaf blades 60–150 cm long,

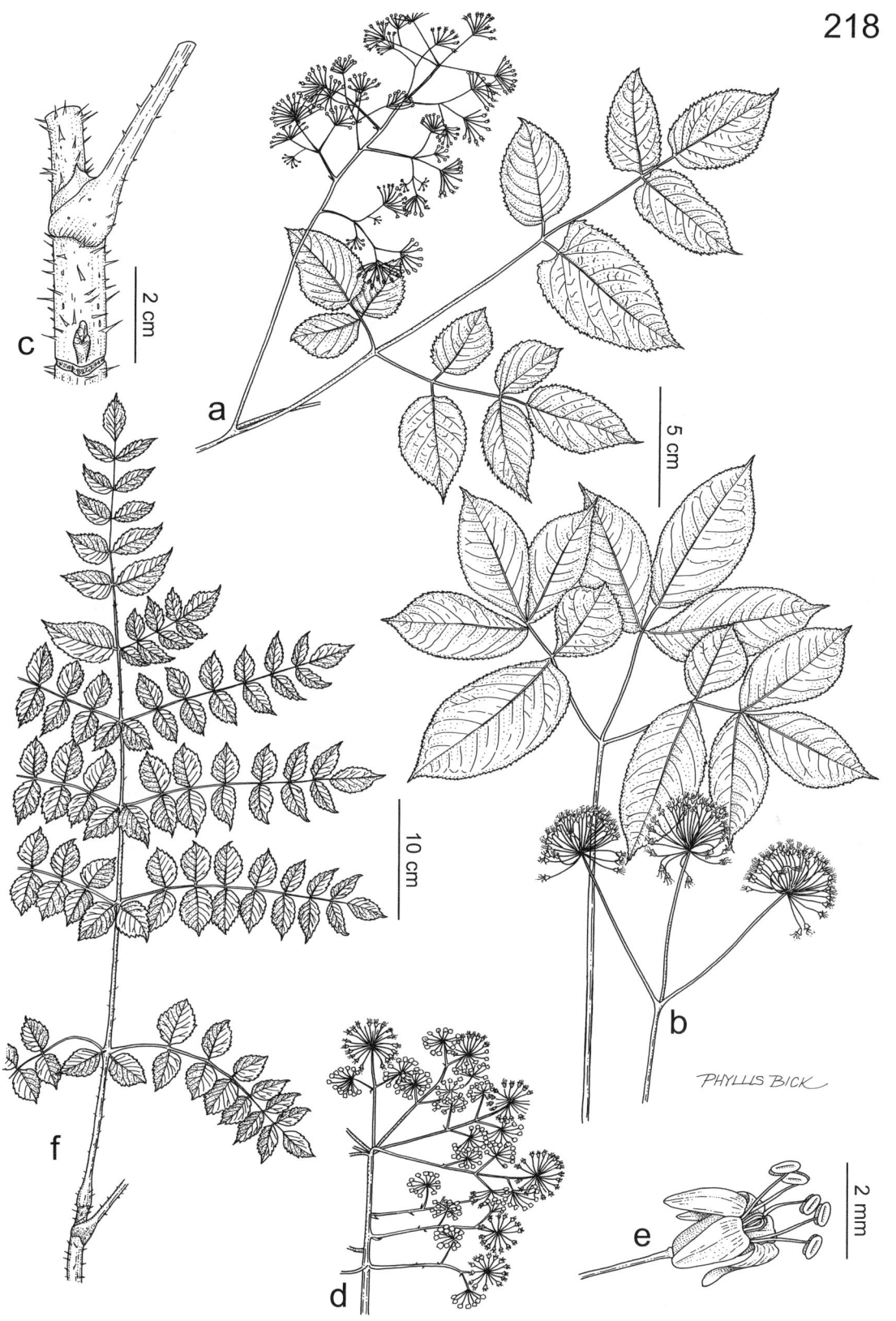

Plate 218. Araliaceae. *Aralia racemosa*, **a)** inflorecence and leaf. *Aralia nudicaulis*, **b)** inflorecence and leaf. *Aralia spinosa*, **c)** node of stem, **d)** portion of inflorescence, **e)** flower, **f)** leaf.

909. Aralia spinosa   910. Hedera helix   911. Panax quinquefolius

2 or 3 times pinnately compound, the ultimate branches with 9–13 leaflets, the leaflets 4–13 cm long, ovate, the margins simply toothed, the upper surface dark green, the undersurface lighter green, often with hairs or minute spines along the midvein. Inflorescences terminal, large, highly branched panicles with numerous umbels, the branches hairy, usually turning red at maturity. Petals 2–3 mm long, white. Fruits 4–6 mm long, black. $2n=24$. June–September.

Uncommon in southeastern Missouri north to St. Louis City and County, mostly in the Mississippi Lowlands Division (southeastern U.S. west to Missouri and Texas, introduced farther north). Mesic upland forests, mostly in ravines, ledges of bluffs, and less commonly bottomland forests in sandy soils.

Steyermark was doubtful that this species is native as far north as St. Louis, although specimens from the Allenton area date back to 1885. Harriman (1969) reported a population in a relatively remote portion of a remnant forest in St. Louis County that he considered to be a native occurrence. Plants presently growing in Forest Park in St. Louis also appear to be well integrated into a remnant mesic upland forest community. Thus, the species is accepted as native in this portion of the state.

The unusual spiny trunks, large compound leaves, and red panicles with white flowers and black fruits make *A. spinosa* an attractive ornamental plant in gardens where there is sufficient space to allow for the suckering of stems. The leaves turn yellow to purplish brown in the autumn. The black fruits with purplish pulp are eaten by birds and other wildlife and have sometimes been used to dye hair black. The relatively soft wood was once used in woodworking to make small items such as pen racks, button boxes, frames for photographs, and small furniture items (Steyermark, 1963). An infusion of the yellow inner bark purportedly was used for toothaches, but the bark and roots cause dermatitis in some individuals.

2. Hedera L. (ivy)

Ten to 15 species, Europe, Asia, Africa.

1. Hedera helix L. (English ivy)

Pl. 219 c; Map 910

Plants lianas, less commonly bushy shrubs. Stems to 20 m or more long, the climbing phase sparsely branched, becoming anchored by adventitious roots, the young stems densely stellate-pubescent. Leaves alternate in the climbing phase, sometimes opposite or whorled in the shrubby phase, evergreen, simple, the blade 2–12 cm long, ovate to nearly orbicular, entire or more commonly palmately 3- or 5-lobed, rounded to pointed at the tip, rounded to cordate at the base, the margins entire or slightly wavy, leathery, dark green on the upper surface, whitened along the veins, yellowish green on the undersurface, stellate-pubescent when young, becoming glabrous or nearly so at maturity. Inflorescences thus far not produced in Missouri material, solitary from the branch tips or in a short raceme, consisting of 1 or more simple umbels each with a stalk 2–8 cm long. Sepals 5 minute teeth or a low, 5-toothed crown. Petals 2.5–3.5 mm long, elliptic-ovate, greenish yellow. Style 1. Fruits 7–10 mm long, globose or somewhat obovoid, black, with 3–5 stones. 2n=48. July–October.

Introduced, uncommon and sporadic (native of Europe; widely cultivated, escaping sporadically in the eastern U.S.). Mesic upland forests; also old homesites and railroads.

English ivy is commonly cultivated as an evergreen ground cover or trellis plant, sometimes also to hide fences and old walls. The heavy growth and the adventitious rootlets by which the plant becomes fastened as it climbs can damage trees, wooden structures, and mortared walls. Plants persist for many years after abandonment and can also occasionally spread from stem pieces that become rooted. The creeping/climbing phase does not produce flowers and generally has lobed leaves, but older plants may sometimes produce stiff, branched stems forming bushy shrubs, which become fertile at the branch tips and have unlobed leaves (Robbins, 1957). This shrubby phase is not known outside cultivation in Missouri. Steyermark (1963) noted that all parts of the plant are poisonous to humans if eaten and can cause dermatitis when handled in some people.

A number of infraspecific taxa have been described for *H. helix;* the few specimens from outside cultivation in Missouri appear to represent the diploid var. *helix* (2n=48), which is characterized by having whitened veins on the upper leaf surface. However, the tetraploid var. *hibernica* G. Kirchn. (2n=96), which is known as Irish ivy or Atlantic ivy and differs most notably in its green veins (Lawrence, 1956; McAllister and Rutherford, 1990), also is cultivated widely and may eventually be recorded as an escape in Missouri. Numerous cultivars are sold commercially as well.

3. Panax L. (ginseng)

Six to 8 species, North America, Asia.

Although much attention is given to *P. quinquefolius* by virtue of its commercial value, it should be noted that a second native species, *P. trifolius* L. (dwarf ginseng), also grows in the northeastern United States (but not as far southwest as Missouri) and is used medicinally to a lesser extent.

1. Panax quinquefolius L. (American ginseng, ginseng)

Pl. 219 d, e; Map 911

Plants perennial herbs with short rhizomes at the tip of the elongate, sometimes branched, fleshy main roots. Aerial stem solitary from the tip of the rhizome, 10–50 cm long, erect or nearly so. Leaves deciduous, in a single whorl of (1–)3(–5) at the tip of a short aerial stem, 1 time palmately compound with 3–5 leaflets. Leaflets 6–15 cm long, oblong-obovate to obovate, tapered to a point at the tip, narrowed to a short stalk at the sometimes slightly asymmetrical base, the margins sharply toothed, glabrous or sparsely hairy along the veins. Inflorescence solitary from the aerial stem tip, a simple umbel from the tip of a stalk 1–12 cm long. Sepals absent or 5 minute teeth. Petals 0.5–1.0 mm long, oblong-elliptic, white to greenish white. Styles 2(3). Fruits 9–10 mm long, somewhat flattened and usually slightly 2-lobed, bright red, shiny, with 2(3) stones. 2n=44. June–July.

Scattered in the Ozark and Ozark Border Divisions, scattered to uncommon in the rest of the state, but apparently absent from most of the Mississippi Lowlands and Unglaciated Plains, also apparently absent from portions of the Glaciated Plains (northeastern U.S. west to south Dakota, south to Georgia; Canada). Mesic upland forests, often in ravines, and ledges of shaded bluffs and rock outcrops.

Evidence that voucher specimens have not been collected in every county in which the species exists comes from annual reports by the Missouri Department of Conservation on the harvest of ginseng per county. Readers also should be aware that the species' apparent range has been altered in some areas by woodland cultivation, the planting by individuals of seeds in natural settings for subsequent commercial harvest of rootstocks.

Roots (and to a lesser extent foliage) of *P. quinquefolius* originally were mostly ground and brewed into a tea but have more recently been incorporated into a variety of products, including tablets and capsules, liquid tonics, soaps and shampoos, lotions and cremes, and even chewing gum. In Chinese herbal medicine and more recently in medicinal practice elsewhere in the world, ginseng and its extracts have been used as a general tonic and stimulant, as a means of lowering blood sugar and cholesterol, for stress relief, and as a sexual stimulant (among a multitude of purported virtues). Although many American medical and pharmaceutical researches remain skeptical of the plant's efficacy, in Europe and Asia it is widely accepted to have therapeutic and stimulant properties.

Carlson (1986) chronicled the fascinating history of the commercialization of *Panax* in the New World, beginning with the discovery in the early 1700s by French missionaries among the Iroquois and other Native American tribes that there existed a native North American counterpart to the Asian species, which had been used medicinally for centuries in China and surrounding countries, and which had by then already been overharvested to near extinction in Asia. Carlson documented that by the 1820s, nearly 3.8 million pounds of dried roots were being exported during the decade, mostly to Asia, with a peak of nearly 6.8 million pounds during the 1880s. Today, American ginseng continues to be the most valuable botanical product wild-harvested commercially in the United States. Of note, one of the nation's largest wholesalers and exporters of ginseng has been based in Eolia (Pike County). Oddly, most American ginseng continues to be exported, whereas most products sold in the United States contain mainly Korean or other Asian species. At present, in Missouri, about 3,000–4,000 pounds of dried roots are harvested each year (mostly from wild-collected plants), which accounts for less than 5 percent of U.S. production (Tim Smith, Missouri Department of Conservation, personal communication). Kentucky (26,000–32,000) and Tennessee (15,000–19,000) lead the nation in pounds harvested annually. Most of the state's harvest is done in the Ozark Division and counties along the Missouri and Mississippi Rivers by root diggers, individuals who supplement their incomes by harvesting a variety of natural products from local areas and selling these to distributors. Less than 10 percent of the state's commercial ginseng production comes from cultivated plants, either grown in beds under netting or other artificial shade or wild-cultivated by planting seeds at natural sites in plant communities where the species can flourish (Lewis, 1989).

The sustained long-term harvest of ginseng for more than 200 years has led to the extirpation or decline of the species in many portions of its range (Yatskievych and Spellenberg, 1993). In some states as well as in Canada, commercial harvest from the wild is presently illegal. Several states and Canada have also designated the species as threatened, endangered, or similar categories, and at one point it was under consideration for listing by the U.S. Fish and Wildlife Service under the Endangered Species Act. In recognition of these conservation concerns, international trade in ginseng is regulated under Appendix II of the Convention on International Trade in Endangered Species of Wild Fauna and Flora (CITES), which mandates that the states monitor harvest and sales within their boundaries. The Missouri Department of Conservation receives and compiles annual reports from all dealers, including detailed information on the quantities and sources of roots introduced into commerce in the state. Collecting is further regulated under the Missouri Wildlife Code, which prohibits wild harvest except during an official season (usually September 1–December 31), coinciding with the months that mature fruits are present on the plants. Diggers are encouraged to squeeze the stones from fruits present into the hole left after the rootstock is excavated, in order to make the natural resource more renewable. Seedlings do not appear until the second spring following planting, and plants apparently do not begin to produce flowers and fruits until they are four years old (Anderson et al., 1993). It is of interest that Lewis (1988) was able to document the gradual recovery of a Missouri population from a seed bank in the soil several years after its decimation by diggers and in the absence of subsequent disruptions.

A number of authors (Gleason and Cronquist, 1991) have continued to treat *Panax* as neuter (as was done by Linnaeus in describing the genus) and therefore to spell the specific epithet *quinquefolium*. However, as noted by Graham (1966), the International Code of Botanical Nomenclature (for most recent edition, see Greuter et al., 1994: Article 62.1a) specifically states that *Panax* and other generic names ending in *-panax* are to be treated as masculine epithets, and the species should thus be spelled *P. quinquefolius*.

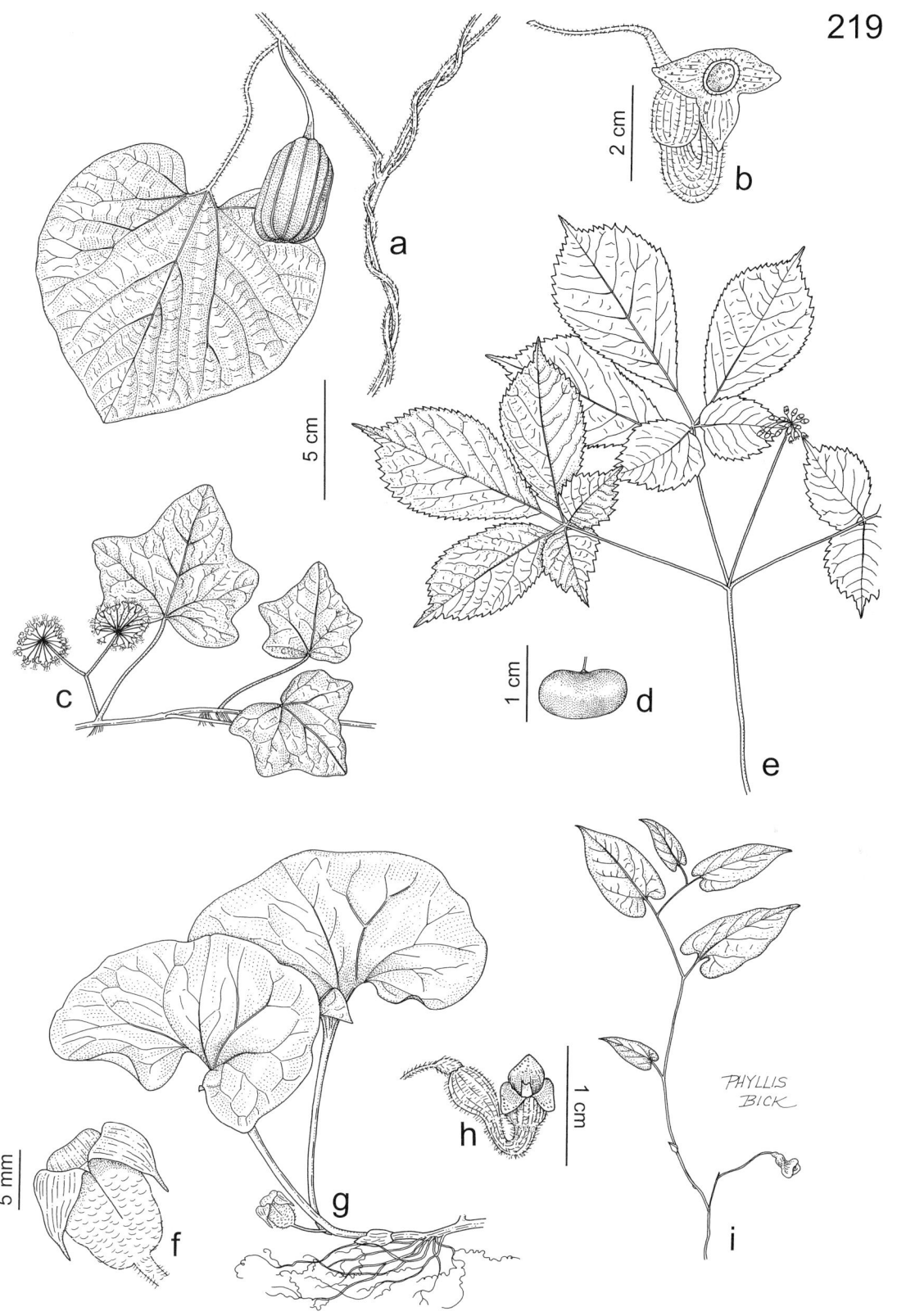

Plate 219. Araliaceae. Aristolochiaceae. *Aristolochia tomentosa*, **a)** fruiting branch, **b)** flower. *Hedera helix*, **c)** flowering branch. *Panax quinquefolius*, **d)** fruit, **e)** habit. *Asarum canadense*, **f)** flower, **g)** habit. *Aristolochia serpentaria*, **h)** flower, **i)** habit.

ARISTOLOCHIACEAE (Birthwort Family)

Plants perennial herbs or lianas (shrubs elsewhere). Leaves alternate, simple, with petioles. Flowers perfect, epigynous. Calyx usually petal-like (not green), often united into a straight or contorted tube toward the base, expanded into 3 shallow or deep lobes at the tip. Corolla absent (reduced to small scales elsewhere). Stamens (5)6 or 12. Pistils 1 per flower. Ovary inferior, 3–6-locular, the ovules numerous. Style 1, the stigmas 3–6. Fruits capsules with numerous seeds. Five genera, about 600 species, nearly worldwide but most diverse in tropical and subtropical regions.

1. Plants with well-developed herbaceous or woody aerial stems; flowers zygomorphic, the calyx tube well developed, hooked or S-shaped .... 1. ARISTOLOCHIA
1. Plants without evident aerial stems, the leaves and flowers arising from widely creeping rhizomes; flowers actinomorphic, the calyx tube absent or very short and straight (note that the overlapping bases of the lobes often appear like a short, straight tube) .... 2. ASARUM

1. Aristolochia L. (birthwort)
(Pfeifer, 1966)

Plants perennial herbs or lianas (shrubs elsewhere). Aerial stems well developed, sometimes twining. Leaves often appearing 2-ranked. Flowers zygomorphic, the calyx tube well developed, hooked or S-shaped, with an expanded ring of tissue at the juncture with the lobes. Stamens 6, the filaments fused with the style into a column, the anthers without a sterile extension between the pollen sacs. Ovary with 3 locules. Stigmas appearing as a low, irregular, 3-lobed crown at the tip of the stylar column. Fruits dehiscing longitudinally into 3 or 6 valves. About 300 species, nearly worldwide but most diverse in tropical and subtropical regions.

The sap of *Aristolochia* species is yellowish and tastes bitter. Various species have been investigated for the possible production of antitumor alkaloids and other compounds. These same compounds, including aristolochic acid and trimethylamine, have also been implicated as having carcinogenic properties. Bruised plant parts emit a faint, unpleasant odor similar to that of turpentine, whereas the flowers of some species sometimes emit an aroma similar to the scent of rotten meat. Pollination in most species is a complex affair involving small flies that become trapped overnight in the expanded basal chamber of the calyx tubes until stiff, downwardly pointing hairs in the narrower portion of the calyx tube relax, releasing the insect, which has been coated with pollen and departs to pollinate a second flower in similar fashion. The leaves of Missouri species are a principal larval food source for the eastern pipevine swallowtail butterfly (*Battus philenor philenor* (L.)).

1. Plants herbaceous perennials with erect or ascending stems ..... 1. A. SERPENTARIA
1. Plants lianas with climbing, twining stems .... 2. A. TOMENTOSA

1. Aristolochia serpentaria L. (Virginia snakeroot)
A. serpentaria var. *hastata* (Nutt.) Duch.
Pl. 219 h, i; Map 912

Plants herbaceous perennials with rhizomes. Aerial stems 15–60 cm long, erect or ascending, sometimes appearing slightly zigzag, glabrous or hairy. Petioles 0.5–3.5 cm long. Leaf blades 5–14 cm long, lanceolate to oblong-ovate or narrowly triangular, tapered to a sharply pointed tip, variously cordate, arrowhead-shaped, or truncate at the base, the main veins pinnate above a somewhat palmate base, the undersurface sometimes hairy. Flowers solitary at the tips of short, scaly branches produced

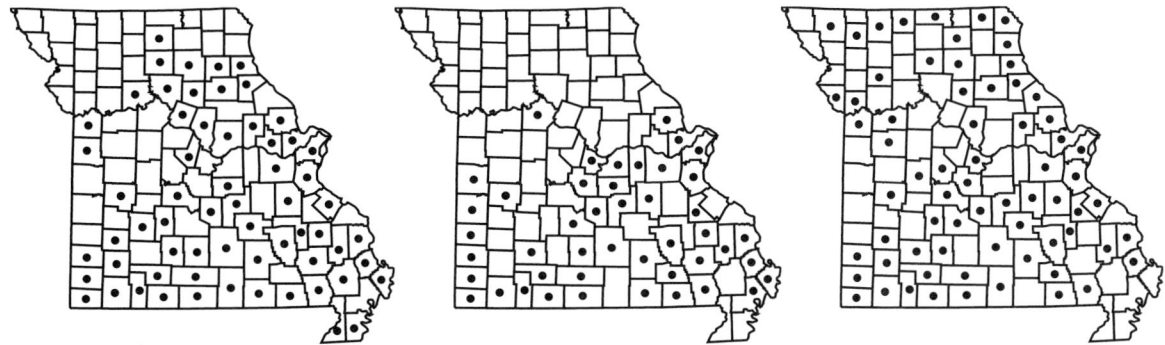

912. Aristolochia serpentaria   913. Aristolochia tomentosa   914. Asarum canadense

near the base of the aerial stem. Calyx 1–3 cm long, hairy on the outer surface, the tube hooked or S-shaped, expanded at both ends, purple to brown above a white to tan base, the lobes ascending to spreading, unequal, shallow, broadly triangular, purple and glabrous on the inner (upper) surface. Fruits 0.8–1.8 cm long, globose or nearly so, 6-ribbed. Seeds 4–5 mm long, concave and with a longitudinal ridge on one side, rounded on the other, ovate in outline, brown, with a lighter pattern of wrinkled ridges and finely pebbled bumps. $2n=28$. May–July.

Scattered to common in southern and central Missouri; rare to absent in the northern third of the state (eastern U.S. west to Iowa, Kansas, and Texas). Bottomland and mesic upland forests, rarely dry upland forests.

The rootstocks of *A. serpentaria* are harvested for the medicinal trade under the names Virginia snakeroot and serpentary. Native Americans, pioneers, and herbalists have made a tonic from the dried rhizomes as treatment for a variety of problems, including general pain, toothache, fevers, colds, rheumatism, worms, snakebite, and as a diuretic. Readers should note the statement above on possible carcinogenic compounds and also that overuse can lead to gastric distress.

Occasional plants may be encountered with somewhat smaller flowers that apparently do not open. Such flowers have been suggested to reproduce cleistogamously. Plants with narrower leaves have been encountered rarely in Dent and Dunklin Counties and have been referred to var. *hastata*, a trivial variant that occurs sporadically nearly throughout the range of the species.

The closely related *A. reticulata* occurs from Texas and Louisiana to Oklahoma and Arkansas and eventually may be found growing in southwestern Missouri. It differs from *A. serpentaria* in its short-petiolate leaves with the bases appearing to clasp the stems and the tips blunt to rounded, and in its slightly smaller seeds (3–4 mm long).

2. Aristolochia tomentosa Sims (woolly pipevine, Dutchman's pipe)

Pl. 219 a, b; Map 913

Plants lianas. Aerial stems to 25 m long, twining and climbing on other vegetation, the younger ones ridged and hairy, the older ones often appearing irregularly flattened, the bark ridged, gray to more commonly brown or blackish brown. Petioles 1–5 cm long. Leaf blades 4–20 cm long, ovate to nearly circular, rounded to pointed at the tip, cordate or less commonly truncate at the base, the main veins palmate, the undersurface hairy. Flowers solitary or less commonly paired, appearing axillary near the tips of young branches or at the nodes opposite the leaves. Calyx 2.5–6.0 cm long, densely hairy on the outer surface, the tube strongly hooked or S-shaped, expanded at both ends, pale yellowish green, purple to maroon on the inner surface at the mouth, the lobes spreading, triangular to nearly oblong, bright greenish yellow and glabrous on the inner (upper) surface. Fruits 5–8 cm long, barrel-shaped or somewhat obovoid, strongly 6-ribbed. Seeds 8–10 mm long, strongly flattened, triangular in outline, brown, smooth. $2n=28$. May–June.

Relatively common, mostly south of the Missouri River (southeastern U.S. west to Kansas and Texas). Bottomland and mesic upland forests, usually associated with banks of streams and rivers, less commonly on open gravel bars, rarely margins of sand prairies; also open, disturbed, floodplain areas.

Aristolochia tomentosa is a characteristic woody climber along streams and rivers in the Ozarks, where hikers and canoeists often overlook the plants because the leaves and flowers can be located high in the adjacent forest canopy.

2. Asarum L.
(Kelly, 2001)

About 10 species, North America, Europe, Asia.

1. Asarum canadense L. (wild ginger)
  A. canadense var. *acuminatum* Ashe
  A. canadense var. *ambiguum* (E.P. Bicknell) Farw.
  A. canadense var. *reflexum* (E.P. Bicknell) B.L. Rob.

Pl. 219 f, g; Map 914

Plants perennial herbs, with widely creeping, often branched rhizomes. Aerial stems not evident. Leaves paired at the tips of rhizome branches. Petiole 8–20 cm long, hairy. Leaf blade 4–15 cm long, broadly kidney-shaped to nearly circular in outline, rounded at the tip, deeply cordate at the base, hairy, the upper surface sometimes sparsely so. Flowers solitary at the rhizome tips, prostrate or somewhat ascending, with a stalk 1–5 cm long, appearing attached between the leaves, actinomorphic. Calyx deeply 3-lobed, sparsely to densely hairy (especially on the outer surface), the tube absent or very short and straight, the lobes 6–24 mm long, narrowly ovate-triangular, erect and slightly overlapping toward the base (appearing tubular), the tips spreading to recurved, narrowed or long-tapered to a point, purplish brown to maroon, the inner surface usually with a well-defined white to light green basal region that is sometimes spotted or mottled with purple, the outer surface often tan toward the base. Stamens 12, the filaments distinct but appressed to the style, the anthers with a minute, sharply pointed, sterile extension between the pollen sacs. Ovary with 6 locules. Stigmas 6, globose or nearly so. Fruits globose to barrel-shaped, somewhat fleshy, dehiscing irregularly, the calyx persistent. Seeds 3.5–4.5 mm long, concave and with a longitudinal ridge on one side, rounded on the other, ovate in outline, olive green to greenish brown, the surface wrinkled, the ridge with a fleshy caruncle toward the end. $2n=26$. April–May.

Common nearly throughout Missouri (eastern U.S. west to North Dakota and Oklahoma; Canada). Bottomland forests, mesic upland forests, banks and terraces of streams and rivers, and bases of bluffs.

Earlier botanists divided this species into several varieties based on differences in calyx lobe length and shape, but as noted by Steyermark (1963) and Kelly (2001) there is too much variation in these characters over too continuous a range to permit formal recognition of discrete entities.

The pungent rootstock is sometimes used as a seasoning or cooked with sugar as a substitute for ginger. Medicinally, the plant has been used as a general stimulant and was once thought to cure a variety of maladies ranging from digestive disorders to colds, asthma, tuberculosis, and venereal disease. Wild ginger is also gaining in popularity among gardeners as a shade-tolerant ground cover. Steyermark (1963) noted that handling the plants has been reported to cause dermatitis in some people.

ASCLEPIADACEAE (Milkweed Family)

Plants perennial herbs (woody elsewhere), lacking tendrils, usually with white latex, the sap thus appearing milky (except in *Cynanchum* and *Asclepias tuberosa*). Stems sometimes twining. Leaves opposite or less commonly alternate or whorled, simple. Stipules absent or minute, linear, and shed during leaf development. Leaf blades variously shaped, the margins entire or somewhat undulate, sometimes curled under. Inflorescences umbels, these solitary or in clusters of 2–4, axillary and/or terminal. Flowers perfect, hypogynous, actinomorphic, without subtending bracts. Calyces 5-lobed nearly to the base, often with minute glandular or scalelike projections near the tip of the tube between the lobes, often persistent at fruiting, the lobes spreading or reflexed. Corollas usually deeply 5-lobed, spreading, reflexed, or less commonly erect, spirally twisted in bud. Stamens 5, the filaments fused to the corolla tube and united into a columnar sheath around the carpels, the anthers fused to and forming a headlike structure with the stigmatic complex, with a corona of 5 variously shaped petaloid outgrowths covering the anther sacs (and often also the rest of the stamens and stigmas). Staminodes absent. Pollen grains of each of the 2 anther locules fused into a saclike mass (pollinium), the rightmost pollinium of each anther united with the lefthand pollinium of the adjacent anther via a short connecting arm (translator). Pistils 2 per flower, each of 1 carpel, these free at the ovary and style, but the stigmas (and anthers) fused into a relatively massive, 5-lobed or angled, headlike structure. Each ovary more or less superior, with 1 locule, the placentation parietal. Ovules numerous. Fruits follicles, these sometimes paired, variously shaped. Seeds variously shaped, usually flattened and winged, usually with a tuft of long, silky hairs at the tip. Fifty to 250 genera, 2,000–3,000 species, worldwide.

The flowers of the Asclepiadaceae are complex and have developed specialized structures to promote outcrossing by various insect pollinators. A specialized terminology has developed to account for these unusual floral structures. The complex of carpels, stamens, and coronas is collectively termed the *gynostegium* (Pl. 220 c). The pollinia from adjacent anthers are connected by an elaborate, acellular, wishbone-shaped complex consisting of 2 threadlike arms, called *translators,* and a central, longitudinally grooved, disclike structure called the *corpusculum*. The complex of 2 pollinia along with the translator apparatus is technically known as a *pollinarium* (Pl. 220 d), but in practice many botanists continue to refer merely to *pollinia* (*pollinium,* singular) in discussing the pollen transfer of asclepiad flowers. The corona may develop from the base of the staminal tube or from the tips of the anthers.

The space between adjacent corona segments, as well as that between the small, outwardly pointed, winglike flaps of adjacent anthers, forms a slot that guides the legs (and sometimes other body parts) of pollinators to become entangled in the translator complex of the pollinarium (with its pair of pollinia). Pollinators often struggle to remove their trapped legs from the flowers. Occasionally, various Asclepiadaceae are encountered with dead insects trapped in the flowers, having been unable to extract body parts from the slots, and moths have even been noted dangling by their proboscises. To effect pollination, the hapless pollinator that wrestles free of a flower must then visit a second flower and have its leg guided into a slot on the next gynostegium, where, if the pollinarium has previously been removed, a sticky stigmatic region in the chamber below the slot traps pollen from the pollinia, leading to pollination. With such a specialized pollination mechanism, it is easy to understand why most asclepiads produce few fruits relative to the large number of flowers.

The families Asclepiadaceae and Apocynaceae are treated here in the traditional sense, as they have been in most floristic works (Steyermark, 1963; Hartman, 1986a, b; Gleason and Cronquist, 1991). However, botanists have long accepted a close relationship between these groups. In recent decades, a number of authors (summarized in Rosatti, 1989; Judd et al., 1994, 2002) have argued that the unique floral characters of the Asclepiadaceae represent a syndrome of specializations within the Apocynaceae and, as such, the Asclepiadaceae should more accurately be classified as a subfamily or tribe of the latter family. Rosatti (1989) noted, however, that even those botanists who combine the two families continue to recognize the milkweeds as distinct at some level and that, especially in temperate floras, there is utility in continuing to recognize two separate families.

The generic classification adopted here follows that of Woodson (1941), who accepted relatively few, broadly circumscribed genera of North American Asclepiadaceae. Although American botanists during the last few decades have, with minor quibbles, almost universally embraced Woodson's concepts, those in other parts of the world have tended to treat the family as it occurs in various regions as consisting of many more, much more narrowly circumscribed genera. The ultimate disposition of most of these segregates awaits future comprehensive investigations involving studies of these complexes on a worldwide basis.

1. Stems erect to spreading, not twining or climbing; calyces and corollas reflexed (except in *A. viridis*) .... 1. ASCLEPIAS
1. Stems twining, usually climbing; calyces and corollas spreading to ascending
    2. Corollas erect or ascending at flowering, the lobes 1.5–6.0 mm long; sap watery (plants with clear latex) .... 2. CYNANCHUM
    2. Corollas spreading or loosely ascending at flowering (erect in bud and sometimes after flowering), the lobes 7–15 mm long; sap milky (plants with milky latex)
        3. Fruits sharply 5-angled, the surface otherwise smooth; calyces glabrous except for sparse hairs at the tips of the lobes; corollas strongly spreading at flowering, yellow to yellowish green, sometimes tinged with purple or brown, glabrous .... 3. GONOLOBUS
        3. Fruits not noticeably angled, covered with slender, warty projections; calyces hairy; corollas more or less spreading to loosely ascending at flowering, white to light cream-colored or dark purple to brownish purple, rarely greenish yellow, hairy on the outer surface .... 4. MATELEA

1. Asclepias L. (milkweed)
(Woodson, 1954)

Plants with white latex and thus milky sap (except in *A. tuberosa*). Stems erect to spreading, not twining or climbing. Leaves alternate, opposite, or whorled. Calyces with the lobes reflexed at maturity (spreading in *A. viridis*). Corollas reflexed at maturity (spreading to loosely ascending in *A. viridis*), white, green, pink, purple, or orange. Gynostegium sessile or appearing short-stalked, the corona modified into 5 lateral segments (hoods, Pl. 220 c), these petaloid or fleshy, tubular to obconical, erect or ascending, often appearing curved or arched, sometimes with an erect or incurved, hairlike or linear appendage (horn, Pl. 220 c) extended from the open tip. Fruits single or in pairs, mostly erect or ascending (sometimes from a deflexed stalk), narrowly elliptic-lanceolate to ovate in outline, circular or slightly flattened in cross-section (not angled), the surface smooth or with warty tubercles. Seeds ovate to broadly ovate in outline, strongly flattened and usually winged, brown to dark brown, with a tuft of long silky hairs at the tip (except in *A. perennis*). About 150 species, mainly North America and Central America; also South America, Caribbean Islands, and Africa.

The genus *Asclepias* is distinct from other North American genera in its nontwining habit and unusual corona morphology. As with other groups of Asclepiadaceae, *Asclepias* in the broad sense apparently consists of several independent lineages whose interrelationships remain poorly understood. Most of the African species have been split into smaller genera, and segregate generic names are available for some North American species groups. The generic concept retained here is that of Woodson (1954), who combined the North American species into a single, broadly circumscribed genus divided into two subgenera and ten total series.

The latex of most species of milkweeds contains a mixture of chemicals, principally cardiac glycosides, that render the plants both unpalatable and poisonous to livestock and humans (although young leaves, flowers, and immature fruits of some species are eaten by wildlife and have been eaten by humans after boiling to leach toxic constituents). Some insects, notably the monarch butterfly (*Danaus plexippus* (L.), also called milkweed butterfly, whose caterpillars eat only Asclepiadaceae, principally species of *Asclepias*), use the plant as a larval food source and sequester the toxic compounds, rendering the larvae and adults unpalatable to predators like birds. Various species of *Asclepias* also have a long history of economic uses (Rosatti, 1989; Moerman, 1998). Native Americans used plants medicinally as an analgesic, cold remedy, respiratory aid, and emetic, among other uses. The latex was allowed to dry fresh or was boiled first for use as a chewing gum. The stem fibers have been used for cordage and weaving, although they are relatively brittle and subject to breakage. The silky hairs on the seeds have been used to stuff pillows and cushions and were harvested commercially during World War II as a substitute for kapok in life jackets. Additionally, a number of species have been investigated as potential sources of hydrocarbons and rubber. Several species also are cultivated as garden ornamentals.

The hoods of *Asclepias* flowers produce nectar, and several species have fragrant flowers. A number of different insects have been documented to visit flowers of various species. Pollination is accomplished mostly by species of wasps, bees, moths, and butterflies, but beetles and flies are important for some species (Woodson, 1954).

In addition to the species treated below, *A. exaltata* L. (poke milkweed) should be searched for in bottomland and mesic upland forests in southeastern Missouri. This northeastern species was mapped erroneously as present in this portion of Missouri by Broyles and Wyatt (1993), but it does occur in adjacent Illinois. Plants would key imperfectly to either *A. amplexicaulis* or *A. incarnata* in the key below, depending on which characters were emphasized. *Asclepias exaltata* tends to be a taller plant than *A. amplexicaulis* and has branched stems and sharply pointed leaf tips, but like *A. amplexicaulis* it has large flowers with more or less tubular hoods. Although similar to *A. incarnata* in general aspect, *A. exaltata* differs in its broader leaves that are glaucous beneath and more gradually tapered to the petiole, as well as its larger flowers with more tubular (vs. more conical) hoods.

1. Many or all of the leaves alternate
    2. Corollas bright orangish yellow to reddish orange; sap clear . . . . . 13. A. TUBEROSA
    2. Corollas pale green to green or light yellow, sometimes purple-tinged; sap milky
        3. Calyx lobes spreading; corolla lobes 10–17 mm long, ascending to spreading . . . . 17. A. VIRIDIS
        3. Calyx lobes reflexed; corolla lobes 4–8 mm long, reflexed
            4. Corolla lobes 6–8 mm long; hoods 4.0–5.5 mm long . . . . 16. A. VIRIDIFLORA
            4. Corolla lobes 4–6 mm long; hoods 1.5–4.0 mm long
                5. Gynostegium appearing stalked (visible between the corona and corolla); corona hoods without horns, the tip entire . . . . 2. A. HIRTELLA
                5. Gynostegium appearing sessile (the corona base touching the corolla or nearly so); corona with a minute horn fused to each hood, the hood thus appearing 3-toothed or shallowly 3-lobed or at the tip . . . . 9. A. STENOPHYLLA
1. Leaves opposite or whorled (rarely the uppermost nodes with only 1 leaf)
    6. Leaves all or nearly all in whorls of 3–6, the blades 0.5–3.0(–4.0) mm wide, linear
        7. Reduced branchlets bearing pairs or clusters of short leaves in the axils of many of the main leaves . . . . 10. A. SUBVERTICILLATA
        7. Axillary branchlets with pairs or clusters of leaves absent or rare
            . . . . 15. A. VERTICILLATA
    6. Leaves all opposite, or, if some of the leaves are in whorls of 3 or 4, then the blades 7–80 mm wide, not linear
        8. Corona hoods without horns . . . . 16. A. VIRIDIFLORA
        8. Corona hoods each with a hairlike or linear horn
            9. Corona hoods shorter than to about as long as the tip of the anther/stigma head, the horns extended conspicuously beyond the tips of the hoods

10. Leaves mostly 2–5 pairs per stem, sessile or nearly so, the blade oblong-ovate, the base truncate to shallowly cordate and frequently overlapping that of the opposite leaf, the tip rounded, sometimes also with an abrupt, short, sharp point; corolla lobes 8–11 mm long . . . . 1. A. AMPLEXICAULIS
10. Leaves 6 to numerous pairs per stem, mostly petiolate, the blade narrowly lanceolate or elliptic to narrowly oblong or less commonly ovate, the base rounded to tapered, occasionally shallowly cordate but not overlapping that of the opposite leaf, the tip narrowed or tapered to a sharp point; corolla lobes 3–6 mm long
    11. Leaf blades abruptly narrowed or rounded at the base, occasionally shallowly cordate; seeds with a tuft of long silky hairs at the tip; corolla lobes 4–6 mm long . . . . . . 3. A. INCARNATA
    11. Leaf blades gradually narrowed or tapered at the base; seeds without a tuft of hairs at the tip; corolla lobes 2.5–4.0 mm long . . . . 5. A. PERENNIS
9. Corona hoods conspicuously longer than the tip of the anther/stigma head, the horns arched over the head but not extended past the tips of the hoods
    12. Leaves moderately to densely hairy on the undersurface
        13. Hoods 9–15 mm long, abruptly narrowed below the middle, the apical half narrowly lanceolate to lanceolate, tapered to the tip . . . . . . 8. A. SPECIOSA
        13. Hoods 3.5–8.0 mm long, gradually narrowed from at or above the middle, the apical portion oblong to ovate
            14. Hoods not noticeably lobed along the margins; corollas reddish purple to dark purple; fruits without warty tubercles on the surface . . . . 6. A. PURPURASCENS
            14. Hoods with a pair of sharply triangular, ascending and incurved lobes at about the middle of the margins; corollas green to lavender, usually tinged with pink and/or white; fruits with narrow, warty tubercles on the surface . . . . 12. A. SYRIACA
    12. Leaves glabrous on the undersurface (except sometimes along the margins) or sparsely hairy, mostly along the midvein
        15. Leaves sessile or nearly so, the blades rounded to shallowly cordate at the base
            16. Leaves mostly in 2–5 pairs per stem, the blades with minute, rough hairs along the margins; corollas white to pale cream-colored or pale green . . . . 4. A. MEADII
            16. Leaves in 7–15 pairs per stem, the blades glabrous along the margins; corollas light pink to purplish-pink, sometimes tinged with green . . . . 11. A. SULLIVANTII
        15. Leaves mostly with noticeable petioles, the blades gradually narrowed or tapered at the base
            17. Usually 1 or more nodes with whorls of 4 leaves, the leaf blades lanceolate to narrowly ovate-elliptic, the tip gradually narrowed or tapered to a sharply point; corolla lobes 4.5–6.0 mm long . . . . 7. A. QUADRIFOLIA
            17. Leaves usually opposite, the blades ovate to broadly oblong-elliptic, the tip rounded or narrowed to a blunt point or rounded but often with an abrupt, short, sharp point; corolla lobes 7–9 mm long . . . . 14. A. VARIEGATA

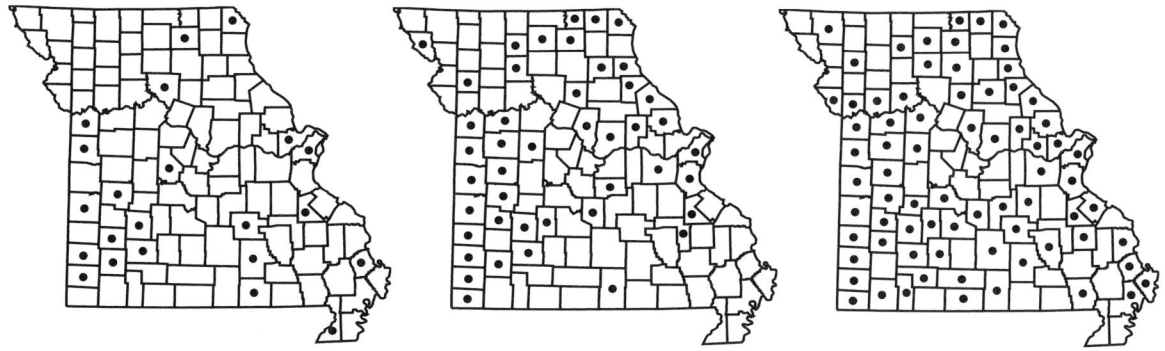

915. Asclepias amplexicaulis   916. Asclepias hirtella   917. Asclepias incarnata

1. Asclepias amplexicaulis Sm. (sand milkweed, bluntleaf milkweed)

Pl. 221 a, b; Map 915

Plants with white latex and deeply seated rhizomes. Stems 20–80 cm long, usually unbranched, mostly erect or ascending, glabrous, glaucous, mostly with 2–5 nodes. Leaves opposite, sessile or nearly so. Leaf blades 4–15 cm long, 2–8 cm wide, oblong-ovate, the base truncate to shallowly cordate and frequently overlapping that of the opposite leaf, the tip rounded, sometimes with an abrupt, minute point, the margins flat or somewhat wavy, usually minutely hairy, the surfaces glabrous, glaucous. Inflorescence 1(2), terminal, with a stout, usually long stalk, with 18–40 flowers. Calyces reflexed, glabrous, the lobes 3.0–5.5 mm long, lanceolate to narrowly ovate. Corollas reflexed, glabrous, green, often pink- or purple-tinged, the lobes 8–11 mm long, lanceolate to elliptic-lanceolate. Gynostegium appearing stalked (the column visible below the bases of the hoods), pink to pale orangish pink, the corona about as long as or slightly longer than the tip of the anther/stigma head. Corona hoods 4.5–5.5 mm long, erect or strongly ascending, attached near their bases, oblong in outline, the tips broadly rounded to truncate, the margins sometimes indistinctly toothed near the tip, the bases slightly pouched. Horns attached near the hood bases, extended conspicuously beyond the tips of the hoods and incurved over the anther/stigma head, linear, not flattened, tapered to a sharp point at the tip. Fruits 9–16 cm long, erect or ascending from usually deflexed stalks, narrowly lanceolate in outline, the surface smooth, glabrous, glaucous. Seeds with the body 6.5–9.0 mm long, the margins mostly narrowly winged, the terminal tuft of hairs white or more commonly light cream-colored. April–July.

Scattered, mostly south of the Missouri River (eastern U.S. west to Nebraska and Texas). Upland prairies, sand prairies, glades, and openings of dry upland forests; also roadsides, railroads, and open, sandy, disturbed areas.

2. Asclepias hirtella (Pennell) Woodson (prairie milkweed, tall green milkweed)

Acerates hirtella Pennell

Asclepias longifolia Michx. ssp. *hirtella* (Pennell) J. Farmer & C.R. Bell

Pl. 221 c–e; Map 916

Plants with white latex and a thickened woody rootstock. Stems 25–110 cm long, sometimes few-branched toward the tip, mostly erect or ascending, moderately to densely short-hairy, sometimes in longitudinal lines, with numerous nodes. Leaves mostly alternate, sessile or short-petiolate. Leaf blades 4–17 cm long, 0.3–1.5 cm wide, linear to narrowly lanceolate, the base narrowed or tapered, the tip tapered gradually to a sharp point, the margins flat, sparsely to moderately rough-pubescent with short, stiff hairs, especially along the veins and margins. Inflorescences (1)2–10 in the leaf axils, short-stalked or sometimes appearing sessile, with 25–90 flowers. Calyces reflexed, hairy on the outer surface, the lobes 1.5–3.0 mm long, lanceolate to ovate. Corollas reflexed, glabrous, green to pale green, usually somewhat purple-tinged, the lobes 4–6 mm long, elliptic-lanceolate. Gynostegium appearing stalked (the column visible below the bases of the hoods), pale green, the corona noticeably shorter than the tip of the anther/stigma head. Corona hoods 1.5–2.5 mm long, erect, attached most of their length, oblong-elliptic in outline, the opening oriented toward the column, the tips broadly rounded to truncate, the margins not toothed, the bases pouched. Horns absent. Fruits 7–14 cm long, erect or ascending from usually deflexed stalks, lanceolate to narrowly ovate in outline, the surface smooth, minutely hairy. Seeds with the body 7–9 mm long, the margins relatively broadly winged,

918. Asclepias meadii   919. Asclepias perennis   920. Asclepias purpurascens

the terminal tuft of hairs white or more commonly light cream-colored to tan. May–August.

Scattered, mostly in the Glaciated Plains and Unglaciated Plains Divisions; absent from the Mississippi Lowlands (West Virginia to Minnesota, Iowa, and Kansas south to Tennessee, Arkansas, and Oklahoma; Canada). Bottomland and upland prairies and glades; also pastures, roadsides, and railroads.

3. Asclepias incarnata L. ssp. **incarnata**
(swamp milkweed)
A. incarnata f. *albiflora* A. Heller
Pl. 220 g, h; Map 917

Plants with white latex and a fibrous rootstock. Stems 50–200 cm long, mostly several-branched, erect or ascending, glabrous or sparsely and minutely hairy in longitudinal lines, with 6 to numerous nodes. Leaves opposite, short-petiolate. Leaf blades 4–15 cm long, 0.5–4.5 cm wide, narrowly lanceolate to lanceolate or less commonly ovate, the base abruptly narrowed or rounded, occasionally shallowly cordate but not overlapping that of the opposite leaf, the tip gradually tapered to a sharp point, the margins flat or slightly curled under, glabrous or more commonly the undersurface sparsely and minutely hairy. Inflorescences 2–12, terminal or occasionally appearing lateral (at the tips of short branches), short- to long-stalked, with 10–40 flowers. Calyces reflexed, short-hairy on the outer surface, the lobes 1.5–2.5 mm long, lanceolate to ovate. Corollas reflexed, glabrous, pink or rarely white, the lobes 4–6 mm long, elliptic to oblanceolate. Gynostegium appearing stalked (the column visible below the bases of the hoods), pale pink, rarely white, the corona slightly shorter than to about as long as the tip of the anther/stigma head. Corona hoods 2.0–2.7 mm long, strongly ascending, attached near their bases, oblong-ovate in outline, the tips broadly rounded, the margins not toothed, the bases not pouched. Horns attached below the middle of the hoods, extended conspicuously beyond the tips of the hoods and incurved over the anther/stigma head, linear, not flattened, tapered to a sharp point at the tip. Fruits 5–9 cm long, erect or ascending from erect or less commonly deflexed stalks, lanceolate in outline, the surface smooth, usually minutely hairy. Seeds with the body 6.5–9.0 mm long, the margins relatively broadly winged, the terminal tuft of hairs white. $2n=22$. June–August.

Scattered nearly throughout Missouri (eastern U.S. west to North Dakota and Arizona; Canada). Swamps, sloughs, marshes, margins of ponds and lakes, banks of streams and rivers, bottomland prairies, and occasionally bottomland forests; also ditches and railroads.

Swamp milkweed is gaining popularity as a garden ornamental; however, the young shoots and leaves are browsed by mammals such as rabbits and deer. Its roots are a food source for muskrats and other wetland mammals. Woodson (1954) separated a series of populations in other states to the east and south of Missouri characterized by less-branched stems, more conspicuous pubescence, and broader, less gradually tapered leaves as ssp. *pulchra* (Ehrh.) Woodson.

4. Asclepias meadii Torr. ex A. Gray (Mead's milkweed)
Pl. 221 j, k; Map 918

Plants with white latex and shallow to deep-set rhizomes. Stems 20–50 cm long, unbranched, erect or ascending with often nodding inflorescence, less commonly spreading, glabrous, often somewhat glaucous, mostly with 2–5 nodes. Leaves opposite, sessile. Leaf blades 3–8 cm long, 1–5 cm wide, lanceolate to broadly ovate, the base rounded or less commonly narrowed or shallowly cordate, the tip narrowed to a usually sharp point, the margins flat,

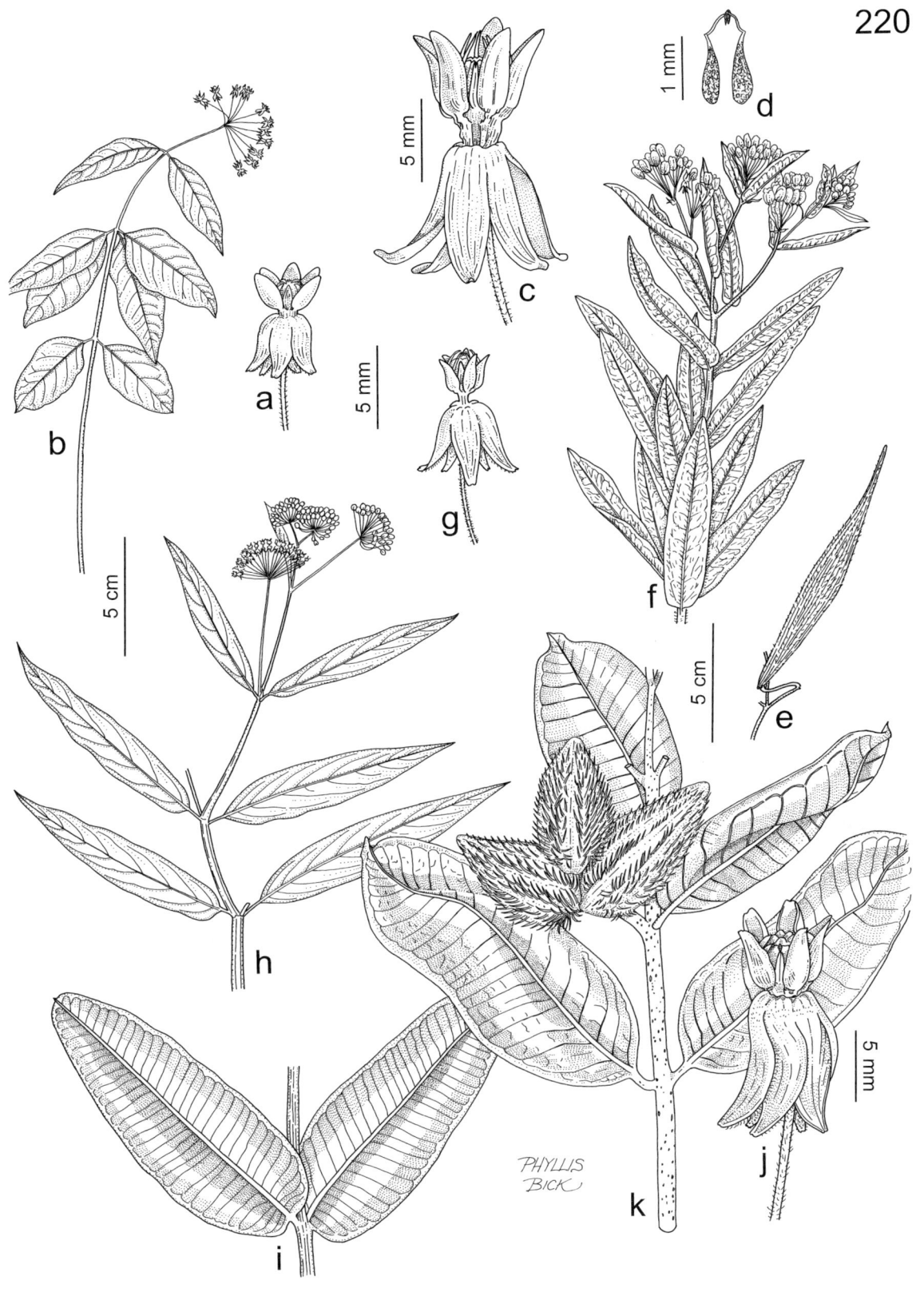

Plate 220. Asclepiadaceae. *Asclepias quadrifolia*, **a)** flower, **b)** habit. *Asclepias tuberosa*, **c)** flower (not gynostegium with hoods and horns), **d)** pollinium (two pollinia with translator), **e)** fruit, **f)** habit. *Asclepias incarnata*, **g)** flower, **h)** flowering branch. *Asclepias sullivantii*, **i)** leaves. *Asclepias syriaca*, **j)** flower, **k)** fruiting branch.

the surfaces glabrous except for minute, rough hairs along the margins. Inflorescence 1 per stem, terminal, long-stalked, with 8–20 flowers. Calyces reflexed, sparsely to moderately short-hairy, the lobes 2.5–4.0 mm long, lanceolate to ovate. Corollas reflexed, glabrous, white to pale cream-colored or pale green, the lobes 9–12 mm long, lanceolate to oblong-lanceolate. Gynostegium appearing short-stalked (the column visible below the bases of the hoods), pale cream-colored to pale green, occasionally tinged with purple, the corona conspicuously longer than the tip of the anther/stigma head. Corona hoods 4.0–5.5 mm long, erect, attached near their bases, ovate to broadly ovate in outline, the tip rounded to broadly rounded, the margins with a pair of blunt, triangular teeth near the middle, the bases pouched. Horns attached toward the hood bases, relatively short, curved inward over the anther/stigma head and not extended past the tips of the hoods, sickle-shaped, relatively stout, not flattened, tapered to a sharp point at the tip. Fruits 8–10 cm long, erect or ascending from deflexed stalks, narrowly lanceolate to narrowly elliptic in outline, the surface smooth, minutely hairy. Seeds with the body 5–7 mm long, the margins narrowly winged, the terminal tuft of hairs white. May–June.

Uncommon in western Missouri, mostly in the Unglaciated Plains Division, and locally farther east (Indiana to Illinois, Wisconsin, and Iowa south to Missouri and Kansas). Upland prairies and igneous glades.

Mead's milkweed has become extirpated from portions of its historical range, primarily because of habitat destruction. Many of the remaining populations continue to decline. It is presently listed as Threatened by the U.S. Fish and Wildlife Service under the federal Endangered Species Act. The Missouri Natural Heritage Program has records in its database of a number of unvouchered occurrences from western Missouri, but populations in this part of the state generally have continued to decline, even at sites where high-quality prairie habitats continue to exist. In contrast, populations in the eastern Ozarks, which were rediscovered in 1991, more than ninety years after the single initial collection was made from this region, appear to be mostly healthy and stable. Paradoxically, the management practices that preserved apparently high-quality plant communities at some prairie sites have been shown to be detrimental to this critically imperilled species. Tecic et al. (1998) and Bowles et al. (1998) completed detailed studies on genetic variation in *A. meadii* and correlated land management histories where the sampled populations occurred. They concluded that at sites in western Missouri and adjacent states where preservation of prairies was accomplished through annual haying, aboveground portions of Mead's milkweed plants were mowed before fruits could mature, resulting in largely clonal (by spreading of rhizomes), genetically depauperate (by lack of successful sexual reproduction and as individuals gradually died without replacement) populations. Sites that had experienced periodic burning and no mowing, however, had much greater genetic variation, presumably because plants were more successful in completing their life cycles during most years. Land managers recently have begun adjusting management strategies at a number of sites to include less frequent haying or haying in conjunction with prescribed burns during times when Mead's milkweed plants are dormant.

5. Asclepias perennis Walter (white milkweed, smoothseed milkweed)

Pl. 221 h, i; Map 919

Plants with white latex and a fibrous or sometimes slightly woody rootstock. Stems 20–50(–80) cm long, unbranched or more commonly branched, erect or ascending, glabrous or sparsely pubescent with minute hairs in longitudinal lines, with 6 to numerous nodes. Leaves opposite, short-petiolate. Leaf blades 4–15 cm long, 0.5–3.5(–6) cm wide, narrowly lanceolate or narrowly elliptic to elliptic, the base gradually narrowed or tapered, the tip gradually narrowed or tapered to a sharp point, the margins flat, glabrous. Inflorescences 1–8, terminal and in the upper leaf axils, short- to long-stalked, with 8–25 flowers. Calyces reflexed, sparsely and minutely hairy on the outer surface, the lobes 1.0–1.5 mm long, oblong-elliptic. Corollas reflexed, glabrous, white, sometimes tinged with pale pink, the lobes 2.5–4.0 mm long, lanceolate to narrowly elliptic. Gynostegium appearing stalked (the column visible below the bases of the hoods), white, the corona slightly shorter than to about as long as the tip of the anther/stigma head. Corona hoods 2.0–2.8 mm long, ascending, attached near their bases, oblong-ovate in outline, the tips broadly rounded, the margins not toothed, the bases not pouched. Horns attached toward the hood bases, extended conspicuously beyond the tips of the hoods and incurved over the anther/stigma head, linear, not flattened, tapered to a sharp point at the tip. Fruits 4–7 cm long, pendant from usually deflexed stalks, ovate to broadly elliptic-ovate in outline, the surface smooth, glabrous. Seeds with the body 12–17 mm long, the margins broadly winged, the terminal tuft of hairs absent. May–September.

921. Asclepias quadrifolia   922. Asclepias speciosa   923. Asclepias stenophylla

Scattered in the Mississippi Lowlands Division and adjacent Ozarks (southeastern U.S. west to Texas, mostly in the Atlantic and Gulf Coastal Plains, north up the Mississippi and Ohio River valleys). Swamps, sloughs, bottomland forests, margins of ponds and lakes, and occasionally banks of streams, often an emergent aquatic; also ditches and moist roadsides.

Edwards et al. (1994) studied the dispersal ecology of A. perennis and concluded that the drooping rather than erect follicles and large, winged seeds lacking terminal tufts of hair are adaptations to seed dispersal by water, in contrast to wind dispersal of the plumed seeds of most other milkweeds.

6. Asclepias purpurascens L. (purple milkweed)

Pl. 221 f, g; Map 920

Plants with white latex and deep-set rhizomes. Stems 35–100 cm long, unbranched or less commonly with a single branch at the tip, erect or ascending, mostly glabrous but often densely short-hairy toward the tip, with 4–10 nodes. Leaves opposite or less commonly 1 node with a whorl of 4, short-petiolate. Leaf blades 4–20 cm long, 2–10 cm wide, lanceolate-elliptic to ovate or oblong-elliptic, the base narrowed or less commonly rounded, the tip narrowed or tapered to a sharp point, less commonly rounded but with a short, sharp point, the margins flat, the upper surface sparsely short-hairy to nearly glabrous with a usually densely hairy midvein, the undersurface moderately to densely short-hairy. Inflorescences 1–5, terminal and sometimes also in the upper leaf axils, usually long-stalked, with 12–50 flowers. Calyces reflexed, sparsely to moderately short-hairy on the outer surface, the lobes 2.5–4.0 mm long, lanceolate to ovate-lanceolate. Corollas reflexed, glabrous or less commonly minutely hairy on the inner surface toward the base, reddish purple to dark purple, the lobes 7–10 mm long, ovate to elliptic-lanceolate. Gynostegium appearing very short-stalked (the column barely visible below the bases of the hoods), pale purple (rarely tan and purple-tinged) to reddish purple, the corona conspicuously longer than the tip of the anther/stigma head. Corona hoods 4–8 mm long, loosely ascending to ascending, attached near their bases, gradually narrowed from at or above the middle, the apical portion oblong to ovate in outline, the tip narrowly rounded, the margins not noticeably lobed, the bases not pouched. Horns attached toward the hood bases, relatively short, bent or curved abruptly inward over the anther/stigma head and not extended past the tips of the hoods, sickle-shaped, relatively stout, slightly flattened, tapered to a more or less sharp point at the tip. Fruits 10–16 cm long, erect or ascending from deflexed stalks, narrowly elliptic to ovate and slightly arched in outline, the surface smooth, minutely hairy. Seeds with the body 5–7 mm long, the margins narrowly winged, the terminal tuft of hairs white. May–July.

Scattered nearly throughout the state (eastern [mostly northeastern] U.S. and adjacent Canada west to South Dakota and Texas). Upland prairies (sometimes in moist swales), glades, savannas, tops of bluffs, mesic to dry upland forests, and less commonly banks of streams; also pastures, roadsides, railroads, and open, disturbed areas.

Where A. purpurascens and A. syriaca grow together, apparent hybrids are found occasionally. Thus far, the only documented occurrence of such putative, morphologically intermediate hybrids is a specimen from a disturbed site in Iron County.

7. Asclepias quadrifolia Jacq. (whorled milkweed, fourleaf milkweed)

Pl. 220 a, b; Map 921

Plants with white latex and slender rhizomes. Stems 20–50 cm long, unbranched or occasionally branched once near the tip, erect or ascending, glabrous or more commonly short-hairy, often in longitudinal lines, with 2–5(–8) nodes. Leaves opposite, usually those at 1 or 2 nodes in whorls of 4 (rarely all opposite), short-petiolate, those at the uppermost and lowermost nodes usually much reduced in size. Leaf blades 2–12 cm long, 1–7 cm wide, lanceolate to narrowly ovate-elliptic, the base gradually narrowed or tapered, the tip gradually narrowed or tapered to a sharp point, the margins flat, sparsely hairy, mostly along the midvein. Inflorescences 1–3, terminal and occasionally in the uppermost leaf axils, short- to more commonly long-stalked, with (8–)15–35 flowers. Calyces reflexed, glabrous, the lobes 1–3 mm long, lanceolate to ovate. Corollas reflexed, glabrous, light pink or rarely cream-colored, the lobes 4.5–6.0 mm long, lanceolate to elliptic-lanceolate. Gynostegium appearing stalked (the column visible below the bases of the hoods), white, the corona conspicuously longer than the tip of the anther/stigma head. Corona hoods 4–5 mm long, spreading to loosely ascending, attached near their bases, oblong-elliptic in outline, the tips broadly rounded, the margins with a pair of triangular teeth or lobes at or below the middle, the bases not pouched. Horns attached in the basal half of the hoods, arched over the anther/stigma head but not extended past the tips of the hoods, linear, not flattened, tapered to a sharp point at the tip. Fruits 8–14 cm long, erect or ascending from mostly erect stalks, narrowly elliptic-lanceolate in outline, the surface smooth, glabrous or minutely hairy. Seeds with the body 7–8 mm long, the margins narrowly winged, the terminal tuft of hairs white or light cream-colored to tan. May–July.

Scattered nearly throughout the state, but apparently absent from the Mississippi Lowlands Division and the western portion of the Glaciated Plains (northeastern U.S. and adjacent Canada west to Iowa, Kansas, and Oklahoma).

8. Asclepias speciosa Torr. (showy milkweed)
Map 922

Plants with white latex and deep-set rhizomes. Stems 40–100 cm long, unbranched or less commonly few-branched, erect or ascending, densely short-hairy, at least toward the tip, with (5–)7 to numerous nodes. Leaves opposite, sessile or short-petiolate. Leaf blades 6–20 cm long, 2.5–10 cm wide, lanceolate to ovate or oblong-elliptic, the base rounded to shallowly cordate, less commonly narrowed, the tip narrowed or tapered to a sharp point, less commonly rounded but with a short, sharp point, the margins flat, the upper surface sparsely to densely short-hairy, especially along the midvein, the undersurface densely short-hairy. Inflorescences 1–6, terminal and in the upper leaf axils, usually long-stalked, with 10–40 flowers. Calyces reflexed, densely short-hairy on the outer surface, the lobes 4.0–6.5 mm long, lanceolate to ovate. Corollas reflexed, moderately to densely short-hairy on the outer surface except for a thin, glabrous, marginal band, purplish pink or greenish purple, the lobes 9–15 mm long, lanceolate to oblong-lanceolate. Gynostegium appearing very short-stalked (the column barely visible below the bases of the hoods), pale pink to cream-colored, the corona conspicuously longer than the tip of the anther/stigma head. Corona hoods 9–15 mm long, spreading to loosely ascending, attached near their bases, abruptly narrowed below the middle, the apical half narrowly lanceolate to lanceolate in outline, the tip tapered to a sharp point, the margins with a pair of triangular teeth or lobes at or below the middle, the bases not pouched. Horns attached toward the hood bases, arched over the anther/stigma head but not extended past the tips of the hoods, linear, not flattened, tapered to a sharp point at the tip. Fruits 7–12 cm long, erect or ascending from deflexed stalks, ovate and slightly arched in outline, the surface with soft, narrow, warty tubercles and densely hairy (woolly). Seeds with the body 6–9 mm long, the margins narrowly to relatively broadly winged, the terminal tuft of hairs white to light cream-colored or tan. $2n=22$. May–August.

Uncommon, known from a single historical collection from St. Louis County (eastern U.S. west to Minnesota and Texas; Canada; introduced eastward to Michigan). Habitat unknown, but presumably the bank of a river or stream or bottomland prairie.

This species was reported for the Missouri flora by Palmer and Steyermark (1935), based on a single specimen collected in St. Louis County in 1894. Later, Steyermark (1963) erroneously excluded it from the flora, stating that the specimen was a misdetermined sample of *A. syriaca*. Woodson (1954) failed to cite the specimen in his monograph of the genus, having in 1951 deaccessioned the sole sheet from the Missouri Botanical Garden and sent it to the University of Minnesota herbarium as part of his infamous purge of supposedly superfluous MO specimens (Solomon, 1998). The specimen fortunately was part of a 75,000-sheet set that was returned to St. Louis in 1993, where it came to be examined during the present study. The St. Louis County record is disjunct from the main western range of *A. speciosa* and the circumstances surrounding its collection are not known. However, the range does approach Missouri in the eastern

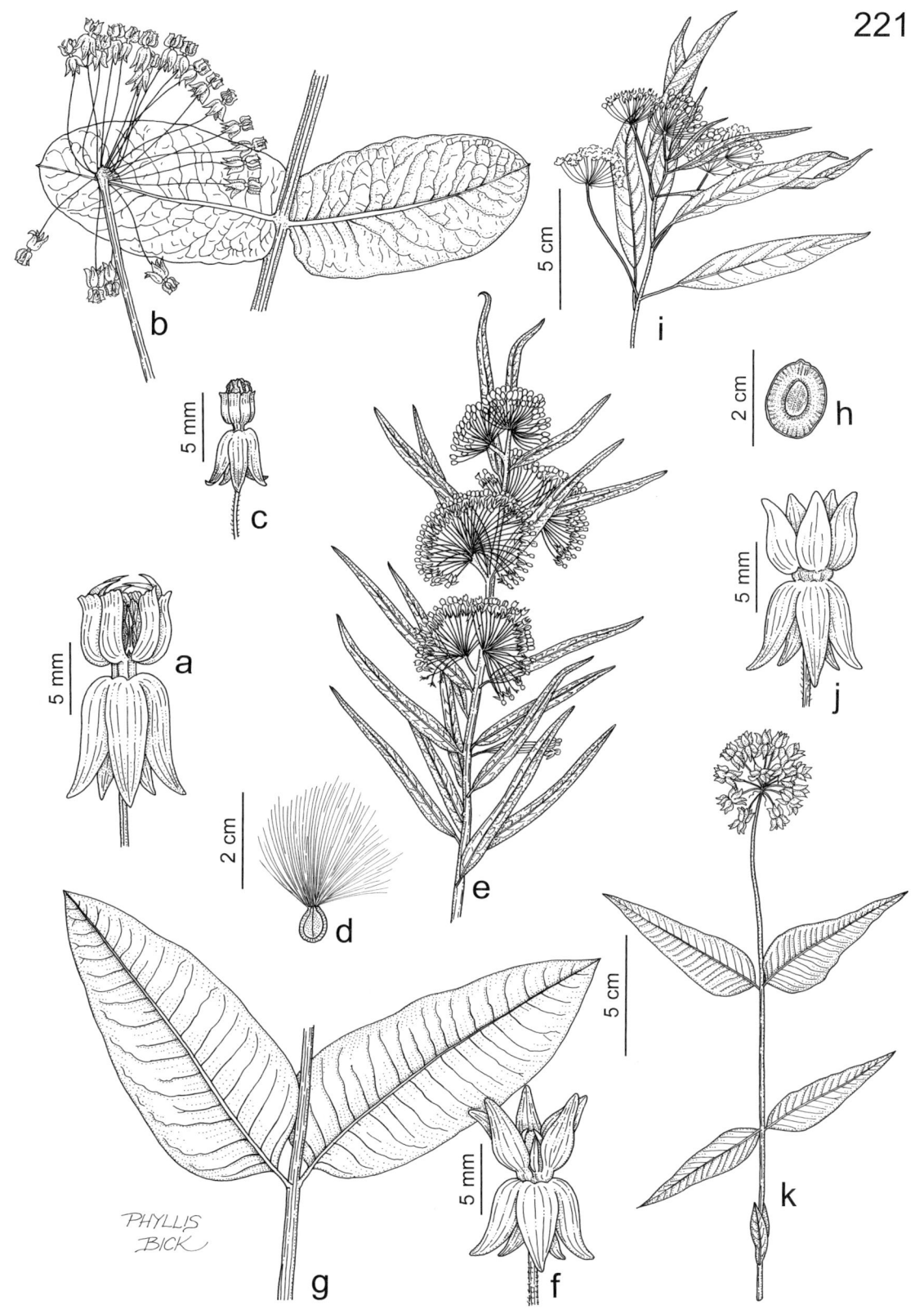

Plate 221. Asclepiadaceae. *Asclepias amplexicaulis*, **a)** flower, **b)** inflorescence and leaves. *Asclepias hirtella*, **c)** flower, **d)** seed, **e)** habit. *Asclepias purpurascens*, **f)** flower, **g)** leaves. *Asclepias perennis*, **h)** seed, **i)** habit. *Asclepias meadii*, **j)** flower, **k)** habit.

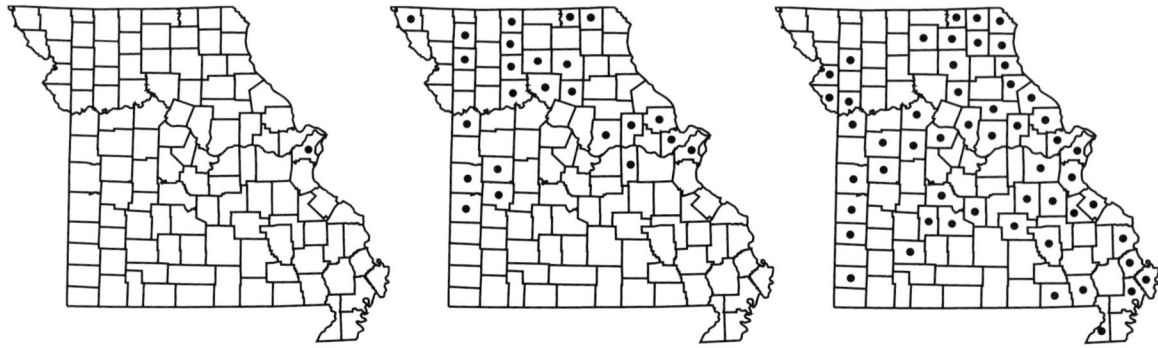

924. Asclepias subverticillata   925. Asclepias sullivantii   926. Asclepias syriaca

portions of Kansas and Nebraska, as well as southwestern Iowa, so it should be searched for in bottomland prairies and other open bottomlands in northwestern Missouri. Where *A. speciosa* grows with *A. syriaca*, hybrids have been reported (Adams et al., 1987; Voss, 1996).

9. Asclepias stenophylla A. Gray (narrow-leaved milkweed)
Polyotus angustifolius Nutt.
Acerates angustifolia (Nutt.) Decne.

Pl. 222 c, d; Map 923

Plants with white latex and a thickened, somewhat tuberous rootstock. Stems 20–100 cm long, sometimes few-branched toward the tip, mostly erect or ascending, moderately to sparsely short-hairy, with numerous nodes. Leaves mostly alternate, sessile or very short-petiolate. Leaf blades 4–18 cm long, 0.1–0.6 mm wide, linear, the base narrowed or tapered, the tip narrowed or tapered to a sharp point, the margins usually curled under, glabrous or more commonly sparsely and minutely hairy along the midvein and margins. Inflorescences (1)2–12 in the leaf axils, short-stalked, usually appearing sessile, with 10–25 flowers. Calyces reflexed, minutely hairy on the outer surface, the lobes 2.0–3.3 mm long, lanceolate. Corollas reflexed, glabrous, pale green to light yellow, sometimes slightly purple-tinged at the tip, the lobes 4.5–6.0 mm long, elliptic-lanceolate. Gynostegium appearing sessile (the corona base touching the corolla or nearly so), pale green to white, the corona noticeably shorter than to slightly longer than the tip of the anther/stigma head. Corona hoods 3–4 mm long, erect, attached toward their bases, narrowly oblong in outline, the tips appearing 3-toothed or shallowly 3-lobed, the margins with a pair of triangular teeth or lobes below the middle, the bases pouched. Horns present but fused nearly the entire length to the inner surface of the hood, visible as a low ridge and extended as the middle tooth or lobe of the hood. Fruits 9–12 cm long, erect or ascending from usually deflexed stalks, narrowly elliptic-lanceolate in outline, the surface smooth, glabrous or more commonly minutely hairy. Seeds with the body 5–6 mm long, the margins narrowly winged, the terminal tuft of hairs light cream-colored or tan. May–July.

Scattered in the western and northern portion of the Ozark Division and in the Ozark Border and Unglaciated Plains (Illinois to Montana south to Arkansas and Texas). Upland prairies, savannas, glades, exposed ledges and tops of bluffs, often on calcareous substrates; also pastures, roadsides, and railroads.

10. Asclepias subverticillata (A. Gray) Vail (poison milkweed)
A. verticillata L. var. *subverticillata* A. Gray

Pl. 222 a; Map 924

Plants with white latex and deep-set rhizomes, the rootstock sometimes woody near the stem bases. Stems 30–90 cm long, occasionally few-branched toward the tip and with axillary branchlets with pairs or clusters of short leaves in the axils of many of the main leaves, mostly erect or ascending, glabrous or sparsely short-hairy in longitudinal lines, with numerous nodes. Leaves of the main stem all or nearly all in whorls of 3–5, sessile or nearly so. Leaf blades 2–12 cm long, 1–3(–4) mm wide, linear, the base narrowed or tapered, the tip narrowed or tapered to a sharp point, the margins curled under, glabrous or sparsely and minutely hairy. Inflorescences (1)2–8 in the leaf axils, mostly short-stalked, with 6–25 flowers. Calyces reflexed, glabrous or sparsely short-hairy on the outer surface, the lobes 1.2–2.5 mm long, narrowly lanceolate to ovate. Corollas reflexed, glabrous, white, occasionally slightly tinged with green or purple, the lobes 3.5–4.5 mm long, elliptic. Gynostegium appearing stalked (the column visible below the bases of the hoods), pale green to white, the corona noticeably shorter than the tip of the anther/stigma head. Corona hoods 1.5–2.0 mm long, erect or ascending, attached toward their bases, broadly oblong in outline, the tips broadly rounded, the margins not toothed, the bases not pouched. Horns attached near the hood bases, extended conspicuously beyond the tips of the hoods and incurved over the anther/stigma head, linear, not flattened, tapered to a sharp point at the tip. Fruits 5–9 cm long, erect or ascending from erect to deflexed stalks, narrowly elliptic-lanceolate in outline, the surface smooth, glabrous or minutely hairy. Seeds with the body 6–8 mm long, the margins narrowly to somewhat more broadly winged, the terminal tuft of hairs white. $2n=22$. June–August.

Introduced, known only from the city of St. Louis (Kansas to Idaho south to Texas and Arizona; Mexico). Railroads.

This southwestern species was first reported for Missouri by Mühlenbach (1979). It is very similar to the native *A. verticillata,* differing most notably in the axillary branchlets, the leaves sometimes more spreading and flexuous, and in details of the morphology of the hoods.

11. Asclepias sullivantii Engelm. ex A. Gray
(smooth milkweed, prairie milkweed)
Pl. 220 i; Map 925

Plants with white latex and deep-set rhizomes. Stems 40–110 cm long, unbranched or less commonly with 1 or 2 branches toward the tip, erect or ascending, glabrous, often somewhat glaucous, with 7–15 nodes. Leaves opposite, sessile or nearly so. Leaf blades 4–16 cm long, 1.5–9.0 cm wide, narrowly oblong or lanceolate to oblong, oblong-elliptic, or broadly ovate, the base rounded to truncate or shallowly cordate, the tip narrowed to a blunt or sharp point or rounded but usually with a short, sharp point, less commonly nearly truncate or very shallowly notched, the margins flat, the upper surface glabrous, the undersurface glabrous or rarely sparsely and minutely hairy along the midvein or toward the base, often somewhat glaucous. Inflorescences 1–6, terminal and in the upper leaf axils, short- or long-stalked, with 15–40 flowers. Calyces reflexed, glabrous, the lobes 4–6 mm long, lanceolate to ovate. Corollas reflexed, glabrous, light pink to purplish-pink, sometimes tinged with green, the lobes 9–12 mm long, lanceolate to oblong-lanceolate. Gynostegium appearing sessile or very short-stalked (the column sometimes barely visible below the bases of the hoods), pale pink to pink, the corona conspicuously longer than the tip of the anther/stigma head. Corona hoods 5.0–6.5 mm long,

loosely ascending to ascending, attached near their bases, oblong to oblong-ovate in outline, the tip rounded to broadly rounded, the margins not toothed, with a pair of small, pouched areas near the base. Horns attached toward the hood bases, relatively short, bent or curved abruptly inward over the anther/stigma head and not extended past the tips of the hoods, sickle-shaped, relatively stout, slightly flattened, tapered to a sharp point at the tip. Fruits 8–10 cm long, erect or ascending from deflexed stalks, narrowly ovate to ovate and slightly arched in outline, the surface with small, warty, spinelike tubercles above the middle, glabrous or minutely hairy. Seeds with the body 7–8 mm long, the margins narrowly or relatively broadly winged, the terminal tuft of hairs white. $2n=22$. June–July.

Scattered in northern and western Missouri, mostly in the Glaciated Plains Division (Ohio to North Dakota and Oklahoma; Canada). Bottomland and upland prairies, margins of lakes and marshes, banks of rivers and streams, and less commonly openings of bottomland and mesic upland forests; also pastures, ditches, roadsides, and railroads.

12. Asclepias syriaca L. (common milkweed)
A. syriaca var. *kansana* (Vail) E.J. Palmer & Steyerm.
Pl. 220 j, k; Map 926

Plants with white latex and deep-set rhizomes. Stems 50–200 cm long, unbranched or less commonly with a single branch at the tip, erect or ascending, sparsely to densely short-hairy, especially toward the tip, with 7–15 nodes. Leaves opposite (rarely the uppermost nodes with only 1 leaf), short-petiolate. Leaf blades 6–30 cm long, 3–11 cm wide, oblong to narrowly or broadly elliptic-ovate, less commonly elliptic-lanceolate, the base narrowed to broadly rounded, less commonly truncate, the tip narrowed to a blunt or sharp point or rounded but usually with a short, sharp point, the margins flat, the upper surface sparsely to moderately short-hairy, especially along the midvein, the undersurface densely short-hairy (felty). Inflorescences (1–)2–10, terminal and in the upper leaf axils, usually long-stalked, with 20 to more than 100 flowers. Calyces reflexed, moderately to densely short-hairy on the outer surface, the lobes 2.5–4.0 mm long, lanceolate to elliptic-lanceolate. Corollas reflexed, moderately to densely short-hairy on the outer surface and usually also on the inner surface near the base, green to lavender, usually tinged with pink and/or white, the lobes 7–10 mm long, elliptic-lanceolate. Gynostegium appearing very short-stalked (the column barely visible below the bases of the hoods), pale pink to pale purple, the

927. Asclepias tuberosa   928. Asclepias variegata   929. Asclepias verticillata

corona conspicuously longer than the tip of the anther/stigma head. Corona hoods 3.5–5.0 mm long, spreading to ascending, attached near their bases, gradually narrowed from at or above the middle, the apical portion oblong to ovate in outline, the tip rounded, the margins with a pair of sharply triangular, ascending and incurved lobes at about the middle, the bases not pouched. Horns attached toward the hood bases, relatively short, bent or curved abruptly inward over the anther/stigma head and not extended past the tips of the hoods, sickle-shaped, relatively stout, slightly flattened, tapered to a sharp point at the tip. Fruits 7–12 cm long, erect or ascending from deflexed stalks, narrowly ovate to ovate and usually noticeably arched in outline, the surface with soft, narrow, warty tubercles, densely hairy (woolly). Seeds with the body 6–8 mm long, the margins narrowly winged, the terminal tuft of hairs white to light cream-colored or tan. $2n=22$. May–August.

Common throughout Missouri (much commoner than the distribution map records) (eastern U.S. and adjacent Canada west to North Dakota and Texas; introduced farther west). Bottomland and upland prairies, openings and edges of mesic upland forests, and banks of streams and rivers; also pastures, old fields, margins of crop fields, ditches, roadsides, railroads, and open, disturbed areas.

This is the most abundant species of milkweed in the state and the one common in the largest variety of disturbed habitats. Forms with white flowers (f. *leucantha* Dore) or fruits lacking tubercles (f. *inermis* Churchill) are encountered locally in other parts of the range but have yet to be reported from Missouri. Rare putative hybrids between *A. syriaca* and *A. amplexicaulis*, *A. speciosa*, *A. sullivantii*, and *A. verticillata*, among other species, have been reported in the literature (Woodson, 1954; Kephart and Heiser, 1980; Adams et al., 1987; Klips and Vulley, 2004), but so far only the hybrid involving *A. purpurascens* has been documented from Missouri, based on a specimen from a mixed population at disturbed site in Iron County.

13. Asclepias tuberosa L. ssp. interior
Woodson (butterfly weed, chigger flower, pleurisy root)
A. tuberosa var. *interior* (Woodson) Shinners
A. tuberosa f. *lutea* (Clute) Steyerm.

Pl. 220 c–f; Map 927

Plants with clear latex and deep-set rhizomes. Stems 20–90 cm long, sometimes few-branched toward the tip, mostly erect or ascending, densely hairy, with numerous nodes. Leaves all or mostly alternate, sessile or nearly so. Leaf blades 2–10 cm long, 0.4–2.3 cm wide, linear to broadly lanceolate, the base usually deeply cordate and clasping the stem, the tip narrowed or tapered gradually to a sharp point, the margins usually somewhat revolute, the surfaces hairy, especially along the veins. Inflorescences 1 to several, terminal and in the upper leaf axils, sessile or short-stalked, with 6–25 flowers. Calyces reflexed, hairy on the outer (under) surface, the lobes 2–4 mm long, linear to elliptic-lanceolate. Corollas reflexed, glabrous, bright orangish yellow to reddish orange, the lobes 6–10 mm long, lanceolate to elliptic. Gynostegium appearing stalked (the column visible below the bases of the hoods), bright yellowish orange to reddish orange, the corona conspicuously longer than the tip of the anther/stigma head. Corona hoods 4.5–6.5 mm long, ascending, attached near their bases, lanceolate in outline, the tips rounded, the margins with a pair of short, triangular teeth or lobes below the middle, the bases not pouched. Horns attached below the middle of the hoods, extended to about the tips of the hoods and angled or somewhat curved inward over the anther/stigma head, linear, not flattened, tapered to a sharp point at the tip. Fruits 8–15 cm long, erect or ascending from

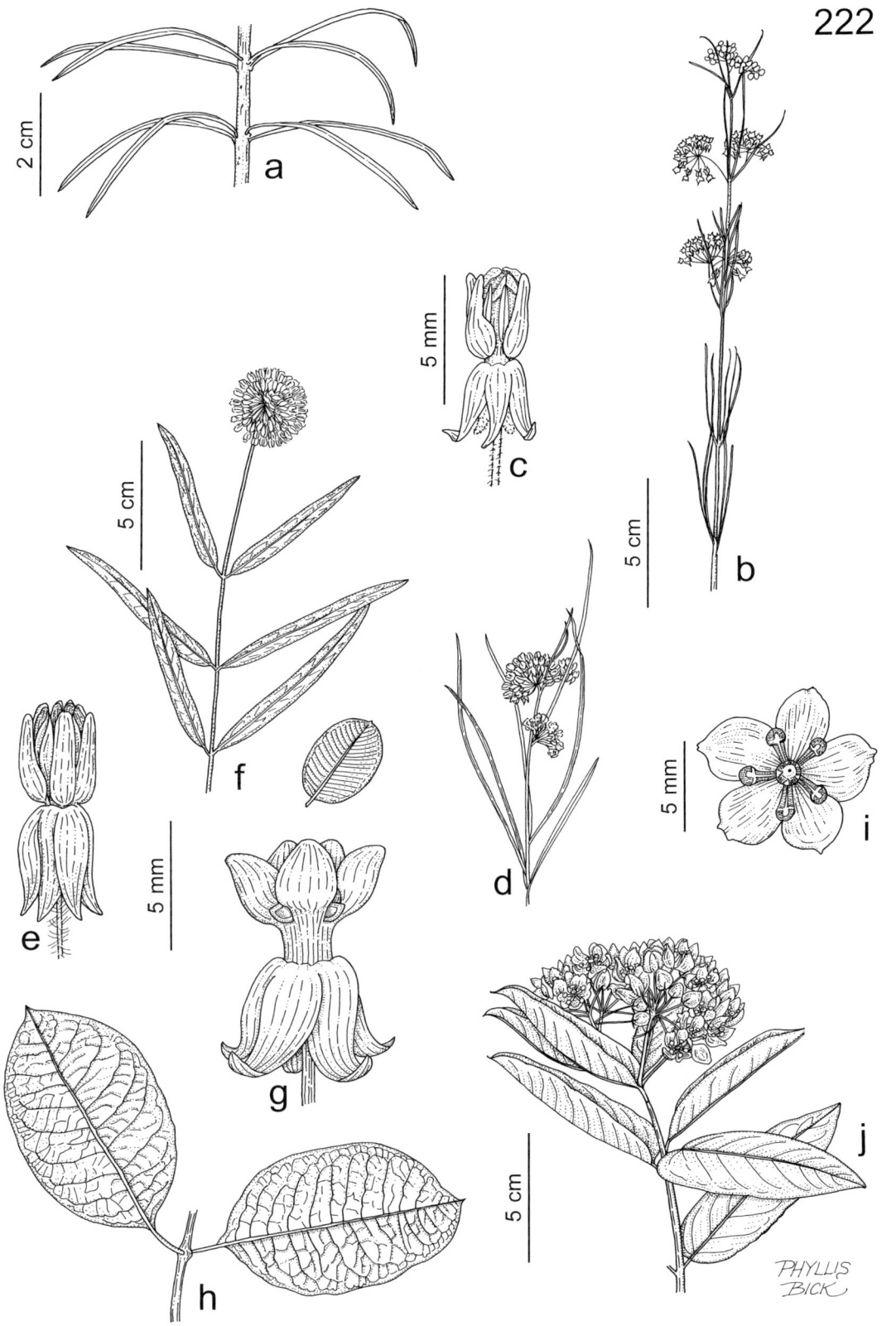

Plate 222. Asclepiadaceae. *Asclepias subverticillata*, **a)** leaves. *Asclepias verticillata*, **b)** habit. *Asclepias stenophylla*, **c)** flower, **d)** habit. *Asclepias viridiflora*, **e)** flower, **f)** narrow-leaved habit and broader leaf. *Asclepias variegata*, **g)** flower, **h)** leaves. *Asclepias viridis*, **i)** flower, **j)** habit.

ascending or deflexed stalks, narrowly lanceolate in outline, the surface minutely hairy. Seeds with the body 5–7 mm long, the margins narrowly winged, the terminal tuft of hairs white. 2n=22. May–September.

Scattered to common nearly throughout Missouri (eastern U.S. [mostly west of the Appalachian Mountains] and adjacent Canada west to South Dakota, Utah, and Arizona). Upland prairies, glades, savannas, and openings of mesic to dry upland forests; also pastures, roadsides, railroads, and open, disturbed areas.

Woodson (1954) concluded from his earlier morphometric studies that *A. tuberosa* could be classified into four more or less geographically distinct subspecies based on differences in leaf shapes, these tending to intergrade in distributional zones of overlap. The widespread eastern ssp. *tuberosa* and ssp. *rolfsii* (Britton) Woodson of the Gulf Coastal Plain are characterized by leaves tending to be widest above the middle. The ssp. *terminalis* Woodson of the southwestern United States and northern Mexico differs in having the leaf bases mostly rounded or truncate. Later, Woodson (1964) abandoned ssp. *terminalis,* placing all western populations into ssp. *interior.* Woodson (1964) also reexamined leaf variation along a 1,200-mile roadside transect and found that during a fourteen-year period between studies ssp. *interior* and ssp. *tuberosa* had expanded their zone of contact westward through the establishment of new populations along disturbed highway margins, with a resultant breakdown in differentiating characters in portions of the intervening area.

Woodson (1962) also studied variation in flower color in the species. The yellow ground color present in all flowers is caused by the presence of carotenoid compounds, which are selectively masked by red pigmentation from anthocyanins. Woodson noted that populations in Missouri and adjacent states tend to have more reddish orange flowers, but that more distant plants in all directions tend to have progressively more yellow flowers, with redder flowers also present locally in some western portions of the range.

Butterfly weed is not weedy. The bright floral displays it provides along roadsides and in glades and prairies are among the showiest of Missouri wildflowers. It has become important enough horticulturally that cultivars are being bred commercially to accentuate particular flower colors and growth forms. It is an important plant in wildlife gardening, particularly in butterfly gardens. However, the deep-set fleshy rhizomes are easily damaged during transplantation, so plants for the garden should be grown from seeds or purchased from reputable nurseries.

14. Asclepias variegata L. (variegated milkweed)

Pl. 222 g, h; Map 928

Plants with white latex and a thickened, somewhat tuberous rootstock. Stems 30–120 cm long, unbranched or occasionally branched once near the tip, erect or ascending, usually glabrous below the middle and short-hairy toward the tip, often in longitudinal lines, with 4–7(–10) nodes. Leaves usually opposite, occasionally those at 1 or 2 nodes in whorls of 4, short- to long-petiolate. Leaf blades 2–15 cm long, 1–8 cm wide, ovate to broadly oblong-elliptic, occasionally narrowly elliptic (uppermost leaves) or nearly circular (lowermost leaves), the base short- to long-tapered, the tip rounded or narrowed to a blunt point or rounded but often with an abrupt, short, sharp point, the margins flat or somewhat wavy, less commonly slightly curled under, short-hairy, the upper surface glabrous, the undersurface sparsely short-hairy, mostly along the midvein, somewhat glaucous. Inflorescences 1–6, terminal and occasionally in the uppermost leaf axils, short- to long-stalked, with 18–45 flowers. Calyces reflexed, moderately to densely hairy on the outer surface, the lobes 1.5–3.0 mm long, lanceolate to triangular-ovate. Corollas reflexed, glabrous, white, occasionally tinged with pale pink, the lobes 7–9 mm long, oblong-elliptic. Gynostegium appearing stalked (the column visible below the bases of the hoods), white with a purple column, the corona conspicuously longer than the tip of the anther/stigma head. Corona hoods 2.0–2.5 mm long, spreading to loosely ascending, attached near their bases, broadly obovate in outline to nearly globose, the tips broadly rounded, the margins sometimes with a pair of indistinct, bluntly triangular teeth or lobes at or below the middle, the bases not pouched. Horns attached in the basal half of the hoods, bent or curved abruptly inward over the anther/stigma head but not extended past the tips of the hoods, sickle-shaped, stout, somewhat flattened above the middle, the basal half somewhat bulbous-inflated, tapered to a sharp point at the tip. Fruits 10–15 cm long, erect or ascending from mostly erect stalks, narrowly elliptic-lanceolate in outline, the surface smooth, minutely hairy, glaucous. Seeds with the body 5–6 mm long, the margins narrowly to moderately broadly winged, the terminal tuft of hairs white. May–July.

Uncommon in the Mississippi Lowlands Division and southern portion of the Ozarks (eastern [mostly southeastern] U.S. west to Oklahoma and

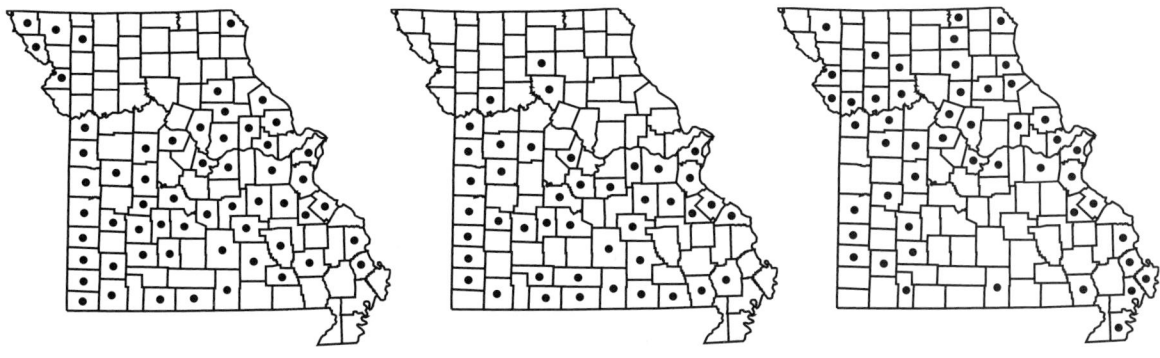

930. Asclepias viridiflora 931. Asclepias viridis 932. Cynanchum laeve

Texas). Openings of mesic to dry upland forests and savannas, often in sandy soils, less commonly bottomland forests; also roadsides and open, sandy, disturbed areas.

15. Asclepias verticillata L. (whorled milkweed, horsetail milkweed)

Pl. 222 b; Map 929

Plants with white latex and a fibrous rootstock. Stems 20–60(–90) cm long, occasionally few-branched toward the tip and with axillary branchlets with pairs or clusters of leaves absent or rare, mostly erect or ascending, sparsely short-hairy in longitudinal lines, with numerous nodes. Leaves all or nearly all in whorls of 3–6, sessile. Leaf blades 2–8 cm long, 0.5–2.5 mm wide, linear, the base narrowed or tapered, the tip narrowed or tapered to a sharp point, the margins curled under, glabrous or sparsely and minutely hairy. Inflorescences (1)2–10 in the leaf axils, mostly short-stalked, with 6–20 flowers. Calyces reflexed, glabrous or sparsely short-hairy on the outer surface, the lobes 1.5–2.5 mm long, narrowly lanceolate to ovate. Corollas reflexed, glabrous, white to pale green (usually with age), sometimes slightly purple-tinged at the tip, the lobes 3.5–5.0 mm long, elliptic. Gynostegium appearing stalked (the column visible below the bases of the hoods), pale green to white, the corona slightly shorter than to about as long as the tip of the anther/stigma head. Corona hoods 1.5–2.0 mm long, erect, attached toward their bases, broadly oblong in outline, the tips broadly rounded, the margins not toothed, the bases not pouched. Horns attached near the hood bases, extended conspicuously beyond the tips of the hoods and incurved over the anther/stigma head, linear, not flattened, tapered to a sharp point at the tip. Fruits 7–10 cm long, erect or ascending from erect to deflexed stalks, narrowly elliptic-lanceolate in outline, the surface smooth, glabrous or minutely hairy. Seeds with the body 5–6 mm long, the margins narrowly to somewhat more broadly winged, the terminal tuft of hairs white. $2n=22$. May–September.

Scattered nearly throughout the state, but nearly absent from the Mississippi Lowlands Division (eastern U.S. and adjacent Canada west to Montana and Arizona). Upland prairies, savannas, glades, exposed ledges and tops of bluffs, and less commonly dry upland forests; also pastures, roadsides, and railroads.

16. Asclepias viridiflora Raf. (green milkweed)
Acerates viridiflora (Raf.) Pursh ex Eaton
Asclepias viridiflora var. *lanceolata* (E. Ives) Torr.
Asclepias viridiflora var. *linearis* (A. Gray) Fernald

Pl. 222 e, f; Map 930

Plants with white latex and a thickened, somewhat woody rootstock. Stems 20–90 cm long, sometimes few-branched toward the tip, erect or ascending to less commonly spreading, sparsely to moderately short-hairy, sometimes in longitudinal lines or stripes, with 8 to numerous nodes. Leaves mostly opposite or subopposite, less commonly mostly alternate, sessile or very short-petiolate. Leaf blades very variable, 2–12 cm long, 0.3–6.0 cm wide, linear to lanceolate, ovate, oval, or nearly circular in outline, the base rounded, narrowed, or tapered, the tip rounded or narrowed or tapered to a blunt or sharp point, the margins flat, minutely hairy along the margins, glabrous or sparsely to moderately pubescent with minute hairs on the undersurface, especially along the midvein. Inflorescences 1–6 in the leaf axils, short-stalked or appearing sessile, with 20–80 flowers. Calyces reflexed, minutely hairy on the outer surface, especially along the margins, the lobes 2.0–3.5 mm long, lanceolate. Corollas reflexed, glabrous or sparsely hairy on the outer surface toward the tip, pale green, the lobes 5–7 mm long, elliptic-lanceolate. Gynostegium appearing sessile (the corona base touching the corolla or nearly so), pale green, the corona shorter than to nearly as long as the tip of the anther/stigma head. Corona hoods 4–5 mm long, erect, attached toward their bases, narrowly oblong-elliptic in outline, the tips rounded to bluntly pointed, the margins with a pair of short, triangular teeth or lobes below the middle, the bases pouched. Horns absent. Fruits 7–15 cm long, erect or ascending from usually deflexed stalks, narrowly elliptic-lanceolate in outline, the surface smooth, glabrous or more commonly minutely hairy. Seeds with the body 5.5–7.0 mm long, the margins narrowly winged, the terminal tuft of hairs light cream-colored or tan. $2n=22$. May–August.

Scattered, mostly south of the Missouri River and in counties adjacent to the Big Rivers Division (eastern U.S. and adjacent Canada and Mexico west to Montana and Arizona). Upland prairies, sand prairies, savannas, glades, and exposed ledges and tops of bluffs, often on calcareous substrates; also roadsides.

Asclepias viridiflora exhibits a near continuum of variation in leaf shape, and although the extremes of long, narrow leaves vs. short, broad leaves appear very dissimilar, separation of the species into discretely defined varieties is not possible (Woodson, 1954).

17. Asclepias viridis Walter (green-flowered milkweed, spider milkweed, Ozark milkweed)
Asclepiodora viridis (Walter) A. Gray
Pl. 222 i, j; Map 931

Plants with white latex and a thickened, somewhat tuberous rootstock. Stems 20–60 cm long, sometimes few-branched toward the tip, ascending to spreading, glabrous or sparsely short-hairy toward the tip, with 12 to numerous nodes. Leaves mostly alternate, sessile or short-petiolate. Leaf blades 2–12 cm long, 1–6 cm wide, narrowly to broadly oblong, elliptic-lanceolate, or ovate, the base rounded, narrowed, or less commonly shallowly cordate, the tip rounded or narrowed to a blunt point, the margins flat, glabrous or sparsely and minutely hairy along the midvein. Inflorescences 1–5, terminal and in the leaf axils, short-stalked, with 3–20 flowers. Calyces spreading, glabrous or sparsely and minutely hairy on the outer surface, the lobes 3–5 mm long, lanceolate to narrowly ovate. Corollas ascending to spreading, glabrous, pale green to green, the lobes 10–17 mm long, elliptic-lanceolate to ovate. Gynostegium appearing sessile (the corona base touching the corolla or nearly so), pale purple to purple, the corona noticeably shorter than the tip of the anther/stigma head. Corona hoods 4–6 mm long, deflexed in the basal $1/3$–$1/2$, with curved, ascending tips, attached in the deflexed portion, the ascending portion narrowly club-shaped, the tips hoodlike, the margins not toothed, the bases not pouched. Horns absent (reduced to a small, platelike appendage near the tip of the inner surface of the hoods). Fruits 6–13 cm long, erect or ascending from usually deflexed stalks, broadly lanceolate to ovoid in outline, the surface smooth, glabrous or more commonly minutely hairy. Seeds with the body 7–8 mm long, the margins narrowly winged, the terminal tuft of hairs white to light cream-colored. May–June.

Scattered mostly south of the Missouri River (southeastern U.S. west to Nebraska and Texas). Upland prairies and glades, usually on calcareous substrates; also roadsides and railroads.

2. Cynanchum L.

Plants with clear latex and thus watery sap (milky or yellow latex elsewhere). Stems twining, usually climbing, branched or unbranched. Leaves opposite, short- to long-petiolate. Leaf blades narrowly to broadly ovate or triangular-ovate, the base rounded, truncate, or cordate, the tip tapered to a sharp point, the margins flat or somewhat wavy, sometimes slightly curled under. Inflorescences solitary in the leaf axils, mostly relatively short-stalked, consisting of sometimes branched or slightly racemose, umbellate clusters. Calyces with the lobes spreading or loosely ascending at maturity, narrowly triangular or lanceolate to ovate, sparsely short-hairy on the outer surface. Corollas erect or ascending at flowering, white to cream-colored or dark purple to nearly black. Gynostegium appearing sessile or nearly so, the corona as long as or usually longer than the anther/stigma head, modified into 5 erect, lateral, petaloid segments or a shallow, cuplike, fleshy disc. Fruits pendant, narrowly elliptic-lanceolate to ovate in outline, circular or slightly flattened in cross-section (not angled), the surface smooth, glabrous or short-hairy. Seeds strongly flattened and narrowly winged, brown to dark brown, with a tuft of long, white, silky hairs at the tip. Two hundred to 400 or more species, nearly worldwide, mostly in tropical and warm-temperate regions.

Generic limits and the circumscriptions and relationships among subgroups in the *Cynanchum* alliance are among the most poorly understood in the Asclepiadaceae. Rosatti (1989) noted that because previous authors included so many species with exceptions to the morphological character states used to define most of the subgroups, comparisons of different classifications have been nearly impossible. Liede (1997) and her colleagues have been studying the infrageneric classification of the genus and its segregates for a number of years, but this classification should still be viewed as preliminary and not fully resolved. The two species present in Missouri have very different coronal structures and probably are not closely related. The North American *Ampelamus* group contains only the species treated here as *Cynanchum laeve*, a relatively isolated taxon whose relationships are not well understood (Sundell, 1981; Liede, 1996, 1997). *Cynanchum louiseae* is part of the *Vincetoxicum* group, a relatively cohesive segregate that comprises up to 100 or more Old World species (Liede, 1996), of which 3 have become introduced in temperate North America as escapes from cultivation (Sheeley and Raynal, 1996). It is likely that ongoing molecular studies will result in the breakup of *Cynanchum* in the broad sense into a number of smaller genera, but until further data become available in the literature it seems premature at the present time to recognize more than a single polymorphic genus.

1. Corollas white to light cream-colored; leaf blades deeply cordate at the base
   .... 1. C. LAEVE
1. Corollas dark purple to nearly black; leaf blades rounded, truncate, or shallowly cordate at the base .... 2. C. LOUISEAE

1. Cynanchum laeve (Michx.) Pers. (sand vine, climbing milkweed, blue vine)
Ampelamus laevis (Michx.) Krings
A. albidus (Nutt.) Britton
Pl. 223 a, b; Map 932

Stems 1–10 m long, sparsely to moderately hairy, at least toward the tip, usually in longitudinal lines. Leaf blades 2–10 cm long, triangular-ovate to broadly ovate, the base deeply cordate, glabrous or sparsely short-hairy, mostly along the veins. Inflorescences with 5–40 flowers. Calyx lobes 1.5–3.0 mm long. Corollas white to light cream-colored, the lobes 4–7 mm long, narrowly oblong to lanceolate, glabrous. Corona modified into 5 erect, lateral, petaloid segments, each segment 4–6 mm long, white, the basal portion oblong-ovate, the apical 1/3–1/2 divided into 2 erect, linear lobes. Fruits usually single, 8–14 cm long, narrowly ovate to ovate in outline. Seeds with the body 7–9 mm long, obovate in outline. July–September.

Common in northern and eastern Missouri, scattered in the remainder of the state (eastern U.S. and adjacent Canada west to Nebraska and Texas). Bottomland forests, banks of rivers and streams, and margins of ponds and lakes; also crop fields, fallow fields, gardens, yards, fencerows, roadsides, railroads, and open, disturbed areas.

Cynanchum laeve is a problem weed of crop fields and gardens, where it can be difficult to eradicate. However, it is cultivated as an ornamental in some places and is recommended by beekeepers as an excellent honey plant.

2. Cynanchum louiseae Kartesz & Gandhi (black swallowwort)
C. nigrum (L.) Pers. (1805), not *C. nigrum* Cav. (1793)
Vincetoxicum nigrum (L.) Moench
Map 933

Stems 1–2 m long, sparsely to moderately hairy, usually in longitudinal lines. Leaf blades 3–10 cm long, narrowly to broadly ovate or triangular-ovate, the base rounded, truncate, or shallowly cordate, sparsely to moderately short-hairy along the veins and margins. Inflorescences with 5–14 flowers. Calyx lobes 1.5–2.0 mm long. Corollas dark purple to nearly black, the lobes 2.0–3.5 mm long, ovate, minutely hairy on the inner surface and margins, glabrous on the outer surface. Corona 0.4–0.6 mm long, modified into a shallow, cuplike, fleshy disc,

160 ASCLEPIADACEAE

933. Cynanchum louiseae

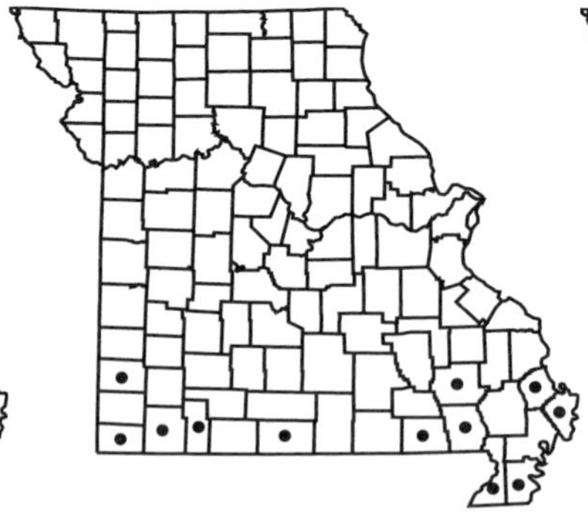

934. Gonolobus suberosus

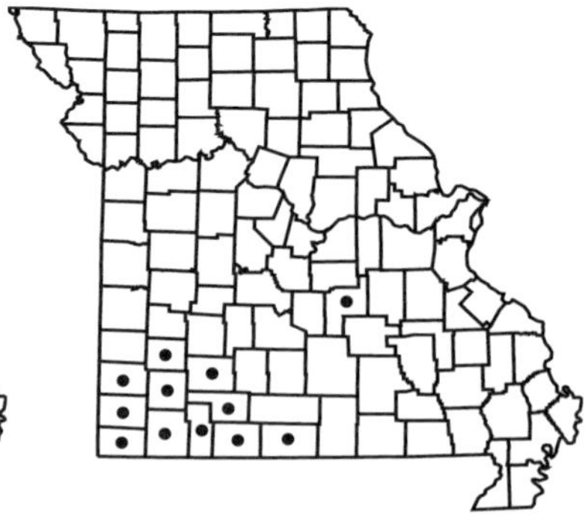

935. Matelea baldwyniana

this shallowly 5-lobed, dark purple to nearly black. Fruits single or paired, 4–7 cm long, narrowly elliptic-lanceolate in outline. Seeds with the body 7–8 mm long, ovate in outline. $2n=22$, 44. June–August.

Introduced, known thus far only from Adair County (native of Europe; escaped from cultivation sporadically in the northeastern U.S. and adjacent Canada west to Minnesota and Kansas, and disjunctly in California). Roadsides, gardens, and brushy, disturbed areas.

Cynanchum louiseae is sometimes cultivated as a garden ornamental. It was discovered growing in the Kirksville area in 1995 by Randy Walker and may be more widespread in northern Missouri than has been documented to date. Sheeley and Raynal (1996) discussed the distributional history of this species (as *Vincetoxicum nigrum*) in North America and indicated that the oldest specimen they were able to locate was collected in 1854 in Massachusetts. They reported a 1910 mixed collection of *C. nigrum* and *C. rossicum* from the city of St. Louis by Earl Sherff, but this collection originated from cultivated plants, and neither species has been recorded from spontaneous occurrences in the St. Louis area thus far.

Cynanchum rossicum (Kleopow) Borhidi (*V. rossicum* (Kleopow) Barbar.) is another native of Europe that has escaped sporadically in the northeastern United States and Canada. It differs from *C. louiseae* in its slightly larger, pink to light purple or maroon corollas and its more deeply lobed coronas. Some authors treat this taxon as a color phase of the closely related *C. vincetoxicum* (L.) Pers. (*V. hirundinaria* Medik.), which has otherwise yellowish white corollas. As noted above, the specimen upon which Sheeley and Raynal (1996) based their report of *C. rossicum* in Missouri consisted of cultivated material, and the species is thus excluded from the flora.

3. Gonolobus Michx. (angle-pod)

One hundred to 150 species, U.S., Mexico, Central America, South America, Caribbean Islands.

A number of authors have treated *Gonolobus* as a subgroup within *Matelea* (Steyermark, 1963; Drapalik, 1970; Hartman, 1986b; Gleason and Cronquist, 1991). Rosatti (1989) defended Woodson's (1941) decision to keep the two genera separate, based on differences in anther structure and fruit morphology, noting that in tropical America, where most of the species grow, the overall distinctions between the two genera are much clearer than they might seem in the limited context of only the few temperate species.

1. Gonolobus suberosus (L.) W.T. Aiton
Matelea suberosa (L.) Shinners
G. gonocarpos (Walter) L.M. Perry
M. gonocarpa (Walter) Shinners
Pl. 223 j–l; Map 934

Plants with milky latex and thus white sap. Stems 1–3(–6) m long, twining, usually climbing, branched or unbranched, sparsely to moderately pubescent with spreading hairs, usually also with sparse to moderate, minute glandular hairs. Leaves

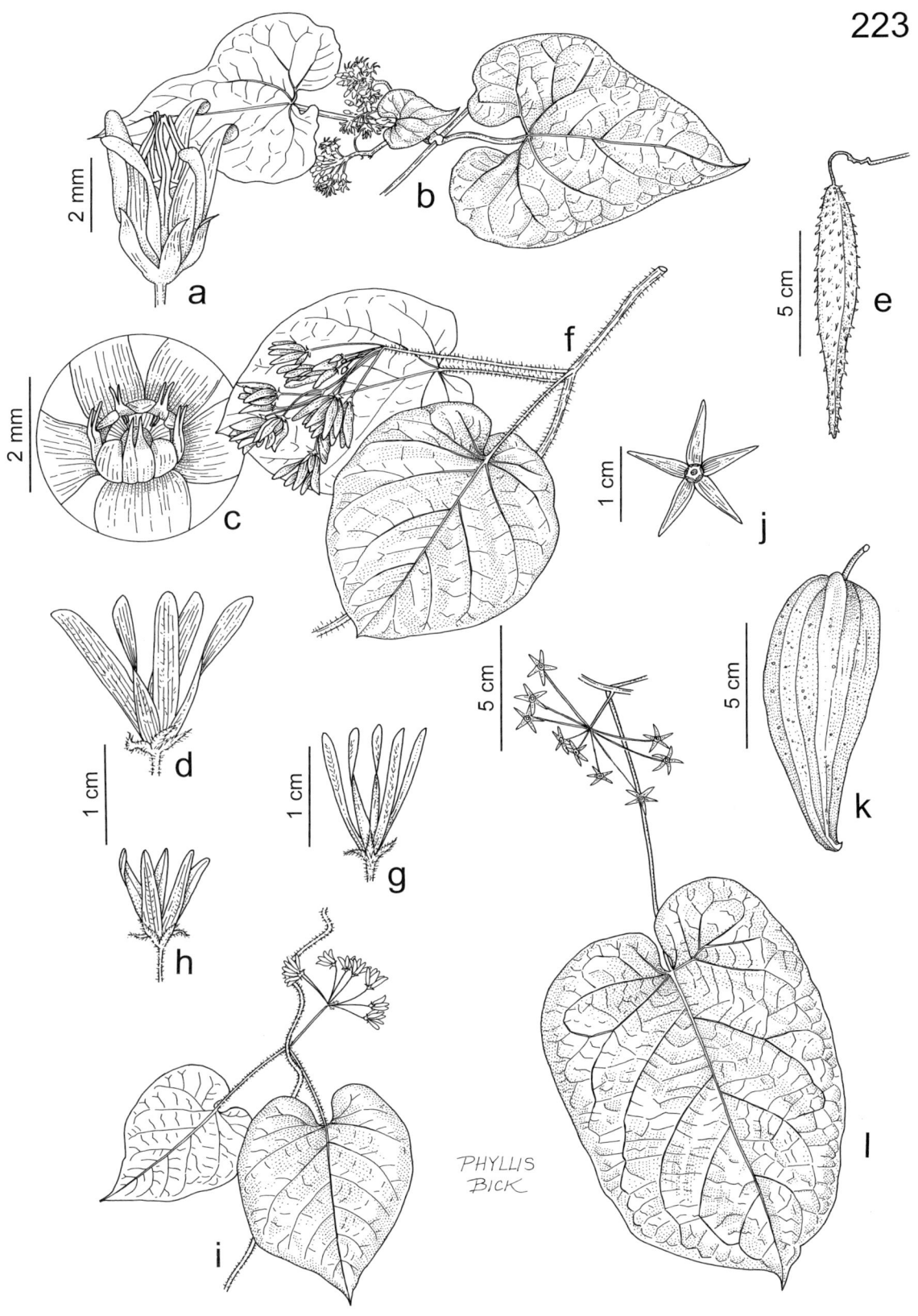

Plate 223. Asclepiadaceae. *Cynanchum laeve*, **a)** flower, **b)** habit. *Matelea decipiens*, **c)** flower detail, **d, g)** flower, **e)** fruit, **f)** habit. *Matelea baldwyniana*, **h)** flower, **i)** habit. *Gonolobus suberosus*, **j)** flower, **k)** fruit, **l)** habit.

opposite, mostly long-petiolate. Leaf blades 4–20 cm long, oblong-ovate to triangular-ovate or broadly ovate, the base deeply cordate, the tip abruptly or gradually tapered to a sharp point, the margins flat, sparsely short-hairy along the veins, usually also with sparse to moderate, minute glandular hairs. Inflorescences solitary in the leaf axils, sessile to long-stalked, consisting of sometimes branched, umbellate clusters with 2–25 flowers. Calyces with the lobes 2.5–6.0 mm long, spreading at maturity, narrowly triangular to narrowly lanceolate, glabrous except for sparse, short hairs along the margins toward the tip. Corollas with the lobes 7–15 mm long, strongly spreading at flowering, narrowly lanceolate, yellow to yellowish green, glabrous. Gynostegium appearing sessile or nearly so, the corona present as a flattened, fleshy, irregularly 5-angled disc shorter than the anther/stigma head, yellow to orangish brown. Fruits 7–15 cm long, pendant, lanceolate to narrowly ovate in outline, sharply 5-angled in cross-section, the surface smooth, glabrous. Seeds with the body 6–10 mm long, elliptic-obovate in outline, strongly flattened and with a narrow to relatively broad, irregularly toothed wing, dark brown, with a tuft of long, white, silky hairs at the tip. June–August.

Uncommon in southern Missouri, mostly in the Mississippi Lowlands Division (southeastern U.S. west to Kansas and Texas). Swamps, bottomland forests, lower ledges of bluffs, margins of lakes and sloughs, and banks of streams and rivers; also disturbed wooded areas.

The nomenclature and taxonomy of this species complex are relatively complicated. The present treatment follows that of Lipow and Wyatt (1998), who summarized problems of nomenclature and typification in the introduction to their study on the reproductive biology and breeding system of the species. Drapalik (1970) presented evidence that *G. gonocarpos* and *G. suberosus* should be treated as part of a single variable species, based on his observation of specimens intermediate between the two traditionally accepted taxa. Reveal and Barrie (1992) studied lectotypification (designation of an official type specimen) of the epithet *suberosus* and concluded that it has priority over *gonocarpos* if the two names are considered to represent a single species. More recently, Krings and Xiang (2005) concluded that the two taxa are more or less distinguishable in the complex, but they emphasized different floral characters to separate the two than those used by most earlier authors. Krings and Xiang concluded that if two entities are to be recognized within *G. suberosus*, then the western taxon (including Missouri) with glabrous, concolorous corollas should be called var. *granulatus* (Scheele) Krings & Q.Y. Xiang.

4. Matelea Aubl. (climbing milkweed)

Plants with white latex and thus milky sap. Stems 1–3 m long, twining, usually climbing, branched or unbranched, moderately pubescent with spreading hairs and also minute glandular hairs (these often dark purple wherever they occur). Leaves opposite, mostly long-petiolate. Leaf blades ovate to broadly ovate or nearly circular in outline, less commonly oblong-ovate, the base deeply cordate, the tip gradually or abruptly tapered to a sharp point, the margins flat or slightly curled under, the surfaces moderately to densely pubescent with short hairs and minute glandular hairs. Inflorescences solitary in the leaf axils, mostly relatively long-stalked, consisting of sometimes branched, umbellate clusters, with 4–25 flowers. Calyces with the lobes spreading at maturity, hairy on the outer surface. Corollas more or less spreading to loosely ascending at flowering (erect in bud and sometimes after flowering), white to light cream-colored or dark purple to nearly black, hairy on the outer surface. Gynostegium appearing sessile or nearly so, the corona longer than the anther/stigma head, modified into a shallow, 5-lobed, cup-shaped crown, with narrow appendages between the lobes. Fruits pendant, lanceolate to narrowly ovate in outline, circular or slightly flattened in cross-section (not noticeably angled), the surface covered with slender, warty, hardened projections, minutely pubescent, usually with mostly glandular hairs. Seeds strongly flattened and winged, brown to dark brown, with a tuft of long, silky hairs at the tip. Two hundred to 250 species, U.S., Mexico, Central America, South America, Caribbean Islands.

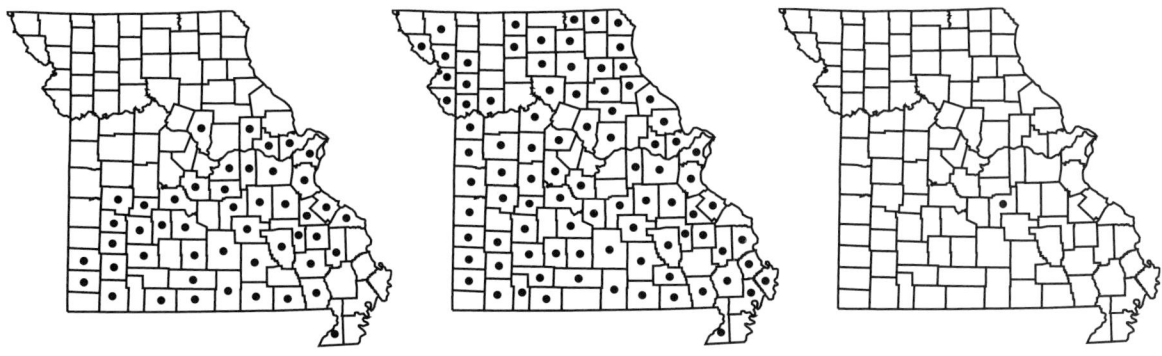

936. Matelea decipiens   937. Achillea millefolium   938. Achillea ptarmica

Vegetatively, Missouri *Matelea* sometimes cannot be distinguished from *Gonolobus,* although in the latter genus the stems often are more sparsely hairy, and the frequency of minute glandular hairs is usually far less than in *Matelea*. Even species within *Matelea* cannot be discriminated reliably based only on vegetative characters.

1. Corollas white to light cream-colored .... 1. M. BALDWYNIANA
1. Corollas dark purple to nearly black, rarely greenish yellow .... 2. M. DECIPIENS

1. Matelea baldwyniana (Sweet) Woodson

Pl. 223 h, i; Map 935

Leaf blades 5–16 cm long. Calyces with the lobes 1.5–3.5 mm long, lanceolate, moderately to densely pubescent on the outer surface with short hairs and usually also minute glandular hairs. Corollas with the lobes 7–13 mm long, narrowly oblong to narrowly oblanceolate or linear, white to light cream-colored, moderately pubescent on the outer surface with short hairs and usually also sparse, minute glandular hairs. Corona white to light cream-colored or light yellow, the lobes bluntly triangular, the appendages about twice as long as the lobes, each deeply divided into 2 narrowly triangular, sharply pointed lobes. Fruits 6.5–9.0 cm long. Seeds 7–9 mm long, ovate, narrowly winged, the terminal tuft of hairs white to light cream-colored or tan. May–early July.

Scattered in the western portion of the Ozark Division, and locally eastward to Phelps County (southeastern U.S. west to Missouri and Oklahoma). Glades, savannas, openings of dry upland forests, and bases of bluffs, on calcareous substrates.

2. Matelea decipiens (Alexander) Woodson

Pl. 223 c–g; Map 936

Leaf blades 4–15 cm long. Calyces with the lobes 2–4 mm long, lanceolate to ovate, moderately to densely pubescent on the outer surface with short hairs and also minute glandular hairs. Corollas with the lobes 7–18 mm long, narrowly oblong to narrowly oblanceolate or linear, dark purple to nearly black, rarely greenish yellow, moderately pubescent on the outer surface with short hairs and usually also sparse, minute glandular hairs (these difficult to see against the dark background color). Corona dark purple to nearly black, rarely greenish yellow, the lobes rounded to broadly and bluntly triangular, the appendages about twice as long as the lobes, each deeply divided into 2 narrowly triangular, sharply pointed lobes. Fruits 8–11 cm long. Seeds 8–9 mm long, ovate, narrowly to relatively broadly winged, the terminal tuft of hairs white to light cream-colored or tan. May–June.

Scattered to common, mostly south of the Missouri River (southeastern U.S. west to Kansas and Texas). Glades, savannas, tops of bluffs, mesic to dry upland forests, banks of rivers and streams, and less commonly bottomland forests; also roadsides.

A specimen from Crawford County documents a single plant with greenish yellow corollas growing within a population of normal purple-flowered plants. Similar rare mutants lacking anthocyanin pigments in the corolla have been documented in the closely related *M. obliqua* (Jacq.) Woodson, another purple-flowered species of *Matelea* (Drapalik, 1970).

Steyermark (1963) included *M. obliqua* in the flora of Missouri based on uncited specimens from four eastern counties. Thus far, all of the specimens located that were labeled as this species are either vegetative gatherings or clearly referable to *M. decipiens*. Drapalik (1970) also mentioned Missouri in the range of *M. obliqua* without citing voucher specimens. The only specimen located that was annotated by him as this species is a historical collection from the state of Mississippi that he thought might possibly have originated from Mississippi County, Missouri. On this basis, the species is excluded from the Missouri flora, but it should be searched for, especially on calcareous substrates in the eastern portion of the Ozarks and Ozark Border, as it has been recorded from Illinois, Kentucky, and Arkansas. *Matelea obliqua* tends to have corolla lobes with a greater length-to-width ratio (mostly 4–6 times as long as wide vs. mostly 2–4 times in *M. decipiens*). The corona tip in *M. obliqua* is a low crown that is only slightly longer than the anther/stigma head and has 5 short (about as long as) 2-lobed appendages between the lobes, giving the corona margin an undulate appearance. In contrast, the corona lobes in *M. decipiens* extend well past the anther/stigma head, with the appendages somewhat strap-shaped and about twice as long as the lobes.

ASTERACEAE (COMPOSITAE) (Sunflower Family)

Plants annual or perennial herbs (shrubs or less commonly trees), rarely scrambling or climbing. Stems branched or unbranched, rarely twining. Leaves alternate, opposite, or less commonly whorled, occasionally all or mostly basal, variously simple or compound, petiolate or sessile, rarely perfoliate, lacking stipules. Inflorescences terminal and/or axillary, composed of heads, these 1 to many per stem, sessile or stalked, solitary or more commonly in secondary inflorescences of clusters, spikes, racemes, or panicles. Heads composed of a flat to conical or rarely concave receptacle surrounded by an involucre of bracts and with few to many dense florets, these sessile or very short stalked on the receptacle, sometimes with subtending receptacular (chaffy) bracts or hairs. Involucral bracts few to many, of various shapes and sizes, arranged in 1 to several series, rarely the inner series fused into a tube or cup. Flowers within a head (collectively referred to as florets) all perfect or sometimes some or all pistillate or staminate (Missouri plants then completely or incompletely monoecious, elsewhere less commonly dioecious), epigynous, actinomorphic (disc florets) or zygomorphic (ray or ligulate florets). Calyx represented by a pappus of hairs, awns, and/or scales, this sometimes reduced to a low crown or lacking. Corollas (3–)5-lobed, the lobes sometimes difficult to interpret in ray and ligulate florets. Stamens (4)5, the filaments usually free but attached to the corolla tube, the anthers usually fused laterally into a tubular ring, dehiscing by vertical slits along the inner side. Pistil 1 per floret, composed of 2 fused carpels, but with only 1 locule and 1 ovule. Placentation basal. Style 1, forked toward the tip, each of the 2 branches with an elongate stigmatic band or pair of lines along the inner side toward the tip (the tip itself often nonstigmatic). Fruits achenes. About 1,535 genera, 23,000–32,000 species, cosmopolitan.

Specialists in the Asteraceae and Orchidaceae have competed for decades for the honor of claiming their group to be the largest family of flowering plants, but in recent years the orchidologists seem to be outpacing the asterologists in the race to describe new species. Regardless, the Asteraceae are one of the largest and most morphologically diverse plant families, found in nearly all habitats and having almost every type of growth form imaginable. Given its size, the family has only moderate economic importance. Numerous species are cultivated as ornamentals and specimen plants, but a number of other species are considered noxious weeds, both in agricultural areas and in natural plant communities. Although several species are used as foods and beverages, the composites overall are less important in this regard than some other relatively large families, such as the grasses and legumes. Sunflower (*Helianthus annuus*), which is grown for its edible seeds and the oil that they contain, has perhaps the greatest economic impact. Natural products produced by members of the family include pyrethrin insecticides (originally from *Tanacetum*), an alternative source of natural rubber (*Parthenium*), resins (several genera, but notably *Grindelia*), and various medicinal remedies. The windborne pollen grains of *Ambrosia, Iva,* and *Xanthium* are strong allergens, contributing heavily to the incidence of hay fever during the late summer and autumn.

The most distinctive feature of the family is the dense grouping of flowers (florets) into a head that functions much like an individual flower in most other plant families. Heads may contain from few to numerous florets. In a few genera, the head may contain only a single floret, and such reduced heads may even be grouped into secondary heads. The involucre associated with each head consists of bracts that are most often green and herbaceous, but they can be papery, leathery, or hardened (occasionally burlike) in a minority of genera. These involucral bracts may occur in a single series and be similar in shape and size, but more commonly they form 2 to several overlapping series, with the outermost bracts sometimes noticeably shorter or longer than the others. In a few groups, an inner series of bracts is fused into a cuplike or cylindrical structure, usually with shorter, free bracts below or outside these. The modified branch tip (receptacle) that forms the center of each head varies from elongate-cylindrical to cone-shaped in some of the coneflowers to hemispherical or nearly spherical in several genera, or strongly concave in a few species of cudweeds (*Gamochaeta*); however, in most species it is relatively flat to slightly convex (Pl. 289 e). Subtending each of the florets in some species is a scalelike bract (Pl. 289 a), in which case the receptacle is said to be chaffy (vs. naked). In a few groups, the floret bases are surrounded by short or long hairs. The bracts subtending individual florets are known as chaffy or receptacular bracts, in contrast to the involucral bracts that subtend an entire head. Chaffy bracts are scalelike and hard or papery in texture. In many cases, the best way to observe the shape of the receptacle and the presence of chaffy bracts is to section a head lengthwise (Pl. 289 e) and to examine the cut surface.

The florets themselves are quite variable and provide characters important in the delineation of tribes, genera, and species. Heads may contain all perfect florets, all staminate or pistillate florets, or a mixture of these types. In some species, some of the florets may be totally neutral, functioning only as a visual attractant to pollinators. When two different types of florets are present, these are usually distinguishable based on differences in size and shape of their component structures. Staminate florets almost always have a slender, nonfunctional ovary and may even produce a style that remains undivided at the tip. Pistillate florets do not produce staminodes.

In all Asteraceae, what one might think of as normal sepals are absent and instead the outermost floral whorl is known as a pappus. Most species have a pappus consisting of numerous capillary bristles (fine hairs), plumose (featherlike) bristles, 1 to several awns (stiff, stout, bristlelike structures visibly tapered from the base), few to several scales, or some combination of these structures. Rarely, the pappus is reduced to a low crown or rim of tissue or is totally absent.

The most visible floral variations are in the corolla are size and shape. Many members of the family produce actinomorphic florets referred to as disc florets. In these, the corolla is usually 5-lobed. It may be a slender tube (as in most thistles), or it may be shorter and differentiated into a narrow basal tube, expanded throat, and erect or spreading lobes. Occasionally, one of the sinuses between the lobes may be deeper than the others. Zygomorphic florets among Missouri genera occur in two distinct types. In the tribe Cichorieae, the florets of a head are all perfect and similar in size and shape, with a single long, strap-shaped lobe that usually has 5 teeth at the tip. Such ligulate florets are unique to this tribe. Ray florets, in contrast, occur in most other tribes and are always positioned along the periphery of a head that also contains a central group of disc florets. They may be pistillate (look for a divided style at flowering) or neutral. Heads with all-ligulate florets are referred to as ligulate, heads with only disc florets are called discoid, and heads with central disc florets and marginal ray florets are referred to as radiate.

Classification within the Asteraceae has remained controversial. Traditionally, the family was divided on a worldwide basis into two subfamilies, Lactucoideae (Liguliflorae) and Asteroideae (Tubuliflorae), with 13–19 total tribes mostly in the Asteroideae. Beginning in the 1970s, a lengthy series of studies by various authors based on different types of data began to dismantle the traditional approach, combining some tribes and segregating some genera into new tribes (for a review, see Bremer, 1994). Molecular studies of the family began with the work of Jansen and Palmer (1987), whose research suggested that the most primitive living Asteraceae were a small group of morphologically anomalous genera of shrubs growing in arid to seasonally dry portions of South America for which a third subfamilial name, Barnadesioideae, was proposed. Subsequent molecular and morphological analyses have broken up the traditional Lactucoideae (now called Cichorioideae) with its single tribe Cichorieae (sometimes referred to by the synonymous Lactuceae) into a series of groups recognized variously as tribes or subfamilies, none of which occur in Missouri except Cichorieae in the strict sense. These studies also have tended to group the traditional Vernonieae and Cynareae (now Cardueae) with the cichorioid groups. Perhaps the greatest disagreement over how many tribes to recognize has focused on the traditional tribes Eupatorieae, Helenieae, and Heliantheae, which various specialists have treated as a single tribe or a multitude, based primarily on molecular evidence (Baldwin et al., 2002). Different authors (Bremer, 1994; Panero and Funk, 2002) presently recognize 3–11 subfamilies and 17–35 total tribes worldwide. The classification of genera of Asteraceae into natural lineages is likely to continue to change as future research incorporates more species samples and additional data sets into the phylogenetic analyses.

The present treatment has borrowed shamelessly from the Flora of North America's Asteraceae volumes (Flora of North America Editorial Committee, 2006), which were under production at the time that the Missouri draft was being prepared. This classification is slightly more conservative than the last major familial summary (Bremer, 1994), wherein the authors of the chapter on Helenieae, Karis and Ryding, continued to accept this tribe while admitting that it was not a natural group. The present treatment does not, however, reflect the current trend toward recognition of many more tribal names to circumscribe all of the fully monophyletic lineages while also preserving all of the traditional tribal epithets (Baldwin et al., 2002; Panero and Funk, 2002). It has the practical advantages of deviating less from the traditional classification in Steyermark (1963) and that users of the present book will be able to locate general information on temperate North American composites more easily within the Flora of North America volumes. Also, many of the larger, traditional genera have been broken up and the tribal grouping accepted here continues to place the segregates relatively closely together in the work. Those interested in comparing the traditional tribal organization of Missouri Asteraceae in Steyermark's (1963) treatment with that accepted here should note that: 1) Steyermark's Inuleae have been split into three tribes, Gnaphalieae, Inuleae, and Plucheeae; 2) Cynareae are now known by the older name Cardueae; and 3) Steyermark's Helenieae and Heliantheae are combined under the latter name.

Because the question of how many subfamilies should be formally recognized taxonomically remains controversial, it seems most useful to deal with the Missouri Asteraceae only at the tribal level. The following key to tribes of Missouri Asteraceae tends to favor relatively easily observed characters, but flowering heads must be present to determine most of the tribes. It draws inspiration mostly from keys in Steyermark (1963), Cronquist (1980), Gleason and Cronquist (1991), and Voss (1996).

1. Florets all ligulate (zygomorphic, perfect, the corollas with the short tube split at the tip into an elongate, flattened, strap-shaped lobe, this usually 5-toothed at the tip); sap milky, usually white, less commonly tan or orange
   . . . . 4. CICHORIEAE, p. 343
1. Some or all of the florets discoid (more or less actinomorphic although sometimes elongate and slender, perfect or imperfect, the corollas with 5 shallow or less commonly deeper lobes), sometimes ray florets also present along the margin of the head (these zygomorphic, pistillate or functionally sterile, the corollas with the short tube split at the tip into a ligulate lobe, this sometimes with 2 or 3 teeth at the tip); sap clear, not appearing milky
   2. Stems and/or leaves appearing spiny or prickly, sometimes only with spine-tipped teeth along the leaf margins . . . . 3. CARDUEAE, p. 315
   2. Stems and leaves not spiny or prickly, but sometimes with roughened texture or stiff, pustular-based hairs
      3. Some or all of the heads with the involucre either burlike and enclosing the florets or having hooks, spines, and/or tubercles on the bracts (sometimes only at the bract tip)
         4. At least some of the involucres burlike, enclosing the florets, the surface and/or tip of the bracts with hooked tubercles (*Xanthium*)
            . . . . 7. HELIANTHEAE, p. 435
         4. Involucres not burlike, but some or all of the bracts with straight spines or tubercles at the tip
            5. Pappus absent (do not confuse pappus with chaffy bracts) (*Ambrosia*) . . . . 7. HELIANTHEAE, p. 435
            5. Pappus present on all or most of the florets . . . . . . 3. CARDUEAE, p. 315
      3. Heads with the involucre neither burlike nor bearing hooks or spines on the involucre (specimens with heads appearing immersed in cottony hairs should be keyed under this lead)
         6. Receptacle with well-developed chaffy bracts or numerous conspicuous bristles subtending or intermixed with all or many of the florets (occasionally only produced toward the central portion or toward the margin of the receptacle)
            7. Individual heads difficult to differentiate, grouped into a dense, more or less spherical, headlike cluster at each stem or branch tip, the entire mass obscured by dense, woolly hairs; true involucral bracts absent, each head merely with chaffy bracts (*Diaperia*)
               . . . . 6. GNAPHALIEAE, p. 425
            7. Individual heads easily differentiated, not grouped into dense, spherical masses (but grouped into hemispherical, headlike clusters subtended by 3 leaflike bracts in *Elephantopus* [Vernonieae]), not obscured by woolly hairs; involucral bracts present and well developed around each head
               8. Involucral bracts with a well-differentiated appendage at the tip, this flattened, abruptly different in color and/or texture from the basal portion of the bract, and often with long-fringed or bristly margins; anthers with the base prolonged into a pair of slender, elongate, tail-like lobes (*Centaurea*)
                  . . . . 3. CARDUEAE, p. 315
               8. Involucral bracts not differentiated into a basal portion and a dissimilar apical appendage; anthers with the base truncate, rounded, or cordate with short, rounded lobes

9. Involucral bracts entirely dry and papery or scalelike, straw-colored to tan or brown, less commonly with a narrow, green central band or small apical region; style branches mostly lacking a well-defined, sterile apical portion beyond the stigmatic lines, often truncate and densely hairy at the tip . . . . 1. ANTHEMIDEAE, p. 171

9. Involucral bracts with a well-developed, green central portion (the green color sometimes partially masked by purplish pigments), with or without thinner membranous to papery margins; style branches mostly with a well-defined (but sometimes short), sterile apical portion beyond the stigmatic lines or bands, this variously shaped and sometimes densely hairy . . . . 7. HELIANTHEAE, p. 435

6. Receptacle naked, sometimes with minute, irregular or toothed ridges around the attachment points of the florets or occasionally with a few short, inconspicuous bristles or bracts toward the margin of the head

10. Pappus absent, a minute crown, or of 1–20 awns or scales, these uncommonly with 1 to several bristlelike tips

11. Disc florets with the corolla bright yellow to orangish yellow or orange

12. Involucral bracts entirely dry and papery or scalelike, straw-colored to tan or brown, less commonly with a narrow, green central band or small apical region; stem leaves alternate

13. Style branches mostly lacking a well-defined, sterile apical portion beyond the stigmatic lines, often truncate and densely hairy at the tip; leaf blades most commonly deeply lobed (unlobed in *Achillea ptarmica* and some *Artemisia* species, all of which are aromatic when bruised or crushed) . . . . 1. ANTHEMIDEAE, p. 171

13. Style branches with a well-defined, sterile apical portion beyond the stigmatic lines, this more or less elongate, usually densely and minutely hairy on the outer surface and glabrous on the inner surface; leaf blades unlobed, the margins entire or toothed; plants not aromatic when bruised or crushed . . . . 2. ASTEREAE, p. 198

12. Involucral bracts with a well-developed green central portion (the green color sometimes partially masked by purplish pigments), with or without thinner membranous to papery margins; stem leaves alternate or opposite (the leaves rarely all or mostly basal)

14. All of the leaves or at least those toward the stem base opposite or, if all alternate, then the stem leaves with the blades lobed or divided . . . . 7. HELIANTHEAE, p. 435

14. All of the leaves alternate and none of the stem leaves lobed or divided

15. Stems wingless (except in *Boltonia decurrens,* with white to pale pink ray corollas); ray florets absent or with the corolla white, pink, or yellow; style branches with a well-defined, sterile apical portion beyond the stigmatic lines, this more or less elongate, usually densely and minutely hairy on the outer surface and glabrous on the inner surface
. . . . 2. ASTEREAE, p. 198

15. Stems appearing narrowly to broadly winged; ray florets with the corolla yellow (sometimes with reddish lines or reddish-tinged at the base); style branches lacking a well-defined, sterile apical portion beyond the stigmatic lines, more or less truncate and densely hairy at the tip (*Helenium*)
.... 7. HELIANTHEAE, p. 435
11. Disc florets with the corolla white, pink, lavender, purple, or blue
    16. Heads radiate (*Helenium*) .... 7. HELIANTHEAE, p. 435
    16. Heads discoid
        17. Primary heads grouped into hemispherical, headlike clusters subtended by an involucre of 3 leaflike bracts (*Elephantopus*)
        .... 11. VERNONIEAE, p. 590
        17. Primary heads sometimes in clusters, but not aggregated into secondary heads subtended by an involucre of bracts
            18. Stem leaves all or mostly opposite, the blade triangular-ovate to ovate (*Ageratum*).................. 5. EUPATORIEAE, p. 393
            18. Stem leaves alternate, the blades linear (*Palafoxia*)
            .... 7. HELIANTHEAE, p. 435
10. Pappus of (10–)20 to numerous slender (capillary) bristles, occasionally also with an outer series of shorter scales
    19. Heads radiate, the ray florets with the corolla well developed and showy
        20. Involucre 2.0–2.5 cm long, 3–5 cm in diameter; anthers with the base prolonged into a pair of slender, elongate, tail-like lobes; style branches lacking a sterile tip, the stigmatic lines continuous around the rounded, more or less glabrous tip (*Inula*) .... 8. INULEAE, p. 572
        20. Involucre 0.3–1.5 cm long, 0.3–3.0 cm in diameter; anthers with the base truncate, rounded, or cordate with short, rounded lobes; style branches with or without a sterile tip, the stigmatic lines not continuous around the tip, this variously truncate to elongate and often minutely hairy
            21. Involucral bracts in 2–9 unequal to subequal noticeably overlapping series; style branches with the sterile apical portion more or less elongate, usually densely and minutely hairy on the outer surface and glabrous on the inner surface................. 2. ASTEREAE, p. 198
            21. Involucral bracts of a single series of bracts of similar size and length, these not or only slightly laterally overlapping, sometimes fused laterally, usually subtended by an outer series of fewer minute, slender, nonoverlapping bracts; style branches lacking a well-defined, sterile apical portion, the tip usually truncate and densely hairy
            .... 10. SENECIONEAE, p. 576
    19. Heads discoid or apparently so, sometimes with pistillate marginal florets but these with relatively short corollas (shorter than to about as long as those of the other disc florets), sometimes completely or incompletely dioecious in Gnaphalieae, the staminate and pistillate flowers then with similar corollas
        22. Involucral bracts white to straw-colored or purplish-tinged, sometimes pale greenish-tinged toward the base when young, this fading to white or tan as the head matures; corollas (except those of the innermost florets in *Pluchea* [Plucheeae]) slender and tubular, with minute, inconspicuous lobes (except sometimes in staminate florets of the dioecious *Antennaria* [Gnaphalieae], with spreading, slightly more conspicuous lobes)

23. Involucral bracts thin and papery throughout, dull or somewhat shiny, glabrous except for the dense, woolly hairs at the base, these partially or nearly completely obscuring the bracts; leaves with densely woolly pubescence, at least on the undersurface, sometimes also with stalked glands . . . . 6. GNAPHALIEAE, p. 425
23. Involucral bracts thin and papery above a thicker, herbaceous to leathery basal portion, minutely hairy and usually also glandular, but not woolly, the hairs not obscuring the bracts; leaves minutely hairy on the undersurface, also with sessile glands (*Pluchea*)
. . . . 9. PLUCHEEAE, p. 573
22. Involucral bracts green, at least toward the tip or along the midnerve, sometimes partially purplish-tinged; corollas of all but the outermost florets (these relatively numerous in *Pluchea* [Plucheeae]) relatively showy, the tube more or less slender but flared or swollen toward the tip, with sometimes small but noticeable, spreading lobes
24. Disc florets with the corollas bright yellow
25. Involucral bracts in 2–9 unequal to subequal, noticeably overlapping series; style branches with the sterile apical portion more or less elongate, usually densely and minutely hairy on the outer surface and glabrous on the inner surface . . . . 2. ASTEREAE, p. 198
25. Involucral bracts of a single series of bracts of similar size and length, these not or only slightly laterally overlapping, sometimes fused laterally, usually subtended by an outer series of fewer minute, slender, nonoverlapping bracts; style branches lacking a well-defined, sterile apical portion, the tip usually truncate and densely hairy
. . . . 10. SENECIONEAE, p. 576
24. Disc florets with the corollas white, pale cream-colored, pink, purple, or blue
26. Involucral bracts of a single series of bracts of similar size and length, these not or only slightly laterally overlapping, sometimes fused laterally, sometimes subtended by an outer series of fewer minute, slender, nonoverlapping bracts . . . . 10. SENECIONEAE, p. 576
26. Involucral bracts in 2–9 unequal to subequal, noticeably overlapping series
27. Stem leaves all or mostly opposite or whorled . . . . . 5. EUPATORIEAE, p. 393
27. Stem leaves alternate
28. Pappus of each disc floret double, of an inner series of numerous capillary bristles and an outer series of minute scales or bristles (*Vernonia*) . . . . 11. VERNONIEAE, p. 590
28. Pappus of each disc floret single, the bristles all similar in length
29. Inflorescences unbranched terminal spikes or spikelike racemes, sometimes leafy and the lowermost heads then appearing axillary (*Liatris*) . . . . 5. EUPATORIEAE, p. 393
29. Inflorescences flat-topped, rounded or elongate panicles (in poorly developed plants sometimes nearly racemose or reduced to a few stalked clusters of heads), the branches sometimes appearing somewhat racemose

30. Involucre 1.5–4.0 mm long; most of the florets pistillate and with the corolla reduced to a minute flap or short fringe on 1 side of the tip of the slender tube, only the central florets of the head truly discoid (perfect and with actinomorphic corollas) (*Conyza*) .... 2. ASTEREAE, p. 198
30. Involucre 4–15 mm long; all of the florets perfect or most of the florets pistillate (only the inner florets perfect), but then with a slender, tubular corolla with 3 minute, equal lobes at the tip
    31. Leaf blades linear to lanceolate or narrowly elliptic; pappus bristles plumose, not fused at the base, white to light tan; florets all discoid, perfect; plants not aromatic nor with a faint, sweet fragrance when bruised or crushed (*Brickellia*) .... 5. EUPATORIEAE, p. 393
    31. Leaf blades ovate to elliptic; pappus bristles minutely toothed or barbed, not plumose, fused into a ring at the base, usually pinkish-tinged; florets of 2 kinds, the innermost florets perfect and discoid, those of the outer several series pistillate, with a slender, tubular corolla with 3 minute, equal lobes at the tip; plants strongly aromatic with a camphorlike or musky odor when bruised or crushed (*Pluchea*) .... 9. PLUCHEEAE, p. 573

1. Tribe Anthemideae Cass.
(Bremer and Humphries, 1993)

Plants annual, biennial, or perennial herbs, sometimes from a woody rootstock, often aromatic, the sap not milky. Stems not spiny or prickly. Leaves alternate (rarely opposite elsewhere), sometimes also in a basal rosette, sometimes appearing fasciculate, sessile or short-petiolate, not spiny or prickly. Leaf blades often lobed, the venation pinnate or palmate, with 1 or few main veins. Inflorescences terminal panicles, clusters of heads, or the heads solitary at the branch tips. Heads discoid or radiate. Involucre of 2 to several series of bracts, not spiny or tuberculate. Receptacle flat to hemispherical or conical, naked or at least some of the florets subtended by chaffy bracts, receptacle rarely long-hairy. Ray florets (when present) pistillate or less commonly sterile, sometimes inconspicuous, the corollas sometimes very short, white, cream-colored, or yellow, rarely pink. Disc florets all perfect or the innermost sometimes staminate or the outermost sometimes pistillate, the corolla yellow to greenish yellow, less commonly white or pink, the 4 or more commonly 5 short lobes spreading to ascending. Pappus of few scales or reduced to a low collar or crown, sometimes absent, when present persistent at fruiting. Stamens with the filaments not fused together, the anthers fused into a tube, each tip without a distinct appendage, each base truncate or with a pair of short lobes. Style branches usually somewhat flattened, each with a stigmatic line along each inner margin, the sterile tip short and usually truncate, often with dense, minute hairs. Fruits sometimes dimorphic (the outer series then thicker-walled and with different surface ornamentation), circular or more commonly angled in cross-section, oblong to slightly wedge-shaped in profile, sometimes ribbed, rarely appearing nearly winged, not beaked. About 110 genera, about 1,750 species, worldwide.

The tribe Anthemidae contains a number of garden ornamentals, including the numerous cultivated members of the *Chrysanthemum* alliance (Soreng and Cope, 1991). Some members of the tribe are pharmacologically or biochemically important and have been used as medicinal herbs, in cosmetics and shampoos, and as a source of insecticides. Similarly, because of their aromatic compounds, some species have been eaten as potherbs and garnishings or used in herbal teas and liquors. However, some members of the tribe are noxious weeds of croplands and pastures.

The taxonomy of the tribe is still not fully understood, and several of the generic complexes require further systematic and phylogenetic study. Although the breakup of the formerly large genus *Chrysanthemum* L. is now widely accepted (for discussion, see the treatment of *Leucanthemum*), the transfer of some of the species groups into *Tanacetum* remains controversial. Similarly, many botanists have not accepted the segregation of *Seriphidium* (Besser ex Less.) Fourr. from *Artemisia*. The generic taxonomy of the chamomiles and their relatives remains especially confusing. The present treatment of the tribe may not be entirely satisfying to readers (for it is less than satisfying to the author), and it should be regarded as an interim classification at best.

Note that the genus *Hymenopappus,* which was included in the Anthemidae in some of the older botanical literature, is here treated in the helenioid group of Heliantheae, following Bremer and Humphries (1993).

1. Heads discoid or appearing so, the marginal florets sometimes pistillate and slightly zygomorphic
   2. Inflorescences spikes or racemes, sometimes grouped into panicles, these usually elongate or pyramidal, the branches spikelike, racemose, or sometimes reduced to small clusters sessile along the main axis . . . . 3. ARTEMISIA
   2. Inflorescences solitary heads or loose clusters of heads at the branch tips, or flat-topped to dome-shaped panicles with solitary heads or clusters of heads at the ultimate branch tips
      3. Receptacles hemispherical or strongly convex, elongating as the fruits mature. . . . . 6. MATRICARIA
      3. Receptacles flat or slightly convex, not elongating as the fruits mature
         . . . . 7. TANACETUM
1. Heads radiate
   4. Rays yellow . . . . 4. COTA
   4. Rays white
      5. Receptacle chaffy, the bracts sometimes absent from the marginal portion
         6. Inflorescences dense panicles, flat-topped to dome-shaped; ray florets with corollas 2–5(–7) mm long; pappus absent; fruits flattened
            . . . . 1. ACHILLEA
         6. Inflorescences of solitary heads at the branch tips; ray florets with corollas 5–15 mm long; pappus a short collar or crown, rarely absent; fruits more or less (8–)10-ribbed. . . . . 2. ANTHEMIS
      5. Receptacle naked
         7. Receptacle flat or slightly convex, not or only slightly elongating as the fruits mature; ultimate segments or lobes of the leaves variously shaped but not narrowly linear or threadlike
            8. Inflorescences of solitary heads at the stem tips; leaf blades narrowly obovate to oblong-oblanceolate in outline (those of the uppermost leaves sometimes linear), pinnately lobed or less commonly only toothed, the lobes or teeth mostly more than 7
               . . . . 5. LEUCANTHEMUM
            8. Inflorescences of flat-topped to dome-shaped terminal panicles; leaf blades ovate to elliptic in outline, pinnately compound or less commonly the uppermost only deeply lobed, the primary leaflets (or lobes) mostly 3 or 5 . . . . 7. TANACETUM

7. Receptacle hemispherical or strongly convex, tending to elongate as the fruits mature; ultimate segments of the leaves narrowly linear or threadlike
    9. Fresh plants strongly aromatic when bruised (with the aroma of pineapple); receptacle at fruiting conical; fruits 5-ribbed .... 6. MATRICARIA
    9. Fresh plants odorless or nearly so when bruised; receptacle at fruiting hemispherical to dome-shaped; fruits 3-ribbed (2 marginal ribs and a winglike rib on 1 of the faces) .... 8. TRIPLEUROSPERMUM

1. Achillea L. (yarrow, milfoil)

Plants perennial herbs (annual elsewhere), with rhizomes, weakly to strongly aromatic with an unpleasant odor. Stems erect or ascending, unbranched below the inflorescence or less commonly few-branched, sometimes with short, leafy branches in the leaf axils, finely ridged to angled, sparsely to densely pubescent with white, woolly hairs. Leaves alternate and basal, the basal leaves sometimes withered by flowering time, sessile or short-petiolate with somewhat broadened, more or less clasping bases. Leaf blades narrowly oblong to lanceolate or oblanceolate, unlobed or deeply 2(3) times pinnately lobed, glabrous or sparsely to densely pubescent with woolly hairs. Inflorescences panicles at the stem tips, rarely reduced to a cluster of heads, dense, flat-topped to dome-shaped, the stalks sparsely to densely hairy, more or less bractless (a few reduced leaflike bracts sometimes present along the branches). Heads radiate. Involucre narrowly nearly cylindrical to hemispherical or broadly cup-shaped, the bracts in more or less 3 overlapping series, the outer ones shorter than the inner ones, elliptic-lanceolate to narrowly oblong or narrowly ovate-triangular, bluntly to sharply pointed at the tip, cobwebby-hairy, often becoming tan to brown with age, usually with a narrow, pale, raised midrib, this surrounded by a narrow to broad, green area, the narrow, paler margins thin and papery. Receptacle slightly convex to hemispheric, solid, chaffy. Ray florets 3–12 (numerous in doubled forms), pistillate, the corolla sometimes glandular or appearing minutely pebbled, white, rarely pink. Disc florets perfect, 10 to numerous, the corolla white to grayish white, sometimes glandular, the 5 lobes without resin canals, persistent, the tube often somewhat flattened toward the tip, slightly swollen at the base. Pappus absent. Fruits 1–2 mm long, oblong-obovate to slightly wedge-shaped in profile, flattened, broadly rounded to nearly truncate at the tip, the margins somewhat thinner and sometimes appearing as blunt wings, the surface otherwise smooth or with several faint, longitudinal lines, glabrous, tan to light brown with lighter margins. About 115 species, North America, Europe, Asia, Africa, most diverse in Europe, Asia.

1. Leaf blades 2(3) times deeply pinnately lobed; involucre narrowly cup-shaped to nearly cylindrical .... 1. A. MILLEFOLIUM
1. Leaf blades unlobed, the margins usually finely toothed; involucre broadly cup-shaped to hemispherical .... 2. A. PTARMICA

1. Achillea millefolium L. (common yarrow, common milfoil)
A. millefolium var. *lanulosa* (Nutt.) Piper ex Piper & Beattie
A. millefolium ssp. *lanulosa* (Nutt.) Piper
Pl. 224 a, b; Map 937

Plants with usually short rhizomes. Stems 20–80 cm long, without short branches in the leaf axils, sparsely to densely pubescent with white, woolly hairs. Leaves sessile or short-petiolate. Leaf blades 1–12 cm long (those of the basal leaves sometimes to 30 cm), narrowly oblong to lanceolate or oblanceolate, deeply 2(3) times pinnately lobed, pinnately veined, glabrous or sparsely to densely pubescent with woolly hairs, the ultimate segments 1–3(–5) mm long, linear to threadlike, sharply pointed at the tip, 1-veined. Inflorescences flat-topped to dome-shaped. Involucre 4–5 mm long, narrowly cup-shaped to nearly cylindrical. Receptacle convex to hemispheric. Ray florets (3–)5(–8), the corolla 2–4 mm long, often with minute, yellow, club-shaped glands, white, rarely pink. Disc florets perfect, 10–20, the corolla 1.8–3.5 mm long, usually with sparse, minute, yellow, club-shaped glands, white to grayish white. $2n=18, 27, 36, 48, 54$ (all or mostly $2n=36$ in Missouri). May–November.

Scattered nearly throughout the state, some populations possibly introduced (North America, Europe, Asia). Upland prairies, glades, openings of mesic to dry upland forest; also pastures, old fields, railroads, roadsides, and open, disturbed areas.

Yarrow is a common garden perennial and is also a component of some wildflower seed mixes. However, it often becomes an aggressive plant that is difficult to control. Steyermark (1963) also mentioned that the species has been used in herbal teas and as a tonic. He noted that one common name, nosebleed, might be attributed to the supposed property of the plants to cause nosebleeds, but after testing this by stuffing foliage up his nostrils, he was able to report that *A. millefolium* caused no irritating or burning sensation and thus was not an effective means of instigating a bloody nose. An alternative explanation comes from other European names for the species such as bloodwort, staunchweed, and soldier's woundwort, which apparently relate to the application in medieval times of fresh leaves to wounds to stop bleeding (Antonio and Masi, 2001).

Among the mutants dignified with names, the form with light pink to rose-pink corollas occurs sporadically in Missouri and has been called f. *roseum* E.L. Rand & Redfield. Many authors (Steyermark, 1963; Gleason and Cronquist, 1991) have attempted to separate *A. millefolium* into two or more varieties, subspecies, or species and to segregate native populations from those introduced from the Old World. Mulligan and Bassett (1959) investigated the situation cytologically and concluded that native North American populations are tetraploids ($2n=36$), whereas plants introduced from the Old World are hexaploid ($2n=48$). However, further published chromosome counts by various workers (for example, Gervais, 1977) and studies by Tyrl (1975) have shown that the situation is not that clear-cut. Whatever the original situation may have been prior to the settlement of North America by Europeans, anthropogenic spread and subsequent hybridization (particularly in the northern United States) between plants with different chromosome numbers have created a confusing circumboreal complex that includes diploid, triploid, tetraploid, pentaploid, hexaploid, and octoploid plants. In Missouri and in the eastern Great Plains, Pireh and Tyrl (1980) showed that most or all of the populations are tetraploid, in spite of various combinations of morphological features. In this region and probably elsewhere, the morphological characters that have been used to attempt to identify native vs. introduced populations are confusing and seem to vary independently of one another and of ploidy. These characters have included density and persistence of the woolly pubescence, color and texture of the margins of the involucral bracts, inflorescence shape (flat-topped vs. rounded), subtle differences in shape of the lobes of the leaves, and whether the divisions of the leaf are positioned in a relatively flat vs. three-dimensional (like a bottle brush) pattern. During the present research, determination of the Missouri specimens using different characters from this array produced strikingly different results and it has not been possible to develop consistent morphological criteria to separate putatively native populations from naturalized ones. The situation warrants further research. Thus no attempt has been made here to formally recognize these variants taxonomically.

2. Achillea ptarmica L. (sneezewort, sneezeweed)

Map 938

Plants with short to often relatively long rhizomes. Stems 20–70 cm long, sometimes with short, leafy branches in the leaf axils (these often appearing as fascicles of leaves), sparsely to densely pubescent with white, woolly to cobwebby hairs, sometimes nearly glabrous toward the base at maturity. Leaves sessile. Leaf blades 1–10 cm long, linear to narrowly lanceolate, unlobed, sharply pointed at the tip, the margins sharply but minutely toothed or rarely nearly entire, glabrous or sparsely to moderately pubescent with short, straight to incurved hairs, the venation of 1 raised midvein and also usually 1 or 2 pairs of nearly parallel, somewhat thinner main veins. Inflorescences usually relatively flat-topped. Involucre 4–5 mm long, broadly cup-shaped to hemispherical. Receptacle slightly to moderately convex. Ray florets 8–10(–12) (or more in doubled forms), the corolla 3–5 mm long (to 7 mm in doubled forms), sometimes appearing pebbled or roughened, occasionally with minute, impressed, yellow glands, white. Disc florets perfect, numerous (or few in doubled forms), the corolla 2.5–4.0 mm long, usually with sparse, minute,

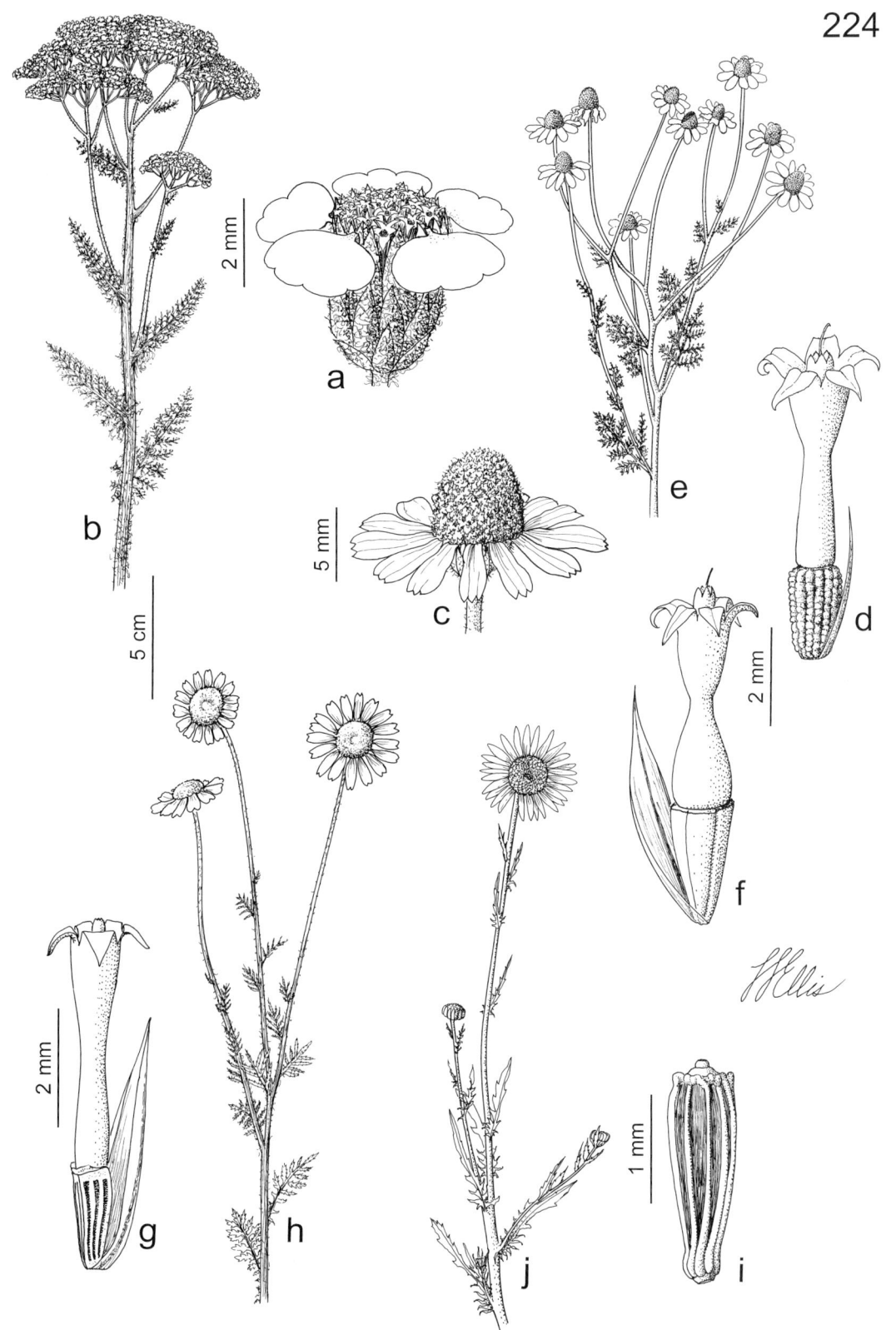

Plate 224. Asteraceae. *Achillea millefolium*, **a)** head, **b)** habit. *Anthemis cotula*, **c)** head, **d)** floret with fruit, **e)** habit. *Anthemis arvensis*, **f)** floret with fruit. *Cota tinctoria*, **g)** floret with fruit, **h)** habit. *Leucanthemum vulgare*, **i)** fruit, **j)** habit.

impressed, yellow glands, white to grayish white. 2n=18. June–September.

Introduced, known thus far only from a historical collection from Phelps County (native of Europe, Asia; introduced widely but sporadically in the northern U.S. and Canada). Roadsides.

This species is cultivated as an ornamental and several cultivars exist, including forms in which most of the disc florets have been transformed into rays. Gleason and Cronquist (1991) and some earlier authors included Missouri in the range of *A. ptarmica,* but Steyermark (1963) excluded it from the Missouri flora, based upon his finding that historical specimens collected in the city of St. Louis by Earl Sherff originated from cultivated rather than escaped plants. Thus, the species is not mapped from the St. Louis area, but a specimen collected by John Kellogg that was overlooked by Steyermark from a roadside near Rolla justifies the inclusion of *A. ptarmica* in the flora.

2. Anthemis L. (chamomile)

Plants annual (perennial elsewhere), with taproots, weakly to more commonly strongly aromatic, sparsely to densely pubescent with somewhat appressed, sometimes 2-branched hairs. Stems erect or ascending, usually branched, finely ridged. Leaves alternate and basal (basal leaves sometimes withered by flowering time), sessile or short-petiolate with winged petioles and slightly broadened and more or less clasping bases. Leaf blades deeply 1–3 times pinnately lobed, hairy and glandular, the ultimate segments mostly linear to threadlike, sharply pointed at the tip, mostly 1-veined. Inflorescences of solitary heads at the branch tips, bractless or with a few reduced leaves well below the head. Heads radiate. Involucre cup-shaped to hemispheric, the bracts more or less in 2 or 3 loosely overlapping series, the outer ones somewhat shorter, elliptic-lanceolate to narrowly oblong or narrowly ovate-triangular, rounded to bluntly pointed at the tip, sparsely to moderately hairy, tan to brown, often with a narrow, green or brown midvein, the midrib not keeled, the margins broad, thin and papery, somewhat irregular to unevenly fringed. Receptacle hemispheric at flowering, elongating to conical or cylindrical at fruiting, solid, chaffy at least toward the center. Ray florets 10–20, pistillate or sterile, becoming reflexed after flowering, white, rarely pinkish-tinged. Disc florets perfect, numerous, the corolla yellow, rarely purplish-tinged, minutely glandular, the 5 lobes without resin canals, persistent, the tube often somewhat flattened toward the tip, becoming swollen at fruiting. Pappus absent or a very short collar or crown. Fruits oblong-obovoid to slightly wedge-shaped in profile, bluntly 4-angled to irregularly polygonal in cross-section, not flattened, truncate at the base, the tip often slightly obliquely truncate, strongly (8–)10-ribbed, the ribs rounded, smooth or appearing cross-wrinkled or with low tubercles, the surface otherwise glabrous or glandular, brown to dark brown. About 175 species, native of Europe, Asia, Africa, introduced nearly worldwide.

The genus *Anthemis* is treated here in a somewhat restricted sense, with the removal of the introduced *A. tinctoria* (and its foreign relatives) to the genus *Cota*. For further discussion, see the treatment of that genus.

1. All of the florets subtended by chaffy bracts; ray florets pistillate; fruits with the ribs smooth or slightly uneven; plants aromatic with a pleasant, sweet to musky odor .... 1. A. ARVENSIS
1. Only the innermost florets subtended by chaffy bracts; ray florets sterile; fruits with the ribs appearing strongly cross-wrinkled or tubercled; plants aromatic with an unpleasant odor .... 2. A. COTULA

939. Anthemis arvensis		940. Anthemis cotula		941. Artemisia absinthium

1. Anthemis arvensis L. (corn chamomile)

Pl. 224 f; Map 939

Plants aromatic with a pleasant, sweet to musky odor. Stems 10–40(–60) cm long, erect, ascending, or spreading with ascending tips, usually branched throughout or mainly below the midpoint. Leaf blades 1–4 cm long, oblanceolate to oblong-elliptic or ovate, deeply 1 or 2 times pinnately lobed, the basal lobes sometimes appearing fascicled, the ultimate segments 0.5–4.0 mm long. Heads mostly long-stalked, the stalks 3–15 cm long at flowering. Involucre 2.5–5.0 mm long. Receptacle with chaffy bracts throughout. Ray florets pistillate, the corolla 5–15 mm long, sometimes inconspicuously glandular. Disc florets with the corolla 1.5–4.0 mm long, the lobes often minutely glandular. Fruits 1.7–2.2 mm long, the ribs smooth or slightly uneven. $2n=18$. May–October.

Introduced, uncommon in the eastern half of Missouri and Jackson and Stone Counties (native of Europe, introduced widely in the U.S. and adjacent Canada, except some southwestern states). Margins of ponds and banks of streams; also railroads, roadsides, and open, disturbed areas.

Plants with chaffy bracts noticeably shorter than the disc florets, which include most North American specimens, have been called var. *agrestis* (Wallr.) DC. Those with the chaff about as long as or longer than the disc florets are var. *arvensis*. Steyermark (1963) included var. *arvensis* for Missouri on the basis of a single specimen collected by Viktor Mühlenbach in the St. Louis railyards; that specimen unfortunately could not be relocated during the present study. This character seems relatively variable, although none of the Missouri materials examined to date have the chaffy bracts distinctly longer than the disc florets. Elsewhere, the overall bract length also more or less correlates with an awnlike tip on the longer bracts and merely a narrowly pointed tip on the shorter ones, but none of the Missouri plants examined appear to possess awns.

2. Anthemis cotula L. (Mayweed, dog fennel, stinking chamomile)

Pl. 224 c–e; Map 940

Plants aromatic with an unpleasant odor. Stems 10–50(–80) cm long, erect or ascending, usually branched above the midpoint. Leaf blades 1–6 cm long, oblanceolate to elliptic or ovate, deeply 2 or 3 times pinnately lobed, the basal lobes sometimes appearing fascicled, the ultimate segments 0.5–4.0(–8.0) mm long. Heads mostly long-stalked, the stalks 3–8 cm long at flowering. Involucre 2.5–5.0 mm long. Receptacle with chaffy bracts confined to the central portion. Ray florets sterile, the corolla 5–9 mm long, sometimes inconspicuously glandular. Disc florets with the corolla 1.5–3.0 mm long, the lobes often minutely glandular. Fruits 1–2 mm long, the ribs appearing strongly cross-wrinkled or tubercled, often also with glands between the ribs. $2n=18$. May–October.

Introduced, scattered nearly throughout the state (native of Europe, introduced widely in North America). Openings of bottomland forests, banks of streams and rivers, and rarely glades; also crop fields, fallow fields, orchards, barnyards, ditches, railroads, roadsides, and open, disturbed areas.

3. Artemisia L. (wormwood, sage, mugwort)

Plants annual, biennial, or perennial herbs (shrubs elsewhere), sometimes from a woody rootstock, weakly to strongly aromatic, glabrous or sparsely to densely hairy. Stems erect or ascending, branched or unbranched, finely ridged. Leaves alternate and sometimes also basal, sessile or short-petiolate (the lowermost leaves usually long-petiolate in *A. absynthium*), sometimes with slightly broadened, more or less clasping bases. Leaf blades entire or variously lobed or compound, glandular and otherwise glabrous to variously hairy, the blade or lobes with 1 to several veins, these sometimes difficult to observe. Inflorescences terminal and often also axillary, spikes or racemes, sometimes grouped into terminal panicles, these usually elongate or pyramidal, the branches spicate, racemose, or sometimes reduced to small clusters sessile along the main axis, bractless or with bracts subtending the heads. Heads discoid, all of the florets perfect (the marginal florets sometimes only pistillate) and potentially producing fruits, or the marginal florets perfect and the central florets only staminate and not producing fruits. Involucre broadly ovoid to nearly cylindrical, the bracts more or less in 2–4 overlapping series, the outer ones somewhat shorter, narrowly ovate to linear, sharply pointed at the tip, variously glabrous or hairy and glandular, often with a green or brown midvein or subapical area, otherwise grayish tan to brown, usually flat, at least the innermost usually with the margins thin and papery. Receptacle usually strongly convex at flowering, not conspicuously elongating at fruiting, solid, naked or less commonly densely bristly-hairy. Disc florets numerous, the corolla yellow or greenish yellow, sometimes purplish-tinged, minutely glandular, the 5 lobes without resin canals, persistent, the tube not flattened toward the tip but sometimes slightly oblique at the tip, not becoming swollen at fruiting. Pappus absent. Fruits ellipsoid to ellipsoid-obovoid, less commonly nearly cylindrical, not or only slightly flattened, often somewhat asymmetric at the base, the tip often slightly obliquely truncate, relatively strongly 5–10-ribbed to inconspicuously nerved or lined, the surface otherwise usually glabrous, tan to brown. About 350 species, nearly worldwide.

The species of sagebrush in the western states are all shrubby members of the genus *Artemisia*. Various species have long been cultivated as ornamentals for their silvery foliage. The name wormwood comes from the use of some species in medieval times to treat intestinal worms. Various other species have been used as spices and flavorings. Because the genus is mainly wind-pollinated, some species, particularly *A. annua,* are considered bad hay fever plants.

1. Leaves glabrous or sparsely hairy at maturity, never appearing woolly, felty, or silky on either surface, the green surfaces exposed
    2. Leaves mostly unlobed or the larger ones sometimes with 1(2) pairs of slender lobes toward the base; plants perennial, with a hard rhizome or branched, woody rootstock, lacking a taproot at flowering . . . . 6. A. DRACUNCULUS
    2. Largest leaves 2 times pinnately compound or lobed, the remaining stem leaves sometimes only once compound or lobed, but then mostly with more than 2 pairs of leaflets or lobes, only the uppermost leaves unlobed or with 1 or 2 pairs of leaflets or lobes; plants annual or biennial, with a taproot at flowering
        3. Ultimate leaflets and/or lobes mostly long and narrowly linear (threadlike), the margins entire; heads with the central florets staminate and not forming fruits, only the marginal florets perfect and developing fruits
         . . . . 4. A. CAMPESTRIS
        3. Ultimate leaflets and/or lobes linear to narrowly elliptic-oblanceolate but not threadlike, the margins noticeably toothed (except on the sometimes threadlike and entire uppermost leaves); heads with all of the florets perfect and potentially developing fruits
            4. Plants strongly aromatic when bruised; inflorescences appearing open-paniculate with the branches loosely flowered, the heads stalked; involucre 1.0–1.5(–2.0) mm long. . . . . 2. A. ANNUA

4. Plants odorless or nearly so when bruised; inflorescences appearing densely and narrowly paniculate or spikelike with short branches densely flowered, the heads sessile or nearly so; involucre 2–3 mm long . . . . 3. A. BIENNIS
1. Leaves densely pubescent with woolly, felty, or silky hairs, at least on the undersurface, the grayish white pubescence hiding the surface
   5. Leaves 0.5–1.5 cm long (including petiole) . . . . 7. A. FRIGIDA
   5. Largest leaves more than 1.5 cm long
      6. Heads relatively large, the involucre 6.0–7.5 mm long; florets with the corolla 3.2–4.0 mm long . . . . 9. A. STELLERIANA
      6. Heads relatively small, the involucre 2–5 mm long; florets with the corolla 1.2–2.8 mm long
         7. Receptacle with relatively long, bristly hairs between the florets
            . . . . 1. A. ABSINTHIUM
         7. Receptacle naked
            8. Leaves mostly less than 3(–5) cm long (including the petiole), with the central axis (and lobes) 0.5–1.0(–1.5) mm wide, often threadlike . . . . 5. A. CARRUTHII
            8. Largest leaves more than 5 cm long (including the petiole), the central axis (and often also the lobes) more than 2 mm wide
               9. Largest leaves entire or toothed to shallowly lobed or, if deeply lobed, then with the lobes either entire (not toothed) or with a pair of deep lobes, lacking stipulelike lobes at the base
                  . . . . 8. A. LUDOVICIANA
               9. Largest leaves with the primary lobes toothed or lobed, often with 1 or 2 pairs of small, stipulelike lobes at the leaf base
                  . . . . 10. A. VULGARIS

1. Artemisia absinthium L. (absinthe, common wormwood)

Pl. 226 a, b; Map 941

Plants perennial herbs, with stout, woody taproots, strongly aromatic when bruised. Stems 35–120 cm long, erect or ascending, moderately to densely pubescent with silky hairs and minutely glandular when young, sometimes becoming nearly glabrous with age. Leaves 1–20 cm long, the basal and lower to median leaves relatively long-petiolate, the upper leaves short-petiolate to sessile, lacking stipulelike lobes or teeth at the base. Leaf blades mostly 2–3 times pinnately compound or deeply lobed and ovate in outline, the uppermost unlobed or few-lobed and linear to narrowly elliptic in outline, the main leaves with 3 or 5 primary lobes, the ultimate segments or lobes 1–3 mm wide, narrowly oblong to linear but not threadlike, rounded to sharply pointed at the tip, both surfaces densely pubescent with silky to woolly hairs when young, the upper surface sometimes becoming nearly glabrous at maturity, also minutely glandular. Inflorescences appearing as open, leafy panicles, the branches spicate or narrowly racemose with more or less loosely spaced heads. Heads with the central florets perfect and the marginal florets pistillate, thus all of the florets potentially producing fruits. Involucre 2–3 mm long, the bracts in 2 or 3 overlapping rows, the main body oblong-elliptic to broadly elliptic, densely silky-hairy and minutely glandular, with broad, thin, transparent margins and tip, these usually cobwebby-hairy along the edge. Receptacle with relatively long, bristly hairs between the florets. Corollas 1.2–1.8 mm long. Fruits 0.7–1.0 mm long, more or less obovoid, faintly lined, tan to grayish brown, shiny. 2n=18. July–September.

Introduced, known thus far only from the city of St. Louis (native of Europe, Asia, introduced widely in northern North America south to Oregon, Utah, and South Carolina). Railroads and open, disturbed areas.

This species was reported for Missouri by Mühlenbach (1979) during his botanical research in the St. Louis railyards. It is cultivated sporadically as a garden ornamental around the state, but it is unlikely to escape widely, as Missouri is at the southern edge of its climatic tolerance.

180 ASTERACEAE

942. Artemisia annua 943. Artemisia biennis 944. Artemisia campestris

Wormwood is an active ingredient in absinthe, an alcoholic beverage that also usually contains flavorings such as anise and fennel (to mask the innate bitterness of the *Artemisia*). This liquor was popular in nineteenth-century Europe (and elsewhere) but has since been banned as a hazardous narcotic in many countries. The reputation of absinthe for inducing drunkenness, euphoria, and hallucinations (sometimes also psychosis and schizophrenia) is in part because keeping compounds in the wormwood extract dissolved requires a high percentage of alcohol in the solution. The principal mind-altering agent, thujone, is a neurotoxic terpenoid that can cause kidney failure in high doses.

2. Artemisia annua L. (sweet wormwood, annual wormwood, sweet sagewort)

Pl. 225 a, b; Map 942

Plants annual, with taproots, strongly aromatic when bruised. Stems 30–200(–300) cm long, erect or ascending, glabrous but minutely glandular. Leaves 1–10 cm long, the basal and lower leaves relatively long-petiolate, often withered by flowering time, the median and upper leaves short-petiolate to sessile, lacking stipulelike lobes or teeth at the base. Leaf blades 1 or 2 times pinnately compound or deeply lobed, lanceolate to broadly ovate in outline, the main leaves with 7–11 primary lobes, the ultimate segments or lobes 0.5–2.0 mm wide, narrowly elliptic-oblanceolate to linear but not threadlike (except sometimes on the uppermost leaves), mostly sharply pointed at the tip, the margins toothed, both surfaces glabrous but minutely glandular. Inflorescences appearing as open, leafy panicles, the branches narrowly racemose with more or less loosely spaced, stalked heads. Heads with the central florets perfect and the marginal florets perfect or pistillate, thus all of the florets potentially producing fruits. Involucre 1.0–1.5(–2.0) mm long, the bracts in 2 or 3 overlapping rows, the main body oblong-elliptic, glabrous but minutely glandular, with broad, thin, transparent margins and tip, these glabrous. Receptacle naked, without bristly hairs. Corollas 0.7–1.0 mm long. Fruits 0.7–0.9 mm long, more or less obovoid, faintly lined, tan to grayish brown, shiny. $2n=18$. August–November.

Introduced, scattered in the southern half of the state and in the Big Rivers Division (native of Asia, widely naturalized in the Americas and Europe). Bottomland forests, mesic upland forests in ravines, banks of streams and rivers, and margins of lakes; also crop fields, fallow fields, old fields, pastures, gardens, railroads, roadsides, and open, disturbed areas.

For many centuries, ingredients in *A. annua* L. were used in Chinese traditional medicine to treat fevers and malaria under the name *qinghaosu*. More recently, Western medicine has come to accept that this species is highly effective against *Plasmodium* species, the microbial parasites that cause malaria. The main active ingredient is a sesquiterpene lactone known as artemisinin, and synthetic derivatives also have been developed from this compound (Klayman, 1985, 1993). In addition to its medical virtues, artemisinin has shown potential as a natural herbicide (Duke et al., 1987).

3. Artemisia biennis Willd. (biennial wormwood)

Pl. 225 c, d; Map 943

Plants annual or biennial, with taproots, not or only slightly aromatic when bruised. Stems 30–150(–300) cm long, erect or ascending, glabrous but minutely glandular. Leaves 1–15 cm long, the basal and lower leaves often withered by flowering time, all the leaves short-petiolate to sessile, often with 2 or 3 pairs of small, linear, stipulelike lobes or leaflets at the base. Leaf blades 1 or 2 times pinnately

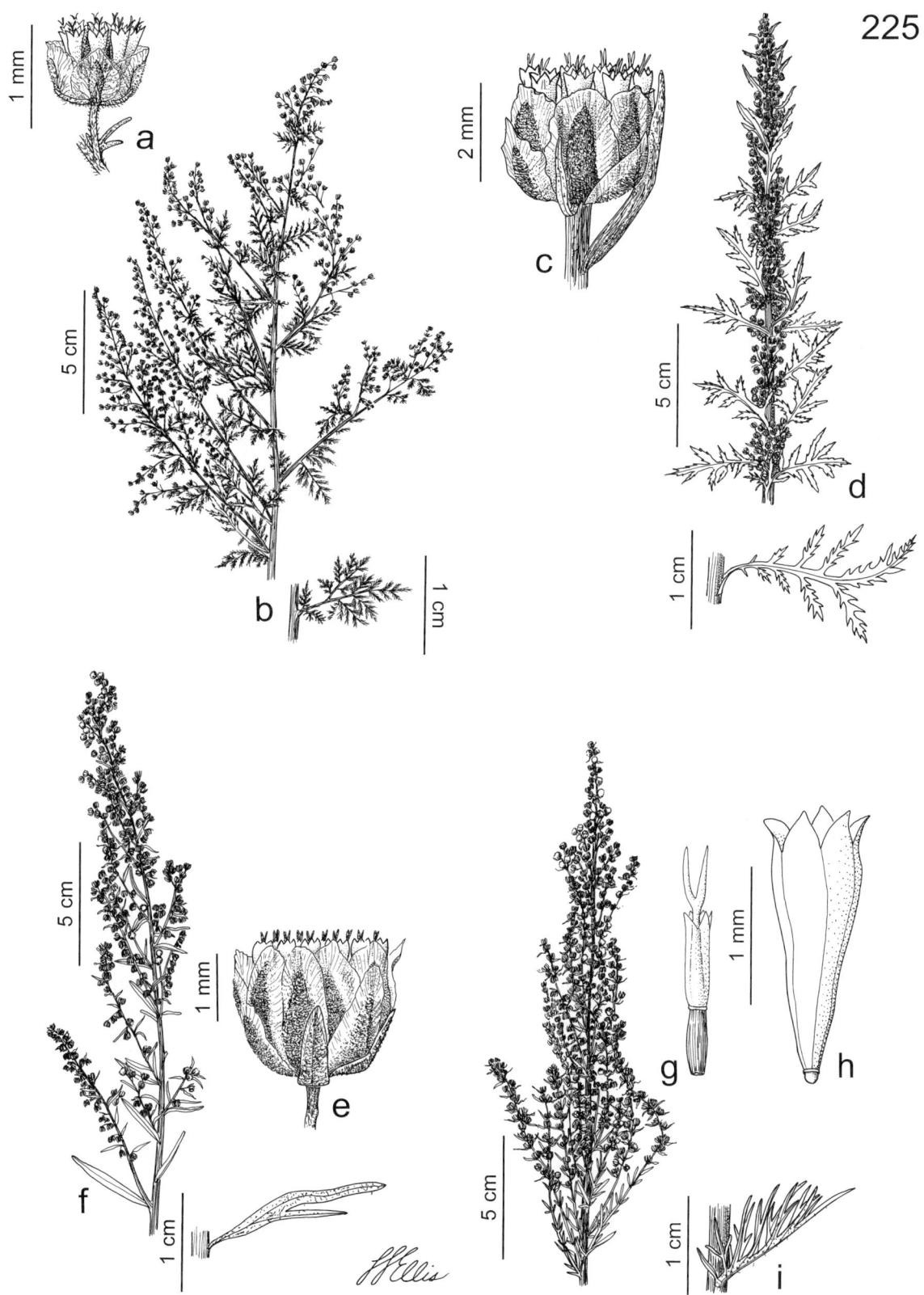

Plate 225. Asteraceae. *Artemisia annua*, **a)** head, **b)** inflorescences and leaves. *Artemisia biennis*, **c)** head, **d)** inflorescences and leaves. *Artemisia dracunculus*, **e)** head, **f)** inflorescences and leaves. *Artemisia campestris*, **g)** pistillate floret, **h)** staminate floret, **i)** inflorescences and leaves.

compound or deeply lobed, lanceolate to ovate or obovate in outline, the main leaves with 7–11 primary lobes, the ultimate segments or lobes 0.5–3.0 mm wide, narrowly elliptic-oblanceolate to linear but not threadlike (except sometimes on the uppermost leaves), mostly sharply pointed at the tip, the margins toothed, both surfaces glabrous but minutely glandular. Inflorescences appearing densely and narrowly paniculate or spikelike with short, densely flowered branches, the heads sessile or nearly so. Heads with the central florets perfect and the marginal florets perfect or pistillate, thus all of the florets potentially producing fruits. Involucre 2–3 mm long, the bracts in 2 or 3 overlapping rows, the main body oblong-elliptic, glabrous but minutely glandular, with broad, thin, transparent margins and tip, these glabrous. Receptacle naked, without bristly hairs. Corollas 0.7–1.1 mm long. Fruits 0.7–0.9 mm long, narrowly ellipsoid-obovoid, mostly 5-nerved or finely 5-ribbed, somewhat flattened, reddish brown to brown, shiny. $2n=18$. June–November.

Introduced, uncommon in Atchison County and the St. Louis and Kansas City metropolitan areas (native of the northwestern quarter of the U.S. and adjacent Canada, introduced widely but sporadically farther east). Banks of rivers; also railroads and open, disturbed areas.

4. Artemisia campestris L. ssp. caudata
(Michx.) H.M. Hall & Clem. (western sagewort, wild wormwood)
A. caudata Michx.
A. campestris L. var. *caudata* (Michx.) E.J. Palmer & Steyerm.
Pl. 225 g–i; Map 944

Plants biennial (rarely short-lived perennial) with taproots, not or only slightly aromatic when bruised. Stems 30–120 cm long, erect or ascending, glabrous at maturity but minutely glandular. Leaves 1–16 cm long, the basal and lowermost leaves usually withered by flowering time, long-petiolate, the main leaves mostly short-petiolate, lacking stipulelike lobes or leaflets at the base. Leaf blades 1–3 times pinnately compound or deeply lobed (those of the uppermost leaves occasionally unlobed), lanceolate to ovate or obovate in outline, the main leaves with 3–7 primary lobes, the ultimate segments or lobes 0.5–1.5(–2.0) mm wide, narrowly linear (threadlike) and often elongate, mostly sharply pointed at the tip, the margins entire, both surfaces glabrous but minutely glandular at maturity, sometimes sparsely to moderately pubescent with fine, cobwebby hairs when young. Inflorescences appearing as open, leafy panicles, the branches narrowly racemose with more or less loosely spaced, stalked heads. Heads with the central florets staminate and not producing fruits, only the marginal florets perfect and developing fruits. Involucre 2.5–4.0 mm long, the bracts in 3 or 4 overlapping rows, the main body elliptic-ovate to ovate, glabrous but minutely glandular, with broad, thin, transparent margins and tip, these glabrous. Receptacle naked, without bristly hairs. Corollas 0.7–1.7 mm long (those of the fertile florets shorter than those of the staminate ones). Fruits 0.7–0.9 mm long, narrowly oblong-obovoid, faintly lined, somewhat flattened, reddish brown to brown, shiny. $2n=18, 36$. July–October.

Uncommon in the east-central portion of the Ozark Division and Jefferson and St. Louis Counties (eastern U.S. west to Wyoming and New Mexico; Canada). Banks of streams and rivers, tops and ledges of sandstone bluffs; also railroads and open, sandy, disturbed areas.

Artemisia campestris is a circumboreal taxon that has been split into a number of intergrading subspecies and varieties (Hall and Clements, 1923; Cronquist, 1955). The ssp. *campestris* is confined to the Old World. The eastern ssp. *caudata* differs from other North American members of the complex in its biennial (vs. perennial) habit, however, a few of the Missouri specimens show development of new overwintering rosettes as offshoots from the base of a flowering stem. Within ssp. *caudata*, plants from the Great Plains tend to be more densely and persistently hairy than those from farther east (including Missouri).

5. Artemisia carruthii A.W. Wood ex Carruth.
Pl. 226 e; Map 945

Plants perennial herbs, with long-creeping rhizomes, sometimes producing short vegetative stems among the taller flowering ones, strongly aromatic when bruised. Stems 10–60 cm long, erect or ascending from sometimes spreading bases, densely pubescent with grayish white, woolly or felty hairs, these hiding the minute glands. Leaves 0.5–2.5(–5.0) cm long, relatively long-petiolate to sessile, often with a pair of slender, stipulelike lobes or leaflets at the base. Leaf blades mostly 1 time pinnately compound or deeply lobed, narrowly oblanceolate to oblong-ovate or obovate in outline, the main leaves with 3–5(–7) lobes, the ultimate segments or lobes 0.5–2.0 mm wide, narrowly linear and often threadlike, mostly sharply pointed at the tip, the margins entire and usually rolled under, both surfaces densely pubescent with woolly or felty hairs, the upper surface sometimes glabrous or nearly so, also minutely glandular. Inflorescences

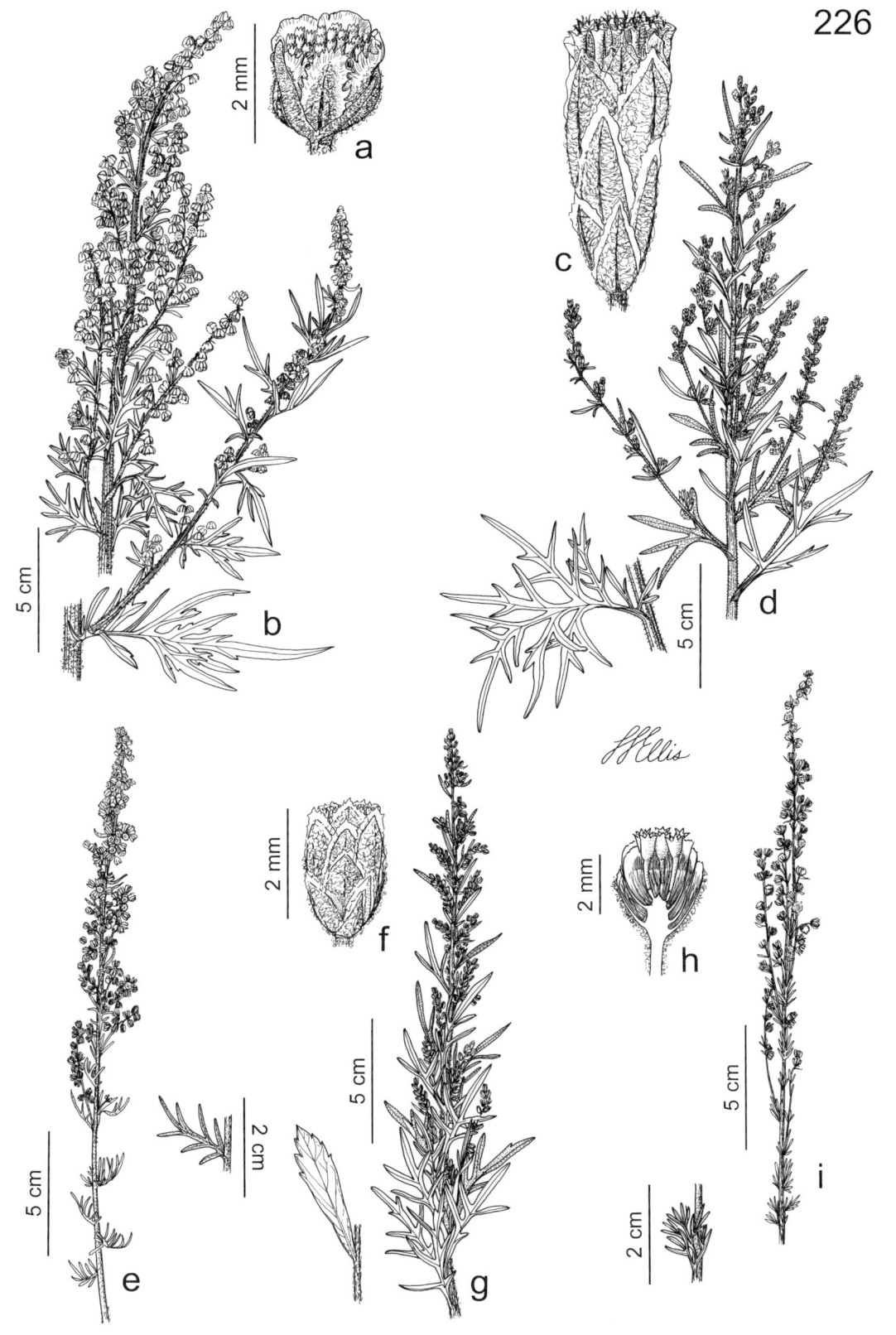

Plate 226. Asteraceae. *Artemisia absinthium*, **a)** head, **b)** inflorescences and leaves. *Artemisia vulgaris*, **c)** head, **d)** inflorescences and leaves. *Artemisia carruthii*, **e)** inflorescences and leaves. *Artemisia ludoviciana*, **f)** head, **g)** inflorescences and leaves. *Artemisia frigida*, **h)** head (longitudinal section), **i)** inflorescences and leaves.

945. Artemisia carruthii   946. Artemisia dracunculus   947. Artemisia frigida

appearing paniculate, ranging from narrow and spikelike with short, densely flowered branches to more open with longer, ascending branches, the heads sessile and/or very short-stalked. Heads with the central florets perfect and the marginal florets usually pistillate, thus all of the florets potentially producing fruits. Involucre 2.5–4.0 mm long, the bracts in 2 or 3 overlapping rows, the often indistinct main body linear to oblong-elliptic, moderately to densely woolly-hairy and minutely glandular, with broad, thin, transparent margins and tip, these usually glabrous but sometimes appearing cobwebby-hairy. Receptacle naked, without bristly hairs. Corollas 1.3–2.0 mm long. Fruits 0.7–1.0 mm long, narrowly ellipsoid-obovoid to nearly cylindrical, faintly lined, somewhat flattened, reddish brown to brown, shiny. $2n=18$. August–October.

Introduced, uncommon, known only from historical collections from Jackson County (native from Nevada to Kansas south to Arizona and Texas; Mexico; introduced eastward to New York). Railroads and open, disturbed areas.

6. Artemisia dracunculus L. (tarragon, silky wormwood)
  A. cernua Rydb.
  A. glauca Pall. ex Willd.
  A. dracunculus ssp. *dracunculina* (S. Watson) H.M. Hall & Clem.
  A. glauca var. *dracunculina* (S. Watson) Fernald
  A. dracunculus ssp. *glauca* (Pall. ex Willd.) H.M. Hall & Clem.
    Pl. 225 e, f; Map 946

Plants perennial herbs, with short rhizomes or somewhat woody rootstocks, variously not or slightly to strongly aromatic when bruised. Stems 40–150 cm long, erect or ascending, glabrous or more commonly sparsely to moderately pubescent with short, curly hairs, also minutely glandular. Leaves 1–8 cm long, short-petiolate to sessile, lacking stipulelike lobes or teeth at the base. Leaf blades mostly unlobed or the larger ones sometimes with 1(2) pair(s) of slender, ascending lobes toward the base, narrowly linear to lanceolate in outline, the ultimate segments or lobes 0.5–2.5(–4.0) mm wide, narrowly linear and threadlike with the margins curled under or slightly broader but still linear and relatively flat, sharply pointed at the tip, both surfaces glabrous or less commonly sparsely pubescent with short, curly hairs, also minutely glandular. Inflorescences appearing as open, leafy panicles, the branches spicate with relatively densely spaced, sessile to very short-stalked heads or less commonly narrowly racemose with more loosely spaced, longer-stalked heads. Heads with the central florets staminate and not producing fruits, only the marginal florets perfect (or rarely only pistillate) and developing fruits. Involucre 2.0–3.5 mm long, the bracts in 2 or 3 overlapping rows, the main body linear to oblong-elliptic, glabrous but minutely glandular, with broad, thin, transparent margins and tip, these glabrous. Receptacle naked. Corollas 1.2–2.0 mm long (those of the fertile florets shorter than those of the staminate ones). Fruits 0.5–0.8 mm long, more or less obovoid, finely and sometimes faintly 8–10-nerved, tan to grayish brown, shiny. $2n=18$. June–October.

Uncommon in northwestern Missouri (Wisconsin to Texas west to Alaska, Washington, and California; Canada, Europe, Asia; introduced sporadically in the northeastern U.S.). Loess hill prairies, margins of lakes, and glades; also roadsides and open, sandy, disturbed areas.

In cultivation, this species is known as tarragon. It has a long history of use as a flavoring in soups and other foods. It also is cultivated occasionally as an ornamental foliage plant in gardens. Steyermark (1963) reported a noncultivated occurrence from Clark County, but the voucher specimen could not be located during the present study. It seems possible that the species might have occurred on the sandy terraces of the Des Moines River, and it should be searched for in that area. Statewide, the species apparently has not been collected since 1934, and it may have become extirpated from the Missouri portion of its range.

The taxonomy of this circumboreal complex needs further study. The names *A. dracunculus* and *A. glauca* both were originally described based upon Siberian materials, and the North American plants were long known under later names like *A. cernua* Rydb. and *A. dracunculoides* Pursh, which were based upon American specimens. The interpretation of Hall and Clements (1923) that plants of the New World and Old World should be classified as a single species was ignored by many later botanists, perhaps in part because Hall and Clements erected an overly complicated infraspecific classification whose subspecies and varieties were not easily separable in the keys. Some of the problems may be due to apparent limited morphological intergradation between *A. dracunculus* and *A. campestris* in the southern Great Plains (Barkley, 1986). Most authors today sidestep the issue by choosing not to treat any of the subspecies or varieties. For those users of the present work who wish to attempt to subdivide *A. dracunculus*, the Missouri materials more or less fall along a gradient between the following two extremes. The rare ssp. *dracunculina* tends to have slightly smaller heads that have longer stalks and are grouped into more open inflorescences, whereas the more common ssp. *glauca* has slightly larger heads that are sessile or short-stalked and grouped into somewhat more congested inflorescences.

7. Artemisia frigida Willd. (prairie sagewort)

Pl. 226 h, i; Map 947

Plants perennial herbs, forming low mounds or mats from woody rootstocks, often producing short vegetative stems in addition to the flowering ones, strongly aromatic when bruised. Stems 8–40(–60) cm long, ascending from spreading bases, densely pubescent with grayish white, woolly or felty hairs, these hiding the minute glands. Leaves 0.5–1.5 cm long, short-petiolate or the uppermost leaves sessile, often with a pair of slender, stipulelike lobes or leaflets at the base. Leaf blades mostly 2 or 3 times ternately compound or deeply lobed, narrowly oblanceolate to broadly obovate in outline, the ultimate segments or lobes 0.5–2.0 mm wide, narrowly linear and often threadlike, mostly sharply pointed at the tip, the margins entire and usually rolled under, both surfaces densely pubescent with woolly or felty hairs, also minutely glandular. Inflorescences appearing paniculate or racemose, ranging from narrow and spikelike with short, densely flowered branches to less commonly more open with longer, ascending branches, the heads sessile and/or very short-stalked. Heads with the central florets perfect and the marginal florets usually pistillate, thus all of the florets potentially producing fruits. Involucre 2–3 mm long, the bracts in 2 or 3 overlapping rows, the often indistinct main body linear to oblong-elliptic, moderately to densely woolly-hairy and minutely glandular, the innermost with broad, thin, transparent margins and tip, these also hairy. Receptacle with relatively long, bristly hairs between the florets. Corollas 1.2–1.6 mm long. Fruits 0.7–1.0 mm long, narrowly ellipsoid-obovoid to nearly cylindrical, faintly lined, somewhat flattened, reddish brown to brown, shiny. $2n=18$. July–September.

Introduced, uncommon and sporadic (western U.S. [including Alaska] west to Wisconsin and Texas; Canada; introduced farther east). Railroads, roadsides, and open, disturbed areas.

8. Artemisia ludoviciana Nutt. (white sage, western mugwort)

Pl. 226 f, g; Map 948

Plants perennial herbs, with rhizomes, strongly aromatic when bruised. Stems 30–100 cm long, erect or ascending from sometimes spreading bases, densely pubescent with woolly or felty hairs, at least toward the tip, also minutely glandular. Leaves 1–11 cm long, short-petiolate to sessile, lacking stipulelike lobes or teeth at the base. Leaf blades unlobed and/or with 1–4 pairs of lobes, linear or lanceolate to oblong-elliptic in outline, the lobes linear to oblong-linear or narrowly oblong-triangular in outline, entire or rarely with a pair of deep lobes, the margins not toothed, the ultimate segments or lobes 1–9 mm wide (mostly more than 2 mm wide), the margins flat or curled under, sharply pointed at the tip, both surfaces densely pubescent with woolly to felty hairs or the upper surface sparsely hairy to glabrous, also minutely glandular. Inflorescences appearing as open, leafy panicles, the branches spicate with usually relatively loosely spaced, sessile to short-stalked heads. Heads with the central florets perfect and the marginal florets pistillate or less commonly perfect, thus all of the florets potentially producing fruits. Involucre 2.5–5.0 mm long, the bracts in 3 or 4 overlapping rows, the often indistinct main body linear to oblong-elliptic, densely woolly-hairy and minutely glandular, the innermost with relatively narrow to broad, thin, transparent margins and tip, these glabrous.

948. Artemisia ludoviciana   949. Artemisia stelleriana   950. Artemisia vulgaris

Receptacle naked. Corollas 1.5–2.8 mm long. Fruits 0.8–1.2 mm long, narrowly oblong-obovoid, faintly lined, tan to yellowish brown, shiny. $2n=18, 36, 72$. June–October.

Scattered in the western and northern halves of the state, uncommon farther south and east (western U.S. east to Illinois and Louisiana; Canada, Mexico, Central America; introduced farther eastward in the U.S.). Glades, ledges and tops of bluffs, openings of mesic to dry upland forest, bottomland and upland prairies, sand prairies, loess hill prairies, banks of rivers, and rarely marshes; also pastures, old fields, fencerows, cemeteries, railroads, roadsides, and open, disturbed areas.

This widespread species was described by one of the first two professional botanists to visit Missouri, Thomas Nuttall, based on plants that he collected along the banks of the Mississippi River near St. Louis during the winter of 1810–1811. Morphological variation across the distributional range is quite complex (Keck, 1946). Most authors have accepted some sort of infraspecific classification, as there appear to be geographic and/or habitat differences between some of the entities, especially in the western United States. However, the characters emphasized to separate taxa have not been applied uniformly by different authors. In Missouri, the two types of plants present appear to sort out reasonably well, although a number of specimens appear intermediate.

1. Largest and/or lowest leaves with the blades entire or with shallow teeth or, if lobed, then the central portion more than 4 mm wide, usually densely hairy on both surfaces, the upper surface sometimes becoming less hairy with age; inflorescence branches with the heads usually dense and overlapping in pressed specimens
.... 8A. VAR. LUDOVICIANA

1. Largest and/or lowest leaves with the blades deeply lobed, the central portion less than 4 mm wide, often glabrous on the upper surface, but sometimes both surfaces densely hairy; inflorescence branches usually more open, with the heads only occasionally overlapping in pressed specimens ... 8B. VAR. MEXICANA

8a. var. ludoviciana

A. ludoviciana var. *gnaphalodes* (Nutt.) Torr. & A. Gray

A. serrata Nutt.

Leaves often relatively thick-textured and relatively stiff. Blades of the largest and/or lowest leaves entire or with shallow teeth or, if lobed, then the central portion more than 4 mm wide and the lobes usually more or less triangular, the margins flat, both surfaces usually densely hairy and appearing gray, the upper surface sometimes becoming less hairy with age but usually still appearing gray. Inflorescences usually appearing relatively narrow and few-leaved, the branches usually with the heads dense and somewhat overlapping in pressed specimens. $2n=18, 36, 72$. June–October.

Scattered, mostly north of the Missouri River (western U.S. east to Illinois and Louisiana; Canada, Mexico, Central America; introduced farther eastward in the U.S.). Glades, openings of mesic to dry upland forest, bottomland and upland prairies, sand prairies, loess hill prairies, banks of rivers, and rarely marshes; also pastures, old fields, cemeteries, railroads, roadsides, and open, disturbed areas.

A few historical specimens from Jackson County were annotated in 1993 by the noted *Artemisia* specialist Ling Yeou-ruenn of the South China Institute of Botany as *A. douglasiana* Besser, a taxon endemic to the far-western United States and adjacent Baja California that differs from

A. ludoviciana primarily in its broader leaves. They are treated as var. *ludoviciana* in the present study.

8b. var. mexicana (Willd. ex Spreng.) Fernald
A. ludoviciana ssp. *mexicana* (Willd. ex Spreng.) D.D. Keck

Leaves sometimes thinner textured and herbaceous. Blades of the largest and/or lowest leaves usually deeply lobed, the central portion less than 4 mm wide, the lobes usually linear-oblong, the margins often somewhat rolled under, the upper surface often glabrous, but sometimes both surfaces densely hairy. Inflorescences usually appearing relatively open and leafy, the branches usually more open, with the heads only occasionally overlapping in pressed specimens. $2n=18, 36$. June–October.

Scattered, mostly south of the Missouri River (Missouri to Texas and California; Mexico; introduced sporadically farther east in the U.S.). Glades, ledges and tops of bluffs, openings of mesic to dry upland forest, bottomland and upland prairies, sand prairies, loess hill prairies, banks of rivers, and rarely marshes; also pastures, old fields, fencerows, cemeteries, railroads, roadsides, and open, disturbed areas.

A few relatively small-headed specimens from disturbed habitats in southern Missouri were annotated in 1993 by Ling Yeou-ruenn as *A. redolens* A. Gray, which most American authors have treated as *A. ludoviciana* var. *redolens* (A. Gray) Shinners or ssp. *redolens* (A. Gray) D.D. Keck. This taxon differs from var. *mexicana* primarily in its unlobed leaves, but the Missouri specimens have the larger leaves deeply lobed and thus cannot represent a state record for this otherwise southwestern taxon. They are treated as var. *mexicana* in the present study.

9. Artemisia stelleriana Besser (beach wormwood, dusty miller)

Map 949

Plants perennial herbs, with long, thin rhizomes, not or only slightly aromatic when bruised. Stems 30–70 cm long, erect or ascending, densely silky- or woolly-hairy and minutely glandular. Leaves 2–10 cm long, the basal and lower to median leaves short- to long-petiolate, the upper leaves short-petiolate to sessile, lacking stipulelike lobes or teeth at the base but sometimes slightly expanded around the stem. Leaf blades mostly 1 or 2 times pinnately or ternately shallowly to deeply lobed, mostly obovate to oblanceolate in outline, the 3 or 5 primary lobes (1–)2–9 mm wide, oblong-triangular to narrowly oblong or less commonly linear but not threadlike, sometimes with a few coarse teeth toward the tip (the lowermost sometimes shallowly lobed a third time), mostly rounded to bluntly pointed at the tip, the margins flat or those of the uppermost leaves slightly curled under, both surfaces densely pubescent with woolly to felty, white or grayish white hairs and minutely glandular. Inflorescences appearing as relatively narrow, mostly leafless panicles, the branches spicate or narrowly racemose with relatively densely spaced heads. Heads with the central florets perfect and the marginal florets pistillate, thus all of the florets potentially producing fruits. Involucre 6.0–7.5 mm long, the bracts in 2 or 3 overlapping rows, the main body indistinguishable, hidden by the densely woolly to cobwebby hairs, also minutely glandular, at least the innermost with narrow to somewhat broader, thin, transparent margins and tip, these glabrous. Receptacle naked. Corollas 3.2–4.0 mm long. Fruits 3–4 mm long, more or less obovoid, slightly flattened, smooth, yellowish brown to dark brown. $2n=18$. June–September.

Introduced, known thus far from a single historical collection from Butler County (native of Alaska, Europe, Asia; introduced widely but sporadically in the U.S. and Canada, mostly in coastal areas). Habitat unknown, but presumably disturbed areas.

This species was first noted as occurring in Missouri by Ling Yeou-ruenn in 1993 during research at the Missouri Botanical Garden Herbarium. It is sometimes cultivated for its attractive white foliage under the name dusty miller. This vernacular name has been applied to various Asteraceae with felty, white foliage that are grown as bedding plants, including selected species of *Artemisia, Centaurea, Senecio,* and *Tanacetum.* The most commonly grown of these is *Senecio cineraria* DC., a native of the Mediterranean region that has a number of cultivars in the horticultural trade.

10. Artemisia vulgaris L. (common mugwort)
A. vulgaris var. *glabra* Ledeb.
A. vulgaris var. *latiloba* Ledeb.

Pl. 226 c, d; Map 950

Plants perennial herbs, with rhizomes, strongly aromatic when bruised. Stems 30–150(–200) cm long, erect or ascending from sometimes spreading bases, glabrous or sparsely hairy toward the tip, also minutely glandular. Leaves 1–10 cm long, short-petiolate to sessile, usually with 1 or 2 pairs of small, stipulelike lobes at the base. Leaf blades 1 or 2 times pinnately lobed with 1–3 pairs of primary lobes, the uppermost merely toothed or entire, linear to ovate or obovate in outline, the primary lobes linear to oblong or narrowly fan-shaped in outline, the margins toothed or lobed (except those of the uppermost leaves), the ultimate segments or lobes 2–11 mm wide (mostly more than 2 mm wide), the margins flat or those of the upper leaves curled under, bluntly to more commonly sharply pointed at the tip, the upper surface glabrous, the undersurface densely pubescent with woolly to felty hairs, also minutely glandular. Inflorescences appearing as open, leafy panicles, the branches spicate with usually relatively densely spaced, sessile to short-stalked heads. Heads with the central florets perfect and the marginal florets pistillate or less commonly perfect, thus all of the florets potentially producing fruits. Involucre 2.5–4.0 mm long, the bracts in 3 or 4 overlapping rows, the often indistinct main body linear to oblong-elliptic, sparsely to densely woolly-hairy and minutely glandular, at least the innermost with relatively broad, thin, transparent margins and tip, these hairy or glabrous toward the tip. Receptacle naked. Corollas 2.0–2.8 mm long. Fruits 0.6–1.2 mm long, narrowly oblong-obovoid, not or very faintly lined, tan to yellowish brown, shiny. $2n=16$ (unusual counts of $2n=18, 34, 36, 40, 54$ also have been reported). July–October.

Introduced, uncommon and widely scattered, mostly in urban areas (northwestern U.S. including Alaska; Canada, Europe, Asia; introduced farther east in North America). Gardens, railroads, roadsides, and open, disturbed areas.

In Europe, this plant was used medicinally to treat intestinal worms and also as a natural insect repellant. Steyermark (1963) reported that overuse could lead to various pains, spasms, and other toxic effects. It has been cultivated in the United States in gardens for its attractive and aromatic foliage. The lower leaves can appear quite similar to those of some of the cultivated mums. Interestingly, some cultivated and escaped plants appear to produce few or no achenes in a given year.

Ling (1995) segregated plants with slightly larger heads and slightly shorter stems as *A. indica* Willd. (*A. vulgaris* var. *kamtschatica* Besser) and annotated some of the Missouri materials as representing this taxon. However, there appears to be too much variation across the range of *A. vulgaris* to allow the segregation of species or even subspecies. Also, the few Missouri specimens in question seem to fit Ling's concept of *A. vulgaris* in the strict sense better than they do *A. indica*.

4. Cota J. Gay ex Guss.

About 40 species, Europe, Asia, Africa.

Recent molecular studies by Oberprieler (2001) have shown that most of the taxa traditionally treated as *Anthemis* subgenus *Cota* (J. Gay ex Guss.) Rouy are more distantly related to the remainder of the *Anthemis* lineage than are species of *Tripleurospermum*. Although the single species of *Cota* in Missouri is easily distinguished from *Anthemis* species, overall the genus *Cota* differs from *Anthemis* in details of fruit shape, ribbing, and wall anatomy, as well as phytochemical and cytogenetic data, which are not popular characters among writers and users of floristic manuals. Nevertheless, its inclusion in a broadly circumscribed *Anthemis* would result in an unnatural assemblage if the morphologically discordant *Tripleurospermum* continued to be recognized as a separate genus.

1. Cota tinctoria (L.) J. Gay ex Guss. **ssp. tinctoria** (yellow chamomile, golden marguerite)
Anthemis tinctoria L.

Pl. 224 g, h; Map 951

Plants short-lived perennial herbs, with short rhizomes and sometimes also stolons. Stems 30–70 cm long, erect or ascending, unbranched or more commonly few-branched toward the tip, inconspicuously ridged, sparsely to densely pubescent with woolly hairs. Leaves alternate, mostly sessile, with somewhat broadened, more or less clasping bases. Leaf blades 1–5 cm long, elliptic to oblong-oblanceolate, deeply 1 time pinnately lobed, the lobes pinnately toothed or again lobed, the basal lobes sometimes appearing fascicled (the leaf then deeply 2 times pinnately lobed), moderately to densely woolly-hairy, especially on the undersurface, the primary lobes 1–20 mm long (the ultimate segments variable but shorter), mostly linear, sharply pointed at the tip, pinnately veined, the ultimate segments 1-veined. Inflorescences of solitary heads at the branch tips, the stalks relatively long, woolly- or cobwebby-hairy, bractless (but with reduced leaves toward the base). Heads radiate. Involucre 4–8 mm long, broadly cup-shaped to bell-shaped, the bracts more or less in 3 or 4 loosely overlapping series, the outer ones somewhat shorter than the inner

951. Cota tinctoria   952. Leucanthemum vulgare   953. Matricaria chamomilla

ones, lanceolate to oblong-lanceolate, bluntly to sharply pointed at the tip, cobwebby-hairy, tan to brown, with a narrow, green midvein, this not keeled, the margins thin and papery to leathery. Receptacle convex to hemispheric, solid, chaffy. Ray florets 20–30, pistillate, the corolla 7–17 mm long, yellow. Disc florets perfect, numerous, the corolla 3.0–3.5 mm long, yellow, glabrous, the 5 lobes without resin canals, persistent, the tube often somewhat flattened toward the tip, becoming swollen at fruiting. Pappus a short collar or crown, less commonly absent. Fruits 1.5–2.0 mm long, nearly rectangular to slightly wedge-shaped in profile, strongly 4-angled, truncate at the base, the tip often slightly obliquely truncate, indistinctly several-nerved, the surface otherwise smooth, glabrous, light brown to brown. $2n=18$. May–September.

Introduced, known thus far only from Cape Girardeau and Greene Counties (native of Europe, introduced widely but sporadically in the northern U.S. and adjacent Canada south to Virginia, Arkansas, and California). Disturbed areas.

Yellow chamomile is grown as an ornamental in gardens. The report of this species from St. Louis by Steyermark (1963) seems reasonable, but no voucher specimen could be located during the present study. In Europe, a complicated infraspecific classification has been erected to account for minor variations in the species. The North American materials seem to correspond best to ssp. *tinctoria*.

5. Leucanthemum Mill.

About 33 species, Europe, Africa, introduced in the New World.

It has now become widely accepted that the former broad circumscription of the genus *Chrysanthemum* involved the recognition of an unnatural group. The most recent classifications separate this unwieldy group into about 38 genera (Soreng and Cope, 1991; Bremer and Humphries, 1993). Unfortunately, the characters supporting this reclassification are mostly microscopic and/or anatomical, which has made it difficult to write keys to the segregate genera. *Chrysanthemum* in the strict sense is now confined to a few annual species native to the Mediterranean region. It and a few other non-Missouri genera are part of a group with florets with two different types of achenes: those of the showy, mostly yellow ray florets strongly 3-ribbed or 3-winged; and those of the disc florets thinner-walled and less strongly angled (sometimes 1 of the angles winged) (Soreng and Cope, 1991). Both types of florets lack a pappus, but this and some other defining characteristics are not individually unique to *Chrysanthemum*, which can only be separated easily by a combination of different characters. Species of *Chrysanthemum* in the strict sense are cultivated under names like garland chrysanthemum and crown daisy (*C. coronarium* L.), corn marigold (*C. carinatum* Schousb.), and tricolor chrysanthemum (*C. segetum* L.).

In the other genera, when both disc and ray florets produce fruits, these tend to be similar in size and morphology. The familiar garden and florist's mums are mostly cultivars and hybrids involving a plant currently known as *Dendranthema* ×*grandiflorum* (Ramat.) Kitam. *Leucanthemum* belongs to a group of about a dozen genera characterized by having resin ducts in the achene wall, among other features. In addition to the species treated below, several other species sometimes are grown as ornamentals. They include the Shasta daisy, *L.* ×*superbum* (Bergmans ex J.W. Ingram) Soreng & E. Cope, which was developed from a cross between the Portuguese chrysanthemum, *L. lacustre* (Brot.) Samp., and the daisy-chrysanthemum, *L. maximum* (Ramond) DC. (Soreng and Cope, 1991).

1. Leucanthemum vulgare Lam. (ox-eye daisy, white daisy)

L. vulgare var. *pinnatifidum* (Lecoq & Lamotte) Moldenke

Chrysanthemum leucanthemum L.

C. leucanthemum var. *pinnatifidum* Lecoq & Lamotte

Pl. 224 i, j; Map 952

Plants perennial herbs, with rhizomes. Stems 20–90 cm long, erect or ascending, unbranched or few-branched from near the base, finely ridged, sometimes finely hairy toward the tip when young, glabrous or nearly so at maturity. Leaves alternate and basal (basal leaves sometimes withered by flowering time), the basal and lowermost stem leaves long-petiolate, grading abruptly to sessile leaves along most of the stem. Leaf blades 1–12 cm long, narrowly obovate to oblong-oblanceolate; the uppermost sometimes linear, usually pinnately lobed (the lobes sometimes few-toothed) but sometimes only with coarse, narrow, rounded teeth; the main leaves usually clasping the stem and often with stipulelike lobes or teeth toward the base, rounded at the tip, the surfaces glabrous or nearly so, the lobes or teeth 7 to numerous, generally with 1 main vein. Inflorescences of solitary heads at the stem tips, the upper portion of the stem leafless. Heads radiate. Involucre 7–9 mm long, cup-shaped to broadly cup-shaped, the bracts more or less in 3 loosely overlapping series, subequal (the innermost slightly elongate), lanceolate to narrowly ovate-triangular, bluntly to sharply pointed (sometimes rounded on the innermost) at the tip, glabrous, green to yellowish green, the midrib not keeled, the margins purple to brown, at least the margins and tip of the innermost bracts also thin and papery. Receptacle somewhat convex to nearly flat, usually hollow, naked. Ray florets 15–35, pistillate, the corolla 10–20 mm long, white. Disc florets perfect, numerous, the corolla 2.5–3.0 mm long, yellow, glabrous, the 5 lobes without resin canals, persistent, the tube not flattened toward the tip or becoming swollen at fruiting. Pappus absent. Fruits 2.0–2.8 mm long, narrowly obovoid to nearly cylindrical, more or less circular in cross-section, truncate at the base and tip, with usually 10 rounded, light tan to white ribs, the surface otherwise glabrous, dark brown to nearly black and sometimes with minute, short, white lines. $2n=18$ (36, 54, 72). May–August.

Introduced, scattered to common throughout the state (native of Europe, Asia, introduced throughout the U.S. and Canada, south to Mexico, Central America, South America). Upland prairies, glades, tops of bluffs, savannas, and openings of mesic to dry upland forests; also pastures, old fields, fallow fields, fencerows, railroads, roadsides, and open, disturbed areas.

Plants with the leaves deeply lobed have been called var. *pinnatifidum*, which is the common phase of the species in Missouri. Uncommonly seen plants with the leaves more entire are var. *vulgare*. However, these morphological extremes appear to grade into one another.

Ox-eye daisy is an aggressive rhizomatous colonizer that is difficult to control in gardens and that escapes readily. Although it is a beautiful wildflower, land managers in many states, particularly in the East and Midwest, consider it a problem invasive exotic species. It appears to increase in abundance in upland prairies that are hayed annually and to decrease when these same prairies are subjected to frequent prescribed burns.

6. Matricaria L. (wild chamomile)
Contributed by John H. Pruski

Plants annual (short-lived perennial herbs elsewhere), with taproots, usually strongly aromatic. Stems erect or ascending, usually branched, finely ridged or lined, glabrous or less commonly hairy. Leaves alternate, sessile or short-petiolate with winged petioles and somewhat broadened, more or less clasping bases. Leaf blades deeply 1–3 times pinnately lobed, glabrous or less commonly hairy, the ultimate segments mostly linear. Inflorescences of solitary heads or loose clusters at the branch tips, sometimes appearing paniculate, the stalks glabrous or minutely hairy, bractless. Heads radiate or discoid. Involucre bell-shaped to broadly ovoid, the bracts mostly in 2 or 3 loosely overlapping series, subequal, elliptic-lanceolate to narrowly oblong or linear-oblanceolate, rounded or angled to a minute, sharp point at the tip, glabrous or sparsely to moderately and minutely hairy, green when young but soon becoming tan to straw-colored, the midrib not keeled, green to reddish brown and often somewhat glandular. Receptacle conical or strongly convex, tending to elongate and become more conical as the fruits mature, hollow, naked. Ray florets (when present) pistillate, the corolla white, not glandular, the upper surface sometimes with minute papillae. Disc florets perfect (the outer series of florets in discoid species occasionally only pistillate), numerous, the corolla yellow or sometimes greenish yellow, occasionally minutely glandular, the 4 or 5 lobes usually lacking resin canals (these appearing as minute, brown lines or streaks), persistent, the tube often somewhat flattened toward the tip, becoming swollen at fruiting. Pappus absent or a short collar or crown. Fruits wedge-shaped in profile, bluntly angled in cross-section, slightly flattened, curved and slightly asymmetrical at the base, the tip often obliquely truncate, 3–5-nerved, the surface otherwise smooth, glabrous or glandular, light brown to brown with lighter nerves. Seven species, worldwide.

The nomenclature and application of the name *Matricaria* and of several species within the genus have remained unstable. Linnaeus (1753) originally treated five species in *Matricaria*, of which only two are still referred to the genus, and he later changed the circumscription and application of the two most widespread of his species (Linnaeus, 1755). Selection among the five species to typify the generic name also has been complex and confusing (Grierson, 1974; Jeffrey, 1979), leading Rauschert (1974) to reject the name *Matricaria* as a nomen confusum in favor of *Chamomilla* Gray. However, this interpretation has been rejected by many subsequent authors (see Gandhi and Thomas, 1991; Applequist, 2002).

The genus is closely related to *Tripleurospermum*, which differs in its solid, more or less dome-shaped receptacles, corolla lobes with resin canals (visible as small, darker lines or streaks), and fruits that are asymmetrically and strongly 3-ribbed (each rib is about as big as the achene body), with the surface between the ridges wrinkled to weakly tuberculate and having 1 or 2 resiniferous or mucilaginous glands. Species of *Tripleurospermum* are often known as scentless chamomiles because the plants are essentially odorless, whereas the two species of *Matricaria* that grow in Missouri are aromatic when bruised or crushed (often strongly so).

1. Heads radiate, the stalks slender, 2–6 cm long; leaf segments finely 3-veined; disc corollas mostly 5-lobed . . . . 1. M. CHAMOMILLA
1. Heads discoid, the stalks stout, 0.3–1.2 cm long; leaf segments 1-veined; disc corollas mostly 4-lobed . . . . 2. M. DISCOIDEA

1. Matricaria chamomilla L. (wild chamomile, false chamomile, German chamomile)

M. chamomilla var. *coronata* (J. Gay) Coss. & Germ.

M. courrantiana DC.

M. recutita L.

Chamomilla recutita (L.) Rauschert

M. suaveolens L.

Pl. 227 h–j; Map 953

Plants aromatic, with a distinct sweet, musky odor when bruised or crushed. Stems 8–60 cm tall, glabrous or less commonly sparsely hairy. Leaf blades 2–7 cm long, elliptic to oblong-obovate in outline, mostly deeply 2 times pinnately lobed, the basal primary lobes often appearing clustered, the ultimate lobes 2–15 mm long, linear to threadlike, finely 3-veined. Heads radiate, the stalks 2–6 cm long, slender. Involucre 2–3 mm long. Ray florets 10–15(–20), often becoming deflexed with age, the corolla 4–10 mm long, white. Disc florets with the corolla 1.5–2.0 mm long, mostly 5-lobed, mostly yellow. Pappus a short collar or crown (0.2–0.6 mm

954. Matricaria discoidea   955. Tanacetum balsamita   956. Tanacetum parthenium

long), less commonly absent. Fruits 0.7–0.9 mm long, usually 5-ribbed, the ribs well developed but not winglike. 2n=18. May–October.

Introduced, scattered, mostly in the eastern half of the state (native of Europe, Africa, Asia; introduced nearly worldwide). Fallow fields, railroads, roadsides, and open, disturbed areas.

This species is commonly grown in gardens and also commercially. It is the common false chamomile that is used medicinally for its anti-inflammatory, antiseptic, and soothing properties. Most commonly dried stems, foliage, and heads are brewed into a tea. An extract of the essential oils is also used as a lotion and an ingredient in some soaps and shampoos.

The pappus in some American plants forms a larger crown than that of plants from elsewhere in North America and Europe. Although some botanists have referred to such plants as var. *coronata*, there appear to be too many intermediates to support formal taxonomic recognition of the extremes. *Matricaria chamomilla* is very similar vegetatively to *Chamaemelum nobile* (L.) All. (true chamomile, English chamomile, Roman chamomile) and *Tripleurospermum maritimum* (L.) W.D.J. Koch (scentless chamomile), and specimens lacking flowers or fruits may be difficult to determine with confidence. *Chamaemelum* differs by having a chaffy receptacle and glandular corolla lobes. It is occasionally cultivated in the Americas. *Tripleurospermum* is discussed further in the treatment of that genus and above after the genus description for *Matricaria*.

2. Matricaria discoidea DC. (pineapple weed)
   M. matricarioides, misapplied
   Chamomilla suaveolens (Pursh) Rydb.
                              Pl. 227 f, g; Map 954
   Plants aromatic, with an often strong odor of pineapple when bruised or crushed. Stems 4–30 cm tall, glabrous or nearly so. Leaf blades 1–3(–7) cm long, elliptic to oblong-obovate in outline, mostly deeply 2 times pinnately lobed, the basal primary lobes usually well spaced and not appearing clustered, the ultimate lobes 2–10 mm long, linear to threadlike, 1-veined. Heads discoid, the stalks 0.3–1.2 cm long, relatively stout. Involucre 2.5–4.0 mm long. Disc florets with the corolla 1.0–1.5 mm long, mostly 4-lobed, greenish yellow. Pappus absent or a minute collar (to 0.1 mm long). Fruits 0.9–1.2 mm long, 3–5-ribbed, the ribs often relatively small. 2n=18. May–October.

Introduced, scattered nearly throughout the state (native of the western U.S.; introduced nearly worldwide). Fallow fields, pastures, barnyards, railroads, roadsides, and open, disturbed areas.

This species was used by Native Americans for its medicinal and aromatic properties (Moerman, 1998). The limits of its native range have been obscured by its widespread weediness. Steyermark (1963) noted that it was well established in the St. Louis area by 1825 and that George Engelmann mistakenly considered it native to Missouri. It appears to have become much more abundant, particularly in the southern half of the state, since Steyermark's (1963) treatment.

Matricaria discoidea formerly was widely known as *M. matricarioides* (Less.) Porter (Steyermark, 1963; Gleason and Cronquist, 1991). However, that name is a taxonomic synonym of *Tanacetum huronense* Nutt., a different species that occurs only to the north of Missouri (Gandhi and Thomas, 1991).

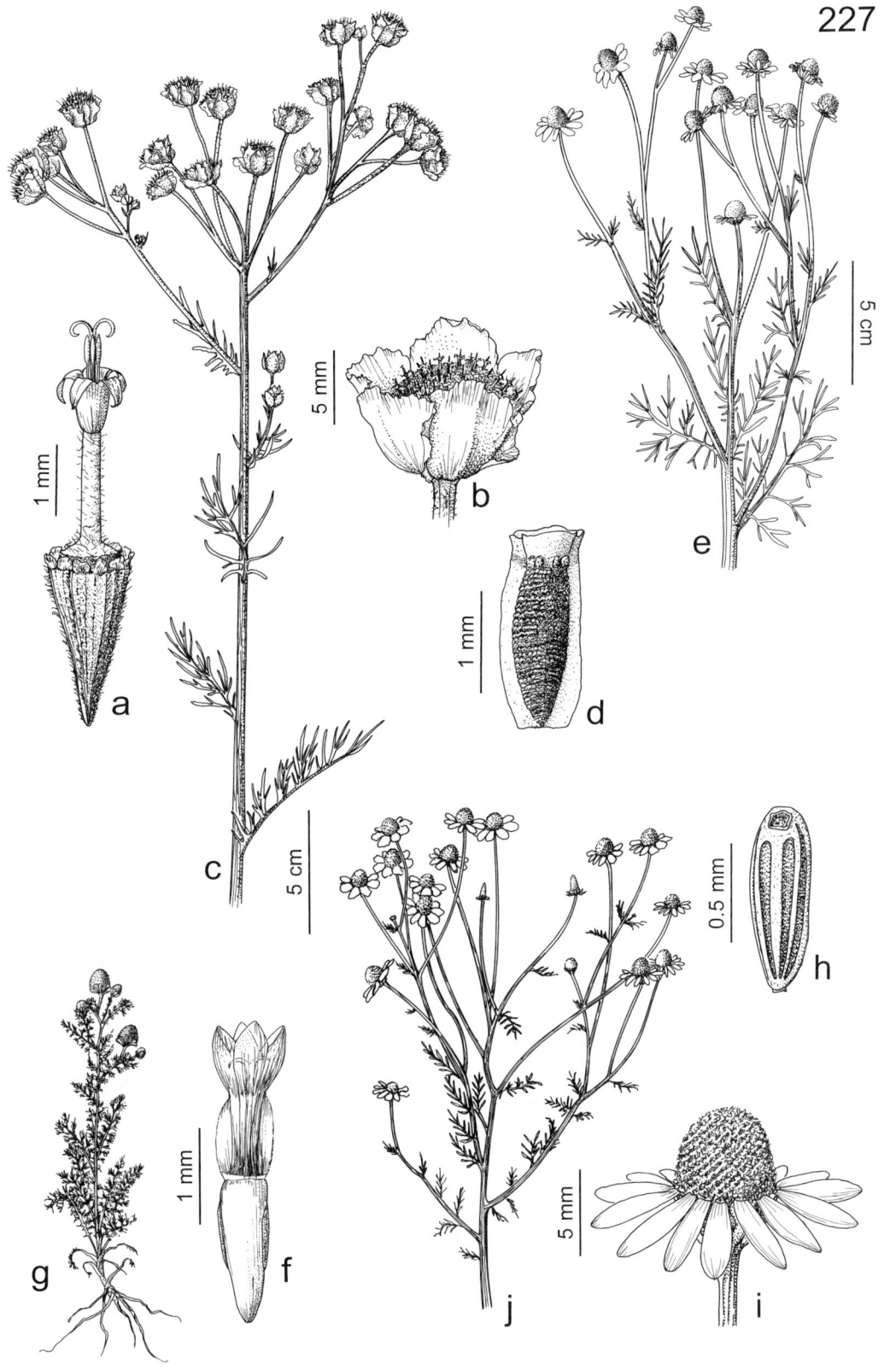

Plate 227. Asteraceae. *Hymenopappus scabiosaeus var. scabiosaeus*, **a)** floret with fruit, **b)** head, **c)** habit. *Tripleurospermum inodorum*, **d)** fruit, **e)** habit. *Matricaria discoidea*, **f)** floret, **g)** habit. *Matricaria chamomilla*, **h)** fruit, **i)** head, **j)** habit.

7. Tanacetum L. (tansy)

Plants perennial herbs (annual elsewhere), sometimes with rhizomes, weakly to more commonly strongly aromatic, glabrous or sparsely to densely hairy. Stems erect or ascending, branched or unbranched, finely ridged. Leaves alternate and basal (basal leaves sometimes withered by flowering time), sessile or short- to long-petiolate, sometimes with slightly broadened, more or less clasping bases. Leaf blades unlobed and bluntly toothed or deeply 1–2 times pinnately compound (sometimes only deeply many-lobed) with the leaflets deeply lobed and the ultimate segments sharply toothed and/or lobed, sharply pointed at the tip, hairy and/or glandular to nearly glabrous, 1- to 5-veined. Inflorescences panicles at the stem tips, these flat-topped or dome-shaped, bractless or with a few reduced leaves toward the base. Heads radiate or appearing discoid. Involucre cup-shaped to broadly cup-shaped, the bracts more or less in 2–5 overlapping series, the outer ones somewhat shorter, oblong-lanceolate to narrowly elliptic, rounded to bluntly or sharply pointed at the tip, sparsely to moderately hairy, tan to brown, sometimes with a green or brown midvein or subapical area, the midrib sometimes keeled, the margins and often also the tip thin and papery, somewhat irregular. Receptacle flat to somewhat convex at flowering, not conspicuously elongating at fruiting, solid, naked. Ray florets (when present) 10–20 or rarely more, pistillate, persistent, becoming reflexed after flowering, white (sometimes yellow when marginal florets are pistillate, not markedly enlarged, and only slightly zygomorphic), sometimes pinkish-tinged. Disc florets perfect (the marginal ones sometimes pistillate), numerous, the corolla yellow, minutely glandular, the 5 lobes without resin canals, persistent, the tube often somewhat flattened toward the tip, becoming swollen at fruiting. Pappus absent or more commonly a very short collar or crown. Fruits oblong-obovoid to slightly wedge-shaped in profile, sharply 5-angled or nearly circular in cross-section, not flattened, truncate at the base, the tip often slightly obliquely truncate, strongly 5-ribbed to slightly 10-ribbed, the ribs slender, smooth, the surface otherwise sparsely glandular, tan to brown. About 160 species, North America, Europe, Asia, Africa, introduced widely.

The taxonomic circumscription of *Tanacetum* remains controversial, especially the inclusion of species formerly classified into the genus *Chrysanthemum*. The genus is quite heterogeneous morphologically. For example, as presently circumscribed, the genus contains a large number of species with white to pink ray florets that formerly were segregated in the genus *Pyrethrum* (Zinn) Rchb. f., whereas *Tanacetum* in the strict sense has been characterized variously as having discoid heads or radiate heads with yellow rays (Bremer and Humphries, 1993).

Several, including all of the escaped Missouri taxa, are cultivated as garden ornamentals. The genus also contains a number of species used medicinally. Heads of the eastern European ornamental species *T. cineariifolium* (Trev.) Sch. Bip. are the primary natural source of the group of monoterpenes called pyrethrins. These compounds are the active ingredient in some tick repellents and also have a long history of other uses in insecticides. Synthetic derivatives of pyrethrins are known as pyrethroids.

1. Leaves simple, unlobed or the basal leaves rarely with a few deep, irregular basal lobes, the margins otherwise bluntly and regularly toothed ... 1. T. BALSAMITA
1. Leaves pinnately compound or deeply pinnately lobed with several pairs of leaflets or lobes throughout the length of the blade
    2. Leaf blades with the primary leaflets or lobes mostly 3 or 5, these further divided; heads radiate with conspicuous, white ray florets 4–8 mm long
       .... 2. T. PARTHENIUM
    2. Leaf blades with the primary leaflets or lobes 9–21, these further divided; heads discoid, the marginal florets sometimes pistillate and slightly zygomorphic, but yellow and not enlarged .... 3. T. VULGARE

1. Tanacetum balsamita L. (costmary, mint geranium)

Chrysanthemum balsamita L.
C. balsamita var. *tanacetoides* Boiss.
Balsamita major Desf.
B. major var. *tanacetoides* (Boiss.) Moldenke

Pl. 228 h, i; Map 955

Plants with rhizomes. Stems 30–120 cm long, glabrous toward the base, sparsely to moderately hairy toward the tip. Leaves 1–27 cm long, the basal and lower stem leaves usually much larger than the others, long-petiolate, grading relatively abruptly into the short-petiolate to sessile median and upper stem leaves. Leaf blades simple and unlobed or the lowermost rarely with a few deep, irregular basal lobes, oblanceolate to elliptic, mostly rounded at the tip, angled to long-tapered at the base, the margins bluntly and evenly toothed, both surfaces moderately to densely glandular and densely pubescent with appressed, grayish, silky hairs when young, sometimes the lowermost becoming glabrous or nearly so at maturity. Heads usually discoid. Involucre 3–7 mm long, cup-shaped, the bracts in 3–5 series, the main body narrowly lanceolate to lanceolate-triangular, tapered to a conspicuous, thin, papery tip, the margins also thin and nearly transparent, the outer surface glandular and hairy. Ray florets absent or the marginal florets sometimes pistillate, occasionally somewhat raylike but inconspicuous and not markedly enlarged, usually white. Disc florets with the corollas 1.5–2.7 mm long. Pappus a short collar or crown. Fruits 1.3–1.6 mm long, more or less circular in cross-section, mostly obscurely 10-ribbed. $2n=18$, 54. August–October.

Introduced, known thus far from a single historical collection from Franklin County (native of Europe, Asia, introduced widely in North America). Fencerows and open, disturbed areas.

Tanacetum balsamita escapes and becomes naturalized sporadically, most commonly in New England and the Great Lakes region. Some botanists have segregated the species into its own genus, *Balsamita* Mill., based primarily on its unlobed leaves, but most botanists do not believe that the differences merit recognition of a separate monotypic genus (Soreng and Cope, 1991; Bremer, 1994). The rayed forms sometimes have been called var. *tanacetoides* Boiss.

2. Tanacetum parthenium (L.) Sch. Bip. (feverfew)

Chrysanthemum parthenium L.

Pl. 228 c, d; Map 956

Plants taprooted, not producing rhizomes. Stems 30–80 cm long, glabrous toward the base, minutely hairy toward the tip. Leaves 1–8(–12) cm long, the basal leaves usually absent by flowering time, short-petiolate to sessile (the basal leaves long-petiolate). Leaf blades pinnately compound, less commonly the uppermost only deeply lobed, ovate to elliptic in outline, the primary leaflets (or lobes) mostly 3 or 5, these deeply pinnately or ternately lobed, oblanceolate to elliptic in outline, rounded to bluntly pointed at the tip, angled or short-tapered at the base, the margins otherwise sharply toothed, both surfaces moderately to densely glandular and sparsely to moderately pubescent with fine, curly hairs. Heads conspicuously radiate. Involucre 2.5–4.0 mm long, broadly and shallowly cup-shaped, the bracts in 2–4 series, the main body narrowly lanceolate to lanceolate or oblong-lanceolate, narrowed to a sharply pointed tip, the margins with a narrow, thin, papery border, the outer surface glandular and somewhat cobwebby-hairy, at least toward the margins. Ray florets 10–21, the corolla 4–8 mm long, white. Disc florets with the corollas 1.5–2.2 mm long. Pappus a short collar or crown, sometimes absent. Fruits 1.3–1.8 mm long, usually fairly strongly 7–10-ribbed. $2n=18$. June–September.

Introduced, uncommon and widely scattered (native of Europe, Asia, introduced widely in North America). Open, disturbed areas.

Feverfew produces an essential oil that is used medicinally as an anti-inflammatory agent and analgesic and also as a natural insect repellant. The oil also sometimes is used in combination with other ingredients in aromatherapy. Side effects that have been reported include decreased blood clotting, swelling of the mouth, and stomachache, especially if the leaves are chewed.

3. Tanacetum vulgare L. (common tansy, golden buttons)

Pl. 228 e–g; Map 957

Plants producing rhizomes. Stems 40–150 cm long, moderately pubescent with short, curly hairs when young, becoming glabrous or nearly so by flowering time. Leaves 3–20 cm long, the basal leaves usually absent by flowering time, short-petiolate to sessile. Leaf blades pinnately compound or deeply pinnately lobed, oblong-obovate to elliptic in outline, the primary leaflets or lobes 9–21 (with short wings or reduced, accessory lobes between them), these pinnately lobed, narrowly elliptic to lanceolate or oblong-lanceolate, rounded to more commonly bluntly pointed at the tip, broadly sessile or short-angled at the base, the margins otherwise sharply toothed, both surfaces moderately to densely glandular but otherwise glabrous or nearly so at maturity. Heads usually discoid. Involucre 3–

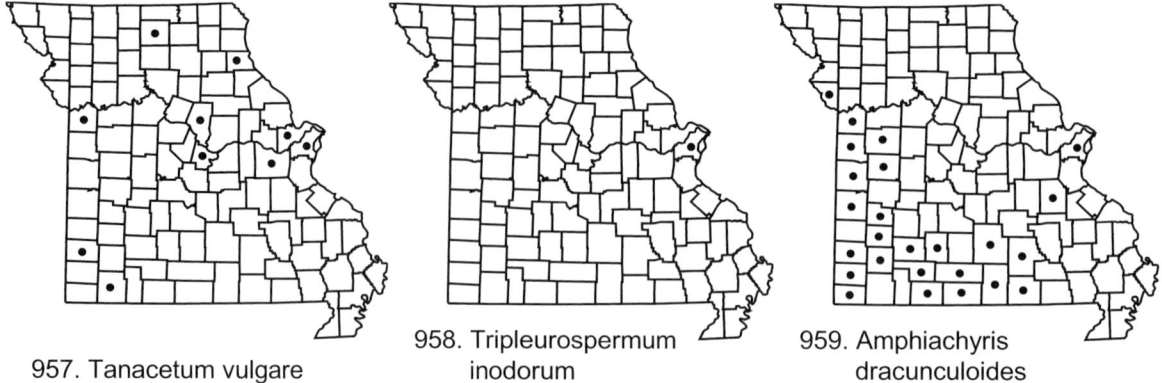

957. Tanacetum vulgare
958. Tripleurospermum inodorum
959. Amphiachyris dracunculoides

7 mm long, broadly and shallowly cup-shaped, the bracts in 3–5 series, the main body narrowly oblong-lanceolate to triangular-lanceolate, tapered to a conspicuous, thin, papery tip, the margins also thin and nearly transparent, the outer surface glandular and hairy. Ray florets absent or the marginal florets rarely pistillate, somewhat zygomorphic, but inconspicuous and not markedly enlarged, yellow. Disc florets with the corollas 1.5–2.5 mm long. Pappus a short collar or crown or absent. Fruits 1.3–1.7 mm long, moderately to strongly 5-angled or 5-ribbed, those of the ray florets sometimes only 3-angled. 2n=18. July–September.

Introduced, scattered (native of Europe, Asia, introduced widely in North America). Tops of bluffs; also fencerows, roadsides, railroads, and open, disturbed areas.

Common tansy has been used medicinally to treat fevers and headaches, and as a tonic. However, Burrows and Tyrl (2001) noted that several cases of fatal overdoses have been documented involving ingestion of either a concentrated extract or a tea brewed from dried plants. An unusual mutant with irregularly undulate (crisped) leaf margins that is sometimes cultivated has been called f. *crispum* (L.) Hayek.

8. Tripleurospermum Sch. Bip. (scentless chamomile)

About 40 species, nearly worldwide, but most diverse in Europe, Asia, Africa.

Species of *Tripleurospermum* formerly were included in *Matricaria* but are accepted as a distinct genus in many recent treatments (Bremer and Humphries, 1993; Kartesz and Meacham, 1999; Applequist, 2002), based on such characters as solid, rounded (vs. hollow, conical) receptacles, the presence of resin canals in the disc corollas, and the unequally but strongly 3-ribbed fruits, as well as differences in phytochemistry.

1. Tripleurospermum inodorum (L.) Sch. Bip.
(scentless mayweed, scentless false chamomile)
Matricaria inodora L.
M. inodora var. *agrestis* (Weiss) Wilmott
M. maritima L. var. *inodora* (L.) Soó
T. maritimum ssp. *inodorum* (L.) Applequist
M. perforata Mérat

Pl. 227 d, e; Map 958

Plants annual, with taproots. Stems 8–40(–70) cm long, erect or ascending, usually branched, finely ridged, glabrous or very sparsely hairy. Leaves alternate and basal (basal leaves usually withered by flowering time), sessile or short-petiolate with winged petioles and somewhat broadened, more or less clasping bases. Leaf blades 1.5–6.0(–10.0) cm long, elliptic to oblong-obovate, deeply 1–3 times pinnately lobed, glabrous, the ultimate segments 3–20 mm long, mostly linear to threadlike, sharply pointed at the tip, mostly 1-veined. Inflorescences of solitary heads or more commonly loose clusters at the branch tips, the stalks glabrous or sparsely hairy, bractless. Heads radiate. Involucre 3–4 mm long, cup-shaped to hemispheric, the bracts more or less in 3 or 4 loosely overlapping series, subequal, elliptic-lanceolate to narrowly oblong or narrowly ovate-triangular, mostly bluntly pointed at the tip, glabrous, tan to brown, often with a narrow, green midvein, the midrib not keeled, the margins thin and papery. Receptacle convex to hemispheric, tending to elongate but remain relatively rounded as the fruits mature, solid, naked. Ray florets 12–25,

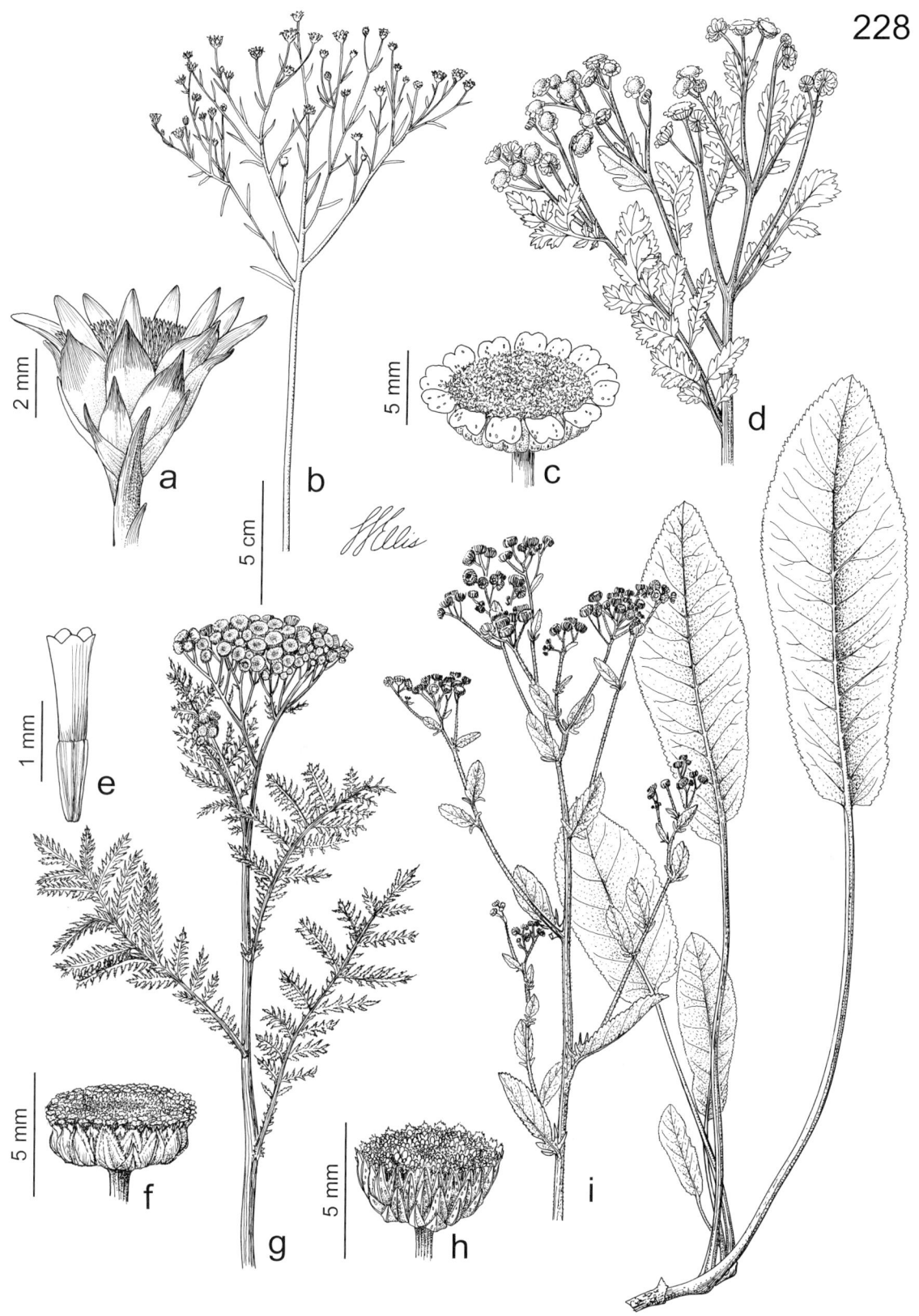

Plate 228. Asteraceae. *Amphiachyris dracunculoides*, **a)** head, **b)** habit. *Tanacetum parthenium*, **c)** head, **d)** habit. *Tanacetum vulgare*, **e)** floret, **f)** head, **g)** habit. *Tanacetum balsamita*, **h)** head, **i)** habit and lower leaves.

pistillate, the corolla 6–12(–16) mm long, white, rarely pinkish-tinged. Disc florets perfect, numerous, the corolla 1–2 mm long, yellow or sometimes greenish yellow, minutely glandular, the 5 lobes with resin canals (appearing as minute, brown lines or streaks), persistent, the tube often somewhat flattened toward the tip, becoming somewhat swollen at fruiting. Pappus a short collar or crown. Fruits 1.5–2.0 mm long, nearly rectangular to slightly wedge-shaped in profile, unevenly triangular in cross-section, slightly flattened, truncate at the base, the tip often slightly obliquely truncate, strongly 3-ribbed, the ribs on the margins and 1 of the faces almost winglike, the surface otherwise finely wrinkled or finely tuberculate, glabrous, brown to dark brown with lighter ribs. 2n=18, 36. May–October.

Introduced, known thus far only from the city of St. Louis (native of Europe; introduced widely in the U.S. [except some southern states], Canada). Railroads and open, disturbed areas.

Some authors prefer to treat this species as a variety or subspecies of the closely related *T. maritimum* (Applequist, 2002), which differs in its biennial to perennial habit, more numerous and more spreading stems, leaves with shorter, thicker lobes, and fruits with thicker ribs. True *T. maritimum* occurs natively mostly in coastal regions in northern North America, Europe, and Asia. Steyermark (1963) used the name *Matricaria maritima* var. *agrestis* (Knaf) Wilmott for this taxon, but this appears to represent an unpublished combination based on the illegitimate earlier name *Dibothrospermum agreste* Knaf (Applequist, 2002).

2. Tribe Astereae Cass.
(Nesom, 2000)

Plants annual, biennial, or perennial herbs (sometimes woody elsewhere), sometimes from a woody rootstock, rarely aromatic, the sap not milky. Stems not spiny or prickly. Leaves alternate (rarely opposite elsewhere), sometimes also in a basal rosette, sessile to long-petiolate, not spiny or prickly. Leaf blades entire to lobed, the venation pinnate or less commonly with 3 main veins. Inflorescences mostly terminal panicles, sometimes reduced to clusters of heads, or the heads solitary at the branch tips. Heads discoid or radiate. Involucre of 2 to more commonly several series of bracts, not spiny or tuberculate. Receptacle flat to slightly convex or less commonly hemispherical to conical, naked (rarely with a few chaffy bracts toward the margin in *Solidago*), sometimes minutely concave around the base of each floret and with irregular low ridges between them, minutely hairy in *Euthamia*. Ray florets (when present) pistillate or less commonly sterile, sometimes inconspicuous, the corolla sometimes very short, white, pink, purple, blue, or yellow. Disc florets all perfect or the outermost sometimes pistillate, the corolla yellow or less commonly white (in some species turning reddish purple after pollen has been shed), the 4 or more commonly 5 lobes spreading to ascending. Pappus most commonly of numerous capillary bristles, these usually with inconspicuous, short barbs, less commonly with an additional outer series of much shorter bristles or scales, in a few genera only of short awns or scales, rarely absent, when present more or less persistent at fruiting. Stamens with the filaments not fused together, the anthers fused into a tube, each tip without an appendage, each base truncate or with a pair of short lobes. Style branches somewhat flattened, each with a stigmatic line along each inner margin, the sterile tip relatively long and tapered, often with dense, minute hairs, at least on the outer surface. Fruits rarely somewhat dimorphic (the outer series then thicker-walled and with different surface ornamentation), often somewhat flattened, oblong-obovoid to slightly wedge-shaped in profile, often somewhat ribbed, not beaked. About 175–200 genera, 2,800–3,020 species, worldwide.

The Astereae contain a number of genera that are commonly cultivated as garden ornamentals, including *Aster, Bellis* L. (English daisy), *Boltonia,* and *Solidago*. As in other tribes of Asteraceae, the current trend among students of the Astereae has been to divide the larger genera taxonomically into smaller taxonomic segregates reflecting monophyletic lineages. For example, users of this volume should not be shocked to see that Missouri now only has one species in the genus *Aster* (a garden escape; see below) and that all of the native asters have been classified into four other genera.

Users should note that *Solidago bicolor* L. (silverrod), a species that may potentially occur in Missouri but has not yet been confirmed for the state, will not key correctly in the key to genera below. This species is unusual in the genus in having yellow disc corollas and white (or pale cream-colored) ray florets. The plants otherwise resemble those of *S. hispida;* see that species further discussion.

1. Heads radiate, the ray florets with yellow corollas (in a few cases, the corolla may be darker-tinged on the undersurface)
    2. Pappus of less than 10 scales or awns, sometimes absent or only a minute, toothed crown in ray florets
        3. Involucre 6–15 mm long; ray florets 14–45, the corolla 8–20 mm long . . . . 21. GRINDELIA
        3. Involucre 3–5 mm long; ray florets 7–15, the corolla 4–7 mm long
            4. Disc florets staminate (style branches not separating at flowering; florets not maturing into fruits), their pappus of 5–9 awns or slender scales, these about as long as the corolla, fused toward the base . . . . 9. AMPHIACHYRIS
            4. Disc florets perfect, their pappus a minute (less than 0.2 mm) crown or rarely an irregularly toothed (scalelike) ring, much shorter than the corolla . . . . 22. GUTIERREZIA
    2. Pappus of numerous capillary or occasionally stouter bristles, sometimes additionally with an outer ring of short bristles
        5. Pappus of the disc florets of 2 types, the inner series of capillary bristles much longer than the outer series
            6. Ray florets lacking a pappus . . . . 23. HETEROTHECA
            6. Ray florets with a well-developed pappus
                7. Plants annual, with usually short, stout taproots . . . . . 14. BRADBURIA
                7. Plants perennial, often with taproots but often also producing rhizomes, the rootstock often somewhat woody . . . . 23. HETEROTHECA
        5. Pappus of the disc florets more or less uniform, not differentiated into distinctly different inner and outer series (some of the bristles sometimes slightly shorter but then more or less intermingled with the longer ones)
            8. Involucre 10–15 mm long; ray florets 25–45, the corolla 10–20 mm long . . . . 21. GRINDELIA
            8. Involucre 2–9 mm long; ray florets (2–)3–25(–35), the corolla 0.5–7.0 mm long
                9. Leaf blades all linear to narrowly linear, the margins all entire, one or both surfaces dotted with impressed resin glands (these sometimes faint); receptacle minutely hairy around the base of each floret; ray florets usually more numerous than the disc florets . . . . 20. EUTHAMIA
                9. Leaf blades not all linear (some or all of the leaves shaped differently) and/or at least some of the blades few- to many-toothed along the margins, lacking impressed dots or resin glands (except in *S. odora*); receptacle not hairy, often with irregular low ridges between the florets; ray florets often fewer than the disc florets . . . . 25. SOLIDAGO

1. Heads discoid (check carefully to rule out the presence of relatively short, inconspicuous ray florets) or, if radiate, then the ray florets with the corollas white, pink, purple, or blue
    10. Disc florets with the corollas white, fading to light brown with age
        .... 25. SOLIDAGO
    10. Disc florets with the corollas yellow, sometimes turning reddish purple after pollen has been shed or fading to brown with age
        11. Pappus of disc florets of less than 10 scales and/or awns, sometimes absent or only a minute ridge or crown
            12. Heads completely discoid, resinous-sticky, especially when young
                .... 21. GRINDELIA
            12. Heads radiate (the ray florets sometimes relatively small and inconspicuous), not resinous or sticky
                13. Plants robust perennials, fibrous-rooted, with rhizomes or basal offshoots; stems 40–250 cm long .... 13. BOLTONIA
                13. Plants annual, taprooted, lacking rhizomes and basal offshoots; stems 3–35(–50) cm long
                    14. Involucre 5–8 mm long, the bracts in 4 or 5 unequal, overlapping series; ray florets 18–45 .... 10. APHANOSTEPHUS
                    14. Involucre 2.5–5.0 mm long, the bracts in 2 or 3 unequal or subequal, overlapping series; ray florets 5–22
                        15. Receptacle flat or slightly convex; pappus of 5 nearly transparent scales 0.2–1.8 mm long, these often alternating with 5 bristlelike awns 1–3 mm long ..... 15. CHAETOPAPPA
                        15. Receptacle hemispherical to conical; pappus essentially absent, represented by a faint line or minute ridge
                            .... 12. ASTRANTHIUM
        11. Pappus of at least the disc florets of relatively numerous (sometimes as few as 8 in *Erigeron*) capillary or occasionally stouter bristles, sometimes additionally with an outer ring of short bristles or scales
            16. Heads discoid or, if radiate, then the ray florets with inconspicuous corollas that are shorter than to slightly longer than the pappus and disc corollas
                17. Involucre 5–11 mm long .... 26. SYMPHYOTRICHUM
                17. Involucre 2–5 mm long
                    18. Perfect disc florets relatively few (less than 30), the head consisting mostly of pistillate ray florets with very short corollas, the strap-shaped portion sometimes reduced to a short fringe at the tip of a slender tube; florets all with a well-developed pappus .... 16. CONYZA
                    18. Perfect disc florets relatively numerous (more than 250), more numerous than the pistillate ray florets, these lacking or more commonly fewer than 100 (but with very short corollas) in 1(2) marginal series; pistillate (outermost) florets with a pappus of reduced bristles and/or scales that are noticeably shorter than the bristles of the other florets
                        .... 18. ERIGERON

16. Heads radiate, the ray florets with conspicuous corollas that are noticeably longer than the pappus and disc corollas
    19. All or most of the basal and lower stem leaves long-petiolate and with the blade truncate to cordate at the base
        20. Inflorescences relatively short and broad in outline, flat-topped or shallowly dome-shaped, the bracts along the branches relatively few and more or less leaflike; outermost involucral bracts 1–2 mm wide and mostly 1.5–3.0 times as long as wide; plants with long-creeping rhizomes . . . . 19. EURYBIA
        20. Inflorescences usually relatively elongate (occasionally short and broad in poorly developed or young plants), not appearing flat-topped or dome-shaped, the bracts along the branches often (but not always) relatively numerous and conspicuously more slender than the leaves; outermost involucral bracts 0.2–0.8 mm wide and mostly 4–8 times as long as wide; plants with the rhizomes often (but not always) relatively short and stout or absent . . . . 26. SYMPHYOTRICHUM
    19. None of the leaves both long-petiolate and with the blade truncate to cordate at the base
        21. Basal leaves very robust, mostly 25–50 cm long and 3–10 cm wide, usually persistent at flowering . . . . 11. ASTER
        21. Basal leaves much smaller, often withered or absent at flowering
            22. Involucral bracts in 1 or 2 series, all similar in length or less commonly the outer series somewhat shorter, the green central portion of each bract relatively uniform from the base to the tip and not conspicuously broadened toward the tip; style branches with the sterile tip (beyond the stigmatic lines) 0.1–0.3 mm long, lanceolate to broadly triangular; plants mostly flowering from spring through early summer . . . . 18. ERIGERON
            22. Involucral bracts in 3–9 series, mostly unequal in length with the outer series usually noticeably shorter (but subequal in *Eurybia hemispherica* and some species of *Symphyotrichum*), the green central portion of each bract absent or very slender toward the base and broadened abruptly; style branches with the sterile tip (beyond the stigmatic lines) (0.2–)0.3–0.5 mm long, linear to lanceolate; plants mostly flowering from late summer through autumn
                23. Pappus of 2 types, the inner 1 or more series of numerous longer bristles, the outer series of few to many much shorter bristles
                    24. Leaves broadly to narrowly elliptic or lanceolate-elliptic; involucre 3–5 mm long; innermost pappus bristles usually slightly broadened toward the tip (narrowly club-shaped); rays with the corolla white . . . . 17. DOELLINGERIA
                    24. Leaves all linear to linear-oblanceolate; involucre 6–9 mm long; innermost pappus bristles all tapered toward the tip; rays with the corolla lavender to purple (rare, white-flowered plants may be found in Missouri in the future) . . . . . 24. IONACTIS
                23. Pappus of 1 or more series of bristles that are all of similar length (the outermost bristles may be slightly shorter)

25. Involucre 9–12 mm long; leaves linear to linear-lanceolate; innermost pappus bristles usually slightly broadened toward the tip . . . . 19. EURYBIA
25. Involucre 2.5–8.0 mm long or, if longer, then all or at least some of the leaves lanceolate to oblanceolate or broader; pappus bristles all tapered toward the tip
. . . . 26. SYMPHYOTRICHUM

9. Amphiachyris (broomweed)
(Lane, 1979)

Two species, U.S.

The decision by Lane (1979, 1982) to accept the segregation of the genus *Amphiachyris* from the closely related *Gutierrezia* based upon a suite of morphological and cytological features was corroborated by Suh and Simpson's (1990) subsequent molecular phylogenetic analysis.

1. Amphiachyris dracunculoides (DC.) Nutt.
(broom snakeroot, broomweed)
Gutierrezia dracunculoides (DC.) S.F. Blake
Xanthocephalum dracunculoides (DC.) Shinners

Pl. 228 a, b; Map 959

Plants annual, with taproots, sometimes appearing slightly woody at the base. Stems solitary, 30–100 cm long, erect or strongly ascending, with numerous ascending to loosely ascending branches from below the midpoint, usually with faint, longitudinal lines (the smaller branches angular), glabrous. Basal leaves absent. Stem leaves sessile or less commonly short-petiolate, 0.5–6.0 cm long, the blade narrowly linear to linear-lanceolate or linear-oblanceolate, mostly sharply pointed at the tip, more or less tapered to the nonclasping base, the margins entire, the surfaces glabrous but with moderate to dense, impressed, resinous dots, often somewhat sticky to the touch. Inflorescences appearing as more or less flat-topped panicles, the heads solitary or in small, loose clusters at the branch tips, the branch tips and stalks short to relatively long, with few leaflike, linear bracts 0.3–1.0 cm long (the heads also usually closely subtended by 2 or 3 small, linear bracts). Heads radiate, sticky, resinous. Involucre 2.5–5.0 mm long, narrowly cup-shaped to obconical. Involucral bracts in 2 or 3 unequal, overlapping series, oblong-lanceolate to oblong-ovate, with a straw-colored to yellowish, hard, shiny basal portion and an ascending, triangular, green tip, glabrous but the green portion finely resin-dotted. Receptacle flat or slightly convex, with noticeable toothlike ridges around the attachment points of the florets. Ray florets 7–10(–12), pistillate, the corolla 4–6 mm long, yellow, sometimes persistent at fruiting. Disc florets 10–25, staminate (style branches not separating at flowering; florets not maturing into fruits), the corolla 2.5–3.2 mm long, yellow. Pappus of the ray florets absent or more commonly a minute crown of scalelike teeth. Pappus of the disc florets of 5–9 minutely toothed awns or slender scales, these about as long as the corolla, fused toward the base, usually white. Fruits 1.5–2.2 mm long, oblong-obovoid, sometimes slightly flattened, with 7–9 fine, green nerves, the surface otherwise densely pubescent with short, white hairs, purplish brown to nearly black. $2n=8$, 10. July–October.

Scattered in the Unglaciated Plains Division and the western portion of the Ozarks, sporadic farther north and east (Nebraska to Ohio south to Arizona and South Carolina; introduced in Pennsylvania and other parts of the eastern portion of the range). Upland prairies, limestone and dolomite glades, tops of bluffs, and banks of streams and rivers; also fallow fields, old fields, railroads, roadsides, and open, disturbed areas.

Steyermark (1963) noted that plants of this species appear densely branched and bushy and are sometimes almost as wide as tall.

10. Aphanostephus DC. (lazydaisy)
(Shinners, 1946b; Turner, 1984)

Four species, U.S., Mexico.

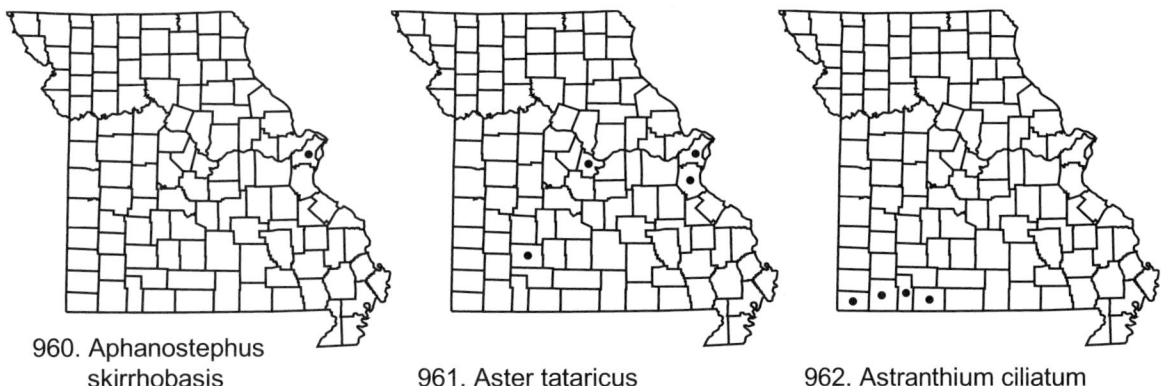

960. Aphanostephus skirrhobasis

961. Aster tataricus

962. Astranthium ciliatum

Preliminary molecular studies (Noyes and Rieseberg, 1999; Noyes, 2000a) have suggested that *Aphanostephus* may represent a specialized species complex within *Erigeron*. In general appearance, it strongly resembles a small *Erigeron*.

1. Aphanostephus skirrhobasis (DC.) Trel.
   var. skirrhobasis (Arkansas lazydaisy)

   Map 960

Plants annual, usually relatively slender, with taproots, the vegetative portions moderately to densely short-hairy (the whole plant often appearing grayish). Stems 1 or few, 5–35(–50) cm long, erect or ascending, unbranched or with few to several loosely to strongly ascending branches from near the base or above the midpoint. Basal leaves sometimes withered by flowering time, 2–7 cm long, 1–2 cm wide, the blade oblanceolate to broadly oblanceolate, rounded to bluntly pointed at the tip, long-tapered to the sometimes short-petiolate base, the margins finely to relatively coarsely scalloped or bluntly toothed from just below the broadest portion to the tip (occasionally appearing lobed). Stem leaves often somewhat reduced toward the tip, 1–5 cm long, rounded to bluntly or sharply pointed at the tip, more or less tapered to a usually sessile, nonclasping base, the margins entire or few-toothed toward the tip. Inflorescences of solitary or rarely paired heads at the branch tips, these relatively long, bractless or nearly so. Heads radiate, not sticky or resinous. Involucre 5–8 mm long, cup-shaped to shallowly cup-shaped. Involucral bracts in 4 or 5 unequal, overlapping series, narrowly lanceolate to narrowly oblong-lanceolate, the tip ascending, with a slender, green or brown central stripe and broad, thin, pale margins. Receptacle conical, with minute, irregular ridges around the concave attachment points of the florets. Ray florets 18–45, pistillate, the corolla 8–15 mm long, white, often with a longitudinal pinkish or purplish central line or band on the undersurface, withered but persistent at fruiting, the base becoming pale, somewhat hardened, swollen, and fused to the tip of the maturing fruit. Disc florets numerous (usually more than 250), perfect, the corolla 2.0–2.5 mm long, yellow (often with faint orange lines or mottling), persistent at fruiting, the base becoming pale, somewhat hardened, swollen, and fused to the tip of the maturing fruit. Pappus of the ray and disc florets similar or absent in the ray florets, when present a low, irregular crown (often appearing somewhat toothed), 0.1–0.3 mm long, white or light tan. Fruits 1.5–2.2 mm long, more or less cylindrical but expanded apically into the persistent corolla base and often somewhat 4-angled in cross-section, the surface with 4–12 rounded ribs (appearing finely 4–12-grooved), sparsely and minutely hairy, the hairs hooked or coiled at the tip, yellowish white to tan. $2n=6$. May–August.

Introduced, known thus far from a single collection from St. Louis (native of New Mexico to Louisiana north to Kansas and Arkansas; introduced in Missouri). Railroads.

This species was collected in 1961 by Viktor Mühlenbach during his botanical surveys of the St. Louis railyards, but originally it was misidentified as an *Astranthium*. The specimen was redetermined in 2002 by Guy Nesom of the Botanical Research Institute of Texas.

Turner (1984) treated *A. skirrhobasis* as consisting of three varieties. The var. *thalassius* Shinners is a low, bushy plant with thickened leaves that grows on sand dunes in the Gulf Coastal Plain of the southeastern U.S. and adjacent Mexico. The var. *kidderi* (S.F. Blake) B.L. Turner, which occurs in Texas and adjacent Mexico, is superficially similar to var. *skirrhobasis* but differs in its pappus of 5 or 10 well-developed scales.

11. Aster L. (aster)

About 180 species, North America (1 native taxon), Europe, Asia, introduced widely.

As has been the case with most of the big genera of Asteraceae, recent studies have caused taxonomists to reevaluate the generic limits and taxonomic interrelationships among species groups. A number of earlier studies of individual species groups had already caused the reclassification of some taxa to other portions of the tribe. For example, the aberrant *A. ptarmicoides* (Nees) Torr. & A. Gray is now generally accepted to belong to a small group of species most closely related to *Solidago* (Brouillet and Semple, 1981). The first comprehensive study of higher-order relationships within the *Aster* group was by Nesom (1994), who concluded, based upon morphological comparisons as well as on biogeographical and cytological data, that the New World and Old World species were not very closely related within the family. His work resulted in a new framework that recognized 7 major lineages of former asters and more than 25 total genera. Since that time, additional morphological studies (Nesom 2000) as well as ongoing research involving molecular data (Noyes and Rieseberg, 1999; Semple et al., 2002) have supported this general classification while fine-tuning the numbers and limits of the various smaller genera. Although *Aster* remains the largest genus in the group, the only species native to North America is the circumboreal *A. alpinus* L. The sole true aster found growing wild in Missouri is an escape from gardens.

1. Aster tataricus L.f. (Tatarian aster)

Map 961

Plants perennial herbs, with stout rootstocks, often densely colonial from relatively thick, somewhat fleshy rhizomes. Stems 1 to several, 50–200 cm long, erect or ascending, often with several to numerous ascending branches above the midpoint, moderately to densely roughened, with short, broad-based hairs toward the tip, glabrous to sparsely hairy toward the base. Basal leaves persistent at flowering, robust, mostly 30–50 cm long, 3–10 cm wide, the blade oblanceolate to narrowly elliptic, narrowed or short-tapered to a sharply pointed tip, long-tapered to the base, the margins shallowly toothed, the surfaces and margins roughened with sparse to moderate, short, broad-based hairs. Stem leaves gradually reduced toward the tip, 3–30 cm long, narrowed or short-tapered to a sharply pointed tip, tapered to a sessile, somewhat sheathing but not strongly clasping base, the margins entire to shallowly toothed, the surfaces and margins roughened with sparse to moderate, short, broad-based hairs. Inflorescences often elongate panicles (rarely reduced to a solitary terminal cluster), the heads in flat-topped to somewhat rounded clusters at the ascending branch tips, the bracts relatively numerous, leaflike, linear to narrowly oblong-elliptic or narrowly oblanceolate. Heads radiate, not sticky or resinous. Involucre 7–12 mm long, cup-shaped to slightly bell-shaped. Involucral bracts in 3 or 4 unequal, overlapping series, linear to narrowly lanceolate, the tip ascending and often somewhat purplish-tinged, poorly differentiated into a green central band and lighter margins, the margins sometimes somewhat irregular, occasionally sparsely hairy toward the tip, the outer surface glabrous or sparsely hairy. Receptacle flat or slightly convex, with minute, irregular ridges around the concave attachment points of the florets. Ray florets 15–30, pistillate, the corolla 10–15 mm long, pale lavender to purple or purplish blue. Disc florets 25–50, perfect, the corolla 4–6 mm long, yellow, often turning purplish after the pollen has been shed, not persistent at fruiting. Pappus of the ray and disc florets similar, of numerous capillary, finely barbed bristles in 2 similar series, 6–8 mm long, white or cream-colored. Fruits 1.5–2.0 mm long, narrowly obconical, not or only slightly flattened, 4–6-ribbed, glabrous to finely hairy, light brown to tan. $2n=54$. September–November.

Introduced, uncommon and sporadic, mostly near urban areas (native of Europe, Asia, introduced sporadically in the eastern U.S. west to Iowa and Missouri). Banks of streams; also gardens and moist, disturbed areas.

Tatarian aster is a striking and robust plant that is grown as an ornamental in gardens. Under some conditions, it can be relatively aggressive, forming dense colonies from the rhizomes. This species was first reported from Missouri by Yatskievych and Summers (1993) based on collections made in the 1970s in Jefferson County by James Solomon of the Missouri Botanical Garden.

12. Astranthium Nutt. (western daisy)
(DeJong, 1965; Nesom, 2005)

About 12 species, U.S., Mexico.

1. Astranthium ciliatum (Raf.) G.L. Nesom
  A. *integrifolium* (Michx.) Nutt. var. *ciliatum* (Raf.) Larsen
  A. *integrifolium* ssp. *ciliatum* (Raf.) DeJong
      Pl. 229 h, i; Map 962

Plants annual, usually relatively slender, with sometimes short taproots. Stems 1 or few, 5–45 cm long, erect or ascending, unbranched or with few to several ascending branches above the midpoint, often 4–6-angled, sparsely to moderately pubescent (especially along the ridges) with strongly to loosely ascending white hairs. Basal leaves sometimes withered by flowering time, 1.5–5.0 cm long, 0.3–1.2 cm wide, the blade oblanceolate to obovate or spatulate, rounded to bluntly pointed at the tip, long-tapered to the mostly short-petiolate base, the margins entire (rarely slightly wavy or scalloped), the surfaces and especially the margins sparsely to more commonly moderately hairy. Stem leaves often somewhat reduced above the lower portion of the stem, 0.5–5.0 cm long, rounded to bluntly or sharply pointed at the tip, more or less tapered to a slightly expanded but not strongly clasping base, the margins entire, the surfaces and especially the margins sparsely to moderately hairy. Inflorescences of solitary heads at the branch tips, these relatively long, bractless or nearly so. Heads radiate, not sticky or resinous. Involucre 2.5–4.5 mm long, cup-shaped to slightly bell-shaped. Involucral bracts in 2 similar to slightly unequal, overlapping series, lanceolate to elliptic or narrowly ovate, the tip ascending, with a relatively broad, green central stripe and broad, thin, pale (often somewhat transparent) margins, glabrous or sparsely hairy. Receptacle hemispherical to conical, with minute, irregular ridges around the concave attachment points of the florets. Ray florets 13–25, pistillate, the corolla 6–12 mm long, white, sometimes purplish-tinged on the undersurface, usually turning purple to purplish blue with age, not persistent at fruiting. Disc florets numerous (usually 120–170), perfect, the corolla 2–3 mm long, yellow, not persistent at fruiting. Pappus of the ray and disc florets essentially absent, represented by a faint line or minute ridge. Fruits 1.0–1.6 mm long, more or less obovoid, somewhat flattened, the angles more or less rounded or slightly ribbed, the surface smooth, usually sparsely to moderately pubescent with minute hairs, the hairs often slightly curved toward the midpoint but straight at the tip, tan to greenish brown or dark brown. $2n=8$. April–June.

Uncommon in the southwestern portion of the Ozark Division (Missouri and Kansas south to Texas; Mexico). Chert, limestone, and dolomite glades, ledges and tops of bluffs, thin-soiled areas in upland prairies, savannas, and banks of streams and rivers.

DeJong (1965) accepted two taxa within *A. integrifolium*. He characterized ssp. *integrifolium* as having fibrous roots, longer achenes (1.6–2.2 mm) with sparser hairs, slightly larger heads (involucre 4–6 mm long) and corollas (disc corollas 2.0–3.2 mm, ray corollas 12–15 mm), and a generally more western distribution (Kentucky to Georgia west to Missouri and Arkansas). DeJong indicated that both of his subspecies were present in Missouri but annotated most of the Missouri specimens as representing intermediates or hybrids between the two subspecies. Yatskievych and Turner (1990) erroneously listed only ssp. *integrifolium* for Missouri. In fact, most of the specimens in question seem to represent ssp. *ciliatum,* even by DeJong's criteria. Nesom (2005) refined the taxon limits slightly and raised ssp. *ciliatum* to species level, while maintaining that the number of intermediate specimens in the region of contact with *A. integrifolium* in the strict sense was many fewer than earlier reports had suggested. Nesom's circumscription of taxa is accepted in the present treatment, which results in the exclusion of true *A. integrifolium* from the Missouri flora.

13. Boltonia L'Hér. (false aster)
(Morgan, 1967)
Contributed by Sarah M. Tofari

Plants robust perennial herbs, fibrous-rooted, with rhizomes or basal offshoots, sometimes somewhat woody at the base. Stems solitary or few, 40–250 cm long, erect or ascending, with usually numerous ascending to loosely ascending branches above the basal $1/3$, with prominent pale ridges, glabrous. Basal leaves absent at flowering, narrowly oblanceolate to oblong-obovate. Stem leaves mostly sessile, progressively reduced toward the stem tip, the blades oblanceolate to elliptic or linear, bluntly to sharply pointed at the tip, tapered to long-tapered at the base, the margins entire or minutely toothed, the surfaces glabrous. Inflorescences panicles, these often large and highly branched, flat-topped or more commonly rounded, the heads solitary at the branch tips, the branches with numerous linear to linear-lanceolate or linear-elliptic, leaflike bracts. Heads radiate, not sticky or resinous. Involucral bracts in 2–5(6) unequal to subequal, overlapping series, linear to lanceolate, oblanceolate, or spatulate, the tip ascending, differentiated into a somewhat thickened, green to yellowish green central band, and narrow to relatively broad, pale margins, these occasionally minutely toothed, glabrous. Receptacle hemispherical or conical, sometimes with inconspicuous, minute, irregular ridges around the concave attachment points of the florets. Ray florets 20–60, pistillate, the corolla white or less commonly pinkish- to purplish-tinged. Disc florets numerous (50–400 or more), perfect, the corolla yellow, not persistent at fruiting. Pappus a short, irregular crown of 4–10(–12) minute awns or narrow scales and 2(–4) longer (0.5–2.0 mm) awns, the longer awns often absent in the ray florets and sometimes also in the disc florets. Fruits of 2 types; those developing from the disc florets wedge-shaped to obovate in outline, relatively strongly flattened, broadly rounded or angled to shallowly notched at the tip, the margins winged, the surface and margins often minutely hairy, tan to grayish brown with lighter wings; fruits developing from ray florets more or less wedge-shaped, 3-angled, with 3 narrow wings. Five species, U.S., Canada.

Boltonia species superficially resemble *Erigeron,* but recent molecular and morphological studies (Noyes and Rieseberg, 1999; Nesom and Noyes, 2000) have suggested that *Boltonia* and two other small groups formerly associated with *Erigeron* are more closely related to *Symphyotrichum* and some other North American genera within the tribe Astereae. The two close relatives of *Boltonia* are the genera *Chloracantha* G.L. Nesom et al., which ranges from Central America to the southwestern United States, and *Batopilasia* G.L. Nesom & Noyes, which is found only in a small area of southwestern Chihuahua, Mexico.

1. Leaf bases decurrent below the attachment point as wings of green tissue along the stem ridges .... 2. B. DECURRENS
1. Leaf bases not decurrent
    2. Flower heads relatively large, the disc usually 6–14 mm in diameter at flowering, the ray corollas 7–15 mm long; lower and median leaves with the blades oblanceolate to elliptic; pappus awns 0.5–2.0 mm long, mostly well developed.. 1. B. ASTEROIDES
    2. Flower heads relatively small, the disc 3–6 mm in diameter at flowering, the ray corollas 5–8 mm long; lower and median leaves with the blades linear to sometimes oblanceolate; pappus awns 0.3–0.7 mm long, mostly poorly developed.. 3. B. DIFFUSA

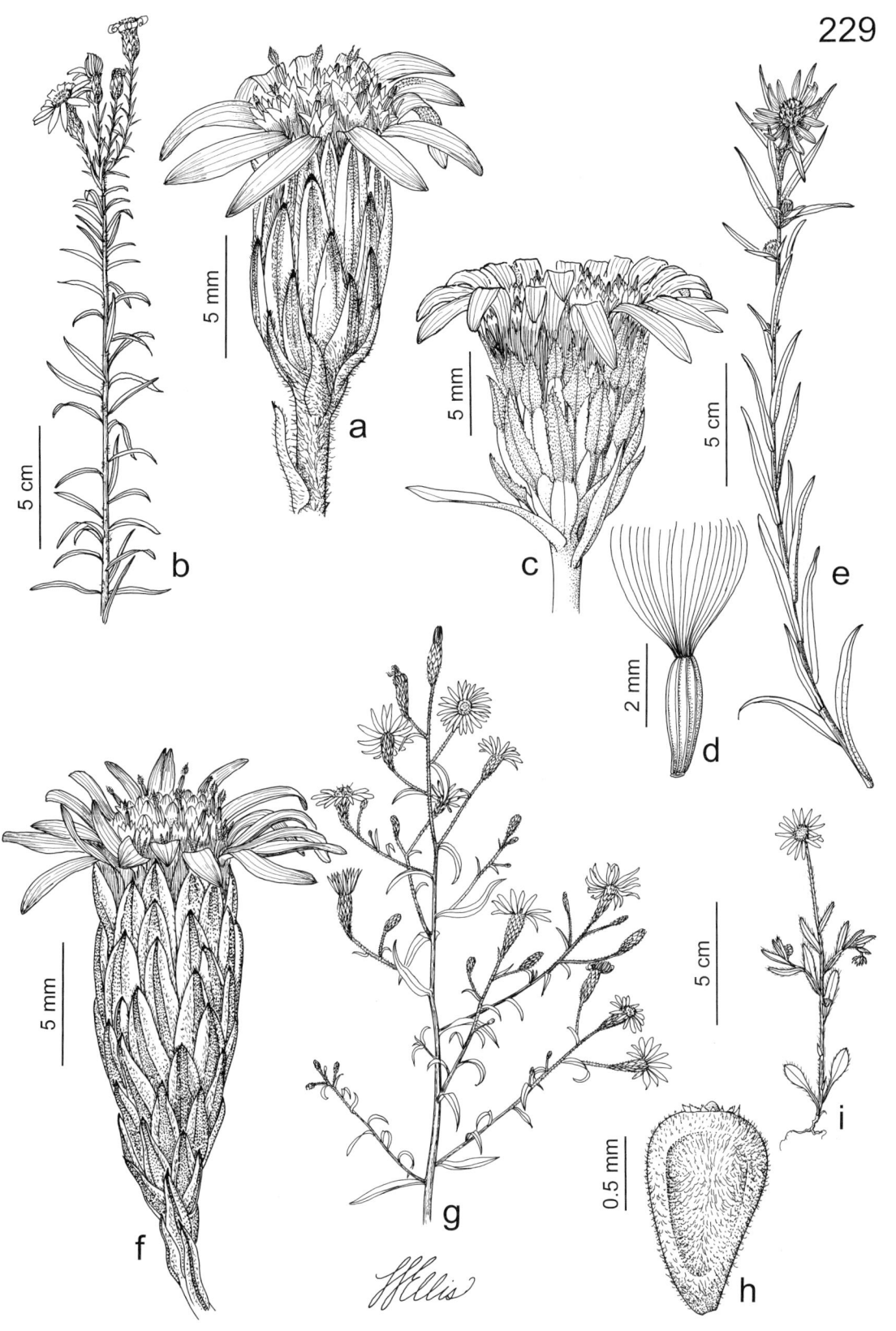

Plate 229. Asteraceae. *Ionactis linariifolius*, **a)** flower, **b)** habit. *Eurybia hemispherica*, **c)** flower, **d)** fruit, **e)** habit. *Symphyotrichum turbinellum*, **f)** flower, **g)** habit. *Astranthium ciliatum*, **h)** fruit, **i)** habit.

1. Boltonia asteroides (L.) L'Hér. (false aster, false starwort)

Pl. 231 f–i; Map 963

Plants producing basal offshoots and often also elongate rhizomes. Stems 40–150 cm long (taller in some cultivated forms). Leaf blades 2–15 cm long, 4–25 mm wide, those of the lower and median leaves oblanceolate to narrowly elliptic, those of the upper leaves narrowly oblanceolate to linear, the base not decurrent below the attachment point (the stems thus unwinged). Inflorescences appearing leafy, the bracts 1.5–5.0 cm long, 2–6 mm wide. Heads relatively large, the receptacle usually 6–14 mm in diameter at flowering. Involucre 3–5 mm long, the bracts in mostly 3 subequal series. Ray florets 25–60, the corolla 7–15 mm long. Disc florets 60–180. Pappus of disc florets a short, irregular crown of awns or narrow scales 0.1–0.4 mm long and 2(–4) awns 0.5–2.0 mm long, the longer awns mostly well developed in the disc florets, often absent in the ray florets. Fruits 1.5–3.0 mm long, the wings 0.1–0.5 mm wide. $2n=18, 36$. July–October.

Scattered nearly throughout the state (eastern U.S. west to North Dakota and Texas; Canada; introduced in Idaho, Oregon). Banks of streams and rivers, margins of ponds and lakes, bottomland prairies, bottomland forests, sloughs, fens, and marshes, also margins of crop fields, fallow fields, levees, banks of ditches, railroads, roadsides, and moist, sandy, disturbed areas.

Boltonia asteroides is the most common species of *Boltonia* in the state. In recent years, it has become popular as an ornamental in native plant gardens throughout the eastern and central United States because of its tolerance to deer browsing and the beautiful white blooms it produces in the autumn. Fruits of this species are commonly eaten by waterfowl.

Currently there are three varieties of this species accepted by most authors, only two of which are found in Missouri. The var. *asteroides* differs in its subequal, relatively narrow involucral bracts and slightly shorter pappus awns. It occurs mainly along the Atlantic and Gulf Coastal Plains.

1. Involucral bracts awl-shaped to oblanceolate, the outer series 0.9–1.6 mm wide, the tips rounded or broadly and bluntly angled, sometimes with an abrupt, minute, sharp point
 .... 1A. VAR. LATISQUAMA
1. Involucral bracts oblong to lanceolate, the outer series 0.4–0.7 mm wide, the tips narrowed or tapered to a sharp point .... 1B. VAR. RECOGNITA

1a. var. latisquama (A. Gray) Cronquist
B. latisquama A. Gray

Pl. 231 f–h

Plants with slender rhizomes and often also basal offshoots. Involucral bracts awl-shaped to oblanceolate, rounded or bluntly and broadly angled at the tip, sometimes with an abrupt, short, sharp point, the outer series 2.5–3.5 mm long, 0.9–1.6 mm wide, the inner series 3.2–4.0 mm long, 0.9–1.3 mm wide. $2n=18$. July–October.

Scattered in the northern and southwestern portions of the state but apparently absent from the western portion of the Glaciated Plains Division and nearly absent from the Ozarks (North Dakota to Oklahoma east to Wisconsin and Arkansas; introduced in New England). Banks of streams and rivers, margins of ponds and lakes, bottomland prairies, bottomland forests, sloughs, fens, and marshes, also margins of crop fields, fallow fields, levees, banks of ditches, railroads, roadsides, and moist, sandy, disturbed areas.

1b. var. recognita (Fernald & Griscom) Cronquist
B. latisquama A. Gray var. *recognita* Fernald & Griscom
B. latisquama var. *microcephala* Fernald & Griscom
B. latisquama var. *occidentalis* Fernald & Griscom

Pl. 231 i

Plants with only slender, underground rhizomes. Involucral bracts oblong to lanceolate, narrowed or tapered to a sharply pointed tip, the outer series 1.5–2.8 mm long, 0.4–0.7 mm wide, the inner series 2.1–3.0 mm long, 0.6–0.9 mm wide. $2n=36$. July–October.

Scattered nearly throughout the state (northeastern U.S. west to North Dakota and Oklahoma; Canada; introduced in Idaho, Oregon). Banks of streams and rivers, margins of ponds and lakes, bottomland prairies, bottomland forests, sloughs, fens, and marshes, also margins of crop fields, fallow fields, levees, banks of ditches, railroads, roadsides, and moist, sandy, disturbed areas.

Plants with the ray corollas pink or pinkish-tinged have been called f. *rosea* Benke.

2. Boltonia decurrens (Torr. & A. Gray) A.W. Wood (decurrent false aster)
B. asteroides var. *decurrens* (Torr. & A. Gray) Fernald & Griscom
B. latisquama var. *decurrens* (Torr. & A. Gray) Fernald & Griscom

Pl. 231 c–e; Map 964

963. Boltonia asteroides   964. Boltonia decurrens   965. Boltonia diffusa

Plants producing basal offshoots but no rhizomes. Stems (50–)120–250 cm long. Leaf blades 5–20 cm long, 5–25 mm wide, those of the lower and median leaves oblanceolate to narrowly elliptic, those of the upper leaves mostly narrowly elliptic to linear-elliptic, the base strongly decurrent below the attachment point as a pair of wings of green tissue along the stem ridges (the stems thus appearing irregularly winged). Inflorescences appearing leafy, the bracts 0.5–5.0 cm long, 1–11 mm wide. Heads relatively large, the receptacle usually 6–12 mm in diameter at flowering. Involucre 3–5 mm long, the bracts in 3–5 subequal to somewhat unequal series, narrowed or tapered to a sharply pointed tip or sometimes rounded to an abrupt, short, sharp point. Ray florets 45–60, the corolla 9–11(–14) mm long. Disc florets 250–400. Pappus of disc florets a short, irregular crown of awns or narrow scales 0.1–0.4 mm long and 2(–4) awns 1–2 mm long, the longer awns well developed in the disc florets, sometimes absent in the ray florets. Fruits 1.5–2.5 mm long, the wings 0.3–0.5 mm wide. $2n=18$. August–October (rarely to December).

Uncommon in the Big Rivers Division from Lincoln to St. Charles Counties with isolated historical collections from the city of St. Louis and Cape Girardeau County; introduced in Howell and St. Louis Counties (endemic to the Illinois and adjacent Mississippi River floodplains in Illinois and Missouri). Banks of rivers, margins of ponds and lakes, and bottomland prairies; also margins of crop fields, fallow fields, and levees.

The present treatment follows that of Schwegman and Nyboer (1985) in recognizing this taxon as a species rather than a variety within *B. asteroides*, as was done by many earlier authors (Steyermark, 1963). *Boltonia decurrens* is heavily dependent on periodic flooding or disturbance to eliminate competing vegetation and to provide the high light and moist soil conditions that the achenes require for germination (Smith and Keevin, 1998).

During the Great Flood of 1993, the entire global range of decurrent false aster was inundated for a period of 8–10 weeks, but in spite of gloomy predictions that the species would be driven toward extinction, it responded relatively well to the flooding, which created large areas of relatively open floodplain habitat. Overall, however, populations have decreased in size and number in recent decades due to the construction of levees and locks and dams along the Illinois and Mississippi Rivers, which have prevented flooding in many areas. In 1988, this globally imperiled species was listed as Threatened by the U.S. Fish and Wildlife Service under the Federal Endangered Species Act.

3. Boltonia diffusa Elliott (doll's daisy)
B. diffusa var. *interior* Fernald and Griscom
Pl. 231 a, b; Map 965

Plants producing basal offshoots and/or elongate rhizomes. Stems 40–150 cm long. Leaf blades 2–12 cm long, 2–18 mm wide, those of the lower and median leaves linear to narrowly lanceolate, those of the upper leaves mostly linear, the base not decurrent below the attachment point (the stems thus unwinged). Inflorescences usually not appearing leafy, the relatively few bracts 0.2–2.5 cm long, 0.5–5.0 mm wide. Heads relatively small, the receptacle usually 3–6 mm in diameter at flowering. Involucre 2.5–3.5 mm long, the bracts in 3–5(6) more or less unequal series, narrowly oblong to nearly linear, narrowed or tapered to a sharply pointed tip or sometimes rounded to an abrupt, short, sharp point. Ray florets 20–50, the corolla 5–8 mm long. Disc florets 55–150. Pappus of disc florets a short, irregular crown of awns or narrow scales 0.1–0.3 mm long and 2(–4) awns 0.3–0.7 mm long, the longer awns mostly poorly developed in the disc florets, usually absent in the ray florets. Fruits 1.5–2.5 mm long, the wings 0.1–0.4 mm wide. $2n=18$. July–October.

Uncommon, mostly in the southeasternmost counties of the Ozark Division and the Mississippi Lowlands (southeastern U.S. west to Missouri, Oklahoma, and Texas). Banks of streams and rivers, swamps, sloughs, bottomland prairies, bottomland forests, and margins of sinkhole ponds; also banks of ditches, roadsides, and moist, disturbed areas.

This species is sometimes difficult to distinguish from smaller-headed variants of *B. asteroides*. In fact, Morgan (1967) hypothesized that some of the Missouri material represents hybrids between the two taxa, although this has not been confirmed by further research. Characters that help to differentiate these species include the apparent leafiness of the inflorescence and the tips of the involucral bracts. The inflorescences of *B. diffusa* have few typically narrow bracts, whereas those of *B. asteroides* var. *recognita* tend to have relatively numerous wider bracts and an overall leafy appearance. The tips of the involucral bracts of *B. diffusa* narrow or taper to a sharp point, but the bracts of *B. asteroides* var. *latisquama* are either rounded or rounded with a short, abrupt point.

Morgan (1967) recognized three varieties of *B. diffusa*; however, the new combination for her var. *caroliniana* was never validly published. Most botanists consider this to represent a separate species, *B. caroliniana* (Walter) Fernald, which is endemic to the Coastal Plain and Piedmont from South Carolina to Georgia and has achenes that are wingless or nearly so. Morgan followed a traditional classification in separating the remaining plants into two varieties that were both attributed to Missouri. The var. *diffusa* was said to produce slender rhizomes more frequently, to produce involucral bracts that are awl-shaped to nearly linear, and to have the stalks of the heads threadlike, with some of the heads sometimes drooping. In contrast, var. *interior* (which occupies parts of the western and northern portion of the species range) supposedly has a nonrhizomatous habit, involucral bracts that are linear-oblong, and slender but not threadlike stalks of the heads, with the heads not drooping. Nearly all of the Missouri specimens are more or less attributable to var. *interior*, but two historical collections from Dunklin and St. Louis Counties appear similar to specimens of var. *diffusa* from the southeastern states. In practice, however, many specimens of *B. diffusa* are not clearly determinable at the infraspecific level, and there is more variation in the shapes of involucral bracts than may have been apparent to earlier workers. The present treatment follows that of Cronquist (1980) in not formally recognizing varieties within this species.

14. Bradburia Torr. & A. Gray

Two species, endemic to the southeastern U.S.

In the past, *Bradburia* has variously been included in *Chrysopsis* and *Heterotheca*. The consensus among recent workers (Nesom, 1991; Semple, 1996) is that the group is closely related to *Chrysopsis,* and the decision to treat *Bradburia* as a separate genus rather than a section of *Chrysopsis* is somewhat arbitrary (Nesom, 1997). Semple (1996) argued that because the two species involved have a suite of minor morphological features not otherwise found in *Chrysopsis,* segregating them into a separate genus makes sense for pragmatic reasons. In Missouri, where there are no other species of true *Chrysopis* present, the decision to follow Semple's classification was made primarily to limit the taxonomic and nomenclatural differences between the present work and the forthcoming volume of the Flora of North America Project.

This genus was named to honor John Bradbury, one of the first two trained botanists to explore in Missouri. Interested readers may consult the introductory chapter on the history of floristic botany in Missouri in the first volume of the present work (Yatskievych, 1999) for a brief biography of this ill-fated early plant explorer.

1. Bradburia pilosa (Nutt.) Semple (soft goldenaster)
Chrysopsis pilosa Nutt.
C. nuttallii Britton
Heterotheca pilosa (Nutt.) Shinners
Pl. 232 h–j; Map 966

Plants annual, with often short, stout taproots. Stems usually solitary, 25–60(–80) cm long, erect or ascending, with several ascending branches above the midpoint, with fine, longitudinal lines or grooves, moderately to densely pubescent with longer and shorter, fine, bent or curled hairs, many

966. Bradburia pilosa   967. Chaetopappa asteroides   968. Conyza canadensis

of the shorter hairs gland-tipped. Basal leaves usually withered by flowering time, oblanceolate, the margins usually with 2 to several pairs of teeth. Stem leaves sometimes somewhat reduced toward the tip of the stem, 1–6 cm long, oblanceolate to narrowly oblong-oblanceolate or linear, rounded to bluntly or sharply pointed at the tip, more or less tapered to a slightly expanded but not strongly clasping base, the margins entire or few-toothed toward the tip (the lowermost stem leaves sometimes with more teeth), the surfaces and especially the margins moderately to densely pubescent with longer and shorter, fine, bent or curled hairs, many of the shorter hairs gland-tipped. Inflorescences appearing as flat-topped to shallowly convex panicles, the heads solitary or in small, loose clusters toward the branch tips, the branch tips and stalks short to relatively long, with usually several leaflike, linear bracts 0.5–1.5 cm long. Heads radiate, not or only slightly sticky, not resinous. Involucre 6–10 mm long, cup-shaped to slightly bell-shaped. Involucral bracts in 3 or 4 unequal, overlapping series, narrowly lanceolate to linear, the slender tip ascending to loosely ascending and often brownish- or purplish-tinged, at least the longer bracts with the green central stripe absent or threadlike toward the base, broadened and narrowly elliptic above the midpoint, the whitish or straw-colored marginal areas becoming correspondingly narrower above the midpoint, the outer surface moderately hairy, at least some of the shorter hairs gland-tipped. Receptacle flat or slightly convex, sometimes with low, toothlike ridges around the attachment points of the florets, especially toward the outer florets. Ray florets 13–25, pistillate, the corolla 6–12 mm long, yellow, not persistent at fruiting. Disc florets 25–60, perfect, the corolla 4.5–6.0 mm long, yellow, not persistent at fruiting. Pappus of the ray and disc florets similar, a low crown of several shorter, white to off-white scales 0.5–1.0 mm long and numerous longer, finely barbed bristles 4–6 mm long, these usually white when young but often turning straw-colored to light orangish brown as the fruits mature. Fruits 1.7–2.5 mm long, more or less wedge-shaped to slightly obovoid, often slightly flattened, finely 10-nerved, the nerves sometimes difficult to observe, the surface moderately pubescent with fine, appressed, silvery hairs, light tan to brown. $2n=8$. June–October.

Scattered in the southern third of the state (Tennessee to Mississippi west to Kansas and Texas; introduced sporadically farther north and east). Upland prairies, sand prairies, glades, savannas, banks of streams and rivers, and rarely margins of fens; also fallow fields, old fields, railroads, and roadsides.

15. Chaetopappa DC.
(Shinners, 1946a; Nesom, 1988)

About 11 species, U.S., Mexico.

1. Chaetopappa asteroides (Nutt.) DC. **var. asteroides** (Arkansas leastdaisy; tiny lazydaisy)

Pl. 232 f, g; Map 967

Plants annual, relatively slender, with short taproots. Stems 1 or less commonly few, 3.5–30.0 cm long, erect or ascending, unbranched or with few to several ascending to spreading branches, often slightly angled, sparsely to moderately pubescent (often appearing somewhat roughened) with appressed to loosely ascending, white hairs, these sometimes broadened at the base. Basal leaves sometimes withered by flowering time, 0.5–2.5 cm long, 0.1–0.5 cm wide, the blade oblanceolate to obovate or spatulate, rounded or less commonly bluntly pointed at the tip, long-tapered to the sometimes more or less short-petiolate base, the margins entire, the surfaces and margins sparsely to moderately hairy. Stem leaves often somewhat reduced toward the tip of the stem, 0.3–1.5 cm long, linear to narrowly oblanceolate, rounded to bluntly or sharply pointed at the tip, more or less tapered to a slightly expanded but not clasping base, the margins entire, the surfaces and especially the margins sparsely to moderately hairy. Inflorescences sometimes appearing as more or less flat-topped panicles after the leaves wither, consisting of solitary heads at the branch or stem tips, these short to relatively long, usually with few to several minute, linear bracts. Heads radiate, not sticky or resinous. Involucre 2.5–4.0 mm long, narrowly obconical. Involucral bracts in 2–4 unequal, overlapping series (the outer 1 or 2 series usually with relatively few bracts), narrowly lanceolate to narrowly oblong-elliptic, the tip ascending, with a relatively broad, green central stripe and narrow to broad, thin, pale (often somewhat transparent) margins, glabrous or sparsely to moderately hairy, at least the inner series finely fringed along the margins toward the tip. Receptacle flat or slightly convex, not noticeably ridged around the attachment points of the florets. Ray florets 5–13, pistillate, the corolla 3–5 mm long, white, sometimes pinkish- or purplish-tinged toward the tip, usually turning purple to purplish blue with age, not persistent at fruiting. Disc florets 6–25, perfect (but the innermost few sometimes not developing fully to flowering), the corolla 2–3 mm long, yellow, not persistent at fruiting. Pappus of the ray and disc florets similar (or that of the ray florets occasionally with the awns shorter or lacking), a low crown of usually (4)5(6) shorter, white to nearly transparent scales 0.1–0.8 mm long that alternate with an equal number of longer, straw-colored to light brown awns 1–3 mm long. Fruits 1.5–2.0 mm long, more or less wedge-shaped to slightly obovoid, 5-angled, the angles occasionally slightly ribbed, the surface smooth, usually moderately pubescent with minute, appressed hairs, brown to dark brown. $2n=16$. April–August.

Scattered in the Unglaciated Plains Division (Kansas and Missouri south to Texas and Louisiana; Mexico). Glades, rock outcrops in upland prairies, and savannas, usually on sandstone and chert substrates.

The other variety of this inconspicuous species, var. *grandis* Shinners, differs in its slightly more numerous disc and ray florets and slightly longer pappus awns. It is endemic to the valley of the Rio Grande in Texas.

16. Conyza Less.

Plants annual, slender or robust, with taproots, the vegetative portions glabrous or more commonly sparsely to densely hairy. Stems 1 or few to several, erect to spreading, usually much-branched either throughout or only above the midpoint, finely to coarsely angled or longitudinally lined. Basal leaves absent at flowering. Stem leaves somewhat reduced above the lower stem or relatively uniform, narrowly linear to oblong-lanceolate or oblanceolate, mostly sharply pointed at the tip, more or less tapered to a sessile or short-petiolate, often slightly expanded but nonclasping base, the margins entire or few-toothed toward the tip. Inflorescences panicles or sometimes appearing as racemes, the heads long-stalked to nearly sessile along the branches and/or in loose clusters at the branch tips, these usually with few to several minute, linear bracts. Heads appearing discoid but actually radiate (see below), not sticky or resinous. Involucre 1.5–4.0 mm long, narrowly to broadly cup-shaped or slightly bell-shaped, occasionally appearing nearly urn-shaped at flowering. Involucral bracts in 2–4 more or less unequal, overlapping series, narrowly lanceolate to linear, the tip ascending, with a slender to relatively broad, green or brown central stripe (usually with a slender, yellowish midvein), this sometimes not extending to the bract tip, and with relatively slender, thin, pale margins. Receptacle flat or nearly so, relatively smooth. Ray florets 18–40 in 1–3 series, pistillate, the corolla inconspicuous, the strap-shaped portion 0.2–0.8 mm long, sometimes reduced to a short fringe at the tip of the slender tube, white or pink to light purple, shed before fruiting. Disc

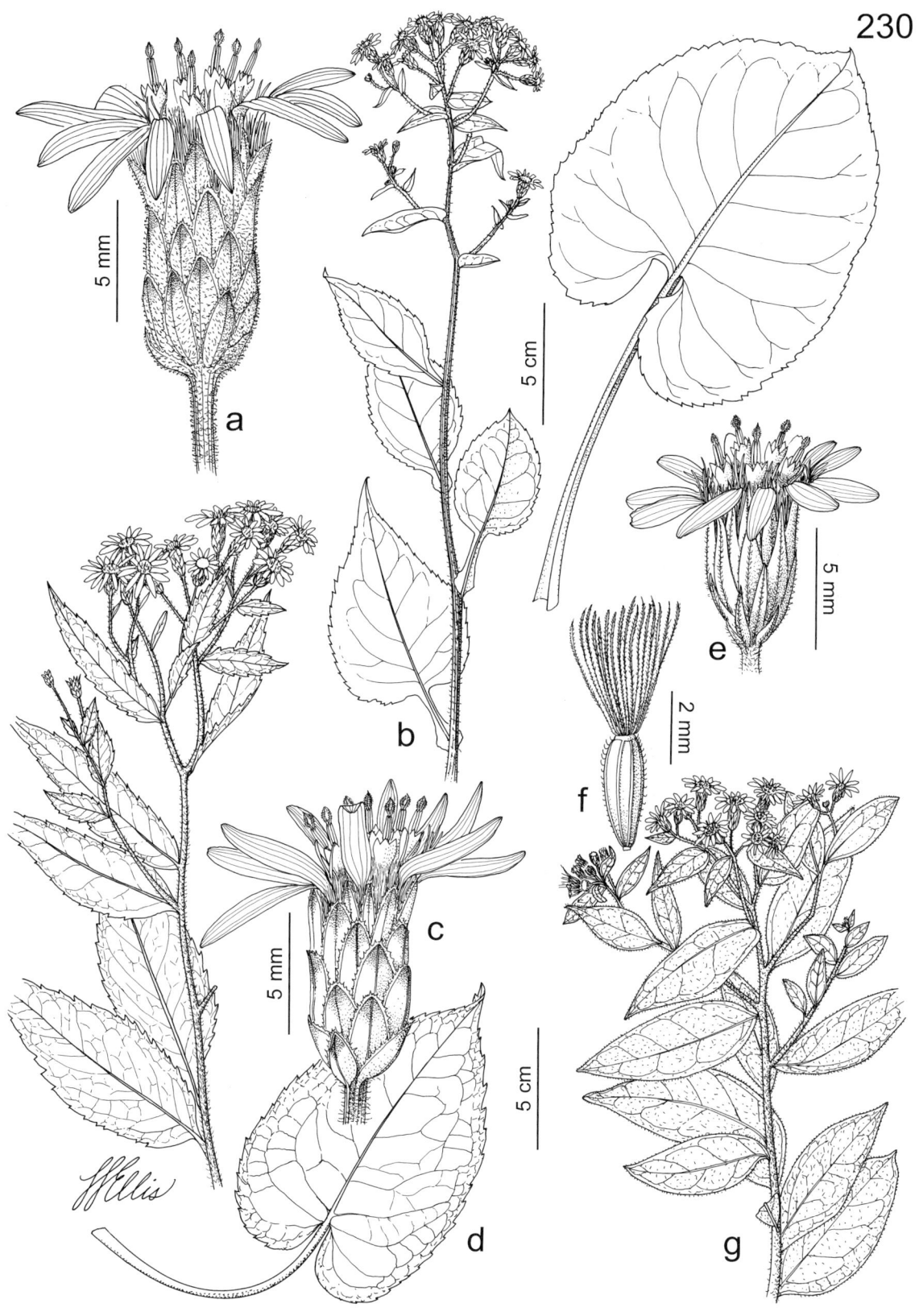

Plate 230. Asteraceae. *Eurybia macrophylla*, **a)** head, **b)** habit and basal leaf. *Eurybia furcata*, **c)** head, **d)** habit and basal leaf. *Doellingeria umbellata*, **e)** head, **f)** fruit, **g)** habit.

florets relatively few (less than 30), perfect, the corolla 1.5–2.5 mm long, yellow, sometimes turning pinkish-tinged after the pollen has been shed, shed before fruiting. Pappus of the ray and disc florets similar, of numerous (15–25) capillary bristles, 2–3 mm long, usually white, less commonly somewhat pinkish- or yellowish-tinged. Fruits 1.0–1.5 mm long, narrowly oblong in outline (slightly tapered at the base), flattened, the angles usually with inconspicuous nerves (these occasionally appearing as narrow wings), the surface glabrous or more commonly sparsely to moderately and minutely hairy, light tan to pale grayish brown. Twenty-five to 40 species, nearly worldwide, mostly in tropical or warm-temperate regions.

Cronquist (1943) segregated *Conyza* from *Erigeron,* based primarily upon its reduced ray corollas and relatively few disc florets. Nesom (1990b) reexamined generic limits in the group and suggested some additional morphological distinctions, including that most species of *Erigeron* have 1-nerved (vs. 3-nerved) involucral bracts (very difficult to observe without the use of a histological clearing agent) and that, in *Conyza,* the pappus bristles tend to elongate slightly as the fruits mature. However, preliminary molecular studies (Noyes and Rieseberg, 1999; Noyes, 2000a) have suggested that *Conyza* represents a mere specialized group within the *Erigeron* lineage that possibly would be better resubmerged within that genus.

1. Main stem solitary, erect, sparsely to densely branched above the midpoint
   .... 1. C. CANADENSIS
1. Main stem 1 to several, spreading, sometimes ascending toward the tip, moderately to densely branched nearly throughout .... 2. C. RAMOSISSIMA

1. **Conyza canadensis** (L.) Cronquist
    (horseweed, Canada fleabane, hogweed)
    Erigeron canadensis L.
    Pl. 232 a, b; Map 968

Stems solitary, very variable from less than 5 cm to more than 250 cm long, erect or strongly ascending, sparsely (in small plants) to densely branched above the midpoint, glabrous or more commonly sparsely to densely pubescent with mostly spreading, often broad-based white hairs. Leaves 0.5–10.0 cm long, sessile or short-petiolate, the blade linear to oblong-lanceolate or oblanceolate, glabrous or more commonly sparsely to moderately hairy, mostly along the margins and midvein. Inflorescences short (in small plants) to elongate (in larger plants) panicles, usually dense and well developed but in small plants appearing as short racemes and/or loose clusters at the branch tips. Involucre 2.5–4.0 mm long, the bracts glabrous or nearly so. Ray florets 20–40, the corolla white or less commonly pinkish-tinged. Disc florets 8–28. $2n=18$. June–November.

Common throughout the state (throughout the U.S.; Canada, Mexico). Banks of streams and rivers, upland prairies, sand prairies, glades, and openings of mesic upland forests; also pastures, old fields, fallow fields, crop fields, gardens, railroads, roadsides, and open, disturbed areas.

Horseweed is an extremely variable plant, having been recorded to flower from a 3 cm stem as well as those nearly 3 m tall. On larger plants, the leaves are usually very numerous. In Missouri, the species mostly has been collected in highly disturbed habitats. It also is a serious crop weed, especially of corn and soy beans. In keeping with its variable and adaptive nature, strains that are resistant to various herbicides have been recorded (see http://www.weedscience.org), including races that are resistant to glyphosate herbicides such as Roundup (Van Gessel, 2001; Rogers, 2003).

1. Stems sparsely to more commonly moderately or densely hairy; involucral bracts with green or white tips
   .... 1A. VAR. CANADENSIS
1. Stems glabrous or nearly so; involucral bracts with dark purple tips
   .... 1B. VAR. PUSILLA

1a. var. canadensis

C. canadensis var. *glabrata* (A. Gray) Cronquist

Stems very variable from less than 5 cm to more than 250 cm long, nearly glabrous or more commonly sparsely to densely hairy. Leaves 0.5–10.0 cm long, glabrous or more commonly sparsely to moderately hairy, mostly along the margins and midvein. Involucral bracts green or white at the tip. $2n=18$. June–November.

Common throughout the state (throughout the U.S.; Canada, Mexico). Banks of streams and rivers, upland prairies, sand prairies, glades, and

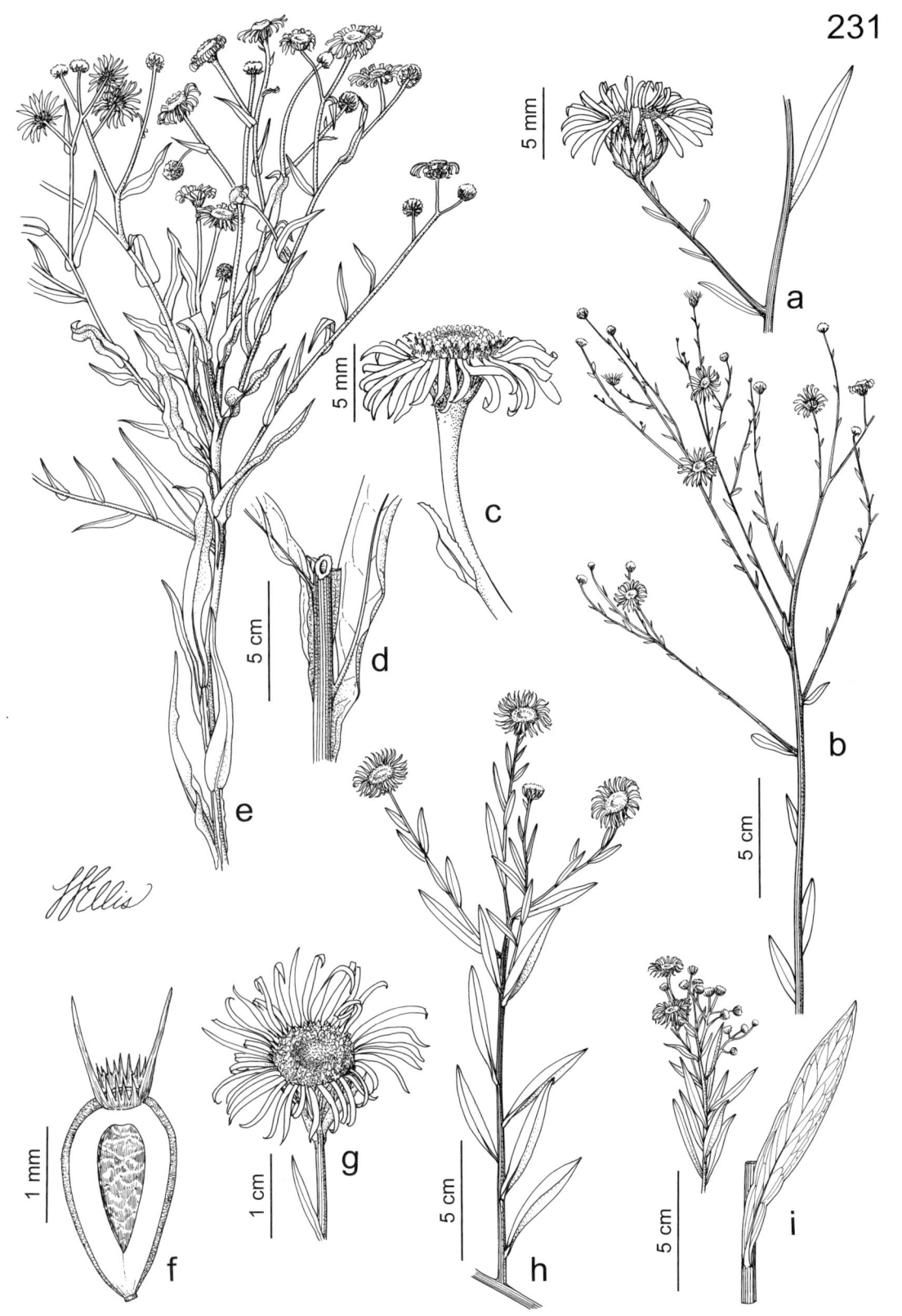

Plate 231. Asteraceae. *Boltonia diffusa*, **a)** head, **b)** habit. *Boltonia decurrens*, **c)** head, **d)** leaf bases, **e)** habit. *Boltonia asteroides* var. *latisquama*, **f)** fruit, **g)** head, **h)** branch with heads. *Boltonia asteroides* var. *recognita*, **i)** heads and leaves.

969. Conyza ramosissima   970. Doellingeria umbellata   971. Erigeron annuus

openings of mesic upland forests; also pastures, old fields, fallow fields, crop fields, gardens, railroads, roadsides, and open, disturbed areas.

Some botanists recognize plants with nearly glabrous stems but lacking purple tips on the involucral bracts as var. *glabrata,* but as noted by Steyermark (1963), the pubescence character alone is too variable to allow formal segregation of varieties. The involucral bracts of this segregate also are supposed to have less-green coloration, but this character also is quite variable and seems to vary independently of pubescence density.

1b. var. pusilla (Nutt.) Cronquist

Erigeron pusillus Nutt.

Stems somewhat less variable, 5–100 cm long, glabrous or nearly so. Leaves 0.5–8.0 cm long, glabrous or more commonly sparsely hairy along the margins and sometimes also the midvein. Involucral bracts mostly with a small area of dark purple at the tip. June–November.

Uncommon in the southeastern portion of the state (southeastern U.S. north locally to New York and west locally to Arizona). Sand prairies and openings of mesic to dry upland forests; also old fields and open, disturbed areas.

2. Conyza ramosissima Cronquist (dwarf fleabane, spreading fleabane)

Erigeron divaricatus Michx.

Pl. 232 c–e; Map 969

Stems 1 to more commonly several, 5–25(–40) cm long, spreading, sometimes ascending toward the tip, moderately to densely branched nearly throughout (except in very small plants, which are branched only 1 or 2 times near the base), moderately to densely pubescent with strongly ascending (abruptly bent just above the stout base), white hairs (the whole plant often appearing grayish). Leaves 0.5–2.5(–4.0) cm long, sessile, the blade narrowly linear, moderately to densely appressed-hairy. Inflorescences relatively short panicles, often dense and well developed but sometimes appearing as short racemes or loose clusters at the branch tips. Involucre 1.5–3.0 mm long, the bracts glabrous or sparsely appressed-hairy. Ray florets 18–30, the corolla pink or light purple. Disc florets 3–8. $2n=18$. May–September.

Scattered nearly throughout the state (North Dakota to Pennsylvania south to New Mexico and Alabama; Canada). Upland prairies, sand prairies, glades, and banks of rivers; also pastures, lawns, fallow fields, sidewalks, railroads, roadsides, and open, disturbed areas.

17. Doellingeria Nees

Three species, U.S., Canada.

Doellingeria was long treated by most North American botanists as a relatively distinctive small subgenus within a broadly circumscribed *Aster*. Molecular analyses have suggested that the group comprises the most primitive lineage of North American asters (Noyes and Rieseberg, 1999; Semple et al., 2002). There has been some controversy over whether eight closely related Asian species should be included in the genus (Nesom, 1993b, 1994, 2000), but the most recent studies indicate that the Asian species are more closely related to other Old World asters.

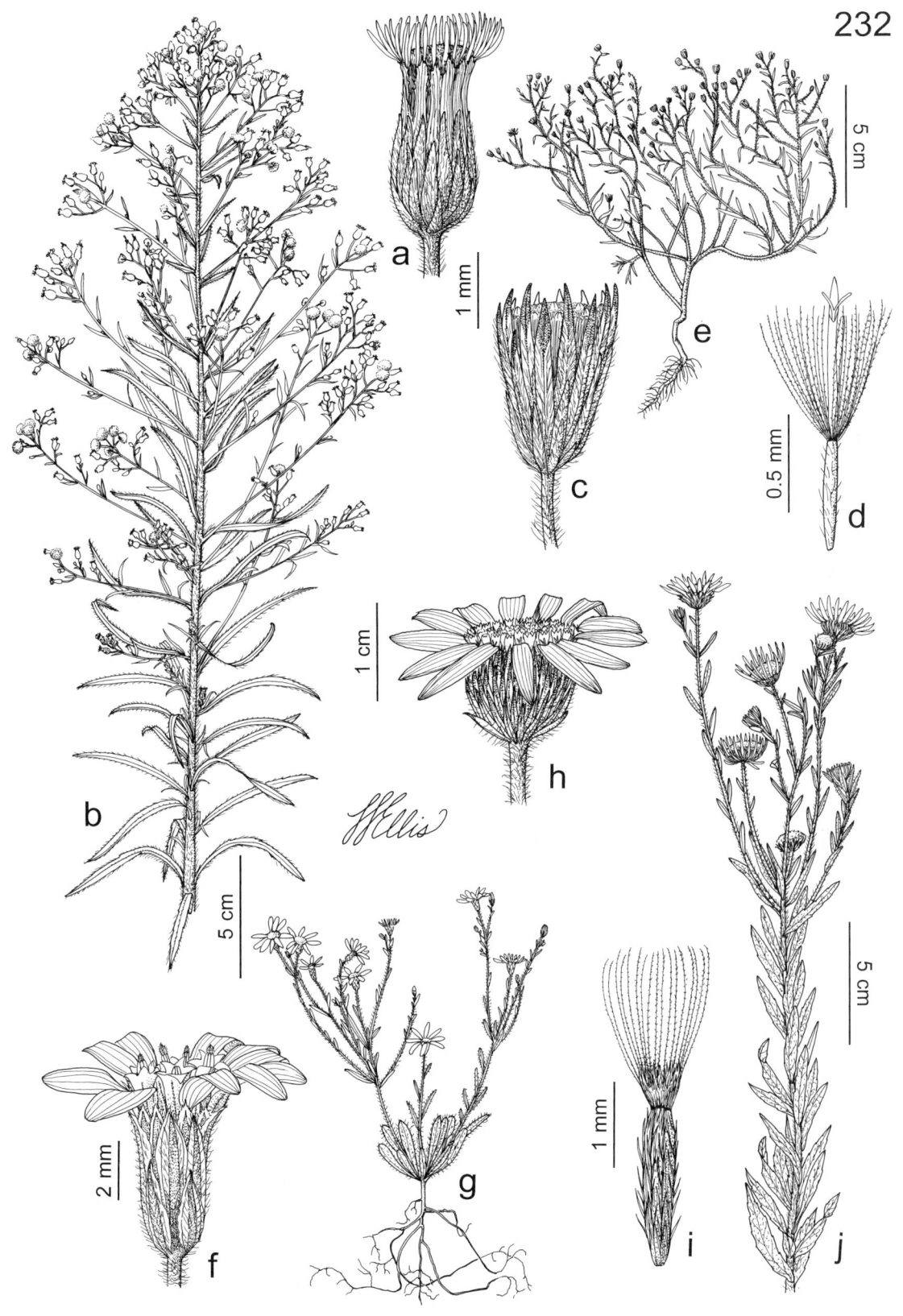

Plate 232. Asteraceae. *Conyza canadensis*, **a)** head, **b)** habit. *Conyza ramosissima*, **c)** head, **d)** ray floret with reduced corolla, **e)** habit. *Chaetopappa asteroides*, **f)** head, **g)** habit. *Bradburia pilosa*, **h)** head, **i)** fruit, **j)** habit.

1. **Doellingeria umbellata** (Mill.) Nees **var. pubens** (A. Gray) Britton (flat-topped white aster)

Aster umbellatus Mill. var. *pubens* A. Gray

A. pubentior Cronquist

Pl. 230 e–g; Map 970

Plants perennial herbs, with slender to thick, sometimes woody rhizomes. Stems 1 or rarely few, 50–150(–200) cm long, erect or ascending, usually branched only in the inflorescence, with fine, longitudinal lines, sparsely to densely roughened with minute, curved, broad-based hairs. Basal leaves absent at flowering. Stem leaves somewhat reduced toward the tip, 3–8(–16) cm long, narrowly to broadly elliptic or lanceolate-elliptic, narrowed or short-tapered to a sharply pointed tip, narrowed or tapered to a sessile, slightly expanded but not clasping base, the margins entire, the surfaces and margins roughened with sparse to moderate, short, curly, broad-based hairs, especially along the veins. Inflorescences flat-topped panicles, the heads in usually dense clusters at the ascending to loosely ascending branch tips, the bracts relatively numerous, 1–6 cm long, leaflike, narrowly elliptic. Heads radiate, not sticky or resinous. Involucre 3–5 mm long, cup-shaped to slightly bell-shaped. Involucral bracts in 3 or 4 unequal, overlapping series, lanceolate to narrowly oblong-lanceolate, the tip ascending, with a narrow, green central band (this sometimes very slightly broadened toward the tip) and broad lighter margins, the margins appearing uneven or finely hairy, especially toward the tip, the outer surface glabrous or sparsely hairy. Receptacle flat or shallowly convex, with minute, irregular ridges around the concave attachment points of the florets. Ray florets 4–12, pistillate, the corolla 5–10 mm long, white. Disc florets 8–30, perfect, the corolla 3.5–6.0 mm long, relatively deeply lobed (0.5–1.3 mm deep), yellow, not persistent at fruiting. Pappus of the ray and disc florets similar, of 2 types, an inner series of numerous finely barbed bristles 3–4 mm long, the longest of these usually slightly broadened toward the tip (narrowly club-shaped), and an outer series of fewer bristles 0.2–0.8 mm long, both series pale cream-colored to light tan. Fruits 1.8–2.5 mm long, oblong-obovoid to somewhat obconical, somewhat flattened, with 4–6 yellowish brown ribs, sparsely to moderately hairy, especially along the ribs, light yellowish brown. $2n=18$. August–October.

Uncommon, known thus far only from Lafayette County (North Dakota to Nebraska east to Michigan and Illinois). Fens.

This taxon was first reported (as *Aster pubentior*) for Missouri by Gremaud (1988) based on his inventories of some isolated deep-muck fens in northwestern Missouri. These unusual shakey springs fen communities also contain populations of some other regionally rare species with a generally more northern provenance, such as *Eupatorium maculatum* (Asteraceae: Eupatorieae), *Berula erecta* (Apiaceae), *Epilobium leptophyllum* (Onagraceae), and *Caltha palustris* (Ranunculaceae). Gremaud interpreted these plants as glacial relicts.

Botanists have disagreed on the classification of taxa within the *D. umbellata* complex. The western populations that include the Missouri plants were initially described as a variety of *Aster umbellatus*. Cronquist (1947b) raised these to species status, but most other botanists have either retained them at the varietal level or have not accepted them as taxonomically distinct at any level. *Doellingeria umbellata* var. *umbellata* occurs in the eastern United States and Canada west to Minnesota, Iowa, and Alabama and differs in its slightly larger heads with up to 14 ray florets and 40 disc florets, as well as its less densely pubescent foliage. The ranges of the two varieties overlap in portions of Minnesota, Iowa, Illinois, Michigan, and Canada. Further intermediate collections that have accumulated since the time of Cronquist's (1947b) research and a detailed morphometric analysis of the complex (Semple et al., 1991) have made it more reasonable to retain the two taxa as varieties rather than species.

18. Erigeron L. (fleabane)

Plants annual or perennial herbs, lacking taproots, sometimes with rhizomes or stolons, variously hairy. Stems 1 or few to several, erect or ascending, unbranched or more commonly branched, finely to coarsely angled or longitudinally lined. Basal leaves absent or more commonly present at flowering. Stem leaves progressively reduced toward the tip, sessile or short-petiolate, the blade linear to lanceolate, oblanceolate, or oblong-lanceolate, the base sometimes somewhat clasping the stem, the margins entire or relatively few-toothed above the midpoint. Inflorescences panicles or sometimes (in *E. pulchellus*) a small cluster or solitary head at the stem or branch tips, the heads long-stalked to nearly sessile, the inflorescence branches with few to several small, leaflike bracts. Heads radiate (rarely discoid in *E. strigosus*), not sticky or resinous. Involucre cup-shaped to broadly cup-shaped or slightly bell-shaped. Involucral bracts in 2 or 3(4) equal or subequal, overlapping series, narrowly oblanceolate to narrowly lanceolate or linear, the tip ascending (sometimes loosely so in *E. pulchellus*), with a relatively broad (slender elsewhere), uniform, green central stripe (often with a slender, yellow or orange midvein) and with relatively narrow, thin, pale margins, sometimes purplish-tinged toward the tip. Receptacle flat or shallowly convex, relatively smooth (the veins leading to the florets often appearing as raised points after the fruits have been shed). Ray florets usually numerous (50–400) in 1–4 series, pistillate, the corolla usually conspicuous (occasionally a few of the innermost ray florets with reduced or absent corollas), relatively slender, white or often tinged with pink, lavender, or blue, especially with age or on the undersurface, shed before fruiting. Disc florets numerous (more than 100), perfect, yellow, shed before fruiting. Pappus of the ray and disc florets similar or of 2 types, of relatively few to numerous (8–30) capillary bristles and usually also an outer series of fewer short bristles or narrow scales, usually white, the pappus of the ray florets sometimes with only the outer, shorter series present. Style branches with the sterile tip (beyond the stigmatic lines) 0.1–0.3 mm long, lanceolate to broadly triangular. Fruits narrowly oblong-obovate (slightly tapered toward the base) in outline, flattened, the angles usually with thickened nerves or ribs, rarely with an additional nerve on each face, the surface sparsely hairy or glabrous, pale tan to light yellowish brown. About 390 species, nearly worldwide, most diverse in temperate montane regions.

Thus far, most botanists have resisted the temptation to subdivide *Erigeron* into smaller genera. In fact, the segregates *Conyza* and *Aphanostephus*, which Cronquist (1943, 1947c) and Shinners (1946b) retained as genera separate from *Erigeron*, should perhaps be reintegrated into the genus, based on molecular data (Noyes, 2000a). In temperate North America, where more than 170 species occur, the greatest diversity is in the western mountains, where numerous narrowly endemic taxa grow. The molecular studies of Noyes (2000a), which involved samples from nearly 70 species spread across the taxonomic and geographic diversity of *Erigeron*, provided support for the natural (monophyletic) circumscription of the genus and suggested a North American origin for it.

That said, the morphological characters separating *Erigeron* from some other genera of the tribe Astereae are rendered less effective by variation within some species in each of the groups. In the eastern half of the United States, the general rule of thumb has been that a plant flowering before July with relatively numerous ray florets and subequal involucral bracts having uniform, green central stripes is likely an *Erigeron*, whereas a plant flowering after July with somewhat fewer ray florets and unequal involucral bracts in which the green stripe is expanded toward the tip is probably an aster (*Doellingeria, Eurybia, Symphyotrichum*). However, in individual species one or more of these characters are difficult to interpret or break down, and the two groups have overlapping flowering periods from about August to October. In the western United States, the differences between the two groups are further blurred by the presence of other related genera, such as *Machaeranthera* Nees and its segregates. In Missouri, care also must be taken to avoid confusion between *Erigeron* and *Boltonia*, which differ primarily in their pappus types and in the generally later flowering period in the latter.

1. Plants with long rhizomes or stolons, often occurring in large colonies; receptacle 12–20 mm in diameter at flowering; disc florets with the corolla 4.5–6.0 mm long
 . . . . 3. E. PULCHELLUS
1. Plants sometimes with few short offsets (in *E. philadelphicus*) but not producing rhizomes or stolons, not colonial; receptacle 6–15 mm in diameter at flowering; disc florets with the corolla 1.5–3.5 mm long

2. Stem leaves with the base rounded to shallowly cordate and more or less clasping the stem; ray florets 120–400; disc corollas 2.5–3.5 mm long
  . . . . 2. E. PHILADELPHICUS
2. Stem leaves with the base tapered to narrowly rounded, not or only slightly clasping the stem; ray florets 50–125; disc corollas 1.5–2.5 mm long
  3. Pappus of the ray and disc florets similar, both with 1 or more inner series of longer bristles and an outer series of shorter bristles or scales; stems 10–45 cm long; leaves 1–5 cm long . . . . 5. E. TENUIS
  3. Pappus of the ray and disc florets dissimilar, the disc florets with 1 or more inner series of longer bristles and an outer series of shorter bristles or scales, the ray florets lacking the inner series of longer bristles; stems 30–150 cm long; leaves 1–15 cm long (species sometimes difficult to distinguish)
    4. Stems (50–)60–150 cm long, the hairs mostly spreading; stem leaves usually relatively numerous, the blade oblanceolate to elliptic or lanceolate, all but the uppermost usually with several sharp teeth on each side, these often produced from below the midpoint to the tip . . . . . 1. E. ANNUUS
    4. Stems 30–70(–90) cm long, the pubescence mostly of appressed to ascending hairs (some of the longer hairs sometimes spreading toward the tip); stem leaves often relatively few, the blade linear to oblanceolate, the margins entire or with few irregular teeth toward the tip . . . . . 4. E. STRIGOSUS

1. Erigeron annuus (L.) Pers. (daisy fleabane, whitetop fleabane, annual fleabane)

Pl. 233 g, h; Map 971

Plants annual or less commonly biennial, with shallow, fibrous roots. Stems 1 to several, (50–)60–150 cm long, usually well branched above the lower ⅓, sparsely to densely roughened with mostly spreading hairs (the hairs sometimes more appressed toward the tip). Basal leaves sometimes withered by flowering time, 3–15 cm long, mostly long-petiolate, the blade broadly oblanceolate to broadly obovate, short- to long-tapered at the base, mostly rounded at the tip, the margins coarsely and sharply toothed, the surfaces and margins sparsely to moderately pubescent with short, relatively stiff, curved hairs. Stem leaves usually relatively numerous, 1–10 cm long, the lower ones short-petiolate, the median and upper ones sessile, the blade oblanceolate to elliptic or lanceolate (the lower leaves rarely obovate), angled or tapered to a mostly sharply pointed tip, angled, tapered, or narrowly rounded at the base, not or only very slightly clasping the stem, the margins of all but the uppermost leaves usually with several sharp teeth on each side, these often produced from below the midpoint to the tip, the surfaces and margins sparsely to moderately hairy. Inflorescences rounded to more or less flat-topped panicles, usually open and often with numerous heads. Involucre 3–5 mm long, the receptacle 6–12 mm in diameter at flowering, the bracts sparsely to moderately pubescent with more or less spreading hairs and often also minutely glandular. Ray florets 80–125, the corolla 4–10 mm long.

Disc florets with the corolla 1.5–2.5 mm long. Pappus of the ray and disc florets of 2 types, an inner series of 10–15 threadlike bristles 1.2–2.2 mm long and an outer series of several shorter bristles or slender scales 0.1–0.4 mm long, the ray florets lacking the longer, inner series. Fruits 0.8–1.0 mm long, sparsely and inconspicuously hairy. $2n=27, 54$. May–November.

Common throughout the state (U.S., Canada; introduced in Europe). Banks of streams and rivers, margins of ponds and lakes, openings of bottomland and mesic upland forests, and bottomland and upland prairies; also pastures, old fields, fallow fields, crop fields, railroads, roadsides, and open, disturbed areas.

Erigeron annuus is a mostly triploid taxon that reproduces apomictically. Using allozyme markers, Hancock and Wilson (1976) showed that multiple genotypes may be present within populations. Stratton (1991) demonstrated that there was considerable morphological variation within experimental populations grown under greenhouse conditions. This genotypic variation and phenotypic plasticity are responsible in part for difficulties that some botanists have had in distinguishing some plants of *E. annuus* from *E. strigosus* in the field and herbarium. Frey et al. (2003) studied introduced populations of these species in Europe and a small number of North American plants, and they concluded that plants attributed to *E. strigosus* var. *septentrionalis* (Fernald & Wiegand) Fernald should more properly be considered a variety of *E. annuus*. In fact, some botanists have suggested that

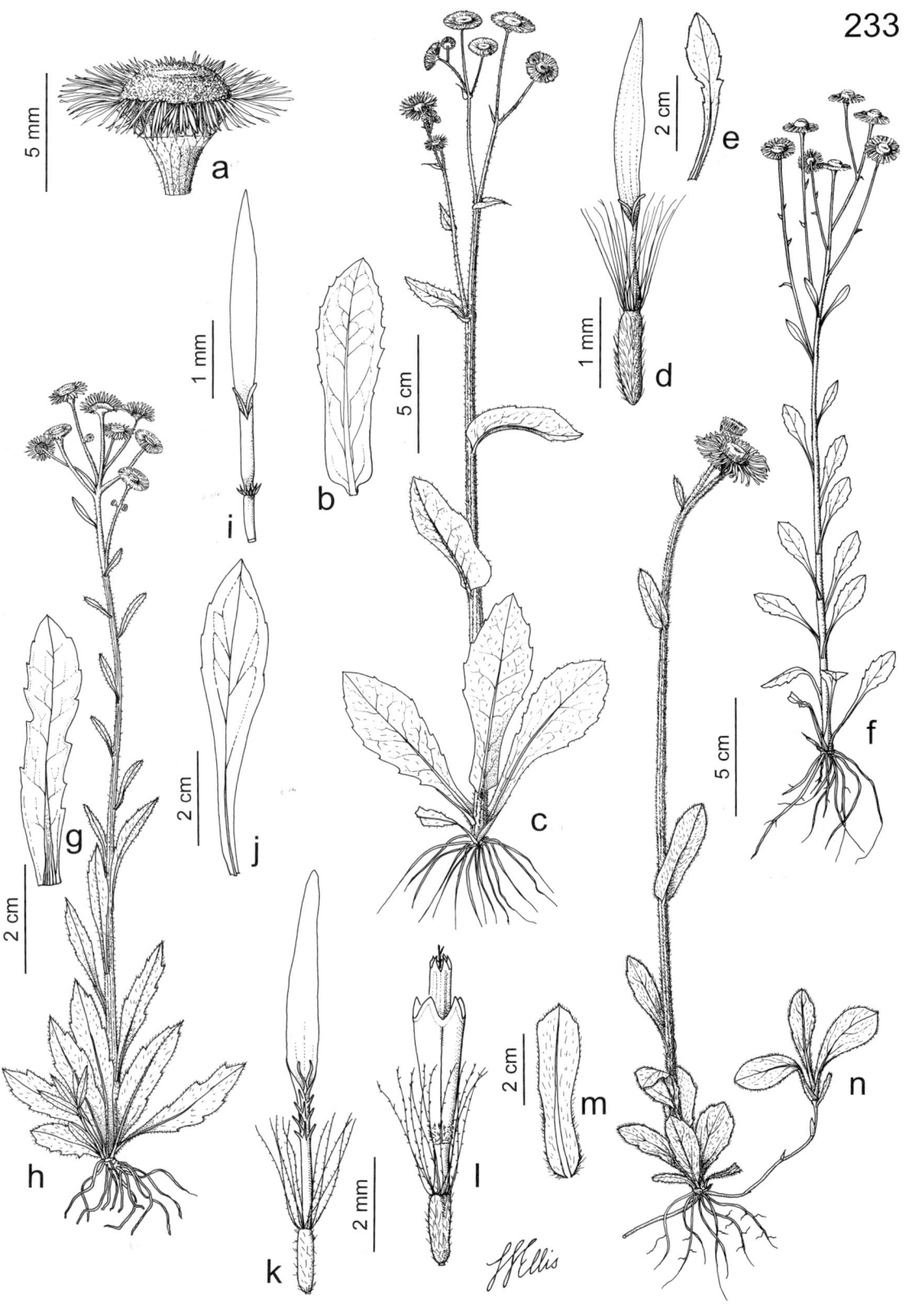

Plate 233. Asteraceae. *Erigeron philadelphicus*, **a)** head, **b)** leaf, **c)** habit. *Erigeron tenuis*, **d)** ray floret, **e)** leaf, **f)** habit. *Erigeron annuus*, **g)** leaf, **h)** habit. *Erigeron strigosus*, **i)** ray floret, **j)** leaf. *Erigeron pulchellus*, **k)** ray floret, **l)** disk floret, **m)** leaf, **n)** habit.

972. Erigeron philadelphicus   973. Erigeron pulchellus   974. Erigeron strigosus

E. strigosus as a whole might be treated better as a subspecies of *E. annuus,* but this approach merely avoids dealing with the problem rather than the more difficult job of reassessing evolutionary lineages within the group. Further research involving more in-depth sampling of North American populations of both species will be necessary to resolve the situation. See the treatment of *E. strigosus* for further discussion.

Cronquist (1947c) called rare plants from Quebec, Canada, with the heads lacking ray florets var. *discoideus* (Vict. & J. Rousseau) Cronquist. This rare mutant should deserve no more recognition than that of a form (if at all).

Steyermark (1963) noted that plants of *E. annuus* have been used medicinally as an astringent, diuretic, and tonic. He also noted that deer frequently browse the foliage. According to Stratton (1991), a single plant of *E. annuus* may produce 10,000–100,000 fruits in one growing season, which helps to account for the success of the species geographically and ecologically.

2. Erigeron philadelphicus L. (Philadelphia fleabane)

Pl. 233 a–c; Map 972

Plants biennial or perennial (often relatively short-lived) herbs, with fibrous roots, occasionally producing short offsets at the end of the growing season, but not producing rhizomes or stolons. Stems 1 to several, 20–80(–150) cm long, usually sparsely to moderately branched above the lower 1/3, sparsely to moderately pubescent with mostly spreading hairs (the hairs sometimes more appressed toward the tip). Basal leaves sometimes withered by flowering time, 2–15 cm long, sessile to short-petiolate, the blade narrowly oblanceolate to oblanceolate or less commonly obovate, mostly long-tapered at the base, mostly rounded at the tip, the margins coarsely and bluntly to sharply toothed or scalloped, rarely shallowly pinnately lobed, the surfaces and margins sparsely to moderately pubescent with short to long, spreading or loosely appressed hairs. Stem leaves usually relatively numerous, 1–10 cm long, sessile, the blade oblanceolate to elliptic or lanceolate (the lower leaves rarely obovate), angled or tapered to a usually sharply pointed tip, rounded to shallowly cordate at the base and more or less clasping the stem, the margins entire, scalloped, or with few to several sharp teeth on each side, these sometimes produced only from above the midpoint, the surfaces and margins sparsely to more commonly moderately hairy. Inflorescences rounded to more or less flat-topped panicles, usually open and often with numerous heads. Involucre 4–6 mm long, the receptacle 6–15 mm in diameter at flowering, the bracts glabrous or more commonly sparsely to moderately pubescent with more or less spreading hairs and sometimes also minutely glandular. Ray florets 120–400, the corolla 5–10 mm long. Disc florets with the corolla 2.5–3.5 mm long. Pappus of the ray and disc florets similar, both with an inner series of (15–)20–30 threadlike bristles 1.2–3.2 mm long and an outer series of usually several shorter bristles or slender scales 0.1–0.4 mm long. Fruits 0.6–1.2 mm long, sparsely and inconspicuously hairy. $2n=18$. April–June.

Scattered to common nearly throughout the state, but uncommon or apparently absent from western portions of the Ozark Division (U.S., Canada; introduced in Europe). Banks of streams and rivers, margins of ponds and lakes, bottomland forests, mesic upland forests, savannas, upland prairies, sand prairies, and ledges and tops of bluffs; also pastures, old fields, fallow fields, gardens, cemeteries, railroads, roadsides, and open, disturbed areas.

Some botanists have attempted to subdivide *E. philadelphicus* by recognizing morphological extremes, but none of the segregates in this very variable species seems worthy of formal taxonomic recognition, and none of them appears to occur in the wild in Missouri. For example, plants with the stem leaves glabrous or nearly so have been called var. *glaber* J.K. Henry and var. *provancheri* (Vict. & J. Rousseau) B. Boivin, whereas robust plants with relatively large, slightly succulent leaves (similar to plants that grow under some garden situations) have been called var. *scaturicola* (Fernald) Fernald (the oldest epithet for this taxon is actually var. *acaulescens,* but Lunell's name was never formally transferred to *E. philadelphicus*). Cronquist (1947c) attributed such unusual morphologies to extreme environmental conditions, such as abundant moisture and nutrient-rich soils or dry, rocky habitats. Morton (1988) made a somewhat stronger case for the recognition of var. *provancheri* by studying plants grown in the greenhouse, but this accounted only for a few populations in the northeastern United States and Canada.

Native Americans used this species medicinally as a cold remedy, analgesic, antidiarrheal agent, and a poultice for sores, and to reduce excessive bleeding following childbirth (Moerman, 1998).

3. Erigeron pulchellus Michx. var. pulchellus
(robin's plantain)

Pl. 233 k–n; Map 973

Plants perennial herbs, with fibrous roots and long, slender rhizomes or stolons, often occurring in large colonies. Stems solitary, 15–40(–60) cm long, unbranched below the inflorescence, moderately to densely pubescent with relatively long, spreading hairs, especially toward the tip. Basal leaves present at flowering, 2–13 cm long, sessile to short-petiolate, the blade oblanceolate to broadly obovate, mostly long-tapered at the base, mostly rounded at the tip, the margins entire or bluntly to less commonly sharply toothed or scalloped, the surfaces and margins moderately to densely pubescent with relatively long, spreading to loosely appressed hairs. Stem leaves usually relatively few, 1–7 cm long, sessile, the blade lanceolate to oblong, oblong-oblanceolate, or ovate, angled or tapered to a bluntly or more commonly sharply pointed tip, rounded to shallowly cordate at the base and more or less clasping the stem, the margins entire or the lowermost leaves with a few teeth toward the tips, the surfaces and margins moderately to densely hairy. Inflorescences of solitary heads or 2–5-headed, more or less flat-topped panicles. Involucre 5–7 mm long, the receptacle 12–20 mm in diameter at flowering, the bracts sparsely to moderately pubescent with more or less spreading hairs and often also minutely glandular. Ray florets 50–80(–100), the corolla 6–10 mm long. Disc florets with the corolla 4.5–6.0 mm long. Pappus of the ray and disc florets similar, both with an inner series of 20–35 threadlike bristles 4–5 mm long and often an outer series of relatively few shorter bristles 0.1–0.4 mm long. Fruits 1.3–2.0 mm long, sparsely and inconspicuously hairy. $2n=18$. April–June.

Scattered nearly throughout Missouri except the western portion of the Glaciated Plains Division (eastern U.S. west to Minnesota and Texas; Canada). Banks of streams and rivers, rocky openings of mesic to dry upland forests, savannas, and ledges and tops of bluffs; also pastures, old fields, and rarely lawns.

This species is relatively uniform morphologically over most of its range. In addition to the widespread var. *pulchellus,* most botanists recognize two other varieties. Essentially glabrous plants that occur mostly in the Appalachian Mountains are called var. *tolsteadtii* Cronquist, and a rare rock ledge specialist in Minnesota with white rays, slightly shorter disc florets, and relatively densely hairy achenes is known as var. *brauniae* Fernald. In var. *pulchellus,* the ray florets are almost always tinged with pink, purple, or blue, and the stems and leaves are pubescent with relatively long, spreading hairs.

Robin's plantain is an attractive wildflower that deserves more widespread cultivation as a groundcover in woodland gardens.

4. Erigeron strigosus Muhl. ex Willd. var. strigosus (daisy fleabane, whitetop fleabane)
E. annuus ssp. *strigosus* (Muhl. ex Willd.) Wagenitz
E. strigosus var. *beyrichii* (Fisch. & C.A. Mey.) Torr. & A. Gray
E. strigosus var. *discoideus* J.W. Robbins ex A. Gray
E. ramosus Britton, Sterns & Poggenb.

Pl. 233 i, j; Map 974

Plants annual or less commonly biennial, with shallow, fibrous roots. Stems 1 to several, 30–70(–90) cm long, often well branched above the lower $1/3$, sparsely to moderately pubescent (occasionally roughened) with appressed to ascending hairs (some of the longer hairs sometimes spreading toward the tip). Basal leaves sometimes withered by flowering time, 3–15 cm long, mostly long-petiolate, the blade oblanceolate to elliptic-oblanceolate, mostly long-tapered at the base, mostly bluntly to sharply pointed at the tip, the margins entire or coarsely and sharply toothed usually above the midpoint, the surfaces and margins sparsely to moderately pubescent with short, relatively stiff

975. Erigeron tenuis   976. Eurybia furcata   977. Eurybia hemispherica

hairs (these often curved or bent toward the base). Stem leaves usually appearing relatively few (this mostly because the leaves are sparser toward the stem tip), 1–10 cm long, the lower ones short-petiolate, the median and upper ones sessile, the blade linear to oblanceolate, angled or tapered to a mostly sharply pointed tip, short- to long-tapered at the base, not clasping the stem, the margins entire or with few irregular teeth toward the tip, the surfaces and margins sparsely to moderately hairy. Inflorescences rounded to more or less flat-topped panicles, usually open, often with numerous heads. Involucre 2–5 mm long, the receptacle 4–12 mm in diameter at flowering, the bracts glabrous or more commonly sparsely pubescent with short, appressed and/or longer, more or less spreading hairs, often also minutely glandular. Ray florets 50–100 or rarely absent, the corolla 4–7 mm long. Disc florets with the corolla 1.5–2.5 mm long. Pappus of the ray florets (when present) and disc florets of 2 types, an inner series of 8–15 threadlike bristles 1.2–2.2 mm long and an outer series of several shorter bristles or slender scales 0.1–0.4 mm long, the ray florets lacking the longer, inner series. Fruits 0.8–1.2 mm long, sparsely and inconspicuously hairy. $2n=18, 27, 36, 54$. May–September.

Common throughout the state (U.S., Canada; introduced in Europe). Banks of streams and rivers, openings of mesic to dry upland forests, savannas, upland prairies, and glades; also pastures, old fields, railroads, roadsides, and open, disturbed areas.

This species exists throughout much of its range as an apomictic polyploid, with sexual diploids apparently uncommon in portions of the southeastern United States (Noyes, 2000b). Cronquist (1947c) treated *E. strigosus* as comprising four varieties. Of these, small-headed plants (var. *beyrichii*) and rayless plants (var. *discoideus*) are rare mutations that occur sporadically within populations (including in Missouri) and probably should be accorded no higher taxonomic rank than that of a form. At the other extreme, two perennial, rhizomatous varieties endemic to dolomite glades in central Alabama, var. *calcicola* J. Allison and var. *dolomiticola* J. Allison, were described relatively recently (Allison and Stevens, 2001) and are so distinctive that they perhaps should be considered varieties of a novel species. The var. *septentrionalis* (Fernald & Wiegand) Fernald refers to plants growing mostly to the north of Missouri that have relatively broad basal leaves and long hairs on the stem, with the pubescence of the involucre unusual in appearing flattened and narrowly ribbonlike when dried. It should be searched for in northern Missouri. As noted above, Frey et al. (2003) considered this variant to be more closely related to *E. annuus* than to *E. strigosus*.

Erigeron strigosus can be difficult to distinguish from the closely related *E. annuus*. Both species are widespread in Missouri, but *E. strigosus* tends to occupy somewhat drier sites and tends not to occur in cropped areas. The fewer narrower leaves give the plants a sparser, more open appearance. The largest leaves of plants of *E. strigosus* tend to be narrower and less toothed along the margins. Morphologically intermediate plants are encountered sporadically but fairly frequently.

Native Americans used this species medicinally for heart ailments and as an analgesic (Moerman, 1998).

5. Erigeron tenuis Torr. & A. Gray

Pl. 233 d–f; Map 975

Plants annual or less commonly biennial, with shallow, fibrous roots. Stems 1 to few, 10–45 cm long, unbranched or sparsely branched above the lower $^{1}/_{3}$, sparsely to moderately pubescent (occasionally roughened) with appressed to ascending hairs (some of the longer hairs sometimes spreading toward the tip). Basal leaves sometimes withered by flowering time, 2–5 cm long, mostly long-petiolate, the blade oblanceolate to obovate or broadly obovate, mostly long-tapered at the base, mostly bluntly to sharply pointed at the tip, the margins entire or coarsely and sharply toothed usually above the midpoint, occasionally 3-lobed at the tip, the surfaces and margins sparsely to moderately (rarely densely) pubescent with short, relatively stiff hairs (these often curved or bent toward the base), occasionally the hairs longer and somewhat spreading toward the stem base. Stem leaves few to occasionally relatively numerous, 1–5 cm long, the lower ones short-petiolate, the median and upper ones sessile, the blade linear to oblanceolate or less commonly oblong-lanceolate, angled or tapered to a bluntly or sharply pointed tip, short- to long-tapered, angled, or narrowly rounded at the base, not or only slightly clasping the stem, the margins entire or with few to several irregular teeth, often only above the midpoint, the surfaces and margins sparsely to moderately hairy. Inflorescences rounded to more or less flat-topped panicles, usually open and with few to numerous heads. Involucre 2.5–4.0 mm long, the receptacle 4–10 mm in diameter at flowering, the bracts glabrous or more commonly sparsely to moderately pubescent with short, appressed or curved hairs, sometimes also minutely glandular. Ray florets 60–120, the corolla 3–5 mm long. Disc florets with the corolla 1.5–2.5 mm long. Pappus of the ray and disc florets similar, both with an inner series of 10–15 threadlike bristles 1.2–2.2 mm long and an outer series of few to several shorter bristles or slender scales 0.1–0.4 mm long. Fruits 0.9–1.2 mm long, sparsely and inconspicuously hairy. $2n=18, 36$. April–June.

Scattered in the southern half of the state (Kansas to Missouri south to Texas and Florida, possibly also North Carolina). Upland prairies, glades, tops of bluffs, and margins of mesic upland forests; also pastures, roadsides, and open, disturbed areas.

Most frequently, *E. tenuis* has the aspect of a small variant of *E. strigosus*. However, a series of historical collections from Jasper County resemble *E. annuus* more closely. These are more foliose plants, with relatively well-developed leaves toward the stem tip, and with lower and median leaves that have more teeth along the margins than is typical for the species in Missouri. Indeed, the inner bristles of the ray pappus constitute the only consistent morphological distinction between *E. tenuis* and dwarf examples of the other two species. Steyermark (1963) noted that the ray corollas of *E. tenuis* are more often pinkish-tinged or bluish purple when fresh rather than white. *Erigeron tenuis* has a much shorter overall flowering period than do *E. strigosus* and *E. annuus,* and in any given year it usually begins to flower a couple of weeks earlier than the other two species.

19. Eurybia (Cass.) S.F. Gray

Plants perennial herbs, usually with elongate, fleshy rhizomes, often forming colonies, the rootstock sometimes somewhat woody in *E. hemispherica*. Stems usually solitary, erect or ascending, unbranched below the inflorescence, with fine, longitudinal lines, glabrous or hairy. Basal leaves present or absent at flowering. Stem leaves gradually reduced toward the tip, variously shaped, glabrous or hairy. Inflorescences either slender racemes or flat-topped to somewhat dome-shaped panicles or loose clusters, rarely of solitary heads, the heads nearly sessile to long-stalked, the bracts relatively few and more or less leaflike. Heads radiate, not sticky or resinous. Involucre 7–12 mm long, broadly cup-shaped to obconical or slightly bell-shaped. Involucral bracts either in 5–7 unequal or in 2–4 nearly equal, overlapping series, variously shaped, the tip ascending or more or less spreading, with a pale, thickened base, this with or without a slender, green midvein, with an ovate to broadly diamond-shaped green portion toward the tip, this sometimes with narrow purple margins. Receptacle flat or shallowly convex, with minute, irregular ridges around the concave attachment points of the florets. Ray florets 9–35, pistillate, the corolla white or lavender to purple. Disc florets 15–80, perfect, the corolla 5.5–8.0 mm long, relatively shallowly lobed, yellow, turning reddish purple to brownish purple after the pollen has been shed, not persistent at fruiting. Pappus of the ray and disc florets similar (the outermost bristles sometimes appearing slightly shorter), apparently of 1 (but often actually 2) series of numerous (60–80) finely barbed bristles 5–8 mm long, the innermost bristles sometimes slightly broadened toward the tip, light tan to pale orangish brown. Style branches with the sterile tip (beyond the stigmatic lines) (0.2–)0.3–0.5 mm long, linear to lanceolate. Fruits narrowly obconical to oblong-ellipsoid, oblong-obovoid, or nearly cylindrical, sometimes slightly flattened, with 7–16 yellowish brown ribs, glabrous or hairy, brown to greenish brown, rarely tan. About 28 species, U.S., Canada, Europe, Asia.

Eurybia is part of an as-yet poorly understood group of genera that includes the large *Aster*-segregate *Symphyotrichum* as well as several genera of western North American plants formerly treated in *Aster* and/or *Haplopappus* and currently known as the *Machaeranthera* Nees alliance (Semple et al., 2002). The generic limits of *Eurybia* require further study, and it is possible that some of the species presently included in the genus are more closely related to those in other genera (Semple et al., 2002). In Missouri, the three species are all morphologically relatively distinct.

1. Basal and lower stem leaves often absent at flowering, when present, sessile or short-petiolate, the blade linear to narrowly oblanceolate, tapered at the base
   .... 2. E. HEMISPHERICA
1. Basal and lower leaves present at flowering, long-petiolate, the blade heart-shaped to triangular-ovate, deeply cordate at the base
   2. Inflorescence branches (and upper portion of stem) sparsely to moderately pubescent with short, nonglandular hairs; ray florets with the corolla white, sometimes turning pinkish- or lavender-tinged with age .... 1. E. FURCATA
   2. Inflorescence branches (and upper portion of stem) moderately to densely pubescent with short, gland-tipped hairs; ray florets with the corolla lavender to purple .... 3. E. MACROPHYLLA

1. Eurybia furcata (E.S. Burgess) G.L. Nesom
(forked aster)
Aster furcatus E.S. Burgess

Pl. 230 c, d; Map 976

Plants with long, sometimes relatively stout, fleshy rhizomes, sometimes forming dense colonies (but usually prevented from doing so by the nature of the rock ledge habitat). Stems 30–120 cm long, often slightly zigzag, sparsely to moderately pubescent with short, nonglandular hairs toward the tip, usually glabrous toward the base. Basal leaves often absent at flowering but lower stem leaves well developed, long-petiolate, the blade 4–15 cm long, heart-shaped to triangular-ovate, short-tapered at the tip, deeply cordate at the base, the margins sharply and usually coarsely toothed, the upper surface sparsely to moderately roughened-hairy with short, stiff, nonglandular hairs, the undersurface sparsely to moderately roughened-hairy with short, stiff, nonglandular hairs and usually also with softer, longer, curled, nonglandular hairs along the veins. Median and upper stem leaves gradually smaller, ovate to narrowly elliptic or narrowly lanceolate, rounded to angled or tapered at the base, the petioles becoming progressively shorter (the uppermost leaves often nearly sessile), slightly expanded at the base but not clasping the stem. Inflorescences flat-topped to somewhat dome-shaped panicles or clusters, the heads in loose to dense clusters or solitary at the branch tips, the branches sparsely to moderately pubescent with short, curved, nonglandular hairs. Involucre 7–10 mm long, the bracts in 4–7 strongly unequal, overlapping series, 1–2 mm wide and mostly 1.5–3.0 times as long as wide, narrowly lanceolate to oblong-lanceolate or narrowly oblong-ovate, rounded to bluntly pointed at the ascending tip, with a narrow (or sometimes absent), green central band toward the base (this sometimes somewhat keeled) and broad, relatively firm, pale yellowish margins, the green area much-broadened toward the tip, the margins otherwise appearing finely and densely hairy and sometimes dark purple or purplish-tinged, the outer surface moderately pubescent with fine, ascending, nonglandular hairs. Ray florets 12–20, the corollas 10–18 mm long, white, sometimes becoming pinkish- or lavender-tinged with age. Disc florets 25–40, the corollas 6–8 mm long, the lobes 0.8–1.2 mm long. Fruits 2.5–4.0 mm long, narrowly oblong-ellipsoid, usually somewhat flattened, with 8–12 ribs, sparsely to moderately pubescent with short, fine hairs. 2n=18. July–October.

Uncommon, mostly in the Ozark and Ozark Border Divisions (Arkansas, Illinois, Indiana, Iowa, Michigan, Missouri, and Wisconsin). Bases and ledges of moist limestone and dolomite bluffs, and occasionally adjacent banks of streams.

This species is uncommon throughout its range and at one time was under consideration for protection under the federal Endangered Species Act. Genetic studies by Les et al. (1991) indicated that although the species is an obligate outcrosser, it has surprisingly low levels of overall genetic variation, which they interpreted to indicate that localized extirpations that reduced the amount of variation further might drive the species toward extinction. Steyermark noted that although forked aster is restricted to seepy, calcareous ledges and adjacent habitats in Missouri, farther north it is found in a larger variety of habitats.

2. Eurybia hemispherica (Alexander) G.L. Nesom (single-stemmed bog aster, southern prairie aster)

Aster hemisphericus Alexander

A. paludosus Aiton ssp. *hemisphericus* (Alexander) Cronquist

A. paludosus Aiton var. *hemisphericus* (Alexander) Waterf.

Heleastrum hemisphaericum (Alexander) Shinners

Pl. 229 c–e; Map 977

Plants with an often stout, woody rootstock and long, scaly rhizomes, sometimes forming loose colonies. Stems 12–120 cm long, usually not zigzag, glabrous or sparsely to moderately pubescent with short, nonglandular hairs toward the tip, glabrous toward the base. Basal leaves often absent at flowering, when present these and the lower stem leaves sessile or short-petiolate, the blade 2–10 cm long, linear to narrowly oblanceolate, angled or tapered to the usually sharply pointed tip, tapered at the base, relatively stiff and leathery, the margins entire but roughened with minute, stiff hairs, the surfaces glabrous, the upper surface often somewhat shiny. Largest leaves above the stem base, the median and upper stem leaves progressively smaller, linear, tapered at the base, sessile or nearly so, slightly expanded at the base and somewhat sheathing the stem. Inflorescences slender racemes, sometimes nearly spicate, sometimes appearing as few-branched, more or less elongate panicles with ascending spicate to racemose branches, occasionally reduced to a solitary terminal head, the stem or branches glabrous or sparsely to moderately pubescent with short, nonglandular hairs. Involucre 9–12 mm long, the bracts in 4–6 somewhat unequal to subequal, overlapping series, 1–2 mm wide and mostly 2–4 times as long as wide, narrowly lanceolate to oblong-lanceolate or narrowly oblong-ovate, sharply pointed at the loosely ascending to spreading or reflexed tip, with a narrow (or sometimes absent) central groove or midvein toward the mostly somewhat thickened base and broad, relatively firm, pale yellowish margins, the green area much-broadened abruptly toward the tip, the margins otherwise appearing minutely roughened or moderately short-hairy and sometimes dark purple or purplish-tinged, the outer surface glabrous. Ray florets 15–35, the corollas 10–25 mm long, purple to bluish purple. Disc florets 40–80(–95), the corollas 5.5–7.0 mm long, the lobes 0.7–1.2 mm long. Fruits 2.5–3.7 mm long, cylindrical or slightly wedge-shaped to narrowly oblong-ellipsoid, sometimes slightly flattened, with 9–16 ribs, glabrous or more commonly sparsely to moderately pubescent with short, fine hairs. $2n=18, 36$. August–October.

Scattered in the Unglaciated Plains Division and locally north and east to Callaway and Howell Counties; introduced in Marion County (Kansas to Texas east to Kentucky and Florida). Upland prairies and less commonly savannas and openings of dry upland forests; also pastures, railroads, and roadsides.

Steyermark (1963) and some other authors (Barkley, 1986) treated this taxon as a subspecies of the closely related *E. paludosa* (*Aster paludosus*). Cronquist (1980) noted that the two taxa (as *A. paludosus* and *A. hemisphericus*) were sometimes difficult to distinguish morphologically but were geographically distinct. *Eurybia paludosa* occurs along the Atlantic Coastal Plain from North Carolina to Florida, whereas *E. hemispherica* tends to be more of an inland plant. The also two differ in that the inflorescence of *E. paludosa* is flat-topped to hemispherical, the stems tend to be hairier (and often with slightly longer hairs) toward the tip, and the involucral bracts are less strongly thickened, with the margins tending to have slightly longer hairs.

Steyermark (1963) noted that this showy species deserved to be cultivated more widely as an ornamental wildflower. In recent years, it has become available through some of the state's wildflower nurseries.

3. Eurybia macrophylla (L.) Cass. (large-leaved aster)

Aster macrophyllus L.

Pl. 230 a, b; Map 978

Plants with long, sometimes relatively stout, fleshy rhizomes, often forming large colonies. Stems 30–120 cm long, often slightly zigzag, moderately to densely pubescent with short, gland-tipped hairs toward the tip, sparsely glandular to nearly glabrous toward the base. Basal and lower stem leaves present, long-petiolate, the blade 5–25 cm long, heart-shaped to irregularly ovate, short-tapered at the tip, deeply cordate at the base, the margins sharply toothed, the upper surface usually sparsely to moderately roughened-hairy with short, stiff, nonglandular hairs (rarely also sparsely glandu-

978. Eurybia macrophylla   979. Euthamia graminifolia   980. Euthamia gymnospermoides

lar), the undersurface sparsely to moderately pubescent with a mixture of gland-tipped and nonglandular hairs, sometimes nearly glabrous. Median and upper stem leaves progressively smaller, ovate to elliptic, truncate to tapered at the base, the petioles becoming progressively shorter and more broadly winged, slightly expanded at the base and sometimes somewhat clasping the stem. Inflorescences flat-topped to somewhat dome-shaped panicles or clusters, the heads in loose to dense clusters or less commonly solitary at the branch tips, the branches moderately to densely pubescent with short, gland-tipped hairs. Involucre 7–11 mm long, the bracts in 5–7 strongly unequal, overlapping series, 1–2 mm wide and mostly 1.5–3.0 times as long as wide, oblong-lanceolate to narrowly oblong-ovate, rounded to bluntly pointed at the ascending tip, with a narrow (or sometimes absent), green central band toward the base (this sometimes somewhat keeled) and broad, relatively firm, pale yellowish margins, the green area abruptly much-broadened toward the tip, the margins otherwise appearing uneven or finely hairy and sometimes dark purple or purplish-tinged, the outer surface glabrous or more commonly moderately to densely pubescent with a mixture of gland-tipped and nonglandular hairs. Ray florets 9–20, the corollas 8–15 mm long, usually lavender to purple, rarely white. Disc florets 20–40, the corollas 6–8 mm long, the lobes 1.0–1.5 mm long. Fruits 2.5–4.5 mm long, narrowly oblong-ellipsoid, usually somewhat flattened, with 7–12 ribs, glabrous or sparsely hairy toward the tip. $2n=72$. August–October.

Uncommon, known thus far only from Howell, Madison, Shannon, and Texas Counties (eastern [mostly northeastern] U.S. west to Minnesota and Missouri; Canada; introduced in Europe). Ledges and tops of bluffs and mesic upland forests on steep slopes, rarely banks of streams.

This species was first reported for Missouri by Summers and Yatskievych (1990). The earliest Missouri specimens were collected in 1970 by the late Art Christ but initially were misdetermined. This may have been because the somewhat disjunct Ozarkian populations usually produce few flowering stems in any given year, existing mostly vegetatively as large colonies of basal rosettes.

20. Euthamia (Nutt.) Cass. (flat-topped goldenrod)
(Sieren, 1970, 1981)

Plants perennial herbs, often colonial from relatively long-creeping, branched rhizomes, often somewhat aromatic, the stem bases sometimes somewhat woody. Stems usually solitary, erect or ascending, with few to many ascending branches above the midpoint, with often relatively coarse, longitudinal ridges, glabrous or less commonly sparsely pubescent with minute, spreading hairs. Basal and lower stem leaves absent at flowering. Stem leaves gradually or relatively abruptly reduced toward the tip of the stem, sessile or nearly so, the blade linear to narrowly lanceolate, sharply pointed at the tip, more or less tapered to a slender or slightly expanded but not clasping base, the margins entire but sometimes roughened with minute, stout, ascending, stiff hairs, the surfaces glabrous or less commonly sparsely to moderately pubescent with minute, spreading hairs mostly along the veins, slightly to strongly resinous with small, impressed or somewhat pustular glandular dots. Inflorescences appearing as relatively dense, flat-topped panicles, the heads solitary or more commonly in small clusters at the branch tips, the stalks mostly short, with relatively few leaflike, linear to elliptic bracts 0.2–1.0 cm long. Heads radiate, slightly to strongly resinous-sticky. Involucre 3–6 mm long, cup-shaped to nearly cylindrical. Involucral bracts in 3–5 unequal, overlapping series, linear to narrowly oblong-lanceolate or lanceolate, the ascending tips rounded to less commonly bluntly or sharply pointed (sometimes in the same head), often somewhat concave or thickened along the midvein, entirely straw-colored to light yellow or some of the bracts with a short, elliptic to obovate green area toward the tip, glabrous, slightly to strongly resinous. Receptacle flat or slightly convex, with low, toothlike ridges around the attachment points of the florets, sometimes also with short, fine hairs. Ray florets 7–35, pistillate, the corolla 1–3 mm long, ascending to somewhat spreading, yellow, not persistent at fruiting (but sometimes trapped irregularly by the pappus and resinous exudate). Disc florets 3–13, perfect, the corolla 1.5–4.0 mm long, yellow, not persistent at fruiting (but sometimes trapped irregularly by the pappus and resinous exudate). Pappus of the ray and disc florets similar, of numerous slender, finely barbed bristles about as long as the corollas, white. Fruits 0.5–1.5 mm long, more or less oblong-elliptic in outline to slightly obovate, circular in cross-section or very slightly flattened, faintly 2–4-nerved, the nerves often difficult to observe, the surface moderately pubescent with fine, ascending, white to somewhat silvery hairs, straw-colored to light tan to light greenish brown, sometimes with slightly darker nerves. Five to 8 species, temperate North America; introduced in Europe, Asia.

Traditionally, *Euthamia* was considered part of *Solidago* (Steyermark, 1963). In his doctoral dissertation, Sieren (1970) presented data suggesting that although the plants have superficial similarities to that genus, there are strong differences in some morphological features of the heads as well as the anatomy of the leaf glands. He suggested a closer relationship between *Euthamia* and a group of genera related to *Gutierrezia*. Anderson and Creech (1975) studied details of leaf anatomy in the group and also concluded that *Euthamia* and *Solidago* should be regarded as distinct genera. More recently, molecular research (summarized by Beck et al., 2004) has supported the hypothesis of separate lineages and generally upheld the notion of a close relationship between *Euthamia* and the *Gutierrezia* group.

At the species level, the differences between the eight species accepted by Sieren (1981) are often fairly subtle and there is a lot of morphological variation within and between populations of each species. Some workers have suggested that there are fewer actual species in the genus (Nesom, 2000). There has not been a recent comprehensive review of the genus, thus the specific and infraspecific taxonomy remain somewhat controversial.

1. Stems and often also leaves sparsely and minutely spreading-hairy; heads with 15–25(–35) ray florets and 4–10(–13) disc florets; leaves all or mostly with 3 main veins, lateral pair sometimes finer than the midvein, the larger leaves sometimes with a second pair of additional main veins (5-veined) . . . . 1. E. GRAMINIFOLIA
1. Stems glabrous (except sometimes a few hairs at the nodes); leaves glabrous except along the margins; heads with 7–15 ray florets and 3–5(–9) disc florets; leaves with 1 main vein or more commonly at least the larger ones with a pair of additional main veins, the lateral pair usually finer than the midvein
    2. Leaves relatively thick in texture, with dense, noticeable impressed or pustular glandular dots; involucre appearing resinous
        . . . . 2. E. GYMNOSPERMOIDES
    2. Leaves relatively thin in texture, with sparse or faint pustular glandular dots (these sometimes only observable through the leaf by holding it up to a strong light); involucre not appearing resinous . . . . 3. E. LEPTOCEPHALA

1. Euthamia graminifolia (L.) Nutt. (common flat-topped goldenrod)
Solidago graminifolia (L.) Salisb.
S. graminifolia var. *nuttallii* (Greene) Fernald
E. graminifolia var. *nuttallii* (Greene) W. Stone

Pl. 234 g–i; Map 979

Stems 40–120 cm long, sparsely pubescent with minute, spreading hairs, at least toward the tip. Leaf blades 1–12 cm long, 1–10 mm wide, relatively thick, the margins moderately roughened with minute, stout, ascending, stiff hairs, the surfaces glabrous or sparsely pubescent with minute, spreading hairs, moderately resinous with impressed or pustular glandular dots, the smaller leaves with 1 midvein, the larger leaves with 3 or occasionally 5 main veins. Involucre 3–5 mm long, relatively resinous, the bracts variously rounded to sharply pointed at the tip. Ray florets 15–25(–35), the corollas 1–2 mm long, the short ligule ascending. Disc florets 4–10(–13), the corolla 1.5–2.0 mm long, the lobes 0.3–0.7 mm long. $2n=18$. August–October.

Uncommon, mostly in counties bordering the Missouri River (northern U.S. south to Wyoming, Oklahoma, and Georgia; Canada; introduced in Europe, Asia). Loess hill prairies, savannas, and bottomland forests; also roadsides.

The present treatment differs markedly from that of Steyermark (1963). Steyermark treated *E. graminifolia* (as *Solidago*) with two varieties, basing his treatment on earlier taxonomic studies summarized in Fernald (1950). He excluded the glabrous var. *graminifolia* from the Missouri flora but accepted a larger-headed variant with somewhat hairy leaves and stems as *S. graminifolia* var. *nuttallii* and a smaller-headed variant with leaves only minutely roughened along the margins as *S. graminifolia* var. *media* (Greene) S.K. Harris. He reported populations of var. *nuttallii* as occurring uncommonly in a few counties along the Missouri River, with var. *media* widespread in the state. Since that time, Missouri botanists have had great difficulty in using this treatment to distinguish the latter variety from *E. gymnospermoides,* as it turns out with just cause. Croat (1970) studied a number of morphological characters throughout the range of the two taxa and observed widespread intergradation between *S. graminifolia* and *S. gymnospermoides* var. *media*. Subsequently, Sieren (1970, 1981) studied plants of the complex in the field and herbarium from Illinois and Wisconsin west into the Great Plains. His research confirmed that plants previously treated as *S. graminifolia* var. *media* were better included within *E. gymnospermoides* and also that there was too much intergradation between different morphotypes within that species to allow formal recognition of varieties within that species.

Within *E. graminifolia,* Sieren (1970, 1981) accepted three strongly intergrading varieties, with the hairy, robust var. *nuttallii* perhaps the most distinctive element in the species. Missouri specimens are less hairy than those from the main portion of the range, and some authors (Cronquist, 1991) have questioned whether the variety is consistently distinguishable. The glabrous taxa var. *graminifolia* and var. *major* (Michx.) Moldenke supposedly differ from one another primarily in the length-to-width ratio of the larger leaves, and both occur mainly to the north of Missouri. Because these variants overlap considerably and do not seem to have much ecological or geographic separation, there seems little point in recognizing them.

2. Euthamia gymnospermoides Greene ex Porter & Britton (viscid bushy goldenrod, Great Plains flat-topped goldenrod)
Solidago gymnospermoides (Greene) Fernald
S. graminifolia var. *media* (Greene) S.K. Harris

Pl. 234 d–f; Map 980

Stems 30–100 cm long, glabrous (except sometimes a few hairs at the nodes). Leaf blades 1–10 cm long, 1–8 mm wide, relatively thick, the margins slightly to moderately roughened with minute, stout, ascending, stiff hairs, the surfaces glabrous, strongly resinous with dense, conspicuous, impressed or less commonly pustular glandular dots, with 1 midvein or more commonly at least the larger leaves often with 3 main veins, the lateral pair usually finer than the midvein. Involucre 4–6 mm long, relatively resinous, the bracts mostly rounded to bluntly pointed at the tip. Ray florets 7–11(–15), the corollas 2–3 mm long, the short ligule ascending to somewhat spreading. Disc florets 3–5(–9), the corolla 2.5–4.0 mm long, the lobes 0.4–0.9 mm long. $2n=18$, 36. August–October.

Scattered nearly throughout the state, but uncommon in the eastern half of the Ozark Division and the Mississippi Lowlands (Ohio to Tennessee west to North Dakota, Colorado, and Texas; Canada). Bottomland and upland prairies, savannas, glades, banks of streams and rivers, and margins of ponds and lakes; also old fields, fallow fields, pastures, ditches, railroads, and roadsides.

As noted above in the treatment of *E. graminifolia,* recent studies have supported the contention of Croat (1970) and Sieren (1970, 1981) that most of the specimens included by Steyermark (1963) under *E. graminifolia* are better classified within *E. gymnospermoides*.

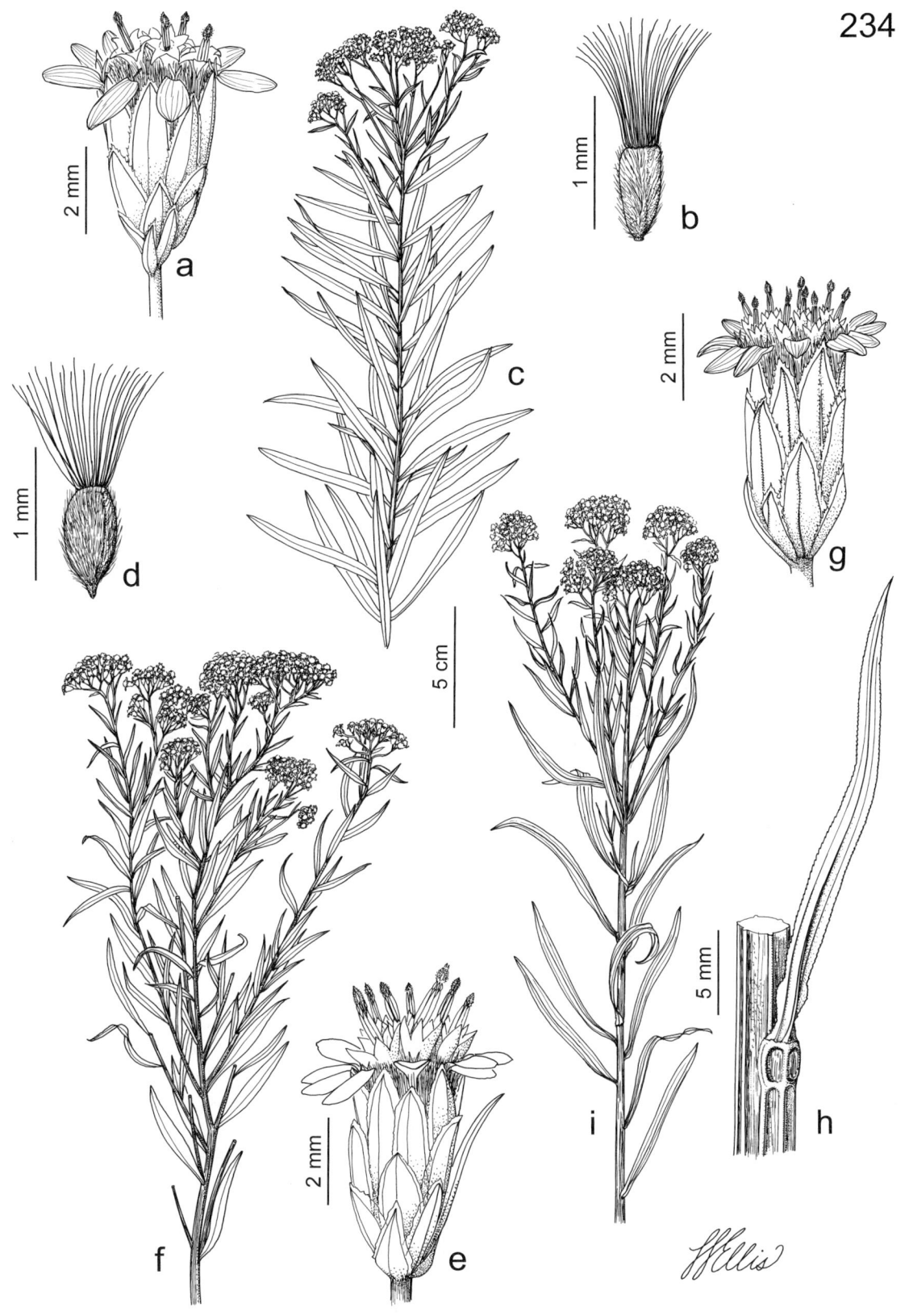

Plate 234. Asteraceae. *Euthamia leptocephala*, **a)** head, **b)** fruit, **c)** habit. *Euthamia gymnospermoides*, **d)** fruit, **e)** head, **f)** habit. *Euthamia graminifolia*, **g)** head, **h)** leaf, **i)** habit.

981. Euthamia leptocephala   982. Grindelia ciliata   983. Grindelia lanceolata

Euthamia gymnospermoides is thus by far the most abundant and widespread member of the genus in Missouri. Croat (1970) further advocated making *E. gymnospermoides* a variety of *E. graminifolia*, but subsequent authors have not followed that classification (Sieren, 1981; Barkley, 1986; Gleason and Cronquist, 1991).

3. Euthamia leptocephala (Torr. & A. Gray) Greene ex Porter & Britton (Mississippi Valley flat-topped goldenrod)

Solidago leptocephala Torr. & A. Gray

Pl. 234 a–c; Map 981

Stems 30–100 cm long, glabrous. Leaf blades 1–10 cm long, 1–8 mm wide, relatively thin in texture, the margins slightly roughened with minute, stout, ascending, stiff hairs, the surfaces glabrous, slightly resinous with sparse or faint, inconspicuous pustular-gland dots (these sometimes only observable through the leaf by holding it up to a strong light), with 1 midvein or more commonly at least the larger leaves often with 3 main veins, the lateral pair usually finer than the midvein. Involucre 4–6 mm long, not or only slightly resinous, the bracts variously rounded to sharply pointed at the tip. Ray florets 7–15, the corollas 2–3 mm long, the short ligule ascending to somewhat spreading. Disc florets 3–5(–9), the corolla 2.5–4.0 mm long, the lobes 0.4–0.9 mm long. $2n=18, 54$. August–October.

Uncommon in the Mississippi Lowlands Division and the adjacent portion of the Ozarks; disjunct in Jasper County (Kentucky, Illinois, and Missouri south to Georgia and Texas). Bottomland forests, swamps, sloughs, and less commonly mesic upland forests and savannas; also ditches, railroads, and roadsides; often in sandy soils.

21. Grindelia Willd. (gumweed, gum plant)

Plants annual, biennial, or perennial herbs, usually with taproots. Stems 1 to several, erect or ascending, usually with few to several ascending branches toward the tip, with fine, longitudinal lines or grooves, glabrous or occasionally sparsely hairy toward the base (variously hairy elsewhere). Basal leaves absent at flowering, similar to the lower stem leaves, short- to long-petiolate. Stem leaves sometimes somewhat reduced toward the tip of the stem, sessile, the blade linear to narrowly oblong, oblong-lanceolate or oblong-obovate, rounded to bluntly or sharply pointed at the tip, tapered to rounded, truncate, or shallowly cordate at the usually slightly to strongly clasping or sheathing base, the margins entire or variously toothed (the lowermost leaves occasionally appearing somewhat lobed), the surfaces glabrous, with moderate to dense, impressed glandular dots. Inflorescences appearing as solitary heads or less commonly loose clusters at the branch tips, occasionally a few heads also in the axils of the adjacent leaves. Heads often relatively short-stalked, with few to several short, inconspicuous, linear bracts that grade into the involucral bracts, radiate or rarely discoid, slightly to strongly resinous-sticky. Involucre 6–15 mm long, cup-shaped to hemispherical or broadly urn-shaped, slightly to strongly resinous-sticky. Involucral bracts in 3–9 unequal to subequal, overlapping series, mostly narrowly lanceolate to linear, thick and yellowish below the midpoint, the tip green and usually with inrolled margins, loosely ascending to recurved or curled, the surface glabrous, with sparse to dense, impressed glandular dots. Receptacle flat or slightly convex, sometimes with low, toothlike ridges around the attachment points of the florets. Ray florets 14–45 or rarely absent, when present pistillate, the corolla 8–20 mm long, yellow, not persistent at fruiting. Disc florets numerous (more than 100), perfect or some of the outer or inner ones sometimes functionally staminate, the corolla 4–8 mm long, yellow, not persistent at fruiting. Pappus of the ray and disc florets similar, either of 2–8 awns or numerous bristles, in either case, these variously slightly shorter than to longer than the disc corollas, smooth or minutely barbed, off-white to straw-colored or light brown, not persistent or more or less persistent at fruiting. Fruits often of 2 types, those of the disc florets somewhat flattened, those of the ray florets more or less 3- or 4-angled in cross-section, both types more or less oblong to slightly wedge-shaped in outline, often more or less truncate at the tip (sometimes with the angles extended into minute teeth), the surface smooth or finely and inconspicuously few-nerved, glabrous. About 30–55 species, North America to South America, introduced in the Old World.

Grindelia was the subject of Julian Steyermark's doctoral research, and his dissertation studies resulted in a lengthy series of papers on various aspects of the ecology, cytology, and systematics of the genus, including a taxonomic monograph of the North American species (Steyermark, 1934). A notable feature of gumweed species is their production of sticky, resinous exudates consisting mostly of complex mixtures of terpenes. These are produced by glands that are usually sunken into the surface of most of a plant's tissues but are especially noticeable on the leaves and heads. Young heads often have the saucerlike space on top of the developing disc florets filled with a characteristic shiny layer of milky-white resin. Resin production in the genus has been studied for possible industrial applications as a substitute for conifer resins (Hoffmann et al., 1984). In particular, the Californian *G. camporum* Greene has been documented to produce resins at about 10 percent of the dry weight of the aboveground portion of the plant (McLaughlin, 1986; Hoffmann and McLaughlin, 1986), making it feasible for potential future development as an arid-adapted crop in the southwestern United States. This same resinous exudate had a long history of use among Native Americans for skin sores and lesions, to bind together edges of wounds, as an inhalant for bronchitis and asthma, and taken internally for coughs and a variety of other ailments (Steyermark, 1963; Moerman, 1998).

1. Pappus of numerous bristles, these fused at the base, persistent or sometimes shed long after the fruit matures . . . . 1. G. CILIATA
1. Pappus of 2–8 slender awns, these not fused, usually shed as the fruit matures
    2. Involucral bracts loosely ascending to slightly curved outward; leaves usually appearing only slightly resinous, the impressed glandular dots moderate to dense but inconspicuous and only slightly differing in color from the surrounding leaf tissue; margins of the leaves sharply toothed (rarely entire), the teeth mostly with a minute, bristlelike extension at the tip
        . . . . 2. G. LANCEOLATA
    2. Involucral bracts strongly curled or recurved; leaves usually appearing strongly resinous, the impressed glandular dots relatively dense and conspicuously darker than the surrounding leaf tissue; margins of the leaves rarely entire or nearly so, more commonly relatively bluntly toothed, the teeth usually with a thickened or glandular tip and lacking a spinelike or bristlelike extension . . . . 3. G. SQUARROSA

1. Grindelia ciliata (Nutt.) Spreng.   Pl. 237 a, b; Map 982
(goldenweed)
Prionopsis ciliatus (Nutt.) Nutt.
Haplopappus ciliatus (Nutt.) DC.
G. papposa G.L. Nesom

Plants annual or less commonly biennial. Stems (10–)40–120 cm long. Stem leaves sessile, the blades 2–8 cm long, oblong to ovate or oblong-obovate, cordate at the base and strongly clasping the stem, rounded or bluntly pointed at the tip, the margins with many narrow, coarse, sharp teeth, these mostly with a spinelike or bristlelike extension at the tip, the surfaces appearing not or only slightly resinous, with usually moderate glandular dots, but these inconspicuous and only slightly differing in color from the surrounding leaf tissue. Inflorescences of solitary heads or loose clusters at the branch tips. Receptacle 1.5–3.0 cm in diameter. Involucre 10–15 mm long, the bracts in 3–5 unequal series, at least those of the outer few series with the tip spreading to recurved or curled outward. Ray florets 25–50, the corolla 10–20 mm long. Disc florets perfect or some of the inner ones functionally staminate, the corollas 6–8 mm long. Pappus of numerous (30–60) bristles, 3–10 mm long, these minutely barbed, fused at the base, persistent at fruiting or sometimes shed tardily as a unit, those of the outer series shorter, stouter, and slightly thickened toward the base (awnlike), straw-colored to light orangish tan. Fruits 2–4 mm long, light gray to whitish gray. 2n=12. August–September.

Possibly introduced, uncommon in southwestern and west-central Missouri, introduced sporadically elsewhere in the northern half of the state and in a few southern counties (Nebraska south to New Mexico and Texas; introduced west to California, east to Iowa, Illinois, Michigan, and Maryland, and south to Louisiana). Glades and upland prairies; also railroads, roadsides, and open, disturbed areas.

Goldenweed presumably is native to portions of the Great Plains, but it has dispersed along railroads and highways to extend its range in all directions. In Missouri, its status is unclear. Arguments in favor of it being a rare native at some sites are the fact that it has been collected in some reasonably undisturbed prairies and glades in the western half of the state, and also that it was collected as early as 1882 in southwestern Missouri (introduced occurrences in most other states were all far more recently documented). On the other hand, the large majority of specimens, even those collected as early as the first decade of the 1900s, clearly document nonnative occurrences in disturbed habitats.

Steyermark (1963) and most earlier authors treated this species as a member of the large and polymorphic genus *Haplopappus* Cass. A series of morphological, anatomical, and molecular studies by various investigators (summarized by Lane and Hartman, 1996) resulted in the gradual dismemberment of that genus in North America into a large number of smaller, more homogeneous groups with diverse affinities in the tribe Astereae. A number of these studies agreed that the species known as *H. ciliatus* was closely related to species of *Grindelia*. Following Correll and Johnston (1970), the generic name *Prionopsis* Nutt. was resurrected to accommodate this one unusual species. Nesom et al. 1993) reviewed the data that had accumulated on these genera and concluded that *P. ciliatus* was best classified as a morphologically atypical member of *Grindelia*, a conclusion that has been followed by recent authors (Kartesz and Meacham, 1999) and also in the present work. However, their renaming of the species as *G. papposa* was superfluous.

2. Grindelia lanceolata Nutt. (spiny-toothed gumweed)

G. lanceolata f. *latifolia* Steyerm.

Pl. 235 f–h; Map 983

Plants biennial or perennial herbs (flowering only once before dying). Stems 20–150 cm long. Stem leaves sessile, the blades 2–11 cm long, linear to lanceolate-oblong (rarely narrowly ovate), angled to rounded or occasionally nearly truncate at the base and slightly clasping to moderately sheathing the stem, tapered to a sharply pointed tip, the margins with sparse to moderate, narrow, fine or less commonly coarse, sharp teeth or less commonly some of the leaves entire, the teeth mostly with a minute, bristlelike extension at the tip, the surfaces appearing not or only slightly resinous, with moderate to dense glandular dots, but these inconspicuous and only slightly differing in color from the surrounding leaf tissue. Inflorescences of solitary heads or loose clusters at the branch tips. Receptacle 1.5–2.0 cm in diameter. Involucre 10–20 mm long, the bracts in 4–6 subequal series, loosely ascending to slightly curved outward. Ray florets 14–30, the corolla 8–16 mm long. Disc florets perfect or some of the inner and/or outer ones functionally staminate, the corollas 4.0–7.5 mm long. Pappus of 2 slender awns, 4–8 mm long, these not barbed, not fused at the base, not persistent at fruiting (usually shed individually as the fruit matures), off-white to straw-colored. Fruits 3–6 mm long, straw-colored to light gray. 2n=12. August–October.

Scattered mostly in the southern half of the state (Kansas to Texas east to Ohio, Virginia, and Alabama). Glades, upland prairies, openings of mesic to dry upland forests, and rarely banks of streams and rivers; also pastures, old fields, railroads, and roadsides; often on calcareous substrates.

Steyermark (1934, 1963) referred to plants with slightly broader upper stem leaves as f. *latifolia*.

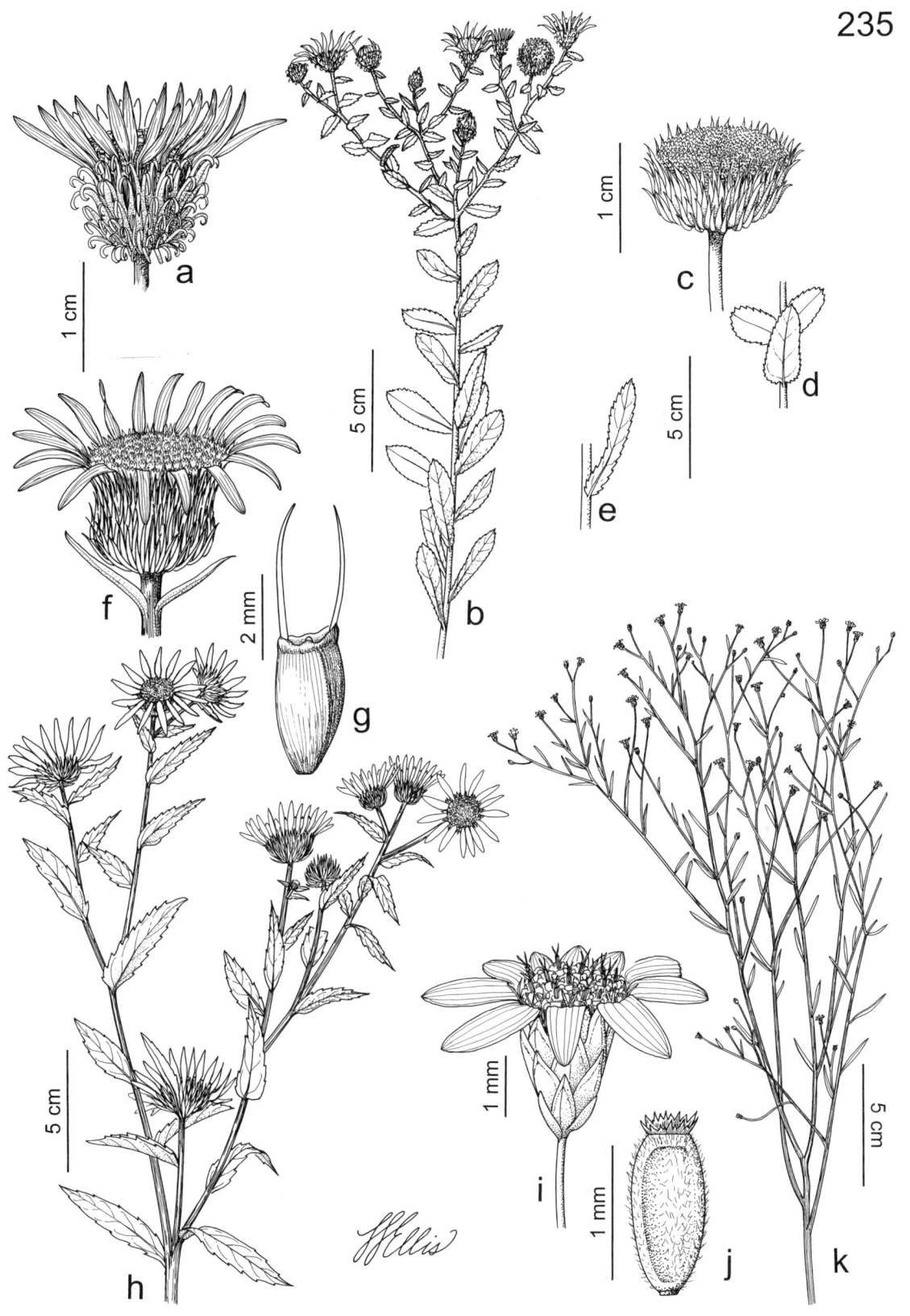

Plate 235. Asteraceae. *Grindelia squarrosa* var. *squarrosa*, **a)** head, **b)** habit. **c)** leaf. *Grindelia squarrosa* var. *nuda*, **d)** head, **e)** leaves. *Grindelia lanceolata*, **f)** head, **g)** fruit, **h)** habit. *Gutierrezia texana*, **i)** head, **j)** fruit, **k)** habit.

984. Grindelia squarrosa   985. Gutierrezia texana   986. Heterotheca camporum

3. Grindelia squarrosa (Pursh) Dunal (curlytop gumweed)

Pl. 235 a–e; Map 984

Plants biennial or rarely short-lived perennials. Stems 10–100 cm long. Stem leaves sessile, the blades 1–7 cm long, oblanceolate to narrowly oblong, oblong, or ovate, more or less truncate to shallowly cordate at the base and slightly to moderately clasping the stem, mostly rounded to a bluntly pointed tip (the uppermost leaves occasionally sharply pointed), the margins with moderate to numerous, narrow, fine or coarse, relatively blunt teeth, rarely sparsely and inconspicuously toothed to entire, the teeth mostly with a thickened or glandular tip and lacking a bristlelike extension, the surfaces appearing strongly resinous with relatively dense glandular dots, these conspicuously darker than the surrounding leaf tissue. Inflorescences of solitary heads or loose clusters at the branch tips, occasionally a few heads also in the axils of the adjacent leaves. Receptacle 1–2 cm in diameter. Involucre 6–11 mm long, the bracts in 5–9 unequal series, strongly curled or recurved. Ray florets 20–40 or rarely absent, when present the corolla 7–15 mm long. Disc florets perfect or some of the inner and/or outer ones functionally staminate, the corollas 3.5–6.5 mm long. Pappus of 2–8 slender awns, 2.5–6.0 mm long, these barbed or less commonly smooth or nearly so, not fused at the base, not persistent at fruiting (usually shed individually as the fruit matures), off-white to straw-colored. Fruits 2–3 mm long, straw-colored to light gray or tan. $2n=12$. July–September.

Introduced, scattered mostly in the northern half of the state (native range unclear; present nearly throughout the U.S. except most of the Southeast; Canada, Mexico; introduced in Europe). Pastures, railroads, roadsides, and open, disturbed areas.

The native range of this species originally probably was restricted mainly to portions of the Great Plains and the eastern edge of the Rocky Mountains (Steyermark, 1934; Cronquist, 1980), but by the early 1900s it had increased its range greatly eastward and westward. Steyermark (1963) stated that it had been documented rarely from prairies and alluvial areas, but the specimens examined during the present study originated from highly disturbed sites.

The taxonomy of the G. squarrosa complex is still not fully understood. Steyermark (1934, 1963) included a series of varieties and forms, and recognized some additional species that have been treated as varieties by some other authors. Steyermark's var. *serrulata* refers to plants that correspond to var. *squarrosa* but tend to have somewhat narrower leaves, and it is not accepted here. The plants treated here as var. *quasiperennis* were treated by Steyermark as a separate species, *G. perennis*. Future populational studies and molecular analyses may help to elucidate the taxonomy of this group.

1. Ray florets absent . . . . 3A. VAR. NUDA
1. Ray florets present, the corolla well developed
    2. Leaf margins mostly with sparse and inconspicuous teeth or entire
        . . . . 3B. VAR. QUASIPERENNIS
    2. Leaf margins mostly with moderate to numerous narrow, fine or coarse, relatively blunt teeth
        . . . . 3C. VAR. SQUARROSA

3a. var. nuda (A.W. Wood) A. Gray

G. nuda A.W. Wood

Pl. 235 d, e

Plants biennial. Leaf blades with the margins mostly with moderate to numerous narrow, fine or coarse, relatively blunt teeth. Ray florets absent. Fruits all similar, somewhat flattened. July–September.

Introduced, uncommon and widely scattered, known thus far from Adair, Dunklin, Jackson, and St. Louis Counties (Colorado, New Mexico, Kansas, Oklahoma, and Texas; introduced sporadically farther east and in Canada). Pastures, railroads, and open, disturbed areas.

Nesom (1990c) advocated the recognition of this taxon as a separate species. However, aside from the lack of ray florets and the associated production of only flattened (rather than 3- or 4-angled) disc achenes, these plants appear to fit within the range of morphological variation of *G. squarrosa*.

3b. var. quasiperennis Lunell
G. perennis A. Nelson

Plants usually short-lived perennials. Leaf blades mostly narrowly oblong to oblong-oblanceolate, the margins mostly with sparse and inconspicuous teeth or entire, the lowermost leaves sometimes with several shallow lobes. Ray florets present, the corolla well developed. Fruits usually of 2 types, those of the disc florets somewhat flattened, those of the ray florets more or less 3- or 4-angled. July–September.

Introduced, known thus far from a single collection from the city of St. Louis (northwestern U.S. east to North Dakota and Colorado; Canada; introduced sporadically farther east). Railroads.

The presence of this taxon was first noted in 1985 by Mark Allen Wetter (now at the University of Wisconsin). The sole, somewhat scrappy specimen was collected by Viktor Mühlenbach in 1955 during his botanical surveys of the St. Louis railyards and was initially misdetermined by Julian Steyermark as var. *serrulata*.

3c. var. squarrosa
G. squarrosa var. *serrulata* (Rydb.) Steyerm.
Pl. 235 a–c

Plants biennial. Leaf blades mostly narrowly to broadly oblong or oblanceolate to ovate, the margins mostly with moderate to numerous narrow, fine or coarse, relatively blunt teeth. Ray florets present, the corolla well developed. Fruits usually of 2 types, those of the disc florets somewhat flattened, and those of the ray florets more or less 3- or 4-angled. $2n=12$. July–September.

Introduced, scattered mostly in the northern half of the state (native range unclear; present in most of the U.S. except most southeastern and a few western states; Canada, Mexico; introduced in Europe). Pastures, railroads, roadsides, and open, disturbed areas.

22. Gutierrezia Lag. (snakeweed)
(Lane, 1985)

About 28 species, North America, South America.

Lane (1982) established the modern classification of *Gutierrezia* and its relatives, recognizing several segregate genera that are widely accepted today and transferring a number of species between genera. For example, the sole native species of *Gutierrezia* treated by Steyermark (1963) is now considered part of the small, closely related genus, *Amphiachyris*. Suh and Simpson (1990) confirmed most of Lane's conclusions using molecular data.

1. Gutierrezia texana (DC.) Torr. & A. Gray
Pl. 235 i–k; Map 985

Plants annual, with taproots. Stems solitary, 10–100 cm long, erect or strongly ascending, with numerous ascending to loosely ascending branches from below the midpoint, longitudinally lined or ridged, glabrous. Basal leaves absent. Stem leaves sessile or less commonly short-petiolate, 0.5–5.0 cm long, the blade narrowly linear to linear-lanceolate or linear-oblanceolate, mostly sharply pointed at the tip, more or less tapered to the nonclasping base, the margins entire, the surfaces glabrous but with moderate to dense, impressed resinous dots, often somewhat sticky to the touch. Inflorescences appearing as more or less flat-topped panicles, the heads solitary or in small, loose clusters along the branches and/or at the branch tips, the branch tips and stalks short to relatively long, with few leaflike, linear bracts 0.3–1.0 cm long. Heads radiate, sticky, resinous. Involucre 2–4 mm long, narrowly cup-shaped to obconical. Involucral bracts in 2–4 unequal, overlapping series, oblong-lanceolate to oblong-ovate, with a straw-colored to yellowish, hard, shiny basal portion (this sometimes with a slender, green midvein) and an ascending, triangular or diamond-shaped, green tip, glabrous but the green portion finely resin-dotted. Receptacle flat or slightly convex, minutely hairy around the attachment points of the florets. Ray florets 5–14, pistillate, the corolla 3–6 mm long, yellow, sometimes persistent at fruiting. Disc florets 7–13, perfect, the corolla 1.5–2.5 mm long, yellow. Pappus of the ray and disc florets similar, a minute, light-colored crown or ridge less than 0.2 mm long, rarely a minute, toothed (appearing scalelike) ring, always much shorter than the corolla. Fruits 0.8–1.5 mm long, oblong-obovoid, sometimes slightly flattened, with 6–9 fine, green nerves, the surface otherwise densely pubescent with short, grayish purple hairs, purplish brown to nearly black. $2n=8, 10, 16$. (July–)October.

Introduced, known only from a single collection from the city of St. Louis (Oklahoma to Texas and adjacent Arkansas and Louisiana; Mexico; introduced in Illinois and Missouri). Railroads.

This species was reported by Steyermark (1963) in the addendum at the end of his *Flora of Missouri*, based on a specimen collected by Viktor Mühlenbach in the St. Louis railyards. A second variety (var. *glutinosa* (S. Schauer) M.A. Lane), with heads having more florets and a better-developed pappus, occurs in eastern Mexico and southern Texas.

23. Heterotheca Cass. (golden aster, camphorweed)

Plants annual or perennial herbs, usually with taproots, sometimes with rhizomes, the stem bases sometimes somewhat woody. Stems 1 to several, erect or ascending, with usually several to numerous ascending to loosely ascending branches above the midpoint, sometimes only few-branched toward the tip, with fine, longitudinal ridges or grooves, moderately to densely hairy, sometimes also glandular. Basal leaves often withered or absent by flowering time, the blade narrowly to broadly oblanceolate, tapered at the base to a winged petiole, the margins entire, shallowly undulate, or variously toothed. Stem leaves slightly to moderately reduced toward the tip of the stem, sessile, the blade narrowly oblanceolate to oblong-lanceolate, oblong-ovate, or ovate, bluntly or sharply pointed at the tip, tapered to shallowly cordate at the base, sometimes clasping the stem, the margins entire or variously toothed, the surfaces and especially the margins moderately to densely hairy, sometimes also glandular. Inflorescences of solitary heads at the branch tips or of small, loose clusters, sometimes appearing paniculate, the branch tips and stalks short to relatively long, usually with few to several linear bracts 0.5–1.5 cm long. Heads radiate, not or only slightly sticky, sometimes aromatic but not resinous. Involucre 4–10 mm long, cup-shaped to slightly bell-shaped. Involucral bracts in 3–6 unequal, overlapping series, narrowly lanceolate or narrowly triangular-lanceolate to linear, tapered to an ascending to loosely ascending, slender, sharply pointed tip, with a green central stripe nearly the entire length or only above the midpoint (sometimes difficult to observe in *H. canescens*), the marginal and basal areas whitish or straw-colored, the outer surface usually densely hairy, sometimes also glandular. Receptacle flat or slightly convex, usually with low, toothlike ridges around the attachment points of the florets. Ray florets 10–35 (absent elsewhere), pistillate, the corolla 4–15 mm long, yellow, not persistent at fruiting. Disc florets 15–65, perfect, the corolla 3–9 mm long, yellow, not persistent at fruiting. Pappus of the ray and disc florets similar or dissimilar (sometimes absent in ray florets), a low crown of several shorter, white to off-white scales or bristles 0.2–1.0 mm long and numerous (25–45) longer, finely barbed bristles 5–9 mm long, these usually white when young but sometimes turning straw-colored to light orangish brown as the fruits mature. Fruits 1.2–4.0 mm long, sometimes of 2 types, those of the disc florets somewhat flattened, those of the ray florets sometimes more or less 3- or 4-angled in cross-section, both types obovate to narrowly obovate in outline, the surface glabrous or moderately to densely pubescent with fine, appressed, silvery hairs, light tan to grayish tan. About 28 species, U.S., Canada, Mexico.

Traditionally, most botanists separated this complex into two genera, *Chrysopsis* (Nutt.) Elliott and *Heterotheca*, with *Chrysopsis* comprising species with a pappus similar in disc and ray florets and producing only one type of fruit and *Heterotheca* in the strict sense confined to species having ray florets lacking a pappus or nearly so and producing angled (vs. flattened) ray fruits. Beginning with the work of Shinners (1951), some botanists developed arguments for treating the entire complex as a single genus under the name *Heterotheca* (Harms, 1963, 1965a, 1968, 1974), citing exceptions to this rule. More recently, Semple (1977, 1996) and Semple et al. (1980) developed cytological, morphological, and molecular data to support the hypothesis that there exist three main lineages within the complex, as well as a few miscellaneous, small segregates. As circumscribed by Semple and his colleagues, *Chrysopsis* consists of about eleven non-Missouri species native to the southeastern United States. Semple (1996) excluded the two species of *Bradburia,* one of which does occur in Missouri, and which some botanists continue to accept as a primitive member of *Chrysopsis* (Nesom, 2000). See the treatment of *Bradburia* for further discussion. Semple et al. (1980) and Semple (1996) further segregated a group of about seven species of so-called grass-leaved golden asters as the genus *Pityopsis* Nutt. These also grow in the southeastern United States, with one species also in the Neotropics, and none occurs in Missouri. The confusing array of treatments for *Heterotheca* and its relatives in the botanical literature is sufficient to cause temporary lightheadedness in botanists, many of whom nevertheless have strong opinions on the appropriate classification.

1. Leaves appearing gray, not roughened, densely pubescent with appressed or curved hairs mostly lacking a bulbous base, the margins entire. . . . 2. H. CANESCENS
1. Leaves appearing green or rarely slightly grayish-tinged, roughened, moderately pubescent with curved and loosely appressed to spreading hairs, at least the longer hairs with an expanded, bulbous, pustular base, the margins of at least some of the leaves toothed
   2. Ray florets with a pappus similar to that of the disc florets; fruits developing from the ray florets somewhat flattened, moderately to densely hairy
    . . . . 1. H. CAMPORUM
   2. Ray florets lacking a pappus; fruits developing from the ray florets 3- or 4-angled, glabrous . . . . 3. H. SUBAXILLARIS

1. Heterotheca camporum (Greene) Shinners
Chrysopsis camporum Greene
C. villosa (Pursh) Nutt. ex DC. var. *camporum* (Greene) Cronquist
Pl. 236 a, b; Map 986

Plants perennial herbs, with stout taproots and rhizomes, the rootstock often somewhat woody. Stems 40–140 cm long, slender to more commonly relatively stout, moderately pubescent (especially toward the tip) with a mixture of minute, appressed or curved, slender-based hairs and longer, spreading hairs with expanded, bulbous, pustular bases, also with moderate to dense, minute, sessile or slightly stalked glands. Stem leaves (1–)3–7 cm long, linear to narrowly oblanceolate or oblanceolate, mostly short-tapered to a sharply pointed-tip, the margins entire or with few to less commonly several fine, sharp teeth, hairy with at least some of the hairs relatively long, stout, and spreading, the surfaces appearing green or rarely slightly grayish-tinged, moderately roughened with loosely appressed to somewhat curved or spreading hairs, these all or mostly with an expanded, bulbous base, not glandular or rarely with sparse to moderate, minute, sessile to slightly stalked glands. Involucre 7–10 mm long, the bracts in 3–6 unequal, overlapping series, narrowly lanceolate to nearly linear, sometimes purplish-tinged at the tip, the green central stripe usually easily observed, not hidden by the sparse to moderate, short, curved hairs, not glandular or sparsely to moderately glandular. Ray florets 15–35, the corollas 11–15 mm long. Disc florets 25–65, the corollas 5.0–6.5 mm long, glabrous or occasionally very sparsely and minutely hairy on the outer surface toward the tip. Pappus of the ray and disc florets similar, consisting of an outer series of several bristlelike scales 0.2–1.0 mm long and an inner series of 25–45 bristles 5–7 mm long. Fruits of the ray and disc florets similar, 1.5–4.0 mm long, somewhat flattened, 3- or 4-nerved on each face, the surface moderately hairy. 2n=36. July–October.

Scattered mostly south of the Missouri River (Virginia to North Carolina west to Illinois, Missouri, Arkansas, and Mississippi). Sand prairies, glades, ledges and tops of bluffs, openings of dry upland forests, banks of streams and rivers, bottomland forests, and swamps; also pastures, old fields, fallow fields, levees, railroads, roadsides, and open, disturbed areas, often in sandy soils.

Steyermark (1963) and most earlier authors treated this taxon as a variety of *H. villosa,* but most recent authors (Gleason and Cronquist, 1991; Semple, 1996) have followed Harms (1963, 1968) in accepting it as a separate species. In a fascinating study, Semple (1983) detailed the apparent differentiation of the species into two varieties toward the beginning of the twentieth century (see further discussion below), both of which occur in Missouri.

1. Stems 40–80(–100) cm long, the branches and upper stem portions sparsely to moderately glandular or lacking glands (the glands minute and in addition to the longer, nonglandular hairs) . . . . 1A. VAR. CAMPORUM
1. Stems (60–)80–140 cm long, the branches and upper stem portions moderately to densely glandular (the glands minute and in addition to the longer, nonglandular hairs) . . . . 1B. VAR. GLANDULISSIMUM

1a. var. camporum

Stems 40–80(–100) cm long, moderately hairy, the branches and upper stem portions with sparse to moderate, minute, sessile or slightly stalked glands or lacking glands. Heads with the stalks and involucral bracts moderately hairy, sparsely to moderately glandular or lacking glands. $2n=36$. July–October.

Scattered mostly south of the Missouri River (Missouri, Arkansas, Illinois, Indiana, possibly also Nebraska). Sand prairies, glades, ledges and tops of bluffs, openings of dry upland forests, banks of streams and rivers, bottomland forests, and swamps; also pastures, old fields, fallow fields, levees, railroads, roadsides, and open, disturbed areas.

The var. *camporum* is the less weedy phase of the species. It tends to have somewhat smaller leaves and hairier stems and foliage.

1b. var. glandulissimum Semple

Chrysopsis camporum var. *glandulissimum* (Semple) Cronquist

Stems (60–)80–140 cm long, moderately hairy, the branches and upper stem portions with moderate to dense, minute, sessile or slightly stalked glands or lacking glands. Heads with the stalks and involucral bracts sparsely to moderately hairy, moderately to densely glandular. $2n=36$. July–October.

Uncommon, known thus far only from Christian and Reynolds Counties (Virginia to North Carolina west to Michigan, Missouri, and Mississippi). Roadsides and open, disturbed areas.

Pursuing clues from earlier studies, Semple (1983) discovered that, beginning in about 1925, specimens began to document a rapid expansion of the range of *H. camporum*, especially eastward. The plants involved in this expansion were generally found growing in highly disturbed environments and generally were more robust, less hairy, and more densely glandular than was typical of the species in natural habitats within its original range. Semple hypothesized that at some point in the early twentieth century some populations underwent morphological differentiation that was correlated with increased fitness in disturbed habitats, leading to an overall expansion of the geographical range. He chose to name these plants var. *glandulissimum*. The earliest collection from Missouri was made by Semple in 1991. This variety is to be expected along roadsides in other southern Missouri counties in the future.

2. Heterotheca canescens (DC.) Shinners

Chrysopsis villosa (Pursh) Nutt. ex DC. var. *canescens* (DC.) A. Gray

C. canescens (DC.) Torr. & A. Gray

Pl. 236 c–e; Map 987

Plants perennial herbs, often with taproots and usually also rhizomes, the rootstock often somewhat woody. Stems 15–60 cm long, relatively slender, moderately to densely pubescent with appressed or curved, silvery hairs, not glandular. Stem leaves 1–3 cm long, linear to narrowly oblanceolate or oblanceolate, mostly short-tapered to a sharply pointed tip, the margins entire and hairy, the surfaces appearing gray, not roughened, densely pubescent with appressed or less commonly also curled, silvery hairs, these all or nearly all lacking an expanded, bulbous base, not glandular. Involucre 5–7 mm long, the bracts in 4–6 unequal, overlapping series, narrowly triangular-lanceolate to nearly linear, often purplish-tinged at the tip, the green central stripe often difficult to observe, hidden by the relatively dense, appressed, silvery hairs, not glandular. Ray florets 10–22, the corollas 4–10 mm long. Disc florets 15–45, the corollas 4.5–6.5 mm long, often sparsely and minutely hairy on the outer surface toward the tip. Pappus of the ray and disc florets similar, consisting of an outer series of relatively few bristlelike scales 0.2–0.5 mm long and an inner series of 25–40 bristles 5.0–7.5 mm long. Fruits of the ray and disc florets similar, 1.2–3.0 mm long, somewhat flattened, 3- or 4-nerved on each face, the surface moderately hairy. $2n=18$, 36. August–October.

Uncommon, known only from historical collections from Holt County (Missouri, Kansas, Oklahoma, Texas, and New Mexico). Loess hill prairies.

The Missouri occurrence of this species is unusual. Not only does it represent an eastern disjunction from the main range of the species in the

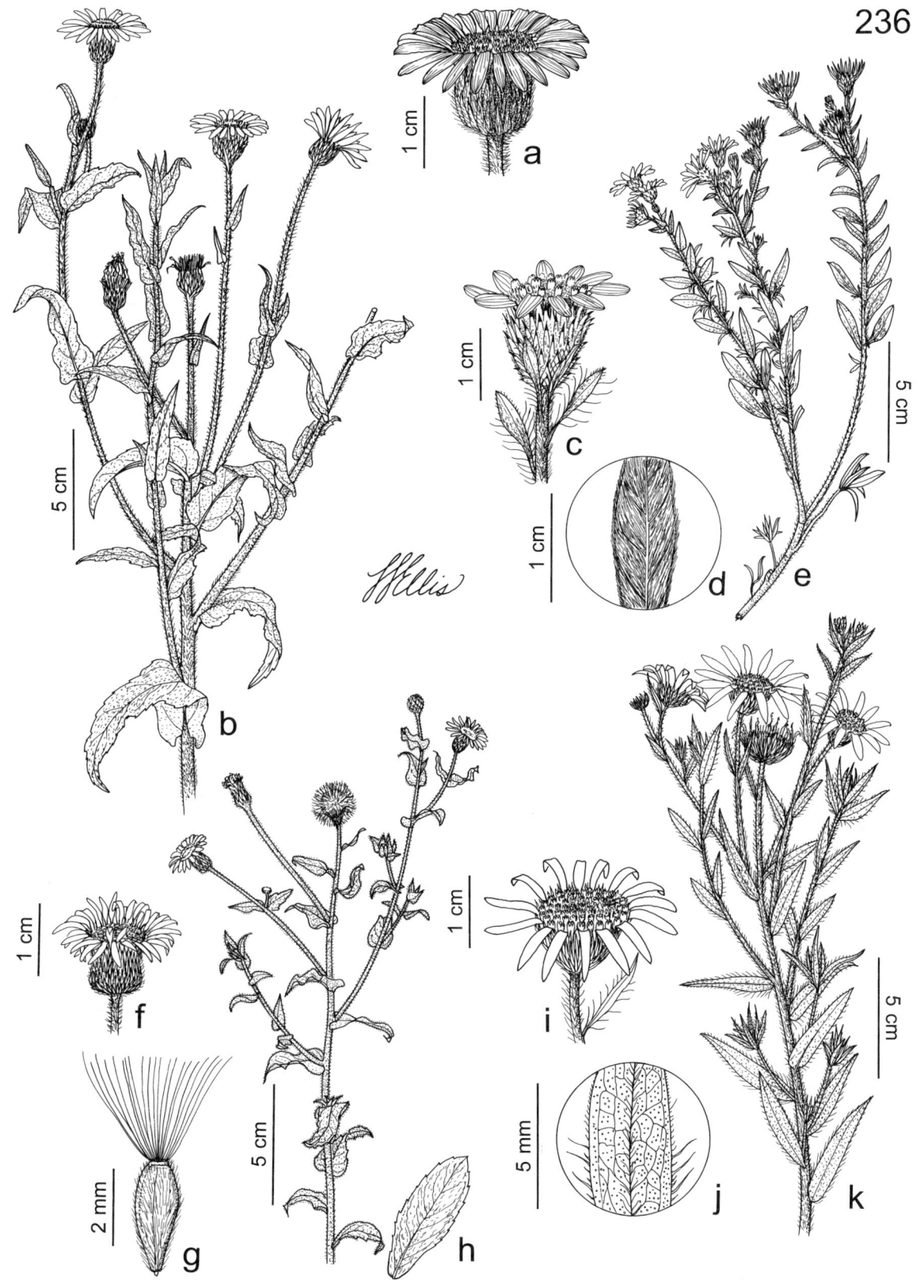

Plate 236. Asteraceae. *Heterotheca camporum*, **a)** head, **b)** habit. *Heterotheca canescens*, **c)** head, **d)** leaf detail, **e)** habit. *Heterotheca subaxillaris*, **f)** head, **g)** fruit, **h)** habit. *Heterotheca stenophylla*, **i)** head, **j)** leaf detail, **k)** habit.

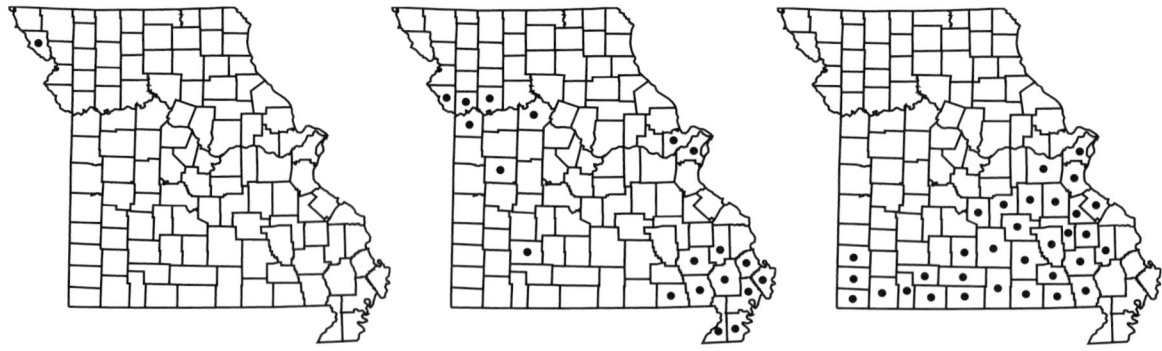

987. Heterotheca canescens   988. Heterotheca subaxillaris   989. Ionactis linariifolius

southern Great Plains, but *H. canescens* has not been reported from elsewhere in the loess hills along the Missouri River or from hill prairies (Semple, 1996). This might lead the reader to believe that the taxon is nonnative in the state, but the brief labels on the more than ample series of specimens examined by Steyermark (1963) and later botanists seem to indicate that it was part of the natural landscape when it was first discovered there by John Kellogg and B. F. Bush in 1931. The species has not been rediscovered in Missouri since that time and has probably become extirpated from the state.

Steyermark (1963) reported *Chrysopsis villosa* var. *angustifolia* (Rydb.) Cronquist as a member of the Missouri flora and cited a single historical specimen accessioned in the herbarium of the New York Botanical Garden from Greene County (Springfield, July 1905, *P.C. Standley s.n.*). He apparently did not personally examine the specimen, stating that the report was based on information provided to him by William J. Dress, who was completing a doctoral dissertation on the group (Dress, 1954). Vernon L. Harms (1963, 1968), who also completed doctoral studies on the group, transferred the taxon into *Heterotheca* (*H. villosa* (Pursh) Shinners var. *angustifolia* (Rydb.) V.L. Harms) but did not cite the specimen or include Missouri in the range of this mostly Great Plains species. Following the systematic studies of John C. Semple (1996), the taxon currently is known as *H. stenophylla* (A. Gray) Shinners var. *angustifolia* (Rydb.) Semple, but Semple also did not include Missouri in its range. Attempts to locate the original voucher specimen during the present work were unsuccessful, leading to the conclusion that this specimen either was miscited by Dress or misdetermined. *H. stenophylla* var. *angustifolia* is similar to *H. canescens,* but it differs in its somewhat larger leaves with less dense, more spreading hairs and heads with slightly larger involucres, among other characters (Pl. 236

i–k). It is not very likely to be found in southwestern Missouri.

3. **Heterotheca subaxillaris** (Lam.) Britton & Rusby (camphorweed, telegraph plant)
H. latifolia Buckley
H. subaxillaris ssp. *latifolia* (Buckley) Semple
H. latifolia var. *arkansana* B. Wagenkn.
Pl. 236 f–h; Map 988

Plants annual or biennial, with taproots. Stems 15–150(–200) cm long, usually relatively stout, sparsely to more commonly moderately to densely pubescent with relatively long, spreading hairs mostly with expanded, bulbous, pustular bases, also with moderate to dense, minute, sessile or slightly stalked glands. Stem leaves 1–9 cm long, lanceolate to elliptic or ovate, some of the lower leaves sometimes narrowed somewhat above the base and expanded again below the midpoint, mostly shallowly cordate at the clasping base (the lowermost leaves occasionally tapered to a short petiole and with a separate pair of auricles clasping the stem), angled or short-tapered to a bluntly or more commonly sharply pointed tip, the margins mostly sharply toothed, those of the upper leaves sometimes only irregularly wavy or entire, also hairy, the surfaces appearing green or slightly grayish-tinged, moderately roughened with loosely appressed to somewhat curved hairs, these all or mostly with an expanded, bulbous base, also with sparse to dense, minute, sessile to slightly stalked glands. Involucre 5–10 mm long, the bracts in 4–6 unequal, overlapping series, narrowly lanceolate to nearly linear, the green central stripe usually easily observed, not hidden by the sparse to moderate, short, curved hairs, these mostly along the midvein, moderately glandular. Ray florets 15–40, the corollas 5–10 mm long. Disc florets 25–60, the corollas 3–9 mm long, glabrous. Pappus of the ray florets absent, that of the disc florets consisting of an outer series of several bristlelike scales 0.2–0.6 mm long and an inner series of 25–45 bristles 4–9 mm long. Fruits 2–4 mm long, obscurely 3- or 4-nerved, those of the ray florets 3- or 4-angled, glabrous, those of the disc florets somewhat flattened, the surface moderately hairy. $2n=18$. July–November.

Scattered, mostly in the Mississippi Lowlands Division and in counties bordering the Missouri and Mississippi Rivers (nearly throughout the U.S. except some northwestern states; Mexico). Sand prairies and loess hill prairies; also fallow fields, margins of crop fields, banks of ditches, railroads, roadsides, and open, disturbed areas, especially in sandy soil.

Some botanists have accepted three species in this complex: true *H. subaxillaris* of the southeastern Coastal Plain, the widespread *H. latifolia,* and the southwestern *H. psammophila* B. Wagenkn. (Wagenknecht, 1960; Steyermark, 1963; Semple, 1996). Other studies have tended to emphasize the overall similarities and innate variability of these plants in various habitats and soil types (Burk, 1961; Harms, 1965b; Nesom, 1990d; Semple, 2004), and a single-species classification without any additional varieties or subspecies is accepted here.

24. Ionactis Greene

About 5 species, U.S., Canada.

Ionactis, with its short, more or less woody rootstocks, narrow, relatively uniform stem leaves, and relatively few heads, is another small, morphologically cohesive taxon that was recognized by most botanists as a section of the genus *Aster* until recent years. Nesom and Leary (1992) discussed the generic delimitation of *Ionactis,* and its distinctness was discussed by Nesom (1994, 2000). Molecular studies (Noyes and Rieseberg, 1999; Semple et al., 2002) also have reaffirmed the distinctness of the group.

1. Ionactis linariifolius (L.) Greene (stiff aster, stiff-leaved aster, flax-leaved aster)

Aster linariifolius L.

Pl. 229 a, b; Map 989

Plants perennial herbs, with a short, thick, somewhat woody rootstock, rarely also with slender rhizomes. Stems 1 to several, 10–50(–70) cm long, erect or ascending from a sometimes somewhat spreading base, usually unbranched, nearly smooth or with fine, longitudinal lines, nearly glabrous or sparsely pubescent with minute, curled hairs toward the base, moderately to densely and minutely hairy toward the tip. Basal leaves absent at flowering. Stem leaves more or less uniform in size and spacing, 1–4 cm long, 0.5–4 mm wide, linear or very narrowly oblong-oblanceolate, stiff and somewhat leathery, narrowed or short-tapered to a sharply pointed tip, abruptly rounded to a sessile, nonclasping base, the margins entire, the surfaces glabrous (the upper surface often somewhat shiny), the margins strongly roughened with moderate to dense, minute, stiff, triangular hairs. Inflorescences of solitary heads or racemes with relatively few heads, but rarely somewhat branched and appearing as rounded to flat-topped panicles of up to 30 heads, the heads nearly sessile to long-stalked, the bracts relatively dense, 0.3–1.2(–2.5) cm long, herbaceous but noticeably shorter than the adjacent leaves, linear. Heads radiate, not sticky or resinous. Involucre 6–9 mm long, cup-shaped to somewhat obconical. Involucral bracts in 4–7 unequal, overlapping series, oblong-lanceolate to lanceolate-triangular, the tip ascending, with a thickened, slender, green central band that is broadened and narrowly elliptic toward the tip (the green color sometimes absent below the midpoint), the broad, lighter margins often with a purple border, especially toward the tip, the margins appearing finely hairy, especially toward the tip, the outer surface glabrous. Receptacle flat or shallowly convex, with minute, irregular ridges around the concave attachment points of the florets. Ray florets 6–20, pistillate, the corolla 5–12 mm long, lavender to purple (rarely white elsewhere). Disc florets 20–40, perfect, the corolla 4–7 mm long, relatively deeply lobed (0.3–0.7 mm), yellow, sometimes turning reddish purple after the pollen has been shed, not persistent at fruiting. Pappus of the ray and disc florets similar, of 2 types, an inner series of numerous finely barbed bristles 4–7 mm long, these not broadened toward the tip, and an outer series of fewer but still relatively numerous bristles 0.2–0.8 mm long, both series straw-colored to light orangish tan. Fruits 2.0–3.5 mm long, narrowly obconical, somewhat 4–6-angled, the angles with yellowish brown ribs, densely silky-hairy, dark brown to purplish brown. $2n=18$. August–November.

Scattered in the Ozark and Ozark Border Divisions (eastern U.S. west to Wisconsin, Kansas, and Texas; Canada). Mesic to dry upland forests (usually on rocky slopes), savannas, ledges of bluffs, and margins of glades; also roadsides.

Steyermark (1963) noted that this attractive clump-forming species was a desirable addition to rock gardens but that it required acidic soils for best growth. In recent years, it has become available commercially through native plant nurseries. White-rayed plants that eventually may be found in Missouri have been called f. *leucactis* Benke.

25. Solidago L. (goldenrod)

Plants perennial herbs, often with short- to long-creeping rhizomes, less commonly with a short, branched, somewhat woody rootstock. Stems 1 to several, erect, ascending, or less commonly spreading to pendant, usually with few to many ascending to spreading branches above the midpoint, with fine to coarse, longitudinal ridges, glabrous or variously hairy. Basal and lower stem leaves either absent at flowering or present (and then the largest leaves on the plant), sessile or short- to long-petiolate. Stem leaves gradually or relatively abruptly reduced toward the tip of the stem either from about the midpoint or from the base, sessile or short-petiolate, the blade variously shaped, usually sharply pointed at the tip, tapered to a slender or slightly expanded but not clasping base or less commonly truncate to rounded and slightly clasping the stem, the margins entire to coarsely toothed, sometimes also roughened with minute, stout, ascending, stiff hairs, the surfaces glabrous or variously hairy. Inflorescences variously consisting of axillary clusters, or terminal, then either narrow and racemelike with clusters along the main axis or open to dense, pyramidal or flat-topped panicles, the heads solitary or more commonly in small clusters, sometimes all or mostly oriented upward (in species with pyramidal panicles), the stalks subtending the heads mostly short, usually with relatively few minute, scalelike, linear to narrowly oblong bracts to 0.4 mm long, the inflorescence branches often also with reduced, leaflike bracts conspicuously shorter than the main foliage leaves. Heads radiate. Involucre 2–8 mm long, cup-shaped to nearly cylindrical. Involucral bracts in 3–6 usually unequal, overlapping series, variously shaped, the ascending tips (the lower series sometimes spreading in *S. petiolaris*) rounded to bluntly or sharply pointed, the midvein 1 or less commonly 3–7 and often slightly thickened and translucent, entirely straw-colored to light yellow or with an often short, elliptic to obovate green area toward the tip, glabrous or hairy, not resinous. Receptacle flat or slightly convex, with low, toothlike ridges around the attachment points of the florets, uncommonly with a few chaffy bracts toward the margins. Ray florets (1)2–15, pistillate, the corolla well developed but relatively short, mostly spreading, yellow (white in *S. ptarmicoides*; see also the discussion of *S. bicolor* at the beginning of the Astereae treatment), not persistent at fruiting. Disc florets 2–35, perfect, the corolla relatively short, yellow (white in *S. ptarmicoides*), not persistent at fruiting. Pappus of the ray and disc florets similar, of numerous (25–45) slender, finely barbed bristles about as long as the corollas, the innermost sometimes slightly thickened toward the tip, white or off-white. Fruits (0.5–)1.0–4.0 mm long, obovoid to more commonly narrowly obovoid to nearly cylindrical, usually slightly flattened, often slightly several-angled in cross-section, 5–8(–10)-nerved, the surface glabrous or hairy, straw-colored to brown. About 100 species, North America, South America, Europe, Asia.

The generic concept of *Solidago* accepted here is broader than that embraced by some other botanists (Nesom, 1993a, 2000). In particular, some authors prefer to segregate *Oligoneuron* Small for six eastern North American species with generally flat-topped inflorescences, relatively broad involucral bracts with more or less parallel veins, and minute differences in achene anatomy. In Missouri, this generic controversy involves three morphologically unusual species,

S. ptarmicoides, S. riddellii, and *S. rigida.* Anderson and Creech (1975) felt that the two groups should be united based on their investigation of leaf anatomy. Nesom's (1993a) morphological analysis of the complex yielded ambiguous results, as did a recent molecular analysis of ITS sequences in the tribe (Beck et al., 2004). However, molecular studies of chloroplast DNA variation by Zhang (1996) supported the hypothesis that the *Oligoneuron* species are members of the *Solidago* lineage, a conclusion accepted in the floristic treatment of Ontario goldenrods by Semple et al. (1999), as well as most other recent floristic manuals for areas surrounding Missouri (Barkley, 1986; Gleason and Cronquist, 1991). The present treatment has benefited from early discussions with John Semple of the University of Waterloo concerning generic, specific, and infraspecific limits.

Two aids to identification in this morphologically variable genus are the inflorescence type and the arrangement of the leaves. Missouri species are separable into three main inflorescence types. A small group that some botanists segregate as the genus *Oligoneuron* (see the preceding paragraph) comprises species with relatively flat-topped or rounded, paniculate inflorescences having the heads in small clusters at the branch tips. Another group has either axillary clusters of heads or narrow, elongate terminal inflorescences that often appear spicate or racemose with the heads in small clusters along the main axis, but can have the heads oriented in several directions along short branches of a narrow panicle. The third group has often dense panicles that have been described in the literature as pyramidal. This term refers to an elongate inflorescence, sometimes nodding at the tip, with at least the lower branches arching outward or nodding, and the longest branches usually toward the inflorescence base with progressively shorter branches toward the tip. The heads are all or mostly oriented toward the upper side of each racemose branch. In a few species (like *S. ulmifolia*), the branches tend to be relatively widely separated along the main axis, and because the axis and branches appear leafy, these lowest branches might be interpreted as being separate inflorescences. Within this volume, to maintain continuity across the treatment of *Solidago,* references to a pyramidal shape apply to the overall branching pattern and disposition of heads on a single main flowering stem. Occasional depauperate individuals with small inflorescences can cause problems with the determination of all of the goldenrod species, but in such cases the orientation and position of the heads relative to the branches can be helpful in discriminating among the groups.

There are two main forms of leaf arrangement, and collectors should be careful either to gather the stem base as part of a specimen or to record the information on basal leaves carefully in the field. In some species, the basal leaves are well represented at flowering (in fact there often are few to several additional basal rosettes in proximity to the flowering stem), the basal and adjacent lower stem leaves the largest leaves on the plant, and the leaves gradually reduced from the stem base toward the tip. Some authors have referred to this leaf arrangement pattern as basally disposed (Cronquist 1980, 1991). In contrast, other species have the basal leaves withered and usually absent at flowering. The largest leaves occur from about $1/3$–$1/2$ the way up the stem. This leaf arrangement is sometimes referred to as chiefly cauline.

Some tribes of Native Americans used goldenrods in the treatment of colds, congestion, fever, cramps, and heart ailments (Moerman, 1998). Because they are showy when in flower and often grow in proximity to wind-pollinated plants like ragweeds (*Ambrosia,* Asteraceae) and pigweeds (*Amaranthus,* Amaranthaceae) that flower at the same time of year, goldenrods have developed an undeserved reputation as causing hay fever. However, all of the goldenrods have sticky pollen that is dispersed by a variety of different insects. A number of species are attractive ornamentals in the garden and are available through wildflower nurseries. Cultivars of a few species, like *S. canadensis,* are more widely available in the nursery trade. Gardeners should note that some of the most widespread Missouri species, like *S. altissima,* can become very aggressive in the garden.

1. Inflorescences terminal panicles (occasionally appearing as a dense terminal cluster), appearing flat-topped or rounded in overall outline, the heads solitary or in small clusters at the branch tips
    2. Disc and ray florets white or less commonly pale cream-colored (note that the disc florets usually appear yellow because of the yellow stamens, but the corollas are white) .... 17. S. PTARMICOIDES
    2. Disc and ray florets yellow
        3. Leaves relatively narrow, the blade linear to narrowly lanceolate or narrowly oblanceolate, the margins entire; stems glabrous below the inflorescence, shiny .... 19. S. RIDDELLII
        3. Leaves relatively broad, the blade broadly oblanceolate to elliptic-obovate, ovate, or oblong-elliptic, the margins of all but the uppermost leaves usually finely scalloped or minutely and bluntly toothed; stems densely short-hairy (except in the rare var. *glabrata*), not shiny .... 20. S. RIGIDA
1. Inflorescences either consisting only of axillary clusters or, if terminal, then elongate and appearing as spikelike racemes or pyramidal panicles; if paniculate, then the heads mostly oriented upward and single or in small clusters along the branches
    4. Inflorescences either consisting of axillary clusters or, if terminal, then narrow with small clusters or spikelike branches along the main axis, the branches not arching or nodding, the heads oriented in several directions
        5. Basal and lowermost stem leaves the longest on the plant, usually persistent and conspicuous at flowering
            6. Stems and leaves conspicuously hairy .... 10. S. HISPIDA
            6. Stems glabrous below the inflorescence or nearly so; leaves glabrous on the surfaces, the margins inconspicuously hairy ...... 22. S. SPECIOSA
        5. Basal and lowermost stem leaves somewhat smaller than the longest ones, which occur about $1/3–1/2$ the way up the stem, the lowermost leaves withered or more commonly absent at flowering
            7. Stems sparsely to moderately short-hairy, at least above the midpoint; at least the outermost involucral bracts usually loosely ascending or with the tips somewhat spreading to recurved
                8. Leaf blades relatively broad (the largest 25–50 mm wide) and thin-textured, the margins usually sharply toothed, the undersurface moderately pubescent with short (0.4–0.7 mm long) hairs along the main veins .... 3. S. BUCKLEYI
                8. Leaf blades relatively narrow (the largest 5–35 mm wide) and thick-textured, the margins usually entire or sparsely and minutely toothed above the midpoint, the undersurface glabrous or sparsely to moderately pubescent with minute (0.1–0.4 mm long) hairs along the main veins .... 16. S. PETIOLARIS
            7. Stems glabrous except sometimes along the inflorescence branches; involucral bracts appressed-ascending
                9. Leaves broadly ovate to broadly elliptic-ovate, rounded then tapered abruptly at the base to a relatively long, winged petiole (except sometimes the uppermost stem leaves); stems often strongly zigzag, especially toward the tip .... 7. S. FLEXICAULIS
                9. Leaves (except sometimes those of the basal and lowermost stem leaves, which may be broader) lanceolate, oblong-lanceolate, elliptic, or narrowly ovate, tapered gradually at the base, sessile or short-petiolate (except sometimes the basal and lowermost stem leaves); stems not or only slightly zigzag

10. Inflorescences axillary clusters (occasionally a few of the
clusters somewhat elongate); stems glaucous . . . . 4. S. CAESIA
10. Inflorescences terminal panicles (sometimes appearing
racemose or spicate); stems not glaucous . . . . 22. S. SPECIOSA
4. Inflorescence a more or less pyramidal terminal panicle (see discussion above),
often somewhat nodding at the tip, at least the lower branches arching or nodding, the heads mostly oriented upward along the racemose branches
11. Basal and lowermost stem leaves the longest on the plant, usually persistent
and conspicuous at flowering
12. Stems moderately to densely pubescent with curved to spreading hairs
(the pubescence sometimes slightly less dense toward the stem base)
13. Stems and leaves densely pubescent with minute (0.1–0.3 mm),
mostly curved hairs . . . . 13. S. NEMORALIS
13. Stems and leaves moderately pubescent with longer (mostly 0.5–1.5
mm), mostly spreading hairs
14. Disc florets 8–20, the corollas 3.5–4.0 mm long . . . . 2. S. ARGUTA
14. Disc florets 4–7, the corollas 2–3 mm long . . . . 23. S. ULMIFOLIA
12. Stems below the inflorescence glabrous or sparsely pubescent with
mostly spreading hairs
15. Leaves with the upper surface strongly roughened (sandpapery) with
minute, stiff, stout, broad-based (pustular) hairs . . . . 15. S. PATULA
15. Leaves with the upper surface glabrous, soft-hairy, or sparsely to
moderately roughened with short, stiff hairs having slender to stout,
nonpustular bases
16. Basal and lower stem leaves with the blade broadly ovate to
broadly elliptic, tapered abruptly to the winged petiole . . . 2. S. ARGUTA
16. Basal and lower stem leaves with the blade narrowly oblanceolate to narrowly elliptic-obovate, tapered gradually to the
winged petiole
17. Leaf blades with only 1 main vein or, if 3-veined, then the
lateral pair poorly developed and only faintly visible near the
leaf base
18. Basal and lower stem leaves tapered gradually to the
petiole; ray florets 7–12, the corolla 2.0–2.5 mm long; disc
florets 8–15 . . . . 11. S. JUNCEA
18. Basal and lower stem leaves tapered relatively abruptly
to the petiole; ray florets 3–5, the corolla 1.5–2.0 mm long;
disc florets 4–7 . . . . 23. S. ULMIFOLIA
17. Leaf blades (at least of the basal and lower stem leaves) with
3 main veins for most of the length of the blade, the lateral
pair well developed but sometimes thinner than the midvein
19. Rootstock short and branched, not producing creeping
rhizomes; ray florets 5–8; receptacle naked; nodes of the
flowering stem usually lacking small clusters of leaves in
the axils of the main leaves . . . . 8. S. GATTINGERI
19. Rootstock producing creeping rhizomes, often also short
and thickened at the stem bases; ray florets 7–13; receptacle frequently with a few slender, chaffy bracts toward
the margin; median and upper nodes of the flowering
stem usually producing small clusters of leaves in the
axils of the main leaves . . . . 12. S. MISSOURIENSIS

11. Basal and lowermost stem leaves somewhat smaller than the longest ones, which occur about 1/3–1/2 the way up the stem, the lowermost leaves withered or more commonly absent at flowering
    20. Stem leaves mostly short-petiolate (the uppermost leaves sometimes sessile or nearly so), the blade 1–2 times as long as wide . . . . 6. S. DRUMMONDII
    20. Stem leaves mostly sessile (the lowermost leaves sometimes short-petiolate), the blade 2–8 times as long as wide
        21. Leaf blades with only 1 main vein or, if 3-veined, then the lateral pair poorly developed and only faintly visible near the base, the secondary veins, if visible, arranged pinnately along the midvein
            22. Leaf blades with minute, impressed, translucent dots (these best observed on the undersurface under magnification while the leaf is held to a strong light), the margins entire . . . . 14. S. ODORA
            22. Leaf blades lacking impressed dots, the margins of some or all of the leaves with at least a few shallow teeth
                23. Rootstock producing creeping rhizomes, often also short and thickened at the stem bases; ray florets 6–11; leaf blades mostly angled at the base, not tapered . . . . 21. S. RUGOSA
                23. Rootstock short and branched, not producing creeping rhizomes; ray florets 3–5; leaf blades mostly fairly abruptly tapered at the base . . . . 23. S. ULMIFOLIA
        21. Leaf blades with 3 main veins, the lateral pair well developed and only slightly thinner than the midvein, conspicuous in at least the basal 1/2 of the length of the blade
            24. Stems glabrous below the inflorescence, usually glaucous . . . . 9. S. GIGANTEA
            24. Stems hairy throughout or sometimes nearly glabrous toward the base
                25. Stem leaves below the inflorescence with the blades relatively broad, 2–6 times as long as wide; ray florets 4–8; stems 40–120 cm long but mostly shorter than 100 cm . . . . 18. S. RADULA
                25. Stem leaves below the inflorescence with the blades relatively narrow, 5–13 times as long as wide; ray florets 6–15; stems 30–250 cm long but usually longer than 100 cm (except sometimes in *S. altissima* var. *gilvocanescens* and in plants growing in extreme habitats such as railroad cinders)
                    26. Involucre (2.5–)3.0–4.5 mm long; ray florets 10–16, the corolla 3–4 mm long; disc florets 3–7, the corolla 3.0–3.5 mm long . . . . 1. S. ALTISSIMA
                    26. Involucre 2–3 mm long; ray florets 6–12, the corolla 2–3 mm long; disc florets 2–5, the corolla 2.3–2.7 mm long . . . . 5. S. CANADENSIS

1. Solidago altissima L. (tall goldenrod)

Pl. 241 a–c, g–j; Map 990

Plants with branched short- to long-creeping rhizomes, often also thickened at the stem bases. Stems 1 to several, 30–250 cm long (but usually longer than 100 cm), erect or ascending, with several fine, longitudinal lines or grooves, moderately to densely pubescent with short, curved hairs, sometimes less densely hairy toward the base, not shiny, not glaucous. Leaves chiefly cauline, the largest leaves in the lower 1/3–1/2 of the stem, the basal leaves usually absent at flowering. Basal and lowermost stem leaves with the blade 3–15 cm long, 0.5–2.5 cm wide, mostly 5–13 times as long as wide, narrowly oblanceolate to narrowly elliptic-lanceolate or narrowly lanceolate, somewhat thickened and stiff, angled or more commonly tapered gradually to a sessile or very short-petiolate base, angled

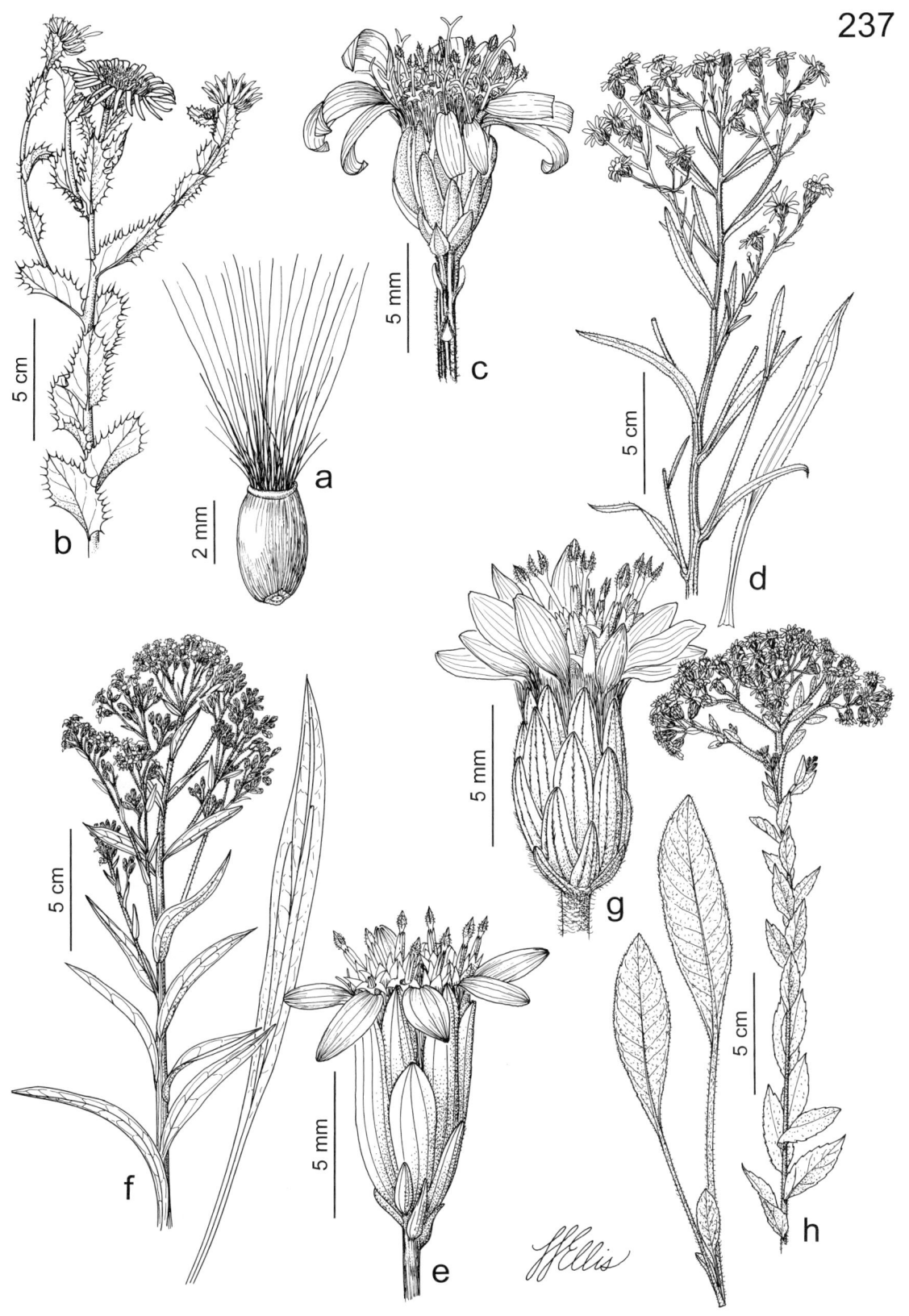

Plate 237. Asteracae. *Grindelia ciliata*, **a)** fruit, **b)** habit. *Solidago ptarmicoides*, **c)** head, **d)** habit and lower leaf. *Solidago riddellii*, **e)** head, **f)** habit and lower leaf. *Solidago rigida*, **g)** head, **h)** habit and lower leaves.

990. Solidago altissima   991. Solidago arguta   992. Solidago buckleyi

or tapered to a sharply pointed tip, the margins sharply toothed to nearly entire and usually microscopically roughened or minutely hairy, sometimes slightly curled under, the upper surface moderately to densely pubescent with short, curved to spreading hairs or sparsely to moderately roughened with minute, bulbous-based hairs, the undersurface moderately to densely pubescent with short, spreading hairs (these also sometimes slightly bulbous-based), with 3 main veins, the lateral pair often originating well above the leaf base, finer than the midvein, the veinlets often difficult to observe, forming an irregular, dense network. Median and upper stem leaves 1–15 cm long, narrowly elliptic to narrowly lanceolate or nearly linear, sessile, the margins of the uppermost leaves often entire, otherwise similar to the lower stem leaves. Inflorescences open to more commonly relatively dense, usually broad, often large, pyramidal panicles, the longer branches and often also the tip arched or nodding, the heads oriented upward along the branches. Involucre (2.5–)3.0–4.5 mm long, the bracts in 3–6 unequal series. Involucral bracts lanceolate to narrowly oblong-lanceolate and bluntly to more commonly sharply pointed at the appressed-ascending tip, the margin sparsely hairy, the outer surface glabrous, entirely yellowish or with a poorly differentiated, narrow, green central region, this tapered gradually to the midvein above or below the bract midpoint, the midvein somewhat thickened and no additional veins present. Receptacle naked. Ray florets 10–16, the corollas 3–4 mm long, yellow. Disc florets 3–7, the corollas 3.0–3.5 mm long, the lobes 0.5–0.9 mm long, yellow. Pappus 2.8–3.5 mm long, some of the bristles slightly thickened toward the tip. Fruits 0.6–1.2 mm long, narrowly obovoid, finely hairy. $2n=18, 36, 54$. August–November.

Common throughout the state (eastern U.S. west to Montana and New Mexico; Canada, Mexico; introduced in the southwestern U.S., Hawaii, Europe, Asia). Upland prairies, loess hill prairies, sand prairies, mesic to dry upland forests, savannas, banks of streams and rivers, and rarely fens, also pastures, old fields, fallow fields, margins of crop fields, fencerows, gardens, railroads, roadsides, and open, disturbed areas.

Solidago altissima is very widespread and common in the state. Some authors (Croat, 1972; Cronquist, 1980; Barkley, 1986; Gleason and Cronquist, 1991) have treated *S. altissima* as part of the variation within a broadly defined *S. canadensis*. In general, that species has slightly smaller heads (this determined by an average involucre length of several heads measured at flowering) and stems that tend to be nearly glabrous toward the base. Because each of the species has been divided into multiple infraspecific taxa and because chromosome counts indicate the presence of multiple ploidies within each species, more detailed taxonomic studies involving molecular and population-genetic data are needed. The treatment here follows that of Semple et al. (1984, 1999) in recognizing two of the apparent main evolutionary lineages within the complex as separate species, but it should be viewed as provisional.

Interestingly, some groups of insects are very adept at discriminating between *S. altissima* and other members of the *S. canadensis* complex. The rich and diverse insect fauna that feeds on these plants has been studied by a number of entomologists (for example, Fontes et al., 1994). Frequently, plants of *S. altissima* are encountered with conspicuous, large, swollen areas along the main stem, these sometimes associated with abnormally developed leaves. Stem galls on this host are mostly produced by small flies (midges) and moths whose larvae mature within the galls (leaving a small, circular opening when the mature insect emerges). The tephritid fly *Eurosta solidaginis* Fitch

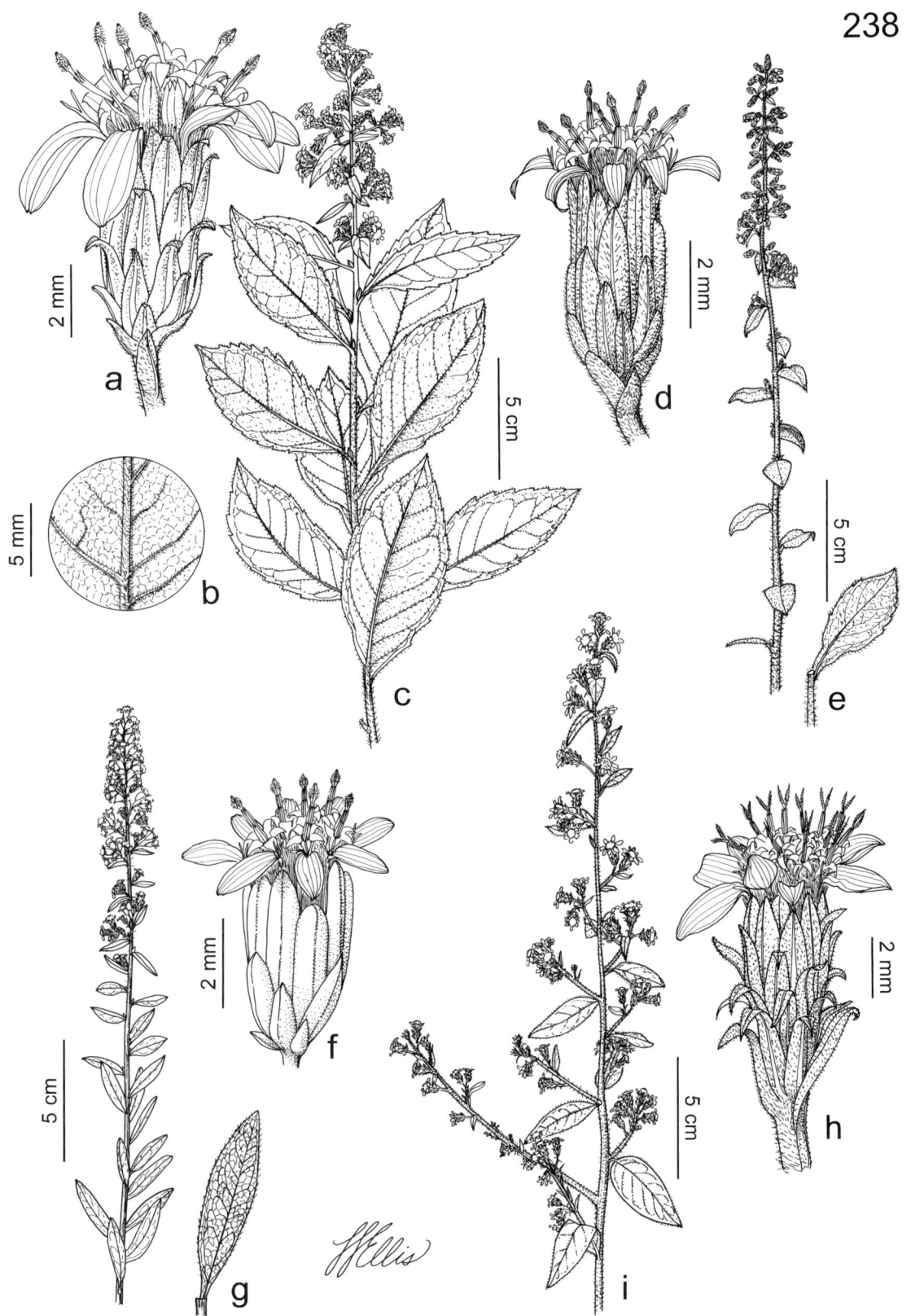

Plate 238. Asteraceae. *Solidago buckleyi*, **a)** head, **b)** leaf detail, **c)** habit and leaf. *Solidago hispida*, **d)** head, **e)** habit and larger leaf. *Solidago speciosa*, **f)** head, **g)** habit and larger leaf. *Solidago petiolaris*, **h)** head, **i)** habit.

(goldenrod stem galler) produces more or less spherical, smooth galls, whereas leafy galls are produced by a different species, the cecidomyid *Rhopalomyia solidaginis* (Loew) (goldenrod bunch galler). The gelechiid moth *Gnorimoschema gallaesolidaginis* (Riley) (goldenrod gall moth) produces relatively leafless galls that are more elliptic (spindle-shaped) in outline. The same insects only rarely create stem galls on the closely related *S. canadensis* and *S. gigantea*. The biology of these parasitic associations has been studied for many years (see Stinner and Abrahamson [1979] and Cronin and Abrahamson [2001] for literature reviews; see Semple et al. [1999] for additional citations) and is a popular topic for student exercises in field ecology classes.

There is some morphological overlap between the two varieties treated here.

1. Leaf blades with the upper surface sparsely to moderately roughened with minute, bulbous-based hairs (stouter and somewhat shorter than those of the undersurface), sometimes only along the veins . . . . 1A. VAR. ALTISSIMA
1. Leaf blades with the upper surface moderately to densely pubescent with short, curved to spreading hairs (similar to those of the undersurface) . . . . 1B. VAR. GILVOCANESCENS

1a. var. altissima

Pl. 241 h–j

S. canadensis L. var. *scabra* (Muhl. ex Willd.) Torr. & A. Gray

Plants usually with long-creeping rhizomes. Stems mostly 100–250 cm long. Largest leaves usually 6–15 cm long. Leaf blades with the margins often finely (and sometimes sparsely) toothed, the upper surface sparsely to moderately roughened with minute, bulbous-based hairs (stouter and somewhat shorter than those of the undersurface), sometimes only along the veins, tending to be relatively thick and stiff. $2n=36, 54$. August–November.

Common throughout the state (eastern U.S. west to Minnesota, Nebraska, and Texas; Canada, Mexico; introduced in the southwestern U.S., Hawaii, Europe, Asia). Upland prairies, loess hill prairies, sand prairies, mesic to dry upland forests, savannas, banks of streams and rivers, and rarely fens, also pastures, old fields, fallow fields, margins of crop fields, fencerows, gardens, railroads, roadsides, and open, disturbed areas.

This is the common phase of the species in Missouri.

1b. var. gilvocanescens (Rydb.) Semple

S. canadensis var. *gilvocanescens* Rydb.

Pl. 241 a–c, g

Plants often with short-creeping rhizomes. Stems 30–120 cm long. Largest leaves usually 3.5–7.5 cm long. Leaf blades with the margins often relatively coarsely and sharply toothed, the upper surface moderately to densely pubescent with short, curved to spreading hairs (similar to those of the undersurface), tending to be less strongly thickened and more pliable. $2n=18, 36$. August–November.

Scattered mostly outside the Ozark Division (Illinois and possibly Kentucky to Montana south to Texas and New Mexico; Canada). Upland prairies, mesic to dry upland forests, savannas, and banks of streams and rivers, also pastures, old fields, margins of crop fields, gardens, railroads, roadsides, and open, disturbed areas.

2. Solidago arguta Aiton

Pl. 239 a, b; Map 991

Plants with the rootstock short, stout, sometimes horizontal, branched or with offsets. Stems 1 or occasionally few, 40–120 cm long, erect or ascending, with several fine, longitudinal ridges or grooves but not noticeably angled, glabrous below the inflorescence or sparsely to moderately pubescent with spreading and/or curved hairs, not or only slightly shiny, not glaucous. Leaves basally disposed and usually persistent at flowering (additional rosettes usually present adjacent to the flowering stem). Basal and lowermost stem leaves with the blade 8–30 cm long, 4–10 cm wide, mostly 2–4 times as long as wide, broadly ovate to broadly elliptic, somewhat thickened, tapered fairly abruptly to the winged petiole at the base, rounded or more commonly tapered to a bluntly or sharply pointed tip, the margins finely to coarsely and sharply toothed, microscopically roughened, the upper surface glabrous or slightly to moderately roughened with forward-angled minute, stiff, stout hairs lacking a pustular base, the undersurface glabrous or moderately pubescent with mostly spreading hairs, with 1 main vein, the fine, pinnate secondary veins easily observed (these usually forming an irregular network). Median and upper stem leaves 1–8 cm long, lanceolate to elliptic or oblanceolate, sessile or short-petiolate, otherwise similar to the lower stem leaves. Inflorescences relatively dense and narrow to more commonly open and broad, pyramidal panicles (the lower branches sometimes elongate), the longer branches and occasionally also the tip usually somewhat arched or nodding, the heads oriented upward along the branches. Involucre 4–6 mm long, the bracts in 3–5 unequal series. Involucral bracts oblong-ovate to narrowly oblong and mostly rounded to bluntly pointed at the appressed-ascending tip (those of the outer series sometimes sharply pointed), the thin, white to yellowish white margins hairy, the outer surface glabrous, with a poorly differentiated green to light green central region mostly above the midpoint, this tapered gradually or not at all to the midvein, the midvein not or only slightly thickened, and no additional veins present. Receptacle naked. Ray florets 2–8, the corollas 4–6 mm long, yellow. Disc florets 8–20, the corollas 3.5–4.0 mm long, the lobes 0.5–1.5 mm long, yellow. Pappus 3.0–3.5 mm long, some of the bristles slightly thickened toward the tip. Fruits 1.5–2.0 mm long, narrowly obovoid, glabrous or finely hairy. $2n=18, 36$. June–October.

Scattered in the Ozark and Ozark Border Divisions (eastern U.S. west to Illinois and Texas; Canada). Mesic to dry upland forests, glades, ledges and bases of bluffs, banks of streams and rivers, and occasionally bottomland forests.

Steyermark (1963) noted that this attractive species is a desirable addition to the wildflower garden. The taxonomy and nomenclature of the *S. arguta* group are complex and still not fully understood. Steyermark (1963) based his treatment of the group on extensive field and herbarium studies in the state, concluding that four elements could be recognized in Missouri, which he treated as varieties. Morton (1974) summarized and amended the taxonomic and nomenclatural conclusions developed in his doctoral research (Morton, 1973), which involved intensive field and herbarium studies across the range of the complex. He argued that the rare *S. arguta* var. *neurolepis* (Fernald) Steyerm. (described from plants collected at Oronogo, Jasper County, but also known from St. Francois County and to be expected elsewhere) actually represents a putative hybrid between *S. arguta* and *S. ulmifolia,* which he called *S.* ×*neurolepis* Fernald. Morton (1973, 1974) also examined the type specimen of *S. boottii* Hook. and concluded that this name had been misapplied. Steyermark (1963) and others had applied this name to plants with glabrous leaves and called plants with the leaf undersurface hairy *S. arguta* var. *strigosa.* However, the type specimen of *S. boottii* is a hairy-leaved plant, and the oldest name for this pubescent phase is var. *boottii.* The next oldest name for the glabrous-leaved plants that Steyermark called var. *boottii* is *S. arguta* var. *caroliniana.* According to Morton (1973), the epithet *S. strigosa* actually applies to a southeastern species, *S. ludoviciana* (A. Gray) Small, which does not occur in Missouri. Other infraspecific taxa recognized by Morton do not occur in Missouri. His taxonomy and nomenclature were adapted by Cronquist (1980) and Gleason and Cronquist (1991), whose treatments are followed here. The differences between the Missouri varieties are summarized in the key below. Note that vegetative material or plants in early flower cannot be determined satisfactorily below the species level in most cases.

1. Fruits glabrous; stems glabrous or sparsely hairy toward the tip
.... 2A. VAR. ARGUTA
1. Fruits hairy; stems glabrous or sparsely to moderately hairy
    2. Leaf blades hairy on the undersurface; stems sparsely to moderately hairy ..... 2B. VAR. BOOTTII
    2. Leaf blades glabrous on the undersurface; stems glabrous below the inflorescence or rarely sparsely hairy toward the tip
.... 2C. VAR. CAROLINIANA

2a. var. arguta

Stems glabrous below the inflorescence or sparsely pubescent with mostly curved hairs toward the tip. Leaf blades with the undersurface glabrous or less commonly sparsely to moderately pubescent with spreading hairs, mostly along the midvein. Fruits glabrous. $2n=18$. June–October.

Scattered in the Ozark and Ozark Border Divisions (northeastern U.S. west to Illinois and Missouri; Canada). Mesic to dry upland forests, ledges and bases of bluffs, banks of streams and rivers, and occasionally bottomland forests.

This variety occupies the northern half of the species range. It is by far the commonest of the three varieties in Missouri.

2b. var. boottii (Hook.) E.J. Palmer & Steyerm.

S. arguta var. *strigosa* (Small) Steyerm., misapplied

S. strigosa Small, misapplied

Stems sparsely to moderately pubescent with spreading and/or curved hairs, sometimes nearly glabrous toward the base. Leaf blades with the undersurface moderately pubescent with spreading hairs, mostly along the veins. Fruits finely hairy. $2n=18$. August–October.

Scattered in mostly the western half of the Ozark Division (southeastern U.S. west to Oklahoma and Texas). Mesic to dry upland forests, glades, ledges and bases of bluffs, banks of streams and rivers, and occasionally bottomland forests.

As noted above, earlier botanists misapplied the epithet var. *boottii* to glabrous-leaved plants that are now called var. *caroliniana*.

2c. var. caroliniana A. Gray

S. arguta ssp. *caroliniana* (A. Gray) G.H. Morton

S. arguta var. *boottii,* misapplied

Stems glabrous below the inflorescence or sparsely pubescent with mostly curved hairs toward the tip. Leaf blades with the undersurface glabrous. Fruits finely hairy. $2n=18$, 36 (apparently $2n=18$ in Missouri). June–October.

Uncommon and widely scattered in the Ozark and Ozark Border Divisions (southeastern U.S. west to Missouri and Louisiana). Mesic upland forests often above bluffs and banks of streams and rivers, often on acidic substrates.

This is the least common of the three varieties in Missouri.

3. Solidago buckleyi Torr. & A. Gray

Pl. 238 a–c; Map 992

Plants with the rootstock short and sometimes branched, not producing rhizomes. Stems 1 to several, 40–120 cm long, erect or ascending, finely ridged, sparsely to moderately pubescent with short, curved or occasionally stiff hairs, sometimes only sparsely hairy toward the base, not shiny, not glaucous. Leaves chiefly cauline, the largest leaves in the lower 1/3 of the stem, the basal leaves absent at flowering. Basal and lowermost stem leaves with the blade 6–12 cm long, 2–5 cm wide, variously 3–5 times as long as wide, oblanceolate to elliptic or obovate, relatively thin, angled or tapered to a sessile or short-petiolate base, angled or more commonly short-tapered to a sharply pointed tip, the margins usually sharply toothed and hairy, the surfaces not sticky or shiny, the upper surface glabrous or sparsely to moderately short-hairy along the main veins, the undersurface moderately pubescent with short (0.4–0.7 mm long) hairs along the main veins, with 1 main vein, the fine, pinnate secondary veins relatively easily observed (these usually forming an irregular network). Median and upper stem leaves 1–15 cm long, the uppermost usually lanceolate to narrowly elliptic, otherwise similar to the lower stem leaves. Inflorescences of axillary clusters grading into a narrow, racemose panicle, the heads oriented in several directions when short ascending branches are present. Involucre 4.5–6.0 mm long, the bracts in 3 or 4 unequal series. Involucral bracts mostly narrowly oblong-lanceolate and sharply pointed at the tip, appressed-ascending or more commonly at least the outermost involucral bracts loosely ascending or with the tips somewhat spreading to recurved, the thin, white to yellowish white margins hairy, the outer surface glabrous or more commonly sparsely to moderately hairy, the hairs minute and sometimes gland-tipped, with a poorly defined, green central region toward the tip, this tapered abruptly to the midvein above or below the bract midpoint, the midvein usually noticeably thickened. Receptacle naked. Ray florets 6–9, the corollas 3.5–5.0 mm long, yellow. Disc florets 8–15, the corollas 4–5 mm long, the lobes 0.9–1.5 mm long, yellow. Pappus 4–5 mm long, a few of the bristles often slightly thickened toward the tip. Fruits 2–3 mm long, narrowly obovoid, glabrous at maturity. September–October.

Scattered in the Ozark and Ozark Border Divisions, and also on Crowley's Ridge (Missouri to Arkansas east to Indiana, Kentucky, and Alabama). Glades, bases, ledges, and tops of bluffs, mesic to dry upland forests, and savannas; also roadsides and open, disturbed areas.

Steyermark (1963) retained this species as doubtfully distinct from *S. petiolaris,* based on the presence of apparently intermediate plants in Missouri. Cronquist (1980) mentioned *S. buckleyi* following his treatment of *S. petiolaris* but did not formally treat it for the southeastern United States, stating that its taxonomic position was uncertain. Nesom (1990e) studied a large series of herbarium specimens of the *S. petiolaris* complex and provisionally maintained *S. buckleyi* as a distinct species. However, he noted differences between the eastern populations that have been called *S. buckleyi* (on which the name originally was based) and the Ozarkian populations, speculating that the two sets of populations may represent separate, morphologically cryptic taxa. As had earlier workers, Nesom concluded that more detailed population-level studies were needed to clarify the taxonomy of the complex.

4. Solidago caesia L. (blue-stemmed goldenrod, wreath goldenrod)

Pl. 239 c–e; Map 993

Plants with the rootstock short, stout, and sometimes branched, sometimes also producing long, slender rhizomes. Stems 1 to several, 30–100 cm long, erect to more commonly arched or pendant, sometimes with fine, inconspicuous, longitudinal lines but not noticeably ridged or grooved, glabrous, not shiny, glaucous. Leaves chiefly cauline, the largest leaves in the lower 1/3 of the stem, the basal and lower stem leaves absent at flowering. Basal and lowermost stem leaves with the blade 6–10 cm long,

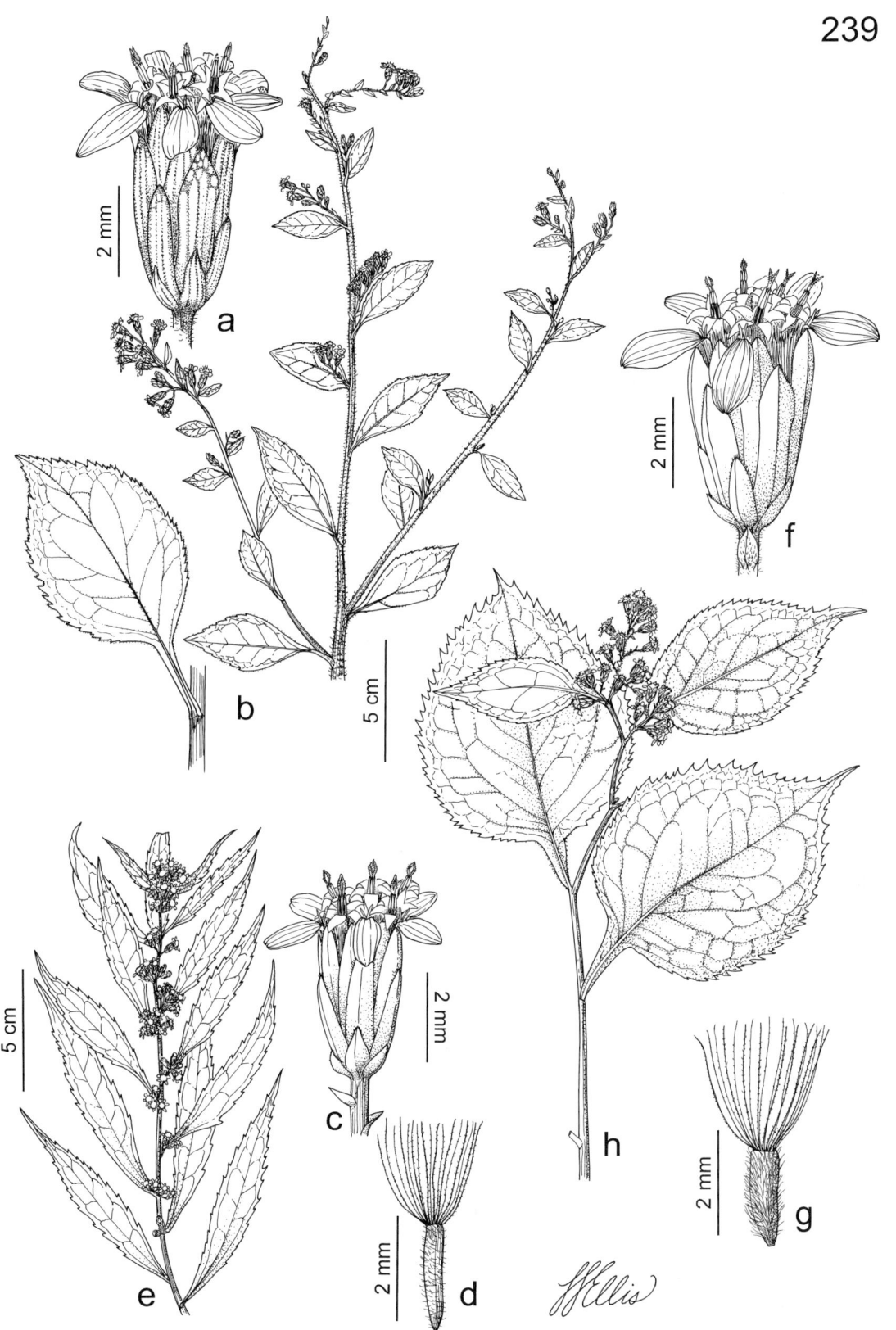

Plate 239. Asteraceae. *Solidago arguta*, **a)** head, **b)** flowering branch and larger leaf. *Solidago caesia*, **c)** head, **d)** fruit, **e)** habit. *Solidago flexicaulis*, **f)** head, **g)** fruit, **h)** habit.

993. Solidago caesia   994. Solidago canadensis   995. Solidago drummondii

1–3 cm wide, mostly 3–8 times as long as wide, narrowly elliptic-oblanceolate to elliptic or elliptic-obovate, relatively thin, tapered to a sessile base, tapered to a sharply pointed tip, the margins sharply toothed and usually inconspicuously hairy, the surfaces glabrous, the undersurface with 1 main vein, the fine, pinnate secondary veins usually relatively easily observed (these usually forming an irregular network). Median and upper stem leaves 2–8 cm long, elliptic-oblanceolate to narrowly elliptic, the margins of the uppermost leaves sometimes entire, otherwise similar to the lower stem leaves. Inflorescences of axillary clusters, the heads oriented in several directions. Involucre 2.5–4.5 mm long, the bracts in 3–5 unequal series. Involucral bracts mostly oblong to narrowly oblong and rounded to bluntly pointed (those of the outer series often oblong-lanceolate and sharply pointed) at the appressed-ascending tip, the thin, white to yellowish white margins hairy toward the tip, the outer surface glabrous, with a poorly defined, green central region toward the tip, this tapered abruptly to the midvein above or below the bract midpoint, the midvein often slightly thickened and sometimes with a faint, additional pair of veins present. Receptacle naked. Ray florets (1–)2–5, the corollas 3.0–3.5 mm long, yellow. Disc florets 5–9, the corollas 3.0–3.5 mm long, the lobes 0.9–1.5 mm long, yellow. Pappus 2.5–3.0 mm long, a few of the bristles often slightly thickened toward the tip. Fruits 1.0–1.8 mm long, narrowly ellipsoid-obovoid, finely hairy. $2n=18$. August–October.

Scattered in the southern quarter of the state, almost entirely within the Ozark Division (eastern U.S. west to Wisconsin and Texas; Canada). Bases and ledges of shaded bluffs, banks of streams and rivers, bottomland forests, and mesic upland forests; also roadsides.

This species is found most frequently along sheltered bluffs and adjacent lower terraces of streams. It does well in full sun or partial shade in the wildflower garden with its graceful, arching, glaucous stems and attractive clusters of heads.

Cook and Semple (2004) recently divided the species into two varieties, based mainly on differences in leaf morphology. They segregated plants with broadly lanceolate to elliptic or elliptic-ovate median leaves 5–9 cm long and less-arched stems as var. *zedia* R.E. Cook & Semple, which they stated to occur from southern Arkansas through Mississippi into northern Florida. They characterized the var. *caesia* as having narrowly lanceolate midstem leaves 5–15 cm long and strongly arched stems. Although Cook and Semple did not include Missouri in the range of var. *zedia*, Rachel Cook annotated about a third of the specimens from throughout the species range in the state as this variety. The determinations of these specimens seem equivocal, with a number of the Missouri specimens seemingly intermediate in leaf shape. None of the specimens in question include observations by the collectors on stem arching. In light of the difficulties in applying the defining features of these infraspecific taxa to plants in the Missouri flora, it seems wisest for the present time not to attempt a formal recognition of varieties.

5. **Solidago canadensis** L. **var. hargeri** Fernald (common goldenrod)

Pl. 241 d–f; Map 994

Plants with branched short- to long-creeping rhizomes, often also thickened at the stem bases. Stems 1 to several, 30–200 cm long (but usually longer than 100 cm), erect or ascending, with several fine, longitudinal lines or grooves, moderately to densely pubescent with short, spreading to somewhat curved hairs for most of the length, sparsely hairy to nearly glabrous toward the base, not shiny, not glaucous. Leaves chiefly cauline, the largest leaves in the lower $1/3$–$1/2$ of the stem, the basal

Plate 240. Asteraceae. *Solidago rugosa*, **a)** head, **b)** habit. *Solidago ulmifolia*, **c)** head, **d)** habit and larger leaf. *Solidago patula*, **e)** head, **f)** habit and larger leaf.

leaves usually absent at flowering. Basal and lowermost stem leaves with the blade 3–15 cm long, 0.5–3.0 cm wide, mostly 5–12 times as long as wide, narrowly oblanceolate to narrowly elliptic-lanceolate or narrowly lanceolate, not or only slightly thick and stiffened, angled or more commonly tapered gradually to a sessile or very short-petiolate base, angled or tapered to a sharply pointed tip, the margins sharply toothed and usually microscopically roughened or minutely hairy, usually flat, the upper surface sparsely to moderately roughened with minute, bulbous-based hairs, the undersurface sparsely to densely pubescent with short, spreading hairs (sometimes only along the veins), with 3 main veins, the lateral pair often originating well above the leaf base, finer than the midvein, the veinlets often difficult to observe, forming an irregular, dense network. Median and upper stem leaves 1–15 cm long, narrowly elliptic to narrowly lanceolate or nearly linear, sessile, the margins of the uppermost leaves often entire, otherwise similar to the lower stem leaves. Inflorescences open to more commonly relatively dense, usually broad, often large, pyramidal panicles, the longer branches and often also the tip arched or nodding, the heads oriented upward along the branches. Involucre 2–3 mm long, the bracts in 3–5 unequal series. Involucral bracts lanceolate to narrowly oblong-lanceolate and bluntly to more commonly sharply pointed at the appressed-ascending tip, the margin sparsely hairy, the outer surface glabrous, entirely yellowish or with a poorly differentiated, narrow, green central region, this tapered gradually to the midvein above or below the bract midpoint, the midvein somewhat thickened and no additional veins present. Receptacle naked. Ray florets 6–12, the corollas 2–3 mm long, yellow. Disc florets 2–5, the corollas 2.3–2.7 mm long, the lobes 0.3–0.7 mm long, yellow. Pappus 1.5–2.1 mm long, some of the bristles slightly thickened toward the tip. Fruits 1.0–1.5 mm long, narrowly obovoid, finely hairy. $2n=18$. July–October.

Scattered in the state but uncommon in much of the Glaciated Plains Division (northeastern U.S. west to North Dakota and Kansas; Canada). Upland prairies, mesic upland forests, savannas, and banks of streams and rivers, also pastures, old fields, margins of crop fields, railroads, roadsides, and open, disturbed areas.

For further discussion on species delimitation in this species complex, see the treatment of *S. altissima*. The other variety of *S. canadensis*, var. *canadensis*, has the stems only densely hairy toward the tip and occurs mainly to the northeast of Missouri. It is cultivated as an ornamental and particularly dwarf forms are popular in European and North American perennial gardens. It has escaped from cultivation sporadically in some portions of the western United States and Europe.

6. Solidago drummondii Torr. & A. Gray
(Ozark goldenrod)

Pl. 241 k–m; Map 995

Plants with the rootstock short and often branched, sometimes also producing stout rhizomes. Stems few to more commonly several, 30–100 cm long, erect to loosely ascending, with several fine, longitudinal ridges or grooves, moderately to densely pubescent with short, mostly spreading hairs 0.1–0.4 mm long (sometimes less densely hairy toward the stem base), not shiny, not glaucous. Leaves chiefly cauline, the largest leaves about $1/3$ of the way up the stem, the basal and lower stem leaves withered or absent at flowering (additional rosettes usually absent). Basal and lowermost stem leaves with the blade 3–8 cm long, 2–7 cm wide, mostly 1–2 times as long as wide, elliptic-ovate to broadly obovate, relatively thick and stiff, tapered relatively abruptly to a short petiole at the base, angled or tapered to a sharply pointed tip, the margins sharply toothed and inconspicuously hairy, the surfaces moderately to densely pubescent with spreading or curved hairs, the upper surface sometimes only sparsely roughened, the undersurface more or less with 3 main veins, but additional pinnate secondary veins usually also well developed and easily observed (the veinlets usually forming an irregular network). Median and upper stem leaves 1–7 cm long, broadly ovate to elliptic, sometimes lanceolate-elliptic toward the tip, the margins toothed or those of the uppermost leaves entire, otherwise similar to the lower stem leaves. Inflorescences open, more or less pyramidal panicles, the branches usually arched or nodding, the lowermost branches often relatively long, the heads oriented upward along the branches (or occasionally apparently all nodding). Involucre 3.0–4.5 mm long, the bracts in 3 or 4 unequal series. Involucral bracts oblong-ovate to narrowly oblong-lanceolate and rounded to bluntly pointed at the appressed-ascending tip (those of the outer series sometimes sharply pointed), the thin, white to yellowish white margins hairy (at least toward the tip), the outer surface glabrous, with an elliptic or narrowly diamond-shaped, green to light green central region mostly above the midpoint, this tapered gradually to the midvein, the midvein often slightly thickened and keeled and no additional veins present. Receptacle naked. Ray florets 3–7, the corollas 2–3 mm long, yellow. Disc florets 4–7, the

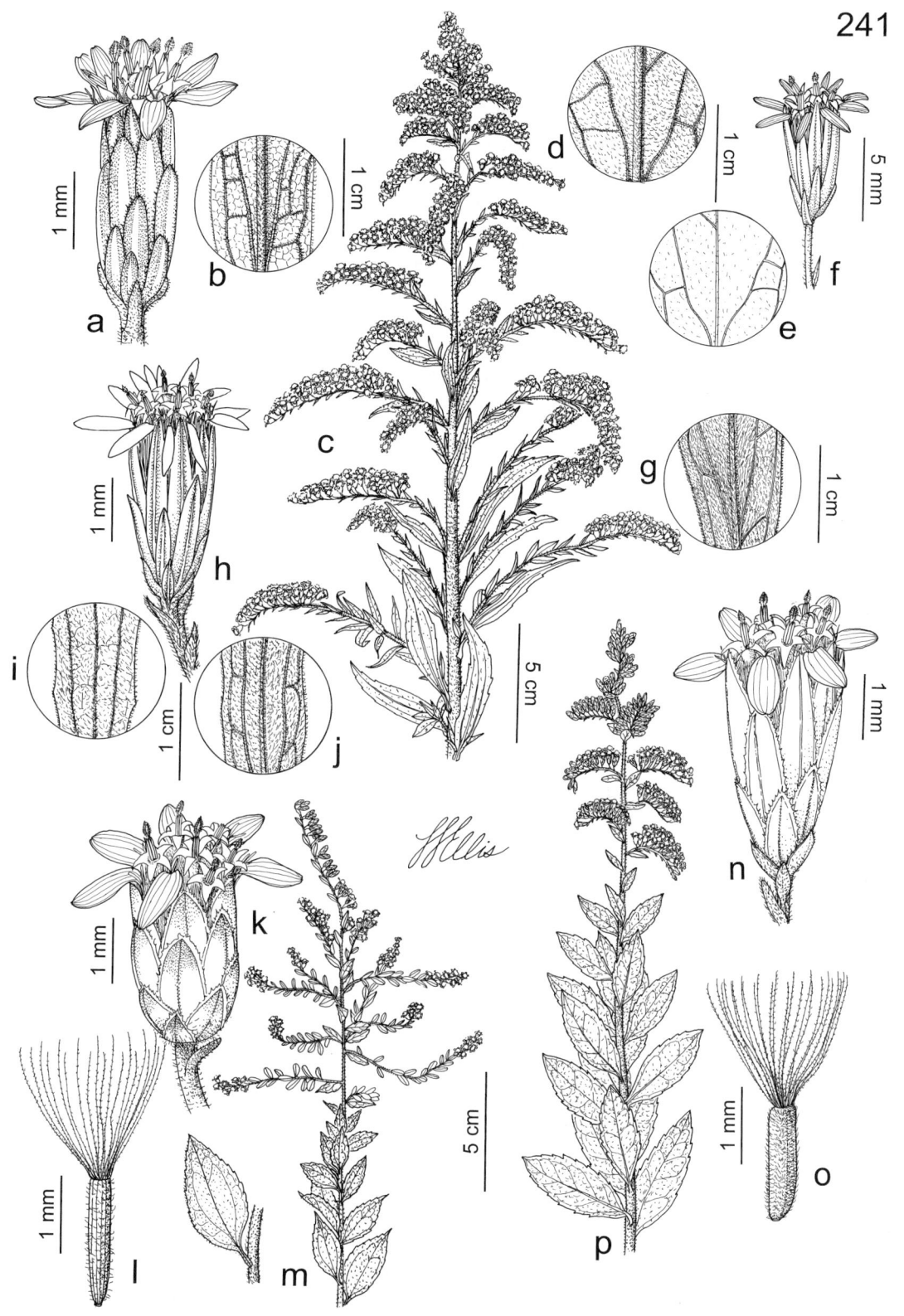

Plate 241. Asteraceae. *Solidago altissima* var. *gilvocanescens*, **a)** head, **b)** abaxial leaf detail, **c)** habit, **g)** abaxial leaf detail. *Solidago canadensis* var. *hargeri*, **d)** abaxial leaf detail, **e)** adaxial leaf detail, **f)** head. *Solidago altissima* var. *altissima* , **h)** head, **i)** adaxial leaf detail, **j)** abaxial leaf detail. *Solidago drummondii*, **k)** head, **l)** fruit, **m)** habit and larger leaf. *Solidago radula*, **n)** head, **o)** fruit, **p)** habit.

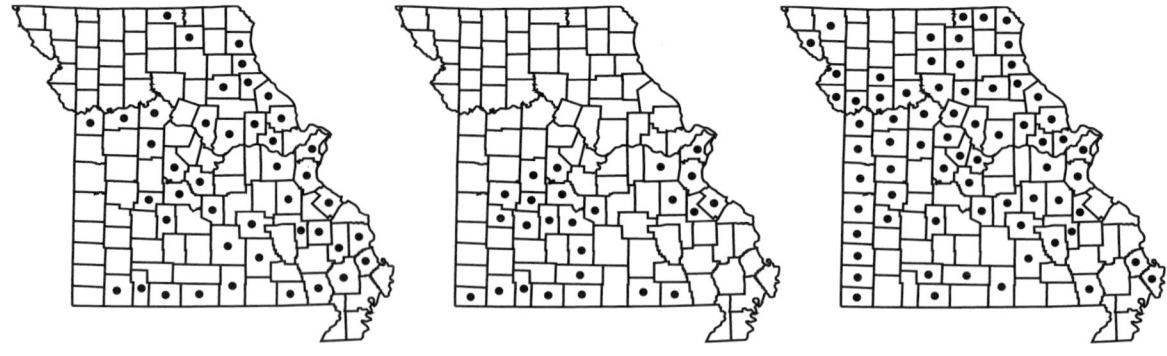

996. Solidago flexicaulis 997. Solidago gattingeri 998. Solidago gigantea

corollas 3.0–3.5 mm long, the lobes 0.5–0.9 mm long, yellow. Pappus 2.0–2.5 mm long, a few of the bristles often slightly thickened toward the tip. Fruits 1.5–2.0 mm long, narrowly obovoid, finely hairy. $2n=18$. August–December.

Scattered mostly in the eastern half of the state (Missouri, Arkansas, and Illinois). Ledges and tops of calcareous bluffs and rarely banks of rivers.

This species might be confused with *S. ulmifolia*, but it is easily differentiated by the key characters. Inexplicably, Kartesz and Meacham (1999) listed it as a synonym of *S. rugosa* ssp. *aspera* (Aiton) Cronquist, even though this was not Cronquist's (1947a) intent, and no other authors have suggested that these two names should refer to the same taxon.

7. Solidago flexicaulis L. (broadleaf goldenrod)
Pl. 239 f–h; Map 996

Plants with long, slender, branched rhizomes. Stems usually solitary, 30–120 cm long, erect to loosely ascending, noticeably few-angled and ridged, often strongly zigzag especially toward the tip, glabrous below the inflorescence (the inflorescence branches moderately pubescent with curved hairs), sometimes slightly shiny, not glaucous. Leaves chiefly cauline, the largest leaves in the lower $^1\!/_3$–$^1\!/_2$ of the stem, the basal leaves absent at flowering. Basal and lowermost stem leaves with the blade 7–15 cm long, 3–10 cm wide, mostly 1–2 times as long as wide, broadly ovate to broadly elliptic-ovate, relatively thin, rounded then tapered abruptly at the base to a relatively long, winged petiole, short-tapered to a sharply pointed tip, the margins sharply toothed and usually inconspicuously hairy, the upper surface glabrous, the undersurface sparsely to moderately pubescent with short, spreading hairs, with 1 main vein, the fine, pinnate secondary veins relatively easily observed (these usually forming an irregular network). Median and upper stem leaves 1–10 cm long, the uppermost leaves sometimes lanceolate to narrowly elliptic and with the margins entire or nearly so, otherwise similar to the lower stem leaves. Inflorescences of axillary clusters grading into a narrow, racemose panicle, the heads oriented in several directions when short ascending branches are present. Involucre 4–6 mm long, the bracts in 3–5 unequal series. Involucral bracts mostly narrowly oblong and rounded to bluntly pointed (those of the outer series often oblong-lanceolate and sharply pointed) at the appressed-ascending tip, the thin, white to yellowish white margins hairy toward the tip, the outer surface glabrous, with a poorly defined, green central region toward the tip, this tapered abruptly to the midvein above or below the bract midpoint, the midvein often slightly thickened. Receptacle naked. Ray florets 3–5, the corollas 2.5–5.0 mm long, yellow. Disc florets 5–9, the corollas 3–5 mm long, the lobes 0.9–1.2 mm long, yellow. Pappus 2–3 mm long, a few of the bristles often slightly thickened toward the tip. Fruits 1.5–2.0 mm long, narrowly obovoid, finely hairy. $2n=18$, 36. July–October.

Scattered in the eastern half of the state, uncommon farther west (eastern U.S. west to North Dakota and Oklahoma; Canada). Bases and ledges of shaded bluffs, banks of streams and rivers, bottomland forests, and mesic upland forests.

8. Solidago gattingeri Chapm.
Pl. 242 c, d; Map 997

Plants with the rootstock usually relatively short, stout, vertical to horizontal, and branched, not producing rhizomes. Stems 1 to few, 30–80 (–100) cm long, erect or ascending, with several fine, longitudinal lines or grooves, glabrous below the inflorescence (sparse, short hairs occasionally present along the inflorescence branches), not shiny, not glaucous, the nodes usually lacking clusters of leaves in the axils of the main leaves. Leaves basally disposed and often persistent at flowering (additional rosettes sometimes present adjacent to the flowering stem). Basal and lowermost stem leaves with the blade 8–17 cm long, 1–2 cm wide, mostly 7–10 times as long as wide, narrowly oblanceolate to oblanceolate, somewhat thickened and stiff, tapered gradually to the winged petiole at the base, angled to a sharply pointed tip, the margins entire or finely and sharply toothed, minutely hairy, the surfaces glabrous, the undersurface with 3 main veins, the lateral pair much finer than the midvein, the veinlets usually easily observed, forming an irregular, dense network. Median and upper stem leaves 1–6 cm long, narrowly oblanceolate to narrowly elliptic or nearly linear, sessile or short-petiolate, often with only 1 main vein, otherwise similar to the lower stem leaves. Inflorescences relatively dense, narrow to more commonly broad, pyramidal panicles, the longer branches and often also the tip arched or nodding, the heads oriented upward along the branches. Involucre 3–5 mm long, the bracts in 3 or 4 unequal series. Involucral bracts oblong-ovate to narrowly lanceolate and rounded or bluntly pointed (those of the outer series often sharply pointed) at the appressed-ascending tip, the margin sparsely hairy toward the tip, the outer surface glabrous, usually entirely yellowish but occasionally with a poorly differentiated, greenish yellow central region above the midpoint, the midvein somewhat thickened and keeled, and no additional veins present. Receptacle naked. Ray florets 5–8, the corollas 2–3 mm long, yellow. Disc florets 3–9, the corollas 3–4 mm long, the lobes 0.5–0.8 mm long, yellow. Pappus 2–3 mm long, some of the bristles slightly thickened toward the tip. Fruits 1.5–2.0 mm long, obovoid, glabrous or less commonly sparsely hairy. $2n=18$. July–October.

Scattered in the Ozark and Ozark Border Divisions (Missouri to Arkansas and Tennessee). Calcareous glades and rock outcrops in upland prairies; rarely also roadsides.

This species has an odd distribution, occurring in glades of the northern half of the Ozarks and in the cedar barrens of Tennessee.

9. Solidago gigantea Aiton (late goldenrod, tall goldenrod)
S. gigantea var. *leiophylla* Fernald
S. gigantea var. *serotina* (Kuntze) Cronquist
Pl. 242 a, b; Map 998

Plants with branched, long-creeping rhizomes, often also thickened at the stem bases. Stems 1 to several, 50–200 cm long, erect or ascending, with several fine, longitudinal lines or grooves, glabrous below the inflorescence (sparse to moderate, short hairs sometimes present along the inflorescence branches), not shiny, usually somewhat glaucous. Leaves chiefly cauline, the largest leaves toward the midpoint of the stem, the basal leaves usually absent at flowering. Basal and lowermost stem leaves with the blade 6–15 cm long, 1–3 cm wide, mostly 4–8 times as long as wide, narrowly oblanceolate to elliptic-lanceolate or narrowly elliptic, not or only slightly thickened (occasionally slightly fleshy when young), angled or more commonly tapered gradually to a sessile or very short-petiolate base, angled or tapered to a sharply pointed tip, the margins sharply toothed and usually microscopically roughened, the surfaces glabrous or more commonly the undersurface sparsely hairy mainly along the main veins, with 3 main veins, the lateral pair often originating well above the leaf base, finer than the midvein, the veinlets usually easily observed, forming an irregular, dense network. Median and upper stem leaves 1–20 cm long, elliptic to lanceolate or narrowly lanceolate, sessile, the margins of the uppermost leaves often entire, otherwise similar to the lower stem leaves. Inflorescences open to more commonly relatively dense, narrow to broad, often large, pyramidal panicles, the longer branches and often also the tip arched or nodding, the heads oriented upward along the branches. Involucre 2.5–4.0 mm long, the bracts in 3–6 unequal series. Involucral bracts lanceolate to narrowly oblong and bluntly to sharply pointed at the appressed-ascending tip, the margin sparsely hairy, the outer surface glabrous, with an often poorly differentiated, green central region of varying width toward the tip, this tapered gradually to the midvein above or below the bract midpoint (the outer series often nearly entirely green), the midvein somewhat thickened and no additional veins present. Receptacle naked. Ray florets 8–17, the corollas 3.5–5.0 mm long, yellow. Disc florets 6–12, the corollas 3.0–3.5 mm long, the lobes 0.5–0.9 mm long, yellow. Pappus 2.0–2.5 mm long, some of the bristles slightly thickened toward the tip. Fruits 1.3–1.7 mm long, obovoid, sparsely and finely hairy. $2n=18, 36, 54$. July–October.

Scattered nearly throughout the state (U.S.; Canada; introduced in Europe). Bottomland forests, mesic to less commonly dry upland forests, banks of streams and rivers, margins of ponds and lakes, fens, bases and ledges of bluffs, bottomland and upland prairies, and rarely glades; also pastures, margins of crop fields, railroads, roadsides, and open, disturbed areas.

A number of varieties and forms have been segregated from typical *S. gigantea* based on differ-

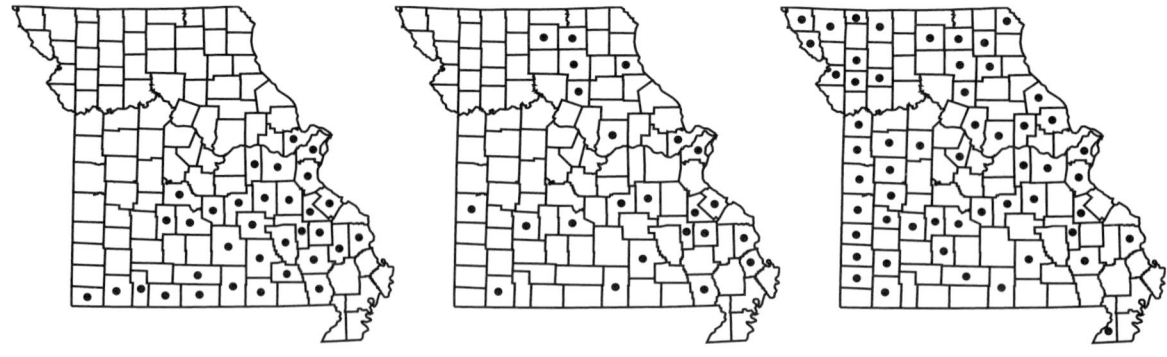

999. Solidago hispida   1000. Solidago juncea   1001. Solidago missouriensis

ences in leaf width and pubescence. Morton (1984) studied the morphology of the complex throughout its range and concluded that there is no practical way of grouping the complex patterns of variation within the species into recognizable infraspecific taxa. His treatment is followed here. *Solidago gigantea* is a relatively glabrous member of the widespread *S. canadensis* complex. For a discussion of stem galls in this and related species, see the treatment of *S. altissima*.

10. Solidago hispida Muhl. ex Willd. **var. hispida** (hairy goldenrod)
S. bicolor L. var. *concolor* Torr. & A. Gray
S. bicolor var. *hispida* (Muhl. ex Willd.) Britton, Sterns & Poggenb.

Pl. 238 d, e; Map 999

Plants with the rootstock short and sometimes branched, usually not producing rhizomes. Stems usually solitary, 20–100 cm long, erect or ascending, with several fine, longitudinal ridges or grooves, moderately to densely pubescent with fine, spreading hairs, not shiny. Leaves basally disposed, usually persistent at flowering (additional rosettes often present adjacent to the flowering stem). Basal and lowermost stem leaves with the blade 6–20 cm long, 1.5–6.0 cm wide, mostly 2–6 times as long as wide, oblanceolate to elliptic or obovate, not or only slightly thickened, tapered to a relatively long, winged petiole at the base, angled to a usually sharply pointed tip, the margins nearly entire to scalloped or toothed, the surfaces moderately to densely pubescent with fine, spreading to loosely appressed hairs, the undersurface with 1 main vein, the fine, pinnate secondary veins usually easily observed (these usually forming an irregular network). Median and upper stem leaves 1–8 cm long, lanceolate to elliptic or oblanceolate, the margins of at least the uppermost leaves entire, otherwise similar to the lower stem leaves. Inflorescences of axillary clusters or short, axillary racemes, these usually appearing as a spicate to racemose, elongate terminal panicle, the heads oriented in several directions when ascending to spreading branches are present. Involucre 4–6 mm long, the bracts in 3–5 unequal series. Involucral bracts mostly oblong to narrowly oblong, rounded to bluntly pointed (occasionally those of the inner series sharply pointed) at the appressed-ascending tip, the thin, white to yellowish white margins hairy (at least toward the tip), the outer surface glabrous or finely hairy, with a green central region of varying width toward the tip, this tapered to the midvein below the bract midpoint, the midvein often slightly thickened and no additional veins present. Receptacle naked. Ray florets 7–14, the corollas 3.5–4.5 mm long, yellow. Disc florets 9–15, the corollas 2.5–3.5 mm long, the lobes 0.6–1.2 mm long, yellow. Pappus 2.5–3.0 mm long, a few of the bristles often slightly thickened toward the tip. Fruits 0.8–1.5 mm long, narrowly obovoid, glabrous. $2n=18$. July–October.

Scattered in the Ozark and Ozark Border Divisions (eastern U.S. west to South Dakota and Oklahoma; Canada). Mesic to dry upland forests and ledges and tops of bluffs; also roadsides; frequently on acidic substrates.

This is a relatively easily recognized species that is variable in pubescence across its overall range. Various authors have accepted four or more varieties based mostly upon various extremes of pubescence. Missouri plants are uniformly var. *hispida*. The other varieties mostly are localized in the northern part of the species range. Because corolla color may fade in pressed specimens, some collectors have confused plants of *S. hispida* with the closely related *S. bicolor* L., and this has even led some botanists to consider the two taxa varieties of a single species. *Solidago bicolor* (silverrod) was reported for Missouri without details by Cronquist (1952,

1980), but it was excluded from the Missouri flora by Steyermark (1963). It is distinguished from other Missouri goldenrods by its white to cream-colored ray corollas and light yellow disc corollas in combination with its elongate inflorescence. It is widespread east of the Mississippi River and may eventually be discovered in eastern Missouri.

11. Solidago juncea Aiton (early goldenrod)
S. juncea f. *scabrella* (Torr. & A. Gray) Fernald

Pl. 243 c–e; Map 1000

Plants with the rootstock usually relatively short, stout, horizontal, branched, and often rhizomatous, sometimes also with deep-set, longer-creeping rhizomes. Stems 1 to few, 30–120 cm long, erect or ascending, with several fine, longitudinal lines or grooves, glabrous below the inflorescence (rarely sparsely and inconspicuously hairy along the inflorescence branches), not shiny, not glaucous. Leaves basally disposed and usually persistent at flowering (additional rosettes often present adjacent to the flowering stem). Basal and lowermost stem leaves with the blade 8–20(–35) cm long, 1.5–4.0(–8.0) cm wide, mostly 5–8 times as long as wide, narrowly oblanceolate to narrowly elliptic-obovate, not or only slightly thickened, tapered gradually to the winged petiole at the base, angled or tapered to a sharply pointed tip, the margins finely to coarsely and sharply toothed, microscopically roughened or minutely hairy, the surfaces glabrous or rarely sparsely roughened or with a few softer hairs, the undersurface with 1 main vein, the fine, pinnate secondary veins usually easily observed (these and the veinlets forming an irregular, dense network). Median and upper stem leaves 2–12 cm long, lanceolate to narrowly elliptic or nearly linear, sessile or short-petiolate, otherwise similar to the lower stem leaves. Inflorescences relatively dense and narrow to relatively open and broad, pyramidal panicles (the lower branches sometimes elongate), the longer branches and often also the tip arched or nodding, the heads oriented upward along the branches. Involucre 3–5 mm long, the bracts in 3 or 4 unequal series. Involucral bracts oblong-ovate to narrowly oblong or narrowly lanceolate and bluntly to less commonly sharply pointed at the appressed-ascending tip, the thin, white to yellowish white margins sparsely hairy toward the tip, the outer surface glabrous, with a poorly differentiated green to light green central region above the midpoint, this tapered gradually to the midvein, the midvein not or only slightly thickened, and no additional veins present. Receptacle usually with a few chaffy bracts (similar to the involucral bracts but somewhat shorter) toward the margin. Ray florets 7–12, the corollas 2.0–2.5 mm long, yellow. Disc florets 8–15, the corollas 2.5–3.5 mm long, the lobes 0.5–0.8 mm long, yellow. Pappus 2–3 mm long, some of the bristles slightly thickened toward the tip. Fruits 1.0–1.5 mm long, obovoid, sparsely hairy or rarely glabrous. $2n=18$. June–October.

Scattered in the southern half of the state and in the northeastern quarter (eastern U.S. west to Minnesota and Louisiana; Canada). Upland prairies, glades, savannas, and openings of mesic to dry forests; also railroads, roadsides, and open, disturbed areas.

12. Solidago missouriensis Nutt. **var. fasciculata** Holz. (Missouri goldenrod)

Pl. 242 e, f; Map 1001

Plants with branched, short- to long-creeping rhizomes, often also thickened at the stem bases. Stems 1 to few, 30–100 cm long, erect or ascending, with several fine, longitudinal lines or grooves, glabrous below the inflorescence (sparse, short hairs occasionally present along the inflorescence branches), not shiny, not glaucous, the median and upper nodes usually producing small clusters of leaves in the axils of the main leaves. Leaves basally disposed but often withered or absent at flowering (additional rosettes sometimes present adjacent to the flowering stem). Basal and lowermost stem leaves with the blade 6–15 cm long, 0.5–1.5 cm wide, mostly 7–10 times as long as wide, narrowly oblanceolate, somewhat thickened and stiff, tapered gradually to the winged petiole at the base, angled to a sharply pointed tip, the margins sharply toothed, minutely hairy, the surfaces glabrous, the undersurface with 3 main veins, the lateral pair finer than the midvein, the veinlets usually easily observed, forming an irregular, dense network. Median and upper stem leaves 1–10 cm long, narrowly oblanceolate to narrowly elliptic or nearly linear, sessile or short-petiolate, the margins toothed to entire, all but the uppermost with 3 main veins, otherwise similar to the lower stem leaves. Inflorescences relatively dense, narrow to broad, pyramidal panicles, the longer branches and often also the tip arched or nodding, the heads oriented upward along the branches. Involucre 2.5–4.5 mm long, the bracts in 3 or 4 unequal series. Involucral bracts oblong-ovate to narrowly lanceolate and rounded or bluntly pointed (those of the outer series often sharply pointed) at the appressed-ascending tip, the margin sparsely hairy toward the tip, the outer surface glabrous, usually entirely yellowish but occasionally with a poorly differentiated, greenish yellow central region above the midpoint,

1002. Solidago nemoralis   1003. Solidago odora   1004. Solidago patula

the midvein somewhat thickened and keeled, and no additional veins present. Receptacle frequently with a few slender, chaffy bracts (similar to the involucral bracts but usually somewhat shorter) toward the margin. Ray florets 7–13, the corollas 2–3 mm long, yellow. Disc florets 6–15, the corollas 2.5–4.0 mm long, the lobes 0.5–0.9 mm long, yellow. Pappus 2.5–3.0 mm long, some of the bristles slightly thickened toward the tip. Fruits 1.0–2.2 mm long, obovoid, glabrous or finely hairy. $2n=18$, 36. July–October.

Scattered nearly throughout the state, but uncommon in the eastern portion of the Ozark Division and the Mississippi Lowlands (Michigan to Tennessee west to Washington and Arizona; Canada). Upland prairies, loess hill prairies, glades, savannas, and rarely openings and margins of mesic to dry upland forests; also old fields, railroads, and roadsides.

This species superficially resembles S. gattingeri but usually is fairly easily distinguished by the key characters. The infraspecific taxonomy is complex and requires further study. Steyermark (1963) treated two varieties as occurring in Missouri: var. *fasciculata* Holz., with broader inflorescences and shorter, less hairy fruits, and var. *missouriensis*, with narrower inflorescences and slightly longer, hairier fruits. Cronquist (1955) reevaluated the complex, recognizing a widespread tall-stemmed, relatively leafy var. *fasciculata* and three shorter-stemmed, less leafy varieties endemic to the northwestern United States differing from each other in details of head size and inflorescence morphology. His treatment has been adopted in several later works (Cronquist, 1980; Barkley, 1986; Gleason and Cronquist, 1991; Semple et al., 1999) and excludes Missouri from the range of var. *missouriensis*. Semple et al. (1999) noted that tetraploid plants ($2n=36$) are known only from the western portion of the species range, but relatively few chromosome counts exist to date for the species. In the northwestern United states, there apparently is some morphological intergradation between plants attributed to var. *fasciculata* and var. *missouriensis*.

13. Solidago nemoralis Aiton (old-field goldenrod, gray goldenrod)

Pl. 243 a, b; Map 1002

Plants with branched short- to moderately creeping rhizomes. Stems solitary or more commonly few to several, 20–100 cm long, erect or ascending, with several fine, longitudinal ridges or grooves, densely pubescent with minute, mostly curved hairs 0.1–0.3 mm long, not shiny, not glaucous, the upper nodes sometimes producing small clusters of leaves in the axils of the main leaves. Leaves basally disposed and persistent at flowering (additional rosettes usually present adjacent to the flowering stem). Basal and lowermost stem leaves with the blade 2–10 cm long, 0.7–2.0 cm wide, mostly 3–8 times as long as wide, nearly linear to narrowly oblanceolate, oblanceolate, or narrowly obovate, relatively thin or sometimes somewhat thickened, tapered gradually to a short to long, winged petiole at the base, rounded or angled to a bluntly or sharply pointed tip, the margins entire to finely scalloped or bluntly to sharply toothed, inconspicuously hairy, the surfaces densely pubescent with minute, mostly curved hairs 0.1–0.3 mm long, the undersurface usually with 1 main vein (a basal or median pair of secondary veins sometimes slightly more prominent toward the base than the others), the fine, pinnate secondary veins usually easily observed (these usually forming an irregular network). Median and upper stem leaves 0.8–6.0(–8.0) cm long, linear to narrowly oblanceolate or lanceolate, the margins mostly entire, otherwise similar to the lower stem leaves. Inflorescences dense, often narrowly pyramidal panicles (broad-

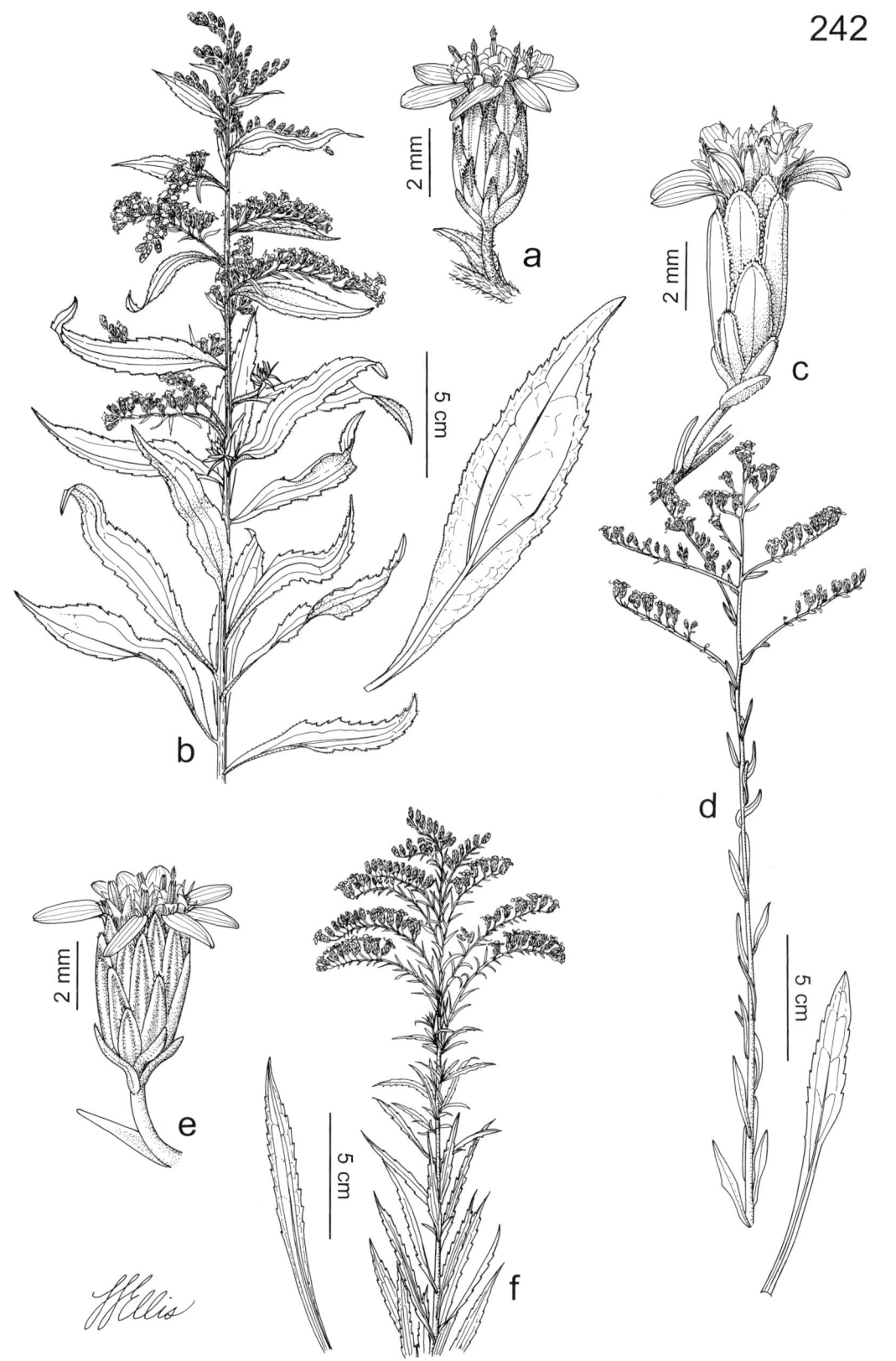

Plate 242. Asteraceae. *Solidago gigantea*, **a)** head, **b)** habit and larger leaf. *Solidago gattingeri*, **c)** head, **d)** habit and larger leaf. *Solidago missouriensis*, **e)** head, **f)** habit and larger leaf.

est near the base), the longer branches and tip usually somewhat arched or nodding, the heads oriented upward along the branches. Involucre 3–6 mm long, the bracts in 3–5 unequal series. Involucral bracts oblong-ovate to narrowly oblong-lanceolate and mostly rounded to bluntly pointed at the appressed-ascending tip (those of the outer series sometimes sharply pointed), the thin, white to yellowish white margins hairy (at least toward the tip), the outer surface glabrous, with a poorly differentiated, green to light green central region above the midpoint, this tapered gradually to the midvein, the midvein often slightly thickened and no additional veins present. Receptacle naked. Ray florets 5–9, the corollas 3.0–5.5 mm long, yellow. Disc florets 3–9, the corollas 2.5–4.5 mm long, the lobes 0.5–0.9 mm long, yellow. Pappus 2–4 mm long, a few of the bristles often slightly thickened toward the tip. Fruits 0.5–2.0 mm long, narrowly obovoid, sparsely to moderately finely hairy. $2n=18, 36$. July–November.

Scattered nearly throughout the state (eastern U.S. west to Montana and New Mexico; Canada). Upland prairies, loess hill prairies, savannas, glades, ledges and tops of bluffs, openings of mesic to dry upland forests, and occasionally banks of streams and margins of ponds and fens; also old fields, pastures, railroads, roadsides, and open, disturbed areas.

Solidago nemoralis is a characteristic species of old fields and other open, grass-dominated habitats. Traditionally, most botanists have divided *S. nemoralis* into three or more infraspecific taxa. Brammall and Semple (1990) and Semple et al. (1990) performed detailed studies of cytological and morphometric variation across the range of the species, concluding that there were only two elements that could be separated consistently. These two subspecies are similar to the ones that Steyermark (1963) recognized as varieties, although Steyermark emphasized differences in leaf morphology, whereas Semple and his colleagues emphasized quantitative features of the heads, florets, and fruits that tend to correlate with ploidy level. Some Missouri specimens are difficult to determine below the species level, and there is some degree of morphological overlap between the two subspecies for every feature cited below.

1. Basal and lower stem leaves mostly narrowly oblanceolate to nearly linear, the margins entire or with sparse, minute, sharp teeth toward the tip; fruits moderately hairy at maturity
 . . . . 13A. SSP. DECEMFLORA

1. Basal and lower stem leaves mostly oblanceolate to narrowly obovate (occasionally narrowly oblanceolate), the margins finely scalloped or with minute to short, usually blunt teeth, especially above the midpoint; fruits sparsely hairy at maturity . . . . 13B. SSP. NEMORALIS

13a. ssp. decemflora (DC.) Brammall ex Semple

S. nemoralis var. *decemflora* (DC.) Fernald
S. nemoralis var. *longipetiolata* (Mack. & Bush) E.J. Palmer & Steyerm.

Basal and lower stem leaves mostly narrowly oblanceolate to nearly linear, the margins entire or with sparse, minute, sharp teeth toward the tip. Involucres mostly 4.5–6.0 mm long. Pappus usually longer than the disc corolla tube. Fruits 1–2 mm long, moderately hairy at maturity. $2n=36$. July–November.

Scattered nearly throughout the state (Indiana to Montana south to Arkansas and New Mexico; Canada). Upland prairies, loess hill prairies, savannas, glades, ledges and tops of bluffs, openings of mesic to dry upland forests, and occasionally banks of streams; also old fields, pastures, and roadsides.

This subspecies is somewhat less abundant in Missouri than ssp. *nemoralis* and occurs scattered through the range of that subspecies. It represents a tetraploid variant of the species that is most abundant in the western half of the species range.

13b. ssp. nemoralis

S. nemoralis var. *haleana* Fernald

Basal and lower stem leaves mostly oblanceolate to narrowly obovate (occasionally narrowly oblanceolate), the margins finely scalloped or with minute to short, usually blunt teeth, especially above the midpoint. Involucres mostly 3.0–4.5 mm long. Pappus usually about as long as the disc corolla tube. Fruits 0.5–1.5 mm long, sparsely hairy at maturity. $2n=18, 36$. July–November.

Scattered nearly throughout the state (eastern U.S. west to North Dakota and Texas; Canada). Upland prairies, loess hill prairies, savannas, glades, ledges and tops of bluffs, openings of mesic to dry upland forests, and occasionally banks of streams and margins of ponds and fens; also old fields, pastures, railroads, roadsides, and open, disturbed areas.

This subspecies represents mostly diploid populations that are widespread in the eastern two-thirds of the species range. Tetraploid plants are relatively uncommon from the eastern seaboard sporadically westward to Michigan and Ohio and have on average slightly larger involucres than do diploid plants (Brammall and Semple, 1990; Semple et al., 1990).

14. Solidago odora Aiton **var. odora** (sweet goldenrod, fragrant goldenrod)

Pl. 243 f–i; Map 1003

Plants with the rootstock short, often horizontal and sometimes branched, not producing rhizomes. Stems 1 to several, 50–120 cm long, erect or ascending, finely ridged, moderately to densely pubescent in longitudinal lines or bands with short, curved hairs toward the tip, often nearly glabrous toward the base, somewhat shiny, not glaucous. Leaves chiefly cauline, the largest leaves in the lower 1/3 of the stem, the basal leaves usually absent at flowering. Basal and lowermost stem leaves with the blade 5–9 cm long, 0.6–2.0 cm wide, variously 4–8 times as long as wide, oblanceolate, somewhat thickened, angled, or tapered to a sessile or short-petiolate base, angled or tapered to a sharply pointed tip, the margins entire and minutely roughened, the surfaces glabrous but with minute, impressed, translucent dots (these best observed on the undersurface under magnification while the leaf is held to a strong light), with 1 main vein or, if 3-veined, then the lateral pair poorly developed and only faintly visible, the fine, pinnate secondary veins often very faint. Median and upper stem leaves 1–12 cm long, nearly linear to narrowly lanceolate (the lowest rarely narrowly obovate), otherwise similar to the lower stem leaves. Inflorescences open to dense, more or less pyramidal panicles, the branches and often also the tip arched or nodding, the lowermost branches short to relatively long, the heads oriented upward along the branches. Involucre 3.5–5.0 mm long, the bracts in 3 or 4 unequal series. Involucral bracts mostly narrowly ovate to narrowly lanceolate and sharply pointed at the appressed-ascending tip, the yellowish white margins glabrous or minutely hairy toward the tip, the outer surface glabrous, with a poorly defined (sometimes nearly absent), green central region toward the tip, this tapered gradually to the midvein above the bract midpoint, the midvein usually noticeably thickened. Receptacle naked. Ray florets 3–5(–6), the corollas 3.5–5.0 mm long, yellow. Disc florets 3–5, the corollas 3–4 mm long, the lobes 0.5–0.9 mm long, yellow. Pappus 2.5–3.5 mm long, a few of the bristles often slightly thickened toward the tip. Fruits 1.5–2.3 mm long, narrowly obovoid, glabrous or finely hairy. $2n=18$. July–September.

Scattered in the southeastern portion of the Ozark Division and also on Crowley's Ridge (eastern [mostly southeastern] U.S. west to Missouri and Texas). Openings of mesic upland forests, margins of spring branches, and savannas; also old fields and roadsides.

The crushed foliage of this species has an aroma reminiscent of anise. The translucent punctations can be difficult to observe in fresh leaves. A second variety, var. *chapmannii* (A. Gray) Cronquist is endemic to Florida and differs in its more even stem pubescence and somewhat shorter, broader leaves.

15. Solidago patula Muhl. ex Willd. **var. patula** (rough-leaved goldenrod)

Pl. 240 e, f; Map 1004

Plants with the rootstock short, stout, sometimes horizontal, branched or with offsets. Stems usually solitary, 50–180 cm long, erect or strongly ascending, with several fine, longitudinal ridges and also relatively strongly 2- or 3-angled (the angles sometimes ridged or narrowly winged), glabrous below the inflorescence or sparsely pubescent with short, spreading hairs toward the tip, not shiny, sometimes slightly glaucous. Leaves basally disposed and usually persistent at flowering (additional rosettes usually present adjacent to the flowering stem). Basal and lowermost stem leaves with the blade 8–30 cm long, 4–10 cm wide, mostly 2–4 times as long as wide, broadly ovate to elliptic, usually somewhat thickened, tapered gradually or fairly abruptly to a long, winged petiole at the base, tapered to a sharply pointed tip, the margins finely toothed, microscopically roughened, the upper surface strongly roughened (sandpapery) with forward-angled, minute, stiff, stout, broad-based (pustular) hairs, the undersurface glabrous, with 1 main vein, the fine, pinnate secondary veins easily observed (these usually forming an irregular network). Median and upper stem leaves 1–9 cm long, lanceolate to elliptic-oblanceolate, sessile or short-petiolate, otherwise similar to the lower stem leaves. Inflorescences relatively dense and narrow to open and broad pyramidal panicles (the lower branches sometimes elongate), the longer branches and occasionally also the tip usually somewhat arched or nodding, the heads oriented upward along the branches. Involucre 3.0–4.5 mm long, the bracts in 3–5 unequal series. Involucral bracts oblong-ovate to narrowly oblong-ovate and mostly bluntly pointed at the appressed-ascending tip (those of the outer series sometimes sharply pointed), the thin, white to yellowish white margins hairy, the outer surface glabrous or more commonly microscopically roughened (often only the tiny, globose, pustular

1005. Solidago petiolaris   1006. Solidago ptarmicoides   1007. Solidago radula

bases visible), with a poorly differentiated, green to light green central region mostly above the midpoint, this tapered fairly abruptly to the midvein (sometimes appearing nearly truncate), the midvein often slightly thickened and keeled, and no additional veins present. Receptacle naked. Ray florets 5–12, the corollas 2.0–3.5 mm long, yellow. Disc florets 5–20, the corollas 2.5–3.0 mm long, the lobes 0.5–1.5 mm long, yellow. Pappus 2–3 mm long, a few of the bristles often slightly thickened toward the tip. Fruits 1.3–1.8 mm long, narrowly obovoid, glabrous or more commonly sparsely and minutely hairy. $2n=18$. August–October.

Scattered in the eastern portion of the Ozark and Ozark Border Divisions and the Mississippi Lowlands (eastern U.S. west to Wisconsin, Missouri, and Mississippi; Canada). Fens, acid seeps, swamps, seepy banks of streams, and ledges of calcareous bluffs.

The var. *strictula* Torr. & A. Gray occurs mostly on the Coastal Plain and Piedmont of the southeastern United States (Cronquist, 1980). It has somewhat narrower, less-toothed leaves, and the plants overall are less robust. This taxon was reported erroneously from southern Missouri by Kartesz and Meacham (1999).

16. Solidago petiolaris Aiton (downy goldenrod)

S. petiolaris var. *angusta* (Torr. & A. Gray) A. Gray

S. petiolaris var. *wardii* (Britton) Fernald

Pl. 238 h, i; Map 1005

Plants with the rootstock short and sometimes branched, occasionally producing slender rhizomes. Stems solitary or more commonly few to several, 40–150 cm long, erect or ascending, finely ridged, sparsely to moderately pubescent with short, curved or occasionally stiff hairs, sometimes only sparsely hairy toward the base, not shiny, not glaucous. Leaves chiefly cauline, the largest leaves in the lower 1/3 of the stem, the basal leaves absent at flowering. Basal and lowermost stem leaves with the blade 6–12 cm long, 0.5–3.5 cm wide, variously 3–8 times as long as wide, narrowly elliptic-oblanceolate to elliptic or obovate, relatively thick and stiff, angled or tapered to a sessile or short-petiolate base, angled or more commonly short-tapered to a sharply pointed tip, the margins entire or sparsely and shallowly toothed above the midpoint and hairy, the surfaces sometimes sticky and somewhat shiny, the upper surface glabrous or somewhat roughened, the undersurface glabrous, somewhat roughened, or sparsely to moderately pubescent with minute (0.1–0.4 mm long) hairs along the main veins, with 1 main vein, the fine, pinnate secondary veins relatively easily observed (these usually forming an irregular network). Median and upper stem leaves 1–15 cm long, nearly linear to narrowly obovate, otherwise similar to the lower stem leaves. Inflorescences of axillary clusters grading into a narrow racemose panicle, the heads oriented in several directions when short ascending branches are present. Involucre 4.5–7.5 mm long, the bracts in 3–5 unequal series. Involucral bracts mostly narrowly oblong-lanceolate and sharply pointed at the tip, appressed-ascending or more commonly at least the outermost involucral bracts loosely ascending or with the tips somewhat spreading to recurved, the thin, white to yellowish white margins hairy, the outer surface glabrous or sparsely to moderately hairy, sometimes glandular and somewhat sticky, with a poorly defined, green central region toward the tip, this tapered abruptly to the midvein above or below the bract midpoint, the midvein usually noticeably thickened. Receptacle naked. Ray florets (5–)7–9, the corollas 3.5–7.5 mm long, yellow. Disc florets (8–)10–16, the corollas 4–5 mm long, the lobes 0.9–1.8 mm long, yellow. Pappus 4.0–4.5 mm long, a few of the bristles often slightly thickened toward the tip. Fruits 3–4 mm long, narrowly obovoid, glabrous or nearly so at maturity. 2n=18, 36. May–November.

Scattered mostly south of the Missouri River (Nebraska to New Mexico east to Illinois, North Carolina, and Florida). Glades, bases, ledges, and tops of bluffs, openings of mesic to dry upland forests, and savannas; also pastures and roadsides.

Steyermark (1963) noted that this attractive species performs well in the wildflower garden and that it is one of the earliest goldenrods to start flowering in Missouri. The heads are relatively large for a goldenrod.

The taxonomy of the *S. petiolaris* complex requires further study. On the one hand, there are problems surrounding the distinctness of *S. buckleyi* from *S. petiolaris* (for further discussion, see the treatment of that species). However, there also are problems of infraspecific variation within the species. Two elements traditionally have been accepted as occurring in Missouri (Steyermark, 1963), but different authors have emphasized different characters to separate the two. As noted by Steyermark (1963) and others (Cronquist, 1980), there is widespread intergradation between these phases. Plants that have been segregated variously as var. *angusta* and var. *wardii* have been characterized variously as having relatively narrow leaves, leaves that are sticky and somewhat shiny, leaves with the undersurface glabrous or slightly roughened, and/or involucres that are glabrous but with sessile glands. In contrast, various authors have characterized var. *petiolaris* as having relatively broad median and lower leaves, leaves that are not sticky or shiny, leaves with the undersurface sparsely to moderately pubescent with minute (0.1–0.4 mm long) hairs along the main veins, and/or involucres that are glabrous to more commonly sparsely to moderately hairy and sometimes somewhat sticky. The last author to study the complex was Nesom (1990e), who examined a large suite of herbarium specimens from throughout the range of the species. He concluded that there was no correlation in variation between any of the aforementioned characters. Plants from the Ozarks were most likely to have sticky, glabrescent leaves, but narrow-leaved and broad-leaved plants were equally likely to have sticky leaves. Similarly, the pubescence of the involucre varied independently of the leaf characters. Nesom concluded that for the present there is no rational way to formally subdivide *S. petiolaris* into varieties and that in fact a more distinctive (but still morphologically confluent) element within the range of variation occurred not in the western portion of the range but in a series of populations in the southeastern states. It remains for future studies of the population genetics of the complex to tease apart the intricate patterns of morphological variation present.

17. Solidago ptarmicoides (Torr. & A. Gray) B. Boivin (white upland aster, sneezewort aster)
Aster ptarmicoides Torr. & A. Gray
Oligoneuron album (Nutt.) G.L. Nesom
S. asteroides Semple

Pl. 237 c, d; Map 1006

Plants with the rootstock short and branched, not producing creeping rhizomes. Stems 1 to several, 10–50 cm long, erect or ascending, with several fine, longitudinal ridges or lines, moderately roughened with stiff, stout, broad-based, upward-curved hairs toward the tip, sparsely roughened to nearly glabrous toward the base, not shiny. Leaves basally disposed, often persistent at flowering (additional rosettes sometimes present adjacent to the flowering stem). Basal and lowermost stem leaves with the blade 6–20 cm long, 0.5–1.0 cm wide, more than 10 times as long as wide, narrowly oblanceolate to nearly linear, relatively thick and stiff, long-tapered to a sessile or short-petiolate base, angled to a sharply pointed tip, the margins entire or with a few shallow, sharp teeth and minutely roughened, the surfaces glabrous or sparsely roughened, the undersurface usually with 3 main veins. Median and upper stem leaves 1.5–8.0 cm long, linear to very narrowly elliptic-oblanceolate, the undersurface with 1 main vein or with a faint pair of lateral main veins, otherwise similar to the lower stem leaves. Inflorescences terminal panicles, appearing flat-topped or shallowly rounded in overall outline, the heads solitary or in small clusters at the branch tips. Involucre 4–7 mm long, the bracts in 4–6 unequal series. Involucral bracts linear to oblong-lanceolate, all but the outer series rounded to bluntly pointed at the appressed-ascending tip, the pale margins slightly irregular to irregularly hairy (but sometimes curled under and not observed), the outer surface glabrous, pale yellow to straw-colored at the base, with an oblong to elliptic or somewhat diamond-shaped green area above the base, the midvein usually thickened or keeled, the 1 or 2 pairs of additional veins usually faint. Receptacle naked. Ray florets 10–25, the corollas 6–9 mm long, white or less commonly pale cream-colored. Disc florets 30–38, the corollas 3.5–4.0 mm long, the lobes 0.5–0.7 mm long, white or less commonly pale cream-colored (usually appearing yellow because of the exserted yellow anthers). Pappus 3.5–4.0 mm long, most of the bristles slightly thickened toward the tip. Fruits 1.0–1.5 mm long, narrowly obovoid, glabrous. $2n=18$. July–September.

Scattered in the Ozark and Ozark Border Divisions (eastern U.S. west to Montana, Colorado, and Oklahoma; Canada). Glades, tops of bluffs, and upland prairies; also pastures, railroads, and roadsides, often on calcareous substrates.

Long classified as an unusual aster (Steyermark, 1963), this taxon was first transferred to *Solidago* by Boivin (1971–1972), based on the observation that the taxon hybridizes with other goldenrods but not with asters. Brouillet and Semple (1981) performed a morphological analysis that supported this reclassification. Zhang's (1996) molecular studies also supported a close relationship between *S. ptarmicoides* and members of the *Oligoneuron* group of *Solidago*. The species nomenclature has been controversial as well, with various opinions on the correct authorship of the epithet *ptarmicoides* and whether that name was legitimately published. However, the name seems to have become stabilized as cited above in the recent botanical literature.

This species differs from all other asters and goldenrods in the state in having both ray and disc florets with the corollas white. In Missouri, no putative hybrids involving *S. ptarmicoides* have been discovered yet, but elsewhere in its range it hybridizes with *S. canadensis*, *S. riddellii*, and *S. rigida*, as well as some non-Missouri species.

18. Solidago radula Nutt. (rough goldenrod)
S. radula var. *laeta* (Greene) Fernald
S. radula var. *stenolepis* Fernald

Pl. 241 n–p; Map 1007

Plants with the rootstock short and somewhat thickened, often also producing branched short- to long-creeping rhizomes. Stems solitary or few to several, 40–120 cm long (but usually shorter than 100 cm), erect to loosely ascending, with several fine, longitudinal ridges or grooves, moderately to densely pubescent with short, stiff, mostly spreading hairs 0.1–0.6 mm long, these often broad-based (the stem roughened to the touch) and sometimes sparse toward the stem base, not shiny, not glaucous. Leaves chiefly cauline, the largest leaves about ⅓ of the way up the stem, the basal and lower stem leaves usually absent at flowering. Basal and lowermost stem leaves with the blade 3–10 cm long, 0.7–2.5 cm wide, mostly 4–6 times as long as wide, elliptic-lanceolate to elliptic or elliptic-oblanceolate, occasionally narrowly elliptic-obovate, somewhat thickened and relatively stiff, tapered gradually to a short to long, winged petiole at the base, angled or tapered to a sharply pointed tip, the margins sharply but finely toothed to shallowly scalloped or nearly entire and microscopically roughened to minutely hairy, the surfaces sparsely to moderately roughened with short, mostly spreading, broad-based hairs, the undersurface with 3 main veins, the lateral pair (and midvein) often somewhat raised, usually only slightly more prominent than the other pinnate secondary veins (these forming an irregular network), the veinlets often difficult to observe. Median and upper stem leaves 1–8 cm long, sessile or very short-petiolate, the blade mostly 2–6 times as long as wide, lanceolate to elliptic or narrowly elliptic-obovate, the margins of at least the uppermost leaves entire, otherwise similar to the lower stem leaves. Inflorescences dense, narrow to broad, more or less pyramidal panicles, the branches and often also the tip usually arched or nodding, the lowermost branches sometimes relatively long, the heads oriented upward along the branches. Involucre 3–5 mm long, the bracts in 3 or 4 unequal series. Involucral bracts oblong-ovate to narrowly oblong-lanceolate and mostly bluntly to sharply pointed at the appressed-ascending tip, the thin, white to yellowish white margins hairy (at least toward the tip), the outer surface glabrous, with an often poorly differentiated, elliptic or narrowly diamond-shaped, green to light green central region mostly above the midpoint, this tapered gradually to the midvein, the midvein often slightly thickened and keeled and no additional veins present. Receptacle naked. Ray florets 4–8, the corollas 2.5–4.0 mm long, yellow. Disc florets 4–6, the corollas 2.5–3.0 mm long, the lobes 0.5–0.9 mm long, yellow. Pappus 2.5–3.0 mm long, a few of the bristles often slightly thickened toward the tip. Fruits 1.5–2.5 mm long, narrowly obovoid, finely hairy (sometimes only sparsely so). $2n=18, 36$. May–October.

Scattered in the southern half of the state north locally to Monroe and Marion Counties, but absent from nearly all of the Mississippi Lowlands Division (southeastern U.S. west to Kansas and Texas). Openings of mesic to dry upland forests, ledges and tops of bluffs, glades, savannas, and rarely upland prairies; also roadsides.

Steyermark (1963) noted that this is one of the earliest goldenrods to start flowering in Missouri. He and some other authors attempted to segregate var. *laeta* and var. *stenolepis* from var. *radula* based on minor differences in involucral bract length and width, but these features overlap too much to make such recognition advisable.

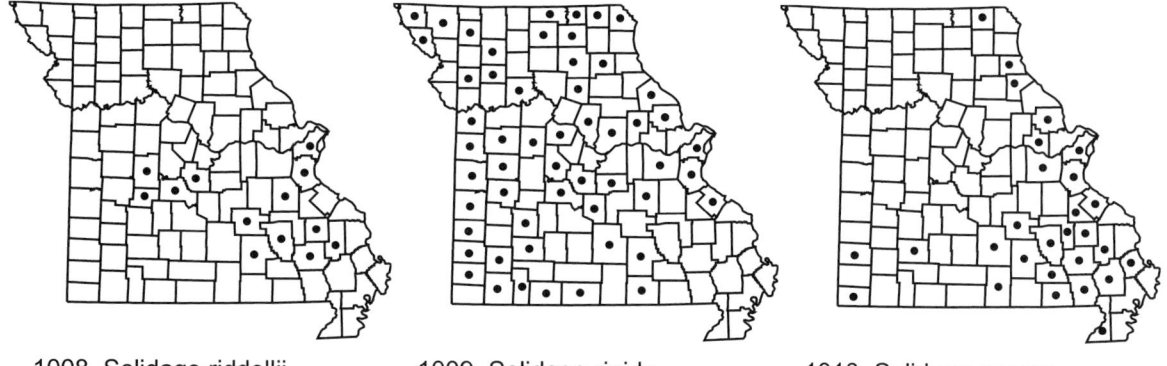

1008. Solidago riddellii   1009. Solidago rigida   1010. Solidago rugosa

19. Solidago riddellii Frank ex Riddell
Oligoneuron riddellii (Frank ex Riddell) Rydb.

Pl. 237 e, f; Map 1008

Plants with the rootstock short and branched, sometimes producing short-creeping, stout rhizomes. Stems 1 to several, 40–100 cm long, erect or ascending, with several fine, longitudinal ridges or grooves, glabrous below the inflorescence (where scattered, inconspicuous, short hairs sometimes occur), somewhat shiny. Leaves basally disposed, often withered at flowering (additional rosettes usually present adjacent to the flowering stem), the papery remains often persistent. Basal and lowermost stem leaves with the blade 8–25 cm long, 0.8–1.5 cm wide, more than 10 times as long as wide, narrowly oblanceolate to linear, relatively stiff and folded longitudinally along the midvein, long-tapered to a more or less long-petiolate, winged base, angled or tapered to a sharply pointed tip, the margins entire but minutely roughened, the surfaces glabrous, the undersurface usually with 3–7 faint to conspicuous main veins. Median and upper stem leaves 2–8(–12) cm long, linear (the uppermost often narrowly lanceolate), sessile or with a short, poorly differentiated petiole, otherwise similar to the lower stem leaves. Inflorescences terminal panicles (or sometimes appearing as a dense terminal cluster), appearing flat-topped or shallowly rounded in overall outline, the heads solitary or in small clusters at the branch tips. Involucre 4.5–6.0 mm long, the bracts in 3–5 unequal series. Involucral bracts linear to narrowly oblong, rounded to broadly and bluntly pointed at the appressed-ascending tip, the pale margins slightly irregular and/or finely hairy (but sometimes curled under), the outer surface glabrous, pale yellow to straw-colored toward the base, with an oblong to elliptic or somewhat diamond-shaped green area toward the tip, the midvein usually thickened or keeled, the 1 or 2 pairs of additional veins often faint. Receptacle naked. Ray florets 7–9, the corollas 5–6 mm long, yellow. Disc florets 6–10, the corollas 4.5–5.5 mm long, the lobes 0.7–1.5 mm long, yellow. Pappus 3.5–4.5 mm long, most of the bristles slightly thickened toward the tip. Fruits 1.5–2.2 mm long, narrowly obovoid, glabrous or with a few hairs along the ribs. $2n=18$. August–October.

Scattered to uncommon in the Ozark and Ozark Border Divisions (North Dakota to Arkansas east to Michigan and Ohio; apparently disjunct in Georgia). Fens and calcareous seeps along streams.

This attractive species is an indicator of high-quality calcareous seepage wetlands, especially fens.

20. Solidago rigida L. (stiff goldenrod, rigid goldenrod)
Oligoneuron rigidum (L.) Small

Pl. 237 g, h; Map 1009

Plants with the rootstock short and branched, often producing short-creeping, stout rhizomes. Stems 1 to several, 30–150 cm long, erect or ascending, with several fine, longitudinal ridges or grooves, usually moderately to densely pubescent with short, curved hairs, rarely sparsely hairy to glabrous toward the base or nearly entirely glabrous below the inflorescence, not shiny. Leaves basally disposed, often persistent at flowering (additional rosettes sometimes present adjacent to the flowering stem). Basal and lowermost stem leaves with the blade 8–20 cm long, 1.8–4.0 cm wide, mostly 3–6 times as long as wide, oblanceolate to elliptic-obovate, ovate, or oblong-elliptic, not or only slightly thickened but often somewhat stiff, tapered to a long petiole at the base, angled to a usually sharply pointed tip, the margins finely scalloped or minutely and bluntly toothed, the surfaces moderately to densely pubescent with short, fine, curved hairs (somewhat roughened to the touch), the undersurface with 1 main vein, the pinnate secondary veins usually faint. Median and upper stem leaves 1.5–11.0 cm long, lanceolate to narrowly elliptic or elliptic, the margins of at least the uppermost leaves usually entire, otherwise similar to the lower stem leaves. Inflorescences terminal panicles, appearing flat-topped or less commonly shallowly rounded in overall outline, the heads solitary or in small clusters at the branch tips. Involucre 5–9 mm long, the bracts in 3–5 unequal series. Involucral bracts mostly oblong and rounded to bluntly pointed at the appressed-ascending tip, those of the outermost and rarely also innermost series sometimes narrowly oblong and bluntly to sharply pointed at the tip, the thin, white to yellowish white margins hairy, the outer surface glabrous or finely hairy, with a green to pale green central region of varying width, the midvein slightly thickened, the 1–3 pairs of additional veins usually easily observed. Receptacle naked. Ray florets 6–14, the corollas 3.0–5.5 mm long, yellow. Disc florets 15–35, the corollas 4.5–6.0 mm long, the lobes 0.6–1.1 mm long, yellow. Pappus 3.0–5.5 mm long, most of the bristles slightly thickened toward the tip. Fruits 0.8–1.7 mm long, obovoid, glabrous or sparsely hairy toward the tip. $2n=18, 36$. August–October.

Scattered nearly throughout the state, but uncommon in the eastern portion of the Ozark Division and apparently absent from the Mississippi Lowlands (eastern U.S. west to Montana, Colorado, and Oklahoma; Canada). Bottomland prairies, upland prairies, loess hill prairies, savannas, glades, openings of mesic to dry upland forests, and uncommonly banks of streams and rivers; also old fields, pastures, railroads, roadsides, and open, disturbed areas.

Stiff goldenrod is an attractive species that is available for use in wildflower gardens at some native plant nurseries. Steyermark (1963) noted that three varieties were traditionally accepted across the range of *S. rigida,* but he was able to locate Missouri specimens to document only var. *rigida.* Heard and Semple (1988) completed detailed morphometric studies on the species and also reported a number of chromosome counts. They validated the three infraspecific taxa accepted by earlier authors but chose to elevate them to subspecies status and also refined the characters separating them. During their examination of more than 1,800 herbarium specimens from throughout the range of *S. rigida,* Heard and Semple discovered a small number of specimens to document native occurrences of ssp. *glabrata* and ssp. *humilis* in Missouri.

1. Involucral bracts all glabrous on the outer surface (hairy along the margins); stems glabrous or sparsely hairy below the midpoint (often more densely hairy toward the tip) ...... 20A. SSP. GLABRATA
1. At least the outer series of involucral bracts finely hairy on the outer surface; stems moderately to densely hairy throughout
    2. Involucral bracts all finely hairy on the outer surface, those of the inner series often relatively narrow and sharply pointed at the tip; stems mostly 30–70 cm long
    .... 20B. SSP. HUMILIS
    2. Inner series of involucral bracts glabrous or nearly so on the outer surface, relatively broad and rounded to bluntly pointed at the tip; stems (40–)60–150 cm long
    .... 20C. SSP. RIGIDA

20a. ssp. glabrata (E.L. Braun) S.B. Heard & Semple
  S. rigida var. *glabrata* E.L. Braun
  Oligoneuron rigidum var. *glabratum* (E.L. Braun) G.L. Nesom

Stems 40–120 cm long, glabrous or sparsely hairy below the midpoint (often more densely hairy toward the tip). Involucral bracts all glabrous on the outer surface (hairy along the margins), those of the inner series relatively broad and rounded to bluntly pointed at the tip. $2n=18$. September–October.

Uncommon, known thus far from a single historical collection from Jefferson County (southeastern U.S. west to Missouri and Texas). Dolomite glades.

This subspecies tends to have slightly thinner, more lax leaves than the others. The report of ssp. *glabrata* from Ozark County by Turner and Yatskievych (1992) was based on a misdetermined specimen of ssp. *rigida.*

20b. ssp. humilis (Porter) S.B. Heard & Semple
  S. rigida var. *humilis* Porter
  Oligoneuron rigidum var. *humile* (E.L. Braun) G.L. Nesom

Stems mostly 30–70 cm long, moderately to densely hairy throughout. Involucral bracts all finely hairy on the outer surface (also along the margins), those of the inner series relatively narrow and sharply pointed at the tip. $2n=18$. August–October.

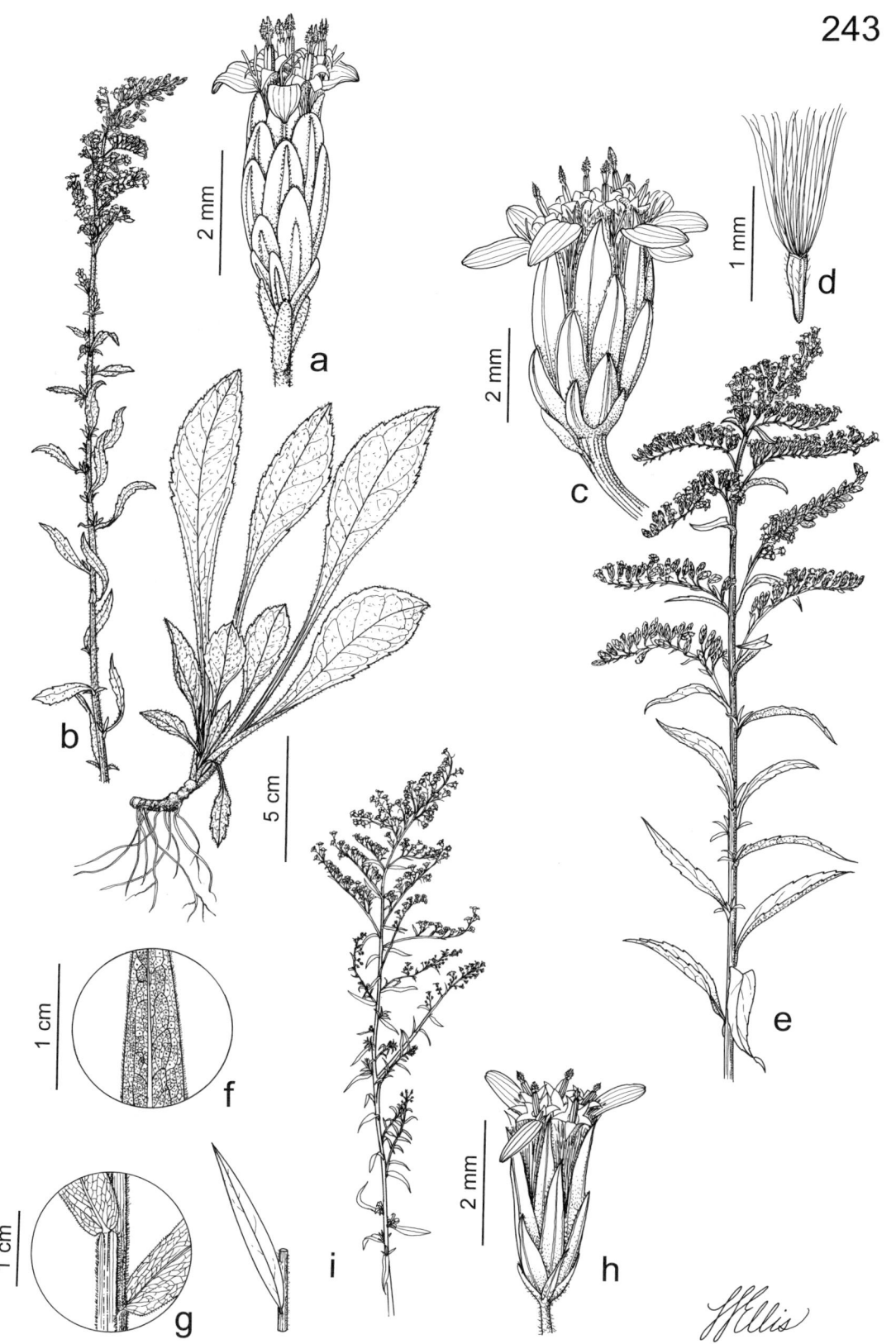

Plate 243. Asteraceae. *Solidago nemoralis*, **a)** head, **b)** habit and basal rosette. *Solidago juncea*, **c)** head, **d)** fruit, **e)** habit. *Solidago odora*, **f)** leaf detail, **g)** leaf base detail, **h)** head, **i)** habit.

Uncommon, known at present only from historical collections from Clark, Jackson, and Jasper Counties (Wisconsin to Montana south to Missouri and New Mexico; Canada; introduced farther east). Upland prairies.

Heard and Semple (1988) mapped an additional occurrence of this subspecies from Atchison County, but the specimen could not be located during the present research. The taxon should be searched for in westernmost Missouri as it occurs in adjacent portions of Nebraska and Kansas.

20c. ssp. rigida

Stems (40–)60–150 cm long, moderately to densely hairy throughout. All but the outermost series of involucral bracts glabrous on the outer surface (hairy along the margins), those of the outer series finely hairy, those of the inner series relatively broad and rounded to bluntly pointed at the tip. $2n=18, 36$. August–October.

Scattered nearly throughout the state, but uncommon in the eastern portion of the Ozark Division and apparently absent from the Mississippi Lowlands (eastern U.S. west to Minnesota and Texas; Canada). Bottomland prairies, upland prairies, loess hill prairies, savannas, glades, openings of mesic to dry upland forests, and uncommonly banks of streams and rivers; also old fields, pastures, railroads, roadsides, and open, disturbed areas.

21. Solidago rugosa Mill. (rough-leaved goldenrod, rough-stemmed goldenrod)

Pl. 240 a, b; Map 1010

Plants with branched, long-creeping rhizomes, often also thickened at the stem bases. Stems 1 to few, 30–150(–200) cm long, erect to loosely ascending, with several fine, longitudinal ridges or grooves, moderately to densely pubescent with short, mostly spreading hairs 0.1–1.2 mm long (sometimes less densely hairy toward the stem base), not shiny, not glaucous. Leaves chiefly cauline, the largest leaves about ⅓ of the way up the stem, the basal and lower stem leaves withered or absent at flowering (additional rosettes usually absent). Basal and lowermost stem leaves with the blade 6–12 cm long, 1–3 cm wide, mostly 3–8 times as long as wide, elliptic to lanceolate, sometimes relatively thick and stiff, angled to a sessile or occasionally very short-petiolate base, angled or tapered to a sharply pointed tip, the margins sharply toothed and inconspicuously hairy, the upper surface glabrous or sparsely to moderately roughened with short, stiff, broad-based hairs (sometimes only along the midvein) or occasionally with somewhat softer pubescence, the undersurface moderately to densely pubescent with spreading or curved hairs, with 1 main vein, the pinnate secondary veins sometimes strongly raised, easily observed (usually forming an irregular network). Median and upper stem leaves 1–12 cm long, ovate to elliptic or lanceolate, the margins toothed or those of the uppermost leaves entire, otherwise similar to the lower stem leaves. Inflorescences narrow to broad, open, more or less pyramidal panicles, the branches usually arched or nodding, the lowermost branches sometimes relatively long, the heads oriented upward along the branches. Involucre 2.5–4.5 mm long, the bracts in 3 or 4 unequal series. Involucral bracts lanceolate to narrowly oblong-lanceolate and bluntly to sharply pointed at the appressed-ascending tip, the thin, white to yellowish white margins hairy (at least toward the tip), the outer surface glabrous, with an elliptic or narrowly diamond-shaped, green to light green central region mostly above the midpoint, this tapered gradually to the midvein, the midvein often slightly thickened and keeled and no additional veins present. Receptacle naked. Ray florets 6–11, the corollas 1.5–3.0 mm long, yellow. Disc florets 3–8, the corollas 2–4 mm long, the lobes 0.7–1.2 mm long, yellow. Pappus 2.0–2.5 mm long, a few of the bristles often slightly thickened toward the tip. Fruits 0.9–1.5 mm long, narrowly obovoid, finely hairy. $2n=18, 36, 54$. August–October.

Scattered mostly in the southeastern quarter of the state (eastern U.S. west to Wisconsin and Texas; Canada). Bottomland forests, banks of streams, rivers, and spring branches, fens, bases and ledges of bluffs, and less commonly mesic to dry upland forests; also pastures and roadsides.

Infraspecific variation in this is complex and poorly understood, and a number of subspecies and varieties have been named. There has not been any attempt to correlate the morphological features with the three ploidy levels recorded thus far. For further discussion see the treatments of the accepted varieties below.

1. Leaf blades relatively thick and stiff, the margins often bluntly toothed, sometimes sparsely so, the undersurface with the secondary veins strongly raised; ray florets 6–8 . . . . 21A. SSP. ASPERA
1. Leaf blades relatively thin and not or only slightly stiffened, the margins usually sharply toothed, usually moderately so, the undersurface with the secondary veins not or only slightly raised; ray florets 8–11
. . . . 21B. SSP. RUGOSA

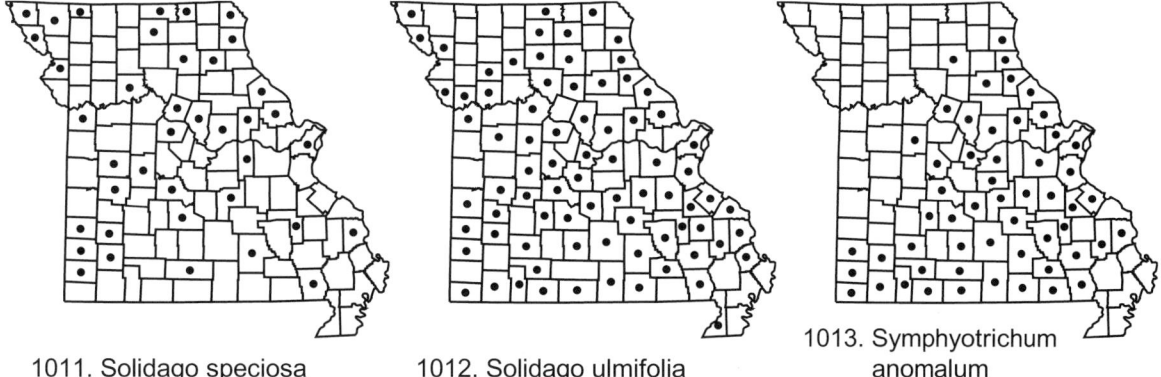

1011. Solidago speciosa   1012. Solidago ulmifolia   1013. Symphyotrichum anomalum

21a. ssp. aspera (Aiton) Cronquist
  S. rugosa var. *aspera* (Aiton) Fernald
  S. rugosa var. *celtidifolia* (Small) Fernald

Leaf blades relatively thick and stiff, mostly angled at the tip, the margins sparsely to moderately toothed, the teeth often blunt, the upper surface glabrous or sparsely to moderately roughened along the main veins, the undersurface with the secondary veins strongly raised. Ray florets 6–8. $2n=18, 36$. August–October.

Scattered mostly in the southeastern quarter of the state (eastern U.S. west to Illinois and Texas; Canada). Bottomland forests, banks of streams, rivers, and spring branches, fens, bases and ledges of bluffs, and less commonly mesic to dry upland forests; also pastures and roadsides.

In addition to var. *rugosa,* Steyermark (1963) accepted var. *aspera* and var. *celtidifolia.* However, he noted that there was morphological overlap between them and that some authors chose to combine the two. He separated var. *celtidifolia* based on its elongate lower inflorescence branches and relatively broad upper stem leaves and reported it as uncommon in five counties in southeastern Missouri. John Semple of the University of Waterloo examined specimens of the complex during his studies of the genus for Ontario (Semple et al., 1999) and for a forthcoming treatment for the Flora of North America. He reached a different conclusion than Steyermark did, annotating two specimens (from Butler and Lincoln Counties) as var. *celtidifolia* and the other pertinent specimens as var. *aspera.* As suggested by these differing interpretations of the Missouri materials, the degree of inflorescence branching and upper leaf shape do not appear to separate Missouri plants into two discrete taxa. Gleason and Cronquist (1991) and Semple et al. (1999) treated the main two variants within *S. rugosa* as subspecies and discussed the segregation of additional varieties within the two.

They noted that this subspecies is more common in the southeastern portion of the species range.

21b. ssp. rugosa

Leaf blades relatively thin and only slightly stiffened, mostly tapered at the tip, the margins usually moderately toothed, the teeth usually sharp, the upper surface sparsely to moderately roughened or occasionally with softer hairs, the undersurface with the secondary veins not or only slightly raised. Ray florets 8–11. $2n=18, 36, 54$. August–October.

Scattered mostly in the southeastern quarter of the state (eastern U.S. west to Wisconsin and Texas; Canada). Bottomland forests, banks of streams, rivers, and spring branches, fens, bases and ledges of bluffs, and less commonly mesic to dry upland forests; also pastures and roadsides.

Gleason and Cronquist (1991) and Semple et al. (1999) noted that this subspecies is more common in the northern portion of the species range. These authors accepted additional varieties within ssp. *rugosa:* the glabrous var. *sphagnophila* Graves, and plants with a narrow, leafy inflorescence called var. *villosa* (Pursh) Fernald. However, those varieties occur far to the east of Missouri and all of our specimens appear referable to var. *rugosa.*

22. Solidago speciosa Nutt. (prairie goldenrod, showy goldenrod)

Pl. 238 f, g; Map 1011

Plants with the rootstock short, stout, and sometimes branched, usually not producing rhizomes. Stems 1 to several, 40–150(–200) cm long, erect or ascending, with several fine, longitudinal ridges or grooves, glabrous or nearly so below the inflorescence (the inflorescence axis and/or branches often minutely hairy), not shiny, not glaucous. Leaves basally disposed or the largest leaves about ⅓ of the way up the stem, absent or persistent at flowering (additional rosettes occasionally present adjacent to the flowering stem). Basal and lowermost stem leaves with the blade 5–30 cm long, 1–10 cm wide, mostly 2–6 times as long as wide, oblanceolate to elliptic or obovate, somewhat thickened and firm, tapered to a relatively short petiole at the base, angled to a bluntly or sharply pointed tip, the margins entire to shallowly scalloped or toothed and inconspicuously hairy, the surfaces glabrous, the undersurface with 1 main vein, the fine, pinnate secondary veins often relatively faint (these usually forming an irregular network). Median and upper stem leaves 1–10 cm long, elliptic to narrowly lanceolate or nearly linear, the margins entire, otherwise similar to the lower stem leaves. Inflorescences of axillary clusters or axillary racemes, these usually appearing as a spicate to racemose or broader, elongate terminal panicle (this not pyramidal or nodding), the heads oriented in several directions when ascending to spreading branches are present. Involucre 3–6 mm long, the bracts in 3–5 unequal series. Involucral bracts mostly oblong to narrowly oblong and rounded to bluntly pointed (those of the outer series often narrowly ovate and sharply pointed) at the appressed-ascending tip, the thin, white to yellowish white margins hairy, the outer surface glabrous and often somewhat sticky, with a poorly defined, pale green to green central region toward the tip, this tapered abruptly to the midvein above the bract midpoint, the midvein often slightly thickened and no additional veins present. Receptacle naked. Ray florets 5–8, the corollas 3.5–5.0 mm long, yellow. Disc florets 7–10, the corollas 2–4 mm long, the lobes 0.5–0.9 mm long, yellow. Pappus 2.0–3.5 mm long, a few of the bristles often slightly thickened toward the tip. Fruits 1.0–1.8 mm long, narrowly obovoid, glabrous. $2n=18$, 36. August–November.

Scattered nearly throughout the state but uncommon in most of the Ozark, Ozark Border, and Mississippi Lowlands Divisions (eastern U.S. west to North Dakota and New Mexico; Canada). Upland prairies, loess hill prairies, ledges and tops of bluffs, openings of mesic to dry upland forests, savannas, glades, banks of streams and rivers, and rarely swamps; also railroads and roadsides.

This is mainly a species of upland sites. It grows well as an ornamental in sunny wildflower gardens and prairie plantings. Moerman (1998) noted that the Chippewa used an extract from the roots and sometimes also stems in treating hemorrhages, sprains, and skin problems, and as a tonic. Several infraspecific taxa have been accepted by various authors, two of which occur in Missouri. In addition to these, var. *pallida* Porter includes plants with relatively pale leaves and glaucous stems that occur in the western portion of the species range, and var. *jejunifolia* (E.S. Steele) Cronquist comprises plants with relatively few stem leaves and slender, persistent, long-petiolate basal leaves that grow in parts of the northern portion of the species range.

1. Basal and lower stem leaves often withered or absent at flowering, relatively slender, 5–12 cm long, 1.0–2.5 cm wide; median stem leaves mostly 0.8–3.0 cm wide; stems mostly 40–100 cm long
 . . . . 22A. VAR. RIGIDIUSCULA
1. Basal and lower stem leaves usually persistent at flowering, relatively broad, 8–30 cm long, 4–10 cm wide; median stem leaves mostly 2.5–5.0 cm wide; stems mostly 60–150(–200) cm long
 . . . . 22B. VAR. SPECIOSA

22a. var. rigidiuscula Torr. & A. Gray

Stems mostly 40–100 cm long, rarely slightly roughened along the ridges. Leaves basally disposed or the largest leaves about ⅓ of the way up the stem. Basal and lower stem leaves often withered or absent at flowering, relatively slender, 5–12 cm long, 1.0–2.5 cm wide. Median stem leaves mostly 0.8–3.0 cm wide. $2n=18$. August–November.

Scattered nearly throughout the state but most commonly in the western half (Ohio to Tennessee west to North Dakota and Texas; Canada). Upland prairies, loess hill prairies, ledges and tops of bluffs, openings of mesic to dry upland forests, savannas, and glades; also railroads and roadsides.

Steyermark (1963) referred to this taxon as var. *angustata*, but from an examination of the type specimens Cronquist (1980) concluded that this name refers to a somewhat narrower-leaved extreme of var. *speciosa*. Plants of var. *rigidiuscula* tend to be less robust and somewhat stiffer-leaved, and to grow at somewhat drier sites, than do those of var. *speciosa*.

22b. var. speciosa

S. speciosa var. *angustata* Torr. & A. Gray

Stems mostly 60–150(–200) cm long, smooth. Leaves basally disposed. Basal and lower stem leaves persistent at flowering, relatively broad, 8–30 cm long, 4–10 cm wide. Median stem leaves mostly 2.5–5.0 cm wide. $2n=18, 36$. August–November.

Scattered nearly throughout the state but most commonly in the northeastern quarter (eastern U.S. west to Wisconsin, Kansas, and Oklahoma).

Upland prairies, ledges and tops of bluffs, openings of mesic to dry upland forests, savannas, banks of streams and rivers, and rarely swamps; also roadsides.

23. Solidago ulmifolia Muhl. ex Willd. (elm-leaved goldenrod)

Pl. 240 c, d; Map 1012

Plants with the rootstock short and often branched, not producing rhizomes. Stems solitary or more commonly few to several, 40–120 cm long, erect to loosely ascending, with several fine, longitudinal ridges or grooves, glabrous below the inflorescence or moderately pubescent with mostly spreading hairs 0.5–1.5 mm long, not shiny, not glaucous. Leaves basally disposed or the largest leaves about ⅓ of the way up the stem, absent or persistent at flowering (additional rosettes usually absent). Basal and lowermost stem leaves with the blade 6–12 cm long, 2–5 cm wide, mostly 2–6 times as long as wide, elliptic to obovate or narrowly obovate, relatively thin, tapered relatively abruptly to a short to long, winged petiole at the base, angled or tapered to a sharply pointed tip, the margins sharply toothed and inconspicuously hairy, the surfaces sparsely to moderately pubescent with spreading or curved hairs, the upper surface often somewhat roughened to the touch, the undersurface with 1 main vein, the fine, pinnate secondary veins usually easily observed (these usually forming an irregular network). Median and upper stem leaves 1–6 cm long, elliptic to narrowly lanceolate, the margins toothed or those of the uppermost leaves entire, otherwise similar to the lower stem leaves. Inflorescences open, more or less pyramidal panicles, the branches usually arched or nodding, the lowermost branches often relatively long, the heads oriented upward along the branches. Involucre 2.5–4.0 mm long, the bracts in 3 or 4 unequal series. Involucral bracts oblong-ovate to narrowly oblong-lanceolate and bluntly to sharply pointed at the appressed-ascending tip, the thin, white to yellowish white margins hairy (at least toward the tip), the outer surface glabrous, with an elliptic or narrowly diamond-shaped, green to light green central region above the midpoint, this tapered gradually to the midvein, the midvein often slightly thickened and no additional veins present. Receptacle naked. Ray florets 3–5, the corollas 1.5–2.0 mm long, yellow. Disc florets 4–7, the corollas 2.5–3.0 mm long, the lobes 0.5–0.9 mm long, yellow. Pappus 2.0–2.5 mm long, a few of the bristles often slightly thickened toward the tip. Fruits 1.0–1.6 mm long, narrowly obovoid, finely hairy. $2n=18$. August–November.

Scattered nearly throughout the state (eastern U.S. west to Wisconsin and Texas; Canada). Mesic to dry upland forests, ledges and tops of bluffs, glades, savannas, margins of ponds and sinkhole ponds, acid seeps, and banks of streams and rivers; also pastures, old fields, dry ditches, and roadsides.

This species usually produces relatively open inflorescences with the lower branches relatively widely spaced, long, and noticeably arching. Although present in a number of habitats, it is a characteristic species of dry bluffs. Most botanists have accepted two varieties, differing in pubescence pattern.

1. Stems moderately pubescent with mostly spreading hairs 0.5–1.5 mm long
.... 23A. VAR. PALMERI
1. Stems glabrous below the inflorescence (often with spreading hairs along the inflorescence main axis and branches)
.... 23B. VAR. ULMIFOLIA

23a. var. palmeri Cronquist

Stems moderately pubescent with mostly spreading hairs 0.5–1.5 mm long. $2n=18$. August–November.

Scattered nearly throughout the state (eastern U.S. west to Wisconsin and Texas; Canada). Mesic to dry upland forests, ledges and tops of bluffs, glades, savannas, margins of ponds and sinkhole ponds, acid seeps, and banks of streams and rivers; also roadsides.

Scattered, most commonly south of the Missouri River (Missouri, Arkansas, Louisiana, and Mississippi). Mesic to dry upland forests, ledges and tops of bluffs, glades, savannas, margins of ponds and sinkhole ponds, acid seeps, and banks of streams and rivers; also roadsides.

In the Arkansas portion of the Ozarks, this variety apparently is the prevalent phase of the species (Cronquist, 1980). Steyermark (1963) knew only one historical collection from Greene County and referred a number of other hairy collections to *S. rugosa*. However, during his studies of the genus for the Flora of North America Project, John Semple of the University of Waterloo transferred these back to *S. ulmifolia* var. *palmeri*. Morphological differences between the two species are summarized in the key to species above. A number of additional specimens of var. *palmeri* also have been collected during the past few decades, including some outside the Ozark Division.

23b. var. ulmifolia

Stems glabrous below the inflorescence (often with spreading hairs along the inflorescence main axis and branches). 2n=18. August–November.

Scattered nearly throughout the state (eastern U.S. west to Wisconsin and Texas; Canada). Mesic to dry upland forests, ledges and tops of bluffs, glades, savannas, margins of ponds and sinkhole ponds, acid seeps, and banks of streams and rivers; also pastures, old fields, dry ditches, and roadsides.

26. Symphyotrichum Nees (aster)

Plants perennial or rarely annual herbs, usually with short and stout rhizomes (elongate in a few species), less commonly with a short, woody rootstock, rarely taprooted. Stems 1 to few, erect or ascending, unbranched or branched, with fine or less commonly coarse, longitudinal lines or ribs, glabrous or variously hairy. Basal leaves present or absent at flowering. Stem leaves usually gradually reduced toward the tip, variously shaped, glabrous or hairy. Inflorescences usually relatively elongate (occasionally short and broad in poorly developed or young plants), not appearing flat-topped or hemispherical, the branches sometimes appearing racemose, the heads short- to long-stalked, the bracts often (but not always) relatively numerous and conspicuously more slender than the leaves. Heads radiate (appearing discoid in *S. ciliatum*), not sticky or resinous. Involucre 3–10 mm long, cup-shaped to obconical or slightly bell-shaped. Involucral bracts in 3–9 subequal or unequal, overlapping series, 0.2–0.8 mm wide (to 1.2 mm in *S. turbinellum*) and mostly 4–10 times as long as wide, linear to narrowly lanceolate or narrowly oblanceolate, the tip ascending or less commonly spreading to reflexed, with a usually white to pale straw-colored, relatively thin base having a very slender, green midvein, this broadened to an ovate to broadly diamond-shaped green portion toward the tip, sometimes also with narrow purple margins. Receptacle flat or shallowly convex, often with minute, irregular ridges around the concave attachment points of the florets. Ray florets 6–75(–100) (or appearing absent in *S. ciliatum*, in which the marginal pistillate florets have inconspicuous, tubular corollas shorter than the pappus and disc corollas), pistillate, the corolla white, pink, lavender to purple, or purplish blue. Disc florets 6–100, perfect, the corolla shallowly to relatively deeply lobed, yellow, usually turning reddish purple to brownish purple after the pollen has been shed, not persistent at fruiting. Pappus of the ray and disc florets similar (the outermost bristles sometimes appearing slightly shorter), of 1(2) series of numerous (mostly 60–90) finely barbed bristles, these all tapered toward the tip, white to grayish white, straw-colored, light tan, or pale orangish brown. Style branches with the sterile tip (beyond the stigmatic lines) (0.2–)0.3–0.5 mm long, linear to lanceolate. Fruits narrowly obovoid, often somewhat flattened, with 2–6 ribs, glabrous or sparsely hairy, variously tan or brown. About 90 species, North America to South America.

Symphyotrichum is the largest segregate of *Aster* in the New World and includes most of the Missouri species formerly placed in that genus. It is less closely related to the Old World species of true asters (see the treatment of *Aster* for further discussion) than it is to the American *Eurybia* and the group of genera related to *Machaeranthera* Nees (Semple et al., 2002). Generic limits and relationships are still somewhat unsettled within this group. *Symphyotrichum* is morphologically variable, and the classification and taxonomic relationships within some groups of closely related taxa also remain controversial.

The change in disc corolla color from yellow to reddish purple in most species of *Symphyotrichum* and in a few related genera of tribe Astereae has been remarked upon by many botanists. Niesenbaum et al. (1999) investigated this color change in *S. racemosum* (as *Aster vimineus*) and determined that the development of reddish pigmentation in disc corollas was correlated with a reduction in the quantity of pollen available in a given floret, but it apparently resulted in increased overall visitation of heads by insects that are potential pollinators.

Steyermark (1963) noted that several asters provide browse for deer (especially the basal rosettes during the winter months) and that the fertile portions are eaten by turkeys and other wildlife. A number of species of *Symphyotrichum* are cultivated as garden ornamentals in the United States and Canada. Some of the species with larger heads and purple rays are choice ornamentals for the garden. However, care should be taken in planting the species with smaller heads and white ray corollas, as most of these can spread aggressively both by fruits and rhizomes. For situations where large displays of white-rayed heads are desired, species and cultivars of *Boltonia* appear to be a more easily managed choice.

1. Basal and lower stem leaves long-petiolate (the petiole sometimes winged) and the blade with a cordate to abruptly rounded or less commonly truncate base
    2. Involucral bracts with the tips spreading to reflexed; ray florets (18–)20–45 .... 1. S. ANOMALUM
    2. Involucral bracts with the tips appressed or strongly ascending; ray florets 10–25
        3. Leaf blades with the margins entire or rarely with a few minute, widely spaced teeth, at least the upper surface strongly roughened (sandpapery); involucral bracts all or mostly with the midvein fairly abruptly expanded into a broadly diamond-shaped (up to 2.5 times as long as wide) green area toward the tip............................. 14. S. OOLENTANGIENSE
        3. Leaf blades (at least those of the lower leaves) with the margins more densely or deeply toothed, the upper surface glabrous or hairy but at most only somewhat roughened, the undersurface with relatively soft hairs; involucral bracts with the midvein gradually expanded into a narrowly diamond-shaped or narrowly elliptic (2.5–5.0 times as long as wide) green area toward the tip (species difficult to distinguish)
            4. Involucral bracts relatively short-tapered or narrowed at the tip, bluntly to sharply pointed or with an abrupt, short, sharp point, the midvein expanded into a well-defined, diamond-shaped or elliptic (mostly 3–5 times as long as wide), green area toward the tip; basal and lower stem leaves with the petiole unwinged or less commonly very narrowly winged .... 3. S. CORDIFOLIUM
            4. Involucral bracts relatively long-tapered at the tip, sharply pointed, the midvein slightly expanded into a sometimes indistinct, narrowly elliptic to nearly linear (mostly 6–10 times as long as wide) green area toward the tip; some or all of the basal and lower stem leaves with the petiole noticeably winged
                5. Stems moderately to densely and evenly hairy, at least above the midpoint; ray florets with the corollas usually purplish blue (sometimes pale), less commonly lavender......... 4. S. DRUMMONDII
                5. Stems glabrous or sparsely to moderately hairy in longitudinal lines or bands toward the tip; ray florets with the corollas usually white, less commonly pale lavender .... 24. S. UROPHYLLUM
1. Basal and lower stem leaves not with both character states as above, either sessile to short-petiolate or with the blade gradually rounded to tapered at the base
    6. Heads appearing discoid, the pistillate marginal florets with the inconspicuous, tubular corolla shorter than the pappus and disc corollas..... 2. S. CILIATUM
    6. Heads radiate, the pistillate marginal florets with the well-developed ligulate corolla noticeably longer than the pappus and the disc corollas

7. Leaves with both surfaces densely pubescent with appressed, silky hairs, appearing grayish or silvery .... 21. S. SERICEUM
7. Leaves glabrous or more commonly with 1 or both surfaces sparsely to moderately pubescent, the hairs not appressed and silky, the leaves not appearing grayish or silvery
    8. Heads with the stalk and involucre with short, gland-tipped hairs, the upper portions of the stem and the leaves sometimes also glandular
        9. Involucral bracts mostly linear, long-tapered to the slender, sharply pointed tip; ray florets 40–100; stem leaves with the base cordate, clasping the stem .... 11. S. NOVAE-ANGLIAE
        9. Involucral bracts mostly narrowly oblong to narrowly oblong-oblanceolate, all but the inner series angled or short-tapered to the sharply pointed tip; ray florets 12–35; stem leaves with the base variously shaped
            10. Stem leaves narrowed or rounded at the sometimes slightly expanded base, occasionally appearing truncate, but not or only slightly clasping the stem; involucral bracts subequal to somewhat unequal, the tips all spreading to recurved .... 12. S. OBLONGIFOLIUM
            10. Stem leaves deeply cordate at the base, strongly clasping the stem; involucral bracts strongly unequal, the tips erect to somewhat loosely ascending, some of them occasionally somewhat spreading .... 16. S. PATENS
    8. Heads with the stalk and involucre glabrous to variously hairy, but not glandular; stems and leaves not glandular
        11. Leaves (except the basal and sometimes the lowermost stem leaves) with the relatively broad base strongly clasping the stem, often deeply cordate
            12. Stems moderately to densely and evenly hairy .... 16. S. PATENS
            12. Stems glabrous below the inflorescence or inconspicuously and sparsely hairy in longitudinal lines toward the tip and occasionally immediately below the upper leaf bases
                13. Involucral bracts subequal or somewhat unequal but not in several well-defined series, often relatively long-tapered at the tip, the midvein expanded below the midpoint into a narrowly elliptic, green apical area; leaves not glaucous ...... 19. S. PUNICEUM
                13. Involucral bracts unequal in several well-defined series, angled or relatively short-tapered at the tip, the midvein expanded near the tip into a diamond-shaped or broadly elliptic, green apical area; leaves glabrous and usually glaucous................... 8. S. LAEVE
        11. Leaves with the base slender or slightly expanded, sometimes somewhat sheathing but not or only slightly clasping the stem, tapered, angled, rounded, or occasionally appearing truncate
            14. All or at least the outermost few series of involucral bracts somewhat inrolled toward the tip, tapered to a somewhat thickened, awl-shaped (outward curved then upward curved), sharply pointed, green tip, with a short, white to yellowish- or purplish-tinged, relatively stout and spinelike or less commonly slender and hairlike point at the very tip
                15. Plants annual, taprooted; involucral bracts entire and glabrous along the margins .... 22. S. SUBULATUM
                15. Plants perennial, the rootstock relatively short and stout, fibrous-rooted, sometimes also with rhizomes; involucral bracts minutely toothed or irregular and often sparsely short-hairy along the margins

16. Involucre 3.0–4.5 mm long, narrowly ellipsoidal to narrowly cup-shaped or nearly cylindrical when fresh (becoming obconical when pressed); disc florets 6–12; ray florets 10–16(–18), the corollas 3.5–6.0 mm long . . . . 15. S. PARVICEPS
16. Involucre 4–8 mm long, urn-shaped to more or less cup-shaped when fresh (becoming bell-shaped to broadly obconical when pressed); disc florets 20–40; ray florets 15–35, the corollas 5–10 mm long . . . . 17. S. PILOSUM
14. Involucral bracts not inrolled toward the tip, angled or tapered to a relatively flat, sharply pointed tip, this sometimes with a minute, white to tan or purplish-tinged, slender (occasionally hairlike) point at the very tip
17. Involucre (6–)7–12 mm long, the bracts in 6–9 unequal series, rounded or angled to a bluntly pointed or sometimes sharply pointed tip; ray florets with the corolla 10–18 mm long, purple to purplish blue . . . . 23. S. TURBINELLUM
17. Involucre 2.5–7.0(–8.0) mm long, the bracts in 3–6 subequal to unequal series, angled or tapered to a sharply pointed tip; ray florets with the corolla 2.5–10.0(–12.0) mm long, usually white when fresh, rarely pinkish-tinged or lavender (often drying purplish or blue)
18. Plants taprooted annuals; stems and leaves glabrous . . . . . 22. S. SUBULATUM
18. Plants fibrous-rooted perennials, the rootstock usually stout and somewhat woody, often also with rhizomes; stems and/or leaves sparsely to moderately hairy (check closely with magnification, as the hairs may be inconspicuous and in lines or only along leaf midveins)
19. At least the outer involucral bracts loosely ascending to spreading, sparsely to moderately hairy on the outer surface (sometimes only along the midvein), relatively densely hairy along the margins, the tip with a minute, bristlelike extension of the midvein or a short, hard, white to yellowish or purple, spinelike point
20. Involucre 3–5 mm long; ray florets 8–20; heads relatively numerous along the inflorescence branches, all or mostly oriented toward 1 side of each branch . . . . 6. S. ERICOIDES
20. Involucre 5–8 mm long; ray florets (15–)20–35; heads appearing fewer, solitary or in small clusters at the tips of the inflorescence branches, oriented in various directions . . . . 7. S. FALCATUM
19. Involucral bracts erect or strongly ascending, the outermost occasionally somewhat loosely ascending, glabrous or sparsely hairy on the outer surface, glabrous (but often somewhat uneven) or sparsely hairy along the margins, the tip bluntly to sharply pointed, lacking a bristlelike extension of the midvein or a short, hard, spinelike point (or, if sometimes with a short, bristlelike tip, in *S. ontarionis*, then the body of the bract flat and not inrolled toward the tip)
21. Disc corollas with the lobes relatively long (0.9–1.7 mm), 45–75 percent of the total length of the expanded upper portion (above the slender basal portion of the tube) of the corolla
22. Stem leaves with the undersurface moderately to densely short-hairy along the midvein and rarely with a few hairs along the lateral veins, otherwise glabrous (except for the minutely hairy margins) . . . . 10. S. LATERIFLORUM

22. Stem leaves (at least the median and upper ones) with the undersurface sparsely to moderately and evenly short-hairy on the undersurface (including the tissue between the veins), sometimes with slightly longer or denser hairs along the midvein .... 13. S. ONTARIONIS
21. Disc corollas with the lobes relatively short (0.4–1.2 mm), 15–45 percent of the total length of the expanded upper portion (above the slender basal portion of the tube) of the corolla
23. Leaf undersurface with the veinlets relatively prominent, forming a network with areoles that are about as long as wide or slightly longer than wide (these often containing free veinlets); ray corollas purple or bluish purple .... 18. S. PRAEALTUM
23. Leaf undersurface with the secondary veins and/or veinlets either faint or, if relatively prominent, then forming a network with areoles that are much longer than wide (these sometimes containing free veinlets); ray corollas mostly white when fresh (often turning somewhat bluish-tinged when dried), less commonly bluish-tinged or lavender (species difficult to distinguish)
24. Involucre 3.5–8.0 mm long; rays 5–12 mm long; heads solitary or clustered at the branch tips or oriented in various directions and more or less racemose along the inflorescence branches; largest stem leaves (3–)6–40 mm wide .... 9. S. LANCEOLATUM
24. Involucre 2.5–4.0 mm long; rays 3–8 mm long; heads solitary at the branch tips or arranged in mostly 1-sided racemes along the inflorescence branches; largest stem leaves 1–7(–11) mm wide
25. Median and inner series of involucral bracts with a relatively short, elliptic to diamond-shaped green tip, this up to $^1\!/_2$ the length of the bract; heads mostly appearing solitary at the ends of inflorescence branches or, if appearing racemose, then the heads mostly relatively long-stalked .... 5. S. DUMOSUM
25. Median and inner series of involucral bracts with a relatively elongate, elliptic green tip, this mostly more than $^1\!/_2$ the length of the bract; heads appearing solitary or in clusters at the ends of inflorescence branches or, if appearing racemose, then the heads mostly relatively short-stalked
26. Heads mostly in small clusters toward the branch tips or appearing racemose, the stalks relatively short and few-bracted .... 9. S. LANCEOLATUM
26. Heads mostly solitary at the branch tips or sometimes in small, loose clusters, the stalks mostly relatively long and many-bracted .... 20. S. RACEMOSUM

1. Symphyotrichum anomalum (Engelm. ex Torr. & A. Gray) G.L. Nesom
Aster anomalus Engelm. ex Torr. & A. Gray
Pl. 244 a, b; Map 1013

Plants perennial herbs, usually from a short, stout, somewhat branched rootstock, this sometimes somewhat woody, occasionally also producing elongate rhizomes. Stems 1 to few, 30–100 cm long, unbranched or with few to several ascending branches above the midpoint, relatively uniformly and moderately to densely roughened with short, spreading hairs, sometimes only sparsely so toward the base. Basal and/or lower stem leaves present at flowering, long-petiolate, the petiole sometimes narrowly winged (often only toward the tip), the blade 4–10 cm long, 2.0–5.5 cm wide, heart-shaped, deeply cordate at the base, angled or tapered to a sharply pointed tip, the margins entire or slightly irregular, rarely with a few small teeth, moderately to densely roughened with minute, stiff hairs on the upper surface, densely pubescent with slightly longer (still somewhat sandpapery) hairs on the

Plate 244. Asteraceae. *Symphyotrichum anomalum*, **a)** head, **b)** habit and lower leaf. *Symphyotrichum cordifolium*, **c)** head, **d)** habit and lower leaf. *Symphyotrichum drummondii*, **e)** head, **f)** habit. *Symphyotrichum oolentangiense*, **g)** head, **h)** habit and lower leaf.

undersurface, the secondary veins on the leaf undersurface faint or sometimes easily observed, often irregularly fused toward their tips, the veinlets often indistinct, forming a dense, irregular network of relatively short areoles. Median and upper stem leaves progressively smaller, with long to short, often winged but not clasping or sheathing petioles, the blades 1–6 cm long, heart-shaped to narrowly ovate (those along the inflorescence axis mostly lanceolate), cordate to truncate, rounded, or short-tapered at the base, otherwise similar to the lower stem leaves. Inflorescences usually panicles with relatively long, loosely ascending, few-headed, racemose branches, sometimes with denser heads along the branches or reduced to a solitary, open raceme, the heads often appearing relatively long-stalked, the bracts along the ultimate branches 0.6–2.5 cm long, leaflike, linear to narrowly lanceolate, more or less grading into the foliage leaves. Heads mostly 2.0–3.5 cm in diameter (including the extended ray corollas) at flowering. Involucre 5–10 mm long, the bracts in 4–7 unequal, overlapping series. Involucral bracts linear to narrowly lanceolate, relatively long-tapered at the sharply pointed tip, the tip spreading to reflexed, the slender midvein broadened gradually in the apical $1/2$–$2/3$ into a narrowly elliptic or narrowly diamond-shaped, green tip, this also sometimes purplish-tinged, the outer surface sparsely to moderately hairy, the margins moderately to densely hairy, especially toward the tip. Ray florets 20–45 in 1 or 2 series, the corollas well developed, 8–18 mm long, purple to blue, rarely white. Disc florets 20–40, the corollas 4.0–5.5 mm long, the slender portion of the tube shorter than the slightly expanded apical portion, the lobes 0.4–0.6 mm long, 20–25 percent of the total length of the expanded portion. Pappus bristles 3.5–5.0 mm long, off-white to straw-colored or light tan, occasionally purplish-tinged. Fruits 2–3 mm long, mostly with 5 or 6 longitudinal ribs, purplish brown with lighter ribs, glabrous. $2n$=16. July–November.

Scattered, mostly in the Ozark and Ozark Border Divisions and the eastern half of the Glaciated Plains (Kansas, Missouri, and Illinois south to Oklahoma and Arkansas). Glades, upland prairies, savannas, mesic to dry upland forests on rocky slopes, and ledges and tops of bluffs; also fencerows and roadsides.

Rare plants with white ray corollas have been called *A. anomalus* f. *albidus* Steyerm. The global distribution of the species appears to be restricted mostly to the Ozark Mountains. Although in Missouri *S. anomalum* is often associated with acidic substrates, in other states the species reportedly grows more commonly at calcareous sites (Gleason and Cronquist, 1991).

2. Symphyotrichum ciliatum (Ledeb.)
G.L. Nesom (rayless alkali aster)
Brachyactis ciliata (Ledeb.) Ledeb.
B. ciliata ssp. *angusta* (Lindl.) A.G. Jones
Aster brachyactis S.F. Blake
Pl. 249 d, e; Map 1014

Plants annual, with a relatively short, sometimes stout taproot. Stem usually solitary, 10–60 cm long, often with several to numerous loosely to strongly ascending branches near the base, but sometimes branched only above the midpoint, glabrous or very sparsely pubescent around the nodes with short, spreading hairs. Basal and lower stem leaves often absent at flowering, sessile or with a short, poorly differentiated petiole, usually slightly succulent, the blade 3–10 cm long, 0.2–1.0 cm wide, linear to narrowly oblanceolate, tapered at the base, tapered to a sharply pointed tip, the margins entire but often with sparse to moderate, short, stout hairs, the surfaces bluish green, glabrous, the secondary veins faint, usually nearly parallel to the more prominent midvein. Median and upper stem leaves somewhat smaller, mostly sessile (note that small clusters of leaves are sometimes produced in the main leaf axils), the base somewhat sheathing the stem, the blades 1–6 cm long, linear, otherwise similar to the lower stem leaves. Inflorescences usually panicles with short to long, ascending, few- to many-headed, racemose branches, rarely reduced to a solitary terminal raceme, the heads appearing short- or long-stalked, the bracts along the ultimate branches 0.3–1.5 cm long, leaflike, mostly linear, often only slightly shorter than the adjacent foliage leaves. Heads mostly 0.4–0.8 cm in diameter (broader when pressed) at flowering. Involucre 5–11 mm long, the bracts in 3 or 4 subequal, overlapping series, those of the outermost series usually somewhat longer and leaflike. Involucral bracts linear to narrowly oblong-oblanceolate, tapered at the sharply pointed tip, the tip erect to loosely ascending (sometimes appearing spreading when pressed), green most of the length, the green color grading indistinctly into $^{1}/_{10}$–$^{1}/_{3}$ of the pale basal area, the outer surface glabrous, the margins sparsely hairy. Ray florets apparently absent but actually 25–60 (more numerous than the disc florets) in several series, the corollas poorly developed, tubular, and lacking a ligulate portion, 1.5–2.2 mm long (shorter than the style), pink to pale purple. Disc florets 15–40, the corollas 3.5–5.0 mm long, the slender portion of the tube about as long to longer than the slightly expanded apical portion, the lobes 0.3–0.4 mm long, 10–15 percent of the total length of the expanded portion. Pappus bristles 4.0–6.5 mm long, noticeably longer than the florets, white to off-white, rarely appearing slightly pinkish-tinged. Fruits 1.5–2.5 mm long, with 2 or 4 longitudinal ribs, gray to light tan, occasionally with purplish streaks, sparsely to moderately hairy. $2n=14$. August–September.

Introduced, known thus far only from a single historical specimen from Clay County (Washington to New Mexico east to Minnesota, Iowa, and Oklahoma; Canada, Siberia; introduced eastward to Pennsylvania). Railroads.

Steyermark (1963) discussed a putatively native occurrence of this taxon from a loess hill prairie in Atchison County. Although this occurrence seems likely, based on the rest of the species range, a voucher specimen could not be located during the present study, nor has the species been relocated in the field in spite of relatively intensive botanizing in the loess hills during recent years by a number of botanists.

3. Symphyotrichum cordifolium (L.) G.L. Nesom (blue wood aster)

Aster cordifolius L.

A. sagittifolius Wedem. ex Willd.

A. cordifolius ssp. *sagittifolius* (Wedem. ex Willd.) A.G. Jones

A. cordifolius var. *polycephalus* Porter

S. cordifolium var. *polycephalum* (Porter) G.L. Nesom

Pl. 244 c, d; Map 1015

Plants perennial herbs, usually from a short, stout, somewhat branched rootstock, this sometimes somewhat woody, often also producing elongate rhizomes. Stems 1 to several, 25–120 cm long, unbranched or with few to several ascending branches above the midpoint, glabrous or sparsely to moderately pubescent toward the tip with short, curled hairs in longitudinal lines or bands. Basal and/or lower stem leaves usually present at flowering, long-petiolate, the petiole unwinged or less commonly very narrowly winged, the blade 4–13 cm long, 1–6 cm wide, narrowly to broadly heart-shaped, deeply cordate at the base, tapered to a sharply pointed tip, the margins sharply and often relatively coarsely toothed, glabrous or sparsely roughened with minute, stiff hairs on the upper surface, sparsely to moderately pubescent with slightly longer hairs mostly along the veins on the undersurface, the secondary veins on the leaf undersurface usually easily observed, often irregularly fused toward their tips, the faint veinlets forming a dense, irregular network of relatively short areoles. Median and upper stem leaves relatively abruptly smaller toward the stem tip, with mostly long (the uppermost ones often short), unwinged to narrowly winged, nonclasping but somewhat sheathing petioles, the blades 1–7 cm long, mostly ovate (the uppermost narrowly ovate to narrowly lanceolate), cordate to short-tapered at the base, otherwise similar to the lower stem leaves. Inflorescences usually panicles with relatively long, loosely ascending, few- to many-headed, racemose branches, rarely reduced to a solitary terminal raceme or cluster, the heads appearing short- or long-stalked, the bracts along the ultimate branches 0.3–1.2 cm long, leaflike, linear or less commonly narrowly oblong-lanceolate, noticeably to only somewhat shorter and narrower than the adjacent foliage leaves. Heads mostly 1–2 cm in diameter (including the extended ray corollas) at flowering. Involucre 3.5–5.5 mm long, the bracts in 4–6 unequal, overlapping series. Involucral bracts linear to narrowly lanceolate or narrowly oblanceolate, angled to short-tapered at the bluntly to sharply pointed tip or with an abrupt, short, sharp point at the tip, the tip erect or ascending (the lowermost bracts occasionally slightly spreading), the slender midvein broadened in the apical $^{1}/_{4}$–$^{1}/_{2}$ into a well-defined, diamond-shaped or elliptic (mostly 3–5 times as long as wide), green area (often also purplish-tinged along the margins), the outer surface glabrous or less commonly sparsely short-hairy, the margins irregularly hairy, especially toward the tip. Ray florets 8–15, usually in 1 series, the corollas well developed, 6–12 mm long, purplish blue to lavender. Disc florets 12–20, the corollas 3.5–5.0 mm long, the slender portion of the tube shorter than the slightly expanded apical portion, the lobes 0.6–0.9 mm long, 20–25 percent of the total length of the expanded portion. Pappus bristles 3–5 mm long, off-white to pale cream-colored or light tan, occasionally pale purplish-tinged. Fruits 1.5–2.5 mm long, mostly with 4 longitudinal ribs, purplish brown to brown, often with lighter ribs, glabrous or sparsely hairy. $2n=16, 32$. August–November.

Scattered nearly throughout the state (eastern U.S. west to North Dakota and Oklahoma; Canada). Bottomland forests, mesic upland forests, banks of streams and rivers, and bases and ledges of bluffs; also railroads and roadsides.

A number of infraspecific taxa have been accepted within this species at various times. As noted in the treatment of *S. urophyllum,* Steyermark (1963) and many other botanists once considered that species as part of the variation within *Aster sagittifolius* until Almut G. Jones (1980) and Jones and Hiepko (1981) established that the type specimen of *A. sagittifolius* instead belongs to *A. cordifolius.* However, the attempts by Jones (1980, 1989) to recognize plants with less rhizome production and somewhat more truncate bases on the upper leaf blades seem ill-advised, given the variability of plants growing in different soils and with different light levels. Similarly, Steyermark (1963) and a few other authors segregated *A. cordifolius* var. *polycephalus* for this same morphological variant. On the other hand, Steyermark (1963) noted that some earlier authors had suggested that var. *moratum* (Shinners) G.L. Nesom (as *A. cordifolius* var. *moratus* (Shinners) Shinners) possibly represented hybrids between *S. cordifolium* and *S. urophyllum* (as *A. sagittifolius*), which seems reasonable. Steyermark (1963) also noted unpublished studies by Edgar Anderson suggesting that apparent widespread hybridization between *S. cordifolium* and *S. drummondii* (as *Aster*) was mostly the result of relatively recent, human-mediated habitat disturbances creating opportunities for them to grow in close proximity more frequently.

4. Symphyotrichum drummondii (Lindl. ex Hook.) G.L. Nesom ssp. drummondii
(Drummond aster)

Aster drummondii Lindl. ex Hook.
A. sagittifolius Wedem. ex Willd. var.
  drummondii (Lindl. ex Hook.) Shinners
Pl. 244 e, f; Map 1016

Plants perennial herbs, usually from a short, stout, somewhat branched rootstock, this sometimes somewhat woody. Stems 1 to few, 30–120 cm long, unbranched or with few to several ascending branches above the midpoint, moderately to densely pubescent above the midpoint with short, curled hairs sometimes in longitudinal lines or bands, sparsely hairy to glabrous toward the base. Basal and/or lower stem leaves usually present at flowering, long-petiolate, the petiole noticeably but sometimes narrowly winged, the blade 4–15 cm long, 2–6 cm wide, heart-shaped, deeply cordate to truncate or abruptly rounded at the base, angled or tapered to a usually sharply pointed tip, the margins sharply and relatively coarsely to less commonly finely toothed, glabrous or more commonly sparsely to moderately pubescent (sometimes somewhat sandpapery) with minute, often stiff hairs on the upper surface, moderately to densely pubescent with slightly longer hairs on the undersurface, the secondary veins on the leaf undersurface often faint, often irregularly fused toward their tips, the veinlets usually very faint, forming a dense, irregular network of relatively short areoles. Median and upper stem leaves progressively smaller (sometimes only toward the stem tip), with long to short, winged, nonclasping but often somewhat sheathing petioles or the uppermost leaves occasionally sessile, the blades 1–10 cm long, ovate to narrowly lanceolate, rounded to tapered at the base, otherwise similar to the lower stem leaves. Inflorescences usually panicles with relatively long, loosely ascending, few- to many-headed, racemose branches, rarely reduced to a solitary terminal raceme or cluster, the heads appearing short- or long-stalked, the bracts along the ultimate branches 0.3–1.2 cm long, leaflike, linear or narrowly elliptic to narrowly oblanceolate, often noticeably shorter and narrower than the adjacent foliage leaves. Heads mostly 1.0–2.2 cm in diameter (including the extended ray corollas) at flowering. Involucre 4–7 mm long, the bracts in 4–6 unequal, overlapping series. Involucral bracts linear to narrowly lanceolate, angled to long-tapered at the sharply pointed tip, the tip erect or ascending (the lowermost bracts occasionally slightly spreading), the slender midvein broadened gradually in the apical $^1/_3$–$^2/_3$ into a usually distinct, narrowly elliptic or narrowly diamond-shaped to occasionally nearly linear, green tip (mostly 6–10 times as long as wide), the outer surface glabrous or less commonly sparsely short-hairy (the inner surface sometimes also sparsely hairy), the margins irregularly hairy toward the tip. Ray florets 8–20, usually in 1 series, the corollas well developed, 5–12 mm long, purplish blue to lavender. Disc

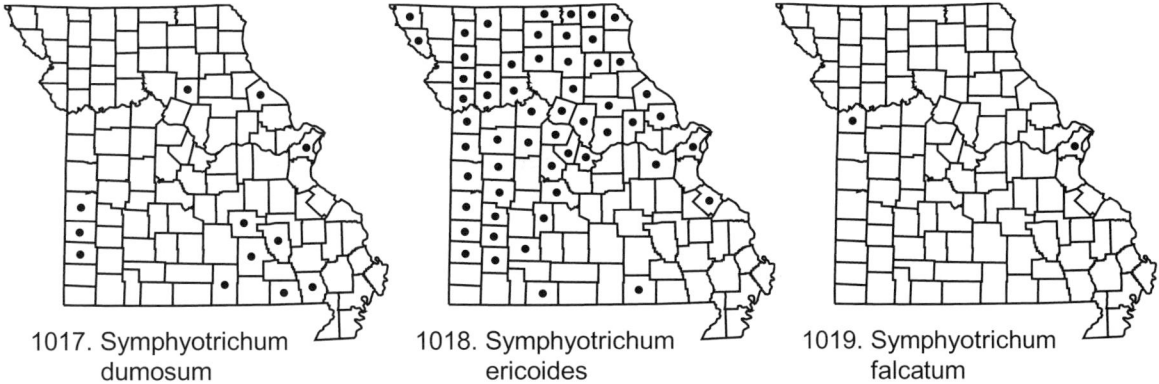

1017. Symphyotrichum dumosum

1018. Symphyotrichum ericoides

1019. Symphyotrichum falcatum

florets 10–18, the corollas 3.5–5.0 mm long, the slender portion of the tube shorter than the slightly expanded apical portion, the lobes 0.5–0.9 mm long, 20–25 percent of the total length of the expanded portion. Pappus bristles 3.0–4.5 mm long, off-white to pale cream-colored or light tan, occasionally pale purplish-tinged. Fruits 2–3 mm long, mostly with 4 longitudinal ribs, purplish brown to brown, often with lighter ribs, glabrous or sparsely hairy. $2n=16, 32$. August–November.

Scattered nearly throughout the state (Minnesota, Nebraska, and Texas east to Pennsylvania and Alabama). Bottomland forests, mesic to dry upland forests, banks of streams and rivers, margins of ponds and lakes, bases, ledges, and tops of bluffs, savannas, upland prairies, glades; also banks of ditches, pastures, railroads and roadsides.

Symphyotrichum drummondii has a relatively broad tolerance for different environmental conditions. At wetter sites it sometimes grows with or near *S. cordifolium*, and at drier sites it sometimes occurs with *S. urophyllum*. For discussions of putative hybrids with these closely related species, see their treatments.

5. **Symphyotrichum dumosum** (L.) G.L. Nesom **var. strictior** (Torr. & A. Gray) G.L. Nesom

Aster dumosus L. var. *strictior* Torr. & A. Gray
A. dumosus var. *dodgei* Fernald
S. dumosum var. *dodgei* (Fernald) G.L. Nesom
Pl. 246 a, b; Map 1017

Plants perennial herbs, often somewhat colonial from relatively long, slender, branched rhizomes. Stems usually solitary, 20–90 cm long, unbranched or with few to most commonly many loosely ascending branches mostly above the midpoint, sparsely pubescent with short, curled hairs toward the tip, these usually in longitudinal lines or bands, usually glabrous toward the base or sometimes nearly throughout. Basal and/or lower stem leaves usually absent from the flowering stems, sessile or with a short, poorly differentiated petiole, the blade 1–5 cm long, 0.4–1.5 cm wide, oblanceolate, tapered at the base, rounded or angled to a bluntly or sharply pointed tip, the margins with spreading to forward-pointing hairs and usually scalloped, the surfaces glabrous or the upper surface sparsely and minutely roughened, the secondary veins on the leaf undersurface often difficult to distinguish from the veinlets, forming an irregular network of elongated areoles. Median and upper stem leaves progressively smaller, the larger ones occasionally withered by flowering time, sessile, the base sometimes slightly expanded but not clasping the stem, the blades 1–10 cm long, linear to narrowly oblanceolate, the margins usually entire and somewhat curled under, angled or tapered at the base, angled or tapered to a sharply pointed tip, otherwise similar to the lower stem leaves. Inflorescences usually appearing as panicles with relatively long, loosely ascending branches (these mostly with the heads solitary or in small, loose clusters toward the branch tips), the heads appearing mostly long-stalked and usually oriented in various directions, the bracts along the ultimate branches 0.1–0.6 cm long, often relatively numerous, more or less leaflike, linear, noticeably shorter than the adjacent foliage leaves. Heads mostly 0.8–1.5 cm in diameter (including the extended ray corollas) at flowering. Involucre 3–5 mm long, cup-shaped to slightly bell-shaped when fresh (sometimes becoming obconical when pressed), the bracts in 4 or 5 unequal, overlapping series. Involucral bracts linear-lanceolate to narrowly oblong-oblanceolate (the median series mostly twice as wide as the outer series), angled or short-tapered at the usually sharply pointed tip, lacking a bristlelike or spinelike point at the ascending tip, the slender midvein broadened abruptly in the apical ¼–½ into a relatively short, oblanceolate or elliptic to diamond-shaped (2–4 times as long as wide), green tip, the outer surface glabrous, the margins often slightly irregular and sparsely to moderately hairy. Ray florets 15–30 in usually 1 or 2 series, the corollas well developed, 4–8 mm long, white to pale pink or bluish-tinged. Disc florets 15–30, the corollas 3.5–4.5 mm long, the slender portion of the tube noticeably shorter than the slightly expanded apical portion, the lobes 0.6–1.1 mm long, 25–35 percent of the total length of the expanded portion. Pappus bristles 3.5–4.5 mm long, white or off-white. Fruits 1.5–2.5 mm long, with 3–5 longitudinal ribs, pinkish gray to tan (sometimes with pinkish lines or streaks), sparsely hairy, sometimes only along the ribs, the hairs lacking swollen bases. $2n=16, 32$. August–November.

Uncommon, mostly in southern Missouri, introduced in the city of St. Louis (northeastern U.S. west to Wisconsin and Missouri; Canada). Fens, margins of sinkhole ponds, and bottomland and less commonly upland prairies; also railroads, moist roadsides, and open, disturbed areas.

Steyermark (1963) recognized two varieties as occurring in Missouri, var. *strictior* and var. *dodgei*. Semple et al. (2002) indicated that the latter was merely a growth form of var. *strictior*. To the east of Missouri, several additional varieties sometimes have been recognized. Jones (1989) suggested that the characteristics that separate var. *strictior* from var. *dumosum* (ascending branches, sparse pubescence, relatively short-stalked heads) may be due to past hybridization between *S. dumosum* var. *dumosum* and *S. racemosum* (as *Aster fragilis*). The taxonomy and relationships of the entire complex deserve more intensive study.

6. Symphyotrichum ericoides (L.) G.L. Nesom (wreath aster)

Aster ericoides L.

Pl. 245 a, b; Map 1018

Plants perennial herbs, often somewhat colonial from relatively long, slender, branched rhizomes. Stems usually solitary, 30–100 cm long, unbranched or with few to most commonly many ascending branches mostly above the midpoint, moderately to densely pubescent with short, appressed to upward-curved or spreading to reflexed hairs, sometimes only sparsely so toward the base, the hairs relatively evenly distributed. Basal and/or lower stem leaves absent from the flowering stems, sessile or with a short, poorly differentiated petiole, the blade 2–6 cm long, 0.3–1.0 cm wide, oblanceolate, tapered at the base, rounded or angled to a minute, abrupt, sharp point at the tip, the margins with spreading to forward-pointing hairs and usually entire, the surfaces glabrous or sparsely pubescent with spreading to somewhat curved hairs, the secondary veins on the leaf undersurface usually a single prominent pair (the leaf thus usually appearing 3-veined) the veinlets often faint and difficult to distinguish, forming an irregular network of elongated areoles. Median and upper stem leaves progressively smaller (the main stem leaves often with clusters of smaller leaves along short, axillary branches), the larger ones often withered by flowering time, sessile, the base sometimes slightly expanded but not clasping the stem, the blades 1–6 cm long, linear or narrowly oblong-oblanceolate, the margins mostly entire, tapered at the base, angled or tapered to a sharply pointed tip (usually with a short, hard point at the tip), usually appearing with 1 main vein, the veinlets usually relatively easily observed, the surfaces sparsely to moderately hairy, otherwise similar to the lower stem leaves. Inflorescences usually appearing as panicles with short to more commonly relatively long, loosely ascending, racemose branches, sometimes reduced to a solitary raceme, the heads appearing mostly long-stalked and oriented upward, usually relatively numerous, the bracts along the ultimate branches 0.2–0.8 cm long, relatively numerous, more or less leaflike, linear, noticeably shorter than the adjacent foliage leaves. Heads mostly 0.7–1.2 cm in diameter (including the extended ray corollas) at flowering. Involucre 3–5 mm long, narrowly ellipsoidal to narrowly cup-shaped or nearly cylindrical when fresh (becoming obconical when pressed), the bracts in 3 or 4 unequal, overlapping series. Involucral bracts linear-lanceolate to oblanceolate, not inrolled but often somewhat thickened toward the tip, rounded to tapered at the bluntly or sharply pointed tip, with a white to yellowish- or purplish-tinged, minute, bristlelike or spinelike point at the very tip, the tip usually somewhat spreading, the outer series with the narrowly elliptic, green portion often reaching nearly to the base, the other series with the slender midvein broadened gradually or relatively abruptly in the apical ⅓–⅔ into an elliptic or diamond-shaped (2–5 times as long as wide), green tip, the outer surface sparsely to moderately hairy (sometimes only along the midvein), the margins usually relatively densely pubescent with spreading hairs. Ray florets 8–20 in usually 1 or 2 series, the corollas well developed, 3–6 mm long, white (rarely lavender to blue). Disc florets 7–15, the corollas 2.5–4.0 mm long, the slender portion of the tube noticeably shorter than the slightly expanded apical portion, the lobes 0.4–0.7 mm long, 20–30 percent of the total length of the expanded portion. Pappus bristles 2–4 mm long, white. Fruits

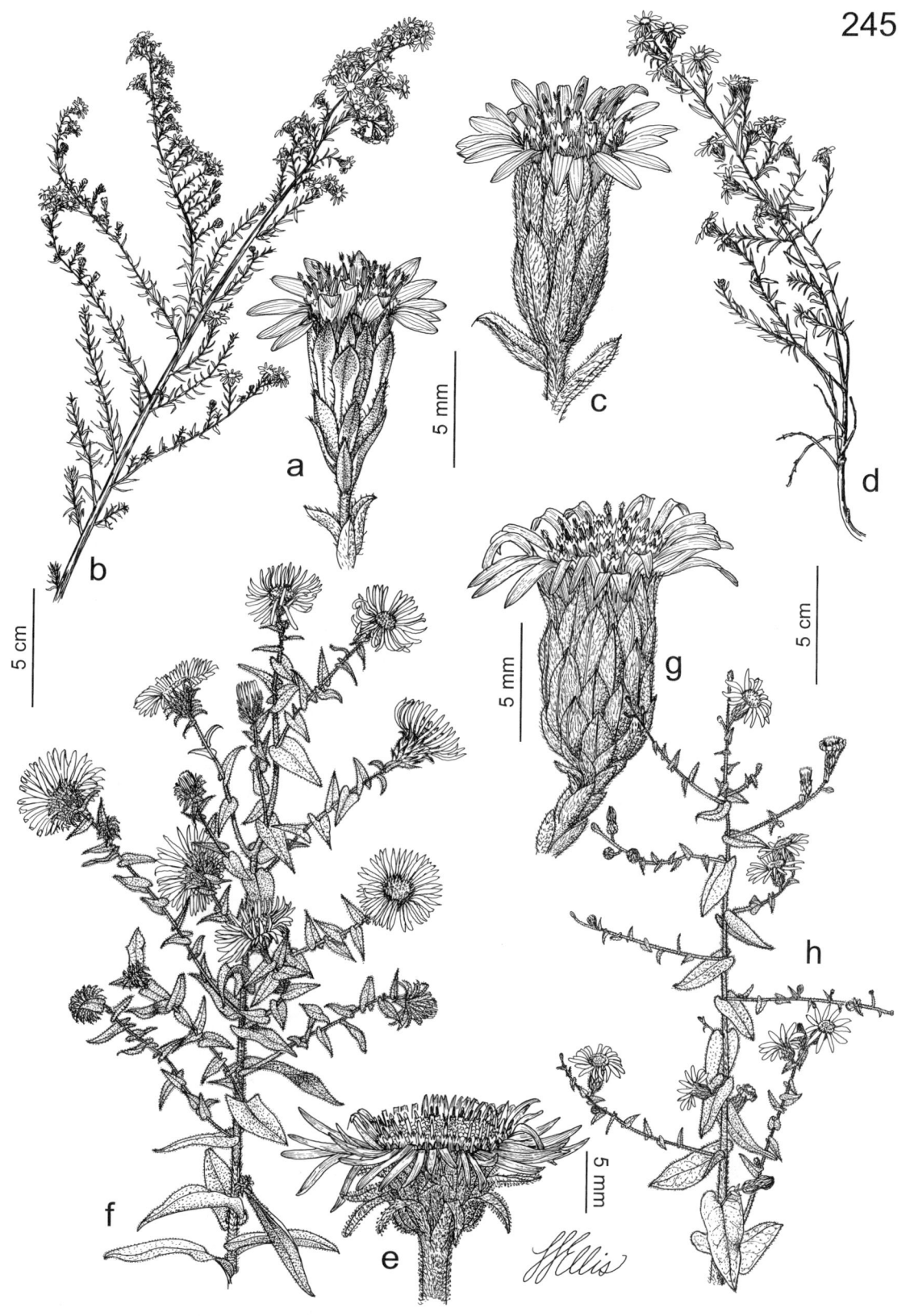

Plate 245. Asteraceae. *Symphyotrichum ericoides*, **a)** head, **b)** habit. *Symphyotrichum falcatum*, **c)** head, **d)** habit. *Symphyotrichum novae-angliae*, **e)** head, **f)** habit. *Symphyotrichum patens*, **g)** head, **h)** habit.

1.0–2.2 mm long, with 7–9 often faint, longitudinal ribs, purplish brown to brown, moderately to densely hairy, the hairs lacking swollen bases. 2n=10, 20. July–October.

Scattered to common in the Glaciated and Unglaciated Plains Divisions, but mostly absent from the rest of the state (Kansas, Oklahoma, Iowa, Missouri, Arkansas, and Illinois; apparently disjunct in North Carolina). Upland prairies, loess hill prairies, glades, and mesic upland forests; also pastures, old fields, railroads, roadsides, and open, disturbed areas.

A rare variant that occurs in northern Missouri with the corollas of the ray florets blue to violet has been called *Aster ericoides* f. *caeruleus* (Benke) S.F. Blake. Steyermark (1963) recorded a number of putative hybrids between this species and *S. pilosum*; however, Jones (1978b) was unable to cross these species artificially and considered similar specimens from Illinois to represent aberrant individuals of *S. ericoides*. An early collection from Audrain County may represent a hybrid with *S. racemosum*. Steyermark also treated a rather well-marked hybrid with *S. novae-angliae* that has been named as *S.* ×*amethystinum* (Nutt.) G.L. Nesom (*Aster* ×*amethystinus* Nutt.). See the treatment of the other parental species for further discussion. Readers should also consult the treatment of *S. parviceps* for further discussion on vegetative similarities between that species and *S. ericoides*.

The last monographer of the group (Jones, 1978a, b) treated *S. ericoides* (as *Aster ericoides*) as comprising an eastern and northwestern subspecies, each with two varieties. In transferring these epithets to *Symphyotrichum*, Nesom (1994) chose to recognize all four taxa at varietal level. The northwestern taxa, var. *pansum* (S.F. Blake) G.L. Nesom and var. *stricticaule* (Torr. & A. Gray) G.L. Nesom, both have a less strongly rhizomatous habit with clustered stems. The two eastern varieties, both of which occur in Missouri, are both relatively strongly rhizomatous with usually solitary stems often appearing in loose colonies.

1. Stems with the pubescence of ascending (sometimes appressed) hairs
 .... 6A. VAR. ERICOIDES
1. Stems with the pubescence of spreading to somewhat reflexed hairs
 .... 6B. VAR. PROSTRATUM

6a. var. ericoides

Stems with the pubescence of ascending (sometimes appressed) hairs, these usually relatively soft to the touch. Stem leaves with the hairs mostly appressed or nearly so. Involucral bracts usually sparsely hairy on the outer surface, often only along the midvein. 2n=10, 20. July–October.

Scattered to common in the Glaciated and Unglaciated Plains Divisions, but mostly absent from the rest of the state (Kansas, Oklahoma, Iowa, Missouri, Arkansas, and Illinois; apparently disjunct in North Carolina). Upland prairies, loess hill prairies, glades, and mesic upland forests; also pastures, old fields, railroads, roadsides, and open, disturbed areas.

6b. var. prostratum (Kuntze) G.L. Nesom

Aster ericoides var. *prostratus* (Kuntze)
 S.F. Blake
A. ericoides f. *prostratus* (Kuntze) Fernald

Stems with the pubescence of spreading to somewhat reflexed hairs, these usually relatively sandpapery to the touch. Stem leaves with the hairs mostly spreading. Involucral bracts sparsely to moderately hairy on the outer surface. 2n=10, 20(?). July–October.

Scattered to common in the Glaciated and Unglaciated Plains Divisions, but mostly absent from the rest of the state (Kansas, Oklahoma, Iowa, Missouri, Arkansas, and Illinois; apparently disjunct in North Carolina). Upland prairies, loess hill prairies, glades, and mesic upland forests; also pastures, old fields, railroads, roadsides, and open, disturbed areas.

This is the less commonly encountered variety in Missouri. Its range is mostly the same as that for var. *ericoides;* however, it appears to be nearly absent from the western portion of the Glaciated Plains Division.

7. Symphyotrichum falcatum (Lindl.) G.L. Nesom ssp. commutatum (Torr. & A. Gray) Semple (white prairie aster)

Aster falcatus Lindl. var. *commutatus* (Torr. & A. Gray) A.G. Jones
S. falcatum var. *commutatum* (Torr. & A. Gray) G.L. Nesom
A. commutatus (Torr. & A. Gray) A. Gray
 Pl. 245 c, d; Map 1019

Plants perennial herbs, often somewhat colonial from relatively long, slender, branched rhizomes. Stems usually solitary, 20–70 cm long, unbranched or with few to most commonly many spreading to ascending branches mostly above the midpoint, moderately to densely pubescent with short, appressed to upward-curved hairs, sometimes only sparsely so toward the base, the hairs relatively evenly distributed. Basal and/or lower stem leaves absent from the flowering stems, sessile or with a short, poorly differentiated petiole, the blade 2–6 cm long, 0.3–1.0 cm wide, oblanceolate, tapered at the base, rounded or angled to a minute, abrupt, sharp point at the tip, the margins with spreading to forward-pointing hairs and usually entire, the surfaces glabrous or sparsely to moderately pubescent with somewhat curved, broad-based hairs, the secondary veins on the leaf undersurface usually a single prominent pair (the leaf thus usually appearing 3-veined), the veinlets often faint and difficult to distinguish, forming an irregular network of elongated areoles. Median and upper stem leaves progressively smaller or not much reduced beyond the median leaves (the main stem leaves often with clusters of smaller leaves along short, axillary branches), the larger ones often persistent at flowering, sessile, the base sometimes slightly expanded but not clasping the stem, the blades 1–6 cm long, linear or narrowly oblong, the margins entire, angled or tapered at the base, rounded to short-tapered to a bluntly or sharply pointed tip (usually with a short, hard point at the tip), the veinlets usually relatively easily observed, the surfaces moderately to densely hairy, otherwise similar to the lower stem leaves. Inflorescences usually appearing as panicles with short to relatively long, loosely ascending branches (these racemose or more commonly with the heads in clusters toward the tip or solitary at the tip), sometimes reduced to a solitary raceme, the heads appearing short- to less commonly long-stalked and oriented in various directions, few to numerous (but fewer than in S. ericoides), the bracts along the ultimate branches 0.3–0.8 cm long, relatively few, more or less leaflike, linear or narrowly oblong-lanceolate, somewhat shorter than the adjacent foliage leaves. Heads mostly 1–2 cm in diameter (including the extended ray corollas) at flowering. Involucre 5–8 mm long, cup-shaped to slightly bell-shaped when fresh (sometimes becoming obconical when pressed), the bracts in 3 or 4 subequal to somewhat unequal, overlapping series. Involucral bracts linear-lanceolate to oblong-oblanceolate, not inrolled but often somewhat thickened toward the base, angled or tapered at the usually sharply pointed tip, with a white to yellowish- or purplish-tinged, minute, bristlelike or spinelike point at the very tip, the tip usually somewhat spreading, the slender midvein broadened relatively abruptly in the apical $1/3$–$1/2$ into an elliptic or diamond-shaped (2–4 times as long as wide), green tip, the outer surface sparsely to moderately hairy (sometimes mostly in the green portion), the margins usually relatively densely pubescent with spreading hairs. Ray florets 20–35 in usually 1 or 2 series, the corollas well developed, 4–9 mm long, white (rarely lavender to blue elsewhere). Disc florets 15–30, the corollas 3.5–5.0 mm long, the slender portion of the tube noticeably shorter than the slightly expanded apical portion, the lobes 0.7–1.2 mm long, 30–45 percent of the total length of the expanded portion. Pappus bristles 3–5 mm long, white. Fruits 1.0–2.2 mm long, with 7–9 very faint, longitudinal ribs, purplish brown to brown, moderately to densely hairy, the hairs lacking swollen bases. $2n=20, 30$. July–October.

Introduced, known thus far only from historical collections from Jackson County and the city of St. Louis (Wisconsin to Texas west to Montana and Arizona; Canada, Mexico). Open, disturbed areas.

The ssp. *falcatum* comprises plants with short, relatively nonrhizomatous rootstocks and occurs from the western United States through Canada to Alaska. The rhizomatous ssp. *commutatum* is mainly a plant of the Great Plains and Rocky Mountains. Its range approaches northwestern Missouri and it may be discovered in the loess hill prairies of that region at some point in the future.

8. Symphyotrichum laeve (L.) Á. Löve & D. Löve (smooth aster)
Aster laevis L.

Pl. 248 e, f; Map 1020

Plants perennial herbs, from short- to long-creeping, relatively stout, woody rhizomes. Stems 1 or few, (30–)50–120 cm long, with few to more commonly several ascending branches above the midpoint, glabrous or very sparsely pubescent toward the tip and around the nodes, usually somewhat glaucous. Basal as well as lower to median stem leaves sometimes absent at flowering, sessile or with short to long, poorly differentiated petioles, the blade 3–10 cm long, 1–3 cm wide, oblanceolate to oblong-ovate or less commonly a few of the leaves narrowly heart-shaped, tapered or rarely cordate at the base, rounded or angled to a bluntly or sharply pointed tip, the margins entire or sparsely to moderately toothed, the surface glabrous, smooth, usually glaucous, the secondary veins on the leaf undersurface relatively easily observed, ascending and sometimes fused toward their tips, the veinlets relatively faint and forming a dense, irregular network of relatively short areoles. Median and upper stem leaves few to relatively numerous, more or less progressively smaller toward the stem tip, sessile or the lowermost short-petiolate, slightly to strongly clasping the stem (the uppermost leaves sometimes somewhat sheathing), the blades 2–15 cm long, lanceolate to oblanceolate (often narrowly so) or oblong-oblanceolate, cordate to truncate at the base, angled or tapered to a sharply pointed tip, the margins entire or shallowly toothed, otherwise similar to the lower stem leaves. Inflorescences usually appearing as open panicles,

1020. Symphyotrichum laeve

1021. Symphyotrichum lanceolatum

1022. Symphyotrichum lateriflorum

sometimes with solitary heads or small clusters at the tips of relatively long, ascending branches, the heads appearing short- to long-stalked, the bracts noticeably shorter than the adjacent foliage leaves, 0.3–0.8 cm long, narrowly lanceolate to linear. Heads 1.5–3.0 cm in diameter (including the extended ray corollas) at flowering. Involucre 5–8 mm long, the bracts in 4–6 unequal, overlapping series. Involucral bracts linear to narrowly oblong-oblanceolate, angled or relatively short-tapered at the sharply pointed, often reddish tip, the tip ascending, the slender midvein broadened abruptly above the often slightly thickened base in the apical ¼– ½ into an elliptic or diamond-shaped (2–5 times as long as wide), green tip, the outer surface glabrous, often somewhat glaucous, the margins sometimes sparsely hairy, especially toward the tip. Ray florets 15–30 in 1 or 2 series, the corollas well developed, 8–15 mm long, lavender to purple to bluish purple. Disc florets 15–40, the corollas 4.5–6.5 mm long, the slender portion of the tube noticeably shorter than the slightly expanded apical portion, the lobes 0.7–1.1 mm long, 18–25 percent of the total length of the expanded portion. Pappus bristles 4.5–6.0 mm long, off-white to pale cream-colored or light tan. Fruits 2–3 mm long, with 4 or 5 longitudinal ribs, purplish brown to brown, glabrous. $2n=48$. August–October.

Scattered nearly throughout the state but apparently absent from most of the Mississippi Lowlands Division (eastern U.S. west to Minnesota, Nebraska, and Louisiana; Canada). Upland prairies, loess hill prairies, glades, ledges and tops of bluffs, openings of mesic to dry upland forests, and less commonly bottomland prairies, fens, and margins of ponds and lakes; also roadsides.

Exceptionally broad-leaved plants have been called *Aster laevis* f. *latifolius* (Porter) Shinners. A number of infraspecific taxa have been described within *S. laeve,* some of which are based on cultivated material and most of which probably are not worthy of formal taxonomic recognition. Steyermark (1963) noted that some Missouri specimens had been referred to var. *geyeri* (A. Gray) G.L. Nesom (as *Aster laevis* var. *geyeri* A. Gray), which supposedly differs from var. *laeve* in its somewhat narrower, more sharply pointed involucral bracts in fewer series. However, he was unable to separate the Missouri materials into two taxa and recommended that this variety not be treated formally. Steyermark (1963) also noted putative hybrids with *S. oolentangiense* (as *Aster azureus*) and *S. cordifolium* (as *A. sagittifolius*).

9. Symphyotrichum lanceolatum (Willd.) G.L. Nesom (tall white aster, panicled aster)
Aster lanceolatus Willd.
A. simplex Willd.

Pl. 246 e, f; Map 1021

Plants perennial herbs, often somewhat colonial from relatively long, slender, branched rhizomes. Stems usually solitary, 20–150 cm long, unbranched or with few to most commonly many spreading to ascending branches mostly above the midpoint, sparsely to moderately pubescent with short, spreading or curled hairs toward the tip, these usually in longitudinal lines or bands, usually glabrous toward the base or sometimes nearly throughout. Basal and/or lower stem leaves usually absent from the flowering stems, sessile or with a short, poorly differentiated petiole, the blade 1–8 cm long, 0.5–2.0 cm wide, oblanceolate, tapered at the base, rounded or angled to a bluntly or sharply pointed tip, the margins with spreading to forward-pointing hairs and usually toothed or scalloped, the surfaces usually glabrous, the secondary veins on the leaf undersurface often difficult to distinguish from the veinlets, forming an irregular network of elongated areoles. Median and upper stem leaves progressively smaller or not much reduced beyond the median leaves, the larger ones occasionally withered by flowering time, sessile, the base

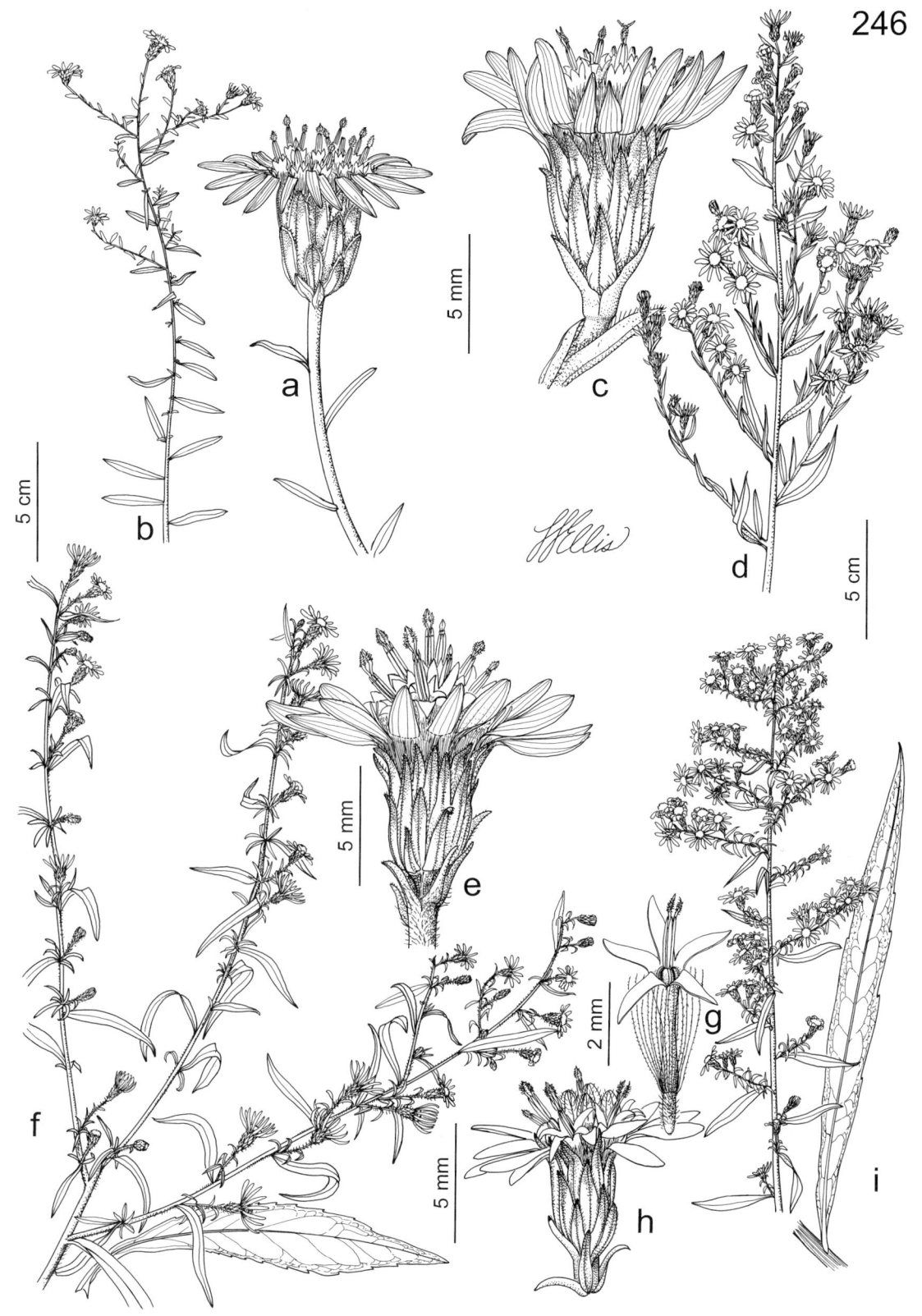

Plate 246. Asteraceae. *Symphyotrichum dumosum*, **a)** head, **b)** habit. *Symphyotrichum racemosum*, **c)** head, **d)** habit. *Symphyotrichum lanceolatum*, **e)** head, **f)** habit. *Symphyotrichum lateriflorum*, **g)** disc floret, **h)** head, **i)** habit and larger leaf.

sometimes slightly expanded but not clasping the stem, the blades 1–14 cm long, linear to narrowly elliptic or oblanceolate to broadly oblanceolate, the margins entire or sparsely toothed, angled or tapered at the base, angled or tapered to a sharply pointed tip, otherwise similar to the lower stem leaves. Inflorescences usually appearing as panicles with short to relatively long, loosely ascending to spreading branches (these racemose or more commonly with the heads solitary or in clusters toward the branch tips), the heads appearing mostly short-stalked and usually oriented in various directions, often relatively few per branch (except sometimes in var. *interior*), the bracts along the ultimate branches 0.2–1.0 cm long, relatively few, more or less leaflike, linear or narrowly oblong-lanceolate, somewhat shorter than the adjacent foliage leaves. Heads mostly 1.0–2.5 cm in diameter (including the extended ray corollas) at flowering. Involucre 3–7 mm long, cup-shaped to slightly bell-shaped when fresh (sometimes becoming obconical when pressed), the bracts in 3–5 more or less unequal, overlapping series. Involucral bracts linear-lanceolate to narrowly oblong-oblanceolate, angled or tapered at the usually sharply pointed tip, lacking a bristlelike or spinelike point at the ascending tip, the slender midvein broadened gradually in the apical $1/2$–$3/4$ into a narrowly elliptic (4–10 times as long as wide), green tip, the outer surface glabrous or sparsely hairy, the margins often slightly irregular and sparsely hairy. Ray florets 20–45 in usually 1 or 2 series, the corollas well developed, 5–12 mm long, white (rarely pink, lavender, or bluish-tinged). Disc florets 15–40, the corollas 3–6 mm long, the slender portion of the tube noticeably shorter than the slightly expanded apical portion, the lobes 0.7–1.2 mm long, 30–45 percent of the total length of the expanded portion. Pappus bristles 3–6 mm long, white or off-white. Fruits 1.2–2.0 mm long, with 4 or 5 longitudinal ribs, gray to tan, moderately hairy, the hairs lacking swollen bases. $2n=32, 40, 48, 56, 64$. August–October.

Scattered nearly throughout the state (U.S., Canada, Mexico). Bottomland forests, mesic upland forests, swamps, bases and ledges of bluffs, banks of streams, rivers, and spring branches, margins of ponds, lakes, and sinkhole ponds, fens, sloughs, and moist depressions in upland prairies; also pastures, fencerows, ditches, railroads, roadsides, and open, disturbed areas.

This is a widespread, variable species. It might be confused with *S. lateriflorum* and *S. ontarionis* at one extreme and with *S. racemosum* at the other. These similarities led Semple and Brammall (1982) to hypothesize that *S. ontarionis* might have evolved from past hybridization involving *S. lanceolatum* and *S. lateriflorum*. Putative hybrids have been recorded from Missouri with *S. lateriflorum* and *S. praealtum*.

Semple and Chmielewski (1987) were the latest to study the taxonomy of the *S. lanceolatum* polyploid complex (as *Aster lanceolatus*), arriving at a complex classification that included two subspecies, one of which was subdivided further into four varieties. Of these, the western ssp. *hesperium* (A. Gray) G.L. Nesom (*A. hesperius* A. Gray, *A. lanceolatus* ssp. *hesperius* (A. Gray) Semple & Chmiel.) was excluded from the Missouri flora by Steyermark (1963), who redetermined the sole historical Jackson County collection upon which earlier reports of this taxon in the state had been based. Barkley (1986), Yatskievych and Turner (1990), and Gleason and Cronquist (1991) have continued to include Missouri in the range of ssp. *hesperium*, but to date no specimens to confirm its presence in Missouri have come to light. It differs from ssp. *lanceolatum* in its subequal involucral bracts and somewhat larger, broader bracts subtending the heads. Of the four named variants within ssp. *lanceolatum*, all but the northern var. *hirsuticaule* (Semple & Chmiel.) G.L. Nesom (with densely and evenly hairy stems) have been reported from Missouri. The three varieties are not strongly marked in Missouri, except in their extremes, and some specimens are difficult to determine below the species level.

1. Heads appearing relatively dense on the inflorescence branches, mostly 1.0–1.5 cm in diameter (including the extended ray corollas) at flowering, the involucre 3–5 mm long; disc florets with the corolla 3.0–3.5 mm long
    .... 9A. VAR. INTERIOR
1. Heads sometimes appearing relatively sparse or loose on the inflorescence branches, mostly 1.8–2.5 cm in diameter (including the extended ray corollas) at flowering, the involucre 4–7 mm long; disc florets with the corolla usually 4–6 mm long
    2. Median and upper stem leaves mostly linear to narrowly elliptic, the margins entire or inconspicuously toothed; inflorescences appearing more or less leafy, but not densely so
        .... 9B. VAR. LANCEOLATUM
    2. Median and upper stem leaves mostly oblanceolate to broadly oblanceolate, the margins noticeably toothed; inflorescences appearing relatively densely leafy
        .... 9C. VAR. LATIFOLIUM

9a. var. interior (Wiegand) G.L. Nesom
Aster lanceolatus var. *interior* (Wiegand) Semple & Chmiel.
A. simplex var. *interior* (Wiegand) Cronquist

Median and upper stem leaves mostly linear to narrowly elliptic, the margins entire or inconspicuously toothed. Inflorescences usually not appearing very leafy (owing to the relatively small size of the upper leaves), the branches with the heads often appearing relatively dense and racemose (occasionally all or mostly oriented toward the upper side of the branches). Heads appearing relatively small, mostly 1.0–1.5 cm in diameter (including the extended ray corollas) at flowering. Involucre 3–5 mm long. Ray florets 20–28, the corollas well developed, 5–9 mm long, white. Disc florets mostly 15–30, the corolla 3.0–3.5 mm long. $2n=48, 64$. August–October.

Scattered sporadically nearly throughout the state but uncommon in the eastern portion of the Ozark Division and the Mississippi Lowlands (northeastern U.S. west to Nebraska and Oklahoma). Bottomland forests, mesic upland forests, swamps, bases and ledges of bluffs, banks of streams and rivers, margins of ponds and lakes, sloughs, and moist depressions in upland prairies; also pastures, fencerows, ditches, railroads, roadsides, and open, disturbed areas.

9b. var. lanceolatum
A. simplex var. *ramosissimus* (Torr. & A. Gray) Cronquist

Median and upper stem leaves mostly linear to narrowly elliptic, the margins entire or inconspicuously toothed. Inflorescences appearing more or less leafy, but not densely so, the branches with the heads often appearing relatively sparse or loose, solitary or in small clusters toward the branch tips, less commonly in short racemes (the heads oriented in several different directions). Heads appearing medium-sized, mostly 1.8–2.5 cm in diameter (including the extended ray corollas) at flowering. Involucre 4–7 mm long. Ray florets 25–45, the corollas well developed, 7–12 mm long, white or less commonly lavender to bluish-tinged. Disc florets mostly 20–40, the corolla (3.5–)4.0–6.0 mm long. $2n=32, 40, 48, 56, 64$. August–October.

Scattered nearly throughout the state (eastern U.S. west to Montana and Texas; Canada). Bottomland forests, mesic upland forests, swamps, bases and ledges of bluffs, banks of streams, rivers, and spring branches, margins of ponds, lakes, and sinkhole ponds, fens, sloughs, and moist depressions in upland prairies; also pastures, fencerows, ditches, railroads, roadsides, and open, disturbed areas.

This is the most widespread and abundant variety in the state.

9c. var. latifolium (Semple & Chmiel.) G.L. Nesom
Aster lanceolatus var. *latifolius* Semple & Chmiel.

Median and upper stem leaves mostly oblanceolate to broadly oblanceolate, the margins noticeably toothed. Inflorescences appearing relatively densely leafy, the branches with the heads often appearing relatively sparse or loose, solitary or in small clusters toward the branch tips, less commonly in short racemes (the heads oriented in several different directions). Heads appearing medium-sized, mostly 1.8–2.5 cm in diameter (including the extended ray corollas) at flowering. Involucre 4–7 mm long. Ray florets 25–45, the corollas well developed, 7–12 mm long, white or rarely pink. Disc florets mostly 25–40, the corolla 4–6 mm long. $2n=48, 64$. August–October.

Scattered, mostly in the southern half of the state and especially in counties bordering the Mississippi and Missouri Rivers (eastern U.S. west to North Dakota and Texas; Canada). Bottomland forests, mesic upland forests, swamps, banks of streams and rivers, and sloughs; also pastures, ditches, railroads, roadsides, and open, disturbed areas.

This variant tends to occur in rich soils of bottomland sites. Although Semple and Chmielewski (1987) were able to distinguish it statistically in their numerical morphological study, there is still some possibility that it represents merely an environmentally induced variant of var. *lanceolatum*.

10. Symphyotrichum lateriflorum (L.) Á. Löve & D. Löve (white woodland aster)
Aster lateriflorus (L.) Britton
Pl. 246 g–i; Map 1022

Plants perennial herbs, with a short, somewhat woody, relatively stout rootstock, sometimes also with stout, somewhat succulent rhizomes. Stems 1 to several, 20–150 cm long, unbranched or with few to many ascending to spreading branches mostly above the midpoint, sparsely to moderately pubescent with short, spreading hairs toward the tip, these usually in longitudinal lines or bands, usually glabrous toward the base. Basal and/or lower stem leaves usually absent at flowering, sessile or with a short, poorly differentiated petiole, the blade 1–12 cm long, 0.5–3.0 cm wide, oblanceolate to obovate, tapered at the base, rounded or angled to a usually bluntly pointed tip, the margins minutely hairy and bluntly to sharply toothed or less commonly entire, glabrous or sparsely pubescent with minute hairs on the upper surface, glabrous on the undersurface except along the midvein (and rarely also on the adjacent bases of the lateral veins), where moderately to densely short-hairy, the secondary veins on the leaf undersurface often indistinguishable from the veinlets or nearly so but the finer veins usually relatively easily observed, forming a dense, irregular network of relatively short areoles. Median and upper stem leaves progressively smaller, sessile, the base sometimes slightly expanded but not clasping the stem, the blades 1–15 cm long, elliptic to elliptic-lanceolate (the uppermost leaves sometimes nearly linear), the margins entire or shallowly toothed (sometimes sparsely so), angled or tapered at the base, angled or tapered to a sharply pointed tip, otherwise similar to the lower stem leaves. Inflorescences usually appearing as panicles with short to more commonly relatively long, loosely ascending to spreading, racemose branches, sometimes reduced to a solitary raceme or with heads clustered toward the branch tips, the heads appearing mostly short-stalked and oriented in several directions, the bracts along the ultimate branches 0.6–2.0 cm long, leaflike, linear to lanceolate or elliptic, mostly only slightly shorter than the adjacent foliage leaves. Heads mostly 0.8–1.5 cm in diameter (including the extended ray corollas) at flowering. Involucre 3.5–5.0 mm long, the bracts in 3–5 unequal, overlapping series. Involucral bracts linear to very narrowly oblong-oblanceolate, angled or short-tapered at the usually sharply pointed tip, the tip erect or ascending (the lowermost bracts occasionally slightly spreading), the outer series with the narrowly elliptic, green portion reaching nearly to the base, the other series with the slender midvein broadened gradually in the apical $1/3$–$2/3$ into a narrowly elliptic (more than 4 times as long as wide), green tip, the surfaces glabrous, the margins sometimes somewhat uneven, sparsely or irregularly short-hairy. Ray florets 8–15(–23) in usually 1 series, the corollas well developed, 4–8 mm long, white (rarely pinkish-tinged or light lavender). Disc florets 8–15, the corollas 2.5–4.5 mm long, the slender portion of the tube slightly shorter than the slightly expanded apical portion, the lobes 0.9–1.7 mm long, 50–75 percent of the total length of the expanded portion. Pappus bristles 2.5–4.5 mm long, white. Fruits 1.5–2.2 mm long, with 3–5 longitudinal ribs, gray to tan, sparsely hairy, the hairs lacking swollen bases. $2n=16, 32$. August–November.

Nearly throughout the state, common south of the Missouri River, scattered farther north (eastern U.S. west to South Dakota and Texas; Canada). Bottomland forests, mesic upland forests, bases and ledges of bluffs, banks of streams, rivers, and spring branches, margins of ponds, lakes, and sinkhole ponds, fens, sloughs, and moist depressions in upland prairies; also pastures, gardens, railroads, roadsides, and open, disturbed areas.

Morphological variation in *S. lateriflorum* is complex across its range and is not well understood. Some authors have recognized several varieties based on differences in pubescence density, leaf width, and other characters (Semple et al., 2002). Jones (1989) noted that plants growing in Illinois were not easily separable into varieties owing to broad intergradation, and this also seems to be the case in Missouri. Most of the material in the state is more or less assignable to var. *lanceolatum*, but occasional plants with a more spreading habit seem referable to var. *horizontale* (Desf.) G.L. Nesom. Plants corresponding to some of the other named variants may also be present, but the situation requires more detailed study. Jones (1989) also noted that part of the problem involved plants somewhat intermediate between *S. lateriflorum* and related species with base chromosome numbers of x=8, such as *S. dumosum*, *S. lanceolatum*, *S. ontarionis*, and *S. racemosum* (as *Aster fragilis*). Semple reported further hybridization with *S. laeve* in southern Canada. Steyermark (1963) reported rare putative hybrids in Missouri between *S. lateriflorum* and *S. puniceum*, *S. lanceolatum* (as *Aster simplex*), and *S. ontarionis*. The inherent weediness (ability to colonize and spread rapidly in disturbed sites) of *S. lateriflorum* no doubt contributes to its promiscuity.

11. Symphyotrichum novae-angliae (L.) G.L. Nesom (New England aster)
Aster novae-angliae L.

Pl. 245 e, f; Map 1023

Plants perennial herbs, with a compact, woody rootstock, usually with 1 or few short, stout, rhizomatous branches, less commonly the rhizomes slender and longer-creeping. Stems 1 to several, 40–170 cm long, with few to more commonly several ascending to spreading branches above the midpoint, densely and evenly pubescent with more or less spreading hairs (less so toward the base), also with moderate to dense, stalked glands toward the tip. Basal and lower stem leaves absent at flowering, sessile or short-petiolate, the blade 2–6 cm long, 0.5–1.5 cm wide, oblanceolate, tapered at the base, rounded or angled to a usually bluntly pointed tip, the margins entire or sparsely toothed and hairy, the surfaces sparsely pubescent with short, loosely ascending to spreading hairs, appearing more or less 3-veined, the secondary veins on the leaf undersurface relatively easily observed, strongly ascending parallel to the midvein, often fused irregularly toward the tip, the veinlets forming a dense, irregular network of relatively short to somewhat elongate areoles. Median and upper stem leaves often relatively numerous, more or less

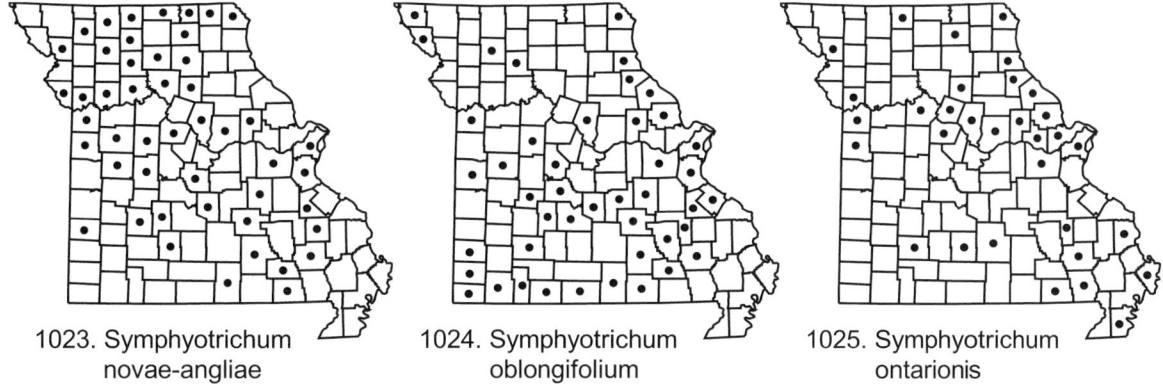

1023. Symphyotrichum novae-angliae

1024. Symphyotrichum oblongifolium

1025. Symphyotrichum ontarionis

progressively smaller toward the stem tip, sessile, strongly clasping the stem, the blades 1–12 cm long, narrowly oblanceolate (but often slightly broadened at the base) to narrowly oblong-lanceolate, rarely broader, cordate to nearly truncate at the base, angled or tapered to a sharply pointed tip, the margins entire, hairy, the surfaces moderately to densely hairy, usually with 3 main veins. Inflorescences usually appearing as open panicles, sometimes with solitary heads or small clusters at the tips of short to long branches, the heads appearing mostly long-stalked, the bracts relatively few, 0.5–1.2 cm long, mostly linear. Heads 2.0–4.5 cm (broader in some cultivars) in diameter (including the extended ray corollas) at flowering. Involucre 6–12 mm long, the bracts in 3–6 subequal, overlapping series. Involucral bracts mostly linear, long-tapered to the slender, sharply pointed tip, the tip spreading or reflexed, those of the outer series more or less leaflike, the green portion extending more than $2/3$ of the way to the base, those of the median and inner series with a progressively shorter, elliptic, green apical area (this mostly $1/2$ the length or more), usually purplish-tinged, the outer surface with relatively dense, stalked glands and often also with sparse, longer, nonglandular hairs (the inner surface sparsely to moderately hairy), the margins with relatively long, spreading hairs. Ray florets 40–100 in usually 2 or 3 series, the corollas well developed, 10–25 mm long, reddish purple to purple (to pink or white elsewhere). Disc florets 50–110, the corollas 4.5–6.5 mm long, the slender portion of the tube shorter than the slightly expanded apical portion, the lobes 0.4–0.7 mm long, 20–25 percent of the total length of the expanded portion. Pappus bristles 4.0–6.5 mm long, mostly pale orangish brown to light tan, occasionally purplish-tinged. Fruits 2–3 mm long, with 7–10 longitudinal ribs, purplish brown to brown, densely hairy and sometimes also sparsely glandular. $2n=10$ (rarely $2n=15$, 20 in Illinois). July–October.

Scattered nearly throughout the state but apparently absent from the Mississippi Lowlands Division and uncommon in portions of the Glaciated and Unglaciated Plains (eastern U.S. west to Wyoming and New Mexico; Canada; introduced farther west). Bottomland prairies, moist depressions of upland prairies, fens, bases of bluffs, banks of streams and rivers, and margins of ponds and lakes; also pastures, fencerows, banks of ditches, railroads, and roadsides.

Symphyotrichum novae-angliae is perhaps the most popular of our native asters in cultivation. A number of cultivars with different growth forms and various ray corolla colors are available commercially. In the garden, this showy species tends to begin flowering earlier than most of the other asters and to stay in flower a longer time.

An unusual hybrid between this species and *S. ericoides* has been recorded thus far from Macon, Randolph, and Washington Counties. This hybrid, known as *S.* ×*amethystinum* (Nutt.) G.L. Nesom, resembles *S. novae-angliae* in its relatively large heads with purple ray corollas and relatively broad leaves. However, its involucral bracts lack the strongly reflexed tips and glandular pubescence. Superficially, this hybrid also resembles *S. oblongifolium*.

12. Symphyotrichum oblongifolium (Nutt.) G.L. Nesom (aromatic aster, oblong-leaved aster)

Aster oblongifolius Nutt.

A. oblongifolius var. *angustatus* Shinners

Pl. 249 f, g; Map 1024

Plants perennial herbs, with a somewhat woody, horizontal rootstock, often also with 1 or more slender and longer-creeping, rhizomatous branches. Stems 1 several, 15–80 cm long, with few to more commonly several ascending to spreading branches above the midpoint, sparsely to moderately and usually evenly pubescent with short, spreading hairs, progressively more glandular toward the tip, the branches and apical portion of the main stem usually with dense, stalked glands. Basal and lower stem leaves absent at flowering, sessile or nearly so, the blade 2–6 cm long, 0.5–1.5 cm wide, oblanceolate, tapered at the base, rounded or angled to a usually bluntly pointed tip (sometimes abruptly tapered to a minute, sharp point), the margins entire and hairy, the surfaces usually sparsely pubescent with short, loosely ascending to spreading hairs and also usually sparsely glandular, appearing more or less 3-veined, the secondary veins on the leaf undersurface often relatively faint, strongly ascending parallel to the midvein, often fused irregularly toward the tip, the veinlets forming a dense, irregular network of relatively short to somewhat elongate areoles. Median and upper stem leaves often relatively numerous, more or less progressively smaller toward the stem tip, sessile, not clasping the stem (a few of the larger leaves sometimes slightly clasping, the blades 1–10 cm long, narrowly oblong-lanceolate to narrowly lanceolate, rarely broader, narrowed or rounded at the sometimes slightly expanded base, occasionally appearing truncate, angled or tapered to a bluntly or sharply pointed tip, the margins entire, hairy, the surfaces moderately to densely hairy, those of the upper leaves also with moderate to dense, stalked glands, the venation often faint, sometimes more or less with 3 main veins. Inflorescences usually appearing as open panicles, sometimes with solitary heads or small clusters at the tips of short to long branches, the heads appearing mostly longstalked, the bracts few to several, 0.3–0.8 cm long, linear to narrowly oblong. Heads 2–3 cm (broader in some cultivars) in diameter (including the extended ray corollas) at flowering. Involucre 5–8 mm long, the bracts in 4–6 subequal to somewhat unequal, overlapping series. Involucral bracts mostly narrowly oblong to narrowly oblong-oblanceolate, all but the inner series angled or short-tapered to the sharply pointed tip, the tip spreading or reflexed, those of the outer series often more or less leaflike, the green portion extending more than 2/3 of the way to the base, those of the other series with a somewhat thickened and keeled base having a slender, green midvein that is expanded abruptly in the apical 1/3–1/2 into an oblong to elliptic, green apical area, sometimes purplish-tinged, the surfaces and margins with relatively dense, stalked glands. Ray florets 15–35 in usually 1 or 2 series, the corollas well developed, 9–15 mm long, reddish purple to bluish purple, rarely pink. Disc florets 30–50, the corollas 4.5–6.0 mm long, the slender portion of the tube usually slightly shorter than the slightly expanded apical portion, the lobes 0.4–0.7 mm long, 18–25 percent of the total length of the expanded portion. Pappus bristles 4–6 mm long, mostly pale orangish brown to light tan, occasionally purplish-tinged. Fruits 2.0–2.5 mm long, with 7–10 longitudinal ribs, purplish brown to brown, sparsely hairy. $2n=10, 20$. July–November.

Scattered, mostly south of the Missouri River, but extending northward locally mostly in counties adjacent to the Mississippi and Missouri Rivers and apparently absent from the Mississippi Lowlands Division (eastern U.S. [except a few southeastern states] west to Montana and New Mexico). Glades, ledges and tops of bluffs, upland prairies, and openings of dry upland forests; also railroads.

Rare plants with pink ray corollas have been called *Aster oblongifolius* f. *roseoligulatus* (Benke) Shinners. Semple and Ford (1981) studied variation in leaf morphology in this species and *S. novae-angliae* (both as species of the segregate genus *Lasallea* Greene) and concluded that there was no statistical basis for the taxonomic recognition of varieties described from leaf size and shape extremes in either species.

A putative hybrid between this species and *S. ericoides* has been called *S.* ×*batesii* (Rydb.) G.L. Nesom (Nesom, 1994) and was originally described from Nebraska. It may be found eventually in Missouri, possibly in the northwestern portion of the state where these two species occur in proximity.

13. Symphyotrichum ontarionis (Wiegand) G.L. Nesom var. ontarionis (Ontario aster)

Aster ontarionis Wiegand

Pl. 247 d, e; Map 1025

Plants perennial herbs, often somewhat colonial from relatively long, slender, branched rhizomes. Stems usually solitary, 30–120 cm long, unbranched or with few to several ascending to spreading branches above the midpoint (some of the branches sometimes relatively short), sparsely to moderately pubescent with short, spreading hairs, these relatively evenly distributed toward the stem tip, often only in longitudinal lines or bands toward the stem midpoint, very sparse (or the stem appearing glabrous) toward the base. Basal and/or lower stem leaves usually absent at flowering, sessile or with a short, poorly differentiated petiole, the blade 2–8 cm long, 0.5–3.5 cm wide, oblanceolate to obovate, tapered at the base, angled or tapered to a usually sharply pointed tip, the margins bluntly to sharply toothed, glabrous or more commonly sparsely pubescent with minute

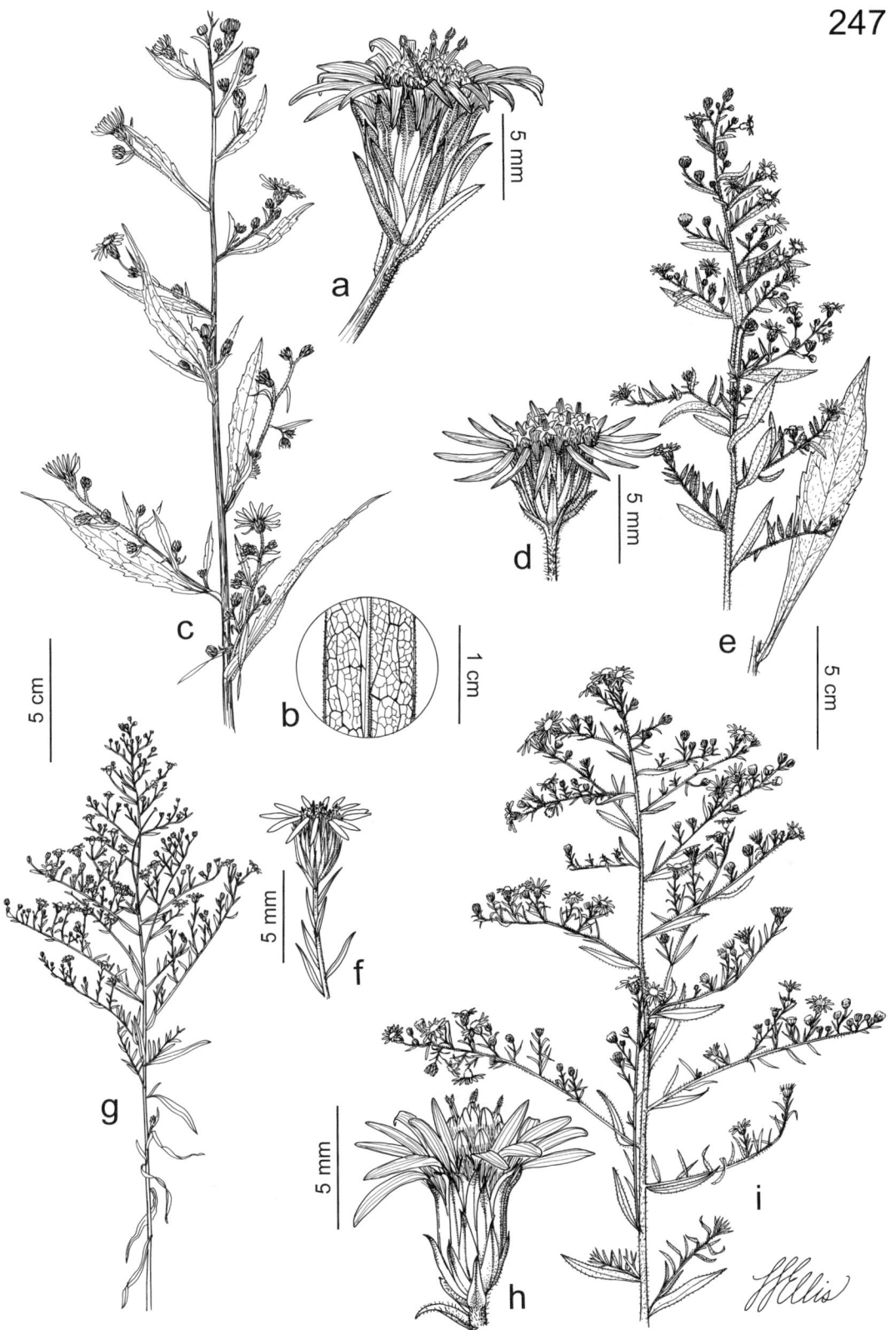

Plate 247. Asteraceae. *Symphyotrichum praealtum*, **a)** head, **b)** leaf detail **c)** habit. *Symphyotrichum ontarionis*, **d)** head, **e)** habit and larger leaf. *Symphyotrichum parviceps*, **f)** head, **g)** habit. *Symphyotrichum pilosum*, **h)** head, **i)** habit.

hairs on the surfaces, the secondary veins on the leaf undersurface fine but usually relatively easily observed, often irregularly fused toward their tips, the veinlets often indistinct, forming a dense, irregular network of relatively short areoles. Median and upper stem leaves progressively smaller, sessile, the base sometimes slightly expanded but not clasping the stem, the blades 1–12 cm long, narrowly elliptic-ovate to lanceolate, oblanceolate, the uppermost leaves usually with entire margins and occasionally linear, angled or tapered at the base, the surfaces sparsely to moderately and relatively evenly pubescent with minute hairs along and in between the veins, sometimes with slightly longer or denser hairs along the midvein, otherwise similar to the lower stem leaves. Inflorescences usually appearing as panicles with short to relatively long, loosely ascending to spreading, racemose branches, sometimes reduced to a solitary raceme or with heads clustered toward the branch tips, the heads appearing mostly short-stalked and oriented in several directions, the bracts along the ultimate branches 0.6–1.5 cm long, leaflike, linear to lanceolate or elliptic, mostly only slightly shorter than the adjacent foliage leaves. Heads mostly 0.8–1.5 cm in diameter (including the extended ray corollas) at flowering. Involucre 3–5 mm long, the bracts in 3–5 unequal, overlapping series. Involucral bracts linear to very narrowly oblong-oblanceolate, angled or short-tapered at the usually sharply pointed tip, the tip erect or ascending (the lowermost bracts occasionally slightly spreading), the outer series with the narrowly elliptic, green portion reaching nearly to the base, the other series with the slender midvein broadened gradually in the apical $1/3$–$2/3$ into a narrowly elliptic (more than 4 times as long as wide), green tip, the surfaces glabrous or sparsely short-hairy, the margins sparsely or irregularly short-hairy. Ray florets 15–25 in 1 or 2 series, the corollas well developed, 4–8 mm long, white (rarely darkening upon drying). Disc florets 12–25, the corollas 2.5–4.5 mm long, the slender portion of the tube slightly shorter than the slightly expanded apical portion, the lobes 0.7–1.1 mm long, 45–65 percent of the total length of the expanded portion. Pappus bristles 3.0–4.5 mm long, white to pale cream-colored. Fruits 1.2–2.0 mm long, with 3–5 longitudinal ribs, gray, moderately hairy, the hairs usually with minutely swollen bases. $2n=32$. August–October.

Scattered nearly throughout the state, most commonly in counties adjacent to the Missouri and Mississippi Rivers, but apparently absent from the Unglaciated Plains Division (New York to Ohio and Georgia west to South Dakota and Texas; Canada).

Bottomland forests, mesic upland forests, swamps, sloughs, banks of streams and rivers, margins of ponds and lakes, bases of bluffs, and rarely tops of bluffs and glades; also pastures, ditches, and roadsides.

Some plants of *S. ontarionis*, particularly specimens lacking the rootstock, can be difficult to distinguish from the closely related *S. lateriflorum*. Jones (1989) suggested intergradation in Illinois with *S. lateriflorum*, *S. lanceolatum*, and *S. racemosum*. Semple and Brammall (1982) hypothesized that *S. ontarionis* may have evolved following past hybridization between *S. lanceolatum* and *S. lateriflorum*. Additionally, occasional plants of *S. ontarionis* may be confused with *S. pilosum* because the involucral bracts may terminate in a short, bristlelike tip. However, in such plants the body of the bract is flat and relatively strongly ascending rather than inrolled and somewhat curved outward at the tip. Some botanists recognize a second variety of *S. ontarionis*, var. *glabratum* (Semple) Brouillet & Labrecque, which is restricted to portions of eastern Canada and differs in being glabrous or nearly so.

14. Symphyotrichum oolentangiense (Riddell) G.L. Nesom (azure aster, blue devil)

Aster oolentangiensis Riddell

A. azureus Lindl. ex Hook.

Pl. 244 g, h; Map 1026

Plants perennial herbs, usually from a short, stout, somewhat branched rootstock, this sometimes somewhat woody, occasionally also producing short rhizomes. Stems 1 to few, 30–120 cm long, unbranched or with few to several ascending branches above the midpoint, glabrous or sparsely to moderately roughened with short, spreading hairs in longitudinal lines or bands, mostly toward the stem tip. Basal and/or lower stem leaves present at flowering, long-petiolate, the petiole sometimes narrowly winged (often only toward the tip), the blade 4–18 cm long, 1–5 cm wide, heart-shaped or occasionally ovate, deeply cordate to less commonly truncate or abruptly rounded at the base, angled or tapered to a sharply pointed tip, the margins entire or with a few minute, widely spaced teeth, moderately to densely roughened with minute, stiff hairs on the upper surface, densely pubescent with slightly longer (still somewhat sandpapery) hairs on the undersurface, the secondary veins on the leaf undersurface faint or sometimes easily observed, often irregularly fused toward their tips, the veinlets often indistinct, forming a dense, irregular network of relatively short areoles.

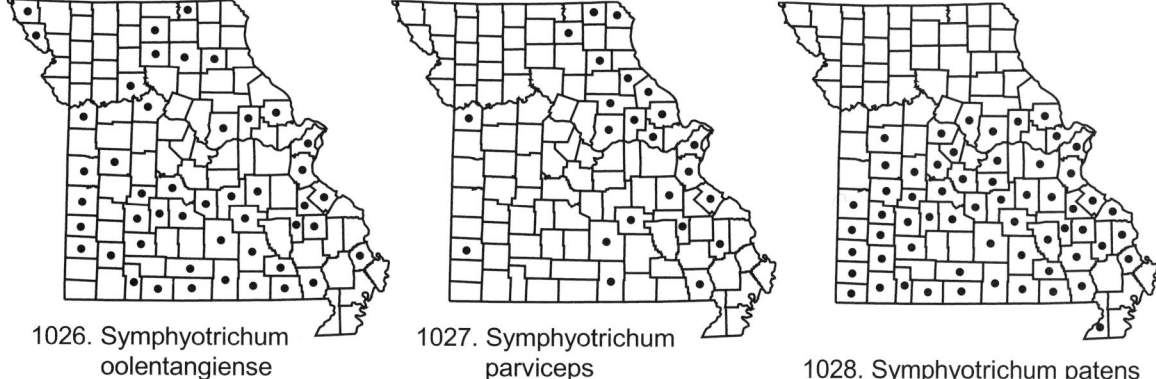

1026. Symphyotrichum oolentangiense

1027. Symphyotrichum parviceps

1028. Symphyotrichum patens

Median and upper stem leaves progressively smaller, with long to short, often winged but not clasping or sheathing petioles or the uppermost leaves sessile, the blades 1–6 cm long, narrowly ovate to narrowly lanceolate, rounded to tapered at the base, otherwise similar to the lower stem leaves. Inflorescences usually panicles with relatively long, loosely ascending, few-headed, racemose branches, sometimes with denser heads along the branches or reduced to a solitary raceme, the heads appearing short- or long-stalked, the bracts along the ultimate branches 0.3–0.8(–1.2) cm long, leaflike, mostly linear, noticeably shorter and narrower than the adjacent foliage leaves. Heads mostly 1.5–2.5 cm in diameter (including the extended ray corollas) at flowering. Involucre 4.5–8.0 mm long, the bracts in 4–6 unequal, overlapping series. Involucral bracts linear to narrowly lanceolate or narrowly oblong, angled or short-tapered at the mostly sharply pointed tip, the tip erect or ascending (the lowermost bracts occasionally slightly spreading), the slender midvein broadened relatively abruptly in the apical $1/5$–$1/3$ (rarely more) into an elliptic or broadly diamond-shaped (up to 2.5 times as long as wide) green tip, the outer surface glabrous or sparsely short-hairy, the margins moderately hairy, especially toward the tip. Ray florets 10–25 in 1 or 2 series, the corollas well developed, 6–12 mm long, lavender to purple to blue, rarely pink or white. Disc florets 15–28, the corollas 4–5 mm long, the slender portion of the tube shorter than the slightly expanded apical portion, the lobes 0.4–0.7 mm long, 20–25 percent of the total length of the expanded portion. Pappus bristles 3.5–5.0 mm long, off-white to pale cream-colored, occasionally purplish-tinged. Fruits 1.5–2.0 mm long, mostly with 4 longitudinal ribs, purplish brown (rarely tan) often with lighter ribs, glabrous or sparsely hairy. $2n=32$. August–November.

Scattered nearly throughout the state (South Dakota to Texas east to New York, Ohio, and Florida; Canada). Glades, upland prairies, sand prairies, savannas, mesic to dry upland forests on rocky slopes, and ledges and tops of bluffs; also roadsides.

This species was long known under the name *Aster azureus* until A. G. Jones (1983) resurrected the slightly older name *A. oolentangiensis*. This epithet represents a noncorrectable misspelling of the place where the species was first discovered, the Olentangy River in Ohio. Several authors have noted that it is closely related to *Symphyotrichum laeve* (Semple et al., 2002), and apparent rare hybrids between the two have been reported (Steyermark, 1963).

Variation within *S. oolentangiensis* is complex and some botanists prefer not to recognize infraspecific taxa within the species (Gleason and Cronquist, 1991). Plants with glabrous stems have sometimes been segregated as *A. oolentangiense* var. *laevicaulis* (a name that has not been transferred to *Symphyotrichum*), but this feature occurs in both of the varieties accepted below. When plants have hairy stems, the density of pubescence in Missouri populations varies from nearly glabrous to fairly densely hairy. It thus seems best not to attempt formal taxonomic recognition of pubescence variants.

1. Median stem leaves mostly 1–3 cm wide, lanceolate to narrowly ovate; heads mostly 2.0–2.5 cm in diameter (including the extended ray corollas) at flowering
   .... 14A. VAR. OOLENTANGIENSE
1. Median stem leaves mostly 0.2–1.0 cm wide, narrowly lanceolate to lanceolate; heads mostly 1.5–2.0 cm in diameter (including the extended ray corollas) at flowering .... 14B. VAR. POACEUM

14a. var. oolentangiense

Aster azureus f. *laevicaulis* Fernald
A. oolentangiensis var. *laevicaulis* (Fernald) A.G. Jones

Basal and lower stem leaves mostly deeply cordate at the base, less commonly truncate. Median stem leaves with the blade mostly 1–3 cm wide, lanceolate to narrowly ovate. Inflorescence branches usually relatively strongly ascending. Heads mostly 2.0–2.5 cm in diameter (including the extended ray corollas) at flowering. Involucre 5–8 mm long. $2n=32$. August–November.

Scattered nearly throughout the state (South Dakota to Texas east to New York, Ohio, and Florida; Canada). Glades, upland prairies, sand prairies, savannas, mesic to dry upland forests on rocky slopes, and ledges and tops of bluffs; also roadsides.

14b. var. poaceum (Burgess) G.L. Nesom

Aster azureus var. *poaceus* (Burgess) Fernald
A. oolentangiensis var. *poaceus* (Burgess) A.G. Jones

Basal and lower stem leaves mostly abruptly rounded at the base, less commonly truncate or shallowly cordate. Median stem leaves with the blade mostly 0.2–1.0 cm wide, narrowly lanceolate to lanceolate. Inflorescence branches usually loosely ascending or spreading. Heads mostly 1.5–2.0 cm in diameter (including the extended ray corollas) at flowering. Involucre 4.5–6.0 mm long. $2n=32$. August–November.

Scattered mostly south of the Missouri River (Texas, Oklahoma, Missouri, Arkansas, and Tennessee). Glades, savannas, and mesic to dry upland forests on rocky slopes; also roadsides.

This variety is less common and less widely distributed in the state than is var. *oolentangiense*.

15. Symphyotrichum parviceps (Burgess) G.L. Nesom (small white aster)

Aster parviceps (Burgess) Mack. & Bush
A. pilosus Willd. ssp. *parviceps* (Burgess) A.G. Jones

Pl. 247 f, g; Map 1027

Plants perennial herbs, with a short, somewhat woody, relatively stout, rhizomatous rootstock. Stems 1 to few, 20–80 cm long, unbranched or with few to many ascending to spreading branches mostly above the midpoint, glabrous or sparsely to moderately pubescent with short, spreading and/or curved hairs, more densely so toward the tip, the hairs relatively evenly distributed or in longitudinal lines or bands. Basal and/or lower stem leaves usually absent from the flowering stems, sessile or with a short, poorly differentiated petiole, the blade 1–4 cm long, 0.3–0.8 cm wide, oblanceolate, tapered at the base, rounded or angled to a minute, abrupt, sharp point at the tip, the margins with spreading to forward-pointing hairs and usually entire, the surfaces glabrous or sparsely pubescent with spreading to somewhat curved hairs, the secondary veins on the leaf undersurface often faint and difficult to distinguish from the veinlets, these forming an irregular network of elongated areoles. Median and upper stem leaves progressively smaller (the main stem leaves often with clusters of smaller leaves along short, axillary branches), the larger ones often withered by flowering time, sessile, the base sometimes slightly expanded but not clasping the stem, the blades 1–8 cm long, linear or narrowly oblong-oblanceolate, the margins entire or shallowly toothed, tapered at the base, angled or tapered to a sharply pointed tip (usually with a short, hard point at the tip), the veinlets usually relatively easily observed, otherwise similar to the lower stem leaves. Inflorescences usually appearing as panicles with short to more commonly relatively long, loosely ascending, racemose branches, sometimes reduced to a solitary raceme, the heads appearing mostly long-stalked and oriented upward, the bracts along the ultimate branches 0.2–0.8 cm long, relatively numerous, more or less leaflike, linear, noticeably shorter than the adjacent foliage leaves. Heads mostly 0.7–1.2 cm in diameter (including the extended ray corollas) at flowering. Involucre 3.0–4.5 mm long, narrowly ellipsoidal to narrowly cup-shaped or nearly cylindrical when fresh (becoming obconical when pressed), the bracts in 3–5 unequal, overlapping series. Involucral bracts linear to narrowly lanceolate, at least the outer few series somewhat inrolled, thickened, and tapered at the sharply pointed, usually awl-shaped tip, with a white to yellowish- or purplish-tinged, minute, bristlelike or spinelike point at the very tip, the tip ascending or somewhat outward- then upward-curved, the outer series with the narrowly elliptic, green portion often reaching nearly to the base, the other series with the slender midvein broadened gradually in the apical $1/4$–$1/2$ into an elliptic or diamond-shaped (3–7 times as long as wide), green tip, the surfaces glabrous, the margins minutely toothed or somewhat irregular and often sparsely short-hairy. Ray florets 10–16(–18) in usually 1 series, the corollas well developed, 3.5–6.0 mm long, white. Disc florets 6–12, the corollas 2–3 mm long, the slender portion of the tube noticeably shorter than the slightly expanded apical portion, the lobes 0.3–0.6 mm long, 20–25 percent of the total length of the expanded portion. Pappus bristles 2–3 mm long, white. Fruits 0.8–1.5 mm long, with 3–4 often faint, longitudinal ribs, pale gray to tan, sparsely to moderately hairy, the hairs lacking swollen bases. 2n=16, 32. August–November.

Scattered, mostly in the eastern half of the state (Kansas, Oklahoma, Iowa, Missouri, Arkansas, and Illinois; apparently disjunct in North Carolina). Upland prairies, mesic to dry upland forests, ledges and tops of bluffs, and less commonly bottomland prairies, margins of ponds, and banks of streams; also pastures, old fields, cemeteries, railroads, roadsides, and open, disturbed areas.

This species is much less common in Missouri than S. pilosum, with which it shares the characteristic inrolled spinelike tips of the involucral bracts. It is easily distinguished from that species when fertile by its smaller, fewer-flowered heads. Steyermark (1963) noted that putative hybrids between the two taxa were not uncommon in northern Missouri. Also, as with S. pilosum, Steyermark knew S. parviceps mainly as an upland species, but it now occurs at some bottomland sites as well. He stated that it occurred most commonly in the Glaciated Plains Division. In discussing their new report of tetraploid (2n=32) and hexaploid (2n=48) plants in the S. parviceps/pilosum complex from dolomite glades in Washington County, Semple et al. (1993) suggested that nearly glabrous plants in these populations were difficult to determine as either S. parviceps or S. pilosum.

Because the main stem leaves often have withered prior to flowering, mature plants of S. parviceps often superficially resemble those of S. ericoides, which also has relatively small heads and some of the involucral bracts with minute, bristlelike tips. However, in S. ericoides the involucral bracts are relatively flat (not inrolled toward the tip) and are moderately hairy along the margins (and also sometimes the outer surface).

16. Symphyotrichum patens (Aiton) G.L. Nesom (spreading aster, purple daisy)
Aster patens Aiton

Pl. 245 g, h; Map 1028

Plants perennial herbs, usually from a short, stout, somewhat branched rootstock, this sometimes somewhat woody, occasionally also producing short rhizomes. Stems 1 to few, (15–)30–120 cm long, with few to several ascending to spreading branches above the midpoint, evenly and moderately to densely pubescent (often somewhat roughened) with short, upward-curved hairs, these sometimes somewhat sparser toward the stem base. Basal as well as lower to median stem leaves absent at flowering, sessile to short-petiolate, the blade 3–7 cm long, 1–3 cm wide, oblanceolate, tapered at the base, rounded or angled to a bluntly pointed tip, the margins entire or sparsely toothed, the margins and surfaces moderately roughened with short, stiff, curved hairs, the secondary veins on the leaf undersurface relatively easily observed, often irregularly fused toward their tips, the veinlets sometimes indistinct, forming a dense, irregular network of relatively short to somewhat elongate (usually wider than long) areoles. Median and upper stem leaves more or less progressively smaller toward the stem tip, sessile, strongly clasping the stem, the blades 2–10 cm long, lanceolate to more or less oblong-lanceolate (frequently the largest leaves somewhat narrowed toward the midpoint and broader toward the base and tip), the uppermost leaves often ovate, deeply cordate at the base, angled or tapered to a sharply pointed tip, the margins entire, otherwise similar to the lower stem leaves. Inflorescences sometimes appearing as open panicles but more often appearing as solitary heads or small clusters at the tips of relatively long branches, the heads appearing short- or long-stalked, the bracts 0.2–0.9 cm long, noticeably shorter than the adjacent foliage leaves, oblong-ovate to linear. Heads mostly 2.0–3.5 cm in diameter (including the extended ray corollas) at flowering. Involucre 5–10 mm long, the bracts in 4–8 unequal, overlapping series. Involucral bracts linear to broadly lanceolate, angled or tapered at the bluntly to sharply pointed tip, the tip erect, ascending, or often somewhat outward-curved, the slender midvein broadened relatively abruptly (mostly in the outer series) or relatively gradually in the apical $1/5$–$1/2$ into an elliptic or diamond-shaped, variable-length green tip, this also often purplish-tinged, the outer surface sparsely to densely pubescent with short, white, mostly appressed hairs and sometimes also with sessile and/or stalked, yellowish glands, the margins usually sparsely to moderately hairy toward the tip. Ray florets 12–25 in usually 1 series, the corollas well developed, 10–16 mm long, purple to bluish purple. Disc florets 20–50, the corollas 5–7 mm long, the slender portion of the tube shorter than the slightly expanded apical portion, the lobes 0.6–0.9 mm long, 18–22 percent of the total length of the expanded portion. Pappus bristles 4.5–7.0 mm long, off-white to cream-colored or tan. Fruits 2.5–3.5 mm long, with 7–10 longitudinal ribs, purplish brown to brown, moderately hairy. 2n=10, 20. August–October.

Scattered mostly south of the Missouri River (eastern U.S. west to Kansas and Texas). Glades, upland prairies, savannas, and openings of mesic to dry upland forests; also old fields, fencerows, railroads, and roadsides; usually on acidic substrates.

Most authors have accepted three varieties in the *S. patens* polyploid complex following the studies of Ronald Jones (1983, 1992). In Missouri, there is intergradation among these varieties, particularly in that many specimens assigned to var. *patentissimus* approach var. *patens* in head size and particularly in having glands on the involucral bracts, and that a few specimens are somewhat intermediate between var. *gracilis* and var. *patens*.

1. Involucre 8–10 mm long, the bracts mostly in 5–8 series, densely hairy, sometimes only toward the tip and along the margins, glandless or with sparse, sessile glands
.... 16C. VAR. PATENTISSIMUM
1. Involucre 5–8 mm long, the bracts mostly in 4–6 series, sparsely to densely hairy, sparsely to densely glandular with sessile and/or stalked glands
   2. Plants relatively short and slender-stemmed and with relatively long branches; involucre 5.0–6.5 mm long, the bracts moderately to densely hairy, often in a central band, mostly relatively blunt at the tip, sparsely glandular, those of the median series mostly 0.7–0.9 mm wide
   .... 16A. VAR. GRACILE
   2. Plants relatively robust (tall or relatively thick-stemmed) with short to long branches; involucre 6–8 mm long, the bracts sparsely to occasionally moderately hairy, relatively densely glandular, at least some of the glands usually stalked, mostly sharply pointed at the tip, those of the median series mostly 1.0–1.2 mm wide .... 16B. VAR. PATENS

16a. var. gracile (Hook.) G.L. Nesom

A. patens var. *gracilis* Hook.

Plants relatively short. Stems 30–70 cm long, slender to relatively thick, moderately hairy. Median and upper stem leaves mostly 2–5 cm long. Involucre 5.0–6.5 mm long, the bracts mostly in 4–6 series, those of the median series mostly 0.7–0.9 mm wide, most of the bracts angled to a relatively bluntly pointed tip, the outer surface moderately to densely hairy, with sparse, mostly sessile glands. $2n=10, 20$. August–October.

Uncommon, known from mostly historical collections from the southwestern portion of the state and a single specimen from Carter County (Kansas and Missouri south to Texas, Alabama, and possibly Florida). Glades, upland prairies, and mesic to dry upland forests.

This variety was first correctly reported for the state from Jasper County by Ronald Jones (1980, 1983). Jones also mistakenly reported the variety from Clay County, based on a specimen from Smithfield (Jasper County). Initially, the taxon was reported from Jasper and McDonald Counties by Palmer and Steyermark (1935, as *Aster patens* var. *gracilis*). However, Steyermark (1963) noted that because of confusion about varietal circumscriptions in the older literature, specimens that he originally had annotated as var. *gracile* were better referred to var. *patentissimum*.

16b. var. patens

Plants relatively stout. Stems (15–)30–120 cm long, usually relatively thick, moderately to densely hairy. Median and upper stem leaves mostly 2–10 cm long. Involucre 6–8 mm long, the bracts mostly in 4–6 series, those of the median series mostly 1.0–1.2 mm wide, all of the bracts angled or tapered to a sharply pointed tip, the outer surface sparsely to occasionally moderately hairy, with relatively dense, sessile and usually also stalked glands. $2n=20$ (occasionally $2n=10$ elsewhere). August–October.

Uncommon, mostly in the Mississippi Lowlands Division but also in counties farther west (eastern U.S. west to Kansas and Texas). Glades, upland prairies, and mesic to dry upland forests; also railroads and roadsides.

A single specimen from Poplar Bluff (Butler County) collected by Stan Hudson has a relatively slender, unbranched stem only about 15 cm long (as in var. *gracile*) but has a pair of terminal heads with bracts (more typical of var. *patens*). It is interpreted to represent a depauperate or damaged specimen of the latter variety.

16c. var. patentissimum (Lindl. ex DC.) G.L. Nesom

A. patens var. *patentissimus* (Lindl. ex DC.) Torr. & A. Gray

Plants short to robust. Stems 30–120 cm long, slender to relatively thick, usually relatively densely hairy. Involucre 8–10 mm long, the bracts mostly in 5–8 series, those of the median series mostly 1.0–1.3 mm wide, all of the bracts mostly angled or tapered to a sharply pointed tip, the outer surface densely hairy, sometimes only toward the tip and along the margins, glandless or with sparse, sessile glands. $2n=20$. August–October.

Scattered mostly south of the Missouri River (Kansas to Illinois and Kentucky south to Texas and Louisiana). Glades, upland prairies, savannas, and openings of mesic to dry upland forests; also old fields, fencerows, railroads, and roadsides.

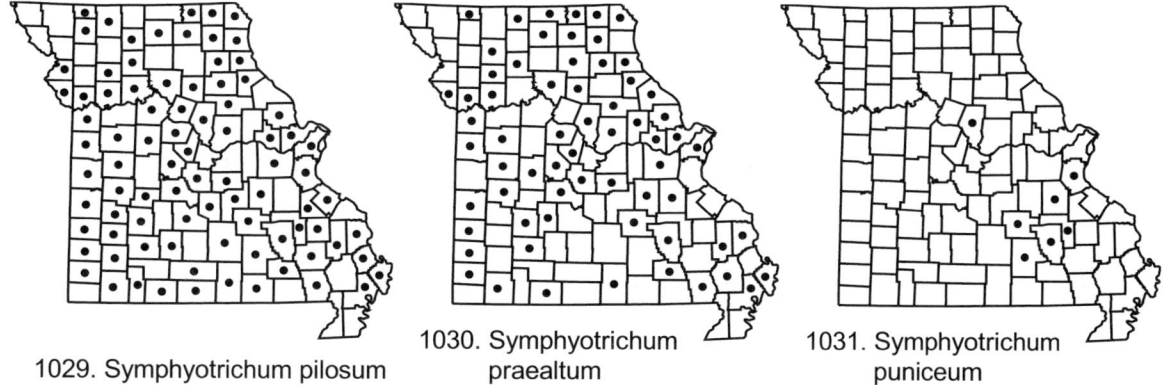

1029. Symphyotrichum pilosum
1030. Symphyotrichum praealtum
1031. Symphyotrichum puniceum

This is by far the most common variety of *S. patens* in Missouri. It is primarily a taxon of the Interior Highlands of the south-central United States.

17. Symphyotrichum pilosum (Willd.) G.L. Nesom (white heath aster)

Aster pilosus Willd.

A. pilosus var. *demotus* S.F. Blake

A. pilosus var. *platyphyllus* (Torr. & A. Gray) S.F. Blake

A. pilosus var. *pringlei* (A. Gray) S.F. Blake

S. pilosum var. *pringlei* (A. Gray) G.L. Nesom

Pl. 247 h, i; Map 1029

Plants perennial herbs, with a short, somewhat woody, relatively stout rootstock, sometimes also with stout, somewhat succulent rhizomes. Stems 1 to several, 20–150 cm long, unbranched or with few to many ascending to spreading branches mostly above the midpoint, glabrous or sparsely to densely pubescent with short, spreading and/or curved hairs, more densely so toward the tip, the hairs usually relatively evenly distributed, occasionally in longitudinal lines or bands. Basal and/or lower stem leaves usually absent from the flowering stems (but new rosettes often produced from the rootstock by flowering time), sessile or with a short, poorly differentiated petiole, the blade 4–8 cm long, 1.0–2.5 cm wide, oblanceolate to obovate, tapered at the base, rounded or angled to a usually bluntly pointed tip, the margins usually with spreading hairs and entire or with sparse, blunt, shallow teeth, the surfaces glabrous or the upper surface sparsely to moderately roughened with short, stiff hairs, the secondary veins on the leaf undersurface often faint and difficult to distinguish from the veinlets, these forming an irregular network of relatively short areoles. Median and upper stem leaves progressively smaller, the larger ones often withered by flowering time, sessile, the base sometimes slightly expanded but not clasping the stem, the blades 1–10(–15) cm long, linear to less commonly elliptic-lanceolate or oblanceolate, the margins entire or shallowly toothed (sometimes somewhat curled under), tapered at the base, angled or tapered to a sharply pointed tip (often with a short, hard point at the tip), the veinlets usually relatively easily observed, otherwise similar to the lower stem leaves. Inflorescences usually appearing as panicles with short to more commonly relatively long, loosely ascending to spreading, racemose branches, sometimes reduced to a solitary raceme, the heads appearing mostly long-stalked and oriented upward, the bracts along the ultimate branches 0.2–1.0 cm long, relatively numerous, more or less leaflike, linear, noticeably shorter and often narrower than the adjacent foliage leaves. Heads mostly 1.4–2.0(–2.5) cm in diameter (including the extended ray corollas) at flowering. Involucre 4–8 mm long, urn-shaped to more or less cup-shaped (becoming bell-shaped to broadly obconical when pressed), the bracts in 4–6 unequal, overlapping series. Involucral bracts narrowly oblong-oblanceolate (sometimes slightly broadened at the base), at least the outer few series somewhat inrolled, thickened, and tapered at the sharply pointed, usually awl-shaped tip, with a white to yellowish- or purplish-tinged, bristlelike, minute spinelike, or rarely softer hairlike point at the very tip, the tip ascending or somewhat outward- then upward-curved, the outer series with the narrowly elliptic green portion reaching nearly to the base, the other series with the slender midvein broadened gradually in the apical $1/4$–$1/2$ into an elliptic or diamond-shaped (3–7 times as long as wide) green tip, the surfaces glabrous or the outer surface sparsely hairy, the margins minutely toothed or somewhat irregular and often sparsely short-hairy. Ray florets 15–35 in usually 1 or 2 series, the corollas well developed, 5–10 mm long, white (rarely pink). Disc florets 20–40, the corollas 2.5–5.0 mm long, the slender portion of the tube noticeably shorter than the slightly expanded apical portion, the lobes 0.5–0.9 mm long, 22–30 percent of the total length of the expanded portion. Pappus bristles 2.5–5.0 mm long, white. Fruits 1.0–1.5 mm long, with 3–4 longitudinal ribs, pale gray to tan, moderately hairy, the hairs lacking swollen bases. $2n=32, 40, 48$. August–November.

Common throughout the state (eastern U.S. west to South Dakota and Texas; Canada; introduced in the Pacific Northwest). Bottomland and upland prairies, glades, bottomland forests, mesic to dry upland forests, savannas, banks of streams and rivers, margins of ponds, lakes, and sinkhole ponds; also fallow fields, pastures, old fields, fencerows, gardens, railroads, roadsides, and open, disturbed areas.

Symphyotrichum pilosum is among the most widespread and weediest of our native asters. Steyermark (1963) knew it primarily as an upland species, but now it is encountered with some frequency at bottomland sites as well. He recommended a form with pink ray corollas (*Aster pilosus* var. *pilosus* f. *pulchellus* Benke) for cultivation as an ornamental, but it can spread aggressively by seed in a garden situation. The species comprises a polyploid complex and also is quite variable in its vegetative morphology. Semple and Chmielewski (1985) studied the cytology of the group across its range, finding both tetraploid ($2n=32$) and hexaploid ($2n=48$) populations in Missouri, with a reported pentaploid ($2n=40$) that may have resulted from hybridization between the other two cytotypes. Steyermark (1963) reported putative hybrids with *S. parviceps*, *S. praealtum*, and *S. racemosum* (as *Aster vimineus*).

Because of the complex cytological and morphological variation within the species, opinions have differed on the merits of providing formal taxonomic recognition to any of the variants (Jones, 1989). Steyermark (1963) treated three varieties, but most subsequent students of the group have treated his *A. pilosus* var. *platyphyllus* as broad-leaved plants showing an environmental response to particular soil and moisture conditions (Jones, 1989). Nesom (1994) and Semple et al. (2002) accepted the other two of Steyermark's varieties, with glabrous-stemmed plants called var. *pringlei* (*Aster pilosus* var. *demotus*) and hairy-stemmed plants called var. *pilosum*. Earlier, Semple (1978) had stated that glabrous-stemmed plants can produce hairy stems in greenhouse culture, suggesting that this feature is at least partially under environmental control. According to Steyermark (1963), both types are equally widespread in the state. In practice, however, a large number of specimens have sparse hairs in lines extending downward from each node and are difficult to categorize into either variant. Semple and Chmielewski (1985) found that plants of var. *pringlei* that they studied were exclusively hexaploids, whereas plants assignable to var. *pilosum* were either tetraploids or hexaploids, but they also suggested that hybridization occurred between the two varieties. Subsequently, Semple et al. (1993) studied tetraploid ($2n=32$) and hexaploid ($2n=48$) plants in the *S. parviceps/pilosum* complex from dolomite glades in Washington County, stating that morphologically some of the plants were unusually glabrescent and thus vegetatively very similar to hexaploid plants that they had determined previously as *S. pilosum* var. *pringlei* but also noting similarities to *S. priceae* (Britton) G.L. Nesom, an unusual octoploid that is endemic to a small region from Kentucky to Alabama and Georgia. In Missouri, plants referable to var. *pringlei* tend to have more slender stems and narrower main stem leaves than do most specimens of var. *pilosus*. In light of the confusion surrounding the infraspecific classification within this species, no varieties are recognized formally in the present treatment.

18. Symphyotrichum praealtum (Poir.) G.L. Nesom (willow-leaved aster)

Aster praealtus Poir.

S. praealtum var. *angustior* (Wiegand) G.L. Nesom

A. praealtus var. *angustior* Wiegand

S. praealtum var. *subasperum* (Lindl. ex Hook.) G.L. Nesom

A. praealtus var. *subasper* (Lindl. ex Hook.) Wiegand

Pl. 247 a–c; Map 1030

Plants perennial herbs, often somewhat colonial from relatively long, slender, branched rhizomes. Stems usually solitary, 30–150 cm long, unbranched or with few to most commonly many ascending branches mostly above the midpoint, sparsely to moderately pubescent with short, spreading or curled hairs toward the tip, these usually in longitudinal lines or bands, usually glabrous toward the base or sometimes nearly throughout, often somewhat glaucous. Basal and/or lower stem leaves usually absent from the flowering stems, sessile or with a short, poorly differentiated petiole, the blade 3–7 cm long, 1.0–2.5 cm wide, oblanceolate, tapered at the base, rounded or angled to a bluntly or sharply pointed tip, the margins with minute, forward-pointing hairs and usually entire (rarely with a few shallow teeth), the upper surface glabrous or sparsely to moderately and minutely roughened, the undersurface glabrous or nearly so, the secondary veins on the leaf undersurface usually difficult to distinguish but the

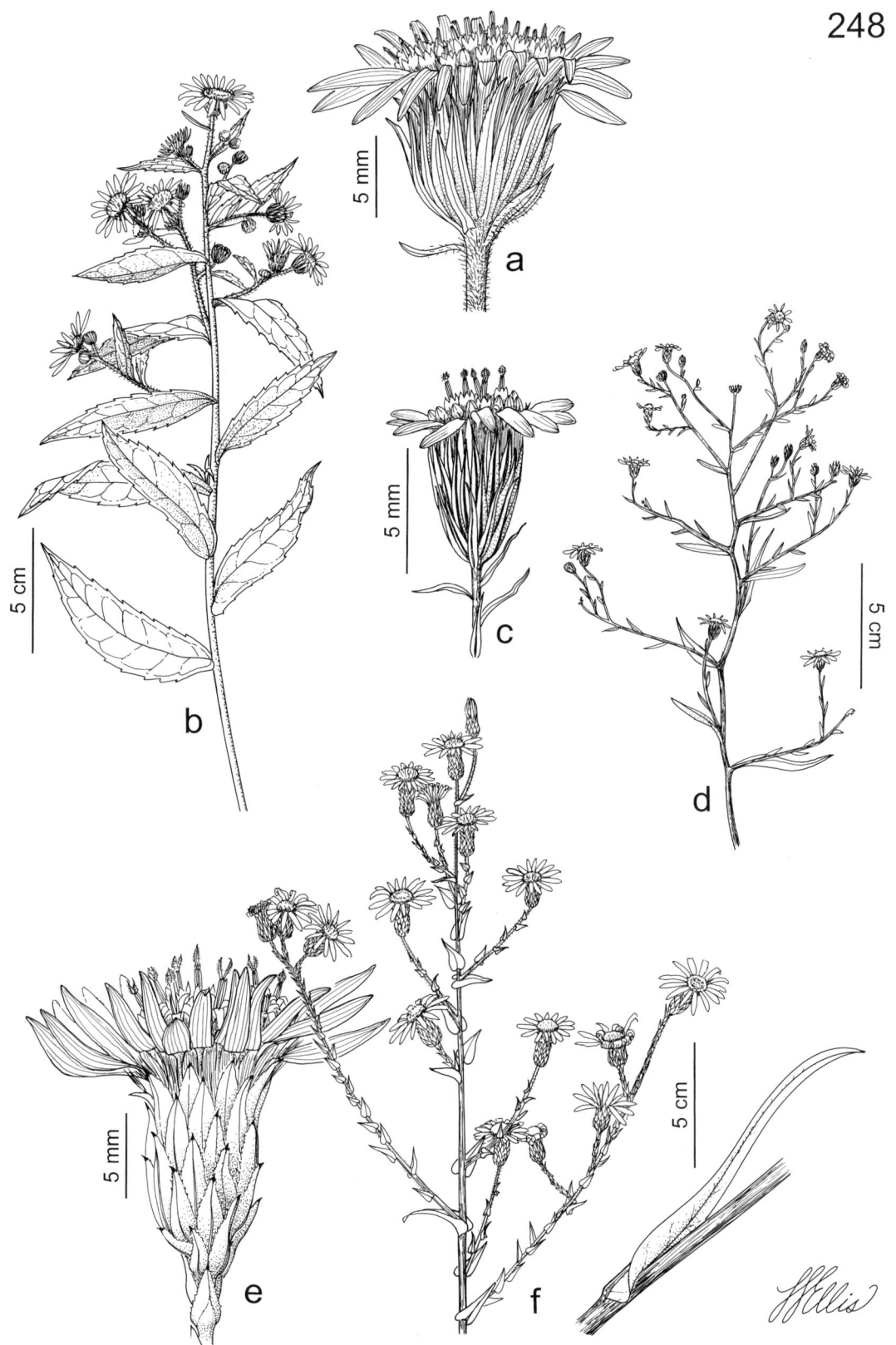

Plate 248. Asteraceae. *Symphyotrichum puniceum*, **a)** head, **b)** habit. *Symphyotrichum subulatum*, **c)** head, **d)** habit. *Symphyotrichum laeve*, **e)** head, **f)** habit and leaf.

veinlets relatively prominent, forming a network with areoles that are about as long as wide or slightly longer than wide (these often containing finer veinlets). Median and upper stem leaves progressively smaller, sessile, the base sometimes slightly expanded but not clasping the stem, the blades 1–15 cm long, linear to narrowly elliptic or oblong-lanceolate, the margins entire or sparsely toothed, angled or tapered at the base, angled or tapered to a sharply pointed tip, otherwise similar to the lower stem leaves. Inflorescences usually appearing as panicles with short to relatively long, loosely ascending to spreading branches (these racemose or less commonly with the heads solitary or in small clusters toward the branch tips), the heads appearing short- to more commonly relatively long-stalked and usually oriented upward, the bracts along the ultimate branches 0.4–1.5 cm long, relatively numerous, more or less leaflike, linear or narrowly oblong-lanceolate, often noticeably shorter than the adjacent foliage leaves. Heads mostly 1.5–2.5 cm in diameter (including the extended ray corollas) at flowering. Involucre 4–7 mm long, cup-shaped to slightly bell-shaped when fresh (sometimes becoming obconical when pressed), the bracts in 4 or 5 unequal, overlapping series. Involucral bracts linear-lanceolate to narrowly oblong-oblanceolate, angled or tapered at the bluntly or sharply pointed tip, lacking a bristlelike or spinelike point at the ascending tip (sometimes with an abrupt, minute, reddish, sharp point at the tip), the slender midvein broadened gradually or more commonly relatively abruptly in the apical $1/4$–$1/2$ into an elliptic, diamond-shaped, or oblanceolate (2–5 times as long as wide) green tip, the outer surface glabrous (the inner surface often sparsely hairy), the margins often slightly irregular and sparsely to moderately hairy especially toward the tip. Ray florets 20–35 in usually 1 or 2 series, the corollas well developed, 5–12 mm long, purple or bluish purple). Disc florets 20–35, the corollas 4–6 mm long, the slender portion of the tube noticeably shorter than the slightly expanded apical portion, the lobes 0.4–1.1 mm long, 18–25 percent of the total length of the expanded portion. Pappus bristles 4.0–6.5 mm long, white or off-white. Fruits 1.5–2.0 mm long, with 4 or 5 longitudinal ribs, purplish brown or tan with purple lines or streaks, sparsely to moderately hairy, the hairs lacking swollen bases. $2n=32, 48$. August–October.

Scattered nearly throughout the state (eastern U.S. west to Minnesota, Colorado, and Texas; Canada). Bottomland prairies, moist depressions of upland prairies, banks of streams, rivers, and spring branches, fens, margins of ponds and lakes, bottomland forests, and mesic upland forests; also ditches, railroads, roadsides, and open, disturbed areas.

Steyermark (1963) recorded this species as nearly absent from the eastern half of the Ozark Division, but more recent collections have served to fill in the range. The infraspecific taxonomy requires more detailed study (Semple et al., 2002). Steyermark accepted three varieties as occurring in the state, but at least two additional varieties have been described from states to the west of Missouri. One of the varieties, var. *subasperum*, originally was described (as *Aster subasper* Lindl. ex Hook.) based on specimens collected by Thomas Drummond near St. Louis. Jones (1989) indicated that at least in Illinois the variants accepted by Steyermark (1963) intergrade too much to make formal taxonomic recognition feasible. In his treatment of *S. praealtum* (as *Aster praealtus*), Steyermark separated var. *angustior* as a rare, narrow-leaved variant that was present sporadically around the state. He treated rare, relatively broad-leaved plants (also with slightly broader involucral bracts) occurring sporadically in the state as var. *subasperum*. The bulk of the Missouri specimens with leaves of intermediate length-to-width ratios were classified in var. *praealtum*. Since Steyermark's (1963) treatment, a number of collections have been made, but these serve to blur the lines between the variants rather than reinforcing them. Because of this, the present treatment does not provide formal taxonomic recognition for varieties within *S. praealtum*.

Steyermark (1963) noted that this species, with its relatively leafy branches and bright, showy flowering heads, does well in the wildflower garden. It is available commercially in some midwestern wildflower nurseries. Steyermark also indicated the presence of rare putative hybrids with *S. pilosum* and *S. lateriflorum*, and putative hybrids with *S. lanceolatum* also have been collected.

19. Symphyotrichum puniceum (L.) Á. Löve & D. Löve (glossy-leaved aster)
Aster puniceus L.
A. firmus Nees
A. puniceus var. *firmus* (Nees) Torr. & A. Gray
A. puniceus ssp. *firmus* (Nees) A.G. Jones
S. firmum (Nees) G.L. Nesom
A. puniceus f. *lucidulus* (A. Gray) Fernald
A. lucidulus (A. Gray) Wiegand
Pl. 248 a, b; Map 1031

Plants perennial herbs, colonial from relatively long-creeping rhizomes. Stems 1 or less commonly few, 50–170 cm long, with few to more commonly several ascending to spreading branches above the midpoint, sparsely pubescent toward the tip with lines or bands of spreading, often stout and stiff (sandpapery) hairs (more densely and evenly hairy elsewhere), glabrous toward the base. Basal as well as lower to median stem leaves absent at flowering, sessile to short-petiolate, the blade 3–12 cm long, 0.5–3.0 cm wide, narrowly oblanceolate to oblanceolate, tapered at the base, rounded or angled to a bluntly or sharply pointed tip, the margins entire to sparsely toothed and relatively strongly roughened, the upper surface often somewhat shiny, glabrous or less commonly somewhat roughened with sparse, short, stout hairs, the undersurface sparsely roughened with short, stout hairs along the midvein, occasionally with a few softer hairs between the veins, the secondary veins on the leaf undersurface relatively easily observed, fused into an irregular network, the veinlets also forming a dense, irregular network of relatively short to somewhat elongate areoles. Median and upper stem leaves often relatively numerous, more or less progressively smaller toward the stem tip, sessile, usually strongly clasping the stem (the uppermost leaves sometimes somewhat sheathing), the blades 2–16 cm long, lanceolate to oblanceolate (often narrowly so), oblong-oblanceolate, or less commonly nearly elliptic, cordate to nearly truncate at the base, angled or tapered to a sharply pointed tip, the margins entire or shallowly toothed, otherwise similar to the lower stem leaves. Inflorescences usually appearing as open panicles, sometimes with solitary heads or small clusters at the tips of relatively long branches, the heads appearing mostly short-stalked, the bracts somewhat shorter than the adjacent foliage leaves, 0.5–1.5 cm long, narrowly lanceolate or oblanceolate to linear. Heads mostly 2–4 cm in diameter (including the extended ray corollas) at flowering. Involucre 6–12 mm long, the bracts in 4–6 subequal, overlapping series. Involucral bracts linear to narrowly oblong-oblanceolate, angled or tapered at the sharply pointed tip, the tip loosely ascending to somewhat spreading, those of the outer series more or less leaflike, the green portion extending more than $2/3$ of the way to the somewhat thickened, pale base, those of the median and inner series with a progressively shorter, elliptic, green apical area (this mostly $1/2$ the length or more), sometimes purplish-tinged along the margins toward the tip, the outer surface glabrous (the inner surface usually sparsely hairy toward the tip), the margins usually finely hairy, especially toward the base. Ray florets 25–45 in 1 or 2 series, the corollas well developed, 12–20 mm long, lavender to purple to bluish purple.

Disc florets 30–70, the corollas 4.5–6.5 mm long, the slender portion of the tube shorter than the slightly expanded apical portion, the lobes 0.4–0.8 mm long, 20–30 percent of the total length of the expanded portion. Pappus bristles 4.5–7.0 mm long, off-white to pale cream-colored. Fruits 2.0–3.5 mm long, with 4 or 5 longitudinal ribs, purplish brown to brown, glabrous or sparsely hairy. $2n=16, 32$. August–October.

Scattered mostly in the eastern half of the Ozark and Ozark Border Divisions with a disjunct population in Boone County (Minnesota, Nebraska, and Missouri east to West Virginia, North Carolina, and Georgia). Fens, calcareous seepage areas along streams, and bottomland prairies.

Steyermark (1963) considered *S. puniceum* (as *Aster puniceus*) in Missouri to represent a relict from Pleistocene glacial times confined to cool, moist microhabitats. He recorded rare putative hybrids between this species and *S. lateriflorum* from Dent and Shannon Counties. Jones (1989) mentioned intergradation in Illinois with other asters having x=8 as their base chromosome number, including *S. lanceolatum* and *S. praealtum*.

The taxonomy of the *S. puniceum* complex requires more study. On the one hand, some authors (A. G. Jones, 1980; Gleason and Cronquist, 1991; Warners and Laughlin, 1999) have recognized two species, *S. puniceum* and *S. firmum* (with a couple of additional morphological extremes sometimes segregated within *S. puniceum* as subspecies). On the other hand, some recent students of the group (Semple et al., 2002) have felt it unwise to provide any formal taxonomic recognition for the two main variants because of a perceived large number of morphological intermediates and widely overlapping ranges. Still other authors (Steyermark, 1963; Jones, 1989) have treated the main variants as varieties or subspecies of *S. puniceum*. This last option has merit, but the epithet *firmus* has not yet been formally transferred to *Symphyotrichum* at the varietal or subspecific level and is therefore not available for use. Thus far, the most detailed study of the complex was that of Warners and Laughlin (1999), who presented evidence that the two taxa do not intergrade for critical characters in a series of populations where they co-occur in Michigan. They characterized *S. firmum* as having a more strongly rhizomatous rootstock with less clustered stems that are less densely hairy and generally smaller basal leaves, among other characters. However, their observation that *S. firmum* has white to pale lavender ray corollas does not hold up in material from other states, where the ray corollas in both morphotypes usually are purple

1032. Symphyotrichum racemosum

1033. Symphyotrichum sericeum

1034. Symphyotrichum subulatum

to bluish purple. In Missouri, as noted by Steyermark (1963), the populations occurring in calcareous seepage communities in the eastern Ozarks and adjacent counties correspond to *S. firmum. Symphyotrichum puniceum* in the strict sense is not yet known from Missouri, but it occurs in the adjacent portions of the Great Plains.

20. **Symphyotrichum racemosum** (Elliott) G.L. Nesom **var. subdumosum** (Wiegand) G.L. Nesom (small white aster, frost flower)

Aster fragilis Willd., misapplied
A. vimineus Lam., misapplied
A. vimineus var. *subdumosus* Wiegand
A. fragilis var. *subdumosus* (Wiegand) A.G. Jones

Pl. 246 c, d; Map 1032

Plants perennial herbs, often somewhat colonial from relatively long, slender, branched rhizomes. Stems usually solitary, 20–100 cm long, unbranched or with few to most commonly many spreading to ascending branches mostly above the midpoint, sparsely pubescent with short, curled hairs toward the tip, these usually in longitudinal lines or bands, usually glabrous toward the base or sometimes nearly throughout. Basal and/or lower stem leaves usually absent from the flowering stems, sessile or with a short, poorly differentiated petiole, the blade 1–4 cm long, 0.4–1.0 cm wide, oblanceolate, tapered at the base, rounded or angled to a bluntly or sharply pointed tip, the margins with spreading to forward-pointing hairs and usually scalloped, the surfaces glabrous, the secondary veins on the leaf undersurface often difficult to distinguish from the veinlets, forming an irregular network of elongated areoles. Median and upper stem leaves progressively smaller, the larger ones occasionally withered by flowering time, sessile, the base sometimes slightly expanded but not clasping the stem, the blades 1–6 cm long, linear to narrowly oblanceolate, the margins entire or sparsely toothed, angled or tapered at the base, angled or tapered to a sharply pointed tip, otherwise similar to the lower stem leaves. Inflorescences usually appearing as panicles with relatively long, loosely ascending to spreading or arched branches (these mostly with the heads solitary or in small, loose clusters toward the branch tips), the heads appearing mostly long-stalked and usually oriented in various directions, the bracts along the ultimate branches 0.1–1.0 cm long, often relatively numerous, more or less leaflike, linear, noticeably shorter than the adjacent foliage leaves. Heads mostly 0.6–1.0 cm in diameter (including the extended ray corollas) at flowering. Involucre 2.5–4.0 mm long, cup-shaped to slightly bell-shaped when fresh (sometimes becoming obconical when pressed), the bracts in 4 or 5 unequal, overlapping series. Involucral bracts linear-lanceolate to narrowly oblong-oblanceolate, angled or tapered at the usually sharply pointed tip, lacking a bristlelike or spinelike point at the ascending tip, the slender midvein broadened gradually in the apical $1/2$–$3/4$ into a narrowly elliptic (4–10 times as long as wide), green tip, the outer surface glabrous, the margins often slightly irregular and sparsely hairy. Ray florets 12–20 in usually 1 or 2 series, the corollas well developed, 3–6 mm long, white (rarely pink). Disc florets 15–25, the corollas 2.5–3.5 mm long, the slender portion of the tube noticeably shorter than the slightly expanded apical portion, the lobes 0.4–0.7 mm long, 30–45 percent of the total length of the expanded portion. Pappus bristles 2.5–3.5 mm long, white or off-white. Fruits 1.0–1.8 mm long, with 4 or 5 faint, longitudinal ribs, gray to tan, sparsely to moderately hairy, the hairs lacking swollen bases. $2n=16, 32$. August–October.

Scattered, mostly in the southern half of the state, most commonly in the Mississippi Lowlands Division (eastern U.S. west to Wisconsin and Texas; introduced in Canada). Bottomland forests, swamps, moist depressions of upland prairies, banks of streams and rivers, margins of ponds, lakes, and sinkhole ponds, fens, and sloughs; also ditches, railroads, roadsides, and moist, open, disturbed areas.

The taxonomy of the *S. racemosum* complex is still not well understood, especially its separation from *S. lanceolatum* var. *interior*. The epithet used for the species by Steyermark (1963) and many earlier botanists, *Aster vimineus,* apparently was based on specimens referable to *S. lateriflorum* (Gleason and Cronquist, 1991; Semple et al., 2002). It has also been called *A. fragilis,* but according to Gleason and Cronquist (1991) that name is based on specimens representing a cultivated hybrid involving *S. lanceolatum* and some unknown second parent. The varietal epithet var. *subdumosum* is derived from the strong superficial similarity of the plants to those of *S. dumosum,* which differs in its involucral bracts (see the key to species). The var. *racemosum,* which occurs to the south and east of Missouri, supposedly differs in having somewhat shorter-stalked heads, but it is not clear that the two varieties really are distinct.

21. **Symphyotrichum sericeum** (Vent.) G.L. Nesom (silky aster)
    Aster sericeus Vent.
    Pl. 249 a–c; Map 1033

Plants perennial herbs, usually with a short, stout, cormlike rootstock, this somewhat woody, sometimes more elongate and rhizomatous. Stems 1 to several, 30–70 cm long, with few to several ascending to loosely ascending branches above the midpoint, ranging from glabrous or nearly so toward the base to densely pubescent with appressed, silky hairs toward the tip. Basal as well as lower to median stem leaves absent at flowering, sessile, the blade 1.5–5.0 cm long, 0.4–1.0 cm wide, mostly oblanceolate to oblong-elliptic, short-tapered to rounded at the base, mostly angled and with a minute, sharp point at the tip, the margins entire, the margins and surfaces moderately to densely pubescent with appressed, silky hairs (the leaves appearing gray or silvery), the secondary veins on the leaf undersurface not visible or faint and few. Median and upper stem leaves relatively uniform, sessile, not clasping or sheathing the stem, the blades 0.7–3.0 cm long, lanceolate or elliptic, angled or rounded at the base, otherwise similar to the lower stem leaves. Inflorescences sometimes appearing as open panicles but more often appearing as solitary heads or small clusters at the branch tips, the heads appearing short- or less commonly long-stalked, the bracts similar to the adjacent foliage leaves. Heads mostly 2–3 cm in diameter (including the extended ray corollas) at flowering. Involucre 6–10 mm long, the bracts in 3–6 unequal, overlapping series. Involucral bracts lanceolate to ovate, angled or short-tapered at the sharply pointed tip, the tip erect or ascending, the green, leaflike portion in the apical $^{1}/_{2}$–$^{2}/_{3}$ abruptly contracted (sometimes only slightly narrower) to the pale, thickened, yellowish basal portion, the outer surface and margins densely appressed-hairy. Ray florets 10–25 in usually 1 series, the corollas well developed, 9–15 mm long, dark purple or nearly blue. Disc florets 15–35, the corollas 5–7 mm long, minutely hairy, the slender portion of the tube much shorter than the slightly expanded apical portion, the lobes 0.7–0.9 mm long, 18–22 percent of the total length of the expanded portion. Pappus bristles 5.5–7.0 mm long, often slightly longer than the disc corollas, off-white to cream-colored or tan. Fruits 2–3 mm long, with 7–10 longitudinal ribs, purplish brown to brown, glabrous. $2n=10$. August–October.

Scattered in the Ozark, Ozark Border, and Unglaciated Plains Divisions, north locally to Lincoln County and along the Missouri River to Atchison County (North Dakota to Texas east to Michigan, Indiana, and Arkansas; disjunct in Georgia). Glades, upland prairies, and loess hill prairies.

This beautiful wildflower is available for sale at some wildflower nurseries and should be used more widely in gardens. The white-rayed *Aster sericeus* f. *albiligulatus* Fassett has not yet been found to grow in Missouri.

22. **Symphyotrichum subulatum** (Michx.) G.L. Nesom **var. ligulatum** (Shinners) S.D. Sundb. (inland saltmarsh aster, freeway aster)
    Aster subulatus Michx. var. *ligulatus* Shinners
    A. exilis Elliott
    S. divaricatum (Nutt.) G.L. Nesom
    Pl. 248 c, d; Map 1034

Plants annual, with a sometimes relatively slender, short taproot. Stem solitary, 10–80 cm long, usually with numerous loosely to strongly ascending branches, glabrous or with a few short, stout hairs toward the tip. Basal and lower stem leaves usually absent at flowering, sessile or with a short, poorly differentiated petiole, this sometimes slightly sheathing the stem, the blade 3–12 cm long, 0.3–1.0 cm wide, linear to narrowly oblanceolate, tapered at the base, tapered to a sharply pointed tip, the margins entire or rarely sparsely and shallowly toothed, the surfaces often somewhat sea green to bluish green, glabrous, the secondary veins and veinlets faint, forming an irregular network of areoles. Median and upper stem leaves progressively smaller, sessile, the base not clasping or sheathing the stem, the blades 0.8–7.0 cm long, linear, otherwise similar to the lower stem leaves. Inflorescences usually panicles with relatively long, ascending to spreading, few- to many-headed branches, these sometimes somewhat racemose, occasionally reduced to a solitary terminal raceme, the heads appearing short- or long-stalked, the bracts along the ultimate branches 0.3–0.8 cm long, sometimes relatively few, leaflike, mostly linear, noticeably shorter than the adjacent foliage leaves. Heads mostly 1.0–1.7 cm in diameter (including the extended ray corollas) at flowering. Involucre 5–8 mm long, the bracts in 3 or 4 unequal, overlapping series. Involucral bracts linear to narrowly lanceolate, tapered at the sharply pointed tip, the tip erect or ascending, the slender midvein broadened gradually usually just above the base (in the apical $^2/_3$–$^9/_{10}$) into a narrowly elliptic green tip with relatively broad, white or purplish-tinged margins, the outer surface and margins glabrous. Ray florets 15–35 in 1 or 2 series, the corollas 4–9 mm long (often tightly inrolled from tip to base if not pressed immediately upon collection), relatively slender (often resembling those of *Erigeron*), white to pale bluish purple. Disc florets 18–35, the corollas 3–4 mm long, the slender portion of the tube shorter than the very slightly expanded apical portion, the lobes 0.2–0.5 mm long, 10–20 percent of the total length of the expanded portion. Pappus bristles 4–5 mm long, slightly to noticeably longer than the florets, white to off-white or light straw-colored. Fruits 2.0–2.5 mm long, with 4–6 sometimes inconspicuous, longitudinal ribs, gray to light tan, moderately to densely hairy. $2n=10$. September–January.

Introduced, scattered in the Mississippi Lowlands Division, as yet uncommon farther north and west (New Mexico to Alabama north to Nebraska, Missouri, and Tennessee; Mexico). Edges of ponds, oxbows, and swamps; also margins of crop fields, fallow fields, lawns, ditches, railroads, roadsides, and open, disturbed areas.

The native range of the widespread *S. subulatum* has been obscured by its relatively recent migration into highly disturbed habitats, particularly along roads that receive applications of salt during the winter. The distribution given above for var. *ligulatum* includes states in which the plants probably did not occur prior to 1900. Sundberg (1986, 2004) recognized five varieties that sometimes have been elevated to species status (Nesom, 1994, 2005). Of these, three are relatively restricted in their distribution in the continental United States: var. *elongatum* (Boss.) S.D. Sundb. in coastal Florida and Georgia; var. *parviflorum* (Nees) S.D. Sundb. in the southwestern United States; and var. *squamatum* (Spreng.) S.D. Sundb., a native of South America that has been introduced sporadically in the southeastern United States. Of the other two varieties, the var. *subulatum,* which has very short, inconspicuous ray corollas, originally was primarily a plant of brackish coastal areas along both the Atlantic Ocean and the Gulf of Mexico, but it has spread inland mainly in the Great Lakes states to Illinois (and Ontario) and also more centrally into southern Arkansas. Eventually some vehicle tire may transport plants of this variety into Missouri.

The var. *ligulatum* originally had more of an inland distribution in the United States from Alabama to New Mexico but probably did not occur natively north of Arkansas. In Missouri, Julian Steyermark collected the first specimens in Dunklin and Pemiscot Counties in 1948 (Steyermark, 1963). Its period of biggest spread in the state apparently began in the late 1980s and continues today. Plants that grow at relatively unmanaged sites generally tend to be upright with ascending branches, but the species responds well to summer mowing, resprouting aggressively into a spreading to low, clump-forming habit. It is usually quite distinctive in its green to light bluish green or sea green (sometimes purplish-tinged in part) stems and leaves, which are essentially glabrous.

23. Symphyotrichum turbinellum (Lindl. ex Hook.) G.L. Nesom (prairie aster)
Aster turbinellus Lindl.

Pl. 229 f, g; Map 1035

Plants perennial herbs, with a compact, woody, sometimes short-branched, usually horizontal rootstock. Stems 1 to several, 40–120 cm long, with few to more commonly several ascending to spreading branches above the midpoint, sparsely pubescent toward the tip, usually in lines or bands, with spreading, often stout and stiff (sandpapery) hairs, glabrous toward the base or sometimes nearly the entire length. Basal as well as lower to median stem leaves often absent at flowering, sessile to short-petiolate, the blade 4–12 cm long, 0.5–2.0 cm wide, narrowly oblanceolate to narrowly oblong-oblanceolate, tapered at the base, rounded or angled to a bluntly or sharply pointed tip, the margins entire or sparsely toothed and relatively strongly roughened, the upper surface often somewhat shiny, glabrous, the undersurface sparsely roughened with short, stout hairs along the midvein, otherwise glabrous, the secondary veins on the leaf undersurface

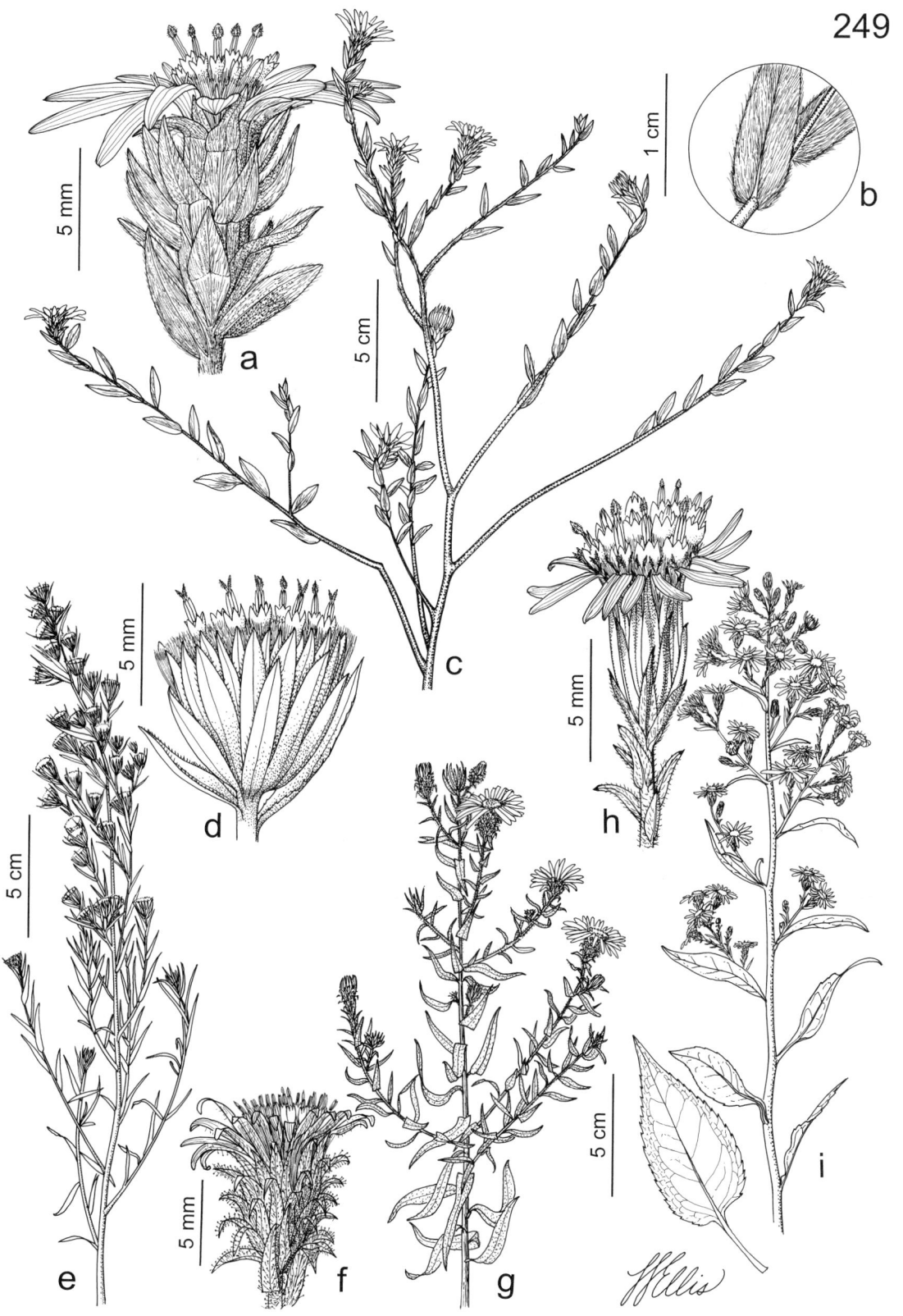

Plate 249. Asteraceae. *Symphyotrichum sericeum*, **a)** head, **b)** leaf detail, **c)** habit. *Symphyotrichum ciliatum*, **d)** head, **e)** habit. *Symphyotrichum oblongifolium*, **f)** head, **g)** habit. *Symphyotrichum urophyllum*, **h)** head, **i)** head and larger leaf.

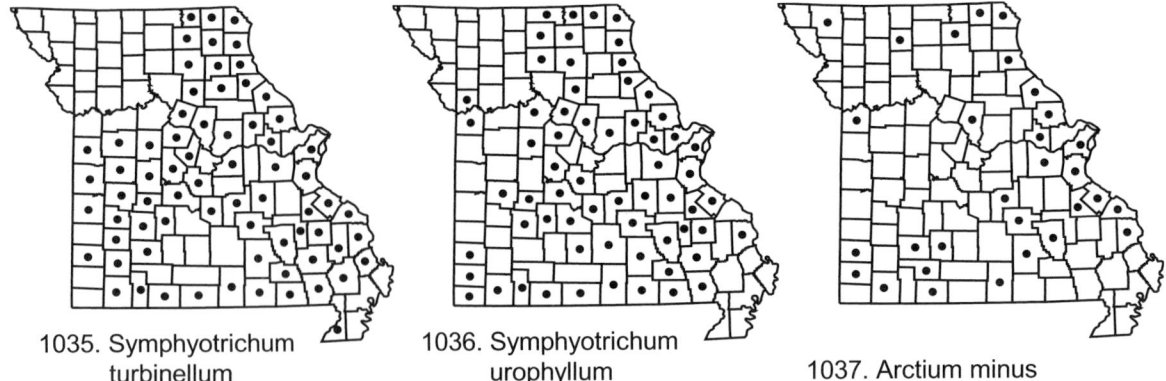

1035. Symphyotrichum turbinellum
1036. Symphyotrichum urophyllum
1037. Arctium minus

usually relatively faint, fused into an irregular network, the veinlets also forming a dense, irregular network of short to elongate areoles. Median and upper stem leaves variable, more or less progressively smaller toward the stem tip, sometimes abruptly smaller in the inflorescence, sessile, not clasping or sheathing the stem, the blades 2–12 cm long, linear to narrowly lanceolate, narrowly elliptic, lanceolate, or elliptic, tapered at the base, angled or tapered to a sharply pointed tip, the margins usually entire, otherwise similar to the lower stem leaves. Inflorescences appearing as open panicles, sometimes with relatively long, loosely racemose branches, the heads appearing mostly long-stalked, the bracts usually noticeably shorter and narrower than the adjacent foliage leaves, 0.2–0.7(–1.2) cm long, narrowly lanceolate or narrowly elliptic to more commonly linear. Heads mostly 2–3 cm in diameter (including the extended ray corollas) at flowering. Involucre 7–12 mm long, the bracts in 6–9 unequal, overlapping series. Involucral bracts narrowly oblong-oblanceolate to oblanceolate, rounded or broadly angled to a bluntly pointed tip (the innermost series sometimes with an abrupt, short, sharp point at the tip), the tip ascending, the base somewhat thickened and keeled, the slender midvein broadened relatively abruptly in the apical $^1/_5$–$^1/_3$ into a broadly elliptic to diamond-shaped or obovate green tip with relatively broad, white to straw-colored margins nearly to the tip, these often also somewhat purplish-tinged toward the tip, the outer surface glabrous or less commonly sparsely hairy toward the tip, the margins usually finely hairy, especially toward the tip. Ray florets 14–20 in 1 series, the corollas well developed, 10–16 mm long, lavender to purple to bluish purple. Disc florets 15–30, the corollas 4.5–6.5 mm long, the slender portion of the tube shorter than the slightly expanded apical portion, the lobes 0.6–0.9 mm long, 17–22 percent of the total length of the expanded portion. Pappus bristles 4.5–6.0 mm long, off-white to light tan, rarely purplish-tinged. Fruits 2–3 mm long, with 4 or 5 longitudinal ribs, purplish brown to brown, sparsely hairy. $2n=96$. August–November.

Scattered nearly throughout the state, but apparently absent from most of the western half of the Glaciated Plains Division (Nebraska to Oklahoma east to Illinois and Louisiana). Openings of mesic to dry upland forests, savannas, glades, and ledges and tops of bluffs; also railroads and roadsides; often on acidic substrates.

Steyermark (1963) noted that this showy species does well in cultivation and should be grown more widely as an ornamental in gardens. Barkley (1986) stated that the colloquial name prairie aster seems inappropriate for this species, as it does not grow in prairies in the Great Plains. The same can be said for Missouri, where this is mostly a woodland species. Presumably, farther east it grows more frequently in grassland communities.

24. Symphyotrichum urophyllum (Lindl. ex DC.) G.L. Nesom
Aster urophyllus Lindl. ex DC.
A. sagittifolius Wedem. ex Willd., misapplied
A. sagittifolius Wedem. ex Willd. f. *hirtellus* (Lindl.) Shinners

Pl. 249 h, i; Map 1036

Plants perennial herbs, usually from a short, stout, somewhat branched rootstock, this sometimes somewhat woody. Stems 1 to few, 30–120 cm long, unbranched or with few to several ascending branches above the midpoint, glabrous or sparsely to moderately pubescent toward the tip with short, curled hairs in longitudinal lines or bands. Basal and/or lower stem leaves usually present at flowering, long-petiolate, the petiole usually narrowly winged, the blade 4–12 cm long, 2–5 cm wide, heart-shaped to ovate, deeply cordate to truncate or abruptly rounded at the base, angled or tapered to a sharply pointed tip, the margins usually sharply but often shallowly toothed, glabrous or sparsely to less commonly moderately pubescent (sometimes moderately sandpapery) with minute, stiff hairs on the upper surface, sparsely to moderately pubescent with slightly longer hairs on the undersurface, the secondary veins on the leaf undersurface usually easily observed, often irregularly fused toward their tips, the veinlets forming a dense, irregular network of relatively short areoles. Median and upper stem leaves progressively smaller, with long to short, often winged but not clasping (often slightly sheathing) petioles or the uppermost leaves sessile, the blades 1–7 cm long, narrowly ovate to narrowly lanceolate, rounded to tapered at the base, otherwise similar to the lower stem leaves. Inflorescences usually panicles with relatively long, loosely ascending, few- to many-headed, racemose branches, rarely reduced to a solitary terminal raceme or cluster, the heads appearing short- or long-stalked, the bracts along the ultimate branches 0.5–1.2 cm long, leaflike, mostly linear, noticeably shorter and narrower than the adjacent foliage leaves. Heads mostly 1.0–1.5 cm in diameter (including the extended ray corollas) at flowering. Involucre 4.5–6.5 mm long, the bracts in 4–6 unequal, overlapping series. Involucral bracts linear to narrowly lanceolate or narrowly oblong, angled to long-tapered at the sharply pointed tip, the tip erect or ascending (the lowermost bracts occasionally slightly spreading or the bract sometimes terminating in a minute, spinelike, hardened, outward-angled point), the slender midvein broadened gradually in the apical $1/3$–$2/3$ into a sometimes indistinct, narrowly elliptic to nearly linear, green tip (mostly 6–10 times as long as wide), the outer surface glabrous or less commonly sparsely short-hairy, the margins irregularly hairy. Ray florets 8–15, usually in 1 series, the corollas well developed, 4.5–9.0 mm long, white or rarely lavender. Disc florets 8–20, the corollas 3.5–5.0 mm long, the slender portion of the tube shorter than the slightly expanded apical portion, the lobes 0.4–0.7 mm long, 20–25 percent of the total length of the expanded portion. Pappus bristles 3.0–4.5 mm long, off-white to pale cream-colored, occasionally pale purplish-tinged. Fruits 1.5–2.5 mm long, mostly with 4 longitudinal ribs, purplish brown to brown, often with lighter ribs, glabrous. $2n=16$ (rarely 32, 48 elsewhere). August–November.

Scattered nearly throughout Missouri, but absent or uncommon in the far northwestern and far southeastern corners of the state (eastern U.S. west to Minnesota, Nebraska, and Oklahoma; Canada). Mesic to dry upland forests, savannas, margins of glades, ledges and tops of bluffs, and banks of streams and rivers; also pastures, old fields, fencerows, railroads, and roadsides.

The nomenclature of this species has a confusing history and remains somewhat controversial. A number of authors (Steyermark, 1963; Gleason and Cronquist, 1963, 1991; Cronquist, 1980; Barkley, 1986) have called the taxon *Aster sagittifolius,* based in large part upon Arthur Cronquist's assessment of the type material in the Willdenow Herbarium in Berlin. However, another aster specialist, Almut Jones, also examined these specimens and had a different interpretation (A. G. Jones, 1980; Jones and Hiepko, 1981). She lectotypified the name based on a specimen that is referable to *A. cordifolius* and applied the name *A. urophyllus* to the present species. In transferring most of the North American *Aster* names to *Symphyotrichum* and other segregate genera, Nesom (1994) unfortunately transferred all of the epithets in question without selecting a particular taxonomic circumscription to follow. More recently, Semple et al. (2002) reviewed the situation and found themselves in agreement with the treatment of Jones, and this interpretation is followed in the present work.

Typically, the ray florets of *S. urophyllum* have white corollas. Rare plants with light lavender ray corollas may represent hybrids with *S. cordifolium,* but this situation has not been studied in detail. *Symphyotrichum cordifolium* tends to occur in moister environments than does *S. urophyllum.* Steyermark (1963) reported the existence of rare putative hybrids between *S. urophyllum* (as *Aster sagittifolius*) and *S. dumosum* (as *A. dumosus*).

3. Tribe Cardueae Cass.

Plants annual, biennial, or perennial herbs (rarely woody elsewhere), the sap not milky. Stems unbranched or branched, sometimes with spiny wings (from decurrent leaf bases). Leaves alternate, sometimes also in a basal rosette, frequently spiny along the margins. Leaf blades simple, commonly with marginal teeth or pinnately lobed, rarely entire. Inflorescences terminal or less commonly axillary, consisting of solitary heads at the stem or branch tips or few- to several-headed clusters (sometimes appearing as small, flat-topped panicles in *Cirsium arvense*). Heads entirely discoid or the marginal florets sometimes sterile, somewhat enlarged, and appearing raylike. Involucre of several overlapping series of bracts, these appressed or with a spreading tip, the tip often spiny or with an elongate bristle or flattened appendage. Receptacle flat to short-conical (fleshy in *Onopordum*), with numerous bristles or less commonly short scales. Disc florets perfect (the plants incompletely dioecious in *Cirsium arvense*) or the outermost ones sometimes sterile, somewhat enlarged, and raylike. Pappus rarely absent, sometimes of several short scales or awns, most commonly of numerous bristles, these often of different lengths, usually finely barbed or plumose (featherlike with numerous long, capillary side branches), persistent at fruiting or more commonly shed before fruiting individually or as a unit. Corollas yellow, white, pink, purple, or blue, the tube usually slender and elongate, with relatively long, sometimes asymmetrically cut, slender lobes. Stamens with the filaments not fused together (fused toward the base in *Silybum*), the anthers fused into a tube, each tip with a short to long, sometimes indistinct appendage, each base prolonged into a pair of slender, elongate (short in *Silybum*), tail-like lobes, these often hairy. Style branches usually somewhat flattened, each with a stigmatic band along the inner face, lacking a sterile tip. Fruits variously shaped, not winged, not beaked but often with a minute crown or conical projection at the tip. About 83 genera, about 2,500 species, nearly worldwide but most diverse in the Old World.

Species are mostly easily recognized as members of the Cardueae, but generic and specific distinctions within the tribe generally are more problematic. The presence of involucral bracts having the tip modified with a spine, bristle, or flattened appendage; the usually long, narrowly tubular corollas with slender lobes; and the bristly (rarely scaly) receptacle are easily observed characters that tend to mark the tribe. Interestingly, at least in the midwestern United States, those taxa having spiny wings on the stems are all nonnative. However, the group lacking this character comprises both native and introduced taxa. In many botanical works (Steyermark, 1963; Gleason and Cronquist, 1963, 1991), this tribe has been treated under the illegitimate name Cynareae (Scott, 1990).

The principal economic importance of the Cardueae is for its noxious weeds of pastures and cropland, as well as species that are invasive exotics in natural communities. However, several species provide positive economic benefits. Various species in several genera are cultivated as ornamentals. *Cynara cardunculus* L. (cardoon) is cultivated for its edible celery-like petioles. The closely related *C. scolymus* L. (artichoke, globe artichoke) is prized for the edible bracts and receptacle of immature heads (the so-called choke consists of the receptacular [chaffy] bristles and the developing florets with abundant pappus bristles).

Carthamus tinctorius L. (safflower) has long been cultivated for a number of uses. Before the development of artificial dyes, it was cultivated mostly in the Old World for its bright flowers, which were used as a dye, a colorant for cosmetics (rouge), and as a food coloring substitute for the unrelated and more costly saffron. Today, safflower is widely grown commercially for its seeds, which provide a polyunsaturated oil popular for cooking and in salad dressings. Safflower also is a constituent of many birdseed mixes, and plants occasionally are reported to the Flora of Missouri Project by curious homeowners who observe them under and near their bird feeders. Thus far, this species has not been documented to persist or reproduce itself in the wild in Missouri, and it is thus not formally treated here. *Carthamus tinctorius* is relatively easily distinguished from other thistles by the lower involucral bracts, which are enlarged and leaflike, and by its corollas, which most commonly are bright orange to reddish orange (less frequently yellow or red).

1. Stems winged, the wings spiny along the margins
   2. Pappus bristles plumose (featherlike with numerous long, capillary side branches) . . . . 30. CIRSIUM
   2. Pappus bristles not plumose, although sometimes with short, ascending barbs
      3. Receptacle with dense bristles, not fleshy, the fruits not embedded in the surface or sunken into pits; leaves glabrous or sparsely prickly-hairy on the upper surface, often densely woolly- or felty-hairy on the undersurface . . . . 28. CARDUUS

3. Receptacle with low, broad scales, fleshy, the fruits embedded in the surface or appearing sunken into pits; leaves densely woolly- or felty-hairy on both surfaces . . . . 31. ONOPORDUM
1. Stems not winged or, if winged, then the wings not spiny along the margins
    4. Leaf surface noticeably white-mottled; stamens with the filaments fused toward the base. . . . . 32. SILYBUM
    4. Leaf surface uniformly green, the coloration sometimes obscured by pubescence; stamens with the filaments not fused
        5. Involucral bracts with a long, stiff bristle at the tip, this hooked at the tip . . . . 27. ARCTIUM
        5. Involucral bracts spiny, with a flattened, fringed appendage, or unmodified at the tip
            6. Fruits asymmetrical at the base, appearing obliquely or laterally attached or the base appearing twisted to the side (only slightly so in *C. repens*); pappus occasionally absent or more commonly of several series of bristles and/or scales, the outermost series shorter than the inner ones, the bristles, when present, with fine, ascending barbs but not plumose . . . . 29. CENTAUREA
            6. Fruits symmetrical at the base, appearing basally attached; pappus of well-developed bristles, these plumose (featherlike with numerous long, capillary side branches) . . . . 30. CIRSIUM

27. Arctium L. (burdock)

Five species, Europe, Asia.

1. Arctium minus (Hill) Bernh. (common burdock)

Pl. 250 a–c; Map 1037

Plants coarse, biennial, with stout taproots. Stems 50–180 cm long, erect or ascending, usually branched, stout, longitudinally ridged, not winged, minutely hairy, at least toward the tip, often reddish- or purplish-tinged. Leaves basal and alternate, long-petiolate (petioles of the basal leaves usually hollow), not decurrent, the blades with the margins irregularly wavy or shallowly lobed and often also toothed, the upper surface glabrous or with scattered hairs along the main veins, the undersurface pale-colored, sparsely to moderately pubescent with minute, few-branched hairs (these sometimes appearing cobwebby), sometimes becoming nearly glabrous with age. Basal leaves with the blades 30–60 cm long, narrowly to broadly ovate, more or less cordate at the base, rounded to bluntly or sharply pointed at the tip. Stem leaves progressively smaller toward the tip, ovate to triangular-ovate, shallowly cordate, truncate, or abruptly tapered at the base, rounded to bluntly or sharply pointed at the tip, sometimes abruptly tapered to a short, sharp point. Inflorescences axillary and terminal, the heads sessile or short-stalked, appearing clustered or more commonly in short, dense racemes. Heads discoid, the involucre broadly ovoid to nearly spherical (sometimes appearing bell-shaped when pressed), 1.2–2.5 cm in diameter, the florets all appearing similar and perfect. Receptacle flat, with numerous bristles. Involucral bracts 4–13 mm long, the body narrowly lanceolate, appressed-ascending, glabrous or somewhat glandular, occasionally cobwebby-hairy, tapered to a long, stiff, ascending bristle, this hooked at the tip (the innermost bracts often with somewhat flattened, hookless bristles). Pappus of several unequal series of short, flattened, bristlelike awns, these 1–3 mm long, with short, ascending barbs, mostly shed individually by fruiting. Corollas 7–9 mm long, pink to purple, often somewhat glandular. Fruits appearing basally attached, 4–6 mm long, oblong or slightly narrower at the symmetrical base, somewhat flattened, the surface finely wrinkled, grayish brown with darker mottling. $2n=32, 36$. July–October.

Introduced, scattered sporadically nearly throughout the state (native of Europe, Asia, introduced widely in North America). Rarely disturbed openings of mesic upland forests and banks of streams; more commonly barnyards, feedlots, old fields, pastures, railroads, and open, disturbed areas.

Young stems of burdock are eaten fresh or more commonly baked or boiled. The young herbage sometimes also is eaten by livestock. The thick roots can be cooked and eaten, and in the past they sometimes were dried and ground for use as a filler in coffee. They also have been used medicinally as a laxative, diuretic, to lessen the symptoms of rheumatism, and in a paste to treat burns and sores (Steyermark, 1963). The burlike heads are dispersed mostly in animal fur, and the individual fruits are shed from the head very tardily.

Most plants have pink to light reddish purple corollas. Rare individuals with darker purple corollas have been called f. *purpureum* (Blytt) A.H. Evans, and white-flowered plants are known as f. *pallidum* Farw.

Steyermark (1963) treated *A. tomentosum* Mill. (cotton burdock) based on historical collections from Jackson County. However, Mühlenbach (1983) concluded that all of these specimens were misdetermined plants of *A. minus*. Arctium tomentosum is another Eurasian species that is weedy in the New World. It differs from *A. minus* in its shorter, more flat-topped inflorescences with mostly long-stalked, generally slightly larger heads, and corollas that are usually somewhat glandular on the outer surface (Moore and Frankton, 1974). The few Missouri specimens in question do have longer stalks, but the heads otherwise resemble those of *A. minus,* and the inflorescences overall are elongate rather than flat-topped.

28. **Carduus** L. (plumeless thistle)

Plants coarse, biennial or less commonly annual, with stout taproots. Stems erect, usually branched, noticeably spiny-winged, glabrous or finely hairy, sometimes felty-hairy toward the tip. Leaves basal and alternate, sessile, the bases decurrent into wings along the stem, these wavy or scalloped to evenly lobed, spiny along the margins. Basal leaves in a dense, overwintering rosette, the blades large, broadly elliptic to lanceolate or oblanceolate, deeply pinnately lobed, the lobes somewhat irregular and also toothed to lobed, the margins spiny, the surfaces glabrous or hairy. Stem leaves shallowly to deeply pinnately lobed, occasionally nearly entire, elliptic to lanceolate or narrowly oblong-elliptic or the margins spiny, the surfaces glabrous or hairy. Inflorescences axillary and terminal, the heads long-stalked to nearly sessile, solitary or in small clusters at the branch tips. Heads discoid, the involucre cup-shaped to somewhat bell-shaped, the florets all appearing similar and perfect. Receptacle flat or slightly convex, with numerous bristles. Involucral bracts (except sometimes the innermost ones) tapered to a spiny tip. Pappus of numerous long capillary bristles, these fused at the base, roughened with minute, ascending barbs or teeth, shed more or less as a unit before fruiting. Corollas reddish purple to purple, rarely white. Fruits appearing basally attached, 2.5–4.0 mm long, oblong or slightly narrower at the symmetrical base, somewhat flattened and sometimes slightly 4-angled in cross-section, the tip usually with a slightly raised rim, the surface somewhat shiny, light brown to grayish brown, with numerous longitudinal, darker brown stripes. About 90 species, Europe, Asia, Africa.

1. Heads erect, sessile or the stalk narrowly spiny-winged; involucral bracts 1–2 mm wide, narrowly lanceolate, the outer and median ones loosely ascending to spreading, not bent or reflexed at the tip .... 1. C. CRISPUS
1. Heads all or mostly nodding, the stalk unwinged; involucral bracts 2–8 mm wide, lanceolate to narrowly ovate (often slightly constricted in the basal ½), the outer and median ones spreading, bent or reflexed at the tip .... 2. C. NUTANS

1. Carduus crispus L. (curly thistle, welted thistle)

Pl. 250 d–f; Map 1038

Stems 40–200 cm long, brittle, usually cobwebby-hairy. Leaves and stem wings armed with relatively weak, slender, straw-colored spines. Basal leaves 15–30 cm long, lanceolate to oblanceolate or elliptic, the upper surface glabrous or cobwebby-hairy, the undersurface cobwebby- to felty-hairy. Stem leaves 3–12 cm long, lanceolate to narrowly oblong-elliptic, the lobes mostly relatively short, broadly triangular to broadly ovate, the upper surface glabrous or cobwebby-hairy, the undersurface cobwebby- to felty-hairy. Heads

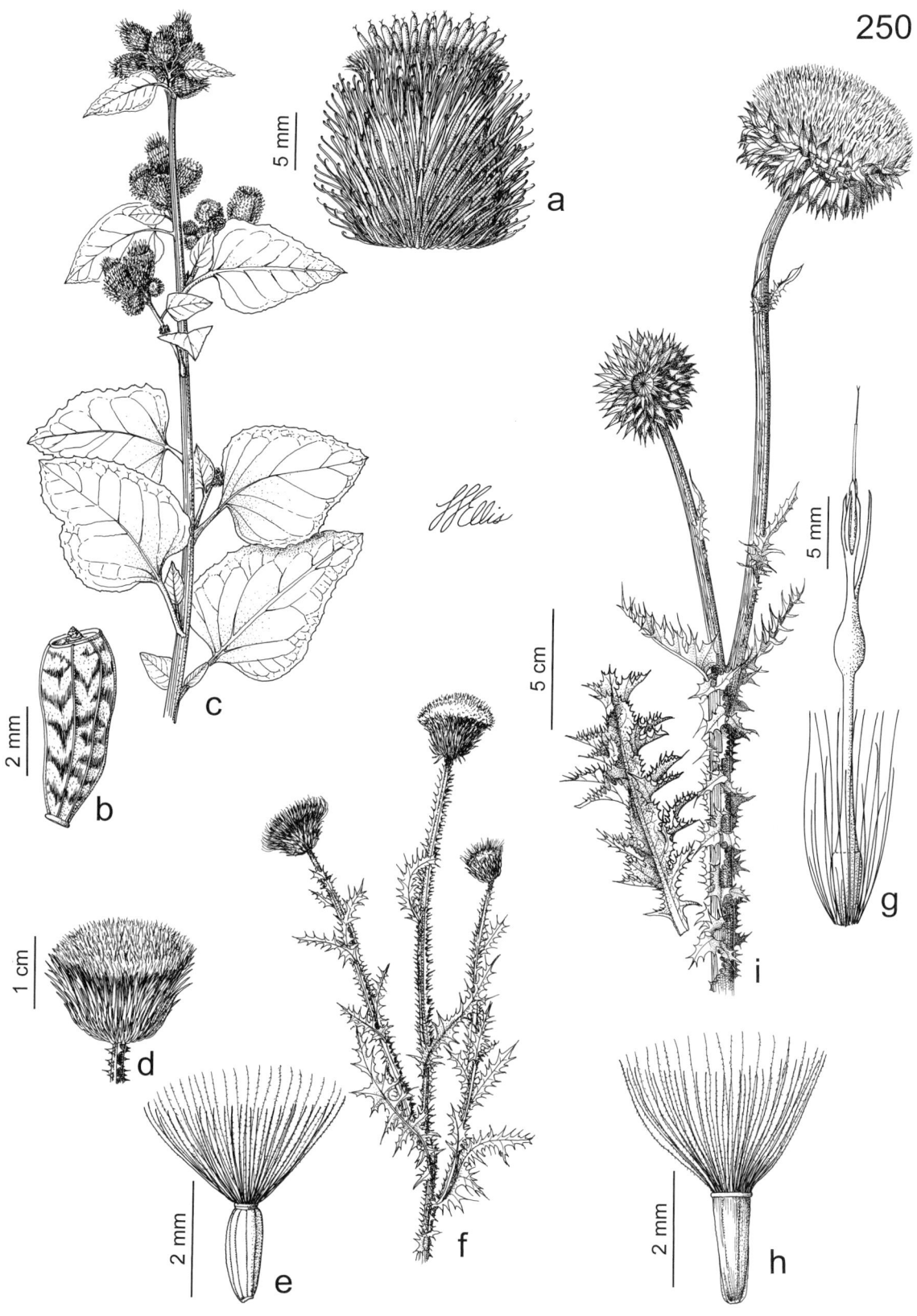

Plate 250. Asteraceae. *Arctium minus*, **a)** head, **b)** fruit, **c)** habit. *Carduus crispus*, **d)** head, **e)** fruit, **f)** habit. *Carduus nutans*, **g)** floret, **h)** fruit, **i)** habit.

1038. Carduus crispus

1039. Carduus nutans

1040. Centaurea americana

solitary or in small clusters, erect, 1.2–2.5 cm in diameter, sessile or the stalk relatively short and spiny-winged to the tip or nearly so, cobwebby- to felty-hairy. Involucral bracts 6–14 mm long (including the spiny tip), 1–2 mm wide, narrowly lanceolate, the outer and median ones loosely ascending to spreading, not bent or reflexed at the tip, gradually tapered to a short, slender, relatively weak, straw-colored or light brown, spiny tip, the surfaces cobwebby-hairy. Pappus 9–12 mm long, white. Corollas 12–15 mm long. $2n=16$. June–September.

Introduced, known thus far from a single historical collection from St. Louis County (native of Europe, Asia, introduced sporadically in the northeastern U.S. and adjacent Canada west to North Dakota and Arkansas). Roadsides and presumably open, disturbed areas.

2. Carduus nutans L. (musk thistle, nodding thistle)

Pl. 250 g–i; Map 1039

Stems 40–300 cm long, not brittle, glabrous or finely hairy between the ribs, usually felty-hairy toward the tip. Leaves and stem wings armed with relatively hard, straw-colored spines. Basal leaves 15–60 cm long, lanceolate to broadly elliptic, glabrous or less commonly sparsely to moderately hairy on both surfaces. Stem leaves 3–30 cm long, lanceolate to broadly elliptic, the lobes mostly triangular to ovate, glabrous or less commonly sparsely to moderately hairy on both surfaces, occasionally somewhat glaucous. Heads solitary, all or mostly nodding, 2–7(–8) cm in diameter, the stalk usually long and relatively naked (a few small, scattered, bracteal leaves often present), less commonly short (normal-sized leaves present nearly to the head), usually densely felty-hairy. Involucral bracts 15–50 mm long (including the spiny tip), 2–8 mm wide, lanceolate to narrowly ovate (often slightly constricted in the basal ½), the outer and median ones spreading to reflexed above the midpoint, gradually or more commonly abruptly tapered to a hard, straw-colored or occasionally purple, spiny tip, the surfaces glabrous or felty- to cobwebby-hairy. Pappus 13–18 mm long, white. Corollas 18–30 mm long. $2n=16$. June–October.

Common nearly throughout the state, but very undercollected (native of Europe, Asia, widely introduced nearly throughout the U.S. and adjacent Canada). Upland prairies and potentially glades; more commonly pastures, crop fields, fallow fields, old fields, fencerows, roadsides, railroads, and open, disturbed areas.

This species was introduced in the mid-nineteenth century in the eastern United States, possibly as a contaminant in ballast dumped from ships (Stuckey and Forsyth, 1971). By the early twentieth century, it was recognized widely as a noxious weed in pastures and crop fields. Its spread into Missouri was slow. Although musk thistle was declared a noxious weed by the state legislature in 1909, the first infestation in Missouri was not recorded until 1941 (Marion County). Steyermark (1963) knew it from only four widely scattered counties, including the St. Louis and Kansas City metropolitan regions. Since that time, it has spread into probably every county using roadsides and railroads as dispersal corridors and as a contaminant in hay. Control of musk thistle involves spraying with a foliar herbicide (best during the rosette stage) or digging up rootstocks in combination with mowing prior to fruiting. The U.S. Department of Agriculture also has attempted biological control of the species. Beginning in 1979 in Missouri (earlier elsewhere), controlled releases were carried out involving two small insects found to be predators of the species in Europe: rosette weevils of the *Trichosirocalus horridus* (Panz.) complex and a flower head weevil, *Rhinocyllus conicus* Froel. (Puttler and Bailey, 2001). Unfortunately, although these insects can be effective in controlling the spread of *Carduus* species, it has since been shown that *Rhinocyllus conicus* also attacks many North American species of the thistle genus *Cirsium* (Louda, 2000). In Missouri, the effects of so-called nontarget infestations have not been studied in detail, but anecdotal observations by a number of botanists suggest that *Cirsium muticum,* which is of restricted distribution in the state, has been impacted the most adversely. The release of *Rhinocyllus* as a biological control agent in North America without sufficient evaluation of its host range has been among the most widely cited examples of the dangers involved in the deliberate introduction of foreign organisms into natural ecosystems in America.

Within its native range, a number of subspecies of *C. nutans* appear to be separable. However, the relationship of these to populations introduced in North America is not clear. The following key serves to separate typical examples of the variation:

1. Involucral bracts tapered relatively evenly to the spiny tip . . . . . . ssp. *macrolepis*
1. Involucral bracts tapered relatively abruptly to the spiny tip
   2. Heads with relatively long, leafless stalks, the stems below these glabrous or sparsely hairy between the ribs; leaves glabrous or both surfaces sparsely hairy . . . ssp. *leiophyllus*
   2. Heads short-stalked, the stems more or less persistently hairy and often densely felty-hairy, at least toward the tip; leaves moderately hairy on both surfaces . . . . ssp. *macrocephalus*

McGregor (1986e), in dealing with populations in the Great Plains, where some of the largest infestations occur, noted that the degree of tapering of the bract tips varies greatly, even on different heads of a single individual, and also was more gradual in plants flowering the first growing season. He therefore chose not to recognize ssp. *macrolepis* (Peterm.) Kazmi within his region. He contrasted the more common ssp. *leiophyllus* (Petrovic) Stoy. & Stef. and the mostly western ssp. *macrocephalus* (Desf.) Nyman, noting that intermediates were relatively uncommon. Musk thistle is greatly underrepresented in the Missouri herbaria, making it difficult to evaluate the infraspecific variation present in the state. Most specimens of well-developed plants appear to represent ssp. *leiophyllus,* as keyed above. However, occasional specimens, especially from the St. Louis and Kansas City metropolitan areas, appear to key better to ssp. *macrocephalus*. Missouri materials mostly have glabrous or very sparsely hairy leaves, but the density and persistence of stem pubescence appear to be more variable than in plants of the Great Plains. Efforts should be made to sample more populations from throughout the state to address the question of whether musk thistle subspecies can be separated in Missouri.

29. Centaurea L. (star thistle, knapweed)

Plants annual, biennial, or perennial herbs, sometimes with rhizomes. Stems erect to loosely ascending, unbranched or branched, usually finely angled or longitudinally ridged, not winged, or with slender, nonspiny wings. Leaves alternate and often also basal, sessile to short-petiolate or the basal ones sometimes long-petiolate, sometimes decurrent into slender wings with the margins unlobed and not spiny, the blades entire to deeply pinnately lobed. Inflorescences terminal, of solitary (clustered elsewhere) heads or panicles, the heads long-stalked to nearly sessile. Heads discoid but often appearing radiate, the involucre variously shaped, the florets all appearing similar and perfect or more commonly the marginal ones sterile, with an enlarged, raylike corolla. Receptacle flat or slightly convex, with numerous bristles. Involucral bracts with the body appressed, glabrous or cobwebby-hairy, the margins variously entire to fringed, the tip loosely ascending to spreading or reflexed, often with a spiny or flattened appendage. Florets numerous (as few as 25 in smaller-headed species, as many as 400 in *C. americana*). Pappus occasionally absent or more commonly of several series of bristles and/or scales, the outermost series shorter than the inner ones, the bristles, when present, with fine, ascending barbs but not plumose, mostly persistent at fruiting. Corollas white, pink, purple, blue, or yellow, often of two types, those of most of the florets discoid, slender, with slender, erect to spreading lobes; those of the marginal florets enlarged, appearing zygomorphic, one of the sinuses between the lobes much deeper than the rest and splitting the upper half of the tube, the portion above the split usually more or less fan-shaped. Fruits more or less oblong in outline, often appearing somewhat narrowed toward the base, somewhat flattened or less commonly somewhat 4-angled in cross-section, the basal portion asymmetrical, often appearing twisted to the side, with a slightly to strongly oblique or lateral attachment scar (only slightly so in *C. repens*), the surface glabrous or finely hairy, somewhat shiny, with longitudinal lines or stripes. Four hundred and fifty to 650 species, nearly worldwide, most diverse in Europe and Asia.

Generic delimitation in the subtribe Centaureinae Dumort. continues to be controversial. Recent molecular work has been correlated with data from cytology, pollen ultrastructure, and morphology to suggest that the genus is unnatural (Susanna et al., 1995; Wagenitz and Hellwig, 1996; Garcia-Jacas et al., 2000, 2001). On the one hand, plants long recognized in other genera, including *Carthamus* L. and *Cnicus* L. (blessed thistle), have been suggested as representing specialized groups nested within the revised concept of *Centaurea*. On the other hand, a number of species groups traditionally treated within *Centaurea* might be treated more properly in several segregate genera. For Missouri, at a minimum this would result in the recognition of two such segregates: 1) The two native North American basket flowers, the relatively widespread *C. americana* and the southwestern *C. rothrockii* Greenm., would be part of an odd, mostly New World group known as *Plectocephalus* D. Don; 2) *Centaurea repens* would be segregated into the monotypic *Acroptilon* Cass. Unfortunately, the dismemberment of *Centaurea* is not without problems. The conclusions of Garcia-Jacas et al. (2001) based on their molecular analyses were weakened by poor resolution at the more basal nodes of their phylogeny and by the failure of the DNA of many species groups to amplify for all of the sequences under study. Nomenclaturally, one problem is that the type species, the African *C. centaurium* L., corresponds to one of the segregate genera, which would require conservation of the name *Centaurea* with a new type species in order to preserve the traditional usage of that name (Greuter et al., 2001). For the present, it seems prudent to maintain a broad view of *Centaurea*, while acknowledging that in the future the taxonomic splitting of the genus probably will gain better support.

1. Involucral bracts with the margins entire and somewhat papery; fruits nearly symmetrical at the base, the attachment scar appearing only slightly oblique .... 8. C. REPENS
1. Involucral bracts with the margins spiny, fringed, or irregular, not papery (except in *C. cyanus*); fruits noticeably asymmetrical at the base, the attachment scar appearing lateral or strongly oblique
    2. Most or all of the involucral bracts with a spiny tip, the terminal spine longer and/or stouter than any lateral spines that may be present
        3. Leaves not or only slightly decurrent, the stems angled but not winged; median and outer involucral bracts with the terminal spine 1–5 mm long; corollas pinkish-purple to reddish purple, rarely white
            4. Pappus absent or of minute, bristly scales to 0.5 mm long; florets appearing all discoid, the marginal florets similar to the discoid florets but functionally sterile, the corolla 9–13 mm long; basal and lower stem leaves 2 times pinnately lobed................ 3. C. DIFFUSA
            4. Pappus of numerous unequal bristles 3–5 mm long; florets of 2 types, the discoid florets with the corolla 18–22 mm long, the raylike florets with the corolla 25–30 mm long; basal and lower stem leaves 1 time pinnately lobed.. 4. C. DILUTA

3. Leaves long-decurrent, the stems noticeably slender-winged; median and outer involucral bracts with the terminal spine 5–30 mm long; corollas yellow
    5. Corollas 10–12 mm long; involucral bracts with the terminal spine 5–10 mm long .... 5. C. MELITENSIS
    5. Corollas 13–20 mm long; involucral bracts with the terminal spine 11–30 mm long .... 9. C. SOLSTITIALIS
2. Involucral bracts without a spiny tip, at most with a short, firm tip, more commonly the tip fringed or irregular
    6. Heads relatively large, the involucre 25–45 cm long, wider than long, broadly bell-shaped; pappus bristles 8–14 mm long .... 1. C. AMERICANA
    6. Heads smaller, the involucre 12–18 cm long, longer than to about as long as wide, narrowly cup-shaped or ovoid to nearly spherical; pappus absent or the bristles 0.5–5.0 mm long
        7. Principal leaves all deeply lobed, the smaller leaves toward the branch tips and occasionally also a few of the basal leaves entire or toothed .... 10. C. STOEBE
        7. Leaves with the margins entire or toothed, occasionally a few of the lowermost ones with the margins lobed
            8. Plants with dense, woolly hairs when young (appearing pale or whitened), the pubescence often partially reduced to woolly or cobwebby tufts at maturity, the leaf undersurface remaining persistently woolly; principal leaves 2–9 mm wide, linear or narrowly lanceolate; involucral bracts not strongly differentiated into a main body with an apical appendage, the broad, white or brownish- to purplish-tinged margins papery and with coarse, ascending, triangular teeth .... 2. C. CYANUS
            8. Plants roughened or hairy with cobwebby hairs when young (not appearing pale or whitened), sometimes glabrous or nearly so at maturity; leaves variable, but at least some of the largest ones 12–40 mm wide; involucral bracts differentiated into a smooth-margined green body and a brown to dark brown or black apical appendage, the appendage margins comblike with a fringe of stiff, spreading or loosely upward-curved, parallel bristles
                9. Involucre about as long as wide; apical appendages of the involucral bracts relatively large, broader than the main body, the intact involucre often appearing solid dark brown to black; florets all discoid .... 6. C. NIGRA
                9. Involucre longer than wide; apical appendages of the involucral bracts relatively small, as wide as or narrower than the main body, the intact involucre with at least some green coloration visible; marginal florets usually raylike .... 7. C. NIGRESCENS

1. Centaurea americana Nutt. (American basket flower)
Plectocephalus americanus D. Don
Pl. 251 a–c; Map 1040

Plants annual, with a taproot, moderately roughened, sometimes smooth toward the stem base, not appearing pale or whitened at maturity. Stems 20–100 cm long, erect or ascending, unbranched or with few ascending branches above the midpoint, somewhat angled or ridged, but not winged. Leaves 3–12 cm long, often dotted with minute, yellow to brown resin glands; basal and lower stem leaves with the blades mostly 15–35 mm wide, narrowly ovate to elliptic-obovate, angled or tapered to a sharply pointed tip, mostly sessile, entire or with few fine teeth; median and upper

1041. Centaurea cyanus   1042. Centaurea diffusa   1043. Centaurea diluta

stem leaves gradually reduced, sessile, the base not decurrent (the elevated midvein usually running into the narrow ridges), median and upper stem leaves somewhat reduced, entire, lanceolate to narrowly lanceolate. Heads solitary at the stem tip or branch tips. Involucre 25–45 mm long, wider than long, broadly bell-shaped. Involucral bracts with the body narrowly to broadly elliptic, the margins entire, the outer surface glabrous or cobwebby-hairy, partially concealed by the appendages, the outer surface with several fine, parallel veins or grooves; the apical appendage well differentiated and appearing jointed to the body, ascending, lanceolate to narrowly lanceolate, mostly narrower than the main body, somewhat overlapping, straw-colored to reddish brown, the involucre with at least some green coloration easily visible, the margins comblike with a fringe of usually stiff, spreading or loosely upward-curved, parallel bristles. Florets discoid, but the marginal florets raylike. Pappus of many unequal bristles, these 8–14 mm long, white, sometimes shed by fruiting. Corollas of discoid florets 14–20 mm long, white or cream-colored, less commonly light pinkish purple, those of raylike florets 20–25 mm long, light pinkish purple to reddish purple or rarely white. Fruits 4–5 mm long, somewhat flattened, the attachment scar appearing lateral or strongly oblique, the surface grayish brown to black with faint, lighter stripes, glabrous or sparsely hairy. $2n=26$. June–July.

Uncommon in the southwestern portion of the Ozark Division; introduced sporadically elsewhere in the state (Missouri to Louisiana west to Kansas and Arizona; introduced sporadically farther north and east). Glades and openings of mesic to dry upland forests; rarely banks of rivers; also pastures, railroads, and roadsides.

This showy native wildflower occasionally is cultivated as a garden ornamental.

2. Centaurea cyanus L. (cornflower, bachelor's button, blue bottle)

Pl. 252 d, e; Map 1041

Plants annual, with a taproot, pubescent with dense, woolly hairs when young (appearing pale or whitened), the pubescence often partially reduced to woolly or cobwebby tufts at maturity, at least the leaf undersurface remaining persistently woolly. Stems 20–120 cm long, erect or ascending, with ascending branches at or above the midpoint, angled and/or ridged but usually not winged. Leaves 4–13 cm long; basal and lower stem leaves with the blades mostly 2–5(–9) mm wide, linear or narrowly lanceolate, tapered to a sharply pointed tip, tapered gradually to a sessile or short-petiolate base, the margins entire or the basal leaves rarely with a few linear lobes (the leaves then wider); median and upper stem leaves somewhat reduced, mostly sessile, the base sometimes narrowly decurrent, the blades linear, entire. Heads solitary at the branch tips. Involucre 11–16 mm long, longer than wide (sometimes about as long as wide when pressed), more or less bell-shaped. Lower and median involucral bracts with the body ovate, the margins entire, the outer surface glabrous or finely woolly, not concealed by the appendages; the apical appendage not strongly differentiated, ascending, the broad, white or brownish- to purplish-tinged margins papery and with coarse, ascending, triangular teeth. Upper involucral bracts similar but oblong-lanceolate. Florets discoid, but the marginal florets raylike. Pappus of many unequal bristles, these 2–4 mm long, straw-colored to brown, usually persistent at fruiting. Corollas of discoid florets 10–15 mm long, those of raylike florets 20–25 mm long, blue or less commonly purple, pink, or white. Fruits 3.5–5.0 mm long, somewhat flattened, the attachment scar appearing lateral, the surface grayish brown to yellowish brown, brown or nearly

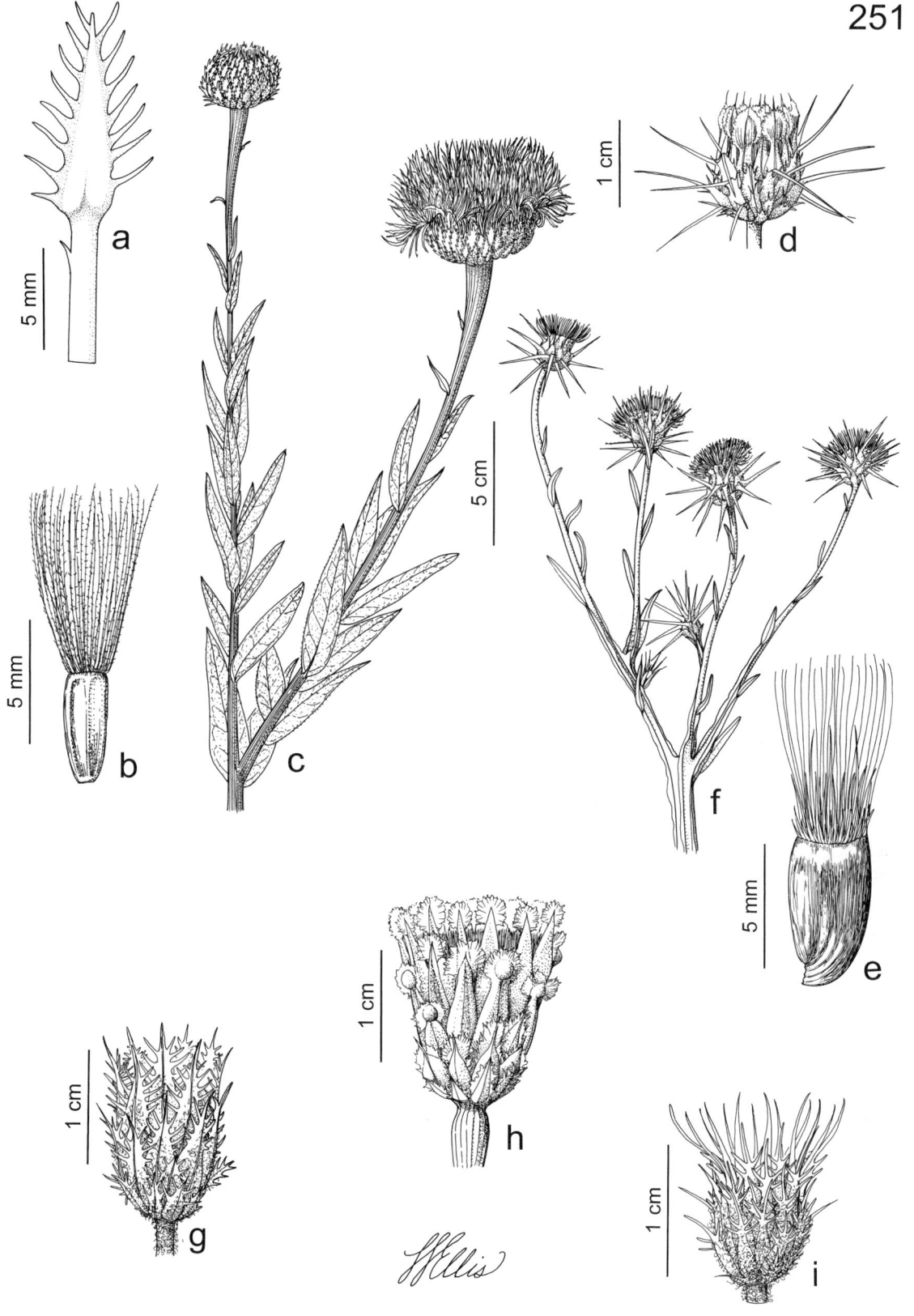

Plate 251. Asteraceae. *Centaurea americana*, **a)** involucral bract, **b)** fruit, **c)** habit. *Centaurea solstitialis*, **d)** head, **e)** fruit, **f)** habit. *Centaurea diffusa*, **g)** head. *Centaurea diluta*, **h)** head. *Centaurea melitensis*, **i)** head.

black, with sometimes faint, lighter stripes, finely hairy. 2n=24. May–September.

Introduced, widely scattered in Missouri (native of Europe, Asia; introduced throughout the U.S. and Canada). Banks of rivers and ledges of bluffs; also fallow fields, railroads, roadsides, and open, disturbed areas.

Centaurea cyanus is a popular garden annual and a component of some wildflower seed mixes. Steyermark (1963) knew it from only ten widely scattered counties but suggested that it actually was present in most counties. Surprisingly, since that time, it has been collected in only a few additional counties.

3. Centaurea diffusa Lam. (diffuse knapweed, tumble knapweed)

Pl. 251 g; Map 1042

Plants annual or less commonly perennial, usually with a taproot, pubescent with cobwebby hairs when young, not appearing pale or whitened, sometimes nearly glabrous at maturity. Stems 20–90 cm long, erect or ascending with loosely ascending or spreading branches at or above the midpoint, somewhat angled but not winged. Leaves 2–20 cm long, often dotted with minute, yellow to brown resin glands; basal and lower stem leaves with the blades mostly 12–80 mm wide, oblanceolate to obovate, rounded or angled to a bluntly pointed tip, tapered gradually to a sessile or short-petiolate, usually somewhat expanded base, 2 times pinnately lobed, the margins otherwise entire or finely toothed; median and upper stem leaves gradually reduced, mostly sessile, the base sometimes slightly clasping the stem, not or only slightly decurrent, the blades linear to oblong-oblanceolate, entire or rarely toothed or lobed, occasionally 2 times pinnately lobed. Heads solitary at the branch tips. Involucre 10–13 mm long, longer than wide, narrowly ovoid to narrowly ovoid-cylindrical. Lower and median involucral bracts with the body ovate to narrowly ovate, the margins fringed with small, slender, straw-colored spines, the outer surface glabrous or slightly cobwebby-hairy, not concealed by the appendages; the apical appendage well differentiated, ascending, narrower than the main body, not or only slightly overlapping, straw-colored or pale greenish-tinged, the involucre often entirely straw-colored, the margins with a fringe of slender, spreading teeth and a short central spine, this 1–3 mm long. Upper involucral bracts lanceolate, the appendages not well differentiated from the relatively broad (at least toward the tip), thin, white, papery margins, the tips usually sharply pointed, irregularly toothed or narrowly lobed. Florets all discoid, the marginal florets appearing discoid but functionally sterile. Pappus absent or of minute, bristly scales, these 0.1–0.5 mm long, white, usually persistent at fruiting. Corollas 9–13 mm long, cream-colored to nearly white, rarely pinkish- or purplish-tinged. Fruits 2.5–3.5 mm long, somewhat flattened, the attachment scar appearing lateral, the surface usually dark brown, usually glabrous and somewhat shiny. 2n=18. June–September.

Introduced, known thus far only from the city of St. Louis (native of Europe; introduced widely in the western and midwestern U.S. and adjacent Canada, also sporadically in the northeastern U.S.). Railroads.

Steyermark (1963) reported this species based on collections made by Viktor Mühlenbach in the St. Louis railyards. However, some of Mühlenbach's collections and also a later one from Howard County are actually the fertile hybrid between *C. diffusa* and *C. stoebe,* which has been called *C. ×psammogena* Gáyer. This hybrid is variable, but generally differs from *C. diffusa* in its less divided leaves with narrower lobes (more like those of *C. stoebe*), slightly broader heads with well-developed marginal raylike florets, and involucral bracts with a shorter central spine at the tip (Ochsmann, 1997, 2001b). True *C. diffusa* is represented by only two of Mühlenbach's several voucher specimens. The species is a widespread weed in the northern and western United States and Canada that may have been spread into Missouri by the transport of hay or livestock.

4. Centaurea diluta Aiton (North African knapweed, lesser star thistle)

Pl. 251 h; Map 1043

Plants annual or less commonly perennial, usually with a taproot, pubescent with cobwebby hairs when young, not appearing pale or whitened, sometimes nearly glabrous at maturity. Stems 30–150 cm long, erect or ascending with loosely ascending or spreading branches at or above the midpoint, somewhat angled but not winged. Leaves 2–15 cm long, often dotted with minute, yellow to brown resin glands; basal and lower stem leaves with the blades mostly 12–80 mm wide, oblanceolate to obovate, rounded or angled to a bluntly pointed tip, tapered gradually to a sessile or short-petiolate, usually somewhat expanded base, mostly deeply pinnately lobed, the margins otherwise entire or finely toothed; median and upper stem leaves gradually reduced, mostly sessile, the base sometimes slightly clasping the stem, not or only slightly decurrent, the blades linear to oblong-oblanceolate, entire or rarely toothed or shallowly lobed. Heads solitary at the branch tips. Involucre 15–18 mm long, longer than to about as long as wide, ovoid

1044. Centaurea melitensis   1045. Centaurea nigra   1046. Centaurea nigrescens

(sometimes appearing narrowly bell-shaped when pressed). Lower and median involucral bracts with the body ovate to narrowly ovate, the margins narrowly thin and papery, entire or somewhat irregular, the outer surface glabrous or nearly so, not concealed by the appendages; the apical appendage well differentiated, ascending, narrower than the main body, not or only slightly overlapping, white to straw-colored or brownish-tinged, the involucre with at least some green coloration visible, the margins with a fringe of slender, spreading teeth and a short central spine, this 1–5 mm long. Upper involucral bracts lanceolate, the appendages not well differentiated from the relatively broad (at least toward the tip), thin, white, papery margins, the tips bluntly or sharply pointed, irregularly toothed or narrowly lobed. Florets discoid, but the marginal florets raylike. Pappus of many unequal bristles, these 3–5 mm long, white, usually persistent at fruiting. Corollas of discoid florets 18–22 mm long, those of raylike florets 25–30 mm long, pinkish purple to reddish purple, rarely white. Fruits 3.0–3.5 mm long, somewhat flattened, the attachment scar appearing lateral, the surface tan to grayish brown with lighter stripes, finely hairy. $2n=20$. June–September.

Introduced, known thus far only from the city of St. Louis (native of Europe, Africa; introduced sporadically in California, Missouri, and New York). Railroads.

This species was first reported for Missouri by Mühlenbach (1979). It is much less common in the United States than *C. diffusa,* with which it sometimes has been confused.

5. Centaurea melitensis L. (Maltese star thistle)

Pl. 251 i; Map 1044

Plants annual, with a taproot, pubescent with cobwebby to woolly hairs, often appearing somewhat pale or whitened, at least when young. Stems 10–60 cm long, erect or ascending, unbranched or with loosely ascending branches, noticeably winged, the wings with irregular and sometimes broadly toothed margins. Leaves 1–15 cm long, often dotted with minute, yellow to brown resin glands; basal and lower stem leaves with the blades mostly 5–35 mm wide, oblong to oblanceolate, usually rounded at the tip, tapered gradually to a sessile or short-petiolate base, deeply pinnately lobed, usually withering by flowering time; median and upper stem leaves gradually reduced, mostly sessile, the base strongly decurrent, the blades linear to oblong-lanceolate, entire or toothed. Heads solitary or in small clusters of 2–4 at the branch tips (rarely on short, axillary branches and thus appearing lateral). Involucre 10–15 mm long (excluding spines), longer than to about as long as wide, broadly ovoid to nearly spherical. Lower and median involucral bracts with the body ovate, the margins entire, the outer surface more or less cobwebby-hairy, not concealed by the appendages; the apical appendage well differentiated, spreading, more or less narrower than the main body, straw-colored and commonly purplish-tinged, the involucre with at least some green coloration easily visible (sometimes becoming entirely straw-colored with age), the margins with 1 or 2 pairs of short, spreading, spinelike, lateral bristles and a central spine, this 5–10 mm long. Upper involucral bracts lanceolate, the appendages papery, tapered, short-spined or merely irregularly toothed at the tip. Florets all discoid and similar or the marginal florets similar in appearance but functionally sterile. Pappus of many unequal bristles, these 2–3 mm long, white, usually persistent at fruiting. Corollas 10–12 mm long, bright yellow. Fruits 2.5–3.5 mm long, somewhat flattened, the attachment scar appearing lateral, the surface yellowish brown to brown with fine, lighter, longitudinal stripes, glabrous or more commonly finely hairy, sometimes somewhat shiny. $2n=24$. June–September.

Introduced, known only from historical collections from Jackson County (native of Europe, Africa; introduced widely in the western U.S. and adjacent Canada, more sporadically farther east). Railroads, roadsides, and open, disturbed areas.

6. Centaurea nigra L. (black knapweed, Spanish buttons)
C. jacea L. ssp. nigra (L.) Bonnier & Layens

Pl. 252 f, g; Map 1045

Plants perennial, often with short rhizomes, pubescent with short, stiff hairs or cobwebby hairs when young, not appearing pale or whitened, sometimes nearly glabrous at maturity. Stems 30–150 cm long, erect or ascending, sometimes from a prostrate base, with loosely ascending or spreading branches at or above the midpoint, somewhat angled but not winged. Leaves 3–25 cm long; basal and lower stem leaves with the blades mostly 12–50 mm wide, oblanceolate to elliptic, angled to a sharply pointed tip, tapered gradually to a sessile or short-petiolate base, the margins entire or few-toothed to shallowly lobed; median and upper stem leaves gradually reduced, mostly sessile, the base not decurrent, the blades linear to lanceolate, entire or toothed. Heads solitary at the branch tips. Involucre 14–18 mm long, about as long as wide, bell-shaped to hemispheric. Lower and median involucral bracts with the body lanceolate to ovate, the margins entire, the outer surface glabrous or cobwebby-hairy, often concealed by the appendages; the apical appendage well differentiated, ascending, broader than the main body, overlapping, brown to dark brown or black, the involucre thus often appearing solid brown or black, the margins comblike with a fringe of stiff, spreading or loosely upward-curved, parallel bristles. Upper involucral bracts similar but the appendages merely irregularly toothed along the margins, the tips rounded to truncate. Florets all discoid and similar. Pappus of many unequal bristles, these 0.5–1.0 mm long, straw-colored, sometimes shed by fruiting. Corollas 15–18 mm long, purple or rarely white. Fruits 2.5–3.0 mm long, somewhat 4-angled, the attachment scar appearing lateral, the surface tan to grayish brown with lighter stripes, finely hairy. $2n=22$. June–September.

Introduced, known only from Boone County and the city of St. Louis (native of Europe; introduced in the northeastern and western U.S. and adjacent Canada). Railroads and roadsides.

Plants that appear to represent fertile hybrids between *C. nigra* and *C. jacea* L. (brown knapweed, not yet reported from Missouri) have been called *C.* ×*pratensis* Thuill. or more correctly *C.* ×*moncktonii* C.E. Britton (meadow knapweed), and were collected by Viktor Mühlenbach during his botanical inventories of the St. Louis railyards.

7. Centaurea nigrescens Willd. (Tyrol knapweed, short-fringed knapweed)
C. vochinensis Bernh. ex Rchb.
C. dubia Suter ssp. vochinensis (Bernh. ex Rchb.) Hayek

Pl. 252 j, k; Map 1046

Plants perennial, often with short rhizomes, pubescent with short, stiff hairs or cobwebby hairs when young, not appearing pale or whitened, sometimes nearly glabrous at maturity. Stems 30–150 cm long, erect or ascending, sometimes from a prostrate base, with loosely ascending or spreading branches at or above the midpoint, somewhat angled but not winged. Leaves 3–25 cm long; basal and lower stem leaves with the blades mostly 12–50 mm wide, oblanceolate to elliptic, angled to a sharply pointed tip, tapered gradually to a sessile or short-petiolate base, the margins entire or few-toothed to shallowly lobed; median and upper stem leaves gradually reduced, mostly sessile, the base not decurrent, the blades linear to lanceolate, entire or toothed. Heads solitary at the branch tips. Involucre 12–16 mm long, longer than wide, narrowly cup-shaped or narrowly bell-shaped to ovoid. Lower and median involucral bracts with the body lanceolate to ovate, the margins entire, the outer surface glabrous or cobwebby-hairy, not concealed by the appendages; the apical appendage well differentiated, ascending, as wide as or narrower than the main body, not or only slightly overlapping, brown to dark brown or black, the involucre with at least some green coloration visible, the margins comblike with a fringe of stiff, spreading or loosely upward-curved, parallel bristles. Upper involucral bracts similar but the appendages merely irregularly toothed along the margins, the tips rounded to truncate. Florets all discoid and similar or the marginal florets raylike. Pappus absent or of a low crown of unequal bristles, these 0.1–0.5 mm long, straw-colored, sometimes shed by fruiting. Corollas of discoid florets 15–18 mm long, those of raylike florets (when present) 20–25 mm long, purple or rarely white. Fruits 2.5–3.0 mm long, somewhat 4-angled, the attachment scar appearing lateral, the surface tan to grayish brown with lighter stripes, finely hairy. $2n=22$. June–September.

Introduced, known thus far only from Boone and Jackson Counties (native of Europe; introduced sporadically in the northeastern and northwestern U.S. and adjacent Canada). Pastures, railroads, and open, disturbed areas.

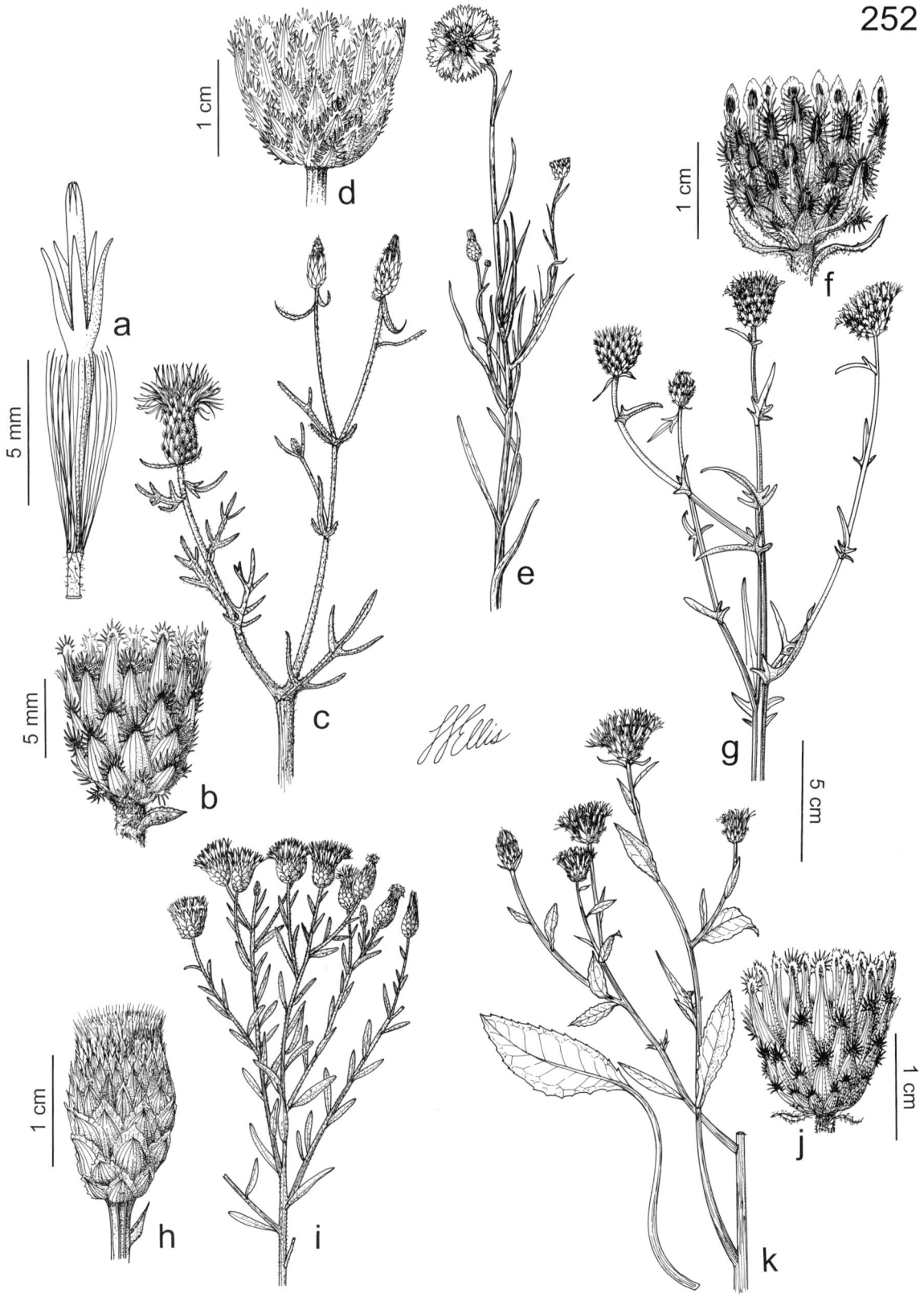

Plate 252. Asteraceae. *Centaurea stoebe*, **a)** floret, **b)** head, **c)** habit. *Centaurea cyanus*, **d)** head, **e)** habit. *Centaurea nigra*, **f)** head, **g)** habit. *Centaurea repens*, **h)** head, **i)** habit. *Centaurea nigrescens*, **j)** head, **k)** habit and lower leaf.

1047. Centaurea repens   1048. Centaurea solstitialis   1049. Centaurea stoebe

Moore and Frankton (1974) discussed the taxonomy and nomenclature of this species. It seems likely that the name *C. dubia* Suter also should be considered a synonym of *C. nigrescens* (Voss, 1996); however, the taxonomic status of native populations in Europe requires further research. Steyermark's (1963) report of Tyrol knapweed (as *C. vochinensis*) from the city of St. Louis could not be confirmed during the present study.

8. Centaurea repens L. (Russian knapweed)
Acroptilon repens (L.) DC.

Pl. 252 h, i; Map 1047

Plants perennial, suckering from deep-set, creeping, dark brown to black roots, pubescent with fine, cobwebby hairs when young (appearing somewhat grayish when young), sometimes appearing nearly glabrous at maturity. Stems 20–100 cm long, erect or ascending, with loosely ascending branches, angled and/or ridged but not winged. Leaves 1–15 cm long; basal and lower stem leaves with the blades mostly 20–40 mm wide, oblanceolate, rounded or angled to a bluntly or sharply pointed tip, tapered to a sessile or short-petiolate base, the margins with coarse teeth or more commonly deeply pinnately lobed with ascending, oblong-triangular lobes; median and upper stem leaves progressively reduced, mostly sessile, the base not or only slightly decurrent, the blades lanceolate to linear, mostly with a few widely spaced teeth, the larger ones sometimes shallowly lobed. Heads solitary at the branch tips. Involucre 9–15 mm long, longer than wide (sometimes about as long as wide when pressed), ovoid. Lower and median involucral bracts with the body broadly ovate to ovate, the margins entire, the outer surface glabrous, becoming straw-colored to light brown at maturity, the apical appendage well differentiated, ascending, as wide as or slightly narrower than the main body, broadly rounded at the tip not or only slightly overlapping, thin, papery, white, the margins entire. Upper involucral bracts narrowly ovate, narrowed to a sharply pointed tip, this sometimes irregularly and finely toothed along the margins, plumose-hairy. Florets all discoid and similar. Pappus of many unequal bristles (the longest often somewhat plumose), these 6–11 mm long, white, shed by fruiting. Corollas 12–14 mm long, pinkish purple to reddish purple or purple. Fruits 3.0–3.5 mm long, somewhat flattened, the attachment scar appearing nearly basal (slightly oblique), the surface pale grayish white, with faint, fine, darker ridges, glabrous. 2n=24. May–September.

Introduced, known only from historical collections from Jackson County (native of Asia; introduced widely in the western U.S. and Canada east to Ohio, Kentucky, Arkansas, and Texas). Railroads and open, disturbed areas.

This species was introduced into the northern states and Canada in the late 1800s as a contaminant in alfalfa seeds. Because of its deep-set rootstock, it is a very difficult species to eradicate. It has allelopathic properties similar to those of *C. stoebe* and is poisonous to horses (see discussion under the treatment of *C. solstitialis*).

9. Centaurea solstitialis L. (yellow star thistle, Barnaby's thistle)

Pl. 251 d–f; Map 1048

Plants annual, with a taproot, pubescent with cobwebby to woolly hairs, appearing somewhat pale or whitened. Stems 20–80 cm long, erect or ascending, with loosely ascending branches, noticeably winged, the wings with irregular and sometimes toothed margins. Leaves 1–20 cm long; basal and lower stem leaves with the blades mostly 5–30 mm wide, oblanceolate, rounded or angled to a bluntly pointed tip, tapered gradually to a sessile or short-petiolate base, deeply pinnately lobed, often withering by flowering time; median and upper stem leaves gradually reduced, mostly sessile, the base strongly decurrent, the blades oblong-linear to narrowly lanceolate, entire or toothed. Heads solitary at the branch tips. Involucre 10–15 mm long (excluding spines), longer than to about as long as wide, broadly ovoid to nearly spherical. Lower and median involucral bracts with the body ovate, the margins entire, the outer surface more or less cobwebby-hairy, not concealed by the appendages; the apical appendage well differentiated, spreading, narrower than the main body, straw-colored, the involucre with at least some green coloration easily visible (sometimes becoming entirely straw-colored with age), the margins with 1 or 2 pairs of short, spreading, spinelike, lateral bristles and a central spine, this 11–30 mm long. Upper involucral bracts lanceolate, the appendages papery, tapered, merely irregularly toothed at the tip. Florets all discoid and similar (but the marginal florets usually lacking a pappus). Pappus (except in marginal florets) of many unequal bristles, these 2–5 mm long, white, usually persistent at fruiting. Corollas 13–20 mm long, bright yellow to orangish yellow. Fruits 2.5–3.5 mm long, somewhat flattened, the attachment scar appearing lateral, the surface straw-colored to yellowish brown with darker brown mottling or (in marginal florets) uniformly dark brown, glabrous, often somewhat shiny. $2n=16$. June–October.

Introduced, known thus far only from Boone and Jackson Counties and the city of St. Louis (native of Europe, Asia; introduced widely in the western U.S. and sporadically elsewhere in the U.S. and Canada). Crop fields, railroads, and roadsides.

In the western states, this species is a severe pest of rangeland. Its spines can cause mechanical injury to the limbs and mouths of animals and deter grazing of mature plants by most livestock. At overgrazed sites, animals will sometimes graze the young plants, but when horses ingest yellow star thistle for lengthy periods, they can develop a usually fatal neurological disorder known as chewing disease and equine nigropallidal encephalomalacia, in which the mouth parts are affected, leading to starvation and dehydration (Burrows and Tyrl, 2001). The compounds implicated in chewing disease are not fully understood but include sesquiterpene lactones, principally repin. *Centaurea repens* is similarly toxic to horses, but neither species appears to affect other livestock.

10. Centaurea stoebe L. **ssp. micranthos**
(S.G. Gmel. ex Gugler) Hayek (spotted knapweed)
C. biebersteinii DC.

Pl. 252 a–c; Map 1049

Plants short-lived perennials, usually with a stout taproot, pubescent with woolly or cobwebby hairs when young, not appearing pale or whitened at maturity (sometimes grayish when young). Stems 20–150 cm long, erect or ascending, with ascending branches at or above the midpoint, somewhat angled or ridged, but not winged. Leaves 3–15 cm long, dotted with minute, yellow to brown resin glands; basal and lower stem leaves with the blades mostly 40–70 mm wide, oblanceolate to elliptic, angled to a usually bluntly pointed tip and often also with a minute, sharp point, sessile or short-petiolate, mostly with 2–4 pairs of deep lobes (the basal leaves rarely entire or few-toothed), the lobes lanceolate to elliptic, entire or few-toothed; median and upper stem leaves gradually reduced, mostly sessile, the base not or only slightly decurrent, the blades of the main stem leaves with 1 to several pairs of mostly linear lobes, those of the uppermost leaves sometimes unlobed. Heads solitary at the branch tips. Involucre 9–13 mm long, narrower than to about as long as wide, ovoid (often appearing bell-shaped when pressed). Lower and median involucral bracts with the body ovate to oblong-ovate, the margins entire, the outer surface glabrous or cobwebby-hairy, not concealed by the appendages, the outer surface with several darker, parallel veins or ribs; the apical appendage more or less well differentiated, ascending, about as wide as the main body, not overlapping, brown to dark brown or black, the involucre with at least some green coloration easily visible, the margins comblike with a fringe of usually stiff, spreading or loosely upward-curved, parallel bristles. Upper involucral bracts with the appendages less differentiated, entire to irregularly toothed along the margins, the tips mostly bluntly pointed. Florets discoid, the marginal florets enlarged and sterile, but usually relatively slender and not greatly expanded laterally. Pappus rarely absent, more commonly of many unequal bristles, these 1.0–2.5 mm long, white, usually persistent at fruiting. Corollas of discoid florets 12–15 mm long, those of raylike florets (when present) 15–25 mm long, light purple to pinkish purple or rarely white. Fruits 2.5–3.5 mm long, somewhat flattened, the attachment scar appearing lateral, the surface grayish brown to brown with lighter stripes, glabrous or finely hairy. $2n=36$. June–September.

Introduced, scattered in eastern and southern Missouri (native of Europe, introduced widely in the U.S. and Canada). Banks of streams, margins of ponds, fens, and sand prairies; also old fields, pastures, ditches, railroads, roadsides, and open, disturbed areas.

Ochsmann (2001a) clarified the taxonomy of the *C. stoebe* complex. The name *C. maculosa* Lam., which had long been used for the tetraploid invasive taxon in North America, is a synonym of *C. stoebe* ssp. *stoebe,* a diploid (2n=18) noninvasive European taxon that is not present in North America. It differs from ssp. *micranthos* in its slightly larger heads with the tips of the involucral bracts having more marginal bristles, its biennial life cycle, and single-stemmed (vs. usually multiple-stemmed) habit. Ochsmann (2001b) also documented the presence in seven states of *C.* ×*psammogena* Gáyer, the fertile hybrid between *C. stoebe* ssp. *micranthos* and *C. diffusa* Lam. (diffuse knapweed). The hybrid has smaller, narrower heads than does *C. stoebe* and has the involucral bracts tipped with short spines (Ochsmann, 1997). This taxon was first reported for Missouri from the St. Louis railyards by Mühlenbach (1979) based on specimens misdetermined as *C. diffusa.* Later it was collected in Howard County. For further discussion of *C. diffusa,* see the treatment of *C. diluta,* which it resembles.

Spotted knapweed is an especially bad problem in overgrazed rangeland in the western states. It apparently was introduced to northwestern North America in the late 1800s as a contaminant in alfalfa seed, but it also became established locally along the eastern seaboard from ship ballast. Cattle generally avoid the plants unless no other food sources are available. LeJeune and Seastedt (2001) reviewed the literature on knapweed invasiveness and concluded that although knapweeds degrade rangeland, the impact of earlier human-mediated perturbations such as grazing and fire suppression contributed to changes in native grasslands that rendered them more susceptible to establishment and spread of knapweeds. They suggested that knapweeds have taken advantage of changes in resource availability (such as water or minerals like nitrogen and phosphorus) to infest millions of acres and in doing so have changed the structure of the prior native grassland communities to mimic more of an old field environment. Spotted knapweed plants are allelopathic; that is, their roots secrete chemicals that inhibit the growth and establishment of other plant species in the vicinity. The chemical basis of this phenomenon is under investigation at Colorado State University for its possible herbicidal applications. An alarming report was circulated on the World Wide Web about the potential of spotted knapweed sap to cause tumors on the limbs of humans working to control infestations by hand-pulling, but to date this has not been substantiated in the medical literature or by further reported cases (see http://tncweeds.ucdavis.edu/news/100297.html).

30. Cirsium Mill. (thistle)

Plants biennial or perennial. Stems erect, branched or less commonly unbranched, in some species noticeably spiny-winged, in unwinged species angled or ridged, variously glabrous or hairy. Leaves basal and alternate, spiny along the margins, usually densely pubescent with white, woolly hairs on both surfaces or only the undersurface, those of the basal rosette with the margins scalloped or toothed but often unlobed, the stem leaves often pinnately lobed, in wing-stemmed species the leaf bases decurrent along the stem, the wings wavy or scalloped to evenly lobed, spiny along the margins. Inflorescences terminal on the branches, the heads long-stalked to nearly sessile, solitary or in small clusters at the branch tips. Heads discoid, the involucre variously shaped, the florets all appearing similar and perfect (or the plants incompletely dioecious in *C. arvense*). Receptacle flat or short-conical, with numerous bristles. Involucral bracts (except sometimes the innermost ones) tapered to a spiny tip (spineless or nearly so in *C. muticum*). Pappus of numerous often unequal, long bristles (the marginal florets sometimes with somewhat fewer, less plumose bristles), these fused at the base, plumose (featherlike with numerous long, capillary side branches), shed more or less as a unit before fruiting. Corollas cream-colored or pink to purple, rarely white. Fruits appearing basally or more commonly somewhat obliquely attached, oblong or slightly narrower at the usually somewhat asymmetrical base, often slightly curved or arched in profile, somewhat flattened and sometimes slightly 4-angled in cross-section, the tip usually with an angular rim or raised crown surrounding a small, knoblike or conical projection, the surface somewhat shiny, straw-colored or light brown, grayish brown or brown. Two hundred to 350 species, widespread in the Northern Hemisphere.

In addition to the species treated below, *C. hillii* (Canby) Fernald (*C. pumilum* ssp. *hillii* (Canby) R.J. Moore & Frankton) should be searched for in eastern Missouri. Steyermark (1963) misdetermined two historical specimens of this taxon from the St. Louis metropolitan region as

C. pumilum (Nutt.) Spreng. but correctly noted that they were actually collected in St. Clair County, Illinois. On that basis, he excluded these specimens from the Missouri flora. In 1979, Marlin L. Bowles of the Morton Arboretum (Lisle, Illinois) photographed a plant of probable *C. hillii* in a small remnant upland prairie strip between a road and a railroad near the town of Ely in Ralls County. However, the photographs were insufficient in detail to confirm with certainty the identity of the species. Perhaps because of construction in the vicinity, repeated attempts by several botanists to relocate Hill's thistle at this site have failed. The main distribution of *C. hillii* is to the north and east of Missouri (Moore and Frankton, 1966), but western Illinois populations are known from sites immediately adjacent to northeastern Missouri. The species will not key well below. *Cirsium hillii* is a perennial with long, thickened, hollow roots giving rise to stout stems. The stem leaves are generally narrowly oblong-elliptic, pinnately lobed with relatively short, irregularly triangular lobes, and with the undersurface finely hairy or cobwebby, but the green color not persistently hidden by the pubescence. The heads are relatively large, with the involucre 3.5–5.0 cm long and the narrow involucral bracts having a slender, sticky dorsal ridge (the outer and median ones are tapered to a short, slender, ascending, spiny tip).

In most of our thistles, the heads are subtended by one to several small leaves and thus appear sessile or short-stalked. Unlike the situation in some western species, these leaves are relatively small and few-lobed in the taxa occurring in Missouri. In *C. carolinianum,* the accessory leaves are highly reduced and widely spaced along the long stalk, and the heads thus appear long-stalked.

1. Leaf bases long-decurrent (more than 1 cm), the stems with spiny-margined wings, at least above the midpoint
    2. Leaf blades with the upper surface pubescent with felty (when young) to cobwebby hairs, not roughened, the undersurface densely pubescent with white, woolly to felty hairs; corollas cream-colored, rarely pale pink
    .... 3. C. CANESCENS
    2. Leaf blades with the upper surface strongly roughened with numerous short, stiff, spinelike or barblike bristles, not cobwebby-hairy, the undersurface thinly or finely pubescent with cobwebby hairs; corollas reddish purple to purple .... 9. C. VULGARE
1. Leaf bases not or only short-decurrent, the stems not winged
    3. None of the involucral bracts spiny at the tip, at most tapered to a minute, sharp point
        4. Heads usually in loose clusters at the branch tips; involucre 10–20 mm long, plants incompletely dioecious (some plants with only pistillate florets, others with a combination of staminate and perfect florets); perennials, colonial from widely creeping, deep-set, black roots
        .... 2. C. ARVENSE
        4. Heads solitary at the branch tips; involucre (17–)20–35 mm long; plants with the florets all perfect; biennial, with a cluster of stout, white to brown roots .... 6. C. MUTICUM
    3. At least the lower and median involucral bracts spiny at the tip
        5. Stem leaves all unlobed or shallowly lobed (less than 1/3 of the way from the margin to midrib), the reduced leaves just below the heads rarely more deeply lobed
            6. Leaves mostly basal or along the lower 1/3 of the stem, few and much-reduced in size toward the stem tip, the heads thus appearing long-stalked; involucre 15–20 mm long .... 4. C. CAROLINIANUM
            6. Leaves well developed along the stem, those toward the branch tips somewhat reduced in size, the heads appearing sessile or very short-stalked; involucre 25–45 mm long

7. Stems glabrous or sparsely pubescent with spreading hairs, sometimes with patches of white, woolly to felty hairs toward the tip; leaves with the upper surface green with sparse to moderate, stiff, straight hairs, often becoming nearly glabrous with age, the undersurface densely and persistently pubescent with white, felty hairs . . . . 1. C. ALTISSIMUM

7. Stems densely pubescent with persistent, white, woolly to felty hairs; leaves with both surfaces appearing grayish or whitish with relatively dense, woolly hairs, the pubescence sometimes becoming thinner on both surfaces with age (the leaves still appearing uniformly gray to grayish green) . . . . 8. C. UNDULATUM

5. Most or all of the leaves deeply lobed (more than ½ of the way from the margin to midrib)
   8. Involucre 25–35 mm long; heads appearing sessile or very short-stalked, the branch tips with reduced leaves . . . . 5. C. DISCOLOR
   8. Involucre 15–22 mm long; heads appearing relatively long-stalked, the leaves strongly reduced in size above the stem midpoint, few and scattered toward the tip of the stem or branches
      9. Largest leaves narrowly lanceolate to linear, at most with a few narrow lobes, long-tapered at the base, not decurrent; plants with a short, inconspicuous taproot in addition to the often somewhat thickened, fibrous roots . . . . 4. C. CAROLINIANUM
      9. Largest leaves narrowly ovate to elliptic or broadly oblanceolate, with several pairs of relatively broad lobes, rounded at the base and somewhat clasping the stem, often short-decurrent; plants with a well-developed taproot, this usually strongly thickened and larger than the fibrous roots . . . . 7. C. TEXANUM

1. Cirsium altissimum (L.) Spreng. (tall thistle, roadside thistle)

Pl. 253 a–c; Map 1050

Plants biennial or short-lived perennials, often with a slightly thickened taproot in addition to the fibrous roots. Stems 100–250 cm long, well branched, glabrous or sparsely pubescent with spreading hairs, sometimes with patches of white, woolly to felty hairs toward the tip, sometimes appearing slightly glaucous, without spiny-margined wings. Basal leaves 10–30 cm long, 4–15 cm wide, narrowly ovate to elliptic or obovate, tapered at the base, rounded to more commonly bluntly angled at the tip, unlobed or rarely with several deep lobes, the margins otherwise toothed or wavy and spiny, the upper surface appearing green, nearly glabrous to moderately pubescent with stiff, straight hairs, the undersurface appearing white, densely pubescent with felty hairs. Stem leaves well developed throughout, the main leaves 4–25 cm long, those toward the branch tips usually somewhat reduced, all unlobed or with shallow (less than ⅓ of the way from the margin to midrib), broad lobes (reduced leaves just below the heads rarely more deeply lobed), tapered to a slightly expanded and sometimes minutely decurrent base, otherwise like the basal leaves. Heads usually relatively numerous, usually solitary at the branch tips, appearing sessile or very short-stalked. Involucre 25–35 mm long, as long as or slightly longer than wide (often appearing broader when pressed), often somewhat cobwebby-hairy, the lower and median bracts tapered to a spreading, spiny tip, this 2–5 mm long, straw-colored to light yellow, usually also somewhat sticky along the midrib. Corollas 22–32 mm long, usually pinkish purple to reddish purple, the lobes 6–9 mm long. Pappus 17–27 mm long, white or occasionally slightly grayish-tinged. Fruits 4.5–6.0 mm long. $2n=18$. July–October.

Scattered to common nearly throughout the state (eastern U.S. west to North Dakota and Texas). Bottomland forests, banks of streams and rivers, and bases of bluffs, less commonly glades, upland prairies, and openings of upland forests; also old fields, railroads, and roadsides.

Rare, white-flowered plants have been called f. *moorei* Steyerm. See the treatment of the closely related *C. discolor* for discussion on hybridization between the two species and plant uses.

1050. Cirsium altissimum   1051. Cirsium arvense   1052. Cirsium canescens

2. Cirsium arvense (L.) Scop. (Canada thistle, field thistle)

C. arvense var. *horridum* Wimm. & Grab.

Pl. 254 c, d; Map 1051

Plants perennial, imperfectly dioecious (some plants with only pistillate florets, others with a combination of staminate and perfect florets), suckering and forming potentially large, clonal colonies from widely creeping, deep-set, black roots. Stems 30–150 cm long, unbranched or more commonly several-branched toward the tip, glabrous or with patches of cobwebby hairs (especially when young), without spiny-margined wings. Basal leaves usually absent at flowering, 8–30 cm long, 2–6 cm wide, narrowly oblong-lanceolate to elliptic or ovate, tapered at the base, sharply pointed at the tip, shallowly to deeply pinnately lobed or occasionally merely coarsely toothed, the lobes otherwise spiny and sometimes irregularly toothed, the upper surface appearing green, glabrous to thinly pubescent with woolly or cobwebby hairs, the undersurface appearing green or more commonly gray, glabrous to densely pubescent with woolly or felty hairs. Stem leaves well developed nearly to the inflorescence, mostly 2–20 cm long, slightly expanded and sometimes minutely decurrent at the base, often somewhat narrower than but otherwise like the basal leaves. Heads usually relatively numerous, usually in loose clusters at the branch tips, mostly appearing short-stalked. Involucre 10–20 mm long, as long as or slightly longer than wide (often appearing broader when pressed), glabrous or more commonly somewhat cobwebby-hairy, the lower and median bracts tapered to an appressed or ascending, nonspiny tip, this often with a minute, sharp point, green or purplish-tinged, usually also somewhat sticky along the midrib. Corollas 12–20 mm long (those of staminate florets slightly shorter than those of pistillate florets), pinkish purple to lavender-purple, rarely white, the lobes 3–5 mm long. Pappus 12–25 mm long at flowering (longer in pistillate than in staminate florets), white or grayish-tinged. Fruits 2.5–4.0 mm long. $2n=34$. June–October.

Introduced, scattered, sporadic, most widespread in the northwestern quarter of the state (native of Europe, Asia, introduced nearly throughout the U.S. and Canada). Crop fields, fallow fields, railroads, roadsides, and open, disturbed areas.

Canada thistle is a bad weed of crop fields and rangeland farther north, but although the Missouri legislature declared it a noxious weed in 1909, the species has not yet become the problem in the state that was predicted. Steyermark (1963) knew it as a rare introduction in only five counties. *Cirsium arvense* apparently was an early introduction into eastern Canada as a contaminant in agricultural seed and was already the subject of a Vermont weed ordinance as early as 1795 (Hansen, 1918). Its southward spread was aided by the transport of contaminated hay during the Civil War era, as well as in impure crop seed harvested from infested sites. Because of its deep, spreading rootstock it is a difficult plant to control once established. Thus far, attempts to develop an effective biological control agent for it have failed. Paradoxically, the species is a good nectar plant for bees and butterflies, and seed-eating birds such as goldfinches (*Carduelis* species) relish the fruits.

Although sometimes described as dioecious, many staminate clones of *C. arvense* produce heads containing at least a few perfect florets. Some authors have recognized several varieties. Among the two most commonly treated, the var. *horridum* is said to differ from var. *arvense* in its more deeply lobed leaves with spinier margins. Both extremes exist in Missouri along with plants with intermediate characters. Morphological variation among North American populations is complex and rendered more confusing by the great amount of vegetative reproduction through root suckers that often results in populations that are relatively

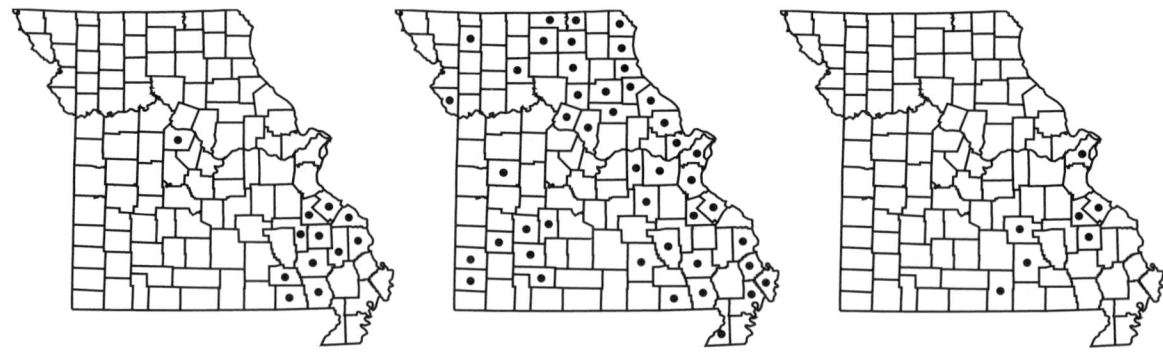

1053. Cirsium carolinianum   1054. Cirsium discolor   1055. Cirsium muticum

uniform morphologically. Thus, it seems inadvisable to attempt the segregation of varieties among the introduced North American populations of this species at the present time.

3. Cirsium canescens Nutt. (Platte thistle)

Map 1052

Plants biennial or short-lived perennials, with a long, slender or thick taproot, not suckering to form clonal colonies. Stems 20–80 cm long, unbranched to several-branched toward the tip, densely pubescent with persistent white, woolly to felty hairs, with spiny-margined wings, at least above the midpoint. Basal leaves 10–30 cm long, 3–6 cm wide, narrowly elliptic to broadly oblanceolate, tapered at the base, bluntly to sharply angled at the tip, usually with several pairs of shallow to deep, relatively broad lobes, the margins otherwise toothed or wavy and spiny, both surfaces appearing grayish or whitish with relatively dense, woolly hairs, the pubescence sometimes becoming thinner on the upper surface with age (which appears gray or green). Stem leaves well developed throughout or progressively reduced above the stem midpoint, the main leaves 4–25 cm long, narrowly oblong to oblong-elliptic, mostly with shallow (less than $1/3$ of the way from the margin to midrib), irregular lobes or wavy, rounded to a clasping and strongly decurrent (more than 1 cm) base, otherwise like the basal leaves. Heads few to several, solitary at the branch tips, appearing sessile or very short-stalked. Involucre (15–)25–40 mm long, as wide as or slightly wider than long, usually cobwebby-hairy (from the bract margins), the lower and median bracts tapered to a loosely ascending to spreading, spiny tip, this 2–4 mm long, straw-colored to light yellow, also sticky along the midrib. Corollas 20–35 mm long, cream-colored to nearly white, rarely pale pink, the lobes 4–8 mm long. Pappus 18–30 mm long, usually white. Fruits 5–7 mm long. $2n=34, 36$. June–October.

Introduced, known from a single historical collection from Jackson County (Idaho to South Dakota south to New Mexico and Nebraska; introduced farther east). Railroads.

The Missouri specimen of *C. canescens* was determined as *C. undulatum* by its collector, B. F. Bush. In 1951, during his taxonomic and nomenclatural studies of these two taxa, Gerald B. Ownbey of the University of Minnesota annotated the sheet as *C. canescens* but did not cite voucher specimens in his publication (Ownbey, 1952). Steyermark (1963) overlooked this specimen during his research for the *Flora of Missouri*. It resurfaced during the research of David J. Keil of California Polytechnic State University toward a treatment of *Cirsium* for the Flora of North America Project (Flora of North America Editorial Committee, in press).

4. Cirsium carolinianum (Walter) Fernald & B.G. Schub. (Carolina thistle, small-headed thistle)

Pl. 253 f, g; Map 1053

Plants biennial or short-lived perennials, with a short, inconspicuous taproot in addition to the often somewhat thickened, fibrous roots. Stems 50–150 cm long, unbranched or more commonly few-branched toward the tip, glabrous or with patches of cobwebby hairs (especially when young), without spiny-margined wings. Basal leaves 8–30 cm long, 1–3 cm wide (to 5 cm wide if lobed), narrowly lanceolate to narrowly oblanceolate or linear, tapered at the base, mostly sharply pointed at the tip, unlobed or less commonly with a few narrow, irregular lobes below the midpoint, the margins otherwise toothed or wavy and spiny, the upper surface appearing green, nearly glabrous to moderately pubescent with stiff, straight hairs, the undersurface appearing white, densely pubescent with felty hairs. Stem leaves progressively reduced from the stem base, those along the upper $1/3$ few,

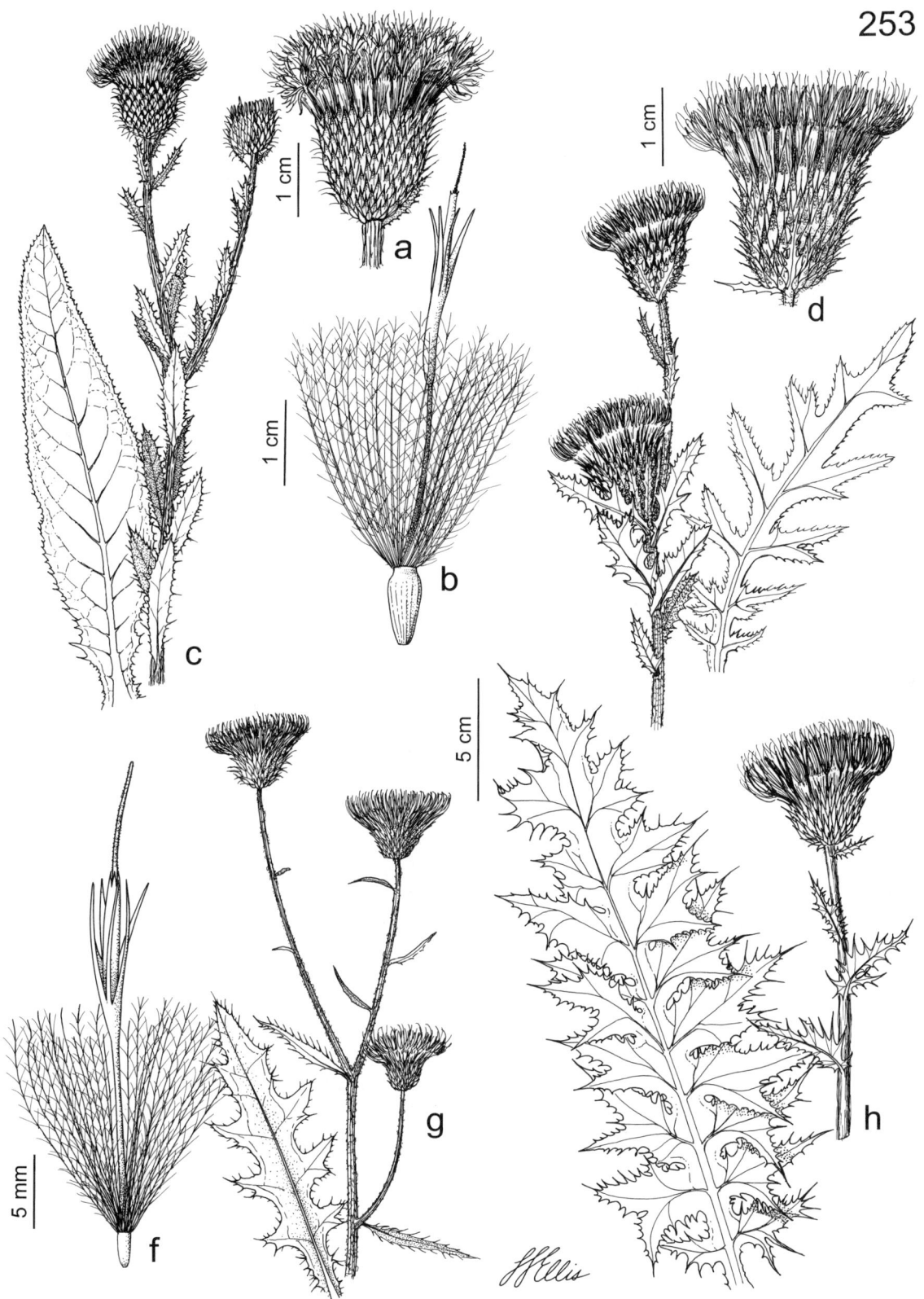

Plate 253. Asteraceae. *Cirsium altissimum*, **a)** head, **b)** floret, **c)** habit and larger leaf. *Cirsium discolor*, **d)** head, **e)** habit and larger leaf. *Cirsium carolinianum*, **f)** floret, **g)** habit and basal leaf. *Cirsium undulatum*, **h)** branch with head and larger leaf.

and widely spaced, mostly 1–15 cm long, all unlobed or the lower ones with a few narrow, irregular lobes below the midpoint, tapered to an often slightly expanded, nondecurrent base, otherwise like the basal leaves. Heads usually relatively few, solitary at the branch tips, appearing long-stalked. Involucre 15–20 mm long, as long as or slightly longer than wide (often appearing broader when pressed), often somewhat cobwebby-hairy, the lower and median bracts tapered to a loosely ascending to spreading, spiny tip, this 1.5–4.0 mm long, straw-colored to light yellow. Corollas 15–24 mm long, pinkish purple to reddish purple, the lobes 4–7 mm long. Pappus 12–17 mm long, white or occasionally slightly grayish-tinged. Fruits 3–4 mm long. $2n=20$. May–June.

Scattered to uncommon in the eastern portion of the Ozark and Ozark Border Divisions, with a single historical specimen from Cooper County (southeastern U.S. west to Missouri and Texas). Openings of mesic to dry upland forests, rarely banks of streams and bottomland forests; also roadsides; usually on acidic substrates.

5. Cirsium discolor (Muhl. ex Willd.) Spreng. (field thistle)

Pl. 253 d, e; Map 1054

Plants biennial or short-lived perennials, often with a slightly thickened taproot in addition to the fibrous roots. Stems 100–250 cm long, well branched, glabrous or sparsely pubescent with spreading hairs, sometimes with patches of white, woolly to felty hairs toward the tip, sometimes appearing slightly glaucous, without spiny-margined wings. Basal leaves 10–50 cm long, 4–25 cm wide, narrowly ovate to elliptic or obovate, more or less tapered at the base, rounded to more commonly bluntly angled at the tip, with several deep lobes, the margins otherwise coarsely toothed and spiny, the upper surface appearing green, nearly glabrous to moderately pubescent with stiff, straight hairs, the undersurface appearing white, densely pubescent with felty hairs. Stem leaves well developed throughout, the main leaves 4–25 cm long, those toward the branch tips usually somewhat reduced, with deep (more than ½ of the way from the margin to midrib), narrow to relatively broad lobes, somewhat clasping and often slightly decurrent at the base, otherwise like the basal leaves. Heads usually relatively numerous, usually solitary at the branch tips, appearing sessile or very short-stalked. Involucre 25–35 mm long, as long as or slightly longer than wide (often appearing broader when pressed), often somewhat cobwebby-hairy, the lower and median bracts tapered to a spreading, spiny tip, this 2–5 mm long, straw-colored to light yellow, usually also somewhat sticky along the midrib. Corollas 25–32 mm long, usually pinkish purple to reddish purple, the lobes 6–9 mm long. Pappus 18–25 mm long, white or occasionally slightly grayish-tinged. Fruits 4.0–5.5 mm long. $2n=20$. July–November.

Uncommon in the Ozark Division, scattered to common elsewhere in the state (eastern U.S. west to North Dakota and Louisiana; Canada). Upland prairies, glades, tops of bluffs, openings of mesic to dry upland forests, and less commonly banks of streams, and bottomland prairies; also fallow fields, old fields, railroads, roadsides, and open, disturbed areas.

Rare, white-flowered plants have been called f. *albiflorum* (Britton) House. This species is closely related to *C. altissimum*, but it usually occurs in more upland habitats than that species and also tends to grow in more disturbed habitats. The two taxa occasionally hybridize at sites where they grow together. Dabydeen (1997) studied one such hybrid in eastern Nebraska and concluded that because the parents have different chromosome base numbers, the hybrids are sterile. Hybrids (mostly sterile) with *C. muticum* can also occur where the two species grow in proximity (Ownbey, 1951), but these have not yet been reported from Missouri. Steyermark (1963) noted that the young shoots and leaves have been cooked and eaten.

6. Cirsium muticum Michx. (swamp thistle)

Pl. 254 a, b; Map 1055

Plants usually biennial, with a well-developed cluster of stout, fleshy, white to brown roots (rarely only a single fleshy taproot) in addition to the fibrous roots. Stems 40–250 cm long, few- to several-branched, glabrous or sparsely pubescent with spreading hairs, rarely with patches of white, woolly to felty hairs toward the tip, not appearing glaucous, without spiny-margined wings. Basal leaves 15–55 cm long, 6–20 cm wide, ovate to elliptic or obovate, usually tapered at the base and tip, with several pairs of deep lobes, the lobes sometimes irregularly lobed again, the margins otherwise toothed and finely spiny, the upper surface appearing green, nearly glabrous to sparsely pubescent with short, curly hairs, the undersurface appearing green, nearly glabrous to sparsely pubescent with cobwebby hairs. Stem leaves progressively reduced from about the stem midpoint, 3–15 cm long, with deep (more than ½ of the way from the margin to midrib), often relatively narrow lobes, sometimes slightly clasping the stem and minutely decurrent at the base, otherwise like the basal leaves. Heads usually relatively numerous, solitary at the branch tips, appearing sessile or very

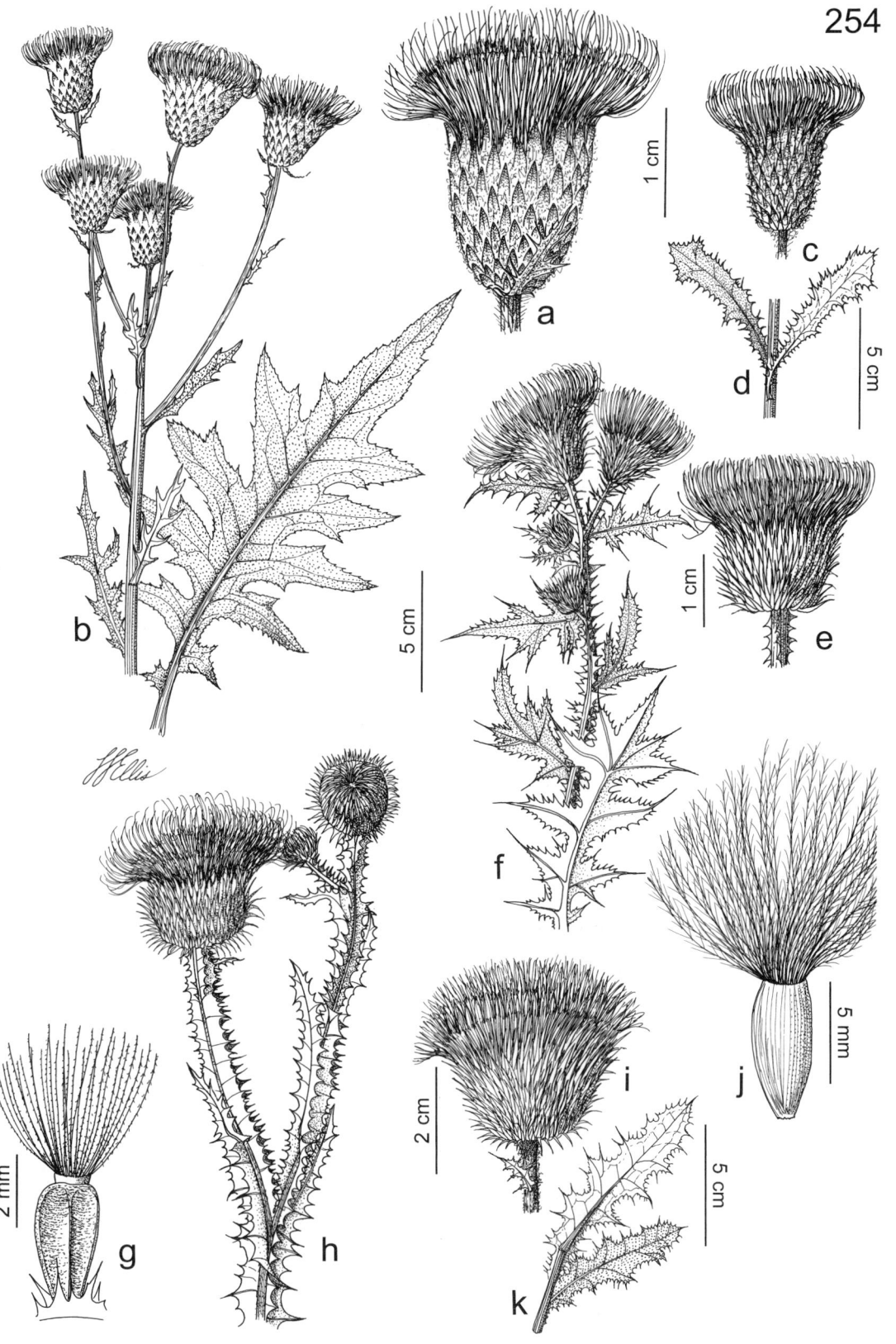

Plate 254. Asteraceae. *Cirsium muticum*, **a)** head, **b)** habit and larger leaf. *Cirsium arvense*, **c)** head, **d)** leaves. *Cirsium texanum*, **e)** head, **f)** habit. *Onopordum acanthium*, **g)** fruit, **h)** habit. *Cirsium vulgare*, **i)** head, **j)** fruit, **k)** leaves.

1056. Cirsium texanum   1057. Cirsium undulatum   1058. Cirsium vulgare

short-stalked. Involucre (17–)20–35 mm long, as long as or slightly longer than wide (often appearing broader when pressed), often somewhat cobwebby-hairy, the lower and median bracts tapered to a bluntly or sharply pointed, appressed or ascending, nonspiny tip, this often with a minute, sharp point, usually also somewhat sticky along the midrib, green with a darker or purplish-tinged area toward the tip. Florets all perfect. Corollas 20–32 mm long, reddish purple to dark purple (rarely white elsewhere), the lobes 4–8 mm long. Pappus 13–20 mm long, white or slightly grayish-tinged. Fruits 4.0–5.5 mm long. $2n=20–23, 30$. July–October.

Scattered to uncommon in the eastern portion of the Ozark and Ozark Border Divisions (eastern U.S. west to North Dakota and Texas; Canada). Fens and less commonly banks of creeks and bases of bluffs, usually in areas of calcareous seepage.

7. Cirsium texanum Buckley **var. texanum**
(Texas thistle)

Pl. 254 e, f; Map 1056

Plants biennial or short-lived perennials, with a short, conspicuously thickened, turnip-shaped taproot in addition to the smaller, fibrous roots, occasionally producing new plants by offsets from the main root. Stems 40–150 cm long, usually well branched, densely pubescent with persistent, white, woolly to felty hairs, the pubescence often becoming patchy with age, without spiny-margined wings. Basal leaves 8–30 cm long, 2–8 cm wide, ovate to elliptic or broadly oblanceolate, slightly tapered to rounded at the base, mostly sharply pointed at the tip, usually with several pairs of shallow to deep, relatively broad lobes, the margins otherwise toothed or wavy and spiny, both surfaces densely pubescent with woolly or felty hairs, the upper surface sometimes appearing green at maturity, nearly glabrous to moderately pubescent, the undersurface appearing persistently gray or white. Stem leaves progressively reduced from the stem base, those along the upper 1/3 few, and widely spaced, mostly 1–15 cm long, all but the uppermost shallowly lobed, rounded to a somewhat clasping, often short-decurrent base, otherwise like the basal leaves. Heads few to numerous, solitary at the branch tips, mostly appearing long-stalked. Involucre 15–22 mm long, about as long as wide (often appearing broader when pressed and at fruiting), usually somewhat cobwebby-hairy, the lower and median bracts tapered to a loosely ascending to spreading, spiny tip, this 1.5–5.0 mm long, straw-colored to light yellow. Corollas 18–25 mm long, pinkish purple to reddish purple, the lobes 4–7 mm long. Pappus 15–18 mm long, white. Fruits 3–5 mm long. $2n=22–24$. May–June.

Introduced, known thus far only from the city of St. Louis (New Mexico, Texas, and Louisiana north to Oklahoma; Mexico; introduced in Missouri). Railroads.

This species was first reported for Missouri by Mühlenbach (1979) from the St. Louis railyards. Superficially it might be mistaken for *C. carolinianum*, but that species has generally narrower leaves and a poorly developed taproot.

8. Cirsium undulatum (Nutt.) Spreng. (wavy-leaved thistle)
C. undulatum var. *megacephalum* (A. Gray) Fernald

Pl. 253 h; Map 1057

Plants perennial (individual stems often appearing biennial), with a relatively short, thick, taproot with few to many spreading main branches, usually suckering from these to form clonal colonies. Stems 30–120 cm long, unbranched or few-branched, densely pubescent with persistent, white, woolly to felty hairs, without spiny-margined wings. Basal leaves 10–30 cm long, 2–8 cm wide, elliptic to broadly oblanceolate, tapered at the base, bluntly to sharply angled at the tip, with several pairs of shallow to deep, relatively broad lobes, the margins otherwise toothed or wavy and spiny, both surfaces appearing grayish or whitish with relatively dense, woolly hairs, the pubescence sometimes becoming thinner on both surfaces with age (the leaves still appearing uniformly gray to grayish green). Stem leaves well developed throughout or progressively reduced above the stem midpoint, the main leaves 4–25 cm long, mostly with shallow (less than 1/3 of the way from the margin to midrib), broad lobes or wavy, angled or rounded to a somewhat clasping and often minutely decurrent base, otherwise like the basal leaves. Heads few to several, solitary at the branch tips, appearing sessile or very short-stalked. Involucre 25–40 mm long, as long as or slightly longer than wide (often appearing broader when pressed or at fruiting), usually cobwebby-hairy (from the bract margins), the lower and median bracts tapered to a loosely ascending to spreading, spiny tip, this 2–5 mm long, straw-colored to light yellow, also sticky along the midrib. Corollas 25–45 mm long, usually light purple to pinkish purple or purple, the lobes 6–10 mm long. Pappus 20–38 mm long, white or slightly grayish-tinged. Fruits 5–7 mm long. $2n=26$. June–October.

Uncommon in Jackson and Atchison Counties, introduced sporadically in eastern Missouri (western U.S. east to Michigan and Texas; Canada, Mexico; introduced farther east). Upland prairies and loess hill prairies; also railroads and roadsides.

Ownbey (1952) studied the nomenclature, typification, and morphology of the *C. undulatum* complex and concluded that the name *C. undulatum* should be restricted to plants treated earlier as *C. undulatum* var. *megacephalum,* with the other varieties formerly treated in a broadly circumscribed concept of the species segregated into other taxa such as *C. canescens. Cirsium undulatum* is unusual within the complex in that the fruits become mucilaginous externally when moistened.

9. Cirsium vulgare (Savi) Ten. (bull thistle)

Pl. 254 i–k; Map 1058

Plants biennial, with a short, thickened taproot in addition to the fibrous roots. Stems 40–200 cm long, usually well branched, sparsely to moderately pubescent with cobwebby hairs, often nearly glabrous toward the base, with sometimes-narrow, spiny-margined wings, at least above the midpoint. Basal leaves 10–40 cm long, 3–15 cm wide, narrowly elliptic to oblanceolate, rarely obovate, more or less tapered at the base, mostly narrowly angled at the tip, with several shallow to deep lobes, the margins otherwise coarsely lobed or toothed and spiny, the upper surface appearing green, strongly roughened with numerous short, stiff, spinelike or barblike bristles, not cobwebby-hairy, the undersurface thinly or finely pubescent with cobwebby hairs. Stem leaves progressively reduced from near the stem base, the main leaves 3–15 cm long, with deep (more than 1/2 of the way from the margin to midrib), narrow lobes, slightly clasping and long-decurrent (more than 1 cm), otherwise like the basal leaves. Heads usually relatively numerous, solitary or in loose clusters at the branch tips, appearing sessile or very short-stalked. Involucre 25–40 mm long, as long as or slightly longer than wide (often appearing broader when pressed or at fruiting), somewhat cobwebby-hairy, the lower and median bracts tapered to a spreading, spiny tip, this 2–5 mm long, straw-colored to light yellow, rarely also slightly sticky along the midrib. Corollas 25–35 mm long, reddish purple to purple, the lobes 5–7 mm long. Pappus 20–28 mm long, white or light tan. Fruits 3.0–4.5 mm long. $2n=68$. June–September.

Introduced, widely scattered in Missouri (native of Europe, Asia, introduced throughout the U.S., Canada). Upland prairies and openings of disturbed, mesic to dry upland forests; also pastures, railroads, roadsides, and open, disturbed areas.

31. Onopordum L.

About 40 species, Asia.

1. Onopordum acanthium L. **ssp. acanthium**
(Scotch thistle, cotton thistle)

Pl. 254 g, h; Map 1059

Plants coarse, biennial, with stout, fleshy taproots, densely pubescent with whitish, woolly to felty hairs. Stems 50–200 cm tall, erect, usually branched, noticeably spiny-winged. Leaves basal and alternate, sharply pointed and spiny at the tip, strongly spiny along the margins. Basal leaves in a dense, overwintering rosette, sessile or short-petiolate, the blades 10–60 cm long, 5–20 cm wide, ovate to lanceolate or narrowly elliptic, irregularly pinnately lobed, the lobes more or less triangular, irregularly lobed. Stem leaves sessile, the blades gradually reduced toward the tip, 5–30 cm long, oblong-ovate to lanceolate or oblanceolate, irregularly toothed or shallowly pinnately lobed, the bases decurrent into wings along the stem, these wavy

1059. Onoportum acanthium   1060. Silybum marianum   1061. Cichorium intybus

or scalloped and with strongly spiny margins. Inflorescences terminal, the heads long-stalked to nearly sessile, solitary or in clusters of 2–5 at the branch tips. Heads discoid, the involucre 25–40 mm long, cup-shaped to nearly spherical, the florets all appearing similar and perfect. Receptacle relatively flat, fleshy, the florets with the ovaries sunken into pits, the margins of the pits with low, broad scales. Involucral bracts lanceolate to narrowly lanceolate, the lower and median bracts tapered to spreading to reflexed, spiny tips, these 3–5 mm long, straw-colored or sometimes purplish-tinged, the upper bracts tapered to ascending, less spinelike, flattened tips. Pappus of numerous unequal capillary bristles, these 6–10 mm long, fused at the base, roughened with minute, ascending barbs or teeth, tan, shed more or less as a unit before fruiting. Corollas 20–26 mm long, reddish purple to purple, rarely pinkish white. Fruits appearing basally attached, 4–5 mm long, oblong-obovate in outline, somewhat flattened and slightly 4-angled in cross-section, the surface glabrous, finely cross-wrinkled, light brownish-grayish with darker brown to nearly black mottling. $2n=34$. July–September.

Introduced, uncommon and sporadic, known only from historical collections in southeastern Missouri (native of Europe, Asia, introduced widely in the U.S. and Canada). Roadsides and open, disturbed areas.

Onopordum acanthium is a striking plant that presumably was introduced into the United States as a garden ornamental in the late 1800s. In Europe it also has a long history of medicinal and food use, and its young herbage and immature heads can be cooked as vegetables (Steyermark, 1963). An oil extracted from the seeds was used as a lamp oil (Moore and Frankton, 1974). However, it is a problem weed of rangeland and crop fields in some parts of the western and northern United States and Canada. Although it was declared a noxious weed by the state legislature in 1909, there is no evidence that this species was ever widespread or even locally abundant in Missouri. Scotch thistle is the national emblem of Scotland.

32. Silybum Adans. (milk thistle)

Two species, Europe, Asia, Africa.

1. Silybum marianum (L.) Gaertn. (milk thistle, blessed milk thistle)

Map 1060

Plants coarse, annual or biennial, with taproots. Stems 60–180 cm tall, erect, usually branched, not winged, finely cobwebby-hairy. Leaves basal and alternate, mostly sharply pointed and spiny at the tip, spiny along the often somewhat corrugated or wavy margins, usually thinly pubescent with cobwebby hairs when young, glabrous at maturity, strongly white-mottled, particularly along the main veins. Basal leaves in a dense, overwintering rosette, sessile or short-petiolate, the blades 15–80 cm long, 5–40 cm wide, elliptic, pinnately lobed, the lobes more or less triangular to hemispherical, irregularly and mostly shallowly lobed. Stem leaves mostly sessile, the blades gradually reduced toward the tip, 1–40 cm long, elliptic to narrowly ovate or the uppermost lanceolate, irregularly toothed or shallowly pinnately lobed, the bases not decurrent, those of all but the lowermost leaves clasping the stem. Inflorescences terminal, the heads mostly long-stalked, solitary at the branch tips, sometimes somewhat nodding. Heads discoid, the involucre

30–45 mm long, broadly ovoid to hemispherical, the florets all appearing similar and perfect. Receptacle flat or slightly convex, with numerous bristles. Involucral bracts glabrous, 25–40 mm long, somewhat thickened and fleshy or leathery; all but the uppermost with the body appressed, oblong-ovate to narrowly ovate, finely and sharply toothed along the margins, the tip with a well-developed appendage, this spreading, lanceolate, the upper surface concave, spiny along the margins, tapered to a stout, straw-colored to light brown, spiny tip; uppermost bracts with the body lanceolate, the appendage less strongly differentiated, ascending, tapered to a sharply pointed or short-spined tips. Pappus of numerous unequal capillary bristles, these 15–20 mm long, fused at the base, roughened with minute, ascending barbs or teeth, white, shed as a unit before fruiting. Corollas 28–35 mm long, reddish purple to purple. Stamens with the filaments fused into a short tube toward the base. Fruits appearing basally or slightly obliquely attached, 6–7 mm long, oblong-obovate in outline, somewhat flattened, with a minute, raised crown at the tip, the surface glabrous, shiny, dark brown to black, often with lighter streaks or mottling, the apical crown yellow. $2n=34$. May–July.

Introduced, known only from historical specimens without further locality data and from the city of St. Louis (native of Europe, Asia, introduced widely in North America). Open, disturbed areas.

Steyermark (1963) overlooked the few historical voucher specimens during his research on the state's flora and did not treat this species. The vernacular and scientific names are derived from Christian folklore, in which the mottling of the leaves was purported to have developed following exposure to a drop of the Virgin Mary's milk. *Silybum marianum* is sometimes grown as a garden ornamental, although in some western states it has become a rangeland weed. The species has a long history of use in Europe and Asia for its food and medicinal value. The immature heads and young herbage can be cooked and eaten as vegetables, and boiled plants were once consumed in the springtime as a blood cleanser (Moore and Frankton, 1974). The fruits (kenguel seed) also were formerly a component of some bird seed mixes and were ground as a substitute for coffee (Mabberley, 1997). The principal medicinal uses date back to the time of Dioscorides and involve the use of a seed extract for treatment of poisoning and ailments of the liver, including jaundice and hepatitis. Such treatment is still popular in Europe, and milk thistle pills also are available commercially in the United States as an herbal dietary supplement to maintain liver health. Studies have shown milk thistle extract to contain flavonoids effective in the treatment of *Amanita* Pers. (death angel mushrooms) poisoning (Mabberley, 1997).

4. Tribe Cichorieae Lam. & DC. (Tribe Lactuceae Cass.)

Plants annual, biennial, or perennial herbs (rarely woody elsewhere), the sap usually milky (white or colored). Stems not spiny or prickly. Leaves alternate and/or in a basal rosette, sessile or petiolate, the base sometimes somewhat sheathing the stem, occasionally spiny or prickly. Leaf blades entire to deeply pinnately lobed, the venation pinnate (or the veins too faint to observe), mostly with 1 main vein (the leaves usually grasslike in *Tragopogon* with few to several parallel veins). Inflorescences mostly terminal; panicles, clusters of heads, or occasionally racemes, or the heads solitary (sometimes appearing as axillary clusters or terminal and spikelike in *Cichorium*). Heads ligulate. Involucre most commonly of a longer, inner series of uniform bracts (these sometimes fused laterally, at least toward the base) and 1 or more shorter series of outer bracts, less commonly with 2 or more overlapping series of more or less similar bracts, the bracts not spiny or tuberculate, often becoming reflexed as or after the fruits mature. Receptacle flat or nearly so, naked or less commonly the florets subtended by chaffy bracts or minute hairs. Ligulate florets perfect, the corollas 5-toothed at the tip, variously colored. Disc and ray florets absent. Pappus usually of numerous capillary bristles (these sometimes plumose), occasionally with an outer, shorter whorl of bristles, hairs, or scales, less commonly only of scales or absent, when present persistent at fruiting (except sometimes in *Picris*). Stamens with the filaments not fused together, the anthers fused into a tube, each tip usually with a short appendage, each base with a pair of auricles or slender lobes. Style branches short or long and usually not or only slightly flattened, each with a stigmatic band along the inner surface, the sterile tip short and rounded or less commonly truncate, the outer surface and tip often with dense, minute hairs. Fruits usually all similar, often flattened or angled in cross-section, variously shaped, sometimes ribbed, not winged, sometimes developing a short and stout or elongate and slender beak at maturity. About 120 genera, about 3,200 species, worldwide.

The Cichorieae are perhaps the most easily recognized tribe of Asteraceae in Missouri, because of their ligulate heads and milky sap. Dried specimens collected in bud occasionally may be confused with Senecioneae because of similarities in involucral bracts and pappus.

The tribe is economically important primarily for the large number of weedy species, although many of these are restricted to highly disturbed areas. Most of the plants are edible, at least when young, and some genera are (or formerly were) of commercial importance for use in food and beverages, such as *Cichorium* (chicory, endive), *Lactuca* (lettuce), and *Tragopogon* (salsify). Relatively few species are cultivated as ornamentals.

Plants of this tribe tend to be quite variable in leaf morphology, so the key to genera below emphasizes pappus characters, which are usually observable (sometimes only with magnification) on fertile specimens, regardless of whether buds, flowers, or fruits are present. In general, *Hypochaeris, Leontodon, Nothocalais, Taraxacum,* and some species of *Hieracium* and *Krigia* have leaves only in a basal rosette (excluding any bracts in the inflorescence), whereas the other genera have stems noticeably leafy above the base. However, young or depauperate plants of species otherwise in the latter group rarely may appear to have only basal leaves, and *Lygodesmia* may appear nearly leafless at maturity. In Missouri, prickly leaves may be found in some species of *Lactuca, Picris,* and *Sonchus.* The key to genera requires both flowers and fruits for a few of the more difficult genera. Fortunately, these groups tend to have long periods of bloom with heads at different stages present on the same plant. Additionally, flowering heads will often continue to develop fruits during the pressing process or if an inflorescence branch is placed in an envelope and allowed to dry.

Missouri botanists should be aware of the following potential future addition to the state's flora, which does not key well in the key to genera below. *Chondrilla juncea* L. (rush skeletonweed, hogbite) is a weed of crop fields, pastures, and natural grasslands that is listed as a noxious weed in several states to the west and north of Missouri. It has also been spreading sporadically in a number of eastern states and may eventually be found in Missouri. This Eurasian perennial can grow to 1.5 m tall. The stems have numerous ascending branches from a usually unbranched basal portion. Although a rosette of pinnately lobed leaves is produced, this usually withers by flowering time and the main stem leaves are inconspicuous, linear, and unlobed. The slender heads have an involucre of 5–9 main bracts 9–12 mm long surrounding 7–15 florets with yellow corollas. The fruits are about 3 mm long, several-nerved, and have a whorl of 3 or 4 tiny scales at the expanded tip of the long, slender beak in addition to the numerous capillary bristles.

1. Pappus absent or of scales (these sometimes short and inconspicuous in *Krigia*), if of scales then sometimes also with bristles (*Leontodon* and *Nothocalais,* which have a pappus of mixed bristles and long, slender [broad-based] scales, should be keyed under this lead)
    2. Pappus a mixture of bristles and scales, the scales sometimes slender and awnlike from an expanded base or short and inconspicuous; plants usually with the leaves all basal (the inflorescence sometimes with a leaflike basal bract in *Krigia*); corollas yellow or orange

3. Pappus mostly of long, slender scales, these mostly bristlelike and plumose from a flattened, expanded base, usually associated with a few somewhat shorter, merely barbed, capillary bristles (note that the outermost florets of the head have a different pappus that is composed of a short, scaly crown, without any longer scales or bristles); some of the hairs, especially on the leaves and involucres, minutely forked at the tip .... 41. LEONTODON
3. Pappus with as many bristles as scales or more bristles than scales, the bristles smooth or minutely barbed, not plumose; hairs all simple or the plants glabrous
    4. Involucre 4–14 mm long; pappus of 5 to numerous bristles but only 5 or (8–)10 scales .... 38. KRIGIA
    4. Involucre 17–25 mm long; pappus of numerous bristles intermixed with usually numerous scales .... 43. NOTHOCALAIS
2. Pappus absent or only of scales; plants usually with at least a few leaves above the base along the stem (except on *Leontodon*) corollas yellow or blue (rarely white or pink)
  5. Pappus present, of minute or slender scales; corollas yellow or blue (rarely white or pink)
    6. Pappus of minute scales; corollas blue (rarely white or pink); leaves usually basal and alternate along the stems (the stem leaves often much smaller than the basal ones) .... 33. CICHORIUM
    6. Pappus of the outermost florets a short, scaly crown, without any longer scales, that of the other florets of slender, awnlike scales, these bristlelike above an expanded, flattened base; corollas yellow; leaves all basal .... 41. LEONTODON
  5. Pappus absent; corollas yellow
    7. Involucre with 1 or 2 series of 4–7(–9) ascending bracts, these all similar in size and shape (no outer, shorter bracts present); fruits 1.4–1.7 mm long .... 38. KRIGIA
    7. Involucre with an inner series of usually 8 longer, ascending bracts and an irregular outer series of minute, spreading bracts; fruits 3–5 mm long .... 40. LAPSANA
1. Pappus only of bristles, these often minutely barbed or plumose (*Pyrrhopappus*, which has a ring of minute, reflexed hairs attached immediately below the pappus bristles, should be keyed under this lead)
  8. Pappus with some or all of the bristles plumose
    9. Leaves all basal
      10. Receptacle with long, slender, chaffy bracts subtending the florets; leaves pubescent with unbranched hairs .... 37. HYPOCHAERIS
      10. Receptacle naked (the florets sunken into minute pits); leaves with at least some of the hairs minutely forked at the tip ...... 41. LEONTODON
    9. Leaves basal and alternate, the basal leaves sometimes withered by flowering time
      11. Leaves mostly grasslike, with few to several main veins, glabrous or with small patches of inconspicuous cobwebby hairs toward the base when young .... 49. TRAGOPOGON
      11. Leaves not grasslike, with 1 main vein and sometimes also a faint network of anastomosing secondary veins, pubescent with some of the hairs barbed at the tip with 2–5 minute, spreading to recurved branches from a knoblike tip

12. Outer involucral bracts 3–5, in 1 series, all similar, slightly shorter than to slightly longer than the inner series, ovate to broadly ovate; fruits with a slender beak 1–2 times as long as the body .... 35. HELMINTHOTHECA
12. Outer involucral bracts 7–13 in 2 or 3 progressively longer series ranging from much shorter than to more than ½ as long as the inner series, narrowly lanceolate to narrowly oblong-lanceolate; fruits sometimes slightly tapered toward the tip but not beaked .... 44. PICRIS
8. Pappus with all of the bristles smooth or with minute, ascending barbs
  13. Leaves all basal
    14. Leaves deeply and irregularly lobed and toothed; fruits long-beaked at maturity .... 48. TARAXACUM
    14. Leaves not lobed, entire or less commonly sometimes slightly wavy or with a few inconspicuous teeth; fruits not beaked at maturity
      15. Inflorescence a panicle, raceme, or cluster of flowers; leaf surfaces and margins with relatively long, spreading hairs having a bulbous or slightly expanded base, often also with minute, stellate hairs (visible only with magnification) .... 36. HIERACIUM
      15. Inflorescence consisting of a solitary long-stalked head; leaves with the surfaces glabrous or sparsely and inconspicuously hairy with short, curled hairs, the margins with a dense fringe of very short, curly hairs .... 43. NOTHOCALAIS
  13. Leaves basal and alternate, the basal leaves sometimes withered by flowering time (*Lygodesmia,* with the stem leaves linear to very narrowly triangular, often scalelike and mostly withered by flowering time, should be keyed under this lead)
    16. Fruits with a dense ring of minute, reflexed hairs attached immediately below the pappus bristles; involucral bracts with an abrupt, small, crestlike thickening just below the tip, because of this often appearing irregular, notched, or jointed near the tip .... 46. PYRRHOPAPPUS
    16. Fruits without a ring of hairs just below the pappus; involucral bracts not thickened at or near the tip, not appearing notched or jointed
      17. Leaves not lobed, the margins entire or less commonly slightly wavy or with a few shallow, inconspicuous teeth
        18. Stem leaves linear to very narrowly triangular, inconspicuous and often scalelike and mostly withering or shed by flowering time; ligulate florets 4–6 per head .... 42. LYGODESMIA
        18. Stem leaves variously shaped but not scalelike or withered by flowering time; ligulate florets 8 to numerous per head (except for *Prenanthes altissima,* with only 5 or 6 florets)
          19. Leaves with relatively long, spreading hairs having a bulbous or slightly expanded base (often also with minute, stellate hairs visible only with magnification) .... 36. HIERACIUM
          19. Leaves glabrous or with sparse to moderate, short, curled, unbranched hairs and/or stouter, spreading hairs on the undersurface along the midvein, these often tapered from a slightly broadened, somewhat flattened, nonbulbous base and sometimes grading into prickles (minute, stellate hairs sometimes also present in *Prenanthes*)

20. Pappus bristles white or nearly so; leaves mostly 4–20 times as long as wide or, if 2–3 times as long as wide (in *L. sativa*), then the surface strongly and irregularly crisped or curled and the blade sessile or nearly so; mature fruits flattened. . . . . 39. LACTUCA

20. Pappus bristles straw-colored to tan, orangish brown, or reddish brown; leaves mostly 1–3 times as long as wide or, if 3–5 times as long as wide, then the blade tapered abruptly to a winged petiole; mature fruits more or less circular in cross-section or 4- or 5-angled, not flattened . . . . 45. PRENANTHES

17. At least the largest leaves deeply lobed and/or coarsely toothed
   21. Pappus bristles straw-colored to tan, orangish brown, or reddish brown; corollas white, pink, purple, or cream-colored, rarely greenish yellow . . . . 45. PRENANTHES
   21. Pappus bristles white; corollas lemon yellow to orangish yellow (purplish blue to blue in some *Lactuca* species)
      22. Mature fruits not flattened, more or less circular or (4)5–10-angled in cross-section . . . . 34. CREPIS
      22. Mature fruits flattened (also with 1 or more nerves on each face)
         23. Mature fruits with a long beak (except in *L. floridana*, which has blue corollas), the pappus attached at an expanded, disclike tip (this more or less visible even at flowering); ligulate florets 9–55 per head; corollas yellow or purplish blue to blue . . . . 39. LACTUCA
         23. Mature fruits not beaked, the pappus attached at an unexpanded, unmodified tip; ligulate florets 80–250 or more per head; corollas yellow . . . . 47. SONCHUS

33. Cichorium L. (chicory)
(Kiers-van der Steen, 2000)

About 6 species, Europe, Asia, Africa, introduced widely in the New World.

The other commercially important member of the genus is *C. endivia* L., endive, which is cultivated as a salad vegetable and also was a popular plant for physiological and developmental research in the past (Vuilleumier, 1973).

1. Cichorium intybus L. (common chicory, blue sailors)

Pl. 255 h; Map 1061

Plants perennial herbs, with a sometimes branched taproot. Latex white. Stems solitary to several, 30–150 cm long, erect or ascending, usually moderately branched, longitudinally ridged, glabrous or sparsely pubescent with curly, white, multicellular hairs, these sometimes with the base slightly broadened and flattened. Leaves alternate and basal, abruptly reduced above the stem base, the basal leaves mostly with a short to long, winged petiole, the stem leaves mostly sessile and clasping the stem. Leaf blades 1–35 cm long, mostly oblanceolate, ranging from pinnately several-lobed with irregular triangular lobes and rounded sinuses toward the stem base to entire toward the stem tip, the margins of the more divided leaves usually also irregularly toothed, glabrous or sparsely to moderately pubescent with minute, curly, white hairs along the margins and sometimes also the undersurface. Venation of 1 main vein and a faint, complex network of anastomosing secondary veins. Inflorescences panicles with ascending, spicate branches, sometimes reduced to a single spike or spikelike raceme or appearing as small, axillary, sessile clusters of heads. Heads sessile or short-stalked. Involucre 10–15 mm long, cup-shaped to broadly cylindrical, the bracts in 2 series, glabrous or sometimes with gland-tipped hairs, those of the outer series 4–6, about ½ as long as the inner series, ovate to narrowly ovate, somewhat thickened, hardened, and pale toward the base, spreading at the tip; those of the inner series 8–12, narrowly lanceolate to linear. Receptacle naked. Ligulate florets 12–30. Corollas 14–25 mm long, purplish blue to blue, rarely pink or white. Pappus a minute, low crown of numerous scales, these 0.1–0.2 mm long, irregularly truncate at the tip, white, persistent at fruiting. Fruits 2–3 mm long, oblong-ovoid in outline, not flattened, bluntly 5-angled, truncate at the tip, the surface sometimes with 8–10 shallow, longitudinal lines, minutely pebbled to finely cross-wrinkled, glabrous, brown to nearly black, sometimes finely mottled. $2n=18$. May–October.

Introduced, common nearly throughout the state (native of Europe, introduced widely nearly worldwide). Pastures, railroads, roadsides, and open, disturbed areas.

The pretty blue flowers of chicory are a common sight along Missouri roadsides in the summer and fall. On sunny days, the heads are said to open in the morning and usually to close by noon, but this behavior is fairly variable. The mapped county voucher specimens grossly underdocument the actual distribution of *C. intybus* in the state. White-flowered (f. *album* Neum.) and pink-flowered (f. *roseum* Neum.) mutants occur sporadically within some populations.

The species has a long history of medicinal use as a laxative, diuretic, and antiseptic, and the young greens have been eaten raw or cooked (Vuilleumier, 1973). However, it is best known for the beverage and coffee adulterant prepared from the ground, roasted roots. More recently, the roots have provided a commercial source for inulin, a fructo-oligosaccharide that is used as a soluble fiber dietary supplement and as a bulking agent and fat substitute in some foods, especially some brands of yogurt (Kiers-van der Steen, 2000).

34. Crepis L. (hawksbeard)
Contributed by David J. Bogler and George Yatskievych

Plants annual, biennial, or perennial herbs, taprooted (with rhizomes or woody rootstocks elsewhere). Latex white. Stems erect to loosely ascending, unbranched or branched, finely ridged, variously hairy or glandular. Leaves basal and alternate, glabrous or variously hairy and sometimes also glandular, sessile or short- to long-petiolate. Basal leaves with the blades (at least the largest) pinnately lobed, variously shaped, the margins otherwise entire or toothed, with 1 main vein visible and sometimes also a faint network of anastomosing secondary veins. Stem leaves similar to the basal ones, but often sessile, smaller, narrower, and generally less divided, the base with a pair of narrowly triangular clasping lobes. Inflorescences mostly terminal panicles, sometimes appearing as solitary heads or loose clusters at the stem or branch tips. Involucre not or only slightly elongating as the fruits mature, cup-shaped or somewhat bell-shaped, the bracts in 1(2) longer, inner series and 1 shorter, outer series, the inner bracts similar in size, mostly lanceolate, the margins sometimes thin and pale, the tip ascending at flowering; the outer bracts unequal in size, linear to broadly ovate, loosely ascending to spreading. Receptacle naked or sometimes with minute hairs around the base of each floret. Ligulate florets 10–70 per head. Corollas lemon yellow to orangish yellow, sometimes purplish-tinged on the undersurface. Pappus of numerous bristles, these white (slightly off-white in *C. pulchra*), often shed irregularly at fruiting. Fruits nearly cylindrical to narrowly oblong-elliptic in outline, beaked or not beaked, not flattened, circular or somewhat 5–10-angled in cross-section, with 10–20 longitudinal ribs, these often minutely roughened or barbed, glabrous, reddish brown to dark brown, the pappus attached to a sometimes expanded tip. About 200 species, North America, South America, Europe, Asia, Africa.

Crepis was the subject of a monumental monographic study by Ernest Babcock (1947a, b) that is still studied by students of plant taxonomy as a landmark publication in biosystematics.

1062. Crepis capillaris   1063. Crepis pulchra   1064. Crepis setosa

1. Stems pubescent with short, stiff, spreading, nonglandular hairs; fruits tapered abruptly to a slender beak .... 3. C. SETOSA
1. Stems glabrous or pubescent with curled, appressed, or gland-tipped hairs; fruits narrowed but not beaked at the tip
    2. Stems sticky, densely pubescent with gland-tipped hairs toward the base, glabrous or nearly so toward the tip; fruits somewhat dimorphic, the inner ones shorter than the outer ones .... 2. C. PULCHRA
    2. Stems not sticky, sparsely to densely pubescent with nonglandular hairs at least toward the base and just below the heads, glandular-hairy only toward the branch tip; fruits all similar or nearly so
        3. Involucral bracts glabrous on the inner surface; fruits 1.5–2.5 mm long, light brown to yellowish brown .... 1. C. CAPILLARIS
        3. Inner series of involucral bracts finely appressed-hairy on the inner surface; fruits 2.5–4.0 mm long, reddish brown to dark brown .... 4. C. TECTORUM

1. Crepis capillaris (L.) Wallr. (smooth hawksbeard)

Pl. 255 c–e; Map 1062

Plants annual or biennial. Stems 10–40(–90) cm long, erect or ascending, unbranched or few- to less commonly many-branched, not sticky, moderately to densely but inconspicuously pubescent with cobwebby hairs (the minute, yellowish hairs sometimes few-branched), at least toward the base and just below the heads, often also with short, spreading, gland-tipped hairs toward the branch tips. Basal leaves 3–20 cm long, sessile to long-petiolate, the blade unlobed or more commonly irregularly pinnately lobed, the lobes spreading, mostly narrow, bluntly to sharply triangular, the surfaces glabrous or sparsely and inconspicuously pubescent with minute, curled to cobwebby, yellowish hairs. Inflorescences sometimes relatively few-flowered, the heads often appearing in loose clusters at the branch tips. Inner series of involucral bracts 8–16, 5–8 mm long, glabrous on the inner surface, the outer surface glabrous or minutely pubescent with minute, cobwebby hairs, often also with a longitudinal band of short, spreading, gland-tipped hairs, these with somewhat broadened bases, dark green to black, the outer series of bracts much shorter than to about ½ as long as the inner series. Receptacle glabrous. Ligulate florets 20–60, the corolla 8–12 mm long. Pappus 3–4 mm long. Fruits all similar or nearly so, 1.5–2.5 mm long, narrowly oblong-elliptic in outline, not beaked, not expanded at the tip, 10(–12)-ribbed, the ribs smooth or minutely roughened or barbed, light brown to yellowish brown. $2n=6$. July–October.

Introduced, known thus far only from St. Louis County (native of Europe; introduced sporadically nearly throughout temperate North America). Open, disturbed areas.

Crepis capillaris is superficially similar to *C. tectorum* but tends to have denser basal rosettes of leaves at flowering time and fewer-headed inflorescences.

2. Crepis pulchra L. (small-flowered hawksbeard)

Pl. 255 a, b; Map 1063

Plants annual. Stems 10–100 cm long, erect or ascending, usually unbranched below the inflores-

1065. Crepis tectorum
1066. Helminthotheca echioides
1067. Hieracium caespitosum

cence, sticky, moderately to densely pubescent with gland-tipped hairs, more or less glabrous toward the tip. Basal leaves 2–20 cm long, mostly short-petiolate, the blade coarsely toothed to irregularly pinnately lobed, the lobes spreading, narrow to broad, sharply triangular, the surfaces and margins moderately to densely pubescent with gland-tipped hairs. Inflorescences usually appearing paniculate, less commonly the heads appearing in loose clusters at the branch tips. Inner series of involucral bracts 10–14, 8–12 mm long, developing a prominent, bluntly keeled midrib (at least toward the base) at fruiting, glabrous on both surfaces, the outer series of bracts much shorter than the inner series. Receptacle glabrous. Ligulate florets 15–30, the corolla 5–12 mm long. Pappus 4–5 mm long. Fruits somewhat dimorphic, the inner ones 4–5 mm long, the outer ones 5–6 mm long, more or less cylindrical, not beaked at the tip, 10(–12)-ribbed, the inner fruits with the ribs usually smooth, the outer fruits with the ribs usually minutely roughened or barbed, yellowish brown to light brown. $2n=8$. May–July.

Introduced, uncommon in St. Charles and St. Louis Counties and the city of St. Louis (native of Europe, Asia; introduced sporadically in the eastern half of the U.S., Oregon, Canada). Railroads, roadsides, and open, disturbed areas.

3. Crepis setosa Haller f. (bristly hawksbeard)

Pl. 255 f, g; Map 1064

Plants annual. Stems 8–50(–80) cm long, erect to loosely ascending, unbranched or few-branched, not sticky, moderately to densely pubescent with stiff, spreading, straw-colored, bristlelike, nonglandular hairs (these mostly with a somewhat flattened, broad base), usually also with inconspicuous, cobwebby to woolly hairs toward the tip. Basal leaves 3–30 cm long, sessile to short-petiolate, the blade unlobed or more commonly coarsely toothed to irregularly pinnately lobed, the lobes spreading or slightly curved toward the leaf base, narrow to broad, mostly sharply triangular, the surfaces and margins moderately to densely pubescent with fine, bristlelike hairs, the midvein often also with flattened, stiff, bristlelike hairs. Inflorescences appearing paniculate or sometimes relatively few-flowered, the heads sometimes appearing in loose clusters at the branch tips. Inner series of involucral bracts 12–16, 6–8 mm long, appressed-hairy toward the tip on the inner surface, the outer surface with a row of flattened, stiff, bristlelike hairs, also cobwebby- to woolly-hairy, the outer series of bracts much shorter than to about ½ as long as the inner series. Receptacle with minute hairs around the base of each floret. Ligulate florets 10–25, the corolla 6–10 mm long. Pappus 3–4 mm long. Fruits all similar or nearly so, with the body 2–4 mm long, narrowly oblong-elliptic in outline, tapered to a slender beak 1–2 mm long, the pappus attached to an expanded, disclike or concave tip, 10(–12)-ribbed, the ribs minutely roughened or barbed, light brown to yellowish brown. $2n=8$. June–August.

Introduced, uncommon and sporadic (native of Europe; introduced sporadically in the U.S., Canada). Crop fields, lawns, and open, disturbed areas.

Several subspecies are sometimes segregated in the European literature, but their application to North American plants is not clear.

4. Crepis tectorum L. (narrow-leaved hawksbeard)

Pl. 255 i, j; Map 1065

Plants annual or biennial. Stems 10–100 cm long, erect or ascending, few- to many-branched, not sticky, sparsely to densely but inconspicuously pubescent with cobwebby hairs (the minute hairs sometimes few-branched), at least toward the base and just below the heads, often also with short, spreading, gland-tipped hairs toward the branch tips. Basal leaves 3–15 cm long, sessile or more

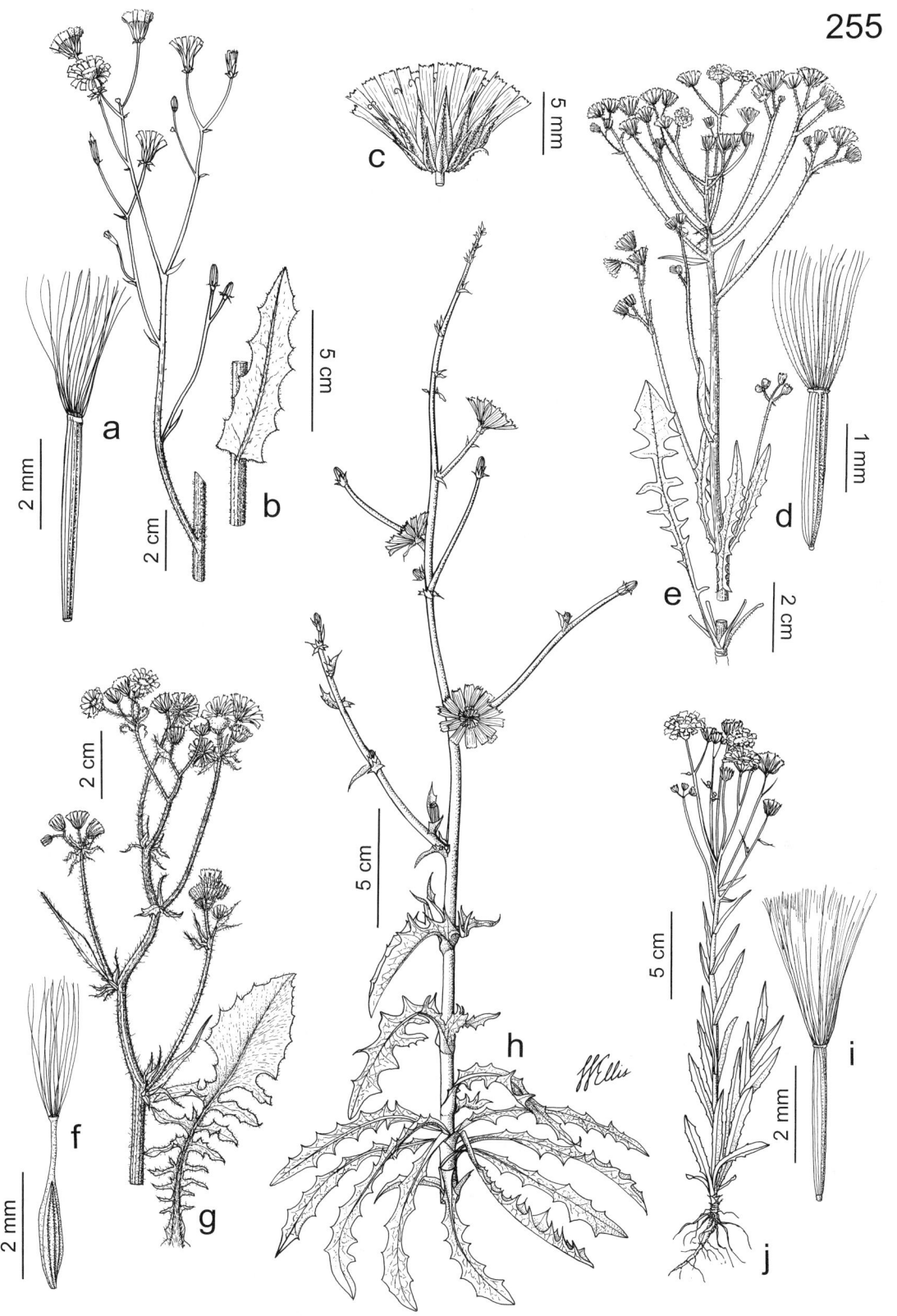

Plate 255. Asteraceae. *Crepis pulchra*, **a)** fruit, **b)** inflorescence branch and leaf. *Crepis capillaris*, **c)** head, **d)** fruit, **e)** habit and larger leaf. *Crepis setosa*, **f)** fruit, **g)** inflorescence and lower leaf. *Cichorium intybus*, **h)** habit. *Crepis tectorum*, **i)** fruit, **j)** habit.

commonly short-petiolate, the blade usually irregularly pinnately lobed, the lobes spreading, mostly narrow, bluntly to sharply triangular, the surfaces glabrous or sparsely and inconspicuously pubescent with minute, curled to cobwebby hairs. Inflorescences usually appearing paniculate, less commonly the heads appearing in loose clusters at the branch tips. Inner series of involucral bracts 12–15, 6–9 mm long, finely appressed-hairy on the inner surface, the outer surface glabrous or more commonly pubescent with minute, cobwebby hairs, often also with a longitudinal band of short, spreading, gland-tipped hairs, these with somewhat broadened bases, nearly transparent to light brownish yellow, the outer series of bracts much shorter than to about ½ as long as the inner series. Receptacle with minute hairs around the base of each floret. Ligulate florets 30–70, the corolla 9–13 mm long. Pappus 4–5 mm long. Fruits all similar or nearly so, 2.5–4.0 mm long, narrowly oblong-elliptic in outline, not beaked or only slightly short-beaked, not expanded at the tip, 10(–12)-ribbed, the ribs usually minutely roughened or barbed, reddish brown to dark brown, with an abrupt, lighter yellowish tip. $2n=8$. June–July.

Introduced, uncommon, known thus far from Jefferson County and the city of St. Louis (native of Europe, Asia; introduced sporadically in the U.S. [mostly in northern states], Canada). Railroads and open, disturbed areas.

This species was first reported by Steyermark (1963) in the supplement to his *Flora of Missouri*, based on a collection by Viktor Mühlenbach from the St. Louis railyards.

35. Helminthotheca Zinn (ox-tongue)

About 5 species, Europe, Africa.

Traditionally, *Helminthotheca* has been treated as a specialized group within *Picris* by most authors (Barkley, 1986; Gleason and Cronquist, 1991), but Lack (1974) demonstrated that it is separable from that genus based primarily upon its unusual involucral bracts and beaked achenes, as well as some micromorphological traits. Preliminary molecular studies (Whitton et al., 1995) have placed *Helminthotheca* as the most primitive member of the *Picris* lineage, but taxon sampling has been far too incomplete for this conclusion to be well supported.

1. Helminthotheca echioides (L.) Holub
(bristly ox-tongue)

Picris echioides L.

Pl. 260 j, k; Map 1066

Plants annual or biennial, usually taprooted. Latex white. Stems solitary or few, 20–80 cm long, erect or ascending, few- to several-branched, finely longitudinally ridged, moderately to densely pubescent with white, spreading hairs, these often somewhat flattened and expanded or swollen at the base, all or most of these barbed at the tip with 2–5 minute, spreading to recurved branches from a knoblike tip. Leaves basal and alternate, not grasslike, the basal leaves often present at flowering and tapered to a sessile or short- to less commonly long-petiolate base, the stem leaves sessile. Leaf blades 2–25 cm long, oblanceolate to elliptic, lanceolate, or narrowly oblong-lanceolate, mostly with rounded basal lobes somewhat clasping the stem, unlobed, the margins entire or with shallow, spreading, broadly triangular teeth and rounded sinuses, both surfaces moderately to densely pubescent with white, spreading hairs, all or most of these barbed at the tip with 2–5 minute, spreading to recurved branches from a knoblike tip, the margins and surfaces also with sparse to moderately dense, short, white to cream-colored prickles. Venation of 1 main vein and sometimes very faint, arching secondary veins. Inflorescences appearing paniculate, the heads mostly in small clusters at the tips of leafy branches, occasionally solitary at the branch tips. Involucre 10–20 mm long at flowering, not or only slightly elongating at fruiting, cup-shaped to broadly urn-shaped, the bracts in 2 series, moderately to densely pubescent with white, spreading hairs (those of the inner series glabrous on the inner surface), all or most of these barbed at the tip with 2–5 minute, spreading to recurved branches from a knoblike tip, the margins, tip, and often midvein also with sparse to moderately dense, short, white to cream-colored prickles; those of the inner series 7–13, lanceolate to narrowly lanceolate, thinner and somewhat papery, ascending at flowering, those of the outer series 3–5, slightly shorter than to slightly longer than those of the inner series, ovate to broadly ovate, thick and leathery, green, loosely ascending at flowering. Receptacle naked, shallowly pitted at the base of each floret.

Ligulate florets 30–80. Corollas 10–17 mm long, bright yellow to deep yellow, often somewhat reddish-tinged on the outer surface. Pappus of numerous plumose bristles, white to pale straw-colored, 4–7 mm long, often somewhat shorter on the marginal florets than on the central florets. Fruits 4–8 mm long (including the beak), the body 2.0–2.7 mm long, narrowly oblong-elliptic in outline, somewhat flattened (oval in cross-section), tapered abruptly to a slender beak 1–2 times as long as the body, the pappus attached to an expanded, disclike tip, the body with 5–8 faint, longitudinal nerves and a network of fine, raised cross-wrinkles, otherwise glabrous or the marginal fruits inconspicuously hairy, orangish brown to yellowish brown. $2n=10$. July–September.

Introduced, known thus far only from St. Louis (native of Europe, Africa; introduced widely in temperate and tropical regions, including sporadically in the western, midwestern, and northeastern U.S., Canada). Railroads.

This species was first reported for Missouri by Mühlenbach (1979) based on his botanical surveys of the St. Louis railyards.

36. Hieracium L. (hawkweed)

Plants perennial herbs, with rhizomes or a short, hard rootstock and fibrous roots. Latex white. Stems usually solitary (sometimes few to several in *H. caespitosum*), erect or strongly ascending, unbranched below the inflorescence, finely ridged, often noticeably spreading-hairy (the hairs with a bulbous or slightly expanded base), at least toward the base, sometimes also with minute, stellate hairs and/or gland-tipped hairs. Leaves basal and sometimes also alternate, hairy or sometimes nearly glabrous, sessile or short- to long-petiolate. Leaf blades unlobed, narrowly oblanceolate to obovate, those of the uppermost leaves sometimes oblong-lanceolate to oblong-ovate, the margins otherwise entire, with 1 main vein visible and often also a pinnate pattern of secondary veins (these sometimes forming a network), occasionally also a faint network of anastomosing tertiary veins. Stem leaves gradually reduced in size, sometimes only produced toward the stem base, without a pair of narrowly triangular, clasping lobes at the base. Inflorescences terminal panicles or racemes, sometimes appearing as loose clusters at the stem tip. Involucre not or only slightly elongating as the fruits mature, narrowly to broadly cup-shaped, the bracts 20–35 or more, arranged variously, generally in 1 or 2 inner series and usually 1 or more additional shorter, outer series, the inner series more or less similar in size, the narrow margins sometimes thin and pale, the tip ascending at flowering. Receptacle with minute, broadly triangular scales around the base of each floret, these fused into an irregular low ridge or wing. Ligulate florets 20 to more than 100 per head. Corollas light yellow to bright yellow (orange to reddish orange elsewhere). Pappus of more or less numerous bristles, these appearing smooth but microscopically barbed, white or straw-colored to light yellowish to orangish brown. Fruits nearly cylindrical to narrowly oblong-elliptic in outline, not beaked, not flattened, circular or finely angled in cross-section, with (8–)10 longitudinal ribs, these appearing smooth or nearly so (microscopically cross-wrinkled), glabrous, purplish brown to nearly black, the pappus attached to a relatively broad but not or only slightly expanded tip. About 100 to many more than 1,000 species, North America to South America, Caribbean Islands, Europe, Asia, Africa.

The species of *Hieracium* native to North America are generally well behaved biologically, aside from occasional interspecific hybridization. However, in the Old World, particularly in Europe, where polyploidy and apomixis are pervasive, the formal taxonomic description of various minor forms and races as so-called microspecies has led to an incredible nomenclatural proliferation involving thousands of published species epithets. This has tended to preclude precise estimates of species numbers in the genus.

Unfortunately, several of the Old World species have become important weeds of pastures and natural grasslands in the United States. Thus far, only one of these has begun to make inroads into Missouri (see below).

1. Stems and leaves with the spreading hairs mostly 10–20 mm long
   .... 3. H. LONGIPILUM
1. Stems and leaves with the spreading hairs 1–9 mm long
    2. Fruits 1.5–2.0 mm long; pappus bristles white; stem leaves only 1 or 2 toward the stem base; plants frequently producing stolons, often appearing in dense patches of rosettes .... 1. H. CAESPITOSUM
    2. Fruits 2–4 mm long; pappus bristles light yellowish to orangish brown; stem leaves often 3 or more, often extending above the stem base for some distance (sometimes few and nearly basal in *H. gronovii*); plants not producing stolons, not appearing colonial
        3. Basal leaves usually present at flowering; ligulate florets 20–40; fruits narrowly ellipsoidal, somewhat tapered to a slightly expanded tip
           .... 2. H. GRONOVII
        3. Basal leaves withered at flowering; ligulate florets 40–100 or more; fruits more or less cylindrical, not tapered or narrowed at the unexpanded tip
           .... 4. H. SCABRUM

1. Hieracium caespitosum Dumort. (yellow king-devil)

Map 1067

Plants with short to long, spreading rhizomes and also usually with stolons, thus frequently forming colonies. Stems solitary or few to several, 25–100 cm long, moderately to densely pubescent toward the base with light orangish brown, spreading to loosely ascending hairs 2–4 mm long having a bulbous or slightly expanded base, these becoming sparse or absent toward the tip, also inconspicuously pubescent with minute, branched hairs, especially toward the tip, also with moderate to dense, dark-colored, gland-tipped hairs toward the tip. Basal leaves persistent at flowering, sessile or with a short, indistinct, winged petiole, the blade 4–25 cm long, narrowly oblanceolate, sharply pointed at the tip, the surfaces and margins pubescent with moderate to dense, spreading, bulbous-based hairs (these often relatively dark-colored) and sparse to moderate, minute, branched hairs. Stem leaves only 1 or 2 toward the stem base, similar to but shorter than the basal leaves, mostly sessile, narrowly oblanceolate to linear, the base not clasping the stem. Inflorescences mostly short, spreading panicles, sometimes reduced to a loose or dense terminal cluster of few to several heads. Involucre 6–9 mm long, the inner series of bracts narrowly oblong-elliptic, pubescent with spreading, dark-colored, gland-tipped hairs and usually also inconspicuous, minute, cobwebby, branched hairs, the outer series variable and grading into the inner series, some of the bracts more than ½ as long as those of the inner series. Ligulate florets 25–70. Corollas 8–14 mm long, bright yellow. Pappus bristles 4–6 mm long, white. Fruits 1.5–2.0 mm long, more or less cylindrical, not tapered at the tip. 2n=18, 27, 36, 45. June–September.

Introduced, known thus far only from Franklin County (native of Europe, introduced widely in the eastern and northwestern U.S., Canada). Open, grassy areas.

This species was first collected in 1993 by Jane Stevens (now curator of insects at the St. Louis Zoo) in a meadow at the Shaw Nature Reserve (then known as the Shaw Arboretum) in Gray Summit, where it had not been planted and was well naturalized. Efforts to control it have since been undertaken.

2. Hieracium gronovii L. (beaked hawkweed)

H. gronovii var. *foliosum* Michx.

H. floridanum Britton

Pl. 256 a, b; Map 1068

Plants with a short, usually erect or ascending rootstock, rarely with a short, spreading rhizome. Stems mostly solitary, 30–85(–120) cm long, moderately to densely pubescent, at least toward the base, with light orangish brown, spreading to loosely ascending hairs 4–9 mm long having a bulbous or slightly expanded base, these sometimes becoming sparse or nearly absent toward the tip, usually also inconspicuously pubescent with cobwebby, minute, branched hairs, often also with sparse, gland-tipped hairs toward the tip. Basal leaves usually persistent at flowering, sessile to short-petiolate, the blade 3–20 cm long, narrowly oblanceolate to obovate, rounded to sharply pointed at the tip, the surfaces and margins pubescent with sparse to moderate, spreading, bulbous-based hairs and often also sparse to moderate, minute, inconspicuous, branched hairs. Stem leaves sometimes few and nearly basal, sometimes several and well spaced, similar to the basal leaves but gradually reduced in size, more often sessile, oblong-obovate

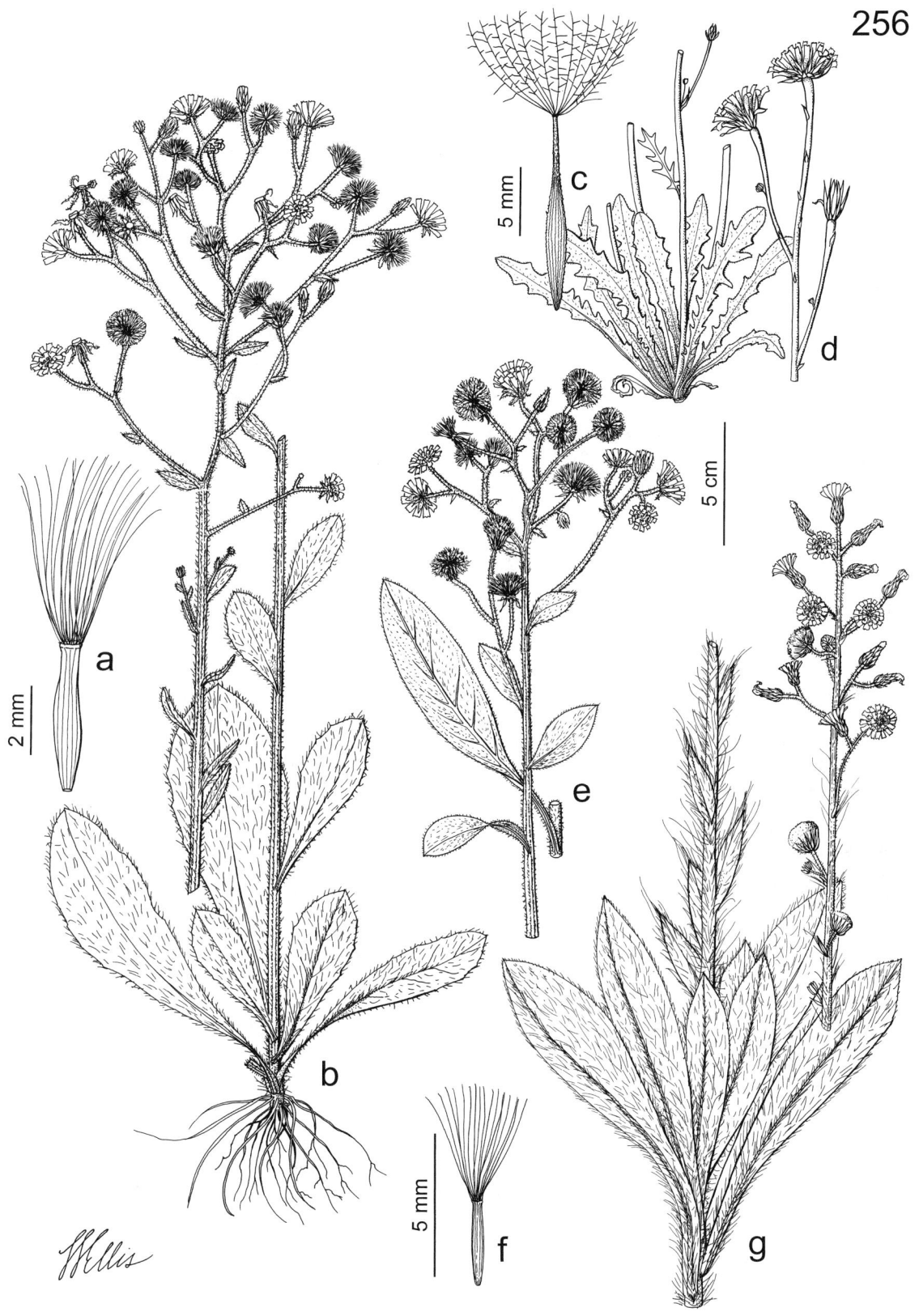

Plate 256. Asteraceae. *Hieracium gronovii*, **a)** fruit, **b)** habit. *Hypochaeris radicata*, **c)** fruit, **d)** habit. *Hieracium scabrum*, **e)** habit. *Hieracium longipilum*, **f)** fruit, **g)** habit.

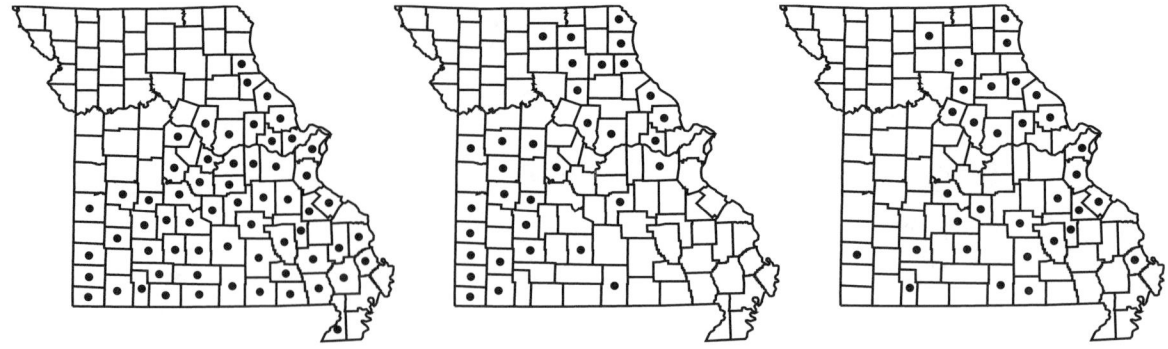

1068. Hieracium gronovii   1069. Hieracium longipilum   1070. Hieracium scabrum

to oblanceolate, the uppermost with the base often slightly clasping the stem. Inflorescences elongate cylindrical panicles or racemes. Involucre 6–9 mm long, the inner series of bracts narrowly oblong-lanceolate, pubescent with inconspicuous, cobwebby, branched hairs and sometimes also longer, spreading, usually dark-colored, gland-tipped hairs, the outer series much less than ½ as long as the inner series. Ligulate florets 20–40. Corollas 7–9 mm long, light yellow to yellow. Pappus bristles 4–5 mm long, light yellowish to orangish brown. Fruits 2.5–4.0 mm long, somewhat tapered to the slightly expanded tip. $2n=18$. May–October.

Scattered, mostly south of the Missouri River, absent from all but the eastern portion of the Glaciated Plains Division (eastern U.S. west to Illinois, Kansas, and Texas; Canada, Mexico, Central America, Caribbean Islands). Mesic to dry upland forests, ledges and tops of bluffs, and borders of glades; also pastures, old fields, and roadsides.

There has been some controversy about the correct application of the name *H. gronovii*, but the epithet was conserved at an international botanical congress with a type specimen that fixes the application to the traditionally accepted widespread species (Greuter et al., 2000). Specimens with more stem leaves and relatively few-flowered inflorescences occasionally have been misdetermined as *H. scabrum*.

Steyermark (1963) corrected earlier reports for Missouri of the eastern *H. venosum* L., annotating the several specimens in question to *H. gronovii* based on their apically somewhat tapered (vs. cylindrical) fruits. This was confirmed during the present study. *Hieracium venosum* should continue to be excluded from the Missouri flora. *Hieracium gronovii* is relatively variable in plant height, as well as number and size of the stem leaves. One of these specimens, a historical collection from Newton County, has unusually narrow leaves and relatively long, spreading hairs. Possibly, it represents a hybrid with *H. longipilum*. In general, the two species do not grow together, but they are known to hybridize regularly in other parts of their ranges.

3. Hieracium longipilum Torr. (long-haired hawkweed)

Pl. 256 f, g; Map 1069

Plants with a short, ascending to spreading, occasionally branched rootstock. Stems mostly solitary, 50–170 cm long, densely pubescent with white to light orangish brown, spreading to loosely ascending hairs (8–)10–20 mm long (rarely longer) having a bulbous or slightly expanded base, these becoming sparser toward the tip, usually also inconspicuously pubescent with cobwebby, minute, branched hairs, often also with sparse, gland-tipped hairs toward the tip. Basal leaves often persistent at flowering, mostly short-petiolate, the blade 5–30 cm long, narrowly oblanceolate to oblanceolate, rarely obovate, rounded to sharply pointed at the tip, the surfaces and margins pubescent with dense, more or less spreading, bulbous-based hairs (these sometimes somewhat shorter than those of the stem) and rarely also sparse, minute, inconspicuous, branched hairs. Stem leaves usually several, similar to the basal leaves but gradually reduced in size, more often sessile, oblanceolate to narrowly oblanceolate, the base usually not clasping the stem. Inflorescences usually elongate cylindrical panicles, occasionally only spikelike racemes. Involucre 7–10 mm long, the inner series of bracts narrowly oblong-lanceolate, pubescent with inconspicuous, cobwebby, branched hairs and longer, spreading, usually dark-colored, gland-tipped hairs, the outer series variable and grading into the inner series, some of the bracts more than ½ as long as those of the inner series. Ligulate florets 40–90. Corollas 7–9 mm long, yellow. Pappus bristles 5–7 mm long, light yellowish to orangish brown. Fruits 3.0–4.5 mm long, somewhat tapered to the slightly expanded tip. $2n=18$. May–October.

Scattered in a broad band from southwestern to northeastern Missouri, absent from most of the northwestern and southeastern quarters of the state (Minnesota to Texas east to Ohio, Tennessee, and Louisiana; Canada). Upland prairies and openings of dry upland forests; also old fields, railroads, and roadsides.

Plants with the involucres mostly or entirely pubescent with inconspicuous, nonglandular hairs have been called f. *eglandulosum* E.J. Palmer & Steyerm., but, as noted by Deardorff (1977) and others, patterns of pubescence are too complex and variable in this genus to provide stable characters for subdivision of species.

4. Hieracium scabrum Michx. (sticky hawkweed)

H. scabrum var. *intonsum* Fernald & H. St. John

Pl. 256 e; Map 1070

Plants with a short, usually erect or ascending rootstock. Stems mostly solitary, 20–80 cm long, moderately to densely pubescent toward the base, with light yellow to orangish brown, spreading to loosely ascending hairs 1–5 mm long having a bulbous or slightly expanded base, these becoming sparse or absent toward the tip, also inconspicuously pubescent with cobwebby, minute, branched hairs especially toward the tip, sometimes also with sparse, gland-tipped hairs toward the tip. Basal leaves withered or absent at flowering, when present (before the flower stem elongates) mostly short-petiolate, the blade 4–20 cm long, oblanceolate to obovate, rounded to broadly pointed at the tip (sometimes with an abrupt, minute, sharp point), the surfaces and margins pubescent with sparse to moderate, spreading, bulbous-based hairs. Stem leaves several and well spaced, similar to the basal leaves but progressively reduced in size, mostly sessile, oblong-obovate to broadly oblanceolate, the base usually not clasping the stem. Inflorescences mostly panicles, these short and few-headed in smaller plants but elongate and cylindrical in larger plants, occasionally only a loose terminal cluster of 2–5 heads. Involucre 6–9 mm long, the inner series of bracts narrowly oblong-lanceolate, pubescent with inconspicuous, cobwebby, branched hairs toward the base and usually also longer, spreading, usually dark-colored, gland-tipped hairs, the outer series variable and grading into the inner series, some of the bracts more than $1/2$ as long as those of the inner series. Ligulate florets 40–100 or more. Corollas 9–11 mm long, yellow. Pappus bristles 6–7 mm long, light yellowish to orangish brown. Fruits 2–3 mm long, more or less cylindrical, not tapered at the tip. $2n=18$. June–September.

Scattered to uncommon, mostly in the eastern half of the state, apparently absent from the Unglaciated Plains Division and the western portion of the Glaciated Plains (eastern U.S. west to Minnesota and Oklahoma). Mesic to dry upland forests, ledges and tops of bluffs, and rarely banks of streams; also rarely roadsides.

Uncommonly encountered plants with slightly longer (3–5 mm) stem hairs have been called var. *intonsum,* but too many intermediates exist to divide the species into varieties based solely on this character.

37. Hypochaeris L. (cat's ear)
Contributed by David J. Bogler and George Yatskievych

Fifty to 60 species, most diverse in South America, but also native to Europe, Asia, Africa. The generic name has been spelled *Hypochoeris* in some of the older literature, based on the inconsistent spellings adopted by Linnaeus in his various publications (Harriman, 1980).

1. Hypochaeris radicata L. (spotted cat's ear)

Pl. 256 c, d; Map 1071

Plants perennial herbs, with a thickened, vertical rootstock and 1 or few often thickened main roots. Latex white. Stems 1 or more commonly few to numerous, 10–60 cm long, erect or ascending, usually few-branched above the midpoint, finely longitudinally ridged, glabrous or sparsely to moderately pubescent with white, somewhat spreading hairs, these often somewhat flattened and gradually expanded at the base. Leaves all basal (except for much-reduced, mostly scalelike, bracteal leaves along the branches), sessile or with a short, winged petiole. Leaf blades 3–35 cm long, narrowly oblong to oblanceolate, ranging from pinnately several-lobed with bluntly triangular to oblong-rounded lobes and rounded sinuses to unlobed but with few to several broad, shallow, spreading teeth, both surfaces moderately to densely pubescent with relatively coarse, white, spreading hairs. Venation of 1 main vein and sometimes very faint, arching secondary veins. Heads solitary at the

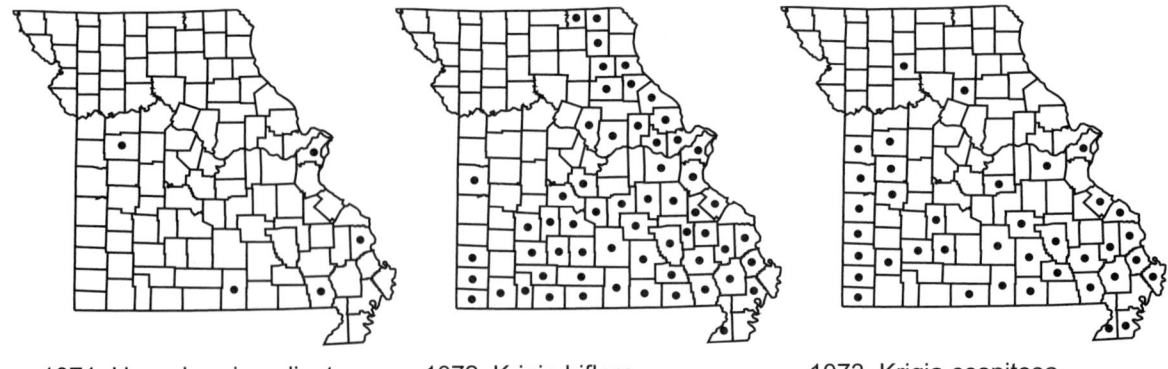

1071. Hypochaeris radicata   1072. Krigia biflora   1073. Krigia cespitosa

branch tips (sometimes appearing loosely paniculate), appearing long-stalked. Involucre 10–15 mm long at flowering, elongating to 17–25 mm at fruiting, cup-shaped to broadly cylindrical, the bracts mostly 20–28 in 4–6 overlapping series, ovate to narrowly lanceolate or oblong-lanceolate, glabrous or sparsely to moderately hairy along the midnerve, green, the margins often thin and pale, the tip appressed-ascending, often with a minute fringe of hairs, sometimes purplish-tinged. Receptacle with a chaffy bract wrapped around the base of each floret, these usually extending just past the pappus at flowering and elongating as the fruits mature, very narrowly triangular, white, papery. Ligulate florets 25–55. Corollas 10–16 mm long, bright yellow, occasionally somewhat grayish-tinged on the outer surface. Pappus 10–12 mm long, of numerous bristles, the outer series shorter and finely barbed, the others longer and plumose, straw-colored, persistent at fruiting. Fruits 8–15 mm long (including the beak), the body narrowly oblong-elliptic in outline, not flattened, 10–14-ribbed, tapered at the tip to a slender beak about as long as the body, the pappus attached to the expanded, disclike tip, the surface roughened with minute, ascending tubercles or barbs, otherwise glabrous, brown to reddish brown. $2n=8$. May–August.

Introduced, uncommon and sporadic in the southern half of the state (native of Europe, Asia, introduced widely in the U.S. [including Alaska, but excluding the Great Plains]; Canada). Lawns, railroads, roadsides, and open, disturbed areas.

A second species has become naturalized sporadically in the southern states and is known from both Arkansas and southern Illinois. *Hypochaeris glabra* L. (smooth cat's ear) is an annual with smaller, more glabrous leaves, finer stems, and smaller, less showy heads than those of *H. radicata*. This European species also has dimorphic fruits, with the outermost florets producing beakless achenes and those produced by the remaining florets having long, slender beaks.

38. Krigia Schreb. (dwarf dandelion)

Plants annual or perennial herbs, taprooted or fibrous-rooted (producing tubers in *K. dandelion*). Latex white. Stems erect or ascending, sometimes loosely ascending from a spreading base, finely ridged, glabrous or hairy. Leaves all basal or basal and alternate, glabrous or hairy, sessile or short- to long-petiolate. Leaf blades entire or pinnately lobed, linear to lanceolate, ovate, or oblong-ovate in outline, the margins entire or with widely spaced, broadly triangular, spreading teeth, with 1 main vein visible and sometimes also a faint network of anastomosing secondary veins. Inflorescences of solitary flowers at the stem or branch tips or open terminal clusters, rarely appearing as small panicles in *K. biflora*. Involucre not or only slightly becoming elongated as the fruits mature, cup-shaped or somewhat bell-shaped, the bracts in 1 or 2 series, all similar in size and shape (no shorter, outer bracts present), narrowly lanceolate to oblong-ovate, often purplish-tinged, the margins sometimes thin and pale, the tip ascending at flowering. Receptacle naked, usually minutely pitted at the base of each floret. Ligulate florets 6–60 per head. Corollas yellow to orange. Pappus of 5 to numerous bristles and/or 5 or (8–)10 scales or absent. Fruits variously shaped, not flattened, sometimes 5-angled in cross-section, often somewhat oblique at the base, with 10–20 fine, longitudinal nerves or ridges, usually also finely cross-wrinkled, glabrous, reddish brown to dark brown, not beaked, the pappus attached to a sometimes slightly expanded tip. Seven species, U.S. and adjacent Canada, Mexico.

1. Involucral bracts 4–7(–9), lanceolate to ovate or oblong-ovate, persistently scalelike and remaining ascending with age (sometimes eventually withering in place); pappus absent or with the longest bristles 1–2 mm long
    2. Leaves basal and alternate along the stems; pappus absent .... 2. K. CESPITOSA
    2. Leaves all basal; pappus present, of 5 short bristles alternating with 5 shorter scales .... 4. K. OCCIDENTALIS
1. Involucral bracts 8–16, narrowly lanceolate to oblong-lanceolate, shriveling and becoming strongly reflexed with age; pappus with the longest bristles 4–8 mm long
    3. Involucre 4–7 mm long; pappus of 5 bristles alternating with 5 shorter scales; plants annual .... 5. K. VIRGINICA
    3. Involucre 7–15 mm long; pappus of 20–45 bristles and 10(–15) short, inconspicuous scales; plants perennial
        4. Stems with at least 1 leaf above the base, few-branched toward the tip; corollas yellowish orange to orange; plants not producing tubers
        .... 1. K. BIFLORA
        4. Leaves all basal, the stems usually unbranched; corollas lemon yellow to orangish yellow; plants producing small, globose to oblong-ovoid tubers
        .... 3. K. DANDELION

1. Krigia biflora (Walter) S.F. Blake **var. biflora** (two-flowered Cynthia, orange dwarf dandelion)

K. biflora f. glandulifera Fernald

Pl. 257 j, k; Map 1072

Plants perennial, the short rootstock with fibrous roots, the main roots often slightly fleshy. Stems 1 to few, 10–60 cm long, erect or ascending, with few to several ascending branches from above the midpoint, glabrous or sparsely to moderately pubescent toward the branch tips with spreading, gland-tipped hairs, occasionally somewhat glaucous. Leaves basal and usually 1–5 alternate along the lower stems, the primary (and sometimes also secondary) branch point with 1 or 2 well-developed, bractlike leaves, the basal and lower stem leaves mostly short- to more commonly long-petiolate (the petiole sometimes winged), those higher on the stem abruptly different, sessile with rounded, clasping bases. Blades of basal and lower stem leaves 2–10 cm long, broadly obovate to oblong-oblanceolate, entire, wavy, toothed, or uncommonly shallowly to deeply pinnately lobed, the teeth spreading, mostly broadly triangular-pointed, the lobes usually oblong-rounded, the leaf tip rounded to bluntly pointed, the surfaces glabrous, usually at least the undersurface glaucous. Blades of median and upper stem leaves similar to the others but linear to narrowly oblong-ovate. Involucral bracts 10–18, 7–11 mm long, lanceolate, flat (not keeled), glabrous, withering and becoming reflexed with age. Ligulate florets 25–60. Corollas 12–25 mm long, yellowish orange to orange. Pappus of 20–40 bristles and 10(–15) short, inconspicuous scales, the bristles 4.0–5.5 mm long, off-white to very light tan, the scales 0.3–0.5 mm long, mostly lanceolate, transparent. Fruits 2–3 mm long, more or less cylindrical (slightly expanded at the tip, slightly tapered at the base), more or less circular in cross-section, with 12–15 blunt, broad ribs, these microscopically roughened or barbed, reddish brown to dark brown. $2n=10, 20$. May–August.

Scattered in the Ozark and Ozark Border Divisions and the eastern portion of the Glaciated Plains; uncommon in the Mississippi Lowlands Division (eastern U.S. and adjacent Canada west to Minnesota, Kansas, and Oklahoma). Mesic to dry upland forests, upland prairies, margins of ponds and sinkhole ponds, and banks of streams; also pastures and roadsides.

Plants that occur disjunctly in portions of Arizona, Colorado, and New Mexico consistently have slightly smaller leaves and shorter stems than those in the main portion of the species' distribution. These have been segregated as var. *viridis* (Standl.) K.-J. Kim, although whether these differences are due to environmental or genetic variation has not been well studied.

1074. Krigia dandelion   1075. Krigia occidentalis   1076. Krigia virginica

2. Krigia cespitosa (Raf.) K.L. Chambers **ssp. cespitosa** (common dwarf dandelion)
Serinia cespitosa Raf.
S. oppositifolia (Raf.) Kuntze (not a valid name)

Pl. 257 f, g; Map 1073

Plants annual, with fibrous roots. Stems 1 to several, 5–45 cm long, erect to loosely ascending, sometimes from a spreading base, with few to several ascending branches, glabrous or rarely sparsely to moderately pubescent toward the branch tips with spreading, gland-tipped hairs, occasionally somewhat glaucous. Leaves basal and alternate along the stems, the uppermost few sometimes appearing opposite, the basal leaves often with short to long, winged petioles, the stem leaves mostly sessile with rounded, clasping bases. Leaf blades 1–15 cm long, linear to narrowly oblanceolate, the first rosette leaves sometimes ovate, entire, wavy, toothed, or shallowly to deeply pinnately lobed, the teeth or lobes spreading, oblong and rounded or triangular and pointed, the leaf tip rounded to more commonly pointed, the surfaces glabrous, occasionally somewhat glaucous. Involucral bracts 4–7(–9), 2–5 mm long, lanceolate to ovate or oblong-ovate, becoming somewhat keeled as the fruits mature, glabrous, persistently scalelike and remaining ascending with age. Ligulate florets 12–25(–35). Corollas 3–7 mm long, light yellow to yellow. Pappus absent. Fruits 1.4–1.7 mm long, narrowly obovoid, more or less circular in cross-section, with usually 15 ribs, these minutely roughened or barbed, reddish brown. $2n=8$. April–June.

Scattered, mostly south of the Missouri River (eastern U.S. west to Nebraska and Texas; Mexico). Bottomland prairies, upland prairies, glades, fens, openings of mesic to dry upland forests, and margins of ponds, lakes, and sloughs; also pastures, fallow fields, old fields, banks of ditches, cemeteries, railroads, roadsides, and open, disturbed areas.

Steyermark and other earlier authors used the name *Serinia oppositifolia* (Raf.) Kuntze for this taxon. Shinners (1947), Vuilleumier (1973), Kim and Turner (1992), and Chambers (2004) all presented evidence that *Serinia* is merely a specialized kind of *Krigia* lacking a pappus. Chambers (1973, 2004) further noted that the nomenclatural basis for this name, *K. oppositifolia* Raf., was not validly published, because Rafinesque did not accept it as a valid name in his own publication. Also, because the species epithet originally was spelled *K. cespitosa,* the later variant *K. caespitosa* that is found in some botanical manuals (Barkley, 1986) currently is considered incorrect.

Some populations in Louisiana, Oklahoma, and Texas with slightly larger heads and slightly longer corollas have been called *K. gracilis* DC., *K. cespitosa* f. *gracilis* (DC.) K.-J. Kim, or *K. cespitosa* ssp. *gracilis* (DC.) K.L. Chambers.

3. Krigia dandelion (L.) Nutt. (potato dandelion)

Pl. 257 h, i; Map 1074

Plants perennial, mostly with fibrous roots and slender, sometimes nearly vertical rhizomes producing 1 or a few globose to ovoid, potato-like tubers 5–15 mm in diameter. Stems 1 to few, 10–50 cm long, erect or ascending, unbranched, glabrous or sparsely to moderately pubescent with spreading, gland-tipped hairs, especially toward the tip. Leaves basal, sessile or with a short to less commonly long, winged petiole. Leaf blades 2–25 cm long, linear to narrowly lanceolate or narrowly oblanceolate, entire, wavy, toothed, or shallowly to deeply pinnately lobed, the teeth or lobes spreading, broadly triangular and pointed or occasionally oblong and rounded, the leaf tip rounded to more commonly sharply pointed, the surfaces glabrous or rarely sparsely pubescent with short, appressed hairs, usually glaucous. Involucral bracts (12–)14–16, 10–15 mm long, narrowly lanceolate to narrowly

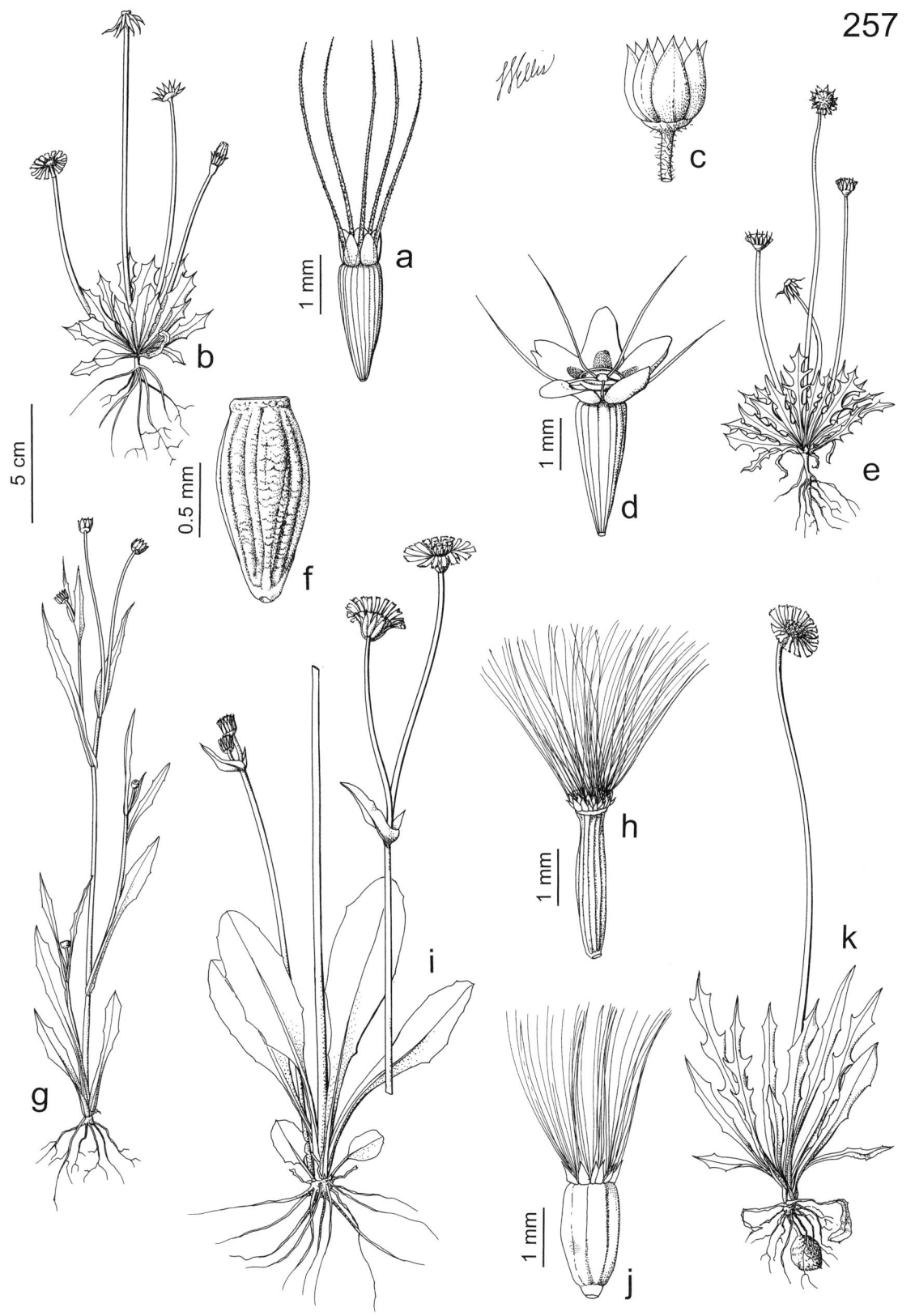

Plate 257. Asteraceae. *Krigia virginica*, **a)** fruit, **b)** habit. *Krigia occidentalis*, **c)** fruiting head, **d)** fruit, **e)** habit. *Krigia cespitosa*, **f)** fruit, **g)** habit. *Krigia dandelion*, **h)** fruit, **i)** habit. *Krigia biflora*, **j)** fruit, **k)** habit.

oblong-lanceolate, flat (not keeled), glabrous, withering and becoming reflexed with age. Ligulate florets 25–35. Corollas 12–25 mm long, lemon yellow to orangish yellow, those of the outer florets sometimes purplish-tinged on the outer surface. Pappus of 25–45 bristles and 10–15 short, inconspicuous scales, the bristles 5–8 mm long, white or nearly so to very light tan, the scales 0.5–1.0 mm long, irregularly lanceolate to oblanceolate (rarely broader), transparent but sometimes with a darker central line or stripe. Fruits 2.2–2.7 mm long, more or less cylindrical (slightly expanded at the tip, slightly tapered at the base), more or less circular in cross-section, with 12–15 blunt, broad ribs, these microscopically roughened or barbed, reddish brown to dark brown. $2n=60$. April–June.

Scattered, mostly south of the Missouri River (eastern U.S. west to Iowa and Texas). Upland prairies, glades, ledges and tops of bluffs, openings of mesic to dry upland forests, margins of sinkhole ponds, and banks of streams and rivers; also rarely disturbed areas.

4. Krigia occidentalis Nutt. (western dwarf dandelion)

Pl. 257 c–e; Map 1075

Plants annual, with fibrous roots. Stems 1 to few, 4–10(–16) cm long, erect or ascending, unbranched, glabrous or sparsely to moderately pubescent with spreading, gland-tipped hairs, especially toward the tip. Leaves basal, sessile or with a short to long, often winged petiole. Leaf blades 0.5–6.0 cm long, linear to oblanceolate or obovate, entire, wavy, toothed, or shallowly to deeply pinnately lobed, the teeth or lobes usually broadly triangular and pointed, the leaf tip rounded to sharply pointed, the surfaces glabrous or the undersurface (and margins) sometimes sparsely pubescent with short, spreading hairs, especially toward the base, these often gland-tipped, sometimes slightly glaucous. Involucral bracts 4–7, 2.5–6.5 mm long, narrowly lanceolate to oblong-lanceolate, fused toward the base, becoming somewhat keeled or developing 2 or 3 ribs as the fruits mature, glabrous, more or less persistently scalelike and remaining ascending with age (eventually withering in place). Ligulate florets 6–25. Corollas 4–9 mm long, yellow to orangish yellow. Pappus of 5 bristles and 5 short, inconspicuous scales, the bristles 1–2 mm long, white or nearly so to pale straw-colored, the scales 0.4–0.6 mm long, broadly oblong with a broadly rounded tip, somewhat transparent, white. Fruits 1.2–1.8 mm long, somewhat obconical, 4- or 5-angled in cross-section, with 10–15 blunt ribs, these microscopically roughened or barbed, reddish brown to dark brown. $2n=12$. April–May.

Uncommon in the southwestern portion of the Ozark Division (Missouri to Kansas south to Louisiana and Texas). Glades and openings of dry upland forests.

This rarely seen species is easily confused with the superficially similar and much more abundant *K. virginica,* which can occur in the same habitat and even at the same site. The fewer, somewhat keeled, persistently ascending involucral bracts are the best way to distinguish the two taxa.

5. Krigia virginica (L.) Willd. (Virginia dwarf dandelion)

Pl. 257 a, b; Map 1076

Plants annual, with fibrous roots. Stems (see discussion below) 1 to numerous, 4–35 cm long, erect or ascending, rarely from a spreading base, unbranched or rarely few-branched from below the midpoint, sparsely to moderately pubescent with spreading, gland-tipped hairs, especially toward the tip. Leaves basal and rarely also 1–4 toward the stem base (see discussion below), sessile or with a short to long, often winged petiole. Leaf blades 0.4–18.0 cm long, linear to lanceolate, oblanceolate or broadly ovate to broadly obovate, entire, toothed, or shallowly to deeply pinnately lobed, the teeth or lobes broadly triangular and pointed, the leaf tip rounded to sharply pointed, the surfaces glabrous or the undersurface (and margins) sometimes sparsely pubescent with short, spreading hairs, these often gland-tipped, sometimes glaucous. Involucral bracts 8–14, 4–7 mm long, narrowly lanceolate to narrowly oblong-lanceolate, flat (not keeled), glabrous, withering and becoming reflexed with age. Ligulate florets 14–35. Corollas 4–12 mm long, yellow to less commonly yellowish orange, those of the outer florets sometimes purplish-tinged on the outer surface. Pappus of 5 bristles and 5 short, inconspicuous scales, the bristles 4–6 mm long, white or nearly so to very light tan or pale straw-colored, the scales 0.5–1.0 mm long, broadly oblong with an irregularly truncate to bluntly angled tip, somewhat transparent, white to light brown (especially toward the base). Fruits 1.5–2.2 mm long, somewhat obconical, 4- or 5-angled in cross-section, with 15–18 blunt ribs, these microscopically roughened or barbed, reddish brown to dark brown. $2n=10, 20$. April–August.

Scattered, mostly south of the Missouri River (eastern U.S. west to Wisconsin, Kansas, and Texas; Canada). Glades, sand prairies, swales in upland prairies, tops and exposed ledges of bluffs, and openings of mesic to dry upland forests; also fallow fields, old fields, roadsides, and open, disturbed areas.

Steyermark (1963) suggested that this plant is an attractive addition rock gardens with acidic soils and noted that it reseeds itself. Shinners (1947) noted that although this species is annual, on some occasions a second growth form is produced after flowering has finished and the original basal rosette leaves have withered. This secondary phase, which is quite rare in Missouri, has erect or strongly ascending stems from a sometimes few-branched, spreading to loosely ascending base that appear somewhat longer than the primary stems. It can produce 1–4 slender or short stem leaves toward the stem base that are alternate or sometimes nearly opposite. The heads produced during this supplementary cycle are usually smaller than those that form earlier in the year and have somewhat fewer florets, flowering from late May to late August in Missouri and even later farther south.

39. Lactuca L. (lettuce)
Contributed by Kuo-Fang Chung

Plants annual, biennial, or perennial herbs, usually taprooted, but often also with fibrous roots at maturity, occasionally with rhizomes. Latex white or light tan to pale orange. Stems erect or ascending, finely to less commonly coarsely ridged, glabrous or hairy. Leaves alternate and often also basal, mostly 4–20 times as long as wide or, if 2–3 times as long as wide (in *L. sativa*), then the blade strongly and irregularly crisped or curled and sessile or nearly so, glabrous or hairy, sessile or short-petiolate, the basal leaves mostly withered before flowering, the stem leaves slightly expanded or with a small pair of lobes and somewhat clasping the stem. Leaf blades unlobed or pinnately lobed, linear to lanceolate or ovate in outline, the margins entire or finely to coarsely toothed, sometimes irregularly so. Inflorescences spikes, racemes, or panicles. Involucre becoming elongated as the fruits mature, narrowly cylindrical to narrowly cup-shaped or urn-shaped, the bracts in 3–5 overlapping series, ovate to lanceolate or narrowly oblong-lanceolate, often purplish- or reddish-tinged, or purple-tipped, glabrous, the margins often thin and pale, the tip appressed-ascending or somewhat outward-curved. Receptacle naked, usually minutely pitted at the base of each floret. Ligulate florets 9–55 per head. Corollas yellow, blue, purple, or rarely white. Pappus of numerous apparently smooth (microscopically barbed) bristles, these white (grayish white elsewhere). Fruits with the body ovate to elliptic, lanceolate, or oblanceolate in outline, flattened, sometimes tapered to a short or long beak at the tip, the pappus attached to an expanded disc at the very tip, with 1 to several longitudinal nerves or ridges on each face, often also finely cross-wrinkled. Fifty to 75 species, North America, Central America, Europe, Asia, Africa.

The generic circumscription of *Lactuca* has been somewhat controversial, with some authors (Bremer, 1994) adopting a narrow generic concept in which (among other segregates) *L. tatarica* and its Old World relatives are classified in *Mulgedium* Cass. Based on earlier biosystematic studies (Dille, 1976), this species and its relatives had been suggested as possible ancestors of the North American allopolyploid species complex that includes *L. canadensis*, *L. floridana*, *L. hirsuta*, and *L. ludoviciana*. A study utilizing molecular data (Koopman et al., 1998) lent support to a relatively broad generic concept of *Lactuca* to include the *L. tatarica* group, which is adopted here.

1. Heads relatively large, the involucre 12–20 mm long at flowering, elongating to 15–23 mm at fruiting
    2. Corollas light purplish blue to blue; fruits with 3–5 nerves or ridges on each face . . . . 7. L. TATARICA
    2. Corollas yellow to orangish yellow, sometimes turning blue with age or upon drying; fruits with 1 nerve or ridge on each face
        3. Leaves best-developed toward the stem base, much smaller above the stem midpoint, the margins of both lobed and unlobed leaves sharply toothed and sparsely short-hairy but not prickly, also hairy but not prickly on the undersurface; florets (12–)15–20(–22) per head . . . . . 3. L. HIRSUTA

3. Leaves well developed along the stem, the margins of both lobed and unlobed leaves sharply toothed and glabrous but somewhat prickly, also glabrous but prickly on the undersurface midvein; florets 20–30(–55) per head . . . . 4. L. LUDOVICIANA
1. Heads relatively small, the involucre 7–10 mm long at flowering, elongating to 10–15(–18) mm at fruiting
    4. Corollas lavender to purplish blue or blue, rarely white; fruits beakless or with a short, stout beak much less than ½ as long as the body . . . . 2. L. FLORIDANA
    4. Corollas light yellow to orangish yellow or orange, sometimes turning blue with age or upon drying; fruits with a noticeable, slender beak somewhat shorter than to twice as long as the body
        5. Corollas yellowish orange to orange; fruits with 1 nerve or ridge on each face, tapered to a beak somewhat shorter than to about as long as the body; sap light tan to pale orange . . . . 1. L. CANADENSIS
        5. Corollas light yellow to bright yellow; fruits with 5–7 nerves or ridges on each face, tapered to a beak 1–2 times as long as the body; sap white
            6. Leaves with the margins and undersurface not prickly, the margins of both lobed and unlobed leaves lacking teeth; inflorescences spikelike panicles . . . . 5. L. SALIGNA
            6. Leaves with prickly margins and sharp teeth, also prickly on the undersurface midvein; inflorescences usually well-branched, open panicles . . . . 6. L. SERRIOLA

1. Lactuca canadensis L. (wild lettuce)

Pl. 258 a, b; Map 1077

Plants annual or biennial. Latex light tan to pale orange. Stems (30–)100–200(–300) cm long, hollow between the nodes, glabrous or rarely pubescent with short, curled hairs, often purple-spotted. Leaves well developed along the stems, extremely variable; the basal and lower stem leaves mostly 20–30 cm long, sessile or more commonly with a winged petiole, narrowly ovate, ovate, or obovate in outline, variously toothed and/or deeply pinnately lobed, the margins minutely hairy, sometimes with a pair of narrowly triangular basal lobes clasping the stem, the undersurface with the midvein often short-hairy; the middle and upper stem leaves mostly linear to lanceolate, ovate, or obovate, pinnately lobed to nearly entire, the margins minutely hairy or rarely glabrous, the base narrowed or tapered, sometimes with a pair of narrowly to broadly triangular basal lobes clasping the stem, the undersurface glabrous. Inflorescences mostly well-branched panicles with 50–100 or more heads. Involucre cylindrical or urn-shaped, 8–10 mm long at flowering, elongating to 10–14 mm at fruiting, the bracts 17(–19). Florets (10–)17–22 (–25). Corollas orangish yellow or orange (yellow elsewhere), occasionally reddish at the tip, sometimes turning blue with age or upon drying (rarely blue at flowering). Pappus 4–7 mm long. Fruits with the body 3–4 mm long, 1.5–2.0 mm wide, dark brown to black, flattened, with prominent lateral wings and a conspicuous ridge on each face, tapered abruptly to a slender beak somewhat shorter than to about as long as the body. $2n=34$. July–September.

Common throughout the state (eastern U.S. west to North Dakota and Texas; Canada; introduced farther west in the U.S.). Bottomland forests, openings and margins of mesic upland forests, savannas, sand savannas, bottomland prairies, upland prairies, sand prairies, banks of streams and rivers, margins of ponds, lakes, and sinkhole ponds; also pastures, fallow fields, old fields, fencerows, railroads, roadsides, and open, disturbed areas.

Corolla color is a relatively easy way to separate *L. canadensis* from the vegetatively similar *L. floridana*, however, readers are cautioned that corollas of *L. canadensis* sometimes darken to blue upon withering or drying, and that a single plant in Scott County was observed during the present study in which a few of the heads were blue at flowering. Barkley (1986) noted that although corolla color is fairly uniformly orange in the western portion of the species' natural range, to the east of St. Louis the color is bright yellow.

Numerous infraspecific taxa of *L. canadensis* have been described based on variations in leaf morphology and pubescence, including f. *angustipes* Wiegand, with entire or shallowly toothed, lanceolate to linear leaves; var. *obovata* Wiegand, with usually finely toothed, oblanceolate to narrowly

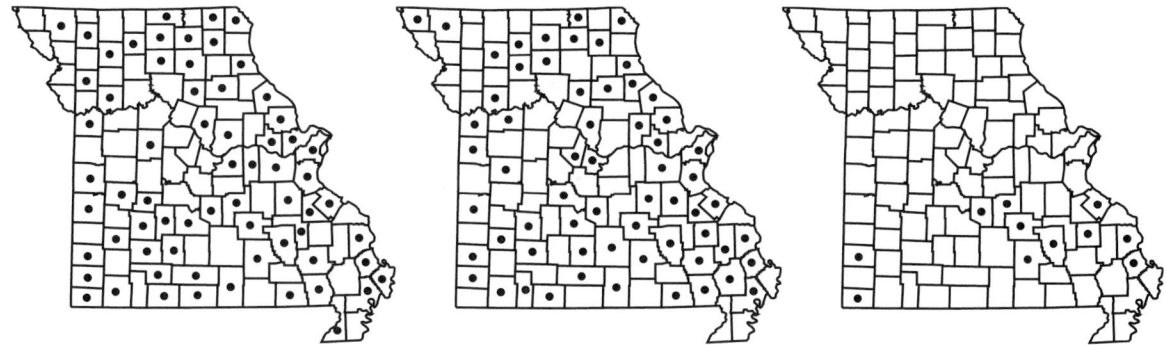

1077. Lactuca canadensis   1078. Lactuca floridana   1079. Lactuca hirsuta

obovate leaves, the base sagittate and clasping the stem; f. *stenopoda* Wiegand, with leaves similar to var. *obovata,* but the base tapered or narrowed and not clasping the stem; var. *longifolia* (Michx.) Farw., with the leaf lobes linear to narrowly triangular, the base sagittate and clasping the stem; f. *angustata* Wiegand, with leaves similar to var. *longifolia* but the base tapered or narrowed; var. *latifolia* Kuntze, with the leaf lobes broadly triangular to obovate, the base sagittate and clasping the stem; f. *exauriculatata* Wiegand, with leaves similar to f. *latifolia* but the bases tapered or narrowed; f. *villicaulis* Fernald, with leaves similar to f. *latifolia* but more or less hairy; and other varieties and forms. Whitaker (1944) studied interspecific hybridization in *Lactuca* and suggested that at least some of these variations were probably due to single-gene differences and are thus unworthy of formal taxonomic recognition.

Steyermark noted an earlier report of *L. graminifolia* Michx. from Butler County but excluded this species from the state's flora based upon his redetermination of the historical specimen as *L. canadensis. Lactuca graminifolia* is a southern species (North Carolina to Arizona) with mostly basal, unlobed leaves that has usually blue corollas (as in *L. floridana*) but fruits resembling those of *L. canadensis* in their beaks and nervation. This specimen also is referred to *L. canadensis* in the present study.

2. Lactuca floridana (L.) Gaertn. (Florida lettuce, woodland lettuce)

Pl. 258 c, d; Map 1078

Plants annual or biennial. Latex plants white. Stems 40–250(–350) cm long, hollow between the nodes, glabrous, often purple-spotted. Leaves well developed along the stems, extremely variable; the basal and lower stem leaves mostly 7–35(–45) cm long, sessile or more commonly with a winged petiole, narrowly ovate, ovate, or obovate in outline, variously toothed and/or deeply pinnately lobed, the margins sometimes minutely hairy, sometimes with a pair of narrowly triangular basal lobes clasping the stem, the undersurface sometimes short-hairy, especially along the midvein; the middle and upper stem leaves mostly lanceolate to ovate, or obovate, pinnately lobed to nearly entire, the margins usually glabrous, sometimes with a pair of narrowly to broadly triangular basal lobes clasping the stem, the undersurface glabrous. Inflorescences mostly well-branched panicles with 50–100 or more heads. Involucre cylindrical or urn-shaped, 8–9 mm long at flowering, elongating to 10–14 mm at fruiting, the bracts 14–17. Florets 10–17(–25). Corollas lavender to purplish blue or blue, rarely white. Pappus 5–7 mm long. Fruits with the body 4–6 mm long, 1.5–2.0 mm wide, brown to dark brown, often mottled, flattened, with somewhat thickened margins and 4 or 5 nerves or ridges on each face, narrowed or tapered abruptly, beakless or with a short, stout beak much less than $^{1}/_{2}$ as long as the body. $2n=34$. July–October.

Common throughout the state (eastern U.S. west to South Dakota and Texas; Canada). Banks of streams and rivers, bottomland forests, mesic upland forests, savannas, sand savannas, glades, bases of bluffs, and margins of ponds and lakes; also railroads, roadsides, and open, disturbed areas.

As in *L. canadensis,* a number of infraspecific variants have been described documenting different leaf morphologies. Of these, f. *villosa* (Jacq.) Cronquist, with unlobed, toothed leaves, occurs fairly commonly in Missouri. White-flowered plants have been called f. *leucantha* Fernald.

3. Lactuca hirsuta Muhl. ex Nutt. (downy lettuce, hairy lettuce)

Pl. 258 e, f; Map 1079

Plants usually biennial. Latex light tan to pale orange. Stems 30–200 cm long, hollow between the

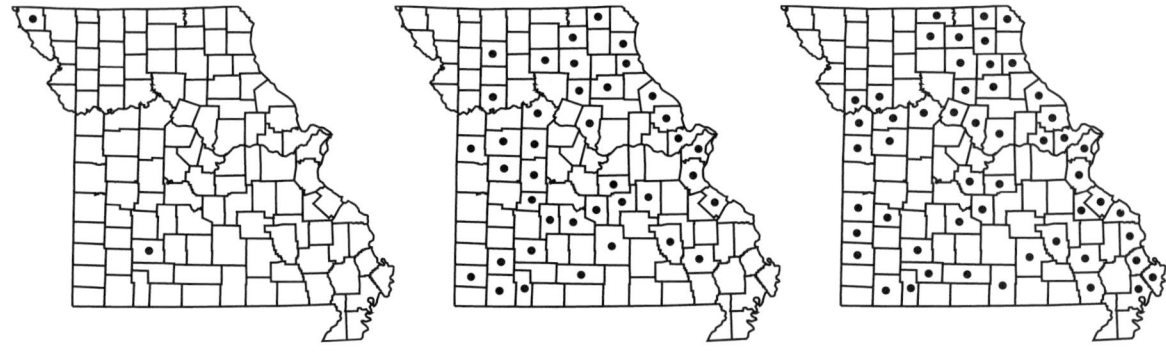

1080. Lactuca ludoviciana 1081. Lactuca saligna 1082. Lactuca serriola

nodes, glabrous to densely pubescent with relatively long, curled hairs, often purple-streaked or purplish-tinged. Leaves best-developed toward the stem base, reduced in size and often unlobed above the stem midpoint, the basal and lower stem leaves mostly 10–20 cm long, with a winged petiole, ovate or obovate in outline, deeply pinnately lobed and sharply toothed, the margins sparsely hairy, sometimes with a pair of narrowly to broadly triangular basal lobes clasping the stem, the surfaces usually short-hairy, less commonly only sparsely hairy on the undersurface midvein. Inflorescences mostly well-branched panicles with (15–)25–100 heads, occasionally appearing nearly racemose. Involucre cylindrical or urn-shaped, 13–20 mm long at flowering, elongating to 15–22 mm at fruiting, the bracts 17(–19). Florets (12–)15–20(–22). Corollas yellow to orangish yellow, sometimes turning blue with age or upon drying. Pappus 7–10(–12) mm long. Fruits with the body 3–5 mm long, 2.0–2.5 mm wide, dark brown to black, flattened, with prominent lateral wings and a conspicuous ridge on each face, tapered abruptly to a slender beak somewhat shorter than to about as long as the body. $2n=34$. June–September.

Uncommon in the Ozark and Ozark Border Divisions (eastern U.S. west to Illinois and Texas; Canada). Mesic to dry upland forests, savannas, sand savannas, and margins of sinkhole ponds; also roadsides.

This species is similar to *L. canadensis* but tends to be hairier and to have larger involucres and fruits. It also has the leaves more basally disposed. Several infraspecific taxa have been described based on the extent of hairiness and the colors of florets. Two phases that occur in the state are var. *hirsuta*, with stems and both sides of the leaves hairy, and var. *sanguinea* (Bigelow) Fernald, with stems glabrous or nearly so and leaves glabrous or sparsely hairy along the midvein on the undersurface. These variations seem unworthy of formal taxonomic recognition, and intermediates exist.

4. Lactuca ludoviciana (Nutt.) Riddell (western wild lettuce)

Pl. 258 g, h; Map 1080

Plants biennial or occasionally short-lived perennials. Latex light tan. Stems (30–)50–150(–200) cm long, hollow between the nodes, glabrous, sometimes purple-streaked or purplish-tinged. Leaves well developed along the stem; the basal and lower stem leaves mostly 10–20(–30) cm long, usually sessile or nearly so, oblanceolate to oblong-ovate in outline, unlobed to deeply pinnately lobed, the margins also sharply toothed and somewhat prickly but glabrous, often with a pair of narrowly to more commonly broadly triangular basal lobes clasping the stem, the surfaces glabrous but the undersurface usually short-prickly along the midvein; the middle and upper stem leaves mostly lanceolate to linear, unlobed or with 1 or 2 pairs of narrowly pinnate lobes, the margins glabrous, usually with a pair of narrowly to broadly triangular basal lobes clasping the stem, the undersurface glabrous but often somewhat prickly along the midvein. Inflorescences mostly well-branched panicles with 50–100 or more heads. Involucre cylindrical, 13–15(–18) mm long at flowering, elongating to 18–20(–23) mm at fruiting, the bracts 17(–19). Florets 20–30(–55). Corollas yellow. Pappus (6–)8–10 mm long. Fruits with the body 4–6 mm long, 2–3 mm wide, dark brown to black, flattened, with prominent lateral wings and a conspicuous ridge on each face, tapered abruptly to a slender beak slightly shorter than to about as long as the body. $2n=34$. June–August.

Uncommon, known only from two historical collections from Atchison and Greene Counties (western U.S. east to Indiana and Mississippi; Canada). Loess hill prairies and savannas.

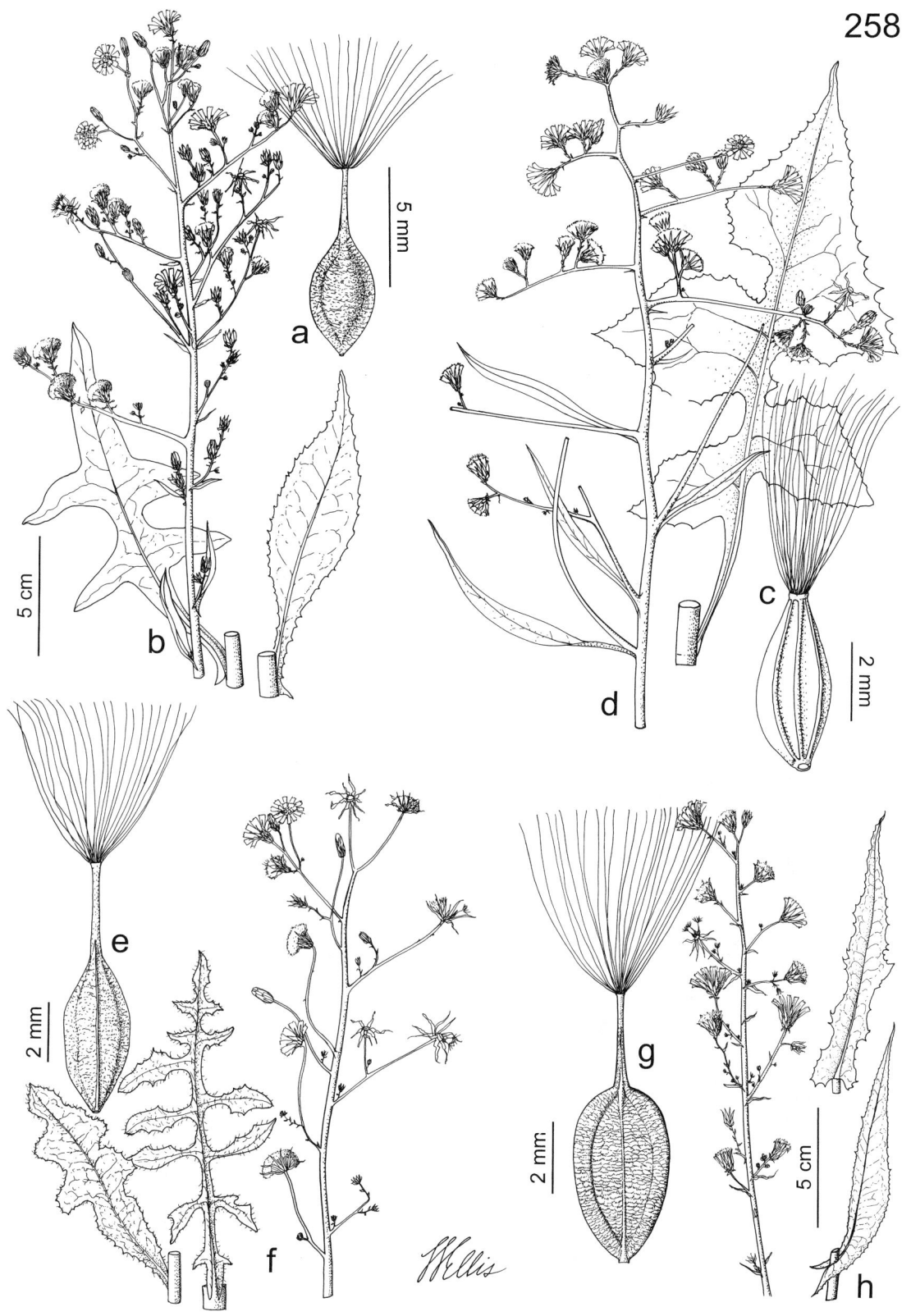

Plate 258. Asteraceae. *Lactuca canadensis*, **a)** fruit, **b)** inflorescence and leaves. *Lactuca floridana*, **c)** fruit, **d)** inflorescence and leaves. *Lactuca hirsuta*, **e)** fruit, **f)** inflorescence and leaves. *Lactuca ludoviciana*, **g)** fruit, **h)** inflorescence and leaves.

The Atchison County specimen is atypical in having the leaves all unlobed, a condition that is comparatively uncommon elsewhere in the range of the species. The Greene County specimen was overlooked by Steyermark (1963).

5. Lactuca saligna L. (willow-leaved lettuce)

Pl. 259 a, b; Map 1081

Plants annual or biennial. Latex white. Stems (30–)50–100 cm long, hollow or more commonly solid between the nodes, glabrous, whitish tan. Leaves well developed along the stems, 2–15(–20) cm long, sessile or occasionally appearing as having a winged petiole, linear and unlobed or narrowly elliptic-lanceolate in outline and with 1 or 2 pairs of linear lobes, the margins otherwise entire, glabrous, all or mostly with a pair of narrowly triangular basal lobes clasping the stem, the surfaces glabrous or the undersurface rarely sparsely short-hairy along the midvein. Inflorescences spikelike panicles with 30–100 or more heads. Involucre cylindrical, 7–9 mm long at flowering, elongating to 10–15(–18) mm at fruiting, the bracts 17(–19). Florets (9–)11–14(–16). Pappus 4–6 mm long. Corollas light yellow to yellow, sometimes some of them bluish- or purplish-tinged on the undersurface. Fruits with the body 3.0–3.5 mm long, about 1.0–1.5 mm wide, greenish brown to dark gray, flattened, with 5–7 conspicuous nerves or ridges on each face, tapered abruptly to a slender beak 1.5–2.0 times as long as the body. $2n=18$. July–October.

Introduced, scattered nearly throughout the state, but less common or absent from the Ozark Border and Mississippi Lowlands Divisions and the eastern half of the Ozarks (native of Europe, Asia; introduced widely in the U.S., Canada). Banks of streams and spring branches, margins and openings of bottomland forests, margins of glades, and disturbed portions of bottomland prairies; also banks of ditches, levees, fallow fields, gardens, pastures, railroads, roadsides, and open, disturbed areas.

This species is relatively distinctive in its narrow leaves and spikelike inflorescences. Two major infraspecific taxa based on leaf morphology have been recognized by some authors, but the application of these names to the plants was confused by Steyermark (1963) and in some of the earlier floristic literature. Plants with unlobed leaves represent the typical form of the species (var. *saligna*). Steyermark (1963) called plants with mostly pinnately lobed leaves f. *ruppiana* (Wallr.) Beck., but that name also refers to entire-leaved plants. The f. *ruppiana* is by far the more common phase in Missouri. Plants with divided leaves are more properly referred to as var. *runcinata* Gren. & Godr. (Feráková, 1977; Yatskievych and Turner 1990). However, Barkley (1986) noted that plants with all entire leaves, those with all divided leaves, and intermediates can occur in the same population, and these variants thus are not treated further in the present study.

6. Lactuca serriola L. (prickly lettuce)

Pl. 259 c, d; Map 1082

Plants annual or biennial. Latex white. Stems 30–150(–200) cm long, hollow between the nodes, glabrous but often short-prickly toward the base, whitish tan. Leaves well developed along the stems, extremely variable; the basal and lower stem leaves mostly (5–)15–20(–30) cm long, sessile or with a winged petiole, narrowly oblong, ovate, or oblong-ovate in outline, unlobed to more commonly deeply pinnately lobed, the margins sharply toothed and prickly but otherwise glabrous, usually with a pair of narrowly to broadly triangular basal lobes clasping the stem, the surfaces glabrous but the undersurface short-prickly along the midvein; the middle and upper stem leaves mostly linear to lanceolate, unlobed, the margins sometimes lacking teeth but with short prickles, the base more or less rounded, with a pair of narrowly triangular basal lobes clasping the stem, the undersurface usually short-prickly along the midvein. Inflorescences mostly well-branched panicles with 50–100 or more heads. Involucre narrowly cup-shaped to cylindrical or urn-shaped, 7–8 mm long at flowering, elongating to 10–15 mm at fruiting, the bracts 17(–19). Florets (12–)18–24(–30). Corollas light yellow to lemon yellow, occasionally bluish- or purplish-tinged on the outer surface, sometimes turning blue with age or upon drying. Pappus 4–5 mm long. Fruits with the body 3–4 mm long, 1.5–2.0 mm wide, yellowish brown to grayish brown, somewhat flattened, hairy toward the tip, angled along the margins and with 5–7 roughened to minutely awned nerves or ridges on each face, tapered to a slender beak 1–2 times as long as the body. $2n=18$. July–October.

Introduced, scattered to common nearly throughout the state (native of Europe, Asia; introduced widely in the U.S., Canada). Banks of streams and rivers, margins and openings of mesic upland forests, and disturbed portions of upland prairies; also crop fields, fallow fields, roadsides, railroads, and open, disturbed areas.

The species epithet has sometimes been spelled *L. scariola* in the older literature (Steyermark, 1963) based on the inconsistent spelling of the name by Linnaeus in his various publications (de Vries and Jarvis, 1987). As in other *Lactuca* species,

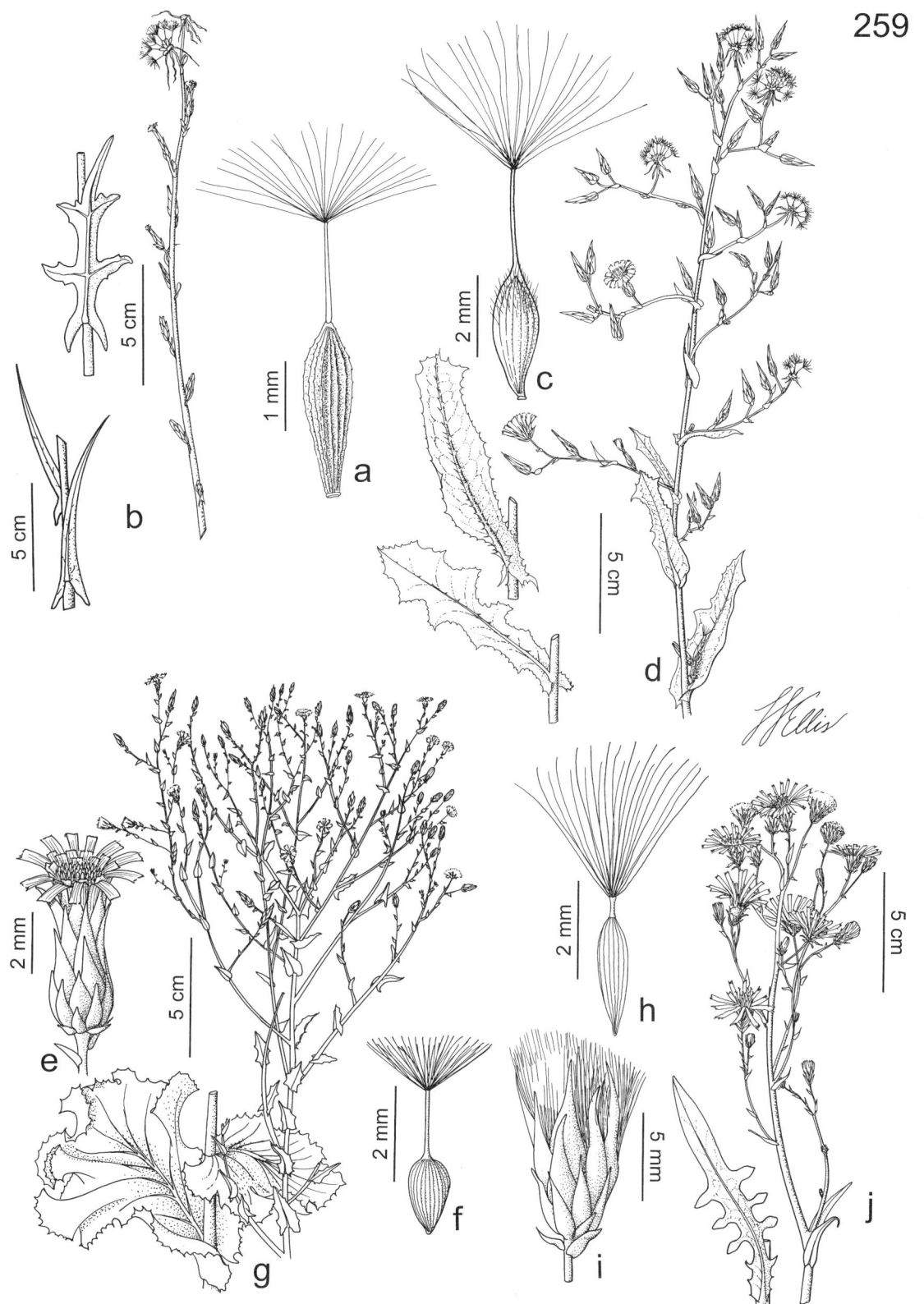

Plate 259. Asteraceae. *Lactuca saligna*, **a)** fruit, **b)** inflorescence and leaves. *Lactuca serriola*, **c)** fruit, **d)** inflorescence and leaves. *Lactuca sativa*, **e)** head, **f)** fruit, **g)** inflorescence and leaves. *Lactuca tatarica*, **h)** fruit, **i)** head, **j)** inflorescence and leaves.

1083. Lactuca tatarica   1084. Lapsana communis   1085. Leontodon taraxacoides

variants have been named to account for phases with different degrees of leaf division. The nominate form has deeply lobed leaves, and plants with narrow, more or less unlobed leaves currently are known as f. *integrifolia* (Gray) S.D. Prince & R.N. Carter. Prince and Carter (1977) speculated that this trait may be under the control of a single major gene. Interestingly, the stem leaves often become twisted at the base during development and assume a vertical position at maturity.

This weedy species is genetically very close to the cultivated lettuce, *L. sativa,* and it has been used as a genetic resource in lettuce breeding. Recent DNA fingerprinting studies (Koopman et al., 2001) have suggested that the closely related *L. sativa* L., *L. serriola, L. dregeana* DC., and *L. altaica* Fisch. & C.A. Mey. are all so closely related genetically that they should be considered part of the same evolutionary gene pool and thus combined into a single species under the name *L. sativa.* However, as the garden lettuce is a cultigen whose origins may have included other ancestral taxa in addition to the preceding list, it seems preferable to maintain it as a separate species for the present time until further, more intensive studies document the genetic limits of this crop plant. *Lactuca sativa* was reported for Missouri from the St. Louis railyards by Mühlenbach (1983), based on a single plant that was uprooted before it could become fully mature. Thus, because it has not been documented to reproduce itself in the wild and has not been seen in the state since the original report, the garden lettuce is not treated fully in the present work. The species has become established outside of cultivation sporadically in the United States and Canada, but it does not appear to persist in the wild anywhere for very long. *Lactuca sativa* is similar to *L. serriola* but differs in its usually leafier stems with broadly ovate to nearly circular leaves that tend to have a corrugated or ruffled appearance along the margins (or to be entire and unlobed with a rounded tip; Pl. 259 e–g). The leaves and stems usually lack prickles.

7. Lactuca tatarica (L.) C.A. Mey. **ssp. pulchella** (Pursh) Stebbins (blue lettuce)
   L. pulchella (Pursh) DC.
   L. oblongifolia Nutt.
   Pl. 259 h–j; Map 1083

Plants perennial, with a long, vertical rootstock and deep-seated, branched rhizomes. Latex white. Stems (20–)30–100 cm long, usually hollow between the nodes, glabrous, often purple-streaked or purplish-tinged. Leaves well developed along the stem; 3–20 cm long, usually sessile or nearly so, linear or narrowly oblong to oblanceolate in outline, unlobed or the lower ones sometimes shallowly to deeply pinnately lobed, the margins otherwise entire, glabrous, sometimes with a pair of small, narrowly to broadly triangular basal lobes clasping the stem, the surfaces glabrous, the undersurface often somewhat glaucous. Inflorescences narrow panicles with 20–50 heads. Involucre narrowly cup-shaped to slightly conical, 12–15 mm long at flowering, elongating to 15–18 mm at fruiting, the bracts 17–19. Florets 18–50. Corollas light purplish blue to blue. Pappus 8–10 mm long. Fruits with the body 3.5–4.0 mm long, 1.0–1.5 mm wide, reddish brown to dark brown, often somewhat mottled, flattened, with somewhat thickened margins and 3–5 nerves or ridges on each face, tapered abruptly to a relatively stout, often lighter beak somewhat shorter than to about as long as the body. $2n=18$. June–August.

Uncommon in northwestern Missouri (western U.S. [including Alaska] east to Michigan and Missouri; Canada, Asia; introduced in the northeastern and south-central U.S.). Loess hill prairies and bottomland prairies; also banks of ditches, roadsides, and open, disturbed areas.

This beautiful wild lettuce has the showiest heads of any *Lactuca* in the state, averaging about 3 cm in diameter (measured across the rays) when in flower. The plants have a long, taprootlike caudex that produces long, deep-set, branched rhizomes, which are easily broken off when a specimen is collected. Stebbins (1939) suggested that American plants were the product of a glacial migration of *L. tatarica* from Siberia across the Bering land bridge during the Pliocene or Pleistocene, and he designated the North American plants ssp. *pulchella*. Feráková (1977) found that this subspecies is also distributed widely in Asia. The mostly European ssp. *tatarica* apparently differs in its slightly shorter stems with noticeably fewer leaves having finely toothed margins.

40. Lapsana L.
Contributed by David J. Bogler and George Yatskievych

One species, Europe, Asia, introduced widely in North America.

Traditionally, this genus was thought to comprise 5–10 mostly Asian species, but Pak and Bremer (1995) reclassified all of the other species into distantly related genera, leaving only a single widespread species in *Lapsana*.

1. Lapsana communis L. (nipplewort)

Pl. 260 a–c; Map 1084

Plants annual, with shallow, fibrous roots. Latex white. Stems 1 or few, 15–80 cm long, erect or ascending, unbranched or few-branched below the midpoint, finely longitudinally ridged, sparsely to moderately pubescent with slender, multicellular, spreading hairs, these often gland-tipped, especially toward the stem tip, occasionally nearly glabrous with age. Leaves basal and alternate, sessile or with a short to long, winged petiole. Leaf blades 1–15 cm long, those of the basal and lower stem leaves obovate to oblanceolate, with a large, ovate terminal lobe abruptly tapered to a narrower, pinnate series of lobes, these sharply triangular with mostly broadly rounded sinuses; those of the upper (and sometimes also median) stem leaves mostly unlobed, entire or with few to several broad, shallow, spreading teeth, both surfaces sparsely to moderately pubescent with short, white, more or less spreading, nonglandular hairs. Venation of 1 main vein and a complex network of anastomosing secondary and tertiary veins. Inflorescences terminal panicles or occasionally loose terminal clusters. Heads appearing mostly long-stalked. Involucre 5–10 mm long, cup-shaped, the bracts in 2 series, glabrous or inconspicuously cobwebby- or glandular-hairy along the margins, those of the outer series 4 or 5, 0.5–2.5 mm long, ovate to narrowly triangular, green; those of the inner series 8–10, narrowly oblong, usually becoming keeled as the fruits mature, sometimes becoming somewhat hardened and pale yellow at maturity. Receptacle naked. Ligulate florets 8–15. Corollas 7–10 mm long, lemon yellow to yellow. Pappus absent. Fruits 3–5 mm long (those of the outer florets longer than those of the inner ones), narrowly oblong in outline, often slightly flattened, with numerous fine ribs, somewhat curved, slightly tapered to a more or less truncate tip, not beaked, the surface otherwise smooth, glabrous, shiny, light brown to yellowish brown. $2n=12, 14, 16$. June–September.

Introduced, uncommon and sporadic in the Ozark and Ozark Border Divisions (native of Europe, Asia, introduced nearly throughout the U.S. [including Alaska], Canada, Greenland). Banks of streams and rivers; also roadsides and open, disturbed areas.

In its native range, this species occasionally is eaten raw or cooked. Some European manuals divide *L. communis* into a number of subspecies based on minor, mostly quantitative variations. Missouri material keys to ssp. *communis* in these treatments.

41. Leontodon L. (hawkbit)
Contributed by David J. Bogler and George Yatskievych

Forty to 50 species, Europe, Asia, Africa.

1. Leontodon taraxacoides (Vill.) Mérat ssp. taraxacoides (lesser hawkbit)

Map 1085

Plants perennial herbs, with a short, thickened rootstock and fibrous roots. Latex white. Stems few to more commonly numerous, 10–40 cm long, ascending, unbranched, finely longitudinally ridged, sparsely to moderately pubescent toward the base with white, somewhat spreading hairs, these often somewhat flattened and gradually expanded at the base, usually glabrous above the midpoint and sparsely hairy near the tip. Leaves all basal, with a mostly short, winged petiole. Leaf blades 2–15 cm long, narrowly oblong to narrowly oblanceolate, mostly shallowly pinnately several-lobed with broadly triangular lobes and rounded sinuses, occasionally unlobed but with few to several broad, shallow, spreading teeth, both surfaces moderately to densely pubescent with relatively coarse, white, spreading hairs, some of these minutely forked at the tip. Venation of 1 main vein and sometimes very faint, arching secondary veins. Head solitary at each stem tip. Involucre 6–11 mm long at flowering, not or only slightly elongating at fruiting, cup-shaped, the bracts in 2 series, glabrous or more commonly hairy (at least some of the hairs minutely forked at the tip), those of the outer series 9–12, 1–3 mm long, narrowly triangular, green or sometimes purplish-tinged; those of the inner series 10–12, narrowly lanceolate. Receptacle naked, shallowly pitted at the base of each floret. Ligulate florets 35–60. Corollas 8–10 mm long, bright yellow, often somewhat grayish- or brownish-tinged on the outer surface. Pappus light yellow to straw-colored, of 2 types: that of the outermost florets a short crown of fused scales 0.5–0.8 mm long, irregularly incised at the tip; that of the other florets mostly of slender scales 5–7 mm long, these mostly bristlelike and plumose from a flattened, expanded base (often weakly fused at the base), usually associated with at least a few somewhat shorter, merely barbed, capillary bristles, more or less persistent at fruiting. Fruits 4–6 mm long, narrowly oblong-elliptic in outline, not flattened, 10–14-ribbed, often slightly curved toward the base, tapered to a slightly flask-shaped tip, not or very short-beaked, the pappus attached to a slightly expanded conical or disclike tip, the surface roughened with minute, ascending cross-wrinkles, otherwise glabrous, brown to reddish brown. $2n=8$. July–September.

Introduced, known thus far only from the city of St. Louis (native of Europe, introduced sporadically in the eastern and western U.S., Canada). Lawns, open, disturbed areas.

Some populations from California appear to be functionally annual and to have slightly longer beaks (to 3 mm long) on the inner fruits (Stebbins, 1993). This variant corresponds to plants that have been called ssp. *longirostris* Finch & Sell in the European literature.

42. Lygodesmia D. Don (Tomb, 1980)
Contributed by David J. Bogler and George Yatskievych

Eight or 9 species, U.S., Mexico.

1. Lygodesmia juncea (Pursh) D. Don ex Hook. (skeleton plant)

Pl. 260 g–i; Map 1086

Plants perennial herbs, with deep-set, woody, usually vertical rootstock and branched rhizomes. Latex cream-colored to yellowish white. Stems 1 or few, 10–70 cm long, erect to loosely ascending, sometimes from a spreading base, highly branched, finely longitudinally ridged, glabrous, grayish green. Leaves basal and alternate, sessile, unlobed, entire, glabrous. Basal leaves relatively few, 15–40 mm long, linear or very narrowly lanceolate, withering before the flowers develop. Stem leaves 1–12(–40) mm long, scalelike, linear to very narrowly triangular, mostly withering or shed before the flowers develop. Venation of 1 or 3 main veins, sometimes not apparent. Heads solitary at the branch tips or rarely lateral along the branches, appearing sessile to long-stalked. Involucre 10–16 mm long at flowering, not or only slightly elongating at fruiting, narrowly cylindrical, the bracts in 1 long, inner series and 3 or 4 shorter, outer series, those of the outer series 9–19, variously 1–4 mm long, narrowly triangular, usually purplish-tinged toward the tip, glabrous or sparsely to moderately hairy along the margins; those of the inner series more or less equal, 5–7, narrowly oblong-lanceolate, the margins often thin and pale, the tip appressed-ascending, often purplish-tinged. Receptacle naked but with minute, scaly ridges between the florets. Ligulate florets 4–6. Corollas 12–20 mm long, light lavender, less commonly pink or nearly white. Pappus 6–9 mm long, of numerous apparently smooth (microscopically barbed) bristles, straw-colored to pale tan, persistent at fruiting. Fruits not produced (see discussion below). $2n=18$. June–September.

Plate 260. Asteraceae. *Lapsana communis*, **a)** fruiting head, **b)** fruit, **c)** inflorescence and leaves. *Nothocalais cuspidata*, **d)** leaf detail, **e)** fruit, **f)** habit. *Lygodesmia juncea*, **g)** head, **h)** fruit, **i)** habit. *Helminthotheca echioides*, **j)** fruit, **k)** inflorescence and leaves. *Picris hieracioides*, **l)** fruit, **m)** inflorescence and leaves. *Picris rhagadioloides*, **n)** fruit, **o)** inflorescence.

1086. Lygodesmia juncea   1087. Nothocalais cuspidata   1088. Picris hieracioides

Uncommon, known only from the western portion of the Glaciated Plains Division from Buchanan to Atchison Counties, and from a presumably introduced historical occurrence in St. Louis County (Washington to Nevada and New Mexico east to Minnesota and Texas). Loess hill prairies and loess bluffs; also rarely open, disturbed areas.

Smith (1988) dismissed an Arkansas specimen of this species as representing a nonpersistent chance introduction. A single nonfertile specimen at the University of Missouri herbarium collected on 23 September 1886 by Henry Eggert in St. Louis County (west of Forest Park) also probably documents a waif that failed to persist in some disturbed habitat. Steyermark (1963) excluded this specimen from his map in the belief that it was mislabeled.

Tomb (1980) noted that although *L. juncea* is widespread, most populations apparently produce few or no viable fruits. This appears to be the case in all of the Missouri specimens examined during the present study. After flowering, the ovary elongates as though the fruit is developing, but at maturity it is abruptly slender and somewhat shrunken below the slightly expanded tip. Tomb (1980) reported that well-formed achenes of *L. juncea* are 6–10 mm long, cylindrical, obscurely longitudinally lined, and glabrous.

Plants of *L. juncea* frequently have spherical galls along the stems. These are caused by a small cynipid wasp, *Antistrophus pisum* Ashmead, whose developing larvae feed on plant tissue inside the galls. The genus *Antistrophus* utilizes only members of the Asteraceae (Krombein et al., 1979). Apparently, *A. pisum* has evolved to parasitize only *L. juncea* and thus requires the skeleton plant to complete its life cycle.

Lygodesmia juncea has a bitter flavor and is thus avoided by livestock. Although there is no evidence that the plants themselves are poisonous, they can accumulate nitrates from the soil in their tissues and thus can become toxic to livestock (Burrows and Tyrl, 2001). Native Americans used the plants medicinally to treat coughs, heartburn, diarrhea, skin problems, kidney problems, and a host of miscellaneous ailments, as well as to increase lactation in nursing mothers (Moerman, 1998).

43. Nothocalais (A. Gray) Greene

Four species, western U.S., Canada.

Chambers (1957) argued for the segregation of the four species of *Nothocalais* from the large, polymorphic genus *Microseris* D. Don, based on their morphological cohesiveness and unique combination of character states. Subsequent molecular studies (Jansen et al., 1991; Whitton et al., 1995) have reinforced the notion that *Nothocalais* is closely related to the segregates *Agoseris* Raf. and *Uropappus* Nutt., which each should be kept separate from *Microseris*.

1. Nothocalais cuspidata (Pursh) Greene
(prairie dandelion)
Agoseris cuspidata (Pursh) D. Dietr.
Microseris cuspidata (Pursh) Sch. Bip.
Pl. 260 d–f; Map 1087

Plants perennial herbs, with a fleshy to somewhat hardened taproot. Latex white. Stems usually solitary, 9–35 cm long, erect, unbranched, finely longitudinally lined or ridged, moderately to densely pubescent with white, curled to woolly hairs near the tip, more sparsely hairy below the apical $1/3$, often glabrous or nearly so toward the base.

Leaves all basal, sessile or tapered to an indistinct, short, winged petiole, usually appearing somewhat sheathing at the base (note that usually 2 or 3 short [1–4 cm], brown, sheathing, scalelike leaves are present outside the normal leaves). Leaf blades 7–35 cm long, linear, unlobed and grasslike, the margins entire, flat or undulate, densely pubescent with short, curled hairs, the surfaces glabrous or sparsely pubescent with somewhat longer, curled hairs, glaucous. Venation of 1 broad, pale main vein and a usually faint network of elongate, anastomosing secondary veins. Head solitary at each stem tip. Involucre 13–27 mm long at flowering, only slightly elongating at fruiting, cup-shaped to bell-shaped, the bracts 13–35, in usually 2 subequal but overlapping series, narrowly lanceolate to lanceolate, mostly long-tapered to a sharply pointed tip, glabrous or minutely hairy toward the tip, green with narrow to broad, thin, pale margins, sometimes purplish-spotted or with a faint purple midvein. Receptacle naked, shallowly pitted at the base of each floret. Ligulate florets 35–80. Corollas 15–27 mm long, bright yellow, often somewhat grayish- or purplish-tinged on the outer surface. Pappus 8–11 mm long, of numerous bristles, these minutely barbed, white, grading into varying numbers (often numerous) of very narrow, awnlike scales similar to and difficult to distinguish from the bristles. Fruits 7–10 mm long, narrowly cylindrical (tapered slightly at the base), not flattened, (8–)10-ribbed, not beaked, not expanded at the tip, the surface minutely roughened, otherwise glabrous, yellowish brown to brown. $2n=18$. April–June.

Uncommon, widely scattered in the eastern and western portions of the state (Wisconsin to Arkansas west to Montana and Texas; Canada). Glades and loess hill prairies.

The Missouri Natural Heritage Database records occurrences for this taxon in Pettis and Ralls Counties, but apparently no vouchers were collected from these populations and the counties thus are not mapped.

44. Picris L. (ox-tongue)

Plants annual, biennial, or perennial herbs. Latex white. Stems solitary to several, often well branched above the midpoint, erect to loosely ascending, finely ridged, pubescent with some or all of the hairs barbed at the tip with 2–4 minute, spreading to recurved branches from a knoblike tip (also usually somewhat enlarged at the base). Leaves alternate and basal, mostly more than 3 times as long as wide, not grasslike, sessile or the basal ones tapered to a short or long petiole, the basal leaves often present at flowering, the stem leaves sessile, with a pair of rounded or pointed basal lobes, more or less clasping the stem. Leaf blades mostly unlobed, those of the basal leaves oblanceolate to oblong-lanceolate or rarely ovate; those of the stem leaves oblong-lanceolate to narrowly lanceolate or linear; the margins entire or with spreading, triangular teeth and rounded sinuses, sometimes somewhat wavy or corrugated, the surfaces pubescent with some or all of the hairs barbed at the tip with 2–4 minute, spreading to recurved branches from a knoblike tip (also usually somewhat enlarged at the base). Venation of 1 main vein and sometimes also a faint network of arching or anastomosing secondary veins. Inflorescences mostly open panicles, the heads solitary or in small clusters at the branch tips. Involucre not or only slightly elongated at fruiting, cup-shaped to bell-shaped or somewhat urn-shaped, the bracts in 1 long, inner series and 2 or 3 shorter, outer series, pubescent with some or all of the hairs barbed at the tip with 2–4 minute, spreading to recurved branches from a knoblike tip, those of the inner series 13–18, all similar, narrowly lanceolate to narrowly oblong-lanceolate; those of the outer series 7–13, ranging from much shorter than to more than ½ as long as the inner series, lanceolate to elliptic or oblong-lanceolate, the margins sometimes thin and pale, the tip mostly loosely ascending at flowering. Receptacle naked, sometimes shallowly pitted at the base of each floret. Ligulate florets 30–50 per head. Corollas bright yellow, sometimes reddish-tinged on the outer surface. Pappus of numerous bristles, these smooth or microscopically barbed, white to straw-colored, often shed as an intact unit after the fruits mature. Fruits with the body nearly narrowly oblong-elliptic to oblong-oblanceolate in outline and at least the outermost often somewhat curved, somewhat flattened in cross-section (irregularly oval in cross-section), not beaked, the pappus attached to an unmodified or only

1089. Picris rhagadioloides   1090. Prenanthes alba   1091. Prenanthes altissima

slightly expanded tip, finely 5–10-nerved, the surface also cross-wrinkled or cross-ridged, yellowish brown to reddish brown or dark brown. About 40 species, Europe, Asia, Africa, introduced widely elsewhere.

1. Fruits 3–5 mm long, not surrounded by the inner involucral bracts, which remain relatively flat, thin, and papery to herbaceous; plants biennial or short-lived perennials, with several slightly thickened main roots . . . . 1. P. HIERACIOIDES
1. Fruits 2.5–3.0 mm long, the outer series partially surrounded by concave bases of the thickened, hard inner involucral bracts; plants annual, taprooted
. . . . 2. P. RHAGADIOLOIDES

1. Picris hieracioides L. ssp. **hieracioides**
(hawkweed ox-tongue, cat's ear)
Pl. 260 l, m; Map 1088

Plants biennial or short-lived perennials, with a fascicle of several slightly thickened main roots. Stems 20–80 cm long, moderately to densely pubescent with spreading, barb-tipped hairs, also sparsely to moderately pubescent with minute, appressed, sometimes cobwebby, branched hairs toward the tip. Basal and lower stem leaves 4–25 cm long, moderately pubescent with spreading, barb-tipped hairs, usually also sparsely to moderately pubescent with minute, appressed, branched hairs. Involucre 8–15 mm long, the inner series of bracts relatively flat, thin, papery to herbaceous, not becoming wrapped around the outer series of fruits, pubescent with a central band of spreading, barb-tipped hairs, usually also pubescent with sparse to moderate, minute, appressed, cobwebby, branched hairs. Corollas 1.2–2.0 cm long. Pappus 5–7 mm long. Fruits 3–5 mm long. $2n=10$. July–September.

Introduced, known thus far only from the city of St. Louis (native of Europe, Africa; introduced). Railroads.

This species was first reported for Missouri by Mühlenbach (1979), based on his botanical inventories of the St. Louis railyards. In Europe, several subspecies have been segregated based on minor differences in involucre size and color, as well as inflorescence shape (Chater, 1976). The application of these names to North American nonnative populations has by no means been firmly established, but the single collection from Missouri to date appears to correspond to ssp. *hieracioides*.

2. Picris rhagadioloides (L.) Desf. (bitterweed)
P. altissima Delile
Pl. 260 n, o; Map 1089

Plants annual, taprooted. Stems 20–100 cm long, moderately to densely pubescent with spreading, barb-tipped hairs, usually also sparsely to moderately pubescent with minute, appressed, branched hairs. Basal and lower stem leaves 3–15 cm long. Involucre 7–12 mm long, the inner series of bracts thickened and hard at fruiting, the basal portion becoming wrapped around the outer series of fruits (these usually becoming dispersed tardily, remaining with the involucre after the inner fruits have been shed), pubescent with a central band of spreading, barb-tipped hairs, usually also pubescent with sparse to moderate, minute, appressed, branched hairs. Corollas 0.8–1.5 cm long. Pappus 4–6 mm long (that of the outer florets sometimes shorter elsewhere). Fruits 2.5–3.0 mm long. $2n=10$. June–September.

Introduced, known thus far only from St. Louis County (native of Europe, Africa; introduced). Railroads.

The correct name for this plant remains somewhat controversial. Traditionally, this species has been called *P. sprengerana* (L.) Poir. in most of the botanical literature (Steyermark, 1963), but as noted by Lack (1974) and Greuter (2003), that name instead applies to an Old World species of *Hieracium*. Lack (1974) referred to the species as *P. integrifolia* Desf., but Holzapfel (1994) suggested that this epithet was not validly published and used the name *P. altissima*. Most recently, Greuter (2003), noted that the name *P. rhagadioloides* is the oldest valid name for the taxon, rejecting Lack's (1974) earlier arguments against the acceptance of this name.

45. Prenanthes L. (rattlesnake root, white lettuce)
(Milstead, 1964)
Contributed by David J. Bogler and George Yatskievych

Plants perennial herbs, usually with somewhat tuberous-thickened rootstocks, often producing offsets and then colonial. Latex white. Stems usually solitary, erect or ascending, finely to less commonly coarsely ridged, glabrous or variously hairy. Leaves alternate and basal, mostly 1–3 times as long as wide, or, if 3–5 times as long as wide, then the blade tapered abruptly to a winged petiole, glabrous or hairy, sessile to long-petiolate, the basal leaves mostly withered before flowering (except in *P. racemosa*), the stem leaves not clasping the stem or the rounded bases somewhat clasping. Leaf blades often quite variable on the plant, unlobed to 3-lobed or shallowly pinnately lobed, variously shaped, the margins entire or finely to coarsely toothed, sometimes irregularly so. Inflorescences consisting of small clusters, these appearing axillary and/or terminal, the terminal ones often arranged into narrow, spicate or broad, open panicles. Involucre not or only slightly elongated at fruiting, narrowly cylindrical to narrowly bell-shaped or urn-shaped, the bracts in 1 long inner and 2 or 3 shorter outer series, those of the inner series all similar, narrowly oblong-elliptic to narrowly oblong-lanceolate, those of the outer series ovate to narrowly lanceolate, sometimes purplish- or reddish-tinged, glabrous or hairy, the margins sometimes thin and pale, the tip appressed-ascending at flowering. Receptacle naked. Ligulate florets 4–35 per head. Corollas white, yellow, pink, or purple. Pappus of numerous bristles, these straw-colored to tan, orangish brown, or reddish brown, often shed irregularly at fruiting. Fruits with the body nearly cylindrical to narrowly oblong-elliptic in outline, more or less circular in cross-section or sometimes 4- or 5-angled, not flattened, not beaked, the pappus attached to an unmodified or only slightly expanded tip, often (8–)10–12-ribbed, the surface otherwise relatively smooth, yellowish brown to reddish brown. Twenty-six to 30 species, North America, Europe, Asia, Africa.

The present treatment uses the traditional broad circumscription of the genus. However, in a preliminary phylogenetic study of molecular data, Kim et al. (1996) found that *Prenanthes* may actually consist of three species groups that are only distantly related within the subtribe Sonchinae K. Bremer. If future research yields similar results, then the name *Prenanthes* must remain with a group of about nine Old World species related to the type species, *P. purpurea* L. The Macaronesian *P. pendula* Sch. Bip. is isolated in the second group. The approximately 14 species in North America form the third group and would have to be restored to a separate genus under the name *Nabalus* Cass.

1. Involucral bracts glabrous (the margins rarely microscopically hairy near the tip)
    2. Involucre with the inner series consisting of 7–11 bracts; florets 7–13 per head . . . . 1. P. ALBA
    2. Involucre with the inner series consisting of 4–6 bracts; florets 4–6 per head
    . . . . 2. P. ALTISSIMA

1. Involucral bracts hairy
   3. Lower and usually also median stem leaves with a winged petiole; inflorescences appearing broadly paniculate or as loose clusters of heads at the end of relatively well-developed branches (small axillary clusters of heads may also be present farther down the stem).................... 4. P. CREPIDINEA
   3. Lower and median stem leaves mostly sessile (the basal and sometimes also lowermost stem leaves usually short-petiolate); inflorescences appearing spicate, as narrowly cylindrical panicles, or as small, dense axillary clusters of heads along an unbranched main stem
      4. Stems and the undersurface of the leaves hairy; corollas cream-colored to pale yellow .... 3. P. ASPERA
      4. Stems and the undersurface of the leaves glabrous below the inflorescence; corollas pink to lavender or sometimes white (then often with a purplish-tinged base)................................. 5. P. RACEMOSA

1. Prenanthes alba L. (white lettuce, rattlesnake root, white snakeroot)

Nabalus albus (L.) Hook.

Pl. 261 a, b; Map 1090

Stems 30–170 cm long, usually relatively stout, glabrous or sparsely and inconspicuously pubescent with short, curled hairs toward the tip, usually purplish-tinged or with dark purple mottling, often somewhat glaucous. Leaves variable, glabrous or the undersurface sparsely and inconspicuously pubescent with short, curled hairs, often somewhat glaucous. Basal and lower stem leaves usually long-petiolate, the blade 4–30 cm long, usually broadly triangular, sometimes with a pair of triangular basal lobes, the margins often with few to several broad, spreading teeth. Median and upper leaves gradually reduced, with progressively shorter petioles (often sessile toward the stem tip), the blade entire or nearly so or finely to coarsely toothed or shallowly pinnately lobed, triangular to ovate or ovate-elliptic. Inflorescences usually elongate panicles, narrow to broad, the heads tending to be clustered toward the branch tips, terminal and often also from the upper leaf axils, sometimes with additional small clusters of heads below the main inflorescence, the heads commonly nodding. Involucre 10–14 mm long, the inner bracts 7–9(–11), rounded to sharply pointed at the often somewhat incurved tip, glabrous (the margins rarely microscopically hairy near the tip), often somewhat purplish-tinged, appearing somewhat glaucous or minutely pebbled (the effect caused by microscopic white papillae on the outer surface). Ligulate florets 7–13. Corollas 9–15 mm long, white to pale pink or lavender, sometimes reddish- or purplish-tinged toward the base. Pappus 6–7 mm long, orangish brown to reddish brown. Fruits 4–6 mm long, light brown to yellowish brown. $2n=32$. July–September.

Uncommon, known mostly from historical collections from the northeastern quarter of the state (northeastern U.S. west to North Dakota and Arkansas; Canada). Bottomland forests, mesic upland forests, ledges of bluffs, and banks of streams and rivers.

As in *P. altissima,* some populations in the eastern portion of the range of *P. alba* have the pappus straw-colored to light yellow.

2. Prenanthes altissima L. (rattlesnake root, tall white lettuce)

P. altissima var. *cinnamomea* Fernald

Nabalus altissimus (L.) Hook.

Pl. 261 e–g; Map 1091

Stems 30–250 cm long, slender to stout, glabrous or sparsely and inconspicuously pubescent with short, curled hairs toward the tip, often purplish-tinged or with dark purple mottling or spots, sometimes somewhat glaucous. Leaves variable, glabrous or the undersurface sparsely and inconspicuously pubescent with short, curled hairs, the undersurface pale but not glaucous. Basal and lower stem leaves long-petiolate, the blade 4–20 cm long, broadly ovate to broadly triangular or less commonly somewhat heart-shaped, usually with 1 or 2 pairs of triangular or ovate basal lobes, the margins often with few to several often irregular, broad, spreading teeth or less commonly shallow lobes. Median and upper leaves gradually reduced, with progressively shorter petioles (usually short-petiolate toward the stem tip), the blade entire or more commonly finely to coarsely toothed or shallowly pinnately lobed, sometimes even those of the upper leaves deeply 3-lobed, triangular to ovate or ovate-elliptic. Inflorescences usually elongate panicles, narrow to broad, the heads tending to be clustered toward the branch tips, terminal and often also from the upper leaf axils, sometimes with

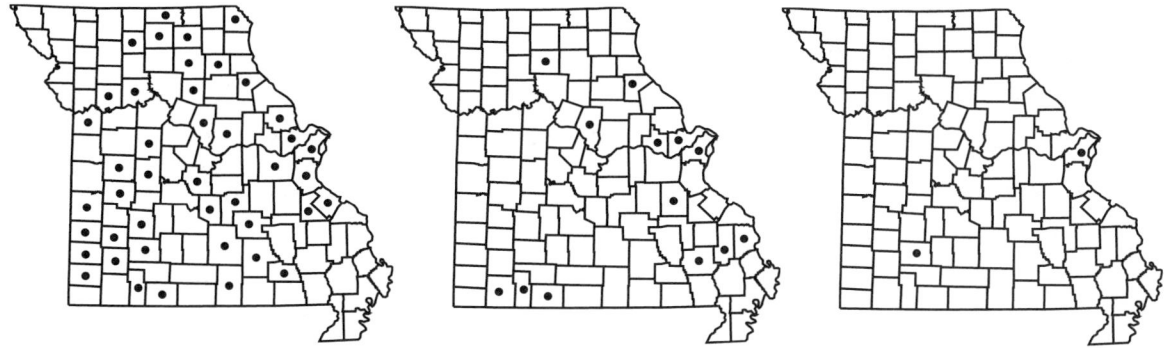

1092. Prenanthes aspera   1093. Prenanthes crepidinea   1094. Prenanthes racemosa

additional small clusters of heads below the main inflorescence, the heads commonly nodding. Involucre 9–14 mm long, the inner bracts 4–6, rounded to sharply pointed at the often somewhat incurved or outcurved tip, glabrous (the margins rarely microscopically hairy near the tip), pale green with a darker green base (this often darkening further upon drying), not glaucous or pebbled. Ligulate florets 4–6. Corollas 7–15 mm long, greenish yellow to cream-colored or pale yellow, sometimes appearing nearly white. Pappus 5–6 mm long, orangish brown. Fruits 4–5 mm long, light brown to yellowish brown. $2n=16$. July–October.

Scattered, mostly in the Ozark and Ozark Border Divisions (eastern U.S. west to Michigan, Missouri, and Texas; Canada). Swamps, bottomland forests, mesic upland forests in ravines, banks of streams and rivers, and bases and ledges of bluffs; also rarely pastures.

This species is quite variable in different portions of its range. Some eastern populations have much more hairy stems and have been called var. *hispidula* Fernald. The pappus is typically straw-colored to light yellow in the eastern portion of the range but gradually deepens in color farther west. In the western part of the range, including Missouri, the pappus is a uniform cinnamon color. Such populations have been treated as var. *cinnamomea* by some botanists (Steyermark, 1963). *Prenanthes altissima* is often observed in the springtime as rosettes of usually 3-lobed leaves. By midsummer, in most cases the leaves will have died back for the year, with only a few individuals producing flowering stems.

3. Prenanthes aspera Michx. (rattlesnake root, rough white lettuce)

Nabalus asper (Michx.) Torr. & A. Gray

Pl. 261 h–j; Map 1092

Stems 35–170 cm long, slender to somewhat stout, moderately to densely pubescent with short, spreading hairs and minute, curled hairs above the midpoint, often glabrous or nearly so toward the base, often purplish-tinged or with dark purple mottling, not glaucous. Leaves with both surfaces minutely roughened, the undersurface also with short, stiff, stout hairs and often also inconspicuously pubescent with short, curled hairs along at least the main veins, not glaucous. Basal and lowermost stem leaves sessile to more commonly short-petiolate, the blade 4–11 cm long, more or less obovate, the margins entire or with few to several often irregular, fine teeth or rarely shallow lobes. Lower, median, and upper leaves gradually reduced, sessile, the blade entire or more commonly finely toothed, rarely more deeply toothed or shallowly pinnately lobed, narrowly lanceolate to lanceolate, oblong-elliptic, or rarely ovate. Inflorescences appearing spicate, as narrowly cylindrical panicles, or as small, dense axillary clusters of heads along an unbranched main stem, the heads mostly ascending. Involucre 12–17 mm long, the inner bracts (6–)8–10, narrowly rounded to more commonly sharply pointed at the tip, the surface pubescent with relatively coarse, stiff, spreading to ascending hairs, the margins frequently pubescent with minute, curled hairs toward the tip, yellowish green to pale green, often with a darker green to dark purplish-tinged base and/or tip (often darkening further upon drying), not glaucous or pebbled. Ligulate florets (8–)11–14(–19). Corollas 9–17 mm long, cream-colored to pale yellow. Pappus 6–8 mm long, usually straw-colored. Fruits 5–6 mm long, usually bright yellowish brown. $2n=32$. August–October.

Scattered nearly throughout the state, but apparently absent from parts of the northwestern and southeastern portions (South Dakota to Pennsylvania south to Oklahoma, Louisiana, and Alabama). Upland prairies, glades, openings of mesic to dry upland forests, and savannas; also railroads and roadsides.

Steyermark (1963) remarked that the florets of this species produce a "pleasantly delicious fragrance resembling that of *Sabatia angularis* (Gentianaceae)." He also noted the existence of a putative hybrid with *P. crepidinea* from Greene County.

4. Prenanthes crepidinea Michx. (rattlesnake root, great white lettuce)
Nabalus crepidineus (Michx.) DC.

Pl. 261 c, d; Map 1093

Stems 100–250 cm long, usually relatively stout, glabrous or sparsely to moderately pubescent with short, curled hairs toward the tip, occasionally also sparsely glandular, often purplish-tinged or with dark purple mottling or spots, not glaucous. Leaves variable, the upper surface often roughened with short, stiff, stout hairs, both surfaces or just the undersurface moderately but inconspicuously pubescent with short, curled hairs along at least the main veins, the undersurface sometimes lighter than the upper surface but not glaucous. Basal and lower stem leaves long-petiolate, the blade 8–30 cm long, broadly ovate to broadly triangular-ovate, usually with 1 or 2 pairs of triangular or ovate basal lobes, the margins usually with few to several often irregular, broad, spreading teeth or less commonly shallow lobes. Median and upper leaves gradually reduced, with progressively shorter petioles (usually short-petiolate toward the stem tip), the blade entire or more commonly finely to coarsely toothed or shallowly pinnately lobed, ovate or oblong-elliptic. Inflorescences usually elongate panicles, relatively broad, the heads tending to be clustered toward the branch tips, terminal and often also from the upper leaf axils, sometimes with additional small clusters of heads below the main inflorescence, the heads commonly nodding. Involucre 12–16 mm long, the inner bracts 10–15, narrowly rounded to more commonly sharply pointed at the tip, the surface pubescent with relatively coarse, stiff, spreading to ascending hairs, the margins frequently pubescent with minute, curled hairs toward the base and occasionally also toward the tip, light green to green, often with a darker green base and/or tip (often darkening further upon drying), not glaucous or pebbled. Ligulate florets (15–)19–27(–38). Corollas 9–15 mm long, greenish yellow to cream-colored or pale yellow, rarely white. Pappus 6–8 mm long, straw-colored to yellowish brown. Fruits 5–6 mm long, yellowish brown to reddish brown. 2n=32. August–October.

Uncommon, known mostly from widely scattered historical collections (northeastern U.S. west to Minnesota and Arkansas). Bottomland forests, mesic upland forests in ravines, banks of streams and rivers, and margins of lakes; also rarely moist depressions along roadsides.

This species can form relatively large colonies of basal rosettes, which are most easily observed in the springtime and that tend to die back by midsummer. Only a small proportion of these rosettes produce flowering stems during any given year. Steyermark (1963) noted the existence of a historical specimen from Greene County that represents a putative hybrid between *P. crepidinea* and *P. aspera*. As the two have different ploidy levels, such a hybrid would be expected to be sterile.

5. Prenanthes racemosa Michx. (rattlesnake root, glaucous white lettuce)
Nabalus racemosus (Michx.) Hook.

Pl. 261 k, l; Map 1094

Stems 35–170 cm long, slender to more commonly stout, glabrous or sparsely to densely pubescent with short, spreading hairs toward the tip, often glabrous below the inflorescence, often purplish-tinged or with dark purple mottling, glaucous. Leaves with both surfaces glabrous, glaucous. Basal and lowermost stem leaves short- to long-petiolate, the blade 4–25 cm long, oblanceolate to obovate, the margins entire or irregularly toothed, often somewhat crisped or corrugated. Median and upper leaves gradually reduced, sessile, the blade entire or finely toothed, obovate to more commonly ovate, oblong-elliptic, or lanceolate. Inflorescences appearing spicate, as narrowly cylindrical panicles, or as small, dense axillary clusters of heads along an unbranched main stem, the heads ascending to spreading or less commonly some of them nodding. Involucre 9–14 mm long, the inner bracts (6–)8(–10), narrowly rounded to more commonly sharply pointed at the tip, the surface pubescent with relatively coarse, stiff, spreading to ascending hairs, the margins frequently pubescent with minute, curled hairs toward the tip, usually dark purple, glaucous, not pebbled. Ligulate florets 9–16. Corollas 7–15 mm long, pink to lavender or sometimes white (then often with a purplish-tinged base). Pappus 5–7 mm long, usually straw-colored. Fruits 5–6 mm long, usually bright yellowish brown. 2n=16. June–September.

Uncommon, known only from historical collections from Greene and St. Louis Counties (northern U.S. south to Colorado, Kentucky, and New Jersey; Canada). Habitat unknown, but (according to Steyermark [1963]) possibly "Wet prairies and low ground bordering streams."

Plants occurring to the north of Missouri with 10–14 (vs. 6–10) longer involucral bracts and

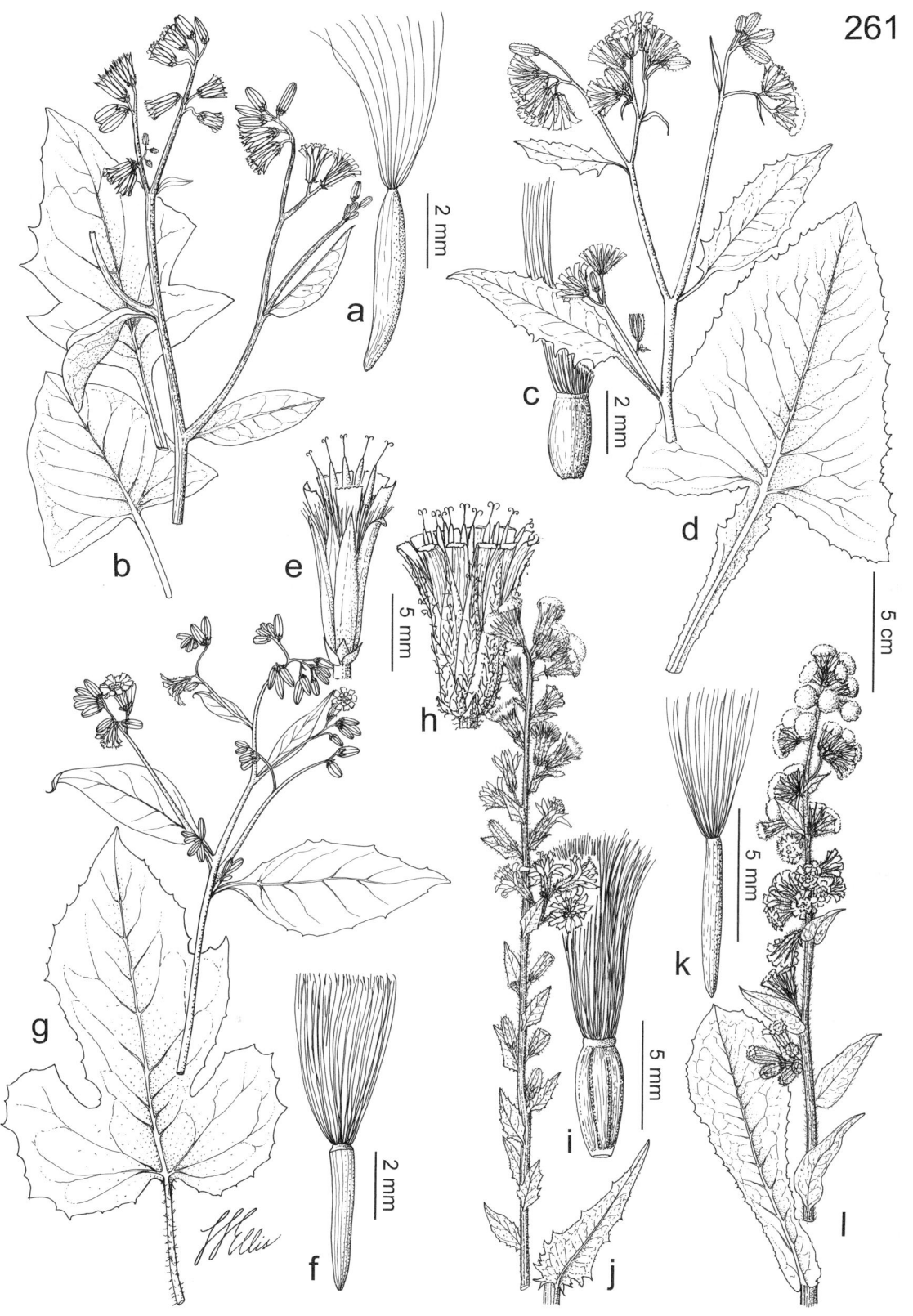

Plate 261. Asteraceae. *Prenanthes alba*, **a)** fruit, **b)** inflorescence and leaves. *Prenanthes crepidinea*, **c)** fruit, **d)** inflorescence and leaves. *Prenanthes altissima*, **e)** head, **f)** fruit, **g)** inflorescence and leaves. *Prenanthes aspera*, **h)** head, **i)** fruit, **j)** inflorescence and leaves. *Prenanthes racemosa*, **k)** fruit, **l)** inflorescence and leaves.

1095. Pyrrhopappus carolinianus 1096. Sonchus arvensis 1097. Sonchus asper

17–26 (vs. 9–16) florets were segregated by Cronquist (1948) as ssp. *multiflora* Cronquist. However, in describing the variation, he noted a broad area of geographic overlap between the two in which plants of both kinds are routinely encountered in the same population. *Prenanthes racemosa* has not been seen in Missouri since the 1890s and presumably has become extirpated from the state. It strongly resembles *P. aspera*. One character that facilitates separation of the two species is that in *P. racemosa* the large basal leaves are often persistent at flowering, whereas in *P. aspera* the basal and lowermost stem leaves usually are withered or absent by flowering time.

46. Pyrrhopappus DC. (false dandelion)
(Northington, 1974; Turner and Kim, 1990)

Five species, U.S., Mexico.

1. Pyrrhopappus carolinianus (Walter) DC.

Pl. 262 g, h; Map 1095

Plants annual, with a taproot. Latex white. Stems 1 or few, 10–75 cm long, erect or ascending, unbranched or usually few-branched, finely longitudinally ridged, moderately to densely and minutely hairy, especially toward the tip, often glabrous or nearly so toward the base. Leaves basal and alternate, sessile or with a short, winged petiole, the base frequently slightly expanded or with a small pair of lobes and somewhat clasping the stem. Leaf blades (1–)3–25 cm long, linear to more commonly oblanceolate, ranging from pinnately several-lobed with usually slender, oblong-triangular lobes and rounded or sometimes angled sinuses to unlobed and entire or nearly so, both surfaces glabrous or the undersurface sparsely and minutely hairy along the midvein. Venation of 1 main vein and a faint network of anastomosing secondary and sometimes also tertiary veins. Heads solitary or rarely paired at the branch tips (sometimes appearing loosely paniculate), appearing long-stalked. Involucre 15–25 mm long at flowering, not or only slightly elongating at fruiting, cup-shaped, the bracts in 1 long inner and 2 or 3 shorter outer series, glabrous or pubescent with dense, minute hairs and sometimes also sparse, longer, cobwebby hairs, those of the outer series 13–17, variously 4–11 mm long, linear to narrowly lanceolate, usually purplish-tinged toward the tip; those of the inner series more or less equal, 17–21, narrowly oblong-lanceolate, with an abrupt, small, purple to brown, crestlike thickening just below the tip, because of this often appearing irregular, notched, or jointed near the tip. Receptacle naked. Ligulate florets 55–140. Corollas 9–22 mm long, bright yellow, occasionally somewhat brownish- or purplish-tinged on the outer surface. Pappus 7–12 mm long, of numerous apparently smooth (microscopically barbed) bristles, straw-colored, tan, or light reddish brown, persistent at fruiting, a dense ring of minute, reflexed, white hairs attached immediately below the pappus bristles. Fruits 11–16 mm long (including the beak), the body narrowly oblong-elliptic in outline, not flattened, deeply 5-grooved, tapered at the tip to a slender beak usually longer than the body, the pappus attached to a more or less unexpanded tip, the surface of the rounded ribs minutely roughened with ascending tubercles or barbs and also appearing faintly or finely cross-wrinkled, otherwise glabrous, brown to reddish brown. $2n=12$. May–October.

Scattered nearly throughout the state (southeastern U.S. west to Nebraska and Texas). Upland prairies, glades, swamps, banks of streams and rivers, and margins of ponds and lakes; also pastures, old fields, fallow fields, lawns, banks of ditches, railroads, roadsides, and open, disturbed areas.

Pyrrhopappus carolinianus is a native species capable of growing in diverse substrate and moisture situations from swamps to dry upland prairies and glades. It has been quite successful in spreading into various disturbed habitats.

Northington (1971) included Missouri in the range of the closely related *P. grandiflorus* (Nutt.) Nutt. based on a single historical collection from Barry County. However, the specimen is actually an atypical sample of *P. carolinianus,* as noted on Northington's annotation of the sheet at the Gray Herbarium of Harvard University. However, *P. grandiflorus* does occur in portions of Oklahoma and Kansas adjacent to southwestern Missouri and may be discovered here in the future. It differs from *P. carolinianus* most notably in its perennial habit, with a slender taproot bearing a tuber 1–2 cm in diameter at its tip, its shorter stems (15–30 cm long), and somewhat larger heads (with the involucre 15–20 mm long).

47. Sonchus L. (sow thistle)
(Boulos, 1972, 1973, 1974a, b; Pons and Boulos, 1972; Roux and Boulos, 1972)

Plants annual, biennial, or perennial herbs, taprooted or with rhizomes. Latex white. Stems 1 or few, erect or ascending, unbranched or branched, finely to relatively coarsely ridged, usually hollow between the nodes, glabrous below the inflorescence (rarely stalked-glandular in *S. asper*). Leaves alternate and basal, mostly 4–20 times as long as wide, glabrous or the undersurface rarely sparsely pubescent with minute, inconspicuous, unbranched hairs, sessile or the basal leaves with short to long, winged petioles, the basal leaves usually persistent at flowering, the stem leaves with a pair of prominent lobes clasping the stem. Leaf blades shallowly to deeply and often irregularly pinnately lobed, narrowly oblong-lanceolate to lanceolate, oblanceolate, or less commonly obovate in outline, the margins with sharp, spreading teeth, these often irregular and prickly at the tips. Inflorescences terminal panicles, the heads solitary or more commonly in loose clusters at the branch tips, sometimes reduced to a solitary terminal cluster of heads. Involucre becoming slightly but not noticeably elongated as the fruits mature, ovoid to pear-shaped at early flowering, usually becoming cup-shaped or somewhat bell-shaped by late flowering or fruiting, the bracts 1 inner and 3 or 4 outer series, often somewhat thickened toward the base, glabrous or sparsely to moderately pubescent with spreading, gland-tipped hairs, occasionally with minute, branched, cobwebby to woolly hairs toward the base, sometimes darkened or purplish-tinged toward the tip, those of the outer series 17–25, overlapping and varying from much shorter than to nearly as long as the inner series, linear to narrowly lanceolate; those of the inner series more or less equal, 15–27, narrowly oblong-lanceolate, tapered to a sharply pointed tip, the tip appressed-ascending to loosely ascending at flowering. Receptacle naked, usually minutely pitted at the base of each floret. Ligulate florets 80–250 or more per head. Corollas light yellow to orangish yellow. Pappus of numerous apparently smooth (microscopically barbed) bristles, these white, often shed irregularly at fruiting. Fruits with the body oblong-elliptic to elliptic or oblanceolate in outline, flattened, not beaked, the pappus attached to an unexpanded, unmodified tip, with 3 to several longitudinal nerves or ridges on each face, sometimes also finely cross-wrinkled. Fifty to 70 species, Europe, Asia, Africa, introduced widely.

1. Involucre (10–)14–22 mm long; flowering heads 2.5–4.5 cm in diameter (measured across the spreading corollas); corollas 12–25 mm long; plants perennial, with deep-set, branched rhizomes .... 1. S. ARVENSIS
1. Involucre 9–13(–15) mm long; flowering heads 1.5–2.7 cm in diameter (measured across the spreading corollas); corollas 8–15 mm long; plants annual, taprooted

2. Stem leaves with the clasping basal lobes all rounded; fruits not cross-wrinkled (although the longitudinal nerves may be microscopically roughened) .... 2. S. ASPER
2. Most of the stem leaves with the clasping basal lobes sharply pointed; fruits noticeably cross-wrinkled (in addition to the longitudinal nerves) .... 3. S. OLERACEUS

1. Sonchus arvensis L. (field sow thistle, perennial sow thistle)

Pl. 262 e, f; Map 1096

Plants perennial, with deep-set, branched rhizomes. Stems 40–150 cm long, often somewhat glaucous. Leaves with the clasping basal lobes rounded or pointed, the margins with the teeth having relatively stiff, short, slender prickles at the tips, the upper surface glabrous, not or only slightly shiny, the undersurface glabrous or rarely sparsely pubescent with minute, inconspicuous, unbranched hairs. Basal and lower stem leaves 6–40 cm long, usually irregularly and deeply lobed. Median and upper stem leaves gradually reduced in size, variously shallowly or deeply lobed, sometimes unlobed and merely toothed. Inflorescence branches glabrous or sparsely to moderately pubescent with spreading, gland-tipped hairs, occasionally with minute, branched, cobwebby to woolly hairs toward the tip. Flowering heads 2.5–4.5 cm in diameter (measured across the spreading corollas). Involucre (10–)14–22 mm long, glabrous or sparsely to moderately pubescent with a central band of spreading, gland-tipped hairs, occasionally with minute, branched, cobwebby to woolly hairs toward the base. Corollas 12–25 mm long, bright yellow to orangish yellow. Pappus 8–14 mm long. Fruits 2.5–3.5 mm long, noticeably 5–8-ribbed on each face, also finely cross-wrinkled, reddish brown to dark brown. $2n=36, 54$. July–October.

Introduced, uncommon, known thus far only from Boone, Marion, and St. Louis Counties and the city of St. Louis (native of Europe, introduced widely in North America). Railroads, gardens, and disturbed areas.

Most botanists recognize two subspecies based mainly upon differences in pubescence patterns of the inflorescence (and underlying differences in ploidy). Both types appear to be widespread weeds in North America, although usually they are not found in mixed populations. The paucity of documented chromosome counts makes it impossible to determine whether the ploidy differences are consistently correlated with the morphological characters of the subspecies.

1. Involucres and at least the upper portions of the inflorescence branches moderately to densely pubescent with spreading, gland-tipped hairs, rarely also sparsely to moderately pubescent with minute, branched, cobwebby hairs
.... 1A. SSP. ARVENSIS
1. Involucres and inflorescence branches glabrous or sparsely to densely pubescent with minute, branched, cobwebby to woolly hairs (toward the base of the involucre and along the adjacent branch tip), rarely also with a few spreading, gland-tipped hairs
.... 1B. SSP. ULIGINOSUS

1a. ssp. arvensis

Branches of the inflorescence moderately to densely pubescent with spreading, gland-tipped hairs toward the tip, rarely also sparsely to moderately pubescent with minute, branched, cobwebby hairs. Involucre with a central stripe of moderate to dense, spreading, gland-tipped hairs, rarely also sparsely to moderately pubescent with minute, branched, cobwebby hairs toward the base. $2n=54$. July–October.

Introduced, uncommon, known only from the city of St. Louis (native of Europe, introduced widely in North America). Gardens.

Plants of ssp. *arvensis* tend to differ from those of ssp. *uliginosus* in being overall somewhat more robust, with slightly larger lower leaves and slightly larger heads, but the differences are not so marked that they can be circumscribed easily in a key. Steyermark (1963) reported this subspecies from Jackson County, but no specimens could be located during the present study to confirm its presence there. The St. Louis occurrence is based on a single historical collection from the grounds of the Missouri Botanical Garden.

1b. ssp. uliginosus (M. Bieb.) Nyman

S. arvensis var. *glabrescens* Günther, Grab. & Wimm.

Branches of the inflorescence glabrous or occasionally sparsely to densely pubescent with minute, branched, cobwebby to woolly hairs toward the tip,

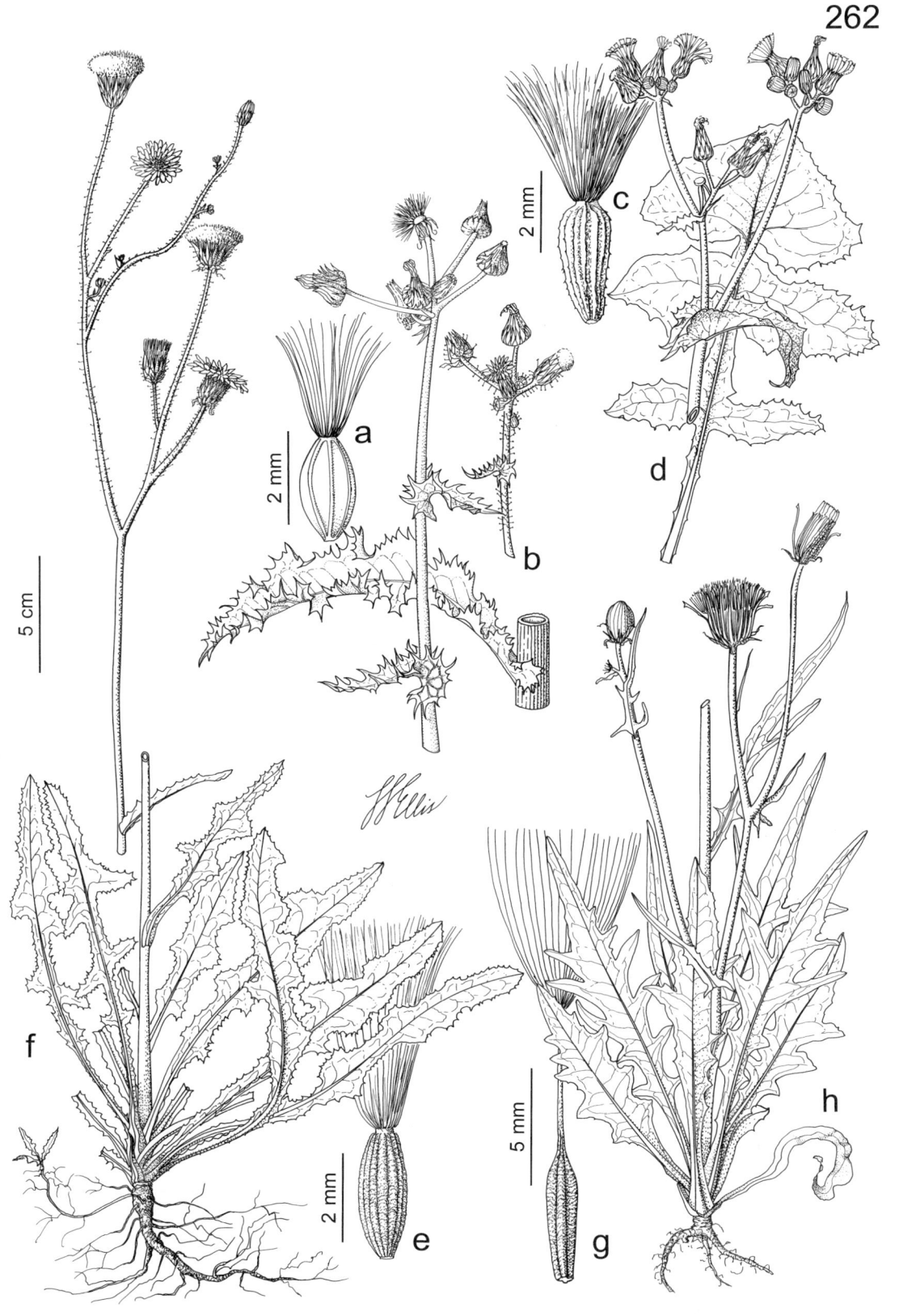

Plate 262. Asteraceae. *Sonchus asper*, **a)** fruit, **b)** inflorescences of two subspecies and leaves. *Sonchus oleraceus*, **c)** fruit, **d)** inflorescence and leaves. *Sonchus arvensis*, **e)** fruit, **f)** habit. *Pyrrhopappus carolinianus*, **g)** fruit, **h)** habit.

1098. Sonchus oleraceus

1099. Taraxacum erythrospermum

1100. Taraxacum officinale

lacking gland-tipped hairs. Involucre glabrous or occasionally sparsely to densely pubescent with minute, branched, cobwebby to woolly hairs toward the base, lacking gland-tipped hairs. $2n=36$. July–October.

Introduced, uncommon, known thus far only from Boone, Marion, and St. Louis Counties and the city of St. Louis (native of Europe, introduced widely in North America). Railroads, gardens, and disturbed areas.

2. Sonchus asper (L.) Hill **ssp. asper** (prickly sow thistle, spiny-leaved sow thistle)
S. asper f. glandulosus Hauman
Pl. 262 a, b; Map 1097

Plants annual, taprooted. Stems 10–150(–200) cm long, glabrous or rarely with sparse, spreading, reddish brown, gland-tipped hairs toward the tip, sometimes somewhat glaucous. Leaves with the clasping basal lobes rounded, the margins with the teeth having relatively stiff, short, slender prickles at the tips, glabrous, the upper surface darker green and shiny, the undersurface pale and sometimes somewhat glaucous. Basal and lower stem leaves 4–40 cm long, usually irregularly and deeply lobed. Median and upper stem leaves gradually reduced in size, variously unlobed to shallowly lobed, less commonly deeply lobed. Inflorescence branches glabrous or rarely sparsely pubescent with spreading, reddish brown, gland-tipped hairs, occasionally with minute, branched, cobwebby to woolly hairs toward the tip when young. Flowering heads 1.5–2.7 cm in diameter (measured across the spreading corollas). Involucre 9–13(–15) mm long, glabrous or occasionally with minute, branched, cobwebby to woolly hairs toward the base when young. Corollas 8–15 mm long, lemon yellow to bright yellow. Pappus 6–9 mm long. Fruits 2–3 mm long, finely 3(5)-nerved on each face, not cross-wrinkled (although the longitudinal nerves may be sparsely and microscopically roughened), yellowish brown to reddish brown. $2n=18$. May–October.

Introduced, scattered nearly throughout the state (native of Europe, introduced widely in North America). Banks of streams and rivers and disturbed bottomland prairies; also pastures, margins of crop fields, fallow fields, cemeteries, gardens, barnyards, railroads, roadsides, and moist, disturbed areas.

Steyermark (1963) noted that this species has been undercollected in Missouri, and it remains so today. The distribution map thus grossly underrepresents its actual distribution in the state. Boulos (1973) accepted two subspecies within S. asper. The Missouri plants represent ssp. asper, comprising annual plants with generally well-developed stem leaves and the fruits not or only slightly roughened on the nerves with sparse, microscopic awns. The name ssp. glaucescens (Jord.) Ball applies to biennial plants with mostly basal leaves and more densely barbed nerves on the achenes.

3. Sonchus oleraceus L. (common sow thistle)
Pl. 262 c, d; Map 1098

Plants annual, taprooted. Stems 10–200 cm long, glabrous or with sparse, spreading, reddish brown, gland-tipped hairs toward the tip, usually somewhat glaucous. Leaves with the clasping basal lobes mostly sharply pointed (less commonly some of them rounded), the margins with the teeth having relatively soft, short, slender prickles at the tips, glabrous, the upper surface somewhat darker green but not shiny, the undersurface pale and usually glaucous. Basal and lower stem leaves 4–40 cm long, usually irregularly and deeply lobed. Median and upper stem leaves gradually reduced in size, variously unlobed to more commonly shallowly or deeply lobed. Inflorescence branches usually sparsely pubescent with spreading, reddish brown, gland-tipped hairs, sometimes with minute, branched, cobwebby to woolly hairs toward the tip. Flowering heads 1.5–2.7 cm in diameter (measured across the spreading corollas). Involucre 9–13(–15) mm long, usually with at least a few spreading, reddish brown, gland-tipped hairs, sometimes also with minute, branched, cobwebby to woolly hairs toward the base when young. Corollas 8–15 mm long, light yellow to lemon yellow. Pappus 5–8 mm long. Fruits 2.5–3.0 mm long, faintly to noticeably 3- or 5-nerved on each face, also finely to relatively coarsely cross-wrinkled, yellowish brown to reddish brown. 2n=32, 36. June–October.

Introduced, scattered nearly throughout the state (native of Europe, introduced widely in North America). Pastures, banks of ditches, gardens, barnyards, railroads, roadsides, and moist, disturbed areas.

Steyermark (1963) noted that this species has been undercollected in Missouri, and it remains so today. The distribution map thus grossly underrepresents its actual distribution in the state.

48. Taraxacum F.H. Wigg. (dandelion)

Plants perennial herbs, with somewhat fleshy taproots. Latex white. Stems 1 to several, continuing to elongate as the heads mature, erect or ascending, unbranched, hollow, finely ridged or nearly smooth, glabrous or with patches of fine, white, cobwebby hairs, sometimes purplish-tinged or purplish-mottled. Leaves all basal (see discussion under *T. erythrospermum*), short- to long-petiolate. Leaf blades shallowly to deeply and irregularly pinnately lobed, sometimes nearly compound, narrowly oblanceolate to obovate or elliptic in outline, the lobes triangular with angled sinuses, irregularly toothed, glabrous or 1 or both surfaces sparsely to moderately pubescent along the veins with fine, irregularly curled, white, sometimes somewhat cobwebby hairs, sometimes pinkish- to purplish-tinged toward the base. Venation of 1 main vein and a network of anastomosing secondary and tertiary veins. Heads solitary at the stem tips. Involucre elongating somewhat as the fruits mature, somewhat urn-shaped to cup-shaped at flowering, the bracts in 1 inner series of 11–23 and 1 or 2 additional outer series of 11–17, usually glabrous, often purplish-tinged, especially toward the tip; those of the inner series similar in size and shape, sometimes more or less fused along the margins toward the base when young, lanceolate, with well-differentiated, thin, pale margins, the tip sharply pointed or minutely notched, ascending at flowering; those of the outer series less than to more than half as long as the inner series, ovate to narrowly ovate, mostly becoming reflexed as the heads first develop. Receptacle naked. Ligulate florets 40–120 or more per head. Corollas bright yellow, sometimes purplish- or grayish-brown-tinged on the outer surface. Pappus of numerous bristles, these smooth or microscopically barbed, white. Fruits with the body oblanceolate in outline, tapered to a slender beak usually more than twice as long as the body, not or only slightly flattened, with 4 or 5 rounded or flattened ribs, these sometimes with 1 or 2 shallow longitudinal grooves, smooth or appearing smooth or more commonly minutely pebbled or roughened, with several rows of prominent barbs toward the tip, glabrous, variously colored, the pappus attached to a relatively broad, expanded, disclike tip. About 60 to more than 2,000 species, North America, South America, Europe, Asia, introduced worldwide.

The taxonomy of the weedy dandelions is in a state of perpetual confusion. As in *Hieracium*, botanists working in the Old World centers of taxonomic diversity for the genus have described large numbers of apomictic polyploid races as microspecies, which has tended to confound application of species-level nomenclature to the introduced populations in North America. The two taxa present in Missouri have been assigned to two different sections, sect. *Erythrosperma* (H. Lindb.) Dahlst. and sect. *Ruderalia* Kirschner, H. Øllg. & Štěpánek, which are characterized by large numbers of weedy polyploid apomictic taxa within their native ranges. In fact, ecological and biochemical studies of nonnative populations in the northwestern United States have indicated that the two taxa present in Missouri may not be particularly distinct morphologically, biochemically, or ecologically (Taylor, 1987). Conversely, genetic studies (King and

Schaal, 1990; King, 1993) have documented surprisingly high genetic diversity among North American dandelions, as well as intertaxon hybridization, due in part to low levels of outcrossing sexual reproduction in these predominantly apomictic plants (Valentine and Richards, 1967). The nomenclature and taxonomy of both of the Missouri dandelions remains controversial, with most recent authors provisionally relying on traditional names and circumscriptions for the species until the taxonomic situation with these complexes within their native ranges becomes stabilized. See further discussion below.

1. Fruits dull brick red to reddish or purplish brown at maturity
 . . . . 1. T. ERYTHROSPERMUM
1. Fruits olive-colored to greenish brown at maturity . . . . 2. T. OFFICINALE

1. Taraxacum erythrospermum Andrz. ex Besser (red-seeded dandelion)

T. laevigatum (Willd.) DC.

T. laevigatum f. *scapifolium* F.C. Gates & S.F. Prince

Pl. 263 a; Map 1099

Stems 2–25 cm long. Leaves (3–)6–25 cm long, almost always very deeply lobed (nearly compound), except on small plants. Involucre 10–20(–25) mm long, the inner series of bracts mostly 11–18, the outer series usually less than half to about half as long as the inner series. Fruits with the body 2.5–4.0 mm long, dull brick red to reddish or purplish brown at maturity. $2n=16, 22, 24, 26, 32$. January–December.

Introduced, scattered nearly throughout the state (native of Europe, introduced nearly throughout temperate North America). Banks of streams and rivers, openings of dry upland forests, margins of glades, and disturbed sand prairies; also lawns, gardens, crop fields, fallow fields, pastures, roadsides, railroads, and open, disturbed areas.

There is widespread disagreement on the correct name for this taxon. Many authors have continued to call these plants *T. laevigatum* (Barkley, 1986), but Shinners (1949) suggested that the original description of that name did not describe the plant that has been naturalized in North America. He suggested instead the epithet *T. erythrospermum,* which has also been adopted in most of the recent literature on dandelions in Europe. However, both of these names are of uncertain application, as type specimens still have not been designated for them. The nomenclatural situation merits further research. Gates and Prince (1938) studied an unusual mutant ascribed to this species from Stone County that they named f. *scapifolium,* which was characterized by leafy stems as well as leaflike bracts subtending the heads. This odd form has not been seen since their initial discovery, and to avoid confusion it has not been included in the descriptions of the genus and species above.

This species is somewhat less commonly encountered in Missouri than is *T. officinale*. Steyermark (1963) suggested the existence of hybrids between the two species of dandelion in Missouri, but this could not be confirmed during the present study. In the absence of mature fruits, Missouri plants of the two dandelions are nearly impossible to distinguish. Individuals of *T. erythrospermum* tend to be slightly smaller plants than those of *T. officinale* and are more likely to have all of the leaves deeply lobed all the way to the rachis. They also are more likely to occur in sandy soils. However, *T. officinale* is so variable in its morphology and habitats that there is almost total overlap between the species.

2. Taraxacum officinale F.H. Wigg. (common dandelion)

T. vulgare Schrank (an illegitimate name)

Pl. 263 b, c; Map 1100

Stems 2–40 cm long. Leaves (3–)8–45 cm long, variably shallowly to very deeply lobed (nearly compound), sometimes more shallowly lobed toward the tip than toward the base. Involucre 12–25 mm long, the inner series of bracts mostly 13–23, the outer series less than to more than half to about half as long as the inner series. Fruits with the body 3–4 mm long, olive-colored to greenish brown at maturity. $2n=16, 18, 22, 24, 26, 32, 33, 37, 46, 47, 48$. January–December.

Introduced, common throughout the state (native of Europe, introduced nearly worldwide). Banks of streams and rivers, ledges of bluffs; also lawns, gardens, cemeteries, crop fields, fallow fields, pastures, roadsides, railroads, and open, disturbed areas.

The name *T. officinale* was lectotypified by Richards (1985) to refer to a very different species, a Scandinavian endemic that is only distantly related to the widespread, weedy taxon found in North America. Unfortunately, it is unclear which of the nearly 50 species names in sect. *Ruderalia* (Kirschner and Štěpánek, 1987) should be used for

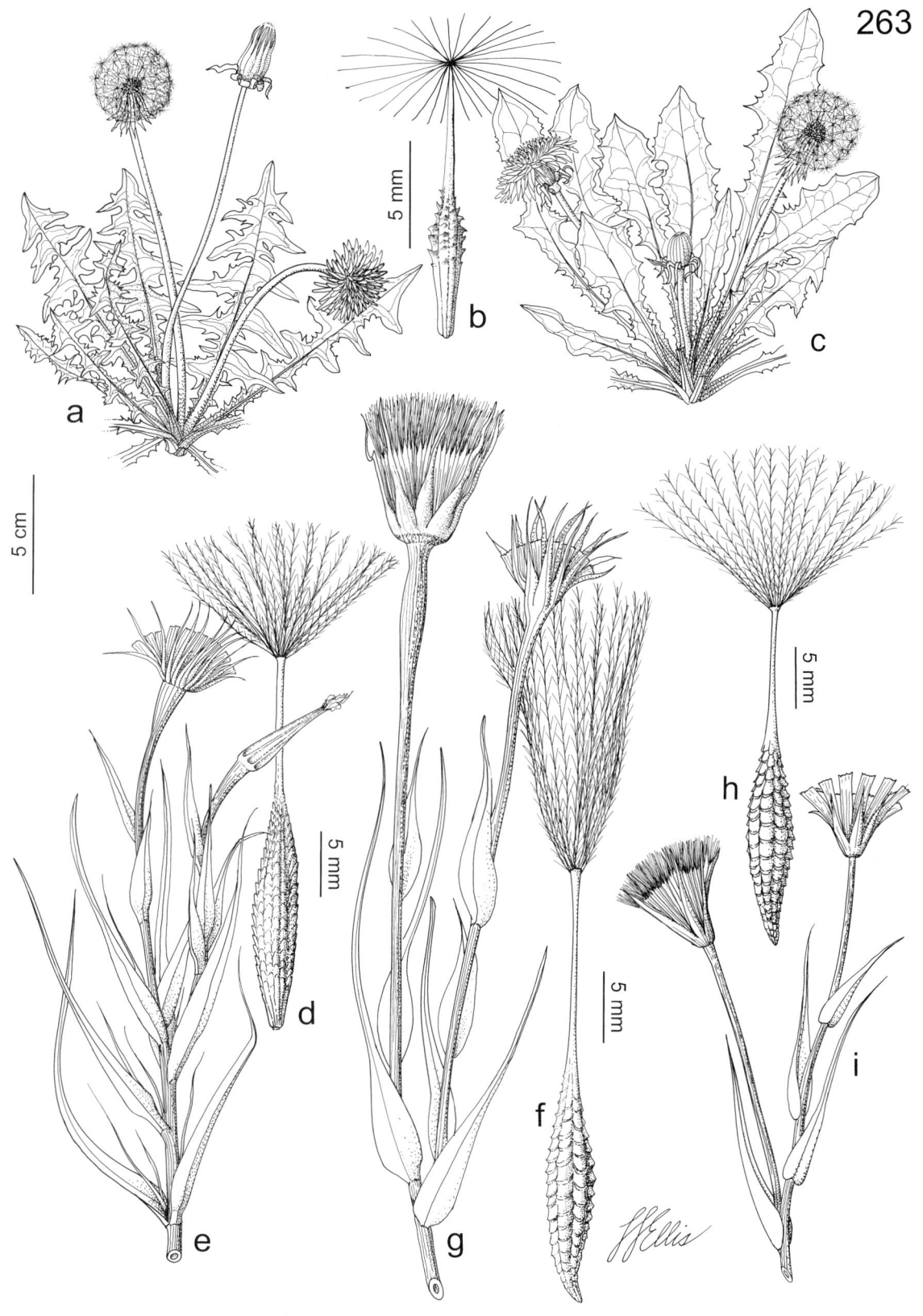

Plate 263. Asteraceae. *Taraxacum erythrospermum*, **a)** habit. *Taraxacum officinale*, **b)** fruit, **c)** habit. *Tragopogon dubius*, **d)** fruit, **e)** habit. *Tragopogon porrifolius*, **f)** fruit, **g)** habit. *Tragopogon pratensis*, **h)** fruit, **i)** habit.

the introduced North American plants, which represent presumed multiple introductions from different Old World sources. The present treatment uses the traditional nomenclature for the common dandelion as broadly circumscribed, until further study clarifies the situation.

Dandelions flowering in lawns are both welcomed by urbanites as a sign of spring and cursed by homeowners, who have helped to fuel the herbicide and lawn care industries through their desire to produce the unblemished great American lawn. However, dandelions also are becoming popular as a field green among gourmets and are slowly expanding from a familiar weed to a minor organic crop in portions of the northern United States. The leaves are highly nutritious and are a good source of various vitamins and iron. They are frequently sold in mesclun salad mixtures, which contain several fancy lettuce cultivars as well as leafy greens from various species of Brassicaceae, Chenopodiaceae, and Asteraceae (Ryder, 2002). Care should be taken that any plants harvested from nature be from areas free of pesticide residues. In addition to its use in salads, the flowering heads or more commonly just the corollas are mixed with sugar and various flavorings, then allowed to ferment, to produce the alcoholic beverage known as dandelion wine. The roots occasionally have been used as a chicory substitute and are also a source of the sugar substitute inulin (see the treatment of *Cichorium* for more information). In Europe, the plants have a history of medicinal use as a laxative, diuretic, and tonic and were also used to cleanse the blood (for liver ailments).

49. Tragopogon L. (goat's beard)

Plants mostly biennial herbs (sometimes short-lived perennials), with somewhat fleshy taproots. Latex white. Stems 1 to several, erect or ascending, unbranched or more commonly with ascending branches, often hollow (at least toward the tip), often swollen toward the tip (at least at fruiting), finely ridged, usually with patches of fine, white, cobwebby to woolly hairs when young, appearing glabrous or nearly so at maturity, often glaucous, sometimes purplish-tinged or purplish-mottled around the lower nodes. Leaves basal and alternate, sessile, the basal leaves sometimes withered at flowering. Leaf blades grasslike, unlobed, linear from a slightly broadened, more or less rounded base (this sometimes purplish-tinged), the margins often slightly corrugated or somewhat incurled and with a narrow, pale band, this often slightly and microscopically roughened, the surfaces sometimes with small patches of inconspicuous, cobwebby hairs toward the base when young, appearing glabrous at maturity, usually glaucous. Venation of few to several apparently parallel main veins, sometimes the central main vein somewhat thicker than the others, and a sometimes faint network of anastomosing secondary veins. Heads solitary at the branch tips. Involucre elongating as the fruits mature, conical to bell-shaped at flowering, the bracts in 1 series of 8–13, all similar in size and shape (no shorter outer bracts present), lanceolate to narrowly lanceolate, long-tapered at the tip, usually with a narrow, pale margin, glabrous or cobwebby-hairy at the base, sometimes purplish-tinged. Receptacle naked, but often with minute, scaly ridges between the florets. Ligulate florets 50–180 or more per head. Corollas yellow or purple. Pappus of numerous plumose bristles, these off-white to straw-colored or tan to light brown, a few usually distinctly longer than the rest. Fruits with the body narrowly lanceolate to nearly linear in outline, often somewhat curved or asymmetrical, tapered to a slender or relatively stout beak 1–2 times as long as the body (rarely some of the outermost fruits with a much shorter beak), not or only slightly flattened, with 5–10 nerves or fine, rounded ribs, the surface with dense tubercles or blunt, ascending barbs, straw-colored to tan or dark brown (often varying in color in a single head), the pappus attached to a relatively broad, expanded, disclike tip (except in *T. porrifolius,* with a narrow, club-shaped tip), this frequently with woolly hairs. About 50 species, Europe, Asia, Africa.

Tragopogon defies conventional definitions of the native vs. nonnative status of species in the North American flora. The three Old World diploid species in Missouri are widely naturalized in temperate North America. Where mixed-species populations occur (first studied in the northwestern United States and not yet noted in Missouri), sterile hybrids are produced frequently. In several independent cases, two of these hybrids have regained their fertility through allopolyploidy (doubling of the chromosomes in an interspecific hybrid), with each tetraploid derivative acting biologically as a full species. This phenomenon in *Tragopogon* formed the basis for a seminal series of studies in plant biosystematics by Marion Ownbey and his colleagues at Washington State University (Ownbey, 1950; Ownbey and McCollum, 1953, 1954; Belzer and Ownbey, 1971). Since then, Ownbey's findings on reticulate evolution in the genus have been confirmed and refined by Pamela and Douglas Soltis (then at Washington State University) and their colleagues (Novak et al., 1991; Soltis and Soltis, 1991; Soltis et al., 1995; Cook et al. 1998). Because the two fertile allotetraploid species, *T. mirus* Ownbey (derived from past hybridization between *T. dubius* and *T. porrifolius*) and *T. miscellus* Ownbey (derived from past hybridization between *T. dubius* and *T. pratensis*), have been documented to have evolved during the twentieth century in the western United States (and apparently have not formed in Europe), they have been treated by most authors as natives of North America, even though their progenitors are not native to the continent.

Heads of *Tragopogon* species open each morning and, except on cloudy days, generally close by noon. The plumose pappus bristles tend to intertwine, and when the involucral bracts become reflexed as the fruits mature, the pappus takes on an intricate globose shape. These fruiting heads are sometimes carefully dried and then sprayed with an aerosol resin (to prevent dispersal of the fruits) for use in dried flower arrangements. Before the heads are produced, plants of *Tragopogon* might be confused with some monocot genera such as *Tradescantia* (Commelinaceae) or some grasses. However, the grasslike leaves in *Tragopogon* do not have sheathing bases as in nearly all monocots.

1. Corollas light purple to bright purple, the tubular basal portion sometimes yellow . . . . 2. T. PORRIFOLIUS
1. Corollas light yellow to bright yellow
   2. Involucral bracts longer than the florets; stems usually swollen toward the tip . . . . 1. T. DUBIUS
   2. Involucral bracts shorter than to as long as the florets; stems not swollen toward the tip at flowering, occasionally becoming slightly inflated at fruiting . . . . 3. T. PRATENSIS

1. Tragopogon dubius Scop. (yellow salsify, western salsify)

Pl. 263 d, e; Map 1101

Stems 30–100 cm long, usually relatively strongly swollen at the tip at flowering and fruiting. Leaves 5–30 cm long, usually occasionally curled or recurved at the tip. Involucre 2.5–4.0 cm long at flowering, longer than the florets, becoming elongated to 4–7 cm long at fruiting, the bracts (8–)13. Ligulate florets 100–180 or more per head, the corollas 21–28 mm long, usually lemon yellow, rarely bright yellow. Pappus 12–29 mm long, off-white to straw-colored or less commonly yellowish tan. Fruits with the body about 10–14 mm long, tapered to a relatively stout beak 12–20 mm long, the expanded disc at the tip densely pubescent with short, woolly hairs, the surfaces straw-colored to tan or yellowish to reddish brown. $2n=12$. May–August.

Introduced, scattered nearly throughout the state (native of Europe, introduced nearly throughout the U.S.; Canada). Disturbed portions of glades, upland prairies, and marshes; also pastures, margins of crop fields, banks of ditches, levees, railroads, roadsides, and open, disturbed areas.

Steyermark (1963) reported this species primarily from the northern half of Missouri. It has greatly increased its distribution south of the Missouri River since that time. He also noted that its culinary uses were similar to those described for *T. porrifolius*. *Tragopogon dubius* is the most robust among the species in Missouri, both in terms of stem size and numbers of leaves.

2. Tragopogon porrifolius L. (salsify, oyster plant, vegetable oyster)

Pl. 263 f, g; Map 1102

Stems 35–100 cm long, usually swollen at the tip at flowering and fruiting. Leaves 3–30 cm long, usually relatively straight at the tip. Involucre

1101. Tragopogon dubius   1102. Tragopogon porrifolius   1103. Tragopogon pratensis

2.5–4.0 cm long at flowering, as long as or more commonly somewhat longer than the florets, becoming elongated to 4–7 cm long at fruiting, the bracts 8(–13). Ligulate florets mostly 85–110 per head, the corollas 16–35 mm long, light purple to bright purple (rarely yellowish toward the base or pinkish purple). Pappus 17–26 mm long, straw-colored to more commonly yellowish brown. Fruits with the body about 10–16 mm long, tapered to a relatively slender beak 15–25 mm long, the narrow, club-shaped tip only slightly and inconspicuously hairy, the surfaces straw-colored to light brown, the outer series sometimes darker brown. 2n=12. May–July.

Introduced, uncommon, widely scattered in the state (native of southern Europe, cultivated widely, introduced nearly throughout the U.S.; Canada, northern Europe, Asia, Africa). Railroads, roadsides, and open, disturbed areas.

Salsify has a long history of cultivation in Europe. The carrot-shaped taproots have an oysterlike flavor when cooked and have been used in soups and stews or as a relish. They also have been dried and ground for use as a substitute for chicory (Vuilleumier, 1973). The young shoots are sometimes cooked and the foliage can be eaten raw as a salad green. The roots also were used medicinally for liver ailments and as a diuretic.

3. Tragopogon pratensis L. (meadow salsify; Jack-go-to-bed-at-noon)

Pl. 263 h, i; Map 1103

Stems 20–80 cm long, not swollen at the tip at flowering, occasionally becoming slightly inflated at fruiting. Leaves 3–30 cm long, usually sometimes curled or recurved at the tip. Involucre 1.2–2.5 cm long at flowering, shorter than to as long as the florets, becoming elongated to 2–4 cm long at fruiting, the bracts 8(–12). Ligulate florets mostly 50–110 per head, the corollas 15–27 mm long, lemon yellow to bright yellow. Pappus 15–27 mm long, off-white to straw-colored or less commonly yellowish tan. Fruits with the body about 10–14 mm long, tapered to a relatively stout beak 4–9 mm long, the expanded disc at the tip densely pubescent with short, woolly hairs, the surfaces straw-colored to tan or yellowish brown. 2n=12. May–August.

Introduced, known thus far only from Nodaway County and the city of St. Louis (native of Europe, introduced nearly throughout the U.S., Canada). Railroads and open, disturbed areas.

Palmer and Steyermark (1935) included *T. pratensis* from fifteen counties, but Steyermark (1963) redetermined all of these records as representing *T. dubius*. Nevertheless, one of the specimens appears to represent a valid record of *T. pratensis*. The St. Louis record, which Steyermark (1963) did accept, resulted from the botanical inventories of the local railyards by Viktor Mühlenbach. The relatively short beak on the fruits is an additional character to help in distinguishing *T. pratensis* from the far more common *T. dubius*.

The taxonomy of the *T. pratensis* complex requires further study. Based on their morphological and cytogenetic studies of materials mostly from the Iberian Peninsula of Spain, Díaz de la Guardia and Blanca (1988) attempted to recognize two taxa within what had been known traditionally as *T. pratensis*, segregating a second species under the name *T. lamottei* Rouy. They subsequently (Díaz de la Guardia and Blanca, 1992) argued that the older name *T. pratensis* should be restricted to plants having only 5 (vs. usually 8) involucral bracts per head, these somewhat shorter and with green (vs. reddish) margins, and the leaves on average shorter and narrower (among other subtle characters). They further characterized *T. pratensis* in the strict sense as having shorter fruits with a shorter beak than is found in *T. lamottei* (Blanca and Díaz de la Guardia, 1997). Because populations in North America tend to have 8 bracts per involucre, there has been a temptation to adopt the name *T. lamottei* for our plants, but in fact, the situation is more complex. North American materials tend to have fruits that are larger and longer-beaked than the sets of measurements given for either taxon by Blanca and Díaz de la Guardia (1997). Ownbey and McCollum's (1954) karyological studies showed more variation in the chromosomal morphology of North American plants than was accounted for by Díaz de la Guardia and Blanca (1988). The Spanish plants apparently all have swollen stem apices but were said to differ in whether the stem tip was contracted (in *T. pratensis*) or not (in *T. lamottei*) just below the base of the head. North American plants have stems that are not or only slightly swollen toward the tip at flowering, sometimes becoming enlarged as the fruits mature, and they appear to be variable for whether there is a constriction at the base of the head. Most recently, preliminary molecular studies of the genus by Mavrodiev et al. (2005) have shown that a sample of *T. lamottei* from Spain and *T. pratensis* from the state of Washington are not the same species. The relationship of the Spanish populations to those in the remainder of the range of the complex, including the North American nonnative occurrences, should be studied in much greater detail before the broad, traditional taxonomic concept of *T. pratensis* is abandoned.

5. Tribe Eupatorieae Cass.
(King and Robinson, 1987)

Plants annual (in *Ageratum*) or perennial herbs (shrubs elsewhere), sometimes from a woody rootstock or corm, occasionally twining climbers, the sap not milky. Stems not spiny or prickly. Leaves alternate, opposite, or whorled, sometimes also in a basal rosette, sessile to long-petiolate, not spiny or prickly. Leaf blades entire to pinnately dissected, the venation mostly pinnate, with 1 or 3(–7) main veins. Inflorescences terminal or less commonly axillary panicles, spikes, or racemes, rarely appearing as solitary or clustered axillary heads. Heads discoid. Involucre of 2 to several series of bracts (1 series in *Mikania*), these overlapping, of more or less equal to strongly unequal lengths, not spiny or tuberculate. Receptacle flat to slightly convex, less commonly conical, naked (chaffy elsewhere). Disc florets all perfect, the corolla white, pink, or purple to nearly blue, the 5 short or less commonly long lobes spreading to ascending. Pappus most commonly of usually numerous capillary bristles, less commonly of relatively few scales or awns, more or less persistent at fruiting. Stamens with the filaments not fused together or fused into a short tube toward the base, the anthers fused into a tube, each tip with a flattened appendage, each base truncate or broadly rounded. Style branches usually not flattened, each with a short stigmatic line along each inner margin, the sterile tip elongate, usually with dense, minute papillae. Fruits monomorphic, mostly several-angled or several-ribbed in cross-section, oblong to slightly wedge-shaped in profile, not beaked. About 170 genera, about 2,400 species, nearly worldwide, but most diverse in the New World.

The modern classification of genera of Eupatorieae largely has been the work of Robert Merrill King and Harold Robinson (both then at the Smithsonian Institution). Starting with a lengthy series of morphological taxonomic studies published over more than two decades and culminating in a comprehensive book-length monograph of the tribe (King and Robinson, 1987), the generic classification of the tribe is among the most thoroughly researched projects in the family. Subsequent molecular studies (Schilling et al., 1999; Schmidt and Schilling, 2000; Ito et al., 2000a, b) have largely supported King and Robinson's insightful taxonomic observations of morphological and anatomical details. As has been the trend in other tribes of Asteraceae, these studies have resulted in the dissection of the large, polymorphic traditional genus *Eupatorium* into a series of smaller, more homogeneous genera, four of which occur in Missouri.

1. Stems twining; heads with 4 involucral bracts, these more or less equal in size
 . . . . 57. MIKANIA
1. Stems not twining (but sometimes reclining on other plants in *Fleischmannia*); heads with more than 4 involucral bracts, these of unequal lengths

2. Leaves mostly in well-separated whorls of 3–7, the uppermost leaves sometimes opposite or alternate . . . . 54. EUPATORIUM
2. Leaves opposite or alternate, the internodes well spaced or sometimes short and the crowded leaves then appearing more or less indefinitely whorled
   3. Stem leaves alternate but sometimes so numerous and dense as to appear indefinitely whorled; fruits finely 10-nerved or 10-ribbed
      4. Basal leaves absent at flowering or when rarely present the basal leaves about as large as or smaller than the lower and median stem leaves; inflorescence a flat-topped or rounded panicle (in poorly developed plants sometimes reduced to few stalked clusters of heads), or appearing as a leafy panicle with racemose branches . . . . 52. BRICKELLIA
      4. Basal leaves often present at flowering, these and the adjacent lowermost stem leaves the largest on the plant; inflorescence an unbranched terminal spike or spikelike raceme, sometimes leafy and the lowermost heads then appearing axillary . . . . 56. LIATRIS
   3. Stem leaves opposite, mostly well spaced; inflorescence a broad panicle (in poorly developed plants sometimes reduced to a few umbellate clusters of heads) or (in *Eupatorium capillifolium*) appearing as a leafy panicle with numerous ascending, spicate branches; fruits 3–5-angled or 3–5-ribbed (but 10-ribbed in *Brickellia*)
      5. Corollas pink to purple or lavender-blue (note that rarely isolated individuals with white flowers may be found scattered within populations of plants otherwise with pigmented corollas)
         6. Pappus of 5(6) scales; plants annual . . . . 51. AGERATUM
         6. Pappus of numerous capillary bristles; plants perennial
            7. Receptacle conical; florets 35–70 per head; plants with rhizomes . . . . 53. CONOCLINIUM
            7. Receptacle flat or slightly convex; florets 13–25 per head; plants without rhizomes . . . . 55. FLEISCHMANNIA
      5. Corollas white or cream-colored, sometimes appearing pale gray; leaf blades variously linear to triangular-ovate
         8. Leaves sessile or short-petiolate . . . . 54. EUPATORIUM
         8. Leaves long-petiolate (petioles at least $1/3$ as along as the blades)
            9. Leaf blades linear to lanceolate, narrowly triangular, or narrowly elliptic or, if broader in outline, then pinnately dissected into narrowly linear lobes . . . . 54. EUPATORIUM
            9. Leaf blades triangular-ovate, unlobed
               10. Leaf blades mostly broadly angled or broadly rounded at the base; fruits 5-angled or 5-ribbed . . . . 50. AGERATINA
               10. Leaf blades mostly truncate to cordate at the base; fruits finely 10-nerved or 10-ribbed . . . . 52. BRICKELLIA

50. **Ageratina** Spach (snakeroot)
(Clewell and Wooten, 1971)

About 250 species, North America, Central America, South America, Caribbean Islands, most diverse in the Neotropics, introduced widely.

1104. Ageratina altissima   1105. Ageratum conyzoides   1106. Brickellia eupatorioides

1. Ageratina altissima (L.) R.M. King & H. Rob. **var. altissima** (white snakeroot)

A. altissima var. *angustatum* (A. Gray) Clewell & Wooten

Eupatorium rugosum Houtt.

E. rugosum f. *villicaule* Fernald

E. rugosum var. *tomentellum* (B.L. Rob.) S.F. Blake

Pl. 267 f, g; Map 1104

Plants perennial, fibrous-rooted. Stems solitary or few, 30–150 cm long, erect or ascending, glabrous or short-hairy, rarely with longer (0.5–1.0 mm) hairs, occasionally somewhat glaucous. Leaves opposite (the nodes well separated), short- to long-petiolate. Leaf blades 2–18 cm long, triangular to triangular-ovate, the uppermost sometimes only lanceolate or narrowly lanceolate, mostly broadly angled or broadly rounded at the base, rarely truncate, tapered to a sharply pointed tip, the margins broadly but often shallowly toothed, the surfaces glabrous or sparsely to moderately pubescent with short, curly hairs, often mostly along the veins, with 3 main veins or the main veins palmate. Inflorescences panicles, occasionally reduced to stalked clusters, flat-topped to dome-shaped. Heads with 9–25 disc florets. Involucre 3–5 mm long, more or less cup-shaped, the bracts 8–14 (the head often also subtended by 1 or 2 other shorter bracts), in usually 2 unequal, overlapping series, linear to more commonly narrowly oblong-elliptic or narrowly oblanceolate, tapered to a bluntly or sharply pointed tip, glabrous or sparsely to moderately and finely short-hairy, sometimes only along the margins. Receptacle flat or nearly so. Corollas white. Pappus of numerous capillary bristles. Fruits 1.7–3.0 mm long, 5-angled, somewhat wedge-shaped in profile (usually slightly and unevenly tapered at the base) to nearly linear, glabrous, brown to dark brown. $2n=34$. July–October.

Common throughout the state (eastern U.S. west to North Dakota and Texas; Canada). Bottomland forests, mesic and rarely dry upland forests, ledges of bluffs, banks of streams and rivers, and rarely glades; also pastures, old fields, railroads, roadsides, and open, disturbed areas.

Clewell and Wooten (1971) accepted var. *angustata* based on somewhat narrower-leaved plants occupying mainly the western portion of the species range (including portions of Missouri). However, these authors also recorded almost complete overlap between the overall distributions of var. *altissima* and var. *angustata* and also concluded that considerable variation in leaf morphology occurs in all portions of the species range. Thus it seems inadvisable to accept this variety in the present treatment. The other accepted variant, the mostly Appalachian var. *roanensis* (Small) Clewell and Wooten, differs in its slightly smaller heads with abruptly sharp-pointed involucral bracts.

White snakeroot is toxic when consumed in quantity or cumulatively in small quantities, causing an often fatal condition called trembles in livestock and called milk sickness in humans. As recently as 1998, horses in northeastern Missouri were reported to have died from white snakeroot poisoning (Anonymous, 1998). The toxic compounds are known as tremetol, which has been discovered to represent a complex mixture of sterols and methyl ketone benzofuran derivatives, notably tremetone (Burrows and Tyrl, 2001). Animals may ingest the toxins from fresh plants or hay and are at most risk when allowed to graze in wooded habitats where *A. altissima* can benefit from disturbance by livestock and form dense stands. The poisons are fat-soluble and can be transmitted in milk products and possibly also in meat from poisoned animals. Milk sickness was responsible for thousands of deaths in the eastern and midwestern

United States in the eighteenth and nineteenth centuries. In some portions of Indiana and Ohio, it apparently accounted for ¼–½ of all fatalities in the early 1800s. The mother of Abraham Lincoln, Nancy Hanks Lincoln, is said to have perished from milk sickness in southern Indiana in October 1818. Symptoms include weakness, muscle spasms, vomiting, constipation, thirst, delirium, and coma.

51. Ageratum L.
(Johnson, 1971)

About 40 species, North America, Central America, South America, Caribbean Islands, most diverse in the Neotropics.

1. Ageratum conyzoides L. **ssp. conyzoides**
(ageratum)
Pl. 264 g, h; Map 1105

Plants annual (occasionally perennial elsewhere), fibrous-rooted. Stems solitary or few, 10–50(–150) cm long, erect or ascending, sometimes from a spreading base, sparsely to densely hairy, sometimes also somewhat glandular. Leaves mostly opposite (the nodes well separated), the uppermost leaves occasionally alternate, short- to long-petiolate. Leaf blades 1–9 cm long, triangular-ovate to ovate, the uppermost sometimes only lanceolate, mostly broadly angled or broadly rounded at the base, rarely truncate, tapered to a bluntly or sharply pointed tip, the margins broadly but often shallowly scalloped or toothed, the upper surface glabrous or sparsely to moderately hairy along the veins, the undersurface moderately to densely hairy, especially along the veins, both surfaces often also somewhat glandular, with mostly 3 main veins. Inflorescences panicles, sometimes reduced to stalked clusters, usually more or less flat-topped. Heads with numerous (more than 25) disc florets. Involucre 3–5 mm long, the bracts 18–30 (the head often also subtended by 1 or a few other narrower bracts), in usually 2 more or less equal, overlapping series, narrowly oblong-elliptic or narrowly oblong-lanceolate, tapered to a sharply pointed tip, noticeably few-nerved or few-ribbed, glabrous or more commonly sparsely short-hairy. Receptacle conical. Corollas purple or lavender-blue (rarely white elsewhere). Pappus of 5(6) more or less lanceolate scales. Fruits 1.3–2.0 mm long, 5-angled, somewhat wedge-shaped in profile (usually slightly and unevenly tapered at the base), glabrous, the angles finely roughened or toothed, dark brown to black. 2n=20, 40. August–October.

Introduced, known thus far only from a single collection from Marion County (native of Mexico, Central America, South America; introduced widely in tropical and warm-temperate regions of the world). Banks of streams.

Ageratums are cultivated commonly as bedding plants in gardens for their beautiful displays of bluish flowers in autumn. A number of cultivars exist. Among native populations in the Neotropics, Johnson (1971) accepted ssp. *latifolium* (Cav.) M.F. Johnson, which differs in its shorter pappus scales (0.2–0.9 vs. 1.5–3.0 mm).

52. Brickellia Elliott

Plants perennial, often with thickened roots. Stems solitary to several, erect or ascending, usually sparsely to densely pubescent with short, curly or curved hairs. Leaves alternate or opposite (the nodes usually well separated), sessile to long-petiolate, the basal leaves usually absent at flowering, when present then about as large as or smaller than the lower and median stem leaves. Leaf blades linear to ovate-triangular or nearly heart-shaped, tapered, angled, truncate, or cordate at the base, usually tapered to a sharply pointed tip, the surfaces variously short-hairy to glabrous or nearly so, at least the undersurface also glandular, with (1)3(5) main veins. Inflorescences panicles, sometimes reduced to stalked clusters at the branch tips, flat-topped to dome-shaped, or less commonly appearing as a leafy panicle with racemose branches. Heads with 6–40 disc florets. Involucre narrowly cup-shaped to narrowly bell-shaped, the bracts 22–35 (the head often also subtended by 1 or more other shorter bracts) in usually several unequal, overlapping series, the bracts linear to narrowly oblong-elliptic or narrowly oblanceolate, those of the outer series sometimes ovate, tapered or narrowed to a bluntly or sharply pointed tip, glabrous or sparsely to moderately and finely short-hairy, also glandular. Receptacle flat or nearly so. Corollas cream-colored to greenish yellow. Pappus of 20–25 bristles, these minutely barbed or plumose. Fruits 1.7–5.0 mm long, (8–)10-nerved or ribbed, slightly wedge-shaped in profile (usually slightly and unevenly tapered at the base) to nearly linear, usually minutely hairy, brown to dark brown. About 100 species, North America, Central America, South America, most diverse in the U.S. and Mexico.

1. Leaves alternate, sessile or very short-petiolate, linear to lanceolate or narrowly elliptic, tapered at the base; heads usually ascending; pappus bristles plumose
 . . . . 1. B. EUPATORIOIDES
1. Leaves all or mostly opposite, long-petiolate, mostly ovate-triangular, truncate to cordate at the base; heads usually nodding; pappus bristles minutely barbed
 . . . . 2. B. GRANDIFLORA

1. Brickellia eupatorioides (L.) Shinners (false boneset)

Kuhnia eupatorioides L.

Pl. 264 a–f; Map 1106

Stems 30–80(–120) cm long, densely and finely hairy to nearly glabrous. Leaves alternate, numerous, usually relatively closely spaced and thus sometimes appearing opposite or whorled at some nodes, short-petiolate or sessile. Leaf blades 0.8–10.0 cm long, linear, narrowly elliptic, or narrowly to broadly lanceolate, short-tapered at the base, tapered to a sharply pointed tip, the margins entire or shallowly to rarely coarsely few-toothed, the upper surface somewhat roughened or glabrous, also with scattered glands, the undersurface sparsely to moderately and finely hairy, sometimes only along the veins, also moderately to densely glandular. Inflorescences small panicles or appearing as stalked clusters at the branch tips, or less commonly appearing as a leafy panicle with loosely racemose branches, the heads usually erect or ascending (except in drought-stressed plants). Involucre 7–15 mm long, narrowly cup-shaped. Disc florets 6–35. Corollas 4–6 mm long. Pappus bristles plumose. Fruits 3–5 mm long. $2n=18$. July–October.

Scattered nearly throughout the state, but absent from most of the Mississippi Lowlands Division (eastern U.S. west to Montana and Arizona). Upland prairies, sand prairies, loess hill prairies, glades, savannas, openings of mesic to dry upland forests, ledges and tops of bluffs, banks of streams and rivers, and rarely fens; also pastures, old fields, fallow fields, and roadsides.

The plants once widely referred to as *Kuhnia eupatorioides* and its allies are now widely accepted as a specialized group within *Brickellia* (Shinners, 1971; King and Robinson, 1987). Shinners (1946c, 1971) and Turner (1989) have slightly different interpretations of the infraspecific classification of this species, but both accept a number of variants within *B. eupatorioides*. These varieties seem reasonably distinct throughout much of their ranges, but in areas of geographic overlap they tend to intergrade freely. This is the case in much of Missouri, where the ranges of the three varieties treated below overlap, particularly those of var. *corymbulosa* and var. *eupatorioides*. Foliage and inflorescence characters generally are too variable to be useful for separating varieties. Although the species itself is quite distinct in the Missouri flora, users will encounter problems in keying out some specimens below the species level.

1. Middle and outer series of involucral bracts tapered to conspicuous, hairlike tips, these frequently twisted or contorted, mostly $1/2$ or more as long as inner series . . . . 1C. VAR. TEXANA
1. Middle and outer series of involucral bracts narrowed or tapered to sharply pointed but not hairlike tips, not contorted, mostly $1/2$ or less as long as the inner series
  2. Involucre 9–15 mm; heads with 15–35 disc florets
   . . . . 1A. VAR. CORYMBULOSA
  2. Involucre 7–11 mm long; heads with 6–18 disc florets
   . . . . 1B. VAR. EUPATOROIDES

1a. var. corymbulosa (Torr. & A. Gray) Shinners

Kuhnia eupatorioides var. *corymbulosa* Torr. & A. Gray

Pl. 264 e, f

Involucre 9–15 mm long, the middle and outer series of bracts mostly less than $1/2$ as long as the inner series, narrowed or tapered to a sharply pointed but not hairlike or contorted tip. Disc florets 15–35. $2n=18$. July–October.

1107. Brickellia grandiflora   1108. Conoclinium coelestinum   1109. Eupatorium altissimum

Scattered in the state, but absent from most of the Ozark and Mississippi Lowlands Division (Ohio to Arkansas west to Montana and Arizona). Upland prairies, sand prairies, loess hill prairies, glades, savannas, openings of mesic to dry upland forests, ledges and tops of bluffs, and banks of streams and rivers; also pastures, old fields, fallow fields, and roadsides.

1b. var. eupatorioides

Pl. 264 c, d

Involucre 7–11 mm long, the middle and outer series of bracts mostly less than ½ as long as the inner series, narrowed or tapered to a sharply pointed but not hairlike or contorted tip. Disc florets 6–18. July–October.

Scattered nearly throughout the state, but absent from most of the Mississippi Lowlands Division (eastern U.S. west to Missouri and Texas). Upland prairies, sand prairies, loess hill prairies, glades, savannas, openings of mesic to dry upland forests, ledges and tops of bluffs, and banks of streams and rivers; also pastures, old fields, fallow fields, and roadsides.

1c. var. texana (Shinners) Shinners

Kuhnia eupatorioides var. *angustifolia* Raf.
K. eupatorioides var. *ozarkana* Shinners
K. eupatorioides var. *texana* Shinners

Pl. 264 a, b

Involucre 7–11(–13) mm long, the middle and outer series of bracts mostly ½ or more as long as the inner series, tapered to a conspicuous, hairlike tip, this frequently twisted or contorted. Disc florets 10–15(–32). July–October.

Scattered, mostly in the Ozark and Ozark Border Divisions (Illinois to Kansas south to Arkansas and Texas). Glades, savannas, openings of mesic to dry upland forests, ledges and tops of bluffs, banks of streams and rivers, and rarely fens; also pastures, old fields, and roadsides.

Shinners (1946c, 1971) treated plants with 18–33 florets per head, which occur mainly in Oklahoma and Texas, as var. *texana,* and he included plants with 10–14 florets per head that are distributed mostly in the Interior Highlands in his var. *ozarkana.* However, he noted the existence of a number of specimens resembling var. *texana* in Arkansas and Missouri, which he stated were probably hybrids between var. *corymbulosa* and var. *ozarkana* rather than true var. *texana.* Turner (1989), who reanalyzed variation within the complex, took the more conservative approach followed here in combining all of these plants under the name var. *texana.* Steyermark (1963) mapped the Missouri plants (as var. *angustifolia*) from relatively few counties in the southern portion of the Ozarks, but since that time populations have been discovered that extend the range as far north as Boone County.

2. Brickellia grandiflora (Hook.) Nutt. (tassel flower)

Pl. 264 i–k; Map 1107

Stems 30–90 cm long, moderately to densely pubescent with minute, sometimes spreading hairs. Leaves all or mostly opposite, the uppermost leaves sometimes alternate, well spaced, long-petiolate. Leaf blades 2–7 cm long, ovate-triangular to triangular, truncate to cordate at the base, narrowed or tapered to a sharply pointed tip, the margins bluntly toothed or scalloped, the surfaces short-hairy and with scattered glands, the undersurface sparsely to moderately and finely hairy, sometimes only along the veins, also moderately glandular. Inflorescences small panicles or appearing as stalked clusters at the branch tips, occasionally also appearing axillary, the heads usually nodding. Involucre 7–12 mm long, narrowly cup-shaped to narrowly bell-shaped. Disc florets 18–45. Corollas 6–8 mm long. Pappus bristles minutely barbed. Fruits 3–5 mm long. $2n=18$. July–October.

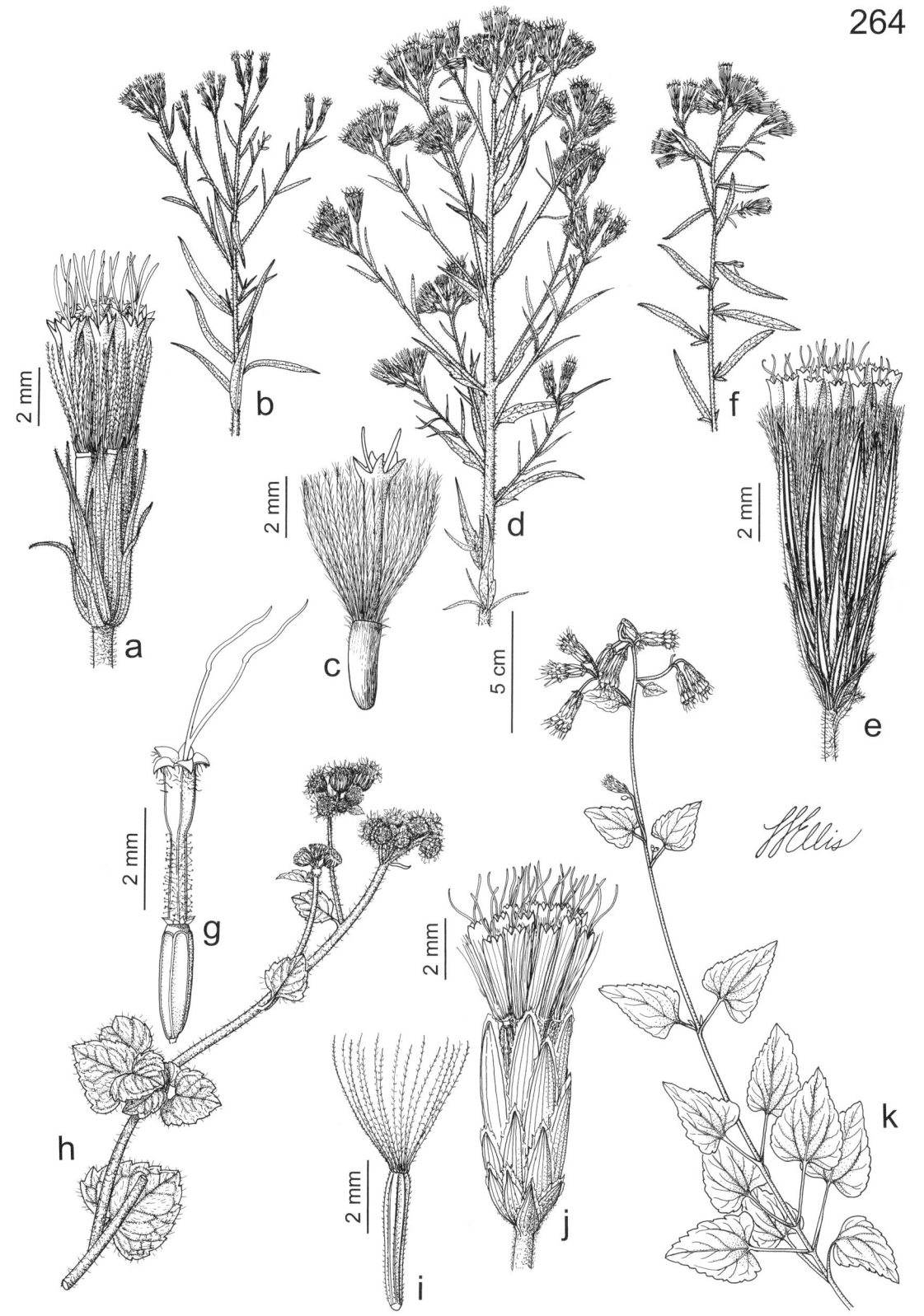

Plate 264. Asteraceae. *Brickellia eupatorioides* var. *texana*, **a)** head, **b)** habit. *Brickellia eupatorioides* var. *eupatorioides*, **c)** floret, **d)** habit. *Brickellia eupatorioides* var. *corymbulosa*, **e)** head, **f)** habit. *Ageratum conyzoides*, **g)** floret, **h)** habit. *Brickellia grandiflora*, **i)** fruit, **j)** head, **k)** habit.

Scattered in the Ozark and Ozark Border Divisions (Washington to California east to Missouri and Texas). Glades, ledges and tops of bluffs, on limestone and dolomite substrates.

Steyermark (1963) noted that plants can have a somewhat fetid odor when bruised or crushed.

53. Conoclinium DC.

Three species, U.S., Mexico.

1. Conoclinium coelestinum (L.) DC. (mistflower, blue boneset, wild ageratum)
Eupatorium coelestinum L.
Pl. 267 a–c; Map 1108

Plants perennial, with rhizomes. Stems 1 or few, unbranched or few- to several-branched below the inflorescence, 25–90 cm long, erect or ascending, moderately to densely pubescent with short, curly hairs. Leaves opposite (the nodes well separated), short- to long-petiolate. Leaf blades 1–10 cm long, triangular-ovate to narrowly triangular, broadly angled to truncate at the base, narrowed or tapered to a usually sharply pointed tip, the margins shallowly scalloped or bluntly toothed, both surfaces glabrous or minutely hairy, usually only along the veins, the undersurface also glandular, with mostly 3 main veins. Inflorescences small panicles or stalked clusters at the branch tips, flat-topped or less commonly dome-shaped. Heads with 35–70 disc florets. Involucre 3–5 mm long, the bracts 18–30 (the head often also subtended by 1 or a few other narrower bracts), in usually 2 or 3 somewhat unequal to nearly equal, overlapping series, narrowly elliptic to narrowly lanceolate, tapered to a sharply pointed, often purple tip, all but the outermost noticeably but finely few-nerved or few-ribbed, sparsely to moderately short-hairy and usually also glandular. Receptacle conical. Corollas purple or lavender-blue (rarely white). Pappus of numerous capillary bristles (although sometimes appearing relatively sparse). Fruits 1.6–2.4 mm long, finely 5-ribbed, slightly wedge-shaped in profile (usually slightly and unevenly tapered at the base), glabrous, the ribs smooth or with a few minute, ascending teeth, lighter colored, the body otherwise irregularly glandular and dark brown to black. $2n=20$. August–October.

Scattered, mostly south of the Missouri River (eastern U.S. west to Nebraska and Texas; Canada). Bottomland forests, swamps, banks of streams and rivers, margins of ponds and lakes, marshes, and fens; also ditches, gardens, railroads, roadsides, and shaded to open, disturbed areas.

Several independent molecular studies have suggested that *Conoclinium* forms a natural group with *Ageratum* and *Fleischmannia* that is only distantly related to *Eupatorium* in the strict sense within the tribe Eupatorieae (Schilling et al., 1999; Schmidt and Schilling, 2000; Ito et al., 2000b). See the treatment of *Fleischmannia* for further discussion. A rare, white-flowered mutant has been called *E. coelestinum* f. *album* Alexander.

54. Eupatorium L., thoroughwort

Plants fibrous-rooted, usually with rhizomes, these sometimes somewhat fleshy. Stems solitary to several, erect or ascending, the nodes often encircled by a low ridge after the leaves are shed. Leaves opposite, whorled, and/or alternate, sessile (perfoliate in *E. perfoliatum*) to long-petiolate, the nodes well spaced or less commonly crowded. Leaf blades variously shaped, simple (pinnately dissected in *E. capillifolium*), the margins usually toothed, usually with 3 main veins. Inflorescences panicles, occasionally reduced to stalked clusters, flat-topped to dome-shaped, rarely appearing as elongate panicles with racemose branches. Heads with 3–23 disc florets. Involucre more or less cup-shaped to nearly cylindrical, the bracts 10–22 (the head often also subtended by 1 or 2 other shorter bracts), in 2 to several usually unequal, overlapping series. Receptacle flat or slightly convex. Corollas white or less commonly pink to lavender or purple. Pappus of numerous capillary bristles. Fruits mostly 5-angled, somewhat wedge-shaped in profile (usually slightly and unevenly tapered at the base) to nearly linear, sparsely to moderately glandular, otherwise glabrous or sparsely and minutely hairy along the angles, yellowish brown to more commonly dark brown or black. About 40 species, North America, Europe, Asia.

Following King and Robinson (1987) and later molecular workers (Schilling et al., 1999; Schmidt and Schilling, 2000; Ito et al., 2000a, b), the genus *Eupatorium* is treated here in a restricted sense. Of the segregate genera, the blue-flowered *Conoclinium* and *Fleischmannia* appear to be more closely related to *Ageratum* than to *Eupatorium* in the strict sense. Various workers have not fully resolved the affinities of the other Missouri segregate, *Ageratina,* but all of the recent students of the tribe have agreed that it is not closely related to the main core of *Eupatorium.*

King and Robinson (1970) segregated the five temperate North American species of Joe-pye weeds from *Eupatorium* as the genus *Eupatoriadelphus,* but later (King and Robinson, 1987) recanted to treat them as a well-marked section within a revised concept of *Eupatorium.* Subsequent molecular studies (Schilling et al., 1999; Schmidt and Schilling, 2000) consistently have placed *Eupatoriadelphus* as the closest relative to the rest of *Eupatorium* in the strict sense. Although these two lineages are separable morphologically, there does not appear to be a good rationale for treating them as separate genera, and the retention of the Joe-pye weeds within *Eupatorium* still allows circumscription of a natural (monophyletic) group. Also, if the Joe-pye weeds are treated as a segregate genus, the name *Eutrochium* Raf. has priority over *Eupatoriadelphus* R.M. King & B. Rob. (Lamont, 2004).

The systematics of *Eupatorium* in North America is complicated by polyploidy and apomixis within some populations of a number of species (Sullivan, 1976) and reports of several interspecific hybrids (Sullivan, 1978; Cronquist, 1980; Gleason and Cronquist, 1991).

1. Leaves mostly pinnately deeply lobed or dissected into threadlike segments 0.2–0.5(–1.0) mm wide, the uppermost simple and threadlike . . . . . . 2. E. CAPILLIFOLIUM
1. Leaves undivided, not threadlike (1 mm or more wide), the margins entire or more commonly toothed
    2. Leaves all or mostly whorled, the uppermost sometimes opposite or alternate
        3. Leaves linear to narrowly oblong-elliptic or narrowly oblanceolate, 1–6 cm long, 0.5–6.0 mm wide . . . . 4. E. HYSSOPIFOLIUM
        3. Leaves narrowly lanceolate or narrowly ovate to broadly ovate, 5–30 cm long, 15–90 mm wide
            4. Heads with 8–22 florets; inflorescences usually flat-topped or nearly so . . . . 5. E. MACULATUM
            4. Heads with 4–8 florets; inflorescences dome-shaped
                5. Stems more or less evenly dark purplish-tinged, glaucous, hollow with a relatively large central cavity (except sometimes toward the tip) . . . . 3. E. FISTULOSUM
                5. Stems dark purple only at the nodes, not or only slightly glaucous, solid or rarely becoming hollow with a slender central cavity (usually toward the base) . . . . 7. E. PURPUREUM
    2. Leaves all or mostly opposite, the uppermost sometimes alternate, rarely 1 or 2 nodes on a stem with a whorl of 3 leaves
        6. Leaves perfoliate, the opposite pair at all but the uppermost nodes with the bases broadly fused around the stem . . . . 6. E. PERFOLIATUM
        6. Leaves not perfoliate, the bases not fused (although sometimes slightly overlapping in *E. sessilifolium*)
            7. Disc florets 9–15 per head; leaves with a well-differentiated petiole mostly 8–30 mm long . . . . 10. E. SEROTINUM

7. Disc florets 5(–7) per head; leaves sessile or with a short petiole 1–3 mm long, sometimes tapered to a short, poorly differentiated petiolar base
    8. Leaves narrowly lanceolate to elliptic or oblanceolate, narrowly tapered at the base
        9. Leaves narrowly lanceolate to lanceolate or less commonly narrowly elliptic, tapered to a sharply pointed tip, the 3 main veins separate to the base; involucre 4.5–7.0 mm long .... 1. E. ALTISSIMUM
        9. Leaves narrowly oblanceolate to oblanceolate or less commonly narrowly elliptic, rounded at the tip or angled to a bluntly pointed tip, the 2 lateral veins branching from the midvein well above the leaf base; involucre 2.5–4.0 mm long .... 9. E. SEMISERRATUM
    8. Leaves lanceolate to broadly ovate or nearly circular, broadly angled, rounded, truncate, or shallowly cordate at the base
        10. Stems densely short-hairy; leaves with 3 main veins, moderately to densely short-hairy, the bases of the pair at each node not overlapping .... 8. E. ROTUNDIFOLIUM
        10. Stems glabrous below the inflorescence; leaves with 1 main vein, glabrous except for the scattered glands; the bases of the pair at all but the uppermost nodes often somewhat overlapping .... 11. E. SESSILIFOLIUM

1. Eupatorium altissimum L. (tall thoroughwort)

Pl. 266 a, b; Map 1109

Stems 50–200 cm long, not hollow, moderately to densely short-hairy above the nearly glabrous basal portion, green to yellowish green, sometimes purplish-tinged or brownish-mottled, often somewhat glaucous, some nodes often with small fascicles of axillary leaves less than ¹/₂ as long as the main stem leaves. Leaves mostly opposite, those of the uppermost nodes sometimes alternate, sessile or with poorly differentiated petioles to 15 mm long. Leaf blades 2–15 cm long, 3–25 mm wide, narrowly lanceolate to lanceolate or less commonly narrowly elliptic, tapered at the base, tapered to a sharply pointed tip, the margins entire or more commonly sharply toothed above the midpoint, the surfaces moderately to densely short-hairy, also densely gland-dotted, with 3 main veins from the base. Inflorescences terminal panicles, flat-topped or broadly dome-shaped. Involucre 4.5–7.0 mm long, more or less cup-shaped, the bracts ovate to narrowly oblong, rounded to bluntly pointed at the tip, the margins thin and pale, mostly faintly 3-nerved, densely short-hairy, green. Disc florets 5. Corollas 4–5 mm long, the surface often somewhat glandular, white. Fruits 1.5–2.5 mm long. $2n=20, 30, 40$. August–October.

Scattered nearly throughout the state (eastern U.S. west to Minnesota, Nebraska, and Texas; Canada). Upland prairies, loess hill prairies, glades, savannas, openings of mesic to dry upland forests, banks of streams and rivers, and ledges and tops of bluffs; also old fields, fallow fields, pastures, ditches, quarries, railroads, roadsides, and open, disturbed areas.

For a discussion of presumed hybrids with *E. hyssopifolium*, see the treatment of that species.

2. Eupatorium capillifolium (Lam.) Small (dog fennel)

Pl. 266 e, f; Map 1110

Stems 50–250 cm long, not hollow, moderately to densely short-hairy above the often nearly glabrous basal portion, green to yellowish green, sometimes purplish-tinged or brownish-mottled, occasionally slightly glaucous, the nodes often with small fascicles of axillary leaves less than ¹/₂ as long as the main stem leaves. Leaves mostly opposite, those of the uppermost nodes sometimes alternate, sessile or with short petioles 1–10 mm long. Leaf blades 2–10 cm long, mostly 1–3 times pinnately or ternately deeply and irregularly lobed or dissected into relatively few long, threadlike segments (the uppermost merely simple and threadlike), these 0.2–0.5(–1.0) mm wide, the margins more or less entire, sometimes rolled under, the surfaces glabrous but sparsely to densely gland-dotted, with 1 main vein. Inflorescences axillary clusters, these appearing as elongate terminal panicles with racemose branches. Involucre 2.0–3.5 mm long, more or less cup-shaped to nearly cylindrical (sometimes bell-shaped after pressing), the bracts linear to narrowly lanceolate or narrowly oblong, sharply pointed at the tip, the narrow margins thin and pale (sometimes only toward the base), faintly

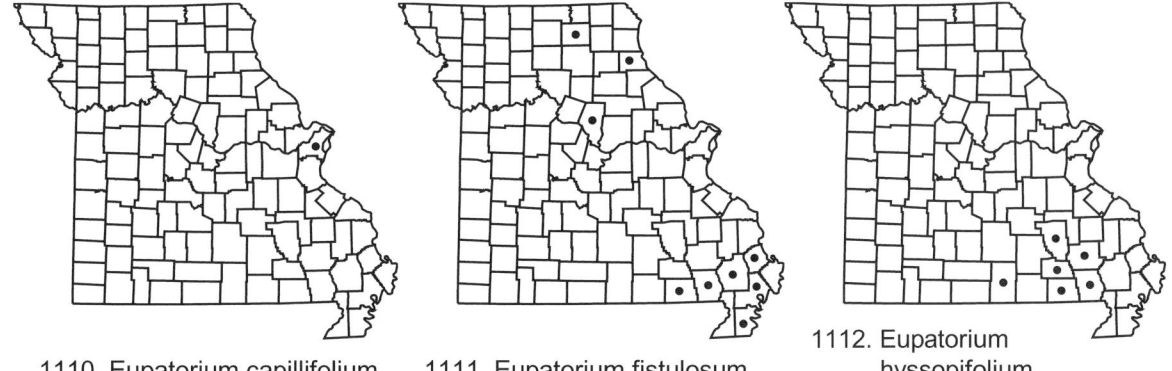

1110. Eupatorium capillifolium   1111. Eupatorium fistulosum   1112. Eupatorium hyssopifolium

1- or 3-nerved, glabrous, light green, sometimes purplish-tinged toward the tip. Disc florets 3–6. Corollas 2.0–2.5 mm long, the surface often inconspicuously glandular, white, greenish white, or pale cream-colored. Fruits 1.0–1.5 mm long. 2n=20. August–October.

Introduced, uncommon, known thus far only from St. Louis City and County (native from New Jersey to Florida west to Texas; introduced farther north and west). Railroads, gardens, and open, disturbed areas.

Eupatorium capillifolium was first reported by Steyermark (1963) in the supplement to his *Flora of Missouri,* based on collections made by Viktor Mühlenbach in the St. Louis railyards. In parts of the southeastern United States, the species is a problem weed in pastures. Apparently it contains low levels of the toxin tremetol (see the treatment of *Ageratina* for further discussion), which can cause dehydration and other symptoms in livestock that ingest the plants in quantity.

3. **Eupatorium fistulosum** Barratt (hollow-stemmed Joe-pye weed)
Eutrochium fistulosum (Barratt) E.E. Lamont
Eupatoriadelphus fistulosus (Barratt) R.M. King & H. Rob.

Pl. 265 e–g; Map 1111

Stems 60–300 cm long, hollow between the nodes, with a relatively large central cavity (except sometimes toward the tip), usually glabrous below the inflorescence, more or less evenly dark purplish-tinged, moderately glaucous, generally lacking small axillary branches or fascicles of axillary leaves. Leaves mostly in whorls of 4–7, the uppermost leaves rarely opposite, the petiole 2–25 mm long. Leaf blades 5–30 cm long, 15–60 mm wide, narrowly to broadly lanceolate or elliptic-lanceolate, tapered at the base, tapered to a sharply pointed tip, the margins sharply toothed, the upper surface glabrous or sparsely short-hairy, the undersurface glabrous to sparsely to moderately short-hairy mostly along the veins, also glandular, with 1 main vein. Inflorescences terminal panicles, often large, usually narrowly dome-shaped. Involucre 6.5–9.0 mm long, slender, the bracts ovate to lanceolate or narrowly oblong-elliptic, rounded or bluntly pointed at the tip, often 3-nerved, often minutely hairy, often somewhat purplish-tinged. Disc florets 4–7(–8). Corollas 4.5–7.5 mm long, the surface often somewhat glandular, pale pink or less commonly somewhat purplish-tinged. Fruits 3.0–4.5 mm long. 2n=20. July–September.

Uncommon, mostly in the eastern half of the state and most frequently recorded from the Mississippi Lowlands Division (eastern Missouri west to Michigan, Missouri, and Texas; Canada). Bottomland forests, mesic upland forests, swamps, and banks of streams and rivers; also roadsides.

Eupatorium fistulosum tends to be the tallest Joe-pye weed in the state, with stems frequently more than 2 m. It also tends to have narrower main stem leaves than do *E. maculatum* and *E. purpureum*. Collectors sampling only the uppermost portion of the plant have been fooled by the stems, which frequently are solid rather than hollow toward the tip.

4. **Eupatorium hyssopifolium** L. var. **calcaratum** Fernald & B.G. Schub. (hyssop-leaved thoroughwort)

Pl. 266 g, h; Map 1112

Stems 30–100 cm long, usually not hollow, moderately to densely short-hairy, dark purple to purplish brown, not glaucous, with small axillary branches or fascicles of axillary leaves at all or most nodes up to ½ as long as the main stem leaves. Leaves mostly in whorls of 4, those of the upper and lower nodes sometimes in whorls of 3 or opposite, rarely alternate, sessile or nearly so. Leaf blades 1–6 cm long, 0.5–6.0 mm wide, linear to narrowly oblong-elliptic or narrowly oblanceolate, more or less tapered at the base, tapered to a sharply pointed tip, the margins entire or less commonly with a few shallow teeth, often somewhat curled under, the surfaces glabrous or sparsely to moderately short-hairy along the midvein, also densely gland-dotted, with 1 main vein. Inflorescences terminal panicles, more or less flat-topped. Involucre 4–6 mm long, cup-shaped, the bracts lanceolate or narrowly oblong, mostly rounded or bluntly pointed at the tip, the margins thin and pale or transparent, faintly 1- or 3-nerved, densely short-hairy, green. Disc florets 5. Corollas 3.5–4.0 mm long, the surface often somewhat glandular, white. Fruits 2.5–3.5 mm long. $2n=20, 30, 40$. August–November.

Uncommon in the eastern portion of the Ozark Division (eastern [mainly southeastern] U.S. west to Missouri and Texas). Savannas, upland prairies, and margins of sinkhole ponds; more commonly old fields, ditch banks, roadsides, and dry, open, disturbed areas.

This taxon was first reported from the open margin of a sinkhole pond in Howell County by Steyermark (1953), who considered it a relictual population disjunct from the Gulf Coastal Plain. More recently, however, it has become apparent that this species is naturally expanding its range northward into the Midwest, making use of disturbance corridors such as highways. It seems unlikely that it was present in Missouri prior to the twentieth century, and thus one might argue that it should not be considered native to the state. There is no evidence to suggest that it was introduced intentionally or unintentionally directly through human activities, and the northward migration of this species and others such as *E. rotundifolium* may have more to do with a warming trend in the region's climate than with environmental perturbations.

Missouri plants described above are assignable to var. *calcaratum*, which is included within var. *hyssopifolium* by some authors (Gleason and Cronquist, 1991). It differs from the nominate variety in its shorter, narrower, mostly entire leaves, but there is a lot of intergradation between the two taxa. The two occupy similar ranges and habitats. More research is necessary to determine whether these morphological trends can be correlated with cytological or molecular characters.

A third variety that is accepted by most botanists is var. *laciniatum* A. Gray, which refers to more robust, apparently polyploid plants with the leaves mainly in whorls of 3 or opposite. Plants corresponding to this morphology are suspected to have arisen through past hybridization between *E. hyssopifolium* and one or more related species, but now reproduce themselves independently as a species with a unique range. They have been reported from as close as central Kentucky (Gleason and Cronquist, 1991). Their relationship to morphologically similar plants discussed in the next paragraph is unclear and requires further study. Sullivan (1978) reported hybrids between *E. hyssopifolium* and *E. semiserratum* from Alabama and Florida, some of which she suggested appeared morphologically similar to var. *laciniatum*. The possibility of these taxa hybridizing in southern Missouri should not be overlooked.

In Wayne County, on property owned by Dr. Robert Cacchione near Williamsville, a large, mixed population of *E. hyssopifolium* and *E. altissimum* in an old field has produced a number of morphologically intermediate hybrid plants. These putative hybrids, which appear to be fertile, will key to *E. hyssopifolium* in the key to species above, as they tend to produce leaves in whorls of 3 at many nodes. They are more robust plants than those of *E. hyssopifolium*, with broader leaves having relatively coarse serrations. These teeth tend to be deeper and more evenly distributed along the margins than in *E. altissimum*, but the leaves are not so large or broad as in that species. This hybrid situation, which is to be expected elsewhere in the southeastern Ozarks, has not been reported previously. It was brought to the attention of the Flora of Missouri Project by Bob Cacchione and Bill Summers, to whom thanks are extended. Voucher specimens are in the herbarium of the Missouri Botanical Garden.

5. Eupatorium maculatum L. var. bruneri
(A. Gray) Breitung (spotted Joe-pye weed)
E. maculatum ssp. *bruneri* (A. Gray) G.W. Douglas
Eutrochium maculatum (L.) E.E. Lamont var. *bruneri* (A. Gray) E.E. Lamont
Eupatoriadelphus maculatus (L.) R.M. King & H. Rob. var. *bruneri* (A. Gray) R.M. King & H. Rob.

Pl. 265 c, d; Map 1113

Stems 60–200 cm long, solid or rarely becoming hollow with a slender central cavity (usually toward the base), usually moderately short-hairy throughout, sometimes becoming sparsely hairy with age, with dark purple mottling or less commonly more evenly dark purplish-tinged, not or only slightly glaucous, generally lacking small axillary branches or fascicles of axillary leaves. Leaves mostly in whorls of (3)4 or 5(6), the uppermost leaves sometimes alternate or opposite, the petiole 2–25 mm long. Leaf blades 5–20(–30) cm long,

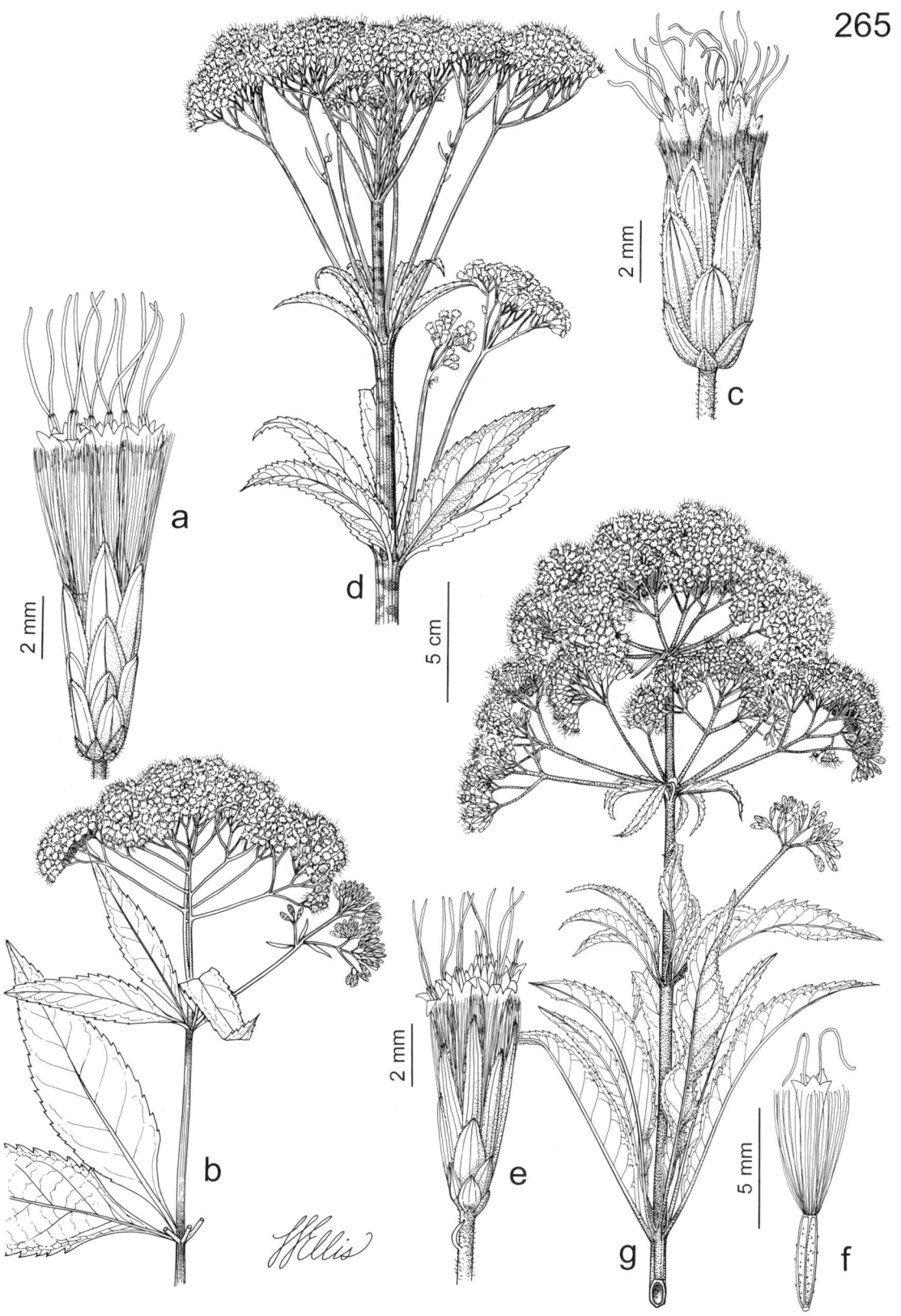

Plate 265. Asteraceae. *Eupatorium purpureum*, **a)** head, **b)** habit. *Eupatorium maculatum*, **c)** head, **d)** habit. *Eupatorium fistulosum*, **e)** head, **f)** floret, **g)** habit.

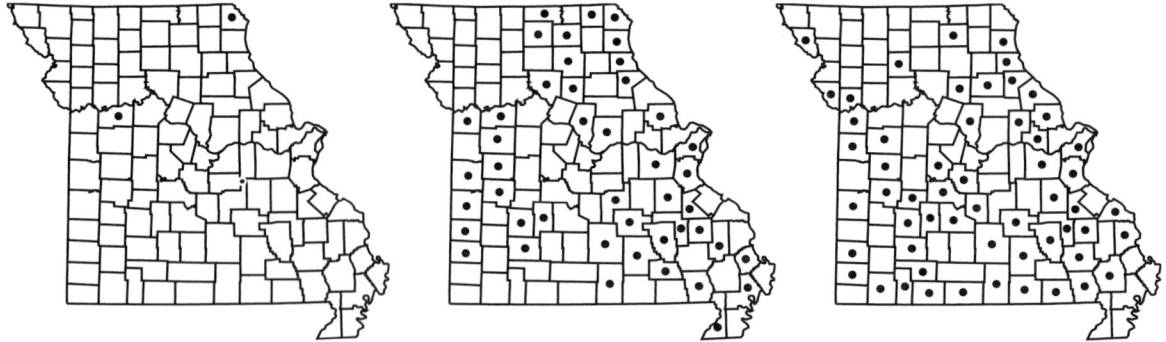

1113. Eupatorium maculatum   1114. Eupatorium perfoliatum   1115. Eupatorium purpureum

15–70 mm wide, lanceolate to elliptic or ovate, narrowed or tapered at the base, tapered to a sharply pointed tip, the margins sharply toothed, the upper surface glabrous or sparsely to moderately short-hairy or roughened, the undersurface densely short-hairy, also glandular, with 1 main vein. Inflorescences terminal panicles, often large, usually flat-topped or nearly so (sometimes the individual branches flat-topped and the entire panicle broadly dome-shaped). Involucre 6.5–9.0 mm long, cup-shaped, the bracts ovate to lanceolate or narrowly oblong-elliptic, rounded to less commonly bluntly pointed at the tip, often 3-nerved, often minutely hairy, usually purplish-tinged to dark purple. Disc florets 8–22. Corollas 4.5–7.5 mm long, the surface often somewhat glandular, mostly pale purple or purplish-tinged. Fruits 3.0–4.5 mm long. $2n=20$. July–September.

Uncommon, known thus far only from Clark and Lafayette Counties (Wisconsin to Missouri west to Washington and Arizona; Canada). Fens and edges of marshes.

This species was first reported for Missouri by Doug Ladd (1990). Across its range, it has long been treated as comprising three infraspecific variants. The Missouri plants appear to be assignable to var. *bruneri*, which is characterized by densely pubescent stems and leaf undersurfaces, and occupies the western portion of the species range. However, the specimen from Clark County is less densely hairy than is typical of the variety. The var. *maculatum* is characterized by sparsely hairy to nearly glabrous leaf undersurfaces and stems (below the inflorescences). It occurs in adjacent northern Illinois and eastern Iowa and is generally distributed in the northeastern United States and adjacent Canada. The far northeastern var. *foliosum* (Fernald) Wiegand has pubescence similar to that of the typical variety but differs in having the uppermost leaves relatively unreduced and mostly surpassing the inflorescence.

6. **Eupatorium perfoliatum** L. var. **perfoliatum** (boneset, thoroughwort)

Pl. 268 a, b; Map 1114

Stems 40–150 cm long, not hollow, densely pubescent with fine, mostly spreading hairs, tan to purple but often appearing gray, not glaucous, not producing small fascicles of axillary leaves. Leaves mostly opposite, rarely 1 or 2 of the lower nodes with a whorl of 3 leaves, sessile, mostly perfoliate, all but the uppermost nodes with the bases of the pair of leaves broadly fused around the stem. Leaf blades 3–20 cm long, 10–450 mm wide, narrowly lanceolate to elongate-triangular, tapered to a sharply pointed tip, the margins bluntly to sharply toothed, the surfaces densely short-hairy (the undersurface usually appearing grayer and more densely hairy), roughened or felty to the touch, also sparsely to moderately gland-dotted, with 1 main vein. Inflorescences terminal panicles, more or less flat-topped. Involucre 4–6 mm long, cup-shaped, the bracts ovate to oblong-lanceolate or narrowly oblong, rounded to bluntly or sharply pointed at the tip, the margins thin and pale or transparent, especially toward the tip, mostly faintly 3-nerved, densely short-hairy, green but appearing mostly gray. Disc florets 9–23. Corollas 2.5–3.5 mm long, the surface often somewhat glandular, white. Fruits 1.5–2.5 mm long. $2n=20$. July–October.

Scattered nearly throughout the state, but apparently absent from much of the western portion of the Glaciated Plains Division (eastern U.S. west to North Dakota and Texas; Canada). Banks of streams and rivers, margins of ponds, lakes, and sinkhole ponds, bases and ledges of bluffs, fens, borders of sloughs, bottomland prairies, moist depressions of upland prairies, marshes, bottomland forests, and rarely openings of mesic upland forests; also margins of crop fields, fallow fields, and moist roadsides.

This species was used medicinally by Native Americans and has a long history of use in folk medicine to treat colds, fevers, and tapeworms and as a tonic, among numerous other uses. The vernacular name boneset refers to the belief by some herbalists that a tea from the plants would help to heal and strengthen bones.

A second variety, with narrower leaves and shorter hairs, is restricted to saline, mostly coastal habitats in the far northeastern United States and adjacent Canada and is called var. *colpophilum* Fernald & Griscom. *Eupatorium perfoliatum* has been suggested to hybridize sporadically with *E. serotinum,* mostly in disturbed, moist habitats. Such plants of intermediate morphology are known as *E.* ×*truncatum* Muhl. ex Willd. but have also been referred to in the past as *E. perfoliatum* var. *cuneatum* (Engelm. ex Torr. & A. Gray) Engelm. ex A. Gray, *E. resinosum* Torr. ex DC. var. *kentuckiense* Fernald, and *E.* ×*polyneuron* (F.J. Herm.) Wunderlin (Wunderlin, 1972; Cronquist, 1980; Tucker and Dill, 1989). Sullivan (1978) reported a putative hybrid between *E. perfoliatum* and *E. sessilifolium* from North Carolina. This hybrid should be searched for in Missouri.

7. Eupatorium purpureum L. (green-stemmed Joe-pye weed)

Eutrochium purpureum (L.) E.E. Lamont
Eupatoriadelphus purpureus (L.) R.M. King & H. Rob.

Pl. 265 a, b; Map 1115

Stems 40–200 cm long (sometimes to 4 m or more in cultivation), solid or rarely becoming hollow with a slender central cavity (usually toward the base), mostly glabrous below the inflorescence, dark purple only at the nodes, not or only slightly glaucous, generally lacking small axillary branches or fascicles of axillary leaves. Leaves mostly in whorls of 3 or 4(5), the uppermost leaves sometimes alternate or opposite, the petiole 2–20 mm long. Leaf blades 5–30 cm long, 25–90 mm wide, narrowly ovate to ovate, elliptic-ovate, or triangular-ovate, tapered at the base, tapered to a sharply pointed tip, the margins sharply toothed, the upper surface glabrous or sparsely to moderately short-hairy, the undersurface glabrous to densely short-hairy, also glandular, with 1 main vein. Inflorescences terminal panicles, often large, broadly to narrowly dome-shaped. Involucre 6.5–9.0 mm long, slender, the bracts ovate to lanceolate or narrowly oblong-elliptic, bluntly to sharply pointed at the tip, often 3-nerved, usually glabrous, usually purplish-tinged to dark purple. Disc florets 4–7 (–8). Corollas 4.5–7.5 mm long, the surface often somewhat glandular, pale pink or somewhat purplish-tinged. Fruits 3.0–4.5 mm long. $2n=20$. July–September.

Scattered nearly throughout the state but more abundant south of the Missouri River (eastern U.S. west to Minnesota, Nebraska, and Oklahoma; Canada). Bottomland forests, mesic upland forests, and banks of streams and rivers; also roadsides.

This is by far the most common of the three Joe-pye weeds in Missouri. It is sometimes grown as an ornamental in gardens, and under some conditions it can grow to over 4 m tall. Native Americans and early European settlers used the three Joe-pye weeds (especially the rhizomes) medicinally to treat kidney ailments and inflammations and as a general tonic (Lamont, 1995).

Putative hybrids between this species and both *E. fistulosum* and *E. maculatum* have been recorded from farther east (Lamont, 1995), but these have not been found in Missouri to date. Two varieties of *E. purpureum* have been recognized by most botanists.

1. Leaf blades with the undersurface densely and relatively evenly short-hairy .... 7A. VAR. HOLZINGERI
1. Leaf blades with the undersurface glabrous (except for the small, spherical glands) or sparsely hairy along the main veins .... 7B. VAR. PURPUREUM

7a. var. holzingeri (Rydb.) E.E. Lamont

Eutrochium purpureum var. *holzingeri* (Rydb.) E.E. Lamont

Petioles often sparsely hairy. Leaf blades with the undersurface densely and relatively evenly short-hairy, also with small, spherical glands.

Uncommon and sporadic (Wisconsin, Illinois, and Arkansas west to Nebraska and Kansas). Bottomland forests, mesic upland forests, and banks of streams and rivers; also roadsides.

7b. var. purpureum

Petioles glabrous. Leaf blades with the undersurface sparsely hairy along the main veins, also with small, spherical glands.

Scattered nearly throughout the state but more abundant south of the Missouri River (eastern U.S. west to Minnesota, Nebraska, and Oklahoma; Canada). Bottomland forests, mesic upland forests, and banks of streams and rivers; also roadsides.

8. Eupatorium rotundifolium L. var. scabridum (Elliott) A. Gray (round-leaved boneset)

E. rotundifolium var. *cordigerum* Fernald

Map 1116

Stems 30–120 cm long, not hollow, densely short-hairy, purplish brown, not glaucous, some nodes

1116. Eupatorium rotundifolium
1117. Eupatorium semiserratum
1118. Eupatorium serotinum

with small fascicles of axillary leaves much shorter than the main stem leaves. Leaves mostly opposite, those of the uppermost nodes rarely alternate, sessile or with petioles to 3 mm long. Leaf blades 1–5 cm long, 8–30 mm wide, ovate to broadly ovate or nearly circular, rounded or broadly angled at the base (the bases of the leaf pair at each node not overlapping), angled to a bluntly or sharply pointed tip, the margins sharply toothed, the surfaces glabrous or sparsely to moderately to densely short-hairy, also densely gland-dotted, with 3 main veins, the lateral pair branched from the midvein 1–2 mm from the base of the blade. Inflorescences terminal panicles, more or less flat-topped. Involucre 4.5–6.5 mm long, cup-shaped, the bracts ovate to lanceolate or narrowly oblong, sharply pointed at the tip, the margins thin and pale or transparent, mostly faintly 3-nerved, densely short-hairy, green. Disc florets 5. Corollas 3–4 mm long, the surface often somewhat glandular, white. Fruits 2.0–3.5 mm long. $2n=20, 30, 40$. August–November.

Uncommon, known thus far only from Butler, Carter, and Ripley Counties (South Carolina to Florida west to Oklahoma and Texas). Savannas and edges of mesic to dry upland forests; also old fields, roadsides, and dry, open, disturbed areas.

Eupatorium rotundifolium is a recent addition to the state's flora that was first collected in 1993 by the late Stanton Hudson in Ripley County. Like *E. hyssopifolium,* it appears to represent a taxon that is naturally expanding its range northward from the Gulf Coastal Plain into the Midwest utilizing disturbance corridors such as highways.

The infraspecific taxonomy of *E. rotundifolium* requires further study, especially in light of the apparent cytological variation within the species. Four or more varieties have been recognized at various times. Overall, the species occurs throughout the southeastern United States and northward to Missouri, Indiana, Ohio, and New England. The varieties differ in details of leaf size, shape, margins, pubescence, and venation. Some of the varieties may have arisen through past hybridization between *E. rotundifolium* and related species (Gleason and Cronquist, 1991).

9. Eupatorium semiserratum DC.
E. cuneifolium Willd. var. *semiserratum* (DC.) Fernald & Griscom

Pl. 266 c, d; Map 1117

Stems 50–120 cm long, not hollow, moderately to densely short-hairy above the sometimes nearly glabrous basal portion, usually purplish-tinged or purplish brown, sometimes somewhat glaucous, some nodes often with small fascicles of axillary leaves less than ½ as long as the main stem leaves. Leaves mostly opposite, those of the uppermost nodes sometimes alternate, sessile or with poorly differentiated petioles to 8 mm long, twisted at the base so that the leaves appear nearly vertically oriented. Leaf blades 1–8 cm long, 2–15(–25) mm wide, narrowly oblanceolate to oblanceolate or less commonly narrowly elliptic, tapered at the base, rounded or angled to a bluntly pointed tip, the margins sharply toothed mostly above the midpoint, the surfaces moderately to densely short-hairy, also densely gland-dotted, with 3 main veins, the 2 lateral veins branching from the midvein 2–12 mm above the blade base. Inflorescences terminal panicles, more or less flat-topped. Involucre 2.5–4.0 mm long (sometimes appearing longer at fruiting), more or less cup-shaped, the bracts ovate to narrowly oblong, rounded to bluntly pointed at the tip, the margins thin and pale, mostly faintly 3-nerved, densely short-hairy, green. Disc florets 5. Corollas 2.5–3.5 mm long, the surface often somewhat glandular, white. Fruits 1.5–2.0 mm long. $2n=20$. August–October.

Uncommon in the Mississippi Lowlands Division and adjacent portions of the Ozarks; also historically disjunct in the St. Louis region (southeastern U.S. west to Missouri and Texas). Edges of mesic

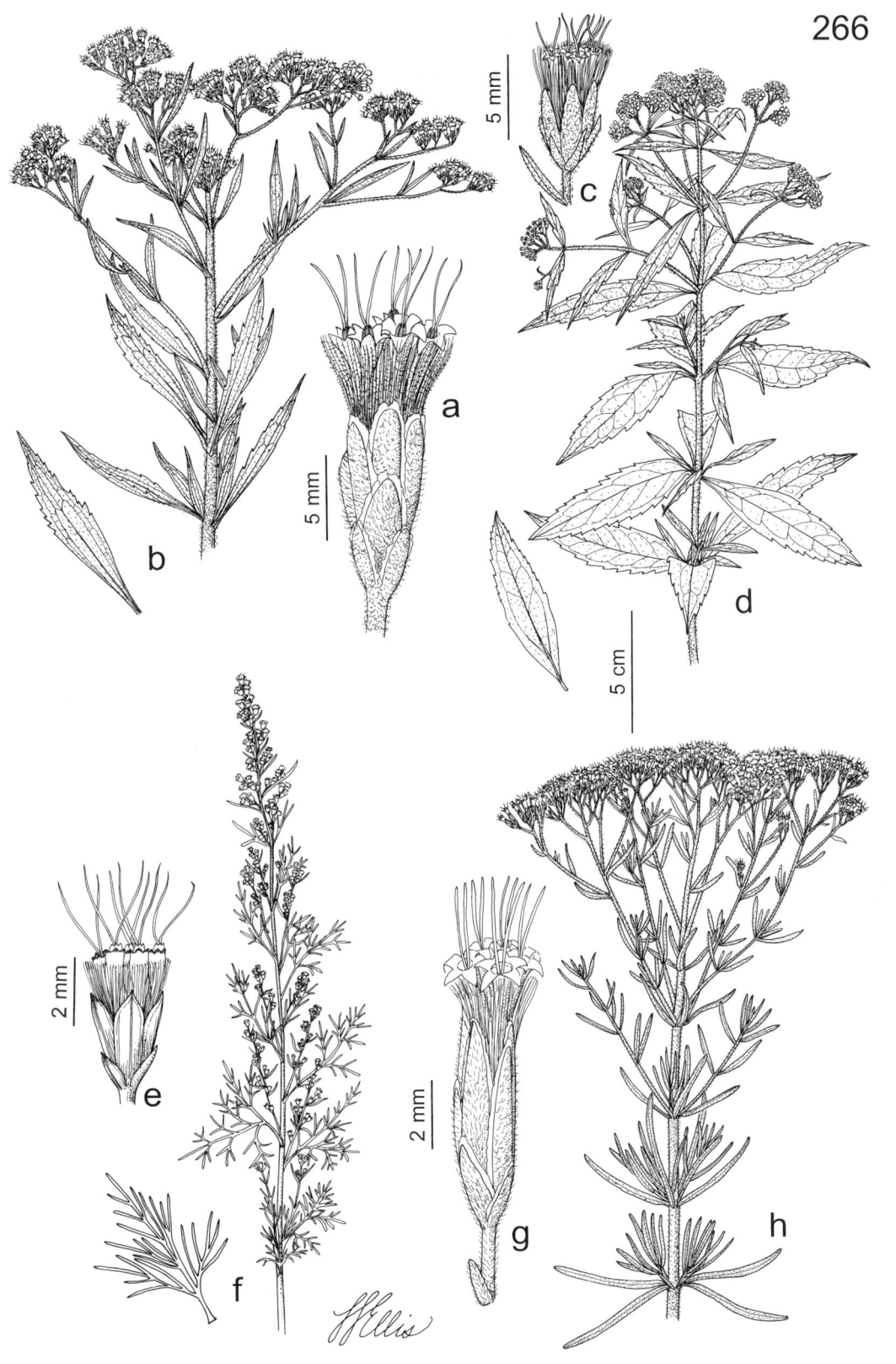

Plate 266. Asteraceae. *Eupatorium altissimum*, **a)** head, **b)** habit and larger leaf. *Eupatorium semiserratum*, **c)** head, **d)** habit and larger leaf. *Eupatorium capillifolium*, **e)** head, **f)** habit and larger leaf. *Eupatorium hyssopifolium*, **g)** head, **h)** habit.

1119. Eupatorium sessilifolium   1120. Fleischmannia incarnata   1121. Liatris aspera

upland forests, sand savannas, bottomland forests, and swamps; also old fields, roadsides, and open, disturbed areas.

For a discussion of possible hybrids with *E. hyssopifolium,* see the treatment of that species. Steyermark (1963) considered *E. semiserratum* (as *E. cuneifolium* var. *semiserratum*) to inhabit only moist, bottomland habitats, but most of the collections since 1990 originated from drier, upland sites.

10. Eupatorium serotinum Michx. (late boneset)

Pl. 268 c, d; Map 1118

Stems 40–200 cm long, not hollow, densely short-hairy above the sometimes nearly glabrous basal portion, tan to grayish purple, not glaucous, often producing small fascicles of axillary leaves much shorter than the main stem leaves, at least at a few nodes. Leaves mostly opposite, those of the uppermost nodes frequently alternate, short-petiolate, the well-differentiated petiole (2–)8–30 mm long. Leaf blades 3–20 cm long, 6–100 mm wide, narrowly lanceolate to ovate, angled or tapered at the base, tapered to a sharply pointed tip, the margins sharply and often coarsely toothed, the upper surface glabrous to moderately short-hairy, the undersurface moderately to densely short-hairy, both surfaces also sparsely to moderately gland-dotted, with 3(5) main veins, the lateral veins branching from at or just above the base of the midvein. Inflorescences terminal panicles, flat-topped or shallowly dome-shaped. Involucre 3–4 mm long, cup-shaped, the bracts oblong-lanceolate to narrowly oblong, rounded to bluntly or less commonly sharply pointed at the tip, the margins thin and pale or transparent, especially toward the tip, mostly faintly 3-nerved, densely short-hairy, green but appearing mostly gray. Disc florets 9–15. Corollas 2.5–3.5 mm long, the surface often somewhat glandular, white. Fruits 1–2 mm long. $2n=20$. August–October.

Scattered nearly throughout the state (eastern U.S. west to Minnesota, Nebraska, and Texas; Canada). Upland prairies, margins of glades, savannas, openings of mesic to dry upland forests, bottomland forests, banks of streams and rivers, margins of ponds and lakes, and ledges and tops of bluffs; also old fields, fallow fields, pastures, gardens, railroads, roadsides, and disturbed areas.

Eupatorium serotinum is among the weediest species of Missouri thoroughworts, with little fidelity to any particular habitat, moisture regime, or substrate type. It spreads aggressively by seeds in gardens. Superficially it bears a resemblance to *Ageratina altissima* (white snakeroot). In addition to the characters in the key to genera above, the involucres of *E. serotinum* appear relatively grayish compared to the greener appearance of those in *A. altissima,* and the leaf blades generally appear somewhat narrower, with more narrowly angled bases. It is important not to confuse the two species, given that *A. altissima* is quite poisonous.

11. Eupatorium sessilifolium L. (upland boneset)

Eupatorium sessilifolium var. *brittonianum* Porter

Pl. 267 h, i; Map 1119

Stems 30–120 cm long, not hollow, glabrous below the inflorescence, tan or less commonly purplish brown, rarely somewhat glaucous, not producing small fascicles of axillary leaves much shorter than the main stem leaves. Leaves mostly opposite, those of the uppermost nodes rarely alternate, sessile or less commonly with petioles to 2 mm long. Leaf blades 3–18 cm long, 8–60 mm wide, lanceolate to narrowly ovate, rounded, truncate or shallowly cordate at the base (the bases of the pair at all but the uppermost nodes often somewhat overlapping), tapered to a sharply pointed tip, the margins sharply toothed, the surfaces glabrous, also sparsely to moderately gland-dotted, with 1 main

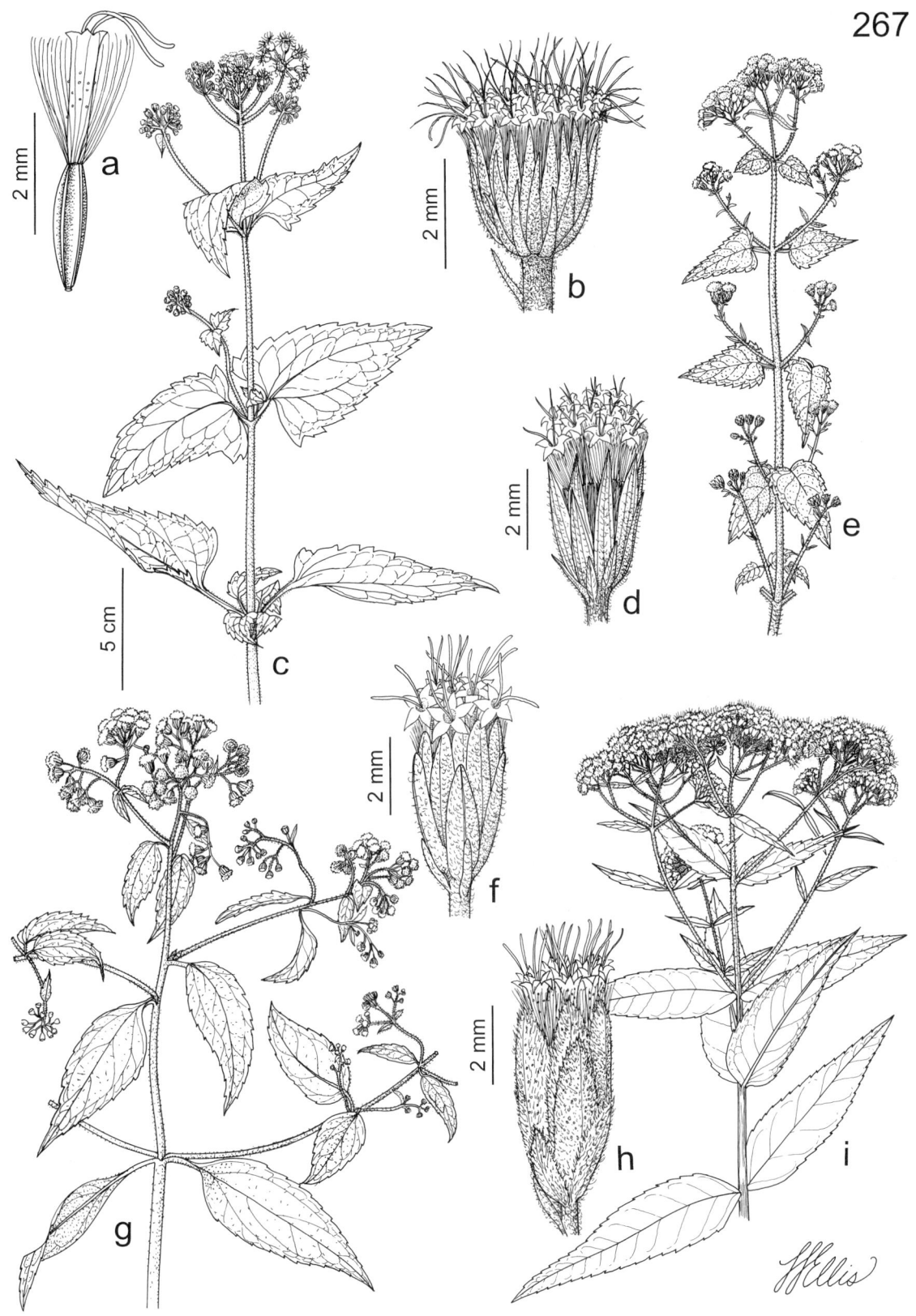

Plate 267. Asteraceae. *Conoclinium coelestinum*, **a)** floret, **b)** head, **c)** habit. *Fleischmannia incarnata*, **d)** head, **e)** habit. *Ageratina altissima*, **f)** head, **g)** habit. *Eupatorium sessilifolium*, **h)** head, **1)** habit.

vein. Inflorescences terminal panicles, more or less flat-topped. Involucre 4.5–6.5 mm long, cup-shaped, the bracts ovate to lanceolate or narrowly oblong, rounded to bluntly or less commonly sharply pointed at the tip, the margins thin and pale or transparent, mostly faintly 3-nerved, densely short-hairy, green. Disc florets 5(–7). Corollas 3–4 mm long, the surface often somewhat glandular, white. Fruits 2–3 mm long. 2n=20, 30. July–September.

Scattered in the Ozark and Ozark Border Divisions and in the eastern half of the Glaciated Plains (northeastern U.S. south to North Carolina and Missouri). Bottomland forests, mesic to dry upland forests, bases and ledges of bluffs, and banks of streams and rivers.

Missouri plants represent a phase with relatively long, slender leaves that has been called var. *brittonianum*. Elsewhere to the east, var. *sessilifolium* tends to have shorter, broader leaves and is sometimes similar in leaf shape to long-leaved variants of *E. rotundifolium*. However, as noted by Steyermark (1963), there is a lot of intergradation between the leaf variants, which argues against formal taxonomic recognition of varieties.

55. Fleischmannia Sch. Bip.

About 80 species, North America, Central America, South America.

1. Fleischmannia incarnata (Walter) R.M. King & H. Rob. (pink thoroughwort)
Eupatorium incarnatum Walter
Pl. 267 d, e; Map 1120

Plants perennial, fibrous-rooted, often reclining on other vegetation. Stems usually few, several-branched, 30–120 cm long, loosely ascending, often from a spreading base, sparsely to moderately pubescent with short, curly hairs. Leaves opposite (the nodes well separated), sessile to short- or long-petiolate. Leaf blades 0.8–7.0 cm long, triangular-ovate to more or less heart-shaped, the uppermost sometimes only triangular-lanceolate, truncate to more commonly shallowly cordate at the base (the uppermost leaves sometimes broadly angled), tapered to a usually sharply pointed tip, the margins shallowly scalloped or bluntly toothed, the upper surface glabrous or minutely hairy, sometimes only along the veins, the undersurface moderately to densely hairy, especially along the veins, usually not glandular, with mostly 3 main veins. Inflorescences small panicles or stalked clusters at the branch tips, usually more or less flat-topped. Heads with 13–25 disc florets. Involucre 3–5 mm long, the bracts 18–30 (the head often also subtended by 1 or a few other narrower bracts), in usually 2 or 3 unequal to subequal, overlapping series, narrowly oblong-elliptic to narrowly lanceolate, tapered to a bluntly or more commonly sharply pointed tip, all but the outermost noticeably but finely few-nerved or few-ribbed, glabrous or sparsely short-hairy. Receptacle flat or slightly convex. Corollas purple or lavender-blue (rarely white elsewhere). Pappus of numerous capillary bristles. Fruits 1.8–2.5 mm long, finely 5(–7)-ribbed, slightly wedge-shaped in profile (usually slightly and unevenly tapered at the base), glabrous, the ribs smooth or with a few minute, ascending teeth, lighter colored, the body otherwise dark brown to black. 2n=20. August–October.

Uncommon in the Mississippi Lowlands Division and also in McDonald and Ozark Counties (southeastern U.S. west to Illinois, Missouri, and Texas; Mexico). Swamps, bottomland forests, bases of bluffs, and banks of streams and rivers; also ditches, railroads, and roadsides.

Molecular studies have suggested that *Fleischmannia* forms a natural group with *Ageratum* and *Conoclinium* that is only distantly related to *Eupatorium* in the strict sense within the tribe Eupatorieae (Schilling et al., 1999; Schmidt and Schilling, 2000; Ito et al., 2000b). This is reflected in the overall similarity in general appearance between *F. incarnata, A. conyzoides,* and *C. coelestinum,* which all have bluish purple flowers and ovate-triangular main leaves. Steyermark (1963) noted that plants of *F. incarnata* (as *Eupatorium incarnatum*) tend to have an odor resembling vanilla when dried and that the leaves generally are thinner, darker green, and more heart-shaped than those of *C. coelestinum*. He also noted that plants of *A. conyzoides* tend to be more densely hairy on the stems and leaves.

56. Liatris Gaertn. ex Schreb. (blazing star)

Plants perennial, the rootstock an ovoid to depressed-globose corm, this usually covered with the persistent, brown, fibrous remains of old leaf bases (except in *L. punctata*, with an elongate rootstock). Stems solitary to several, erect or ascending. Leaves alternate but sometimes so numerous and dense as to appear indefinitely whorled, the basal and lower stem leaves short-petiolate, grading abruptly into the sessile or very short-petiolate stem leaves; the basal leaves often present at flowering, these and the adjacent lowermost stem leaves the largest on the plant. Blades of the basal leaves usually oblanceolate to narrowly elliptic, grading abruptly into the linear or less commonly narrowly elliptic to narrowly oblong stem leaves, all tapered at the base, usually tapered to a sharply pointed tip, the margins entire, the surfaces variously short-hairy to glabrous or nearly so, often also gland-dotted, with 1 or 3(5) main veins. Inflorescences unbranched terminal spikes or spikelike racemes (sometimes somewhat branched elsewhere), sometimes leafy and the lowermost heads then appearing axillary. Heads with 4–80 disc florets. Involucre cylindrical to broadly cup-shaped or bell-shaped, the bracts mostly 18–40, in usually 3 to several unequal, overlapping series, the bracts lanceolate to broadly ovate, tapered (often abruptly so) to a sharply pointed tip, glabrous or short-hairy, often also glandular. Receptacle flat or nearly so. Corollas pink to purple, rarely white, often sparsely glandular on the outer surface, sometimes hairy on the inner surface. Pappus of 12–40 bristles, these minutely barbed (the short barbs mostly 0.1–0.3 mm long) or plumose (the feathery barbs mostly 0.5–1.0 mm long). Fruits (8–)10-nerved or ribbed, slightly wedge-shaped in profile (usually slightly and unevenly tapered at the base) to nearly linear, usually minutely hairy and glandular, brown to dark brown. About 45 species, U.S., Canada, Mexico, Caribbean Islands.

Liatris species are among the showier plants to grace native wildflower gardens. A number of species are in cultivation (Dress, 1959), and most of the Missouri taxa can be purchased at native plant specialty nurseries in the state. *Liatris spicata* has become more widely available at home supply centers and general plant nurseries and also is coming into broader use as a cut flower at florists and even grocery stores. In the garden, these plants attract a host of animals, including a wide variety of insects visiting the flowers and birds feeding on the achenes, but the sweet, thickened rootstocks are a magnet for voles and other herbivorous mammals. Rootstocks of various species also were consumed raw or baked by early settlers and by Native Americans. Various tribes also used some *Liatris* species medicinally as a general analgesic, to settle upset stomachs, for urinary tract problems, and in poultices to ease skin inflammations, among other uses (Moerman, 1998). The common name snakeroot, applied to some species, apparently is in reference to a belief in their efficacy in the treatment of snakebite. Horses sometimes were fed a decoction of the rootstocks as a stimulant before races (Moerman, 1998).

Liatris is a taxonomically difficult genus that requires much further research. For some of the complexes, species limits still are not well understood, and the impact of hybridization and polyploidy upon the recognition of species has not been well studied for most of our taxa. In perhaps the only detailed study of hybridization in the group, Levin (1968) examined a mixed population of *L. aspera*, *L. cylindracea*, *L. ligulistylis* (probably actually *L. scariosa*), and *L. spicata* in northern Illinois, documenting an extensive swarm of mostly fertile hybrids. Steyermark (1963) noted that the last monographer of the genus, Lulu O. Gaiser (1946), did not examine specimens from any of the herbaria having large holdings of Missouri materials.

1. Pappus bristles plumose
    2. Leaves mostly with 3(5) main veins; heads with 10–60 disc florets; corolla lobes hairy on the inner surface
        3. Involucre with the outer series of bracts not appearing noticeably longer than the other series, the bracts appressed or strongly ascending, broadly rounded or broadly angled to a bluntly pointed tip, or more commonly short-tapered to an abrupt, sharp point at the tip . . . . 2. L. CYLINDRACEA
        3. Involucre with the outer series of bracts usually appearing noticeably longer than the other series (especially on the terminal head), the bracts abruptly spreading to recurved above the basal portion, mostly long-tapered to a sharply pointed tip . . . . 8. L. SQUARROSA
    2. Leaves mostly with 1 main vein; heads with 3–7 disc florets; corolla lobes glabrous; species difficult to distinguish
        4. Rootstock a corm, more or less globose; leaves green, glabrous; involucre 7–12 mm long . . . . 3. L. MUCRONATA
        4. Rootstock more or less a thickened, elongate taproot, vertical or occasionally somewhat spreading; leaves grayish green, at least some of them hairy along the margins (note that the hairs break off with age, leaving minute, stubby bases); involucre 10–14 mm long . . . . 4. L. PUNCTATA
1. Pappus bristles merely barbed
    5. Heads with (3)4–9(–14) disc florets; involucre narrowly cup-shaped to nearly cylindrical; basal and lower stem leaves (and often most other stem leaves) with 3 or 5 main veins
        6. Involucral bracts tapered to long, sharply pointed tips, these spreading or recurved . . . . 5. L. PYCNOSTACHYA
        6. Involucral bracts rounded or broadly angled to a bluntly pointed tip, appressed . . . . 7. L. SPICATA
    5. Heads with (11–)14–80 disc florets; involucre broadly cup-shaped to bell-shaped; leaves with 1 main vein; species difficult to distinguish
        7. Involucral bracts mostly with broad, thin, pale to transparent margins, these sometimes strongly purplish-tinged, appearing irregularly torn or strongly uneven (irregularly wavy or scalloped), the main body appearing swollen or pouched . . . . 1. L. ASPERA
        7. Involucral bracts with green (unmodified) or narrow, thin, pale to transparent margins, these sometimes strongly purplish-tinged, entire or sometimes appearing wavy or shallowly scalloped (sometimes also minutely irregular), the main body sometimes recurved at the tip but appearing not or only slightly swollen or pouched
            8. Heads with 28–80 disc florets, relatively long-stalked, the stalks 12–40(–150) mm long, often with several small bracts; all or nearly all of the involucral bracts erect or ascending, the lowermost sometimes spreading or reflexed . . . . 6. L. SCARIOSA
            8. Heads with 11–26(–28) disc florets, sessile or relatively short-stalked, the stalks to 15 mm long, the longer ones sometimes with a few small bracts; nearly all of the involucral bracts with the tip spreading or reflexed, only the innermost series usually ascending
                . . . . 9. L. SQUARRULOSA

1122. Liatris cylindracea   1123. Liatris mucronata   1124. Liatris punctata

1. **Liatris aspera** Michx. (rough gayfeather)
   L. sphaeroidea Michx.

   Pl. 269 f–h; Map 1121

Rootstock a more or less globose to ovoid corm, sometimes appearing somewhat erect and angular or irregular. Stems (30–)50–180 cm long, moderately to densely pubescent with short, curled hairs, sometimes nearly glabrous toward the base. Basal and adjacent lower stem leaves short- to long-petiolate, the blades 6–25 cm long, 4–25 mm wide, oblanceolate to narrowly elliptic (rarely broader), the surfaces glabrous to densely short-hairy, green, with 1 main vein, grading abruptly to the shorter stem leaves, these mostly sessile, 2–10 cm long, narrowly oblanceolate to more commonly linear. Inflorescences elongate spicate racemes, the heads mostly relatively loosely to more densely spaced (the axis usually visible between heads), sessile or with stalks to 8 mm long, these usually with only 1 basal bract. Heads with 14–30 disc florets, the terminal head not noticeably larger than the others. Involucre 9–16 mm long, broadly cup-shaped to broadly bell-shaped, with 4 or 5 unequal, overlapping series of bracts (the outer series appearing progressively shorter). Involucral bracts broadly obovate to oblong-spatulate, all but the innermost series spreading or recurved at the tip, mostly with broad, thin, pale to transparent margins, the margins or entire bracts sometimes strongly purplish-tinged, appearing irregularly torn or strongly uneven (irregularly wavy or scalloped), the main body appearing swollen or pouched toward the tip. Corollas 8–11 mm long, the tube hairy on the inner surface. Pappus bristles barbed. Fruits 4–6 mm long. $2n=20$. August–November.

Scattered nearly throughout the state, although apparently absent from most of the Mississippi Lowlands Division (eastern U.S. west to North Dakota and Texas; Canada). Upland prairies, loess hill prairies, glades, exposed ledges and tops of bluffs, savannas, openings of mesic to dry upland forests, and rarely banks of streams; also pastures, railroads, and roadsides.

Rare, white-flowered individuals have been called f. *benkei* (J.F. Macbr.) Fernald. A putative hybrid with *L. squarrosa* is known from Barry County, and a possible instance of hybridization with *L. squarrulosa* has been collected in Texas County. Gaiser (1946) named a putative hybrid between *L. aspera* and *L. pycnostachya* from Minnesota as *L. ×frostii* Gaiser. A single historical collection from Pettis County appears intermediate between these two species. Gaiser had annotated this sheet as *L. sphaeroidea,* which she thought might be a hybrid involving *L. aspera* and one or more other unknown parents, but that name apparently instead is merely a synonym of *L. aspera.*

2. **Liatris cylindracea** Michx.
   L. squarrosa (L.) Michx. var. *intermedia* (Lindl.) DC.

   Pl. 268 g, h; Map 1122

Rootstock a more or less globose corm. Stems 20–60 cm long, glabrous or sparsely hairy toward the base. Basal and adjacent lower stem leaves sessile to short-petiolate, the blades 8–25 cm long, 2–6(–12) mm wide, linear, the margins usually light, hard, and thickened, often curled under, the surfaces glabrous or the undersurface sparsely to moderately pubescent with short, spreading hairs, green, with 3(5) main veins, grading to the shorter stem leaves, these mostly sessile, 2–12 cm long, linear. Inflorescences elongate racemes, the heads loosely spaced (the axis easily visible between heads), sessile or more commonly with stalks 2–35(–65) mm long, these with 1 or few basal bracts. Heads with 10–35 disc florets, the terminal head usually slightly longer than the others. Involucre 11–20 mm long, narrowly cup-shaped to nearly cylindrical, with 5–7 unequal, overlapping series of bracts (the outer series appearing progressively shorter). Involucral bracts broadly ovate-triangular to oblanceolate-oblong, the outer series broadly rounded or narrowed to a sharply pointed, appressed or strongly ascending tip, grading to the inner series with a truncate to abruptly rounded tip, often with an abrupt, minute, sharp point, mostly with narrow, thin, pale to transparent margins, these sometimes slightly to strongly purplish-tinged, entire to minutely irregular and sometimes also with minute, irregular, hairlike processes, the main body appearing flat below the tip. Corollas 12–14 mm long, glabrous or the tube hairy on the inner surface. Pappus bristles plumose. Fruits 5–7 mm long. $2n=20$. July–September.

Scattered in the Ozark and Ozark Border Divisions north locally in the eastern portion of the state to Lewis County (Minnesota to New York south to Kansas, Arkansas, and Alabama; Canada). Glades, ledges and tops of bluffs, openings of dry upland forests, and upland prairies; also pastures and roadsides.

White-flowered individuals (f. *bartelii* Steyerm.) have not yet been recorded from Missouri. A surprising number of putative hybrids of intermediate morphology with *L. squarrosa* (apparently mostly with var. *hirsuta*) have been collected in Missouri. These require further study.

3. Liatris mucronata DC. (narrow-leaved gayfeather)

Pl. 269 i, j; Map 1123

Rootstock a more or less globose corm. Stems 20–90 cm long, glabrous. Basal and adjacent lower stem leaves sessile to short-petiolate, the blades 7–15 cm long, 1–4 mm wide, linear, the narrow, light margins occasionally curled under, the surfaces glabrous, green, with 3(5) main veins, grading toward the stem tip to shorter leaves, these mostly sessile, 1.5–12.0 cm long, linear. Inflorescences elongate spikes, the heads densely spaced (the axis mostly not visible between heads), sessile or nearly so, with 1 basal bract. Heads with 3–6 disc florets, the terminal head sometimes slightly longer than the others. Involucre 7–12 mm long, narrowly cup-shaped to nearly cylindrical, with 4–6 unequal, overlapping series of bracts (the outer series appearing somewhat shorter). Involucral bracts broadly lanceolate to narrowly oblong-obovate, tapered to a sharply pointed, ascending to somewhat spreading tip, mostly with narrow, thin, pale to transparent margins, these sometimes slightly to strongly purplish-tinged, more or less entire but sometimes cobwebby-hairy, the main body appearing flat below the tip. Corollas 9–11 mm long, glabrous or the tube sparsely hairy on the inner surface toward the base. Pappus bristles plumose. Fruits 5.5–7.5 mm long. July–October.

Uncommon in the western portion of the Ozark Division, mostly in the watershed of the White River and its tributaries (Kansas and Missouri south to Texas and possibly Louisiana). Glades and ledges and tops of bluffs, on limestone and dolomite substrates.

The aboveground portions of plants of *L. mucronata* and *L. punctata* can be surprisingly difficult to distinguish. In Missouri, where the ranges of the two do not overlap, it has proven easier to separate specimens lacking rootstocks by geography than morphology. The combination *L. punctata* var. *mucronata* (DC.) B.L. Turner is available if future research should show that the apparent morphological differences in rootstock are due more to environmental variation than to taxonomy. One qualitative difference between them is that in *L. mucronata* the leaves are described as green, relatively soft-herbaceous, mostly arched or curved outward, and glabrous, whereas in *L. punctata* the leaves are said to be relatively thick and leathery, mostly straight or the largest ones only slightly curved outward, and at least some of them hairy along the margins (note that the hairs break off with age, leaving minute, stubby bases).

These plants are part of a polyploid complex that is widespread in the center of the United States and needs more intensive biosystematic study. A recent segregate in this complex from northern Texas and Oklahoma was described as *L. aestivalis* G.L. Nesom & O'Kennon and differs from typical *L. mucronata* in its fewer longer involucral bracts, slightly more pronounced leaf venation, and earlier flowering period, among other subtle characters. However, Nesom and O'Kennon (2001) noted that some plants attributed to *L. mucronata* from northern Arkansas, southern Missouri, and adjacent Kansas may represent an undescribed taxon more closely related to *L. aestivalis* than to *L. mucronata* in the strict sense. Further studies of population-level variation and cytology are needed.

4. Liatris punctata Hook. **var. punctata**
(prairie snakeroot)

Pl. 270 f, g; Map 1124

Rootstock a more or less thickened, elongate taproot, vertical or occasionally somewhat spreading. Stems 20–90 cm long, glabrous. Basal and adjacent lower stem leaves sessile to short-petiolate, the blades 7–15 cm long, 1–4 mm wide, linear, the narrow, light margins occasionally curled under, the surfaces glabrous but the margins of at least some

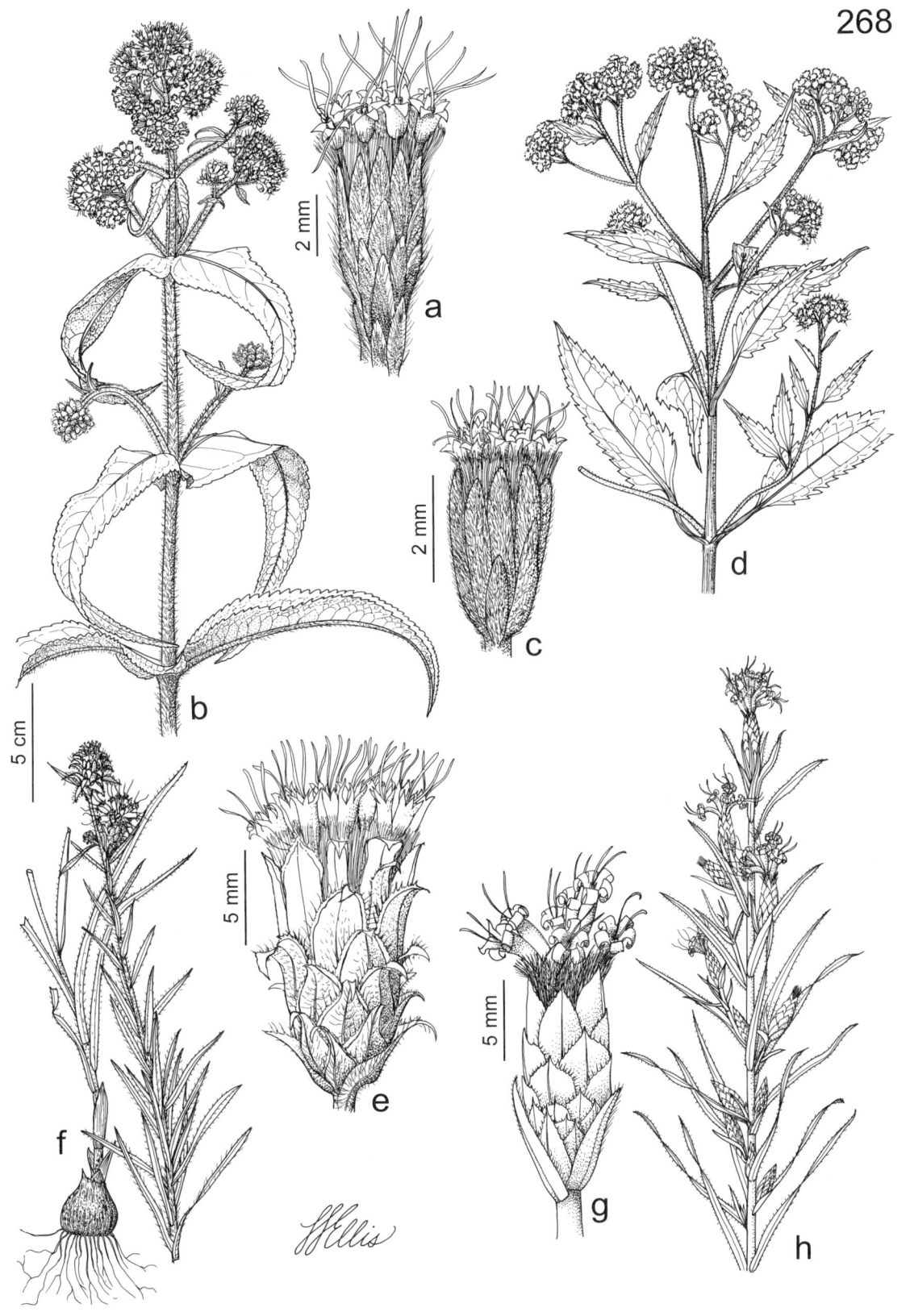

Plate 268. Asteraceae. *Eupatorium perfoliatum*, **a)** head, **b)** habit. *Eupatorium serotinum*, **c)** head, **d)** habit. *Liatris squarrosa*, **e)** head, **f)** habit. *Liatris cylindracea*, **g)** head, **h)** habit.

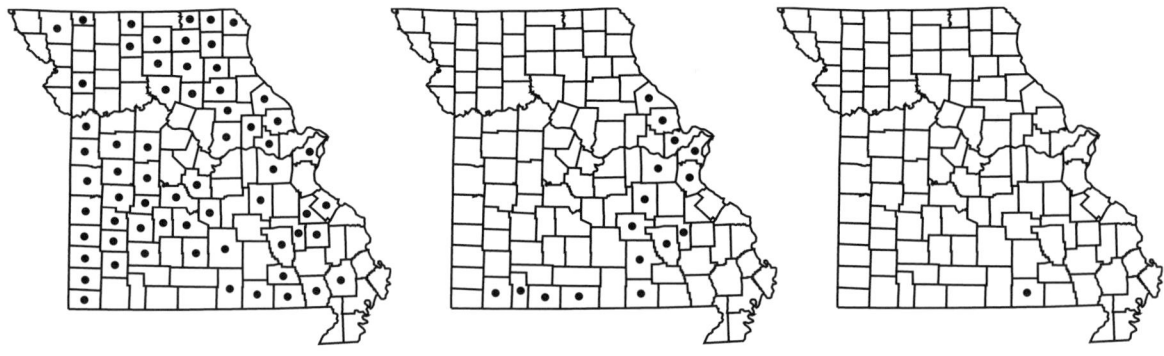

1125. Liatris pycnostachya   1126. Liatris scariosa   1127. Liatris spicata

leaves hairy (note that the hairs break off with age, leaving minute, stubby bases), grayish green, with 3(5) main veins, grading toward the stem tip to shorter leaves, these mostly sessile, 1.5–12.0 cm long, linear. Inflorescences elongate spikes, the heads densely spaced (the axis mostly not visible between heads), sessile or nearly so, with 1 basal bract. Heads with 3–7 disc florets, the terminal head sometimes slightly longer than the others. Involucre 10–14 mm long, narrowly cup-shaped to nearly cylindrical, with 4 or 5 unequal, overlapping series of bracts (the outer series appearing somewhat shorter). Involucral bracts broadly lanceolate to narrowly oblong-obovate, tapered to a sharply pointed, ascending to somewhat spreading tip, mostly with narrow, thin, pale to transparent margins, these sometimes slightly to strongly purplish-tinged, more or less entire but usually cobwebby-hairy, the main body appearing flat below the tip. Corollas 9–11 mm long, glabrous. Pappus bristles plumose. Fruits 5.5–7.5 mm long. August–October.

Uncommon along the western border of the state, mostly in the Glaciated Plains Division (South Dakota to New Mexico east to Michigan, Tennessee, and Louisiana). Loess hill prairies and rarely upland prairies.

Steyermark (1963) followed Gaiser (1946) in accepting Missouri plants as part of var. *nebraskana* Gaiser based upon minor differences in leaf length and width, as well as involucral bract shape and pubescence. However, he questioned whether these taxa should be formally recognized, as specimens from Missouri and adjacent states are somewhat intermediate between var. *nebraskana* and var. *punctata*. Further study of additional collections has reinforced the futility of attempting to separate populations into two varieties based upon these minor and inconsistent variations. The wisdom in recognizing somewhat better-marked populations from Texas, New Mexico, and adjacent Mexico that have a more loosely flowered inflorescence as var. *mexicana* Gaiser requires further study. See the treatment of *L. mucronata* for further discussion of problems of species recognition in this complex. The specimen of *L. punctata* cited by Steyermark (1963) as having been collected in Bates County could not be located during the present study.

5. Liatris pycnostachya Michx. **var. pycnostachya** (gayfeather, button snakeroot)

Pl. 270 a–c; Map 1125

Rootstock a more or less globose corm. Stems 50–150 cm long, moderately to densely pubescent with short, curled hairs, sometimes nearly glabrous toward the base. Basal and adjacent lower stem leaves mostly short-petiolate, the blades 8–40 cm long, 3–13 mm wide, linear or narrowly oblanceolate, the surfaces glabrous to densely short-hairy, green, with 3 or 5 main veins, grading to the shorter stem leaves, these mostly sessile, 1.5–15.0 cm long, linear or very narrowly oblanceolate to very narrowly lanceolate. Inflorescences elongate spicate racemes, the heads densely spaced (the axis mostly not visible between heads), sessile or with stalks to 1 mm long, these with only 1 basal bract. Heads with 4–9(–12) disc florets, the terminal head not or only slightly larger than the others. Involucre 7–11 mm long, narrowly cup-shaped to nearly cylindrical, with 4 or 5 unequal, overlapping series of bracts (the outer series appearing progressively shorter). Involucral bracts broadly lanceolate to oblong-lanceolate, tapered to a long, sharply pointed, spreading or recurved tip, mostly with thin, pale to transparent margins, the margins or entire bract sometimes strongly purplish-tinged, minutely irregular or more commonly with irregular, hair-like processes, the main body appearing flat below the tip. Corollas 7–10 mm long, glabrous. Pappus bristles barbed. Fruits 3.5–5.0 mm long. $2n=20, 40$. July–October.

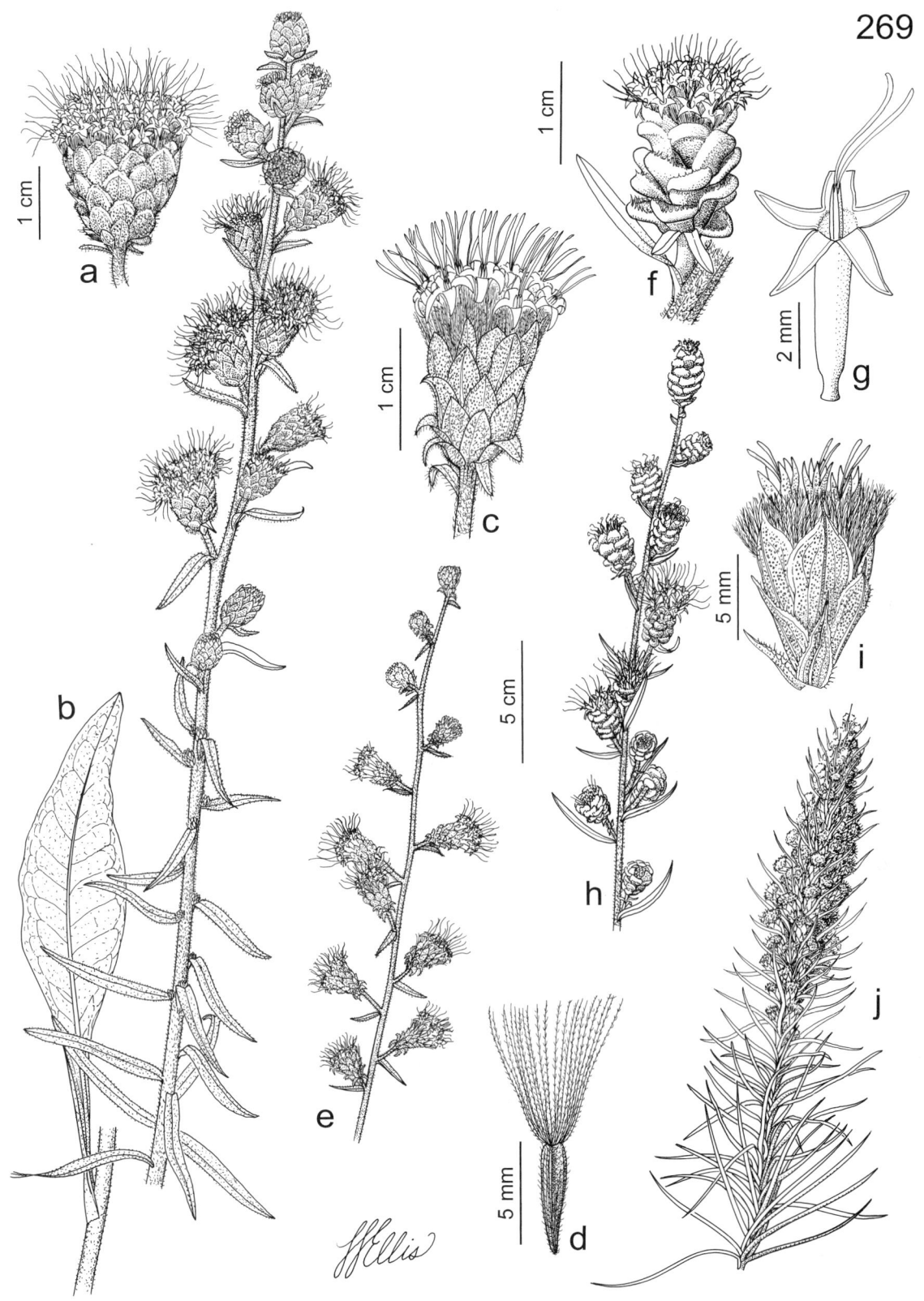

Plate 269. Asteraceae. *Liatris scariosa*, **a)** head, **b)** habit and larger leaf. *Liatris squarrulosa*, **c)** head, **d)** fruit, **e)** inflorescence. *Liatris aspera*, **f)** head, **g)** floret (ovary not drawn), **h)** inflorescence. *Liatris mucronata*, **i)** head, **j)** habit.

Scattered nearly throughout the state (North Dakota to Texas east to Massachusetts, Kentucky, and Mississippi). Glades, upland prairies, ledges and tops of bluffs, savannas, openings of mesic to dry upland forests, and rarely banks of streams; also ditch banks, fencerows, pastures, railroads, and roadsides.

Rare, white-flowered plants have been called f. *hubrichtii* E.S. Anderson. Another variety, var. *lasiophylla* Shinners, occurs sporadically from Texas to Louisiana and is a hairier taxon with an often more elongate rootstock. For a discussion of possible hybridization of *L. pycnostachya* with *L. aspera* and *L. spicata,* see the treatments of those species.

6. Liatris scariosa (L.) Willd. var. nieuwlandii E.G. Voss

Lacinaria scariosa (L.) Hill var. *nieuwlandii* Lunell

Liatris novae-angliae (Lunell) Shinners var. *nieuwlandii* (Lunell) Shinners

Liatris ×*nieuwlandii* (Lunell) Gaiser

Pl. 269 a, b; Map 1126

Rootstock a more or less globose to ovoid corm. Stems 40–200 cm long, moderately to densely pubescent with short, curled hairs, sometimes nearly glabrous toward the base. Basal and adjacent lower stem leaves short- to long-petiolate, the blades mostly 10–50 cm long, 25–55 mm wide, oblanceolate to elliptic, the surfaces glabrous to sparsely short-hairy along the midvein, also usually minutely hairy along the margins, green, with 1 main vein, grading gradually to the shorter stem leaves, these mostly sessile, 3–15 cm long, narrowly oblanceolate to more commonly narrowly oblong-elliptic, the uppermost often linear. Inflorescences elongate, often somewhat open racemes, the heads mostly relatively loosely spaced (the axis visible between heads), with stalks 12–40(–150) mm long, these usually with several small, loosely spaced bracts. Heads with 29–80 disc florets, the terminal head almost always somewhat larger than the others. Involucre 11–17 mm long, broadly cup-shaped to broadly bell-shaped, with 4–6 unequal, overlapping series of bracts (the outer series appearing progressively shorter). Involucral bracts broadly obovate to oblong-obovate, all appressed-ascending or the outermost series sometimes with somewhat spreading tips, mostly with narrow, thin, pale to dark purple margins (occasionally green throughout), these nearly entire to shallowly scalloped (sometimes also minutely irregular), the main body appearing flat or very slightly swollen just below the tip. Corollas 9–11 mm long, the tube glabrous or hairy on the inner surface. Pappus bristles barbed. Fruits 4.5–6.5 mm long. July–October.

Scattered in the Ozark Division north locally to Pike County (northeastern U.S. west to Wisconsin, Missouri, and Arkansas). Upland prairies, glades, exposed ledges and tops of bluffs, savannas, and openings of mesic to dry upland forests; also pastures and open, grassy, disturbed areas.

The taxonomy of this species and its variants is still not well understood. As currently accepted (Cronquist, 1980; Gleason and Cronquist, 1991), it comprises three varieties, each of which has been treated as a separate species at some point in the past. The mostly midwestern var. *nieuwlandii* has the largest heads, tallest stems, and most leaves; var. *scariosa* is mostly Appalachian and has much-fewer-flowered (19–33) heads and many fewer leaves; and var. *novae-angliae* occurs mostly in New England, is glabrous or nearly so, and has an intermediate number of disc florets per head.

In her monograph, Gaiser (1946) suggested that *L. scariosa* is a stabilized hybrid involving past hybridization between *L. ligulistylis* (A. Nelson) K. Schum. and some unknown parental taxon. Steyermark (1963) and Bowles et al. (1988) rejected this treatment, suggesting that the midwestern plants are not of hybrid origin. Steyermark (1963), however, included Missouri plants in *L. ligulistylis* rather than *L. scariosa,* apparently following Gaiser's suggestion (since disproven) that the corollas of the latter should always be hairy internally (Missouri plants vary from glabrous to hairy). *Liatris ligulistylis* presently is understood to be distributed from the eastern Rocky Mountains through the western and northern Great Plains, and it differs from *L. scariosa* mainly in its involucral bracts with relatively broad, irregular, scarious margins.

The application of names to Missouri plants is more complicated than even the confusing nomenclatural history of the complex suggests. Specimens annotated by Steyermark as *L. ligulistylis* are by no means all assignable to *L. scariosa*. A significant proportion of these are treated here as part of the closely related *L. squarrulosa*. Gaiser (1946) classified *L. squarrulosa* as a variety of *L. scariosa* but also recognized as a separate species something that she called *L. scabra*. Steyermark (1963) accepted *L. scabra* as a species without further comment, but later authors (Cronquist, 1980; Gleason and Cronquist, 1991) have tended to consider *L. scabra* and *L. squarrulosa* as a single species that is separable from *L. scariosa* by its relatively short-stalked heads with the involucral bracts having reflexed tips. In practice, Missouri plants of *L.*

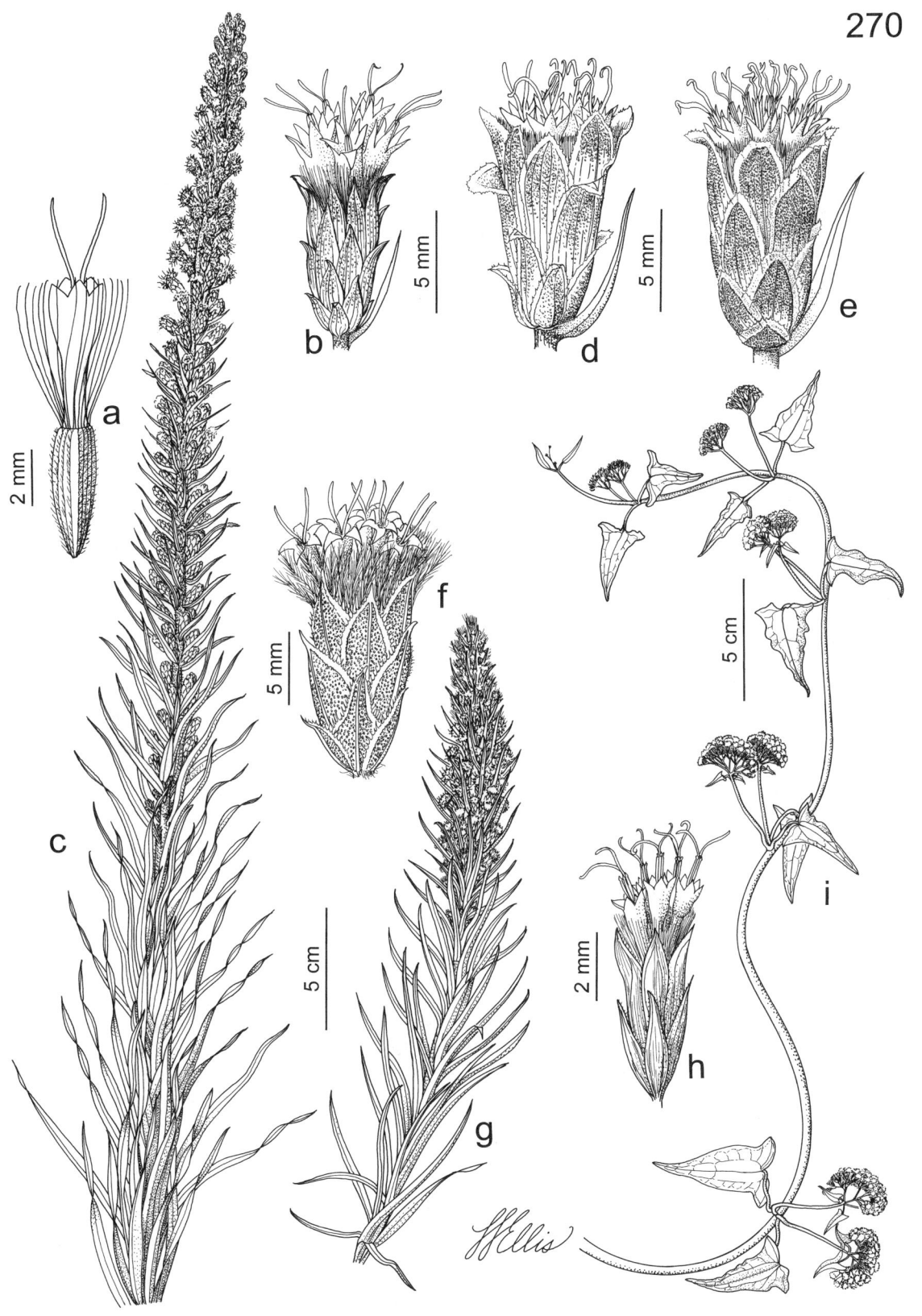

Plate 270. Asteraceae. *Liatris pycnostachya*, **a)** floret, **b)** head, **c)** habit. *Liatris ×bebbiana*, **d)** head. *Liatris spicata*, **e)** head. *Liatris punctata*, **f)** head, **g)** habit. *Mikania scandens*, **h)** head, **i)** habit.

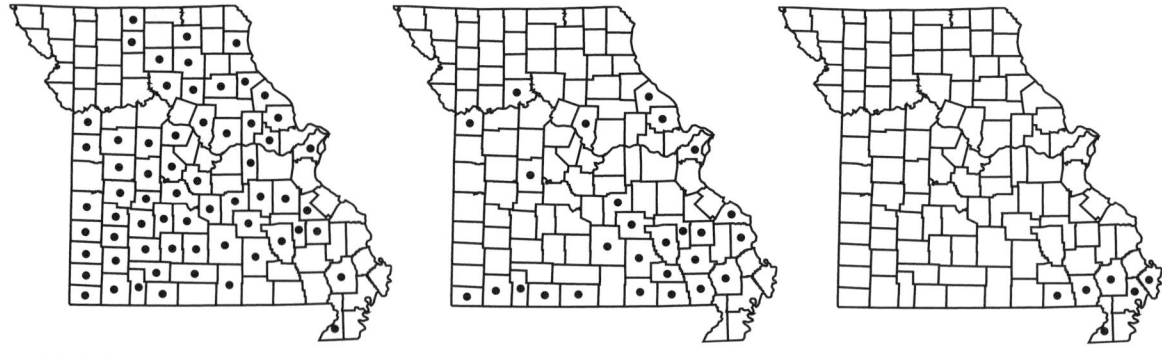

1128. Liatris squarrosa   1129. Liatris squarrulosa   1130. Mikania scandens

aspera and *L. scariosa* are relatively easily separated, but plants assignable to *L. squarrulosa* are variable and are sometimes difficult to separate from particularly *L. aspera*.

7. Liatris spicata (L.) Willd. (button snakeroot)
Pl. 270 e; Map 1127

Rootstock a more or less globose corm. Stems 40–150 cm long, glabrous (sometimes hairy elsewhere). Basal and adjacent lower stem leaves mostly short-petiolate, the blades 8–40 cm long, 3–12(–20) mm wide, linear or narrowly oblanceolate, the surfaces glabrous (occasionally sparsely hairy elsewhere), green, with 3 or 5 main veins, grading to the shorter stem leaves, these mostly sessile, 1.5–15.0 cm long, linear or very narrowly oblanceolate to very narrowly lanceolate. Inflorescences elongate spicate racemes, the heads densely or rarely more loosely spaced (the axis mostly not visible between heads), sessile or with stalks to 1 mm long, these with only 1 basal bract. Heads with 4–8(–14) disc florets, the terminal head not or only slightly larger than the others. Involucre 7–11 mm long, narrowly cup-shaped to nearly cylindrical, with 4 or 5 unequal, overlapping series of bracts (the outer series appearing progressively shorter). Involucral bracts ovate to narrowly oblong-ovate, rounded or broadly angled to a bluntly pointed, appressed tip, mostly with thin, pale to transparent margins, the margins or entire bract sometimes strongly purplish-tinged, entire or with few minute, irregular, hairlike processes, the main body appearing flat below the tip. Corollas 7–11 mm long, glabrous. Pappus bristles barbed. Fruits 4–6 mm long. $2n=20$. July–October.

Uncommon, known thus far from a single site in Oregon County (eastern U.S. west to Wisconsin, Missouri, and Mississippi; Canada). Upland prairies.

Because this species is relatively widespread in cultivation, it may become recorded as an escape in other counties in Missouri at some point in the future. The Oregon County site for this eastern species was studied by Steyermark (1963) in the early 1930s. He found plants locally common in open, gravelly ground with upland prairie vegetation near the town of Bardley. In 1986, Tim Nigh of the Missouri Department of Conservation relocated this station. However, in the intervening decades, *L. pycnostachya* became common at the site and hybridized extensively with the original population of *L. spicata,* with the result that genetically pure *L. spicata* apparently no longer exists there. Plants of intermediate morphology apparently are referable to *L. ×bebbiana* Rydb. (Pl. 270 d) and are now the most abundant type there. Interestingly, another putative hybrid was collected by Bill Summers in 1986 in adjacent Howell County at a roadside station from which *L. spicata* has never been documented.

The Missouri specimens of true *L. spicata* are referable to var. *spicata*. A dwarf variant that grows along the southeastern Coastal Plain is known as var. *resinosa* (Nutt.) Gaiser.

8. Liatris squarrosa (L.) Michx.
Pl. 268 e, f; Map 1128

Rootstock a more or less globose corm. Stems 20–75 cm long, glabrous or moderately hairy. Basal and adjacent lower stem leaves sessile to short-petiolate, the blades 7–20 cm long, 2–12 mm wide, linear to narrowly lanceolate or narrowly oblanceolate, the margins usually light, hard, and thickened, sometimes also sparsely hairy, flat, the surfaces glabrous or sparsely to moderately pubescent with spreading, curly hairs, green, with 3–5 main veins, grading toward the stem tip to shorter leaves, these mostly sessile, 3–15 cm long, linear. Inflorescences relatively short to more elongate spicate racemes (rarely reduced to a single terminal head), the heads loosely spaced (the axis easily visible between heads), sessile or with stalks 1–10 mm long, these with usually 1 basal bract. Heads with 15–45(–60) disc florets, the terminal head usually slightly larger than the others. Involucre 11–25 mm long, narrowly cup-shaped to cup-shaped, with 5–7 weakly unequal, overlapping series of bracts (the outer series usually appearing noticeably longer than the other series, especially on the terminal head). Involucral bracts broadly ovate to oblong-lanceolate, long-tapered to a sharply pointed, abruptly spreading to recurved tip, usually with narrow, slightly thickened, pale margins, these sometimes slightly to strongly purplish-tinged, entire but with relatively dense, spreading hairs (the surface sometimes also sparsely to moderately hairy), the main body appearing flat below the tip. Corollas 12–14 mm long, the lobes with often dense, short, stiff hairs on the inner (upper) surface. Pappus bristles plumose. Fruits 4–6 mm long. $2n=20$. July–September.

Scattered nearly throughout the state, but absent from most of the western half of the Glaciated Plains Division (eastern U.S. west to South Dakota, Colorado, and Texas). Glades, ledges and tops of bluffs, openings of dry upland forests, savannas, and upland prairies; also pastures, fencerows, railroads, and roadsides.

For discussion of putative hybrids with *L. cylindracea,* see the treatment of that species. In Missouri, there is some morphological overlap between the three varieties, which are reasonably distinct elsewhere in their ranges.

1. Stems and leaves glabrous below the inflorescence . . . . 8A. VAR. GLABRATA
1. Stems and leaves hairy
    2. All but the outermost involucral bracts relatively short-tapered to an abruptly spreading or reflexed tip; stems and leaves pubescent with relatively straight, spreading hairs . . . . 8B. VAR. HIRSUTA
    2. Most or all of the involucral bracts relatively long-tapered to a gradually or less commonly abruptly spreading or reflexed tip; stems and leaves pubescent with bent or curled, appressed to loosely matted hairs . . . . 8C. VAR. SQUARROSA

8a. var. glabrata (Rydb.) Gaiser

Stems glabrous below the inflorescence. Leaves glabrous. Most or all of the involucral bracts relatively long-tapered to a gradually or less commonly abruptly spreading or reflexed tip, glabrous (except for the hairy margins), the outer series usually noticeably longer than the others. July–September.

Scattered mostly in the Unglaciated Plains Division but occasionally also in the Ozark, Ozark Border, and Glaciated Plains (South Dakota, Colorado, and Texas east to Iowa and Arkansas). Glades, ledges and tops of bluffs, openings of dry upland forests, savannas, and upland prairies; also roadsides.

8b. var. hirsuta (Rydb.) Gaiser

L. hirsuta Rydb.

Stems and leaves pubescent with relatively long, straight, spreading hairs. All but the outermost involucral bracts relatively short-tapered to an abruptly spreading or reflexed tip, mostly hairy on the surface and margins, the outer series usually only somewhat longer than the others. July–September.

Scattered nearly throughout the state but absent from most of the western half of the Glaciated Plains Division and the Mississippi Lowlands (Nebraska to Texas east to North Carolina and Georgia). Glades, ledges and tops of bluffs, openings of dry upland forests, savannas, and upland prairies; also pastures, fencerows, railroads, and roadsides.

This is by far the commonest variety of *L. squarrosa* in the state.

8c. var. squarrosa

Stems and leaves pubescent with short to long, bent or curled, appressed to loosely matted hairs. Most or all of the involucral bracts relatively long-tapered to a gradually or less commonly abruptly spreading or reflexed tip, glabrous (except for the hairy margins) or sparsely hairy, mostly along the midvein, the outer series usually noticeably longer than the others. $2n=20$. July–September.

Scattered mostly in the Ozark and Ozark Border Divisions, but uncommonly also in the Glaciated Plains and Mississippi Lowlands (Texas to Florida north to Missouri, Michigan, Maryland, and Delaware). Glades, ledges and tops of bluffs, openings of dry upland forests, and savannas; also roadsides.

The inflorescence of var. *squarrosa* tends to be denser and to consist of more heads than in the other two varieties in Missouri, but this feature is too variable to include in the key.

9. Liatris squarrulosa Michx.

L. scabra (Greene) K. Schum.

Pl. 269 c–e; Map 1129

Rootstock a more or less globose to ovoid corm, sometimes appearing somewhat erect and angular or irregular. Stems 25–150 cm long, moderately to densely pubescent with short, curled hairs, sometimes roughened to the touch. Basal and adjacent lower stem leaves short- to long-petiolate, the blades 8–25 cm long, 9–45 mm wide, oblanceolate to narrowly elliptic, the surfaces glabrous to densely short-hairy, green, with 1 main vein, grading abruptly to the shorter stem leaves, these mostly sessile, 2–15 cm long, narrowly oblanceolate to narrowly oblong-elliptic or sometimes linear. Inflorescences short to elongate spicate racemes, the heads mostly relatively loosely spaced (the axis visible between heads), sessile or with stalks to 15 mm long, these with only 1 or few bracts. Heads with 11–26(–28) disc florets, the terminal head not or sometimes slightly larger than the others. Involucre 7–14 mm long, broadly cup-shaped to broadly bell-shaped, with 4–6 unequal, overlapping series of bracts (the outer series appearing progressively shorter). Involucral bracts broadly obovate to oblong-spatulate, all but the innermost series spreading or recurved at the tip, mostly with relatively narrow, thin, pale to transparent margins, the margins or entire bract sometimes strongly purplish-tinged, nearly entire to shallowly scalloped (sometimes also minutely irregular), the main body appearing flat or very slightly swollen just below the tip. Corollas 9–12 mm long, the tube hairy on the inner surface. Pappus bristles barbed. Fruits 3.5–5.5 mm long. $2n=20$. August–November.

Scattered mostly in the Ozark and Ozark Border Divisions (Ohio to Florida west to Missouri and Texas). Glades, bases, ledges, and tops of bluffs, savannas, openings of mesic to dry upland forests, and rarely upland prairies; also ditch banks, fencerows, pastures, railroads, and roadsides.

Liatris squarrulosa is part of a taxonomically difficult species complex that also includes *L. aspera* and *L. scariosa*. For further discussion, see the treatment of the latter species. Most of the specimens of *L. squarrulosa* from Missouri are relatively short plants with relatively few heads. However, at scattered sites, especially in sandy soils of Crowley's Ridge, populations occur in which the plants are much taller, with elongate, many-headed inflorescences. The relationship between these different growth forms and their intermediates is not presently understood.

57. Mikania Willd.
(Holmes, 1981)

About 400 species, North America, Central America, South America, Caribbean Islands, Africa, Asia south to New Guinea, most diverse in the Neotropics.

1. Mikania scandens (L.) Willd. (climbing hempweed)

M. pubescens Nutt.

Pl. 270 h, i; Map 1130

Plants perennial twining herbs, with fleshy, clustered roots. Stems prostrate or more commonly climbing in other vegetation, 1–5 m long, densely and minutely hairy to nearly glabrous, often also sparsely glandular. Leaves opposite (the nodes well separated), long-petiolate. Leaf blades 2–12 cm long, triangular to triangular-ovate, truncate to cordate at the base, tapered to a sharply pointed tip, often with a pair of short, bluntly triangular, spreading basal lobes, the margins otherwise entire to somewhat wavy, scalloped or bluntly toothed, the surfaces sparsely to moderately pubescent with short, curved hairs, also sparsely to moderately glandular, the main veins palmate. Inflorescences axillary panicles or stalked clusters, dome-shaped. Heads with 4 disc florets. Involucre 4–6 mm long, more or less cylindrical, the bracts 4 (the head often also subtended by 1 or 2 other shorter bracts), more or less equal in size, narrowly oblong-elliptic to narrowly elliptic-lanceolate, tapered to a sharply pointed tip, glabrous or sparsely to moderately and finely short-hairy. Receptacle flat or nearly so. Corollas white to pale pink or rarely pale lavender, glandular. Pappus of numerous capillary bristles. Fruits 1.5–2.5 mm long, 5-angled, somewhat wedge-shaped in profile (usually slightly and unevenly tapered at the base) to nearly linear, densely glandular, dark brown to black. $2n=38$. July–October.

Scattered in the Mississippi Lowlands Division (eastern [mostly southeastern] U.S. west to Michigan, Missouri, and Texas; Canada, Mexico, Caribbean Islands). Swamps, bottomland forests, banks of oxbows and sloughs, and margins of lakes and ponds; also ditches, levees, fencerows, and roadsides.

6. Tribe Gnaphalieae Benth.

Plants annual or perennial herbs (woody elsewhere), sometimes entirely or incompletely dioecious, fibrous-rooted or taprooted, sometimes with rhizomes or stolons, usually densely pubescent with white to grayish white woolly hairs, the sap not milky. Stems unbranched or branched, not spiny or prickly. Leaves basal and alternate, not spiny or prickly, sessile or with short, poorly differentiated petioles. Leaf blades simple, the margins entire or with a few small teeth (more divided elsewhere). Inflorescences terminal and sometimes also axillary spikes, racemes, or flat-topped to headlike clusters, these sometimes grouped into small panicles. Heads appearing discoid, with usually numerous florets. Involucre cylindrical to ovoid or hemispherical, with few to several unequal series of overlapping bracts, these appressed or spreading at maturity, at least partially white to straw-colored or purplish-tinged, scalelike or papery, not spiny or tuberculate, often all or partially obscured by dense pubescence. Receptacle usually flat, less commonly convex or concave, naked or with chaffy scales in *Diaperia*. Central florets (except in dioecious taxa) perfect or staminate (the style branches not or only slightly spreading at maturity). Marginal florets (except in staminate plants of dioecious taxa) pistillate (lacking stamens). Corollas very slender, tubular, with very short lobes. Pappus absent (in *Diaperia*) or more commonly of capillary bristles (awns or scales elsewhere), these finely toothed or barbed (plumose elsewhere), sometimes somewhat expanded and narrowly club-shaped at the tip, free or sometimes fused at the base, usually shed before fruiting. Stamens with the filaments not fused together, the anthers fused into a tube, each tip with a short, sometimes indistinct appendage, each base prolonged into a pair of short lobes. Style branches usually somewhat flattened, each with a stigmatic line along each inner margin, the sterile tip rounded or more commonly truncate, hairy. Fruits relatively small, variously shaped, not winged, not beaked, glabrous, hairy, or with minute papillae. About 180 genera, about 2,000 species, worldwide.

Traditionally, the genera treated here as tribe Gnaphalieae were included in an expanded concept of Inuleae (Steyermark, 1963; Cronquist, 1980; Gleason and Cronquist, 1963, 1991; Barkley, 1986). Anderberg (1991, 1994), using morphological and anatomical data, and Panero and Funk (2002), using molecular data, have shown that this group has closer affinities to the Astereae and Anthemideae than to the Inuleae and Plucheeae. In spite of this, the limits of the tribe Gnaphalieae remain less than perfectly understood, and some generic limits within the group also require further study. In the absence of data to the contrary, the present treatment of generic limits follows that of the forthcoming Flora of North America treatment more for expediency than out of full confidence that the relatively finely split genera in that volume will withstand the test of time.

1. Receptacle with chaffy bracts; pappus absent; individual heads grouped into a dense, more or less spherical, headlike cluster at each stem or branch tip, the individual heads difficult to differentiate . . . . 60. DIAPERIA
1. Receptacle naked; pappus of numerous capillary bristles; inflorescences various, but the heads not grouped into dense, headlike clusters or, if so, then not restricted to the stem tip
    2. Heads short-stalked; plants dioecious or nearly so, perennial, with rhizomes or stolons
        3. Pistillate heads usually with a few central staminate flowers (plants incompletely dioecious); plants with rhizomes, lacking a basal rosette of large leaves at flowering time . . . . 58. ANAPHALIS
        3. Pistillate heads lacking central staminate florets (plants completely dioecious); plants with stolons, forming colonies of rosettes of basal leaves, the basal leaves much larger than those of the flowering stems
         . . . . 59. ANTENNARIA

2. Heads sessile or short-stalked; plants not dioecious, the heads with the marginal florets pistillate, the central florets perfect, annual or biennial, with taproots
    4. Inflorescences narrow, spikelike; heads sessile or minutely stalked; involucre 3–5 mm long, usually brownish- and/or purplish-tinged, the bracts in 3–5 overlapping series; pappus bristles fused at the base, shed intact as a ring; plants not or only slightly aromatic . . . . 61. GAMOCHAETA
    4. Inflorescences relatively broad panicles; heads sessile or more commonly short-stalked; involucre 5–8 mm long, usually white to straw-colored, rarely faintly pinkish- or purplish-tinged, the bracts in 5–7 overlapping series; pappus bristles mostly free at the base, shed individually or in small groups; plants moderately to strongly aromatic when crushed or bruised
. . . . 62. PSEUDOGNAPHALIUM

58. Anaphalis DC. (everlasting)

About 110 species, North America, South America, Asia south to New Guinea.

1. Anaphalis margaritacea (L.) Benth. ex C.B. Clarke. (pearly everlasting)

Pl. 293 a–d; Map 1131

Plants perennial, incompletely dioecious, often branched, with rhizomes. Stems 25–80 cm long, erect or ascending, densely woolly, the hairs sometimes becoming thinner or reddish tan (rusty) with age. Basal leaves usually absent at flowering, not noticeably larger than the lower stem leaves. Stem leaves numerous, sessile, 2–12 cm long, linear to elliptic-lanceolate or oblanceolate, bluntly to sharply pointed at the tip, sometimes slightly expanded and clasping at the base, the margins entire and sometimes curled under, both surfaces densely white-woolly, the upper surface sometimes becoming nearly glabrous with age. Inflorescences rounded to more or less flat-topped, often relatively dense panicles, the individual heads mostly short-stalked. Heads with all staminate or mostly pistillate florets, the pistillate heads usually with 2–4 staminate central florets. Involucre 5–8 mm long, broadly ovoid to cup-shaped, the bracts in 7–12 overlapping series, mostly loosely appressed when young, spreading with age or upon drying, lanceolate to ovate, mostly bluntly pointed at the tip, woolly at the base, bright white (sometimes darker at the base), usually slightly shiny. Receptacle flat or somewhat convex, naked. Corollas 3.5–4.5 mm long, yellow to greenish yellow. Pappus of numerous capillary bristles, these free and shed individually, minutely toothed. Fruits 0.7–1.0 mm long, narrowly ellipsoid-obovoid, strongly flattened, the surface appearing pebbled or roughened with minute papillae, brown to olive brown. $2n=28$. July–September.

Introduced, uncommon, known thus far only from Boone and St. Louis Counties (native of the western and northern U.S. south to Virginia, Nebraska, Arizona, and California; Canada, Asia; introduced sporadically farther southeast in the U.S.). Bases of bluffs and disturbed openings of mesic upland forests.

This species is cultivated as an ornamental in gardens and is popular both as a fresh-cut flower and in dried flower arrangements. Native Americans used the plants medicinally for a variety of treatments, including headaches, tuberculosis, colds, coughs, infections, and so on, and also in various religious ceremonies (Moerman, 1998).

59. Antennaria Gaertn. (pussytoes)
(Bayer, 1993)

Plants perennial, dioecious, fibrous-rooted, with more or less leafy stolons (with rhizomes elsewhere), often forming colonies. Flowering stems erect or ascending, moderately to densely woolly, the hairs sometimes becoming thinner with age. Basal leaves persistent at flowering, much larger than the lower stem leaves. Stem leaves several, sessile, the margins entire, both

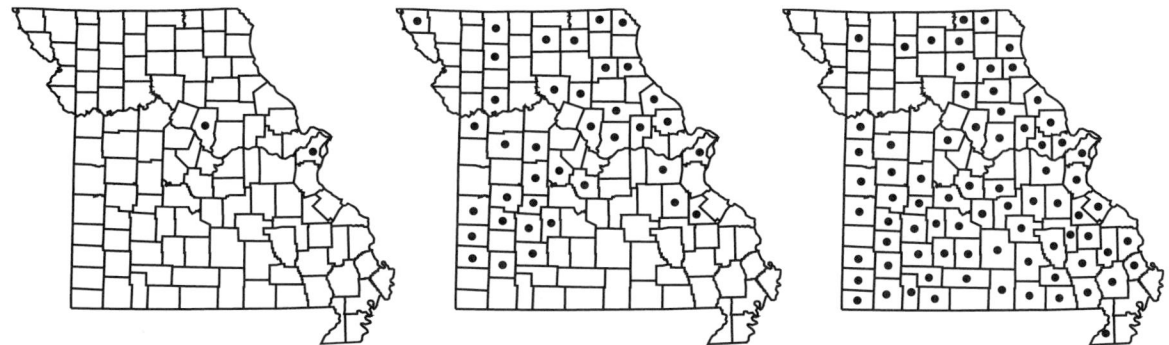

1131. Anaphalis margaritacea   1132. Antennaria neglecta   1133. Antennaria parlinii

surfaces densely white-woolly or the upper surface sparsely hairy to glabrous. Inflorescences terminal, relatively dense clusters, occasionally appearing as small panicles, flat-topped to more commonly rounded, the individual heads mostly short-stalked. Heads with all staminate or all pistillate florets, the pistillate heads very rarely with a few central staminate florets, the staminate heads with the involucre usually slightly shorter than those of the pistillate ones. Involucre narrowly ovoid to narrowly bell-shaped at flowering, becoming broadly bell-shaped or funnelform at fruiting or when pressed, the bracts in 5–8 overlapping series, appressed when young, often somewhat spreading with age or upon drying, those of the outer few series shorter and ovate, densely woolly, with short, broadly pointed tips; those of the inner few series noticeably longer and narrowly lanceolate, sparsely woolly, with elongate, tapered, thin tips, green (turning straw-colored or brown with age) toward the base, white to translucent toward the tip, usually at least somewhat purplish-tinged. Receptacle flat or convex, naked. Corollas slender, those of the staminate florets shorter than those of the pistillate florets, tubular or those of the staminate florets narrowly trumpet-shaped, white or yellow, sometimes reddish- or purplish-tinged. Pappus of the pistillate florets of numerous slightly longer capillary bristles, that of the staminate florets of fewer slightly shorter bristles, the bristles all or mostly free and shed individually or in small groups, minutely toothed and sometimes slightly expanded and narrowly club-shaped at the tip. Fruits 1.0–1.5 mm long, narrowly elliptic-obovoid, not or only slightly flattened, the surface appearing pebbled or roughened with minute papillae, brown to olive brown. Seventy to 100 species, North America, South America, Europe, Asia.

Species recognition in *Antennaria* is difficult because hybridization, polyploidy, and apomixis are widespread among the North American taxa. Fortunately, relatively few taxa have been documented thus far from Missouri. The classification of Bayer and Stebbins (1982) represents a practical compromise between the need to provide taxonomic recognition of the polyploid hybrid derivatives as separate species from their diploid progenitors and the problems for identification that would be created if each of the races resulting from similar but independent hybridization events were attempted to be recognized. See Cronquist (1946) for a more taxonomically conservative interpretation.

Bayer and Stebbins (1982) mapped the occurrence of *A. solitaria* Rydb. from southeasternmost Missouri without citation of locality or specimens. Randall Bayer (then of the University of Alberta) kindly checked his notes and was unable to substantiate the report, thus it is excluded from the Missouri flora for now. It does occur in adjacent portions of Arkansas, Kentucky, and Tennessee, so it should be searched for in Missouri. *Antennaria solitaria* differs from other members of the genus in consistently having a single relatively large head (involucre 8–14 mm long) at the tip of each flowering stem. Vegetatively, it resembles *A. parlinii* in having broad basal leaves with multiple main veins, and, in fact, *A. solitaria* is one of the diploid sexual progenitors in the *A. parlinii* complex.

1. Basal leaves with the blades less than 2 cm wide, with 1 main vein, occasionally with an additional faint pair of main veins. . . . . 1. A. NEGLECTA
1. Basal leaves with the blades mostly 2.0–4.5 cm wide, with 3 or 5 main veins
. . . . 2. A. PARLINII

1. Antennaria neglecta Greene (field pussytoes)

A. *neglecta* var. *campestris* (Greene) Steyerm.
A. *longifolia* Greene

Pl. 293 e–h; Map 1132

Stolons frequently relatively long, slender, densely woolly, leafy. Flowering stems (1–)3–20(–34) cm long, densely woolly, sometimes becoming glabrous in patches with age. Basal leaves 1.5–5.0 cm long, 4–15(–18) mm wide, narrowly oblanceolate to narrowly obovate or narrowly spatulate, rounded to broadly and abruptly pointed at the tip, tapered at the base, the upper surface moderately to densely woolly, often becoming glabrous or nearly so with age, the undersurface densely woolly, with 1 main vein, occasionally with an additional faint pair of main veins. Stem leaves 0.8–2.5 cm long, linear to narrowly oblong-lanceolate, the lowermost often narrowly oblanceolate, mostly sharply pointed at the tip, the median and upper leaves with a short, hairlike extension of the midvein, truncate or slightly tapered at the base, the blade tissue sometimes extending along the stem as 2 narrow wings below the main attachment point, densely woolly on both surfaces. Involucre 5–10 mm long. Corollas 2.5–6.5 mm long. 2n=28. April–June.

Scattered in the Glaciated Plains and Unglaciated Plains Divisions, but mostly absent from the Ozarks and Mississippi Lowlands (northeastern U.S. west to Wyoming and Colorado; Canada). Bottomland and upland prairies, openings of mesic to dry upland forests, and rarely banks of streams; also pastures, lawns, cemeteries, railroads, and roadsides.

Antennaria neglecta is a diploid sexual species that is one of several parental taxa in the *A. howellii* Greene (*A. neodioica* Greene) polyploid complex in the northern United States (Bayer, 1985a). Plants often form relatively extensive, dense mats of basal rosettes, which are most easily observed in the spring before grasses and other dense growth obscure the ground from view.

2. Antennaria parlinii Fernald (plainleaf pussytoes, ladies' tobacco, Indian tobacco)

Pl. 293 i–m; Map 1133

Stolons short to relatively long, slender, densely woolly, leafy. Flowering stems (4–)9–28(–40) cm long, densely woolly, sometimes becoming glabrous in patches with age. Basal leaves 2.5–9.5 cm long, (8–)18–45 mm wide, oblanceolate to more commonly broadly obovate or circular-obovate, rounded to broadly pointed at the tip, often with a minute, abrupt, sharp point, tapered at the base, the upper surface glabrous to densely woolly, the undersurface densely woolly, with 3 or 5 main veins. Stem leaves 0.5–4.5 cm long, linear to narrowly oblong-lanceolate, the lowermost often narrowly oblanceolate, mostly sharply pointed at the tip, the upper leaves often with a short, hairlike extension of the midvein, truncate or somewhat rounded at the base, the blade tissue not extending along the stem below the attachment point, densely woolly on both surfaces or the upper surface sometimes only moderately hairy. Involucre 7–13 mm long. Corollas 3.5–7.0 mm long. 2n=56, 84, 70, 112. April–June.

Common nearly throughout the state (eastern U.S. west to North Dakota and Texas; Canada). Mesic to dry upland forests, upland prairies, savannas, and ledges and tops of bluffs, less commonly glades, ditches, banks of streams and rivers, and margins of ponds and lakes; also pastures, roadsides, and open, disturbed areas.

Antennaria parlinii is one of the few flowering plants in Missouri that grows well in dry, shaded habitats (although it can tolerate sun), and it can be cultivated as a groundcover or rock garden plant in areas where gardeners may find it otherwise difficult to grow plants without supplemental irrigation. Native Americans used the species medicinally to treat gastrointestinal and gynecological pain (Moerman, 1998). It has been shown to represent a polyploid complex formed from past hybridization of diploid progenitor species followed by polyploidy and apomixis (Bayer, 1985b). The main diploids that have hybridized repeatedly in the creation of various subspecies and races of *A. parlinii* include *A. plantaginifolia* (L.) Richardson, *A. solitaria* (see above), and *A. racemosa* Hook. (presently restricted to the western United States and adjacent Canada). None of these has been documented from Missouri. However, Bayer and Stebbins (1982) mapped the occurrence of *A. plantaginifolia* (Pl. 293 n, o) from the far southeastern portion of the state. Randall Bayer (then of the University of Alberta) kindly checked his notes and was unable to substantiate the report, thus it is excluded from the Missouri flora for now.

It does occur in adjacent portions of Arkansas, Kentucky, and Tennessee, so it should be searched for in Missouri in wooded uplands in the southeastern portion of the state. *Antennaria plantaginifolia* will key out to *A. parlinii* ssp. *fallax* in the key to subspecies below. Its basal leaves tend to be persistently hairy on the upper surface. Its stolons also tend to be somewhat shorter. Additionally, its heads (particularly the pistillate heads) are somewhat smaller than those of *A. parlinii* (pistillate involucre 5–7 mm vs. 7–10 mm long). In staminate plants, the corollas tend to be slightly shorter (2–4 mm vs. 3–5 mm long).

Bayer and Stebbins (1982) and Bayer (1985b) recognized two subspecies within *A. parlinii,* both of which occur in Missouri. Unfortunately the distinctions between the two are not always obvious. Assignment of many specimens to one subspecies or the other seems relatively arbitrary and various authors of recent floristic manuals have applied the names somewhat differently. The leaves of plants that start out hairy tend to become more glabrous as the season progresses. Also, the broader basal leaves tend to become more glabrous with age than the narrower ones. In Missouri, there are very few specimens in which the young basal leaves are totally glabrous.

1. Basal leaves moderately to densely woolly or felty on the upper surface, usually becoming less hairy at maturity .... 2A. SSP. FALLAX
1. Basal leaves sparsely hairy or glabrous on the upper surface, often appearing cobwebby rather than woolly, even when young, becoming glabrous or nearly so at maturity .... 2B. SSP. PARLINII

2a. ssp. fallax (Greene) R.J. Bayer & Stebbins
   A. plantaginifolia (L.) Hook. var. *ambigens* (Greene) Cronquist

Stems usually lacking purplish hairs. Basal leaves moderately to densely woolly or felty on the upper surface, usually becoming less hairy at maturity. April–June.

Common nearly throughout the state (eastern U.S. west to North Dakota and Texas; Canada). Mesic to dry upland forests, upland prairies, savannas, and ledges and tops of bluffs, less commonly glades, ditches, banks of streams and rivers, and margins of ponds and lakes; also pastures, roadsides, and open, disturbed areas.

This appears to be the somewhat less common phase of the species in Missouri, except in portions of western Missouri.

2b. ssp. parlinii
   A. plantaginifolia var. *arnoglossa* (Greene) Cronquist

Stems usually sparsely to moderately pubescent with purplish glandular hairs toward the tip. Basal leaves sparsely hairy or glabrous on the upper surface, often appearing cobwebby rather than woolly, even when young, becoming glabrous or nearly so at maturity. April–June.

Scattered nearly throughout the state (eastern U.S. west to Minnesota and Texas; Canada). Mesic to dry upland forests, upland prairies, savannas, and ledges and tops of bluffs, less commonly glades, ditches, banks of streams and rivers, and margins of ponds and lakes; also pastures, roadsides, and open, disturbed areas.

As interpreted here, this appears to be the more common phase of the species in Missouri. However, the two types do not seem to segregate by region or habitat in Missouri and do sometimes grow in proximity to one another.

60. Diaperia Nutt.

Three species, southwestern U.S. and adjacent Mexico.

The generic classification of the so-called *Filago* group is still unsettled, and the small segregate *Diaperia* has been resurrected only recently. *Diaperia* often has been included in *Evax* Gaertn. (Cronquist, 1980), which often has been included in *Filago* L. (Anderberg, 1991, 1994; Arriagada, 1998). Morefield (1992, 2004) took a cladistic approach and attempted to recognize a number of smaller, monophyletic groups as genera. According to his studies, the alternative to splitting the subtribe Filagininae Bentham & Hook. f. into numerous small genera that are very difficult to circumscribe morphologically is to combine virtually the entire group into a single, morphologically heterogeneous *Filago,* an approach that Morefield (1992) considered relatively uninformative.

1134. Diaperia prolifera   1135. Gamochaeta argyrinea   1136. Gamochaeta pensylvanica

1. Diaperia prolifera (Nutt. ex DC.) Nutt. **var. prolifera** (rabbit tobacco)

Evax prolifera Nutt. ex DC.

Pl. 294 i, j; Map 1134

Plants annual, often branched at the base and sometimes also the stem tip. Stems 2–10(–15) cm long, erect or ascending, densely woolly. Basal leaves absent at flowering. Stem leaves numerous, sessile, 0.3–1.5 cm long, linear to oblanceolate or narrowly spatulate, rounded at the tip, tapered at the base, the margins entire, both surfaces densely white-woolly. Inflorescences consisting of more or less globose, headlike, dense clusters of 4 to numerous heads, these 5–20 mm wide, the individual heads difficult to differentiate, each head closely subtended by a leaf that is somewhat exserted from the cluster and the whole cluster associated with an apparent involucre of leaves; the cluster terminal on the stem and in larger plants closely subtended by 1–5 spreading to loosely ascending, naked branches that have similar terminal clusters of heads. Heads narrowly ellipsoid to nearly conical, with mostly pistillate florets, the 2–5 central florets staminate. Involucre technically absent but an apparent involucre formed by chaffy bracts subtending the outermost florets, these 2–4 mm long, in a single series, oblong-ovate, the outer surface densely woolly, straw-colored. Receptacle somewhat convex to short-conical, chaffy, the bracts oblong-oblanceolate, with a woolly patch near the tip, straw-colored. Corollas 1.2–2.0 mm long, pale yellow to greenish yellow. Pappus absent. Fruits 0.7–1.0 mm long, narrowly ellipsoid to oblong-ovoid, somewhat flattened, the surface appearing smooth or pebbled with minute papillae, light brown. May–June.

Introduced, uncommon, known thus far from a single site in Ozark County (Alabama to New Mexico north to Arkansas, North Dakota, and Montana, introduced in California [historically], Missouri, and South Carolina). Roadsides.

This species has an unusual distribution, being found in the southeastern United States, mostly in chalk prairies along the Gulf Coastal Plain (Cronquist, 1980), but also ranging widely through the Great Plains (McGregor, 1986f). It was first reported for Missouri by Sullivan (1991). Arriagada (1998) incorrectly considered the species to be introduced from Europe. The recently named var. *barnebyi* Morefield occurs in New Mexico, western Texas, and western Oklahoma and differs (among other characters) in its silvery (vs. grayish white) pubescence and inflorescence clusters with only 1–3 heads.

61. Gamochaeta Wedd. (cudweed, everlasting)

Plants annual or biennial (perennial elsewhere), unbranched or few-branched from the base, fibrous-rooted or with slender taproots, not or only slightly aromatic when crushed or bruised. Stems erect or loosely to strongly ascending, sometimes ascending from a spreading base, moderately to densely pubescent with woolly hairs. Basal leaves sometimes present at flowering, not noticeably larger than the lower stem leaves. Stem leaves several to numerous, sessile, linear to oblanceolate or narrowly spatulate, rounded to bluntly or less commonly sharply pointed at the tip, mostly tapered at the base, the margins entire and sometimes finely wavy, the upper surface sparsely to densely woolly, sometimes appearing nearly glabrous, the undersurface densely white-woolly. Inflorescences narrow, often appearing as interrupted leafy spikes, with small clusters of heads axillary in the uppermost leaves and a short to somewhat elongate terminal spike, this sometimes reduced to a dense, conical or headlike mass in poorly developed plants, the individual heads sessile or minutely and inconspicuously stalked. Heads with the marginal florets pistillate, the central florets perfect. Involucre 3–5 mm long, narrowly ovoid to cup-shaped, the bracts in 3–5 overlapping series, mostly appressed, lanceolate to ovate or triangular, mostly sharply pointed at the tip, with dense, woolly hairs that obscure all or most of the involucre, ranging from white to more commonly straw-colored, usually brownish- and/or purplish- to pinkish-tinged, often slightly shiny. Receptacle flat, slightly convex, or concave, naked. Corollas 2.5–3.5 mm long, usually white, often purple at the tip. Pappus of numerous capillary bristles, these fused at the base and shed intact as a ring, minutely toothed. Fruits 0.4–0.7 mm long, narrowly oblong-obovoid, slightly flattened, the surface appearing pebbled or roughened with minute papillae, tan to yellowish brown, sometimes somewhat shiny. About 50 species, North America to South America, Caribbean Islands, introduced in the Old World.

Gamochaeta is one of the more easily recognizable segregates of *Gnaphalium* L. in temperate North America (although perhaps not elsewhere), even though no shared derived morphological character unique to the group has been identified to date. The name is accepted by many recent authors (Cabrera, 1961; Nesom, 1990a; Anderberg, 1991, 1994; Arriagada, 1998). The North American species of *Gamochaeta* have been studied intensively by Guy L. Nesom of the Botanical Research Institute of Texas in conjunction with his forthcoming treatment of the group for the Flora of North America Project. He has uncovered several morphologically cryptic taxa that formerly were included in *G. purpurea* and has done an excellent job of diagnosing these. The present treatment was improved following discussions with Dr. Nesom and his willingness to share unpublished data on the group. However, the three Missouri species remain very difficult to distinguish.

1. Receptacle becoming deeply concave (cuplike) at fruiting; leaves (except sometimes the upper ones) appearing only slightly to moderately bicolorous, the upper surface moderately to densely woolly but still appearing somewhat darker or greener than the undersurface. . . . . 2. G. PENSYLVANICA
1. Receptacle becoming flat to slightly concave at fruiting; leaves appearing strongly bicolorous, the upper surface sparsely woolly or with patches of cobwebby hairs, often appearing nearly glabrous
   2. Hairs not swollen at the base; involucre 3.0–3.5 mm long . . . . . . 1. G. ARGYRINEA
   2. Most or all of the foliar hairs (most easily observed on the more sparsely pubescent upper surface) with a minute, swollen or expanded basal cell (requires magnification to observe); involucre 4.0–4.5 mm long . . . . 3. G. PURPUREA

1. Gamochaeta argyrinea G.L. Nesom

Pl. 295 d; Map 1135

Plants usually fibrous-rooted, rarely with slender taproots. Stems 10–35(–45) cm long. Basal leaves usually present at flowering. Leaves 1–6 cm long, oblanceolate to oblong-oblanceolate, the upper ones sometimes linear, strongly bicolorous, the upper surface sparsely woolly or with patches of cobwebby hairs, sometimes appearing nearly glabrous, the undersurface densely woolly, the undersurface hairs with a slender, unexpanded basal cell (even with magnification). Involucre 3.0–3.5 mm long, the outermost bracts ovate to narrowly ovate with sharply pointed tips, the innermost oblong to oblong-elliptic, rounded to truncate at the tip, sometimes with a minute, abrupt, sharp point. Receptacle flat or slightly convex at flowering, usually becoming shallowly concave at fruiting. April–June.

Uncommon, known thus far only from Howell County (southeastern U.S. west to Kansas and Texas; Puerto Rico). Pastures.

This species was described only recently (Nesom, 2004a) and is known thus far from one site in Missouri. Because it has not been searched for elsewhere in southern Missouri, its range and native habitat are not yet known. The stem leaves tend to be narrower and less spatulate than those of the morphologically similar *G. purpurea*.

1137. Gamochaeta purpurea
1138. Pseudognaphalium helleri
1139. Pseudognaphalium micradenium

2. **Gamochaeta pensylvanica** (Willd.) Cabrera
   Gnaphalium pensylvanicum Willd.
   Gnaphalium purpureum L. var. spathulatum (Lam.) Ahles

   Pl. 295 e; Map 1136

   Plants usually with slender taproots, less commonly fibrous-rooted. Stems 10–45 cm long. Basal leaves usually present at flowering. Leaves 1–7 cm long, narrowly obovate to oblanceolate or spatulate, the upper ones sometimes linear, slightly to moderately bicolorous, the upper surface moderately to densely woolly but still appearing somewhat darker or greener than the undersurface, the undersurface densely woolly, the hairs with a slender, unexpanded basal cell (even with magnification). Involucre 3.0–3.5 mm long, the outermost bracts ovate-triangular with sharply pointed tips, the innermost lanceolate-triangular, tapered to a sharply pointed or less commonly bluntly pointed tip. Receptacle flat or slightly concave at flowering, becoming deeply concave (cuplike) at fruiting. $2n=28$. April–June.

   Possibly introduced, uncommon, widely scattered, mostly south of the Missouri River (southeastern U.S. north to Pennsylvania and west to Oklahoma and Texas, disjunct in California; Mexico, Central America, South America; introduced widely in the Old World). Upland prairies and glades; also pastures, old fields, roadsides, and open, disturbed areas.

   This species can be difficult to distinguish from the more widely distributed G. purpurea based on foliage characters. The strongly concave (craterlike), mature receptacles are most easily observed after the fruits have dispersed. Nesom (2004b) has discussed the possibility that this taxon represents an early introduction into North America, possibly from tropical America, but the evidence is far from conclusive, and the native range of this widespread, weedy species may never be known with certainty.

3. **Gamochaeta purpurea** (L.) Cabrera (purple cudweed, early cudweed)
   Gnaphalium purpureum L.

   Pl. 295 a–c; Map 1137

   Plants usually with slender taproots, less commonly fibrous-rooted. Stems 7–45 cm long. Basal leaves present or sometimes withered by flowering time. Leaves 1–6(–8) cm long, oblanceolate to spatulate, the upper ones sometimes linear, strongly bicolorous, the upper surface sparsely woolly or with patches of cobwebby hairs, sometimes appearing nearly glabrous, the undersurface densely woolly, most or all of the hairs with a minute, swollen or expanded, transparent basal cell (requires 10× magnification to observe). Involucre 4.0–4.5 mm long, the outermost bracts ovate-triangular with sharply pointed tips, the innermost lanceolate-triangular, tapered to a sharply pointed tip. Receptacle flat or slightly convex at flowering, usually becoming shallowly concave at fruiting. $2n=14, 28$. April–June.

   Scattered, mostly south of the Missouri River (eastern U.S. west to Wisconsin, Kansas, and Texas; disjunct in Arizona; Mexico, Central America, South America, Caribbean Islands; introduced widely in the Old World). Upland prairies, sand prairies, glades, savannas, openings of mesic to dry upland forests, tops of bluffs, and banks of streams and rivers; also pastures, old fields, fallow fields, roadsides, open, disturbed areas, and rarely lawns.

   The small, bulbous, basal cells of the hairs on the leaves can only be seen under a hand lens or dissecting microscope and are best observed on the upper surface. They appear similar to minute grains of sand or small glands where they persist even after the elongate portions of the hairs have been shed or abraded away. The stem leaves of G. purpurea tend to be somewhat broader and less oblong (more strongly oblanceolate to spatulate) than those of G. argyrinea.

62. Pseudognaphalium Kirp. (cudweed, everlasting)

Plants annual or biennial, unbranched or more commonly few-branched from the base and few-branched to moderately branched toward the tip, with taproots, usually slightly to strongly aromatic with a resinous odor when crushed or bruised. Stems erect or ascending, densely pubescent with woolly hairs, at least toward the tip, sometimes also glandular. Basal leaves occasionally present at flowering, broader but not noticeably longer than the lower stem leaves, oblanceolate to spatulate, rounded to broadly and bluntly pointed at the tip, often with 3 main veins. Stem leaves numerous, sessile, linear to narrowly oblanceolate or narrowly lanceolate, mostly sharply pointed at the tip, truncate or slightly tapered at the base, the margins entire and sometimes finely wavy, the upper surface appearing green, sparsely pubescent with cobwebby to woolly hairs, also with scattered, minute, stalked glands, the undersurface densely white-woolly. Inflorescences relatively broad panicles, the individual heads sessile or more commonly and noticeably short-stalked. Heads with the marginal florets pistillate, the central florets perfect. Involucre 5–8 mm long, narrowly ovoid to cup-shaped, often appearing bell-shaped when pressed, the bracts in 5–7 overlapping series, appressed to loosely appressed, oblong-lanceolate to lanceolate or ovate, rounded to bluntly or sharply pointed at the tip, the inner few series sometimes irregularly truncate, with dense, woolly hairs toward the base, usually white to straw-colored, rarely faintly pinkish- or purplish-tinged, shiny. Receptacle flat or somewhat convex, naked. Corollas white to more commonly yellow, sometimes purplish-tinged at the tip. Pappus of numerous capillary bristles, these mostly free at the base and shed individually or in small groups, minutely toothed. Fruits 0.6–0.9 mm long, narrowly oblong-obovoid, slightly flattened, the surface appearing smooth, yellowish brown to greenish brown, sometimes somewhat shiny. About 80 species, nearly worldwide, but most diverse in Central America and South America.

The segregation of *Pseudognaphalium* from *Gnaphalium* has been somewhat controversial (Arriagada, 1998). The modern concept of these genera initially was suggested by Hilliard and Burtt (1981) and subsequently was refined by Anderberg (1991, 1994). However, there is no single derived character that separates the two, and it has been argued that the details of involucral bract morphology and other minor features that have been used in combination to separate the genera are insufficient for the purpose. In the belief that further, more detailed study probably will show increased support for *Pseudognaphalium* and because the genus will be accepted in the forthcoming treatment for the Flora of North America Project, it is accepted here as well.

The basal rosettes of particularly *P. obtusifolium* may be confused with those of *Antennaria parlinii*. Although in both species the upper surface may appear glabrous or nearly so at maturity, the leaves of *Pseudognaphalium* differ in their scattered, minute, stalked glands. Also, a fourth species of *Pseudognaphalium* eventually may be collected in Missouri, as it has been documented in Illinois (historically) and Tennessee. *Pseudognaphalium macounii* (Greene) Kartesz differs from the species known thus far from Missouri in having its leaves extending down the stem from the main attachment point as two short, narrow wings of tissue.

Steyermark (1963) noted that plants of this genus provide food for Missouri's turkeys and deer, and that in folk medicine the leaves were boiled in milk as a remedy for flux (the abnormal bodily discharge of fluid; diarrhea).

1. Stems moderately to densely woolly, the pubescence sometimes becoming abraded in small patches with age, not appearing glandular (but sparse glands sometimes present toward the stem base and hidden under the woolly hairs) . . . . 3. P. OBTUSIFOLIUM

1. Stems sometimes somewhat woolly when young, but not appearing woolly below the inflorescence at maturity, instead moderately to densely glandular or glandular-hairy
    2. Stems with minute to short, stalked glands (appearing glandular-hairy), these variously 0.2–1.0 mm long on the same stem; largest stem leaves linear to narrowly lanceolate or narrowly oblong-lanceolate . . . . 1. P. HELLERI
    2. Stems with minute, stalked glands, these 0.1–0.2 mm long; largest stem leaves lanceolate or oblong-oblanceolate . . . . 2. P. MICRADENIUM

1. Pseudognaphalium helleri (Britton)
 Anderb. (rabbit tobacco)
Gnaphalium helleri Britton
Pl. 294 g, h; Map 1138

Plants usually strongly aromatic when bruised or crushed. Stems 25–75 cm long, sometimes somewhat woolly when young, but not appearing woolly below the inflorescence at maturity, instead moderately to densely glandular-hairy, the stalked glands variously 0.2–1.0 mm long on the same stem. Leaves 1–9 cm long, linear to narrowly lanceolate or narrowly oblong-lanceolate, the upper surface with moderate to dense, minute, stalked glands. August (July–November elsewhere).

Possibly introduced, known from a single historical collection from St. Louis County (southeastern U.S. west to Oklahoma and Texas). Habitat unknown.

Steyermark (1963) overlooked the single specimen from Missouri, and it remained hidden in the Missouri Botanical Garden Herbarium under the name *Gnaphalium polycephalum* Michx. until its rediscovery in 2000 by Guy Nesom of the Botanical Research Institute of Texas. The closest known native populations in central Arkansas, northern Mississippi, and eastern Kentucky and Tennessee are not particularly close to the Missouri record from Allenton. Although it is plausible that the historical St. Louis County collection represents a native occurrence, it seems more likely that the plant originated from a disturbed habitat such as a railroad embankment.

2. Pseudognaphalium micradenium (Weath.)
 G.L. Nesom (rabbit tobacco)
Gnaphalium obtusifolium var. *micradenium* Weath.
G. helleri var. *micradenium* (Weath.) Mahler
P. helleri var. *micradenium* (Weath.) Kartesz
Map 1139

Plants usually strongly aromatic when bruised or crushed. Stems 25–75 cm long, sometimes somewhat woolly when young, but not appearing woolly below the inflorescence at maturity, instead moderately to densely glandular, the minute, stalked glands 0.1–0.2 mm long. Leaves 1–9 cm long, linear to narrowly lanceolate or narrowly oblong-lanceolate, the upper surface with moderate to dense, minute, stalked glands. August–October.

Uncommon, known thus far only from Shannon and Ste. Genevieve Counties (eastern U.S. west to Minnesota and Louisiana). Savannas and openings of dry upland forests.

Steyermark (1963) noted that this taxon occurs in acidic soils on rocky slopes and ridgetops. Some authors have treated it as a variety of *Gnaphalium obtusifolium* (Steyermark, 1963) or the closely related *G. helleri* (Mahler, 1975). The present treatment follows that of Nesom (2001), who noted that the ranges of *P. helleri* and *P. micradenium* are largely distinct and that hybrids appear to be absent from the zone of overlap between the two.

3. Pseudognaphalium obtusifolium (L.)
 Hilliard & B. L. Burtt (sweet everlasting, fragrant cudweed, old-field balsam, catfoot)
Gnaphalium obtusifolium L.
Pl. 294 d–f; Map 1140

Plants usually only slightly aromatic when bruised or crushed. Stems 15–100 cm long, moderately to densely woolly, the pubescence sometimes becoming abraded in small patches with age, not appearing glandular, but with sparse glands 0.2–1.0 mm long, sometimes present toward the base and hidden under the woolly hairs. Leaves 1–10 cm long, linear to narrowly oblanceolate or narrowly lanceolate, the upper surface usually with sparse, stalked glands and sometimes also with sparse, woolly to cobwebby hairs along the midvein. $2n=14$. July–November.

Scattered nearly throughout the state (eastern U.S. west to Minnesota, Nebraska, and Texas; Canada). Upland prairies, openings of mesic to dry upland forests, savannas, glades, tops of bluffs, and banks of streams and rivers; also pastures, old fields, railroads, roadsides, and open, disturbed areas.

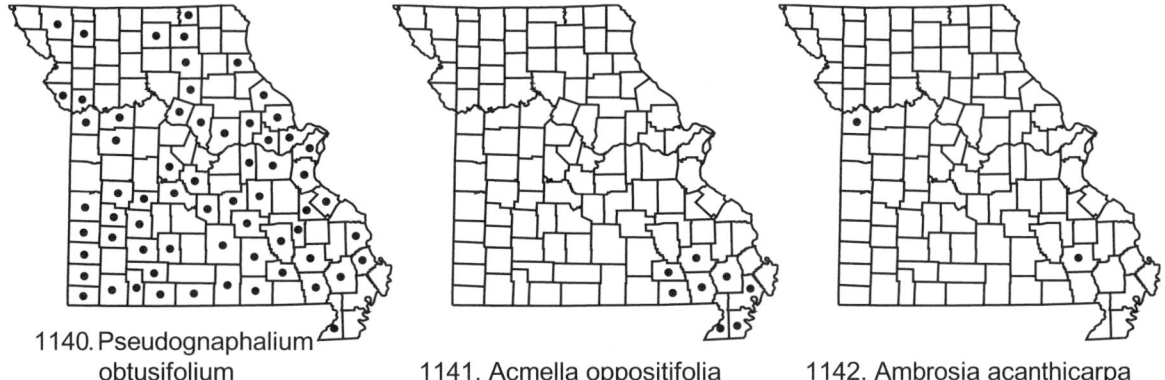

1140. Pseudognaphalium obtusifolium
1141. Acmella oppositifolia
1142. Ambrosia acanthicarpa

7. Tribe Heliantheae Cass.

Plants annual, biennial, or perennial herbs (sometimes woody elsewhere), sometimes from a woody rootstock, sometimes aromatic, occasionally monoecious, the sap not milky. Stems not spiny or prickly. Leaves alternate and/or opposite (whorled elsewhere), rarely only in a basal rosette, sessile to long-petiolate, not spiny or prickly. Leaf blades entire to lobed or pinnately compound, the venation pinnate or less commonly with 3 or more main veins. Inflorescences terminal panicles, spikes, or racemes, sometimes reduced to clusters of heads, or the heads solitary at the branch tips. Heads discoid or radiate. Involucre of 1 or 2 to more commonly several series of bracts, in a few genera these spiny or tuberculate. Receptacle flat to hemispherical or conical, naked or with chaffy bracts or hairs subtending all or some of the florets. Ray florets (when present) pistillate or sterile, sometimes inconspicuous, the corolla showy or sometimes very short or rarely absent, when present white, pink, purple, blue, yellow, or orange. Disc florets all perfect or functionally staminate, the corolla absent or highly reduced in a few monoecious genera, when present yellow, purple, purplish brown, the 4 or more commonly 5 lobes spreading to ascending. Pappus variously of numerous capillary bristles or 1 to several awns or scales, occasionally reduced to a minute crown or absent, when present persistent or not persisting at fruiting. Stamens with the filaments not fused together and the anthers fused into a tube or in a few genera with the filaments more or less fused into a tube and the anthers more or less free, each anther tip with a short to long appendage, each anther base truncate or less commonly with a pair of short lobes. Style branches somewhat flattened (the style undivided at maturity in functionally staminate flowers), each with a stigmatic line along each inner margin or more commonly with a stigmatic band along each inner surface, the sterile tip usually relatively long and tapered, less commonly short and truncate, usually with dense, minute hairs, at least on the outer surface. Fruits sometimes dimorphic (the outer series then thicker-walled and with different surface ornamentation), in a few genera enclosed in the persistent hardened involucre (forming a bur), sometimes somewhat flattened or angled to ribbed, variously shaped in profile, not beaked. About 300 genera, about 3,300 species, worldwide.

The Heliantheae are one of the largest and morphologically most diverse tribes in the family, which makes description of the tribe a challenge. The circumscription adopted here parallels that in the upcoming Asteraceae volumes of the Flora of North America series and differs from that adopted by Steyermark (1963) most notably in the inclusion of the genera that he separated as the tribe Helenieae Benth. Other recently proposed classifications, based mostly on molecular data, have varied widely, from those advocating the inclusion of the genera of Eupatorieae into a very broadly defined Heliantheae (Jansen and Kim, 1996; Bayer and Starr,

1998) to those who have suggested dissection of the Helenieae into six or more tribes in order to preserve the more traditional taxonomic circumscriptions of Heliantheae and Eupatorieae while providing formal recognition to each of the groups identified as monophyletic from the molecular data (Baldwin et al., 2002; Panero and Funk, 2002). Ultimately further research will be necessary to stabilize the limits of the Heliantheae and its sister tribes, and the present classification undoubtedly will have to be changed in the future to better reflect evolutionary relationships in the family.

A number of genera of Heliantheae contain members that are commonly cultivated as garden ornamentals and/or cut flowers, including *Bidens, Coreopsis, Cosmos, Echinacea, Gaillardia, Helenium, Heliopsis, Rudbeckia, Tagetes,* and *Zinnia* L. *Helianthus* is the most important crop genus, including *H. annuus,* the common sunflower, which is grown for its oil and seeds, as well as *H. tuberosus,* Jerusalem artichoke, which produces an edible tuberous root. *Guizotia* is the source of Niger thistle, which is a popular source of bird seed. On the negative side, a few genera contain species that are aggressive weeds of agricultural lands and producers of allergenic wind-borne pollen, including *Ambrosia, Iva,* and *Xanthium.*

1. Heads discoid, none of the florets producing a ligulate corolla (note that some radiate genera with minute rays are keyed under both leads to minimize confusion)
    2. Staminate and pistillate florets in different heads (none of the heads with perfect florets both staminate and pistillate florets), the staminate heads in terminal spikes or spikelike racemes, the pistillate heads appearing axillary below the staminate spikes; pistillate florets tightly enclosed by the involucral bracts, these persistent and with straight or hooked to curled spines or blunt tubercles on the outer surface
        3. Staminate heads with the involucral bracts fused; pistillate heads with the involucral bracts having straight spines or blunt tubercles on the outer surface . . . . 64. AMBROSIA
        3. Staminate heads with the involucral bracts free; pistillate heads with the involucral bracts having sharp spines on the outer surface that are hooked or curled at the tip . . . . 95. XANTHIUM
    2. Heads either with all of the florets perfect or with the marginal florets pistillate and the central florets perfect or functionally staminate; inflorescences various, but the heads all more or less similar; involucral bracts not enclosing the florets, the outer surface not spiny or tuberculate
        4. Involucral bracts with prominent yellow to yellowish brown glandular dots or lines embedded in the outer surface; plants strongly aromatic when bruised or crushed . . . . 70. DYSSODIA
        4. Involucral bracts without glandular dots or lines embedded in the surface, either nonglandular or with inconspicuous, minute, sessile to stalked glands; plants not or only slightly aromatic when bruised or crushed
            5. Receptacle naked, none of the florets subtended by chaffy bracts (note that minute, irregular ridges may surround the attachment points of the florets)
                6. Leaves opposite; florets with the corolla yellow; involucral bracts apparently 3 per head, but with 2 additional minute bracts at the base of the head . . . . 74. FLAVERIA
                6. Leaves alternate; florets with the corolla white to pale cream-colored or pink to purplish pink; involucral bracts more than 5 per head

7. Florets with the corolla white to pale cream-colored; inflorescence branches finely hairy, not glandular .. 81. HYMENOPAPPUS
7. Florets with the corolla pink to purplish pink; inflorescence branches with conspicuous, black, tack-shaped glands, otherwise glabrous................................ 85. PALAFOXIA
5. Receptacle chaffy, all or at least the outer florets subtended by chaffy bracts (the bracts appearing as slender, flattened bristles in *Eclipta*)
8. Pappus absent or a minute rim or crown
9. None of the florets perfect, the outer ones pistillate (lacking stamens), the inner ones functionally staminate with a small, stalklike ovary and an undivided style; stamens with the filaments fused into a tube, the anthers more or less free
10. Heads not subtended by bracts (small, leaflike bracts present only at the inflorescence branch points); involucral bracts free to the base; pistillate florets lacking a corolla .... 69. CYCLACHAENA
10. Each head subtended by a conspicuous bract; involucral bracts fused toward the base; pistillate florets with a tubular corolla 1.0–1.5 mm long .... 82. IVA
9. Florets mostly perfect, some of the outer ones only pistillate; stamens with the filaments free and the anthers fused into a tube
11. Leaf blades not lobed, finely toothed, narrowly lanceolate to elliptic in outline .... 72. ECLIPTA
11. Leaf blades pinnately lobed and irregularly toothed, broadly oblong to ovate in outline .... 87. POLYMNIA
8. Pappus present, of awns, scales, or bristles
12. Florets with the corolla yellow
13. Involucral bracts all free to the base............. 66. BIDENS
13. Inner series of involucral bracts fused $1/4$–$2/3$ of the way to the tip .... 93. THELESPERMA
12. Florets with the corolla white to off-white or pale pink
14. Florets all perfect; leaves mostly basal...... 83. MARSHALLIA
14. None of the florets perfect, most of the florets staminate, a few of the outer ones pistillate; leaves basal and also well distributed along the stem .... 86. PARTHENIUM
1. Heads radiate, the rays well developed but sometimes with relatively short corollas
15. Receptacle naked, none of the florets subtended by chaffy bracts (note that minute, irregular ridges or hairs may surround the attachment points of the florets), or with slender bristles fused irregularly at the base
16. Involucral bracts with prominent yellow to yellowish brown glandular dots or lines embedded in the outer surface; plants strongly aromatic when bruised or crushed
17. Ray florets inconspicuous, the corolla 2–4 mm long; involucral bracts fused only at the base .... 70. DYSSODIA
17. Ray florets showy, the corolla 10–30 mm long; involucral bracts fused to well above the midpoint .... 92. TAGETES

16. Involucral bracts without glandular dots or lines embedded in the surface, either nonglandular or with inconspicuous, minute, sessile to stalked glands; plants not or only slightly aromatic when bruised or crushed
    18. Ray floret 1 per head, inconspicuous, the corolla 1.5–2.5 mm long; disc florets 3 or 4; involucral bracts in an inner series of 3 and an outer series of 2 much shorter bracts .... 74. FLAVERIA
    18. Ray florets 5–21 per head, usually showy, the corolla 5–30 mm long; disc florets numerous; involucral bracts usually 14–20 in 2 or 3 series
        19. Receptacle with numerous slender bristles fused irregularly at the base or these reduced to a network of low, irregular teeth; ray florets with the corolla red, brownish red, or reddish purple, at least toward the base (entirely yellow in G. aestivalis); disc florets with the corolla lobes woolly-hairy, not glandular; style branches with a relatively long, tapered, sterile tip .... 75. GAILLARDIA
        19. Receptacle naked, without bristles; ray florets yellow; disc florets with the corollas glandular; style branches with a relatively short, slightly expanded, more or less truncate, sterile tip .... 78. HELENIUM
15. Receptacle chaffy, all or at least the outer florets subtended by chaffy bracts
    20. Leaves all or mostly alternate (rarely appearing opposite or nearly so at a few nodes), occasionally all or mostly basal
        21. None of the florets perfect, the disc florets staminate (with a small ovary and undivided style), the ray florets pistillate (lacking stamens)
            22. Ray florets usually 5, the corolla 1–3 mm long, white or off-white .... 86. PARTHENIUM
            22. Ray florets 7–35, the corolla 10–50 mm long, yellow
                23. Ray florets (8–)13–35 in 2 or 3 overlapping marginal series, the corolla 15–50 mm long; involucre 15–40 mm long .... 90. SILPHIUM
                23. Ray florets (5–)8(–13) in 1 marginal series, the corolla 9–17 mm long; involucre 6–10 mm long
                    24. Leaf blades coarsely toothed or scalloped but not lobed; involucral bracts mostly broadly obovate to broadly rhombic, usually uniformly green, with a bluntly to sharply pointed tip; disc corollas brownish red to dark reddish purple .... 65. BERLANDIERA
                    24. Leaf blades pinnately deeply lobed, the slender lobes often toothed or lobed; involucral bracts with a linear to narrowly oblong-lanceolate, green tip above a broader, thickened, yellowish base; disc corollas yellow .... 73. ENGELMANNIA
        21. Disc florets all perfect, the ray florets pistillate (lacking stamens or sterile) (producing neither stamens nor a style exserted from the corolla tube)
            25. Leaf blades all or mostly deeply lobed .... 88. RATIBIDA
            25. Leaf blades with the margins entire or variously scalloped or toothed, but not lobed
                26. Receptacle noticeably conical or columnar
                    27. Chaffy bracts noticeably longer than the disc florets (including the corolla), tapered to a hard, spinelike tip; disc florets with the corolla slightly bulbous-thickened at the base .... 71. ECHINACEA

27. Chaffy bracts shorter than to slightly longer than the disc florets (including the corolla), truncate or rounded to angled or tapered to a sharply pointed tip, this unawned or with a soft, bristlelike awn; disc florets with the corolla not thickened at the base .... 89. RUDBECKIA
26. Receptacle flat or convex, but not noticeably elongate
28. Disc florets with the corollas dark reddish purple to purplish brown; ray florets with the corollas yellow ...... 79. HELIANTHUS
28. Disc and ray florets with the corollas all yellow (sometimes greenish yellow toward the base of the disc corollas) or all white (in *Verbesina virginica*)
29. Leaf blades linear to narrowly lanceolate, the margins curled under, entire .... 79. HELIANTHUS
29. Leaf blades lanceolate to elliptic-lanceolate, ovate, or triangular, the margins relatively flat, toothed .... 94. VERBESINA
20. Leaves all or mostly opposite (sometimes alternate at the uppermost few nodes)
30. None of the florets perfect, the disc florets staminate (with a small ovary and undivided style), the ray florets pistillate (lacking stamens)
31. Ray florets with the corolla 2–10 mm long, white or pale cream-colored; involucre 3–8 mm long
32. Outer series of involucral bracts fused in the basal $1/6$–$1/3$; inner involucral bracts folded around the ovary or fruit of a ray floret .... 84. MELAMPODIUM
32. Outer series of involucral bracts free to the base; inner involucral bracts relatively flat and not folded around the ray florets ... 87. POLYMNIA
31. Ray florets with the corolla 12–50 mm long, yellow; involucre 10–40 mm long
33. Leaf blades with the margins entire or toothed but not lobed; ray florets (8–)13–35 in 2 or 3 overlapping marginal series ..... 90. SILPHIUM
33. Leaf blades with 3 or less commonly 5 broad, irregular, triangular lobes, the margins also irregularly toothed; ray florets 7–13 in 1 marginal series .... 91. SMALLANTHUS
30. Disc florets all perfect, the ray florets pistillate or sterile (producing neither stamens nor a style exserted from the corolla tube)
34. Ray florets with the corolla white or pink
35. Leaf blades deeply lobed or compound; ray florets with the corolla 5–40 mm long
36. Leaf blades mostly 1 time pinnately compound (the uppermost leaves sometimes only deeply lobed), the leaflets oblong-lanceolate to ovate, toothed; fruits narrowed or somewhat tapered at the tip but not beaked...................... 66. BIDENS
36. Leaf blades 1 or 2 times pinnately dissected, the ultimate segments linear to threadlike, mostly entire; fruits tapered to a noticeable beak .... 68. COSMOS
35. Leaf blades finely toothed but not lobed; ray florets with the corolla 1–3 mm long
37. Pappus absent or a minute rim or crown; involucral bracts 10–12 .... 72. ECLIPTA
37. Pappus of 12–20 fringed scales, sometimes highly reduced or absent in the ray florets; involucral bracts 6–9 .... 76. GALINSOGA

34. Ray florets with the corolla yellow to orange, occasionally somewhat reddish-tinged toward the base
    38. Disc florets with the corolla densely pubescent with tangled, more or less spreading hairs toward the base .... 77. GUIZOTIA
    38. Disc florets with the corolla glabrous or rarely sparsely to densely pubescent with more or less appressed, straight hairs or glandular
        39. Inner series of involucral bracts fused $1/4$–$2/3$ of the way to the tip .... 93. THELESPERMA
        39. Involucral bracts all free or fused at the very base
            40. Ray florets pistillate (with an exserted, forked style and an ovary similar in size to those of the disc florets at flowering, developing into a fruit)
                41. Receptacle flat or broadly convex to shallowly conical; pappus of 2 slender to stout, stiff awns; fruits strongly flattened, the margins narrowly to broadly winged .... 94. VERBESINA
                41. Receptacle noticeably conical or somewhat columnar; pappus absent or rarely of 1 or 2 minute, slender awns; fruits 3- or 4-angled or, if flattened, then the margins usually hairy but not winged
                    42. Stems spreading with ascending tips and branches, often rooting at the lower nodes; involucre 4–6 mm long; ray corollas not persistent at fruiting, the corolla 3–9 mm long; fruits flattened or those developing from the ray florets often somewhat 3-angled, the margins sparsely hairy .... 63. ACMELLA
                    42. Stems erect or ascending, not rooting at the nodes (note that short rhizomes may be present); involucre 6–16 mm long; ray corollas becoming papery and persistent at fruiting, the corolla 15–40 mm long; fruits of the disc florets 4-angled, glabrous .... 80. HELIOPSIS
            40. Ray florets sterile (lacking a style at flowering and with an ovary that is shorter and thinner than those of the disc florets, not developing into a fruit)
                43. Leaf blades deeply lobed or compound
                    44. Leaflets or lobes with the margins toothed .... 66. BIDENS
                    44. Leaflets or lobes with the margins entire
                        45. Plants submerged aquatics with only the stem tips emergent .... 66. BIDENS
                        45. Plants terrestrial, at most growing in moist soil along stream banks but often in drier habitats
                            46. Ray florets with the corolla yellow to occasionally orangish yellow; pappus absent or of a few short bristles but more commonly of 2 short, slender scales or awns (these smooth or with sparse, upward-angled barbs); fruits 2–7 mm long, flattened, more or less truncate at the tip, not beaked .... 67. COREOPSIS
                            46. Ray florets with the corolla yellowish orange to reddish orange; pappus of 2–4 short awns (these with downward-angled barbs) or rarely absent; fruits 7–30 mm long (including the beak), not flattened, 4-angled, tapered to a noticeable beak at the tip .... 68. COSMOS

43. Leaf blades simple, unlobed, the margins entire to variously toothed or scalloped
   47. Involucral bracts of 2 distinct types, the outer series green and more or less herbaceous, ascending or spreading, the inner series membranous to scale-like, reddish-, brownish-, or yellowish-tinged and with numerous fine longitudinal nerves or lines, erect or strongly ascending
      48. Leaf blades with the margins toothed; outer series of involucral bracts as long as or mostly longer than the inner series . . . . 66. BIDENS
      48. Leaf blades with the margins entire; outer series of involucral bracts shorter than the inner series . . . . 67. COREOPSIS
   47. Involucral bracts in 1 to several series of similar or different lengths, but all similar in color and texture, green and herbaceous from a sometimes yellowish base, without numerous longitudinal nerves or lines, erect to spreading
      49. Pappus of 2 awns with broad, flattened bases (or less commonly scales) that usually are shed prior to fruit maturity; fruits 4-angled or somewhat flattened with blunt to rounded margins, not winged . . . . 79. HELIANTHUS
      49. Pappus of 2 awns that are persistent at fruiting; fruits strongly flattened, the margins sharply angled or more commonly winged . . . . . . 94. VERBESINA

63. Acmella Rich. ex Pers.
(Jansen, 1985)

About 30 species, North America to South America, Caribbean Islands, Africa, Asia south to Australia.

Jansen (1981) discussed the variety of morphological features separating *Acmella* from the closely related genus *Spilanthes* Jacq., to which the Missouri species was assigned by most earlier authors (Steyermark, 1963).

1. Acmella oppositifolia (Lam.) R.K. Jansen
   var. repens (Walter) R.K. Jansen
   A. repens (Walter) Rich.
   Spilanthes americana (L.f.) Hieron. var.
      repens (Walter) A.H. Moore
                              Pl. 272 a, b; Map 1141

Plants perennial herbs, with fibrous roots and sometimes also rhizomes. Stems 30–100 cm long, few- to several-branched, spreading with ascending tips and branches, often rooting at the lower nodes, finely ridged or grooved, glabrous to sparsely or moderately pubescent with soft, curved hairs. Leaves opposite or the uppermost few alternate, short- to long-petiolate. Leaf blades 1–8(–10) cm long, narrowly lanceolate to broadly ovate, tapered at the base, angled or tapered to a bluntly or more commonly sharply pointed tip, the margins finely to coarsely and bluntly to sharply toothed and sometimes microscopically hairy, the surfaces glabrous or sparsely pubescent along the veins with soft, curved hairs. Inflorescences of solitary heads at the branch tips, the heads mostly with long, bractless, sparsely to densely hairy stalks. Heads radiate. Involucre 4–6 mm long, 3–6 mm in diameter, cup-shaped, the bracts in 2 more or less similar (the inner series slightly more membranous) series. Involucral bracts 6–16, ascending, lanceolate, sharply pointed at the tip, green (the inner series sometimes somewhat yellowish-tinged) with slightly irregular to fringed margins, the outer surface glabrous or sparsely hairy, with usually 3 fine nerves. Receptacle conical, elongating greatly and sometimes becoming somewhat cylindrical as the fruits mature, with chaffy bracts subtending the ray and disc florets, these 2.5–4.5 mm long, oblong-lanceolate, folded longitudinally and more or less wrapped around the florets. Ray florets 5–15, pistillate (with a 2-branched style exserted from the short tube at flowering), the corolla 3–9 mm long, relatively broad, yellow to orangish yellow, sparsely hairy toward the base, not persistent at fruiting. Disc florets numerous, perfect, the corolla 1.5–3.0

mm long, yellow, glabrous, not expanded at the base or persistent at fruiting. Style branches with the sterile tip very short and more or less truncate. Pappus of the ray and disc florets absent or the disc florets rarely with 1 or 2 short, slender awns. Fruits 1.0–2.5 mm long, narrowly oblong-obovate to somewhat wedge-shaped in outline, flattened (those of the ray florets somewhat 3-angled), smooth or more commonly irregularly warty, the surface otherwise glabrous, dark brown to black, the margins and warts straw-colored to yellowish brown. $2n=52$. July–October.

Scattered in the Mississippi Lowlands Division and the adjacent southeastern portion of the Ozarks (southeastern U.S. west to Missouri and Texas; Mexico). Swamps, bottomland forests, banks of streams and rivers, and rarely margins of sinkhole ponds; also ditches; sometimes emergent aquatics.

Jansen (1985) noted that the name *Spilanthes americana* was based on the illegitimate name *Anthemis americana* L. f. and that the next-oldest specific epithet available for the taxon was *A. oppositifolia* Lam. He separated the species into the widespread Latin American var. *oppositifolia* (with somewhat broader involucral bracts, usually a pappus of 2 or 3 slender, bristlelike awns, and the fruits either glabrous or with relatively long hairs along the margins) and the var. *repens* of mostly the southeastern United States (with somewhat narrower involucral bracts, the pappus absent, and the margins of the fruits with short hairs). Farther southeast in its range, the var. *repens* is more variable in its hairiness than the above description suggests, but Missouri plants all appear to vary toward the more glabrous end of the spectrum for the taxon.

64. Ambrosia L. (ragweed)
(Payne, 1964)

Plants annual or perennial herbs (shrubs elsewhere), sometimes with taproots, rhizomes, or a branched, woody rootstock. Stems few- to many-branched, mostly erect or ascending, finely ridged or grooved, sparsely to densely pubescent with often pustular-based hairs. Leaves alternate and/or opposite, variously sessile to long-petiolate. Leaf blades variously shaped, simple and entire or more commonly 1–3 times pinnately or palmately lobed, variously sparsely to densely hairy. Inflorescences of separate staminate and pistillate heads, the staminate heads in spikes or spikelike racemes terminal on the branch tips, these sometimes appearing paniculate, the pistillate heads sessile, solitary or in clusters toward the base of the staminate spikes or in the adjacent upper leaf axils. Heads discoid (but this not evident in pistillate heads), the staminate heads pendant. Involucre of the staminate heads cup-shaped to saucer-shaped, often somewhat asymmetrical, the 5–12 involucral bracts in 1 series, fused irregularly well above the base, green. Involucre of the pistillate heads with the main body globose to ovoid or somewhat pear-shaped, the involucral bracts closely enclosing the florets and fused into a bur, the outer surface with straight spines or tubercles, more or less beaked at the tip (where an opening allows exsertion of the stigmas), green or sometimes purplish-tinged. Receptacle flat (difficult to observe in pistillate heads), not elongating as the fruits mature, that of the staminate heads with chaffy bracts subtending the florets, these narrowly linear to narrowly lanceolate, usually hairy and sometimes also glandular, not wrapped around the florets. Staminate heads with 10–150 disc florets, these with a minute, nonfunctional ovary and undivided style, the stamens with the filaments more or less fused into a tube and the anthers free but positioned closely adjacent to one another in a ring, the corolla 2–4 mm long, narrowly bell-shaped, white to pale yellow, sometimes purplish-tinged toward the tip, usually minutely hairy and often also glandular. Pistillate heads with 1 or 2 florets, the corolla absent. Pappus of the staminate and pistillate florets absent. Fruits 3.0–4.5 mm long, more or less globose to ovoid, not flattened (or less commonly somewhat flattened), not angled, grayish tan to nearly black, glabrous, completely enclosed in the persistent pistillate involucre and dispersed as an intact bur. About 46 species, nearly worldwide, but most diverse in North America.

Ambrosia and its relatives form a specialized group within the Heliantheae that at one time was treated as a separate tribe, Ambrosieae Cass., by some botanists. Payne (1964) justified the inclusion of species formerly separated as *Franseria* Cav. (Steyermark, 1963) within *Ambrosia*. More recent molecular studies (Miao et al. 1995c) generally supported this conclusion and resulted in the additional lumping of the four southwestern species of the genus *Hymenoclea* Torr. & A. Gray ex A. Gray into *Ambrosia*. The ragweeds are wind-pollinated, and especially the species that form large stands in disturbed areas are a leading cause of hay fever during the late summer.

1. Leaf blades unlobed or with 3 or 5 lobes
    2. Leaves all or mostly alternate, the main stem leaves 0.4–1.0 cm wide, lanceolate to narrowly oblong-lanceolate, unlobed or with a pair of small basal lobes; staminate heads sessile along the axis in a solitary terminal spike
    .... 3. A. BIDENTATA
    2. Leaves mostly opposite (except for the uppermost ones), the main stem leaves usually much more than 1 cm wide (often 10–20 cm wide), ovate to broadly elliptic, with 3 or 5 conspicuous lobes (less commonly unlobed); staminate heads short-stalked along the axis in a solitary, terminal, spikelike raceme or sometimes in 3 to several racemes grouped into a panicle .... 6. A. TRIFIDA
1. Leaf blades (except sometimes those of the uppermost leaves) deeply 1 or 2 times pinnately lobed with more than 5 lobes
    3. Pistillate involucre with 2 florets and 2 beaks; leaves with the undersurface densely white-hairy, appearing somewhat whitened .... 5. A. TOMENTOSA
    3. Pistillate involucre with 1 floret and 1 tip or beak; leaves variously hairy but not appearing whitened
        4. Pistillate involucre with several series of relatively long, strongly flattened spines scattered across the surface (these developing as the fruits mature) .... 1. A. ACANTHICARPA
        4. Pistillate involucre with 1 series of relatively short, not or only slightly flattened spines in a ring toward the tip (these developing as the fruits mature)
            5. Plants annual, with taproots; leaves short- to long-petiolate, the blades mostly 2 times pinnately lobed (the smaller leaves sometimes only 1 time lobed)........................... 2. A. ARTEMISIIFOLIA
            5. Plants perennial, with long-creeping rhizomes; leaves sessile or nearly so, the blade mostly only 1 time pinnately lobed
            .... 4. A. PSILOSTACHYA

1. Ambrosia acanthicarpa Hook. (annual bursage)

Franseria acanthicarpa (Hook.) Coville

Pl. 271 f–h; Map 1142

Plants annual, with taproots. Stems 10–80 (–120) cm long, variously with short and/or longer hairs. Leaves opposite toward the stem base, alternate toward the stem tip, mostly short- to long-petiolate (the uppermost ones sessile or nearly so). Leaf blades 2–8 cm long, ovate to lanceolate in outline, irregularly 1 or 2 times pinnately lobed (at least the larger leaves with 7 or 9 primary lobes), the lobes linear to ovate, entire or few-toothed, the surfaces sparsely to moderately pubescent with more or less appressed, white hairs, the undersurface sometimes appearing somewhat pale or whitened. Staminate heads in spikelike racemes, these frequently in paniculate clusters, the staminate involucre 3–12 mm wide, with 3–9 shallow to relatively deep lobes, glabrous but usually with 1 or few black lines or streaks. Pistillate heads at the base of the staminate racemes (or less commonly the racemes mostly pistillate), the involucre enclosing 1 floret and with 1 beak, 5–10 mm long at fruiting, more or less ovoid, with several series of relatively long, strongly flattened spines scattered across the surface, otherwise glabrous. $2n=36$. August–October.

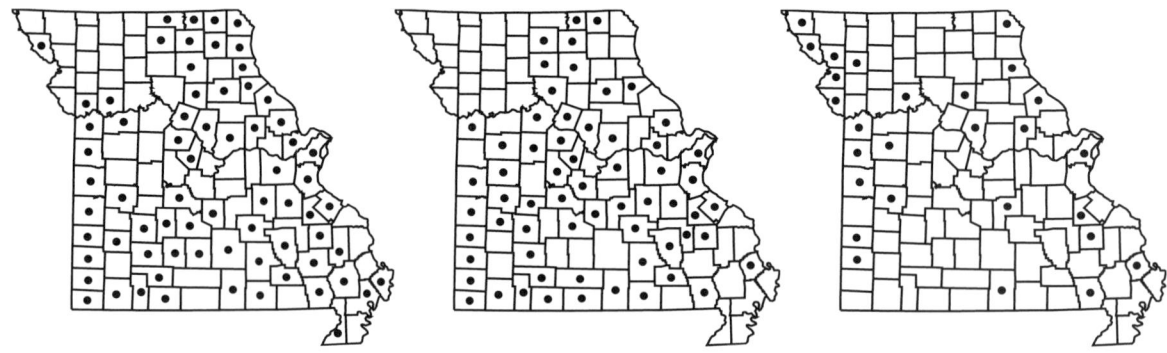

1143. Ambrosia artemisiifolia   1144. Ambrosia bidentata   1145. Ambrosia psilostachya

Introduced, uncommon, known thus far only from historical collections from Jackson and Wayne Counties (western U.S. east to North Dakota and Texas; Canada; introduced sporadically farther east). Glades; also railroads and disturbed areas.

2. Ambrosia artemisiifolia L. (common ragweed, bitterweed, Roman wormwood, hogweed)

A. artemisiifolia var. *elatior* (L.) Descourt.
A. artemisiifolia f. *villosa* Fernald & Griscom
Pl. 271 d, e; Map 1143

Plants annual, with taproots. Stems 30–120 cm long, sparsely to densely pubescent with relatively long, spreading hairs usually with minute, pustular bases and/or shorter, appressed hairs. Leaves opposite toward the stem base, alternate toward the stem tip, with short to long, narrowly winged petioles. Leaf blades 3–10 cm long, ovate to broadly ovate in outline (the uppermost leaves sometimes lanceolate to linear), 2–3 times pinnately lobed with more than 5 primary lobes (the uppermost leaves sometimes few-lobed to nearly entire), the ultimate lobes lanceolate to narrowly oblong, entire or few-toothed, the surfaces sparsely to moderately pubescent with short, somewhat broad-based hairs and sometimes appearing somewhat grayish, the undersurface also usually with longer hairs along the main veins, not or only slightly paler than the upper surface. Staminate heads in spikelike racemes, these usually not in paniculate clusters, the staminate involucre 2–4 mm wide, with 3–9 shallow lobes, glabrous or sparsely hairy. Pistillate heads in small axillary clusters (or sometimes solitary), the involucre enclosing 1 floret and with 1 beak, 3–5 mm long at fruiting, more or less ovoid, with 1 series of not or only slightly flattened, short spines in a ring toward the tip, sparsely to moderately hairy. $2n=34, 36$. July–November.

Common nearly throughout the state (U.S., Canada; introduced in Hawaii, Europe). Upland prairies, savannas, glades, tops of bluffs, banks of streams, rivers, and spring branches, marshes, margins of ponds, lakes, and sinkhole ponds, and openings of bottomland to mesic or dry upland forests; also pastures, old fields, fallow fields, crop fields, levees, ditches, farmyards, railroads, roadsides, and open, disturbed areas.

This is a variable species well adapted to disturbed sites. Steyermark (1963) noted that putative hybrids with *A. trifida* (*A.* ×*helenae* Rouleau) occur in surrounding states and eventually may be discovered in Missouri. To the north of Missouri, hybrids with *A. psilostachya* (*A.* ×*intergradiens* W.H. Wagner & Beals) also have been documented (Wagner and Beals, 1958). Steyermark (1963) also noted that although the fruits provide food for wild turkey and other wildlife, grazing of the plants apparently causes a nauseous effect in cattle. The species is a problem agricultural weed, and anecdotal reports indicate that a strain resistant to glyphosate-based herbicides such as Roundup has evolved in central Missouri (Bradley, 2005).

3. Ambrosia bidentata Michx. (lanceleaf ragweed, southern ragweed)

Pl. 271 a–c; Map 1144

Plants annual, with taproots. Stems 30–100 cm long, moderately roughened-pubescent with a mixture of short-ascending to spreading and longer, stiffly spreading hairs with pustular bases, sometimes appearing nearly glabrous toward the base and often somewhat shiny. Leaves mostly alternate (the lowermost ones occasionally opposite), sessile. Leaf blades 1–7 cm long, 0.4–1.0 cm wide, lanceolate to narrowly oblong-lanceolate, unlobed or with a pair of small lobes toward the base, the lobes triangular, the margins otherwise entire, the surfaces sparsely to moderately roughened-pubescent with short, loosely ascending, pustular-based hairs and stiff, longer hairs (mostly along the midvein), the upper surface sometimes only sparsely hairy, the

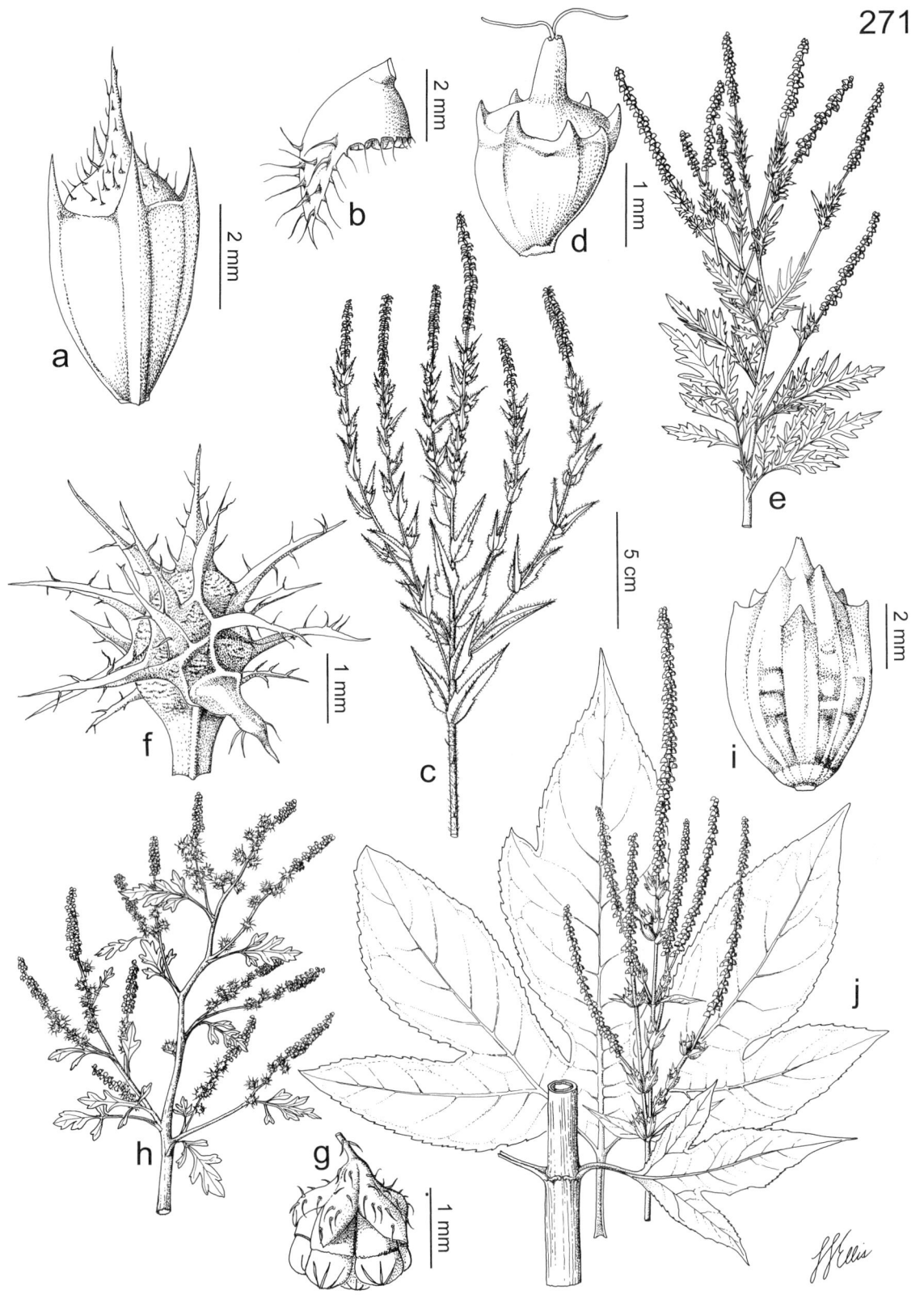

Plate 271. Asteraceae. *Ambrosia bidentata*, **a)** fruiting involucre, **b)** staminate involucre, **c)** habit. *Ambrosia artemisiifolia*, **d)** fruiting involucre, **e)** habit. *Ambrosia acanthicarpa*, **f)** fruiting involucre, **g)** staminate head, **h)** habit. *Ambrosia trifida*, **i)** fruiting involucre, **j)** inflorescence, stem with node, and larger leaf.

undersurface often with pronounced, small, sessile glands (slightly sticky to the touch), not or only slightly paler than the upper surface. Staminate heads sessile in relatively dense, solitary spikes, the staminate involucre 2.5–4.0 mm wide, with 3(5) lobes, the lateral ones small, the terminal lobe elongate, the outer surface with minute, sessile glands, the terminal lobe with a patch of relatively long, stiff hairs. Pistillate heads in small axillary clusters (or occasionally solitary), the involucre enclosing 1 floret and with 1 stout, conical beak, 5–8 mm long at fruiting, more or less ovoid, with usually 4 longitudinal angles or ridges, each terminating in a short spine, moderately hairy. $2n=34$. July–October.

Scattered nearly throughout the state, but uncommon or absent from portions of the northwestern quarter (Ohio to Louisiana west to Nebraska and Texas; introduced farther northward and eastward). Upland prairies, savannas, glades, tops of bluffs, banks of streams and rivers, and openings of mesic or dry upland forests; also old fields, fallow fields, railroads, roadsides, and open, disturbed areas.

Sterile hybrids between *A. bidentata* and *A. trifida* have been collected sporadically in Missouri, mostly in the western half of the state. These plants resemble *A. bidentata* in stature but have the leaves irregularly and mostly deeply 3- or 5-lobed with relatively narrow, pinnate lobes.

4. Ambrosia psilostachya DC. (western ragweed)

A. coronopifolia Torr. & A. Gray

Pl. 272 f, g; Map 1145

Plants perennial, usually colonial from deep, often widely creeping rhizomes. Stems 30–70(–100) cm long, moderately to densely pubescent with relatively short, appressed-ascending hairs usually with minute, pustular bases, sometimes also with longer, spreading hairs. Leaves opposite, the uppermost sometimes alternate, sessile or with very short, narrowly winged petioles. Leaf blades 3–10 cm long, lanceolate to ovate in outline, mostly 1 time pinnately lobed with more than 5 lobes (the uppermost leaves sometimes few-lobed to nearly entire), the lobes narrowly lanceolate to narrowly triangular, entire or few-toothed, the surfaces moderately to densely pubescent with somewhat pustular-based hairs and often appearing somewhat grayish, the undersurface not or only slightly paler than the upper surface. Staminate heads in spikelike racemes, these usually not in paniculate clusters, the staminate involucre 2–3 mm wide, with 3–9 shallow lobes, usually moderately hairy. Pistillate heads in small axillary clusters (or sometimes solitary), the involucre enclosing 1 floret and with 1 beak, 2.5–3.5 mm long at fruiting, more or less ovoid, with 1 series of not or only slightly flattened, short tubercles in a ring toward the tip, these sometimes absent, moderately to densely hairy, especially above the midpoint. $2n=36, 72, 108, 144$. August–October.

Scattered, mostly in the Unglaciated Plains Division and the western portion of the Glaciated Plains (western U.S. east to Michigan and Louisiana; Canada; introduced eastward to Maine and Florida). Upland prairies, loess hill prairies, and sand prairies; also railroads, roadsides, and sandy, open, disturbed areas.

Many earlier botanists segregated midwestern plants from those farther south in the range as *A. coronopifolia* based on slight morphological differences as well as different ploidy (Steyermark, 1963). Following more detailed investigations of morphological variation and chromosome number, Payne (1970) concluded that the complex is best treated as a single variable species.

5. Ambrosia tomentosa Nutt. (perennial bursage)

Franseria discolor Nutt.

Pl. 272 d, e; Map 1146

Plants perennial, colonial from deep, often widely creeping rhizomes. Stems 10–35 cm long, sparsely to densely pubescent with short, appressed and sometimes also longer, woolly hairs. Leaves alternate except sometimes toward the stem base, mostly short- to long-petiolate (the uppermost ones sometimes nearly sessile). Leaf blades 2–10 cm long, ovate to lanceolate in outline, irregularly 1–3 times pinnately lobed (at least the larger leaves with 7–15 primary lobes), the lobes irregularly elliptic to ovate, the margins mostly toothed and often slightly thickened or curled under, the upper surface moderately roughened-pubescent with minute, stiff, more or less appressed, white hairs, the undersurface more densely hairy, pale or whitened, the hairs sometimes appearing more or less tangled. Staminate heads in spikelike racemes, these usually not in paniculate clusters, the staminate involucre 4–7 mm wide, with 7–12 shallow to moderately deep lobes, densely hairy. Pistillate heads in small axillary clusters (or sometimes solitary), the involucre enclosing 2 florets and with 2 stout beaks, 4–6 mm long at fruiting, ovoid to ellipsoid, with 1 or few series of relatively short, sometimes slightly flattened spines scattered across the surface (these rarely absent or nearly so), otherwise minutely hairy. July–September.

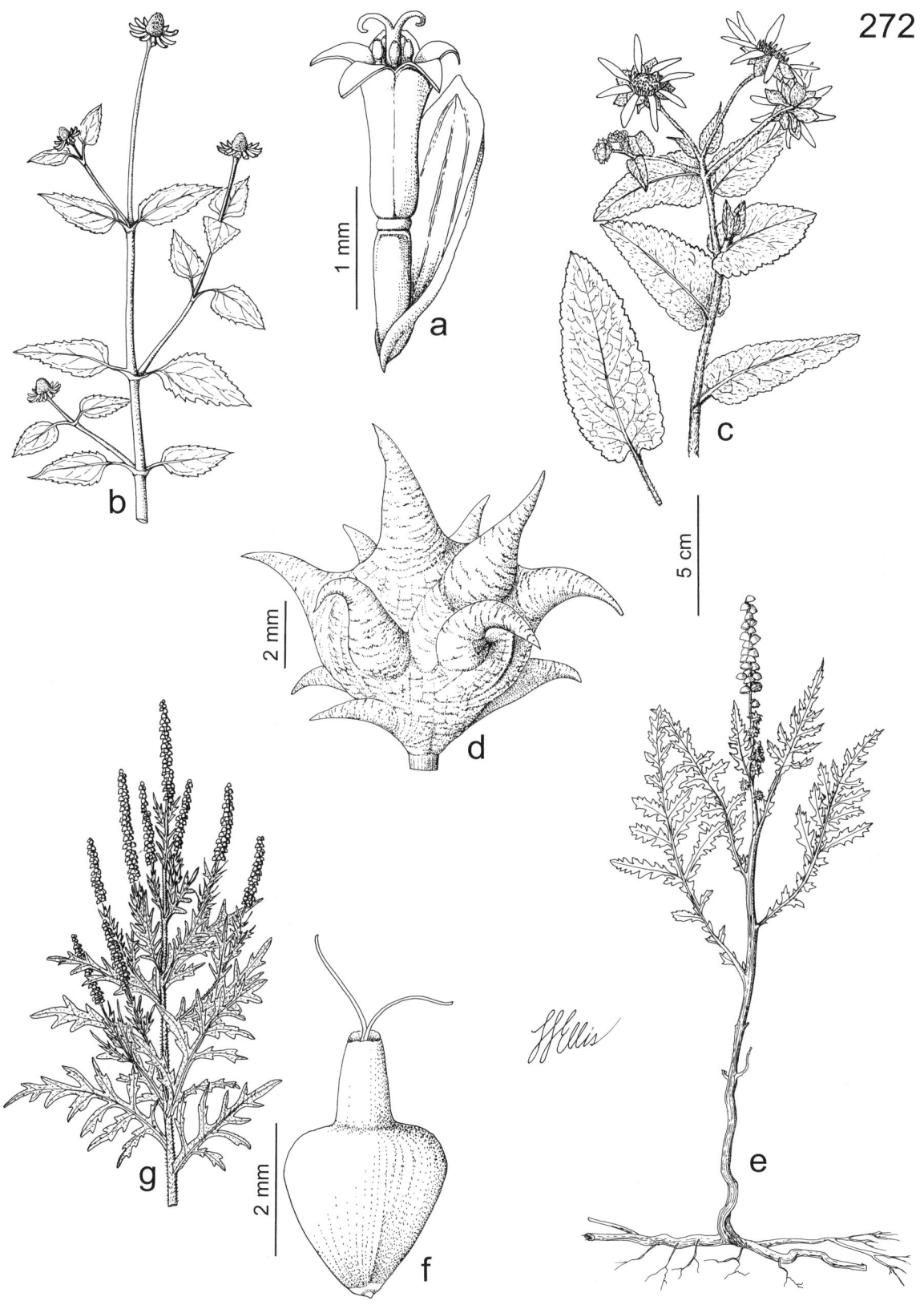

Plate 272. Asteraceae. *Acmella oppositifolia*, **a)** habit, **b)** disc floret. *Berlandiera texana*, **c)** habit with larger leaf. *Ambrosia tomentosa*, **d)** fruiting involucre, **e)** habit. *Ambrosia psilostachya*, **f)** fruiting involucre, **g)** habit.

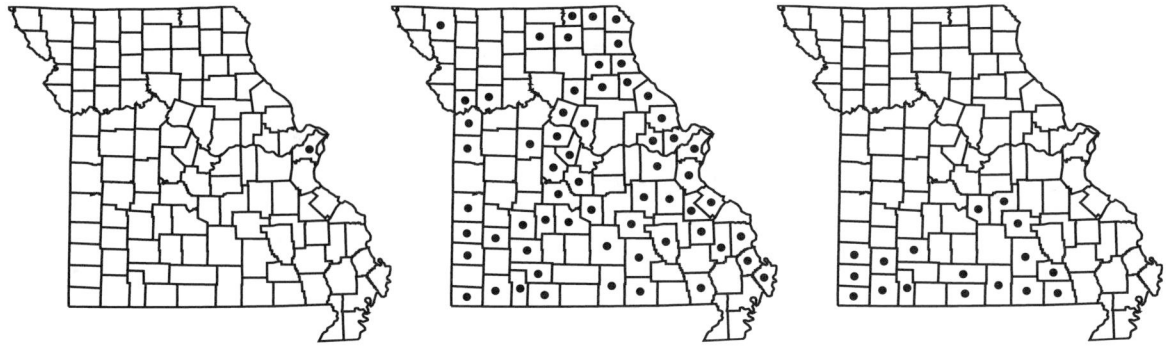

1146. Ambrosia tomentosa   1147. Ambrosia trifida   1148. Berlandiera texana

Introduced, uncommon, known thus far only from the city of St. Louis (Wyoming to Arizona east to South Dakota and Nebraska; introduced sporadically elsewhere in the U.S.). Railroads.

Mühlenbach (1979) observed the colony that he discovered in the St. Louis railyards for fourteen years and remarked that it persisted in spite of repeated sprayings with herbicides.

6. Ambrosia trifida L. (giant ragweed, great ragweed, horseweed, buffalo weed)
  A. trifida var. *texana* Scheele
  A. trifida f. *integrifolia* (Muhl.) Fernald
     Pl. 271 i, j; Map 1147

Plants annual, with taproots. Stems 30–500 cm long, sparsely to moderately roughened with short, ascending to loosely appressed hairs with pustular bases, sometimes also with scattered, longer, spreading hairs toward the tip. Leaves mostly opposite (the uppermost ones often alternate), with short to more commonly long petioles, those of the upper leaves often narrowly winged. Leaf blades 3–30 cm long, the main stem leaves usually much more than 1 cm wide (often 10–20 cm wide), ovate to broadly elliptic or nearly circular in outline, unlobed or more commonly with 3 or 5 deeply palmate lobes, occasionally with an additional pair of smaller lobes, the lobes oblong-elliptic to more or less elliptic or oblong-triangular, usually with many fine or coarse teeth, the upper surface sparsely to moderately roughened with short, stout, pustular-based hairs and sometimes with sparse, longer hairs along the main veins, the undersurface similarly but more sparsely roughened, not or only slightly paler than the upper surface. Staminate heads in spikelike racemes, these sometimes 3 to several in paniculate clusters, the staminate involucre 1.5–3.0 mm wide, with 5–8 shallow to moderately deep lobes, minutely hairy and usually with pronounced black lines from the attachment point to the tips of the lobes. Pistillate heads in small axillary clusters (or occasionally solitary), the involucre enclosing 1 floret and with 1 stout, conical beak, 5–10 mm long at fruiting, more or less ovoid or somewhat pear-shaped, with 4–8 longitudinal angles or ridges, each terminating in a short, stout tubercle or spine, glabrous or sparsely hairy, the spines and beak often more densely hairy. $2n=24$. July–September.

Common nearly throughout the state (U.S., Canada; introduced in Europe). Bottomland and upland prairies, banks of streams and rivers, sloughs, marshes, margins of ponds and lakes, and bottomland forests; also pastures, old fields, fallow fields, crop fields, levees, ditches, railroads, roadsides, and open, disturbed areas.

Giant ragweed is an exceedingly variable species. Plants varying in characters such as degree of leaf lobing and size of the fruiting involucre have been segregated as forms or varieties by some botanists, but Payne (1970) argued against such splitting of the species based on his observations of plants from throughout the range. For a discussion of uncommon putative hybrids with *A. bidentata*, see the treatment of that species. Hybridization also occurs with *A. artemisiifolia* in other states, but it has not yet been reported from Missouri.

Ambrosia trifida forms extensive colonies in disturbed bottomland and agricultural areas and is a leading cause of hay fever in the late summer. It is a problem weed in crop fields, particularly in soybean fields, and anecdotal evidence suggests that in Illinois a strain has evolved that is resistant to glyphosate-based herbicides such as Roundup (Associated Press, 2002). Payne and Jones (1962) studied archaeological remains of *A. trifida* with unusually large fruiting involucres from prehistoric bluff shelters in the Ozark Plateau and concluded that the species was cultivated by aboriginal Americans for its edible fruits. Moerman

(1998) noted that Native Americans used the plant ceremonially and medicinally, principally to treat stings, hives, infections, fever, and pneumonia. Steyermark (1963) stated that the species was used medicinally as an astringent. He also noted that the plants are a source of food for wildlife.

65. Berlandiera DC. (green eyes)
(Pinkava, 1967)

Five species, U.S., Mexico.

1. Berlandiera texana DC. (Texas green eyes)

Pl. 272 c; Map 1148

Plants perennial herbs, with fleshy taproots, the stem bases often somewhat woody. Stems 50–120 (–200) cm long, usually several-branched toward the tip, erect or ascending, finely ridged or grooved, moderately to densely pubescent with soft, loosely tangled hairs. Leaves alternate, sessile or very short-petiolate. Leaf blades 1–15 cm long, the uppermost often narrowly lanceolate, the main stem leaves ovate to ovate-triangular, shallowly cordate to truncate or rounded at the often somewhat clasping base, angled or tapered to a bluntly or more commonly sharply pointed tip, the margins mostly coarsely and bluntly toothed or scalloped and short-hairy, the upper surface moderately pubescent with short, curved hairs, the undersurface densely hairy with short, soft, curved hairs and often appearing grayish green. Inflorescences of solitary heads or small clusters at the branch tips, sometimes appearing as open, more or less flat-topped panicles, the heads with short to relatively long, bractless, densely hairy stalks. Heads radiate. Involucre 6–10 mm long, 9–15 mm in diameter, mostly broadly cup-shaped to slightly bell-shaped, the bracts in 2 or 3 somewhat unequal, overlapping series. Involucral bracts 14–22, loosely ascending or the tip somewhat reflexed or spreading, those of the outer series oblong-lanceolate to narrowly ovate, bluntly to sharply pointed at the tip, those of the inner series broadly obovate to broadly rhombic, sharply pointed at the tip, both series green with entire, hairy margins, the surfaces densely hairy and with minute, orange glands, with 5–11 fine nerves (these sometimes obscured by the hairs). Receptacle shallowly convex to low-conical but usually with a flat or slightly concave central portion, not noticeably elongating as the fruits mature, with chaffy bracts subtending the disc florets, these 4–5 mm long, linear to narrowly oblong, hairy, especially along the margins, relatively flat and not wrapped around the florets. Ray florets 5–11, pistillate (with a 2-branched style exserted from the short tube at flowering), the corolla 10–20 mm long, relatively broad, yellow to orangish yellow, the tube densely hairy and glandular, the ligule with the undersurface sparsely to moderately hairy, glandular, and with green veins, not persistent at fruiting. Disc florets numerous, staminate (the nonfunctional ovary slender and hairy), the corolla 3–4 mm long, brownish red to dark reddish purple, sparsely glandular and with the lobes usually densely short-hairy on the upper surface, not expanded at the base or persistent at fruiting. Style branches with the sterile tip very short and more or less truncate. Pappus of the ray and disc florets absent or occasionally of a few minute, papery teeth. Fruits 4.5–6.0 mm long, obovate to broadly oblong-obovate, strongly flattened, black, the inner surface densely hairy but difficult to observe because obscured by the basally fused sterile ovaries of the adjacent disc florets and chaffy bracts, the outer surface fused with the adjacent involucral bract (this tan to brown at fruiting), the whole assemblage dispersed as an intact unit. $2n=30$. June–October.

Scattered in the Ozark Division (New Mexico to Texas north to Kansas and Missouri). Upland prairies, glades, ledges and tops of bluffs, and openings of mesic to dry upland forests; also pastures, old fields, and roadsides.

This attractive species deserves to be cultivated more widely as an ornamental in sunny gardens.

66. Bidens L. (beggar-ticks)

Plants annual or perennial herbs (shrubs elsewhere). Stems erect to loosely ascending (except in submerged aquatics), unbranched or more commonly few- to numerous-branched, with several fine to coarse longitudinal lines or ridges, sometimes relatively strongly 4- or 5-angled, glabrous or sparsely to moderately pubescent. Leaves opposite or less commonly in whorls of 3 and sometimes also basal, variously sessile to long-petiolate, the bases slightly expanded and wrapping around the stem. Leaf blades simple or 1–3 times pinnately lobed, dissected, or compound (repeatedly dichotomously dissected in submerged leaves of *B. beckii*), variously shaped, the margins otherwise usually toothed, glabrous or variously hairy, not glandular. Inflorescences of solitary terminal heads or appearing as loose, open clusters or small panicles, occasionally some of the heads appearing axillary, the heads with short to long, bractless stalks or rarely the stalk with 1 or 2 inconspicuous, minute bracts. Heads radiate or discoid. Involucre broadly to narrowly cup-shaped to slightly bell-shaped, the bracts in 2(3) dissimilar overlapping series. Involucral bracts free at the base, those of the outer series 2–25, mostly as long as or longer than the others, variously shaped, green, glabrous or sparsely to moderately hairy, not glandular, usually inconspicuously 1- or 3-nerved; those of the inner series 5–8(–12), variously shaped, yellowish brown to yellowish green, more or less scalelike with narrow to broad, lighter, thinner margins, glabrous (inconspicuously hairy in *B. alba*), not glandular, with several to numerous conspicuous nerves or longitudinal lines. Receptacle flat or slightly convex, not elongating as the fruits mature, with chaffy bracts subtending the disc florets, these linear to narrowly oblong or narrowly lanceolate, relatively flat to somewhat concave, sometimes slightly wrapped around the florets toward the base. Ray florets absent or 1–8(–12), when present sterile (lacking stamens and style at flowering and with an ovary that is shorter and thinner than those of the disc florets, not developing into a fruit; pistillate elsewhere), the corolla inconspicuous or showy, 2–30 mm long, relatively broad, yellow or white, not persistent at fruiting. Disc florets 10–100(–150), perfect, the corolla usually 5-lobed (sometimes 4-lobed in *B. tripartita*), yellow to orangish yellow, not thickened at the base, not persistent at fruiting. Style branches with the sterile tip slightly to moderately elongate and tapered to a sharply pointed tip. Pappus of the disc florets of (1)2–4(–8) short or long awn(s) (these smooth or with sparse, upward- or downward-angled barbs), rarely absent, when present usually persistent at fruiting. Fruits 3–18 mm long, sometimes dissimilar in the same head (the outer ones grading into the inner ones) narrowly oblong to linear in outline, strongly flattened or 4-angled, not appearing curved or curled, more or less truncate at the tip, not beaked, the angles lacking wings or less commonly with slender wings, the surfaces often angled or with 1 or more longitudinal grooves, glabrous or variously hairy, tan to dark brown or nearly black, sometimes mottled, the wings (if present) usually pale, sometimes slightly shiny. More than 200 species, nearly worldwide.

As noted in the treatment of the genus *Coreopsis,* that genus is difficult to separate from *Bidens,* and further research undoubtedly will result in the revision of generic circumscriptions in the subtribe Coreopsidinae. In addition, some of the species of *Bidens* are difficult to distinguish from one another. Barkley (1986) advised studying several specimens when attempting to identify *Bidens* species.

1. Plants submerged aquatics with 2 dissimilar kinds of leaves, the submerged portion with the leaves repeatedly dichotomously dissected into narrowly linear (threadlike) segments, grading abruptly into the emergent leaves, which are unlobed; fruits more or less circular in cross-section; pappus 13–30 mm long
 . . . . 3. B. BECKII
1. Plants terrestrial or less commonly emergent aquatics (occasionally epiphytic), the leaves all unlobed or more or less similar in lobing or dissection; fruits flattened or 4-angled in cross-section; pappus absent or more commonly 2–8 mm long
    2. Ray florets present, the corolla white . . . . 1. B. ALBA
    2. Ray florets absent or, if present, then the corolla yellow

3. Leaf blades all unlobed or, if lobed or divided, then at most with 3–5 divisions, none of the divisions of the lobed leaves stalked at the base
   4. Outer series of involucral bracts loosely ascending to spreading; heads not nodding at fruiting; heads discoid or less commonly radiate, when ray florets are present then 1–5, the corollas 3–8 mm long . . . . 10. B. TRIPARTITA
   4. Outer series of involucral bracts spreading or more commonly reflexed; heads usually nodding at fruiting; heads radiate or rarely discoid, when ray florets are present then 6–8, the corollas (2–)10–30 mm long
      5. Chaffy bracts straw-colored to yellowish-tinged at the tip; ray florets (when present) with the corolla (2–)10–15 mm long; angles of the fruit pale and somewhat thickened or sometimes narrowly winged . . . . . 5. B. CERNUA
      5. Chaffy bracts reddish brown to orangish brown at the tip; ray florets (when present) with the corolla (10–)15–30 mm long; angles of the fruit not paler than the body, not thickened or winged . . . . 8. B. LAEVIS
3. At least some of the leaves with the blades 1–3 times pinnately or ternately lobed, divided, or compound, the divisions or leaflets more than 5 or, if only 3, then at least the central segment or leaflet usually tapered to a stalked base
   6. Heads radiate, the ray florets 5–8, showy, the corollas 10–25 mm long
      7. Fruits mostly 1.5–2.5 times as long as wide, wedge-shaped to obovate in outline; disc florets with the corolla 2.0–3.5 mm long . . . . 2. B. ARISTOSA
      7. Fruits mostly 2.5–5.0 times as long as wide, narrowly wedge-shaped to oblong-oblanceolate or nearly linear in outline; disc florets with the corolla 3–5 mm long . . . . 9. B. TRICHOSPERMA
   6. Heads discoid or less commonly radiate, when ray florets are present then 1–5, inconspicuous, the corollas 2–4 mm long
      8. All except sometimes the uppermost leaves with the blades 2 or 3 times pinnately lobed, divided, and/or compound into 7 to numerous ultimate lobes or segments; fruits linear in outline, strongly 4-angled (more or less square in cross-section) . . . . 4. B. BIPINNATA
      8. All of the leaves with the blades 1 time ternately or pinnately divided or compound into 3 or 5 divisions or leaflets; fruits wedge-shaped to narrowly oblong-obovate in outline, flattened, often slightly 3- or 4-angled (1 or both faces sometimes with a broad, low longitudinal angle or ridge)
         9. Involucral bracts of the outer series 3–5, the margins glabrous; disc florets 10–20, the corolla 1.5–2.0 mm long; pappus rarely absent, usually of 2 awns 0.5–2.0 mm long; fruits 3–6 mm long . . . . 6. B. DISCOIDEA
         9. Involucral bracts of the outer series 5–21, the margins with spreading hairs, at least toward the base; disc florets 20–100(–150), the corolla 2.5–4.0 mm long; pappus of 2 awns 2–7 mm long; fruits 5–12 mm long
            10. Involucral bracts of the outer series mostly 5–8 . . . . . . 7. B. FRONDOSA
            10. Involucral bracts of the outer series mostly 10–21 . . . . 11. B. VULGATA

1. Bidens alba (L.) DC. **var. radiata** (Sch. Bip.) Ballard
   B. pilosa L. var. *radiata* (Sch. Bip.) J.A. Schmidt

Pl. 274 e, f; Map 1149

Plants annual, terrestrial, with taproots. Stems 30–150 cm, erect or ascending, the upper portion usually 4-angled, glabrous (sparsely hairy elsewhere). Leaves all more or less similar, short- to long-petiolate, opposite, the blade 2–10 cm long, mostly broadly ovate to ovate-triangular in outline, deeply 1 time ternately or pinnately divided or compound into 3–5(–7) segments or leaflets, these lanceolate to ovate, angled or short-tapered (rarely rounded to shallowly cordate) to a stalked or stalklike, winged base, tapered to a sharply pointed

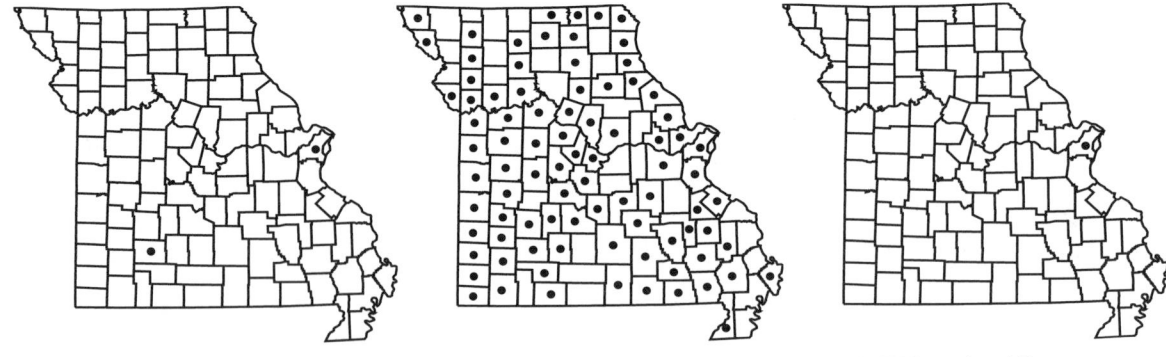

1149. Bidens alba   1150. Bidens aristosa   1151. Bidens beckii

tip, the margins sharply toothed, the surfaces glabrous or the undersurface sparsely short-hairy, especially along the veins. Inflorescences of solitary terminal heads or appearing in loose, open clusters, the heads radiate, not nodding at fruiting. Involucre with the outer series of 8–12 bracts 2–5 mm long, ascending to loosely ascending, narrowly oblong to oblanceolate, the margins with spreading hairs, the outer surface glabrous or sparsely short-hairy; the inner series of 7–13 bracts 4–7 mm long, lanceolate to oblong-oblanceolate, inconspicuously pubescent with short, appressed hairs. Chaffy bracts narrowly oblong, usually purplish-tinged at the tip. Ray florets present, 5–8, the corolla showy, 7–15 mm long, white. Disc florets 20–40(–80), the corollas 3–5 mm long, yellow. Pappus absent or more commonly of 2 or 3 awns 2–4 mm long, these with downward-pointed barbs, usually erect at fruiting. Fruits 4–12 mm long, linear, more or less flattened (outer florets) or 4-angled (inner florets) in cross-section, not winged, the faces each with a pair of longitudinal grooves, reddish brown to black, with short, upward angled, usually pustular-based hairs. $2n=48$. July–October.

Introduced, uncommon, known thus far only from the city of St. Louis (southwestern U.S. and Florida; Mexico, Central America, South America, Caribbean Islands; introduced sporadically farther north in the U.S., also Canada, Europe, Asia, Africa). Railroads.

This species was first reported for Missouri by Mühlenbach (1979, as *B. pilosa*). The present work follows the careful study by Ballard (1986), who separated *B. alba*, *B. odorata*, and *B. pilosa* based on a suite of subtle morphological characters, differences in ploidy, and phytochemistry. True *B. pilosa* tends to have fewer outer-series involucral bracts, absent or inconspicuous ray florets, and fewer pappus awns, among other differences, and also is a hexaploid ($2n=72$) taxon.

2. **Bidens aristosa** (Michx.) Britton (tickseed sunflower)
   B. polylepis S.F. Blake
   B. aristosa var. *fritcheyi* Fernald
   B. aristosa var. *mutica* (A. Gray) Gatt. ex Fernald
   B. aristosa f. *mutica* (A. Gray) Wunderlin
   B. aristosa var. *retrorsa* (Sherff) Wunderlin
   B. polylepis var. *retrorsa* Sherff
   B. aristosa f. *involucrata* (Nutt.) Wunderlin
   Pl. 273 g–i; Map 1150

Plants annual or biennial, terrestrial, sometimes with taproots. Stems 15–100(–150) cm, erect or ascending, glabrous or sparsely pubescent with short, ascending hairs. Leaves all more or less similar, short- to less commonly long-petiolate, opposite, the blade 2–15 cm long, lanceolate to ovate or triangular-ovate in outline, all but occasionally those of the uppermost leaves 1 or 2 times pinnately deeply divided or compound into 3–7 segments or leaflets, these linear to lanceolate or narrowly elliptic, angled or more commonly tapered at the base, the middle one frequently with a stalklike base, tapered to a sharply pointed tip, the margins sharply and finely to coarsely toothed, also minutely hairy, the surfaces glabrous or sparsely hairy. Inflorescences of solitary terminal heads or appearing in loose, open clusters or small panicles, the heads radiate, not nodding at fruiting. Involucre with the outer series of 8–25 bracts 4–12(–25) mm long, spreading to more commonly reflexed, mostly not leaflike, linear, the margins entire but usually with short, spreading hairs, the outer surface glabrous or with short, spreading hairs; the inner series of 6–8(–12) bracts 4–12 mm long, narrowly lanceolate to narrowly ovate, glabrous. Chaffy bracts narrowly lanceolate, usually with broad, yellowish margins and tip, occasionally purplish-tinged at the tip. Ray florets 5–8, the corolla showy, 10–25 mm long, yellow. Disc florets 20–80(–120), the corollas

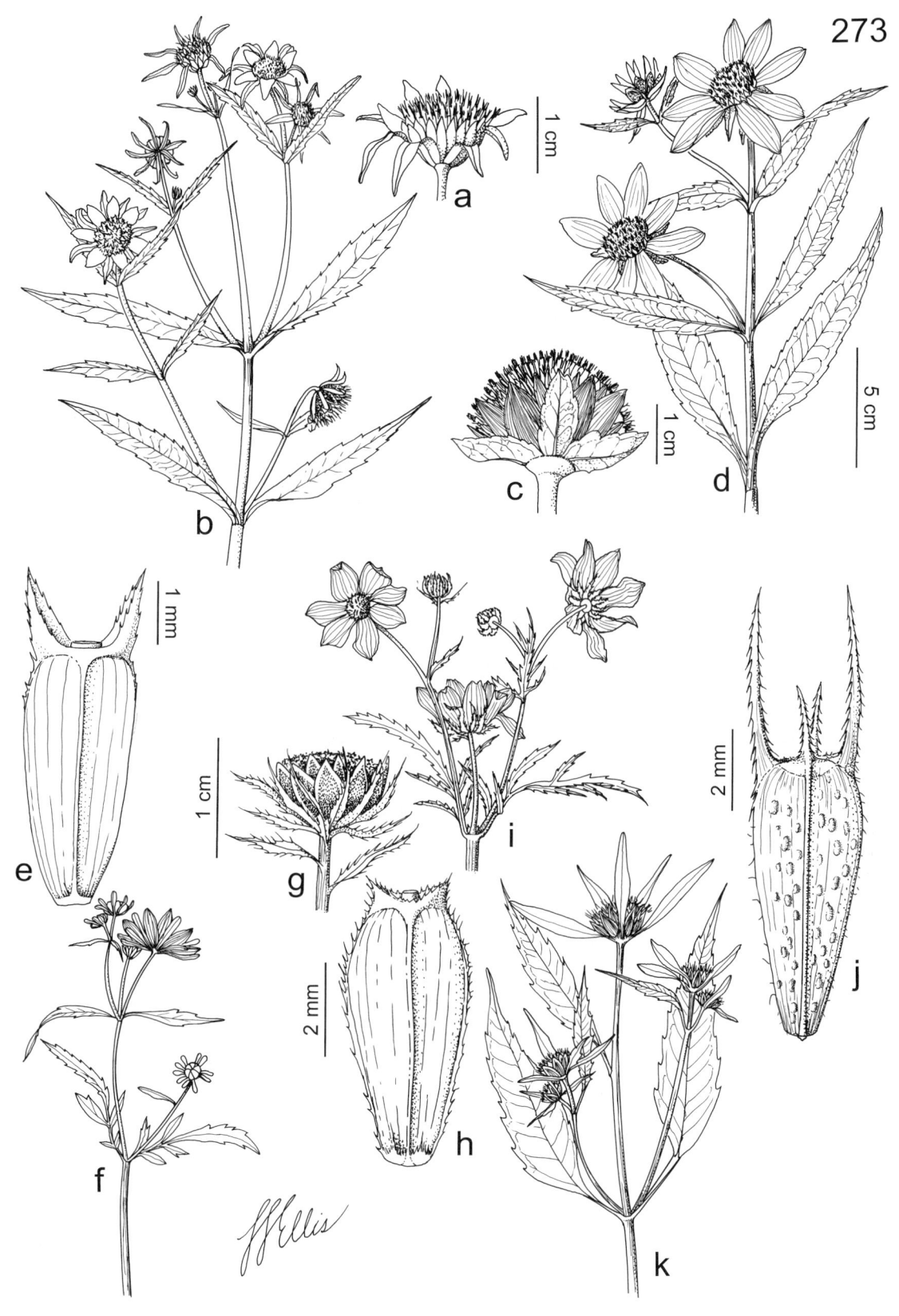

Plate 273. Asteraceae. *Bidens cernua*, **a)** head, **b)** habit. *Bidens laevis*, **c)** head, **d)** habit. *Bidens trichosperma*, **e)** fruit, **f)** habit. *Bidens aristosa,* **g)** head, **h)** fruit, **i)** habit. *Bidens tripartita,* **j)** fruit, **k)** habit.

2.0–3.5 mm long, yellow. Pappus absent, reduced to a minute crown with 2 teeth, or more commonly of 2 awns variously 0.5–4.0 mm long, these with upward- or downward-pointed barbs, erect to somewhat spreading at fruiting. Fruits 4–8 mm long, wedge-shaped to obovate (mostly 1.5–2.5 times as long as wide), more or less flattened and somewhat 3-angled in cross-section, the angles roughened with short, stiff, ascending, pustular-based hairs, the faces sometimes with few to several faint longitudinal lines, dark brown to black, sparsely to moderately pubescent with short, fine, mostly appressed hairs. 2n=24. August–October.

Scattered to common throughout the state (eastern U.S. west to Minnesota, Colorado, and New Mexico; Canada). Bottomland prairies, upland prairies, bottomland forests, margins of ponds and lakes, sloughs, and fens; also crop fields, fallow fields, railroads, roadsides, and open, disturbed areas.

Bidens aristosa is highly variable morphologically. The present treatment accepts a broader species circumscription for the species than that found in much of the North American botanical literature. Most authors have treated the complex as comprising two species, *B. aristosa* and *B. polylepis,* with a number of additional infraspecific taxa (Sherff, 1955). *Bidens polylepis* has been said to differ from *B. aristosa* in its involucral bracts of the outer series more numerous, larger, and hairier, as well as its slightly smaller fruits. However, most authors who have studied the group have conceded that there is strong overlap between the two extremes (Steyermark, 1963). In a short note preliminary to an account of Illinois Asteraceae, Wunderlin (1972) reduced the main varieties of *B. polylepis* to forms of *B. aristosa* var. *retrorsa* (although current rules would require that the as-yet unpublished name var. *polylepis* be used instead). Subsequently, Lipscomb and Smith (1977) completed a detailed numerical analysis of morphological variation in some Arkansas populations, concluding that there is no rational basis for the continued recognition of two entities at any taxonomic level. Their treatment is followed here. Occasional plants with the pappus awns having downward-pointed (vs. upward-pointed in f. *aristosa*) barbs have been called f. *fritcheyi,* var. *fritcheyi,* or var. *retrorsa,* whereas plants with the awns absent or highly reduced have been called f. *mutica* or var. *mutica.*

3. Bidens beckii Torr. ex Spreng. (water marigold)

Megalodonta beckii (Torr. ex Spreng.) Greene

Pl. 284 g, h; Map 1151

Plants perennial herbs, submerged aquatics (except for the stem tips and heads). Stems 50–200 cm or more long, glabrous. Leaves of two dissimilar kinds, both kinds sessile or nearly so, glabrous, the submerged leaves opposite or sometimes in whorls of 3, the blade 1.5–3.5 cm long, fan-shaped to nearly circular in outline, repeatedly dichotomously dissected into narrowly linear (threadlike) segments, grading abruptly into the emergent leaves, which are 0.5–4.0 cm long, lanceolate to narrowly oblong-elliptic, unlobed, tapered to a sharply pointed tip, the margins finely to coarsely and sharply toothed. Inflorescences of solitary terminal heads, occasionally with 1 or a few additional heads from the uppermost leaf axils, the heads radiate, not nodding at fruiting. Involucre with the outer series of 5 or 6 bracts 5–8 mm long, spreading to loosely ascending, oblong to obovate, glabrous; the inner series of 7–9 bracts 7–10 mm long, ovate, glabrous. Chaffy bracts narrowly lanceolate, usually purplish-tinged at the tip. Ray florets present, 6–8, the corolla showy, 10–15 mm long, yellow. Disc florets 10–40, the corollas 5–6 mm long, light yellow. Pappus of (2–)3–6 awns 13–30 mm long (elongating as the fruit matures), these with downward-pointed barbs toward the tip, more or less spreading at fruiting. Fruits 10–15 mm long, linear, more or less circular in cross-section, brownish yellow to yellowish brown, the surfaces smooth, glabrous. 2n=36. July–October.

Uncommon, known only from a single historical collection from the city of St. Louis (northern U.S. south to Oregon, Illinois, and Maryland; Canada). Lakes, submerged aquatics.

Steyermark (1963) discussed the ambiguity in whether George Engelmann's 1846 St. Louis record was collected in Missouri or whether it actually originated from just across the Mississippi River in the American Bottoms area of St. Clair County, Illinois. In the absence of data to the contrary, the record is accepted here. Regardless of its historical status, the species appears to have become extirpated from the St. Louis metropolitan region. This unusual species is relatively uncommon throughout its range and apparently disappears from ponds and lakes quickly following hydrological disturbances. Roberts (1985) noted that the species produces specialized modified stem segments known as turions that facilitate vegetative dispersal of plants and also act as a mechanism for overwintering. He discussed that these are morphologically similar to turions produced in another genus of submerged aquatics, *Myriophyllum* (Haloragaceae). Vegetatively, however, *B. beckii* resembles more closely the genus *Ceratophyllum* (Ceratophyllaceae), which is common in still waters in Missouri. Members of *Ceratophyllum* tend to have the

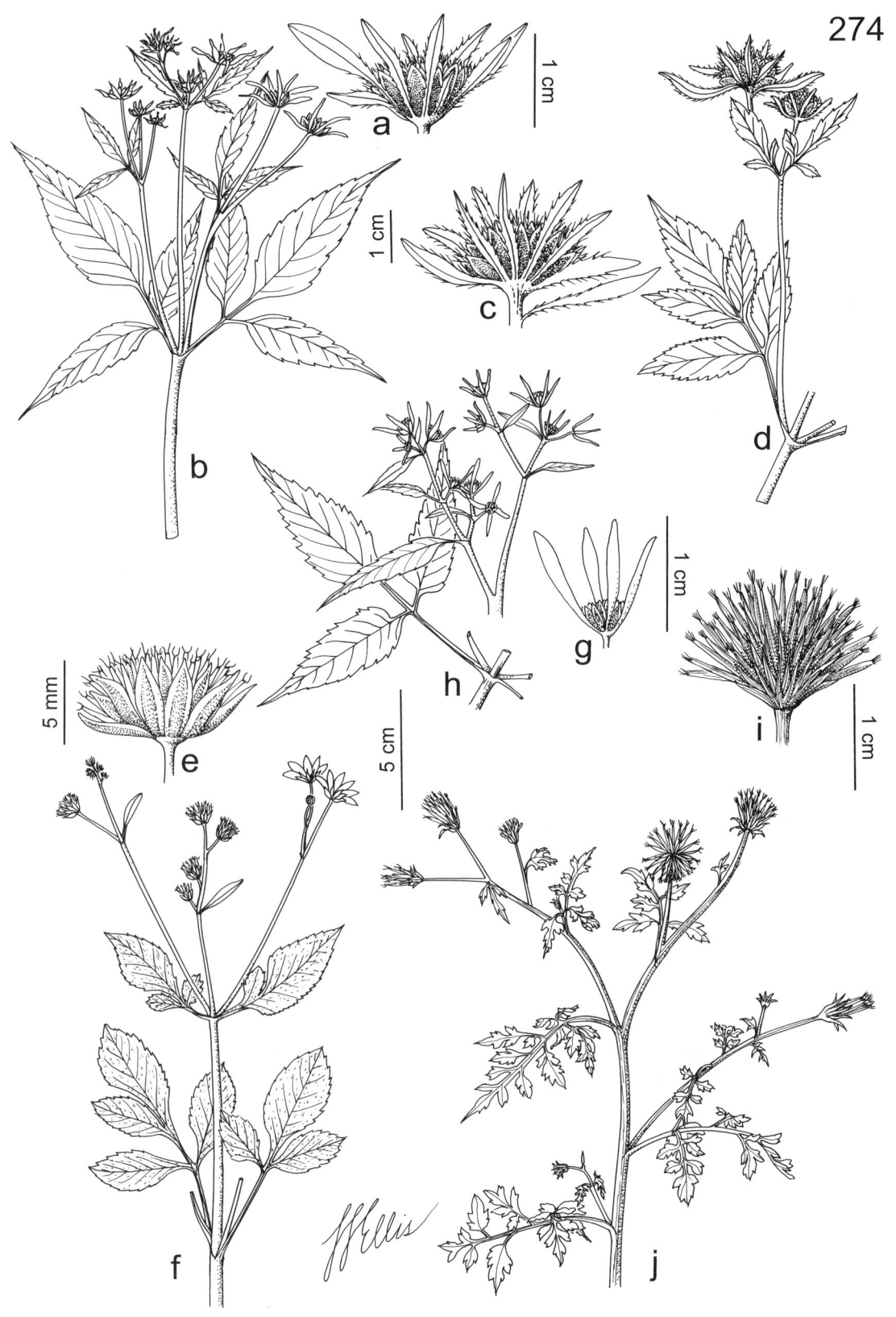

Plate 274. Asteraceae. *Bidens frondosa*, **a)** head, **b)** habit. *Bidens vulgata*, **c)** head, **d)** habit. *Bidens alba*, **e)** head, **f)** habit. *Bidens discoidea*, **g)** head, **h)** habit. *Bidens bipinnata*, **i)** head, **j)** habit.

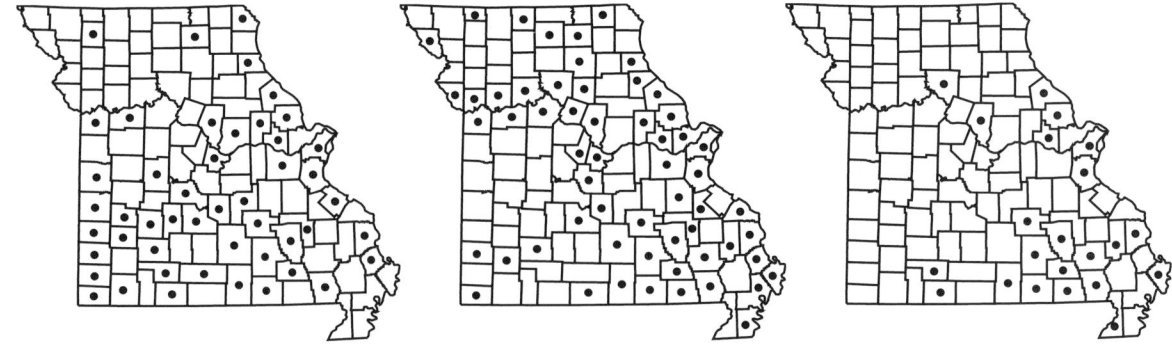

1152. Bidens bipinnata   1153. Bidens cernua   1154. Bidens discoidea

dichotomously parted leaves divided fewer times than in *B. beckii,* and the margins tend to have minute teeth. *Ceratophyllum* also produces minute axillary flowers that develop into individual achenes with a spiny tip and margins.

Roberts (1985) followed Sherff (1955) in segregating the species into its own genus, *Megalodonta* Greene, based on its morphological adaptations to an aquatic lifestyle, differences in phytochemistry, and an unusual chromosome number. Preliminary molecular phylogenetic studies (Kimball and Crawford, 2004) did not uphold this generic segregation, although they failed to resolve details of the relationships among *B. beckii* and its relatives in *Bidens* and *Coreopsis*.

4. Bidens bipinnata L. (Spanish needles)

Pl. 274 i, j; Map 1152

Plants annual, terrestrial, usually with taproots. Stems 15–60(–150) cm, erect or ascending, glabrous or sparsely to moderately pubescent with minute, more or less spreading hairs. Leaves all more or less similar, short- to long-petiolate, opposite, the blade 3–20 cm long, lanceolate to ovate in outline, all except rarely those of the uppermost leaves 2 or 3 times pinnately lobed, divided, and/or compound into 7 to numerous ultimate lobes or segments, these lanceolate to obovate, mostly angled or tapered at the base, without a stalklike base, tapered to a sharply pointed tip, the margins otherwise entire or coarsely few-toothed, sometimes minutely hairy, the surfaces glabrous or the undersurface sparsely and minutely hairy along the main veins. Inflorescences of solitary terminal heads or appearing in loose, open clusters, the heads discoid or appearing discoid, not nodding at fruiting. Involucre with the outer series of 7–10 bracts 3–5 mm long, ascending to more commonly spreading, mostly not leaflike, linear to narrowly oblong-oblanceolate, the margins entire but usually with minute, ascending hairs, the outer surface glabrous; the inner series of 8–12 bracts 4–9 mm long, narrowly lanceolate to linear, glabrous. Chaffy bracts narrowly oblong to linear (elongating somewhat as the fruits mature), usually with a minute fringe of white hairs at the otherwise greenish tip. Ray florets absent or 1–5, the corolla inconspicuous, 2–4 mm long, yellow. Disc florets 12–27, the corollas 1–2 mm long, yellow. Pappus of 2–4 awns 1–4 mm long, these with downward-pointed barbs, erect to slightly spreading at fruiting. Fruits 10–18 mm long, linear, strongly 4-angled (more or less square in cross-section), the angles glabrous or with a few minute, stiff, ascending hairs (these denser in immature fruits), the faces each with a pair of slender longitudinal grooves, dark brown to black, often somewhat mottled, glabrous. $2n=24, 72.$ August–October.

Scattered, mostly south of the Missouri River (eastern U.S. west to Nebraska and Arizona; Canada, Mexico, Central America, South America, Asia, Madagascar, Malesia, Australia). Upland prairies, glades, openings of mesic to dry upland forests, bottomland forests, and banks of streams and rivers; also ditches, pastures, fallow fields, gardens, railroads, roadsides, and open, disturbed areas.

The natural range of this species is not well understood as it long ago became established as a weed in many parts of the world. Sherff (1955) speculated that the native range prior to the period of European expansionism was in the eastern United States and eastern Asia. The distributional summary above does not attempt to discriminate native vs. adventive components of the overall range of the species.

5. Bidens cernua L. (sticktight, nodding bur marigold)

B. cernua var. *elliptica* Wiegand
B. cernua var. *integra* Wiegand

Pl. 273 a, b; Map 1153

Plants annual, terrestrial or occasionally

emergent aquatics, sometimes with taproots. Stems 10–80(–250) cm, erect to loosely ascending, sometimes from a spreading base, occasionally rooting at the lower nodes, glabrous or rarely somewhat roughened with sparse, stiff, ascending hairs. Leaves all more or less similar, sessile, opposite or rarely the lowermost leaves in whorls of 3, the blade 2–20 cm long, oblong-lanceolate to oblanceolate in outline, unlobed, narrowed or tapered at the base, tapered to a sharply pointed tip, the margins sharply and finely to coarsely toothed, rarely nearly entire, also minutely roughened, the surfaces glabrous. Inflorescences of solitary terminal heads or appearing in loose, open clusters, the heads radiate or rarely discoid, usually nodding at fruiting. Involucre with the outer series of 5–9 bracts 6–30(–45) mm long, spreading to more commonly reflexed, mostly leaflike, narrowly lanceolate to oblong-lanceolate or linear, the margins entire but usually minutely roughened, the outer surface glabrous or sparsely short-hairy toward the base; the inner series of 6–9 bracts 4–10 mm long, lanceolate to narrowly ovate or ovate, glabrous. Chaffy bracts narrowly oblong, straw-colored to yellowish-tinged at the tip. Ray florets rarely absent, when present then 6–8, the corolla usually showy, (2–)10–15 mm long, yellow. Disc florets 40–100(–150), the corollas 3–4 mm long, yellow to orangish yellow. Pappus of (2–)4 awns mostly 2–4 mm long, these with downward-pointed barbs, erect at fruiting. Fruits 4–8 mm long, narrowly wedge-shaped, more or less flattened and somewhat 4-angled in cross-section, the angles pale and somewhat thickened or sometimes narrowly winged, the faces sometimes with several faint longitudinal lines, dark brown to purplish black, glabrous or with sparse, short, pustular-based hairs. $2n=24, 48$. August–October.

Scattered nearly throughout the state but absent or uncommon in the Unglaciated Plains Division (U.S. [including Alaska], Canada, Europe, Asia). Banks of streams and rivers, margins of ponds and lakes, sloughs, swamps, bottomland forests, and fens; also ditches and railroads.

Steyermark (1963) noted that the species has been used medicinally to treat urinary tract infections. Rare plants with discoid heads have been called f. *discoidea* (Wimm. & Grab.) Briq. & Cavill. A number of infraspecific taxa have been named, but these appear to represent either ecological variants or minor variants unworthy of formal taxonomic recognition.

6. Bidens discoidea (Torr. & A. Gray) Britton
(few-bracted bur marigold)
Pl. 274 g, h; Map 1154
Plants annual, terrestrial or epiphytic, usually with taproots. Stems 15–80(–150) cm, erect or ascending, glabrous. Leaves all more or less similar, short- to long-petiolate, opposite, the blade 3–15 cm long, broadly ovate-triangular in outline, all except rarely those of the uppermost leaves 1 time ternately (pinnately) compound with 3 discrete leaflets, these lanceolate to obovate, mostly angled at the base, each with a well-developed stalk, tapered to a sharply pointed tip, the margins usually sharply and relatively coarsely toothed, sometimes minutely hairy, the surfaces glabrous or the undersurface sparsely to moderately pubescent with minute, fine hairs. Inflorescences of solitary terminal heads or appearing in loose, open clusters or small panicles, the heads discoid, sometimes nodding or spreading at fruiting. Involucre with the outer series of 3–5 bracts 8–25 mm long, ascending to spreading, leaflike, linear to narrowly oblong-oblanceolate or narrowly oblanceolate, the margins entire and glabrous (rarely with a few hairs toward the base), the outer surface glabrous or sparsely and minutely hairy, especially toward the base; the inner series of 5–7 bracts 4–7 mm long, oblong to narrowly lanceolate, glabrous. Chaffy bracts narrowly oblong to oblong, usually with broad, yellowish margins but purplish-tinged at the tip. Ray florets absent. Disc florets 10–20, the corollas 1.5–2.0 mm long, yellow. Pappus rarely absent, usually of 2 awns 0.5–2.0 mm long, these with upward-pointed barbs, erect to slightly spreading at fruiting. Fruits 3–6 mm long, wedge-shaped to narrowly oblong-obovate, often slightly 3- or 4-angled (1 or both faces sometimes with a broad, low longitudinal angle or ridge), the angles with minute, stiff, ascending hairs, the faces reddish brown to black, moderately to densely pubescent with fine, more or less appressed hairs. $2n=24$. August–October.

Scattered, mostly in the southern half of the Ozark Division and in counties bordering the Mississippi and Missouri Rivers (eastern U.S. west to Minnesota and Texas; Canada). Oxbows, swamps, bottomland forests, and margins of sinkhole ponds, sometimes epiphytic (see below).

This species frequently is found growing on mossy hummocks, rotting logs, and stumps, or as an epiphyte on the bases of *Cephalanthus occidentalis* (buttonbush, Rubiaceae), *Nyssa aquatica* (water tupelo, Cornaceae), *Populus heterophylla* (swamp cottonwood, Salicaceae), and *Taxodium distichum* (bald cypress, Cupressaceae). Steyermark (1963) noted two records from the Unglaciated Plains Division, collected at oxbow lakes in Bates and Vernon Counties. The specimens supporting these reports could not be located during the present research.

1155. Bidens frondosa   1156. Bidens laevis   1157. Bidens trichosperma

7. Bidens frondosa L. (beggar-ticks, sticktight)
Pl. 274 a, b; Map 1155

Plants annual, terrestrial, usually with taproots. Stems 15–80(–120) cm, erect or ascending, glabrous or inconspicuously pubescent with sparse, cobwebby hairs, mostly at the nodes. Leaves all more or less similar, short- to more commonly long-petiolate, opposite, the blade 3–12 cm long, broadly ovate-triangular in outline, all except rarely those of the uppermost leaves 1 time ternately or pinnately divided or compound with 3–5 segments or discrete leaflets, these lanceolate to narrowly ovate, angled or tapered at the base, each with a well-developed stalk (this sometimes narrowly winged), tapered to a sharply pointed tip, the margins usually sharply and finely to coarsely toothed, sometimes minutely hairy, the surfaces glabrous or the undersurface sparsely to moderately pubescent with minute, fine hairs. Inflorescences of solitary terminal heads or appearing in loose, open clusters or small panicles, the heads discoid or apparently discoid, rarely nodding at fruiting. Involucre with the outer series of 5–8 bracts 5–25(–50) mm long, ascending to spreading, leaflike, linear to narrowly oblong-oblanceolate or narrowly oblanceolate, the margins entire but with short, spreading hairs, at least toward the base, the outer surface glabrous or sparsely and minutely hairy, especially toward the base; the inner series of 6–12 bracts 5–9 mm long, oblong to narrowly ovate or ovate, glabrous. Chaffy bracts narrowly oblong to oblong-lanceolate, with narrow to broad, yellowish margins but sometimes purplish-tinged at the tip. Ray florets absent or 1–5, the corolla inconspicuous, 2–4 mm long, yellow. Disc florets 20–100(–150), the corollas 2.5–4.0 mm long, yellow. Pappus of 2 awns 2–7 mm long, these with downward-pointed barbs (upward-pointed elsewhere), erect to somewhat spreading at fruiting. Fruits 5–12 mm long, wedge-shaped to oblong-obovate, often slightly 3- or 4-angled (1 or both faces sometimes with a broad, low longitudinal angle or ridge), the angles with minute, stiff, usually ascending hairs, the faces dark brown to black, moderately to densely pubescent with fine, more or less appressed hairs. $2n=24$, 48, 72. August–October.

Scattered nearly throughout the state (U.S. [including Alaska], Canada; introduced in Europe). Banks of streams and rivers, bottomland forests, margins of ponds, lakes, and sinkhole ponds, swamps, sloughs, marshes, and fens; also ditches, fallow fields, railroads, roadsides, and moist, open, disturbed areas.

This variable species can be difficult to distinguish from *B. vulgata*.

8. Bidens laevis (L.) Britton, Sterns & Poggenb. (showy bur marigold)
Pl. 273 c, d; Map 1156

Plants annual (sometimes perennial farther south), terrestrial or occasionally emergent aquatics, sometimes with taproots. Stems 20–80(–150) cm, erect to loosely ascending, sometimes from a spreading base, occasionally rooting at the lower nodes, glabrous. Leaves all more or less similar, sessile, opposite or rarely the lowermost leaves in whorls of 3, the blade 2–15 cm long, oblong-lanceolate to oblanceolate in outline, less commonly obovate or linear, unlobed, narrowed or tapered at the base, tapered to a sharply pointed tip, the margins usually sharply and finely to coarsely toothed, rarely nearly entire, also minutely roughened, the surfaces glabrous. Inflorescences of solitary terminal heads or appearing in loose, open clusters, the heads radiate (rarely discoid elsewhere), usually nodding at fruiting. Involucre with the outer series of 5–9 bracts 6–16 mm long, spreading to more commonly reflexed, somewhat leaflike, narrowly lanceolate to narrowly oblong or linear, the margins entire but usually minutely roughened, the outer surface glabrous or sparsely short-hairy toward the base; the inner series of 6–9 bracts 4–10 mm long, lanceolate to narrowly ovate or ovate, glabrous. Chaffy bracts narrowly oblong, reddish brown to orangish brown at the tip. Ray florets 6–8, the corolla showy, (10–)15–30 mm long, yellow. Disc florets 40–100(–150), the corollas 3–4 mm long, yellow to orangish yellow. Pappus of (2–)4 awns mostly 2–4 mm long, these with downward-pointed barbs, erect at fruiting. Fruits 4–8 mm long, narrowly wedge-shaped, more or less flattened and somewhat 4-angled in cross-section, the angles not pale and not thickened or winged, the faces sometimes with several faint longitudinal lines, dark brown to purplish black, glabrous or with sparse, short, pustular-based hairs. $2n=22$, 24. August–October.

Uncommon, known thus far only from a single historical collection from Dunklin County (southern U.S. north to California, Missouri, and Maine; Mexico, Central America, South America; introduced in Hawaii). Habitat unknown, but presumably banks of streams and rivers or bottomland forests.

Steyermark (1963) noted that a number of additional specimens originally determined as *B. laevis* had been redetermined during his studies as the closely related *B. cernua*. He also discussed that the two species tend to intergrade and perhaps should be considered varieties of a single species.

9. Bidens trichosperma (Michx.) Britton
Coreopsis trichosperma Michx.
B. coronata (L.) Britton (1913), not *B. coronata* Fisch. ex Colla (1834)
Pl. 273 e, f; Map 1157

Plants annual or biennial, terrestrial, sometimes with taproots. Stems 15–80(–150) cm, erect or ascending, glabrous or nearly so. Leaves all more or less similar, short-petiolate, opposite, the blade 2–12 cm long, mostly ovate or triangular-ovate in outline, 1 or 2 times deeply pinnately divided or compound into 3–7(–9) segments or leaflets, these linear to more commonly lanceolate to oblanceolate, angled or tapered at the base, the middle one frequently with a stalklike base, tapered to a sharply pointed tip, the margins usually sharply and finely to coarsely toothed, rarely entire or nearly so, the surfaces glabrous or sparsely hairy. Inflorescences of solitary terminal heads or appearing in loose, open clusters or small panicles, the heads radiate, not nodding at fruiting. Involucre with the outer series of (6–)8(–11) bracts 4–8 mm long, ascending to spreading, somewhat leaflike, linear to narrowly oblanceolate, the margins entire but sometimes with minute, ascending hairs, the surfaces glabrous; the inner series of 6–8 bracts 4–8 mm long, oblong-elliptic, glabrous. Chaffy bracts narrowly oblong-lanceolate, usually with broad, yellowish margins and tip, occasionally purplish-tinged at the tip. Ray florets (6–)8(–9), the corolla showy, 10–25 mm long, yellow. Disc florets 40–80(–120), the corollas 3–5 mm long, yellow. Pappus of 2 awns 1–4 mm long (rarely shorter and scalelike), these with upward barbs, erect to somewhat spreading at fruiting. Fruits 4–8 mm long, narrowly wedge-shaped to oblong-oblanceolate or nearly linear (mostly 2.5–5.0 times as long as wide), more or less flattened and somewhat 4-angled in cross-section, the angles roughened with short, stiff, ascending, pustular-based hairs, the faces sometimes with 1 or few faint longitudinal line(s), dark brown to black, sparsely to moderately pubescent with short, fine, mostly appressed hairs. $2n=24$. August–October.

Introduced, known thus far only from the city of St. Louis (eastern [mostly northeastern] U.S. west to Minnesota, Nebraska, and Mississippi). Railroads.

This species was first reported for Missouri by Mühlenbach (1979).

10. Bidens tripartita L. (beggar-ticks, swamp beggar-ticks)
B. comosa (A. Gray) Wiegand
B. connata Muhl. ex Willd.
B. connata var. *petiolata* (Nutt.) Farw.
Pl. 273 j, k; Map 1158

Plants annual, terrestrial, with taproots. Stems 10–90(–200) cm, erect or ascending, sometimes straw-colored or purplish-tinged, glabrous. Leaves all more or less similar, sessile or short- to long-petiolate, the petiole often partially winged, opposite, the blade 2–8(–15) cm long, lanceolate to elliptic (in unlobed leaves) or broadly ovate to ovate-triangular (in divided leaves) in outline, mostly unlobed, but occasionally those of the larger leaves deeply and sometimes irregularly 1 time ternately or pinnately lobed into 3(5) lobes or segments or leaflets, these lanceolate to elliptic, angled or tapered but not stalked at the base, tapered to a sharply pointed tip, the margins sharply and usually coarsely toothed, rarely nearly entire, the surfaces glabrous or sparsely to moderately pubescent with fine, short hairs. Inflorescences of solitary terminal heads or appearing in loose, open clusters, the heads discoid or appearing discoid, not nodding at fruiting. Involucre with the outer series of 4–9 bracts 7–35(–70) mm long, loosely ascending to spreading, mostly leaflike, oblanceolate to elliptic,

1158. Bidens tripartita   1159. Bidens vulgata   1160. Coreopsis grandiflora

oblong-lanceolate, or linear, the margins entire or finely toothed, often also with spreading hairs, the outer surface glabrous or sparsely short-hairy toward the base; the inner series of 7 or 8(–12) bracts (4–)7–12 mm long, narrowly ovate to ovate, glabrous. Chaffy bracts narrowly lanceolate, usually purplish-tinged at the tip. Ray florets absent or less commonly present, if present then 1–5, the corolla inconspicuous, 3–8 mm long, yellow. Disc florets 20–40(–80), the corollas 3–4 mm long, sometimes only 4-lobed, yellow to orangish yellow. Pappus absent or more commonly of (2–)3 or 4 awns mostly 2–3 mm long, these with upward- or more commonly downward-pointed barbs, erect or somewhat spreading at fruiting. Fruits 3–11 mm long, linear to narrowly wedge-shaped, more or less flattened and somewhat 3- or 4-angled in cross-section, not winged, the faces each with a longitudinal nerve, dark brown to purplish black, glabrous or with sparse, short, upward angled, fine hairs, sometimes also with minute tubercles. $2n=24, 48, 60, 72$. July–October.

Scattered nearly throughout the state but absent or uncommon in the southern portion of the Ozark Division (U.S., Canada, Europe, Asia, Africa; introduced in the Pacific Islands, Australia). Banks of streams and rivers, margins of ponds and lakes, sloughs, swamps, bottomland forests, and fens; also ditches and railroads.

The *B. tripartita* complex comprises both Old World plants and those native to the New World that have been segregated variously under the names *B. comosa* and *B. connata,* as well as a number of infraspecific names (Hall, 1967; Cronquist, 1980). Morphological differences between populations that various authors have used to separate taxa include stem color, leaf divisions, number of disc corolla lobes, relative size and shape of the involucral bracts, and pappus details, but these do not seem to correlate well enough to allow the segregation of discrete species or varieties based on present knowledge. However, not all of the plants currently called *B. tripartita* may represent the same biological entity, as evidenced by the broad geographic range, great morphological variation, and multiple ploidy levels reported in the literature. Crowe and Parker (1981) studied this and related species of *Bidens* in Ontario and suggested that populations they called *B. connata* probably arose through past hybridization between *B. cernua* and *B. frondosa,* which had regained fertility by switching to an apogamous life cycle. Further studies are needed involving genetic variation in plants from throughout the range of the complex.

11. Bidens vulgata Greene (beggar-ticks, sticktight)
   B. vulgata f. *puberula* (Wiegand) Fernald
   Pl. 274 c, d; Map 1159

Plants annual, terrestrial, usually with taproots. Stems 30–150 cm, erect or ascending, glabrous or inconspicuously pubescent with sparse, cobwebby hairs, mostly at the nodes. Leaves all more or less similar, short- to more commonly long-petiolate, opposite, the blade 3–15 cm long, broadly ovate-triangular in outline, all except rarely those of the uppermost leaves 1 time ternately or pinnately divided or compound with 3–5 segments or discrete leaflets, these lanceolate to narrowly ovate, angled or tapered at the base, each with a well-developed stalk (this sometimes narrowly winged), tapered to a sharply pointed tip, the margins usually sharply and finely to coarsely toothed, sometimes minutely hairy, the surfaces glabrous or the undersurface sparsely to moderately pubescent with minute, fine hairs. Inflorescences of solitary terminal heads or appearing in loose, open clusters or small panicles, the heads discoid or apparently discoid, rarely nodding at fruiting. Involucre with the outer series of 10–21 bracts 10–20(–40) mm long, ascending to spreading, leaflike, linear to narrowly oblong-oblanceolate or narrowly oblanceolate, the margins entire but with short, spreading hairs, at least toward the base, the outer surface glabrous or sparsely to moderately pubescent with minute, fine hairs, especially toward the base; the inner series of 8–14 bracts 5–9 mm long, oblong to narrowly ovate or ovate, glabrous. Chaffy bracts narrowly oblong to oblong-lanceolate, with narrow to broad, yellowish margins but sometimes purplish-tinged at the tip. Ray florets absent or 1–5, the corolla inconspicuous, 2–4 mm long, yellow. Disc florets 40–100(–150), the corollas 2.5–4.0 mm long, yellow. Pappus of 2 awns 3–7 mm long, these with downward-pointed barbs, erect to somewhat spreading at fruiting. Fruits 5–12 mm long, wedge-shaped to oblong-obovate, 1 or both faces sometimes with a fine longitudinal nerve or ridge, the angles with minute, stiff, usually ascending hairs, the faces dark brown to black, sometimes with minute, purplish spots, glabrous or sparsely pubescent with fine, more or less appressed hairs. $2n=24, 48$. August–October.

Scattered to uncommon nearly throughout the state (U.S. [except some southwestern states], Canada; introduced in Europe). Banks of streams and rivers, margins of ponds and lakes, and sloughs; also ditches, fallow fields, railroads, roadsides, and moist, open, disturbed areas.

This variable species can be difficult to distinguish from *B. frondosa*. It tends to be more robust than that species. Some botanists separate plants with more finely divided leaves occurring to the north of Missouri as var. *schizantha* Lunell, and plants with hairy leaves and outer involucral bracts (which occur sporadically in Missouri) have been called f. *puberula* (Wiegand) Fernald.

67. Coreopsis L. (tickseed, coreopsis)
(Smith, 1976)

Plants annual or perennial herbs (shrubs elsewhere). Stems erect or ascending, unbranched or more commonly few- to numerous-branched, with several fine to coarse longitudinal lines or ridges, sometimes relatively strongly 4- or 5-angled, glabrous or sparsely to moderately pubescent. Leaves opposite and sometimes also basal, variously sessile to long-petiolate, the bases slightly expanded and wrapping around the stem. Leaf blades simple or 1 or 2 times pinnately or less commonly palmately lobed, dissected, or compound, variously shaped, the margins otherwise entire, glabrous or variously hairy, not glandular. Inflorescences of solitary terminal heads or appearing as loose, open clusters or panicles, the heads with long, bractless stalks or the stalk with 1 or 2 inconspicuous, minute bracts. Heads radiate. Involucre cup-shaped, the bracts in 2 dissimilar overlapping series. Involucral bracts fused at the base, those of the outer series (5–)8(–12), shorter and narrower than the others, ovate-triangular to narrowly triangular, tapered to the sharply pointed tip, green, sometimes reddish-tinged, the margins sometimes thin and white, often with spreading hairs, the surfaces glabrous (the outer surface sometimes sparsely to moderately hairy in *C. tripteris*), not glandular, usually inconspicuously 3-nerved; those of the inner series usually 8, oblong-elliptic to oblong-ovate, rounded or angled to a bluntly or less commonly sharply pointed tip, yellowish brown to yellowish green, sometimes reddish-tinged, more or less scalelike with usually broad, lighter, thinner margins, glabrous, not glandular, inconspicuously several-nerved. Receptacle flat or slightly convex, not elongating as the fruits mature, with chaffy bracts subtending the disc florets, these variously shaped, relatively flat, not or only slightly wrapped around the florets. Ray florets (5–)8, sterile (lacking stamens and style at flowering and with an ovary that is shorter and thinner than those of the disc florets, not developing into a fruit; pistillate elsewhere), the corolla showy, 12–30 mm long, relatively broad, yellow, or occasionally orangish yellow (usually with a reddish brown to brownish purple base in *C. tinctoria*), not persistent at fruiting. Disc florets 40–80 (–150), perfect, the corolla yellow to orangish yellow or reddish purple, not thickened at the base, not persistent at fruiting. Style branches with the sterile tip slightly to moderately elongate and tapered to a bluntly or sharply pointed tip. Pappus of the disc florets of (1)2 short awn(s) (these smooth or with sparse, upward-angled barbs), sometimes reduced to a low rim or crown with 1 or 2 short teeth or absent, when present usually persistent at fruiting. Fruits 2–7 mm long, narrowly oblong to oblong-ovate or oblong-obovate in outline, strongly flattened, flat or more commonly the base and tip appearing curled or arched inward at maturity, more or less truncate at the tip, not beaked, the angles usually with slender to broad wings, the surfaces relatively flat to slightly convex and with a faint longitudinal angle, the inner face sometimes with a bulbous thickening at 1 or both ends, glabrous, sometimes appearing finely pebbled or with numerous small tubercles, dark brown to black with pale wings, often slightly shiny. About 140 species (depending on the generic circumscription), mainly North America, but also Central America, South America, Africa, Pacific Islands.

Several species of *Coreopsis* are cultivated as garden ornamentals and for cut flowers, and doubled cultivars with numerous ray florets have been developed for some of these species. The fruits also provide food for birds and small mammals. Some authors have noted that the genera *Bidens* and *Coreopsis* are difficult to separate morphologically (Sherff, 1955; Barkley, 1986). Recent molecular phylogenetic studies (Kimball and Crawford, 2004) also found that, as traditionally circumscribed, the two genera are not natural units. Further research undoubtedly will result in the revision of generic circumscriptions in the subtribe Coreopsidinae.

1. Disc corollas mostly 4-lobed; ray corollas with a well-differentiated region of reddish brown to brownish purple toward the base, otherwise yellow; style branches with a short, bluntly pointed, sterile tip .... 5. C. TINCTORIA
1. Disc corollas 5-lobed; ray corollas lacking reddish brown or brownish purple color, uniformly yellow to orangish yellow; style branches with a sharply pointed, sterile tip
    2. Ray corollas entire or with 2 or 3(4) minute teeth in the center of the otherwise more or less rounded tip; chaffy bracts more or less linear, with a slender base, sometimes slightly thickened toward the rounded or angled, bluntly to sharply pointed tip
        3. Disc corollas yellow, sometimes with yellowish orange lobes; leaf blades 3-lobed from well above the base, the lobes sometimes lobed again but all of the ultimate segments narrowly oblong, more or less the same width throughout, sometimes very slightly narrowed toward the base .... 3. C. PALMATA
        3. Disc corollas reddish purple, sometimes with a yellow tube; leaf blades deeply divided or compound from the base, the divisions or leaflets angled or tapered to a slender attachment point at the base .... 6. C. TRIPTERIS
    2. Ray corollas with 3–5 deep, sometimes irregular (appearing jagged) teeth or lobes around the tip; chaffy bracts narrowly triangular, long-tapered from an abruptly broadened, flat basal portion to a sharply pointed tip
        4. Leaves confined to 1–3(–5) nodes (occasionally more in cultivated forms) mostly in the lower ½ of the plant .... 2. C. LANCEOLATA
        4. Leaves at (5)6–12 nodes along usually ⅔ or more of the length of the stems
            5. Leaf blades (except sometimes for the unlobed lowermost few stem leaves) mostly 1 or 2 times deeply pinnately or palmately divided into (3–)5–9 relatively slender, narrowly linear to narrowly lanceolate lobes or divisions .... 1. C. GRANDIFLORA
            5. Leaf blades unlobed or 1 time deeply pinnately lobed with 1 or 2(3) relatively broad, lanceolate to ovate or elliptic lobes toward the base .... 4. C. PUBESCENS

1. Coreopsis grandiflora Hogg ex Sweet
(bigflower coreopsis)

C. grandiflora var. *harveyana* (A. Gray) Sherff

Pl. 275 d; Map 1160

Plants perennial, with a short, sometimes branched rootstock. Stems 30–100 cm long, glabrous (hairy elsewhere). Leaves distributed at (5)6–10 nodes along (½–)⅔ of the length of the stems, sessile or short-petiolate, the lowermost leaves sometimes with relatively long petioles. Leaf blades (2–)3–9 cm long, narrowly oblanceolate or narrowly elliptic (in undivided leaves) to oblong-ovate or broadly ovate (in divided leaves) in outline, those of the basal and lowermost stem leaves sometimes unlobed, but those of most leaves 1 or 2 times deeply pinnately or palmately divided into (3–)5–9 relatively slender, narrowly linear to narrowly lanceolate lobes or divisions, in entire leaves the blade angled or tapered at the base; in divided leaves, the lateral lobes or divisions shorter than to longer than the terminal lobe or division, 0.5–3.0(–5.0) mm wide, narrowly linear to narrowly oblong-lanceolate, angled or tapered at the base, rounded or more commonly angled or tapered to a usually sharply pointed tip, glabrous or the margins sometimes with a few spreading hairs at the leaf base. Inflorescences of solitary heads or appearing as loose, open clusters, the heads with the stalk mostly 8–20 cm long. Involucre with the outer series of bracts 4–10 mm long; the inner series of bracts 7–12 mm long. Chaffy bracts narrowly triangular, long-tapered from an abruptly broadened, flat basal portion to a sharply pointed tip. Ray florets with the corolla 12–25 mm long, with 3–5 deep, sometimes irregular (appearing jagged) teeth or lobes around the tip, uniformly yellow to orangish yellow. Disc florets with the corollas 3.5–5.0 mm long, 5-lobed, yellow, sometimes with yellowish orange lobes. Style branches tapered abruptly to a sharply pointed, sterile tip. Pappus of 1 or 2 scalelike teeth 0.1–0.3 mm long. Fruits 2.0–3.5 mm long, the base and tip appearing curled or arched inward at maturity, the angles with broad, pale wings having entire or occasionally slightly irregular margins, the inner face with a bulbous thickening at 1 or both ends, dark brown to black, 1 or both surfaces smooth to minutely pebbled, sometimes with few to numerous small, lighter-colored tubercles, especially on the inner surface. $2n=26$. April–July.

Scattered mostly in the Unglaciated Plains Division and the eastern portion of the Ozarks (southeastern U.S. west to Kansas and Texas; introduced farther north and west). Upland Prairies, glades, ledges and tops of bluffs, and openings of dry upland forests; also railroads and roadsides.

Morphological variation within and between populations of *C. grandiflora* is complex. Sherff (1955) and Smith (1976) attempted to segregate a series of varieties and forms based on supposed differences in leaf segment width, pubescence, and the size of the pappus teeth. Steyermark (1963) separated Missouri materials into two of these infraspecific taxa, with var. *grandiflora* differing from var. *harveyana* based on its somewhat broader leaf lobes and a distribution in the Unglaciated Plains Division (vs. the eastern Ozarks). More recent collections have shown that plants with both narrower and broader leaf segments occur throughout the range in Missouri. Although there may be some basis in recognizing some of the other infraspecifc taxa from other states that differ in pubescence and pappus characters, the separation of Missouri materials into varieties seems arbitrary and without merit.

2. Coreopsis lanceolata L. (tickseed coreopsis)
C. lanceolata var. *villosa* Michx.

Pl. 275 a, b; Map 1161

Plants perennial, with a short, horizontal rootstock. Stems 20–60 cm long (taller in cultivated races), glabrous or sparsely to moderately pubescent with spreading hairs, especially toward the base. Leaves confined to 1–3(–5) nodes (occasionally more in cultivated forms) mostly in the lower half of the plant, the uppermost leaves sessile, the basal and lowermost stem leaves mostly long-petiolate. Leaf blades 2.5–12.0 cm long, narrowly oblanceolate or narrowly elliptic to ovate or oblong-ovate in outline, unlobed or with 1 or 2(–5) deep basal (pinnate) lobes or divisions, in entire leaves the blade angled to long-tapered at the base, in divided leaves, the lateral lobes or divisions much shorter than the terminal lobe or division, 2–9 mm wide, linear, narrowly oblong, or elliptic, more or less narrowed toward the base, angled or tapered to a usually sharply pointed tip, the surfaces glabrous or sparsely to moderately pubescent with short, spreading hairs. Inflorescences of solitary heads or appearing as loose, open clusters, the heads with the stalk mostly 8–40 cm long. Involucre with the outer series of bracts 5–10 mm long; the inner series of bracts 9–12 mm long. Chaffy bracts narrowly triangular, long-tapered from an abruptly broadened, flat basal portion to a sharply pointed tip. Ray florets with the corolla 15–30 mm long, with 3–5 deep, sometimes irregular (appearing jagged) teeth or lobes around the tip, uniformly yellow. Disc florets with the corollas 3.5–5.0 mm long, 5-lobed, yellow, sometimes with yellowish orange lobes. Style branches tapered abruptly to a sharply

1161. Coreopsis lanceolata   1162. Coreopsis palmata   1163. Coreopsis pubescens

pointed, sterile tip. Pappus of 1 or 2 scalelike teeth 0.3–0.8 mm long. Fruits 2.5–4.0 mm long, the base and tip appearing curled or arched inward at maturity, the angles with broad, pale wings having entire or more commonly somewhat irregular margins, the inner face with a bulbous thickening at 1 or both ends, dark brown to black, 1 or both surfaces with numerous small, lighter-colored tubercles. $2n=26$. April–July.

Scattered, mostly in the Ozark and Ozark Border Divisions; introduced elsewhere in the state (eastern U.S. west to Wisconsin and Texas; Canada; introduced in the western U.S.).

Steyermark (1963) maintained that plants encountered outside the Ozarks represented escapes from cultivation. The species also sometimes is planted along highways for roadside beautification. Smith (1976) examined specimens from throughout the range of this species and determined that none of the infraspecific taxa based on differences in leaf dissection and pubescence that had been accepted by some earlier authors were worthy of continued recognition.

3. Coreopsis palmata Nutt. (finger coreopsis)

Pl. 275 e, f; Map 1162

Plants perennial, with sometimes long-creeping rhizomes. Stems 40–90 cm long, glabrous or sparsely pubescent with short, spreading hairs, mostly around the nodes. Leaves distributed at 6–12 nodes along ⅔ or more of the length of the stems, sessile or occasionally minutely petiolate. Leaf blades 3–8 cm long, mostly broadly obovate in outline, 3-lobed from well above the base, the lobes sometimes lobed again, the 3–7(–9) ultimate segments 2–7 mm wide, narrowly oblong, more or less the same width throughout, sometimes very slightly narrowed toward the base, the surfaces glabrous (but the margins occasionally with a few hairs at the leaf base). Inflorescences of solitary heads or appearing as loose, open clusters, the heads with the stalk mostly 1–5 cm long. Involucre with the outer series of bracts 3–8 mm long; the inner series of bracts 6–10 mm long. Chaffy bracts more or less linear, with a slender base, often slightly thickened toward the bluntly to sharply pointed tip. Ray florets with the corolla 15–30 mm long, entire or with 2 or 3(4) minute teeth in the center of the otherwise more or less rounded tip, uniformly yellow to orangish yellow. Disc florets with the corollas 5.0–6.5 mm long, 5-lobed, yellow, sometimes with yellowish orange lobes. Style branches tapered abruptly to a sharply pointed, sterile tip. Pappus absent or of 1 or 2 teeth to 0.2 mm long. Fruits 4.5–6.5 mm long, the base and tip appearing slightly arched inward at maturity, the angles unwinged or more commonly with narrow, pale wings having entire margins, the inner face not thickened at the ends, dark brown to black, the surfaces smooth. $2n=26$. May–September.

Scattered nearly throughout the state but uncommon in the Mississippi Lowlands Division and portions of the Glaciated Plains (Michigan to South Dakota south to Louisiana and Oklahoma). Upland prairies, glades, ledges and tops of bluffs, savannas, and openings of dry upland forests; also fallow fields, old fields, railroads, and roadsides.

4. Coreopsis pubescens Elliott **var. pubescens** (star tickseed)

Pl. 275 c; Map 1163

Plants perennial, with a short, sometimes branched rootstock. Stems 25–100 cm long, sparsely to moderately pubescent with more or less spreading hairs. Leaves distributed at (5)6–12 nodes along ⅔ or more of the length of the stems, short-petiolate. Leaf blades 1–8(–10) cm long, elliptic to ovate in outline or those of the uppermost leaves sometimes narrowly elliptic to lanceolate, unlobed or with 1 or 2(3) deep basal lobes or divisions (rarely fully compound), in entire leaves the blade angled or tapered at the base, in divided leaves, the

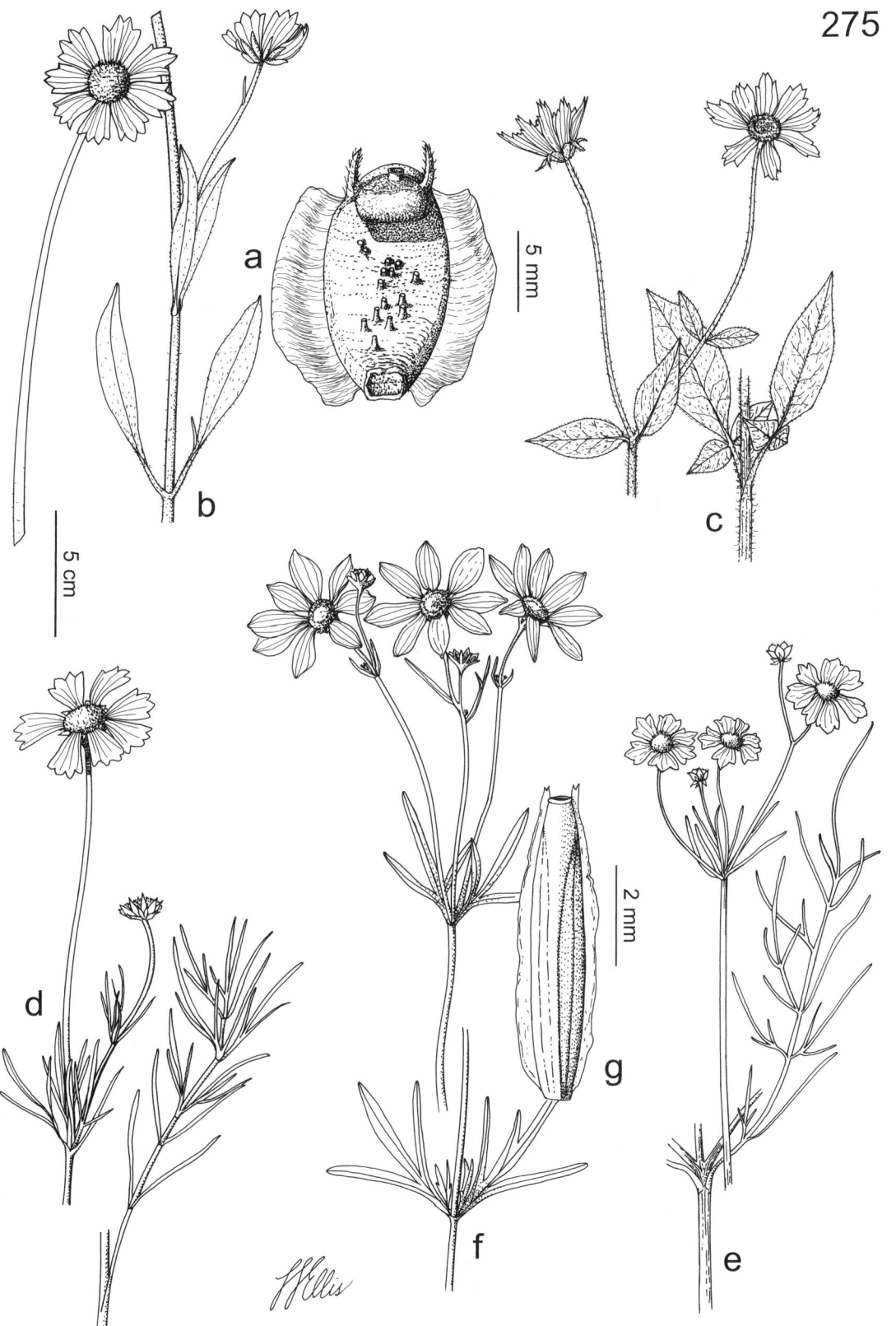

Plate 275. Asteraceae. *Coreopsis lanceolota*, **a)** fruit, **b)** habit. *Coreopsis pubescens*, **c)** habit. *Coreopsis grandiflora*, **d)** habit. *Coreopsis palmata*, **e)** habit, **f)** fruit. *Coreopsis tinctoria*, **g)** habit.

1164. Coreopsis tinctoria   1165. Coreopsis tripteris   1166. Cosmos bipinnatus

lateral lobes or divisions much shorter than the terminal lobe or division, 4–12 mm wide, lanceolate to ovate or elliptic, rounded or angled at the base, angled or tapered to a usually sharply pointed tip, the upper surface glabrous or with sparse, minute hairs along the midvein, the undersurface moderately pubescent with short, spreading hairs, the sometimes narrowly whitened margins relatively densely pubescent with minute, loosely appressed hairs. Inflorescences of solitary heads or appearing as loose, open clusters, the heads with the stalk mostly 5–14 cm long. Involucre with the outer series of bracts 3–7 mm long; the inner series of bracts 9–12 mm long. Chaffy bracts narrowly triangular, long-tapered from an abruptly broadened, flat basal portion to a sharply pointed tip. Ray florets with the corolla 12–25 mm long, with 3–5 deep, sometimes irregular (appearing jagged) teeth or lobes around the tip, uniformly yellow. Disc florets with the corollas 3.5–4.5 mm long, 5-lobed, yellow, sometimes with yellowish orange lobes. Style branches tapered abruptly to a sharply pointed, sterile tip. Pappus of 1 or 2 scalelike teeth 0.1–0.4 mm long. Fruits 2.5–3.5 mm long, the base and tip appearing curled or arched inward at maturity, the angles with broad, pale wings having entire or more commonly somewhat irregular margins, the inner face with a bulbous thickening at 1 or both ends, dark brown to black, 1 or both surfaces with numerous small, lighter-colored tubercles. $2n=26$. May–September.

Scattered, mostly in the Ozark and Ozark Border Divisions (eastern [mostly southeastern] U.S. west to Kansas and Texas). Banks of streams and rivers, bottomland forests, mesic upland forests in ravines, and bases and ledges of bluffs; also railroads and roadsides.

Smith (1976) followed earlier authors in accepting three varieties within *C. pubescens*. Missouri plants are all the widespread var. *pubescens*, which is hairy and has relatively broad, elliptic leaf divisions with angled or short-tapered tips. Scattered populations from Louisiana, Mississippi, Alabama, and possibly Tennessee that are glabrous and have wiry, much-branched stems and narrowly oblanceolate, often undivided leaves have been called var. *debilis* (Sherff) E.B. Sm. Some southern Appalachian populations comprising glabrous plants with narrowly elliptic leaves or terminal leaf divisions with relatively tapered tips are recognized by some botanists as var. *robusta* A. Gray ex Eames.

5. Coreopsis tinctoria Nutt. (plains coreopsis, calliopsis)
  C. tinctoria var. *similis* (F.E. Boynton) H.M. Parker ex E.B. Sm.
  C. cardaminifolia (DC.) Torr. & A. Gray
                                    Pl. 275 g; Map 1164

Plants usually annual, with taproots. Stems (10–)40–120 cm long, glabrous. Leaves distributed at (5)6–12 nodes along $^2/_3$ or more of the length of the stems, sessile or short-petiolate. Leaf blades 1.5–10.0 cm long, oblong to elliptic or obovate in outline, 1 or 2 times pinnately dissected, the mostly 5–25 ultimate segments 1–4 mm wide, narrowly linear to linear-lanceolate, usually somewhat tapered at the base and tip, the surfaces glabrous. Inflorescences mostly appearing as panicles, the heads with the stalk mostly 2–5 cm long. Involucre often reddish-tinged, the outer series of bracts 1–4 mm long; the inner series of bracts 4–9 mm long. Chaffy bracts linear, not widened at the base or tip. Ray florets with the corolla 12–20 mm long, with 3(4) deep, sometimes irregular (appearing rounded or less commonly somewhat jagged) teeth or lobes around the tip, yellow with a well-differentiated region of reddish brown to brownish purple toward the base. Disc florets with the corollas 2.5–3.5 mm long, mostly 4-lobed, reddish purple, sometimes with a yellow tube. Style branches with a short, bluntly pointed, sterile tip. Pappus absent or of 1 or 2 teeth or awns to 1.5 mm long. Fruits

Plate 276. Asteraceae. *Coreopsis tripteris*, **a)** habit. *Rudbeckia amplexicaulis*, **b)** habit. *Dyssodia papposa*, **c)** head, **d)** flowering branch. *Cosmos bipinnatus*, **e)** flowering branch. *Cosmos sulphureus*, **f)** flowering branch. *Cosmos parviflorus*, **g)** fruit, **h)** habit.

1.5–3.0 mm long, the base and tip appearing curled or arched inward at maturity, the angles usually with narrow to broad, pale wings having entire to slightly irregular margins, the inner face slightly thickened at 1 or both ends, dark brown to black, 1 or both surfaces usually with numerous small, lighter-colored tubercles. $2n=24$. June–September.

Scattered, mostly south of the Missouri River (U.S., Canada, Mexico). Glades and banks of streams and rivers; also ditches, old fields, railroads, roadsides, and open, disturbed areas.

The native range of this species has been obscured by its natural spread along disturbance corridors such as highways and railroads and also its escape from cultivation in gardens. Some authors have suggested that its original range included at least the southern Great Plains and Missouri, Oklahoma, Texas, Arkansas, and Louisiana (Barkley, 1986; Gleason and Cronquist, 1991).

Steyermark (1963), in the appendix to his manual, added *C. cardaminifolia* based on a collection made by Viktor Mühlenbach in the St. Louis railyards. Smith and Parker (1971) studied the relationship of this taxon to *C. tinctoria,* noting the presence of additional material attributable to *C. tinctoria* and intermediate specimens from Missouri, but they concluded that there was no taxonomic or genetic basis for maintaining two species in the complex. Missouri plants appear to be relatively uniform in having the basal ¼–½ of the corolla reddish brown to brownish purple. Elsewhere in the range of the species, a number of variants occur with ray corollas ranging from entirely yellow (f. *tinctoria*) to entirely reddish (f. *atropurpurea* (Hook.) Fernald). Native Americans derived a dark red dye from the heads of this species. It was also used medicinally as a tonic and for treatment of venereal diseases (Moerman, 1998).

6. Coreopsis tripteris L. (tall tickseed, tall coreopsis)

C. tripteris var. *deamii* Standl.

Pl. 276 a; Map 1165

Plants perennial, with short, stout rhizomes. Stems 60–180(–250) cm long, usually glabrous, sometimes somewhat glaucous. Leaves distributed at mostly 10 or more nodes along ⅔ or more of the length of the stems, short-petiolate, the uppermost leaves sometimes nearly sessile. Leaf blades 3–12 cm long, mostly broadly obovate in outline, those of the uppermost leaves simple and unlobed but most of the leaves with the blades ternately or palmately deeply divided or compound with 3 or less commonly 5 divisions or leaflets, these entire (in some ternately compound leaves the middle leaflet deeply divided into 3 divisions), 5–25(–30) mm wide, lanceolate to narrowly elliptic, angled or tapered to a slender attachment point at the base, angled or tapered to a bluntly or sharply pointed tip, the surfaces glabrous or sparsely to densely hairy with short, curved hairs. Inflorescences mostly appearing as loose, open clusters or panicles, the heads with the stalk mostly 2–6 cm long. Involucre with the outer series of bracts 2–5 mm long, glabrous or sparsely to moderately hairy; the inner series of bracts 6–9 mm long. Chaffy bracts more or less linear, with a slender base, sometimes slightly thickened toward the rounded or angled, bluntly to sharply pointed tip. Ray florets with the corolla 12–25 mm long, entire or with 2 or 3(4) minute teeth in the center of the otherwise more or less rounded tip, uniformly yellow (sometimes with fine, purplish veins). Disc florets with the corollas 5–6 mm long, 5-lobed, reddish purple, sometimes with a yellow tube. Style branches tapered abruptly to a sharply pointed, sterile tip. Pappus absent or more commonly of 2 awns to 0.5 mm long. Fruits 4–7 mm long, the base and tip appearing slightly arched inward at maturity, the angles with narrow, pale wings having entire margins and occasionally with a few hairs at the tip, the inner face not thickened at the ends, dark brown to black, the surfaces smooth. $2n=26$. July–September.

Scattered nearly throughout the state but apparently absent from the Mississippi Lowlands Division (eastern U.S. west to Iowa and Texas; Canada). Upland prairies, savannas, openings of mesic to dry upland forests, ledges and tops of bluffs, and banks of streams and rivers; also old fields, railroads, and roadsides.

Smith (1976) examined specimens from throughout the range of this species and determined that none of the infraspecific taxa based on differences in leaf dissection and pubescence that had been accepted by some earlier authors were worthy of continued recognition.

68. Cosmos Cav. (cosmos)

Plants annual (perennial herbs elsewhere), with taproots. Stems erect or ascending, unbranched or few- to several-branched, with several fine longitudinal lines or ridges, glabrous or sparsely to moderately pubescent, sometimes minutely roughened. Leaves opposite (basal leaves absent except in seedlings), variously sessile to long-petiolate, the bases slightly expanded and wrapping around the stem. Leaf blades 1 or 2 times pinnately lobed or dissected, all but those of the uppermost leaves oblong-ovate to oblong-elliptic in outline, the ultimate segments linear or oblong-triangular to oblong-lanceolate, tapered to a sharply pointed tip, the margins otherwise all or mostly entire, glabrous, minutely roughened, or sparsely pubescent with short, spreading hairs, the surfaces glabrous or sparsely and inconspicuously hairy or roughened, sometimes also dotted with sparse, sessile or impressed, yellow glands. Inflorescences of solitary terminal heads or appearing as loose, open clusters, the heads with long, bractless stalks. Heads radiate. Involucre narrowly cup-shaped to cup-shaped, the bracts in 2 dissimilar overlapping series. Involucral bracts fused at the base, variously shaped, glabrous but sometimes sparsely dotted with minute, impressed glands, inconspicuously 3- or 5-nerved, those of the outer series of (5–)8 green, slightly narrower and often slightly shorter than those of the often more scalelike inner series of (5–)8. Receptacle flat, not elongating as the fruits mature, with chaffy bracts subtending the ray and disc florets, these narrowly oblong to narrowly oblong-lanceolate, somewhat concave, sometimes wrapped around the florets toward the base. Ray florets (5–)8, sterile (lacking stamens and style at flowering and with an ovary that is shorter and thinner than those of the disc florets, not developing into a fruit), the corolla showy (except in *C. parviflorus*), 5–40 mm long, relatively broad, white, pink, purple, yellow, or orange, not persistent at fruiting. Disc florets 10–20 (–80 elsewhere), perfect, the corolla yellow to less commonly orange, not thickened at the base, not persistent at fruiting. Style branches with the sterile tip somewhat elongate and tapered. Pappus of the disc florets of 2–4 short awns (–8 elsewhere), these with downward-angled barbs, rarely very short or absent, when present then mostly persistent and somewhat spreading at fruiting. Fruits 7–30 mm long (including the beak), linear in outline (usually slightly curved and asymmetrically slightly thickened toward the midpoint), not or only slightly flattened, the body 4-angled in cross-section, tapered at the tip into an elongate, slender beak, each face with a slender longitudinal groove, the surface glabrous or sparsely hairy, sometimes appearing finely pebbled or roughened, dark brown to black, somewhat shiny. About 26 species, southwestern U.S., Mexico, Central America, South America, introduced nearly worldwide.

Several species of *Cosmos* are commonly cultivated as ornamentals in gardens, including two of the species in Missouri, *C. bipinnatus* and *C. sulphureus*. These two species also are components of so-called native wildflower seed mixes that were planted for roadside beautification in the past by the Missouri Department of Transportation and other public agencies (see also under *Gaillardia pulchella,* below). They are thus undoubtedly more widespread in the state than the few specimens in herbaria indicate.

1. Ray florets with the corollas yellow to reddish orange; leaf blades lobed, the ultimate segments not threadlike, mostly 2–5 mm wide . . . . 3. C. SULPHUREUS
1. Ray florets with the corolla white, pink, or purple; leaf blades deeply divided, the ultimate segments often threadlike, 0.5–1.5 mm wide
   2. Ray florets with the corolla 1.5–4.0 cm long; fruits smooth or appearing minutely pebbled . . . . 1. C. BIPINNATUS
   2. Ray florets with the corolla 0.5–1.5 cm long; fruits appearing minutely roughened (with minute, ascending teeth or hairs) . . . . 2. C. PARVIFLORUS

1. Cosmos bipinnatus Cav. (garden cosmos, Mexican aster)

Pl. 276 e; Map 1166

Stems 25–200 cm long, glabrous or sparsely pubescent with short, appressed-ascending hairs, rarely minutely roughened to the touch. Leaves sessile or with a short petiole to 10 mm long. Leaf blade 1–12 cm long, 1 or 2 times deeply pinnately divided, the ultimate segments narrowly linear, often threadlike, 0.5–1.5 mm wide. Involucre 6–15

1167. Cosmos parviflorus   1168. Cosmos sulphureus   1169. Cyclachaena xanthiifolia

mm long, 7–15 mm in diameter, the outer series of bracts 6–13 mm long, spreading to loosely ascending, linear to narrowly triangular or narrowly lanceolate, tapered to a sharply pointed tip, mostly with slender, white margins; the inner series of bracts 7–13 mm long, erect or strongly ascending, lanceolate to narrowly ovate, rounded or more commonly angled to a bluntly or sharply pointed tip, mostly with relatively broad, white margins. Ray florets with the corolla 1.5–4.0 cm long, most commonly pink, less commonly white or purplish pink. Disc florets with the corolla 5–7 mm long, yellow. Fruits 7–16 mm long (including the beak), the surface glabrous but often dotted with scattered impressed glands, smooth or appearing minutely pebbled. $2n=24$. July–October.

Introduced, uncommon and sporadic (native of Arizona, Mexico; introduced widely in the U.S., Canada). Railroads, roadsides, and open, disturbed areas.

2. Cosmos parviflorus (Jacq.) Pers.

Pl. 276 g, h; Map 1167

Stems 25–90 cm long, glabrous or sparsely to moderately pubescent with minute, mostly spreading hairs, rarely minutely roughened to the touch. Leaves sessile or with a short petiole to 5 mm long. Leaf blade 1–7 cm long, 1 or 2 times deeply pinnately divided, the ultimate segments narrowly linear, often threadlike, 0.5–1.5 mm wide. Involucre 7–12 mm long, 5–10 mm in diameter, the outer series of bracts 5–9 mm long, spreading to loosely ascending, lanceolate, tapered to a sharply pointed tip, rarely with slender, white margins; the inner series of bracts 6–10 mm long, erect or strongly ascending, oblong to oblong-ovate, rounded or more commonly angled to a bluntly or sharply pointed tip, mostly with relatively broad, white margins. Ray florets with the corolla 0.5–1.5 cm long, most commonly pink, less commonly white or purple.

Disc florets with the corolla 4–5 mm long, yellow. Fruits 7–16 mm long (including the beak), the surface glabrous or appearing minutely roughened (with minute, ascending teeth or hairs). $2n=24$. July–October.

Introduced, known thus far only from the city of St. Louis (Arizona, New Mexico, Colorado, and Texas; Mexico; introduced sporadically east to Maine). Railroads.

This species was first reported for Missouri by Mühlenbach (1979), based on plants found during his surveys of the St. Louis railyards. It is not widely cultivated.

3. Cosmos sulphureus Cav. (sulphur cosmos)

Pl. 276 f; Map 1168

Stems 15–200 cm long, glabrous or sparsely pubescent with short, appressed-ascending and/or longer, spreading hairs, rarely minutely roughened to the touch. Leaves sessile or with a petiole to 70 mm long. Leaf blade 1–15 cm long, 1 or 2 times pinnately lobed, the ultimate segments oblong-triangular to oblong-lanceolate, not threadlike, mostly 2–5 mm wide. Involucre 8–14 mm long, 6–10 mm in diameter, the outer series of bracts 4–8 mm long, spreading to loosely ascending, linear to narrowly triangular or narrowly lanceolate, tapered to a sharply pointed tip, sometimes with slender, white margins; the inner series of bracts 7–12 mm long, erect or strongly ascending, oblong-lanceolate to narrowly oblong-ovate, rounded or more commonly angled to a bluntly or sharply pointed tip, mostly with relatively broad, white margins. Ray florets with the corolla 1.5–3.0 cm long, yellow to orange or reddish orange. Disc florets with the corolla 5–7 mm long, yellow to yellowish orange. Fruits 15–30 mm long (including the beak), the surface minutely pubescent with sparse to moderate, ascending hairs or rarely glabrous. $2n=24, 48$. August–October.

Introduced, uncommon and sporadic (native of Mexico, Central America, South America, Caribbean Islands; introduced widely in the eastern U.S. west to Missouri and Texas, also California). Railroads, roadsides, and open, disturbed areas.

Sherff (1955) separated native Latin American populations of this species into three varieties, based on minor differences in pubescence and persistence of the pappus. The introduced Missouri plants appear to represent var. *sulphureus* (with nearly glabrous stems and mostly persistent pappus) if these taxa are accepted. However, the variability of plants in nature and the effects of plant breeding on the cultivated races has obscured the supposed differences among the varieties, making it imprudent to attempt their formal taxonomic recognition.

69. Cyclachaena Fresen.

One species, western U.S., Canada; introduced elsewhere in temperate North America, Europe.

Cyclachaena is a recently resurrected genus formerly included in *Iva*. See the treatment of that genus for further discussion.

1. Cyclachaena xanthiifolia (Nutt.) Fresen.
(big marsh elder, careless weed)
Iva xanthiifolia Nutt.
Pl. 284 a, b; Map 1169

Plants annual, with taproots. Stems 25–150 (–200) cm long, unbranched or more commonly several-branched, erect or ascending, finely ridged or grooved, sparsely to moderately pubescent with fine, curved or curled, slender hairs, often nearly glabrous toward the base, sometimes also with scattered, longer, stiff, spreading, pustular-based hairs toward the tip. Leaves opposite or the uppermost few alternate, mostly long-petiolate (the uppermost leaves sometimes short-petiolate). Leaf blades 3–25 cm long, simple or shallowly 3-lobed (usually with 3 main veins), lanceolate to more commonly ovate or broadly ovate, angled or short-tapered at the base, mostly tapered at the sharply pointed tip, the margins irregularly and often coarsely toothed and minutely hairy, the upper surface nearly glabrous to sparsely or moderately roughened with short, more or less appressed hairs, sometimes only along the main veins, the undersurface moderately to densely pubescent with appressed hairs, often appearing pale or whitened, both surfaces also with minute, sessile, yellow glands. Inflorescences panicles at the branch tips and from the upper leaf axils, the heads sessile or very short-stalked, solitary or in small clusters of 2–4 at the nodes, the inflorescence branches mostly with short, leaflike bracts, but the individual heads or clusters not subtended by bracts. Heads discoid, spreading in several directions. Involucre 1.5–3.0 mm long, cup-shaped to broadly cup-shaped, not or only slightly asymmetrical, the 5 involucral bracts in 1 series, free to the base and overlapping, green, sparsely pubescent with somewhat tangled hairs and also sessile glands. Receptacle flat, not elongating as the fruits mature, with chaffy bracts subtending the florets, those subtending the pistillate florets obovate, hairy along the margins (also glandular) and wrapped around the florets, those subtending the staminate florets linear to narrowly oblanceolate, glandular, not wrapped around the florets. Central staminate florets 8–20, these with a minute, nonfunctional ovary and undivided style, the stamens with the filaments more or less fused into a tube and the anthers free but positioned closely adjacent to one another in a ring, the corolla 2.0–2.5 mm long, funnelform to narrowly bell-shaped, 5-lobed, white to pale yellow, sometimes purplish-tinged toward the tip, glandular. Marginal pistillate florets usually 5, the corolla absent (a persistent, minute, raised collar sometimes present at the tip of the ovary and fruit). Pappus of the staminate and pistillate florets absent. Fruits 2–3 mm long, obovoid, sometimes somewhat flattened, not angled or very bluntly angled on 1 face, otherwise appearing minutely pebbled, dark brown to nearly black, glabrous. $2n=36$. June–October.

Uncommon, known only from historical collections from northwestern Missouri and Lincoln County; possibly both native and introduced in Jackson County, and introduced in St. Louis City and County (Wisconsin to Washington south to North Dakota and Texas; Canada; introduced eastward to the East Coast and in Europe). Floodplains of rivers and margins of lakes; also gardens, railroads, and open, disturbed areas.

This species is an important cause of hay fever in regions where it is abundant, but it is too uncommon in Missouri to present a severe prob-

1170. Dyssodia papposa   1171. Echinacea angustifolia   1172. Echinacea pallida

lem. Pruski (2005) has argued that the correct author citation for this taxon should be C. xanthiifolia Fresen. Additionally, many authors have followed older misspellings of the name as C. xanthifolia, but compound epithets intended to convey the meaning "leaves like *Xanthium*" should have *ii* as the combining vowels.

70. Dyssodia Cav.
(Strother, 1969, 1986)

Four species, North America, Central America.

Strother (1969) monographed *Dyssodia* as a genus of 32 species but subsequently adopted a narrower set of generic circumscriptions of the group, recognizing six total genera, with only 4 retained in *Dyssodia*.

1. Dyssodia papposa (Vent.) Hitchc. (fetid marigold)

Pl. 276 c, d; Map 1170

Plants annual, aromatic, with taproots. Stems 10–50 cm long, erect to loosely ascending from a spreading base, several- to more commonly many-branched (especially toward the tip), with several inconspicuous, fine longitudinal ridges and grooves toward the tip (occasionally also somewhat angled), sparsely to moderately pubescent with short, curved or curled hairs, sometimes nearly glabrous toward the base. Leaves opposite or the uppermost few alternate, mostly sessile, the bases of each pair slightly expanded and wrapped around the stem. Leaf blades 1–5 cm long, elliptic to ovate in outline, 1 or 2 times pinnately dissected with 7–15 primary lobes, the ultimate segments linear to narrowly linear, the margins otherwise entire or occasionally with a few irregular teeth, the surfaces glabrous or sparsely short-hairy and dotted with scattered, brownish yellow, sessile oil glands. Inflorescences of solitary, short-stalked to nearly sessile heads at the branch tips. Heads radiate. Involucre 6–10 mm long, more or less urn-shaped to narrowly bell-shaped, the disc 2–4 mm in diameter, the bracts in 2 unequal series. Involucral bracts 10–21, the surface glabrous but dotted or lined with conspicuous, sessile, yellowish brown oil glands, those of the outer series 4–9, 2–5 mm long, linear, free, green, the margins usually noticeably fringed; those of the inner series 6–12, 6–9 mm long, oblong-obovate to elliptic, somewhat thickened along the midnerve toward the base, fused at the base, greenish yellow and usually pinkish-tinged, the margins inconspicuously hairy. Receptacle shallowly convex, not elongating as the fruits mature, with minute, irregular ridges around the attachment points of the florets. Ray florets 4–8, pistillate (with a 2-branched style exserted from the short tube at flowering), the corolla 2–4 mm long, inconspicuous, yellow to orangish yellow. Disc florets 12–50, perfect, the corolla 2.5–3.5 mm long, brownish yellow, the tube not expanded at the base or persistent at fruiting, glabrous. Style branches with the sterile tip elongate and truncate at the tip. Pappus of 18–20 slender scales, 2.0–3.5 mm long, white with purplish-tinged tips or orangish yellow, each scale dissected irregularly into 5–10 ascending bristles. Fruits 3.0–3.5 mm long, narrowly oblanceolate to narrowly wedge-shaped in outline, 3–5-angled and slightly flattened, the surface moderately pubescent with silky, ascending hairs, especially along the angles, dark gray to black. 2n=26. May–October.

Scattered in most of the state but uncommon in the eastern half of the Ozark Division and apparently absent from the Mississippi Lowlands (U.S., Canada, Mexico, Central America; introduced in South America). Upland prairies, loess hill prairies, glades, and banks of streams; also pastures, levees, roadsides, and open, disturbed areas.

Plants of fetid marigold often form a dense, narrow band at the edge of the asphalt along highways. Their strong disagreeable odor and bad flavor cause livestock to avoid the species.

71. Echinacea Moench (coneflower)
(McGregor, 1968; Binns et al., 2002)

Plants perennial herbs, with a usually elongated, vertical rootstock and often somewhat tuberous main roots (merely fibrous-rooted in *E. purpurea*), sometimes also with short, stout rhizomes. Stems erect or ascending, unbranched or few- to several-branched, with several longitudinal lines or ridges, variously hairy (glabrous elsewhere), usually roughened to the touch. Leaves basal and alternate, the basal and lower stem leaves long-petiolate, the petioles progressively shorter up the stem, the upper stem leaves sometimes sessile or nearly so, the bases usually only slightly expanded, those of the basal and lower stem leaves usually somewhat wrapping around the stem. Leaf blades simple, linear to lanceolate, elliptic, or ovate, tapered at the base (those of the basal leaves rounded to heart-shaped in *E. purpurea*), mostly tapered to a sharply pointed tip, the margins entire or irregularly toothed, the surfaces roughened-hairy, not glandular, with 3 or 5 main veins. Inflorescences of solitary terminal heads, the heads with long, bractless stalks. Heads radiate. Involucre broadly cup-shaped or saucer-shaped, the bracts in mostly 2–4 subequal, overlapping series. Involucral bracts about 17–30 (the innermost bracts grading into the chaffy bracts), narrowly lanceolate to lanceolate, the outermost bracts occasionally narrowly ovate, spreading to reflexed above the midpoint, green, the margins and outer surface roughened-hairy (glabrous elsewhere), not glandular (rarely a few sessile, yellow glands present in *E. paradoxa*), the midnerve inconspicuous. Receptacle strongly convex to conical, usually elongating somewhat as the fruits mature (also broadening somewhat as the fruits mature), with chaffy bracts subtending the ray and disc florets, these concave and wrapped around the florets, the sharply pointed, spinelike tips noticeably longer than the tips of the disc corollas, hardened (somewhat softer and leathery in *E. purpurea*), the apical portion orange to dark purple, persistent at fruiting. Ray florets 8–21, sterile (lacking stamens and style at flowering and with an ovary that is shorter and thinner than those of the disc florets, not developing into a fruit) or rarely a few pistillate, the corolla showy, relatively slender, spreading to drooping at flowering, pink, purple, or yellow, rarely white, not persistent at fruiting. Disc florets numerous (more than 200), perfect, inconspicuous (because of the overtopping chaffy bracts), the corolla pink, purple, yellow, or occasionally green, slightly bulbous-thickened at the base, not persistent at fruiting (but sometimes trapped by the subtending bract). Style branches with the sterile tip somewhat elongate and tapered. Pappus of the disc florets of a low rim or crown similar in color and texture to the fruit body, the margin slightly irregular and sometimes with 2–4 triangular teeth (at the angles of the fruit), persistent at fruiting. Fruits wedge-shaped in outline, slightly flattened and usually somewhat 4-angled in cross-section (3-angled in rare fruits of ray florets), more or less smooth, the surface glabrous (hairy elsewhere), tan to nearly white, usually with an abrupt, brown to dark brown region toward the tip, sometimes slightly shiny. About 9 species, U.S., Canada.

The rootstocks of *Echinacea* species are commercially important as medicinals. Native Americans used the plants to treat a variety of ailments ranging from snakebites to toothaches, burns, arthritis, rheumatism, swollen glands, and other pains (Moerman, 1998). McGregor

(1968) discussed the more recent history of medicinal use, beginning with Meyer's Blood Purifier, a tonic prepared from *E. angustifolia* by a Nebraska doctor named Meyer, who learned of the plant from local North American Indians during the 1880s. Because medical science was unable to substantiate the curative properties of the genus, the American Medical Association discouraged the use of *Echinacea,* but its use in homeopathic medicine flourished in Europe, particularly in Germany (Hobbs, 1989; Foster, 1991). Beginning in the 1940s, new research began to hint at various antibacterial and antiviral properties of the root extract. Today the plants are used to provide a nonspecific stimulant to the immune system, both as a general tonic and to ward off colds and other illnesses, and *Echinacea* extract can be found in many health-food shops and even grocery stores. Much of the research on the efficacy of the plants continues to occur in Germany (Bauer and Wagner, 1990), but clinical studies in the United States continue to offer somewhat contradictory findings. All five of the species present in Missouri have been utilized, but apparently *E. pallida* is the preferred species.

The international market for *Echinacea* roots has led to intense demand for wild-collected roots. In Missouri, the genus ranks among the most-collected by so-called root-diggers (individuals who supplement their incomes by harvesting a variety of natural products from local areas and selling these to distributors). The gradual elimination of these beautiful wildflowers from mile upon mile of the state's roadsides was one of the main reasons for the enactment of laws restricting the collecting of plants from public highways in Missouri. Although much of the collection of rootstocks continues to take place legally on private property with landowner permission, managers of public lands such as state conservation areas and the state parks system have recorded numerous instances of unscrupulous individuals who have vandalized high-quality glades and prairies that were protected by state law for the preservation of native ecosystems and their enjoyment by all Missourians. Perhaps in part because of restrictions placed upon where *Echinacea* can be collected legally, some wholesalers have reported counterfeit coneflower roots collected from *Parthenium* species, which superficially resemble those of true *Echinacea.*

Species of *Echinacea* also are cultivated as ornamentals. All of the species growing in Missouri are available through wildflower nurseries to some extent, but only *E. purpurea* (and its cultivars and hybrids) currently is sold more widely in the general nursery trade. The fruiting heads also sometimes are used in dried flower arrangements and other craft projects.

1. Leaf blades narrowly to broadly ovate, 1.5–5.0 times as long as wide, those of the stem leaves mostly short-tapered to more or less rounded at the base, those of the basal leaves often rounded or heart-shaped, the margins usually irregularly toothed . . . . 4. E. PURPUREA
1. Leaf blades linear to lanceolate or elliptic, mostly 5–20 times as long as wide, long-tapered to narrowly angled at the base, the margins entire
    2. Ray florets with the corolla yellow; stems pubescent with appressed or ascending hairs . . . . 3. E. PARADOXA
    2. Ray florets with the corollas pink to purplish pink (rarely white); stems pubescent with spreading to loosely ascending hairs (species difficult to distinguish)
        3. Ray florets with the corolla 2.0–3.5(–4.0) cm long, mostly spreading at flowering; pollen yellow when fresh (sometimes faded to pale yellow on herbarium specimens) . . . . 1. E. ANGUSTIFOLIA
        3. Ray florets with the corolla (3–)4–9 cm long, reflexed or drooping at flowering; pollen yellow when fresh or white
            4. Pollen white . . . . 2. E. PALLIDA
            4. Pollen yellow when fresh (sometimes faded to pale yellow on herbarium specimens) . . . . 5. E. SIMULATA

1. **Echinacea angustifolia** DC. (narrow-leaved coneflower)
Brauneria angustifolia (DC.) A. Heller
E. angustifolia var. *strigosa* McGregor
E. pallida var. *angustifolia* (DC.) Cronquist
Pl. 277 a, b; Map 1171

Plants with a usually elongated, vertical rootstock and often somewhat tuberous main roots, sometimes also with short, stout rhizomes. Stems 10–50(–70) cm long, mostly unbranched, sparsely to densely pubescent with stiff, spreading (loosely ascending elsewhere), pustular-based hairs. Leaves with the margins entire and usually pubescent with spreading hairs, the surfaces moderately to densely pubescent with stiff, mostly spreading (loosely ascending elsewhere), pustular-based hairs, strongly roughened to the touch, with 3(5) main veins. Basal leaves 5–25 cm long, the blade narrowly elliptic to lanceolate, mostly 5–20 times as long as wide, long-tapered or narrowly angled at the base. Stem leaves 3–10 cm long, linear to narrowly elliptic or lanceolate, otherwise similar to the basal leaves. Involucral bracts 6–12 mm long, the outer surface moderately pubescent with mostly pustular-based hairs, not glandular. Receptacle 2.0–3.5 cm in diameter, the chaffy bracts 9–14 mm long, hardened, usually dark purple toward the tip. Ray florets with the corolla 2.0–3.5(–4.0) cm long, 5–8 mm wide, mostly spreading at flowering, pink to purplish pink (rarely white elsewhere). Disc florets with the corolla 6–8 mm long, the tube yellow to green, the lobes usually dark purple. Pollen bright yellow when fresh (sometimes faded to pale yellow on herbarium specimens). Fruits 4–5 mm long. $2n=22$ ($2n=44$ elsewhere). May–July.

Introduced, known only from the city of St. Louis (Montana to Wisconsin south to Texas). Railroads.

A number of historical specimens from Missouri attributed to this species were redetermined as *E. pallida* during Ronald McGregor's (1968) research. These mostly represented fruiting specimens, immature plants in which the ray corollas were not fully elongated or drooping, or generally incomplete specimens. Steyermark (1963) cited a single specimen from a prairie in Shelby County to document the presence of *E. angustifolia* in Missouri, but this collection could not be located during the present study and likely was misdetermined. However, a collection made by Viktor Mühlenbach during his botanical survey of the St. Louis railyards does document the species in Missouri. In 1990, Michael Skinner of the Missouri Department of Conservation discovered plants on a dolomite glade in Ozark County that had somewhat shorter ray corollas than is usual for *E. pallida,* and for a time this population was thought to represent an extant native occurrence of *E. angustifolia*. However, recent critical examination of the sheet at the Missouri Botanical Garden Herbarium by Craig Freeman of the University of Kansas revealed that because of its drooping rays and pollen color, this population is better considered to represent atypical plants of *E. pallida*. The natural range of *E. angustifolia* approaches the Missouri border in adjacent Iowa, Nebraska, and Kansas, and it is in northwestern Missouri that this species should be sought in the future.

McGregor (1968) segregated var. *strigosa* for plants from Kansas, Oklahoma, and Texas with more appressed pubescence. In their morphometric analysis of the genus, Binns et al. (2002) were unable to discriminate these plants from those attributed to var. *angustifolia* in the portion of the species range where the two types co-occur. Cronquist (1955, 1980; Gleason and Cronquist, 1991) and Binns et al. (2002) also treated *E. angustifolia* and several other species as varieties of *E. pallida,* but this obscures the biosystematic differences between the taxa (McGregor, 1968; Baskin et al., 1993).

2. **Echinacea pallida** (Nutt.) Nutt. (pale purple coneflower)
Brauneria pallida (Nutt.) Britton
Pl. 277 c, d; Map 1172

Plants with a usually elongated, vertical rootstock and often somewhat tuberous main roots, sometimes also with short, stout rhizomes. Stems (40–)60–150 cm long, mostly unbranched, sparsely to moderately pubescent with stiff, spreading, minutely pustular-based hairs. Leaves with the margins entire and usually pubescent with spreading hairs, the surfaces moderately to densely pubescent with stiff, mostly spreading, mostly minutely pustular-based hairs, moderately to strongly roughened to the touch, with 3(5) main veins. Basal leaves 8–35 cm long, the blade narrowly elliptic to narrowly lanceolate or lanceolate, mostly 5–20 times as long as wide, long-tapered or narrowly angled at the base. Stem leaves 4–25 cm long, linear to narrowly elliptic or narrowly lanceolate, otherwise similar to the basal leaves. Involucral bracts 7–15 mm long, the outer surface moderately pubescent with mostly pustular-based hairs, not glandular. Receptacle 2–4 cm in diameter, the chaffy bracts 9–14 mm long, hardened, usually dark purple toward the tip. Ray florets with the corolla (3–)4–9 cm long, 5–8 mm wide, reflexed or drooping at flowering, pale pink to pink (rarely white elsewhere). Disc florets with the corolla 6–8 mm long, the tube

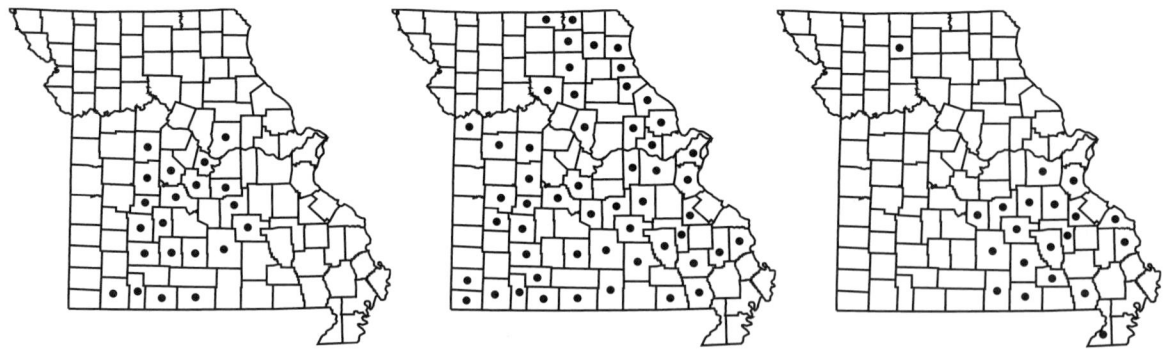

1173. Echinacea paradoxa 1174. Echinacea purpurea 1175. Echinacea simulata

yellow to green, the lobes pink to dark purple. Pollen white when fresh. Fruits 2.5–5.0 mm long. $2n=44$. May–July.

Scattered to common nearly throughout the state but apparently absent from the Mississippi Lowlands Division (Nebraska to Texas east to Indiana and Georgia; Canada; introduced sporadically elsewhere in the eastern U.S.). Upland prairies, glades, savannas, and openings of dry upland forests, also pastures, railroads, and roadsides.

McGregor (1968) suggested that *E. pallida* is an allopolyploid derived from past hybridization between *E. simulata* and the closely related *E. sanguinea* Nutt. (of Oklahoma, Arkansas, Texas, and Louisiana). Binns et al. (2002) chose to interpret all three of these taxa as varieties of *E. pallida*. Rare plants with white ray corollas occur as isolated individuals within some populations and have been called f. *albida* Steyerm.

3. **Echinacea paradoxa** (Norton) Britton **var. paradoxa** (yellow coneflower)
   Brauneria paradoxa Norton
   E. atrorubens (Nutt.) Nutt. var. *paradoxa* (Norton) Cronquist
   Pl. 277 g; Map 1173

Plants with a usually elongated, vertical rootstock and often somewhat tuberous main roots, sometimes also with short, stout rhizomes. Stems 40–90 cm long, unbranched or rarely with 1 or 2 ascending branches, sparsely to moderately pubescent with stiff, appressed or ascending, broad-based hairs. Leaves with the margins entire and usually pubescent with loosely appressed hairs, the surfaces moderately to densely pubescent with stiff, appressed to loosely appressed, sometimes minutely pustular-based hairs, slightly to moderately roughened to the touch, with 3 or 5 main veins. Basal leaves 8–45 cm long, the blade narrowly elliptic to narrowly lanceolate, oblong-lanceolate, or lanceolate, mostly 5–20 times as long as wide, long-tapered or narrowly angled at the base. Stem leaves 4–35 cm long, linear to narrowly elliptic or narrowly lanceolate, otherwise similar to the basal leaves. Involucral bracts 7–12 mm long, the outer surface glabrous (except along the margins) or sparsely pubescent with mostly appressed hairs, not glandular. Receptacle 2.0–3.5 cm in diameter, the chaffy bracts 9–14 mm long, hardened, usually reddish purple or orange toward the tip. Ray florets with the corolla 3–7 cm long, 5–8 mm wide, reflexed or drooping at flowering, yellow. Disc florets with the corolla 4–6 mm long, the tube pale yellow, the lobes yellow or pinkish-tinged. Pollen yellow when fresh. Fruits 4.0–5.5 mm long. $2n=22$. May–June.

Scattered mostly in the western half of the Ozark and Ozark Border Divisions (Missouri, Arkansas). Limestone and dolomite glades, upland prairies, and savannas; also roadsides.

This species is nearly endemic to the Ozarks. McGregor (1968) noted the occurrence of putative hybrids between this species and both *E. pallida* and *E. simulata*. A single individual that apparently represents *E. paradoxa* × *simulata* was observed but not collected by a survey team from the Missouri Department of Conservation at a glade restoration site in Ozark County. Such hybrids should be looked for elsewhere in Missouri where *E. paradoxa* grows in proximity to other species of *Echinacea*. The suspected hybrids usually are easily distinguished from either parent by the mottled or variegated ray corollas, which sometimes are a mixture of pink and purple and other times may be orangish. Researchers at the Chicago Botanic Garden currently are breeding various artificial hybrid combinations involving yellow- and pink-rayed coneflowers and recently released a bright-orange-rayed cultivar into commercial horticulture for the first time.

Plants with pink to purple (rarely white) ray corollas and somewhat hairy fruits that are endemic

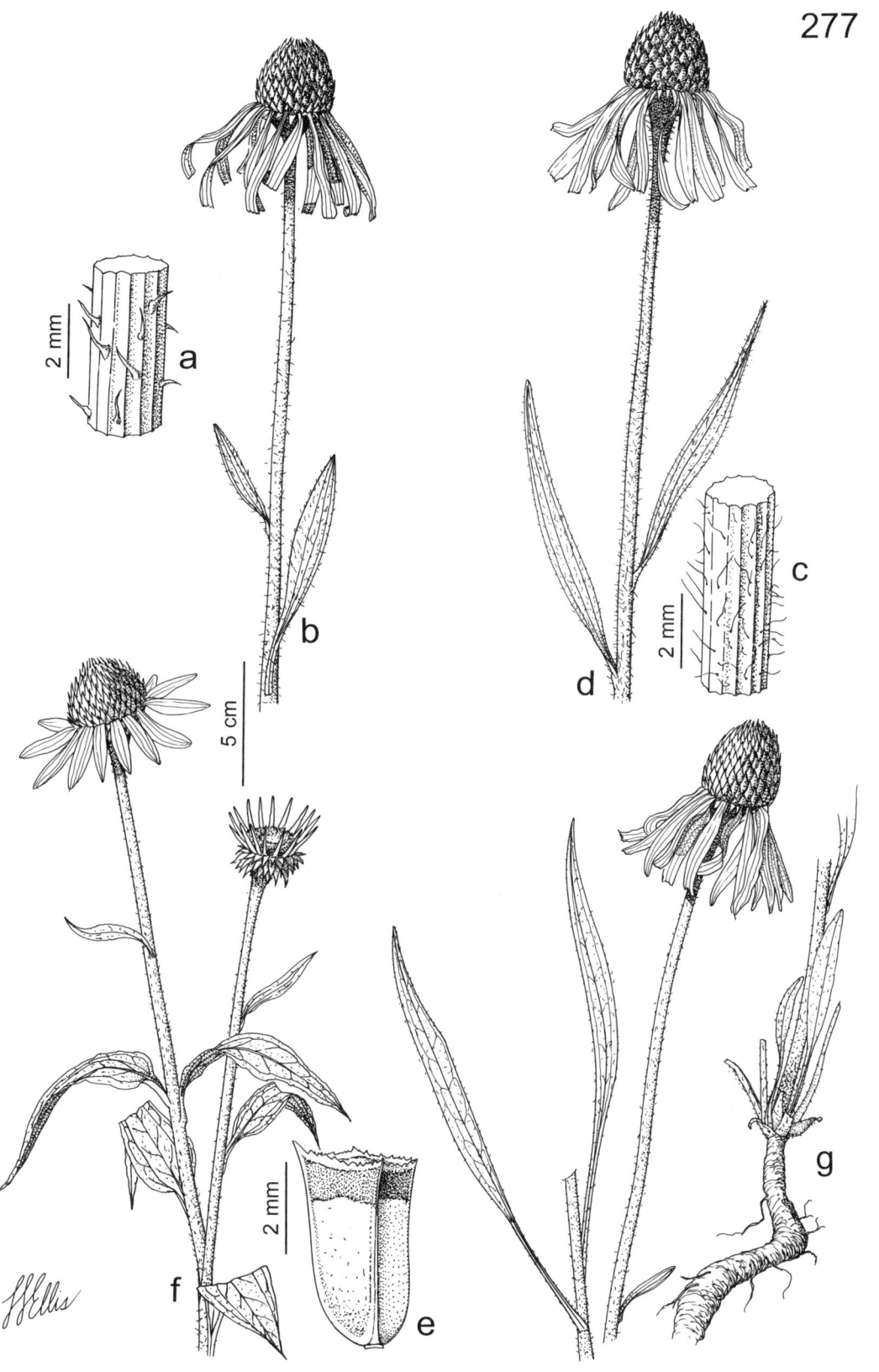

Plate 277. Asteraceae. *Echinacea angustifolia*, **a)** stem detail, **b)** stem tip. *Echinacea pallida*, **c)** stem detail, **d)** stem tip. *Echinacea purpurea*, **e)** fruit, **f)** stem tip. *Echinacea paradoxa*, **g)** stem tip, leaves, and rootstock.

to portions of Oklahoma and Texas were segregated by McGregor (1968) as var. *neglecta* McGregor. Binns et al. (2002) chose to emphasize the close relationship of the two varieties of *E. paradoxa* by treating them as varieties of *E. atrorubens,* a relatively uncommon species with the ray corollas short, spreading, and dark red that is restricted to Kansas, Oklahoma, and Texas.

4. **Echinacea purpurea** (L.) Moench (purple coneflower)

Pl. 277 e, f; Map 1174

Plants with a short rootstock and fibrous roots. Stems 50–150 cm long, unbranched or with few to several ascending branches, sparsely to moderately pubescent with stiff, appressed or ascending, broad-based (often pustular-based) hairs. Leaves with the margins usually irregularly toothed and pubescent with ascending hairs, the surfaces moderately to densely pubescent with stiff, appressed to loosely appressed, usually minutely pustular-based hairs, slightly to more commonly moderately roughened to the touch, with (3)5 main veins. Basal leaves 8–45 cm long, the blade narrowly ovate to broadly ovate, mostly 1.5–5.0 times as long as wide, often rounded or cordate at the base. Stem leaves 4–35 cm long, mostly narrowly ovate to broadly ovate, short-tapered to more or less rounded at the base, otherwise similar to the basal leaves. Involucral bracts 7–12 mm long, the outer surface glabrous (except along the margins) or sparsely to moderately pubescent with loosely appressed hairs, not glandular. Receptacle 2–4 cm in diameter, the chaffy bracts 9–15 mm long, somewhat hardened and leathery, usually orange or reddish-tinged toward the tip. Ray florets with the corolla 3–8 cm long, 7–14(–19) mm wide, spreading to somewhat drooping at flowering, pink to purple (rarely white). Disc florets with the corolla 4.5–6.0 mm long, the tube green, the lobes green or purplish-tinged. Pollen yellow when fresh. Fruits 3–5 mm long. $2n=22$. May–October.

Scattered nearly throughout the state but uncommon or absent from the Mississippi Lowlands Division and the western portion of the Glaciated Plains (eastern U.S. west to Kansas and Texas; introduced sporadically farther west and north). Mesic to dry upland forests, savannas, upland prairies, ledges and tops of bluffs, banks of streams, and rarely fens and sinkholes; also pastures, old fields, railroads, and roadsides.

The name *E. purpurea* as used here was officially conserved at the 2005 International Botanical Congress. Binns et al. (2001a, b) had pointed out that when Linnaeus originally described *Rudbeckia purpurea* L. he was referring to plants from the eastern United States that most modern botanists have called *E. laevigata* (C.L. Boynton & Beadle) S.F. Blake.

This species is commonly cultivated as an ornamental in gardens, and a number of hybrids and cultivars are available commercially. Rare plants with white ray corollas occur as isolated individuals within some populations and have been called f. *liggettii* Steyerm. Occasionally, plants are observed in gardens and in the wild with abnormal heads of two main types. Plants exposed to excess water through flooding or overwatering early in the growing season sometimes develop crested heads in which the disc becomes irregularly enlarged, broadened, and somewhat flattened. It is not known whether this condition is caused by direct injury to the meristems of the developing flowering stems or whether waterlogging leaves the plants susceptible to infection by some fungus or other microorganism. Another problem that occasionally afflicts *E. purpurea* plants (and also a variety of other unrelated crop and wild plants) is known as aster yellows and is caused by a group of prokaryotic microorganisms called aster yellows phytoplasmas (*Candidatus* Phytoplasma), which reside in the phloem of infected plants and are spread between plants by foraging leafhoppers of the genus *Macrosteles* Fieber (Delahaut, 1997; Lee et al., 2004; IRPCM Phytoplasma/Spiroplasma Working Team—Phytoplasma Taxonomy Group, 2004). Diseased plants generally have stunted or twisted, often chlorotic foliage, and the corollas often are malformed and greenish. Frequently the disc in infected plants has irregularly elongate stalks with small, headlike clusters of sterile disc florets at the tips. As there is no cure for this condition, plants in a garden should be removed to minimize the risk of spreading the infection.

5. **Echinacea simulata** McGregor (glade coneflower, pale purple coneflower)
E. speciosa McGregor (1968), not *E. speciosa* (Wender.) Paxton (1849)
E. pallida var. *simulata* (McGregor) Binns, B.R. Baum & Arnason

Map 1175

Plants with a usually elongated, vertical rootstock and often somewhat tuberous main roots, sometimes also with short, stout rhizomes. Stems (40–)60–120 cm long, mostly unbranched, sparsely to moderately pubescent with stiff, spreading, minutely pustular-based hairs. Leaves with the margins entire and usually pubescent with loosely appressed hairs, the surfaces moderately to densely pubescent with stiff, mostly spreading, mostly minutely pustular-based hairs, moderately to strongly

1176. Eclipta prostrata   1177. Engelmannia peristenia   1178. Flaveria campestris

roughened to the touch, with 3(5) main veins. Basal leaves 8–35 cm long, the blade narrowly elliptic to narrowly lanceolate or lanceolate, mostly 5–20 times as long as wide, long-tapered or narrowly angled at the base. Stem leaves 4–25 cm long, linear to narrowly elliptic or narrowly lanceolate, otherwise similar to the basal leaves. Involucral bracts 7–15 mm long, the outer surface moderately pubescent with mostly pustular-based hairs, not glandular. Receptacle 2–4 cm in diameter, the chaffy bracts 9–14 mm long, hardened, usually dark purple toward the tip. Ray florets with the corolla 4–9 cm long, 4–7 mm wide, reflexed or drooping at flowering, pale pink to purplish pink (rarely white elsewhere). Disc florets with the corolla 5–7 mm long, the tube yellow to green, the lobes pink to dark purple. Pollen bright yellow to lemon yellow when fresh. Fruits 3.0–4.5 mm long. $2n=22$. May–July.

Scattered mostly in the eastern half of the Ozark and Ozark Border Divisions (Missouri, Arkansas, Kentucky, Tennessee, Alabama, and Georgia; apparently introduced in Illinois, North Carolina, and Virginia). Limestone and dolomite glades, tops of bluffs, savannas, and edges and openings of dry upland forests; also ditches and roadsides.

As noted above, *E. simulata* apparently is one of the parental taxa that gave rise to *E. pallida*. It can be very difficult to distinguish from that species when pollen is not being shed. McGregor noted that the pollen in *E. pallida* is larger (24.0–28.5 vs. 22.5–24.5 μm in diameter), as are the guard cells of the stomates. Anatomically, in *E. simulata* the median vascular bundle in the petiole is fan-shaped (vs. circular in cross-section in *E. pallida*). Additionally, in *E. simulata* the ray corollas tend to be a slightly darker pink, and whereas *E. pallida* occurs both in prairie and glade habits, *E. simulata* has not been recorded growing in any prairies thus far in Missouri. Hybrids between the tetraploid ($2n=44$) *E. pallida* and its diploid ($2n=22$) parent, *E. simulata*, are sterile triploid individuals. These potentially occur in southern Missouri but have not yet been documented from the state.

72. Eclipta L.

Two to 4 species, nearly worldwide in tropical and warm-temperate regions.

1. Eclipta prostrata (L.) L. (yerba de tajo)
E. alba (L.) Hassk.

Pl. 278 a–c; Map 1176

Plants annual, sometimes forming mats, with small taproots. Stems 5–75 cm long, spreading to erect, sometimes ascending from a spreading base, sometimes rooting at the nodes, few- to many-branched, with fine longitudinal ridges and grooves, sparsely to densely pubescent with appressed-ascending hairs, especially toward the tip, sometimes glabrous toward the base. Leaves opposite, sessile or with a short, poorly differentiated petiole, the base slightly expanded and wrapping around the stem. Leaf blades 1–10 cm long, narrowly lanceolate or narrowly elliptic to lanceolate or elliptic, unlobed, tapered or narrowly angled at the base, tapered to a sharply pointed tip, the margins sparsely to densely and finely toothed and with short, stiff, appressed hairs, the surfaces sparsely to moderately pubescent with short, stiff, appressed hairs, these usually with a minutely expanded, somewhat resinous, pustular base. Inflorescences of solitary terminal and/or axillary heads or sometimes in small clusters of 2 or 3, the heads appearing short- to long-stalked (the stalks bractless). Heads radiate but sometimes appearing discoid. Involucre 2–5 mm long, 3–5 mm in diameter (becoming broader as the fruits mature), cup-shaped to broadly cup-shaped or slightly bell-shaped, the bracts in usually 2 subequal series (the inner series sometimes slightly shorter than the others). Involucral bracts 10–12, green but with the basal half lighter and prominently several-nerved, mostly ascending at the tip (the outer series sometimes loosely ascending at the tip), narrowly lanceolate to ovate. Receptacle flat to shallowly convex, broadening as the fruits mature, the florets subtended by chaffy bracts, these narrowly linear, awnlike, and shed with the florets. Ray florets 20 to numerous in 2 or 3 series, pistillate (with a 2-branched style exserted from the short tube at flowering), the corolla 1–3 mm long, relatively slender, white, glabrous, not persistent at fruiting. Disc florets 15 to numerous (more than 100), perfect, the corolla 1.0–1.5 mm long, white, the short tube not expanded at the base or persistent at fruiting, glabrous, the mostly 4 minute lobes glabrous. Style branches with the sterile tip short-tapered to a bluntly pointed tip. Pappus absent or a minute rim or crown. Fruits 2.0–2.5 mm long, more or less wedge-shaped to somewhat triangular in outline, 3- or 4-angled, sometimes slightly flattened, the surface usually appearing wrinkled or warty, straw-colored to brown or less commonly black, the tip minutely hairy and usually darker than the rest of the fruit (green when immature). $2n=22$. July–October.

Scattered to common throughout the state (eastern U.S. west to South Dakota, Texas, and California, Mexico, Central America, South America, Caribbean Islands, Asia south to Australia; introduced in Hawaii, Canada, and Europe). Banks of streams, rivers, and spring branches, marshes, seeps, and margins of ponds, lakes, sinkhole ponds, and sloughs; also crop fields, fallow fields, ditches, gardens, railroads, roadsides, and moist, open, disturbed areas.

The native distribution of this species is poorly understood as it is a widespread colonizer of disturbed habitats. *Eclipta prostrata* was long known as *E. alba* in the literature on New World plants. Koyama and Boufford (1981) discussed the taxonomy of this complex in the context of a proposal to make a minor change in the wording of an article in the International Code of Botanical Nomenclature. They noted that plants growing in the Old World could not be distinguished from those in the New World and that plants from throughout the combined range should be united under the name *E. prostrata*.

73. Engelmannia Torr. & A. Gray ex Nutt.

One species, southwestern U.S., Mexico.

1. Engelmannia peristenia (Raf.) Goodman & C.A. Lawson (Engelmann's daisy)
 E. pinnatifida A. Gray ex Nutt.
 Pl. 278 d, e; Map 1177

Plants perennial herbs, with somewhat woody taproots. Stems 30–100 cm long, usually few-branched toward the tip, erect or ascending, finely ridged or grooved, moderately to densely pubescent with short and long, spreading and ascending, mostly pustular-based hairs. Leaves alternate and basal, the basal ones short- to long-petiolate, the stem leaves mostly sessile (the lowermost sometimes short-petiolate). Leaf blades 1–30 cm long, the uppermost linear to narrowly lanceolate and sometimes entire, the main stem leaves ovate to oblong-ovate in outline, deeply pinnately lobed, the lobes mostly narrowly oblong to linear (less commonly narrowly lanceolate), the basal lobes sometimes slightly clasping the stem, rounded or angled or tapered to a bluntly or sharply pointed tip, the margins often few-toothed to deeply lobed again, somewhat thickened, and strongly roughened with short, stout, curved hairs, the surfaces strongly roughened with short, stout, curved hairs. Inflorescences of solitary heads or small clusters at the branch tips, sometimes appearing as open panicles, the heads with short to relatively long, mostly bractless, densely hairy stalks. Heads radiate. Involucre 6–10 mm long, 7–12 mm in diameter, cup-shaped, the bracts in 3 slightly unequal, overlapping series. Involucral bracts 21–29, with a linear to narrowly oblong-lanceolate, green, spreading to loosely ascending tip (this often very short in the inner series) above a broader, thickened, yellowish, appressed base, the green portion roughened with short, stout, pustular-based hairs, the basal portion with finer, more or less appressed hairs, mostly with a single midnerve visible, the basal portion sometimes with a faint pair of additional nerves. Receptacle flat or shallowly convex, not

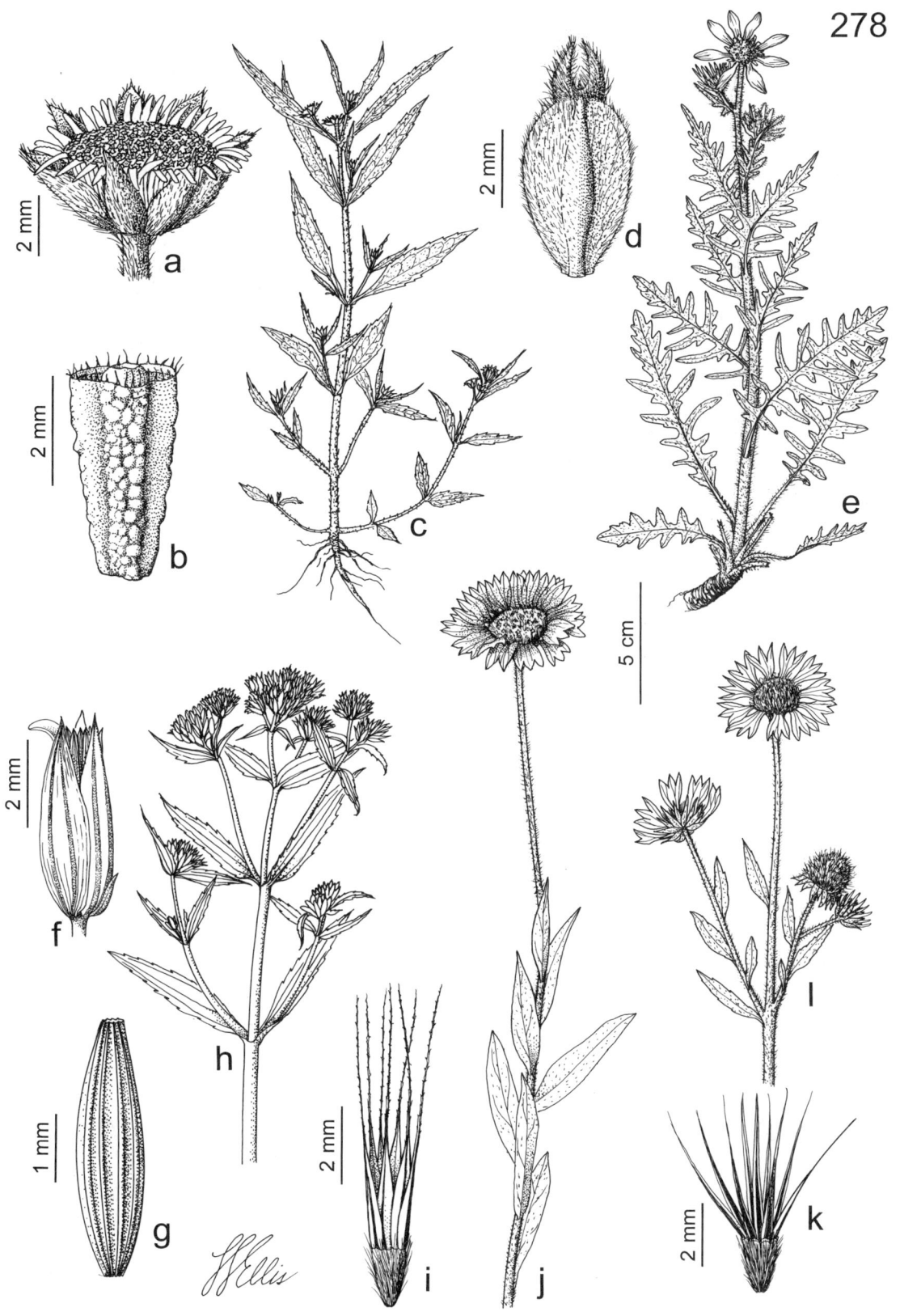

Plate 278. Asteraceae. *Eclipta alba*, **a)** head, **b)** fruit, **c)** habit. *Engelmannia peristenia*, **d)** fruit, **e)** habit. *Flaveria campestris*, **f)** head, **g)** fruit, **h)** habit. *Gaillardia pulchella*, **i)** fruit, **j)** habit. *Gaillardia aestivatlis*, **k)** fruit, **l)** habit.

elongating as the fruits mature, with chaffy bracts subtending the disc florets, these 5–6 mm long, linear to narrowly oblong, bristly-hairy, especially toward the tip, relatively flat and not wrapped around the florets. Ray florets 7–9, pistillate (with a 2-branched style exserted from the short tube at flowering), the corolla 10–18 mm long, relatively broad, yellow, the tube glabrous or sparsely hairy, the ligule with the undersurface sparsely to moderately hairy, not persistent at fruiting. Disc florets 25–50, staminate (the nonfunctional ovary slender and hairy), the corolla 3.5–5.0 mm long, yellow, glabrous or sparsely hairy toward the tip, not expanded at the base or persistent at fruiting. Style branches with the sterile tip very short and more or less truncate. Pappus of the ray and disc florets of 2–4 small, hairy scales. Fruits 3.0–4.5 mm long, obovate, flattened but somewhat 4-angled, tan to brown, moderately to densely and minutely hairy, more or less fused at the base with the adjacent disc florets, chaffy bracts, and adjacent involucral bract (this tan to brown at fruiting), the whole assemblage often dispersed as an intact unit. $2n=18$. June–August.

Introduced, known only from historical collections from Jackson County (Arizona to Texas north to Colorado, South Dakota, and Arkansas; Mexico). Habitat unknown but possibly railroads.

At first glance, the occurrence of this species in Missouri would seem to be native. The labels on the specimens collected by B. F. Bush in 1892 contain no data other than the collection date, county name, and "rare." However, Steyermark (1963) indicated that it was nonnative and found along train tracks. Perhaps he had the opportunity early in his career to discuss this species with Bush during a visit. Furthermore, in his flora of Jackson County, Mackenzie (1902) indicated that the plant had been collected by Bush as a waif in the Sheffield area (along the Big Blue River in northeastern Kansas City).

Engelmannia peristenia is an attractive species that is sometimes cultivated as an ornamental.

74. Flaveria Juss.
(Powell, 1978)

Twenty-one species, North America to South America, Caribbean Islands, Africa, Asia, Australia.

1. Flaveria campestris J.R. Johnst. (alkali goldentops)

Pl. 278 f–h; Map 1178

Plants annual, with taproots. Stems 20–75 cm long, erect or ascending, several-branched or sometimes unbranched below the inflorescence, with several usually relatively conspicuous longitudinal ridges and grooves, glabrous except for small tufts of spreading hairs at some of the nodes. Leaves opposite, sessile, the bases of each pair slightly expanded and wrapped around the stem. Leaf blades 3–7 cm long, narrowly lanceolate to lanceolate or narrowly elliptic, simple, tapered to a sharply pointed tip, angled or tapered at the base, the margins finely toothed to nearly entire, the teeth sometimes minutely spinelike or bristlelike at the tip, the surfaces glabrous, the larger leaves usually with 3(5) main veins. Inflorescences of dense, sessile to very short-stalked clusters of 12–40 or more heads subtended by a pair of leaflike bracts, these at the branch tips and on short branches from the upper leaf axils, sometimes appearing as small, open, more or less flat-topped panicles. Heads radiate but the ray florets inconspicuous or occasionally absent from some heads. Involucre 5–7 mm long, urn-shaped to more or less cylindrical, the receptacle 1–3 mm in diameter, the bracts in 2 unequal series. Involucral bracts 5, those of the outer series 2, 1–3 mm long, narrowly lanceolate, those of the inner series 3, 5–7 mm long, elliptic to oblong-obovate, bluntly angled along the midnerve, both series glabrous (not aromatic or glandular), yellowish green. Receptacle shallowly convex, not elongating as the fruits mature, naked. Ray floret absent or more commonly 1 per head, pistillate (with a 2-branched style exserted from the short tube at flowering), the corolla 1.5–2.5 mm long, yellow. Disc florets 3–8 (but often only 1–3 producing mature fruits), perfect, the corolla 2.5–3.5 mm long, yellow, the tube not expanded at the base or persistent at fruiting, glabrous. Style branches with the sterile tip somewhat elongate and tapered to a bluntly pointed tip. Pappus absent. Fruits 2.5–3.5 mm long, narrowly oblanceolate to narrowly oblong-oblanceolate in outline, somewhat flattened, the surface with 10 slender longitudinal ridges, glabrous, dark gray to black. $2n=36$. July–September.

Introduced, uncommon, known only from historical collections from Jackson and Marion Counties (Utah and Arizona east to Kansas and Texas; introduced in Missouri, Indiana). Floodplains, also railroads and disturbed areas.

75. Gaillardia Foug. (gaillardia, blanketflower)

Plants annual or perennial herbs, often with taproots. Stems erect or ascending, few- to many-branched, with fine longitudinal ridges and grooves, moderately to densely pubescent with curved or curled hairs, some of the hairs sometimes gland-tipped. Leaves alternate and sometimes also basal (all basal elsewhere), long-petiolate to sessile, the base sometimes slightly expanded or clasping the stem. Leaf blades oblanceolate to oblong or lanceolate, unlobed or with rounded pinnate lobes, angled or tapered to more or less rounded at the base, rounded or angled or tapered to a sharply or bluntly pointed tip, the margins otherwise entire, wavy or toothed, the surfaces densely pubescent with short, curved or curled hairs, also dotted with sessile to impressed glands, usually somewhat roughened to the touch. Inflorescences of solitary heads terminal on the branches, the heads appearing long-stalked. Heads radiate. Involucre cup-shaped to shallowly cup-shaped, the bracts in 2 or 3 subequal series. Involucral bracts 15–28, green with a sometimes yellowish base, inconspicuously 1-nerved, spreading to reflexed at flowering, lanceolate to ovate, sometimes somewhat concave, the surfaces and margins densely hairy and usually also glandular. Receptacle strongly convex, often somewhat enlarging as the fruits mature, the florets usually subtended by numerous bristles, these sometimes fused irregularly at the base, usually straw-colored, occasionally reduced to a network of low, irregular teeth. Ray florets 6–15 in 1 series (absent elsewhere), sterile (lacking stamens and style at flowering and with an ovary that is shorter and thinner than those of the disc florets, not developing into a fruit), the corolla relatively broad above a slender base, yellow or red, brownish red, or reddish purple, at least toward the base, the tubular portion and often also the ligule with dense, somewhat tangled hairs and also glandular, not persistent at fruiting. Disc florets 40 to numerous (more than 100), perfect, the corolla 4–7 mm long, yellow, orangish red, purple, or purplish brown, the tube not expanded at the base or persistent at fruiting, usually woolly toward the tip, the 5 sharply pointed lobes densely woolly. Style branches with the sterile tip long-tapered to a sharply pointed tip. Pappus of 6–10 scales, these with a somewhat thickened, orangish brown midnerve and thin, nearly transparent margins, the lanceolate to ovate basal portion tapered to a relatively long, awned tip. Fruits somewhat wedge-shaped, more or less 4-angled, the surface densely pubescent with relatively long, yellowish, ascending hairs, black. Fifteen to 18 species, North America, South America.

1. Ray florets with the corollas yellow; disc florets with the corollas yellow; receptacle with the florets subtended by a network of low, irregular teeth
 . . . . 1. G. AESTIVALIS
1. Ray florets with the corollas red, brownish red, or reddish purple, sometimes yellowish toward the tip; disc florets with the corollas orangish red, purple, or purplish brown; receptacle with the florets subtended by numerous bristles
 . . . . 2. G. PULCHELLA

1. Gaillardia aestivalis (Walter) H. Rock **var. flavovirens** (C. Mohr) Cronquist (prairie blanketflower; prairie gaillardia)
 G. *lutea* Greene

Pl. 278 k, l; Map 1179

Plants short-lived perennial herbs but frequently flowering the first growing season, sometimes with slender, deep-set rhizomes. Stems 20–60 cm long. Basal leaves usually absent at flowering. Stem leaves mostly sessile, the lowermost usually short-petiolate. Leaf blades 1–6 cm long, narrowly oblong-elliptic to lanceolate or nearly linear, tapered at the base, those of the median and upper leaves more or less rounded and somewhat

1179. Gaillardia aestivalis   1180. Gaillardia pulchella   1181. Galinsoga parviflora

clasping the stem, angled or tapered to a sharply pointed tip, unlobed, the margins entire or toothed. Involucre 8–12 mm long, 18–25 mm in diameter, the bracts lanceolate to narrowly ovate. Receptacle with the florets subtended by a network of low, irregular, papery teeth. Ray florets with the corollas 12–25 mm long, yellow, the ligule frequently with both surfaces hairy and glandular. Disc florets with the corollas yellow. Pappus of 8–10 scales, these 5–7 mm long. Fruits 1.5–2.0 mm long. 2n=17, 34 (other varieties counted). May–October.

Uncommon, known thus far only from historical collections from Dunklin, Mississippi, and Scott Counties (Florida to Texas north to Illinois and Missouri). Sand prairies; also roadsides and open, sandy, disturbed areas.

This taxon is known historically from the Crowley's Ridge Section of the Mississippi Lowlands, especially in the northeastern portion in the Sikeston Ridge extension where most of the state's sand prairies occurred. Interestingly, the name *G. lutea,* which was applied to this taxon for many years (Biddulph, 1944; Steyermark, 1963), originally was described based on a specimen collected by B. F. Bush in Dunklin County. Turner (1979) discussed the three varieties of *G. aestivalis,* each with a different corolla color combination: var. *aestivalis,* with orangish brown disc corollas and reddish orange ray corollas; var. *flavovirens,* with yellow disc and ray corollas; and var. *winkleri,* with white disc and ray corollas. The var. *aestivalis* is widespread in the southeastern United States west to Kansas and Texas, but the white-flowered variety is endemic to southeastern Texas.

2. Gaillardia pulchella Foug. **var. pulchella**
(Indian blanket, fire-wheels)
Pl. 278 i, j; Map 1180

Plants annual. Stems 30–60 cm long. Basal leaves often present at flowering, long-petiolate, the blades oblong-elliptic to obovate, usually with several blunt, shallow to deep, pinnate lobes, tapered at the base, rounded or broadly angled to a bluntly pointed tip. Stem leaves short-petiolate to sessile, the lowermost sometimes petiolate. Leaf blades 1–8(–12) cm long, narrowly oblong-elliptic to lanceolate (those of the lowermost leaves sometimes narrowly obovate and tapered at the base), those of the median and upper leaves more or less rounded and clasping the stem, rounded or more commonly angled to a bluntly or sharply pointed tip, unlobed or with few to occasionally several shallow to deep pinnate lobes, the margins otherwise entire, wavy, or toothed. Involucre 8–16 mm long, 18–28 mm in diameter, the bracts lanceolate to narrowly lanceolate or narrowly triangular. Receptacle with the florets subtended by numerous bristles, these 2–3 mm long, stiff, sometimes fused irregularly at the base. Ray florets with the corollas 15–28 mm long, red, brownish red, or reddish purple, sometimes yellowish toward the tip, the ligule usually with only the outer surface hairy and glandular. Disc florets with the corollas orangish red, purple, or purplish brown. Pappus of 6–8 scales, these 4–7 mm long. Fruits 2.0–2.5 mm long. 2n=34. May–September.

Introduced, uncommon and widely scattered in the state although undoubtedly more widespread than collections indicate (southwestern U.S. east to Nebraska and Texas, east along the Gulf Coastal Plain to Florida; Mexico; introduced farther north). Fallow fields, railroads, roadsides, and open, disturbed areas.

This attractive exotic species regrettably was seeded along some highway medians and roadsides in the past by the Missouri Department of Transportation and other public agencies along with other native and nonnative species (see also under the genus *Cosmos,* above) in so-called native wildflower plantings. Some of the plants that have become naturalized as a result of this practice are atypical and may represent cultivars or results of past hybridization with some other species of *Gaillardia*. Turner and Whalen (1975) segregated two additional geographically restricted varieties: Missouri plants correspond to the widespread var. *pulchella,* and plants along Gulf Coastal sand dunes with somewhat spreading stems and thickened leaves are referred to var. *picta* (Sweet) A. Gray. The same researchers named a series of populations in southern Texas and adjacent Mexico with most of the stem leaves lobed as var. *australis* B.L. Turner & M.A. Whalen.

76. Galinsoga Ruiz & Pav. (quickweed)
(Canne, 1977)

Plants annual, with small taproots. Stems 8–70 cm long, erect or ascending, few- to many-branched, with fine longitudinal ridges and grooves, glabrous or sparsely to moderately pubescent with appressed or spreading hairs, the hairs sometimes gland-tipped toward the stem tip. Leaves opposite, long-petiolate to nearly sessile, the base slightly expanded and wrapping around the stem. Leaf blades lanceolate to broadly ovate, unlobed, mostly angled to short-tapered at the base, angled or tapered to a sharply pointed tip, the margins entire or more commonly sparsely to moderately toothed and with short, slender, more or less spreading hairs, the surfaces sparsely to densely pubescent with short, slender, more or less spreading hairs, usually with 3 main veins. Inflorescences of irregular, open, terminal panicles or sometimes merely loose clusters, these sometimes also from the upper leaf axils, rarely reduced to solitary heads, with small, leaflike bracts at the branch points, the heads appearing mostly long-stalked. Heads radiate. Involucre 2.5–5.0 mm long, 2.5–6.0 mm in diameter, cup-shaped to somewhat bell-shaped, the bracts in 2 somewhat unequal series. Involucral bracts 5–9, the outer series of 1–3 bracts somewhat shorter and narrower than the 4–6 inner bracts, green but relatively thin, noticeably several-nerved, ascending at the tip, oblong-elliptic to broadly ovate, glabrous or sparsely hairy. Receptacle convex or short-conical, not or only slightly elongating as the fruits mature, the florets subtended by chaffy bracts, these narrowly lanceolate to oblong-obovate, concave and more or less wrapping around the florets. Ray florets (3–)5(–8) in 1 series, pistillate (with a 2-branched style exserted from the short tube at flowering), the corolla 1.5–3.0 mm long, relatively broad, white (sometimes light pink elsewhere), the tubular portion with minute, dense glandular hairs, not persistent at fruiting. Disc florets 5–50, perfect, the corolla 0.8–1.5 mm long, yellow, the tube not expanded at the base or persistent at fruiting, glabrous, the 5 minute lobes glabrous. Style branches with the sterile tip short-tapered to a sharply pointed tip. Pappus of 12–20 slender, fringed scales, sometimes highly reduced or absent in the ray florets. Fruits 1.5–2.5 mm long, narrowly wedge-shaped to wedge-shaped in outline, 4-angled, those of the ray florets usually slightly flattened, the surface usually with minute, yellowish, ascending hairs, black, those of the disc florets shed individually with their subtending chaffy bract, those of the ray florets either shed separately or as an intact unit with the basally fused adjacent 2 or 3 chaffy bracts and the adjacent inner involucral bract. About 15 species, nearly worldwide.

1. Ray florets with the pappus absent or reduced (much shorter than the corolla tube); disc florets with the pappus scales narrowed to a bluntly pointed, awnless tip; upper portions of the stem and the inflorescence branches with nonglandular hairs, the involucral bracts usually glabrous .... 1. G. PARVIFLORA
1. Ray florets with a well-developed pappus about as long as the corolla tube; disc florets with the pappus scales tapered to a sharply pointed, often minutely awned tip; upper portions of the stem, inflorescence branches, and often also the involucral bracts with gland-tipped hairs .... 2. G. QUADRIRADIATA

1182. Galinsoga quadriradiata   1183. Guizotia abyssinica   1184. Helenium amarum

1. Galinsoga parviflora Cav. (gallant soldier)

Pl. 279 a, b; Map 1181

Stems glabrous to moderately pubescent with more or less spreading, nonglandular hairs. Leaf blades 2–11 cm long, lanceolate to broadly ovate. Involucre 2.5–4.0 mm long, 2.5–5.0 mm in diameter, persistent at fruiting, usually glabrous, with an outer series of 1 or 2 bracts, these often with a thin white margin, and an inner series of 4–6 bracts, these fused basally with 2 or 3 adjacent chaffy bracts but not shed with the fruits developing from the ray florets. Chaffy bracts subtending the disc florets mostly deeply 3-lobed for more than 1/3 of their length, often persisting after the fruits have been shed. Ray florets with the pappus scales absent or reduced (much shorter than the corolla tube). Disc florets with the pappus scales narrowed to a bluntly pointed, awnless tip. $2n=16$. May–November.

Introduced, uncommon, known only from historical collections from Jackson County and the city of St. Louis (probably native from the southwestern U.S. to South America; introduced nearly worldwide). Roadsides and open, disturbed areas.

Steyermark (1963) included a dot for Jasper County on the map in his treatment of the species but did not note an occurrence from the county in his text. This apparently represents a specimen that he originally determined as *G. parviflora* but which subsequently was redetermined as *G. quadriradiata*. Canne (1977) discussed two intergrading morphotypes, with some plants from Mexico and the southwestern United States tending to have shorter stems, less-toothed to entire leaves, and denser clusters of heads. Some earlier authors had accepted these under the name var. *semicalva* A. Gray, but Canne argued that too much intergradation exists to allow formal recognition of infraspecific taxa.

2. Galinsoga quadriradiata Ruíz & Pav. (fringed quickweed)

G. ciliata (Raf.) S.F. Blake

Pl. 279 c, d; Map 1182

Stems moderately to densely pubescent with more or less spreading hairs, those toward the stem tip (and on the inflorescence branches) with minute, dark, glandular tips. Leaf blades 2–7 cm long, lanceolate to ovate. Involucre 3–5 mm long, 3–6 mm in diameter, not persistent at fruiting, usually sparsely pubescent with spreading, gland-tipped hairs, with an outer series of 2 or 3 bracts and an inner series of 4–6 bracts, these fused basally with 2 or 3 adjacent chaffy bracts and this group shed as an intact unit with the fruits developing from the ray florets. Chaffy bracts subtending the disc florets mostly unlobed or shallowly 2- or 3-lobed for less than 1/5 of their length, shed with the fruits. Ray florets with a well-developed pappus about as long as the corolla tube. Disc florets with the pappus scales tapered to a sharply pointed, often minutely awned tip. $2n=32, 48, 64$. May–November.

Introduced, widely scattered in the state although undoubtedly more widespread than collections indicate (probably native to Central America, South America; introduced nearly worldwide). Banks of rivers; also gardens, farmyards, railroads, roadsides, and open, disturbed areas.

Canne (1977) noted that in parts of tropical America a variant occurs with entirely nonglandular pubescence.

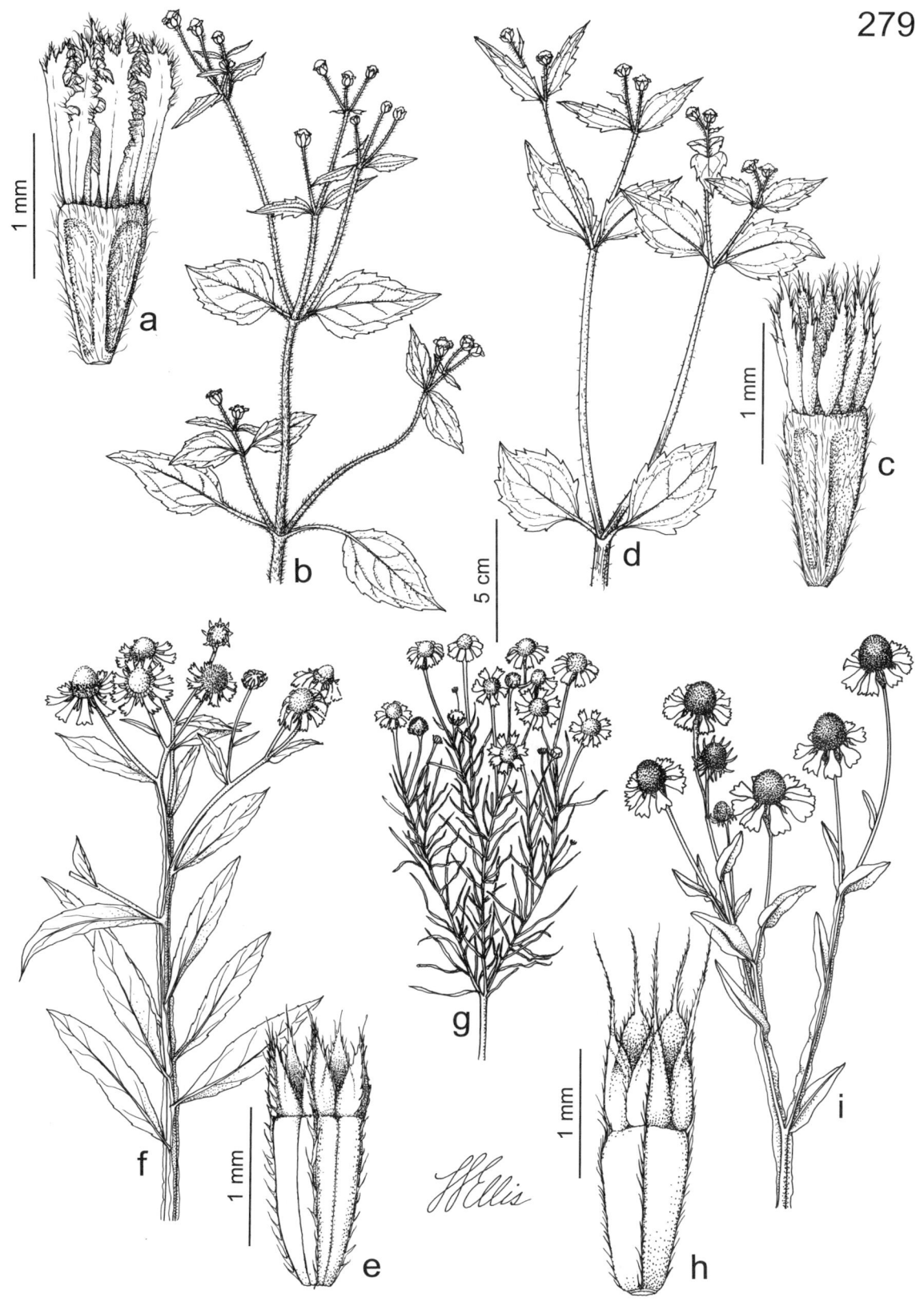

Plate 279. Asteraceae. *Galinsoga parviflora*, **a)** fruit, **b)** habit. *Galinsoga quadriradiata*, **c)** fruit, **d)** habit. *Helenium autumnale*, **e)** fruit, **f)** habit. *Helenium amarum*, **g)** habit. *Helenium flexuosum*, **h)** fruit, **i)** habit.

77. Guizotia Cass.
(Baagøe, 1974)

Six species, Africa.

1. Guizotia abyssinica (L. f.) Cass. (Niger thistle, Niger seed)

Map 1183

Plants annual, with taproots. Stems 40–100 (–150) cm long, loosely ascending to erect, several-branched, with several longitudinal lines or rounded ridges, the branches often strongly angled, sparsely to moderately pubescent with soft, broad-based hairs, often glabrous toward the base. Leaves opposite or the uppermost few alternate, sessile, the bases somewhat expanded and wrapping around the stem. Leaf blades 3–14 cm long, oblong-lanceolate to oblong-elliptic, rounded to shallowly cordate at the clasping base (sometimes tapered above the base and somewhat expanded again at the very base), tapered to a sharply pointed tip, the margins sharply toothed and short-hairy, the surfaces nearly glabrous to moderately or strongly roughened with minute, broad-based hairs, the undersurface also with scattered, sessile, spherical, yellow glands. Inflorescences of solitary heads or small, loose clusters at the branch tips, the heads mostly with relatively long, bractless, woolly stalks. Heads radiate. Involucre 7–10 mm long, 10–18 mm in diameter, broadly cup-shaped to somewhat bell-shaped, the bracts in 2 dissimilar series. Involucral bracts of the outer series 5 or 6, ascending, slightly longer than those of the inner series, ovate to elliptic-obovate, rounded to bluntly pointed at the tip, green, with thin, yellowish, slightly irregular or sparsely hairy margins, the outer surface sparsely to moderately pubescent with short, curved hairs, also sparsely glandular, with 5–11 fine nerves; involucral bracts of the inner series 8–11, ascending, narrowly oblong-obovate, rounded at the tip, straw-colored to light yellow, the thin margins with a fringe of short hairs toward the tip, the outer surface usually sparsely glandular, with usually 5 nerves. Receptacle conical, often elongating somewhat as the fruits mature, with chaffy bracts subtending the ray and disc florets, these similar to the inner series of involucral bracts, 5–7 mm long, relatively flat and not wrapped around the florets. Ray florets 6–13, pistillate (with a 2-branched style exserted from the short tube at flowering), the corolla 12–20 mm long, relatively broad, yellow, with a short band of dense, tangled, more or less spreading hairs toward the base, not persistent at fruiting. Disc florets numerous, perfect, the corolla 4–5 mm long, yellow, with a short band of dense, tangled, more or less spreading hairs toward the base, sparsely hairy toward the tip, not expanded at the base or persistent at fruiting. Style branches with the sterile tip somewhat elongate and tapered. Pappus of the ray and disc florets absent. Fruits 4–5 mm long, narrowly oblanceolate to nearly narrowly wedge-shaped in outline, somewhat flattened and bluntly 3- or 4-angled, smooth or sometimes with a few very fine, lighter lines or nerves, the surface otherwise glabrous, dark brown to black, shiny. $2n=30$. September–October.

Introduced, known thus far only from Howell and St. Louis Counties (native of Africa, introduced nearly worldwide). Roadsides and open, disturbed areas.

In parts of Asia and Africa, an oil extracted from the seeds is used similarly to that produced by the cultivated sunflower (*Helianthus annuus*). In the United States and Europe, Niger thistle is a popular component of birdseed and is often used pure to attract various species of finches and siskins in special tube-feeders with smaller-than-usual holes. The common name is a misnomer, as *Guizotia* is not a member of the thistle tribe, Cardueae, and the plants do not resemble thistles morphologically. The birdseed is sometimes sold under the name Nyjer seed, which was trademarked by the Wild Bird Feeding Industry trade association to differentiate their heat-sterilized product from true thistles and to eliminate possible offensive mispronunciation of the common name. Although most of the supply is imported from India and Africa (Purseglove, 1969), there has been interest among some farmers in the Midwest in domestic cultivation of *Guizotia* as a crop plant. Although the achenes of Niger thistle are supposed to be heat-sterilized prior to sale in this country, a percentage of any given supply apparently retains viability (Denison, 1977), and the species is thus expected to become established sporadically at disturbed sites around the state, especially in affluent suburban areas.

78. Helenium L. (sneezeweed)

Plants annual or perennial herbs. Stems erect or ascending, sometimes from a spreading base, unbranched or few- to many-branched mostly above the midpoint, with fine longitudinal ridges and grooves, sometimes appearing winged, glabrous or variously hairy, sometimes also with small, sessile, spherical, yellow glands. Leaves alternate and sometimes also basal, sessile (the basal leaves occasionally with a short, winged petiole), the base extended downward along the stems as narrow wings of green tissue (except in *H. amarum*). Leaf blades linear to elliptic or narrowly ovate, unlobed, shallowly pinnately lobed, or (in *H. amarum*) deeply pinnately divided, mostly angled or tapered at the base, angled or tapered to a sharply or less commonly bluntly pointed tip, the margins otherwise entire, wavy or toothed, the surfaces variously glabrous or hairy, also dotted with relatively dense, sessile to impressed, yellow to yellowish brown glands, smooth or very slightly roughened to the touch. Inflorescences of solitary heads terminal on the branches or appearing as loose, open clusters or open, leafy panicles, the heads appearing mostly long-stalked. Heads radiate (discoid elsewhere). Involucre more or less saucer-shaped, the bracts in 2(3) unequal to subequal series, those of the outer series sometimes fused at the base. Involucral bracts 15–21, green with sometimes thinner, white margins, 1-nerved, spreading to more commonly reflexed at flowering, linear to narrowly triangular or narrowly lanceolate, the surfaces and margins glabrous to variously hairy and dotted with relatively dense, sessile to impressed glands. Receptacle strongly convex (hemispherical to broadly conical or nearly globose), often slightly enlarging as the fruits mature, naked. Ray florets 5–21 in usually 1 series (absent elsewhere), pistillate or sterile, the corolla relatively broad above a slender base, yellow, occasionally with reddish streaks or reddish-tinged toward the base, the tubular portion and the undersurface of the ligule with sparse to moderate, minute, curled hairs and also with moderate to dense, sessile and spherical or somewhat impressed, yellow glands, not persistent at fruiting. Disc florets 75 to numerous (more than 500), perfect, the corolla 1.5–4.0 mm long, yellow or reddish brown to dark purple, the tube not expanded at the base (but somewhat expanded above the base) or persistent at fruiting, glandular, the 4 or 5 sharply pointed lobes also glandular on the outer surface. Style branches with the sterile tip slightly expanded and more or less truncate. Pappus of 5–8 scales, these papery and white or thinner and nearly transparent, lanceolate to narrowly ovate, tapered to a short- or relatively long-awned or sharply pointed tip, the margins usually slightly irregular (toothed or fringed elsewhere). Fruits wedge-shaped to narrowly wedge-shaped in outline, 4–8-angled or ribbed, the surface moderately to densely pubescent with ascending hairs, at least on the angles or ribs, often also glandular, brown. About 35 species, North America to South America, Caribbean Islands.

Species of *Helenium* contain helenanolide sesquiterpene lactones, especially helenalin and tenulin, which render the plants bitter and toxic. Ingestion of significant quantities causes a condition known as spewing disease, first recognized in cattle grazing on other species in high-elevation pastures in the western states (Burrows and Tyrl, 2001). Symptoms include vomiting, diarrhea, and death. The bitter flavor of tenulin (the main compound in *H. amarum*) can also flavor milk from cattle that have grazed on these plants. Because livestock avoid ingesting *Helenium* species unless no other food sources are available, the genus (particularly *H. amarum*) is an indicator of overgrazed pastures.

1. Leaf bases not decurrent, the stems not appearing winged; all linear to narrowly linear, unlobed or the lower leaves rarely deeply pinnately divided into linear segments; plants annual, with a taproot . . . . 1. H. AMARUM
1. Leaf bases decurrent as narrow wings downward along the stem; leaves narrowly lanceolate to elliptic-ovate, unlobed or the basal leaves sometimes with shallow pinnate lobes (these rounded, not linear); plants perennial, with fibrous roots

2. Ray florets sterile; disc florets with the corollas reddish brown to dark purple
.... 3. H. FLEXUOSUM
2. Ray florets pistillate; disc florets with the corollas yellow
   3. Leaves basally disposed, the basal and lower stem leaves significantly larger than the median and upper stem leaves, usually persistent at flowering
.... 2. H. AUTUMNALE
   3. Leaves well developed along the stem, the basal and lower stem leaves usually somewhat smaller than the median ones, absent at flowering
.... 4. H. VIRGINICUM

1. Helenium amarum (Raf.) H. Rock **var. amarum** (bitterweed, bitter sneezeweed, yellow dog fennel)

H. tenuifolium Nutt.

Pl. 279 g; Map 1184

Plants annual, with taproots. Stems erect or ascending, 10–40(–80) cm long, few- to many-branched usually toward the tip, with fine, rounded longitudinal ridges, not appearing winged, glabrous or sparsely pubescent with short, slender, ascending hairs, also moderately to densely dotted with sessile to impressed, yellow glands. Leaves glabrous but densely dotted with sessile to impressed, yellow glands. Basal and lowermost stem leaves absent or withered at flowering, somewhat smaller than the median stem leaves, the blade linear or rarely ovate and then deeply pinnately divided into narrowly linear segments. Median and upper stem leaves 1–8 cm long, linear to narrowly linear, unlobed, the margins entire, tapered at the base, not decurrent along the stem, tapered to a sharply pointed tip. Involucre 5–9 mm long, 6–12 mm in diameter, the bracts all free, usually with a bluntly thickened, sometimes pinkish- or purplish-tinged midnerve, glabrous or the outer surface sparsely pubescent with minute, ascending hairs, both surfaces also densely gland-dotted. Ray florets 5–10, pistillate (with a 2-branched style exserted from the short tube at flowering and a well-developed ovary that potentially develops into a fruit), the corolla 5–12 mm long, yellow. Disc florets with the corolla 1.5–3.0 mm long, yellow, 5-lobed. Pappus of mostly 6–8 scales, 1.0–1.8 mm long, the awned tip relatively long. Fruits 0.7–1.5 mm long, wedge-shaped, with 4 or 5 angles or slender ribs, the surface brown, relatively densely pubescent with ascending, yellowish hairs. $2n=30$. June–November.

Introduced, scattered to common, mostly south of the Missouri River (native of Texas, Louisiana, introduced widely in the eastern U.S. and California). Banks of streams and rivers, openings of dry upland forests, and disturbed portions of upland prairies; also pastures, farmyards, railroads, roadsides, and open, disturbed areas.

Stanford and Turner (1988) discussed the historical spread of this taxon from its original range in Texas and Louisiana. The earliest Missouri collection dates to about 1879 from Poplar Bluff (Butler County), but it reached the St. Louis and Kansas City areas, as well as southwestern Missouri, quickly during the following twenty years. A second variety, var. *badium* (A. Gray ex S. Watson) Waterf., is said to differ in its purple disc corolla lobes and more persistent and sometimes pinnately divided basal and lower stem leaves (Bierner, 1989). However, some botanists believe that there is too much overlap to allow recognition of infraspecific taxa within the species (Stanford and Turner, 1988).

This species can form dense populations along mowed roadsides and in overgrazed pastures. It is the weediest species of *Helenium* in the state. For a discussion of its poisonous and bitter properties, see the generic treatment.

2. Helenium autumnale L. (common sneezeweed, autumn sneezeweed)

H. autumnale var. *canaliculatum* (Lam.) Torr. & A. Gray

H. autumnale var. *parviflorum* (Nutt.) Fernald

Pl. 279 e, f; Map 1185

Plants perennial herbs, with fibrous roots. Stems erect or ascending, 30–150 cm long, few- to many-branched above the midpoint, narrowly several-winged, glabrous or sparsely to densely pubescent with short, sometimes curved or curled, more or less spreading to loosely ascending hairs, also moderately dotted with sessile to impressed, yellow glands. Leaves glabrous or more commonly moderately to densely pubescent with short, sometimes curved, mostly spreading hairs, also densely dotted with sessile to impressed, yellow glands. Basal and lowermost stem leaves absent or withered at flowering, somewhat smaller than the median stem leaves, the blade oblanceolate to obovate, unlobed or with a few shallow, rounded, pinnate lobes. Median and upper stem leaves 4–15 cm long, oblanceolate to elliptic, less commonly lanceolate,

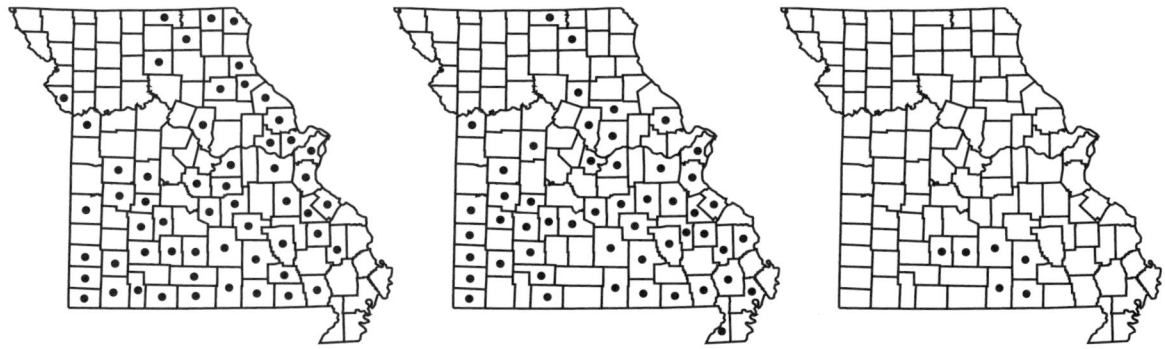

1185. Helenium autumnale   1186. Helenium flexuosum   1187. Helenium virginicum

oblong, or obovate, unlobed, the margins entire or finely to coarsely toothed (often only above the midpoint), tapered at the base, long-decurrent as narrow wings of green tissue along the stem, angled or tapered to a usually sharply pointed tip. Involucre 8–20 mm long, 8–23 mm in diameter, the outer series of involucral bracts fused at the base, the midnerve inconspicuous, not thickened, the outer surface moderately to densely pubescent with minute, curved hairs, both surfaces also moderately gland-dotted. Ray florets 8–21, pistillate (with a 2-branched style exserted from the short tube at flowering and a well-developed ovary that potentially develops into a fruit), the corolla (3–)8–25 mm long, yellow. Disc florets with the corolla 2.5–3.5 mm long, yellow, 5-lobed. Pappus of mostly 5–7 scales, 0.5–0.9 mm long, the awned tip relatively short. Fruits 1–2 mm long, narrowly wedge-shaped, with (5–)8 often lighter-colored ribs, the surface brown, moderately to densely pubescent with white to straw-colored hairs, mostly along the ribs. $2n=32, 34, 36$. August–November.

Scattered nearly throughout the state, but uncommon or absent in the Mississippi Lowlands Division and the western portion of the Glaciated Plains (nearly throughout the U.S.; Canada). Banks of streams, rivers, and spring branches, margins of ponds, lakes, and sinkhole ponds, sloughs, fens, and calcareous seeps, marshes, bottomland prairies, moist depressions of upland prairies, and bottomland forests; also pastures, ditches, railroads, roadsides, and moist, open, disturbed areas.

This widespread species comprises populations exhibiting complex variation for an array of morphological characters that seem to vary independently, and several infraspecific taxa have been described. Steyermark (1963) accepted three varieties as occurring in Missouri: the smaller-headed var. *parviflorum,* the narrow-leaved var. *canaliculatum,* and the var. *autumnale* for plants not fitting either of the preceding extremes. There seems little purpose served in attempting to recognize formal infraspecific taxa in Missouri.

For a discussion of putative hybrids between this species and other perennial sneezeweeds, see the treatments of *H. flexuosum* and *H. virginicum*. Native Americans used a powder from the dried heads of this and other species of *Helenium* as snuff to induce sneezing as a treatment for headaches, colds, blocked sinuses, and so on (Moerman, 1998). An infusion of the stems was applied to the skin for fevers.

3. Helenium flexuosum Raf. (southern sneezeweed, purple-headed sneezeweed)
Pl. 279 h, i; Map 1186

Plants perennial herbs, with fibrous roots. Stems erect or ascending, 20–120 cm long, few- to many-branched above the midpoint, narrowly several-winged, glabrous or sparsely to moderately pubescent with short, sometimes curved or curled, more or less spreading to loosely ascending hairs, also moderately dotted with sessile to impressed, yellow glands. Leaves glabrous or sparsely to moderately pubescent with short, sometimes curved, mostly spreading hairs, also moderately dotted with sessile to impressed, yellow glands. Basal and lowermost stem leaves absent or withered at flowering, not or only slightly smaller than the median stem leaves, the blade narrowly oblanceolate to less commonly narrowly obovate, unlobed or with few to several shallow, rounded, pinnate lobes. Median and upper stem leaves 3–12 cm long, narrowly oblanceolate to oblong-oblanceolate, unlobed, the margins entire or few-toothed (often only above the midpoint), somewhat tapered at the base, long-decurrent as narrow wings of green tissue along the stem, angled or tapered to a usually sharply pointed tip. Involucre 6–18 mm long, 8–18 mm in diameter, the outer series of involucral bracts fused at the base, the midnerve inconspicuous or sometimes somewhat thickened (keeled), the outer surface moderately to densely pubescent with minute, curved hairs, also moderately gland-dotted. Ray florets 8–13 (occasionally absent elsewhere), sterile (lacking stamens and style at flowering and with an ovary that is shorter and thinner than those of the disc florets, not developing into a fruit), the corolla 5–20 mm long, yellow, occasionally with reddish streaks or reddish-tinged toward the base. Disc florets with the corolla 2.5–4.0 mm long, reddish brown to dark purple, usually 4-lobed. Pappus of 5(6) scales, 0.8–2.0 mm long, the awned tip relatively long. Fruits 1.0–1.7 mm long, wedge-shaped, with 5 ribs, the surface brown but this obscured by the often dense, sessile, yellow glands, the ribs moderately to densely pubescent with white hairs. $2n=28$. June–November.

Scattered, mostly south of the Missouri River (eastern U.S. west to Wisconsin and Texas; Canada). Banks of streams, rivers, and spring branches, margins of sinkhole ponds, sloughs, swamps, bottomland prairies, moist depressions of upland prairies, bottomland forests, and seepy ledges of bluffs; also pastures, old fields, ditches, railroads, roadsides, and moist, open, disturbed areas.

Although no voucher specimens have been collected thus far, John Knox (personal communication) of Washington and Lee University has reported observing rare hybrids between *H. flexuosum* and *H. autumnale* in 1994 from a pasture adjacent to a stream in Shannon County. These hybrids had well-developed but relatively narrow stem leaves and dark-colored disc corollas, and they were sterile. The two species grow in proximity often enough that such hybrids should be searched for in the future. *Helenium flexuosum* frequently also grows in proximity to *H. virginicum*. A possible hybrid between these two species was collected in 2003 by Bill Summers from a mixed population in Howell County. This unusual specimen had somewhat reduced median and upper leaves and disc corollas with a greenish tube and 4 brownish purple lobes. However, it appeared to be developing at least some fully formed fruits, so it may represent merely a depauperate individual of *H. flexuosum*.

4. Helenium virginicum S.F. Blake (Virginia sneezeweed)

Map 1187

Plants perennial herbs, with fibrous roots. Stems erect or ascending, 10–150 cm long, few- to many-branched above the midpoint, narrowly several-winged, glabrous or sparsely to moderately pubescent with short, sometimes curved or curled, more or less spreading to loosely ascending hairs, also moderately dotted with sessile to impressed, yellow glands. Leaves glabrous or sparsely to moderately pubescent with short, sometimes curved, mostly spreading hairs, also moderately to densely dotted with sessile to impressed, yellow glands. Basal and lowermost stem leaves present at flowering, significantly larger than the median stem leaves (to 18 cm long), the blade narrowly oblanceolate to oblanceolate, unlobed or with few to several shallow, rounded, pinnate lobes. Median and upper stem leaves 2–12 cm long, narrowly oblong to narrowly oblanceolate or narrowly lanceolate, unlobed, the margins entire or few-toothed, tapered at the base, long-decurrent as narrow wings of green tissue along the stem, angled or tapered to a sharply pointed tip. Involucre 6–15 mm long, 8–16 mm in diameter, the outer series of involucral bracts fused at the base, the midnerve inconspicuous or sometimes somewhat thickened (keeled), the outer surface sparsely to moderately pubescent with minute, curved hairs, also moderately gland-dotted. Ray florets 8–13, pistillate (with a 2-branched style exserted from the short tube at flowering and a well-developed ovary that potentially develops into a fruit), the corolla 5–20 mm long, yellow, rarely brownish- or reddish-tinged toward the base. Disc florets with the corolla 2.5–3.5 mm long, yellow, 5-lobed. Pappus of 5(6) scales, 0.6–1.5 mm long, the awned tip relatively long. Fruits 1.8–2.5 mm long, narrowly wedge-shaped, with (5–)8 blunt, orangish tan ribs, the surface light brown, moderately to densely pubescent with white to straw-colored hairs, mostly along the ribs, also with often inconspicuous, moderate, yellow to orange, sessile glands. $2n=28$. July–October.

Uncommon in the south-central portion of the Ozark Division, most abundant in Howell County (Missouri, Virginia). Sinkhole ponds and moist swales of upland prairies; also pastures, ditches, roadsides, and moist, open, disturbed areas.

For many years, this species was thought to be endemic to a series of sinkhole pond margins in and adjacent to the Piedmont of Virginia. Steyermark (1960) first observed the Missouri plants at a single site in Howell County, but he considered them to represent a putative hybrid between *H. autumnale* and *H. flexuosum*, which were both growing in the immediate vicinity. That hybrid, however, differs in a number of morphological features and is sterile. Beginning in the mid-1980s, John Knox of Washington and Lee University began studying the ecology, breeding system, and taxonomy of *H. virginicum* and stumbled upon Steyermark's specimen during the course of herbarium searches. In a series of studies (Knox, 1987, 1997; Knox et al., 1995; Messmore and Knox, 1997; Simurda and Knox, 2000; Simurda et al., 2005), Knox and his colleagues established that *H. virginicum* is a species separate from *H. autumnale* and *H. flexuosum,* that it is not a hybrid involving those two species, and that the Missouri populations appear to represent the same species as the populations in Virginia. In 1998, *H. virginicum* was listed as Threatened under the Federal Endangered Species Act, and Missouri was added to the designated range of the species at about the time of publication of the Draft Recovery Plan developed by a committee under the authority of the U.S. Fish and Wildlife Service (Van Alstine, 2000). One of the actions recommended in this plan was for further inventory. Subsequent fieldwork in southern Missouri by Rhonda Rimer and Bill Summers of the Missouri Department of Conservation beginning in 2003 resulted in the discovery of about 30 additional populations in south-central Missouri. Thus far, no plants have been discovered in states between Missouri and Virginia, but inventory efforts will likely continue in the future.

79. Helianthus L. (sunflower)
(Heiser et al., 1969)

Plants annual or perennial herbs, the annuals with taproots, the perennials usually with rhizomes. Stems erect or ascending, unbranched or more commonly several-branched, mostly above the midpoint, with several longitudinal lines or ridges, variously glabrous to roughened or hairy. Leaves opposite and/or alternate, in a few species mostly basal, sessile or petiolate, the bases sometimes somewhat expanded and wrapping around the stem. Leaf blades simple, narrowly linear to nearly circular, tapered to shallowly cordate at the base, mostly tapered or less commonly broadly angled to a sharply pointed tip (rarely rounded), the margins entire to variously sharply toothed, the surfaces variously glabrous to roughened or hairy, sometimes also with scattered, sessile, spherical, yellow glands. Inflorescences most commonly panicles (sometimes few-flowered and appearing racemose) or in some species loose axillary clusters or of solitary axillary or terminal heads, the heads with short to more commonly relatively long, usually bractless stalks. Heads radiate. Involucre cup-shaped to broadly cup-shaped or somewhat bell-shaped, the bracts in mostly 2–4 unequal to subequal, overlapping series. Involucral bracts 12–40, variously narrowly lanceolate to ovate, ascending or with the tips loosely spreading, green with often hairy margins, the outer surface glabrous or more commonly variously roughened or hairy, often also with scattered, sessile, spherical, yellow glands, with 1 to several fine nerves. Receptacle flat to convex or slightly conical, usually not elongating as the fruits mature (occasionally broadening somewhat as the fruits mature), with chaffy bracts subtending the ray and disc florets, these somewhat concave and wrapped around the florets. Ray florets sterile (lacking stamens and style at flowering and with an ovary that is shorter and thinner than those of the disc florets, not developing into a fruit), the corolla showy, relatively broad, yellow (often pale yellow in *H. decapetalus*), not persistent at fruiting. Disc florets numerous, perfect, the corolla yellow or purple to reddish brown, not expanded at the base or persistent at fruiting (but often with a swollen portion toward the midpoint of the tube). Style branches with the sterile tip somewhat elongate and tapered. Pappus of the ray and disc florets of 2 short awns 1–5 mm long, sometimes also with 1–6 additional inconspicuous scales or awns 0.2–2.0 mm long, not persistent at fruiting. Fruits obovate to somewhat wedge-shaped in outline, somewhat flattened and often somewhat 4-angled in cross-section, the surface smooth or sometimes with several fine lines or nerves, glabrous or hairy, purplish black, sometimes with lighter mottling, usually not shiny. About 51 species, U.S., Canada, Mexico.

The sunflowers are a variable group, but most species are relatively easily recognized as members of the genus. Sometimes specimens of *Helianthus* are confused with *Heliopsis* (see the treatment of that genus for further discussion) or with *Silphium* (whose species tend to be more resinous and to have larger, broader involucral bracts). Within *Helianthus,* delimitation of particularly the perennial species is complicated by the presence of polyploidy and hybridization. Determination of specimens is complicated by the widespread incidence of hybridization among several of the common species, but in Missouri the hybrids generally are encountered as solitary or few individuals within populations of one or both putative parents. The exception to this rule is *H.* ×*laetiflorus,* which occasionally occurs as small populations separate from either parent. In the Ozarks, the main problems of species-level recognition appear to center around *H. divaricatus, H. hirsutus,* and *H. strumosus,* although smaller plants of *H. tuberosus* also can be problematic to identify. See the treatments of those species for further discussion.

1. Leaves all or mostly alternate (sometimes appearing relatively crowded along the stem)
    2. Leaf blades very narrow (mostly 7–20 times as long as wide), those of the largest leaves 0.1–1.0(–1.5) cm wide, linear to narrowly lanceolate
        3. Leaves mostly narrowly lanceolate, folded longitudinally along the midvein at maturity but with relatively flat margins; disc florets with the corollas yellow .... 7. H. MAXIMILIANII
        3. Leaves all or mostly linear, not folded longitudinally but with the margins curled under; disc florets with the corollas reddish brown to dark purple (at least the lobes and the upper portion of the tube)
            4. Stems sparsely to moderately hairy, especially toward the base, not glaucous; plants with rhizomes absent or very short, not occurring as colonies of stems .... 1. H. ANGUSTIFOLIUS
            4. Stems glabrous and often somewhat glaucous; plants with long-creeping, branched rhizomes, often colonial .... 13. H. SALICIFOLIUS
    2. Leaf blades broad to moderately narrow (mostly 1.2–10.0 times as long as wide), those of the largest leaves 1–35 cm wide, lanceolate to broadly ovate-triangular
        5. Plants annual, with taproots; disc florets with the corollas reddish brown to dark purple (at least the lobes and the upper portion of the tube); receptacle flat to slightly convex; largest leaves with the blade usually ovate to triangular-ovate or broadly ovate, less commonly oblong-lanceolate
            6. Involucral bracts with relatively long, spreading hairs along the margins and often also on the outer surface; chaffy bracts with inconspicuous, short hairs toward the tip; fruits glabrous or more often densely and minutely hairy when young, but usually appearing glabrous or nearly so at maturity .... 2. H. ANNUUS
            6. Involucral bracts with minute, ascending to sometimes more or less spreading hairs along the margins and the outer surface; chaffy bracts with the middle lobe densely pubescent with conspicuous, white hairs at the tip; fruits usually persistently moderately short-hairy, at least along the margins .... 12. H. PETIOLARIS
        5. Plants perennial, with a coarse, sometimes woody rootstock and short to long rhizomes; disc florets with the corollas yellow; receptacle convex to short-conical; largest leaves with the blade usually lanceolate to narrowly oblong-elliptic or narrowly ovate

7. Stems glabrous below the midpoint, often sparsely to moderately pubescent with short, ascending hairs toward the tip; leaf blades flat or only shallowly concave, not folded longitudinally, the upper surface sparsely to moderately pubescent with minute, broad-based hairs, usually not or only slightly roughened to the touch .... 5. H. GROSSESERRATUS
7. Stems moderately roughened-pubescent with short, ascending hairs throughout, more densely so toward the tip; leaf blades folded longitudinally along the midvein at maturity, the upper surface moderately to densely pubescent with short, pustular-based hairs, strongly roughened to the touch .... 7. H. MAXIMILIANII
1. Leaves all or mostly opposite (sometimes appearing all or mostly basal in *H. occidentalis*)
   8. Disc florets with the corollas reddish brown to dark purple (at least the lobes and the upper portion of the tube); involucral bracts in 3 or 4 noticeably unequal, overlapping series, tightly appressed at flowering
      9. Leaf blades lanceolate to narrowly ovate, (2.0–)2.5–8.0 times as long as wide, tapered to a sessile base or to a short, mostly winged petiole, tapered gradually to a sharply pointed tip or angled more abruptly to a sharply or bluntly pointed tip .... 11. H. PAUCIFLORUS
      9. Leaf blades ovate to broadly ovate or nearly circular, 1.0–1.7(–2.0) times as long as wide, rounded or abruptly short-tapered to an unwinged, short petiole, rounded or broadly angled to a bluntly pointed tip .... 14. H. SILPHIOIDES
   8. Disc florets with the corollas yellow; involucral bracts in 2–4 subequal, more or less overlapping series, loosely appressed and sometimes with spreading tips at flowering (except in *H. occidentalis,* with often unequal, sometimes more or less appressed bracts)
      10. Leaves mostly basal, the 3–8 pairs of stem leaves much smaller than those of the basal rosette (occasionally the lowermost pair of stem leaves nearly as large as the basal ones); involucral bracts in 3 or 4 noticeably unequal, overlapping series, usually appressed at flowering .... 10. H. OCCIDENTALIS
      10. Leaves well distributed along the stems, gradually reduced toward the stem tip, the stem leaves usually 8–15 pairs (except rarely in depauperate plants); involucral bracts in 2–4 subequal, more or less overlapping series, the tips usually at least somewhat spreading at flowering
         11. Heads relatively small, the involucre 5–7 mm long, 4–10 mm in diameter; ray florets 5–8, the corolla 1.0–1.5 cm long .... 8. H. MICROCEPHALUS
         11. Heads relatively large, the involucre 5–12 mm long, 15–30 mm in diameter; ray florets (8–)10–30, the corolla (1.5–)2.0–4.0 cm long
            12. Stem leaves all sessile or with a minute petiole less than 5 mm long, the blade rounded or shallowly cordate at the base
               13. Stems (at least above the midpoint) and leaves moderately to more commonly densely pubescent with short, spreading hairs and usually also shorter, ascending hairs, these mostly not pustular-based, usually appearing uniformly grayish, slightly to moderately roughened to the touch .... 9. H. MOLLIS

13. Stems sparsely to moderately pubescent (at least above the midpoint) with short, stiff, loosely ascending to spreading, pustular-based hairs; leaves moderately pubescent, the upper surface with short, stiff, loosely ascending to spreading, pustular-based hairs, not appearing uniformly grayish, strongly roughened to the touch (the undersurface somewhat lighter in color and sometimes with somewhat softer hairs than the upper surface) (species sometimes difficult to distinguish)
    14. Stems glabrous or hairy only toward the tip and along the inflorescence branches, sometimes somewhat glaucous; involucre 10–15 mm in diameter; disc florets with the corollas 4.0–5.5 mm long .... 4. H. DIVARICATUS
    14. Stems hairy throughout or at least above the midpoint, not glaucous; involucre (10–)15–20 mm in diameter; disc florets with the corollas (5.0–)5.5–6.5 mm long
    .... 6. H. HIRSUTUS
12. At least the largest stem leaves short- to long-petiolate, the petiole more than 5 mm long or, if appearing nearly sessile, then the blade angled or tapered at the base to a poorly defined, winged petiole
    15. Leaf blades with a single midvein; stems often with 20–25 pairs of leaves
    .... 5. H. GROSSESERRATUS
    15. Leaf blades with 3 main veins, the lateral pair arching upward from at or near the blade base; stems usually with 8–20 pairs of leaves (species sometimes difficult to distinguish)
        16. Stems glabrous or hairy only toward the tip and along the inflorescence branches, sometimes somewhat glaucous
            17. Leaf blades relatively thin-textured, those of at least the larger leaves with the margins usually coarsely toothed; uppermost stem leaves usually alternate; petioles of at least the larger leaves 2–5 cm long; involucral bracts extending conspicuously beyond the disc florets (sometimes difficult to observe in pressed specimens)
            .... 3. H. DECAPETALUS
            17. Leaf blades relatively thick-textured, the margins entire or finely toothed; uppermost stem leaves usually opposite; petioles of the larger leaves 1–3 cm long; involucral bracts extending to about the tips of the disc florets .... 15. H. STRUMOSUS
        16. Stems sparsely to moderately hairy, at least above the midpoint, not glaucous
            18. Uppermost stem leaves usually opposite; largest leaves with the blade 0.7–9.0 cm wide, usually rounded or less commonly abruptly short-tapered to a relatively well-differentiated petiole 0.5–1.5 cm long; rhizomes not producing tubers .... 6. H. HIRSUTUS
            18. Uppermost stem leaves usually alternate; largest leaves with the blade 6–15 cm wide, tapered at the base to a partially winged, sometimes poorly differentiated petiole (1.5–)2.0–8.0 cm long; rhizome branches usually with small tubers at the tip .... 16. H. TUBEROSUS

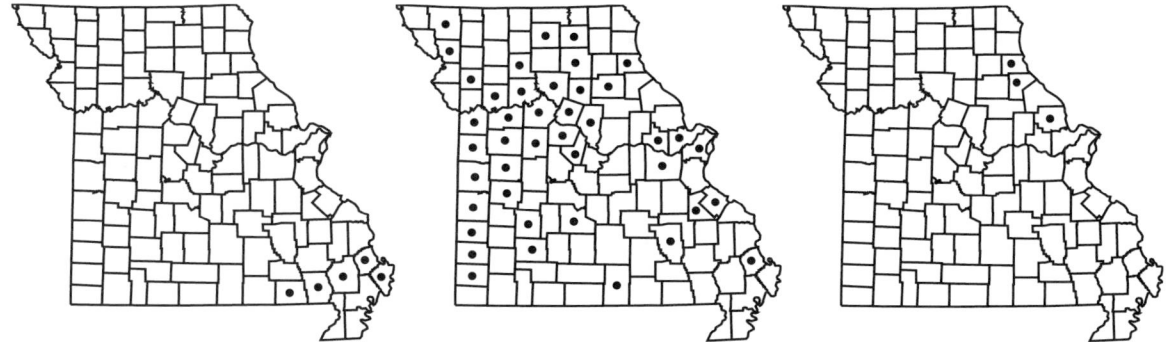

1188. Helianthus angustifolius   1189. Helianthus annuus   1190. Helianthus decapetalus

1. Helianthus angustifolius L. (narrow-leaved sunflower)

Pl. 280 a, b; Map 1188

Plants perennial herbs, with a short, clumped rootstock, with rhizomes absent or very short, not occurring as colonies of stems. Stems usually solitary, 50–170 cm long, moderately pubescent with slender, mostly loosely ascending, minutely pustular-based hairs, these often sparser above the stem midpoint, not glaucous. Leaves relatively numerous and well developed along the stem (usually more than 30 nodes), mostly alternate, sessile or nearly so (sometimes also with petiolate basal leaves to 2 cm wide). Blades of the stem leaves 3–20 cm long, 0.1–1.0(–1.5) cm wide, linear or those of the lowermost leaves sometimes narrowly lanceolate (mostly 7–20 times as long as wide), not folded longitudinally, mostly short-tapered at the base, tapered to a sharply pointed tip, the margins entire, mostly curled under, the upper surface moderately roughened with stiff, pustular-based hairs (these often shed, leaving only the expanded base), the undersurface usually pale green, moderately to densely pubescent with somewhat tangled, soft, easily shed hairs, both surfaces often also with scattered, sessile, yellow glands, with 1 main vein. Inflorescences of solitary terminal heads or appearing as open, few-headed clusters, the heads mostly long-stalked. Involucre 8–12 mm long, 12–20 mm in diameter, extending to about the tips of the disc corollas, the bracts in 2 or 3 subequal series, lanceolate to narrowly lanceolate, tapered to a sharply pointed, slender, loosely ascending to more commonly spreading or recurved tip, the margins and outer surface sparsely to moderately roughened with minute, ascending, pustular-based hairs and also with sessile, yellow glands. Receptacle shallowly convex, the chaffy bracts 5.5–6.5 mm long, oblong-oblanceolate, tapered to a sharply pointed, purplish-tinged, glabrous tip, the outer surface minutely hairy. Ray florets 10–21, the corolla 1.2–3.0 cm long, minutely hairy and with sessile, yellow glands on the outer surface. Disc florets with the corolla 4.0–4.5 mm long, the lobes and tip of the tube reddish brown to dark purple above a sometimes yellowish basal portion. Pappus of 2 scales 1.5–2.0 mm long, these lanceolate, tapered to a sharply pointed, minute, awnlike tip, occasionally with an additional pair of thin, irregular apical lobes. Fruits 2–3 mm long, more or less wedge-shaped, flattened but 4-angled in cross-section, the surface glabrous, finely mottled with dark brown to nearly black and lighter brown patches. $2n=34$. August–October.

Uncommon in the Mississippi Lowlands Division (southeastern U.S. west to Missouri, Oklahoma, and Texas). Upland prairies, sand prairies, and savannas; also pastures, ditches, and roadsides.

Steyermark (1963) noted that this attractive species does well in cultivation. Although it has fewer heads per stem and fewer stems per plant than does *H. salicifolius,* its poorly developed rhizome system makes it less aggressive in the garden.

2. Helianthus annuus L. (common sunflower)

H. annuus var. *lenticularis* (Douglas) Steyerm.

H. annuus var. *macropocarpus* (DC.) Cockerell

Pl. 281 g, h; Map 1189

Plants annual, with taproots. Stems solitary, (25–)50–300 cm long, stout, moderately pubescent with short, stiff, ascending, pustular-based hairs, these often breaking off toward the stem base, leaving the persistent expanded base. Leaves well developed along the stem (usually with 8–25 nodes), mostly alternate, long-petiolate. Blades of the stem leaves 7–40 cm long, 3–35 cm wide, ovate to triangular-ovate or broadly ovate-triangular (mostly

1.2–2.5 times as long as wide), flat or sometimes slightly drooping, not folded longitudinally, cordate to truncate, broadly rounded, or short-tapered at the base, tapered to a usually sharply pointed tip, the margins finely to coarsely and often somewhat irregularly toothed (rarely nearly entire), flat, the surfaces moderately to densely roughened, pubescent with minute, loosely appressed, pustular-based hairs, also with moderate to dense, sessile, yellow glands, more or less with 3 main veins, the lateral pair branching from the midnerve at the base of the blade. Inflorescences rarely of solitary terminal heads, more commonly appearing as open panicles. Involucre 10–30 mm long, (15–)20–50 mm in diameter (longer and much wider in cultivated forms), mostly shorter than or extending about to the tips of the disc corollas, the bracts in 3 or 4 subequal to somewhat unequal series, narrowly ovate to ovate, tapered to a sharply pointed, slender, loosely ascending to more commonly spreading or recurved tip, the margins with relatively long, stiff, spreading hairs, the outer surface moderately to densely pubescent with short (or occasionally longer), stiff, loosely ascending to spreading, pustular-based hairs, both surfaces also usually with small, sessile, yellow glands. Receptacle flat or slightly convex, the chaffy bracts 9–12 mm long (slightly longer in cultivated forms), narrowly oblong-triangular, usually 3-lobed above the midpoint, the lobes tapered to sharply pointed, straw-colored to dark purple, inconspicuously short-hairy tips, the outer surface usually glabrous below the tip. Ray florets 17–40 (more in cultivated forms), the corolla 2.5–5.0 cm long, variously glabrous or both surfaces inconspicuously hairy toward the base, or the outer surface minutely hairy and occasionally also with minute, sessile, yellow glands. Disc florets with the corolla 5–8 mm long, reddish brown to dark purple (at least the lobes and the upper portion of the tube; rarely yellow elsewhere). Pappus of 2 scales 2.0–3.5 mm long, these narrowly lanceolate-triangular, tapered to a sharply pointed, often minutely awnlike tip, papery, often also with 1–4 additional oblong scales 0.2–1.0 mm long. Fruits 3–7 mm long (longer in cultivated forms), narrowly wedge-shaped to obovate, flattened but usually more or less 4-angled in cross-section, the surface glabrous or more commonly densely and minutely hairy when young, but usually appearing glabrous or nearly so at maturity, uniformly black to variously with gray, brown, or white stripes or mottling. $2n=34$. July–November.

Mostly absent from the Ozark and Mississippi Lowlands Divisions but scattered elsewhere in the state (nearly throughout the U.S.; Canada, Mexico). Upland prairies, openings of bottomland to mesic upland forests, and banks of streams and rivers; also crop fields, fallow fields, old fields, ditches, railroads, roadsides, and open, disturbed areas.

The common sunflower is the most important crop plant to have originated from a species native to the United States. It is the state flower of Kansas, where sunflower fields are a common sight. For a thorough account of this plant and its uses, see Heiser's (1976) enjoyable book on the genus. *Helianthus annuus* is cultivated nearly worldwide for its edible seeds and for the oil produced from the seeds, which is one of the most important of the world's vegetable oils for food production and also industrial and other uses. In recent years, the market for sunflower seeds for wild bird feeding also has become increasingly important in the United States. The species is cultivated as an ornamental, and plants also are used to a limited extent for fodder for livestock and for wildlife food plantings. At one time, Missouri was among the largest producers of commercial sunflowers in the country, but production has shifted primarily to other states, particularly North Dakota.

Heiser (1951, 1976) discussed the domestication of common sunflower. According to him, the species originally was native to the western states. However, the remains of fruits have been found by archaeologists in scattered prehistoric cliff shelters as far east as South Carolina (including sites in Missouri). Heiser noted that Native Americans had a long history of harvesting the plant from the wild for food, oil, and a yellow dye, which led to its establishment as a weed in and around sites of habitation and to its cultural dissemination among tribes during pre-Columbian times. The plant became cultivated at some point, which led to selection for traits such as fewer but larger heads and larger fruits. As early as the 1500s, the species was brought to Europe by early explorers and was cultivated there. It eventually became an important crop plant in Russia, where the 'Mammoth Russian' type, with its unbranched stem, greatly enlarged head, and large fruits, was bred. Cultivated forms of the common sunflower subsequently were reimported into the United States as a crop plant during the second half of the 1800s. Continued breeding both in Europe and in this country have resulted in a large number of commercially available cultivars adapted to different climates and for various uses. For example, Steyermark (1963) recorded a few collections from Holt County and the city of St. Louis of an ornamental dwarf form with the florets mostly converted to ray florets, which has been referred to under the name cv. 'Nanus

Plate 280. Asteraceae. *Helianthus angustifolius*, **a)** fruiting head, **b)** habit. *Helianthus salicifolius*, **c)** fruiting head, **d)** habit. *Helianthus grosseserratus*, **e)** fruiting head, **f)** habit. *Helianthus maximilianii*, **g)** fruiting head, **h)** habit.

Florepleno.' Interestingly, the large-headed sunflower that was independently domesticated by Native Americans survives in the American Southwest in a race known as the Hopi sunflower, which is cultivated on a very limited basis in modern times. The commercially important plants grown around the world today first originated as a result of efforts by plant breeders in Europe.

Although at times up to five taxonomic varieties of *H. annuus* have been recognized, a formal infraspecific classification mostly has been abandoned, owing to frequent hybridization that has blurred the boundaries between cultivated and wild plants (Heiser et al., 1969). The original, wild plants of the southwestern United States and adjacent Mexico at one time were called var. *lenticularis* and differ in their more numerous, smaller heads with narrower (4–7 mm) involucral bracts, smaller (2.0–3.5 cm in diameter) receptacles, fewer (17–26) ray florets with shorter (2.5–4.0 cm) corollas, and small (3–5 mm) fruits. Steyermark (1963) cited a single historical Missouri occurrence of this taxon from Jackson County (but see below). At the other extreme, the commonly cultivated plants (which rarely escape) were known in the past as var. *macropocarpus* and have an unbranched stem with usually a solitary terminal head having big (8.5–15.0 mm or more wide) involucral bracts, large (5.5–60.0 cm or more in diameter) receptacles, more numerous (30–70) ray florets with longer (3.5–10.0 cm) corollas, and large (6.5–15.0 mm) fruits. The sunflower that is widely distributed outside the original native range of the species has been called var. *annuus* and is more or less intermediate between the other two main variants for virtually every character. Heiser (1954, 1976) hypothesized that this weedy variety either arose through introgressive hybridization between the cultivated sunflower and the original wild type or that it evolved from the so-called camp-follower weed that became established around prehistoric habitations as an indirect result of harvesting of the species from the wild.

Given the complex biogeographical and ethnobotanical situation surrounding the species, it is not possible to address the issue of whether to consider the species native in Missouri with any confidence. The oldest Missouri herbarium specimen of common sunflower examined during the present study was collected in Cass County by G. C. Broadhead in 1876 and represents the var. *lenticularis* phase rather than the weedy phase. However, archaeological remains from digs at prehistoric dwellings indicate that a probably cultivated variant was present in Missouri hundreds or thousands of years ago (Heiser, 1978). Based on the herbarium record, the weedy variant moved into the state relatively early, reaching St. Louis County by 1896, Jasper County by 1905, and Marion County by 1914. It seems unlikely that *H. annuus* was ever a natural component of a native plant community prior to European settlement of the state, but the spread of the weedy variant into and across the state could be interpreted to have occurred without any direct human assistance.

Heiser (1947) studied natural and artificial hybridization between *H. annuus* and *H. petiolaris*. He noted that the hybrids have reduced fertility but are not totally sterile. Such hybrids have been reported from Cole and Jackson Counties and the city of St. Louis and can be very difficult to distinguish morphologically from one or the other parent.

3. Helianthus decapetalus L. (thin-leaved sunflower, pale sunflower)

Pl. 282 e, f; Map 1190

Plants perennial herbs, with relatively slender, long-creeping, branched rhizomes (some of the branches rarely with small tubers at the tip), usually occurring in dense colonies. Stems usually appearing loosely clumped, less commonly solitary, 60–200 cm long, glabrous below the inflorescence or the uppermost portion of the stem sparsely pubescent with short, stiff, loosely ascending, pustular-based hairs, often somewhat glaucous. Leaves well developed along the stem (usually with 8–20 nodes), mostly opposite (often alternate in the upper third), short- to long-petiolate (the petioles of the larger leaves usually 2–5 cm long and somewhat winged toward the tip). Leaf blades 6–20 cm long, (1.5–)3.0–10.0 cm wide, relatively thin-textured, lanceolate to ovate (mostly 2–5 times as long as wide), flat, not folded longitudinally, rounded to tapered at the base, tapered to a sharply pointed tip, the margins finely to more commonly (at least on larger leaves) coarsely toothed, flat, the upper surface strongly roughened with moderate, minute, stout, pustular-based hairs, the undersurface glabrous (and usually pale green to silvery) or sparsely to moderately pubescent with stiff or somewhat softer, loosely appressed hairs and with sparse, sessile, yellow glands, more or less with 3 main veins, the lateral pair sometimes relatively thin, branching from the midnerve at or near the base of the blade, arching upward. Inflorescences of solitary terminal heads or appearing as open clusters or occasionally open panicles. Involucre 12–18 mm long, 12–25 mm in diameter, conspicuously longer than the tips of the disc corollas, the bracts in 3 or

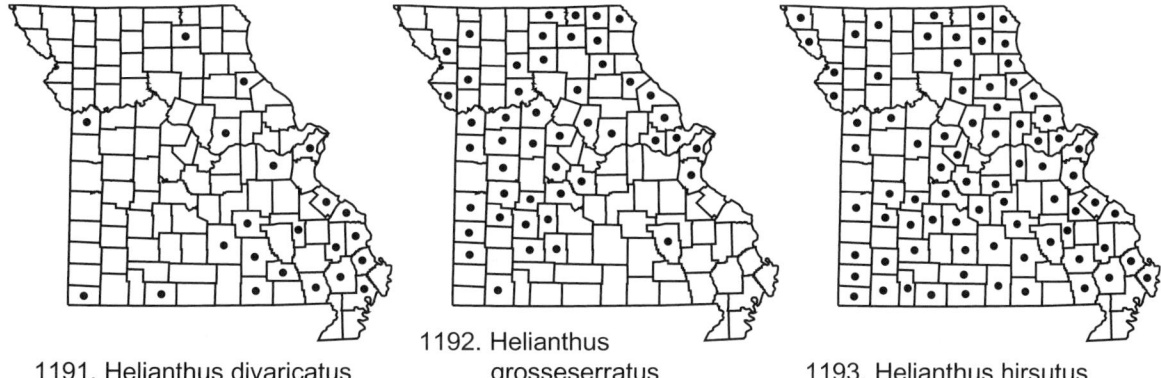

1191. Helianthus divaricatus
1192. Helianthus grosseserratus
1193. Helianthus hirsutus

4 subequal, overlapping series, narrowly lanceolate, tapered to a sharply pointed, spreading or recurved tip, the margins with a fringe of short, spreading to ascending hairs, the outer surface glabrous or more commonly sparsely to moderately pubescent with short, stout, ascending, often pustular-based hairs, usually lacking glands. Receptacle convex to short-conical, the chaffy bracts 8–10 mm long, narrowly oblong to narrowly oblong-oblanceolate, with 3 short-tapered, sharply pointed lobes at the tip, these straw-colored, the outer surface minutely hairy toward the tip. Ray florets 8–15, the corolla 1.5–3.5 cm long, yellow to pale yellow, the outer surface usually with sparse, minute hairs. Disc florets with the corolla 6.5–7.5 mm long, the corollas yellow, the lobes often minutely hairy on the outer surface. Pappus of 2 scales 3–4 mm long, these lanceolate to narrowly triangular, tapered to a sharply pointed, often minutely awnlike tip. Fruits 3.5–5.0 mm long, wedge-shaped to narrowly obovate, somewhat flattened and more or less bluntly 4-angled in cross-section, the surface glabrous or with a few minute hairs at the tip, uniformly brown or with fine, darker and lighter brown mottling. 2n=34, 68. July–October.

Uncommon in the northeastern portion of the state (eastern [mostly northeastern] U.S. west to Wisconsin, Iowa, and Missouri). Banks of streams and rivers, bottomland forests, mesic upland forests in ravines, and bases and ledges of bluffs.

Steyermark's (1963) treatment of *H. decapetalus* differed sharply from that of Heiser et al. (1969). He accepted the species as occurring in eleven counties, mostly in a diagonal band from the northeastern to southwestern corners of the state, but also in the Bootheel and in Jackson County. All but one of these specimens subsequently were annotated by Heiser and his colleagues as either *H. divaricatus* or *H. strumosus,* and they did not include Missouri in their distribution map of the species. In its typical phase, the thin leaves with well-developed petioles and relatively coarse, blunt teeth are good morphological markers for *H. decapetalus.* The confusion among the taxa involved may be blamed in part on the presence of tetraploid populations, which apparently are inherently more variable morphologically than are the diploids. As noted by Steyermark (1963), further studies are still needed to accurately assess the status and range of *H. decapetalus* in Missouri.

4. Helianthus divaricatus L. (woodland sunflower)

Pl. 282 c, d; Map 1191

Plants perennial herbs, with relatively slender, long-creeping, branched rhizomes, usually occurring in colonies. Stems usually solitary, 50–150 cm long, glabrous below the inflorescence or the uppermost portion of the stem sparsely to moderately pubescent with short, stiff, loosely ascending, pustular-based hairs, sometimes somewhat glaucous. Leaves well developed along the stem (usually with 8–20 nodes), opposite, sessile or with a minute petiole less than 5 mm long. Leaf blades 4–15 cm long, 1–5 cm wide, relatively thick-textured, lanceolate to narrowly triangular or triangular-ovate (mostly 3–7 times as long as wide), flat, not folded longitudinally, rounded to truncate or shallowly cordate at the base, tapered to a sharply pointed tip, the margins finely toothed to nearly entire, flat, the upper surface strongly roughened with moderate, minute, stout, broad-based hairs, the undersurface moderately to densely pubescent with somewhat softer, more or less spreading hairs but usually lacking sessile, yellow glands, with 3 main veins, the lateral pair branching from the midnerve at the base of the blade, arching upward. Inflorescences of solitary terminal heads or appearing as open clusters or occasionally open panicles. Involucre 8–12 mm long, 10–15 mm in diameter, about as long as or slightly longer than the tips of the disc corollas, the bracts in 3 or 4 subequal, overlapping series, lanceolate, tapered to a sharply pointed, loosely ascending to somewhat spreading or recurved tip, the margins with a fringe of short, spreading to ascending hairs, the outer surface glabrous or more commonly sparsely to moderately pubescent with short, stout, ascending, often pustular-based hairs, usually lacking glands. Receptacle convex to short-conical, the chaffy bracts 5–8 mm long, narrowly oblong to narrowly oblong-oblanceolate, with 3 short-tapered, sharply pointed lobes at the tip, these straw-colored or rarely purplish-tinged, the outer surface minutely hairy toward the tip. Ray florets 5–15, the corolla 1.5–3.0 cm long, the outer surface usually with sparse, minute hairs. Disc florets with the corolla 4.0–5.5 mm long, the corollas yellow, the lobes often minutely hairy on the outer surface. Pappus of 2 scales 2.0–2.5 mm long, these lanceolate to narrowly triangular, tapered to a sharply pointed, often minutely awnlike tip. Fruits 3–4 mm long, wedge-shaped to narrowly obovate, somewhat flattened and more or less bluntly 4-angled in cross-section, the surface glabrous or with a few minute hairs at the tip, uniformly brown or with fine, darker and lighter brown mottling. $2n=34$. July–October.

Scattered in the Ozark, Ozark Border, and Mississippi Lowlands Divisions, uncommon and sporadic in the Glaciated Plains (eastern U.S. west to Wisconsin, Oklahoma, and Louisiana; Canada). Glades, savannas, openings of dry upland forests, tops of bluffs, and rarely banks of streams and rivers; also fencerows and roadsides.

In its typical phase, *H. divaricatus* is easily recognized by its thick, flat, sessile leaves that usually spread from the stem at about a 90-degree angle. However, as noted by Heiser et al. (1969), plants in the Ozarks are relatively diverse morphologically and can be difficult to distinguish from some other perennial woodland sunflowers. Heiser and his colleagues suggested that because plants growing in the Ozarks tend to have hairier stems and leaves and more numerous ray florets and involucral bracts than is typical of the species elsewhere in the Midwest, plants in the region may have hybridized with *H. mollis* in the past. This seems less likely than an interpretation of potential hybridization with other species of woodland sunflowers of the Ozarks. In Missouri, the principal difficulties are in distinguishing *H. divaricatus* from *H. hirsutus* and *H. strumosus,* and occasional specimens (especially those in which only the top of the plant was collected) are difficult to separate from *H. tuberosus.* Heiser et al. noted that in particular the problems of identification involve tetraploid plants of *H. hirsutus* and *H. strumosus,* which both tend to have slightly larger disc corollas and fruits. They suggested that these polyploid races may have arisen following past interspecific hybridization with *H. divaricatus,* but this hypothesis still needs to be tested.

The distribution of *H. divaricatus* in Missouri shown in the present work is broader than that in Steyermark's (1963) map. Oddly, the specimens from western Missouri added since 1963 do not represent populations newly discovered in the field, but rather are mostly older specimens that Steyermark (1963) had misdetermined as *H. decapetalus.* See the treatment of that species for further discussion.

5. Helianthus grosseserratus M. Martens (sawtooth sunflower)

H. grosseserratus ssp. *maximus* R.W. Long

Pl. 280 e, f; Map 1192

Plants perennial herbs, with short-creeping or sometimes longer, thick, branched rhizomes, sometimes occurring as dense colonies of stems. Stems often appearing somewhat clumped, 50–300 cm long, glabrous below the midpoint, often sparsely to moderately pubescent with short, ascending hairs toward the tip, often somewhat glaucous. Leaves relatively numerous and well developed along the stem (usually with 20–25 nodes), mostly alternate or occasionally most of the leaves opposite and only the uppermost ones alternate, all or nearly all short-petiolate. Blades of the stem leaves 5–30 cm long, 1–9 cm wide, narrowly lanceolate to narrowly ovate (mostly 5–10 times as long as wide), flat or only shallowly concave, not folded longitudinally, tapered at the base, tapered to a sharply pointed tip, the margins finely to coarsely and sharply toothed (rarely entire), flat, the upper surface sparsely to moderately pubescent with minute, broad-based hairs, usually not or only slightly roughened to the touch, the undersurface densely pubescent with minute, soft, appressed hairs, both surfaces also with sparse to moderate, sessile, yellow glands, more or less with 1 main vein but the lowermost pair of lateral veins usually slightly more prominent than the other pinnate lateral veins. Inflorescences of solitary terminal heads or more commonly appearing as open clusters or open panicles. Involucre 10–15 mm long, 15–25 mm in diameter, mostly extending slightly beyond the tips of the disc corollas, the bracts in 2 or 3 subequal series, narrowly lanceolate to nearly linear, tapered to a sharply pointed, slender, loosely ascending to more commonly spreading or recurved tip, the margins with short, ascending to occasionally spreading hairs, at least toward the base, the outer surface glabrous or sparsely hairy toward the base but usually lacking glands. Receptacle convex, the chaffy bracts 7–9 mm long, narrowly oblong-triangular to nearly linear, angled or short-tapered to a sharply pointed, green, minutely hairy tip, the outer surface also minutely hairy. Ray florets 10–25, the corolla 2.0–4.5 cm long, glabrous. Disc florets with the corolla 5–6 mm long, yellow throughout. Pappus of 2 scales 2.0–2.5 mm long, these narrowly triangular, tapered abruptly to a sharply pointed, often minutely awnlike tip. Fruits 3–4 mm long, narrowly wedge-shaped, flattened but more or less 4-angled in cross-section, the surface glabrous, often finely mottled with dark brown to nearly black and lighter brown patches. $2n=34$. July–October.

Scattered in the Glaciated and Unglaciated Plains Divisions and the western portion of the Ozark and Ozark Border Divisions; absent from most of the southeastern quarter of the state (eastern U.S. west to North Dakota and Texas; Canada; introduced sporadically in the northwestern U.S.). Bottomland and upland prairies; bases of bluffs, banks of streams and rivers, fens, and margins of ponds and lakes; also ditches, margins of cultivated fields, pastures, railroads, roadsides, and open, disturbed areas.

Rarely collected plants with most of the disc florets converted to rays have been called f. *pleniflorus* Wadmond. It is not clear whether this represents some form of genetic mutation or a symptom of a disease. Steyermark (1963) noted that hybrids have been recorded between *H. grosseserratus* and *H. maximilianii* (*H.* ×*intermedius* R.W. Long) in adjacent states and should be searched for in Missouri. Thus far a single collection made by B. F. Bush in 1916 in Jackson County (and labeled as introduced there) appears to represent this hybrid.

6. Helianthus hirsutus Raf. (hairy sunflower, bristly sunflower)

H. hirsutus var. *stenophyllus* Torr. & A. Gray
H. hirsutus var. *trachyphyllus* Torr. & A. Gray
Pl. 282 a, b; Map 1193

Plants perennial herbs, with slender to stout, usually long-creeping, branched rhizomes (the branches sometimes slightly thickened at the tip but not with well-defined tubers), sometimes occurring in dense colonies. Stems usually solitary, 60–150(–200) cm long, sparsely to more commonly moderately or densely pubescent with short, stiff, loosely ascending, sometimes pustular-based hairs, sometimes nearly glabrous toward the base, not glaucous. Leaves well developed along the stem (usually with 8–20 nodes), usually all opposite, mostly with a well-differentiated, short petiole (the petioles of the larger leaves 0.5–1.5 cm long). Leaf blades 4–18 cm long, 0.7–6.0(–9.0) cm wide, relatively thick-textured, narrowly lanceolate to ovate (mostly 2–7 times as long as wide), flat, not folded longitudinally, rounded to angled or short-tapered at the base, tapered to a sharply pointed tip, the margins finely toothed to entire, flat, the upper surface moderately to strongly roughened with moderate to dense, minute to short, stout, sometimes pustular-based hairs, the undersurface moderately to densely pubescent with somewhat softer, loosely appressed to more or less spreading hairs, also with sparse to moderate, sessile, yellow glands, with 3 main veins, the lateral pair branching from the midnerve at or near the base of the blade, arching upward. Inflorescences of solitary terminal heads or appearing as open clusters or rarely open panicles. Involucre 8–14 mm long, 10–25 mm in diameter, about as long as or slightly longer than the tips of the disc corollas, the bracts in 3 or 4 subequal, overlapping series, narrowly lanceolate to lanceolate, tapered to a sharply pointed, loosely ascending to spreading or recurved tip, the margins with an irregular fringe of short, spreading to ascending hairs, the outer surface glabrous or sometimes sparsely to moderately pubescent with short, stout, ascending hairs, but usually lacking minute, sessile, yellow glands. Receptacle convex to short-conical, the chaffy bracts 7–9 mm long, narrowly oblong to narrowly oblong-oblanceolate, usually with 3 short, sharply pointed lobes at the tip, these green or straw-colored, the outer surface minutely hairy, especially toward the tip. Ray florets 10–15, the corolla 1.5–3.0 cm long, the outer surface usually with sparse, minute hairs and scattered, minute, sessile, yellow glands. Disc florets with the corolla 5.5–6.5 mm long, the corollas yellow, the lobes often minutely hairy on the outer surface. Pappus of 2 scales 2.5–3.5 mm long, these lanceolate to narrowly triangular, tapered to a sharply pointed, often minutely awnlike tip. Fruits 3.5–4.5 mm long, wedge-shaped to obovate, somewhat flattened and more or less bluntly 4-angled in cross-section, the surface glabrous or with a few minute hairs at the tip, uniformly brown or with fine, darker and lighter brown mottling. $2n=68$. July–October.

Scattered nearly throughout the state (eastern U.S. west to Minnesota, Nebraska, and Texas; Canada, Mexico). Mesic to dry upland forests, savannas, upland prairies, sand prairies, ledges and tops of bluffs, and banks of streams, rivers, and spring branches; also pastures, quarries, ditches, railroads, roadsides, and disturbed areas.

1194. Helianthus maximilianii

1195. Helianthus microcephalus

1196. Helianthus mollis

Heiser et al. (1969) stated that this is a morphologically variable species that some earlier authors had divided into three varieties, based on supposed differences in pubescence patterns and leaf shape. They mentioned that *H. hirsutus* may have been involved in the parentage of *H. tuberosus* and discussed the problems of distinguishing some specimens of *H. hirsutus* from *H. divaricatus* and *H. strumosus*. In Missouri, especially in the Ozarks, many specimens collected in relatively dry, partially sunny sites appear somewhat intermediate between *H. hirsutus* and especially *H. strumosus*. In fact, *H. hirsutus* appears to some extent to be the name of last resort applied by botanists to perplexing plants that do not key well. A detailed investigation of genetic and cytological variation within and among populations of upland perennial sunflowers in the Midwest will be necessary to tease apart the confusing patterns of morphological variation that have confounded specimen identifications for many plants within portions of the region.

7. Helianthus maximilianii Schrad.
(Maximilian sunflower)

Pl. 280 g, h; Map 1194

Plants perennial herbs, with relatively short-creeping, thick, branched rhizomes and often somewhat succulent roots, often occurring as dense colonies of stems. Stems often appearing somewhat clumped, 50–200(–250) cm long, moderately to densely pubescent with more or less stiff, short, ascending, often pustular-based hairs throughout, not glaucous. Leaves relatively numerous and well developed along the stem (usually more than 30 nodes), mostly alternate, mostly short-petiolate, often appearing arched. Blades of the stem leaves 4–30 cm long, 0.5–5.5 cm wide, narrowly lanceolate to lanceolate or narrowly elliptic-lanceolate (mostly 7–20 times as long as wide), folded longitudinally along the midvein, tapered at the base, tapered to a sharply pointed tip, the margins entire or less commonly with minute, widely spaced teeth, flat, the surfaces moderately to more commonly densely roughened-pubescent with minute, usually stout, loosely appressed, pustular-based hairs (usually appearing grayish green), also with moderate to dense, sessile, yellow glands, with 1 main vein. Inflorescences of solitary terminal heads or small terminal clusters, also usually with axillary single or clustered heads present from the upper leaves, commonly appearing overall spicate or racemose, the heads short- to long-stalked. Involucre 12–25 mm long, 15–28 mm in diameter, mostly extending beyond the tips of the disc corollas, the bracts in 2 or 3 subequal series, narrowly lanceolate to nearly linear, tapered to a sharply pointed, slender, loosely ascending to more commonly spreading or recurved tip, the margins with a dense fringe of short hairs, at least toward the base, the surfaces moderately to densely roughened-hairy and often also with scattered, sessile, yellow glands. Receptacle convex, the chaffy bracts 7–11 mm long, narrowly oblong-triangular to nearly linear, angled or short-tapered to a sharply pointed, green, minutely hairy tip, the outer surface also minutely hairy. Ray florets 10–25, the corolla 2.5–4.0 cm long, glabrous but often with scattered, sessile, yellow glands. Disc florets with the corolla 5–7 mm long, yellow throughout. Pappus of 2 scales 3–4 mm long, these oblong-lanceolate, tapered abruptly to a sharply pointed, often minutely awnlike tip. Fruits 3–4 mm long, narrowly wedge-shaped, flattened but more or less 4-angled in cross-section, the surface glabrous, often finely mottled with dark brown and lighter brown patches. $2n=34$. July–October.

Scattered nearly throughout the state (most of the U.S. [except some western and southeastern states]; Canada). Calcareous glades, ledges and tops of bluffs, upland prairies, loess hill prairies, and savannas; also old fields, railroads, roadsides, and open, disturbed areas.

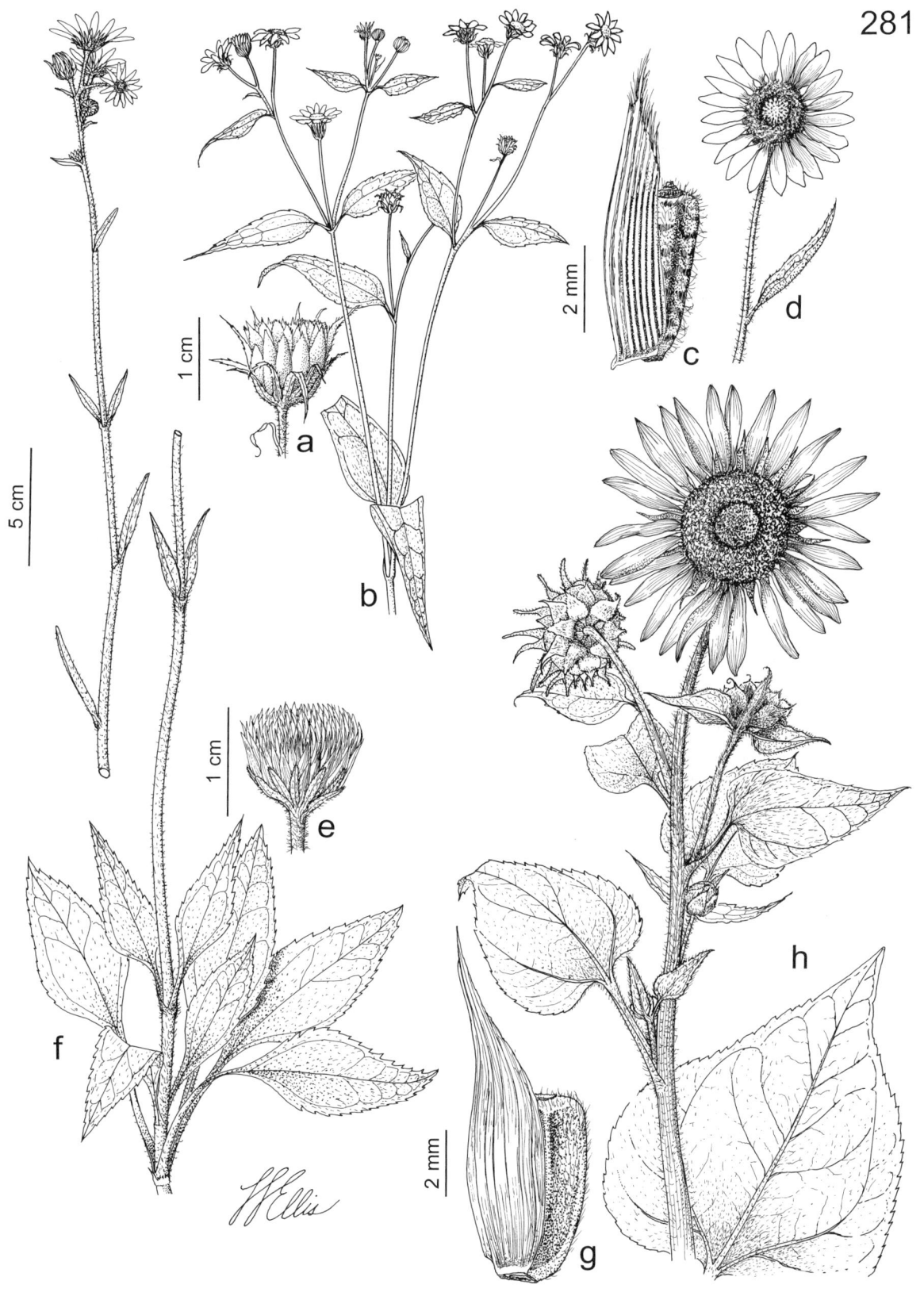

Plate 281. Asteraceae. *Helianthus microcephalus*, **a)** fruiting head, **b)** habit. *Helianthus petiolaris*, **c)** fruit with bract, **d)** flowering stalk. *Helianthus occidentalis*, **e)** fruiting head, **f)** habit. *Helianthus annuus*, **g)** fruit with bract, **h)** habit.

Maximilian sunflower is a strikingly beautiful plant that makes an excellent addition to the wildflower garden where there is room for it to grow. In cultivation, plants usually develop into large, dense clumps of many stems to 3 m tall, increasing in diameter slowly each year. Steyermark (1963) noted that the form usually observed in disturbed habitats in eastern Missouri is similar to a cultivar developed in the St. Louis area with more numerous heads having longer ray corollas. He considered the species native only from the western portion of the Ozarks northward to the loess hills of northwestern Missouri. Today, it has become difficult to assess the eastern edge of the species' native distribution as it has been collected more frequently away from roadsides in the eastern half of the state. Gleason and Cronquist (1991) considered *H. maximilianii* to occur natively at least as far eastward as Ohio.

8. Helianthus microcephalus Torr. & A. Gray
(small woodland sunflower)

Pl. 281 a, b; Map 1195

Plants perennial herbs, lacking rhizomes or with short, thick rhizomes, often occurring as clumps. Stems usually few to several, 50–250(–450) cm long, glabrous below the inflorescence, often somewhat glaucous. Leaves well developed along the stem (usually with 8–20 nodes), gradually reduced toward the stem tip, all or mostly opposite, mostly with a short petiole less than 4 cm long. Leaf blades 4–18 cm long, 1–6 cm wide, relatively thick-textured, lanceolate to narrowly ovate (mostly 4–9 times as long as wide), flat, not folded longitudinally, rounded to short-tapered at the base, tapered to a sharply pointed tip, the margins finely toothed to nearly entire, flat, the upper surface strongly roughened with moderate, minute, stout, broad-based hairs, the undersurface moderately to densely pubescent with short, softer, loosely appressed hairs and also with sparse to moderate, sessile, yellow glands, with 3 main veins, the lateral pair branching from the midnerve well above the base of the blade, arching upward. Inflorescences appearing as open clusters or more commonly open panicles. Involucre 5–7 mm long, 8–10 mm in diameter, about as long as or slightly longer than the tips of the disc corollas, the bracts in 3 or 4 more or less subequal, overlapping series, narrowly lanceolate, tapered to a sharply pointed, loosely ascending to spreading or recurved tip, the margins with a fringe of short, spreading to ascending hairs, the outer surface glabrous or sparsely hairy toward the base but usually lacking glands. Receptacle convex to short-conical, the chaffy bracts 5–7 mm long, narrowly oblong to narrowly oblong-oblanceolate, with 3 short-tapered, sharply pointed lobes at the tip, these straw-colored or rarely purplish-tinged, the outer surface minutely hairy toward the tip. Ray florets 5–8, the corolla 1.0–1.5 cm long, the outer surface usually with sparse, sessile, yellow glands. Disc florets with the corolla 4.0–5.5 mm long, the corollas yellow, the lobes minutely hairy on the outer surface. Pappus of 2 scales 1.5–2.5 mm long, these lanceolate to narrowly triangular, tapered to a sharply pointed, often minutely awnlike tip. Fruits 3.5–4.5 mm long, wedge-shaped to narrowly obovate, only slightly flattened but more or less bluntly 4-angled in cross-section, the surface glabrous, with fine, darker and lighter brown mottling. $2n=34$. August–September.

Uncommon in the Crowley's Ridge Section of the Mississippi Lowlands Division and the adjacent eastern portion of the Ozarks (eastern U.S. west to Minnesota and Louisiana). Mesic upland forests, banks of streams, and margins of acid seeps; rarely also old fields.

Heiser et al. (1969) noted that this species hybridizes with *H. divaricatus* (such hybrids have been called *H.* ×*glaucus* Small); such hybrids have not been reported for Missouri but might be found in the state in the future.

9. Helianthus mollis Lam. (ashy sunflower)
H. mollis f. *flavidus* Steyerm.

Pl. 282 g, h; Map 1196

Plants perennial herbs, with relatively thick, long-creeping, branched rhizomes, usually occurring in colonies. Stems solitary, 50–120 cm long, moderately to more commonly densely pubescent with short, spreading hairs and usually also shorter, ascending hairs, these mostly not pustular-based, usually appearing uniformly grayish, slightly to moderately roughened to the touch, occasionally nearly glabrous toward the base with age. Leaves well developed along the stem (usually with 8–17 nodes), all or mostly opposite, sessile or with a minute petiole less than 5 mm long. Leaf blades 3–15 cm long, 1–7 cm wide, broadly lanceolate to broadly ovate, rounded or shallowly cordate at the base, tapered to a usually sharply pointed tip, the margins entire or less commonly finely toothed, flat, the surfaces moderately to more commonly densely pubescent with short, slender, curved or ascending hairs, usually appearing uniformly grayish, slightly to moderately roughened to the touch, often also with sparse to dense, sessile, yellow glands, with 3 main veins, the lateral pair branching from the midnerve well above the base of the blade, arching upward. Inflorescences of solitary terminal heads

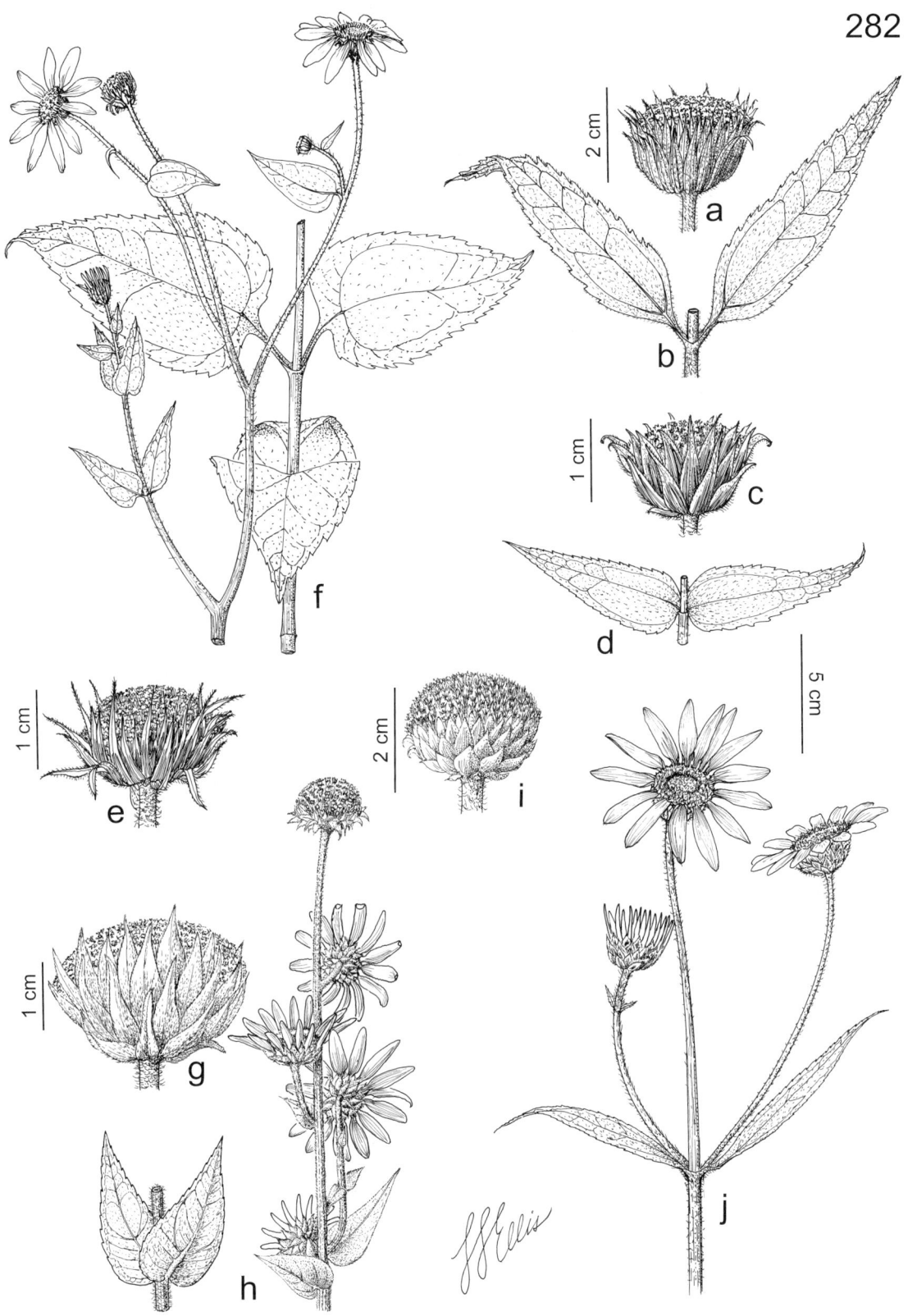

Plate 282. Asteraceae. *Helianthus hirsutus*, **a)** fruiting head, **b)** node with leaves. *Helianthus divaricatus*, **c)** fruiting head, **d)** node with leaves. *Helianthus decapetalus*, **e)** fruiting head, **f)** habit. *Helianthus mollis*, **g)** fruiting head, **h)** inflorescence and node with leaves. *Helianthus pauciflorus*, **i)** fruiting head, **j)** habit.

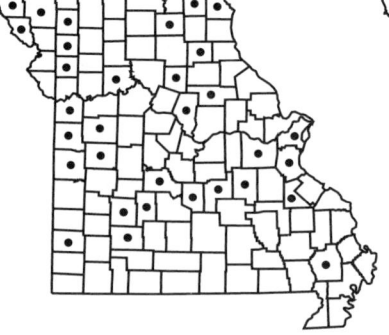

1197. Helianthus occidentalis   1198. Helianthus pauciflorus   1199. Helianthus petiolaris

or appearing as leafy spikes or narrow racemes. Involucre 7–12 mm long, 20–30 mm in diameter, about as long as or slightly longer than the tips of the disc corollas, the bracts in 2–4 somewhat unequal, overlapping series, narrowly lanceolate to lanceolate, tapered to a sharply pointed, appressed-ascending or somewhat spreading tip, the margins and outer surface densely pubescent with slender, ascending to somewhat spreading or tangled hairs, also with moderate to dense, minute, sessile, yellow glands. Receptacle convex, the chaffy bracts 9–11 mm long, narrowly oblong to linear, tapered to a sharply pointed, green or straw-colored tip, the outer surface densely short-hairy and glandular. Ray florets 17–30, the corolla 2.0–3.5 cm long, glabrous or the outer surface with sparse to dense, short, slender hairs and often dense, minute, sessile, yellow glands. Disc florets with the corolla 6.0–7.5 mm long, the corollas yellow or rarely pale yellow, the lobes usually minutely hairy on the outer surface (sometimes only along the margins). Pappus of 2 scales 2.5–3.5 mm long, these lanceolate to narrowly triangular, tapered to a sharply pointed, often minutely awnlike tip. Fruits 3–4 mm long, wedge-shaped to obovate, only slightly flattened but usually bluntly 4-angled in cross-section, moderately to densely pubescent with short, ascending hairs when young, only those at the tip persistent at maturity, uniformly dark brown or with fine, darker and lighter brown mottling. $2n=34$. July–October.

Common in the Unglaciated Plains Division, scattered elsewhere in the state but uncommon or absent from most of the Glaciated Plains (Wisconsin to Nebraska south to Louisiana and Texas; introduced in the eastern U.S. and adjacent Canada). Upland prairies and glades; also pastures, old fields, fencerows, margins of ditches, railroads, and roadsides.

Helianthus ×*cinereus* Torr. & A. Gray var. *sullivantii* Torr. & A. Gray is a rare morphologically intermediate hybrid between *H. mollis* and *H. occidentalis* ssp. *occidentalis*. Steyermark (1963) reported it from a mixed population of the parental species from Laclede County, and it also has been collected more recently in Howell County. Because *H. occidentalis* has the leaves greatly reduced above the stem base, hybrids derived from it are fairly easily recognizable in the field. Heiser et al. (1969) also stated that *H. mollis* hybridizes with a number of other sunflower species, including a few that occur in Missouri: *H. divaricatus*, *H. grosseserratus*, *H. maximilianii*, and *H. microcephalus*. A single late-season collection made by B. F. Bush in 1893 in Shannon County may represent a hybrid with *H. divaricatus*, but otherwise none of these hybrids has been discovered yet in Missouri, although *H. mollis* can be found in proximity to all but the last species in the state.

10. Helianthus occidentalis Riddell ssp. **occidentalis** (naked-stemmed sunflower)
Pl. 281 e, f; Map 1197

Plants perennial herbs, with slender, long-creeping, branched rhizomes and sometimes also stolons, often occurring in colonies. Stems solitary, 50–100 (–150) cm long, sparsely to densely pubescent with short-ascending or longer and tangled, slender hairs toward the base, glabrous or nearly so above the midpoint (often for most of the length). Leaves mostly basal (usually with 3–8 nodes), those of the stem much smaller than those of the basal rosette (rarely some of the stem leaves nearly as large as the basal ones), all or mostly opposite, the basal leaves usually long-petiolate, the stem leaves sessile or the lowermost stem leaves with a winged, short petiole. Blades of the basal leaves 6–15(–25) cm long, 1–7(–10) cm wide, narrowly to broadly elliptic or ovate, tapered to long-tapered at the base, tapered to a usually sharply pointed tip; those of the stem leaves 1–9(–12) cm long, narrowly oblong-lanceolate to ovate (mostly 2–7 times as long as

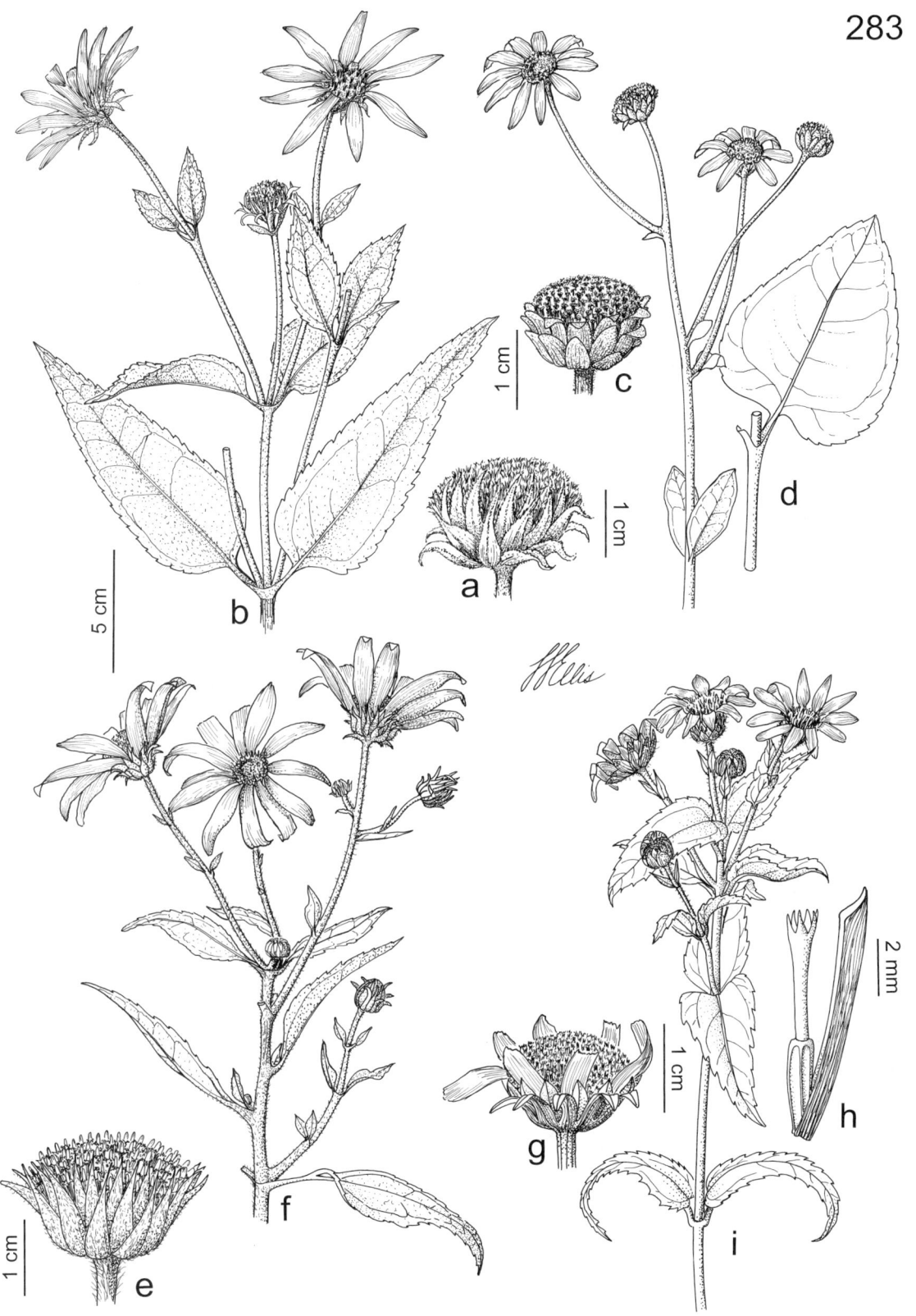

Plate 283. Asteraceae. *Helianthus strumosus*, **a)** fruiting head, **b)** habit. *Helianthus silphioides*, **c)** fruiting head, **d)** stem tip and larger leaf. *Helianthus tuberosus*, **e)** fruiting head, **f)** habit. *Heliopsis helianthoides*, **g)** head, **h)** disc floret, **i)** habit.

wide), flat or somewhat cupped or folded longitudinally, angled or tapered at the base, rounded or broadly angled to a bluntly pointed tip, all of the leaves with the margins entire or less commonly finely toothed, flat, the surfaces strongly roughened-pubescent with moderate to dense, short, stiff, broad-based hairs, often also with sparse, sessile, yellow glands, with 3 main veins, the lateral pair branching from the midnerve usually well above the base of the blade, arching upward. Inflorescences of solitary terminal heads or open clusters, occasionally with a few additional heads on branches from the upper leaf axils, then appearing somewhat paniculate. Involucre 7–12 mm long, 9–15 mm in diameter, slightly shorter than to less commonly slightly longer than the tips of the disc corollas, the bracts in 3 or 4 noticeably unequal, overlapping series, lanceolate, tapered to a sharply pointed, usually appressed-ascending tip (at least at flowering), the margins with a fringe of short, spreading hairs, the outer surface glabrous or sparsely pubescent with short, stout, ascending, sometimes broad-based hairs, especially toward the base, often also with sparse, minute, sessile, yellow glands. Receptacle convex, the chaffy bracts 5–7 mm long, narrowly oblong to narrowly oblong-oblanceolate, angled or short-tapered to a sharply pointed, green or straw-colored tip with hairy margins, occasionally with a pair of additional short, broad lobes toward the tip, the outer surface also usually glabrous. Ray florets 8–15, the corolla 1.5–3.0 cm long, glabrous or the outer surface with inconspicuous, short, slender hairs and scattered, minute, sessile, yellow glands. Disc florets with the corolla 4.5–5.5 mm long, the corollas yellow, glabrous. Pappus of 2 scales 1.5–3.0 mm long, these lanceolate to narrowly triangular, tapered to a sharply pointed, often minutely awnlike tip, also usually with 1–4 additional shorter, ovate scales 0.3–1.0 mm long. Fruits 3–4 mm long, wedge-shaped to obovate, only slightly flattened but usually bluntly 4-angled in cross-section, moderately to densely pubescent with short, ascending hairs when young, only those at or toward the tip persistent at maturity, with fine, darker and lighter brown mottling. $2n=34$. July–September.

Scattered mostly in the Ozark and Ozark Border Divisions (eastern U.S. west to Minnesota and Texas). Upland prairies, glades, savannas, and openings of dry upland forests; also pastures, old fields, fencerows, and roadsides.

This species is easily recognized by its highly reduced stem leaves and relatively big basal rosettes. Rare specimens with basal leaves absent or withered at flowering and the stem leaves less abruptly reduced might seem like hybrids between *H. occidentalis* and some other species, but these do not appear to differ from typical plants in their overall leaf shape and pubescence or in their involucral bract morphology. These plants are similar to those described as *H. dowellianus* M.A. Curtis from the southeastern United States, which Heiser et al. (1969) placed in synonymy under *H. occidentalis*. They are interpreted here as atypical specimens of *H. occidentalis* perhaps growing in response to unusual soil or moisture conditions. Heiser et al. (1969) reported hybrids between *H. occidentalis* and several other species. For discussion of a putative hybrid with *H. mollis,* see the treatment of that species. The only other putative hybrid involving a parental taxon that occurs in Missouri is *H. occidentalis* × *grosseserratus,* but although the parents sometimes are found growing in proximity, this hybrid has not yet been found in Missouri. The other subspecies of *H. occidentalis,* ssp. *plantagineus* (Torr. & A. Gray) Heiser, D.M. Sm., Clev. & W.C. Martin, occurs in Texas, Arkansas, and Louisiana and differs in its leaves with softer, appressed hairs.

11. Helianthus pauciflorus Nutt. (stiff sunflower, prairie sunflower)
H. rigidus Desf.
H. laetiflorus Pers. var. *rigidus* (Desf.) Fernald
Pl. 282 i, j; Map 1198

Plants perennial herbs, with relatively long-creeping, thick, branched rhizomes, often occurring as colonies of stems. Stems usually appearing solitary, 30–150(–200) cm long, sparsely to densely roughened-pubescent with short, loosely ascending to more or less spreading, short, stiff, pustular-based hairs. Leaves well developed along the stem (usually with 5–15 nodes), all or mostly opposite, sessile or with a poorly differentiated, winged petiole to about 1 cm long. Leaf blades 4–25 cm long, 2–6 cm wide, relatively thick-textured, lanceolate to elliptic-ovate or those of the uppermost leaves sometimes narrowly oblanceolate (mostly [2.0–]2.5–8.0 times as long as wide), flat, not folded longitudinally, tapered at the base, tapered gradually to a sharply pointed tip or angled more abruptly to a bluntly or sharply pointed tip, the margins finely toothed to nearly entire, flat, the surfaces strongly roughened with moderate to dense, minute, stout, pustular-based hairs, also with sparse to moderate, sessile, yellow glands, with 3 main veins, the lateral pair branching from the midnerve well above the base of the blade, arching upward. Inflorescences of solitary terminal heads or more commonly appearing as open clusters or open panicles. Involucre 12–20 mm long, 15–25 mm in diameter, shorter than the tips of the disc corollas, the bracts in 3 or

4 noticeably unequal, overlapping series, elliptic to oblong-ovate, narrowed to a bluntly or sharply pointed, tightly appressed tip, the margins with a fringe of short, spreading hairs, the outer surface glabrous or sparsely hairy but usually lacking glands. Receptacle convex, the chaffy bracts 8–10 mm long, narrowly oblong to narrowly oblong-oblanceolate, angled or short-tapered to a sharply pointed, green or purplish-tinged, minutely hairy tip, the outer surface also usually minutely hairy above the midpoint. Ray florets 10–21, the corolla 2.0–3.5 cm long, glabrous or the outer surface microscopically roughened. Disc florets with the corolla 6.0–7.5 mm long, the corollas reddish brown to dark purple (at least the lobes and the upper portion of the tube). Pappus of 2 scales 4–5 mm long, these lanceolate to narrowly triangular, tapered to a sharply pointed, often minutely awnlike tip, also usually with 2–8 additional minute, oblong scales 0.3–2.0 mm long. Fruits 5–6 mm long, narrowly wedge-shaped to narrowly obovate, flattened but more or less 4-angled in cross-section, the surface often with fine, ascending hairs when young but usually glabrous or nearly so at maturity, uniformly dark brown to black or sometimes with lighter brown streaks or mottling. 2n=102. August–October.

Scattered in the Glaciated and Unglaciated Plains Divisions, uncommon in the Ozarks and Mississippi Lowlands (Montana to New Mexico east to Maine and Georgia; Canada). Upland prairies, loess hill prairies, glades, and openings of dry upland forests; also pastures, railroads, and roadsides.

Specimens from 15 (so far) counties scattered nearly throughout the state are referable to *Helianthus* ×*laetiflorus* Pers., which comprises a series of hexaploid (2n=102) hybrids and backcrosses between *H. pauciflorus* and *H. tuberosus*. Steyermark (1963) and some other earlier authors considered this a true species, but its hybrid origins have now been confirmed with data from morphology, artificial crosses, and chemical compounds. It has a relatively broad distribution in the midwestern states (sometimes forming populations in the absence of both parents) and also occurs sporadically farther east. The natural range of the taxon is not known with certainty, as it has been cultivated as a garden ornamental for a long time and subsequently escaped from cultivation. Fertility apparently varies a great deal from plant to plant (Heiser et al., 1969). It differs most notably from *H. pauciflorus* in its yellow disc corollas and from *H. tuberosus* in its more strongly unequal, relatively short involucral bracts.

Helianthus pauciflorus was long known as *H. rigidus,* but the former is the older of the two available names (Spring and Schilling, 1990; Kartesz and Gandhi, 1990). Most botanists currently accept two morphologically overlapping subspecies within *H. pauciflorus*.

1. Stems mostly 80–150(–200) cm long, with mostly 8–15 nodes below the inflorescence; uppermost leaves usually alternate; median and lower leaves mostly 8–25 cm long, oblong-lanceolate to narrowly ovate, tapered gradually to a sharply pointed tip
  . . . . 11A. SSP. PAUCIFLORUS
1. Stems mostly 30–120 cm long, with mostly 5–10 nodes below the inflorescence; uppermost leaves usually opposite; median and lower leaves mostly 5–12 cm long, lanceolate to elliptic-ovate, angled to a bluntly or sharply pointed tip . . . . 11B. SSP. SUBRHOMBOIDEUS

11a. ssp. pauciflorus

Stems mostly 80–150(–200) cm long, with mostly 8–15 nodes below the inflorescence. Uppermost stem leaves usually alternate. Median and lower stem leaves mostly 8–25 cm long, oblong-lanceolate to narrowly ovate, tapered gradually to a sharply pointed tip. 2n=102. August–October.

Scattered in the Glaciated and Unglaciated Plains Divisions, uncommon in the Ozarks and Mississippi Lowlands (Montana to Texas east to Maine and Georgia; Canada). Upland prairies, loess hill prairies, glades, and openings of dry upland forests; also railroads and roadsides.

11b. ssp. subrhomboideus (Rydb.) O. Spring & E.E. Schill.

H. pauciflorus ssp. *subrhomboideus* (Rydb.) Kartesz & Gandhi (a later publication)

H. pauciflorus var. *subrhomboideus* (Rydb.) Cronquist

H. subrhomboideus Rydb.

H. rigidus ssp. *subrhomboideus* (Rydb.) Heiser, D.M. Sm., Clev. & W.C. Martin

Stems mostly 30–120 cm long, with mostly 5–10 nodes below the inflorescence. Uppermost stem leaves usually opposite. Median and lower stem leaves mostly 5–12 cm long, lanceolate to elliptic-ovate, angled to a bluntly or sharply pointed tip. 2n=102. August–October.

Introduced, uncommon and sporadic (Montana to Wisconsin south to New Mexico, Texas, Iowa, and Illinois; Canada; introduced farther south and east). Pastures, railroads, and roadsides.

The status of this taxon in Missouri is unclear, but it seems likely that it should not be considered native. It was first reported as occurring in the state by Heiser et al. (1969) without citation of a voucher specimen (but likely based on a historical specimen from a pasture in St. Francois County). Mühlenbach (1983) provided a second report based on an atypical specimen confirmed by Charles Heiser of Indiana University from railroad tracks in the city of St. Louis. The only other specimen discovered thus far was collected by Norlan Henderson in 1994 along a highway embankment in Clinton County. Clevenger and Heiser (1963), Heiser et al. (1969), and Gleason and Cronquist (1991) indicated that the main native range of the taxon is to the north and west of the state, with scattered, introduced occurrences farther east.

12. Helianthus petiolaris Nutt. ssp. petiolaris (prairie sunflower, Kansas sunflower, plains sunflower)

Pl. 281 c, d; Map 1199

Plants annual, with taproots. Stems solitary, (25–)40–100(–150) cm long, usually relatively stout, moderately pubescent with short, stiff, ascending hairs. Leaves well developed along the stem (usually with 8–25 nodes), mostly alternate, long-petiolate. Blades of the stem leaves 4–15 cm long, 1–8 cm wide, lanceolate to triangular-ovate (mostly 2–5 times as long as wide), flat, not folded longitudinally, tapered to truncate at the base, tapered to a usually sharply pointed tip, the margins entire or finely and sometimes somewhat irregularly toothed, flat, the surfaces moderately to densely pubescent with short, straight, appressed, somewhat pustular-based hairs (somewhat roughened to the touch), sometimes also with sparse, sessile, yellow glands, more or less with 3 main veins, the lateral pair branching from the midnerve at the base of the blade. Inflorescences rarely of solitary terminal heads, more commonly appearing as open panicles. Involucre 8–14 mm long, (10–)15–25 mm in diameter, mostly shorter than or extending about to the tips of the disc corollas, the bracts in 2–4 subequal to somewhat unequal series, lanceolate to narrowly ovate, tapered to a sharply pointed, slender, loosely ascending to more commonly spreading or recurved tip, the margins with minute, stiff, ascending to more or less spreading hairs, the outer surface moderately to densely pubescent with short, stiff, loosely ascending to spreading, pustular-based hairs (the inner surface often microscopically roughened or hairy), both surfaces sometimes also with small, sessile, yellow glands. Receptacle flat or slightly convex, the chaffy bracts 5–8 mm long, narrowly oblong-triangular, usually 3-lobed above the midpoint, the lateral lobes tapered to sharply pointed, usually straw-colored, glabrous tips, the middle lobe somewhat differentiated and tapered to a sharply pointed, usually purplish-tinged, densely and conspicuously white-hairy tip, the outer surface usually glabrous below the tip. Ray florets 12–30, the corolla 1.7–2.5 cm long, glabrous or the upper surface inconspicuously hairy toward the base, both surfaces occasionally also with minute, sessile, yellow glands. Disc florets with the corolla 4.5–6.0 mm long, reddish brown to dark purple (at least the lobes and the upper portion of the tube). Pappus of 2 scales 1.5–3.0 mm long, these narrowly lanceolate-triangular, tapered to a sharply pointed, often minutely awnlike tip, more or less papery, rarely also with 1 or 2 additional oblong scales 0.2–0.5 mm long. Fruits 3.0–4.5 mm long, narrowly wedge-shaped to obovate, flattened but usually more or less 4-angled in cross-section, the surface moderately densely and usually persistently (at least along the margins) pubescent with fine, ascending hairs, finely mottled with dark brown and lighter brown patches. $2n=34$. May–October.

Uncommon, mostly south of the Missouri River (Montana to New Mexico east to Wisconsin and Missouri; Canada; introduced in the western and eastern U.S.). Upland prairies, sand prairies, and tops of bluffs; also quarries, railroads, roadsides, and open, disturbed areas, often in sandy soil.

Some plants from the southwestern United States have been segregated into one or two other subspecies, based on differences in pubescence patterns and quantitative features of the heads and florets. Missouri plants are all referable to ssp. *petiolaris*. For a discussion of hybridization between this species and *H. annuus*, see the treatment of that species.

13. Helianthus salicifolius A. Dietr. (willow-leaved sunflower)

Pl. 280 c, d; Map 1200

Plants perennial herbs, with long-creeping, branched rhizomes, often occurring as colonies of stems. Stems not densely clumped, 50–200(–250) cm long, glabrous, often somewhat glaucous. Leaves numerous (sometimes appearing in irregular fascicles) and well developed along the stem (usually many more than 30 nodes), mostly alternate, sessile or nearly so, often appearing arched or drooping. Blades of the stem leaves 4–20 cm long, 0.1–1.0 cm wide, linear or those of the lowermost leaves sometimes narrowly lanceolate (mostly 12–20 times as long as wide), not folded longitudinally, tapered at the base, tapered to a sharply pointed tip, the margins entire, mostly curled under, the surfaces

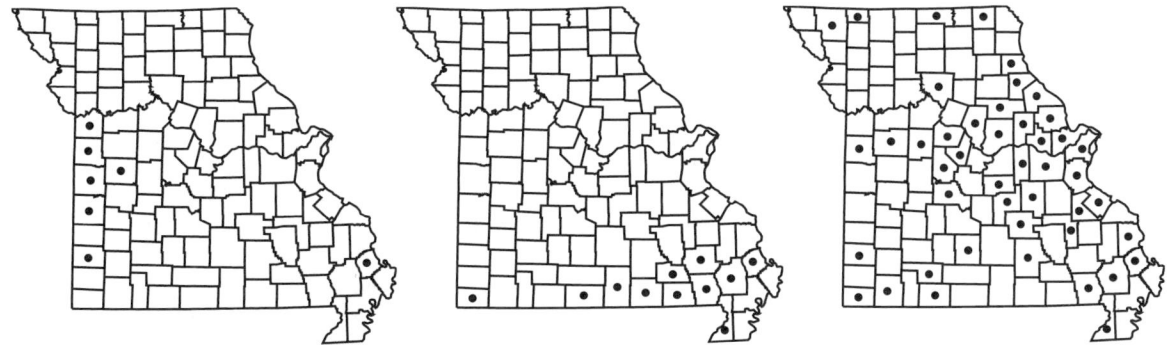

1200. Helianthus salicifolius   1201. Helianthus silphioides   1202. Helianthus strumosus

glabrous or sparsely pubescent with minute, slender, soft, appressed hairs, also with moderate to dense, sessile, yellow to brown glands, with 1 main vein. Inflorescences of solitary terminal heads or more commonly appearing as open, few-headed clusters or open panicles, the heads mostly long-stalked. Involucre 9–14 mm long, 10–18 mm in diameter, mostly extending beyond the tips of the disc corollas, the bracts in 2 or 3 subequal series, narrowly lanceolate to nearly linear, tapered to a sharply pointed, slender, loosely ascending to more commonly spreading or recurved tip, the margins with a dense fringe of short hairs, the surfaces glabrous or inconspicuously and minutely hairy and often also with scattered, sessile, yellow glands. Receptacle shallowly convex, the chaffy bracts 6–9 mm long, oblong-oblanceolate, tapered to a sharply pointed, green, minutely hairy tip, the outer surface also minutely hairy. Ray florets 10–21, the corolla 2.5–3.5 cm long, glabrous. Disc florets with the corolla 5–6 mm long, the lobes and sometimes also the tip of the tube reddish brown to dark purple above a yellow basal portion. Pappus of 2 scales 3.0–3.5 mm long, these oblong-lanceolate, tapered abruptly to a sharply pointed, minutely awnlike tip, often with an additional pair of thin, irregular apical lobes, also usually with 2–6 additional minute, oblong scales 0.2–0.5 mm long. Fruits 4–6 mm long, narrowly wedge-shaped, flattened but 4-angled in cross-section, the surface glabrous, often finely mottled with dark brown and lighter brown patches. $2n=34$. August–October.

Scattered in the Unglaciated Plains Division north locally to Jackson County; introduced in Scott County (Texas, Oklahoma, Kansas, and Missouri; introduced sporadically farther north and east). Upland prairies and rarely sand prairies and calcareous glades; also roadsides.

In recent years, this species has become available as an ornamental through native plant nurseries. In its native habitat, the plants generally grow in rather poor, dry soils. Gardeners who grow willow-leaved sunflowers under typical garden conditions may be surprised when the plants produce numerous stems more than 4 m long. The species also may spread aggressively by rhizomes and seeds. Finches and other seed-eating birds relish the achenes.

14. Helianthus silphioides Nutt. (rosinweed sunflower)

Pl. 283 c, d; Map 1201

Plants perennial herbs, lacking rhizomes or with very short, thick rhizomes, often occurring as clumps. Stems solitary or more commonly few to several, 80–250(–300) cm long, sparsely to densely pubescent with short, ascending, slender hairs and/or more or less spreading, short, stiff, pustular-based hairs toward the base, usually glabrous above the midpoint. Leaves well developed along the stem (usually with 8–15 nodes), all or mostly opposite, mostly with an unwinged, short petiole 1–4 cm long. Blades of the stem leaves 3–15 cm long, 1–15 cm wide, ovate to broadly ovate or nearly circular (mostly 1.0–1.7[–2.0] times as long as wide), the uppermost leaves occasionally somewhat narrower, flat, not folded longitudinally, rounded or abruptly short-tapered at the base, rounded or broadly angled to a bluntly pointed tip, the margins finely toothed or scalloped to nearly entire, flat, the surfaces strongly roughened with moderate, minute, stout, pustular-based hairs, but lacking sessile, yellow glands, with 3 main veins, the lateral pair branching from the midnerve at or just above the base of the blade, arching upward. Inflorescences open clusters or open panicles. Involucre 8–15 mm long, 10–20 mm in diameter, slightly shorter than to about as long as the tips of the disc corollas, the bracts in 3 or 4 noticeably unequal, overlapping series, oblong to oblong-ovate, rounded or narrowed to a bluntly pointed, tightly appressed tip, the margins often with a fringe of minute, spreading hairs, the outer surface glabrous or sparsely hairy toward the base but lacking glands. Receptacle convex, the chaffy bracts 8–10 mm long, narrowly oblong to narrowly oblong-oblanceolate, angled or short-tapered to a sharply pointed, green, straw-colored, or purplish-tinged, glabrous tip, occasionally with a pair of additional short, broad lobes toward the tip, the outer surface also usually glabrous. Ray florets (8–)12–15, the corolla 1.5–2.0 cm long, glabrous or the outer surface with inconspicuous, short, slender hairs along the veins. Disc florets with the corolla 6–7 mm long, the corollas reddish brown to dark purple (at least the lobes and the upper portion of the tube), the lobes and tip of the tube often minutely hairy on the outer surface. Pappus of 2 scales 2.5–3.0 mm long, these lanceolate to narrowly triangular, sometimes with an irregular pair of small basal lobes, tapered to a sharply pointed, often minutely awnlike tip. Fruits 3.0–4.5 mm long, narrowly wedge-shaped to narrowly obovate, flattened but sometimes more or less 4-angled in cross-section, the tip and margins moderately to densely pubescent with short, ascending hairs, the surface otherwise glabrous, uniformly dark brown or more commonly with fine, darker and lighter brown mottling. $2n=34$. August–October.

Scattered in the Mississippi Lowlands Division and westward along the southern portion of the Ozarks to McDonald County (Oklahoma, Missouri, Illinois, and Kentucky south to Arkansas, Tennessee, Louisiana, Mississippi, and Alabama). Banks of streams, sand prairies, and openings of mesic to dry upland forests; also fallow fields, fencerows, and roadsides.

Steyermark (1963) noted that this showy species does well in the wildflower garden.

15. Helianthus strumosus L. (pale-leaved sunflower)

Pl. 283 a, b; Map 1202

Plants perennial herbs, with slender to stout, usually long-creeping, branched rhizomes (the branches sometimes with small tubers at the tip), usually occurring in dense colonies. Stems usually solitary, less commonly appearing in small clumps, 80–200 cm long, glabrous below the inflorescence or the uppermost portion of the stem sparsely pubescent with short, stiff, loosely ascending, pustular-based hairs, often somewhat glaucous. Leaves well developed along the stem (usually with 8–20 nodes), all or less commonly mostly opposite, mostly short-petiolate (the petioles of the larger leaves 1–2[–3] cm long). Leaf blades 4–20 cm long, 1–10 cm wide, relatively thick-textured, lanceolate to ovate (mostly 2–5 times as long as wide), flat, not folded longitudinally, rounded to broadly angled or short-tapered at the base, tapered to a sharply pointed tip, the margins finely toothed to entire, flat, the upper surface strongly roughened with moderate to dense, minute, stout, pustular-based hairs, the undersurface sparsely to densely pubescent with somewhat softer, loosely appressed to more or less spreading hairs, less commonly glabrous or nearly so and pale green to silvery (this phase more common elsewhere), also with sparse to moderate, sessile, yellow glands, with 3 main veins, the lateral pair branching from the midnerve at or just above the base of the blade, arching upward. Inflorescences of solitary terminal heads or appearing as open clusters or occasionally open panicles. Involucre 7–12 mm long, 8–20 mm in diameter, about as long as or slightly longer than the tips of the disc corollas, the bracts in 3 or 4 subequal, overlapping series, lanceolate, tapered to a sharply pointed, loosely ascending to spreading or recurved tip, the margins with an irregular fringe of short, spreading to ascending hairs, the outer surface glabrous or more commonly sparsely to moderately pubescent with short, stout, ascending, often pustular-based hairs, sometimes also with minute, sessile, yellow glands. Receptacle convex to short-conical, the chaffy bracts 5.0–6.5 mm long, narrowly oblong to narrowly oblong-oblanceolate, often with 3 short-tapered, sharply pointed lobes at the tip, these green or straw-colored, the outer surface minutely hairy, especially toward the tip. Ray florets 8–15, the corolla 1.5–3.0(–4.0) cm long, the outer surface usually with sparse, minute hairs and scattered, minute, sessile, yellow glands. Disc florets with the corolla 5.5–6.5 mm long, the corollas yellow, the lobes often minutely hairy on the outer surface. Pappus of 2 scales 1.5–2.5 mm long, these lanceolate to narrowly triangular, tapered to a sharply pointed, often minutely awnlike tip. Fruits 4.0–5.5 mm long, wedge-shaped to obovate, somewhat flattened and more or less bluntly 4-angled in cross-section, the surface glabrous or with a few minute hairs at the tip, uniformly brown or with fine, darker and lighter brown mottling. $2n=68$, 102. July–September.

Scattered nearly throughout the state (eastern U.S. west to Wisconsin, Kansas, and Texas; Canada). Bottomland forests, mesic upland forests, banks of streams and rivers, bases and ledges of bluffs, fens, and upland prairies; also railroads and roadsides.

Although apparently reasonably distinct elsewhere in its range, this species can be very difficult to distinguish from *H. hirsutus* in Missouri,

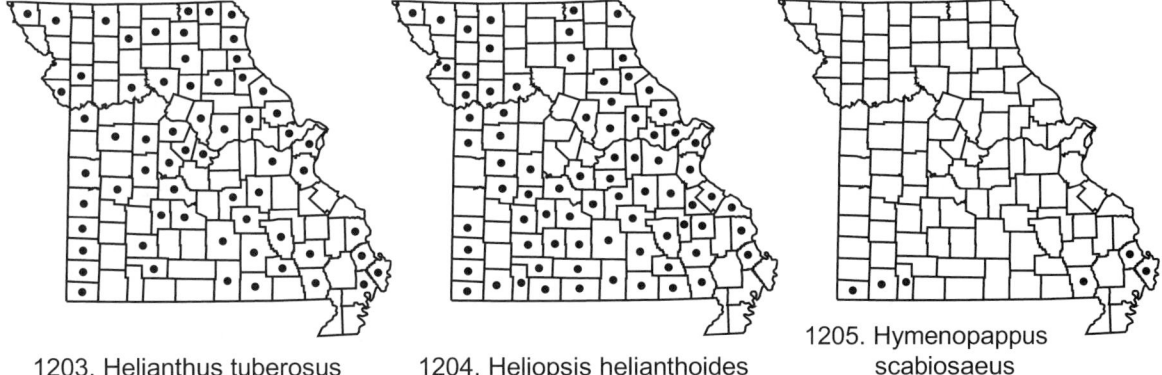

1203. Helianthus tuberosus   1204. Heliopsis helianthoides   1205. Hymenopappus scabiosaeus

especially in the Ozarks. Plants with slightly hairy, nonglaucous stems and the leaf undersurface relatively hairy are encountered with some frequency. How much of this variation represents the results of introgressive hybridization between the taxa is not known and requires further detailed genetic study. Heiser et al. (1969) thought that these plants, which they considered a tetraploid (2n=64) race, might as easily be accommodated in *H. hirsutus*. It is possible that future studies may show that the apparent separation of *H. strumosus* from *H. hirsutus* is an illusion and that these should be combined as a single species, perhaps with two or more varieties to account for the various races.

Occasional specimens also can be difficult to distinguish from *H. divaricatus* and *H. tuberosus*. Heiser et al. (1969) discussed a hexaploid (2n=102) race of *H. strumosus* from Illinois, Missouri, Arkansas, and Oklahoma, which had been named *H. formosus* S. Watson. These plants were characterized as having the leaf undersurface strongly glaucous and the rhizomes with well-defined, small tubers.

16. Helianthus tuberosus L. (Jerusalem artichoke)

H. tuberosus var. *subcanescens* A. Gray

Pl. 283 e, f; Map 1203

Plants perennial herbs, with slender, long-creeping, branched rhizomes (the branches usually with small tubers at the tip), usually occurring in dense colonies. Stems usually solitary, less commonly appearing in small clumps, 80–300 cm long, sparsely to more commonly moderately pubescent with short, stiff, loosely ascending, usually pustular-based hairs, not glaucous, sometimes nearly glabrous toward the base. Leaves well developed along the stem (usually with 8–20 nodes), mostly opposite (usually alternate in the upper third), with a short to less commonly long, sometimes poorly differentiated, partially winged petiole (the petioles of the larger leaves 1.5–8.0 cm long). Leaf blades 6–25 cm long, 3–15 cm wide, relatively usually moderately thick-textured, oblong-lanceolate to lanceolate or ovate (mostly 2–4 times as long as wide), flat, not folded longitudinally, short- to more commonly long-tapered at the base, tapered to a sharply pointed tip, the margins finely to coarsely toothed, flat, the upper surface strongly roughened with moderate to dense, minute, stout, pustular-based hairs, the undersurface moderately to densely pubescent with somewhat softer, loosely appressed to more or less spreading hairs, also with sparse to moderate, sessile, yellow glands, with 3 main veins, the lateral pair branching from the midnerve well above the base of the blade, arching upward. Inflorescences of solitary terminal heads or more commonly appearing as open clusters or open panicles. Involucre 8–12 mm long, 10–25 mm in diameter, about as long as or slightly longer than the tips of the disc corollas, the bracts in 3 or 4 subequal, overlapping series, narrowly lanceolate to lanceolate, often darkened or with a dark stripe toward the base, tapered to a sharply pointed, loosely ascending to spreading or recurved tip, the margins with an irregular fringe of short, spreading to ascending hairs, the outer surface glabrous or more commonly sparsely to moderately pubescent with short, stout or slender, ascending hairs, but usually lacking sessile, yellow glands. Receptacle convex to short-conical, the chaffy bracts 8–9 mm long, narrowly oblong to narrowly oblong-oblanceolate, usually with 3 short, sharply pointed lobes at the tip, these green or straw-colored, the outer surface minutely hairy, especially toward the tip. Ray florets 10–20, the corolla 2–4 cm long, the outer surface with sparse to moderate, minute hairs and usually also scattered, minute, sessile, yellow glands. Disc florets with the corolla 6–7 mm long, the corollas yellow, the lobes glabrous or with a few inconspicuous, minute hairs on the outer surface. Pappus of 2 scales 2–3 mm long, these lanceolate to narrowly triangular, tapered to a sharply pointed, often minutely awnlike tip, occasionally also with an additional minute, oblong scale 0.4–0.8 mm long. Fruits 5–7 mm long, wedge-shaped to narrowly obovate, somewhat flattened and more or less 4-angled in cross-section, the surface glabrous or with a few minute hairs at the tip, uniformly brown or with fine, darker and lighter brown mottling, sometimes purplish-tinged. $2n=102$. August–October.

Scattered to common throughout the state (eastern U.S. west to North Dakota and Texas; Canada; introduced westward to the Pacific Coast). Banks of streams, rivers, and spring branches, bottomland forests, mesic upland forests, sloughs, margins of ponds and lakes, and moist depressions of upland prairies; also pastures, fencerows, railroads, roadsides, and disturbed areas.

Heiser et al. (1969) stated that this species is exceedingly variable morphologically, but that most plants are fairly easily recognized by the alternate upper leaves, leaf blades that are relatively coarsely toothed and longer-tapered at the base, and involucral bracts that often are darkened toward the base, among other features. They discussed, however, difficulties in separating some specimens from *H. strumosus*. Heiser and his colleagues treated var. *subcanescens* as part of the environmentally based variation in the species but noted that plants determined by Steyermark (1963) as this variety actually represented misdetermined specimens of *H. strumosus*. The origins of this hexaploid ($2n=102$) species are not fully understood, and the degree to which it hybridizes with other sunflower species in nature requires further study. For discussion of putative hybrids with *H. rigidus* (known as *H.* ×*laetiflorus*), see the treatment of that species.

The seasonally produced fleshy tubers produced by this species are edible and have a crisp texture and a flavor reminiscent of nuts and artichokes; they can be eaten raw, cooked, or pickled. They contain the storage-carbohydrate inulin (instead of starch), which is a fructose-based oligosaccharide with a slightly sweet flavor that passes through the human digestive tract largely undigested and is thus tolerated well by persons with diabetes. Inulins occur in a large number of plants (see also the treatment of *Cichorium* [Asteraceae tribe Cichorieae]) and are extracted commercially from a few species for use as a soluble fiber dietary supplement and as a bulking agent and fat substitute in some foods, especially some brands of yogurt.

Heiser (1976) reviewed the history of Jerusalem artichoke cultivation. Although there is no archaeological data to document the antiquity of the food use of this species in North America, the historical use of both cultivated and wild-collected plants by various tribes of Native Americans has been well documented (Moerman, 1998). The plant reached Europe in the early 1600s, where it was cultivated and became a fashionable food item for a time, but it was never economically of major importance. Russian plant breeders used *H. tuberosus* as a source of disease resistance in breeding programs to improve strains of the giant annual sunflower. An artificial hybrid between *H. tuberosus* and *H. annuus* also has been bred in the United States with improved tuber production, and it has become available commercially under the name sunchoke. Today, both Jerusalem artichokes and sunchokes are both available in various forms in some grocery stores, particularly health food stores, although not as commonly as they were during the 1980s.

Amato (1993) chronicled the remarkable account of the brief and unfortunate attempt to develop Jerusalem artichoke (which is neither native to the Middle East nor closely related to artichokes) as an alternative crop for ethanol production as a source of alternative energy. During a period of about 18 months beginning in 1981, a Minnesota company called American Energy Farming Systems convinced farmers from agriculturally depressed areas of 31 states and 3 Canadian provinces to invest more than 25 million dollars in Jerusalem artichoke seeds for this purpose. Through a combination of bad politics and financial mismanagement, this enterprise failed spectacularly, and the company owners subsequently were prosecuted as con artists. A regrettable consequence of this series of events was a loss of confidence in the overall crop potential of *H. tuberosus* and a sharp decline in its cultivation.

80. Heliopsis Pers. (oxeye)
(Fisher, 1957)

About 16 species, North America to South America, most diverse in Mexico.

1. Heliopsis helianthoides (L.) Sweet (oxeye, false sunflower, sunflower heliopsis)

Pl. 283 g–i; Map 1204

Plants perennial herbs, with fibrous roots and sometimes a short, stout rhizome. Stems 40–150 cm long, erect or ascending, usually several-branched, with fine longitudinal lines or ridges, glabrous to sparsely to moderately pubescent with slender, ascending hairs (these rarely rough to the touch) or more commonly moderately roughened with short, stout, ascending, broad-based hairs (and sometimes also scattered, longer hairs). Leaves opposite or the uppermost few alternate, sessile or relatively short-petiolate, the lowermost leaves sometimes with longer petioles, the bases minutely expanded and wrapping around the stem (leaving a low ridge around the stem after the leaves are shed), the petiole sometimes with sparse to moderate, long, spreading hairs. Leaf blades 3–15 cm long, oblong-lanceolate to ovate or ovate-triangular, short-tapered or angled to more or less truncate at the base, tapered to a sharply pointed tip, the margins sharply and finely to coarsely toothed and minutely hairy, the upper surface glabrous to minutely pubescent with slender, sometimes broad-based hairs or more commonly strongly roughened with short, stout, broad-based hairs, the undersurface glabrous or minutely pubescent with slender hairs (occasionally with a few longer hairs along the midvein) or more commonly strongly roughened with short, stout, broad-based hairs, both surfaces also with scattered, sessile, spherical, yellow glands. Inflorescences of solitary or less commonly 2–4 heads at the branch tips and from the upper leaf axils, the heads usually with long, bractless, glabrous or roughened stalks. Heads radiate. Involucre 6–16 mm long, 10–22 mm in diameter, cup-shaped to broadly bell-shaped, the bracts in 2 or 3 subequal series. Involucral bracts 17–29, lanceolate to ovate, rounded to sharply pointed at the tip, the outer surface usually minutely hairy or roughened, those of the outer series often somewhat longer than the others and with loosely ascending to spreading tips, green, those of the inner series slightly shorter, usually somewhat yellowish green and somewhat scalelike, more strongly ascending at the tip. Receptacle conical, elongating somewhat as the fruits mature, with chaffy bracts subtending the ray and disc florets, these 8.0–8.5 mm long, oblong-lanceolate, concave and wrapped around the florets. Ray florets 8–16(–35), pistillate (with a 2-branched style exserted from the short tube at flowering), the corolla 15–40 mm long, relatively broad, pale yellow to orangish yellow, glabrous or nearly so, persistent and turning papery at fruiting. Disc florets 10–80, perfect, the corolla 4–5 mm long, greenish yellow to brownish yellow, glabrous, not expanded at the base or persistent at fruiting. Style branches with the sterile tip somewhat elongate and tapered. Pappus of the ray and disc florets absent or the disc florets with 2–4 minute teeth. Fruits 3.0–3.5 mm long, narrowly rectangular to slightly wedge-shaped in outline, strongly 3-angled (ray florets) or 4-angled (disc florets), the surface usually appearing finely pebbled, glabrous, dark brown to black, somewhat shiny. $2n=28$. May–September.

Scattered nearly throughout the state but apparently absent from most of the Mississippi Lowlands Division (eastern U.S. west to North Dakota and New Mexico; Canada). Mesic to dry upland forests, savannas, glades, upland prairies, sand prairies, banks of streams and rivers, seeps, and ledges and tops of bluffs; also margins of crop fields, pastures, fencerows, railroads, and roadsides.

Oxeye is sometimes cultivated as an ornamental in gardens. Examination of specimens in herbaria suggests that many students and botanists have difficulty in distinguishing this species from some members of *Helianthus*. This may be because of difficulties in interpretation of the character of persistent papery ray corollas. More reliably, the flowering heads of *Heliopsis* contain ray florets with a well-developed ovary and an exserted, branched style, whereas those of *Helianthus* contain ray florets with a slender, nonfunctional, stalklike ovary and the style apparently absent (very short, included in the short, tubular portion of the ray corolla, and undivided at the tip). The ray florets of *Heliopsis* develop well-formed, conspicuously 3-angled achenes, whereas in *Helianthus* the ovary portion of the ray florets is frequently shed with the corolla or, when persistent, is very difficult to observe.

Fisher (1957, 1958) treated *H. helianthoides* as comprising three morphologically overlapping subspecies with different centers of distribution. In particular, his ssp. *occidentalis* and ssp. *scabra* were noted to intergrade for most of the characters separating them. In fact, Fisher's mapped ideograms representing character coding for features such as petiole length and head stalk length seem to show a clinal north-to-south variation in the central portion of the United States with Missouri indicated to occupy part of the main geographic zone of intermediacy. Other authors, such as Steyermark (1963) and Turner (1988a), also commented on the large number of specimens intermediate for one or several of the purported differences between ssp. *occidentalis* and ssp. *scabra*. More recent collections continue to reinforce the problems in separating Missouri material of these two variants. Accordingly, they have been reduced to a single taxon in the present treatment. Steyermark (1963) and Turner (1988a) also chose to treat infraspecific variation within *H. helianthoides* at the varietal rather than subspecific level, which seems reasonable in light of the overlapping ranges of the taxa and the lack of cytological differences between them.

1. Leaves glabrous or minutely pubescent with slender hairs (occasionally with a few longer hairs on the undersurface midvein), those of the upper surface sometimes broad-based and rough to the touch; stems glabrous or sparsely to moderately pubescent with slender, ascending hairs, these rarely rough to the touch . . . . . 1A. VAR. HELIANTHOIDES
1. Leaves strongly roughened with short, stout, broad-based hairs on both surfaces; stems also moderately roughened with short (and sometimes also longer), stout, broad-based hairs
. . . . 1B. VAR. SCABRA

1a. var. helianthoides

Stems glabrous or sparsely to moderately pubescent with slender, ascending hairs, rarely rough to the touch. Leaf blades glabrous or minutely pubescent with slender hairs (occasionally with a few longer hairs on the undersurface midvein), those of the upper surface sometimes broad-based and rough to the touch.

Uncommon, known thus far from Crawford and Ralls Counties and possibly from the city of St. Louis (eastern U.S. west to Wisconsin, Missouri, and Louisiana). Banks of streams and rivers and saline seeps.

Steyermark (1963) did not accept the occurrence of var. *helianthoides* in Missouri. He suggested that two historical specimens collected by Sherff in the city of St. Louis and annotated by Fisher as this taxon probably were based on cultivated material and should be excluded from the flora. However, Barkley (1986) continued to include Missouri in the range of this variety. Examination of specimens during the present research documented a few widely scattered recent collections.

1b. var. scabra (Dunal) Fernald (rough oxeye)
H. helianthoides ssp. *scabra* (Dunal) T.R. Fisher
H. helianthoides ssp. *occidentalis* T.R. Fisher
H. helianthoides var. *occidentalis* (T.R. Fisher) Steyerm.

Stems moderately roughened with short (and sometimes also longer), stout, broad-based hairs. Leaf blades strongly roughened with short, stout, broad-based hairs on both surfaces, sometimes with longer, spreading hairs toward the base and along the petiole.

Scattered nearly throughout the state but apparently absent from most of the Mississippi Lowlands Division (eastern U.S. west to North Dakota and New Mexico; Canada). Mesic to dry upland forests, savannas, glades, upland prairies, sand prairies, banks of streams and rivers, and ledges and tops of bluffs; also margins of crop fields, pastures, fencerows, railroads, and roadsides.

This is the common variant present in Missouri. As noted above, Fisher (1957, 1958) applied the name ssp. *scabra* to an Ozarkian element with relatively narrow, longer-petiolate leaves and smaller heads with shorter ray corollas and treated plants from the western and the northern portion of the species range with broader leaves, shorter petioles, and larger heads as ssp. *occidentalis*. He indicated that much of Missouri was in the range of geographic and morphological overlap between his two subspecies, and more recent studies of Missouri specimens have confirmed that there is nearly continuous variation for every character said to distinguish the two taxa.

81. Hymenopappus L'Hér.
(Turner, 1956)

About 12 species, U.S., Mexico.
Some authors have included this genus in the tribe Anthemideae.

1. Hymenopappus scabiosaeus L'Hér. **var. scabiosaeus** (old plainsman)

Pl. 227 a–c; Map 1205

Plants biennial, with taproots. Stems 40–150 cm long, erect or ascending, several-branched or sometimes unbranched below the inflorescence, with several usually relatively conspicuous longitudinal ridges and grooves, moderately to densely pubescent with fine, woolly to cobwebby hairs, often becoming glabrous in patches with age. Leaves basal and alternate, of two kinds: leaves of the first-season basal rosette mostly short- to long-petiolate, the blade 8–25 cm long, oblanceolate to elliptic or lanceolate in outline, 1 or 2 times deeply pinnately lobed (rarely fully compound), the lobes narrowly triangular to narrowly oblong or oblong-elliptic, the upper surface bright green and glabrous or nearly so, the undersurface densely pubescent with white, felty or woolly hairs; basal and stem leaves of the second season short- to long-petiolate or the median and upper leaves sessile, the blade 2–18 cm long, lanceolate to elliptic or ovate in outline, 1–4 times deeply pinnately lobed, the lobes mostly linear (sometimes slightly broader) to threadlike, the upper surface bright green and glabrous or nearly so, the undersurface moderately (on upper leaves) to densely pubescent with white, cobwebby to felty or woolly hairs. Inflorescences more or less flat-topped, open terminal panicles with broad, irregularly shaped, white, membranous bracts at the branch points, the heads usually relatively long-stalked. Heads discoid. Involucre 7–12 mm long, cup-shaped to somewhat bell-shaped, the receptacle 5–9 mm in diameter, the bracts in 2 or 3 subequal series. Involucral bracts 8–14, glabrous or sparsely cobwebby-hairy toward the base and also with minute, sessile glands, green toward the base, white apically to well below the midpoint, those of the outer series ovate to nearly circular, those of the inner series often lanceolate to narrowly elliptic. Receptacle flat or slightly convex, not elongating as the fruits mature, naked, with minute ridges around the attachment points of the florets. Disc florets 20–60, perfect, the corolla 3.5–5.0 mm long, white to pale cream-colored above the slender, greenish tube, the tube not expanded at the base or persistent at fruiting, glandular-hairy, the abruptly expanded throat and the relatively long lobes mostly with scattered, sessile glands on the outer surface. Style branches with the sterile tip somewhat elongate and tapered to a bluntly pointed tip. Pappus of 14–18 scales, these 0.2–1.0 mm long, obovate, membranous. Fruits 3–5 mm long, more or less obovate to wedge-shaped in outline, somewhat 4-angled, the surface with 8–14 blunt longitudinal ridges, moderately pubescent with fine, more or less spreading hairs, dark brown to black. $2n=34$. April–July.

Uncommon in the southwestern portion of the Ozark Division and along the northern edge of the Mississippi Lowlands (southeastern U.S. west to Missouri and Oklahoma, north disjunctly to Illinois and Indiana). Glades, tops of bluffs, rocky portions of upland prairies, and sand prairies; also fencerows, cemeteries, and roadsides.

This attractive species deserves to be cultivated more widely for its striking white and green color combination and for the sweet scent of its flowers. A second variety, var. *corymbosus* (Torr. & A. Gray) B.L. Turner, occurs from Nebraska to Texas and adjacent Mexico. It differs in having the inflorescence stalks bractless or with small, scalelike bracts and the involucral bracts somewhat shorter and with a somewhat larger proportion of green coloration.

82. Iva L. (marsh elder)

About 9 species, North America.

Until recently, *Iva* was considered by most botanists to be a nearly cosmopolitan genus of about 15 species (Jackson, 1960). Bolick (1983) performed a cladistic analysis of Heliantheae subtribe *Ambrosiinae* Less. based on morphological, palynological, and phytochemical characters and suggested that some of the traditional generic circumscriptions did not describe natural phylogenetic lineages. In particular, she noted that some species of *Iva* apparently were more closely related to species of *Ambrosia* than to the remainder of *Iva*. Karis (1995) expanded on this study by sampling additional species and characters, and also concluded that the traditional circumscription of *Iva* included members of more than one evolutionary lineage. Molecular phylogenetic research by Miao et al. (1995a, b) provided further evidence that *Iva* in the broad sense is not a natural group. Strother (2000) reviewed the data from these earlier studies and concluded that a number of generic segregations were necessary in order to reflect the phylogeny of the group, resulting in the recognition of six total genera for the North American species formerly included in *Iva*. For Missouri, the practical consequence has been that *I. xanthiifolia* is now treated as the monotypic segregate, *Cyclachaena*. *Cyclachaena* differs from *Iva* in a number of gross morphological features, including its more strongly

1206. Iva annua

1207. Marshallia caespitosa

1208. Melampodium cinereum

paniculate inflorescences with small, bractless clusters of heads at the nodes and its pistillate florets lacking a corolla, as well as differences in secondary compounds and pollen ultrastructure.

Species of *Iva* shed abundant wind-borne pollen and are important causes of hay fever in areas where plants are abundant. Some species also can cause contact dermatitis in susceptible individuals.

1. Iva annua L. (marsh elder, sump weed)
I. ciliata Willd.
I. annua var. *caudata* (Small) R.C. Jacks.

Pl. 284 c, d; Map 1206

Plants annual, with taproots. Stems 30–120 (–200) cm long, unbranched or more commonly several-branched, erect or ascending, finely ridged or grooved, sparsely to moderately roughened with short, ascending, pustular-based hairs, often also with longer, spreading hairs toward the tip, usually glabrous or nearly so toward the base. Leaves opposite or the uppermost few alternate, variously sessile to long-petiolate. Leaf blades 2–15 cm long, simple, lanceolate to broadly ovate, angled or short-tapered at the base, tapered at the sharply pointed tip, the margins entire or more or less toothed and hairy, the surfaces nearly glabrous to sparsely or moderately roughened with short, more or less appressed hairs, often also with sparse, longer, spreading hairs along the main veins and/or near the base, usually also sparsely to moderately gland-dotted. Inflorescences of spikelike racemes terminal on the branch tips, these sometimes appearing paniculate, the heads very short-stalked, solitary at the nodes, subtended by ovate to narrowly lanceolate, leaflike bracts. Heads discoid, pendant. Involucre 2–4 mm long, more or less cup-shaped, somewhat asymmetrical, the 3–5 involucral bracts in 1 series, fused irregularly at or just above the base, green, sparsely long-hairy, especially along the margins. Receptacle flat, not elongating as the fruits mature, with chaffy bracts subtending the florets, these narrowly linear to narrowly oblanceolate, usually glandular along the margins, not wrapped around the florets. Central florets 9–16, staminate, with a minute, nonfunctional ovary and undivided style, the stamens with the filaments more or less fused into a tube and the anthers free but positioned closely adjacent to one another in a ring, the corolla 2.0–2.5 mm long, funnelform to narrowly bell-shaped, 5-lobed, white to pale yellow, sometimes purplish-tinged toward the tip, usually glabrous. Marginal florets 3–5, pistillate, the corolla 1.0–1.5 mm long, narrowly tubular, often slightly oblique at the tip (but unlobed), white to pale yellow, often persistent at fruiting. Pappus of the staminate and pistillate florets absent. Fruits 2.5–4.0 mm long, broadly obovoid to somewhat pear-shaped, somewhat flattened, not angled or at most very bluntly angled on 1 face, otherwise appearing smooth or with numerous faint, fine longitudinal lines, brown to dark brown, glabrous but with minute, sessile glands. 2n=34. July–October.

Scattered nearly throughout the state, but most commonly in the Unglaciated Plains Division and counties adjacent to the Missouri and Mississippi Rivers (Indiana to North Dakota south to Mississippi and New Mexico; Mexico; introduced eastward to Maine and Florida). Banks of streams and rivers, margins of ponds and lakes, swamps, sloughs, bottomland prairies, bottomland forests, and rarely moist depressions of upland prairies; also fallow fields, crop fields, ditches, pastures, railroads, roadsides, and open, disturbed areas.

Jackson (1960) recognized plants with the bracts of the inflorescence narrowly lanceolate, with

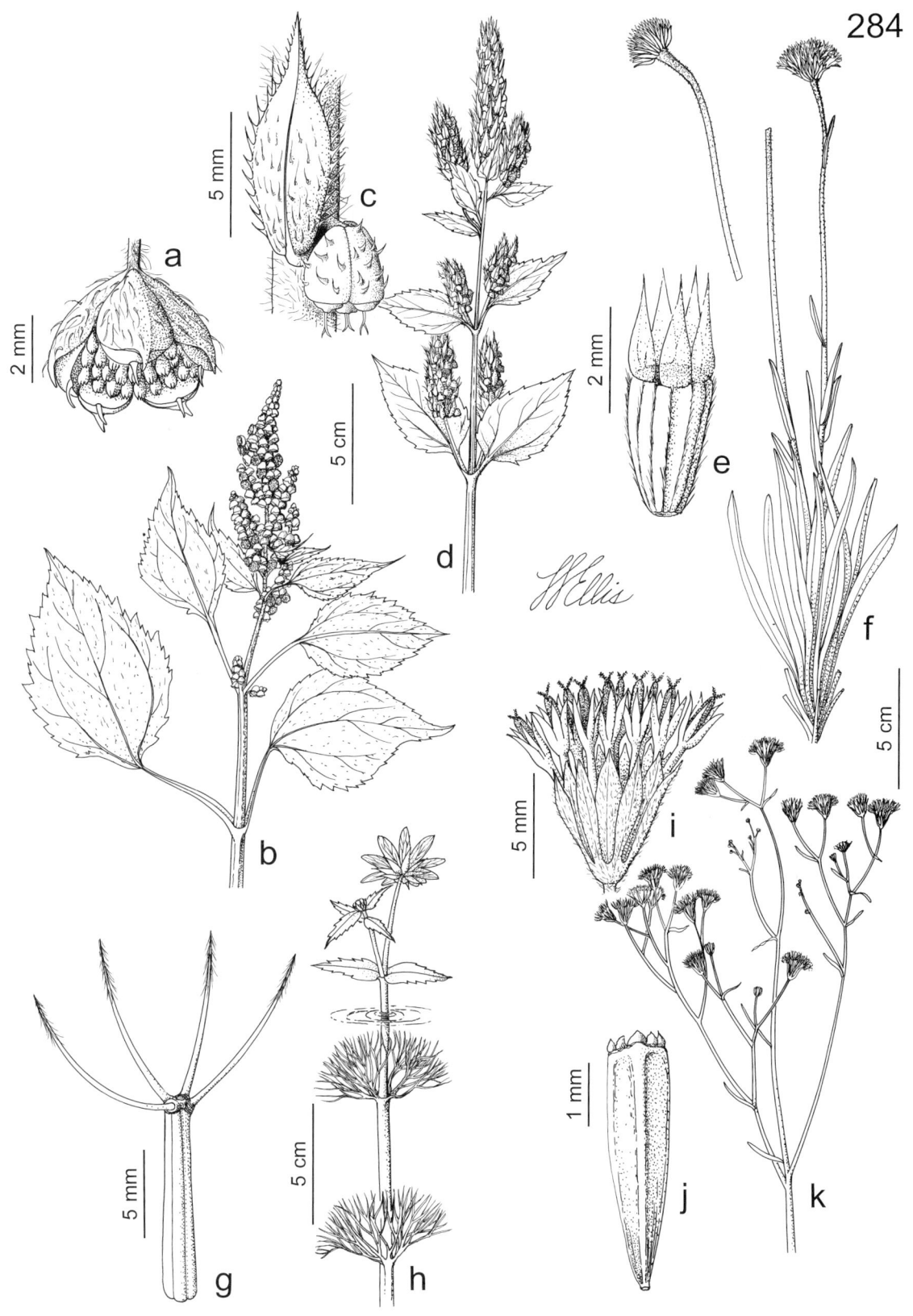

Plate 284. Asteraceae. *Cyclachaena xanthiifolia*, **a)** head, **b)** habit. *Iva annua*, **c)** head at node, **d)** habit. *Marshallia caespitosa*, **e)** fruit, **f)** habit. *Bidens beckii*, **g)** fruit, **h)** habit. *Palafoxia callosa*, **i)** head, **j)** fruit, **k)** habit.

slender, long-tapered tips (vs. ovate to lanceolate and more gradually tapered) as var. *caudata*. In the Midwest, these differences appear to form a more or less continuous spectrum of variation, and segregation of varieties to document the extreme cases seems ill-advised. Blake (1939) described a form of the species based on relatively large fruits found by archaeologists at rock shelter sites in Arkansas, Kentucky, and Missouri. Jackson (1960) included this taxon in his monograph of the genus as *I. annua* var. *macrocarpa* (S.F. Blake) R.C. Jacks. (*I. ciliata* Willd. var. *macrocarpa* S.F. Blake). This fossil marsh elder appears to be extinct in modern times and to have existed only as a food or medicinal plant cultivated by prehistoric Native Americans.

83. Marshallia Schreb.

Seven species, south-central and southeastern U.S.

1. Marshallia caespitosa Nutt. ex DC.
(Barbara's buttons)
Pl. 284 e, f; Map 1207

Plants perennial herbs, with a short rootstock and fibrous roots. Stems (6–)20–50 cm long, erect or ascending, unbranched or with few ascending branches toward the tip, with fine longitudinal ridges and grooves, sparsely to densely pubescent with short, curved, ascending hairs, especially toward the tip, usually glabrous toward the base. Leaves basal and sometimes also alternate, the median and upper stem leaves then sessile and much shorter than the basal and lower stem leaves (which have long or short, poorly differentiated petioles that more or less wrap around the stem), all relatively thick and stiff. Leaf blades 1–15 cm long, linear to narrowly elliptic or narrowly lanceolate, unlobed, the margins entire, the surfaces glabrous, dotted or impressed with small, yellow to yellowish brown, sessile glands. Inflorescences of solitary heads terminal on the stem or branches, the heads appearing long-stalked (the stalks bractless). Heads discoid. Involucre 6–12 mm long, 12–25 mm in diameter, cup-shaped to somewhat bell-shaped, the bracts in 1 or 2 subequal series. Involucral bracts 12–21, concave and somewhat cupped around the outer florets, glabrous but sometimes with a few sessile to impressed glandular dots, green, sometimes with thin, white margins especially toward the base, otherwise relatively thick and stiff, ascending at the tip, narrowly oblong to narrowly lanceolate or narrowly elliptic. Receptacle convex to short-conical, elongating slightly as the fruits mature, sometimes hollow, naked, with chaffy bracts subtending the florets, these linear, green but relatively thick and stiff, somewhat concave and cupped around the florets, often slightly incurved at the sharply pointed tip. Disc florets 40–90, perfect, the corolla 10–12 mm long, white to off-white or pale pink above the slender, slightly darker pink tube, the tube not expanded at the base or persistent at fruiting, minutely hairy, the abruptly expanded, minute throat and the relatively long lobes mostly with scattered, minute hairs on the outer surface and margins. Style branches with the sterile tip truncate or short-tapered to a more or less rounded tip. Pappus of 5(6) scales, these 1.5–2.5 mm long, ovate, papery, white, with irregular margins. Fruits 2.5–4.0 mm long, narrowly wedge-shaped in outline, 5-angled, the surface with 10 fine longitudinal ridges, densely pubescent with silvery, ascending hairs especially along the angles, minutely gland-dotted between the ribs, dark brown to black. $2n=18, 36$. April–June.

Uncommon in the southwestern portion of the Ozark Division and the adjacent Unglaciated Plains (Kansas and Missouri south to Texas and Louisiana). Glades on limestone, dolomite, and chert substrates, upland prairies, and banks of streams.

Steyermark (1963) noted that this species is an attractive ornamental in the wildflower garden. Most botanists have followed Channell (1957) in recognizing two varieties. Watson and Estes (1990) suggested that var. *caespitosa* is mostly diploid ($2n=18$) and that var. *signata* is tetraploid ($2n=36$).

1. Leaves all basal or nearly so; stems unbranched, the head 1 per flowering stem . . . . 1A. VAR. CAESPITOSA
1. Leaves basal and noticeably alternate up the flowering stem (but the median and upper leaves much shorter than the basal and lower stem leaves); stems with ascending branches toward the tip each with a terminal head, the heads 2–5(–12) per flowering stem . . . . 1B. VAR. SIGNATA

1a. var. caespitosa

Stems branched. Leaves all basal or the stems with a few alternate leaves in the basal third, these not extending significantly above the tips of the basal leaves. Head 1 per flowering stem. $2n=36$. April–June.

Uncommon in the southern portion of the Unglaciated Plains Division and eastward locally in Howell and Oregon Counties (Kansas and Missouri south to Texas and Louisiana). Glades on dolomite and chert substrates, upland prairies, and banks of streams.

Steyermark (1963) knew this variety only from the Unglaciated Plains Division, but recent fieldwork by Bill Summers has resulted in the first collections from the southern portion of the Ozark Division.

1b. var. signata Beadle & F.E. Boynton

Stems with ascending branches toward the tip. Leaves basal and the stems with few to several leaves nearly throughout their length, the stem leaves extending well beyond the tips of the basal leaves, the median and upper stem leaves sessile and much shorter than the basal and lower stem leaves. Heads 2–5(–12) per flowering stem. $2n=18$, 36. April–June.

Uncommon, known thus far only from a single site in Ozark County (Texas, Louisiana, Arkansas, and Missouri). Dolomite glades.

Steyermark (1963) noted that the original collection from Ozark County had been misdetermined earlier (Palmer and Steyermark, 1935) as *M. obovata* (Walt.) Beadle and F.E. Boynton, a species restricted to areas in and around the Piedmont region in the southeastern states. The Missouri plants are the northernmost in the distribution of the variety and are somewhat disjunct from the nearest population, in central Arkansas.

84. Melampodium L.

About 36 species, U.S., Mexico, Central America, South America; introduced in the Old World.

1. Melampodium cinereum DC. **var. ramosissimum** (DC.) A. Gray (hoary blackfoot)

Map 1208

Plants perennial herbs (somewhat shrubby elsewhere), the rootstock sometimes somewhat spreading. Stems 8–20 cm long, loosely ascending to strongly ascending, few- to several-branched, with fine longitudinal ridges and grooves, moderately to densely pubescent with short, stiff, ascending hairs. Leaves opposite, sessile. Leaf blades 0.7–3.5 cm long, linear to narrowly oblong, unlobed or with 2–10 blunt, pinnate lobes (sometimes appearing merely wavy), the margins otherwise entire and on the largest leaves sometimes curled under, angled at the base, rounded or narrowed to a bluntly pointed tip, the surfaces moderately to densely pubescent with short, appressed hairs toward the base. Inflorescences of solitary, relatively long-stalked heads at the branch points and sometimes also the branch tips. Heads radiate. Involucre 3–8 mm long, 4–12 mm in diameter, cup-shaped, the bracts in 2 dissimilar series. Involucral bracts 12–18; those of the outer series 5, 3–7 mm long, ovate, fused in the basal $^1/_6$–$^1/_3$, green, the outer surface and margins moderately pubescent with short, white, appressed hairs; those of the inner series 7–13, straw-colored or light yellowish brown and hardened or leathery at fruiting, each wrapped tightly around a ray floret and expanded at the ovary tip into a cup-shaped or hooded asymmetrical extension (as long as or longer than the basal portion), this rounded or abruptly tapered to a slender point on the outer side, the outer surface of the basal portion with moderate to dense, usually sharply pointed tubercles. Receptacle more or less flat but usually minutely elevated above the base of the involucre, not elongating as the fruits mature, with chaffy bracts subtending the disc florets, these wrapped around the disc florets, oblong-lanceolate, straw-colored, shiny, somewhat expanded and fringed at the tip. Ray florets 7–13, the corolla 2–10 mm long, relatively broad and somewhat wedge-shaped, white or pale cream-colored. Disc florets 25–40, staminate (with a slender, stalklike ovary and a more or less undivided style), the corolla 1.5–2.5 mm long, yellow, the tube not expanded at the base, glabrous, the relatively short lobes usually minutely hairy on the outer surface. Style branches with the sterile tip elongate and tapered to a bluntly pointed tip. Pappus absent. Fruits 1.3–2.0 mm long, asymmetrically obovate in outline (somewhat D-shaped) and often minutely beaked, slightly flattened, the surface otherwise smooth, glabrous, purplish black, shiny, more or less hidden within the enveloping bract and dispersed with it as a unit. $2n=20$. July–September.

Introduced, known only from a single historical collection from Jasper County (Texas; Mexico). Gardens.

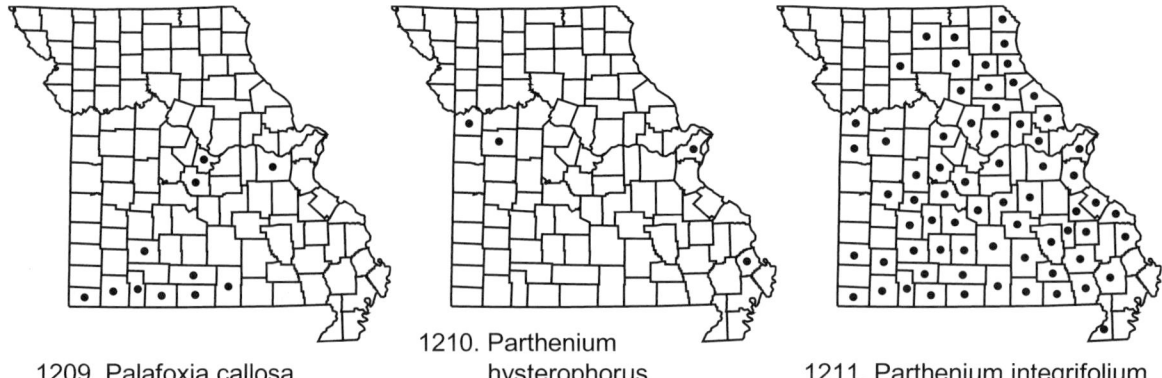

1209. Palafoxia callosa
1210. Parthenium hysterophorus
1211. Parthenium integrifolium

This species was not treated by Steyermark (1963), apparently because of doubts that it was a member of the flora. The label on the voucher specimen indicates that plants were weedy adventives persisting in E. J. Palmer's garden from seeds of Texas origin that he had discarded earlier. Stuessy (1972) recognized three varieties in the species, with var. *ramosissimum* differing from var. *cinereum* in its narrower leaves, and var. *hirtellum* Stuessy differing from the other two varieties in its longer hairs toward the leaf base.

85. Palafoxia Lag. (Spanish needles)
(Turner and Morris, 1976)

Twelve species, U.S., Mexico.

1. Palafoxia callosa (Nutt.) Torr. & A. Gray

Pl. 284 i–k; Map 1209

Plants annual, with taproots. Stems 15–60 cm long, erect or ascending, few- to several-branched, with fine longitudinal ridges and grooves, sparsely to densely pubescent with short, stiff, ascending hairs, these replaced abruptly with dark, tack-shaped glands toward the tip. Leaves alternate, sessile or the largest leaves with a short, poorly differentiated petiole. Leaf blades 2–7 cm long, linear, relatively thick (the smaller leaves often nearly as thick as wide), unlobed, the margins entire and on the largest leaves sometimes curled under, tapered at the base, tapered or narrowed to a sharply pointed tip, the surfaces sparsely to densely roughened-pubescent with short, loosely appressed, pustular-based hairs, the bases sometimes conspicuously darkened. Inflorescences open, more or less flat-topped panicles with short, leaflike bracts at the branch points, the branches with moderate to dense, dark, tack-shaped glands, the heads usually solitary and long-stalked at the branch tips. Heads discoid. Involucre 3–6 mm long, 2–5 mm in diameter, conical, the bracts in 2 subequal series. Involucral bracts 9–12, linear, ascending at the tip, the outer surface and margins moderately pubescent with short, white, appressed hairs and/or dark, more or less tack-shaped glands. Receptacle flat, not elongating as the fruits mature, naked. Disc florets 5–30, perfect, the corolla 5–6 mm long, pink to purplish pink (rarely white elsewhere) above the slender, slightly darker purple tube, the tube not expanded at the base or persistent at fruiting, usually sparsely hairy, the minute, abruptly expanded throat and the relatively long, slender lobes mostly glabrous. Style branches with the sterile tip elongate and tapered to a usually sharply pointed tip. Pappus of usually 8 scales, these 0.3–1.0 mm long, oblanceolate to obovate, papery, straw-colored. Fruits 3–5 mm long, narrowly wedge-shaped in outline, strongly 4-angled, the surface otherwise smooth, moderately pubescent with fine, appressed-ascending hairs (these sometimes produced in fascicles of 2 or 3 hairs), purplish black to black, dull or slightly shiny. $2n=20$. August–October.

Scattered in the southwestern portion of the Ozark Division and introduced sporadically farther north and east (Missouri south to Texas and Mississippi; Mexico). Glades and tops of bluffs on calcareous substrates, less commonly banks of streams and rivers; also quarries, roadsides, and dry, open, disturbed areas.

Steyermark (1963) noted that this attractive annual performed well in a sunny portion of his garden.

86. Parthenium L.
(Rollins, 1950)

Plants annual or perennial herbs (shrubs elsewhere), often somewhat aromatic when bruised or crushed, sometimes with rhizomes or a tuberous rootstock. Stems erect or ascending, unbranched below the inflorescence to many-branched, with fine longitudinal ridges, glabrous or sparsely to densely hairy and sometimes also with minute, sessile, spherical, yellow glands. Leaves basal and alternate, the basal and lower leaves short- to more commonly long-petiolate, the median and upper leaves mostly sessile. Leaf blades variously shaped, unlobed or 1 or 2 times deeply pinnately lobed, the margins otherwise entire or more commonly toothed or scalloped, also usually minutely hairy, the surfaces variously hairy, sometimes roughened to the touch, also usually with minute, sessile, spherical, yellow glands. Inflorescences usually small, more or less flat-topped terminal panicles, sometimes reduced to a small, loose terminal cluster, subtended by small, leaflike bracts at the branch points, the heads mostly with relatively short, densely hairy stalks. Heads radiate but sometimes appearing discoid. Involucre cup-shaped, the bracts in 2–4 subequal, overlapping series, those of the outer series usually somewhat narrower than the others. Involucral bracts ascending, straw-colored, sometimes greenish-tinged toward the tip, mostly hardened and leathery, the outer surface densely hairy, those of the inner series about as long as the outermost chaffy bracts. Receptacle convex to short-conical, not elongating as the fruits mature, with chaffy bracts subtending the ray and disc florets, these densely hairy on the outer surface toward the tip, those of the disc florets concave and wrapped around the florets. Ray florets (4–)5(–7), pistillate (with a 2-branched style exserted from the short tube at flowering), the corolla with a short (1–2 mm) or rarely absent ligule (then reduced to a minute tube), when present the ligule relatively broad, white or off-white, the short, tubular base (and the outer surface of the ligule) densely short-hairy, usually persistent at fruiting. Disc florets about 15–65, staminate (with a small, stalklike ovary and an undivided style), all but the outermost florets usually shed as an intact unit at fruiting, the corolla 1.2–2.0 mm long (only slightly surpassing the chaffy bract), off-white to pale cream-colored, minutely hairy on the outer surface of the lobes, not expanded at the base. Style branches with the sterile tip broad and bluntly pointed to rounded. Pappus of the disc florets usually absent, that of the ray florets of 2 scales or 2 or 3 slender awns. Fruits obovate to narrowly obovate in outline, flattened, the surface minutely hairy, dark gray to black, fused basally to the subtending chaffy bract as well as to the adjacent 2(3) disc florets and their chaffy bracts, the whole group shed intact as a unit at fruiting. About 16 species, North America to South America, Caribbean Islands.

Parthenium argentatum A. Gray is a shrubby member of the genus native to the Chihuahuan Desert region that is known as guayule. It has long been known to contain polyisoprenoid natural rubber in its tissues. During World War II, when the supply of commercial rubber (mostly produced from the latex of the rubber tree, *Hevea brasiliensis* (A. Juss.) Müll. Arg. [Euphorbiaceae]) from plantations in the Philippines and other Malesian islands was cut off, interest in guayule and its relatives was rekindled. In fact, the monograph of the genus by Rollins (1950) was an indirect result of his involvement with the search for alternative sources of rubber during the early 1940s. However, after the war ended, interest in research and cultivation of guayule lagged because extraction of the rubber requires the harvest of entire plants (as opposed to sustainable harvest from *Hevea* rubber by tapping latex from the tree's trunk), which makes the process relatively costly and inefficient. On the other hand, the rubber refined from *Parthenium* supposedly causes fewer allergic reactions than does the rubber from *Hevea*, so there may be a market for rubber from *Parthenium* for the manufacture of items such as surgical gloves and condoms (Cornish and Siler, 1996). Currently, research on a stable domestic supply of natural rubber involving *Parthenium* and other genera continues in the United States, but at a level somewhat below that in the early 1940s.

Plants of *Parthenium* have a bitter flavor and usually are avoided by livestock and other grazing mammals.

1. Leaves 1 or 2 times deeply pinnately lobed; plants annual, with taproots
 .... 1. P. HYSTEROPHORUS
1. Leaves unlobed, the margins coarsely toothed; plants perennial, with short and somewhat tuberous or long-rhizomatous rootstocks .... 2. P. INTEGRIFOLIUM

1. Parthenium hysterophorus L. (Santa Maria)

Pl. 285 e, f; Map 1210

Plants annual, with taproots. Stems 30–120 cm long, usually much-branched above the midpoint, moderately to densely pubescent with short, stiff, more or less ascending hairs, usually also with minute, sessile, spherical, yellow glands. Basal leaves short- to long-petiolate, the petioles progressively shorter up the stem, the uppermost leaves usually sessile or nearly so. Leaf blades mostly 3–20 cm long, ovate to elliptic, 1 or 2 times deeply pinnately lobed, the lobes oblong-triangular to narrowly oblong or linear, the margins otherwise entire or with few coarse teeth, the surfaces sparsely to densely roughened-pubescent with short, stiff, loosely appressed hairs, often also with minute, sessile, spherical, yellow glands. Blades of uppermost leaves 1–3 cm long, usually linear to narrowly oblong, unlobed, narrowed or tapered at the base and not clasping the stem. Involucre 2–4 mm long, 3–4 mm in diameter, the bracts of the outer series slightly shorter than the others, lanceolate to narrowly elliptic, those of the inner series ovate to broadly ovate. Pappus of 2 ovate scales 0.7–1.5 mm long, these petaloid in color and texture, appressed to and often difficult to distinguish from the corolla. Fruits 1.5–3.0 mm long. 2n=34. August–October.

Introduced, uncommon and sporadic (native of Mexico, Central America, South America, Caribbean Islands; introduced in the eastern U.S. west to Kansas and Texas; also Europe, Asia, Pacific Islands).

Although apparently not a problem in North America, plants introduced and naturalized in India have poisoned cattle and buffalo and caused contact dermatitis in some humans. The agents responsible apparently are sesquiterpene lactones, especially parthenin (Burrows and Tyrl, 2001).

2. Parthenium integrifolium L. (American feverfew, wild quinine)

Pl. 285 a–d; Map 1211

Plants perennial herbs, with short to more commonly long-creeping rhizomes or a somewhat thickened, tuberous rootstock. Stems 30–100 cm long, usually unbranched below the inflorescence, moderately to densely pubescent with short, stiff, spreading to ascending hairs toward the tip, sometimes also with minute, sessile, spherical, yellow glands, glabrous or sparsely to densely pubescent with stiff, spreading or ascending hairs toward the base. Basal and lower stem leaves long-petiolate, the blades 12–30 cm long, elliptic to ovate, long-tapered to the petiole at the base, angled or tapered to a bluntly or sharply pointed tip, unlobed or those of the largest leaves rarely with a few short lobes toward the base, the margins otherwise coarsely toothed or scalloped, the upper surface sparsely to moderately roughened-pubescent with short, stiff hairs, the undersurface moderately to densely pubescent with short or longer, stiff, spreading hairs, both surfaces also usually with minute, sessile, spherical, yellow glands. Median and upper stem leaves short-petiolate to sessile, the blade mostly 2–15 cm long, lanceolate to ovate, angled or tapered to the petiole or, in sessile leaves, often rounded to shallowly cordate and sometimes somewhat clasping the stem, the margins and surfaces more or less like those of the lower leaves. Involucre 3–6 mm long, 4–10 mm in diameter, the bracts of the outer series slightly shorter than the others, lanceolate to broadly ovate, those of the inner series broadly ovate to nearly circular. Pappus of 2 or 3 slender awns 0.2–0.5 mm long. Fruits 3–5 mm long. 2n=72. May–October.

Scattered nearly throughout the state but uncommon or absent from the northwestern quarter (eastern U.S. west to Minnesota and Texas; Canada). Glades, upland prairies, savannas, openings of mesic to dry forests, and ledges and tops of bluffs; also pastures, railroads, and roadsides.

Wild quinine was a minor medicinal plant for some tribes of Native Americans, who applied poultices of the leaves to treat burns (Moerman, 1998). Although it occurs in other vegetation types, it is a characteristic species of high-quality upland prairie plant communities.

Botanists in Missouri apparently have had trouble distinguishing the two native taxa accepted by Steyermark (1963), *P. hispidum* and *P. integrifolium,* as evidenced by the number of

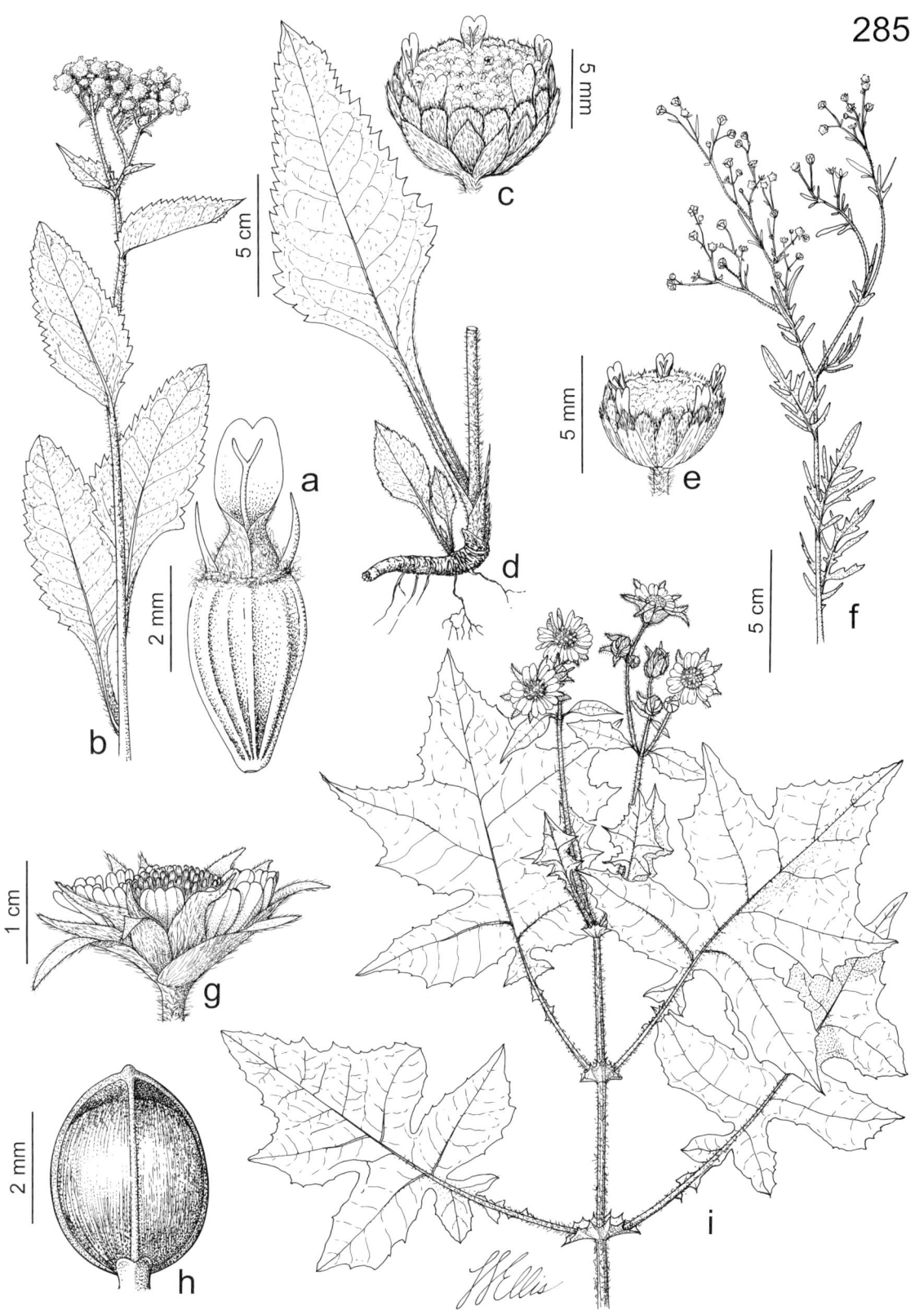

Plate 285. Asteraceae. *Parthenium integrifolium* var. *integrifolium*, **a)** ray flower, **b)** habit. *Parthenium integrifolium* var. *hispidum*, **c)** head, **d)** basal portion of plant. *Parthenium hysterophorus*, **e)** head, **f)** habit. *Polymnia canadensis*, **g)** head, **h)** fruit, **i)** habit.

specimens with multiple annotations back and forth. Superficially, the two would seem amply distinct, but in fact a number of seemingly intermediate specimens exist for each of the characters used by Steyermark (1963) and Rollins (1950) to separate them. In his monograph of the genus, Rollins noted the broad geographic overlap between the two taxa in Arkansas and Missouri and postulated that hybridization between them probably was occurring, based on his inability to assign some specimens collected in the region to one or the other taxon with certainty. Mears (1975), in a preliminary report to a more detailed monograph of the complex that regrettably was never published, chose to treat the two members of the *P. integrifolium* complex as varieties (he recognized five total varieties in *P. integrifolium*). Mears noted that var. *integrifolium* occurs virtually throughout the species range, with his other four varieties circumscribing more geographically localized variants, sometimes adapted to somewhat different habitats than those of the typical variety. Clearly, the situation requires more detailed biosystematic study. Whether the Missouri plants represent two species whose boundaries are slightly blurred by occasional hybridization or a widespread species with an incompletely distinct variety adapted to somewhat drier sites cannot be determined without more intensive study in the field, herbarium, and laboratory. For the present, the two are accepted provisionally as varieties, in recognition of the difficulties involved in the determination of some specimens.

1. Plants usually colonial, with scattered stems from a branched rhizome; stems moderately to more commonly densely pubescent with short, stiff, spreading hairs toward the base; leaf blades with relatively dense, spreading hairs along the undersurface midvein
. . . . 2A. VAR. HISPIDUM
1. Plants usually not colonial, the rootstock often somewhat thickened and tuberous, occasionally with a short rhizome; stems glabrous or sparsely to moderately pubescent with short, soft or relatively stiff, loosely ascending hairs; leaf blades with moderate, more or less appressed hairs along the undersurface midvein
. . . . 2B. VAR. INTEGRIFOLIUM

2a. var. **hispidum** (Raf.) Mears
P. hispidum Raf.
P. repens Eggert

Pl. 285 c, d

Plants usually colonial, with scattered stems from a branched, long-creeping rhizome. Stems moderately to more commonly densely pubescent with short, stiff, spreading hairs toward the base, the hairs becoming shorter and more appressed toward the stem tip. Leaf blades with relatively dense, spreading hairs along the undersurface midvein, otherwise moderately to densely pubescent with shorter, spreading hairs, those of the upper leaves often shallowly cordate and clasping the stem. Involucre 4–6 mm long, 6–10 mm in diameter. Fruits 3–5 mm long. $2n=72$. May–October.

Scattered in the Ozark and Ozark Border Divisions, extending slightly into the Glaciated Plains, and historically in Dunklin County (Kansas to Texas east to Wisconsin, Illinois, and Tennessee; introduced in Michigan). Glades, upland prairies, savannas, openings of mesic to dry forests, and ledges and tops of bluffs; also railroads and roadsides.

2b. var. **integrifolium**

Pl. 285 a, b

Plants usually not colonial, the rootstock often somewhat thickened and tuberous, occasionally with a short rhizome. Stems glabrous or sparsely to moderately pubescent with short, soft or relatively stiff, loosely ascending hairs toward the base, the hairs becoming more uniformly appressed and often denser toward the stem tip. Leaf blades relatively uniformly moderately to densely pubescent with shorter, spreading hairs on the undersurface, those of the upper leaves often rounded to nearly truncate at the base but usually not or only slightly clasping the stem. Involucre 3–5 mm long, 4–7 mm in diameter. Fruits 3–4 mm long. $2n=72$. May–September.

Scattered nearly throughout the state but uncommon or absent from the northwestern quarter (eastern U.S. west to Minnesota and Texas; Canada). Glades, upland prairies, savannas, openings of mesic to dry forests, and ledges and tops of bluffs; also pastures, railroads, and roadsides.

Plants of var. *integrifolium* often have somewhat more open inflorescences than those of var. *hispidum* and are somewhat less frequently encountered in glade habitats.

87. Polymnia L. (leaf cup)

Plants perennial herbs (annuals elsewhere), usually colonial from rhizomes. Stems erect or ascending, usually several-branched, with fine longitudinal lines or ridges, occasionally bluntly angled, the internodes of the main stems often hollow, glabrous or moderately to densely pubescent with nonglandular or gland-tipped hairs toward the tip. Leaves opposite or the uppermost few alternate, sessile or short-petiolate, the bases mostly slightly or greatly expanded and wrapping around the stem. Leaf blades ovate to broadly oblong-ovate or broadly elliptic, more or less pinnately 3–11-lobed, tapered or more or less truncate at the base, relatively thin in texture, the main lobes tapered to a sharply pointed tip, the margins finely to coarsely toothed and hairy, the surfaces glabrous or sparsely to moderately pubescent with short, soft, curved hairs, the undersurface pale green and usually with minute, sessile, spherical, yellow glands. Inflorescences of relatively open, irregular, often spreading to nodding clusters at the branch tips, these sometimes appearing as small panicles, subtended by leaflike bracts at the branch points, the heads mostly with relatively short, glabrous or more commonly hairy and/or glandular stalks. Heads radiate but sometimes appearing discoid. Involucre cup-shaped, the bracts in 2 dissimilar series, those of the outer series usually narrower and slightly shorter or longer than the others. Involucral bracts loosely ascending, those of the inner series slightly longer than the chaffy, outermost bracts, oblong-lanceolate to elliptic or broadly ovate, mostly narrowed or tapered to a sharp point at the tip, the margins with minute or spreading hairs, the outer surface glabrous or sparsely to densely pubescent with gland-tipped hairs, the outer series of bracts green and somewhat leaflike, at least toward the tip. Receptacle flat or shallowly convex, not elongating as the fruits mature, with chaffy bracts subtending the ray and disc florets, narrowly ovate to obovate, thin and papery, concave and wrapped around the florets. Ray florets 4–7 (2 or 3 elsewhere), pistillate (with a 2-branched style exserted from the short tube at flowering), the corolla sometimes with a short or absent ligule (then reduced to a minute tube), when present the ligule relatively broad, white or pale cream-colored, the short, tubular base moderately to densely short-hairy and/or glandular, withered and often not persistent at fruiting. Disc florets 12–40, staminate (with a small, stalklike ovary and a more or less undivided style), the corolla light yellow to pale yellow, often sparsely glandular toward the tip, not expanded at the base. Style branches with the sterile tip somewhat elongate and tapered. Pappus of the ray and disc florets absent. Fruits 3–4 mm long, elliptic-obovate to slightly pear-shaped in outline, slightly flattened, with 3–6 blunt angles or less commonly ribs, the surface otherwise glabrous, dark brown to black, sometimes with lighter brown mottling, somewhat shiny. Three species, eastern U.S., Canada.

Most authors (Steyermark, 1963; Wells, 1965; Barkley, 1986; Gleason and Cronquist, 1991) have included the species of *Smallanthus* in *Polymnia,* but Robinson (1978) noted differences in morphology and anatomy of the achenes, pubescence patterns of the florets, and chromosomal base number that appear to justify separating the two genera. The third species of *Polymnia* in the restricted sense is *P. cossatotensis* Pittman & V.M. Bates, which is endemic to the Ouachita Mountain region of west-central Arkansas. This taxon is distinguished readily from the rest of the genus by its annual habit, mostly unlobed leaves that are cordate at the base, and relatively small heads with only 2 or 3 pistillate florets.

1. Leaves and stems (at least toward the tip) moderately to densely pubescent with noticeable, often gland-tipped hairs; involucre 5–8 mm long, 6–13 mm in diameter; fruits with 3 blunt angles or less commonly ribs . . . . 1. P. CANADENSIS
1. Leaves and stems glabrous or inconspicuously pubescent with inconspicuous, nonglandular hairs; involucre 3–4 mm long, 3–6 mm in diameter; fruits with 4–6 angles or ribs . . . . 2. P. LAEVIGATA

1212. Polymnia canadensis   1213. Polymnia laevigata   1214. Ratibida columnifera

1. Polymnia canadensis L. (pale-flowered leaf cup)

Pl. 285 g–i; Map 1212

Plants usually not fragrant when crushed or bruised. Stems 50–180 cm long, moderately to densely pubescent with mostly gland-tipped or sticky hairs toward the tip, glabrous or nearly so toward the base. Leaves sessile or short-petiolate, the larger leaves often with rounded appendages of green tissue at the petiole base, these wrapping around the stem, the appendages of the adjacent leaves at a given node sometimes more or less fused into a short cup around the stem. Leaf blades 3–40 cm long, mostly broadly tapered at the base, mostly with 3–7 more or less pinnate lobes (the smaller, upper leaves sometimes unlobed), the lobes tapered to a sharply pointed tip, the margins otherwise finely to coarsely toothed, the surfaces glabrous or sparsely to moderately pubescent with more or less spreading, sometimes somewhat sticky or gland-tipped hairs, especially along the veins. Involucre 5–9 mm long, 6–13 mm in diameter, the bracts moderately to densely pubescent with sticky or gland-tipped hairs on the outer surface. Outer series of 2–4 involucral bracts somewhat longer and narrower than the others, lanceolate to narrowly ovate, the inner series ovate to broadly obovate. Ray florets (4)5, the corolla reduced to a minute tube and lacking a ligule or with a short, broad ligule 2–10 mm long. Disc florets 26–40, the corolla 3–4 mm long. Fruits with 3 blunt angles or less commonly ribs, dark brown to black, usually with some reddish brown mottling. $2n=30$. May–October.

Scattered in the Ozark and Ozark Border Divisions and in some additional counties bordering the Mississippi and Missouri Rivers (eastern U.S. west to Minnesota and Oklahoma; Canada). Bottomland forests, mesic upland forests in ravines, bases and ledges of bluffs, and banks of streams and rivers; also fencerows, pastures, railroads, and roadsides.

This species frequently occurs in large colonies along stream terraces and also is a characteristic component of the vegetation along moist, shaded talus areas at the bases of steep, rocky slopes or bluffs. The form with relatively well-developed ray corollas has been called f. *radiata* (A. Gray) Fassett and is about as common as the nearly rayless form.

2. Polymnia laevigata Beadle

Pl. 286 g, h; Map 1213

Plants usually somewhat aromatic when crushed or bruised. Stems 50–200 cm long, glabrous or sparsely to moderately pubescent with inconspicuous, loosely appressed, sometimes somewhat sticky hairs toward the tip. Leaves short-petiolate, lacking appendages at the petiole base, these wrapping around the stem, the appendages of the adjacent leaves at a given node sometimes more or less fused into a short cup around the stem. Leaf blades 3–40 cm long, broadly rounded to shallowly cordate but most commonly more or less truncate at the base, mostly with 5–11 deep, pinnate lobes (the smaller, upper leaves sometimes unlobed or shallowly 3-lobed), the lobes tapered to a sharply pointed tip, the margins otherwise finely to coarsely toothed, the surfaces glabrous or sparsely to moderately pubescent with minute, sometimes somewhat sticky hairs along the veins. Involucre 3–4 mm long, 3–6 mm in diameter, the bracts glabrous or minutely hairy on the outer surface toward the base, minutely hairy along the margins. Outer series of 3–5 involucral bracts somewhat shorter and narrower than the others, lanceolate to narrowly oblong, the inner series ovate to broadly ovate. Ray florets 4–6, the corolla with a short, broad ligule 2.5–3.5 mm long. Disc florets 12–18, the corolla 2–3 mm long. Fruits with 4–6 angles or ribs, dark brown. $2n=30$. July–October.

Uncommon, known only from a single collection from Pemiscot County (Missouri to Kentucky south to Florida). Bottomland forests.

This species appears to be relatively infrequently encountered throughout most of its range. In Missouri, searches have failed to relocate the plant in the vicinity of Steyermark's original 1953 collection site near Portageville.

88. Ratibida Raf. (prairie coneflower)
(Richards, 1968)

Plants perennial herbs, with a taproot or a stout, horizontal rootstock (often a short rhizome) and fibrous roots. Stems erect or ascending, unbranched or few- to several-branched, with several longitudinal lines or ridges, moderately to densely pubescent with short, stiff, strongly to occasionally loosely ascending hairs, sometimes nearly glabrous toward the base, also with scattered, minute, sessile, spherical, yellow glands. Leaves alternate, the lower stem leaves short- to long-petiolate, the petioles progressively shorter up the stem, the upper stem leaves usually sessile or nearly so, the bases usually only slightly expanded, not or only slightly wrapping around the stem. Leaf blades 1 or 2 times deeply divided or compound, those of the uppermost leaves occasionally unlobed, variously shaped, the divisions or leaflets linear to lanceolate, mostly tapered at the base, tapered to a sharply pointed tip, the margins otherwise usually entire and with minute, appressed hairs, the surfaces moderately to densely pubescent with short, stiff, appressed to more or less spreading, sometimes pustular-based hairs (sometimes roughened to the touch), usually also dotted with scattered, minute, sessile to impressed glands, with 1 or 3 main vein(s). Inflorescences of solitary terminal heads, occasionally appearing in loose, open terminal clusters, the heads with long, usually bractless stalks. Heads radiate. Involucre narrowly saucer-shaped, the bracts in 2 somewhat similar overlapping series (the outer series longer than the inner series). Involucral bracts about 5–15, those of the outer series narrowly ovate-lanceolate, those of the inner series narrowly lanceolate to nearly linear, usually reflexed at flowering, green, the margins and outer surface moderately to densely hairy, the inner surface not or only sparsely hairy but often with scattered, minute, sessile, yellow glands, the midnerve usually inconspicuous. Receptacle strongly convex to nearly spherical, ovoid, or cylindrical, elongating as the fruits mature, with chaffy bracts subtending the disc florets, these concave and folded around the florets, oblong to oblong-obovate, the abruptly sharply pointed tips somewhat incurved and slightly concave, moderately to densely white-hairy along the margins and toward the tip, also with a large glandular spot on each face, persistent at fruiting. Ray florets 4–15, sterile (lacking stamens and style at flowering and with an ovary that is shorter and thinner than those of the disc florets, not developing into a fruit), the corolla showy, relatively broad or somewhat narrower (in *R. pinnata*), moderately to strongly drooping at flowering, the short tube densely hairy, the ligule glandular and hairy, at least on the outer surface, yellow, sometimes with a well-differentiated region of reddish brown to brownish purple toward the base (less commonly yellow only near the tip or the yellow color completely absent), withered but sometimes more or less persistent at fruiting. Disc florets 50 to numerous (more than 200), perfect, the corolla yellow to yellowish green, sometimes purplish-tinged toward the tip, not bulbous-thickened at the base, not persistent at fruiting (but sometimes trapped by the subtending bract), the 5 lobes with the outer surface glandular and sometimes also hairy. Style branches with the sterile tip short or elongate and rounded to sharply pointed. Pappus of the disc florets absent or either a low rim or crown or of 1 or 2 minute teeth, when present persistent at fruiting. Fruits oblong in outline, slightly to moderately oblique at the base, flattened (biconvex), 1 or both of the angles (also the tip) often hairy or minutely fringed, the surface usually glabrous, dark brown to black, with fine, sometimes faint longitudinal lines or grooves, sometimes slightly shiny. Seven species, U.S., Canada, Mexico.

All three of the species found in Missouri are cultivated as garden ornamentals.

1. Ray florets with the corolla 4–9 mm long; pappus a minute rim or crown .... 3. R. TAGETES
1. Ray florets with the corolla (8–)10–60 mm long; pappus absent or of 1 or 2 minute teeth
    2. Pappus of 1 or 2 minute teeth; receptacle 1–5(–7) cm long, columnar; plants with a taproot .... 1. R. COLUMNIFERA
    2. Pappus absent; receptacle 1.0–2.5 cm long, oblong-ovoid to nearly spherical; plants with a stout, horizontal rootstock (often a short rhizome) and fibrous roots .... 2. R. PINNATA

1. Ratibida columnifera (Nutt.) Wooton & Standl. (longhead prairie coneflower, Mexican hat)

R. columnaris Pursh

Pl. 286 a, b; Map 1214

Plants with a taproot. Stems 30–60(–100) cm long, solitary or clustered. Leaf blades 2–15 cm long, mostly oblong-elliptic to oblong-obovate in outline, 1 or 2 times deeply pinnately divided, the divisions 3–15, 0.1–6.0 cm long, linear to narrowly oblong or oblong-oblanceolate, with 1 inconspicuous vein. Heads positioned mostly well above the leaves, the stalks to 45 cm long. Involucral bracts 5–14, those of the outer series 4–15 mm long, linear, those of the inner series 1–3 mm long, mostly narrowly ovate. Receptacle columnar, 1–5(–7) cm long. Ray florets 4–11, the corolla (8–)10–35 mm long, yellow or less commonly with a well-defined zone of reddish brown to brownish purple toward the base, this occasionally entirely masking the yellow coloration, the outer surface moderately to densely but inconspicuously short-hairy, both surfaces usually with scattered, minute, sessile, spherical, yellow glands. Disc florets numerous, the corolla 1.5–2.5 mm long, yellow to yellowish green, sometimes purplish-tinged toward the tip. Style branches with the sterile tip short and blunt. Pappus of 1 or 2 minute, triangular, brownish teeth. Fruits 1.5–3.0 mm long, somewhat obliquely oblong, the tip and the angle opposite the chaffy bract with a minute, orangish brown fringe. $2n=28$. June–September.

Possibly introduced, widely scattered, mostly in the western half of the state (Idaho to Arizona east to Minnesota and Louisiana; Canada, Mexico; introduced elsewhere nearly throughout the U.S.). Upland prairies; also pastures, railroads, roadsides, and open, disturbed areas.

The native status of *R. columnifera* in Missouri is questionable. Steyermark (1963) treated the species as native at least at some sites in Missouri. Earlier, however, Palmer and Steyermark (1935) had stated, "Generally and perhaps everywhere introduced from farther west." Nearly all of the herbarium specimens, even those from the western portion of the state, are indicated on their labels as representing introduced plants. The few occurrences at upland prairie sites are from remnant prairies adjacent to roads or railroads. A complicating factor has been that the species sometimes is a component of wildflower seed mixes sown along public roadways for roadside beautification, which has increased the spread and abundance of *R. columnifera* in recent years. On the other hand, Missouri is contiguous with the main portion of the species range, and the species has been noted as a native colonizer of disturbed habitats.

Uncommon plants with the ray corollas entirely or partially reddish brown to brownish purple have been called f. *pulcherrima* (DC.) Fernald.

2. Ratibida pinnata (Vent.) Barnhart (grayheaded coneflower, drooping coneflower)

Pl. 286 e, f; Map 1215

Plants with a stout, horizontal rootstock (often a short rhizome) and fibrous roots. Stems 40–150 cm long, solitary or few to several, sometimes appearing clustered. Leaf blades 4–40 cm long, mostly broadly oblong-ovate to oblong-obovate or oblong-elliptic in outline (those of the undivided leaves usually lanceolate), 1 time deeply pinnately divided or compound (those of the uppermost leaves sometimes undivided), the divisions or leaflets 3–9, 1–15 cm long, entire or deeply toothed or pinnately lobed, narrowly lanceolate to ovate, with 1 or 3 main veins. Heads positioned mostly well above the leaves, the stalks to 30 cm long. Involucral bracts 10–15, those of the outer series 3–15 mm long, linear, those of the inner series 3–6 mm long, mostly narrowly ovate. Receptacle oblong-ovoid or less commonly nearly spherical, 1.0–2.5 cm long. Ray florets 6–15, the corolla (20–)30–60 mm long, yellow, the outer surface usually densely but inconspicuously short-hairy, the inner surface often sparsely hairy toward the base, both surfaces usually with scattered, minute, sessile, spherical, yellow glands. Disc florets numerous, the corolla 2–4 mm long, greenish yellow to yellowish green, sometimes purplish-tinged toward the tip. Style

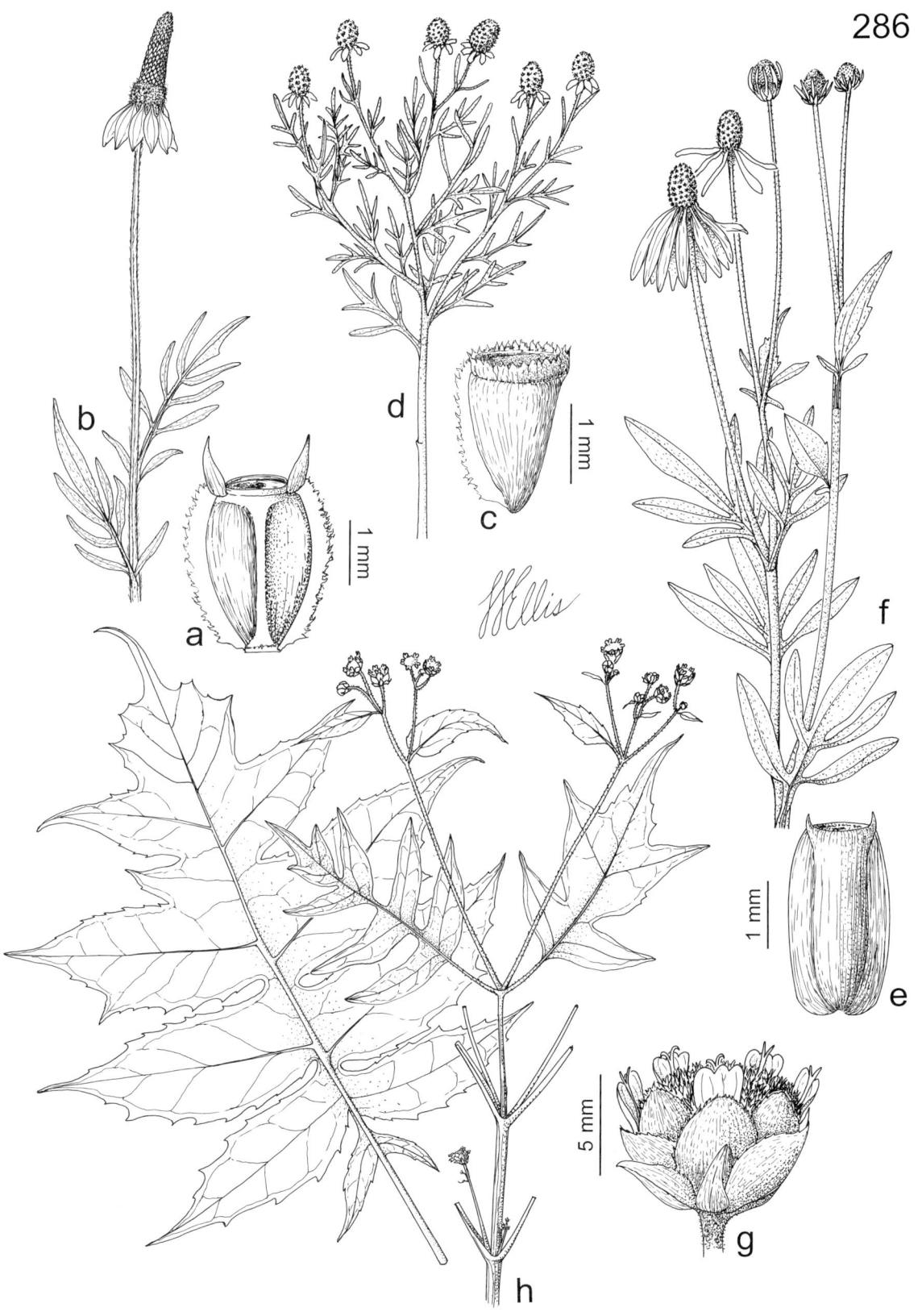

Plate 286. Asteraceae. *Ratibida columnifera*, **a)** fruit, **b)** habit. *Ratibida tagetes*, **c)** fruit, **d)** habit. *Ratibida pinnata*, **e)** fruit, **f)** habit. *Polymnia laevigata*, **g)** head, **h)** habit.

1215. Ratibida pinnata　　　1216. Ratibida tagetes　　　1217. Rudbeckia amplexicaulis

branches with the sterile tip slender and sharply pointed. Pappus absent. Fruits 2–4 mm long, slightly obliquely oblong, the angle opposite the chaffy bract sometimes sparsely hairy or with a narrow, lighter-colored wing. $2n=28$. May–September.

Scattered nearly throughout the state but apparently absent from the Mississippi Lowlands Division and the adjacent southeastern counties of the Ozarks (eastern U.S. west to South Dakota to Oklahoma; Canada). Upland prairies, glades, savannas, openings and edges of mesic to dry upland forests, and rarely banks of streams; also fencerows, pastures, railroads, and roadsides.

3. **Ratibida tagetes** (E. James) Barnhart (short-ray prairie coneflower, green prairie coneflower)

Pl. 286 c, d; Map 1216

Plants with a taproot. Stems 30–60(–100) cm long, solitary (but then often several-branched from near the base) or clustered. Leaf blades 0.5–9.0 cm long, mostly oblong-elliptic to oblong-obovate in outline (those of the undivided leaves linear to narrowly oblong), 1 or 2 times deeply pinnately divided (those of the uppermost and/or lowermost leaves sometimes undivided), the divisions (2)3–7, 0.3–4.0 cm long, linear to narrowly oblong or narrowly oblong-lanceolate, with 1 inconspicuous vein. Heads positioned mostly shortly above the leaves, the stalks to 6.5 cm long. Involucral bracts 10–12, those of the outer series 1.5–6.0 mm long, linear to lanceolate, those of the inner series 1–4 mm long, lanceolate to narrowly ovate. Receptacle oblong-ovoid to short-columnar or less commonly nearly spherical, 0.8–1.5 cm long. Ray florets 5–10, the corolla 4–9 mm long, yellow or more commonly with a well-defined zone of reddish brown to brownish purple toward the base, this occasionally entirely masking the yellow coloration, the outer surface densely short-hairy, both surfaces usually with scattered, minute, sessile, spherical, yellow glands. Disc florets 50 to numerous, the corolla 1.5–2.5 mm long, greenish yellow to yellowish green, sometimes purplish-tinged toward the tip. Style branches with the sterile tip slender and sharply pointed. Pappus a minute rim or crown. Fruits 1.9–2.8 mm long, somewhat obliquely oblong, the angle opposite the chaffy bract often with a minute, orangish brown fringe, the tip and the apical portion of the winged angle sometimes also with a minute, orangish tan fringe. $2n=32$. August–October.

Introduced, known thus far only from the city of St. Louis (Wyoming to Arizona east to Nebraska and Texas; Mexico; introduced in Missouri). Railroads.

89. Rudbeckia L. (coneflower)

Plants annual or more commonly perennial herbs, mostly fibrous-rooted, less commonly with rhizomes or taproots. Stems erect or ascending, unbranched or few- to several-branched toward the tip, with several longitudinal lines or ridges, glabrous or variously hairy, slightly roughened to the touch. Leaves basal and alternate, the basal and lower stem leaves long-petiolate (except in *R. missouriensis*), the petioles progressively shorter up the stem, the upper stem leaves usually sessile or nearly so, the bases usually moderately expanded and more or less clasping or wrapping around the stem. Leaf blades simple, sometimes deeply ternately or pinnately lobed, those of the basal leaves usually ovate to elliptic-ovate (oblanceolate in *R. missouriensis*), those of the stem leaves variously linear to broadly ovate, tapered or rounded at the base, mostly angled or tapered to a sharply pointed tip, the margins otherwise entire or toothed, the surfaces glabrous or more commonly hairy (usually roughened), sometimes with sessile or impressed glands, with 1–5(–7) main veins. Inflorescences of solitary terminal heads, these sometimes appearing as loose, open clusters or leafy panicles, the heads mostly with relatively long stalks, these often with 1 or 2 bracts similar to the involucral bracts at or near the tip. Heads radiate (discoid elsewhere). Involucre broadly more or less saucer-shaped, the bracts in 1 or 2 subequal, overlapping series. Involucral bracts 5–25, linear to narrowly lanceolate or narrowly ovate, spreading to reflexed, green, the margins and outer (usually also the inner) surface roughened-hairy, not glandular, the midnerve inconspicuous. Receptacle nearly spherical to conical or less commonly somewhat cylindrical, often elongating somewhat as the fruits mature, with chaffy bracts subtending the disc florets (also the ray florets in *R. amplexicaulis*), these shorter than to slightly longer than the disc florets (including the corolla), concave and wrapped around the florets, truncate or rounded to angled or tapered to a sharply pointed tip, this unawned or with a soft, bristlelike awn, the apical portion green to dark purple, persistent at fruiting. Ray florets 5–21, sterile (lacking stamens and style at flowering and with an ovary that is shorter and thinner than those of the disc florets, not developing into a fruit), the corolla showy, relatively slender to somewhat broadened, spreading to drooping at flowering, yellow to orangish yellow, sometimes strongly reddish- or orangish-tinged toward the base, not persistent at fruiting. Disc florets 50 to numerous (more than 200), perfect, the corolla yellow, yellowish green, purple, or purplish brown, not thickened at the base, not persistent at fruiting. Style branches with the sterile tip somewhat elongate, tapered, and bluntly or sharply pointed. Pappus of the disc florets absent, a low rim or crown, or of 2–6 minute, unequal scales, usually persistent at fruiting. Fruits narrowly wedge-shaped to nearly oblong in outline, usually strongly 4-angled in cross-section, each face with several faint to more prominent but slender lines, grooves, or ribs, the angles sometimes very narrowly winged, the surface glabrous, brown to black, sometimes somewhat shiny. About 23 species, U.S., Canada; introduced in Europe.

A number of species of *Rudbeckia* are cultivated as ornamentals in gardens (Dress, 1961). A few species, such as *R. hirta,* also are sold as cut flowers. Native Americans used several species both topically and internally for a variety of ailments ranging from sores and burns to worms, snakebites, kidney disease, and heart problems. The foliage and stems of *R. laciniata* also were cooked and eaten by some tribes (Moerman, 1998). Species of *Rudbeckia,* especially *R. laciniata,* have been implicated in livestock poisoning. The active agents apparently belong to a group of sesquiterpene lactones. Burrows and Tyrl (2001) noted that experimental efforts to reproduce symptoms in pigs and sheep first reported anecdotally were only partially successful, with depression, loss of appetite, loss of coordination, and increased respiration apparently passing away relatively quickly without long-term effects. They also noted that because the plants have a strongly disagreeable flavor, livestock avoid eating them unless no alternative food plants are present.

Abrahamson and McCrea (1977) researched different patterns of absorption and reflectance of light in the ultraviolet portion of the spectrum on the corollas of *R. hirta, R. laciniata,* and *R. triloba.* A number of insects (such as bees) are able to see light in the ultraviolet portion of the spectrum, which is invisible to humans. The ultraviolet corolla patterns, which have been studied in a variety of plants, act as visual cues to attract appropriate pollinators.

1. At least some of the leaves (at least the lower ones) with the blade ternately or pinnately lobed
   2. Chaffy bracts as long as or slightly longer than the disc florets (including the corolla), tapered to a slender, sharply pointed, somewhat awnlike tip, glabrous . . . . 9. R. TRILOBA
   2. Chaffy bracts shorter than to nearly as long as the disc florets (including the corolla), truncate, rounded, or broadly angled to a short, bluntly or sharply pointed tip, the outer surface and margins with dense, short, often somewhat matted hairs
      3. Stems glabrous, sometimes glaucous; leaf blades with the upper surface glabrous or sparsely hairy, smooth to the touch; disc corollas (and chaffy bracts) dull yellow to yellowish green . . . . 5. R. LACINIATA
      3. Stems moderately to densely pubescent with short, spreading hairs, at least above the midpoint; leaf blades with the upper surface moderately pubescent with short, spreading, pustular-based hairs, roughened to the touch; disc corollas (and chaffy bracts) dark purple toward the tip (rarely greenish yellow) . . . . 8. R. SUBTOMENTOSA
1. All of the leaves with the blade unlobed, the margins entire, scalloped, or toothed
   4. Stems and leaves glabrous, smooth to the touch
      5. Stems 20–70(–90) cm long; stems and leaves not or only slightly glaucous, appearing green to somewhat bluish green when fresh; all but the lowermost leaves sessile and deeply cordate-clasping at the base; receptacle 10–30 mm long, 8–15 mm in diameter, hemispherical to ovoid or conical . . . . 1. R. AMPLEXICAULIS
      5. Stems (60–)80–250 cm long; stems and leaves strongly glaucous, appearing gray to bluish green when fresh; the median and uppermost leaves sessile and rounded to shallowly cordate-clasping at the base; receptacle 35–80 mm long, 15–35 mm in diameter, ovoid to conical or somewhat cylindrical . . . . 6. R. MAXIMA
   4. Stems and/or leaves sparsely to densely hairy, especially toward the stem tip, at least the leaves usually somewhat roughened to the touch
      6. Plants often with taproots (sometimes with only fibrous roots), usually annual (occasionally resprouting the second year); pappus absent; stigma lobes elongate and more or less sharply pointed at the tip; chaffy bracts relatively long-fringed with spreading, bristly hairs toward the tip, the very tip with 1 or less commonly 2 bristles . . . . 4. R. HIRTA
      6. Plants without taproots, fibrous rooted (sometimes with somewhat fleshy roots), often with rhizomes or stolons, perennial; pappus present but sometimes minute; stigma lobes short and rounded to bluntly pointed at the tip; chaffy bracts glabrous or short-hairy, never with bristles at the tip
         7. Chaffy bracts with the outer surface and margins densely pubescent with short, sometimes glandular or somewhat matted hairs toward the tip; ray florets with the corolla 30–50(–70) mm long, reflexed or strongly drooping at flowering; receptacle (10–)14–30 mm long, ovoid to conical . . . . 3. R. GRANDIFLORA
         7. Chaffy bracts with the outer surface glabrous (rarely with a few short hairs), the margins glabrous or with a fringe of minute hairs toward the tip; ray florets with the corolla 10–40 mm long, spreading to slightly drooping at flowering; receptacle 6–16 mm long, hemispherical to spherical, ovoid, or short-conical

8. Chaffy bracts as long as or slightly longer than the disc florets (including the corolla), tapered to a slender, sharply pointed, somewhat awnlike tip, the margins glabrous .... 9. R. TRILOBA
8. Chaffy bracts shorter than to nearly as long as the disc florets (including the corolla), rounded or short-tapered to a broadly triangular, bluntly or sharply pointed tip
   9. Leaf blades relatively broad, those of the basal leaves 10–110 mm wide, long-petiolate and lanceolate to ovate or somewhat heart-shaped, those of the stem leaves (5–)12–50(–70) mm wide, lanceolate to elliptic or ovate; plants with stolons, the new basal rosettes occurring at the stolon tips and often at some distance from (not immediately adjacent to) the older stems .... 2. R. FULGIDA
   9. Leaf blades relatively narrow, those of the basal leaves 5–20 mm wide, more or less sessile to short- or less commonly long-petiolate and broadly linear to oblanceolate or narrowly spatulate, those of the stem leaves 4–10 mm wide, linear to narrowly oblong-lanceolate; plants with rhizomes, the new basal rosettes occurring close (immediately adjacent) to the older stems .... 7. R. MISSOURIENSIS

1. Rudbeckia amplexicaulis Vahl (clasping coneflower)

Dracopis amplexicaulis (Vahl) Cass.

Pl. 276 b; Map 1217

Plants annual, with taproots. Stems 20–70(–90) cm long, glabrous, occasionally slightly glaucous. Leaves all unlobed, all (except sometimes the basal ones) strongly cordate at the base and clasping the stem, the margins entire or bluntly and usually finely toothed, the surfaces glabrous, smooth, usually somewhat glaucous, appearing bluish green when fresh. Basal and lowermost stem leaves usually absent at flowering, long-petiolate to nearly sessile, the blade 2–15 cm long, 6–15 mm wide, oblanceolate to oblong-oblanceolate, rounded to sharply pointed at the tip. Median and upper stem leaves sessile, the blade 1–10 cm long, 8–40 mm wide, progressively shorter and broader, oblong-ovate to oblong-obovate, ovate or heart-shaped, tapered to a sharply pointed tip. Inflorescences mostly appearing as loose, open clusters of heads, sometimes of solitary heads. Involucral bracts 7–12, 3–10 mm long, linear to lanceolate, glabrous. Receptacle 10–30 mm long, 8–15 mm in diameter, hemispherical to somewhat ovoid at the start of flowering, then elongating and becoming ovoid to conical. Chaffy bracts subtending the ray and disc florets, shorter than the disc florets, broadly angled or abruptly short-tapered to a usually sharply pointed tip, the outer surface glabrous, the margins with a fringe of minute, spreading hairs. Ray florets 6–10(–12), the corolla 12–30 mm long, relatively broad, strongly reflexed at flowering, yellow, sometimes strongly reddish- or orangish-tinged toward the base, the outer surface moderately to densely short-hairy. Disc florets numerous, the corolla 2.5–3.5 mm long, greenish yellow to yellow toward the base, dark purple to brownish purple toward the tip, the lobes usually strongly curled downward at flowering. Stigma lobes elongate and more or less sharply pointed at the tip. Pappus absent or occasionally a minute rim or crown. Fruits 1.8–2.5 mm long. $2n=32$. May–July.

Uncommon in Jasper and Newton Counties, also introduced there and elsewhere in western Missouri as well as the city of St. Louis (southern [mostly southeastern] U.S. north to Kansas and Missouri; introduced sporadically farther north). Upland prairies and glades; also edges of crop fields, railroads, roadsides, and open, disturbed areas.

Steyermark (1963) treated *R. amplexicaulis* (as *Dracopis*) as native in Missouri, and the oldest specimens, from Jasper and Newton Counties (collected by E. J. Palmer in 1909), apparently were collected from native plant communities. The species also grows natively in adjacent portions of Oklahoma and Kansas. However, most of the historical materials and all of the more recent collections undoubtedly represent introduced populations. Clasping coneflower is cultivated as an ornamental in gardens and can escape to form extensive populations in disturbed habitats. It also has been planted along some highways as a component of wildflower seed mixes for so-called roadside beautification projects.

The classification of this species remains somewhat controversial. Steyermark (1963) and most earlier authors maintained it as the only species of

1218. Rudbeckia fulgida   1219. Rudbeckia grandiflora   1220. Rudbeckia hirta

Dracopis Cass. This interpretation was supported by the phylogenetic analysis of morphological and anatomical characters in the various coneflower genera of Cox and Urbatsch (1990), who concluded that the genus is a relative of *Ratibida*. Molecular phylogenetic studies of the coneflower genera involving restriction site variation within the chloroplast genome was inconclusive but weakly favored a similar interpretation (Urbatsch and Jansen, 1995). However, an expanded molecular study that integrated the earlier chloroplast genomic data with nuclear sequence data resolved *Dracopis* as a distinct lineage nested within *Rudbeckia* (Urbatsch et al., 2000). Urbatsch and his colleagues favored a classification in which *Dracopis* is treated as one of three subgenera within *Rudbeckia*. This classification is followed in the present treatment.

2. Rudbeckia fulgida Aiton (orange coneflower)
Pl. 288 f, g; Map 1218

Plants perennial, with fibrous roots and often stolons (new basal rosettes occurring at the stolon tips and often not immediately adjacent to the older stems). Stems 30–120 cm long, sparsely pubescent with short, spreading to ascending hairs, often glabrous toward the base, not glaucous. Leaves all unlobed, variously tapered, angled, rounded, truncate, or cordate at the base, conspicuously clasping to inconspicuously wrapping around the stem, the margins entire or bluntly to sharply and finely to coarsely toothed, the surfaces sparsely to moderately hairy with spreading to loosely ascending, minutely pustular-based hairs, usually slightly to moderately roughened to the touch, not glaucous, green when fresh. Basal and lowermost stem leaves often present at flowering (or present on nearby rosettes), long-petiolate, the blade 5–30 cm long, 10–110 mm wide, narrowly to broadly ovate or elliptic, sometimes more or less heart-shaped, angled or tapered to a sharply pointed tip. Median and upper stem leaves sessile or with a short or rarely long, winged petiole, the blade (1–)2–20 cm long, (5–)12–50(–70) mm wide, lanceolate to oblanceolate or more commonly elliptic to ovate, angled or tapered to a sharply pointed tip. Inflorescences of solitary heads or appearing as loose, open clusters. Involucral bracts 8–14, 8–22 mm long, lanceolate to linear, the outer surface glabrous or sparsely hairy, the margins usually with moderate to dense, ascending hairs. Receptacle 10–16 mm long, 10–18 mm in diameter, hemispherical at the start of flowering, then elongating somewhat and often becoming ovoid. Chaffy bracts subtending only the disc florets, shorter than to nearly as long as the disc florets (including the corolla), rounded or short-tapered to a broadly triangular, bluntly or sharply pointed tip, the outer surface glabrous (rarely with a few short hairs) and often somewhat shiny, the margins with a fringe of moderate to dense, minute, spreading hairs. Ray florets 8–15, the corolla 10–40 mm long, relatively slender, spreading to slightly drooping at flowering, yellow or less commonly the basal portion or the entire corolla orange, the outer surface sparsely short-hairy and sometimes also minutely gland-dotted. Disc florets 50 to numerous, the corolla 3–4 mm long, yellowish green toward the base, dark purple to purplish brown toward the tip, the lobes ascending at flowering. Stigma lobes relatively short and bluntly pointed at the tip. Pappus a minute rim or crown. Fruits 2–4 mm long. $2n=38, 76$. July–October.

Scattered in the Ozark, Ozark Border, and Unglaciated Plains Divisions (eastern U.S. west to Wisconsin, Missouri, and Texas; Canada). Banks of streams, rivers, and spring branches, fens and calcareous seeps, marshes, bases and ledges of bluffs, bottomland forests, and rarely moist depressions of dolomite glades; also roadsides.

Some authors have divided this species into as many as seven varieties (Perdue, 1957). The var.

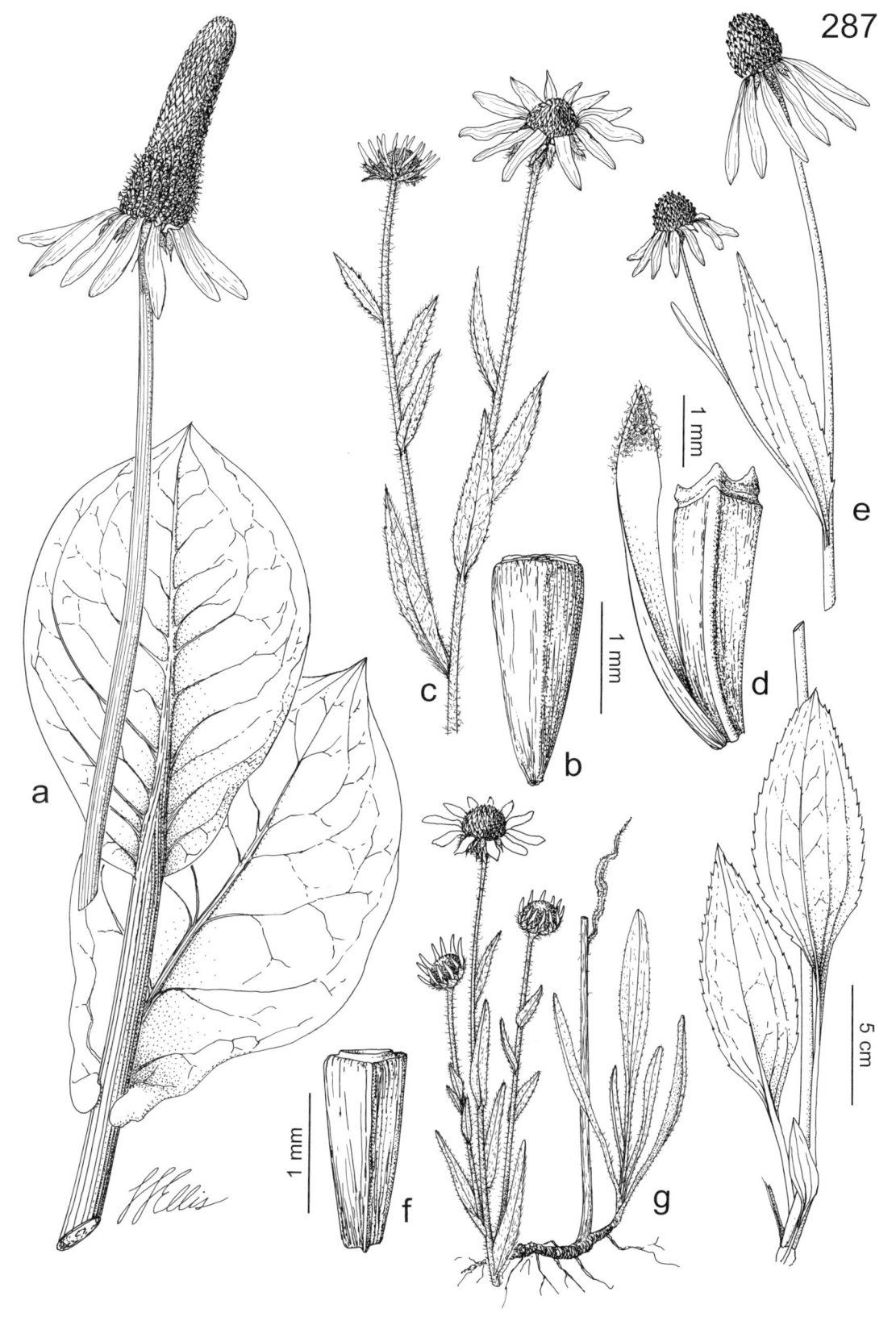

Plate 287. Asteraceae. *Rudbeckia maxima*, **a)** habit. *Rudbeckia hirta*, **b)** fruit, **c)** habit. *Rudbeckia grandiflora*, **d)** fruit with bract, **e)** inflorescence. *Rudbeckia missouriensis*, **f)** fruit, **g)** habit.

fulgida occurs to the east of Missouri and is characterized by its relatively narrow basal and stem leaves with the latter mostly narrowed to short, winged petioles. Steyermark (1963) reluctantly accepted two of the other varieties as occurring in Missouri, but he noted the innate variability of the plants, lack of correlation between some of the distinguishing characters, and large number of specimens intermediate between the two varieties (as well as between these and some other varieties). Cronquist (1980) and Gleason and Cronquist (1991) recognized only three overlapping varieties across the species range and also considered two of these to occur in Missouri. The two varieties are accepted here with some hesitation in part because they tend to occupy somewhat different ranges in Missouri. Users should note that some plants cannot be determined below the species level satisfactorily.

1. Ray florets 10–21, the corolla mostly 2.5–4.0 cm long; median and upper stem leaves with the blade lanceolate, oblanceolate, or elliptic, the margins usually sharply toothed
.... 2A. VAR. SULLIVANTII
1. Ray florets 8–15, the corolla 1.0–2.5 (–3.0) cm long; median and upper stem leaves with the blade mostly broadly ovate to elliptic, the margins entire to bluntly or less commonly sharply toothed .... 2B. VAR. UMBROSA

2a. var. sullivantii (C.L. Boynton & Beadle) Cronquist (showy coneflower)
R. speciosa Wender. var. *sullivantii* (C.L. Boynton & Beadle) B.L. Rob.
R. fulgida var. *speciosa* (Wender.) Perdue

Stems sparsely hairy with short, usually spreading to ascending hairs, often glabrous toward the base. Basal leaves with the blade mostly narrowly to broadly ovate or elliptic, seldom cordate at the base. Median and upper stem leaves with the blade lanceolate, oblanceolate, or elliptic, often tapered or angled at the base, only slightly expanded and inconspicuously wrapping around the stem, the margins usually sharply toothed. Ray florets 10–21, the corolla mostly 2.5–4.0 cm long. $2n=38$, about 76. July–October.

Scattered in the Ozark, Ozark Border, and Unglaciated Plains Divisions (eastern U.S. west to Wisconsin and Arkansas; introduced in Canada). Banks of streams, rivers, and spring branches, fens and calcareous seeps, marshes, and bottomland forests; also roadsides.

2b. var. umbrosa (C.L. Boynton & Beadle) Cronquist (orange coneflower)
R. fulgida var. *palustris* (Eggert ex C.L. Boynton & Beadle) Perdue

Stems sparsely hairy with short, spreading to ascending hairs, often glabrous toward the base. Basal leaves with the blade mostly broadly ovate or elliptic to sometimes more or less heart-shaped, often cordate at the base. Median and upper stem leaves with the blade mostly broadly ovate to elliptic, broadly rounded, truncate, or shallowly cordate at the base, often noticeably clasping the stem, the margins usually entire to bluntly or less commonly sharply toothed. Ray florets 8–15, the corolla mostly 1.0–2.5(–3.0) cm long. $2n=38$. August–October.

Scattered in the Ozark and Ozark Border Divisions (southeastern U.S. west to Missouri and Texas). Banks of streams, rivers, and spring branches, fens and calcareous seeps, bases and ledges of bluffs, and rarely moist depressions of dolomite glades.

This is the more abundant of the two varieties in Missouri, but its range apparently does not extend out of the Ozarks. Perdue (1957) recognized var. *palustris* for plants with the leaf bases truncate to broadly angled, restricting the name var. *umbrosa* to plants with the leaf bases broadly rounded to cordate. He included Missouri in the range of var. *palustris* (Missouri to Texas) but attributed a range to var. *umbrosa* from Ohio to Alabama and Georgia. In fact, plants with both morphologies occur in Missouri, sometimes in the same population, and there is a near continuum of leaf base morphologies represented in the herbaria.

3. Rudbeckia grandiflora (Sweet) J.F. Gmel. ex DC. **var. grandiflora** (rough coneflower)
Pl. 287 d, e; Map 1219

Plants perennial, the rootstock usually short and somewhat woody, with fibrous roots, sometimes with stout rhizomes. Stems 50–120 cm long, moderately to densely pubescent with short, spreading to loosely ascending hairs, often also with scattered, minute, sessile glands toward the tip, sometimes nearly glabrous toward the base, not glaucous. Leaves all unlobed, tapered at the base, only slightly expanded at the base and inconspicuously wrapping around the stem, the margins entire or more commonly relatively sharply but finely toothed, the surfaces moderately to densely hairy with short, spreading to loosely ascending, minutely pustular-based hairs and also dotted with minute, sessile, yellowish to orange glands, slightly to moderately roughened to the touch, not glaucous, green when fresh. Basal and lowermost stem leaves

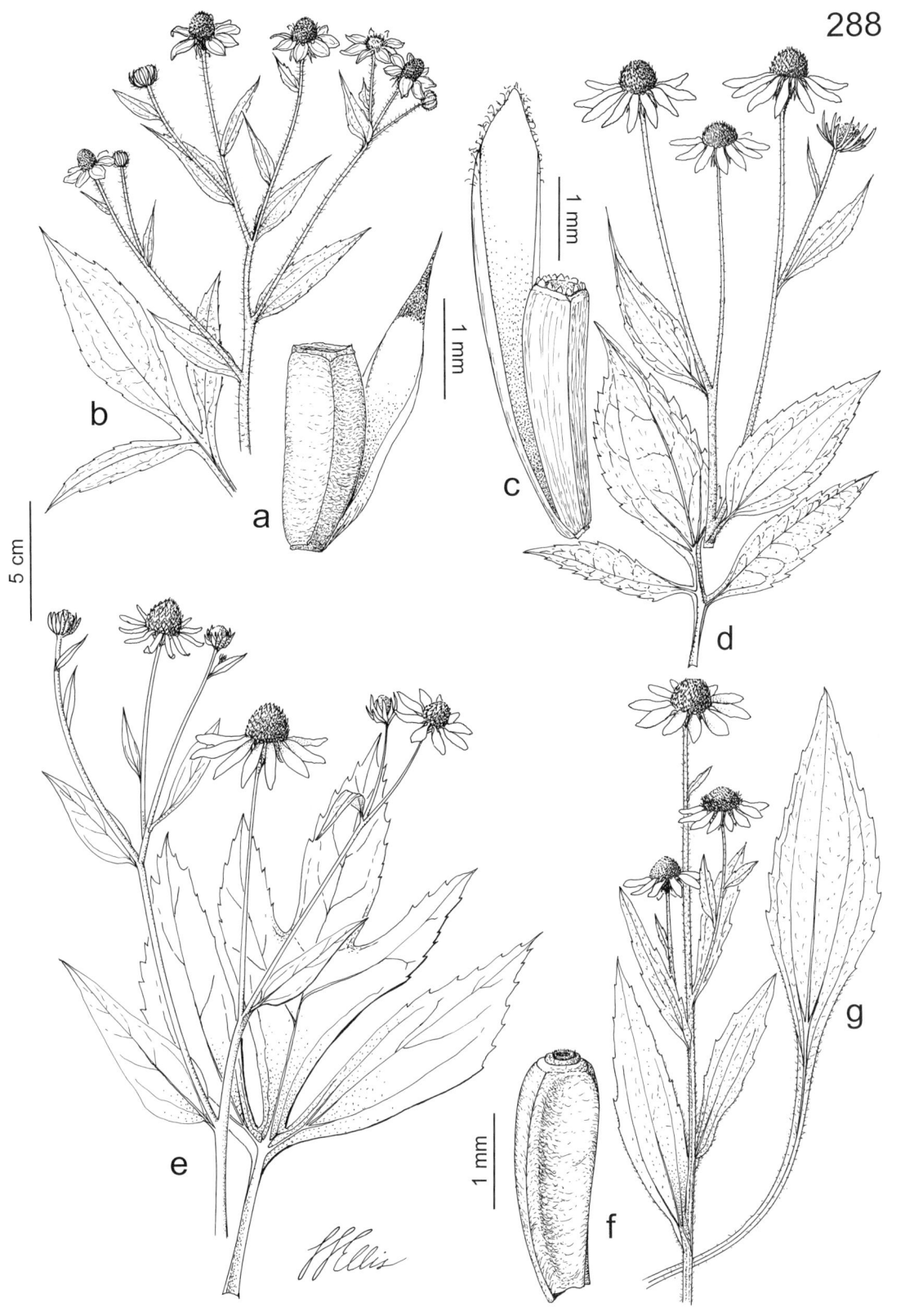

Plate 288. Asteraceae. *Rudbeckia triloba*, **a)** fruit with bract, **b)** inflorescence and larger leaf. *Rudbeckia subtomentosa*, **c)** fruit with bract, **d)** inflorescence and larger leaf. *Rudbeckia laciniata*, **e)** inflorescence and larger leaf. *Rudbeckia fulgida*, **f)** fruit, **g)** inflorescence and larger leaf.

usually present at flowering, long-petiolate, the blade 10–35 cm long, 40–90 mm wide, ovate to elliptic or narrowly ovate, angled or tapered to a sharply pointed tip. Median and upper stem leaves sessile to short-petiolate, the blade 4–15 cm long, 15–70 mm wide, ovate to elliptic or narrowly ovate, angled or tapered to a sharply pointed tip. Inflorescences mostly appearing as solitary heads, a second head rarely produced. Involucral bracts 12–20, 6–15 mm long, narrowly lanceolate to linear, the outer surface and margins moderately to densely roughened-hairy. Receptacle (10–)14–30 mm long, 15–25 mm in diameter, hemispherical or somewhat ovoid at the start of flowering, then elongating and becoming ovoid to conical. Chaffy bracts subtending only the disc florets, shorter than the disc florets, rounded or broadly angled to a bluntly pointed tip, the outer surface and margins densely pubescent with short, sometimes glandular or somewhat matted hairs toward the tip. Ray florets 10–21, the corolla 30–50(–70) mm long, relatively slender, reflexed or strongly drooping at flowering, yellow, the outer surface moderately short-hairy, both surfaces usually minutely gland-dotted. Disc florets numerous, the corolla 3.5–5.0 mm long, greenish yellow to yellow toward the base, dark purple to purplish brown toward the tip, the lobes usually strongly curled downward at flowering. Stigma lobes relatively short and rounded to bluntly pointed at the tip. Pappus a minute, irregularly toothed crown 0.2–0.5 mm long. Fruits 3–4 mm long. $2n=38$. July–August.

Possibly introduced, uncommon and sporadic (Kansas to Texas east to Ohio and Georgia; Canada). Openings of mesic upland forests; also railroads, roadsides, and open, disturbed areas.

Although Missouri is encompassed in the overall distribution of *R. grandiflora,* there is some doubt as to whether it should be considered native to the state's flora. Steyermark (1963) considered it native, but he listed no natural plant communities in his summary of its habitats. Although not all of the material that Steyermark apparently examined was relocated during the present study (he also reported it from Shannon County), the labels on the majority of the specimens that he studied either state that the plant was introduced where collected or originated from disturbed habitats such as railroads. The most recent collection, however, was made in Phelps County by the legendary Missouri botanist Bill Summers, who documented a relatively large population growing with other native species in openings of an upland forest made by a gas pipeline.

Plants with minutely hairy stems and leaves and the stems more or less glabrous toward the base have been called var. *alismifolia* (Torr. & A. Gray) Cronquist. These occur mostly in the southwestern portion of the range of var. *grandiflora* in Arkansas, Texas, Kentucky, Louisiana, and Mississippi (Perdue, 1957).

4. Rudbeckia hirta L. var. pulcherrima Farw. (black-eyed Susan)
R. hirta f. *flavescens* Clute
R. hirta f. *homochroma* Steyerm.
R. bicolor Nutt.
R. serotina Nutt.

Pl. 287 b, c; Map 1220

Plants usually annual (occasionally resprouting the second year), usually with taproots but sometimes with only fibrous roots. Stems 10–100 cm long, sparsely to densely pubescent with short, spreading to ascending hairs, not glaucous. Leaves all unlobed, tapered or angled (in lower leaves) to rounded or truncate (sometimes in upper leaves) at the base, mostly inconspicuously wrapping around the stem, the upper leaves sometimes more conspicuously clasping the stem, the margins entire or sometimes with widely spaced, usually fine, sharp teeth, the surfaces sparsely to moderately hairy with spreading to loosely ascending, often minutely pustular-based hairs, usually slightly to moderately roughened to the touch, not glaucous, green when fresh. Basal and lowermost stem leaves sometimes present at flowering, mostly with a long, often winged petiole, the blade 5–30 cm long, 5–30 mm wide, mostly oblanceolate to elliptic, rounded or angled to a bluntly pointed tip. Median and upper stem leaves sessile or with a short, winged petiole, the blade 2–20 cm long, 3–40 mm wide, lanceolate to elliptic to ovate, occasionally nearly linear, angled or tapered to a bluntly or sharply pointed tip. Inflorescences of solitary heads or more commonly appearing as loose, open clusters. Involucral bracts 12–20, 8–30 mm long, lanceolate to linear, the outer surface moderately to densely hairy, the margins usually with moderate to dense, more or less spreading hairs. Receptacle 10–22 mm long, 10–20 mm in diameter, hemispherical at the start of flowering, then elongating somewhat and often becoming ovoid. Chaffy bracts subtending only the disc florets, shorter than to nearly as long as the disc florets (including the corolla), mostly angled to a triangular, sharply pointed tip, the outer surface glabrous or with short, stiff hairs toward the tip, the margins with a relatively long fringe of spreading, bristly hairs toward the tip, the very tip with 1 or less commonly 2 bristles. Ray florets

1221. Rudbeckia laciniata 1222. Rudbeckia maxima 1223. Rudbeckia missouriensis

8–21, the corolla 15–40 mm long, relatively slender, spreading to somewhat drooping at flowering, yellow to orangish yellow, sometimes the basal portion reddish- or purplish-tinged, the outer surface sparsely to densely short-hairy (especially toward the tip) and usually also minutely gland-dotted. Disc florets numerous, the corolla 3.0–4.5 mm long, greenish yellow to nearly white toward the base, dark purple to purplish brown toward the tip, the lobes ascending to spreading at flowering. Stigma lobes elongate and more or less sharply pointed at the tip. Pappus absent. Fruits 1.5–3.0 mm long. $2n=38$. May–October.

Scattered to common throughout the state (U.S., Canada; introduced in Europe). Bases of bluffs, openings of mesic to dry upland forests, upland prairies, and glades; also pastures, old fields, railroads, roadsides, and open, disturbed areas.

This is the most abundant species of *Rudbeckia* in the state and the one best adapted to growing in disturbed habitats. In addition to its native occurrences, the species also is widely cultivated both in gardens and as a component of wildflower seed mixes for roadside beautification projects, from which it occasionally becomes naturalized. A number of cultivars exist, including ones with especially long ray corollas or ray corollas with more reddish coloration than is usually found in natural populations. The var. *pulcherrima* is the most widespread variety, growing throughout the species range. It is one of up to five semidistinct varieties maintained by some authors (Perdue, 1957) that differ in degree and position of stem branching, degree of leaf toothing, leaf shape, and length of the stalks subtending the heads. The species as a whole is quite variable morphologically, and most of the named taxa reflect regional extremes in variation for one or more characters. Steyermark (1963) noted that var. *hirta* may be discovered in Missouri in the future (although such plants may represent escapes from cultivation). This variety is widespread east of the Mississippi River and differs from var. *pulcherrima* in its relatively broad, coarsely toothed leaves.

5. Rudbeckia laciniata L. **var. laciniata** (wild goldenglow, cutleaf coneflower)

Pl. 288 e; Map 1221

Plants perennial, with fibrous roots and often a somewhat woody rootstock. Stems 50–250 cm long, glabrous, sometimes glaucous. Leaves mostly deeply 1 or 2 times 3(5 or 7)-lobed or deeply divided, the basal and lowermost stem leaves sometimes fully compound, the uppermost stem leaves sometimes unlobed, mostly angled or tapered at the base, only slightly expanded at the base and inconspicuously wrapping around the stem, the margins otherwise entire or sharply finely to coarsely toothed, the surfaces glabrous or sparsely pubescent with short, stiff hairs, at least the upper surface smooth to the touch, both surfaces sometimes somewhat glaucous, green to grayish green when fresh. Basal and lowermost stem leaves often absent at flowering, long-petiolate, the blade 15–50 cm long, 100–300 mm wide, ovate to broadly ovate in outline, the lobes ovate to elliptic or obovate, tapered to a sharply pointed tip. Median and upper stem leaves short- to long-petiolate, the uppermost leaves occasionally sessile, the blade 3–40 cm long, 1–20 mm wide, the unlobed leaf blade or the lobes of a divided blade lanceolate to ovate to elliptic, tapered to a sharply pointed tip. Inflorescences mostly appearing as loose, open clusters or leafy panicles. Involucral bracts 8–15, 3–15 mm long, oblong-ovate to oblong-lanceolate or narrowly oblong, the outer surface and margins glabrous or moderately to densely and finely hairy, especially toward the base. Receptacle 15–20 mm long, 10–20 mm in diameter, spherical to broadly ovoid at the start of flowering, then elongating somewhat but remaining spherical to ovoid. Chaffy bracts subtending only the disc florets, shorter than to nearly as long as the disc florets (including the corolla), truncate, rounded or broadly angled to a short, bluntly or sharply pointed tip, the outer surface and margins with dense, short, often somewhat matted hairs, dull yellow to yellowish green. Ray florets 8–13, the corolla 20–45 mm long, usually relatively slender, angled downward or drooping at flowering, yellow, the outer surface sparsely short-hairy. Disc florets numerous, the corolla 3.5–4.5 mm long, dull yellow to yellowish green, the lobes ascending at flowering. Stigma lobes relatively short and bluntly to less commonly sharply pointed at the tip. Pappus a short, toothed crown 0.4–0.8 mm long. Fruits 2.0–3.5 mm long. $2n=38, 54, 72, 102+$. July–September.

Scattered nearly throughout the state, but apparently absent from the Mississippi Lowlands Division and uncommon in the western portion of the Glaciated Plains (U.S. [except a few western states]; Canada; introduced in Europe). Bottomland forests, mesic upland forests in ravines, banks of streams and rivers, margins of ponds and lakes, and sloughs; also roadsides.

Most authors have accepted between two and five varieties within *R. laciniata*. The most distinctive of these is the western var. *ampla* (A. Nelson) Cronquist (excluded from the Missouri flora by Steyermark [1963]), which has receptacles to 3 cm long and to 2 cm in diameter, as well as bigger disc florets and fruits, and a number of subtle vegetative differences. Some authors have argued that it is better regarded as a distinct species (Jones, 1957). Among the eastern varieties, none is particularly distinct, although the Florida endemic var. *heterophylla* (Torr. & A. Gray) Fernald & B.G. Schub., with its unlobed lower leaves and relatively densely hairy leaves and stem tips, is striking in its extreme form. Missouri plants uniformly are referable to var. *laciniata*. It should be noted that at least the plants with triploid and pentaploid chromosome numbers ($2n=54, 102+$) apparently are apomictic, a condition not uncommon in the variety (Cronquist, 1980).

One of the most common forms of the species in cultivation has been called var. *hortensis* L.H. Bailey or cv. 'Hortensis,' the garden goldenglow, and it has all or most of the disc florets converted to ray florets. Some of the other plants sold as cultivars of *R. laciniata* represent variants developed following hybridization of *R. laciniata* with related species such as *R. nitida* Nutt. (Dress, 1961).

6. Rudbeckia maxima Nutt. (great coneflower, cabbage coneflower)

Pl. 287 a; Map 1222

Plants perennial, with rhizomes. Stems (60–)80–250 cm long, glabrous, glaucous. Leaves all unlobed, all except the basal and lower stem leaves rounded to shallowly and somewhat obliquely cordate at the base, more or less clasping the stem, the margins entire or shallowly scalloped to bluntly toothed, the surfaces glabrous, smooth, strongly glaucous, appearing gray to bluish green when fresh. Basal and lowermost stem leaves often present at flowering, usually long-petiolate, the blade 10–35 cm long, 40–120 mm wide, narrowly obovate to broadly elliptic or broadly elliptic-obovate, rounded to sharply pointed at the tip. Median and upper stem leaves sessile, the blade 3–20 cm long, 20–150 mm wide, broadly oblong-ovate to oblong-circular, the uppermost sometimes ovate, abruptly tapered to a bluntly or sharply pointed tip. Inflorescences mostly appearing as solitary heads, sometimes 2- or 3-headed open clusters. Involucral bracts 12–20, 8–22 mm long, narrowly ovate to lanceolate or narrowly oblong-lanceolate, usually glabrous and somewhat glaucous, the margins sometimes slightly thickened and yellowish. Receptacle 35–80 mm long, 15–35 mm in diameter, ovoid at the start of flowering, then elongating and becoming conical or somewhat cylindrical. Chaffy bracts subtending only the disc florets, shorter than the disc florets, rounded or broadly angled to a bluntly or sharply pointed tip, the outer surface mostly moderately to densely pubescent with short, tan hairs but glabrous at the tip, the margins similarly hairy but glabrous toward the tip. Ray florets 10–21, the corolla 30–80 mm long, relatively broad, reflexed or strongly drooping at flowering, yellow, the outer surface sparsely short-hairy to nearly glabrous. Disc florets numerous, the corolla 4–6 mm long, greenish yellow to yellow toward the base, dark purple to purplish brown toward the tip, the lobes ascending at flowering. Stigma lobes relatively short and rounded to bluntly pointed at the tip. Pappus of 4–6 scales 0.8–1.8 mm long. Fruits 5–7 mm long. $2n=36$. July–August.

Introduced, known thus far only from historical collections from Jackson County (Texas to Louisiana north to Oklahoma and Arkansas; introduced in Missouri and South Carolina). Habitat unknown, but according to Steyermark (1963) railroads.

This species is occasionally cultivated as an ornamental in gardens for its striking grayish foliage and large heads. Cox and Urbatsch (1994) noted that in Texas the plants can become aggressive pasture weeds and that cattle avoid grazing on them.

7. **Rudbeckia missouriensis** Engelm. ex C.L. Boynton & Beadle (Missouri coneflower)
R. *fulgida* Aiton var. *missouriensis* (Engelm. ex C.L. Boynton & Beadle) Cronquist
Pl. 287 f, g; Map 1223

Plants perennial, with fibrous to somewhat fleshy roots and stout rhizomes (new basal rosettes occurring immediately adjacent to the older stems). Stems 20–50(–80) cm long, moderately to densely pubescent with relatively long, spreading hairs, not glaucous. Leaves all unlobed, tapered or angled at the base, only slightly expanded at the base and inconspicuously wrapping around the stem, the margins entire or with a few inconspicuous, sharp teeth, the surfaces moderately to densely hairy with mostly spreading minutely pustular-based hairs, usually moderately roughened to the touch, not glaucous, green to grayish green when fresh. Basal and lowermost stem leaves often present at flowering (or present on adjacent rosettes), more or less sessile to short- or less commonly long-petiolate, the blade 5–20 cm long, 5–20 mm wide, broadly linear to oblanceolate or narrowly spatulate, rounded or more or less angled to a bluntly pointed tip. Median and upper stem leaves sessile, the blade 2–15 cm long, 4–10 mm wide, linear to narrowly oblong-lanceolate, angled or short-tapered to a bluntly or sharply pointed tip. Inflorescences usually appearing as strongly ascending clusters. Involucral bracts 10–20, 6–14 mm long, ovate to lanceolate or occasionally linear, both surfaces sparsely to densely hairy, the margins usually with moderate to dense, ascending hairs. Receptacle 8–15 mm long, 10–17 mm in diameter, hemispherical at the start of flowering, then elongating somewhat and sometimes becoming ovoid. Chaffy bracts subtending only the disc florets, shorter than the disc florets (including the corolla), short-tapered to a broadly triangular, usually sharply pointed tip, the outer surface glabrous (sometimes sparsely hairy toward the base) and often somewhat shiny, the margins usually glabrous. Ray florets 9–15, the corolla 10–25 mm long, relatively slender, spreading to slightly drooping at flowering, yellow or less commonly the basal portion orangish- or reddish-tinged, the outer surface sparsely to moderately short-hairy. Disc florets numerous, the corolla 3.5–4.5 mm long, greenish yellow toward the base, dark purple to purplish brown toward the tip, the lobes ascending at flowering. Stigma lobes relatively short and bluntly pointed at the tip. Pappus a minute rim or crown. Fruits 1.5–3.0 mm long. $2n=38$. June–October.

Scattered in the Ozark and Ozark Border Divisions (Oklahoma to Texas east to Illinois, Kentucky, and Louisiana). Glades, ledges and tops of bluffs, rock outcrops in upland prairies, and rocky openings of dry upland forests, rarely banks of streams; also roadsides, usually on limestone and dolomite substrates.

This species is recognized by its relatively low stature, often large colonies of dense rosettes, narrow leaves, strongly ascending branches, and relatively small heads.

8. **Rudbeckia subtomentosa** Pursh (sweet coneflower)
Pl. 288 c, d; Map 1224

Plants perennial, with fibrous roots and stout rhizomes. Stems 50–200 cm long, moderately to densely pubescent with short, spreading hairs, at least above the midpoint, not glaucous. Leaves mostly (or occasionally only the lowermost leaves) deeply 3-lobed or deeply divided, mostly angled or tapered at the base, only slightly expanded at the base and inconspicuously wrapping around the stem, the margins otherwise sharply and finely to coarsely toothed, the surfaces (especially the undersurface) moderately to densely pubescent with short, spreading, minutely pustular-based hairs, often also minutely gland-dotted, moderately to strongly roughened to the touch, green to grayish green when fresh. Basal and lowermost stem leaves often absent at flowering, long-petiolate, the blade 15–30 cm long, 30–100 mm wide, ovate to broadly ovate or elliptic in outline, the lobes ovate to narrowly elliptic or lanceolate, tapered to a sharply pointed tip. Median and upper stem leaves short- to long-petiolate, the uppermost leaves occasionally sessile, the blade 2–20 cm long, 1–120 mm wide, the unlobed leaf blade or the lobes of a divided blade lanceolate to ovate to elliptic, tapered to a sharply pointed tip. Inflorescences mostly appearing as loose, open clusters or leafy panicles. Involucral bracts 15–25, 5–15 mm long, lanceolate to nearly linear, the outer surface and margins densely pubescent with short, usually minutely pustular-based hairs. Receptacle 10–17 mm long, 8–16 mm in diameter, usually hemispherical at the start of flowering, then elongating somewhat and becoming more or less conical. Chaffy bracts subtending only the disc florets, shorter than to nearly as long as the disc florets (including the corolla), angled to a short, usually sharply pointed tip, the outer surface and margins with dense, short, often somewhat matted hairs, dark purple toward the tip or rarely dull greenish yellow (see discussion below). Ray florets 12–20, the corolla 20–40 mm long, usually relatively slender, spreading or slightly drooping at flowering, yellow to orangish

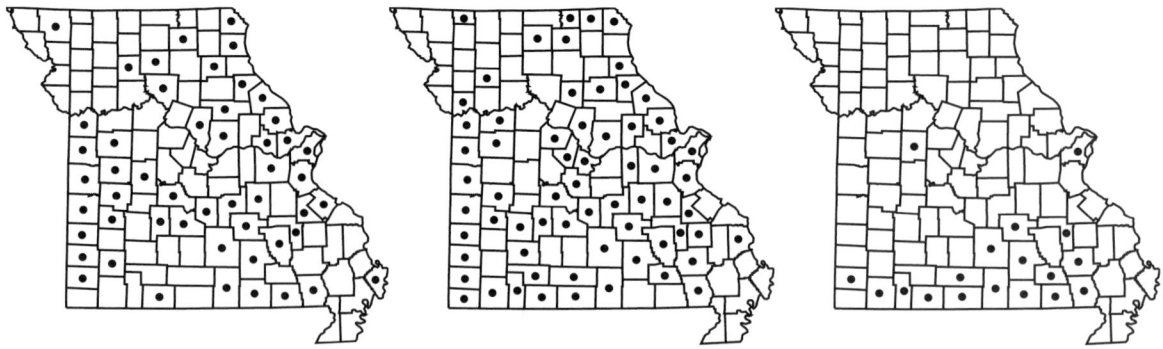

1224. Rudbeckia subtomentosa   1225. Rudbeckia triloba   1226. Silphium asteriscus

yellow, the outer surface sparsely short-hairy and moderately minutely gland-dotted. Disc florets numerous, the corolla 3.0–4.5 mm long, yellowish green toward the base (or rarely throughout), normally dark purple to purplish brown toward the tip, the lobes spreading to curled downward at flowering. Stigma lobes relatively short and usually sharply pointed at the tip. Pappus a minute rim or crown. Fruits 2.0–3.5 mm long. 2n=38. July–October.

Scattered nearly throughout the state (Kansas to Texas east to Michigan and Alabama, possibly also North Carolina; introduced in New England). Bottomland forests, edges of mesic upland forests, upland prairies, banks of streams and rivers, margins of ponds and lakes, and fens; also pastures, ditches, railroads, and roadsides.

Rare plants with the disc corollas and chaffy bracts dull yellow to yellowish green have been called *R. subtomentosa* f. *craigii* (Sherff) Fernald. This form originally was described based on plants collected in 1910 by Moses Craig in St. Louis County. It may still be known only from the original site, where plants historically grew with more typical plants having purple disc florets and chaffy bracts.

9. Rudbeckia triloba L. var. triloba (brown-eyed Susan)

Pl. 288 a, b; Map 1225

Plants perennial, with fibrous roots and rhizomes. Stems 30–150 cm long, sparsely to moderately hairy with short, mostly spreading hairs, sometimes glabrous toward the base, not glaucous. Leaves all unlobed or the basal and/or larger stem leaves deeply (2)3-lobed, tapered at the base, only slightly expanded at the base and inconspicuously wrapping around the stem, the margins entire or more commonly relatively sharply and finely to coarsely toothed, the surfaces moderately hairy with spreading, pustular-based hairs and usually also dotted with minute, sessile to impressed, yellowish glands, slightly to moderately roughened to the touch, not glaucous, green when fresh. Basal and lowermost stem leaves often absent at flowering, with long, mostly winged petioles, the blade 10–30 cm long, 30–100 mm wide, ovate to broadly ovate in outline, the lobes (when present) elliptic, ovate, or oblong-lanceolate, angled or tapered to a sharply pointed tip. Median and upper stem leaves sessile to short-petiolate, the blade 1–12 cm long, 6–30 mm wide (to 90 mm in lobed leaves), the unlobed leaf blade or the lobes of a divided blade narrowly elliptic-lanceolate to ovate to elliptic, angled or tapered to a sharply pointed tip. Inflorescences mostly appearing as loose, open clusters or leafy panicles. Involucral bracts 6–12, 5–15 mm long, lanceolate to linear, the outer surface and margins moderately roughened-hairy. Receptacle 6–16 mm long, 5–15 mm in diameter, hemispherical at the start of flowering, then elongating and often becoming ovoid to short-conical. Chaffy bracts subtending only the disc florets, as long as or slightly longer than the disc florets (including the corolla), tapered to a slender, sharply pointed, somewhat awnlike tip, the outer surface glabrous (rarely with a few short hairs) and often somewhat shiny, the margins glabrous. Ray florets 10–16, the corolla 10–20(–40) mm long, broad or slender, spreading to slightly drooping at flowering, yellow or less commonly the basal portion or the entire corolla orange, the outer surface sparsely short-hairy and sometimes also minutely gland-dotted. Disc florets 50 to numerous, the corolla 3–4 mm long, yellowish green toward the base, dark purple to purplish brown toward the tip, the lobes ascending at flowering. Stigma lobes relatively short and bluntly pointed at the tip. Pappus a minute rim or crown. Fruits 2.0–3.5 mm long. 2n=38, about 57. June–November.

Scattered nearly throughout the state, but apparently absent from the Mississippi Lowlands Division and uncommon in the western portion of the Glaciated Plains (eastern U.S. west to Minnesota, Utah, and Texas; introduced in Canada).

This is among the smallest-headed species of *Rudbeckia* in Missouri and tends to have the shortest ray corollas. Many plants do not have 3-lobed leaves at flowering, either because they did not produce any or because the larger or lower leaves were shed earlier in the growing season. Most botanists divide *R. triloba* into three varieties (Perdue, 1957). Missouri plants represent the widespread var. *triloba*, with relatively small receptacles, relatively few ray florets, and leaves unlobed or sometimes 3-lobed. Scattered populations mostly in North Carolina, Kentucky, and Tennessee differ in their larger receptacles (14–20 mm wide) and longer ray corollas (18–30 mm) and have been called var. *rupestris* (Chick.) A. Gray. The var. *pinnatiloba* differs from both of the preceding in its usually more delicate habit and at least some of the leaves pinnately 5- or 7-lobed. It occurs sporadically in the southeastern United States north to Kentucky and Virginia.

90. **Silphium** L. (rosinweed)

Plants perennial herbs, with rhizomes or taproots. Stems erect or ascending, unbranched or more commonly few- to several-branched toward the tip, with several longitudinal lines or ridges, strongly 4-angled (square in cross-section) in *S. perfoliatum,* variously glabrous to roughened or hairy. Leaves opposite or alternate (rarely a few nodes with whorls of 3), and sometimes also basal (nearly entirely basal in *S. terebinthinaceum*), sessile or the lowermost leaves short- to long-petiolate, the bases sometimes slightly to relatively strongly expanded and wrapping around or clasping the stem, often strongly perfoliate (leaves opposite and with the bases of each pair fused into a leafy cup around the stem) in *S. perfoliatum.* Leaf blades unlobed or deeply pinnately lobed or divided, lanceolate to ovate, elliptic, or occasionally nearly circular in outline, tapered to deeply cordate at the base, angled or short- to long-tapered to a usually sharply pointed tip, the margins otherwise entire or finely to coarsely sharply toothed, the surfaces variously glabrous to roughened or hairy. Inflorescences panicles or loose, open clusters (rarely of solitary heads) or narrow racemes or racemelike panicles, the heads with short to long, usually bractless stalks. Heads radiate. Involucre 15–40 mm long, cup-shaped to broadly cup-shaped or somewhat bell-shaped, the bracts in mostly 2–4 unequal to subequal, overlapping series. Involucral bracts 11–45, variously shaped, ascending or those of the outer series sometimes with the tips loosely spreading, green and somewhat leaflike to purplish brown or yellowish brown and somewhat hardened or leathery, the outer surface glabrous or variously roughened or hairy, with several fine, often inconspicuous nerves. Receptacle flat to slightly convex, not elongating as the fruits mature (occasionally broadening somewhat as the fruits mature), with chaffy bracts subtending the ray and disc florets, these narrowly oblong to linear, relatively flat with the margins not or only slightly curled partially around the florets. Ray florets (8–)13–35 in 2 or 3 overlapping marginal series, pistillate (with a 2-branched style exserted from the short tube at flowering), the corolla showy, 15–50 mm long, relatively slender, yellow, spreading to slightly drooping at flowering, not persistent at fruiting. Disc florets 40 to numerous (more than 150), staminate (with a small, stalklike ovary and an undivided style), the corolla yellow (rarely white elsewhere), slender throughout. Style branches with the sterile tip somewhat elongate and tapered. Pappus absent or that of the ray florets of 2 short, triangular, awnlike extensions of the winged angles of the fruit 1–5 mm long (the tip of the fruit then appearing deeply notched), persistent at fruiting. Fruits obovate in outline, strongly flattened, the surface smooth but usually with several fine, faint lines, glabrous or less commonly finely hairy, brown to black, sometimes with minute resinous dots. About 12 species, U.S., Canada.

Some species of *Silphium* resemble members of the genus *Helianthus* but differ in their staminate (vs. perfect) disc florets, pistillate (vs. sterile) ray florets, and frequently in their relatively large, broad outer involucral bracts. The name rosin weed refers to the gummy resin that oozes from damaged tissues. This exudate was used by Native Americans and pioneers as a kind of chewing gum. Most of the species also were utilized in various ways medicinally for urinary tract infections and as a general analgesic, among other uses (Moerman, 1998).

Several species of *Antistrophus* Walsh, a genus of small, gall-forming wasps in the family Cynipidae, are obligate parasites on the stems of various *Silphium* species (Krombein et al., 1979; Tooker and Hanks, 2004). The female wasp oviposits eggs into various portions of the developing *Silphium* stem (depending on the wasp and host species) and the larvae feed on a specialized lining of nutritious cells within their chamber, which is visible externally as an ellipsoidal to spherical swelling of the stem. In contrast, the larvae of most other *Silphium* stem-boring insects, such as the beetle *Mordellistema aethiops* Smith (Mordellidae), form galleries within the stem that do not result in externally evident abnormalities like galls (Tooker and Hanks, 2004). At maturity, the adult *Antistrophus* wasps bore through the stem, leaving a small hole. Complicating the situation, other wasps in various families, such as the Eurytomidae, Ormyridae, Eupelmidae, and Pteromalidae, oviposit their eggs into existing stem galls, and their larvae parasitize the gall-forming *Antistrophus* larvae (Tooker and Hanks, 2004). Apparently, in some cases, these secondary gall inhabitants may themselves be parasitized by yet other wasps.

1. Blades of all or most of the leaves deeply (rarely shallowly) lobed or divided
   .... 3. S. LACINIATUM
1. Leaf blades unlobed, the margins entire or finely to coarsely toothed
   2. Main leaves basal, often 1 or a few attached just above the stem base, long-petiolate, the median and upper portions of the stem naked or with widely spaced, much-reduced, sessile, bractlike leaves ...... 6. S. TEREBINTHINACEUM
   2. Leaves well distributed along the stems
      3. Stems strongly 4-angled (square in cross-section); leaves opposite, most or rarely only the uppermost leaves perfoliate (the bases of adjacent leaves in each pair fused into a leafy cup around the stem)
         .... 4. S. PERFOLIATUM
      3. Stems not or only slightly angled (more or less circular in cross-section); leaves alternate and/or opposite, the bases not perfoliate, although sometimes cordate and clasping the stems
         4. Stems glabrous or moderately to densely hairy or roughened-hairy, the hairs very short (mostly less than 0.5 mm long); ray florets (17–)20–35 .... 2. S. INTEGRIFOLIUM
         4. Stems moderately to densely hairy or roughened-hairy, most of the hairs longer (at least some of the hairs 0.8–2.0 mm long, especially toward the stem tip); ray florets (8–)13–35
            5. Ray florets (8–)13–17(–20), the corolla 15–30 mm long
               .... 1. S. ASTERISCUS
            5. Ray florets 20–35, the corolla 30–45 mm long .... 5. S. RADULA

1. Silphium asteriscus L. **var. asteriscus**
   (starry rosinweed)
   S. asteriscus var. *scabrum* Nutt.
   S. asperrimum Hook.
   S. gatesii C. Mohr

Pl. 289 a, b; Map 1226

Plants with short, stout rhizomes. Stems 30–100(–150) cm long, solitary or appearing clustered, more or less circular in cross-section (often finely many-angled), moderately to densely roughened-hairy (many of the hairs 0.8–2.0 mm long, especially toward the stem tip). Leaves relatively thin to relatively thick but not leathery, the surfaces moderately to strongly roughened or short-hairy,

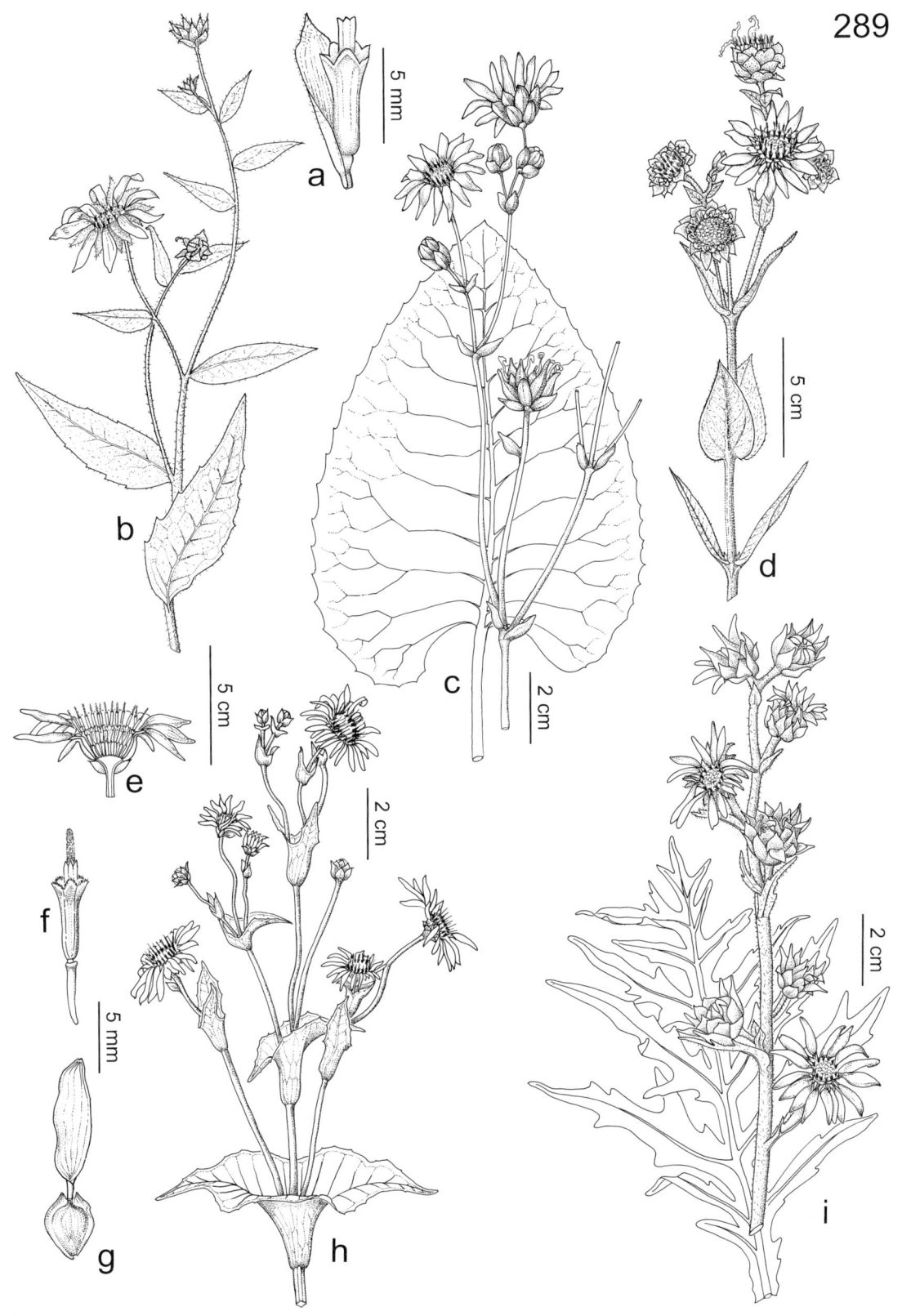

Plate 289. Asteraceae. *Silphium asteriscus*, **a)** disk floret with bract, **b)** habit. *Silphium terebinthinaceum*, **c)** inflorescence and basal leaf. *Silphium integrifolium*, **d)** inflorescence. *Silphium perfoliatum*, **e)** head (longitudinal section), **f)** disc floret, **g)** ray floret, **h)** inflorescence. *Silphium laciniatum*, **i)** inflorescence and basal leaf.

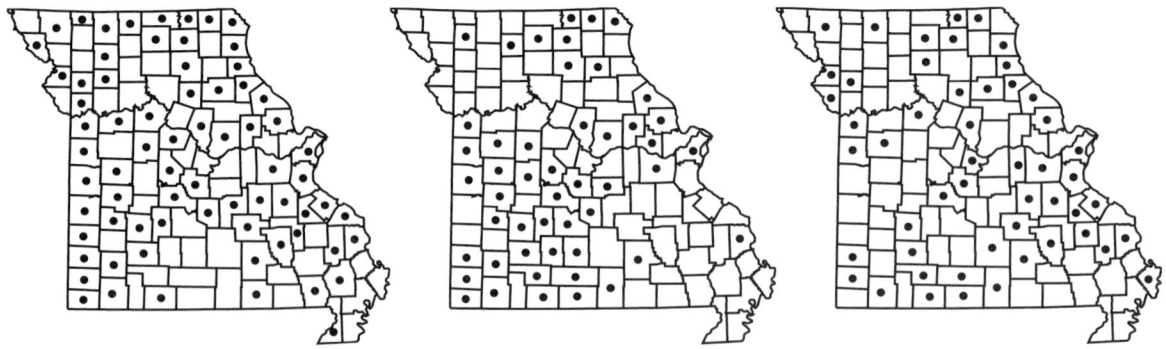

1227. Silphium integrifolium   1228. Silphium laciniatum   1229. Silphium perfoliatum

the hairs often pustular-based. Basal and lowermost stem leaves sometimes absent or withered at flowering, long-petiolate, the bases slightly expanded and somewhat wrapped around the stem, the blade 12–25 cm long, elliptic to narrowly ovate or lanceolate, unlobed, tapered to a usually sharply pointed tip, tapered at the base, the margins entire or finely to coarsely toothed and usually with minute, spreading to loosely appressed hairs. Stem leaves somewhat reduced toward the tip of the stem, alternate to less commonly mostly opposite, 2–15 cm long, sessile or the lower leaves short- to less commonly long-petiolate, lanceolate or narrowly elliptic to ovate or broadly ovate, rarely somewhat heart-shaped, angled, tapered, rounded, or occasionally shallowly cordate at the base, occasionally clasping the stem (but not perfoliate), angled or tapered to a sharply pointed tip. Inflorescences loose, open clusters or panicles, rarely of solitary heads, the heads long-stalked to short-stalked or nearly sessile. Involucral bracts 12–26, 10–20 mm long, elliptic to ovate, spreading to somewhat recurved at the sharply pointed tip, the surfaces moderately to densely hairy or roughened with often pustular-based, more or less spreading hairs, often also with stalked and/or sessile glands, the margins fringed with short, spreading to loosely ascending hairs. Receptacle 1.0–2.2 cm in diameter. Ray florets (8–)13–17(–20), the corolla 15–30 mm long. Disc florets 40 to numerous, the corolla 5–6 mm long, slightly shorter than or extending slightly beyond the tips of the chaffy bracts. Fruits 5–15 mm long, the surfaces glabrous or with fine, curled hairs at the tip, the angles with narrow to broad, lighter wings, each wing rounded or tapered at the tip, the fruit usually with a shallow, semicircular to broadly U-shaped apical notch. $2n=14$. May–September.

Scattered, mostly in the Ozark Division (southeastern U.S. west to Oklahoma and Texas). Glades, tops of bluffs, sinkholes, and openings of dry upland forests; also pastures, ditches, railroads, and roadsides.

Cronquist (1980) segregated the Ozarkian plants as var. *scabrum,* based on the observation that some specimens have chaffy bracts that are stalked-glandular on the outer surface in addition to having nonglandular hairs. Clevinger (2004), in her broader view of species limits in rosinweeds, reduced four other *Silphium* species to varieties of *S. asteriscus* but did not specifically address the epithet var. *scabrum*. However, it is clear from her annotations on the Missouri collections that she intended to include them in her concept of var. *asteriscus*. Examination of the Missouri materials reveals that many specimens do have at least a few stalked, reddish brown glands on the chaffy bracts, although other specimens have only short, white, nonglandular hairs toward the bract tips. Clevinger attributed such ambiguities to introgressive hybridization between the varieties in their zones of geographic overlap followed by a gradual diffusion of genetic traits acquired from a different infraspecific taxon through the range of a given variety. The situation requires further study.

2. Silphium integrifolium Michx. (rosinweed)

Pl. 289 d; Map 1227

Plants with short, stout rhizomes. Stems 40–200 cm long, solitary or more commonly appearing clustered, more or less circular in cross-section (often finely many-angled), glabrous or moderately to densely pubescent (variously velvety to roughened) with relatively short (mostly 0.1–0.5 mm), slender to stout hairs, if glabrous then sometimes strongly glaucous. Leaves usually relatively thick and sometimes somewhat leathery, the upper surface glabrous or sparsely to strongly roughened, the undersurface glabrous or moderately to densely roughened or hairy with short, sometimes pustular-based hairs, when glabrous both surfaces sometimes moderately to strongly glaucous. Basal and lowermost stem leaves absent or withered at flowering, with a winged, more or less petiolar base, the bases slightly expanded and somewhat clasping or wrapped around the stem, the blade 7–15 cm long, ovate to elliptic-ovate or narrowly ovate, unlobed, tapered to a usually sharply pointed tip, tapered at the base, the margins entire or finely to occasionally coarsely toothed and usually with minute, more or less appressed hairs. Stem leaves somewhat reduced toward the tip of the stem, opposite (rarely alternate or in whorls of 3), 3–17 cm long, sessile, lanceolate to ovate, broadly ovate, or somewhat heart-shaped, angled to tapered or more commonly rounded to shallowly cordate at the base, sometimes clasping the stem (but not perfoliate), angled or tapered to a sharply pointed tip. Inflorescences loose, open clusters or panicles, rarely of solitary heads, the heads long-stalked to short-stalked or nearly sessile. Involucral bracts 25–38, 10–22 mm long, elliptic to ovate, spreading at the sharply pointed tip, the surfaces glabrous or moderately to densely short-hairy, sometimes also with stalked and/or sessile glands, the margins glabrous or more commonly with short, ascending to spreading hairs. Receptacle 1.5–2.5 cm in diameter. Ray florets (17–)20–35, the corolla 20–50 mm long. Disc florets numerous, the corolla 4.5–5.5 mm long, slightly shorter than or extending slightly beyond the tips of the chaffy bracts. Fruits 8–12 mm long, the surfaces glabrous or inconspicuously but densely hairy, the angles with relatively broad, lighter wings, each wing rounded to somewhat angled at the tip, the fruit usually with a narrowly U-shaped apical notch. $2n=14$. July–September.

Scattered nearly throughout the state (Wyoming to New Mexico east to Michigan and Alabama; Canada). Upland prairies, loess hill prairies, openings of mesic to dry upland forests, savannas, tops of bluffs, glades, banks of streams and rivers, and rarely margins of ponds and lakes; also pastures, edges of crop fields, ditches, railroads, and roadsides.

Apparent hybrids between *S. integrifolium* and *S. perfoliatum* have been collected in Bates and Warren Counties and are to be expected sporadically at other sites where the two putative parents grow in proximity.

1. Stems and leaves not glaucous; leaf blades with the undersurface roughened, hairy, or rarely glabrous; involucral bracts with the outer surface and margins hairy
.... 2A. VAR. INTEGRIFOLIUM

1. Stems and often also the leaves glaucous; leaf blades with the undersurface glabrous or sparsely to moderately hairy only along the midvein; involucral bracts with the outer surface glabrous, the margins sometimes minutely hairy
.... 2B. VAR. LAEVE

2a. var. integrifolium

S. integrifolium var. *deamii* L.M. Perry

Stems variously hairy or rarely glabrous, but not glaucous, appearing green, purplish-tinged, or brownish-tinged. Leaf blades with the upper surface sparsely to strongly roughened, the undersurface roughened, hairy, or rarely glabrous, not glaucous. Involucral bracts with both surfaces moderately to densely roughened or hairy, the margins with minute, loosely ascending hairs. $2n=14$. July–September.

Scattered nearly throughout the state (South Dakota to Texas east to Michigan and Alabama; Canada). Upland prairies, openings of mesic to dry upland forests, savannas, glades, banks of streams and rivers, and rarely margins of ponds and lakes; also pastures, edges of crop fields, ditches, railroads, and roadsides.

Plants with the leaves tending to be densely velvety and the involucral bracts more or less glandular have been called f. *deamii* (L.M. Perry) Steyerm. Settle and Fisher (1970) upheld earlier treatments of this taxon as a variety, but it does not seem distinct enough morphologically nor to have a sufficiently discrete range to warrant taxonomic recognition.

2b. var. laeve Torr. & A. Gray

S. speciosum Nutt.

Stems glabrous, strongly glaucous, appearing silvery or bluish green. Leaf blades with the upper surface glabrous, the undersurface glabrous or sparsely to moderately hairy only along the midvein, both surfaces moderately to strongly glaucous. Involucral bracts with both surfaces glabrous, the margins with minute, loosely ascending hairs. $2n=14$. July–September.

Scattered in counties along the western margin of the state and uncommon sporadically farther east (Wyoming to New Mexico east to Wisconsin and Arkansas). Loess hill prairies, openings of mesic to dry upland forests, and tops of bluffs; also railroads.

This taxon has been regarded as a separate species by some botanists (Settle and Fisher, 1970). As noted by Steyermark (1963), it is striking and distinctive, but at least a few intermediates with var.

integrifolium exist in Missouri for every character said to distinguish the two taxa.

3. Silphium laciniatum L. (compass plant)

Pl. 289 i; Map 1228

Plants with woody taproots. Stems (40–)80–200(–250) cm long, usually solitary, more or less circular in cross-section (often finely many-angled), moderately roughened-pubescent with relatively long (mostly 1–4 mm), slender, pustular-based hairs, often also with minute, spreading hairs, some of these usually gland-tipped. Leaves thick and leathery, the surfaces sparsely to moderately roughened-pubescent with slender, spreading, pustular-based hairs, often also dotted with scattered, sessile to impressed glands. Basal leaves present at flowering, long-petiolate, the blade 30–60 cm long, ovate in outline, 1 time or more commonly 2 times pinnately deeply (rarely shallowly) lobed or divided, the lobes or divisions narrowly oblong to oblong-triangular, mostly tapered to a sharply pointed tip, broadly attached at the base, the margins otherwise entire or few-toothed and with minute, appressed hairs. Stem leaves progressively reduced from the stem base, alternate, the lowermost leaves similar to the basal leaves, the uppermost leaves 4–15 cm long, mostly short-petiolate to nearly sessile, the blade 1 time pinnately lobed, rarely merely toothed. Inflorescences racemes or more commonly narrow, racemelike, short-branched panicles, the heads short-stalked to nearly sessile. Involucral bracts 25–45, 20–40 mm long, ovate, often spreading or recurved at the sharply pointed tip, the outer surface sparsely to densely hairy and usually also glandular, the margins with a conspicuous fringe of dense, spreading hairs. Receptacle 2–3 cm in diameter. Ray florets 20–35, the corolla 20–50 mm long. Disc florets numerous, the corolla 6–7 mm long, usually extending beyond the tips of the chaffy bracts. Fruits 8–14 mm long, the surfaces glabrous, but usually strongly gland-dotted, the angles with relatively broad, slightly lighter wings, each wing incurved and abruptly angled at the tip, the fruit with a relatively deep, sometimes narrowly U-shaped to nearly square apical notch. $2n=14$. July–September.

Scattered nearly throughout the state, but nearly absent from the southeastern quarter of Missouri (South Dakota to New Mexico east to New York and Alabama; Canada). Glades, upland prairies, savannas, and openings of dry upland forests; also railroads and roadsides.

The name compass plant refers to the habit of the large, relatively flat, strongly ascending basal leaves to orient themselves with the surfaces facing east and west (the lateral margins north and south), taking maximum advantage of the sun's rays. Fisher (1959) characterized rare sterile hybrids of intermediate morphology between *S. laciniatum* and *S. terebinthinaceum* from mixed populations in Illinois and Indiana. Subsequently, Fisher (1966) reported a more widespread pattern of hybridization between these species in Ohio. Steyermark (1963) remarked upon this phenomenon but did not locate any instances of such hybridization in Missouri. However, Redfearn (1980) reported such hybrids from Dallas and Taney Counties, as well as from northern Arkansas.

4. Silphium perfoliatum L. (cup plant, cup rosinweed)

Pl. 289 e–h; Map 1229

Plants with rhizomes. Stems 100–250 cm long, solitary or more commonly appearing clustered, square in cross-section (strongly 4-angled), glabrous or rarely sparsely pubescent with relatively short (mostly 0.2–0.5 mm), slender hairs toward the base (at a few of the lowermost nodes), occasionally slightly glaucous. Leaves usually relatively thick but not or only slightly leathery, the upper surface moderately roughened with small pustules and sometimes also with sparse, minute hairs, the undersurface moderately to densely roughened-pubescent with a mixture of short and minute, spreading, mostly pustular-based hairs. Basal and lowermost stem leaves absent or withered at flowering, long-petiolate, the petioles often expanded into a pair of basal auricles, these wrapped around the stem and those of the stem leaves often more or less perfoliate, the blade 10–30 cm long, ovate to triangular-ovate, unlobed, tapered to a usually sharply pointed tip, tapered to angled at the base, the margins otherwise finely to coarsely toothed and with minute, more or less appressed hairs. Stem leaves progressively reduced from about the midpoint of the stem, opposite (the uppermost leaves rarely alternate), 3–35 cm long, the largest pairs of leaves with short to long, broadly winged petioles, these expanded toward the base and strongly perfoliate (fused into a leafy cup around the stem) or rarely not perfoliate, the median and upper leaves mostly short-petiolate to sessile. Inflorescences loose, open clusters or panicles, the heads long-stalked to short-stalked or nearly sessile. Involucral bracts 25–38, 12–27 mm long, elliptic to ovate, loosely ascending to somewhat spreading at the bluntly to sharply pointed tip, the outer surface usually glabrous, the margins with minute, ascending hairs. Receptacle 1.5–2.5 cm in diameter. Ray florets 18–35, the corolla 15–40 mm long. Disc florets numerous, the corolla 6–7 mm long, usually extending slightly beyond the tips of

1230. Silphium radula

1231. Silphium terebinthinaceum

1232. Smallanthus uvedalius

the chaffy bracts. Fruits 10–15 mm long, the surfaces glabrous, the angles with relatively broad, lighter wings (tapered toward the fruit base), each wing irregularly rounded and minutely hairy at the tip, the fruit with a broadly rounded apical notch. $2n=14$. July–September.

Scattered nearly throughout the state but uncommon in the Mississippi Lowlands Division (eastern U.S. west to North Dakota and Oklahoma; Canada; introduced in Europe). Banks of streams and rivers, bottomland forests, and margins of ponds and lakes; also edges of crop fields, railroads, and roadsides.

Some botanists have divided S. perfoliatum into two varieties, but the utility in maintaining these is not evident. The var. connatum (L.) Cronquist (ssp. connatum (L.) Cruden), differs from var. perfoliatum in having at least the upper portion of the stem and the stalks of the heads with spreading hairs, sometimes longer (1–2 mm) hairs on the leaf undersurface, more of the leaves sessile, and often fewer (8–13) ray florets. It is said to occur in the mountains and Piedmont portions of North Carolina, Virginia, and West Virginia. However, Cruden (1962) reported slightly hairy (intermediate) plants from the extreme western and eastern portions of the species range and called these f. *hornemannii* (Schrad.) Cruden. Additionally, Steyermark (1963) included a rare Missouri variant from Stone County with only the uppermost leaves perfoliate, which he called f. *petiolatum* E.J. Palmer & Steyerm. Patterns of overall morphological variation within the species appear to be more complex than earlier authors have understood.

5. Silphium radula Nutt. var. radula (rough-leaved rosinweed)

Map 1230

Plants with short, stout rhizomes. Stems 40–250 cm long, solitary or more commonly appearing clustered, more or less circular in cross-section (often finely many-angled), moderately to densely roughened-hairy (many of the hairs 0.8–2.0 mm long, especially toward the stem tip). Leaves relatively thin to relatively thick but not leathery, the surfaces moderately to strongly roughened or less commonly short-hairy, the hairs often pustular-based. Basal and lowermost stem leaves usually absent or withered at flowering, with a more or less winged, short to moderately long petiole, the bases slightly expanded and somewhat clasping or wrapped around the stem, the blade 7–18 cm long, ovate to elliptic-ovate or narrowly ovate, unlobed, tapered to a usually sharply pointed tip, tapered at the base, the margins entire or finely to occasionally coarsely toothed and usually with minute, more or less appressed hairs. Stem leaves somewhat reduced toward the tip of the stem, alternate to mostly opposite, 2–22 cm long, sessile, lanceolate to ovate, broadly ovate, or somewhat heart-shaped, angled to tapered or more commonly rounded to shallowly cordate at the base, sometimes clasping the stem (but not perfoliate), angled or tapered to a sharply pointed tip. Inflorescences loose, open clusters or panicles, rarely of solitary heads, the heads long-stalked to short-stalked or nearly sessile. Involucral bracts 18–38, 12–25 mm long, elliptic to ovate, spreading to somewhat recurved at the sharply pointed tip, the surfaces moderately to densely roughened with short, pustular-based, more or less spreading hairs, the margins glabrous or more commonly with short, ascending hairs. Receptacle 1.5–2.8 cm in diameter. Ray florets 20–35, the corolla 30–45 mm long. Disc florets numerous, the corolla 5.5–6.5 mm long, slightly shorter than or extending slightly beyond the tips of the chaffy bracts. Fruits 10–16 mm long, the surfaces glabrous or finely but densely hairy toward the tip, the angles with relatively broad, lighter wings, each wing sharply angled at the tip, the fruit usually with a deep, semicircular to U-shaped apical notch. $2n=14$. June–September.

Possibly introduced, uncommon, known thus far from a single collection from Vernon County (Kansas to Texas east to Georgia). Roadsides, but presumably also upland prairies.

The sole collection thus far from Missouri was made by Norlan Henderson in 1965 along a highway near Nevada, in a region where frequent remnant patches of upland prairie continue to persist along roadsides. The specimen label regrettably does not contain information on the plant community present along the roadside at the collection site, and the population has not been rediscovered in recent years. This specimen initially was determined as *S. integrifolium,* but it was annotated correctly as *S. radula* var. *radula* by Jennifer Clevinger in 1999 during her doctoral studies on the genus at the University of Texas. Clevinger (2004) treated *S. radula* as comprising two varieties, var. *radula* and var. *gracile* (A. Gray) J.A. Clevinger, the latter of which is restricted to portions of Oklahoma, Texas, Arkansas, and Louisiana and differs in its shorter stems, fewer (12–18) ray florets, and more persistent basal leaves.

At one time, the name *S. asperrimum* Hook. was misapplied to this species, but that epithet is more properly regarded as a synonym of *S. asteriscus* (Cronquist, 1980). Perry (1937) reported *S. asperrimum* from Missouri based on a historical collection by B. F. Bush from Dunklin County. Steyermark (1963) redetermined this specimen as *S. asteriscus* and excluded *S. radula* (as *S. asperrimum*) from the Missouri flora. His redetermination of this specimen appears correct. The 1965 record from southwestern Missouri has not previously been reported in the literature.

6. Silphium terebinthinaceum Jacq. (prairie dock)

Pl. 289 c; Map 1231

Plants with woody taproots. Stems (40–)80–200(–250) cm long, usually solitary, more or less circular in cross-section (often finely many-angled), glabrous, sometimes slightly glaucous. Leaves thick and leathery, the surfaces glabrous or more commonly moderately roughened-pubescent with slender, spreading, strongly pustular-based hairs. Basal leaves present at flowering, long-petiolate, the blade (10–)15–50 cm long, oblong-ovate, ovate-triangular, more or less heart-shaped, or nearly circular, unlobed, tapered, rounded, truncate, or more commonly shallowly to deeply cordate at the base, rounded to angled or short-tapered to a bluntly or sharply pointed tip, the margins otherwise coarsely toothed (rarely shallowly toothed) and with minute, appressed hairs. Stem leaves mostly restricted to 1 or a few leaves similar to the basal leaves and positioned near the stem base, the median and upper portions of the stem naked or with widely spaced, much-reduced, sessile, bractlike leaves, these alternate, 1–3 cm long, sessile, ovate, the base wrapped around the stem, the margins usually entire. Inflorescences loose, open clusters or panicles, the heads short- to long-stalked or sometimes nearly sessile. Involucral bracts 21–35, 13–25 mm long, those of the shorter outer series mostly broadly elliptic, those of the longer inner series oblong-elliptic, usually ascending at the rounded or broadly angled tip, the outer surface glabrous, the margins sometimes with a minute fringe of dense, spreading hairs. Receptacle 1.5–2.5 cm in diameter. Ray florets (13–)15–21(–29), the corolla 17–30 mm long. Disc florets numerous, the corolla 6–7 mm long, slightly shorter than to about as long as the chaffy bracts. Fruits 7–13 mm long, the surface glabrous or finely hairy, the angles with relatively broad, lighter wings, each wing incurved but abruptly truncate at the tip, the fruit with a U-shaped to semicircular apical notch. $2n=14$. July–October.

Scattered, mostly in the Ozark and Ozark Border divisions (Iowa to Arkansas east to Ohio, Virginia, and Georgia; Canada). Glades, upland prairies, tops of bluffs, savannas, openings of dry upland forests, and rarely banks of streams; also old fields, railroads, and roadsides.

Steyermark (1963) noted that colonies of vegetative rosettes with few or no flowering stems are frequently encountered. He also mentioned that grazing mammals find the foliage palatable. Some botanists divide *S. terebinthinaceum* into two varieties, with the widespread (including Missouri) var. *terebinthinaceum* having unlobed, toothed leaves vs. the deeply pinnately lobed or divided leaves of var. *pinnatifidum* (Elliott) A. Gray, which occurs sporadically to the east of Missouri. However, Fisher and Speer (1978) defended this taxon as a separate species, rather than a variety of *S. terebinthinaceum* or a fertile hybrid between that species and *S. laciniatum,* as suggested by some earlier authors. For further discussion of this hybrid, see the treatment of *S. laciniatum.*

91. Smallanthus Mack.

About 20 species, North America to South America.

Most authors (Steyermark, 1963; Wells, 1965; Barkley, 1986; Gleason and Cronquist, 1991) have included the species of *Smallanthus* in *Polymnia,* but Robinson (1978) noted differences in morphology and anatomy of the achenes, pubescence patterns of the florets, and chromosomal base number that appear to justify separating the two genera.

1. Smallanthus uvedalius (L.) Mack.
(bearsfoot, yellow-flowered leaf cup)
Polymnia uvedalia (L.) L.
P. uvedalia var. *densipilis* S.F. Blake
Pl. 290 g, h; Map 1232

Plants perennial herbs, with fibrous or sometimes somewhat tuberous roots. Stems 40–150(–250) cm long, erect or ascending, usually several-branched, with fine longitudinal lines or ridges, moderately to densely pubescent with more or less spreading, nonglandular and/or gland-tipped hairs toward the tip, often glabrous or nearly so toward the base. Leaves opposite or the uppermost few alternate, sessile or short-petiolate (below the tapered, winglike blade base), the bases slightly expanded and wrapping around the stem. Leaf blades 3–35(–55) cm long, ovate to broadly ovate or broadly obovate, more or less palmately 3- or 5-lobed, tapered (abruptly long-tapered on larger leaves) at the base, the main lobes tapered to a sharply pointed tip, the margins coarsely and irregularly few-toothed or shallowly few-lobed and hairy, the upper surface sparsely to moderately roughened-pubescent with slender, stiff, somewhat curved hairs, the undersurface with moderate to dense, short, somewhat spreading hairs, especially along the veins, also with sessile, spherical, yellow glands. Inflorescences of relatively open, irregular clusters at the branch tips, these sometimes appearing as small panicles, subtended by leaflike bracts mostly in pairs at the branch points, the heads mostly with relatively short, hairy and/or glandular stalks. Heads radiate. Involucre 10–20 mm long, 8–15 mm in diameter, cup-shaped, the bracts in 1 series. Involucral bracts 4–6, loosely ascending, slightly longer than the outermost chaffy bracts, broadly lanceolate to more commonly elliptic or broadly ovate, rounded or narrowed to an abrupt, minute, blunt or sharp point at the tip, green and somewhat leaflike, the margins with spreading hairs, the outer surface glabrous or sparsely to moderately pubescent with short, fine hairs, often also sparsely to moderately glandular. Receptacle flat or slightly convex, not elongating as the fruits mature, with chaffy bracts subtending the ray and disc florets, those subtending the ray florets slightly longer than the others, 6–14 mm long, lanceolate to ovate, thin and papery but usually with a greenish-tinged apical portion, concave and wrapped around the florets. Ray florets 7–13, pistillate (with a 2-branched style exserted from the short tube at flowering), the corolla 12–25(–30) mm long, relatively broad and often slightly cupped, yellow, the short, tubular base moderately short-hairy and/or glandular, the outer surface of the ligule often minutely glandular, withered and often not persistent at fruiting. Disc florets 45–80, staminate (with a small, stalklike ovary and a more or less undivided style), the corolla 7–10 mm long, yellow, sparsely glandular or hairy, not expanded at the base. Style branches with the sterile tip somewhat elongate and tapered. Pappus absent. Fruits 5.5–6.5 mm long, elliptic-obovate in outline, slightly flattened, with numerous shallow longitudinal grooves, the surface otherwise glabrous, dark brown to purplish black, somewhat shiny. $2n=32$. July–September.

Scattered in the Ozark, Ozark Border, Big Rivers, and Mississippi Lowlands Divisions (eastern U.S. west to Illinois, Kansas, and Texas; Caribbean Islands). Bottomland forests, mesic upland forests in ravines, bases and ledges of bluffs, and banks of streams and rivers; also fencerows, pastures, railroads, and roadsides.

Steyermark (1963) recognized two varieties, which differ in whether the inflorescence branches had glandular (var. *uvedalia*) or mostly nonglandular (var. *densipilis*) hairs. Wells (1969) studied variation within the species and concluded that, at least in part, the pubescence patterns said to differentiate the three named variants that he had accepted earlier (Wells, 1965) are under environmental rather than strictly genetic control. He concluded that recognition of infraspecific taxa within the species was unwarranted. Turner's (1988b) research on Mexican populations supported this view.

92. Tagetes L. (marigold)

Plants annual (perennials elsewhere), aromatic, with taproots. Stems erect or ascending, unbranched to several-branched, with several fine longitudinal ridges and grooves, glabrous. Leaves mostly alternate, the lowermost few sometimes opposite, mostly sessile, the bases of each pair slightly expanded and wrapped around the stem. Leaf blades 3–12 cm long, elliptic to ovate in outline, 1 or 2 times pinnately dissected with 9–25 primary lobes, the ultimate segments linear to narrowly lanceolate, the margins otherwise entire or with a few irregular teeth, the surfaces glabrous and dotted with scattered, brownish yellow, sessile oil glands. Inflorescences of solitary, mostly long-stalked heads at the branch tips or less commonly axillary from the uppermost leaves. Heads radiate (discoid elsewhere). Involucre cup-shaped to slightly bell-shaped, the disc 5–12 mm in diameter, the bracts in 1 series. Involucral bracts 7–10, fused more than ⅔ of the way to the tip, with triangular teeth, the surface glabrous but dotted or lined with conspicuous, sessile, yellowish brown oil glands, somewhat thin and membranous, green but often pinkish- or purplish-tinged. Receptacle convex or somewhat conical, not or only slightly elongating as the fruits mature, with minute, irregular ridges around the attachment points of the florets. Ray florets 5–8 (commonly numerous in doubled horticultural forms), pistillate, showy, the corolla 10–30 mm long, relatively broad, yellow to orange, sometimes red or reddish brown toward the base. Disc florets 40–120, perfect, the corolla greenish yellow to orange, the tube slender, not expanded at the base nor persistent at fruiting, glabrous, the lobes sometimes hairy along the margins. Style branches with the sterile tip elongate and with a small, more or less spherical, expanded portion at the very tip. Pappus of 5–12 slender, unequal scales, these sometimes fused at the base, straw-colored or orangish brown, the margins irregular or more commonly with ascending barbs or teeth. Fruits linear to very narrowly wedge-shaped in outline, 4- or 5-angled and slightly flattened, the surface usually moderately pubescent with short, ascending hairs, especially along the angles, black. About 40 species, North America to South America; introduced in the Old World.

The two species of marigold in the Missouri flora are both escapes from cultivation and are sometimes difficult to differentiate (Neher, 1966). Some botanists have treated them as components of a single species. A third, very different, cultivated species of *Tagetes* is likely to be encountered in the future as an escape. *Tagetes minuta* L. (variously known as chinchilla, anisillo, tall khakiweed, Mexican marigold, and stinkweed, among other names) is widely grown as a seasoning for meats, soups, and vegetables. It is a densely branched annual with erect stems 30–100 cm long and pinnately dissected leaves. However, it differs from the other cultivated marigolds in Missouri in its more numerous heads with the slender, nearly cylindrical involucre 8–12 mm long, only 1–3 ray florets with inconspicuous corollas 1–2 mm long, only 3–5 disc florets, and somewhat shorter pappus and fruits.

1. Involucre 17–22 mm long; ray florets with the corolla 15–25 mm long, yellow or orange; stalks of the heads relatively strongly inflated or swollen toward the tip
.... 1. T. ERECTA
1. Involucre 12–16 mm long; ray florets with the corolla 8–15 mm long, yellow or orange but often with a red or reddish brown region toward the base (or nearly entirely so); stalks of the heads not or only slightly inflated or swollen toward the tip... 2. T. PATULA

1. Tagetes erecta L. (common marigold, African marigold)

Pl. 290 e, f; Map 1233

Stems 10–50(–120) cm long. Heads mostly 6–10 cm in diameter (including the spreading ray corollas), the stalk relatively strongly inflated or swollen toward the tip, often hollow. Involucre 17–22 mm long. Ray florets with the corolla 15–25 mm long, yellow or orange. Disc florets with the corolla 10–16 mm long. Pappus scales variously 5–12 mm long. Fruits 7–11 mm long. $2n=24$. July–October.

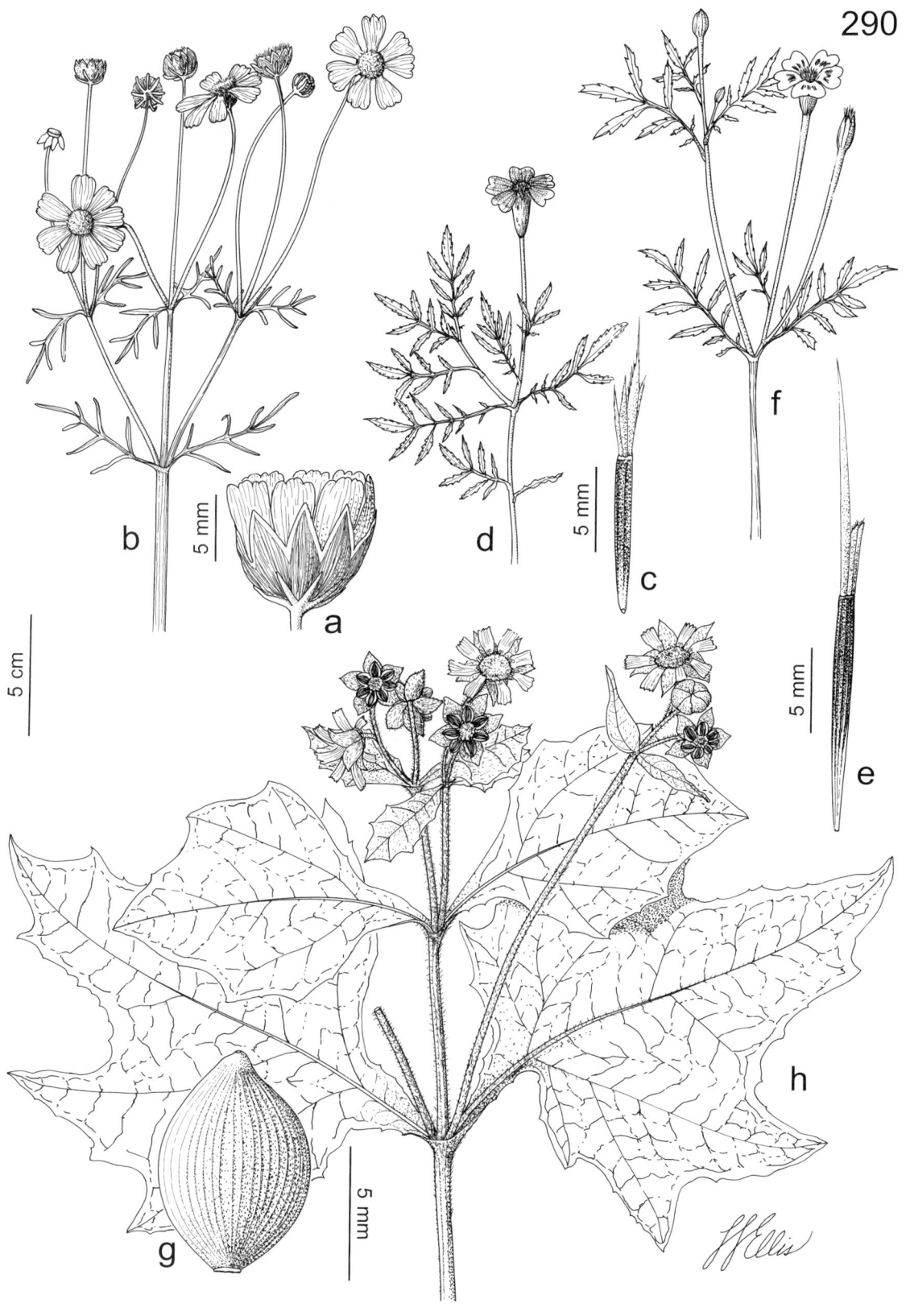

Plate 290. Asteraceae. *Thelesperma ambiguum*, **a)** head, **b)** habit. *Tagetes patula*, **c)** fruit, **d)** habit. *Tagetes erecta*, **e)** fruit, **f)** habit. *Smallanthus uvedalius*, **g)** fruit, **h)** habit.

1233. Tagetes erecta   1234. Tagetes patula   1235. Thelesperma filifolium

Introduced, uncommon and sporadic (native of Mexico, widely introduced in tropical and temperate regions nearly worldwide). Railroads and open, disturbed areas.

The common marigold is cultivated widely as a bedding plant and as a cut flower. It has a number of ceremonial and religious uses in Mexico and has also been used medicinally for a number of ailments and as an insect and tick repellant. A yellow dye can be extracted from the flowers. *Tagetes erecta* is one of four closely related taxa that have been bred and selected intensively in cultivation. Because it is so widely cultivated in Latin America, its native range is poorly understood. This taxon is a diploid ($2n=24$) that may have been crossed with another diploid cultivated species, *T. tenuifolia* Cav., long ago to produce the fertile tetraploid ($2n=48$) taxon *T. patula* (Neher, 1966).

2. Tagetes patula L. (French marigold)

Pl. 290 c, d; Map 1234

Stems 10–80(–120) cm long. Heads mostly 3.5–5.0 cm in diameter (including the spreading ray corollas), the stalk not or only slightly inflated or swollen toward the tip, hollow or less commonly solid. Involucre 12–16 mm long. Ray florets with the corolla 8–15 mm long, yellow or orange but often with a red or reddish brown region toward the base, sometimes nearly uniformly reddish brown. Disc florets with the corolla 8–13 mm long. Pappus scales variously 5–10 mm long. Fruits 7–9 mm long. $2n=48$. July–October.

Introduced, uncommon, known thus far only from the city of St. Louis (probably native of Mexico and Central America, widely introduced in tropical and temperate regions nearly worldwide). Railroads.

This species was first reported for the state by Mühlenbach (1979), based on his botanical surveys of the St. Louis railyards. Its uses are similar to those of *T. erecta*, discussed above. Because it is a tetraploid ($2n=48$), presumably derived following past hybridization between diploid progenitors, it is unclear whether this species ever had a native distribution in Latin America (Neher, 1966).

93. Thelesperma Less.

Plants annual or perennial herbs. Stems erect or ascending, unbranched or few- to several-branched, with several fine longitudinal lines or ridges, glabrous or sparsely pubescent with short, more or less spreading hairs at the nodes. Leaves opposite and usually also basal, the uppermost few leaves rarely alternate, short-petiolate to nearly sessile, the petiole sometimes with short, sparse, spreading hairs toward the base, the bases slightly expanded and wrapping around the stem. Leaf blades mostly 1–3 times pinnately dissected, those of the uppermost leaves rarely simple, obovate to fan-shaped in outline (narrowly linear in simple leaves), the ultimate segments narrowly linear (threadlike) or rarely very narrowly oblanceolate, the margins otherwise entire and usually appearing somewhat curled under, glabrous, not glandular. Inflorescences of solitary terminal heads, the heads with long, bractless stalks. Heads radiate or discoid. Involucre cup-shaped to somewhat urn-shaped, the bracts in 2 dissimilar overlapping series. Involucral bracts of the outer series 3–10, shorter than the others, free except sometimes fused at the base, spreading to somewhat ascending, variously shaped, green, usually with thin, pale margins, glabrous, not glandular, inconspicuously 1(3)-nerved; those of the inner series usually 8, fused 1/4–2/3 of the way from the base to the tip, the free portion lanceolate to ovate-triangular, bluish green or yellowish green, membranous to leathery with narrow to broad, lighter, thinner margins, glabrous or minutely fringed at the tip, not glandular, with several to numerous usually conspicuous nerves or longitudinal lines. Receptacle flat or slightly convex, not elongating as the fruits mature, with chaffy bracts subtending the disc florets, these oblong or narrowly ovate, usually somewhat concave and wrapped around the florets, at least toward the base, white or nearly so with usually 2 purplish or reddish nerves. Ray florets absent or present, when present usually 8, sterile (lacking stamens and style at flowering and with an ovary that is shorter and thinner than those of the disc florets, not developing into a fruit), the corolla showy (sometimes inconspicuous elsewhere), 10–22 mm long, relatively broad, yellow, occasionally reddish brown toward the base, not persistent at fruiting. Disc florets 20–100(–120), perfect, the corolla usually 5-lobed, reddish brown or yellow to orangish yellow with reddish brown veins, not thickened at the base, not persistent at fruiting. Style branches with the sterile tip slightly to moderately elongate and tapered to a usually sharply pointed tip. Pappus of the disc florets of usually 2 short awns, these with downward-angled barbs or fine hairs, persistent at fruiting. Fruits 3–8 mm long, sometimes dissimilar in the same head (the shorter, sometimes somewhat curved outer ones grading into the inner ones) narrowly oblong to narrowly oblong-oblanceolate or nearly linear in outline, circular to slightly (and often unevenly) flattened, more or less truncate at the tip, not beaked, the angles lacking wings, the surfaces smooth or 1 or both with irregular wrinkles, ridges, or rounded tubercles (appearing lumpy), otherwise glabrous, brown or black, the black ones somewhat shiny (and usually smooth). About 15 species, U.S., Mexico.

Thelesperma is morphologically similar to some species groups in *Coreopsis,* differing mainly in the partially united inner series of involucral bracts. Native Americans used species in the genus (mostly *T. megapotamicum*) medicinally to treat tuberculosis, toothache, and various pediatric ailments. They also derived various dyes from the flowering heads and rootstocks and boiled the young foliage to make a tea (Moerman, 1998).

1. Heads radiate, the ray corollas showy; involucre with the outer series of 7–10 bracts 4–8 mm long, linear to narrowly lanceolate-triangular ...... 1. T. FILIFOLIUM
1. Heads discoid (radiate but with inconspicuous ray corollas 3–6 mm long elsewhere); involucre with the outer series of 3–5(6) bracts 1–3 mm long, lanceolate to oblong or ovate .... 2. T. MEGAPOTAMICUM

1. Thelesperma filifolium (Hook.) A. Gray (stiff greenthread)

Pl. 292 a, b; Map 1235

Plants annual or occasionally short-lived perennial herbs, with taproots or less commonly a somewhat branched, vertical rootstock. Stems 20–70 cm long. Leaves well distributed along the stems. Leaf blades 3–12 cm long, all but occasionally those of the uppermost leaves 2 or 3 times pinnately dissected, the ultimate segments 5–30 mm long. Heads radiate. Involucre with the outer series of 7–10 bracts 4–8 mm long, linear to narrowly triangular with a long-tapered tip; the inner series of bracts 7–10 mm long, mostly fused in the basal 1/4–1/3, the free portion lanceolate to narrowly ovate. Ray florets usually 8, the corolla 10–22 mm long, yellow, occasionally tinged reddish brown toward the base. Disc florets with the corollas 5–7 mm long, yellow to orangish yellow with reddish brown veins (entirely reddish brown elsewhere). Pappus absent or of 2 short, stout awns 0.5–1.0(–2.0) mm long. Fruits 3–6 mm long. $2n=16, 18, 20, 22$. May–August.

Uncommon, known thus far only from Greene County, also introduced in Jackson County and the city of St. Louis (Wyoming to New Mexico east to Missouri and Mississippi; Mexico). Glades and tops of bluffs, on limestone substrates; also railroads and open, disturbed areas.

1236. Thelesperma megapotamicum   1237. Verbesina alternifolia   1238. Verbesina encelioides

Steyermark (1963) and some other earlier authors called this taxon *T. trifidum* (Poir.) Britton, but Shinners (1950) explained that the name is of uncertain application and cannot refer to the plants most recent authors have called *T. filifolium,* as Poiret's original description noted alternate leaves. Alexander (1955) similarly excluded the name *T. trifidum,* but without explanation.

Steyermark's (1963) report of this species from Lawrence County could not be substantiated during the present research. Some authors (Barkley, 1986) have accepted var. *intermedium* (Rydb.) Shinners, a weakly separable variant in the western portion of the species range characterized by relatively short stems, relatively long outer involucral bracts, and somewhat lighter yellow ray corollas. Our plants correspond to var. *filifolium,* but the number of intermediate specimens in the region of overlap to the southwest of Missouri makes it impractical to recognize these varieties.

2. Thelesperma megapotamicum (Spreng.) Kuntze (Colorado greenthread, Navajo tea, Hopi tea)

Pl. 292 c, d; Map 1236

Plants perennial herbs, with short, somewhat woody rhizomes. Stems 30–80 cm long. Leaves sometimes more or less crowded below the stem midpoint, in other plants more evenly distributed along most of the stem. Leaf blades 3–12 cm long, all but occasionally those of the uppermost leaves 1 or less commonly 2 times pinnately dissected, the ultimate segments 5–50 mm long. Heads discoid (radiate elsewhere). Involucre with the outer series of 3–5(6) bracts 1–3 mm long, lanceolate to oblong or ovate; the inner series of bracts 7–10 mm long, fused in the basal $1/3$–$2/3$, the free portion narrowly ovate to ovate-triangular. Ray florets absent (rarely present elsewhere but with inconspicuous, yellow corollas 3–6 mm long). Disc florets with the corollas 7–9 mm long, yellow to orangish yellow, usually with reddish brown veins. Pappus absent or of 2 short, stout awns 1–2(–3) mm long. Fruits 5–8 mm long. $2n=22, 44$. May–July.

Introduced, known thus far from Jackson County and the city of St. Louis (southwestern U.S. east to South Dakota and Texas; introduced farther east). Railroads.

Steyermark (1963) reported a record of *T. ambiguum* A. Gray (Pl. 290 a, b) in the appendix to his manual, based on a collection made by Viktor Mühlenbach in the St. Louis railyards. However, Mühlenbach (1979) later reported that the material had been redetermined as *T. filifolium* (as *T. trifidum*). *Thelesperma ambiguum,* which probably is better treated as *T. megapotamicum* var. *ambiguum* (A. Gray) Shinners (Greer, 1997), is native to Texas and Mexico, differing from *T. megapotamicum* mainly in its shorter stems, usually radiate heads, and reddish brown disc corollas.

94. Verbesina L. (crownbeard, wingstem)

Plants annual or perennial herbs (shrubs or trees elsewhere). Stems erect or ascending, unbranched or more commonly several- to many-branched, mostly above the midpoint, with several longitudinal lines or ridges, these often winged, sparsely to densely hairy. Leaves opposite or alternate, sessile or petiolate, the bases sometimes somewhat expanded and wrapping around the stem or decurrent below the attachment point as wings of green tissue. Leaf blades simple, lanceolate to elliptic-lanceolate, ovate, or triangular, tapered to shallowly cordate at the base, angled or tapered to a sharply pointed tip, the margins relatively flat, variously toothed and usually minutely hairy, the surfaces variously hairy, sometimes roughened, not glandular. Inflorescences of solitary terminal heads or more commonly loose, open clusters or panicles, the heads with short to long, usually bractless stalks. Heads radiate (discoid elsewhere). Involucre cup-shaped to somewhat conical or saucer-shaped, the bracts in mostly 1 or 2 (more elsewhere) subequal (the outer series often somewhat shorter), overlapping series. Involucral bracts 9–21, linear to oblanceolate or narrowly ovate, ascending or spreading to reflexed, more or less green, the outer surface variously hairy, not glandular, with usually 1 inconspicuous nerve. Receptacle flat or broadly convex to shallowly conical, not elongating as the fruits mature, with chaffy bracts subtending the ray and disc florets, these narrowly oblong to narrowly oblong-oblanceolate (linear in *V. encelioides*), somewhat concave or folded, and wrapped around the florets. Ray florets 1–15, sterile or pistillate, the corolla usually showy, relatively broad, yellow or white, not persistent at fruiting. Disc florets 8 to numerous (more than 150), perfect, the corolla yellow or white, not expanded at the base or persistent at fruiting. Style branches with the sterile tip somewhat flattened, elongate and tapered to a sharply pointed tip. Pappus of the disc florets of 2 short awns 0.5–3.5 mm long, that of the rays similar (absent in *V. encelioides*), more or less persistent at fruiting (usually shed in *V. encelioides*). Fruits oblanceolate to broadly obovate in outline, flattened, the margins sharply angled or more commonly narrowly to broadly winged, the surface usually with a longitudinal angle or ridge, sometimes also with 1 or 2 additional inconspicuous nerves, finely hairy, dark brown to black, sometimes with fine, lighter mottling, the wings lighter colored, usually not shiny. About 300 species, North America, Central America, South America, Caribbean Islands.

Verbesina is a morphologically variable genus for which a comprehensive modern monograph has yet to be published. Most of the species occur in the American tropics. Our species (especially *V. alternifolia* and *V. virginica*) are noted as frequently producing spectacular formations called frost flowers at the end of the growing season. These are created when sudden overnight freezing temperatures cause the stems to burst and release quantities of sap, which freezes into intricate, layered, petal-like shapes of ice sometimes more than 10 cm long (Steyermark, 1963; Swihart, 2000).

1. Ray and disc florets with the corollas white; disc florets 8–12 . . . . 4. V. VIRGINICA
1. Ray and disc florets with the corollas yellow; disc florets 40 to numerous (more than 150)
    2. Leaves mostly with a well-defined, relatively long petiole; plants annual, with taproots; stems not winged . . . . 2. V. ENCELIOIDES
    2. Leaves sessile or tapered to an often short, poorly differentiated, winged petiole; plants perennial, with fibrous roots and often also rhizomes; stems usually more or less winged, sometimes only toward the base
        3. Heads 8–100 per stem; involucral bracts 8–12, spreading to reflexed at flowering; ray florets 2–10, the corollas strongly drooping at flowering; fruits spreading in all directions, the fruiting head appearing nearly spherical . . . . 1. V. ALTERNIFOLIA
        3. Heads 1–10 per stem; involucral bracts 16–21, ascending at flowering; ray florets 8–15, the corollas spreading at flowering; fruits mostly ascending, the fruiting head appearing more or less hemispherical
            . . . . 3. V. HELIANTHOIDES

1. Verbesina alternifolia (L.) Britton ex Kearney (yellow ironweed)
Actinomeris alternifolia (L.) DC.

Pl. 291 a–c; Map 1237

Plants perennial herbs with fibrous, sometimes slightly fleshy roots and often stout rhizomes. Stems 40–250 cm long, narrowly and sometimes incompletely winged, often only below the midpoint, sparsely to moderately pubescent with short, spreading hairs, especially toward the tip. Leaves alternate or the lowermost leaves sometimes opposite, less commonly all or nearly all of the leaves opposite (see discussion below), sessile or with a short, often poorly differentiated, winged petiole, the base usually minutely to strongly decurrent below the attachment point as a pair of wings. Leaf blades (2–)6–25 cm long, lanceolate to elliptic or occasionally ovate, tapered at the base, tapered to a usually sharply pointed tip, the margins coarsely to finely toothed or sometimes nearly entire, the upper surface strongly roughened with sparse to moderate, short, spreading, pustular-based hairs, the undersurface similarly hairy, but often somewhat less roughened to the touch. Inflorescences usually terminal panicles with 8–100 heads. Involucre 10–14 mm in diameter, more or less saucer-shaped, with 8–12 bracts. Involucral bracts 3–8 mm long, linear to narrowly lanceolate or narrowly oblanceolate, spreading to reflexed at flowering, the outer surface moderately hairy. Chaffy bracts narrowly lanceolate to narrowly oblong-oblanceolate, sparsely to moderately hairy. Ray florets 2–10, sterile (lacking stamens and style at flowering and with an ovary that is shorter and thinner than those of the disc florets, not developing into a fruit), the corolla 10–25(–30) mm long, strongly drooping, yellow. Disc florets 40–80, the corolla 3.5–4.5 mm long, yellow. Pappus of the ray and disc florets similar (but that of the ray florets often slightly shorter and more flattened), of 2 slender awns 1.5–2.0 mm long, smooth or with fine, upward-pointed barbs, more or less persistent at fruiting. Fruits spreading in all directions at maturity (forming a nearly spherical mass), 4–5 mm long, the body usually oblanceolate to broadly obovate, usually relatively broadly winged, less commonly narrowly winged or wingless, the surface glabrous or sparsely to moderately pubescent with fine, slightly pustular-based hairs. $2n=68$. August–October.

Scattered nearly throughout the state (eastern U.S. west to Nebraska and Texas; Canada). Banks of streams and rivers, bases of bluffs, sloughs, bottomland forests, and occasionally mesic upland forests; also pastures and roadsides.

Gleason and Cronquist (1963, 1991) and Cronquist (1980) included Missouri in their distributional summaries of *V. occidentalis* (L.) Walter without further data or discussion. Those reports may have been based on historical specimens collected by John Davis in Marion, Pike, and Ralls Counties and originally determined as that species. Examination of these specimens has resulted in their redetermination as *V. alternifolia,* based on their relatively narrow, spreading involucral bracts, sterile ray florets, and (where visible) partially spreading to reflexed, winged developing fruits. The two species appear superficially similar, based on leaf shape, degree of stem wingedness, size of the flowering heads, and number, size, and color of the ray florets. Part of the confusion may be because published descriptions of *V. occidentalis* indicate the species to have opposite leaves, whereas *V. alternifolia* has been characterized as having alternate leaves. In eastern Missouri there are a number of populations in which the stems have mostly opposite leaves, especially toward the tips (which are all that many botanists collect in making specimens). In the 1990s, Carl Darigo and James Sullivan of the Webster Groves Nature Study Society assembled an excellent series of collections (housed at the Missouri Botanical Garden Herbarium) to document the morphology of these plants. Although their samples have all or mostly opposite leaves, they all have heads typical of *V. alternifolia*. Careful field studies may yet confirm the mostly southeastern *V. occidentalis* as occurring in Missouri, as it grows in adjacent portions of Illinois, Kentucky, and Tennessee (as well as Oklahoma), but for now it must be excluded from the state's flora.

2. Verbesina encelioides (Cav.) Benth. & Hook. f. ex A. Gray (golden crownbeard)
Ximenesia encelioides Cav.
V. encelioides ssp. *exauriculata* (B.L. Rob. & Greenm.) J.R. Coleman
V. encelioides var. *exauriculata* B.L. Rob. & Greenm.

Pl. 291 d, e; Map 1238

Plants annual with taproots. Stems 7–80(–150) cm long, not winged, densely pubescent with short, mostly straight, loosely appressed hairs, especially toward the tip. Leaves alternate or the lowermost leaves sometimes opposite, short-petiolate or more commonly with a well-defined, long petiole, the base of most or sometimes only the uppermost leaves with small to conspicuous, oblong or obovate to nearly circular auricles of leafy tissue (these rarely absent). Leaf blades 1–14 cm long, narrowly

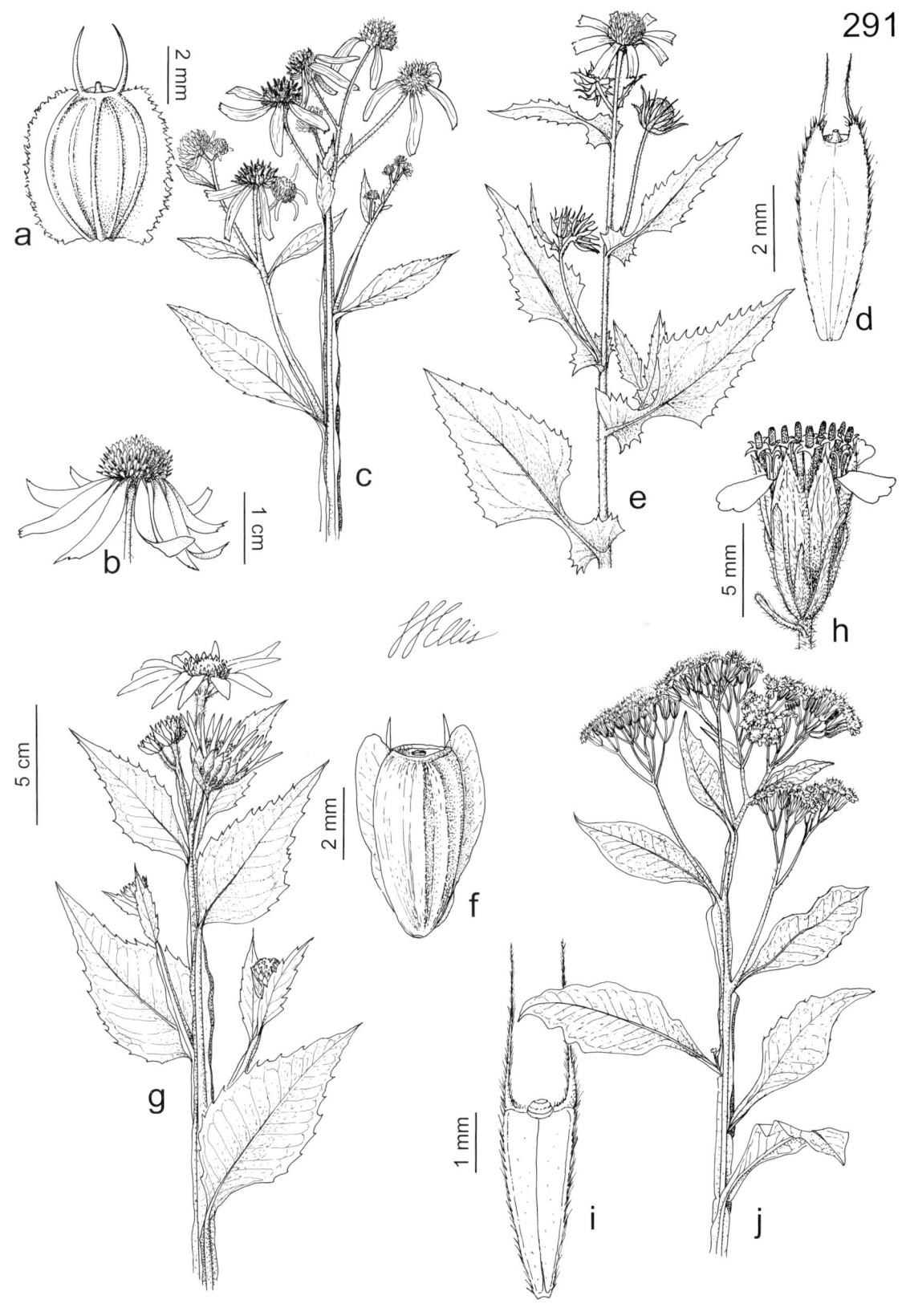

Plate 291. Asteraceae. *Verbesina alternifolia*, **a)** fruit, **b)** head, **c)** habit. *Verbesina encelioides*, **d)** fruit, **e)** habit. *Verbesina helianthoides*, **f)** fruit, **g)** habit. *Verbesina virginica*, **h)** head, **i)** fruit, **j)** habit.

1239. Verbesina helianthoides   1240. Verbesina virginica   1241. Xanthium spinosum

lanceolate (on small plants) or more commonly lanceolate to ovate or ovate-triangular, short-tapered to more or less truncate at the base, angled or tapered to a usually sharply pointed tip, the margins usually coarsely toothed, less commonly finely toothed to nearly entire, the upper surface sparsely to moderately pubescent with appressed hairs, usually appearing green, the undersurface densely appressed-hairy, usually appearing silvery. Inflorescences loose, open clusters with 2–8 heads, sometimes only a solitary head. Involucre 10–20 mm in diameter, broadly cup-shaped, with 12–21 bracts. Involucral bracts 6–23 mm long, narrowly ovate to lanceolate, narrowly oblong-lanceolate, or occasionally linear, ascending to loosely ascending at flowering, the outer surface densely hairy. Chaffy bracts linear, moderately to densely hairy toward the tip. Ray florets 10–15, pistillate (with a 2-branched style exserted from the short tube at flowering), the corolla 10–20 mm long, spreading, yellow. Disc florets 80 to numerous (more than 150), the corolla 2.5–3.5 mm long, yellow. Pappus of the ray florets absent, that of the disc florets of 2 more or less slender awns 0.5–2.0 mm long, usually with fine, upward-pointed barbs, usually shed as the fruit matures. Fruits ascending to somewhat spreading at maturity (forming a more or less hemispherical mass), 4.0–6.5 mm long, the body usually narrowly oblong-obovate, relatively broadly winged, the surface sparsely to moderately pubescent with fine, often pustular-based hairs. $2n=34$. May–October.

Introduced, uncommon, sporadic (southwestern U.S. east to Nebraska and Texas; Mexico; introduced sporadically farther north and east in the U.S., also Hawaii, South America, Caribbean Islands, Asia, Mauritius, Australia). Crop fields, railroads, roadsides, and open, disturbed areas.

Coleman (1966) treated *V. encelioides* as comprising two weakly separable subspecies. He indicated that their ranges in the United States differed, with ssp. *encelioides* along the Atlantic and Gulf Coastal Plains from eastern Texas to Florida and north to North Carolina, and ssp. *exauriculata* occupying a range in the southwestern United States. However, in doing so, he (and some subsequent authors) apparently confused the native range of ssp. *encelioides*, which appears to be native in southernmost Texas but has been treated as introduced in the major floristic works involving the southeastern states (for example, Cronquist, 1980). Both subspecies have a large overlapping range in Mexico. The ssp. *encelioides* was first reported for Missouri by Coleman (1966), and its distribution and spread into Missouri were discussed by Wagner (1979). It supposedly differs in having most of the petioles with conspicuous, obovate to nearly circular auricles at the base and slightly larger heads, whereas ssp. *exauriculata* supposedly is characterized by petioles lacking auricles or those of only the upper leaves with relatively small, oblong auricles and slightly smaller heads. Examination of specimens in the Missouri Botanical Garden Herbarium from throughout the range of the species do not support any correlation between these characters, nor do the specimens suggest that there are strong differences in the natural ranges. The size and shape of the leaf auricles is often not possible to ascertain, as many specimens comprise only the upper portions of the plant, and these characters appear to vary more continuously than has been suggested in the literature. The characters involved probably should be regarded merely as part of the overall species variation.

3. Verbesina helianthoides Michx. (yellow crownbeard)
Phaethusa helianthoides (Michx.) Britton
Pterophyton helianthoides (Michx.) Alexander
Pl. 291 f, g; Map 1239

Plants perennial herbs with fibrous, sometimes

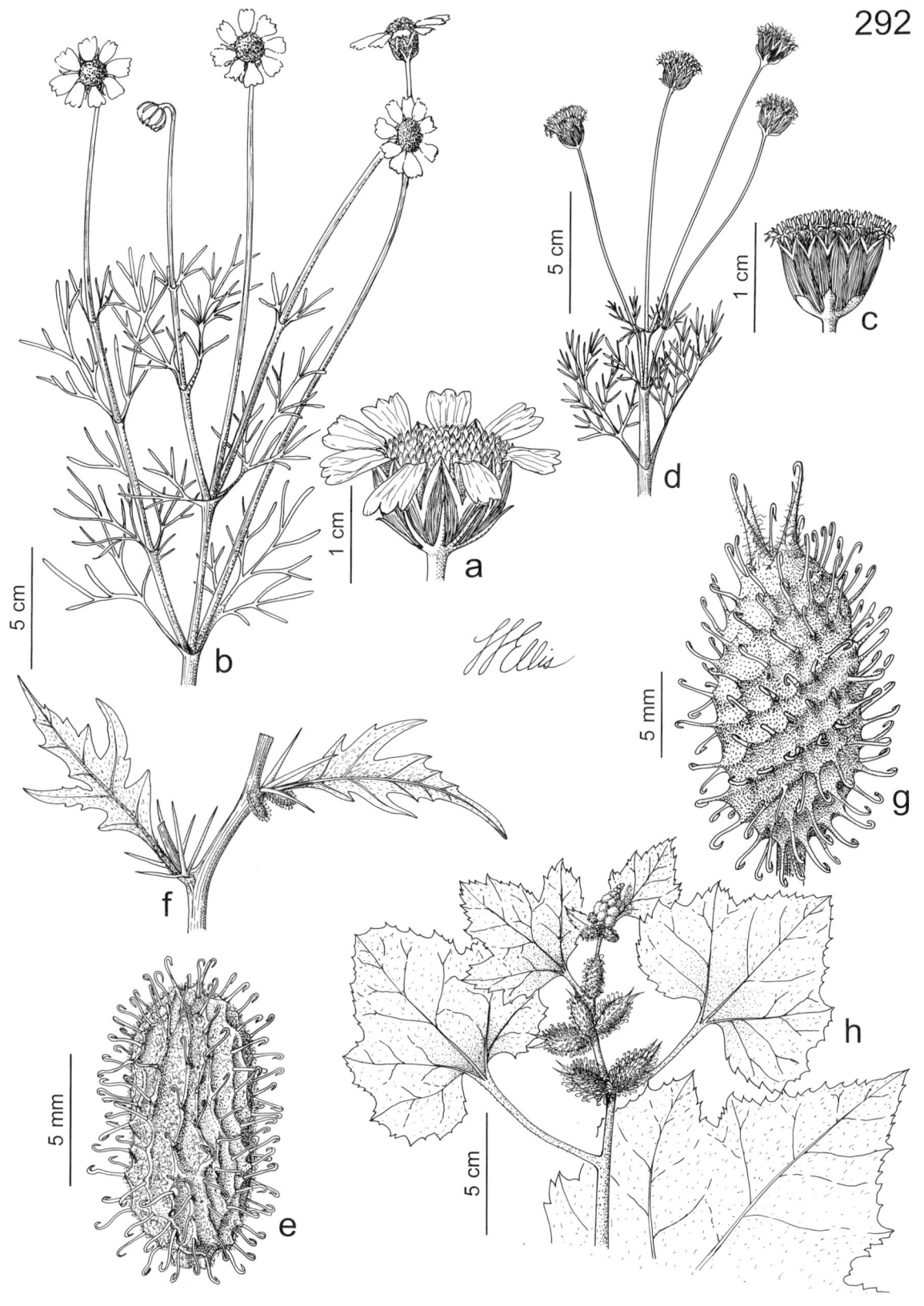

Plate 292. Asteraceae. *Thelesperma filifolium*, **a)** head, **b)** habit. *Thelesperma megapotamicum*, **c)** head, **d)** habit. *Xanthium spinosum*, **e)** bur, **f)** node with leaves. *Xanthium strumarium*, **g)** bur, **h)** habit.

slightly fleshy roots and usually short, more or less stout rhizomes. Stems (20–)50–120 cm long, narrowly and sometimes incompletely winged, moderately to densely pubescent with short, fine, loosely ascending hairs. Leaves alternate or the lowermost leaves sometimes opposite, sessile or the lowermost leaves with a short, usually poorly differentiated, winged petiole, the base usually strongly decurrent below the attachment point as a pair of wings. Leaf blades (1–)5–55 cm long, lanceolate to elliptic or narrowly ovate, tapered at the base, angled or tapered to a sharply pointed tip, the margins usually finely toothed, occasionally nearly entire, the upper surface moderately to densely pubescent (and sometimes slightly roughened to the touch) with relatively soft, spreading, pustular-based hairs, the undersurface densely pubescent with appressed to somewhat spreading hairs (and usually felty to the touch). Inflorescences usually loose, open clusters with 2–10 heads, sometimes with only a solitary head. Involucre 9–16 mm in diameter, cup-shaped, with 16–21 bracts. Involucral bracts 6–9 mm long, narrowly lanceolate or narrowly oblong-lanceolate to lanceolate, ascending to loosely ascending at flowering, the outer surface moderately to densely hairy. Chaffy bracts narrowly lanceolate to narrowly oblong-oblanceolate, moderately to densely hairy, especially toward the tip. Ray florets 8–15, pistillate or sterile, the corolla 10–30 mm long, spreading, yellow. Disc florets 40–80(–120), the corolla 3.5–4.5 mm long, yellow. Pappus of the ray and disc florets usually similar (but that of the ray florets usually somewhat broader, flattened, and scalelike, sometimes absent in sterile florets), of 2 slender to relatively stout awns 0.5–1.5 mm long, smooth or with fine, upward-pointed barbs, more or less persistent at fruiting. Fruits ascending to somewhat spreading at maturity (forming a more or less hemispherical mass), 4–5 mm long, the body usually oblanceolate to narrowly obovate, narrowly to broadly winged, occasionally wingless, the surface glabrous or more commonly moderately to densely pubescent with short, stout, pustular-based hairs. $2n=68$. May–October.

Scattered south of the Missouri River and north locally to Linn, Macon, and Ralls Counties (Kansas to Texas east to Ohio, North Carolina, and Georgia). Upland prairies, savannas, glades, and mesic to dry upland forests; also railroads and roadsides.

4. Verbesina virginica L. **var. virginica** (white crownbeard, frostweed, tickweed)

Phaethusa virginica (L.) Britton

Pl. 291 h–j; Map 1240

Plants perennial herbs with often somewhat fleshy roots. Stems 50–200(–250) cm long, narrowly winged, often only below the midpoint, densely pubescent with short, curved, often somewhat felty or woolly hairs, especially toward the tip. Leaves alternate or the lowermost leaves sometimes opposite, sessile or with an often short, poorly differentiated, winged petiole, the base minutely to strongly decurrent below the attachment point as a pair of wings. Leaf blades 5–20 cm long, narrowly elliptic or lanceolate-elliptic to ovate, tapered at the base, tapered to a sharply pointed tip, the margins coarsely to finely toothed or sometimes nearly entire, the upper surface sparsely to moderately roughened-pubescent with short, fine, broad-based hairs, the undersurface moderately to densely pubescent with short, fine, sometimes felty hairs. Inflorescences usually terminal panicles with (20–)40–200 heads, rarely reduced to an open cluster. Involucre 3–7 mm in diameter, more or less conical, with 8–13 bracts. Involucral bracts 2.5–5.5 mm long, oblong-oblanceolate to oblanceolate, erect or ascending at flowering, the outer surface densely hairy. Chaffy bracts narrowly oblong to narrowly oblong-oblanceolate, densely hairy toward the tip. Ray florets 1–5, pistillate (with a 2-branched style exserted from the short tube at flowering), the corolla 3–9 mm long, spreading to somewhat drooping, white. Disc florets 8–12, the corolla 2.5–3.5 mm long, white. Pappus of the ray and disc florets similar, of 2 slender awns 1.5–3.0 mm long, with fine, upward-pointed barbs, more or less persistent at fruiting. Fruits ascending to somewhat spreading at maturity (forming a more or less hemispherical mass), 3.5–6.0 mm long, the body usually oblanceolate, usually relatively broadly winged, less commonly narrowly winged or wingless, the surface sparsely to moderately pubescent with fine, pustular-based hairs. $2n=32, 34$. August–October.

Scattered south of the Missouri River (eastern U.S. west to Kansas and Texas). Banks of streams and rivers, bases of bluffs, bottomland forests, and mesic upland forests; also pastures, railroads, and roadsides.

Olsen (1979) treated *V. virginica* as comprising two varieties, the widespread var. *virginica*, with all of the leaves unlobed, and var. *laciniata*, which differs in having some or all of the leaves lobed or divided and is endemic to the Atlantic Coastal Plain from Florida to North Carolina.

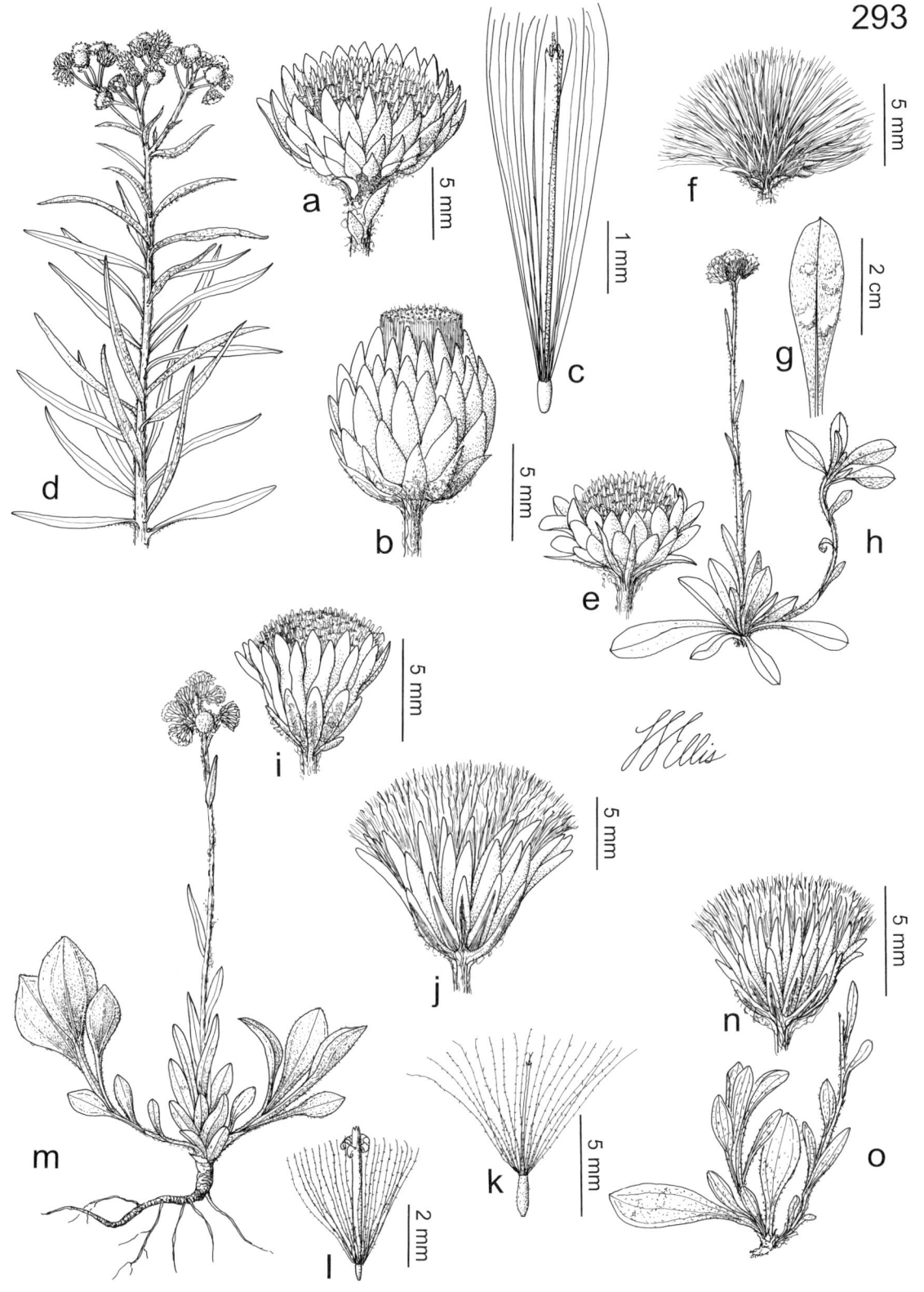

Plate 293. Asteraceae. *Anaphalis margaritacea*, **a)** staminate head, **b)** pistillate head, **c)** pistillate flower, **d)** habit. *Antennaria neglecta*, **e)** staminate head, **f)** pistillate head, **g)** leaf, **h)** habit. *Antennaria parlinii*, **i)** staminate head, **j)** pistillate head, **k)** pistillate flower, **l)** staminate flower, **m)** habit. *Antennaria plantaginifolia*, **n)** pistillate head, **o)** leaf base.

95. Xanthium L. (cocklebur)

Plants annual, with taproots. Stems usually few- to many-branched, erect or ascending, sometimes from a spreading base, finely ridged or grooved, glabrous or sparsely to moderately roughened or short-hairy. Leaves mostly alternate, sometimes opposite at the basal few nodes, sessile to long-petiolate. Leaf blades variously shaped, unlobed or more commonly with 3–7 main lobes, the margins entire to somewhat wavy or coarsely and irregularly toothed, the surfaces glabrous to densely hairy. Inflorescences of separate staminate and pistillate heads in small, dense axillary spikes (the uppermost spikes sometimes entirely staminate, the lowermost spikes occasionally entirely pistillate), the staminate heads usually several toward the tip of the spike, the pistillate heads usually solitary or in a small cluster of 2–4 at the base of the spike, each subtended by a small bract. Heads discoid (but this not evident in pistillate heads), the staminate heads spreading in several directions. Involucre of the staminate heads 2–4 mm in diameter, saucer-shaped, symmetrical, the 6–18 involucral bracts in 1–3 series, free to the base, green. Involucre of the pistillate heads with the main body ellipsoid to oblong-ellipsoid, the involucral bracts several to numerous in several overlapping series, closely enclosing the florets and all but the outermost few fused into a bur, the outer surface with sharp spines that are hooked or curled at the tip, often more or less 1- or 2-beaked at the tip (where an opening allows exsertion of the stigma), green, turning brown to orangish brown or brownish yellow at fruiting. Receptacle of the staminate heads conical (impossible to observe in pistillate heads), somewhat elongating as the florets mature, with chaffy bracts subtending the florets, these narrowly linear to narrowly lanceolate, usually hairy and sometimes also glandular, not wrapped around the florets. Staminate heads with 20–60 disc florets, these with a minute, nonfunctional ovary and undivided style, the stamens with the filaments more or less fused into a tube and the anthers free but positioned closely adjacent to one another in a ring, the corolla 1–2 mm long, narrowly bell-shaped to nearly tubular, shallowly 5-lobed, white to pale yellow, sometimes purplish-tinged toward the tip, usually minutely hairy and often also glandular. Pistillate heads with 2 florets in separate chambers, the corolla absent. Pappus of the staminate and pistillate florets absent. Fruits 8–11 mm long, narrowly elliptic to narrowly oblanceolate in outline, somewhat flattened with rounded angles, grayish brown to dark gray or black, nearly smooth or more commonly with noticeable branched veins, glabrous, often somewhat shiny, completely enclosed in the persistent pistillate involucre and dispersed as an intact bur. Three species, nearly worldwide, but probably introduced in the Old World.

Based on a cladistic analysis of morphological data, Karis (1995) suggested that *Xanthium* might best be treated as a specialized subgroup within *Ambrosia*. However, this was in disagreement with the results of earlier, similar studies (Bolick, 1983), and molecular studies of chloroplast DNA variation (Miao et al., 1995b) also did not confirm the hypothesis.

The species-level taxonomy of *Xanthium* remains somewhat controversial. More than 50 species have been accepted by various authors in different parts of the world (over 20 in North America), differentiated mostly on subtle details of bur morphology. Beginning with Wiegand (1926) and Cronquist (1945), some North American authors began questioning the validity of recognizing most of the taxa present on the continent at other than infraspecific levels. The classification followed here is adapted from that proposed by Löve and Dansereau (1959), who accepted only two well-marked species originally described by Linnaeus in 1753. Steyermark (1963) discussed the confusing taxonomic situation within the genus and preferred to recognize eight species in Missouri, seven of which are here considered part of the complex pattern of ecological and morphological variation within *X. strumarium*. The third species in the genus is *X. ambrosioides* Hook. and Arn., an Argentinean taxon related to *X. spinosum* but differing in its more or less spreading stems, more finely divided leaves, and densely hairy burs. Löve and Dansereau (1959) suggested that this taxon might better be treated as a variety of *X. spinosum* but did not discuss their reasons for doing so.

The burs of *Xanthium* can be hazardous to livestock, causing mechanical injury and sores to mouth parts and other soft tissues, as well as blockages in the throat and intestines. Additionally, the embryonic tissue in the seed and the cotyledons of the seedling contain highly toxic diterpene glycosides such as carboxyatractyloside (gummiferin), which are poisonous to mammals and birds when ingested at even low concentrations (Burrows and Tyrl, 2001). Ecologically, *Xanthium* is a colonizer of disturbed open-soil habitats with an interesting behavior in that one of the two fruits within each bur generally germinates quickly and readily, with the other fruit waiting in reserve in case the first seedling fails to survive. *Xanthium strumarium* (in the broad sense) has been a model species for physiological research on seed germination and plant phenology. In particular, the physiological and biochemical mechanisms that result in the onset of flowering in response to shortening day length have been well documented (see Löve and Dansereau [1959] for a review of the older literature) in various populations at different latitudes.

1. Stems with prominent, slender, 3-parted spines at the nodes; leaves lanceolate to more or less elliptic; burs 9–13 mm long, the beak absent or 1, short and straight . . . . 1. X. SPINOSUM
1. Stems lacking spines; leaves broadly ovate to more or less kidney-shaped or nearly circular; burs 15–30 mm long, the beaks 2, relatively prominent and somewhat incurved at the tips . . . . 2. X. STRUMARIUM

1. Xanthium spinosum L. (spiny cocklebur)
Pl. 292 e, f; Map 1241

Stems 10–120 cm long, sparsely to moderately pubescent with short, relatively soft, loosely appressed hairs and sometimes also with scattered, longer spreading hairs, with prominent, 3-parted axillary spines at the nodes, these 5–20 mm long, slender, straw-colored to dull yellow. Leaves mostly sessile to short-petiolate. Leaf blades 2–6 cm long, lanceolate to more or less elliptic, mostly angled or tapered at the base, mostly tapered to a sharply pointed tip, unlobed or with 3–7 narrowly to broadly triangular, pinnate lobes, the margins otherwise entire to somewhat wavy, the upper surface nearly glabrous or sparsely to moderately pubescent with short, straight, loosely appressed, silvery white hairs, especially along the midvein (this sometimes appearing whitened), the undersurface similarly but densely hairy, appearing whitened. Pistillate heads 1 or 2 per spike, the bur 9–13 mm long, the beak absent or 1, when present short and straight, the surface with short, fine, somewhat cobwebby hairs and numerous slender spines, these usually hooked at the tip. $2n=36$. October–November.

Introduced, uncommon, known only from historical collections from the city of St. Louis (native of tropical America, introduced nearly worldwide). Roadsides and open, disturbed areas.

2. Xanthium strumarium L. (common cocklebur)
X. strumarium var. *canadense* (Mill.) Torr. & A. Gray
X. italicum Moretti
X. pensylvanicum Wallr.
X. speciosum Kearney
X. varians Greene
X. wootonii Cockerell
X. strumarium var. *glabratum* (DC.) Cronquist
X. chinense Mill.
X. inflexum Mack. & Bush
Pl. 292 g, h; Map 1242

Stems 10–150(–200) cm long, sparsely to moderately roughened-pubescent with short, stout, broad-based, ascending hairs and often also with minute glandular hairs or inconspicuous, sessile glands, spineless. Leaves all or mostly long-petiolate. Leaf blades 2–18 cm long, broadly ovate or ovate-triangular to more or less kidney-shaped or nearly circular, mostly shallowly to deeply cordate at the base (often broadly short-tapered to the petiole within the notch), variously rounded to bluntly or sharply pointed at the tip, unlobed or with 3 or 5 usually shallow, broad, palmate main lobes, the margins otherwise coarsely and irregularly toothed, the surfaces sparsely to moderately roughened with short, stout, loosely appressed, broad-based hairs, sometimes only along the veins, usually also glandular, the undersurface not appearing whitened. Pistillate heads mostly 2–4 per spike, the bur 15–30(–35) mm long, the beaks 2, usually relatively prominent and incurved, the surface variously (see below) nearly glabrous or with sessile glands, stalked glands, and/or short, spreading to loosely appressed hairs, also with numerous slender to slightly broad-based spines, these hooked or curled at the tip. $2n=36$. July–November.

1242. Xanthium strumarium   1243. Inula helenium   1244. Pluchea camphorata

Scattered to common throughout the state (nearly worldwide, but probably introduced in the Old World). Banks of streams, rivers, and spring branches, swamps, sloughs, bottomland prairies, bottomland forests, and margins of ponds, lakes, sinkhole ponds, marshes, fens, and seeps; also fallow fields, crop fields, pastures, barnyards, ditches, railroads, roadsides, and open, disturbed areas.

Patterns of morphological variation in *X. strumarium* are very subtle. Characters of the burs used to differentiate variants within the complex have included size, color, relative length and shape of the beaks, spine density and pubescence, and pubescence of the bur surface. None of these features can be observed if mature burs are absent from a given sample. Löve and Dansereau (1959) suggested that recent expansion of the ranges of formerly more isolated entities followed by hybridization in mixed or adjacent populations accounts in part for the difficulties in distinguishing taxa within the group. Wiegand (1926) was the first botanist to combine the various North American taxa into a single species (under the name *X. orientale* L.) but listed four informal subgroups occurring in the northeastern United States. Cronquist (1945) formalized Wiegand's treatment into three varieties of *X. strumarium,* combining two different hairy-burred entities listed by Wiegand into a single variety and noting the presence of common intermediates between his taxa. Löve and Dansereau (1959) treated *X. strumarium* in a provisional (informal) sense hierarchically to include two major subunits (which they suggested might be subspecies) divided into nine provisional varieties and with three additional putative intervarietal hybrids. Infraspecific combinations for most of their taxa have yet to be validly published. In Missouri, for those specimens with mature burs, there appear to be more specimens intermediate for any given differentiating character than there are morphological extremes. Thus no attempt has been made in the present treatment to segregate varieties. For those who enjoy tormenting themselves with attempts to assign infraspecific names to cocklebur specimens, the following key has been adapted from that of Cronquist (1945). Of the Missouri specimens that can be keyed successfully, there are no discernable differences in abundance or distribution between the two supposedly distinct varieties. In Cronquist's treatment, var. *strumarium,* which is most common in Europe but occurs sporadically in the United States (not reported from Missouri), differs from the two widespread varieties in its yellowish green burs mostly 15–20 mm long with straight beaks and minute pubescence. For convenience, in the list of synonyms of *X. strumarium* above, the specific epithets applied to Missouri plants by Steyermark (1963) are in two groups under the corresponding varietal names in Cronquist's classification.

1. Burs mostly 20–30 mm long, the base of the spines with short, more or less spreading hairs and stalked glands, the surface between the spines glabrous or with similar glands and hairs .... var. *canadense*
1. Burs often 15–20 mm long, the base of the spines with sessile glands, the surface between the spines glabrous, with sessile glands, or with minute glandular hairs .... var. *glabratum*

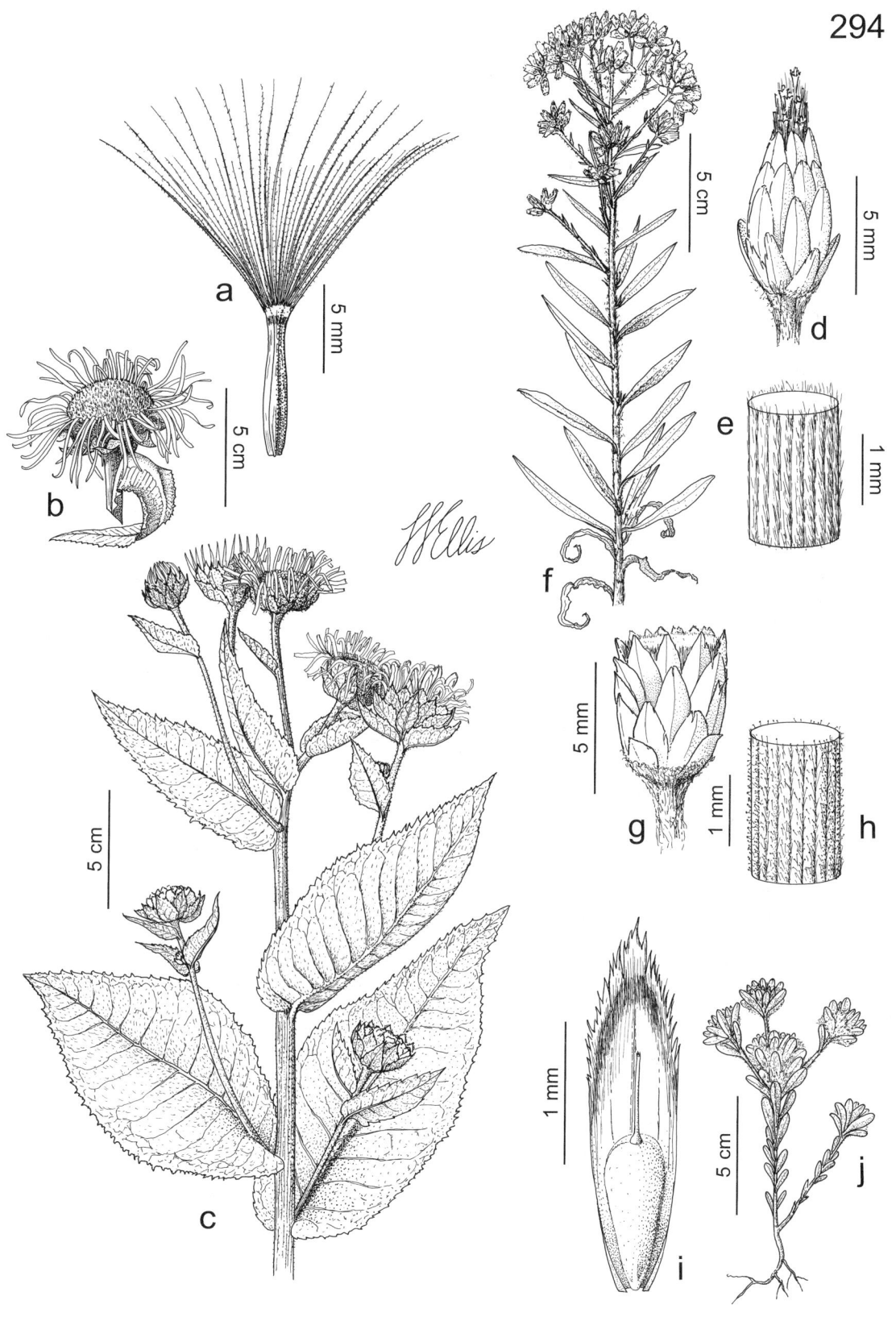

Plate 294. Asteraceae. *Inula helenium*, **a)** disk flower, **b)** head, **c)** habit. *Pseudognaphalium obtusifolium*, **d)** head, **e)** stem, **f)** habit. *Pseudognaphalium helleri*, **g)** head, **h)** stem. *Diaperia prolifera*, **i)** bract, **j)** habit.

8. Tribe Inuleae Cass.

About 38 genera, about 480 species, Europe, Asia, Africa.

As currently circumscribed, most of the genera of the traditional Inuleae that were included in North American floras are classified into the tribes Gnaphalieae and Plucheeae (see those tribes for further discussion). Inuleae in the strict sense is an Old World group represented in North America only by garden escapes. The larger heads with usually well-developed ray florets serve to separate this tribe from the North American members of the Gnaphalieae and Plucheeae.

In addition to *Inula,* the other main genus of Inuleae that is cultivated for its horticultural and medicinal value is *Pulicaria* Gaertn. (pulicaria, false fleabane). Species of *Pulicaria* are grown infrequently in Missouri, have not been documented outside cultivation in the state, and have only rarely escaped in California and the northeastern states.

96. Inula L.

About 100 species, Europe, Asia, Africa.

1. Inula helenium L. (elecampane)

Pl. 294 a–c; Map 1243

Plants perennial herbs from a thickened rootstock. Stems 40–120(–200) cm long, erect or ascending, usually unbranched below the inflorescence, with fine, spreading hairs, not spiny or prickly. Leaves alternate and sometimes also basal, simple, not spiny or prickly, the blade with the margins irregular and finely toothed, the upper surface with sparse, spreading hairs, the undersurface densely pubescent with fine, velvety hairs. Lowermost leaves long-petiolate, the blade 25–50 cm long, 8–20 cm wide, elliptic, tapered at the base, tapered to a sharply pointed tip, grading to the upper leaves, these sessile, clasping the stem, ovate, the base cordate, the tip narrowed or tapered to a sharp point. Inflorescences relatively few-headed, axillary and terminal, the heads solitary or in small, loose clusters, sometimes appearing as small panicles. Heads radiate. Involucre 2.0–2.5 cm long, 3–5 cm in diameter, bell-shaped to more or less hemispherical, with 2–4 series of overlapping bracts, these loosely appressed to somewhat spreading, all but the innermost series narrowly ovate to elliptic-ovate or oblong-ovate, not spiny or tuberculate, the outer surface and margins with fine, velvety hairs, green, but often with lighter, papery margins, these sometimes becoming reddish brown, those of the innermost series linear to narrowly oblong-lanceolate, glabrous, membranous to papery, reddish brown to purple, rarely lighter colored. Receptacle flat to slightly convex, naked. Ray florets numerous, pistillate, slender, the corolla 1.5–2.5 cm long, yellow. Disc florets numerous, perfect, the corolla 8–12 mm long, yellow. Pappus (of ray and disc florets) of a single series of numerous capillary bristles, these 6–9 mm long, more or less fused at the base, with short, fine, ascending awns. Stamens with the filaments not fused together, the anthers fused into a tube, each tip with a short, often indistinct appendage, each base prolonged into a pair of slender, tail-like lobes. Style branches flattened, each with a stigmatic line along each inner margin that is continuous around the rounded tip, hairy. Fruits 4–5 mm long, narrowly oblong in outline, more or less 4-angled or square in cross-section, not winged, not beaked, tan to light brown, somewhat shiny, glabrous. $2n=20$. June–September.

Introduced, uncommon and sporadic (native of Europe, widely but sporadically escaped from cultivation in the northeastern and western U.S. and adjacent Canada). Old fields, roadsides, and open, disturbed areas.

Elecampane is not as frequently grown as it once was. Early settlers used an extract of the mucilaginous, thickened root to treat pulmonary diseases and various other ailments, including upset stomach, diarrhea, worms, sciatica, and skin ailments. The rootstock, which is rich in inulin (a source of fructose) also has been cooked and candied and used to flavor vermouth and absinthe (Steyermark, 1963; Arriagada, 1998). Native Americans, who got plants from settlers, used the species as a gynecological aid, as heart medicine, and to treat domesticated animals (Moerman, 1998). The roots contain alantolactone, an analgesic with sedative properties.

The cultivated *I. britannica* L. (British yellowhead) has become established outside cultivation sporadically in New York and adjacent Canada. This aggressive garden perennial was collected recently as a weed in a flower bed in St. Louis County. Eventually it may become naturalized in the state. *Inula britannica* is a biennial or short-lived perennial with shorter stems and smaller leaves and heads than those found in *I. helenium*.

9. Tribe Plucheeae Anderb.

About 28 genera, about 220 species, nearly worldwide, but most diverse in tropical regions.

This relatively small but widely distributed tribe contains species ranging from annual herbs to trees. Steyermark (1963), Cronquist (1980), Gleason and Cronquist (1963, 1991), Barkley (1986), and most earlier authors treated the Plucheeae as a subtribe of a broadly circumscribed Inuleae along with genera here classified in the Gnaphalieae, although some of the tropical genera in the currently circumscribed Plucheeae originally were classified in the Vernonieae. The decision to dismember the Inuleae into three smaller tribes has been supported by recent analyses of molecular data sets (Panero and Funk, 2002) but was suggested originally based on differences in subtle morphological details, such as stylar ultrastructure (Bremer, 1987, 1994). This classification remains somewhat controversial and at least one recent molecular study (Anderberg et al., 2005) suggested that the Plucheeae may best be considered a subtribe of the Inuleae.

97. Pluchea Cass. (marsh fleabane)

Plants annual or perennial herbs (woody elsewhere), fibrous-rooted, sometimes with short rhizomes, usually glandular, the sap not milky. Stems erect or ascending, hairy, unbranched or few-branched toward the tip, not spiny or prickly. Leaves alternate, not spiny or prickly, sometimes slightly succulent, sessile or short-petiolate. Leaf blades simple, the margins toothed. Inflorescences mostly terminal panicles or the stem branches with a small cluster of heads at the tip. Heads appearing discoid. Involucre urn-shaped to hemispherical or somewhat bell-shaped, with several series of more or less overlapping bracts, the inner series progressively longer, these usually appressed, not spiny or tuberculate, usually glandular. Receptacle flat, naked. Inner florets few, appearing perfect but mostly functionally staminate (the style branches usually not spreading at maturity), the corolla with short lobes. Outer florets more numerous, pistillate (lacking stamens), the corolla very slender and shorter than the style, the lobes very short. Pappus a ring of numerous capillary bristles, these finely toothed, persistent at fruiting. Stamens with the filaments not fused together, the anthers fused into a tube, each tip with a short, often indistinct appendage, each base prolonged into a pair of slender lobes. Style branches not flattened, each with a stigmatic line along each inner margin, the sterile tip rounded, hairy. Fruits 0.6–1.0 mm long, cylindrical, with 4–6 ribs, not winged, not beaked, pinkish tan, hairy, at least along the ribs. 50–80 species, nearly worldwide.

Plants of *Pluchea* usually are quite aromatic when crushed or bruised, with a camphorlike or musky odor. Steyermark (1963) likened the odor of *P. camphorata* to that of a skunk. Native Americans used some species medicinally for diarrhea, fever, skin ailments, and hemorrhoids, and also as a sedative (Moerman, 1998).

1. Leaves short-petiolate or, if sessile, then tapered at the base and not clasping the stem; corollas pink to pinkish purple .... 1. P. CAMPHORATA
1. Leaves sessile, clasping the stem; corollas light cream-colored .... 2. P. FOETIDA

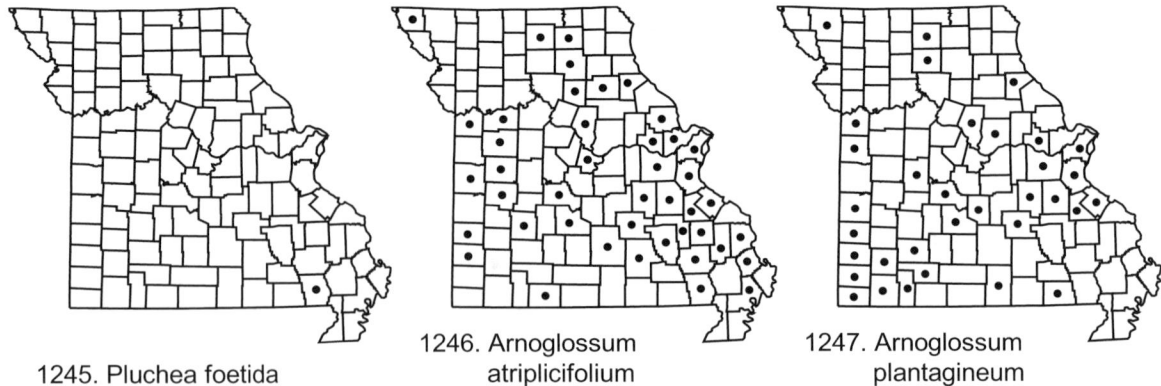

1245. Pluchea foetida
1246. Arnoglossum atriplicifolium
1247. Arnoglossum plantagineum

1. **Pluchea camphorata** (L.) DC. (stinkweed, camphorweed, inland marsh fleabane)

Pl. 295 g–i; Map 1244

Plants annual or short-lived perennial. Stems 40–100(–200) cm long, minutely hairy, sometimes nearly glabrous toward the base, occasionally also with sparse, longer hairs. Leaves short-petiolate or, if sessile, then not clasping the stem. Leaf blades 4–15 cm long, lanceolate to ovate or elliptic, angled or tapered at the base, angled or tapered to a sharply pointed tip, the margins with widely spaced teeth or more or less scalloped, rarely nearly entire, both surfaces with sparse to moderate, sessile, spherical glands, the undersurface also minutely hairy. Inflorescences of relatively dense clusters at the branch tips, these sometimes paniculate, appearing flat-topped or more commonly rounded in profile. Heads 5–9 mm in diameter. Involucre 4–6 mm long, the outer bracts triangular-ovate, grading to the inner, oblong-lanceolate bracts, angled to tapered at the tip, papery and white to tan, often pinkish- or purplish-tinged, at least toward the tip, the outer surface with sessile, spherical glands and often also minute hairs, rarely nearly glabrous. Corollas 3–6 mm long, pink to pinkish purple. Pappus bristles 3–5 mm long, white or pinkish-tinged. $2n=20$. August–October.

Scattered in the southern portion of the state north disjunctly to St. Louis County; apparently introduced in Jackson County (Pennsylvania to Florida west to Kansas and Texas). Bottomland forests, swamps, sloughs, banks of streams and rivers, margins of lakes, ponds, and sinkhole ponds, bottomland prairies, and marshes; also ditches and roadsides.

2. **Pluchea foetida** (L.) DC. **var. foetida** (marsh fleabane, stinking fleabane)

Pl. 295 f; Map 1245

Plants perennial, sometimes with short rhizomes. Stems 40–100 cm long, minutely glandular-hairy, especially toward the tip, usually also with fine, cobwebby hairs, at least toward the tip. Leaves sessile, clasping the stem, often somewhat purplish-tinged. Leaf blades 4–10(–13) cm long, oblong to elliptic or less commonly narrowly ovate to ovate, cordate at the base, rounded or angled to short-tapered to a bluntly or sharply pointed tip, the margins finely toothed, both surfaces with sparse, sessile, spherical glands, the undersurface also minutely hairy. Inflorescences of relatively dense clusters at the branch tips, these sometimes paniculate, usually appearing flat-topped in profile. Heads 6–12 mm in diameter. Involucre 4–8 mm long, the bracts oblong-lanceolate, mostly tapered at the tip, green or more commonly papery and straw-colored, the inner bracts sometimes pinkish- or purplish-tinged, the outer surface with sessile, spherical glands and often also minute hairs. Corollas 4–7 mm long, light cream-colored. Pappus bristles 3–6 mm long, white or pinkish-tinged. $2n=20$. August.

Known only from a single historical collection from Butler County (Florida to New Jersey west to Missouri and Texas, mostly along the southeastern Coastal Plain; Caribbean Islands). Habitat unknown, but presumably bottomland forests, swamps, or sloughs.

The var. *imbricata* Kearney, which some botanists choose not recognize, occurs in southern Georgia and Florida. It differs from var. *foetida* in its somewhat taller stems and slightly larger heads.

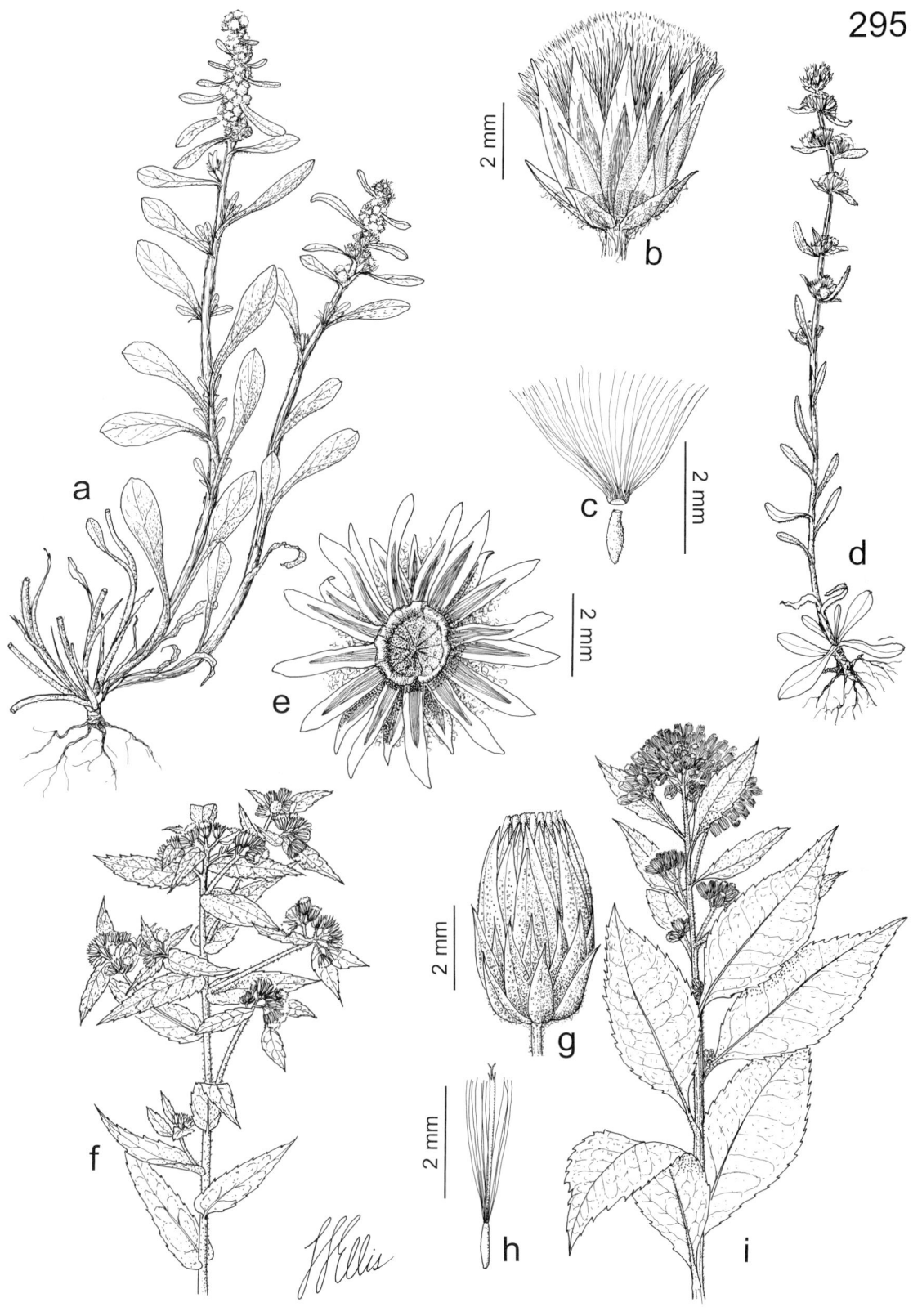

Plate 295. Asteraceae. *Gamochaeta purpurea*, **a)** habit, **b)** head, **c)** disk flower. *Gamochaeta argyrinea*, **d)** habit. *Gamochaeta pensylvanica*, **e)** concave receptacle after fruits have been dispersed. *Pluchea foetida*, **f)** habit. *Pluchea camphorata*, **g)** head, **h)** disk flower, **i)** habit.

10. Tribe Senecioneae Cass.

Plants annual, biennial, or perennial herbs (woody or succulent elsewhere), the sap not milky. Stems not spiny or prickly. Leaves alternate, often also with a basal rosette, not spiny or prickly. Leaf blades simple or less commonly compound, entire to variously toothed and/or lobed. Inflorescences mostly terminal panicles or the stem branches with a small cluster or a solitary head at the tip. Heads entirely discoid or radiate. Involucre of a single series of bracts of similar size and length, these usually appressed, sometimes fused laterally, not spiny or tuberculate, usually subtended by a group of often shorter, narrow bracts. Receptacle flat to slightly convex (sometimes with a minute, nipplelike or beaklike central outgrowth), naked. Ray florets (when present) pistillate; the pappus of numerous capillary bristles, these occasionally finely barbed, usually shed individually before the fruit is dispersed; the corollas yellow. Disc florets perfect (the outermost ones in discoid taxa occasionally only staminate); the pappus of numerous capillary bristles, these occasionally finely barbed, usually shed individually before the fruit is dispersed; the corollas yellow, white, cream-colored, or pink, the tube usually expanded above the midpoint, the 5 short lobes spreading to ascending. Stamens with the filaments not fused together, the anthers fused into a tube, each tip with a short, often indistinct appendage, each base truncate or with a pair of short lobes. Style branches usually somewhat flattened, each with a stigmatic line along each inner margin or these sometimes fused into a single band, the sterile tip usually truncate, minutely hairy. Fruits mostly angled or with longitudinal lines, not winged, not beaked. About 120 genera, about 3,200 species, worldwide.

A number of members of the Senecioneae are cultivated as garden ornamentals or for cut flowers. In addition to the genera growing wild in Missouri, examples include some species of *Pericallis* D. Don (cineraria), *Ligularia* Cass. (ligularia, leopard plant), *Petasites* Mill. (butterbur, winter heliotrope), and *Tussilago* L. (coltsfoot). Most members of the tribe produce pyrrolizidine alkaloids and are considered poisonous to humans and livestock.

Circumscriptions of the two larger, traditionally recognized genera represented in the Missouri flora, *Cacalia* and *Senecio,* have been altered in recent years following studies showing that some species groups within each of these genera were less closely related to each other than they were to other genera of Senecioneae (Bremer, 1994; Barkley, 1999). Application of the generic name *Cacalia* L. has been particularly problematic, and its use for any group of species has been officially rejected under the International Code of Botanical Nomenclature (Greuter et al., 2000). The Missouri species are now treated under the names *Arnoglossum* and *Hasteola*. The limits of *Senecio,* formerly the largest genus of Asteraceae with about 3,000 species, are still subject to reinterpretation. Presently, the genus is thought to include about 1,300 mostly Old World species, but it still comprises a large number of morphologically diverse species groups. The native Missouri species are treated within the segregate genus, *Packera*.

1. Heads radiate (ray florets present)
    2. At least the basal leaves with a long, well-defined petiole; stem leaves usually pinnately lobed .... 101. PACKERA
    2. Leaves all sessile or the lowermost long-tapered to a poorly defined petiole; stem leaves with the margins shallowly toothed to nearly entire, not lobed
    .... 102. SENECIO
1. Heads discoid (ray florets absent)
    3. Corollas bright yellow; outer series of involucral bracts with black tips
    .... 102. SENECIO
    3. Corollas white, cream-colored, or pink; outer series of involucral bracts absent or not darkened at the tips

4. Leaves with 3–10 palmate or more or less parallel main veins; heads with 5 florets, the inner series of 5 involucral bracts 7–10 mm long . . . 98. ARNOGLOSSUM
4. Leaves with 1 or 3 main veins, the venation pinnate or the pair of lateral veins spreading into the basal lobes; heads with numerous (more than 20) florets, the inner series of 9 to numerous involucral bracts 10–17 mm long
   5. Venation pinnate, with 1 main vein; largest leaves elliptic to ovate, the margins toothed, the upper leaves grading to narrowly elliptic-lanceolate and often with several irregular lobes . . . . 99. ERECHTITES
   5. Venation palmate, with 1 main vein and a pair of spreading main veins; largest leaves strongly triangular to hastate, with a pair of spreading, triangular basal lobes, the margins also toothed, the uppermost leaves usually narrowly lanceolate and unlobed . . . . 100. HASTEOLA

98. Arnoglossum Raf. (Indian plantain)

Plants perennial, the rootstock with somewhat fleshy roots. Stems erect or strongly ascending, usually unbranched below the inflorescence, sometimes lined, grooved, or angled, glabrous, sometimes glaucous. Leaves in a basal rosette and alternate, progressively reduced in size from the stem base to tip, glabrous (except sometimes along portions of the margins), sometimes glaucous. Basal and lowermost stem leaves long-petiolate, the blades simple, entire to palmately lobed, the margins otherwise entire or shallowly, irregularly, and bluntly toothed, the venation palmate or appearing more or less parallel, with 3–10 main veins. Inflorescences panicles, terminal and axillary from the uppermost leaves, usually flat-topped in profile. Heads discoid, short- to long-stalked, with 5 florets. Involucre 7–10 mm long, more or less cylindrical, the bracts in 1 or less commonly 2 series, glabrous, the inner series of 5 bracts, these rounded dorsally or sharply keeled, uniformly greenish white or green centrally with white margins; the outer series absent or of few minute bracts, these appressed, green or greenish white. Corollas white, cream-colored, or rarely somewhat pinkish-tinged. Style branches with a stigmatic band along each inner surface. Fruits 3.5–5.0 mm long, narrowly oblong to narrowly oblong-elliptic in outline, not flattened, 10-ribbed (12–15-ribbed in *A. reniforme*), glabrous, dark brown. Eight species, eastern U.S. and adjacent Canada.

1. Leaf blades ovate to elliptic, the margins entire or shallowly toothed, not lobed, the main veins appearing more or less parallel (diverging from the base, converging toward the tip); involucral bracts with a sharp, winglike keel
   . . . . 2. A. PLANTAGINEUM
1. Blades of at least the lower leaves broadly triangular-ovate to kidney-shaped or heart-shaped, the margins irregularly toothed and usually shallowly palmately lobed, the main veins palmate; involucral bracts rounded dorsally, not keeled
   2. Stems glaucous, circular in cross-section, rounded or with fine longitudinal grooves; leaf blades glaucous on the undersurface . . . . 1. A. ATRIPLICIFOLIUM
   2. Stems not glaucous, angled in cross-section, with conspicuous longitudinal grooves; leaf blades not or only slightly glaucous on the undersurface
      . . . . 3. A. RENIFORME

1. Arnoglossum atriplicifolium (L.) H. Rob.
(pale Indian plantain)
Cacalia atriplicifolia L.
Pl. 296 a, b; Map 1246
Rootstock not tuberous-thickened, but with somewhat fleshy roots. Stems 1.2–3.0 m tall, circular in cross-section, rounded or with fine longitudinal grooves, glaucous. Leaves herbaceous in texture. Basal and lower leaves long-petiolate, the blades 15–45 cm long, broadly triangular-ovate to

kidney-shaped or heart-shaped, often shallowly palmately lobed, the tip and lobes usually broadly pointed, usually broadly cordate at the base, the margins otherwise irregularly toothed, the venation palmate with mostly 7–10 main veins, glabrous, the undersurface conspicuously glaucous. Upper leaves short-petiolate, ovate to ovate-triangular (the uppermost sometimes narrowly obovate), coarsely toothed, tapered at the base. Involucre 7–9 mm long, the inner bracts rounded dorsally, uniformly greenish white, rarely darker along the margins, the outer series usually absent (rarely with 1 or 2 minute bracts toward the tip of the stalk of the head). Corollas 8–10 mm long, white, cream-colored, or rarely somewhat pinkish-tinged. $2n=50$, 52, 54, 56. June–October.

Scattered nearly throughout the state (eastern U.S. west to Nebraska and Oklahoma). Bottomland forests, mesic upland forests, bases and ledges of bluffs, and banks of streams and rivers; also occasionally pastures, railroads, and roadsides.

This species, with its interesting foliage, would make an attractive addition to the garden, but it appears not to be widely available at nurseries. Steyermark (1963) suggested that it prefers neutral to somewhat limy soils.

The leaves of *A. atriplicifolium* tend to be somewhat more angular than those of *A. reniforme*. The lower leaves are often more coarsely toothed, with lobes that are usually somewhat deeper and more angular. They are, however, quite variable in size and lobing. *Arnoglossum atriplicifolium* grows in the shade, as does *A. reniforme,* but it also seems to tolerate sunnier sites than those in which that species is found. Coleman (1965) studied a sterile hybrid between *A. atriplicifolium* and *A. reniforme* (as *Cacalia atriplicifolia* and *C. muhlenbergii*) from a site in southern Indiana where the two parents were growing in proximity. Gleason and Cronquist (1991) also mentioned hybridization between these two species. Although not yet confirmed from Missouri, this hybrid, which is morphologically intermediate between the two parental species, eventually may be discovered in the state as isolated individuals at sites where the parents are growing mixed or adjacent to one another.

2. Arnoglossum plantagineum Raf. (Indian plantain)
Cacalia plantaginea (Raf.) Shinners
C. tuberosa Nutt.
Pl. 296 h, i; Map 1247

Rootstock sometimes somewhat tuberous-thickened, also with somewhat fleshy roots. Stems 0.5–1.6 m tall, angled in cross-section, usually with conspicuous, reddish purple longitudinal lines, not glaucous. Leaves thickened and somewhat leathery or papery in texture. Basal and lower leaves long-petiolate, the blades 8–20 cm long, elliptic to ovate, unlobed, the tip rounded to bluntly pointed, tapered at the base, the margins entire or less commonly shallowly toothed, the venation appearing more or less parallel with 7–10 main veins diverging from the base and converging toward the tip, glabrous, not glaucous. Upper leaves short-petiolate, ovate to narrowly obovate, the margins usually entire, tapered at the base. Involucre 7–10 mm long, the inner bracts with a sharp, winglike, dorsal keel, green, usually with a white dorsal band and tip, the outer series of a few minute, ascending and incurved bracts toward the tip of the stalk of the head. Corollas 8–11 mm long, white or cream-colored. $2n=54$. May–August.

Scattered, but apparently absent from portions of the Glaciated Plains Division and the Mississippi Lowlands (Ohio to Louisiana west to South Dakota and Texas; Canada). Upland and occasionally bottomland prairies, calcareous glades, tops of bluffs, savannas, and openings of mesic to dry upland forests; also pastures, railroads, and roadsides.

3. Arnoglossum reniforme (Hook.) H. Rob. (great Indian plantain)
Cacalia muhlenbergii (Sch. Bip.) Fernald
Pl. 296 c–e; Map 1248

Rootstock not tuberous-thickened, but with somewhat fleshy roots. Stems 1–3 m tall, angled in cross-section, with conspicuous longitudinal grooves, not glaucous. Leaves herbaceous in texture. Basal and lower leaves long-petiolate, the blades 25–65 cm long, broadly triangular-ovate to kidney-shaped, sometimes shallowly palmately lobed, the tip and lobes rounded or very bluntly pointed, usually broadly cordate at the base, the margins otherwise irregularly toothed, the venation palmate with mostly 7–10 main veins, glabrous except along the margins in the sinuses between the lobes or larger teeth, where often minutely hairy, the undersurface not or only slightly glaucous. Upper leaves short-petiolate, ovate to more or less kidney-shaped (the uppermost sometimes narrowly obovate), finely or occasionally relatively coarsely toothed, broadly angled, truncate, or shallowly cordate at the base. Involucre 7–9 mm long, the inner bracts rounded dorsally, uniformly greenish white, sometimes darker along the margins, the outer series usually absent (rarely with 1 or 2 minute bracts toward the tip of the stalk of the head). Corollas 7–10 mm long, white, cream-colored, or rarely somewhat pinkish-tinged. $2n=50$. May–September.

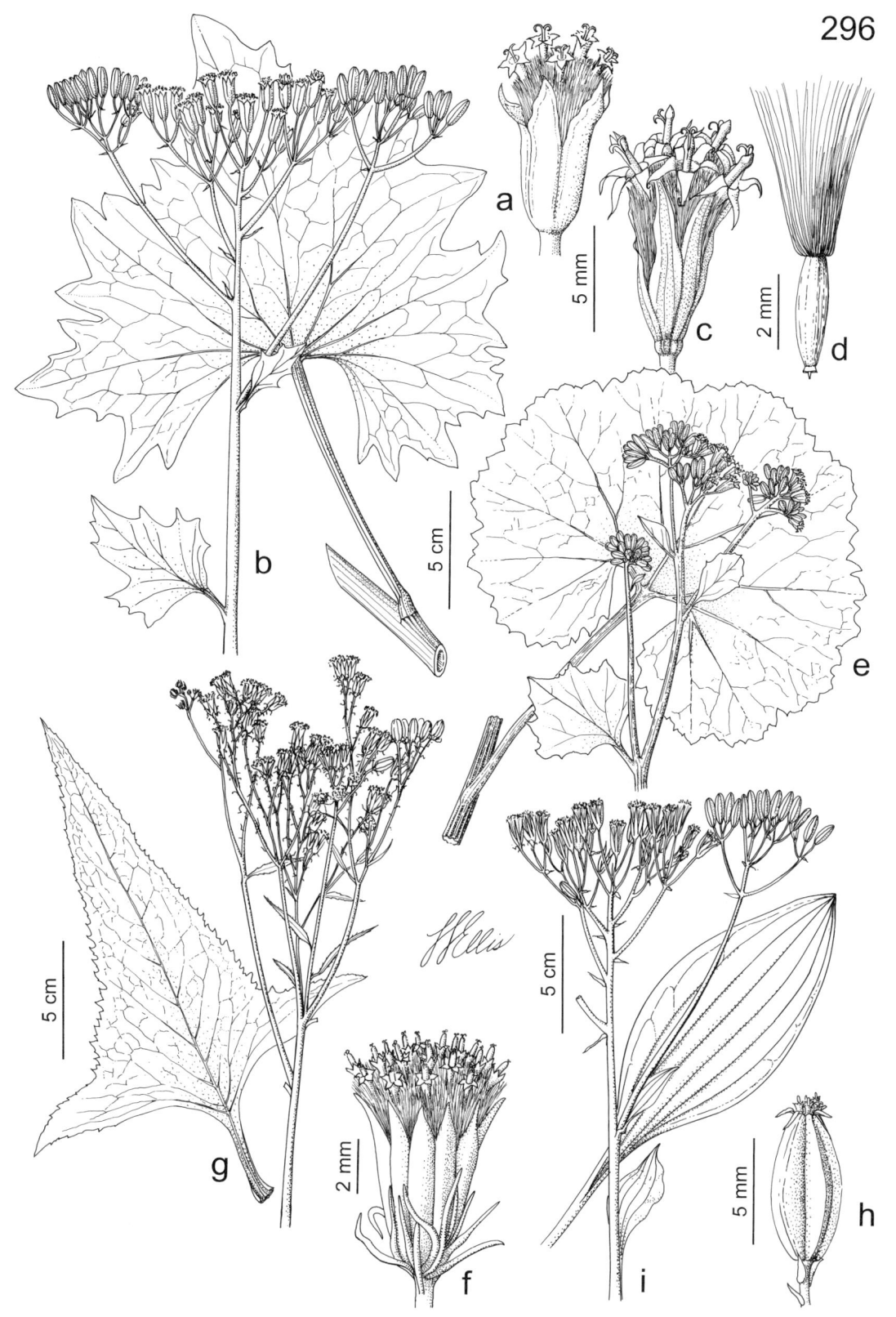

Plate 296. Asteraceae. *Arnoglossum atriplicifolium*, **a)** head, **b)** inflorescence and larger leaf. *Arnoglossum reniforme*, **c)** head, **d)** fruit, **e)** inflorescence and larger leaf. *Hasteola suaveolens*, **f)** head, **g)** inflorescence and larger leaf. *Arnoglossum plantagineum*, **h)** head, **i)** inflorescence and larger leaf.

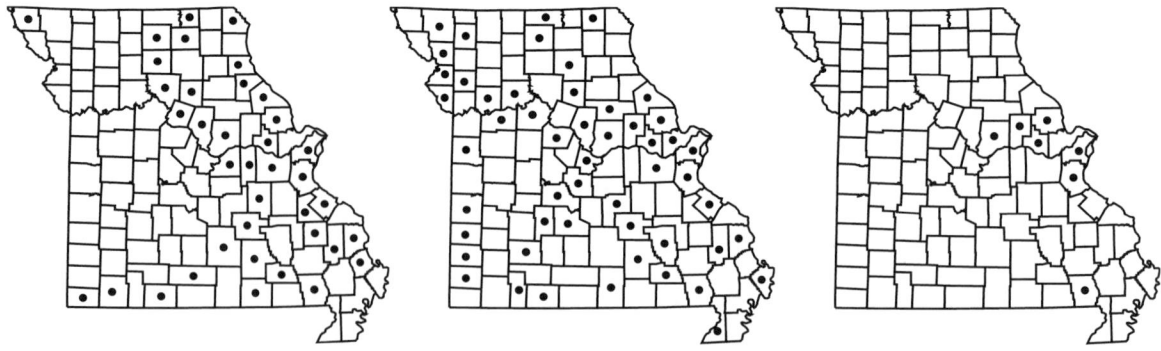

1248. Arnoglossum reniforme   1249. Erechtites hieracifolius   1250. Hasteola suaveolens

Scattered, mostly in the Ozark, Ozark Border, and Glaciated Plains Divisions (eastern U.S. west to Wisconsin and Oklahoma). Bottomland forests, mesic upland forests, bases and ledges of bluffs, and banks of streams and rivers; also occasionally along roadsides.

Steyermark (1963) noted that this species can spread aggressively by seeds in the garden and suggested that it prefers neutral to somewhat limy soils. For notes on distinguishing this species from the closely related *A. atriplicifolium* and a discussion of hybridization between the two, see the treatment of that species.

99. Erechtites Raf.
(Belcher, 1956)

About 12 species, North America to South America, introduced in the Old World.

Most botanists have treated the name *Erechtites* as feminine (*E. hieracifolia*), but the International Code of Botanical Nomenclature (Greuter et al., 2000) has clarified that generic names ending in *-ites* must be treated as masculine.

1. Erechtites hieracifolius (L.) Raf. ex DC.
var. hieracifolius (fireweed)
E. hieracifolius var. *intermedius* Fernald

Pl. 297 c–e; Map 1249

Plants annual, taprooted. Stems 0.1–2.5 m long, erect or strongly ascending, usually unbranched below the inflorescence, lined and/or angled, glabrous to moderately pubescent with more or less spreading hairs. Leaves in a basal rosette (this sometimes absent at flowering) and alternate, reduced in size from the stem base to tip, glabrous to moderately hairy, at least on the undersurface. Basal and lowermost stem leaves mostly short-petiolate, the blades simple, 3–20 cm long, elliptic to narrowly ovate, tapered to a sharply pointed tip, occasionally irregularly and shallowly pinnately lobed, the base tapered to the petiole, the margins sharply toothed, the venation pinnate. Median and upper leaves narrower with progressively shorter petioles and usually deeper lobes, sometimes sessile and with clasping bases, the very uppermost leaves often narrowly elliptic-lanceolate. Inflorescences panicles, terminal and axillary from the uppermost leaves, elongate to rounded in profile. Heads discoid, short- to long-stalked, mostly with 20–45 florets. Involucre 10–17 mm long, cylindrical above a somewhat swollen, disclike base, the bracts in 2 series, usually glabrous, the inner series of 9–15 bracts, these relatively flat dorsally and uniformly green; the outer series of 4–7 bracts, these minute, ascending, and incurved, uniformly green. Corollas 7–13 mm long, pale cream-colored. Style branches with a stigmatic line along each inner margin. Fruits 2–3 mm long, more or less linear in outline, not flattened, 10–12-ribbed, glabrous, light brown. $2n=40$. July–November.

Scattered nearly throughout the state (eastern U.S. west to South Dakota and Texas; Canada, Mexico, West Indies, Central America; introduced in the western U.S., Hawaii). Openings of mesic to dry upland forests, savannas, and banks of streams and rivers; also pastures, gardens, fallow fields, roadsides, railroads, and open, disturbed areas.

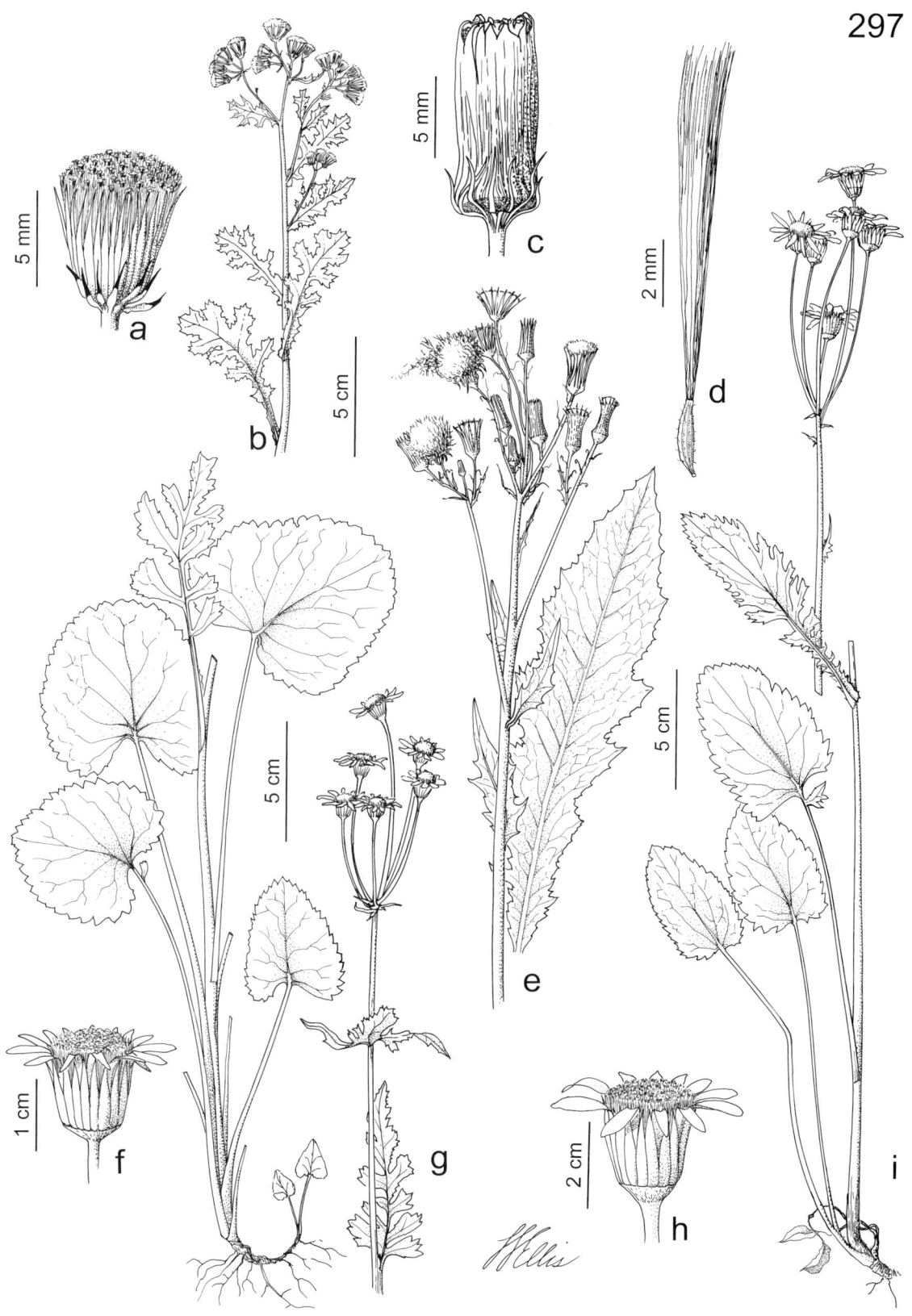

Plate 297. Asteraceae. *Senecio vulgaris*, **a)** head, **b)** habit. *Erechites hieracifolius*, **c)** head, **d)** fruit, **e)** inflorescence and larger leaf. *Packera aurea*, **f)** head, **g)** habit. *Packera pseudaurea*, **h)** head, **i)** habit.

This early successional species is well adapted to colonize open, disturbed soil, such as following fires or construction. It disappears as other plants grow more densely and outcompete it. Most botanists subdivide the species into a series of varieties. The other variety occurring in the United States is var. *megalocarpus* (Fernald) Cronquist, which is endemic to coastal salt marshes along the Atlantic seaboard. It has somewhat succulent foliage, somewhat broader heads, and slightly larger achenes with 16–20 ribs.

100. Hasteola Raf.
(Anderson, 1994)

Two species, eastern U.S.

A suite of superficially similar species occurs in Eurasia, but these are currently treated in a different genus, *Parasenecio* W.W. Sm. & Small. The other species of *Hasteola*, *H. robertiorum* L.C. Anderson, was first described in 1994 and is endemic to a small portion of Gulf-coastal northern Florida.

1. Hasteola suaveolens (L.) Pojark. (false Indian plantain)
Cacalia suaveolens L.
Synosma suaveolens (L.) Britton
Pl. 296 f, g; Map 1250

Plants perennial, the rootstock with somewhat fleshy roots and slender, fleshy rhizomes. Stems 0.4–1.8 m long, erect or strongly ascending, usually unbranched below the inflorescence, usually finely lined or grooved, glabrous or nearly so. Leaves in a basal rosette and alternate, progressively reduced in size from the stem base to the tip, glabrous or nearly so. Basal and lowermost stem leaves long-petiolate, the blades simple, 15–35 cm long, sharply triangular to hastate, tapered to a sharply pointed tip, with a pair of spreading, triangular basal lobes, the base otherwise truncate to nearly cordate, tapering abruptly to the petiole, the margins sharply toothed, the venation palmate with 1 main vein and a pair of spreading main veins running into the lobes. Upper leaves with progressively shorter petioles and shorter lobes, the uppermost leaves usually narrowly lanceolate and unlobed. Inflorescences panicles (occasionally reduced to small, loose clusters), terminal and axillary from the uppermost leaves, broadly rounded to flat-topped in profile. Heads discoid, short- to long-stalked, mostly with 20–45 florets. Involucre 10–14 mm long, narrowly bell-shaped (cylindrical to narrowly ovoid in bud), the bracts in 2 series, sometimes sparsely and minutely hairy near the base and/or tip, the inner series of 10–15 bracts, these relatively flat dorsally and uniformly green; the outer series of 5–11 bracts, these $^{1}/_{2}$ or more as long as the inner series, spreading, uniformly green. Corollas 8–12 mm long, white, cream-colored, or rarely somewhat pinkish-tinged. Style branches with a stigmatic line along each inner margin. Fruits 5–8 mm long, more or less linear in outline, not flattened, 8–12-ribbed, glabrous, light brown to pale green. $2n=40$. July–September.

Uncommon in east-central Missouri and also in Butler County (Maine to Georgia west to Wisconsin and Missouri). Bottomland forests, bases of sheltered bluffs, margins of sloughs, and banks of streams and rivers.

This species, with its unusual leaves, is an attractive ornamental, forming colonies of stems in the garden when sufficient moisture is present. However, as noted by Steyermark (1963), it spreads both by rhizomes and seeds and can become an aggressive problem plant.

101. Packera Á. Löve & D. Löve (ragwort)

Plants annual, biennial, or perennial, sometimes with rhizomes or stolons. Stems erect or strongly ascending, usually unbranched below the inflorescence, sometimes lined or angled, glabrous or with tangled or woolly (cobwebby) hairs. Leaves in a basal rosette (this sometimes absent at flowering) and alternate, progressively reduced in size from the stem base to tip, glabrous or with tangled or woolly (cobwebby) hairs, sessile or the lowermost short- to often long-petiolate, the bases weakly to strongly clasping the stem. Leaf blades unlobed to weakly or strongly pinnately lobed, less commonly pinnately compound, the terminal lobe or leaflet then usually larger than the lateral ones, the margins otherwise toothed or scalloped to nearly entire, the venation pinnate, the undersurface often dark purple. Inflorescences terminal and axillary from the uppermost leaves, panicles consisting of loose clusters to less commonly solitary heads at the branch tips, broadly rounded to more or less flat-topped in profile. Heads radiate (discoid elsewhere), short- to long-stalked, with numerous florets. Involucre cylindrical to slightly wedge-shaped or somewhat hemispherical, the bracts in 2 series, glabrous (cobwebby-hairy in *P. tomentosa*), those of the inner series 13–21, more or less flat dorsally, those of the outer series usually 3–7 (sometimes obscured by hairs in *P. tomentosa*) minute, ascending and usually incurved. Ray florets mostly 8–13, the corolla bright yellow, the lobe usually minutely 3-toothed at the tip. Disc corollas 4.5–7.0 mm long (including the lobes), bright yellow. Style branches with a stigmatic line along each inner margin. Fruits narrowly oblong to narrowly oblong-elliptic in outline, not flattened, 5–10-ribbed, minutely hairy or glabrous, brown to dark brown. About 60 species, North America, Asia.

A number of authors have noted the problems of delimiting species in this group and noted how, in some portions of their ranges, individual species can be difficult to determine with confidence. In Missouri, the main problem appears to be in distinguishing *P. paupercula* from *P. plattensis,* although specimens of *P. aurea, P. obovata,* and *P. pseudaurea* also are sometimes misdetermined in herbaria. Steyermark (1963) recommended that basal leaves and rootstocks should be collected to facilitate identification of species.

Barkley (1962, 1978) stated that putative hybrids tend to occur at sites where two or more species of *Packera* grow together. He also suggested that the changing climate at the close of the Pleistocene ice age brought together taxa that formerly were relatively isolated geographically. Various experts over the years have annotated Missouri specimens representing the following putative hybrids: *P. aurea* × *obovata, P. aurea* × *paupercula, P. obovata* × *paupercula,* and *P. obovata* × *plattensis*. Other hybrid combinations are to be expected, but they may be difficult to detect morphologically, for example *P. paupercula* × *plattensis*.

1. Basal and lower stem leaves all deeply pinnately lobed or appearing pinnately compound; plants annual or perennial, the rootstock various
   2. Plants annual or less commonly biennial, lacking rhizomes or stolons, glabrous or inconspicuously hairy; basal leaves usually absent at flowering; lower stem leaves with the terminal lobe or leaflet broadly wedge-shaped to nearly circular, shorter than to about as wide as long . . . . 2. P. GLABELLA
   2. Plants perennial or occasionally biennial, sometimes producing stolons, densely pubescent with felty hairs when young, sometimes becoming nearly glabrous by flowering time except for patches of dense, cobwebby or woolly hairs at the leaf bases and/or inflorescence branch points; basal leaves usually present at flowering; lower stem leaves with the terminal lobe ovate to elliptic-obovate (rarely much narrower), longer than wide . . . . 5. P. PLATTENSIS
1. At least some of the basal and often also lower stem leaves unlobed or at most with a pair of reduced basal lobes, the basal leaves well developed at flowering; plants perennial (rarely biennial in *P. plattensis*), usually producing short rhizomes (often appearing as leafy tufts or rosettes adjacent to the flowering stem) or stolons
   3. Plants more or less persistently pubescent with felty, woolly, or cobwebby, dense hairs on the stems, leaves, inflorescence branches, involucre, and leaf undersurface

4. Stem leaves all or mostly with the blade deeply lobed, the margins otherwise with relatively sharp, sometimes irregular teeth; plants more or less evenly pubescent with dense, felty hairs when young, usually becoming nearly glabrous by flowering time except for patches of dense, cobwebby or woolly hairs at the leaf bases and/or inflorescence branch points... 5. P. PLATTENSIS

4. Lower stem leaves with the blade unlobed or with few irregular lobes toward the base, the margins otherwise scalloped or relatively evenly and bluntly toothed; plants more or less evenly and persistently pubescent with dense, felty hairs, the leaf blades (particularly the upper surface) and upper portion of stems sometimes becoming nearly glabrous by flowering time.. 7. P. TOMENTOSA

3. Plants mostly glabrous, sometimes sparsely to densely hairy along lower portion of the stems, sparsely hairy on the leaf undersurface, cobwebby-hairy in small patches at the leaf bases, or with sparse pubescence along the inflorescence branches

5. Basal leaves with the blades mostly tapered at the base, if truncate to shallowly cordate, then with a pair of narrow wings of tissue extending along most of the petiole

6. Plants producing well-developed, slender stolons; basal leaves with the blade base tapered to truncate or slightly cordate, the tissue extending along most of the petiole as a pair of narrow wings .... 3. P. OBOVATA

6. Plants not producing stolons or rarely producing short stolons; basal leaves with the blades mostly truncate to heart-shaped, the tissue not extending along the petiole or extending along only the terminal portion of the petiole .... 4. P. PAUPERCULA

5. Basal leaves with the blades truncate to cordate at the base, if somewhat tapered, then with blade tissue not or only slightly extending along the terminal portion of the petiole

7. Basal leaves with the blade oblong-ovate to nearly circular, truncate or cordate at the base, at least some deeply cordate .... 1. P. AUREA

7. Basal leaves with the blade oblong-ovate to broadly lanceolate, abruptly short-tapered to more commonly truncate or shallowly cordate at the base .... 6. P. PSEUDAUREA

1. Packera aurea (L.) Á. Löve & D. Löve (golden ragwort, squaw weed)

Senecio aureus L.

S. aureus var. *gracilis* (Pursh) Hook.

S. aureus var. *intercursus* Fernald

Pl. 297 f, g; Map 1251

Plants perennial, from a short, stout to slender, ascending to horizontal, often branched rootstock, sometimes producing a few stolons. Stems mostly 1, occasionally 2 or 3, 20–80 cm long, sometimes pubescent with felty or cobwebby hairs toward the base when young, but usually glabrous or nearly so at flowering. Basal leaves usually present at flowering, long-petiolate, the petioles sometimes woolly or cobwebby when young, usually glabrous or nearly so at flowering, the blades 1–11 cm long, unlobed or rarely with few narrow, irregular lobes toward the base, oblong-ovate to nearly circular, truncate or cordate at the base, at least some deeply cordate, the tissue not extending along the petiole or extending only along the terminal portion of the petiole, rounded at the tip, the margins with blunt to sharp teeth, the surfaces glabrous. Stem leaves gradually reduced toward the stem tip, sessile or nearly so, the blades mostly deeply pinnately lobed, sometimes irregularly so, occasionally pinnately compound, the margins otherwise relatively sharply toothed, the surfaces glabrous. Involucre 4–8 mm long, glabrous or very sparsely hairy near the base. Ray florets usually 7–13, the lobe 6–13 mm long. Fruits 2.5–3.0 mm long, glabrous. $2n=44, 66$, about 132. April–June.

Scattered, mostly in the Ozark Division, but also sporadically in the Glaciated Plains (eastern U.S.

Plate 298. Asteraceae. *Packera tomentosa*, **a)** head, **b)** stem leaf, **c)** habit. *Packera plattensis*, **d)** head, **e)** stem leaf, **f)** habit. *Packera obovata*, **g)** head, **h)** habit. *Packera paupercula*, **i)** head, **j)** habit.

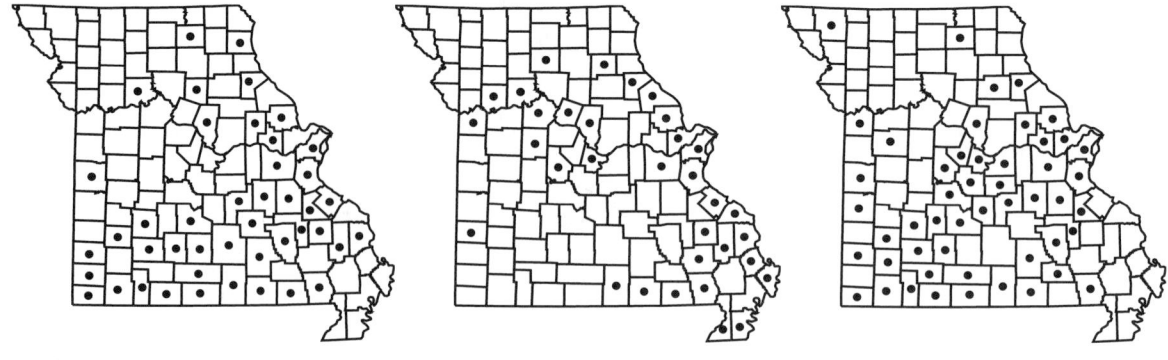

1251. Packera aurea   1252. Packera glabella   1253. Packera obovata

west to Minnesota and Texas; Canada). Banks of streams, rivers, and spring branches, fens, seepy ledges of bluffs, and occasionally bottomland forests; also rarely roadside ditches and depressions in pipeline or transmission line corridors.

For a discussion of difficulties in separating this species from *P. pseudaurea,* see the treatment of that species.

2. Packera glabella (Poir.) C. Jeffrey (butterweed)

Senecio glabellus Poir.

Pl. 299 i, j; Map 1252

Plants annual or less commonly biennial, lacking rhizomes or stolons. Stems mostly 1, rarely 2, 15–80 cm long, sometimes appearing somewhat inflated, glabrous or inconspicuously hairy. Basal leaves usually absent at flowering, short- to less commonly long-petiolate, the petioles rarely cobwebby when young, glabrous at flowering, the blades 3–20 cm long, pinnately compound with several (rarely few) pairs of lateral pinnae, less commonly only pinnately lobed, oblanceolate to elliptic-obovate in outline, the lobes or leaflets rounded at the tip, sometimes deeply few-lobed, the margins otherwise scalloped or with blunt to sharp teeth, the surfaces glabrous. Stem leaves gradually reduced toward the stem tip, sessile or nearly so, the blades mostly deeply pinnately lobed, the lower leaves sometimes with the blades pinnately compound, the lobes or leaflets sometimes deeply lobed, the terminal lobe or leaflet broadly wedge-shaped to nearly circular, shorter than to about as wide as long, the margins otherwise scalloped or bluntly to sharply toothed, the surfaces glabrous. Involucre 4–7 mm long, glabrous. Ray florets usually 7–13, the lobe 5–12 mm long. Fruits 2.5–3.0 mm long, glabrous or more commonly hairy, especially along the ribs. $2n=46$. April–June.

Scattered, principally in the Big Rivers and Mississippi Lowlands Divisions, but also sporadically elsewhere in the eastern half of the state (southern U.S. west to South Dakota and Texas; extirpated from Canada). Bottomland forests, swamps, banks of streams, rivers, and sloughs, bottomland prairies, and less commonly moist depressions of upland prairies and sand prairies, also crop fields, fallow fields, railroads, roadsides, and moist, open, disturbed areas.

3. Packera obovata (Muhl. ex Willd.) W.A. Weber & Á. Löve (roundleaf groundsel, squaw weed)

Senecio obovatus Muhl. ex Willd.

S. obovatus var. *umbratilis* Greenm.

Pl. 298 g, h; Map 1253

Plants perennial, from a short, relatively slender, erect to horizontal rootstock, often branched and producing well-developed, slender stolons. Stems mostly 1, occasionally 2 or 3, 20–70 cm long, often pubescent with dense, felty hairs in the axils of the basal leaves, the portion above the base thinly pubescent to glabrous, but sometimes with cobwebby patches of dense hairs in the leaf axils. Basal leaves usually present at flowering, long-petiolate, the petioles glabrous or sparsely hairy above the woolly or cobwebby base, the blades 1–8(–14) cm long, unlobed or less commonly with few to several narrow, irregular lobes toward the base, narrowly obovate to oblanceolate, oblong-ovate, or nearly circular, tapered to truncate or slightly cordate at the base, the tissue extending along most of the petiole as a pair of narrow wings, usually rounded at the tip, the margins scalloped or more commonly with blunt to sharp, sometimes irregular teeth, the undersurface usually glabrous, sometimes sparsely hairy, the upper surface glabrous. Stem leaves gradually reduced toward the stem tip, sessile or nearly so, the blades mostly shallowly to deeply pinnately lobed, sometimes irregularly so, the margins otherwise relatively sharply toothed, the surfaces glabrous except sometimes the lowermost

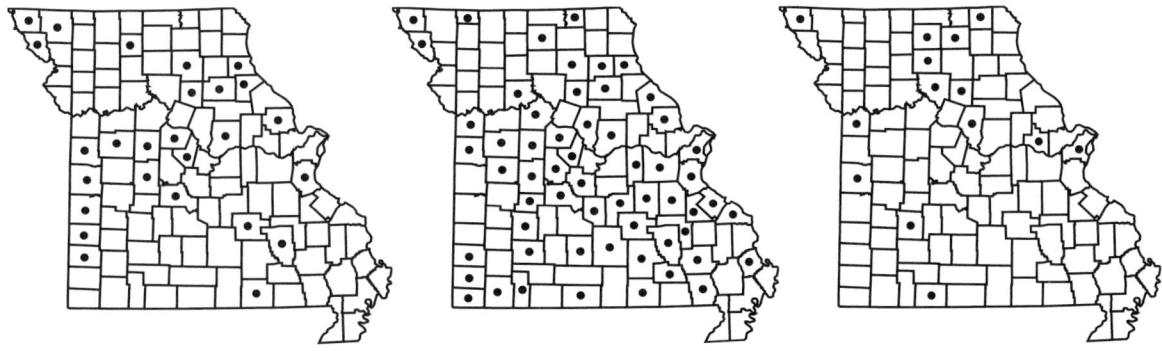

1254. Packera paupercula   1255. Packera plattensis   1256. Packera pseudaurea

leaves with patches of dense, cobwebby hairs at the very base. Involucre 4–7 mm long, glabrous or somewhat cobwebby-hairy near the base. Ray florets usually 7–13, the lobe 5–10 mm long. Fruits 2.5–3.0 mm long, glabrous or hairy along the ribs. $2n=44$. April–June.

Scattered nearly throughout the state, but apparently absent from the Mississippi Lowlands Division and portions of the Glaciated Plains (eastern U.S. west to Illinois, Kansas, and Texas; Canada). Bottomland forests, mesic upland forests, banks of streams, ledges of bluffs, and rarely moist depressions of glades and upland prairies; also roadsides and open, disturbed areas.

4. **Packera paupercula** (Michx.) W.A. Weber & Á. Löve (northern ragwort)
Senecio pauperculus Michx.
S. pauperculus var. *balsamitae* (Muhl. ex Willd.) Fernald
Pl. 298 i, j; Map 1254

Plants perennial, from a short, stout to slender, erect to horizontal rootstock, often producing a short to moderately creeping rhizome, rarely producing short stolons. Stems mostly 1, occasionally 2 or 3, 10–60 cm long, often pubescent with dense, felty hairs in the axils of the basal leaves, the portion above the base thinly pubescent to glabrous, but sometimes with cobwebby patches of dense hairs in the leaf axils. Basal leaves usually present at flowering, long-petiolate, the petioles glabrous or sparsely hairy above the usually woolly or cobwebby base, the blades 1–8 cm long, unlobed or less commonly with few narrow, irregular lobes toward the base, narrowly ovate-elliptic to oblanceolate, mostly truncate to cordate at the base (rarely narrowed), the tissue not extending along the petiole or else extending along only the terminal portion of the petiole, rounded to bluntly pointed at the tip, the margins with relatively sharp, sometimes irregular teeth (rarely scalloped), the undersurface glabrous or sparsely to moderately (but inconspicuously) hairy, the upper surface glabrous. Stem leaves gradually reduced toward the stem tip, sessile or nearly so, the blades mostly deeply pinnately lobed, sometimes irregularly so, the margins otherwise relatively sharply toothed, the surfaces glabrous except sometimes for patches of dense, cobwebby hairs at the very base. Involucre 4–7 mm long, glabrous or somewhat cobwebby-hairy near the base. Ray florets usually 7–13, the lobe 5–10 mm long. Fruits 2.5–3.0 mm long, glabrous or hairy along the ribs. $2n=44, 88$. April–August.

Scattered, mostly in the northern half of the state (eastern U.S. west to Washington and Arizona [but absent from most of the Great Plains]; Canada). Upland prairies, loess hill prairies, openings of mesic to dry upland forests, ledges of bluffs, and rarely bottomland forests; also pastures, railroads, and roadsides.

Packera paupercula varies relatively greatly in overall size, leaf shape, and pubescence density. For some specimens, it is virtually impossible to distinguish this species from the closely related *P. plattensis*. Barkley (1963) detailed the situation for plants from Wisconsin, where the problems are similar to those in Missouri. In general, both species usually may start out relatively evenly and densely hairy, but plants of *P. plattensis* tend to remain more persistently woolly or cobwebby-hairy than do plants of *P. paupercula*. Also, plants of *P. paupercula* in Missouri tend to produce short rhizomes more frequently, whereas those of *P. plattensis* tend to produce slender stolons more regularly. It should be noted that Allison Mahoney of Minnesota State University and Robert Kowal of the University of Wisconsin have annotated some Missouri specimens of stoloniferous, relatively small-leaved plants previously referred to *P. plattensis* by Barkley (1962, as *Senecio*) as representing an as-yet unpublished variety of *P. paupercula*. The name *Senecio pseudotomentosus* may in fact apply to this variant, but Barkley referred that name to synonymy under *P. plattensis*. Evidently there is still substantial disagreement on species limits within the complex among specialists on the group.

5. Packera plattensis (Nutt.) W.A. Weber & Á. Löve (prairie ragwort)

Senecio plattensis Nutt.

S. pseudotomentosus Mack. & Bush

Pl. 298 d–f; Map 1255

Plants perennial or sometimes biennial, from a short, usually stout, erect to horizontal rootstock, sometimes producing well-developed stolons. Stems mostly 1, occasionally 2 or 3, 20–70 cm long, more or less evenly pubescent with dense, felty hairs when young, the portion above the base sometimes becoming more or less glabrous at flowering, except sometimes for cobwebby patches in the leaf axils or at the inflorescence branch points. Basal leaves usually present at flowering, long-petiolate, the petioles with dense, felty hairs when young, the blades 1–9 cm long, unlobed or less commonly with few narrow, irregular lobes toward the base, ovate-elliptic to oblanceolate or rarely nearly circular, tapered to nearly truncate at the base, rounded to bluntly pointed at the tip, the margins with relatively sharp, sometimes irregular teeth (rarely scalloped or nearly entire), the surfaces evenly and more or less persistently pubescent with dense, felty hairs when young, usually becoming nearly glabrous by flowering time, except toward the base of the petiole. Stem leaves gradually reduced toward the stem tip, sessile or nearly so, the blades entire or more commonly irregularly and usually deeply pinnately lobed, the terminal lobe ovate to elliptic-obovate (rarely much narrower), longer than wide, the margins otherwise relatively sharply toothed, the surfaces evenly and more or less persistently pubescent with dense, felty hairs when young, usually becoming nearly glabrous by flowering time. Involucre 4–7 mm long, glabrous or somewhat cobwebby-hairy near the base. Ray florets usually 7–9, the lobe 5–10 mm long. Fruits 2.5–3.0 mm long, usually hairy along the ribs, occasionally glabrous or the surface evenly hairy. $2n=46, 92$. April–June.

Scattered nearly throughout the state (Pennsylvania to North Carolina west to Montana and Arizona; Canada). Upland prairies, loess hill prairies, glades, ledges and tops of bluffs, openings of mesic to dry upland forests, and rarely stream banks; also pastures, railroads, roadsides, and open, disturbed areas.

For a discussion of difficulties in distinguishing this species from the closely related *P. paupercula*, see the treatment of that species.

6. Packera pseudaurea (Rydb.) W.A. Weber & Á. Löve var. semicordata (Mack. & Bush) Trock & T.M. Barkley (western golden ragwort, squaw weed)

Senecio aureus L. var. *semicordatus* (Mack. & Bush) Greenm.

S. semicordatus Mack. & Bush

Pl. 297 h, i; Map 1256

Plants perennial, from a short, stout to slender, erect (but sometimes branched) or rarely horizontal rootstock, not producing stolons. Stems mostly 1, occasionally 2 or 3, 10–60 cm long, often pubescent with felty or cobwebby hairs toward the base when young, but usually glabrous or nearly so at flowering. Basal leaves usually present at flowering, long-petiolate, the petioles sometimes woolly or cobwebby when young, usually glabrous or nearly so at flowering, the blades 1–6(–8) cm long, unlobed or rarely with few narrow, irregular lobes toward the base, oblong-ovate to broadly lanceolate, abruptly short-tapered to more commonly truncate or shallowly cordate, the tissue not extending along the petiole or extending along only the terminal portion of the petiole, rounded at the tip, the margins with blunt to sharp teeth, the surfaces glabrous. Stem leaves gradually reduced toward the stem tip, sessile or nearly so, the blades mostly deeply pinnately lobed, sometimes irregularly so, the margins otherwise relatively sharply toothed, the surfaces glabrous. Involucre 3–7 mm long, glabrous or very sparsely hairy near the base. Ray florets usually 7–13, the lobe 6–13 mm long. Fruits 2.5–3.0 mm long, glabrous or hairy along the ribs. $2n=46$. April–June.

Scattered in western and northern Missouri (Indiana to Wisconsin west to North Dakota and Kansas; Canada). Bottomland prairies, bottomland forests, mesic upland forests in ravines, banks of streams and rivers, and occasionally ledges of bluffs; also pastures and ditches.

The var. *pseudaurea* and var. *flavula* (Greene) D.K. Trock occur to the west of the Great Plains. Barkley (1962, 1978) noted that populations in the eastern and southern portion of the species range are often difficult to separate from the closely related *P. aurea*, which is certainly the case in Missouri. In addition to the key characters, plants of *P. pseudaurea* tend to have basal leaves with ascending blades, whereas those of *P. aurea* frequently tend to droop or spread. Also, the involucral bracts in *P. pseudaurea* frequently have relatively little purplish coloration, whereas those of *P. aurea* often are noticeably purple, at least toward the tip.

1257. Packera tomentosa

1258. Senecio ampullaceus

1259. Senecio vulgaris

7. Packera tomentosa (Michx.) C. Jeffrey
Senecio tomentosus Michx.

Pl. 298 a–c; Map 1257

Plants perennial, from a short, stout, erect to horizontal rootstock, rarely producing a few stolons. Stems 1 to several, 15–60 cm long, evenly and persistently pubescent with dense, felty hairs, the apical portions sometimes becoming glabrous in patches. Basal leaves usually present at flowering, long-petiolate, the petioles with dense, felty hairs, the blades 2–14 cm long, unlobed or uncommonly with few narrow, irregular lobes toward the base, oblong-ovate to narrowly ovate, ovate, or oblong-elliptic, tapered abruptly to nearly truncate at the base, rounded to bluntly and broadly pointed at the tip, the margins scalloped or with relatively even, shallow, blunt teeth, sometimes nearly entire, the surfaces evenly and more or less persistently pubescent with dense, felty hairs, particularly the upper surface sometimes becoming nearly glabrous by flowering time. Stem leaves conspicuously shorter than the basal ones, mostly short-petiolate or sessile, the blades entire or irregularly pinnately lobed, the margins otherwise toothed or scalloped, the surfaces evenly and more or less persistently pubescent with dense, felty hairs, both surfaces or sometimes only the upper surface sometimes becoming nearly glabrous by flowering time. Involucre 4–7 mm long, cobwebby-hairy, at least near the base. Ray florets usually 11 or 13, the lobe 5–9 mm long. Fruits 2.5–3.0 mm long, hairy along the ribs. $2n=40, 46$. March–May.

Known from a single historical collection from Barry County (New Jersey to Florida west to Missouri, Oklahoma, and Texas). Banks of rivers.

Steyermark (1963) overlooked the single specimen collected at Eagle Rock and correctly determined by B. F. Bush in 1898. The species was first mapped from Missouri by Barkley (1980), who reported that his specimen notes for the species were lost and therefore cited no specimens. The specimen was rediscovered at the Missouri Botanical Garden Herbarium during the present study.

102. Senecio L. (groundsel)

Plants annual. Stems erect or ascending, unbranched or few-branched, sometimes lined or angled, glabrous or with patches of fine, woolly (cobwebby) hairs, sometimes with scattered, minute hairs toward the tip. Leaves in a basal rosette and alternate, progressively reduced in size from the stem base to tip, cobwebby-hairy when young, often becoming glabrous or nearly so at maturity, sessile or the lowermost long-tapered to a poorly defined petiole, the bases strongly clasping the stem. Leaf blades sometimes pinnately lobed, the margins otherwise shallowly or irregularly toothed or wavy to nearly entire, the venation pinnate. Inflorescences terminal and axillary from the uppermost leaves, panicles consisting of loose clusters to less commonly solitary flowers at the branch tips, rounded to broadly rounded in profile. Heads discoid or radiate, short- to long-stalked, with numerous florets. Involucre cylindrical to slightly wedge-shaped, the bracts in 2 series, glabrous or nearly so, those of the inner series more or less flat dorsally, those of the outer series minute, ascending and usually incurved. Disc corollas bright yellow. Style branches with a stigmatic line along each inner margin. Fruits narrowly oblong to narrowly oblong-elliptic in outline, not flattened, 5–10-ribbed, minutely hairy to less commonly nearly glabrous, brown. About 1,300 species, nearly worldwide, but most diverse in the Old World.

The description above applies only to the two species present in Missouri. Even in the restricted sense, *Senecio* on a worldwide basis includes plants with a bewildering array of growth forms and morphologies. Note that the native Missouri species formerly classified in *Senecio* have been transferred to the genus *Packera*.

1. Ray florets present, conspicuous; leaves unlobed, the margins shallowly toothed to nearly entire . . . . 1. S. AMPULLACEUS
1. Rays absent; leaves mostly pinnately lobed . . . . 2. S. VULGARIS

1. Senecio ampullaceus Hook. (Texas groundsel)

Pl. 299 c, d; Map 1258

Plants with thin, fibrous roots. Stems 20–70 cm long, when young with patches of fine, woolly (cobwebby) hairs, usually becoming glabrous or nearly so with age. Basal and lower leaves 3–10 cm long, the blades narrowly ovate, unlobed, bluntly to sharply pointed at the tip, tapered at the base to a poorly defined petiole, the margins otherwise shallowly toothed to nearly entire, moderately to densely pubescent with cobwebby hairs, often becoming glabrous or nearly so at maturity. Upper leaves sessile, narrowly ovate to narrowly lanceolate-triangular, unlobed, the margins shallowly toothed to nearly entire. Heads mostly erect or ascending. Involucre 7–11 mm long, the inner bracts 11–14, uniformly green, the outer series extending down the stalk, linear, uniformly green. Disc corollas 7–10 mm long. Ray florets mostly 8, the corolla lobe 11–17 mm long, the tip rounded, entire, minutely notched, or minutely 3-toothed at the tip. Fruits 2–3 mm long. April–May.

Introduced, known only from St. Louis (endemic to Texas). Railroads.

2. Senecio vulgaris L. (common groundsel)

Pl. 297 a, b; Map 1259

Plants more or less taprooted. Stems 8–40 cm long, when young with fine, woolly (cobwebby) hairs, sometimes becoming nearly glabrous with age, often reddish- or purplish-tinged. Basal and lower leaves 2–10 cm long, the blades narrowly elliptic to narrowly obovate, irregularly pinnately lobed, mostly bluntly pointed at the tip, tapered at the base to a poorly defined petiole, the margins otherwise irregularly few-toothed or with a few secondary lobes, sparsely to moderately pubescent with cobwebby hairs or scattered, minute, curled hairs. Upper leaves sessile, narrowly ovate to narrowly oblong-elliptic, mostly pinnately lobed, the margins irregularly few-toothed, wavy, or rarely with a few secondary lobes. Heads often somewhat nodding. Involucre 4–8 mm long, the inner bracts 18–22, green with minute, black tips, the outer series occasionally extending slightly down the stalk, linear, green with minute, black tips. Disc corollas 4–6 mm long. Ray florets absent. Fruits 2–3 mm long. $2n=40$. April–May.

Introduced, uncommon and sporadic (native of Eurasia, introduced widely in North America). Greenhouses, gardens, railroads, and open, disturbed areas.

Most of the specimens until now have originated from the St. Louis metropolitan region. This species sometimes is spread as weeds in bedding plants. As such, it eventually may become established in other urban areas of the state.

11. Tribe Vernonieae Cass.
(Robinson, 1999)

Plants perennial herbs (annual or woody elsewhere), the sap not milky. Stems not spiny or prickly. Leaves alternate, sometimes also in a basal rosette, not spiny or prickly. Leaf blades simple, unlobed, the venation pinnate, with 1 main vein. Inflorescences mostly terminal panicles. Heads discoid (radiate elsewhere). Involucre of 2 to several series of bracts, not spiny or tuberculate. Receptacle flat to slightly convex, naked. Disc florets all perfect. Pappus of bristles and/or scales, these occasionally finely barbed, usually shed individually before the fruit is dispersed. Corollas white, cream-colored, or pink, the tube usually expanded above the midpoint, the 5 short lobes spreading to ascending. Stamens with the filaments not fused together, the anthers fused into a tube, each tip with a short, often indistinct appendage, each base truncate or with a pair of short lobes. Style branches relatively long and usually somewhat flattened, each with a stigmatic line along each inner margin or sometimes these fused into a single band, the sterile tip usually truncate, minutely hairy. Fruits mostly angled or with longitudinal lines, not winged, not beaked. About 98 genera, 1,300–1,530 species, nearly worldwide, but most diverse in tropical and subtropical regions; apparently absent from Europe.

In this tribe, the genus *Stokesia* L'Hér., with its sole species, *S. laevis* (Hill) Greene of the southeastern United States, is gaining popularity as a garden perennial and cut flower. It usually is sold under the name Stokes aster.

1. Primary heads grouped into dense, headlike clusters, each cluster subtended by 3 leaflike bracts; pappus of a single series of 5 flattened, narrow, awnlike scales, these long-tapered to bristlelike tips . . . . 103. ELEPHANTOPUS
1. Heads separate, not grouped into secondary headlike clusters; pappus of an inner series of numerous capillary bristles and an outer series of minute scales or bristles . . . . 104. VERNONIA

103. Elephantopus L. (elephant's foot)
(Clonts, 1972; Clonts and McDaniel, 1978)

About 30 species, nearly worldwide, except Europe and northern Asia.

In addition to the single species recorded thus far from Missouri, two additional taxa of the southeastern United States have been collected in adjacent counties of Arkansas. Both differ from *E. carolinianus* in having the leaves mostly basal and stem leaves much smaller than the basal ones. *Elephantopus nudatus* A. Gray has slightly smaller heads and achenes, and the pappus bristles are flattened and broadened to a triangular base. *Elephantopus tomentosus* L. has relatively large heads and achenes and has usually woolly pubescence on the involucral bracts (and sometimes also the leaf undersurface). These species might possibly be discovered in the southernmost tier of Missouri counties in the future.

1. Elephantopus carolinianus Raeusch.
(Carolina elephant's foot)
Pl. 299 a, b; Map 1260

Plants with a relatively stout, spreading, somewhat rhizomatous rootstock. Stems 1 or less commonly 2 to several, 30–100 cm long, erect or ascending, sometimes from a spreading base, unbranched or branched toward the tip, moderately hairy, but often becoming nearly glabrous toward the tip at flowering. Leaves basal and alternate, sessile or with a short, indistinct petiole, the basal leaves sometimes absent at flowering, somewhat larger than the stem leaves. Stem leaves gradually reduced toward the stem tip, the blade 2–18(–23) cm long, elliptic-obovate to ovate or lanceolate, tapered or narrowed to a bluntly or more commonly sharply pointed tip, usually long-tapered at the base, the margins shallowly scalloped or toothed, rarely nearly entire, the surfaces sparsely to moderately hairy. Inflorescences terminal panicles, the primary heads grouped into dense, headlike clusters at the branch tips, each cluster of 3–20 heads subtended by 3 leaflike bracts, these 1–3 cm long, ovate-triangular, more or less folded lengthwise, overlapping, with cordate bases. Heads sessile or nearly so, with 4 florets. Involucre 6–10 mm long, cylindrical, the bracts in 2 alternating series with the outer series about half as long as the inner one, narrowly elliptic to narrowly lanceolate, somewhat rounded dorsally, sparsely to moderately hairy and with minute, impressed resin glands toward the tip, membranous toward the margins, the central portion green, sometimes purplish-tinged toward the tip, the basal portion often pale or whitened. Pappus of a single series of 5 flattened, narrow, awnlike scales, persistent at fruiting, these 4–5 mm long, tapered to bristlelike tips and gradually broadened toward the base, with minute, ascending barbs. Corollas 7–9 mm long, reddish purple to purple or less commonly white, relatively deeply

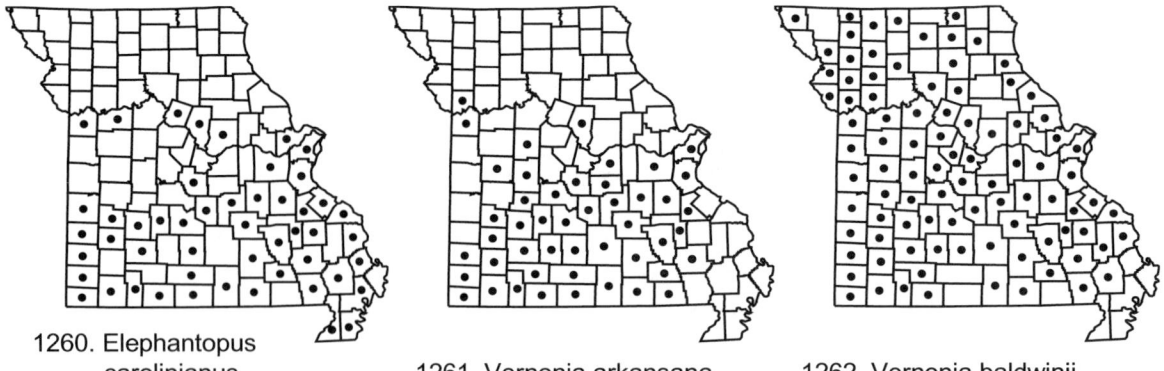

1260. Elephantopus carolinianus
1261. Vernonia arkansana
1262. Vernonia baldwinii

lobed, 1 of the sinuses between the lobes deeper than the other 4. Fruits (excluding the pappus) 3–4 mm long, narrowly oblong-obtriangular in outline, not flattened, with 8–10 relatively narrow ribs, hairy, light brown to brown. $2n=22$. August–October.

Scattered, mostly south of the Missouri River (Pennsylvania to Kansas south to Florida and Texas). Bottomland forests, mesic upland forests, and banks of streams and rivers; also pastures and roadsides.

104. Vernonia Schreb. (ironweed)
(Jones and Faust, 1978)

Plants with a stout, often short-rhizomatous rootstock. Stems 1 to several, several-branched toward the tip, erect or ascending, usually with fine longitudinal lines or ridges. Leaves alternate (basal leaves usually absent at flowering), sessile or short-petiolate, the margins toothed (occasionally entire in *V. arkansana*), the undersurface sometimes dotted with minute, impressed resin glands. Inflorescences irregularly branched terminal panicles, sometimes appearing somewhat flat-topped, the heads not grouped into secondary headlike clusters at the branch tips. Heads sessile to long-stalked, with 9 to more commonly numerous florets. Involucre urn-shaped to short-cylindrical, hemispherical or somewhat bell-shaped, the bracts in several overlapping series, the inner series progressively longer, variously shaped, flattened or with a slightly raised midvein dorsally, green or more commonly purplish-tinged to uniformly dark purple. Pappus of an inner series of numerous capillary bristles and an outer series of minute scales or bristles, persistent at fruiting, with minute, ascending barbs. Corollas reddish purple to purple or rarely white, relatively deeply lobed, the sinuses between the lobes all similar. Fruits narrowly oblong to narrowly oblong-triangular in outline, not flattened, with 8–10 relatively narrow ribs, usually hairy, at least along the ribs, often with minute resin dots between the ribs, grayish brown to brown. About 500 species, North America to South America, West Indies, Asia, Africa.

The taxonomy of *Vernonia* has not yet stabilized. Robinson (1999) studied the American species and confirmed the conclusions of earlier workers that the traditional, broad concept of *Vernonia* was unnatural. He suggested that *Vernonia* in the strict sense should be restricted to about 20 species of North America and the Bahamas plus 2 temperate South American species, but he noted that further studies were still necessary to assess the numerous Old World taxa.

The name ironweed may be attributable to the grayish cast of plants in some of the species, which is caused by the relatively dense pubescence on the leaves and sometimes also the stems. In the Midwest, ironweeds are a familiar sight in overgrazed pastures, presumably because the plants are unpalatable to cattle. Most species of *Vernonia* produce toxic sesquiterpene

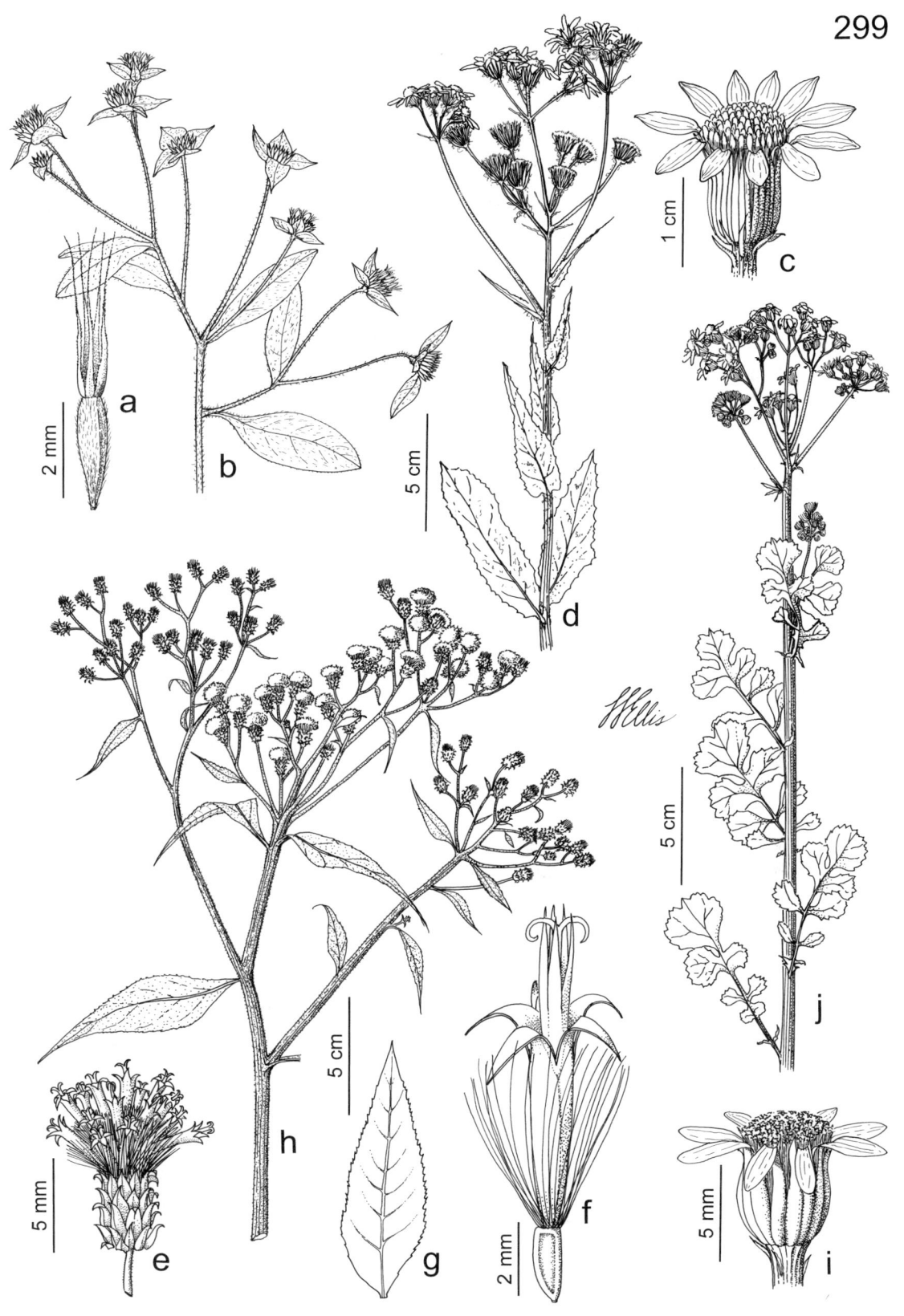

Plate 299. Asteraceae. *Elephantopus carolinianus*, **a)** fruit, **b)** habit. *Senecio ampullaceus*, **c)** head, **d)** habit. *Vernonia baldwinii* ssp. *interior*, **e)** head, **f)** floret, **g)** leaf, **h)** habit. *Packera glabella*, **i)** head, **j)** habit.

lactones, although the midwestern species have not yet been implicated directly in livestock or human poisoning. Lactones (vernoniosides, especially vernolepin) extracted from the African *V. hymenolepis* A. Rich. have been shown to inhibit tumor formation (Burrows and Tyrl, 2001). Some of the North American species were used medicinally by Native Americans, principally as pain relievers (Moerman, 1998).

Readers should note that the North American ironweeds have a notorious reputation for promiscuity, and that the numerous hybrids likely to be encountered in nature are fertile. In Missouri, all possible parental combinations have been recorded among the five species growing in the state, although some hybrids are less frequent than others.

1. Involucral bracts (except the outermost ones) linear to narrowly lanceolate, long-tapered and somewhat curled to a threadlike, sharply pointed tip
 . . . . 1. V. ARKANSANA
1. Involucral bracts ovate to oblong-obovate or oblong-lanceolate, rounded, angled or short-tapered to an appressed to spreading (but not appearing curly), rounded to sharply pointed tip
    2. Leaf blades with the undersurface glabrous, appearing dotted with minute, impressed resin glands (these sometimes difficult to observe in fresh material, but darkening and becoming more noticeable in dried leaves)
     . . . . 3. V. FASCICULATA
    2. Leaf blades with the undersurface moderately to densely but sometimes minutely hairy, the hairs obscuring any glandular dots
        3. Involucral bracts angled or more commonly short-tapered to a sharply pointed tip, this loosely ascending or recurved-spreading, usually with abundant minute, impressed resin glands on both sides of the noticeably keeled midvein . . . . 2. V. BALDWINII
        3. Involucral bracts rounded or broadly angled to a bluntly pointed tip, sometimes abruptly short-tapered to a minute, sharp point, glandless or with sparse, minute, impressed resin glands, the midvein not or only inconspicuously keeled
            4. Disc florets 13–30 per head; stems minutely hairy, sometimes becoming nearly glabrous toward the base; leaf blades minutely hairy on the undersurface, especially along the veins (occasionally sparse, longer hairs also present along the veins) . . . . 4. V. GIGANTEA
            4. Disc florets 32–60 per head; stems with longer, spreading or often bent to tangled hairs toward the tip, sometimes minutely hairy toward the base; leaf blades with longer, often bent or tangled hairs, more or less evenly distributed on the undersurface . . . . . 5. V. MISSURICA

1. Vernonia arkansana DC.
V. crinita Raf.

Pl. 300 g–j; Map 1261

Stems 70–160 cm long, glabrous or nearly so, sometimes minutely hairy toward the tip, often appearing somewhat glaucous. Leaf blades 8–20 cm long, linear to narrowly elliptic-lanceolate, tapered at both ends, the margins sharply toothed or less commonly entire or nearly so, usually appearing somewhat turned under, both surfaces glabrous to sparsely hairy and appearing dotted with minute, impressed resin glands (these sometimes difficult to observe in fresh material, but darkening and becoming more noticeable in dried leaves). Heads with 50–120 florets. Involucre 9–16 mm long, hemispherical or somewhat bell-shaped, the bracts 6–14 mm long, all but the outermost ones linear to narrowly lanceolate, long-tapered and somewhat curled to a threadlike, sharply pointed tip, cobwebby-hairy along the margins and sometimes also minutely hairy on the outer surface, green, the midvein indistinct. Pappus brownish purple, the inner bristles 6–7 mm long, the outer scales 0.7–1.0 mm long. Corollas 10–12 mm long. Fruits 4–5 mm long. $2n=34$. July–October.

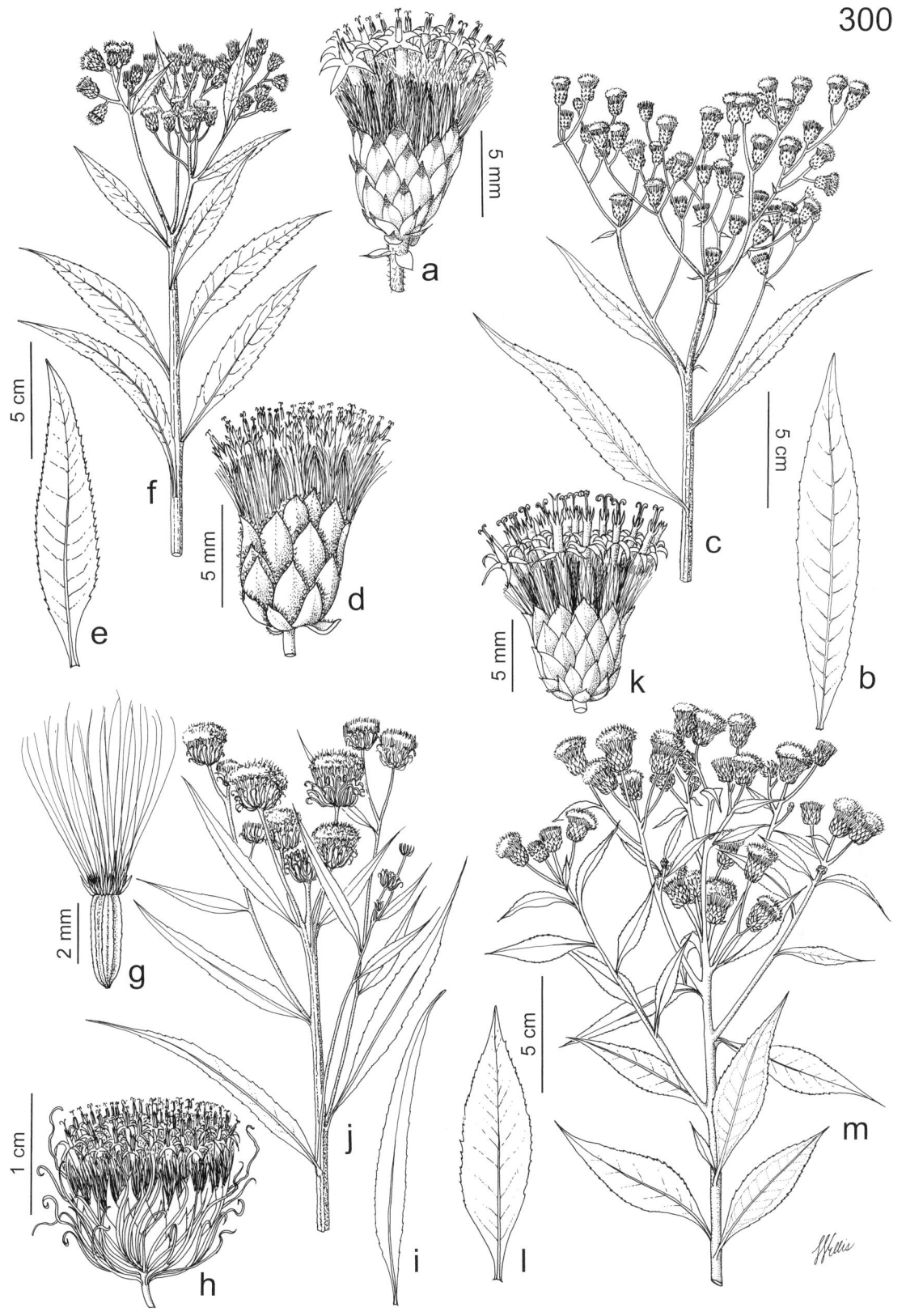

Plate 300. Asteraceae. *Vernonia fasciculata*, **a)** head, **b)** leaf, **c)** habit. *Vernonia gigantea*, **d)** head, **e)** leaf, **f)** habit. *Vernonia arkansana*, **g)** fruit, **h)** head, **i)** leaf, **j)** habit. *Vernonia missurica*, **k)** head, **l)** leaf, **m)** habit.

Scattered in the Ozark and Ozark Border Divisions, and uncommon in the Big Rivers (Kansas and Oklahoma east to Missouri and Arkansas; introduced north- and eastward to Wisconsin and New York). Banks of streams, margins of sloughs, fens, openings of bottomland forests, mesic upland forests, bottomland prairies, and rarely glades; also pastures and roadsides.

Vernonia arkansana is a characteristic species of stream banks and gravel bars in the Ozarks, but it can occur in a variety of other habitats. Hybrids with other Missouri species of *Vernonia* are relatively easily noted, because *V. arkansana* is relatively distinctive in its larger heads and elongate involucral bracts. Cora Steyermark (1939) grew seeds harvested from natural Ozarkian populations of *V. arkansana* (as *V. crinita*) in a garden, and she concluded that hybridization involving this species was relatively common, as the progeny often exhibited variable and intermediate morphologies. Hybrids with *V. baldwinii* are especially common (Jones et al., 1970) and often occur where a road near a stream brings drier roadside and moister streamside habitats into close proximity. Harms (1969) studied ironweeds in southeasternmost Kansas and documented a population showing introgression between *V. arkansana*, *V. baldwinii*, and *V. missurica*.

2. Vernonia baldwinii Torr. (western ironweed)
Pl. 299 e–h; Map 1262

Stems 60–150 cm long, with relatively long, spreading or often bent hairs, occasionally with only minute hairs toward the tip. Leaf blades 6–18 cm long, lanceolate to ovate or elliptic-ovate, tapered to angled at the base, tapered at the tip, the margins finely to less commonly coarsely toothed, the upper surface minutely hairy, the undersurface with more or less evenly distributed, relatively long, often bent or tangled hairs. Heads with 17–34 florets. Involucre 4.5–8.0 mm long, short-cylindrical to somewhat hemispherical or bell-shaped, the bracts 4–7 mm long, lanceolate to ovate, angled or more commonly short-tapered to a sharply pointed tip, loosely ascending or the tip recurved-spreading, glabrous or minutely hairy, usually with abundant minute, impressed resin glands on both sides of the midvein, the margins usually cobwebby-hairy, brownish green to dark purple, the midvein usually noticeably keeled. Pappus tan to brown, often purplish-tinged, the inner bristles 5.0–6.5 mm long, the outer scales 0.5–0.8 mm long. Corollas 8–10 mm long. Fruits 2.5–4.0 mm long. $2n=34$. May–September.

Scattered to common nearly throughout Missouri, but only scattered in portions of the Glaciated Plains and Mississippi Lowlands (Colorado and New Mexico east to Minnesota, Illinois, and Louisiana). Upland prairies, glades, tops of bluffs, savannas, mesic to dry upland forests, less commonly banks of streams and margins of ponds; also pastures, fencerows, old fields, railroads, roadsides, open, disturbed areas, and occasionally banks of ditches.

Vernonia baldwinii is by far the most abundant species of ironweed in Missouri and is thus most frequently involved in hybridization with the other species. Faust (1972) studied morphological, cytological, and phytochemical variation in *V. baldwinii* and concluded that it comprises two reasonably distinct subspecies, the eastern ssp. *baldwinii* and the widespread but mostly western ssp. *interior*, which had been treated as separate species by most earlier authors. Where the ranges of these two taxa overlap, including much of Missouri, some populations occur that are intermediate between the two. The key and descriptions below do not accommodate these intermediates and the distributions of the subspecies within the state similarly are unclear.

1. Middle and upper involucral bracts loosely ascending, usually hairy on the inner surface . . . . 2A. SSP. BALDWINII
1. Middle and upper involucral bracts with the tip recurved-spreading, usually glabrous on the inner surface
. . . . 2B. SSP. INTERIOR

2a. ssp. baldwinii
V. duggeriana Daniels
V. flavipapposa Daniels
V. parthenioides Daniels
V. interior Small var. *baldwinii* (Torr.) Mack. & Bush

Stems with relatively long, spreading or often bent hairs, these sometimes tangled. Heads with 23–34 florets. Involucre 4.5–6.7 mm long. Middle and upper involucral bracts loosely ascending, usually hairy on the inner surface. $2n=34$. May–September.

Common in the Ozark, Ozark Border, and Unglaciated Plains Divisions, but only scattered in the Glaciated Plains and Mississippi Lowlands (Kansas to Texas east to Illinois, and Louisiana). Upland prairies, glades, tops of bluffs, savannas, mesic to dry upland forests, less commonly banks of streams and margins of ponds; also pastures, fencerows, old fields, railroads, roadsides, and open, disturbed areas.

A hybrid with *V. missurica* has been called *V.* ×*peralta* Daniels (*V. pseudobaldwinii* Daniels,

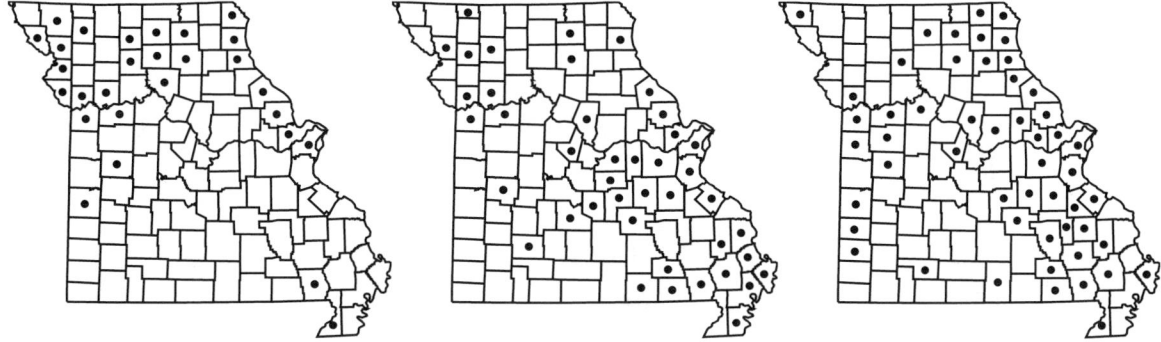

1263. Vernonia fasciculata   1264. Vernonia gigantea   1265. Vernonia missurica

V. pseudodrummondii Daniels) (Faust, 1972; Yatskievych and Turner, 1990). White-flowered plants, which occur sporadically within populations, have been called f. *albiflora* E.J. Palmer & Steyerm.

2b. ssp. interior (Small) W.Z. Faust
   V. interior Small
   V. baldwinii var. *interior* (Small) B.G. Schub.
   Pl. 299 e–h

Stems with relatively long, spreading or often bent hairs or sometimes with only minute hairs toward the tip. Heads with 17–27 florets. Involucre 5.5–8.0 mm long. Middle and upper involucral bracts with the tip recurved-spreading, usually glabrous on the inner surface. $2n=34$. May–September.

Common nearly throughout Missouri, but only scattered in the eastern portion of the Glaciated Plains, Mississippi Lowlands, and southeastern portions of the Ozarks (Colorado and New Mexico east to Minnesota, Illinois, and Louisiana). Upland prairies, glades, tops of bluffs, savannas, mesic to dry upland forests, less commonly banks of streams and margins of ponds; also pastures, fencerows, old fields, railroads, roadsides, open, disturbed areas, and occasionally banks of ditches.

White-flowered plants, which occur rarely within populations, have been called *V. baldwinii* f. *alba* Waterf.

3. Vernonia fasciculata Michx. **ssp. fasciculata** (prairie ironweed)
   Pl. 300 a–c; Map 1263

Stems 50–140 cm long, glabrous or nearly so, sometimes minutely hairy toward the tip. Leaf blades 4–18 cm long, narrowly to broadly lanceolate, tapered at both ends, the margins sharply toothed, the upper surface glabrous or occasionally sparsely and minutely hairy toward the margins, sometimes with sparse, minute, impressed resin glands, the undersurface glabrous, appearing dotted with minute, impressed resin glands (these sometimes difficult to observe in fresh material, but darkening and becoming more noticeable in dried leaves). Heads with 10–26 florets. Involucre 4–9 mm long, short-cylindrical to somewhat bell-shaped, the bracts 3–7 mm long, ovate to oblong-ovate, rounded or broadly angled to a bluntly pointed tip, the innermost bracts sometimes with a minute, sharp point, appressed, glabrous, the margins occasionally minutely hairy, green or more commonly purplish-tinged, especially along the midvein and toward the tip, the midvein not keeled or only slightly so toward the tip. Pappus tan to brownish purple, the inner bristles 5–6 mm long, the outer scales 0.5–1.0 mm long. Corollas 9–11 mm long. Fruits 2.8–3.5 mm long. $2n=34$. July–September.

Scattered in the Glaciated Plains and Unglaciated Plains Divisions, and uncommon in the Ozark and Big Rivers Divisions (South Dakota to Oklahoma east to Massachusetts, Kentucky, and Mississippi). Bottomland prairies, marshes, fens, margins of lakes, and banks of streams and rivers; also fallow fields, railroads, and roadsides.

The *V. fasciculata* complex was studied by Jones (1972). Although *V. fasciculata* has been documented to hybridize with the other four species of *Vernonia* in Missouri, hybrids with *V. arkansana* are uncommon as the two taxa rarely grow in proximity to one another. Steyermark noted that some of the Missouri hybrids with *V. gigantea* were misdetermined by earlier botanists as representing records of *V. fasciculata* ssp. *corymbosa* (Schwein. ex Keating) S.B. Jones (var. *nebraskensis* Gleason, var. *corymbosa* (Schwein. ex Keating) B.G. Schub.). This western subspecies, which occurs from Kansas to Minnesota west to Colorado and Montana (also adjacent Canada), differs from ssp. *fasciculata* in its generally shorter stems and shorter, broader leaves with the upper surface usually somewhat roughened.

4. Vernonia gigantea (Walt.) Trel. ex Branner & Coville **ssp. gigantea** (tall ironweed)
V. altissima Nutt.
V. altissima var. *taeniotricha* S.F. Blake
Pl. 300 d–f; Map 1264

Stems 80–200(–300) cm long, minutely hairy, sometimes becoming nearly glabrous toward the base. Leaf blades 6–30 cm long, lanceolate to oblong-lanceolate and narrowly ovate, occasionally elliptic-oblanceolate, tapered at both ends, the margins sharply toothed, the upper surface glabrous or occasionally somewhat roughened toward the margins, the undersurface minutely hairy, especially along the veins, occasionally sparse, longer hairs also present along the veins. Heads with 13–30 florets. Involucre 3–7 mm long, short-cylindrical to somewhat hemispherical or bell-shaped, the bracts 2.0–5.5 mm long, ovate to oblong-ovate or oblong-lanceolate, rounded or broadly angled to a bluntly pointed tip, the innermost bracts rarely abruptly tapered to a minute, sharp point, appressed, glabrous or sparsely and minutely hairy, the margins occasionally also minutely hairy, purplish-tinged to uniformly dark purple, the midvein not keeled or only slightly so toward the tip. Pappus tan to brownish purple, the inner bristles 5–6 mm long, the outer scales 0.2–0.8 mm long. Corollas 9–11 mm long. Fruits 2.8–3.5 mm long. $2n=34$. August–October.

Scattered nearly throughout the state, but apparently absent from the Unglaciated Plains Division and portions of the Ozarks and Glaciated Plains (eastern U.S. west to Iowa and Texas; Canada). Banks of streams, rivers, and spring branches, margins of ponds and lakes, bottomland forests, swamps, bottomland prairies, and fens; also pastures, ditches, and roadsides.

This taxon is the most widely distributed member of a small species complex otherwise restricted mainly to the southeastern United States (Urbatsch, 1972). The species consists of two subspecies. The ssp. *ovalifolia* (Torr. & A. Gray) Urbatsch of southern Georgia and Florida differs in its generally shorter, broader leaves and narrower heads. Hybrids between *V. gigantea* and the other four Missouri species of *Vernonia* are relatively uncommon. The hybrid with *V. baldwinii* has been called *V. ×sphaeroidea* Nutt. Hybrids with *V. missurica* are known as *V. ×illinoensis* Gleason (Faust, 1972, 1977), but the older name *V. drummondii* Shuttlew. ex H. Werner (which has not been typified and is thus of somewhat ambiguous application) also may represent this hybrid combination (Faust, 1977). A white-flowered variant that occurs sporadically within populations has been called *V. altissima* f. *alba* Moldenke.

5. Vernonia missurica Raf. (Missouri ironweed)
V. reedii Daniels
Pl. 300 k–m; Map 1265

Stems 75–200 cm long, with relatively long, spreading or often bent to tangled hairs toward the tip, sometimes minutely hairy toward the base. Leaf blades 6–20 cm long, lanceolate to narrowly ovate, tapered to angled at the base, tapered at the tip, the margins coarsely toothed to nearly entire, the upper surface minutely hairy or roughened, sometimes nearly glabrous, the undersurface with more or less evenly distributed, relatively long, often bent or tangled hairs. Heads with 32–60 florets. Involucre 7.0–10.5 mm long, short-cylindrical to somewhat hemispherical or bell-shaped, the bracts 4–10 mm long, ovate to oblong-ovate, rounded or broadly angled to a bluntly pointed tip, the innermost bracts often abruptly tapered to a minute, sharp point, appressed, glabrous or minutely hairy, glandless or with sparse, minute, impressed resin glands toward the tip, the margins usually cobwebby-hairy, usually uniformly dark purple, sometimes green along the midvein, the midvein not keeled or only slightly so toward the tip. Pappus tan to brownish purple, the inner bristles 6–8 mm long, the outer scales 0.6–1.0 mm long. Corollas 9–11 mm long. Fruits 4–6 mm long. $2n=34$. July–September.

Scattered in the eastern half of the state, uncommon farther west (Michigan and Ohio to Louisiana and Florida west to Nebraska and New Mexico; Canada; possibly introduced in Massachusetts). Banks of streams, rivers, and spring branches, bottomland forests, swamps, fens, margins of sinkhole ponds, bottomland to upland prairies; also pastures, ditches, old fields, fencerows, railroads, roadsides, and open, disturbed areas.

Gleason and Cronquist (1991) suggested that in the eastern portion of the species range, many of the populations of *V. missurica* show the effects of introgression with *V. gigantea*. In Missouri, hybridization with *V. baldwinii* appears to be more common. Faust (1972) noted that *V. missurica* tends to inhabit wetter sites than does *V. baldwinii*, but both species occur in a variety of disturbed habitats. Rare, white-flowered plants have been called f. *swinkii* Steyerm., but these have not yet been recorded from Missouri.

BALSAMINACEAE (Touch-Me-Not Family)
Contributed by Alan E. Brant

Two genera, 400–850 species, North America, Central America, Caribbean Islands, Europe, Africa, Asia, south to Java.

1. Impatiens L. (jewelweed, touch-me-not)

Plants annual herbs (perennial elsewhere). Stems ascending, often somewhat succulent, simple or branched, hollow between the nodes, the nodes usually somewhat swollen. Leaves alternate (opposite or whorled elsewhere), mostly petiolate, the blades simple, the margins finely toothed or scalloped. Stipules absent. Inflorescences axillary, the flowers solitary or 2–5 in small clusters or panicles, some of the flowers usually small and nonopening (cleistogamous). Flowers zygomorphic, perfect, hypogynous, stalked, twisted at the base during development so that the top of the flower is oriented toward the bottom at maturity (resupinate). Sepals 3, the apparent lowest one petaloid and inflated into a conical pouch, this tapered to a slender nectar-producing spur that is usually recurved or strongly bent at maturity, the two lateral sepals 2–7 mm long, free, somewhat cupping the flower, broadly and obliquely ovate, tapered abruptly to a sharply pointed tip, and green or pale-colored. Petals 5, but appearing as 3, each of the 2 apparent lateral ones lobed, representing a fused pair, the single apparently upper petal free, wider than long, and keeled. Stamens 5, the filaments short and flat, fused toward the tips, each with a scalelike appendage on the inner side, these appendages fused into a cap over the pistil, the short, stout anthers often also somewhat fused, the whole complex of stamens and cap shed as a unit before the stigmas mature. Pistil 1 per flower, of 5 fused carpels, the ovary superior, 5-locular, the placentation axile. Ovules 3 to many per locule. Style absent or very short, the stigmas 5, minute. Fruits capsules, the 5 fleshy valves elastic, coiling violently from the base to the tip at dehiscence, scattering the seeds. Four hundred to 850 species, North America, Central America, Caribbean Islands, Europe, Africa, Asia, south to Java.

The common name touch-me-not refers to the intriguing explosive dehiscence of the capsules when touched, a trait that makes a patch of fruiting plants irresistible to children of all ages; hence, they are among our most familiar native wildflowers. The name jewelweed apparently refers to the shiny, silvery appearance of the wet foliage. The juice from crushed plants is reputed to counteract the itching effects of chiggers, mosquito bites, and poison ivy when rubbed on the skin. Native Americans had multiple medicinal uses for species of *Impatiens*, including as treatments for dermatitis, burns, fever, and gastrointestinal ailments; they also boiled plants for use as dyes (Moerman, 1998). Several species are cultivated as ornamentals. The native species are easily grown in moist soil of sunny locations but can spread aggressively by seeds.

1. Plants hairy, at least when young; leaf blades with the margins sharply and finely toothed; flowers purple to red, pink, or white . . . . 1. I. BALSAMINA
1. Plants glabrous, the stems and leaves usually somewhat glaucous; leaf blades with the margins coarsely to finely scalloped or bluntly toothed; flowers orange or yellow
   2. Flowers orange with reddish brown spots (rarely pale yellow, with or without spots); spurred sepal with the conical pouched portion usually about twice as long as wide, the slender spur (6–)7–10(–12) mm long . . . . 2. I. CAPENSIS
   2. Flowers lemon yellow, with or without red spots; spurred sepal with the conical, pouched portion about as long as wide or slightly longer than wide, the slender spur 4–6 mm long . . . . 3. I. PALLIDA

1266. Impatiens balsamina 1267. Impatiens capensis 1268. Impatiens pallida

1. Impatiens balsamina L. (balsam, garden balsam)

Pl. 301 a; Map 1266

Plants (stems, leaves, sepals, and fruits) pubescent with short, curved, somewhat glandular hairs, at least when young. Stems 30–80 cm long. Leaves 3–10(–15) cm long, the blade oblanceolate to narrowly elliptic, tapered gradually to the short petiole, tapered to the sharply pointed tip, the margins sharply and finely toothed, the basal portion and petiole usually with several small, hemispheric to stalked or tack-shaped, dark-colored glands. Inflorescences of solitary or less commonly pairs of axillary flowers, these purple to red, pink, white, or combinations thereof. Spurred sepal with the pouched portion 6–14 mm long, broadly conical, wider than long, the spur 1.3–2.2 cm long, gradually recurved. Fruits 1.2–2.0 cm long, asymmetrically elliptic in outline. Seeds 2.5–4.0 mm long, oblong-ovate in outline, rounded at the tip, bluntly 4-angled, the surface pebbled to finely warty, dark brown, with minute, lighter flecks. $2n=14$. July–September.

Introduced, known only from St. Louis County and City (native of Asia, widely cultivated and escaping sporadically; in the eastern U.S. west to Wisconsin and Louisiana). Railroads, creek beds, and moist, disturbed areas.

2. Impatiens capensis Meerb. (jewelweed, spotted touch-me-not)

I. biflora Walter
I. fulva Nutt.

Pl. 301 b–d; Map 1267

Plants glabrous. Stems 30–150 cm long, often slightly glaucous. Leaves 2–22 cm long, the blade ovate to elliptic, narrowed or tapered to the short or long petiole, narrowed to the bluntly or sharply pointed tip, the margins coarsely to finely scalloped or bluntly toothed, the teeth usually ending in minute, sharp points, the basal portion (but not the petiole) sometimes with few to several small-stalked, dark-colored glands, the undersurface (but usually not the upper surface) somewhat glaucous. Inflorescences of solitary flowers or more commonly of small panicles of 2–5 flowers, these orange or rarely cream-colored to white, with abundant red to reddish brown spots. Spurred sepal with the pouched portion 12–22 mm long, conical, usually about twice as long as wide, the slender spur (6–)7–10(–12) mm long, strongly recurved, with most of the length appearing parallel to the sepal body. Fruits 1.4–2.0 cm long, elliptic to oblanceolate in outline. Seeds 4–5 mm long, elliptic-ovate in outline, the tip nipplelike or beaklike, strongly and irregularly 4-angled, the surface with a network of low ridges in several irregular columns, sometimes appearing irregularly warty, dark brown. $2n=20$. June–September.

Scattered to common throughout Missouri, possibly in every county (eastern U.S. west to North Dakota, Colorado, and Texas, also Idaho, Oregon, and Washington; Canada; introduced in Europe). Banks of streams, rivers, spring branches, and sloughs, margins of ponds and sinkhole ponds, swamps, bottomland forests, and bases of bluffs; also moist, disturbed areas.

The two native species (*I. capensis* and *I. pallida*) cannot be distinguished reliably without flowers, although *I. capensis* generally tends to be somewhat less glaucous on the stem and upper leaf surface. The two jewelweeds often grow together, but there seems to be no evidence of hybridization between them (Wood, 1975). Also, the effective pollinators of the two appear to be different. The rather narrow saccate sepal and orange and red color of *I. capensis* favor the ruby-throated hummingbird (*Archilochus colubris*), which seems to be the only legitimate visitor to the flowers both before and after the stamens fall (various bees collect pollen,

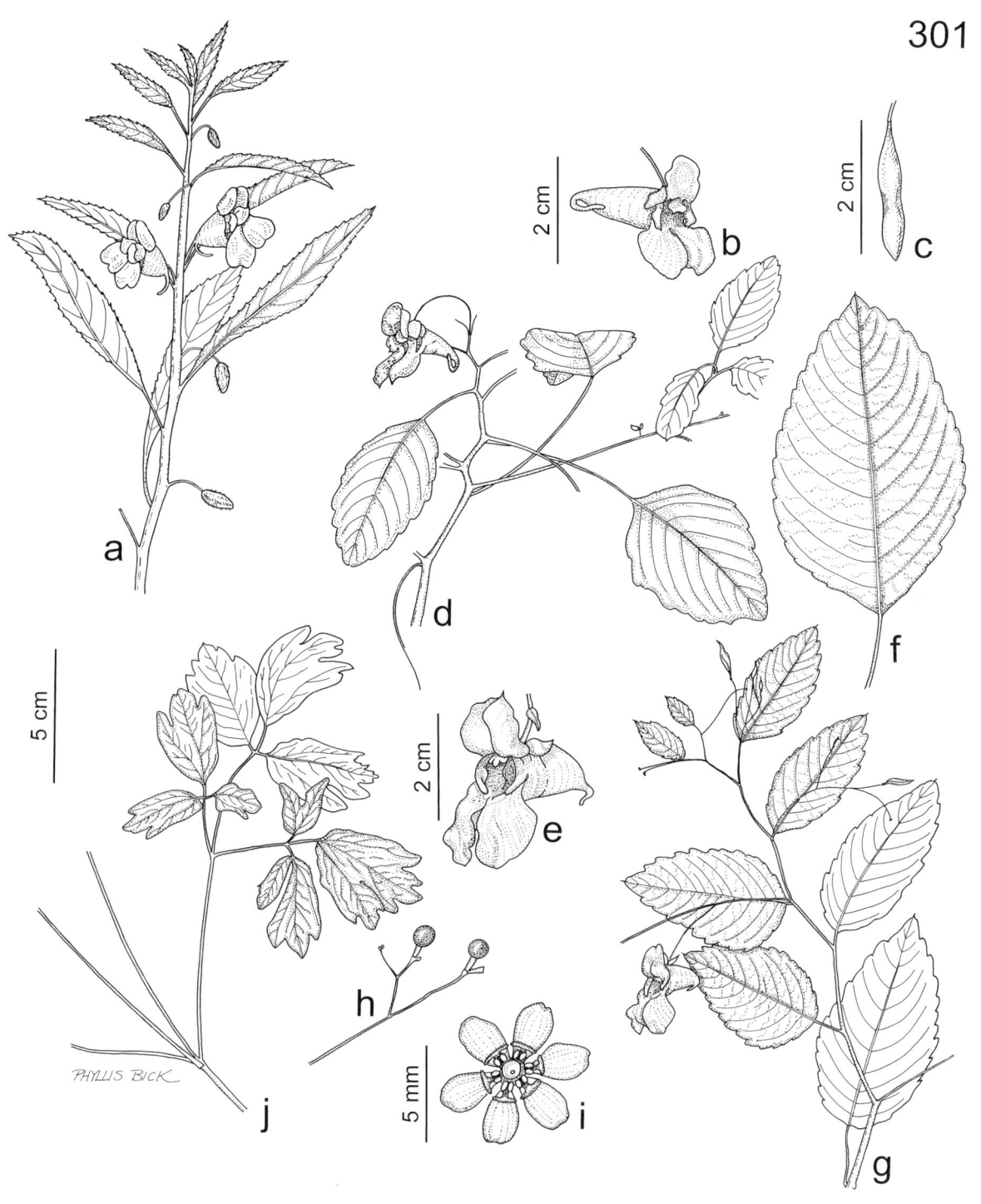

Plate 301. Balsaminaceae, Berberidaceae. *Impatiens balsamina*, **a)** habit. *Impatiens capensis*, **b)** flower, **c)** fruit, **d)** habit. *Impatiens pallida*, **e)** flower, **f)** leaf, **g)** habit. *Caulophyllum thalictroides*, **h)** fruiting branch, **i)** flower, **j)** node with leaf.

and bees and wasps bite holes in the nectar spur, but these do not pollinate the flowers). *Impatiens pallida,* with its yellow flowers and larger, saccate sepal, is visited by species of bumblebees (*Bombus* spp.), which Robertson (1889) noted were more constant in their visits to flowers of this species than to those of *I. capensis.*

A relatively small proportion of the showy, fully opening flowers of our two native species produce seed. The bulk of seed production appears to be from cleistogamous flowers (see Wood, 1975, for a summary with bibliography). These closed flowers appear morphologically similar to the developing buds of the other flower type, but the perianth never expands, and the stamens do not develop the cap of tissue that would otherwise isolate the stigmas from the anthers. The perianth is forced to drop off as the developing fruit elongates.

A number of rare color forms have been described for *I. capensis,* only one of which has been discovered thus far in Missouri: f. *albiflora* (E.L. Rand & Redfield) Fernald & B.G. Schub., which has white or cream-colored flowers with pink or brownish red spots and has been collected thus far in Washington County. Elsewhere, an unspotted form is f. *immaculata* (Weath.) Fernald & B.G. Schub.; plants having lemon yellow flowers and red spots are f. *citrina* (Weath.) Fernald & B.G. Schub.; and those having cream-colored flowers with pink coloring over most of the surface and with pink spots are f. *peasei* (A.H. Moore) Fernald & B.G. Schub.

3. Impatiens pallida Nutt. (jewelweed, pale touch-me-not)

Pl. 301 e–g; Map 1268

Plants glabrous. Stems 30–150 cm long, usually somewhat glaucous. Leaves 2–22 cm long, the blade ovate to elliptic, narrowed or tapered to the short or long petiole, narrowed to the bluntly or sharply pointed tip, the margins coarsely to finely scalloped or bluntly toothed, the teeth usually ending in minute, sharp points, the basal portion (but not the petiole) sometimes with few to several small-stalked, dark-colored glands, the leaf surfaces somewhat glaucous. Inflorescences of solitary flowers or more commonly of small panicles of 2–5 flowers, these lemon yellow, with or without red spots. Spurred sepal with the pouched portion 10–19 mm long, broadly conical, about as long as wide or slightly longer than wide, the slender spur 4–6 mm long, bent abruptly to somewhat recurved, usually appearing to spread at about a right angle to the sepal body. Fruits 1.4–2.0 cm long, elliptic to oblanceolate in outline. Seeds 5–6 mm long, elliptic-ovate in outline, the tip nipplelike or beaklike, strongly and irregularly 4-angled, the surface with a network of low ridges, sometimes appearing irregularly warty, dark brown. $2n=20$. June–September.

Scattered to common nearly throughout Missouri, probably more common than specimens indicate (eastern U.S. west to North Dakota and Oklahoma; Canada). Banks of streams, rivers, spring branches, and sloughs, margins of ponds and sinkhole ponds, swamps, bottomland forests, and bases of bluffs.

For notes on the pollination biology and relationships to *I. capensis,* see the treatment of that species. Rare color forms of *I. pallida* include f. *dichroma* Steyerm., first described from Missouri, with the saccate sepal mostly yellow and the petals white. Plants with creamy-white flowers (f. *speciosa* Jenn.) have not been recorded from Missouri thus far.

BERBERIDACEAE (Barberry Family)
Contributed by Alan Whittemore

Shrubs or rhizomatous perennial herbs, stems sometimes spiny. Leaves basal, alternate, or subopposite, simple or compound. Flowers actinomorphic, perfect, hypogynous; sometimes with 3–4 bractlets adjacent to the calyx. Calyces of 6 free sepals, sometimes falling as flowers open. Corollas of 6–9 free petals, these showy or inconspicuous. Stamens 6–18, free, the anthers opening by apical flaps or longitudinal slits. Pistil 1 per flower, of apparently 1 carpel. Ovary superior, with 1 locule, the placentation basal. Style short or absent, the stigma appearing sessile. Fruits berries or capsules, or the fruit wall rupturing early in development and the seeds then exposed at maturity. Seeds 1–50 per fruit. Fifteen genera, about 650 species, North America, Europe, Asia, and the mountains of South America and east Africa.

The structure of the pistil in plants of this family is very simple, with a single unlobed stigma and an ovary having a single locule containing basal ovule(s). It is not clear whether the pistil is composed of a single carpel, like the pistils of Ranunculaceae and Menispermaceae, or of several fused carpels, as in Papaveraceae and Fumariaceae.

In addition to the five species included here, there are unconfirmed reports of *Jeffersonia diphylla* (L.) Pers. (twinleaf) from Missouri. Steyermark (1963) excluded the species from the flora and did not accept an anecdotal report by B. F. Bush of a population in a creek bottom in Taney County. More recently, members of the Webster Groves Nature Study Society have suggested that the species has escaped from plantings at an estate in Jefferson County. *Jeffersonia diphylla,* which superficially resembles *Stylophorum diphyllum* (Papaveraceae), is a rhizomatous perennial herb to 30 cm tall with basal leaves having the blades divided into two asymmetrically ovate or kidney-shaped leaflets 7–12 cm long. The solitary, long-stalked flowers open while the leaves are expanding, have white to light pink petals 11–23 mm long, and have stamens with the anthers attached at the base and opening by longitudinal slits. The fruit is a tough-walled, ellipsoidal capsule that opens by the loss of a terminal lid. Twinleaf is known from Illinois and Iowa, and botanists should continue to search for it in Missouri.

1. Shrubs with stems usually spiny . . . . 1. BERBERIS
1. Rhizomatous perennial herbs lacking spines
    2. Leaves compound; inflorescences of 4–16 flowers; fruit wall rupturing early in development, the seeds exposed at maturity . . . . 2. CAULOPHYLLUM
    2. Leaves simple, often deeply lobed; flowers solitary; fruit a berry
    . . . . 3. PODOPHYLLUM

1. Berberis L. (barberry)

Plants shrubs. Stems with the leaves mostly replaced by spines, the normal foliage leaves clustered on very short shoots in the axils of the spines. Leaves many per plant, simple and unlobed, pinnately veined. Inflorescences axillary racemes or umbels, or flowers rarely solitary in leaf axils. Flowers with 3 small bractlets immediately below the calyx. Sepals 6, 3–5 mm long, petaloid, yellow. Petals 6, 2.5–3.0 mm long, yellow, bearing nectaries. Stamens 6, the anthers attached at the middle, opening by 2 apical flaps. Fruits berries. Seeds 4–6 mm long, narrowly to broadly ovoid, the seed coat hard, tan or more commonly reddish brown to black. About 500 species, North America, Europe, Asia, and the mountains of South America and east Africa.

The Missouri species of *Berberis* are deciduous shrubs with spiny stems, simple leaves, lax inflorescences, and red berries, but many species from other parts of the world are quite different. One such species, *B. aquifolium* Pursh (*Mahonia aquifolium* (Pursh) Nutt.), is widely grown as an ornamental in Missouri under the name Oregon grape. It is an unbranched evergreen shrub with spineless stems, pinnately compound leaves with spiny hollylike leaflets, dense racemes 3–11 cm long with 30–60 flowers, and glaucous blue berries. It has been reported to escape from cultivation occasionally in Michigan and eastern Canada, and it might be capable of persisting outside of cultivation in Missouri, at least briefly.

Many species of *Berberis* are grown as ornamental shrubs. However, some species serve as the alternate host for *Puccinia graminis* Pers., the fungus that causes black stem rust of wheat and other grain crops. *Puccinia graminis* has a two-phase life cycle, with the dikaryotic (telial and uredinial) stage attacking grains and wild grasses and the homokaryotic (aecial) stage attacking some species of *Berberis* (Alexopoulos et al., 1996). The fungus can overwinter on grasses only in areas with mild winters, but it overwinters on *Berberis* species even in very cold climates, so the presence of susceptible *Berberis* species can greatly increase losses from this fungus in nearby grain fields. Farming is only affected when susceptible barberries grow in open areas adjacent to grain fields; plants in wooded areas or areas remote from farmland have no significant impact on crops. The sale, transport, or cultivation of susceptible or untested species of *Berberis* is now banned by state and federal quarantine regulations, and extensive eradication programs have almost eliminated susceptible barberries (primarily *B. vulgaris*) in many areas where they once grew (Fulling, 1943; Roelfs, 1982). Because of this, barberries no longer affect grain growing in the United States, but quarantine laws are still enforced in order to prevent susceptible species from becoming reestablished in farming areas in the future (Roelfs, 1982).

Aside from the berries of some species, barberry plants are variously considered poisonous. Decoctions of a few species were used by Native Americans to treat sore throats, infected gums, jaundice, and diarrhea (Moerman, 1998).

1. Flowers solitary or in umbels; leaves 3–10 mm wide, the margins entire
   .... 2. B. THUNBERGII
1. Flowers in racemes; leaves 8–33 mm wide, the margins toothed
   2. Second-year bark purple or brown; each leaf margin with 3–12 teeth; racemes with 3–12 flowers .... 1. B. CANADENSIS
   2. Second-year branches gray; each leaf margin with (8–)16–30 teeth; racemes with 10–20 flowers .... 3. B. VULGARIS

1. Berberis canadensis Mill. (American barberry)

Pl. 302 b, c; Map 1269

Shrubs 40–200 cm tall. Second-year twigs with the bark purple or brown; spines simple or pinnately 3(–7)-branched. Petioles 2–8(–13) mm long. Leaf blades 1.8–7.5 cm long, 0.8–3.3 cm wide, oblanceolate or sometimes narrowly elliptical, long-attenuate at the base, rounded or bluntly pointed at the tip, each margin with 3–12 bristle-tipped teeth. Inflorescences lax racemes 2.0–5.5 cm long, with 3–12 flowers. Fruits 10 mm long, red, oblong-ellipsoidal. $2n=28$. May.

Uncommon, restricted to the south-central portion of the Ozark Division (Pennsylvania to Missouri south to Georgia and Tennessee). Ledges of dolomite bluffs and blufftop glades, occasionally on sandstone near the contact zone with dolomite strata.

Berberis canadensis is uncommon enough to be considered imperiled in Missouri and elsewhere in its range, but it is susceptible to infection by *Puccinia graminis* and is legally subject to eradication. Because it is uncommon and grows in areas remote from farms, it has no impact on grain farming in the state, so this native species should be exempted from eradication. Despite its scientific name, *B. canadensis* is not found in Canada.

2. Berberis thunbergii DC. (Japanese barberry)

Pl. 302 a; Map 1270

Shrubs 30–300 cm high. Second-year twigs with the bark purple or brown; spines simple or 3-branched. Petioles absent or 1–8 mm long. Leaf blades obovate to spatulate, 1.2–2.4 cm long, 0.3–1.0 cm wide, long-attenuate at the base, rounded or bluntly pointed at the tip, the margins entire. Inflorescences umbels 1.0–1.5 cm long, with 2–5 flowers, sometimes reduced to a solitary flower. Fruits 9–10 mm long, red, ellipsoidal or spherical. $2n=28$. April–May.

Introduced, uncommon and widely scattered (native of Japan; widely naturalized in eastern and central North America west to Wyoming). Bottomland forests and mesic upland forests; also disturbed areas.

Berberis thunbergii is very commonly grown as an ornamental in Missouri, and occasionally escapes. It is resistant to infection by *Puccinia graminis,* although the hybrid *B. thunbergii* × *vulgaris* (not yet found in Missouri) is susceptible.

3. Berberis vulgaris L. (common barberry)

Pl. 302 d, e; Map 1271

Shrubs 100–300 cm high. Second-year twigs with the bark gray; spines simple or 3-branched. Petioles 2–8 mm long. Leaf blades obovate to oblanceolate or almost elliptical, 2–5 cm long, 0.9–2.3

Plate 302. Berberidaceae. *Berberis thunbergii*, **a)** fruiting branch. *Berberis canadensis*, **b)** flower, **c)** fruiting branch. *Berberis vulgaris*, **d)** fruits, **e)** flowering branch. *Podophyllum peltatum*, **f)** flower, **g)** leaf blade (top view) and habit.

1269. Berberis canadensis   1270. Berberis thunbergii   1271. Berberis vulgaris

cm wide, short- to long-attenuate at the base, rounded or bluntly pointed at the tip, each margin with (8–)16–30 spine- or bristle-tipped teeth. Inflorescences lax racemes 2–6 cm long, with 10–20 flowers. Fruits 10–11 mm long, red or purple, ellipsoidal. $2n=28$. May–June.

Introduced, uncommon in the eastern half of the state (native of Europe; widely naturalized across temperate North America). Mesic to dry upland forests, often in rocky places; also dry pastures and roadsides.

During the eighteenth and nineteenth centuries, *B. vulgaris* was very commonly cultivated in North America for thorn hedges and as a source of jam, jelly, flavoring, beverage, medicine, and yellow dye. It sometimes escaped from cultivation and became widely naturalized in portions of the United States and Canada, including Missouri. Prior to 1920, as an important winter host of black stem rust of wheat, *B. vulgaris* was responsible for heavy losses to grain growers. However, it has been the subject of a vigorous eradication program by the federal and state governments, and it is now seldom seen in Missouri.

2. Caulophyllum Michx. (blue cohosh)

Three species, eastern North America, eastern Asia.

1. Caulophyllum thalictroides (L.) Michx.
(blue cohosh)

Pl. 301 h–j; Map 1272

Plants perennial herbs, with rhizomes 5–7 mm thick. Aerial stems 40–60 cm tall, without spines. Leaves cauline, 2 per plant, often not fully expanded at flowering time; lower leaf 3 times ternately compound, 20–32 cm long, usually sessile, occasionally with a petiole to 4.5 cm long; upper leaf twice ternate, 8.5–14.0 cm long. Leaflets obovate, sometimes 2- or 3-parted, and generally with a few rounded teeth, the upper surface green, the lower surface glaucous. Inflorescences terminal, open clusters or small panicles of 4–16 flowers, 13–60 mm long, sometimes with a second inflorescence in the axil of the upper leaf. Flowers with 3 or 4 sepal-like bracts immediately below the calyx. Sepals 6, 3–5 mm long, yellow, purple, or green. Petals 6, 1–2 mm long, yellow, purple, or green, bearing nectaries. Stamens 6, anthers attached at the base, opening by 2 apical flaps. Fruit wall rupturing early in development to expose the developing seeds. Seeds 1 or 2 per fruit, 6–9 mm long, spherical, each on a thick stalk 4–7 mm long, the seed coat fleshy, blue, glaucous. $2n=16$. April–May.

Scattered nearly throughout Missouri, but less common in the western half of the state (New Brunswick south to northern Georgia, west to Arkansas and southern Manitoba). Mesic upland forests, often on rich slopes in ravines, bottomland forests, and shaded ledges of bluffs.

Rootstocks of blue cohosh were used medicinally by Native Americans for a variety of treatments, including rheumatism, toothaches, menstrual problems, and various ailments of the digestive and urinary tracts (Moerman, 1998). The berrylike seeds are considered poisonous and the plants rarely can cause dermatitis in some people.

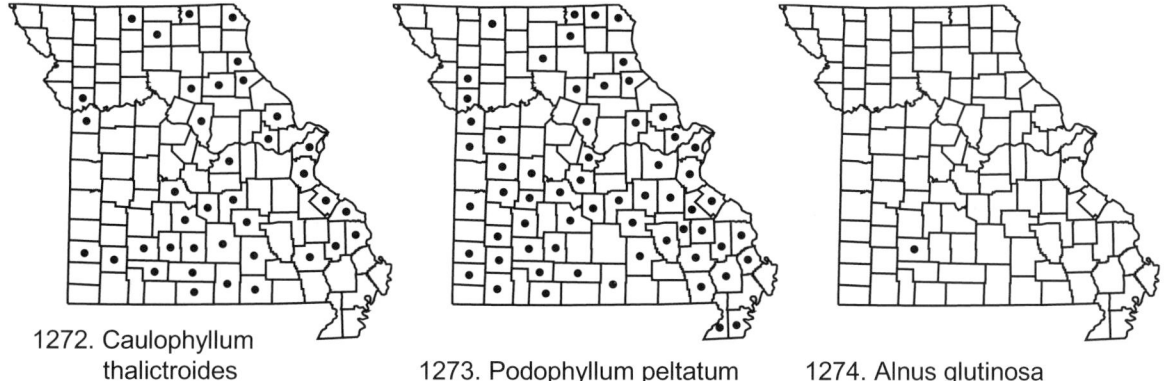

1272. Caulophyllum thalictroides

1273. Podophyllum peltatum

1274. Alnus glutinosa

3. Podophyllum L. (mayapple, mandrake)

Two species, eastern North America and eastern Asia.

1. Podophyllum peltatum L. (mayapple)

Pl. 302 f, g; Map 1273

Plants perennial herbs, 20–50 cm tall, with long creeping branched rhizomes 4–7 mm thick. Aerial stems without spines. Vegetative portions with one leaf attached directly near the tip of the rhizome, fertile portions with 2(3) leaves attached at the tip of an aerial stem. Leaves simple, the blade palmately veined, often perfoliate, 18–35 cm in diameter, deeply 5–9-lobed, the lobes oblanceolate, sometimes further 2-lobed, the margins toothed, the upper surface green, the lower surface glaucous. Flowers solitary, appearing from the angle between the leaves, without bracts. Sepals 6, falling as the flowers open, green (rarely pink). Petals 6–9, 15–33 mm long, white (rarely pink), without nectaries. Stamens 12–18, the anthers attached at the base, opening by longitudinal slits. Fruits ellipsoid berries, 25–45 mm long, 20–36 mm in diameter, yellow (rarely orange or purple). Seeds enclosed in a yellow (rarely purple) aril. $2n=12$. Late March–June.

Common throughout the state (southern Quebec south to northern Florida, west to eastern Nebraska and eastern Texas). Mesic upland forests, bottomland forests, ledges of bluffs; also pastures, roadsides, and railroads.

The ripe berries (mayapples) of *P. peltatum* are edible and were an important food for Native Americans. They are picked by native-food enthusiasts to be eaten raw or prepared into beverages, jellies, and preserves. However, all other parts of the plant are poisonous, and handling of the rootstocks also causes dermatitis in some individuals. Etoposide, a semisynthetic derivative of a lignan isolated from *P. peltatum,* is used as a chemotherapy agent in treating several types of cancer. Rare plants with pink sepals and petals and strong purple pigmentation of the leaves, stems, and rhizomes have been called f. *deamii* Raymond.

BETULACEAE (Birch Family)
Contributed by Alan Whittemore

Plants trees or shrubs, monoecious. Leaves alternate, short- to less commonly long-petiolate. Leaf blades simple, pinnately veined, the margins toothed. Stipules scarious or somewhat hardened, broadly ovate, mostly shed during leaf development. Inflorescences of separate staminate and pistillate catkins (the pistillate ones in globose clusters in *Corylus*). Flowers actinomorphic, imperfect. Staminate flowers with calyces absent or of (1–)4(–6) minute, scalelike sepals; corollas absent; stamens mostly 4–6, with a reduced, nonfunctional pistil sometimes present. Pistillate flowers with calyces rudimentary or absent; corollas absent;

stamens and staminodes absent; ovary inferior (but often appearing naked in the absence of perianth), with 2 locules toward the base but often appearing 1-locular toward the tip, the placentation axile. Styles 2 (sometimes united basally), each with a linear stigmatic region toward the tip. Fruits nuts, nutlets, or samaras, surrounded by bracts and often grouped into conelike infructescences. Six genera, 125–150 species, North America to South America, Europe, Asia.

Species of Betulaceae form an important part of the woody vegetation in many parts of Missouri. Many Betulaceae are tolerant of flooding and waterlogging, and they are a common component of the woody cover on stream banks and wet bottomlands. Other species dominate the woody understory in some Missouri forests. The seeds of all of our species are eaten by birds and other wildlife.

1. Leaf blades with the tip broadly or bluntly pointed to shallowly notched; fruits in dense aggregates of 50–120, separated by persistent woody bracts, the infructescence thus resembling a small woody cone; young growth (twigs, leaves, and inflorescences) sticky or resinous . . . . 1. ALNUS
1. Leaf blades with the tip narrowed or tapered to a sharp point; fruits either in clusters of 5 or fewer, or if more per cluster then the bracts shed after flowering; young growth not sticky or resinous
    2. Bark peeling in thin, papery sheets; undersurface of the leaf blade glaucous; fruits samaras 2–3 mm long, more or less hidden between overlapping bracts to form a smooth, conelike infructescence that disintegrates at maturity
       . . . . 2. BETULA
    2. Bark smooth, ridged, grooved, or scaly, not peeling; undersurface of the leaf blade not glaucous; fruits nutlets or nuts 3–15 mm long, surrounded by bracts that do not overlap tightly, clustered or loosely overlapping to form an infructescence that is not smooth or conelike
        3. Fruits spherical or depressed-spherical nuts 10–15 mm long, in clusters of 1–3(–5); leaf blades broadly ovate to broadly elliptic, with 6–9 veins on each side of the midrib . . . . 4. CORYLUS
        3. Fruits ovoid or flat-ovate nutlets (mostly 2 or 3 per flower), 2–6 mm long, in clusters of mostly 10–25; leaf blades narrowly ovate or elliptic to obovate, with 10–16 veins on each side of the midrib
            4. Bracts lobed, relatively flat, and not surrounding the fruits; bark smooth . . . . 3. CARPINUS
            4. Bracts unlobed, inflated, and surrounding the fruits; bark scaly
                . . . . 5. OSTRYA

1. Alnus Mill. (alder)
(Furlow, 1979)

Plants shrubs or trees, sometimes forming thickets. Young growth (twigs, leaves, and inflorescences) with a sticky or resinous coating. Twigs 1.5–2.0 mm thick, dark purplish brown, usually hairy, the pith more or less triangular in cross-section. Buds stalked, with 2 or 3 scales. Leaf blades elliptic, rhombic, or very broadly obovate to almost circular, the tip broadly or bluntly pointed to shallowly notched, the undersurface green, hairy (at least along the veins), but not or scarcely felty to the touch, the lateral veins 6–11 on each side of the midrib and sometimes branched. Stamens 4, undivided. Styles persistent at fruiting. Fruits samaras with 2 small lateral wings, brown, arranged in conelike aggregates of mostly 50–120 fruits. Bracts fused into a relatively flat structure, unlobed or very shallowly 5-lobed woody scales that overlap and more or less conceal the fruits, the scales persistent on the axis after the fruits are dispersed. About 25 species, North America to South America, Europe, Asia.

Alders are very important ecologically for their ability to fix nitrogen, that is, to transform inert gaseous nitrogen from the atmosphere into nitrates and other compounds useful to living things. Like other nitrogen-fixing plants, the chemical reactions occur in root nodules that contain symbiotic microorganisms—in the case of *Alnus,* bacteria (actinomycetes) in the genus *Frankia*. The leaves and bark are rich in tannins, so alders have been used for tanning leather; in traditional medicine they have been used to treat various kinds of infections and inflammations, especially of the skin (Moerman, 1998). The foliage and bark also yield dyes; depending on the parts of the plant used and the method of preparing and applying the dye, they can be yellow, red, or black. The conelike infructescences of some species have been marketed in handicrafts and jewelry as miniature pinecone substitutes.

1. Plants trees 4–15 m tall; bark medium gray, smooth on younger branches but fissured with age; leaf blades usually nearly circular, occasionally very broadly obovate, the tip shallowly notched, truncate, or occasionally broadly rounded, the margins coarsely toothed and sometimes also with small, shallow lobes
. . . . 1. A. GLUTINOSA
1. Plants shrubs or small trees 1.5–3.0(–6.0) m tall; bark light gray, smooth; leaf blades elliptic to rhombic, the tip broadly or bluntly pointed to rounded, the margins finely toothed, not lobed. . . . . 2. A. SERRULATA

1. Alnus glutinosa (L.) Gaertn. (black alder, European alder)

Map 1274

Plants trees 4–15 m tall, the bark medium gray, smooth on younger branches but fissured with age. Twigs with the buds 5–7 mm long (excluding the stalk). Petioles 11–32 mm long (the longest petioles on a branch always longer than 2 cm). Leaf blades usually almost circular, occasionally very broadly obovate, 4–12 cm long, 4–10 cm wide, the base rounded to broadly narrowed, the tip shallowly notched, truncate, or occasionally broadly rounded, the margins coarsely toothed and sometimes also with small, shallow lobes, the largest teeth 0.6–7.0 mm long, each side of the midrib with 6–8 strong secondary veins. Conelike infructescences 1.5–3.0 cm long, the fruits 3–5 mm long (including the styles), narrowly winged. 2n=28. March–April.

Introduced, known thus far only from a single specimen from Greene County (native of Europe; naturalized in the northeastern U.S. from Massachusetts and Tennessee west to Minnesota and Missouri). Margins of a lake.

Black alder is tolerant of waterlogged soil and is sold as an ornamental for wet or boggy soil, especially in the cooler parts of the northeastern United States. It has escaped over a wide region in the northeastern states and has become a pest at a few sites, but is known as an escape thus far at only one locality in Missouri, along the shore of Lake Springfield in Springfield.

The wood of black alder is durable when continually wet or submerged, and in Europe it was traditionally the wood of choice for pilings and foundations in saturated or submerged soil. The pilings that support the old city of Venice and parts of Amsterdam are mostly black alder. It is a poor conductor of heat and is the wood of choice for wooden shoes. It also has been used for furniture and small carved or turned items.

2. Alnus serrulata (Aiton) Willd. (common alder, smooth alder, tag alder)

Pl. 303 e–g; Map 1275

Plants shrubs or small trees 1.5–3.0(–6.0) m tall, the bark light gray, smooth. Twigs with the buds 4–5 mm long (excluding the stalk). Petioles 5–18 mm long. Leaf blades elliptical to rhombic, 5–13 cm long, 3.0–7.5 cm wide, the base rounded to narrowed, the tip broadly or bluntly pointed to rounded, the margins finely toothed, not lobed, the teeth 0.2–0.5 mm long, each side of the midrib with 7–11 strong secondary veins. Conelike infructescences 1–2 cm long, the fruits 2.5–4.5 mm long (including the styles), narrowly winged. 2n=28. March–April.

Scattered, mostly south of the Missouri River (eastern U.S. west to Kansas and Texas; Canada). Banks of streams, spring branches, and rivers, fens, acid seeps, and less commonly bottomland forests and margins of lakes and sloughs.

The leaves stay green later than most of our deciduous species. The underside of the blade

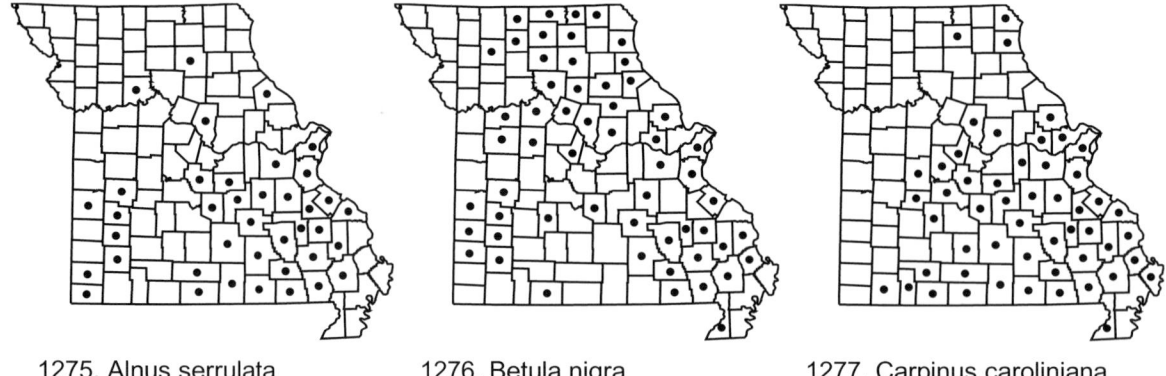

1275. Alnus serrulata   1276. Betula nigra   1277. Carpinus caroliniana

varies from almost glabrous (only scattered hairs on the main veins) to conspicuously pubescent; the latter form has been called f. *noveboracensis* (Britton) Fernald.

2. Betula L. (birch)

About 35 species, throughout the cool and temperate parts of the Northern Hemisphere.

1. Betula nigra L. (river birch, red birch)

Pl. 303 a–d; Map 1276

Plants trees 10–15 m tall, the bark peeling in thin, papery sheets brown to reddish brown. Young growth not sticky or resinous. Twigs 1–2 mm thick, dark purplish brown, initially cobwebby-hairy, but glabrous or nearly so at maturity, the buds sessile, with 2 or 3 scales. Petioles 7–14 mm long. Leaf blades 3.5–14.0 cm long, 2–10 cm wide, triangular-ovate, the undersurface glaucous, sparsely or moderately soft-hairy, smooth or felty to the touch, the tip narrowed or short-tapered to a sharp point, the base broadly rounded to nearly truncate, the margins finely toothed and sometimes also shallowly lobed, the lateral veins 6–10 on each side of the midrib, seldom branched. Stamens 2 or 3, each divided almost to the base. Fruits samaras, 2–3 mm long, smooth, brown, discoid, arranged in conelike aggregates with mostly 30–70 samaras. Bracts fused into a relatively flat structure, deeply 3-lobed, green, hairy scales, leaflike in texture, that overlap and almost conceal the samaras, these falling from the axis with the fruits. $2n=28$. April–May.

Scattered to common nearly throughout the state (eastern U.S. west to Wisconsin and Texas). Banks of streams and rivers, margins of ponds and lakes, and bottomland forests.

The slender, often leaning or twisted trunk and the pendant masses of foliage give this tree a graceful appearance, and it has long been used in horticultural plantings, especially in places with damp, poorly drained soil. It only does well in low pH soils. The wood has been used for furniture, handicrafts, and specialty items (artificial limbs and wooden toys), as well as for pulp for paper manufacturing. The foliage turns pale to bright yellow in the autumn. Steyermark (1963) noted that the staminate catkins produce copious airborne pollen that can cause hay fever.

3. Carpinus L. (hornbeam)
(Furlow, 1987)

About 25 species, North America, Central America, Europe, Asia.

1. Carpinus caroliniana Walter (blue beech, hornbeam, musclewood)

Pl. 303 h, i; Map 1277

Plants small trees 4–10 m tall, the bark smooth. Young growth not sticky or resinous. Twigs about 1 mm thick, medium to dark brown, pubescent with appressed or spreading hairs, the buds sessile, with more than 10 scales. Petioles 6–13 mm long. Leaf

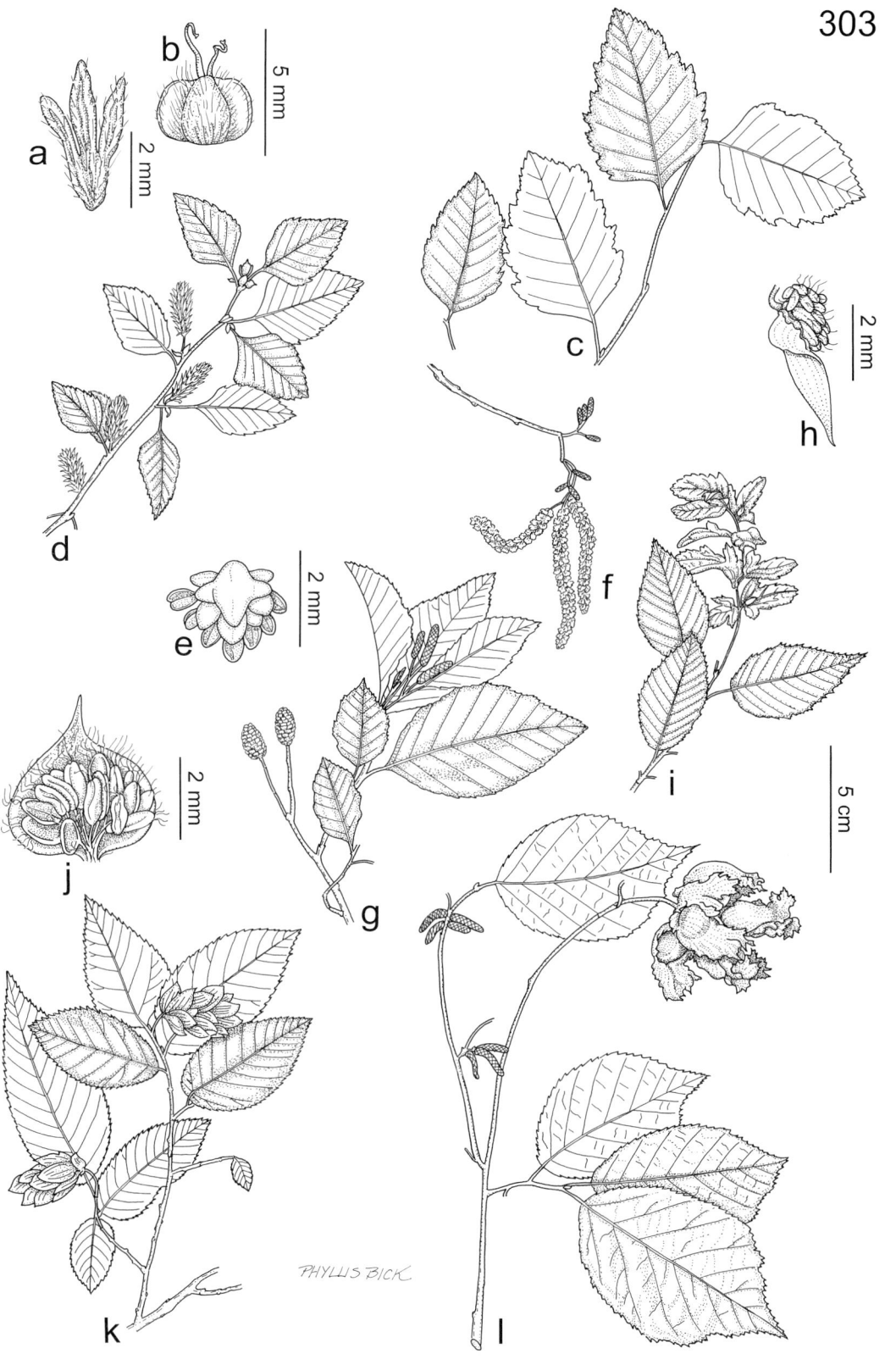

Plate 303. Betulaceae. *Betula nigra*, **a)** scale, **b)** pistillate flower, **c)** branch with leaves, **d)** flowering branch. *Alnus serrulata*, **e)** staminate flowers, **f)** flowering twig, **g)** fruiting branch. *Carpinus caroliniana*, **h)** staminate flowers, **i)** fruiting branch. *Ostrya virginiana*, **j)** staminate flowers, **k)** fruiting branch. *Corylus americana*, **l)** habit.

blades 5–10 cm long, 2.0–4.5 cm wide, narrowly elliptic to narrowly ovate or obovate, the undersurface green, pubescent with scattered long, weak hairs, smooth or slightly felty to the touch, the tip narrowed or tapered to a sharp point, the base broadly narrowed or rounded to weakly cordate, the margins sharply toothed to the base, the lateral veins 10–16 on each side of the midrib, rarely branched. Stamens 3, each divided almost to the base. Fruits nutlets (mostly 2 or 3 per flower), 2–5 mm long, 2–5 mm wide, ovoid, somewhat flattened, the shell thin, longitudinally veined, brown, arranged in elongate spikes with mostly 10–20 nutlets. Bracts 18–27 mm long, fused into a relatively flat structure, deeply lobed, herbaceous, strongly veined, green, bluntly lanceolate or narrowly oblong with 1 or more short basal lobes, sparsely hairy primarily on the veins, bracts not concealing the nutlets but falling with them. $2n=16$. March–May.

Scattered nearly throughout the state, but uncommon in the Glaciated Plains Division and apparently absent from the Unglaciated Plains (eastern U.S. west to Minnesota and Texas; Canada). Bases and ledges of sheltered bluffs, banks of streams and rivers, margins of sinkhole ponds, and mesic upland forests in ravines.

The trunks of *C. caroliniana* are usually conspicuously fluted, a feature that is made more apparent by the smooth, light gray bark. Their form is reminiscent of the limbs of a sinewy, muscular person, hence the common name musclewood. The species performs well as a landscape plant and is becoming common in the nursery trade. The wood is almost as hard and strong as that of *Ostrya virginiana* and has been put to similar uses, but as with the latter species, its usefulness is limited by the small size of the trees.

1. Leaf blades 5–8 cm long, 2.0–3.3 cm wide, narrowed or rarely slightly tapered at the tip, the undersurface not glandular .... 1A. VAR. CAROLINIANA

1. Leaf blades 6.5–10.0 cm long, 3.0–4.5 cm wide, tapered at the tip, the undersurface often glandular
.... 1B. VAR. VIRGINIANA

1a. var. caroliniana

Bark medium or brownish gray. Leaf blades 5–8 cm long, 2.0–3.3 cm wide, narrowly ovate or narrowly oblong-ovate, narrowed or rarely slightly tapered at the tip, the undersurface not glandular, the marginal teeth fine to rather coarse. Bracts 18–25 mm long. March–April.

Uncommon in the Mississippi Lowlands Division (southeastern U.S. west to Missouri and Texas). Bottomland forests.

The two varieties of *C. caroliniana* are distinctive over most of their range, but intermediate populations are fairly common in and near the area where their ranges overlap. The var. *caroliniana* is primarily a plant of the Atlantic and Gulf Coastal Plains and the Mississippi Embayment, barely entering Missouri in Dunklin County. A few populations from other parts of southern Missouri are more or less intermediate between the two varieties (Furlow, 1987).

1b. var. virginiana (Marshall) Fernald
C. caroliniana ssp. *virginiana* (Marshall) Furlow

Bark bluish gray. Leaf blades 6.5–10.0 cm long, 3.0–4.5 cm wide, narrowly ovate or narrowly obovate, tapered at the tip, the undersurface often with tiny dark brown glands, the marginal teeth moderate to coarse. Bracts 22–27 mm long. $2n=16$. March–May.

Scattered nearly throughout the state, but uncommon in the Glaciated Plains Division and apparently absent from the Unglaciated Plains (eastern U.S. west to Minnesota and Oklahoma; Canada). Bases and ledges of sheltered bluffs, banks of streams and rivers, margins of sinkhole ponds, and mesic upland forests in ravines.

4. Corylus L. (hazel)
(Drumke, 1964)

About 15 species, North America, Europe, Asia.

1. Corylus americana Walter (hazelnut, American hazelnut)
C. americana var. *indehiscens* E.J. Palmer & Steyerm.

Pl. 303 l; Map 1278

Plants shrubs 1.5–3.0 m tall, sometimes forming thickets, the bark smooth or sometimes becoming finely grooved or ridged with age, grayish brown. Young growth not sticky or resinous. Twigs 1–2 mm thick, tan to dark brown, pubescent with long, spreading hairs, sometimes nearly glabrous late in the growing season. Buds sessile, with 4–6

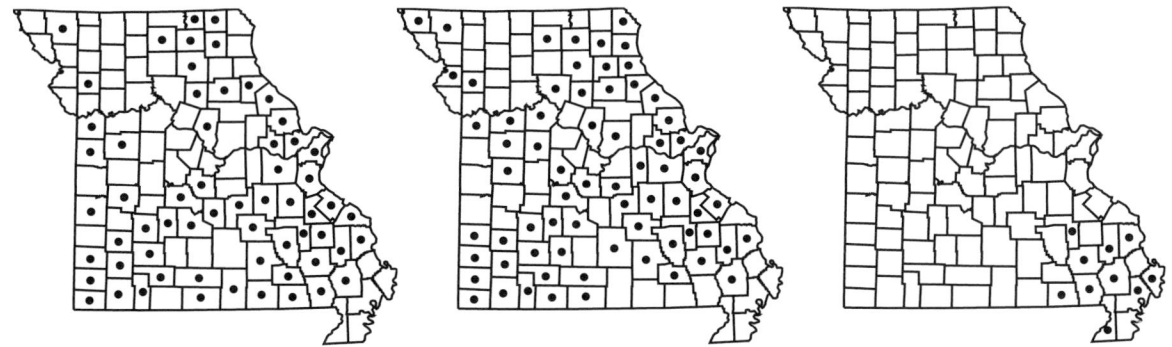

1278. Corylus americana   1279. Ostrya virginiana   1280. Bignonia capreolata

scales. Petioles 7–30 mm long. Leaf blades 7–17 cm long, 4–13 cm wide, broadly ovate to broadly elliptic, the undersurface green, pubescent with spreading hairs, felty and soft to the touch, the tip abruptly short-tapered, the base cordate, the margins sharply toothed to the base and also with a few coarse, shallow, blunt lobes, the lateral veins 6–9 on each side of the midrib, sometimes branched. Stamens 4, each divided almost to the base. Pistillate flowers in globose terminal clusters. Fruits nuts, 10–15 mm long, spherical or depressed-spherical, 11–18 mm wide, the shell thick and hard, arranged in clusters of 1–3(–5) nuts. Bracts more or less flat, unlobed, herbaceous to papery in texture, forming an involucre that completely encloses the nut and falls with it. Involucre 20–32 mm long, projecting far beyond the nut, not contracted to a beak, strongly veined, more or less felty-hairy, the outer edge sharply toothed. 2n=22, 28. February–April.

Scattered nearly throughout the state (eastern U.S. west to North Dakota, Oklahoma, and Louisiana). Upland prairies, savannas, glades, mesic to dry upland forests, bases, ledges, and tops of bluffs, banks of streams and rivers, and margins of ponds and lakes; also roadsides and railroads.

The nuts of this species have long been prized as a tasty and nutritious food, similar to *C. avellana* L., the European hazel or filbert. They are an important food for birds and other wildlife, and also are gathered from the wild locally by natural-food enthusiasts and other nut lovers. However, Furlow (1990) noted that the windborne pollen can cause hay fever in areas where the plants are common. Data from early literature and specimens indicate that in presettlement times, hazel thickets were a common component of Missouri prairies. Except for a few highly localized examples in the Glaciated Plains Division, hazel prairies, as they were known, essentially have disappeared from the landscape.

Gleason and Cronquist (1963, 1991) reported *C. cornuta* Marshall (beaked hazel) from Missouri, but the presence of this species could not be confirmed during the present research, nor was it mapped from anywhere close to Missouri by Drumke (1964). This widespread eastern and northern species differs from *C. americana* in its larger involucres (40–70 mm long) with the bracts fused together and narrowed into a tubular beak. Its presence in Missouri seems unlikely. The bracts around the fruit of *C. americana* are normally separate almost to the base, but plants are occasionally found in which these are fused for most of their length, at least on one side; these have been called var. *indehiscens*. However, Drumke (1964) observed that on some individuals free-bracted involucres and those with fused bracts can occur in the same cluster of flowers. The hairs on twigs, petioles, and involucres are usually gland-tipped; plants that lack glandular hairs have been called f. *missouriensis* (A. DC.) Fernald.

5. Ostrya Scop. (hop hornbeam)

About 5 species, North America, Central America, Europe, Asia.

1. Ostrya virginiana (Mill.) K. Koch (eastern hop hornbeam, ironwood)
O. virginiana var. *lasia* Fernald

Pl. 303 j, k; Map 1279

Plants small trees (3.5–)5.0–10.0 m tall, the bark scaly, light reddish gray to grayish brown. Young growth not sticky or resinous. Twigs grayish brown, about 1 mm thick, pubescent with long, soft, loosely appressed hairs. Buds sessile, with mostly 5–7 scales. Petioles 2–9 mm long. Leaf blades 6–13 cm long, 3.0–7.5 cm wide, narrowly elliptic or narrowly obovate, the undersurface green, pubescent with spreading hairs, felty and soft to the touch, the tip tapered to a sharp point, the base shallowly cordate to rounded, the margins sharply toothed to the base, without lobes, the lateral veins 10–16 on each side of the midrib, sometimes branched. Stamens 3, each divided almost to the base. Fruits nutlets, 5–6 mm long, ovate in outline, strongly flattened, the shell thin, smooth, whitish green, arranged in dense, elongate spikes 3–6 cm long with mostly 10–25 nutlets. Bracts 12–25 mm long, papery, inflated, lightly but distinctly veined, tan or pale green, hairy, completely enclosing the nutlets and falling with them. $2n=16$. April–May.

Scattered nearly throughout the state (eastern U.S. west to North Dakota, Wyoming, and Texas; Canada). Borders of mesic to dry upland forests, glades, savannas, ledges and tops of bluffs, and rarely margins of fens and banks of streams.

The wood of the hop hornbeam is extremely hard and strong, and it was once widely used for small items such as tool handles. The usefulness of hop hornbeam for lumber is limited by the small size of the trees, and it has been used almost exclusively by local craftsmen and artisans. It is seldom seen in the nursery trade, perhaps because it grows fairly slowly, although it has performed well when tried.

Pubescence of the twigs is variable. Plants with unusually dense hairs have been called var. *lasia,* but these appear to grade into those with more typical pubescence. Individuals with stout, long-stalked glands on the twigs, petioles, and midribs have been called f. *glandulosa* (Spach) J.F. Macbr. The glandular form is common across the northern edge of the species' range, but it is rare in Missouri.

BIGNONIACEAE (Trumpet Creeper Family)

Plants trees, shrubs, or lianas, sometimes with tendrils. Leaves opposite or whorled, lacking stipules, simple or compound. Inflorescences various, axillary or terminal. Flowers perfect, hypogynous, without subtending bracts. Calyces actinomorphic or zygomorphic, splitting irregularly into 2 lobes or 5-lobed, the lobes sometimes reduced to teeth or nearly absent. Corollas zygomorphic, 5-lobed, sometimes appearing 2-lipped. Stamens 2 or 4, the filaments fused to the corolla tube, the anther sacs often appearing relatively distinct from each other and spreading, sometimes attached asymmetrically with one sac appearing terminal and the other appearing somewhat lateral. Staminodes sometimes also present, 1–3, particularly in some genera with only 2 functional stamens. Pistil 1 per flower, of 2 fused carpels. Ovary superior, with 2 locules, the placentation usually axile. Style 1 per flower, the stigma 1, with 2 deep flaplike lobes. Ovules numerous. Fruits capsules, linear-elongate, 2-valved, longitudinally dehiscent. Seeds variously shaped, flattened, winged. One hundred to 120 genera, about 800 species, cosmopolitan, but most diverse in tropical and subtropical regions, especially northern South America.

In the tropics, the Bignoniaceae are among the most important families of lianas. A number of lianas, shrubs, and trees in several genera are cultivated as ornamentals, especially in the southern states.

The genus *Paulownia,* which was included in the Scrophulariaceae by Steyermark (1963) and in the Bignoniaceae by Cronquist (1981, 1991), is segregated into its own family in the present work, for reasons discussed in the treatment of Paulowniaceae. *Paulownia* is distinguished most easily from the superficially similar *Catalpa* by its bluish purple corollas and ovoid fruits.

1. Trees or less commonly shrubs; leaves simple .... 3. CATALPA
1. Lianas; leaves compound

2. Leaves with 2 lateral leaflets and a terminal, branched tendril . . . . . 1. BIGNONIA
2. Leaves with 7–13 pinnate leaflets, the terminal leaflet well developed; tendrils absent . . . . 2. CAMPSIS

1. Bignonia L. (cross vine, quarter vine)

One species, southeastern U.S. west to Missouri, Oklahoma, and Texas.

1. Bignonia capreolata L. (cross vine, quarter vine)

Anisostichus capreolata (L.) Bureau

Pl. 304 f–h; Map 1280

Plants lianas, creeping or climbing, with leaf tendrils, lacking adventitious rootlets. Stems to 25 m long, glabrous or inconspicuously hairy at the nodes, finely ridged, the older ones often angled, the dark grayish brown bark becoming wrinkled and sometimes peeling in thin, papery strips. Leaves opposite, compound with 2 lateral leaflets and a terminal, branched tendril, relatively short-petiolate. Leaflets 4–18 cm long, lanceolate to elliptic or oblong-ovate, shallowly cordate at the base, rounded or more commonly tapered to a sharply pointed tip, the margins entire or slightly irregular, the surfaces glabrous or sparsely pubescent with minute, unbranched hairs along the main veins. Inflorescences axillary clusters. Calyces 6–9 mm long, shallowly 5-lobed or without lobes, glabrous or minutely hairy along the margin, reddish green, the lobes (when present) much shorter than the tube, broadly and bluntly triangular. Corollas 4–5 cm long, somewhat zygomorphic, glabrous or more commonly minutely hairy on the outer surface, somewhat thickened, red to reddish orange on the outer surface, yellow to yellowish orange on the inner surface, 5-lobed, only slightly 2-lipped, the tube narrowly bell-shaped and slightly bent toward the middle, the lobes much shorter than the tube, the margins entire or slightly irregular. Stamens 4. Staminodes absent. Fruits 10–20 cm long, flattened, narrowly elliptic in cross-section, the valves glabrous, with a leathery texture, and tan to brown at maturity. Seeds with the body 7–10 mm long, flattened, the body elliptic in outline, 2-lobed, brown, with a wing around the middle (longest at each end) or less commonly only at the ends, the wings papery, light tan, with irregular margins. $2n=40$. April–June.

Uncommon, southeastern Missouri in the Mississippi Lowlands Division and adjacent Ozarks (southeastern U.S. west to Oklahoma and Texas). Bottomland forests, swamps, and banks of streams and rivers; also roadsides, fencerows, fallow fields, and wet, disturbed areas.

The leaves of *B. capreolata* have been called semievergreen. Farther south in the range, they remain evergreen, but in southern Missouri, at the northern edge of the species' climatic tolerance, they sometimes are shed during the coldest winter temperatures. The unusual foliage and brightly colored, sweetly scented flowers make this an attractive garden ornamental for fences and trellises. However, in plants that climb on trees or poles, the flowers generally become restricted to the upper portion of the plant, too high to be appreciated without binoculars. The common name cross vine refers to the cross-shaped pattern visible in a cross-section of older stems, which develops as a result of anomalous growth in the wood.

2. Campsis Lour. (trumpet creeper)

Two species, U.S., Canada, Asia.

1. Campsis radicans (L.) Seem. (trumpet creeper, trumpet vine, devil's shoelaces, devil's shoestrings, hell vine)

Pl. 304 i, j; Map 1281

Plants lianas, lacking tendrils, creeping or climbing, with aerial rootlets along the stems. Stems to 20 m long, glabrous or inconspicuously hairy at the nodes, usually finely ridged, the older ones often angled or somewhat flattened, the yellowish brown bark often peeling in thin, tangled strips. Leaves opposite, pinnately compound with (5–)7–11(–13) leaflets, petiolate. Leaflets 2–8 cm long, lanceolate to ovate, rounded or narrowed to a short, winged stalk at the base, long-tapered to a

1281. Campsis radicans   1282. Catalpa bignonioides   1283. Catalpa ovata

short, winged stalk at the base, long-tapered to a sharply pointed tip, the margins coarsely toothed, the upper surface glabrous, the undersurface sparsely pubescent with minute, unbranched hairs along the main veins. Inflorescences short, terminal panicles, appearing as clusters. Calyces 15–22 mm long, 5-lobed, glabrous (but with scattered inconspicuous glands toward the base of the lobes), reddish green, the lobes shorter than the tube, triangular. Corollas 6–9 cm long, somewhat zygomorphic, glabrous, thickened, somewhat waxy-textured, red to reddish orange, yellowish orange toward the base, 5-lobed, only slightly 2-lipped, the tube narrowly bell-shaped to nearly cylindrical, the lobes much shorter than the tube, somewhat overlapping, the upper 2 lobes slightly smaller than the others, the margins entire or slightly irregular. Stamens 4. Staminodes absent. Fruits 10–28 cm long, slightly flattened, elliptic in cross-section with noticeable ridges along the sutures between the valves, the valves glabrous, with a leathery texture, and tan to brown at maturity. Seeds with the body 6–9 mm long, flattened, the body elliptic in outline, 2-lobed, brown, with a wing at each end, the wings papery, light tan, with irregular margins. $2n=40$. May–August.

Common throughout the state (eastern U.S. west to South Dakota and Texas). Bottomland forests, mesic upland forests, and banks of streams and rivers; also fencerows, pastures, fallow fields, old fields, roadsides, railroads, and open, disturbed areas.

Although it is found in a number of natural habitats, trumpet creeper is a disturbance-adapted species. It frequently is encountered along roadsides, where it often climbs on other vegetation and telephone poles. It is cultivated as an ornamental for trellises and fences, but care must be taken as mature plants can become massive enough that their weight will cause insufficiently supported structures to collapse. Handling plants can also cause dermatitis in some individuals (Steyermark, 1963). The exterior of the flowers is usually bright orangish red to red, but occasional plants may have orange or peach-colored flowers. Flowers of trumpet creeper are pollinated primarily by hummingbirds and secondarily by bumblebees and honeybees (Bertin, 1982). In addition to nectar produced in the flowers, *C. radicans* also has small nectar-producing glands on the outside of the flowers and fruits, as well as along the petioles. These nectaries, which appear as inconspicuous pustular or platelike glands, attract ants, which usually are abundant on the plants and use the exudate for food. However, Stevens (1990) was unable to document an obligate ant-plant relationship, in spite of indications from her data that patrolling by ants may reduce the incidence of fruit infestation by parasitic moth species.

3. Catalpa (catalpa)

Plants trees or less commonly shrubs, lacking tendrils. Twigs stout, glabrous or sometimes hairy when young, yellowish to reddish brown, with prominent white lenticels, the leaf scars prominent. Leaves opposite or whorled, simple, long-petiolate. Leaf blades ovate, sometimes shallowly 3-lobed or 3-angled toward the base, cordate or less commonly truncate at the base, narrowed or tapered to a sharply pointed tip, the margin otherwise entire. Inflorescences large,

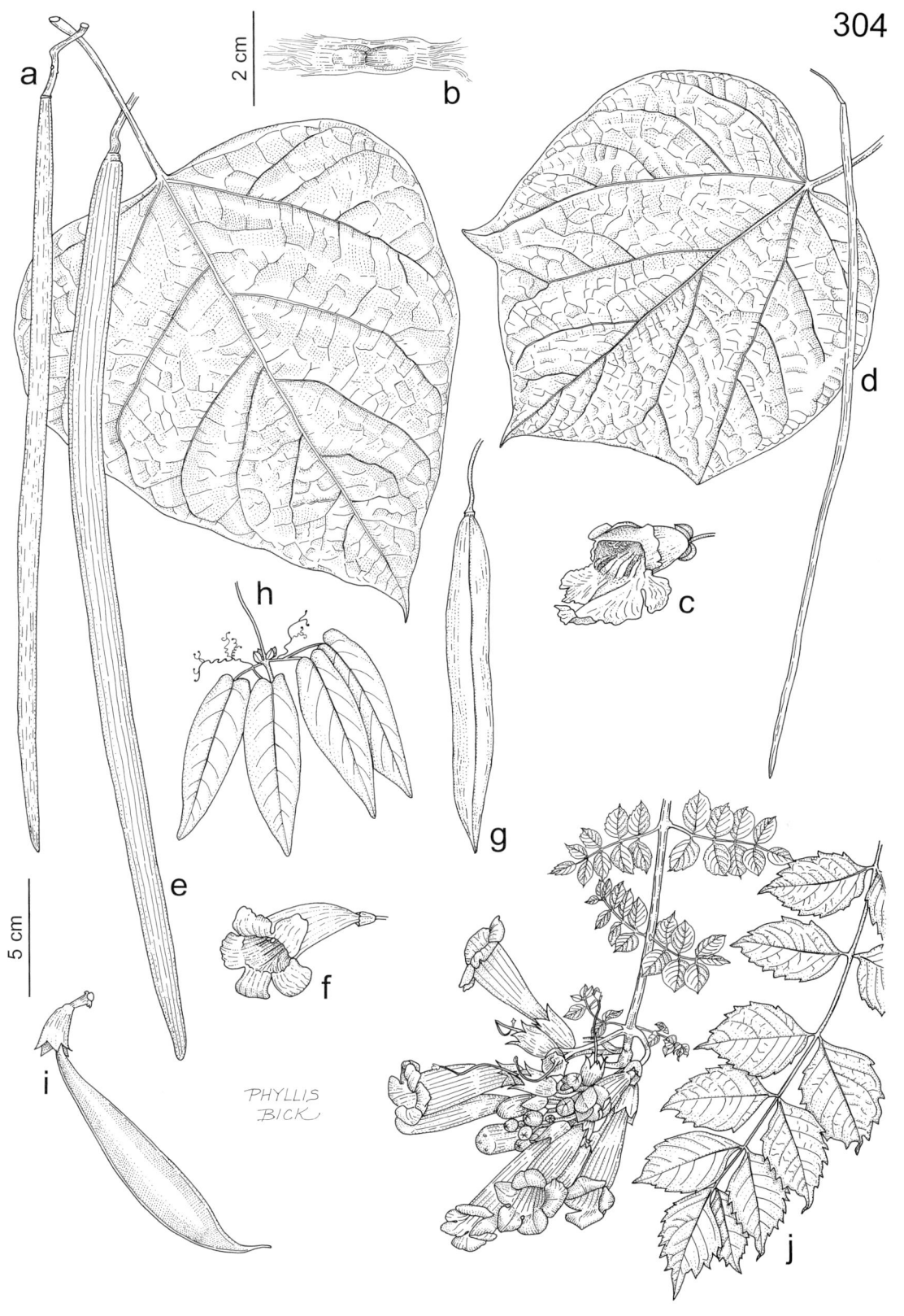

Plate 304. Bignoniaceae. *Catalpa speciosa*, **a)** fruit and leaf, **b)** seed, **c)** flower. *Catalpa ovata*, **d)** fruit and leaf. *Catalpa bignonioides*, **e)** fruit. *Bignonia capreolata*, **f)** flower, **g)** fruit, **h)** branch with leaves. *Campsis radicans*, **i)** fruit, **j)** flowering branch and larger leaf.

terminal panicles. Calyces splitting deeply into two irregular lobes at maturity, glabrous, usually purplish-tinged, the lobes broadly ovate, pointed at the tip. Corollas zygomorphic, glabrous, 5-lobed, appearing obliquely 2-lipped, white or light yellow with 2 longitudinal yellow to orange lines and a pattern of purple to brownish purple spots and short lines in the throat, the tube bell-shaped, the lobes shorter than the tube, the margins irregular and appearing somewhat crinkled. Stamens 2 or rarely 4. Staminodes 3 (or 1 in flowers with 4 stamens), minute, fused to the corolla tube. Fruits cylindrical, circular in cross-section, glabrous, brown at maturity. Seeds flattened, elliptic in outline, 2-lobed, light brown, with a long tuft of dense hairs at each end, each tuft fused toward the base into a papery wing. Ten species, U.S., Caribbean Islands, Asia.

Catalpas are the required food source for caterpillars of the catalpa sphinx moth (*Ceratomia catalpae* Boisd.), which can entirely defoliate trees during peak years but are usually heavily parasitized themselves by small wasps. The flowers are pollinated primarily by carpenter bees (*Xylocopa* spp.) and secondarily by bumblebees and honeybees. The North American catalpas are frequently cultivated as shade trees and ornamentals. The wood is a minor source of fence posts, rails, poles, and lumber for furniture, and it is sometimes mixed with other woods to make pulp for paper. The leaves turn yellow or brown in the autumn.

Cox and Dunn (1973–1974) and Cox (1973–1974) studied catalpa trees growing under cultivation and as escapes in Columbia (Boone County), where they documented putative introgression between the three species treated below. Based on a morphological analysis and the electrophoretic study of seed proteins, they identified individuals of *C. bignonioides* × *C. ovata*, *C. bignonioides* × *C. speciosa*, and *C. ovata* × *C. speciosa*, as well as putative backcrosses to the parental taxa. Hybrids were noted to form viable seeds but had greatly reduced fruit set. The hybrid between *C. bignonioides* and *C. ovata* originally was developed in the horticultural trade through artificial crosses and has been named *C.* ×*erubescens* Carrère (*C.* ×*hybrida* Späth). In nature, these trees rarely if ever grow together, at least in Missouri. Thus, there is little likelihood of hybridization in more natural settings.

1. Leaves often shallowly lobed, glabrous or becoming glabrous or nearly so at maturity; corollas 2.0–2.5 cm long, light yellow . . . . 2. C. OVATA
1. Leaves mostly unlobed, the undersurface persistently and densely hairy; corollas 2.5–5.0 cm long, white
    2. Corollas 2.5–4.0 cm long, the middle lobe of the lower lip not notched; fruits relatively thin-walled, the valves becoming flattened after dehiscence; bark light brown and divided into thin, scaly plates on older trunks
    . . . . 1. C. BIGNONIOIDES
    2. Corollas 4–5 cm long, the middle lobe of the lower lip shallowly notched; fruits relatively thick-walled, the valves remaining concave after dehiscence; bark reddish brown and divided into thick, scaly plates or furrows on older trunks . . . . 3. C. SPECIOSA

1. Catalpa bignonioides Walter (southern catalpa, common catalpa)
Pl. 304 e; Map 1282

Plants trees to 15 m tall. Bark light brown, divided into thin, scaly plates on older trunks. Leaf blades 8–22 cm long, entire or less commonly shallowly 3-lobed or 3-angled toward the base, narrowed or short-tapered at the tip, the upper surface glabrous, the undersurface persistently pubescent with short, curly, unbranched to stellate hairs, especially along the veins. Calyces 9–12 mm long. Corollas 2.5–4.0 cm long, white, the middle lobe of the lower lip not notched. Fruits 20–45 cm long, 0.6–1.0 cm in diameter, relatively thin-walled, the valves becoming flattened after dehiscence. Seeds with the body 12–16 mm long, 2.5–4.5 mm wide, the hairs of the tufts converging to a more or less pointed tip. $2n=40$. May–June.

Introduced, widely scattered, mostly in southern and central Missouri (native of the southeastern U.S. from Florida to Texas; introduced farther north). Margins of streams and rivers; also roadsides, railroads, and moist, disturbed areas.

The two North American catalpas can be difficult to distinguish. In addition to the key characters separating them, *C. bignonioides* tends to be a somewhat smaller tree, with more numerous, slightly smaller flowers per inflorescence than *C. speciosa*. Also, the corollas have a denser pattern of lines and spots in the throat, and the leaves have a strong unpleasant odor when crushed or bruised, whereas those of *C. speciosa* are less strongly and more pleasantly scented.

2. Catalpa ovata G. Don (Chinese catalpa)
Pl. 304 d; Map 1283

Plants eventually becoming trees to 15 m tall, but already flowering when still a shrub 1 m tall. Bark gray to grayish brown, divided into shallow furrows and thin plates on older trunks. Leaf blades 8–25 cm long, entire or more commonly shallowly 3-lobed or 3-angled toward the middle, noticeably tapered at the tip, the surfaces glabrous or minutely pubescent with short, straight hairs along the veins when young and then becoming glabrous or nearly so at maturity. Calyces 6–9 mm long. Corollas 2.0–2.5 cm long, light yellow, the middle lobe of the lower lip not notched. Fruits 20–35 cm long, 0.5–0.7 cm in diameter, relatively thin-walled, the valves becoming flattened after dehiscence. Seeds with the body 6–8 mm long, 2.5–4.5 mm wide, the hairs of the tufts more or less parallel. $2n=40$. May–June.

Introduced, known thus far only from Boone and Crawford Counties (native of China; introduced sporadically in the eastern U.S.). Roadsides and open, disturbed areas.

This species is far less commonly cultivated in the United States than are the two North American species. As in *C. bignonioides*, the leaves have a strong unpleasant odor when crushed or bruised.

3. Catalpa speciosa Warder ex Engelm.
(northern catalpa, catawba tree, cigar tree, hardy catalpa, Indian bean)
Pl. 304 a–c; Map 1284

Plants trees to 30 m tall. Bark reddish brown, divided into thick, scaly plates or furrows on older trunks. Leaf blades 15–30 cm long, entire or less commonly shallowly 3-lobed or 3-angled toward the base, noticeably tapered at the tip, the upper surface glabrous, the undersurface persistently pubescent with short, curly, unbranched to stellate hairs, especially along the veins. Calyces 9–12 mm long. Corollas 4–5 cm long, white, the middle lobe of the lower lip shallowly notched. Fruits 20–50 cm long, 1.0–1.5 cm in diameter, relatively thick-walled, the valves remaining concave after dehiscence. Seeds with the body 8–15 mm long, 4–6 mm wide, the hairs of the tufts more or less parallel. $2n=40$. May–June.

Native in the Mississippi Lowlands Division, introduced widely elsewhere in the state, mostly south of the Missouri River (native in the southeastern U.S. west to Missouri and Texas; introduced farther north, west, and east). Bottomland forests, mesic upland forests, margins of streams and rivers, and bases of bluffs; also roadsides, railroads, and moist, disturbed areas.

During the 1930s and 1940s, *C. speciosa* was widely planted in groves for windbreaks and fence posts (Settergren and McDermott, 1962), which resulted in trees becoming naturalized outside the native range. It is still popular as an ornamental and shade tree. Some individuals develop dermatitis after handling the flowers, and even the scent of the flowers is reportedly poisonous to some individuals when inhaled (Steyermark, 1963).

BORAGINACEAE (Borage Family)
Contributed by David J. Bogler and George Yatskievych

Plants annual, biennial, or perennial herbs (shrubs or trees elsewhere). Stems unbranched or branched, often hairy, sometimes appearing bristly, the hairs often with persistent pustular bases. Leaves alternate and sometimes also basal, well developed (except sometimes toward the stem base in *Lithospermum* and a few other genera), the margins entire. Stipules absent. Leaf blades simple, sessile or tapered to an often winged petiole, the surfaces usually stiffly hairy, the hairs often with persistent pustular bases (with calcified or silicified walls, known as cystoliths) and roughened to the touch, less commonly glabrous. Inflorescences mostly of terminal racemes or spikes, these often appearing coiled (scorpioid) and uncoiling as the flowers develop, sometimes panicles, seldom of solitary axillary flowers, the flowers most commonly subtended by bracts. Flowers actinomorphic (the corolla zygomorphic in *Echium,* the calyx zygomorphic in some *Myosotis* species), hypogynous, usually perfect; cleistogamous flowers sometimes present. Calyces usually deeply 5-lobed, the lobes equal or unequal, persistent and sometimes becoming enlarged at fruiting. Corollas (short and inconspicuous in cleistogamous flowers) usually shallowly 5-lobed, often trumpet-shaped, less commonly saucer-shaped, funnelform, narrowly bell-shaped, or tubular, the inside of the throat sometimes blocked by small appendages. Stamens usually 5, the filaments attached in the corolla tube, short or long, the anthers most commonly not exserted, rarely exserted, attached at their base, usually yellow. Pistil 1 per flower (but sometimes appearing as 2–4), of 2 fused carpels. Ovary usually deeply 4-lobed, usually 4-locular and with 1 ovule per locule, the placentation axile or sometimes appearing nearly basal. Style 1, usually attached in the pit between the ovary lobes, sometimes on a short, upward projection of the receptacle between the lobes, entire, usually not persistent at fruiting, the stigma sometimes 2-lobed, capitate or discoid. Fruits usually schizocarps splitting into (1–)4 nutlets, these 1-seeded, indehiscent, with a hardened, sometimes bony outer wall, less commonly a drupe with 4 (1-seeded) or 2 (2-seeded) stones. One hundred to 130 genera, about 2,500 species, nearly worldwide.

In the strict sense, the Boraginaceae are recognized by their alternate leaves, stiff hairs with pustular bases (cystoliths), tightly coiled scorpioid inflorescences, actinomorphic flowers, 4-lobed ovaries with a gynobasic style, and fruits consisting of 4 nutlets. Interestingly, gynobasic styles and nutlets apparently evolved independently in the Verbenaceae-Lamiaceae alliance (which usually have zygomorphic corollas). The Boraginaceae in the strict sense are related to a series of taxa variously recognized as subfamilies or separate families, including the Hydrophyllaceae, Heliotropiaceae, Cordiaceae, Ehretiaceae, Lennoaceae, and Wellstediaceae. The Hydrophyllaceae mostly share the distinctive scorpioid inflorescence and sometimes are combined into Boraginaceae, but they differ in having a capsular fruit and parietal placentation. They are treated as a separate family in the present work. *Heliotropium* is the only Missouri genus of Heliotropiaceae, a group of 3 genera that is distinguished by a terminal style, usually dry but drupaceous fruit, and the frequent presence of a stigmatic appendage. Ehretiaceae (11 genera, about 170 species) and Cordiaceae (3 genera, about 325 species), which some botanists combine under the former name, are mostly woody plants with divided styles and 2 stigmas in the former group and 4 stigmas in the latter group. They are most diverse in tropical and warm-temperate regions. The Cordiaceae do not occur in Missouri, but the introduced genus *Tiquilia* is here treated in the segregate family Ehretiaceae. The Lennoaceae (2 genera, 4 species) are a small group of succulent root parasites lacking chlorophyll and occur from the southwestern United States to northern South America. The small African genus *Wellstedia* Balf. f. (3 species) usually is treated as a subfamily of the Boraginaceae and is unique within the complex in its somewhat flattened capsular fruits.

The number of families to be recognized in this complex is still somewhat controversial. Molecular studies using ITS rDNA sequences and secondary structures (Gottschling et al., 2001) and *ndhF* sequence data (Ferguson, 1998) indicate these groups are indeed related and support the traditional classification, with Boraginaceae in the strict sense as the sister group to the other taxa. The decision to split the Boraginaceae into several related families that are easily separable morphologically avoids the problem of having to combine the Hydrophyllaceae into a broadly circumscribed concept of Boraginaceae. See the treatments of Ehretiaceae, Heliotropiaceae, and Hydrophyllaceae (the last two in the forthcoming Volume 3) for further discussion.

Although the Boraginaceae usually are easy to identify at the family level, the characters used to identify genera and species are rather technical and may require observation at relatively high magnification. The leaves provide only a few characters, differing mainly in size and shape. The flowers vary in size, shape, and color. Some genera have characteristic appendages or flaps of tissue (fornices, faucal appendages) inside the tube of the corolla or blocking the throat. The size and shape of these appendages are important characters but are difficult to observe, especially in pressed specimens. The length of the style relative to the anthers is also variable and sometimes an important character. Several genera of Boraginaceae are heterostylous, in which populations contain individuals with long styles (pin flowers) and other individuals with short styles (thrum flowers), relative to the position of the anthers. The insertion of the stamens is sometimes variable, and even the pollen is sometimes dimorphic. Fruits vary in the size and shape of the nutlets, how they are attached to the receptacle, and surface ornamentation. Mature fruits sometimes are needed to identify the species with confidence.

Members of this family are of relatively little economic importance, although many species are cultivated as ornamentals and a few species are used medicinally. Boraginaceae produce a variety of pyrrolizidine alkaloids and other chemical compounds that may be toxic or carcinogenic (Burrows and Tyrl, 2001). Livestock can be poisoned by eating the plants directly or as contaminants in their feed. The symptoms are cumulative and include loss of coordination, delirium, depression, and liver necrosis. The bark of the root in some species produces a purple napthaquinone pigment (alkannin), which was used as a paint and dye by Native Americans. The purple pigment sometimes stains herbarium sheets as well. The family is named after the Mediterranean genus *Borago* L., a name derived from the Latin word *burra,* meaning rough-hairy. *Borago officinalis* L. (borage) is among the few species in the family to be used as a culinary herb and is also grown for its ornamental and medicinal value.

Key to Flowering Material
1. Stems and leaves completely glabrous .... 10. MERTENSIA
1. Stems and leaves more or less hairy
    2. Some or all of the leaves with the blade tissue extending down the stem as narrow wings .... 13. SYMPHYTUM
    2. Leaf blades not decurrent (but occasionally clasping the stem), the stems not winged
        3. Corollas yellow or orange
            4. Stems and leaves with conspicuous, long, bristly hairs; uppermost (and often median) flowers lacking subtending bracts; plants annual, with taproots; corollas orange......................... 1. AMSINCKIA
            4. Stems and leaves with shorter or less conspicuous, spreading or ascending hairs; all of the flowers subtended by bracts; plants perennial, not taprooted (except in *L. incisum,* which has yellow corollas); corollas yellow or orange .... 9. LITHOSPERMUM
        3. Corollas white, greenish white, cream-colored, lavender, blue, or maroon
            5. Corollas zygomorphic, the upper lobes much longer then the lower ones, blue .... 6. ECHIUM
            5. Corollas actinomorphic, the lobes all similar, white, greenish white, cream-colored, lavender, maroon, or blue
                6. Style relatively long, conspicuously exserted from the corolla; corollas tubular, white to cream-colored or greenish white
                    .... 12. ONOSMODIUM
                6. Style relatively short, much shorter than the corolla, not exserted; corollas variously shaped (but not tubular), white, lavender, blue, or maroon, not greenish-tinged

7. Corollas white
    8. Flowers solitary in the axils of the upper leaves
        9. Leaves linear to narrowly lanceolate, without noticeable lateral veins; plants annual, with a slender taproot .... 4. BUGLOSSOIDES
        9. Leaves lanceolate to narrowly ovate, with conspicuous lateral veins; plants perennial, with fibrous roots .... 9. LITHOSPERMUM
    8. Flowers in sometimes paired inflorescences at the branch tips, these sometimes on short lateral branches and thus appearing axillary, appearing as clusters at the start of flowering, then becoming elongated into usually scorpioid spikelike racemes
        10. Inflorescences with a bract subtending each flower; leaves all sessile
            .... 8. LAPPULA
        10. Inflorescences with a bract subtending the lower flowers, the upper flowers bractless, or the bracts few and alternating with the flowers or absent altogether; at least the lower leaves petiolate
            11. Calyces lobed nearly to the base, actinomorphic, the lobes all similar in appearance; leaves (2–)5–30 cm long, (8–)20–100 mm wide
                .... 7. HACKELIA
            11. Calyces lobed 1/3–2/3 of the way to the base, zygomorphic, more or less 2-lipped, the upper 3 lobes shorter than the lower 2 lobes; leaves 1–8 cm long, 5–16 mm wide .... 11. MYOSOTIS
7. Corollas lavender, blue, or purplish red
    12. Calyx lobed 1/3–2/3 of the way to the base .... 11. MYOSOTIS
    12. Calyx lobed to the base or nearly so
        13. Inflorescences without bracts subtending the flowers
            14. Basal leaves with the blade more or less heart-shaped, cordate at the base, the petiole unwinged or winged only at the tip; corollas 2.5–3.5 mm long, bright blue .... 3. BRUNNERA
            14. Basal leaves with the blade oblanceolate to elliptic or oblong-elliptic, tapered at the base, the petiole winged all or most of its length; corollas 4–7 mm long, purplish red or pale blue to light blue
                .... 5. CYNOGLOSSUM
        13. Inflorescences with a bract subtending each flower
            15. Leaves with the blade 10–30 mm wide, lanceolate to oblanceolate, those of the lower leaves tapered to a winged petiole (the median and upper leaves usually sessile), corollas 10–20 mm long .... 2. ANCHUSA
            15. Leaves with the blade 3–10 mm wide, linear to linear-oblong, or narrowly oblanceolate, sessile; corollas 2–4 mm long .... 8. LAPPULA

Key to Fruiting Material
1. Stems and leaves completely glabrous .... 10. MERTENSIA
1. Stems and leaves more or less hairy
    2. Some or all of the leaves with the blade tissue decurrent at the base, the stems thus appearing more or less narrowly winged .... 13. SYMPHYTUM
    2. Leaf blades not decurrent (but occasionally clasping the stem), the stems not winged
        3. Flowers and fruits solitary in the axils of the upper leaves
            4. Nutlets wrinkled or somewhat warty, brown or dull gray
                .... 4. BUGLOSSOIDES
            4. Nutlets smooth or nearly so, white or ivory-colored ... 9. LITHOSPERMUM

3. Flowers and fruits in sometimes paired, more or less elongated, usually scorpioid, spikelike racemes at the branch tips, these sometimes on short lateral branches and thus appearing axillary, occasionally a few solitary axillary flowers also present
    5. Nutlets with hooked or apically barbed tubercles
        6. Leaves 1–5(–8) cm long, 3–10 mm wide, linear to linear-oblong, or narrowly oblanceolate, all sessile; inflorescences with a bract subtending each flower and fruit; fruits with the stalk ascending . . . . 8. LAPPULA
        6. Leaves 4–30 cm long, 15–100 mm wide, at least the lowermost leaves with the blade tapered to a winged petiole (these sometimes absent at fruiting); inflorescences with a bract subtending all or most of the lower flowers and fruits, the median and upper flowers and fruits bractless; fruits with the stalk spreading to drooping or recurved
            7. Nutlets 5–7 mm long, asymmetrically broadly obovoid; calyx lobes 4–10 mm long at fruiting . . . . 5. CYNOGLOSSUM
            7. Nutlets 2–3 mm long, more or less ovoid; calyx lobes 2–3 mm long at fruiting . . . . 7. HACKELIA
    5. Nutlets smooth, wrinkled, warty, or with blunt tubercles
        8. Nutlets attached laterally, well above the base . . . . 1. AMSINCKIA
        8. Nutlets attached basally but usually off-center (toward the inner edge of the basal region)
            9. Style elongate, extending conspicuously past the nutlets, mostly persistent at fruiting (check several fruits)
                10. Nutlets 3-angled in cross-section, the surface roughened or wrinkled, brown to brownish gray . . . . 6. ECHIUM
                10. Nutlets rounded in cross-section, the surface smooth or faintly pitted, white or ivory-colored . . . . 12. ONOSMODIUM
            9. Style short, shorter than or extending only slightly past the nutlets, rarely persistent at fruiting
                11. Nutlets with the surface smooth
                    12. Nutlets 3–4 mm long, not flattened, with a slightly raised ridge along most of the ventral surface, but otherwise rounded, white or ivory-colored . . . . 9. LITHOSPERMUM
                    12. Nutlets 1–2 mm long, somewhat flattened, with a longitudinal rim or keel all the way around, brown to black
                        . . . . 11. MYOSOTIS
                11. Nutlets with the surface roughened to pebbled or wrinkled
                    13. Basal leaves with the blade lanceolate to oblanceolate, tapered to a winged petiole; upper leaves lanceolate to oblanceolate; inflorescences with a bract subtending each flower . . . 2. ANCHUSA
                    13. Basal leaves with the blade more or less heart-shaped, cordate at the base, the petiole unwinged or winged only at the tip; upper leaves ovate; inflorescences without bracts subtending the flowers
                        14. Leaves oblong to lanceolate, veins inconspicuous; inflorescence with bracts at the base of each flower; nutlets roughened, the base with a thickened annular ring and stalklike extension . . . . 2. ANCHUSA
                        14. Leaves ovate to cordate, conspicuously veined; inflorescence lacking bracts at the base of each flower; nutlets longitudinally wrinkled and roughened, lacking a stalklike extension at the base . . . . 3. BRUNNERA

1. Amsinckia Lehm.

Plants annual, with slender taproots. Stems erect or ascending, unbranched or branched, pubescent with sparse to moderate, spreading, bristly, pustular-based hairs and usually also sparse to dense, fine, short, more or less appressed, sometimes somewhat matted hairs. Leaves alternate and often also basal, sessile, gradually reduced toward the stem tip. Leaf blades linear to narrowly oblong or oblong-lanceolate, the uppermost sometimes narrowly ovate, narrowed to somewhat rounded at the base, mostly angled to a bluntly or sharply pointed tip, the lowermost leaves tapered at the base and often rounded at the tip, the surfaces and margins moderately to densely pubescent with more or less spreading, bristly, pustular-based hairs. Inflorescences not paired, at first appearing as dense, terminal and apparently axillary (terminal on short axillary branches) clusters, subsequently elongating into ascending, scorpioid, spikelike racemes, occasionally appearing as few-branched panicles, the flowers short-stalked, usually lacking bracts (rarely a few present at the base of the inflorescence). Calyces more or less actinomorphic, 5-lobed nearly to the base, occasionally some of the lobes fused (the calyx then appearing 3- or 4-lobed), the lobes mostly linear to narrowly lanceolate-triangular, with bristly, pustular-based hairs, persistent and ascending at fruiting. Corollas more or less trumpet-shaped, actinomorphic, yellow or orange, sometimes with reddish streaks in the throat, the throat sometimes with small, scalelike appendages, the lobes more or less spreading, rounded. Stamens attached above or less commonly below the midpoint of the corolla tube, the filaments very short, the anthers oblong, not exserted from the corolla. Ovary deeply 4-lobed, the style slender, not exserted from the corolla, often not persistent at fruiting, the stigma capitate, unlobed. Fruits usually dividing into 4 nutlets, these erect, angular-ovoid, attached to the pyramidal gynobase toward the basal portion of a ventral keel, the relatively small scar usually surrounded by a small outgrowth of tissue, the surface appearing variously roughened, warty, or somewhat wrinkled, usually at least in part with small, blunt tubercles, white to grayish white or occasionally tan. Fifteen to 50 species, North America, South America.

The taxonomy of *Amsinckia* is complex and not well understood, with species boundaries difficult to evaluate because of confusing patterns of morphological variation between populations. Many new species were described by Suksdorf (1931), but the majority of them now seem unworthy of recognition. The nomenclature consequently is confusing, with some species treated under different names in various state floristic manuals. In the present work, the nomenclatural suggestions of Ray and Chisaki (1957) are followed in dealing with these taxa.

In general, *Amsinckia* species are weedy and are found in open, disturbed sites in a variety of soil types. All the taxa found in Missouri are considered toxic to livestock, especially to horses (Burrows and Tyrl, 2001). Collectors should also exercise caution as the bristly hairs found in abundance on the plants can cause itching and/or dermatitis.

1. Calyx lobes mostly 3 or 4, with 1 or 2 pairs of the developmentally 5 lobes usually fused and thus appearing unequal in width; corolla tube faintly 20-nerved below the midpoint; nutlets with dense, rounded tubercles, these often fused irregularly into blunt ridges .... 3. A. TESSELLATA
1. Calyx lobes 5, all distinct and more or less equal in width; corolla faintly 10-nerved below the midpoint; nutlets warty or somewhat wrinkled
    2. Corolla 6–8 mm long, orange to orange-yellow.............. 1. A. LYCOPSOIDES
    2. Corolla 4–6 mm long, light yellow......................... 2. A. MENZIESII

1. Amsinckia lycopsoides Lehm. ex Fisch. & C.A. Mey. (tarweed fiddleneck; bugloss fiddleneck)
    A. barbata Greene
    A. hispida (Ruiz & Pav.) I.M. Johnst.
    A. idahoensis M.E. Jones
    A. parviflora A. Heller

Pl. 305 a–c; Map 1285

1284. Catalpa speciosa   1285. Amsinckia lycopsoides   1286. Amsinckia menziesii

Stems 15–60 cm long, densely pubescent, with the longer, bristly hairs 2–3 mm long. Leaves 2–10 cm long, 4–12 mm wide. Inflorescences elongating to 3–15 cm long. Flower stalks 1–2 mm long. Calyx lobes 5, 4–8 mm long, 1–2 mm wide, more or less similar in width. Corollas 6–8 mm long, yellowish orange to orange, the tube 5–6 mm long, with 10 faint nerves below the midpoint, the throat with hairy, scalelike appendages, the lobes 1.0–1.5 mm long. Stamens attached above or below the middle of the tube, the filaments 0.3–0.5 mm long, the anthers 0.8–1.1 mm long. Style 2–3 mm long. Nutlets 1.8–2.0 mm long, 1.0–1.4 mm wide, the surface warty or somewhat wrinkled. 2n=30. May–July.

Introduced, known thus far only from Jackson County and the city of St. Louis (western U.S. east to North Dakota and Oklahoma, Alaska; Canada; introduced sporadically east to Pennsylvania and New Hampshire). Railroads.

This species sometimes is cultivated as an ornamental. Although Steyermark (1963) cited a number of other historical collections from Jackson County, at least some of these subsequently were redetermined as *A. menziesii*.

2. Amsinckia menziesii (Lehm.) A. Nelson & J.F. Macbr. (small-flowered fiddleneck)
Echium menziesii Lehm.
A. retrorsa Suksd.
A. micrantha Suksd.

Pl. 305 d–f; Map 1286

Stems 10–60 cm long, densely pubescent, with the longer, bristly hairs 1.5–2.5 mm long. Leaves 1–7 cm long, 3–8 mm wide. Inflorescences elongating to 3–12 cm long. Flower stalks 1–2 mm long. Calyx lobes 5, 3–5 mm long, 1.0–1.5 mm wide, more or less similar in width. Corollas 4–6 mm long, light yellow, the tube 3–5 mm long, with 10 faint nerves below the midpoint, the throat without scalelike appendages, often with reddish lines or streaks, the lobes 0.8–1.2 mm long. Stamens attached near the tip of the tube, the filaments 0.2–0.3 mm long, anthers 0.4–0.6 mm long. Style 2.0–2.5 mm long. Nutlets 2–3 mm long, 1–2 mm wide, the surface warty or somewhat wrinkled. 2n=16, 26, 34. May–July.

Introduced, uncommon and sporadic (western U.S. east to Idaho and Nevada, Alaska; Canada; introduced sporadically east to Maine and South Carolina). Railroads and open, disturbed areas.

The treatment of this species also includes plants reported by Mühlenbach (1979) as *A. hispida* (Ruiz & Pav.) I.M. Johnst., a South American taxon that Mühlenbach equated with *A. micrantha*.

3. Amsinckia tessellata A. Gray (devil's lettuce)
Pl. 305 j–l; Map 1287

Stems 15–60 cm long, densely pubescent, with the longer, bristly hairs 2–4 mm long. Leaves 2–8 cm long, 6–20 mm wide. Inflorescences elongating to 5–20 cm long. Flower stalks 1–2 mm long. Calyx lobes mostly 3 or 4, with 1 or 2 pairs of the developmentally 5 lobes usually fused and thus appearing unequal in width, these usually notched at the tip, 6–8 mm long, 1–2 mm wide. Corollas 8–12 mm long, orangish yellow to orange, the tube 5–9 mm long, with 20 faint nerves below the midpoint, the throat without scalelike appendages, sometimes with scattered hairs or glands, the lobes 1.0–1.5 mm long. Stamens attached above the midpoint of the tube, the filaments 0.2–0.4 mm long, anthers 0.7–1.0 mm long. Style 3–4 mm long. Nutlets 2.5–4.0 mm long, 1.5–3.0 mm wide, the surface with dense, broadly rounded tubercles, these often fused irregularly into blunt ridges. 2n=24. May–July.

Introduced, uncommon, known thus far only from the city of St. Louis (western U.S. east to Idaho and New Mexico; Canada, Mexico; introduced in Missouri). Railroads.

2. Anchusa L. (bugloss)

Plants perennial herbs (annual or biennial elsewhere), with sometimes somewhat woody, stout taproots. Stems erect or ascending, unbranched or branched, pubescent with sparse to moderate, spreading to loosely ascending, bristly, pustular-based hairs. Leaves alternate and basal, the basal and lower stem leaves with a winged petiole, the median and upper leaves sessile. Leaf blades linear to narrowly oblong, narrowly lanceolate, or narrowly oblanceolate, those of the basal and lower stem leaves long-tapered at the base, those of the median and upper leaves usually angled or rounded at the base, sometimes clasping the stem, angled or short-tapered to a bluntly or more commonly sharply pointed tip, the surfaces and margins moderately to densely pubescent with more or less spreading, bristly, pustular-based hairs. Inflorescences panicles with ascending to spreading or arched branches, these scorpioid, spikelike racemes, the flowers short-stalked, each subtended by a bract. Calyces actinomorphic, 5-lobed, the lobes linear to narrowly triangular, with bristly, pustular-based hairs, persistent and ascending at fruiting. Corollas funnelform to trumpet-shaped, actinomorphic (zygomorphic elsewhere), blue, purple, or purplish red, the throat with small, scalelike appendages, the lobes loosely ascending or spreading, rounded. Stamens attached above the midpoint of the corolla tube, the filaments very short, the anthers oblong, positioned at about the level of the scales, not exserted from the corolla. Ovary deeply 4-lobed, the style slender, not exserted from the corolla, often more or less persistent at fruiting, the stigma capitate, somewhat 2-lobed. Fruits dividing into 2–4 nutlets, these erect or angled obliquely, ovoid or oblong-obovoid, attached to the relatively flat gynobase at the base, the attachment scar with a short, stalklike projection and surrounded by a collarlike ring, the surface appearing variously wrinkled, ridged, and/or tuberculate, white to grayish white. Thirty-five to 40 species, Europe, Asia, Africa.

Steyermark (1963) overlooked the presence of both *Anchusa* species as rare escapes from cultivation in Missouri.

1. Calyces divided to above or slightly below the midpoint at flowering, the lobes 4–6 mm long at flowering, elongating to 9–12 mm at fruiting; nutlets 7–9 mm long, erect, oblong-obovoid, rounded at the tip .... 1. A. AZUREA
1. Calyx lobes divided nearly to the base, the lobes 6–10 mm long at flowering, elongating to 12–18 mm at fruiting; nutlets 3–4 mm long, angled obliquely, ovoid, bluntly pointed at the tip .... 2. A. OFFICINALIS

1. Anchusa azurea Mill. (showy bugloss, Italian bugloss, alkanet)

Map 1288

Stems 30–80 cm long, mostly solitary and usually unbranched below the inflorescence. Leaves 1–25 cm long, 10–30 mm wide, the basal leaves much longer than the stem leaves, these gradually reduced toward the stem tip, the basal and lower stem leaves narrowly oblanceolate to narrowly oblong-oblanceolate and long-tapered to a winged petiole, those of the median and upper stem leaves mostly lanceolate, sessile and rounded at the base, somewhat clasping the stem. Inflorescence bracts 4–12 mm long, 1–3 mm wide, linear to narrowly lanceolate or narrowly triangular, moderately to densely bristly-hairy. Flower stalks 1–3 mm long at flowering, elongating to 7–10 mm at fruiting, ascending. Calyces 8–10 mm long at flowering, divided to above or slightly below the midpoint, the lobes 4–6 mm long at flowering, elongating to 9–12 mm at fruiting. Corollas 10–20 mm long, the tube 6–10 mm long, bright blue to purplish blue, the throat with the scales white and densely hairy, the lobes 2–3 mm long. Nutlets 7–9 mm long, 2–3 mm wide, erect, oblong-obovoid, rounded at the tip, the surface with a coarse network of ridges, these appearing longitudinal toward the nutlet base. $2n=32$. April–June.

Introduced, known thus far from a single historical collection from Jefferson County (Native of Europe, introduced sporadically, in the U.S. mostly in western and northeastern states). Pastures and roadsides.

1287. Amsinckia tessellata   1288. Anchusa azurea   1289. Anchusa officinalis

2. Anchusa officinalis L. (common bugloss)

Map 1289

Stems 20–100 cm long, solitary or several, unbranched below the inflorescence or branched toward the base. Leaves 1–20 cm long, 6–20 mm wide, the basal leaves much longer than the stem leaves, these gradually reduced toward the stem tip, the basal and lower stem leaves narrowly oblanceolate to narrowly oblong-oblanceolate and long-tapered to a winged petiole, those of the median and upper stem leaves linear to narrowly lanceolate or lanceolate, sessile and angled to rounded at the base, sometimes slightly clasping the stem. Inflorescence bracts 6–12 mm long, 0.8–1.2 mm wide, linear, moderately to densely bristly-hairy. Flower stalks 1–2 mm long at flowering, elongating to 3–5 mm at fruiting, ascending. Calyces 6–10 mm long at flowering, divided nearly to the base, the lobes elongating to 12–18 mm at fruiting. Corollas 12–17 mm long, the tube 7–10 mm long, bright blue to purple or reddish purple, the throat with the scales white to lavender or pale blue and usually with dense, short, club-shaped hairs, the lobes 3–5 mm long. Nutlets 3–4 mm long, 2–3 mm wide, angled obliquely, ovoid, bluntly pointed at the tip, the surface more or less wrinkled and/or with blunt tubercles. $2n=16$. April–June.

Introduced, uncommon, known thus far from a single historical collection from Greene County (native of Europe, Asia, Africa, introduced sporadically, in the U.S. mostly in the northern half of the country). Habitat unknown, but presumably open, disturbed areas.

3. Brunnera Steven

Three species, Europe, Asia.

1. Brunnera macrophylla (Adams) I.M. Johnst. (Siberian bugloss)

Pl. 305 g–i; Map 1290

Plants perennial herbs, with stout, horizontal rhizomes. Stems 15–50 cm long, erect or ascending, solitary or few to several, mostly unbranched below the inflorescence, sparsely to moderately pubescent with patches of minute, appressed hairs and also with scattered, short, stiff, spreading, minutely pustular-based hairs. Leaves alternate and basal. Basal leaves with a long petiole 8–12 cm long, this unwinged or winged only at the tip, the blade 4–20 cm long, 25–120 mm wide, more or less heart-shaped, shallowly to deeply cordate at the base, angled to a sharply pointed tip, the surfaces and margins moderately pubescent with minute, appressed hairs and also with scattered, short, stiff, spreading, minutely pustular-based hairs. Stem leaves long- to short-petiolate, the uppermost often sessile, the blade 1–8 cm long, 5–50 mm wide, ovate to somewhat heart-shaped, shallowly cordate to rounded at the base, otherwise similar to those of the basal leaves. Inflorescence panicles with ascending to spreading branches, these scorpioid, spikelike racemes, the flowers with stalks 2–8 mm long, lacking bracts. Calyces actinomorphic, 5-lobed nearly to the base, the lobes 1.5–2.0 mm long, not becoming noticeably enlarged at fruiting, narrowly triangular, minutely appressed-hairy, occasionally with a few longer, bristly hairs, persistent and ascending at fruiting. Corollas 3–4 mm long, trumpet-shaped to nearly saucer-shaped, actinomorphic,

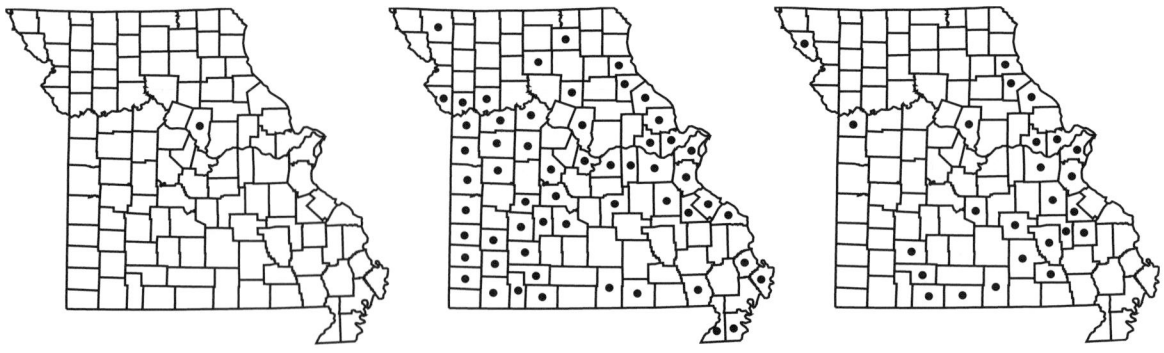

1290. Brunnera macrophylla   1291. Buglossoides arvense   1292. Cynoglossum officinale

bright blue, the tube 0.8–1.5 mm long, the throat with small, scalelike appendages (these often 2-lobed, with dense, minute papillae, white or yellow), the lobes 1.0–1.5 mm long, spreading, rounded. Stamens attached toward the midpoint of the corolla tube, the filaments very short, the anthers oblong, positioned just below the scales, not exserted from the corolla. Ovary deeply 4-lobed, the style short, not exserted from the corolla, often more or less persistent at fruiting, the stigma capitate, somewhat 2-lobed. Fruits dividing into mostly 4 nutlets, these 2.5–3.5 mm long, erect to slightly oblique, oblong-ovoid, attached to the relatively flat gynobase at the base, the attachment scar surrounded by a collarlike ring, bluntly pointed at the tip, the surface irregularly wrinkled and roughened, the wrinkles appearing longitudinal toward the nutlet base, white to grayish white. $2n=12$. May–June.

Introduced, uncommon, known thus far from a single collection from Boone County (native of Europe and Asia, introduced uncommonly in Missouri, possibly also Ohio, New York). Alleys.

This species was first reported for Missouri by Dunn (1982). It is a common garden perennial. A number of cultivars are available commercially, including several with variegated foliage.

4. Buglossoides Moench

About 7 species, Europe, Asia, introduced nearly worldwide.

Johnston (1954) resurrected this segregate of *Lithospermum* based on differences in flower morphology between the two, especially the glandular nectar guides (vs. scales) extending from the corolla throat to the stamens in *Buglossoides*. Elsewhere, the genus includes both annuals and perennials, with blue or white corollas.

1. Buglossoides arvense (L.) I.M. Johnst. (corn gromwell, bastard alkanet)
Lithospermum arvense L.

Pl. 307 g–i; Map 1291

Plants annual, with taproots. Stems 10–50 cm long, erect or ascending, solitary, unbranched or few- to several-branched at the base and above, densely pubescent with short, appressed-ascending, pustular-based hairs. Leaves alternate and basal, but the basal leaves often withered or absent at flowering. Basal leaves shorter than those of the lower stem, with a petiole 1.5–2.0 cm long, this unwinged or winged toward the tip, the blade 2–3 cm long, 3–9 mm wide, oblanceolate to narrowly oblanceolate, tapered at the base, rounded or angled to a bluntly or sharply pointed tip, the surfaces and margins moderately to densely pubescent with short, loosely appressed, pustular-based hairs. Stem leaves short-petiolate to sessile, the blade 1–7 cm long, 2–8 mm wide, linear to narrowly lanceolate, angled or tapered at the base, without noticeable lateral veins, otherwise similar to those of the basal leaves. Inflorescences of solitary flowers in the axils of the upper leaves, eventually appearing as leafy, spikelike racemes, the flowers with stalks 0.5–1.5 mm long. Calyces actinomorphic, 5-lobed nearly to the base, the lobes 6–8 mm long at flowering, becoming elongated to 8–13 mm at fruiting, linear to narrowly triangular, appressed-hairy, persistent and ascending at fruiting. Corollas 5–8 mm long, narrowly funnelform, actinomorphic, white or occasionally bluish-tinged, the tube 4–6

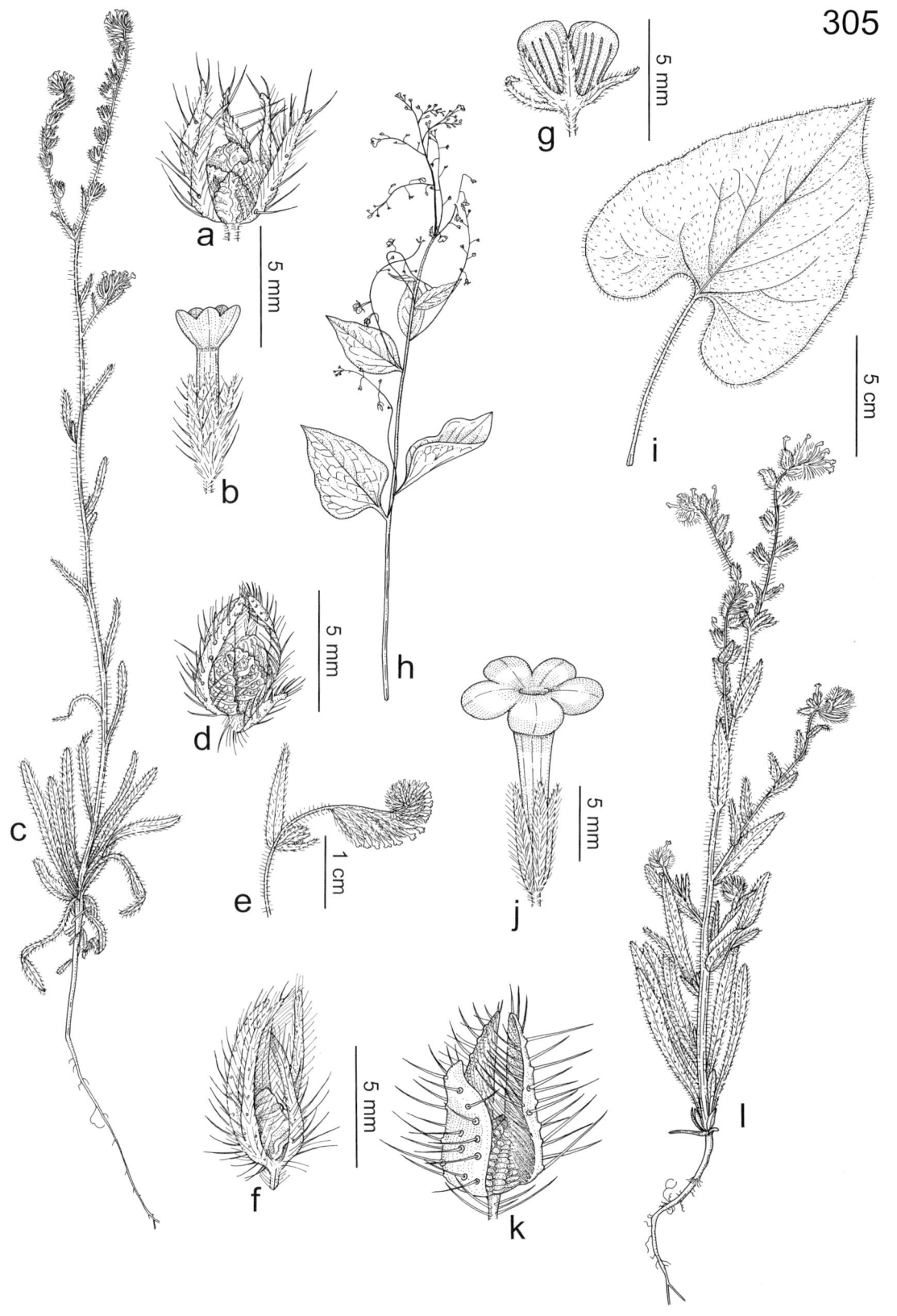

Plate 305. Boraginaceae. *Amsinckia lycopsoides*, **a)** fruit, **b)** flower, **c)** habit. *Amsinckia menziesii*, **d, f)** fruit, **e)** inflorescence. *Brunnera macrophylla*, **g)** fruit, **h)** habit. **i)** leaf. *Amsinckia tessellata*, **j)** flower, **k)** fruit, **l)** habit.

mm long, the throat lacking scales but with 5 lines or bands of minute glandular hairs extending into the tube, the lobes 1.0–1.5 mm long, more or less ascending, rounded to bluntly pointed. Stamens attached near the base of the corolla tube, the filaments short, the anthers oblong, positioned near the tube midpoint, not exserted from the corolla. Ovary deeply 4-lobed, the style short, not exserted from the corolla, often not persistent at fruiting, the stigma capitate, 2-lobed. Fruits dividing into mostly 4 nutlets, these 2.5–3.5 mm long, erect to slightly oblique, angular-ovoid with a blunt ventral keel, attached to the relatively flat gynobase at the base or nearly so, the attachment scar surrounded by a low, collarlike ring, bluntly pointed at the tip, the surface strongly longitudinally wrinkled and pitted, tan to grayish brown with darker depressions. $2n=28$. April–June.

Introduced, scattered nearly throughout the state (native of Europe, introduced nearly throughout the U.S., Canada). Glades, ledges and tops of bluffs, and savannas; also pastures, old fields, fallow fields, margins of crop fields, railroads, roadsides, and open, disturbed areas.

Al-Shehbaz (1991) noted that the several varieties of this species recognized by some European authors should be reduced to synonymy within the species.

5. Cynoglossum L. (hound's tongue)

Plants biennial or perennial herbs, with stout taproots or less commonly short, stout rhizomes, the rootstock sometimes somewhat woody. Stems erect, usually solitary, unbranched below the inflorescence or few- to several-branched toward the tip, variously hairy. Leaves alternate and basal, the basal leaves well developed at flowering, relatively large (10–30 cm long, 20–50 mm wide), the blade oblanceolate to elliptic or oblong-elliptic, tapered at the base into a long, winged petiole, grading abruptly into the stem leaves. Stem leaves sessile or the lowermost with short, winged petioles, the blade variously shaped, the surfaces moderately to densely hairy. Inflorescences sometimes paired, of short to elongate, ascending to arched, terminal, scorpioid, spikelike racemes, these usually grouped into panicles, the flowers with stalks 2–5(–12) mm long at flowering, usually elongating to 5–12 mm at fruiting, spreading to drooping or recurved at fruiting, the flowers lacking bracts but the inflorescence branch points sometimes with leaflike bracts. Cleistogamous flowers not produced. Calyces actinomorphic, 5-lobed nearly to the base, the lobes 2–5 mm long, oblong-elliptic to lanceolate, becoming somewhat elongated to 4–10 mm at fruiting and elliptic to ovate, hairy, persistent and spreading to loosely ascending or occasionally reflexed at fruiting. Corollas 4–7 mm long, broadly funnelform to trumpet-shaped, actinomorphic, purplish red or pale blue to light blue, the tube relatively short, the throat with 5 small, hairy, scalelike appendages, the lobes 3–5 mm long. Stamens attached near the midpoint of the corolla tube, the filaments short, the anthers oblong to elliptic, not exserted from the corolla. Ovary deeply 4-lobed, the style relatively short and not exserted from the corolla, usually persistent and sometimes becoming stout and tapered from the base at fruiting, the stigma capitate, shallowly 2-lobed. Fruits dividing into usually 4 nutlets, these 5–7 mm long, technically spreading but this not apparent at maturity, asymmetrically broadly obovoid, attached to the narrowly pyramidal gynobase at the tip (but appearing to be attached laterally toward the base), the attachment scar large and sometimes surrounded by a rounded, puckered rim, rounded or bluntly pointed at the tip, the surface with dense, stout, tapered, apically barbed tubercles and sometimes also finely wrinkled, pale gray to nearly white or tan to dull brown. About 75 species, nearly worldwide.

1. Stems densely pubescent with slender, relatively soft, sometimes somewhat woolly hairs, leafy to the tip; inflorescences with leaflike bracts at the branch points; corollas purplish red . . . . 1. C. OFFICINALE
1. Stems moderately to densely pubescent with stiff, straight hairs and also much shorter, appressed hairs toward the tip, often also with scattered patches of short, woolly hairs when young, leafless toward the tip; inflorescences bractless; corollas pale blue to light blue . . . . 2. C. VIRGINIANUM

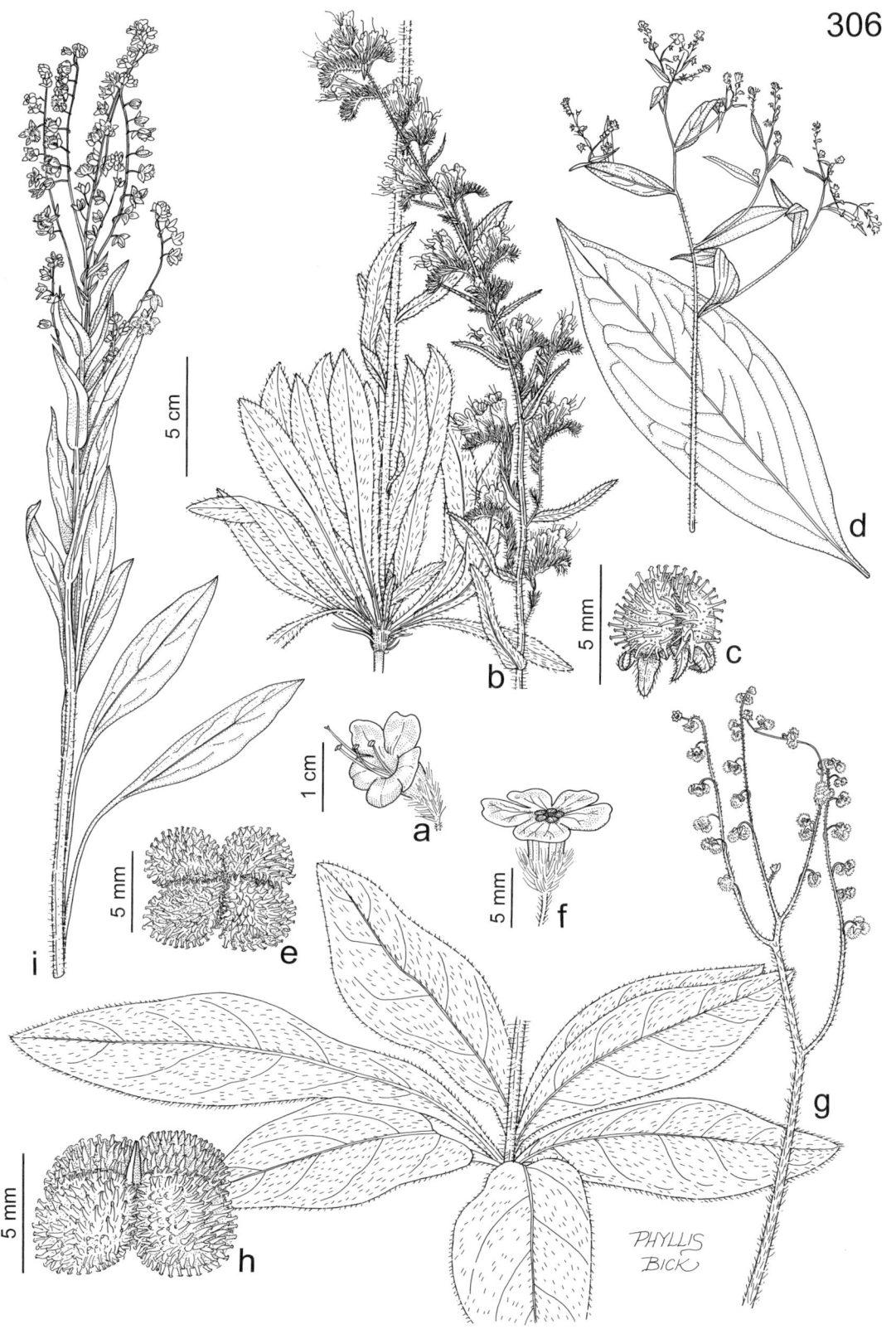

Plate 306. Boraginaceae. *Echium vulgare*, **a)** flower, **b)** inflorescence and base. *Hackelia virginiana*, **c)** fruit, **d)** inflorescence and leaf. *Cynoglossum virginianum*, **e)** fruit, **f)** flower, **g)** inflorescence and base. *Cynoglossum officinale*, **h)** fruit, **i)** inflorescence.

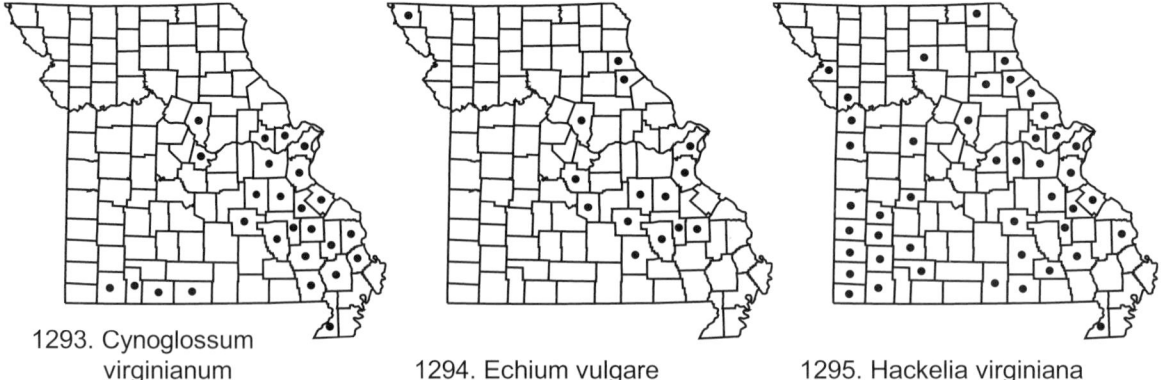

1293. Cynoglossum virginianum
1294. Echium vulgare
1295. Hackelia virginiana

1. **Cynoglossum officinale** L. (common hound's tongue)

Pl. 306 h, i; Map 1292

Plants biennial, with stout, woody taproots, producing an unpleasant musky odor when bruised or crushed. Stems 30–60 cm long, densely pubescent with slender, relatively soft, sometimes somewhat woolly hairs, leafy to the tip. Stem leaves oblanceolate to narrowly elliptic, narrowly oblong-elliptic, or narrowly lanceolate, tapered to rounded at the base, sometimes slightly clasping the stem, angled or tapered to a sharply pointed tip, grading into the bracts, the upper surface moderately roughened-pubescent with stiff, pustular-based hairs, the undersurface moderately to densely hairy with fine, softer hairs. Inflorescences usually not paired, with leaflike bracts 1–5 cm long at all or most of the branch points. Calyces 6–10 mm long at fruiting, spreading or loosely cupped around the fruit, broadly elliptic to broadly ovate. Corollas purplish red. Nutlets somewhat flattened along the dorsal surface, the attachment point usually with a well-developed rim. $2n=24$. May–July.

Introduced, scattered, mostly in the Ozark and Ozark Border Divisions (native of Europe, Asia, introduced nearly throughout the U.S. and Canada). Banks of streams and rivers; also pastures, railroads, roadsides, and open, disturbed areas.

This species is considered a noxious weed in some western states, and toxicity to livestock due to pyrrolizidine alkaloids is a hazard (Burrows and Tyrl, 2001). Aside from its purple corollas and leafy inflorescences, it tends to have narrower stem and basal leaves than does *C. virginianum*.

2. **Cynoglossum virginianum** L. (wild comfrey, giant forget-me-not)

Pl. 306 e–g; Map 1293

Plants perennial, with stout, woody taproots or short, stout rhizomes, usually not producing an odor when bruised or crushed. Stems 20–90 cm long, moderately to densely pubescent with stiff, straight hairs and also much shorter, appressed hairs toward the tip, sometimes also with scattered patches of short, woolly hairs when young, leafless toward the tip. Stem leaves oblanceolate to oblong or narrowly lanceolate, the lowermost leaves tapered at the base, the median leaves rounded to cordate at the base and usually strongly clasping the stem, rounded or angled to tapered to a usually sharply pointed tip, the surfaces moderately roughened-pubescent with stiff, more or less spreading, pustular-based hairs. Inflorescences sometimes paired or appearing dichotomously branched, bractless. Calyces 4–7 mm long at fruiting, spreading or somewhat reflexed, oblong-elliptic to broadly lanceolate. Corollas pale blue to light blue. Nutlets rounded along the dorsal surface, the attachment point with a usually poorly developed rim. $2n=24, 28$. April–June.

Bottomland forests, mesic upland forests, and banks of streams and rivers; also pastures.

Open rocky woods, wooded slopes, roadsides. Native to much of the eastern and central U.S., from New England to Florida, west to Kansas and Texas. This species was sometimes used medicinally by Native Americans for a variety of ailments (Moerman, 1998). It sometimes has been confused by herbalists with the European comfrey (*Symphytum officinale*).

6. Echium L.

About 60 species, Europe, Asia, Africa, Macaronesia.

1. Echium vulgare L. (blueweed, blue devil, blue thistle)

Pl. 306 a, b; Map 1294

Plants biennial, with taproots. Stems 30–80 cm long, erect or ascending, solitary or occasionally few, usually unbranched below the inflorescence, densely pubescent with minute, usually downward-pointed hairs and scattered to dense, longer, stiff, strongly pustular-based hairs. Leaves alternate and basal, the basal leaves sometimes withered at flowering and with the blades tapered at the base to a short or long petiole, the stem leaves mostly sessile and progressively reduced toward the stem tip. Leaf blades 2–25 cm long, 5–30 mm wide, those of the basal and lower stem leaves oblanceolate to narrowly oblanceolate, those of the median and upper leaves narrowly oblong to narrowly lanceolate or occasionally nearly linear, tapered to angled or rounded at the base, mostly angled or short-tapered to a sharply pointed tip, the surfaces and margins moderately to densely pubescent with stiff, bristly, spreading to somewhat ascending, pustular-based hairs. Stem leaves short-petiolate to sessile, the blade 1–7 cm long, 2–8 mm wide, linear to narrowly lanceolate, angled or tapered at the base, without noticeable lateral veins, otherwise similar to those of the basal leaves. Inflorescences spikelike, scorpioid spikes, these aggregated into a short or elongate panicle, the branch points with leaflike bracts 1–6 cm long, the flowers with short, leaflike bracts, sessile or nearly so. Cleistogamous flowers not produced. Calyces more or less actinomorphic (2 of the lobes sometimes slightly reduced), 5-lobed nearly to the base, the lobes 5–7 mm long at flowering, sometimes becoming elongated to 6–10 mm at fruiting, narrowly triangular to narrowly lanceolate, bristly-hairy, persistent and ascending at fruiting. Corollas 12–20 mm long, more or less funnelform, zygomorphic, the upper 2 lobes longer than the other 3, bright blue (pink in bud), the tube 8–12 mm long, the throat lacking scales, the lobes 2–5 mm long, more or less ascending, rounded to bluntly pointed, hairy on the outer surface. Stamens attached at different levels below the midpoint of the corolla tube, the filaments elongate (1 shorter than the others), the anthers oblong to somewhat heart-shaped, long-exserted from the corolla (the shortest stamen only slightly exserted). Ovary deeply 4-lobed, the style long-exserted from the corolla, finely hairy, usually withered at fruiting, the stigma capitate, strongly 2-lobed. Fruits dividing into mostly 4 nutlets, these 2.0–2.8 mm long, erect to slightly oblique, angular-ovoid with a relatively sharp ventral keel, attached to the relatively flat gynobase at the base or nearly so, the attachment scar surrounded by a collarlike ring, bluntly pointed at the slightly oblique tip, the surface strongly longitudinally wrinkled and warty or tuberculate, dark brown to nearly black with tan to white raised areas. $2n=16, 32$. May–September.

Introduced, scattered, mostly in the eastern half of the Ozark and Ozark Border Divisions (native of Europe, introduced nearly throughout the U.S. [including Alaska] and Canada). Banks of streams and rivers, upland prairies, and openings of mesic upland forests; also pastures, railroads, roadsides, and open, disturbed areas.

Echium is distinguished from other Missouri genera of Boraginaceae by its irregular, oblique corollas and unequal, purple stamens. Pusateri and Blackwell (1979) studied morphological variation in introduced North American populations and concluded that it is impossible to delineate infraspecific taxa within the species. As with many other members of the Boraginaceae, the species contains toxic pyrrolizidine alkaloids (Burrows and Tyrl, 2001) and might become a problem for livestock if it were to spread into an overgrazed pasture from an adjacent roadside.

7. Hackelia Opiz (stickseed)
(Gentry and Carr, 1976)

About 45 species, North America to South America, Europe, Asia.

1. Hackelia virginiana (L.) I.M. Johnst.
Myosotis virginiana L.
Lappula virginiana (L.) Greene

Pl. 306 c, d; Map 1295

Plants biennial, with short taproots. Stems 40–120 cm long, erect, usually solitary, few- to several-branched above the midpoint, moderately pubescent (sometimes somewhat roughened) with short, fine hairs, these downward-angled toward the stem base, spreading to upward-curved above the stem midpoint. Leaves alternate and basal, but the basal leaves usually withered or absent at flowering, long-petiolate, similar to the lower stem leaves but shorter. Leaf blades (2–)5–30 cm long, (8–)20–100 mm wide, those of the lower leaves elliptic to ovate, becoming progressively narrower (to narrowly lanceolate) toward the stem tip, angled or tapered to a sessile base or those of the lower leaves with a short, winged petiole, tapered to a sharply pointed tip, the upper surface moderately roughened with minute, stiff, pustular-based hairs, the undersurface moderately pubescent with fine, minute, and somewhat longer, mostly minutely pustular-based hairs, especially along the veins, with 3–7 noticeable pairs of lateral veins. Inflorescences usually paired, terminal on the branches, sometimes also on short lateral branches and thus appearing axillary, appearing as dense clusters at the start of flowering, then becoming elongated into usually scorpioid, spikelike racemes, these usually appearing aggregated into leafy panicles, the flowers with stalks 1–4(–10) mm long, these elongating slightly and spreading to drooping or recurved at fruiting, the lower and median flowers subtended by small, progressively reduced, lanceolate to linear bracts, the upper flowers bractless. Cleistogamous flowers not produced. Calyces actinomorphic, 5-lobed nearly to the base, the lobes 1.0–1.5 mm long at flowering, elongating to 2–3 mm long at fruiting, lanceolate, moderately to densely short-hairy, persistent and spreading to reflexed at fruiting. Corollas 1.7–2.5 mm long, more or less trumpet-shaped, actinomorphic, white (rarely pale blue), the tube 1.0–1.4 mm long, the throat with small, scalelike appendages, the lobes 0.5–0.8 mm long. Stamens inserted at about the midpoint of the tube, the filaments very short, the anthers oblong or elliptic, not exserted from the corolla. Ovary deeply 4-lobed, the style very short, not exserted from the corolla, usually persistent but inconspicuous at fruiting, the stigma capitate, shallowly 2-lobed. Fruits dividing into mostly 4 nutlets, these 2–3 mm long, erect, more or less angular-ovoid with a relatively sharp ventral keel, attached to the narrowly pyramidal gynobase at about the midpoint of the ventral keel, the attachment scar relatively broad, bluntly to sharply pointed at the tip, the surface with dense, tapered, apically barbed tubercles and minute warts or tubercles, brown. $2n=24$. June–September.

Scattered nearly throughout the state (eastern U.S. west to North Dakota and Texas; Canada). Bottomland forests, mesic upland forests, and banks of streams and rivers; also edges of pastures and roadsides.

8. Lappula Moench (stickseed, beggar's lice)

Plants annual or rarely biennial, with short taproots. Stems 5–60 cm long, erect or ascending, solitary or few to several, usually moderately branched above the midpoint, moderately to densely pubescent with shorter and longer, fine, spreading to ascending, often minutely pustular-based hairs. Leaves alternate and basal (the lower leaves sometimes withered at flowering), sessile, the blade 1–8 cm long, 3–10 mm wide, linear to narrowly oblong or narrowly oblanceolate, short-tapered to rounded at the base, rounded or angled to short-tapered to a bluntly or sharply pointed tip, the surfaces moderately to densely hairy, without noticeable lateral veins. Inflorescences usually paired, terminal on the branches, sometimes also on short lateral branches and thus appearing axillary, appearing as dense clusters at the start of flowering, then becoming elongated into usually scorpioid, spikelike racemes, these sometimes appearing aggregated into leafy panicles, the flowers usually with stalks 0.5–2.0 mm long, these not elongating noticeably at fruiting, erect to loosely ascending, each flower subtended by a small, lanceolate to linear bract. Calyces 5-lobed nearly to the base, the lobes 1.5–2.5 mm long at flowering, becoming elongated to 2–4 mm at fruiting, linear to oblong-lanceolate, moderately to densely pubescent with stiff, ascending, pustular-based hairs. Corollas small, broadly funnelform to somewhat trumpet-shaped, white or pale blue, the tube about as long as the calyx, the throat with scalelike appendages. Stamens inserted at about the midpoint of the tube, the filaments very short, the anthers oblong, not exserted from the corolla. Ovary deeply 4-lobed, the style very short, not exserted from the corolla, usually persistent but inconspicuous at fruiting, the stigma capitate, unlobed. Fruits dividing into mostly 4 nutlets, these 2–3 mm long, erect, more or less angular-ovoid with a relatively sharp ventral keel, attached to the very narrowly pyramidal gynobase along the ventral keel, the attachment scar slender and elongate, bluntly to sharply pointed at the tip, the dorsal surface with 1 or 2 rows of tapered,

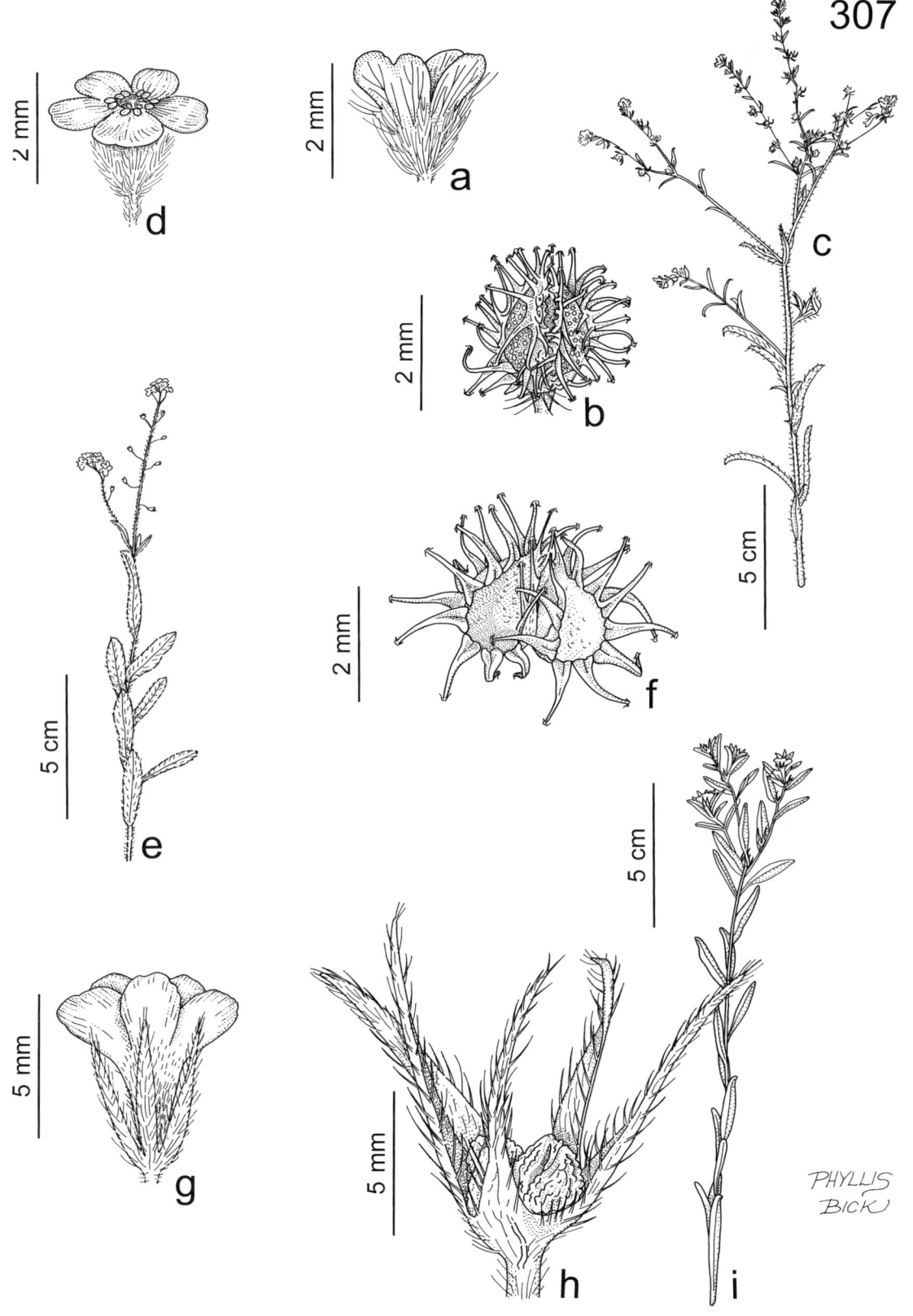

Plate 307. Boraginaceae. *Lappula squarrosa*, **a)** flower, **b)** fruit, **c)** habit. *Myosotis scorpioides*, **d)** flower, **e)** habit. *Lappula redowski*, **f)** fruit. *Buglossoides arvense*, **g)** flower, **h)** fruit, **i)** habit.

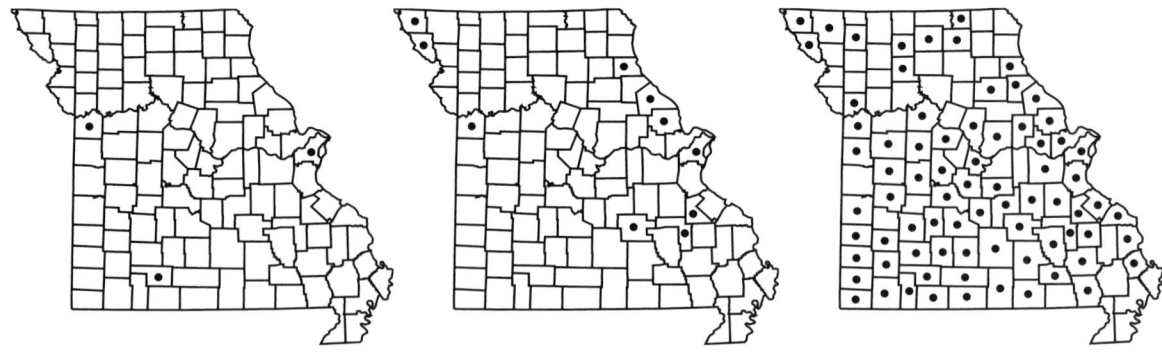

1296. Lappula redowskii   1297. Lappula squarrosa   1298. Lithospermum canescens

apically barbed tubercles, these sometimes expanded and fused into a narrow, irregular ridge at the base, otherwise with minute warts or tubercles, brown. Twelve to 40 species, North America, Europe, Asia; introduced nearly worldwide.

The two species of *Lappula* in Missouri are very similar in overall morphology. As discussed by Higgins (1979) and Al-Shehbaz (1991), the innate morphological variability of populations has resulted in the segregation of numerous species and infraspecific taxa from both of the species treated below. The presence of many intermediate plants for all of the characters said to delineate these taxa argues against their recognition. The genus is closely related to *Hackelia,* and the barbed nutlets in both genera facilitate dispersal by hooking into the fur (or trousers) of creatures coming into contact with fruiting plants.

1. Hairs of stem more or less spreading; hairs of leaves soft; dorsal surface of the nutlets with 1 row of apically barbed tubercles around the margin . . . 1. L. REDOWSKII
1. Hairs of stem mostly appressed; hairs of leaves rough; dorsal surface of the nutlets with 2(3) rows of apically barbed tubercles around the margin
. . . . 2. L. SQUARROSA

1. Lappula redowskii (Hornem.) Greene
(western stickseed)
L. occidentalis (S. Watson) Greene
L. redowskii var. *occidentalis* (S. Watson) Rydb.
L. texana (Scheele) Britton
L. texana var. *cupulata* (A. Gray) M.E. Jones
Pl. 307 f; Map 1296

Stems with the longer hairs mostly spreading or loosely ascending. Leaf blades slightly roughened, the pubescence mostly of relatively soft, slender hairs, these often minutely pustular-based. Corollas 2.5–4.0 mm long, the tube 2–3 mm long, the lobes 0.7–1.0 mm long, rounded. Nutlets with the dorsal surface having 1 row of apically barbed tubercles around the margin, these often more or less fused basally into a narrow, irregular ridge. $2n=48$. May–September.

Introduced, uncommon, known thus far from Christian and Jackson Counties and the city of St. Louis (native of western North America, Europe, Asia; introduced farther east in the U.S. sporadically to the East Coast). Railroads.

2. Lappula squarrosa (Retz.) Dumort. (two-row stickseed)
Myosotis squarrosa Retz.
L. echinata Gilib.
L. lappula (L.) Karst.
Pl. 307 a–c; Map 1297

Stems with the hairs mostly strongly ascending to appressed. Leaf blades strongly roughened, the pubescence mostly of more or less stiff, slender hairs, these strongly pustular-based. Corollas 2.5–3.0 mm long, the tube 1.5–2.0 mm long, the lobes 0.7–1.0 mm long, rounded. Nutlets with the dorsal surface having 2(3) rows of apically barbed tubercles around the margin, these distinct at the base and not forming a ridge. $2n=48$. May–September.

Introduced, scattered and sporadic (native of Europe, Asia; introduced nearly throughout the U.S. [absent from most southeastern states], also Canada, Africa). Banks of streams and rivers; also pastures, railroads, roadsides, and open, disturbed areas.

9. Lithospermum L. (puccoon)

Plants perennial herbs, with rhizomes or taproots, the rootstock sometimes somewhat woody, often with a purple pigment that leaches out when the tissue is bruised or in pressed specimens. Stems erect or ascending, sometimes from a spreading base, solitary or few to several, unbranched or more commonly few- to several-branched, moderately to densely pubescent with relatively soft and/or stiff, sometimes pustular-based hairs. Leaves alternate and sometimes also basal, but the basal leaves usually withered or absent at flowering, in some species the lower stem leaves reduced and scalelike. Stem leaves sessile or very short-petiolate, the blade variously shaped, the surfaces moderately to densely hairy, in *L. latifolium* the lateral veins prominent. Inflorescences not paired, of solitary flowers in the axils of the upper leaves or appearing as dense terminal clusters, these sometimes subsequently elongating into ascending, scorpioid, spikelike racemes, the flowers with stalks 1–4 mm long, each subtended by a leaflike bract. Cleistogamous flowers sometimes present, produced after the open flowers, solitary or in small clusters at the tips of short axillary branches. Calyces more or less actinomorphic but 1 or 2 of the lobes usually somewhat longer and/or wider than the others, 5-lobed nearly to the base, the lobes becoming somewhat elongated at fruiting, linear to narrowly triangular or narrowly lanceolate, hairy, persistent and ascending at fruiting. Corollas narrowly funnelform to trumpet-shaped, actinomorphic, yellow to orange or white to greenish white, the tube conspicuous, the throat usually with 5 small, scalelike appendages but these sometimes inconspicuous, hairy on the outer surface, sometimes only in lines. Stamens variously attached in the corolla tube (correlated with short-styled vs. long-styled flowers), the filaments short, the anthers oblong, not or only slightly exserted from the corolla. Ovary deeply 4-lobed, the styles often of different lengths in different plants of the same species (heterostylous), not or rarely slightly exserted from the corolla, often not persistent at fruiting, the stigma capitate, 2-lobed. Fruits dividing into usually 4 nutlets (in some species routinely only 1 or 2 nutlets), these erect to slightly oblique, ovoid with a blunt ventral keel, attached to the relatively flat gynobase at the base, the attachment scar surrounded by a low, collarlike ring, bluntly pointed at the tip, the surface smooth or pitted (mostly near the keel), white to ivory-colored or pale yellowish brown, shiny. About 45 species, nearly worldwide except Australia, most diverse in North America.

Eight of the species currently included in *Lithospermum* exhibit the phenomenon known as heterostyly, in which different plants in a given population produce flowers that are either short-styled or long-styled (Johnston, 1952; Al-Shehbaz, 1991). Correlated with this is usually a difference in the position of the stamens, with short-styled flowers having stamens attached higher in the tube and anthers positioned above the stigma, whereas in long-styled flowers the stamens are attached below the midpoint of the tube and the anthers are positioned below the stigma. Additionally, some species regularly produce cleistogamous flowers from short axillary branches in the leaf axils later in the growing season. Such flowers have small corollas that remain closed over the stamens and ovary and are shed as the fruits mature.

Species of *Lithospermum* have a long history of medicinal and ceremonial use by various cultures in North America and Asia (Al-Shehbaz, 1991; Moerman, 1998). Various species with roots that produce a purple or red dye have been used to color fabrics, for body decoration, and as food colorants. The common name puccoon apparently was derived from a Native American word for a dye plant. Nowadays, species of *Lithospermum* mainly are cultivated as garden ornamentals.

1. Leaf blades lanceolate to narrowly ovate; flowers solitary in the axils of the upper leaves; corollas white to greenish white, the tube 3–5 mm long, about as long as the calyx lobes . . . . 4. L. LATIFOLIUM
1. Leaf blades (except those of the reduced, scalelike lower leaves) linear to narrowly oblong or narrowly lanceolate; flowers crowded into dense terminal clus-

ters or spikelike racemes; corolla yellow or orange, the tube 8–30 mm long, surpassing the calyx lobes

2. Corollas with the tube 15–30 mm long, the lobes unevenly toothed to nearly fringed along the margin and often also somewhat corrugated ..... 3. L. INCISUM
2. Corollas with the tube 7–15 mm long, the lobes entire

3. Leaves densely and relatively softly pubescent with mostly nonpustular-based hairs; calyx lobes 3–6 mm long at flowering; corollas 12–18 mm long; nutlets 2.5–3.0 mm long, pale yellowish brown .... 1. L. CANESCENS
3. Leaves moderately to densely and roughly pubescent with stiff, pustular-based hairs; calyx lobes 7–12 mm long at flowering; corollas 14–25 mm long; nutlets 3–4 mm long, white to ivory-colored .... 2. L. CAROLINIENSE

1. Lithospermum canescens (Michx.) Lehm.
(hoary puccoon, orange puccoon)
Batschia canescens Michx.
Anchusa canescens (Michx.) Muhl.
L. canescens f. *pallida* Steyerm.
Pl. 308 a–c; Map 1298

Plants perennial herbs, with rhizomes or a short, woody rootstock. Stems 10–40 cm long, erect or ascending, solitary or few to less commonly several, unbranched or sparsely branched toward the tip, densely pubescent with loosely ascending to spreading, sometimes somewhat tangled, soft, slender, mostly nonpustular-based hairs. Stem leaves 1–5 cm long, 3–12 mm wide, narrowly oblong to lanceolate, relatively thick, angled to narrowly rounded at the base, rounded or angled to a usually bluntly pointed tip, sometimes with an abrupt, minute, sharply pointed tip, the surfaces densely pubescent with short, loosely ascending, sometimes minutely pustular-based hairs, grayish green, without noticeable lateral veins. Inflorescences dense terminal clusters, these subsequently elongating into ascending, scorpioid, spikelike racemes 3–10 cm long. Cleistogamous flowers absent. Calyces 3–6 mm long at flowering, elongating to 6–9 mm at fruiting. Corolla 12–18 mm long, broadly funnelform, yellow to orange, rarely pale yellow, the tube 7–10 mm long, the lobes 3–6 mm long, entire. Stamens attached just below the midpoint of the corolla tube in long-styled flowers, just above the midpoint in short-styled flowers. Style 5–9 mm long in long-styled flowers, 1–2 mm long in short-styled flowers. Nutlets 2.5–3.0 mm long, the tip more or less symmetrical, usually with a short, scoop-shaped appendage at the attachment point, smooth, yellowish white to pale yellowish brown. $2n=14$. March–June.

Scattered to common nearly throughout the state (eastern U.S. west to North Dakota and Texas; Canada). Glades, savannas, upland prairies, loess hill prairies, ledges and tops of bluffs, openings of mesic upland forests, dry upland forests, and rarely margins of lakes; also pastures, railroads, roadsides, and open, disturbed areas.

2. Lithospermum caroliniense (Walter ex J.F. Gmel.) MacMill. (plains puccoon)
Batschia caroliniense Walter ex J.F. Gmel.
L. croceum Fernald
Pl. 308 d–f; Map 1299

Plants perennial herbs, with stout, woody taproots or occasionally short rhizomes. Stems 25–60 cm long, erect or ascending, solitary or more commonly few to several, unbranched or more commonly sparsely branched toward the tip, roughened-pubescent with moderate to dense, short, fine, loosely ascending but stiff hairs and sparse to moderate longer, stouter hairs, both kinds minutely pustular-based. Stem leaves 2–6 cm long, 3–12 mm wide, linear to narrowly oblong or lanceolate, relatively thick, angled to narrowly rounded at the base, rounded or angled to a usually bluntly pointed tip, sometimes with an abrupt, minute, sharply pointed tip, the surfaces moderately to densely pubescent with short, stiff, loosely ascending, pustular-based hairs, grayish green to green, without noticeable lateral veins. Inflorescences dense terminal clusters, these subsequently elongating into ascending, scorpioid, spikelike racemes 5–25 cm long. Cleistogamous flowers absent. Calyces 7–12 mm long at flowering, elongating to 10–18 mm at fruiting. Corolla 14–25 mm long, broadly funnelform, yellow to orange, the tube 7–14 mm long, the lobes 5–8 mm long, entire. Stamens attached near the tip of the corolla tube in long-styled flowers, near the midpoint in short-styled flowers. Style 5–12 mm long in long-styled flowers, 2–5 mm long in short-styled flowers. Nutlets 3–4 mm long, the tip more or less symmetrical, without an appendage at the attachment point, smooth, white to ivory-colored. $2n=24$. April–June.

Widely scattered, mostly in the Mississippi Lowlands Division (eastern U.S. west to South Dakota and Texas; Canada). Sand prairies, upland prairies, and occasionally margins of lakes; also pastures, levees, roadsides, and open, disturbed areas, usually in sandy soils.

This attractive species tends to be more robust than the closely related *L. canescens* and has

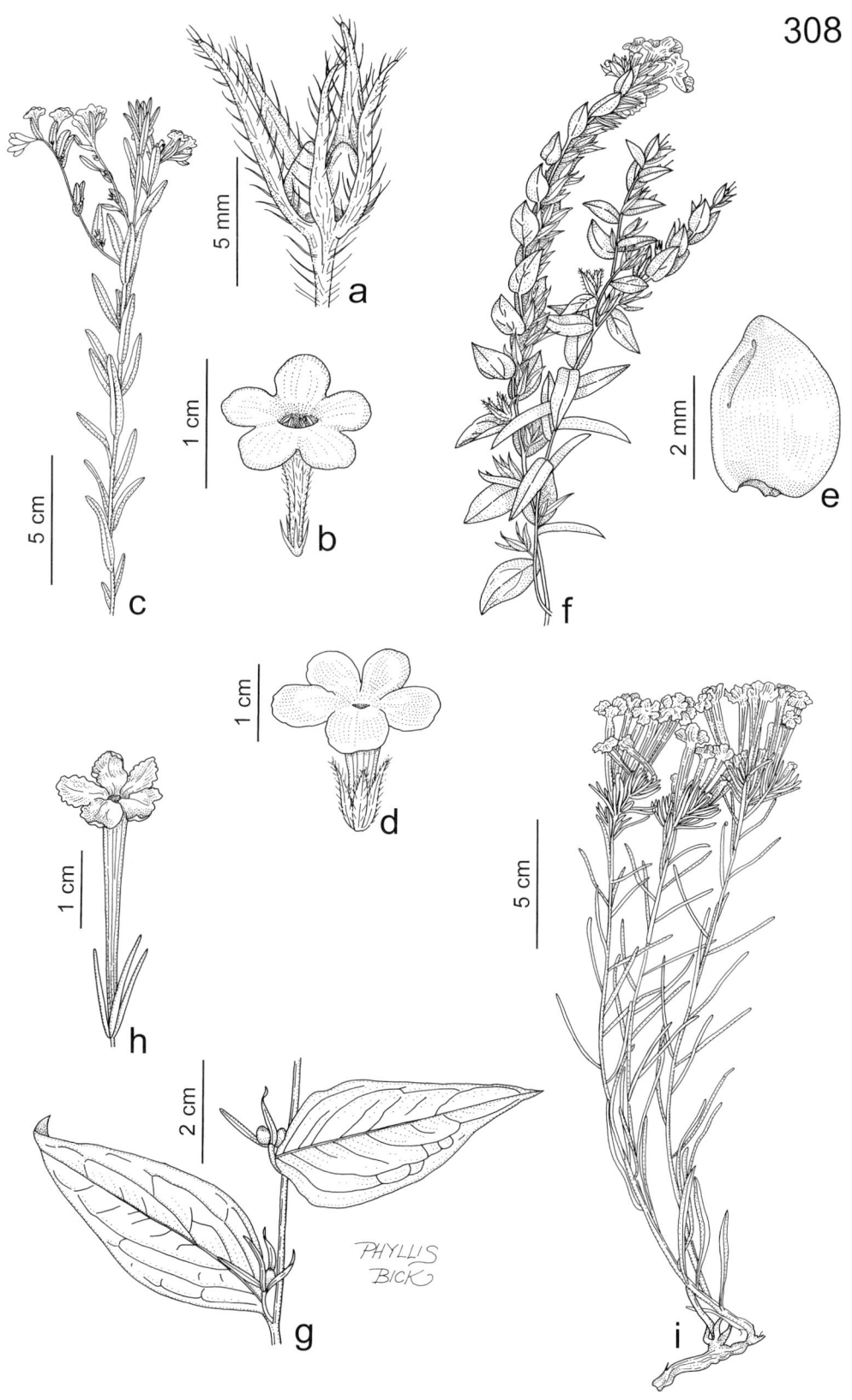

Plate 308. Boraginaceae. *Lithospermum canescens*, **a)** fruit, **b)** flower, **c)** habit. *Lithospermum caroliniense*, **d)** flower, **e)** fruit, **f)** habit. *Lithospermum latifolium*, **g)** fruiting branch. *Lithospermum incisum* **h)** flower, **i)** habit.

1299. Lithospermum caroliniense

1300. Lithospermum incisum

1301. Lithospermum latifolium

rougher pubescence. Some botanists recognize two varieties, with var. *caroliniense* the southern phase having flat calyx lobes and somewhat fewer leaves, and var. *croceum* having slightly keeled calyx lobes and somewhat more leaves per stem. Missouri plants appear to correspond more or less to var. *croceum* (Fernald) Cronquist, but the characters are too variable to allow consistent recognition of infraspecific taxa.

3. Lithospermum incisum Lehm. (yellow puccoon, narrow-leaved puccoon, plains stoneseed)

L. *angustifolium* Michx.
L. *linearifolium* Goldie

Pl. 308 g, h; Map 1300

Plants perennial herbs, with stout, woody taproots. Stems 10–40 cm long, erect or ascending, solitary or few, at first unbranched but usually becoming sparsely to moderately branched later in the season, densely pubescent with short, stiff, appressed-ascending, minutely pustular-based hairs. Stem leaves 2–7 cm long, 3–6 mm wide, linear to narrowly lanceolate, relatively thick, tapered at the base, tapered to a sharply pointed tip, the surfaces moderately to densely pubescent with short, stiff, appressed-ascending, minutely pustular-based hairs, grayish green, without noticeable lateral veins. Inflorescences dense terminal clusters, these not becoming elongated. Cleistogamous flowers solitary in the axils of the leaves along the later-produced branches. Calyces 6–12 mm long at flowering (shorter in cleistogamous flowers), elongating to 10–15 mm at fruiting. Corolla 25–40 mm long, trumpet-shaped (2–6 mm long and ovoid in cleistogamous flowers), lemon yellow to bright yellow, the tube 15–30 mm long, the lobes 3–8 mm long, unevenly toothed to nearly fringed along the margin and often also somewhat corrugated. Stamens inserted near the tip of the corolla tube. Style 20–32 mm long (much shorter in cleistogamous flowers).

Nutlets 3–4 mm long, the tip slightly oblique, without an appendage at the attachment point, smooth or with scattered pits or short grooves, mostly toward the base and along the keel, white or ivory-colored. $2n=24, 28, 36$. April–June.

Scattered in the Unglaciated Plains and portions of the Glaciated Plains, Ozark, and Ozark Border Division (western U.S. to Michigan, Louisiana, and Florida; Canada, Mexico). Limestone and dolomite glades, upland prairies, loess hill prairies, and ledges and tops of bluffs; also pastures, railroads, and roadsides.

The early open-flowering phase of this species is recognized easily by the very long corolla tube and ruffled lobes. The nutlets are white, plump, shiny, constricted at the base, and resemble small teeth. The morphological disparity between the early open-flowering and later cleistogamous phases is remarkable, and they could easily be mistaken for different species. The open-flowering phase apparently produces relatively few fruits.

4. Lithospermum latifolium Michx. (American gromwell)

L. *officinale* L. var. *latifolium* (Michx.) Lehm.

Pl. 308 i; Map 1301

Plants perennial herbs, with stout, somewhat woody taproots or short rhizomes. Stems 40–100 cm long, erect or ascending, solitary or few, unbranched or branched, moderately to densely pubescent with short, stiff, appressed-ascending, slender, usually minutely pustular-based hairs. Stem leaves 2–14 cm long, 8–60 mm wide, lanceolate to narrowly ovate, thin, angled or tapered at the base, tapered to a sharply pointed tip, the upper surface moderately roughened with minute, stiff, loosely appressed, pustular-based hairs and usually also sparse, longer hairs, dark green, the undersurface moderately to densely pubescent with short, fine, softer, loosely appressed hairs, grayish green, with 3–5 pairs of prominent, arched lateral veins. Inflo-

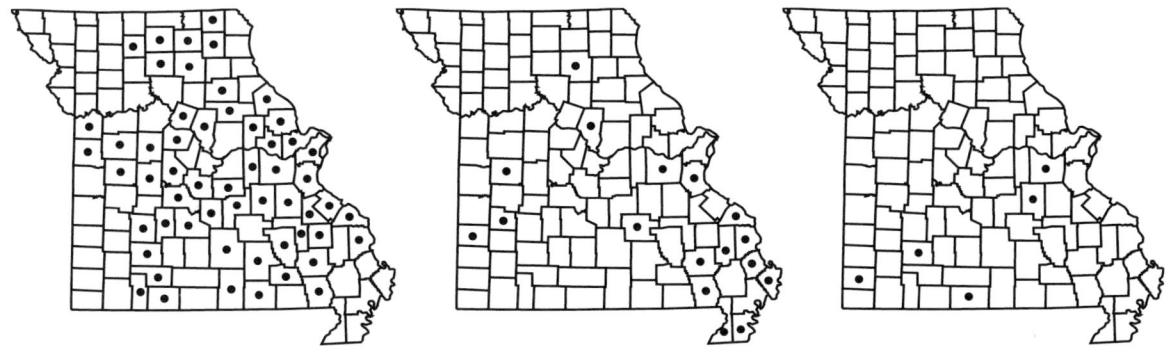

1302. Mertensia virginica   1303. Myosotis macrosperma   1304. Myosotis scorpioides

rescences of solitary flowers in the upper leaf axils. Cleistogamous flowers not produced. Calyces 4–6 mm long at flowering, elongating to 5–8 mm at fruiting. Corolla 5–8 mm long, narrowly funnelform, white to greenish white, the tube 4–6 mm long, the lobes 1.0–1.5 mm long, entire. Stamens inserted near the middle of the corolla tube. Style 1–2 mm long. Nutlets 3.5–5.0 mm long, the tip more or less symmetrical, without an appendage at the attachment point, smooth or faintly pitted, white. $2n=28$. May–June.

Uncommon in the northern half of the state (eastern U.S. west to South Dakota and Kansas; Canada). Mesic upland forests, mostly on north-facing slopes; also roadsides.

This species apparently is relatively uncommon throughout its range. Steyermark (1963) noted its superficial resemblance to *Onosmodium molle* ssp. *occidentale,* which differs in its usually paired scorpioid inflorescences and leaf venation strongly impressed on the upper surface and strongly ridged on the undersurface, in addition to its quite different flowers.

10. Mertensia Roth

About 45 species, North America, Europe, Asia.

1. Mertensia virginica (L.) Pers. ex Link
   (bluebells, Virginia cowslip)
   Pulmonaria virginica L.
   M. pulmonarioides Roth
                            Pl. 309 f, g; Map 1302

Plants perennial herbs, with a stout, woody, branched, often somewhat rhizomatous rootstock. Stems 25–70 cm long, ascending or arched, solitary or more commonly few to several, unbranched or few-branched below the inflorescence, glabrous, usually somewhat glaucous. Leaves alternate and basal, the basal leaves mostly with long, usually winged petioles, the stem leaves mostly short-petiolate, the upper leaves sometimes sessile, the bases usually extending down the stem as narrow wings or ridges. Leaf blades (2–)4–35 cm long, 15–130 mm wide, broadly ovate to elliptic, obovate, or oblanceolate, tapered at the base or those of the upper leaves sometimes rounded, rounded or angled to a bluntly or less commonly sharply pointed tip, those of the largest leaves occasionally shallowly notched, the surfaces glabrous or nearly so, particularly the undersurface often glaucous, with 3–6 noticeable pairs of lateral veins (these arched, forming a prominent network). Inflorescences not paired, terminal on the branches, appearing as dense clusters at the start of flowering, then becoming slightly elongated into short, more or less scorpioid, spikelike racemes with often pendant flowers, these continuing to elongate as the fruits mature, often appearing aggregated into leafy panicles, the flowers with stalks 2–5 mm long, these elongating to 4–12 mm at fruiting, usually spreading to drooping, the flowers lacking bracts. Cleistogamous flowers not produced. Calyces actinomorphic, 5-lobed about the midpoint, the lobes 2–5 mm long at flowering, elongating slightly at fruiting, lanceolate to oblong-ovate, glabrous, often somewhat glaucous, persistent and ascending at fruiting. Corollas 18–28 mm long, broadly funnelform, actinomorphic, usually pink in bud and lavender blue to blue at flowering, rarely persistently pink or white, the tube 10–14 mm long, the throat with minute, inconspicuous, scalelike appendages, the broad lobes 1–2 mm long. Stamens inserted near the tip of the corolla tube, the filaments relatively short (usually longer than the anthers), the anthers oblong, usually slightly exserted from the corolla. Ovary deeply 4-lobed, the style elongate, exserted from the corolla, usually somewhat withered but persistent at fruiting, the stigma capitate, unlobed. Fruits dividing into mostly 4 nutlets, these 3–4 mm long, erect or nearly so, more or less angular-ovoid with a relatively sharp ventral keel, attached to the mound-shaped to shallowly pyramidal gynobase at or just below the base of the ventral keel, the attachment scar relatively small, mostly bluntly pointed at the usually slightly oblique tip, the surface wrinkled, dull brown to dark brown. $2n=24$. March–June.

Scattered nearly throughout the state but uncommon or absent from most of the northwestern quarter (eastern U.S. west to Minnesota, Nebraska, and Arkansas; Canada). Bottomland forests, mesic upland forests in ravines, swamps, bases and ledges of bluffs, and banks of streams and rivers.

This beautiful spring wildflower is commonly cultivated as an ornamental in shade gardens. It is easily transplanted and has become a target for unscrupulous collectors who sometimes remove entire populations from the wild, leaving only ugly craters under the trees. Pringle (2004) discussed the correct use of the name *M. virginica* for this species, as opposed to *M. pulmonarioides*, an epithet accepted in some of the horticultural literature. The corollas usually change color from pink to blue as the buds mature. Uncommon plants with persistently pink or white corollas occur within otherwise blue-flowered populations and have been called f. *rosea* Steyerm. and f. *berdii* Moldenke, respectively. The aboveground parts of the plant wither and disappear soon after the fruits mature.

11. Myosotis L. (forget-me-not, scorpiongrass)

Plants annual or perennial herbs. Stems erect or ascending to arched or occasionally more or less spreading, solitary or few to several, unbranched or branched, variously hairy. Leaves alternate and usually also basal, the basal and lower stem leaves with a winged petiole, the median and upper stem leaves sessile. Leaf blades 1–8 cm long, 5–16 mm wide, variously shaped, variously hairy, the hairs sometimes hooked at the tip, often without noticeable lateral veins. Inflorescences sometimes paired, appearing as dense clusters at the start of flowering, then becoming elongated into usually sometimes scorpioid, spikelike racemes, these sometimes appearing aggregated into panicles, the flower stalks variously elongating or not as the fruits mature, erect to spreading, drooping, or reflexed, the lower flowers subtended by small, progressively reduced, slender bracts or the bracts few and alternating with the flowers or absent altogether. Cleistogamous flowers not produced. Calyces 5-lobed 1/3–2/3 of the way to the base, actinomorphic or zygomorphic (then somewhat 2-lipped with the 3 upper lobes shorter than the 2 lower lobes), the lobes usually elongating somewhat at fruiting, variously hairy, the hairs sometimes hooked at the tip. Corollas trumpet-shaped to funnelform or broadly trumpet-shaped to nearly saucer-shaped, blue or white, sometimes with a yellow spot in the throat or pink in bud, the throat with small, scalelike appendages, these often hairy, the lobes rounded. Stamens attached variously in the corolla tube, the filaments short, the anthers oblong to ovate, not exserted from the corolla. Ovary deeply 4-lobed, style short, either shorter than or slightly longer than the nutlets, not exserted from the corolla, the stigma capitate (sometimes minutely so), unlobed or 2-lobed. Fruits usually dividing into 4 nutlets, these more or less flattened-ovoid, the ventral side often angled slightly toward the tip, the lateral margin with a longitudinal rim or keel all the way around, attached to the flat or low mound-shaped gynobase at the base, the attachment scar relatively small, bluntly to broadly but sharply pointed at the tip, the surface smooth, brown to black, shiny. About 100 species, nearly worldwide, most diverse in temperate regions.

1. Plants perennial, with fibrous roots; stems spreading with ascending tips, often rooting at the lower nodes; calyces with the hairs appressed and not hooked at the tip; corollas broadly trumpet-shaped to nearly saucer-shaped, showy, the spreading portion 5–9 mm in diameter (measured across the tips of the lobes) .... 2. M. SCORPIOIDES

1. Plants annual, with slender taproots; stems erect or ascending, not rooting at the nodes; calyces with some or all of the hairs spreading and hooked at the tip; corollas funnelform to more or less trumpet-shaped, not showy, 1.0–2.5 mm in diameter (measured across the tips of the lobes)
   2. Undersurface of the leaf blades with the hairs hooked at the tip; calyces actinomorphic, the lobes all similar in appearance; corollas blue
      .... 3. M. STRICTA
   2. Undersurface of the leaf blades with the hairs not hooked at the tip; calyces zygomorphic, more or less 2-lipped, the 3 upper lobes shorter than the 2 lower lobes; corollas white or rarely pale blue
      3. Stems 20–60 cm long; leaves 2–8 cm long; flower stalks loosely ascending, angled away from the axis at fruiting; calyces 5–8 mm long at fruiting, the hairs all hooked at the tip; nutlets 1.5–2.2 mm long .... 1. M. MACROSPERMA
      3. Stems 8–35 cm long; leaves 1–5 cm long; flower stalks erect or strongly ascending, mostly appressed against the axis at fruiting; calyces 3–6 mm long at fruiting, the hairs mostly not hooked at the tip; nutlets 1.2–1.5 mm long .... 4. M. VERNA

1. Myosotis macrosperma Engelm. (big-seeded scorpiongrass)
  M. virginica var. *macrosperma* (Engelm.) Fernald
  M. verna var. *macrosperma* (Engelm.) Chapm.
      Pl. 309 a–c; Map 1303

Plants annual, with slender taproots. Stems 20–60 cm long, erect or ascending, not rooting at the lower nodes, solitary or occasionally few, usually unbranched below the inflorescence, moderately to densely pubescent with fine, loosely ascending to spreading, usually minutely pustular-based hairs, these not hooked at the tip. Leaf blades 2–8 cm long, 6–16 mm wide, lanceolate to narrowly oblong-elliptic or oblanceolate, rounded or angled to a bluntly or sharply pointed tip, the surfaces and margins densely pubescent with fine, loosely ascending to spreading, minutely pustular-based hairs, these not hooked at the tip. Inflorescences not paired, the spikelike racemes sometimes aggregated into few-branched panicles, the flowers with stalks 0.5–2.0 mm long at flowering, elongating to 2–4 mm at fruiting and loosely ascending (angled away from the axis) at fruiting with a noticeable bend or curve at the tip, the inflorescence with linear to narrowly oblong, leaflike bracts at the branch points and lowermost flowers. Calyces 1.5–2.5 mm long at flowering, elongating to 5–8 mm at fruiting, slightly zygomorphic at flowering but becoming nearly 2-lipped at fruiting, 5-lobed slightly less than (shorter teeth) to slightly more than (longer teeth) 1/2 of the way to the base, the 3 upper lobes shorter than the 2 lower lobes, especially at fruiting, triangular to narrowly triangular, densely pubescent with spreading, stiff hairs that are hooked at the tip. Corollas 2–3 mm long, broadly funnelform to trumpet-shaped, the tube 1.4–2.0 mm long, the spreading portion 1–2 mm in diameter (measured across the tips of the lobes), white or rarely pale blue. Stamens inserted below the midpoint of the corolla tube. Style 0.2–0.3 mm long, shorter than the nutlets. Nutlets 1.5–2.2 mm long, greenish brown to dark brown. April–June.

Scattered in the southeastern and southwestern portions of the state, uncommon north of the Missouri River (eastern U.S. west to Missouri and Texas; Canada). Bottomland prairies, upland prairies, bases and ledges of bluffs, swamps, bottomland forests, and mesic upland forests; also pastures, fallow fields, railroads, roadsides, and open, disturbed areas.

Some botanists have considered *M. macrosperma* a variety of the closely related *M. verna* (Steyermark, 1963), but Al-Shehbaz (1991) argued forcefully against this interpretation. In Missouri, it is the less common of the two and apparently is unable to colonize drier sites as efficiently as *M. verna*. Although the two taxa can be difficult to distinguish at flowering, the pubescence of the calyx appears to separate them reliably, as noted in the key to species above. At fruiting, *M. macrosperma* tends to be a more robust plant than is *M. verna* and has more open, elongated racemes (the fruits spaced 10–30 vs. 5–9 mm apart, as measured between the bases of adjacent stalks). The orientation of the stalks at fruiting also is a reliable character, but care must be taken not to misinterpret it in poorly pressed or immature specimens.

1305. Myosotis stricta   1306. Myosotis verna   1307. Onosmodium molle

2. Myosotis scorpioides L. (forget-me-not,
    water scorpiongrass)
  M. scorpioides var. *palustris* L.
  M. palustris (L.) L.

Pl. 307 d, e; Map 1304

Plants perennial herbs, with fibrous roots. Stems 15–60 cm long, spreading with ascending tips, often rooting at the lower nodes, solitary or more commonly few to several, usually few- to several-branched, sparsely to moderately pubescent with fine, stiff, mostly appressed-ascending, sometimes minutely pustular-based hairs. Leaf blades 2–8 cm long, 5–15 mm wide, narrowly oblong to narrowly oblong-elliptic or oblanceolate, rounded or angled to a bluntly pointed tip, the surfaces and margins moderately pubescent with short, stiff, appressed, pustular-based hairs. Inflorescences often appearing paired at the branch tips, the spikelike racemes sometimes aggregated into small panicles, the flowers with stalks 3–5 mm long, these elongating to 5–8 mm and spreading to slightly drooping at fruiting, all or nearly all lacking bracts. Calyces 2.5–4.0 mm long, actinomorphic, 5-lobed less than 1/2 of the way to the base, the lobes all more or less similar in appearance, broadly triangular, moderately to densely pubescent with short, appressed hairs, these straight at the tip. Corollas 4–10 mm long, broadly trumpet-shaped to nearly saucer-shaped, the tube 2.0–3.5 mm long, the spreading portion 5–9 mm in diameter (measured across the tips of the lobes), light blue to blue with a yellow spot in the throat, usually pink in bud. Stamens inserted near the tip of the corolla tube. Style 2.0–2.5 mm long, equal to or extending slightly beyond the tips of the nutlets. Nutlets 1–2 mm long, dark brown to black. $2n=22, 44, 64, 66$. April–October.

Introduced, sporadic, mostly in the southern half of the state (native of Europe, introduced nearly throughout temperate North America). Banks of streams, rivers, and spring branches and margins of lakes; also ditches.

This species is one of three or more Old World species cultivated in gardens under the name forget-me-not. It rarely becomes established outside of cultivation in the Midwest. A morphologically similar relative, *M. laxa* Lehm. (smaller forget-me-not), is native and widespread in both the eastern and western United States but has not been collected yet in Missouri or most of the adjacent states. It differs from *M. scorpioides* in its stems mostly not rooting at the lower nodes, smaller corollas (the spreading portion 2–5 mm in diameter), and styles shorter than the nutlets.

3. Myosotis stricta Link ex Roem. & Schult.
    (small-flowered forget-me-not, blue
    scorpiongrass)
  M. micrantha Pall. ex Lehm., misapplied

Pl. 309 h, i; Map 1305

Plants annual, with slender taproots. Stems 2–20 cm long, erect or ascending, not rooting at the lower nodes, solitary or few to several, unbranched or with few ascending branches, densely pubescent with fine, loosely ascending to spreading, usually minutely pustular-based hairs, at least some of these hooked at the tip. Leaf blades 1–2 cm long, 4–7 mm wide, lanceolate to narrowly oblong-elliptic or oblanceolate, rounded or broadly angled to a bluntly pointed tip, the surfaces and margins densely pubescent with fine, loosely ascending to spreading, minutely pustular-based hairs, at least some of these hooked at the tip. Inflorescences solitary at the stem tips, the spikelike racemes usually with additional solitary flowers in the axils of the median (and sometimes even lower) leaves, the flowers with stalks lacking or nearly so at flowering, elongating to 0.6–1.0 mm at fruiting and strongly to loosely ascending at fruiting, those above the apparent foliage leaves lacking bracts. Calyces 1.5–4.0 mm long, actinomorphic, 5-lobed about 1/2 of the way to the base, the lobes all more or less similar in appearance, narrowly triangular, densely pubescent with short, appressed hairs, these

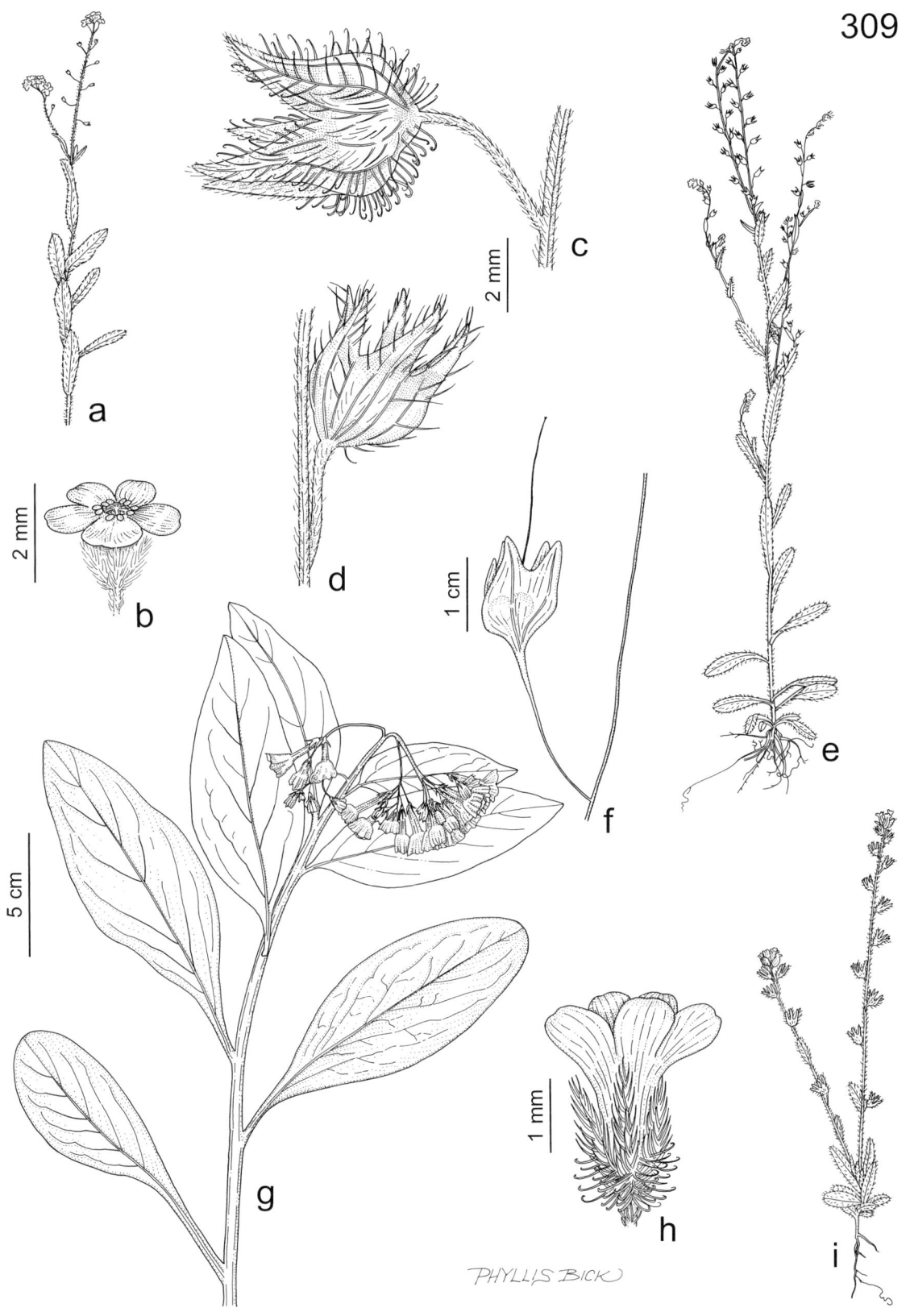

Plate 309. Boraginaceae. *Myosotis macrosperma*, **a)** habit, **b)** flower, **c)** fruit. *Myosotis verna*, **d)** fruit, **e)** habit. *Mertensia virginica*, **f)** fruiting calyx, **g)** habit. *Myosotis stricta*, **h)** flower, **i)** habit.

straight at the tip, the tube also with sparse to moderate spreading hairs that are hooked at the tip. Corollas 1.5–2.5 mm long, broadly funnelform to trumpet-shaped, the tube 0.6–1.0 mm long, the spreading portion 0.8–1.2 mm in diameter (measured across the tips of the lobes), light blue to blue, sometimes with a yellow spot in the throat, the outer surface often yellowish below the lobes. Stamens inserted toward the base of the corolla tube. Style 0.2–0.3 mm long, shorter than the nutlets. Nutlets 0.5–1.0 mm long, greenish brown. $2n=24$, 36, 48. April–June.

Introduced, uncommon, known thus far only from Linn, Montgomery, and St. Louis Counties (native of Europe, introduced widely but sporadically in mostly the northern U.S., Canada). Tops of bluffs; also lawns and open, disturbed areas.

The nomenclature of this taxon remains controversial because of ambiguities in the application of names. Gleason and Cronquist (1991) called the species *M. micrantha* Pall. ex Lehm., which is the oldest name potentially applied to it. This was in agreement with the analysis of Stroh (1935), who suggested that the names *M. micrantha* and *M. stricta* applied to the same taxon based on his analysis of the literature and possible type specimens in the herbarium of the Royal Botanical Garden in Berlin. Other authors, however, have doubted that the two names in fact refer to the same taxon. For example, Al-Shehbaz (1991) suggested that, based on the published characters of yellow axillary flowers and wrinkled nutlets, the name *M. micrantha* should refer to a taxon in some genus other than *Myosotis*. Because the original publication of *M. micrantha* cited no type specimens and is therefore of somewhat uncertain application, the present treatment follows most of the recent European floristic literature in accepting *M. stricta* as the correct name for the taxon in question.

Myosotis stricta was first reported for Missouri by Christ (1984) from a site in Bee Tree County Park (St. Louis County) and has been found only rarely in the state.

4. Myosotis verna Nutt. (early scorpiongrass)

M. virginica (L.) Britton, Sterns & Poggenb.

Pl. 309 d, e; Map 1306

Plants annual, with slender taproots. Stems 8–35 cm long, erect or ascending, not rooting at the lower nodes, solitary or occasionally few, usually unbranched below the inflorescence, moderately to densely pubescent with fine, loosely ascending to spreading, usually minutely pustular-based hairs, these not hooked at the tip. Leaf blades 1–5 cm long, 3–9 mm wide, lanceolate to narrowly oblong-elliptic or oblanceolate, rounded or angled to a bluntly or sharply pointed tip, the surfaces and margins densely pubescent with fine, loosely ascending to spreading, minutely pustular-based hairs, these not hooked at the tip. Inflorescences not paired, the spikelike racemes sometimes aggregated into few-branched panicles, the flowers with stalks 0.5–1.5 mm long at flowering, elongating to 2–4 mm at fruiting and erect to strongly ascending (mostly appressed to the axis) at fruiting with a strong bend at the tip, the inflorescence with linear, leaflike bracts at the branch points and lowermost flowers. Calyces 1.5–2.0 mm long at flowering, elongating to 3–6 mm at fruiting, slightly zygomorphic at flowering but becoming nearly 2-lipped at fruiting, 5-lobed slightly less than (shorter teeth) to slightly more than (longer teeth) 1/2 of the way to the base, the 3 upper lobes shorter than the 2 lower lobes, especially at fruiting, triangular, densely pubescent with short, appressed, straight hairs and scattered, longer, spreading hairs that are hooked at the tip. Corollas 2–3 mm long, broadly funnelform to trumpet-shaped, the tube 1.4–2.0 mm long, the spreading portion 1–2 mm in diameter (measured across the tips of the lobes), white or rarely pale blue. Stamens inserted toward the base of the corolla tube. Style 0.2–0.3 mm long, shorter than the nutlets. Nutlets 1.2–1.5 mm long, greenish brown to dark brown. April–June.

Scattered nearly throughout the state (nearly throughout the U.S.; Canada). Bottomland prairies, upland prairies, glades, ledges and tops of bluffs, savannas, bottomland forests, and mesic to dry upland forests; also pastures, railroads, roadsides, and open, disturbed areas.

For a discussion of the separation of this species from the closely related *M. macrosperma*, see the treatment of that species.

12. Onosmodium Michx. (false gromwell)

About 7 species, North America.

An interesting characteristic of *Onosmodium* species that has resulted in underestimation of perianth size in some floristic accounts is that the flowers tend to appear precociously fertile. That is, the style becomes long-exserted from the bud and the stamens shed their pollen before the corolla has become fully enlarged (Al-Shehbaz, 1991).

1. Onosmodium molle Michx. (western false gromwell)

O. bejariense A. DC.

Pl. 310 c–j; Map 1307

Plants perennial herbs, with a stout, woody rootstock. Stems 30–120 cm long, erect or ascending to somewhat arched, solitary or few, unbranched or few-branched below the inflorescence, nearly glabrous to densely hairy. Leaves alternate, sessile, the basal leaves present only in seedlings, the lowermost leaves reduced and usually withered or shed at flowering. Leaf blades 2–14 cm long, 10–40 mm wide, narrowly lanceolate to lanceolate, elliptic, or ovate, angled or tapered at the base, angled or tapered to a sharply pointed tip, the upper surface densely pubescent with appressed to spreading, stiff, pustular-based hairs, the undersurface moderately to densely pubescent with softer, sometimes nonpustular-based hairs, with 2–4 prominent pairs of lateral veins, these usually strongly impressed on the upper surface and strongly ridged on the undersurface. Inflorescences often paired, terminal, appearing as dense clusters at the start of flowering, then becoming elongated into scorpioid, spikelike racemes, the flowers with stalks 0.3–1.0 mm long at flowering, these elongating to 1.5–3.0 mm at fruiting, ascending, each flower subtended by a leaflike bract 10–35 mm long. Cleistogamous flowers not produced. Calyces actinomorphic, 5-lobed nearly to base, the lobes 3.5–9.0 mm long at flowering, elongating slightly at fruiting, linear to narrowly triangular or narrowly oblanceolate, densely hairy on both surfaces, persistent and ascending at fruiting. Corollas 7–16 mm long at full flowering (see discussion above), tubular with a small bulge in the throat, actinomorphic, white to cream-colored or greenish yellow, hairy on the outer surface, the tube 5–12 mm long, the throat lacking scalelike appendages, the lobes 2–4 mm long. Stamens inserted near the tip of the corolla tube, the filaments very short, the anthers lanceolate to slightly arrowhead-shaped, not or only slightly exserted from the corolla. Ovary deeply 4-lobed, the style elongate, strongly exserted from the corolla, persistent at fruiting, stigma capitate, shallowly 2-lobed. Fruits dividing into mostly 1 or 2 nutlets, these 2.5–5.0 mm long, ovoid to broadly ovoid, sometimes with a shallow ventral groove or an indistinct, blunt ventral ridge, white to ivory-colored or brownish-tinged, attached to the flat gynobase at the base, the attachment scar relatively large, the surface smooth or rarely minutely pitted, sometimes shiny. $2n=24, 28$. May–August.

Scattered, mostly south of the Missouri River (eastern U.S. west to Montana and New Mexico; Canada). Glades, savannas, bottomland forests, mesic to dry upland forests, ledges and tops of bluffs, upland prairies, and loess hill prairies; also old fields, railroads, and roadsides; often on calcareous substrates.

Onosmodium molle is a widespread species that is polymorphic in size, branching, pubescence, and fruit characters. Correlation of the variation with geographic regions has led to the recognition of several taxa as varieties (Cronquist, 1959), subspecies (Cochrane, 1976; Turner, 1995), or species (Mackenzie, 1906; Steyermark, 1963; Turner, 1995) within the complex. Intermediates between the taxa are commonly encountered, and some characters used by previous authors to separate them, such as nutlet size, pitting, and luster, do not hold up under careful scrutiny (Cochrane, 1976). Most recently, Turner (1995) recognized the Missouri material as varieties of *O. bejariense* and restricted *O. molle* to cedar glades in Tennessee and adjacent states, but this treatment is not strongly supported. We follow the nomenclature of Cochrane (1976) and recognize the Missouri material as subspecies of *O. molle*. Subspecific status is supported by the correlation of variation with geographic distribution as mapped by Turner (1995). Two other subspecies do not occur in Missouri. *Onosmodium molle* ssp. *molle* appears to be restricted to Tennessee, Kentucky, northern Alabama, and southern Illinois (Baskin et al., 1983) and differs in its slightly narrower leaves with relatively soft, dense pubescence on both surfaces. The ssp. *bejariense* (A. DC.) Cochrane occupies the southwestern portion of the species range in portions of Texas, Oklahoma, Arkansas, and Louisiana and has somewhat longer (2–4 mm vs. 1.0–1.5 mm) stem hairs than does ssp. *occidentale*, as well as slightly smaller nutlets (Turner, 1995). The entire complex would benefit from more detailed biosystematic study.

In Missouri, a small number of specimens exist that are intermediate between all combinations of pairs of the three subspecies treated below. These specimens represent putative hybrids, but to date there is no experimental evidence to support this hypothesis. It also is not known whether such specimens were collected from mixed populations of the putative parental subspecies or whether these morphological intermediates form their own uniform populations. The plants in question appear to produce fully formed nutlets.

1. Stems glabrous or nearly so below the midpoint, becoming progressively hairier toward the inflorescence
 . . . . 1C. SSP. SUBSETOSUM
1. Stems moderately to densely hairy throughout

2. Upper surface of the leaf blades with the longer and shorter hairs more or less spreading; nutlets often flared basally into a small collar around the attachment point
. . . . 1A. SSP. HISPIDISSIMUM

2. Upper surface of the leaf blades with the hairs appressed and all of similar length; nutlets more or less rounded at the base, usually without a collar around the attachment point
. . . . 1B. SSP. OCCIDENTALE

1a. ssp. hispidissimum (Mack.) B. Boivin
   O. hispidissimum Mack.
   O. molle Michx. var. *hispidissimum* (Mack.) Cronquist
   O. bejariense A. DC. var. *hispidissimum* (Mack.) B.L. Turner

Pl. 310 c–e

Stems relatively robust (to 120 cm long), often few-branched toward the tip, moderately to densely pubescent throughout with spreading, shorter and longer hairs, the longest hairs about 2 mm long. Leaf blades 2–14 cm long, lanceolate to elliptic or ovate, the upper surface densely pubescent with more or less spreading, longer and shorter, strongly pustular-based hairs, the undersurface moderately to densely pubescent with softer, longer and shorter, minutely pustular-based hairs. Nutlets often flared basally into a small collar around the attachment point, smooth or occasionally minutely pitted. $2n=24$. May–July.

Scattered, sporadic within the range of ssp. *subsetosum* (eastern U.S. [mostly from the Appalachian Mountains westward] west to Minnesota and Louisiana; Canada). Glades, savannas, dry upland forests, upland prairies, and loess hill prairies; also railroads and roadsides.

In Missouri, this subspecies is known almost entirely from older collections, and since the 1950s it appears to have become less common in the state than in previous decades.

1b. ssp. occidentale (Mack.) Cochrane
   O. occidentale Mack.
   O. molle Michx. var. *occidentale* (Mack.) I.M. Johnst.
   O. bejariense A. DC. var. *occidentale* (Mack.) B.L. Turner

Pl. 310 i, j

Stems less robust (mostly 30–80 cm long), often unbranched below the inflorescence, moderately to densely pubescent throughout with appressed-ascending hairs of more or less uniform length and usually also with scattered, loosely ascending to spreading hairs, the longest hairs about 1.5 mm long. Leaf blades 2–10 cm long, narrowly lanceolate to lanceolate or narrowly elliptic, both surfaces densely pubescent with appressed to loosely appressed, usually minutely pustular-based hairs all of similar length. Nutlets rounded basally, lacking a collar around the attachment point (rarely with a small collar), smooth. $2n=28$. June–August.

Scattered, sporadic within the range of ssp. *subsetosum* (Montana to New Mexico east to Wisconsin and Georgia; Canada). Glades, savannas, dry upland forests, tops of bluffs, upland prairies, and occasionally bottomland forests; also old fields, railroads, and roadsides.

1c. ssp. subsetosum (Mack. & Bush ex Small) Cochrane
   O. subsetosum Mack. & Bush ex Small
   O. molle Michx. var. *subsetosum* (Mack. & Bush ex Small) Cronquist
   O. bejariense A. DC. var. *subsetosum* (Mack. & Bush ex Small) B.L. Turner

Pl. 310 f–h

Stems variously more or less robust (mostly 40–120 cm long), often unbranched below the inflorescence, glabrous or nearly so below the midpoint, becoming progressively more pubescent toward the inflorescence with mostly appressed-ascending hairs of more or less uniform length, the longest hairs about 1.0 mm long. Leaf blades 2–12 cm long, narrowly lanceolate to lanceolate or elliptic, the upper surface moderately to densely pubescent with appressed, strongly pustular-based hairs, all of similar length, the undersurface densely pubescent with softer, loosely appressed to curved, often nonpustular-based hairs. Nutlets rounded basally, lacking a collar around the attachment point, smooth. May–July.

Scattered, mostly south of the Missouri River (Oklahoma, Missouri, Arkansas, and Tennessee). Glades, savannas, bottomland forests, mesic to dry upland forests, ledges and tops of bluffs, and upland prairies; also railroads and roadsides.

This is by far the most commonly encountered subspecies in Missouri. Its global distribution is mostly restricted to the Ozark Mountains.

13. **Symphytum** L. (comfrey) (Gadella, 1984)

About 35 species, Europe, Asia.

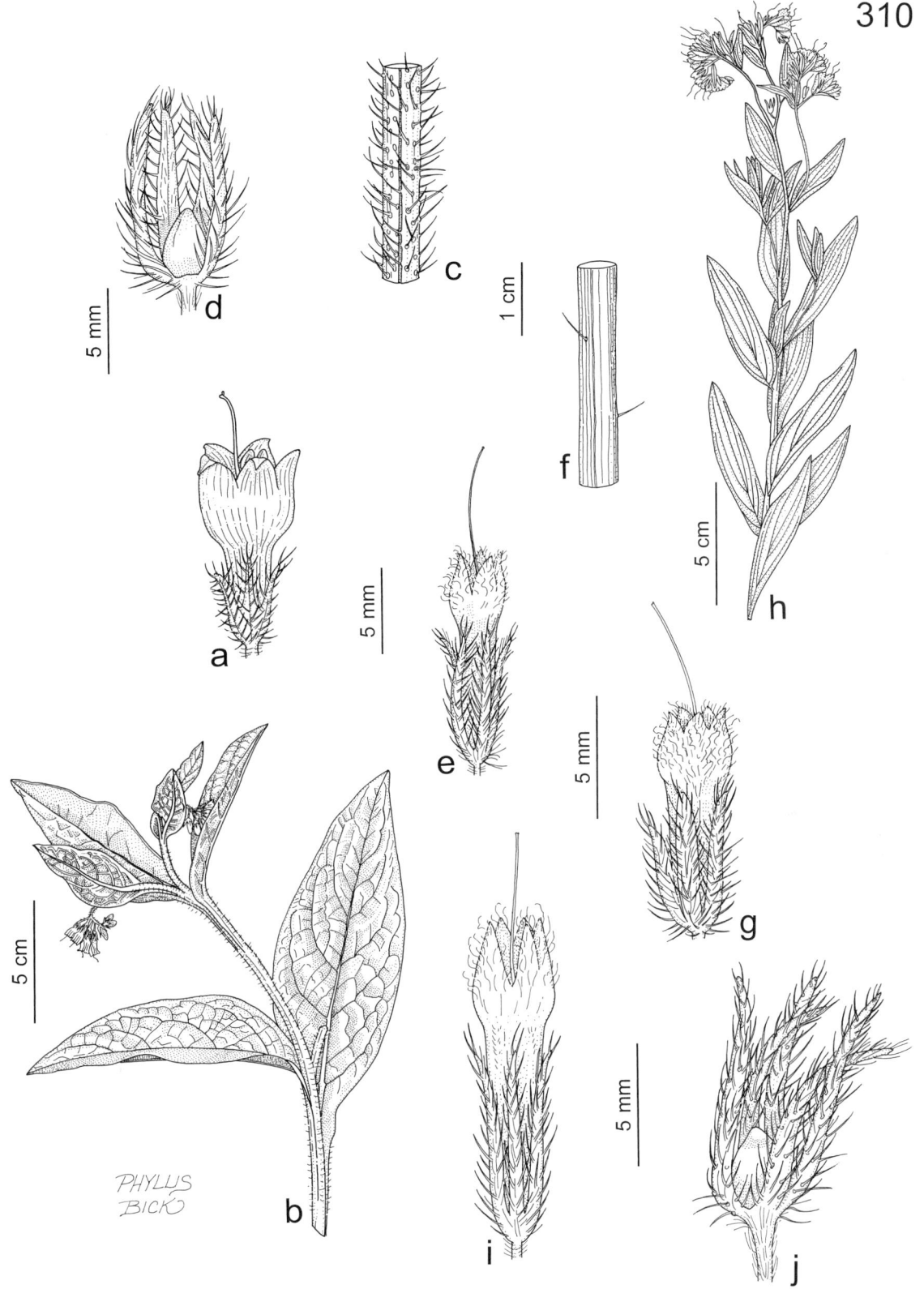

Plate 310. Boraginaceae. *Symphytum officinale*, **a)** flower, **b)** habit. *Onosmodium molle* ssp. *hispidissmum*, **c)** stem, **d)** fruit, **e)** flower. *Onosmodium molle* ssp. *subsetosum*, **f)** stem, **g)** flower, **h)** habit. *Onosmodium molle* ssp. *occidentale*, **i)** flower, **j)** fruit.

1308. Symphytum officinale   1309. Alliaria petiolata   1310. Alyssum alyssoides

1. Symphytum officinale L. (common comfrey)
Pl. 310 a, b; Map 1308

Plants perennial herbs, with stout, somewhat fleshy taproots. Stems 30–1200 cm long, ascending or arched, solitary or more commonly few to several, unbranched or few- to several-branched below the inflorescence, moderately to densely pubescent with minute, spreading hairs often having microscopically hooked tips and also sparse to moderate longer, stiff, spreading to somewhat downward-curved, usually strongly pustular-based hairs. Leaves alternate and basal, the basal leaves mostly with long, usually winged petioles, the stem leaves mostly short-petiolate, the upper leaves sometimes sessile, the bases mostly extending down the stem as wings, these often narrow. Leaf blades 4–35 cm long, 15–150 mm wide, ovate to elliptic, those of the uppermost leaves sometimes lanceolate, tapered at the base, tapered to a sharply pointed tip, the upper surface roughened-pubescent with shorter and longer, stiff, spreading, usually strongly pustular-based hairs, the shortest of these often minutely hooked at the tip, the undersurface moderately to densely hairy with short, soft, curved hairs, with usually 4–7 noticeable pairs of lateral veins (these arched, forming a faint network). Inflorescences not paired, terminal on the branches, appearing as dense clusters at the start of flowering, then becoming slightly elongated into short, more or less scorpioid, spikelike racemes with often pendant flowers, these continuing to elongate as the fruits mature, sometimes appearing aggregated into leafy panicles, the flowers with stalks 2–10 mm long, these elongating to 4–15 mm at fruiting, usually spreading to drooping, the flowers lacking bracts. Cleistogamous flowers not produced. Calyces actinomorphic or nearly so, 5-lobed about 3/4 of the way to the base, the lobes 3–6 mm long at flowering, elongating to 5–10 mm at fruiting, narrowly triangular to nearly linear, densely pubescent with minute, spreading hairs having microscopically hooked tips and scattered, longer, stiff, spreading, pustular-based hairs, persistent and ascending at fruiting. Corollas 12–18 mm long, narrowly funnelform to narrowly bell-shaped, actinomorphic, purple to dull purplish blue (rarely cream-colored to pale lemon yellow elsewhere), the tube 6–12 mm long, the throat with small, scalelike appendages, the lobes often spreading to outward-curled, 1–2 mm long. Stamens inserted near the midpoint of the corolla tube, the filaments relatively short (shorter than the anthers), the anthers oblong, not exserted from the corolla. Ovary deeply 4-lobed, the style elongate, twisted just below the tip, exserted slightly from the corolla, usually somewhat withered but persistent at fruiting, the stigma capitate, unlobed. Fruits dividing into mostly 4 nutlets, these 4–6 mm long, erect or ascending, more or less angular-ovoid with a blunt ventral keel, attached to the relatively flat gynobase at the base or nearly so, the attachment scar surrounded by a low, collarlike ring with an irregularly toothed margin, mostly bluntly pointed at the somewhat oblique tip, the surface smooth, black, shiny. $2n=24$, 32, 40–48, 56. June–August.

Introduced, uncommon and sporadic, thus far mostly in the eastern half of the state (native of Europe, introduced nearly throughout the U.S., Canada). Roadsides and disturbed areas.

Comfrey once was grown widely for animal feed, being very rich in minerals and protein, and the young leaves have been used by humans in teas and salads. It also has a long history of medicinal use, often taken in the form of teas, extracts, and poultices. The lengthy register of putative uses includes treatment in various types of applications for lung diseases, diarrhea, colds, gangrene, burns, anemia, ulcers, wounds, headaches, tuberculosis, bee stings, insect bites, and broken bones (Al-Shehbaz, 1991). However, the presence of pyrrolizidine alkaloids and potential liver toxicity makes *Symphytum* species a potential health risk for human consumption (Burrows and Tyrl, 2001).

BRASSICACEAE (CRUCIFERAE) (Mustard Family)
Contributed by Ihsan A. Al-Shehbaz and George Yatskievych

Plants annual or perennial herbs, rarely woody at the base or shrubby. Leaves alternate or basal, rarely opposite or whorled, lacking stipules, entire to deeply lobed or compound. Inflorescences terminal (except in some species of *Lepidium*), short to elongate racemes or panicles, or reduced to single, long-stalked flowers (in *Leavenworthia*). Flowers mostly actinomorphic, perfect. Calyces of 4 free or rarely united sepals. Corollas of 4 free petals, these uncommonly reduced or absent, often narrowed to stalklike bases. Stamens (2, 4)6, often the outer 2 shorter than the inner 4. Ovary 1 per flower, superior, of 2 fused carpels, usually with 2 locules. Style 1 per flower, persistent in the fruits, the stigma 1, entire or 2-lobed. Ovules 1 to numerous. Fruits uncommonly indehiscent and achenelike or more commonly 2-valved capsules that dehisce longitudinally leaving a persistent *replum* (the thin, placental band of tissue around the periphery of the *septum,* the partition between the 2 locules), these arbitrarily referred to as *siliques* when more than 3 times as long as wide or *silicles* when less than 3 times as long as wide. Seeds variously shaped, with curved embryos. About 350 genera, about 3,500 species, worldwide, but most diverse in temperate and alpine regions and dry areas.

The petals of most species of Brassicaceae are arranged in the shape of a cross, leading to the common name crucifer and the familial name Cruciferae. The family contains a large number of economically important species, both beneficial plants cultivated for food and oils and detrimental weeds. Although separate keys to flowering and fruiting material are given, most species flower for long enough that flowers and fruits are present at the same time. It is recommended that both keys are used, and a more reliable determination to the genus is achieved when both keys are successfully used to reach to the same genus.

An important character of the fruits is whether they are circular in cross-section, 4-angled, or slightly to strongly flattened. If flattened, they can be flattened parallel or at a right angle to the septum. In parallel-flattened fruits, the septum is a broad band of tissue between the 2 faces (valves) extending the full width of the fruit, and the replum is visible as a line along the edge of each face. In fruits flattened at a right angle, the septum is a narrow line of tissue bisecting each face, and the replum is visible as a line along the middle of the face. Such fruits sometimes also have lines or wings along the margins and are almost always shorter than 3 times as long as wide.

Key to Fruiting Material
1. Fruits at most 3 times longer than wide
    2. Upper stem leaves perfoliate, clasping, or with auricles at the base
        3. Seeds 1 or 2 per fruit
            4. Fruits 2 to 3 times longer than wide, with 1 seed in the middle, borne on reflexed stalks . . . . 24. ISATIS
            4. Fruits about as long as wide, often with 2 seeds (1 in each locule), borne on erect or spreading stalks
                5. Plants with branched hairs; fruits with a coarse network or honeycomb-like pattern of ridges and pits . . . . 31. NESLIA
                5. Plants glabrous or with unbranched hairs; fruits with smooth walls . . . . 26. LEPIDIUM
        3. Seeds (3)4 or more per fruit
            6. Fruits not flattened; flowers (when present) with yellow or cream-colored petals
                7. Plants with branched hairs; fruits pear-shaped or obovoid, the valve tip extending into a beak in the stylar area . . . . . . 11. CAMELINA

7. Plants glabrous or with unbranched hairs; fruits oblong or rarely ovoid, the valve tip not extending into a beak . . . . 36. RORIPPA
6. Fruits flattened at a right angle to the septum; flowers (when present) with white petals
    8. Fruits triangular, not winged; plants with branched hairs . . . . 12. CAPSELLA
    8. Fruits obovate or circular in outline, winged, at least at the tip; plants glabrous or with simple hairs
        9. Seed surface with a series of concentric, arched ribs or with a coarse, honeycomb-like pattern of ridges and pits; leaf margin often toothed . . . . 40. THLASPI
        9. Seed surface smooth; leaf margin entire . . . . . . 29. MICROTHLASPI
2. Upper stem leaves petiolate or sessile but not clasping or auriculate at base, sometimes the stems leafless and only rosette leaves present
  10. Plants glabrous or with exclusively unbranched hairs
    11. All of the fruit stalks subtended by bracts . . . . 37. SELENIA
    11. Almost all of the fruit stalks lacking bracts
      12. Fruits 2.0–4.5 cm long, the stalk above the attachment of the perianth 7–18 mm long; styles 4–10 mm long . . . . 28. LUNARIA
      12. Fruits at most 0.8 cm long, the stalk above the attachment of the perianth absent or rarely to 1 mm long; styles absent or to 3 mm long
        13. Fruits clearly constricted at the middle into upper and lower segments; styles 1–3 mm long. . . . . 35. RAPISTRUM
        13. Fruits not constricted into segments; styles absent or rarely to 0.3 mm long
          14. Fruits strongly flattened at a right angle to the septum; seeds 1 or 2 per fruit . . . . 26. LEPIDIUM
          14. Fruits circular in cross-section or only slightly flattened; seeds rarely produced but the undeveloped ovules numerous per fruit
            15. Plants aquatic; styles 2–4 mm long. . . . . 36. RORIPPA
            15. Plants terrestrial; styles to 1 mm long . . . . 5. ARMORACIA
  10. Plants with at least some branched hairs
    16. Plants uniformly pubescent with appressed hairs, each with 2 opposite branches, thus appearing as a straight line . . . . 27. LOBULARIA
    16. Plants with other types of hairs, often more than 1 kind present
      17. Styles absent or rarely to 0.7 mm long
        18. Fruits circular in profile; seeds 1–4 per fruit . . . . 2. ALYSSUM
        18. Fruits oblong, ovate, or elliptic; seeds more than 4 (to 80) per fruit . . . . 18. DRABA
      17. Styles (1–)2–8 mm long
        19. Fruits glabrous, circular to obovate, elliptic, or pear-shaped in profile, circular in cross-section . . . . 32. PHYSARIA
        19. Fruits pubescent, elliptic to oblong in profile, flattened parallel to the septum
          20. Plants perennial; styles 4–8 mm long; fruits ascending or spreading . . . . 6. AUBRIETA
          20. Plants annual; styles 1–3(–4) mm long; fruits erect . . . . 8. BERTEROA

1. Fruits more (often much more) than 3 times longer than wide
    21. At least some fruits solitary on long stalks originating from the basal rosette
        .... 25. LEAVENWORTHIA
    21. All fruits borne in racemes or panicles
        22. Upper stem leaves perfoliate, clasping, or with distinct auricles at the base
            23. Fruits and their stalks erect, appressed to the infructescence axis
                24. Plants glabrous or with simple hairs near the base; stems angled
                    .... 7. BARBAREA
                24. Plants with at least some branched hairs near the base, stems circular in cross-section
                    25. Plant glabrous and glaucous in the upper half; fruits circular or slightly 4-angled in cross-section; seeds in 2 rows per locule
                        .... 41. TURRITIS
                    25. Plants not glaucous, rarely glabrous in the upper half; fruits flattened; seeds in 1 row per locule .... 4. ARABIS
            23. Fruits and their stalks not erect, not appressed to the inflorescence rachis
                26. Leaves compound, with 3–9(–13) leaflets; seed surface with a coarse, netlike or honeycomb-like pattern of 25–60 ridges and pits on each side; plants rooting from most nodes .... 30. NASTURTIUM
                26. Leaves simple; seed surface with a minute, netlike pattern of more than 60 ridges and pits on each side; plants not rooting from the nodes
                    27. Seeds in 2 rows per locule .... 36. RORIPPA
                    27. Seeds in 1 row per locule
                        28. Fruits flattened parallel to the septum; plants with branched hairs (if glabrous or with only simple hairs, then the fruits strongly recurved above the middle)
                            .... 9. BOECHERA
                        28. Fruits circular or 4-angled in cross-section; plants glabrous or with exclusively simple hairs
                            29. Seeds globose .... 10. BRASSICA
                            29. Seeds oblong or oblong-elliptic in profile
                                30. Fruits 5.0–13.5 cm long, strongly 4-angled
                                    .... 15. CONRINGIA
                                30. Fruits 1–4 cm long, circular in cross-section or weakly 4-angled
                                    31. Stems angled along the entire length; inflorescences mostly panicles .... 7. BARBAREA
                                    31. Stems circular in cross-section; inflorescences mostly racemes .... 23. IODANTHUS
        22. Upper stem leaves petiolate or sessile but not clasping or auriculate at base, sometimes absent
            32. Lower portion of the plant or at least the leaf margins with branched hairs
                33. All leaves basal .... 18. DRABA
                33. At least some leaves on the stem
                    34. Leaves 2 or 3 times pinnately divided .... 16. DESCURAINIA
                    34. Leaves entire or toothed

35. Fruits 2.5–4.0 mm wide, strongly flattened parallel to the septum, borne on strongly reflexed stalks ...... 9. BOECHERA
35. Fruits less than 1.5 mm wide, circular, 4-angled, or slightly flattened parallel to the septum, borne on erect, ascending, or spreading stalks
    36. All of the branched hairs sessile .... 21. ERYSIMUM
    36. All of the branched hairs short-stalked
        37. Seeds in 2 rows per locule; styles 4–8 mm long; stellate hairs present.................. 6. AUBRIETA
        37. Seeds in 1 row per locule; styles 0.1–4.0 mm long; stellate hairs absent
            38. Upper stem leaves lanceolate or ovate-lanceolate, the margins toothed; styles 3–4 mm long; fruits (4–)6–10(–14) cm long; plants often with stalked glands .... 22. HESPERIS
            38. Upper stem leaves linear, the margins entire; styles 0.1–1.5 mm long; fruits 0.8–4.5 cm long; plants not glandular .... 3. ARABIDOPSIS
32. Plants glabrous or with exclusively unbranched hairs
    39. Plants with large, stalked glands .... 14. CHORISPORA
    39. Plants not glandular
        40. Upper stem leaves compound
            41. Seeds in 1 row per locule; fruit valves coiled during dehiscence; plants not rooting from the lower nodes .... 13. CARDAMINE
            41. Seeds in 2 rows per locule; fruit valves not coiled during dehiscence; plants rooting from the lower nodes................. 30. NASTURTIUM
        40. Upper stem leaves simple or pinnately lobed
            42. Seeds in 2 rows per locule
                43. Fruits with a flattened beak 4–11 mm long .... 19. ERUCA
                43. Fruits not beaked, or the beaklike style less than 4 mm long and not flattened
                    44. Leaves mostly basal; fruits 2.5–4.5 cm long ...... 17. DIPLOTAXIS
                    44. Leaves mostly on the stem; fruits less than 2.5 cm long .... 36. RORIPPA
            42. Seeds in 1 row per locule
                45. At least lower part of the inflorescence bracteate at fruiting .... 20. ERUCASTRUM
                45. Inflorescence not bracteate at fruiting
                    46. Fruits flattened parallel to the septum
                        47. Fruit valves becoming coiled during dehiscence; replum flattened .... 13. CARDAMINE
                        47. Fruit valves not coiled during dehiscence; replum not flattened .... 33. PLANODES
                    46. Fruits circular or 4-angled in cross-section
                      48. Fruit valves obscurely veined or only with the midnerve prominent
                        49. Fruit not corky, the lower segment seeded, the beak 0.1–1.0 cm long........................ 10. BRASSICA
                        49. Fruit often corky, the lower segment seedless, minute, stalklike, the beak 1–5 cm long.......... 34. RAPHANUS

48. Fruit valves often with 3–7 prominent nerves
   50. Seed surface with prominent, longitudinal ribs; fruit stalk as thick as the fruit; stigma entire ..... 1. ALLIARIA
   50. Seed surface smooth or with a honeycomb pattern; fruit stalk narrower than the fruit, or if as wide as the fruit, then the stigmas 2-lobed
      51. Seeds globose, 2–12 per fruit; fruits beaked; stigmas entire .... 38. SINAPIS
      51. Seeds oblong, 10–120 per fruit; fruits beakless; stigmas 2-lobed .... 39. SISYMBRIUM

Key to Flowering Material
1. All or nearly all of the flowers solitary on long stalks originating from the basal rosette .... 25. LEAVENWORTHIA
1. All flower stalks arranged in racemes or panicles
   2. Stamens 2 .... 26. LEPIDIUM
   2. Stamens (4)6
      3. Petals absent
         4. Plants with at least some branched hairs .... 18. DRABA
         4. Plants glabrous .... 36. RORIPPA
      3. Petals present
         5. Petals white, pink, or purple
            6. Petals deeply 2-lobed at the tip
               7. Stem leaves many........................... 8. BERTEROA
               7. Stem leaves absent .... 18. DRABA
            6. Petals not lobed, entire or slightly notched at the tip
               8. Stems rooting from most nodes .... 30. NASTURTIUM
               8. Stems not rooting from nodes
                  9. Upper stem leaves clasping or auriculate
                     10. Plants with at least some branched hairs
                        11. Ovaries triangular to obcordate in profile; stellate hairs sessile .... 12. CAPSELLA
                        11. Ovaries linear in outline; stellate hairs absent or with a short-stalked base
                           12. Stellate hairs stalked; young fruits appressed to the axis of the inflorescence .... 4. ARABIS
                           12. Stellate hairs absent, or if present, then the young fruits spreading in relationship to the axis of the inflorescence .... 9. BOECHERA
                     10. Plants glabrous or exclusively with simple hairs
                        13. Ovaries and young fruits linear in profile
                           .... 23. IODANTHUS
                        13. Ovaries and young fruits ovate to circular in profile
                           14. Ovules 2 per ovary .... 26. LEPIDIUM
                           14. Ovules at least 4 per ovary
                              15. Stem leaves with the margins entire; crushed or bruised plants with neither an unpleasant aroma nor an odor of garlic
                                 .... 29. MICROTHLASPI

>   15. Stem leaves with the margins toothed; crushed or bruised plants usually with an unpleasant aroma or an odor of garlic .... 40. THLASPI
> 9. Upper stem leaves neither clasping nor auriculate, sometimes absent
>   16. Plants with at least some branched hairs
>   17. Plants uniformly pubescent with appressed hairs each having 2 opposite branches, thus appearing as a straight line with an attachment point near the center .... 27. LOBULARIA
>   17. Plants with more than one kind of hair
>   18. Ovaries and young fruits linear
>   19. Uppermost stem leaves linear, the margins entire .... 3. ARABIDOPSIS
>   19. Uppermost stem leaves ovate to lanceolate, the margins usually toothed
>   20. Petals purple or pink, rarely white; stigmas 2-lobed; plants sparsely to moderately glandular .... 22. HESPERIS
>   20. Petals white; stigmas entire; plants not glandular .... 9. BOECHERA
>   18. Ovaries and young fruits oblong, elliptic, or ovate
>   21. Petals purple or pink; plants perennial; styles 4–8 mm long .... 6. AUBRIETA
>   21. Petals white; plants annual; styles absent or to 0.2 mm long .... 18. DRABA
>   16. Plants glabrous or exclusively with unbranched hairs
>   22. Plants with stalked glands .... 14. CHORISPORA
>   22. Plants without glands
>   23. Petals purple, or if pink or white, then with darker veins
>   24. Petal veins distinctly darker than the rest of the petal; ovaries and young fruits sessile......................... 34. RAPHANUS
>   24. Petal veins not darker than the rest of the petal; ovaries and young fruits with a stalk above the receptacle (attachment point of the calyx and corolla)............................ 28. LUNARIA
>   23. Petals white or if pink then the veins not darker than the rest of the petal
>   25. Ovaries and young fruits circular or ovate in profile
>   26. Ovules 2 per ovary .... 26. LEPIDIUM
>   26. Ovules many per ovary........................ 5. ARMORACIA
>   25. Ovaries and young fruits linear in profile
>   27. Plants perennial with rhizomes or tubers........ 13. CARDAMINE
>   27. Plants annual or biennial
>   28. Upper stem leaves broadly ovate to triangular, the margins coarsely toothed but not lobed; plants with the odor of garlic when crushed or bruised .... 1. ALLIARIA
>   28. Upper stem leaves pinnately divided
>   29. Young fruits with the replum winged; seeds wingless .... 13. CARDAMINE
>   29. Young fruits with the replum wingless; seeds winged .... 33. PLANODES

5. Petals yellow or orange, rarely pale yellow
   30. Inflorescences bracteate, at least toward the base
      31. Ovaries linear in profile; petals 4–7 mm long; styles 1.5–3.0 mm long
         .... 20. ERUCASTRUM
      31. Ovaries oblong or elliptic in profile; petals 8–11 mm long; styles 5–10 mm long .... 37. SELENIA
   30. Inflorescences not bracteate
      32. Petal veins distinctly darker than the rest of blade
         33. Stigma 2-lobed; pistil not segmented .... 19. ERUCA
         33. Stigma not lobed; pistil segmented above the base into a sterile lower segment and ovulate upper segment .... 34. RAPHANUS
      32. Petal veins not darker than the rest of blade
         34. Upper stem leaves perfoliate, clasping, or auriculate
            35. Plants with at least some branched hairs
               36. Ovaries and young fruits linear in profile.......... 41. TURRITIS
               36. Ovaries and young fruits ovate, pear-shaped, or circular in profile
                  37. Petals 1.5–2.5 mm long; ovules 4 per ovary .... 31. NESLIA
                  37. Petals 3.0–5.5 mm long; ovules 6–12 per ovary
                     .... 11. CAMELINA
            35. Plants glabrous or with exclusively simple hairs
               38. Ovules 1 or 2 per ovary
                  39. Lowermost stem leaves entire or toothed .... 24. ISATIS
                  39. Lowermost stem leave 2 or 3 times pinnately divided
                     .... 26. LEPIDIUM
               38. Ovules more than 8 per ovary
                  40. Upper stem leaves entire, often glaucous
                     41. Pistils and young fruits beaked............ 10. BRASSICA
                     41. Pistils and young fruits not beaked ...... 15. CONRINGIA
                  40. Upper stem leaves toothed or pinnately lobed, not glaucous
                     42. Stems angled; ovules in 1 row per locule .... 7. BARBAREA
                     42. Stems not angled; ovules in 2 rows per locule
                        .... 36. RORIPPA
         34. Upper stem leaves petiolate or sessile, not clasping or auriculate
            43. Plants with at least some branched hairs
               44. Leaves 2 or 3 times pinnately divided; branched hairs treelike (with a stalked base and somewhat ascending branches)
                  .... 16. DESCURAINIA
               44. Leaves entire or toothed; branched hairs stellate, sessile
                  45. Ovaries and young fruits linear in profile ..... 21. ERYSIMUM
                  45. Ovaries and young fruits circular or nearly so in profile
                     46. Petals 2–4 mm long; simple hairs present in addition to the branched hairs; style less than 1 mm long
                        .... 2. ALYSSUM
                     46. Petals 5–11 mm long; simple hairs absent, the hairs all branched; styles 2–5 mm long .... 32. PHYSARIA
            43. Plants glabrous or with simple hairs only
               47. Most of the leaves basal .... 17. DIPLOTAXIS
               47. Most of the leaves on the stem

48. Sepals spreading or reflexed at flowering . . . . 38. SINAPIS
48. Sepals erect or ascending at flowering
    49. Ovules 1–4 per ovary . . . . 35. RAPISTRUM
    49. Ovules more than 10 per ovary
        50. Plants aquatic or of wet areas; ovary and young fruits oblong in profile, rarely linear; ovules in 2 rows per locule . . . . 36. RORIPPA
        50. Plants terrestrial; ovary and young fruits linear in profile; ovules in 1 row per locule
            51. Stigmas 2-lobed; uppermost leaves pinnately divided . . . 39. SISYMBRIUM
            51. Stigmas entire; uppermost leaves often toothed or entire
            . . . . 10. BRASSICA

1. Alliaria Heist. ex Fabr.

Two species, Europe, Asia, introduced widely in North America.

1. Alliaria petiolata (M. Bieb.) Cavara & Grande (garlic mustard)
A. officinalis Andrz.

Pl. 311 i, j; Map 1309

Plants biennial, terrestrial, glabrous or with sparse, unbranched, nonglandular hairs, with the odor of garlic when fresh plants are crushed or bruised. Stems 30–120 cm long, erect or ascending, unbranched or branched from the base. Leaves alternate and basal, 3–12 cm long, the uppermost sessile, the lower ones with progressively longer petioles, the bases not clasping the stem. Leaf blades triangular to broadly ovate, less commonly kidney-shaped, the margins shallowly to coarsely toothed and wavy. Inflorescences racemes, sometimes few-branched panicles with racemose branches, without bracts. Sepals (2–)3–4 mm long, ascending, oblong-elliptic, shed soon after the flower opens. Petals 4–7(–9) mm long, unlobed, white. Styles 1–2(–3) mm long. Fruits spreading, straight or nearly so, (2–)3–7(–8) cm long, more than 10 times as long as wide, linear, somewhat 4-angled in cross-section, not beaked, dehiscing longitudinally, each valve with a midnerve and 2 lateral, longitudinal nerves. Seeds in 1 row in each locule, 2.0–4.5 mm long, oblong or narrowly elliptic, the margins not winged, the surface with a pattern of longitudinal ribs, black. $2n=42$. May–June.

Introduced, scattered in the state, mostly in urban areas and in the Big Rivers Division (native of Europe and Asia, widely naturalized in the U.S. and southern Canada). Bottomland and mesic upland forests in valleys and floodplains of creeks and rivers, often in soils derived from calcareous substrates; also in disturbed, mostly shaded sites.

This species originally escaped during the 1800s in the northeastern states, probably from gardens. In Europe, it has a long history of use as a potherb, salad green, garlic substitute, and source of fatty oils (from the seeds), and medicinally as a diuretic, diaphoretic, expectorant, and treatment for asthma and dropsy (Al-Shehbaz, 1988b). It is extremely invasive in moist, shaded habitats and has become a serious threat to natural plant communities in much of the eastern half of North America (Nuzzo, 1991, 1993). Cattle that graze on it produce garlic-flavored milk, and deer usually avoid browsing on it. Characteristics of the species that promote its ability to spread and become naturalized include self-compatibility of the flowers and seeds that may persist in the soil for several years. Once established, it rapidly replaces the native ground flora.

In Missouri, garlic mustard apparently was absent until recently. The earliest reports for the state were from eastern counties (Mohlenbrock, 1979; Wiese, 1979), followed quickly by reports from the western and southwestern areas (Henderson, 1980; Nightingale, 1980). The plant has become widely established in floodplains of the Missouri River and other rivers and is expected to continue its spread in the state.

2. Alyssum L. (alyssum)

Plants annual (perennial elsewhere), terrestrial, with stellate, mostly appressed, nonglandular hairs. Stems 5–25 cm long, erect to spreading, usually branched near the base.

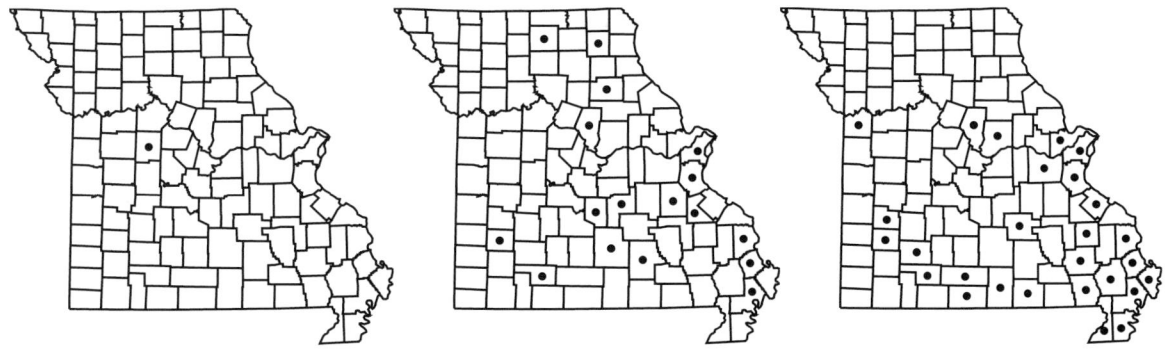

1311. Alyssum desertorum   1312. Arabidopsis lyrata   1313. Arabidopsis thaliana

Leaves alternate, 6–15(–30) mm long, the bases not clasping the stem, the margins entire. Inflorescences racemes, without bracts. Sepals erect or ascending, 1.5–2.0 mm long, oblong-elliptic. Petals shallowly notched at the tip, pale yellow or cream-colored. Styles 0.3–0.7 mm long (longer elsewhere). Fruits spreading or ascending, 3–5 mm long, about as wide as long or less than 2 times as long as wide, broadly oval to circular in outline, strongly flattened parallel to the septum, usually shallowly notched at the tip, sometimes with a minute style, dehiscing longitudinally, the valves lacking midnerves. Seeds 1 or 2 per locule, 1.0–1.5 mm long, obovate to broadly elliptic in outline, somewhat flattened, the margins narrowly winged, the surface minutely roughened, orange. About 170 species, Europe, Asia, Africa, Alaska, Canada.

Some species of *Alyssum* are cultivated as garden ornamentals.

1. Sepals persistent through fruit maturity; fruits hairy at maturity
   .... 1. A. ALYSSOIDES
1. Sepals deciduous soon after the flower opens; fruits glabrous at maturity
   .... 2. A. DESERTORUM

1. Alyssum alyssoides (L.) L. (pale alyssum)

Pl. 311 g, h; Map 1310

Plants pubescent with 6–10-rayed stellate hairs mixed with simple and forked ones on the flower stalks and sepals. Stems 5–35(–50) cm long. Leaves short-petiolate. Leaf blades 3.0–4.5 cm long, oblong, oblanceolate, to linear, sometimes spatulate or obovate. Sepals persistent through fruit maturity, (1.5–)2.0–3.0 mm long. Petals 2–4 mm long, whitish or pale yellow. Filaments slender, neither appendaged nor toothed. Styles 0.3–0.6 mm long. Fruits circular, (2–)3–4(–5) mm long, inflated at the center, flattened at the margins, uniformly pubescent with minute, stellate hairs at maturity. Seeds oblong or ovate, 1 or 2 per locule, 1–2 mm long, narrowly margined all around. 2n=32. April–July.

Introduced, known thus far only from western Missouri (native of Europe, Asia, naturalized in North America). Railroads, roadsides, and disturbed areas.

This species was first reported for Missouri by Henderson (1980).

2. Alyssum desertorum Stapf

Pl. 311 c; Map 1311

Plants pubescent with 8–20-rayed stellate hairs. Stems (2–)5–18(–28) cm long. Leaves short-petiolate or more commonly sessile, linear to narrowly oblanceolate, 0.5–2.5(–3.0) cm long. Sepals deciduous soon after the flower opens. Petals 2–3 mm long, pale yellow. Median filaments narrowly winged at the base, lateral filaments with a broadly winged appendage apically notched into 2 teeth. Styles 0.3–0.7 mm long. Fruits circular, 2.5–4.5 mm long, inflated at the center, broadly flattened at margin, glabrous and glaucous at maturity. Seeds ovate, 2 per locule, 1.2–1.5 mm long, slightly margined or not margined. 2n=32. April–June.

Introduced, known thus far only from Pettis County (native of Europe, extensively naturalized in western North America). Disturbed areas.

The Missouri population, located at the state fairgrounds in Sedalia, is disjunct from the closest naturalized populations in western Nebraska. It was first reported by Castaner (1984).

3. Arabidopsis Heynh.
(O'Kane and Al-Shehbaz, 1997)

Plants annual or perennial herbs, terrestrial, with simple and 2- or 3-branched, stalked hairs. Stems 2–45(–80) cm long, erect to loosely ascending from a spreading base, branched at the base or unbranched. Leaves alternate and basal, 1–6(–10) cm long, the bases not clasping, the margins entire, toothed, or pinnately divided. Inflorescences racemes, without bracts. Sepals 1.0–3.5 mm long, erect or ascending, oblong. Petals entire or notched at the tip, white, pink, or purple. Styles 0.5–1.5 mm long. Fruits spreading or ascending, 1.0–4.5 cm long, to 50 times as long as wide, linear, circular in cross-section or flattened parallel to the septum, dehiscing longitudinally, the valves with or without a midnerve. Seeds 8–35 per locule, 0.3–1.4 mm long, oblong or elliptic in outline, sometimes slightly flattened, brown. Nine species, North America, Europe, Asia.

1. Plants biennial or perennial; fruits flattened parallel to the septum; petals 4–8 mm long . . . . 1. A. LYRATA
1. Plants annual; fruits circular in cross-section; petals 2.0–3.5 mm long
. . . . 2. A. THALIANA

1. Arabidopsis lyrata (L.) O'Kane & Al-Shehbaz ssp. lyrata (sand cress)
Arabis lyrata L.
Arabis lyrata f. *parvisiliqua* M. Hopkins
Cardaminopsis lyrata (L.) Hiitonen

Pl. 311 d–f; Map 1312

Plants biennial or perennial herbs. Stems (5–)10–30(–45) cm long, erect or ascending, branched at the base, usually sparsely pubescent with unbranched and 2- or 3-branched, stalked hairs toward the base. Basal leaves (1–)2–4(–8) cm long, spatulate to oblanceolate, subentire or toothed to more commonly pinnately divided into 3–9 segments, narrowed at the base to a short petiole, usually pubescent on both surfaces with unbranched and 2- or 3-branched, stalked hairs. Stem leaves (0.5–)1.0–3.0(–4.0) cm long, narrowed at the base, not clasping, mostly sessile, the lower leaves often pinnately divided, the upper leaves entire to toothed, mostly glabrous. Sepals 2–4 mm long, oblong-elliptic, green. Petals 4–8 mm long, white, sometimes tinged with light pink or purple. Styles 1.0–1.5 mm long. Fruits ascending at maturity, straight or slightly arched upward, 1.5–4.5 cm long, flattened parallel to the septum, the valves without nerves. Seeds in 1 row in each locule, (0.7–)0.9–1.2 mm long, oblong-elliptic in outline, flattened, sometimes narrowly winged at the tip, the surface with a netlike or honeycomb-like pattern of ridges and pits, orangish yellow. $2n=16$, 32. April–May.

Scattered in the eastern half of the state, mostly in the Ozark and Ozark Border Divisions (eastern U.S. and adjacent Canada west to Wisconsin and Missouri). On and around dolomite and sandstone bluffs, sometimes among rocks in the adjacent mesic upland forest.

The f. *parvisiliqua* was based on plants with shorter fruits, but there appears to be total intergradation in fruit lengths in some populations. The arctic ssp. *kamchatika* (Fisch. ex DC.) O'Kane & Al-Shehbaz, which is distributed from Washington to Alaska and eastern Asia, is characterized by stouter stems, thicker leaves, and obtuse (vs. pointed) fruits. Recent molecular studies (O'Kane and Al-Shehbaz, 2003) have suggested that *A. lyrata* is more at home in the genus *Arabidopsis* than in *Arabis*. It was placed in *Arabidopsis* by O'Kane and Al-Shehbaz (1997, 2002c) and is now widely recognized as a member of this genus.

2. Arabidopsis thaliana (L.) Heynh. (mouse-eared cress, thale cress)

Pl. 311 a, b; Map 1313

Plants annual, terrestrial. Stems (2–)5–30(–50) cm long, erect or ascending, branched or unbranched below the inflorescence, often glabrous in the apical half, pubescent toward the base with mostly unbranched hairs intermixed with 2- or 3-branched stalked hairs. Leaves mostly basal, those of the stems alternate, 1–5 cm long, sessile or the leaf blades narrowed to short petioles, the bases not clasping. Leaf blades oblong to spatulate, progressively reduced and narrowed to linear toward the stem tip, the margins entire or with few shallow teeth, sparsely to densely pubescent with unbranched, forked, and/or rarely 2- or 3-branched stalked hairs. Inflorescences panicles or racemes. Sepals 1.0–2.0(–2.5) mm long, ascending, oblong,

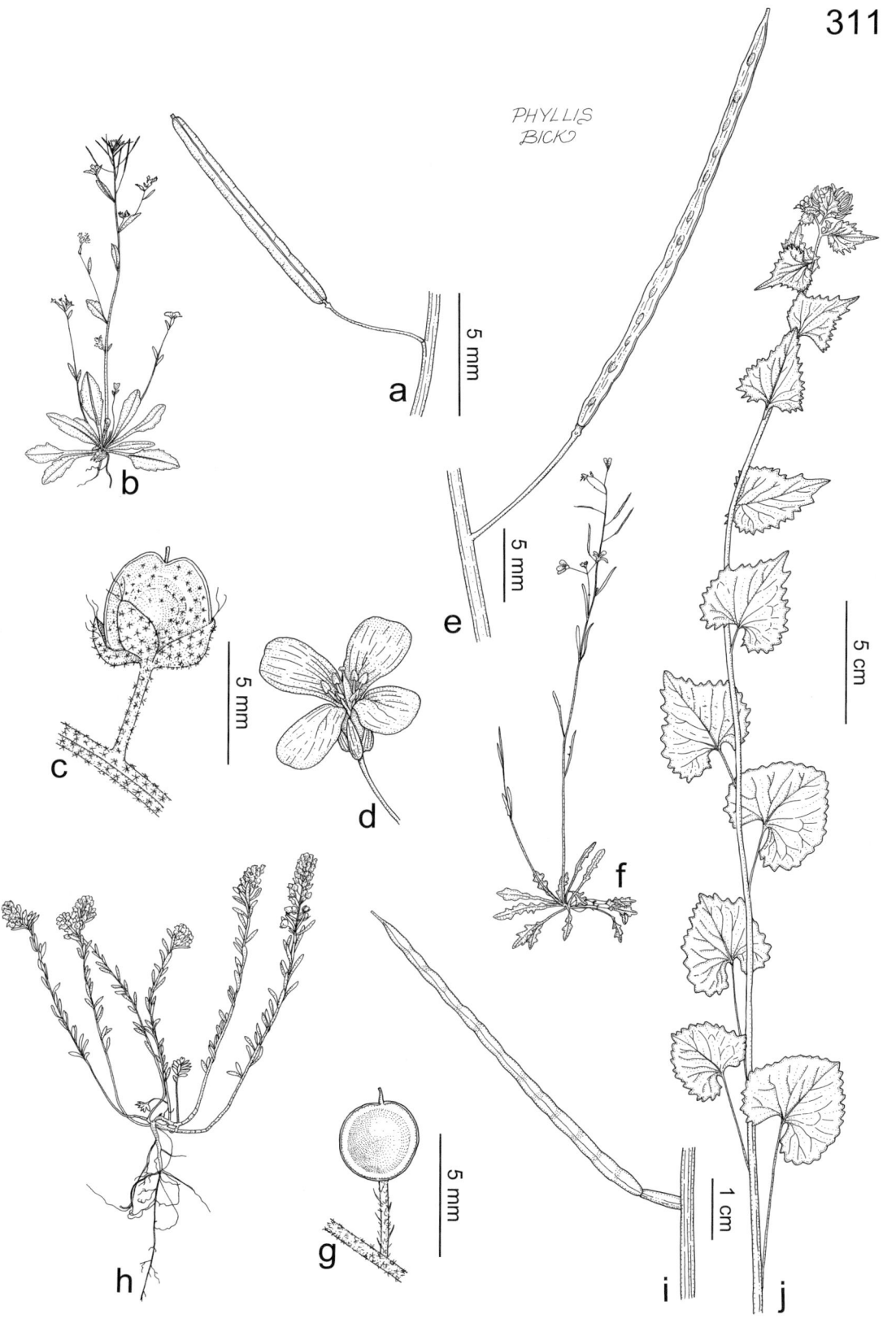

Plate 311. Brassicaceae. *Arabidopsis thaliana*, **a)** fruit, **b)** habit. *Alyssum desertorum*, **c)** fruit. *Arabidopsis lyrata*, **d)** flower, **e)** fruit, **f)** habit. *Alyssum alyssoides*, **g)** fruit, **h)** habit. *Alliaria petiolata*, **i)** fruit, **j)** habit.

1314. Arabis hirsuta   1315. Armoracia rusticana   1316. Aubrieta deltoidea

hairy. Petals 2.0–3.5 mm long, unlobed, white. Styles 0.1–0.5 mm long. Fruits spreading to ascending, straight or nearly so, (0.8–)1.0–1.8 cm long, to 20 times as long as wide, circular in cross-section, not beaked, each valve with a midnerve. Seeds in 1 row in each locule, 0.3–0.5 mm long, oblong in outline, the surface with a fine, netlike or honeycomb-like pattern of ridges and pits to nearly smooth, orange. 2n=10. April–May.

Introduced, scattered in eastern and southern Missouri (native of Europe and Asia, widely naturalized in temperate areas of the world). Disturbed ground in pastures, fallow fields, lawns, waste areas, railroads, and roadsides.

Arabidopsis thaliana is an inconspicuous weed of disturbed areas. In recent years, however, it has become an extremely important tool in scientific studies on various aspects of plant physiology, biochemistry, pathology, ecology, development, and genetics. As a laboratory organism, it has the advantages of relatively few chromosomes, a relatively simple genetic system, a short life cycle that can be completed in a few weeks, and small plant size, making it amenable to cultivation in greenhouses, growth chambers, and in test tubes under sterile conditions.

4. Arabis L. (rock cress)

About 80 species, North America (about 12 species), Europe, Asia, and northern and alpine tropical Africa.

The genus *Arabis,* as traditionally interpreted by North American botanists (Steyermark, 1963), recently has been found to represent an unnatural group of superficially similar species many of whose relationships are with other genera of Brassicaceae. See the treatments of *Arabidopsis, Boechera,* and *Turritis* for further discussion.

1. Arabis hirsuta (L.) Scop. (hairy rock cress)
Pl. 312 a, b; Map 1314

Plants biennial or short-lived perennial herbs, terrestrial. Stems 20–80(–110) cm long, erect, often few-branched from the base and in the upper half, pubescent with stellate, stalked hairs, rarely glabrous in the upper half. Basal leaves 2–8(–10) cm long, oblong to broadly oblanceolate in outline, entire to toothed or irregularly pinnately divided, narrowed at the base to a short petiole, pubescent with stellate, stalked hairs. Stem leaves appressed to spreading, (1–)2–5(–7) cm long, lanceolate, oblong, to oblanceolate in outline, sessile, clasping the stem, usually with pointed auricles, entire or coarsely few-toothed, glabrous or hairy. Sepals 2.5–4.5 mm long, oblong-elliptic, green. Petals 3–5(–6) mm long, white to pink. Fruits 1.5–5.5(–7)cm long, erect at maturity, appressed to the inflorescence axis, straight or nearly so, flattened parallel to the septum. Seeds in 1 row in each locule, (0.8–)1.1–1.4(–1.7) mm long, oblong-elliptic in outline, somewhat flattened, usually winged at both ends, the surface with a netlike or honeycomb-like pattern of ridges and pits, yellow to brown. 2n=8, 14, 16, 28, 30, 32, 64. May–June.

Scattered nearly throughout Missouri, mostly absent from the northwestern quarter of the state (North America, Europe, Asia; see discussion be-

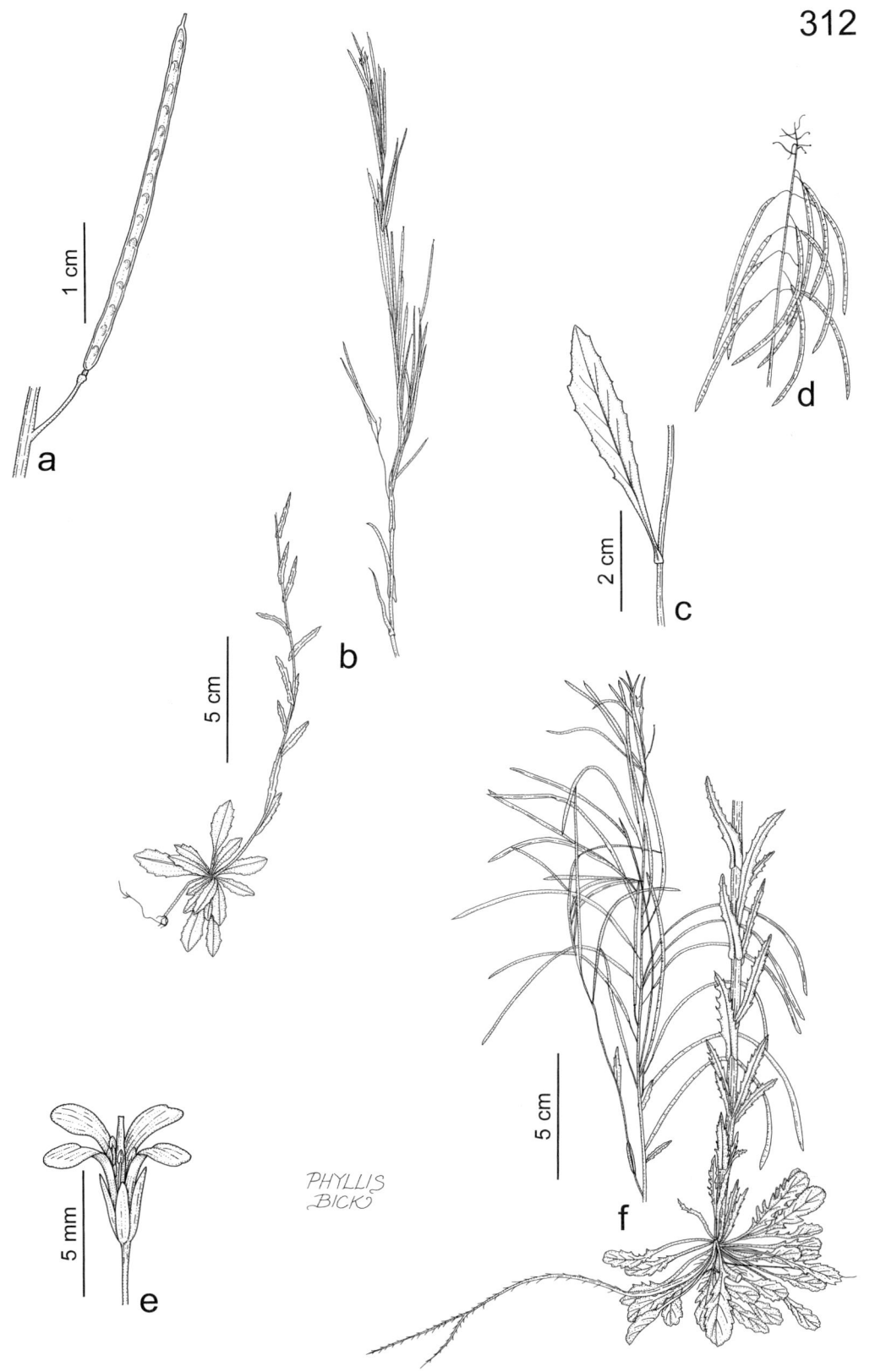

Plate 312. Brassicaceae. *Arabis hirsuta*, **a)** fruit, **b)** fruiting branch and base. *Boechera canadensis*, **c)** leaf, **d)** fruiting branch. *Boechera missouriensis*, **e)** flower, **f)** fruiting branch and base.

low). Ledges of bluffs and rock outcrops, mostly on calcareous substrates (rarely on sandstone).

Arabis hirsuta comprises a circumboreal complex that is represented in North America by four varieties (Al-Shehbaz, 1988b; Rollins, 1993). The typical variety grows in Europe. Two western varieties, var. *glabrata* and var. *eschscholtziana* (Andrz.) Rollins, differ from those found in Missouri in their larger petals (5–9 mm), which are white to pinkish white, and in their more widely spaced leaves on the stems, among other characters. The two varieties found in Missouri differ mainly in their pubescence patterns.

1. Stems with the hairs mostly tightly appressed, centrally attached, and with 2 opposite branches, thus appearing as a straight line .... 1A. VAR. ADPRESSIPILIS
1. Stems with the hairs spreading, mostly unbranched .... 1B. VAR. PYCNOCARPA

1a. var. adpressipilis (M. Hopkins) Rollins

Stems moderately pubescent with the hairs mostly tightly appressed, centrally attached, and with 2 opposite branches, thus appearing as a straight line. Median and upper leaves glabrous, the lowermost leaves sparsely to moderately hairy with hairs similar to those of the stems. $2n=8, 14, 16, 28, 30, 32$. May–June.

Scattered nearly throughout Missouri, mostly absent from the northwestern quarter of the state (eastern U.S. and adjacent Canada west to Iowa and Arkansas). Ledges of bluffs and rock outcrops, mostly on calcareous substrates (rarely on sandstone).

1b. var. pycnocarpa (M. Hopkins) Rollins

Stems sparsely to moderately pubescent with the hairs spreading, basally attached, and mostly unbranched. Stem leaves sparsely to moderately pubescent with spreading, mostly unbranched hairs. $2n=32, 64$. May–June.

Uncommon and widely scattered in the state (northern U.S. south to Georgia, Arkansas, Arizona, and California; Canada). Ledges of bluffs and rock outcrops, mostly on calcareous substrates (rarely on sandstone).

5. Armoracia P. Gaertn., B. Mey., & Scherb.

Three species, Europe, Asia, introduced widely.

1. Armoracia rusticana (Lam.) P. Gaertn., B. Mey. & Scherb. (horseradish)

Pl. 313 c–e; Map 1315

Plants perennial herbs with thick, fleshy roots, lacking rhizomes, terrestrial. Stems (40–)50–120 (–200) cm long, erect, unbranched below the inflorescence, glabrous. Leaves alternate and basal. Leaf blade (5–)10–45(–60) cm long, glabrous, the basal and lower stem leaves with petioles to 60 cm long, the median and upper leaves progressively reduced and short-petiolate or sessile, the bases not clasping, the leaf blades of the lower leaves broadly oblong or ovate with irregularly toothed or scalloped margins, those of the upper leaves lanceolate to linear, entire to lobed. Inflorescences panicles, the lower branches subtended by small, leaflike bracts. Sepals 2–4 mm long, ascending, elliptic to obovate, glabrous. Petals 5–8 mm long, unlobed, white. Styles absent or to 0.5 mm long. Fruits ascending, often aborting prior to maturity, 5–8 mm long, about as long as wide or less than 2 times as long as wide, globose to ovoid, circular in cross-section or slightly flattened at a right angle to the septum, dehiscing longitudinally, each valve with a faint midnerve. Seeds in 2 rows in each locule, often aborting prior to maturity, 1.0–1.5 mm long, nearly circular in outline, the surface with a netlike or honeycomb-like pattern of ridges and pits, brownish orange to brown. $2n=32$. May–July.

Introduced, widely scattered in the state, mostly in and around the Missouri River floodplain (native of Europe and Asia, introduced sporadically in North America). Roadsides, margins of cultivated fields, and disturbed areas; sometimes a weed in flower beds.

Horseradish is widely cultivated for its roots, which are used as a flavoring and ingredient in various dishes.

6. Aubrieta Adans.

About 15 species, Europe, Asia.

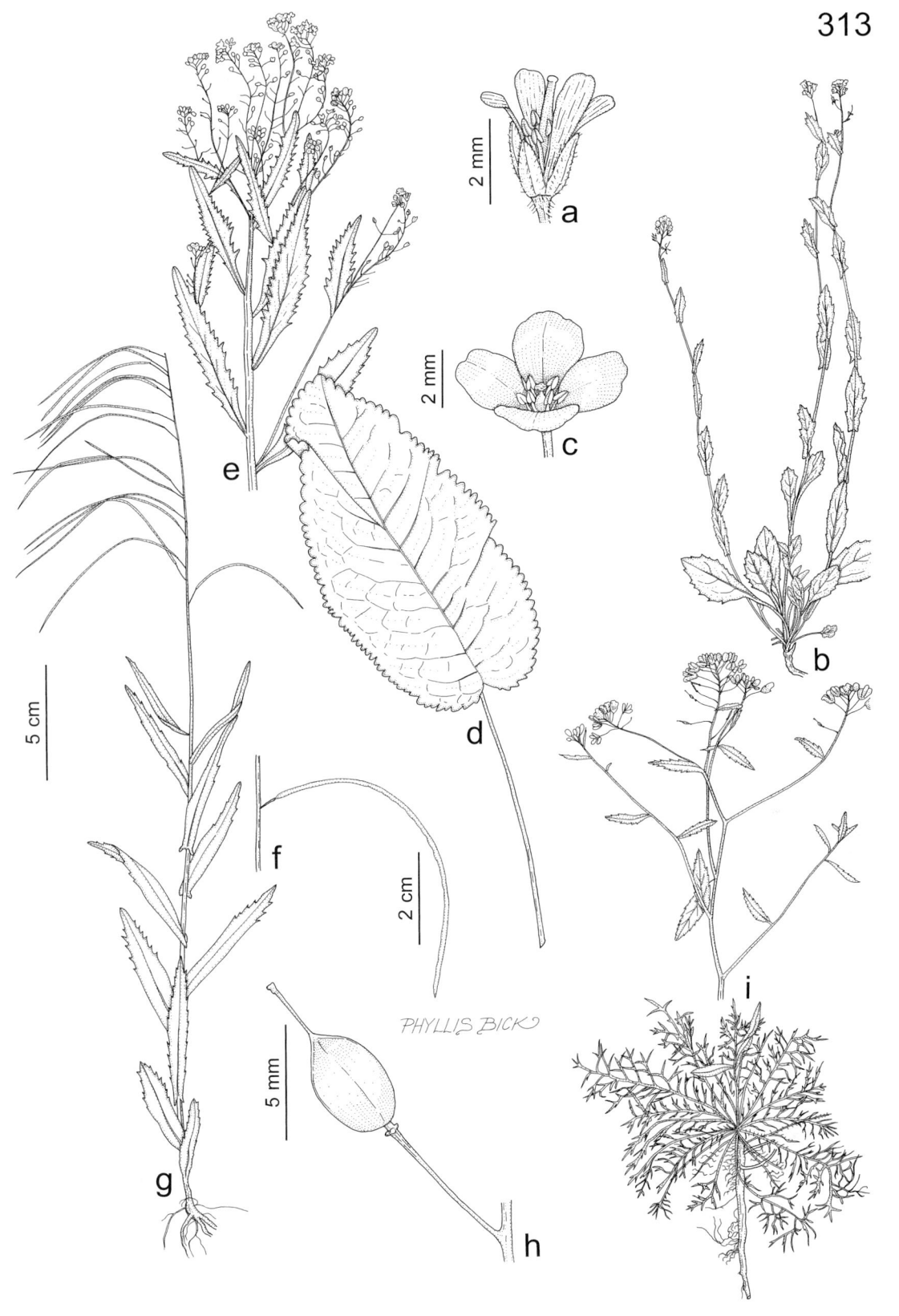

Plate 313. Brassicaceae. *Boechera shortii*, **a)** flower, **b)** habit. *Armoracia rusticana*, **c)** flower, **d)** leaf, **e)** inflorescence. *Boechera laevigata*, **f)** fruit, **g)** habit. *Rorippa aquatica*, **h)** fruit, **i)** inflorescence and leaf base.

1. Aubrieta deltoidea (L.) DC. (purple alyssum, aubrietia)

Pl. 314 f, g; Map 1316

Plants perennial herbs, terrestrial. Stems (5–)10–30 cm long, spreading or creeping, usually somewhat branched from the base, sometimes forming clumps or mats, sparsely to densely pubescent with unbranched and stalked, branched, and/or stellate hairs. Leaves alternate, (0.5–)1.0–3.0 cm long, sessile, the bases not clasping, broadly spatulate, oblanceolate, to narrowly obovate, the margins entire or coarsely few-toothed, sparsely to densely pubescent with branched and stellate hairs mixed on the lower surface with much fewer unbranched hairs. Inflorescences racemes or few-branched panicles, the lower branches subtended by reduced leaves. Sepals 6–10 mm long, ascending, oblong, hairy. Petals 12–28 mm long, unlobed, reddish purple. Styles 4–8 mm long. Fruits spreading to ascending, 6–18 mm long, 2–5 times as long as wide, narrowly oblong to elliptic, somewhat flattened parallel to the septum, hairy, dehiscing longitudinally, each valve with a midnerve. Seeds in 2 rows in each locule, 1.2–1.5 mm long, ovoid, the surface roughened, black. $2n=16$. April–June.

Introduced, known only from Boone County (native of Europe, rarely escaped in the U.S.). Disturbed ground.

Aubrieta deltoidea is commonly cultivated in flower beds and rock gardens, and various hybrids and cultivars exist. It is unknown whether plants have become naturalized at the single locality reported by Dunn (1982), and this species perhaps should not be regarded as a member of the flora.

7. Barbarea R. Br. (winter cress)

Plants biennial or rarely annual or short-lived perennial herbs (woody elsewhere), terrestrial, glabrous or less commonly the stem with very sparse unbranched, nonglandular hairs toward the base. Stems erect, usually unbranched below the inflorescence, more or less angled in cross-section. Leaves alternate and basal, the basal leaves usually long-petiolate, the stem leaves progressively reduced and mostly sessile, the bases usually rounded and clasping the stem, the blades entire to pinnately divided and toothed. Inflorescences panicles with racemose branches (rarely reduced to a solitary raceme), the lower branches rarely subtended by reduced leaves, the flowers bractless. Sepals erect or ascending, oblong-ovate to oblong-lanceolate, glabrous. Petals unlobed, yellow, without conspicuously darkened veins. Fruits (7–)15–70(–80) mm long, mostly more than 10 times as long as wide, loosely ascending to nearly erect, straight or slightly arched upward, circular or somewhat 4-angled in cross-section, not beaked except for the persistent style, dehiscing longitudinally, each valve with a midnerve and usually a more or less distinct pair of lateral, longitudinal nerves. Seeds more or less in 1 row in each locule, 1.2–2.5 mm long, circular to somewhat oblong in outline, sometimes somewhat flattened, the margins not winged, the surface with a fine to coarse, netlike or honeycomb-like pattern of ridges and pits, dark brown to grayish black. About 22 species, North America, Europe, Asia, Africa, a few species introduced nearly worldwide.

1. Fruits (4.5–)5.5–7.0(–8.0) cm long; ovules or seeds (34–)38–48(–52) per ovary or fruit; fruit stalks nearly as thick as the fruits . . . . 1. B. VERNA
1. Fruits (0.7–)1.5–3.0 cm long; ovules or seeds 18–24(–28) per ovary or fruit; fruit stalks narrower than fruits . . . . 2. B. VULGARIS

1. Barbarea verna (Mill.) Asch. (early winter cress, American cress)

Map 1317

Stems (10–)25–80 cm long. Leaves 2–15 cm long, the basal and lowermost stem leaves with the blade oblanceolate, deeply pinnately divided with (3–)6–10 lateral segments or lobes and a large, circular to oval, terminal lobe, the median and upper stem leaves oblanceolate, irregularly lobed, toothed, or nearly entire. Sepals 3–5 mm long. Petals 5–8 mm long. Styles 0.2–2.0 mm long. Fruits (4.5–)5.5–7.0(–8.0) cm long, the stalks 3–6 mm long, nearly as thick as the fruits. Ovules or seeds (34–)38–48(–52) per ovary or fruit, the seeds 1.5–2.5 mm long, more or less oblong in outline, broadly elliptic in cross-section, slightly flattened. $2n=16$. April–June.

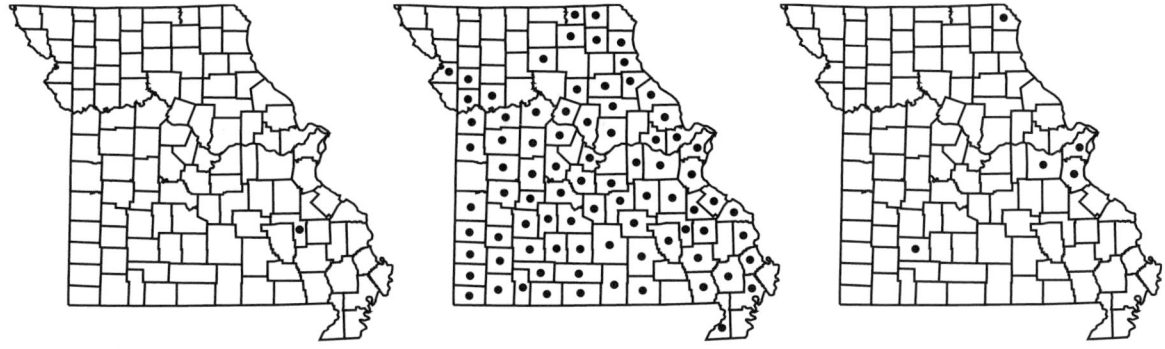

1317. Barbarea verna   1318. Barbarea vulgaris   1319. Berteroa incana

Introduced, uncommon, known thus far only from a single historical collection from Iron County (native of Europe, Asia; introduced widely in North America, Africa, Australia). Habitat unknown, but presumably open, disturbed areas.

The Iron County record originally was misdetermined as *B. vulgaris* and was discovered during the present research. The species occurs in most surrounding states and thus will likely be found elsewhere in Missouri in the future.

2. Barbarea vulgaris R. Br. (yellow rocket, common winter cress)

B. vulgaris var. *arcuata* (Opiz ex J. Presl & C. Presl) Fr.

Pl. 314 h–j; Map 1318

Stems (20–)30–80(–100) cm long. Leaves 2–15 cm long, the basal and lowermost stem leaves with the blade oblanceolate, deeply pinnately divided with 2–8 small, lateral segments or lobes and a large, circular to oval, terminal lobe, the median and upper stem leaves oblanceolate, irregularly lobed, toothed, or nearly entire. Sepals 3–4 mm long. Petals 5–9 mm long. Styles 1.5–3.0 mm long. Fruits (0.7–)1.5–3.0 cm long, the stalks 3–7 mm long, narrower than the fruits. Ovules or seeds 18–24(–28) per ovary or fruit, the seeds 1.0–1.5 mm long, circular in outline, oblong in cross-section, somewhat flattened. $2n=16$. April–June.

Introduced, common throughout Missouri (native of Europe, Asia, widely naturalized in North America). Banks of streams; also pastures, fields, roadsides, railroads, and open, disturbed areas.

The young leaves of *B. vulgaris* sometimes are eaten cooked or raw.

8. Berteroa DC.

Five species, Europe, Asia, introduced in North America.

1. Berteroa incana (L.) DC. (hoary alyssum)

Pl. 314 d, e; Map 1319

Plants annual (biennial or perennial elsewhere), terrestrial. Stems (20–)35–80(–110) cm long, erect, usually branched, densely pubescent with mostly stellate, appressed hairs mixed with unbranched ones. Leaves numerous, alternate, the basal leaves usually withering by flowering, sessile or the lowermost leaves short-petiolate, the bases not clasping, 1–5 cm long, the blades oblanceolate, entire, densely pubescent with mostly stellate, appressed hairs. Inflorescences panicles, the lower branches subtended by reduced leaves. Sepals 1–2 mm long, ascending, elliptic to lanceolate, stellate-pubescent, becoming detached soon after the flower opens. Petals (4.0–)5.0–6.5(–8.0) mm long, deeply 2-lobed at the tip, white. Styles 1–3(–4) mm long. Fruits erect, straight or arched, (4–)5–8(–10) mm long, 2–3 times as long as wide, elliptic, broadly elliptic in cross-section and somewhat flattened, stellate-pubescent, eventually dehiscing longitudinally, each valve with a faint midnerve near the base. Seeds in 2 rows in each locule, 1.5–2.3 mm long, circular to broadly obovate in outline, flattened, the margin narrowly winged, the surface with a fine, netlike or honeycomb-like pattern of ridges and pits, brown to dark brown. $2n=16$. May–September.

Introduced, uncommon and widely scattered (native of Europe and Asia, widely naturalized in North America, mostly to the north and east of Missouri). Railroads, roadsides, and open, disturbed areas.

9. Boechera Á. Löve & D. Löve

Plants biennial or perennial herbs (sometimes woody at the base elsewhere), terrestrial. Stems erect, ascending or rarely spreading, usually branched. Leaves alternate and basal, the basal ones often forming rosettes, sometimes absent at flowering time, variously shaped and divided, the bases of the stem leaves clasping or not, rarely petiolate, often pubescent with unbranched or forked hairs (with stalked, more or less stellate hairs elsewhere), rarely glabrous or with unbranched hairs only. Inflorescences racemes or panicles, the flowers not subtended by bracts. Sepals erect or ascending. Petals not lobed; white, pink, or purple. Styles absent or 0.5–2.0 mm long (rarely longer elsewhere). Stigma entire. Fruits more than 10 times as long as wide, slightly to strongly flattened parallel to the septum, dehiscing longitudinally by the valves breaking off without coiling, each valve usually with a midnerve extending the full length or rarely only near the base. Seeds in 1 or 2 rows in each locule, winged or rarely wingless. About 65 species, North America.

Extensive molecular and cytological studies (summarized in Al-Shehbaz, 2003) strongly support the recognition of *Boechera* as a genus not closely related to *Arabis*. All species of *Arabis* have chromosome numbers based on x=8, whereas all species of *Boechera* have x=7. The morphological differences between the two genera are discussed in that paper.

1. Leaves not clasping the stem, narrowed at the base; stalks of the fruits reflexed or arched downward; fruits 2.5–4.0 mm wide .... 1. B. CANADENSIS
1. Some or all of the leaves clasping the stem, with auricles at the base; stalks of the fruits erect to spreading; fruits 0.7–2.0(–2.5) mm wide
    2. Fruits 1.5–3.0(–4.0) cm long; petals 2–3 mm long; stems ascending or spreading, usually several from the base .... 4. B. SHORTII
    2. Fruits 5–10 cm long; petals 3–8 mm long; stems erect and solitary
        3. Petals 3–5 mm long, as long as or slightly longer than the sepals; basal leaves entire or shallowly toothed .... 2. B. LAEVIGATA
        3. Petals (5–)6–9 mm long, about twice as long as the sepals; basal leaves pinnately lobed to coarsely toothed .... 3. B. MISSOURIENSIS

1. Boechera canadensis (L.) Al-Shehbaz (sicklepod)
Arabis canadensis L.

Pl. 312 c, d; Map 1320

Plants biennial. Stems 30–100 cm long, erect, sometimes few-branched in the upper half, sparsely pubescent with mostly unbranched hairs toward the base. Basal leaves often withered by flowering time, 2.5–12.0 cm long, obovate to lanceolate, entire to sharply toothed, narrowed at the base to a short petiole, pubescent on both surfaces with unbranched and much fewer, stalked, forked hairs. Stem leaves usually overlapping, 2–10 cm long, narrowed at the base, not clasping, the lower leaves sometimes short-petiolate, entire to toothed, the upper leaves often glabrous, the lower leaves hairy. Sepals 2–4 mm long, oblong-elliptic, often yellow- or purple-tinged. Petals 3–5 mm long, white. Fruits (4–)5–11 cm long, 2.5–4.0 mm wide, spreading horizontally to strongly recurved at maturity, on pedicels arched downward, strongly flattened. Seeds in 1 row in each locule, 1.3–1.6 mm long, oval to nearly circular in outline, flattened, strongly winged, the surface with a netlike or honeycomb-like pattern of ridges and pits, orange. $2n=14$. April–June.

Scattered throughout Missouri (eastern U.S. and adjacent Canada west to Minnesota and Texas). Mesic to dry upland forests, particularly in openings or on rocky slopes, and along the edges of bluffs, on both acidic and calcareous substrates.

2. Boechera laevigata (Muhl. ex Willd.) Al-Shehbaz (smooth rock cress)
Arabis laevigata (Muhl. ex Willd.) Poir.

Pl. 313 f, g; Map 1321

Plants biennial, glabrous (seedlings sometimes with scattered, unbranched hairs), pale green, glaucous. Stems 30–100 cm long, erect, often few-branched in the upper or lower half. Basal leaves

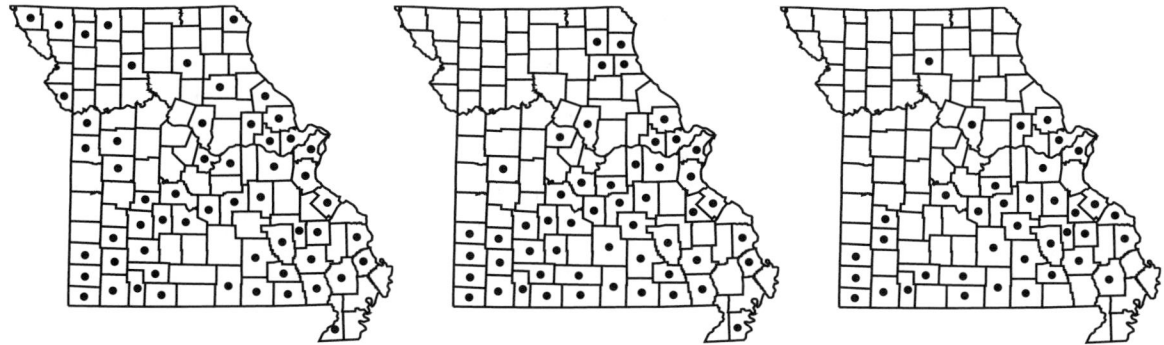

1320. Boechera canadensis   1321. Boechera laevigata   1322. Boechera missouriensis

often withered by flowering time, 3–11 cm long, obovate to narrowly oblanceolate in outline, usually toothed or entire, narrowed at the base to a short petiole. Stem leaves mostly spreading, sometimes overlapping, 3–15 cm long, lanceolate to linear in outline, sessile, clasping the stem, usually with pointed auricles, entire or toothed. Sepals 2.5–4.0(–5.0) mm long, oblong-elliptic, light green. Petals 3–5 mm long, white. Fruits 5–10 cm long, 1.0–2.0(–2.5) mm wide, widely ascending to spreading horizontally along the lower half, usually arched downward, somewhat flattened. Seeds in 1 row in each locule, 1.3–1.8 mm long, oblong in outline, flattened, winged, the surface with a netlike or honeycomb-like pattern of ridges and pits, orange. $2n=14$. April–June.

Scattered nearly throughout Missouri, although nearly absent from the northwestern quarter (eastern U.S. and adjacent Canada west to Minnesota and Oklahoma). Bottomland and mesic upland forests, often in drainages, banks of streams, and ledges of sheltered bluffs, often on calcareous substrates.

3. Boechera missouriensis (Greene) Al-Shehbaz (rock cress, Missouri rock cress)

Arabis missouriensis Greene

A. laevigata (Muhl. ex Willd.) Poir. var. *missouriensis* (Greene) H.E. Ahles

A. missouriensis var. *deamii* (M. Hopkins) M. Hopkins

Pl. 312 e, f; Map 1322

Plants biennial, glabrous or less commonly sparsely pubescent with mostly unbranched hairs, not glaucous. Stems 20–60 cm long, erect, often few-branched in the upper or lower half. Basal leaves usually present at flowering, 2–9 cm long, lanceolate to spatulate, sharply toothed to pinnately divided, narrowed at the base to a short petiole. Stem leaves mostly appressed or strongly ascending, 1–8 cm long, lanceolate to linear-lanceolate in outline, sessile, clasping the stem, with pointed auricles, the lower leaves pinnately divided or coarsely toothed, the upper leaves mostly entire. Sepals 3–5 mm long, oblong-elliptic, green. Petals (5–)6–9 mm long, milky white. Fruits (5–)6–9 cm long, 1.7–2.0 mm wide, widely ascending to spreading horizontally or recurved at maturity, usually arched downward along upper half, strongly flattened. Seeds in 1 row in each locule, 1.6–1.9 mm long, oblong to broadly elliptic in outline, flattened, winged, the surface roughened and/or irregularly patterned, orange. April–June.

Scattered in the Ozark and Ozark Border Divisions, locally north to St. Charles County, and also on Crowley's Ridge (eastern U.S. west to Wisconsin and Oklahoma). Mesic to dry upland forests, on rocky or sandy slopes, mostly on acidic substrates such as chert, sandstone, granite, and sand.

The uncommonly collected var. *deamii,* consisting of plants with scattered pubescence, appears as a polymorphism within populations of otherwise glabrous plants. As such, it seems unworthy of formal taxonomic recognition (Rollins, 1993).

4. Boechera shortii (Fernald) Al-Shehbaz

Arabis shortii (Fernald) Gleason

A. shortii var. *phalacrocarpa* (M. Hopkins) Steyerm.

A. dentata (Torr.) Torr. & A. Gray

Pl. 313 a, b; Map 1323

Plants biennial. Stems 20–50(–60) cm long, spreading or ascending, usually several-branched at the base, sometimes also few-branched in the upper half, sparsely pubescent with mostly 2–4-branched hairs. Leaves pubescent with simple hairs on the upper surface and with mostly 3- or 4-branched hairs on the undersurface. Basal leaves 4–15 cm long, obovate to oblanceolate in outline, irregularly toothed, narrowed at the base to a short

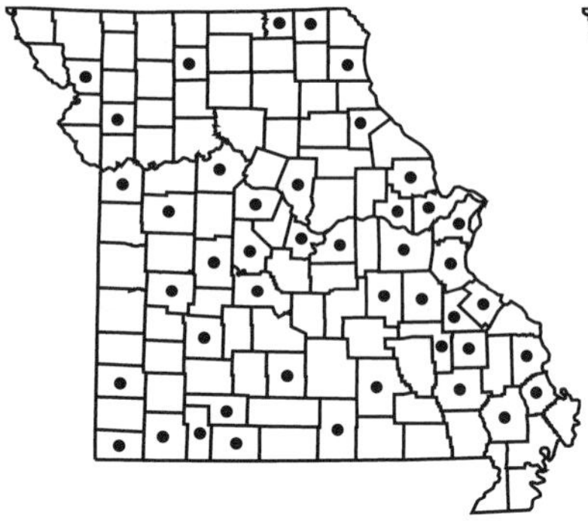

1323. Boechera shortii

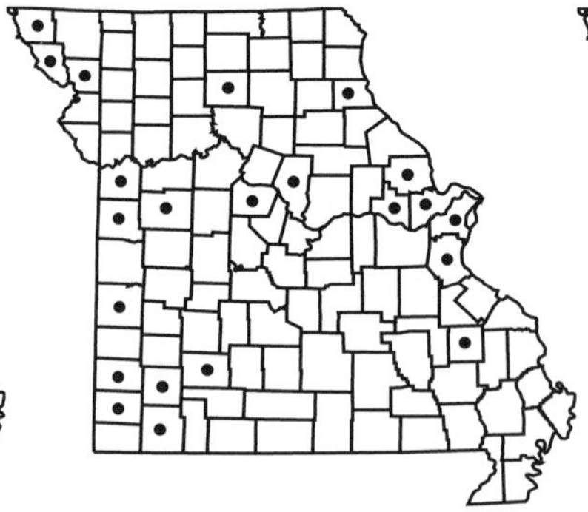

1324. Brassica juncea

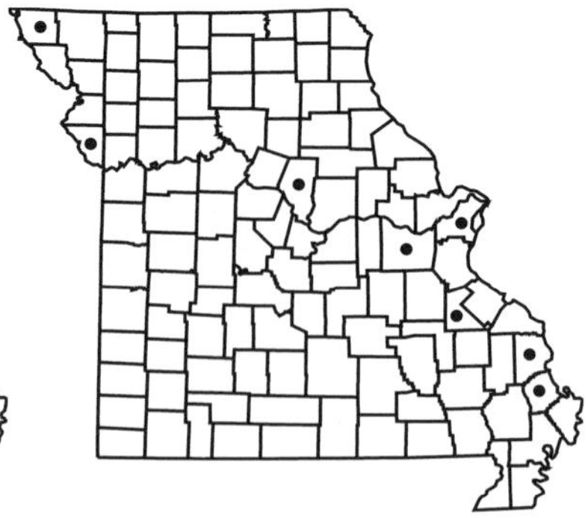

1325. Brassica napus

or long petiole. Stem leaves 1–6 cm long, narrowly oblanceolate to nearly linear in outline, sessile, clasping the stem, with pointed auricles, sharply and irregularly toothed. Sepals 1.5–2.5 mm long, narrowly oblong, green. Petals 2–3 mm long, white. Fruits 1.5–3.0(–4.0) cm long, spreading, straight or slightly arched upward, strongly flattened, glabrous or pubescent with 3- or 4-branched hairs. Seeds in 1 row in each locule, 1.0–1.2 mm long, elliptic to narrowly oblong in outline, not flattened or winged, the surface roughened, orange. $2n=12$. April–June.

Scattered nearly throughout Missouri, but apparently absent from the northeastern quarter (eastern U.S. west to South Dakota and Oklahoma). Bottomland and mesic upland forests, banks of streams, open floodplains of rivers, and ledges of sheltered bluffs, most commonly on calcareous substrates.

Glabrous-fruited plants, which are more common in Missouri than those with hairy fruits, have been called var. *phalacrocarpa* by some botanists. However, the degree of pubescence varies from very sparse to moderately dense, and there are no other morphological characters to correlate with this variation. Thus, no infraspecific taxa are recognized for this species in the present treatment (see also Rollins, 1993).

10. Brassica L. (mustard)

Plants annual or biennial (woody-based perennials or shrubs elsewhere), terrestrial, glabrous or with unbranched, frequently coarse, spreading hairs. Stems erect, usually branched. Leaves alternate and basal, the lower leaves usually relatively long-petiolate, the upper leaves progressively reduced and short-petiolate or sessile, the bases clasping or not clasping, the leaf blades entire to pinnately divided and toothed. Inflorescences panicles or racemes, the lower branches rarely subtended by reduced leaves, the flowers bractless. Sepals erect or ascending, mostly narrowly oblong or linear-lanceolate. Petals unlobed, yellow, without conspicuously darkened veins. Fruits 10–70 mm long, mostly more than 10 times as long as wide, spreading, ascending, or erect (reflexed elsewhere), straight or slightly arched upward, circular or somewhat 4-angled in cross-section, short- to long-beaked with a distinct, tapering, usually seedless area in addition to the style, the portion below the beak dehiscing longitudinally, each valve with a midnerve. Seeds in 1 row in each locule, 1.2–1.7 mm long, globose, the surface with a fine to coarse, netlike or honeycomb-like pattern of ridges and pits, reddish brown to gray or black. About 40 species, Europe, Asia, Africa, a few species introduced nearly worldwide.

The genus *Brassica* is of tremendous economic importance for its agricultural crop species, with major uses ranging from vegetables to seed oils. One species, *B. oleracea* L., has numerous economically important cultivars, including broccoli, Brussels sprouts, cabbage, cauliflower, collards, kale, and kohlrabi, the so-called kohl crops that have been advocated as potentially reducing the risk of heart disease when eaten regularly. Several species have been used phar-

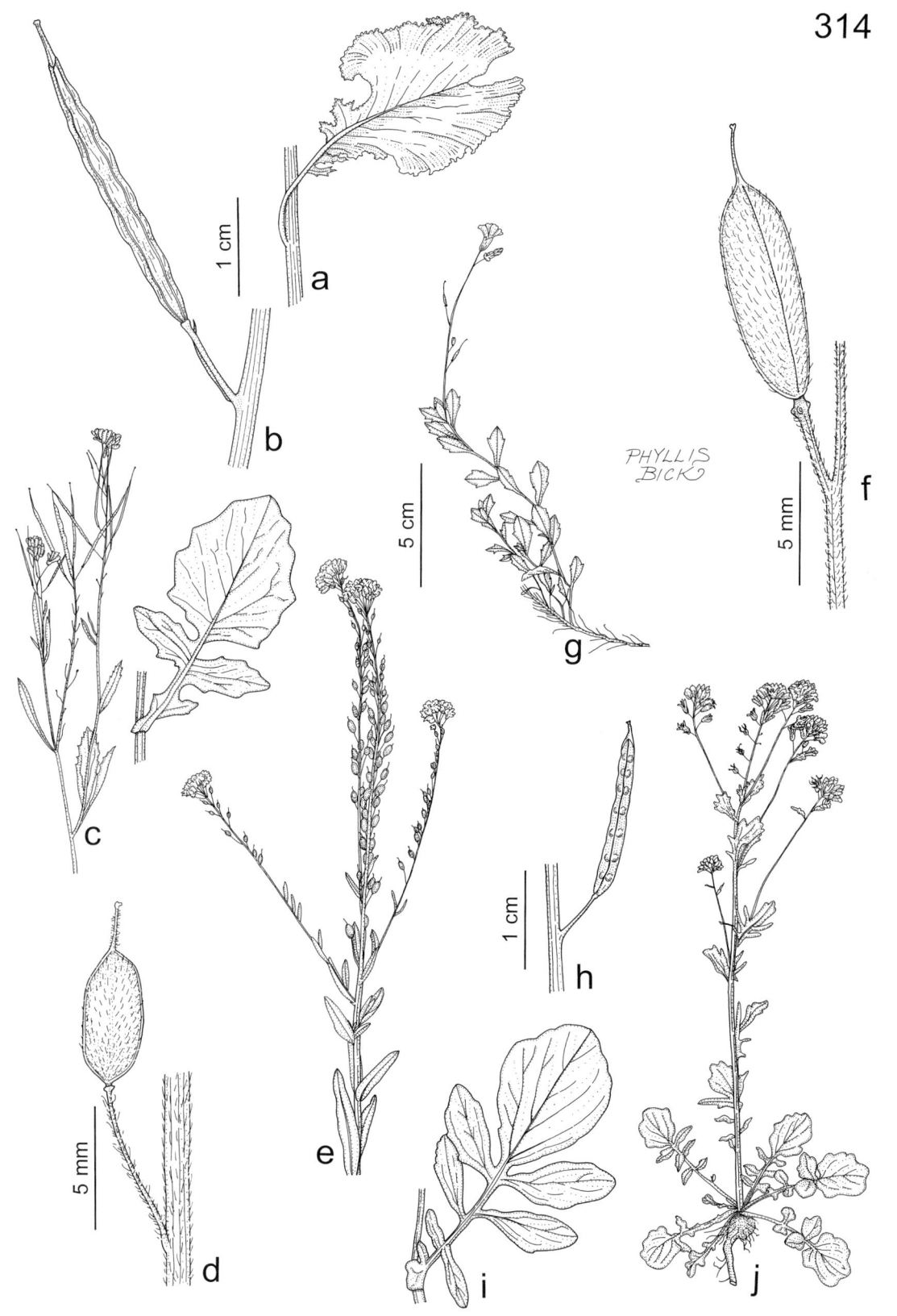

Plate 314. Brassicaceae. *Brassica juncea*, **a)** leaf, **b)** fruit, **c)** inflorescence and leaf. *Berteroa incana*, **d)** fruit, **e)** inflorescence. *Aubrieta deltoidea*, **f)** fruit, **g)** habit. *Barbarea vulgaris*, **h)** fruit, **i)** leaf, **j)** habit.

maceutically and medicinally, and some of the weedy species have been shown to be poisonous to humans and livestock when ingested in large quantities (Al-Shehbaz, 1985).

Mühlenbach (1983) reported *B. oleracea* as a member of the synanthropic railroad flora of the St. Louis area; however, this report was based upon misdetermined specimens of *B. juncea*. *Brassica oleracea* rarely if ever becomes established outside of cultivation and was not accepted as naturalized in the recent treatments of Al-Shehbaz (1985) for the southeastern United States and Rollins (1993) for North America.

1. Upper leaves sessile, the bases clasping the stems and with auricles
   2. Plants glabrous or nearly so, glaucous; petals (0.9–)1.0–1.6(–1.8) mm long, usually pale yellow; flowers not overtopping the buds . . . . 2. B. NAPUS
   2. Plants hairy (at least on the developing leaves) or nearly glabrous, but not glaucous; petals 0.7–1.0(–1.3)mm long, usually bright yellow; flowers overtopping the buds . . . . 4. B. RAPA
1. Upper leaves petiolate, or if sessile, then tapered at the base and not clasping
   3. Fruits (5–)10–25(–27) mm long, erect, appressed to the infructescence axis, the beak and style (1–)2–5(–6) mm long; petals usually bright yellow
      . . . . 3. B. NIGRA
   3. Fruits (20–)30–50(–60) mm long, spreading to ascending, not appressed to the infructescence axis, the beak and style (4–)5–10(–15) mm long; petals pale yellow . . . . 1. B. JUNCEA

1. Brassica juncea (L.) Czern. (brown mustard, leaf mustard, Chinese mustard, Indian mustard)

Pl. 314 a–c, 315 a–c; Map 1324

Plants annual, glabrous or nearly so, glaucous. Stems (20–)30–100(–180) cm long. Basal and lower leaves (4–)10–30(–80) cm long, pinnately divided into 5–9 irregularly toothed divisions, petiolate, elliptic to obovate in outline. Stem leaves progressively reduced toward the tip, the uppermost 2–5(–10) cm long, short-petiolate or sessile with nonclasping bases, oblanceolate to elliptic or nearly linear in outline. Flowers usually not overtopping the buds. Sepals 4–8 mm long. Petals 7–13 mm long, pale yellow. Fruits (20–)30–50(–60) mm long, spreading to ascending, not appressed to the inflorescence axis, circular in cross-section or nearly so, the beak and style (4–)5–10 mm long. Seeds 12–30(–40) per fruit, globose, 1.0–1.7 mm in diameter. 2n=36. April–September.

Introduced, widely scattered in Missouri (native of Europe, Asia, widely naturalized in the New World). Pastures, margins of crop fields, roadsides, railroads, and open, disturbed areas.

This species is occasionally cultivated as a leafy, green vegetable. In Asia, the seeds are sometimes used as a spice. They are also occasionally used in the preparation of massage oils. It is thought to have originated in Asia through past hybridization between *B. nigra* and *B. rapa* (Al-Shehbaz, 1985). Numerous leafy forms are cultivated in China and cooked as a green vegetable.

A related species that should be searched for in Missouri is *B. tournefortii* Gouan (Sahara mustard), an aggressive weed that was first reported as naturalized in the southwestern United States and that has spread at least as far east as Texas within the past few decades (Rollins, 1993). It will probably arrive in the state at some point either along railroad tracks or roadsides. This species differs from *B. juncea* in having hairy rather than glabrous lower leaves and stem bases, more persistent rosettes of basal leaves with more numerous divisions (15–30), and shorter petals (4–6 mm).

2. Brassica napus L. (rape, rapeseed, rutabaga, Swedish turnip)

Pl. 315 g–i; Map 1325

Plants annual or biennial, glabrous or nearly so, glaucous, the rootstock sometimes thickened and somewhat fleshy. Stems 30–150 cm long. Basal and lower leaves often absent at flowering, 5–25(–40) cm long, irregularly pinnately divided or lobed into 5–13 irregularly toothed divisions, petiolate but sometimes with rounded auricles at the base, mostly obovate in outline. Stem leaves progressively reduced toward the tip, the uppermost 2–5 cm long, sessile, the bases clasping and with rounded auricles of tissue, linear to narrowly oblanceolate in outline. Flowers not overtopping the buds. Sepals (5–)6–10 mm long. Petals (0.9–)1.0–1.6(–1.8) mm long, usually pale yellow. Fruits (3.5–)5.0–9.5(–11.0) mm long, ascending or spreading, circular in cross-

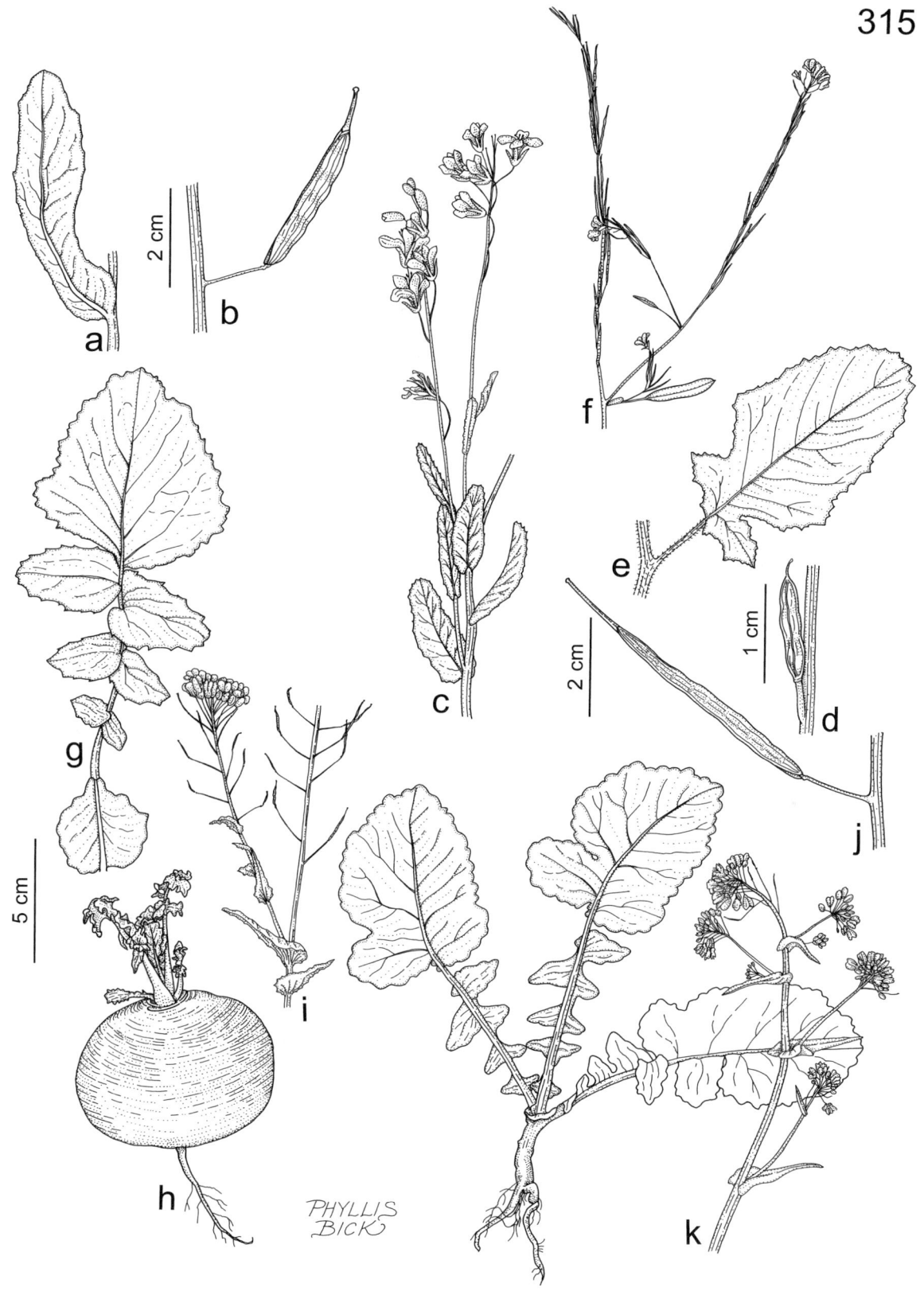

Plate 315. Brassicaceae. *Brassica juncea*, **a)** leaf, **b)** fruit, **c)** inflorescence. *Brassica nigra*, **d)** fruit, **e)** leaf, **f)** inflorescence. *Brassica napus*, **g)** leaf, **h)** fruit, **i)** inflorescence. *Brassica rapa*, **j)** fruit, **k)** inflorescence and base.

1326. Brassica nigra   1327. Brassica rapa   1328. Camelina microcarpa

section or nearly so, the slender, tapered beak and style (5–)9–15 mm long. Seeds 24–40(–60) per fruit, globose, (1.2–)1.5–2.5(–3.0) mm in diameter. $2n=38$. April–September.

Introduced, uncommon in northwestern and eastern Missouri (not known from native populations, but presumed to have originated in Europe; widely cultivated and sporadically escaped in North America). Roadsides, disturbed ground.

Brassica napus originated from past hybridization between *B. oleracea* ($2n=18$) and *B. rapa* ($2n=20$). It can be difficult to distinguish from the latter parent (see key). Some botanists separate this species into two varieties, based upon differences in the rootstocks. In var. *napus*, popularly known as rape or rapeseed, the rootstock is an unthickened taproot, whereas in var. *napobrassica* (L.) Peterm., the rutabaga, the rootstock is thickened and somewhat succulent. When plants become established outside cultivation, these differences tend to disappear.

This species is presently very uncommon as an escape from cultivation, but it is expected to become more common as commercial production of rapeseed increases in the state. The specimen from railroads in St. Louis reported by Steyermark (1963) as the original voucher for this species was a misdetermined collection of *B. rapa*.

Rapeseed is the source of canola oil, which has become very popular as a cooking oil in recent years, as well as having pharmaceutical, chemical, and industrial uses. Increasingly, it is being grown as an alternative crop regionally by farmers in the United States and Canada.

3. Brassica nigra (L.) W.D.J. Koch (black mustard)

Pl. 315 d–f; map 1326

Plants annual, sparsely to densely pubescent, at least near the base, often somewhat glaucous above. Stems 30–200(–310) cm long. Basal and lower leaves 6–25(–40) cm long, irregularly pinnately divided or lobed into 3–7 irregularly toothed divisions, sometimes unlobed, petiolate, elliptic to obovate in outline. Stem leaves progressively reduced toward the tip, the uppermost 1–5 cm long, petiolate with nonclasping bases, oblanceolate to elliptic in outline. Flowers usually not overtopping the buds. Sepals (3–)4–6(–7) mm long. Petals (5–)7–11(–13) mm long, usually bright yellow. Fruits (5–)10–25(–27) mm long, erect, appressed to the inflorescence axis, somewhat 4-angled in cross-section, abruptly narrowed to a linear beak and style (1–)2–5(–6) mm long. Seeds 4–10(–16) per fruit, globose, 1.2–2.0 mm in diameter. $2n=16$. April–November.

Introduced, widely scattered in Missouri, mostly north of the Missouri River (native of Europe, Asia, widely naturalized in North America). Pastures, margins of crop fields, roadsides, railroads, and open, disturbed areas.

Until recently replaced by *B. juncea*, *B. nigra* was the chief source of seed used in making table mustard, which also contains extracts from the seeds of white mustard, *Sinapis alba*. Extracts from the seeds are also used medicinally and in the preparation of some scented soaps.

4. Brassica rapa L. (field mustard, turnip, bird's rape)

Pl. 315 j–k; Map 1327

Plants annual or biennial, hairy (at least on the developing leaves) or nearly glabrous, but not glaucous, the rootstock sometimes thickened and somewhat fleshy. Stems 20–120(–190) cm long. Basal and lower leaves often absent at flowering, 15–40(–60) cm long, irregularly pinnately divided or lobed into 3–9 irregularly toothed divisions, petiolate but sometimes with rounded auricles of tissue at the base, mostly obovate in outline. Stem leaves progressively reduced toward the tip, the uppermost 2–5 cm long, sessile, the bases clasping and with rounded auricles, linear to narrowly oblanceolate in outline. Flowers overtopping the buds. Sepals 3–6 mm long. Petals 0.7–1.0(–1.3) mm long, usually bright yellow. Fruits (20–)30–80(–110) mm long, ascending or spreading, circular in cross-section or nearly so, the slender, tapered beak (3–)10–25(–35) mm long. Seeds 16–30 per fruit, globose, 1.0–1.8 mm in diameter. $2n=20$. April–September.

Introduced, widely scattered in Missouri (native of Europe, Asia, widely cultivated and commonly escaped in North America). Pastures, margins of crop fields, roadsides, railroads, and open, disturbed areas.

This species is one of the parents of *B. napus*, and it can be very difficult to distinguish from that taxon (see key). Many botanists divide this species into two varieties. The var. *rapa*, commonly known as turnip, is a biennial with a thickened, somewhat fleshy rootstock. In contrast, var. *oleifera* DC. (var. *campestris* (L.) Koch, *B. campestris* L.) is an annual with an unthickened taproot. As noted by Al-Shehbaz (1985), these differences tend not to be apparent in uncultivated plants.

11. Camelina Crantz (false flax)

Plants annual, terrestrial, pubescent with both unbranched and stalked forked and stellate hairs, rarely nearly glabrous. Stems 30–100 cm long, erect or ascending, sometimes few-branched in the upper half. Leaves alternate and usually also a few basal at flowering, 2–7 cm long, the lower leaves short-petiolate, the upper leaves sessile, clasping with prominent, pointed auricles, the leaf blades oblanceolate to lanceolate or nearly linear in outline, the margins entire, wavy, or with shallow, widely spaced teeth. Inflorescences racemes or few-branched panicles, the flowers not subtended by bracts. Sepals narrowly oblong to oblanceolate, erect or ascending. Petals not lobed, but sometimes shallowly notched at the tip, light yellow or yellow. Styles 1.0–3.5 mm long. Fruits ascending, less than 2 times as long as wide, obovoid or pear-shaped, tapered at the base to a short stalk above the attachment point of the perianth, elliptic in cross-section, slightly flattened parallel to the septum, the valves with an inconspicuous nerve in the middle, sometimes also with a faint network of smaller veins, the edges slightly raised but not winged, the valve tip extending into a beak into the persistent style, tardily dehiscent. Ovules 4–12 (more elsewhere) per locule. Seeds in 2 rows in each locule, usually 4 per locule (6–12 per fruit). Six species, Europe, Asia.

The two species found in Missouri can be difficult to differentiate, in spite of the apparent differences noted in the key to species.

1. Stems and leaves pubescent with rough, unbranched, spreading hairs 1.0–2.5 mm long in addition to the shorter, stellate hairs; fruits (2.5–)3.5–6.0 mm long; petals pale yellow; seeds 0.8–1.4(–1.5) mm long . . . . 1. C. MICROCARPA
1. Stems and leaves sparsely pubescent, with the minute, unbranched hairs not longer than the stellate hairs, sometimes nearly glabrous; fruits 7–9(–10) mm long; petals yellow; seeds 1.5–1.7(–3.0) mm long . . . . 2. C. SATIVA

1. Camelina microcarpa Andrz. ex DC.
(littlepod false flax, flaxweed, gold-of-pleasure, Dutch flax)
C. sativa (L.) Crantz ssp. *microcarpa* (Andrz. ex DC.) Em. Schmid

Pl. 316 c, d; Map 1328

Stems (8–)20–60(–80) cm long; stems and leaves pubescent with rough, unbranched, spreading hairs 1.0–2.5 mm long in addition to the shorter, forked and stellate hairs. Stem leaves (0.8–)1.5–5.5(–7.0) cm long, margins entire or minutely toothed, the base arrowhead-shaped or minutely auriculate. Sepals 2.0–2.5 mm long. Petals pale yellow, 3–4 mm long. Fruits (2.5–)3.5–6.0 mm long. Styles 1.0–3.5 mm long. Seeds 0.8–1.4(–1.5) mm long, oblong-elliptic in outline, slightly longer than wide, the surface with a fine, netlike or honeycomb-like pattern of ridges and pits, reddish brown. $2n=16, 32, 40$. April–September.

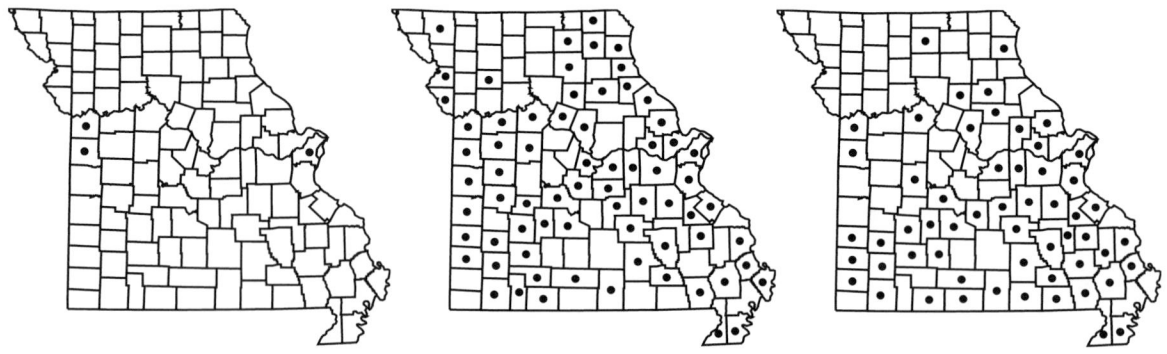

1329. Camelina sativa   1330. Capsella bursa-pastoris   1331. Cardamine bulbosa

Introduced, scattered nearly throughout Missouri (native of Europe, Asia; widely introduced in North America). Pastures, margins of crop fields, roadsides, railroads, and open, disturbed areas.

This species is probably more common in Missouri than herbarium specimens indicate.

2. Camelina sativa (L.) Crantz (common false flax, false flax)

Pl. 316 e, f; Map 1329

Stems (12–)30–80(–100) cm long; stems and leaves sparsely pubescent, with the minute, unbranched hairs not longer than the stellate hairs, sometimes nearly glabrous. Stem leaves (1–)2–7(–9) cm long, margins commonly wavy and/or with shallow, widely spaced teeth, base sagittate to strongly auriculate. Sepals 2–3 mm long. Petals yellow, (3.0–)4.0–5.5 mm long. Fruits 7–9(–10) mm long. Styles 1.0–2.5 mm long. Seeds 1.5–1.7(–3.0) mm long, narrowly elliptic in outline, about 2 times as long as wide, the surface roughened, light orange to pale yellowish brown. $2n=26, 28, 40$. April–August.

Introduced, uncommon and widely scattered in Missouri (native of Europe, Asia; sporadically introduced in North America). Margins of crop fields, railroads, and open, disturbed areas.

McGregor (1985a) reported a third species for Missouri, *C. alyssum* (Mill.) Thell., based on a single historical collection from Cass County. However, the specimen has been redetermined as *C. sativa*. McGregor suggested that this European species was an early introduction in the Great Plains that became extirpated from the region by about 1950. It differs from *C. sativa* in its lobed or toothed leaves, nodding inflorescences, relatively broad fruits, and slightly larger seeds. *Camelina alyssum* apparently still persists locally in southern Canada (Gleason and Cronquist, 1991), but it should be considered excluded from the Missouri flora.

12. Capsella Medik.

One species, Europe, Asia, introduced nearly worldwide.

1. Capsella bursa-pastoris (L.) Medik.
(shepherd's purse)

Pl. 316 i–k; Map 1330

Plants annual, terrestrial. Stems (2–)10–50(–70) cm long, erect, usually few-branched, pubescent near the base with sessile stellate, stalked forked, and unbranched hairs. Leaves (1–)2–10(–15) cm long, sparsely to moderately pubescent with sessile stellate and unbranched hairs, the basal leaves usually numerous, petiolate, the leaf blades oblanceolate, toothed to deeply pinnately lobed with numerous irregular lobes, the stem leaves few, progressively reduced, alternate, mostly sessile, the bases clasping with prominent, rounded to somewhat pointed auricles, linear to lanceolate, entire or shallowly toothed. Inflorescences panicles, the lower branches subtended by reduced leaves. Sepals 1–2 mm long, ascending, elliptic, usually sparsely pubescent, sometimes with white or reddish purple margins. Petals 2–4(–5) mm long, not lobed, white. Styles 0.2–0.7 mm long. Fruits spreading to ascending, (3–)4–8(–10) mm long, about as long as wide, obtriangular, flattened at a right angle to the septum, the tip usually slightly concave, dehiscing longitudinally, each valve with a network of nerves. Seeds (6–)8–20 in each locule, 0.8–1.1

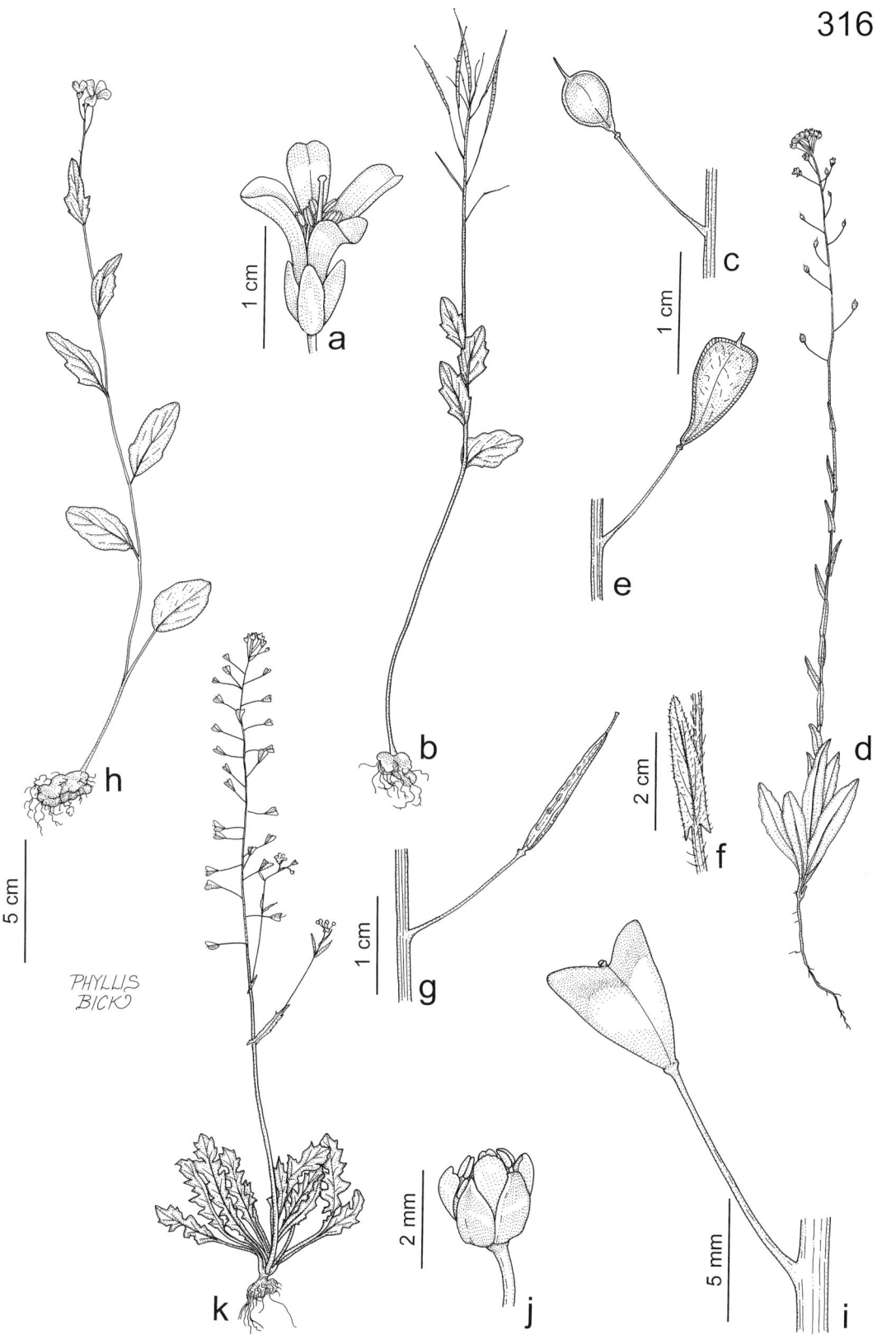

Plate 316. Brassicaceae. *Cardamine douglassii*, **a)** flower, **b)** habit. *Camelina microcarpa*, **c)** fruit, **d)** habit. *Camelina sativa*, **e)** fruit, **f)** leaf. *Cardamine bulbosa*, **g)** fruit, **h)** habit. *Capsella bursa-pastoris*, **i)** fruit, **j)** flower, **k)** habit.

mm long, oblong in outline, the surface with a fine, netlike or honeycomb-like pattern of ridges and pits, orangish yellow. 2n=16, 32. January–December.

Introduced, common throughout the state (native of Europe, western Asia; introduced nearly worldwide). Pastures, fields, lawns, flower beds, roadsides, railroads, and open, disturbed areas.

This weedy species has several adaptations that help to account for its worldwide distribution, including partially self-fertile flowers and the ability to continue blooming nearly throughout the year, with a single plant capable of producing up to 90,000 seeds in a year (Al-Shehbaz, 1986). The seeds have a mucilaginous sheath and become sticky when wet, aiding in their dispersal by shoe soles, animals, and other items that they come into contact with.

The young rosettes and fruits can be eaten raw and are rich in vitamin C. A variety of medicinal uses have also been attributed to the plants (reviewed by Al-Shehbaz, 1986).

13. Cardamine L. (bitter cress)

Plants annual, biennial, or perennial herbs, terrestrial or occasionally emergent aquatics (in *C. bulbosa*), glabrous or pubescent with unbranched hairs. Stems 10–30 cm long (longer elsewhere), erect or ascending, unbranched or few-branched. Leaves alternate, opposite, or whorled and often also basal, simple, trifoliate, or pinnately or palmately compound, the margins entire or toothed, the bases not clasping. Inflorescences racemes or less commonly few-branched panicles, the flowers not subtended by bracts. Sepals lanceolate to narrowly elliptic or oblong, erect or ascending. Petals 7–19 mm (or more) long, not lobed, white, pink, or purple. Stamens 6, rarely reduced to 4. Fruits erect or ascending, usually more than 10 times as long as wide, straight, flattened parallel to the septum, usually tapered toward the tip into the style, the replum narrowly winged, the valves unveined or with a faint midnerve, longitudinally dehiscent. Seeds in 1 row in each locule. About 200 species, worldwide.

The present circumscription of the genus includes species formerly segregated from *Cardamine* as the genus *Dentaria* by many North American authors, including Steyermark (1963). However, within the context of variation in *Cardamine* throughout the world, none of the characters thought to separate the two groups can be applied consistently or reliably (Al-Shehbaz, 1988a, b). Thus, although these species appear amply distinct when compared to other Missouri taxa, they are best maintained taxonomically as part of an expanded concept of *Cardamine*. This expanded concept has become widely accepted in recent years (Rollins, 1993; Appel and Al-Shehbaz, 2002) and is supported by extensive molecular data (Franzke et al., 1998; Sweeney and Price, 2000).

1. Leaves simple, entire or toothed
    2. Stems glabrous or rarely with sparse, minute hairs to 0.1 mm long; petals white, sometimes faintly pinkish-tinged . . . . 1. C. BULBOSA
    2. Stems with spreading hairs (0.2–)0.3–0.6(–0.8) mm long, at least in the apical half; petals pink or purplish pink . . . . 4. C. DOUGLASSII
1. Leaves pinnately or palmately compound or divided
    3. Leaves palmately compound or divided into 3 or 5 leaflets or main divisions; petals 7–20 mm long; plants perennial, with fleshy rhizomes
        4. Rhizomes beadlike, breaking easily at the constrictions; stem leaves in a whorl of (2)3(4), the leaflets linear to oblanceolate or lanceolate, sharply and usually coarsely toothed to irregularly incised . . . . . . 2. C. CONCATENATA
        4. Rhizomes not constricted, uniform in diameter, not fragile; stem leaves 2 (rarely 3 elsewhere), opposite or subopposite, the leaflets ovate to broadly elliptic, with more or less regularly spaced, usually blunt or rounded teeth . . . . 3. C. DIPHYLLA

3. Leaves pinnately compound or divided, with more than 3 leaflets or divisions; petals 1.5–5.0 mm long; plants annual, without rhizomes
   5. Basal leaves numerous at flowering time, their petioles with spreading hairs; fruits erect and appressed to the axis of the inflorescence; stamens 4(6) .... 6. C. HIRSUTA
   5. Basal leaves few at flowering time, glabrous; fruits ascending to spreading; stamens (4)6
      6. Lateral leaflets linear to narrowly wedge-shaped or narrowly oblanceolate, 0.3–3.0 mm wide, tapered to the base and not flared along the rachis; terminal leaflet about as wide as or slightly wider than the lateral leaflets; fruits 0.6–0.9 mm wide; petals 1.3–3.0 mm long .... 7. C. PARVIFLORA
      6. Lateral leaflets oblanceolate to obovate, often more than 4 mm wide, broadened at the very base and extending along the leaf rachis, often connected by the flared leaflet bases; terminal leaflet usually broadly obovate, wider than the lateral leaflets; fruits 1.0–1.5 mm wide; petals 2.5–5.0 mm long.
         7. Axis of the inflorescence wavy or somewhat zigzag, at least at fruiting; lateral leaflets stalked .... 5. C. FLEXUOSA
         7. Axis of the inflorescence straight or nearly so (occasionally somewhat arched at fruiting); lateral leaflets sessile .... 8. C. PENSYLVANICA

1. Cardamine bulbosa (Schreb. ex Muhl.) Britton, Sterns & Poggenb. (spring cress)
C. bulbosa f. *fontinalis* E.J. Palmer & Steyerm.
C. rhomboidea (Pers.) DC.

Pl. 316 g, h; Map 1331

Plants perennial herbs, occasionally emergent aquatics, with short, tuberous rhizomes, these unsegmented or occasionally irregularly constricted into 2 or 3 segments. Stems (15–)20–60 cm long, glabrous or rarely with sparse, minute hairs to 0.1 mm long (visible only with strong magnification) in the apical half. Leaves 2–7 cm long, simple, entire, wavy-margined, or with few, shallow, widely spaced teeth, glabrous; the basal leaves usually withered by flowering time, long-petiolate, the leaf blades ovate to cordate; the stem leaves 4–14, mostly sessile, ovate to lanceolate or narrowly oblong. Sepals 2.5–5.0 mm long, greenish yellow. Petals (6–)7–12(–16) mm long, white, sometimes faintly pinkish-tinged. Styles 2–3(–5) mm long. Fruits 20–30 mm long, sometimes aborting before maturity. Seeds 1.7–2.1 mm long, irregularly oblong to circular in outline, the surface slightly roughened, orange to greenish yellow. $2n=64, 80, 96$. March–June.

Scattered in the southern, central, and northeastern portions of the state (eastern U.S. west to South Dakota and Texas; Canada). Bottomland forests, banks of streams and spring branches, fens, and less commonly seepy bluffs and acid seeps.

There is controversy surrounding the validity of the name *C. bulbosa*. Some authors have maintained that the basionym *Arabis bulbosa* Schreb. ex Muhl. was not validly published, and they use the later name *C. rhomboidea* for this species. The present treatment follows the recommendation of Merrill and Hu (1949), who investigated the names published by Henry Muhlenberg in detail.

Cardamine bulbosa and *C. douglassii* are both complex, morphologically and cytologically variable species. They are not entirely distinct morphologically in all parts of their ranges, and they have sometimes been treated as varieties of *C. bulbosa*. Naturally occurring hybrids have been documented from Ohio by Hart and Eshbaugh (1976), who also documented several biochemical and morphological races within each taxon. In Missouri, *C. bulbosa* flowers on average 2 weeks later than does *C. douglassii* (Steyermark, 1963), and the characters presented in the key to species above seem to separate the two species adequately.

The rhizomes and aboveground portions of *C. bulbosa* have a flavor reminiscent of horseradish (*Armoracia rusticana*) and have been used as a substitute for it in salads and condiments. The species is commonly but not exclusively associated with calcareous substrates.

1332. Cardamine concatenata   1333. Cardamine diphylla   1334. Cardamine douglassii

2. Cardamine concatenata (Michx.) O. Schwarz (cut-leaved toothwort)
Dentaria laciniata Muhl. ex Willd.

Pl. 317 i, j; Map 1332

Plants perennial herbs with elongate rhizomes consisting of thickened, beadlike, tuberous portions connected by thinner, threadlike portions. Stems 15–40(–55) cm long, glabrous toward the base, usually somewhat pubescent with spreading hairs toward the tip. Basal leaves usually absent on flowering plants, 13–30 cm long, long-petiolate, the leaf blades trifoliate, the central leaflet sometimes divided or lobed into 3 divisions, the lateral leaflets sometimes unevenly 2(3)-lobed; the leaflets 2.5–10.0 cm long, linear to oblanceolate or lanceolate, glabrous or sometimes with minute, spreading, marginal hairs, the margins sharply and usually coarsely toothed to irregularly incised, rarely nearly entire. Stem leaves (2)3, in a single whorl, 4–12 cm long, short-petiolate, linear to oblanceolate or lanceolate, glabrous or sometimes with minute, spreading, marginal hairs, the margins sharply and usually coarsely toothed to irregularly incised, rarely nearly entire. Sepals (4–)5–10 mm long, green, sometimes pinkish-tinged, the margins usually pale. Petals (8–)10–20 mm long, white to light pink. Styles (4–)5–8 mm long. Fruits 20–40 mm long. Seeds 1.9–2.7 mm long, broadly oblong to ovate in outline, the surface with a netlike or honeycomb-like pattern of ridges and pits, orange to greenish orange. $2n=128, 240, 256$. March–May.

Common nearly throughout Missouri, possibly in every county (eastern U.S. and adjacent Canada west to Minnesota and Oklahoma). Mesic bottomland and upland forests, occasionally on drier slopes.

Toothwort is a conspicuous member of the herbaceous spring flora in rich woodlands of valleys and moist ravines. The segmented, spicy rhizomes are sometimes eaten raw in salads or dried and ground as a horseradish substitute.

3. Cardamine diphylla (Michx.) A.W. Wood (crinkleroot, broad-leaved toothwort)
Dentaria diphylla Michx.

Pl. 317 g, h; Map 1333

Plants perennial herbs with elongate, unsegmented rhizomes uniform in diameter and not fragile. Stems 15–40 cm long, glabrous. Basal leaves usually present on flowering plants, 12–25 cm long, long-petiolate, the leaf blades trifoliate, the leaflets 4–8 cm long, broadly ovate-elliptic, glabrous or sometimes with minute, appressed, marginal hairs, the margins with more or less regularly spaced, blunt or rounded teeth. Stem leaves 2(3), opposite or subopposite, 3–14 cm long, short-petiolate, the leaflets ovate to broadly elliptic, glabrous or sometimes with minute, appressed, marginal hairs, the margins with more or less regularly spaced, usually blunt or rounded teeth. Sepals (4–)5–8 mm long, green or pinkish white. Petals (7–)11–17 mm long, white to pink or light purple, sometimes with darker veins. Styles 5–7 mm long. Fruits 15–40 mm long, often aborting before maturity. Seeds usually not produced. $2n=96$. April–June.

Uncommon, known from a single historical collection from Chariton County (eastern U.S. and adjacent Canada west to Minnesota and Alabama; apparently disjunct in Arkansas and Missouri). Habitat in Missouri unknown, but elsewhere bottomland and mesic upland forests, banks of streams, and ledges of moist bluffs, on both calcareous and acidic substrates.

This species is hesitantly included in the flora based upon a single collection lacking habitat and precise locality data. This specimen may have been cultivated, but in the absence of evidence to the contrary, it cannot be excluded from the flora, particularly in light of the existence of a similarly ambiguous historical collection from Arkansas. The Arkansas and Missouri collections are disjunct from the main portion of the species range (to the north

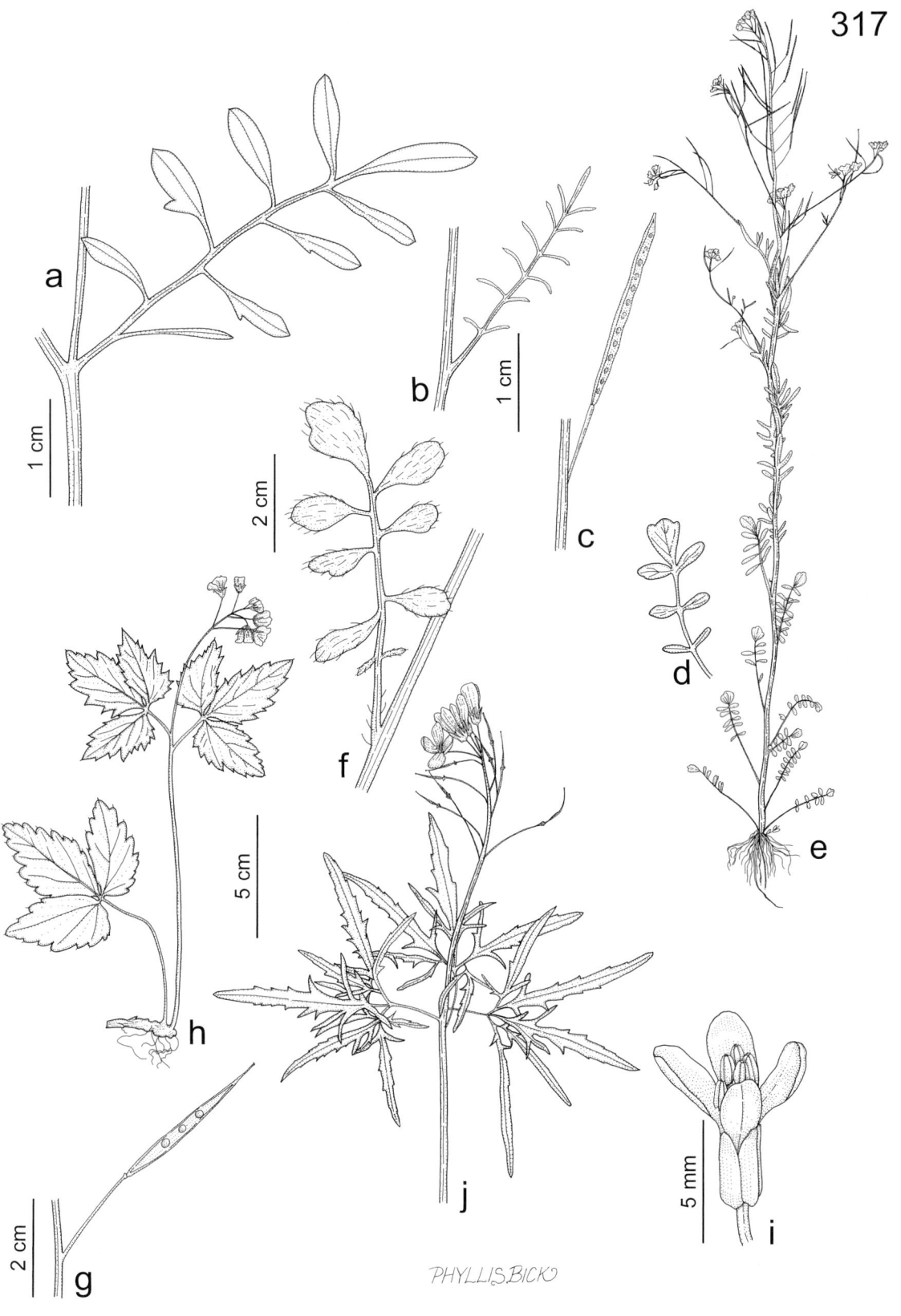

Plate 317. Brassicaceae. *Cardamine parviflora*, **a)** leaf, **b)** leaf, **c)** fruit. *Cardamine pensylvanica*, **d)** leaf, **e)** habit. *Cardamine hirsuta*, **f)** leaf. *Cardamine diphylla*, **g)** fruit, **h)** habit. *Cardamine concatenata*, **i)** flower, **j)** inflorescence.

1335. Cardamine flexuosa   1336. Cardamine hirsuta   1337. Cardamine parviflora

and east). This species reproduces exclusively asexually throughout much of its range (Harriman, 1965; Al-Shehbaz, 1988b), and its wide distribution in eastern North America is not easily explained. It should be searched for in remnant forests in northern Missouri.

4. Cardamine douglassii (Torr.) Britton (purple cress, northern bitter cress)

C. bulbosa (Schreb. ex Muhl.) Britton, Sterns & Poggenb. var. *purpurea* (Torr.) Britton, Sterns & Poggenb.

Pl. 316 a, b; Map 1334

Plants perennial herbs with short, tuberous, unsegmented rhizomes. Stems (7–)10–30 cm long, with spreading hairs (0.2–)0.3–0.6(–0.8) mm long, especially in the apical half. Leaves 2–7 cm long, simple, entire, wavy-margined, or with few, shallow, widely spaced teeth, often sparsely hairy; the basal leaves usually withered by flowering time, long-petiolate, the leaf blades ovate to cordate; the stem leaves mostly 3–5, mostly sessile, ovate to lanceolate or narrowly elliptic. Sepals (2.5–)3.0–6.0 mm long, reddish purple. Petals (7–)8–14 mm long, pink to purplish pink. Styles 3–4 mm long. Fruits (15–)25–40 mm long, sometimes aborting before maturity. Seeds 1.7–2.1 mm long, irregularly oblong to circular in outline, the surface slightly roughened, orange to greenish yellow. $2n=64, 96, 144$. March–April.

Scattered in northeastern Missouri and disjunctly in Cape Girardeau County (eastern U.S. and adjacent Canada west to Minnesota, and Missouri). Mesic to wet bottomland forests.

For a discussion of the separation between this species and the closely related *C. bulbosa*, see the treatment of that species.

5. Cardamine flexuosa With.

Map 1335

Plants annual or biennial. Stems (6–)10–50 cm long, glabrous or sparsely to densely pubescent with spreading hairs near the base (and often also along the petioles). Basal leaves often withered at flowering time, (2–)4–10(–14) cm long, short-petiolate, pinnately compound with 5–13(–14) glabrous, stalked leaflets, the lateral leaflets circular to obovate or obovate to oblong, entire or with 1–5 irregular teeth, the terminal leaflet often broader than the lateral ones. Stem leaves usually 3–15, (1–)2–5 cm long, pinnately compound with 5–11 short-stalked leaflets, these linear to oblanceolate or ovate to circular, glabrous, the stalk broadened at the very base and extending slightly along the leaf rachis. Sepals 1.5–2.5 mm long, green. Petals 2.5–4.0(–5.0) mm long, white. Stamens (4)6. Styles 0.3–1.0(–1.5) mm long. Fruits (8–)12–28 mm long. Seeds 0.9–1.5 mm long, oval to oblong in outline, the surface with a fine, netlike or honeycomb-like pattern of ridges and pits, orange. $2n=32$. March–July.

Introduced, uncommon and sporadic (native of Europe, introduced widely but sporadically in the eastern and midwestern U.S. and California). Greenhouses, gardens, and open, disturbed areas.

This species is spread primarily as a seed contaminant in the soil of bedding plants and other cultivated species distributed through the horticultural industry. It can be found as a weed in and around plant nurseries and also in flower beds, but it has not been documented to invade any natural habitats as yet. This is the first report of the species in Missouri.

6. Cardamine hirsuta L. (hoary bitter cress)

Pl. 317 f; Map 1336

Plants annual. Stems (3–)10–35(–45) cm long, glabrous or with sparse, spreading hairs near the base. Basal leaves numerous at flowering time, (1.5–)3.0–10.0(–13.0) cm long, short-petiolate, the petioles with spreading hairs, pinnately compound with 3–15(–23) glabrous leaflets, the lateral leaflets circular to ovate or obovate, entire or with 1–3 irregular teeth, tapered to a sessile or short-stalked base, this not expanded along the rachis, the terminal leaflet usually broader than the lateral ones. Stem leaves 2–5(6), 1–5(–7) cm long, pinnately compound with 5–15(–21) leaflets, these linear to oblanceolate or oblong to ovate, sometimes hairy on the upper surface. Sepals 1.0–2.5 mm long, green. Petals 1.5–4.0(–5.0) mm long, white. Stamens 4(6). Styles 0.1–0.6(–1.0) mm long. Fruits (9–)15–25(–28) mm long. Seeds 0.9–1.3(–1.5) mm long, oval to oblong in outline, the surface with a fine, netlike or honeycomb-like pattern of ridges and pits, orange. $2n=16$. March–April.

Introduced, uncommon and widely scattered in Missouri (native of Europe, Asia, naturalized nearly worldwide). Sandy banks of streams and moist depressions of sandstone glades; also lawns and pastures.

This uncommon weed sometimes is confused with *C. pensylvanica* and *C. parviflora,* but it may be distinguished by characters in the key to species above. For a discussion of differences between this complex and *Planodes virginica,* see the treatment of that species.

7. Cardamine parviflora L. var. arenicola
(Britton) O.E. Schulz (small-flowered bitter cress)

C. arenicola Britton

Pl. 317 a–c; Map 1337

Plants annual, glabrous. Stems 5–30(–35) cm long. Basal leaves usually few at flowering time, (2–)4–10 cm long, short-petiolate, pinnately compound with 5–13 leaflets 0.3–3.0 mm wide, the lateral leaflets linear to narrowly wedge-shaped or oblanceolate, entire or with 1–2 irregular teeth, tapered to a sessile or short-stalked base, this not expanded along the rachis, the terminal leaflet about as wide as or somewhat wider than the lateral leaflets. Stem leaves usually 5–8, 2–6(–7) cm long, pinnately compound with 5–13 leaflets, these linear to oblanceolate or narrowly wedge-shaped, tapered to the base and not expanded along the rachis, the terminal leaflet about as wide as or slightly wider than the lateral ones. Sepals 1.0–1.5(–2.0) mm long, green. Petals 1.3–3.0 mm long, white. Styles 0.3–0.7(–1.0) mm long. Fruits (5–)10–20(–30) mm long, 0.6–0.9 mm wide. Seeds 0.6–0.9 mm long, oblong in outline, the surface with a netlike or honeycomb-like pattern of ridges and pits, orange. $2n=16$. March–July.

Scattered throughout Missouri (eastern U.S. and adjacent Canada west to Minnesota and Texas). Mesic to dry upland forests, particularly on rocky ridges and slopes, glades, ledges of bluffs, fallow fields, margins of crop fields, and disturbed ground, often on acidic substrates.

This species is quite variable in size and leaflet shape. Plants growing in moister situations may sometimes closely resemble *C. pensylvanica* but can usually be distinguished by the nondecurrent bases of the lateral leaflets and the terminal leaflets that are about the same width as the lateral ones. *Cardamine parviflora* also tends to grow in somewhat drier sites than does *C. pensylvanica.* For a discussion of differences between this species and *Planodes virginica,* see the treatment of that species.

The other variety, var. *parviflora,* is widespread in Europe, Asia, and northern Africa, but it has not been recorded from North America yet. It differs in its narrower, more numerous leaflets. The relationship between the two varieties requires further study.

8. Cardamine pensylvanica Muhl. ex Willd.
(bitter cress, Pennsylvania bitter cress)

Pl. 317 d, e; Map 1338

Plants annual, biennial, or rarely short-lived perennial herbs, glabrous or the stem bases (but not the petioles) sparsely hairy. Stems (5–)10–70 cm long. Basal leaves usually few at flowering time, 4–15 cm long, short-petiolate, pinnately compound with 5–13 leaflets, the lateral leaflets oblanceolate to obovate, entire or with 1–2 irregular teeth, tapered to a sessile base, this broadened at the very base and extending along the leaf rachis, the leaflets sometimes more or less connected by the flared leaflet bases, the terminal leaflet somewhat wider than the lateral leaflets, obovate to nearly circular. Stem leaves usually 5–8, mostly 4–8 cm long, pinnately compound with 5–13 leaflets, these oblanceolate or narrowly wedge-shaped to obovate, tapered to the base, this broadened at the very base and extending along the leaf rachis, the leaflets sometimes more or less connected by the flared leaflet bases, the terminal leaflet somewhat wider than the lateral leaflets, oblanceolate to obovate or nearly circular. Sepals 1.5–2.3 mm long, green. Petals 1.5–4.0 mm long, white. Styles 0.5–1.0 mm long. Fruits 15–25(–30) mm long. Seeds 0.7–0.9 mm long, oval to oblong in outline, the margins sometimes slightly winged, the surface finely roughened, orangish tan. $2n=32, 64$. March–July.

Scattered nearly throughout Missouri, but apparently more common south of the Missouri River (U.S., Canada). Mesic bottomland and upland forests, ledges of sheltered bluffs, on banks and in

1338. Cardamine pensylvanica   1339. Chorispora tenella   1340. Conringia orientalis

shallow water of streams and spring branches, margins of crop fields, and open, disturbed areas.

Cardamine pensylvanica is perhaps the most morphologically variable species of bittercress in Missouri. Depauperate plants can be very difficult to distinguish from *C. parviflora* (see the treatment of that species for further discussion) and *C. flexuosa*. The potential confusion is heightened by the fact that the leaflets of upper leaves tend to be narrower than those of lower leaves in most plants, and the character of decurrent leaflet bases is not always easily observed on every leaf of a given plant.

Plants of *C. pensylvanica* growing in the moist sand of overhung sandstone ledges or in shallow, running water sometimes have less strongly ascending stems and large, membranous leaves with very broad terminal leaflets. In these respects they are superficially similar to *C. flexuosa* With. However, that species has ciliate petiole bases similar to those found in *C. hirsuta*. For a discussion of differences between this species and *Planodes virginica*, see the treatment of that species.

Young herbage of this species is sometimes eaten raw in salads.

14. Chorispora R. Br. ex DC.

Thirteen species, Europe, Asia.

1. Chorispora tenella (Pall.) DC. (blue mustard)

Pl. 318 j–l; Map 1339

Plants annual (perennial elsewhere), terrestrial. Stems (5–)20–40(–56) cm long, erect to spreading, branched from the base and above, with stalked glands usually mixed with sparse, unbranched hairs. Leaves alternate and basal, (1.5–)3.0–8.0(–13.0) cm long, sparsely covered with stalked glands (rarely also a few unbranched hairs), usually short-petiolate, not clasping, the leaf blades oblanceolate to lanceolate or narrowly elliptic, shallowly and broadly toothed, those of the lower leaves less commonly deeply pinnately divided. Inflorescences usually racemes, the lower branches subtended by reduced leaves. Sepals (3–)4–6(–7) mm long, narrowly ascending, linear-lanceolate, sparsely covered with stalked glands. Petals 6–10(–12) mm long, not lobed, purple to bluish purple. Styles minute or absent at flowering, becoming elongated as a beak as the fruit matures. Fruits ascending, curved upward, (1.5–)3.0–5.0 cm long, more than 10 times as long as wide, linear, circular in cross-section, usually somewhat constricted between the seeds, the tip with a tapered beak 1–2 cm long, the valves lacking well-defined midnerves, not dehiscing, but breaking transversely into 2-seeded segments, glabrous or with stalked glands. Seeds in 1 row in each locule, imbedded in the thick, corky fruit-wall, 1.4–1.6 mm long, broadly ovate-elliptic in outline, flattened, the margins not winged, the surface smooth to minutely roughened, greenish yellow. $2n=14$. April–June.

Introduced, uncommon and widely scattered (native of Asia, widely naturalized in North America). Railroads.

Plate 318. Brassicaceae. *Lepidium draba*, **a)** fruit, **b)** habit. *Lepidium chalapense*, **c)** fruit. *Lepidium appelianum*, **d)** leaf, **e)** fruit. *Conringia orientalis*, **f)** fruit, **g)** habit. *Lepidium didymum*, **h)** fruit, **i)** habit. *Chorispora tenella*, **j)** fruit, **k)** flower, **l)** habit.

15. Conringia Heist. ex Fabr.

Six species, Europe, Asia.

1. Conringia orientalis (L.) Dumort. (hare's-ear mustard)

Pl. 318 f, g; Map 1340

Plants annual, terrestrial, glabrous, glaucous. Stems (10–)30–60(–80) cm long, erect, unbranched or few-branched from the base. Leaves alternate and basal, 3–9 cm long, the margins entire, the basal leaves obovate to oblanceolate, tapered at the base, the stem leaves narrowly elliptic to oblong-lanceolate, the base rounded and strongly clasping. Inflorescences panicles, the lower branches subtended by reduced leaves. Sepals 6–8 mm long, ascending, linear. Petals (5.5–)7.0–10.0(–12.0) mm long, not lobed, pale yellow. Styles 0.5–1.0 mm long. Fruits ascending, straight, (5.0–)7.5–11.0(–13.5) cm long, more than 10 times as long as wide, linear, strongly 4-angled in cross-section, not beaked except for the persistent style, each valve with a more or less well-defined midnerve, dehiscing longitudinally. Seeds in 1 row in each locule, 1.9–2.5 mm long, narrowly oblong-elliptic in outline, the margins not winged, the surface with a fine, netlike or honeycomb-like pattern of ridges and pits, reddish brown. $2n=14$. May–August.

Introduced, widely scattered, especially in northern and eastern Missouri (native of Europe, Asia, widely introduced in North America). Railroads, roadsides, and open, disturbed areas.

16. Descurainia Webb & Berthel.

Plants annual or biennial (perennials or shrubs elsewhere), terrestrial, pubescent with short-stalked, branched hairs occasionally mixed with unbranched hairs and frequently also with stalked glands (less commonly nearly all of the hairs replaced with stalked glands). Stems erect, usually branched in the upper half. Leaves alternate and usually also a few basal at maturity, short-petiolate or sessile, not clasping, the lower leaves ovate to obovate or oblanceolate in outline, 1–3 times pinnately divided, with toothed, oblanceolate to linear leaflets and/or divisions, the upper leaves progressively reduced and less divided, ovate to lanceolate or linear in outline, 1 time pinnately compound or divided, the leaflets or divisions linear, toothed. Inflorescences racemes or less commonly panicles, the flowers often not subtended by bracts. Sepals lanceolate to elliptic or ovate, erect or ascending, green, often reddish-tinged. Petals not lobed, pale yellow to yellow or greenish yellow. Styles absent or less than 0.5 mm long. Fruits ascending to spreading, 4–30 times as long as wide, circular in cross-section or nearly so, dehiscent longitudinally. Seeds in 1 or 2 rows in each locule, 0.8–1.1 mm long, elliptic to narrowly ovate in outline, the margins not winged, the surface with a fine, netlike or honeycomb-like pattern of ridges and pits, orange to reddish brown. About 40 species, North America to South America, Europe, Asia, Africa.

The seeds of this genus have been used as a substitute for mustard seeds. Massive ingestion of the plants can poison livestock. The plants also have been used for various medicinal purposes.

1. Fruits 5–10(–15) mm long, 1–2 mm wide, blunt or rounded at the tip; seeds at least in part in 2 rows in each locule; septum of the fruit not veined . . . 1. D. PINNATA
1. Fruits (12–)15–25(–30) mm long, 0.5–1.2 mm wide, tapered and pointed at the tip; seeds in 1 row in each locule; septum of the fruit 2- or 3-veined . . . . . 2. D. SOPHIA

1. Descurainia pinnata (Walter) Britton (tansy mustard)

Pl. 319 a–c; Map 1341

Stems (10–)20–50(–80) cm long. Leaves 1–10 cm long, with or without stalked glands, the lower leaves 2 times pinnately dissected, the upper ones reduced in size and often 1 time pinnately dissected. Sepals 1–2 mm long. Petals 1.0–3.5 mm long, light yellow, sometimes fading to white upon drying. Fruits club-shaped or oblong, 5–10(–15) mm long,

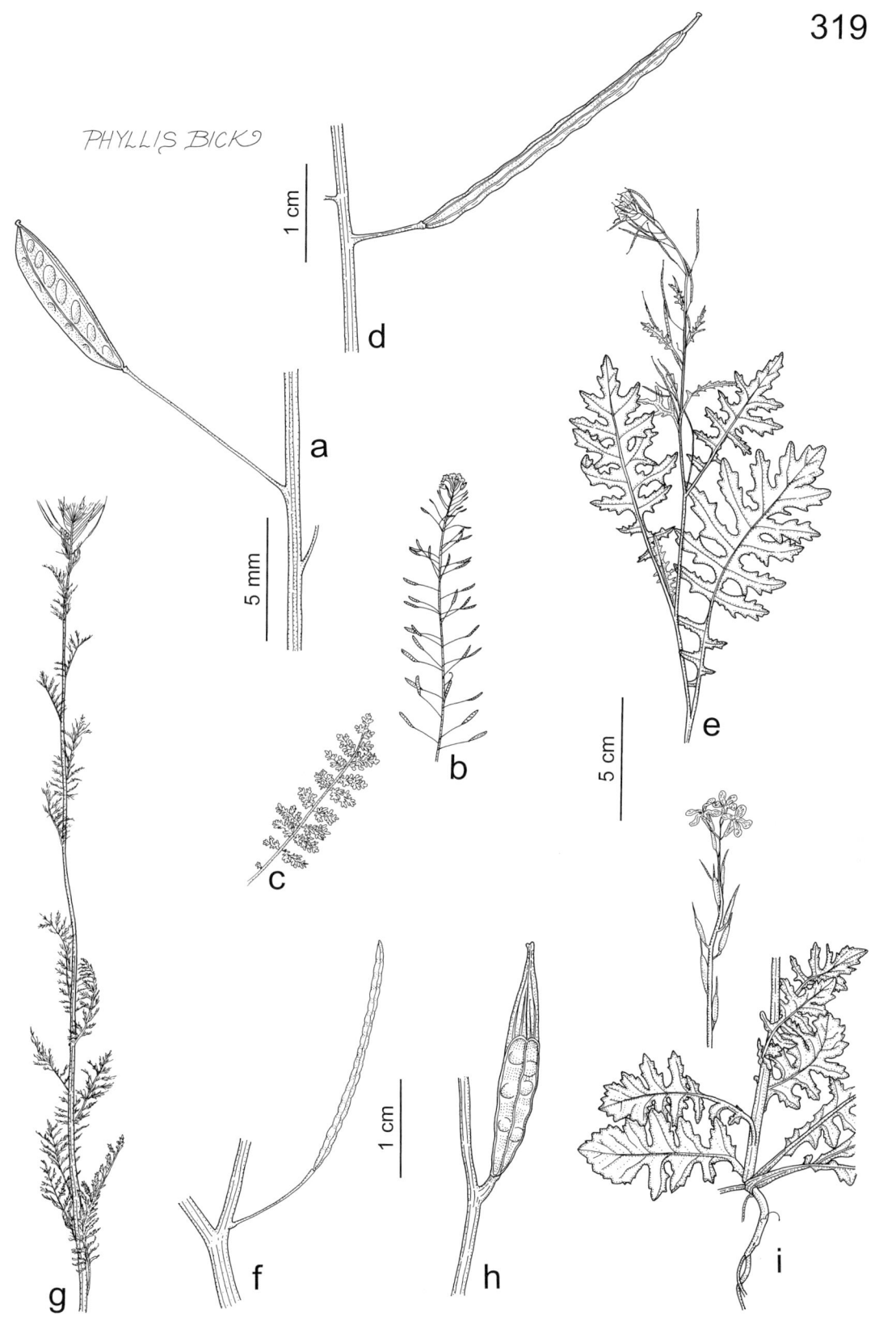

Plate 319. Brassicaceae. *Descurainia pinnata*, **a)** fruit, **b)** inflorescence, **c)** leaf. *Erucastrum gallicum*, **d)** fruit, **e)** inflorescence and leaves. *Descurainia sophia*, **f)** fruit, **g)** habit. *Eruca vescicaria*, **h)** fruit, **i)** inflorescence and base.

1341. Descurainia pinnata   1342. Descurainia sophia   1343. Diplotaxis muralis

1–2 mm wide, blunt or rounded at the tip, the valves with 1 midnerve. Styles 0.1–0.2 mm long. Seeds 10–40 per fruit, at least in part in 2 rows in each locule. 2n=14, 28, 42. March–July.

Common throughout Missouri (North America). Banks of streams and rivers, openings of bottomland and mesic upland forests, ledges of calcareous bluffs, glades, and prairies; also fallow fields, railroads, roadsides, and open, disturbed areas.

1. Stalks of the fruits mostly ascending; stems and leaves densely pubescent with stalked glands, the plants appearing green . . . . 1A. SSP. BRACHYCARPA
1. Stalks of the fruits widely spreading; stems and leaves densely pubescent with mostly many-branched, nonglandular hairs, the plants appearing grayish green . . . . 1B. SSP. PINNATA

1a. ssp. brachycarpa (Richardson) Detling
    D. pinnata var. *brachycarpa* (Richardson) Fernald

Stems and leaves densely pubescent, but with many or most of the hairs replaced with stalked glands, the plants appearing green. Stalks of fruits mostly ascending at about a 45° angle from the infructescence axis. Fruits mostly 5–10 mm long. 2n=14, 28. March–July.

Common throughout Missouri (U.S., Canada). Banks of streams and rivers, openings of bottomland forests, ledges of calcareous bluffs, glades, and prairies; also fallow fields, railroads, roadsides, and open, disturbed areas.

This is by far the more common of the subspecies in Missouri. Occasional plants have branched hairs in sufficient numbers to appear intermediate between the two subspecies.

1b. ssp. pinnata

Stems and leaves densely pubescent with mostly many-branched, nonglandular hairs, the plants appearing grayish green. Stalks of fruits broadly spreading at about a 75° angle from the infructescence axis. Fruits mostly 6–13 mm long. 2n=14, 28, 42. March–May.

Uncommon and widely scattered in Missouri (southeastern U.S. north to Virginia and west to Texas). Openings of bottomland and mesic upland forests, banks of rivers; also fallow fields and roadsides.

2. Descurainia sophia (L.) Webb ex Prantl
    (flixweed, tansy mustard)
                                        Pl. 319 f–g; Map 1342

Stems (10–)25–80(–100) cm long, moderately to usually densely pubescent with many-branched, nonglandular hairs. Leaves 1–15 cm long, moderately to usually densely pubescent with many-branched hairs, 2 or 3 times pinnately dissected. Sepals 2.0–2.5 mm long. Petals 2.0–2.5 mm long, yellow. Fruits linear, (12–)15–25(–30) mm long, 0.5–1.2 mm wide, tapered and pointed at the tip, the valves with a prominent midnerve, the septum with 2 or 3 longitudinal veins, the fruits and their stalks ascending. Styles absent to 0.2 mm. Seeds 20–40 per fruit, in 1 row in each locule. 2n=20, 28, 38. May–July.

Introduced, uncommon and widely scattered in Missouri (native of Europe, Asia, widely introduced in the U.S., Canada). Railroads, roadsides, and open, disturbed areas.

17. Diplotaxis DC.

About 30 species, Europe, Asia, Africa.

1. Diplotaxis muralis (L.) DC. (sand rocket, stinking wall-rocket)

Map 1343

Plants annual or rarely perennial herbs, terrestrial. Stems (5–)20–40(–60) cm long, erect to spreading, few-branched from the base, glabrous or sparsely pubescent with unbranched hairs. Leaves mostly basal, 2–10 cm long, the margins wavy, toothed or pinnately lobed, the basal leaves petiolate, the leaf blades oblong to oblanceolate, the few stem leaves alternate, sessile, not clasping, the leaf blades oblong, tapered at the base. Inflorescences panicles, the lower branches subtended by reduced leaves. Sepals 3.0–5.5 mm long, ascending, narrowly oblong. Petals (4.5–)6.0–8.0(–10.0) mm long, not lobed, yellow. Styles stout and beaklike, (1.0–)1.5–3.0(–3.5) mm long. Fruits ascending, straight, 2.5–4.5 cm long, more than 10 times as long as wide, linear, somewhat flattened parallel to the septum, not beaked except for the persistent style, each valve with a midnerve, dehiscing longitudinally. Seeds in 2 rows in each locule, 0.9–1.3 mm long, oblong to nearly circular in outline, the surface smooth or nearly so, sometimes minutely roughened, orange or gray. 2n=42. May–September.

Introduced, known thus far only from the city of St. Louis (native of Europe, introduced sporadically in North America). Railroads.

18. Draba L. (whitlow wort, whitlow grass)

Plants annual or biennial (perennial or woody elsewhere), terrestrial, variously pubescent with unbranched to branched and/or stellate hairs. Stems erect or ascending. Leaves basal and/or alternate, sessile or short-petiolate, not clasping (clasping elsewhere), simple, the margins entire or shallowly toothed. Inflorescences racemes, few-branched panicles, rarely appearing nearly umbellate, the flowers not subtended by bracts (with bracts elsewhere). Sepals lanceolate to elliptic or ovate, erect or spreading, green. Petals shallowly to deeply 2-lobed at the tip or unlobed and rounded, white (yellow, lavender, or purple elsewhere), rarely absent. Styles absent or less than 0.3 mm long (to 17.0 mm elsewhere). Fruits ascending, 2–5 times (to more than 40 times elsewhere) as long as wide, flattened parallel to the septum (circular in cross-section elsewhere), dehiscent longitudinally. Seeds in 2 rows in each locule, 0.5–1.5 mm long (longer elsewhere), ovate to broadly elliptic in outline, the margins not winged, the surface finely pebbled or warty, sometimes faintly reticulate, yellow to light orange. About 350 species, nearly worldwide, except for the Australian region, mostly temperate to arctic.

Draba is the largest genus in the family Brassicaceae. Rollins (1993) included more than 100 species from North America, most of these from the western and northern portions of the continent.

1. Petals deeply 2-lobed at the tip; leaves all in a basal rosette, the stems leafless
 .... 5. D. VERNA
1. Petals unlobed or at most only slightly notched at the tip, rarely absent; stems with leaves, these sometimes few and mostly basal (basal rosettes often also present)
    2. Stem leaves more or less evenly distributed along the stem; fruits 2.5–6.0 mm long
        3. Fruits hairy; stems 1(–3) per plant; lateral branches of the inflorescence frequently short and densely flowered, the flowers sometimes appearing fascicled .... 1. D. APRICA
        3. Fruits glabrous; stems usually several per plant; branches of the inflorescence all elongate, the flowers scattered along the axes
 .... 2. D. BRACHYCARPA
    2. Stem leaves mostly near the stem base, the stems mostly leafless; fruits 5–22 mm long

4. Leaf margins noticeably toothed; axis of the inflorescence hairy ... 3. D. CUNEIFOLIA
4. Leaf margins entire; axis of the inflorescence glabrous .... 4. D. REPTANS

1. Draba aprica Beadle
D. brachycarpa Nutt. ex Torr. & A. Gray var. fastigiata Nutt. ex Torr. & A. Gray

Pl. 320 a, b; Map 1344

Plants annual. Stems usually 1(–3) per plant, unbranched below the inflorescence, 7–40 cm long, hairy. Leaves basal and also more or less evenly distributed along the stems, 0.4–2.0 cm long, sessile or the basal leaves often short-petiolate, linear to obovate or broadly elliptic, the margins entire or less commonly few-toothed, hairy on both surfaces. Inflorescences usually several-branched panicles, the branches often short and densely flowered. Flowers dense, those of the lateral branches often appearing fascicled. Sepals 0.8–1.5 mm long. Petals 2–3 mm long (sometimes absent), white, rounded or very slightly indented at the tip. Styles absent or to 0.2 mm long. Fruits 3–6 mm long, linear to narrowly elliptic in outline, hairy. Seeds 4–8 per fruit. $2n=16, 24$. April–May.

Uncommon and local in the southwestern and eastern Ozarks (Georgia to Missouri and Oklahoma). Banks of streams, openings of mesic to dry upland forests, rock outcrops and ledges of shut-ins, glades, on both calcareous and acidic substrates; also roadsides.

Draba aprica was once thought to be merely a variety of the more widespread *D. brachycarpa*. In addition to the characters in the key to species above, *D. aprica* may also be distinguished by its fewer (4–8 vs. 8–15), larger seeds (1.0–1.5 mm vs. 0.5–0.8 mm long), and its generally larger fruits (3–6 mm vs. 2.5–4.5 mm long). It also begins to bloom about the time that *D. brachycarpa* finishes flowering. The species sometimes grow together, but no intermediates have been found at these sites.

This species is quite local in its distribution, although at least in eastern Missouri it has proven to be somewhat more common than was thought initially. In southwestern Missouri, it is associated mostly with calcareous glades and rock outcrops, but in the eastern Ozarks it is associated primarily with igneous glades, rock outcrops, and shut-ins of the St. Francois Mountains, less commonly on sandstone or chert. In this region, plants also tend to colonize the gravel piles left by graders at the edges of unpaved roads.

2. Draba brachycarpa Nutt. ex Torr. & A. Gray
(whitlow grass, shortpod draba)

Pl. 320 g, h; Map 1345

Plants annual. Stems usually several per plant, unbranched below the inflorescence, 4–20 cm long, hairy. Leaves basal and also more or less evenly distributed along the stems, 0.3–2.0 cm long, sessile or the basal leaves sometimes short-petiolate, linear to obovate, the margins entire, densely hairy on both surfaces. Inflorescences few-branched panicles or less commonly racemes. Flowers dense, but scattered along the inflorescence axis. Sepals 0.5–2.0 mm long. Petals 2–3 mm long (sometimes absent), white, rounded or very slightly indented at the tip. Styles absent or to 0.1 mm long. Fruits 2.5–4.5 mm long, narrowly elliptic to lanceolate or narrowly oblong in outline, glabrous. Seeds 8–15 per fruit. $2n=16, 24$. February–April.

Common nearly throughout Missouri, less commonly in the northern third of the state (southeastern U.S. north to Virginia and Indiana, west to Kansas and Texas). Mesic to dry upland forests, glades, prairies, and banks of streams and rivers, frequently on acidic substrates; also crop fields, fallow fields, pastures, railroads, roadsides, and open, disturbed areas.

For a discussion of the separation of this species from *D. aprica*, see the treatment of that species.

3. Draba cuneifolia Nutt. ex Torr. & A. Gray
var. cuneifolia (whitlow grass, wedgeleaf draba)

Pl. 320 i, j; Map 1346

Plants annual. Stems 1 per plant, usually few-branched from near the base, 5–30 cm long, hairy from the base to the tip of the inflorescence axis. Leaves in a basal rosette and also a few to several toward the bases of the stem branches, 0.5–5.0 cm long, sessile or nearly so, spatulate to oblanceolate, the margins noticeably few-toothed, especially above the middle, densely hairy on both surfaces. Inflorescences unbranched racemes or rarely few-branched panicles. Flowers almost always scattered along the inflorescences. Sepals 1–2 mm long. Petals 2–5 mm long (less commonly absent), white, rounded or more usually with a broad, shallow notch at the tip. Styles absent or to 0.1 mm long. Fruits 6–15 mm long, linear to narrowly oblong in outline, hairy. Seeds 20–80 per fruit. $2n=32$. February–May.

Scattered to common, mostly south of the Missouri River (southern U.S. north to Ohio and Nevada, Mexico). Ledges and tops of bluffs, glades, and openings of mesic to dry upland forests, mostly on calcareous substrates, but also on other substrates; also old fields and roadsides.

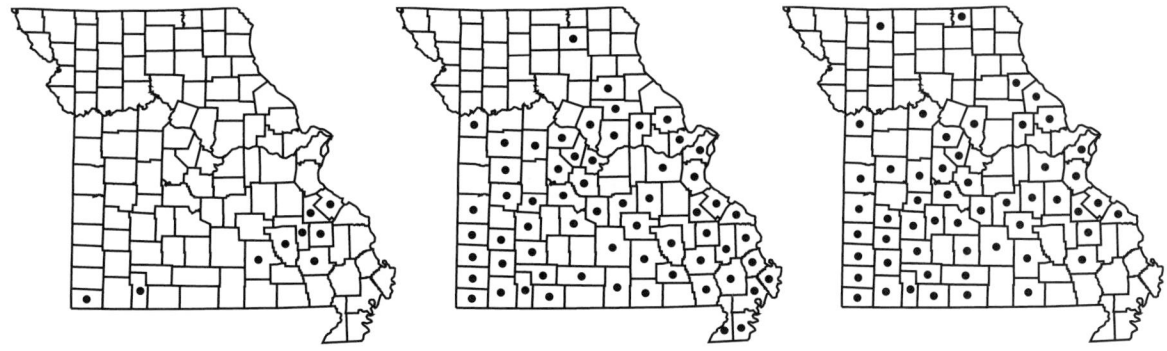

1344. Draba aprica   1345. Draba brachycarpa   1346. Draba cuneifolia

The two other varieties of this species are restricted to the southwestern United States and adjacent Mexico and differ in the size and type of pubescence of their fruits. Those flowers lacking petals in this species are cleistogamous; they produce fruits through a process of self-fertilization without opening (Al-Shehbaz, 1987).

4. Draba reptans (Lam.) Fernald (white whitlow wort)
   D. reptans var. *micrantha* (Nutt. ex Torr. & A. Gray) Fernald
   D. reptans var. *stellifera* (O.E. Schulz) C.L. Hitchc.

Pl. 320 e, f; Map 1347

Plants annual. Stems 1 or few per plant, unbranched or few-branched from near the base, (2–)4–15 cm long, hairy toward the base, the inflorescence axis glabrous. Leaves in a basal rosette and also a few to several toward the bases of the stems, 0.5–3.0 cm long, sessile or nearly so, narrowly elliptic to obovate, the margins entire, densely hairy on both surfaces. Inflorescences unbranched racemes or rarely few-branched panicles, sometimes appearing subumbellate. Flowers dense near the tips of the inflorescences or branches. Sepals 1–2 mm long. Petals 2–5 mm long or less commonly absent, white, usually rounded at the tip. Styles absent or to 0.1 mm long. Fruits 5–22 mm long, linear to narrowly oblong in outline, glabrous or hairy. Seeds 15–80 per fruit. $2n=16, 30, 32$. February–May.

Scattered nearly throughout Missouri (U.S. and adjacent Canada). Glades, upland prairies, and openings of mesic to dry upland forests, on both calcareous and acidic substrates; also in fallow fields, pastures, railroads, roadsides, and open, disturbed areas.

Those flowers lacking petals in this species are cleistogamous; they produce fruits through a process of self-fertilization without opening (Al-Shehbaz, 1987).

5. Draba verna L. (vernal whitlow grass)
   D. verna var. *boerhaavii* H. Hall
   Erophila verna (L.) Chev.
   Erophila verna ssp. *praecox* (Steven) Walters

Pl. 320 c, d; Map 1348

Plants annual. Stems usually several per plant, 5–20 cm long, hairy near the base, glabrous along the inflorescence axis. Leaves all in a basal rosette, 1–2 cm long, sessile, spatulate to oblanceolate, the margins entire or few-toothed, glabrous or sparsely hairy on the upper surface, hairy on the undersurface. Inflorescences unbranched racemes. Flowers scattered along the inflorescences. Sepals 1.0–2.0(–2.5) mm long. Petals (1.5–)2.0–4.5(–6) mm long, white, deeply 2-lobed from the tip to about the middle. Styles absent or to 0.1 mm long. Fruits (2.5–)4.0–9.0(–12.0) mm long, elliptic to oblong in outline, glabrous. Seeds 40–60 per fruit. $2n=14, 16, 24, 30, 32, 34, 36, 38, 40, 52, 54, 58, 60, 64$. February–April.

Introduced, common, mostly in eastern and central Missouri (native of Europe and Asia, widely introduced in the New World). Lawns, crop fields, fallow fields, pastures, roadsides, and disturbed areas.

A large number of segregates and infraspecific taxa have been named; these apparently maintain themselves primarily through high levels of inbreeding. There are no correlations between the slight, individual characters that define these various segregates, and there is no useful purpose in recognizing them taxonomically. There are over 200 names associated with this single species in the European literature (Al-Shehbaz, 1987).

Draba verna is one of the earliest species to flower in the spring. It was once thought to cure whitlow, an inflammation of horses' hooves, hence the common name.

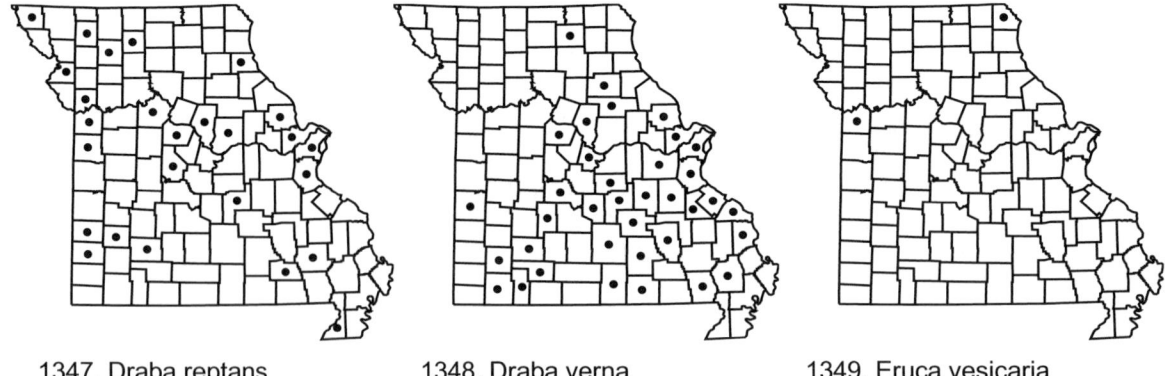

1347. Draba reptans   1348. Draba verna   1349. Eruca vesicaria

19. Eruca Mill.

One species, Europe, Asia, Africa

1. Eruca vesicaria (L.) Cav. **ssp. sativa** (Mill.)
Thell. (garden rocket, arugula)
E. sativa Mill.
Brassica eruca L.

Pl. 319 h, i; Map 1349

Plants annual, terrestrial. Stems (10–)20–90(–100) cm long, erect to spreading, usually few-branched from the base, sparsely pubescent with unbranched hairs. Leaves alternate and sometimes basal, (2–)4–15(–20) cm long, the lower leaves petiolate, the upper leaves sessile, not clasping, the leaf blades elliptic to oblanceolate in outline, irregularly pinnately lobed or divided with a large, terminal segment and 4–10 smaller, lateral segments, the upper leaves sometimes merely toothed, glabrous. Inflorescences panicles, the lower branches subtended by reduced leaves. Sepals (6–)7–10(–12) mm long, erect, narrowly oblong, becoming detached as the flower opens. Petals (12–)15–20(–26) mm long, not lobed, light yellow to nearly white with dark purple to brown veins. Styles 2–5 mm long at flowering, becoming elongated into a beak as the fruits mature, the stigma lobes decurrent along the upper portion of the style. Fruits ascending, straight, (11–)15–40 mm long, 5–8 times as long as wide, transversely segmented into 2 distinctly dissimilar parts, the lower portion narrowly ovoid to oblong, circular in cross-section, containing the seeds, each valve with a prominent midnerve, dehiscing longitudinally, the upper portion beaklike, (4–)5–10(–11) mm long, somewhat shorter than the lower portion, narrowly triangular, strongly flattened, seedless, indehiscent. Seeds in 2 rows in each locule, 1.7–1.9(–2.5) mm in diameter, broadly ovoid to globose, the surface with a fine, netlike or honeycomb-like pattern of ridges and pits, orange or grayish brown. $2n=22$. May–October.

Introduced, uncommon and sporadic in Missouri (native of Europe, Asia, widely introduced in North America). Railroads.

20. Erucastrum C. Presl

About 20 species, Europe, Asia, Africa.

1. Erucastrum gallicum (Willd.) O.E. Schulz
(dog mustard, rocketweed)
Sisymbrium gallicum Willd.

Pl. 319 d, e; Map 1350

Plants annual or biennial, terrestrial. Stems 20–75 cm long, erect, branched from the base, pubescent with unbranched hairs. Leaves alternate and basal, 3–15(–25) cm long, the upper leaves progressively reduced, petiolate, not clasping, the leaf blades oblong to oblanceolate, 1 or 2 times pinnately divided, the divisions usually irregularly toothed, pubescent with unbranched hairs. Inflorescences panicles, the branches and most of the flowers subtended by reduced, leaflike bracts. Sepals 2–4(–5) mm long, ascending, narrowly oblong. Petals 4–7 mm long, not lobed, pale yellow. Styles 1.5–3.0 mm

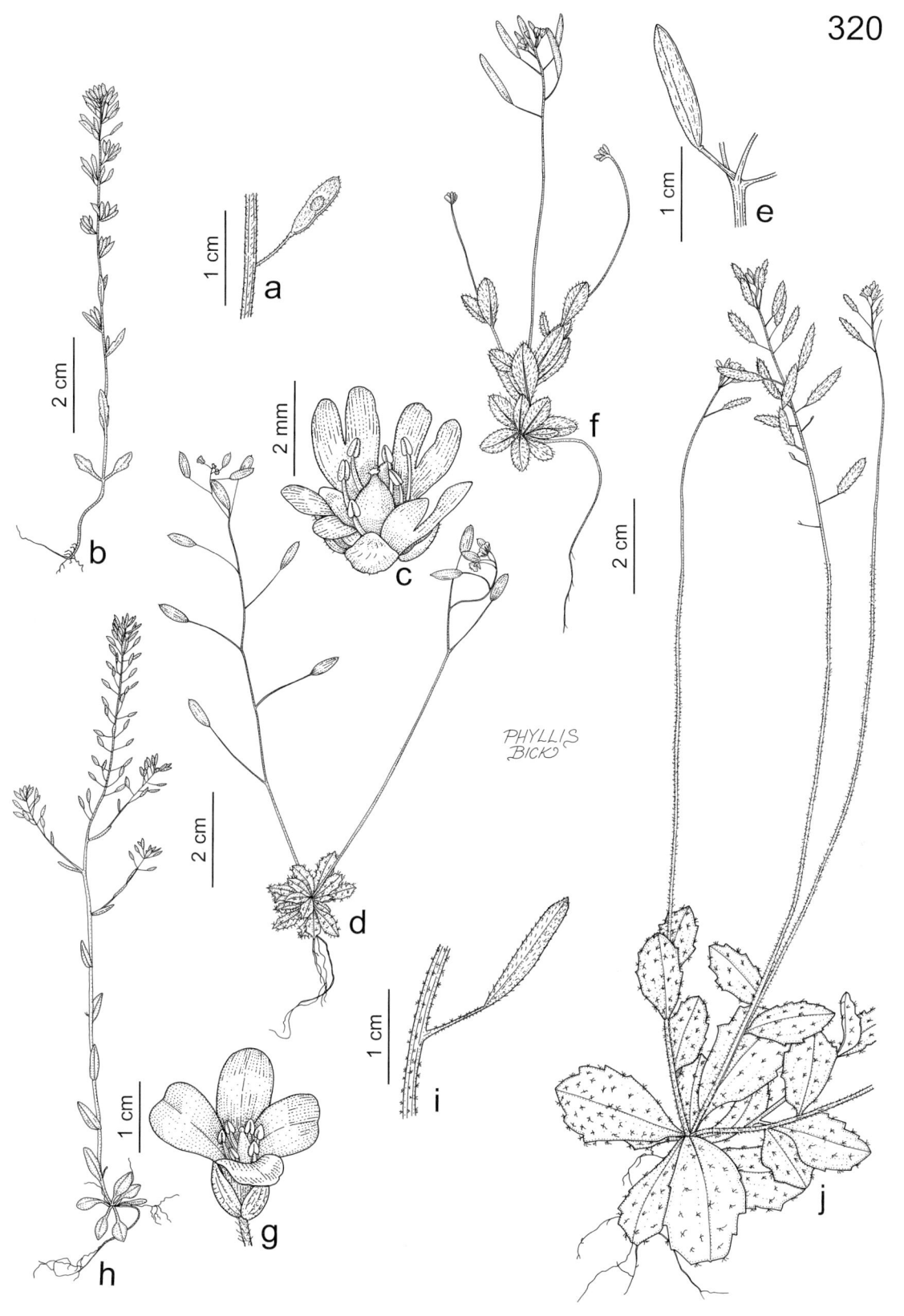

Plate 320. Brassicaceae. *Draba aprica*, **a)** fruit, **b)** habit. *Draba verna*, **c)** flower, **d)** habit. *Draba reptans*, **e)** fruit, **f)** habit. *Draba brachycarpa*, **g)** flower, **h)** habit. *Draba cuneifolia* **i)** fruit, **j)** habit.

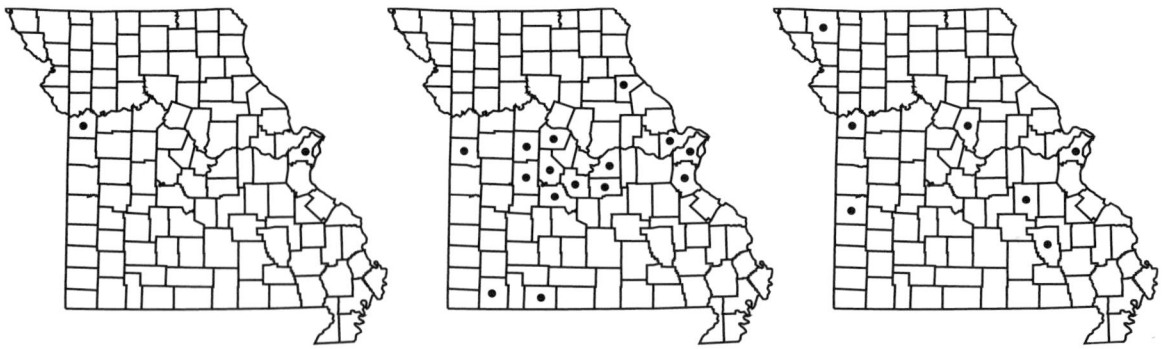

1350. Erucastrum gallicum   1351. Erysimum capitatum   1352. Erysimum cheiranthoides

long. Fruits ascending, straight, 2–4 cm long, more than 10 times as long as wide, linear, somewhat 4-angled in cross-section, not beaked except for the persistent style, each valve with a raised, sometimes winglike midnerve, dehiscing longitudinally. Seeds in 1 row in each locule, 1.2–1.3 mm long, oblong in outline, not winged, the surface with a fine, netlike or honeycomb-like pattern of ridges and pits, reddish orange. $2n=30$. May–September.

Introduced, Jackson County and the city of St. Louis (native of Europe, Asia, widely introduced in North America). Ditches and railroads.

Plants of *Erucastrum* bear a strong resemblance to some *Brassica* species but differ in several subtle morphological features of the inflorescences, flowers, and fruits (Al-Shehbaz, 1985; Rollins, 1993), in addition to those presented in the key to genera above.

The spread of *E. gallicum* in North America was documented by Luken et al. (1993). The species was first documented in the United States from collections made along railroads in Wisconsin in 1903, and it was first collected in Missouri in 1918. It is presently known from at least 29 states and all of the Canadian provinces and territories. However, because of heavy herbicide use along railroads during the past few decades, it probably no longer occurs at many of its former sites.

21. Erysimum L. (wallflower)

Plants annual, biennial, or perennial herbs (shrubs elsewhere), terrestrial, pubescent with sessile, 2- or more branched hairs, the hairs often positioned flat along the stem or leaves and with 2 opposite branches, thus appearing as a straight line. Stems erect or ascending, unbranched below the inflorescence or several-branched. Leaves basal and alternate, sessile or short-petiolate, not clasping, simple, the margins entire, wavy, toothed, or pinnately lobed. Inflorescences racemes or panicles, the flowers not subtended by bracts. Sepals linear to narrowly oblong, erect, green, hairy. Petals unlobed, light yellow to yellow or orange (white, pink, or purple elsewhere). Styles variously absent or to 5 mm long (to 12 mm elsewhere), the stigma more or less 2-lobed. Fruits erect, ascending, or horizontally spreading, more than 10 times as long as wide, linear in outline, circular or 4-angled in cross-section, dehiscent longitudinally. Seeds in 1 row in each locule, the surface with a fine, netlike or honeycomb-like pattern of ridges and pits, sometimes also finely roughened, yellowish orange to reddish brown. About 150 species, North America, Europe, Asia, Africa.

1. Petals (12–)15–30 mm long, bright yellow to deep orange; seeds (1.5–)2.0–3.0 (–3.4) mm long . . . . 1. E. CAPITATUM
1. Petals 3.5–10.0 mm long, light to bright yellow; seeds 1.0–1.7(–2.0) mm long
    2. Fruit stalk 2–4 mm long, about as thick as fruit . . . . 4. E. REPANDUM
    2. Fruit stalk 5–13 mm long, considerably narrower than fruit

3. Sepals 2.0–3.5 mm long; petals 3–5 mm long; fruit valves pubescent on the inside
.... 2. E. CHEIRANTHOIDES
3. Sepals 4–7 mm long; petals 6–10 mm long; fruit valves glabrous on the inside
.... 3. E. INCONSPICUUM

1. Erysimum capitatum (Douglas ex Hook.) Greene **var. capitatum** (western wallflower)

Pl. 321 c, d; Map 1351

Plants biennial or perennial herbs. Stems 1 or few per plant, (10–)20–90(–100) cm long, usually unbranched below the inflorescence, pubescent with mostly 2-branched hairs. Leaves 2–12(–27) cm long, narrowly elliptic to linear-lanceolate, the margins entire or shallowly and broadly toothed, pubescent with 2- and 3-branched hairs. Inflorescences mostly racemes. Sepals 7–15 mm long. Petals (12–)15–30 mm long, bright yellow to deep orange. Styles 0.5–5.0 mm long. Fruits erect or ascending, 3–9(–12) cm long, usually somewhat 4-angled in cross-section, rarely slightly flattened toward the tip, pubescent with 2–5-branched hairs, the stalks relatively stout and about as wide as or slightly narrower than the fruits. Seeds (1.5–)2.0–3.0(–3.4) mm long, oblong-elliptic in outline, somewhat flattened, sometimes with a narrow wing at the tip. $2n=36$. April–July.

Scattered in a broad band through the central portion of the state, uncommon farther south, mostly associated with major river drainages (western U.S. east to Ohio and Tennessee; Mexico). Calcareous bluffs and glades, less commonly road cuts.

Erysimum capitatum is a complex, morphologically variable species that has been divided into several varieties, all but one of which occur to the west of Missouri. The other varieties differ in a number of characters relating to pubescence type and density, fruit shape, seed morphology, and petal size and color (Rollins, 1993). Reports in some floras of the related *E. asperum* (Nutt.) DC. from Missouri are based upon misdetermined specimens of *E. capitatum*.

This showy species is also cultivated as a garden ornamental, and it is possible that some of the few collections from roadside embankments represent escapes.

2. Erysimum cheiranthoides L. (wormseed mustard, wormseed wallflower)

Pl. 321 e, f; Map 1352

Plants annual or biennial. Stems 1 or few per plant, 15–75(–100) cm long, usually unbranched or less commonly few-branched in the inflorescence, pubescent with 2-branched hairs. Leaves (1–)2–9(–11) cm long, linear to lanceolate or oblanceolate, the margins usually entire or less commonly shallowly and broadly toothed, pubescent with mostly 3-branched hairs. Inflorescences mostly racemes. Sepals 2.0–3.5 mm long. Petals 3–5 mm long, light yellow to yellow. Styles 0.5–1.0(–1.5) mm long. Fruits erect or ascending, (1.0–)1.5–2.5(–4.0) cm long, circular to somewhat 4-angled in cross-section, the valves pubescent on the inside and outside with mostly 3-branched hairs, the stalks slender and noticeably narrower than the fruits. Seeds 0.8–1.1(–1.5) mm long, oblong-elliptic in outline, somewhat flattened, often pointed at the tip, not winged. $2n=16$. May–September.

Introduced, uncommon and widely scattered, mostly in central and northwestern Missouri (native of Europe, Asia, and possibly western North America, widely introduced in the U.S.). Roadsides, railroads, and open, disturbed areas.

Rollins (1993) has noted the controversy over whether this species is indigenous in some parts of Alaska, Canada, and the Rocky Mountain states. The application of subspecific names to North American plants also is problematic. The other subspecies, ssp. *altum* Ahti, originally was designated for plants from far northern Europe and Asia that are taller than plants of ssp. *cheiranthoides* and have more internodes, as well as long-tapered leaves appressed to the stems. Rollins (1993) pointed out that patterns of north-to-south variation in this country approximate those in the Old World, however, and that the American plants are not easily subdivided into discrete taxa. Most authors treat all North American materials as belonging to ssp. *cheiranthoides*.

3. Erysimum inconspicuum (S. Watson) MacMill. (smallflower wallflower)

Pl. 321 a, b; Map 1353

Plants mostly perennial herbs. Stems 1 or few per plant, 20–60 cm long, usually unbranched below the inflorescence, pubescent with 2-branched hairs. Leaves 2–8 cm long, linear to narrowly oblanceolate, the margins usually entire or less commonly shallowly and broadly toothed, pubescent with 2- and 3-branched hairs. Inflorescences racemes or less commonly panicles. Sepals 4–7 mm long. Petals 6–10 mm long, light yellow to yellow. Styles 1–2 mm long. Fruits erect or ascending,

1353. Erysimum inconspicuum **1354. Erysimum repandum** **1355. Hesperis matronalis**

(2–)3–5 cm long, circular to slightly 4-angled in cross-section, pubescent with 2-branched hairs, the stalks slender and noticeably narrower than the fruits. Seeds (1.0–)1.5–1.7(–2.0) mm long, oblong in outline, somewhat flattened, not winged. $2n=54$. May–June.

Introduced, uncommon and widely scattered in Missouri (Alaska and Canada south to Nevada and Michigan, adventive farther south and east). Roadsides, railroads, and open, disturbed areas.

The var. *coarcticum* (Fernald) Rossbach of eastern Canada supposedly differs from var. *inconspicuum* in its denser infructescences, shorter styles, and slightly broader fruits, but these varieties do not appear to be worthy of formal taxonomic recognition.

4. Erysimum repandum L. (bushy wallflower, treacle mustard)

Pl. 321 i–k; Map 1354

Plants mostly perennial herbs. Stems 1 per plant, (4–)7–50(–70) cm long, usually unbranched below the inflorescence, pubescent with 2- and 3-branched hairs. Leaves (1–)2–8(–10) cm long, linear to narrowly oblanceolate, the margins usually entire or the lower leaves commonly shallowly and broadly toothed, rarely pinnately lobed, pubescent with 2- and 3-branched hairs. Inflorescences mostly panicles, the branches often spreading. Sepals 4–6 mm long. Petals 6–8(–10) mm long, light yellow to yellow. Styles 0.5–2.0 mm long. Fruits loosely ascending or more commonly spreading horizontally, (2–)3–8(–12) cm long, usually noticeably 4-angled in cross-section, pubescent with 2-branched hairs, the stalks 2–4 mm long, stout and about as wide as the fruits. Seeds 0.9–1.1(–1.5) mm long, oblong-elliptic in outline, somewhat flattened, not winged. $2n=14, 16$. March–June.

Introduced, scattered to common nearly throughout Missouri, but apparently absent from most of the southwestern portion of the Ozark Division and the Unglaciated Plains (native of Europe, widely naturalized in North America). Roadsides, railroads, fallow fields, open margins of ponds, and open, disturbed areas.

Where present, this species sometimes occurs in large masses that turn roadsides and other areas into blankets of pale yellow during the flowering season. In the early spring, young plants tend to have more erect branches than the dense, spherical inflorescences of spreading branches found in older plants later in the season.

22. Hesperis L.

About 25 species, Europe, Asia, Africa.

1. Hesperis matronalis L. (dame's rocket)

Pl. 322 f, g; Map 1355

Plants biennial or perennial herbs, terrestrial. Stems 50–120 cm long, erect, branched from the base and usually also in the upper half, pubescent with 2-branched and unbranched hairs, sometimes also with very sparse glands. Leaves alternate and basal, (2–)4–20 cm long, the lower leaves petiolate, the upper ones often sessile, not clasping, the leaf blades lanceolate to ovate-lanceolate, the margins toothed, pubescent on the upper surface with unbranched hairs and on the undersurface mostly with 2-branched hairs. Inflorescences panicles, the lower branches subtended by reduced leaves.

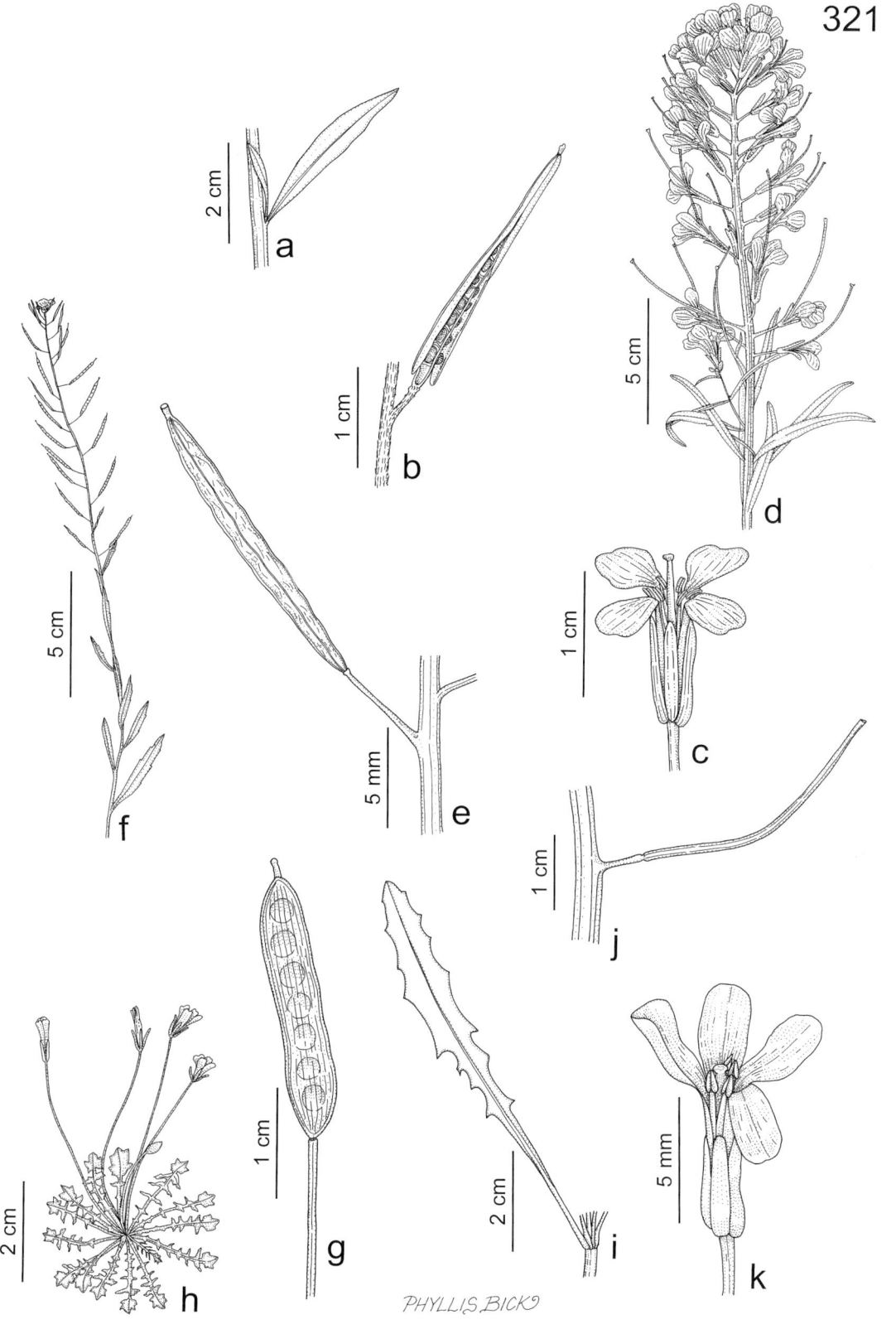

Plate 321. Brassicaceae. *Erysimum inconspicuum*, **a**) leaf, **b**) dehiscing fruit. *Erysimum capitatum*, **c**) flower, **d**) habit. *Erysimum cheiranthoides*, **e**) fruit, **f**) habit. *Leavenworthia uniflora*, **g**) fruit, **h**) habit. *Erysimum repandum*, **i**) leaf, **j**) fruit, **k**) flower.

1356. Iodanthus pinnatifidus   1357. Isatis tinctoria   1358. Leavenworthia torulosa

Sepals 5–8(–10) mm long, ascending, oblanceolate. Petals (11–)15–25 mm long, not lobed, pink to purple, rarely white. Styles 3–4 mm long. Stigma lobes decurrent. Fruits ascending, straight to slightly arched upward, (4–)6–10(–14) cm long, more than 10 times as long as wide, linear, circular in cross-section, not beaked except for the persistent style, each valve with a midnerve and sometimes 2 lateral, longitudinal nerves, dehiscing longitudinally. Seeds in 1 row in each locule, 2.1–3.0 (–4.0) mm long, oblong-elliptic in outline, somewhat flattened, usually with narrow wings at both ends, the surface with a fine, netlike or honeycomb-like pattern of ridges and pits, sometimes roughened, reddish brown. $2n$=12, 14, 16, 24, 26, 28, 32, but counts other than $2n$=24 may be erroneous (Al-Shehbaz, 1988b). May–June.

Introduced, widely scattered in the state (native of Europe, introduced widely in North America, mostly east of the Rocky Mountains). Railroads, roadsides, and open, disturbed areas.

This attractive plant has a long history of cultivation for its fragrant, showy flowers. In recent years, dense colonies have become more common along roadsides in the state. Plants with the petals nearly white occasionally grow intermixed with pink- or purple-flowered plants.

23. Iodanthus Torr. & A. Gray ex Steud.
(Rollins, 1942)

One species, U.S.

1. Iodanthus pinnatifidus (Michx.) Steud.
(purple rocket, violet rocket)
Pl. 322 c–e; Map 1356

Plants perennial herbs, terrestrial. Stems 30–80 cm long, erect, usually unbranched, densely to sparsely pubescent with unbranched hairs. Leaves alternate and sometimes basal, 4–15 cm long, the lower leaves petiolate, the upper ones often sessile, sometimes clasping the stems with rounded auricles, the leaf blades lanceolate to ovate-lanceolate, the lower leaves frequently pinnately lobed or divided, the margins toothed, glabrous. Inflorescences racemes or rarely few-branched panicles, the lower branches subtended by reduced leaves. Sepals 4–6 mm long, ascending, oblong. Petals 8–14 mm long, not lobed, pale pink to light purple. Styles 1.5–2.0 mm long. Fruits ascending to spreading, straight to slightly arched upward, 2–4 cm long, more than 10 times as long as wide, linear, circular in cross-section or nearly so, not beaked except for the persistent style, each valve with an indistinct midnerve, usually covered with minute, transparent papillae, dehiscing longitudinally. Seeds in 1 row in each locule, 1.0–1.5 mm long, oblong in outline, somewhat flattened, narrowly winged at both ends, the surface with a netlike or honeycomb-like pattern of ridges and pits, reddish brown. May–June.

Widely scattered in the state, but absent from much of the Ozark Division and the western portion of the Glaciated Plains (Pennsylvania to Alabama west to Minnesota, Kansas, and Texas). Bottomland forests and banks of streams and rivers.

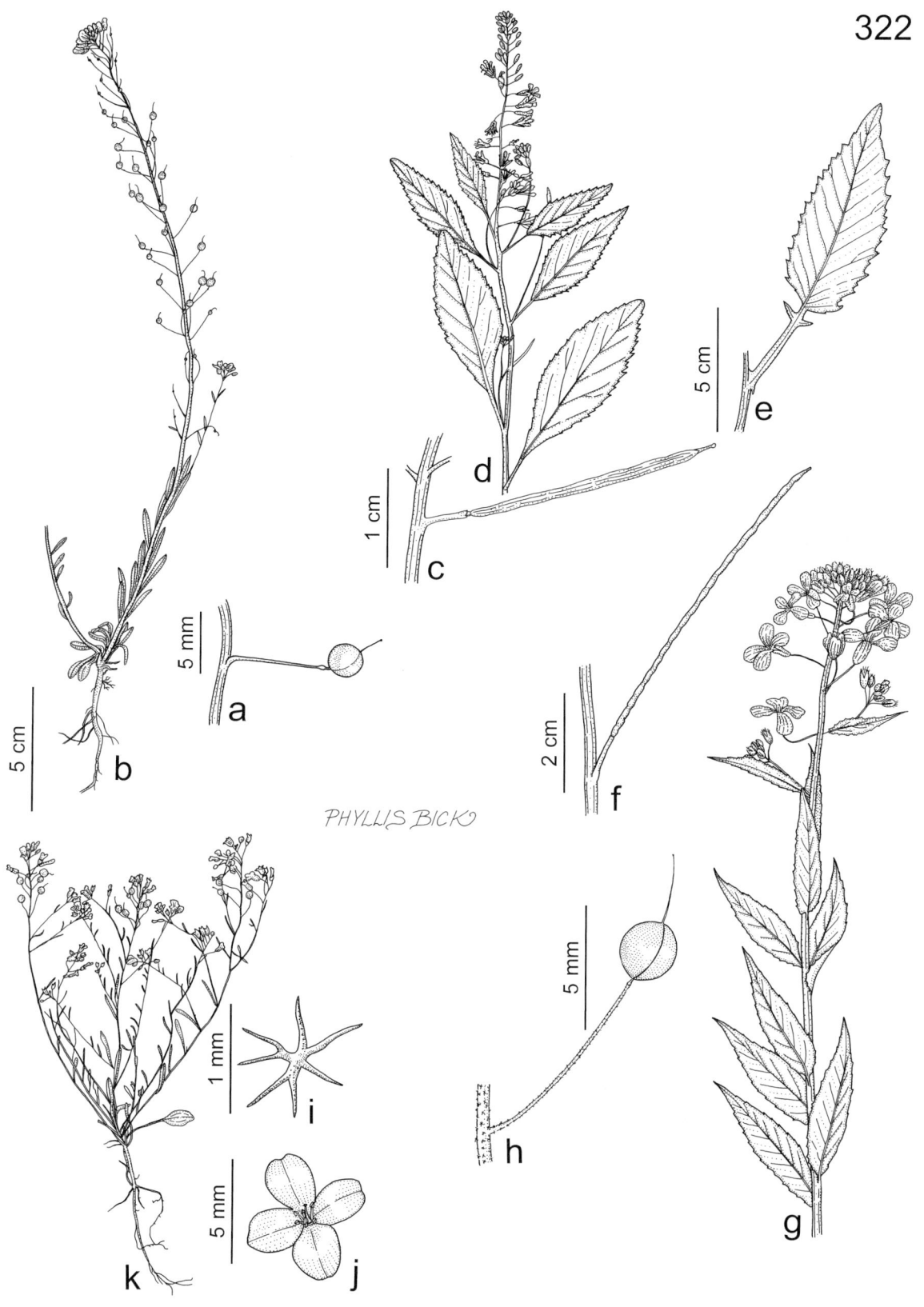

Plate 322. Brassicaceae. *Physaria gracilis*, **a)** fruit, **b)** habit. *Iodanthus pinnatifidus*, **c)** fruit, **d)** habit, **e)** leaf. *Hesperis matronalis*, **f)** fruit, **g)** habit. *Physaria filiformis*, **h)** fruit, **i)** stellate hair, **j)** flower (top view), **k)** habit.

24. Isatis L.

About 50 species, Europe, Asia, Africa.

1. Isatis tinctoria L. (woad)

Map 1357

Plants perennial herbs, terrestrial. Stems (30–)40–100(–150) cm long, erect, usually unbranched at the base, many-branched above, glabrous and often glaucous, sometimes sparsely pubescent with unbranched hairs toward the stem base. Leaves mostly 2–15 cm long, the basal and lower stem leaves short-petiolate, the median and upper leaves sessile, clasping the stems, with rounded or acute auricles. Leaf blades oblong or oblanceolate, unlobed, the margin entire or toothed. Inflorescences often several-branched panicles, the lower branches subtended by reduced leaves. Sepals 1.5–2.8 mm long, oblong. Petals 2.5–4.0 mm long, not lobed, yellow. Styles absent. Fruits oblong to oblanceolate or sometimes elliptic-obovate, indehiscent, winged all around, (9–)11–20(–27) mm long, about 3 times as long as wide or less, the seed-bearing portion thickened but somewhat flattened, with a distinct midnerve, the replum absent at maturity, the stalks slender, reflexed, thickened and club-shaped at the tip. Seed 1 per fruit, 2.3–3.5(–4.5) mm long, oblong, light brown. $2n=28$. April–June.

Introduced, known thus far only from Jackson County (native of Europe, introduced widely in North America, most abundantly in the western U.S.). Gardens.

This species was used traditionally as the source of woad, a blue dye extracted from the fermented plants. It can become a persistent weed in gardens.

25. Leavenworthia Torr.
(Rollins, 1963)

Plants annual, terrestrial, glabrous. Leaves all or nearly all basal and forming rosettes, petiolate, the blades pinnately lobed, the margins otherwise entire to minutely toothed or wavy, the terminal lobe distinctly or slightly larger than the 2–10 pairs of lateral lobes. Inflorescences usually of solitary, long-stalked flowers arising from the rosette leaves, rarely a short, reduced, few-flowered raceme with long-stalked flowers. Sepals narrowly oblong, spreading or erect. Petals unlobed or distinctly notched at the apex, white, yellow, or lavender toward the tip and yellow to orange toward the base. Styles 1.5–5.0 mm long, unlobed. Fruits linear or narrowly oblong (elliptic or nearly circular in outline elsewhere), more than 10 times (1 or 2 times elsewhere) as long as wide, flattened (circular in cross-section elsewhere), somewhat fleshy, dehiscent longitudinally. Seeds in 1 row in each locule, circular, flattened, broadly winged all around or nearly wingless, the surface with a prominent honeycomb-like pattern, reddish brown. Eight species, southeastern U.S.

1. Fruits constricted along the replum between the seeds; petals 3.5–6.0 mm wide, deeply notched at the tip; terminal leaf lobe considerably larger than the lateral lobes; seeds nearly wingless . . . . 1. L. TORULOSA
1. Fruit not constricted between the seeds; petals 2.5–3.5 mm wide, not notched at the tip; terminal leaf lobe only slightly larger than the lateral lobes; seeds broadly winged . . . . 2. L. UNIFLORA

1. Leavenworthia torulosa A. Gray (necklace gladecress)

Map 1358

Plants annual. Stems absent. Leaves in a basal rosette, (2–)3–8(–9) cm long, petiolate, the leaf blades oblanceolate in outline, pinnately divided with 12–18 lateral lobes and 1 considerably larger, terminal lobe, these toothed or lobed. Inflorescences of single, long-stalked flowers arising from the rosette leaves. Sepals 3.5–5.5 mm long, nearly erect, oblong, greenish or lavender. Petals 6–10 mm long, 3.5–6.0 mm wide, deeply notched at the tip, white, lavender, or yellow. Styles 2.5–5.0 mm long. Fruits somewhat fleshy, erect, 1.5–3.0 cm long, 5–8 times as long as wide, linear, somewhat flattened parallel to the septum, constricted along the replum be-

1359. Leavenworthia uniflora   1360. Lepidium appelianum   1361. Lepidium campestre

tween the seeds. Seeds in 1 row in each locule, 2.5–3.5 mm in diameter, circular in outline, flattened, the margin nearly wingless, the surface with a netlike or honeycomb-like pattern of ridges and pits, dark brown. $2n=30$. March–April.

Uncommon, known thus far only from St. Louis County (Kentucky, Tennessee, Missouri). Habitat unknown but probably limestone or dolomite glades.

A single historical specimen at the Missouri Botanical Garden Herbarium labeled as having been made by George Letterman at Allenton in 1882 consists of three plants of *L. uniflora* mounted on the same sheet with a single small specimen of *L. torulosa*. *Lesquerella torulosa* was once considered for listing under the federal Endangered Species Act because of its rarity and the perceived risk of extinction, but it subsequently was judged still to be too widespread in Kentucky and Tennessee to warrant listing. The plant labeled as having originated from Missouri represents a significant disjunction in the overall range of the species and perhaps was the result of some sort of mixup when the specimens were mounted.

2. Leavenworthia uniflora (Michx.) Britton
(Michaux's gladecress, leavenworthia)
Pl. 321 g, h; Map 1359

Plants annual. Stems absent or rarely present but highly reduced. Leaves in a basal rosette, 2–10 cm long, petiolate, the leaf blades oblanceolate in outline, pinnately divided with 4–18 lateral lobes and 1 slightly larger, terminal lobe, the lobes toothed or lobed. Inflorescences usually of single, long-stalked flowers arising from the rosette leaves, rarely a short, reduced, few-flowered raceme of long-stalked flowers. Sepals 3.5–5.0 mm long, ascending, oblong, often turning purple with age. Petals 5–7 mm long, 2.5–3.5 mm wide, rounded at the tip, white. Styles 1.5–3.0 mm long. Fruits somewhat fleshy, erect, 1.5–3.0 cm long, 5–8 times as long as wide, linear, somewhat flattened parallel to the septum, not constricted between the seeds. Seeds in 1 row in each locule, 3–4 mm in diameter, circular in outline, flattened, the margin broadly winged, the surface with a netlike or honeycomb-like pattern of ridges and pits, dark brown. $2n=30$. March–April.

Scattered in the Ozark and Ozark Border Divisions, locally north to Montgomery and St. Louis Counties (Alabama to Arkansas north to Ohio, Indiana, and Missouri). Limestone and dolomite glades.

This diminutive species is one of the smallest members of the Brassicaceae in Missouri. It is referred to as a winter annual by some botanists, because of its life cycle. Seeds germinate in the fall and the plants overwinter as rosettes, flowering early the following spring and disappearing soon after the fruits mature in late spring. *Leavenworthia uniflora* is the most widespread species in a genus noted for its narrow endemics. This is in part because plants of this species are self-fertile. The populations in Arkansas and Missouri are somewhat disjunct from the remainder of the species range.

26. Lepidium L. (pepper grass)

Plants annual, biennial, or perennial herbs (shrubby elsewhere), sometimes with rhizomes but the roots all slender and almost always nonfleshy, terrestrial, glabrous or pubescent with unbranched hairs. Stems mostly erect or ascending, unbranched below the inflorescence or several-branched from the base. Leaves alternate and often also basal, sessile or short-petiolate, not clasping or rarely clasping, sometimes perfoliate, simple or pinnately lobed or divided, the margins entire or toothed. Inflorescences panicles or less commonly racemes, the flowers not subtended by bracts. Sepals erect to spreading, green. Petals 1–6 mm long, unlobed, white, less commonly yellow, sometimes absent. Stamens 2 or 6, less commonly 4. Styles absent or very short (rarely as long as the fruit elsewhere). Fruits ascending or spreading, 1–2 times as long as wide, circular to ovate or obovate to oblong in outline, flattened at a right angle to the septum or circular (4-angled elsewhere) in cross-section, the margins entire, but the tip frequently winged and with a notch, dehiscent or indehiscent, the valves thin and smooth or thick and with warts. Seeds 1 per locule. About 220 species, nearly worldwide.

Recent molecular studies, summarized in Mummenhoff et al. (2001) and Al-Shehbaz et al. (2002), have shown that *Cardaria* and *Coronopus* are nested within *Lepidium* and that the various species groups of *Coronopus* are less closely related to one another than to other species of *Lepidium*. The broad generic concept of *Lepidium* adopted in this flora basically follows the same circumscription recognized in several other floristic works in progress, especially the forthcoming volume in the Flora of North America series.

The seeds of *Lepidium* species have been used as a flavoring for meat, salads, and soups. They also are relished by seed-eating pet birds, such as canaries. A number of medicinal and herbal uses have been attributed to the genus, but most are probably not efficacious (Al-Shehbaz, 1986).

1. At least the upper stem leaves clasping the stem with auricles or perfoliate; stamens 6
    2. Corollas yellow; leaves glabrous or nearly so, the basal and lower stem leaves 2 or 3 times deeply pinnately divided, the median and upper leaves perfoliate; seeds narrowly winged all around .... 8. L. PERFOLIATUM
    2. Corollas white; leaves densely hairy, never pinnately divided, the upper leaves clasping the stem with auricles; seeds wingless
        3. Plants annual; fruits broadly winged at the tip, dehiscent, the valves with minute papillae .... 2. L. CAMPESTRE
        3. Plants perennial with rhizomes; fruits not winged, indehiscent, the valves without papillae
            4. Fruits densely or rarely sparsely short-hairy, globose . . 1. L. APPELIANUM
            4. Fruits glabrous, obovate to ovate in outline, flattened or circular in cross-section
                5. Fruit flattened, the base cordate, the valves with a relatively prominent, fine, netlike or honeycomb-like pattern of ridges and pits .... 6. L. DRABA
                5. Fruit usually inflated, the base rounded or obtuse to truncate, the valves with an inconspicuous, very fine, netlike or honeycomb-like pattern of ridges and pits or nearly smooth ... 3. L. CHALEPENSE
1. Stem leaves not clasping the stem, lacking auricles; stamens 2 (6 in *L. latifolium*)
    6. Plants perennial with long-creeping rhizomes; stamens 6 ..... 7. L. LATIFOLIUM
    6. Plants annual or biennial with taproots; stamens 2
        7. Fruit not winged, notched at the tip and base, the 2 halves wrinkled or with a network of ridges and pits, not releasing the seeds, the two lobes eventually dispersed intact individually .... 5. L. DIDYMUM
        7. Fruits winged and notched only at the tip, smooth, the 2 halves readily dehiscing to release the seeds

8. Petals present and often much longer than the sepals; fruits circular; axis of the inflorescence pubescent with curved hairs .... 10. L. VIRGINICUM
8. Petals absent or present but shorter than the sepals; fruits broadly obovate or elliptic; axis of the inflorescence with minute, straight hairs (these often appearing as papillae)
   9. Fruits broadly obovate, widest above the middle; fruit stalks usually hairy only on the upper surface; basal leaves toothed or 1 time pinnately divided
   .... 4. L. DENSIFLORUM
   9. Fruits elliptic, widest at the middle; fruit stalks pubescent all around; basal leaves 2 or 3 times pinnately divided .... 9. L. RUDERALE

1. Lepidium appelianum Al-Shehbaz (globe-podded hoary cress, white top)
Cardaria pubescens (C.A. Mey.) Jarm.
Hymenophysa pubescens C.A. Mey., not *Lepidium pubescens* Desv.
C. pubescens var. *elongata* Rollins
Pl. 318 d, e; Map 1360

Plants perennial herbs, with long-creeping, branched rhizomes, sometimes forming dense colonies. Stems (10–)15–35 cm long, erect or ascending, branched in the upper third, densely short-hairy. Leaves densely short-hairy, the lower leaves petiolate, 2–8 cm long, obovate to oblanceolate, the median and upper leaves sessile, clasping the stem with rounded auricles, 1–5(–8) cm long, oblong to oblanceolate, irregularly toothed or nearly entire. Inflorescences flat-topped panicles or racemes, not elongating as the fruits mature. Sepals 1.5–2.0 mm long, oblong, short-hairy. Petals white, 2.5–4.0 mm long. Stamens 6. Styles 0.7–1.5 mm long. Fruits (2–)3–4(–5) mm in diameter, globose or nearly so, rounded or slightly cordate at the base, rounded at the tip, the valves densely or rarely sparsely short-hairy, not winged, obscurely veined or not veined, the stalk 3–6(–10) mm long. Seeds 1.5–2.0 mm long, not winged, the surface with a fine, netlike or honeycomb-like pattern of ridges and pits, brown or dark brown. $2n=16$. May–September.

Introduced, known thus far only from the city of St. Louis (native of Asia, widely introduced in the U.S. and Canada). Railroads.

The hoary cresses (*L. appelianum, L. chalepense,* and *L. draba*) are considered noxious weeds in some western states, where they are weeds of crop fields and pastures and also invasive in natural plant communities.

2. Lepidium campestre (L.) R. Br. (field cress, cow cress, field pepper grass)
Thlaspi campestre L.
Pl. 323 c, d; Map 1361

Plants annual. Stems (8–)12–50(–60) cm long, erect, usually unbranched below the inflorescence, densely pubescent with short, spreading hairs. Leaves (1–)2–7(–8) cm long, sessile, clasping the stem with pointed auricles, the blades narrowly oblong to oblanceolate in outline, the margins entire to shallowly toothed (the basal leaves less commonly 1 time pinnately lobed), densely short-hairy. Sepals (1.0–)1.3–1.8 mm long, narrowly elliptic. Petals (1.5–)1.8–2.5(–3.0) mm long, white. Stamens 6. Styles 0.1–0.5 mm long. Fruits (4–)5–6 mm long, ovate to broadly oblong in outline, the tip shallowly notched and relatively broadly winged, flattened, the valves with minute papillae. Seeds 2.0–2.3 (–2.8) mm long, obovate in outline, not winged, the surface minutely papillate, reddish gray to dark brown or black. $2n=16$. April–June.

Introduced, scattered nearly throughout Missouri (native of Europe and Asia, widely naturalized in the U.S. and Canada). Roadsides, railroads, fallow fields, pastures, and open, disturbed areas.

Superficially, this species resembles the species of *Thlaspi* found in Missouri. However, in addition to having only 1 seed per locule (vs. 3–8), *L. campestre* is densely hairy, whereas species of *Thlaspi* are glabrous.

3. Lepidium chalepense L. (lens-podded hoary cress)
Cardaria chalepensis (L.) Hand.-Mazz.
Cardaria draba (L.) Desv. ssp. *chalepensis* (L.) O.E. Schulz
Pl. 318 c; Map 1362

Plants perennial herbs, with long-creeping, branched rhizomes, often forming dense colonies. Stems (10–)12–55(–85) cm long, erect or ascending, branched in the upper third, densely short-hairy, sometimes nearly glabrous toward the tip. Leaves densely short-hairy, the lower leaves petiolate, 1.5–7.5 cm long, obovate to oblanceolate, the median and upper leaves sessile, clasping the stem with rounded auricles, 1.5–8.5(–12.5) cm long, ovate or oblong to oblanceolate, irregularly toothed or nearly entire. Inflorescences flat-topped panicles or racemes, not elongating as the fruits mature.

1362. Lepidium chalepense   1363. Lepidium densiflorum   1364. Lepidium didymum

Sepals 1.5–2.2 mm long, oblong, glabrous or nearly so. Petals white, 2.5–4.5 mm long. Stamens 6. Styles (0.7–)1.0–1.8(–2.0) mm long. Fruits (2.5–)3.0–5.5(–6.5) mm in diameter, obovate or ovate in outline, not or only slightly flattened, sometimes slightly grooved along the replum, rounded to truncate at the base, the valves glabrous or rarely sparsely and minutely hairy when young, not winged, obscurely veined or not veined, the fruit stalk 5–10(–15) mm long. Seeds 1.5–2.0 mm long, not winged, the surface with an inconspicuous, very fine, netlike or honeycomb-like pattern of ridges and pits or nearly smooth, dark brown. $2n=48, 80, 128$. April–July.

Introduced, sporadic (native of Europe, Asia, widely introduced in the U.S. and Canada). Railroads, roadsides, and open, disturbed areas.

The fruits of *L. chalepense* are inflated or only slightly flattened compared to the somewhat more flattened fruits of *L. draba*. They also exhibit greater size variation. The hoary cresses (*L. appelianum*, *L. chalepense*, and *L. draba*) are considered noxious weeds in some western states, where they are weeds of crop fields and pastures and also invasive in natural plant communities.

4. Lepidium densiflorum Schrad. (pepper grass, green-flowered pepper grass)

Pl. 323 e, f; Map 1363

Plants annual or biennial. Stems 10–50(–65) cm long, erect or ascending, these and rachis of inflorescence pubescent with minute, straight, spreading hairs with rounded tips (visible with magnification, appearing as papillae). Leaves (1.0–)2.5–8.0(–11.0) cm long, linear to oblanceolate or elliptic in outline, the bases not clasping, glabrous or less commonly minutely hairy on the undersurface, the lower and basal leaves often 1 time pinnately lobed, the margins entire to coarsely toothed, the upper leaves reduced, linear, the margins usually entire. Sepals 0.5–0.8(–1.0) mm long, linear to narrowly elliptic. Petals absent or less than 1 mm long, white. Stamens 2. Styles 0.1–0.2 mm long. Fruits (2.0–)2.5–3.0(–3.5) mm long, broadly obovate, widest above the middle, the tip shallowly notched and sometimes narrowly winged, strongly flattened, glabrous, the stalks pubescent on the upper surface. Seeds 1.1–1.4(–1.5) mm long, narrowly obovate to elliptic in outline, wingless or obscurely winged around most of the margin, the surface minutely roughened, light orange. $2n=32$. April–November.

Scattered nearly throughout Missouri (Canada, U.S., widely introduced in Europe, Asia). Glades, bluff tops, prairies, rocky openings of dry upland forests, pastures, crop fields, fallow fields, old fields, roadsides, railroads, and open, disturbed areas.

Lepidium densiflorum and *L. virginicum* are closely related and can be difficult to distinguish from each other. Some botanists have resorted to obscure characters of the embryos to separate them. Some authors have suggested that part of the problem may be because of hybridization between them, but this has not been documented cytologically (Al-Shehbaz, 1986). A few historical specimens from Jackson County may represent such hybrids. Morphological variation within the species also may account for some of the confusion. Several varieties have been named that are said to differ in position and density of the hairs and in other minor, mostly quantitative characters of the fruits. The Missouri plants have always been placed into var. *densiflorum*, but the varieties seem unworthy of taxonomic recognition.

5. Lepidium didymum L. (wart cress, swine cress)

Coronopus didymus (L.) Sm.
Carara didyma (L.) Britton

Pl. 318 h, i; Map 1364

Plants annual or rarely biennial, terrestrial,

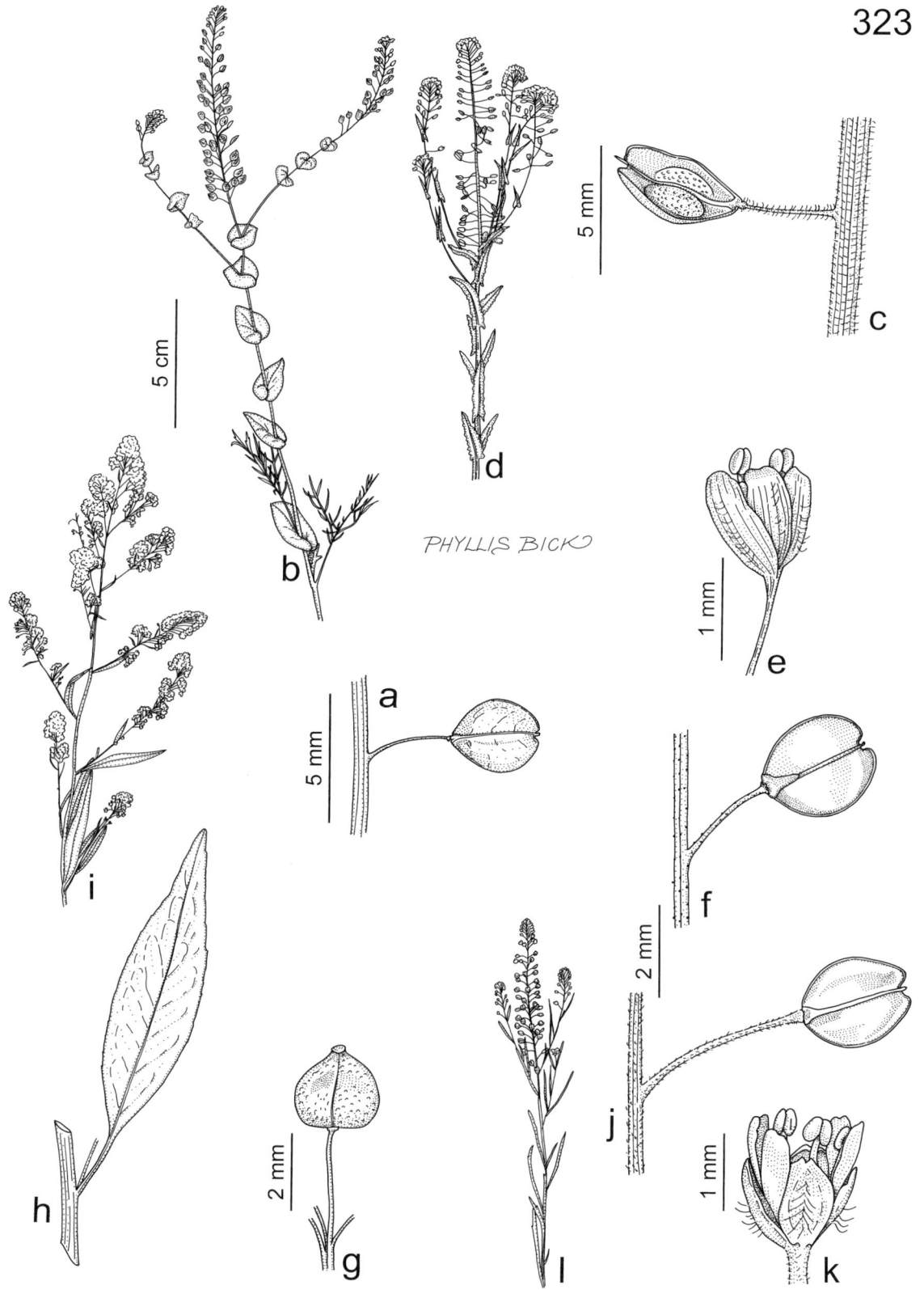

Plate 323. Brassicaceae. *Lepidum perfoliatum*, **a)** fruit, **b)** habit. *Lepidum campestre*, **c)** fruit, **d)** habit. *Lepidum densiflorum*, **e)** flower, **f)** fruit. *Lepidum latifolium*, **g)** fruit, **h)** leaf, **i)** habit. *Lepidum virginicum*, **j)** fruit, **k)** flower, **l)** habit.

1365. Lepidium draba   1366. Lepidium latifolium   1367. Lepidium perfoliatum

sparsely to moderately pubescent with short, straight hairs, less commonly glabrous, usually with a somewhat fetid odor when bruised or crushed. Stems 10–50(–70) cm long, spreading to ascending or erect, branched, sometimes forming mats. Leaves alternate and basal, (1.5–)3.0–10.0 cm long, short-petiolate, not clasping, the blades oblong to ovate in outline, deeply 1 time pinnately divided, the divisions often toothed or pinnately lobed. Inflorescences short, dense racemes, terminal and lateral in the leaf axils (or sometimes opposite the leaves). Sepals 0.5–1.0 mm long, spreading, elliptic to oval. Petals 0.4–0.5 mm long, inconspicuous, not lobed, white. Stamens 2. Styles absent. Fruits spreading, 1.5–2.0 mm long, about as long as wide or wider, with 2 lateral, kidney-shaped or oblong to circular and slightly flattened lobes, notched at the tip and base, the surfaces wrinkled or with a network of ridges and pits, the 2 lobes breaking apart at maturity and shed intact, but not releasing the seeds. Seeds 1 per locule (lobe), 1.0–1.2 mm long, kidney-shaped, appearing coiled, the surface with an indistinct, fine, netlike or honeycomb-like pattern of ridges and pits, light yellow or light brown. $2n=32$. April–October.

Introduced, uncommon, known thus far only from the city of St. Louis (probably native to South America, but introduced nearly worldwide). Sidewalk cracks, lawns, railroads, roadsides, and open, disturbed areas.

The native range of this species is not well understood. Al-Shehbaz (1986) noted that although some botanists consider it a Eurasian native, its closest relatives grow in South America. This worldwide weed is sometimes eaten as a salad green or used as a wound disinfectant. When eaten by cattle, the fetid-smelling plants impart an unpleasant flavor to the milk.

6. Lepidium draba L. (heart-podded hoary cress, heart-podded white top)
Cardaria draba (L.) Desv.

Pl. 318 a, b; Map 1365

Plants perennial herbs, with long-creeping, branched rhizomes, often forming dense colonies. Stems (8–)20–65(–90) cm long, erect or ascending, branched in the upper third, densely pubescent, sometimes the upper portion becoming nearly glabrous at maturity. Lowermost leaves petiolate, (1.5–)3.0–10.0(–15.0) cm long, obovate to oblanceolate, the middle and upper leaves sessile, clasping the stem with rounded auricles, densely pubescent, ovate or oblong to oblanceolate, (1–)2–8(–10) cm long, dentate or nearly entire. Inflorescences flat-topped panicles or racemes, not elongated in fruit. Sepals 1.5–2.2 mm long, oblong. Petals white, (2.5–)3.0–4.0(–5.0) mm long. Stamens 6. Styles (0.8–)1.0–1.8(–2.0) mm long. Fruits (2.5–)3.0–5.0 mm long, cordate or ovate, strongly indented or grooved at the replum, rounded or obtuse to truncate at the base, the valves not inflated, flattened, glabrous, keeled, not winged, prominently veined, the fruit stalk 4.5–11.0(–14.5) mm long. Seeds 1.5–2.0 mm long, not winged, the surface with a relatively prominent but fine, netlike or honeycomb-like pattern of ridges and pits, dark brown. $2n=32$, 64. April–July.

Introduced, sporadic (native of Europe, Asia, widely introduced in the U.S. and Canada). Roadsides, railroads, and open, disturbed areas.

Lepidium draba and *L. chalepense* differ only in fruit size and shape. Partially fertile hybrids have been said to occur in Canada, but no experimental documentation has been published to date. The two are often treated as subspecies of a single species. The hoary cresses (*L. appelianum*, *L. chalepense*, and *L. draba*) are considered noxious weeds in some western states, where they are weeds of crop fields

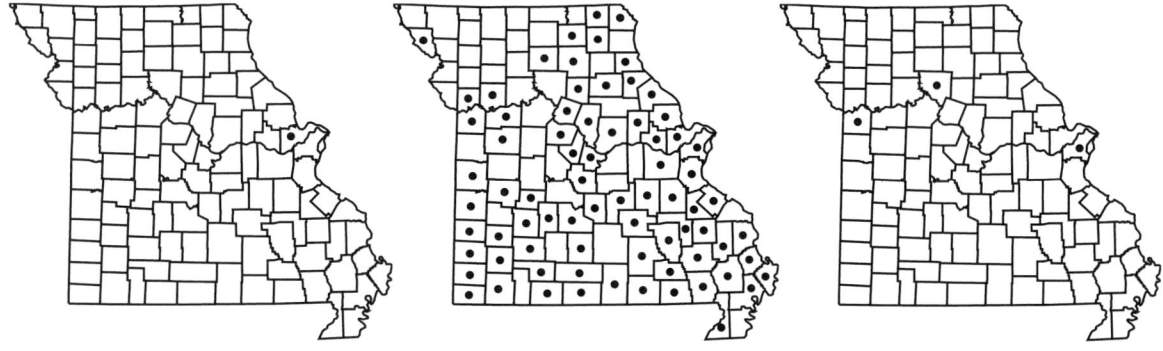

1368. Lepidium ruderale   1369. Lepidium virginicum   1370. Lobularia maritima

and pastures and also invasive in natural plant communities.

7. Lepidium latifolium L. (perennial pepper grass, broad-leaved pepper grass)

Pl. 323 g–i; Map 1366

Plants perennial herbs, with rhizomes. Stems (20–)35–120(–150) cm long, erect or ascending, mostly unbranched below the inflorescence, glabrous or nearly so. Leaves 1–30 cm long, linear to oblanceolate or elliptic in outline, the base not clasping, glabrous, the margins entire to finely toothed, the lower and middle ones petiolate, the upper leaves progressively reduced and sessile. Sepals 0.8–1.5 mm long, broadly elliptic to ovate. Petals 1.5–2.5 mm long, white. Stamens 6. Styles absent. Fruits 1.8–2.2(–2.7) mm long, broadly ovate to nearly circular in outline, the tip rounded, not notched or winged, flattened, sparsely hairy or glabrous. Seeds (0.8–)1.0–1.3 mm long, oblong-elliptic in outline, not winged, the surface with a very fine, netlike or honeycomb-like pattern of ridges and pits, brownish orange. $2n=24, 48$. June–August.

Introduced, known only from a single collection from Jackson County (native of Europe and Asia, widely naturalized in the U.S. and adjacent Canada). Railroads.

This species was first collected in Missouri in 1975 from an area between a road and a railroad in the Kansas City area (Henderson, 1980). Its distribution in the United States is sporadic, but it may be expected to appear along other railroads in Missouri in the future.

8. Lepidium perfoliatum L. (shield cress, perfoliate pepper grass)

Pl. 323 a, b; Map 1367

Plants annual or biennial. Stems (7–)10–40(–65) cm long, erect, usually unbranched below the inflorescence, usually sparsely pubescent with short, spreading hairs toward the base, glabrous and glaucous toward the tip. Leaves of 2 types, the basal and often lower stem leaves (2–)4–13(–19) cm long, 2 or 3 times pinnately divided with linear divisions, minutely hairy, the median and upper stem leaves 0.7–3.0(–4.0) cm long, ovate to circular, the bases perfoliate or strongly clasping the stems with rounded, strongly overlapping auricles, the margins entire to minutely toothed, glabrous and glaucous. Sepals 0.8–1.0(–1.3) mm long, broadly elliptic. Petals 1.0–1.5(–1.9) mm long, yellow. Stamens 6. Styles 0.1–0.3 mm long. Fruits 3.0–4.0(–4.8) mm long, ovate to nearly circular in outline, flattened, the tip with a minute notch and short, narrow wings. Seeds 1.6–2.0(–2.3) mm long, elliptic in outline, narrowly winged all around, the surface with a fine, netlike or honeycomb-like pattern of ridges and pits, reddish or dark brown. $2n=16$. April–June.

Introduced, widely scattered in Missouri (native of Europe and Asia, widely naturalized in the western U.S. and adjacent Canada east to Ohio). Railroads and less commonly roadsides and open, disturbed areas.

9. Lepidium ruderale L. (stinking pepperweed)

Map 1368

Plants annual or biennial, pubescent with papillate hairs, usually with a fetid odor when fresh. Stems (5–)10–35(–55) cm long, erect or ascending, usually branched below the inflorescence, these and the axis of the inflorescence pubescent with minute, straight, spreading hairs with rounded tips (visible with magnification, appearing as papillae). Leaves of 2 types, the basal and lowermost stem leaves (1.5–)3.0–5.0(–7.0) cm long, 2 or 3 times pinnately divided with oblong divisions, the upper stem leaves (0.5–)1.0–2.0(–3.0) cm long, linear, sessile, the bases not auriculate, the margins usually entire, sparsely pubescent. Sepals 0.5–0.9(–1.0) mm long, oblong. Petals absent or present and 0.1–0.4 mm long, white. Stamens 2. Styles to 0.1 mm long. Fruits (1.5–)1.8–2.5(–3.0) mm long, elliptic, widest at the middle, the tip shallowly notched and relatively narrowly winged, flattened, glabrous, the stalks pubescent all around. Seeds 1.0–1.5 mm long, oblong, not winged, the surface with a minute, netlike pattern, brown. $2n=16, 32$. April–May.

Introduced, known thus far only from St. Charles County (native of Europe, introduced sporadically in the eastern and western U.S. and Canada). Open, disturbed areas.

This species was first collected in Missouri in 2002 by James Sullivan from the parking lot of a truck stop.

10. Lepidium virginicum L. (poor man's pepper grass, pepper grass, Virginia pepper grass)

Pl. 323 j–l; Map 1369

Plants annual or biennial. Stems (6–)15–55(–70) cm long, erect or ascending, pubescent especially in the upper parts with minute, curved, mostly ascending hairs with pointed tips (visible with magnification). Leaves (1.0–)2.5–10.0(–15.0) cm long, linear to oblanceolate or elliptic to obovate in outline, the bases not clasping, glabrous or less commonly minutely hairy on the undersurface, the lower and basal leaves often 1 time pinnately lobed, rarely 2 times lobed, the margins entire to coarsely toothed, petiolate, the upper leaves reduced, linear, sessile or nearly so, the margins usually entire. Sepals (0.5–)0.7–1.0(–1.1) mm long, linear to narrowly elliptic. Petals 1.5–2.5 mm long, white, rarely rudimentary. Stamens 2. Styles 0.1–0.2 mm long. Fruits 2–4 mm long, circular, widest at the middle, the tip shallowly notched and sometimes narrowly winged, flattened, glabrous. Seeds 1.1–1.7(–1.9) mm long, narrowly obovate to elliptic in outline, usually winged around most of the margin or at least at the tip, the surface minutely roughened, light orange. $2n=32$. April–November.

Common nearly throughout Missouri (Canada, U.S., widely introduced in South America, Europe, Asia). Glades, tops of bluffs, prairies, and rocky openings of dry upland forests; also pastures, crop fields, fallow fields, old fields, railroads, roadsides, and open, disturbed areas.

For a discussion of the separation of this species from the closely related *L. densiflorum,* see the treatment of that species. As in *L. densiflorum,* several varieties have been named that appear to be based mostly upon trivial characters of pubescence density and distribution, as well as slight differences in the shape and size of the fruits and their stalks. Missouri plants have always been attributed to var. *virginicum,* but the varieties seem unworthy of taxonomic recognition.

27. Lobularia Desv.

Four species, Europe, Asia, Africa (Mediterranean region).

1. Lobularia maritima (L.) Desv. (sweet alyssum)

Alyssum maritimum L.

Pl. 324 a, b; Map 1370

Plants perennial herbs, terrestrial, usually appearing grayish-tinged, uniformly pubescent with appressed hairs, each hair sessile and with 2 opposite branches, thus appearing as a straight line. Stems (5–)12–24(–40) cm long, spreading to ascending, branched. Leaves alternate, (1–)2–5 cm long, sessile or short-petiolate, not clasping, the blades linear to narrowly oblanceolate, the margins entire. Inflorescences racemes or few-branched panicles, the lower branches subtended by reduced leaves. Sepals 1.5–2.0(–2.4) mm long, ascending, oblong to ovate. Petals 2–4 mm long, not lobed, white to purple. Styles 0.5–1.0 mm long. Fruits spreading, (2.0–)2.5–3.5(–4.2) mm long, about as long as wide or less than 2 times as long as wide, elliptic in outline, somewhat flattened parallel to the septum, each valve with an obscure midnerve, sparsely hairy, dehiscing longitudinally. Seeds 1 per locule, 1.0–1.5(–2.0) mm long, circular to broadly elliptic in outline, flattened, the margin wingless or narrowly winged, the surface with minute papillae, orange. $2n=24$. April–October.

Introduced, uncommon and sporadic (native of the western Mediterranean region, cultivated worldwide, and widely introduced in North America). Railroads and disturbed areas.

This species is perhaps the most widely cultivated of the ornamental crucifers (Al-Shehbaz, 1987), and its small, white to purple flowers are quite fragrant. It has escaped from cultivation and become naturalized at a number of sites in the eastern United States, and it is to be expected elsewhere in Missouri.

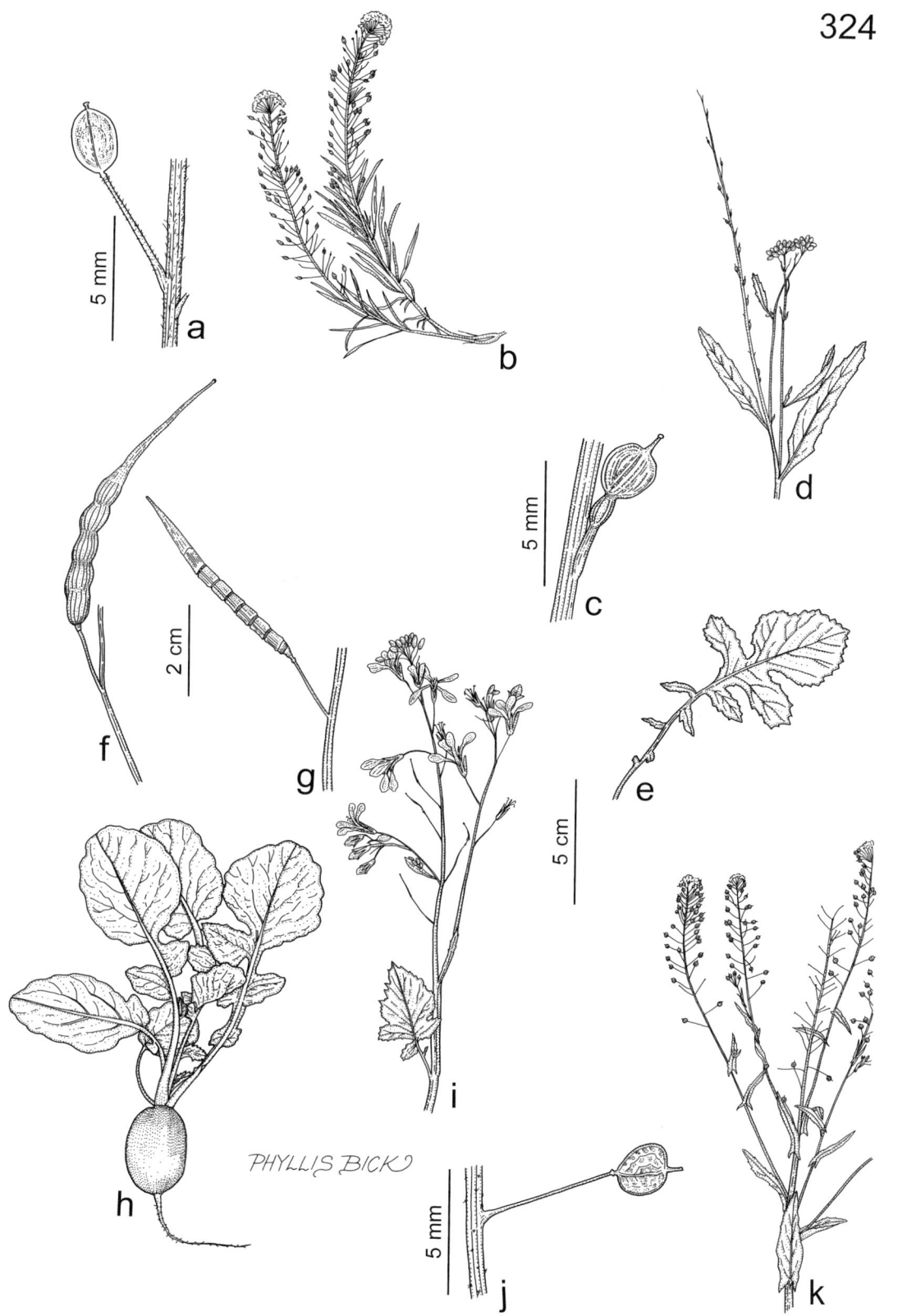

Plate 324. Brassicaceae. *Lobularia maritima*, **a)** fruit, **b)** habit. *Rapistrum rugosum*, **c)** fruit, **d)** inflorescence, **e)** lower leaf. *Raphanus raphanistrum*, **f)** fruit. *Raphanus sativus*, **g)** fruit, **h)** base, **i)** inflorescence. *Neslia paniculata*, **j)** fruit, **k)** inflorescence.

1371. Lunaria annua 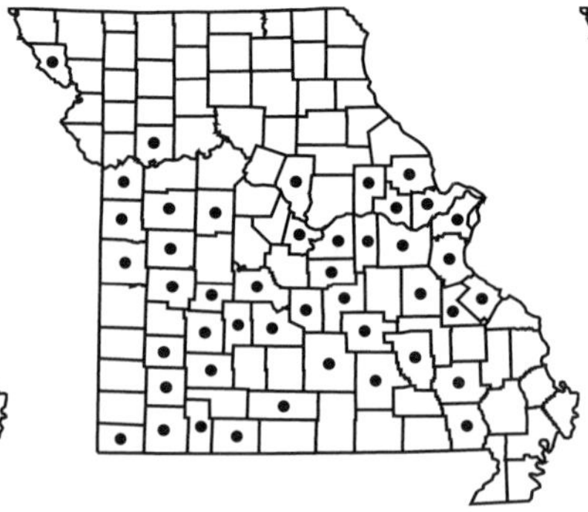   1372. Microthlaspi perfoliatum 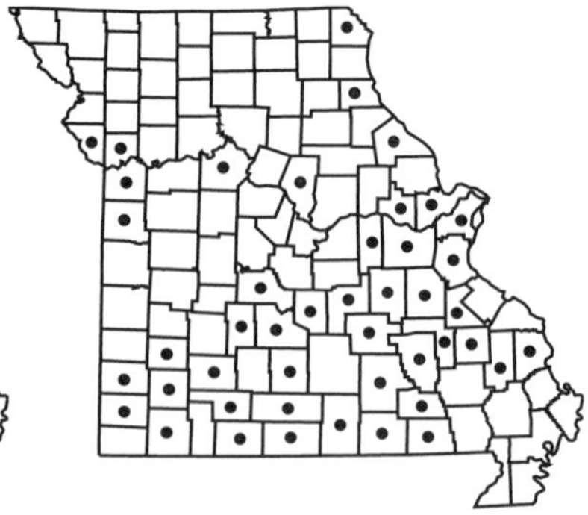   1373. Nasturtium officinale

28. Lunaria L.

Three species, Europe.

1. Lunaria annua L. (honesty, money-plant, silver dollar)

Map 1371

Plants biennial or rarely annual, terrestrial, pubescent with unbranched hairs. Stems (20–)50–120 cm tall, erect, branched, sparsely to moderately pubescent with short, spreading hairs, the upper portion sometimes nearly glabrous. Leaves 2–15 cm long, gradually reduced in size toward the stem tip, alternate or sometimes lowermost opposite, petiolate or uppermost sometimes sessile, cordate at the base and slightly clasping the stem, the margins toothed, the surfaces sparsely hairy. Inflorescences racemes or panicles, the lower branches subtended by reduced leaves. Sepals 6–10 mm long, erect, oblong, the inner pair often pouched at the base. Petals broadly obovate, (15–)17–30 mm long, not lobed, purple to lavender. Styles 4–10 mm long. Fruits (2.0–)2.5–4.5 cm long, about as long as wide or less than 2 times as long as wide, broadly oblong to ovate, strongly flattened parallel to the septum, valves with an obscure midnerve, often glabrous, the fruit stalk above the attachment of the perianth slender and 7–18 mm long, the septum membranous, dehiscing longitudinally. Seeds in 2 rows in each locule, 4–8 per locule, 7–10 mm long, kidney-shaped, winged all around, strongly flattened, grayish brown. $2n=30$. April–June.

Introduced, known thus far only from Cape Girardeau County (native of Europe, introduced sporadically in North America). Margins and openings of mesic upland forests; also gardens and open, disturbed areas.

This species is widely cultivated as an ornamental for its attractive flowers, but especially for the attractive infructescences, which are used in dried arrangements and bouquets after the removal of the fruit valves and seeds. The seeds are sometimes used in Europe as a condiment.

29. Microthlaspi F.K. Mey.

Four species, Europe, Asia, Africa, introduced widely.

1. Microthlaspi perfoliatum (L.) F.K. Mey. (penny cress, perfoliate penny cress)
Thlaspi perfoliatum L.

Pl. 327 g, h; Map 1372

Plants annual, terrestrial, glabrous throughout, not producing an odor when bruised or crushed, usually somewhat glaucous. Stems (3–)5–35(–45) cm long. Leaves alternate and with usually few to several basal leaves at flowering, the stem leaves 1–4 cm long, sessile, clasping the stem with mostly rounded auricles, ovate to narrowly oblong, entire. Sepals (0.8–)1.0–1.5(–1.8) mm long, ovate. Petals 1.5–3.0 mm long. Style absent. Fruits 3.0–5.5(–7.0) mm long, obovate to cordate or circular, the margins broadly winged toward the tip, these narrowed toward the fruit base, the apical notch shallow and wider than deep. Seeds 2–4 per locule, (0.9–)1.3–1.5 mm long, obovate in outline, the surface light orange to tan, smooth or nearly so. 2n=14, 28, 42. March–May.

Introduced, scattered, mostly south of the Missouri River, but apparently absent from the Mississippi Lowlands Division (native of Europe; widely introduced in North America). Banks of streams, disturbed glades; also crop fields, fallow fields, pastures, lawns, railroads, roadsides, and open, disturbed areas.

Extensive molecular studies (summarized in Koch and Al-Shehbaz, 2004) support Meyer's (1973, 1979, 2003) recognition of *Microthlaspi* as a genus separate from *Thlaspi*. The latter has strongly ornamented seed coats with very different anatomy from those of *Microthlaspi*.

30. Nasturtium R. Br.

Five species, North America, Europe, Asia, northern Africa.

1. Nasturtium officinale R. Br. (watercress)
Sisymbrium nasturtium-aquaticum L.
Rorippa nasturtium-aquaticum (L.) Hayek
Nasturtium officinale R. Br. var. *siifolium* (Rchb.) W.D.J. Koch

Pl. 325 i–k; Map 1373

Plants perennial herbs, with rhizomes, usually emergent or floating aquatics, glabrous. Emergent, submerged, and/or floating stems 10–65(–200) cm long, rooting at most nodes. Leaves alternate (basal leaves absent except in seedlings), 2–10 cm long, petiolate, the bases usually clasping the stem with small, rounded auricles, pinnately compound with 3–9(–13) leaflets or less commonly simple, especially when plants occur in relatively deep water, the leaflets linear to irregularly ovate or nearly circular, the margins entire, wavy, or with few, shallow, blunt teeth. Sepals 1.5–2.5(–3.5) mm long. Petals 2.5–4.8(–6.0) mm long, white. Styles absent or 0.1–0.5 mm long. Fruits 10–15(–20) mm long, straight or slightly arched upward. Seeds mostly 20–50 per fruit, in 2 rows in each locule, 0.9–1.3 mm long, nearly circular in outline, the surface with a coarse, netlike or honeycomb-like pattern of 25–60 ridges and pits on each side, reddish orange to reddish brown. 2n=32. April–October.

Introduced, scattered, mostly in the Ozark and Ozark Border Divisions (native of Europe, Asia; widely naturalized in the U.S. and adjacent Canada). Emergent, submerged, and/or floating aquatic in spring branches and streams, less commonly terrestrial on banks of streams or stranded by a receding waterline, and occasionally fens and marshes; also ditches.

This species is a conspicuous member of the aquatic flora in spring branches and streams in the Ozarks. Plants often are encountered as large, nonflowering, submerged colonies. Pieces from these are easily dispersed by water currents, and much of the species' distribution in southern Missouri is probably due to vegetative reproduction. Steyermark (1963) suggested that watercress was native in Missouri, based upon its occurrence in relatively remote spring branches in the Ozarks. Unfortunately, these localities are somewhat less pristine than Steyermark thought, and the native range of *N. officinale* is certainly confined to the Old World. There are relatively few Missouri specimens in herbaria that were collected prior to 1890, and the labels on these indicate that the plants collected then were introduced at the collection sites. Voss (1985) discussed a similar situation in Michigan and also concluded that watercress is not native there.

Watercress is maintained as a separate genus, *Nasturtium,* based on extensive molecular data (Les, 1994; Bleeker et al., 1999, 2002; Sweeney and Price, 2000) and critical morphological comparison with *Rorippa* (Al-Shehbaz and Price, 1998), in which it has been maintained previously (Al-Shehbaz, 1988b). These studies demonstrate that *Nasturtium* is much more closely related to *Cardamine* than to *Rorippa*.

All of the Missouri specimens examined thus far are referable to the diploid (2n=32) cytotype, *N. officinale,* which is characterized by slightly broader and shorter fruits with seeds in 2 rows in each locule, as well as seeds with a coarser pattern of reticulation. The tetraploid (2n=64) cytotype is known as *N. microphyllum* Boenn. ex Rchb.; it also is widely naturalized in the United States and Canada, but it tends to have a slightly more northern distribution. It has been reported from as far south as Nebraska and Kentucky (Barker, 1986; Rollins, 1993), but because most collections do not include mature fruits, its actual distribution in this country is not known well. This taxon should be looked for in Missouri.

1374. Neslia paniculata   1375. Physaria filiformis   1376. Physaria gracilis

Watercress has a long history of use as a salad green. It currently is cultivated to a limited extent in the United States for the gourmet food market. Plants collected from the wild should be washed carefully prior to consumption to avoid accidental ingestion of microscopic parasites, such as the protozoan *Giardia*, that may be present in untreated water.

31. Neslia Desv.

Two species, Europe, Asia, Africa.

1. Neslia paniculata (L.) Desv. (ball mustard)

Pl. 324 j, k; Map 1374

Plants annual, terrestrial, pubescent with 2- to several-branched hairs. Stems 25–70(–90) cm long, erect, usually few-branched toward the tip. Leaves alternate and usually also basal, 3–6 cm long, sessile, those of the stems clasping with prominent, pointed or narrowly rounded auricles, the blades oblanceolate to lanceolate in outline, the margins entire. Inflorescences racemes at the tips of the branches, the flowers not subtended by bracts. Sepals 1.0–1.5 mm long, oblong, ascending. Petals 1.5–2.5 mm long, not lobed, light yellow. Styles 0.5–1.0 mm long. Fruits spreading, 2.0–2.5 mm long, about as long as wide or slightly longer than wide, globose to depressed-globose, each valve with a coarse, netlike or honeycomb-like pattern of ridges and pits, indehiscent. Ovules 2 per locule. Seeds usually 1 per locule, 1.9–2.2 mm long, circular to broadly obovate in outline, the margins not winged, the surface with a coarse, netlike or honeycomb-like pattern of ridges and pits and also finely roughened, tan to dark brown. $2n=14, 28$. May–July.

Introduced, uncommon, sporadic (native of Europe, sporadically introduced in the U.S. and Canada). Railroads.

32. Physaria (Nutt.) A. Gray (bladderpod)

Plants annual or biennial (perennial elsewhere), terrestrial, with sessile, stellate hairs. Stems ascending from a spreading base to erect (creeping elsewhere), usually few- to several-branched from the base. Leaves alternate and basal, the basal leaves usually with petioles, the upper leaves usually narrower and short-petiolate or sessile, the bases not clasping. Inflorescences racemes, the flowers bractless. Sepals mostly ascending. Petals unlobed, yellow (white to light purple elsewhere), without conspicuously darkened veins. Styles 2–5 mm long. Fruits with the body about as long as wide or less than 2 times as long as wide, usually spreading, globose or broadly ellipsoid to obovoid or pear-shaped, not flattened (flattened elsewhere), the slender style often persistent. Seeds in 2 rows in each locule, 1.6–2.2 mm long, circular in outline or nearly so, slightly flattened, the margins rounded, not winged, the surface minutely roughened, reddish brown (variously sized, shaped, or winged elsewhere). About 105 species, North America, South America, eastern Asia (1 species).

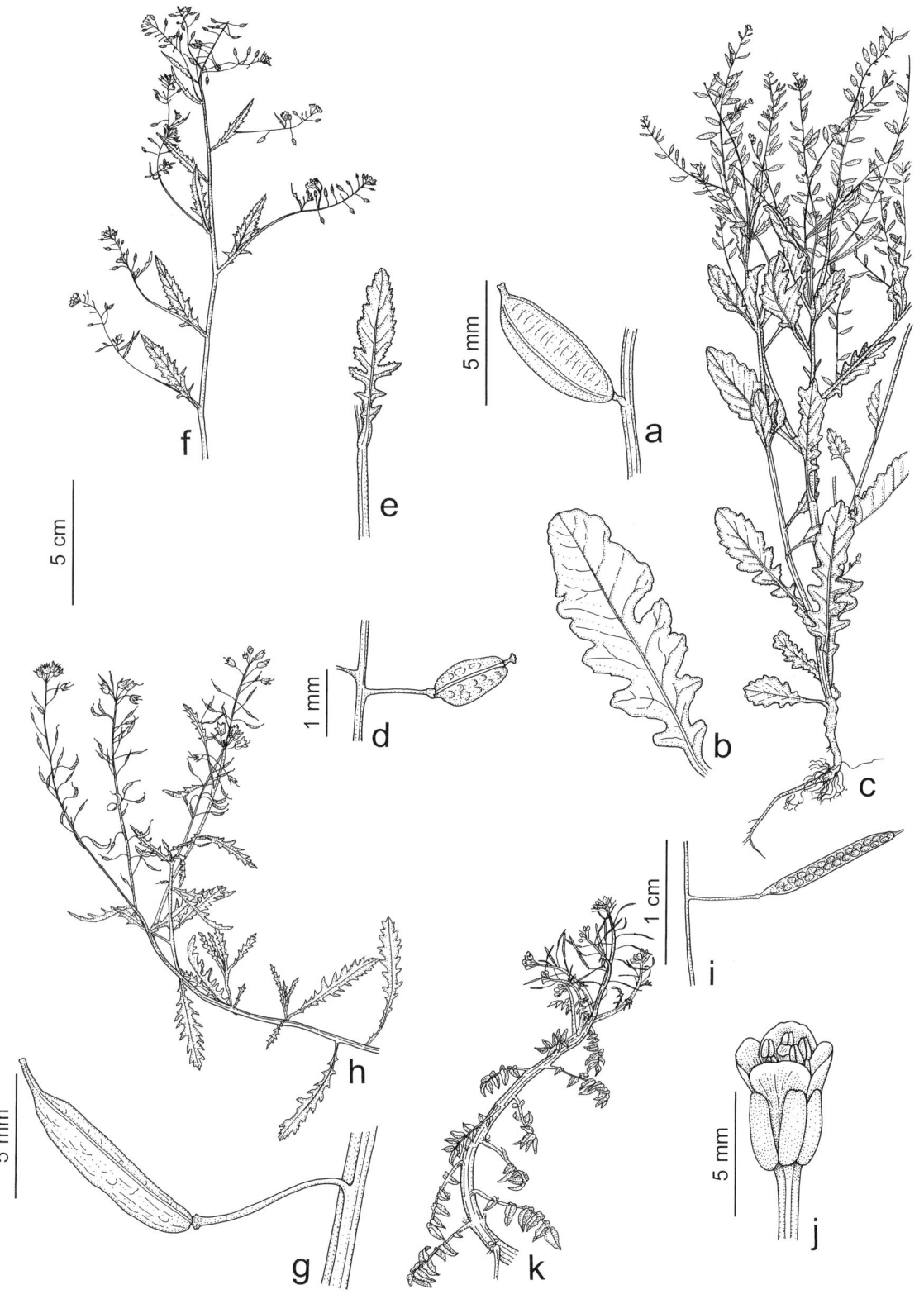

Plate 325. Brassicaceae. *Rorippa sessiliflora*, **a)** fruit, **b)** leaf, **c)** habit. *Rorippa palustris*, **d)** fruit, **e)** leaf, **f)** habit. *Rorippa sinuata*, **g)** fruit, **h)** habit. *Nasturtium officinale*, **i)** fruit, **j)** flower, **k)** habit.

Extensive molecular studies (summarized in O'Kane and Al-Shehbaz, 2002a, b) have indicated that species ascribed to *Physaria* as traditionally circumscribed evolved more than once from different lineages within *Lesquerella* S. Watson. Attempts to conserve *Lesquerella,* the much larger but more recently published genus, against *Physaria* (O'Kane et al., 1999) did not succeed, and when the two genera are united (O'Kane and Al-Shehbaz, 2002a) the resultant larger group must be called *Physaria.*

1. Plants appearing silvery gray, the hairs dense and overlapping; fruits without a stalk above the attachment point of the perianth; seeds 2 per locule
 . . . . 1. P. FILIFORMIS
1. Plants appearing grayish green to green, the hairs sparse to moderately dense, but not overlapping; fruits with a stalk 1–2 mm long above the attachment point of the perianth; seeds 4–10(–12) per locule . . . . 2. P. GRACILIS

1. Physaria filiformis (Rollins) O'Kane & Al-Shehbaz (Missouri bladderpod)
Lesquerella filiformis Rollins
Pl. 322 h–k; Map 1375

Plants annual, appearing silvery gray, the stellate hairs dense and overlapping. Stems 8–25 cm long. Basal leaves 1.0–2.4 cm long, petiolate, the leaf blades circular to spatulate, the margins entire or slightly wavy, both surfaces densely and evenly pubescent. Stem leaves 0.5–2.0 cm long, gradually reduced toward the stem tip; spatulate, oblanceolate, or linear; the lower leaves short-petiolate, the upper leaves sessile; the margins entire. Sepals 2.5–4.5 mm long, oblong or elliptic, ascending. Petals 5–9 mm long, yellow. Styles 3–5 mm long. Fruits 3–4 mm long, globose, glabrous, without a stalk above the attachment point of the perianth. Seeds 2 per locule. 2n=14. April–May.

Uncommon in the southwestern portion of the Ozark Division (Missouri, Arkansas). Limestone glades.

No specimens have been found to verify Steyermark's (1963) assertion that this species occurs in Jasper County. Steyermark and most other botanists considered this species to be endemic to Missouri. The Arkansas report by Smith (1994) was overlooked for a number of years, perhaps because no specimen or even county information was cited, until the species was rediscovered by Bill Summers in Izard County in 1997 (McKenzie, 2003). More recently, botanists at the University of Arkansas herbarium have discovered additional populations as far south as the Ouachita Mountains of west-central Arkansas.

Populations of this attractive annual undergo wide fluctuations from year to year in the number of plants present. Many of the glades in which Missouri bladderpod grows have been heavily impacted by quarrying, grazing, construction, and/or fire suppression, and although a number of populations are known to be extant, several of these are in danger of extirpation. The species originally was listed as Endangered in 1987 by the U.S. Fish and Wildlife Service under the federal Endangered Species Act, but with the discovery of additional populations growing on protected public lands, it later was reclassified as Threatened (McKenzie, 2003).

2. Physaria gracilis (Hook.) O'Kane & Al-Shehbaz **ssp. gracilis** (spreading bladderpod)
Lesquerella gracilis (Hook.) S. Watson
Pl. 322 a, b; Map 1376

Plants annual or biennial, appearing grayish green to green, the stellate hairs sparse to moderately dense, but not overlapping. Stems 10–70 cm long. Basal leaves 1.5–9.0 cm long, petiolate, the leaf blades oblanceolate to narrowly elliptic, the margins entire or more commonly coarsely toothed to pinnately divided, the undersurface more densely pubescent than the upper surface. Stem leaves 1–7 cm long, gradually reduced toward the stem tip, obovate to narrowly elliptic or linear, the lower leaves short-petiolate, the upper leaves sessile, the margins entire or rarely sparsely toothed. Sepals 3.0–6.5 mm long, elliptic to ovate, ascending. Petals 6–11 mm long, bright yellow to nearly orange. Styles 2.0–4.5 mm long. Fruits 3–10 mm long, globose, ellipsoid, or obovoid to pear-shaped, glabrous, with a stalk 1–2 mm long above the attachment point of the perianth. Seeds 4–10(–12) per locule. 2n=12. March–June.

Introduced, uncommon, known only from widely scattered historical collections (Alabama and Mississippi west to Arkansas, Texas, Oklahoma, Kansas, and Nebraska; introduced sporadically farther to the northeast). Railroads and disturbed areas.

1377. Planodes virginica   1378. Raphanus raphanistrum   1379. Raphanus sativus

Although this species is known from several counties in southeastern Kansas and eastern Oklahoma, it has not been found at any localities in Missouri where it might be suspected of being native. It is a coarser, taller plant than *P. filiformis* and is less densely pubescent. A second subspecies, ssp. *nuttallii* (Torr. & A. Gray) O'Kane & Al-Shehbaz, differs from ssp. *gracilis* in its slightly larger, more obovoid fruits and usually shorter stems.

33. **Planodes** Greene

One species, North America.

1. **Planodes virginica** (L.) Greene (Virginia rock cress)
   Cardamine virginica L.
   Arabis virginica (L.) Poir.
   Sibara virginica (L.) Rollins
   Pl. 326 i–k; Map 1377

Plants annual, terrestrial, mostly glabrous above, usually pubescent with mostly unbranched hairs near the base. Stems 1 to several per plant, 10–40 cm long, erect or ascending, sometimes branched in the upper half. Leaves alternate and basal, the basal leaves petiolate, not clasping the stem, the blades oblong in outline, 2–8 cm long, pinnately divided into 5–21 narrow, entire or few-toothed lobes, those of the stems similar but gradually reduced in size and number of lobes. Inflorescences racemes, the flowers not subtended by bracts. Sepals 1.2–1.5 mm long, narrowly oblong-elliptic, erect, usually purple-tinged. Petals 2–3 mm long, not lobed, white. Styles 0.2–0.5 mm long. Fruits ascending, straight, 15–25 mm long, more than 10 times as long as wide, flattened parallel to the septum, the valves extending to the margins, the replum not winged, dehiscing longitudinally, the valves breaking off without coiling and each with a faint midnerve near the base. Seeds in 1 row in each locule, 1.1–1.5 mm long, broadly elliptic to nearly circular in outline, somewhat flattened, the margins winged, the surface with a netlike or honeycomb-like pattern of ridges and pits, reddish orange. $2n=16$. March–May.

Common nearly throughout Missouri, except in the northernmost counties (eastern U.S. west to Kansas and Texas, also California; Mexico). Banks of streams and rivers; also moist pastures, fallow fields, roadsides, and open, disturbed areas.

This common member of the spring flora is occasionally misdetermined as *Cardamine hirsuta, C. parviflora,* or *C. pensylvanica,* but it can be distinguished from these species by its broader fruits with wingless replum, fruit valves not coiled upon fruit dehiscence, and strongly winged seeds.

This species has been known as *Sibara virginica* to most Missouri botanists (Steyermark, 1963; Rollins, 1993), but its generic classification has long been controversial. Ongoing, as-yet unpublished molecular systematic studies by Mark Bielstein, a student in the Missouri Botanical Garden's cooperative graduate studies program with Washington University, indicate that *Sibara* Greene is an unnatural genus because *S. virginica* forms a distinct lineage unrelated to the other *Sibara* species, that this species is more closely related to *Cardamine, Nasturtium,* and *Rorippa,* and that it should be recognized in an independent genus, *Planodes.*

34. Raphanus L. (radish)

Plants annual or biennial, terrestrial, usually sparsely pubescent with unbranched hairs. Stems erect or ascending, usually branched. Leaves alternate and basal, nearly all petiolate, the upper leaves progressively reduced, the bases not clasping the stem, pinnately lobed or divided, the 5–15 lobes progressively larger toward the leaf tip, the margins otherwise toothed or lobed. Inflorescences panicles, the lower branches subtended by reduced leaves, the flowers bractless. Sepals erect or ascending, narrowly oblong to oblanceolate or nearly linear. Petals unlobed, purple, pink, white, or light yellow, with dark purple veins. Stigma not lobed. Fruits more than 5 times as long as wide or rarely less, ascending, straight or nearly so, circular in cross-section, often corky in texture, segmented into apical and basal portions, the basal portion relatively short, seedless, and stalklike, the apical portion 4 or more times as long as the basal portion, tapering in the apical 1/4–1/2 to a distinct, conical, seedless beak in addition to the style, indehiscent or eventually breaking between the seeds into 1-seeded segments. Seeds in 1 row in each locule, 2.0–3.5 mm long, oblong to nearly circular in outline, the surface with a netlike or honeycomb-like pattern of ridges and pits, reddish brown. Three species, Europe, Asia, Africa, introduced nearly worldwide.

The fruits of *Raphanus* are unusual in that the seeds are effectively in a single vertical rank and are immersed in corky tissue. Thus, there are no locules in the mature fruit. The reduced basal segment usually contains vestiges of the replum.

1. Petals pale yellow (sometimes fading to white), with dark purple veins; fruits 3–6 mm wide, strongly narrowed between the seeds, the surface strongly ribbed, especially when dry . . . . 1. R. RAPHANISTRUM
1. Petals purple, pink, or rarely nearly white, with dark purple veins; fruits 5–10 mm wide, not strongly narrowed between the seeds, the surface smooth or faintly ribbed . . . . 2. R. SATIVUS

1. Raphanus raphanistrum L. (wild radish, jointed charlock)

Pl. 324 f; Map 1378

Roots not noticeably thickened. Stems (20–)30–80 cm long. Basal leaves (4–)6–30 cm long, the stem leaves 2–8 cm long. Sepals 8–11 mm long. Petals 16–25 mm long, pale yellow (sometimes fading to white), with dark purple veins. Fruits (2–)4–11(–14) cm long, 3–6(–11) mm wide, narrowly oblong to narrowly lanceolate in outline, somewhat corky, the beak 1–2(–5) cm long, strongly narrowed or constricted between the seeds, the surface strongly ribbed, especially when dry, at maturity breaking transversely into 1-seeded segments, the stalk 0.7–2.5 cm long. Seeds 4–12 per fruit, 2.5–3.5 mm long, oblong-ellipsoid to ovoid. 2n=18, 32. May–November.

Introduced, known thus far only from Christian and Jackson Counties and the city of St. Louis (native of Europe, Asia, Africa, introduced widely but sporadically in North America and Central America). Railroads.

A number of infraspecific taxa, most of questionable validity, have been named for this species in the Old World. The Missouri specimens all appear to represent typical *R. raphanistrum*.

2. Raphanus sativus L. (radish, garden radish)

Pl. 324 g–i; Map 1379

Roots noticeably thickened, fleshy. Stems (10–)30–80(–130) cm long. Basal leaves 5–25 cm long, the stem leaves 2–10 cm long (to 100 cm elsewhere). Sepals (5.5–)8.0–11.0 mm long. Petals 15–20 mm long, purple, pink, or rarely nearly white, with dark purple veins. Fruits (1–)3–6 cm long, 5–15 mm wide, corky, the beak 1–3(–4) cm long, not or only slightly narrowed between the seeds, the surface smooth or faintly ribbed, at maturity remaining indehiscent, the stalk 0.5–4.0 cm long. Seeds 1–4 per fruit, 2.5–4.0 mm in diameter, globose or nearly ovoid. 2n=18. May–August.

Introduced, uncommon and sporadic (native of Europe, Asia, Africa, introduced widely but sporadically in North America and Central America). Gardens, railroads, and roadsides.

The cultivated radish is quite variable morphologically, and a number of cultivars and infraspecific taxa have been named. Outside of Missouri, it hybridizes with *R. raphanistrum*.

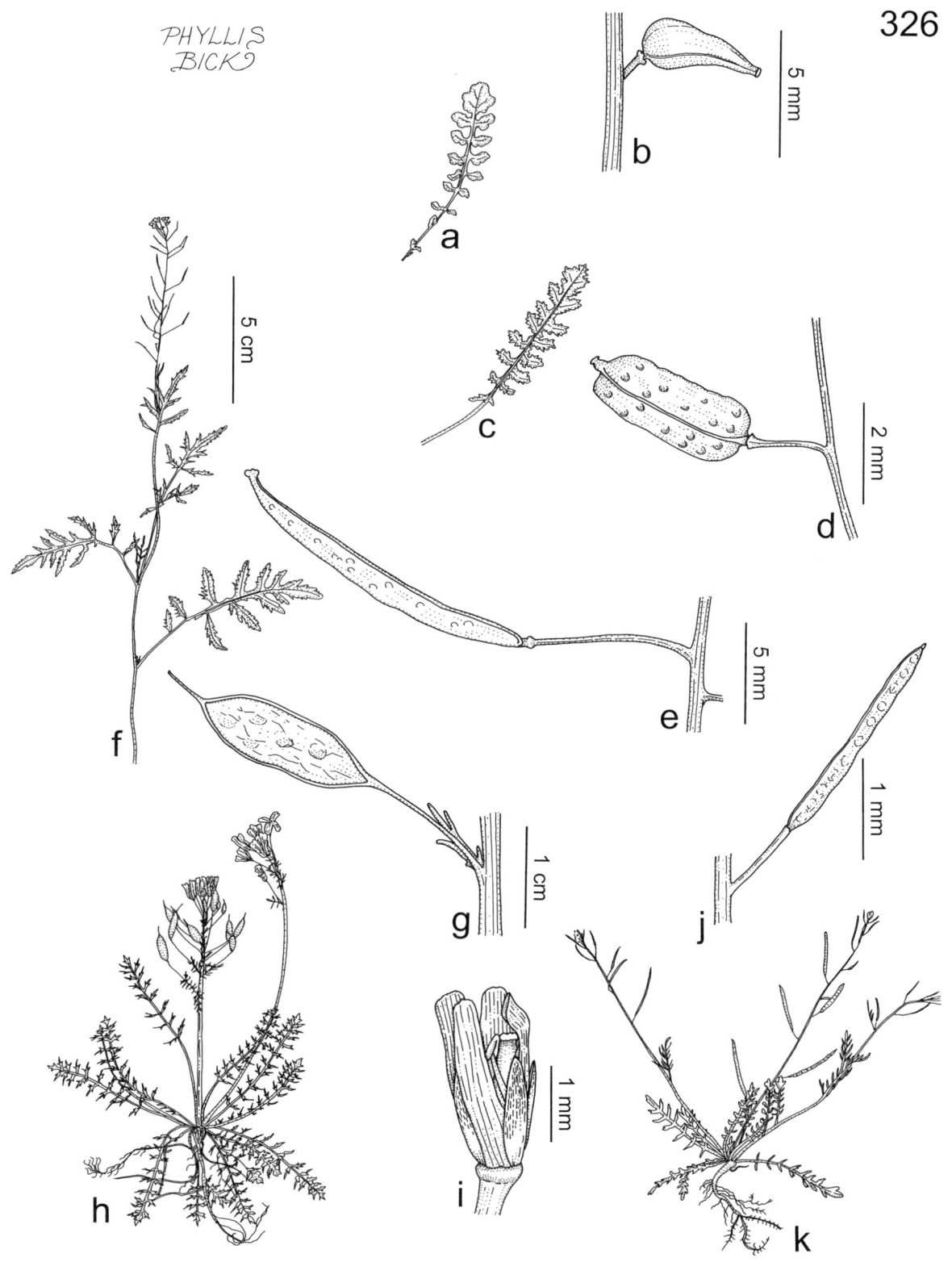

Plate 326. Brassicaceae. *Rorippa tenerrima*, **a)** leaf, **b)** fruit. *Rorippa curvipes*, **c)** leaf, **d)** fruit. *Rorippa sylvestris*, **e)** fruit, **f)** habit. *Selenia aurea*, **g)** fruit, **h)** habit. *Planodes virginica*, **i)** flower, **j)** fruit, **k)** habit.

1380. Rapistrum rugosum   1381. Rorippa aquatica   1382. Rorippa curvipes

35. Rapistrum Crantz

Two species, Europe, Asia.

1. Rapistrum rugosum (L.) All. (wild turnip, turnip weed)

R. rugosum var. *venosum* (Pers.) DC.

Pl. 324 c–e; Map 1380

Plants annual, terrestrial, rough-hairy mostly in the basal portion with simple, coarse hairs. Stems 20–100(–150) cm long, erect, branched in the upper half. Leaves alternate and basal, 1–20(–30) cm long, gradually reduced toward the stem tip, the bases not clasping the stem, the blades oblanceolate to elliptic in outline, the basal leaves petiolate, pinnately divided and coarsely toothed, those of the stems sessile or nearly so, unlobed but toothed. Inflorescences racemes at the tips of the branches, the flowers not subtended by bracts. Sepals 2.5–4.0(–5.0) mm long, oblong, erect or ascending. Petals 5–11 mm long, not lobed, yellow. Styles 1–3 mm long. Fruits erect or ascending, 5–7 mm long, segmented into apical and basal portions, the apical portion (1.5–)2.5–4.0 mm long, globose or nearly so, somewhat 5-ribbed, indehiscent, with a single seed, beaked with the persistent style, the basal portion thinner, 0.7–3.0 mm long, cylindrical to obovoid, dehiscing longitudinally, seedless or with one seed. Seeds 1.5–1.8 mm long, oblong in outline, the surface with a netlike or honeycomb-like pattern of ridges and pits, light orange. $2n=16$. May–July.

Introduced, known thus far from a single site in Barry County (native of southern Europe, sporadically adventive nearly worldwide). Open, disturbed areas.

This species was first reported for Missouri by Rebman (1989) but is expected to appear in disturbed areas elsewhere in the future. Al-Shehbaz (1985) noted that although some European botanists subdivide this species into three subspecies, the almost total intergradation among these and the sporadic nature of the North American collections preclude the formal recognition of infraspecific taxa, at least for the adventive New World populations.

36. Rorippa Scop. (cress)

Plants annual, biennial, or perennial herbs, terrestrial or aquatic, glabrous or pubescent with unbranched hairs. Stems erect to spreading, unbranched or branched. Leaves alternate and sometimes also basal, short-petiolate or sessile, not clasping the stem or the petiole base with small, rounded auricles. Leaf blades simple or pinnately divided or compound. Inflorescences racemes or panicles with the lower branches sometimes subtended by reduced, leaflike bracts, the flowers bractless (with bracts elsewhere). Sepals ovate to narrowly oblong, erect or ascending, often green or yellowish. Petals not lobed, yellow (white in *R. aquatica*), sometimes absent. Stamens (4)6. Styles absent or 0.5–4.0 mm long. Fruits spreading or ascending, 1 or more times as long as wide, ovate to oblong or linear (circular elsewhere) in outline, circular in cross-section or nearly so or somewhat flattened at a right angle to the septum, straight or curved upward, the valves not veined or with a single, indistinct midnerve, dehiscent longitudinally. Ovules mostly in 2 rows in each locule. Seeds 5–100(–300) per locule. Seventy-five to 80 species, nearly worldwide.

Some species of *Rorippa* have been collected for use as a salad green or vegetable. While in Missouri in 1803 during the early days of their epic voyage, the Lewis and Clark Expedition collected some species of *Rorippa* along the Missouri River both for food and as a pressed specimen (Meehan, 1898; see also the introductory chapter on the history of floristic botany in Missouri in the first volume of the present work [Yatskievych, 1999]).

1. Aerial (and/or submerged) stems rooting at most nodes; styles 2–4 mm long
 . . . . 1. R. AQUATICA
1. Stems not rooting at the nodes or rooting sporadically only at the lowermost few nodes; styles absent or rarely 0.5–2.0 mm long
    2. Petals 4–8 mm long, noticeably longer than the sepals; plants perennial, with creeping rhizomes
        3. Leaf bases more or less clasping the stem, with small rounded auricles, the margins of the lobes or divisions entire or with a few blunt teeth; fruits usually with well-developed seeds, these pale yellow at maturity
         . . . . 5. R. SINUATA
        3. Leaf bases not clasping the stem, the margins of the lobes with several sharp teeth; fruits usually not producing viable seeds, when produced, these reddish brown . . . . 6. R. SYLVESTRIS
    2. Petals absent or up to 3.5 mm long, shorter than to about as long as the sepals; plants annual, biennial, or short-lived perennials, taprooted, without creeping rhizomes
        4. Petals absent; seeds mostly 120–200 per fruit . . . . 4. R. SESSILIFLORA
        4. Petals present; seeds mostly 10–80 per fruit
            5. Petals 0.5–0.8 mm long; fruits with minute papillae (visible only with magnification) on the surface . . . . 7. R. TENERRIMA
            5. Petals 1.0–3.5 mm long; fruits smooth on the surface
                6. Stalks of fruits 1–3(–4) mm long; stems mostly spreading with loosely ascending tips, usually several-branched from the base
                 . . . . 2. R. CURVIPES
                6. Stalks of at least some of the fruits 3–7 mm long; stems usually erect or strongly ascending, branched only in the upper half or less commonly few-branched from the base . . . . 3. R. PALUSTRIS

1. Rorippa aquatica (Eaton) E.J. Palmer & Steyerm. (lake cress)
Armoracia aquatica (Eaton) Wiegand
A. lacustris (A. Gray) Al-Shehbaz & V.M. Bates
Neobeckia aquatica (Eaton) Greene
Pl. 313 h, i; Map 1381

Plants perennial herbs, with slender roots, lacking rhizomes, but the stem bases often becoming horizontal and buried, rooting at most nodes. Stems 30–85 cm long, spreading to ascending, unbranched below the inflorescence or branched with age, glabrous. Leaves alternate (and basal when young), 2–7 cm long, glabrous, the submerged or lowermost leaves irregularly pinnately dissected into numerous linear lobes or threadlike segments, oblong in outline, the emergent leaves unlobed, entire or shallowly toothed, lanceolate. Sepals 2–4 mm long, ascending, elliptic to obovate, glabrous. Petals 4–8 mm long, unlobed, white. Styles 2–4 mm long. Fruits spreading, often aborting prior to maturity, 3.5–7.0 mm long, about as long as wide or less than 2 times as long as wide, elliptic to narrowly obovate in outline, circular in cross-section or slightly flattened at a right angle to the septum, the stalks 6–12 mm long. Seeds (when rarely present) 12–30 per fruit, in 2 rows in each locule, 0.7–1.4 mm long, ovoid to subglobose, the surface with a netlike or honeycomb-like pattern of ridges and pits or minutely roughened, brownish orange. $2n=24$. May–August.

Uncommon and widely scattered, mostly south of the Missouri River (eastern U.S. and adjacent Canada west to Minnesota and Texas). Swamps, sloughs, streams, and spring branches; also ditches, often emergent aquatics in open areas with still or slow-moving water, rarely on mud.

This unusual species has become uncommon throughout much of its range. The highly dissected, submerged leaves may sometimes be mistaken for those of *Myriophyllum* (or occasionally those of other genera of submerged aquatics). Some of the leaves become detached during the summer and fall, and these leaves are capable of forming plantlets that root and develop into rosettes that presumably overwinter and flower the following year (reviewed by Al-Shehbaz, 1988b; Les, 1994). The species also is unusual in that the septum between the carpels is incomplete (perforated), and the fruits thus are functionally 1-locular.

The generic placement of lake cress remains controversial, and the species has been classified in six different genera. Les (1994) presented morphological and molecular evidence that it is more closely related to *Rorippa* than to *Armoracia*, although he concluded that it should be treated in a separate genus as *Neobeckia aquatica*.

2. Rorippa curvipes Greene (blunt-leaved yellow cress)

R. truncata (Jeps.) Stuckey

Pl. 326 c, d; Map 1382

Plants annual or biennial, with taproots, not rooting at the nodes or rooting sporadically only at the lowermost few nodes. Stems 6–50 cm long, spreading or less commonly erect or ascending, usually several-branched from the base, glabrous. Leaves basal and alternate, 2–12 cm long, the basal and lowermost stem leaves mostly short-petiolate, the bases usually somewhat clasping the stem with small, rounded auricles, pinnately divided with 3–17 divisions, the lobes linear to oblong or ovate, the margins irregularly toothed, glabrous or the lower leaves sparsely hairy on the petioles and larger veins. Sepals 0.8–1.5 mm long. Petals 1.0–1.5 mm long, yellow. Styles not apparent or up to 1 mm long. Fruits 2.5–6.5 mm long, 1–2 mm wide, ovoid or oblong, straight, usually slightly constricted at about the midpoint, the surface smooth. Seeds mostly 20–80 per fruit, in 2 rows in each locule, 0.5–0.7 mm long, circular in outline, usually with a small notch at the base, the surface slightly uneven or finely bumpy, brown. May–September.

Uncommon, sporadic along the floodplains of the Mississippi and Missouri Rivers (western U.S. and adjacent Canada east to Missouri and Texas). Banks of rivers and margins of ponds.

Steyermark (1963) included *R. obtusa* (Nutt.) Britton in his treatment of the genus. Stuckey (1972) untangled the several taxa in North America that had been referred to collectively under that name. Stuckey noted that *R. teres* (Michx.) Stuckey is the correct name for *R. obtusa,* and that this taxon occurs primarily along the Atlantic Coastal Plain and in Mexico, not in Missouri. According to him, the specimens from Missouri constitute a mixture of two different species, *R. tenerrima* and *R. truncata*. The taxonomic disposition of these segregates from *R. teres* has not been without controversy, however. Gleason and Cronquist (1991) accepted only a single segregate under the name *R. tenerrima* but also suggested that true *R. teres* occurred in the St. Louis area. Examination of specimens has not supported this claim. Rollins (1993) also accepted *R. tenerrima* but treated *R. truncata* as part of another segregate species, *R. curvipes* var. *truncata* (Jeps.) Rollins. Rollins further suggested that the western var. *curvipes* probably was introduced in Missouri, based apparently upon collections from St. Louis of plants with a growth form typical of this taxon. The two growth forms also were noted by Stuckey (1972). The association between these two taxa as components of *R. curvipes* appears quite sound; however, there is so much overlap in the fruit and other characters Rollins used to distinguish his varieties that there is no utility in attempting to formally recognize them taxonomically. For further discussion on the separation of *R. curvipes* from *R. tenerrima,* see the treatment of that species.

3. Rorippa palustris (L.) Besser **var. fernaldiana** (Butters & Abbe) Stuckey (marsh yellow cress, bog yellow cress, yellow watercress)

R. islandica (Oeder ex Murray) Borbás var. *fernaldiana* Butters & Abbe

Pl. 325 d–f; Map 1383

Plants annual or biennial, with taproots. Stems (10–)35–120(–140) cm long, erect or ascending, not rooting at the nodes or rooting sporadically only at the lowermost few nodes, much-branched in the upper half, unbranched or few-branched from the base, glabrous or the lower portion sparsely hairy. Leaves basal and alternate, 2–22(–30) cm long, the basal and lowermost stem leaves sessile to short-

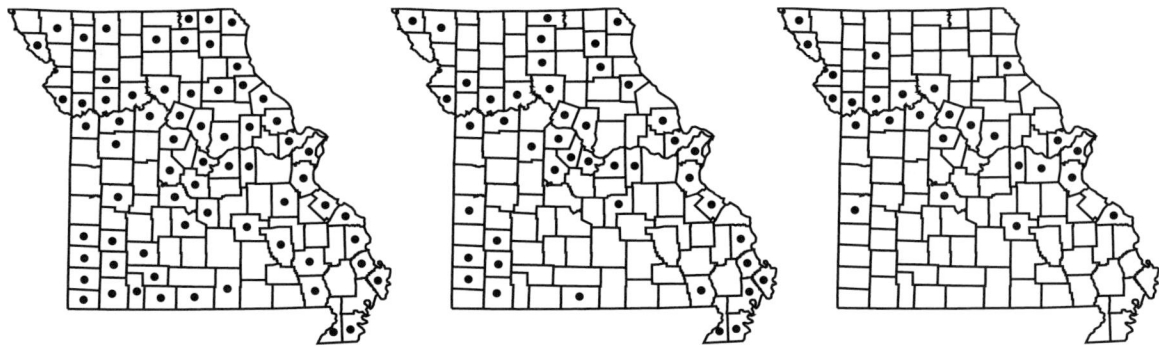

1383. Rorippa palustris   1384. Rorippa sessiliflora   1385. Rorippa sinuata

petiolate, the base usually somewhat clasping the stem, with small, rounded auricles, the blades simple to pinnately divided or compound with 3–17 divisions or leaflets, the lobes linear to oblong or irregularly ovate, the margins sharply and irregularly toothed, glabrous or the lower leaves sparsely hairy on the petioles and larger veins. Sepals 1.5–3.0 mm long. Petals 1.0–3.5 mm long, yellow. Styles absent or 0.5–1.0 mm long. Fruits 2.5–9.0(–14.0) mm long, 1.0–2.5 mm wide, ovoid or oblong, straight or slightly arched upward, usually slightly constricted at about the midpoint, the surface smooth, at least some of the stalks 3–7 mm long. Seeds mostly 20–80 per fruit, in 2 rows in each locule, 0.5–1.0 mm long, circular in outline, usually with a small notch at the base, the surface slightly uneven or finely bumpy, brown or reddish brown. $2n=32$. May–October.

Scattered to common throughout Missouri (eastern U.S. and Canada west to Texas, Colorado, and Montana; possibly introduced in Mexico, Central America, the West Indies, and the Old World). Openings of bottomland forests, banks of streams and rivers, sloughs, and margins of ponds and lakes; also levees, wet pastures, fallow fields, railroads, roadsides, and moist, open, disturbed areas.

This species was known as *R. islandica* (Oeder ex Murray) Borbás in much of the North American floristic literature until Stuckey (1972), following Jonsell (1968), indicated that the epithet *islandica* should be restricted to a different species that is confined to Europe and Greenland. He further noted that plants identified as var. *hispida* (Desv.) Jonsell from Missouri were misdetermined, and that the only variety of *R. palustris* present in the state is var. *fernaldiana*.

This is the commonest species of *Rorippa* in Missouri. The erect habit and relatively long fruit stalks are good characters to distinguish it from other annual species in the state. Overall, this morphologically variable species is circumboreal in its native distribution, but it has become naturalized nearly worldwide. Stuckey's complex infraspecific classification involved recognition of four subspecies with eleven total varieties. The var. *fernaldiana* was classified as part of ssp. *glabra* (O.E. Schulz) Stuckey. Stuckey's subspecies and varieties are distinguished by a number of minor, doubtfully useful features ranging from pubescence types to plant size and color, and fruit length and shape. They are probably oversplit, based on the large variation found in many of the critical characters (Al-Shehbaz, 1988).

For a discussion of putative hybrids between this species and *R. sinuata,* see the treatment of that species.

4. Rorippa sessiliflora (Nutt. ex Torr. & A. Gray) Hitchc. (sessile-flowered cress, marsh cress, yellow cress)
Nasturtium limosum Nutt.

Pl. 325 a–c; Map 1384

Plants annual, with taproots. Stems (10–)20–50 cm long, erect or ascending, not rooting at the nodes or rooting sporadically only at the lowermost few nodes, glabrous. Leaves basal and alternate, 1.5–10.0 cm long, glabrous, the basal and lowermost stem leaves petiolate, the base not or only slightly clasping the stem, simple and entire or wavy-margined to pinnately lobed with 3–15 blunt, irregular lobes, the lobes linear to irregularly ovate, the margins entire, wavy, or with few, shallow, blunt teeth. Sepals 1–2 mm long. Petals absent. Styles absent or less than 0.5 mm long. Fruits 5–10 mm long, 1.8–3.0 mm wide, oblong, straight or slightly arched upward, the surface smooth or slightly roughened with minute, light-colored ridges, the stalk up to 2.5 mm long. Seeds mostly 100–200 per fruit, in 2 rows in each locule, 0.4–0.5 mm long, circular or somewhat cordate in outline, the surface with a fine, netlike or honeycomb-like pattern of ridges and pits, tan to light yellowish brown. $2n=16$. April–October.

Scattered nearly throughout Missouri but absent or very uncommon in most of the Ozark Division (eastern U.S. west to South Dakota and Texas). Bottomland forests, banks of streams and rivers, and sloughs; also levees, railroads, and roadsides.

Rorippa sessiliflora is distinguished from other Missouri species of *Rorippa* by its relatively large, broad fruits that contain large numbers of very small seeds and the absence of petals. The stalks of the fruits are also quite short in this species, usually less than 3 mm long, which distinguishes it from the vegetatively similar *R. palustris,* which has stalks 3–7 mm long.

This species is mainly distributed from the Atlantic and Gulf Coastal Plains north through the Mississippi Embayment. Stuckey (1972) noted that it is not common north of the southern limits of Pleistocene glaciation. Its absence from much of the Ozarks may be a result of its preference for muddy sites, and this may change over time if plants are able to colonize the disturbed margins of livestock ponds.

5. Rorippa sinuata (Nutt. ex Torr. & A. Gray) Hitchc. (spreading yellow cress)

Pl. 325 g, h; Map 1385

Plants perennial herbs, with rhizomes. Stems 10–40(–45) cm long, spreading to ascending, sometimes rooting at the lower nodes, sparsely to densely pubescent with minute, hemispherical hairs, these appearing scalelike after collapsing upon drying. Leaves basal and alternate, 2–8(–10) cm long, the lowermost petiolate, the base mostly clasping the stem with small, rounded auricles, simple and wavy-margined to pinnately lobed or divided with 3–13 blunt lobes or divisions, the lobes linear to irregularly ovate, the margins entire, wavy, or with few, shallow, blunt teeth, the upper surface glabrous, the undersurface sparsely to densely pubescent with minute, hemispherical hairs, these appearing scalelike after collapsing upon drying. Sepals 2–4 mm long. Petals 4–6(–7) mm long, yellow. Styles 1–2 mm long, the stigma no wider than the style. Fruits 5–12 mm long, 1.5–2.0 mm wide, oblong, straight or slightly arched upward. Seeds mostly 25–80 per fruit, in 2 rows in each locule, 0.9–1.1 mm long, angular, ovate, or nearly circular in outline, the surface with a fine, netlike or honeycomb-like pattern of ridges and pits, light yellow. $2n=16$. April–September.

Scattered mostly along the floodplains of the Mississippi and Missouri Rivers (western U.S. and adjacent Canada east to Illinois and Arkansas). Banks of rivers; also edges of crop fields, railroads, roadsides, and open, disturbed areas.

The odd hemispherical hairs, which are visible only with magnification, are a distinguishing feature of this species. The stems frequently form loose mats. Stuckey (1972) hypothesized that the species originated in the Rocky Mountains and subsequently migrated down major river drainages eastward through the Great Plains.

A small number of historical collections from St. Louis and Jefferson Counties appear somewhat intermediate between *R. sinuata* and *R. palustris.* Because these have relatively large petals and appear rhizomatous, they would key to the former species, but they are atypical in their pubescence and leaf division pattern. The lack of extant sites for plants with this morphology has hampered a more detailed study of their putative hybrid origin.

6. Rorippa sylvestris (L.) Besser (creeping yellow cress)

Pl. 326 e, f; Map 1386

Plants perennial herbs, with rhizomes, glabrous or the stem bases sparsely hairy. Stems (5–)20–80 (–100) cm long, erect or ascending, rarely rooting at the lowermost nodes. Leaves basal and alternate, (2.0–)3.5–15.0(–20.0) cm long, petiolate, the bases not clasping the stem, pinnately lobed or divided, with 5–13 lobes or divisions, these linear to oblong or irregularly obovate to ovate, the margins with several sharp teeth. Sepals 2–4 mm long. Petals (3–)4–6(–8) mm long, yellow. Styles 0.5–1.0 mm long, the stigma capitate and wider than the style. Fruits 10–20(–25) mm long, 1–2 mm wide, linear, often slightly arched upward. Seeds produced uncommonly, mostly 25–80 per fruit, in 2 rows in each locule, 0.6–0.7 mm long, oblong in outline, the surface faintly and finely bumpy, reddish brown. $2n=32, 40, 48$. May–September, rarely also January.

Introduced, scattered in southwestern Missouri and along the floodplains of the Mississippi and Missouri Rivers (native of Europe, Asia; introduced widely in eastern and northwestern U.S., Canada). Bottomland forests and sloughs; also edges of crop fields, pastures, ditches, railroads, roadsides, and open, disturbed areas.

The lack of seed set in many populations of *R. sylvestris* may be attributed to the self-incompatibility of the plants and the relatively clonal nature of many populations. The plants are easily spread by rhizome fragments. Crosses between different clones frequently produce copious seeds

1386. Rorippa sylvestris   1387. Rorippa tenerrima   1388. Selenia aurea

(Mulligan and Munro, 1984). Mulligan and Munro also discovered a naturally occurring hybrid between this species and *R. palustris* in Quebec.

7. Rorippa tenerrima Greene

Pl. 326 a, b; Map 1387

Plants annual or biennial, taprooted. Stems 10–35(–40) cm long, spreading or less commonly ascending, not rooting at the nodes or rooting sporadically only at the lowermost few nodes, several-branched from the base, glabrous. Leaves basal and alternate, 2–9(–11) cm long, the basal and lowermost stem leaves mostly short-petiolate, the bases usually not clasping, pinnately divided with 3–17 divisions, the lobes linear to oblong or irregularly ovate, the margins entire or with few, blunt teeth, glabrous or the lower leaves with a few hairs on the petioles. Sepals 0.7–1.2 mm long. Petals 0.5–0.8 mm long, yellow. Styles 0.5–1.0 mm long. Fruits 3–7 mm long, 1.0–1.5 mm wide, oblong to linear, straight or nearly so, usually slightly constricted about the middle, the surface with tiny bumps or papillae (visible only with magnification). Seeds mostly 20–80 per fruit, in 2 rows in each locule, 0.5–0.7 mm long, circular in outline, usually with a small notch at the base, the surface slightly uneven or finely bumpy, brown. May–September.

Uncommon, sporadic along the floodplains of the Mississippi and Missouri Rivers (western U.S. east to Missouri and Texas; Canada). Banks of rivers and bottomland forests.

See the treatment of *R. curvipes* for a discussion of the taxonomic relationships within the *R. teres* species complex. This species is morphologically quite similar to *R. curvipes* and may be distinguished best by the fruit shape (not constricted about the middle and tapering [vs. abruptly rounded] at the tip), presence of minute papillae on the fruits, somewhat shorter petals, and less sharply toothed leaf margins (Harms et al., 1986).

37. Selenia Nutt.

Four species, U.S., Mexico.

1. Selenia aurea Nutt. (selenia, golden selenia)

Pl. 326 g, h; Map 1388

Plants annual, terrestrial, glabrous. Stems 5–30(–40) cm long, erect or ascending, sometimes branched from the base. Leaves alternate and basal, 2–6(–9) cm long, the basal leaves short-petiolate, the stem leaves sessile, the bases not clasping the stem, the blades oblanceolate to broadly elliptic in outline, pinnately divided into 5–29 narrow, entire or toothed lobes. Inflorescences racemes, the flowers subtended by reduced, leaflike bracts. Sepals 4–6 mm long, narrowly oblong to ovate, erect. Petals 8–11 mm long, not lobed, yellow. Styles 5–10 mm long. Fruits erect or ascending, 10–25 mm long, oblong or elliptic, less than 2 times as long as wide, flattened parallel to the septum, the margins entire to slightly wavy, dehiscing longitudinally, the valves lacking midnerves, the stalk 1.0–2.5 cm long. Seeds 6–12 per fruit, in 2 rows in each locule, 2.7–3.1 mm long, circular in outline or nearly so, flattened, the margins broadly winged, the surface with a netlike or honeycomb-like pattern of ridges and pits, orange. $2n=46, 138$. April–May.

Scattered in the Ozark and Unglaciated Plains Divisions (Missouri, Kansas, Oklahoma). Glades, rocky prairies, thin-soiled areas of pastures, and open, disturbed areas, on sandstone and chert.

This showy species has a relatively restricted range and is the only polyploid taxon thus far documented in a genus whose base numbers appear to include both x=7 and x=12 (Al-Shehbaz, 1988). Its origin and relationships to the other members of this small genus are poorly understood.

38. Sinapis L. (charlock)

Plants annual, terrestrial, usually with unbranched, frequently coarse, spreading or recurved hairs, sometimes nearly glabrous above. Stems erect, usually branched. Leaves alternate and basal, the lower leaves usually with petioles, the upper leaves progressively reduced and short-petiolate or sessile, the bases not clasping. Inflorescences panicles or racemes, the lower branches subtended by reduced leaves, the flowers bractless. Sepals spreading or reflexed, narrowly oblong to linear. Petals unlobed, yellow, without conspicuously darkened veins. Fruits 20–45 mm long, mostly more than 10 times as long as wide, spreading or ascending, straight or slightly arched upward, circular or slightly 4-angled in cross-section, prominently beaked with a distinct, conical or flattened, often seedless area in addition to the style, the portion below the beak dehiscing longitudinally, each valve with 3–7 distinct nerves. Seeds in 1 row in each locule, globose. Seven species, native to Europe, Asia, Africa.

Some North American botanists have treated *Sinapis* as part of *Brassica* (Steyermark, 1963), although most European authorities have maintained these groups as separate genera. The two seem quite distinct. The sepals of *Brassica* species are erect and somewhat pouched at the base (saccate), whereas those of *Sinapis* species are spreading to reflexed and not concave. The beaks of the fruits of *Sinapis* species are far more strongly differentiated from the lower portions than in *Brassica* species, and the valves of the lower portions have more veins. The chemistry of the mustard oils and seed proteins in the two groups also is different (Al-Shehbaz, 1985).

1. Beak of the fruit triangular, strongly flattened; lower portion of the fruit with dense, spreading, coarse hairs mixed with much shorter ones . . . . 1. S. ALBA
1. Beak of the fruit conical, circular or angled in cross-section; lower portion of the fruit glabrous or less commonly sparsely pubescent with short hairs fairly uniform in size. . . . . 2. S. ARVENSIS

1. Sinapis alba L. **ssp. alba** (white mustard, yellow mustard)

Brassica alba (L.) Rabenh.
B. hirta Moench

Pl. 327 e, f; Map 1389

Stems (15–)25–100(–220) cm long. Basal and lower leaves 5–20 cm long, obovate in outline, usually pinnately divided into 3–7 irregular, toothed lobes, sometimes only toothed. Upper leaves progressively reduced, the smallest 1–2 cm long, oblanceolate to oblong in outline, short-petiolate or sessile, the margins toothed. Sepals 4–7(–8) mm long. Petals (7–)8–12(–14) mm long. Fruits with the beaks 1.0–2.5(–3.0) cm long, usually somewhat longer than the lower portion, narrowly triangular, strongly flattened, the lower portion of the fruit 0.5–2.0 cm long, with dense, coarse, spreading hairs mixed with much shorter ones, usually somewhat constricted between the seeds. Seeds (1–)2–5 per locule, (1.7–)2.0–3.0(–3.5) mm in diameter, the surface with a netlike or honeycomb-like pattern of ridges and pits, usually light yellow. 2n=24. April–July, rarely also December–January.

Introduced, widely scattered in Missouri (native of Europe, Asia, commonly cultivated and sporadically introduced in North America). Edges of crop fields, railroads, and open, disturbed areas.

White mustard is commonly cultivated for its seeds, which, along with those of *Brassica nigra* (black mustard) and *B. juncea* (brown or leaf mustard), are among the principal ingredients in table mustard. These also contain oils that are used industrially as lubricants, in cooking oils, and in the manufacture of soaps and mayonnaise, as well as in medical research (Al-Shehbaz, 1985). The young foliage sometimes is eaten raw in salads.

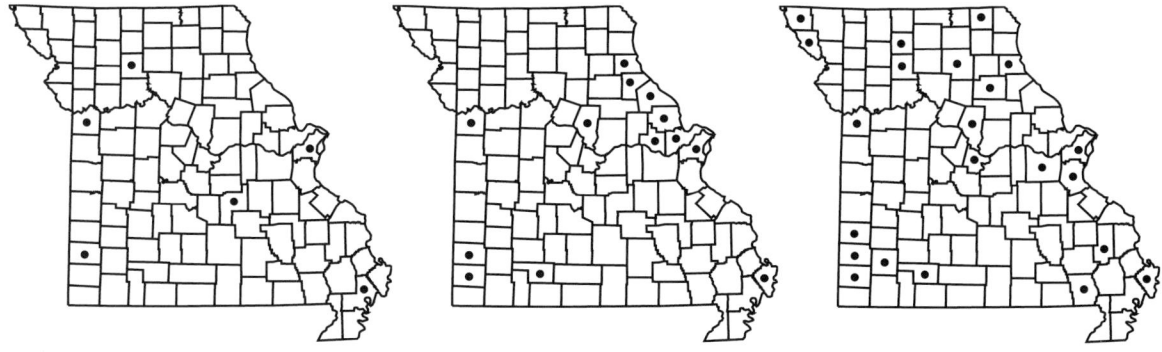

1389. Sinapis alba   1390. Sinapis arvensis   1391. Sisymbrium altissimum

The species is often divided into two subspecies, of which ssp. *dissecta* (Lag.) Bonnier is a widespread weed in flax fields that does not occur in Missouri. It differs from ssp. *alba* in having more-dissected leaves and in its seeds, which have evolved to resemble those of cultivated flax (*Linum usitatissimum* L., Linaceae).

2. **Sinapis arvensis** L. (wild mustard, charlock, crunchweed)

Brassica arvensis (L.) Rabenh.
B. kaber (DC.) L.C. Wheeler
B. kaber var. *pinnatifida* (Stokes) L.C. Wheeler

Pl. 327 a–d; Map 1390

Stems (5–)20–100(–210) cm long. Basal and lower leaves 5–20(–25) cm long, obovate in outline, usually pinnately divided with 3–11 irregular, toothed lobes, sometimes only toothed. Upper leaves progressively reduced, the smallest 1–2 cm long, oblanceolate to oblong in outline, usually sessile, the margins toothed. Sepals (4–)5–6(–8) mm long. Petals (8–)9–12(–17) mm long. Fruits with the beaks 0.8–1.6 cm long, usually about half as long as the lower portion, narrowly conical, circular or angled in cross-section, the lower portion of the fruit (0.6–)1.2–3.5(–4.3) cm long, glabrous or sparsely pubescent with short hairs fairly uniform in size, usually slightly constricted between the seeds. Seeds (2–)4–8(–12) per locule, (1.0–)1.4–2.0 mm in diameter, the surface with a fine, indistinct, netlike or honeycomb-like pattern of ridges and pits, reddish brown to black. $2n=18$. April–July.

Introduced, widely scattered in Missouri (native of Europe, Asia, introduced nearly worldwide). Edges of crop fields, railroads, roadsides, and open, disturbed areas.

This species is more common in Missouri than *S. alba* and is almost certainly more common in Missouri than present herbarium specimens would indicate. It is a worldwide weed of crop fields and disturbed ground that is very difficult to eradicate because of its high production of seeds that are quite long-lived (Al-Shehbaz, 1985).

Although generally considered an introduced species in North America, archaeological studies document that *S. arvensis* was present in the northeastern United States as early as 8,000 years ago, and it was relatively abundant and widespread by 2,000 years before the present (Jacobson et al., 1988). However, it is highly doubtful that seeds of this species can be determined with certainty from other cultivated mustards especially in archaeological remains. It is not known whether the plants were grown by Native Americans at that time or were components of the indigenous flora. Whichever the case, the species appears to have become extirpated from North America prior to European colonization, and the earliest collections found in herbaria are consistent with its status as a weedy introduction, rather than a native species (Rollins, 1993).

39. Sisymbrium L.

Plants annual (perennial elsewhere), terrestrial, glabrous or pubescent with unbranched hairs. Stems erect to spreading, unbranched or branched. Leaves alternate and basal, short-petiolate or sessile, not clasping, the leaf blades mostly pinnately lobed or divided. Inflorescences panicles or rarely racemes, the flowers not subtended by bracts (with bracts elsewhere).

Sepals linear-lanceolate to narrowly oblong, erect or ascending, green. Petals not lobed, yellow. Stamens 6. Styles 0.5–2.0 mm long. Fruits erect to spreading or ascending, more than 10 times as long as wide, linear in outline, circular in cross-section or nearly so, straight or slightly curved upward, not beaked (except for the style), the valves with a distinct midnerve and 2 less-distinct, lateral, longitudinal nerves, dehiscent longitudinally. Ovules in 1 row in each locule. Seeds 0.7–1.5 mm long, ovate to oblong in outline, somewhat flattened, not winged, the surface nearly smooth to slightly roughened or with a netlike or honeycomb-like pattern of ridges and pits, grayish yellow to orange or brown. About 40 species, Europe, Asia, Africa, introduced nearly worldwide.

1. Fruits erect, appressed to the inflorescence axis, 0.7–1.8 cm long; petals 2.5–4.0 mm long; seeds 10–20 per fruit . . . . 3. S. OFFICINALE
1. Fruits spreading to ascending, not appressed to the inflorescence axis, 2–12 cm long; petals (5–)6–8(–10) mm long; seeds 40–120 per fruit
    2. Fruits (5–)6–9(–12) cm long, the stalk relatively stout, about as wide as the fruit . . . . 1. S. ALTISSIMUM
    2. Fruits 2.0–3.5(–5.0) cm long, the stalk slender, noticeably thinner than the fruit . . . . 2. S. LOESELII

1. Sisymbrium altissimum L. (tumble mustard, Jim Hill mustard)

Pl. 328 a–c; Map 1391

Stems (20–)40–120(–160) cm long, usually hairy near the base, usually glabrous in the apical half. Leaves 2–40 cm long, petiolate, lanceolate to oblanceolate in outline, pinnately lobed or divided into 7–17 lobes, these entire to lobed, those of the lowermost leaves more numerous and lanceolate, hairy, those of the uppermost leaves fewer and linear, glabrous. Sepals 4–6 mm long. Petals (5–)6–8(–10) mm long. Style 0.5–2.0 mm long at fruiting. Fruits (5–)6–9(–12) cm long, spreading to loosely ascending at maturity, not appressed to the inflorescence axis, the stalk relatively stout, about as wide as the fruit. Seeds 90–120 per fruit, 0.8–1.0 mm long. $2n=14$. May–August.

Introduced, widely scattered in Missouri (native of Europe, Asia; introduced widely in North America). Banks of streams and rivers; also fallow fields, pastures, railroads, roadsides, and open, disturbed areas.

2. Sisymbrium loeselii L. (tall hedge mustard)

Pl. 328 h–j; Map 1392

Stems (20–)35–120(–175) cm long, usually sparsely to densely hairy, especially near the base, sometimes glabrous near the tip. Leaves 2–15 cm long, petiolate, lanceolate to lanceolate-triangular in outline, pinnately lobed or divided into 3–9 lobes, these entire to irregularly toothed or lobed, hairy or those of the uppermost leaves glabrous. Sepals 3–4 mm long. Petals 6–8 mm long. Style 0.5–0.7 mm long at fruiting. Fruits 2.0–3.5(–5.0) cm long, spreading or ascending at maturity, not appressed to the inflorescence axis, the stalk slender, noticeably narrower than the fruit. Seeds 40–60 per fruit, 0.7–1.0 mm long. $2n=14$. May–October.

Introduced, widely scattered in mostly central Missouri (native of Europe, Asia; introduced widely in the U.S., Canada). Banks of streams and rivers; also levees, pastures, railroads, roadsides, and open, disturbed areas.

This species has become increasingly common during the past 35 years, but it remains restricted to mostly along railroads.

3. Sisymbrium officinale (L.) Scop. (hedge mustard)

S. officinale var. *leiocarpum* DC.

Pl. 328 f, g; Map 1393

Stems 25–75(–110) cm long, sparsely to usually densely hairy, especially near the base. Leaves 2–20 cm long, hairy or the uppermost leaves sometimes nearly glabrous, the middle and lowermost leaves petiolate, lanceolate to lanceolate-triangular in outline, pinnately lobed or divided into 3–11 lobes, these entire to irregularly toothed or lobed, the uppermost leaves often sessile, triangular to narrowly lanceolate in outline, entire or pinnately 3–5 lobed, the lobes entire or few-toothed. Sepals 1.0–2.0(–2.5) mm long. Petals 2.5–4.0 mm long. Style (0.8–)1.0–1.5(–2.0) mm long at fruiting. Fruits 0.7–1.8 cm long, erect, appressed to the inflorescence axis, the stalk 1.5–3.0(–4.0) mm long, relatively stout, about as wide as the fruit. Seeds 10–20 per fruit, 1.0–1.3 mm long. $2n=14$. May–October.

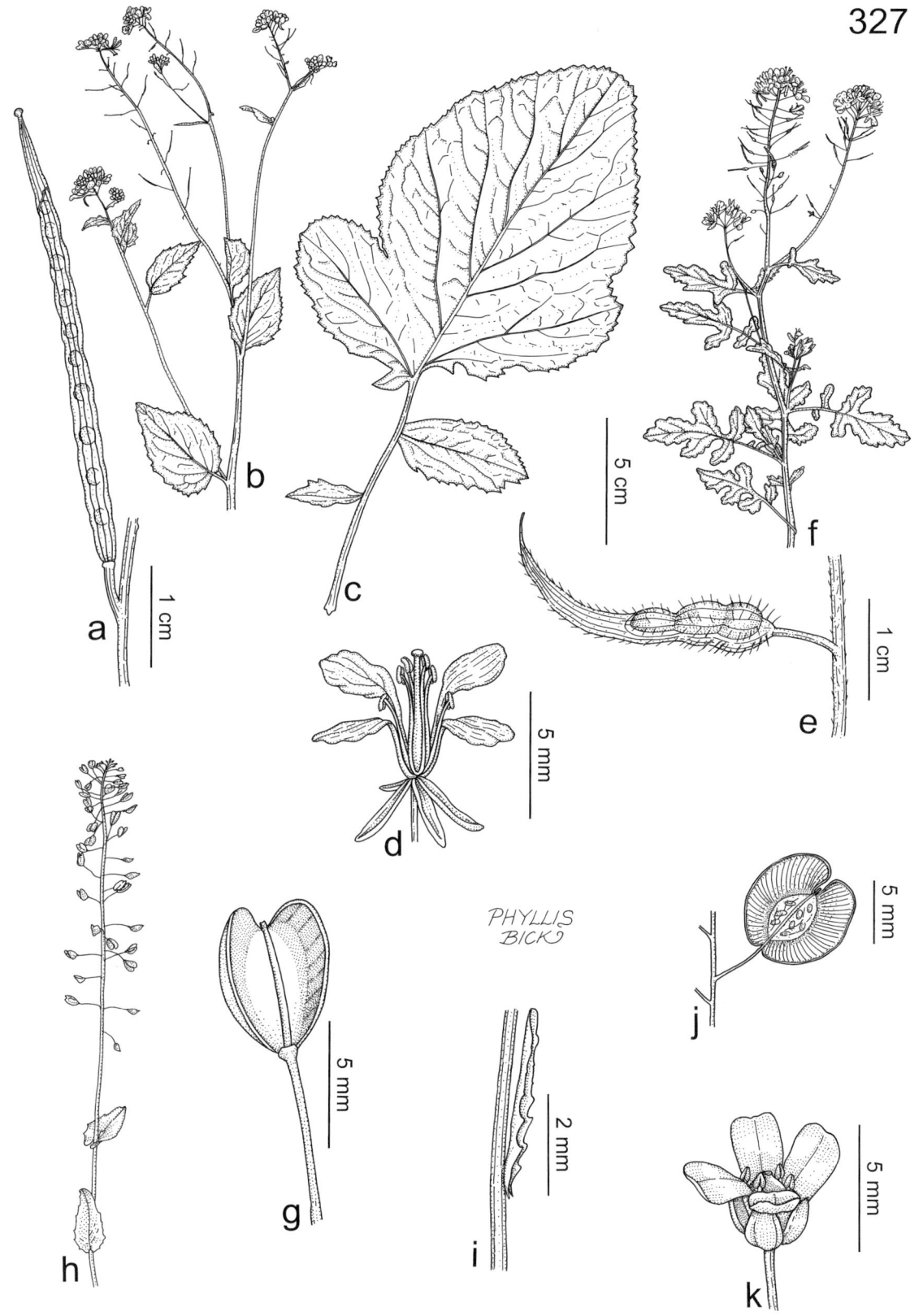

Plate 327. Brassicaceae. *Sinapis arvensis*, **a)** fruit, **b)** inflorescence, **c)** lower leaf, **d)** flower. *Sinapis alba*, **e)** fruit, **f)** habit. *Microthlaspi perfoliatum*, **g)** fruit, **h)** habit. *Thlaspi arvense*, **i)** leaf, **j)** fruit, **k)** flower.

1392. Sisymbrium loeselii   1393. Sisymbrium officinale   1394. Thlaspi alliaceum

Introduced, scattered nearly throughout Missouri (native of Europe, Asia; introduced widely in North America). Banks of streams and rivers; also fallow fields, pastures, farmyards, railroads, roadsides, and open, disturbed areas.

Sisymbrium officinale is the most widespread species of the genus in the state. Missouri plants of this species mostly have glabrous or nearly glabrous fruits. These have been referred to as var. *leiocarpum*. However, presence and density of pubescence varies greatly in the species, and these variants are not worthy of formal taxonomic recognition.

40. Thlaspi L. (penny cress)

Plants annual, terrestrial, glabrous or nearly so, usually with an unpleasant odor when bruised or crushed. Stems mostly erect or ascending, unbranched below the inflorescence or several-branched from the base. Leaves alternate and sometimes also basal, sessile or short-petiolate, the bases clasping the stems with rounded or pointed auricles of tissue, simple, ovate, broadly elliptic, lanceolate, oblanceolate, or narrowly oblong, the margins entire or toothed. Inflorescences racemes or less commonly panicles, the flowers not subtended by bracts. Sepals erect or ascending, 1.0–3.5 mm long, oblong to ovate or elliptic, green, usually with white margins. Petals 2–5 mm long, unlobed or very shallowly notched at the tip, white. Stamens 6. Styles 0.1–0.3 mm long. Fruits spreading to broadly ascending, less than 3 times as long as wide, obovate to cordate or nearly circular in outline, strongly flattened at a right angle to the septum and sometimes inflated, the margins entire, but winged, the tip with a notch, dehiscent. Seeds 3–8 per locule. Six species, Europe, Asia.

Species of *Thlaspi* superficially resemble *Lepidium campestre,* especially when not in fruit. However, that species has densely hairy stems, whereas the *Thlaspi* species known to occur in the state thus far are glabrous. The species that Steyermark (1963) and many other North American botanists have called *T. perfoliatum* recently has been segregated into the genus *Microthlaspi.* See the treatment of that genus for further discussion.

1. Fruits 5–10 mm long, narrowly winged below, slightly more broadly winged above, inflated; seeds with a netlike or honeycomb-like pattern of ridges and pits
 . . . . 1. T. ALLIACEUM
1. Fruits 8–20 mm long, broadly winged all around, flattened; seeds with a series of concentric, arched ribs . . . . 2. T. ARVENSE

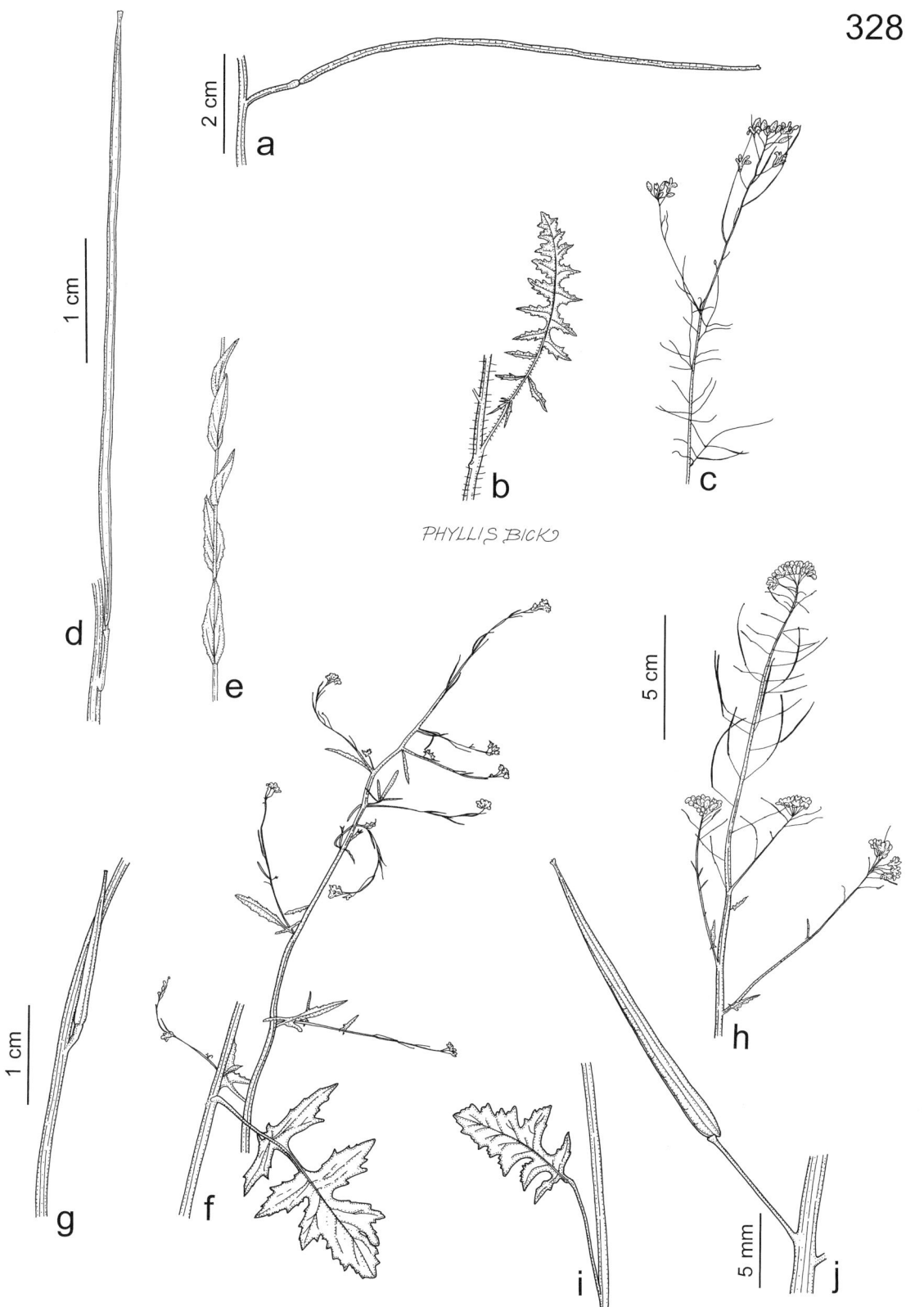

Plate 328. Brassicaceae. *Sysimbrium altissimum*, **a)** fruit, **b)** leaf, **c)** inflorescence. *Turritis glabra*, **d)** fruit, **e)** fruiting branch and leaves. *Sysimbrium officinale*, **f)** leaf and inflorescence, **g)** fruit. *Sysimbrium loesilii*, **h)** inflorescence, **i)** leaf, **j)** fruit.

1395. Thlaspi arvense 1396. Turritis glabra 1397. Pachysandra procumbens

1. Thlaspi alliaceum L. (roadside penny cress)

Map 1394

Plants with a strong smell of garlic when crushed, often glaucous. Stems 20–60(–80) cm long, pubescent near the base. Leaves 1–5 cm long, narrowly elliptic to oblong or ovate, sessile, the basal auricles pointed, the margin toothed, the basal leaves few or absent at flowering time. Sepals 1–2 mm long. Petals 2.5–4.0(–5.0) mm long. Styles 0.2–0.3 mm long. Fruits 5–10 mm long, obovate, inflated, narrowly winged below, slightly more broadly winged above, the apical notch shallow. Seeds mostly 3–5 per locule, 2–4 mm long, broadly ellipsoid, the surface with a netlike or honeycomb-like pattern of ridges and pits, dark brown to black. $2n=14$. March–May.

Introduced, uncommon, known thus far from a single specimen from Platte County (native of Europe; introduced sporadically in the northeastern U.S. and Louisiana). Disturbed floodplains of rivers.

This species was first reported for Missouri by Tenaglia and Yatskievych (2002). As noted by them, the species was first collected in the United States in the early 1960s in North Carolina. It has spread quickly and is expected to be found elsewhere in Missouri in the future.

2. Thlaspi arvense L. (field penny cress, stinkweed)

Pl. 327 i–k; Map 1395

Plants with a sometimes faint fetid odor when bruised or crushed, sometimes glaucous. Stems (9–)15–55(–80) cm long, glabrous. Leaves 0.5–7.0(–8.0) cm long, linear to ovate or broadly elliptic or oblong, the basal auricles pointed or rarely rounded, the margin often toothed, the basal leaves few or absent at flowering time. Sepals (1.5–)2.0–3.0(–3.3) mm long. Petals 3–4(–5) mm long. Styles 0.1–0.3 mm long. Fruits 8–20 mm long, broadly elliptic or obovate to nearly circular in outline, flattened, the margins broadly winged their entire length, the apical notch narrowly U-shaped and longer than deep. Seeds mostly 3–8 per locule, 1.6–2.0(–2.3) mm long, broadly ellipsoid to obovoid, the surface with a series of concentric, arched ribs, dark gray to black. $2n=14$. April–June.

Introduced, common throughout Missouri (native of Europe, widely naturalized in North America). Moist, disturbed depressions of upland prairies and banks of streams; also crop fields, fallow fields, old fields, pastures, lawns, railroads, roadsides, and open, disturbed areas.

When not in fruit, this species can be difficult to distinguish from *Microthlaspi perfoliatum*. It differs in lacking glaucousness, in having slightly larger flowers, and in having usually somewhat more toothed middle and upper leaves. In Missouri, *T. arvense* is more variable morphologically than *M. perfoliatum* and can be far more robust than that species.

41. Turritis L. (tower mustard)

Two species (1 endemic to the Middle East), Europe, Asia, North America.

1. Turritis glabra L. var. glabra (tower mustard)
Arabis glabra (L.) Bernh.

Pl. 328 d, e; Map 1396

Plants biennial or rarely short-lived perennial herbs, terrestrial. Stems (30–)40–120(–150) cm long, erect, often few-branched in the upper or lower half, usually pubescent with mostly unbranched and fewer forked, stalked hairs near the base, glabrous and glaucous in the upper half, rarely entirely glabrous. Basal leaves (4–)6–12(–15) cm long, oblong to broadly oblanceolate or spatulate in outline, entire to toothed or irregularly pinnately divided, narrowed at the base to a short petiole, usually coarsely pubescent with stalked, 2-branched hairs. Stem leaves appressed to spreading, sometimes overlapping, (2–)3–15 cm long, lanceolate to ovate in outline, sessile, clasping the stem, usually with pointed auricles, entire or toothed, glabrous or the lowest ones sometimes sparsely hairy. Inflorescences racemes or panicles, the flowers without bracts. Sepals 3–5 mm long, oblong-elliptic, glabrous, often purple-tinged. Petals (3.0–)5.0–8.5 mm long, cream-colored to pale yellow, rarely tinged with purple. Fruits (3–)5–10 cm long, erect at maturity, appressed to the inflorescence axis, straight or nearly so, nearly circular or slightly 4-angled in cross-section, dehiscing longitudinally by the valves breaking off without coiling, each valve with a midnerve extending the full length or nearly so. Seeds usually in 2 rows in each locule, 0.6–0.9(–1.2) mm long, irregularly angular-elliptic in outline, somewhat flattened, usually irregularly winged at both ends, the surface with a netlike or honeycomb-like pattern of ridges and pits, orange. $2n=12, 16, 32$. May–June.

Uncommon in eastern Missouri (Alaska and Canada south to North Carolina, Arkansas; Mexico, Europe). Bottomland forests.

Turritis glabra comprises a circumboreal complex that has been subdivided variously by European authors. In addition to var. *glabra,* the only other taxon indigenous to North America is var. *furcatipilis* M. Hopkins, which grows in California and Utah and differs from var. *glabra* in having the lower portions of the stems pubescent with appressed, several-branched hairs. Although var. *glabra* originally was reported for Missouri from Perry County (Mohlenbrock, 1979), apparently no voucher specimens have been collected from that county as yet.

Turritis glabra was placed by Rollins (1993) and other North American authors in *Arabis.* However, recent molecular studies (summarized in Al-Shehbaz, 2003) indicate that *Turritis* is unrelated to *Arabis.*

BUXACEAE (Boxwood Family)

Five genera, about 60 species, nearly worldwide.

In temperate regions, the main use of members of the Buxaceae is for cultivation as ornamentals. Species of *Pachysandra* are grown as groundcovers, and various species of *Buxus* L. (box, boxwood), especially *B. sempervirens* L., are common evergreen shrubs also used in hedges.

1. Pachysandra Michx.
(Robbins, 1968; Boufford and Xiang, 1992)

Three species, eastern North America, Asia.

1. Pachysandra procumbens Michx. (Allegheny spurge)

Map 1397

Plants perennial herbs, monoecious, with long-creeping, branched rhizomes, more or less evergreen. Aerial stems 10–40 cm long, spreading to ascending, sparsely to densely (especially toward the tip) pubescent with short, curled hairs, green to nearly white, sometimes purplish-tinged. Leaves alternate, mostly toward the stem tips, mostly long-petiolate. Stipules absent. Leaf blades 3–9 cm long, simple, ovate to broadly obovate or nearly circular, rounded to more commonly sharply pointed at the tip, tapered at the base, the margins with relatively few coarse teeth above the midpoint and also minutely hairy, the surfaces sparsely to moderately and minutely hairy along the veins, the upper surface dark green but often with lighter mottling, the undersurface lighter green. Inflorescences relatively dense spikes or spikelike racemes (the staminate flowers sessile, the pistillate flowers sometimes short-stalked), 5–12 cm long, axillary from the lowermost aerial stem nodes. Flowers actinomorphic, hypogynous, the staminate flowers numerous and positioned toward the inflorescence tip (sometimes also 1 or 2 at the very base), each subtended by 1 small sepal-like bract; the pistillate flowers usually 1–7 and positioned toward the inflorescence base, each subtended by a series of 7 or more small scalelike to sepal-like bracts. Calyces of 4 free sepals (sometimes more in pistillate flowers), 3.5–5.0 mm long, ovate to broadly ovate, rounded to sharply pointed at the tip, usually reddish- or purplish-tinged, the margins minutely hairy. Corollas absent. Staminate flowers with usually 4 stamens, these opposite the sepals and exserted from the calyx, the filaments white, the anthers red to reddish purple, attached below the midpoint of the outer side. Pistillate flowers with 1 pistil, this superior, usually with 3 carpels and apically 3-lobed, each carpel with 2 locules, the surface densely and minutely hairy. Ovules 6, the placentation more or less axile. Styles usually 3, each with a linear stigmatic region along the inner side toward the tip.

Fruits capsules, 12–16 mm long, ovoid to nearly globose, apically 3-lobed, the surface densely and minutely hairy, dehiscing circumscissilely near the base. Seeds 1 per locule, 3.0–4.5 mm long, ovate in outline, triangular in cross-section, with a small outgrowth (caruncle) at one end, the surface smooth, black, shiny. $2n=24$. March–May.

Introduced, known thus far from a single collection from Warren County (southeastern U.S. west to Kentucky and Louisiana; introduced farther north and west). Mesic upland forests.

The presence of the Warren County population was long known to members of the Webster Groves Nature Study Society and monitored regularly by Midge Tooker, who graciously showed the site to the author. The occurrence is seemingly natural, but the presence in the vicinity of a few other nonnative species, such as *Tsuga canadensis,* suggests the former presence of an old homesite.

The Asian *P. terminalis* Siebold & Zucc., a close relative, also is commonly cultivated as a groundcover. It differs from *P. procumbens* in its somewhat narrower leaves, terminal inflorescences, and 2-carpellate flowers. *Pachysandra terminalis* has escaped from cultivation sporadically in the northeastern United States and adjacent Canada, and eventually it may be located in Missouri.

CABOMBACEAE (Water Shield Family)
Contributed by Alan E. Brant

Plants perennial herbs, with short rhizomes (seldom collected) giving rise to long (to 2 m or more) branched stems potentially rooting at the nodes. Leaves opposite and/or alternate, with well-developed, mostly long petioles. Stipules absent. Leaf blades simple and entire or highly dissected. Flowers solitary in the leaf axils, mostly long-stalked, hypogynous, perfect, actinomorphic. Perianth showy, appearing free, the sepals and petals fused only at the very base, persistent at fruiting. Sepals 3 or 4. Petals 3 or 4. Stamens 3–6 or numerous, free, the filament slender and flattened, attached at the anther base. Pistils (1–)2–18, each with 1 carpel, the ovary superior, with 1–5 ovules. Style 1, the stigma capitate or a linear region toward the style tip. Fruits in a ring or cluster, achenelike, indehiscent, leathery. Seeds 1–3. Two genera, 6 species, worldwide.

Genera of Cabombaceae were included in the Nymphaeaceae by Steyermark (1963) and other earlier authors, but most botanists now agree that *Cabomba* and *Brasenia* constitute an independent family, based on a variety of morphological and biochemical characters. Both groups comprise aquatics with submerged and/or floating leaves.

The interesting floral biology of the two species present in Missouri has been studied in detail (Schneider and Jeter, 1982; Osborn and Schneider, 1988). *Brasenia schreberi* is primarily wind-pollinated, and although a number of bees and wasps visit the flowers of *Cabomba caroliniana,* it is mainly fly-pollinated. In both species, individual flowers are open for only

1398. Brasenia schreberi   1399. Cabomba caroliniana   1400. Opuntia humifusa

two days. On the first day, the stigmas become receptive as the flower opens at mid-morning; in the late afternoon, the flower closes and the flower stalk bends, resulting in the submergence of the flower overnight. On the second morning, the flower emerges from the water again, and although the stigmas are no longer receptive, the stamens elongate slightly and shed their pollen. The second evening, the flower again closes and become submerged, with fruit development occurring underwater. Thus, first-day flowers are cross-pollinated with pollen from second-day flowers.

1. Submerged leaves absent, or if present in young plants, then entire and peltate (similar to floating leaves); floating leaves present during entire growing season, 5–11 cm long, conspicuous, peltate, not dissected; submerged parts covered with a thick layer of a jellylike mucilaginous substance . . . . 1. BRASENIA
1. Submerged leaves deeply and finely dissected, not peltate; floating leaves present only at flowering, 0.6–2.0 cm long, relatively inconspicuous, peltate, not dissected; submerged parts only slightly if at all coated with a mucilaginous substance . . . . 2. CABOMBA

1. Brasenia Schreb. (water shield)

One species, North America, Central America, Caribbean Islands, Asia, Africa, Australia.

1. Brasenia schreberi J.F. Gmel. (water shield, snot plant)

Pl. 329 a, b; Map 1398

Plants apparently glabrous (although the mucilage is secreted by minute glandular hairs, these are not apparent on mature herbage), the submerged portion, including the undersurface of the leaf blades, covered with a thick layer of a jellylike mucilaginous substance. Leaves monomorphic, sometimes submerged in young plants, but mostly with floating blades, long-petiolate (to 30 cm or more). Leaf blades 3.5–13.5 cm long, 2–8 cm wide, peltate, broadly elliptic, the margins entire, the undersurface dark reddish purple. Flowers mostly long-stalked, held just above the water at flowering, the stalk bending to become submerged as the fruits develop. Perianth curled outward and downward at flowering, later becoming erect and more or less enfolding the developing fruits. Sepals 3 or less commonly 4, 13–15 mm long, strap-shaped, green on the outer surface, reddish purple to maroon on the inner surface. Petals 3 or less commonly 4, slightly longer and narrower than the sepals, strap-shaped, reddish purple to maroon. Stamens 18–36 (rarely more). Pistils 4–18, the style with a linear stigmatic region along a side toward the tip. Fruits 6–10 mm long, club-shaped to peanut-shaped, tapered to the persistent style. Seeds 2.5–4.0 mm long, oblong to oblong-ovate in outline, the surface finely pebbled, grayish brown to yellowish brown. $2n=80$. May–September.

Scattered, mostly in southern and eastern Missouri (nearly throughout the U.S., except for some western states; Canada, Mexico, Central America, Caribbean Islands, Asia, Africa, Australia). Floating-leaved aquatics in ponds, lakes, and sloughs.

Water shield is of food value to waterfowl, and the tuberous roots reportedly were eaten by Native Americans in California. The thick, jellylike coating of the submerged parts presumably retards drying out of the plants when water levels drop during droughts and may also help to deter mammals from eating the herbage. The young leaves and petioles, before the gelatinous covering becomes prominent, also have served as salad greens in parts of Japan. The floating leaves afford shade and shelter for fish and other aquatic organisms. However, the species can become a nuisance, especially in shallow ponds and lakes, where it can cover the water surface so thickly as to preclude light and gas exchange. It undoubtedly is more widespread in Missouri than the specimen data suggest.

2. Cabomba Aubl. (fanwort)
(Ørgaard, 1991)

Five species, nearly worldwide.

1. Cabomba caroliniana A. Gray **var. caroliniana** (fanwort, Carolina water shield)

Pl. 329 c, d; Map 1399

Plants inconspicuously hairy on young and emergent parts, the submerged portion not mucilaginous or the stems inconspicuously so. Leaves dimorphic, mostly submerged, the floating leaves present only at flowering. Submerged leaves opposite, mostly long-petiolate (1–2 cm), the blades 1.5–4.0 cm long and wide, attached at the base, fan-shaped, palmately deeply divided into 5–7 parts, each of these further 2–6 times dichotomously (rarely trichotomously) deeply dissected 2–6 times into numerous slender ultimate segments. Floating leaves alternate, mostly long-petiolate (1–5 cm), the blades 0.6–2.0 cm long, 0.1–0.2 cm wide, peltate, narrowly oblong to narrowly elliptic, sometimes shallowly notched on the sides or at one or both ends, the margins otherwise entire. Flowers short- or long-stalked, held just above the water at flowering, the stalk bending to become submerged as the fruits develop. Perianth ascending to spreading, flat or curved slightly inward and upward, later becoming erect and more or less enfolding the developing fruits. Sepals 3, 5–10 mm long, petaloid, elliptic to obovate, sometimes tapered to a short, stalklike base, white. Petals 3, 5–10 mm long, elliptic to obovate, sagittate toward the base and thus appearing short-stalked, white with the basal auricles each with a yellow nectary. Stamens (3)6. Pistils 2–4, the stigma capitate. Fruits 4–7 mm long, narrowly ovoid to somewhat peanut-shaped, tapered to the persistent style. Seeds 1.5–3.0 mm long, oblong-ovate in outline, the surface finely pebbled, with 4 longitudinal rows of dense tubercles visible when moistened, greenish brown to brown. $2n=39$, ca. 78, ca. 96, 104. May–September.

Scattered, mostly in southeastern Missouri; introduced in Boone County (southeastern U.S. west to Kansas and Texas; also disjunct in southeastern South America; introduced northward to Michigan and New Hampshire, and disjunctly in Oregon and Washington; Canada). Mostly submerged aquatic in still or slow-moving water of ponds, lakes, swamps, sloughs, streams, rivers, and ditches.

Cabomba caroliniana is commonly cultivated as an aquarium plant and has escaped from cultivation or been deliberately introduced at some sites within and outside of its native range. Fanwort provides food and cover for fish and aquatic invertebrates. Vegetative plants of *Cabomba* might possibly be confused with those of *Ceratophyllum* species or *Ranunculus aquatilis*, both of which have dichotomously and/or palmately highly divided, fan-shaped leaves. *Cabomba* differs from both of these alternate-leaved genera in its opposite submerged leaves.

Two other weakly defined varieties of *C. caroliniana* have been accepted by some authors (Ørgaard, 1991), based on variations in the color of shoots and flowers. The var. *pulcherrima* R.M. Harper has purple-tinged stems and perianth and is distributed sporadically along the Atlantic and Gulf Coastal Plains from North Carolina to northern Florida, whereas var. *flavida* Ørgaard has a pale yellow perianth and occurs in the South American portion of the species' range.

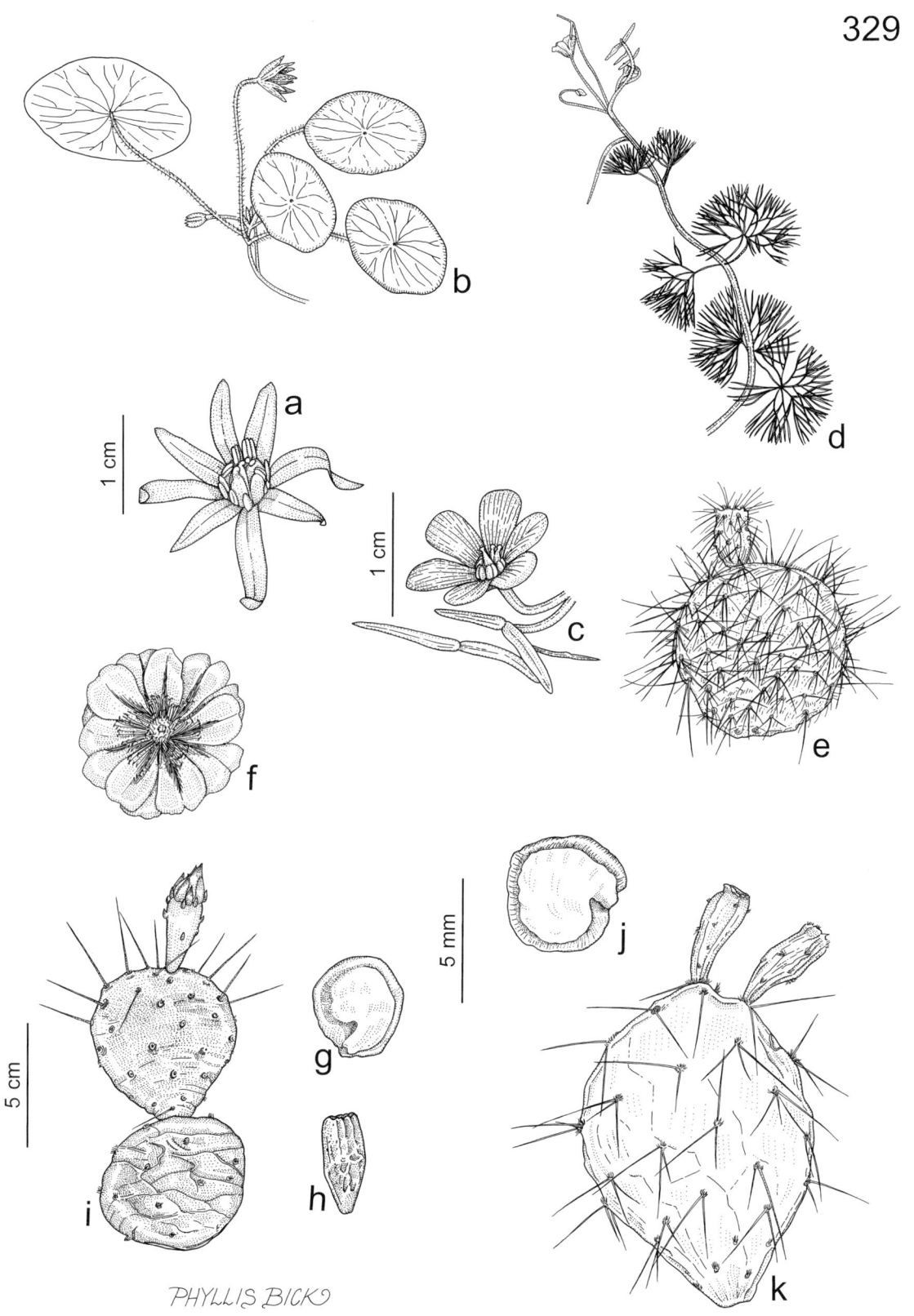

Plate 329. Cactaceae, Cabombaceae. *Brasenia shreberi*, **a)** flower, **b)** habit. *Cabomba caroliniana*, **c)** flower, **d)** inflorescence. *Opuntia polyacantha*, **e)** habit. *Opuntia humifusa*, **f)** flower, **g)** seed, **h)** fruit, **i)** habit. *Opuntia macrorhiza*, **j)** seed, **k)** habit.

CACTACEAE (Cactus Family)
(Benson, 1982; Anderson, 2001)

Ninety-three to 125 genera, 1,400–1,850 species, North America to South America, Caribbean Islands; 1 species in Africa, Madagascar.

The morphology of cactus plants requires some explanation. Except for the relatively primitive genus *Pereskia* Mill., which includes more or less normal-looking spiny shrubs and trees with well-developed leaves, in cacti the stems have become modified and thickened to serve in photosynthesis and water storage, whereas the leaves are essentially absent or very short-lived. There is a bewildering diversity of stem shapes, sizes, surface textures, and branching patterns in the family. The spines of cacti are produced on areoles, which can be thought of as extremely short, highly modified branches. Spines actually develop from the axillary buds associated with the numerous closely spaced nodes of the areole and often appear to have a more or less radial distribution on the areole. Flowers also are produced in association with areoles and developmentally are thus considered to be axillary and solitary (except in *Pereskia*). Most species produce flowers with numerous perianth parts that grade continuously from sepaloid to petaloid morphology along the densely spiraled series and thus are referred to as tepals. Taxonomically, the family is considered to represent a morphologically specialized offshoot of the Portulacaceae (Hershkovitz and Zimmer, 1997; Applequist and Wallace, 2001) and to have experienced an explosive radiation of species in the New World, with the result that many of the species are difficult to distinguish from others within a given complex. This has led to great controversy as to species numbers and delimitation in most of the larger genera (generic delimitation has been equally contentious). In an effort to stabilize the nomenclature and taxonomy of the family, since 1984 an International Cactaceae Systematics Group of specialists has worked to produce a consensus classification and checklist (Hunt, 1999) for use by horticulturalists, conservation officers, and others interested in the family.

With their unusual stems and bright flowers, cacti are popular both among amateur enthusiasts and botanists. In fact, there is a huge horticultural market for cacti, which also have economic importance as landscape plants in dry and seasonally dry regions of the United States and other countries. Overcollection from the wild for horticultural purposes has led to the endangerment of numerous species, thus international trade in most cacti is closely regulated by the Convention on International Trade in Endangered Species of Wild Fauna and Flora (CITES). Some species of cacti, particularly members of *Opuntia* and related genera, also are cultivated for their edible fruits and stems and as hosts for homopteran cochineal insects (*Dactylopius* spp.), which produce a beautiful red dye. Other genera, particularly peyote (*Lophophora williamsii* (Lam. ex Salm-Dyck) J.M. Coult.), have been cultivated and/or wild-harvested for their hallucinogenic properties. The wood of various cacti sometimes is used in handicrafts. Cactus spines can cause damage to the legs, feet, and mouthparts of livestock, and some cacti have been considered pest plants in pastures. This is especially true in Australia, where imported *Opuntia* species rendered millions of acres of rangeland unfit for livestock and other uses until biological controls involving stem-boring larvae of the South American moth genus *Cactoblastus* proved effective.

In addition to the numerous species cultivated in homes, greenhouses, and conservatories, several nonnative cacti are cold-hardy in Missouri's climate and are cultivated outdoors as specimen plants in well-drained soils. These generally have not escaped into the wild, but they occasionally persist at old homesites. The best example of this is a tree cholla, *Cylindropuntia imbricata* (Haw.) F.M. Knuth (*Opuntia imbricata* (Haw.) DC., *O. arborescens* Engelm.), which is native from Colorado to Kansas south to Texas and Mexico. A single individual of this species was located in 1998 by an amateur botanist, T. Owens, in Laclede County growing on a dry ridgetop overlooking the Niangua River. This species has cylindrical stems to 2 m tall and 3 cm in diameter that are covered with coarse, elongate tubercles and dense clusters of spines. The stems are jointed every 5–35 cm and have whorled branches. The flowers have bright pink to reddish purple tepals to 5 cm long, and the broadly obovoid fruits are spineless and more or less yellow at maturity. This species eventually may need to be added to the roster of the state's flora and should be searched for, especially in the Unglaciated Plains Division.

Beginning with an unsubstantiated listing from Pulaski County (Palmer and Steyermark, 1935), there also have been persistent anecdotal reports (mostly from western Missouri) of another regionally native cactus species, *Escobaria missouriensis* (Sweet) D.R. Hunt (*Mammillaria missouriensis* Sweet, *Coryphantha missouriensis* (Sweet) Britton & Rose, *Neobesseya missouriensis* (Sweet) Britton & Rose, *Neomammillaria missouriensis* (Sweet) Britton & Rose, *Neomammillaria similis* (Engelm.) Britton & Rose), which is known by various common names, including ball cactus, beehive cactus, cream cactus, and Missouri pincushion cactus. Its distributional range stretches from Idaho to Arizona east to North Dakota, Kansas, and Texas (also adjacent Mexico). Steyermark (1963) excluded it from the Missouri flora, but searches of thin-soil areas of prairies, glades, and bluff tops may result in its eventual documentation from the state. *Escobaria missouriensis* produces small, globose to broadly obovoid stems, these 2–5 cm long, solitary or forming small, clustered mounds, and covered with nipplelike tubercles tipped with a dense cluster of short, straight spines. The flowers have narrow tepals 1–2 cm long that are cream-colored to light yellow or greenish yellow, often reddish- or pinkish-tinged toward the base. The globose to obovoid fruits are 1–2 cm long, spineless, and bright red at maturity.

1. **Opuntia** L. (prickly pear)

Plants atypical succulent shrubs, spiny, glabrous except for woolly multicellular hairs in the young areoles. Rootstocks somewhat woody, the roots often also tuberous. Stems modified into thick, flattened pads, the surfaces and margins with numerous areoles, green to dark green or bluish green, the surface with a thick, waxy epidermis, often somewhat glaucous, smooth or becoming strongly wrinkled during times of drought or freezing temperatures. Spines of 2 types in each areole, both usually with minute, downward-pointing barbs: numerous fine glochids, these easily shed, 1–3 mm long, straight, yellow to brown; also often 1 to several true spines, these persistent, 1–6 cm long, straight or slightly curved, white to gray or brown at maturity, sometimes reddish-tinged. Leaves present only on new growth, shed soon after development, 4–9 mm long, narrowly conical, narrowed to a sharply pointed tip. Flowers usually solitary from areoles along the margin of the pad, actinomorphic, perfect, epigynous, with usually numerous small, leaflike bracts at the base and on the surface of the ovary, these shed during bud development. Hypanthium a short, broadly conical crown at the tip of the ovary. Perianth of numerous free parts in a dense spiral series, these grading from outer sepaloid tepals, these shorter, ovate, and green, to inner petaloid tepals, these longer, obovate-spatulate, and brightly colored. Stamens numerous, the filaments relatively long and slender, the small anthers attached toward their midpoint, yellow. Pistil 1 per flower, of usually 5–10 fused carpels. Ovary inferior, with 1 locule, the placentation parietal, the outer surface with few to numerous areoles. Style 1 per flower, relatively stout, the stigmas 4–9, fused along their inner margins to form a hemispheric to broadly conical structure. Ovules numerous. Fruits berries, broadly club-shaped to obovoid, frequently long-persistent on the plant. Seeds broadly kidney-shaped to broadly oval or nearly circular in outline, flattened laterally, tan to straw-colored, the margin with a well-differentiated blunt rim, often lighter colored before drying, bony or corky in texture. About 150 species, North America to South America, Caribbean Islands; widely introduced elsewhere.

In recent years, there has been a growing trend to split *Opuntia,* as traditionally recognized (with about 350 species), into several smaller genera (Anderson, 2001; Hunt and Taylor, 2002; Pinkava, 2003). The prickly pears form the largest of these groups and continue to be treated in a more narrowly circumscribed version of *Opuntia,* with several splinter genera, such as *Cylindropuntia* (Engelm.) F.M. Knuth for the approximately 33 cholla species, elevated from their earlier subgeneric status.

Seedlings of prickly pear species or suppressed plants growing in shady areas or dense plant associations frequently have tubular stems and sometimes have been misdetermined as chollas. The vegetative portions of *Opuntia* plants are quite variable, which can make them difficult to identify. In the southwestern United States, the situation is exacerbated by the presence of hybrids where two or more species grow in proximity.

Field botanists generally tend to avoid collecting specimens of *Opuntia,* because of difficulties in drying them and the nuisance that glochids create when plants are handled. A number of strategies have been suggested to facilitate drying. Carefully slicing pads, flowers, and fruits lengthwise (wear thick gloves, or immobilize the sample in thick layers of newspaper first) doubles the effective number of specimens that one plant can provide and also allows moisture to escape more readily. If the plant press cannot be placed in a drying oven or other situation where hot air is passing through the press, it becomes necessary to change the blotters frequently to avoid mold contamination of the specimens. Some collectors also advocate adding salt or silica gel to the cut surfaces to facilitate drying, but this tends to be a messy procedure.

The Missouri species of *Opuntia* are easily cultivated at sunny sites in well-drained soils. Individual pads that are removed and allowed to air-dry for a week or more generally produce new roots quickly if the bases are placed in soil. In pastured areas where cattle spread pads by kicking or stepping on plants, or in highway median areas where mowing equipment achieves a similar result, the rerooting of pads can create large, dense populations in relatively few years.

1. Spines mostly 6–10 per areole; glochids 1.0–1.5 mm long; fruits becoming dry and papery with age, with both glochids and relatively dense spines on the surface; stigmas green . . . . 3. O. POLYACANTHA
1. Spines absent or 1–5(–6) per areole; glochids 2–3 mm long; fruits remaining fleshy with age, with only glochids on the surface; stigmas light yellow to nearly white
    2. Spines absent or 1(–3) per areole, often restricted to near the margin of the pads; seeds with the marginal rim relatively short (extending about 0.5 mm beyond the seed body) and smooth . . . . 1. O. HUMIFUSA
    2. Spines (1–)3–5(–6) per areole, usually present on the surface and margins of the pad; seeds with the marginal rim relatively large (extending about 1 mm beyond the seed body) and often somewhat irregular . . . . 2. O. MACRORHIZA

1. Opuntia humifusa (Raf.) Raf. (eastern prickly pear)

O. compressa (Salisb.) J.F. Macbr.

Pl. 329 f–i; Map 1400

Plants prostrate or less commonly forming low mounds, the main roots sometimes tuberous-thickened. Pads 4–8(–12) cm long, obovate to nearly circular, dark or bright green, often somewhat shiny at maturity, often somewhat glaucous when young. Areoles mostly 10–30 mm apart, 2–4 mm in diameter, oval to circular. Glochids 2–3 mm long, yellowish brown to orangish brown. Spines absent or 1(–3) per areole, often restricted to near the margin of the pads, 1–5 cm long, spreading or deflexed, straight, not flattened, white to gray or brown, often purplish toward the base. Tepals to 3 cm long, the petaloid ones yellow, sometimes orange or pinkish-tinged toward the base. Stigma lobes white or pale yellow. Fruits 2–5 cm long, the surface with glochids but no spines, the areoles sometimes restricted to the apical rim, remaining fleshy with age, green, sometimes becoming reddish- or purplish-tinged with age. Seeds 3.5–4.5 mm long, the raised rim relatively short (extending about 0.5 mm

1401. Opuntia macrorhiza　　1402. Opuntia polyacantha　　1403. Calycanthus floridus

beyond the seed body) and smooth. $2n=22, 44$. May–July.

Scattered nearly throughout the state (eastern U.S. west to South Dakota and New Mexico, disjunct in Montana; Mexico). Rocky areas of upland prairies, sand prairies, glades, tops and exposed ledges of bluffs, and rocky stream terraces; also pastures, roadsides, railroads, and open, disturbed areas.

Benson (1982) treated *O. humifusa* as comprising three varieties, the widespread var. *humifusa* and the southeastern Coastal Plain endemics var. *ammophila* (Small) L.D. Benson and var. *austrina* (Small) Dress. Many subsequent authors have rejected the last two as mere morphological variants within the species that tend to differ in their more upright growth form and longer, narrower pads. These taxa require further study. For a discussion of difficulties in separating *O. humifusa* from *O. macrorhiza*, see the treatment of that species.

2. Opuntia macrorhiza Engelm. (plains prickly pear)

O. compressa var. *macrorhiza* L.D. Benson

Pl. 329 j, k; Map 1401

Plants prostrate or more commonly forming low mounds, the main roots usually tuberous-thickened. Pads 5–14 cm long, obovate to nearly circular, dark green to bluish green, usually not shiny, usually moderately glaucous. Areoles mostly 10–30 mm apart, 2–4 mm in diameter, oval to circular. Glochids 2–3 mm long, yellowish brown to orangish brown. Spines (1–)3–5(–6) per areole, usually present on both the surface and margins of the pad but denser toward the margins, 1.0–5.5 cm long, spreading or deflexed, straight or rarely slightly curved, not or only slightly flattened, white to gray or rarely brown, often purplish toward the base. Tepals to 3 cm long, the petaloid ones yellow, usually orange or red toward the base. Stigma lobes pale yellow to light yellow. Fruits 2–5 cm long, the surface with glochids but no spines, the areoles sometimes restricted to the apical rim, remaining fleshy with age, green to bluish green, sometimes becoming purple or reddish purple with age. Seeds 3.5–5.0 mm long, the raised rim (best seen in fresh, undried seeds) relatively large (extending about 1 mm beyond the seed body) and often somewhat irregular. $2n=22, 44$. May–July.

Uncommon and widely scattered, mostly in southwestern Missouri (western U.S. and adjacent Mexico east to Michigan and Louisiana, introduced in Ohio). Glades, tops and exposed ledges of bluffs, and rocky openings of dry upland forests; also unconfirmed reports from degraded sand prairies (Adair County).

This taxon is accepted at the species level with extreme reluctance in the present treatment. It represents a western element in the *O. humifusa* complex, but there is a large area of geographic overlap from the eastern Great Plains through the midwestern states between it and true *O. humifusa*. Comparing typical *O. humifusa* plants of the eastern United States with typical *O. macrorhiza* from the western states, the two would appear to be amply distinct in growth form, spine morphology, and tepal color. However, although many populations of typical *O. humifusa* exist in Missouri, there are practically no individuals of typical *O. macrorhiza*. A number of populations that yielded specimens determined as *O. macrorhiza* during one growing season seem to key better to *O. humifusa* upon a second visit. Spine length, spine number, tepal color, and pad coloration all do not serve to divide Missouri plants easily into two discrete taxa. Although plants of *O. macrorhiza* routinely produce tuberous-thickened main roots, contrary to statements in some earlier literature some individuals of *O. humifusa* also produce such thickened roots. Benson (1982) discussed differences in the shape and texture of the rim of the seeds, but these characters are difficult to interpret in dried material and also do not appear to correlate entirely with spine, pad, and tepal characters. Benson himself at one time reduced *O. macrorhiza* to a variety under *O. compressa,* a synonym of *O. humifusa,* but a valid varietal combination has not been published for it under the latter name. The current expert on North American *Opuntia,* Donald Pinkava of Arizona State University, has had similar difficulties in distinguishing these two taxa in the broad zone of overlap in their ranges. Thus users should expect to find frustration in applying the above key to species to some plants in the state.

The closely related *O. pottsii* Salm-Dyck is sometimes recognized as *O. macrorhiza* var. *pottsii* (Salm-Dyck) L.D. Benson. It differs from *O. macrorhiza* primarily in its more slender spines and more reddish tepals and grows in the southwestern United States and adjacent Mexico.

3. Opuntia polyacantha Haw. var. polyacantha (starvation cactus)

Pl. 329 e; Map 1402

Plants prostrate or more commonly forming low mounds, the main roots not tuberous-thickened. Pads 4–10(–14) cm long, obovate to nearly circular, dark green to bluish green, usually not shiny, sometimes slightly glaucous. Areoles mostly 6–9 mm apart, 2–3 mm in diameter, oval to circular. Glochids 1.0–1.5 mm long, straw-colored to yellow. Spines (4–)6–10 per areole, relatively dense on both the surface and margins of the pad, 0.6–4.0(–5.5) cm long, spreading and deflexed, straight or rarely slightly curved, not or only slightly flattened, straw-colored to tan, gray, or less commonly reddish brown. Tepals to 4 cm long, the petaloid ones yellow, sometimes pinkish- or reddish-tinged toward the base. Stigma lobes green. Fruits 2–4 cm long, the surface with both glochids and relatively dense spines, the areoles relatively dense on the rim and surface, becoming dry and papery with age, tan to dull brown at maturity. Seeds 3–5 mm long, the raised rim relatively broad (1–2 mm thick) and often somewhat irregular. $2n=44, 66$. May–July.

Uncommon, known only from a single historical collection from Jasper County (western U.S. east to North Dakota and Texas; Canada). Habitat unknown, but presumably glades or upland prairies.

Most authors treat this species as a complex of four or five varieties differing in subtle details of areole and spine distribution and morphology. The Missouri specimen appears to represent the widespread var. *polyacantha.*

Steyermark (1963) excluded *O. polyacantha* from the Missouri flora, having searched without success for a voucher specimen to support earlier literature reports for the state. Benson (1982) included Missouri in the range of the species, although a typographical error in the specimen citation contributed to the confusion surrounding whether the species had ever existed in the state. During the present research, the Jasper County collection made by E. J. Palmer in 1909 was located at the U.S. National Herbarium and its identity verified. The area around Webb City was strip-mined extensively for lead in the early twentieth century, and recent searches have failed to relocate plants in southwestern Missouri. Thus, it seems likely that *O. polyacantha* became extirpated from the state long ago.

CALYCANTHACEAE (Strawberry Shrub Family)
(Nicely, 1965)

Two to 4 genera, 8–10 species, U.S., China, Australia.

1. Calycanthus L.

Two species, U.S.

1. Calycanthus floridus L. var. floridus
(Carolina allspice, sweetshrub)

Map 1403

Plants shrubs, the stems 1–3 m long, erect to ascending. Twigs hairy, purplish brown, the buds naked, usually hidden by the petioles during the growing season. Leaves opposite, simple, short-petiolate, lacking stipules. Leaf blades 5–15 cm long, ovate to oblong-elliptic, tapered to the sharply pointed tip, narrowed to rounded at the base, the margins entire, the upper surface roughened, the undersurface hairy and sometimes somewhat glaucous. Flowers solitary at the tips of branches, actinomorphic, perfect, perigynous, the receptacle expanded and somewhat cup-shaped, with several small bracts on the outer surface. Perianth of numerous free tepals of varying lengths, these 1.5–4.0 cm long, narrowly elliptic to strap-shaped, hairy on the outer surface, red to maroon, the innermost tepals usually white-tipped. Stamens numerous, the outer 10–15 fertile and the inner 10–25 reduced to linear, hairy staminodes, the fertile stamens with a short, broad, hairy filament extended between the anthers into a short, blunt, succulent tip. Pistils 10–35 per flower, each composed of 1 carpel. Ovary with 1 locule and 2 ovules (1 abortive), hairy, especially at the base and tip. Style slender, becoming elongated in fruit, attached off-center at the tip of the ovary, the stigma minute, decurrent. Fruits enclosed within the greatly expanded, flask-shaped receptacle, this 2–6 cm long, with a small opening at the tip, hairy, becoming dry and brown at maturity, persistent.

Individual fruits achenes, 10–12 mm long, plump, oblong-elliptic in outline, with a finely wavy, longitudinal seam around the circumference, the surface finely hairy, somewhat shiny, dark brown. $2n=22, 33$. April–June.

Introduced, uncommon and sporadic (southeastern U.S. west to Kentucky and Mississippi). Margins of mesic upland forests and roadsides.

Calycanthus floridus is widely cultivated in the eastern United States for its ornamental foliage and large, fragrant, unusual flowers, which are pollinated by beetles. The plants (especially the seeds) are poisonous to humans and livestock, containing a number of alkaloids, including calycanthine, whose action is similar to that of strychnine (Nicely, 1965; Wood, 1958). The other variety, var. *glaucus* (Willd.) Torr. & A. Gray, occupies nearly the same range, extending westward into southern Ohio, and differs in its glabrous or nearly glabrous twigs and leaves. The St. Louis County population, which undoubtedly became naturalized from plantings at a former house site, was first discovered by the Botany Group of the Webster Groves Nature Study Society in 1995.

CAMPANULACEAE (Bellflower Family)
Contributed by David J. Bogler

Plants mostly perennial herbs (sometimes shrubby elsewhere), usually with milky sap. Leaves usually alternate, simple, the margins entire, toothed, or lobed, lacking stipules. Inflorescences terminal spikes or racemes, axillary clusters or solitary flowers, or the flowers rarely terminal and solitary. Flowers perfect, epigynous or deeply perigynous. Calyces 5-lobed (sometimes 3- or 4-lobed in cleistogamous flowers), usually actinomorphic, fused to the ovary, the tube sometimes extending slightly past the ovary, persistent, sometimes with a short, reflexed appendage between each of the lobes. Corollas more or less 5-lobed, actinomorphic and funnelform to bell-shaped or saucer-shaped, or zygomorphic and strongly 2-lipped, and usually split along the apparently upper side (through which the stamens and style are exserted), white to purple, blue, or red. Stamens 5, the filaments distinct or united into a tube toward the tip. Anthers attached at their bases, distinct or fused into a tube surrounding the style. Pistil 1 per flower, of 2–5 fused carpels. Ovary inferior or less commonly only partially inferior, with 1 or 2–5 locules, the placentation parietal (if 1 locule) or axile. Style 1 per flower, elongate, the stigma 2–5-lobed. Ovules numerous. Fruits capsules, dehiscing by apical or lateral pores, slits, or longitudinally. Seeds numerous, minute. Sixty to 70 genera, about 2,000 species, nearly worldwide.

The Campanulaceae are recognized by the combination of alternate leaves, milky latex, well-developed corolla with united petals, inferior to half-inferior ovary, often specialized pollination mechanisms, and capsular fruits with numerous seeds. The family comprises two distinct subfamilies that some authors have maintained as separate families (Rosatti, 1986).

The subfamily Campanuloideae has actinomorphic flowers, nontwisted flower stalks, most commonly three carpels and stigma lobes, free filaments and anthers, and pollen grains with spiny ornamentation. The Lobelioideae have strongly zygomorphic flowers borne on a stalk that is twisted so that the upper lip of the flower is positioned basally at flowering (resupinate, as in most orchids), most commonly 2 carpels and stigma lobes, the filaments fused into a tube toward the tip, the anthers fused into a ring, and pollen grains with a network of ridges rather than spines. Recent molecular studies (Cosner et al., 1994) have supported maintaining both groups as a single family. The flowers of species in the Lobelioideae are functionally similar to those of the Asteraceae in that the stigma elongates through the center of the ring of fused anthers, with a brush of hairs near the style tip pushing the pollen out of the flower, and the stigma lobes do not spread and become receptive until after the pollen has been released. This mechanism promotes outcrossing between flowers.

The genus *Sphenoclea* is sometimes included in the Campanulaceae. The present treatment follows that of Rosatti (1986) in treating this group as a separate family, Sphenocleaceae (to be included in the third volume of the present work). Steyermark (1963) noted the existence of a specimen of the commonly cultivated ornamental, *Platycodon grandiflorum* (Jacq.) A. DC. (balloon flower), at the herbarium of William Jewell College, Liberty, Missouri, but excluded the species from the Missouri flora in the belief that it was an isolated, nonpersistent escape from an adjacent planting. This specimen was not examined during the present study. *Platycodon* resembles *Campanula* but differs in its larger flowers (corollas to 45 mm long) and strongly inflated buds. It has been recorded as an escape sporadically from New York to North Carolina, but it has not persisted outside of cultivation yet in any midwestern state.

1. Flowers zygomorphic, bilabiate, the corolla split along the apparent upper side; filaments united above, anthers united in a ring surrounding the style; carpels 2; capsules opening longitudinally by apical valves . . . . 2. LOBELIA
1. Flowers actinomorphic, bell-shaped to saucer-shaped, the fused portion not split along the upper side; filaments free, anthers free; carpels 3–5; capsules opening by lateral pores or slits
    2. Flowers mostly stalked from the axil of a reduced (much shorter and narrower than the leaves) bract, all flowers similar and open-flowering; style curved; fruits obconical to subglobose . . . . 1. CAMPANULA
    2. Flowers sessile in the axil of a leaflike bract, lowermost flowers usually cleistogamous; style straight; fruits ellipsoid to narrowly ellipsoid or narrowly cylindrical . . . . 3. TRIODANIS

1. Campanula L. (bellflower)

Plants annual or perennial herbs. Stems erect or loosely ascending, unbranched or few-branched. Leaves sessile or petiolate, the margins entire or sharply toothed. Inflorescences spikes, racemes, or panicles, the flowers from the axils of mostly reduced (much shorter and narrower than the leaves) bracts. Flowers epigynous, not cleistogamous. Calyces actinomorphic, all 5-lobed, often with a short, reflexed appendage between each of the lobes, persistent at fruiting. Corollas actinomorphic, funnelform to bell-shaped or saucer-shaped, 5-lobed, but usually not divided below the middle, commonly blue or white. Stamens 5, attached to the base of the corolla, the filaments dilated and hairy at the base, the anthers distinct, elongate. Pistil with 3–5 carpels. Ovary totally inferior, with 3–5 locules. Style elongate, the stigma 3–5-lobed. Fruits obconical to nearly globose capsules, often longitudinally ribbed, dehiscent by 3–5 lateral pores or slits. Seeds ellipsoid, sometimes somewhat flattened, the surface brown, shiny.

1404. Campanula americana   1405. Campanula aparinoides   1406. Campanula rapunculoides

About 300 species, nearly worldwide, but most diverse in temperate portions of the Northern Hemisphere.

1. Stem leaves lanceolate, more than 4 mm wide; inflorescences dense racemes, erect or strongly ascending, relatively stout
    2. Leaf blades with the margins evenly toothed; corolla saucer-shaped; style of open flowers S-shaped, strongly exserted from the corolla; fruits erect or ascending, dehiscing by apical pores . . . . 1. C. AMERICANA
    2. Leaf blades with the margins coarsely and somewhat unevenly toothed; corolla bell-shaped; style of open flowers straight, included within or barely exserted from the corolla; fruits nodding, dehiscing by basal pores
       . . . . 3. C. RAPUNCULOIDES
1. Stem leaves linear or narrowly lanceolate, mostly 1–4 mm wide; inflorescences of solitary flowers or in open, few-flowered clusters, spreading to loosely ascending
    3. Stems slender, loosely spreading and often reclining on other plants; leaf blades with recurved hairs having swollen bases along the margins and midvein; corollas 3–8 mm long, pale blue to nearly white . . . . . 2. C. APARINOIDES
    3. Stems relatively robust, erect to loosely ascending, but not reclining on adjacent plants; leaves glabrous; corollas 10–20 mm long, blue
       . . . . 4. C. ROTUNDIFOLIA

1. Campanula americana L. (tall bellflower)
Campanulastrum americanum (L.) Small
Pl. 330 f–h; Map 1404

Plants robust annuals or biennials, with a short taproot. Stems 50–200 cm long, erect or strongly ascending, unbranched or occasionally sparsely branched toward the tip, glabrous or less commonly sparsely to moderately hairy toward the tip. Leaves gradually reduced from near the stem base to the tip, with a slender, often winged petiole 1–5 cm long, the uppermost leaves usually sessile. Leaf blades 5–16 cm long, 2–6 cm wide, lanceolate to oblong-ovate, usually tapered at the base, tapered to a sharply pointed tip, the margins evenly, relatively finely, and sharply toothed, inconspicuously hairy, the surfaces sparsely hairy, especially along the veins. Inflorescences relatively dense spikelike racemes, the flowers solitary or in clusters of 3 at each node, the lowermost bracts more or less leaflike but most of the bracts reduced (much shorter and narrower than the leaves). Calyx tube 3–4 mm long at flowering, elongating as fruit matures, the lobes 7–12 mm long, 0.8–1.2 mm wide, hairlike to very narrowly triangular. Corollas saucer-shaped, the tube 3–4 mm long, the lobes 6–15 mm long, hairy at the tip, blue. Style 12–15 mm long, strongly exserted from the corolla, elongating and becoming S-shaped (curved from near the base toward the bottom of the flower and arched upward toward the tip) in open flowers, the stigma 3-lobed. Fruits 5–12 mm long, 3–4 mm in diameter, obconic, 3-locular, erect or ascending, dehiscing by 3 round pores near the tip; seeds 0.8–1.2 mm long, 0.8–1.0 mm wide, flattened, often with a thin, winglike margin. $2n=58, 102$. July–September.

Forest edges, open, mesic woods, bluffs, disturbed areas, roadsides. Endemic to eastern North America, from southern Canada south to Florida, west to Minnesota and Oklahoma. Abundant in Missouri, this species is easily identified by its stout, tall habit, shortly rotate corolla, and exserted, curved style. It does not appear to be closely related to the other species in the genus and has been considered a separate genus, *Campanulastrum,* by some authors (Small, 1903). However, critical characters such as the rotate corolla are found in other species of *Campanula* (Rosatti, 1986).

Pollination in *C. americana* was described by Shetler (1962). The flowers are protandrous, that is, the pollen matures and is dispersed prior to the maturation and receptivity of the stigmas. Pollen matures in the bud, with the 5 anthers forming a tight column around the terminal part of the style. The anthers shed their pollen on the style, which is equipped with bristly hairs. Once the bud opens, the anthers wither very quickly. A variety of insects, including bees, flies, and butterflies, visit the flowers, attracted by nectar produced at the base of the filaments. When the flower opens, these insects attempt to get at the nectar, rubbing against the pollen-bearing hairs of the style. At about the same time, the style elongates rapidly, develops an S-shaped curve, and the stigmas become receptive to pollen carried from other flowers. Self-pollination might be possible if the unfolding stigma lobes touch the style.

2. Campanula aparinoides Pursh var. aparinoides (marsh bellflower)

Pl. 330 c; Map 1405

Plants slender perennial herbs, with very shallow roots and sometimes slender rhizomes. Stems 15–50 cm long, loosely spreading, often reclining on adjacent plants, often somewhat 3-angled, roughened with short, recurved hairs, sparingly branched toward the tip. Leaves gradually reduced from near the stem base to the tip, sessile. Leaf blades 0.6–2.5(–5.0) cm long, 1–5 mm wide, narrowly lanceolate to linear, angled or tapered at the base, shot-tapered at the tip, with recurved hairs having swollen bases along the margins and midvein, the margins sometimes also with a few minute teeth. Inflorescences of solitary flowers at the branch tips or in open, few-flowered clusters, spreading to loosely ascending. Calyx tube 1–2 mm long, the lobes 1–2 mm long, triangular-ovate. Corolla funnelform to more or less bell-shaped, the tube 1–2 mm long, lobes 3–5 mm long, pale blue to nearly white. Style 1.5–2.0 mm long at flowering, mostly enclosed in the corolla, rarely slightly exserted, not elongating markedly as the fruits mature, the stigma usually 3-lobed. Fruits 2–3 mm long, 1.5–2.5 mm in diameter, obovoid to nearly globose, spreading to more or less pendant, dehiscing by usually 3 pores, these lateral, usually near the base. Seeds 0.8–1.2 mm long, ellipsoid. $2n=34$, 68, 136, 170. June–July.

Uncommon and widely scattered (eastern U.S. west to South Dakota, Nebraska, and Missouri; Canada). Fens and calcareous swamps.

Another variety, var. *grandiflora* Holz., with generally narrower longer leaves and somewhat larger corollas, occurs in the northeastern United States and Canada south to Pennsylvania and Iowa.

3. Campanula rapunculoides L. var. rapunculoides (creeping bellflower, rover bellflower, false rampion)

Pl. 330 d, e; Map 1406

Plants robust perennial herbs, colonial from long-creeping rhizomes, with fleshy roots. Stems 40–100 cm long, erect or strongly ascending, unbranched or occasionally sparsely branched toward the tip, glabrous or sparsely hairy toward the tip. Basal leaves with a long, slender petiole 4–10 cm long, the blade 2–7 cm long, 1.5–3.5 cm wide, broadly ovate to cordate, rounded, or cordate at the base, angled or slightly tapered to a sharply pointed tip, the margins somewhat unevenly, relatively coarsely, and sharply toothed, the surfaces minutely roughened, especially at and near the margins. Stem leaves gradually reduced from near the stem base to the tip, short-petiolate to sessile, the blade 2–5 cm long, those of the lower leaves ovate to ovate-triangular, those of the upper leaves narrowly lanceolate, otherwise like the basal leaves. Inflorescences relatively dense racemes, the flowers mostly oriented toward 1 side, solitary at each node and usually nodding, the lowermost bracts more or less leaflike but most of the bracts reduced (much shorter and narrower than the leaves). Calyx tube 4–5 mm long at flowering, elongating as fruit matures, the lobes 5–7 mm long, 1.5–2.0 mm wide, lanceolate. Corollas bell-shaped, the tube 9–13 mm long, the lobes 7–10 mm long, glabrous at the tip, blue to purple. Style 18–20 mm long, included within or barely exserted from the corolla, straight or nearly so in open flowers, the stigma 3-lobed. Fruits 5–8 mm long, 5–6 mm in diameter, nearly spherical, 3-locular, nodding, dehiscing by 3 round pores near the base; seeds 0.8–1.2 mm long, 0.8–1.0 mm wide, flattened, usually with a thickened margin. $2n=68$, 102. June–October.

Introduced, scattered, mostly in the northwestern quarter of the state (native of Europe, Asia;

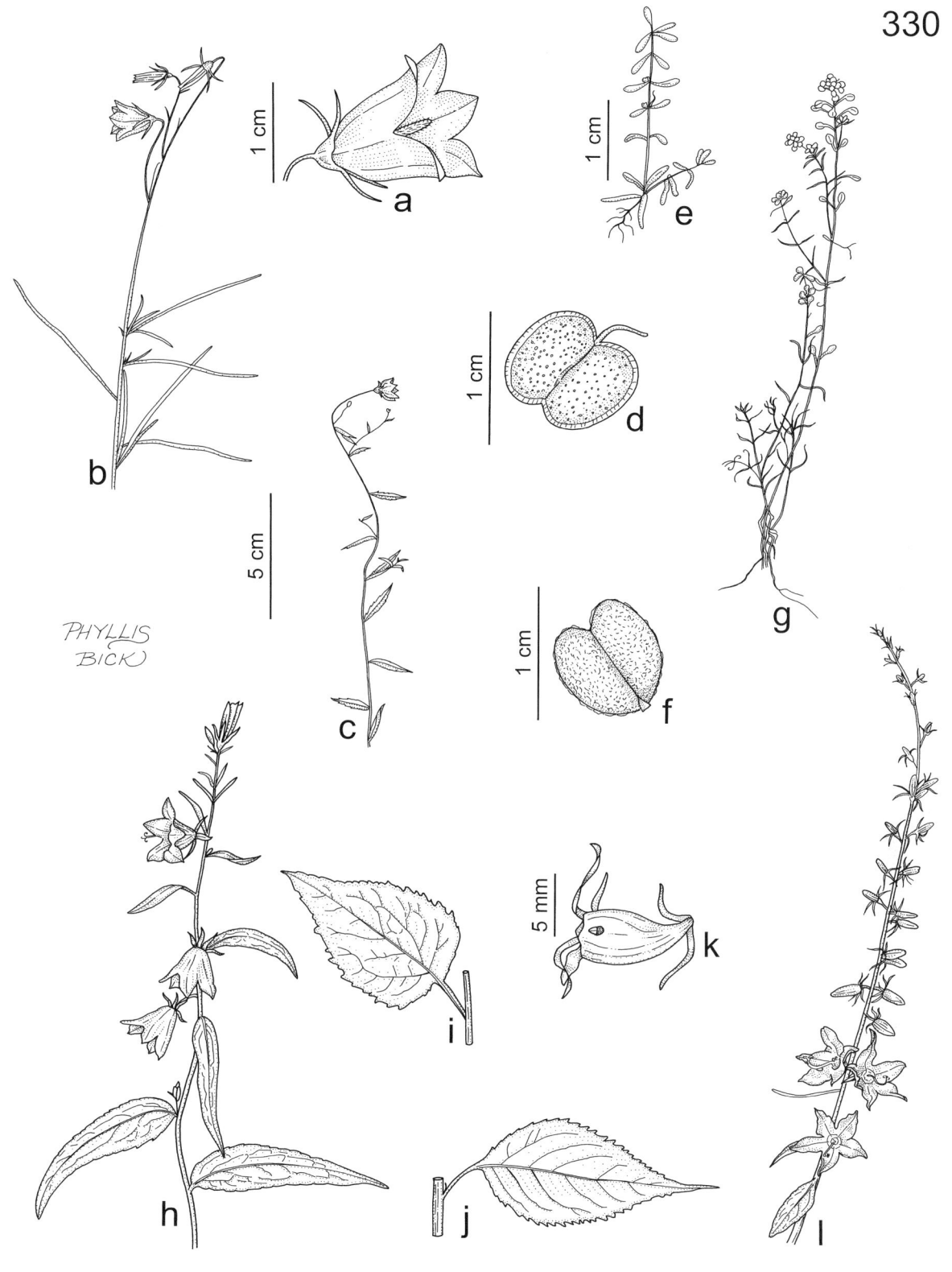

Plate 330. Campanulaceae. *Campanula rotundifolia*, **a)** flower, **b)** inflorescence. *Campanula aparinoides*, **c)** inflorescence. *Campanula rapunculoides*, **d)** leaf, **e)** inflorescence. *Campanula americana*, **f)** leaf, **g)** fruit, **h)** inflorescence.

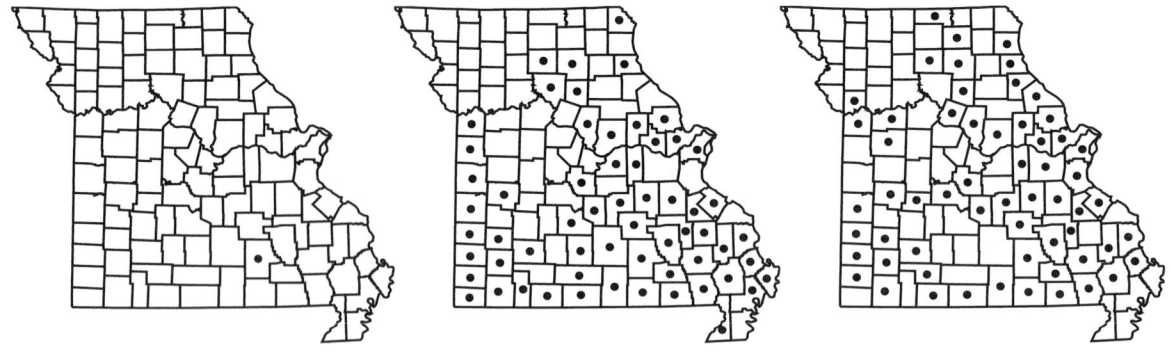

1407. Campanula rotundifolia   1408. Lobelia cardinalis   1409. Lobelia inflata

introduced widely from southern Canada south to Kentucky and Missouri). Railroads.

Creeping bellflower has a long history of cultivation in gardens. However, its widely creeping rhizomes allow it to spread aggressively in some situations, and the fleshy roots make the plant difficult to eradicate from unwanted sites.

4. Campanula rotundifolia L. (harebell, bluebell)

Pl. 330 a, b; Map 1407

Plants slender perennial herbs, with shallow roots and slender rhizomes and/or stolons. Stems 20–70 cm long, erect to loosely ascending, but not reclining on adjacent plants, often slightly 3-angled, glabrous or inconspicuously hairy in longitudinal lines, unbranched or sparingly branched toward the tip. Basal leaves often absent at flowering, with a long, slender petiole 2–3 cm long, the blade 0.5–1.0 cm long, 0.5–0.8 cm wide, broadly ovate to nearly circular, shallowly cordate, rounded, or broadly angled at the base, rounded or angled to a bluntly or sharply pointed tip, the margins entire or few-toothed, the surfaces glabrous. Stem leaves gradually reduced from near the stem base to the tip, long-petiolate to sessile, the blade 3–6 cm long, those of the lower leaves narrowly elliptic to oblanceolate, those of the upper leaves narrowly lanceolate to linear, angled or short-tapered at the base, tapered at the tip, the margins entire, the surfaces glabrous. Inflorescences of 3–8-flowered, open, nodding racemes, less commonly paniculate (of solitary flowers elsewhere). Calyx tube 2–3 mm long, the lobes 4–5 mm long, linear. Corolla bell-shaped, the tube 6–10 mm long, lobes 3–4 mm long, light blue to blue. Style 10–11 mm long at flowering, enclosed in the corolla, not elongating markedly as the fruits mature, the stigma usually 3-lobed. Fruits 4–5 mm long, 2.5–3.0 mm in diameter, obconic to narrowly obovoid, pendant, dehiscing by 3 basal pores. Seeds 0.6–0.9 mm long, narrowly ellipsoid. $2n=34, 56, 68, 102$. June–July.

Uncommon, known only from a portion of the Jacks Fork River in Shannon County (North America, Europe, Asia). Crevices and ledges of tall, north-facing dolomite bluffs.

Steyermark (1963) considered the Missouri populations of the circumboreal *C. rotundifolia* to represent relicts from the Pleistocene epoch. He noted that a small set of species exists in the cool, moist microclimate in the area where the species occurs that all have their present-day main ranges to the north of Missouri and were stranded when the surrounding climate became warmer as the glaciers receded. These include *Campanula rotundifolia, Galium boreale, Trautvetteria caroliniensis,* and *Zigadenus elegans,* among others. See the introductory section on the origins of the Missouri flora in Yatskievych (1999) for further discussion.

2. Lobelia L. (lobelia)

Plants annual or perennial herbs (woody elsewhere) usually with milky juice. Leaves sessile to shortly petiolate, lanceolate to oblanceolate or ovate, the margins finely to coarsely toothed. Inflorescences terminal spikes, racemes or panicles, the flowers from the axils of mostly reduced (shorter and narrower than the leaves) bracts. Flowers more or less epigynous, resupinate (inverted by twisting of the stalk), not cleistogamous. Calyces all 5-lobed, more or less actinomorphic, the tube usually ribbed, the lobes usually longer than the tube, sometimes with short, reflexed appendages (auricles) toward the base alternating with the lobes. Corollas zygomorphic, strongly 2-lipped with spreading lobes, the apparent lower lip 3-lobed, the upper lip 2-lobed and split nearly to the base through which the stamens and style are exserted, the tube sometimes with slitlike openings (fenestrate), red, or blue to white (rarely purple). Stamens 5, free from the corolla, the filaments free at the very base but mostly fused into a tube, the anthers also united into a ring around the style, the lower 2 anthers usually shorter than the others and with white tufts of hair. Pistil with 2 carpels. Ovary half to totally inferior, with 2 locules. Style elongating through the filament tube, stigma lobes 2, protruding and expanding through the anther tube after the pollen has been shed. Fruits hemispherical capsules, usually longitudinally ribbed, crowned by the persistent calyx lobes, withered corolla, stamen tube, and style, longitudinally dehiscent by 2 apical pores. Seeds oblong to oblong-elliptic, the surface yellowish brown, with minute wrinkles or tubercles. About 365 species, nearly worldwide, most diverse in tropical and warm temperate regions.

The pyridine alkaloid lobeline is extracted from several species of *Lobelia*. This alkaloid is similar to nicotine and has been used as an ingredient in antismoking medications (Rosatti, 1986). *Lobelia* extracts have been used medicinally to treat asthma and bronchitis, but in large doses they can cause nausea, paralysis, and even death (Steyermark, 1963).

1. Corollas 4–10 mm long, the tube lacking longitudinal slits; stamens with the filaments 2–4 mm long
   2. Stems usually branched above the midpoint, the apical portion and the inflorescence axis moderately to densely hairy; calyx tube 2–3 mm long; capsules 6–10 mm long, greatly inflated at maturity . . . . 2. L. INFLATA
   2. Stems usually unbranched (occasionally with a few branches near the base), the apical portion and the inflorescence axis glabrous or sparsely hairy; calyx tube 0.5–1.5 mm long; capsules 3–5 mm long, not inflated at maturity
       . . . . 5. L. SPICATA
1. Corollas 15–35 mm long, the tube with longitudinal slits; stamens with the filaments 6–30 mm long
   3. Corollas bright red; stamens with the filaments (15–)18–30 mm long
       . . . . 1. L. CARDINALIS
   3. Corollas blue or rarely white; stamens with the filaments 6–15 mm long
       4. Stems and leaves densely short-hairy; calyces with short, inconspicuous auricles; stamens with the filaments 6–10 mm long . . . . 3. L. PUBERULA
       4. Stems and leaves glabrous or sparsely hairy, the undersurface of the leaves rarely roughened along the veins; calyces with relatively long, conspicuous auricles; stamens with the filaments 12–15 mm long
           . . . . 4. L. SIPHILITICA

1. Lobelia cardinalis L. (cardinal flower)
L. cardinalis ssp. *graminea* (Lam.) McVaugh
L. splendens Humb. & Bonpl. ex Willd.
Pl. 331 a, b; Map 1408

Plants perennial herbs, with fibrous roots. Stems 40–150 cm long, erect or strongly ascending, unbranched or occasionally few-branched toward the tip, glabrous or rarely with sparse, short hairs. Leaves with petioles 1–2 cm long, gradually reduced from about the stem midpoint to the stem tip. Leaf blades 2–15 cm long, 0.5–4.0 cm wide, narrowly lanceolate to oblanceolate or narrowly ovate, angled or tapered at the base, angled or tapered to a sharply pointed tip, the margins shallowly or irregularly toothed, the surfaces glabrous or sparsely and inconspicuously pubescent with short, curved hairs. Inflorescences densely to loosely flowered racemes with 10–40 flowers, the bracts of the lower flowers leaflike, the upper bracts much shorter (each flower also with a pair of minute bracts immediately below the calyx). Calyces 9–15 mm long, the tube 3–5 mm long at flowering, becoming somewhat enlarged as the fruit matures, the slender lobes 6–12 mm long, becoming noticeably elongated as the fruit matures, lacking auricles. Corollas 25–45 mm long, bright red, the outer surface glabrous, the tube 15–30 mm long, with slitlike openings (fenestrate), the lobes 14–20 mm long. Filament tube (15–)18–30 mm long, red, the anther tube 3–5 mm long. Fruits 6–10 mm long. Seeds 0.6–1.0 mm long, ovoid to ellipsoid, the surface with a series of minute pits and ridges, sometimes also with minute, irregular tubercles, yellowish brown. 2n=14. July–October.

Scattered nearly throughout the state but absent or uncommon in the western portion of the Glaciated Plains Division (U.S., Canada, Mexico, Central America, South America). Banks of streams, rivers, and spring branches, openings of bottomland forests, swamps, and sloughs; also ditches and wet roadsides.

Because of its striking red flowers, *L. cardinalis* was introduced very early into cultivation in Europe. It has been painstakingly subdivided into a number of species, subspecies, and varieties (Bowden, 1982), but none of these taxa are supported by recent quantitative morphological analyses (Thompson and Lammers, 1997). About the most that can be said is that there is a trend for populations in the southwestern U.S. to have narrower leaves. Individuals with white or pink corollas are collected occasionally in Missouri and have been called f. *alba* (J. McNab) H. St. John and f. *rosea* H. St. John respectively. *Lobelia cardinalis* is most closely related to *L. siphilitica,* with which it occasionally hybridizes, as discussed under that species. The red flowers of *L. cardinalis* are attractive to hummingbirds, which are probably the major pollinators.

2. Lobelia inflata L. (Indian tobacco, inflated lobelia)

Pl. 331 c, d; Map 1409

Plants annual, with taproots. Stems 20–90 cm tall, erect or ascending, unbranched or more commonly moderately branched toward the tip, not winged, moderately to densely pubescent with short, curved to more or less spreading hairs. Leaves sessile or the lowermost leaves sometimes with a short, winged petiole, gradually reduced above the stem base. Leaf blades 2–11 cm long, 1.0–3.5 cm wide, oblong-elliptic to elliptic, ovate, or obovate, tapered at the base, the margins finely to relatively coarsely and sometimes irregularly scalloped or bluntly toothed, the upper surface glabrous or with sparse, more or less appressed hairs, the undersurface moderately pubescent with curved hairs. Inflorescences loosely flowered narrow racemes, the axis moderately to densely hairy, the bracts similar to the adjacent leaves toward the base, the upper bracts much shorter (each flower also with a pair of minute bracts immediately below the calyx). Calyces 5–9 mm long, the tube 2–3 mm long at flowering, becoming noticeably enlarged and inflated as the fruit matures, the slender lobes 3–6 mm long, lacking auricles. Corollas 5–8 mm long, white, sometimes bluish-tinged or with bluish purple lobes, the tube 3–5 mm long, without longitudinal slits, the lobes 2–3 mm long, the lower lip with a beard of dense hairs on the inner surface toward the base. Filament tube 2–3 mm long, the anther tube 1.2–1.8 mm long. Fruits 6–10 mm long. Seeds 0.5–0.8 mm long, ellipsoid, the surface with a series of minute pits and ridges, reddish brown. 2n=14. June–October.

Scattered nearly throughout the state (eastern U.S. west to Minnesota, Nebraska, and Oklahoma; Canada). Bottomland forests, openings of mesic to dry upland forests, banks of streams and rivers, margins of ponds and lakes, and rarely marshes; also pastures, fallow fields, gardens, railroads, roadsides, and moist, open, disturbed areas.

Lobelia inflata is an aggressive, sometimes weedy species often found in disturbed areas.

3. Lobelia puberula Michx. (purple dewdrop, big blue lobelia, blue cardinal flower)

Pl. 331 g, h; Map 1410

Plants perennial herbs with fibrous roots. Stems 40–120 cm long, erect or strongly ascending, usually unbranched, densely short-hairy throughout. Leaves sessile, gradually reduced toward the stem tip. Leaf blades 3–10 cm long, 1–2 cm wide, angled or tapered at the base, those of the lower leaves oblanceolate and rounded or angled to a bluntly pointed tip, those of the median and upper leaves lanceolate and angled or tapered to a sharply pointed tip, the margins sharply toothed, the surfaces densely short-hairy. Inflorescences dense spikes or spikelike racemes (the flower stalks to 4 mm long), the bracts of the lower flowers more or less leaflike, the upper bracts gradually reduced (each flower also with a pair of minute bracts immediately below the calyx). Calyces 7–10 mm long, the tube 1.5–2.0 mm long at flowering, becoming somewhat enlarged as the fruit matures, the slender lobes 5–8 mm long, becoming somewhat elongated and purplish-tinged as the fruit matures, with short, inconspicuous auricles. Corollas 14–22 mm long, blue to white, the outer surface with short, spreading hairs, the tube 7–12 mm long, with slitlike openings (fenestrate), the lobes 6–10 mm long. Filament tube 6–10 mm long, white to light blue, the anther tube 3–4 mm long. Fruits 5–8 mm

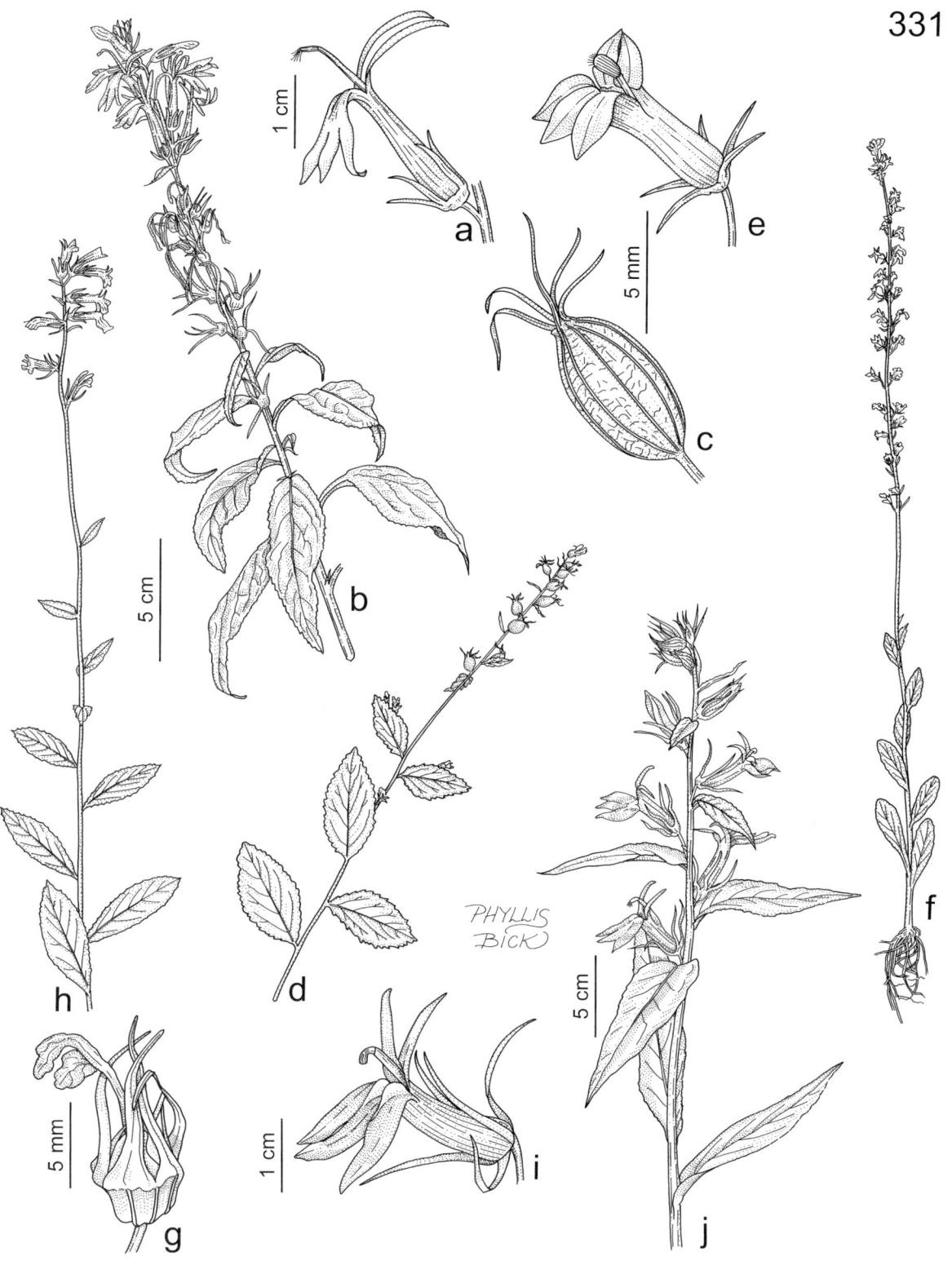

Plate 331. Campanulaceae. *Lobelia cardinalis*, **a)** flower, **b)** habit. *Lobelia inflata*, **c)** fruit, **d)** habit. *Lobelia spicata*, **e)** flower, **f)** habit. *Lobelia puberula*, **g)** fruit, **h)** habit. *Lobelia siphilitica*, **i)** flower, **j)** habit.

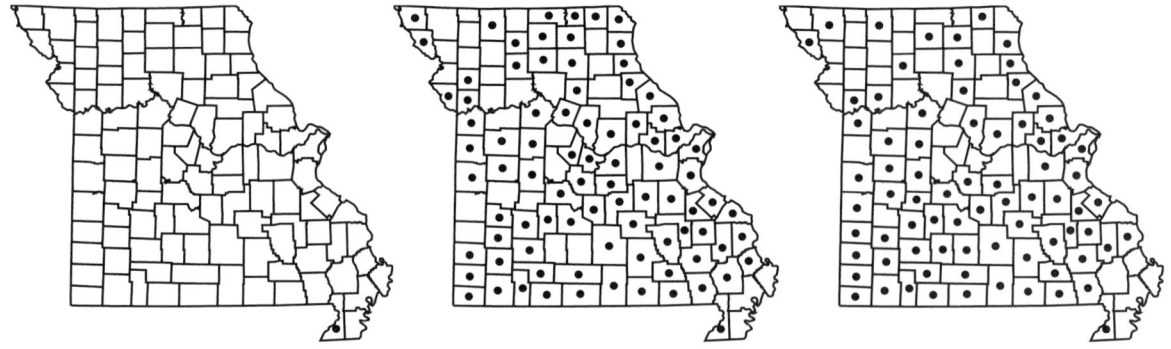

1410. Lobelia puberula   1411. Lobelia siphilitica   1412. Lobelia spicata

long. Seeds 0.5–0.8 mm long, ovoid to ellipsoid, the surface with a series of minute pits and ridges, yellowish brown. 2n=14. August–October.

Uncommon, known thus far from a single historical collection from Dunklin County (eastern U.S. west to Illinois and Texas; Canada). Habitat unknown, but Steyermark (1963) recorded "moist sandy open ground and low woodland."

Lobelia puberula is relatively widespread, and several varieties and forms have been described based on minor components of morphological variation. Steyermark determined the Missouri specimen as var. *mineolana* E. Wimm., a mostly southern element distinguished by its more or less leaflike bracts and relatively broad calyx lobes. McVaugh (1936) chose to describe these variants as unofficial forms of the species, without resorting to formal taxonomic designations, which seems a sensible approach.

4. Lobelia siphilitica L. (blue cardinal flower, great lobelia, blue lobelia)
L. siphilitica var. *ludoviciana* A. DC.

Pl. 331 i, j; Map 1411

Plants perennial herbs with fibrous roots. Stems (10–)30–100(–150) cm long, erect or strongly ascending, unbranched or less commonly few-branched toward the tip, often slightly winged, glabrous or sparsely pubescent with short, loosely ascending hairs. Leaves sessile or short-petiolate, gradually reduced from about the stem midpoint to the stem tip. Leaf blades 2–15 cm long, 0.5–6.0 cm wide, lanceolate to elliptic, narrowly oblong, or oblanceolate, angled or tapered at the base (those of the uppermost leaves sometimes more or less rounded), the base usually somewhat decurrent below the attachment point as a pair of narrow wings of green tissue along the stem, rounded or more commonly angled or tapered to a bluntly or sharply pointed tip, the margins finely and bluntly to sharply toothed, the surfaces glabrous or less commonly the undersurface with sparse, short hairs or minutely roughened along the veins. Inflorescences dense racemes (the flower stalks mostly 3–9 mm long), the bracts of the lower flowers more or less leaflike, the upper bracts gradually reduced (each flower also with a pair of minute bracts immediately below or farther below the calyx). Calyces 8–18 mm long, the tube 2–4 mm long at flowering, becoming somewhat enlarged as the fruit matures, the slender lobes 6–14 mm long, becoming somewhat elongated as the fruit matures, with relatively long, conspicuous auricles. Corollas 16–32 mm long, blue to bluish purple, sometimes with white, longitudinal striping, rarely completely white, the outer surface glabrous or with short, spreading hairs along the veins, the tube 11–20 mm long, with slitlike openings (fenestrate), the lobes 5–12 mm long. Filament tube 12–15 mm long, blue, the anther tube 3–5 mm long. Fruits 5–9 mm long. Seeds 0.6–0.8 mm long, ellipsoid, the surface with a series of minute pits and ridges, yellowish brown. 2n=14. August–October.

Scattered nearly throughout the state (eastern U.S. west to North Dakota, Wyoming, and Texas; Canada). Banks of streams, rivers, and spring branches, margins of ponds and lakes, bottomland forests, moist depressions of upland prairies, sloughs, swamps, fens, and moist ledges of bluffs; also pastures, ditches, and roadsides.

Two varieties of *L. siphilitica* were recognized by McVaugh (1936) and were included for Missouri by Steyermark (1963). *Lobelia siphilitica* var. *siphilitica* is the typical form of the species east of the Mississippi River, with hairy, relatively broad leaves 2–6 cm wide, a densely flowered inflorescence, and hairy calyces and flower stalks. *Lobelia siphilitica* var. *ludoviciana* is more common west of the Mississippi River and is said to be smaller, with glabrous leaves mostly less than 2 cm wide, a more sparsely flowered inflorescence, and glabrous calyces and flower stalks. Although these varieties may be distinguishable in other states, almost all the Missouri material lies somewhere in between these extremes. Plants with entirely white corollas have been called f. *albiflora* Britton.

Lobelia siphilitica and *L. cardinalis* are closely related. Ordinarily, the two species are isolated by differences in floral morphology and pollinator preferences, but occasional natural hybrids are produced. In Missouri, hybrids have been collected most often along the Current, Jacks Fork, and Eleven Point Rivers. The hybrids are intermediate between the parents for most characters, and they usually have distinctive reddish purple or less commonly somewhat blue-, purple-, and white-variegated corollas. They are uncommon and apparently sterile (Bowden, 1982). Such plants were first reported from Missouri by Steyermark (1952, 1963) and rediscovered by Witherspoon (1974). The correct name for these hybrids is *Lobelia* ×*speciosa* Sweet, although Steyermark (1963) referred to them by the incorrect name *L.* ×*siphilitica* var. *hybrida* Hook.

5. Lobelia spicata Lam. (spiked lobelia, palespike lobelia)

Pl. 331 e, f; Map 1412

Plants perennial herbs, with fibrous roots. Stems 20–80(–100) cm tall, erect or ascending, usually unbranched (occasionally with a few branches near the base), narrowly winged, glabrous or sparsely to moderately pubescent with short, curved to more or less spreading hairs mostly toward the base. Leaves sessile, gradually reduced above the stem base. Leaf blades 1–7 cm long, 1.0–2.5 cm wide, obovate to oblanceolate, narrowly oblong, or lanceolate, the base strongly decurrent below the attachment point as a pair of wings of green tissue along the stem, rounded or angled to a bluntly pointed tip, the margins entire or less commonly with sparse, irregular teeth, also usually with short, fine, spreading hairs, the surfaces glabrous or with sparse hairs. Inflorescences densely to loosely flowered spikelike racemes, the axis glabrous or sparsely hairy, the lowermost bracts lanceolate, the upper bracts much shorter and more or less linear (each flower also with a pair of minute bracts immediately below the calyx). Calyces 4–7 mm long, the tube 0.5–1.5 mm long at flowering, not becoming inflated as the fruit matures, the slender lobes 3–5 mm long, often with auricles 0.2–1.5 mm long. Corollas 4–10 mm long, pale blue or white, the tube 3–5 mm long, not slitted, the lobes 2–5 mm long, the lower lip with a beard of dense hairs on the inner surface toward the base. Filament tube 2.3–3.5 mm long, the anther tube 1.4–2.0 mm long. Fruits 3.5–5.0 mm long. Seeds 0.4–0.8 mm long, ovoid to ellipsoid, the surface with a series of minute pits and ridges, reddish brown. $2n=14$. April–August.

Scattered nearly throughout the state (eastern U.S. west to Montana and Texas). Upland prairies, glades, savannas, ledges and tops of bluffs, openings of mesic to dry upland forests, banks of streams and rivers, and fens; also pastures, old fields, railroads, roadsides, and open, disturbed areas.

Several varieties of *L. spicata* have been recognized in Missouri based on supposed differences in pubescence, calyx, and anther characters (McVaugh, 1936; Steyermark, 1963). The most significant character is found in the presence, size, and shape of the auricles at the base of the calyx lobes, which is correlated with differences in size and pubescence of the calyx tube, flowering times, and habitats. However, as noted by McVaugh, a number of collections from Missouri, Illinois, and Iowa are intermediate for all of the features upon which he based his varieties. The present treatment recognizes the two seemingly most stable entities (at least in Missouri), but users will still encounter difficulty in determining some specimens.

McVaugh (1936) noted a historical collection from near Asbury (Jasper County) as possibly representing an aberrant example of the closely related *L. appendiculata* A. DC., which otherwise is known to occur from Kansas to Texas east to Kentucky and Alabama and thus plausibly could grow in Missouri. Steyermark (1963) excluded this species from the state's flora on the basis that it lacks the conspicuous, flattened calyx auricles typical of that species and instead referred the specimen to *L. spicata* var. *leptostachys*. As discussed by Rosatti (1986), *L. appendiculata* also differs from *L. spicata* in its more rounded leaf bases, hairier calyx lobes, and sparser inflorescences that do not have a tapered appearance in profile. In 2004, Tim Smith of the Missouri Department of Conservation rediscovered plants matching the morphology of E. J. Palmer's original collection in a hardpan prairie on the south side of Asbury. Examination of this additional material confirms that although plants from this population have a morphology that is somewhat aberrant for either *L. spicata* or *L. appendiculata,* they agree better with McVaugh's (1936) and Rosatti's (1986) descriptions of the former. *Lobelia appendiculata* thus continues to be excluded from the state's flora, but further fieldwork in southern Missouri may yet turn up more typical examples of this species. Further research is needed to test the relationships among the taxa described in the *L. spicata* complex.

1. Inflorescence axis with an apparent stalked portion 2–5 cm long below the lowermost flower; calyx tube 0.5–1.0 mm long, glabrous, auricles at the base of the calyx lobes well developed, 1.0–1.5 mm long, tapered to a sharply pointed tip . . . . 5A. VAR. LEPTOSTACHYS
1. Inflorescence axis with an apparent stalked portion 4–15 cm long below the lowermost flower; calyx tube 1.0–1.5 mm long, often finely hairy between the ribs, auricles at the base of the calyx lobes absent or greatly reduced, 0.2–0.5 mm long, rounded to bluntly pointed at the tip . . . . 5B. VAR. SPICATA

5a. Lobelia spicata var. **leptostachys** (A. DC.) Mack. & Bush

L. leptostachys A. DC.

Basal leaves absent or few at flowering. Inflorescence axis with an apparent stalk 2–5 cm long below the lowermost flower and above the uppermost leaf. Calyces with the tube 0.5–1.0 mm long, glabrous, the lobes glabrous. Auricles at the base of the calyx lobes well developed, 1.0–1.5 mm long, slender, with a more or less rounded base, abruptly long-tapered to a sharply pointed tip. $2n=14$. June–August, with the peak of flowering in mid-July.

Scattered, most commonly in the Ozark and Ozark Border Divisions (eastern [mostly southeastern] U.S. west to Iowa and Oklahoma). Glades, upland prairies, savannas, openings of dry upland forests, and rarely banks of streams and rivers; also railroads and roadsides.

5b. Lobelia spicata var. **spicata**

L. spicata var. *campanulata* McVaugh

L. spicata var. *hirtella* A. Gray

L. spicata var. *parviflora* A. Gray

Basal leaves absent or present at flowering. Inflorescence axis with an apparent stalked portion 4–15 cm long below the lowermost flower and above the uppermost leaf. Calyces with the tube 1.0–1.5 mm long, often finely hairy or with papillae between the ribs. Auricles at the base of the calyx lobes absent or greatly reduced, 0.2–0.5 mm long, oblong, rounded or bluntly pointed at the tip. $2n=14$. April–July, with the peak of flowering in mid-June.

Scattered nearly throughout the state (eastern U.S. west to Montana and Oklahoma; Canada). Upland prairies, glades, savannas, ledges and tops of bluffs, openings of mesic to dry upland forests, banks of streams and rivers, and fens; also pastures, old fields, railroads, roadsides, and open, disturbed areas.

In typical var. *spicata,* the anthers are blue. White-anthered plants have been called f. *campanulata* (McVaugh) Bowden. These occur sporadically within populations of both varieties of *L. spicata* and produce only aborted pollen (Bowden, 1959; McGregor, 1985b). This mutant form thus has functionally pistillate flowers and can only produce fruits by cross-pollination with other so-called normal plants in the same population.

Steyermark (1963) followed McVaugh (1936) in recognizing var. *hirtella* for Missouri based on plants with stiff, spreading hairs on the stems, bracts, and calyx, as well as the leaves mostly below the midpoint of the stem. Pubescence on the stems and bracts is sparse in most of the Missouri specimens and not closely correlated with other characters (including leaf arrangement). Thus, var. *hirtella* is treated as a synonym of var. *spicata* in the present work, in agreement with the careful analysis of the complex by McGregor (1985b).

3. Triodanis Raf.

Plants annual. Stems erect or loosely ascending, unbranched or less commonly few-branched, mostly from the basal half. Basal leaves usually absent at flowering. Stem leaves sessile or the lowermost leaves less commonly short-petiolate, the margins entire or bluntly to sharply toothed and often inconspicuously hairy. Inflorescences of solitary axillary flowers or small axillary clusters of 2 or 3(–5), mostly very short-stalked, the leaves subtending flowers essentially indistinguishable from the often relatively few foliage leaves, the whole inflorescence appearing spicate. Flowers epigynous, those at the lower nodes cleistogamous (obligately self-fertilizing and probably apogamous), these smaller than the open-flowering ones, the calyx most often with usually only 3 or 4(–6) lobes, the corolla reduced to short flaps of tissue, the stamens variously reduced but when present failing to dehisce or release pollen, and the style highly reduced and nonfunctional. Calyces in normal flowers actinomorphic, (2–)5(6)-lobed, without appendages. Corollas in normal flowers actinomorphic, broadly bell-shaped to saucer-shaped, (3–)5(6)-lobed, the lobes longer than the tube, usually relatively slender (lanceolate to elliptic), blue, purple, or white. Stamens in normal flowers (3–)5(6), attached to the base of the corolla, the filaments short, dilated and hairy at the base, the anthers distinct, elongate. Pistil with 3 carpels. Ovary totally inferior, with 1–3 locules. Style relatively short, straight, the stigma (2)3-lobed. Fruits ellipsoid to narrowly ovoid, narrowly ellipsoid, or narrowly cylindrical capsules, usually with longitudinal nerves, these sometimes minutely ridged toward the tip, dehiscent by 1–3 lateral pores or slits, glabrous or minutely hairy or roughened along the nerves. Seeds ellipsoid, sometimes somewhat flattened, the surface tan to dark brown, shiny (except often in *T. perfoliata*). Seven or 8 species, North America, Europe, Asia, Africa.

The center of diversity for *Triodanis* is in the south-central portion of the United States. Only one species (*T. falcata* (Ten.) McVaugh) occurs in the Old World. All of the species are winter annuals, germinating in the autumn, overwintering as basal rosettes, and flowering the following spring. The regular production of cleistogamous flowers in *Triodanis* is distinctive but also occurs in a few species of non-Missouri *Campanula*. The distinction between *Triodanis* and *Specularia* Heister ex A. DC., and whether these should be viewed as separate from *Campanula*, has been somewhat controversial. The species here recognized as *Triodanis* originally were included in *Campanula* and subsequently segregated into *Specularia*. *Triodanis* was described as a mostly New World segregate of an otherwise Eurasian *Specularia*. However, the latter name has disappeared from the botanical literature on Old World plants, as it is a nomenclatural synonym of *Legousia* J.F. Durande. See McVaugh (1945, 1948), Bradley (1968), and Rosatti (1986) for further discussion. The cultivated Venus' looking-glass, *Legousia speculum* Fisch. ex A. DC. (*Specularia speculum* A. DC.), sometimes is grown in Missouri gardens and differs from *Triodanis* in its more-branched stems, inflorescences in terminal clusters, and lack of cleistogamous flowers.

1. Stem leaves linear to narrowly elliptic or lanceolate, 5–10 times as long as wide; fruits 1(2)-locular, those that develop from normal flowers straight and strongly ascending but those developing from cleistogamous flowers twisted and/or arched away from the stem, dehiscing by 1(2) pore(s) at or near the fruit tip or by longitudinal slits above the fruit midpoint . . . . 4. T. LEPTOCARPA
1. Stem leaves elliptic to narrowly ovate or ovate, less than 3 times as long as wide; fruits 2- or more commonly 3-locular, all similar in size and shape, straight and strongly ascending, dehiscing by (2)3 pores
    2. Pores positioned at about the midpoint of the fruit, linear to narrowly oblong
        . . . . 2. T. HOLZINGERI
    2. Pores positioned either near the base or near the tip of the fruit, oval to broadly oblong-elliptic or nearly circular
        3. Seeds 0.8–1.0 mm long, strongly flattened (narrowly biconvex); pores positioned at or near the tip of the fruit . . . . 3. T. LAMPROSPERMA
        3. Seeds 0.4–0.7 mm long, slightly flattened (relatively plump); pores positioned either near the tip or below the midpoint of the fruit
            4. Stem leaves 1.5–3.0 times as long as wide, sessile but not or only slightly clasping the stem; flowers mostly cleistogamous, the normal open-flowering one(s) usually solitary (rarely 2 or 3) at the stem tip; pores positioned at or near the tip of the fruit . . . . 1. T. BIFLORA
            4. Stem leaves mostly as long as wide or wider than long, clasping the stem; flowers mostly cleistogamous but normal, open flowers usually 1 per node along the upper ⅓–⅔ of the stem; pores positioned below the midpoint of the fruit . . . . 5. T. PERFOLIATA

1413. Triodanis biflora   1414. Triodanis holzingeri   1415. Triodanis lamprosperma

1. **Triodanis biflora** (Ruíz & Pav.) Greene
(small Venus' looking-glass)
Campanula biflora Ruíz & Pav.
Specularia biflora (Ruíz & Pav.) Fisch. & C.A. Mey
T. perfoliata (L.) Nieuwl. var. *biflora* (Ruíz & Pav.) T.R. Bradley

Pl. 332 g; Map 1413

Stems 10–40(–50) cm long, erect or nearly so, roughened with minute, recurved hairs along the angles, at least toward the base. Basal leaves elliptic to obovate, tapered to a sessile or short-petiolate base, rounded or bluntly pointed at the tip. Stem leaves 5–20 mm long, 2–10 mm wide, 1.5–3.0 times as long as wide, elliptic to narrowly ovate, sessile but not or only slightly clasping the stem, mostly bluntly pointed at the tip, the margins finely scalloped or bluntly toothed to nearly entire, the upper surface glabrous or nearly so, the undersurface finely roughened or with relatively soft, inconspicuous hairs, mostly along the veins. Flowers 1–3 per node at most nodes of the stem, mostly cleistogamous, the normal open-flowering one(s) usually solitary (rarely 2 or 3) at the stem tip. Calyces with the tube 3–7 mm long, usually appearing slightly inflated, the lobes in normal flowers 5–8 mm long, narrowly triangular to lanceolate, those in cleistogamous flowers 0.7–2.0 mm long, narrowly triangular. Corollas in normal flowers purple to lavender, the lobes 5–9 mm long, 2–3 mm wide. Fruits all similar in size and shape, straight and strongly ascending, 4.5–8.0 mm long, 1.2–2.0 mm wide, the (usually) 3 pores 1.0–1.2 mm long, oval to nearly circular, positioned at or near the tip of the fruit. Seeds 0.4–0.6 mm long, elliptic to broadly oblong-elliptic, slightly flattened (relatively plump), the surface smooth, shiny. $2n=56$. May–June.

Scattered mostly south of the Missouri River (southern U.S. north to Oregon, Nebraska, Illinois, and New York). Upland prairies, glades, ledges and tops of bluffs, openings of mesic to dry upland forests, fens, margins of ponds, lakes, and sinkhole ponds, and banks of streams and rivers; also old fields, fallow fields, pastures, ditches, railroads, roadsides, and open, disturbed areas.

This species is closely related to *T. perfoliata*, and the two species sometimes are found in mixed colonies. The two taxa can be crossed relatively easily (Bradley, 1968, 1975). The hybrids are fertile, robust, and intermediate between the parental forms in most characters, except that the inflorescence always has more than a few open flowers, as in *T. perfoliata*. Bradley felt that because the two taxa form viable hybrids both in the greenhouse and in nature, they should be considered varieties of a single species. The alternative interpretation was presented by Ward (1978), who noted that most plants can be assigned to one of the parental taxa without difficulty, that several morphological characters correlate to separate the two taxa, and that cleistogamous flowers help to maintain the integrity of each taxon by providing a partial genetic isolating mechanism. For these reasons, the two taxa are maintained as separate species in the present treatment, but users should be aware that a certain proportion of the plants in the complex encountered in the field and herbarium cannot be determined to species with confidence.

2. **Triodanis holzingeri** McVaugh (Holzinger's Venus' looking-glass)
Specularia holzingeri (McVaugh) Fernald

Pl. 332 f; Map 1414

Stems 20–50(–90) cm long, erect or ascending, sometimes from a spreading base, often short-hairy along the angles, sometimes roughened with minute, recurved hairs toward the tip. Basal leaves broadly elliptic to broadly ovate, angled or tapered to a sessile or short-petiolate base, rounded or bluntly pointed at the tip. Stem leaves 12–25 mm

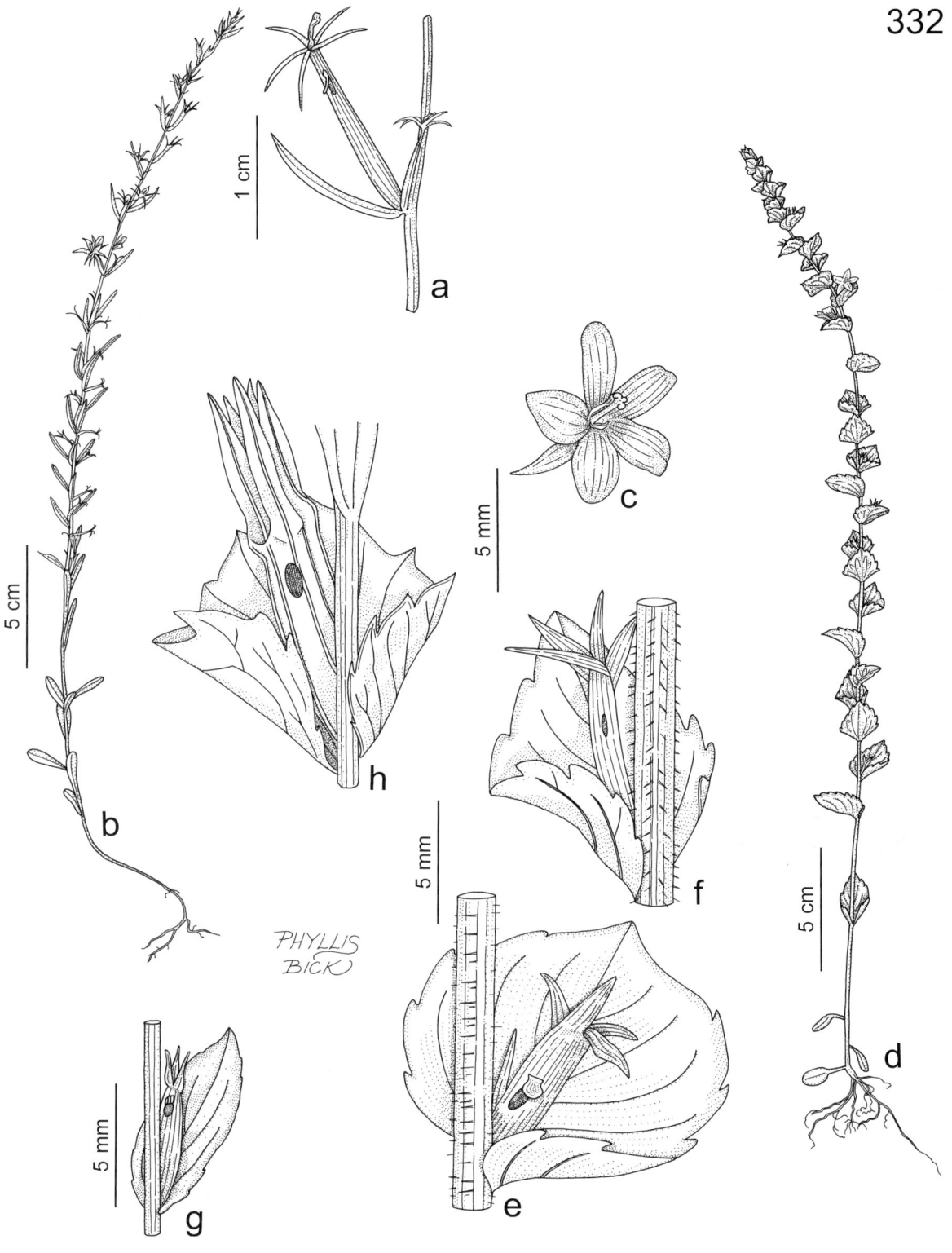

Plate 332. Campanulaceae. *Triodanis leptocarpa*, **a)** fruit, **b)** habit. *Triodanis perfoliata*, **c)** flower, **d)** habit, **e)** fruit and leaf at node. *Triodanis holzingeri*, **f)** fruit and leaf at node. *Triodanis biflora*, **g)** fruit and leaf at node. *Triodanis lamprosperma*, **h)** fruit and leaf at node.

long, 6–15 mm wide, as long as wide or 1.5 times as long as wide, ovate to broadly ovate, sessile, rounded to somewhat cordate and clasping the stem, rounded to bluntly or sharply pointed at the tip, the margins finely and bluntly to occasionally sharply toothed, the upper surface glabrous or nearly so, the undersurface finely roughened or with relatively soft, short hairs. Flowers 1–3 per node at most nodes of the stem, mostly cleistogamous but normal, open flowers usually 1 per node along the upper 1/3–1/2 of the stem. Calyces with the tube 3–5 mm long, usually appearing slightly inflated, the lobes in normal flowers 4–9 mm long, narrowly triangular to lanceolate, those in cleistogamous flowers 0.7–2.5 mm long, narrowly triangular. Corollas in normal flowers bluish purple to purple, the lobes 4–7 mm long, 2–3 mm wide. Fruits all similar in size and shape, straight and strongly ascending, 5–9 mm long, 0.7–2.0 mm wide, the (usually) 3 pores 1.4–2.5 mm long, linear to narrowly oblong, positioned at about the midpoint of the fruit. Seeds 0.4–0.7 mm long, broadly elliptic to nearly circular, somewhat flattened (moderately biconvex), the surface with longitudinal lines of minute tubercles, shiny. April–August.

Uncommon, known thus far from Jackson, Newton, and Osage Counties (Idaho to Arizona east to Tennessee). Chert glades and sandy banks of rivers.

The diagnostic character of *T. holzingeri* is the narrow, slitlike pores positioned midway on the capsule. This species otherwise is very similar to *T. perfoliata*. Bradley (1968) recorded putative hybrids between the two species from Oklahoma.

3. Triodanis lamprosperma McVaugh (slimpod Venus' looking-glass)

Specularia lamprosperma (McVaugh) Fernald

Pl. 332 h; Map 1415

Stems 20–50(–100) cm long, erect or ascending, sometimes from a spreading base, often short-hairy along the angles, sometimes roughened with minute, recurved hairs toward the tip. Basal leaves broadly elliptic to broadly ovate, angled or tapered to a sessile or short-petiolate base, rounded or bluntly pointed at the tip. Stem leaves 5–15 mm long, 3–15 mm wide, mostly as long as wide or slightly wider than long (the uppermost ones sometimes slightly narrower), ovate to broadly ovate or somewhat kidney-shaped, sessile, rounded to somewhat cordate and clasping the stem, rounded to bluntly or occasionally sharply pointed at the tip, the margins finely and bluntly to occasionally sharply toothed, the upper surface glabrous or nearly so, the undersurface finely roughened or with relatively soft, short hairs. Flowers 1 or 2(3) per node at most nodes of the stem, mostly cleistogamous but normal, open flowers usually 1 per node along the upper 1/3–2/3 of the stem. Calyces with the tube 3–5 mm long, usually appearing slightly inflated, the lobes in normal flowers 4–8 mm long, narrowly triangular to lanceolate, those in cleistogamous flowers 2–3 mm long, narrowly triangular. Corollas in normal flowers bluish purple to purple, the lobes 4–7 mm long, 2–3 mm wide. Fruits all similar in size and shape, straight and strongly ascending, 4–10 mm long, 2–3 mm wide, the (usually) 3 pores 1.4–2.0 mm long, elliptic to broadly elliptic, positioned at or near the tip of the fruit. Seeds 0.8–1.0 mm long, broadly elliptic to nearly circular, strongly flattened (narrowly biconvex), the surface smooth and shiny. April–August.

Uncommon in the Unglaciated Plains Division and sporadically eastward in the Ozarks to St. Francois County (Kansas to Texas east to Missouri and Louisiana). Glades and upland prairies; also old fields, railroads, and roadsides.

The most distinctive feature of *T. lamprosperma* is the round, flattened, very shiny seeds that are larger than those of most other *Triodanis* species. Otherwise it is very similar to *T. perfoliata*. An apparent hybrid between *T. lamprosperma* and *T. leptocarpa* from a chert glade in Newton County was reported by Palmer and Steyermark (1955), but Bradley (1968) suggested that this might instead represent *T. leptocarpa* × *perfoliata*, based on his inability to artificially cross *T. lamprosperma* and *T. leptocarpa* and the apparent absence of other examples of this hybrid in herbaria except for one apparently sterile specimen from Texas.

4. Triodanis leptocarpa (Nutt.) Nieuwl. (prairie Venus' looking-glass, slender-leaved Venus' looking-glass)

Specularia leptocarpa (Nutt.) A. Gray

Pl. 332 a, b; Map 1416

Stems 10–70 cm long, erect, ascending, or more commonly spreading to sprawling on surrounding vegetation, short-hairy or roughened along the angles. Basal leaves narrowly elliptic to oblanceolate, tapered to a short-petiolate base, rounded or bluntly pointed at the tip. Stem leaves 8–25 (–30) mm long, 1–6 mm wide, 5–10 times as long as wide, linear to narrowly elliptic or lanceolate, sessile or very short-petiolate, tapered at the base, bluntly to sharply pointed at the tip, the margins entire or very finely scalloped or bluntly toothed, the upper surface glabrous or nearly so, the undersurface finely roughened or with stiff, short hairs. Flowers 1–3(–5) per node (sometimes some

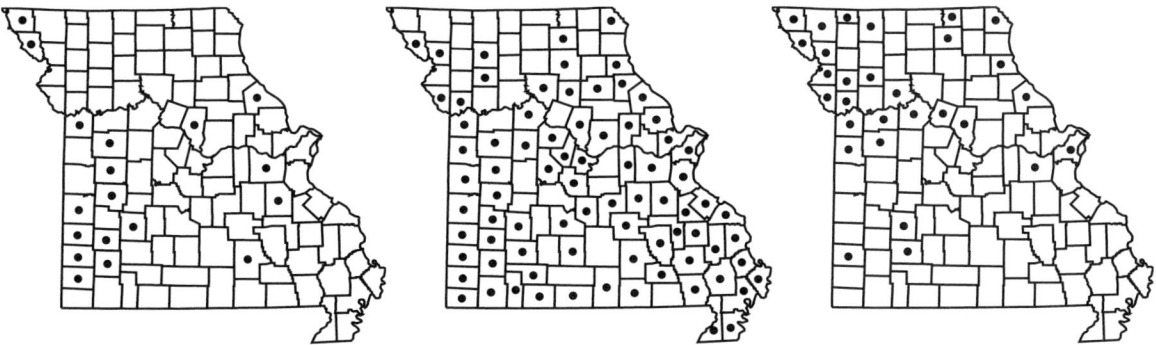

1416. Triodanis leptocarpa   1417. Triodanis perfoliata   1418. Cannabis sativa

of the flowers aborting during development) at most nodes of the stem, mostly cleistogamous but normal, open flowers usually 1 or more per node along the upper 1/3–1/2 of the stem. Calyces with the tube 8–14 mm long, not inflated, the lobes in normal flowers 5–10 mm long, narrowly triangular to narrowly lanceolate or linear, those in cleistogamous flowers 2–5 mm long, narrowly triangular to linear. Corollas in normal flowers bluish purple to purple, the lobes 4–7 mm long, 1.5–3.5 mm wide. Fruits of two different types, those developing from the normal, open flowers straight and strongly ascending, 15–25 mm long, 1.3–1.8 mm wide, the 1(2) pore(s) 1.4–2.5 mm long, oblong-elliptic to narrowly oval, positioned at the tip of the fruit; those developing from the cleistogamous flowers twisted and/or arched away from the stem, 5–18 mm long, 0.7–2.5 mm wide, with slender, longitudinal slits (alternating with the calyx lobes) above the midpoint or 1 pore 1.4–2.5 mm long, oblong-elliptic to narrowly oval, positioned at the tip of the fruit. Seeds 0.8–1.2 mm long, elliptic, somewhat flattened (moderately biconvex), the surface smooth, shiny. May–August.

Scattered in the Unglaciated Plains Division, uncommon farther north and east (Montana to Texas east to Minnesota and Arkansas; introduced east to Indiana). Bottomland and upland prairies, glades, ledges and tops of bluffs, and banks of streams and rivers; also pastures, roadsides, and open, disturbed areas.

This species is distinctive within the genus in its narrow leaves and mostly 1-locular capsules. Frequently there are only 2 stigma lobes. Capsules produced from the cleistogamous flowers are twisted or arched away from the stem. Bradley (1968) recorded an apparent hybrid between this species and *T. perfoliata* from a roadside in Cedar County and also noted an older herbarium specimen from Mercer County that apparently represents this hybrid. McVaugh (1945) suggested that at least some of the populations in eastern Missouri are nonnative occurrences.

5. Triodanis perfoliata (L.) Nieuwl. (clasping Venus' looking-glass)
Campanula perfoliata L.
Specularia perfoliata (L.) A. DC.
Pl. 332 c–e; Map 1417

Stems 10–80(–100) cm long, erect or ascending, sometimes from a spreading base, often short-hairy along the angles, sometimes roughened with minute, recurved hairs toward the tip. Basal leaves broadly elliptic to broadly ovate, angled or tapered to a sessile or short-petiolate base, rounded or bluntly pointed at the tip. Stem leaves 5–20(–25) mm long, 4–20(–27) mm wide, mostly as long as wide or slightly wider than long (the uppermost ones sometimes slightly narrower), broadly ovate to somewhat kidney-shaped, sessile, rounded to cordate and clasping the stem, rounded to bluntly pointed at the tip, the margins finely and bluntly to sharply toothed, the upper surface glabrous or nearly so, the undersurface finely roughened or with relatively soft, short hairs. Flowers 1–3 per node at most nodes of the stem, mostly cleistogamous but normal, open flowers usually 1 per node along the upper 1/3–2/3 of the stem. Calyces with the tube 3–5 mm long, usually appearing slightly inflated, the lobes in normal flowers 4–9 mm long, narrowly triangular to lanceolate, those in cleistogamous flowers 2–3 mm long, narrowly triangular. Corollas in normal flowers purplish blue to purple, rarely with white streaks or entirely white, the lobes 4–7 mm long, 2–3 mm wide. Fruits all similar in size and shape, straight and strongly ascending, 5–10 mm long, 2–3 mm wide, the (usually) 3 pores 1.3–2.0 mm long, elliptic, positioned below the midpoint of the fruit. Seeds 0.4–0.7 mm long, elliptic to broadly oblong-elliptic, slightly flattened (relatively plump), the surface minutely wrinkled or tuberculate, dull or only slightly shiny, rarely smooth and shinier. 2n=56. May–June.

Scattered nearly throughout the state (U.S., Canada, Mexico, South America, Caribbean Islands). Upland prairies, glades, ledges and tops of bluffs, openings of mesic to dry upland forests, oxbows, marshes, margins of lakes, and banks of streams and rivers; also old fields, fallow fields, pastures, ditches, railroads, roadsides, and open, disturbed areas.

Rare individuals with white corollas have been called f. *alba* (J. Voigt) Steyerm. *Triodanis perfoliata* is by far the most widespread species in the genus and is found in the broadest variety of habitats. It is closely related to *T. biflora* and hybridizes readily with that species (Bradley, 1968). See the treatment of *T. biflora* for further discussion. Bradley also discussed possible hybridization with *T. leptocarpa* in southwestern Missouri and Mercer County (and in other states), as well as rare putative hybrids with *T. holzingeri* from Oklahoma.

McVaugh (1945) suggested that smooth-seeded plants (rare in Missouri) are polyploids that have spread north and south from the original distribution, but to date there exist no counts to differentiate these populations, and the species appears to be uniformly tetraploid (2n=56). Interestingly, Trent (1942) recorded that seed viability in populations of *T. perfoliata* that he studied was relatively low and different for normal and cleistogamous flowers: normal flowers produced about 47.5 percent viable seed, whereas in cleistogamous flowers only about 0.07 percent of the seed was viable. His anatomical studies documented that for both flower types nonviable seeds lacked embryos. The cause for this unusual situation has not been determined.

CANNABACEAE (Hemp Family)
Contributed by David J. Bogler and George Yatskievych

Plants annual or perennial herbs, dioecious, sometimes aromatic, the sap not milky. Stems erect or twining, often with glandular hairs. Leaves opposite but usually alternate toward the stem tip, mostly long-petiolate. Stipules small, usually herbaceous, lanceolate to narrowly triangular, sometimes fused laterally, persistent. Leaf blades simple (and often lobed) or palmately compound, the margins sharply toothed, the surfaces with a variety of hairs and glands, cystoliths (calcium carbonate inclusions) commonly present at the base of hairs. Staminate inflorescences open axillary or less commonly terminal panicles with numerous flowers. Pistillate inflorescences dense axillary clusters or spikes, the flowers or pairs of flowers subtended by a bract and an additional closely surrounding smaller bract, these brown and scalelike at maturity, hairy and glandular. Flowers imperfect, incomplete, small and inconspicuous, hypogynous. Calyx of staminate flowers of 5 free sepals, these green or greenish white; that of pistillate flowers saclike, shorter than to nearly as long as and closely surrounding the ovary and fruit, unlobed, membranous to papery. Petals absent. Stamens 5 (absent in pistillate flowers), the filaments short, attached at or near the base of the anthers, the anthers dehiscing longitudinally. Pistil 1 per flower (absent in staminate flowers), the ovary superior, consisting of 2 fused carpels, with 1 locule, the placentation usually nearly apical. Style 1, very short, the stigmas 2, relatively long and slender, shed soon after flowering. Ovule 1. Fruits small achenes, not winged at the tip, surrounded by the persistent calyx. Seed 1, more or less spherical or nearly so (the embryo appearing curved or coiled but not always easily observed). Two genera, 4 species, North America, Europe, Asia; cultivated and introduced nearly worldwide.

Cannabis and *Humulus* form a natural lineage that generally is considered closely related to the Moraceae and Urticaceae, and often has been included in Moraceae. All species are wind-pollinated with corresponding morphological adaptations (Miller, 1970). Staminate flowers release large quantities of wind-borne pollen, which can cause hay fever. The stigmas of pistillate flowers are elongate, exserted from the bracts, and equipped with dense, papillose hairs for catching the pollen. Dioecy in these plants is said to be regulated by inheritance of distinctive sex chromosomes, but it also is influenced by environmental conditions and seasonality, and staminate and/or bisexual flowers are produced occasionally on otherwise pistillate plants under certain conditions or toward the end of the growing season. The family is of considerable economic importance as a source of fiber from the stems, oils from the fruits, and drugs and flavorings from the glandular trichomes of the leaves and bracts.

1. Stems erect, not twining; leaves palmately compound with 3–9 leaflets; stems and petioles lacking 2-armed hairs; fruiting clusters erect, not conelike
.... 1. CANNABIS
1. Stems (and sometimes also petioles) twining; leaves simple or palmately lobed; stems and petioles strongly roughened with 2-armed hairs; fruiting clusters pendant, appearing conelike with enlarged, overlapping bracts .... 2. HUMULUS

1. Cannabis L.

One species, probably native to central Asia but now naturalized nearly worldwide.

1. Cannabis sativa L. (hemp, marijuana)
  C. sativa var. *spontanea* Vavilov
  C. *ruderalis* Janish.

Pl. 333 a, b; Map 1418

Plants annuals, with long taproots, aromatic, the staminate plants mostly taller, more slender, and with sparser leaves than the pistillate ones. Stems 50–500 cm tall, erect, usually with numerous branches, usually coarsely ridged, often hollow at maturity, moderately pubescent with unbranched appressed-ascending hairs, also sparsely dotted with resinous glands. Leaves with the petioles 2–7 cm long, pubescent with unbranched hairs. Stipules narrowly triangular, not fused. Leaf blades palmately compound with mostly 5–9 unequal leaflets, the leaflets sessile or nearly so, 3–17 cm long, 0.3–2.0 cm wide, linear to narrowly lanceolate or narrowly elliptic, tapered at the base, tapered to a sharply pointed tip, the margins coarsely and sharply toothed, the upper surface sparsely pubescent with unbranched hairs and dotted with yellowish brown resinous glands, the hairs with inflated bases and cystoliths, dark green, the undersurface moderately to densely appressed-hairy and with scattered resinous gland-dots and sometimes also stalked glands, pale green. Staminate inflorescences appearing as short panicles consisting of small flower clusters on nearly leafless branches, the flowers with stalks 0.5–3.0 mm long. Pistillate inflorescences consisting of small clusters on short, leafy, spikelike branches, ascending at maturity, not conelike, sparsely to densely covered with stalked or nearly sessile glands, the flowers sessile or nearly so, the bracts lanceolate. Staminate flowers 2.5–4.0 mm long, the sepals lanceolate to ovate, with pale, thin margins, minutely hairy. Pistillate flowers with the ovary 2–3 mm long at flowering. Fruits 3–4 mm long, ovoid, somewhat flattened, enveloped by the persistent glandular bracts and membranous calyx, pale green to light brownish green, often somewhat purplish-mottled. $2n=20$. July–October.

Introduced, scattered, mostly in the northern half of the state, locally common in some portions of northwesternmost Missouri (cultigen presumably native to western Asia; cultivated and introduced nearly worldwide). Banks of streams and rivers, edges of marshes, and disturbed portions of bottomland, upland, and loess hill prairies; also levees, ditches, fencerows, gardens, margins of crop fields, railroads, roadsides, and open, disturbed places.

Cannabis sativa is a remarkable plant with many uses and a fascinating history (Dewey, 1914; Miller, 1970; Abel, 1980; Brown, 1998). Some races have been bred for fiber production, for the long fibers of the stem are used to make rope, twine, bags, nets, cloth, and paper. Some races have been selected for their fruits, which can be eaten (usually roasted) by humans and are often used in bird seed mixtures. The seeds contain an oil that can be used in the manufacture of paints, varnish, lubricants, and soaps, and as a fuel for lamps or even diesel engines. In other races, the leaves and pistillate inflorescences produce a resin used medicinally to treat a wide range of ailments (Brown, 1998; Grotenhermen and Russo, 2002), and that in more recent years has been adopted by modern medicine for the treatment of glaucoma and to ameliorate the side effects of chemotherapy in the treatment of cancer. When used as a drug, marijuana consists of plant fragments (mostly of pistillate inflorescences and associated leaves) that are burned and the smoke inhaled with various narcotic and hallucinogenic effects. Hashish is a more purified resin from the same source that also usually is smoked. The use of these substances as recreational drugs is illegal in most of the developed world, including the United States, although in a few states the possession of small amounts of marijuana has become decriminalized in recent years. The resin involved is most concentrated in the pistillate flower buds and contains terpenoids known as cannabinols, the most potent of which are isomers of tetrahydrocannabinol (THC).

Cannabis sativa appears to be one of mankind's oldest cultivated plants. It apparently originated in central Asia north of the Himalayas and was cultivated in China for thousands of years. It also was used by the ancient Assyrians, Scythians, Indians, and Greeks. *Cannabis* spread to Africa very early, where it assumed great importance in many cultures. It was brought to Europe by 1500 BC and was widely grown for fiber by AD 500. Spanish conquistadors and English pilgrims brought *Cannabis* to the New World. Prior to the Civil War, hemp was a major crop in the United States for the rope industry. In Missouri, it was a leading crop in Saline, Lafayette, and other counties along the Missouri River. Hemp also was grown during World War II when the Japanese armed forces cut off access to supplies of Manila rope.

Cannabis sativa is a highly variable species, due to a natural genetic plasticity, long selection by humans, a wide distribution, and its response to varied environmental factors. Several additional species have been described, and there has been much debate about whether these taxa should be accepted or not (Small, 1979; Small and Cronquist, 1976). *Cannabis indica* Lam. was based on plants from the East Indies that are smaller than typical *C. sativa,* with somewhat firmer, narrower, mostly alternate leaves. In his original description, Lamarck (1785) mentioned its intoxicating properties, and the name is associated to the present day with low-growing plants with small seeds selected for their use as drugs. The names *C. sativa* var. *spontanea* and *C. ruderalis* were described for plants from Russia with so-called wild characteristics such as smaller, harder, and more readily disarticulating fruits. Some botanists believe that at least three species of *Cannabis* exist (Emboden, 1974) or maintain that they might exist (Schultes and Hofmann, 1980). Alternatively, Small and Cronquist (1976) argued for the existence of a single species, *C. sativa,* and provided an infraspecific classification with two subspecies, ssp. *sativa* and ssp. *indica* (Lam.) E. Small & Cronquist, plus two varieties of each. These authors pointed out that variation is continuous, there are no barriers to interbreeding, and that these taxa are maintained through continued natural and artificial selection. Their two subspecies are separated by the percentage (based on dry weight) of THC in the upper portions of plants (0.3 percent being the arbitrary separation), rather than morphological features. Within each subspecies, Small and Cronquist segregated a cultivated variety, with larger, longer-persistent fruits with a somewhat less-persistent calyx, from a variety having more so-called wild-type characteristics. Several authors (Barker and Brooks, 1986) have noted that cultivated hemp plants that have been naturalized in the wild begin to revert back to the wild type as selection favors plants with more easily dispersed fruits. This is true in Missouri populations, many of which have fruits smaller than the arbitrary 3.8 mm length used by Small and Cronquist to distinguish varieties. Although a formal infraspecific classification has use for plant breeders and law enforcement officials, it has limited practical value for botanists dealing with plants growing outside of cultivation in Missouri.

Wild populations found in Missouri more or less correspond to ssp. *sativa,* the taxon that once was widely grown for fiber production and is now a common weed. There are only sporadic occurrences of one or a few plants (mostly in urban areas) of the drug plant, ssp. *indica,* which apparently does not reproduce itself sufficiently outside of cultivation to form persistent populations. Currently, it is illegal to grow or possess any species of *Cannabis* in Missouri, and state law enforcement officials eradicate numerous wild and cultivated plants each year. Although hemp cultivation is legal in many countries, attempts to legalize it in the United States for industrial hemp production have met with stiff resistance.

2. **Humulus** L. (hops)

Plants annual or perennial herbs, with long taproots or stout rhizomes respectively, not strongly aromatic, the staminate and pistillate plants appearing similar vegetatively. Stems twining, branched (except in very depauperate individuals), usually finely ridged or angled, solid at maturity, sparsely to densely rough-pubescent with stiff, prickly 2-branched hairs. Leaves with the petioles shorter than to longer than the blades, sometimes twining, pubescent with short-stalked, 2-branched hairs. Stipules lanceolate, sometimes fused laterally and then appearing solitary rather than paired. Leaf blades ovate to nearly circular in outline, unlobed or palmately 3–9-lobed, deeply cordate at the base, the lobes mostly somewhat narrowed toward the base, tapered to a sharply pointed tip, the margins sharply toothed and sometimes hairy, the surfaces variously hairy and glandular. Staminate inflorescences appearing as open axillary or less commonly terminal panicles with small flower clusters at the branch tips, the flowers with stalks 0.5–3.5 mm long, the branch points and flowers subtended by small bracts. Pistillate inflorescences short, dense spikes continuing to elongate as the fruits mature, pendant at maturity, conelike, the solitary or paired flowers sessile or nearly so, the bracts elliptic or ovate to broadly ovate, sometimes hairy on the surface and along the margins, also with glandular dots and/or short-stalked glands cup-shaped at the tip. Staminate flowers 1.5–3.0 mm long, the sepals lanceolate to ovate or oblong-ovate, with pale, thin margins, glabrous or less commonly sparsely hairy, also with glandular dots. Fruits broadly ovoid to nearly spherical, sometimes slightly flattened, enveloped by the persistent enlarged bracts and membranous calyx. Three species, North America, Europe, Asia, widely cultivated and introduced nearly worldwide.

The branched hairs consist of two short, rigid branches spreading in opposite directions. These are raised from the surface on short, multicellular bases containing cystoliths (calcium carbonate inclusions) that appear as small tubercles. This arrangement apparently is an adaptation to help anchor the plants as they climb; they impart a strong, prickly roughness to the surfaces of the stems and leaves.

1. Plants annual; leaves mostly with the petiole longer than the blade; leaf blades 5–9-lobed, the undersurface roughened with stiff, prickly hairs along the veins; pistillate bracts hairy along the margins, the outer surface not glandular; anthers not glandular . . . . 1. H. JAPONICUS
1. Plants perennial; leaves mostly with the petiole shorter than blade; leaf blades unlobed or with 3(–5) lobes, the undersurface glabrous or soft-hairy; pistillate bracts glabrous along the margins, the outer surface glandular; anthers usually glandular . . . . 2. H. LUPULUS

1. Humulus japonicus Siebold & Zucc. (Japanese hops)

Pl. 333 f, g; Map 1419

Plants annuals, with taproots. Stems 0.5–5.0 m or more long, with dense, 2-armed hairs on the ridges, rough and prickly to the touch, minutely hairy or glabrous between them. Leaves with petioles 4–20 cm long, mostly longer than the blades, usually sparsely to moderately pubescent with stiff, 2-armed hairs. Leaf blades 3–15 cm long, 4–18 cm wide, broadly ovate to depressed-ovate in outline, with 5–9 relatively deep palmate lobes (except on very young stems), the margins with stiff, bulbous-based, prickly hairs, the upper surface sparsely roughened with stiff, bulbous-based, prickly hairs, the undersurface roughened with stiff, spreading, bulbous-based, prickly hairs mostly along the veins and with yellowish sessile glands. Staminate panicles 15–30 cm long, 2–4 cm wide. Pistillate spikes 1–2 cm long at flowering, elongating to 1.5–3.0 cm long at fruiting, the bracts 7–12 mm long, ovate to broadly ovate, the margins densely hairy, the outer surface sparsely to moderately hairy, not glandular. Sepals 2–3 mm long, lanceolate, hairy. Stamens with the anthers lacking glands. Fruits 3–4 mm long, 2.5–4.0 mm wide, the surface smooth, light brown to yellowish brown, the persistent calyx often darker-mottled. $2n=20$. July–October.

Introduced, widely scattered, locally common in some counties along the Missouri and Mississippi Rivers (native of Asia; introduced in the eastern U.S. west to North Dakota, Kansas, and Arkansas; Canada). Bottomland prairies, bottomland forests, and banks of streams and rivers; also ditches, roadsides, railroads, and moist, open, disturbed areas.

The bracts of the conelike fruiting structures of *H. japonicus* do not possess glands, and consequently the species is of no use in brewing. It has

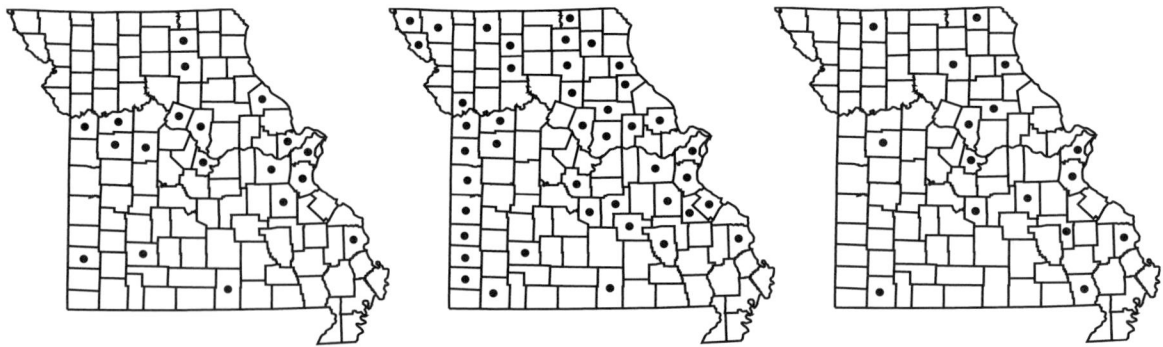

1419. Humulus japonicus   1420. Humulus lupulus   1421. Lonicera ×bella

occasionally been grown as an ornamental climber (variegated forms exist), but it can spread aggressively in the garden. It appears to be increasing its range in the floodplains of some streams and rivers and has become a problem weed in some constructed or restored bottomland prairies.

2. Humulus lupulus L. (common hops)

Pl. 333 e; Map 1420

Plants perennial herbs, with stout rhizomes. Stems 1–6 m or more long, sparsely to moderately pubescent with 2-armed hairs on the ridges, rough and prickly to the touch, minutely hairy or glabrous between them. Leaves with petioles 1–8 cm long, mostly shorter than the blades, usually sparsely to moderately pubescent with stiff, 2-armed hairs. Leaf blades 3–14 cm long, 2–13 cm wide, narrowly to broadly ovate in outline, unlobed or with 3(–5) shallow to relatively deep lobes, the margins sometimes with short, soft hairs, the upper surface sparsely roughened with stiff, bulbous-based, prickly hairs, the undersurface not roughened, glabrous or more commonly sparsely to moderately pubescent with short, fine hairs, also with yellowish, sessile glands. Staminate panicles 3–15 cm long, 2–3 cm wide. Pistillate spikes 0.5–1.5 cm long at flowering, elongating to 1–5 cm long at fruiting, the bracts 10–18 mm long, ovate to elliptic-ovate, the margins glabrous, the outer surface glabrous or sparsely to moderately hairy, also with yellowish to orangish, stalked glands, especially near the base. Sepals 1.5–2.5 mm long, lanceolate to ovate or oblong-ovate, glabrous or hairy, also with yellowish to orangish, stalked glands. Stamens with the anthers usually having orangish glands. Fruits 2.0–2.7 mm long, 2.0–2.5 mm wide, the surface smooth, yellowish brown, the persistent calyx occasionally darker-mottled. $2n=20, 40$. July–October.

Scattered nearly throughout the state but apparently absent from the Mississippi Lowlands Division and some western portions of the Ozarks (U.S., Canada, Mexico, Europe, Asia). Banks of streams and rivers, margins of lakes, bottomland forests, and moist ledges of bluffs; also fencerows, railroads, roadsides, and disturbed areas.

The pistillate inflorescences, which in the brewing industry are referred to as cones, are harvested for beer-making, medicinals, and flavorings. The bitter flavor and antibacterial properties are derived from the yellow resinous exudate from the cup-shaped glands, which contain essential oils known collectively as lupulin. Tannins in the extract also improve the clarity of beer after boiling during the production process. The origin of hops as a crop is obscure. They were known and cultivated in Europe at a very early date, but they might have been brought in from China, where all three species of *Humulus* occur today. How and when hops came to be used to impart flavor and aroma in beer is unknown, but the practice probably originated in Germany (Barth et al., 1994). Prior to the use of hops, beer was preserved with oak leaves, bark, and bitter herbs such as wormwood. Hops were grown in European monasteries in the eighth and ninth centuries and were used to flavor and preserve beverages by the twelfth century. Hops were first brought to England around 1500 by Flemish settlers. The traditional English ale was brewed without hops, but the use of hops eventually caught on. British settlers brought hops with them to New England, where hops became a large and important crop, especially in New York. Hops were well established before the great waves of German immigration to the New World. Prohibition and a fungal disease brought an end to hop production in the east. For a while, hops were widely grown in Wisconsin, California, and other places, but pres-

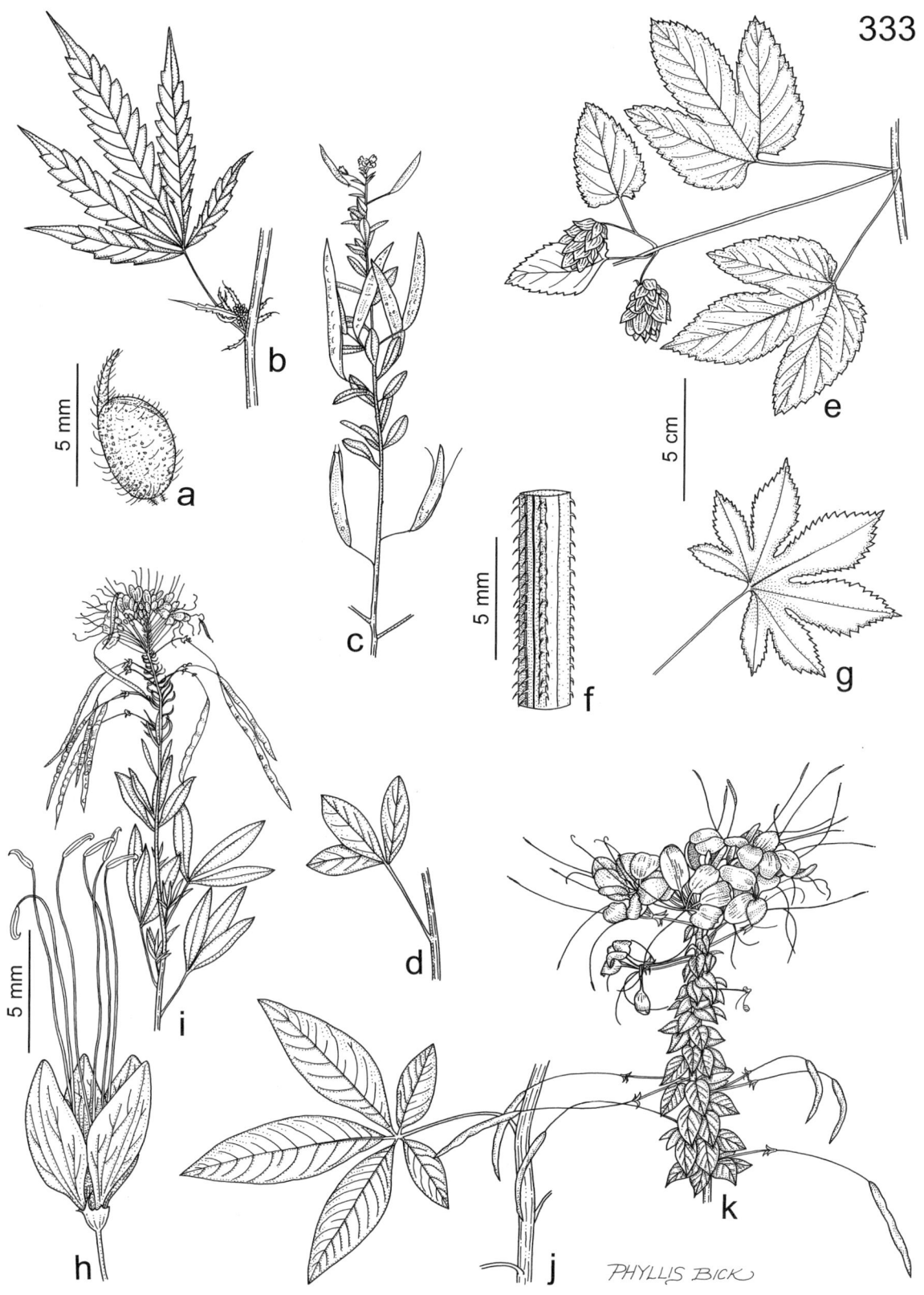

Plate 333. Cannabaceae, Cleomaceae. *Cannabis sativa*, **a**) fruit, **b**) node with leaf and flowers. *Polanisia dodecandra*, **c**) inflorescence, **d**) leaf. *Humulus lupulus*, **e**) inflorescence and leaves. *Humulus japonicus*, **f**) stem detail, **g**) leaf. *Cleome serrulata*, **h**) flower, **i**) inflorescence. *Cleome hassleriana*, **j**) leaf, **k**) inflorescence.

ently the largest producers are Oregon and Washington (Barth et al., 1994).

In addition to their use in the beer industry, hops have been used medicinally, especially by Native Americans, both as a stimulant and mild sedative, for urinary and reproductive problems, to ease fevers and pains, to treat pneumonia and coughs, and as a tonic (Moerman, 1998).

Populations in various parts of North America have been named as distinct species, but the differences are slight and they are now all considered varieties of *H. lupulus*. Potentially making matters more difficult is that varieties of *H. lupulus* native to North America possibly have interbred with introductions from Europe (Small, 1997). A statistical analysis of patterns of variation in eleven morphological characters in *H. lupulus* revealed five intergrading groups roughly correlated with geographic distribution and recognized as varieties (Small, 1978, 1997). Beer is made using only var. *lupulus,* and the North American varieties are avoided because they impart objectionable flavor and aroma to the brew. However, some native North American populations have been a source of germplasm for breeding programs to improve various aspects of cultivated hops (Hampton et al., 2001).

1. Largest leaf blades mostly 5-lobed, the smaller leaves with 5 main veins from the base . . . . 2B. VAR. NEOMEXICANUS
1. Largest leaf blades unlobed or with 3(4) lobes, all the leaves with 3 main veins from the base
    2. Leaf blades sparsely to moderately pubescent and glandular on the undersurface, the midvein with mostly less than 20 hairs per cm, the surface between the veins mostly with less than 25 glands per square cm . . . . 2A. VAR. LUPULUS
    2. Leaf blades densely pubescent and glandular on the undersurface, the midvein with mostly more than 100 hairs per cm, the surface between the veins mostly with more than 25 glands per square cm
    . . . . 2C. VAR. PUBESCENS

2a. var. lupulus

H. americanus Nutt.

Stems relatively sparsely hairy at the nodes. Leaf blades tending to have fewer lobes, those of the larger leaves unlobed or more commonly 3-lobed, all of the leaves with 3 main veins from the base, the undersurface sparsely to moderately pubescent along the veins, glabrous between them, the midvein with mostly fewer than 20 hairs per cm, also moderately glandular with stalked glands, the surface between the veins mostly with fewer than 25 glands per square cm. July–October.

Introduced, uncommon and widely scattered in the state (native of Europe; introduced in the northeastern U.S. west to Wisconsin and Missouri, disjunct in California and possibly Oregon; Canada, Asia). Banks of streams; also railroads and disturbed areas.

This variety was cultivated widely in the eastern United States from the 1600s to the Prohibition era, and many of the herbarium collections from the northeastern United States are said to be of this variety (Small, 1978).

2b. var. neomexicanus A. Nelson & Cockerell

Stems relatively hairy at the nodes. Leaf blades tending to have more lobes, those of the larger leaves mostly 5-lobed, even those of smaller leaves palmately veined with 5 main veins from the base, the undersurface moderately to densely pubescent along the veins and often also between them, the midvein with mostly more than 20 hairs per cm, also densely glandular with stalked glands, the surface between the veins mostly with more than 35 glands per square cm. July–October.

Uncommon, known thus far only from Vernon County (western U.S. east to North Dakota, Kansas, and New Mexico; Canada, Mexico). Banks of streams and edges of bottomland forests.

The Missouri plants were first collected in 1998 by Stephen Timme of Pittsburg State University. They are the easternmost occurrence of the variety and somewhat disjunct from the closest populations in western Kansas.

2c. var. pubescens E. Small

Stems relatively hairy at the nodes. Leaf blades tending to have fewer lobes, those of the larger leaves unlobed or more commonly 3(4)-lobed, all of the leaves with 3 main veins from the base, the undersurface densely pubescent along the veins and usually also between them, the midvein with mostly more than 100 hairs per cm, also moderately to densely glandular with stalked glands, the surface between the veins mostly with more than 25 glands per square cm. July–October.

Scattered nearly throughout the state but apparently absent from the Mississippi Lowlands Division and some western portions of the Ozarks (Nebraska to Arkansas east to Pennsylvania and

North Carolina). Banks of streams and rivers, margins of lakes, bottomland forests, and moist ledges of bluffs; also fencerows, railroads, roadsides, and disturbed areas.

This is by far the most common variety of *H. lupulus* in the state. The mostly northern var. *lupuloides* E. Small occurs immediately to the north of Missouri and should be searched for in the northernmost counties. It differs from var. *lupulus* in being somewhat less densely pubescent on the leaf undersurface, with hairs usually absent between the veins and more than 20 but less than 100 per cm along the midvein. It also tends to have more of the leaves 3-lobed, including most of the smaller leaves, whereas in var. *pubescens* the smaller leaves generally are unlobed and some of the larger leaves also can be unlobed.

CAPRIFOLIACEAE (Honeysuckle Family)
Contributed by Alan Whittemore

Plants shrubs, small trees, lianas, or perennial herbs. Leaves opposite, simple or pinnately compound but infrequently lobed. Stipules absent or, if present, relatively inconspicuous and scalelike or glandular. Inflorescences consisting of solitary, paired, or densely clustered flowers in the leaf axils or of terminal clusters or panicles. Flowers perfect (in paniculate species, occasionally some of the marginal flowers sterile), epigynous. Calyces 5-lobed, actinomorphic, fused to the ovary, the tube sometimes extending slightly past the ovary, usually persistent at fruiting. Corollas 5-lobed, actinomorphic and funnelform to cup-shaped, bell-shaped, or saucer-shaped, or strongly zygomorphic and strongly 2-lipped, white, cream-colored, yellow, pink, or red. Stamens 5, the filaments distinct, attached to the base of the corolla. Anthers oblong to oblong-elliptic or nearly linear, attached above the base on the dorsal side. Pistil 1 per flower (but the ovaries of adjacent flowers sometimes fused), of 2–5 fused carpels. Ovary inferior, with 2–5 locules, the placentation axile. Style 1 per flower with the stigma capitate and sometimes shallowly 2–5-lobed or absent and the sessile stigma deeply 3–5-lobed (usually appearing as 3–5 separate stigmas). Ovules 1 to numerous per locule. Fruits berries or berrylike drupes with one to several seeds or seedlike stones. Fifteen genera, about 400 species, nearly worldwide, most diverse in temperate and montane regions.

Many species of Caprifoliaceae are popularly cultivated as ornamentals. In addition to the genera treated here, two groups of mostly Asian bush honeysuckles, *Abelia* R. Br. and *Weigelia* Schreb., are commonly seen in gardens. They differ from the genera present in the wild in Missouri by their sepals, which are large and prominent, and their fruits, which are dry and capsular.

The circumscription of Caprifoliaceae in the present treatment is the traditional one (Steyermark, 1963; Cronquist, 1981, 1991). However, phylogenetic studies in recent years have suggested that a broad renovation of familial limits in the order Dipsacales is necessary (Donoghue et al., 1992, 2001; Judd et al., 1994; Bell et al., 2001). Both morphological and molecular data provide evidence that the broad traditional circumscription of Caprifoliaceae does not accurately reflect the phylogeny of the group, because the core of the family (including the Missouri genera *Lonicera, Symphoricarpos,* and *Triosteum*) is more closely related to members of the Dipsacaceae and Valerianaceae than to other genera traditionally included in the family. The two Missouri genera with sessile stigmas and saucer- or cup-shaped corollas, *Sambucus* and *Viburnum,* along with three small herbaceous genera, *Adoxa* L., *Sinadoxa* C.Y. Wu, Z.L. Wu & R.F. Huang, and *Tetradoxa* C.Y. Wu, probably are better treated in a separate family, Adoxaceae. Some authors have gone so far as to submerge Dipsacaceae and

Valerianaceae into a redefined Caprifoliaceae (Judd et al., 2002), but further details of familial limits within the order remain to be elucidated through broader taxon sampling and more detailed studies within the major branches of the overall phylogenetic tree.

1. Leaves pinnately compound .... 2. SAMBUCUS
1. Leaves simple
    2. Plants perennial herbs .... 4. TRIOSTEUM
    2. Plants shrubs, lianas, or small trees
        3. Style absent or very short, the stigmas sessile or nearly so; corollas about 2 mm long, saucer-shaped or bell-shaped, white; well-developed leaves toothed .... 5. VIBURNUM
        3. Stigmas raised on a well-developed style 3–50 mm long; corollas 3–48 mm long, white, yellow, pink, or red, cup-shaped to bell-shaped or narrowly funnelform, the lobes often flaring; leaves almost always entire (the first-produced leaves of the season sometimes bluntly lobed)
            4. Corollas 13–48 mm long; fruits berries, red, black, or rarely yellow, with several seeds; shrubs with flowers in pairs in the leaf axils, or lianas .... 1. LONICERA
            4. Corollas 3–9 mm long; fruits berrylike drupes, white or pink to red or purple, with 2 seedlike nutlets; shrubs with flowers in small clusters at the branch tips and in the upper leaf axils .... 3. SYMPHORICARPOS

1. Lonicera L. (honeysuckle)

Plants shrubs or lianas. First-year twigs 1–2 mm thick, the pith solid or hollow. Winter buds more or less ovoid to conical, with 2 to several overlapping scales. Leaves sessile or with short, unwinged petioles, the uppermost leaves sometimes strongly perfoliate (fused at the base to form a single blade surrounding the stem). Stipules absent. Leaf blades simple, unlobed or in *L. japonica* the first leaves of the season sometimes irregularly lobed, elliptic to ovate, oblong, or oblanceolate, the margins entire. Flowers in pairs in the leaf axils, or in 1–4 whorls terminal on the branches, or in *L.* ×*purpurea* in 1 or 2 sessile pairs from buds on second-year branches, the individual flowers usually subtended by small, paired bractlets, the inflorescences subtended by bracts. Corollas 13–48 mm long, actinomorphic and narrowly tubular-funnelform or zygomorphic with spreading to recurved lobes, sometimes the lobes similar in size and sometimes shape but positioned zygomorphically, white, pink, yellow, orange, or red. Style 8–50 mm long. Fruits berries, more or less spherical (oblong-ellipsoid in *L.* ×*purpusii*), red, black, or rarely yellow at maturity. Seeds few, 2.5–6.0 mm long, oblong-ovate to oblong-elliptic in outline, sometimes irregularly so, often somewhat flattened, one or both sides sometimes slightly angled or with a pair of shallow longitudinal grooves, the surface smooth or more commonly with a minute network of low ridges, reddish brown to dark brown. About 180 species, North America, Europe, Asia.

Honeysuckles have long been very popular in horticulture, and numerous species and cultivars exist in cultivation. The plants are easy to grow, tolerant of difficult conditions, and their flowers are attractive and (in many species) fragrant. The berries attract birds in the autumn. However, some exotic species of *Lonicera* are very invasive in natural plant communities, and the genus includes two of the worst invasive exotics in Missouri, *L. japonica* and *L. maackii*.

In Missouri, two escaped taxa of shrubby honeysuckles exist that arose through past interspecific hybridization, but whose parents do not occur in the state. Because these fertile hybrids are not easily related to any of the other species in the state, they are accorded full treatments below.

1. Plants shrubs 1.5–4.0 m tall, the main stems erect or ascending, not twining; flowers in pairs in the leaf axils; corollas white or pink (often changing to cream-colored or yellow after pollination); fruits red, rarely yellow
    2. Twigs with solid white pith; flowers sessile from buds produced on previous year's wood (second-year portions of branches); paired ovaries fused for about ½ of their length . . . . 6. L. ×PURPUSII
    2. Twigs with hollow pith; flowers in the axils of leaves on current year's growth (first-year wood); paired ovaries separate
        3. Leaf blades tapered at the tip; stalk below the paired flowers 2–5(–8) mm long, the fruits appearing sessile or nearly so . . . . 5. L. MAACKII
        3. Leaves rounded or broadly angled to a bluntly or sharply pointed tip, sometimes tapered abruptly to a minute, sharp point; stalk below the paired flowers 5–19 mm long, the fruits appearing noticeably stalked at maturity . . . . 1. L. ×BELLA
1. Plants lianas, the main stems twining around adjacent trees or shrubs or trailing on the ground
    4. Flowers in pairs in the leaf axils; leaves never fused; corollas white, becoming cream-colored or yellow after pollination; fruits black . . . . 4. L. JAPONICA
    4. Flowers clustered in 1–4 whorls at the branch tips, usually with 1 or 2 pairs of leaves below the inflorescence strongly perfoliate (fused at the base to form a single blade surrounding the stem); corollas pale yellow to orange, red, or purplish-tinged, not changing color after pollination; fruits red
        5. Corollas 38–48 mm long, divided less than ⅕ of the way to the base into more or less similar ascending lobes, bright red or orangish red . . . . 8. L. SEMPERVIRENS
        5. Corollas 15–35 mm long, strongly 2-lipped, divided ⅓–½ of the way to the base into spreading to recurved lobes, pale yellow to orange or pink
            6. Perfoliate upper leaves 1.0–1.5 times as long as wide, the upper surface usually strongly glaucous in patches near the center; flowers in usually 2–4 whorls; corollas cream-colored or pale yellow, not pinkish-tinged . . . . 7. L. RETICULATA
            6. Perfoliate upper leaves 1.2–2.2 times as long as wide, the upper surface green; flowers in 1 or 2 whorls; corollas yellow or orange, sometimes pinkish-tinged
                7. Corollas white to lemon yellow and pinkish-tinged, the base of the tube weakly swollen or pouched on one side; undersurface of the leaves very strongly glaucous, often almost white . . . . 2. L. DIOICA
                7. Corollas bright yellow to orange, the base of the tube slender, symmetrical, not pouched; undersurface of the leaves only weakly glaucous, pale green . . . . 3. L. FLAVA

1. Lonicera ×bella Zabel

Pl. 335 c, d; Map 1421

Plants shrubs 1.5–2.5(–4.0) m tall, the main stems erect or ascending, self-supporting. Twigs sparsely to densely pubescent with short, curved, sometimes more or less tangled, unbranched hairs, the pith hollow. Winter buds ovoid or conical, hairy or glandular. Leaf blades 1.7–5.5 cm long, 0.8–2.5 cm wide, oblong to oblong-ovate or less commonly ovate, those of the uppermost leaves occasionally oblong-lanceolate, broadly angled to rounded or more or less truncate at the base, rounded or more commonly angled or short-tapered to a bluntly or sharply pointed tip, the surfaces (and margins) sparsely to moderately pubescent with fine, curved hairs, at least along the main veins, not glaucous, but the undersurface light green to pale green. Flowers in pairs in the axils of the leaves on current year's growth (first-year wood), each pair at the tip of a stalk 5–19 mm long, the 2 bracts each 2–15 mm long, free, linear to narrowly elliptic, glabrous or sparsely hairy, the pair of bractlets on opposite sides of each flower 1/2–2/3 as long as the ovary, free, obliquely oblong to oblong-ovate. Calyces glabrous or with stalked glands and/or scattered, long, straight hairs, the lobes 0.8–1.0 mm long, triangular. Corollas 13–16(–20) mm long, strongly zygomorphic, divided 1/2–2/3 of the way to the base into 5 more or less spreading lobes of about equal length, the upper portion with the 4 lobes slightly less deeply cut than the 1 lobe of the lower portion, the tube noticeably swollen or pouched on the lower side near the base, white or pink, turning yellow or yellowish orange after pollination. Stamens and style exserted from the corolla, shorter than to about as long as the corolla lobes, the style hairy. Ovaries free. Fruits 5–8 mm in diameter, orangish red to red. $2n=18$. April–June.

Scattered nearly throughout the state (artificially derived hybrid between two Eurasian species; introduced widely in the northeastern and central U.S., Canada). Mesic upland forests and banks of streams; also fencerows, railroads, roadsides, and disturbed areas.

This taxon represents a variable group of fertile hybrids, including several genetically distinct forms that have been cultivated widely and occasionally escape. All were derived from hybridization between two commonly cultivated species of Eurasian origin, *L. tatarica* and *L. morrowii*. In recent decades, members of the group have been largely replaced by *L. maackii* (see treatment below) at plant nurseries, although as the aggressive nature of all of the Asian bush honeysuckles has become apparent, even *L. maackii* has become less commonly sold for gardens.

Three other Eurasian species in the *L. tatarica* complex have been reported from Missouri: *L. tatarica* (Palmer, 1961; Steyermark, 1963), *L. morrowii* (Steyermark, 1963), and *L. xylosteum* (Mühlenbach, 1979). These species are closely related and similar to one another. Reexamination of all of the available specimens from Missouri suggests that so far *L. ×bella* is the only entity that can be confirmed as growing in the state. Distinguishing among the species and the various cultivars and hybrids usually requires close examination of the pair of minute bractlets at the base of each ovary, as well as the calyces, especially the type of hairs and/or glands (if any) on them. Further, in some situations in other midwestern states where two or more of these bush honeysuckles grow in mixed populations, spontaneous hybridization has occurred with varying frequency. The illustrations, descriptions, and keys in Green (1966) are very helpful for separating species and hybrids in the complex. Hansen (1966) also provided useful information on the group. The species can be recognized as follows.

Lonicera morrowii A. Gray (Morrow honeysuckle; Pl. 335 a, b). Stems hairy. Winter buds broadly ovoid, rounded, the scales not keeled, glabrous. Leaf blades 10–22 mm wide, oblong or narrowly elliptic, the undersurface hairy. Bractlets as long as the ovary or nearly so, usually more or less obscuring it, the outer surface and margins (also the bracts and calyx lobes) with a few long hairs, never glandular. Corollas strongly 2-lipped, initially white, turning yellow after pollination or with age.

Lonicera tatarica L. (Tatarian honeysuckle, twin sisters). Stems glabrous. Winter buds broadly ovoid, rounded, the scales not keeled, glabrous. Leaf blades 22–27 mm wide, oblong to oblong-ovate, the undersurface glabrous. Bractlets less than 1/2 as long as the ovary, glabrous or sometimes with a few stalked glands on the margins (also on the calyx lobes). Corollas deeply cleft into five more or less equal lobes (usually only slightly zygomorphic), pink or white, not turning yellow with age.

Lonicera xylosteum L. (European fly honeysuckle). Stems hairy. Winter buds conical, pointed, the scales keeled, with a few hairs near their tips. Leaf blades 25–32 mm wide, broadly elliptic or broadly ovate (sometimes almost circular), the undersurface hairy. Bractlets less than 1/2 as long as the ovary, with long, stiff hairs on the surface and margins and many conspicuous stipitate glands. Calyx lobes conspicuously glandular, sometimes also hairy. Corollas strongly 2-lipped, initially white or pink, turning yellow or yellowish orange after pollination or with age.

Some Missouri specimens closely resemble *L. tatarica*, but they have pubescent leaves, and the flowers tend to fade yellow with age, indicating crossing with *L. morrowii*. A few of the Missouri escapes have large bractlets and key to *L. ×minutiflora* Zabel in Green (1966), but they do not have the small flowers of the latter and are more likely a *L. morrowii*-like recombinant or backcross of *L. ×bella*. These plants have stalked glands on the bractlets and calyx and some have pink flowers, so they cannot be genetically pure *L. morrowii*. *Lonicera xylosteum* was reported from the outskirts of Joplin (Jasper County) by Palmer (1961), based

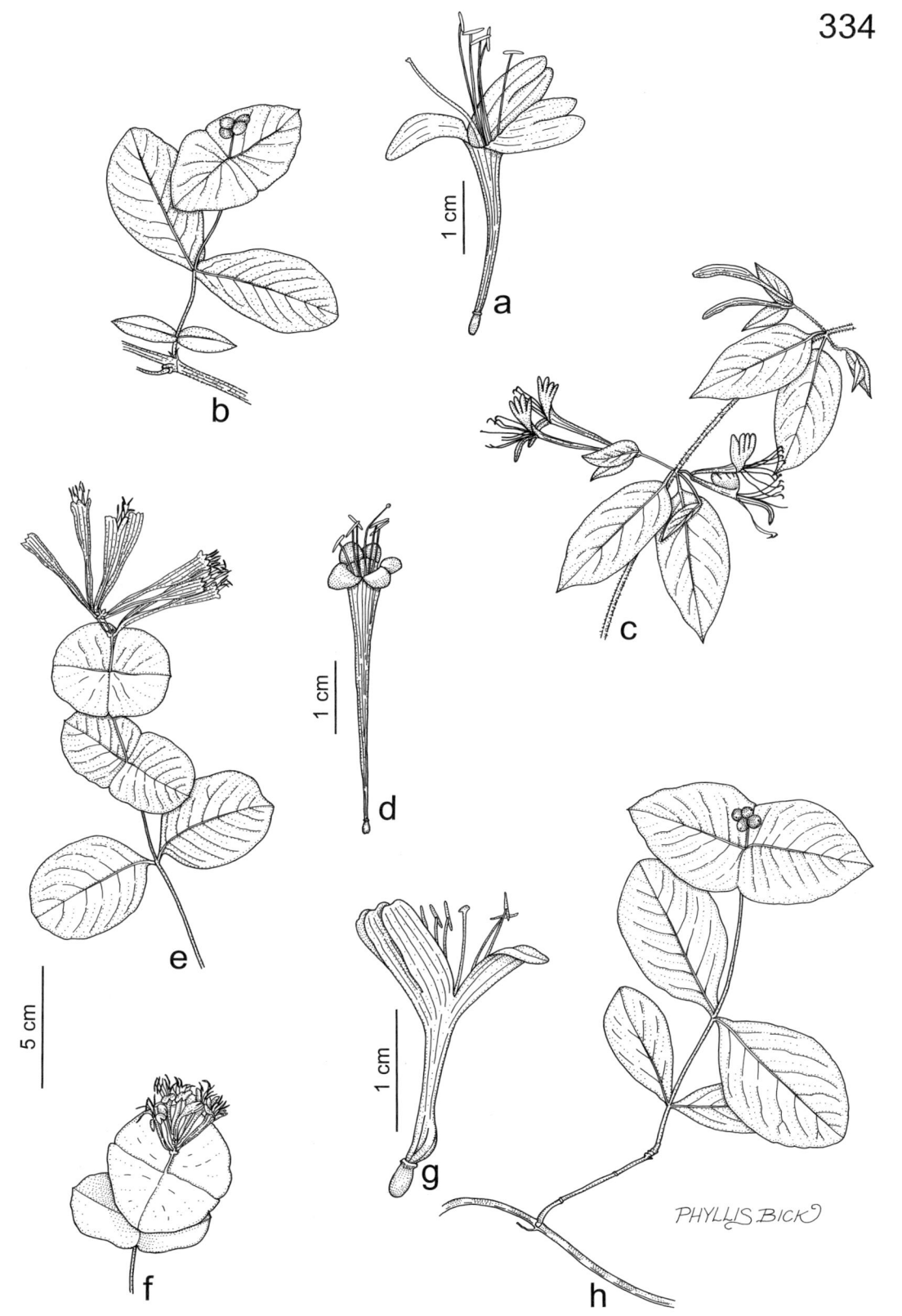

Plate 334. Caprifoliaceae. *Lonicera flava*, **a)** flower, **b)** fruiting branch. *Lonicera japonica*, **c)** flowering branch. *Lonicera sempervirens*, **d)** flower, **e)** inflorescence. *Lonicera reticulata*, **f)** inflorescence. *Lonicera dioica*, **g)** flower, **h)** fruiting branch.

1422. Lonicera dioica

1423. Lonicera flava

1424. Lonicera japonica

on a single collection. This specimen (at the University of Missouri's Dunn-Palmer Herbarium) shows none of the characters of *L. xylosteum* but is typical *L. ×bella*.

2. Lonicera dioica L. (limber honeysuckle, wild honeysuckle, red honeysuckle)

L. dioica var. *glaucescens* (Rydb.) Butters

Pl. 334 g, h; Map 1422

Plants lianas to 4 m or more long, the main stems loosely twining toward the tips, climbing on adjacent vegetation (sometimes twining on themselves and forming loose mounds) or more or less trailing on the ground. Twigs glabrous, the pith hollow, the bark of older branches becoming shredded. Winter buds ovoid but somewhat flattened, glabrous. Leaf blades mostly 3–9 cm long, 2.0–6.5 cm wide, mostly elliptic, rounded or more commonly angled at the base, rounded or more commonly angled or tapered to a bluntly or sharply pointed tip, those of the uppermost 1 or few pairs strongly perfoliate, 1.2–2.2 times as long as wide, the pair broadly elliptic to oblong-elliptic in overall outline, rounded or broadly angled to bluntly pointed tips, sometimes abruptly tapered to minute, sharp points, the upper surface glabrous and bright green (that of the perfoliate leaves sometimes slightly glaucous toward the center), the undersurface glabrous or sparsely to moderately and evenly pubescent with soft, more or less spreading hairs, very strongly glaucous (sometimes almost white). Flowers in 1 or 2 whorls of 6 at the branch tips, the flowers sessile, the 2 bracts each 4–6 mm long, free, very broadly triangular, glabrous, the pair of bractlets on opposite sides of each flower minute (0.3–0.5 mm long), free, oblong to broadly ovate. Calyces glabrous, the lobes 0.2–0.4 mm long, semicircular to broadly oblong-rounded, often pale or whitened. Corollas 20–30 mm long, strongly zygomorphic, divided 1/3–1/2 of the way to the base into 2 recurved-curled lips of about equal length, the upper lip shallowly (3)4-lobed, the lower lip with 1 lobe, the tube weakly swollen or pouched on the lower side near the base, white to lemon yellow and pinkish-tinged (rarely nearly red), not changing color after pollination. Stamens and style exserted from the corolla, slightly longer than the corolla lobes, the style sparsely to densely hairy. Ovaries free. Fruits 5–10 mm in diameter, orangish red to red. $2n=18$. April–June.

Uncommon, widely scattered in the state (eastern [mostly northeastern] U.S. west to North Dakota, Wyoming, and Oklahoma; Canada). Bases and ledges of bluffs, mesic upland forests, and banks of streams and rivers; rarely also fencerows.

Plants with pubescent leaves and densely pubescent styles have been called var. *glaucescens*, but Perino (1978) studied patterns of morphological variation across the species range and concluded that the pubescence patterns showed complete intergradation.

3. Lonicera flava Sims (yellow honeysuckle)

L. flava var. *flavescens* Gleason

Pl. 334 a, b; Map 1423

Plants lianas to 4 m or more long, the main stems loosely twining toward the tips, climbing on adjacent vegetation (sometimes twining on themselves and forming loose mounds) or occasionally more or less trailing on the ground. Twigs glabrous, usually glaucous, the pith hollow, the bark of older branches sometimes becoming shredded. Winter buds conical but somewhat flattened, glabrous. Leaf blades mostly 3–9 cm long, 2.0–6.5 cm wide, most elliptic, rounded or more commonly angled at the base, rounded or more commonly angled or tapered to a bluntly or sharply pointed tip, those of the uppermost 1 or few pairs strongly perfoliate, 1.2–2.2 times as long as wide, the pair broadly elliptic to oblong-elliptic in overall outline, rounded or broadly

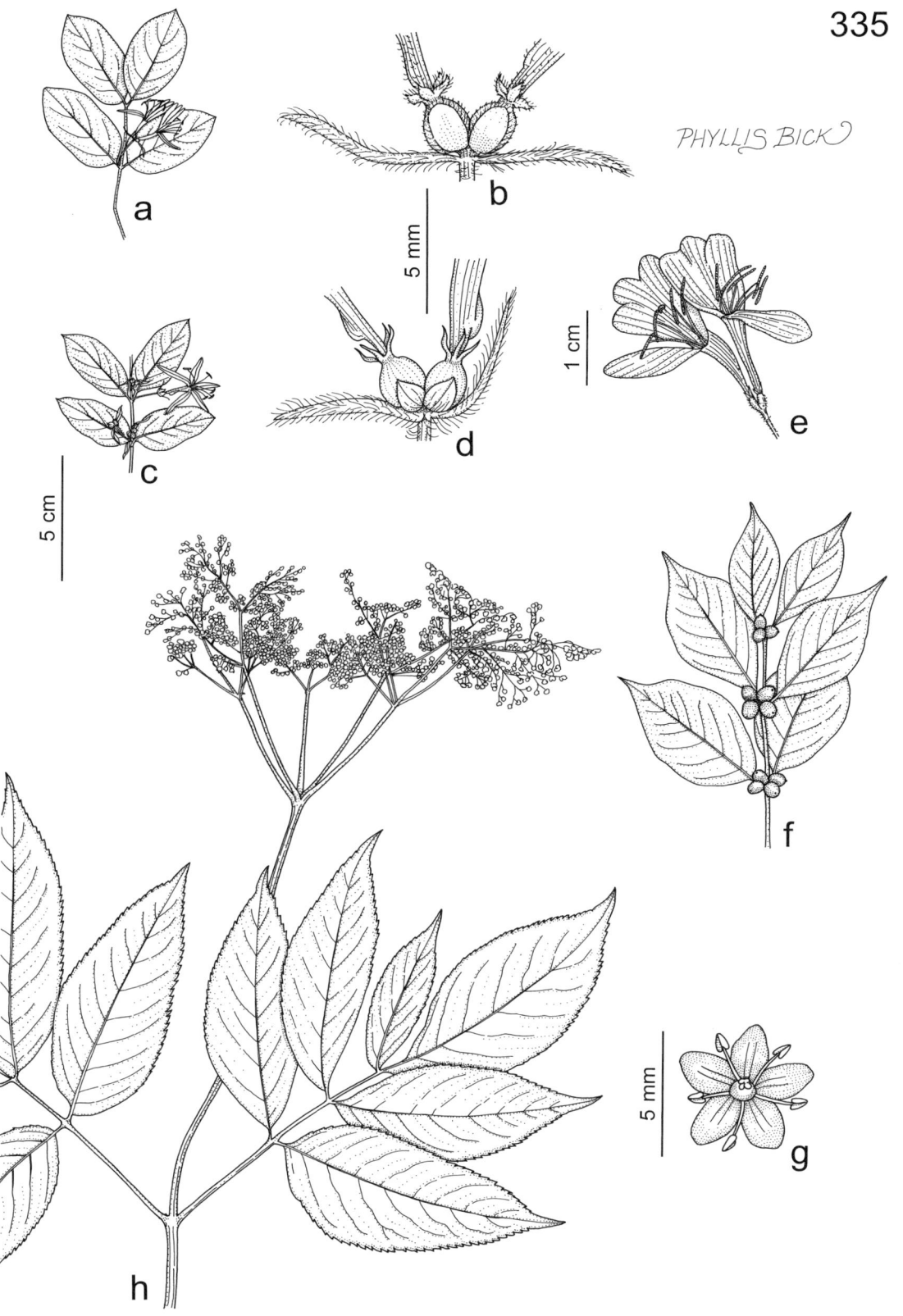

Plate 335. Caprifoliaceae. *Lonicera morrowii*, **a)** branch with flowers, **b)** base of flower pair. *Lonicera ×bella*, **c)** branch with flowers, **d)** base of pair of flowers. *Lonicera maackii*, **e)** flower pair, **f)** fruiting branch. *Sambucus canadensis*, **g)** flower (top view), **h)** inflorescence and leaves.

angled to bluntly pointed tips, sometimes abruptly tapered to minute, sharp points, the upper surface glabrous and bright green (that of the perfoliate leaves sometimes slightly glaucous toward the center), the undersurface glabrous or sparsely to moderately pubescent with soft, more or less spreading hairs along the midvein, slightly to moderately glaucous (usually pale green). Flowers in 1 or 2 whorls of 6 at the branch tips, the flowers sessile, the 2 bracts each 0.8–3.0 mm long, free, oblanceolate to bluntly triangular, glabrous, the pair of bractlets on opposite sides of each flower minute (0.4–1.2 mm long), free, oblong to broadly ovate. Calyces glabrous, the lobes 0.2–0.4 mm long, broadly oblong-rounded to broadly triangular, often pale or whitened. Corollas 20–35 mm long, strongly zygomorphic, divided $1/3$–$1/2$ of the way to the base into 2 recurved-curled lips of about equal length, the upper lip shallowly (3)4-lobed, the lower lip with 1 lobe, the tube slender and not swollen or pouched near the base, bright yellow to orange, sometimes strongly reddish-tinged along the tube, not changing color after pollination. Stamens and style exserted from the corolla, slightly longer than the corolla lobes, the style usually glabrous. Ovaries free. Fruits 5–10 mm in diameter, orangish red to red. $2n=18$. April–June.

Scattered in the Ozark and Ozark Border Divisions (Kansas and Oklahoma east to Ohio, North Carolina, and Georgia). Mesic to dry upland forests, edges of glades, savannas, ledges and tops of bluffs, and banks of streams and rivers; rarely also roadsides.

This species tends to be more robust and colorful than the other native honeysuckles in the state. It is an attractive arbor plant in gardens and recently has become available commercially through a few specialty native plant nurseries.

4. Lonicera japonica Thunb. ex Murray
(Japanese honeysuckle)

Pl. 334 c; Map 1424

Plants lianas to 5 m or more long, the main stems loosely twining, climbing on adjacent vegetation or more or less trailing on the ground. Twigs sparsely to moderately pubescent with spreading hairs, sometimes some of these with minutely glandular tips, the pith hollow, the bark of older branches becoming shredded. Winter buds conical, densely hairy. Leaf blades mostly 3–9 cm long, 1.5–4.5 cm wide, ovate to oblong-elliptic, rounded or angled to short-tapered at the base, angled or tapered to a sharply pointed tip, none perfoliate, the upper surface glabrous or more commonly short-hairy along the midvein, bright green to olive green, the undersurface sparsely to moderately pubescent with short, soft hairs mostly along the main veins, occasionally also with scattered stiffer hairs, not glaucous (pale green). Flowers in pairs in the axils of the leaves on current year's growth (first-year wood), each pair at the tip of a stalk 1–25 mm long, the 2 bracts each 3–15 mm long, free, ovate to oblong-elliptic (leaflike), hairy, the pair of bractlets on opposite sides of each flower minute (0.7–1.2 mm long), free, oblong-ovate to nearly circular, hairy along the margins. Calyces hairy along the margins, the lobes 0.7–1.5 mm long, triangular to narrowly triangular, green, sometimes purplish-tinged. Corollas 30–47 mm long, strongly zygomorphic, divided about $1/2$ of the way to the base into 2 recurved-curled lips of about equal length, the upper lip shallowly 3-lobed, the lower lip with 2 deeper lobes, the tube slender and not swollen or pouched near the base, white, turning cream-colored or pale yellow after pollination, usually hairy and stalked-glandular on the outer surface. Stamens and style exserted from the corolla, slightly longer than the corolla lobes, the style glabrous. Ovaries free. Fruits 5–8 mm in diameter, black. $2n=18$. April–June.

Scattered nearly throughout the state (native of Asia, introduced widely in the U.S., Canada). Bottomland forests, mesic to dry upland forests, banks of streams, rivers, and spring branches, and margins of ponds, lakes, and sinkhole ponds; also fencerows, old fields, old homesites, railroads, roadsides, and mostly shaded disturbed areas.

Lonicera japonica was once common in horticulture because of its easy culture, rapid growth, numerous fragrant flowers, and attractive twining habit. It has proven to be aggressively invasive in a number of native plant communities in the state. In addition to the fruits, which are dispersed by birds, the species can spread by woody underground stems, and the aerial stems can root where they touch the ground. Japanese honeysuckle forms dense mounds and smothers other plants, and it sometimes girdles shrubs and small trees by twisting around their trunks. It may be semievergreen in mild winters.

5. Lonicera maackii (Rupr.) Maxim. (Amur honeysuckle, bush honeysuckle)

Pl. 335 e, f; Map 1425

Plants shrubs 1.5–4.0(–5.0) m tall, the main stems erect or ascending, self-supporting. Twigs moderately to densely pubescent with short, curved, sometimes more or less tangled, unbranched hairs, the pith hollow. Winter buds ovoid, hairy. Leaf blades 3.5–9.5 cm long, 1.5–4.0 cm wide, elliptic to ovate-elliptic, angled or rounded to more commonly

1425. Lonicera maackii   1426. Lonicera ×purpusii   1427. Lonicera reticulata

tapered at the base, tapered to a sharply pointed tip, the surfaces (and margins) sparsely to moderately pubescent with fine, mostly curved hairs, at least along the main veins, not glaucous, but the undersurface light green to pale green. Flowers in pairs in the axils of the leaves on current year's growth (first-year wood), each pair at the tip of a stalk 2–5(–8) mm long (often appearing sessile at fruiting), the 2 bracts each 1–4 mm long, free, linear, hairy, the pair of bractlets on opposite sides of each flower ½ as long to about as long as the ovary, free, oblong-obovate to nearly circular. Calyces with stalked glands and long, straight hairs, the lobes 0.2–0.5 mm long, triangular. Corollas 15–25 mm long, strongly zygomorphic, divided about ½ of the way to the base into 5 more or less spreading lobes of about equal length, the upper lip shallowly 4-lobed, the lower lip 1-lobed, the tube not or very slightly swollen or pouched on the lower side near the base, white (sometimes pinkish-tinged toward the base of the tube), turning yellow or orangish yellow after pollination or with age. Stamens and style exserted from the corolla, shorter than the corolla lobes, the style hairy. Ovaries free. Fruits 4–7 mm in diameter, orangish red to red. $2n=18$. April–June.

Introduced, scattered, mostly around urban areas (native of eastern Asia, introduced widely in the eastern U.S. west to North Dakota and Texas, Canada). Bottomland forests, mesic upland forests, bases and ledges of bluffs, and banks of streams and rivers; also fencerows, gardens, railroads, roadsides, and shaded, disturbed areas.

Luken and Thieret (1995, 1996) reviewed the spread of Amur honeysuckle in North America, from its first introduction in the late 1800s to its establishment as a serious pest. So-called improved cultivars of *L. maackii* were developed by the USDA Soil Conservation Service beginning in 1960 and the species was planted widely for erosion control, as a hedge or screen, and for ornamental purposes through the mid-1980s, when its invasive potential was first realized. In the Midwest, *L. maackii* largely replaced other bush honeysuckles in the horticultural industry and it remains available through some specialty nurseries today. *Lonicera maackii* was first reported for Missouri by Mühlenbach (1983) from the St. Louis railyards, but it remains undercollected in the state (for example, it is a problem plant in the Kansas City metropolitan area but is still undocumented from those counties). It has proven to be aggressively invasive in most midwestern states and has become the dominant species in the understory of many remnant wooded areas in and around cities. More recently it has used rivers and highways as dispersal corridors and is beginning to invade more rural portions of Missouri. Dense stands of Amur honeysuckle leaf out earlier in the spring than other understory plants, reducing sunlight required by spring ephemerals and germinating seeds of other species. Similarly, the species remains green until after other woody species have already lost their leaves. Because the plants strongly impact the structure of forest understories, they have been implicated in changes in nesting patterns of some native bird species, rendering the eggs and chicks more prone to predation. Luken and Thieret (1995) further noted that the berries are mildly toxic to humans but are strongly unpalatable.

6. Lonicera ×purpusii Rehder

Map 1426

Plants shrubs 1.5–2.5 m tall, the main stems erect or ascending, self-supporting. Twigs glabrous or inconspicuously hairy with minute, scurfy hairs, the pith solid. Winter buds narrowly ovoid, glabrous except for the finely hairy margins of the scales. Leaf blades 3.5–9.5 cm long, 1.5–5.0 cm wide, elliptic to ovate-elliptic, angled or tapered at the base, angled or tapered to a sharply pointed tip, the upper surface glabrous, the undersurface (and margins) sparsely to moderately pubescent with stiff, more or less appressed hairs along the midvein, not glaucous, but the undersurface light green. Flowers in pairs, 1 or 2 in the axils of the leaves on previous year's growth (second-year wood), each pair sessile, the 2 bracts each 3–9 mm long, fused at the base, ovate to ovate-triangular, hairy along the margins, the pair of bractlets on opposite sides of each flower somewhat longer than the ovary, usually fused at the base, oblong-ovate to nearly circular. Calyces glabrous, the lobes 0.1–0.5 mm long, broadly semicircular or the margin appearing merely undulate. Corollas 12–18 mm long, somewhat zygomorphic, divided ½ or slightly more than ½ of the way to the base into 5 more or less spreading lobes of about equal length, the lower lobe with deeper sinuses than those between the other 4 lobes, the tube noticeably swollen or pouched on the lower side near the base, white, turning cream-colored or pale yellow after pollination or with age. Stamens and style exserted from the corolla, about as long as the corolla lobes, the style glabrous. Ovaries fused to about the midpoint. Fruits 6–10 mm long (oblong-ellipsoid), red. January–February.

Introduced, uncommon, known thus far from a single site in Cape Girardeau County (artificially derived hybrid between two Asian species; introduced in Missouri). Disturbed mesic upland forests.

This is an artificial hybrid between two Chinese species, *L. fragrantissima* Lindl. & Paxton and *L. standishii* Jacques. Its leaves are semievergreen. The intensely fragrant flowers appear very early in the season. The first Missouri collections were made by Mark Basinger in 2000 at a remnant woodlot within the Cape Girardeau city limits.

7. Lonicera reticulata Raf. (grape honeysuckle)
L. prolifera (G. Kirchn.) Booth ex Rehder

Pl. 334 f; Map 1427

Plants lianas to 4 m or more long, the main stems loosely twining toward the tips, climbing on adjacent vegetation (sometimes twining on themselves and forming loose mounds) or occasionally more or less trailing on the ground. Twigs glabrous, the pith hollow, the bark of older branches becoming shredded. Winter buds conical but somewhat flattened, glabrous. Leaf blades mostly 3–9 cm long, 2.0–8.0 cm wide, elliptic to broadly elliptic or obovate, usually tapered at the base, rounded (occasionally shallowly notched) or angled or tapered to a usually bluntly pointed tip, those of the uppermost 1 or few pairs strongly perfoliate, 1.0–1.5 times as long as wide, the pair broadly elliptic to oblong-elliptic or nearly circular in overall outline, rounded (sometimes shallowly notched) or broadly angled to usually bluntly pointed tips, sometimes abruptly tapered to minute, sharp points, the upper surface usually bright green (that of the perfoliate leaves usually strongly glaucous in patches and irregularly pale green), the undersurface glabrous or sparsely to moderately pubescent with soft, more or less appressed hairs and sometimes also scattered, spreading hairs along the midvein, slightly to strongly glaucous (pale green). Flowers in usually 2–4 whorls of 6 at the branch tips, the flowers sessile, the 2 bracts each 0.5–2.0 mm long, free, linear to bluntly triangular, glabrous, the pair of bractlets on opposite sides of each flower minute (0.2–0.5 mm long), free, oblong to broadly ovate. Calyces glabrous, the lobes 0.2–0.4 mm long, broadly oblong-rounded to broadly triangular, often pale or whitened. Corollas 18–28 mm long, strongly zygomorphic, divided ⅓–½ of the way to the base into 2 recurved-curled lips of about equal length, the upper lip shallowly (3)4-lobed, the lower lip with 1 lobe, the tube swollen or pouched on the lower side near the base, cream-colored or pale yellow, not pinkish-tinged, not changing color after pollination. Stamens and style exserted from the corolla, slightly longer than the corolla lobes, the style sparsely hairy. Ovaries free. Fruits 5–10 mm in diameter, orangish red to red. $2n=18$. April–June.

Scattered, mostly in the northern half of the state (northeastern U.S. west to Minnesota and Oklahoma). Bases, ledges, and tops of bluffs, glades, savannas, mesic to dry upland forests, and banks of streams and rivers; rarely also fencerows and railroads.

This species was long known as *L. prolifera*, but the older name *L. reticulata* has to replace it (Perino, 1978). *Lonicera reticulata* Raf. should not be confused with the later homonym *L. reticulata* Champ. ex Benth., which is a synonym of the Asian species *L. rhytidophylla* Hand.-Mazz.

8. Lonicera sempervirens L. (trumpet honeysuckle, coral honeysuckle)

Pl. 334 d, e; Map 1428

Plants lianas to 4 m or more long, the main stems loosely twining toward the tips, climbing on adjacent vegetation (sometimes twining on themselves and forming loose mounds) or occasionally more or less trailing on the ground. Twigs glabrous, the pith hollow, the bark of older branches becoming shredded. Winter buds conical, glabrous. Leaf blades mostly 3.0–7.5 cm long, 1.5–4.5 cm wide, elliptic to oblong-elliptic, oblanceolate, or occasionally ovate, angled or tapered at the base, rounded

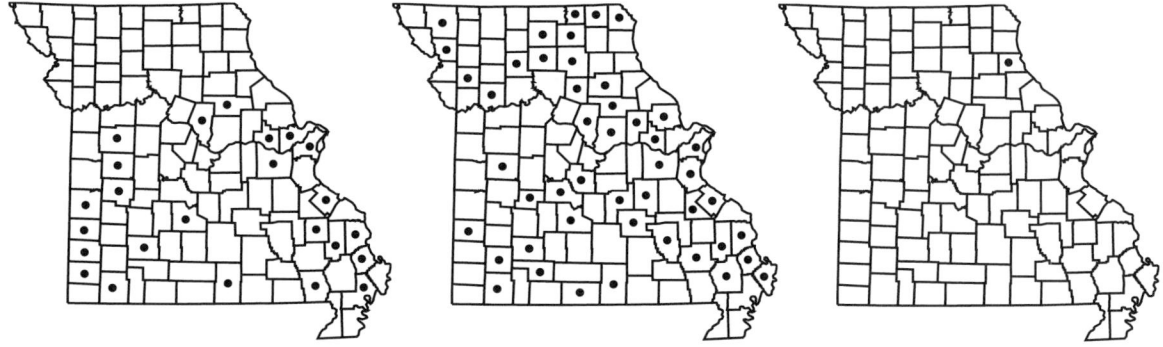

1428. Lonicera sempervirens   1429. Sambucus canadensis   1430. Sambucus pubens

to angled or tapered to a usually bluntly pointed tip, those of the uppermost 1 or few pairs strongly perfoliate, 1.2–2.8 times as long as wide, the pair broadly elliptic to oblong-elliptic in overall outline, rounded or broadly angled to usually bluntly pointed tips, sometimes abruptly tapered to minute, sharp points, the upper surface usually bright green (that of the perfoliate leaves usually bright green), the undersurface glabrous, moderately to strongly glaucous (pale green). Flowers in 1–4 whorls of 6 at the branch tips, the flowers sessile, the 2 bracts each 0.5–1.5 mm long, free or occasionally appearing fused at the base, broadly triangular, glabrous, the pair of bractlets on opposite sides of each flower minute (0.2–0.5 mm long), free, oblong to broadly ovate. Calyces glabrous, the lobes 0.1–0.4 mm long, broadly oblong-rounded to broadly triangular, often pale or pinkish-tinged. Corollas 38–48 mm long, nearly actinomorphic, divided less than $1/5$ of the way to the base into 5 loosely ascending to more or less spreading, similar lobes, the sinuses on either side of 1 lobe slightly deeper than those between the others, the tube with an elongate, slightly swollen area on the lower side near the base, bright red to orangish red, not changing color after pollination. Stamens and style exserted from the corolla, somewhat longer than the corolla lobes, the style glabrous. Ovaries free. Fruits 6–10 mm in diameter, orangish red to red. $2n=36$. April–July.

Introduced, scattered, mostly south of the Missouri River (Maine to Florida west to Oklahoma and Texas; Canada; introduced northwest to Iowa and Kansas). Mesic upland forests, banks of streams and rivers, margins of sinkhole ponds; also fencerows, old homesites, and roadsides.

This beautiful plant deserves more attention from gardeners. It is not particularly invasive in native plant communities. The flowers have no scent, but they are very attractive to hummingbirds. Some botanists have called plants from a series of populations in the southeastern states with hairy upper leaf surfaces and somewhat glandular ovaries var. *hirsutula* Rehder.

2. Sambucus L. (elderberry)

Plants shrubs or small trees with soft wood. First-year twigs 3–8 mm thick, the pith large, solid but soft or spongy. Winter buds more or less ovoid, with several overlapping scales. Leaves with well-developed, unwinged petioles, none perfoliate. Stipules absent or, if present, then small, usually shed early, herbaceous or glandular, linear (note that the primary leaf divisions also frequently have linear, stipulelike outgrowths at their bases). Leaf blades pinnately usually 1 time compound, but rarely (in horticultural forms) the primary divisions again irregularly 1 or 2 times lobed, dissected, or compound, the leaflets otherwise lanceolate to narrowly oblong or elliptic, the margins sharply toothed. Flowers in dense panicles, terminal on the branches, the branch points with inconspicuous, minute, ovate to triangular bracts, the individual flowers bractless. Calyx lobes inconspicuous, 0.2–0.8 mm long, oblong to ovate-triangular. Corollas 2–3 mm long, actinomorphic, more or less bell-shaped with a cup-shaped tube about 1 mm long and a spreading, 5-lobed portion 3–5 mm in diameter (measured across the top of the flower), the lobes rounded, white or pale yellow. Style absent or nearly so, the deeply 3–5-lobed stigma appearing sessile. Fruits berrylike drupes, 4–6(–7) mm in diameter, more or less spherical, orangish red to red or bluish black at maturity. Seedlike nutlets (also called pyrenes or stones) 3–5, 2.5–3.0 mm long, more or less obovate to elliptic in outline, somewhat flattened or bluntly 3-angled, the surface roughened with irregular, fine cross-wrinkles or blunt, low ridges, yellowish brown to brownish yellow. Nine to 20 species, nearly worldwide, most diverse in temperate and montane regions.

The stems, leaves and roots of *Sambucus* species are poisonous, and ingestion can cause vomiting and diarrhea. They have been used medicinally in the past, taken as strong purgatives or applied externally for skin disorders.

1. Inflorescences more or less flat-topped, lacking an elongate main axis, instead with usually 5 primary branches; fruits bluish black to purplish black (rarely red), palatable; stems with the pith white .... 1. S. CANADENSIS
1. Inflorescences ovoid to more or less pyramidal, the solitary main axis elongate; fruits usually bright red (rarely yellow elsewhere), unpalatable; stems with the pith brown .... 2. S. PUBENS

1. Sambucus canadensis L. (common elderberry)

S. *nigra* ssp. *canadensis* (L.) Bolli
S. *canadensis* var. *laciniata* A. Gray
S. *canadensis* var. *submollis* Rehder
S. *canadensis* f. *rubra* E.J. Palmer & Steyerm.

Pl. 335 g, h; Map 1429

Plants shrubs or small trees 1–8 m tall, usually with stout, spreading rootstocks and suckering from the roots. Bark yellowish brown, tight, lacking ridges but appearing roughened or warty. Twigs 3–5 mm thick, the pith white. Leaves with the petiole 3–7 cm long, glabrous except in the ventral groove, where densely pubescent with minute, scurfy hairs. Leaflets (5–)7–9 per leaf, 5–12(–14) cm long, 2–6(–9) cm wide, lanceolate to narrowly oblong or elliptic, the upper surface glabrous, the undersurface usually minutely hairy along the veins, rarely also hairy on the tissue between the veins. Inflorescences more or less flat-topped, lacking an elongate main axis, instead with usually 5 primary branches (these repeatedly branched). Corollas 3–5 mm wide, white. Fruits bluish black to purplish black (rarely red), edible (except when fresh and eaten in quantity) and palatable, with a pleasant flavor. $2n=36$. May–July.

Common throughout the state (eastern U.S. west to North Dakota and Texas; Canada, Mexico, Caribbean Islands). Bottomland forests, mesic upland forests, banks of streams and rivers, margins of ponds, lakes, and sinkhole ponds, and upland prairies; also pastures, old fields, gardens, railroads, roadsides, and disturbed areas.

Sambucus canadensis is closely related to the European S. *nigra* L., and Bolli (1994) treated it as one of six subspecies in his very broadly circumscribed vision of that species. The two taxa differ in leaflet number (almost always five in S. *nigra*, almost always seven in S. *canadensis*), fruit color, and pubescence, and there seems little justification for uniting them. It should be noted that rare plants of S. *canadensis* are found with red berries; these have been called f. *rubra*. Another uncommon form has the leaflets 1 or 2 times deeply lobed or dissected. These plants have been called var. *laciniata* and are popular horticulturally for their lacy foliage. Steyermark (1963) also noted the presence of scattered plants with the undersurface of the leaflets hairy both along the veins and on the tissue between the veins. Such plants have been called var. *submollis*. All of these variants appear to occur as rare mutants that are formed independently and repeatedly within populations of otherwise typical individuals and thus are not worthy of formal taxonomic recognition. Instead, they might better be treated in horticulture as cultivars.

Elderberries have many uses in food, medicine, and winemaking, and as a colorant. Many parts of the plant are toxic, but the toxins may be broken down by proper cooking or (under some circumstances) drying and storage, so it is best to work from proven recipes rather than experimenting with the plant. However, the berries, which ripen between August and October, are relatively edible and palatable, tart but tasty. They can be gathered in quantity in midsummer, but they are rapidly eaten by birds if left on the shrub. They are sometimes eaten fresh, though they contain mild toxins which can cause discomfort if too many are eaten at one time. These compounds apparently break down if the fruits are cooked or dried for the winter. Elderberries make excellent jelly, but they contain no pectin, so they must be mixed with other fruit or else have pectin added. The flowers are used in perfumery and as a flavoring in confections. A tea made from the flowers has been used medicinally, and they may be fried in batter or added to muffins. Both flowers and fruit have often been used to make wine. The fruits provide a magenta colorant used to alter the color of some other beverages and foods, and a black dye has been produced from the bark.

2. Sambucus pubens Michx. (red-berried elderberry, stinking elderberry)

S. racemosa ssp. *pubens* (Michx.) House

Pl. 336 h; Map 1430

Plants shrubs or small trees 1–4 m tall (to 8 m elsewhere), usually with stout, spreading rootstocks and suckering from the roots. Bark grayish brown to greenish brown, tight, lacking ridges but appearing roughened or warty. Twigs 4–8 mm thick, the pith brown. Leaves with the petiole 2.5–7.0 cm long, sparsely to moderately short-hairy. Leaflets 5–7 per leaf, 5–19 cm long, 2.5–6.0 cm wide, lanceolate to narrowly oblong or elliptic, the upper surface glabrous, the undersurface usually moderately to densely pubescent with short, spreading hairs, sometimes nearly glabrous with age. Inflorescences ovoid to more or less pyramidal, the solitary main axis elongate (the side branches repeatedly branched). Corollas 3–4 mm wide, white to pale yellow. Fruits bright red, unpalatable, with an unpleasant flavor. $2n=36$. April–May.

Uncommon, known thus far only from Marion County (northern U.S. [including Alaska] south to California, Missouri, and Georgia; Canada). Ledges and tops of bluffs and mesic upland forests.

This species is closely related to *S. racemosa* L., a species that is widespread in Eurasia in similar habitats. *Sambucus pubens* usually has much larger leaves, with more oblong lateral leaflets, and larger, more open inflorescences than *S. racemosa*. All of these characters are variable, however. Occasional specimens from the two continents are very similar, and the two taxa are sometimes considered subspecies of one widespread species.

The fruits are unpalatable, and some references list them as mildly poisonous.

3. **Symphoricarpos** Duhamel (snowberry)

Plants shrubs, often colonial from rhizomes. First-year twigs 1–3 mm thick, usually short-hairy but often becoming glabrous or nearly so later in the growing season, the pith relatively small, white to tan, often hollow in larger twigs. Winter buds more or less ovate, flattened, with several overlapping scales. Leaves with short, unwinged petioles, none perfoliate. Stipules absent. Leaf blades simple, unlobed or those of the first leaves of the season sometimes varying from undulate to irregularly and bluntly lobed, ovate to oblong or elliptic, the margins otherwise entire. Flowers in dense, small clusters or dense, short spikes or spikelike racemes, these in the axils of the upper leaves and usually also terminal on the branches, the spikes with a pair of small, scalelike bracts at the nodes, these free or fused at the base, ovate to broadly triangular, the individual flowers subtended by a pair of small bractlets, these shorter than to about as long as the ovaries. Calyx lobes 0.5–0.8 mm long, triangular to narrowly triangular. Corollas 3–9 mm long, actinomorphic, bell-shaped with a cup-shaped tube about 2–5 mm long and a spreading, 5-lobed portion 2–9 mm in diameter (measured across the top of the flower), the lobes rounded, pink or greenish and purplish. Style 3–7 mm long. Fruits berrylike drupes, 4–6(–7) mm in diameter, more or less spherical, white to greenish white or pink, red, or purplish red at maturity. Seedlike nutlets (also called pyrenes or stones) 2, 2.5–5.0 mm long, more or less ovate to elliptic in outline, flattened on 1 side and rounded on the other side, the surface smooth or nearly so, white or dull yellow. Nine to 17 species, North America, Asia.

Species of *Symphoricarpos* often spread by underground runners and form dense colonies in dry, lightly shaded places. Some species are cultivated as ornamentals. The berries are not edible and are said to cause vomiting and diarrhea if eaten in large quantities. Birds will eat them during the winter months, but often not until other food sources have been exhausted.

1431. Symphoricarpos albus

1432. Symphoricarpos occidentalis

1433. Symphoricarpos orbiculatus

1. Fruits 3–4 mm in diameter, pink to purple, rarely white; corollas 3–4 mm long, greenish and purplish; styles hairy .... 3. S. ORBICULATUS
1. Fruits 5–15 mm in diameter, white (sometimes drying blackish or bluish); corollas 5–9 mm long, pink or pink and white; styles glabrous or nearly so
   2. Flowers short-stalked; corollas lobed about ⅓ of the way to the base; anthers 1.0–1.5 mm long, not exserted from the corolla .... 1. S. ALBUS
   2. Flowers sessile; corollas lobed about ½ of the way to the base; anthers 1.8–2.2 mm long, exserted from the corolla .... 2. S. OCCIDENTALIS

1. Symphoricarpos albus (L.) S.F. Blake (white coralberry)

Pl. 336 g; Map 1431

Plants shrubs 0.1–1.0 m tall. Bark gray, thin, tending to become shredded. Petioles 1–5 mm long. Leaf blades 2–4 cm long, 12–25 mm wide, oblong-elliptic to ovate, rounded or broadly angled at the base, rounded or angled to a bluntly pointed tip, sometimes abruptly tapered to a minute, sharp point, the upper surface bright green to dark green, the undersurface pale green, glabrous (inconspicuously hairy elsewhere), somewhat glaucous. Flowers short-stalked, mostly appearing in clusters. Corollas 5–7 mm long, lobed about ⅓ of the way to the base, pale pink. Stamens with the anthers 1.0–1.5 mm long, not exserted from the corolla. Styles glabrous. Fruits 7–15 mm in diameter, white. Nutlets 4–5 mm long, elliptic in outline, more or less pointed at each end. 2n=54. May–July.

Introduced, uncommon, known thus far from a single specimen from the city of St. Louis (northern U.S. [including Alaska] south to Virginia, Illinois, and California; Canada; introduced farther south). Railroads.

This species sometimes is grown in Missouri gardens, but it is known outside of cultivation from a single specimen that consists of a depauperate, unhealthy twig, dying back from the tip, with no flowers or fruit, bearing only a few tiny leaves (to 9 mm long) near its base. It was collected and reported by Mühlenbach (1983), but the identification based on such poor material is somewhat uncertain. The description and flowering time above are taken from specimens within the native range of the species. Most material of the species has the leaves pubescent beneath, but our specimen has glabrous leaves. It thus is referable to var. *laevigatus* (Fernald) S.F. Blake, a trivial segregate that occurs natively in the western portion of the species range but is commonly cultivated farther east.

2. Symphoricarpos occidentalis Hook. (wolfberry, western snowberry)

Pl. 336 c, d; Map 1432

Plants shrubs 0.3–1.0 m tall. Bark gray to grayish brown, thin, tending to become shredded. Petioles 3–7 mm long. Leaf blades 2–4 cm long, 12–25 mm wide, oblong-elliptic to broadly ovate, broadly rounded or broadly angled to nearly truncate at the base, rounded or angled to a bluntly pointed tip (those of the lobed first leaves of the season sometimes appearing sharply pointed), the upper surface bright green to dark green, the undersurface pale green, short-hairy, at least along the veins, usually not glaucous. Flowers sessile, appearing in clusters or more commonly in short spikes. Corollas 5–9 mm long, lobed about ½ of the way to the base, pale pink or occasionally greenish white. Stamens with the anthers 1.8–2.2 mm long, exserted

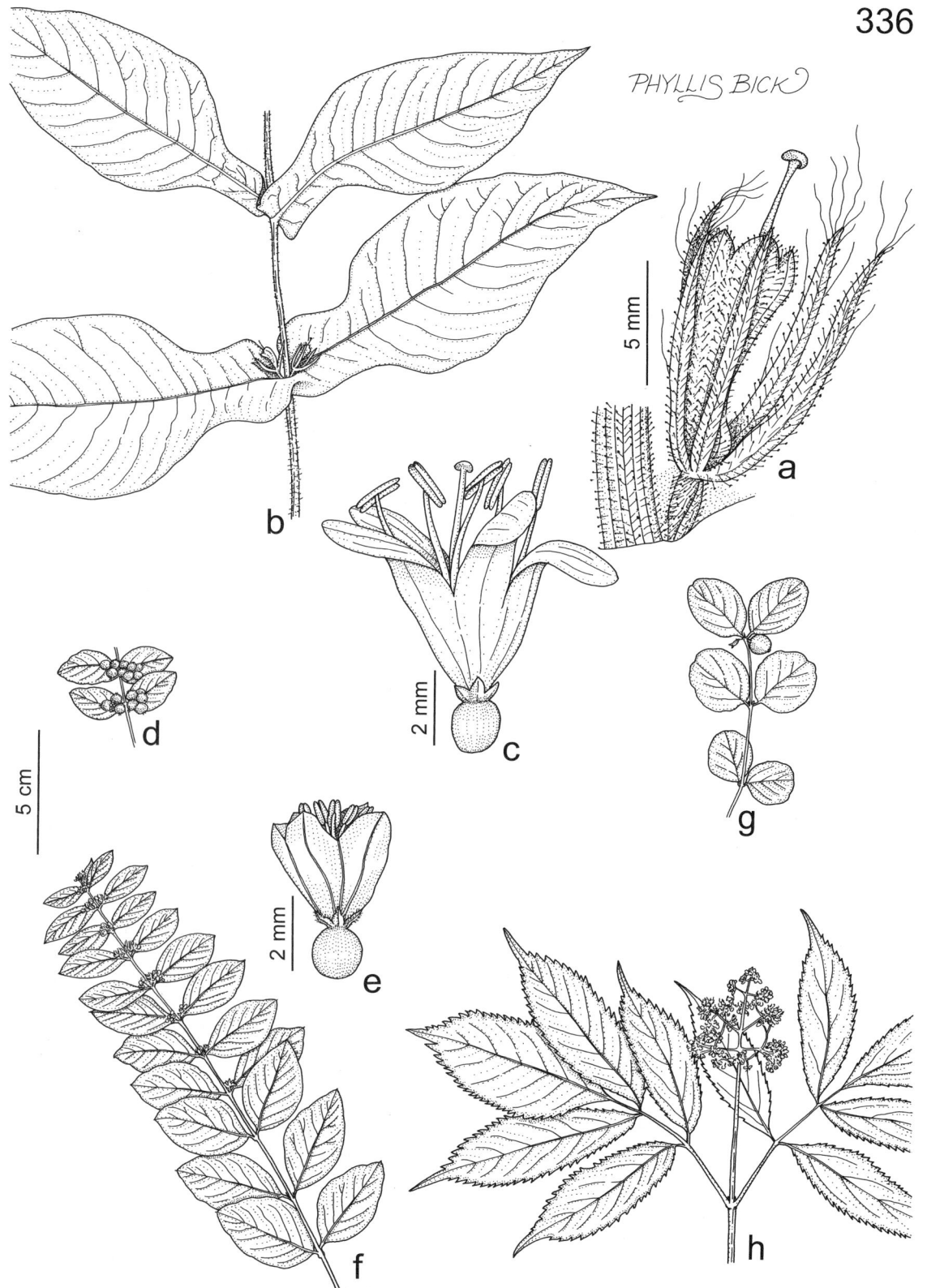

Plate 336. Caprifoliaceae. *Triosteum perfoliatum*, **a)** flower, **b)** median stem portion. *Symphoricarpos occidentalis*, **c)** flower, **d)** fruiting branch. *Symphoricarpos orbiculatus*, **e)** flower, **f)** flowering branch. *Symphoricarpos albus*, **g)** fruiting branch. *Sambucus pubens*, **h)** inflorescence.

from the corolla. Styles glabrous. Fruits 5–9 mm in diameter, white or greenish white. Nutlets 3.0–3.5 mm long, elliptic in outline, rounded to bluntly pointed at each end. June–August.

Uncommon, northwesternmost Missouri (Michigan to Missouri west to Washington and New Mexico; Canada; introduced elsewhere in the U.S.). Mesic upland forests and margins of loess hill prairies.

Steyermark (1963) mapped a presumably nonnative occurrence of this species from Adair County, but this could not be confirmed during the present study. The species is cultivated as an ornamental and border plant and in some western states has been planted for erosion control (Kurz, 1997). In *S. occidentalis* the first leaves produced by elongating twigs each season frequently are bluntly toothed to lobed, a feature usually not found in the more commonly encountered *S. orbiculatus*.

3. Symphoricarpos orbiculatus Moench
(coralberry, buckbrush, Indian currant)
Pl. 336 e, f; Map 1433

Plants shrubs 0.5–1.2(–2.0) m tall. Bark gray to grayish brown, thin, tending to become shredded or occasionally peeling in thin plates. Petioles 2–3(–5) mm long. Leaf blades 1.0–4.5 cm long, 8–28 mm wide, elliptic to ovate or oblong, broadly rounded or broadly angled at the base, rounded or angled to a bluntly or less commonly sharply pointed tip, the upper surface bright green to dark green, the undersurface pale green, short-hairy, usually not glaucous. Flowers mostly sessile, appearing in clusters or short spikes. Corollas 3–4 mm long, lobed about $1/3$ of the way to the base, greenish white and usually purplish-tinged. Stamens with the anthers 0.9–1.2 mm long, not or only slightly exserted from the corolla. Styles hairy. Fruits 3–4 mm in diameter, pink to reddish purple or purple, rarely white. Nutlets 2.5–3.2 mm long, elliptic to oblong-elliptic in outline, more or less rounded at each end. $2n=18$. July–August.

Common throughout the state (eastern U.S. west to South Dakota and Texas; Canada, Mexico; introduced in New Mexico, Utah). Bottomland forests, mesic to dry upland forests; bases, ledges, and tops of bluffs, banks of streams and rivers, and margins of upland prairies; also pastures, old fields, fencerows, railroads, and roadsides.

This species is planted as an ornamental, desirable because of its striking red berries and tolerance of drought, cold, and heavy soils. However, *S. orbiculatus* is an indicator of degraded forests and woodlands that tends to form dense thickets in response to disturbance from grazing. Rare plants with white fruits have been called f. *leucocarpus* (D. Andrews) Rehder. Steyermark (1963) noted that in some southern states, the stems have been used in basketry.

4. Triosteum L. (horse gentian)

Plants perennial herbs. Stems erect or ascending, 2–8 mm in diameter, solitary or few to several, unbranched, sometimes hollow between the nodes, moderately to densely pubescent with shorter and/or longer, spreading hairs, some of these sometimes gland-tipped. Leaves sessile or with indistinct, broadly winged petioles, perfoliate or those of each pair connected by a ridge around the stem. Stipules absent. Leaf blades simple, unlobed or the largest leaves occasionally irregularly scalloped, with shallow, rounded lobes, or somewhat fiddle-shaped, oblanceolate to elliptic, obovate, or oblong-obovate, the margins otherwise entire. Flowers solitary or in dense clusters of 2–6 in the axils of the upper leaves, clusters on each side of the stem sometimes with a leaflike bract, the individual flowers subtended by a pair of leaflike bractlets, these longer than the ovary. Calyx lobes 7–15 mm long, linear to narrowly triangular or narrowly lanceolate. Corollas 9–20 mm long, zygomorphic, somewhat funnelform with a curved tube, a slightly to strongly oblique mouth, and 5 relatively shallow lobes, these oblong to nearly circular and sometimes somewhat overlapping, the overall corolla yellow or red. Style 10–23 mm long. Fruits drupes, appearing berrylike but relatively dry and mealy, 5–10 mm in diameter, more or less spherical, orangish yellow to orangish red or red and occasionally greenish-tinged at maturity. Nutlets usually 3, 4.5–9.5 mm long, more or less elliptic in outline, somewhat 3-angled, with a prominent, blunt keel on the ventral side and rounded with 4 or 5 prominent, rounded longitudinal ridges on the dorsal side, the surface appearing somewhat encrusted or fibrous between the ribs, tan to yellowish brown or reddish brown. Five or 6 (possibly more) species, eastern North America, Asia.

It should be noted that some Asian species of *Triosteum* have terminal inflorescences and that various non-Missouri species differ in fruit color (Gould and Donoghue, 2000). Ripe fruits of *Triosteum* spp. have been dried and roasted for use as a coffee substitute (Ferguson, 1966a).

1. Calyx lobes with the outer surface glabrous or with scattered, short, soft hairs, the margins moderately to densely pubescent with at least some of the hairs longer and stiffer (bristly); leaf blades 2.0–5.5 cm wide, appearing sessile (sometimes tapered to an indistinct, broadly winged petiole) but not perfoliate; corollas pale yellow to yellow, rarely orange or red .... 1. T. ANGUSTIFOLIUM
1. Calyx lobes with the outer surface variously hairy with shorter and/or longer hairs, the margins similarly pubescent; leaf blades 4–9 cm wide, often at least a few perfoliate; corollas red
    2. Leaves not perfoliate or 1–3 median pairs weakly perfoliate, these 1–2 cm wide at the base; stems sparsely pubescent with shorter, mostly gland-tipped hairs 0.3–0.5 mm long (these rarely absent or denser) and moderate to dense, longer hairs 1.5–3.0 mm long; corollas (9–)13–19 mm long, the mouth relatively strongly oblique; style not or only slightly exserted (up to 3 mm beyond the corolla lobes); fruits orange to red.................... 2. T. AURANTIACUM
    2. Usually 3–5 median leaf pairs perfoliate, these 1–6 cm wide at the base; stems densely pubescent with shorter, mostly gland-tipped hairs 0.3–0.5 mm long and also sparse, longer hairs 1.0–1.5 mm long; corollas 9–14 mm long, the mouth only slightly oblique; style weakly to strongly exserted (to 5 mm beyond the corolla lobes); fruits yellowish orange or orange .... 3. T. PERFOLIATUM

1. Triosteum angustifolium L. (yellow-flowered horse gentian)

Pl. 337 a, b; Map 1434

Stems 0.2–0.7 m long, moderately to densely pubescent with straight, spreading to somewhat downward-angled, stiff, bristly hairs 1.5–3.0 mm long, these sometimes mixed with shorter, softer hairs that are all or mostly minutely gland-tipped. Leaf pairs not perfoliate, the bases joined only by a small ridge around the stem. Leaf blades 10–19 cm long, 2.0–5.5 cm wide, oblanceolate to narrowly rhombic or narrowly elliptic, tapered at the base, sometimes to an indistinct, broadly winged petiolar base, tapered to a sharply pointed tip, the margins with relatively dense, stiff, ascending hairs, the upper surface moderately pubescent with long, straight, appressed hairs, the undersurface sparsely to moderately pubescent with stiff, spreading hairs along the veins or less commonly moderately to densely and uniformly pubescent with short, soft hairs. Flowers 1 per leaf axil (2 per node). Paired bracts subtending each flower about as long as to somewhat longer than the flower (including the calyx and corolla), narrowly lanceolate to narrowly elliptic. Calyx lobes 9–12 mm long, the margins with dense, stiff, bristly, longer and shorter hairs, the inner and outer surfaces glabrous or with scattered, short hairs. Corollas 13–17 mm long, pale yellow to yellow, rarely orange or red, narrowly funnelform, the mouth noticeably oblique, the outer surface with gland-tipped hairs. Styles not or only slightly exserted (less than 2 mm beyond the corolla lobes). Fruits 5–7 mm in diameter, orangish yellow to pale orange, moderately to densely hairy at maturity. $2n=18$. April–May.

Scattered in the Ozark and Ozark Border Divisions (eastern U.S. west to Kansas and Texas; Canada). Bottomland forests, mesic upland forests, bases and ledges of bluffs, and banks of streams and rivers.

Rare plants with the undersurface of the leaves evenly pubescent with soft hairs along and between the veins have been called var. *eamesii* Wiegand, in contrast to the leaves in var. *angustifolium*, which have short, stiff hairs along the veins. The differences in hairiness seem trivial. Occasional mutants with red flowers have been called f. *rubrum* F. Lane and occur sporadically within otherwise yellow-flowered populations.

2. Triosteum aurantiacum E. P. Bickn. **var. illinoense** (Wiegand) E.J. Palmer & Steyerm. (red-fruited horse gentian)
Triosteum illinoense (Wiegand) Rydb.

Pl. 337 f, g; Map 1435

Stems 0.3–1.0 m long, moderately to densely pubescent with straight, more or less spreading, bristly hairs 1.5–3.0 mm long, these sometimes mixed with sparse or rarely denser, short (0.3–0.5 mm), softer hairs that are mostly minutely gland-

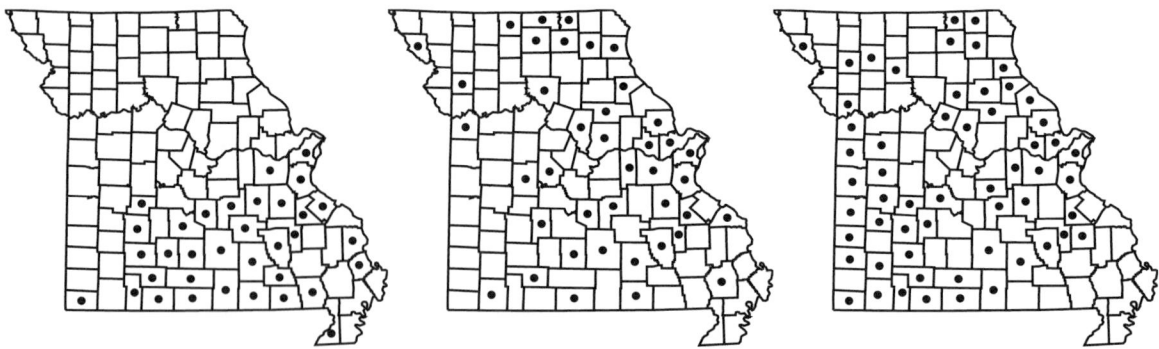

1434. Triosteum angustifolium 1435. Triosteum aurantiacum 1436. Triosteum perfoliatum

tipped. Leaf pairs not perfoliate or 1–3 median pairs weakly perfoliate, these 1–2 cm wide at the base. Leaf blades 9–20 cm long, 4–9 cm wide, oblanceolate to obovate or elliptic, those of the largest leaves occasionally slightly fiddle-shaped, the nonperfoliate leaves tapered at the base, sometimes to an indistinct, broadly winged petiolar base, tapered to a sharply pointed tip, the margins with relatively dense, stiff, ascending hairs, the upper surface moderately pubescent with short, straight, appressed hairs, the undersurface moderately to densely pubescent with short, soft, spreading hairs mostly along the veins. Flowers 1 or less commonly 2 or 3 per leaf axil (2–6 per node). Paired bracts subtending each flower sometimes longer than the ovary but shorter than the flower (including the calyx and corolla), linear to narrowly lanceolate. Calyx lobes 8–15 mm long, the margins with dense, stiff, bristly, longer and shorter hairs, the inner and outer surfaces with scattered, short hairs. Corollas 8–15 mm long, dull red to purplish red, narrowly funnelform, the mouth relatively strongly oblique, the outer surface with gland-tipped hairs. Styles not or only slightly exserted (less than 3 mm beyond the corolla lobes). Fruits 7–10 mm in diameter, red (may be orange when not fully mature), usually densely hairy at maturity. April–July.

Scattered, mostly in the eastern half of the state, and uncommon in the northwestern quarter mostly in counties along the Missouri River (Nebraska to Oklahoma east to Pennsylvania and West Virginia). Bottomland forests, mesic upland forests, and bases and ledges of bluffs.

Triosteum aurantiacum and *T. perfoliatum* differ in several subtle characters, but all of these are variable, and specimens (especially vegetative ones) are sometimes impossible to identify with certainty. Some authors have treated the former taxon as *T. perfoliatum* var. *aurantiacum* (E.P. Bickn.) Wiegand. It is not clear whether the difficulty in distinguishing the taxa is merely due to the variability of the characters, or whether there is actual intergradation due to natural hybridization between them. Lane (1954) suggested that a few specimens collected by Julian Steyermark in Chariton and Howard Counties might possibly represent hybrids between *T. aurantiacum* (as *T. illinoense*) and *T. perfoliatum*, but these have been redetermined in the present work as merely examples of the latter species. Steyermark (1963), who collected the two species occasionally in mixed populations, nevertheless felt that the degree of intergradation in the field was very small.

Most authors have recognized three variants within the *T. aurantiacum* complex, either as varieties of *T. aurantiacum* or as separate species. Because there is considerable morphological overlap between the phases in some eastern portions of the species range, they are best treated as varieties until more intensive studies can be carried out. Plants of var. *illinoense* are the only phase found in Missouri, with the remainder of the complex occurring to the east of the state. They are unique in their stems having mostly relatively long (to 3 mm), shaggy hairs, in contrast to the other varieties, which have mainly shorter (0.5–1.5 mm) hairs. Among eastern North American plants, those assigned to var. *aurantiacum* have the leaves hairy on the undersurface, whereas in var. *glaucescens* Wiegand the leaves are glabrous or nearly so on the undersurface (Gleason and Cronquist, 1991).

3. Triosteum perfoliatum L. (common horse gentian, wild coffee, tinker's weed, feverwort)

Pl. 336 a, b; Map 1436

Stems 0.3–1.2 m long, densely pubescent with short (0.3–0.5 mm), softer hairs that are mostly minutely gland-tipped and also sparse straight, more or less spreading, longer, bristly hairs 1.0–

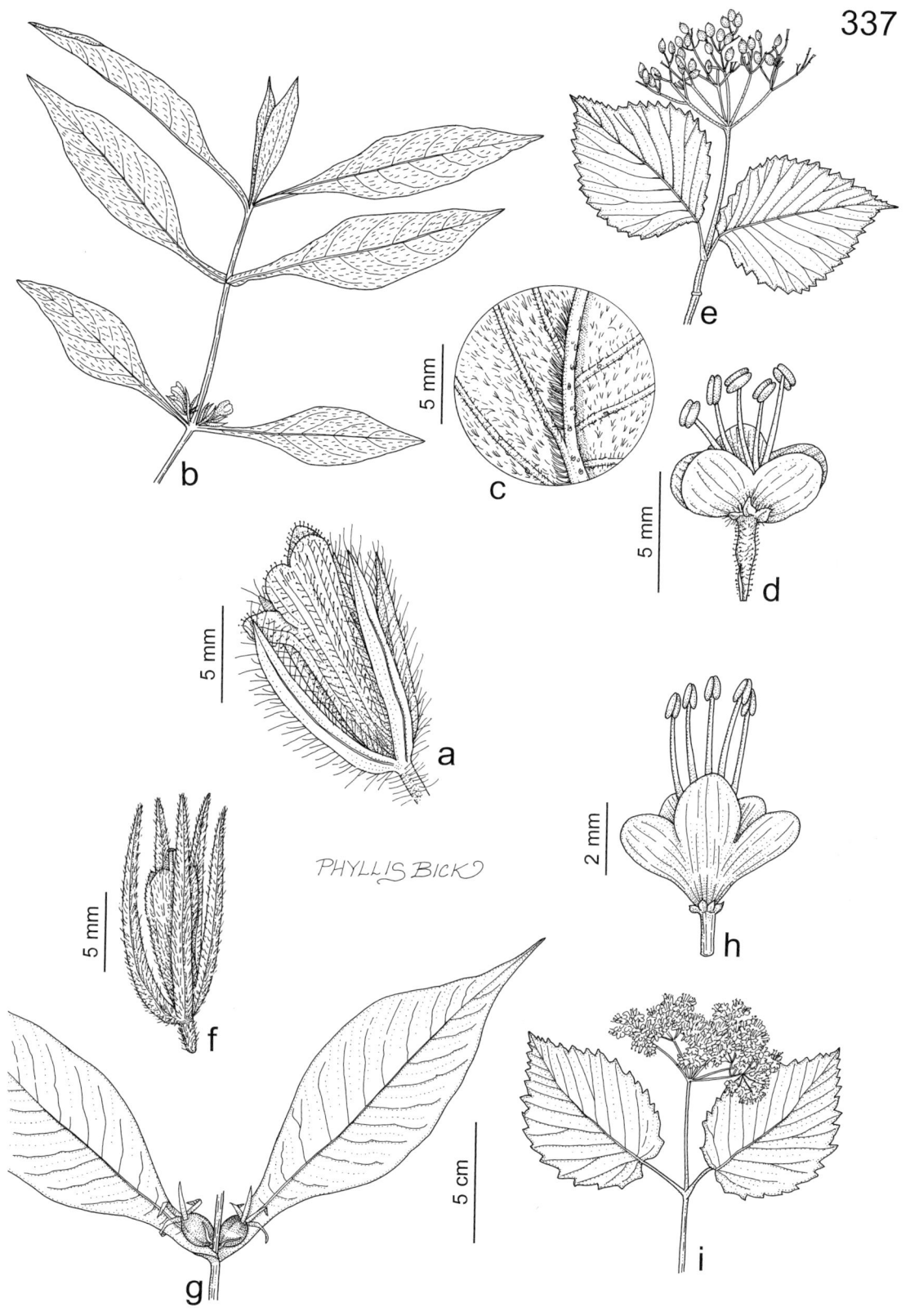

Plate 337. Caprifoliaceae. *Triosteum angustifolium*, **a)** flower, **b)** flowering stem. *Viburnum dentatum*, **c)** abaxial leaf surface detail, **d)** flower, **e)** fruiting branch. *Triosteum aurantiacum*, **f)** flower, **g)** flowering median node of stem. *Viburnum recognitum*, **h)** flower, **i)** flowering branch.

1.5 mm long. Median 3–5 leaf pairs usually moderately to strongly perfoliate, these 1–6 cm wide at the base. Leaf blades 9–20 cm long, 4–9 cm wide, oblanceolate to obovate or elliptic, those of the largest leaves occasionally slightly fiddle-shaped, the nonperfoliate leaves tapered at the base, sometimes to an indistinct, broadly winged petiolar base, tapered to a sharply pointed tip, the margins with relatively dense, stiff, ascending hairs, the upper surface moderately pubescent with short, straight, appressed hairs, the undersurface moderately to densely pubescent with short, soft, spreading hairs mostly along the veins. Flowers 1 or more commonly 2 or 3(4) per leaf axil (2–6[8] per node). Paired bracts subtending each flower sometimes longer than the ovary but shorter than the flower (including the calyx and corolla), linear to narrowly lanceolate. Calyx lobes 10–18 mm long, the margins with dense, stiff, bristly, longer and shorter hairs, the inner and outer surfaces with scattered, short hairs, some of these often minutely gland-tipped. Corollas 8–17 mm long, dull red to purplish red, often mottled with small, cream-colored spots near the base and on the inner surface, occasionally entirely greenish yellow, narrowly funnelform, the mouth only slightly oblique, the outer surface with gland-tipped hairs. Styles usually slightly to moderately exserted (to 5 mm beyond the corolla lobes). Fruits 7–10 mm in diameter, orangish yellow to orange, sparsely hairy at maturity. May–July.

Scattered nearly throughout the state, most abundantly in the Ozark and Ozark Border Divisions (eastern U.S. west to Minnesota, Nebraska, and Texas; Canada). Bottomland forests, mesic upland forests, bases and ledges of bluffs, edges of glades, and banks of streams and rivers; also roadsides.

This is the most abundant and widespread member of the genus in Missouri. See the treatment of *T. aurantiacum* for a discussion of problems in discriminating that species from *T. perfoliatum*.

5. Viburnum L. (viburnum, blackhaw, arrowwood)

Plants shrubs or small trees. First-year twigs 1–3 mm thick, the pith solid. Winter buds various, appearing naked or more commonly with 2 valvate (the margins touching but not overlapping) or 4 overlapping scales. Leaves with short or relatively long, unwinged or narrowly winged petioles, none perfoliate. Stipules absent or present, if present then 2–10 mm long, usually slender, sometimes partially fused to the basal portion of the petiole. Leaf blades simple, elliptic, oblong, ovate, or almost circular, the margins finely to coarsely toothed (3-lobed in *V. opulus*). Flowers in dense, flat-topped panicles, usually appearing as compound umbels, terminal on the branches, occasionally reduced to a loose, umbellate cluster, the branch points with small, linear to narrowly triangular bracts that are mostly shed by flowering, the individual flowers usually subtended by minute, paired bractlets that are mostly shed early in flower development. Calyx lobes 0.3–3.0 mm long, variously shaped. Corollas of the marginal (sterile) flowers of the inflorescence sometimes enlarged and slightly zygomorphic, those of the fertile flowers 2–4 mm long, actinomorphic, more or less saucer-shaped or rarely somewhat bell-shaped with a broadly cup-shaped tube 1–2 mm long and a spreading, (4)5-lobed portion 3–8 mm in diameter (measured across the top of the flower), the lobes rounded, white, sometimes fading to pale yellow. Style absent or nearly so, the ovary often broadly conical-tapered at the tip, the deeply 3-lobed stigma appearing sessile. Fruits berrylike drupes, more or less spherical to more commonly oblong-ellipsoid, red to bluish black. Nutlet 1, 6–11 mm long, oblong-ovate to elliptic in outline, flattened, one or both sides often ridged or with a pair of longitudinal grooves, sometimes appearing longitudinally folded, the surface variously smooth to somewhat warty, yellowish brown to reddish brown or nearly black. About 175 species, North America to South America, Europe, Asia to Java.

Species of *Viburnum* are important for wildlife food, and the fruits of some species also have been eaten by man, raw, cooked, or processed into preserves and jellies. Many Asian species of *Viburnum* have been introduced into horticulture, and several have been reported to escape in states to the east of us. However, only two Old World species, *V. lantana* and *V. opulus*, are known to have become established in Missouri. Some of the native species are planted widely as well, and they sometimes escape in areas outside their native ranges.

The sections of *Viburnum* are well defined (at least in Missouri), but the species within the sections often are difficult to distinguish. For this reason, the sections are noted in the key to species below. Users should note that leaf descriptions in the species treatments below apply to most of the stems on a given plant, including flowering/fruiting stems. However, leaves of vigorous leading shoots, rapidly elongating root suckers, or juvenile plants sometimes differ strikingly in their longer, narrower, and more acute appearance, often with fewer veins and fewer, smaller marginal teeth.

1. Leaves lobed; marginal flowers of the inflorescence sterile, their corollas greatly enlarged, somewhat zygomorphic (section *Opulus* DC.) . . . . 5. V. OPULUS
1. Leaves unlobed; marginal flowers of the inflorescence fertile, similar in size and appearance to the more centrally positioned flowers
   2. Secondary veins curved or arched toward the tip, looping and joined into a network with adjacent veins, not reaching the leaf margin; leaf blades with the undersurface glabrous or with red, woolly hairs, the margins with 5–11 small teeth per cm; winter buds with 1 pair of scales, these valvate (with the margins touching but not overlapping) (section *Lentago* DC.)
      3. Petioles with well-developed, undulate or irregularly curled wings; leaf blades abruptly contracted to a slender, long-tapered tip . . . . . . 3. V. LENTAGO
      3. Petioles with moderately well or more commonly poorly developed, narrow, straight (not undulate or curled) wings; leaf blades rounded or broadly angled to a usually bluntly pointed tip, or contracted to a short-tapered or rarely longer-tapered, sharply pointed tip
         4. Petioles, main veins of the leaf undersurface, and winter buds glabrous or with sparse, red, stellate hairs; leaf blades relatively thin and papery, the upper surface dull . . . . 7. V. PRUNIFOLIUM
         4. Petioles, main veins of the leaf undersurface, and winter buds densely pubescent with minute, red to reddish brown, branched, woolly hairs; leaf blades relatively thick and somewhat leathery, the upper surface shiny . . . . 10. V. RUFIDULUM
   2. Secondary veins relatively straight above the base, usually branched, extending to the leaf margin without looping or joining, each branch ending in a tooth; leaf blades with the undersurface glabrous or with white or gray hairs, the margins variously densely or sparsely toothed; winter buds naked (lacking scales) or with 2 pairs of overlapping scales
      5. Leaves with the petiole and undersurface densely pubescent with small, gray, stellate hairs, the margins with (3–)5–15 small teeth per cm; winter buds naked (lacking scales); fruits red, sometimes turning dark bluish purple with age (section *Viburnum*) . . . . 2. V. LANTANA
      5. Leaves with the petiole and undersurface glabrous or pubescent with all or mostly unbranched hairs, the margins with 1–3 relatively large teeth per cm; winter buds with 2 pairs of overlapping scales; fruits bluish black (section *Odontotinus* Rehder)
         6. Leaf blades mostly deeply cordate at the base, the margins with 18–35 teeth on each side; petioles glabrous or with sparse to moderate minute, stalked glands; bark peeling in papery sheets . . . . 4. V. MOLLE
         6. Leaf blades rounded to truncate or shallowly cordate at the base, the margins with 7–25 teeth on each side; petioles hairy, at least in and around the ventral groove; bark smooth or fissured (rarely peeling in *V. rafinesquianum*)

7. Petioles 3–15(–22) mm long; leaf blades with 7–17 teeth on each side of the margin, usually with 3–5 main veins from the base, the lateral veins otherwise usually relatively evenly distributed along midvein; ovaries densely glandular at flowering; bark peeling or not
    8. Leaf blades 6–10(–14) cm wide, with (10–)12–17 teeth on each side of the margin. Bark not peeling, smooth or appearing somewhat roughened or warty, not fissured .... 6. V. OZARKENSE
    8. Leaf blades 3–7 cm wide, with 7–12(–14) teeth on each side of the margin; bark on young stems firm, sometimes peeling in papery sheets, becoming fissured on older stems........................... 8. V. RAFINESQUIANUM
7. Petioles 16–36 mm long; leaf blades with 12–25 teeth on each side of the margin, usually with 3 or 4 secondary veins on each side crowded near the base of midvein; ovaries glabrous or sparsely glandular at flowering; bark not peeling
    9. Petioles moderately to densely hairy, especially on the upper surface; leaf blades with both surfaces hairy along the veins and on the tissue between the veins; inflorescence branches hairy and glandular .... 1. V. DENTATUM
    9. Petioles glabrous or with scattered hairs along the ventral groove; leaf blades with the upper surface glabrous (rarely with scattered hairs), the undersurface hairy only along the main veins and in small patches in the axils of the veins; inflorescence branches glabrous, sometimes glandular .... 9. V. RECOGNITUM

1. Viburnum dentatum L. (southern arrowwood)

V. dentatum var. deamii (Rehder) Fernald

Pl. 337 c–e; Map 1437

Plants shrubs 1–3 m tall. Bark firm, not peeling, relatively smooth to finely roughened or warty on younger branches, gray or grayish brown to reddish brown. Winter buds ovoid-conical, slightly flattened, with 2 pairs of overlapping scales, glabrous, not or only slightly sticky. Stipules sometimes present but usually shed early, partially fused to the basal portion of the petiole, linear, hairy and glandular. Petioles 16–28 mm long, unwinged, moderately pubescent with short, pale, branched to stellate hairs, especially on the upper surface, lacking prominent glandular swellings near the tip. Leaf blades (5–)7–11 cm long, (5.5–)6.5–10.0 cm wide, unlobed, relatively thin and papery, broadly elliptic to broadly ovate or nearly circular, rounded to truncate or shallowly cordate at the base, rounded or broadly angled to a usually bluntly pointed tip, occasionally abruptly contracted to a short-tapered, more or less sharply pointed tip, the margins coarsely toothed with the teeth 1–3 per cm, 14–21 on each side, the upper surface moderately pubescent along the veins and on the tissue between the veins with mostly unbranched or few-branched hairs, occasionally some of the hairs appearing stellate, the undersurface moderately to densely pubescent along the veins and on the tissue between the veins with mostly branched or stellate hairs, often with 3 or 4 secondary veins on each side crowded near the base of the midvein, pinnately veined above the base, the secondary veins straight, often dichotomously branched but not forming a network, extending to the leaf margin, each branch ending in a tooth. Inflorescences short- to long-stalked, with (3–)5–7 primary branches, these moderately hairy and with minute, stalked glands, the marginal flowers fertile and similar to the other flowers. Ovaries glabrous or occasionally with a few scattered, minute, stalked glands, the tapered, stylelike tip densely short-hairy. Fruits 7–10 mm long, ellipsoid, oblong-ovoid, or nearly spherical, purplish blue to bluish black, not glaucous. Nutlet 6.0–7.5 mm long, yellowish brown. $2n=36, 54, 72$. May–June.

Uncommon, known thus far only from Shelby County (eastern U.S. west to Iowa and Texas). Bottomland forests and mesic upland forests.

Viburnum section *Odontotinus* comprises about 45 species that are widespread in eastern North America and Eurasia. The approximately six species native to North America are mostly part of a taxonomically confusing polyploid complex centered on *V. dentatum*. Most of these taxa have been called arrowwood because the straight branches were used in the past to make arrow shafts. They are difficult to distinguish, especially when mature fruit and flowers are lacking, and they have been treated variously as 6–10 or more species or as about 4 species with several additional varieties. The whole

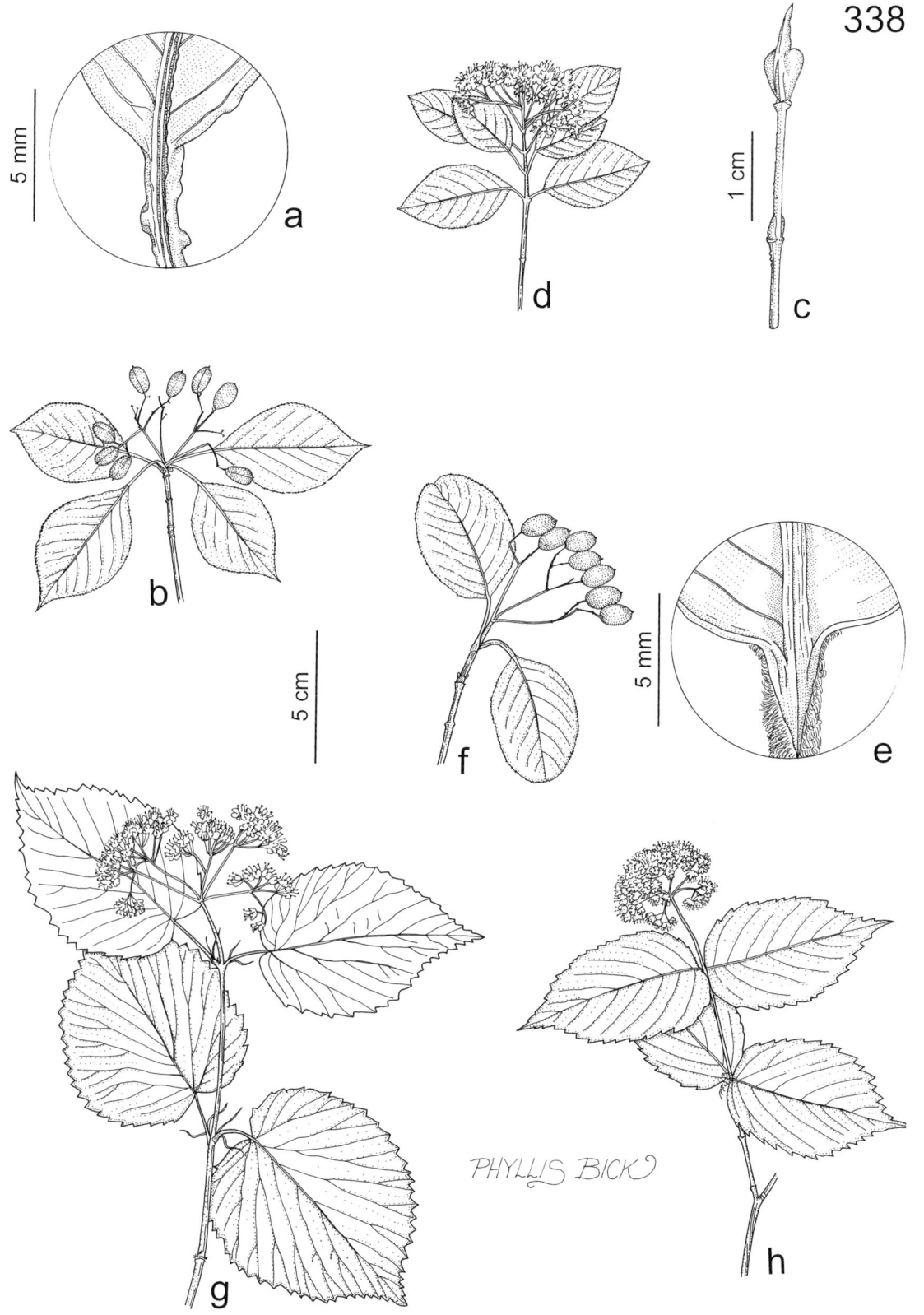

Plate 338. Caprifoliaceae. *Viburnum lentago*, **a)** petiole detail, **b)** fruiting branch. *Viburnum prunifolium*, **c)** winter bud, **d)** flowering branch. *Viburnum rufidulum*, **e)** petiole detail, **f)** fruiting branch. *Viburnum molle*, **g)** flowering branch. *Viburnum rafinesquianum*, **h)** flowering branch.

1437. Viburnum dentatum 1438. Viburnum lantana 1439. Viburnum lentago

complex deserves more detailed biosystematic study. Missouri plants have been called var. *deamii*, a name applied to a series of morphologically poorly differentiated octoploid (2n=72) populations from western Pennsylvania to Missouri that differs from var. *dentatum* in the frequent production of stipules and the presence of scattered stipitate glands on the undersurface of the leaves.

Recently collected specimens from the margins of a disturbed woodland in Columbia (Boone County) were referred to this species by their collectors (Paul McKenzie and Tim Smith). These shrubs have the pubescence of the leaves mostly confined to the veins and vein axils (but with sparse hairs on the intervening tissue) and the inflorescence branches only sparsely pubescent. Plants with similar morphology have been called *V. dentatum* var. *indianense* (Rehder) Gleason, a name that has been applied to uncommon plants occurring in southern Illinois and Indiana (Gleason and Cronquist, 1991). It is possible that the Boone County materials represent an escaped cultivar or hybrid involving *V. dentatum*, but in their overall morphology they also seem close to *V. recognitum*. They are included in the concept of that species in the present treatment but might in fact be treated better as a hybrid derived from these two taxa (see Egolf [1962] for further discussion). They require further study.

2. Viburnum lantana L. (wayfaring tree, twistwood)

Map 1438

Plants shrubs 2–3(–5) m tall. Bark firm, not peeling, usually relatively smooth to finely roughened or warty on younger branches, occasionally becoming somewhat fissured on older stems, gray or grayish brown to reddish brown. Winter buds more or less narrowly oblong, slightly flattened, naked (lacking scales), densely pubescent with minute, stellate hairs, not sticky. Stipules usually absent, when present minute and shed early. Petioles 5–20 mm long, unwinged or less commonly with minute, winglike ridges toward the tip (these usually obscured by the pubescence), densely pubescent with minute, gray, stellate hairs, lacking prominent glandular swellings near the tip. Leaf blades 5–10 cm long, 2.5–5.5 cm wide, unlobed, thin to relatively thick but soft and not leathery, ovate to elliptic or broadly ovate, rounded to truncate or shallowly cordate at the base, angled or short-tapered to a bluntly or sharply pointed tip, rarely rounded at the tip, the margins finely toothed with the teeth (3–)5–12 per cm, 30–45 on each side, the surfaces densely pubescent with minute, gray, stellate hairs or the upper surface sparsely to moderately hairy at maturity, sometimes with 3–5 main veins from the base, otherwise pinnately veined, the secondary veins straight to somewhat arched or curved, dichotomously branched 1–4 times but not forming a network, extending to the leaf margin, each branch ending in a tooth. Inflorescences sessile or short-stalked, with usually 7 primary branches, these with dense, minute, gray, stellate hairs at flowering, the marginal flowers fertile and similar to the other flowers. Ovaries glabrous. Fruits 8–10 mm long, ellipsoid to nearly spherical, red, sometimes turning dark bluish purple with age, not glaucous. Nutlet 7–9 mm long, yellowish brown to reddish brown. 2n=18. March–April.

Introduced, uncommon, known thus far from a single specimen from Cape Girardeau County (native of Europe, Asia; introduced sporadically in the northeastern U.S. west to Iowa and Missouri, also Montana to Colorado; Canada). Mesic upland forests.

Steyermark (1963) was unable to examine the voucher specimen documenting this species from Missouri, which was collected in 1957 in Cape Girardeau, and thus failed to include it in his treatment. The wayfaring tree of Eurasia has long been cultivated as an ornamental in North America, and it occasionally escapes. *Viburnum lantana* is a member of section *Viburnum,* which includes about 18 species, all native to Europe and Asia. A number of *Viburnum* species commonly cultivated in the United States are members of this morphologically diverse section.

3. Viburnum lentago L. (nannyberry, sheepberry, wild raisin)

Pl. 338 a, b; Map 1439

Plants shrubs or small trees 2–6(–10) m tall. Bark firm, not peeling, relatively smooth to finely roughened or warty on younger branches, sometimes forming small plates or flakes on older trunks, gray to grayish black. Winter buds oblong-conical, often somewhat flattened, with a pair of scales, these valvate (the margins touching but not overlapping), glabrous, somewhat sticky. Stipules absent. Petioles 14–25 mm long, with well-developed wings 0.5–1.0 mm wide flanking the ventral groove, especially toward the tip, these undulate or irregularly curled, moderately to densely pubescent with microscopic, red, stellate hairs when young, becoming glabrous or nearly so at maturity, lacking prominent glandular swellings near the tip. Leaf blades 6.5–10.5 cm long, 3.5–7.5 cm wide, unlobed, relatively thin and papery, broadly ovate-elliptic to elliptic or oblong-elliptic, rounded to short-tapered at the base, abruptly contracted to a slender, long-tapered, sharply pointed tip, the margins finely toothed with the teeth 5–11 per cm, numerous on each side, the surfaces glabrous or with scattered microscopic, red, stellate hairs when young, pinnately veined, the secondary veins evenly spaced, curved or arched toward the tip, looping and joined with adjacent veins into a network, not reaching the margin. Inflorescences sessile, with 3–5(–7) primary branches, these glabrous or with scattered microscopic, red, stellate hairs, the marginal flowers fertile and similar to the other flowers. Ovaries glabrous. Fruits 7–18 mm long, ellipsoid to nearly spherical, bluish black, glaucous. Nutlet 8–11 mm long, yellowish brown. $2n=18$. May–June.

Uncommon, known thus far natively only with surety from Schuyler County; introduced sporadically elsewhere (northeastern U.S. west to Montana and Colorado, south locally to Alabama; Canada). Bottomland forests, mesic upland forests, banks of streams and rivers, and fens.

Steyermark (1963) regarded this species as occurring in native populations only in Schuyler County. He discussed historical collections from Franklin and St. Louis Counties, which he dismissed as either misdeterminations or representing cultivated plants. However, he overlooked an additional historical collection from St. Louis County that may represent a native occurrence, although the specimen collected in 1891 by Noah Glatfelter at Meramec Highlands bears no label data to address this situation.

Viburnum lentago and its close relatives *V. prunifolium* and *V. rufidulum* are members of section *Lentago,* which comprises six total species distributed in the U.S., Canada, and Mexico. Incomplete specimens can be difficult to determine to species within the section. Steyermark (1963) stated that *V. lentago* differs from *V. prunifolium* and *V. rufidulum* in having teeth that point straight out from the leaf margin (vs. angled toward the apex). However, serration in *V. lentago* and its relatives is more variable than he indicated, and this character should not be used to discriminate among taxa in the section.

Steyermark (1963) discussed the fall foliage, which turns a dull or deep purplish red with dull green to rose-colored highlights. He noted that the species can spread by root suckers to form thickets, and that in cultivation it is prone to attack by aphids, scales, and leaf hoppers, but he recommended it for cultivation anyway.

4. Viburnum molle Michx. (Missouri arrowwood, Kentucky viburnum)

Pl. 338 g; Map 1440

Plants shrubs 2–6 m tall. Bark peeling in papery sheets, gray or grayish brown to yellowish brown. Winter buds ovoid-conical, slightly flattened, with 2 pairs of overlapping scales, glabrous, not or only slightly sticky. Stipules usually present but often shed early, partially fused to the basal portion of the petiole, linear, glandular. Petioles 23–50 mm long, unwinged, with sparse, minute, stalked glands, lacking prominent glandular swellings near the tip. Leaf blades 5–14 cm long, 5–12 cm wide, unlobed, relatively thin and papery, broadly ovate or broadly triangular-ovate to broadly heart-shaped or nearly circular, mostly deeply cordate at the base, angled or short-tapered to a bluntly or more commonly sharply pointed tip, the margins coarsely toothed with the teeth 1–3 per cm, 18–35 on each side, the upper surface glabrous or more commonly inconspicuously and minutely glandular along the main veins, the undersurface sparsely to moderately pubescent with short, unbranched or few-branched hairs, most abundantly along the main veins, sometimes also sparsely and minutely glandular, often with 3–5 secondary veins on each side crowded near the base of the midvein, pinnately

1440. Viburnum molle   1441. Viburnum opulus   1442. Viburnum ozarkense

veined above the base, the secondary veins straight, often dichotomously branched but not forming a network, extending to the leaf margin, each branch ending in a tooth. Inflorescences short- to more commonly long-stalked, with 5–7 primary branches, these with moderate to dense, minute glands at flowering, sometimes also with scattered, unbranched hairs, the marginal flowers fertile and similar to the other flowers. Ovaries with dense, minute glands. Fruits 7–11 mm long, ellipsoid to oblong-ellipsoid and slightly flattened, bluish black, not glaucous. Nutlet 6–10 mm long, yellowish brown to reddish brown. 2n=18, 36. May–June.

Scattered in the Ozark and Ozark Border Divisions north locally to Lincoln County (Pennsylvania to Tennessee west to Iowa and Arkansas). Bases and ledges of limestone and dolomite bluffs and adjacent bottomland and mesic upland forests.

This species often spreads by underground runners to form clonal colonies. It can become a dominant species on talus slopes and ledges of wooded bluffs. Plants with the pubescence of the leaf undersurface confined to the major veins have been called f. *leiophyllum* Rehder.

5. Viburnum opulus L. var. **opulus** (Guelder rose)

Map 1441

Plants shrubs 1–4 m tall. Bark firm, not peeling, relatively smooth, occasionally becoming somewhat fissured on older stems, gray or grayish brown. Winter buds oblong-ovoid, slightly flattened, with 1 pair of scales, these fused along the margins for part of their length and with the free apical portions overlapping, glabrous, not sticky. Stipules usually present but often shed early, partially fused to the basal portion of the petiole, narrowly linear, glabrous or with a few microscopic glands. Petioles 10–25 mm long, unwinged or more commonly with a pair of slender ridges or narrow wings 0.2–0.5 mm wide flanking the ventral groove, glabrous or with scattered hairs along the ventral groove, these straight (not undulate or curled), glabrous, also with 1 or more pairs of prominent, lozenge-shaped glands 1–2 mm in diameter at and/or near the tip. Leaf blades 1.5–8.0 cm long, 2–9 cm wide, shallowly to moderately 3-lobed, relatively thin but firm, broadly oblong-ovate to broadly ovate in outline, broadly rounded to truncate at the base, the lobes angled or short-tapered to a usually sharply pointed tip, the margins otherwise coarsely toothed with the teeth 1–3 per cm, 3–7(–9) per lobe, also moderately short-hairy, the upper surface glabrous, the undersurface sparsely to moderately pubescent, sometimes only along the veins, with mostly unbranched hairs, palmately 3-lobed at the base, each main vein with pinnate secondary veins, these straight or slightly arched, often dichotomously branched but not forming a network, most but not all of the branches extending to the leaf margin and then ending in a tooth. Inflorescences short- to more commonly long-stalked, with (3–)5–7 primary branches, these glabrous or with sparse, minute, stalked glands, the marginal flowers sterile, with enlarged, showy corollas that are trumpet-shaped from a slender tube and somewhat zygomorphic. Ovaries glabrous. Fruits 8–15 mm long, nearly spherical, red to reddish orange, not glaucous. Nutlet 7–10 mm long, yellowish brown to brownish yellow. 2n=18. April–June.

Introduced, uncommon, known thus far from Iron, St. Charles, and St. Louis Counties (native of Europe, Asia, Africa; introduced in the northeastern U.S. west to Iowa and Missouri; Canada). Bottomland forests and mesic upland forests; also old fields.

Viburnum opulus, with a broad distribution in Eurasia and northern Africa, has long been cultivated in North America for its showy flowers. It sometimes escapes into disturbed habitats and has

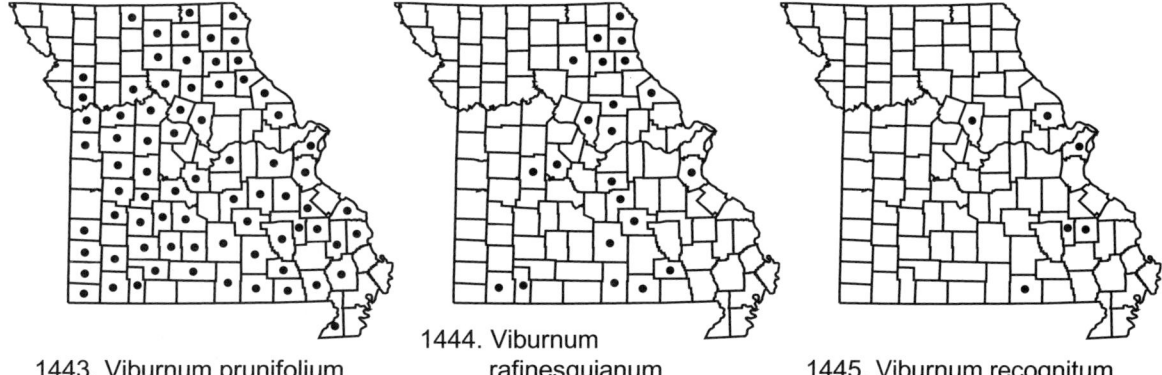

1443. Viburnum prunifolium   1444. Viburnum rafinesquianum   1445. Viburnum recognitum

become abundant around some urban areas, for example in the Chicago region (Swink and Wilhelm, 1994). The species was first reported for Missouri by Ladd (1994).

Plants with all of the flowers sterile and enlarged are known as snowball bush; they are often cultivated for their globose clusters of showy flowers, but they do not set seed. The Guelder rose is closely related to the highbush cranberry of the northern United States and Canada, which is known as *V. opulus* var. *americanum* Aiton or *V. trilobum* Marshall. This taxon differs in its narrowly oblanceolate stipules and short-stalked, somewhat barrel-shaped petiolar glands. The five or six species belonging to section *Opulus* mostly are native to Asia and eastern North America, with the exception of *V. opulus*, whose range extends into Europe.

6. Viburnum ozarkense Ashe (Ozark arrowwood)

Map 1442

Plants shrubs 2–4 m tall. Bark firm, not peeling, usually relatively smooth to finely roughened or warty, not fissured, gray or grayish brown to reddish brown. Winter buds ovoid-conical, slightly flattened, with 2 pairs of overlapping scales, glabrous, not or only slightly sticky. Stipules usually present but often shed early, partially fused to the basal portion of the petiole, linear, glandular. Petioles 5–22(–28) mm long, unwinged, with sparse to moderate minute, stalked glands and usually also scattered, unbranched hairs, lacking prominent glandular swellings near the tip. Leaf blades 7–18 cm long, 6–10(–14) cm wide, unlobed, relatively thin and papery, broadly ovate to nearly circular, rounded to truncate or shallowly cordate at the base, angled or short-tapered to a bluntly or more commonly sharply pointed tip, the margins coarsely toothed with the teeth 1–3 per cm, (10–)12–17 on each side, the surfaces moderately pubescent with short, unbranched hairs, most abundantly along the main veins, sometimes also sparsely and minutely glandular, often with 3–5 secondary veins on each side crowded near the base of the midvein, pinnately veined above the base, the secondary veins straight, often dichotomously branched but not forming a network, extending to the leaf margin, each branch ending in a tooth. Inflorescences short- to more commonly long-stalked, with 5–7 primary branches, these with moderate to dense, minute glands at flowering, sometimes also with scattered, unbranched hairs, the marginal flowers fertile and similar to the other flowers. Ovaries with dense, minute glands. Fruits 8–12 mm long, ellipsoid to oblong-ellipsoid and slightly flattened, bluish black, not glaucous. Nutlet 7–10 mm long, yellowish brown to reddish brown. 2n=36. May–June.

Uncommon, known thus far only from Howell and Oregon Counties (Missouri, Arkansas, Oklahoma). Ledges of dolomite bluffs and adjacent mesic upland forests.

This Ozark endemic often has been listed as a synonym of *V. molle*, but it is actually more similar to *V. rafinesquianum*. Weckman (2002), who resurrected the name as a segregate of *V. molle*, presented a great deal of comparative information on the morphology, range, and habitat of *V. ozarkense*.

7. Viburnum prunifolium L. (blackhaw)
V. bushii Ashe
V. prunifolium var. *bushii* (Ashe) E.J. Palmer & Steyerm.

Pl. 338 c, d; Map 1443

Plants shrubs or small trees 2–5(–10) m tall. Bark firm, not peeling, relatively smooth to finely roughened or warty on younger branches, sometimes forming small plates or furrows on older trunks, gray to grayish brown. Winter buds oblong-conical, often somewhat flattened, with a pair of scales, these valvate (the margins touching but not overlapping), glabrous or with scattered, minute, red to reddish brown, stellate hairs, somewhat sticky. Stipules absent. Petioles 8–16 mm long, with poorly to moderately well-developed wings 0.2–1.0 mm wide flanking the ventral groove, especially toward the tip, these straight (not undulate or curled), glabrous or with scattered microscopic, red to reddish brown, stellate hairs, lacking prominent glandular swellings near the tip. Leaf blades 1.5–8.0 cm long, 1.0–4.5 cm wide, unlobed, relatively thin and papery, broadly to less commonly narrowly elliptic to oblong-ovate, rounded to short-tapered at the base, rounded or broadly angled to a usually bluntly pointed tip, or contracted to a short-tapered or rarely longer-tapered, sharply pointed tip, the margins finely toothed with the teeth 5–11 per cm, numerous on each side, the surfaces glabrous or with scattered microscopic, red to reddish brown, stellate hairs when young, pinnately veined, the secondary veins evenly spaced, curved or arched toward the tip, looping and joined with adjacent veins into a network, not reaching the margin. Inflorescences sessile or short-stalked, with (3)4(5) primary branches, these glabrous or with scattered microscopic, red to reddish brown, stellate hairs, the marginal flowers fertile and similar to the other flowers. Ovaries glabrous. Fruits 7–16 mm long, ellipsoid, oblong-ovoid, or nearly spherical, bluish black, slightly glaucous. Nutlet 9–15 mm long, yellowish brown to dark brown. $2n=18$. April–May.

Scattered nearly throughout the state (eastern U.S. west to Iowa and Texas). Bottomland forests, mesic upland forests, banks of streams and rivers, and bases and ledges of bluffs; also fencerows, pastures, railroads, and roadsides.

Steyermark (1963) described the fall foliage of this species, which turns to a deep lavender, maroon-purple, or rosy purplish red. He also noted that the bark was used medicinally as an astringent, nerve tonic, and antispasmodic, among other uses.

Viburnum prunifolium is closely related to *V. rufidulum,* and some immature plants have pubescence somewhat intermediate between the two species.

8. Viburnum rafinesquianum Schult. (downy arrowwood)

V. rafinesquianum var. *affine* House

Pl. 338 h; Map 1444

Plants shrubs 1–2 m tall. Bark firm, not peeling, usually relatively smooth to finely roughened or warty on younger branches, rarely somewhat peeling in papery sheets, becoming somewhat fissured on older stems, gray or grayish brown to reddish brown. Winter buds ovoid-conical, slightly flattened, with 2 pairs of overlapping scales, glabrous, not or only slightly sticky. Stipules often present but usually shed early, partially fused to the basal portion of the petiole, linear, hairy and usually also glandular. Petioles 3–15(–22) mm long, unwinged, sparsely to moderately hairy, mostly along the ventral groove, lacking prominent glandular swellings near the tip. Leaf blades (4–)6–10 cm long, 3–7 cm wide, unlobed, relatively thin and papery, ovate to broadly oblong-ovate or occasionally nearly circular, rounded to truncate or shallowly cordate at the base, angled or short-tapered to a bluntly or more commonly sharply pointed tip, the margins coarsely toothed with the teeth 1–3 per cm, 7–12(–14) on each side, the upper surface glabrous or more commonly with scattered, unbranched hairs, the undersurface sparsely to moderately pubescent along the main veins with mostly unbranched hairs and usually also with small patches of densely woolly hairs in the axils of the main veins, often with 3–5 secondary veins on each side crowded near the base of the midvein, pinnately veined above the base, the secondary veins straight, often dichotomously branched but not forming a network, extending to the leaf margin, each branch ending in a tooth. Inflorescences sessile or short- to long-stalked, with (3–)5–7 primary branches, these with dense, minute glands at flowering, the marginal flowers fertile and similar to the other flowers. Ovaries with dense, minute glands. Fruits 6–9 mm long, ellipsoid, oblong-ovoid, or nearly spherical, bluish purple to purplish black, not glaucous. Nutlet 6.0–7.5 mm long, yellowish brown to dark reddish brown. $2n=36$. May–June.

Scattered mostly in the eastern half of the state, uncommon or absent from most of western Missouri (eastern U.S. west to North Dakota and Oklahoma; Canada). Bottomland forests, mesic to less commonly dry upland forests, banks of streams and rivers, and bases and ledges of bluffs.

Plants with leaf pubescence absent or limited to the veins have been called var. *affine* and appear to be the more common phase in Missouri. However, because of the large number of specimens with seemingly intermediate levels of hairiness, it has not been possible to accept the varieties in the present treatment. Steyermark (1963) noted that the fall foliage of this species is attractive, varying from dull purplish red to dull purple with reddish highlights.

9. Viburnum recognitum Fernald (northern arrowwood)

V. dentatum var. *lucidum* Aiton

Pl. 337 h, i; Map 1445

Plants shrubs 1–3 m tall. Bark firm, not peel-

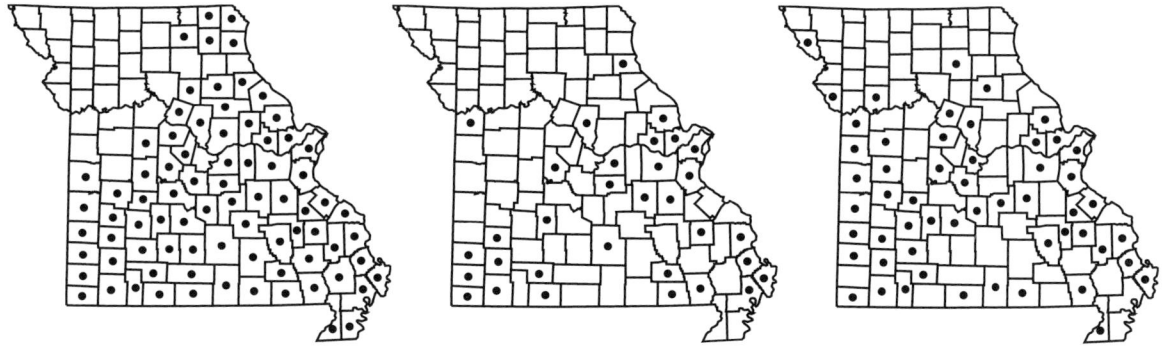

1446. Viburnum rufidulum   1447. Agrostemma githago   1448. Arenaria serpyllifolia

ing, relatively smooth to finely roughened or warty on younger branches, occasionally becoming somewhat fissured on older stems, gray or grayish brown to reddish brown. Winter buds ovoid-conical, slightly flattened, with 2 pairs of overlapping scales, glabrous, not or only slightly sticky. Stipules usually absent. Petioles 18–36 mm long, unwinged, glabrous or with scattered hairs along the ventral groove, lacking prominent glandular swellings near the tip. Leaf blades (5–)7–15 cm long, 4.5–10.0 cm wide, unlobed, relatively thin and papery, broadly elliptic to broadly ovate or nearly circular, rounded to truncate or shallowly cordate at the base, angled to a bluntly or sharply pointed tip or short-tapered to a sharply pointed tip, the margins coarsely toothed with the teeth 1–3 per cm, 12–25 on each side, the upper surface glabrous or rarely with scattered, mostly unbranched hairs, the undersurface sparsely or rarely moderately pubescent along the main veins with mostly unbranched hairs and usually also with small patches of densely woolly hairs in the axils of the main veins, often with 3 or 4 secondary veins on each side crowded near the base of the midvein, pinnately veined above the base, the secondary veins straight, often dichotomously branched but not forming a network, extending to the leaf margin, each branch ending in a tooth. Inflorescences short- to more commonly long-stalked, with (3–)5–7 primary branches, these glabrous or rarely with a few unbranched hairs, occasionally also with sparse, minute, stalked glands, the marginal flowers fertile and similar to the other flowers. Ovaries glabrous, the tapered, stylelike tip densely short-hairy. Fruits 7–10 mm long, ellipsoid, oblong-ovoid, or nearly spherical, purplish blue to bluish black, not glaucous. Nutlet 6–8 mm long, yellowish brown. 2n=36. May–June.

Uncommon in the eastern half of the Ozark Division, locally north to Lincoln County and perhaps west to Boone County (eastern U.S. west to Wisconsin, Missouri, and Arkansas; Canada). Banks of streams and rivers, bottomland forests, and mesic upland forests; also old fields and roadsides; often on acidic substrates.

Steyermark (1963) knew this species only from two sites in Oregon County, but in recent years several additional populations elsewhere in the Ozarks have been discovered. The plants at Cuivre River State Park (Lincoln County) and Forest Park (St. Louis City) are from disturbed successional habitats and may represent escapes from cultivation. For further discussion of anomalous plants from Boone County, see the treatment of the closely related *V. dentatum*.

10. Viburnum rufidulum Raf. (southern blackhaw, wild raisin)

Pl. 338 e, f; Map 1446

Plants shrubs or small trees 2–6(–10) m tall. Bark firm, not peeling, relatively smooth to finely roughened or warty on younger branches, occasionally forming small plates or furrows on older trunks, gray to grayish brown or purplish gray. Winter buds oblong-conical, often somewhat flattened, with a pair of scales, these valvate (the margins touching but not overlapping), densely pubescent with minute, red to reddish brown, stellate hairs, not sticky. Stipules absent. Petioles 7–16 mm long, with poorly to moderately well-developed wings 0.2–1.0 mm wide flanking the ventral groove, especially toward the tip, these straight (not undulate or curled), densely pubescent with microscopic, red to reddish brown, stellate hairs, sometimes appearing less pubescent late in the season, lacking prominent glandular swellings near the tip. Leaf blades 1.5–10.0 cm long, 1.0–6.5 cm wide, unlobed, relatively thick and somewhat leathery, elliptic to oblong-elliptic or oblong-ovate, rounded to angled or occasionally short-tapered at the base, rounded or broadly angled to a usually bluntly pointed tip, rarely contracted to a short-tapered, sharply pointed tip, the margins finely toothed with the teeth 5–11 per cm, numerous on each side, the upper surface glabrous or with scattered microscopic, red to reddish brown, stellate hairs mostly along the midvein, shiny (this sometimes not apparent in dried specimens), the undersurface moderately to densely pubescent with scattered microscopic, red to reddish brown, stellate hairs when young, these becoming more scattered and mostly along the main veins with age, pinnately veined, the secondary veins evenly spaced, curved or arched toward the tip, looping and joined with adjacent veins into a network, not reaching the margin. Inflorescences sessile or short-stalked, with 3 or 4(5) primary branches, these sparsely pubescent with scattered microscopic, red to reddish brown, stellate hairs, usually also with scattered longer, branched or tangled hairs, at least when young, the marginal flowers fertile and similar to the other flowers. Ovaries glabrous. Fruits 8–18 mm long, ellipsoid, oblong-ovoid, or nearly spherical, bluish purple to blue or bluish black, glaucous. Nutlet 7–15 mm long, yellowish brown to dark brown. $2n=18$. April–May.

Scattered south of the Missouri River and in northeastern Missouri (eastern [mostly southeastern] U.S. west to Kansas and Texas). Bottomland forests, mesic to dry upland forests, banks of streams, rivers, and spring branches, edges of glades, and bases and ledges of bluffs; also fencerows, pastures, railroads, and roadsides.

CARYOPHYLLACEAE (Pink Family)
(Rabeler and Hartman, 2005)
Contributed by Richard K. Rabeler and Ronald L. Hartman

Plants annual, biennial, or perennial herbs, rarely dioecious, glabrous or pubescent with unbranched hairs or stalked glands. Stems simple or branched, erect to spreading, often somewhat thickened at the nodes. Leaves opposite, sometimes appearing whorled, the pair of leaves at each node equal in size or nearly so, often petiolate, the bases of adjacent leaves at a node sometimes fused around the stem. Stipules mostly absent, when present then scarious, white or silvery, slender to broadly triangular. Leaf blades simple, linear to broadly ovate, often 1-veined, the margins entire. Inflorescences terminal or axillary clusters or panicles, sometimes appearing headlike or the flowers solitary at the stem or branch tips. Bracts often present at the inflorescence branch points, these opposite or rarely whorled, herbaceous or papery; 1–3 pairs (sometimes appearing whorled) of sepaloid bracts (epicalyx) sometimes also present immediately subtending the flower. Flowers actinomorphic, perfect or rarely imperfect, hypogynous (technically perigynous in a few genera). Calyx of 4 or 5 sepals, these separate or fused, sometimes only at the base, the tips sometimes awned, persistent at fruiting. Corolla absent or more commonly the petals 4 or 5 (rarely 1), distinct, sometimes notched or 2-lobed at the tip, sometimes tapered to a slender, stalklike base, rarely (in *Saponaria*) with a small appendage on the upper surface at the base of the spreading apical portion (blade). Stamens mostly 4, 5, or 10, in 1 or 2 whorls, the filaments distinct or fused basally into a short tube (in *Silene*). Staminodes usually absent (minute and scalelike in *Geocarpon*). Pistil 1 per flower, the ovary superior, of 2–5 fused carpels, with 1 locule (sometimes appearing 2-locular toward the base) or less commonly 3–5 locules. Styles mostly 2–5, sometimes fused toward the base, the stigmas mostly linear. Fruits capsules, many-seeded and dehiscing longitudinally from the tip usually into teeth, or achenelike, 1-seeded and indehiscent or dehiscing irregularly with age. Seeds kidney-shaped to globose, sometimes more or less oblong and flattened, rarely with marginal wings or terminal appendages, the embryo appearing curved or coiled. Eighty-two to 88 genera, about 3,000 species, worldwide, especially in north-temperate, alpine, and Mediterranean areas.

Members of this family and Molluginaceae are exceptional within the order Caryophyllales because they possess anthocyanin pigments rather than betalains (as in the related families

Aizoaceae, Amaranthaceae, Cactaceae, Chenopodiaceae, Nyctaginaceae, Phytolaccaceae, and Portulacaceae). The number of genera recognized within the family depends mainly on whether one adopts a broad (Greuter, 1995; Morton, 2005c) or narrow (Oxelman et al., 2001) circumscription of *Silene,* the largest genus in the family. The number of genera may increase to over 120 if *Silene* is further subdivided (as has been done by some Old World authors) and if other genera, such as *Minuartia* and *Stellaria,* are similarly divided. Additional molecular work may assist in providing evidence for the recognition (or not) of such segregate genera.

This family includes many taxa cultivated as ornamentals, including *Dianthus, Gypsophila* L., and *Silene.* There also are many weedy taxa, mostly introductions from Eurasia. Of the 21 genera found in Missouri, 13 are entirely nonnative: *Agrostemma, Arenaria, Atocion, Dianthus, Holosteum, Lychnis, Myosoton, Petrorhagia, Saponaria, Scleranthus, Spergula, Spergularia,* and *Vaccaria.*

1. Stipules present, linear and bristlelike to ovate-triangular, mostly scalelike, sometimes fused
   2. Fruit 1-seeded, dehiscing irregularly with age; petals absent; styles 2 .... 12. PARONYCHIA
   2. Fruit many-seeded, dehiscing longitudinally from the tip; petals present; styles 3 or 5
      3. Leaves appearing whorled; styles and fruit valves 5 .... 18. SPERGULA
      3. Leaves appearing opposite (but axillary leaf clusters often present); styles and fruit valves 3 .... 19. SPERGULARIA
1. Stipules absent
   4. Sepals fused for at least ¼ of their length, the calyx often tubular or flask-shaped; petals white to pink or red, or absent
      5. Flowers closely subtended by (sometimes partially enclosed in) an epicalyx of 2–6 sepaloid bracts
         6. Calyces with 20–45 nerves or ribs, the tube herbaceous between the teeth or lobes .... 5. DIANTHUS
         6. Calyces with 15 nerves or ribs, the tube thin, papery, and white to tan or pinkish-tinged between the green teeth .... 13. PETRORHAGIA
      5. Flowers not subtended by an epicalyx of sepaloid bracts
         7. Plants densely pubescent with relatively long, usually woolly hairs; styles (4)5; flowers perfect
            8. Calyx lobes longer than the tube, broadly linear, often equaling or exceeding the petals .... 1. AGROSTEMMA
            8. Calyx lobes shorter than the tube, lanceolate, shorter than the petals .... 8. LYCHNIS
         7. Plants glabrous (sometimes glaucous) or, if pubescent, then not with woolly hairs; styles 2 or 3(4) or, if 5, then the flowers imperfect and the plants dioecious
            9. Styles 2(3); fruits dehiscing apically by 4(6) teeth
               10. Inflorescences very dense, the flower stalks 0.1–0.5 cm long; calyces rounded, not angled in cross-section; petals with a pair of small appendages on the upper surface at the base of the expanded, spreading apical portion .... 15. SAPONARIA
               10. Inflorescences open, the flower stalks 0.5–4.0 cm or more; calyces 5-angled or winged in cross-section; petals lacking appendages .... 21. VACCARIA
            9. Styles 3–5 (absent in staminate flowers); fruits dehiscing apically by 3–6 or 10 teeth

11. Plants somewhat succulent; stems 0.7–4.0 cm long; leaves 0.2–0.4 cm long; petals absent .... 6. GEOCARPON
11. Plants not succulent; stems (8–)10–110 cm long; leaves 0.5–14.0 mm long; petals present (rarely absent in *Silene*)
    12. Calyces glabrous, narrowly funnelform, dilated above the midpoint; petals pink .... 3. ATOCION
    12. Calyces glabrous or pubescent, tubular to flask-shaped; petals white or pink (rarely absent) .... 17. SILENE
4. Sepals free or fused only at the base, the calyces not appearing tubular or flask-shaped (note the presence of a hypanthium surrounding and enclosing the ovary in *Scleranthus;* this thick-walled and hardened, different in texture from the sepals); petals white or absent
    13. Fruits 1-seeded and dehiscing irregularly with age; flowers perigynous, the hypanthium deeply cup-shaped to somewhat urn-shaped, surrounding and enclosing the ovary and fruit; flowers inconspicuous, sessile or with a minute, tapered stalk; petals absent............................ 16. SCLERANTHUS
    13. Fruits many-seeded, dehiscing from the tip, flowers appearing hypogynous (the sepals sometimes fused into a shallow, saucer-shaped cup at the base); flowers mostly conspicuous and stalked; petals present or absent
        14. Petals absent or minute and inconspicuous
            15. Fruits cylindrical, often curved, dehiscing by 10 apical teeth; styles 5 .... 4. CERASTIUM
            15. Fruits ovoid to globose, dehiscing longitudinally by valves; styles 3–5
                16. Styles 4 or 5; fruits dehiscing apically by 4 or 5 teeth; stamens 4–8 .... 14. SAGINA
                16. Styles 3; fruits dehiscing apically by 6 teeth; stamens 1–3 .... 20. STELLARIA
        14. Petals present, well developed, and often showy
            17. Petals deeply 2-lobed from the tip, often divided nearly to the base
                18. Fruits cylindrical, often curved, dehiscing by 8 or 10 apical teeth; styles (4)5 or 6 .... 4. CERASTIUM
                18. Fruits ovoid to globose, dehiscing longitudinally by valves; styles 3 or 5(6)
                    19. Inflorescences glandular-pubescent; styles 5(6); fruits dehiscing by 5(6) valves that are 2-lobed or notched toward the tip .... 11. MYOSOTON
                    19. Inflorescences glabrous or nonglandular-pubescent; styles 3; fruits dehiscing by 6 valves that are unlobed ...... 20. STELLARIA
            17. Petals entire, irregularly toothed, or notched at the tip
                20. Styles as many as the 3–5 valves of the fruit
                    21. Styles 3; valves or teeth of the fruits 3; petals much longer than the sepals, relatively showy .... 9. MINUARTIA
                    21. Styles 4 or 5; valves of the fruit 4 or 5; petals shorter than to about as long as the sepals, relatively inconspicuous .. 14. SAGINA
                20. Styles half as many as the 6–10 valves or apical teeth of the fruit
                    22. Fruits cylindrical, dehiscing by 8 or 10 apical teeth
                        23. Inflorescences paniculate clusters; petals shallowly notched at the tip but otherwise entire; seeds kidney-shaped .... 4. CERASTIUM

23. Inflorescences umbellate clusters; petals irregularly toothed or lobed along the margins; seeds more or less oblong .... 7. HOLOSTEUM

22. Fruits ovoid to subglobose, opening by 6 valves or teeth
    24. Leaves 2–8 mm long, angled or tapered to a sharply pointed tip; bracts with herbaceous margins; petals shorter than the sepals; seeds without an appendage .... 2. ARENARIA
    24. Leaves 7–35 mm long, rounded or angled to a bluntly pointed tip; bracts with thin, whitened margins; petals longer than the sepals; seeds with a spongy appendage .... 10. MOEHRINGIA

1. Agrostemma L.
(Hammer et al., 1982)

Two species, Europe, Asia, introduced worldwide.

1. Agrostemma githago L. **var. githago** (corn cockle)

Pl. 339 a, b; Map 1447

Plants annual. Stems 30–100 cm long, erect, branched or unbranched, densely pubescent with long, spreading, white, nonglandular hairs. Leaves opposite, fused basally into a sheath, sessile, lacking axillary clusters of leaves. Stipules absent. Leaf blades 4–15 cm long, linear to narrowly lanceolate, not fleshy, tapered at the base, angled or tapered to a sharply pointed tip. Flowers mostly solitary at the branch tips, sometimes appearing in small, open panicles or axillary, with stalks 5–20 cm long, erect or strongly ascending, the bracts resembling small leaves. Epicalyx absent. Sepals 5, 25–62 mm long, fused basally into a tube 10–17 mm long, this 10-nerved, green and herbaceous between the sepals, the lobes linear or linear-lanceolate, longer than the tube, sharply pointed at the tip, not appearing hooded or awned, the margins green and herbaceous. Petals 5, conspicuous and showy, 15–40 mm long, oblanceolate to spatulate, tapered to a stalklike base, entire or shallowly notched at the tip, purplish red or pink, rarely white, lacking appendages. Stamens 10, the filaments distinct. Staminodes absent. Pistil with 1 locule, sessile. Styles (4)5, each with a subterminal stigmatic area. Fruits capsules, 16–24 mm long, dehiscing apically by (4)5 ascending teeth. Seeds mostly 30–60, 3–4 mm wide, kidney-shaped, the surface tuberculate, black, lacking wings or appendages. $2n=48$. May–September.

Introduced, scattered, mostly in the Ozark, Ozark Border, and Mississippi Lowlands Divisions (native of Europe, Asia; introduced throughout the U.S. except some southwestern states, also Canada). Sand prairies, tops of bluffs, and edges of mesic to dry upland forests; also crop fields, fallow fields, railroads, roadsides, and open, disturbed areas.

This species formerly was a common crop weed that was spread as a seed contaminant. In recent decades, populations have been declining, in part due to a combination of improved grain handling and the more widespread application of herbicides. The large seeds of corn cockle contain toxic steroidal saponins and have been linked to cases of poisoning in birds and cattle (Burrows and Tyrl, 2001). *Agrostemma* is occasionally cultivated as an ornamental for its bright flowers and attractive foliage; seeds were sold in Massachusetts as early as 1845 (Mack, 1991).

2. Arenaria L.

About 210 species, North America, South America, Europe, Asia, introduced nearly worldwide.

The present treatment follows that of McNeill (1962) in segregating *Minuartia* and *Moehringia* from *Arenaria,* rather than the broad, inclusive concept of *Arenaria* often followed by some American authors (Steyermark, 1963; Wofford, 1981).

1. Arenaria serpyllifolia L. (thyme-leaved sandwort)

Pl. 339 c, d; Map 1448

Plants annual. Stems 3–45 cm long, erect or ascending to sprawling or spreading, usually branched, moderately pubescent, with minute, downward-curved hairs, sometimes also stipitate-glandular. Leaves opposite, fused basally into a sheath, short-petiolate (lower leaves) or often sessile, lacking axillary clusters of leaves. Stipules absent. Leaf blades 0.2–0.7 cm long, elliptic to broadly ovate or rarely nearly circular, not fleshy, tapered at the base, angled or tapered to a sharply pointed tip. Flowers in terminal clusters or panicles, with stalks 0.1–1.2 cm long, erect or ascending, the bracts paired and resembling small leaves. Epicalyx absent. Sepals 5, 2–3 mm long, becoming elongated to 4 mm long at fruiting, distinct or fused at the very base, lanceolate to narrowly ovate, narrowly angled or tapered to a sharply pointed tip, not appearing hooded nor awned, the margins thin and whitened. Petals 5, inconspicuous or conspicuous but usually shorter than the sepals, 0.6–2.7 mm long, oblong, not tapered to a stalklike base, entire at the tip, white, lacking appendages. Stamens 10, the filaments distinct. Staminodes absent. Pistil with 1 locule, sessile. Styles 3, each with the stigmatic area along the inner surface. Fruits capsules, 3.0–3.5 mm long, dehiscing apically by 6 ascending to recurved teeth. Seeds 10–15, 0.3–0.6 mm wide, kidney-shaped, the surface tuberculate, black, lacking wings or appendages. $2n=20, 40$. April–August.

Introduced, scattered nearly throughout the state but uncommon or apparently absent from portions of the Glaciated Plains Division (native of Europe, Asia, introduced throughout the U.S., Canada). Glades, ledges and tops of bluffs, and banks of streams and rivers; also lawns, gardens, railroads, roadsides, and open, disturbed areas.

The infraspecific classification of *A. serpyllifolia* accepted here follows that in the recent treatment for the Flora of North America Project (Hartman et al., 2005). Taxonomic authorities from the native range of this polyploid complex have treated it in various ways, accepting varying numbers of taxa and sometimes recognizing these as subspecies or even species (Chater and Halliday, 1993; Abuhadra, 2000). Both of the two infraspecific taxa recorded for North America occur in Missouri, but they can only be distinguished if mature fruits and seeds are present, or by counting their chromosomes.

1. Fruits ovoid to ovoid-conical, broadened or somewhat swollen toward the base; seeds 0.5–0.6 mm wide
 . . . . 1A. VAR. SERPYLLIFOLIA
1. Fruits cylindrical to cylindrical-ovoid, not or only slightly broadened at the base; seeds 0.3–0.4 mm wide
 . . . . 1B. VAR. TENUIOR

1a. var. serpyllifolia (thyme-leaved sandwort)

Sepals 2.5–4.0 mm long. Fruits 2.5–4.0 mm long, ovoid to ovoid-conical, broadened or somewhat swollen toward the base, the main portion with microscopic papillae, the apical teeth usually smooth. Seeds 0.5–0.6 mm wide. $2n=40$. April–August.

Introduced, scattered mostly in the Ozark, Ozark Border, and Mississippi Lowlands Divisions (native of Europe, Asia; introduced nearly throughout the U.S., Canada). Roadsides, railroads, sandy fields, and glades.

1b. var. tenuior Mert. & W.D.J. Koch (slender sandwort)

A. leptoclados (Rchb.) Guss.
A. serpyllifolia ssp. *leptoclados* (Rchb.) Nyman

Sepals 2.5–4.0 mm long. Fruits 2.5–4.0 mm long, cylindrical to cylindrical-ovoid, not or only slightly broadened at the base, the main portion and the apical teeth with microscopic papillae. Seeds 0.3–0.4 mm wide. $2n=20$. April–August.

Introduced, scattered nearly throughout the state but uncommon or apparently absent from portions of the Glaciated Plains Division (native to Eurasia; introduced in the eastern U.S. west to Missouri and Louisiana). Roadsides, sandy fields.

Slender sandwort was first collected in Missouri in 1951 in Joplin County.

3. Atocion Raf.

Five species, Europe, Asia, introduced widely.

Although this group of species has more often been treated as *Silene* section *Compactae* (Boiss.) Schischk., the present treatment follows that of Oxelman et al. (2001) in recognizing *Atocion* as a distinct genus. This is consistent with evidence from ongoing molecular studies that have shown *Atocion* to be one of several genera forming a monophyletic group separate from the core of *Silene* within the tribe Sileneae (Oxelman and Lidén, 1995; Oxelman et al., 1997).

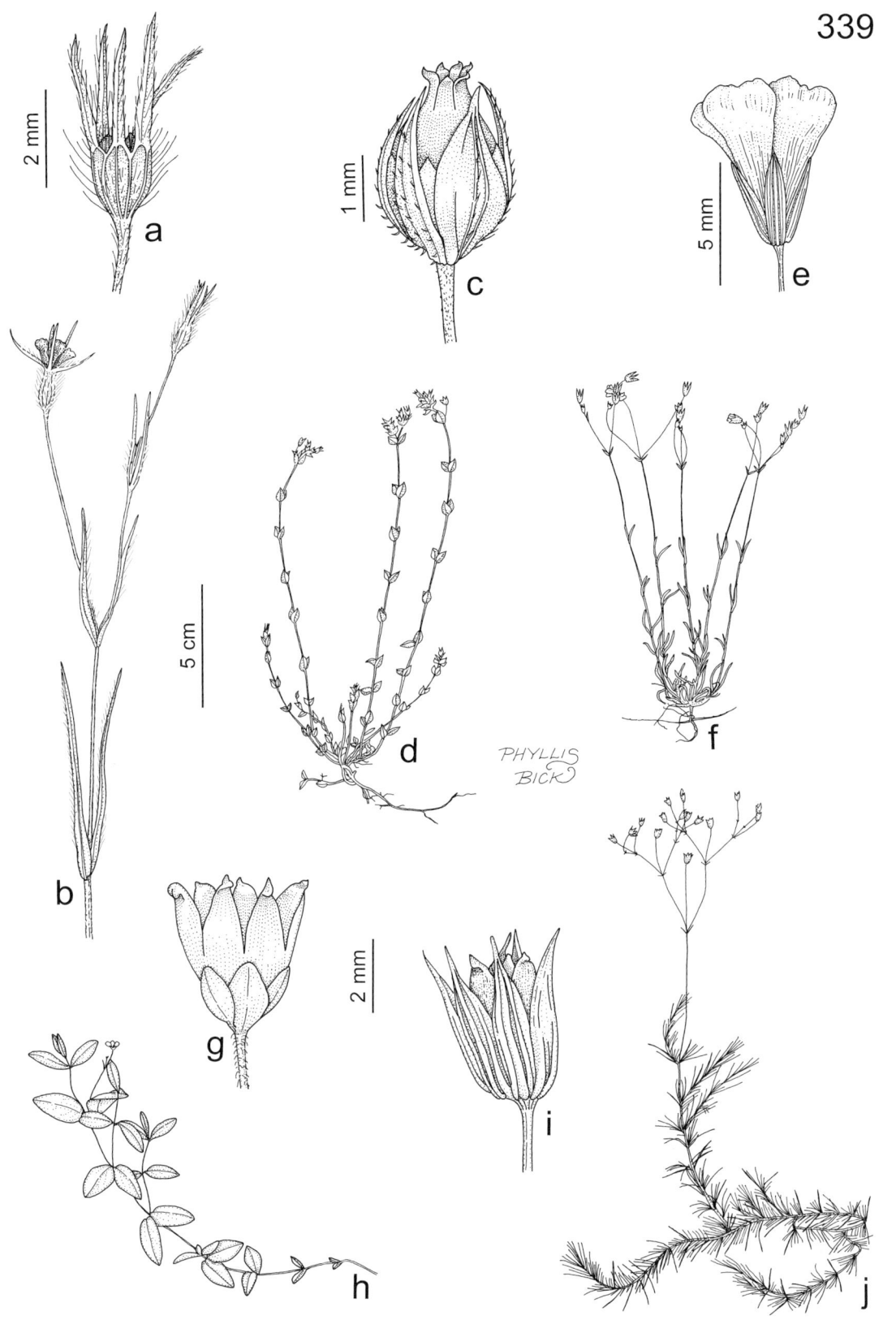

Plate 339. Caryophyllaceae. *Agrostemma githago*, **a)** fruit, **b)** habit. *Arenaria serpyllifolia*, **c)** fruit, **d)** habit. *Minuartia patula*, **e)** flower, **f)** habit. *Moehringia lateriflora*, **g)** flower, **h)** habit. *Minuartia michauxii*, **i)** fruit, **j)** habit.

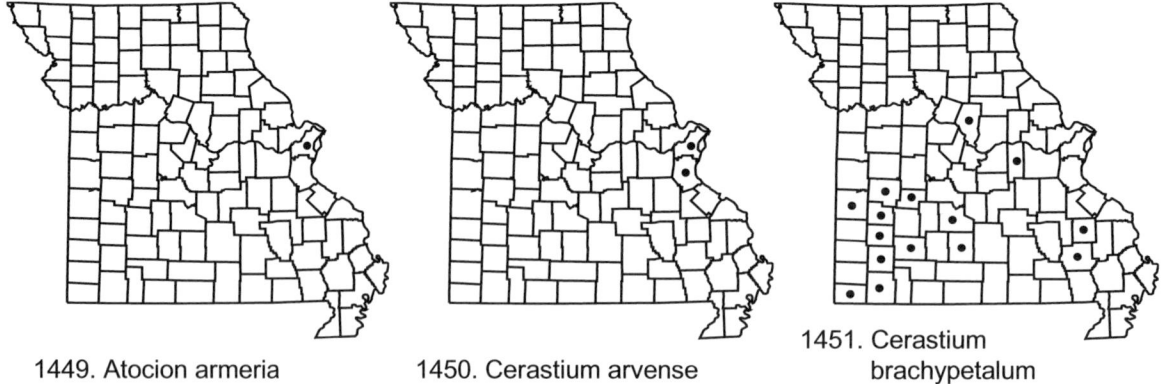

1449. Atocion armeria   1450. Cerastium arvense   1451. Cerastium brachypetalum

1. Atocion armeria (L.) Raf. (sweet William catchfly)

Silene armeria L.

Pl. 345 f; Map 1449

Plants annual or biennial. Stems (10–)20–40 cm long, erect or ascending, usually unbranched below the inflorescence, glabrous or rarely sparsely to moderately stipitate-glandular toward the tip. Leaves opposite, fused basally into a sheath, sessile and sometimes clasping the stem, lacking axillary clusters of leaves. Stipules absent. Leaf blades 1–6 cm long, oblong-lanceolate to elliptic or ovate, not fleshy, tapered (lower leaves) or truncate to cordate at the base, angled to a bluntly or more commonly sharply pointed tip. Flowers in terminal clusters, these grouped into small panicles, the stalks 0.1–0.5 cm long, erect or spreading, the bracts scalelike or leaflike. Epicalyx absent. Sepals 5, 13–17 mm long, fused into a narrow tube, the tube 10-nerved, purplish, and slightly thinner between the lobes, the lobes ovate to triangular, shorter than the tube, angled to a bluntly pointed tip, not appearing hooded or awned, the margins thin and white or purplish-tinged. Petals 5, 10–15 mm long, oblanceolate to spatulate, tapered to a stalklike base, the blade 0.5–0.7 cm long, entire or sometimes slightly notched at the tip, pink or purple, with two small, scalelike appendages on the upper surface at the base of the expanded portion. Stamens 10, the filaments distinct. Staminodes absent. Pistil with 3 locules, the ovary with a stalk 6–7 mm long. Styles 3, each with a stigmatic area along the inner surface. Fruits capsules, 7–10 mm long, dehiscing apically by 6 ascending to somewhat recurved teeth. Seeds 20–30(–45), 0.5–0.7 mm wide, kidney-shaped, the surface with minute papillae, dark brown, lacking wings or appendages. $2n=24$. May–October.

Introduced, uncommon, known thus far only from St. Louis County and City (native to Europe and Asia; probably introduced in the eastern U.S. west to Minnesota, Missouri, and Kentucky, and in the western U.S. from Washington to California and Utah). Railroads and disturbed areas.

4. **Cerastium** L. (mouse-ear chickweed)

Plants annual or perennial. Stems erect or ascending, often from spreading, somewhat matted, sprawling bases, often branched toward the tip, less commonly unbranched below the inflorescence, sparsely to densely pubescent with sometimes woolly hairs, some of the hairs sometimes gland-tipped. Leaves opposite, fused basally into a sheath, short-petiolate (basal leaves) or sessile and sometimes slightly clasping (stem leaves), in a few species with axillary clusters of leaves. Stipules absent. Leaf blades linear-lanceolate to oblong-spatulate or elliptic to broadly ovate, not fleshy, tapered at the base, angled to a bluntly or sharply pointed tip. Flowers in terminal open or relatively dense clusters or panicles, with stalks 0.1–2.5(–4.0) cm long, erect or ascending to somewhat spreading at flowering, sometimes appearing hooked or downward-angled at fruiting, the bracts paired and resembling small leaves, often with thin, whitened margins. Epicalyx absent. Sepals (4)5, usually fused at the base, the lobes narrowly elliptic to lanceolate or broadly lanceolate, green or sometimes reddish-tinged toward the tip, tapered or angled to a bluntly or more commonly sharply pointed tip, not appearing hooded or

Plate 340. Caryophyllaceae. *Cerastium brachypodum*, **a)** fruit, **b)** habit. *Cerastium diffusum*, **c)** fruit, **d)** habit. *Cerastium velutinum*, **e)** fruit, **f)** habit. *Cerastium brachypetalum*, **g)** fruiting branch. *Cerastium fontanum*, **h)** fruit, **i)** habit.

awned, the margins thin and white or translucent. Petals (4)5 or rarely absent, oblanceolate to spatulate, tapered to a stalklike base, shallowly notched or more commonly deeply 2-lobed (to about the petal midpoint) at the tip, white or rarely tinged with purple, lacking appendages. Stamens (4)5 or more commonly 10, the filaments distinct. Staminodes usually absent. Pistil with 1 locule, sessile. Styles (4)5 (rarely 3 elsewhere), distinct, each with a stigmatic area along the inner surface or subterminal. Fruits capsules, often curved, dehiscing apically by (8)10 (occasionally 6 elsewhere) erect to spreading, often somewhat inrolled teeth. Seeds 15–30 or more, wedge-shaped to kidney-shaped, the surface tuberculate or with minute papillae, orangish brown to brown, lacking wings or appendages. About 180 species, nearly worldwide, but most diverse in Europe and Asia, introduced nearly worldwide.

In addition to the ten species treated below, *C. dubium* (Bastard) Guépin (anomalous mouse-ear chickweed) should be expected in Missouri. First collected from eastern North America in central Illinois in 1980 (Shildneck and Jones, 1986), it is now known from at least one population in all states adjacent to Missouri except Iowa; those in Arkansas (Crittenden and Mississippi Counties; Rabeler, 1993) and Kansas (Labette County) are particularly close to our borders. Although the short, sparse glandular hairs on entirely herbaceous bracts would place *C. dubium* closest to *C. nutans* or *C. brachypodum* in the key below, *C. dubium* can be distinguished readily by its slender, almost linear leaves and shallowly notched petals. It also is distinct in having 3(4) styles and a straight (rather than curved) capsule that opens by 6(8) teeth.

Steyermark (1963) noted that the annual species of *Cerastium* sometimes are collected when young in the spring and cooked as vegetables.

1. At least the upper inflorescence bracts with narrow to broad, thin, white to translucent margins
    2. Petals showy, 2–3 times as long as the sepals
        3. Stems usually less than 20 cm long; sepal pubescence nonglandular; petals 7.5–9.0 mm long .... 1. C. ARVENSE
        3. Stems usually 20–40 cm long; sepal pubescence stipitate-glandular; petals 10–15 mm long .... 10. C. VELUTINUM
    2. Petals shorter than to slightly longer than the sepals
        4. Petals deeply lobed (often to the midpoint); sepals densely pubescent with usually nonglandular hairs, rarely some of the hairs stipitate-glandular; stamens 10; fruits curved, (6–)8–11 mm long.............. 5. C. FONTANUM
        4. Petals notched or shallowly lobed at the tip; sepals densely pubescent with stipitate-glandular hairs; stamens usually 5; fruits mostly straight, 4.0–7.5(–8.0) mm long
            5. Bracts all with broad, thin, white to translucent margins (sometimes the entire apical half thin and white to translucent); petal veins not branched; seed tubercles indistinct or low and rounded .... 9. C. SEMIDECANDRUM
            5. Uppermost bracts with narrow, thin, white to translucent margins; petal veins branched; seed tubercles conspicuous, bluntly or sharply pointed .... 8. C. PUMILUM
1. All of the inflorescence bracts with green and herbaceous margins (thin and whitened at the very tip in *C. diffusum*)
    6. Pubescence nonglandular and sometimes also stipitate-glandular, that of the sepals extending beyond and somewhat obscuring the tip
        7. Plants hairy, the hairs not long and silvery; flowers in a dense, crowded or slightly open cluster, the stalk shorter than the sepals .... 6. C. GLOMERATUM

7. Plants covered with long, silvery hairs; flowers in open panicles, the stalk about 2 times as long as the sepals . . . . 2. C. BRACHYPETALUM
6. Pubescence glandular, that of the sepals not extending beyond the tip
   8. Capsules about as long as to twice as long as the sepals; stamens 4 or 5; sepals with dense, relatively long glandular hairs . . . . 4. C. DIFFUSUM
   8. Capsules twice as long or longer than the sepals; stamens 10; sepals with sparse, short glandular hairs
      9. Flower stalks 2–5(–8) times as long as the sepals, at fruiting appearing hooked near the tip, not downward-angled from the base . . . . . . 7. C. NUTANS
      9. Flower stalks 1.0–1.5 times as long as the sepals, at fruiting not appearing hooked near the tip, but often downward-angled from the base
      . . . . 3. C. BRACHYPODUM

1. Cerastium arvense L. ssp. **strictum** Gaudin
(field chickweed)

Map 1450

Plants perennial, with a taproot or short rhizomes. Stems 2–20(–30) cm long, ascending, from a spreading base, branched, the pubescence variable, but often spreading or reflexed, nonglandular hairs mixed with stalked glands toward the tip. Leaves sessile, with axillary clusters of leaves usually present. Leaf blades 0.5–2.5 cm long, linear to narrowly lanceolate or narrowly oblong-spatulate, angled to a bluntly or sharply pointed tip. Flowers in open panicles or clusters, the stalks 1.0–1.7 cm long, about 2 times as long as the sepals, at fruiting ascending, densely pubescent with glandular and nonglandular hairs, the bracts all with narrow, thin, white to translucent margins. Sepals 5, 3.5–6.0 mm long, narrowly elliptic, tapered or angled to a sharply pointed tip, green, pubescent with glandular and nonglandular hairs, these not extending past the sepal tips. Petals 5, 7.5–9.0 mm long, about 2 times the length of the sepals, shallowly 2-lobed at the tip, the veins not apparent or faint and usually branched. Stamens 10. Styles 5. Fruits 7.5–11.0 mm long, less than 2 times as long as the sepals, curved. Seeds 0.6–1.1 mm wide, the surface tuberculate, brown. $2n=36$. May–June.

Uncommon, known thus far only from historical collections from Jefferson and St. Louis Counties (nearly throughout the U.S. except some southeastern states; Canada). Habitat unknown but presumably bluffs or openings of mesic to dry upland forests.

Although most of the specimens included in this showy species by Steyermark (1963) have been redetermined as *C. velutinum,* at least two specimens collected in the late 1880s are still referable to *C. arvense.* Additional efforts should be made to determine if this species is still extant in Missouri. The European ssp. *arvense* has been introduced sporadically into the northeastern states and adjacent Canada and differs in its well-developed rhizomes, more robust stems that are mostly nonglandular-hairy, slightly longer (5–7 mm) sepals, and larger anthers (1.0–1.1 mm vs. 0.8–0.9 mm long).

2. Cerastium brachypetalum Pers. (gray mouse-ear chickweed)
C. brachypetalum ssp. *tauricum* (Spreng.) Murb.
C. brachypetalum f. *glandulosum* W.D.J. Koch.

Pl. 340 g; Map 1451

Plants annual. Stems 5–35 cm long, erect, usually branched toward the tip, moderately to densely pubescent with long, silvery, nonglandular hairs, these sometimes mixed with or replaced by stalked glands toward the tip. Leaves sessile, lacking axillary clusters of leaves. Leaf blades 0.5–2.5 cm long, spatulate (some basal leaves) or elliptic to ovate, angled to a bluntly or sharply pointed tip. Flowers in open panicles, the stalks 0.6–1.6(–2.5) cm long, these (1–)2–3(–6) times as long as the sepals, erect or spreading, at fruiting sometimes appearing hooked near the tip, densely pubescent with glandular hairs, the bracts with herbaceous, green margins. Sepals 5, 3–5 mm long, lanceolate, green, angled to a bluntly or sharply pointed tip, densely pubescent with nonglandular and sometimes also glandular hairs, these extending past and somewhat obscuring the sepal tips. Petals 5, 2.0–3.5 mm long, about $^2/_3$–$^3/_4$ as long as the sepals, shallowly 2-lobed at the tip, the veins usually not apparent. Stamens 10. Styles 5. Fruits 6.0–8.5 mm long, about 1.5 times as long as the sepals, slightly curved. Seeds 0.4–0.5 mm wide, the surface tuberculate, light brown. $2n=88, 90$. April–May.

Introduced, scattered in the western half of the Ozark Division, uncommon and sporadic farther north and east (native of Europe; introduced in the eastern U.S. west to Kansas and Texas, also Idaho

1452. Cerastium brachypodum   1453. Cerastium diffusum   1454. Cerastium fontanum

and Oregon). Glades, ledges of bluffs, upland prairies, and margins of sinkhole ponds; also pastures, roadsides, and open, disturbed areas.

Cerastium brachypetalum is the largest and usually the most diffusely branched of the introduced annual chickweeds. Although most of the Missouri specimens are covered with distinctive long, silvery, nonglandular hairs, the stems, flower stalks, and sepals of some plants are mostly stipitate-glandular. Steyermark (1963) called such plants f. *glandulosum,* but some European botanists have used this as one of the characters used in distinguishing among as many as eight subspecies (Sell and Whitehead, 1993). The present treatment follows that of Morton (2005a) in not formally recognizing this distinction.

3. Cerastium brachypodum (Engelm. ex A. Gray) B.L. Rob. (short-stalked mouse-ear chickweed)

C. nutans Raf. var. *brachypodum* Engelm. ex A. Gray

Pl. 340 a, b; Map 1452

Plants annual. Stems 4–30 cm long, erect, unbranched or sometimes branched below the midpoint, moderately pubescent with stalked glands, these longer toward the base. Leaves sessile, or the lower leaves sometimes short-petiolate, lacking axillary clusters of leaves. Leaf blades 0.5–3.0 cm long, spatulate (some basal leaves) or oblong-lanceolate to narrowly obovate, angled or tapered to a sharply pointed tip. Flowers in narrow, somewhat crowded panicles or clusters, the stalks 0.3–1.0 cm long, 1.0–1.5 times as long as the sepals, mostly ascending, at fruiting often downward-angled from the base, densely pubescent with glandular hairs, the bracts with herbaceous, green margins. Sepals 5, 3.0–4.5 mm long, lanceolate, angled to a sharply pointed tip, green but often with narrow to broad, thin, white to translucent margins, sparsely pubescent with short glandular hairs, these not extending past the sepal tips. Petals 5, 3–5(–6) mm long, 1.0–1.5(–2.0) times as long as the sepals, shallowly 2-lobed at the tip, the veins usually not apparent. Stamens 10. Styles 5. Fruits (5–)8–12 mm long, about 1.0–1.5 times as long as the sepals, curved. Seeds 0.4–0.7 mm wide, the surface tuberculate, yellowish brown. 2n=34. March–June.

Scattered, mostly south of the Missouri River (western U.S. east to Wisconsin, Virginia, and Georgia; Canada; Mexico). Glades, upland prairies, ledges and tops of bluffs, openings of mesic to dry upland forests, savannas, banks of streams and rivers, and margins of ponds and lakes; also pastures, ditches, railroads, roadsides, and open, disturbed areas.

Cerastium brachypodum sometimes has been treated as a variety of *C. nutans*. Although the two taxa share the sparse sepal pubescence that separates them from all other *Cerastium* species in Missouri, they are otherwise distinct (Steyermark, 1963).

4. Cerastium diffusum Pers. (dark green mouse-ear chickweed)

C. tetrandrum Curtis

Pl. 340 c, d; Map 1453

Plants annual. Stems 12–30 cm long, ascending, sometimes from a spreading base, usually branched toward the tip, densely pubescent with short glandular hairs. Leaves sessile, lacking axillary clusters of leaves. Leaf blades 0.5–1.2 cm long, spatulate (some basal leaves) or ovate-elliptic, angled to a bluntly or sharply pointed tip. Flowers in open panicles or clusters, the stalks 0.4–0.8 (–1.2) cm long, 1–3 times as long as the sepals, erect or ascending at flowering and fruiting, stipitate-glandular, the bracts with herbaceous, green margins, but rarely thin and whitened at the very tip. Sepals 4 or 5, (3–)4–7 mm long, lanceolate, angled

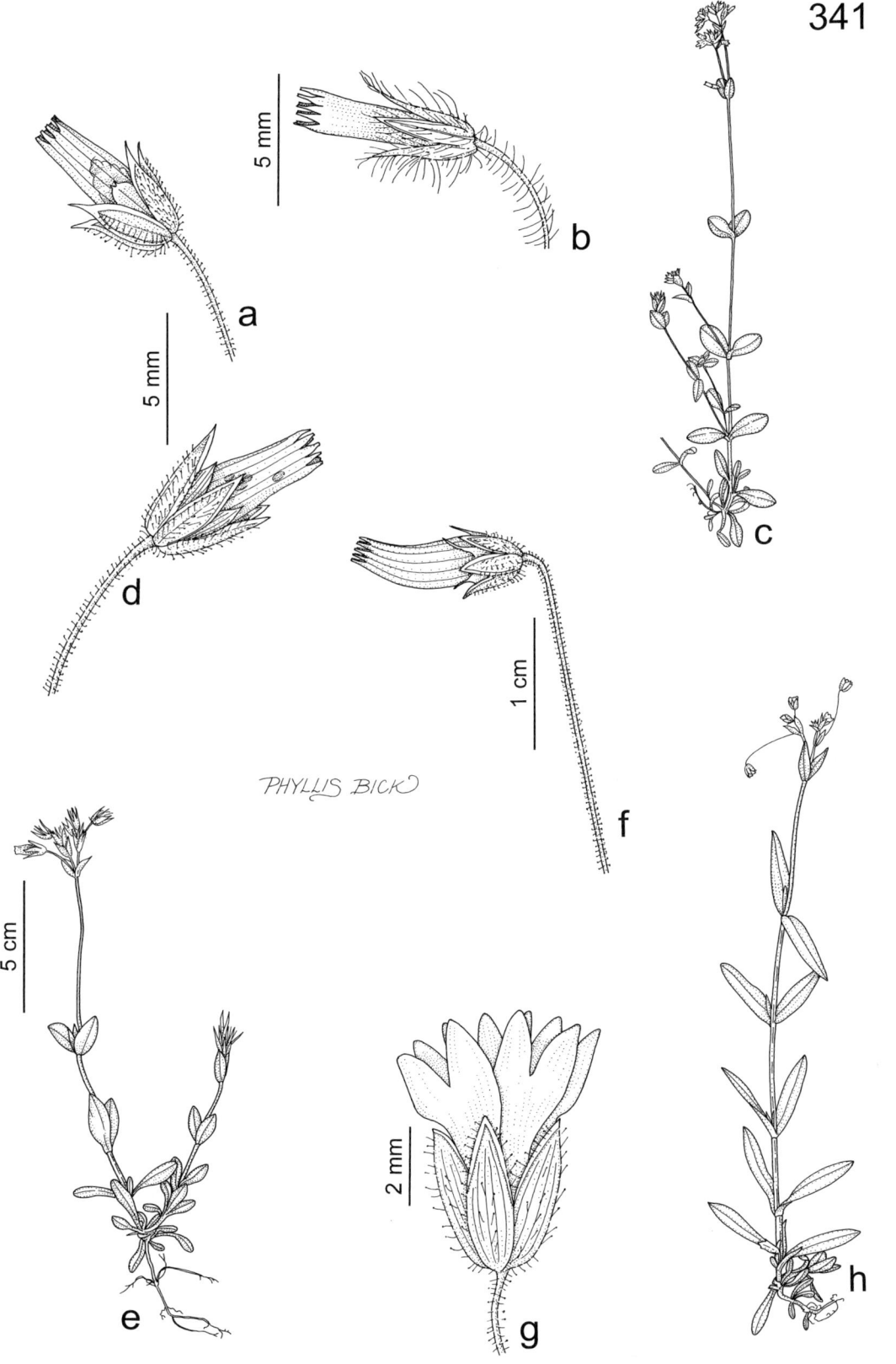

Plate 341. Caryophyllaceae. *Cerastium semidecandrum*, **a)** fruit. *Cerastium glomeratum*, **b)** fruit, **c)** habit. *Cerastium pumilum*, **d)** fruit, **e)** habit. *Cerastium nutans*, **f)** fruit, **g)** flower, **h)** habit.

or tapered to a sharply pointed tip, green, densely pubescent with short glandular hairs, these not extending past the sepal tips. Petals 4 or 5, 3–5 mm long, 0.7–1.0 times as long as the sepals, shallowly 2-lobed at the tip, the veins usually not apparent. Stamens 4 or 5. Styles 4 or 5. Capsules 6.5–8.0 mm long, 1.0–1.2 times as long as the sepals, slightly curved. Seeds 0.5–0.7 mm wide, the surface tuberculate, reddish brown. $2n=72$. March–May.

Introduced, uncommon, known thus far only from Crawford and Pemiscot Counties (native of Europe; introduced sporadically in the southern U.S. north to California, Nebraska, Illinois, and Alabama). Lawns, roadsides, and open, disturbed areas.

Apparently first collected in Crawford County in 1957, *C. diffusum* may be difficult to distinguish from *C. pumilum*. Both species belong to the *C. semidecandrum* complex, a group of 5–8 morphologically similar species native to Europe; see Karlsson (2001) for an extremely detailed treatment of part of the complex and for comments on various ways the taxa can be defined. Although Morton (2005a) included Steyermark's (1963) record of *C. diffusum* in his concept of *C. pumilum,* the present treatment maintains *C. diffusum,* in part to draw attention to plants with some or many 4-parted flowers. Most treatments of *C. pumilum* show that species to have strictly 5-parted flowers.

5. Cerastium fontanum Baumg. **ssp. vulgare** (Hartman) Greuter & Burdet (common mouse-ear chickweed)

C. vulgatum L. f. *glandulosum* (Boenn.) Druce
C. holosteoides Fr. f. *glandulosum* (Boenn.) Möschl

Pl. 340 h, i; Map 1454

Plants usually perennial, with a taproot. Stems (3–)10–65 cm long, erect, often from a spreading or sprawling base, usually branched above the midpoint, moderately to densely pubescent with usually straight, nonglandular hairs, occasionally also stipitate-glandular in the inflorescence. Leaves sessile, rarely with axillary clusters of leaves. Leaf blades 0.5–4.0 cm long, spatulate (some basal leaves) or oblong to oval, angled to a bluntly pointed tip. Flowers in open panicles, the stalks 0.4–1.0 (–1.5) cm long, 1–2 times as long as the sepals, ascending but usually somewhat curved above the midpoint at flowering and fruiting, densely pubescent with nonglandular hairs or occasionally also stalked glands, the bracts green or the upper bracts sometimes with narrow, thin, white to translucent margins. Sepals 5, 4–7 mm long, ovate-lanceolate,

angled or tapered to a sharply pointed tip, green, densely pubescent with nonglandular hairs, rarely also stipitate-glandular, the hairs usually not extending past the sepal tips. Petals 5, 4–5 mm long, 0.6–1.0 times as long as the sepals, deeply 2-lobed (often to the midpoint) at the tip, the veins usually not apparent. Stamens 10. Styles 5. Fruits (6–)8–11 mm long, 1–2 times as long as the sepals, curved. Seeds 0.6–0.9 mm wide, the surface tuberculate, brown. $2n=144$. April–October.

Introduced, scattered nearly throughout the state (native of Europe; introduced nearly worldwide). Sloughs, margins of ponds and lakes, banks of streams, rivers, and spring branches, ledges of bluffs, and sand prairies; also pastures, fallow fields, old fields, ditches, lawns, gardens, railroads, roadsides, and open, disturbed areas.

Cerastium fontanum is a very common species in disturbed areas. It was long treated under the name *C. vulgatum,* a name that has been officially rejected as having ambiguous application (Turland and Wyse Jackson, 1997). The ssp. *fontanum* occurs in Europe and Greenland but apparently is not encountered widely outside its native range. It differs in its petals noticeably longer than the sepals, fruits 11–17 mm long, and seeds 0.9–1.2 mm wide (Morton, 2005a). Two other minor variants sometimes are recognized as subspecies in Europe. Plants with at least some of the pubescence glandular have been called f. *glandulosum,* but this character is expressed in more than one of the subspecies in Europe.

6. Cerastium glomeratum Thuill. (clammy chickweed)

C. viscosum f. *apetalum* (Dumort.) Mert. & W.D.J. Koch

Pl. 341 b, c; Map 1455

Plants annual. Stems 3.5–30.0 cm long, erect or ascending, sometimes branched above the midpoint, densely pubescent with glandular (especially toward the tip) and nonglandular hairs. Leaves sessile, lacking axillary clusters of leaves. Leaf blades 0.5–3.0 cm long, spatulate (some basal leaves) or elliptic to narrowly oblong-obovate, abruptly short-tapered to a minute, sharp point at the tip. Flowers in dense, crowded clusters or later slightly open clusters or small panicles, the stalks 0.1–0.3 cm long, 0.5–1.0 times as long as the sepals, usually longest in the first-opening flowers, erect or spreading at flowering and fruiting, densely pubescent with mostly short glandular hairs, the bracts with herbaceous, green margins. Sepals 5, (2–)3–5 mm long, lanceolate, green or occasionally reddish-tinged at the tip, angled or tapered to a

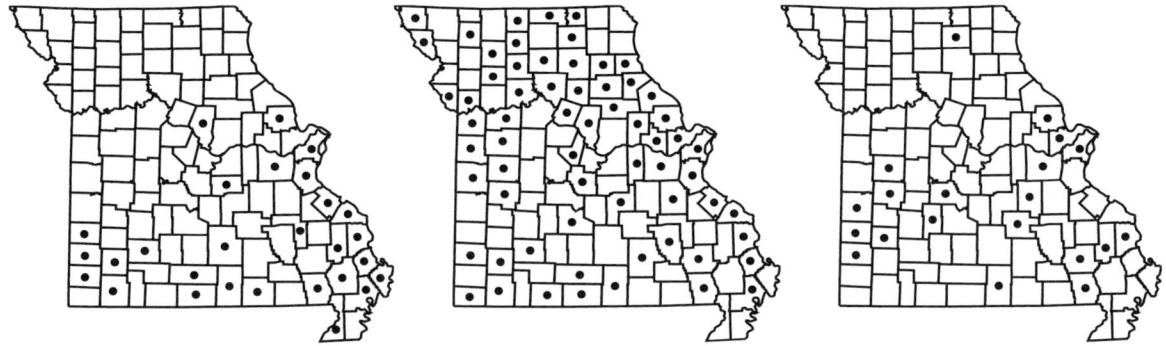

1455. Cerastium glomeratum   1456. Cerastium nutans   1457. Cerastium pumilum

sharply pointed tip, moderately to densely pubescent with glandular and nonglandular hairs, these extending past and somewhat obscuring the sepal tips. Petals 5 (rarely absent), 3–5 mm long, 0.7–1.2 times as long as the sepals, shallowly 2-lobed at the tip, the veins usually not apparent. Stamens 10. Styles 5. Capsules (3.5–)5.0–9.0 mm long, 1–2 times as long as the sepals, slightly curved. Seeds 0.5–0.6 mm wide, the surface tuberculate, brown. $2n=72$. March–July.

Introduced, scattered mostly south of the Missouri River (native of Europe; introduced in the eastern U.S. west to Michigan and Arizona, also Idaho and Nevada west to Washington, Oregon, and California; Canada). Bottomland forests, banks of streams and rivers, ledges and tops of bluffs, and occasionally glades; also crop fields, fallow fields, old fields, ditches, railroads, roadsides, and open, disturbed areas.

Cerastium glomeratum is a common species in disturbed areas, especially in areas to the south and east of Missouri. It was long known as *C. viscosum* (Steyermark, 1963), a name that has been officially rejected as having ambiguous application (Turland and Wyse Jackson, 1997). The rare form with flowers lacking petals has been called f. *apetalum*.

7. Cerastium nutans Raf. ssp. nutans (nodding chickweed, powderhorn)

Pl. 341 f–h; Map 1456

Plants annual. Stems 5–60 cm long, erect or ascending, unbranched below the inflorescence or branched toward the base, moderately to densely pubescent with short glandular hairs, often also with longer glandular hairs at the lower nodes. Leaves sessile or sometimes the lowermost leaves short-petiolate, lacking axillary clusters of leaves. Leaf blades 0.5–8.0 cm long, spatulate (some basal leaves) or oblong-lanceolate to narrowly obovate, angled or tapered to a sharply pointed tip. Flowers in loose, open panicles, the stalks (0.8–)1.5–2.7 (–4.0) cm long, 2–5(–8) times as long as the sepals, ascending, at fruiting appearing hooked near the tip, densely pubescent with glandular hairs, the bracts with herbaceous, green margins. Sepals 5, 2.0–5.5 mm long, lanceolate, angled to a bluntly or sharply pointed tip, green but often with narrow to broad, thin, white to translucent margins, sparsely pubescent with short glandular hairs, these not extending past the sepal tips. Petals 5 (sometimes absent), 5–7 mm long, about 1.0–1.5 times as long as the sepals, shallowly 2-lobed at the tip, the veins usually not apparent. Stamens 10. Styles 5. Fruits (5–)8–13 mm long, 2–3 times as long as the sepals, curved. Seeds 0.5–0.8 mm wide, the surface tuberculate, yellowish brown. $2n=34, 36$. March–June.

Scattered to common throughout the state (nearly throughout the U.S. [including Alaska]; Canada). Banks of streams and rivers, bases and ledges of bluffs, bottomland forests, mesic upland forests, and rarely glades; also crop fields, fallow fields, pastures, gardens, railroads, roadsides, and open, disturbed areas.

Plants with uniformly longer hairs and more persistent but often withered lower leaves that grow in the southwestern United States and adjacent Mexico have been called var. *obtectum* Kearney & Peebles.

8. Cerastium pumilum Curtis (Curtis's mouse-ear chickweed)

C. pumilum f. *medium* Möschl

Pl. 341 d, e; Map 1457

Plants annual. Stems 4–12(–20) cm long, erect, unbranched or branched toward the tip, densely pubescent with short glandular and nonglandular hairs. Leaves sessile, lacking axillary clusters of leaves. Leaf blades 0.3–2.0 cm long, spatulate (especially toward the stem base) or oblong to ovate,

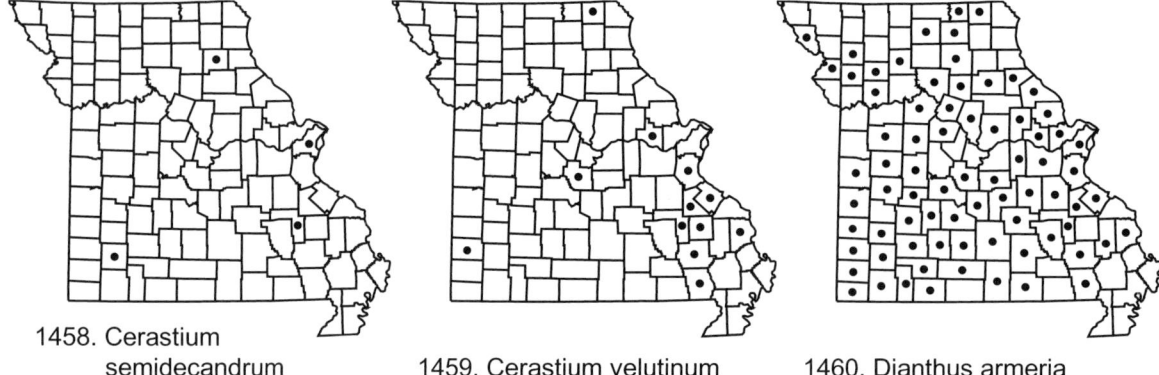

1458. Cerastium semidecandrum

1459. Cerastium velutinum

1460. Dianthus armeria

angled to a bluntly or sharply pointed tip. Flowers in open panicles, the stalks 0.3–0.6(–1.0) cm long, 1.0–1.5 times as long as the sepals, erect or ascending at flowering and fruiting, densely pubescent with short glandular hairs, the bracts green or the upper bracts sometimes with narrow, thin, white to translucent margins. Sepals 5, 3–5 mm long, lanceolate, angled to a sharply pointed tip, green or sometimes reddish-tinged toward the tip, densely pubescent with short glandular hairs, these not extending past the sepal tips. Petals 5, 3–5 mm long, about as long as or slightly longer than the sepals, notched or shallowly lobed at the tip, the veins branched. Stamens mostly 5. Styles 5. Fruits 4.0–7.5(–8.0) mm long, 1.0–1.5(–2.0) times as long as the sepals, straight or slightly curved. Seeds 0.6–0.7 mm wide, the surface with bluntly to sharply pointed tubercles, brown. $2n=72$, 90–95, about 100 (most commonly $2n=72$). April–May.

Introduced, scattered in the southern half of the state, most abundantly in the Unglaciated Plains Division (native of Europe, Asia; introduced in the eastern U.S. west to Michigan, Nebraska, and Texas, also in Oregon and Washington; Canada). Glades, margins of sinkhole ponds, and tops of bluffs; also pastures, crop fields, fallow fields, lawns, railroads, roadsides, and open, disturbed areas.

Cerastium pumilum is abundant in some urban areas and is probably far more common than the relatively few collections suggest. Resemblance to the larger, more common *C. fontanum,* as well as to the early flowering of all of the introduced annual *Cerastium* species, has no doubt contributed to it being undercollected. Plants of *C. pumilum* can be most easily confused with *C. semidecandrum* (all bracts with very broad, thin margins), *C. fontanum* (stamens 10 and all bracts with thin margins), *C. glomeratum* (sepal hairs longer than the sepal tips and entirely herbaceous bracts), and *C. diffusum* (often with 4-parted flowers and only rarely with very slender, thin margins in the uppermost bracts).

9. Cerastium semidecandrum L. (small mouse-ear chickweed, five-stamen mouse-ear chickweed)

Pl. 341 a; Map 1458

Plants annual. Stems 10–25 cm long, erect, unbranched or branched toward the tip, densely pubescent with stalked glands. Leaves sessile, lacking axillary clusters of leaves. Leaf blades 0.3–2.0 cm long, spatulate (some basal leaves) or elliptic to broadly ovate, angled to a bluntly pointed tip. Flowers in panicles, the stalks 0.3–0.9 cm long, 1–2 times as long as the sepals, erect or ascending, at fruiting often downward-angled from the base, densely pubescent with stalked glands, the bracts all with broad, thin, white to translucent margins (sometimes the entire apical half thin and white to translucent). Sepals 5, 3–4 mm long, lanceolate, angled to a sharply pointed tip, densely pubescent with stalked glands, these not extending past the sepal tips. Petals 5, 3–4 mm long, about as long as the sepals, notched or shallowly lobed at the tip, the veins unbranched. Stamens usually 5, occasionally 10. Styles 5. Fruits 4.0–7.5(–8.0) mm long, 1–2 times as long as the sepals, straight or slightly curved. Seeds 0.4–0.6 mm wide, the surface indistinctly roughened or with low, rounded tubercles, brown. $2n=36$. April–May.

Introduced, uncommon and sporadic (native of Europe, Asia; introduced in the eastern U.S. west to Nebraska and Kansas, also in Idaho, Oregon, Washington; Canada). Glades; also pastures, lawns, railroads, roadsides, and open, disturbed areas.

This species was first reported for the state by Mühlenbach (1979). It was collected in St. Louis County in 1971 during his botanical inventories of the St. Louis railyards. *Cerastium semidecandrum* apparently is much less common than *C. pumilum,* the species that it most closely resembles. The broad, thin, white margins of the bracts are unique among the introduced annual *Cerastium* species in Missouri.

Steyermark (1963) excluded *C. semidecandrum* from the Missouri flora, noting that a 1936 collection made by B. F. Bush from Bates County was misdetermined as this species. He stated that the species was likely to be found in the state eventually and further noted that some specimens attributed to *C. glomeratum* appeared to him to be somewhat morphologically intermediate with *C. pumilum*.

10. Cerastium velutinum Raf. **ssp. velutinum**
(field chickweed, prairie chickweed)
C. arvense f. *oblongifolium* (Torr.) Pennell
C. arvense var. *oblongifolium* (Torr.) Hollick & Britton
C. arvense var. *villosum* (Muhl.) Hollick & Britton

Pl. 340 e, f; Map 1459

Plants perennial, usually with a taproot, sometimes with short rhizomes. Stems (15–)20–40 cm long, ascending from an often spreading or sprawling base, branched, moderately to densely pubescent usually with a mixture of stalked glands and spreading, nonglandular hairs. Leaves sessile, with axillary clusters of leaves. Leaf blades 2.0–6.5 cm long, oblong-spatulate to narrowly ovate, angled to a usually bluntly pointed tip. Flowers in open panicles, the stalks 1.2–3.4 cm long, 2–4(–5) times as long as the sepals, erect or ascending, at fruiting usually hooked near the tip, densely pubescent with a mixture of gland-tipped and nonglandular hairs, only the uppermost bracts with narrow, thin, white to translucent margins, the others green and herbaceous throughout. Sepals 5, 4.5–8.5 mm long, narrowly elliptic, angled to a sharply pointed tip, green, densely pubescent with stalked glands, these shorter than the sepal tips. Petals 5, 10–15 mm long, 2(–3) times as long as the sepals, deeply 2-lobed ($^{1}/_{4}$–$^{1}/_{3}$ of the way) at the tip, the veins not apparent or faint and usually branched. Stamens 10. Styles 5. Fruits 10–14 mm long, about 2 times as long as the sepals, curved. Seeds 0.8–1.2 mm wide, the surface tuberculate, brown. $2n=72$. April–June.

Scattered mostly in the eastern portion of the Ozark and Ozark Border Divisions disjunctly westward to Miller and Jasper Counties (northeastern U.S. west to Minnesota and Missouri; Canada). Banks of streams and rivers and openings of mesic to dry upland forests; also pastures and roadsides.

Although it was long considered a variety or form of *C. arvense,* Morton (2005a) and others have recognized this taxon as a separate species. The two species apparently do not hybridize in the wild and for the most part are easily differentiated. Most of the Missouri reports of *C. arvense* and all of the recently collected specimens originally determined as that species are referable to *C. velutinum* Morton (2005a). Morton treated the species as comprising two subspecies. The ssp. *villosissimum* (Pennell) J.K. Morton is endemic to serpentine outcrops in Maryland, New Jersey, and Pennsylvania. It differs in its denser, broader, more persistent leaves that tend to have dense, shaggy pubescence.

5. Dianthus L. (pink, carnation)

Plants annual, biennial, or perennial. Stems erect or ascending, unbranched or branched, sometimes only in the inflorescence, glabrous or variously pubescent, sometimes glaucous. Leaves opposite, fused basally into a sheath, short-petiolate (basal leaves) or sessile (stem leaves), lacking axillary clusters of leaves. Stipules absent. Leaf blades linear to lanceolate or ovate, not fleshy, tapered at the base, angled or tapered to a bluntly (some basal leaves) or sharply pointed tip. Flowers in terminal, open to dense clusters, densely bracted heads, or solitary at the stem or branch tips (rarely axillary in *D. barbatus*), the stalks erect or the flowers nearly sessile, the bracts absent or more commonly paired and resembling smaller leaves. Epicalyx present of 1 or 2(3) pairs of bracts, these resembling small leaves, sometimes with thin, white to translucent margins. Sepals 5, fused into a slender tube, the tube 20–45-veined, herbaceous, green or reddish-tinged between the lobes, the lobes linear to lanceolate or ovate, shorter than the tube, angled or tapered to a bluntly or sharply pointed tip, not appearing hooded nor awned, the margins thin and white or reddish-tinged. Petals 5, conspicuous and showy, oblanceolate to spatulate, tapered to a stalklike base, toothed or deeply divided into an irregular fringe of slender lobes at the tip, white, pink, purple, or red, sometimes spotted, lacking appendages but often hairy on the upper surface, especially toward the base. Stamens 10, the filaments distinct. Staminodes absent. Pistil with 1 locule, the ovary appearing short-stalked. Styles 2, distinct, each with a stigmatic area along the inner surface. Fruits capsules, dehiscing by 4 short teeth. Seeds 40–100 or more, more or less oblong, flattened, the surface with minute papillae or a fine network of ridges, blackish brown, lacking wings or appendages. About 320 species, North America, Europe, Asia, Africa, introduced nearly worldwide.

Each of the *Dianthus* species included here is native to Europe or western Asia and, with the exception of *D. armeria*, will most likely be encountered in Missouri as an escape from cultivation. Carnations have been popular garden plants and cut flowers for many years; however most species do not escape or become established in the wild.

1. Flowers solitary or more commonly 3–20 or more in dense heads or dense clusters, nearly sessile, the stalks 0.1–0.3 cm long; bracts of the epicalyx ³⁄₄–1¹⁄₄ as long as the sepals
   2. Leaf blades linear to narrowly oblanceolate; flowers solitary or more commonly 3–6 in relatively dense clusters; sepals sparsely to densely hairy on the outer surface . . . . 1. D. ARMERIA
   2. Leaf blades lanceolate to ovate; flowers 4–20 or more in a dense head; sepals glabrous on the outer surface but usually hairy along the margins
      . . . . 2. D. BARBATUS
1. Flowers solitary at the stem or branch tips or rarely 2–4 in open clusters; long-stalked, the stalks 0.4–2.5(–3.0) cm long; bracts of the epicalyx ¹⁄₄–¹⁄₂ as long as the sepals
   3. Petals with the expanded portion 4–9 mm, toothed at the tip; bracts of the epicalyx ¹⁄₃–¹⁄₂ as long as the sepals . . . . 3. D. DELTOIDES
   3. Petals with the expanded portion 8–15 mm, deeply fringed at the tip; bracts of the epicalyx ¹⁄₄–¹⁄₃ as long as the sepals . . . . 4. D. PLUMARIUS

1. Dianthus armeria L. **ssp. armeria** (Deptford pink)

Pl. 342 c, d; Map 1460

Plants annual or biennial. Stems (4–)10–88 cm long, erect, unbranched or branched at the base, glabrous or moderately pubescent with curled hairs, mostly at the nodes and toward the tip. Leaf blades 2.0–10.5 cm long, linear to narrowly oblanceolate, angled to a bluntly (most basal leaves) or sharply pointed tip. Flowers solitary or 3–6 in relatively dense clusters, appearing nearly sessile, the stalks 0.1–0.3 cm long. Inflorescence bracts present. Epicalyx of 4 bracts, these ³⁄₄–1 times as long as the sepals, the margins herbaceous and green. Sepals 12–19 mm long, the tube 20–25-veined, sparsely to densely pubescent with minute, more or less appressed hairs; the lobes linear, tapered to a sharply pointed tip. Petals 15–27(–30) mm long, the expanded portion 3–8(–10) mm long, irregularly toothed at the tip, pinkish red to purplish red, spotted with white (rarely entirely white). Fruits 10–16 mm long. Seeds 1.1–1.4 mm long. $2n=30$. May–October.

Introduced, scattered nearly throughout the state (native of Europe, Asia; introduced nearly throughout the U.S. [including Hawaii], Canada, South America). Glades, upland prairies, tops of bluffs, and openings of mesic upland forests; also pastures, old fields, fencerows, railroads, roadsides, and open, disturbed areas.

Unlike the other *Dianthus* species found in Missouri, *D. armeria* is not cultivated very commonly today, although that may have been the original source of its introduction into the wild in the United States. In eastern and southern Europe, plants with some of the bracts of the epicalyx ovate and the calyces dark purple have been called ssp. *armeriastrum* (Wolfner) Velen., but this subspecies apparently has not made it to the New World.

2. Dianthus barbatus L. **ssp. barbatus** (sweet William)

Pl. 342 a, b; Map 1461

Plants perennial, forming clumps. Stems 10–60(–70) cm long, erect, unbranched, glabrous. Leaf blades (1.5–)2.5–10.0 cm long, lanceolate to ovate, angled or tapered to a bluntly (basal leaves) or sharply pointed tip. Flowers 5–20, in dense terminal heads, rarely solitary and axillary in young plants, appearing nearly sessile, the stalks 0.1–0.2

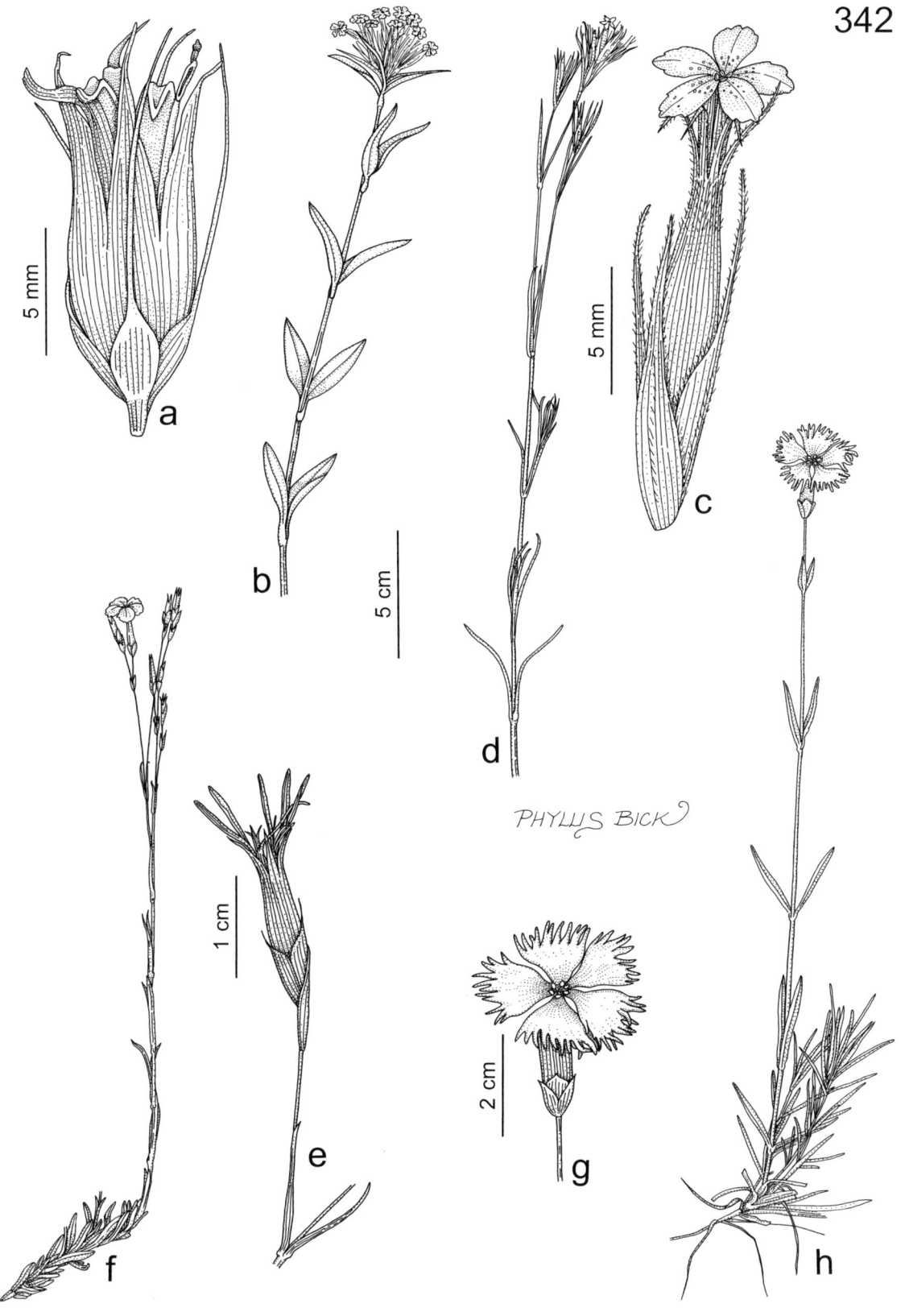

Plate 342. Caryophyllaceae. *Dianthus barbatus*, **a)** fruit, **b)** habit. *Dianthus armeria*, **c)** fruit, **d)** habit. *Dianthus deltoides*, **e)** fruit, **f)** habit. *Dianthus plumarius*, **g)** flower, **h)** habit.

1461. Dianthus barbatus   1462. Dianthus deltoides   1463. Dianthus plumarius

cm long. Inflorescence bracts present. Epicalyx of 4(6) bracts, these ³/₄–1¹/₄ times as long as the sepals, the margins herbaceous and green. Sepals (10–)12–19(–21) mm, the tube about 40-nerved, glabrous, the lobes ovate, tapered to a sharply pointed tip; the margins sometimes hairy. Petals (14–)16–28(–30) mm long, the expanded portion 4–9 mm long, irregularly toothed at the tip, purple, red spotted with white, pink, or white, with (or without) a dark basal portion. Fruits 10–13 mm long. Seeds 2.0–2.5(–2.7) mm long. 2n=30. June–August.

Introduced, uncommon and sporadic (native of Europe, Asia; introduced in the eastern U.S. west to Minnesota and Texas, also Montana to Washington south to California; Canada, Mexico, Central America, South America, Java). Cemeteries, old homesites, and open, disturbed areas.

Small plants from the mountains of central and southern Europe with relatively short bracts and longer-petiolate lower leaves have been called ssp. *compactus* (Kit.) Heuffel. However, these plants are not cultivated and thus have not become established outside of their native range.

3. Dianthus deltoides L. ssp. deltoides
(maiden pink)

Pl. 342 e, f; Map 1462

Plants perennial, appearing clumped or colonial with matted sterile branches or stems. Stems (5–)10–40 cm long, ascending from a spreading basal portion, usually with spreading, sterile basal branches, branched toward the tip, densely pubescent with minute, more or less spreading hairs. Leaf blades 0.6–2.6 cm long, linear to linear-lanceolate (flowering stems) or oblanceolate (sterile branches or stems), angled or tapered to a bluntly (sterile branches or stems) or sharply pointed tip. Flowers solitary at the branch tips or less commonly 2–4 in open clusters, the stalks 0.4–2.5(–3.0) cm long. Inflorescence bracts absent (but the uppermost leaves often appearing as bracts). Epicalyx of 2(4) bracts, these ¹/₃–¹/₂ as long as the sepals, the margins more or less thin and whitened. Sepals 10–15(–17) mm, the tube 25–30-nerved, glabrous or minutely and inconspicuously hairy, the lobes narrowly lanceolate to linear, tapered to a sharply pointed tip. Petals 14–24(–26) mm long, the expanded portion 4–9 mm long, irregularly toothed at the tip, light or dark pink to purple (rarely white), often with a darker band toward the base. Fruits 11–14 mm long. Seeds 1.0–1.2 mm long. 2n=30. May–August.

Introduced, uncommon and sporadic (native of Europe, Asia; introduced in the northeastern U.S. west to Minnesota and Arkansas, also sporadically in the western states; Canada). Roadsides and open, disturbed areas.

Dwarf plants with small, elliptic, glabrous leaves that are endemic to mountains in the western portion of the Balkan Peninsula have been called ssp. *degenii* (Bald.) Strid., but these plants are not cultivated and thus have not become established outside of their native range.

4. Dianthus plumarius L. ssp. plumarius
(cottage pink, garden pink)

Pl. 342 g, h; Map 1463

Plants perennial, forming clumps. Stems 13–40 cm long, erect from an often spreading base, unbranched, glabrous and often glaucous. Leaf blades 2.0–7.5 cm long, linear, tapered to a sharply pointed tip. Flowers solitary or 2–4 in open clusters. Inflorescence bracts absent (but the uppermost leaves often appearing as bracts), the stalks 0.8–2.5 cm long. Epicalyx of usually 4 bracts, these ¹/₄–¹/₃ as long as the sepals, the margins herbaceous and green. Sepals 14–22 mm long, the tube 40–45-nerved, glabrous, the lobes ovate, tapered to a sharply pointed tip. Petals 22–37 mm long, the expanded portion 8–15 mm long, deeply divided into an irregular fringe of slender lobes at the tip, white

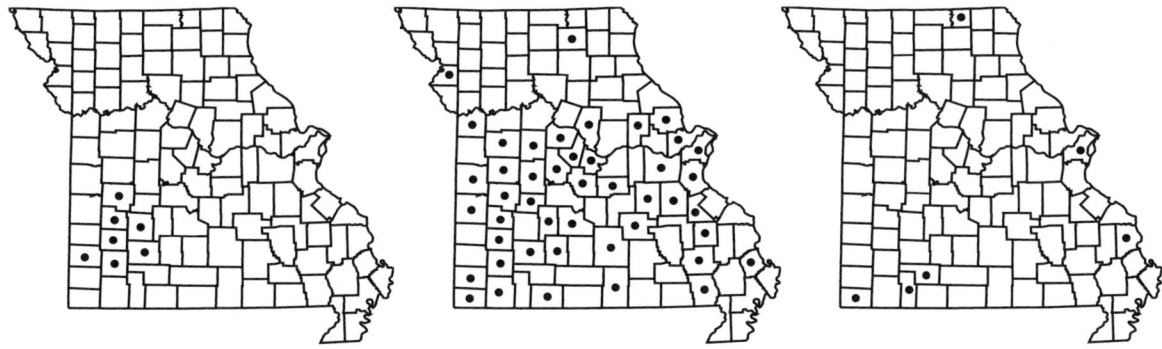

1464. Geocarpon minimum   1465. Holosteum umbellatum   1466. Lychnis coronaria

or pale pink, often with a darker band toward the base. Fruits 2.3–2.7 cm long. Seeds 2.4–3.0 mm long. $2n=30, 60, 90$. May–July.

Introduced, uncommon, known thus far only from Camden County (native of Europe; introduced sporadically in the eastern U.S. west to Wisconsin and Missouri, also California; Canada). Roadsides.

The only Missouri collection thus far was made in 1975 by Scott McReynolds of Louisiana State University and was reported by Rabeler and Thieret (1988). *Dianthus plumarius* is often cultivated and could be expected elsewhere. Some European authors have attempted to recognize infraspecific taxa within this species based on minor variants endemic to portions of the native range of the species (Tutin and Walters, 1993), but the cultivated escapes are all derived from ssp. *plumarius*.

6. Geocarpon Mack.

One species, endemic to Missouri, Arkansas, Louisiana, and Texas.

Although several studies have confirmed Palmer and Steyermark's (1950) conclusion that *Geocarpon* is a member of the Caryophyllaceae (see Nienaber [2005] for a summary), the relationship of *Geocarpon* to other members of the family has yet to be fully elucidated. It should be noted that ontogenetically the flowers are perigynous and that the fused portion of the apparent calyx is a hypanthium with the stamens and staminodes attached along the inner surface. However, unlike *Scleranthus* (see treatment below), in *Geocarpon* the hypanthium tissue is undifferentiated from that of the sepals. To avoid confusing users of the present volume who attempt to identify a specimen of this group, the key to genera of Caryophyllaceae above and the species description below treat the flowers as though they were hypogynous with fused sepals, which is how they would likely be interpreted by all but the most observant biologists.

1. Geocarpon minimum Mack. (geocarpon, tiny Tim)

Pl. 343 a, b; Map 1464

Plants annual. Stems 1–4 cm long, erect or loosely ascending, unbranched or few-branched, often appearing somewhat zigzag, glabrous. Leaves opposite, fused basally into a cuplike sheath, sessile, lacking axillary clusters of leaves. Stipules absent. Leaf blades 0.2–0.4 cm long, narrowly oblong to oblong-triangular, fleshy and often somewhat concave on the upper surface, truncate to slightly expanded toward the base, rounded or angled to a bluntly pointed tip. Flowers appearing solitary or paired at the branch tips and solitary in the leaf axils, the stalks absent, the bracts apparently absent or (in terminal flowers) resembling leaves. Epicalyx absent. Sepals 5, 3.0–3.5(–4.0) mm long, fused into a cup-shaped or slightly flask-shaped tube (see discussion under the generic treatment above), the tube 10-nerved, herbaceous to slightly succulent, green or reddish-tinged between the lobes, the lobes triangular-ovate, shorter than the tube, angled to a sharply pointed tip, often appearing slightly concave at the tip but not noticeably hooded or awned, the narrow margins thin and white. Petals absent. Stamens 5, the filaments distinct, attached to the calyx tube. Staminodes 5, minute and scalelike, alternating with the stamens. Pistil with 1 locule, sessile. Styles actually 3 but apparently absent, each with a stigmatic area along all or most of the inner surface. Fruits capsules, (3.0–)3.5–4.0 mm long, dehiscing apically by 3 minutely notched teeth. Seeds about 30(–60), 0.3–0.5 mm wide, kidney-shaped, the surface with minute papillae or low tubercles, yellowish green to brown, lacking wings or appendages. March–April.

Uncommon in the western portion of the Ozark Division and the adjacent part of the Unglaciated Plains (endemic to Missouri, Arkansas, Louisiana, and Texas). Sandstone glades and sandstone outcrops in upland prairies.

Geocarpon minimum was listed as Threatened in 1987 under the federal Endangered Species Act. The genus and species were first described from a single population in Jasper County (Mackenzie, 1914). Steyermark (1958) discovered the second population more than four decades later in St. Clair County, and others soon followed (Steyermark et al., 1959). It was first reported from Arkansas by Moore (1958), from Louisiana by McInnis et al. (1993), and from eastern Texas by Keith et al. (2004). Most of the populations outside of Missouri are found in a habitat known to land managers as slick spots, which are patches of sparsely vegetated soil rich in magnesium and sodium that are associated with saline prairies. In Missouri, however, the species apparently is restricted to sandstone outcrops of the Channel Sands formation, where plants occur in shallow pockets of the bedrock containing deposits of loose sand grains and with minimal competition from other plants and lichens (Thurman and Hickey, 1989).

7. Holosteum L. (jagged chickweed)
(Shinners, 1965)

Three or 4 species, Europe, Asia, Africa, introduced widely.

1. Holosteum umbellatum L. ssp.
umbellatum (jagged chickweed)
Pl. 343 g, h; Map 1465

Plants annual. Stems 4–35 cm long, erect or ascending, unbranched or branched at the base, glabrous or sparsely to moderately pubescent with short, stalked glands, especially toward the base. Leaves opposite, fused basally into a sheath, short-petiolate (basal rosette leaves) or sessile (stem leaves), lacking axillary clusters of leaves. Stipules absent. Leaf blades 0.2–3.0 cm long, oblanceolate to narrowly oblanceolate (basal leaves) or elliptic to ovate or lanceolate (stem leaves), somewhat thickened and sometimes slightly succulent, tapered to short-tapered at the base, angled or tapered to a sharply pointed tip. Flowers in terminal, long-stalked, umbellate clusters, the stalks 0.5–2.5 cm long, ascending to spreading at flowering, becoming downward-angled from the base after flowering, and becoming erect or ascending again at maturity of the fruit, the bracts whorled, minute, herbaceous with thin, translucent margins to completely papery and white to translucent. Epicalyx absent. Sepals 5, 2.5–4.5 mm long, distinct, lanceolate to ovate, angled to a bluntly or sharply pointed tip, not appearing hooded or awned, the margins thin and white to translucent. Petals 5, conspicuous or not, 3–5 mm long, oblanceolate to spatulate, tapered to a stalklike base, irregularly toothed or shallowly several-lobed at the tip, white or rarely pink, lacking appendages. Stamens 3–5, the filaments distinct. Staminodes absent. Pistil with 1 locule, sessile. Styles 3(–5), each with the stigmatic area subterminal or along the inner surface. Fruits capsules, 4–8 mm long, opening by 6(8 or 10) outward-curled teeth. Seeds 35–60, 0.5–1 mm long, oblong, flattened, the surface appearing pebbled or with broad, low papillae, orange to brown, lacking wings or appendages. $2n=20, 40$. March–May.

Introduced, scattered mostly south of the Missouri River (native of Europe, Asia, Africa; introduced nearly throughout the U.S.; Canada). Glades, sand prairies, and openings of dry upland forests, and ledges of bluffs; also pastures, lawns, ditches, railroads, roadsides, and open, disturbed areas.

Steyermark (1963) noted that at that time *H. umbellatum* was a recent introduction into Missouri, the earliest specimens having been collected in 1950 in Washington County. He reported that it was spreading rapidly. Today, the species probably is present in every county south of the Missouri River, but it has been undercollected because the plants dry and disappear very quickly after the fruits mature, often by the end of May.

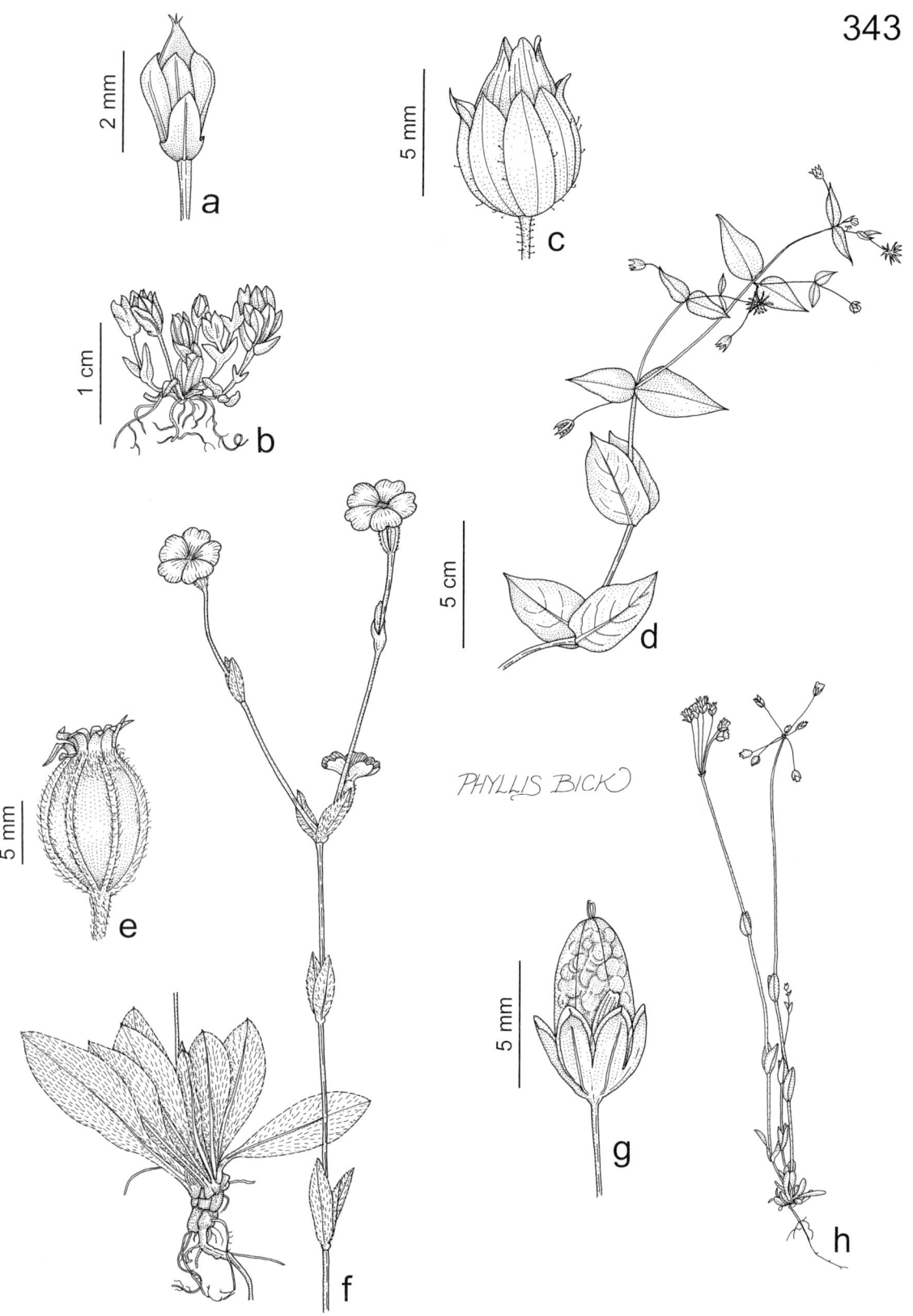

Plate 343. Caryophyllaceae. *Geocarpon minimum*, **a)** fruit, **b)** habit. *Myosoton aquaticum*, **c)** fruit, **d)** habit. *Lychnis coronaria*, **e)** fruit, **f)** habit. *Holosteum umbellatum*, **g)** fruit, **h)** habit.

8. Lychnis L.

About 30 species, Europe, Asia, Africa, introduced widely.

The delimitation of *Lychnis* has remained somewhat controversial, with some botanists classifying all of the species as members of *Silene* (Morton, 2005c) and others circumscribing *Lychnis* in a narrower sense than was done traditionally, with only some of its former members transferred to *Silene* (reviewed by McNeill, 1978). The present treatment follows that of Oxelman et al. (2001), who concluded from morphological and molecular data that some of the subgenera of *Lychnis* form a group distinct from the core lineage of *Silene* and should continue to be recognized as a separate genus. Both of the species treated under the genus *Lychnis* by Steyermark (1963) are included in *Silene* in the present work.

1. Lychnis coronaria (L.) Desr. (mullein-pink, rose-campion)

Pl. 343 e, f; Map 1466

Plants perennial, often with a woody rootstock, densely pubescent throughout with woolly, grayish white hairs. Stems 40–100 cm long, erect or ascending from a sometimes spreading base, unbranched or more commonly branched above the midpoint. Leaves opposite, short-petiolate (basal leaves) or sessile and usually somewhat clasping at the base (stem leaves), lacking axillary clusters of leaves. Stipules absent. Leaf blades 5–10 cm long, oblanceolate to spatulate, not fleshy, rounded to tapered at the base, angled or short-tapered to a sharply pointed tip. Flowers in open terminal panicles, the stalks 5–10 cm long, erect or ascending at flowering and fruiting, the bracts paired and resembling small leaves. Epicalyx absent. Sepals 5, 12–15 mm long, fused basally into a tube 8–12 mm long, the tube 10-nerved, green, and herbaceous between the sepals, the lobes lanceolate to narrowly lanceolate, shorter than the tube, tapered to a sharply pointed tip, not appearing hooded or awned, the margins thin and white. Petals 5, conspicuous and showy, 2.2–3.0 cm long, oblanceolate to broadly spatulate, tapered to a stalklike base, the expanded portion 1.0–1.5 cm long, notched at the tip, deep magenta or rarely white, with a pair of small, scalelike appendages on the upper surface at the base of the expanded portion. Stamens 10, the filaments fused into a short tube toward the base. Staminodes absent. Pistil with 1 locule, the ovary with a short stalk 1–2 mm long. Styles (4)5, each with a stigmatic area along the inner surface. Fruits capsules, 12–15 mm long, dehiscing apically by (4)5 spreading teeth. Seeds 20–30 or more, 1.0–1.5 mm wide, kidney-shaped, the surface tuberculate, grayish black, lacking wings or appendages. $2n=24$. June–September.

Introduced, uncommon and sporadic (native of Europe, Asia; introduced in the eastern U.S. west to Wisconsin and Louisiana, also from Idaho and Utah west to Washington and California; Canada). Old homesites, fencerows, and roadsides.

This species was first reported for the state by Dunn (1982) from a collection made in 1974 in Stone County. *Lychnis coronaria* is a popular garden plant with the striking combination of magenta flowers and grayish white, densely woolly foliage.

9. Minuartia L. (sandwort)

Plants annual or perennial. Stems erect or ascending, sometimes with ascending branches from a spreading, somewhat matted branch system, usually branched, glabrous or sparsely to moderately pubescent with short, stalked glands. Leaves opposite, fused below into a usually cuplike sheath, sessile, with axillary clusters of leaves sometimes present. Stipules absent. Leaf blades narrowly linear to narrowly lanceolate and stiff or linear to oblanceolate and soft, not fleshy, tapered at the base, angled or tapered to a bluntly or sharply pointed tip, this sometimes with a short, fine, spinelike appendage, the surfaces glabrous. Flowers in terminal, open or dense panicles, the stalks erect to arched or spreading, the bracts paired and resembling small leaves, herbaceous or with thin, white to translucent margins. Epicalyx absent. Sepals 5, fused at the base, ovate to broadly lanceolate, green or sometimes purplish-tinged at the tip, narrowly angled or tapered to a sharply pointed tip, not appearing hooded or awned, the margins thin and white. Petals 5, obovate to oblong-obovate, tapered but not to a stalklike base,

1467. Minuartia michauxii   1468. Minuartia muscorum   1469. Minuartia patula

rounded to more or less truncate or notched at the tip, white, lacking appendages. Stamens 10, the filaments distinct, attached to the calyx tube. Staminodes absent. Pistil with 1 locule, sessile. Styles 3, distinct, each with a stigmatic area along the inner surface. Fruits capsules, dehiscing apically by 3 incurved to recurved valves or teeth. Seeds 1–25, nearly globose, the surface tuberculate or with papillae, sometimes appearing pebbled, brown or black, lacking wings or appendages. About 175 species, North America, Europe, Asia.

Steyermark (1963) and many earlier North American authors included the species of *Minuartia* in a broadly circumscribed concept of *Arenaria*. McNeill (1962, 1980) presented compelling morphological evidence to support the recognition of two genera and published many of the new combinations necessary to transfer species into *Minuartia*.

1. Plants perennial; leaf blades stiff and leathery, often spinelike at the tip; axillary clusters of leaves present at most nodes; flower stalks glabrous . . . . . 1. M. MICHAUXII
1. Plants annual; leaf blades soft and herbaceous, soft at the tip; axillary clusters of leaves absent; flower stalks glabrous or more commonly sparsely to moderately pubescent with stalked glands
    2. Sepals 3-nerved, angled to a sharply pointed tip; leaves narrowly lanceolate to narrowly oblanceolate, (0.6–)1.5–3.2 mm wide; seeds 0.6–0.8 mm wide, the surface appearing pebbled or with short papillae, black . . . . 2. M. MUSCORUM
    2. Sepals 5-nerved or 5-ribbed, angled or tapered to a sharply pointed tip; leaves narrowly linear to linear, 0.5–1.5(–1.8) mm wide; seeds 0.4–0.6 mm wide, the surface tuberculate, reddish brown to black . . . . 3. M. PATULA

1. Minuartia michauxii (Fenzl) Farw. (rock sandwort, Michaux's stitchwort)
Arenaria stricta Michx.
A. stricta ssp. *texana* (B.L. Rob.) Maguire
A. stricta var. *texana* (Britton) B.L. Rob.
Sabulina stricta (Michx.) Small

Pl. 339 i, j; Map 1467

Plants perennial, often with a somewhat woody, branched rootstock. Stems 8–40 cm long, often with erect or ascending branches from a spreading, somewhat matted branch system, occasionally the ascending stems originating directly from the rootstock, the ascending portions unbranched or less commonly few-branched, glabrous. Leaves dense and overlapping below the stem midpoint, sparse and widely spaced above, with axillary clusters of leaves at most nodes. Leaf blades 0.8–3.0 cm long, 0.5–1.8 mm wide, narrowly linear to narrowly lanceolate, stiff and leathery, angled or tapered to a bluntly or sharply pointed tip, often with a fine, spinelike appendage at the very tip, green to grayish green. Flowers in open or dense panicles, the stalks 0.3–6.0 cm long, glabrous, the bracts with thin, white margins. Sepals 3–6 mm long (not becoming elongated at fruiting), the lobes lanceolate to ovate, angled or more commonly tapered to a sharply pointed tip, the surface with 3 prominent nerves or ridges. Petals 4–9 mm long, rounded to more or less truncate at the tip. Fruits 3–4 mm long, usually shorter than the calyx. Seeds 0.8–0.9 mm wide, the surface tuberculate, black, not shiny. $2n=30, 44$. May–July.

Scattered in the Ozark and Ozark Border Divisions, north locally to Monroe County (eastern U.S. west to South Dakota and New Mexico). Glades and ledges and tops of bluffs, usually on limestone and dolomite substrates.

Plants with shorter, stiffer stems and somewhat shorter leaves that are crowded into the lower third of the stem have been called *Arenaria stricta* ssp. *texana* (or var. *texana*) and usually are found in the southern part of the species range. Steyermark (1963) studied morphological variation among populations in Missouri and concluded that although most of the plants corresponded with ssp. *texana*, both morphological extremes were present in the state. He thought that the differences were not sharply expressed in Missouri plants and suggested that the morphological variation might be habitat-related, with plants in open, dry glades having shorter, more crowded leaves and plants in semishaded bluff habitats having longer, less-dense leaves. The situation deserves further study.

This attractive species should be investigated for its ornamental potential in rock gardens.

2. Minuartia muscorum (Fassett) Rabeler
   (Dixie stitchwort)
   Arenaria muriculata Maguire
   A. patula Michx. f. *robusta* Steyerm.
   A. patula var. *robusta* (Steyerm.) Maguire
   M. patula (Michx.) Mattf. var. *robusta* (Steyerm.) McNeill
   M. muriculata (Maguire) McNeill

Map 1468

Plants annual. Stems 10–55 cm long, erect or ascending, not matted, unbranched or branched, glabrous or sparsely pubescent with short, stalked glands toward the tip. Leaves widely spaced and not overlapping, lacking axillary clusters of leaves. Leaf blades (0.5–)1.0–3.5(–5.0) cm long, (0.6–)1.5–3.2 mm wide, narrowly lanceolate to narrowly oblanceolate, soft and herbaceous, angled to a sharply pointed but not spinelike tip, green or occasionally somewhat purplish-tinged. Flowers in open panicles, the stalks 0.6–5.5 cm long, sparsely to moderately pubescent with stalked glands, the bracts with green margins. Sepals 3–4 mm long (becoming elongated to 5 mm at fruiting), the lobes lanceolate, narrowly angled to a sharply pointed tip, the surface with 3 prominent nerves or ribs. Petals 5–8 mm long, broadly notched at the tip. Fruits 5–7 mm long, longer than the calyx. Seeds 0.6–0.8 mm wide, the surface appearing pebbled or with short papillae, black, shiny. May–July.

Uncommon, known thus far only from Butler County and historical specimens from Dunklin and St. Louis Counties (Alabama to Texas north to Tennessee and Kansas). Sandy openings of mesic upland forests; also open, disturbed areas.

From the synonymy shown above, it is clear that the relationship between *M. muscorum* and *M. patula* has been expressed at various levels in the past seventy years. Rabeler (1992) considered the two taxa best treated as separate species, separated by characters in the key above.

3. Minuartia patula (Michx.) Mattf. (sandwort, Pitcher's stitchwort)
   Arenaria patula Michx.
   A. patula f. *pitcheri* (Nutt. ex Torr. & A. Gray) Steyerm.
   A. patula f. *media* Steyerm.
   Alsinopsis patula (Michx.) Small

Pl. 339 e, f; Map 1469

Plants annual. Stems 5–30 cm long, erect or ascending, not matted, unbranched or branched, glabrous or sometimes sparsely pubescent with short, stalked glands. Leaves widely spaced and not overlapping or the lowermost nodes sometimes short near the stem base, lacking axillary clusters of leaves. Leaf blades 0.2–2.0 cm long, 0.5–1.5(–1.8) mm wide, narrowly linear to linear, soft and herbaceous, angled to a bluntly or sharply pointed but not spinelike tip, green or purplish-tinged. Flowers in open panicles; the stalks 0.3–3.0 cm long, glabrous or sparsely pubescent with short, stalked glands, the bracts with green margins. Sepals 4.0–5.5 mm long (not becoming elongated at fruiting), the lobes narrowly to broadly lanceolate, narrowly angled or tapered at the tip, the surface with 5 prominent nerves or ridges. Petals 5–9 mm long, broadly notched at the tip. Fruits 3.0–4.2 mm long, shorter than the calyx. Seeds 0.4–0.6 mm wide, the surface tuberculate, reddish brown to black, not shiny. March–May.

Scattered, mostly south of the Missouri River (eastern U.S. west to Wisconsin, Kansas, and Texas). Glades, ledges and tops of bluffs, sandy or rocky openings of mesic to dry upland forests, and upland prairies; also roadsides.

Steyermark (1941, 1963) treated three forms of this species based mainly on differences in the presence and density of pubescence on the calyces and flower stalks. Further collecting throughout the species range has resulted in recognition that variation for this character forms a continuum.

This is a characteristic species of limestone and dolomite glades in Missouri, although it also occurs on other substrates. With its slender leaves and stems almost invisible against the background substrate, the flowers sometimes appear suspended in open space when in dense patches and viewed from above.

1470. Moehringia lateriflora   1471. Myosoton aquaticum   1472. Paronychia canadensis

10. Moehringia L.

Twenty-five species, North America, Europe, Asia.

McNeill (1962) segregated *Moehringia* from *Arenaria* based on differences in chromosome base number (x=12 vs. x=10 or 11) and the white, spongy appendage on the seed.

1. Moehringia lateriflora (L.) Fenzl (grove sandwort)
Arenaria lateriflora L.

Pl. 339 g, h; Map 1470

Plants perennial, with slender rhizomes. Stems 5–30 cm long, ascending, often from a spreading base, usually branched, moderately to densely pubescent with short, downward-curved, nonglandular hairs. Leaves opposite, not fused basally, sessile or nearly so, with axillary clusters of leaves rarely present. Stipules absent. Leaf blades 0.6–3.0 cm long, broadly elliptic to oblong-elliptic or oblanceolate, not fleshy, tapered at the base, rounded or angled to a bluntly pointed tip, the margins and undersurface midvein moderately hairy. Flowers in open terminal clusters or solitary at the branch tips, the stalks 0.3–3.0 cm long, erect, ascending or somewhat curved in flower and fruit, the bracts paired and resembling small leaves, with thin, white margins. Epicalyx absent. Sepals 5, 1.7–2.8 mm long, fused at the base or appearing distinct, ovate to obovate, rounded or angled to a bluntly pointed tip, not appearing hooded or awned, the margins thin and white. Petals 5, conspicuous and somewhat showy, 3–6 mm long (ca. 2 times as long as the sepals), oblanceolate to narrowly oblong-obovate, tapered but not to a stalklike base, rounded at the tip, white, lacking appendages. Stamens 10, the filaments distinct, attached along the margin of a small nectar disc. Staminodes absent. Pistil with 1 locule, sessile. Styles 3, each with a stigmatic area along the inner surface. Fruits capsules, 3–5 mm long, dehiscing apically by 6 recurved teeth. Seeds 2–6, 0.9–1.2 mm wide, more or less kidney-shaped, the surface smooth, reddish brown to black, shiny, not winged, with a white, spongy, elliptic appendage. $2n=48$. May–August.

Scattered in the northeastern corner of the state (northern U.S. [including Alaska] south to Nevada, New Mexico, Missouri, and Virginia; Canada, Europe, Asia). Mesic upland forests and occasionally bottomland forests.

11. Myosoton Moench (giant chickweed)

One species, Europe, Asia, introduced in North America.

Although sometimes included in the genus *Stellaria,* the combination of five (sometimes six) styles and capsule valves with the valves notched at the tip is not found in other species of that genus. The chromosome number of $2n=28$ is very rarely encountered in *Stellaria*.

1. Myosoton aquaticum (L.) Moench (giant chickweed)

Stellaria aquatica (L.) Scop.

Pl. 343 c, d; Map 1471

Plants perennial, with slender rhizomes, the stems also frequently rooting at the lower nodes. Stems 10–100 cm long, sprawling or ascending from a spreading base, unbranched or more commonly branched, moderately to densely pubescent with minute, stalked glands, often glabrous or nearly so toward the base. Leaves opposite, fused basally into a sheath, long-petiolate (basal leaves) or sessile (stem leaves), the uppermost leaves somewhat clasping, lacking axillary clusters of leaves. Stipules absent. Leaf blades 2.0–3.5(–8.5) cm long, ovate to broadly elliptic, not fleshy, angled to rounded or shallowly cordate at the base, angled or tapered to a sharply pointed tip, glabrous or the undersurface glandular-hairy along the midvein. Flowers in axillary and/or terminal clusters, 1–2(–3) cm long, erect or ascending at flowering, downward-angled from the base in fruit, the bracts paired and resembling small leaves. Epicalyx absent. Sepals 5, 4–6 mm long (becoming elongated to 9 mm at fruiting), distinct, ovate, angled or tapered to a bluntly or more commonly sharply pointed tip, not appearing hooded or awned, the margins herbaceous and green or thin and white, glandular-pubescent. Petals 5 (but usually appearing as 10), 4–7 mm long, more or less obovate, angled or tapered but not to a stalklike base, the tips deeply 2-lobed (nearly to the base), white (rarely pink elsewhere), lacking appendages. Stamens (5–)10, the filaments distinct, attached along the margin of a small nectar disc. Staminodes absent or rarely 1–5 and linear. Pistil with 1 locule, sessile. Styles 5(6), each with stigmatic area along the inner surface. Fruits capsules, 5–10 mm long, opening by 5 (rarely 6) somewhat recurved valves or teeth, these narrowly notched at the tip. Seeds 50–100, 0.8–1.0 mm wide, kidney-shaped, the surface with papillae, dark brown, lacking wings or appendages. $2n=28$. May–September.

Introduced, known thus far from a single collection from Clay County (native of Europe and Asia; introduced sporadically in the northeastern U.S. west to Minnesota and Kansas). Bottomland forests.

The only collection from the state thus far was made in 1985 along the Missouri River by Jay Raveill and was reported by Rabeler and Thieret (1988).

12. Paronychia Mill. (nailwort, whitlow wort)

Plants annual or perennial (biennial elsewhere). Stems erect or loosely ascending from a spreading base, often appearing many-branched, glabrous or pubescent with nonglandular hairs. Leaves opposite, fused basally with the accompanying stipules, short-petiolate (some lower leaves) or more commonly all sessile, lacking axillary clusters of leaves. Stipules 2 per node, thin and papery, white or silvery, narrowly triangular to lanceolate (to ovate elsewhere), tapered at the tip and entire (sharply pointed), fringed, or deeply notched. Leaf blades narrowly linear to linear or oblanceolate to obovate or elliptic, thin and herbaceous or somewhat leathery but not fleshy, tapered at the base, angled or tapered to a bluntly or sharply pointed tip, sometimes rounded or abruptly tapered to a minute, sharp point or with a minute, awnlike appendage. Flowers in open or dense terminal clusters or panicles, but sometimes appearing solitary and axillary, the stalks absent or minute and erect or ascending at flowering and fruiting, the bracts paired (absent in axillary flowers), conspicuous, resembling the leaves and/or stipules. Epicalyx absent. Sepals 5 (3 or 4 elsewhere), fused at the base, the lobes ovate to lanceolate, oblong-lanceolate, or narrowly oblong, rounded or angled to a bluntly pointed tip, appearing hooded, with a dorsal extension behind the tip into either a stiff awn or a minute, conical to triangular, blunt to sharp point, green or yellowish-tinged to orange or brown, the margins thin or papery and white or translucent. Petals absent. Stamens 5, the filaments distinct or fused at the base. Staminodes absent or 5 and narrowly linear to hairlike. Pistil with 1 locule, sessile. Style(s) 1 and 2-lobed toward the tip or 2 and distinct, with the stigmatic area subterminal or along the inner surface. Fruits achenelike, 1-seeded and indehiscent. Seeds 0.7–1.1 mm wide, ellipsoid to nearly globose, the surface smooth and shiny, brown or black, lacking wings or appendages. About 110 species, North America, Europe, Asia, Africa.

1473. Paronychia fastigiata   1474. Paronychia virginica   1475. Petrorhagia prolifera

1. Plants perennial, forming dense clumps, the stems usually several, ascending from a usually spreading basal portion; leaf blades narrowly linear to linear; sepals with a short, awnlike dorsal appendage from just behind the tip, this 0.4–1.1 mm long .... 3. P. VIRGINICA
1. Plants annual, the stem usually solitary, erect toward the base; leaf blades narrowly oblanceolate to obovate or elliptic; sepals with a minute, conical to triangular dorsal appendage from just behind the tip, this 0.05–0.3 mm long
   2. Stems glabrous; flowers appearing solitary and axillary; sepals prominently 3-nerved .... 1. P. CANADENSIS
   2. Stems minutely hairy, mostly on one side; flowers in dense terminal clusters; sepals nerveless .... 2. P. FASTIGIATA

1. Paronychia canadensis (L.) A.W. Wood (smooth nailwort, forked chickweed)

Anychia canadensis (L.) Ell.

Pl. 344 j, k; Map 1472

Plants annual. Stem solitary, 3–40 cm long, erect, many-branched, glabrous. Stipules 0.5–4.0 mm long, triangular to lanceolate, long-tapered to a sharply pointed, entire tip. Leaves thin and herbaceous. Leaf blades 0.3–3.0 cm long, elliptic to obovate (occasionally narrowly so), tapered at the base, angled to a bluntly pointed tip to rounded or abruptly tapered to a minute, sharp point. Flowers appearing solitary in the leaf axils (but actually in open terminal clusters with leaflike bracts). Sepals 0.5–1.0 mm long, ovate to oblong-lanceolate, usually rounded at the tip, appearing hooded, with a dorsal extension behind the tip into a minute, conical to triangular, blunt point, green to brown, prominently 3-nerved, the margins thin and white or translucent. Staminodes absent. Styles 2, distinct. Fruits 0.5–1.0 mm long, the surface appearing minutely roughened or pebbled toward the tip, otherwise smooth. Seed 0.6–0.8 mm wide, black. May–October.

Scattered nearly throughout the state but apparently absent from the Mississippi Lowlands Division and uncommon in portions of the Glaciated Plains (eastern U.S. west to Minnesota, Nebraska, Oklahoma, and Louisiana; Canada). Bottomland forests, mesic to dry upland forests, ledges and tops of bluffs, and glades.

2. Paronychia fastigiata (Raf.) Fernald **var. fastigiata** (hairy nailwort, forked chickweed)

Anychia fastigiata Raf.
P. fastigiata var. *paleacea* Fernald

Pl. 344 g–i; Map 1473

Plants annual. Stems 4–30 cm long, erect, many-branched, sparsely to moderately pubescent with minute, spreading to downward-curved hairs, these mostly on one side of the stem. Stipules 0.5–4.5 mm long, narrowly triangular to lanceolate, long-tapered to a sharply pointed, entire or sometimes fringed tip. Leaf blades 0.2–2.5 cm long, narrowly oblanceolate to obovate or elliptic, tapered at the base, angled to a bluntly pointed tip to rounded or abruptly tapered to a minute, sharp point. Flowers in open to dense terminal clusters (occasionally appearing solitary and axillary). Sepals 0.8–1.2 mm long, narrowly oblong, usually rounded at the tip, appearing hooded, with a dorsal extension behind the tip into a minute, conical to triangular, blunt point, green to brown, nerveless, the margins thin and white or translucent. Staminodes absent. Styles 2, distinct. Fruits 0.7–1.0 mm long, the surface with minute papillae. Seed 0.7–0.9 mm wide, brown. $2n=32, 36$. May–October.

Scattered nearly throughout the state (eastern U.S. west to Minnesota, Kansas, Oklahoma, and Texas; Canada). Bottomland forests, mesic to dry upland forests, bottomland and upland prairies, banks of streams and rivers, ledges and tops of bluffs, and glades; also railroads and roadsides.

Some botanists have separated *P. fastigiata* into as many as four varieties differing in details of the bracts, sepals, and styles. Steyermark (1963) accepted two of these as more or less co-occurring in Missouri: var. *paleacea* (with stipular bracts as long as or longer than the sepals) and var. *fastigiata* (with somewhat shorter bracts). Examination of Missouri specimens reveals that too many intermediates exist to justify the segregation of var. *paleacea*. However, the validity of the other named variants, which occur to the east of Missouri, requires further study.

3. Paronychia virginica Spreng. (Virginia whitlow wort, Appalachian whitlow wort, yellow nailwort)

P. virginica var. *scoparia* (Small) Cory

Pl. 344 a, b; Map 1474

Plants perennial, forming dense clumps from a branched, somewhat woody rootstock. Stems usually several, 7–45 cm (flowering) or 3–10 cm (vegetative) long, ascending from a spreading base, unbranched or branched at the base, glabrous or sparsely to moderately pubescent with minute, spreading to loosely ascending hairs. Stipules 6–13 mm long, narrowly lanceolate, long-tapered to a sharply pointed tip, this often deeply and narrowly notched. Leaves stiff and somewhat leathery. Leaf blades 1.0–3.0 cm long, narrowly linear to linear, inconspicuously short-tapered at the base, angled or tapered to a sharply pointed tip, this often with a minute, spinelike appendage. Flowers in open to dense terminal clusters or panicles. Sepals 2.0–2.9 mm long, lanceolate to oblong-lanceolate, yellow to orange or brown, prominently 3-nerved, the margins papery and white, with a short, curved, awnlike dorsal appendage from just behind the tip, this 0.4–1.1 mm long. Staminodes 5, narrowly linear to hairlike. Style 1, 2-lobed toward the tip. Fruits 1.8–2.0 mm long, the surface uniformly smooth. Seed 1.0–1.2 mm wide, black. July–August.

Uncommon, known thus far only from McDonald County (Maryland to North Carolina, Tennessee, and Alabama, also Missouri, to Arkansas, Oklahoma, and Texas). Ledges and faces of limestone bluffs.

This distinctive species was first collected in the state in 1988 by Nina Bicknese (then with the Missouri Department of Conservation). The Missouri populations are about 180 km north of the closest Arkansas population and were first reported by Yatskievych and Summers (1993). They are part of a disjunct set of western populations that some botanists have called var. *scoparia* because of their more ascending inflorescence branches, but this character is too variable to allow segregation of varieties (Core, 1941).

The inflorescences of *P. virginica* are yellow at flowering and gradually turn orange as the fruits mature and brown with age. These colors provide a striking contrast to the bright green, dense, needlelike leaves. The horticultural potential of this species merits investigation.

13. Petrorhagia (Sér.) Link
(Rabeler, 1985)

Thirty-three species, Europe, Asia, Africa; introduced in North America, South America, Pacific Islands, and Australia.

Steyermark (1963) followed many earlier North American botanists in including *Petrorhagia* in a broadly circumscribed concept of *Dianthus*. Rabeler (1985) reviewed the classification of the group as followed here.

1. Petrorhagia prolifera (L.) P.W. Ball & Heyw. (childing pink)

Dianthus prolifer L.
Tunica prolifera (L.) Scop.

Pl. 344 e, f; Map 1475

Plants annual. Stems (6–)20–30(–60) cm long, erect, unbranched or branched toward the tip, glabrous or the median portion slightly roughened. Leaves opposite, fused basally into a sheath, sessile, lacking axillary clusters of leaves. Stipules absent. Leaf blades 1–3 cm long, linear to linear-lanceolate, not fleshy, tapered at the base, angled or tapered to a sharply pointed tip. Flowers in dense, headlike clusters or solitary, the stalks minute (0.5–1.5 mm

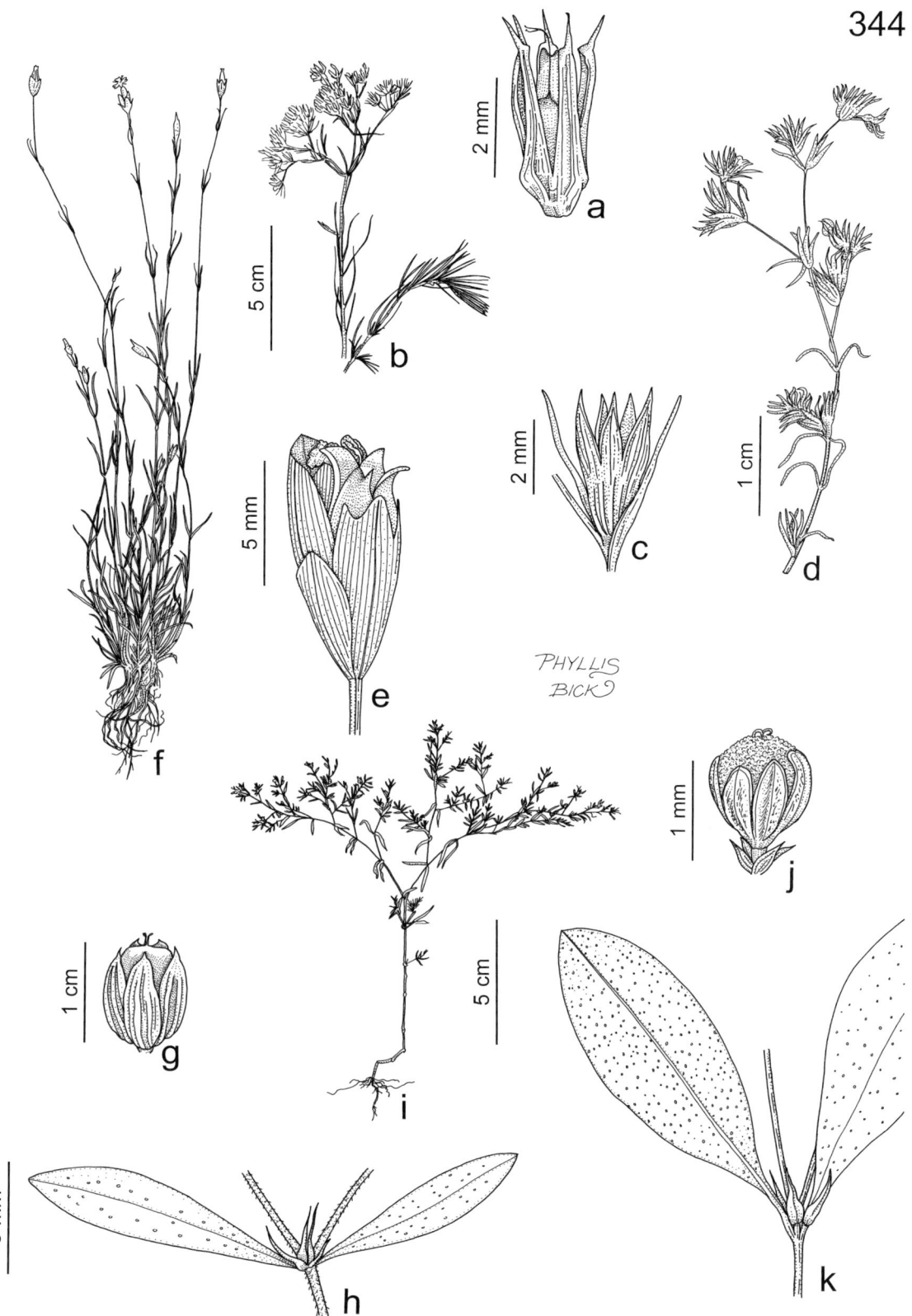

Plate 344. Caryophyllaceae. *Paronychia virginica*, **a)** fruit, **b)** fertile and vegetative branches. *Scleranthus annuus*, **c)** fruit, **d)** habit. *Petrorhagia prolifera*, **e)** fruit, **f)** habit. *Paronychia fastigiata*, **g)** fruit, **h)** node with leaves, **i)** habit. *Paronychia canadensis*, **j)** fruit, **k)** node with leaves.

long), erect at flowering and fruiting, the bracts overlapping, papery and brown, broadly ovate, more or less enclosing the flowers. Epicalyx of 1–3 pairs of bracts closely subtending the sepals, these usually appearing whorled, similar to the inflorescence bracts in size and texture. Sepals 5, (7–)10–12 mm long, fused basally into a tube, the tube 15-nerved, papery and white or translucent between the green to purplish-tinged sepals, the lobes oblong-lanceolate to narrowly lanceolate, shorter than the tube, tapered to a sharply pointed tip, not appearing hooded or awned, the margins herbaceous and green. Petals 5, conspicuous and more or less showy, 12–15 mm long, narrowly obovate to spatulate, tapered to a long, stalklike base, the expanded portion 2–4 mm long, truncate or shallowly and broadly notched at the tip, pink or purplish pink, rarely pale pink to white, lacking appendages. Stamens 10, the filaments distinct. Staminodes absent. Pistil with 1 locule, the ovary minutely stalked at the base. Styles 2, each with a stigmatic area along the inner surface. Fruits capsules, 6–8 mm long, dehiscing apically by 4 straight or slightly recurved teeth. Seeds 8–15, 1.1–1.6(–1.8) mm long, oblong, usually somewhat concave on one side, the surface with a usually fine network of ridges enclosing shallow, laterally elongate pits, black, lacking wings or appendages. $2n=30$. May–October.

Introduced, uncommon and sporadic in the southern portion of the Ozark Division and disjunctly in Pike County (native of Europe, Asia, Africa; introduced sporadically in the eastern U.S. west to Michigan, Arkansas, and Oklahoma, also Idaho; Canada). Glades, tops of bluffs, and banks of streams and rivers; also pastures, old fields, roadsides, and open, disturbed areas.

This species was first collected in 1956 in Stone County by Julian Steyermark.

14. Sagina L. (pearlwort)
(Crow, 1978)

Plants annual or perennial. Stems erect or ascending to sprawling or spreading, usually branched, sometimes matted, glabrous or sparsely to moderately pubescent with stalked glands. Leaves opposite, fused basally, into an often cup-shaped, thin sheath, sessile, sometimes with axillary clusters of leaves at some nodes. Stipules absent. Leaf blades linear to narrowly triangular, somewhat stiff and leathery or sometimes softer and slightly succulent, tapered at the base, angled or short-tapered to an abrupt, minute, sharp point. Flowers in terminal or axillary open clusters or appearing solitary, the stalks erect to spreading at flowering, at fruiting sometimes hooked near the tip, the bracts of terminal inflorescences paired and resembling small leaves. Epicalyx absent. Sepals 4 or 5, distinct, ovate or elliptic to nearly circular, green or purplish-tinged, angled to a bluntly or sharply pointed tip, not appearing hooded or awned, the margins thin and translucent, white, or more commonly purplish-tinged, the surfaces usually glabrous. Petals absent or (1–)4 or 5 and inconspicuous, elliptic to oblong-obovate, angled or tapered but not to a stalklike base, the tips entire (slightly notched elsewhere), white, lacking appendages. Stamens 4, 5(8), or 10, the filaments distinct. Staminodes absent. Pistil with 1 locule, sessile. Styles 4 or 5, distinct, each with the stigmatic area subterminal or along the inner surface. Fruits capsules, dehiscing apically to below the midpoint by 4 or 5 valves. Seeds numerous (more than 100), obliquely triangular with a distinct groove along the dorsal margin (elsewhere sometimes plump and lacking the groove), the surface smooth, pebbled, or tuberculate, light tan to brown, lacking wings or appendages. About 20 species, North America, Central America, South America, Europe, Asia, Africa; most diverse in north-temperate regions.

In addition to the two species treated here, *S. japonica* (Sw.) Ohwi (Japanese pearlwort) has been spreading westward since the first collections in eastern North America were made in the 1940s, and eventually it may be found in Missouri. It was collected in Ohio in 1987 (Rabeler, 1996), in Ontario, Canada, in 1995 (Rabeler, 1996), and in Illinois originally in 1951 and more recently in 1997 and 1999 (Tucker, 2000). *Sagina japonica* would key closest to *S. decumbens*, but it is distinguished by its seeds, which are plump, obovoid to globose, lack a distinct groove, and have a dark brown surface that is either tuberculate or slightly pebbled.

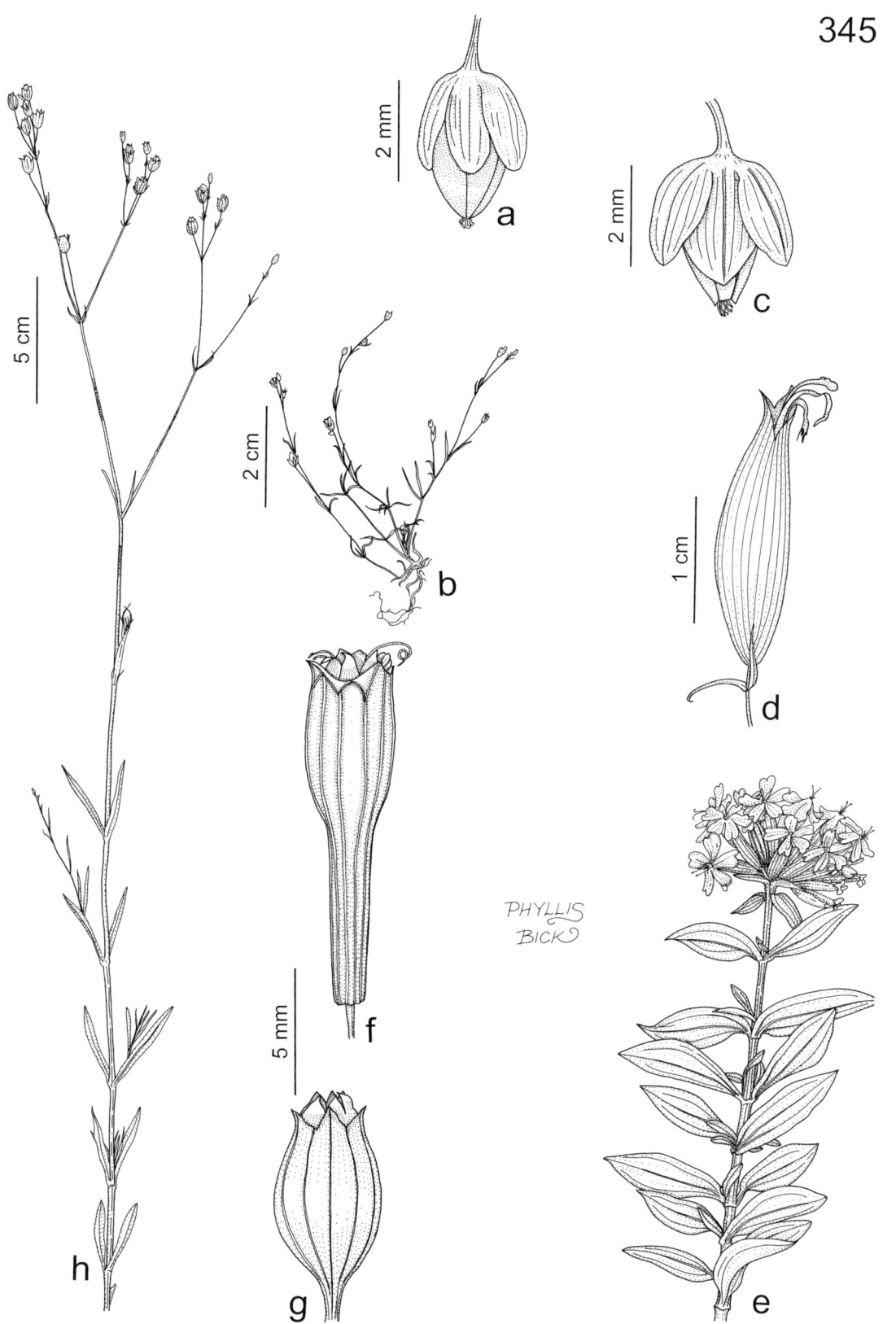

Plate 345. Caryophyllaceae. *Sagina decumbens*, **a)** fruit, **b)** habit. *Sagina procumbens*, **c)** fruit. *Saponaria officinalis*, **d)** fruit, **e)** habit. *Atocion armeria*, **f)** fruit. *Silene antirrhina*, **g)** fruit, **h)** habit.

1476. Sagina decumbens   1477. Sagina procumbens   1478. Saponaria officinalis

1. Plants annual, stems ascending to sprawling; flowers usually with a 5-parted perianth, rarely with 4 sepals and lacking petals, the stalks and sepal bases often glandular-hairy; sepals appressed to the mature fruit, the tips and/or margins often purplish-tinged (rarely white or translucent) .... 1. S. DECUMBENS
1. Plants perennial, stems often spreading; flowers mostly with a 4-parted perianth, the stalks and sepals glabrous; sepals diverging from mature capsules, the tips green and the margins white or less commonly translucent ...... 2. S. PROCUMBENS

1. Sagina decumbens (Elliott) Torr. & A. Gray
   ssp. decumbens (trailing pearlwort)
   Pl. 345 a, b; Map 1476

Plants annual. Stems 3–10(–20) cm long, ascending to sprawling but usually not fully spreading, not rooting at the lower nodes, glabrous or sparsely to moderately glandular-hairy, often purplish-tinged. Axillary clusters of leaves absent; basal rosettes present in young plants. Leaf blades 0.1–2.3 cm long. Flower stalks 0.5–1.0(–1.9) cm long, glabrous or glandular-hairy, erect to spreading at fruiting, at fruiting remaining relatively straight. Sepals (4)5, (1.0–)1.5–2.0(–3.0) mm long, ovate or elliptic, often sparsely glandular-hairy toward the base, the tip and/or margins often purplish-tinged (rarely white or translucent). Petals 5, (1.0–)1.5–2.0(–2.3) mm long, rarely absent in flowers with 4 sepals. Stamens 5(8) or 10. Fruits 2.0–3.0(–3.5) mm long, the sepals appressed to the mature fruits. Seeds (0.2–)0.3–0.4 mm wide, the surface smooth, pebbled, or tuberculate. 2n=36. March–August.

Scattered mostly south of the Missouri River (eastern U.S. west to Kansas and Texas; Canada). Glades, rock outcrops in upland prairies, openings of mesic to dry upland forests; also margins of crop fields, fallow fields, pastures, roadsides, cracks in pavement, and open, disturbed areas.

Sagina decumbens is very variable morphologically. The phase with flowers having mostly four sepals and no petals has been called var. *smithii* (A. Gray) S. Watson, but most botanists currently consider it a minor variant unworthy of formal taxonomic recognition. Along the West Coast, disjunct populations with more ovate, somewhat darker, generally smoother seeds and broader sepals and fruits are known as ssp. *occidentalis* (S. Watson) G.E. Crow.

2. Sagina procumbens L. (matted pearlwort)
   Pl. 345 c; Map 1477

Plants perennial, the stems often rooting at the lower nodes. Stems 3–5(–10) cm long, mostly spreading and matted, glabrous, green. Axillary clusters of leaves often present at some nodes; basal rosettes present in young plants. Leaf blades 0.2–1.7 cm long. Flower stalks 0.5–1.0(–1.4) cm long, glabrous, often hooked near the tip at fruiting. Sepals 4(5), 1.5–2.5 mm long, elliptic to nearly circular, glabrous throughout, the tip green, the margins white or translucent. Petals (1–)4(–5), 0.8–1.0(–1.5) mm long, occasionally absent. Stamens 4 or 5(8). Fruits (1.5–)2.0–2.5(–3.0) mm long, the sepals somewhat arched outward from the mature fruits. Seeds (0.3–)0.4(–0.5) mm wide, the surface smooth or pebbled. 2n=22. June.

Introduced, uncommon, known thus far with certainty only from a single historical collection from the city of St. Louis (native of Europe, Asia; introduced in the northeastern U.S west to Minnesota and Arkansas, and in the western U.S. east to Montana and Colorado; Canada). Habitat unknown but presumably roadsides or open, disturbed areas.

In his revision of *Sagina* in North America, Crow (1978) mentioned Missouri in the range of *S. procumbens* based on a historical specimen in the herbarium of the New York Botanical Garden lacking county name or other more detailed locality data or even the collector's name, and with only the month of June for a collection date. However, a second specimen (also at the New York Botanical Garden) subsequently was discovered during the present study, collected by Henry Eggert on 3 June 1875 in the city of St. Louis. The two sheets possibly are duplicates of the same collection, but the data are insufficient to verify this. Turner and Yatskievych (1992) cited a specimen from Ripley County collected in 1989, but this collection represents misdetermined material of *S. decumbens*. It seems likely that *S. procumbens* eventually will be rediscovered in Missouri, as it is often found in sidewalk cracks, lawn edges, and other highly disturbed habitats where inconspicuous plants frequently are overlooked by botanists.

15. Saponaria L. (soapwort)

About 40 species, Europe, Asia, Africa, introduced nearly worldwide.

1. Saponaria officinalis L. (bouncing-bet, soapwort)

Pl. 345 d, e; Map 1478

Plants perennial, colonial from long-creeping, branched rhizomes. Stems 30–90 cm long, erect or ascending, occasionally from a spreading base, unbranched or branched, glabrous. Leaves opposite, fused basally into a sheath, sessile or with a poorly differentiated, winged petiole to 1.5 cm long, lacking axillary clusters of leaves. Stipules absent. Leaf blades 3–11(–15) cm long, elliptic to ovate or oblanceolate, not fleshy, tapered at the base, rounded or angled to a bluntly or sharply pointed tip. Flowers in dense terminal clusters (those on short lateral branches sometimes appearing axillary), the stalks 0.1–0.5 cm long, erect to spreading at flowering and fruiting, the bracts paired and resembling small leaves. Epicalyx absent. Sepals 5, 15–20 mm long, becoming elongated to 25 mm long at fruiting, fused basally into a slender tube, this 15–25-nerved, herbaceous and green or reddish-tinged between the sepals, the lobes narrowly triangular, shorter than the tube, angled or tapered to a sharply pointed tip, not appearing hooded or awned, the margins papery and white. Petals 5 (rarely 10 or more in doubled horticultural forms), showy, 23–35 mm long, oblanceolate to spatulate, tapered to a stalklike base and this more or less fused to the ovary stalk, the expanded portion 0.8–1.5 cm long, rounded to shallowly notched at the tip, pink to white, with a pair of small appendages on the upper surface at the base of the expanded, spreading apical portion. Stamens 10, the filaments fused at the base. Staminodes absent (present in some horticultural forms). Pistil with 1 locule, the ovary with a stalk 2–5 mm long. Styles 2(3), distinct, each with a stigmatic area along the inner surface. Fruits capsules, 15–20 mm long, opening by 4(6) ascending to recurved teeth. Seeds mostly 30–75, 1.6–2.0 mm wide, kidney-shaped, the surface with minute, broad papillae or appearing pebbled, dark brown, lacking wings or appendages. $2n=28$. June–October.

Introduced, scattered to common nearly throughout the state, but less abundant in the northern counties (native of Europe, Asia; introduced throughout the U.S., Canada, Mexico, Central America, South America, Africa, Australia). Banks of streams, rivers, and spring branches; also old fields, pastures, fencerows, old homesites, gardens, railroads, roadsides, and open, disturbed areas.

Saponaria officinalis is a familiar nonnative roadside wildflower in most of the state. It often forms large colonies that are difficult to eradicate. The plants have tough rhizomes from which many flowering stems arise; new plants can easily arise from small portions of a rhizome. Several cultivars, including ones with doubled petals, staminodes replacing the stamens, and dark pink flowers, are commonly grown in gardens; they may persist at old homesites. The scientific name refers to the high concentration of saponins in the plant; the leaves often were used as a source for soap. The seeds of *Saponaria* also contain saponins and are poisonous (Burrows and Tyrl, 2001).

16. Scleranthus L. (knawel)

About 10 species, Europe, Africa, Asia, south to Australia, introduced nearly worldwide.

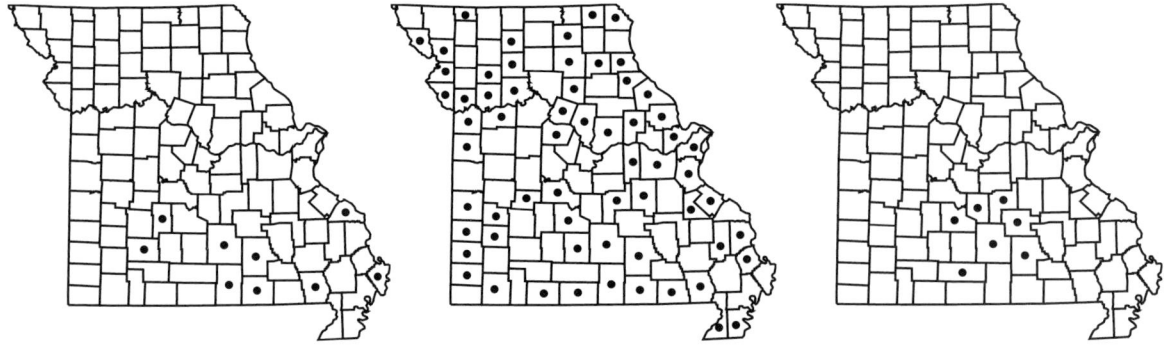

1479. Scleranthus annuus   1480. Silene antirrhina   1481. Silene caroliniana

1. Scleranthus annuus L. ssp. annuus
(knawel, annual knawel)
Pl. 344 c, d; Map 1479

Plants annual or biennial. Stems 2–25 cm long, erect to spreading, usually many-branched, glabrous or moderately to densely pubescent with minute, stalked glands, often mostly on one side of the stem. Leaves opposite, fused basally into a small sheath, sessile, lacking axillary clusters of leaves. Stipules absent. Leaf blades 0.3–2.4 cm long, narrowly triangular to more commonly linear, not fleshy, slightly expanded to more or less tapered at the base, angled or tapered to a bluntly or more commonly sharply pointed tip. Flowers in open to dense terminal and axillary clusters, sometimes also with solitary axillary flowers, deeply perigynous, sessile or appearing with minute, tapered stalks, the bracts paired and resembling the leaves. Epicalyx absent. Hypanthium present, 1.2–2.0 mm long, thick-walled, hardened, and enclosing the ovary, deeply cup-shaped to somewhat urn-shaped with an incurved, white rim along the upper margin, persistent and becoming strongly 10-ribbed at fruiting. Sepals 5, 1.5–4.0 mm long, free and attached along the upper margin of the hypanthium, angled somewhat outward, lanceolate to awl-shaped, leathery, angled to a usually sharply pointed tip, usually somewhat concave but not hooded or awned, the margins often somewhat inrolled, thin and white. Petals absent. Stamens 2–10, the filaments distinct, attached along the inner rim of the calyx tube. Staminodes absent or 5–8, linear or narrowly triangular. Pistil with 1 locule, the ovary minutely stalked. Styles 2, each with a capitate terminal stigmatic area. Fruits achenelike, dehiscing irregularly with age, 3.2–5.0 mm long, enclosed in and shed with the persistent hypanthium and sepals. Seed 1, 0.8–1.0 mm wide, more or less spherical, the surface smooth, yellowish brown, lacking wings or appendages. $2n=22$. March–October.

Introduced, scattered in the Ozark, Ozark Border, and Mississippi Lowlands Divisions (native of Europe, Asia, Africa; introduced in the eastern U.S. west to Minnesota, Nebraska, Oklahoma, and Louisiana, and in the western U.S. from Washington and Idaho south to California; Canada). Sandstone glades and banks of streams and rivers; also pastures, cemeteries, lawns, roadsides, and open, disturbed areas.

Steyermark (1963) knew this species only from Greene County, where he first discovered it in 1958. *Scleranthus annuus* can form large populations in sandy areas and is expected to continue its spread in southern Missouri. European botanists have divided the species into five or more subspecies differing in stem orientation and branching density, leaf shape, and fruit size (Sell, 1993). North American materials all appear to be referable to ssp. *annuus*.

17. Silene L. (campion, catchfly)
(Oxelman et al. 2001; Morton 2005c)

Plants annual, biennial, or perennial, sometimes dioecious (in Missouri, only *S. dioica* and *S. latifolia*). Stems erect to spreading, sometimes reclining, branched or unbranched, glabrous or sparsely to densely pubescent with nonglandular hairs and/or stalked glands, sometimes glaucous. Leaves opposite or whorled (in *S. stellata*), fused basally into a sheath or sometimes clasping the stem, petiolate (basal leaves) or sessile (most stem leaves), usually lacking axillary clusters of leaves (these occasionally present at few to several nodes in some species). Stipules absent. Leaf blades linear to lanceolate, elliptic, obovate, or spatulate, not succulent (except in *S. csereii*), tapered to rounded or cordate at the base, angled or tapered to a bluntly or sharply pointed tip, occasionally tapered abruptly to a minute, sharp point. Flowers in terminal, open panicles or clusters, occasionally solitary and terminal, the stalks erect or ascending at flowering and fruiting or sometimes nearly absent, the bracts paired and resembling small leaves, with herbaceous and green or less commonly thin and white margins. Epicalyx absent. Flowers perfect or all staminate or pistillate (in *S. dioica* and *S. latifolia*). Sepals 5, fused most of their length into a tube, this (8–)10–30-nerved, these green or sometimes whitish, reddish, or greenish purple, sometimes inflated and/or papery, variously herbaceous and green or membranous to papery and translucent or white to purplish-tinged between the nerves, the lobes linear to lanceolate, oblong, or broadly triangular, much shorter than the tube, angled or tapered to a bluntly or sharply pointed tip, not appearing hooded or awned, the margins white or purplish-tinged. Petals 5 or rarely absent, when present oblanceolate to narrowly obovate or spatulate, tapered to a stalklike base, rounded to more or less truncate at the tip, sometimes notched or 2(4–12)-lobed, otherwise irregularly toothed or rarely entire, white, creamy white, pink, purplish pink, or red, sometimes with a pair of small appendages on the upper surface at the base of the expanded portion. Stamens 10 (absent in pistillate flowers), the filaments distinct or fused into a short tube basally. Staminodes absent (rarely 1–10 and linear to hairlike in pistillate flowers). Pistil with 1 locule or appearing 3–5-locular toward the base, the ovary sessile or appearing stalked (small and nonfunctional in staminate flowers). Styles 3(4) or 5 (absent in staminate flowers), when present distinct, each with a stigmatic area along the inner side. Fruits capsules, opening by 3(4) or 5 valves (sometimes each split at the tip) or by 6(8) or 10 spreading or outward-curled teeth. Seeds 15–100(–500 or more), kidney-shaped to nearly globose, the surface usually tuberculate or with small papillae, reddish brown to gray, dark brown, or black, lacking wings or appendages. About 650 (or more) species, North America, South America, Europe, Asia, Africa, introduced nearly worldwide.

The present circumscription of *Silene* follows that of Oxelman et al. (2001), who defined the tribe Sileneae as comprising eight genera. Four of these are represented in Missouri: *Agrostemma, Atocion, Lychnis,* and *Silene.* Most of the nonnative species encountered in Missouri originally were brought into the United States as garden ornamentals or as contaminants in other seeds. A number of species are still cultivated in gardens today.

1. Flowers either staminate or pistillate, the plants dioecious; pistillate flowers with (4)5 styles, the fruits dehiscing by (8)10 teeth
    2. Petals deep pink, the flowers opening during the day; lower portion of the stem pubescent with long, spreading hairs .... 6. S. DIOICA
    2. Petals white, the flowers opening at night; lower portion of the stem pubescent with a mixture of short, stiff, nonglandular hairs (somewhat roughened to the touch) and stalked glands .... 8. S. LATIFOLIA
1. Flowers all perfect, the plants not dioecious; styles 3(–5), the fruits dehiscing by 6(8 or 10) teeth or rarely by 3 valves that may split to produce 6 teeth
    3. Petals bright red
        4. Petals entire or with a few fine teeth toward the tip, rarely notched (but then lacking additional lobes); stem leaves in (6–)10–20 pairs, these larger than the basal leaves, which are withered or absent at flowering, the blades ovate to lanceolate; stems (35–)50–120 cm tall .... 11. S. REGIA
        4. Petals with 2 tapered lobes at the tip, usually also with a pair of smaller lateral lobes (often appearing somewhat fringed); stem leaves in 2–4 pairs, these much smaller than the prominent basal leaves, the blades broadly oblanceolate to narrowly elliptic; stems 20–80 cm tall
         .... 13. S. VIRGINICA

3. Petals white, creamy white, or pink, rarely red or absent
   5. Stem leaves in whorls of 4 .... 12. S. STELLATA
   5. Stem leaves opposite or the leaves mostly basal
      6. Sepal tube with 20–30 parallel nerves, occasionally with a network of nerves between them
         7. Plants glabrous, strongly glaucous; sepal tube 7–10 mm long, glabrous, the 20 nerves not prominent .... 4. S. CSEREII
         7. Plants hairy, not glaucous; sepal tube (15–)20–25 mm long, pubescent with stalked glands, the 25–30 nerves prominent .... 3. S. CONOIDEA
      6. Sepal tube with (8)10 parallel nerves, with or without a network of nerves between them, or sometimes the nerves obscure
         8. Sepal tube uniformly green or purplish, not clearly pale between the mostly obscure nerves; stems glabrous or at most somewhat roughened toward the base; petals white
            9. Stems green, not glaucous; inflorescence bracts slightly smaller than the foliage leaves, the margins herbaceous and green; flowers often appearing solitary; sepal tube glabrous or at most roughened toward the base, the nerves parallel, without a network of nerves between them .... 9. S. NIVEA
            9. Stems appearing gray, glaucous; inflorescence bracts much smaller than the foliage leaves, at least some with the margins thin and white; inflorescences well-developed clusters; sepal tube glabrous, the nerves parallel and occasionally with a network of veins between them
               10. Sepal tube 11–15 mm long, bell-shaped, ascending to somewhat spreading at the tip and not tightly enclosing the fruit, the nerves mostly anastomosing; inflorescences open clusters or panicles .... 14. S. VULGARIS
               10. Sepal tube 7–10 mm long, more or less ellipsoid, narrowed at both ends and tightly enclosing the fruit, the nerves mostly parallel; inflorescences slender panicles, but often forked at the base .... 4. S. CSEREII
         8. Sepal tube nerves conspicuous and green, the tissue between the nerves pale; stems roughened-hairy or with stalked glands, rarely glabrous; petals white, creamy white, pink, red, or rarely absent
            11. Sepal tube 15–25 mm long; petals creamy white or pink
               12. Stems 25–75 cm tall, the stems leafy; flowers opening at night; petals creamy white, often faintly pinkish-tinged, the tip 2-lobed .... 10. S. NOCTIFLORA
               12. Stems 8–20(–25) cm tall, the leaves mostly basal; flowers opening during the day; petals pink, the tip entire or with wavy margins .... 2. S. CAROLINIANA
            11. Sepal tube 5–12(–15) mm long; petals mostly white, sometimes tinged with pink or red, rarely entirely red or absent
               13. Stems glabrous or with short, soft, downward-angled hairs toward the base, often with reddish brown to purplish black, sticky patches along the upper internodes; flower stalks (5–)10–25(–35) cm long; petals mostly white, tinged with pink or red, rarely entirely red, rarely entirely red or absent .... 1. S. ANTIRRHINA

13. Stems roughened with downward-angled hairs throughout or roughened toward the base and with stalked glands toward the tip, lacking sticky patches; flower stalks absent or to 0.5 cm long; petals white
   14. Inflorescences often forked basally; sepal tube with long, spreading hairs; petals 2-lobed at the tip .... 5. S. DICHOTOMA
   14. Inflorescences usually not forked basally; sepal tube with a mixture of long, nonglandular hairs and short, stalked glands; petals entire or notched at the tip .... 7. S. GALLICA

1. Silene antirrhina L. (sleepy catchfly)

S. *antirrhina* f. *apetala* Farw.
S. *antirrhina* f. *bicolor* Farw.
S. *antirrhina* f. *deaneana* Fernald

Pl. 345 g, h; Map 1480

Plants annual. Stems (10–)25–80 cm long, erect, unbranched or branched, glabrous or more commonly with dense, short, soft, downward-angled hairs toward the base, often with reddish brown to purplish black, sticky bands encircling the upper internodes. Basal leaves usually present at flowering, usually shorter than the largest stem leaves, short- to long-petiolate. Stem leaves opposite, in mostly 6–15 pairs, sessile. Leaf blades 1–5(–9) cm long, not succulent, oblanceolate to spatulate (basal leaves) or oblanceolate to linear (stem leaves), tapered at the base, angled to a bluntly or sharply pointed tip, the surfaces glabrous or minutely hairy, not glaucous. Flowers perfect, in open terminal panicles or clusters, rarely solitary, the stalks (5–)10–25(–35) cm long, glabrous, the bracts paired and resembling small leaves, some of the bracts with thin, white margins and a small, papery, often reddish-tinged tip. Sepals 5–9 mm long, the tube with 10 parallel nonanastomosing nerves, ovoid to broadly ellipsoid, the nerves green, pale between the nerves, glabrous, the lobes triangular, often reddish purple, sharply pointed at the tip, the margins thin and white or purplish-tinged. Petals 5 or rarely absent, 8–9 mm long, the expanded portion 2–3 mm long, 2-lobed at the tip, mostly white, occasionally tinged with pink or red, rarely entirely red, with a pair of minute appendages on the upper surface at the base of the expanded portion. Styles 3. Fruits 5–7 mm long, dehiscing apically by 6 teeth, with a basal stalklike portion 0.5–1.0 mm long. Seeds 0.5–0.8 mm wide, kidney-shaped, the surface with minute papillae, grayish black. $2n=24$. April–September.

Scattered nearly throughout the state (nearly throughout North America; South America; introduced in Europe). Glades, upland prairies, openings of mesic to dry upland forests, tops of bluffs, and margins of ponds; also pastures, ditches, levees, fallow fields, old fields, railroads, roadsides, and open, disturbed areas.

Silene antirrhina is extremely variable, and various workers have attempted to classify the variants as either varieties or forms. Steyermark (1963) placed Missouri specimens into four forms, based on petal presence/absence and color, as well as presence/absence of the sticky patches along the upper stems. The present treatments takes a broader approach and follows that of Morton (2005c) in not formally recognizing these taxa.

2. Silene caroliniana Walter **var. wherryi** (Small) Fernald (wild pink, Wherry's pink)

S. *wherryi* Small
S. *caroliniana* ssp. *wherryi* (Small) R.T. Clausen

Pl. 346 a, b; Map 1481

Plants perennial, with a branched, somewhat woody rootstock. Stems 8–20(–25) cm long, ascending, unbranched or less commonly branched, moderately to densely pubescent with soft, nonglandular hairs. Basal leaves well developed at flowering, longer than the relatively few stem leaves, with a relatively long, broadly winged petiole. Stem leaves opposite, 2–4 pairs, short-petiolate to sessile. Leaf blades 3–9 cm long, not succulent, oblanceolate to broadly oblanceolate or spatulate, tapered at the base, angled to a bluntly or sharply pointed tip, more or less glabrous except for the margins. Flowers perfect, in open terminal clusters, sometimes solitary, the stalks 0.2–0.8(–1.5) cm long, densely pubescent with mostly nonglandular hairs, the bracts paired and resembling small leaves, with green margins. Sepals 15–22 mm long, the tube with (8–)10 parallel nonanastomosing nerves, tubular, becoming club-shaped at fruiting, the nerves green, pale between the nerves, densely pubescent with nonglandular hairs, the lobes more or less oblong, green, rounded or bluntly pointed at the tip, the margins thin and white or reddish-tinged. Petals 5, 25–35 mm long, the expanded portion 9–12 mm long, entire or with wavy margins toward the tip, pink, with a pair of small appendages on the upper surface at the base of the expanded por-

1482. Silene conoidea 1483. Silene csereii 1484. Silene dichotoma

tion. Styles 3(4). Fruits 9–12 mm long, dehiscing apically by 6(8) teeth, with a basal stalklike portion 5–8 mm long. Seeds 1.3–1.5 mm wide, kidney-shaped, the surface with minute papillae, reddish brown. $2n=48$. April–May.

Uncommon in the central portion of the Ozark Division (Virginia, Ohio, Kentucky, Missouri, Kansas, and Alabama). Mesic upland forests on rocky slopes; ledges and tops of bluffs; also roadsides; usually on acidic substrates.

Silene caroliniana var. *wherryi* is one of three infraspecific taxa within *S. caroliniana* that have been treated as either varieties or subspecies. The other two varieties both have glandular hairs on the calyces and occur to the east of Missouri. The var. *pensylvanica* (Michx.) Fernald differs from var. *caroliniana* in its basal leaves that are glabrous except along the margins and undersurface main veins (vs. moderately to densely pubescent with glandular and nonglandular hairs), and tapered to a relatively slender petiole (vs. with a winged petiole).

Steyermark (1963) reported putative hybrids between *S. caroliniana* var. *wherryi* and *S. virginica* from Shannon County. Although the two parental taxa usually occupy sunny and shaded habitats respectively and thus seldom grow in close proximity, both were present at Steyermark's site as well as several plants with a mixture of parental traits. These included taller individuals with mostly pink petals shaped as in *S. virginica* and other plants with the short habit of *S. caroliniana* but with the bright red petals normally found in *S. virginica*. Mitchell and Uttall (1969) subsequently studied a similar instance of apparent hybridization in Virginia.

Steyermark (1963) considered *S. caroliniana* one of the most beautiful native wildflowers in the Ozarks and strongly recommended its use in wildflower gardens. It is now available commercially from native plant nurseries in the state.

3. Silene conoidea L. (large sand catchfly)

Pl. 348 d; Map 1482

Plants annual. Stems 12–30 cm long, erect, unbranched or branched, moderately to densely pubescent with stalked glands. Basal leaves usually absent at flowering, when present usually shorter than the largest stem leaves, short- to long-petiolate. Stem leaves opposite, mostly 6–12 pairs, sessile. Leaf blades 2–5(–8) cm long, not succulent, oblanceolate or narrowly lanceolate, tapered at the base, angled to a bluntly or sharply pointed tip. Flowers perfect, in open terminal clusters or panicles, the stalks (1–)2–3 cm long, densely pubescent with stalked glands, the bracts paired and resembling small leaves, with green margins. Sepals (15–)20–25 mm long, the tube with 25–30 parallel nonanastomosing nerves, conical at flowering, becoming flask-shaped and inflated toward the base at fruiting, the nerves green, pale or somewhat translucent between the nerves, densely pubescent with stalked glands, the lobes narrowly lanceolate, green, tapered to a sharply pointed tip, the margins thin and white. Petals 5, 28–40 mm long, the expanded portion 8–12 mm long, entire, irregularly scalloped, or notched at the tip, pink, with a pair of small appendages on the upper surface at the base of the expanded portion. Styles 3. Fruits 11–16 mm long, dehiscing apically by 6 teeth, with a basal stalklike portion 0.5–2.0 mm long. Seeds 1.2–1.7 mm wide, kidney-shaped, the surface tuberculate, brown. $2n=20, 24$. May–July.

Introduced, uncommon, known thus far only from the city of St. Louis (native of Europe, Asia; introduced in the northwestern U.S. sporadically east to Texas and Missouri; Canada). Railroads.

This showy species was first collected from Missouri in 1957 by Viktor Mühlenbach during his botanical inventories of the St. Louis railyards. Although he collected it several times, it has not been encountered elsewhere in the state.

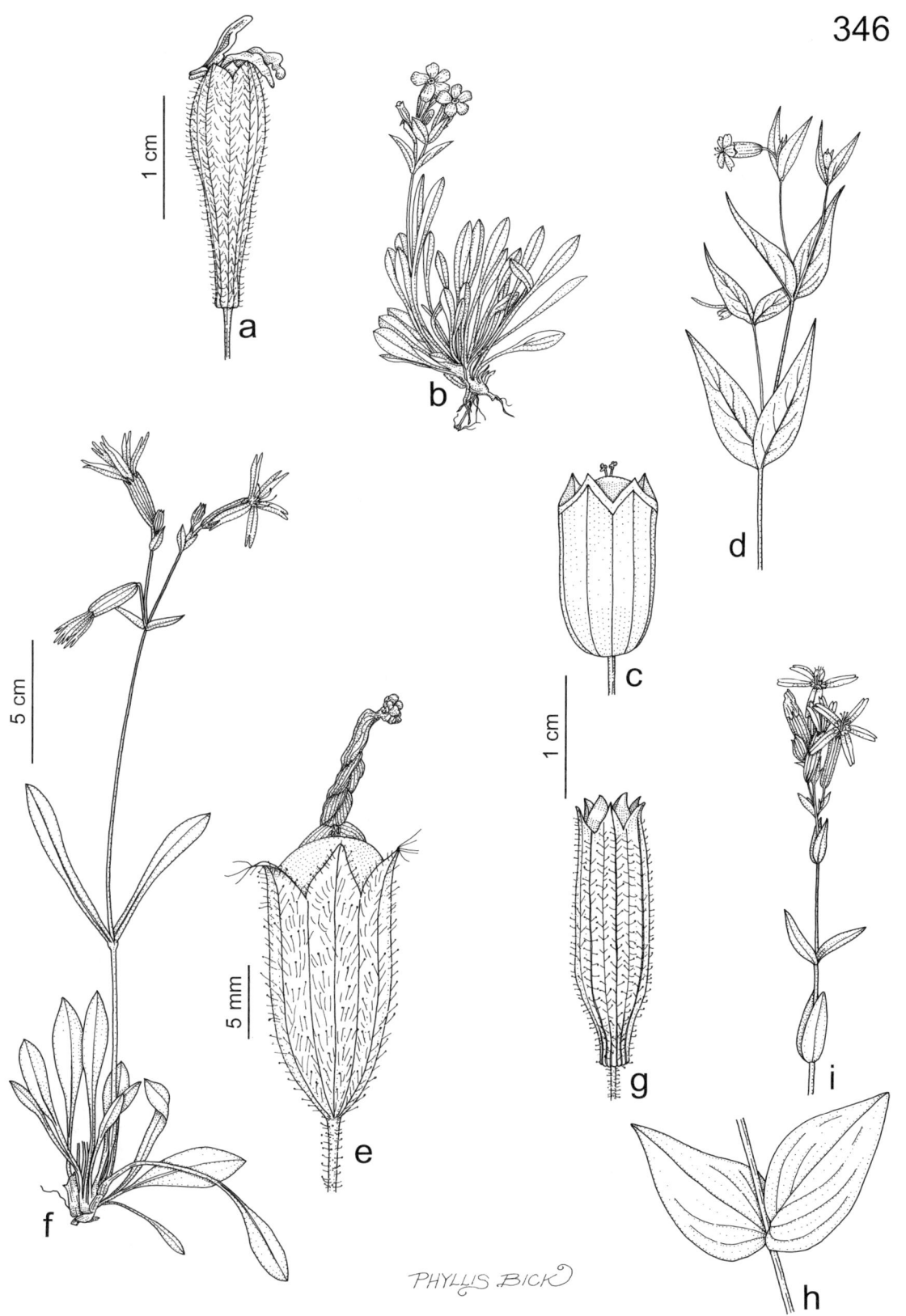

Plate 346. Caryophyllaceae. *Silene caroliniana*, **a)** fruit, **b)** habit. *Silene nivea*, **c)** fruit, **d)** habit. *Silene virginica*, **e)** fruit, **f)** habit. *Silene regia*, **g)** fruit, **h)** node, **i)** habit.

4. Silene csereii Baumg. (smooth catchfly, glaucous campion)

Pl. 347 e, f; Map 1483

Plants annual or biennial. Stems 40–75 cm long, erect, unbranched or few-branched toward the base, glabrous and glaucous. Basal leaves usually beginning to wither at flowering, when present mostly shorter than the stem leaves, usually with a short, poorly differentiated petiole. Stem leaves opposite, mostly 4–10 pairs, sessile. Leaf blades 3.0–7.5 cm long, somewhat thickened and succulent, broadly oblanceolate to obovate or occasionally lanceolate to ovate, tapered at the base, angled or abruptly short-tapered to a sharply pointed tip, glabrous, usually strongly glaucous. Flowers perfect, in terminal, slender, usually elongated panicles, often forked basally, the stalks (0.7–)1.0–2.5(–3.5) cm long, glabrous, the bracts paired and resembling highly reduced leaves, with thin, white margins. Sepals 7–10 mm long, the tube with 20 indistinct, parallel, rarely anastomosing nerves, somewhat inflated, oblong-ovoid to ellipsoid, constricted at both ends and tightly enclosing the fruit, pale green or faintly purplish-tinged, not lighter between the nerves, glabrous, the lobes triangular, pale green or somewhat purplish-tinged, angled or short-tapered to a sharply pointed tip, the margins thin and white. Petals 5, 10–15 mm long, the expanded portion 3–5 mm long, 2-lobed at the tip, white, with a pair of small appendages on the upper surface at the base of the expanded portion. Styles 3. Fruits 7–10 mm long, dehiscing apically by 6 teeth, with a basal stalklike portion 0.7–1.0 mm long. Seeds 0.6–1.0 mm wide, kidney-shaped and plump, the surface with minute papillae, grayish brown. $2n=24$. May–September.

Introduced, uncommon and sporadic (native of Europe; introduced in the northern U.S. south to North Carolina, Missouri, Colorado, Idaho, and Washington; Canada). Dolomite glades; also railroads, roadsides, and open, disturbed areas.

Silene csereii sometimes is mistaken for *S. vulgaris*. Distinguishing features of *S. csereii* include the thicker, somewhat succulent leaves, the narrow, often basally forked inflorescences and the smaller flowers with a less inflated calyx that tightly encloses the fruit. The basal rosettes of thick, gray leaves superficially are similar to those of some species of Crassulaceae that are cultivated by succulent plant enthusiasts. The species epithet was spelled *S. cserei* in the original publication, and that spelling is still used in many recent floristic works. However, *S. csereii* appears to have the correct termination when the name of Johann Baumgarten's patron, Wolfgang von Cserei, is commemorated in a botanical name.

5. Silene dichotoma Ehrh. ssp. dichotoma (forked catchfly)

Pl. 347 a, b; Map 1484

Plants annual. Stems 20–50(–100) cm long, erect, branched, roughened with relatively long, downward-angled hairs. Basal leaves usually withered at flowering, when present usually shorter than the largest stem leaves, with a poorly developed, short petiole. Stem leaves opposite, mostly 5–10 pairs, with a poorly differentiated, short petiole or sessile. Leaf blades 2.0–8.5 cm long, oblanceolate to narrowly lanceolate or narrowly ovate, tapered at the base, angled or tapered to a sharply pointed tip, sparsely to moderately pubescent with more or less spreading hairs. Flowers perfect, in open, terminal, more or less 1-sided panicles, these often forked 1 or more times toward the base, occasionally appearing as leafy branches with solitary axillary flowers, the stalks absent or to 0.3(–1.5) cm long, roughened-hairy and sometimes also with stalked glands, the bracts paired and resembling small or nearly full-sized leaves, at least the uppermost with thin, white margins. Sepals 8–12 (–15) mm long, the tube with 10 parallel nonanastomosing nerves, ellipsoid, not inflated, the nerves green or purplish-tinged, pale between the nerves, sparsely to moderately pubescent with relatively long, spreading hairs, the lobes narrowly triangular, green, tapered to a sharply pointed tip, the margins thin and white. Petals 5, 13–20 mm long, the expanded portion 4–7 mm long, 2-lobed at the tip, white or rarely pink, with a pair of small appendages on the upper surface at the base of the expanded portion. Styles 3. Fruits 7–10 mm long, dehiscing apically by 6 teeth, with a stalklike base 1–3 mm long. Seeds 0.8–1.0 mm wide, kidney-shaped, the surface tuberculate, grayish brown. $2n=24$. June–September.

Introduced, uncommon, known thus far only from Cape Girardeau County and the city of St. Louis (native of Europe, Asia; introduced in the northern U.S. south to Georgia, Missouri, and California, also Texas; Canada). Old fields and railroads.

Nonweedy plants from southern Europe with well-developed rosettes of basal leaves that are densely hairy and relatively dense inflorescences have been called ssp. *racemosa* (Otth) Graebn. Some European botanists recognize one or more additional infraspecific taxa, but none of these is weedy or has become introduced in North America.

Plate 347. Caryophyllaceae. *Silene dichotoma*, **a)** fruit, **b)** stem with leaves and inflorescence. *Silene gallica*, **c)** fruit, **d)** stem with leaves and inflorescence. *Silene csereii*, **e)** fruit, **f)** stem with leaves and inflorescence. *Silene stellata*, **g)** flower, **h)** stem with leaves and inflorescence.

1485. Silene dioica   1486. Silene gallica   1487. Silene latifolia

6. **Silene dioica** (L.) Clairv. (red campion)
Lychnis dioica L.

Pl. 348 c; Map 1485

Plants perennial, dioecious, usually with short rhizomes. Stems 15–80 cm long, ascending, often from a spreading base, branched, moderately to densely pubescent with long, spreading hairs, sometimes also with sparse, stalked glands toward the tip. Basal leaves often present at flowering, shorter than the largest stem leaves with a poorly developed, short or long petiole. Stem leaves opposite, mostly 5–10 pairs, short-petiolate (lower leaves) or sessile. Leaf blades 3–9 cm long, ovate to broadly ovate or elliptic, tapered at the base, angled or short-tapered to a sharply pointed tip, the surfaces moderately to densely short-hairy, especially along the midvein. Flowers imperfect (either all staminate or all pistillate), in open to slender terminal panicles or clusters, often forked basally, the stalks 0.2–2.0 cm long, with long, spreading, nonglandular hairs, rarely also with sparse, stalked glands, the bracts paired and resembling small leaves, with green margins. Sepals 9–17 mm long, the tube with 10 parallel nonanastomosing nerves, more or less bell-shaped, not inflated, the nerves green or often purplish-tinged, pale between the nerves, long-hairy, the lobes triangular to lanceolate, green or purplish-tinged, sharply pointed at the tip, the margins herbaceous and green or thin and white. Petals 5, 18–25 mm long, the expanded portion 7–10 mm long, entire to 2-lobed at the tip, deep pink to pink, with a pair of small appendages on the upper surface at the base of the expanded portion. Styles 5. Fruits 10–12 mm long, dehiscing apically by 10 teeth, lacking a stalklike base. Seeds 1.0–1.6 mm wide, kidney-shaped and plump, the surface with small papillae, dark brown to black. $2n=24$. May–September.

Introduced, uncommon, known thus far from St. Louis County (native of Europe; introduced in the northern U.S. south to Pennsylvania, Missouri, and Oregon; Canada). Habitat unknown but presumably open, disturbed areas.

This species is accepted as part of the Missouri flora with some reservations. Palmer and Steyermark (1935) and Steyermark (1963) cited Jackson County as the source for their inclusion of *S. dioica* (as *Lychnis dioica*) in the Missouri flora. However, no specimens from that county could be located during the present research, and it seems likely that the original determination was in error. Subsequently, however, a specimen was collected by Bill Bauer that possibly originated from St. Louis County (the label has no further locality data, but Bauer was living in Webster Groves at the time). The species is included here based on that record. *Silene dioica* is completely interfertile with the white-flowered *S. latifolia,* and Morton (2005c) stated that hybrids are formed readily where the two species grow in proximity. He also noted that such hybrids are very difficult to distinguish from *S. dioica* from herbarium specimens on which pale pink corollas or a mix of pink and white flowers cannot be observed with confidence. Nevertheless, Morton annotated a specimen from St. Louis County as representing this hybrid, which suggests that both parents were growing somewhere in the vicinity (further support for the dot in St. Louis County in Map 1485).

7. **Silene gallica** L. (small-flowered catchfly)
S. anglica L.

Pl. 347 c, d; Map 1486

Plants annual. Stems 10–45 cm long, erect, unbranched or branched, roughened with dense, short, stiff, downward-angled hairs, also with sparse to moderate longer, softer, more or less crinkled hairs, often also with stalked glands toward the tip. Basal leaves usually withered at flowering, when present shorter than the largest stem leaves, short- to long-petiolate. Stem leaves opposite, usually 6–10 pairs, short-petiolate (lower leaves) or sessile. Leaf blades 0.5–5.0 cm long, spatulate or oblanceolate to lanceolate, tapered at the base, angled or abruptly short-tapered to a sharply pointed tip. Flowers perfect, in open terminal panicles, often appearing as leafy branches with solitary axillary flowers, the stalks 0.1–0.5 cm long, with stalked glands, the bracts paired and resembling small leaves, with green margins. Sepals 8–10 mm long, the tube with 10 parallel nonanastomosing nerves, tubular to ovoid, constricted toward the tip, the nerves green or purplish-tinged, pale or yellowish white between the nerves, pubescent with sparse long, nonglandular hairs and moderate short, stalked glands, the lobes narrowly lanceolate, green or purple, sharply pointed at the tip, the margins green. Petals 5, 11–14 mm long, the expanded portion 3–5 mm long, entire or notched at the tip, white, with a pair of small appendages on the upper surface at the base of the expanded portion. Styles 3. Fruits 6–8 mm long, dehiscing apically by 6 teeth, with a stalklike base 0.8–1.0 mm long. Seeds 0.5–0.8 mm wide, kidney-shaped with concave sides, the surface with a fan-shaped pattern of many fine ridges and low tubercles, dark reddish brown to black. $2n=24$. April–September.

Introduced, uncommon, known thus far only from Jackson County and the city of St. Louis (native of Europe, Asia; introduced in the eastern U.S. west to Missouri and Texas and in the western U.S. from Washington and Idaho south to California and Arizona; Canada). Railroads and open, disturbed areas.

Steyermark (1963) reported a specimen of this species from Cape Girardeau County, but this was redetermined as S. dichotoma in 1998 by John K. Morton of the University of Waterloo, Canada, during his research for the Flora of North America Project (Morton, 2005c).

8. Silene latifolia Poir. (white campion, evening campion, white cockle)

S. pratensis (Rafn) Gren. & Godr.
S. latifolia ssp. alba (Mill.) Greuter & Burdet
Lychnis alba Mill.

Pl. 348 g, h; Map 1487

Plants annual or short-lived perennials, dioecious, with a sometimes woody taproot. Stems 30–100 cm long, erect, sometimes from a spreading base, branched, pubescent with stalked glands and also roughened with short, stiff, downward-angled hairs toward the base. Basal leaves usually withered at flowering, when present shorter than to about as long as the largest stem leaves, mostly long-petiolate. Stem leaves opposite, usually 6–12 pairs, sessile. Leaf blades 3–12 cm long, elliptic to lanceolate, tapered at the base, angled or short-tapered to a sharply pointed tip, the surfaces finely pubescent with sometimes somewhat tangled, nonglandular hairs. Flowers imperfect (either all staminate or all pistillate), in open terminal panicles or clusters, the stalks 0.5–4.0 cm long, those of the pistillate flowers usually longer than those of the staminate flowers, with stalked glands, the bracts paired and resembling small leaves, with green margins. Sepals 10–20 mm long, in pistillate flowers becoming enlarged to 30 mm at fruiting, the tube with 10 (staminate flowers) or 20 (pistillate flowers) parallel nerves, with a network of sparse to moderate, anastomosing, angled cross-nerves, tubular to slightly conical and not inflated at flowering, in pistillate flowers becoming ovoid and inflated as the fruit matures, the nerves green to reddish-tinged, pale to nearly translucent between the nerves, pubescent with a mixture of nonglandular hairs and stalked glands, the lobes ovate to narrowly triangular, green to purplish, angled or short-tapered to a sharply pointed tip, the margins green. Petals 5, 24–32 mm long, the expanded portion 8–12 mm long, entire to 2-lobed at the tip, white, with a pair of small appendages on the upper surface at the base of the expanded portion. Styles (4)5. Fruits 11–18 mm long, dehiscing apically by (8)10 teeth, with a stalklike base 0.8–1.5 mm long. Seeds 1.0–1.3(–1.5) mm wide, kidney-shaped and plump, the surface coarsely tuberculate, dark grayish brown to dark grayish black. $2n=24$. May–September.

Introduced, scattered nearly throughout the state (native of Europe, Asia, Africa; introduced nearly throughout the U.S. [including Alaska] except some southeastern states, Oklahoma, and Texas). Pastures, fallow fields, old fields, old homesites, ditches, railroads, roadsides, and open, disturbed areas.

First collected in Jackson County in 1915, S. latifolia sometimes has been confused with S. noctiflora, a plant with perfect flowers, creamy white or pinkish petals, and 3 styles. The proper name for this taxon has been the subject of intense nomenclatural research (McNeill and Prentice, 1981; Greuter and Burdet, 1982). For a discussion of putative hybrids with S. dioica, see the treatment of that species.

The attractive flowers of S. latifolia open in the early evening and are sweetly fragrant. Dean (1963) studied more than 50,000 individual flowers of this species from Iowa and Minnesota populations. She

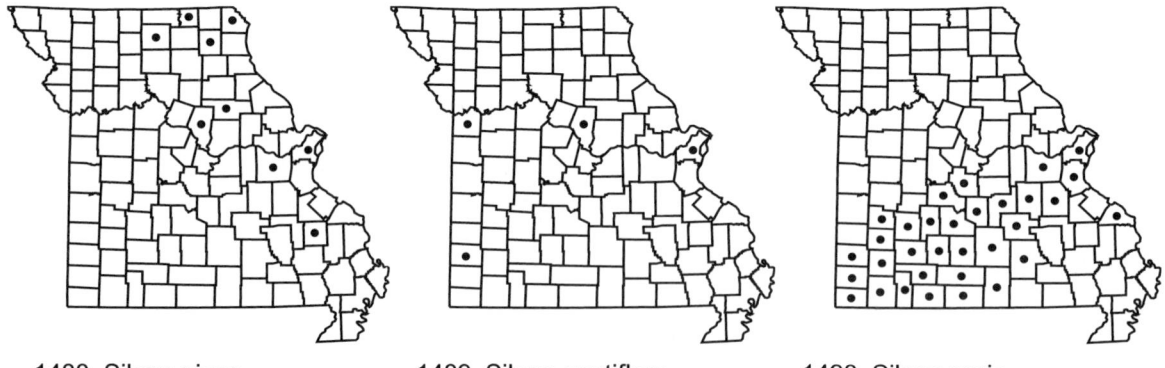

1488. Silene nivea   1489. Silene noctiflora   1490. Silene regia

found that although *S. latifolia* pistillate flowers usually had 5 or less often 4 styles, about 1 percent of her samples had 6–8, 0–3, or 9–13 styles. Morton (2005c) observed that many North American specimens of *S. latifolia* appear to be morphologically intermediate between the two most common subspecies recognized by some European authors, ssp. *latifolia* and ssp. *alba* (Mill.) Greuter & Burdet and thus chose not to formally recognize these taxa. His treatment is followed here.

9. Silene nivea (Nutt.) Muhl. ex Otth (snowy campion)

Pl. 346 c, d; Map 1488

Plants perennial, with rhizomes. Stems 20–90 cm long, erect or ascending, unbranched or occasionally few-branched, glabrous or less commonly moderately pubescent with short, downward-curved hairs toward the base. Basal leaves usually absent at flowering, when present shorter than the largest stem leaves, short-petiolate. Stem leaves opposite, usually 6–10 pairs, short-petiolate (lower leaves) or sessile. Leaf blades 3–11 cm long, elliptic-lanceolate to lanceolate, rounded, angled, or tapered at the base, angled or tapered to a sharply pointed tip, the surfaces glabrous or occasionally short-hairy along the midvein toward the base. Flowers perfect, in open terminal clusters or solitary, occasionally appearing solitary and axillary from the upper leaves, the stalks 1.2–3.0 cm long, glabrous to short-hairy, sometimes becoming hooked toward the tip at fruiting, the bracts paired and resembling small or more commonly nearly full-sized leaves, with herbaceous, green margins. Sepals 14–17 mm long, the tube with 10 faint, parallel nerves, tubular, becoming somewhat inflated and broadly urn-shaped at fruiting, narrowed toward the base, green, not pale between the nerves, glabrous or occasionally sparsely to densely short-hairy, the lobes triangular, green, bluntly pointed at the tip, the margins thin and white. Petals 5, 19–24 mm long, the expanded portion 5–7 mm long, 2-lobed at the tip, white, with a pair of small appendages on the upper surface at the base of the expanded portion. Styles 3. Fruits 8–10 mm long, dehiscing apically by 3 valves that sometimes are split into 6 teeth, with a basal stalklike portion 3–4 mm long. Seeds 0.7–1.0 mm wide, kidney-shaped, the surface finely tuberculate, dark brown to black. $2n=48$. June–August.

Uncommon, sporadic in the eastern half of the state (northeastern U.S. west to Minnesota and Missouri). Banks of streams and rivers, bottomland forests, sloughs; also railroads, roadsides, and disturbed areas.

10. Silene noctiflora L. (night-flowering catchfly, sticky cockle)

Pl. 348 e, f; Map 1489

Plants annual. Stems 25–75 cm long, erect or ascending, unbranched or occasionally branched toward the base, densely pubescent with longer, spreading to somewhat crinkled, nonglandular hairs and shorter, stalked glands. Basal leaves usually withered or absent at flowering, when present shorter than to about as long as the largest stem leaves, short- to occasionally long-petiolate. Stem leaves opposite, usually 6–10 pairs, short-petiolate (lower leaves) or sessile. Leaf blades 1–10 cm long, oblanceolate to elliptic or narrowly lanceolate, tapered at the base, angled or tapered to a sharply pointed tip, the surfaces moderately to densely short-hairy, especially the undersurface also with stalked glands. Flowers perfect, in open terminal clusters or panicles, the stalks 0.3–2.0(–3.0) cm long, with stalked glands, the bracts paired and resembling small leaves, with herbaceous, green margins. Sepals 20–25 mm long, the tube with 10 parallel nerves, with a network of sparse to moderate, anastomosing, angled cross-nerves, ellipsoid to slightly ovoid, narrowed toward the base, green, pale to white between the nerves, the lobes linear

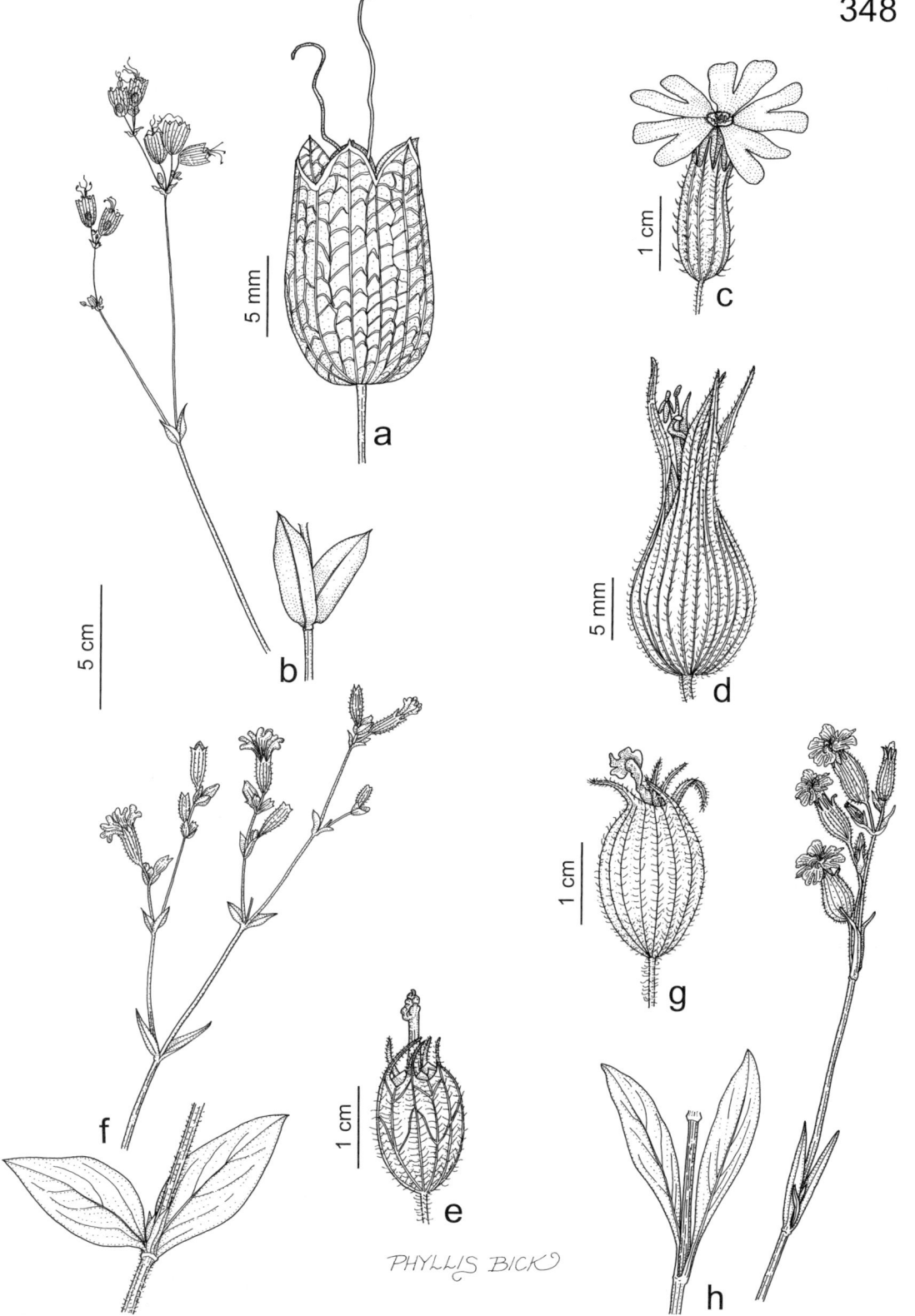

Plate 348. Caryophyllaceae. *Silene vulgaris*, **a)** fruit, **b)** stem with leaves and inflorescence. *Silene dioica*, **c)** flower. *Silene conoidea*, **d)** fruit. *Silene noctiflora*, **e)** fruit, **f)** stem with leaves and inflorescence. *Silene latifolia*, **g)** fruit, **h)** stem with leaves and inflorescence.

to narrowly triangular, green or reddish-tinged, long-tapered to a sharply pointed tip, the margins herbaceous and green or reddish-tinged. Petals 5, 22–28 mm long, the expanded portion 4–6 mm long, 2-lobed at the tip, creamy white or sometimes pinkish-tinged, with a pair of small appendages on the upper surface at the base of the expanded portion. Styles 3. Fruits 11–18 mm long, dehiscing apically by 6 teeth, with a stalklike basal portion 1–3 mm long. Seeds 0.8–1.0 mm wide, kidney-shaped, the surface tuberculate, dark brown to black. $2n=24$. May–September.

Introduced, uncommon and sporadic, mostly in and around urban areas (native of Europe, Asia; introduced nearly throughout the U.S.; Canada). Fallow fields, roadsides, railroads, and open, disturbed areas.

This species has flowers that open in the evening and thus may be confused with *S. latifolia*, but its smaller, creamy white or pinkish-tinged, perfect flowers are distinctive.

11. Silene regia Sims (royal catchfly)

Pl. 346 g–i; Map 1490

Plants perennial, with a stout, fleshy taproot. Stems (35–)50–120 cm long, erect, unbranched below the inflorescence, moderately roughened with short, stiff, downward-angled hairs toward the base and with stalked glands toward the tip, rarely glabrous or nearly so. Basal leaves usually absent at flowering, when present shorter than the largest stem leaves, sessile or with a short, poorly differentiated petiole. Stem leaves opposite, in (6–)10–20 pairs, sessile. Leaf blades 4–14 cm long, ovate to lanceolate, tapered to rounded or somewhat cordate at the base, angled or tapered to a bluntly or sharply pointed tip, the surfaces moderately to densely roughened-hairy or rarely glabrous. Flowers perfect, in open terminal clusters or panicles, the stalks 5–15 mm long, with stalked glands, the bracts paired and resembling small leaves, with herbaceous, green margins. Sepals 17–22 mm long, the tube with 10 parallel, nonanastomosing nerves, tubular to narrowly conical at flowering, narrowed toward the base, becoming somewhat inflated and ellipsoid at fruiting, green, pale between the nerves, with stalked glands, the lobes triangular, green, bluntly pointed at the tip, the margins thin and white or reddish-tinged. Petals 5, 30–40 mm long, the expanded portion 10–16 mm long, entire or with a few fine teeth toward the tip, rarely notched at the tip (but lacking additional lobes), bright red, with a pair of small appendages on the upper surface at the base of the expanded portion. Styles 3 (–5). Fruits 14–16 mm long, opening by 6(8 or 10) teeth, with a stalklike basal portion 3–5 mm long.

Seeds 1.5–2.0 mm wide, kidney-shaped, the surface finely tuberculate, dark reddish brown. $2n=48$. May–October.

Scattered in the Ozark and Ozark Border Divisions and uncommon along the southern margin of the Unglaciated Plains (Ohio to Florida west to Kansas and Oklahoma). Upland prairies, glades, tops of bluffs, savannas, and rocky openings of mesic to dry upland forests; also fencerows, railroads, and roadsides.

Silene regia is a spectacular plant of the tallgrass prairie, with its many flowers and bright red corollas crowning stems sometimes over 1 m tall. Populations have been lost or are threatened by habitat destruction and degradation throughout its range. Missouri has by far the majority of extant populations, but many of these are small and along roadsides or in other degraded habitats. King (1981) mapped several additional counties within the presently circumscribed Missouri range for this species, but voucher specimens for these were not relocated during the present study. *Silene regia* was advocated for use in gardens by Steyermark (1963). It is relatively easily propagated from seed and also is available for sale at most native plant nurseries in the state.

12. Silene stellata (L.) W.T. Aiton (starry campion)

S. stellata var. *scabrella* (Nieuwl.) E.J. Palmer & Steyerm.

Pl. 347 g, h; Map 1491

Plants perennial, with a thick, branched rootstock. Stems 30–110 cm long, erect or ascending, occasionally from a spreading base, unbranched below the inflorescence, moderately to densely pubescent with short, spreading to downward-curled, soft hairs, sometimes nearly glabrous toward the base. Basal leaves usually absent at flowering, when present shorter than the largest stem leaves, sessile or short-petiolate. Stem leaves in whorls of 4, mostly 6–12 pairs, short-petiolate to more commonly sessile. Leaf blades 3–10 cm long, lanceolate, angled or tapered at the base, tapered to a sharply pointed tip. Flowers perfect, in open terminal clusters or panicles, the stalks 0.5–2.5 cm long, glabrous or more commonly densely short-hairy, the bracts paired and resembling very small leaves, with green, herbaceous margins. Sepals 7–12 mm long, the tube with 10 faint nerves and sometimes with a very faint network of fine, irregularly anastomosing veins, deeply cup-shaped to bell-shaped, green, pale between the nerves, glabrous or minutely hairy, the lobes broadly triangular, green, bluntly to sharply pointed at the tip, with herbaceous and green or less commonly thin and white

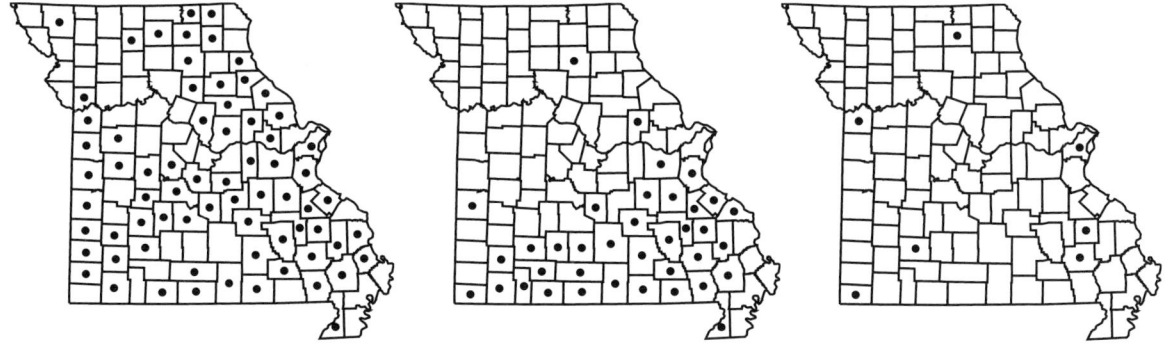

1491. Silene stellata   1492. Silene virginica   1493. Silene vulgaris

margins. Petals 5, 13–16 mm long, the expanded portion 5–8 mm long, irregularly 4–12-lobed (appearing more or less fringed) at the tip, white, lacking appendages. Styles 3. Capsules 6–8 mm long, dehiscing apically by 6 teeth, with a stalklike basal portion 2–3 mm long. Seeds 1.0–1.5 mm wide, kidney-shaped, the surface with fine papillae, dark brown to grayish black. $2n=48$. June–September.

Scattered nearly throughout the state (eastern U.S. west to South Dakota and Texas). Bottomland forests, mesic to dry upland forests, savannas, bottomland prairies, banks of streams and rivers, and margins of ponds and lakes; also fencerows, railroads, and roadsides.

13. Silene virginica L. (fire pink)

Pl. 346 e, f; Map 1492

Plants perennial, with a branched rootstock. Stems 20–70 cm long, erect, unbranched or branched at the base, moderately to densely pubescent with stalked glands, sometimes glabrous or nearly so toward the base. Basal leaves usually numerous at flowering (vegetative rosettes from offsets frequently also present adjacent to the flowering stems), the largest much longer than the largest stem leaves, short- to more commonly long-petiolate. Stem leaves opposite, in 2–4 widely spaced pairs, sessile. Leaf blades 1–12 cm long, those of the basal leaves narrowly to broadly oblanceolate or spatulate, those of the stem leaves oblanceolate to narrowly oblong-elliptic or lanceolate, tapered at the base, angled or tapered to a usually sharply pointed tip (a few of the smaller basal leaves sometimes rounded or bluntly pointed), the surfaces glabrous or short-hairy and/or with stalked glands. Flowers perfect, in open terminal clusters or panicles, the stalks 0.4–2.0 cm long, often angled downward from the base at fruiting, with stalked glands, the bracts paired and resembling small leaves, with herbaceous, green margins. Sepals 16–21 mm long, the tube with 10 parallel, nonanastomosing nerves, more or less tubular at flowering, becoming somewhat inflated and broadly club-shaped at fruiting, green or purplish-tinged, slightly paler and sometimes slightly translucent (at fruiting) between the nerves, with stalked glands, the lobes lanceolate to triangular, green to purple, bluntly or sharply pointed at the tip, the margins herbaceous and green or thin and white to reddish-tinged. Petals 5, 30–45 mm long, the expanded portion 10–19 mm long, with 2 larger and 2 smaller, slender lobes at the tip (appearing irregularly fringed, bright red, with a pair of small appendages on the upper surface at the base of the expanded portion. Styles 3(4). Capsules 14–17 mm long, dehiscing apically by 6(8) teeth, with a stalklike basal portion 2–3 mm long. Seeds 1.0–1.2 mm wide, kidney-shaped, the surface with large, somewhat bulbous papillae, gray. $2n=48$. April–June.

Scattered in the Ozark and Ozark Border Divisions, uncommon and sporadic in the Unglaciated Plains, Glaciated Plains, and Mississippi Lowlands (eastern U.S. west to Iowa, Kansas, Oklahoma, and Louisiana). Bottomland forests, mesic upland forests, banks of streams and rivers, and bases and ledges of bluffs; also edges of pastures and shaded roadsides.

For a discussion of hybridization between *S. virginica* and *S. caroliniana,* see the treatment of that species.

14. Silene vulgaris (Moench) Garcke **ssp. vulgaris** (bladder campion, bladder catchfly)

S. cucubalus Wibel

S. latifolia (Mill.) Rendle & Britten (1907), not Poir. (1789)

Pl. 348 a, b; Map 1493

Plants short-lived perennial from a woody rootstock. Stems 20–80 cm long, erect or ascending, sometimes from a spreading base, few-branched

1494. Spergula arvensis   1495. Spergularia salina   1496. Stellaria graminea

toward the base, glabrous or rarely sparsely pubescent with short, downward-curled hairs toward the base, glaucous. Basal leaves usually withered or absent at flowering, when present shorter than the largest stem leaves, short-petiolate or sessile. Stem leaves opposite, sessile. Leaf blades 2–6 cm long, oblong or broadly elliptic to narrowly lanceolate, tapered at the base (basal leaves) or somewhat clasping (most stem leaves) tapered to a sharply pointed tip, the surfaces glabrous, usually somewhat glaucous. Flowers perfect, in open terminal clusters or panicles, the stalks 0.5–2.0(–3.5) cm long, glabrous, the bracts paired and resembling small leaves, with herbaceous, green margins. Sepals 11–15 mm long, the tube with many faint parallel and anastomosing nerves, narrowly bell-shaped at flowering, becoming inflated and papery at fruiting, ascending to somewhat spreading at the tip and not tightly enclosing the fruit, light green to pale green or purplish-tinged, not paler between the nerves, glabrous, the lobes broadly triangular, green or purplish-tinged, sharply pointed at the tip, the margins herbaceous and green at flowering, similar in texture to the rest of the calyx at fruiting. Petals 5, 14–19 mm long, the expanded portion 4–7 mm long, entire to 2-lobed at the tip, white, with a pair of small appendages on the upper surface at the base of the expanded portion. Styles 3. Capsules 6–9 mm long, dehiscing apically by 6 teeth, with a stalklike basal portion 2–3 mm long. Seeds 1.0–1.5 mm wide, kidney-shaped, the surface finely tuberculate, black. 2n=24. May–August.

Introduced, uncommon and sporadic (native of Europe, Asia, Africa; introduced nearly throughout the U.S. [including Alaska] except some southeastern states, Nevada, Utah, Oklahoma, and Texas; Canada). Disturbed portions of mesic upland forests; also fallow fields, crop fields, railroads, roadsides, and open, disturbed areas.

Botanists in Europe have recognized up to five subspecies of *S. vulgaris* differing in details of leaf shape, capsule size, and habit. However, apparently only ssp. *vulgaris* has been introduced in North America (Morton, 2005c).

18. Spergula L. (spurrey)

Five species, Europe, Asia, introduced nearly worldwide.

As in *Geocarpon,* the flowers in *Spergula* technically are perigynous with a shallow hypanthium having the perianth and stamens attached along its rim. However, this is difficult to observe in most flowers, and these are described below as though they were hypogynous with a short calyx tube.

1. Spergula arvensis L. (corn spurrey)

Pl. 349 f, g; Map 1494

Plants annual. Stems 10–50 cm or more long, erect or ascending to spreading, unbranched or branched (especially above the midpoint), glabrous or moderately to densely pubescent with short, stalked glands. Leaves technically opposite but appearing whorled, with 2 axillary clusters of 8–15 each per node, these all fused basally into an inconspicuous ridge around the stem, sessile. Stipules arising from the basal ridge, 4 per node, thin and white, broadly ovate to triangular, angled or tapered to a bluntly or sharply pointed tip, sometimes narrowly notched at the tip. Leaf blades 1.5–3.0(–5.0) cm long, hairlike to linear, sometimes succulent, more or less tapered at the base, angled to a bluntly or sharply pointed tip, glabrous or with sparse, short, stalked glands. Flowers in open terminal clusters, 0.4–1.5(–3.0) cm long, erect to ascending, at fruiting angled downward from the base, the bracts paired, resembling minute leaves. Epicalyx absent. Sepals 5, fused toward the base, the fused portion short and saucer-shaped, the lobes 3.5–5.0 mm long, elliptic to ovate-elliptic, green or purplish-tinged, rounded or angled to a bluntly pointed tip, not appearing hooded or awned, the margins thin and white, the outer surface glabrous or glandular-hairy. Petals 5, relatively inconspicuous, 2–3(–4) mm long, shorter than to about as long as the sepals and attached along the rim of the calyx tube, more or less ovate, narrowed or short-tapered but not to a stalklike base, rounded to bluntly pointed at the tip, white, lacking appendages. Stamens (5)10, the filaments distinct, attached to the rim of the calyx tube. Staminodes absent. Pistil with 1 locule, the ovary sessile. Styles 5, each with a stigmatic area along the inner surface. Fruits capsules, 3.5–5.0 mm long, dehiscing apically to near the midpoint by 5 spreading or recurved valves. Seeds 5–25, 1.0–1.1 mm wide, subglobose, the surface often appearing pebbled or with minute, white papillae, black, often with a 0.1 mm wide, white wing, lacking an appendage. $2n=18, 36$. May–October

Introduced, uncommon, known thus far with certainty only from a historical collection from Jasper County (native of Europe, Asia; introduced in the eastern U.S. west to Minnesota and Texas, and in the western U.S. from Alaska and Washington to Montana south to California and Wyoming; also south to South America, Africa, Australia). Open, disturbed areas.

Palmer and Steyermark (1935) and Steyermark (1963) mentioned occurrences of this species in Greene and Jackson Counties and suggested that it also grows in agricultural areas and along railroads. However, no specimens to support these occurrences were located during the present research. Contrary to Steyermark's (1963) assertion that the species may be more common in western Missouri than the specimen record indicates, instead it seems likely that corn spurrey has not persisted to modern times in the state. *Spergula arvensis* is known in North America mostly as a weed in sandy fields and disturbed areas, especially in some eastern states and in the Pacific Northwest. However, in Michigan it was grown for testing as a forage crop on nutrient-poor, sandy soil (Clute and Palmer, 1893).

19. **Spergularia** (Persoon) J. Presl & C. Presl (sandspurrey)

About 60 species, North America, South America, Europe, Africa, particularly in saline coastal areas; introduced widely.

1. Spergularia salina J. Presl & C. Presl
   (saltmarsh sandspurrey)
   S. marina (L.) Griseb.

Map 1495

Plants annual. Stems 8–20(–30) cm long, erect or ascending to spreading, usually many-branched above the midpoint, moderately to densely pubescent with short, stalked glands, sometimes nearly glabrous toward the base. Leaves opposite, not fused basally, sessile, with axillary clusters of leaves rarely present. Stipules 2 per node, papery and dull white, broadly triangular, angled or short-tapered to a sharply pointed tip. Leaf blades (0.8–)1.5–4.0 cm long, linear, succulent, tapered at the base, angled or short-tapered to a bluntly or sharply pointed tip, the surfaces with short, stalked glands. Flowers in terminal, sometimes compound, open clusters or solitary and axillary, sometimes appearing racemose, the stalks 0.1–0.7 cm long, erect or ascending, at fruiting angled downward from the base, the bracts usually paired, resembling small leaves, with green or less commonly papery and white margins. Epicalyx absent. Sepals 5, fused basally for about ⅕ the length, the lobes 2–3 mm long (becoming elongated to 4.0 mm at fruiting), elliptic to ovate, rounded or angled to a bluntly or sharply pointed tip, not appearing hooded or awned, the margins thin or papery and white. Petals 5, usually relatively inconspicuous, 1.5–2.0 mm long, shorter than the sepals, ovate to oblong-elliptic, narrowed or short-tapered but not to a stalklike base, rounded to bluntly pointed at the tip, pink (less commonly all or partially white elsewhere), lacking appendages. Stamens (1–)2–3(–5), the filaments distinct, attached near the base of the fused portion of the calyx. Staminodes absent. Pistil with 1 locule, the ovary sessile. Styles 3, each with a stigmatic area along the inner surface. Fruits capsules, 2.8–4.5(–6.4) mm long, opening by 3 spreading to recurved valves. Seeds 20–30 or more, 0.5–0.7 mm wide, broadly ovoid to nearly globose, the surface smooth or often with minute, gland-tipped papillae, light or reddish brown, sometimes with an incomplete, brownish tan wing, lacking appendages. 2n=36. August.

Introduced, uncommon, known thus far only from a single collection from St. Charles County (native of Europe, Asia, and possibly western North America; introduced nearly worldwide). Banks of rivers; presumably also open, disturbed areas.

Spergularia salina was first collected by Neil Snow in 1994 on a floodplain terrace of the Missouri River under a highway bridge. It should be expected elsewhere in the state, especially around metropolitan areas, along highways that are salted during winter, and in the floodplains of large rivers. The inland plants tend to be somewhat smaller in stature and floral dimensions than some populations from coastal salt marshes.

20. Stellaria L. (stitchwort, chickweed)

Plants annual or perennial. Stems erect or ascending to reclining or spreading, unbranched or more commonly branched, often above the midpoint, glabrous, sparsely papillate, or with hairs in 2 longitudinal lines. Leaves opposite, fused basally into a small sheath, short-petiolate (some basal leaves) or sessile, lacking axillary clusters of leaves. Stipules absent. Leaf blades linear, lanceolate, oblong, ovate, triangular-ovate, or elliptic, not fleshy, tapered to rounded or nearly truncate at the base, angled or tapered to a sharply pointed tip. Flowers in terminal or axillary open panicles or clusters or sometimes solitary, the stalks ascending or arched to curved, at fruiting sometimes angled downward from the base, the bracts paired and resembling small leaves and green or much reduced, scalelike, and translucent to white with at most a green midrib. Epicalyx absent. Sepals (4)5, distinct or fused at the base, lanceolate or triangular to ovate-elliptic, green, the margins green or thin and white to translucent, bluntly or sharply pointed at the tip, not hooded or awned. Petals 5 (often appearing as 10) or absent, when present variously shaped, angled or tapered but not to a stalklike base, deeply lobed at the tip to $^2/_3$–$^4/_5$ of the way to the base, white, lacking appendages. Stamens 1–10 (most commonly 5 or 10), the filaments distinct, attached along a small nectar disc. Staminodes usually absent. Pistil with 1 locule, the ovary sessile. Styles 3 (2, 4, or 5 elsewhere), distinct, each with a subterminal or terminal stigmatic area. Fruits capsules, dehiscing apically by 6 ascending to recurved valves or teeth. Seeds 10–20 or more, kidney-shaped to circular in outline, the surface smooth, tuberculate, or coarsely wrinkled, yellowish brown to brown or nearly black, lacking wings or appendages. About 190 species, nearly worldwide, most diverse in temperate regions; introduced nearly worldwide.

In addition to the species treated below, one other species should be mentioned. *Stellaria pubera* Michx. (great chickweed, star chickweed) is a plant of rich woodlands in many areas to the east and southeast of Missouri. Steyermark (1963) included this species with some reservations, based on a single historical collection from Franklin County. However, Yatskievych and Turner (1990) excluded it from the state's flora because John Kellogg's collection in 1927 almost certainly originated from plants cultivated at what was then the Missouri Botanical Garden's recently acquired property in Gray Summit (now Shaw Nature Reserve). *Stellaria pubera* would key imperfectly to either *S. media* or *S. neglecta* in the key to species below. It differs from both of these species most noticeably in its showier flowers with the petals 4–8 mm long and distinctly longer than the sepals. Although no populations have been discovered in the state in spite of Steyermark's treatment of the species, perhaps it will be found somewhere in eastern Missouri in the future.

1. Leaf blades linear, lanceolate, or narrowly elliptic; lower leaves sessile (rarely short-petiolate on vegetative shoots); inflorescence bracts much reduced, membranous and white to translucent with at most a green midrib
   2. Leaf blades lanceolate to narrowly lanceolate or narrowly elliptic, the margins not roughened (but may be inconspicuously hairy near the blade base);

seeds coarsely wrinkled; bracts and often also the sepals with the margins finely short-hairy .... 1. S. GRAMINEA
2. Leaf blades linear, occasionally very narrowly elliptic, the margins roughened with microscopic papillae (this easier to feel than to see); seeds smooth or nearly so; bracts and sepals with the margins glabrous or the bracts (but not the sepals) very rarely finely hairy .... 2. S. LONGIFOLIA
1. Leaf blades oblong, ovate, triangular-ovate, or elliptic; lower leaves rounded to subtruncate at the base with distinct petioles; inflorescence bracts leaflike, herbaceous
3. Plants usually yellowish green when fresh (this sometimes difficult to observe in dried material); flowers usually cleistogamous and lacking petals; sepals 2.5–3.0(–4.0) mm long, usually with an inconspicuous reddish band at the base; stamens 1–3; seeds 0.5–0.8 mm wide, yellowish brown .... 5. S. PALLIDA
3. Plants green or dark green, rarely yellow-green; flowers not cleistogamous, usually with petals; sepals (3–)4–6 mm long, lacking a reddish band at the base; stamens 3–5 or 8–10; seeds (0.8–)0.9–1.7 mm wide, brown to dark brown (sometimes yellowish brown before fully mature)
4. Sepals (3.0–)4.0–4.5(–6.0) mm long; stamens 3–5(–8); seeds brown, (0.8–)0.9–1.4 mm wide, at least the tubercles along the margin broader than tall, more or less hemispherical, blunt or rounded at the tip ...... 3. S. MEDIA
4. Sepals 5–6 mm long; stamens (5–)8–10; seeds brown to dark brown, 1.1–1.7 mm wide, at least the tubercles along the margin taller than broad, conical, sharply pointed .... 4. S. NEGLECTA

1. Stellaria graminea L. (common stitchwort)
S. graminea var. *latifolia* Peterm.

Pl. 349 a, b; Map 1496

Plants perennial, with rhizomes, green to dark green. Stems 10–50 cm long, ascending or sprawling, branched, especially toward the tip, glabrous. Leaves sessile, or the lower leaves on late-season vegetative shoots rarely short-petiolate. Leaf blades 1.5–5.0 cm long, lanceolate to narrowly lanceolate or narrowly elliptic, tapered to short-tapered at the base, angled or tapered to a sharply pointed tip, the margins sometimes inconspicuously hairy near the blade base. Flowers not cleistogamous, in open terminal clusters or panicles, the stalks 1–3 cm long, arched or curved to erect at flowering and fruiting, the bracts small, membranous, white to translucent with at most a green midrib, the margins finely hairy, sometimes sparsely so. Sepals 5, 3.5–5.0 mm long, lanceolate to narrowly triangular, lacking a reddish band at the base, sharply pointed at the tip, the margins thin and white to translucent, often finely hairy. Petals 5, 3–7 mm long, as long as or more commonly slightly longer than the sepals. Stamens 10. Fruits 5–6 mm long, the valves ascending at dehiscence. Seeds 0.9–1.1 mm wide, the surface coarsely wrinkled, black. $2n=52$.

Introduced, uncommon and sporadic, mostly in urban areas (native of Europe; introduced in the eastern U.S. west to Minnesota and Kansas and in the western U.S. from Washington to California east to Montana and Colorado; Canada). Old fields, lawns, and disturbed areas. May–October.

First collected in Boone County in 1929, *S. graminea* is sometimes confused with *S. longifolia* but is seldom found in the wet, native habitats where *S. longifolia* grows. In addition to the characters in the key to species above, it differs in its strongly angled stems that are smooth along the angles and its more strongly 3-veined, stiffer sepals.

2. Stellaria longifolia Muhl. ex Willd. (long-leaved stitchwort)

Pl. 349 c; Map 1497

Plants perennial, with rhizomes, green to dark green. Stems 15–45 cm long, erect or ascending to sprawling, branched especially toward the tip, glabrous or minutely roughened along the angles. Leaves sessile. Leaf blades 2–5 cm long, linear or occasionally very narrowly elliptic, tapered to short-tapered at the base, angled or tapered to a sharply pointed tip, the margins roughened with microscopic papillae (this noticeable by touch but not visible without strong magnification). Flowers not cleistogamous, in terminal or often apparently lateral (at the tips of branches produced from the upper leaf axils) clusters or panicles, 0.3–3.0 cm long, often arched upward at flowering and fruiting, the bracts small, membranous, white to translucent with at most with a green midrib, the margins glabrous or very rarely finely hairy. Sepals 5, 2.5–4.0

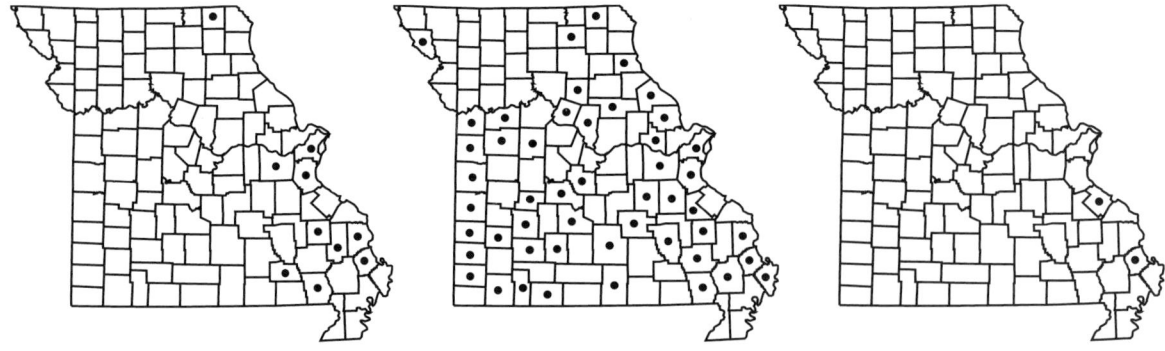

1497. Stellaria longifolia   1498. Stellaria media   1499. Stellaria neglecta

mm long, lanceolate, lacking a reddish band at the base, sharply pointed at the tip, the margins thin and white or translucent, glabrous. Petals 5, 2.5–3.5 mm long, about as long as the sepals. Stamens 10. Fruits 4–6 mm long, the valves ascending at dehiscence. Seeds 0.7–0.8 mm wide, the surface smooth or nearly so, brown. $2n=26$. May–July.

Uncommon in the eastern half of the state (northern U.S. south to South Carolina, Missouri, Arizona, and California; Canada). Banks of streams and rivers and bottomland forests; also margins of crop fields, pastures, and roadsides.

3. Stellaria media (L.) Vill. (common chickweed)
Pl. 349 d, e; Map 1498

Plants annuals or short-lived perennials, green to dark green or rarely yellowish green. Stems 10–80 cm long, erect or ascending to spreading, branched, usually short-hairy in longitudinal lines, rarely nearly glabrous. Leaves petiolate (basal and lower stem leaves) or sessile (median and upper stem leaves). Leaf blades 0.3–4.0 cm long, ovate to elliptic, rounded to nearly truncate at the base, angled or slightly tapered at the tip, the margins glabrous or inconspicuously hairy. Flowers not cleistogamous, in terminal clusters or sometimes solitary, the stalks 0.3–4.0 cm long, ascending at flowering, often angled downward from the base at fruiting, the bracts herbaceous and resembling small leaves. Sepals 5, (3.0–)4.0–4.5(–6.0) mm long, oblong-lanceolate, lacking a reddish band at the base, bluntly pointed at the tip, the margins thin and white, finely hairy. Petals 5 or occasionally absent, when present 1–4 mm long, shorter than to about as long as the sepals. Stamens 3–5(–8). Fruits 4–6 mm long, the valves ascending or rarely recurved at dehiscence. Seeds (0.8–)0.9–1.4 mm wide, the surface tuberculate, the tubercles along the marginal portion broader than tall, more or less hemispherical, blunt or rounded at the tip, brown to reddish brown. $2n=40, 42, 44$. January–December.

Introduced, scattered to common throughout the state, more abundantly south of the Missouri River (native of Europe; introduced throughout the U.S. [including Alaska, Hawaii]; Canada, also nearly worldwide). Banks of streams, rivers, and spring branches, bases and ledges of bluffs, sloughs, and bottomland forests; also pastures, crop fields, fallow fields, ditches, lawns, gardens, roadsides, and disturbed areas.

Stellaria media is one of the most widely distributed weeds in the world. This species is morphologically quite varied and specimens collected in the spring can be difficult to distinguish from the closely related *S. neglecta* and *S. pallida*. Small plants flowering early in the growing season can be confused easily with *S. pallida*, and robust plants can be very similar to *S. neglecta*. In both cases, seed size and ornamentation are important diagnostic features.

Steyermark (1963) noted that in its native range, young herbage of this species sometimes was gathered for use as a spinach substitute. He also discussed that birds (including domesticated songbirds) relish the seeds and that livestock sometimes eat the plants. He advised caution, however, citing anecdotal reports of poisoning of lambs that had ingested large quantities of this species.

4. Stellaria neglecta Weihe (greater chickweed)
S. media (L.) Vill. ssp. *neglecta* (Weihe) Murb.
Map 1499

Plants annual, green or dark green or rarely yellowish green. Stems 35–60(–80) cm long, erect or ascending to less commonly sprawling or spreading, branched, with a longitudinal line of short, stalked glands. Leaves petiolate (basal and lower stem leaves) or sessile (median and upper stem leaves). Leaf blades 0.8–5.0 cm long, broadly ovate to elliptic, rounded to nearly truncate at the base, angled or slightly tapered to a sharply pointed tip, the margins glabrous or inconspicuously hairy.

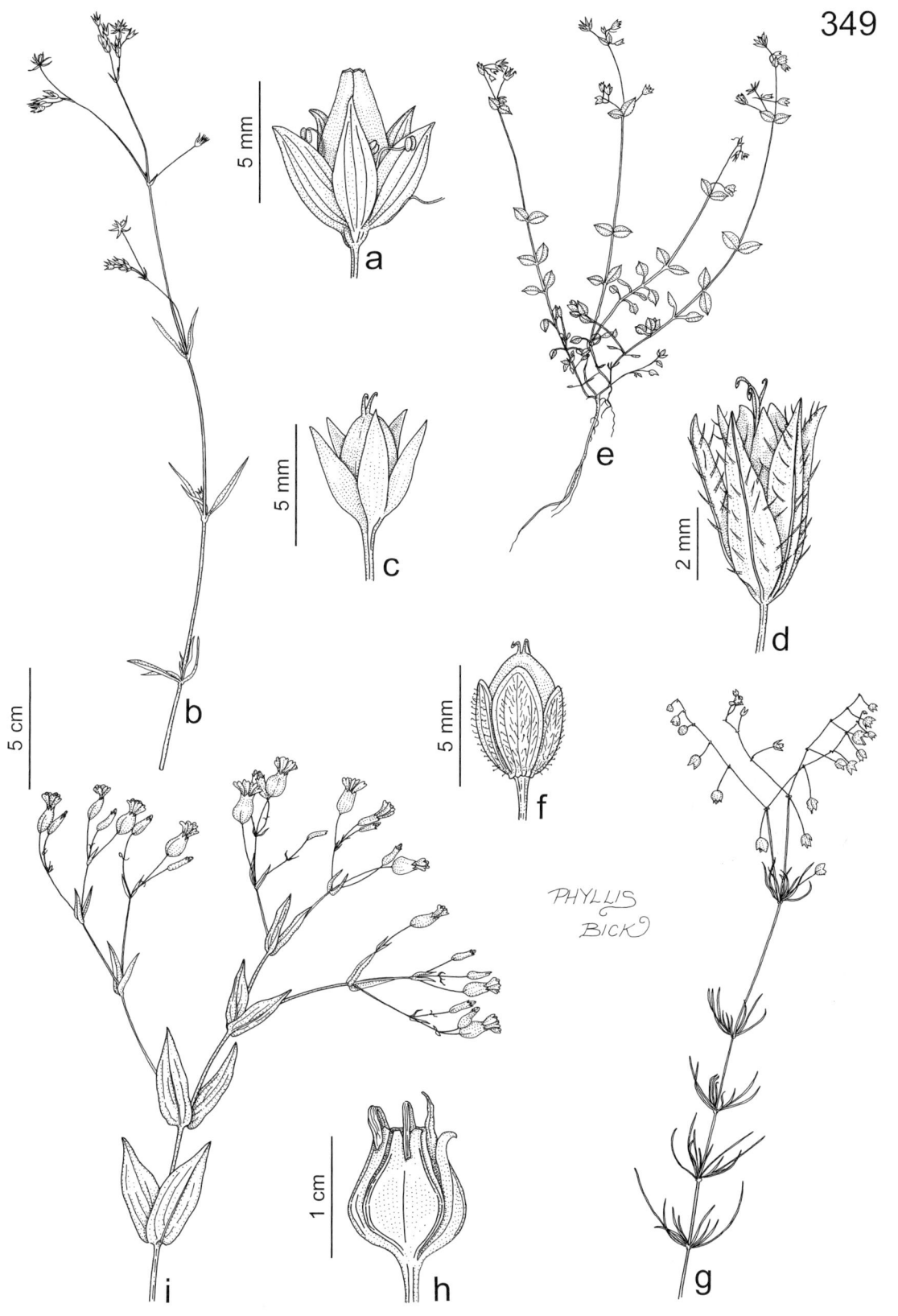

Plate 349. Caryophyllaceae. *Stellaria graminea*, **a)** fruit, **b)** habit. *Stellaria longifolia*, **c)** fruit. *Stellaria media*, **d)** fruit, **e)** habit. *Stellaria arvensis*, **f)** fruit, **g)** habit. *Vaccaria hispanica*, **h)** fruit, **i)** habit.

1500. Stellaria pallida   1501. Vaccaria hispanica   1502. Celastrus orbiculatus

Flowers not cleistogamous, in terminal clusters or sometimes solitary, the stalks 1.0–4.5 cm long, ascending at flowering, often angled downward from the base at fruiting, the bracts herbaceous and resembling small leaves. Sepals 5, 5–6 mm long, oblong-lanceolate, lacking a reddish band at the base, mostly sharply pointed at the tip, the margins thin and white, sparsely and finely hairy. Petals 5 or occasionally absent, 3–5 mm long, shorter than to about as long as the sepals. Stamens (5–)8–10. Fruits 5–6 mm long, the valves ascending with usually recurved tips at dehiscence. Seeds 1.1–1.7 mm wide, the surface tuberculate, the tubercles along the marginal portion taller than broad, conical, sharply pointed at the tip, brown to dark brown (sometimes yellowish brown before full maturity). $2n=22$. April–June.

Introduced, uncommon, known thus far only from Scott and Ste. Genevieve Counties (native of Europe; introduced sporadically in the southeastern U.S. west to Oklahoma and Louisiana, also in California). Banks of streams and bases of bluffs; also lawns and disturbed areas.

First collected in Missouri from Ste. Genevieve County by George Yatskievych in 1993, *S. neglecta* rarely has been reported in North America. Morton (2005b) suggested that the increase in North American populations has occurred since about 1990. The present treatment follows that of Morton in including plants with the petals shorter than the sepals in this species, although most European treatments (for example, Nilsson, 2001) restrict the name *S. neglecta* to plants in which the petals are equal to or often longer than the sepals. North American plants do produce seeds with conical tubercles that are consistent with descriptions and illustrations from the European literature (for example, Berggren, 1981).

5. **Stellaria pallida** (Dumort.) Crép. (lesser chickweed)
   S. media (L.) Vill. ssp. *pallida* (Dumort.) Asch. & Grabn.

Map 1500

Plants annual, usually yellowish green. Stems 10–25 cm long, ascending to spreading, branched or unbranched, glabrous or sparingly short-hairy, the hairs sometimes not in lines. Leaves petiolate (basal and lower stem leaves) or sessile (median and upper stem leaves). Leaf blades 0.2–0.9 cm long, ovate to elliptic, rounded to nearly truncate at the base, usually slightly tapered to a sharply pointed tip, the margins glabrous. Flowers all or mostly cleistogamous, in terminal clusters, the stalks 0.1–1.0 cm long, erect or ascending at flowering and fruiting, the bracts herbaceous and resembling small leaves. Sepals (4)5, 2.5–3.0(–4.0) mm long, oblong-lanceolate, usually with an inconspicuous reddish band around the base, sharply pointed at the tip, the margins herbaceous and green, glabrous. Petals absent or, if occasionally present, then minute (to 0.5 mm long). Stamens 1–3. Fruits 3–4 mm long, the valves ascending with usually recurved tips at dehiscence. Seeds 0.5–0.8 mm wide, the surface tuberculate, the tubercles along the marginal portion broader than tall, more or less hemispherical, blunt or rounded at the tip, yellowish brown. $2n=22$. March–May.

Introduced, uncommon, known thus far sporadically from portions of southern and central Missouri (native of Europe; introduced in the eastern U.S. west to Nebraska, Colorado, and Texas, also Arizona, California, and Washington; Canada, Mexico). Margins of sinkhole ponds; also pastures, lawns, roadsides, and disturbed areas.

This species was first collected from Missouri in Boone County by David Dunn in 1957, but its presence in North America was not reported for another fifteen years (Morton, 1972). It undoubtedly is far more abundant in Missouri than the meager herbarium record indicates. Most specimens of S. *pallida* initially were identified as S. *media*, mostly over confusion with small S. *media* plants that are collected in the early spring and because the two species may be found growing together. It appears that populations of S. *pallida* have increased dramatically in recent years; for example, see the report by Hyatt (2001) for Arkansas. Although the seeds offer the most consistent feature for separating the plants, the reddish band that is usually found at the base of sepals in S. *pallida* plants combined with the erect fruit stalks that often detach at the base are usually good characters for recognizing S. *pallida* (Rabeler, 1988).

The cleistogamous flowers of the lesser chickweed require some explanation. They are obligately self-pollinated and appear more or less closed and somewhat flask-shaped at flowering. The sepals generally are cupped tightly around the ovary and stamens, but with their tips somewhat curved outward above the ovary. The closed nature of the flower ensures that the stamens remain in proximity to the stigmatic areas of the styles as the pollen is shed. However, after pollination, as the ovary enlarges during the development of the fruit, the sepals are forced apart and the flower then appears more or less open.

21. Vaccaria Wolf

One or 4 species, Europe, Asia; introduced nearly worldwide.

1. Vaccaria hispanica (Mill.) Rauschert (cow herb, cow soapwort, cow cockle)
V. pyramidata Medik.
V. hispanica ssp. *pyramidata* (Medik.) Holub
Saponaria vaccaria L.

Pl. 349 h, i; Map 1501

Plants annual. Stems 20–100 cm long, erect, branched above the midpoint, glabrous and glaucous. Leaves opposite, clasping or fused basally into a sheath, short-petiolate (basal leaves) or sessile (stem leaves), lacking axillary clusters of leaves. Stipules absent. Leaf blades 2–10 cm long, lanceolate to ovate-lanceolate or oblong, not fleshy, angled to cordate at the base, angled or tapered to a bluntly or more commonly sharply pointed tip, the surfaces glabrous, somewhat glaucous. Flowers in terminal open, often flat-topped panicles, the stalks (0.5–)1.0–3.0(–5.5) cm long, erect or ascending at flowering and fruiting, the bracts paired and resembling small leaves. Epicalyx absent. Sepals 5, 9–17 mm long, fused most of their length into a tube, this more or less cylindrical to narrowly flask-shaped and strongly 5-angled or winged, each angle or wing with a prominent nerve, whitish green, herbaceous between the sepals, the lobes obovate to broadly triangular, much shorter than the tube, angled or short-tapered to a sharply pointed tip, not appearing hooded or awned, the margins herbaceous and green or more commonly thin and translucent or purplish-tinged. Petals 5, showy, 1.1–2.2 cm long, oblanceolate to spatulate, tapered to a relatively broad, membranous, straplike or stalklike base, the expanded portion 0.3–0.8 cm long, rounded, truncate, or shallowly and broadly notched at the tip, sometimes appearing slightly uneven or ruffled, pink or purplish pink, lacking appendages. Stamens 10, the filaments distinct, attached to the basal portion of the petals. Staminodes absent. Pistil with 1 locule or rarely appearing 2-locular toward the base, with a short stalk 0.5–1.5 mm long. Styles 2(3), each with a stigmatic area along the inner surface. Fruits capsules, (8–)10–13 mm long, the outer wall layer dehiscing apically by 4(6) ascending or slightly spreading teeth, the inner layer dehiscing irregularly. Seeds 8–12, 2.0–2.5 mm wide, subglobose, the surface with papillae, black, lacking wings or appendages. $2n=30$. May–September.

Introduced, uncommon and sporadic (native of Europe, Asia; introduced nearly worldwide). Railroads, roadsides, and open, disturbed areas.

The taxonomy of *Vaccaria* is still not fully resolved. Botanists have treated it as comprising either four species or a single species (*V. hispanica*) with up to four subspecies (of which ssp. *hispanica* is the widespread weed). In the United States, this taxon formerly was more abundant, especially in the plains and midwestern states, as a crop weed. In recent decades, populations have been declining, in part due to a combination of improved grain handling and the more widespread application of herbicides. However, in Missouri it does not appear ever to have been very common. Mühlenbach (1979) noted that it was rarely observed in his botanical inventory of the St. Louis railyards, often as only 1–3 individuals at a site. The seeds of *Vaccaria* contain saponins and are poisonous (Burrows and Tyrl, 2001).

CELASTRACEAE (Staff-Tree Family)
(Brizicky, 1964b)
Contributed by David J. Bogler

Plants trees, shrubs, or lianas, glabrous or nearly so, sometimes incompletely or completely dioecious, occasionally with milky sap. Leaves alternate or opposite, simple, pinnately veined, the margins entire or bluntly toothed, sometimes partially or fully evergreen. Stipules absent or small, scalelike, and shed early. Inflorescences axillary or terminal clusters (sometimes appearing as short racemes or small panicles in *Celastrus*). Flowers perfect or imperfect, actinomorphic, hypogynous or perigynous, small, the stalks jointed. Sepals 4 or 5, small, usually fused toward the base. Petals 4 or 5, separate. Stamens 4 or 5, opposite the sepals, the filaments short, usually distinct, the short, broad anthers attached at or near the base. A nectar disc is present between the stamens and ovary, usually conspicuous, the petals and stamens sometimes appearing attached to the disc. Pistil of 2–5 fused carpels, the ovary superior or appearing partially inferior (the basal portion sometimes appearing sunken into the disc), 1–5-locular, the ovules 2 to several per locule, the placentation axile. Style 1, the stigma capitate or 3-lobed. Fruits capsules (drupes or berries elsewhere), sometimes strongly lobed, dehiscent longitudinally by 3–5 valves. Seeds 3–5 (more elsewhere), covered by a brightly colored, red or orange, fleshy aril. Fifty-five to 90 genera, 850–1,300 species, nearly worldwide, but most diverse in tropical and subtropical regions.

The Celastraceae are of limited economic importance, although a number of species are cultivated as ornamentals, and some taxa are considered invasive exotics. *Euonymus* is used in Chinese herbal medicine. The leaves of *Catha edulis* (Vahl) Endl., called khat, are chewed or brewed into tea for their stimulant effect in East Africa and the Middle East.

1. Leaves alternate; plants lianas with twining stems . . . . 1. CELASTRUS
1. Leaves opposite; plants shrubs or small trees, or if lianas then the stems not twining (adhering to substrate by modified roots) . . . . 2. EUONYMUS

1. Celastrus L.
(Hou, 1955)

Plants twining lianas, incompletely or rarely completely dioecious, sometimes spreading by stolons and root suckers. Stems 1–20 m or more long, the branches circular in cross-section, not winged. Leaves alternate, deciduous, short-petiolate. Leaf blades variable in shape (even on the same plant), the margins finely toothed. Inflorescences axillary or terminal clusters, sometimes appearing as short racemes or small panicles. Flowers usually imperfect. Sepals 5. Petals 5. Staminate flowers with 5 stamens, these inserted under the margin of the often lobed nectar disc, the filaments 1.5–2.0 mm long. Pistillate flowers with minute staminodes, the ovary usually with 3 locules and 2 ovules per locule. Style short, stout, the stigma deeply 3-lobed. Fruits more or less globose, 3-lobed, orange to yellow, dehiscent by 3 valves. Seeds 4–5 mm long, ovoid to ellipsoid, 3–6, each enclosed in a fleshy red to orangish red aril. About 30 species, North America to South America, Asia to Australia, Madagascar.

This genus is closely related to the large pantropical genus *Maytenus* Molina (Hou, 1955). The sexuality of the flowers is variable. Some individuals have only staminate flowers and never produce fruit, some have pistillate flowers with abortive stamens, and some are mostly unisexual but produce a few perfect flowers. Birds are attracted to and eat the bright red arillate seeds, but these are poisonous to humans. The boiled bark has a sweet flavor and was used as a famine food by various Indian tribes (Dillingham, 1907). The fruits of bittersweet commonly are used in wreaths and other winter decorations.

In his original description of the genus, Linnaeus (1753) treated *Celastrus* as masculine and used the *-us* ending for the species name *C. orbiculatus*. However, the classical Greek root for the generic epithet is feminine, and many of the species names have been spelled inconsistently in the botanical literature. Paclt (1998a) made a formal proposal to conserve the name *Celastrus* as feminine, but the Committee for Spermatophyta of the International Association of Plant Taxonomy, which must approve such proposals before they can be voted upon at an International Botanical Congress, chose to table this proposal until a later date (Brummitt, 2000), and no formal ruling on the proper gender of the name has been published to date. As this situation is identical to that found in *Euonymus* (see below) and in that case the committee did rule against changing the original usage of Linnaeus from masculine to feminine, the name *Celastrus* is here treated as masculine.

1. Leaves often obovate to suborbicular; flowers in axillary clusters of 2–4; mature fruits with the valves yellow .... 1. C. ORBICULATUS
1. Leaves more or less elliptic; flowers numerous in terminal clusters; mature fruits with the valves orange .... 2. C. SCANDENS

1. Celastrus orbiculatus Thunb. (Oriental bittersweet)

C. articulatus Thunb.

Pl. 350 d, e; Map 1502

Bark dark brown, not exfoliating. Leaves with petioles 1–2 cm long. Leaf blades 3–8 cm long, 2–5 cm wide, obovate to ovate, broadly elliptic, or nearly circular, those of the flowering branches mostly broadly elliptic to nearly circular, the margins bluntly toothed, rounded, angled, or short-tapered at the base, rounded to strongly tapered at the tip, the upper surface glabrous, green to dark green, the undersurface glabrous or rarely sparsely hairy along the veins, light green to pale green. Flowers in clusters of 2–5(–7), axillary, the staminate flowers also in terminal clusters. Sepals 1.0–1.5 mm long. Petals 3–5 mm long, 1.0–1.5 mm wide, narrowly oblong, rounded at the tip, greenish white. Fruits 6–8 mm long, 8–9 mm in diameter, the valves with the outer surface yellow at maturity, the inner surface orangish yellow. $2n=46$. May–June.

Introduced, uncommon and sporadic but often locally abundant (native of eastern Asia; introduced in the eastern U.S. west to Wisconsin, Nebraska, and Arkansas; Canada). Bottomland forests and mesic upland forests; also old fields.

Celastrus orbiculatus was first reported for Missouri from Boone County by Yatskievych and Summers (1993). It is distinguished from the very similar *C. scandens* by having flowers produced in small axillary clusters, in contrast to the terminal clusters in *C. scandens*. The leaves of *C. orbiculatus* tend to be more rounded on the flowering branches, but both species can vary greatly in leaf shape on a given individual. The two species have the same chromosome number and can be artificially crossed, but the offspring are not vigorous and only sparingly fertile (White and Bowden, 1947).

Celastrus orbiculatus is fast becoming a serious weed in the eastern United States. It is reported to be more vigorous than the native *C. scandens*, with higher seed viability (Dreyer et al., 1987). The fruits and seeds are attractive to birds, which disperse the seeds widely. *Celastrus orbiculatus* proliferates by root suckers, grows fast, and can form a smothering blanket over whole plant communities. The tightly twining stems can constrict the phloem of other plants on which they twine and can kill young trees (Lutz, 1943). These plants are invasive and highly tolerant of low light conditions. In some areas, *C. orbiculatus* appears to be replacing *C. scandens*, although the native taxon appears to tolerate somewhat drier conditions than the invader.

2. Celastrus scandens L. (American bittersweet)

Pl. 350 f, g; Map 1503

Bark light gray, that of older branches smooth. Leaves with petioles 1–2 cm long. Leaf blades 5–12 cm long, 2–7 cm wide, mostly elliptic or obovate, the margins bluntly toothed, rounded to angled or short-tapered at the base, rounded to more commonly sharply tapered at the tip, both surfaces glabrous, the upper surface green to dark green, the undersurface light green to pale green. Flowers in clusters of 12–40, terminal on the branches. Sepals 1.0–1.5 mm long. Petals 3–4 mm long, 1.0–1.5 mm

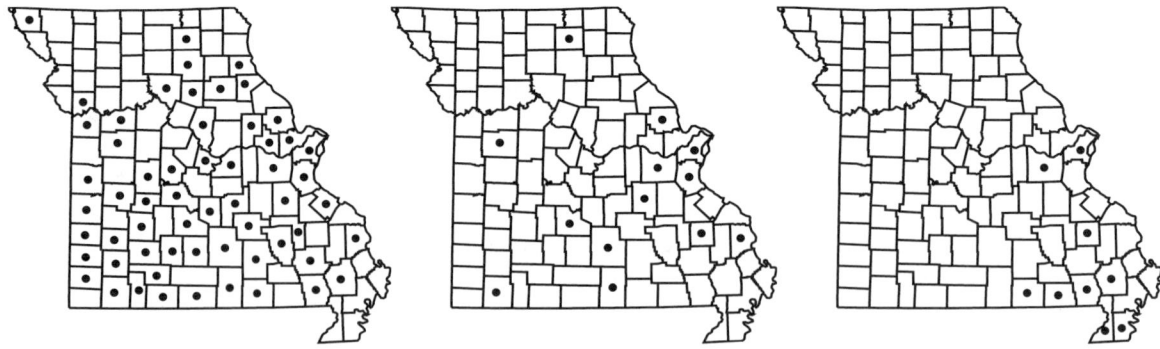

1503. Celastrus scandens   1504. Euonymus alatus   1505. Euonymus americanus

wide, narrowly oblong, rounded at the tip, greenish white to white. Fruits 6–8 mm long, 8–10 mm in diameter, the valves with the outer surface orange at maturity, the inner surface pale yellow to nearly white. $2n=46$. May–June.

Scattered nearly throughout the state (eastern U.S. to Montana and Texas; Canada). Bottomland forests, mesic upland forests, bases and ledges of bluffs, and margins of upland prairies, loess hill prairies, and glades; also fencerows, railroads, and roadsides.

This is the only species of Celastrus native to North America. Bees are probably the major pollinators, although wind pollination also may occur.

2. Euonymus L.
(Ma, 2001)

Plants small trees or shrubs, sometimes trailing or climbing but not twining (anchored to the substrate by adventitious roots). Twigs 4-angled or sometimes circular in cross-section, sometimes with corky wings. Leaves opposite, deciduous or partially to completely evergreen, short-petiolate. Leaf blades variously ovate to elliptic or obovate, the margins entire or finely toothed, the surfaces glabrous or inconspicuously hairy. Inflorescences axillary clusters or small panicles or the flowers occasionally solitary in the leaf axils. Flowers usually perfect (rarely functionally staminate or pistillate). Sepals 4 or 5, fused toward the base. Petals 4 or 5. Stamens 4 or 5, these inserted along the margin of the nectar disc, the filaments minute (0.1–0.5 mm). Ovary usually with 1–5 locules and 2–6 ovules per locule. Style short, stout, the stigma entire or shallowly 3-lobed. Fruits ovoid or more or less globose, sometimes (2)3–5-lobed, pink to red, dehiscent by (2)3–5 valves. Seeds ovoid to ellipsoid, 1 or 2(–6) per locule, brown, each enclosed in a fleshy red to orange aril. About 130 species, nearly worldwide.

The taxonomy of Euonymus has been somewhat confused. There is considerable variation in growth habit and leaf characters, with many variants sometimes receiving separate names and numerous species described from horticultural selections of Asian origin. Many of these problems, especially those concerning the Oriental species, were addressed by Ma (2001), who reduced the number of accepted species from 200 or more to 129.

The generic name Euonymus has engendered a lot of controversy. Linnaeus (1753) spelled it "Evonymus" in his Species Plantarum and "Euonymus" in his Genera Plantarum (Linnaeus, 1737). After decades of debate (Zijlstra and Tolsma, 1991), the spelling Euonymus was conserved at the International Botanical Congress in Tokyo in 1993. Also, Linnaeus (1753) treated Euonymus and its species with a masculine (-us) ending, but the classical Greek root is considered feminine. After decades of inconsistent species-epithet endings, Paclt (1998b) made a formal proposal to conserve the name as feminine (with -a rather than -us) for the species epithets. However, the Committee for Spermatophyta of the International Association of Plant

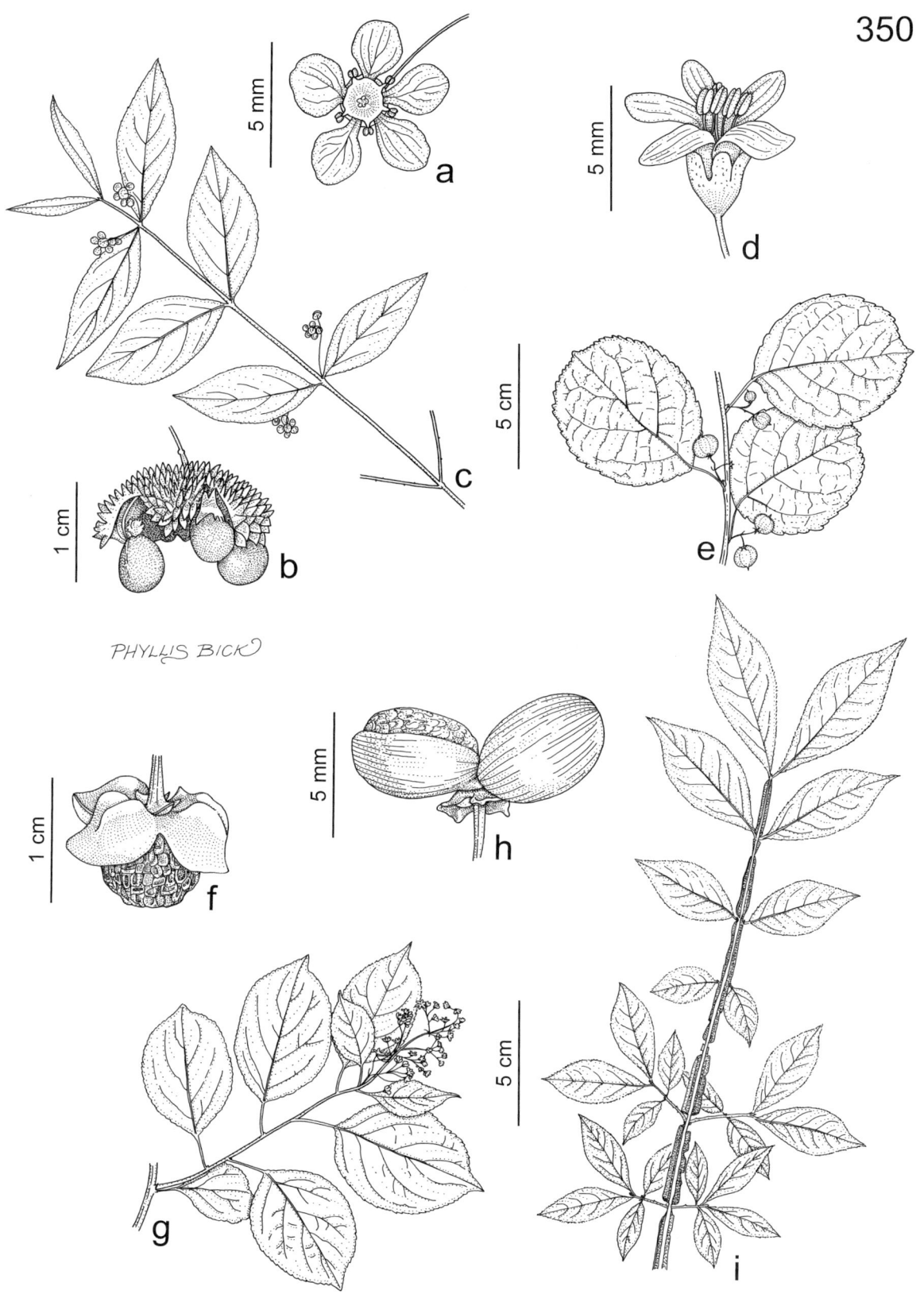

Plate 350. Celastraceae. *Euonymus americanus*, **a)** flower, **b)** fruit, **c)** flowering branch. *Celastrus orbiculatus*, **d)** flower, **e)** fruiting branch. *Celastrus scandens*, **f)** fruit, **g)** flowering branch. *Euonymus alatus*, **h)** fruit, **i)** branch.

Taxonomy, which must approve such proposals before they can be voted upon at an International Botanical Congress, chose not to favor Paclt's view (Brummitt, 2000). The generic name (and thus specific epithets) therefore must be treated as masculine.

Polyembryony has been reported in *E. americanus* and *E. alatus* (Brizicky, 1964a). The seeds of these species contain up to nine embryos, many of them small and apomictic. Only two chromosome levels, $2n=32$ and $2n=64$, have been reported for *Euonymus,* the latter possibly arising following past interspecific hybridization (Nath and Clay, 1972). Several species of *Euonymus* are prized for their colorful fall foliage and attractive fruits with bright red arillate seeds. Although sometimes used in folk medicine, the bark and fruits generally are considered poisonous.

1. Branches with broad, corky ridges or wings . . . . 1. E. ALATUS
1. Branches lacking corky wings
    2. Stems erect or ascending, not climbing
        3. Leaves sessile or nearly so, the petiole to 1 mm long, the undersurface glabrous; petals 5, greenish yellow, sometimes tinged with purple or brown greenish-yellow; capsule strongly warty, the lobes extending the full length . . . . 2. E. AMERICANUS
        3. Leaves with the petiole 1–2 cm long, the undersurface finely hairy; petals 4, dull purple to brownish purple; capsule smooth, the lobes extending to the midpoint . . . . 3. E. ATROPURPUREUS
    2. Stems spreading, trailing, or climbing
        4. Leaves thick, leathery, evergreen, the blade lanceolate or elliptic to broadly ovate; valves of the fruit smooth, straw-colored or occasionally slightly pinkish-tinged; plants often climbing into other shrubs and trees; sepals and petals 4 . . . . 4. E. HEDERACEUS
        4. Leaves relatively thin, herbaceous, deciduous, the blade obovate or less commonly elliptic; valves of the fruit strongly warty, pinkish purple; plants trailing, occasionally with a few ascending branches, but not climbing; sepals and petals 5 . . . . 5. E. OBOVATUS

1. Euonymus alatus (Thunb.) Siebold (burning bush, winged euonymus)

Celastrus alatus Thunb.

Pl. 350 h, i; Map 1504

Plants erect or ascending shrubs with spreading to ascending branches, 1–3 m tall, sometimes spreading by runners. Twigs green, 4-angled with 2–4 conspicuous corky wings 1–4 mm wide. Leaves thin to relatively thick but herbaceous, deciduous, subsessile, the petiole to 2 mm long. Leaf blades 2–6 cm long, 1–2 cm wide, elliptic to obovate, narrowed at the base, narrowed or tapered to a usually sharply pointed tip, the margins finely and usually sharply toothed. Inflorescences axillary clusters of 2 or 3 or solitary flowers. Sepals 4, 0.5–1.0 mm long. Petals 4, 2–3 mm long, 2–3 mm wide, broadly spatulate with a minute, stalklike base, greenish yellow, the margins irregular. Fruits 6–8 mm long, 2–4-lobed their entire length, the valves smooth, dark brown to dull purple. Seeds usually only 1 or 2 per fruit, 4.0–4.5 mm long. $2n=64$. April–June.

Introduced, uncommon and sporadic but becoming more widespread (native of eastern Asia; escaped from cultivation in the northeastern U.S. and adjacent Canada west to Iowa, Missouri, and Montana). Mesic to dry upland forests and banks of streams; also disturbed areas.

The leaves of this species turn brilliant red or purple in autumn. The curious corky wings vary in size with the age of the twig. Cork development begins in a shallow groove on the current year's growth. A phellogen layer originates around the inner side of the first-formed cork cells. The wings continue to grow for several years, but they eventually are shed and are replaced by other layers of bark tissue (Bowen, 1962).

2. Euonymus americanus L. (strawberry bush, brook euonymus)

Pl. 350 a–c; Map 1505

Plants erect or ascending shrubs with often spreading branches, 1–2 m tall, the lowermost branches occasionally rooting. Twigs green, 4-

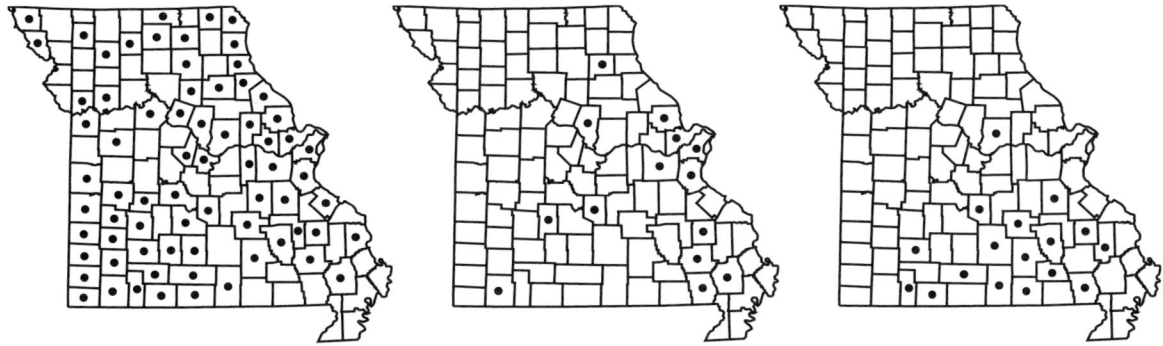

1506. Euonymus atropurpureus   1507. Euonymus hederaceus   1508. Euonymus obovatus

angled but not winged. Leaves relatively thin and herbaceous, deciduous, subsessile, the petiole to 1 mm long. Leaf blades 2–9 cm long, 1–4 cm wide, lanceolate to elliptic or ovate, narrowed or tapered at the base, narrowed or tapered to a usually sharply pointed tip, the margins finely and usually bluntly toothed. Inflorescences axillary clusters of 2 or 3 or solitary flowers, often all but 1 flower aborting during development. Sepals 5, 1–2 mm long. Petals 5, 2–4 mm long, 2–3 mm wide, broadly spatulate with a short, stalklike base, greenish yellow, sometimes tinged with purple or brown, the margins mostly entire. Fruits 14–16 mm long, 3–5-lobed most of their length, the valves strongly warty, pink to red or purplish red. Seeds 1–6 per locule, 4–5 mm long. $2n=64$. April–June.

Uncommon, mostly in the Mississippi Lowlands Division and eastern portion of the Ozarks (eastern U.S. west to Missouri and Texas). Swamps, bottomland forests, mesic upland forests in ravines, and bases of sheltered bluffs.

This species is distinguished by its erect habit, more or less elliptic leaves, and strongly tuberculate, usually red fruits that, when ripe, bear a superficial resemblance to strawberries. Steyermark (1963) noted that the leaves tend to be a darker green than those of the other native species and that it is an attractive shrub in the garden. It is becoming increasingly available in plant nurseries. As discussed below, *E. americanus* is very similar to *E. obovatus,* and the two taxa were considered one species by Ma (2001).

3. Euonymus atropurpureus Jacq. (wahoo, spindle tree)

Pl. 351 h–j; Map 1506

Plants erect or ascending shrubs or small trees with ascending or spreading branches, 2–6 m tall. Twigs green, gray, or less commonly brownish purple, circular in cross-section. Leaves relatively thin and herbaceous, deciduous, short-petiolate, the petiole 10–20 mm long. Leaf blades 4–14 cm long, 2–7 cm wide, elliptic to narrowly ovate or ovate, narrowed or tapered at the base, narrowed or tapered to a usually sharply pointed tip, the margins finely and usually sharply toothed. Inflorescences small axillary panicles with 6–24 flowers. Sepals 4, 1.0–1.5 mm long. Petals 4, 2–3 mm long, 1.5–2.5 mm wide, ovate to broadly ovate, dull purple to brownish purple, the margins often thinner, pale, and slightly irregular. Fruits 6–10 mm long, 2–4-lobed from the tip to about the midpoint, the valves smooth, pink to purple, fading to tan. Seeds 1–6 per locule, 4–5 mm long. $2n=32$. April–June.

Scattered nearly throughout the state (eastern U.S. west to Montana and Texas; Canada). Bottomland forests, mesic upland forests, bases and ledges of bluffs, banks of streams and rivers, and margins of glades, upland prairies, and loess hill prairies; also pastures, railroads, and roadsides.

This species is recognized by the relatively long petioles, large compound inflorescences, and smooth, rosy-red capsules.

Euonymus europaeus L., the European spindle tree, is sometimes cultivated in Missouri and can persist at old homesites. Although the label on a specimen from St. Louis County indicates the potential for localized spread, this species has yet to be documented as fully established and reproducing outside cultivation in the state. *Euonymus europaeus* would key to *E. atropurpureus* in the key to species above but differs in its totally glabrous (vs. often sparsely hairy) leaves that are often larger (to 15 cm long), as well as its fewer-flowered inflorescences (3–8 flowers), greenish white corollas, and seeds with a more orange aril. This native of Europe has become naturalized sporadically in the eastern United States.

4. **Euonymus hederaceus** Champ. ex Benth.
(wintercreeper)
E. fortunei (Turcz.) Hand.-Mazz.
E. kiautschovicus Loes.

Pl. 351 a, b; Map 1507

Plants spreading, trailing, or climbing lianas (rarely appearing as mounding shrubs), the stems to 15 m or more long, dimorphic, the spreading/trailing phase exclusively vegetative, relatively slender, rooting at nodes, the climbing phase vegetative and often also reproductive, slender to more commonly relatively stout, adhering to the substrate by adventitious roots produced at and between nodes. Twigs green to grayish brown, circular in cross-section (but the older climbing stems often somewhat flattened or bluntly 3-angled). Leaves thick and leathery, evergreen, short-petiolate, the petiole 5–10 mm long. Leaf blades dimorphic, those of the spreading, trailing phase 1–6 cm long, 0.5–3.0 cm wide, lanceolate to elliptic or ovate, angled or tapered at the base, angled or tapered to a bluntly or sharply pointed tip, the margins finely and bluntly to less commonly sharply toothed, the upper surface green to dark green, often with pale mottling along the main veins, the undersurface lighter green; those of the climbing, fertile phase changing to somewhat more herbaceous, 4–9 cm long, 2–5 cm wide, elliptic to broadly obovate-elliptic, tapered at the base, tapered to a usually sharply pointed tip, the margins finely and usually sharply toothed, the upper surface green to yellowish green and only occasionally slightly lighter along the main veins, the undersurface light green. Flowers in small axillary panicles of 5–15. Sepals 4, 1.0–1.5 mm long. Petals 4, 3–4 mm long, 2–3 mm wide, nearly circular, white to pale greenish yellow, the margins usually entire. Fruits 6–8 mm long, 2–4-lobed nearly their entire length, the valves smooth, straw-colored or occasionally slightly pinkish-tinged. Seeds 1–4(–6) per locule, 4–6 mm long. $2n=32$. June–August.

Introduced, scattered, mostly in the eastern half of the state thus far but continuing to spread (native of eastern Asia; widely naturalized in the eastern U.S. west to Wisconsin, Kansas, and Mississippi; Canada). Bottomland forests, mesic upland forests, and banks of streams and rivers; also gardens, fencerows, railroads, roadsides, and disturbed areas.

This species is said to be the most common and widespread species in the genus, and it is also one of the most taxonomically confusing (Ma, 2001). It exhibits a variety of growth forms and leaf characters, and each of the different forms were at one time considered to be separate species. The vegetative phase is a creeping shrub with relatively small, dark green leaves. Old plants sometimes produce ascending branches that can take on a moundlike habit. When encountering a vertical surface such as a fence, house, tree, or telephone pole, the plant climbs upward, adhering to the substrate by means of dense patches of adventitious roots, and alters its morphology. The reproductive phase has stouter branches and larger, lighter green leaves. Wintercreeper is widely available horticulturally and is usually sold under the name *E. fortunei*. It is used primarily as an evergreen ground cover, especially in areas where dense shade or dry soil prevent landowners from growing most other species. The brightly arillate seeds are attractive to birds and sometimes also are eaten by small mammals.

The characters that make this species desirable for cultivation—fast, aggressive, dense growth; evergreen leaves; and wide climatic tolerance—also make it a potentially devastating weed. When it escapes, it soon forms a dense mat over other vegetation, sometimes causing tree trunks and branches to break from the weight. This aggressive liana can grow so densely as to choke out most other species.

Euonymus hederaceus is frequently confused with *E. japonicus* Thunb., but that species is nonclimbing (Ma, 2001) and has not been reported as an escape from cultivation in Missouri.

5. **Euonymus obovatus** Nutt. (running strawberry bush)
E. americanus var. *prostratus* E.J. Palmer & Steyerm.

Pl. 351 k, l; Map 1508

Plants spreading or trailing shrubs, occasionally with a few ascending branches, but not climbing, the stems 0.3–1.2 m long, often rooting at the nodes. Twigs green, gray, or less commonly brownish purple, 4-angled but not winged. Leaves relatively thin and herbaceous, deciduous, subsessile, the petiole to 2 mm long. Leaf blades 3–7 cm long, 1.5–3.5 cm wide, obovate or less commonly elliptic, narrowed or tapered at the base, narrowed or tapered to a usually sharply pointed tip, the margins finely and usually sharply toothed. Inflorescences axillary clusters of 2 or 3 or solitary flowers. Sepals 5, 1.0–1.5 mm long. Petals 5, 2–3 mm long, 2–3 mm wide, broadly spatulate with a minute, stalklike base, greenish yellow and tinged with purple, the margins slightly irregular. Fruits 6–8 mm long, 2- or 3-lobed nearly their entire length, the valves strongly warty, pinkish purple. Seeds 1–6 per locule, 3–4 mm long. April–June.

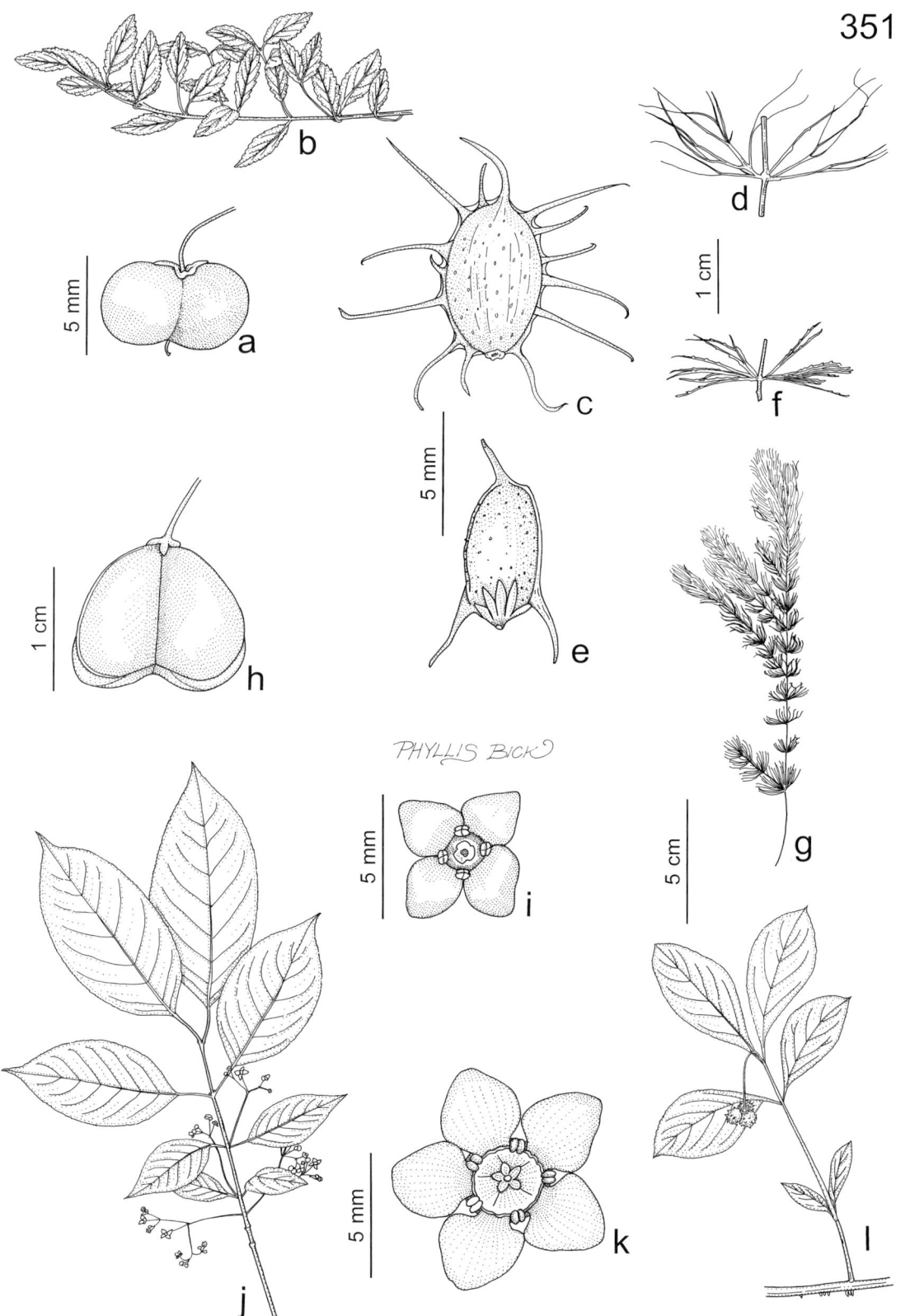

Plate 351. Celastraceae. Ceratophyllaceae. *Euonymus hederaceus*, **a)** fruit, **b)** branch. *Ceratophyllum echinatum*, **c)** fruit, **d)** node. *Ceratophyllum demersum*, **e)** fruit, **f)** node, **g)** leaves. *Euonymus atropurpureus*, **h)** fruit, **i)** flower, **j)** flowering branch. *Euonymus obovatus*, **k)** flower, **l)** fruiting branch.

Scattered in the Ozark and Ozark Border Divisions (eastern U.S. west to Missouri and Arkansas; Canada). Bases and ledges of sheltered bluffs and mesic upland forests; also rarely roadsides.

This species has a distinctive trailing growth habit. The stems spread out, producing many adventitious roots, but they never climb into trees. It is a shade-loving species, and flowers are produced sparingly. In most other respects, *E. obovatus* is very similar to *E. americanus*. In her monograph, Ma (2001) considered *E. obovatus* to be merely a growth form of *E. americanus* that is displayed under poor growing conditions, and she could find no clear way to separate the two taxa. There may be some merit to this assessment. However, in light of the tetraploid chromosome number and minor meiotic irregularities documented by Nath and Clay (1972) for *E. americanus*, further biosystematic investigation is called for. Many specimens of *E. obovatus* found along Ozark streams have characteristic obovate leaves, but a few have elliptical leaves like *E. americanus*. The lower stems on some specimens of *E. americanus* are trailing and rooting like *E. obovatus*. In general, *E. americanus* appears to produce more flowers and fruits than does *E. obovatus*.

CERATOPHYLLACEAE (Hornwort Family)
Contributed by Alan E. Brant

One genus, 6 species, worldwide.

The Ceratophyllaceae have one of the longest fossil histories of any angiosperm family, with fossilized fruits from the Cretaceous period, about 115 million years ago (Dilcher, 1989), and extant species apparently present as long as 45 million years ago (Herendeen et al., 1990). Formerly, the group was considered to be among the most primitive of extant flowering plants, but recent molecular studies have discounted this hypothesis.

1. Ceratophyllum (hornwort, coontail)

Plants perennial herbs, monoecious, lacking roots. Stems to 1 m or more long, freely branched. Leaves whorled, sessile or nearly so. Leaf blades divided 1 or more times dichotomously into 3–10 narrowly linear segments, rarely undivided, the margins with small, broadly triangular teeth, especially toward the segment tip, each with a minute, spinelike or bristlelike tip. Stipules absent. Inflorescences of solitary or less commonly 2 or 3 flowers in the leaf axils (staminate and pistillate flowers usually at different nodes), these sessile or very short-stalked, each subtended by a small involucre (interpreted as a perianth by some authors). Involucres of 8–12 bracts, these 0.5–1.5 mm long, linear to narrowly obovate, similar in appearance to the tips of the leaf divisions. Flowers hypogynous, the sepals and petals absent (but see above). Staminate flowers with 3 to numerous stamens, the filaments short, the anthers with 2 or 3 short teeth at the tip. Pistillate flowers with 1 pistil composed, the ovary superior, with 1 ovule. Style 1, short, tapered, grooved longitudinally along one side, the stigmatic region a small pouch at the base of the groove. Fruits achenes, the body 3.5–5.5 mm long, 2.5–3.5 mm wide, somewhat flattened, oblong-ovate in outline, green to olive-green or dark green, with a slender terminal spine (the elongated stigma) and 2 or more additional spreading spines along the rim, these shorter than to longer than the body. Six species, worldwide.

The taxonomy of *Ceratophyllum* has received three modern studies (Lowden, 1978; Wilmot-Dear, 1984; Les, 1986), which have recognized three, two, and six species respectively. The present treatment follows that of Les (summarized in Les, 1993), whose analyses include the largest and most diverse data set, including morphometric, developmental, chemosystematic, and molecular studies.

1509. Ceratophyllum demersum
1510. Ceratophyllum echinatum
1511. Atriplex argentea

Species of *Ceratophyllum* are unrooted, submerged aquatics that are suspended below the water surface. In Missouri, reproduction is mostly by fragmentation of the branching stems, and fertile specimens are encountered rarely. The genus is unusual in that staminate flowers shed anthers, which trap small exuded air bubbles in their apical teeth and have about the same buoyant density as the surrounding water. These reach the pistillate flowers for pollination seemingly by accident as they are carried in and around the plants by small currents. Coontails have been used in folk medicine as purgatives, diuretics, and as treatments for jaundice, rheumatism, and skin disorders. The fruits and foliage are consumed by waterfowl, which help to disperse the plants, and the plants also provide valuable cover for small fish and aquatic invertebrates. Some species are sold for horticultural use in pools and aquaria, but the plants can become pests, growing so quickly and densely that they crowd out other plants and interfere with fishing and boating.

1. Leaf segment margins with the teeth relatively prominent, the spinelike or bristlelike point ascending or somewhat spreading on the broadly triangular base of tissue; fruits 3-spined (2 basal and 1 terminal stylar spine), the surfaces smooth or slightly warty . . . . 1. C. DEMERSUM
1. Leaf segment margins with the teeth absent or inconspicuous, not raised on a broad base; fruit warty, with marginal spines in addition to the terminal and basal spines . . . . 2. C. ECHINATUM

1. Ceratophyllum demersum L.

Pl. 351 e–g; Map 1509

Leaf blades 1.5–2.5 cm long, divided 1 or 2(3) times dichotomously into 3 or 4(–6) segments, rarely undivided, the marginal teeth relatively prominent, with the spinelike or bristlelike point ascending or somewhat spreading on the broadly triangular base of tissue, giving the leaves a minutely spiny appearance, even to the naked eye. Fruits with the surfaces smooth or slightly warty, the rim 3-spined (1 terminal stylar spine and 2 basal spines). $2n=24$. June–September.

Common nearly throughout the state (worldwide). Submerged aquatic in slow-moving waters of streams, rivers, and spring branches, and still waters of ponds, lakes, and sloughs.

2. Ceratophyllum echinatum A. Gray

Pl. 351 c, d; Map 1510

Leaf blades 0.6–1.5 cm long, divided 3 or more times dichotomously into 4–8(–10) segments, rarely undivided, the marginal teeth absent or more commonly present but relatively inconspicuous, with the spinelike or bristlelike point appressed or ascending on the broadly triangular base of tissue, visible only with magnification. Fruits with the surfaces minutely warty, the rim with up to 10 marginal spines in addition to the terminal and basal spines. $2n=24$. June–September.

Uncommon, widely scattered in the state (eastern U.S. west to Minnesota and Texas; also Oregon and Washington; Canada, Mexico). Submerged aquatic in spring branches, ponds, and sinkhole ponds.

CHENOPODIACEAE (Goosefoot Family)
(Judd and Ferguson, 1999)

Plants annual or perennial herbs (often shrubs or small trees elsewhere), sometimes monoecious or dioecious, sometimes slightly to strongly succulent, often with a taproot, glabrous or hairy, sometimes mealy (with short, white, inflated hairs that collapse and appear lozenge-shaped with age or upon drying), often tinged with pink to purple. Stems usually with fine longitudinal angles, stripes, or ridges. Leaves alternate or occasionally some leaves opposite, sometimes reduced to inconspicuous scales, simple but sometimes lobed, the margins variously entire, toothed, or wavy. Stipules absent. Inflorescences axillary and/or terminal, spikes, spike-like racemes, or panicles, often reduced to axillary clusters or solitary axillary flowers. Flowers sessile or very short-stalked, usually with 1–3 small, herbaceous bracts (usually 1 outer bract and 2 inner, often smaller bracteoles), imperfect or more commonly perfect, hypogynous (perigynous in *Beta*). Calyx absent or more commonly of 1–5 sepals, these free or fused toward the base, usually green at flowering, persistent and sometimes becoming hardened or papery at fruiting. Petals absent. Stamens 1–5, absent or reduced to minute staminodes in pistillate flowers, the filaments sometimes fused toward the base, the anthers attached basally or more or less toward their midpoints, usually yellow. Pistil 1 per flower (absent in staminate flowers), the ovary superior (partly inferior in *Beta*), consisting of 2 or 3 fused carpels, with 1 locule, the placentation usually basal. Styles 1–3, often very short, sometimes fused toward the base, the stigmas 1–3 (4 or 5 in the cultivated *Spinacia*), slender or capitate, occasionally lobed. Ovule 1 per flower. Fruits achenelike or less commonly capsules, sometimes winged, sometimes beaked, indehiscent or more commonly with irregular or circumscissile dehiscence. Seed 1, often somewhat flattened and lens-shaped, circular in outline or nearly so (the embryo appearing curved or coiled but not always easily observed). About 100 genera, about 1,500 species, worldwide.

The Chenopodiaceae are here treated in the traditional sense as a family separate from the Amaranthaceae. However, a number of morphological and molecular phylogenetic studies (Manhart and Rettig, 1993; Rodman, 1990, 1993; Downey et al., 1997; see also Judd et al., 1999) have presented evidence to suggest that the family as thus circumscribed is paraphyletic; that is, that the genera of Amaranthaceae represent a specialized subgroup within the lineage of Chenopodiaceae rather than a separate sister clade. Because some of the conclusions of these papers are contradictory and a few relationships among genera are yet controversial (such as the placement of *Spinacia* L.), it seems premature to combine these families in a floristic treatment until more detailed studies can be completed. The morphological features that generally separate the Chenopodiaceae from Amaranthaceae include their stamens with free (vs. basally fused) filaments and herbaceous (vs. papery) perianth and bracts, but numerous exceptions exist.

The family is economically important primarily as a source of food plants, including beets and Swiss chard (*Beta*), spinach (*Spinacia*), and grains (*Chenopodium*). An extract prepared from beets sometimes is used as a food coloring, and the genus is also a source of processed sugar. The family contains a number of noxious weeds of crop fields and sometimes also native plant communities, particularly Russian thistle (*Salsola*) and the pigweeds (*Chenopodium*). Members of the Chenopodiaceae are nearly all wind-pollinated and have been cited as causing hay fever. Pollen grains of most Amaranthaceae and Chenopodiaceae are virtually indistinguishable morphologically, and the two families are usually lumped into a single pollen class in projects that monitor airborne spores and pollen for air quality and hay fever reports.

Species of Chenopodiaceae have a well-deserved reputation for being difficult to determine. Users are cautioned that in this family mature fruits are usually necessary for identification of genera and species.

1. Stems succulent, appearing jointed; leaves reduced to tiny, opposite scales; flowers sunken into the stem .... 9. SALICORNIA
1. Stems not or only slightly succulent, not appearing jointed; leaves all or mostly herbaceous and well developed; flowers not sunken into the stem
   2. Stems, leaves, and/or bracts sparsely to densely pubescent with small, stellate hairs
      3. Leaf blades elliptic or narrowly to broadly lanceolate; flowers all either staminate or pistillate (plants monoecious); fruits sometimes with a small, 2-lobed wing at the tip .... 2. AXYRIS
      3. Leaf blades linear to narrowly lanceolate; flowers all or nearly all perfect (occasional pistillate flowers sometimes present); fruits usually with a narrow wing around the rim (except sometimes in *C. villosum*)
      .... 5. CORISPERMUM
   2. Plants glabrous or hairy, the pubescence, if present, variously glandular, mealy, and/or of unicellular or multicellular hairs, but the hairs always unbranched
      4. Leaves and bracts having hard, spinelike tips .... 10. SALSOLA
      4. Leaves and bracts without spinelike tips, sometimes sharply pointed or with a short, soft, threadlike extension of the midvein at the tip
         5. Flowers all either staminate or pistillate (plants monoecious); pistillate flowers lacking a calyx (except in *A. hortensis*), the bracts 2, enlarged, more or less fused basally, and enclosing the fruit .... 1. ATRIPLEX
         5. Flowers all or nearly all perfect (occasional pistillate flowers sometimes present); pistillate flowers with a calyx (reduced to 1 sepal in *Monolepis*), the bracts absent or mostly 3, not becoming enlarged, fused, or enclosing the fruit
            6. Leaves sessile, the leaf blades linear to narrowly lanceolate (rarely the largest leaves oblanceolate), the margins entire
               7. Stems and/or leaves sparsely to moderately hairy, sometimes becoming nearly glabrous at maturity (if in doubt, check the leaf margins for more persistent hairs) .... 7. KOCHIA
               7. Stems and leaves glabrous, sometimes glaucous or mealy (with minute, white, lozenge-shaped glands)
                  8. Leaves not fleshy, not or only slightly thickened, relatively flat in cross-section .... 4. CHENOPODIUM
                  8. Leaves fleshy, circular or elliptic in cross-section .... 11. SUAEDA
            6. Leaves variously shaped, at least the largest ones petiolate, the leaf blades broadly lanceolate or oblong to elliptic, ovate, or triangular, the margins often toothed or lobed, less commonly entire
               9. Main root with a globose to top-shaped tuberous thickening; ovary partly inferior; fruit dehiscence circumscissile ...... 3. BETA
               9. Roots not tuberous-thickened; ovary superior; fruits indehiscent or irregularly dehiscent
                  10. Calyx of 1 sepal positioned along one side of the ovary; stamen 1 per flower; largest leaves with a pair of prominent, spreading basal lobes, the margins otherwise entire
                  .... 8. MONOLEPIS
                  10. Calyx of (3–)5 sepals or (3–)5-lobed, the sepals or lobes spaced around the ovary; stamens 3–5 (absent in rare pistillate flowers); leaves entire to variously toothed or lobed, but not with a single pair of spreading basal lobes

11. Calyx at fruiting unwinged or with a low longitudinal ridge along the midvein of each sepal; flowers rarely solitary at the inflorescence nodes, usually in small, dense clusters .... 4. CHENOPODIUM
11. Calyx at fruiting with a prominent, papery, continuous, transverse wing; flowers solitary at the inflorescence nodes............................. 6. CYCLOLOMA

1. Atriplex L. (saltbush, orach, orache)

Plants annual, monoecious (dioecious perennial herbs or shrubs elsewhere), the roots not tuberous-thickened. Stems prostrate to erect, sometimes weak and supported by surrounding plants, not succulent, not appearing jointed, usually much-branched, glabrous or sparsely to moderately mealy, at least when young. Leaves alternate or opposite toward the stem base, well developed, not or only slightly succulent, mostly petiolate. Leaf blades narrowly lanceolate to ovate or triangular, often with a pair of pronounced, spreading basal lobes, flattened in cross-section, not clasping the stem, rounded or narrowed to a sharply pointed tip, the margins entire or sometimes slightly wavy, the surfaces glabrous or sparsely to densely mealy. Inflorescences axillary and/or terminal, consisting of small flower clusters (sometimes solitary flowers), these often appearing as interrupted spikes, or small panicles of interrupted spikes, the staminate flowers intermingled with the pistillate flowers or positioned more terminally along the inflorescence, the flowers not sunken into the axis. Bracts of staminate flowers absent (or sometimes minute bracts subtending flower clusters); bracts of pistillate flowers 2, longer than the flower (absent in some flowers of dimorphic species), more or less fused basally, enclosing the fruit, broadly triangular to rhombic or nearly circular, sometimes strongly veined or ornamented with flaplike or hornlike projections, the margins entire or toothed. Calyx absent in pistillate flowers (present sometimes in dimorphic species, then similar to that of staminate flowers), calyx in staminate flowers 1.5–2.5 mm long, deeply (3–)5-lobed, the lobes spaced evenly around the flower, the tips erect or incurved. Staminate flowers with (3–)5 stamens. Pistillate flowers with the ovary superior. Styles 2, short, sometimes fused at the very base, the stigmas 1 per style, linear. Fruits broadly obovate to nearly circular in outline, somewhat flattened laterally, usually indehiscent, the wall thin and papery to membranous. Seed adhering loosely to the fruit wall, usually positioned vertically, sometimes dimorphic, somewhat flattened, the surface smooth, brown or black, shiny, the coiled embryo usually apparent. About 300 species, nearly worldwide.

Some shrubby species of *Atriplex* (saltbushes) are major components of scrubby vegetation types in the arid portions of the western United States. Nearly all of the species are adapted to grow well in saline soils. The eastern species are annuals of disturbed habitats and for the most part are closely related components of a circumboreal polyploid complex. Species determinations among the taxa present in Missouri are difficult, and mature fruits are usually necessary for certain identification.

1. Pistillate flowers with the bracts fused to well above the base, at maturity the fused portion becoming hardened and somewhat bony, the fruit difficult to separate from the bracts
   2. Leaf blades with the margins all irregularly wavy or more commonly with irregular blunt teeth... 6. A. ROSEA
   2. Some or all of the leaves with the margins entire, sometimes most of the leaves with a single pair of spreading basal lobes
      3. Lowermost leaves opposite or subopposite; leaf blades linear to oblanceolate or elliptic, mostly more than 4 times as long as wide and widest at or above the midpoint, appearing green on the upper surface, silvery gray on the undersurface .... 8. A. WRIGHTII

3. Leaves all alternate, the blades ovate to ovate-triangular, less than 3 times as long as wide and widest toward the base, appearing silvery gray on both surfaces (this sometimes hard to see on older specimens)
    4. Bracts at fruiting 3–6 mm long, obovate to depressed-circular in outline, the margins with several coarse teeth or lobes to below the midpoint . . . . 1. A. ARGENTEA
    4. Bracts at fruiting 2–3 mm long, obtriangular to wedge-shaped, the margins entire except for usually 3 teeth along the truncate tip (sometimes also slightly irregular between the teeth) . . . . 7. A. TRUNCATA
1. Pistillate flowers with the bracts fused only toward the base, at maturity becoming more or less uniformly papery or leathery (somewhat spongy-thickened toward the base in *A. prostrata*), the fruit easily separable from the bracts
    5. Pistillate flowers of 2 kinds on the same plant (best seen at fruiting), differing in the size or presence of the 2 bracts, these (when present) elliptic-ovate to broadly ovate or nearly circular at fruiting
        6. Some pistillate flowers lacking bracts but with a minute, 5-lobed calyx, other pistillate flowers with 2 bracts but lacking a calyx, the bracts enlarging to 6–18 mm long at fruiting . . . . 2. A. HORTENSIS
        6. All pistillate flowers lacking a calyx, both kinds of flowers with 2 bracts; at fruiting some flowers with the bracts 1.5–2.5 mm long, others with the bracts enlarging to 5–6 mm long at fruiting . . . . 3. A. MICRANTHA
    5. Pistillate flowers all similar (but sometimes producing fruits with 2 different kinds of seeds), with 2 bracts, these triangular to broadly ovate-triangular or rhombic at fruiting
        7. Bracts not spongy-thickened, more or less evenly herbaceous; leaves mostly without a pair of basal lobes, when present these loosely angled toward the leaf tip . . . . 4. A. PATULA
        7. Bracts more or less spongy-thickened toward the base; leaves all or mostly with a pair of basal lobes, these spreading . . . . 5. A. PROSTRATA

1. Atriplex argentea Nutt. **var. argentea**
(silver scale)

Map 1511

Stems 15–80 cm tall, erect or strongly ascending, the branches ascending. Leaves all alternate, mostly short-petiolate. Leaf blades 0.5–5.0 cm long, less than 3 times as long as wide and widest toward the base, ovate-triangular to ovate, angled at the base, rounded to bluntly pointed at the tip, sometimes with a pair of short, blunt, spreading basal lobes, the margins otherwise entire or with shallow, irregularly wavy teeth, appearing silvery gray on both surfaces (the mealiness sometimes difficult to discern on older specimens). Staminate flowers appearing axillary among the uppermost leaves and/or in short, dense terminal spikes, these occasionally grouped into small panicles. Pistillate flowers in axillary clusters, all similar, lacking a perianth. Bracts at fruiting fused to about the midpoint, 3–6 mm long, obovate to depressed-circular in outline, appearing sessile or narrowed to a short, stalklike base, the free portions of the margins with several coarse teeth or lobes, the fused portion becoming hardened and somewhat bony, the surfaces sometimes with irregular tubercles or crests. Fruits difficult to separate from the bracts. Seeds all similar in size and color, 1.5–2.0 mm long, brown, more or less shiny, the tip of the radicle (seedling root) positioned above the remaining body of the seed. $2n=18$. July–September.

Introduced, uncommon, known thus far only from the St. Louis and Kansas City metropolitan areas (western U.S. east to North Dakota and Texas; Canada; introduced very sporadically farther east). Railroads and open, disturbed areas.

Mühlenbach (1979) redetermined as *A. truncata* the initial Missouri collections upon which Steyermark (1963) based his inclusion of *A. argentea*, but subsequent herbarium research by Hilda Flores Olvera of the National Autonomous University in Mexico City and others has uncovered bona fide specimens to document the presence of *A. argentea* in the state. *Atriplex argentea* is a morphologically variable polyploid complex that has

1512. Atriplex hortensis   1513. Atriplex micrantha   1514. Atriplex patula

been divided into a number of subspecies and varieties based upon subtle differences in leaf and bract morphology. The Missouri specimens correspond morphologically more or less to the typical variety.

2. Atriplex hortensis L. (garden orach)

Pl. 352 b, c; Map 1512

Stems 15–300 cm tall, erect or ascending, the branches ascending. Leaves mostly alternate, those toward the stem base sometimes opposite or subopposite, sessile to short- or less commonly long-petiolate. Leaf blades 1–7 cm long (those of basal leaves to 15 cm), mostly 2–3 times as long as wide and mostly widest toward the base, lanceolate to ovate or triangular, shallowly cordate to truncate or broadly angled at the base, rounded to bluntly or sharply pointed at the tip, often with a pair of short, blunt, spreading basal lobes, the margins otherwise entire or with few to several irregular wavy teeth, silvery gray on the undersurface or on both surfaces when young, the mealiness usually disappearing by flowering and the surfaces then uniformly green (yellowish green or reddish- to purplish-tinged in some cultivars). Staminate flowers axillary and terminal, appearing as elongate spikes with clusters of flowers and/or panicles with short, spicate branches. Pistillate flowers intermingled with the staminate ones, of 2 kinds on the same plant (best seen at fruiting), differing in the presence or absence of the bracts, the bracted flowers without a calyx, the bractless flowers with a minute, 5-lobed calyx. Bracts (in bracted flowers) at fruiting fused only at the very base, enlarging variously to 6–18 mm long, elliptic-ovate to broadly ovate or nearly circular in outline, appearing sessile (sometimes shallowly cordate), the free portions of the margins entire, at maturity becoming more or less uniformly papery or leathery, the surfaces prominently veined. Fruits easily separable from the bracts. Seeds usually of 3 kinds, those of larger-bracted flowers 3.5–4.5 mm long, positioned vertically, yellowish brown, dull; those of the smaller-bracted flowers sometimes not produced, 1.7–2.0 mm long, positioned vertically, black or rarely yellowish brown, usually shiny; those of the bractless flowers 1.4–1.7 mm long, positioned horizontally, black, shiny; in all types, the tip of the radicle (seedling root) positioned below the remaining body of the seed. $2n=18$. July–September.

Introduced, uncommon, known thus far only from the St. Louis and Kansas City metropolitan areas (origin unknown but presumably a native of Asia; introduced widely in the western and northern U.S. and Canada). Railroads and open, disturbed areas.

This taxon has a long history of cultivation as a garden ornamental (particularly the cultivars with reddish- or purplish-tinged foliage). Its leaves can be cooked and eaten like spinach. In the former Soviet Republics, a blue dye is extracted from the seeds.

3. Atriplex micrantha Ledeb. (two-seeded orach, Russian atriplex)

A. heterosperma Bunge

Pl. 352 d; Map 1513

Stems 30–150 cm tall, erect, unbranched or the branches ascending. Leaves mostly alternate, those toward the stem base often opposite or subopposite, sessile to short- or less commonly long-petiolate. Leaf blades 1–7 cm long (those of basal leaves occasionally to 10 cm), 2–4 times as long as wide and mostly widest toward the base, oblong-ovate to triangular, truncate to angled at the base, rounded to bluntly or sharply pointed at the tip, often with a pair of short, blunt, spreading basal lobes, the margins otherwise entire or with few to many irregular teeth, silvery gray on the undersurface when young, the mealiness usually disappearing by flowering and the surfaces then uniformly green. Stami-

1515. Atriplex prostrata   1516. Atriplex rosea   1517. Atriplex truncata

nate flowers axillary and terminal, appearing as elongate spikes with clusters of flowers and/or panicles with short, spicate branches. Pistillate flowers intermingled with the staminate ones, of 2 kinds on the same plant (best seen at fruiting), differing in the size of the bracts, all lacking a perianth. Bracts at fruiting fused only at the very base, in some flowers 1.5–2.5 mm long, in others enlarging to 5–6 mm long, broadly ovate to nearly circular in outline, appearing sessile or nearly so, the free portions of the margins entire, at maturity becoming more or less uniformly papery or leathery, the surfaces prominently veined. Fruits easily separable from the bracts. Seeds of 2 kinds, those of larger-bracted flowers 2–3 mm long, yellowish brown, dull; those of the smaller-bracted flowers 1.2–1.5(–2.0) mm long, black, shiny; in both types, the tip of the radicle (seedling root) positioned below the remaining body of the seed. $2n=36$. July–September.

Introduced, uncommon, known thus far only from the city of St. Louis (native of Europe, Asia; introduced widely but sporadically in the western U.S. and Canada). Railroads.

This species was first reported for the state by Mühlenbach (1979) under the name *A. heterosperma*. That name has been placed in synonymy under *A. micrantha* in most of the recent floristic literature on eastern Europe and Asia, as well as by Kartesz and Meacham (1999), in their checklist of North American vascular plants.

4. Atriplex patula L. (spear-scale, fat hen saltbush)

Pl. 352 a; Map 1514

Stems 8–150 cm tall, prostrate to erect, often climbing over other vegetation, the branches mostly spreading. Leaves alternate, sessile to more commonly short-petiolate. Leaf blades 1–12 cm long, 2–6 times as long as wide and mostly widest toward the base, linear to triangular-ovate, tapered or sometimes angled at the base, mostly sharply pointed at the tip, mostly unlobed but sometimes the largest leaves with a pair of short, blunt basal lobes, these loosely angled toward the leaf tip, the margins otherwise entire or with few to several irregular wavy teeth, silvery gray especially on the undersurface when young, the mealiness usually disappearing by flowering and the surfaces then uniformly green (rarely persistent to fruiting). Staminate flowers axillary and terminal, appearing as elongate spikes with clusters of flowers, these occasionally appearing paniculate. Pistillate flowers intermingled with the staminate ones, all similar, lacking a perianth. Bracts at fruiting fused to below the midpoint, 3–7 mm long, broadly triangular-ovate or rhombic, appearing sessile or nearly so, the margins entire or with a few teeth around the midpoint, at maturity becoming more or less uniformly papery or leathery, not spongy-thickened, the surfaces occasionally with a pair of narrow, irregular tubercles. Fruits easily separable from the bracts. Seeds usually of 2 kinds, some 2.5–3.5 mm long, brown, dull or shiny; others 1.5–2.5 mm long, black, shiny; in both types, the tip of the radicle (seedling root) positioned below or less commonly alongside the remaining body of the seed. $2n=36$. July–September.

Introduced, uncommon in the northern half of the state (native of Europe, Asia; introduced widely in the U.S. and Canada). Fallow fields, roadsides, railroads, and open, disturbed areas.

For a discussion of differences between this species and the closely related *A. prostrata*, see the treatment of that species.

5. Atriplex prostrata Boucher ex DC. (spear-scale)
A. patula L. var. *hastata* (L.) A. Gray
A. patula var. *triangularis* (Willd.) K.H. Thorne & S.L. Welsh

A. *subspicata* (Nutt.) Rydb.
A. *triangularis* Willd.

Pl. 352 e; Map 1515

Stems 15–150 cm tall, prostrate to erect, often climbing over other vegetation, the branches mostly spreading. Leaves alternate, sessile or short- to more commonly long-petiolate. Leaf blades 1–10 cm long, 1.5–4.5 times as long as wide and mostly widest at or near the base, those of the uppermost leaves narrowly lanceolate, but those of most leaves ovate-triangular to triangular, shallowly cordate to truncate or broadly angled at the base, mostly sharply pointed at the tip, mostly with a pair of short basal lobes, these spreading or slightly recurved, the margins otherwise entire or with few to several irregular wavy teeth, silvery gray especially on the undersurface when young, the mealiness often disappearing (at least on the upper surface) by flowering and the surfaces then more or less uniformly green (sometimes reddish-tinged). Staminate flowers axillary and terminal, appearing as elongate spikes with clusters of flowers, these occasionally appearing paniculate. Pistillate flowers intermingled with the staminate ones, all similar, lacking a perianth. Bracts at fruiting fused toward the base, 2–6 mm long, triangular to broadly ovate-triangular, appearing sessile or nearly so, the margins entire, irregular, or with few to several shallow teeth above the base, at maturity becoming more or less uniformly papery or leathery, somewhat spongy-thickened toward the center (appearing slightly inflated), the surfaces occasionally with a pair of narrow, irregular tubercles. Fruits easily separable from the bracts. Seeds usually of 2 kinds, some 1.5–2.5 mm long, brown, dull or shiny; others 1.0–1.8 mm long, black, shiny; in both types, the tip of the radicle (seedling root) positioned below or less commonly alongside the remaining body of the seed. $2n=18$. July–September.

Introduced, uncommon and sporadic, mostly in counties adjacent to the Missouri River (native of Europe, Asia, Africa; introduced widely in the U.S. and Canada). Roadsides, railroads, and open, disturbed areas.

The nomenclature and taxonomy of the *A. patula* polyploid complex require further investigation. The name *A. hastata* L., which had been used for this taxon in much of the older American literature, was officially rejected from use as an ambiguous epithet at the 1999 International Botanical Congress (Turland, 1996). Many authors have treated *A. prostrata* as a variety of the closely related *A. patula* (Steyermark, 1963). The two differ markedly in fruiting bract morphology, those of *A. prostrata* being ovate-triangular to triangular and somewhat spongy-thickened toward the center (therefore usually appearing slightly inflated), whereas those of *A. patula* are triangular-ovate to rhombic and lack spongy tissue. However, the two taxa are less distinct vegetatively, which has led to some confusion in determination of immature specimens. In general, all or most of the leaves in *A. prostrata* are ovate-triangular to triangular, shallowly cordate to truncate or broadly angled at the base, and with a pair of spreading basal lobes. In contrast, the main leaves of plants of *A. patula* are linear to lanceolate, narrowed or angled at the base, and either unlobed or with the basal lobes ascending. Sterile triploid hybrids between the two have been encountered only rarely (Judd and Ferguson, 1999) and have yet to be found in Missouri, where neither taxon is very common.

Seed dimorphism in a Nebraskan population of this taxon was studied by Ungar (1971), who determined that the thicker-walled black seeds were viable after six years of storage, but they required scarification prior to germination, whereas the thinner-walled yellowish green seeds did not retain their viability but germinated without pretreatment. In the *A. patula* complex and in other *Atriplex* species having dimorphic seeds, the larger, lighter-colored seeds with thin walls apparently are adapted for quick restocking of the local population and short-term spread. The smaller, darker seeds with thick walls are adapted for longer-range dispersal and allow the species to survive at sites following periods of hostile environmental conditions.

6. Atriplex rosea L. (red scale, tumbling orach)

Pl. 352 g; Map 1516

Stems 25–100 cm tall, erect or strongly ascending, the branches ascending or spreading. Leaves all alternate or the lowermost leaves occasionally opposite or nearly so, sessile or short-petiolate. Leaf blades 1–7 cm long, less than 3 times as long as wide and widest toward the base, ovate-triangular to ovate, angled at the base, rounded to sharply pointed at the tip, sometimes with 1 or 2 pairs of short, blunt, spreading basal lobes, the margins otherwise irregularly wavy or more commonly with irregular blunt teeth, moderately mealy and sometimes appearing silvery gray on both surfaces (sometimes becoming nearly glabrous with age). Staminate flowers appearing axillary among the uppermost leaves and/or in short, dense terminal spikes. Pistillate flowers in axillary clusters, all similar, lacking a perianth. Bracts at fruiting fused to about the midpoint, 3–6(–10) mm long, rhombic

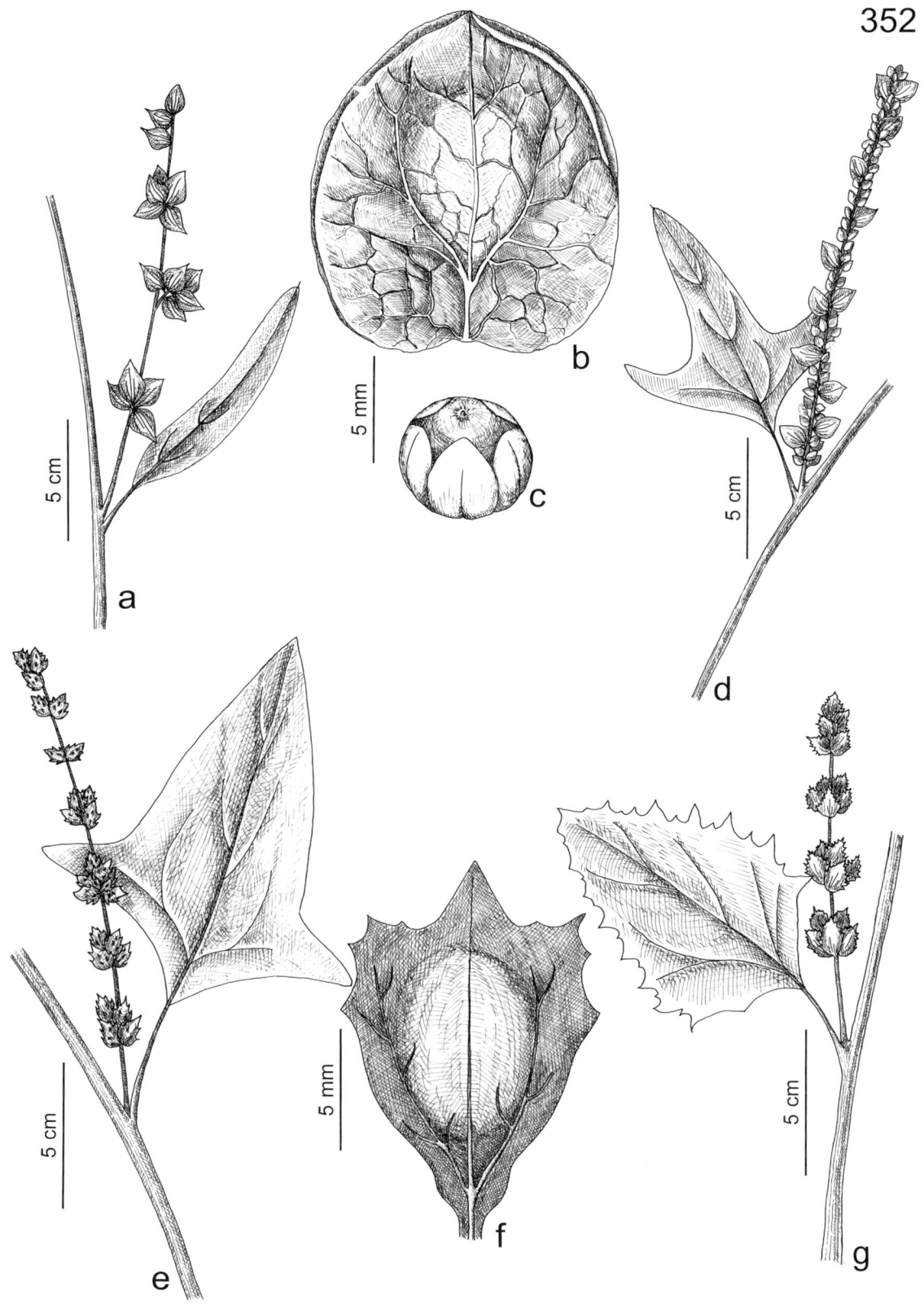

Plate 352. Chenopodiaceae. *Atriplex patula*, **a)** fruiting branch. *Atriplex hortensis*, **b)** fruit in scale, **c)** fruit without scale. *Atriplex micrantha*, **d)** fruiting branch. *Atriplex prostrata*, **e)** fruiting branch. *Atriplex truncata*, **f)** fruit. *Atriplex rosea*, **g)** fruiting branch.

1518. Atriplex wrightii   1519. Axyris amaranthoides   1520. Beta vulgaris

to ovate-triangular in outline, appearing sessile or narrowed to a short, stalklike base, the free portions of the margins with several fine teeth, the fused portion becoming hardened and somewhat bony, the surfaces sometimes with a patch of short, narrow tubercles or a low crest. Fruits difficult to separate from the bracts. Seeds all similar in size and color, 2–3 mm long, brown, dull, the tip of the radicle (seedling root) positioned alongside the remaining body of the seed. $2n=18$. July–September.

Introduced, uncommon, known only from the St. Louis and Kansas City metropolitan areas and Howard County (native of Europe, Asia, Africa; widely introduced in North America and Australia). Saline seeps and banks of streams; also railroads.

In the western United States, red scale has been used for animal feed, and in portions of Europe, it is used to make potash.

7. Atriplex truncata (Torr.) A. Gray (silver scale, wedge-leaved orache)

Pl. 352 f; Map 1517

Stems 15–60 cm tall, erect or strongly ascending, the branches ascending. Leaves all alternate, sessile to short-petiolate. Leaf blades 0.5–3.0 cm long, less than 3 times as long as wide and widest toward the base, ovate-triangular to ovate, short-tapered or angled at the base, the uppermost leaves sometimes shallowly cordate and clasping the stems, bluntly to sharply pointed at the tip, sometimes with a pair of short, blunt, spreading basal lobes, the margins otherwise entire or with a few shallow, irregular teeth, appearing silvery gray on both surfaces (the mealiness sometimes difficult to discern on older specimens). Staminate flowers appearing axillary among the uppermost leaves. Pistillate flowers in axillary clusters, all similar, lacking a perianth. Bracts at fruiting fused to far above the midpoint, 2–3 mm long, obtriangular to wedge-shaped in outline, appearing sessile or narrowed to a short, stalklike base, the more or less truncate free apical portion of the margins with usually 3 teeth (sometimes also slightly irregular between the teeth), the fused portion becoming hardened and somewhat bony, the surfaces usually lacking tubercles or crests. Fruits difficult to separate from the bracts. Seeds all similar in size and color, 1–2 mm long, brown, more or less shiny, the tip of the radicle (seedling root) positioned above the remaining body of the seed. $2n=18$. July–September.

Introduced, uncommon, known thus far only from the city of St. Louis (western U.S. east to Montana and New Mexico; Canada). Railroads.

This species was first reported for Missouri by Frankton and Bassett (1970) based on collections from the St. Louis railroad yards by Viktor Mühlenbach, who also discussed the occurrences (Mühlenbach, 1979).

8. Atriplex wrightii S. Watson (Wright's saltbush)

Map 1518

Stems 15–60(–100) cm tall, erect or strongly ascending, the branches ascending. Leaves mostly alternate, the lowermost leaves opposite or nearly so, sessile to short-petiolate. Leaf blades 0.5–7.0 cm long, mostly more than 4 times as long as wide and widest at or above the midpoint, linear to oblanceolate or elliptic, long-tapered or angled at the base, rounded (sometimes with a minute, sharp point) to sharply pointed at the tip, the margins entire or with shallow, irregular, sometimes wavy teeth, appearing green on the upper surface at maturity (often sparsely mealy when young), silvery gray (mealy) on the undersurface. Staminate flowers in short, clusterlike spikes, these axillary among the uppermost leaves and forming narrow, elongate, terminal, spikelike panicles. Pistillate flowers in axillary clusters, all similar, lacking a perianth. Bracts at fruiting fused to far above the midpoint, 2–3 mm long, depressed-ovate to irregularly circular in outline, appearing sessile or narrowed to a short, stalklike base, the margins irregularly toothed, the fused portion becoming hardened and somewhat bony, the surfaces lacking tubercles or crests. Fruits difficult to separate from the bracts. Seeds all similar in size and color, 1.0–1.5 mm long, light brown to brown, more or less shiny, the tip of the radicle (seedling root) positioned above the remaining body of the seed. $2n=18$. July–September.

Introduced, uncommon, known thus far only from the city of St. Louis (Arizona to Texas). Railroads.

The inclusion of this species in the Missouri flora is based upon a specimen collected in the St. Louis railyards that Viktor Mühlenbach was unable to determine with confidence in 1956, which finally was identified in 1996 by Hilda Flores Olvera of the National Autonomous University in Mexico City.

2. Axyris L.

Seven species, Europe, Asia.

1. Axyris amaranthoides L. (Russian pigweed)

Pl. 357 e, f; Map 1519

Plants monoecious, annual, the taproot not tuberous-thickened. Stems 15–90 cm long, erect or nearly so, not succulent, not appearing jointed, usually with many relatively short, ascending branches, sparsely to densely pubescent with small, stellate hairs. Leaves alternate, well developed, not succulent, mostly short-petiolate. Leaf blades 2–9 cm long, elliptic or narrowly to broadly lanceolate, flattened in cross-section, not clasping the stem, narrowed to a bluntly or sharply pointed tip, narrowed at the base, the margins entire or shallowly and broadly few-toothed, those of the smaller leaves sometimes curled under, the upper surface sparsely pubescent with small, stellate hairs to nearly glabrous, the undersurface moderately to densely pubescent with small, stellate hairs. Inflorescences axillary and terminal, mostly panicles of short spikes (sometimes reduced to a single short, terminal spike); the flowers not sunken into the axis; the staminate flowers at the tips of the panicles and/or tips of the individual pistillate spikes; the pistillate flowers in axillary clusters and/or along the spicate panicle branches. Bracts 3 per flower, 1.5–3.5 mm long, leaflike, lanceolate to ovate, tapered to a sharply pointed tip. Calyx of 3(4) sepals, these tiny (those of staminate flowers 0.4–0.7 mm long, those of pistillate flowers 1.5–2.5 mm long), scalelike, irregularly oblong-ovate, rounded at the tip, persistent at fruiting, more or less concealing the fruit, rounded on the back, not winged, pubescent with dense, shorter, stellate, and usually also scattered longer, unbranched hairs. Staminate flowers with the stamens 3. Pistillate flowers with the ovary superior. Style absent or 1 and very short, the stigmas 2, linear. Fruits 2–3 mm long, obovate or nearly circular in outline, sometimes with a small, 2-lobed wing at the tip, flattened vertically, indehiscent, the wall papery to membranous, usually mottled with reddish brown spots and occasionally a few small, whitish, warty outgrowths. Seed adhering tightly to the fruit wall, positioned vertically, 2–3 mm long, obovate in outline, flattened, the surface smooth, brown, not or only slightly shiny, the embryo appearing U-shaped. $2n=18$. May–October.

Introduced, uncommon, known thus far only from a historical collection from Jackson County (northern U.S. south to Colorado and Missouri; Canada). Open, disturbed areas.

3. Beta L. (beet)

About 12 species, Mediterranean region.

1. Beta vulgaris L. **ssp. vulgaris** (garden beet)

Map 1520

Plants annual or biennial, the taproot with a globose to top-shaped, tuberous, thickened portion. Stems 30–120 cm long, erect or ascending, not succulent, not appearing jointed, usually relatively few-branched, glabrous. Leaves alternate and basal, well developed, progressively reduced toward the stem tip, not succulent, glabrous, often strongly reddish-tinged, sessile to short-petiolate, the basal leaves long-petiolate. Blades of basal leaves 10–40 cm or more long, unlobed, flattened in cross-section, elliptic-ovate to ovate or oblong-ovate, rounded at the tip or less commonly narrowed to a bluntly or sharply pointed tip, cordate at the base with a short-tapered extension of tissue into the petiole, the margins entire or irregularly wavy or scalloped, often appearing somewhat corrugated or crisped. Blades of the stem leaves 1–15 cm long, elliptic to ovate, unlobed, flattened in cross-section, not clasping the stem, narrowed to a sharply or bluntly pointed tip, mostly short-tapered at the base, the margins entire or irregularly wavy or scalloped. Inflorescences terminal panicles consisting of spikes, occasionally reduced to a single spike, the flowers not sunken into the axis. Flowers perfect, perigynous. Bracts 1 per flower, 8–15 mm long, leaflike, linear to narrowly ovate. Calyx 5-lobed to about the midpoint, the basal portion fused to the ovary (and sometimes also to the calyces of adjacent flowers), persistent at fruiting, more or less enclosing the fruit, the lobes 1.0–1.5 mm long, narrowly oblong, with a raised midrib. Stamens 5. Ovary partly inferior. Style absent, the stigmas 2 or 3, narrowly oblong-ovoid. Fruits 0.8–1.6 mm long, circular or nearly so in cross-section, oblong-obovate to obovate in outline, somewhat flattened laterally, indehiscent or more commonly circumscissilely dehiscent, the wall leathery to somewhat succulent toward the tip, thinner and herbaceous toward the base. Seed adhering loosely to the fruit wall, positioned horizontally, 1.5–2.0 mm long, broadly circular to somewhat kidney-shaped in outline, somewhat flattened, the surface smooth, black, shiny, the coiled embryo usually apparent. $2n=18$. July–September.

Introduced, uncommon, known thus far only from the city of St. Louis (originally developed in Europe, cultivated widely, escaped sporadically in the U.S. and Canada). Railroads.

This species was first reported for Missouri by Mühlenbach (1979). The ssp. *vulgaris,* with a tuberous-thickened taproot, more or less erect stems, and larger flower clusters, is a cultigen derived from ssp. *maritima* (L.) Arcang. (sea beet) and has been cultivated for a very long time. This species is an important source of table sugar, and various cultivars are used for livestock feed (fodder beet, mangel-wurzel), as a source for a reddish purple dye (from betalain pigments), as garden ornamentals with colorful foliage, as salad greens and potherbs (Swiss chard), and as an edible root. Extracts of the betalain compounds also have been used medicinally as an herbal treatment for some cancers.

4. Chenopodium L. (goosefoot, pigweed)

Plants usually annual (perennial elsewhere), the roots not tuberous-thickened. Stems prostrate to erect, not succulent, not appearing jointed, few- to much-branched, glabrous, glandular-hairy (sometimes also with nonglandular hairs), or sparsely to moderately mealy. Leaves alternate, well developed, not or only slightly succulent, petiolate or nearly sessile. Leaf blades linear to ovate or triangular, sometimes with a pair of spreading basal lobes, flattened in cross-section, not clasping the stem, rounded or narrowed to a sharply pointed tip, the margins entire, wavy, toothed, or lobed, the surfaces glabrous, glandular- and/or nonglandular-hairy, or sparsely to densely mealy. Inflorescences axillary and/or terminal, consisting of small flower clusters, these often appearing as interrupted spikes or panicles (occasionally a few nodes then with solitary flowers), the flowers not sunken into the axis. Flowers perfect (rarely a few pistillate flowers also present). Bracts absent or rarely 3 and minute. Calyx deeply 5-lobed, the lobes persistent at fruiting and all or partially enclosing the fruit, not winged, spaced evenly around the flower, often keeled or with a low, vertical ridge, rarely fleshy (in *C. capitatum*). Stamens 3–5 (usually only 1 in *C. pumilio*). Ovary superior. Styles 2 or 3, short, sometimes fused toward base, the stigmas 1 per style, linear or capitate. Fruits elliptic to circular in cross-section, depressed-globose to ovate in outline, somewhat flattened laterally or vertically, indehiscent or irregularly dehiscent, the wall thin and papery to membranous. Seed adhering loosely or tightly to the fruit wall, positioned vertically or horizontally, the surface smooth or less commonly roughened or with a low network of fine, rounded ridges, dark brown to black, shiny or less commonly dull, the coiled embryo usually apparent. About 150 species, nearly worldwide.

Immature specimens of *Chenopodium* often are difficult or impossible to identify and should not be collected. Fruits will, however, sometimes mature in the plant press while samples are drying. For large plants, it may be important to note the morphology of the larger, lower leaves when collecting samples for later determination. The adherence of the thin, membranous to papery wall of the fruit to the seed usually can be tested by rubbing a few mature fruits lightly between thumb and forefinger. In species with loosely adhering fruit walls this action will suffice to remove them, but in species with tightly adhering walls it will serve only to strip away the calyx.

Some species of *Chenopodium*, notably *C. album*, are cooked and eaten as a spinach substitute or (in combination with *Amaranthus, Lactuca, Sonchus, Rumex*, etc.) as field greens in the springtime when tender young growth is present (Steyermark, 1963). Native Americans also ground the seeds for flour (Moerman, 1998).

Crawford and Wilson (1986) included Missouri in their synopsis of the range of the western species *C. atrovirens* Rydb. However, in response to a query during the present research, Daniel J. Crawford (then of Ohio State University) indicated that this was a typographical error and was intended to read MT (Montana), rather than MO (Missouri).

Chenopodium bonus-henricus L. (good King Henry) is a perennial herb that is cultivated occasionally for its spinachlike leaves. It may be encountered in the future as an escape in Missouri. Aside from the thick, woody rootstock, this species may be recognized easily by the triangular leaves often with a pair of spreading basal lobes (similar in appearance to those of some Missouri species of *Atriplex*) and the oblong calyx lobes only partially enclosing the fruit, with the tips truncate to rounded or broadly angled and usually somewhat irregular margins.

1. Plants with the leaves and/or calyx pubescent with glandular or nonglandular hairs, and/or with minute, sessile, yellowish resinous glands (these globose when fresh, often flattening when dried); plants often with a resinous or unpleasant odor
    2. Leaves (except the uppermost) mostly more than 5 cm long, lacking hairs but with sparse to moderate sessile, yellowish resinous glands . . . . 2. C. AMBROSIOIDES
    2. Leaves (except the lowermost) mostly less than 4 cm long, lacking sessile glands but with moderate to dense, short glandular and/or nonglandular hairs
        3. Calyces with short glandular hairs; well-developed inflorescences consisting of small axillary panicles; seeds mostly positioned horizontally, the surface roughened or finely wrinkled, often mottled . . . . 4. C. BOTRYS
        3. Calyces with sessile glands and sometimes also short, nonglandular hairs; well-developed inflorescences consisting of short axillary spikes; seeds positioned vertically, the surface smooth . . . . 16. C. PUMILIO
1. Plants with the leaves and/or calyx glabrous or sparsely to densely white-mealy (the mealy hairs more or less bladder-shaped when fresh, lozenge-shaped when dry); plants lacking a pronounced odor (except in the mealy *C. watsonii* and sometimes *C. berlandieri*)
    4. Leaf blades entire; blades of main leaves linear to narrowly oblong, narrowly lanceolate, or less commonly narrowly ovate, mostly more than 3 times as long as wide
        5. Leaf blades linear or very narrowly oblong, with only a single midvein and no other veins visible . . . . 14. C. PALLESCENS
        5. Leaf blades (except sometimes toward the branch tips) narrowly lanceolate to oblong-lanceolate, lanceolate, or less commonly narrowly ovate; venation more complex, with secondary veins branching from the main vein(s); if no secondary veins present, then the blades with 3 main veins from the base

6. Calyx only partially covering the fruit (the lobes extending past the widest part but not reaching the stylar area, leaving much of the portion above the rim exposed); calyx lobes flat or slightly rounded dorsally, not keeled . . . . 19. C. STANDLEYANUM
6. Calyx covering the fruit (the lobes extending to the stylar region or nearly so, obscuring all or nearly all of the fruit surface) (usually somewhat spreading in *C. pratericola* and then more or less exposing the fruit); calyx lobes keeled dorsally
   7. Fruit wall membranous, smooth or often faintly granular, adhering relatively tightly to the seed; seed relatively strongly flattened horizontally, lens-shaped (narrowly biconvex) . . . . . . 1. C. ALBUM
   7. Fruit wall membranous to somewhat papery, smooth, adhering loosely to the seed; seed only somewhat flattened horizontally, depressed-ovoid
      8. Stems prostrate to loosely ascending; leaves mostly relatively thick and sometimes slightly fleshy, moderately to densely mealy on both surfaces; calyx lobes not or only slightly spreading at fruiting, covering the fruit . . . . . . 7. C. DESICCATUM
      8. Stems erect or ascending; leaves relatively thin and herbaceous, moderately to densely mealy on the undersurface, glabrous or sparsely mealy on the upper surface; calyx lobes usually somewhat spreading at fruiting, exposing the tip of the fruit . . . . 15. C. PRATERICOLA
4. At least a few of the leaf blades toothed or lobed; blades of main leaves variable but mostly triangular to rhombic-ovate, ovate, elliptic, or less commonly broadly lanceolate, mostly less than 3 times as long as wide
   9. Calyx 3(–5)-lobed; some or all of the seeds positioned vertically
      10. Leaves densely white-mealy on the undersurface, glabrous or sparsely mealy on the upper surface . . . . 9. C. GLAUCUM
      10. Leaves glabrous on both surfaces, not mealy, green
         11. Flower clusters relatively large, becoming 5–15 mm in diameter at fruiting, the calyx turning fleshy, red . . . . 6. C. CAPITATUM
         11. Flower clusters relatively small, becoming 3–5 mm in diameter at fruiting, the calyx remaining herbaceous to scalelike, green
            . . . . 17. C. RUBRUM
   9. Calyx (4)5-lobed; all of the seeds positioned horizontally
      12. Plants short, bushy, the stems 5–25 cm long, much-branched from the base
         13. Fruit wall appearing membranous and more or less transparent at maturity, easily separated from the seed; plants lacking an odor
            . . . . 10. C. INCANUM
         13. Fruit wall appearing thicker and opaquely white at maturity, difficult to separate from the seed; plants with a pronounced unpleasant odor
            . . . . 22. C. WATSONII
      12. Plants taller, the stems to 200 cm long, if less than 25 cm long then with few to no branches
         14. Fruits with the surface appearing honeycombed (often visible only with magnification), finely pitted, the pits usually more or less rectangular, separated by a network of thin ridges (do not confuse fruit wall with seed surface)

15. Largest leaves with the middle lobe elongate, appearing narrowly oblong, more or less parallel sided for most of its length................................... 8. C. FICIFOLIUM
15. Largest leaves with the middle lobe not elongate, appearing ovate to triangular, angled or tapered from near the base
    16. Calyx lobes with a pronounced broad keel or raised area along the midvein; fruits 1.0–1.5 mm wide, the fruit wall difficult to separate from the seed except for a loose area surrounding the style base, this usually appearing lighter colored and yellowish at fruit maturity ... 3. C. BERLANDIERI
    16. Calyx lobes only slightly and narrowly keeled or raised along the midvein; fruits 1.5–2.2 mm wide, the fruit wall uniformly adhering to the seed, without a loose, lighter-colored apical portion.................... 5. C. BUSHIANUM
14. Fruits with the surface smooth or finely roughened, not appearing honeycombed
    17. Calyx lobes covering the entire fruit, extending to the stylar region and tightly enclosing the fruit except for a minute area surrounding the style, occasionally somewhat spreading in a few flowers of a given inflorescence
        18. Largest leaves with the middle lobe elongate, appearing narrowly oblong, more or less parallel sided for most of its length.......... 8. C. FICIFOLIUM
        18. Largest leaves with the middle lobe not elongate, appearing ovate to triangular, angled or tapered from near the base
            19. Calyx lobed to about the midpoint, the fused portion extending beyond the broadest portion of the fruit........... 13. C. OPULIFOLIUM
            19. Calyx lobed nearly to the base
                20. Fruits 1.2–1.5 mm wide; largest (lowermost) leaves with the blades mostly 1.5–3.0 times as long as wide; stems often reddish-tinged or with reddish purple stripes, but usually lacking a pronounced reddish purple area at the base of each leaf .. 1. C. ALBUM
                20. Fruits 0.9–1.1(–1.2) mm wide; largest (lowermost) leaves with the blades mostly 1.2–1.5 times as long as wide; stems almost always reddish purple at each node (on the side on which the leaf is attached) .... 11. C. MISSOURIENSE
    17. Calyx lobes not covering the entire fruit, either not extending to the stylar region or only loosely enclosing the fruit (or both)
        21. Leaves with the margins entire above the base, at most with a single pair of basal (or nearly basal) lobes
            22. Calyx lobes keeled dorsally, extending to the stylar region or nearly so, but usually somewhat spreading, loosely covering most of the fruit
            .... 15. C. PRATERICOLA
            22. Calyx lobes flat, somewhat rounded, or pouched, or at most narrowly and shallowly keeled dorsally, extending past the widest part of the fruit but not reaching the stylar area, leaving much of the portion above the rim exposed .... 19. C. STANDLEYANUM
        21. At least some of the leaves with the margins toothed above the base
            23. Calyx lobes keeled dorsally along the midrib, the flower appearing more or less pentagonal when viewed from above
                24. Leaf blades mostly with entire margins, only those of the largest (lowermost) leaves shallowly toothed; lowermost leaves with the blades mostly oblong-ovate .... 20. C. STRICTUM

24. Leaf blades all or mostly with wavy to irregularly toothed margins, only those of the smallest (uppermost) leaves with the blades shallowly toothed; lowermost leaves with the blades triangular-ovate to rhombic

25. Calyx lobes relatively short, leaving most of the upper portion of the fruit exposed, the lobes glabrous or rarely sparsely mealy .... 21. C. URBICUM

25. Calyx lobes relatively long, covering most of the fruit, the lobes sparsely to densely mealy

26. Fruits 1.2–1.5 mm wide; largest (lowermost) leaves with the blades mostly 1.5–3.0 times as long as wide; stems often reddish-tinged or with reddish purple stripes, but usually lacking a pronounced reddish purple area at the base of each leaf .... 1. C. ALBUM

26. Fruits 0.9–1.1(–1.2) mm wide; largest (lowermost) leaves with the blades mostly 1.2–1.5 times as long as wide; stems almost always reddish purple at each node (on the side to which the leaf is attached) ..... 11. C. MISSOURIENSE

23. Calyx lobes flat or only slightly rounded or raised dorsally along the midrib, the flower appearing more or less circular in cross-section

27. Largest leaves 7–18 cm long, the margins of the blades with relatively few broad, coarse teeth; seeds 1.5–2.5 mm wide .... 18. C. SIMPLEX

27. Largest leaves 2–5(–8) cm long, the margins of the blades wavy or with few to many relatively narrow, irregular teeth; seeds 0.5–1.5 mm wide

28. Seeds and fruits sharply angled along the margin (rim), the seeds dull .... 12. C. MURALE

28. Seeds and fruits rounded or at most bluntly angled along the margin, the seeds shiny

29. Fruits (1.0–)1.2–1.5 mm wide; leaf blades mostly linear to lanceolate or oblong-lanceolate, those of only the lowermost leaves ovate to ovate-triangular........................... 19. C. STANDLEYANUM

29. Fruits 0.9–1.2 mm wide; leaf blades mostly ovate-triangular, those of only the uppermost leaves linear to lanceolate .... 21. C. URBICUM

1. Chenopodium album L. (pigweed, lamb's quarters)

Pl. 353 f–h; Map 1521

Plants annual, without an odor. Stems 10–150 cm long, erect or ascending, usually few- to several-branched above the base and below the inflorescence, glabrous or more commonly sparsely to moderately white-mealy, sometimes reddish-tinged or reddish purple–striped, but usually lacking a pronounced reddish purple area at the base of each leaf. Leaves mostly long-petiolate. Leaf blades 1–6(–12) cm long, mostly 1–3 times as long as wide (1–4 cm wide), often more than 1.5 times in the largest (lowermost) leaves, rhombic to ovate-rhombic, ovate-triangular, or lanceolate, the uppermost usually linear to narrowly lanceolate, angled or tapered to a bluntly or sharply pointed tip, the middle lobe not appearing unusually elongate, angled at the base, green or reddish-tinged, thin and herbaceous to thickened, somewhat leathery, and slightly succulent in texture, the margins entire to wavy or irregularly several-toothed (the basal pair of teeth usually larger than the others, sometimes appearing shallowly lobed), the upper surface glabrous or sparsely to moderately mealy at maturity, not shiny, the undersurface moderately to more commonly densely white-mealy. Venation noticeably branched, with 1 or 3 main veins. Inflorescences axillary and terminal, consisting of short spikes with small clusters of flowers, the terminal ones usually grouped into small to relatively large panicles. Flowers not all maturing at the same time.

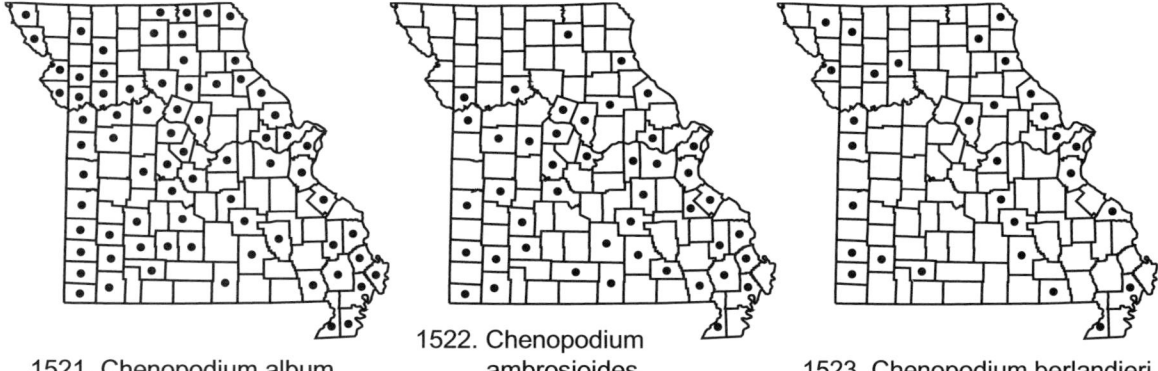

1521. Chenopodium album
1522. Chenopodium ambrosioides
1523. Chenopodium berlandieri

Calyx 5-lobed nearly to the base, usually covering the entire fruit except sometimes for a minute area surrounding the style, occasionally somewhat spreading in a few flowers of a given inflorescence, the lobes 0.7–1.2 mm long, ovate to triangular-ovate, bluntly pointed at the tip, usually with a relatively pronounced broad keel or raised area along the midvein dorsally, moderately to densely white-mealy. Stamens 5. Stigmas 2. Fruits 1.2–1.5 mm wide, depressed-ovoid, the seeds positioned horizontally, the wall thin, membranous, and somewhat translucent, smooth or finely roughened, not appearing honeycombed, usually difficult to separate from the seed. Seeds black, shiny, smooth or nearly so, rounded to very bluntly angled along the rim. $2n=54$. May–October.

Introduced, common nearly throughout the state (nearly worldwide, probably of Eurasian origin). Banks of streams, rivers, and spring branches; also crop fields, fallow fields, gardens, roadsides, railroads, and open, disturbed areas.

Chenopodium album has been interpreted taxonomically in a variety of ways. On the one hand, it has been split into numerous subspecies, varieties, and forms (see Wahl [1952–1953] for synonymy), some of which have been recognized as separate species (Mohlenbrock, 2001). At the other extreme, some authors have circumscribed *C. album* as a highly polymorphic cosmopolitan species, reducing to synonymy or to infraspecific rank taxa treated here as *C. berlandieri*, *C. bushianum*, *C. missouriense*, and *C. opulifolium*, along with some others not yet found growing in Missouri. Although the temptation in any group of plants in which taxa are difficult to identify is to lump them together into a single species, this probably does not reflect the actual taxonomy of the *C. album* complex and it certainly is not the approach taken by those monographers who have had experience with the group on a worldwide basis. Further biosystematic and molecular studies may help to resolve classification in these pigweeds.

Steyermark (1963) separated var. *lanceolatum* (Muhl. ex Willd.) Coss. & Germ. from var. *album* based upon its more spreading branches, somewhat narrower and less-toothed leaves, and more discontinuous flower clusters in the inflorescence. Mohlenbrock (2001) and some other authors have treated this as a separate species, *C. lanceolatum* Muhl. ex Willd., and have noted other morphological distinctions, including less mealy calyces with whitened (vs. yellowish) lobe margins and supposedly slightly smaller seeds. The status of this relatively uncommon taxon is unclear, especially as its fruits are sometimes faintly reticulate on the surface. A number of specimens exist that are intermediate for one or more of the characters. Perhaps var. *lanceolatum* is merely a component of the morphological variation in *C. album*, but it also may represent the effects of past hybridization between this species and *C. berlandieri* var. *boscianum*. The specimens in question appear somewhat different in leaf and fruit morphology from the rare putative hybrids between *C. album* and *C. berlandieri* var. *zschackei*, which are discussed under the treatment of the latter species. Further studies are needed.

2. Chenopodium ambrosioides L. (Mexican tea, wormseed)
Ambrina ambrosioides (L.) Spach
Dysphania ambrosioides (L.) Mosyakin & Clemants

Pl. 353 i–l; Map 1522

Plants annual (biennial or short-lived perennial herbs farther south), usually with an unpleasant odor. Stems 25–100 cm long, spreading to ascending, usually much-branched, with sparse to moderate, sessile, yellowish resin glands, sometimes reddish-tinged or reddish-striped. Leaves sessile (uppermost) to long-petiolate (lowermost). Leaf blades 1–14 cm long (mostly over 5 cm long), those of well-developed leaves mostly 2–5 times as long as wide, linear (uppermost leaves) to oblong, lanceolate, or ovate, bluntly to sharply pointed at the tip, rounded or angled at the base, entire to irregularly lobed, yellowish green to green and herbaceous in texture, the margins often also somewhat wavy or with shallow, irregular, narrow teeth, the surfaces lacking hairs and mealiness, but with sparse to moderate, sessile, yellowish resin glands. Venation noticeably branched, often with 3 main veins from the base. Inflorescences axillary and terminal spikes, the terminal ones sometimes arranged into small panicles, the spikes relatively short and dense, with small clusters of flowers. Flowers not all maturing at the same time. Calyx (4)5-lobed to below the midpoint, covering the fruit at maturity, the lobes 0.7–1.0 mm long, ovate, bluntly pointed at the tip, flat to rounded dorsally, glabrous or with sparse to moderate, short, fine, nonglandular hairs. Stamens (4)5. Stigmas 3. Fruits 0.6–1.0 mm long, ovoid, the seeds positioned horizontally or vertically, the wall thin, papery, and somewhat translucent, smooth to finely wrinkled, easily separated from the seed. Seeds reddish brown to dark brown, shiny, smooth to faintly wrinkled, rounded along the rim. $2n=16, 32, 36, 48(?), 64$. July–November.

Introduced, scattered nearly throughout the state, but more common south of the Missouri River (native of tropical America; naturalized widely in the U.S. and Canada). Banks of rivers; also crop fields, fallow fields, gardens, roadsides, railroads, and open, disturbed areas.

The resinous odor exuded by this species has been described as somewhat reminiscent to that of kerosene. Plants have been used medicinally for a variety of ailments, but principally to treat worms. The active ingredient, a volatile oil called ascaridol, is considered poisonous in larger doses. In portions of Latin America, the species is also cultivated for use as a flavoring for other foods and to prepare a tea. The oils in and on the foliage are allelopathic, retarding the germination and growth of other plant species (Clemants, 1992). Some botanists refer the roughly 32 species of *Chenopodium* with resinous glands rather than farina to the genus *Dysphania* R. Br. in the belief that these species are more closely related to *Atriplex* than the remainder of *Chenopodium* (Mosyakin and Clemants, 2002). In Missouri, these include *C. ambrosioides*, *C. botrys*, and *C. pumilio*. The situation requires further study.

The infraspecific taxonomy of this polyploid complex requires further study. Several variants have been given formal taxonomic recognition in some older literature. Two of these that are treated in many floristic manuals are accepted here, but readers should note the presence of both varieties in some populations and the existence of occasional morphologically intermediate plants.

1. Spikes with leaflike bracts at most of the nodes . . . . 2A. VAR. AMBROSIOIDES
1. Spikes bractless or with only a few small, leaflike bracts
   . . . . 2B. VAR. ANTHELMINTICUM

2a. var. ambrosioides

Spikes with leaflike bracts 0.3–2.5 cm long at most of the nodes. Calyx lobes rounded to very slightly keeled dorsally. July–November.

Introduced, scattered nearly throughout the state, but more common south of the Missouri River (native of tropical America; naturalized widely in the U.S. and Canada). Banks of rivers; also crop fields, fallow fields, gardens, roadsides, railroads, and open, disturbed areas.

2b. var. anthelminticum (L.) A. Gray

C. anthelminticum L.
Ambrina anthelmintica (L.) Spach
Dysphania anthelmintica (L.) Mosyakin & Clemants

Spikes bractless or with only a few small, leaflike bracts to 0.3–1.0 cm long. Calyx lobes rounded to somewhat flattened dorsally. August–November.

Introduced, uncommon and sporadic (native of tropical America; naturalized widely in the U.S. and Canada). Banks of rivers; also crop fields, fallow fields, gardens, roadsides, railroads, and open, disturbed areas.

3. Chenopodium berlandieri Moq. (pitseed goosefoot)

C. album L. var. *berlandieri* (Moq.) Mack. & Bush

Pl. 354 g, h; Map 1523

Plants annual, sometimes with an unpleasant odor. Stems 20–120(–200) cm long, erect or ascending, usually few- to several-branched above the base and below the inflorescence, glabrous or more commonly sparsely to moderately white-mealy, sometimes reddish-tinged or reddish-striped. Leaves mostly long-petiolate. Leaf blades 1–12(–15) cm long, mostly 1–3 times as long as wide (1–5 cm wide), mostly rhombic to ovate-rhombic, less commonly ovate-triangular, the uppermost usually linear to narrowly lanceolate, angled or tapered from below the midpoint to a usually sharply pointed tip, the middle lobe not appearing unusually elongate, angled at the base, green or reddish-tinged, thin and herbaceous to thickened, somewhat leath-

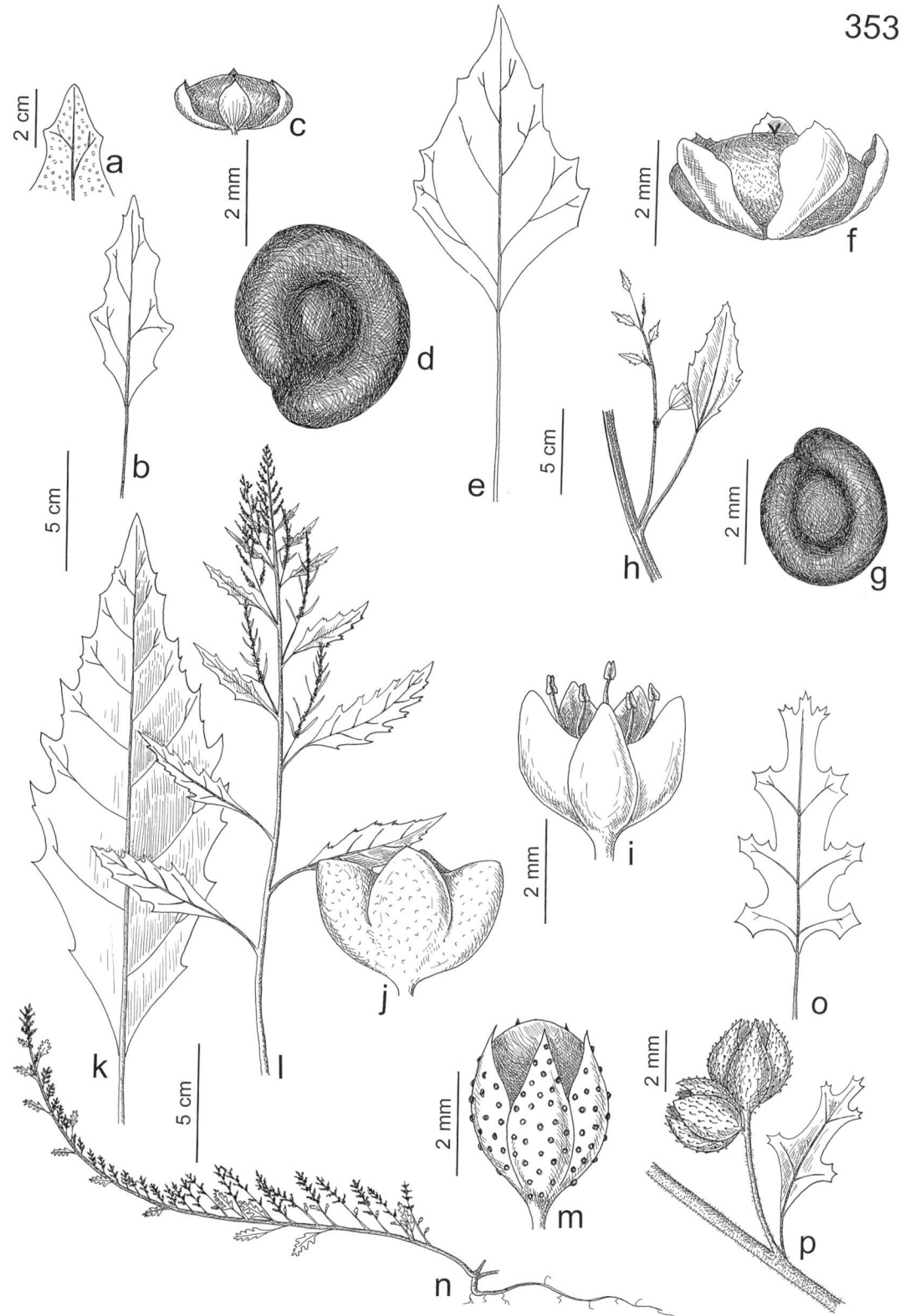

Plate 353. Chenopodiaceae. *Chenopodium glaucum*, **a)** leaf tip, **b)** leaf, **c)** fruit. *Chenopodium bushianum*, **d)** fruit, **e)** leaf. *Chenopodium album*, **f)** fruit, **g)** leaf, **h)** seed. *Chenopodium ambrosioides*, **i)** flower, **j)** fruit with surrounding calyx, **k)** leaf, **l)** inflorescence. *Chenopodium pumilio*, **m)** leaf, **n)** habit. *Chenopodium botrys*, **o)** leaf, **p)** fruits with surrounding calyces.

ery, and slightly succulent in texture, the margins entire to irregularly several-toothed (the basal pair of teeth usually larger than the others), the upper surface glabrous or sparsely to moderately mealy at maturity, not shiny, the undersurface usually densely white-mealy. Venation noticeably branched, with 1 or 3 main veins. Inflorescences axillary and terminal, consisting of short spikes with small clusters of flowers, the terminal ones usually grouped into small to relatively large panicles. Flowers not all maturing at the same time. Calyx 5-lobed nearly to the base, covering the entire fruit except sometimes for a minute area surrounding the style, the lobes 1.0–1.5 mm long, ovate to triangular-ovate, rounded or bluntly pointed at the tip, with a relatively pronounced broad keel or raised area along the midvein dorsally, densely white-mealy. Stamens 5. Stigmas 2. Fruits 1.0–1.5 mm wide, depressed-ovoid, the seeds positioned horizontally, the wall thin, membranous, and somewhat translucent, appearing honeycombed (often visible only with magnification), finely pitted, the pits usually more or less rectangular, separated by a network of thin ridges, the wall usually difficult to separate from the seed (except around the style, where the fruit wall then often appears yellowish and distinctly lighter than the surrounding tissue). Seeds reddish brown to black, shiny, finely wrinkled, rounded to very bluntly angled along the rim. $2n=36$. July–October.

Scattered nearly throughout the state (U.S., Canada, Mexico; introduced in Europe). Banks of streams, rivers, and spring branches, less commonly openings of mesic to dry upland forests and upland prairies; also crop fields, roadsides, railroads, and open, disturbed areas.

Chenopodium berlandieri and *C. bushianum* can be difficult to separate morphologically, which has led some botanists to treat them as varieties of a single species. Variation within the widespread *C. berlandieri* also has led to the recognition by some earlier monographers of numerous intergrading subspecies, varieties, and forms (Wahl, 1952–1953). Only two of the varieties are accepted as distinct in Missouri, and there is considerable morphological overlap in some populations even for these. The var. *berlandieri* refers to plants primarily of the western United States similar to var. *boscianum*, but with leaf blades having a somewhat elongate central lobe.

1. Largest leaves mostly 2–4 cm long, relatively thin and herbaceous in texture, the largest ones usually with conspicuous teeth above the widest point; seeds 1.0–1.3 mm wide
 . . . . 3A. VAR. BOSCIANUM

1. Largest leaves often but not always 4–12(–15) cm long, usually thickened, somewhat leathery, and slightly succulent in texture, the margins often wavy, few-toothed or nearly entire above the widest point; seeds 1.2–1.5 mm wide
 . . . . 3B. VAR. ZSCHACKEI

3a. var. boscianum (Moq.) Wahl
C. boscianum Moq.

Leaf blades 1–4 cm long, those of the largest leaves mostly 2–4 cm long, relatively thin and herbaceous in texture, the largest ones usually with conspicuous teeth above the widest point. Seeds 1.0–1.3 mm wide. July–October.

Uncommon in southern and western Missouri (southeastern U.S. west to Missouri and Texas, Colorado; Mexico). Open, disturbed areas.

Plants of var. *boscianum* are in some ways intermediate in vegetative characters between *C. bushianum* and *C. berlandieri* var. *berlandieri*. However, they have seeds that are smaller than in the former species and sepals that are more strongly keeled than in the latter. The main distribution of var. *boscianum* is along the Atlantic and Gulf Coastal Plains, and the few populations in Missouri may represent nonnative occurrences.

3b. var. zschackei (Murr) Murr ex Asch.
C. berlandieri f. *angustius* (A. Ludw. ex Asch. & Graebn.) Aellen
C. berlandieri f. *latifolium* (A. Ludw. ex Asch. & Graebn.) Aellen
C. berlandieri f. *neglectum* (A. Ludw. ex Asch. & Graebn.) Aellen
C. berlandieri f. *pedunculare* Aellen
C. berlandieri var. *farinosum* (A. Ludw. ex Asch. & Graebn.) Aellen
C. berlandieri var. *foetens* (A. Ludw. ex Asch. & Graebn.) Aellen
C. berlandieri ssp. *zschackei* (Murr) Zobel
C. dakoticum Standl.

Leaf blades 1–12(–15) cm long, those of the largest leaves mostly 4–12 cm long, usually thickened, somewhat leathery, and slightly succulent in texture, the margins often wavy, few-toothed or nearly entire above the widest point. Seeds 1.2–1.5 mm wide. $2n=36$. July–October.

Scattered nearly throughout the state (U.S., Canada, Mexico; introduced in Europe). Banks of streams, rivers, and spring branches, less commonly openings of mesic to dry upland forests and upland prairies; also crop fields, roadsides, railroads, and open, disturbed areas.

A putative hybrid between *C. album* and *C. berlandieri* var. *zschackei* (*C.* ×*variabile* Aellen var.

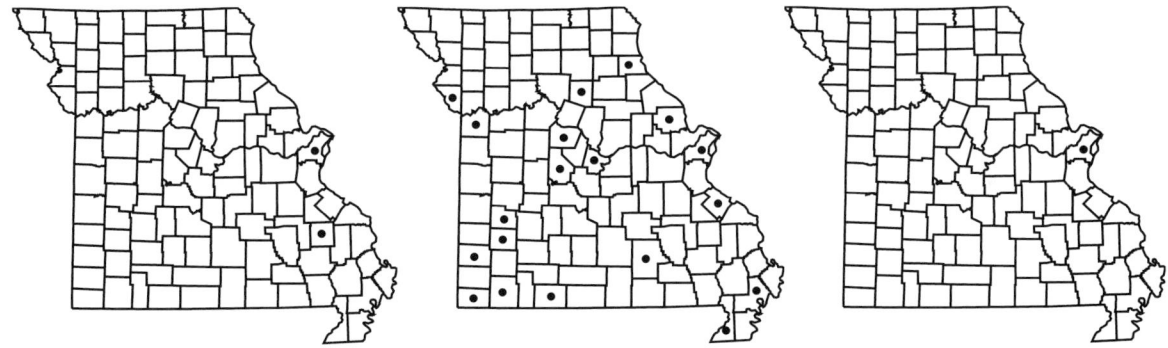

1524. Chenopodium botrys   1525. Chenopodium bushianum   1526. Chenopodium capitatum

murrii Aellen) has been collected in Jackson County and St. Louis City. Differentiating these plants from *C. berlandieri* requires a great deal of imagination, although at least the St. Louis specimen appears to be mostly sterile.

4. Chenopodium botrys L. (Jerusalem oak, feather geranium)

Dysphania botrys (L.) Mosyakin & Clemants

Pl. 353 o, p; Map 1524

Plants annual, with a slight unpleasant odor. Stems 10–60 cm long, erect or ascending, few- to more commonly much-branched, with dense, short glandular hairs, usually not reddish-tinged or reddish-striped. Leaves sessile (uppermost) to long-petiolate (lowermost). Leaf blades 0.5–4.0 cm long (the lowermost rarely to 8 cm long), those of well-developed leaves mostly 1.5–3.0 times as long as wide, oblong, lanceolate, or ovate, mostly bluntly pointed at the tip, rounded or angled to truncate at the base, entire or with few to several rounded lobes, green and herbaceous in texture, the margins usually also wavy or with shallow, blunt, narrow teeth, the surfaces lacking mealiness but with dense, short glandular hairs. Venation noticeably branched, usually with 1 main vein. Inflorescences axillary and terminal, mostly small panicles, some of the axillary ones occasionally reduced to small, sessile clusters of flowers. Flowers not all maturing at the same time. Calyx 5-lobed to below the midpoint, somewhat spreading at maturity and exposing the fruit, the lobes 0.7–0.9 mm long, elliptic-ovate to oblong-elliptic, sharply pointed at the tip, flat or nearly so dorsally, densely pubescent with short glandular hairs. Stamens 5. Stigmas 2. Fruits 0.5–0.8 mm long, ovoid to depressed-ovoid, the seeds mostly positioned horizontally, the wall thin, membranous, and somewhat translucent, smooth, difficult to separate from the seed. Seeds dark brown to nearly black, often somewhat mottled, dull, roughened to finely wrinkled, rounded along the rim. $2n=18$. July–October.

Introduced, uncommon in eastern Missouri but to be expected elsewhere (native of Europe, Asia; introduced widely in the U.S. and Canada). Gardens, railroads, and open, disturbed areas.

5. Chenopodium bushianum Aellen

C. berlandieri var. *bushianum* (Aellen) Cronquist

C. paganum Rchb.

Pl. 353 d, e; Map 1525

Plants annual, without an odor. Stems (20–)40–200 cm long, erect or ascending, usually few- to several-branched above the base and below the inflorescence, glabrous or more commonly sparsely to moderately white-mealy, sometimes reddish-tinged or reddish-striped. Leaves mostly long-petiolate. Leaf blades 1–12 cm long, mostly 1–3 times as long as wide (1–5 cm wide), mostly rhombic to ovate-rhombic, less commonly ovate-triangular, the uppermost usually linear to narrowly lanceolate, angled or tapered from below the midpoint to a usually sharply pointed tip, the middle lobe not appearing unusually elongate, angled at the base, green or reddish-tinged, relatively thin and herbaceous in texture, the margins entire to wavy or irregularly several-toothed (the basal pair of teeth usually larger than the others), sometimes with a pair of shallow lobes and otherwise entire, the upper surface glabrous or sparsely to moderately mealy at maturity, not shiny, the undersurface sparsely to densely white-mealy. Venation noticeably branched, with 1 or 3 main veins. Inflorescences axillary and terminal, consisting of short spikes with small clusters of flowers, the terminal ones usually grouped into small to relatively large panicles, these sometimes drooping at maturity. Flowers not all maturing at the same time. Calyx 5-lobed nearly to the base, covering the entire fruit,

1527. Chenopodium desiccatum
1528. Chenopodium ficifolium
1529. Chenopodium glaucum

the lobes 1.4–2.0 mm long, ovate to triangular-ovate, rounded or bluntly to sharply pointed at the tip, slightly and narrowly keeled or raised along the midvein dorsally, sparsely to less commonly densely white-mealy. Stamens 5. Stigmas 2. Fruits 1.5–2.2 mm wide, depressed-ovoid, the seeds positioned horizontally, the wall thin, membranous, and somewhat translucent, appearing honeycombed (often visible only with magnification), finely pitted, the pits usually more or less rectangular, separated by a network of thin ridges, usually uniformly difficult to separate from the seed. Seeds reddish brown to more commonly black, dull, finely wrinkled, rounded to very bluntly angled along the rim. $2n=36$. August–October.

Scattered but sporadic nearly throughout the state (eastern U.S. west to North Dakota and Kansas, Canada). Banks of streams, rivers, and spring branches, and mesic upland forests; also open, disturbed areas.

Steyermark (1963) noted that in addition to the large, reticulate fruits and large, thin leaves, other characteristic features of this species include the relatively pale green color of the foliage and the often lead-colored, drooping inflorescences produced in the autumn when most of the leaves have already been shed. He predicted that *C. bushianum* eventually would be found in nearly every county of Missouri, but its distribution has remained sporadic in the state.

6. Chenopodium capitatum (L.) Asch. (strawberry blite, strawberry spinach)
Blitum capitatum L.

Pl. 354 e, f; Map 1526

Plants annual, without an odor. Stems 10–60 cm long, erect or ascending, few- to much-branched from near the base, glabrous, often somewhat reddish-tinged or reddish-striped. Leaves short- to long-petiolate. Leaf blades 1–10 cm long, mostly 1.0–2.5 times as long as wide, triangular to ovate-triangular, sharply pointed at the tip, broadly angled to truncate at the base, sometimes with a pair of spreading triangular basal lobes, green and slightly fleshy in texture, the margins also wavy or with sharp, irregular teeth, occasionally entire, the surfaces glabrous. Venation noticeably branched, with 1 or 3 main veins. Inflorescences terminal and sometimes also axillary, consisting of dense clusters of flowers (becoming enlarged to 5–15 mm in diameter at fruiting), these often arranged into spikes. Flowers often all or mostly maturing at more or less the same time. Calyx 3(–5)-lobed nearly to the base, becoming enlarged, red, and fleshy at maturity and covering the fruit, the lobes 0.6–0.7 mm long at flowering, oblong-elliptic to oblong-obovate, rounded to bluntly pointed at the tip, rounded dorsally, glabrous. Stamens usually 3 or 4. Stigmas 2. Fruits 0.7–1.2 mm long, ovoid, the seeds positioned mostly vertically, the wall thin, membranous, and somewhat translucent, smooth, difficult to separate from the seed. Seeds black, dull, finely roughened, angled along the rim. $2n=18$. May–August.

Introduced, known thus far only from the city of St. Louis (native of Europe, Asia; introduced widely in the northern and western U.S., Canada). Open, disturbed areas.

Strawberry blite is cultivated as an ornamental. The red, berrylike fruits are formed from the enlarged calyces of a flower cluster, which turn bright red as the fruits mature and become fused into a fleshy mass. These so-called fruits sometimes have been eaten raw or cooked or used as a coloring for beverages or makeup. The young foliage also sometimes is eaten cooked like spinach.

7. Chenopodium desiccatum A. Nelson

Pl. 354 a, b; Map 1527

Plants annual, without an odor. Stems 10–50 cm long, prostrate to loosely ascending, several- to much-branched from the base, moderately to

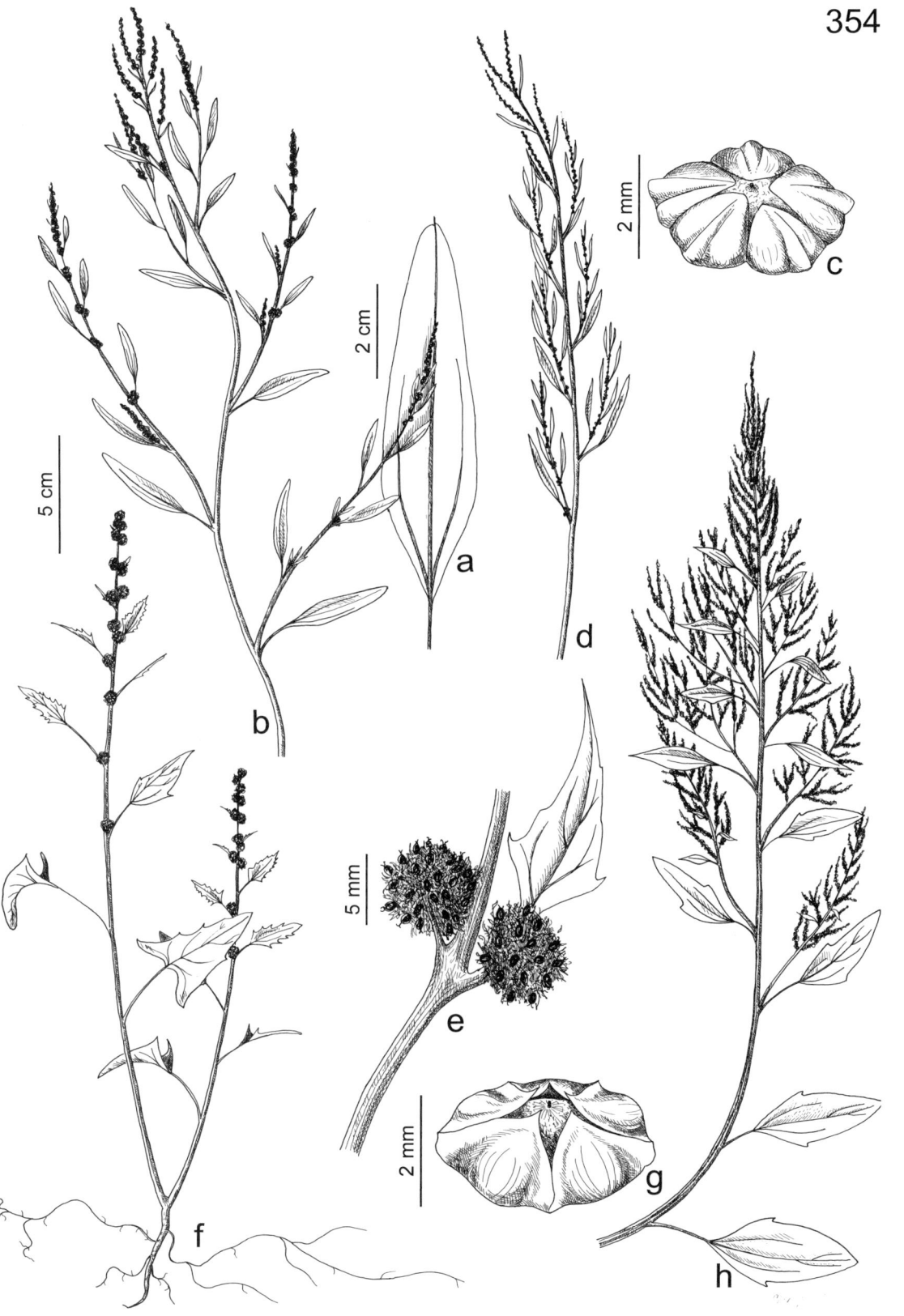

Plate 354. Chenopodiaceae. *Chenopodium desiccatum*, **a)** leaf, **b)** branch. *Chenopodium pratericola*, **c)** fruit, **d)** branch. *Chenopodium capitatum*, **e)** fruit, **f)** node at fruiting. *Chenopodium berlandieri*, **g)** fruit, **h)** branch.

densely white-mealy, sometimes slightly reddish-tinged or reddish-striped. Leaves sessile or more commonly short-petiolate. Leaf blades 1–3(–4) cm long, mostly 3–8 times as long as wide (2–10 mm wide), linear to oblong-lanceolate or narrowly oblong-elliptic, sharply pointed at the tip, angled at the base, unlobed, grayish green, relatively thick and sometimes slightly fleshy in texture, the margins entire, moderately to densely mealy on both surfaces. Venation of 3 main veins, usually with no other branching apparent. Inflorescences axillary and terminal, consisting of short spikes with small clusters of flowers, the terminal spikes usually grouped into small panicles. Flowers not all maturing at the same time. Calyx 5-lobed nearly to the base, extending to the stylar area, not or only slightly spreading at maturity, covering the entire fruit except sometimes for a minute area surrounding the style, the lobes 0.7–1.0 mm long, broadly lanceolate to ovate, bluntly pointed at the tip, strongly keeled dorsally, moderately to densely white-mealy. Stamens (4)5. Stigmas 2. Fruits 0.9–1.1 mm wide, depressed-ovoid, the seeds positioned horizontally, the wall thin, membranous to somewhat papery, somewhat translucent, smooth, difficult to separate from the seed. Seeds black, shiny, finely pebbled or minutely roughened, bluntly angled along the rim. $2n=18$. July–November.

Introduced, uncommon and widely scattered in eastern and western Missouri (native of the western U.S. and adjacent Canada east to Wyoming and New Mexico; introduced sporadically eastward). Sand prairies and banks of rivers; also fallow fields, gardens, railroads, and open, disturbed areas.

8. Chenopodium ficifolium Sm. (fig-leaved goosefoot)

Pl. 355 a, b; Map 1528

Plants annual, without an odor. Stems 20–100 cm long, erect or ascending, unbranched or more commonly few- to much-branched above the base, glabrous or nearly so, often somewhat reddish-tinged or reddish-striped. Leaves short- to more commonly long-petiolate. Leaf blades 2–7 cm long, mostly 2–5 times as long as wide (0.5–3.0 cm wide), oblong to ovate or narrowly rhombic, rounded to bluntly pointed at the tip, angled, tapered, or occasionally rounded at the base, with a conspicuous pair of triangular lobes toward the base and the middle lobe elongate, appearing narrowly oblong, more or less parallel sided for most of its length, green to grayish green, relatively thin and herbaceous in texture, the margins otherwise wavy to irregularly several-toothed, the upper surface glabrous or sparsely white-mealy, the undersurface sparsely to densely white-mealy. Venation noticeably branched, with 1 or 3 main veins. Inflorescences axillary and terminal, consisting of short spikes with small clusters of flowers, these often grouped into small panicles. Flowers not all maturing at the same time. Calyx 5-lobed to about the midpoint, covering the entire fruit except sometimes for a minute area surrounding the style, the lobes 0.5–0.8 mm long, ovate, rounded or bluntly pointed at the tip, rounded and often narrowly and inconspicuously keeled, moderately to densely white-mealy. Stamens 5. Stigmas 2. Fruits 0.9–1.2 mm wide, depressed-ovoid, the seeds positioned horizontally, the wall thin, membranous, and somewhat translucent, smooth or finely but distinctly pitted (the pits usually more or less rectangular, separated by a network of thin ridges), difficult to separate from the seed. Seeds black, shiny, finely pitted or wrinkled at maturity, rounded along the rim. $2n=18$. July–November.

Introduced, uncommon, known thus far only from the city of St. Louis (native of Europe, Asia; introduced in Missouri, Pennsylvania, Canada). Open, disturbed areas.

This species was first reported from Missouri by Mühlenbach (1983), who noted that plants appeared for a period of several years in various portions of the Missouri Botanical Garden following construction of the John G. Lehmann Building. However, the species has not been found in the state since 1976.

9. Chenopodium glaucum L. (oak-leaved goosefoot)

Pl. 353 a–c; Map 1529

Plants annual, without an odor. Stems 5–40 cm long, prostrate to erect, unbranched or more commonly much-branched from near the base, glabrous or sparsely white-mealy, often strongly reddish-tinged or reddish-striped. Leaves short- to long-petiolate. Leaf blades 0.5–4.0 cm long, mostly 1–3 times as long as wide, lanceolate to oblong or ovate, bluntly pointed at the tip, angled at the base, herbaceous in texture, the margins entire, wavy or with few to several sharp or blunt, irregular teeth, the most basal ones occasionally deeper and appearing as a pair of shallow lobes, the upper surface glabrous or nearly so, the undersurface densely white-mealy. Venation noticeably branched (best observed from upper surface), with 1 or more commonly 3 main veins. Inflorescences axillary and terminal, consisting of small clusters of flowers, these sometimes arranged into spikes, the terminal ones sometimes grouped into small panicles. Flowers not all maturing at the same time. Calyx

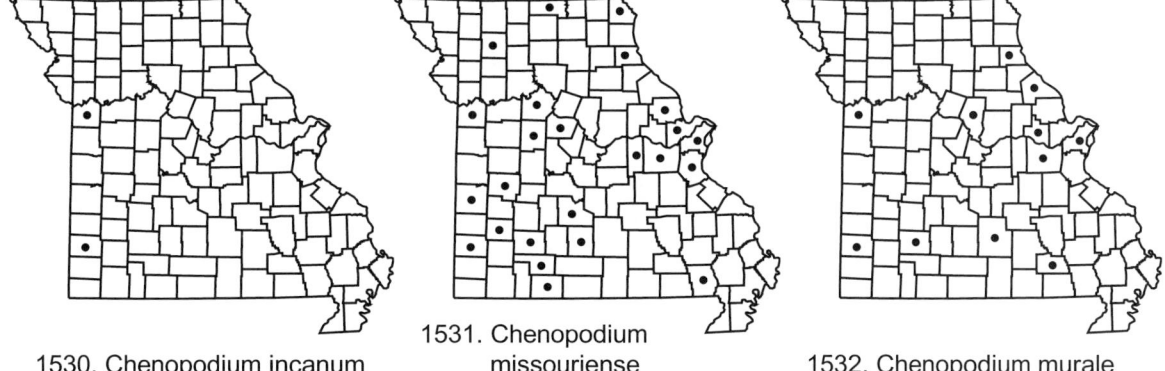

1530. Chenopodium incanum
1531. Chenopodium missouriense
1532. Chenopodium murale

3–5-lobed nearly to the base, shorter than and only partially covering the fruit at maturity, the lobes 0.4–0.6 mm long, oblong to oblong-obovate, rounded to bluntly pointed at the tip, flat to somewhat pouched dorsally, glabrous. Stamens 3–5. Stigmas 2, minute. Fruits 0.6–1.0 mm long, ovoid to depressed-ovoid, the seeds mostly positioned mostly horizontally, but a few vertically oriented seeds also present, the wall thin, membranous, and somewhat translucent, smooth, easily separated from the seed. Seeds reddish brown to dark brown, shiny, smooth or finely roughened, rounded along the rim. $2n=18$. July–November.

Introduced, uncommon and local in northwestern Missouri and the city of St. Louis (native of Europe; introduced widely in the U.S. and Canada). Margins of lakes; also railroads and open, disturbed areas.

As noted by McGregor (1986a), variation within the *C. glaucum* complex requires further study. The present treatment tentatively accepts two varieties.

1. Leaf tips and marginal teeth rounded to bluntly pointed; spikes mostly bractless
   . . . . 9A. VAR. GLAUCUM
1. Leaf tips and marginal teeth sharply pointed; spikes with scattered leaflike bracts . . . . 9B. VAR. SALINUM

9a. var. glaucum

Leaf blades with the tip and marginal teeth rounded to bluntly pointed. Spikes mostly bractless. Fruits 0.6–0.9 mm long. $2n=18$. July–November.

Introduced, uncommon and local in northwestern Missouri and the city of St. Louis (native of Europe; introduced widely in the U.S. and Canada). Margins of lakes; also railroads and open, disturbed areas.

9b. var. salinum (Standl.) B. Boivin

C. salinum Standl.

Leaf blades with the tip and marginal teeth sharply pointed. Spikes with scattered leaflike bracts. Fruits 0.8–1.0 mm long. July–November.

Introduced, uncommon, known thus far from a single collection from the city of St. Louis (native of Europe; introduced widely in the U.S. and Canada). Railroads.

Although Steyermark (1963) treated this taxon as a variety of *C. glaucum,* other North American authors have submerged it entirely within that species (Gleason and Cronquist, 1991; Clemants, 1992) or separated it at the species level (Bassett and Crompton, 1982). Steyermark (1963) noted that the St. Louis railyard collection originated from a population where both varieties were represented. There is some controversy as to the extent of the native range of this taxon. Some authors (Bassett and Crompton, 1982) believe that it occurs natively in portions of the western United States and Canada.

10. Chenopodium incanum (S. Watson) A. Heller var. incanum

C. fremontii S. Watson var. *incanum* S. Watson

Pl. 355 h, i; Map 1530

Plants annual, without an odor. Stems 5–25 cm long, spreading to ascending, bushy, much-branched from the base, moderately to densely white-mealy, sometimes appearing striped. Leaves short- to long-petiolate. Leaf blades 0.7–1.5 cm long, mostly 1.0–1.5 times as long as wide (mostly 5–12 mm wide, the uppermost often 3 mm or narrower), narrowly triangular to rhombic, those of the uppermost leaves often narrowly lanceolate-elliptic, sharply pointed at the tip, angled or tapered at the base, with 1 or rarely 2 pairs of sharply triangular lobes below the midpoint, green to yellowish green, thick and leathery in texture, the margins otherwise entire or rarely shallowly few-toothed, the upper surface sparsely to moderately white-mealy, the undersurface moderately to densely white-mealy. Venation mostly obscured, with usually 3 main veins. Inflorescences axillary and terminal, consisting of clusters or short spikes with small clusters of flowers, the terminal spikes usually grouped into small panicles. Flowers not all maturing at the same time. Calyx 5-lobed to below the midpoint, covering the entire fruit except sometimes for a minute area surrounding the style, the lobes 0.7–1.0 mm long, ovate, bluntly to sharply pointed at the tip, somewhat rounded or pouched, sometimes keeled dorsally, moderately to densely white-mealy. Stamens 5. Stigmas 2. Fruits 0.9–1.2 mm wide, depressed-ovoid, the seeds positioned horizontally, the wall thin, membranous, and somewhat translucent, smooth, easily separated from the seed. Seeds black, shiny, finely and usually faintly wrinkled, bluntly angled along the rim. $2n=18$. June–September.

Introduced, uncommon, known only from historical collections from Jackson and Jasper Counties (Arizona to Texas north to Utah and South Dakota; Canada; introduced eastward sporadically to Maine and South Carolina). Open, disturbed areas.

Crawford (1977) treated *C. incanum* as comprising three varieties. The other two varieties, var. *elatum* D.J. Crawford and var. *occidentale* D.J. Crawford, occur to the southwest and west of var. *incanum* respectively and differ in subtle characters of seed, leaf, and plant size, as well as having somewhat less crowded inflorescences.

11. Chenopodium missouriense Aellen
C. album L. var. *missouriense* Bassett & Crompton

Pl. 355 f, g; Map 1531

Plants annual, without an odor. Stems 20–150 cm long, erect or ascending, usually few- to several-branched above the base and below the inflorescence, glabrous or more commonly sparsely to moderately white-mealy, sometimes reddish-tinged or reddish purple–striped, almost always reddish purple at each node (on the side to which the leaf is attached). Leaves mostly long-petiolate. Leaf blades 1–6 cm long, mostly 1–3 times as long as wide (1–4 cm wide), usually 1.2–1.5 times in the largest (lowermost) leaves, ovate-rhombic or ovate-triangular, the uppermost usually linear to narrowly lanceolate, angled or tapered to a rounded or more commonly bluntly to sharply pointed tip, the middle lobe not appearing unusually elongate, broadly angled at the base, green or reddish-tinged, thin and herbaceous to thickened, somewhat leathery, and slightly succulent in texture, the margins entire to wavy or irregularly and often relatively coarsely several-toothed (the basal pair of teeth sometimes larger than the others, sometimes appearing shallowly lobed), the upper surface glabrous or sparsely to moderately mealy at maturity, not shiny, the undersurface sparsely to moderately white-mealy. Venation noticeably branched, with 1 or 3 main veins. Inflorescences axillary and terminal, consisting of short spikes with often well-separated small clusters of flowers, the terminal ones usually grouped into small to relatively large panicles. Flowers all maturing at more or less the same time. Calyx 5-lobed nearly to the base, usually covering the entire fruit except sometimes for a minute area surrounding the style, occasionally somewhat spreading in a few flowers of a given inflorescence, the lobes 0.7–1.0 mm long, ovate to triangular-ovate, bluntly pointed at the tip, usually with a sometimes poorly developed narrow keel or raised area along the midvein dorsally, sparsely to moderately white-mealy. Stamens 5. Stigmas 2. Fruits 0.9–1.1(–1.2) mm wide, depressed-ovoid, the seeds positioned horizontally, the wall thin, membranous, and somewhat translucent, smooth or finely roughened, not appearing honeycombed, sometimes difficult to separate from the seed. Seeds black, shiny, smooth or nearly so, rounded to very bluntly angled along the rim. $2n=54$. September–October.

Scattered nearly throughout the state (eastern U.S. west to North Dakota and Texas; introduced farther west to California and Alaska; Canada). Banks of streams, rivers, spring branches, and sloughs, less commonly in bottomland forests and mesic upland forests; also crop fields, fallow fields, roadsides, railroads, and open, disturbed areas.

The inflorescences of *C. missouriense* tend to be more open and have more spreading, relatively slender branches than those of *C. album*. Relatively synchronous autumnal flowering and small fruits also are characteristic of *C. missouriense*. The reddish patches at the stem nodes occasionally are poorly developed.

12. Chenopodium murale L. (nettle-leaved goosefoot)

Pl. 355 c–e; Map 1532

Plants annual, without an odor. Stems 10–80 cm long, erect or ascending, few- to much-branched above the base, glabrous or sparsely white-mealy, green to pale green. Leaves mostly long-petiolate. Leaf blades 2–5(–8) cm long, mostly 2.0–2.5 times as long as wide (1–3 cm wide), ovate to ovate-trian-

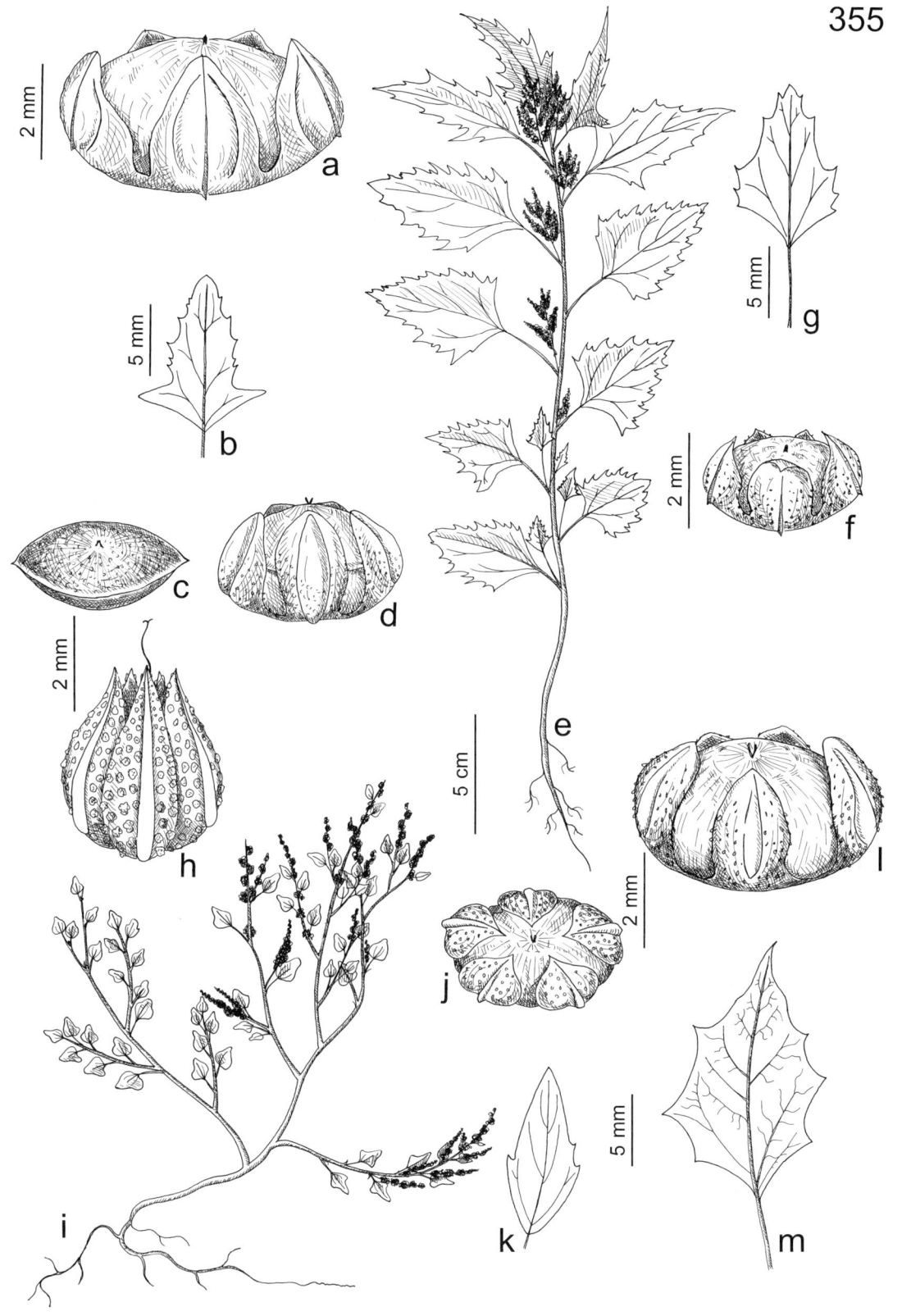

Plate 355. Chenopodiaceae. *Chenopodium ficifolium*, **a)** fruit, **b)** leaf. *Chenopodium murale*, **c)** fruit without calyx, **d)** fruit with calyx, **e)** habit. *Chenopodium missouriense*, **f)** fruit, **g)** leaf. *Chenopodium incanum*, **h)** fruit, **i)** habit. *Chenopodium standleyanum*, **j)** fruit, **k)** leaf. *Chenopodium simplex*, **l)** fruit, **m)** leaf.

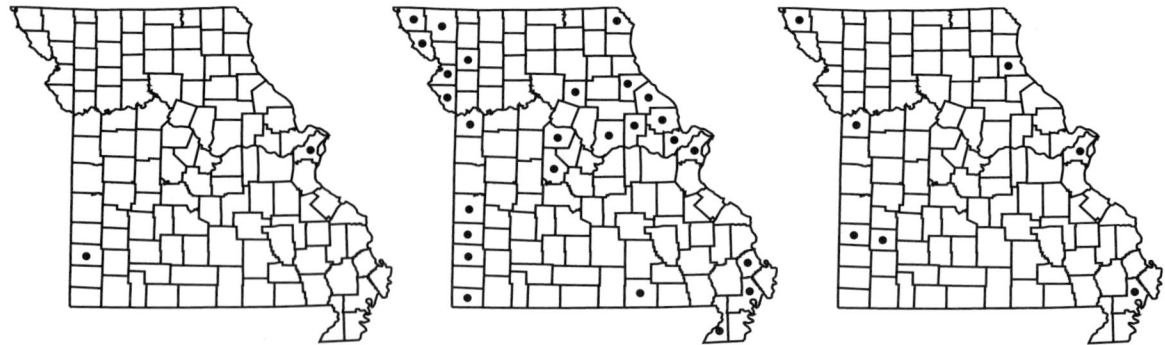

1533. Chenopodium opulifolium   1534. Chenopodium pallescens   1535. Chenopodium pratericola

gular or occasionally rhombic, bluntly to sharply pointed at the tip, broadly angled, truncate or shallowly cordate at the base, green, relatively thin and herbaceous in texture, the margins irregularly several-toothed, the upper surface glabrous or sparsely white-mealy, the undersurface sparsely to moderately white-mealy. Venation noticeably branched, with 1 or 3 main veins. Inflorescences axillary and terminal, consisting of short spikes with small clusters of flowers, these usually grouped into small to relatively large panicles. Flowers not all maturing at the same time. Calyx 5-lobed to below the midpoint, extending past the widest part of the fruit but not reaching the stylar area, leaving much of the portion of the fruit above the rim exposed at maturity, the lobes 0.5–0.8 mm long, ovate, rounded or bluntly pointed at the tip, keeled (often narrowly so) dorsally, glabrous or sparsely white-mealy. Stamens 5. Stigmas 2. Fruits 1.1–1.5 mm wide, depressed-ovoid, the seeds positioned horizontally, the wall thin, membranous, and somewhat translucent, finely wrinkled or roughened at maturity, but occasionally becoming smooth with age, difficult to separate from the seed. Seeds black, dull, finely wrinkled or roughened at maturity, sharply angled along the rim. $2n=18$. June–November.

Introduced, uncommon and sporadic in Missouri (native of Europe, Asia; introduced widely in the U.S. and Canada). Banks of rivers; also pastures, gardens, railroads, and open, disturbed areas.

13. Chenopodium opulifolium Schrad. ex W.D.J. Koch & Ziz

Pl. 356 c–e; Map 1533

Plants annual, without an odor. Stems 20–100 cm long, erect or ascending, unbranched or more commonly few- to much-branched above the base, glabrous or sparsely white-mealy, often somewhat reddish-tinged or reddish-striped. Leaves sessile to more commonly short-petiolate. Leaf blades 1–4 cm long, mostly 1.0–2.5 times as long as wide (1–3 cm wide), ovate to ovate-triangular or rhombic, bluntly to sharply pointed at the tip, angled to nearly truncate at the base, usually with a pair of triangular lobes below the midpoint, green to grayish green, relatively thin and herbaceous in texture, the margins otherwise entire or wavy to irregularly several-toothed, the upper surface glabrous or nearly so, the undersurface sparsely to densely white-mealy. Venation noticeably branched, with usually 3 main veins. Inflorescences axillary and terminal, consisting of short spikes with small clusters of flowers, the terminal ones usually grouped into small to relatively large panicles. Flowers not all maturing at the same time. Calyx 5-lobed to about the midpoint, the fused portion extending beyond the rim of the fruit, covering the entire fruit except sometimes for a minute area surrounding the style, the lobes 0.5–0.8 mm long, broadly ovate, rounded or bluntly pointed at the tip, broadly keeled and often appearing somewhat pouched dorsally, moderately to densely white-mealy. Stamens 5. Stigmas 2. Fruits 1.0–1.4 mm wide, depressed-ovoid, the seeds positioned horizontally, the wall thin, membranous, and somewhat translucent, smooth, difficult to separate from the seed or less commonly easily separable. Seeds black, shiny, smooth at maturity, rounded along the rim. $2n=54$. July–November.

Introduced, uncommon, known thus far only from Jasper County and the city of St. Louis (native of Europe, Asia; introduced sporadically in the U.S. and Canada). Railroads and open, disturbed areas.

This species was first reported for Missouri by Mühlenbach (1979). The determination of Missouri materials is somewhat controversial. Sergei Mosyakin of the Institute of Botany in Kiev, Ukraine, who has been studying North American populations of the *C. album* complex, has annotated some of the Missouri materials as *C.* ×*borbasii* Murr, a fertile putative hybrid between *C. album* and *C. opulifolium*. Dvořák (1991) documented the subtle morphological variation in the complex, noting that various populations of the hybrid could only be distinguished from one or the other parent by careful measurement of quantitative features involving pollen grains and/or leaf blade dimensions and lobing patterns.

14. Chenopodium pallescens Standl. (slim-leaved goosefoot)

Pl. 356 f, g; Map 1534

Plants annual, without an odor. Stems 10–60 cm long, erect or ascending, few- to several-branched, glabrous or sparsely white-mealy, sometimes slightly reddish-tinged or reddish-striped, often appearing pale green. Leaves sessile or short-petiolate. Leaf blades 1–4 cm long, mostly 6–10 times as long as wide (1–6 mm wide), linear to very narrowly oblong, bluntly to sharply pointed at the tip, angled at the base, grayish green and thickened-leathery in texture, the margins entire, the surfaces glabrous or more commonly (especially the undersurface) sparsely white-mealy at maturity. Venation unbranched, only 1 midvein apparent. Inflorescences axillary and terminal, consisting of short spikes with dense, small clusters of flowers, the terminal spikes sometimes grouped into small panicles. Flowers not all maturing at the same time. Calyx 5-lobed to below the midpoint, covering the fruit at maturity, the lobes 1.3–1.6 mm long, oblong-elliptic, rounded to bluntly pointed at the tip, keeled dorsally, sparsely white-mealy. Stamens 5. Stigmas 2. Fruits 1.3–1.6 mm wide, depressed-ovoid, the seeds positioned horizontally, the wall thin, membranous, and somewhat translucent, finely roughened, difficult to separate from the seed. Seeds black, shiny, smooth, rounded along the rim. $2n=18$. June–October.

Uncommon, widely scattered in the state (Indiana to Wyoming south to Arkansas, Texas, and New Mexico). Glades, tops of bluffs, upland prairies, and sand prairies; also roadsides, railroads, and open, rocky, disturbed areas.

The taxonomy of the narrow-leaved species of *Chenopodium* is complex and has been confused in some of the older botanical literature. Crawford (1975) studied the morphology, flavonoid chemistry, and cytology of the group and annotated many of our specimens. Some of the specimens noted and mapped by Steyermark (1963) as this species have been redetermined as *C. desiccatum* or *C. pratericola*, and *C. pallescens* is generally less common in Missouri than he understood it to be.

15. Chenopodium pratericola Rydb. (desert goosefoot)

C. desiccatum A. Nelson var. *leptophylloides* (Murr) Wahl

Pl. 354 c, d; Map 1535

Plants annual, without an odor. Stems 10–100 cm long, usually erect, unbranched or more commonly with to several ascending branches above the base, moderately to densely white-mealy, sometimes slightly reddish-tinged or reddish-striped. Leaves sessile or more commonly short-petiolate. Leaf blades 1–4 cm long, mostly 3–7 times as long as wide (1–15 mm wide), linear to lanceolate or narrowly oblong-elliptic, sharply pointed at the tip, angled at the base, the largest leaves sometimes with 1 or 2 shallow basal lobes, green to grayish green, relatively thin and herbaceous in texture, the margins otherwise entire or inconspicuously wavy to shallowly few-toothed, the upper surface glabrous or sparsely white-mealy, the undersurface moderately to densely white-mealy. Venation of 3 main veins, often with no other branching apparent. Inflorescences axillary and terminal, consisting of short spikes with small clusters of flowers, the terminal spikes usually grouped into small panicles. Flowers not all maturing at the same time. Calyx 5-lobed nearly to the base, extending nearly to the stylar area, usually somewhat spreading at maturity to expose the apical portion of the fruit, the lobes 0.7–1.0 mm long, broadly lanceolate to ovate, bluntly pointed at the tip, strongly keeled dorsally, moderately to densely white-mealy. Stamens (4)5. Stigmas 2. Fruits 1.0–1.4 mm wide, depressed-ovoid, the seeds positioned horizontally, the wall thin, membranous to somewhat papery, somewhat translucent, smooth, easily separated from the seed. Seeds black, shiny, finely pebbled or minutely roughened, bluntly angled along the rim. $2n=18$. July–November.

Introduced, uncommon and widely scattered in eastern and western Missouri (native of the western U.S. and adjacent Canada east to North Dakota and Texas; introduced sporadically eastward to the eastern seaboard). Sand prairies; also railroads and open, disturbed areas.

Most recent authors have followed Crawford (1975) in treating *C. pratericola* as a species separate from *C. desiccatum* and the other members of the so-called narrow-leaved complex within *Chenopodium*. See the treatment of *C. pallescens* for further discussion. The eastern limits of the native range of *C. pratericola* remain controversial. Botanists in some states outside the native range indicated here treat the species as native, and it is considered of conservation concern in Ohio, as well as in Ontario.

1536. Chenopodium pumilio   1537. Chenopodium rubrum   1538. Chenopodium simplex

16. Chenopodium pumilio R. Br. (clammy goosefoot)

Dysphania pumilio (R. Br.) Mosyakin & Clemants

Pl. 353 m, n; Map 1536

Plants annual, with a slight unpleasant odor. Stems 10–40 cm long, prostrate to loosely ascending, few- to much-branched from near the base, with dense, short glandular hairs, usually not reddish-tinged or reddish-striped, often appearing pale green to nearly white. Leaves short- to long-petiolate. Leaf blades 0.5–3.0 cm long, those of well-developed leaves mostly 1.5–3.0 times as long as wide, oblong-elliptic to ovate, mostly bluntly pointed at the tip, rounded or angled at the base, mostly with few rounded lobes, occasionally nearly entire (uppermost leaves), yellowish green and herbaceous in texture, the margins usually also wavy or with shallow, blunt, narrow teeth, the surfaces lacking mealiness, but with dense, sessile, yellowish resin glands. Venation noticeably branched, usually with 1 main vein. Inflorescences axillary, consisting of short spikes or small clusters of flowers. Flowers not all maturing at the same time. Calyx 5-lobed nearly to the base, somewhat spreading at maturity and exposing the fruit at maturity, the lobes 0.5–0.7 mm long, elliptic-ovate to oblong-elliptic, usually sharply pointed at the tip, flat or more commonly rounded to somewhat pouched dorsally, moderately to densely pubescent with dense, sessile, yellowish resin glands, sometimes also with sparse to moderate, short, nonglandular hairs. Stamen usually 1. Stigmas 3. Fruits 0.5–0.7 mm long, ovoid, the seeds positioned vertically, the wall thin, membranous, and somewhat translucent, smooth to finely and faintly wrinkled, difficult to separate from the seed. Seeds reddish brown to dark brown, shiny, smooth, rounded along the rim. $2n=16$. August–October.

Introduced, scattered, mostly in the southern third of the state (native of Australia; introduced widely in the U.S., uncommonly in Canada). Banks of streams and rivers, margins of sinkhole ponds; also crop fields, fallow fields, pastures, farmyards, roadsides, railroads, and open, disturbed areas.

17. Chenopodium rubrum L. (coast blite, red goosefoot)

C. rubrum var. *humile* (Hook.) S. Watson

Pl. 356 l, m; Map 1537

Plants annual, without an odor. Stems 20–80 cm long, erect or ascending, few- to much-branched from near the base, glabrous, often somewhat reddish-tinged or reddish-striped. Leaves short- to long-petiolate. Leaf blades 1–10 cm long, mostly 1–2 times as long as wide, ovate-triangular to somewhat rhombic, bluntly to sharply pointed at the tip, angled or tapered at the base, sometimes with a pair of more or less ascending triangular basal lobes, green and slightly fleshy in texture, the margins otherwise entire, wavy or with few to several blunt, irregular teeth, the surfaces glabrous. Venation noticeably branched, with 1 or 3 main veins. Inflorescences axillary and sometimes also terminal, consisting of dense clusters of flowers (remaining 3–5 mm in diameter at fruiting), the terminal ones usually arranged into spikes, these sometimes grouped into small panicles. Flowers not all maturing at the same time. Calyx 3(–5)-lobed nearly to the base, remaining small, green, and herbaceous to scalelike at maturity, more or less covering the fruit at maturity, the lobes 0.9–1.2 mm long, oblong-elliptic to oblong-lanceolate, rounded to bluntly pointed at the tip, rounded or somewhat pouched dorsally, rarely slightly keeled, glabrous or sparsely white-mealy. Stamens usually 3–5. Stigmas 2. Fruits 0.7–1.2 mm long, ovoid, the seeds positioned mostly vertically, the wall thin, membranous, and somewhat translucent, smooth, easily separated from the seed. Seeds reddish brown to dark brown,

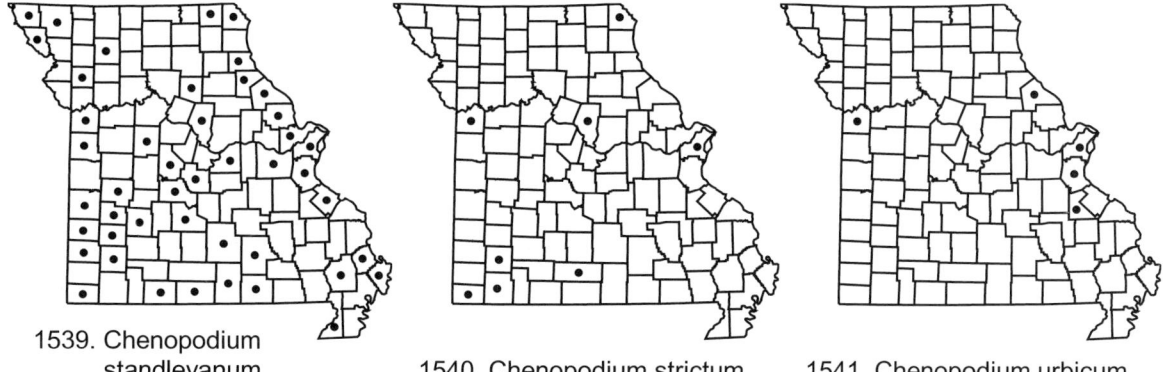

1539. Chenopodium standleyanum
1540. Chenopodium strictum
1541. Chenopodium urbicum

shiny, finely roughened and/or with a faint network of fine, low ridges, angled along the rim. $2n=36$. July–November.

Introduced, uncommon and sporadic (native of eastern North America, Europe, Asia; introduced sporadically elsewhere in the U.S.). Roadsides and open, disturbed areas.

Several variants have been named within this species. Bassett and Crompton (1982), who studied *C. rubrum* in Canada, suggested that these varieties represented merely ecological forms unworthy of formal taxonomic recognition. The few Missouri specimens key to var. *rubrum,* because of their relatively small seeds, ascending stems, well-developed inflorescences, and toothed leaves. The limits of the range of *C. rubrum* are not understood fully, and it is unclear how far west in the United States the species grows natively. However, the few specimens from Missouri all clearly all originated from introduced populations.

18. Chenopodium simplex (Torr.) Raf. (maple-leaved goosefoot)

C. gigantospermum Aellen

Pl. 355 l, m; Map 1538

Plants annual, without an odor. Stems 30–150 cm long, erect or ascending, several-branched above the base, glabrous or sparsely white-mealy, green. Leaves short- to long-petiolate. Leaf blades 3–15 cm long (the largest 7–18 cm), mostly 1.2–2.5 times as long as wide (2–9 cm wide), ovate to broadly ovate or triangular-ovate, sharply pointed at the tip, mostly rounded, truncate, or shallowly cordate at the base, green, relatively thin and herbaceous in texture, the margins with relatively few broad, coarse teeth or (in the uppermost leaves) entire to slightly wavy, the surfaces glabrous or less commonly sparsely white-mealy. Venation noticeably branched, with mostly 3 or 5 main veins. Inflorescences axillary and terminal, consisting of short spikes with small clusters of flowers, these mostly grouped into relatively large, bractless panicles. Flowers not all maturing at the same time. Calyx 5-lobed to below the midpoint, extending past the widest part of the fruit but not reaching the stylar area, leaving much of the portion of the fruit above the rim exposed at maturity, the lobes 0.7–1.0 mm long, lanceolate to ovate, rounded or bluntly pointed at the tip, flat to slightly rounded or narrowly and shallowly keeled dorsally, glabrous or sparsely white-mealy. Stamens 5. Stigmas 2. Fruits 1.5–2.5 mm wide, depressed-ovoid, the seeds positioned horizontally, the wall thin but papery, and somewhat yellowish, smooth, usually easily separated from the seed. Seeds black, shiny, smooth, bluntly angled along the rim. $2n=36$. June–October.

Scattered nearly throughout the state, but absent or uncommon in portions of the southeastern quarter (U.S., Canada). Bases and ledges of bluffs, bottomland forests, mesic upland forests in ravines, and edges of loess hill prairies; also old fields, fencerows, roadsides, railroads, and shaded, disturbed areas.

The prevalence of this species along bluff bases at or near prehistoric shelter sites has given rise to speculation that the relatively large fruits were harvested for food by aboriginal midwesterners. The species apparently is part of a circumboreal complex, with the Old World component referred to as *C. hybridum* L. (Bassett and Crompton, 1982; McGregor, 1986a).

19. Chenopodium standleyanum Aellen (woodland goosefoot)

Pl. 355 j, k; Map 1539

Plants annual, without an odor. Stems 20–120 (–200) cm long, erect or ascending (sometimes arched), usually much-branched above the base, glabrous, sometimes slightly reddish-tinged or reddish-striped. Leaves sessile or short-petiolate. Leaf blades 1–5(–8) cm long, mostly 3–6 times as long as wide (3–15 mm wide), narrowly lanceolate to oblong-lanceolate or lanceolate, those of the lowermost leaves occasionally narrowly ovate to ovate-triangular, sharply pointed at the tip, angled at the base, rarely with a shallow pair of basal lobes, green to dark green, relatively thin and herbaceous in texture, the margins otherwise entire or shallowly few- to less commonly several-toothed, the upper surface glabrous or sparsely white-mealy, the undersurface sparsely to moderately white-mealy. Venation noticeably branched, with 1 or 3 main veins. Inflorescences axillary and terminal, consisting of short spikes with small clusters of flowers, the terminal spikes usually grouped into small panicles. Flowers not all maturing at the same time. Calyx 5-lobed to below the midpoint, extending past the widest part of the fruit but not reaching the stylar area, leaving much of the portion of the fruit above the rim exposed at maturity, the lobes 0.5–0.7 mm long, oblong-obovate, rounded at the tip, flat, somewhat rounded or pouched, rarely narrowly and shallowly keeled dorsally, glabrous or sparsely white-mealy. Stamens 5. Stigmas 2. Fruits (1.0–)1.2–1.5 mm wide, depressed-ovoid, the seeds positioned horizontally, the wall thin, membranous, and somewhat translucent, smooth, easily separated from the seed. Seeds black, shiny, smooth to slightly roughened, rounded along the rim. $2n=18$. June–October.

Scattered nearly throughout the state (eastern U.S. and adjacent Canada west to Montana and New Mexico). Upland forests, tops and ledges of bluffs, banks of streams and rivers, and rarely bottomland forests and edges of upland prairies; also roadsides, railroads, and rarely crop fields.

20. Chenopodium strictum Roth var. glaucophyllum (Aellen) Wahl

Pl. 356 a, b; Map 1540

Plants annual, without an odor. Stems 20–100 cm long, erect or ascending, few- or more commonly much-branched at and above the base, glabrous or sparsely white-mealy, green or reddish-tinged or reddish-striped. Leaves short- to long-petiolate. Leaf blades 1.5–6.0 cm long, mostly 1.5–3.0 times as long as wide (1–2 cm wide, but the uppermost leaves often only 0.5 cm), those of the lower leaves mostly oblong-ovate, those of the upper leaves narrowly oblong-elliptic to lanceolate, bluntly to sharply pointed at the tip, angled at the base, grayish green, relatively thin and herbaceous in texture, the margins mostly entire, often slightly wavy, only those of the largest (lowermost) leaves shallowly and irregularly toothed, the upper surface glabrous or sparsely white-mealy, the undersurface sparsely to moderately white-mealy. Venation noticeably branched, with 1 or 3 main veins. Inflorescences axillary and terminal, consisting of small clusters of flowers, these sometimes grouped into short to relatively long spikes, the terminal ones rarely grouped into small panicles. Flowers not all maturing at the same time. Calyx 5-lobed nearly to the base, extending past the widest part of the fruit and nearly reaching the stylar area, but spreading at fruiting and thus leaving much of the portion of the fruit above the rim exposed at maturity, the lobes 0.7–1.0 mm long, ovate, rounded or bluntly pointed at the tip, relatively narrowly keeled dorsally, sparsely white-mealy. Stamens 5. Stigmas 2. Fruits 0.9–1.2 mm wide, depressed-ovoid, the seeds positioned horizontally, the wall thin, membranous, and somewhat translucent, smooth or finely and faintly roughened at maturity, usually difficult to separate from the seed. Seeds black, shiny, smooth, rounded along the rim. $2n=36$. July–October.

Uncommon and widely scattered in the state (northern U.S. south to California, Arkansas, and South Carolina; Canada, Europe, Asia). Bottomland forests and banks of rivers; also pastures, railroads, and open to shaded disturbed areas.

The var. *strictum* is confined to the Old World and differs in its narrower leaves with more teeth. There has been some debate as to whether var. *glaucophyllum* is native in the New World or represents an early introduction, although most authors have treated it as a native taxon. Wahl (1952–1953) stated that plants in the northern portion of the North American range behaved ecologically as adventives, but some populations from the southern portion behaved similarly to other native taxa of *Chenopodium*. Wahl (1952–1953) also mentioned that some populations from the southern portion of the range (including some plants from Missouri) differed somewhat morphologically from the typical phase of the variety and might represent a different taxon. There has been no further research to follow up on this suggestion.

21. Chenopodium urbicum L. (city goosefoot)

Pl. 356 j, k; Map 1541

Plants annual, without an odor. Stems 20–80 cm long, erect or ascending, usually not or few-branched above the base, glabrous or sparsely white-mealy, green to pale green. Leaves mostly long-petiolate. Leaf blades 2–11 cm long, mostly 1–2 times as long as wide (1–10 cm wide), mostly ovate-triangular, sometimes ovate or rhombic, the uppermost lanceolate to narrowly triangular,

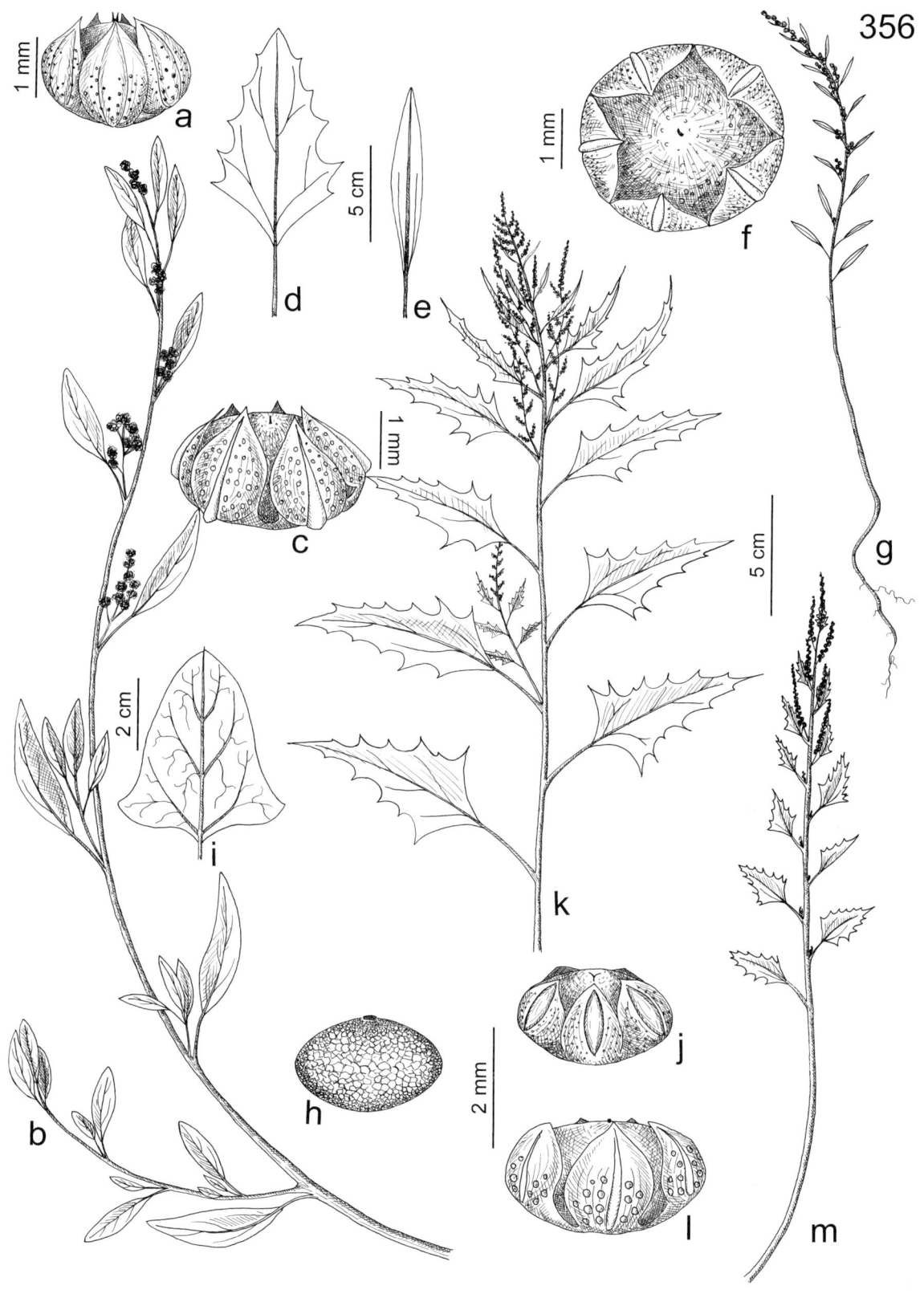

Plate 356. Chenopodiaceae. *Chenopodium strictum*, **a)** fruit, **b)** habit. *Chenopodium opulifolium*, **c)** fruit, **d, e)** leaves. *Chenopodium pallescens*, **f)** fruit (top view), **g)** habit. *Chenopodium watsonii*, **h)** fruit without calyx, **i)** leaf. *Chenopodium urbicum*, **j)** fruit, **k)** fruiting branch. *Chenopodium rubrum*, **l)** fruit, **m)** habit.

1542. Chenopodium watsonii
1543. Corispermum americanum
1544. Corispermum pallasii

bluntly to sharply pointed at the tip, angled to truncate at the base, green, relatively thin and herbaceous in texture, the margins wavy or irregularly several-toothed (the basal pair of teeth usually larger than the others), the upper surface glabrous and usually shiny, the undersurface sparsely to moderately white-mealy. Venation noticeably branched, with 1 or 3 main veins. Inflorescences axillary and terminal, consisting of short spikes with small clusters of flowers, the terminal ones usually grouped into small to relatively large panicles. Flowers not all maturing at the same time. Calyx 5-lobed nearly to the base, extending past the widest part of the fruit but not reaching the stylar area, leaving much of the portion of the fruit above the rim exposed at maturity, the lobes 0.6–0.8 mm long, elliptic to broadly ovate, rounded or bluntly pointed at the tip, flat or somewhat rounded dorsally, glabrous or sparsely white-mealy. Stamens 5. Stigmas 2. Fruits 0.9–1.2 mm wide, depressed-ovoid, the seeds positioned horizontally, the wall thin, membranous, and somewhat translucent, smooth or finely roughened at maturity, usually easily separated from the seed. Seeds reddish brown to black, shiny, finely wrinkled or appearing nearly smooth at maturity, rounded to very bluntly angled along the rim. $2n=36$. July–October.

Introduced, uncommon and sporadic (native of Europe, Asia; introduced widely in portions of the northern and western U.S., Canada). Bottomland forests and banks of rivers; also pastures, railroads, and open to shaded disturbed areas.

The reports by Steyermark (1963) from Boone and Greene Counties could not be confirmed during the present research.

22. Chenopodium watsonii A. Nelson

Pl. 356 h, i; Map 1542

Plants annual, with a pronounced unpleasant odor. Stems 2–15(–30) cm long, spreading to more commonly erect or ascending, bushy, usually much-branched from the base, sparsely to densely white-mealy, sometimes appearing striped. Leaves mostly long-petiolate. Leaf blades 1–4 cm long, mostly 1.0–1.5 times as long as wide (mostly 6–20 mm wide), ovate to broadly ovate or rhombic, rounded or bluntly pointed at the tip, angled or tapered at the base, with 1 or rarely 2 pairs of triangular lobes below the midpoint, green to yellowish green, thick and leathery in texture, the margins otherwise entire or rarely shallowly few-toothed, the upper surface sparsely to moderately white-mealy, the undersurface densely white-mealy. Venation noticeably branched, with 1 or 3 main veins. Inflorescences axillary and terminal, consisting of clusters or short spikes with dense clusters of flowers, the terminal spikes usually grouped into small panicles. Flowers not all maturing at the same time. Calyx 5-lobed to below the midpoint, covering the entire fruit except sometimes for a minute area surrounding the style, the lobes 0.7–1.0 mm long, ovate to oblong-ovate, mostly bluntly pointed at the tip, strongly keeled dorsally, moderately to densely white-mealy. Stamens 5. Stigmas 2. Fruits 0.9–1.2 mm wide, depressed-ovoid, the seeds positioned horizontally, the wall relatively thick, papery and opaque–white at maturity, finely roughened at maturity, difficult to separate from the seed. Seeds black, shiny, wrinkled, bluntly angled along the rim. $2n=18$. June–September.

Introduced, uncommon, known only from the city of St. Louis (Arizona to Montana east to South Dakota and Kansas; Canada; introduced in Missouri and Maine). Railroads.

This species was first reported from Missouri by Mühlenbach (1979).

5. Corispermum L. (bugseed)

Plants annual, the taproot not tuberous-thickened, the aboveground portions sparsely to densely pubescent with small, stellate hairs, sometimes becoming nearly glabrous at maturity. Stems erect or ascending, rarely spreading, not succulent, not appearing jointed, few- to much-branched. Leaves alternate, well developed, not succulent, sessile or nearly so. Leaf blades linear to narrowly lanceolate, flattened in cross-section, not clasping the stem, narrowed to a sharply pointed tip, narrowed at the base, the margins entire. Inflorescences terminal on the branches, slender to somewhat club-shaped spikes, the flowers solitary, not sunken into the axis. Flowers all or nearly all perfect (a few pistillate flowers occasionally present). Bract 1 per flower, somewhat leaflike, cupping the fruit, lanceolate to ovate, tapered to a sharply pointed tip. Calyx of 1 sepal (absent or 3 elsewhere), this tiny (0.5–1.0 mm long), scalelike, irregularly oblong-ovate, rounded at the tip, persistent at fruiting, not concealing the fruit, rounded on the back, not winged. Stamens 1–3(–5). Ovary superior. Style absent or 1 and very short, the stigmas 2, linear. Fruits unequally elliptic in cross-section (the inner surface flattened to slightly concave), elliptic to obovate or nearly circular in outline, flattened vertically, indehiscent, the wall papery to somewhat leathery or hardened, usually with reddish brown spots and occasionally a few small, whitish, warty outgrowths. Seed adhering more or less tightly to the fruit wall, positioned vertically, 1.5–4.0 mm long, elliptic to oval in outline, flattened, the surface smooth, dark brown to nearly black, shiny, the embryo appearing more or less ring-shaped. About 60 species, North America, Europe, Asia.

Species of *Corispermum* are very difficult to distinguish, especially if mature fruits are not present. Perhaps because of this, the taxonomy of the genus in North America is controversial. Some authors (Maihle and Blackwell, 1978; Brooks, 1986) have contended that the three or four species present in North American are introduced from the Old World. However, Mosyakin (1995) accepted eleven species as occurring in the region, with at least eight of these said to be native. Paleobotanical evidence (Betancourt et al., 1984) documents the presence of one or more members of the genus in western North America (at least some of these similar in fruit morphology to modern *C. villosum*) over a period between more than 38,000 and less than 1,500 years ago, so perhaps the modern materials should correctly be considered all or in part indigenous to the New World. However, Mosyakin's (1995) narrow taxonomic circumscriptions (which he states to be preliminary pending more detailed research) have been challenged by some recent authors (Judd and Ferguson, 1999). The bulk of the morphological variation that has given rise to this controversy occurs in plants to the west and/or north of Missouri. Because Mosyakin (1995 and pers. comm.) accepted the same number of taxa for Missouri as did earlier authors (Steyermark, 1963; Maihle and Blackwell, 1978), the present treatment tentatively accepts his newer nomenclature.

In regions to the west of Missouri where *Corispermum* is more abundant, the genus is considered a good forage for livestock.

1. Fruits 1.8–3.0 mm long, elliptic to ovate-elliptic or obovate-elliptic in outline, the wing absent or very narrow (to 0.15 mm wide); inflorescences mostly narrowly club-shaped, conspicuously denser toward the tip; bracts as wide as or wider than the fruits; leaves linear . . . . 3. C. VILLOSUM
1. Fruits 3.5–4.5 mm long, elliptic-obovate to nearly circular in outline, the wing relatively well developed (0.3–0.5 mm wide); inflorescences slender to club-shaped, the flowers relatively evenly spaced; leaves linear to narrowly lanceolate
   2. Inflorescences mostly slender and linear, the flowers only slightly overlapping; fruits broadly obovate to nearly circular in outline, the tip usually

rounded; bracts narrower than to about as wide as the fruits; leaves mostly linear . . . . 1. C. AMERICANUM
2. Inflorescences mostly stouter and club-shaped, the flowers strongly overlapping except at the very base; fruits obovate to elliptic-obovate in outline, the tip usually broadly angled; bracts mostly wider than the fruits; leaves mostly narrowly lanceolate . . . . 2. C. PALLASII

1. Corispermum americanum (Nutt.) Nutt.
(American bugseed)

Pl. 357 a–c; Map 1543

Plants nearly glabrous to sparsely stellate-hairy, sometimes also with a few unbranched hairs. Stems 10–60 cm long. Leaves 1–5 cm long, linear or sometimes the smaller ones narrowly lanceolate. Inflorescences mostly slender, linear, the flowers more or less evenly spaced, only slightly overlapping. Bracts 2–10 mm long, ovate to narrowly ovate, those of the median and upper flowers about as wide as the fruits, those of the lowermost flowers conspicuously narrower than the fruits. Fruits 3.5–4.5 mm long, broadly obovate to nearly circular in outline, usually rounded at the tip, the wing relatively well developed (0.3–0.5 mm wide), the body light brown to dark brown or dark olive green. August–October.

Known only from historical collections from Clark and Jackson Counties (western U.S. east to New York, Ohio, and Texas; Canada, Mexico). Disturbed, sandy areas.

Plants from Clark County were labeled as being introduced, but the status of the Jackson County populations was not indicated. *Corispermum americanum* is a morphologically variable species. Mosyakin (1995) treated some populations from the southwestern United States and adjacent Mexico with slightly larger fruits and somewhat more slender, interrupted spikes as var. *rydbergii* Mosyakin, but these grade into the typical phase too much to warrant formal taxonomic recognition. Although many of the North American specimens of this species originally were determined as *C. hyssopifolium* L., Steyermark (1963) treated the few Missouri collections as *C. nitidum* Kit. ex Schult. In the strict sense, *C. nitidum* supposedly is restricted to eastern Europe (Mosyakin, 1995), but occasional North American specimens seem virtually indistinguishable from those collected in the Old World, even using the characters provided by Mosyakin.

2. Corispermum pallasii Steven (common bugseed)

Pl. 357 g–i; Map 1544

Plants sparsely to densely stellate-hairy. Stems 10–60 cm long. Leaves 1–7 cm long, linear or more commonly narrowly lanceolate. Inflorescences mostly stout, narrowly club-shaped, the flowers more or less evenly spaced, densely overlapping above the basal portion. Bracts 4–10 mm long, ovate, mostly wider than the fruits. Fruits 3.5–4.5 mm long, obovate to elliptic-obovate in outline, usually broadly angled at the tip, the wing relatively well developed (0.3–0.5 mm wide), the body yellowish brown or greenish brown to brown. August–October.

Uncommon in the Missouri and Mississippi River floodplains (North Dakota and South Dakota to Missouri, Michigan, and Illinois; Europe, Asia). Banks of rivers; also disturbed, open, sandy areas.

Specimens of *C. pallasii* correspond to those treated as the Eurasian native *C. hyssopifolium* L. by Steyermark (1963). The few Missouri specimens are mostly immature and somewhat atypical in the morphology of their developing fruits. In his studies of Chenopodiaceae for the Flora of North America Project, Sergei Mosyakin of the Institute of Botany in Kiev, Ukraine, has annotated these as "transitional toward *C. americanum*."

3. Corispermum villosum Rydb. (hairy bugseed)

C. orientale Lam. var. *emarginatum* (Rydb.) J.F. Macbr.

Pl. 357 d; Map 1545

Plants often moderately to densely stellate-hairy. Stems 5–50 cm long. Leaves 1.0–3.5 cm long, linear. Inflorescences mostly stout, narrowly club-shaped, conspicuously denser toward the tip, the flowers densely overlapping but becoming more widely spaced below the midpoint. Bracts 5–15 mm long, ovate to narrowly ovate, as wide as or wider than the fruits. Fruits 1.8–3.0 mm long, elliptic to ovate-elliptic or obovate-elliptic in outline, broadly angled at the tip, the wing absent or very narrow (to 0.15 mm wide), the body yellowish brown, light brown, or dark brown, usually with reddish brown spots and occasionally a few whitish, low, warty outgrowths. August–October.

Uncommon in the Missouri River floodplain (Idaho, Oregon, and Nevada east sporadically to Illinois and Missouri; Canada). Banks of rivers; also open, disturbed, sandy areas.

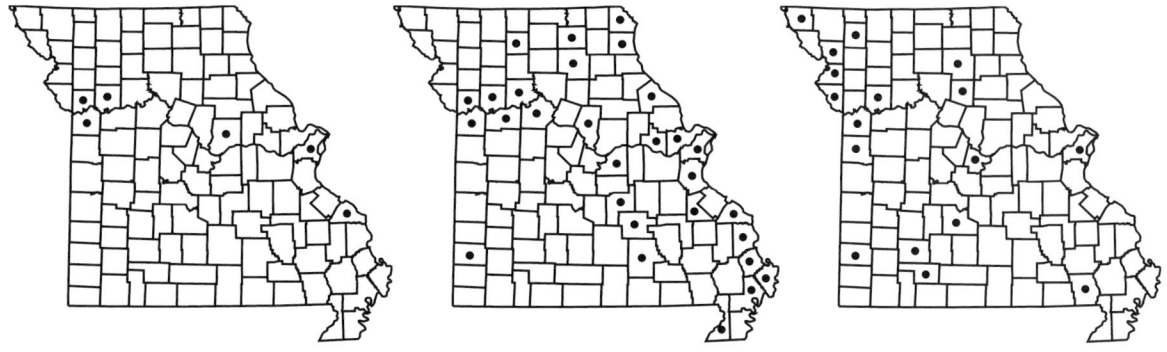

1545. Corispermum villosum   1546. Cycloloma atriplicifolium   1547. Kochia scoparia

6. Cycloloma Moq.

One species, western U.S. east to Indiana, Arkansas, and Texas; Canada, Mexico, introduced farther east.

1. Cycloloma atriplicifolium (Spreng.) J.M. Coult. (winged pigweed, tumble ringwing)
Pl. 357 j, k; Map 1546

Plants annual, the taproot not tuberous-thickened. Stems 10–80 cm long, erect to loosely ascending, not succulent, not appearing jointed, much-branched, moderately to densely pubescent with woolly hairs when young, becoming nearly glabrous at maturity. Leaves alternate, well developed, not succulent, sessile to short-petiolate, shed early. Leaf blades 1–8 cm long, lanceolate to oblong or ovate, flattened in cross-section, not clasping the stem, narrowed to a sharply pointed tip, tapered at the base, the margins coarsely and irregularly wavy and/or toothed, the surfaces hairy when young, becoming glabrous at maturity. Inflorescences terminal, consisting of loose, interrupted spikes usually appearing as irregular panicles, the flowers solitary at the nodes, not sunken into the axis. Flowers perfect or pistillate, green, red, or purple. Bract 1, 3–10 mm long, linear to narrowly elliptic, the margins entire or sparsely toothed. Calyx 5-lobed to about the midpoint, persistent at fruiting, enclosing the fruit, longitudinally angled or ridged, at fruiting the entire calyx developing a prominent, continuous, papery, transverse wing with 5 shallow lobes and an otherwise irregular margin, the lobes 0.4–0.6 mm long, triangular. Stamens 5 (sometimes absent). Ovary superior. Style absent or 1 and very short, the stigmas (2)3, linear. Fruits 0.5–1.0 mm long, 2–4 mm in diameter (including the wing), circular to somewhat 5-angled in cross-section, depressed-elliptic in outline, flattened vertically, indehiscent, the wall thin and papery, glabrous. Seed adhering loosely to the fruit wall, positioned horizontally, 1.3–1.7 mm in diameter, nearly circular in outline, strongly flattened, the surface smooth, black, shiny, the embryo appearing more or less ring-shaped. 2n=36. June–October.

Scattered, mostly in counties with large rivers, nearly absent from the Ozark Division (western U.S. east to Indiana, Arkansas, and Texas; Canada, Mexico, introduced eastward to South Carolina and Massachusetts). Banks of rivers and less commonly streams, and sand prairies; also fallow fields, railroads, and open, disturbed areas, in sandy soil.

The unique equatorial wing in this species, which develops as the fruit matures, makes it one of the easiest members of the family to recognize. The plants often form dense, irregularly spherical masses that break off with age and act as tumbleweeds, dispersing fruits as the wind blows them across the substrate. The wing of the fruit presumably also aids in dispersal by wind, although it also helps the fruit to float during floods.

7. Kochia Roth
(Blackwell et al., 1978)

About 30 species, North America, Europe, Asia, Africa.

The genus *Kochia* is closely related to *Bassia* All. and also to *Cycloloma*. All three genera have outgrowths of various sorts on the sepals. Some botanists have combined *Kochia* and *Bassia* under the latter name (Scott, 1978; Clemants, 1992; Judd and Ferguson, 1999). However, the phylogenetic relationships among these groups require more detailed study, and the three genera (only two of which occur in Missouri) are here maintained as separate.

1. Kochia scoparia (L.) Schrad. (summer cypress, fire bush)
  K. scoparia var. *culta* Farw.
  K. scoparia var. *pubescens* Fenzl
  K. sieversiana (Pall.) C.A. Mey.
  Bassia scoparia (L.) A.J. Scott

Pl. 358 d–f; Map 1547

Plants annual, the taproot not tuberous-thickened. Stems 30–150 cm long, erect, not succulent, not appearing jointed, much-branched, glabrous or sparsely to moderately pubescent with spreading, curled hairs. Leaves alternate, well developed, sessile or short-petiolate, not succulent. Leaf blades 0.5–8.0 cm long, those subtending flowers shorter than the others, linear to lanceolate, oblanceolate, or narrowly obovate, flattened in cross-section, not clasping the stem, narrowed to a sharply pointed tip, narrowed at the base, the margins entire but with relatively long, spreading hairs, the surfaces glabrous or sparsely to moderately pubescent with shorter, somewhat curled hairs. Inflorescences axillary toward the branch tips, sometimes appearing as terminal, interrupted spikes, the flowers paired or less commonly solitary or in small clusters, not sunken into the axis. Flowers perfect and/or functionally staminate and pistillate, the base with a tuft of hairs. Bracts absent (not counting the subtending leaf). Calyx 5-lobed to above the midpoint, persistent at fruiting, enclosing the fruit, indistinctly longitudinally angled, at fruiting each sepal developing an inconspicuous projection or prominent, spreading, papery, transverse wing, the lobes 0.3–0.6 mm long, oblong-triangular. Stamens 5 (sometimes absent). Ovary superior. Style 1, short, the stigmas 2(3), linear. Fruits 1.0–1.6 mm long, 1.5–3.0 mm in diameter (including the wings), circular or nearly so in cross-section, depressed-elliptic in outline, flattened vertically, indehiscent or irregularly dehiscent, the wall thin and papery to membranous. Seed adhering loosely to the fruit wall, positioned horizontally, 1.4–2.0 mm long, oblong-obovate in outline, flattened, the surface smooth to finely granular, brown to black, not shiny, the embryo appearing more or less ring-shaped. $2n=18$. July–October.

Introduced, scattered, mostly in counties with large rivers, nearly absent from the Ozark Division (native of Europe, Asia; introduced nearly worldwide, nearly throughout the U.S.). Banks of rivers; more commonly fallow fields, railroads, roadsides, and open, disturbed areas.

Summer cypress was introduced into North America as an ornamental because its foliage turns bright red in the autumn. A number of cultivars exist, some of which have received formal taxonomic recognition. Steyermark (1963) and some other earlier botanists separated var. *culta* based on its threadlike leaves and var. *pubescens* based on its denser pubescence, but Blackwell et al. (1978), McGregor (1986a), and Judd and Ferguson (1999) all indicated that these variants grade into one another completely in naturalized populations.

According to Mühlenbach (1983), plants that he had previously (Mühlenbach, 1979) attributed to *K. iranica* (Hausskn. & Bornm.) Litv. are actually *K. sieversiana*, a taxon usually considered a synonym of *K. scoparia*. Blackwell et al. (1978) excluded *K. iranica* from the North American flora.

8. Monolepis Schrad.

Four species, western North America, South America, Asia.

1. Monolepis nuttalliana (Roem. & Schult.) Greene (poverty weed)

Pl. 358 i, j; Map 1548

Plants annual, the taproot not tuberous-thickened. Stems 8–35 cm long, loosely to strongly ascending, slightly succulent, not appearing jointed, few- to much-branched, sparsely to moderately mealy when young, becoming glabrous or nearly so at maturity. Leaves alternate, well developed, progressively reduced toward the stem tip, somewhat succulent, short- to long-petiolate. Leaf blades 0.5–4.0 cm long, linear to elliptic-triangular or ovate-triangular, at least the largest ones with a pair of prominent, spreading to loosely ascending

Plate 357. Chenopodiaceae. *Corispermum americanum*, **a)** fruit, **b)** portion of inflorescence, **c)** habit. *Corispermum villosum*, **d)** fruit. *Axyris amaranthoides*, **e)** fruit, **f)** habit. *Corispermum pallasii*, **g)** fruit, **h)** inflorescence, **i)** habit. *Cycloloma atriplicifolium*, **j)** fruit (top view), **k)** habit.

1548. Monolepis nuttalliana   1549. Salicornia europaea   1550. Salsola collina

lobes below the midpoint, flattened in cross-section, not clasping the stem, narrowed to a sharply or less commonly bluntly pointed tip, mostly tapered at the base, the margins otherwise entire, the surfaces sparsely to moderately mealy when young, becoming glabrous or nearly so at maturity. Inflorescences axillary, consisting of small flower clusters (rarely solitary flowers), sometimes appearing as an interrupted terminal spike, the flowers not sunken into the axis. Flowers perfect (often a few pistillate flowers also present in each cluster). Bracts absent, but occasional flowers toward the base of each cluster lacking stamens and pistils, the remaining sepals thus appearing bractlike. Calyx of 1 sepal, this persistent at fruiting, not enclosing the fruit, not winged, positioned along one side of the ovary, 0.7–1.2 mm long, narrowly oblanceolate to spatulate. Stamen 1 (sometimes absent). Ovary superior. Styles 2, short, fused toward the base, the stigmas 1 per style, linear. Fruits 1.0–1.5 mm long, broadly elliptic in cross-section, broadly ovate in outline, somewhat flattened laterally, indehiscent or irregularly dehiscent, the wall thin and papery to membranous, with a faint network of ridges and pits. Seed adhering loosely to the fruit wall, positioned vertically, 0.8–1.4 mm long, broadly ovate in outline, somewhat flattened, the surface smooth, dark brown to black, shiny, the coiled embryo usually apparent. $2n=18$. April–September.

Introduced, uncommon and sporadic (western U.S. east to Minnesota and Texas; Canada; introduced sporadically farther east). Pastures, railroads, and open, disturbed areas.

Steyermark (1963) considered this species native in Missouri, which is along the eastern edge of its natural distribution. However, to date no plants have been discovered growing in a natural habitat, and Bush's early collections from the Kansas City area, where native occurrences might have been expected, are indicated as introduced.

9. Salicornia L. (glasswort)

Fifteen to 28 species, nearly worldwide.

1. Salicornia europaea L. (glasswort, samphire)

Map 1549

Plants annual, the taproot not tuberous-thickened. Stems 10–50 cm long, erect with ascending to spreading branches, succulent, appearing cylindrical and jointed, usually much-branched, glabrous. Leaves opposite, fused into a small sheath, the free portion reduced to tiny scales, these 1.0–1.5 mm long, broadly triangular, rounded to bluntly pointed at the tip, glabrous. Inflorescences terminal, dense spikes with tiny clusters of 3 flowers in pairs at the nodes, these sunken into the axis. Flowers perfect, the central flower of each cluster positioned slightly above the lateral pair. Bracts reduced to tiny, broadly triangular scales, these fused basally into a low cup, bluntly pointed at the tip. Calyx 0.5–0.7 mm long, obpyramidal, splitting across the tip at maturity into 3 more or less truncate lobes, persistent at fruiting, mostly enclosing the fruit, not winged. Stamens 2. Ovary superior. Styles 2, short, fused toward base, the stigmas 1 per style, linear. Fruits 1.0–1.5 mm long, broadly elliptic in cross-section, ovate in outline, slightly flattened laterally, indehiscent or irregularly dehiscent, the wall thin and papery to membranous. Seed adhering loosely to the fruit wall, positioned vertically, 1.0–1.5 mm long, ovate to oblong in outline, somewhat flattened, the surface with scattered minute, curved hairs, greenish brown to brown, not shiny, the coiled embryo usually not apparent. $2n=36$. July–September.

Introduced, known from a single historical collection from St. Louis County (coastal areas of North America, Europe, Asia, Africa; introduced sporadically farther inland). Open, disturbed areas, in sandy soil.

This distinctive species frequently grows in coastal salt marshes and other places with saline soils. In the Midwest it is an uncommon introduced species known from few populations, most of which probably did not persist for very long. The specimen collected in Valley Park in 1927 was overlooked by Steyermark (1963).

Salicornia europaea is treated here in the broad sense to comprise a circumboreal polyploid complex. The North American plants belong to a tetraploid ($2n=36$) variant that has sometimes been treated as a separate species, *S. virginica* L., with the name *S. europaea* then restricted to a diploid ($2n=18$) taxon restricted to Europe. Further studies are needed to resolve the subtle patterns of morphological variation in the group and to assess the relationships of the polyploid taxa to the diploids (Clemants, 1992; Judd and Ferguson, 1999).

10. Salsola L. (Russian thistle, tumbleweed)
(Mosyakin, 1996; Rilke, 1999)

Plants annual (perennial herbs or shrubs elsewhere), the roots not tuberous-thickened. Stems erect to loosely ascending, not succulent, not appearing jointed, much-branched, sparsely pubescent with stiff, short, unbranched, pustular-based hairs, occasionally appearing glabrous at maturity. Leaves alternate or occasionally opposite toward the stem base, well developed, succulent, sessile. Leaf blades linear, unlobed, circular to elliptic in cross-section, the base clasping the stem, the tip with a slender, somewhat spinelike extension of the midvein, the margins entire but pubescent with minute, spreading, unbranched, pustular-based hairs (thus sometimes appearing finely toothed), the surfaces glabrous or sparsely hairy, sometimes slightly glaucous. Inflorescences spikes, terminal on the branches, the flowers solitary (rarely 2 or 3) per node, not sunken into the axis. Bracts longer than and more or less enclosing the flower, the middle bract only slightly longer than the lateral pair, narrowly triangular, spreading or ascending, tapered to a stiff, hard, spinelike tip. Flowers perfect. Calyx deeply 5-lobed or of 5 free sepals spaced around and tightly enclosing the ovary, the tips sometimes erect or somewhat spreading, becoming papery or somewhat hardened at fruiting, with a continuous horizontal ridge or wing above the midpoint (the fruit including calyx thus appearing concave at the tip). Stamens 3–5. Ovary superior. Style 1, short, the stigmas 2(3), linear to narrowly club-shaped. Fruits broadly obovoid to obconic, usually indehiscent, the wall somewhat fleshy, becoming membranous at maturity. Seed adhering loosely to the fruit wall, positioned horizontally, 1.5–2.0 mm in diameter, somewhat flattened, the surface smooth, black, shiny, the strongly coiled embryo usually apparent. About 130 species; native to Europe, Asia, Africa; introduced nearly worldwide.

The vernacular name tumbleweed refers to the habit of the spherical plants dispersing their seeds by breaking off at the base after fruits have matured and rolling across the countryside in response to breezes. Russian thistles are among the worst range weeds in North America. The primary problem species is *S. tragus*, which is present in nearly every state and throughout southern Canada, but *S. collina* occurs sporadically in the central half of the United States as well. *Salsola tragus* was present along the eastern seaboard as early as 1788, but its development into a noxious weed occurred only after a second introduction into South Dakota around 1873, possibly as a contaminant in flax seed from Russia (Rilke, 1999). From there it spread rapidly in all directions, expanding its range through the Great Plains and subsequently crossing the Rockies into the western states. *Salsola collina* apparently was a later introduction and was first recorded from Minnesota in 1938 (Rilke, 1999). Steyermark (1963) noted that the

1551. Salsola tragus 1552. Suaeda calceoliformis 1553. Helianthemum bicknellii

plants can be harvested for hay during times of drought and also have helped to stabilize eroded soils in some regions, but Russian thistle is blamed for the degradation of millions of hectares of rangeland in this country. Thus the tumbleweed of Western novels and films is actually *Salsola,* an invasive exotic species. In addition to lowering the productivity of rangelands, tumbleweeds also impact fields when the dry, spiny plants roll through them (Karpiscak and Grosz, 1979), depositing seeds that sprout along the apparent furrows into linear arrays of plants the following year. In the southwestern United States, tumbleweed trails are sometimes so noticeable that they can be mapped using aerial photography.

1. Bracts ascending and more or less appressed to the flower; inflorescences relatively slender at fruiting, dense and continuous .... 1. S. COLLINA
1. Bracts spreading at maturity; inflorescences relatively stout at fruiting, dense toward the tip but often interrupted toward the base .... 2. S. TRAGUS

1. Salsola collina Pall. (slender Russian thistle)

Pl. 358 b, c; Map 1550

Stems 15–100 cm long, the usually single main stem erect or strongly ascending, usually developing dense, spreading branches before flowering. Leaves 8–40(–60) mm long, about 1 mm wide. Inflorescences 7–40 cm long, relatively slender, dense and continuous, sometimes arched or nodding toward the tip. Bracts 4–9 mm long, ascending and more or less appressed to the flower. Calyx 2.5–3.5 mm long, with a narrow, horizontal ridge or wing at maturity, becoming papery to somewhat hardened below the ridge, the sepal tips soft and ascending, sometimes somewhat twisted. Fruits 1.5–2.5 mm in diameter. $2n=18$. July–October.

Introduced, uncommon and widely scattered, mostly in the St. Louis area (native of Europe, Asia; introduced in the U.S. and adjacent Canada from Montana to Arizona east to Vermont, Kentucky, and Texas). Fallow fields, crop fields, roadsides, railroads, and open, disturbed areas, often in sandy soils.

This species was first reported for Missouri by Mühlenbach (1979). It probably has been undercollected both in the state and elsewhere in the country.

2. Salsola tragus L. (Russian thistle, tumbleweed, saltwort)

S. iberica (Sennen & Pau) Botsch. ex Czerep.
S. kali L. var. *tenuifolia* Tausch
S. pestifer A. Nelson
S. ruthenica Iljin

Pl. 358 a; Map 1551

Stems 20–100 cm long, the usually single main stem erect or strongly ascending, usually developing dense, spreading branches with ascending tips before flowering. Leaves 8–40(–60) mm long, 0.4–0.7 mm wide. Inflorescences 4–15 cm long, relatively stout, at least at fruiting, dense toward the tip but often interrupted toward the base, seldom arched or nodding toward the tip. Bracts 4–12 mm long, spreading at maturity. Calyx 2.5–3.5 mm long, developing a broad, papery, horizontal wing at maturity, becoming hardened below the wing, the sepal tips usually soft and ascending, occasionally somewhat twisted. Fruits 3–9 mm in diameter (including the wings). $2n=36$. July–October.

Introduced, scattered north of the Missouri River, uncommon and sporadic farther south (native of Europe, Asia; introduced nearly worldwide). Banks of rivers, tops of bluffs, and disturbed por-

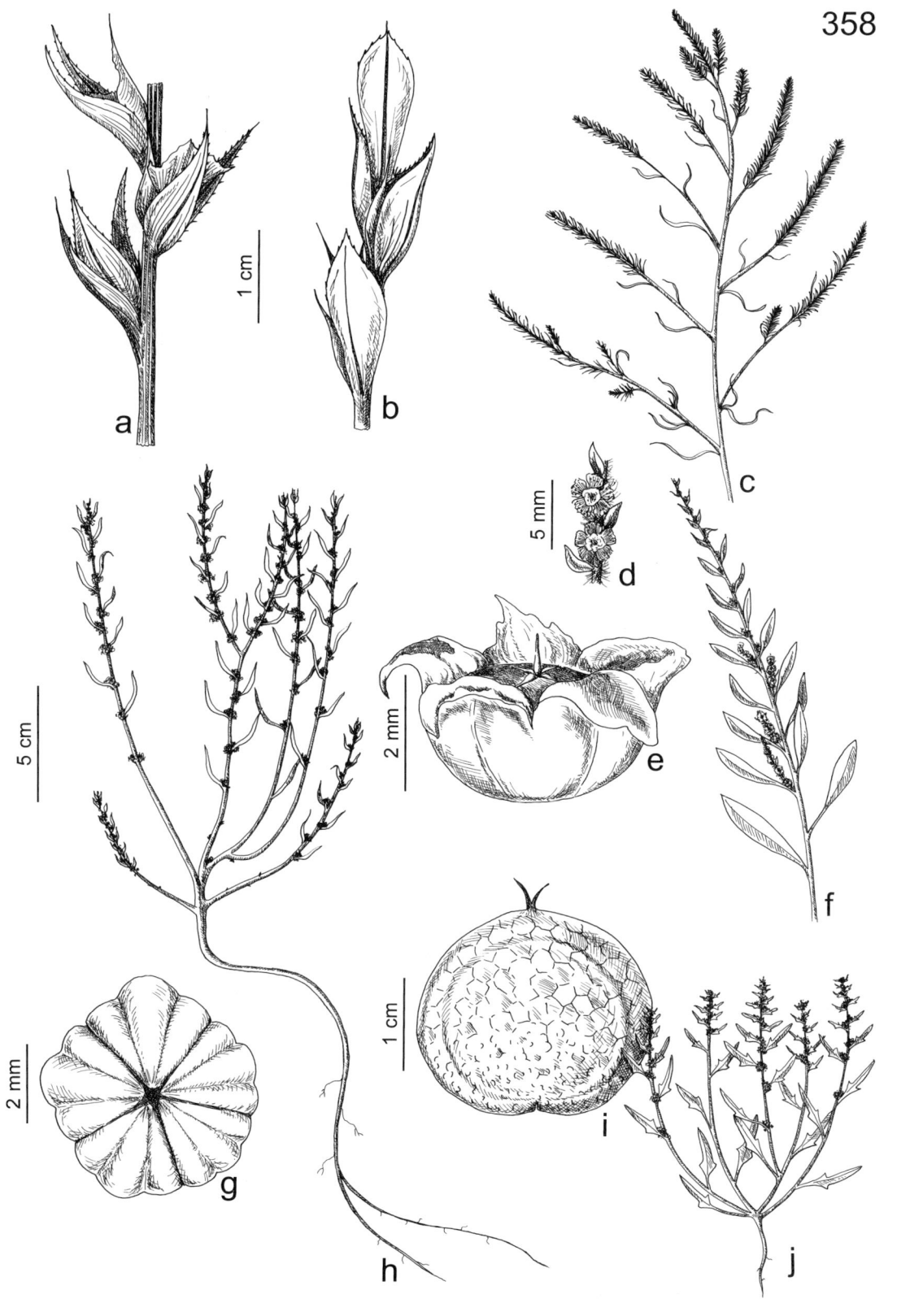

Plate 358. Chenopodiaceae. *Salsola tragus*, **a)** stem with fruits. *Salsola collina*, **b)** stem with fruits, **c)** habit. *Kochia scoparia*, **d)** flowers, **e)** fruit, **f)** habit. *Suaeda calceoliformis*, **g)** fruit (top view), **h)** habit. *Monolepis nuttalliana*, **i)** fruit, **j)** habit.

tions of loess hill prairies; also fallow fields, crop fields, roadsides, railroads, and open, disturbed areas, often in sandy soils.

The taxonomy of this species is complex and still not fully understood. Plants in the United States were long referred to as *S. kali* L. (Steyermark, 1963), a different species restricted to mostly coastal areas with thicker, more strongly succulent leaves having stouter spiny tips and sepal tips stiff and somewhat spiny. The two species are closely related and there is still some disagreement about the classification of certain infraspecific variants, such as *S. tragus* ssp. *pontica* (Pall.) Rilke (*S. kali* ssp. *pontica* (Pall.) Mosyakin).

11. Suaeda Forssk. (seepweed)
(Hopkins and Blackwell, 1977)

About 100 species, nearly worldwide, except in arctic regions.

1. Suaeda calceoliformis (Hook.) Moq. (sea blite)

Pl. 358 g, h; Map 1552

Plants annual, the taproot not tuberous-thickened. Stems 10–80 cm long, loosely to strongly ascending or less commonly erect, not succulent, not appearing jointed, few- to much-branched, glabrous, glaucous. Leaves mostly alternate (the lowermost sometimes opposite), well developed, succulent, sessile. Leaf blades 0.5–4.0 cm long, those subtending flowers shorter than the others, linear, circular or elliptic in cross-section, not clasping the stem, narrowed to a sharply pointed tip, narrowed at the base, the margins entire, the surfaces glabrous, usually somewhat glaucous. Inflorescences axillary toward the branch tips, appearing as terminal, interrupted spikes, the flowers solitary or in small clusters, not sunken into the axis. Flowers perfect or pistillate. Bract 1(2), scalelike. Calyx of 5 sepals, these fused toward the base, persistent at fruiting, more or less enclosing the fruit, 1–3 of them somewhat broader and noticeably hooded or with a small, hornlike projection from about the midpoint, the others rounded or angled on the back, the lobes 0.5–1.0 mm long, ovate to ovate-triangular. Stamens 5 (sometimes absent). Ovary superior. Styles absent or nearly so, the stigmas 2(–5), linear. Fruits 0.5–0.6 mm long, 1.0–1.5 mm in diameter, circular or nearly so in cross-section, depressed-elliptic in outline, flattened vertically, indehiscent or irregularly dehiscent, the wall thin and papery to membranous. Seed adhering loosely to the fruit wall, positioned horizontally, 1.0–1.5 mm long, broadly ovate in outline, flattened, the surface smooth to finely and obscurely pebbled, reddish brown to black, more or less shiny, the coiled embryo usually apparent. $2n=90$. July–October.

Introduced, known only from historical collections from Carter and Jackson Counties (native of the western U.S. and adjacent Canada east to Minnesota and Texas; introduced sporadically in the midwestern and northeastern U.S. and Canada). Habitat unknown, but presumably open, disturbed areas.

Steyermark (1963) noted that where it is abundant this species is cooked and eaten as a vegetable. McNeill et al. (1977) showed that the name *S. depressa* (Pursh) S. Watson, which was used by most earlier botanists for our species (Steyermark, 1963), had been misapplied and was instead referable to a different Eurasian species.

CISTACEAE (Rockrose Family)

Plants perennial herbs (small shrubs elsewhere), sometimes woody at the base. Stems creeping to erect. Leaves alternate or less commonly opposite or whorled, relatively small, sessile or short-petiolate. Stipules absent. Leaf blades simple, the margins entire, sometimes somewhat curled under, with pinnate veins or sometimes only a midvein. Inflorescences terminal or appearing axillary, of solitary flowers or more commonly ranging from small clusters to racemes or panicles, the branches sometimes appearing racemose and with the flowers mainly along the upper side. Flowers actinomorphic, perfect, hypogynous, often subtended by bracts (these usually shed as the fruit matures). Cleistogamous flowers sometimes present (in *Helianthemum*). Calyces of 5 sepals, these free or fused at the base, in 2 whorls, the outer 2 narrower, shorter than, and sometimes partially fused to the inner 3, all persistent at fruiting. Corollas of 3 or 5 free petals. Stamens 3 to numerous, the anthers attached at their bases, dehiscent by longitudinal slits. Pistil 1 per flower, of 3 fused carpels. Ovary superior, with 1 locule (incompletely 3-locular, the locule thus appearing lobed in cross-section), the placentation parietal. Style absent or 1 per flower and short, the stigmas 1 or 3. Fruits capsules, 3-valved, dehiscent longitudinally. Eight genera, about 200 species, nearly worldwide.

Some members of the rockrose genus *Cistus* are cultivated as garden ornamentals, but these generally are not winter-hardy as far north as Missouri. A few species are grown or wild-harvested commercially for the bitter, gummy resins extracted from their twigs and used as scent agents in soaps and deodorants.

1. Leaves with stellate and sometimes also simple hairs; plants not producing specialized overwintering offshoots, the stems all similar; petals 5, showy but withering and shed soon after the flower opens, yellow; style short; stigma 1, capitate, often 3-lobed.................................... 1. HELIANTHEMUM
1. Leaves with simple hairs; plants producing spreading, densely leafy, overwintering offshoots toward the end of the growing season; petals 3, minute, withering but persistent at fruiting, dark red; style absent; stigmas 3, plumose ..... 2. LECHEA

1. Helianthemum Mill. (rockrose, frostweed)
(Daoud and Wilbur, 1965)

Plants not producing specialized overwintering offshoots, the stems all similar. Stems loosely ascending to erect, pubescent with stellate hairs, sometimes becoming nearly glabrous with age. Leaves alternate and sometimes also basal, the blade elliptic-oblanceolate to narrowly elliptic or elliptic, pubescent with stellate and sometimes also simple hairs, usually with pinnate venation. Inflorescences first appearing as a solitary petaliferous flower or a small, dense cluster of petaliferous flowers at the tip of each nearly unbranched stem, later in the growing season developing short, racemose branches with dense clusters of nearly sessile, cleistogamous flowers. Cleistogamous flowers appearing similar to but smaller than mature open-flowering buds and usually with shorter stalks, lacking petals. Sepals becoming slightly enlarged as the fruit matures. Outer 2 sepals somewhat shorter than to nearly as long as the inner ones, linear to narrowly lanceolate, sharply pointed at the tip, those of cleistogamous flowers partially fused to the inner ones. Inner 3 sepals ovate to ovate-elliptic, sharply pointed at the tip. Petals 5, showy but withering and shed soon after the flower opens, obovate-obtriangular, broadly rounded to more or less truncate or shallowly concave at the tip, yellow. Stamens of petaliferous flowers numerous; reduced to 3(5) in cleistogamous flowers. Style 1 per flower, short, the stigma 1, capitate, often 3-lobed. Ovules 9 (3 per carpel) to numerous. Fruits shorter than the persistent sepals, ovoid to ellipsoid, somewhat 3-angled, glabrous. Seeds 1 to numerous, those of the open-flowering fruits slightly larger than those developing from the cleistogamous flowers, variable in shape (dependent on number in fruit), globose or ovoid to strongly 3-angled or more or less trapezoidal, the surface dark brown, often covered with an inconspicuous, thin, membranous outer coat, this sometimes turning gelatinous when moistened. About 110 species, North America to South America, Caribbean Islands, Europe, Asia, Africa.

The so-called normal flowers are produced earlier in the growing season (as indicated in the flowering months cited after the species descriptions) and occur singly or in small clusters toward the stem tips. Lateral branches elongate at or after the first round of flowering. These

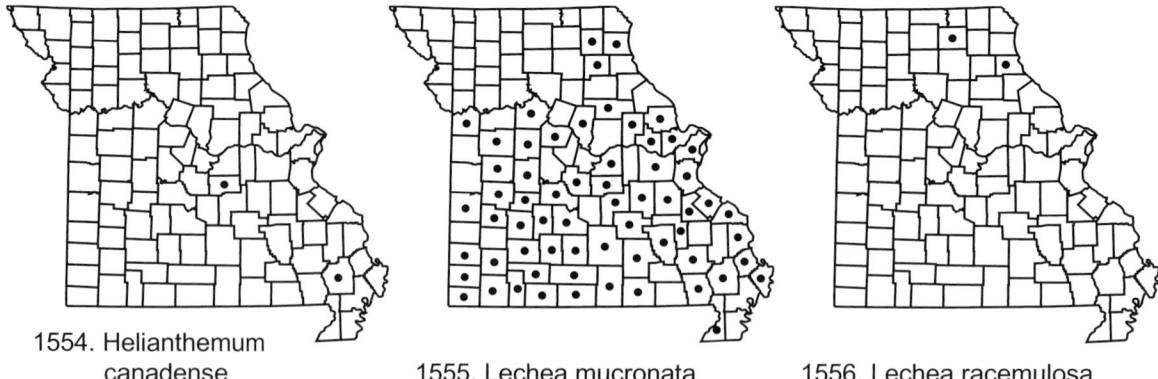

1554. Helianthemum canadense

1555. Lechea mucronata

1556. Lechea racemulosa

bear small clusters of cleistogamous flowers at the nodes. The cleistogamous flowers are much smaller, lack petals, and do not open. As these flowers are often similar in size to those of *Lechea*, care must be taken not to misdetermine plants encountered later in the growing season.

1. Petal-bearing flowers 2–10 at each stem tip; sepals pubescent only with stellate hairs; fruits developing from cleistogamous flowers with 1–3 seeds; seeds with the surface with rows of minute pits or an obscure network of ridges .... 1. H. BICKNELLII
1. Petal-bearing flowers occurring singly or rarely in pairs at each stem tip; sepals of petaliferous flowers with longer, simple hairs intermixed with the shorter, stellate ones (those of cleistogamous flowers usually only stellate-hairy); fruits developing from cleistogamous flowers with 5–12 seeds; seeds with the surface pebbled to finely tubercled .... 2. H. CANADENSE

1. Helianthemum bicknellii Fernald (hoary frostweed)

Pl. 359 e, f; Map 1553

Stems 20–50(–60) cm long. Leaves 4–30 mm long, those of the main stems longer than those of the branches, the upper surface appearing green, but sparsely to moderately stellate-hairy, the undersurface obscured by the dense stellate hairs. Petaliferous flowers 2–10 at each stem tip. Sepals all pubescent with only stellate hairs, lacking longer simple hairs; those of petaliferous flowers 4–8 mm long; those of cleistogamous flowers 1.5–3.5 mm long. Petals 8–12 mm long. Fruits maturing from petaliferous flowers 3.5–5.5 mm long, with 12 to numerous seeds; those developing from cleistogamous flowers 1.5–2.5 mm long, with 1–3 seeds. Seeds 1.0–1.5 mm long, the surface with rows of minute pits or an obscure network of ridges. April–July.

Scattered nearly throughout the state but apparently absent from the Mississippi Lowlands Division and adjacent portions of the Ozarks (eastern [mainly northeastern] U.S. west to North Dakota and Colorado; Canada). Openings of mesic to dry upland forests, upland prairies, and glades; also old fields and roadsides; usually in acidic soils.

2. Helianthemum canadense (L.) Michx.

Pl. 359 g–i; Map 1554

Stems 10–40 cm long. Leaves 4–30 mm long, those of the main stems longer than those of the branches, the upper surface appearing green, but sparsely to moderately stellate-hairy and sometimes also with a few simple hairs, the undersurface obscured by the dense stellate hairs. Petaliferous flowers 1(2) at each stem tip. Sepals of petaliferous flowers 5–9 mm long, with longer, simple hairs intermixed with the shorter, stellate ones; those of cleistogamous flowers 2–4 mm long, only stellate-hairy. Petals 8–15 mm long. Fruits maturing from petaliferous flowers 4–7 mm long, with numerous seeds; those developing from cleistogamous flowers 2–3 mm long, with 5–12 seeds. Seeds 1.0–1.5 mm long, the surface pebbled to finely tubercled. $2n=20$. April–July.

Uncommon, known thus far only from Maries and Stoddard Counties (eastern U.S. west to Minnesota, Missouri, and Alabama; Canada). Ledges of sandstone bluffs and sand blowouts.

Steyermark (1963) excluded *H. canadense* from the flora based on the likelihood that the only specimen that he encountered attributed to this species (a vegetative collection from Barry County) could not be identified reliably to species. In the research toward their revision of the genus, Daoud and Wilbur (1965) annotated this specimen as *H. bicknellii* but instead cited one of Steyermark's own collections (from Maries County) as confirmation that *H. canadense* is present in Missouri. Since then it has been collected only rarely in the state, presumably in part because plants are so infrequently encountered with open flowers.

2. Lechea L. (pinweed)
(Wilbur and Daoud, 1961)

Plants producing spreading, densely leafy, overwintering offshoots toward the end of the growing season, these usually shorter than the flowering stems. Flowering stems ascending to erect, pubescent with simple hairs (stellate hairs absent) or nearly glabrous. Leaves alternate or sometimes partially opposite or whorled (especially on the overwintering offshoots), variously shaped, usually with only a midvein. Inflorescences dense, terminal panicles with leaflike bracts along the branches, sometimes appearing as axillary racemes with clusters of flowers. Flowers all appearing similar in size and morphology, but frequently not appearing to open. Sepals not becoming enlarged as the fruit matures. Outer 2 sepals shorter than to longer than the inner ones, linear to narrowly oblong-elliptic, sharply pointed at the tip. Inner 3 sepals ovate to obovate, often strongly cupping the fruit, rounded to bluntly pointed at the tip, usually with broad, thin, white margins. Petals 3, 0.7–1.2 mm long, shorter than the sepals, withering but persistent at fruiting, oblong-oblanceolate, rounded at the tip, dark red. Stamens (3–)5 to numerous. Style absent, the stigmas 3, plumose with dense, feathery hairs, dark red. Ovules 6 (2 per carpel). Fruits circular in cross-section, glabrous. Seeds 1–6, variable in shape (dependent on number in fruit), globose or ovoid to strongly 3-angled, the surface light brown to dark brown, the outer coat hard and not membranous, usually somewhat shiny. Seventeen to 20 species, North America, Central America, Caribbean Islands.

The flowers of *Lechea* are seldom observed fully opened, and then apparently only during morning hours in bright sunlight. Because of this, self-pollination has been hypothesized as the predominant reproductive mode. This is supported by the fact that most of the flowers appear to mature into fruits, but Hodgdon (1938) and Wilbur and Daoud (1961) observed apparent interspecific hybrids between various species (none has been recorded thus far from Missouri), so cross-pollination must occur to some extent. The feathery stigmas suggest that wind-pollination may occur in the group.

1. Stems moderately to densely pubescent with conspicuous spreading hairs; leaves of the overwintering offshoots ovate-elliptic, the largest ones 8–15 mm long
 . . . . 1. L. MUCRONATA
1. Stems glabrous or sparsely to moderately pubescent with inconspicuous, appressed or ascending hairs; leaves of the overwintering offshoots linear to oblong-elliptic, the largest ones 4–6 mm long
    2. Leaves of the overwintering offshoots lanceolate-elliptic to oblong-elliptic; leaves of the flowering stems narrowly lanceolate to oblanceolate; outer sepals noticeably shorter than the inner ones; fruits narrowly ovoid to narrowly obovoid-ellipsoid, slightly longer than the persistent sepals
 . . . . 2. L. RACEMULOSA
    2. Leaves of the overwintering offshoots linear; leaves of the flowering stems linear; outer sepals somewhat longer than the inner ones; fruits broadly ovoid to globose, slightly shorter than the persistent sepals . . . . 3. L. TENUIFOLIA

1. Lechea mucronata Raf. (hairy pinweed)
  L. villosa Elliott

Pl. 359 b–d; Map 1555

Flowering stems 20–80 cm long, moderately to densely pubescent with conspicuous spreading hairs (these sometimes worn away, but leaving persistent, slightly pustular bases). Leaves with the upper surface glabrous or sparsely hairy, the lower surface moderately to densely hairy, especially along the margins and midvein, those of the overwintering offshoots 4–15 mm long, the largest 8–15 mm long, ovate-elliptic to broadly ovate-elliptic, bluntly to sharply pointed at the tip; those of the flowering stems 5–30 mm long, lanceolate to elliptic or oblanceolate, sharply pointed to less commonly rounded at the tip. Calyx without a differentiated basal portion. Outer 2 sepals 1.9–2.2 mm long, slightly shorter than to slightly longer than the inner ones. Inner 3 sepals 1.5–2.2 mm long, broadly ovate to nearly circular, deeply cupped, strongly angled along the midrib, glabrous or hairy along the midrib. Fruits 1.4–2.1 mm long, usually slightly shorter than the persistent sepals, more or less globose, mostly 3-seeded. Seeds 0.9–1.2 mm long, brown to yellowish brown. July–November.

Scattered nearly throughout the state, but apparently absent from most of the western half of the Glaciated Plains Division (eastern U.S. west to Nebraska and Texas; Canada). Glades, savannas, upland prairies, and dry upland forests; also fallow fields and old fields, mostly in sandy or other acidic soils.

This species was long known as *L. villosa* until Wilbur (1966) resurrected the name *L. mucronata* during his studies of the correct application of species epithets in *Lechea* published by the early botanist Constantine Rafinesque, many of whose vast array of names generally have been difficult to apply. Hodgdon (1938) separated the species into three varieties based on a number of subtle differences in flower and fruit size and leaf pubescence, but, as noted by Wilbur (1969), there is far too much variation within the species to allow recognition of infraspecific taxa.

2. Lechea racemulosa Michx. (Illinois pinweed)

Pl. 359 a; Map 1556

Flowering stems 10–40 cm long, moderately to densely pubescent with inconspicuous, appressed or ascending hairs. Leaves with the upper surface glabrous, the lower surface sparsely to moderately hairy, especially along the midvein and margins, those of the overwintering offshoots 2–6 mm long, the largest 4–6 mm long, lanceolate-elliptic to oblong-elliptic, sharply pointed at the tip; those of the flowering stems 5–20 mm long, narrowly lanceolate to oblanceolate, sharply pointed at the tip. Calyx with a minute, nipplelike base, noticeably differentiated from the main portion of the sepals, at least at fruiting. Outer 2 sepals 1.2–1.6 mm long, noticeably shorter than the inner ones. Inner 3 sepals 1.8–2.2 mm long, narrowly elliptic to obovate, deeply cupped, the midrib indistinct, not angled, sparsely to densely and more or less uniformly hairy on the surface. Fruits 1.6–1.9 mm long, slightly longer than the persistent sepals, narrowly ovoid to narrowly obovoid-ellipsoid, mostly 1–3-seeded. Seeds 1.0–1.3 mm long, usually dark brown. June–November.

Known thus far only from Adair and Marion Counties (eastern U.S. west to Iowa and Louisiana). Habitat unknown, but presumably ledges of bluffs or glades.

The Iowa, Louisiana, and Missouri occurrences are somewhat disjunct from the main range of this species, which is most common in the Appalachians.

Steyermark (1963) noted the existence of a historical specimen of *L. intermedia* Liggett that was collected in Adair County and deposited at the University of Missouri Herbarium. He excluded the species from the flora, however, noting that A. R. Hodgdon, who annotated it during his taxonomic revision of *Lechea,* wrote on the sheet, "Probably not from Missouri." The closest sites for this species to Missouri are in northeastern Illinois, and the present work has accepted Steyermark's suggestion that the original collection was mislabeled. *Lechea intermedia* superficially resembles *L. racemulosa,* differing in its more globose fruits.

3. Lechea tenuifolia Michx. (narrow-leaved pinweed)

Pl. 359 j–l; Map 1557

Flowering stems 10–35 cm long, glabrous or sparsely to moderately pubescent with inconspicuous, appressed or ascending hairs. Leaves with the upper surface glabrous, the lower surface sparsely to moderately hairy, especially along the midvein, those of the overwintering offshoots 2–6 mm long, the largest 4–6 mm long, linear, sharply pointed at the tip; those of the flowering stems 5–20 mm long, linear, sharply pointed at the tip. Calyx without a differentiated basal portion. Outer 2 sepals 2–3 mm long, somewhat longer than the inner ones. Inner 3 sepals 1.6–2.2 mm long, ovate to broadly oblong-ovate, deeply cupped, the midrib prominent but not angled, sparsely to densely and more or less uniformly hairy on the surface. Fruits 1.4–2.1 mm long, slightly shorter than the persistent sepals, broadly ovoid to globose, mostly 2–5-seeded. Seeds 0.9–1.2

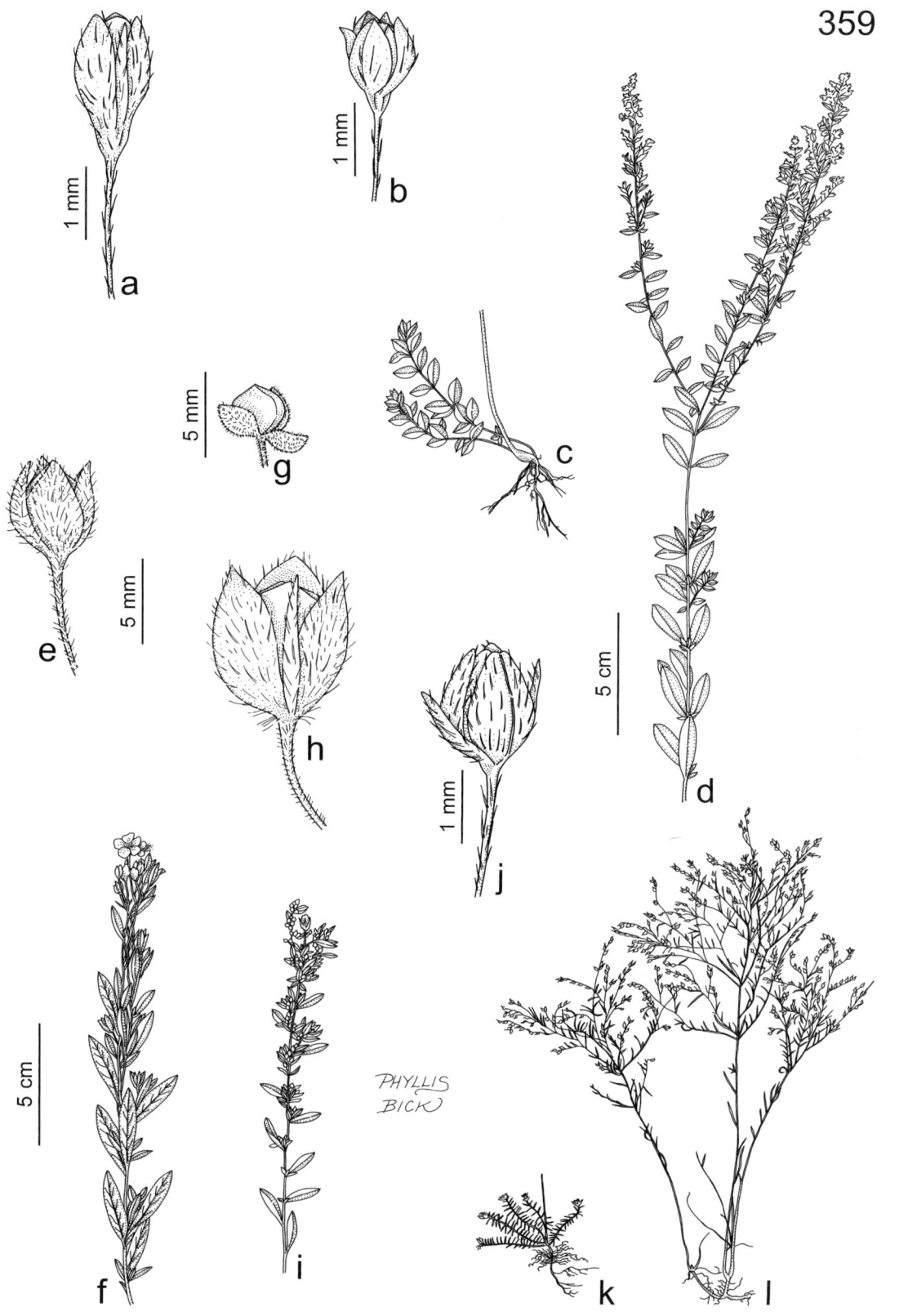

Plate 359. Cistaceae. *Lechea racemulosa*, **a)** fruit. *Lechea mucronata*, **b)** fruit, **c)** base of plant with overwintering branches, **d)** inflorescence. *Helianthemum bicknellii*, **e)** fruit, **f)** inflorescence. *Helianthemum canadense*, **g)** fruit, **h)** fruit, **i)** inflorescence. *Lechea tenuifolia*, **j)** fruit, **k)** base of plant with overwintering branches, **l)** habit.

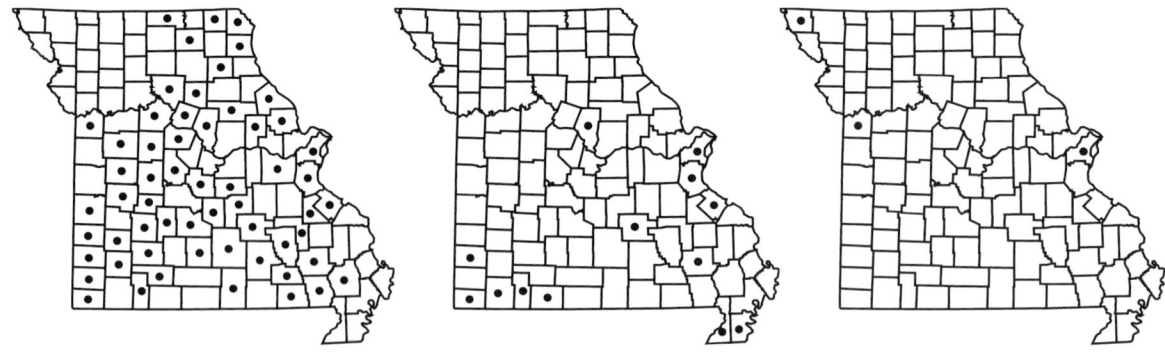

1557. Lechea tenuifolia   1558. Cleome hassleriana   1559. Cleome serrulata

mm long, usually reddish brown, sometimes yellow or nearly black. June–November.

Scattered nearly throughout the state, but apparently absent from most of the western half of the Glaciated Plains Division (eastern U.S. west to Minnesota and Texas; Canada). Glades, savannas, dry upland forests, upland prairies, and tops and ledges of bluffs; also fallow fields, pastures, roadsides, and open, disturbed areas, usually on acidic substrates.

Hodgdon (1938) segregated var. *occidentalis* Hodgdon, based on relatively robust plants with the inflorescence branches more erect than in the typical variety. He cited specimens of this mostly western taxon from Illinois, Kansas, and Nebraska, so its presence in Missouri would not be surprising. However, as noted by Wilbur (1969), the supposed differences between the two varieties are not convincing, with too many intermediate plants to permit the recognition of infraspecific taxa in this species.

CLEOMACEAE (Cleome Family)
Contributed by David J. Bogler

Plants annual (perennial herbs, shrubs, or trees elsewhere). Leaves alternate, palmately compound or simple. Stipules absent or present, sometimes thornlike. Inflorescences terminal racemes (flowers solitary in the leaf axils elsewhere), the flowers subtended by bracts. Flowers perfect or imperfect, actinomorphic to slightly zygomorphic, hypogynous. Calyx usually with 4 sepals, distinct or fused toward the base. Corolla with 4 to numerous separate petals, these ovate to spatulate, equal or often unequal, usually tapered to a stalklike base. Nectar-producing glands present between the corolla and stamens. Stamens 6 to numerous, usually exserted and showy, equal or unequal but not arranged in clusters of 4 and 2, the filaments slender, the anthers attached at or above the base, dehiscing by longitudinal slits. Pistil usually borne on a short to long, stalklike extension of the receptacle (gynophore) above the calyx, cylindrical, of 2 fused carpels, 1-locular, the placentation parietal. Style 1, often very short and inconspicuous, the stigma capitate or a somewhat concave disc, often shallowly 2-lobed. Fruits capsules, dehiscent by 2 valves, a persistent replum (the thin, placental band of tissue visible as a longitudinal line or nerve around the fruit) often present. Seeds numerous, more or less circular in outline, usually somewhat flattened, appearing folded or curled, with a broadly angled furrow or depression running partway through each face and usually with a small notch where it meets the margin. About 8 genera, 275 species, nearly worldwide.

Until recently, most botanists treated the Cleomaceae as a subfamily of the Capparaceae. Both groups have long been regarded as close relatives of the Brassicaceae, but the phylogenetic relationships between the groups were not well understood. Beginning in the early 1990s, several studies using different sources of data produced similar results, that the largely herbaceous Brassicaceae and Cleomaceae are sister groups and that this assemblage shared a common ancestor with the Capparaceae in the strict sense (Rodman et al., 1993; Judd et al., 1994; Hall et al., 2002). Subsequent authors have wrestled with whether to classify the entire group as one heterogeneous family (Brassicaceae) or to recognize two or three separate families (Hall et al., 2002). What has become apparent is that if *Cleome* and its relatives are to be combined with some other family, then they must become a subfamily of the Brassicaceae, not the Capparaceae.

Morphologically and phytochemically, the Brassicaceae in the restricted sense are a relatively easily recognized group distinguished by the presence of mustard oils, specialized capsular fruits with valves separating from persistent placenta tissues (replum) often divided by a false septum, six stamens positioned in groups of four and two (tetradynamous), and a folded embryo (Judd et al., 1994). Capparaceae and Cleomaceae have some of these features as well, but they have six or more nontetradynamous stamens. Capparaceae in the strict sense are usually trees or shrubs and have fruits that are berries or capsules lacking a replum. The mostly herbaceous Cleomaceae have capsules with a replum but lacking a false septum. The present treatment deviates from that of Cronquist (1981, 1991) in recognizing the Cleomaceae as a distinct family. This seems the most pragmatic solution to the problem, one that is phylogenetically sound while preserving two morphologically distinct groups in the Missouri flora. The true Capparaceae, comprising about 25 genera and 440 species, are not represented in the Missouri flora, other than by the edible pickled flower buds of the Mediterranean *Capparis spinosa* L. (capers) included by gourmet chefs in some tasty dishes.

1. Petals similar in size, entire at the tip; stamens 6, all of equal length; buds with the petals well developed, overlapping and covering stamens; fruits on long gynophores.. 1. CLEOME
1. Petals unequal in size, notched at the tip; stamens 8–20, unequal in length; buds with the petals small, not overlapping or covering the stamens; fruits sessile or on short gynophores .... 2. POLANISIA

1. Cleome L.

Leaves short- to long-petiolate. Leaf blades trifoliate or palmately compound, the leaflets with the margins entire or finely toothed. Stipules absent or minute, or spines. Inflorescences elongate bracteate terminal racemes or the flowers solitary in the leaf axils. Buds with the petals well developed, overlapping and wrapped around the young stamens. Flowers zygomorphic, but the petals similar in size, entire and more or less rounded at the tip, purple or less commonly white (rarely yellow elsewhere). Stamens 6, the filaments equal in length, the anthers elongate, attached at the base. Fruits on a long gynophore, usually spreading or pendant, elongate, with a persistent replum, the valves shed at dehiscence and readily releasing the seeds. About 150 species, nearly worldwide, most diverse in seasonally arid portions of tropical America and Africa.

About 58 species are recognized in the New World (Iltis, 1952). Two subgenera are recognized (Iltis, 1952), one in the Old World, and one in the New World. Some species of *Cleome* are cultivated as ornamentals. Several species also are favorites among beekeepers.

1. Leaf blades palmately compound with 5 or 7 leaflets; stipules a pair of conspicuous, stout spines; petals 18–28 mm long .... 1. C. HASSLERIANA
1. Leaf blades trifoliate, stipules absent or minute, lacking stipular spines; petals 7–12 mm long .... 2. C. SERRULATA

1. Cleome hassleriana Chodat (spider flower, pink queen)

C. spinosa Jacq., misapplied

C. houtteana Schltdl., misapplied

Pl. 333 j, k; Map 1558

Stems 30–150 cm tall, unbranched or sparsely branched above, densely glandular and hairy. Leaves palmately compound with 5 or 7 leaflets, densely glandular and sparsely hairy, the petioles 6–15 cm long. Stipules relatively stout golden brown spines 1–4 mm long. Leaflets 3–10 cm long, elliptic to oblanceolate, narrowed or tapered to a sharply pointed base and tip, the margins entire or shallowly toothed, the undersurface sometimes with sparse, short prickles similar to the stipular spines along the midvein. Inflorescences to 35 cm long, relatively dense. Bracts 1–4 cm long, 0.5–2.0 cm wide, simple, short-stalked to nearly sessile, ovate, tapered to a sharply pointed tip, densely glandular. Flower stalks 4–6 cm long, glandular. Sepals fused at the very base, 6–7 mm long, narrowly lanceolate, densely glandular and sometimes also sparsely hairy. Petals all oriented to one side of the flower, similar in size and shape, 18–28 mm long, 6–8 mm wide, abruptly narrowed to a stalklike base 6–12 mm long, pink to purple, rarely white. Glandular disc fleshy, conical, 1–3 mm long. Stamens with the filaments 3–4 cm long, the anthers 7–9 mm long, linear. Gynophore relatively short at flowering (0.5–0.9 cm) but elongating rapidly as the fruit matures to 3–8 cm long. Fruits 2–7 cm long, 3–4 mm wide, linear, usually slightly constricted between seeds, glabrous. Seeds 1.6–2.0 mm in diameter, the surface appearing wrinkled or warty, dark brown with lighter-colored warts and faint concentric lines. 2n=20. August–October.

Introduced, uncommon and sporadic in the southern half of the state (native of tropical America, mainly Brazil; widely planted and often escaping throughout the eastern U.S.). Banks of streams and rivers and rarely glades; also railroads, roadsides, and open, disturbed areas.

This showy species is popular in gardens throughout the world and often is sold under the name *C. spinosa*. There are many varieties available. The plants are very colorful and popular with hummingbirds. The common name spider flower comes from the long, waving stamens that somewhat resemble spider legs. *Cleome hassleriana* has been cultivated and has escaped from gardens in Missouri for a long time. Through the courtesy of Dr. Hugh Iltis of the University of Wisconsin at Madison, a specimen accessioned at the National Natural History Museum in Paris came to light that demonstrates this. In June 1848, the French botanist Auguste Trécul (see the introductory chapter on the history of Missouri botany in Yatskievych, 1999) collected a plant on the bank of the Mississippi River near Ste. Genevieve.

2. Cleome serrulata Pursh (Rocky Mountain bee plant, stinking clover)

Peritoma serrulatum (Pursh) DC.

P. integrifolia Nutt.

Pl. 333 h, i; Map 1559

Stems 30–200 cm tall, unbranched or sparsely branched above, glabrous or with sparse, cobwebby hairs when young, especially around the nodes. Leaves trifoliate with 3 leaflets, glabrous or sparsely to moderately pubescent with cobwebby hairs, the petioles 3–4 cm long. Stipules herbaceous or scalelike, inconspicuous, linear, 0.2–0.8 mm long, withering as the leaves mature. Leaflets mostly 2–5 cm long, narrowly elliptic to lanceolate, narrowed or tapered to a sharply pointed base and tip, the margins entire, the surfaces lacking prickles. Inflorescences to 25 cm long, relatively dense. Bracts 8–12 mm long, 1–2 mm wide, simple, short-stalked to sessile, narrowly elliptic to lanceolate, tapered to a hairlike tip, usually glabrous. Flower stalks 10–12 mm long, glabrous. Sepals fused to at or above the midpoint, 2.5–3.0 mm long, the free portion triangular-ovate, glabrous. Petals spreading asymmetrically, similar in size and shape, 7–12 mm long, 2–4 mm wide, abruptly narrowed to a stalklike base 1–2 mm long, pink to purple. Glandular disc fleshy, short-cylindrical, 1–4 mm long, with a scalelike appendage ascending from the tip. Stamens with the filaments 1.2–1.8 cm long, the anthers 1.5–2.0 mm long, linear. Gynophore short at flowering (0.2–0.4 cm) but elongating as the fruit matures to 1.0–2.5 cm long. Fruits 2–5 cm long, 2–3 mm wide, linear, usually slightly constricted between seeds, glabrous. Seeds 2.5–4.0 mm in diameter, the surface appearing wrinkled or warty, dark brown to black with lighter-colored warts and faint concentric lines. 2n=32, 34, 60. July–August.

1560. Polanisia dodecandra   1561. Hypericum adpressum   1562. Hypericum ascyron

Uncommon in Atchison County, introduced uncommonly and sporadically farther south (western U.S. east to Minnesota and Texas; Canada; introduced sporadically farther east). Loess hill prairies; also railroads and open, disturbed areas.

This species was commonly consumed and used as a poultice for various ailments by Native Americans (Moerman, 1998) and is today sometimes planted as an ornamental. Bees are attracted to the plant, and it is a good source of honey.

2. Polanisia Raf.
(Iltis, 1958, 1966)

Four or 5 species, Canada to Mexico.

1. Polanisia dodecandra (L.) DC. (clammy weed)

Cleome dodecandra L.
P. graveolens Raf.

Pl. 333 c, d; Map 1560

Plants pubescent with dense, minute, spreading glandular hairs and moderate to dense, longer, matted, often gland-tipped hairs usually with a disagreeable odor when bruised or crushed, sticky to the touch. Stems 10–60 cm tall, erect, sparsely to moderately branched. Leaves with petioles 1–6 cm long or sometimes nearly sessile. Stipules hairlike, inconspicuous, 0.2–0.9 mm long, withering quickly as the leaves mature and often visible only on very young growth. Leaf blades trifoliate, the leaflets 0.8–6.0 cm long, ovate to oblanceolate, narrowed or tapered at the base, rounded or bluntly to less commonly sharply pointed at the tip, the margins entire. Inflorescences to 20 cm long, sparsely flowered to more commonly relatively dense. Bracts 0.8–1.5 cm long, 2–5 mm wide, simple, mostly short-stalked, lanceolate-elliptic to narrowly elliptic. Buds with the petals relatively short and maturing late, not overlapping or covering the stamens. Flowers zygomorphic with unequal petals, the stalks 1.0–2.5 cm long. Sepals 2–7 mm long, 1.4–2.4 mm wide, lanceolate to obovate, fused at the base. Petals 4–10 mm long, the lower pair longer than the upper pair, spatulate, narrowed to a stalklike base 2–5 mm long, the tip shallowly and broadly notched (occasionally the shorter pair merely truncate at the tip), white with a pink to purple basal portion. Gland an oblique truncate mass, 1–2 mm long, fleshy. Stamens 8–20, the filaments 5–20 mm long, unequal in length with staggered maturation, the anthers 0.8–1.4 mm long, narrowly elliptic, attached above the base. Gynophore short at flowering (0.2–0.8 cm), not elongating as the fruit matures. Fruits strongly ascending to erect, 2–6 cm long, 4–10 mm wide, not constricted between seeds, dehiscing from the tip, the valves tending to persist and the fruit thus shedding the seeds tardily by shaking in the wind. Seeds 2.2–2.4 mm in diameter, the surface minutely and inconspicuously wrinkled or pebbled, dark brown to reddish brown. $2n=20$. May–October.

Scattered nearly throughout the state (nearly throughout the U.S.; Canada, Mexico). Banks of streams and rivers, glades, ledges and tops of bluffs, and sand prairies; also railroads, roadsides, and open, disturbed areas.

The name *Polanisia* is derived from Greek words meaning "many unequal" in reference to the stamens. The common name clammy weed refers to the stickiness associated with the glandular pubescence. Some Native American tribes ate the young plants in soups and as potherbs and also used this plant in religious ceremonies (Moerman, 1998).

This species is extremely variable in leaf, flower, and fruit size. There is a general continuum in flower size from longer in the south to shorter in the north (Iltis, 1958). Two subspecies are recognized in Missouri, based on the size of the flower parts. Although generally separable by flower size, ssp. *trachysperma* intergrades with ssp. *dodecandra* from Missouri to Minnesota (Iltis, 1966). Fortunately, the number of intermediates in Missouri is not excessive. There is considerable variation in leaf size that does not seem to correlate with flower size. In particular, some collections of *P. dodecandra* ssp. *dodecandra* have very small leaves, but others have large leaves. Fruit size is overlapping and is probably determined in part by environmental factors.

1. Petals 4–6(–7) mm long; filaments 5–9 mm long; styles 1–3 mm long; fruits 2.5–5.0 cm long ...... 1A. SSP. DODECANDRA
1. Petals 6–12 mm long; filaments 9–20 mm long; styles (4–)5–10 mm long; fruits 4–6 cm long .... 1B. SSP. TRACHYSPERMA

1a. ssp. dodecandra

Stems 10–40 cm tall. Petioles 1.0–1.5 cm long. Leaflets 0.8–4.0 cm long, 4–12 mm wide. Flower stalks 0.5–2.0 cm long, purple. Sepals 2–4 mm long, 1.4–2.4 mm wide. Petals 4–6(–7) mm long, 2–3 mm wide, the stalklike base 1.5–3.0 mm long. Stamens 6–10, the filaments 5–9 mm long, usually much longer than the petals, the anthers 0.8–1.0 mm long. Styles 1–3 mm long. Fruits 2.5–5.0 cm long, 4–6 mm wide. May–September.

Scattered nearly throughout the state (nearly throughout the U.S., but uncommon or absent from some of the southeastern states; Canada, Mexico). Banks of streams and rivers, glades, and ledges and tops of bluffs, and sand prairies; also railroads, roadsides, and open, disturbed areas.

This is the more commonly encountered and widespread subspecies in Missouri.

1b. ssp. trachysperma (Torr. & A. Gray) H.H. Iltis

P. trachysperma Torr. & A. Gray

P. dodecandra var. *trachysperma* (Torr. & A. Gray) H.H. Iltis

Stems 30–60 cm tall. Petioles 3.0–5.5 cm long. Leaflets 2–6 cm long, 8–30 mm wide. Flower stalks 1.5–2.5 cm long. Sepals 4–7 mm long, 1.2–2.0 mm wide. Petals 6–12 mm long, the stalklike base 3–5 mm long. Stamens 12–18, the filaments 9–20 mm long, often only slightly longer than the petals, the anthers 1.0–1.4 mm long. Styles (4–)5–10 mm long. Fruits 4–6 cm long, 5–11 mm wide. May–October.

Scattered, mostly south of the Missouri River (nearly throughout the U.S., but uncommon or absent from some of the southeastern and northeastern states; Canada, Mexico). Banks of streams and rivers, glades, and ledges and tops of bluffs; also railroads, roadsides, and open, disturbed areas.

CLUSIACEAE (GUTTIFERAE) (St. John's Wort Family)

Plants annual or perennial herbs or shrubs (trees and lianas elsewhere), sometimes more or less evergreen, sometimes with rhizomes, usually glabrous, the tissues with clear to yellowish to dark green or black resinous secretory cavities, these appearing as dots, lines, or streaks on stems, leaves, and often also floral parts. Leaves opposite, simple, sessile or short-petiolate. Leaf blades simple, the margins entire. Stipules absent. Inflorescences terminal and/or axillary, consisting of clusters of flowers, these often grouped into panicles, sometimes reduced to single flowers, the branch points and flowers often subtended by small, leaflike bracts. Flowers actinomorphic (except in a few *Hypericum*), perfect, hypogynous. Calyces of 4 or 5 free sepals, usually persistent at fruiting. Corollas of 4 or 5 free petals, these mostly spreading, sometimes withered but persistent at fruiting. Stamens 5 to numerous, often in groups by basal fusion of the filaments, the long, slender filaments occasionally fused to the petal bases, sometimes fused into a short ring around the ovary base, the anthers attached toward the base or more or less medially, yellow or occasionally orange. Staminodes absent (in *Hypericum*) or glandular (in *Triadenum*). Pistils of usually 2–5 fused carpels. Ovary superior, with 1–5 locules, with axile (when 2–5-locular) or parietal (when 1-locular) placentation. Styles 1–5, persistent at fruiting, the stigmas capitate or minute. Ovules numerous. Fruits capsules, the body narrowly to broadly ovoid or nearly globose, usually tapered to a stylar beak(s), sometimes appearing somewhat woody at maturity, dehiscing longitudinally. Forty-five to 50 genera, 900–1,350 species, nearly worldwide.

Some authors have treated the group of mostly temperate, herbaceous to shrubby species with perfect flowers, relatively short secretory cavities, glandular-punctate leaves, and seeds lacking arils as a separate family, Hypericaceae, restricting the Clusiaceae to a mostly tropical group of shrubs and trees with usually imperfect flowers, long secretory canals, leaves lacking glandular punctations, and seeds often enclosed in fleshy arils. Studies of comparative anatomy, especially of flower vascular patterns, have offered evidence that the entire set of genera is best treated as a single family (summarized by Wood and Adams, 1976) comprising two or three subfamilies. There is general agreement, however, that *Hypericum* and *Triadenum* are closely related and form a natural group within this family.

1. Petals yellow to orangish yellow or pale yellow; stamens 5 to more commonly numerous, the filaments free, fused basally into a short ring around the ovary base, or those of varying numbers of stamens fused basally into 3–5 sometimes indistinct groups; staminodes absent .... 1. HYPERICUM
1. Petals pink or less commonly flesh-colored; stamens 9, in 3 groups of 3, the filaments within a group noticeably fused toward the base; staminodes 3, alternating with the groups of stamens .... 2. TRIADENUM

1. Hypericum L. (St. John's wort)

Plants annual or perennial herbs or shrubs. Young stems or twigs often angled, sometimes narrowly 2-winged. Leaves sessile, the blades sometimes clasping at the base. Flowers actinomorphic (except in *H. hypericoides*). Calyces of 4 or 5 sepals, these often persistent at fruiting. Corollas of 4 or 5 petals, these yellow, less commonly orangish yellow, usually somewhat asymmetric at the tip, sometimes withered but persistent at fruiting. Stamens 5 to more commonly numerous, the filaments free, fused basally into a short ring around the ovary base, or those of varying numbers of stamens fused basally into 3–5 sometimes indistinct groups. Staminodes absent. Pistils of 2–5 fused carpels. Ovary 1-locular with parietal placentation, or completely or incompletely 3–5-locular and then with more or less axile placentation. Style 1–5, if more than 1 then the styles sometimes fused basally or closely appressed to each other most of their length, the stigmas capitate or minute. Fruits capsules, narrowly to broadly ovoid, ellipsoid, conical or nearly globose. Seeds numerous (except in *H. sphaerocarpum*), tiny, ellipsoid-cylindrical to oblong-cylindrical. Three hundred to 400 species, nearly worldwide, but most diverse in temperate regions of the Northern Hemisphere.

1. Plants shrubs, noticeably woody above the base
   2. Sepals 2 or 4, the inner pair absent or much smaller than the outer pair; petals 4, pale yellow to lemon yellow .... 6. H. HYPERICOIDES
   2. Sepals 5, all of similar size and shape; petals 5, mostly bright yellow to orangish yellow
      3. Fruits 5–7(–8) mm long, lobed longitudinally; styles mostly 4 or 5; ovaries and fruits appearing mostly 4- or 5-locular .... 7. H. LOBOCARPUM
      3. Fruits 7–14 mm long, not lobed, circular or bluntly triangular in cross-section; styles 3(4); ovaries and fruits appearing 3(4)-locular
         .... 11. H. PROLIFICUM

1. Plants annual or perennial herbs, sometimes with a somewhat woody rootstock or with stems woody at the very base
   4. Largest leaves scalelike or needlelike, 0.5–1.5 mm wide, with only a midvein visible
      5. Leaves 5–22 mm long, linear, needlelike; fruits as long as or slightly longer than the sepals at maturity . . . . 3. H. DRUMMONDII
      5. Leaves 1–4 mm long, triangular to ovate, scalelike; fruits 2–3 times as long as the sepals at maturity . . . . 4. H. GENTIANOIDES
   4. Largest leaves variously shaped, but not scalelike or needlelike, 2–20 mm wide, if linear then with (3)5 or 7 main veins from the leaf base
      6. Petals 25–30 mm long; styles 5; ovary and fruit 5-locular, the fruits 15–35 mm long; seeds winged; stems (50–)70–200 cm long . . . . 2. H. ASCYRON
      6. Petals 1.5–25.0 mm long; styles 1 or 3(4); ovary and fruit 1- or 3(4)-locular, the fruits 2–12 mm long; seeds not winged; stems 10–80 cm long
         7. Sepals and/or petals with noticeable yellowish brown to black dots, lines, and/or streaks (these sometimes difficult to observe in fruiting material); petals withered and inconspicuous but usually persistent at fruiting; fruits completely 3-locular (sometimes incompletely so at the very tip)
            8. Upper portion of the stem sharply ridged below each leaf; sepals usually with few yellowish brown to black dots; petals with black dots usually relatively few and restricted to at or near the margins . . . . 10. H. PERFORATUM
            8. Stem rounded or bluntly and inconspicuously angled below each leaf; sepals and petals with yellowish brown to black dots, lines, and/or streaks abundant and occurring irregularly across the entire surface
               9. Sepals 5–7 mm long; petals 8–12 mm long; uppermost leaves ovate to broadly triangular-ovate (those farther down the stem narrowly ovate to narrowly elliptic), mostly sharply pointed at the tip . . . . 12. H. PSEUDOMACULATUM
               9. Sepals 2.5–4.0 mm long; petals 4–8 mm long; leaves oblong-elliptic to narrowly oblong or oblanceolate, rounded to less commonly bluntly pointed at the tip . . . . 13. H. PUNCTATUM
         7. Sepals and/or petals lacking yellowish brown to black dots, lines, or streaks; petals usually shed by fruiting; fruits 1-locular or appearing partially 3-locular by intrusion of the marginal placentae into the locule
            10. Petals 5–9 mm long; stamens 45–85; styles erect, appressed to each other and sometimes fused toward the base, the stigmas minute
               11. Plants with long-creeping rhizomes bearing scattered aerial stems, the rootstocks entirely herbaceous; fruits (below the persistent styles) ellipsoid to ovoid; seeds numerous, 0.6–0.8 mm long . . . . 1. H. ADPRESSUM
               11. Plants with short, poorly developed rhizomes or these more commonly lacking, the aerial stems occurring singly or more commonly 2 or more together, the rootstocks and stem bases often somewhat woody; fruits (below the persistent styles) broadly ovoid to more or less globose; seeds 4–8 per fruit, 2.0–2.7 mm long . . . . 14. H. SPHAEROCARPUM

10. Petals 1.5–5.0 mm long; stamens 5–20; styles spreading to ascending, not appressed or fused except sometimes at the base, the stigmas capitate
   12. Fruits ellipsoid, widest at about the midpoint; sepals linear-lanceolate to narrowly oblong-lanceolate; petals 1.5–2.5 mm long; leaves rounded to bluntly pointed at the tip .... 9. H. MUTILUM
   12. Fruits ovoid or conical, widest well below the midpoint; sepals lanceolate; petals 3.5–5.0 mm long; at least the uppermost leaves sharply pointed at the tip
      13. Leaves lanceolate-triangular to ovate-triangular, broadly rounded to shallowly cordate at the base; plants annual, lacking rhizomes or the remains of previous years' growth; stamens 10–15; fruits 3–5 mm long, narrowly ovoid .... 5. H. GYMNANTHUM
      13. Leaves lanceolate to narrowly elliptic or narrowly oblong, narrowed or narrowly rounded at the base; plants perennial, sometimes with short rhizomes, the rootstock usually with the persistent remains of previous years' growth; stamens 14–20; fruits 5–7 mm long, ovoid-conical
      .... 8. H. MAJUS

1. Hypericum adpressum Raf. ex W.P.C. Barton (creeping St. John's wort)

Map 1561

Plants perennial herbs, the rootstock and stem bases occasionally somewhat spongy but not woody, with well-developed, long-creeping rhizomes. Stems scattered along the rhizomes, occurring singly, 30–80 cm long, erect or ascending, angled or slightly ridged below each leaf toward the tip, rounded toward the base, reddish brown, the surface usually not peeling with age. Leaves not jointed at the base. Leaf blades 20–80 mm long, 2–12 mm wide, narrowly oblong to narrowly elliptic, bluntly to sharply pointed at the tip, tapered or narrowed at the base, the margins somewhat rolled under at maturity, herbaceous to somewhat leathery in texture, with 3 main veins usually visible toward the base, the surfaces lacking noticeable black dots, lines, or streaks but usually with minute, faint, pale dots visible, the upper surface green, the undersurface usually pale green, but not glaucous. Inflorescences appearing as panicles of 13–80 flowers, rounded to more or less flat-topped in outline. Flowers actinomorphic. Sepals 5, all more or less similar in size and shape, 3–7 mm long, not becoming enlarged at fruiting, lanceolate to ovate, the margins occasionally slightly curled, lacking noticeable black dots, lines, or streaks. Petals 5, 6–8 mm long, broadly oblanceolate to obovate, bright yellow, usually shed before fruiting. Stamens 60–80, the filaments not fused into groups. Ovary 1-locular or appearing partially 3-locular by intrusion of the parietal placentae into the locule. Styles 3(4), sometimes fused toward the base, erect and more or less appressed at flowering, persistent and usually separating somewhat as the fruit matures, the stigmas minute. Fruits 3.5–6.0 mm long, ellipsoid to ovoid, widest at or slightly below the midpoint, tapered abruptly to the minute beak (this sometimes absent), more or less circular in cross-section. Seeds numerous, 0.6–0.8 mm long, the surface with a network of ladderlike columns of fine ridges and pits, dark brown to nearly black. $2n=18$. July–August.

Uncommon, known only from Mississippi and Scott Counties (eastern U.S. west to Illinois, Missouri, and Alabama). Moist depressions in sand prairies; also sandy banks of ditches.

Adams (1962) reported the presence of this species in southeastern Missouri, but it was overlooked by Steyermark (1963) and Yatskievych and Turner (1990). John Kartesz (Biota of North America Program) was the first to note this omission from the Missouri literature.

2. Hypericum ascyron L. ssp. **pyramidatum** (Aiton) N. Robson (great St. John's wort, giant St. John's wort)

H. pyramidatum Aiton

Pl. 361 c, d; Map 1562

Plants perennial herbs, the rootstock and stem bases often somewhat woody, sometimes with short, poorly developed rhizomes. Stems occurring singly or more commonly 2 or more together, (50–)70–200 cm long, erect or ascending, the upper portion angled or slightly ridged below each leaf, green, sometimes reddish-tinged, the surface often peeling in thin strips with age. Leaves not jointed at the base. Leaf blades 40–100 mm long, 15–45 mm wide, lanceolate to more commonly elliptic or oblong-ovate, rounded to more commonly bluntly

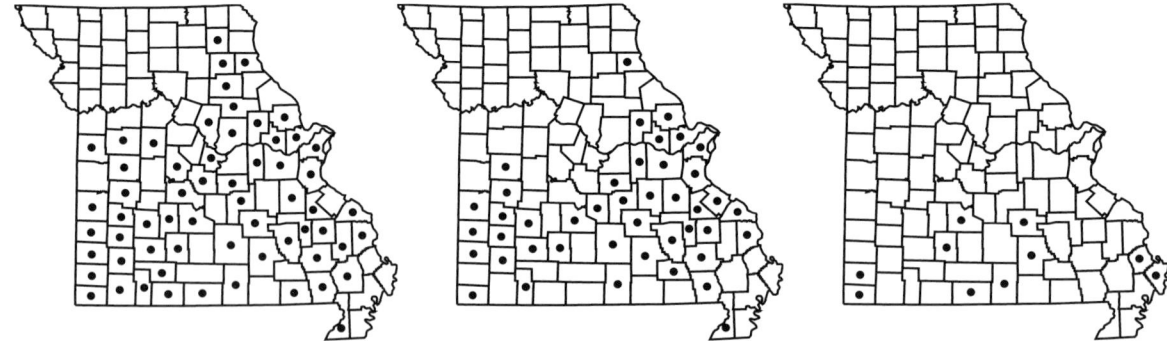

1563. Hypericum drummondii   1564. Hypericum gentianoides   1565. Hypericum gymnanthum

pointed at the tip, narrowed to more commonly rounded or shallowly cordate at the base, often somewhat clasping the stem, the margins flat, herbaceous in texture, with 1 main vein visible toward the base, the surfaces usually lacking noticeable black dots, lines, or streaks but usually with minute, faint, pale to yellowish brown dots visible, the upper surface green, sometimes somewhat glaucous, the undersurface pale green, glaucous. Inflorescences of solitary flowers at the branch tips, or sometimes appearing as pairs or small clusters of up to 5 flowers. Flowers actinomorphic. Sepals (4)5, all more or less similar in size and shape, 8–13 mm long, usually becoming slightly enlarged at fruiting, oblong-lanceolate to narrowly ovate, lacking noticeable yellowish brown or black dots, lines, or streaks. Petals (4)5, 25–35 mm long, narrowly to broadly obovate, bright yellow, withered and inconspicuous but usually persistent at fruiting. Stamens about 120–160, the filaments fused into 5 groups toward the base. Ovary 5-locular, the placentation axile. Styles 5, free above the base, erect and more or less appressed for most of their length at flowering, often persistent and usually separating as the fruit matures, the stigmas capitate. Fruits 15–35 mm long, ovoid, widest well below the midpoint, tapered to the short beak (this sometimes absent), shallowly 5-lobed in cross-section. Seeds numerous, 1.3–1.6 mm long, the surface with a narrow, longitudinal wing along one side, otherwise with a network of fine ridges and pits, often appearing inconspicuously longitudinally ribbed, brown. $2n=18$. June–September.

Uncommon and sporadic in northernmost Missouri and in counties along the western portion of the Missouri River floodplain (northeastern U.S. west to Minnesota, Nebraska, and Kansas; Canada). Bottomland forests, mesic upland forests in ravines, and bottomland prairies.

The taxonomic treatment of the *H. ascyron* complex has been controversial. Gillett and Robson (1981), in their study of the Canadian Clusiaceae, concluded that the characters supposedly separating North American plants (*H. pyramidatum*) from Old World populations (*H. ascyron* L.) were too variable to allow formal segregation of two taxa and recognized a single species under the latter name. Some botanists (Voss, 1985; Kartesz and Meacham, 1999) followed their suggestion, but adoption of this classification has not been universal (Kaul, 1986; Cooperrider, 1989; Gleason and Cronquist, 1991). Most recently, Robson (2001) regarded the North American plants as a subspecies of the otherwise Asian *H. ascyron,* separating it from ssp. *ascyron* based upon its sharply pointed (vs. rounded to bluntly pointed) sepals and nearly free (vs. more or less fused) styles. Another Asian taxon, ssp. *gebleri* (Ledeb.) N. Robson, is characterized by its more slender fruits, shorter sepals, and narrower leaves with more-angled bases. Robson's treatment is followed here.

This showy species deserves to be cultivated more widely as an ornamental in gardens. Curiously, large populations are seldom encountered anywhere in its range. Gillett and Robson (1981) quoted the Canadian botanist, W. G. Dore, who speculated that the distributional pattern of sporadic small populations might have resulted from dispersal of plants by Native Americans, who may have eaten the immature fruits.

3. Hypericum drummondii (Grev. & Hook.) Torr. & A. Gray (nits and lice)
Sarothra drummondii Grev. & Hook.

Pl. 360 d–f; Map 1563

Plants annuals, with taproots, usually with numerous ascending branches. Stems 10–40(–80) cm long, erect, angled or slightly ridged below each leaf toward the tip, with minute, yellowish brown to dark green or black resinous dots, green to reddish brown, sometimes turning orangish brown with age, sometimes peeling in thin strips with age. Leaves not jointed at the base, strongly ascending. Leaf blades 5–22 mm long, 0.5–1.5 mm wide, linear, needlelike, bluntly to sharply pointed at the tip, tapered or narrowed at the base, the margins somewhat rolled under at maturity, somewhat leathery in texture, with 1 main vein visible, the surfaces with minute, yellowish brown to dark green or black resinous dots, the upper surface green, the undersurface usually somewhat paler, but not glaucous. Inflorescences of mostly solitary flowers in the leaf axils, sometimes with small, loose clusters of 3 or 5 flowers at the branch tips. Flowers actinomorphic. Sepals 5, all more or less similar in size and shape, 3–7 mm long, not becoming enlarged at fruiting, narrowly lanceolate to narrowly oblong-lanceolate, with minute, yellowish brown to black dots, lines, and/or streaks. Petals 5, 4–7 mm long, oblong-obovate, orangish yellow, withered and inconspicuous but usually persistent at fruiting. Stamens 10–22, the filaments sometimes irregularly spaced but usually not fused into groups. Ovary 1-locular, with parietal placentation. Styles 3, free above the base, more or less spreading, the stigmas capitate. Fruits 4.5–7.0 mm long, at maturity as long as or slightly longer than the sepals, narrowly ovoid, widest slightly below the midpoint, tapered to the persistent styles, more or less circular in cross-section. Seeds numerous, 0.9–1.1 mm long, the surface with a coarse network of ridges and pits, light brown to dark brown. $2n=24$. June–September.

Scattered nearly throughout the southern half of the state and in the eastern portion of the Glaciated Plains Division (eastern U.S. west to Iowa, Kansas, and Texas). Glades, upland prairies, ledges and tops of bluffs, openings of dry upland forest, savannas, and less commonly banks of streams; also fallow fields and old fields, on acidic substrates.

4. Hypericum gentianoides (L.) Britton, Sterns & Poggenb. (pineweed, orange grass)

Sarothra gentianoides L.

Pl. 360 a–c; Map 1564

Plants annuals, with taproots, usually with numerous ascending branches. Stems 8–50 cm long, erect, angled or slightly ridged below each leaf toward the tip, green to light brown, turning orange to orangish brown with age, sometimes peeling in thin strips with age. Leaves not jointed at the base, strongly ascending. Leaf blades 1–4 mm long, 0.5–1.0 mm wide, triangular to ovate, scalelike, rounded to bluntly or less commonly sharply pointed at the tip, truncate or broadly angled at the base, the margins somewhat rolled under at maturity, papery to somewhat leathery in texture, with 1 main vein visible, the surfaces with inconspicuous, minute, yellowish brown to dark green or black resinous dots (usually in a single row on each side of the midvein), the upper surface green, the undersurface green, not glaucous. Inflorescences of mostly solitary flowers in the leaf axils, sometimes with small, loose clusters of 3 or 5 flowers at the branch tips. Flowers actinomorphic. Sepals 5, all more or less similar in size and shape, 1.5–2.5 mm long, not becoming enlarged at fruiting, narrowly lanceolate to narrowly oblong-lanceolate, usually with inconspicuous, minute, yellowish brown to black dots, lines, and/or streaks. Petals 5, 2–4 mm long, narrowly oblong, orangish yellow to yellow, sometimes reddish-tinged, withered and inconspicuous but usually persistent at fruiting. Stamens 5–11, the filaments sometimes irregularly spaced but usually not fused into groups. Ovary 1-locular, with parietal placentation. Styles 3, free above the base, more or less spreading, the stigmas capitate. Fruits 4–7 mm long, at maturity 2–3 times as long as the sepals, narrowly conical to nearly cylindrical, widest near the base, tapered to the persistent styles, triangular in cross-section. Seeds numerous, 0.4–0.8 mm long, the surface with a coarse network of ridges and pits, sometimes appearing longitudinally ribbed, light brown to dark brown. $2n=24$. June–September.

Scattered, mostly south of the Missouri River (eastern U.S. and adjacent Canada west to Minnesota and Texas; apparently introduced in South America, Caribbean Islands, and Europe). Glades, upland prairies, ledges and tops of bluffs, openings of dry upland forest, and savannas; also old fields, on acidic substrates.

5. Hypericum gymnanthum Engelm. & A. Gray (clasping-leaved St. John's wort, small St. John's wort)

Pl. 360 g–i; Map 1565

Plants annuals, with taproots. Stems 15–70 cm long, erect, rounded or inconspicuously angled or ridged below each leaf, green. Leaves not jointed at the base, spreading to loosely ascending. Leaf blades 5–30 mm long, 3–12 mm wide, lanceolate-triangular to ovate-triangular, rounded or bluntly or occasionally sharply pointed at the tip, broadly rounded to shallowly cordate at the base, often somewhat clasping the stem, the margins flat, herbaceous to slightly leathery in texture, with mostly 5 or 7 main veins from the base, the surfaces with

1566. Hypericum hypericoides   1567. Hypericum lobocarpum   1568. Hypericum majus

inconspicuous, minute, yellowish brown to dark green or black resinous dots, the upper surface green, the undersurface somewhat paler, not glaucous. Inflorescences appearing as panicles of 5–50 flowers, usually more or less flat-topped in outline. Flowers actinomorphic. Sepals 5, all more or less similar in size and shape, 3.5–5.0 mm long, usually becoming slightly enlarged at fruiting, lanceolate, lacking noticeable yellowish brown or black dots, lines, and/or streaks. Petals 5, 3–4 mm long, oblanceolate, bright yellow to lemon yellow, lacking noticeable yellowish brown or black dots, lines, and/or streaks, usually shed by fruiting. Stamens 10–15, the filaments sometimes irregularly spaced but usually not fused into groups. Ovary 1-locular or appearing partially 3-locular by intrusion of the parietal placentae into the locule. Styles 3, free above the base, more or less spreading, the stigmas narrowly capitate. Fruits 3–5 mm long, at maturity about as long as the sepals, narrowly ovoid, widest well below the midpoint, tapered to the persistent styles, more or less circular in cross-section. Seeds numerous, 0.5–0.6 mm long, the surface with a faint network of ridges and pits, appearing inconspicuously longitudinally ribbed to nearly smooth, light brown to brown. $2n=16$. June–September.

Uncommon and widely scattered in the southern third of the state (eastern U.S. west to Kansas and Texas; disjunct in Guatemala, introduced in Europe). Margins of ponds and sinkhole ponds, banks of streams, and upland prairies and sand prairies, usually in moist depressions; also pastures, fallow fields, and ditches.

Steyermark (1963), Robson (1990), and others have noted the presence of putative hybrids between *H. gymnanthum* and *H. mutilum* where these two species occur together. Such hybrids have been recorded thus far from Shannon and Texas Counties and are to be expected nearly everywhere that *H. gymnanthum* is found in Missouri. Hybrid plants, which are difficult to distinguish from the parents, tend to have smaller flowers with less-tapered sepals than those of *H. gymnanthum* and less strongly triangular leaves.

6. Hypericum hypericoides (L.) Crantz (St. Andrew's cross)

Ascyrum hypericoides L.

Pl. 361 i, j; Map 1566

Plants shrubs to 10–80(–150) cm tall. Stems spreading or ascending. Bark reddish brown, smooth, peeling in thin strips or flakes. Twigs angled or ridged below each leaf, reddish brown, the older stems usually angled below each leaf. Leaves obscurely jointed just above the base (the bases frequently persistent after the leaves have been shed), more or less evergreen. Leaf blades 5–35 mm long, 2–8 mm wide, oblanceolate to narrowly oblong or linear, rounded to bluntly pointed at the tip, tapered or narrowed at the base (sometimes with a pair of minute, rounded lobes at the jointed base), the margins often slightly rolled under at maturity, herbaceous to somewhat leathery in texture, with 1 main vein visible (sometimes difficult to observe), the surfaces with minute resinous dots, these yellowish brown to dark green or nearly black, the upper surface green, the undersurface slightly paler, not glaucous. Inflorescences usually of solitary flowers at the branch tips and upper leaf axils, occasionally with small clusters of 3 flowers at the branch tips. Flowers somewhat zygomorphic. Sepals 2 or 4, usually with minute, yellowish brown to dark green or black resinous dots, the outer pair 5.0–12.5 mm long, becoming somewhat enlarged at fruiting, broadly ovate to narrowly elliptic, rounded to shallowly cordate at the base; the inner pair absent or 1–4 mm long, lanceolate to narrowly lanceolate. Petals 4, grouped into 2 pairs, 6–11 mm long, narrowly oblong-elliptic to narrowly obovate

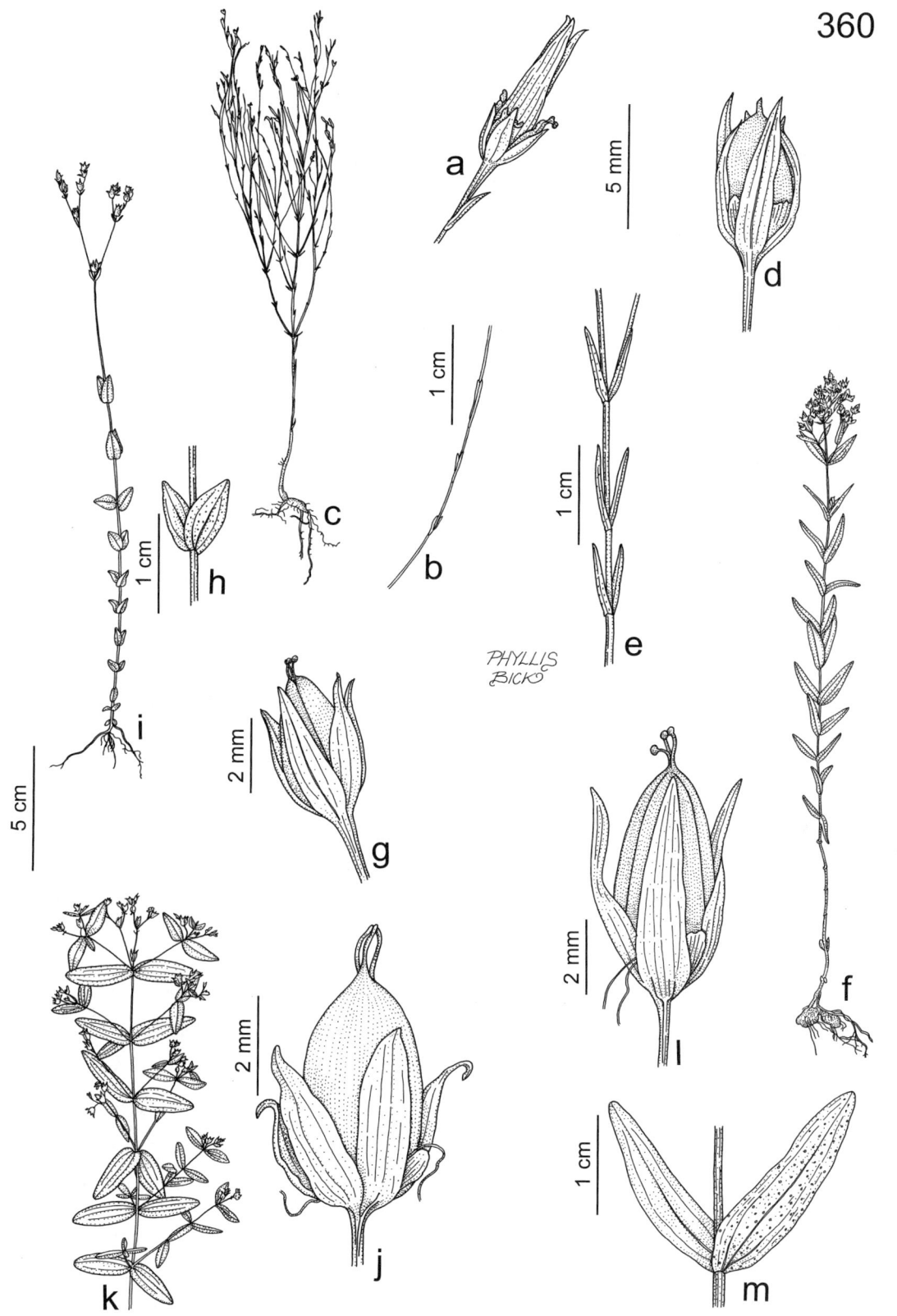

Plate 360. Clusiaceae. *Hypericum gentianoides*, **a)** fruit, **b)** leaves, **c)** habit. *Hypericum drummondii*, **d)** fruit, **e)** leaves, **f)** habit. *Hypericum gymnanthum*, **g)** fruit, **h)** leaves, **i)** habit. *Hypericum mutilum*, **j)** fruit, **k)** habit. *Hypericum majus*, **l)** fruit, **m)** leaves.

or less commonly obovate, pale yellow to lemon yellow, shed before fruiting. Stamens 35–50, the filaments not fused into groups. Ovary from 2 carpels but 1-locular, the placentation parietal. Styles 2, free above the base, somewhat spreading or curved outward, the stigmas minute. Fruits 5–9 mm long, elliptic-ovoid to oblong-elliptic-ovoid, widest at or slightly below the midpoint, tapered abruptly to the short beaks, flattened and thus narrowly elliptic in cross-section. Seeds numerous, 0.6–0.8 mm long, the surface with fine longitudinal lines, these sometimes forming a faint network, dark brown to purplish brown. $2n=18$. July–October.

Scattered in the Ozark, Ozark Border, Glaciated Plains, and Mississippi Lowlands Divisions (eastern U.S. west to Kansas and Texas; Mexico, Central America, Caribbean Islands). Bottomland forests, mesic to dry upland forests, banks of streams and rivers, and occasionally margins of ponds and lakes; also roadsides and open, disturbed areas, usually on acidic substrates.

The flowers of St. Andrew's cross often are neither particularly abundant nor very showy, and the species is thus sometimes overlooked in the field. Steyermark (1963) noted that the species is browsed by deer. He also suggested that it makes a desirable cultivated plant, particularly in rock gardens, but the species apparently is not winter-hardy in the northern half of the state.

Hypericum hypericoides and four or five closely related species sometimes have been segregated into the genus *Ascyrum* L. (Steyermark, 1963; Kaul, 1986), based primarily on their flowers with two pairs of unequal sepals and the four petals strongly spreading and grouped into two pairs. Most botanists currently believe that this group is best treated as a section or subsection within the classification of the large and morphologically diverse *Hypericum* (Adams and Robson, 1961; Robson, 1996). Circumscription of the *H. hypericoides* complex also has been controversial. The two morphological entities in Missouri are relatively easily distinguished and tend to maintain their leaf shape and growth-form characteristics, even when growing side by side. They have been treated variously as separate varieties (Adams, 1959; Steyermark, 1963; Cooperrider, 1989), subspecies (Robson, 1980, 1996), or species (Adams and Robson, 1961; Adams, 1962; Wood and Adams, 1976; Gleason and Cronquist, 1991). A third variant endemic to the Dominican Republic has been described as ssp. *prostratum* N. Robson and differs in details of its strongly prostrate stems, stem-branching pattern, and generally smaller leaves. Because the Caribbean taxon has only been described at the subspecies level, the present treatment of the complex follows the most recent taxonomic revision of the group (Robson, 1996) in recognizing the Missouri taxa as subspecies.

1. Plants 30–80(–150) cm tall, the usually single main stem erect or ascending; leaf blades linear to narrowly oblong-elliptic
 .... 6A. SSP. HYPERICOIDES
1. Plants 10–20(–30) cm tall, the usually several main stems spreading or loosely ascending; leaf blades oblanceolate
 .... 6B. SSP. MULTICAULE

6a. ssp. hypericoides

Plants 30–80(–150) cm tall, usually not branching at the base (the numerous spreading to ascending branches originating well above ground level), the single main stem erect or ascending, not rooting at the nodes. Leaf blades linear to narrowly oblong-elliptic, mostly widest at about the middle. $2n=18$. July–October.

Scattered in the Mississippi Lowlands Division and the southern portion of the Ozarks (southeastern U.S. west to Missouri and Texas; Mexico, Central America, Caribbean Islands). Bottomland forests and banks of streams; also roadsides and open, disturbed areas, often in sandy soils.

6b. ssp. multicaule (Michx. ex Willd.) N. Robson

H. hypericoides var. *multicaule* (Michx. ex Willd.) Fosberg

Ascyrum hypericoides var. *multicaule* (Michx. ex Willd.) Fernald

H. stragulum W.P. Adams & N. Robson

Plants 10–20(–30) cm tall, few- to several-branched at the base (also well branched above the base, the branches frequently ascending), the main stems spreading or loosely ascending, frequently rooting at the nodes. Leaf blades oblanceolate, widest above the middle. $2n=18$. July–October.

Scattered in the Ozark, Ozark Border, Glaciated Plains, and Mississippi Lowlands Divisions (eastern U.S. west to Kansas and Texas). Bottomland forests, mesic to dry upland forests, banks of streams and rivers, and occasionally margins of ponds and lakes; also roadsides, usually on acidic substrates.

7. Hypericum lobocarpum Gatt.

H. densiflorum Pursh var. *lobocarpum* (Gatt.) Svenson

Pl. 362 g, h; Map 1567

Plants shrubs, sometimes woody only toward

the base, to 1.5 m tall. Stems erect or ascending. Bark reddish brown, smooth, usually peeling in thin strips. Twigs angled or slightly ridged below each leaf, reddish brown, the older stems usually rounded but often with longitudinal lines below the leaves. Leaves jointed at the base. Leaf blades 15–50 mm long, 3–12 mm wide, narrowly oblong to oblanceolate or less commonly linear, rounded to bluntly pointed at the tip (often with a minute, sharp point), tapered or narrowed to a petiole-like base, the margins rolled under at maturity, herbaceous to somewhat leathery in texture, with 1 main vein sometimes faintly visible, the surfaces lacking noticeable black dots, lines, or streaks but usually with minute, faint, pale dots visible, the upper surface green, the undersurface pale green and often somewhat glaucous. Inflorescences clusters of mostly 3–17 flowers, at the branch tips and from the axils of the uppermost leaves, appearing paniculate. Flowers actinomorphic. Sepals 5, all similar in size and shape, 3.5–4.5 mm long, not becoming enlarged at fruiting, narrowly elliptic to narrowly oblong-oblanceolate, the margins often somewhat curled, lacking noticeable yellowish brown or black dots, lines, or streaks. Petals 5, 6–8 mm long, oblanceolate to obovate, yellow to orangish yellow, usually shed before fruiting. Stamens 100–150, the filaments not fused into groups. Ovary incompletely mostly 4- or 5-locular, the placentation more or less axile. Styles mostly 4 or 5 (some 3-styled flowers present in Missouri material), free above the base, but erect and more or less appressed at flowering, persistent and usually separating somewhat as the fruit matures, the stigmas minute. Fruits 5–7(–8) mm long, narrowly ovoid or somewhat conical, widest below the midpoint, tapered to the beak(s), noticeably lobed longitudinally and appearing mostly 4- or 5-lobed in cross-section. Seeds numerous, 1.2–1.5 mm long, the surface with fine longitudinal lines, these sometimes forming a faint network, dark brown to nearly black. $2n=18$. June–September.

Uncommon, known only from historical collections from Howell and Ripley Counties (Illinois, Kentucky, and Tennessee west to Oklahoma and Texas). Openings of sandy upland forests and fens.

The situation with this taxon in Missouri is complicated. Steyermark (1963) reported both *H. lobocarpum* (as *H. densiflorum* var. *lobocarpum*) and the closely related *H. densiflorum* for Missouri. However, *H. densiflorum*, which differs in its mostly 3-locular ovaries, unlobed capsules, and details of sepal venation, occurs only to the southeast and east of Missouri. Nowhere does its range overlap with that of *H. lobocarpum* (Adams, 1962; Robson, 1996). Steyermark noted, however, that Preston Adams, who was then preparing a monograph of the group (Adams, 1962), had suggested that the single collection attributable to *H. densiflorum* was in fact *H. lobocarpum*. *Hypericum densiflorum* subsequently was excluded from the flora by Yatskievych and Turner (1990), as well as in the present treatment. Adams, however, further noted in his annotations of the Missouri specimens that although they were closest to *H. lobocarpum*, they possibly represented hybrids between that species and *H. prolificum*. Adams (1972) discussed hybridization between *H. lobocarpum* and *H. prolificum*, for which the binomial *H.* ×*dawsonianum* Rehder was published describing hybrids that arose spontaneously from adjacent plantings of the two parental taxa at the Arnold Arboretum. The Missouri specimens are intermediate insofar as they possess some 3-styled flowers in each inflorescence (a character of *H. prolificum* and *H. densiflorum*), but they otherwise resemble *H. lobocarpum*. The status of *H. lobocarpum* in Missouri thus requires further study to resolve whether the species is truly a member of the Missouri flora or whether the few historical collections represent isolated instances of hybrids spreading into southernmost Missouri from adjacent mixed parental populations in Arkansas. A historical specimen from Dunklin County reported by Steyermark (1963) could not be located during the present study and was not mapped by Adams (1962) or Robson (1996).

8. Hypericum majus (A. Gray) Rusby (greater St. John's wort)

H. canadense L. var. *majus* A. Gray

Pl. 360 l, m; Map 1568

Plants perennial herbs, sometimes with short rhizomes. Stems 10–60 cm long, erect, angled or narrowly ridged below each leaf, at least toward the tip, green, sometimes purplish-tinged. Leaves not jointed at the base, ascending or loosely ascending. Leaf blades 10–45 mm long, 3–10 mm wide, lanceolate to narrowly elliptic or narrowly oblong, rounded or bluntly to more commonly (in at least the uppermost leaves) sharply pointed at the tip, narrowed or narrowly rounded at the base, the uppermost leaves sometimes slightly clasping the stem, the margins flat, herbaceous to slightly leathery in texture, with mostly 5 or 7 main veins from the base, the surfaces with inconspicuous, minute, yellowish brown to dark green or black resinous dots, the upper surface green, the undersurface green, not glaucous. Inflorescences appearing as panicles of 5–30 flowers, rounded to more or less flat-topped in outline. Flowers actinomorphic. Se-

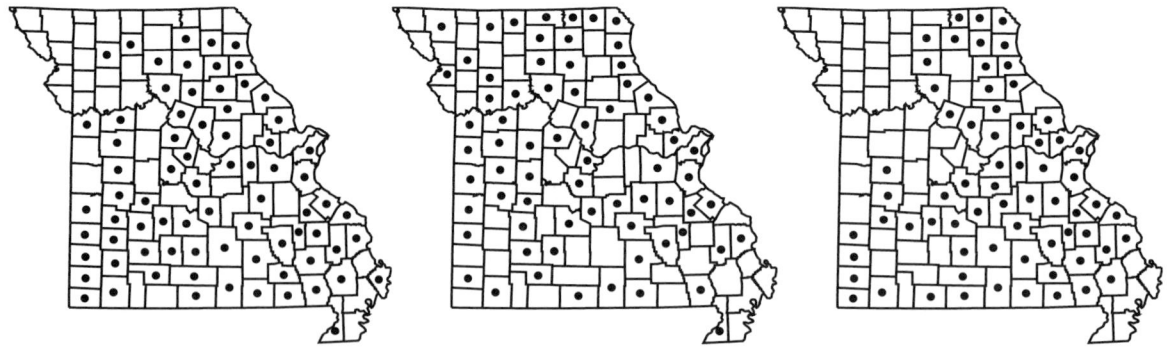

1569. Hypericum mutilum   1570. Hypericum perforatum   1571. Hypericum prolificum

pals 5, all more or less similar in size and shape, 4.0–6.5 mm long, usually becoming slightly enlarged at fruiting, lanceolate, lacking noticeable yellowish brown or black dots, lines, and/or streaks. Petals 5, 3.5–5.5 mm long, oblanceolate, bright yellow, occasionally with reddish veins, lacking noticeable yellowish brown or black dots, lines, and/or streaks, usually shed by fruiting. Stamens 14–20, the filaments sometimes irregularly spaced but usually not fused into groups. Ovary 1-locular or appearing partially 3-locular by intrusion of the parietal placentae into the locule. Styles 3, free above the base, more or less spreading, the stigmas narrowly capitate. Fruits 5–7 mm long, at maturity about as long as the sepals, ovoid-conical, widest well below the midpoint, tapered to the persistent styles, more or less circular in cross-section, usually maroon at maturity. Seeds numerous, 0.5–0.7 mm long, the surface with a network of fine to coarse ridges and pits, appearing longitudinally ribbed, light brown to brown. $2n=16$. July–September.

Uncommon, known thus far only from a single specimen from Sullivan County (northern U.S. south to Pennsylvania, Oklahoma, and Washington; Canada). Open slopes in ravines.

As noted by Yatskievych (1990), Steyermark (1963) incorrectly determined the Missouri plants as *H. canadense*, a closely related species that occurs only to the north and east of the state. It is most easily distinguished from *H. majus* by its 1- or 3-nerved, linear to narrowly oblanceolate leaves that are tapered at the base.

9. Hypericum mutilum L. ssp. mutilum
(dwarf St. John's wort, small-flowered St. John's wort)
H. mutilum var. *parviflorum* (Willd.) Fernald
Pl. 360 j, k; Map 1569

Plants annuals or short-lived perennials, with fibrous roots, sometimes with a rhizomelike creeping basal portion of the stem, this rooting at the nodes. Stems 10–90 cm long, loosely ascending to erect, rounded or inconspicuously angled or ridged below each leaf, green. Leaves not jointed at the base, mostly loosely ascending. Leaf blades 4–40 mm long, 1–15 mm wide, lanceolate to ovate, elliptic, or rarely nearly circular, the lowermost leaves sometimes varying to obovate, rounded or bluntly pointed at the tip, broadly angled to rounded or shallowly cordate at the base, often somewhat clasping the stem, the margins flat, membranous to herbaceous in texture, with mostly 3 or 5 main veins from the base, the surfaces with inconspicuous, minute, yellowish brown to dark green or black resinous dots, the upper surface green, the undersurface somewhat paler, not glaucous. Inflorescences appearing as panicles of 5–60 flowers, usually more or less rounded in outline, sometimes reduced to small clusters of flowers at the branch tips. Flowers actinomorphic. Sepals 5, all more or less similar in size and shape, 2–4 mm long, usually becoming slightly enlarged at fruiting, linear-lanceolate to narrowly oblong-lanceolate, lacking noticeable yellowish brown or black dots, lines, and/or streaks. Petals 5, 1.5–2.5 mm long, narrowly oblong to oblong-oblanceolate, bright yellow to lemon yellow, lacking noticeable yellowish brown or black dots, lines, and/or streaks, usually shed by fruiting. Stamens 5–16, the filaments sometimes irregularly spaced but usually not fused into groups. Ovary 1-locular, with parietal placentation. Styles 3, free above the base, more or less spreading, the stigmas capitate. Fruits 2–4 mm long, at maturity usually slightly longer than the sepals, ellipsoid, widest at about the midpoint, narrowed or short-tapered to the persistent styles, more or less circular in cross-section. Seeds numerous, 0.4–0.7 mm long, the surface with a faint network of fine ridges and pits, appearing inconspicuously longitudinally

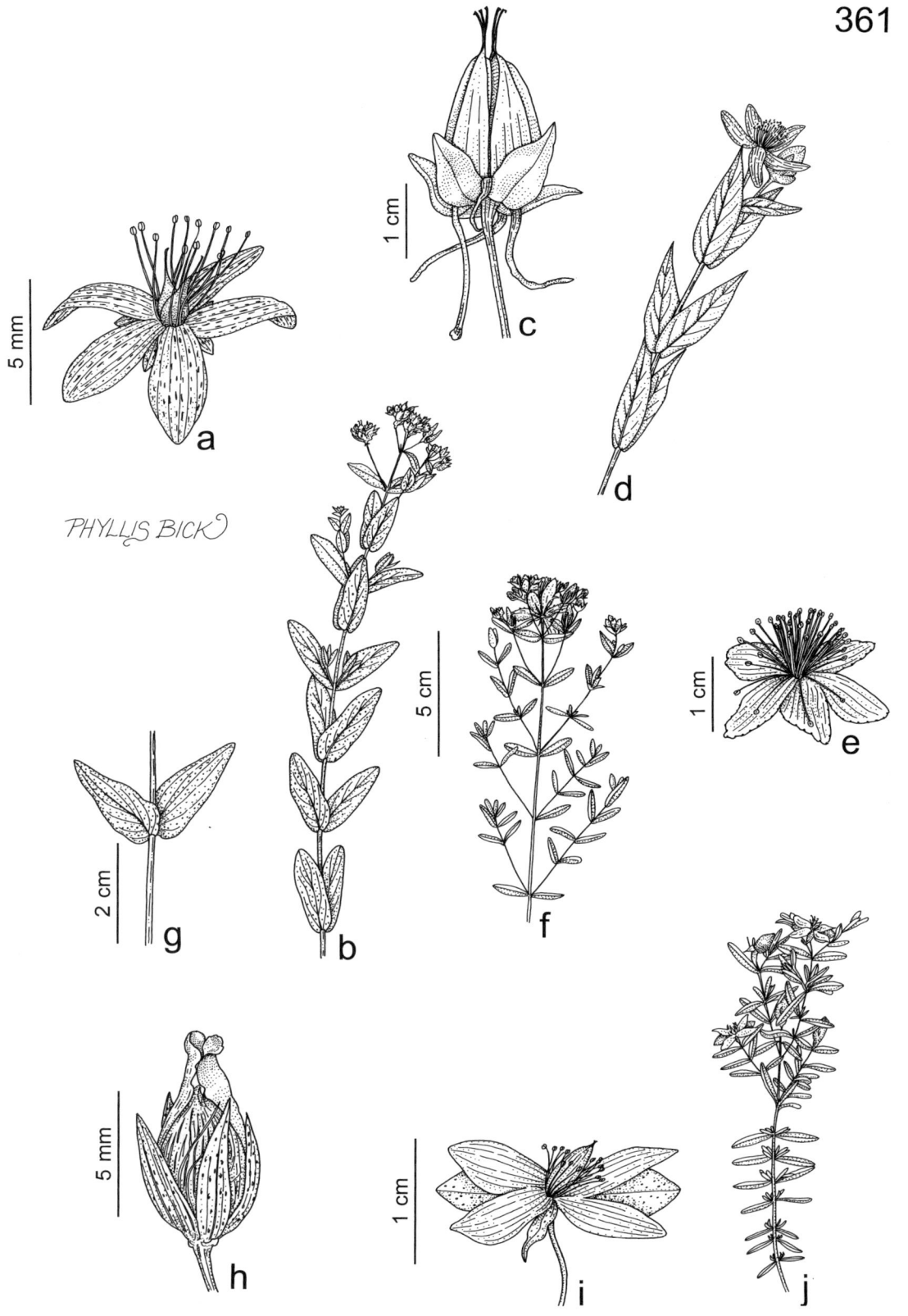

Plate 361. Clusiaceae. *Hypericum punctatum*, **a)** flower, **b)** habit. *Hypericum ascyron*, **c)** fruit, **d)** flowering branch. *Hypericum perforatum*, **e)** flower, **f)** habit. *Hypericum pseudomaculatum*, **g)** leaves, **h)** fruit. *Hypericum hypericoides*, **i)** flower, **j)** habit.

ribbed to nearly smooth, light brown to brown. 2n=16. July–October.

Scattered nearly throughout the state (eastern U.S. west to Minnesota and Texas; Canada, Mexico, Central America, South America, Caribbean Islands; introduced in western North America, Hawaii, Europe, and New Zealand). Banks of streams, rivers, sloughs, and spring branches, margins of ponds, lakes, and sinkhole ponds, fens, acid seeps, bottomland forests, swamps, moist depressions of glades and upland prairies, and rarely openings of mesic upland forests; also ditches, fallow fields, and moist, open, disturbed areas.

Robson (1990) treated *H. mutilum* as comprising three overlapping subspecies, with only ssp. *mutilum* present in Missouri. The ssp. *boreale* (Britton) J.M. Gillett, which has often been treated as a separate species (*H. boreale* (Britton) E.P. Bickn.) in the botanical literature (Gleason and Cronquist, 1991), is characterized by broad, leaflike (vs. linear and inconspicuous) bracts in the inflorescence. It occurs only to the north and east of Missouri. A taxon mostly confined to the southeastern Coastal Plain, ssp. *latisepalum* (Fernald) N. Robson, differs from ssp. *mutilum* in its oblanceolate sepals and in its stems with a more elongate uppermost internode. Fernald (1950), Steyermark (1963), and some other authors separated var. *parviflorum,* based on subtle differences in leaf shape, but Webb (1980) found that this character was too variable and too inconsistent in its distribution to allow formal taxonomic recognition of this minor variant.

For a discussion of putative hybrids between *H. mutilum* and *H. gymnanthum,* see the treatment of that species.

10. Hypericum perforatum L. (common St. John's wort, Klamath weed)

Pl. 361 e, f; Map 1570

Plants perennial herbs, the rootstock and stem bases often somewhat woody, often with short to long rhizomes. Stems occurring singly or less commonly 2 or more together, 25–75(–100) cm long, erect or ascending, the upper portion sharply ridged below each leaf, green to reddish brown, the surface often peeling in thin strips with age. Leaves not jointed at the base. Leaf blades 4–20(–30) mm long, 2–8 mm wide, linear to narrowly oblong-elliptic, rounded to bluntly pointed or occasionally shallowly notched at the tip, narrowed or rounded at the base, the margins flat or somewhat rolled under at maturity, herbaceous in texture, with 1 or less commonly 3 main veins usually visible toward the base, the surfaces with few to many variously noticeable or inconspicuous yellowish brown to dark green or black dots or less commonly lines, or streaks, the upper surface green, the undersurface paler but only rarely slightly glaucous. Inflorescences appearing as panicles of 30–200 flowers, rounded to more or less flat-topped in outline. Flowers actinomorphic. Sepals 5, all more or less similar in size and shape, 3–6 mm long, not becoming enlarged at fruiting, lanceolate to narrowly lanceolate, the margins flat, with usually few noticeable black dots, lines, or streaks (additional inconspicuous yellowish brown or green dots, lines, or streaks may be present), these sometimes difficult to observe in fruiting material. Petals 5, 7–12 mm long, oblanceolate to obovate or elliptic, bright yellow, with yellowish brown or black dots present but usually relatively few and restricted to at or near the margins, withered and inconspicuous but usually persistent at fruiting. Stamens 45–85, the filaments fused into 3 groups toward the base. Ovary completely 3-locular (sometimes incompletely so at the very tip), the placentation axile. Styles 3, free above the base, more or less spreading, persistent, the stigmas capitate. Fruits 3.5–7.0 mm long, ovoid, widest below the midpoint, tapered to the minute beaks, bluntly triangular to more or less circular in cross-section. Seeds numerous, 0.9–1.2 mm long, the surface with a coarse network of ridges and pits, dark brown to black. 2n=32. May–September.

Introduced, scattered nearly throughout the state (native of Europe; introduced nearly throughout the U.S. and adjacent Canada). Glades, upland prairies, and occasionally margins of mesic upland forests; also fallow fields, old fields, pastures, levees, ditches, roadsides, railroads, and open, disturbed areas.

This species has considerable economic importance. Although in Missouri *H. perforatum* mainly colonizes roadsides and other disturbed areas, in the western states, especially in northern California, it is a major weed of rangelands. The plants contain a phototoxic red-fluorescing pigment, hypericin (as well as pseudohypericin and their precursors), which, when ingested, causes livestock to develop swollen muzzles and ears, as well as sores on their bodies, following exposure to sunlight. These pigments account for the dark color of the punctations on the surface of vegetative and/or floral parts in most species of *Hypericum* (Burrows and Tyrl, 2001). European chrysolinid beetles that forage on Klamath weed have proven to be a fairly successful biological control for dense infestations of the species (Wood and Adams, 1976). *Hypericum perforatum* is also one of the biggest success sto-

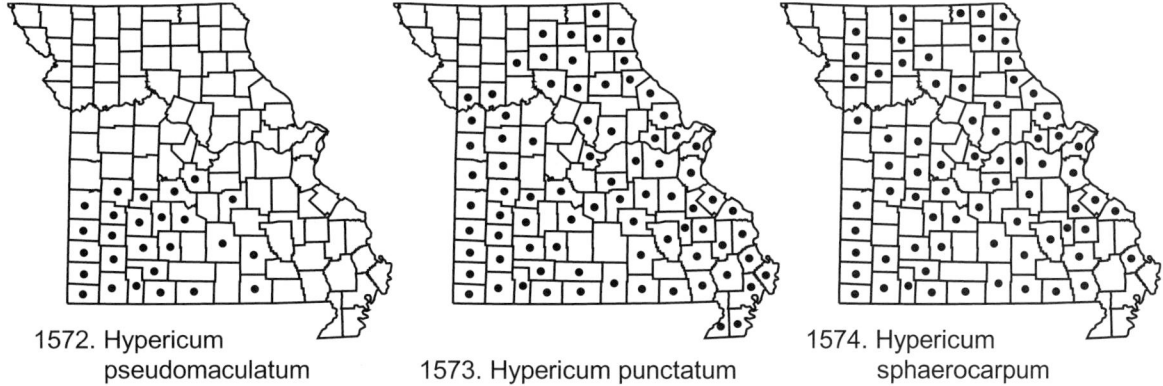

1572. Hypericum pseudomaculatum   1573. Hypericum punctatum   1574. Hypericum sphaerocarpum

ries in modern herbal medicine. The species had a long history of use medicinally as an astringent, diuretic, and sedative to combat mood swings and sleep disorders. Clinical studies have shown that the active ingredient, hypericin, is effective in treating mild to moderate depression without the potential side effects of other mood-altering drugs (Burrows and Tyrl, 2001).

11. Hypericum prolificum L. (shrubby St. John's wort)

H. spathulatum (Spach) Steud.

Pl. 362 c, d; Map 1571

Plants shrubs to 2 m tall. Stems ascending. Bark gray, smooth, usually peeling in thin strips or flakes, with an orange underlayer. Twigs angled or slightly ridged below each leaf, reddish brown, the older stems usually rounded but often with longitudinal lines below the leaves, tan to gray. Leaves jointed at the base. Leaf blades 10–75 mm long, 4–15 mm wide, narrowly oblong to narrowly elliptic-oblanceolate, rounded to bluntly pointed at the tip (rarely with a minute, sharp point or a small notch), tapered or narrowed to a petiole-like base, the margins often somewhat rolled under at maturity, herbaceous to somewhat leathery in texture, with 1 main vein visible, the surfaces lacking noticeable yellowish brown or black dots, lines, or streaks but occasionally with minute, faint, pale dots visible, the upper surface green, the undersurface pale green and often somewhat glaucous. Inflorescences small clusters of mostly 3–7 flowers, often with an additional pair of flowers in the axils of the uppermost leaves. Flowers actinomorphic. Sepals 5, all similar in size and shape, 4–8 mm long, becoming somewhat enlarged at fruiting, obovate to oblanceolate or broadly to narrowly elliptic, lacking noticeable yellowish brown or black dots, lines, or streaks. Petals 5, 7–15 mm long, broadly oblanceolate to obovate, yellow to orangish yellow, usually shed before fruiting. Stamens 150–500, the filaments not fused into groups. Ovary incompletely 3(4)-locular, the placentation more or less axile. Styles 3(4), free above the base, but erect and more or less appressed at flowering, persistent and usually separating somewhat as the fruit matures, the stigmas minute. Fruits 7–14 mm long, ovoid to elliptic-ovoid or somewhat conical, widest below the midpoint, tapered to the beak, not lobed, circular to more commonly bluntly triangular in cross-section. Seeds numerous, 1.5–2.0 mm long, the surface with a network of fine ridges and pits, appearing inconspicuously longitudinally ribbed, dark brown to nearly black. $2n=18$. June–September.

Scattered nearly throughout the state, but apparently absent from portions of the Mississippi Lowlands and Unglaciated Plains Divisions and the western portion of the Glaciated Plains (eastern U.S. and adjacent Canada west to Minnesota and Texas). Mesic to dry upland forests, margins of glades, banks of streams and rivers, ledges, tops, and bases of bluffs, and rarely bottomland forests; also fallow fields and old fields.

Hypericum prolificum is quite variable in its vegetative features. Difficulties in the interpretation of the fragmentary type specimen of this epithet in the Linnaean herbarium caused some earlier authors to adopt the later name *H. spathulatum* for the species (Steyermark, 1963), but Adams (1959, 1962) established the correct nomenclature for this taxon. Shrubby St. John's wort has showy flowers and is becoming more widely cultivated as an ornamental.

12. Hypericum pseudomaculatum Bush ex Britton (spotted St. John's wort)

H. punctatum Lam. var. *pseudomaculatum* (Bush ex Britton) Fernald

H. punctatum f. *flavidum* Steyerm.

Pl. 361 g, h; Map 1572

Plants perennial herbs, the rootstock and stem bases often somewhat woody, often with short to long rhizomes. Stems occurring singly or less commonly 2 or more together, 40–100 cm long, erect or ascending, rounded or bluntly and inconspicuously angled below, green to more commonly reddish brown and usually with noticeable yellowish brown or black dots, lines, and/or streaks, the surface rarely peeling in thin strips with age. Leaves not jointed at the base. Leaf blades 10–60 mm long, 4–20 mm wide, those of the uppermost leaves ovate to broadly triangular-ovate (those farther down the stem narrowly ovate to narrowly elliptic), mostly sharply pointed at the tip, rounded to shallowly cordate at the base, often somewhat clasping the stem, the margins flat or less commonly slightly rolled under, herbaceous in texture, with 1 or less commonly 3 main veins visible toward the base, the surfaces with usually many noticeable yellowish brown to dark green or black dots, the upper surface green, the undersurface paler but only rarely slightly glaucous. Inflorescences appearing as panicles of 20–200 flowers, rounded to more or less flat-topped in outline. Flowers actinomorphic. Sepals 5, all more or less similar in size and shape, 5–7 mm long, not becoming enlarged at fruiting, elliptic-lanceolate, the margins flat, with abundant noticeable black dots, lines, or streaks (additional inconspicuous yellowish brown to dark green dots, lines, or streaks may be present). Petals 5, 8–12 mm long, narrowly obovate, orangish yellow to bright yellow or less commonly lemon yellow or pale yellow, with abundant yellowish brown to black dots, lines, and/or streaks occurring irregularly across the entire surface, withered and inconspicuous but usually persistent at fruiting. Stamens 40–60, the filaments not fused into noticeable groups or occasionally fused inconspicuously into 3 or 5 groups toward the base. Ovary completely 3-locular (sometimes incompletely so at the very tip), the placentation axile. Styles 3, free above the base, more or less spreading, persistent, the stigmas capitate. Fruits 5–7 mm long, ovoid, widest below the midpoint, tapered to the minute beaks, bluntly triangular to more or less circular in cross-section, the surface usually with abundant yellowish brown dots and/or lines. Seeds numerous, 0.7–0.9 mm long, the surface with a fine network of ridges and pits, sometimes appearing inconspicuously longitudinally ribbed, light brown to brown. $2n=16$. June–September.

Scattered in the Unglaciated Plains Division and all but the easternmost portion of the Ozarks (southeastern U.S. west to Oklahoma and Texas). Upland prairies, glades, ledges and tops of bluffs, dry upland forests, and savannas; also ditches and roadsides.

This species was originally described by Britton (1901) based upon B. F. Bush's manuscript and his collections from the vicinity of Swan, in Taney County (Mackenzie and Bush, 1902). Steyermark (1963) followed Fernald (1935) in treating it as a variety of the closely related *H. punctatum,* citing the existence of intermediate specimens in the state. However, Culwell (1970), who studied the systematics of this complex, found very limited morphological overlap between *H. pseudomaculatum* and *H. punctatum* and further suggested that opportunities for hybridization between them were limited by apparent high rates of inbreeding in both taxa. Occasional Missouri specimens may have somewhat intermediate leaf shapes or flower sizes (possibly indicative of hybridization), but the vast majority of specimens can be determined reliably to one or the other species.

13. Hypericum punctatum Lam. (spotted St. John's wort)

H. punctatum f. *subpetiolatum* (E.P. Bickn. ex Small) Fernald

Pl. 361 a, b; Map 1573

Plants perennial herbs, the rootstock and stem bases often somewhat woody, often with short to long rhizomes. Stems occurring singly or less commonly 2 or more together, 35–105 cm long, erect or ascending, rounded or bluntly and inconspicuously angled below, green to more commonly reddish brown and usually with noticeable yellowish brown or black dots, lines, and/or streaks, the surface rarely peeling in thin strips with age. Leaves not jointed at the base. Leaf blades 10–55 mm long, 4–20 mm wide, oblong-elliptic to narrowly oblong or oblanceolate, rounded to less commonly bluntly pointed or rarely shallowly notched at the tip, narrowed or rounded to shallowly cordate at the base, sometimes somewhat clasping the stem, the margins flat or less commonly slightly rolled under, herbaceous in texture, with mostly 1 main vein visible toward the base, the surfaces with usually many noticeable yellowish brown to dark green or black dots, the upper surface green, the undersurface paler but only rarely slightly glaucous. Inflorescences appearing as panicles of 10–200 flowers, rounded to more or less flat-topped in outline. Flowers actinomorphic. Sepals 5, all more or less similar in size and shape, 2.5–4.0 mm long, not becoming enlarged at fruiting, elliptic-lanceolate, the margins flat, with abundant noticeable black dots, lines, or streaks (additional inconspicuous yellowish brown to dark green dots, lines, or streaks may be present). Petals 5, 4–8 mm long, oblanceolate to oblong-elliptic, orangish yellow to

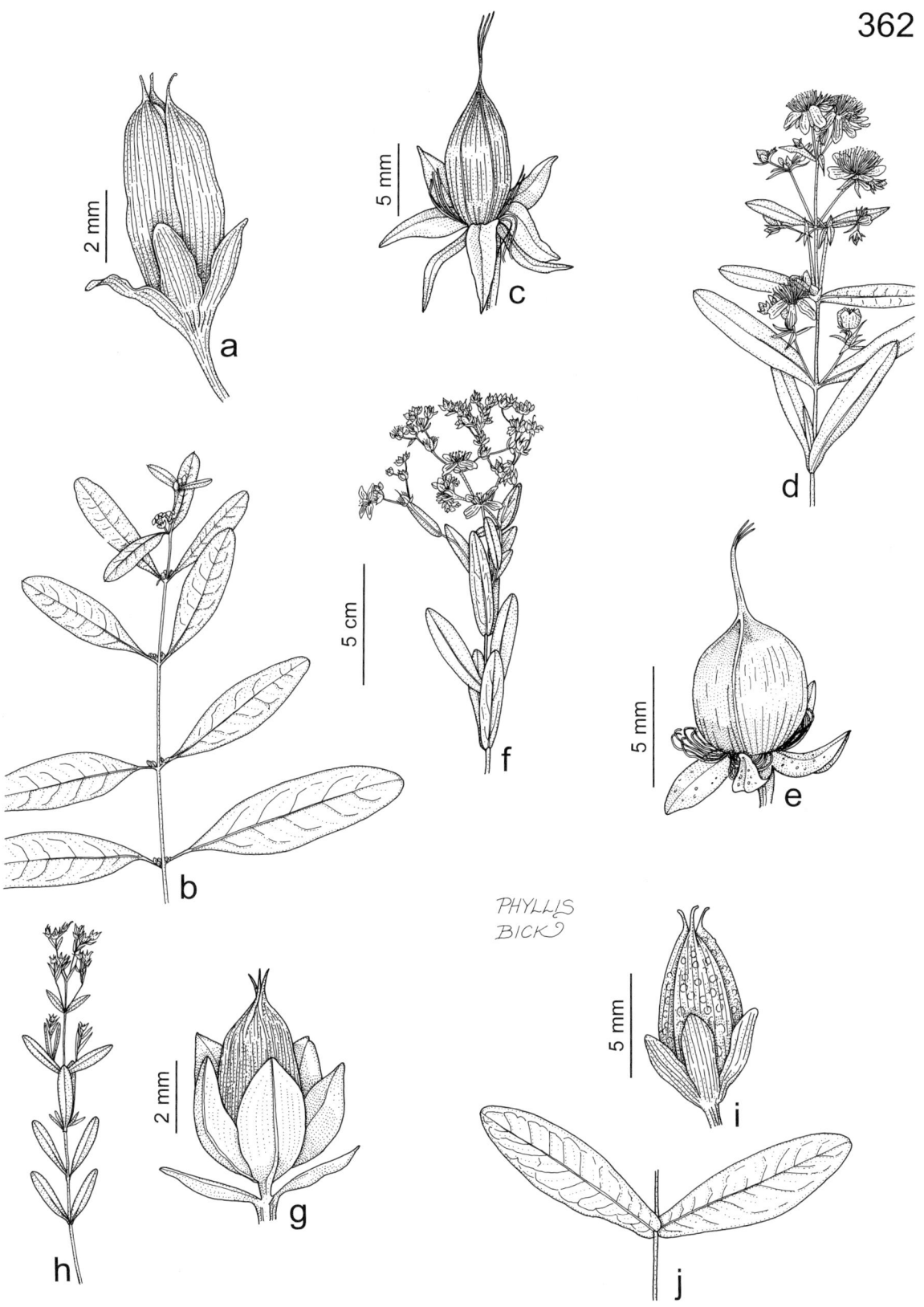

Plate 362. Clusiaceae. *Triadenum walteri*, **a)** fruit, **b)** habit. *Hypericum prolificum*, **c)** fruit, **d)** habit. *Hypericum sphaerocarpum*, **e)** fruit, **f)** habit. *Hypericum lobocarpum*, **g)** fruit, **h)** habit. *Triadenum tubulosum*, **i)** fruit, **j)** leaves.

bright yellow or less commonly lemon yellow, with abundant yellowish brown to black dots, lines, and/or streaks occurring irregularly across the entire surface, withered and inconspicuous but usually persistent at fruiting. Stamens 25–55, the filaments not fused into noticeable groups or occasionally fused inconspicuously into 3 or 5 groups toward the base. Ovary completely 3-locular (sometimes incompletely so at the very tip), the placentation axile. Styles 3, free above the base, more or less spreading, persistent, the stigmas capitate. Fruits 4–6 mm long, ovoid, widest below the midpoint, tapered to the minute beaks, bluntly triangular to more or less circular in cross-section, the surface usually with abundant yellowish brown dots and/or lines. Seeds numerous, 0.6–0.9 mm long, the surface with a fine network of ridges and pits, sometimes appearing inconspicuously longitudinally ribbed, light brown to brown. $2n=16$. June–September.

Scattered nearly throughout the state (eastern U.S. west to Minnesota and Texas; Canada). Bottomland forests, mesic upland forests, banks of streams, rivers, and spring branches, bottomland prairies, upland prairies, glades, and ledges and tops of bluffs; also fallow fields, old fields, pastures, ditches, railroads, roadsides, and disturbed, open areas.

14. Hypericum sphaerocarpum Michx.
(round-fruited St. John's wort)
H. sphaerocarpum var. *turgidum* (Small) Svenson

Pl. 362 e, f; Map 1574

Plants perennial herbs, the rootstock and stem bases often somewhat woody, sometimes with short, poorly developed rhizomes. Stems occurring singly or more commonly 2 or more together, 20–70 cm long, erect or ascending, angled or slightly ridged below each leaf, reddish brown, the surface often peeling in thin strips with age. Leaves not jointed at the base. Leaf blades 20–70 mm long, 3–15 mm wide, narrowly oblong to narrowly elliptic, rounded to bluntly pointed at the tip, tapered or narrowed at the base, the margins flat or somewhat rolled under at maturity, herbaceous to somewhat leathery in texture, with 3 main veins usually visible toward the base, the surfaces lacking noticeable black dots, lines, or streaks but usually with minute, faint, pale to yellowish brown dots visible, the upper surface green, the undersurface pale green and sometimes somewhat glaucous. Inflorescences appearing as panicles of 7–70 flowers, rounded to more or less flat-topped in outline. Flowers actinomorphic. Sepals 5, all more or less similar in size and shape, 2.5–5.0 mm long, not becoming enlarged at fruiting, lanceolate to broadly ovate, the margins occasionally slightly curled, lacking noticeable yellowish brown or black dots, lines, or streaks. Petals 5, 5–9 mm long, oblanceolate to elliptic, bright yellow, usually shed before fruiting. Stamens 45–85, the filaments not fused into groups. Ovary 1-locular or appearing partially 3-locular by intrusion of the parietal placentae into the locule. Styles 3(4), sometimes fused toward the base, erect and more or less appressed at flowering, persistent and usually separating somewhat as the fruit matures, the stigmas minute. Fruits 4.5–8.0 mm long, ovoid to more or less globose, widest at or slightly below the midpoint, tapered abruptly to the minute beak (this sometimes absent), more or less circular in cross-section. Seeds 4–8 per capsule, 2.0–2.7 mm long, the surface with a coarse network of ridges and pits, dark brown to nearly black. May–September.

Scattered nearly throughout the state (eastern U.S. [but mostly absent from the eastern seaboard] west to Nebraska and Texas; Canada). Glades, ledges and tops of bluffs, openings of mesic to dry upland forests, bottomland and upland prairies, banks of streams and rivers, fens, and margins of ponds and lakes.

Steyermark (1963) and other earlier authors recognized a narrow-leaved variant as var. *turgidum*, but Adams (1962) and Robson (1996) both believed that leaf variation was more or less continuous within the species. Numerous historical specimens of *H. sphaerocarpum* in various herbaria were originally misdetermined as *H. cistifolium* Lam., a taller shrubby species with narrower capsules endemic to the southeastern Coastal Plain. Steyermark (1963) also excluded *H. dolabriforme* Vent. from the Missouri flora, which had been reported by Fernald (1950) based upon a misdetermined historical specimen of *H. sphaerocarpum* from southern Missouri. This shrubby species occurs to the east of the state and differs in its loosely ascending stems and unequal sepals.

2. Triadenum Raf. (marsh St. John's wort)

Plants perennial herbs, with rhizomes, glabrous. Young stems or twigs rounded or slightly angled, not winged. Leaves sessile or short-petiolate, the blades sometimes clasping at the

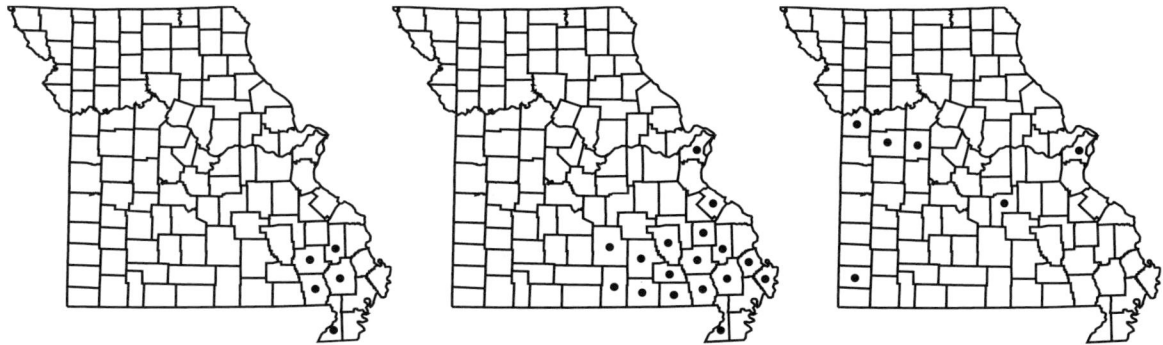

1575. Triadenum tubulosum   1576. Triadenum walteri   1577. Calystegia macounii

base. Flowers often relatively few, actinomorphic. Calyces of 5 sepals. Corollas of 5 petals, these pink, less commonly flesh-colored, not persistent at fruiting. Stamens 9, in 3 groups of 3, the filaments within a group noticeably fused toward the base. Staminodes 3, alternating with the groups of stamens, appearing as ellipsoid, yellow to orange glandular bodies attached at the base of the ovary. Pistils of 3 fused carpels. Ovary 3-locular, with axile placentation. Styles 3, free to the base, loosely ascending at flowering, the stigmas capitate. Fruits capsules, narrowly oblong-ovoid, 2–3 times as long as the sepals. Seeds numerous, 0.8–1.2 mm long, ovoid-cylindrical, not flattened, rounded to bluntly pointed at the ends, the surface with a network of fine ridges and pits, dark brown. Eight to 10 species, eastern U.S. and Canada, Asia.

Steyermark (1963) and various other earlier botanists treated *Triadenum* as a section within *Hypericum*, but Gleason (1947) argued persuasively that the odd staminal characteristics of the species and the petal color warranted the group's recognition as a separate genus. Subsequent workers (see Wood and Adams, 1976) noted differences in vascular patterns of the flowers and corolla positions in the buds. Thus, most recent authors of floristic manuals (Voss, 1985; Kaul, 1986; Gleason and Cronquist, 1991) have maintained *Triadenum* as a separate genus.

1. Leaf undersurface lacking glandular dots; uppermost leaves of plants at flowering time broadly rounded to cordate at the base (note that this character cannot be evaluated on immature vegetative plants) .... 1. T. TUBULOSUM
1. Leaf undersurface with clear to light yellowish resinous secretory cavities (visible with magnification), the punctations appearing as pale to nearly clear (rarely darker) glandular dots; leaves all narrowed or tapered at the base ..... 2. T. WALTERI

1. Triadenum tubulosum (Walter) Gleason
Hypericum tubulosum Walter
Pl. 362 i, j; Map 1575

Stems 45–100 cm long. Leaves sessile (uppermost) to short-petiolate (lowermost). Leaf blades 4–15 cm long, oblong-elliptic or oblong-ovate to elliptic-oblanceolate, the tip rounded to bluntly pointed, the base broadly rounded to cordate in most or only the uppermost leaves, narrowed or tapered in the lowermost leaves (sometimes all of the leaves in juvenile plants), the upper surface green to olive green, the undersurface pale green, sometimes somewhat glaucous, lacking glandular dots. Sepals 4–6 mm long, oblong to oblong-lanceolate, usually sharply pointed at the tip. Petals 5–8 mm long. Stamens in each group with the filaments united to at or above the midpoint. Fruits 9–11 mm long. August–September.

Uncommon in the Mississippi Lowlands Division (eastern [mostly southeastern] U.S. west to Illinois, Oklahoma, and Texas). Swamps and bottomland forests.

In Missouri, *T. tubulosum* is much less common than the closely related *T. walteri*. The two taxa sometimes have been treated as varieties of *T. tubulosum* (Fernald, 1950; Cooperrider, 1989). Records of *T. tubulosum* from the Ozark and Ozark Border Divisions (and some of those from the Mississippi Lowlands) have been found to represent misdetermined specimens of *T. walteri* and usually include immature plants lacking leaves with cordate bases, which tend to develop only later in the season. Steyermark (1963) corrected earlier reports of the closely related *T. virginicum* (L.) Raf. from southeastern Missouri, which were based on misdetermined specimens of *T. tubulosum*. *Triadenum virginicum* has leaves similar in shape to those of *T. tubulosum*, but with the undersurface having resinous punctations, as in *T. walteri*, as well as slightly larger flowers than in either of the Missouri species.

2. Triadenum walteri (J.G. Gmel.) Gleason
  Hypericum walteri J.G. Gmel.
  H. tubulosum Walter var. *walteri* (J.G. Gmel.) Lott
  T. tubulosum Walter var. *walteri* (J.G. Gmel.) Cooperr.

Pl. 362 a, b; Map 1576

Stems 40–100 cm long. Leaves short-petiolate, the uppermost leaves occasionally sessile or nearly so. Leaf blades 3–15 cm long, oblong-elliptic to oblanceolate, the tip rounded to less commonly bluntly pointed, the base narrowed or tapered, the upper surface green to olive green or yellowish green, the undersurface pale green, sometimes somewhat glaucous, with clear to light yellowish resinous secretory cavities, the punctations appearing as pale to nearly clear (rarely darker) glandular dots. Sepals 3–5 mm long, narrowly elliptic, bluntly pointed at the tip. Petals 5–7 mm long. Stamens in each group with the filaments united only toward the base. Fruits 7–12 mm long. July–September.

Uncommon in the southeastern quarter of the state (eastern [mostly southeastern] U.S. west to Illinois, Oklahoma, and Texas). Swamps, bottomland forests, margins of ponds and sinkhole ponds, and less commonly on ledges of sandstone bluffs.

The punctations on the leaf undersurface in this species can be translucent or nearly so and thus difficult to observe. This character should be evaluated carefully to avoid misdetermination of specimens. On any given leaf, at least some of the punctations will be visible under magnification either as colored dots or as minute changes in the surface topography.

CONVOLVULACEAE (Morning Glory Family)

Plants annual or perennial herbs (shrubs and trees elsewhere), often climbing or scrambling but lacking tendrils, sometimes parasitic (in *Cuscuta*). Stems often twining, usually branched, sometimes with milky sap. Leaves alternate, well developed (reduced to small scales in *Cuscuta*). Stipules absent. Leaf blades simple (pinnately dissected and appearing compound in *Ipomoea quamoclit*), entire or lobed, the main venation pattern often palmate. Inflorescences axillary, sessile, of stalked clusters, sometimes appearing as small panicles, sometimes solitary flowers. Flowers actinomorphic, hypogynous, perfect, usually subtended by bracts. Calyces deeply (3–)5(6)-lobed or of (4)5 free sepals, often at fruiting. Corollas shallowly (3–)5(6)-lobed (deeply lobed in *Dichondra*), pleated and spirally twisted in bud (except in *Cuscuta*). Stamens (3–)5(6), alternating with the corolla lobes, the filaments attached in the corolla tube (each subtended by a small scale in *Cuscuta*), the anthers exserted or more commonly not exserted, often linear, attached toward their midpoints (or at least above the base), yellow. Pistil 1 per flower, of 2 fused carpels. Ovary superior, 2(3)-locular, sometimes incompletely so or appearing 4-locular, with usually 2 ovules per locule, the placentation axile or appearing more or less basal. Styles 1 or 2(3), if solitary then sometimes 2-lobed, sometimes persistent at fruiting, the stigmas 1 or 2, disc-shaped or capitate to linear, sometimes shallowly lobed. Fruits capsules, ovoid to globose or depressed-globose (2-lobed in *Dichondra*), variously dehiscent. Seeds 1–4 per locule. Fifty to 56 genera, about 1,600–2,000 species, nearly worldwide, most diverse in tropical and subtropical regions.

Two genera that occur in Missouri sometimes have been treated as separate families, but neither appear to warrant recognition (Wilson, 1960; Cronquist 1981, 1991; Stefanović et al., 2002). The genus *Cuscuta* is often segregated into the Cuscutaceae, but the group seems clearly derived from ancestors within the Convolvulaceae, and the differences between the genus and others in the family can mostly be attributed to structural modifications accompanying the shift to a parasitic habit. *Dichondra,* which is sometimes treated in the Dichondraceae, also appears to represent a mere specialization within the Convolvulaceae. Except for *Cuscuta,* an interesting feature shared by all Missouri genera of Convolvulaceae is the often deeply 2-lobed cotyledons of the seedlings.

1. Leaves reduced to small scales; plants parasitic on the aboveground portions of other plants; stems usually yellow or orange when fresh .... 3. CUSCUTA
1. Leaves well developed; plants not parasitic (but sometimes twining on other plants); stems green, sometimes white or reddish-tinged
    2. Leaf blades pinnately dissected into linear lobes, appearing nearly pinnately compound .... 6. IPOMOEA
    2. Leaf blades entire or with a pair of broad basal lobes, sometimes shallowly notched at the tip (in *Dichondra*) and appearing slightly 2-lobed
        3. Leaf blades linear to elliptic, narrowed or tapered at the base
            4. Styles 2, each 2-lobed toward the tip, the stigmas thus 4 per flower; corollas 3–6 mm long, lavender to blue or less commonly white .... 5. EVOLVULUS
            4. Style 1, unequally 2-lobed toward the tip, the stigmas thus 2 per flower; corollas 12–18 mm long, white .... 8. STYLISMA
        3. Leaf blades ovate to triangular or nearly circular, cordate or less commonly more or less truncate at the base
            5. Ovary and fruit deeply 2-lobed; styles 2, attached basally in the notch between the ovary lobes; corollas 1.5–2.5 mm long; leaf blades 0.7–2.0 cm long, kidney-shaped to less commonly nearly circular .... 4. DICHONDRA
            5. Ovary and fruit unlobed; styles 1 or 2, attached at the tip of the ovary; corolla 5 mm or more long; leaf blades 1–15 cm long, triangular to ovate or oblong-ovate
                6. Inflorescences dense, headlike clusters; stigmas 2, oblong to elliptic in outline, flattened .... 7. JACQUEMONTIA
                6. Inflorescences of solitary or paired flowers, or loose clusters, not forming heads
                    7. Flowers with a pair of conspicuous, more or less leaflike bracts, these closely subtending, longer than, and more or less hiding the calyx .... 1. CALYSTEGIA
                    7. Flowers with inconspicuous scalelike bracts at the inflorescence branch points and/or along the flower stalks, these occasionally absent, usually distant from, always much shorter than, and not covering the calyx
                        8. Stigmas 2, linear; calyx 3–5 mm long ...... 2. CONVOLVULUS
                        8. Stigma 1, capitate, sometimes 2-lobed; calyx 9–25 mm long .... 6. IPOMOEA

1. Calystegia R. Br. (hedge bindweed)

Plants perennial herbs, usually scrambling or twining (except in *C. spithamea*), with usually deep-set rhizomes and root systems. Stems sometimes somewhat angular, glabrous or finely hairy. Leaves mostly relatively long-petiolate (except in *C. spithamea*). Leaf blades triangular to ovate or oblong-ovate, often with 1 or 2 pairs of triangular lobes at the base (then appearing sagittate or hastate), rounded to sharply pointed at the tip, truncate to more commonly deeply cordate at the base, the margins otherwise entire. Inflorescences axillary, the flowers solitary or paired, long-stalked. Bracts 2, closely subtending the flower, leaflike, longer than and more or less hiding the calyx, slightly to strongly overlapping, ovate to oblong-ovate, persistent at fruiting. Calyx of free sepals, these similar in size and shape, narrowly ovate to ovate or oblong, membranous, at least toward the margins, glabrous or nearly so. Corollas very shallowly 5-lobed, funnelform, white or rarely pink. Stamens lacking subtending scales, not exserted. Ovary 2-locular, usually appearing 1-locular toward the tip, with 4 ovules. Style 1, the stigmas 2, oblong to ovate in outline, somewhat flattened. Fruits globose to ovoid, 1-locular, dehiscing longitudinally, the wall separating into 4 segments. Seeds 1–4, oblong-ovate to ovate in outline, somewhat longitudinally angled on the inner face, the surface smooth to very finely granular, dark brown to more commonly black, glabrous. About 30 species, nearly worldwide.

Calystegia is now accepted by most botanists as a genus separate from *Convolvulus*, although emerging molecular data suggests that it represents merely a specialized offshoot of that genus (Brummitt, 2002). Lewis and Oliver (1965) discussed the morphological characters separating the two groups, including differences in pollen morphology, stigma shape, and often locule number. The present treatment was improved greatly following helpful discussion and advice from Richard K. Brummitt (Royal Botanic Gardens, Kew) and Daniel F. Austin (Arizona-Sonora Desert Museum, Tucson).

Steyermark (1963) noted that some species of what is now considered *Calystegia* apparently are poisonous to livestock, but Burrows and Tyrl (2001) do not discuss any poisonous properties. Steyermark also noted that the group can become noxious weeds in crop fields and disturbed sites. Hedge bindweeds do not produce as extensive or deep a rootstock as does true bindweed *(Convolvulus arvensis),* making them less significant agricultural weeds. Some species of *Calystegia* are cultivated as ornamental plants on fences and trellises.

1. Corollas 2.5–4.0 cm long, doubled (at least in Missouri material) .... 2. C. PUBESCENS
1. Corollas 4–7 cm long, not doubled
    2. Petioles less (usually much less) than ½ as long as the midvein of the accompanying leaf blades; stems 7–50 cm long, erect or ascending, not twining
    .... 5. C. SPITHAMAEA
    2. Petioles more than (usually much more than) ½ as long as the midvein of the accompanying leaf blades; stems mostly 40–300 cm long, scrambling or trailing, twining, at least toward the tip (except occasionally in *C. macounii*)
        3. Stems and leaves moderately to densely pubescent with short, velvety hairs; leaf blades with the basal lobes entire (without additional lobes), not spreading, broadly rounded to less commonly bluntly pointed
        .... 1. C. MACOUNII
        3. Stems and leaves glabrous or nearly so; leaf blades with the basal lobes angular or with an additional pair of lobes, often somewhat spreading, bluntly to sharply pointed

4. Leaf blades with the basal sinus quadrate (squared-off, with 2 parallel sides at abrupt right angles to the base; note that the shape is sometimes distorted in pressed specimens); bracts usually strongly overlapping, rounded to bluntly pointed . . . . 4. C. SILVATICA
4. Leaf blades with the basal sinus U-shaped or V-shaped; bracts usually overlapping only toward the base, bluntly to more commonly sharply pointed . . . . 3. C. SEPIUM

1. Calystegia macounii (Greene) Brummitt

Pl. 363 f; Map 1577

Stems mostly 40–90 cm long, scrambling or trailing, usually twining only toward the tip (rarely not twining), moderately to densely pubescent with short, velvety hairs. Leaves mostly long-petiolate, moderately to densely pubescent with short, velvety hairs, the petiole more than ½ as long as the midvein of the accompanying leaf blade. Leaf blades 2–6(–8) cm long, oblong-ovate to ovate, broadly rounded to less commonly bluntly pointed at the tip, deeply cordate at the base, the sinus usually V-shaped, the basal lobes entire (lacking additional shallow lobes along the upper portion), not spreading (oriented more or less toward the leaf base), broadly rounded to less commonly slightly angled and bluntly pointed. Flowers solitary in the axils of leaves, positioned mostly above the stem midpoint. Bracts 18–25 mm long, usually strongly overlapping, often strongly inflated, ovate to oblong-ovate, bluntly or sharply pointed at the tip, sparsely to moderately pubescent with short hairs. Sepals 10–12 mm long, ovate to elliptic. Corollas not doubled, 4–6 cm long, white or rarely pinkish purple. Fruits 10–15 mm long. Seeds 4–6 mm long. May–July.

Uncommon and widely scattered in the state. (Montana to Arizona east to Minnesota and Louisiana and locally east to Virginia and Georgia; Canada). Banks of streams; also pastures, fallow fields, and gardens.

Plants of *C. macounii* were known as *Convolvulus sepium* f. *malacophyllus* or *C. spithameus* in much of the older botanical literature (Steyermark, 1963). At the species and subspecies level, the epithet *malacophylla* applies to a different taxon endemic to California, *Calystegia malacophylla* (Greene) Munz. Records from the eastern half of the state may represent introduced populations. The main range of *C. macounii* is in the central third of the country. However, the range of this species in Missouri is not well understood, and it has probably been undercollected by botanists who have mistakenly assumed plants to be *C. sepium*. Austin (1997) noted that this species tends to be the earliest in the complex to bloom.

2. Calystegia pubescens Lindl. (Japanese bindweed)

Convolvulus pellitus Ledeb. f. *anestius* Fernald

Map 1578

Stems mostly 30–200 cm long, scrambling or trailing, twining, at least toward the tip, glabrous or more commonly sparsely pubescent with short hairs. Leaves short- to less commonly long-petiolate, usually sparsely pubescent with short hairs, the petiole usually less than ½ as long as the midvein of the accompanying leaf blade. Leaf blades 2–8 cm long, oblong to narrowly ovate-triangular, narrowed to a bluntly or more commonly sharply pointed tip, deeply cordate at the base, the sinus U-shaped or occasionally V-shaped, the basal lobes often each with 1 or 2 additional shallow lobes along the upper portion, spreading, angular, mostly sharply pointed. Flowers solitary or less commonly paired in the axils of leaves, positioned mostly above the stem midpoint. Bracts 15–21 mm long, usually overlapping only toward the base, not strongly inflated (usually appearing somewhat angular), ovate to oblong-ovate, rounded to bluntly pointed at the tip, usually glabrous. Sepals 8–10 mm long, ovate to oblong-ovate. Corollas doubled (at least in Missouri material), 2.5–4.0 cm long, pink. Fruits not produced. $2n=22$. May–September.

Introduced, uncommon and widely scattered, mostly in and around urban areas (native of Asia; introduced sporadically in the eastern U.S. west to Iowa, Idaho, and Canada). Fallow fields, gardens, roadsides, railroads, and open, disturbed areas.

Missouri plants were included in the circumscription of *Convolvulus pellitus* f. *anestius*, which was based upon a sterile cultivar with doubled flowers (sometimes known as California rose). Presumably they escape from ornamental plantings via fragments of roots or stems. The correct species determination of this cultivar has been controversial. It has been classified variously under *Calystegia pellita* (Ledeb.) G. Don, *C. hederacea* Wall., *C. dahurica* (Herb.) Choisy, and *C. pubescens*, and under *Convolvulus japonicus* Thunb. (non Choisy) and *C. wallichianus* Spreng. The present assignment follows that of Fang and Brummitt

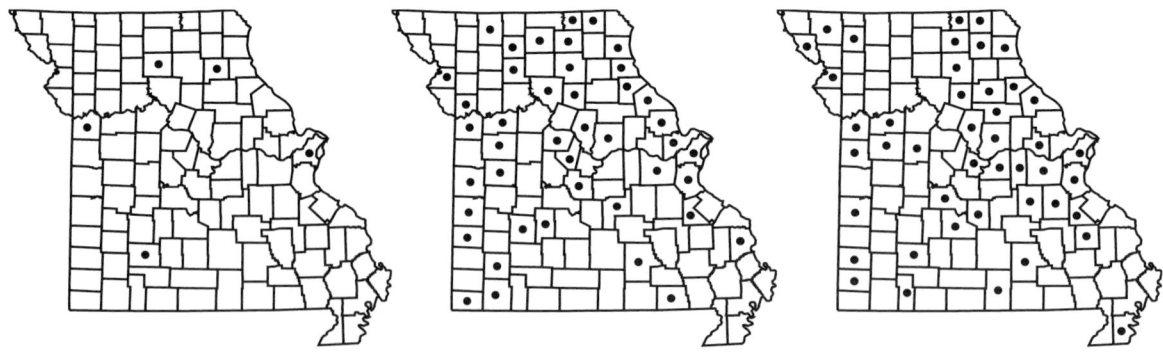

1578. Calystegia pubescens   1579. Calystegia sepium   1580. Calystegia silvatica

(1995) in the recent Convolvulaceae treatment for the Flora of China series.

3. Calystegia sepium (L.) R. Br. (hedge bindweed, wild morning glory, Rutland beauty)
Convolvulus sepium L.
Convolvulus sepium var. *repens* (L.) A. Gray
Convolvulus sepium L. f. *malacophyllus* Fernald
Convolvulus sepium f. *coloratus* Lange

Pl. 363 b, c; Map 1579

Stems mostly 50–300 cm long, scrambling or trailing, twining, at least toward the tip, glabrous or nearly so (hairy elsewhere). Leaves long-petiolate, glabrous or nearly so, the petiole more than (usually much more than) ½ as long as the midvein of the accompanying leaf blade. Leaf blades 2–9 (–15) cm long, narrowly ovate-triangular to oblong-ovate or triangular, narrowed to a bluntly or more commonly sharply pointed tip, deeply cordate at the base, the sinus U-shaped or less commonly V-shaped, the basal lobes often each with 1 or 2 additional shallow lobes along the upper portion, frequently somewhat spreading, angular, bluntly to sharply pointed. Flowers solitary or less commonly paired in the axils of leaves, positioned mostly above the stem midpoint. Bracts 14–25(–32) mm long, usually overlapping only toward the base, not strongly inflated (usually appearing somewhat angular), ovate to triangular-ovate, bluntly to more commonly sharply pointed at the tip, glabrous or sparsely hairy along the margins. Sepals 11–15 mm long, elliptic to narrowly ovate. Corollas not doubled, 4–7 cm long, white or rarely pink. Fruits 10–15 mm long. Seeds 4–5 mm long. 2n=22. May–September.

Scattered, mostly north and west of the Ozark Division (North America, Europe, Asia; in North America, nearly throughout the U.S., the southern half of Canada, and northern Mexico). Borders of bottomland and mesic upland forests, banks of streams and rivers, margins of ponds, lakes, and sloughs, and disturbed portions of upland prairies; also fallow fields, crop fields, pastures, fencerows, gardens, railroads, roadsides, and open, disturbed areas.

According to Brummitt (1980), one name used by Steyermark (1963) and some other authors, *Convolvulus sepium* var. *repens* (L.) A. Gray, is actually a synonym of the unrelated pantropical species, *Ipomoea imperati* (Vahl) Griseb. (cited by Brummitt under another synonym, *I. stolonifera* (Cirillo) J.F. Gmel. [La Valva and Sabato, 1983]). Brummitt (1965, 1980) treated *Calystegia sepium* as consisting of a complex series of about a dozen intergrading subspecies, but in the absence of a comprehensive taxonomic revision of the group, his concepts are merely summarized here and not treated fully. The ssp. *sepium* is native to the Old World and has somewhat smaller flowers than the New World taxa. Nearly all of the Missouri specimens are referable to ssp. *angulata* Brummitt, which occurs throughout the species' range but is especially prominent in the western states and the Great Plains. It is characterized by usually white corollas, glabrous stems and leaves, and leaf blade base with the lobes relatively spreading and the sinus frequently broadly U-shaped. The few pink-flowered specimens collected in Missouri appear otherwise to be referable to ssp. *angulata*. Two other taxa that are present in the eastern United States should be searched for in Missouri. Both have flowers with usually pink or pinkish-tinged corollas. The ssp. *americana* (Sims) Brummitt was erroneously reported for Missouri by Yatskievych and Turner (1990) and has yet to be confirmed as occurring in the state. It differs in being sparsely to moderately hairy and having the leaf blade base with a V-shaped sinus. A single aberrant specimen from Ralls County is possibly referable to ssp. *erratica*

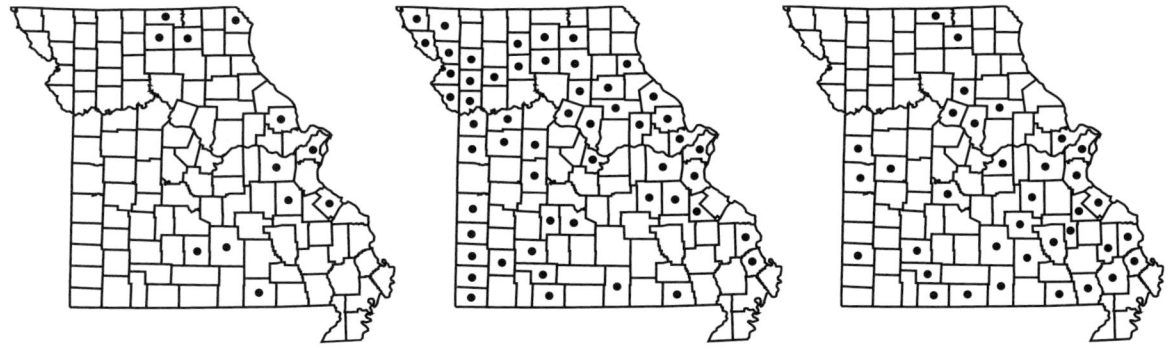

1581. Calystegia spithamaea   1582. Convolvulus arvensis   1583. Cuscuta campestris

Brummitt, a usually glabrous taxon with the bracts grading into and not sharply distinguished from the sepals, as well as the leaf blade base with a narrow sinus and often slightly overlapping lobes. The specimen from Ralls County is unusual in having bracts that are somewhat larger than is typical for the species.

For a discussion of hybridization with *C. silvatica*, see the treatment of that species.

4. Calystegia silvatica (Kit.) Griseb. ssp. fraterniflora (Mack. & Bush) Brummitt
Convolvulus sepium var. *fraterniflorus* Mack. & Bush
Calystegia fraterniflora (Mack. & Bush) Brummitt

Pl. 363 d; Map 1580

Stems mostly 50–300 cm long, scrambling or trailing, twining, at least toward the tip, glabrous. Leaves long-petiolate, glabrous, the petiole more than (usually much more than) ½ as long as the midvein of the accompanying leaf blade. Leaf blades 2–8(–15) cm long, narrowly ovate-triangular to oblong-ovate or triangular, narrowed to a bluntly or more commonly sharply pointed tip, deeply cordate at the base, the sinus quadrate (squared-off, with 2 parallel sides at abrupt right angles to the base; note that the shape is sometimes distorted in pressed specimens) but often with small, angular bands of tissue tapering along the apical portion of the petiole, the basal lobes often each with 1 or 2 additional shallow lobes along the upper portion, more or less spreading, angular, bluntly to sharply pointed. Flowers usually paired in the axils of leaves, less commonly solitary, positioned mostly above the stem midpoint. Bracts 16–28 mm long, usually strongly overlapping, often strongly inflated, ovate, rounded to bluntly pointed at the tip, glabrous or sparsely and minutely hairy along the margins. Sepals 12–16 mm long, ovate. Corollas not doubled, 4–6 cm long, white. Fruits 10–15 mm long. Seeds 4–6 mm long. $2n=20$. May–September.

Scattered nearly throughout the state (eastern U.S. west to Kansas and Texas). Borders of bottomland and mesic upland forests, banks of streams and rivers, margins of ponds, lakes, and sloughs, and disturbed portions of upland prairies; also fallow fields, crop fields, pastures, fencerows, gardens, railroads, roadsides, and open, disturbed areas.

Taxon limits and relationships in the *C. sepium* / *silvatica* complex have been confused historically. The present circumscription of species follows that of Brummitt (1980) and Austin (1997). Brummitt (1980) characterized *C. silvatica* (sometimes spelled *C. sylvatica* in the botanical literature) as comprising three or more subtly different subspecies. The other two taxa, ssp. *silvatica* and ssp. *orientalis* Brummitt, are native to Europe and Asia. In Missouri, most plants are readily assignable to either *C. sepium* or *C. silvatica* based on examination of their leaf bases and floral bracts, and the species normally are not encountered growing together. However, a small number of specimens, mostly from disturbed sites in urban areas, are somewhat intermediate morphologically and have been annotated by Brummitt as representing putative hybrids between the two species. The presence of such apparent hybrids in low numbers in the eastern United States may in part account for the difficulties encountered by some earlier authors (Tryon, 1939) in delimiting taxa within the complex.

5. Calystegia spithamaea (L.) Pursh ssp. spithamaea (low bindweed, dwarf morning glory)
Convolvulus spithamaeus L.

Pl. 363 a; Map 1581

Stems 7–50 cm long, erect or ascending (rarely becoming prostrate with age), not twining, glabrous or sparsely to moderately pubescent with short hairs. Leaves mostly short-petiolate, sparsely to moderately pubescent with short hairs, the petiole less (usually much less) than $^1/_2$ as long as the midvein of the accompanying leaf blade. Leaf blades 2–6(–10) cm long, oblong to ovate, broadly rounded to less commonly bluntly pointed at the tip, truncate or more commonly shallowly cordate at the base, the sinus U- or V-shaped, the basal lobes usually entire (lacking additional shallow lobes along the upper portion), oriented more or less toward the leaf base or occasionally somewhat spreading, broadly rounded to somewhat angled and bluntly pointed. Flowers solitary in the axils of leaves, often positioned below the stem midpoint. Bracts 15–22 mm long, usually strongly overlapping, usually not inflated, ovate to oblong-ovate, sharply pointed at the tip, sparsely to moderately pubescent with short hairs. Sepals 10–12 mm long, ovate to elliptic. Corollas not doubled, 4–6 cm long, white. Fruits 10–15 mm long. Seeds 4–6 mm long. $2n=22$. May–August.

Uncommon in the eastern half of the state but apparently absent from the Mississippi Lowlands Division (northeastern U.S. west to Iowa and Missouri; Canada). Openings of mesic to dry upland forests, exposed ledges and tops of bluffs, and upland prairies; also roadsides.

This species is similar in appearance to *C. macounii* but has more erect stems, longer petioles, denser pubescence, less strongly inflated floral bracts, and a generally eastern distribution. As he did with most other North American *Calystegia* species, Brummitt (1965) treated *C. spithamaea* as a series of morphologically overlapping subspecies. The other two subspecies, ssp. *purshiana* (Wherry) Brummitt of the eastern seaboard and ssp. *stans* of the northeastern United States, differ in subtle details of leaf shape and hairiness, and they have not been reported from Missouri.

2. Convolvulus L. (bindweed)

About 200 species, worldwide, most diverse in the Old World.

1. Convolvulus arvensis L. (field bindweed, small bindweed)
C. arvensis f. *cordifolius* Lasch

Pl. 363 e; Map 1582

Plants perennial herbs, scrambling or twining, with deep-set, somewhat fleshy rhizomes and root systems. Stems 10–100(–200) cm long, sometimes somewhat angular, glabrous or sparsely and minutely hairy. Leaves mostly relatively short-petiolate. Leaf blades 1–5(–10) cm long, narrowly to broadly ovate, oblong-ovate, or triangular, sometimes with a pair of triangular lobes at the base (then appearing sagittate or hastate), rounded to sharply pointed at the tip, truncate to deeply cordate at the base, the margins otherwise entire or with few shallow teeth on the basal lobes, the surfaces glabrous or less commonly minutely hairy. Inflorescences axillary, the flowers solitary or in loose clusters of 2 or 3, usually long-stalked. Bracts 2, 2–8 mm long, usually distant from the flower, scalelike, much shorter than and not hiding the calyx, not overlapping, linear to elliptic or obovate, persistent or shed before fruiting. Calyx of free sepals, 3–5 mm long, similar in size and shape or the outer 3 slightly shorter and narrower than the inner 2, elliptic to oblong, obovate, or nearly circular, herbaceous, the margins usually hairy, the surfaces glabrous or finely hairy. Corollas 1.2–2.5 cm long, very shallowly 5-lobed, broadly funnelform to nearly bell-shaped, white, often pinkish-tinged, especially with age, rarely all pink. Stamens lacking subtending scales, not exserted. Ovary 2-locular, with 4 ovules. Style 1, the stigmas 2, linear in outline, somewhat flattened. Fruits 5–7 mm long, globose to ovoid, 2-locular, dehiscing longitudinally, the wall separating into 4 segments. Seeds 1–4, 3–4 mm long, oblong-ovate to ovate in outline, somewhat longitudinally angled on the inner face, the surface with small, dense tubercles, dark brown to more commonly black, glabrous. $2n=48, 50$. May–September.

Common nearly throughout the state (native probably of Europe, possibly also Asia and Africa, but now naturalized worldwide). Banks of streams and rivers, disturbed portions of upland prairies, and disturbed margins of upland forests and glades; also crop fields, fallow fields, pastures, gardens, fencerows, roadsides, railroads, and open, disturbed areas.

Field bindweed is considered to be among the world's worst agricultural weeds. Because of the extensive deep root system of the plants, they are extremely difficult to eradicate. The spreading, branched rootstocks may reach depths of 4 m or

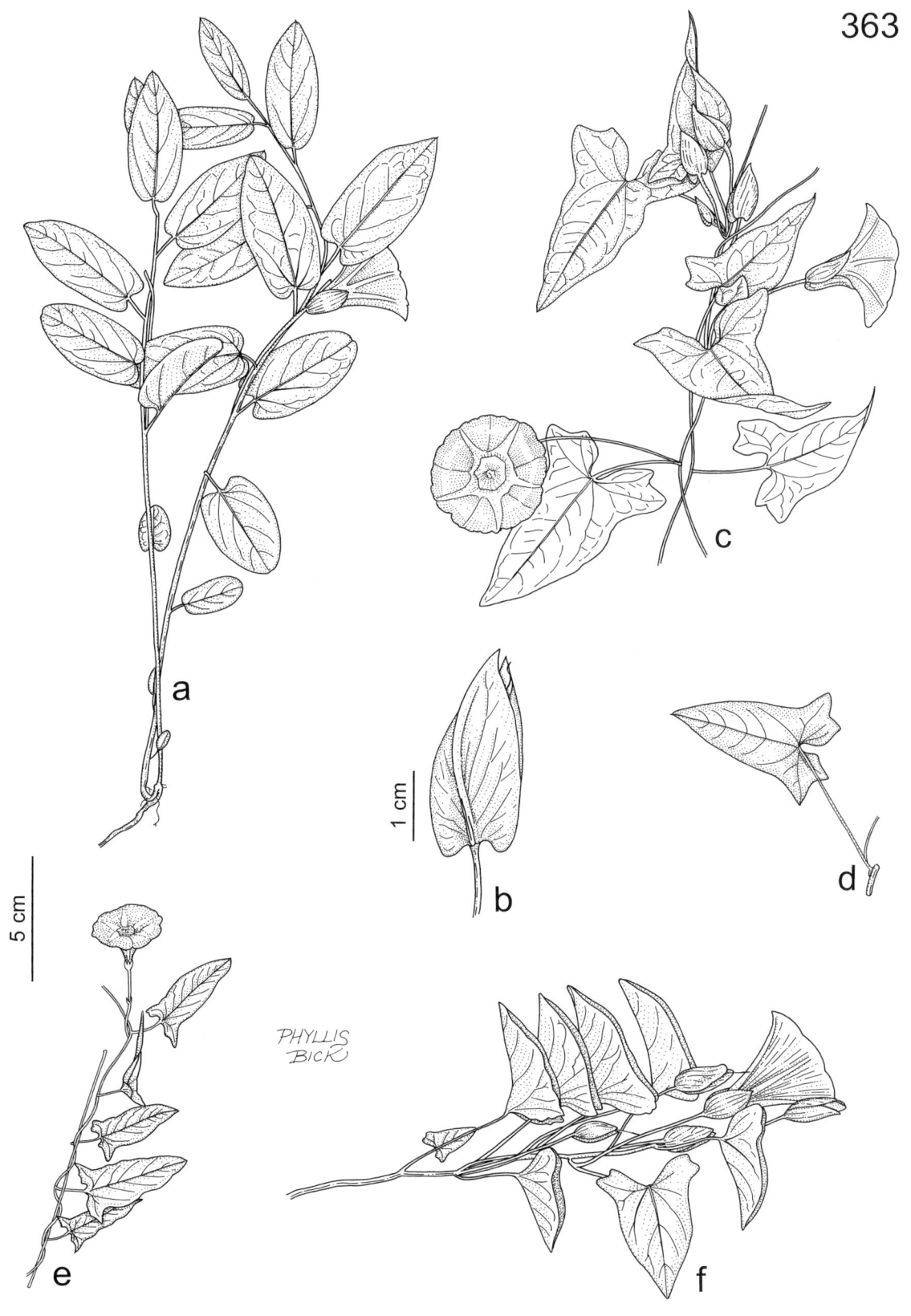

Plate 363. Convolvulaceae. *Calystegia spithameus*, **a)** habit. *Calystegia sepium*, **b)** fruiting bracts, **c)** flowering branch. *Calystegia silvatica*, **d)** leaf. *Convolvulus arvensis*, **e)** flowering branch. *Calystegia macounii*, **f)** flowering branch.

more. Austin (2000) discussed the history of *C. arvensis* in the United States, dating its original introduction to before 1739. According to his review of early literature, it was sold commercially as a medicinal herb and garden ornamental in the eastern states by the early nineteenth century and began to spread outside cultivation in New England soon thereafter. By 1850, the species had reached California, and by the 1880s, it was discussed as a severe weed in the agricultural literature. Today it is regulated by law as a noxious weed in most states, including Missouri. However, such legislation has not resulted in the control of the plants or diminished their spread.

Austin (2000) discussed the long history of medicinal use of *C. arvensis,* both in Europe and North America, mainly for its laxative properties. Steyermark (1963) noted that the roots apparently are poisonous to some livestock.

3. Cuscuta L. (dodder, love vine)
(Yuncker, 1943)

Plants parasitic on the aboveground portions of other plants. Stems twining, glabrous, greenish yellow, yellow, or orange, often forming tangled mats and attaching to host tissues by suckerlike haustoria. Leaves alternate, consisting of small, lanceolate or ovate, sessile scales 1–4 mm long. Inflorescences sessile or stalked clusters along stems, sometimes appearing as small panicles. Flowers with the surface smooth or densely and minutely papillose (warty or bumpy), sometimes also with scattered resinous (pellucid-glandular) cells. Calyces 3–5(6)-lobed, sometimes deeply divided to form separate sepals. Corollas white, rarely greenish, 3–5(6)-lobed, the lobes erect to recurved, the tips sometimes incurved. Stamens with a scalelike appendage (infrastaminal scale) attached to the corolla tube below the attachment point of the filament, this infrastaminal scale with a toothed or fringed margin. Ovary 2-locular. Styles 2, each with a capitate (linear elsewhere) stigma. Fruits papery-walled capsules, usually with an aperture between stigmas, breaking open irregularly (with circumscissile dehiscence elsewhere). Seeds usually 2–4 per fruit, brown. About 145 species. Nearly worldwide.

Contrary to some reports, *Cuscuta* species do produce chlorophyll, although in reduced quantities. Some species of dodder are important agricultural pests and have been spread as contaminants in crop seeds. Species identification in the genus is challenging and requires a hand lens. Details of the flowers are easiest to observe in fresh specimens or in plants that have been dried without pressing. Most dodders parasitize a wide variety of host species, and determination of the host is usually not an aid in *Cuscuta* identification.

1. Calyces deeply divided to form separate or nearly distinct sepals; each flower subtended by 2 to several scalelike bracts
    2. Flowers stalked, in paniculate clusters; bracts oval to ovate, bluntly pointed at the tip .... 5. C. CUSPIDATA
    2. Flowers sessile, in dense clusters along the stem
        3. Bracts appressed to base of flower or erect, oval to orbicular and rounded at the tip .... 3. C. COMPACTA
        3. Bracts spreading to recurved, narrowly lanceolate to oblanceolate and pointed at the tip .... 6. C. GLOMERATA
1. Calyces lobed ½–⅔ of the way to base, a distinct fused portion observable; flowers not regularly subtended by scalelike bracts (occasionally 1 bract present at base of some flowers)
    4. Surface of the flowers with dense, minute papillae; corolla lobes sharply pointed at the tip

5. Calyces and corollas mostly 4-lobed; infrastaminal scales not reaching filament bases, reduced to 2 narrow, toothed, lobes of tissue ..... 4. C. CORYLI
5. Calyces and corollas mostly 5-lobed; infrastaminal scales reaching filament bases, well developed and fringed along the margins ..... 8. C. INDECORA
4. Surface of the flowers not papillose (sometimes slightly irregular in texture when dried); corolla lobes sharply to bluntly pointed at the tip
   6. Calyces and corollas mostly 3- or 4-lobed
      7. Corolla lobes rounded or bluntly pointed at the tip; corolla tube much longer than the calyx and extending past tips of calyx lobes ..... 2. C. CEPHALANTHI
      7. Corolla lobes sharply pointed at the tip; corolla tube shorter than to as long as the calyx and not extending past tips of calyx lobes ..... 10. C. POLYGONORUM
   6. Calyces and corollas mostly 5-lobed
      8. Calyx appearing 5-angled by overlapping lobes that project where they overlap; corolla lobes sharply pointed and often appearing tapered at the tip ..... 9. C. PENTAGONA
      8. Calyx appearing rounded in cross-section (although the lobes may overlap basally); corolla lobes either sharply pointed or rounded to bluntly pointed
         9. Calyx lobes usually extending to top of corolla tube or nearly so; corolla lobes sharply pointed, often appearing tapered at the tip; fruits depressed-globose, the wall not thickened at the tip ..... 1. C. CAMPESTRIS
         9. Calyx lobes usually extending to middle of corolla tube; corolla lobes rounded or bluntly pointed; fruits globose to globose-conical, the wall usually thickened at the tip ..... 7. C. GRONOVII

1. Cuscuta campestris Yunck. (field dodder)

Pl. 365 h–k; Map 1583

Stems relatively slender, usually less than 0.6 mm in diameter. Flowers 2.0–2.5 mm long, with smooth to slightly irregular surfaces, subtended by at most 1 lanceolate to ovate bract (usually none), in dense clusters on short side branches, the pedicels usually shorter than the flowers. Calyces about as long as the corolla tube, 5-lobed $^{1}/_{2}$–$^{2}/_{3}$ of the way to the base, the lobes triangular to broadly ovate, rounded to bluntly pointed at the tip, overlapping basally, but not angled. Corollas narrowed or tapered to 5 sharply pointed lobes, these spreading to recurved, with straight to slightly incurved tips. Infrastaminal scales reaching filament bases, oval, densely fringed along the margins. Fruits depressed-globose, the wall not thickened at the tip. Seeds 1.4–1.6 mm long. June–October.

Scattered throughout the state (U.S., Canada, Mexico, naturalized nearly worldwide). Mostly in moist habitats, but also at margins of fields and in other disturbed areas. Parasitic on a large number of mostly herbaceous hosts, including species in such genera as *Asclepias, Bidens, Euphorbia, Oenothera, Perilla, Pilea, Polygonum, Salix, Saururus,* and *Xanthium.*

2. Cuscuta cephalanthi Engelm. (buttonbush dodder)

Pl. 365 d–f; Map 1584

Stems relatively slender, usually less than 1 mm in diameter. Flowers 2.0–2.5 mm long, with smooth to slightly irregular surfaces, subtended by at most 1 lanceolate to ovate bract (usually none), in spikelike or cymose clusters on short side branches, the pedicels absent or very short. Calyces about $^{1}/_{2}$ as long as the corolla tube, 3- or 4-lobed $^{1}/_{2}$–$^{2}/_{3}$ of the way to the base, the lobes ovate, rounded at the tip, not or slightly overlapping basally, not angled. Corollas with 3 or 4 rounded lobes, these erect to spreading, with incurved tips. Infrastaminal scales usually not quite reaching filament bases, narrowly oblong, deeply toothed to fringed along the margins. Fruits globose to depressed-globose, the wall not thickened at the tip. Seeds 1.5–2.0 mm long. $2n=60$. July–September.

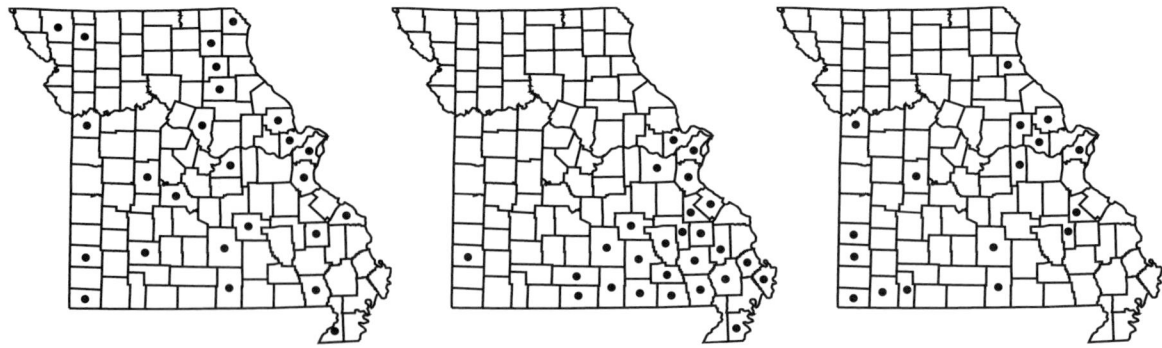

1584. Cuscuta cephalanthi   1585. Cuscuta compacta   1586. Cuscuta coryli

Scattered nearly throughout the state (U.S., Canada). Stream banks, bottomland forests, and wet prairies. Parasitic on both herbaceous and woody host species, including species of *Boehmeria, Cephalanthus, Cynanchum, Eupatorium, Justicia, Lycopus, Lysimachia, Polygonum, Salix, Saururus, Symphyotrichum,* and *Vernonia.*

In this species, frequently only 1 or 2 seeds mature in each fruit, giving the capsules a somewhat lopsided appearance. The papery remains of the corollas frequently cap the fruits.

3. Cuscuta compacta Juss. (compact dodder)

Pl. 364 a–c; Map 1585

Stems relatively stout, usually 1–2 mm in diameter. Flowers 4–5 mm long, with smooth to slightly irregular surfaces, subtended by 2–5 overlapping, ovate to orbicular bracts, forming dense, sessile, ropelike clusters along main stems, occasionally also with tight clusters on short side branches. Calyces about 1/2 as long as the corolla tube, mostly hidden by the bracts, deeply 5-lobed into separate or nearly distinct sepals, the sepals ovate to orbicular, rounded at the tip, strongly overlapping basally, but not angled. Corollas with 5 rounded lobes, these spreading, with straight tips. Infrastaminal scales reaching filament bases, narrowly oval, densely fringed along the margins. Fruits globose to conical, the wall slightly thickened at the tip. Seeds 2.0–2.6 mm long. July–October.

Scattered, mostly in the southeastern quarter of the state (eastern U.S. west to Texas). Occurs in bottomland forests, swamps, and on the banks of streams and sinkhole ponds. Parasitic on both woody and herbaceous hosts, including species of *Alnus, Apios, Aralia, Campsis, Cephalanthus, Corylus, Decodon, Diospyros, Equisetum, Hydrangea, Lindera, Nyssa, Rosa, Rudbeckia, Saururus, Toxicodendron, Triadenum, Verbesina, Vernonia,* and *Vitis.*

4. Cuscuta coryli Engelm. (hazel dodder)

Pl. 364 d; Map 1586

Stems relatively slender, usually less than 1 mm in diameter. Flowers 1.5–2.0 mm long, with succulent, strongly granular to papillate surfaces, subtended by at most 1 lanceolate to ovate bract (usually none), in dense to loose cymose clusters on short side branches, the pedicels usually shorter than the flowers. Calyces about as long as the corolla tube, 3- or 4(5)-lobed 1/2–2/3 of the way to base, the lobes triangular-ovate, pointed at the tip, slightly overlapping basally, but not angled. Corollas narrowed or tapered to 3 or 4 sharply pointed lobes, these erect, with incurved tips. Infrastaminal scales not reaching filament bases, reduced to 2 narrow, toothed lobes along the vein below each filament. Fruits globose to depressed-globose, the wall thickened around the aperture between style bases. Seeds 1.4–1.6 mm long. $2n=30$. July–September.

Relatively uncommon and widely scattered (eastern and southwestern U.S.). Stream banks, bottomland forests, and prairies. Parasitic on various woody and herbaceous hosts, including species of *Campsis, Corylus, Desmodium, Eupatorium, Helianthus, Iresine, Justicia, Salix, Solidago, Symphyotrichum,* and *Toxicodendron.*

5. Cuscuta cuspidata Engelm. (cusp dodder)

Pl. 364 e; Map 1587

Stems relatively slender, usually less than 1 mm in diameter. Flowers 2.5–3.5 mm long, with smooth to slightly irregular surfaces, subtended by 2–4 loosely overlapping, ovate bracts with pointed, erect tips, in loose, paniculate clusters on short side branches, the pedicels usually shorter than the flowers. Calyces 1/2–2/3 as long as the corolla tube,

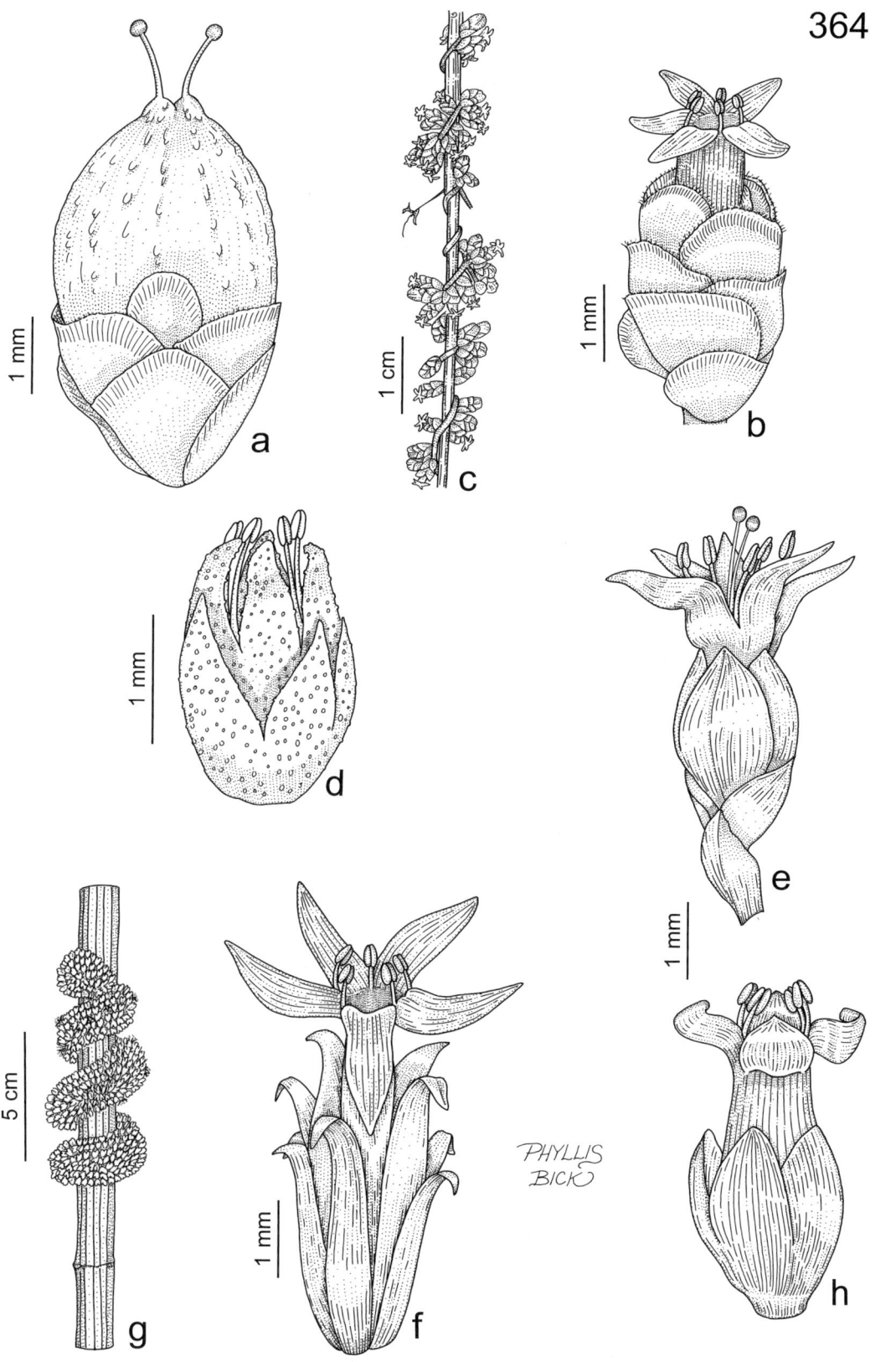

Plate 364. Convolvulaceae. *Cuscuta compacta*, **a)** fruit, **b)** flower, **c)** habit. *Cuscuta coryli*, **d)** flower. *Cuscuta cuspidata*, **e)** flower. *Cuscuta glomerata*, **f)** flower, **g)** habit. *Cuscuta polygonorum*, **h)** flower.

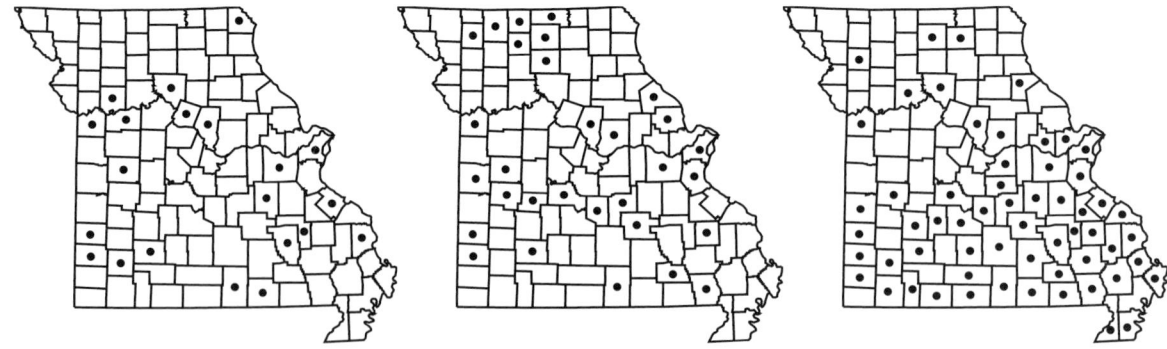

1587. Cuscuta cuspidata   1588. Cuscuta glomerata   1589. Cuscuta gronovii

not hidden by the bracts, deeply 5-lobed into separate or nearly distinct sepals, the sepals ovate to nearly orbicular, with a bluntly pointed tip, strongly overlapping basally, but not angled. Corollas narrowed to 5 sharply or rarely bluntly pointed lobes, these spreading to reflexed, with straight tips. Infrastaminal scales not quite reaching filament bases, narrowly oval, densely fringed along the margins. Fruits globose or nearly so, thickened at the tip. Seeds 1.4–1.5 mm long. July–October.

Scattered in the southern half of the state (Indiana and Louisiana west to Utah). Mostly on the banks of rivers, streams, and ponds, but also in wet prairies; sometimes a weed in fields. Parasitic on a wide variety of herbaceous species (including *Ambrosia, Erigeron, Impatiens, Iva, Lycopus, Symphyotrichum,* and *Vernonia*), mostly in the Asteraceae.

6. Cuscuta glomerata Choisy (rope dodder)
C. coryli Engelm. var. *stylosa* Engelm.
Pl. 364 f, g; Map 1588

Stems relatively slender, usually 1.0–1.5 mm in diameter, often withered or absent by flowering time. Flowers 4–5 mm long, with smooth to slightly irregular surfaces, subtended by several overlapping, lanceolate bracts with recurved, boat-shaped tips, forming dense, sessile, ropelike clusters along main stems. Calyces about as long as the corolla tube, mostly hidden by the bracts, deeply 5-lobed into separate or nearly distinct sepals, the sepals lanceolate, with a spreading, tapered, boat-shaped tip, not overlapping basally or angled. Corollas narrowed to 5 sharply pointed lobes, these spreading to reflexed, with straight tips. Infrastaminal scales reaching filament bases, narrowly oval, densely fringed along the margins. Fruits globose to conical, with a thickened collar at the tip. Seeds 1.5–1.9 mm long. $2n=30$. July–September.

Scattered counties (Michigan to Mississippi west to South Dakota and Texas). Found in a variety of habitats, including wet and dry prairies, various forest types, fens, and roadsides. Parasitic on herbaceous hosts, including species of *Asclepias, Helianthus, Solidago,* and *Vernonia*.

In this dodder, frequently only 1 or 2 seeds mature in each fruit, giving the capsules a somewhat lopsided appearance. The papery remains of the corollas frequently cap the fruits.

7. Cuscuta gronovii Willd. (common dodder)
C. gronovii var. *latiflora* Engelm.
C. vulgivaga Engelm.
Pl. 365 a, b; Map 1589

Stems relatively slender, usually less than 1 mm in diameter. Flowers 2.0–3.5 mm long, with smooth to slightly irregular surfaces, subtended by at most 1 lanceolate to ovate bract (usually none), in dense to loose cymose clusters on short side branches, the pedicels shorter than to longer than the flowers. Calyces about ½ as long as the corolla tube, (4)5(6)-lobed ½–⅔ of the way to base, the lobes ovate, rounded at the tip, overlapping basally, but not angled. Corollas narrowed to (4)5(6) blunt or rounded lobes, these spreading to recurved, with straight to slightly incurved tips. Infrastaminal scales usually reaching filament bases, oval, densely fringed along the margins, especially near the tips. Fruits globose to globose-conical, the wall usually thickened at the tip. Seeds 1.4–1.6 mm long. $2n=60$. July–October.

Scattered to common throughout the state, except in the northernmost counties (U.S., Canada). Bottomland forests, swamps, fens, gravel bars, banks of streams, and along the margins of ponds, lakes, and sinkhole ponds. Also sometimes found along railroad tracks, as a weed in fields, and in other disturbed habitats. Parasitic on a wide variety of herbaceous and woody hosts, including species

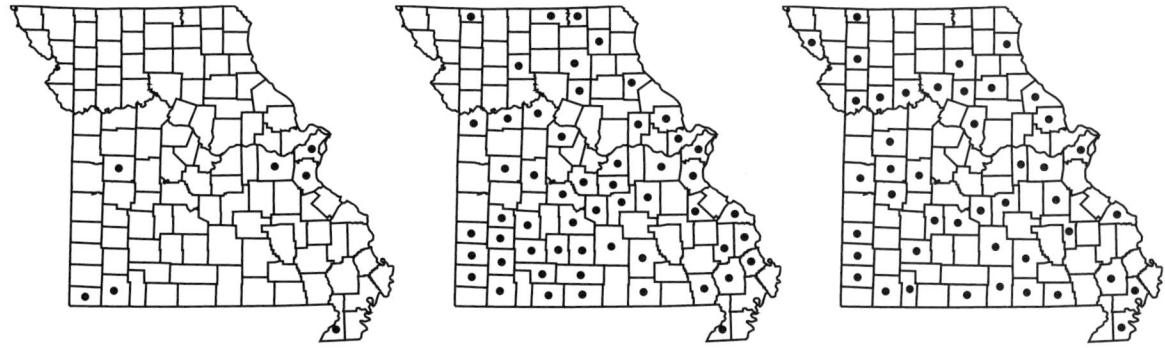

1590. Cuscuta indecora   1591. Cuscuta pentagona   1592. Cuscuta polygonorum

of *Acalypha, Agrimonia, Amphicarpaea, Apocynum, Bidens, Boehmeria, Campsis, Cephalanthus, Eupatorium, Glycine, Hibiscus, Hypericum, Impatiens, Justicia, Laportea, Lespedeza, Mikania, Penthorum, Perilla, Pilea, Polygonum, Rubus, Salix, Saururus, Solanum, Solidago, Symphyotrichum, Verbena, Verbesina, Vernonia,* and *Vitis.*

This is one of the most common species of dodder in the state, and the flowers are quite variable in size.

8. Cuscuta indecora Choisy (large alfalfa dodder, pretty dodder)

C. indecora Choisy var. *neuropetala* (Engelm.) Hitchc.

Pl. 365 c; Map 1590

Stems relatively slender, usually less than 1 mm in diameter. Flowers 2–3 mm long, with succulent, strongly granular to papillose surfaces, subtended by at most 1 lanceolate to ovate bract (usually none), in dense to loose cymose clusters on short side branches, the pedicels shorter than to longer than the flowers. Calyces shorter than to nearly as long as the corolla tube, 5-lobed $1/2$–$2/3$ of the way to the base, the lobes triangular-ovate, pointed at the tip, slightly overlapping basally, but not angled. Corollas narrowed or tapered to 4(5) sharply pointed lobes, these erect to somewhat spreading, with incurved tips. Infrastaminal scales reaching filament bases, obovate, densely fringed along the margins. Fruits depressed-globose, the wall thickened around the aperture between the style bases. Seeds 1.6–1.8 mm long. $2n=30$. June–September.

Introduced along railroad tracks in the city of St. Louis, but apparently also native in the state, based upon historical collections from "low ground" and "rocky areas" in Barry, Franklin, and McDonald Counties (U.S., south to South America). Parasitic on a wide variety of hosts (most commonly on herbaceous plants), including species of *Amphicarpaea, Ceanothus, Convolvulus, Helianthus, Pluchea, Solidago,* and *Symphyotrichum.*

This species is quite variable in flower size and in the relative length of the corolla tube. Small-flowered individuals might be mistaken for the closely related *C. coryli,* which may be distinguished by its 3- or 4-parted flowers and reduced infrastaminal scales.

9. Cuscuta pentagona Engelm. (field dodder)

Pl. 365 g; Map 1591

Stems relatively slender, usually less than 0.6 mm in diameter. Flowers 1.5–2.0 mm long, with smooth to slightly irregular surfaces, subtended by at most 1 lanceolate to ovate bract (usually none), in dense clusters on short side branches, the pedicels usually shorter than the flowers. Calyces about as long as the corolla tube, 5-lobed to halfway to base, the lobes broadly ovate, rounded at the tip, with overlapping bases forming angles. Corollas narrowed or tapered to 5 sharply pointed lobes, these spreading to recurved, with straight to slightly incurved tips. Infrastaminal scales reaching filament bases, oval, densely fringed along the margins. Fruits depressed-globose, the wall not thickened at the tip. Seeds 0.8–1.0 mm long. $2n=56$. June–October.

Scattered throughout the state (throughout the U.S.). Stream banks, swamps, and a variety of dry and wet prairie, glade, and forest types. Also frequently encountered as a weed in fields and along railroad tracks. Parasitic on a wide variety of herbaceous and woody hosts, including species of *Achillea, Amaranthus, Ambrosia, Betula, Bidens, Bromus, Calamintha, Ceanothus, Chaerophyllum, Chenopodium, Convolvulus, Conyza, Dalea, Erigeron, Euphorbia, Froelichia, Grindelia, Hedyotis, Helianthus, Justicia, Lespedeza, Medicago, Oenothera, Penstemon, Phleum, Plantago, Polygonum, Pycnanthemum, Ruellia, Salix,*

1593. Dichondra carolinensis   1594. Evolvulus alsinoides   1595. Evolvulus nuttallianus

Salsola, Sonchus, Stachys, Strophostyles, Symphoricarpos, Symphyotrichum, Thelesperma, Tragia, Trifolium, and *Verbena.*

This is one of the most common species of dodder in the state, and it is probably the most commonly encountered member of the genus in prairies and glades.

10. Cuscuta polygonorum Engelm. (smartweed dodder)

Pl. 364 h; Map 1592

Stems relatively slender, usually less than 0.6 mm in diameter. Flowers 1.5–2.0 mm long, with smooth to slightly irregular surfaces, subtended by at most 1 lanceolate to ovate bract (usually none), in dense clusters on short side branches, the pedicels absent or very short. Calyces as long as or slightly longer than the corolla tube, 3- or 4-lobed $\frac{1}{2}$–$\frac{2}{3}$ of the way to base, the lobes triangular to broadly ovate, rounded at the tip, not overlapping basally, not angled. Corollas narrowed or tapered to 3 or 4 sharply pointed lobes, these usually erect, with straight to slightly incurved tips. Infrastaminal scales usually not reaching filament bases, usually 2-lobed, toothed along the margins, usually slightly fringed near the tips. Fruits globose to depressed-globose, the wall not thickened at the tip. Seeds 1.3–1.5 mm long. July–September.

Scattered throughout the state (eastern U.S. and adjacent Canada west to Texas). Streams, ponds, swamps, bottomland forests, mesic upland forest slopes, and wet areas in prairies. Parasitic on a variety of mostly herbaceous hosts, including species of *Amaranthus, Bromus, Cephalanthus, Justicia, Lycopus, Penthorum, Polygonum, Saururus, Symphyotrichum,* and *Xanthium.*

4. Dichondra (dichondra)

About 15 species, nearly worldwide, most diverse in tropical and warm-temperate regions.

1. Dichondra carolinensis Michx. (pony-foot)

Pl. 366 e, f; Map 1593

Plants perennial herbs, scrambling but not twining. Stems 5–40 cm long, not angled, sparsely to moderately and minutely hairy. Leaves mostly relatively long-petiolate. Leaf blades 0.7–2.0 cm long, kidney-shaped to nearly circular, rounded or shallowly notched at the tip, shallowly to deeply cordate at the base, the margins otherwise entire or nearly so, the surfaces usually finely hairy. Inflorescences axillary, the flowers solitary, short-stalked. Bracts absent. Calyx of free or nearly free sepals, 2–3 mm long, similar in size and shape, obovate to spatulate, herbaceous, finely hairy on the outer surface. Corollas 1.2–2.5 mm long, relatively deeply 5-lobed, broadly funnelform to bell-shaped, pale yellow. Stamens lacking subtending scales, not exserted. Ovary deeply 2-lobed, 2-locular, with 4 ovules. Styles 2, attached in the notch between the carpels, the stigmas capitate. Fruits 2–3 mm long, each carpel developing into a thin-walled, ovoid, indehiscent or irregularly dehiscent, achenelike fruit. Seeds usually 1 per carpel, 1.5–2.5 mm long, ovate in outline, somewhat flattened on the inner face, the surface smooth, dark brown to more nearly black, often mottled, glabrous. $2n=30$. May–November.

Uncommon in southern Missouri (southeastern U.S. west to Missouri and Texas; Caribbean Is-

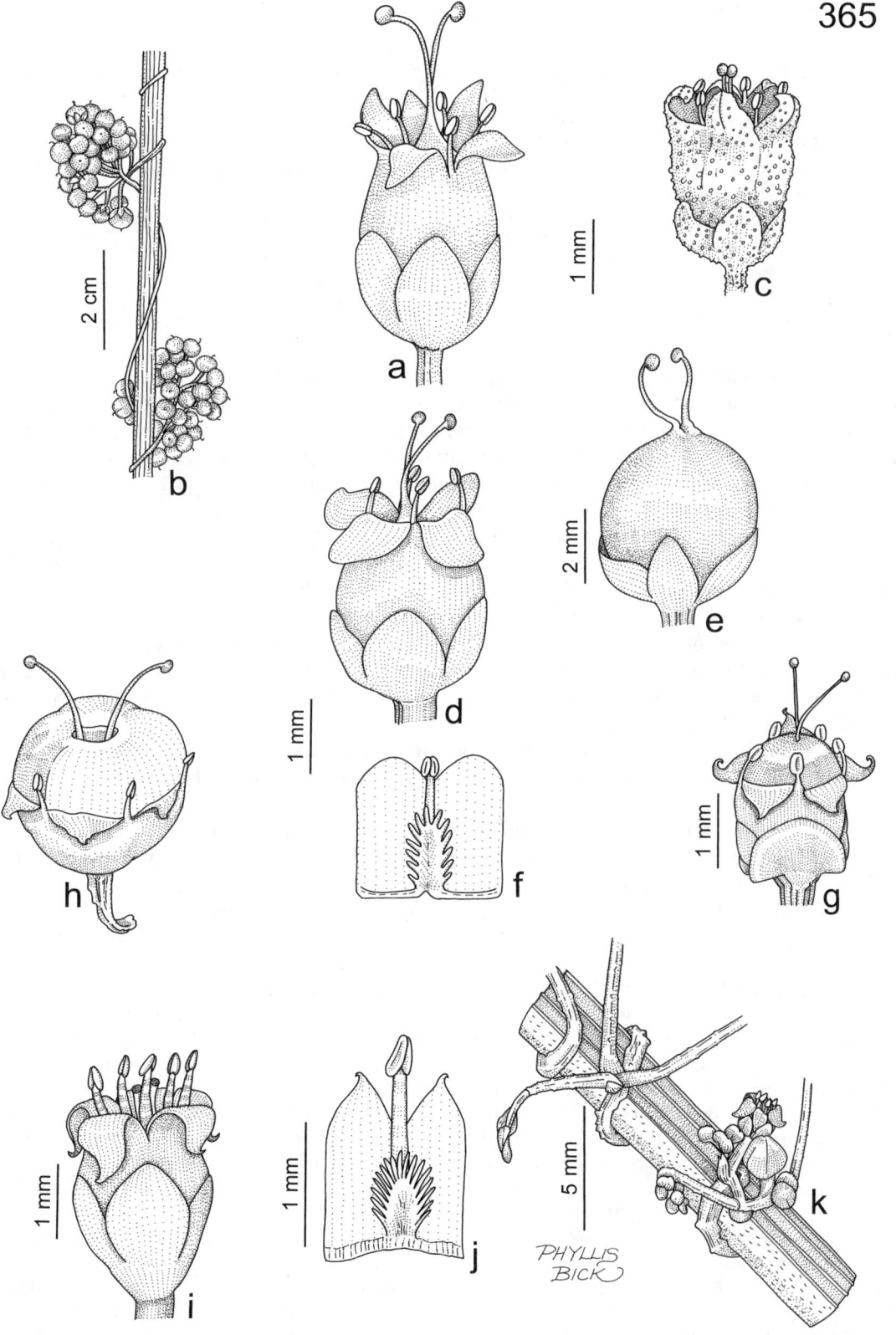

Plate 365. Convolvulaceae. *Cuscuta gronovii*, **a)** flower, **b)** fruit. *Cuscuta indecora*, **c)** flower. *Cuscuta cephalanthi*, **d)** flower, **e)** fruit, **f)** inner surface of corolla with stamen and infrastaminal scale. *Cuscuta pentagona*, **g)** flower. *Cuscuta campestris*, **h)** fruit, **i)** flower, **j)** inner surface of corolla with stamen and infrastaminal scale, **k)** habit.

lands). Pastures, cemeteries, and open, disturbed areas.

This species was first reported outside of cultivation in Missouri by Yatskievych and Summers (1991). The initial Howell County population discovered apparently represents a native occurrence, but other plants in southern Missouri are escapes from cultivation. This species and a few others in the genus are cultivated in lawns in the southeastern United States. Drew (1944) discussed field trials of the closely related *D. repens* Forst. (with which *D. carolinensis* sometimes has been combined) as a turfgrass substitute in central Missouri, concluding that plants could not survive freezing winter temperatures in that region.

5. Evolvulus (evolvulus)

Plants perennial herbs, sometimes woody at the base, not twining, sometimes with somewhat thickened root systems. Stems prostrate to erect, often somewhat angular, moderately to densely pubescent with appressed and/or spreading simple hairs. Leaves sessile or very short-petiolate. Leaf blades linear to elliptic, bluntly to sharply pointed at the tip, narrowed or tapered at the base, the margins entire, the surfaces moderately to densely hairy. Inflorescences axillary, the flowers solitary or rarely in small clusters of 2 or 3, sessile to long-stalked. Bracts 2 per flower, 1–4 mm long, distant from the flower, leaflike, shorter than the flower stalk and calyx and not hiding the calyx, not overlapping, linear, usually persistent at fruiting. Calyx of free sepals, similar in size and shape, herbaceous, relatively densely long-hairy on the outer surface. Corollas 3–6 mm long, very shallowly 5-lobed, broadly funnelform to nearly bell-shaped, lavender to blue or less commonly white. Stamens lacking subtending scales, more or less exserted. Ovary 2-locular, with 4 ovules, densely hairy. Styles 2, each 2-lobed toward the tip, the stigmas thus 4 per flower, capitate. Fruits 1- or 2-locular, dehiscing longitudinally, the wall separating into 2–4 segments. Seeds mostly 1 or 2, ovate to nearly circular in outline, sometimes somewhat longitudinally angled on the inner face, glabrous. About 100 species, nearly worldwide, most diverse in tropical and warm-temperate portions of the New World.

1. Flowers long-stalked, the stalk extending beyond the subtending leaf; sepals 2.0–2.5 mm long; corollas 3–4 mm long, blue or rarely white .... 1. E. ALSINOIDES
1. Flowers sessile or short-stalked, the stalk shorter than the subtending leaf; sepals 4–5 mm long; corollas 5.0–7.5 mm long, pale pinkish purple to nearly white .... 2. E. NUTTALLIANUS

1. Evolvulus alsinoides L.

Pl. 366 a, b; Map 1594

Stems 10–45 cm long, prostrate to more commonly erect or ascending, densely pubescent with both shorter, appressed hairs and longer, spreading hairs. Leaf blades 3–20 mm long, lanceolate to oblong-elliptic or elliptic, both surfaces moderately to densely pubescent with relatively long, appressed hairs, the upper surface sometimes only sparsely hairy at maturity. Flowers long-stalked, the stalk longer than the subtending leaf. Sepals 2.0–2.5 mm long, lanceolate, moderately to densely hairy. Corollas 3–4 mm long, blue or rarely white. Fruits 2.5–3.0 mm long, ovoid to nearly globose, glabrous. Seeds 1.5–2.5 mm long, dark brown to black. 2n=22, 26. June–August.

Introduced, known thus far from a single historical collection from St. Louis County (nearly worldwide in tropical and warm-temperate regions). Habitat unknown, but probably railroads or open, disturbed areas.

Various authors have attempted to subdivide this species into a confusing series of varieties, but the morphology of the species is too variable to warrant this action.

2. Evolvulus nuttallianus Roem. & Schult.

E. pilosus Nutt.

Pl. 366 g–i; Map 1595

Stems 7–15(–35) cm long, prostrate to rarely ascending, densely pubescent with relatively long, loosely appressed to somewhat spreading hairs. Leaf blades 6–20 mm long, oblong-linear to nar-

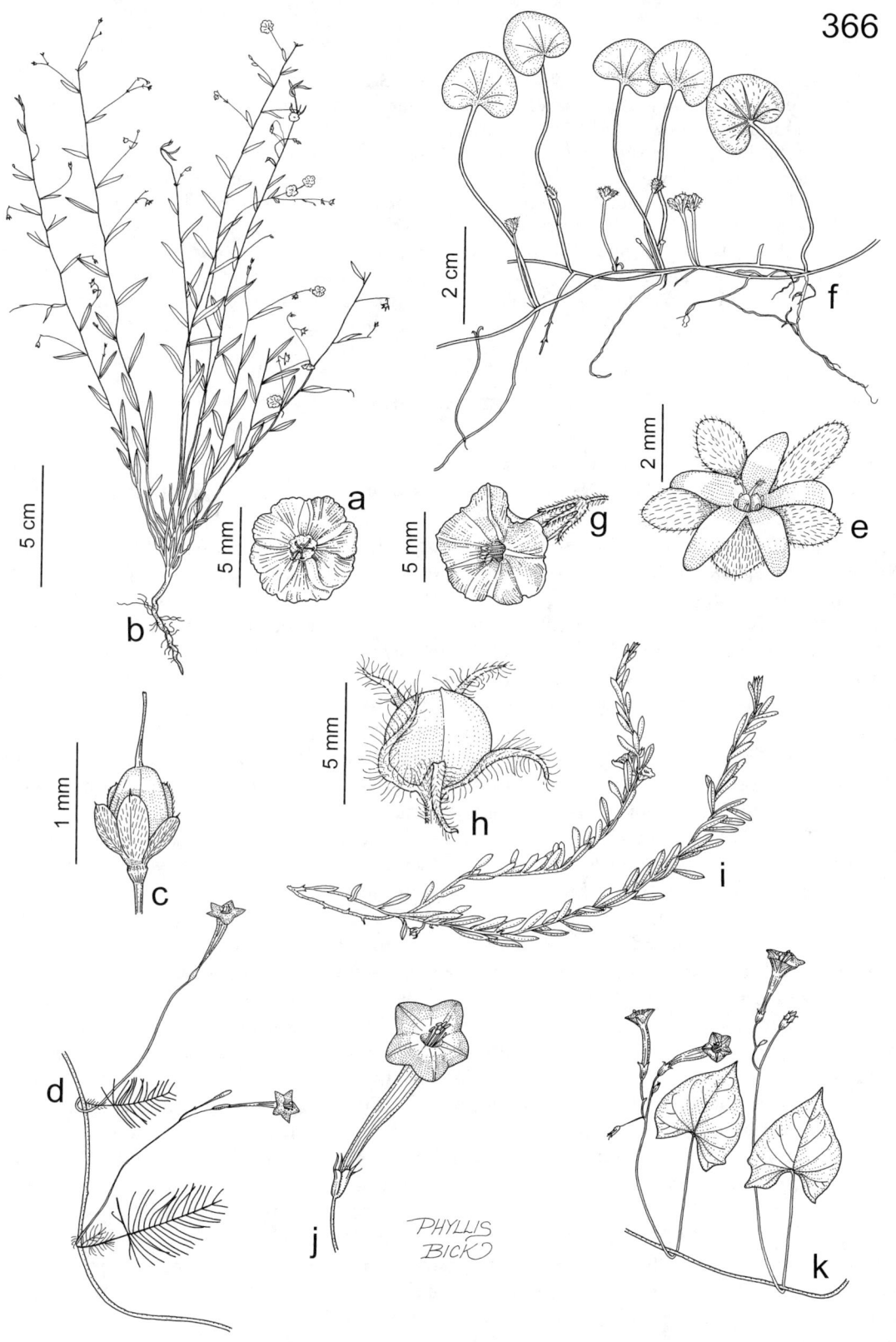

Plate 366. Convolvulaceae. *Evolvulus alsinoides*, **a)** corolla (top view), **b)** habit. *Ipomoea quamoclit*, **c)** fruit, **d)** habit. *Dichondra carolinensis*, **e)** flower, **f)** habit. *Evolvulus nuttallianus*, **g)** flower, **h)** fruit, **i)** habit. *Ipomoea coccinea*, **j)** flower, **k)** habit.

rowly lanceolate or narrowly oblanceolate, both surfaces densely pubescent with relatively long, loosely appressed hairs. Flowers sessile or short-stalked, the stalk shorter than the subtending leaf. Sepals 4–5 mm long, narrowly lanceolate to lanceolate, densely hairy. Corollas 5.0–7.5 mm long, pale pinkish purple to nearly white (pale blue elsewhere). Fruits 3.5–5.0 mm long, usually ovoid, glabrous. Seeds 2.5–3.5 mm long, dark brown to purplish brown. April–June.

Scattered in the Ozark and Ozark Border Divisions (Illinois and Tennessee west to Montana and Arizona). Glades, savannas, and openings of dry upland forests.

Although this species has relatively uniform morphology in Missouri, plants farther west tend to be more variable, having often strongly ascending stems, variable density and coloration of the hairs, and sometimes pale blue corollas. However, the patterns of morphological variation are too complex to permit the recognition of varieties.

6. Ipomoea (morning glory)

Plants annual or perennial herbs (woody elsewhere), usually scrambling or twining, sometimes with tuberous root systems. Stems sometimes somewhat angular, glabrous or hairy. Leaves short- to long-petiolate. Leaf blades variously shaped, most commonly triangular-ovate or heart-shaped, often with 1 pair of triangular lobes at the base (pinnately dissected into numerous linear lobes in *I. quamoclit;* palmately lobed or compound elsewhere), bluntly or sharply pointed at the tip, truncate to more commonly deeply cordate at the base, the margins otherwise entire or less commonly somewhat wavy or few-toothed. Inflorescences axillary, the flowers solitary or in loose clusters, long-stalked. Bracts variable, sometimes absent, often only at the inflorescence branch points, when present inconspicuous and scalelike, usually distant from, always much shorter than, and not covering the calyx, usually not overlapping, linear to ovate, often shed before fruiting. Calyx of free sepals, these similar in size and shape or unequal, 9–25 mm long, often overlapping, variously shaped, herbaceous or thickened and somewhat leathery, glabrous or variously hairy. Corollas very shallowly 5-lobed, funnelform or trumpet-shaped, white to pink, red, purple, or blue. Stamens lacking subtending scales, sometimes somewhat exserted. Ovary 2–4-locular, with 4 ovules. Style 1, the stigma 1, capitate, sometimes 2- or 3-lobed. Fruits globose to ovoid, 2(4)-locular, dehiscing longitudinally, the wall separating into usually 4 segments. Seeds 1–4, oblong-ovate to ovate in outline, somewhat longitudinally angled on the inner face, the surface smooth to very finely granular, tan to dark brown or black, glabrous or hairy. Five hundred to 650 species, nearly worldwide.

Ipomoea is most diverse in tropical and warm-temperate areas. The economically most important member of the genus is *I. batatas* L. (sweet potato), a cultigen of tropical American origin that is grown as a starchy vegetable in warmer regions around the world for its sweet, tuberous roots. A number of species also are cultivated as ornamentals, in Missouri usually as annuals on fences and trellises. Several species are important agricultural weeds. The seeds of various species contain significant quantities of hallucinogenic ergoline alkaloids; those of some species also have been used medicinally for their purgative properties.

Steyermark (1963) reported an introduced occurrence of *I. cairica* (L.) Sweet, based on a single collection by Viktor Mühlenbach (1979) from the St. Louis railyards. This native of Africa differs from other Missouri morning glories in its leaves, which are deeply palmately lobed or compound with 3–7 lobes or leaflets (Pl. 367 h). However, the specimen documenting this find could not be located during the present research and may have been discarded. Because the species has not been rediscovered in Missouri and the original find remains undocumented, this species is excluded from the state's flora for the present.

1. Leaf blades pinnately dissected into numerous linear lobes, appearing nearly pinnately compound .... 7. I. QUAMOCLIT
1. Leaf blades unlobed or with 1 or less commonly 2 pairs of triangular lobes, not appearing compound
    2. Sepals and flower stalks conspicuously pubescent on the surface and margins with spreading to downward-angled hairs; stigmas 3-lobed
        3. Sepals 12–25 mm long, long-tapered to a sharply pointed, linear tip .... 3. I. HEDERACEA
        3. Sepals 10–17 mm long, narrowed or tapered to a sharply pointed, triangular tip .... 6. I. PURPUREA
    2. Sepals and flower stalks glabrous or (in *I. lacunosa*) the sepals inconspicuously pubescent along the margins with spreading hairs; stigmas 2-lobed
        4. Corollas red to orangish red, trumpet-shaped, the tube slender, widened abruptly at the tip; stamens exserted at flowering .... 2. I. COCCINEA
        4. Corollas variously white, pink, purple, and/or blue but not red, the tube funnelform or slightly bell-shaped, widened gradually toward the tip; stamens not exserted at flowering
            5. Sepals narrowed or tapered to a sharply pointed tip, the margins sparsely to moderately hairy; corollas 1.2–2.2 cm long, all white or (in hybrids) pale pink; fruits hairy .... 4. I. LACUNOSA
            5. Sepals truncate, rounded or very bluntly pointed at the tip, sometimes tapered abruptly to a short, sharp point, the margins glabrous; corollas 2–8 cm long, white with a reddish purple center; fruits glabrous
                6. Sepals 13–20 mm long, the outer sepals noticeably shorter than the inner ones; corollas 5–8 cm long .... 5. I. PANDURATA
                6. Sepals 3.5–5.0 mm long, all similar in size; corollas 2–9 cm long
                    7. Sepals 3.5–5.0 mm long; corollas 2–4 cm long .... 1. I. AMNICOLA
                    7. Sepals 4–7 mm long; corollas 5–9 cm long .... 8. I. TRICOLOR

1. Ipomoea amnicola Morong (red-centered morning glory)

Pl. 367 i–k; Map 1596

Plants perennial, with somewhat fleshy, branched rhizomes and root systems. Stems 40–200 cm long, glabrous. Leaves long-petiolate. Leaf blades 2–7 cm long, unlobed, broadly ovate in outline, tapered to a sharply pointed tip, deeply cordate at the base, glabrous, the margins entire. Flowers in loose clusters of 3–8(–12), the stalks glabrous. Sepals similar in size and shape, the main body 3.5–5.0 mm long, broadly elliptic to nearly circular, rounded or very bluntly pointed at the tip, sometimes tapered abruptly to a short, sharp point or shallowly notched, glabrous. Corollas 2–4 cm long, funnelform to slightly bell-shaped, the tube widened gradually toward the tip, white with a reddish purple center. Stamens not exserted. Ovary usually 2-locular, the stigma 2-lobed. Fruits ovoid, the main body 7–10 mm long, the persistent style 0.5–1.0 mm long, glabrous. Seeds 4.5–5.5 mm long, the surface densely pubescent with minute, matted hairs, the angles with a crest of dense long hairs. September–October.

Introduced, known only from a single historical collection from Jackson County (native of South America; introduced uncommonly in Texas, Missouri, and northern Mexico). Disturbed areas.

This species was reported for Missouri by Shinners (1965a) and has not been rediscovered in recent years.

2. Ipomoea coccinea L. (red morning glory, scarlet starglory)

Quamoclit coccinea (L.) Moench

Pl. 366 j, k; Map 1597

Plants annual. Stems 40–300 cm long, glabrous or nearly so. Leaves short- to long-petiolate. Leaf blades 2–12 cm long, usually with a pair of short, downward-pointing or somewhat spreading basal lobes, broadly ovate to somewhat sagittate in outline, tapered to a sharply pointed tip, glabrous or inconspicuously short-hairy toward the base, the lobes triangular to narrowly triangular, sharply

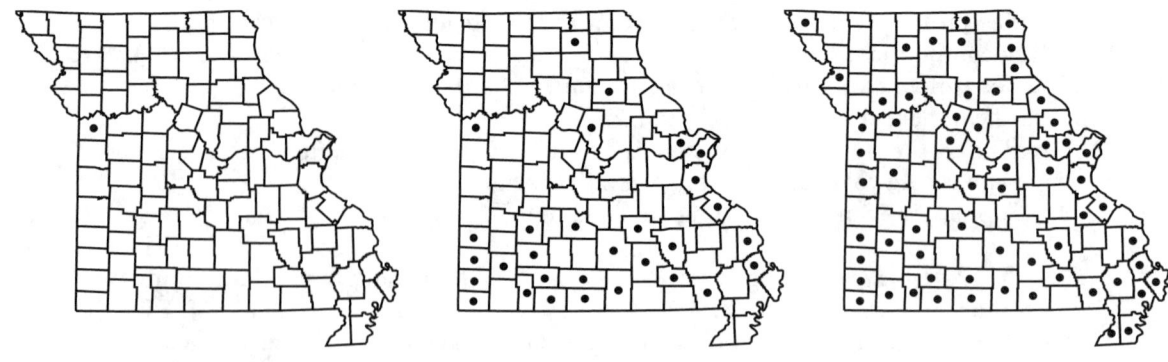

1596. Ipomoea amnicola   1597. Ipomoea coccinea   1598. Ipomoea hederacea

pointed at the tip, the margins sometimes also with a few short teeth toward the base. Flowers rarely solitary or more commonly in loose clusters of 2–8, the stalks glabrous. Sepals similar in size and shape or the outer 2 slightly shorter and narrower, the main body 3–7 mm long, broadly oblong-elliptic, rounded or truncate at the tip but with a slender, tapered awn 2–6 mm long from just below the tip, glabrous. Corollas 2.0–3.5 cm long, trumpet-shaped, the tube slender, widened abruptly at the tip, orangish red to red, with the tube and throat often yellow to yellowish orange. Stamens exserted. Ovary 4-locular, the stigma 2-lobed. Fruits globose or nearly so, the main body 5–7 mm long, the persistent style 3–4 mm long, glabrous. Seeds 3.2–3.6 mm long, the surface densely pubescent with minute, curly hairs. $2n=28$. June–October.

Scattered, mostly south of the Missouri River (eastern [mostly southeastern] U.S. west to Kansas and Texas). Banks of streams and rivers; also pastures, fencerows, ditches, crop fields, fallow fields, railroads, roadsides, and moist, open, disturbed areas.

Many earlier botanists confused *I. coccinea* with the closely related *I. hederifolia* L. (Wilson, 1960). This mostly tropical American species occurs natively as far north as Arizona to Alabama and has only rarely been recorded farther north as an introduction. With the separation of the two taxa (O'Donell, 1959) came the recognition that *I. coccinea* is a native colonizer of stream banks and other periodically disturbed, moist habitats. *Ipomoea hederifolia* differs from *I. coccinea* in its slightly longer and broader inner sepals, as well as the tendency of its leaves to have deeper lobes and sometimes to be palmately 5-lobed. This species has not yet been reported from Missouri.

Ipomoea coccinea occasionally is cultivated as an ornamental for its red flowers, which attract hummingbirds. The self-fertile flowers produce large quantities of seed, and the plant can sometimes become a nuisance in gardens. Hybrids with *I. quamoclit* are discussed under the treatment of that species.

3. Ipomoea hederacea Jacq. (blue morning glory, ivy-leaved morning glory)
I. hederacea var. *integriuscula* A. Gray
Pl. 368 h–j; Map 1598

Plants annual. Stems 30–250 cm long, moderately to densely pubescent with relatively long, spreading to downward-angled hairs. Leaves long-petiolate. Leaf blades 2–12 cm long, unlobed or more commonly deeply 3(5)-lobed, the lobes triangular, broadly ovate to ovate-triangular in overall outline, tapered to a sharply pointed tip, shallowly to more commonly deeply cordate at the base, both surfaces moderately pubescent with straight, appressed to spreading hairs, the margins otherwise entire. Flowers solitary or more commonly in loose clusters of 2 or 3(–6), the stalks moderately to densely pubescent with relatively long, spreading to downward-angled hairs. Sepals similar in size and shape, 12–25 mm long, with a short, ovate basal portion and an outward-curled, long-tapered, sharply pointed, linear tip, the surface and margins moderately to densely pubescent with relatively long, spreading to downward-angled hairs. Corollas 2.5–5.0 cm long, funnelform to slightly bell-shaped, the tube widened gradually toward the tip, purple or light blue with a white or yellowish white center. Stamens not exserted. Ovary 3-locular, the stigma 3-lobed. Fruits globose or slightly depressed-globose, the main body 8–12 mm long, the persistent style 4–15 mm long, glabrous. Seeds 4–5 mm long, the surface moderately to densely minutely hairy. $2n=30$. July–October.

Introduced, common throughout the state (native probably of tropical America, now widely introduced in the U.S. and adjacent Canada [and

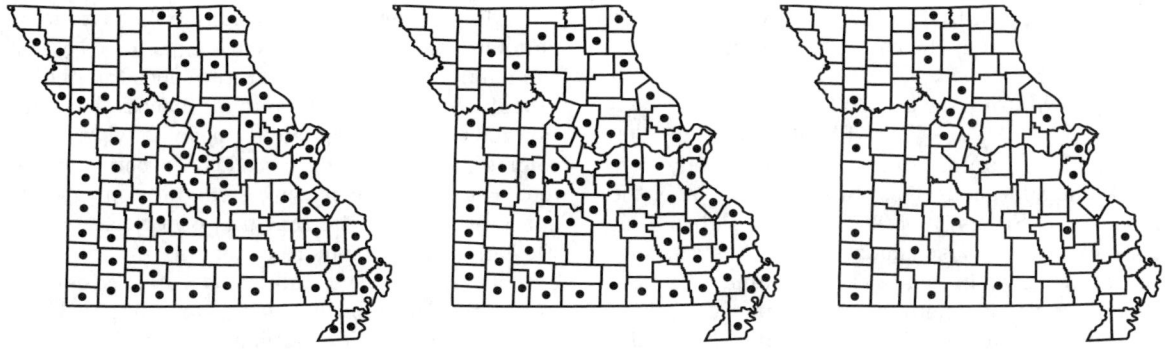

1599. Ipomoea lacunosa 1600. Ipomoea pandurata 1601. Ipomoea purpurea

other warm-temperate to tropical portions of the world] west to North Dakota and Arizona). Banks of streams and rivers; also crop fields, fallow fields, gardens, pastures, fencerows, ditches, railroads, roadsides, and open, disturbed areas.

Ipmoea hederacea and *I. purpurea* were confused in some of the older botanical literature but were treated correctly by Steyermark (1963). Both are weedy species with variable leaf shapes. Elmore (1986) performed controlled crosses between entire-leaved and 3-lobed (ivy-leaved) lineages of *I. hederacea*. He concluded that leaf lobing is controlled by a single gene, with the lobed genotype (which is more common in Missouri) dominant over the unlobed one. The recognition of varieties based upon this minor variation is thus unwarranted. There is some controversy as to the natural range of the species, with most authors postulating a neotropical origin. However, Austin (1990) and others have suggested that *I. hederacea* originally was endemic to the southeastern United States. The species' spread as a contaminant in agricultural products was early enough that it may not be possible to discern its native provenance with any certainty.

4. Ipomoea lacunosa L. (small white morning glory)

Pl. 367 a–d; Map 1599

Plants annual. Stems 10–300 cm long, glabrous or sparsely and inconspicuously hairy. Leaves long-petiolate. Leaf blades 2–8 cm long, unlobed or shallowly to deeply 3(5)-lobed, the lobes triangular, broadly ovate to ovate-triangular in overall outline, tapered to a sharply pointed tip, shallowly to deeply cordate at the base, glabrous or both surfaces sparsely to moderately short-hairy, the margins otherwise entire. Flowers solitary or more commonly in loose clusters of 2–6, the stalks glabrous, often appearing somewhat warty when dry. Sepals similar in size and shape or the outer 2 slightly shorter and narrower, (8–)10–14 mm long, lanceolate to narrowly ovate, or narrowed or tapered to a sharply pointed tip, the margins sparsely to moderately pubescent with spreading hairs. Corollas 1.2–2.2 cm long, funnelform to slightly bell-shaped, the tube widened gradually toward the tip, all white (for pink-flowered individuals, see the discussion of hybrids below). Stamens not exserted. Ovary usually 2-locular, the stigma 2-lobed. Fruits globose or slightly depressed-globose, the main body 10–14 mm long, the persistent style 0.8–1.5 mm long, moderately to densely pubescent with relatively long, often curly, spreading hairs. Seeds 5–6 mm long, the surface glabrous. $2n=30$. August–October (rarely as early as June).

Common nearly throughout the state (eastern U.S. west to Iowa, Kansas, and Texas; Canada). Banks of streams and rivers, margins of ponds and lakes, moist depressions of upland prairies, and fens; also crop fields, fallow fields, old fields, pastures, fencerows, ditches, railroads, roadsides, and open, disturbed areas.

Austin (1978) and Abel and Austin (1981) interpreted plants assignable to *I. lacunosa* f. *purpurata* Fernald as representing introgressive hybrids between *I. lacunosa* and *I. cordato-triloba* Dennst. (*I. trichocarpa* Elliott), a southeastern species that has not been documented from Missouri. These fertile hybrids, known as *I.* ×*leucantha* Jacq., are common in the southeastern United States, and the sporadic occurrences nearly throughout the Missouri range of *I. lacunosa* may originally have been introduced from there. They differ from *I. lacunosa* in having slightly larger flowers with light pink corollas.

5. **Ipomoea pandurata** (L.) G. Mey. (wild potato vine, man-of-the-earth, bigroot morning glory)

I. pandurata f. *leviuscula* Fernald

Pl. 367 e–g; Map 1600

Plants perennial, with a somewhat woody rootstock and a large, deep-set tuberous portion of the main root. Stems 40–500 cm long, glabrous or sparsely and inconspicuously hairy. Leaves long-petiolate. Leaf blades 2–12 cm long, unlobed, broadly ovate or sometimes pear-shaped in outline, tapered to a sharply pointed tip, shallowly to more commonly deeply cordate at the base, glabrous or the undersurface sparsely to moderately short-hairy, some of the hairs sometimes glandular, the margins entire. Flowers solitary or more commonly in loose clusters of 2–7(–13), the stalks glabrous. Sepals not similar in size and shape, the outer sepals noticeably shorter and slightly narrower than the inner ones, 13–20 mm long, oblong-elliptic to less commonly oblong-ovate, rounded or very bluntly pointed at the tip, occasionally tapered abruptly to a short, sharp point or shallowly notched, glabrous or less commonly minutely hairy toward the margins. Corollas 5–8 cm long, funnelform to slightly bell-shaped, the tube widened gradually toward the tip, white with a reddish purple center. Stamens not exserted. Ovary 2- or 4-locular, the stigma 2-lobed. Fruits ovoid, the main body 10–16 mm long, the persistent style 0.5–35.0 mm long and frequently becoming irregularly curled, glabrous. Seeds 7–9 mm long, the surface densely pubescent with minute, matted hairs, the angles with a crest of dense, long hairs. $2n=30$. May–September.

Scattered to common nearly throughout the state (eastern U.S. west to Nebraska and Texas; Canada). Banks of streams and rivers, margins of ponds and lakes, and less commonly glades; also crop fields, fallow fields, old fields, ditches, railroads, roadsides, and open, disturbed areas.

The large, vertical, fleshy, tuberous root of this species, which can reach lengths of 0.6 m or more and can weigh more than 10 kg (Steyermark, 1963), is deep-set and difficult to excavate, but it is edible. Native Americans cooked and ate it as a starchy vegetable. However, the rootstock also has been said to have mild purgative properties.

6. **Ipomoea purpurea** (L.) Roth (common morning glory)

I. purpurea var. *diversifolia* (Lindl.) O'Donell
I. hirsutula Jacq.

Pl. 368 e–g; Map 1601

Plants annual. Stems 30–400 cm long, moderately to densely pubescent with relatively long, spreading to downward-angled hairs. Leaves long-petiolate. Leaf blades 2–12 cm long, unlobed or less commonly deeply 3(5)-lobed, the lobes triangular, broadly ovate to ovate-triangular in overall outline, tapered to a sharply pointed tip, shallowly to more commonly deeply cordate at the base, both surfaces moderately pubescent with straight, appressed to spreading hairs, the margins otherwise entire. Flowers solitary or more commonly in loose clusters of 2 or 3(–6), the stalks moderately to densely pubescent with relatively long, spreading to downward-angled hairs. Sepals similar in size and shape or the outer 2 slightly longer and broader than the inner ones, 10–17 mm long, oblong-lanceolate to narrowly oblong-ovate, narrowed or short-tapered to a sharply pointed, somewhat outward-curved, triangular tip, the surface and margins moderately to densely pubescent with relatively long, spreading to downward-angled hairs. Corollas 2.5–5.0 cm long, funnelform to slightly bell-shaped, the tube widened gradually toward the tip, pink, purple, or less commonly white or light blue with a white or yellowish white center. Stamens not exserted. Ovary 3-locular, the stigma 3-lobed. Fruits globose or slightly depressed-globose, the main body 8–10 mm long, the persistent style 2–5 mm long, glabrous. Seeds 3.8–4.5 mm long, the surface moderately to densely minutely hairy. $2n=30$. July–October.

Scattered sporadically nearly throughout the state (native of tropical America; widely naturalized in the U.S. and Canada, and other warm-temperate to tropical portions of the world). Banks of streams and rivers; also crop fields, fallow fields, gardens, ditches, railroads, roadsides, and open, disturbed areas.

As in *I. hederacea*, differences in leaf lobing are not thought to be worthy of formal taxonomic recognition. Whereas specimens of *I. hederacea* from Missouri mostly have 3-lobed leaves, those of *I. purpurea* mostly have unlobed leaves.

7. **Ipomoea quamoclit** L. (cypress vine)

Quamoclit vulgaris Choisy

Pl. 366 c, d; Map 1602

Plants annual. Stems 100–500 cm long, glabrous. Leaves mostly short-petiolate. Leaf blades 2–9 cm long, pinnately dissected into 9–19 pairs of lobes, appearing nearly pinnately compound, broadly ovate in outline, the lobes linear, sharply pointed at the tip, glabrous. Flowers solitary or in loose clusters of 2–5, glabrous. Sepals similar in size and shape or the outer 2 slightly shorter and narrower, 4–7 mm long, oblong-elliptic, rounded or narrowed to a bluntly pointed tip but with a minute, sharp point from just below the tip, glabrous. Co-

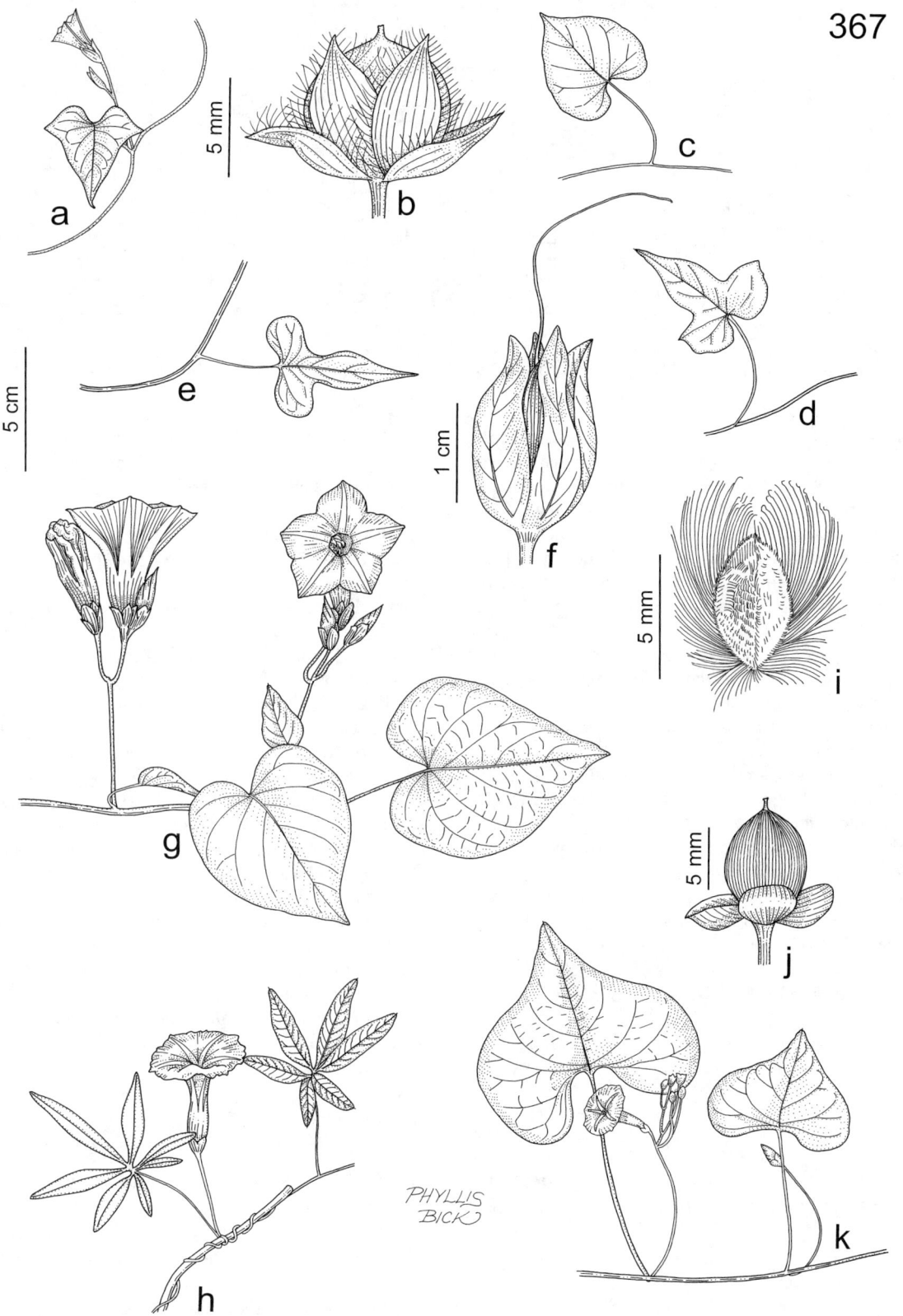

Plate 367. Convolvulaceae. *Ipomoea lacunosa*, **a)** node with flowers and 3-lobed leaf, **b)** fruit, **c)** unlobed leaf, **d)** 3-lobed leaf. *Ipomoea pandurata*, **e)** 3-lobed leaf, **f)** fruit, **g)** habit with unlobed leaves. *Ipomoea cairica*, **h)** flowering branch. *Ipomoea amnicola*, **i)** seed, **j)** fruit, **k)** habit.

1602. Ipomoea quamoclit   1603. Ipomoea tricolor   1604. Jacquemontia tamnifolia

rollas 2.2–3.5 cm long, trumpet-shaped, the tube slender, widened abruptly at the tip, scarlet red (rarely white elsewhere) with the throat usually yellow or white. Stamens exserted. Ovary 4-locular, the stigma 2-lobed. Fruits ovoid, the main body 7–10 mm long, the persistent style 5–9 mm long, glabrous. Seeds 4.5–5.5 mm long, the surface sparsely to moderately minutely hairy. $2n=30$. June–October.

Introduced, uncommon and widely scattered (native of tropical America; widely but sporadically introduced in the eastern U.S.). Banks of streams; also fencerows, roadsides, railroads, and open, disturbed areas.

This species is cultivated for its large displays of brilliant scarlet flowers that attract hummingbirds. Once allowed to fruit, these self-fertile plants will tend to regrow from seed in future years. Fertile hybrids of intermediate morphology with *I. coccinea* have been developed horticulturally and are named *I.* ×*multifida* (Raf.) Shinners, but these have not yet escaped from cultivation in Missouri.

8. Ipomoea tricolor Cav. (morning glory)

Map 1603

Plants annual (sometimes perennial farther south). Stems 40–300 cm long, glabrous. Leaves long-petiolate. Leaf blades 2–12 cm long, unlobed, broadly ovate in outline, tapered to a sharply pointed tip, shallowly to deeply cordate at the base, glabrous, the margins entire. Flowers in loose clusters of 3–8(–15), rarely solitary, the stalks glabrous. Sepals similar in size and shape, 4–7 mm long, narrowly ovate to lanceolate, tapered to a sharply pointed tip, glabrous, with relatively broad, pale margins. Corollas 5–9 cm long, funnelform, the tube widened gradually toward the tip, blue, purple, reddish purple, or white. Stamens not exserted. Ovary usually 2-locular, the stigma 2-lobed. Fruits ovoid, the main body 9–14 mm long, the persistent style 5–9 mm long, usually hairy. Seeds 5–7 mm long, the surface glabrous. September–October.

Introduced, known thus far from a single collection from St. Francois County (native of Mexico, Central America; introduced sporadically in mostly the southern U.S.). Roadsides.

This cultivated morning glory is grown widely on fences and trellises in the United States. A number of cultivars are available commercially, especially cv. 'Heavenly Blue,' which has large, blue corollas. Plants are propagated from seeds and reseed themselves readily.

7. Jacquemontia

Eighty to 120 species, nearly worldwide in tropical and warm-temperate regions, most diverse in the New World.

1. Jacquemontia tamnifolia (L.) Griseb. (tie vine, smallflower morning glory)

Pl. 368 a, b; Map 1604

Plants annuals, scrambling and twining. Stems 40–200 cm long, frequently erect when young, but eventually becoming loosely ascending to prostrate, not angled, sparsely to densely pubescent with relatively long, loosely spreading hairs. Leaves short- to long-petiolate. Leaf blades 3–12 cm long, ovate-triangular to elliptic-ovate, narrowed or tapered to a usually sharply pointed tip, cordate or less commonly rounded at the base, the margins entire and

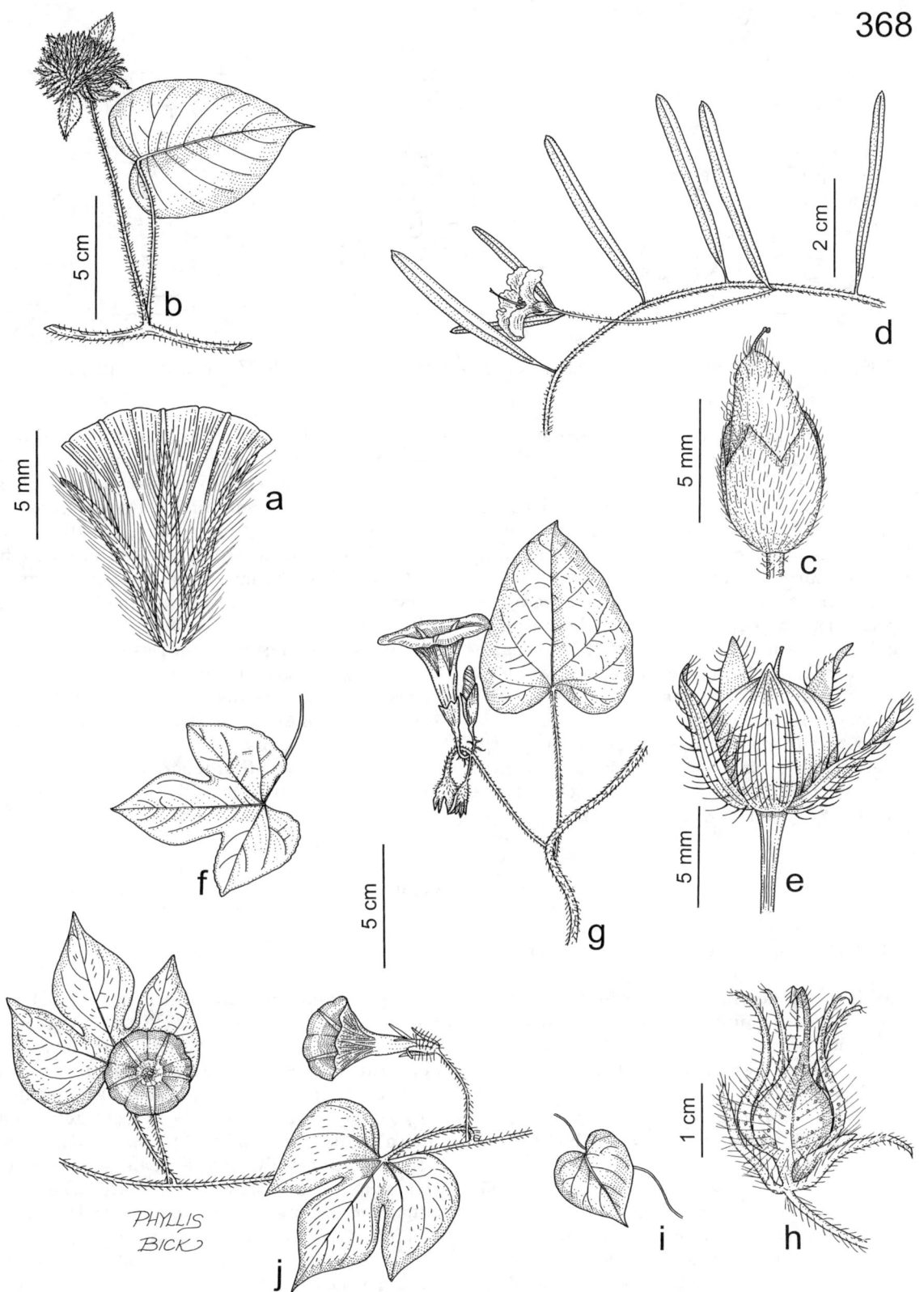

Plate 368. Convolvulaceae. *Jacquemontia tamnifolia*, **a)** flower, **b)** node with leaf and inflorescence. *Stylisma pickeringii*, **c)** fruit, **d)** habit. *Ipomoea purpurea*, **e)** fruit, **f)** 3-lobed leaf, **g)** node with flowers and unlobed leaf. *Ipomoea hederacea*, **h)** fruit, **i)** unlobed leaf, **j)** habit with 3-lobed leaves.

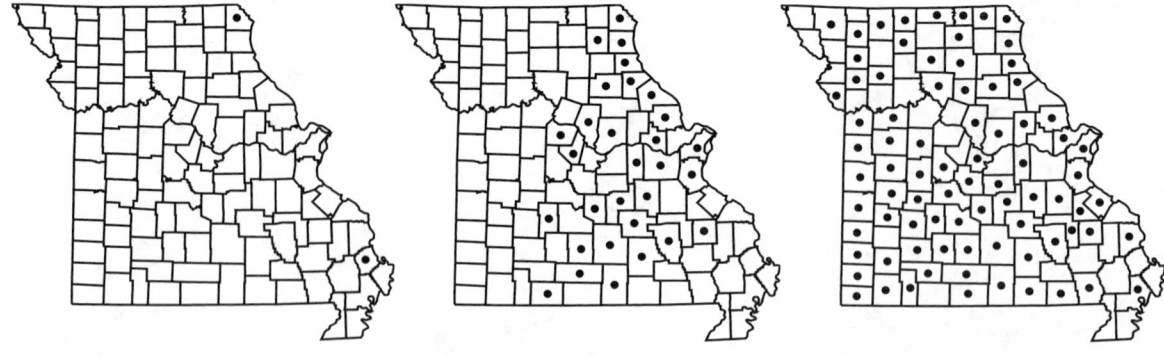

1605. Stylisma pickeringii   1606. Cornus alternifolia   1607. Cornus amomum

hairy, the surfaces glabrous or sparsely hairy. Inflorescences axillary, short- to long-stalked, the flowers sessile or short-stalked in dense, headlike clusters. Bracts usually numerous subtending each flower cluster, forming an involucre, each 9–26 mm long, herbaceous, usually about as long as and slightly obscuring the calyx, often somewhat overlapping, narrowly lanceolate to elliptic, moderately to densely long-hairy, especially along the margins, usually persistent at fruiting. Calyx of free sepals, 7–12 mm long, similar in size and shape, narrowly lanceolate to linear-triangular, herbaceous, densely pubescent with spreading to loosely appressed hairs on the outer surface. Corollas 1.2–1.6 cm long, very shallowly 5-lobed, funnelform to nearly bell-shaped, blue. Stamens lacking subtending scales, not exserted. Ovary 2-locular, with 4 ovules, densely hairy.

Style 1, 2-lobed toward the tip, the stigmas thus 2 per flower, oblong to elliptic in outline, flattened. Fruits 4–7 mm long, ovoid to nearly spherical, 2-locular, dehiscing longitudinally, the wall separating into usually 4 segments. Seeds mostly 4, 2.5–3.0 mm long, ovate in outline, somewhat longitudinally angled on the inner face, the surface tan to brown, pebbled to minutely warty, glabrous. $2n=18$. July–October.

Introduced, uncommon in Butler County and the city of St. Louis (southeastern U.S. west to Texas, introduced farther north; Mexico, Central America, South America). Railroads, gardens, and open, disturbed areas.

This species was first reported for Missouri by Mühlenbach (1979). It is cultivated occasionally in the Midwest as an ornamental.

8. Stylisma

Six species, eastern and central U.S.

1. Stylisma pickeringii (Torr. ex M.A. Curtis) A. Gray **var. pattersonii** (Fernald & B.G. Schub.) Myint

S. pattersonii (Fernald & B.G. Schub.) G.N. Jones

Breweria pickeringii (Torr. ex M.A. Curtis) A. Gray var. *pattersonii* Fernald & B.G. Schub.

Pl. 368 c, d; Map 1605

Plants perennial herbs, scrambling but not twining, with deep-set root systems. Stems usually prostrate, 80–200 cm long, not angled, minutely pubescent with appressed, usually 2-branched hairs. Leaves sessile or very short-petiolate. Leaf blades 2–6 cm long, linear, bluntly to sharply pointed at the tip, narrowed or tapered at the base, the margins entire, the surfaces glabrous to moderately and minutely hairy. Inflorescences axillary, the flowers solitary or in loose clusters of 2–5, usually long-stalked. Bracts 2 per flower, but often appearing clustered at the tip of the inflorescence stalk, 5–25 mm long, herbaceous, usually about as long as the flower stalk and calyx but not hiding the calyx, not overlapping, linear, usually persistent at fruiting. Calyx of free sepals, 4–6 mm long, similar in size and shape, oblong-ovate to oblong-elliptic, herbaceous to somewhat thickened, densely and finely hairy on the outer surface. Corollas 1.2–1.8 cm long, very shallowly 5-lobed, broadly funnelform to nearly bell-shaped, white. Stamens lacking subtending scales, more or less exserted. Ovary 2-locular, with 4 ovules, densely hairy. Style 1, unequally 2-lobed toward the tip, the stigmas thus 2 per flower, capitate. Fruits 5–9 mm long, ovoid, 1-

or 2-locular, dehiscing longitudinally, the wall separating into 2–4 segments. Seeds mostly 1 or 2, 3.5–4.5 mm long, ovate in outline, somewhat longitudinally angled on the inner face, the surface brown to yellowish brown, glabrous. $2n=28$. June–September.

Uncommon, known thus far only from Clark and Scott Counties (Illinois to Kansas south to Louisiana and Texas). Sand prairies; also levees and open, disturbed areas on sandy substrates.

Myint (1966) treated the genus *Stylisma* as separate from *Bonamia* Thouars and *Breweria* R. Br., two mostly tropical groups with which it had been combined by some earlier authors. Myint also treated *S. pickeringii* as comprising two varieties. The var. *pickeringii*, which includes plants segregated as two other varieties by some earlier authors, occurs almost entirely in the southeastern Coastal Plain. It differs in its longer (2–3 mm) subequal style branches and usually broadly rounded sepal tips. In contrast, the scattered, more inland populations ascribed to var. *pattersonii* have shorter (1.0–1.5 mm), usually unequal style branches, and mostly bluntly pointed sepal tips.

This rare taxon was first reported from northeastern Missouri by Conrad (1989) but was subsequently discovered in southeastern Missouri by Holmes (1995). It is an indicator of sand prairie habitats, but apparently it can persist in disturbed, sandy areas where the original prairie has been destroyed.

CORNACEAE (Dogwood Family)
Contributed by David J. Bogler

Plants trees or shrubs (rhizomatous herbs elsewhere), sometimes monoecious or dioecious, variously pubescent, often with branched hairs. Leaves opposite or alternate, simple, usually petiolate, lacking stipules, the blades with the margins entire or less commonly minutely wavy or with a few coarse teeth, the venation pinnate, the secondary veins often arching. Inflorescences terminal on the branches or axillary (in *Nyssa*), short, broad panicles, compound umbels, short racemes, dense or nearly umbellate clusters, heads, or of solitary flowers. Flowers perfect or imperfect, small, actinomorphic, epigynous. Calyces actinomorphic, fused to the ovary, the tube not or only slightly extending past the ovary, the free portion with 4 or 5 minute, toothlike lobes or reduced to a minute rim. Petals 4–10, free (occasionally fused toward the base elsewhere), often relatively small. Stamens 4 or 5, in staminate flowers often 8–10, usually alternate with the petals, the filaments free, the anthers attached at the base or near the midpoint. Staminodes absent (rarely a few stamens nonfunctional in staminate flowers). Pistil 1 per flower (highly reduced and nonfunctional in staminate flowers), composed of usually 2 fused carpels (to 9 carpels elsewhere) but sometimes 1 of the carpels abortive and not apparent at maturity, the inferior ovary with a nectar disc at the tip (this also present in staminate flowers), the styles 1 or 2, the stigma(s) capitate. Ovules 1 per locule, the placentation apical or axile. Fruits drupes (rarely berries elsewhere), with 1 stone, this often with several longitudinal ridges and grooves. Seeds 1 or 2 (1 per locule). Ten to 14 genera, about 120 species, nearly worldwide, but most diverse in temperate portions of the Northern Hemisphere.

Within the Cornaceae there are 2 major groups: *Nyssa* and two related genera with mostly imperfect flowers having mostly 5-parted perianth whorls; and the remaining genera that are allied to *Cornus* with mostly perfect flowers having mostly 4-parted perianth whorls. Some botanists have treated *Nyssa* and its relatives as a separate family, Nyssaceae (Eyde, 1966; Ferguson, 1966b; Cronquist, 1981), but strong similarities in floral morphology, phytochemistry, and chromosome numbers argue for their inclusion in the Cornaceae. The broad circumscription of Cornaceae also is supported by molecular data (Xiang et al., 1998).

Both of the Missouri genera contain some species that are cultivated as ornamentals. A few of the other genera also occasionally are cultivated as hedges, ornamentals, or specimen plants, often in greenhouses and conservatories. These include the Asian genera *Helwingia*

Willd., with the inflorescence appearing to originate from the middle of a foliage leaf, and *Davidia* Baill. (dove tree, ghost tree), with the inflorescences consisting of a small head subtended by a pair of large, showy petaloid bracts.

1. Plants shrubs or small trees; leaves opposite (alternate in *C. alternifolia*); secondary veins strongly arched toward the leaf tip; flowers perfect; petals and stamens 4 .... 1. CORNUS
1. Plants medium-sized to large trees; leaves alternate (but sometimes appearing nearly whorled on short shoots); secondary veins straight or only slightly arched; flowers all or mostly imperfect, the staminate and pistillate flowers often on different plants; petals and stamens 5–10 .... 2. NYSSA

1. Cornus L. (dogwood)

Plants shrubs or small trees (occasionally rhizomatous herbs elsewhere). Leaves usually opposite (alternate in *C. alternifolia*). Winter buds ovoid, with usually 2 scales. Twigs with white to brown pith lacking diaphragms (cross-partitions). Leaf blades with the margins entire or less commonly minutely wavy, the upper surface green to dark green, the undersurface usually pale green (except sometimes in *C. foemina*), the secondary veins strongly arched toward the leaf tip, becoming irregularly fused toward the leaf margin. Inflorescences terminal on the branches, short, broad panicles or usually compound umbels, in *C. florida* dense heads surrounded by 4 showy petaloid bracts. Flowers perfect (imperfect elsewhere). Calyces with the free portion consisting of 4 small, triangular lobes 0.1–2.0 mm long or a minute, low rim. Petals 4, inserted along the margin of the nectar disc, white, cream-colored, or greenish yellow. Stamens 4, the slender filaments 3–4 mm long, attached along the margin of the nectar disc, the anthers 1.0–1.5 mm long, narrowly oblong, attached toward the midpoint. Pistil of 2 fused carpels but 1 carpel sometimes aborting during development, the ovary frequently hairy, with 1 or 2 locules. Style 1, stout or slender, the stigma often 2-lobed. Ovule(s) 1 or 2 (1 per locule). Fruits ovoid to spherical, red, white, or dark blue. Stone 1- or 2-seeded, the seeds oblong, flattened. Forty to 65 species, North America, South America, Europe, Asia, Africa.

In the past, some botanists have treated *Cornus* in a narrower sense, with various species groups segregated into 7–9 small genera (Murrell, 1993). Because these groups are more closely related to each other than to anything outside the genus, most botanists now regard them as subgenera of *Cornus* (Eyde, 1987). There are two major groups within *Cornus,* one with red fruits and large petaloid bracts subtending the inflorescence, and the other with blue or white fruits and the inflorescence bracts minute or absent (Eyde, 1988; Fan and Xiang, 2001). Taxonomic relationships within the blue- to white-fruited lineage are complicated by morphological intergradation and apparent hybridization, in addition to many of the original taxon descriptions having very brief, incomplete diagnoses and lacking designated type specimens. Furthermore, the size and shape of the leaves, degree of pubescence, and color of the twigs and fruits may be affected by exposure to sun and shade or other environmental factors. Flower morphology is more or less uniform, but the inflorescences vary from open panicles to compact, headlike clusters. For this group, the present treatment tentatively follows that of Wilson (1965).

Cornus is prized for its hard wood, which has been used in spears, daggers, and weaving shuttles. Additionally, several species are cultivated in the Midwest as hedges, specimen plants, and as ornamentals, especially cultivars with red or yellow stems. Dogwood fruits also provide food for various mammals and birds and are recommended for wildlife plantings. Various species are larval food plants for several groups of moths. The European species *C. mas* L. (Cornelian cherry) produces delicious fruits that are often prepared into jams and syrups. Steyermark (1963) noted that Native Americans prepared a tobacco substitute from strips of the bark of several species under the name kinnikinnick.

1. Leaves alternate, tending to be clustered toward the tips of branches; petioles 3–5 cm long .... 1. C. ALTERNIFOLIA
1. Leaves opposite, usually relatively evenly dispersed along the branches, petioles 0.5–1.5 cm long
   2. Inflorescences dense heads subtended by 4 showy petaloid bracts; fruit ovoid to ellipsoid, red .... 4. C. FLORIDA
   2. Inflorescences panicles or compound umbels, the bracts absent or rarely a few at the branch points and these minute and scalelike; fruit more or less spherical, blue to dark blue or white
      3. Young twigs densely hairy; leaves with mostly 5 or 6 pairs of lateral veins; sepals 1 mm long or longer; style club-shaped, expanded at the tip; fruits dark blue, stones longitudinally grooved .... 2. C. AMOMUM
      3. Young twigs glabrous; leaves with mostly 3 or 4 pairs of lateral veins; sepals less than 1 mm long; style relatively slender, not expanded at the tip; fruits dark blue or white, stones not grooved
         4. Leaves with the upper surface moderately to strongly roughened (also with minute, appressed hairs), the undersurface moderately pubescent with more or less spreading, often somewhat woolly, unbranched and more or less basally branched (V-shaped or Y-shaped) hairs, the secondary veins mostly arising from the basal half of the blade .... 3. C. DRUMMONDII
         4. Leaves with the upper leaf surface smooth (but often with scattered minute, appressed hairs), the undersurface glabrous or sparsely pubescent with minute, straight, appressed hairs attached at their midpoints (T-shaped hairs), the secondary veins evenly spaced
         .... 5. C. FOEMINA

1. Cornus alternifolia L.f. (pagoda dogwood, alternate-leaved dogwood, green osier, pigeonberry)
Svida alternifolia (L.f.) Small

Pl. 369 d, e; Map 1606

Plants shrubs or small trees 2–6 m tall, usually occurring as solitary individuals, the trunk often forked near the base. Twigs greenish yellow, glabrous, the pith white. Bark smooth or with shallow fissures, that of older stems usually thick and somewhat corky, reddish brown to grayish brown. Leaves alternate, tending to be clustered toward the tips of branches, the petiole 3–5 cm long. Leaf blades 5–10 cm long, 3–6 cm wide, ovate to broadly elliptic or obovate, angled or short-tapered at the base, conspicuously tapered to a sharply pointed tip, the surfaces sparsely to moderately pubescent with mostly appressed, straight, unbranched and T-shaped hairs, the lateral veins 4–6 pairs, these relatively evenly spaced. Inflorescences appearing paniculate, mostly hemispherical, the bracts absent, the flower stalks 3–4 mm long, glabrous, becoming red as the fruits mature. Sepals 0.1–0.2 mm long. Petals 2.5–3.0 mm long, oblong, white to cream-colored. Style 2.0–2.4 mm long, slender, not broadened toward the tip. Fruits 4–5 mm in diameter, spherical, dark blue to bluish black. Stone deeply pitted at the tip. $2n=20$. May–July.

Scattered in the Ozark and Ozark Border Divisions north locally in eastern Missouri to Clark County (eastern U.S. west to Minnesota, Arkansas, and Mississippi; Canada). Bottomland forests, mesic upland forests, bases and ledges of bluffs, and bottoms of forested sinkholes.

This species is relatively widespread in the understories of deciduous forests in the northeastern United States. It is commonly cultivated as an ornamental shrub for its interesting growth form, attractive foliage, and conspicuous infructescences. The leaves turn yellow to red in the autumn and the fruits are eaten by a wide variety of birds and mammals.

2. Cornus amomum Mill. ssp. **obliqua** (Raf.) J.S. Wilson (swamp dogwood, pale dogwood)
C. obliqua Raf.
C. amomum var. *schuetziana* (C.A. Mey.) Rickett

Pl. 369 i; Map 1607

Plants shrubs 1–4 m tall, usually occurring as solitary individuals. Twigs reddish brown to purplish brown, densely hairy when young, becoming glabrous or nearly so with age, the pith brown. Bark smooth or with shallow fissures, not becoming corky, reddish brown, usually with small, slightly

1608. Cornus drummondii   1609. Cornus florida   1610. Cornus foemina

raised to somewhat warty, lighter dots. Leaves opposite, usually relatively evenly dispersed along the branches, the petiole 0.5–1.5 cm long. Leaf blades 4–9 cm long, 1–4 cm wide, narrowly lanceolate or narrowly elliptic to ovate, angled or tapered at the base, angled or more commonly tapered to a bluntly or sharply pointed tip, the surfaces sparsely to moderately pubescent with appressed, straight, T-shaped hairs and somewhat spreading, V-shaped or Y-shaped hairs, the lateral veins (4)5 or 6 pairs, these relatively evenly spaced. Inflorescences appearing more or less umbellate, flat-topped to shallowly convex, the bracts absent or rarely a few at the branch points and these minute and scalelike, the flower stalks 2–3 mm long, hairy, becoming reddish brown as the fruits mature. Sepals 0.6–1.2 mm long. Petals 3.5–5.0 mm long, narrowly oblong-lanceolate, white to cream-colored. Style 2–4 mm long, relatively stout, broadened toward the tip. Fruits 5–8 mm in diameter, spherical, dark blue, sometimes with white blotches. Stone with 6–9 sharply angled longitudinal ridges. $2n=22$. May–July.

Scattered throughout the state (northeastern U.S. west to South Dakota and Oklahoma; Canada). Banks of streams and rivers, margins of ponds and lakes, fens, bottomland forests, moist depressions of upland prairies, and rarely mesic to dry upland forests; also pastures, fencerows, railroads, and roadsides.

This species is distinguished by its silky pubescent, often maroon twigs, smooth leaves, dilated style, relatively long sepals, and blue fruits. *Cornus amomum* has been divided into two subspecies, which many botanists have treated as separate species (Steyermark, 1963; Murrell, 1992). Wilson (1965) recorded putative hybrids between the two subspecies with apparently viable pollen, thus strengthening his argument that the two taxa are components of a single biological species. Missouri plants are ssp. *obliqua,* which is recognized by its narrower leaves with a more tapered base and the undersurface with minute, raised punctations at the bases of the hairs, vs. broadly ovate leaves with a broadly angled to rounded base and a smooth undersurface in ssp. *amomum.* The latter subspecies is widespread to the east of Missouri. A historical specimen of ssp. *amomum* in the herbarium of the Field Museum was collected by Earl E. Sherff in the city of St. Louis. However, Steyermark (1963) did not treat this taxon as part of the state's flora and in his introductory chapters discussed how most of Sherff's collections from the area represent cultivated material.

Wilson (1965) reported putative hybrids between *C. amomum* ssp. *obliqua* and *C. foemina* ssp. *racemosa,* as well as with *C. drummondii.* He noted the high percentages of inviable pollen in such interspecific hybrids. Although he cited historical material collected in Jackson County as a hybrid with *C. drummondii,* he discovered no specimens to document hybridization with *C. foemina* in Missouri. Wilson noted, however, that such hybrids are possible wherever the ranges of the species overlap. Such hybrids may exist where the species grow in proximity, but distinguishing them from their parents will prove very difficult for most botanists.

3. Cornus drummondii C. A. Mey. (rough-leaved dogwood)
C. asperifolia Michx.
Svida asperifolia (Michx.) Small
Pl. 369 a–c; Map 1608

Plants shrubs or small trees 2–6 m tall, sometimes colonial. Twigs green to reddish brown, densely hairy when young, becoming glabrous or nearly so with age, the pith white to brown. Bark usually with shallow fissures, the ridges becoming divided into thin, irregular plates, grayish brown, usually with small, slightly raised, lighter dots. Leaves opposite, usually relatively evenly dispersed along the branches, the petiole 1.0–1.5 cm long. Leaf blades (4–)6–11 cm long, 3–7 cm wide, lanceolate

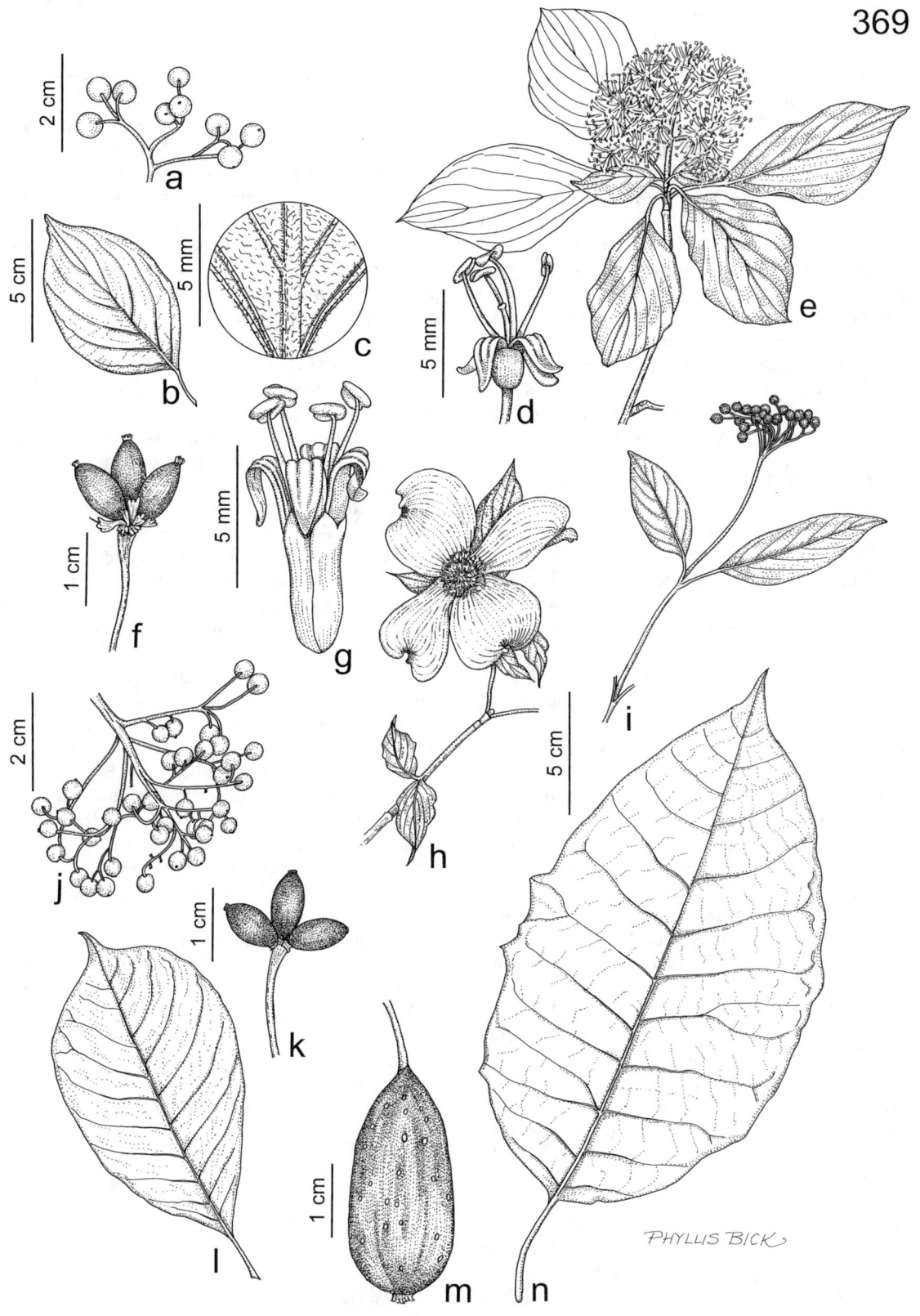

Plate 369. Cornaceae. *Cornus drummondii*, **a)** fruits, **b)** leaf, **c)** detail of leaf undersurface. *Cornus alternifolia*, **d)** flower, **e)** flowering branch. *Cornus florida*, **f)** fruits, **g)** flower, **h)** flowering branch. *Cornus amomum*, **i)** fruiting branch. *Cornus foemina*, **j)** fruits. *Nyssa sylvatica*, **k)** fruits, **l)** leaf. *Nyssa aquatica*, **m)** fruit, **n)** leaf.

to elliptic or broadly ovate, rounded or angled at the base, tapered to a sharply pointed tip, the surfaces moderately pubescent with mostly somewhat spreading, V-shaped or Y-shaped hairs, the upper surface moderately to strongly roughened to the touch, the undersurface softer-hairy and usually with microscopic white papillae, the lateral veins 3 or 4(5) pairs, these mostly arising from the basal half of the blade. Inflorescences appearing more or less umbellate, flat-topped to shallowly convex, the bracts absent or rarely a few at the branch points and these minute and scalelike, the flower stalks 2–8 mm long, hairy, becoming reddish brown to purplish brown as the fruits mature. Sepals 0.4–1.0 mm long. Petals 2.5–4.0 mm long, narrowly oblong-lanceolate, white to cream-colored. Style 2.5–3.5 mm long, relatively slender, not broadened toward the tip. Fruits 4–7 mm in diameter, spherical, white, occasionally mottled with dark blue. Stone smooth or inconspicuously grooved. $2n=22$. May–June.

Common nearly throughout the state (New York to Kentucky and Georgia west to South Dakota and Texas). Banks of streams, rivers, and spring branches, margins of ponds and lakes, bottomland forests, mesic upland forests in ravines, bases, ledges, and tops of bluffs, margins of glades and upland prairies, marshes, and fens; also fencerows, old fields, margins of crop fields, railroads, and roadsides.

Cornus drummondii is perhaps the most common species of dogwood in disturbed habitats in Missouri and also tolerates somewhat drier conditions than do other members of the genus present in the state. It is unique among Missouri *Cornus* taxa in its roughened upper leaf surface, the secondary veins somewhat crowded toward the leaf base, and its white fruits. However, there is considerable variation in the degree of leaf roughening, which has led some botanists to speculate that hybrids between *C. drummondii* and other Missouri dogwoods are common. Such hybrids do occur but apparently are partially sterile and generally are confined to individual plants or small colonies (when the hybrid spreads vegetatively by root sprouts). Hybrids between *C. drummondii* and *C. foemina* ssp. *racemosa* were reported from southern Missouri north to Jackson County by Wilson (1965). Such hybrids have leaves that are only slightly roughened, with many of the hairs T-shaped and appressed rather than Y-shaped and spreading, and fruits (when produced) that are either white or light blue. For a brief discussion of putative hybridization between *C. drummondii* and *C. amomum*, see the treatment of that species.

4. Cornus florida L. (flowering dogwood)
Cynoxylon floridum (L.) Raf.
Pl. 369 f–h; Map 1609

Plants small trees or rarely shrubs 3–15 m tall, usually occurring as solitary individuals. Twigs reddish brown to reddish gray or green, moderately to densely appressed-hairy, the pith usually tan. Bark relatively deeply fissured, the ridges becoming divided into irregularly polygonal to oblong plates, dark gray to brown. Leaves opposite, positioned mostly toward the tips of the branches, the petiole 0.3–1.2 cm long. Leaf blades 5–12 cm long, 3–8 cm wide, ovate to broadly elliptic or occasionally elliptic-obovate, rounded, angled, or short-tapered at the base, angled or more commonly tapered to a sharply pointed tip, the surfaces sparsely to densely pubescent with mostly appressed, straight, T-shaped hairs when young, the undersurface sometimes also with longer, more or less spreading, V-shaped or Y-shaped hairs along the main veins, sometimes becoming glabrous or nearly so at maturity, the lateral veins (4)5–7 pairs, these relatively evenly spaced. Inflorescences dense heads subtended by 4 showy petaloid bracts, these 2.5–5.0 cm long, broadly obovate to nearly circular or somewhat heart-shaped, shallowly notched at the tip, white or rarely pink, the flowers sessile or nearly so. Sepals fused below the midpoint, the lobes 0.5–1.0 mm long. Petals 3–4 mm long, narrowly oblong-lanceolate, greenish yellow. Style 3–4 mm long, relatively slender, not broadened toward the tip. Fruits 7–14 mm long, ovoid to ellipsoid, red (sometimes turning black with age). Stone strongly longitudinally veined but smooth or shallowly and inconspicuously grooved. $2n=22$. April–May.

Scattered to common in the Ozark and Ozark Border Divisions north locally to Shelby and Marion Counties; also in the Crowley's Ridge Section of the Mississippi Lowlands Division (eastern U.S. west to Kansas and Texas; Canada, Mexico). Mesic to dry upland forests, savannas, edges of glades, bases, ledges, and tops of bluffs; less commonly banks of streams and rivers, margins of sinkhole ponds, and bottomland forests; also pastures, old fields, and roadsides.

Cornus florida is the official state tree of Missouri (and also Virginia). It is among the showiest of early spring-flowering trees in the understories of Missouri forests, beginning to flower slightly later than but overlapping the pink flowers of redbuds (*Cercis canadensis*, Fabaceae). The red fruits are an important autumn and winter food source for migrating songbirds and small mammals. Steyermark (1963) noted that Native Americans prepared a red dye from the roots and that the bitter inner bark was used as a quinine substitute to treat malaria. The wood has been shaped into many products, including shuttles, golf clubs, mallets, tool handles, wedges, spindles, pulleys, knitting needles, roller skates, and engraving blocks.

Flowering dogwood also is one of the most popular of ornamental trees in gardens, prized for its showy bracts before the leaves appear, checkered bark, graceful habit, and red fall foliage. Rare pink-bracted individuals have been called f. *rubra* (Weston) E.J. Palmer & Steyerm. and are popular in horticulture. Other cultivars, including some with the bracts pinkish- to reddish-tinged toward the margins, also have been developed by plant breeders.

Unfortunately, flowering dogwood is threatened throughout much of its range by dogwood anthracnose, a disease caused by the fungus *Discula destructiva* Redlin, which attacks some members of the group of dogwoods with showy inflorescence bracts. The disease was first diagnosed in the 1970s in both the northwestern (on *C. nuttallii* Audubon) and northeastern (on *C. florida*) United States, and by the 1990s it had spread throughout the range of the host species. Symptoms include blotches and small spots with purple margins on the leaves and twigs, worsening in wet years and developing into branch diebacks. The tree weakens and dies after usually 1–3 years. The geographic origins of this disease are not known, and it was first described as new to science as recently as 1991. Attempts to link it to *C. kousa* F. Buerger ex Miq. trees imported from China have not been successful, and speculation that a shift in some aspect of the continent's climate might have caused an expansion in the range of a previously undiscovered native fungus also has not been substantiated. For a summary of the pathology of this disease and its affect on *C. florida*, see the excellent review by Daughtrey (1993). Dogwood anthracnose has yet to be documented from wild dogwood populations in Missouri, but it has been recorded from south-central and west-central Illinois (Schwegman et al., 1998), and unconfirmed reports of trees with anthracnose-like symptoms have been recorded from elsewhere in the Midwest.

5. Cornus foemina Mill. (gray dogwood, stiff dogwood)

Pl. 369 j; Map 1610

Plants shrubs or small trees 2–5 m tall, often colonial. Twigs reddish brown to tan or brown when young, becoming olive green, grayish brown, or gray with age, glabrous or sparsely appressed-hairy, the pith white or tan. Bark smooth or with shallow fissures, the ridges becoming divided into thin, irregular plates, gray to grayish brown, sometimes with small, slightly raised, lighter dots. Leaves opposite, usually relatively evenly dispersed along the branches, the petiole 0.5–1.0 cm long. Leaf blades (3–)5–12 cm long, 2–7 cm wide, broadly lanceolate to elliptic or ovate, rounded, angled, or short-tapered at the base, tapered to a sharply pointed tip, the surfaces glabrous or sparsely to moderately pubescent with mostly appressed, straight, T-shaped hairs, the undersurface sometimes with microscopic white papillae and slightly roughened to the touch, the lateral veins 3 or 4(5) pairs, these relatively evenly spaced. Inflorescences either hemispherical to pyramidal panicles or appearing more or less umbellate and flat-topped to shallowly convex, the bracts absent, the flower stalks 2–5 mm long, glabrous or sparsely hairy, becoming reddish brown to purplish brown as the fruits mature. Sepals 0.4–1.0 mm long. Petals 2–4 mm long, narrowly oblong-lanceolate, white to cream-colored, rarely greenish-tinged. Style 2–3 mm long, relatively slender, not broadened toward the tip. Fruits 4–7 mm in diameter, spherical, white or light blue, rarely white-and-blue mottled. Stone strongly longitudinally veined but smooth or shallowly and inconspicuously grooved. $2n=22$. May–July.

Scattered nearly throughout the state but absent from most of the Unglaciated Plains Division (eastern U.S. west to North Dakota and Texas; Canada). Swamps, bottomland forests, mesic upland forests in ravines, banks of streams and rivers, margins of ponds, lakes, and sinkhole ponds, bases of bluffs, fens, acid seeps, and edges of bottomland and upland prairies; also fencerows, old fields, ditches, railroads, and roadsides.

Wilson (1965) treated the *C. foemina* complex as comprising three subspecies that some other botanists have considered separate species. The two Missouri taxa, ssp. *foemina* and ssp. *racemosa,* are southern and northern analogs whose ranges overlap in southern Missouri, where Wilson recorded apparently fertile putative hybrids. The third subspecies, ssp. *microcarpa* (Nash) J.S. Wilson (*C. asperifolia* Michx.), inhabits portions of the Coastal Plain from Florida north to North Carolina and west to Alabama and differs in its relatively strongly roughened twigs and leaves and often slightly smaller fruits. Wilson also reported apparently sterile hybrids between *C. foemina* and both *C. amomum* and *C. drummondii,* which are discussed briefly in the treatments of those species. Steyermark (1963), who also noted possible hybridization between *C. foemina* ssp. *racemosa* (as *C. racemosa*) and *C. amomum* ssp. *obliqua* (as *C.*

obliqua), referred to such plants under the name *C.* ×*arnoldiana* Rehder.

Some botanists (Gleason and Cronquist, 1991; Murrell, 1992) have used the name *C. stricta* in place of *C. foemina* because the original description of *C. foemina* was ambiguous, possibly referring to ssp. *foemina* or to ssp. *racemosa*. However, Wilson (1965) argued that the original materials from which the name was described were cultivated from plants growing in a region and habitat where only ssp. *foemina* grows today, and he accepted the name *C. foemina* for the species.

1. Leaf blades with the undersurface somewhat lighter green than the upper surface but not appearing pale or whitened; young twigs reddish brown, bark relatively smooth; inflorescences flat-topped or slightly convex; fruits light blue or rarely blue-and-white mottled .... 5A. SSP. FOEMINA
1. Leaf blades with the undersurface appearing pale or whitened; young twigs brown, tan, or occasionally pinkish-tinged, bark usually with small, slightly raised, lighter dots; inflorescences hemispherical to pyramidal; fruits white .... 5B. SSP. RACEMOSA

5a. ssp. foemina (stiff dogwood)

C. stricta Lam.

Svida foemina (Mill.) Rydb.

Twigs mostly reddish brown when young, becoming olive green or gray with age, relatively smooth and lacking small, light, raised dots, the pith white. Leaf blades with the upper surface dark green to green, the undersurface somewhat lighter green than the upper surface but not appearing pale or whitened, lacking microscopic white papillae, smooth to the touch. Inflorescences more or less umbellate, flat-topped to shallowly convex. Fruits light blue or rarely blue-and-white mottled. 2*n*=22. May–June.

Uncommon in the Mississippi Lowlands Division and the adjacent southeastern portion of the Ozarks (southeastern U.S. west to Missouri and Texas). Swamps, bottomland forests, banks of streams and rivers, and acid seeps; also fencerows and ditches.

Steyermark (1963) remarked on the horticultural value of this taxon, with its bright blue fruits and brown branches. He also noted that the leaves of ssp. *foemina* tend to be narrower and more tapered at the base than are those of ssp. *racemosa*, which he treated as a separate species.

5b. ssp. racemosa (Lam.) J.S. Wilson (gray dogwood)

C. racemosa Lam.

Twigs brown, tan, or occasionally pinkish-tinged when young, becoming olive green, grayish brown, or gray with age, usually somewhat roughened with small, slightly raised, lighter dots, the pith white or more commonly tan. Leaf blades with the upper surface dark green to green, the undersurface appearing pale or whitened, with microscopic white papillae, slightly roughened to the touch. Inflorescences paniculate, hemispherical to pyramidal. Fruits usually white. 2*n*=22. May–June.

Scattered nearly throughout the state but absent from most of the Unglaciated Plains Division (northeastern U.S. west to North Dakota and Arkansas; Canada). Swamps, bottomland forests, mesic upland forests in ravines, banks of streams and rivers, margins of ponds, lakes, and sinkhole ponds, bases of bluffs, fens, and edges of bottomland and upland prairies; also fencerows, old fields, railroads, and roadsides.

Steyermark (1963) remarked on the ornamental merits of *C. foemina* ssp. *racemosa* because of its attractive white fruits contrasted with the reddish stalks and inflorescence branches, and also its fall foliage, which turns dull purple with yellow to pink and green highlights. He also noted that plants of this taxon tend to produce flowers at a relatively early age, often when merely a meter tall.

2. Nyssa L. (tupelo, sour gum)
(Eyde, 1963; Burckhalter, 1992)

Plants medium-sized to large trees, dioecious, sometimes incompletely so. Winter buds ovoid, with several scales. Twigs with white pith having diaphragms (solid but with numerous differentiated cross-partitions). Leaves alternate, often crowded toward the tips of branches, sometimes appearing nearly whorled on short shoots. Leaf blades with the margins entire or with a few coarse teeth, the secondary veins straight or only slightly arched toward the leaf tip, becoming fine and inconspicuously fused toward the leaf margin. Inflorescences axillary (sometimes appearing terminal or clustered when on short shoots), those of the staminate flowers short, dense racemes or headlike to nearly umbellate clusters, the pistillate flowers solitary or in small clusters of 2–4(5). Flowers mostly imperfect, the staminate and pistillate flowers mostly on separate plants (few to several apparently perfect flowers usually produced per branch on otherwise staminate or pistillate trees). Calyces with the free portion consisting of 5 small, triangular or oblong lobes 0.5–2.0 mm long or more commonly a minute, low rim. Petals 5–10, inserted along the margin of the nectar disc, green to greenish yellow. Stamens 8–15, the slender filaments 3–5 mm long, attached along the margin of the nectar disc, the anthers 0.8–1.2 mm long, oblong, attached at the base. Pistil of 2 fused carpels but 1 carpel usually aborting during development, the ovary glabrous or hairy, with usually 1 locule. Style 1 (sometimes with a short, rudimentary, second style or rarely with a second, fully formed style), relatively stout, bent or reflexed toward the tip, the stigma(s) unlobed. Ovule 1(2) (1 per locule). Fruits ovoid to ellipsoid, dark blue to bluish purple or red. Stone 1(2)-seeded, the seed ovoid or ellipsoid, often somewhat flattened. Five to 10 species, North America, Central America, Asia.

Nyssa sometimes has been treated in a separate family Nyssaceae along with two small Chinese genera, *Davidia* Baill. and *Camptotheca* Decne. (Eyde, 1966). The main character separating these genera from the remainder of Cornaceae is the larger number of petals and stamens. The circumscription of some species of *Nyssa* also has been somewhat controversial (Burckhalter, 1992; Wen and Stuessy, 1993). *Nyssa* was widely distributed in the Northern Hemisphere during the early Tertiary Period (Eyde, 1988).

Nyssa species are economically important as timber trees, primarily in the southeastern states. The wood is used for plywood, boxes, pallets, furniture, flooring, paper pulp, fishing floats, and handicrafts. The honey from tupelo flowers also is highly esteemed.

1. Leaves 10–30 cm long, the blade often with a few coarse, broadly triangular teeth along the margin, the petiole 3–6 cm long; glabrous to sparsely hairy staminate flowers sessile, grouped into headlike clusters; fruits 2–3 cm long, the stone with sharply angled longitudinal ridges .... 1. N. AQUATICA
1. Leaves (2–)4–15 cm long, the blade usually lacking teeth along the margin, the petiole 0.5–2.0 cm long; staminate flowers stalked, grouped into short racemes or umbellate clusters; fruits 0.8–1.2 cm long, the stone with rounded longitudinal ridges
   2. Leaf blades lanceolate to elliptic or oblanceolate, mostly rounded or angled to a bluntly pointed tip; pistillate flowers and fruits 1 or 2 per inflorescence
      .... 2. N. BIFLORA
   2. Leaf blades elliptic to obovate, occasionally narrowly elliptic or oblanceolate, mostly short-tapered to a sharply pointed tip; pistillate flowers and fruits mostly 2–4(5) per inflorescence .... 3. N. SYLVATICA

1. Nyssa aquatica L. (swamp tupelo, water tupelo, cotton gum, tupelo, tupelo gum)
N. uniflora Wangenh.

Pl. 369 m, n; Map 1611

Plants large trees to 35 m tall, the trunk usually tapered from a swollen and/or buttressed basal portion, the bark relatively thin, finely fissured, the ridges sometimes broken into small, scaly plates, dark brown or gray. Twigs reddish brown to brown, relatively stout. Leaves with petioles 3–6 cm long, these usually densely pubescent with spreading, sometimes tangled, mostly 2-branched hairs. Leaf blades 10–30 cm long, 5–12 cm wide, ovate, elliptic, or obovate, the margins entire or with 1 to few, coarse, broadly triangular, spreading teeth and usually also hairy, tapered, angled, or occasionally shallowly cordate at the base, tapered to a sharply pointed tip, the upper surface glabrous, not shiny, the undersurface glabrous or sparsely hairy along the main veins, pale green and glaucous. Staminate flowers in dense, headlike clusters 1.0–1.5 cm in diameter, the inflorescence stalk 1.0–1.5 cm long. Pistillate flower 1 per inflorescence, the inflorescence stalk 2–5 cm long. Petals 2–3 mm long, oblong, usually rounded at the tip. Fruits 2–3 cm long, oblong-ellipsoid, dull yellow to olive green, turning purplish black, with scattered minute, white spots, glaucous, bitter, the stone with 8–10 sharply angled longitudinal ridges. April–May.

1611. Nyssa aquatica   1612. Nyssa biflora   1613. Nyssa sylvatica

Uncommon in the Mississippi Lowlands Division and adjacent southeastern portion of the Ozarks (southeastern U.S. west to Missouri and Texas). Swamps, bottomland forests, sloughs, banks of streams and rivers, and sinkhole ponds, often emergent aquatics.

Tupelo gum is common in the southeastern states in areas that are periodically flooded, in habitats where many other species cannot survive. In Missouri it was a dominant species in the swampy forests of the Mississippi Lowlands Division before the nearly complete clearing and draining of that portion of the state for agriculture. Its disjunct presence in a couple of upland sinkhole ponds in the Ozarks is remarkable.

2. **Nyssa biflora** Walter (swamp black gum, swamp tupelo)
   N. sylvatica Marshall var. biflora (Walter) Sarg.

Map 1612

Plants large trees to 25 m tall, the trunk usually tapered from a swollen and/or buttressed basal portion, the bark relatively thick, finely to more commonly deeply fissured, the ridges often breaking into irregular plates, gray to brown or black. Twigs reddish brown to gray, relatively slender to moderately stout. Leaves with petioles 0.5–1.5 cm long, these usually densely pubescent with spreading, sometimes tangled, mostly 2-branched hairs, sometimes only on the upper surface. Leaf blades (2–)4–8 cm long, 1.5–3.5 cm wide, lanceolate to elliptic or oblanceolate, the margins entire and sometimes also hairy, often minutely curled under, mostly tapered at the base, rounded or angled to short-tapered to a bluntly or less commonly sharply pointed tip, the upper surface glabrous, often somewhat shiny, the undersurface sparsely to moderately hairy, especially along the main veins, pale green but not glaucous. Staminate flowers in short, dense racemes or appearing as dense, subumbellate clusters 0.7–1.2 cm in diameter, the inflorescence stalk 1.5–2.0 cm long. Pistillate flower(s) 1 or 2 per inflorescence, the inflorescence stalk 2–5 cm long. Petals 0.5–1.5 mm long, oblong, usually rounded at the tip. Fruits 0.8–1.2 cm long, ovoid to ellipsoid, purplish blue to nearly black, with scattered minute, white spots, glaucous, bitter, the stone with 8–12 broad, rounded, shallow longitudinal ridges. $2n=44$. April–May.

Uncommon, known thus far from a single historical collection from Dunklin County (southeastern U.S. west to Missouri and Texas). Habitat unknown but presumably swamps.

In most leaf and floral characters *N. biflora* resembles *N. sylvatica,* and some botanists have treated it as a variety of that species (Eyde, 1963, 1966; Gleason and Cronquist, 1991; Wen and Stuessy, 1993). Burckhalter (1992) studied taxonomic relationships in temperate North American *Nyssa* and concluded, based on morphological variation in various habitats and differences in flavonoid chemistry, that the two taxa should be considered separate species.

3. **Nyssa sylvatica** Marshall (black gum, sour gum, black tupelo)
   N. sylvatica var. carolina (Poir.) Fernald
   N. sylvatica var. dilatata Fernald

Pl. 369 k, l; Map 1613

Plants large trees to 30 m tall, the trunk not or only slightly swollen and never buttressed toward the base, the bark relatively thick, finely to more commonly deeply fissured, the ridges usually breaking into irregularly hexagonal plates on older trunks, gray to brown or black. Twigs reddish brown to gray, usually relatively slender. Leaves with petioles 0.5–2.0 cm long, these moderately to densely pubescent with spreading, sometimes tangled, mostly 2-branched hairs, sometimes only on the upper surface or undersurface. Leaf blades 4–15 cm long, 2–6 cm wide, elliptic to obovate, occasionally narrowly elliptic or oblanceolate, the margins entire or rarely with 1 to few, coarse, broadly triangular, spreading teeth, sometimes also hairy, often minutely curled under, broadly angled or tapered at the base, rounded or angled to more commonly short-tapered to a usually sharply pointed tip, the upper surface glabrous, somewhat shiny, the undersurface sparsely to moderately hairy, especially along the main veins, pale green but not glaucous. Staminate flowers in short, dense racemes or appearing as umbellate clusters 0.7–2.5 cm in diameter, the inflorescence stalk 1–2 cm long. Pistillate flowers 2–4(5) per inflorescence, the inflorescence stalk 2–5 cm long. Petals 0.5–1.5 mm long, oblong, usually rounded at the tip. Fruits 0.8–1.2 cm long, ovoid, purplish blue to nearly black, with scattered, minute, white spots, glaucous, bitter, the stone with 8–12 broad, rounded, shallow longitudinal ridges. $2n=44$. April–May.

Scattered to common in the Ozark, Ozark Border, and Mississippi Lowlands Divisions (eastern U.S. west to Wisconsin and Texas; Mexico). Bottomland forests, mesic to dry upland forests, savannas, swamps, sloughs, banks of streams, rivers, and spring branches, and margins of sinkhole ponds; also old fields, railroads, and roadsides; usually in acidic soils.

Black gum trees are grown as ornamentals for their shiny foliage, which turns various shades of purple, red, orange, and yellow in the autumn. Various authors have treated *N. sylvatica* as comprising two to several varieties. On the one hand, some of these (var. *caroliniana,* var. *dilatata*) are merely morphological extremes for various characters. Some others probably are best treated as separate species (Burkhalter, 1992). The latter group includes *N. biflora* (treated above) and *N. ursina* Small (*N. sylvatica* var. *ursina* (Small) J. Wen & Stuessy; shrubs or dwarf trees endemic to northwestern Florida).

CRASSULACEAE (Stonecrop Family)

Thirty to 35 genera, 900–1,500 species, nearly worldwide.

The genus *Penthorum,* which is included in the Crassulaceae by some authors, is here treated in its own family, Penthoraceae. See the treatments of Penthoraceae and Saxifragaceae for further discussion.

1. Sedum L. (stonecrop, orpine)
(Clausen, 1975)

Plants annual, biennial, or perennial herbs, sometimes with thickened roots or a somewhat woody rootstock, sometimes colonial by rhizomes, mat-forming, or becoming propagated vegetatively from stem fragments or leaves that form adventitious roots, glabrous, sometimes somewhat glaucous, sometimes with minute papillae. Stems erect to spreading, usually somewhat succulent. Leaves basal and usually also alternate or whorled, less commonly opposite, sessile or short-petiolate. Stipules absent. Leaf blades simple, thickened and succulent, variously shaped, the margins entire to coarsely scalloped or toothed. Inflorescences panicles, variously branched and shaped, less commonly reduced to spikelike racemes. Flowers usually perfect, actinomorphic, hypogynous. Calyces of 4–7 sepals, these occasionally fused at the base, often somewhat fleshy. Corollas of 4–7 petals (these fused at the very base in *S. acre*). Stamens mostly 8 or 10 (usually twice the number of sepals), the filaments slender, attached at the base of the ovary (less commonly attached at the petal bases), the anthers attached at their bases. Staminodes absent. Pistil of 4 or 5(–7) carpels, these free or fused at the base, more or less erect at flowering, sometimes spreading at fruiting, each with a tiny, scalelike nectary at the base. Ovaries superior, green to greenish white, each with 1 locule, with numerous ovules, the placentation more or less parietal. Styles 1 per carpel, the ovary tip tapered into the tapering style, usually persistent at fruiting, the stigmas terminal, poorly differentiated.

Fruits follicles, longer than wide, dehiscing longitudinally on the inner side along the suture, usually brown, with numerous seeds. Seeds (when produced) tiny, ellipsoid to obovoid, the surface often with fine longitudinal ribs, yellowish brown to dark brown. Three hundred to 600 species, North America to South America, Caribbean Islands, Europe, Asia, Africa, Madagascar.

Results of recent molecular studies (Van Ham and 't Hart, 1998; Mort et al., 2001) suggest that the genus *Sedum* is polyphyletic; that is, the traditionally circumscribed genus contains a collection of several only distantly related species groups, each more closely related to other groups of genera within the family. However, the details (especially the morphological basis) of a new classification still await resolution, and the genus continues to be treated in the broad sense in the present treatment.

A number of hardy *Sedum* species are popular horticulturally as garden ornamentals and cemetery plantings. Some of these taxa sometimes persist at old homesites, but none have become naturalized in Missouri as yet.

1. Leaves sessile or short-petiolate, attached at the very base; blades of largest leaves 12–40 mm wide, relatively flat; flowering stems 25–60 cm long, erect or ascending
    2. Leaves only slightly and gradually reduced in size toward the stem tip, the margins entire or bluntly and irregularly scalloped; petals white to light pink
    . . . . 2. S. ERYTHROSTICTUM
    2. Leaves relatively abruptly reduced in size toward the stem tip, the margins more or less sharply toothed; petals reddish pink to purplish pink
    . . . . 5. S. PURPUREUM
1. Leaves sessile or nearly so, attached slightly above the base, thus with a small pouched or flattened plate or pair of spurs of tissue below the attachment point; blades of largest leaves 1–12 mm wide, circular, angled, or flat in cross-section; stems with the flowering branches 1–20 cm long, erect to loosely ascending
    3. Leaves mostly opposite or in whorls of 3
        4. Leaf blades narrowly oblanceolate to narrowly oblong-elliptic or nearly linear, sharply pointed at the tip; petals yellow . . . . 7. S. SARMENTOSUM
        4. Leaf blades obovate to broadly elliptic or spatulate, rounded to truncate or shallowly notched at the tip, occasionally very bluntly pointed or with an abrupt, minute, sharp point; petals white . . . . 8. S. TERNATUM
    3. Leaves alternate, sometimes appearing densely spiraled along the stem
        5. Plants perennial, the main stems creeping (flowering branches erect or ascending), rooting at some nodes, usually with at least some green leaves at fruiting time
            6. Leaves 2–5 mm long, ovate, bluntly pointed at the tip; sepals (4)5; petals (4)5; fruits spreading at maturity . . . . 1. S. ACRE
            6. Leaves 5–12 mm long, linear to narrowly oblong-lanceolate, sharply pointed at the tip; sepals 5–7; petals 5–7; fruits erect or strongly ascending at maturity . . . . 6. S. REFLEXUM
        5. Plants annual or biennial (rarely perennial in cultivation), the main stems erect or ascending from a compact, fibrous-rooted base, the leaves usually withered by fruiting time
            7. Leaves 4–6(–12) mm long; petals yellow; fruits spreading at maturity
            . . . . 3. S. NUTTALLIANUM
            7. Leaves (5–)9–25 mm long; petals pink, rarely white; fruits erect or strongly ascending at maturity . . . . 4. S. PULCHELLUM

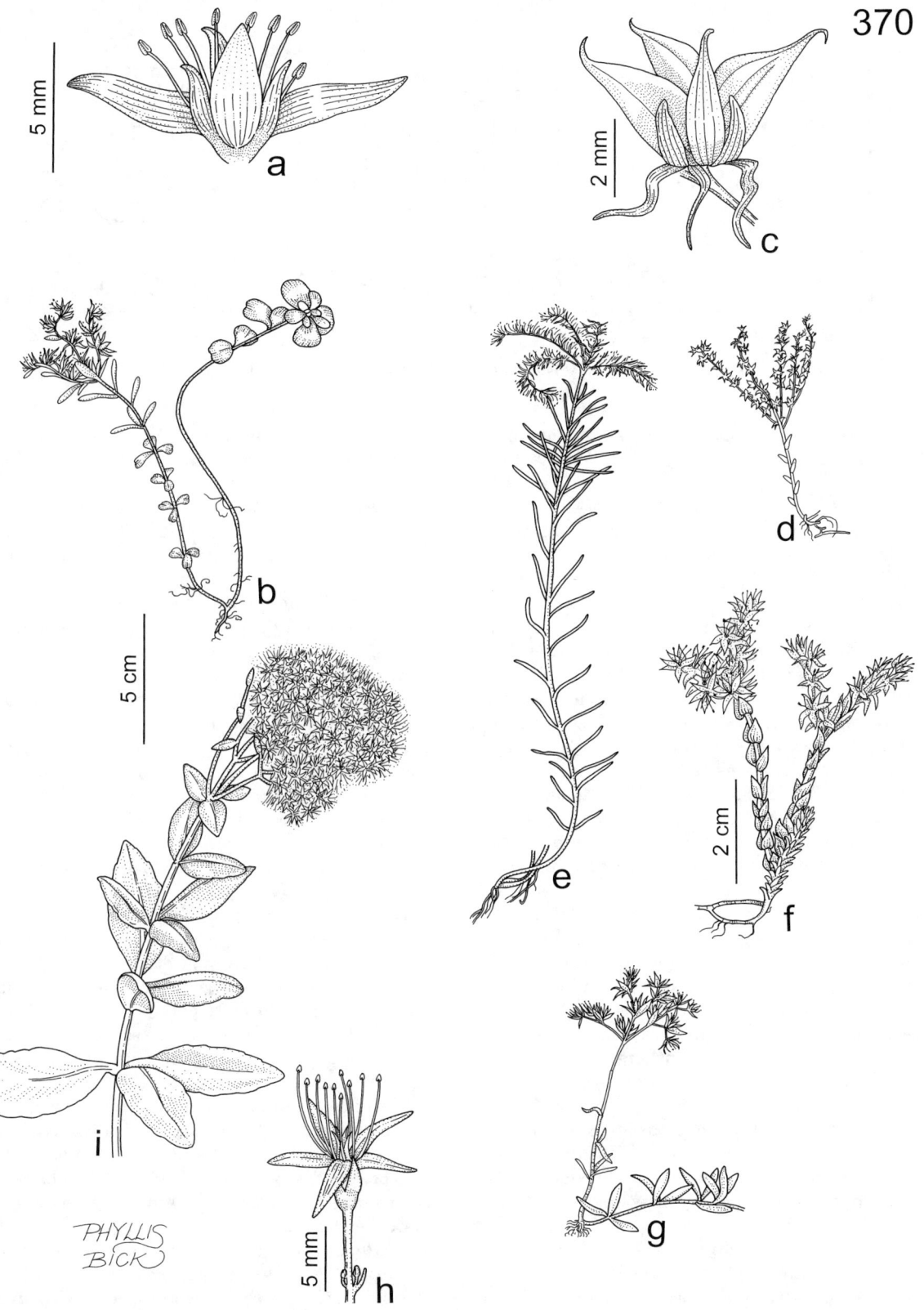

Plate 370. Crassulaceae. *Sedum ternatum*, **a)** fruits with remains of flower, **b)** habit. *Sedum nuttallianum*, **c)** fruits with remains of flower, **d)** habit. *Sedum pulchellum*, **e)** habit. *Sedum acre*, **f)** habit. *Sedum sarmentosum*, **g)** habit. *Sedum erythrostictum*, **h)** flower, **i)** habit.

1614. Sedum acre   1615. Sedum erythrostictum   1616. Sedum nuttallianum

1. Sedum acre L. ssp. **acre** (mossy stonecrop, golden carpet)

Pl. 370 f; Map 1614

Plants perennial from fibrous roots, the foliage and flowers sometimes with resinous lines or dots visible upon drying. Stems green, sometimes becoming white with age, not glaucous, the main stems 2–10 cm or more long (rarely absent in young plants), prostrate, often forming loose or dense mats, the flowering and vegetative branches 2–5(–10) cm long, erect or ascending. Leaves alternate (often absent from the horizontal main stems by flowering), sometimes appearing densely spiraled along the stem, those of the lower portions of the flowering branches usually becoming white and papery by flowering time, 2–5 mm long, sessile, attached just above the minutely pouched base, only slightly and gradually reduced in size toward the stem tip. Leaf blades ovate, more or less elliptic in cross-section, the margins entire, the surfaces yellowish green, not glaucous, bluntly pointed at the tip. Inflorescences erect spikes or spikelike racemes, these sometimes paired and ascending at the branch tips. Sepals (4)5, 2.5–4.0 mm long, ovate, thickened, bluntly pointed at the tip. Petals (4)5, 6–8 mm long, fused at the very base, lanceolate, yellow. Stamens with the anthers yellow, the filaments attached at the petal bases. Nectaries about as long as wide, oblong, yellowish green. Carpels (4)5, greenish yellow. Fruits 2.5–3.5 mm long, spreading at maturity. Seeds 0.5–0.8 mm long, obovoid to oblong-ellipsoid, with a finely honeycombed pattern of ridges and pits, yellowish brown. $2n=80$. June–July.

Introduced, uncommon, known thus far only from Iron County and St. Louis City (native of Europe, Asia; widely but sporadically introduced in the U.S. and Canada). Cemeteries and dry, open, disturbed areas.

This species was first reported for Missouri by Clausen (1975).

2. Sedum erythrostictum Miq. (live-forever)

Hylotelephium erythrostictum (Miq.) H. Ohba

S. alboroseum Baker

S. telephium L. ssp. *alboroseum* (Baker) Fröd.

Pl. 370 h, i; Map 1615

Plants perennial from a stout, spongy to slightly woody rootstock, the roots somewhat thickened and fleshy, the foliage and flowers with tiny, reddish brown, resinous lines or dots visible upon drying. Stems 25–60 cm long, erect or ascending, glaucous. Leaves alternate or rarely opposite, sessile or short-petiolate, attached at the very base, only slightly and gradually reduced in size toward the stem tip. Leaf blades (18–)25–90 mm long, (8–)12–40 mm wide, thickened but flat in cross-section, the margins entire or bluntly and irregularly scalloped, the surfaces pale green, sometimes yellowish- or pinkish-tinged, glaucous. Inflorescences panicles, more or less flat-topped, the branches numerous and ascending. Sepals (4)5(6), 1–2 mm long, lanceolate to narrowly ovate, sharply pointed at the tip. Petals (4)5(6), 4.0–5.5 mm long, narrowly elliptic to lanceolate, white to light pink. Stamens with the anthers pink. Nectaries longer than wide, more or less strap-shaped, yellow. Carpels (4)5(6), pink. Fruits 4–6 mm long, erect or strongly ascending at maturity. Seeds aborting during development. $2n=48, 50$. July–November.

Introduced, uncommon and widely scattered (native of Asia; widely but sporadically introduced in the eastern U.S. and adjacent Canada west to Wisconsin, Iowa, and Louisiana). Rocky banks of rivers; also roadsides, cemeteries, and rocky, open, disturbed areas.

Sedum erythrostictum is a member of the *S. telephium* complex, which consists of about 28 closely related mostly Eurasian species with broad, flattened leaves. The members of this complex sometimes are treated as a separate genus, *Hylotelephium* H. Ohba. Meiosis is irregular in this taxon (which may be of hybrid origin), and plants

1617. Sedum pulchellum   1618. Sedum purpureum   1619. Sedum reflexum

often have a reduced number of stamens and do not produce viable seeds. The only member of the group native to North America is *S. telephioides* Michx. (*Hylotelephium telephioides* (Michx.) H. Ohba), which grows on rock ledges and bluff crevices in portions of the eastern United States and Ontario westward locally to Louisiana and southern Illinois. Botanists should search for this species in adjacent southeastern Missouri on bluffs overlooking the Mississippi River. It differs from *S. erythrostictum* in its nectaries that are only slightly longer than wide and white to pale yellow, and its often pale pink petals. Because it is mostly diploid ($2n=24$, rarely 48), *S. telephioides* regularly produces viable seeds.

3. Sedum nuttallianum Raf. (Nuttall's sedum)
Pl. 370 c, d; Map 1616

Plants annual from fibrous roots, the main stems erect or ascending, the foliage and flowers lacking noticeable resinous lines or dots visible upon drying. Stems 2–12(–18) cm long, green, occasionally becoming reddish- or pinkish-tinged at maturity, not glaucous. Leaves usually withered by fruiting time, in a basal rosette and alternate, sometimes appearing densely spiraled along the stem, 4–6(–12) mm long, sessile, attached just above the minutely pouched base, only slightly and gradually reduced in size toward the stem tip. Leaf blades linear to narrowly oblong-lanceolate, more or less circular in cross-section, the margins entire, the surfaces light to dark green or bluish green, rarely reddish-tinged at maturity, not or only slightly glaucous, rounded to bluntly pointed at the tip. Inflorescences panicles with 2 or 3(4) branches, these spreading to loosely ascending spikes or spikelike racemes, the flowers oriented along the upper side of each branch, with leaflike bracts on the underside. Sepals (4)5(6), 1.5–3.0 mm long, narrowly lanceolate to oblong lanceolate, sharply pointed at the tip. Petals (4)5(6), 2.5–3.5 mm long, oblong-elliptic to oblong-lanceolate, yellow. Stamens with the anthers yellow. Nectaries about as long as wide, oblong to broadly obovate, yellow to milky white. Carpels 4 or 5, greenish yellow. Fruits 1.5–3.0 mm long, spreading at maturity. Seeds 0.5–0.9 mm long, obovoid, very faintly wrinkled or with a few faint, rounded ridges, yellowish brown. $2n=20$. May–June.

Uncommon, restricted to the southwestern portion of the Ozark Division (Missouri and Kansas south to Louisiana and Texas). Chert and sandstone glades; rarely rocky roadsides.

4. Sedum pulchellum Michx. (widow's cross)
Pl. 370 e; Map 1617

Plants annual or biennial (rarely perennial in cultivation) from fibrous roots, the main stems erect or ascending, occasionally from a horizontal base, the foliage and flowers lacking noticeable resinous lines or dots visible upon drying. Stems 2–20 cm long, green, often becoming reddish- or pinkish-tinged at maturity, not glaucous. Leaves usually withered by fruiting time, in a basal rosette and alternate, sometimes appearing densely spiraled along the stem, (5–)9–25 mm long, sessile, attached at the base, with a pair of small, sharply pointed basal spurs of tissue that partially encircle the stem, only slightly and gradually reduced in size toward the stem tip. Leaf blades linear to narrowly oblong-lanceolate, more or less circular in cross-section, the margins entire, the surfaces light to dark green, sometimes reddish-tinged at maturity, not glaucous, rounded to bluntly pointed at the tip. Inflorescences panicles with 3 or 4(–8) branches, these spreading spikes or spikelike racemes, sometimes somewhat scorpioid at the tips, the flowers oriented along the upper side of each branch, with leaflike bracts on the underside. Sepals 4(–7), 2.0–3.5 mm long, narrowly lanceolate, sharply pointed at the tip. Petals 4(–7), 4–6 mm long, narrowly elliptic-lanceolate to narrowly lanceolate, pink, rarely white. Stamens with the anthers red or purplish red. Nectaries about as long as wide, oblong, yellow to white, sometimes pinkish-tinged. Carpels 4 or 5, white to light pink. Fruits 4–7 mm long, erect or strongly ascending at maturity. Seeds 0.8–1.2 mm long, obovoid, finely longitudinally ribbed, light brown to yellowish brown. 2n=22, 44, 66. May–July.

Scattered, mostly south of the Missouri River, but absent from the Mississippi Lowlands Division (Ohio to Georgia west to Kansas and Texas). Glades, ledges and tops of bluffs, rock outcrops in mesic to dry upland forests, and occasionally rocky banks of streams; also roadsides and open, disturbed areas, on both acidic and calcareous substrates.

5. Sedum purpureum (L.) Schult. (live-forever)
S. telephium L. ssp. purpureum (L.) Schinz & R. Keller

Map 1618

Plants perennial from a stout, sometimes slightly woody rootstock, the roots somewhat tuberous-thickened, the foliage and flowers with tiny, reddish brown, resinous lines or dots visible upon drying. Stems 25–60 cm long, erect or ascending, not or only slightly glaucous, sometimes reddish-tinged. Leaves alternate or rarely opposite, sessile or short-petiolate, attached at the very base, relatively abruptly reduced in size toward the flowering stem tip. Blades of main leaves 25–60 mm long, 12–25 mm wide, those of reduced apical leaves 6–35 mm long, 2–12 mm wide, thickened but flat in cross-section, the margins more or less sharply toothed, the surfaces green to dark green, not glaucous. Inflorescences panicles, more or less rounded, the branches usually numerous and ascending. Sepals (4)5(6), 1.0–1.5(–6.0) mm long, lanceolate to narrowly ovate, sharply pointed at the tip. Petals (4)5(6), 4–5 mm long, narrowly elliptic to oblong-lanceolate, reddish pink to purplish pink. Stamens with the anthers red. Nectaries longer than wide, more or less strap-shaped, yellow. Carpels (4)5(6), pink. Fruits 4.0–4.5 mm long, erect or strongly ascending at maturity. Seeds usually aborting during development. 2n=36 (2n=22–24, 48 elsewhere). July–October.

Introduced, uncommon, mostly in counties bordering the Mississippi River (native of Europe; widely but sporadically introduced in the U.S. and Canada). Banks of rivers; also roadsides, cemeteries, and open, disturbed areas.

Steyermark (1963) mentioned the existence of S. fabaria W.D.J. Koch (as S. telephium ssp. fabaria (W.D.J. Koch) Schinz & R. Keller) in Missouri gardens, but he did not record it as an escape from cultivation. As with the other hardy cultivated live-forevers, this taxon may persist at old homesites and eventually may be found to have become naturalized in Missouri. It differs from S. purpureum in its more angled leaf bases and in its regular production of viable seeds. Other specimens from the state originally determined as S. telephium are referable to S. purpureum.

6. Sedum reflexum L. (Jenny's stonecrop)

Map 1619

Plants perennial from fibrous roots, the foliage and flowers sometimes with resinous lines or dots visible upon drying. Stems light green, sometimes reddish- or purplish-tinged and/or becoming white with age, often somewhat glaucous, the main stems 10–25 cm or more long, prostrate, forming loose mats, the flowering and vegetative branches 5–22(–35) cm long, erect or ascending. Leaves alternate (often absent from the horizontal main stems by flowering), appearing densely spiraled along the stem, persistent at fruiting, 5–12 mm long, sessile, attached just above the base, with a small, truncate plate of tissue below the attachment point, only slightly and gradually reduced in size toward the stem tip. Leaf blades linear to narrowly oblong-lanceolate, more or less elliptic in cross-section (sometimes somewhat angular), the margins entire, the surfaces grayish green, glaucous, sharply pointed at the tip. Inflorescences small panicles, often condensed and appearing as clusters at flowering, branched few to several times mostly dichotomously, the branches ascending, sometimes somewhat scorpioid at the tips, the flowers oriented along the upper side of each branch, with small, leaflike bracts. Sepals 5–7, 2.0–3.5 mm long, ovate to narrowly ovate, somewhat thickened, bluntly to sharply pointed at the tip. Petals 5–7, 5.0–6.5 mm long, oblong-lanceolate, yellow. Stamens with the anthers yellow, the filaments attached at the petal bases. Nectaries about as long as wide, oblong, greenish yellow. Carpels mostly 4 or 5(–7), yellowish green. Fruits 4–6 mm long, erect or strongly ascending at maturity. Seeds 1.2–1.6 mm long, narrowly obovoid, prominently ribbed, dark brown to nearly black. 2n=108. May–July.

Introduced, uncommon, known thus far only from Hickory and Phelps Counties (native of Europe, Africa; introduced sporadically in the eastern U.S. and Canada). Cemeteries and roadsides.

This species has not previously been reported as an escape in Missouri. Plants usually produce numerous, shorter, ascending vegetative branches from the main prostrate stems and only sporadic, longer, flowering branches.

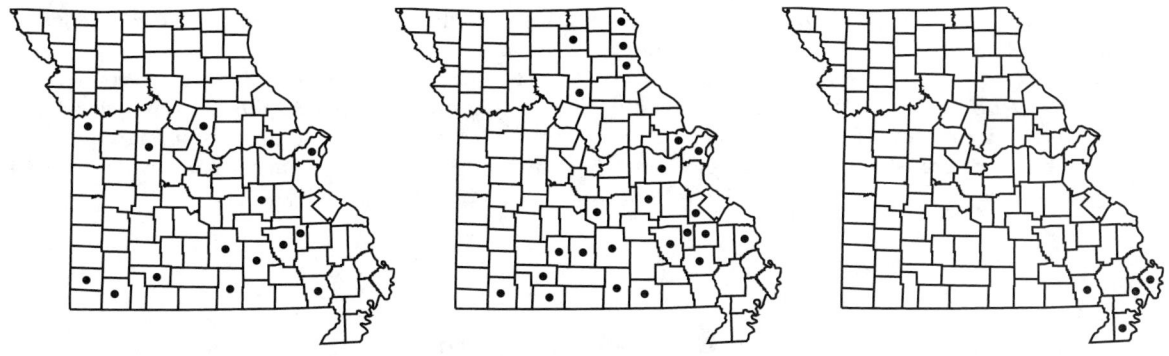

1620. Sedum sarmentosum   1621. Sedum ternatum   1622. Cayaponia quinqueloba

7. Sedum sarmentosum Bunge (yellow stonecrop)

Pl. 370 g; Map 1620

Plants perennial from fibrous roots, the stems rooting at the nodes, the foliage and flowers often with faint, tiny, reddish brown resinous lines or dots visible upon drying. Stems light green to white, rarely reddish- or pinkish-tinged, not glaucous, the main stems 5–30 cm long, prostrate, often forming loose mats, the flowering branches 5–20 cm long, spreading to loosely ascending. Leaves mostly in whorls of 3, less commonly opposite, sessile (rarely a few leaves appearing minutely petiolate), attached just above the minutely pouched base, only slightly and gradually reduced in size toward the stem tip. Leaf blades 10–30 mm long, persistent at fruiting, thickened but somewhat flattened, narrowly depressed-oblong or somewhat rectangular in cross-section, narrowly oblanceolate to narrowly oblong-elliptic or nearly linear, sharply pointed at the tip, the margins entire, the surfaces green, not glaucous. Inflorescences panicles with 2 or 3 branches, these loosely ascending, somewhat zigzag, few-flowered spikes. Sepals (4)5, 3.5–5.0 mm long, oblong-lanceolate, bluntly pointed at the tip, the base with a minute, pouchlike spur of tissue. Petals (4)5, 5–8 mm long, narrowly lanceolate, yellow. Stamens with the anthers yellow, often reddish-tinged, the filaments attached at the petal bases. Nectaries about as long as wide, oblong, yellowish orange to orange. Carpels (4)5, greenish yellow. Fruits usually withering before full maturity, 4.5–6.0 mm long, spreading after flowering. Seeds aborting during development. $2n=36, 72$. May–September.

Introduced, uncommon and widely scattered (native of Asia; widely introduced in the eastern U.S. and adjacent Canada west to Wisconsin, Missouri, and Louisiana). Rocky banks of streams and rivers, glades, and ledges and tops of bluffs; also cemeteries, roadsides, and open to partially shaded disturbed areas, on both acidic and calcareous substrates.

8. Sedum ternatum Michx. (three-leaved stonecrop)

Pl. 370 a, b; Map 1621

Plants perennial from fibrous roots, the stems sometimes rooting at the nodes, the foliage and flowers often with faint, tiny, reddish brown resinous lines or dots visible upon drying. Stems green, sometimes reddish- or pinkish-tinged, not or only slightly glaucous, the main stems 10–40 cm long, prostrate with ascending tips, often forming loose mats, the flowering and vegetative branches 5–15 cm long, erect or ascending. Leaves mostly opposite or in whorls of 3, less commonly in whorls of 4, rarely alternate, sessile, attached just above the base, with a small, truncate plate of tissue below the attachment point, only slightly and gradually reduced in size toward the stem tip. Leaf blades thickened but flat in cross-section, the margins entire or inconspicuously scalloped, the surfaces dark green, sometimes yellowish- or pinkish-tinged, not or only slightly glaucous, those of the main stems and vegetative branches 4–25 mm long, 4–15 mm wide, broadly obovate to broadly elliptic or spatulate, rounded to truncate or shallowly notched at the tip, occasionally very bluntly pointed or with an abrupt, minute, sharp point; those of the flowering branches often slightly shorter or narrower. Inflorescences panicles with (2)3(4) branches, these loosely ascending spikelike racemes. Sepals (4)5(6), 2.5–5.5 mm long, oblong-lanceolate to elliptic, bluntly pointed at the tip. Petals (4)5(6), 4.5–7.0 mm long, narrowly elliptic-lanceolate, white. Stamens with the anthers red or purplish red. Nectaries somewhat longer than wide, oblong, yellow to nearly white. Carpels (4)5(6), white. Fruits 3.5–6.0 mm long, spreading at maturity. Seeds 0.7–1.0 mm long, narrowly asymmetrically ovoid, finely longitudinally ribbed, dark brown. $2n=16, 24, 32, 48$. April–June.

Scattered in the Ozark and Ozark Border Divisions, and apparently disjunct in Marion County (eastern U.S. and adjacent Canada west to Iowa and Arkansas). Bottomland forests, mesic upland forests, rock outcrops, rocky spring outlets, banks of streams and rivers, shaded margins of glades, bases and ledges of bluffs; also railroads and partially shaded disturbed areas; on both acidic and calcareous substrates.

In recent years, *S. ternatum* has become more widely grown as an ornamental groundcover and rock garden plant. In addition to its native distribution, it has escaped from cultivation at scattered sites in the state.

CUCURBITACEAE (Gourd Family)
Contributed by David J. Bogler

Plants annual or perennial vines (woody elsewhere), climbing or trailing with coiled tendrils from the nodes, usually monoecious or dioecious. Stems often 5-angled, often hollow between the nodes, usually hairy. Leaves alternate, often long-petiolate, simple but palmately (rarely pinnately) lobed or divided, lacking stipules. Inflorescences axillary, of solitary flowers or in clusters, sometimes racemes or panicles, sometimes the staminate flowers more numerous and positioned toward the stem tip, the pistillate flowers then fewer and toward the stem midpoint and/or base. Flowers staminate or pistillate, epigynous, with a hypanthium extending above the ovary as a tube. Calyces 5- or 6-lobed, sometimes deeply so, actinomorphic. Corollas 5- or 6-lobed, sometimes deeply so and appearing as free petals, actinomorphic, bell-shaped, saucer-shaped, or less commonly trumpet-shaped, usually yellow or white, those of staminate and pistillate flowers sometimes appearing somewhat different. Stamens 5, but often appearing fewer because of fusion of 1 or 2 pairs of stamens (appearing as 3 stamens in all Missouri genera), attached at or toward the base of the inner side of the hypanthium, the filaments sometimes fused into a column, the anthers facing outward, attached at or toward the base of the inner surface, sometimes fused, often bent or contorted, reduced to staminodes in pistillate flowers. Pistils 1 per flower, of 3 fused carpels, highly reduced or absent in staminate flowers. Ovary inferior (incompletely so elsewhere), with 1 locule, the 2–5 areas of placentation parietal (1 terminal placenta and 1 ovule in *Sicyos*). Style 1 per flower, the stigma 1- or 3-lobed, the lobes sometimes each shallowly 2-lobed. Ovules most commonly numerous. Fruits berries, sometimes with a hardened or leathery rind (then known as a pepo), less commonly dry and capsulelike, indehiscent or bursting irregularly in *Echinocystis*. Seeds numerous, less commonly few or only 1, often somewhat flattened, the seed coat usually of several layers, the outer layer sometimes fleshy. Ninety to 130 genera, 800–900 species, nearly worldwide, but most diverse in tropical and subtropical regions.

All members of the family are sensitive to frost. The cultivated species are annual vines that die back each year, and the temperate species often are perennial vines with a deep, usually massive rootstock. Cucurbitaceae generally are noted for their vining habit, coiled tendrils, palmately lobed leaves, inferior ovaries, imperfect flowers, and often large, berrylike fruits known as pepos. They are further characterized by the presence of a very bitter class of tetracyclic triterpenoids known as cucurbitacins. The family once was thought to be closely related to the Passifloraceae because of their similar tendrils, inferior ovaries, and parietal placentation, but now the Cucurbitaceae are thought to be closer to the Begoniaceae, another family with inferior ovaries having deeply intruding parietal placentae (Angiosperm Phylogeny Group, 1998; Judd et al., 2002).

Many species of Cucurbitaceae have been domesticated and are grown for food, as ornamentals, or for use in handicrafts, including various kinds of melons, squashes, cucumbers, and gourds. Some species of the genus *Luffa* Mill. are cultivated for the network of fibers in their fruits that are dried, cleaned, shaped, and sold commercially as sponges. Seeds from some of these domesticated taxa often wind up in waste areas, where they germinate and persist for many years before ultimately succumbing to competition, herbivory, freezing temperatures, or disease. If the progenitors of the cultivated variety were adapted to local conditions and disturbed habitats, the plants may become feral and occupy habitats similar to those of their ancestors. In the St. Louis area, several species of *Cucurbita* and *Cucumis,* described below, have been found growing untended along railroads and in waste areas.

Coccinia grandis (L.) J. Voigt. (ivy gourd, scarlet gourd) is a dioecious species native to Africa, southern Asia, and Malesia that is cultivated in its native range for its edible fruits and young foliage. It is an aggressive invasive exotic in portions of Australia and on a number of Pacific islands, including Hawaii. It also has been reported sporadically as an introduction in Texas and Florida. This species was discovered in 1996 growing in a fencerow near a motel in Clinton (Henry County) by members of the Missouri Native Plant Society. It is not known whether the plants encountered were planted deliberately as an ornamental or for food, or whether the plants were a chance introduction. The species is a strong perennial with a large, tuberous rootstock and rapidly growing stems. It has unbranched tendrils, glabrous leaves, solitary or paired short-stalked flowers with white corollas 15–20 mm long, and large, fleshy, bright red, ellipsoid to ovoid fruits 2–5 cm long. For now, *C. grandis* has been excluded from formal treatment in the flora, but it should be monitored for potential spread in southern Missouri.

1. Leaf blades deeply pinnately divided, the divisions again deeply lobed .... 2. CITRULLUS
1. Leaf blades shallowly to moderately palmately lobed, but not deeply divided, rarely unlobed
    2. Stems and leaves with dense, soft, sticky hairs; petioles with a pair of disc-shaped glands at the tip .... 6. LAGENARIA
    2. Stems glabrous or sparsely to densely hairy, sometimes roughened but not sticky (except in *Sicyos,* which has somewhat sticky, minutely gland-tipped hairs); petioles lacking glands
        3. Stems relatively stout, 2–5 mm in diameter, coarsely roughened; fruits large, more than 5 cm long; seeds numerous, more than 20
            4. Tendrils unbranched; corollas less than 5 cm wide; ovary and immature fruit pubescent (but the fruit usually glabrous at maturity) .... 3. CUCUMIS
            4. Tendrils branched; corollas 5–10 cm wide or wider; ovary and fruit glabrous .... 4. CUCURBITA
        3. Stems slender, mostly 1–2 mm in diameter, glabrous or hairy, not coarsely roughened; fruits small, less than 5 cm long; seeds few, less than 20
            5. Tendrils branched; staminate flowers in racemes; stamens united; fruits with bristly spines or prickles
                6. Leaf blades with ovate to oblong-triangular lobes, the angle between the major lobes 90° or less; corollas 6-lobed; fruits 2.0–3.5 cm long, solitary, inflated, dehiscing irregularly at the tip, usually 4-seeded .... 5. ECHINOCYSTIS
                6. Leaf blades with broadly triangular lobes, the angle between the major lobes more than 90°; corollas 5-lobed, fruits 1.2–1.8 cm long, in small clusters, not inflated, indehiscent, 1-seeded .... 8. SICYOS
            5. Tendrils unbranched (rarely 2-branched in *Cayaponia*); staminate flowers in clusters in the leaf axils, these either short-stalked to nearly sessile or long-stalked; stamens distinct; fruits smooth, lacking spines or prickles

7. Leaf blades 5–11 cm long, hairy; fruits 1.2–1.8 cm long, the stalk 2–3 mm long, red (rarely yellow); seeds 1–3 per fruit. . . . . 1. CAYAPONIA
7. Leaf blades 2–6 cm long, glabrous; fruits 0.7–1.2 cm long, the stalk 30–50 mm long, mottled with darker and lighter green, becoming black with age; seeds 10–14 per fruit. . . . . 7. MELOTHRIA

1. Cayaponia S. Manso

About 50 species, North America, Central America, South America, Caribbean Islands, Africa.

The tuberous roots of the South American *C. tayuya* (Vell.) Cogn. are used in herbal medicine, primarily as an analgesic and anti-inflammatory agent, but purportedly to cure a variety of ailments.

1. Cayaponia quinqueloba (Raf.) Shinners (melonleaf)
Arkezostis quinqueloba Raf.
Bryonia boykinii Torr. & A. Gray
C. boykinii (Torr. & A. Gray.) Cogn.
C. grandifolia (Torr. & A. Gray) Small
Melothria grandifolia Torr. & A. Gray

Pl. 372 a; Map 1622

Plants monoecious perennial vines with slender rhizomes. Stems to 4 m or more long, slender (1–2 mm in diameter), glabrous or sparsely (more densely on young growth) pubescent with short, nonsticky, spreading hairs, not strongly roughened, the tendrils usually unbranched. Leaves long-petiolate, the petiole 3–5 cm long, lacking glands at the tip, with sparse to moderate spreading hairs. Leaf blades 5–11 cm long, 6–13 cm wide, broadly ovate to nearly circular in outline, palmately moderately 5-lobed with 3 major lobes and 2 minor lobes, the lobes broadly triangular to triangular or ovate, with sharply pointed tips and broadly to narrowly rounded (mostly more than 90°) sinuses, cordate at the base, the margins otherwise sparsely toothed, the surfaces sparsely to moderately roughened with inconspicuous, short, nonsticky, pustular-based hairs. Flowers solitary or in small clusters in the leaf axils, the main stalk 3–5 mm long, the clustered flowers with individual stalks 1–2 mm long, those of the pistillate flowers elongating to 3 mm at fruiting. Calyx lobes 1–3 mm long. Corollas 4–6 mm wide, saucer-shaped to broadly bell-shaped, the usually 5 lobes 3–5 mm long, greenish white to white. Staminate flowers with the stamens distinct. Pistillate flowers with 3 staminodes, the ovary with 1 or 2 ovules per placenta, the stigma 3-lobed. Fruits solitary or paired, juicy berries, 1.2–1.8 cm long, thin-walled, indehiscent, ovoid or ellipsoid, with a stalk 2–3 mm long, the surface red (rarely yellow), smooth (not spiny), glabrous, glossy. Seeds 1–3, 6–9 mm long, 4–6 mm wide, more or less oblong in outline, somewhat flattened, pointed at the tip, with a pair of shallow, blunt lobes toward the base, black. June–August.

Uncommon in the Mississippi Lowlands Division (Florida to Texas north to Missouri). Swamps, bottomland forests, and banks of streams and rivers, usually climbing on shrubs and forming mats.

Two species of *Cayaponia* have been described in the southeastern United States. *Cayaponia grandifolia* was distinguished by having fruits 16–20 mm long and the largest leaf blades more than 10 cm wide, and *C. quinqueloba* (*C. boykinii*) was distinguished by having fruits 12–14 mm long and leaf blades less than 10 cm wide. Although there is considerable variation in leaf size and lobing, this is not correlated with fruit size or other characters. Thus, only one species is recognized here. The name *C. quinqueloba* has priority over *C. boykinii* and *C. grandifolia* (Shinners, 1957). However, critical examination of type collections of the names in question may lead to further reevaluation of the nomenclature and taxonomy of the complex in the future.

2. Citrullus Schrad.

About 4 species, native to Africa, Asia, introduced widely.

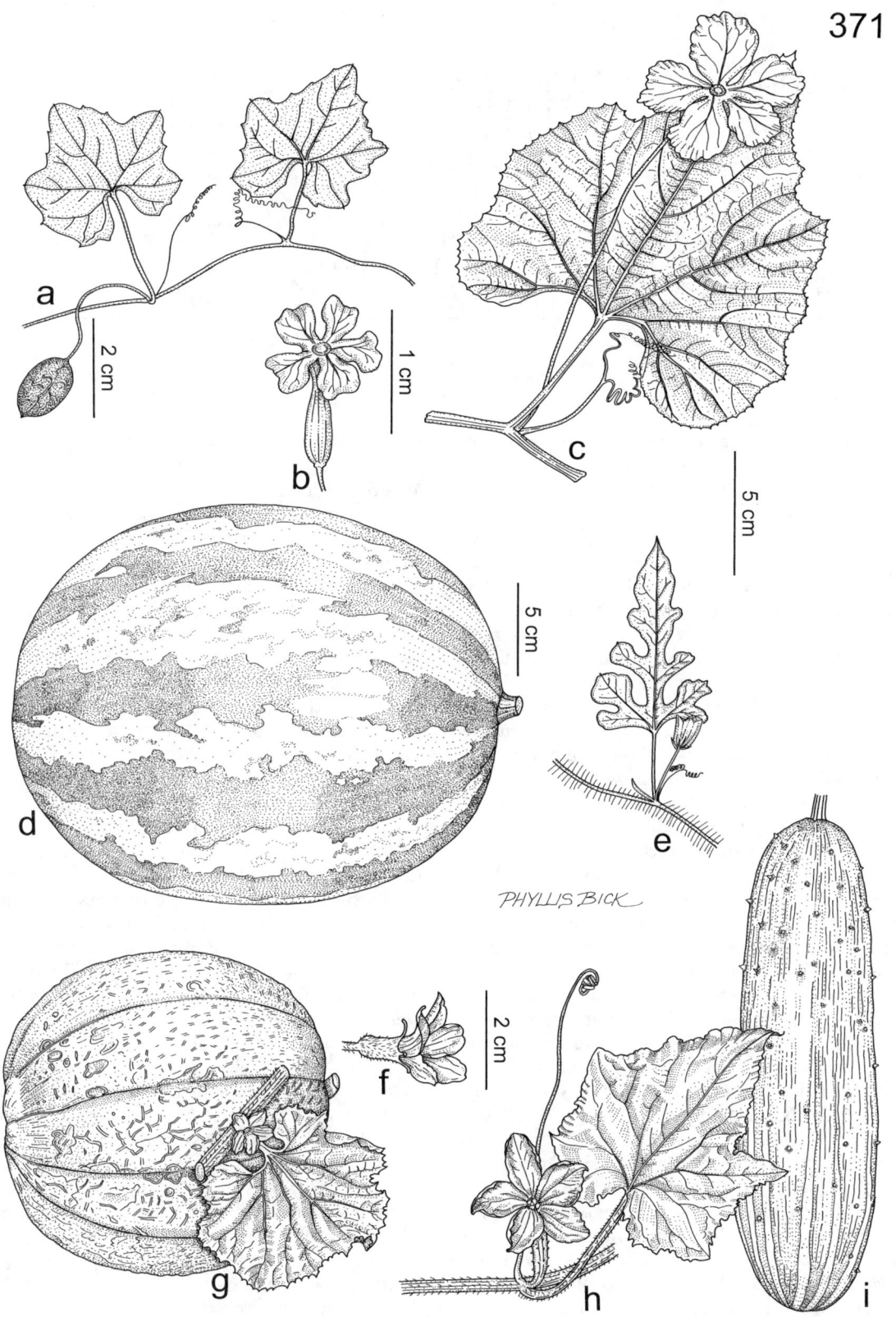

Plate 371. Cucurbitaceae. *Melothria pendula*, **a)** fruiting branch, **b)** flower. *Lagenaria siceraria*, **c)** flowering branch. *Citrullus lanatus*, **d)** fruit, **e)** flowering branch. *Cucumis melo*, **f)** flower, **g)** fruit with flowering branch. *Cucumis sativus*, **h)** flowering branch, **i)** fruit.

1623. Citrullus lanatus   1624. Cucumis melo   1625. Cucumis sativus

1. Citrullus lanatus (Thunb.) Matsum. & Nakai **var. lanatus** (watermelon)

Momordica lanata Thunb.
C. vulgaris Schrad.

Pl. 371 d, e; Map 1623

Plants monoecious annual vines with taproots. Stems to 4 m or more long, relatively stout (2–5 mm in diameter, at least toward the base), densely pubescent with relatively long, sometimes somewhat matted, nonsticky, spreading hairs (these often sparse on older growth), not roughened, the tendrils branched. Leaves short- to long-petiolate, the petiole 1.5–15.0 cm long, lacking glands at the tip, with moderate to dense, spreading hairs. Leaf blades 4–20 cm long, 3–18 cm wide, triangular-ovate to more or less heart-shaped in outline, deeply and irregularly pinnately divided with (3–)5–11 main lobes, at least the lowermost divisions again deeply lobed, the ultimate lobes obovate to oblong or semicircular, with rounded to bluntly or sharply pointed tips and usually narrowly rounded (less than 90°) sinuses, cordate at the base, the margins otherwise irregularly and finely scalloped or toothed, the surfaces slightly to moderately roughened with shorter and longer, nonsticky, mostly pustular-based hairs (these sometimes relatively soft), 1 or both surfaces often appearing gray or grayish-tinged. Flowers solitary in the leaf axils, the stalks of the pistillate flowers 2–12 mm long, those of the staminate flowers 6–30 mm long. Calyx lobes 2–12 mm long. Corollas 15–30 mm wide, broadly bell-shaped to nearly saucer-shaped, the 5 lobes 7–15 mm long, yellow. Staminate flowers with the stamens distinct. Pistillate flowers with 3 small staminodes, the ovary with numerous ovules per placenta, the stigma 3-lobed or more or less 6-lobed. Fruits solitary, juicy berries 5–50 cm long (much larger in some cultivated plants), the rind usually relatively thick, somewhat leathery (but not hardened), indehiscent, oblong-elliptic to nearly spherical, with a stalk 12–35 mm long, the surface usually green with lighter green, irregular longitudinal stripes, occasionally white or yellow, smooth (not spiny), glabrous, usually dull. Seeds numerous, 10–15 mm long, 6–9 mm wide, elliptic-obovate to obovate in outline, usually with a pair of slightly swollen patches at the base, flattened, mostly rounded at the tip, the surface smooth, black or brown with black mottling or spots, rarely ivory-colored, usually shiny. $2n=22$. May–October.

Introduced, scattered and sporadic, mostly south of the Missouri River (cultigen of African origin, introduced nearly worldwide, widely but sporadically in the southern U.S. north to California, Illinois, and Maine). Banks of streams and rivers; also railroads, roadsides, and open, disturbed areas.

Watermelons are grown for their large, juicy fruits and sometimes for their edible seeds. The cultivated watermelon has been classified as var. *lanatus* and originally was domesticated in central or southern Africa, where wild plants referable to var. *citroides* (L.H. Bailey) Mansf. occur (these are also the source of the citron, a bland pale-fleshed fruit used primarily to make pickles and conserves, and for its oily seeds). By 2000 B.C. watermelons were an important crop in Egypt, and by A.D. 1000 they were grown in India, China, and Russia (Robinson and Decker-Walters, 1997). In these places the seeds are sometimes eaten or crushed for their oil. Watermelons were introduced into Spain by the Moors, where the Arabic name became *sandia*. The Spanish introduced watermelons into North America in the 1500s, where they became popular with Native Americans and spread rapidly (Blake, 1981). By the time later European explorers arrived on the scene, watermelons were already cultivated widely. Today, many cultivars have been developed that are disease resistant, early maturing, seedless (these are sterile triploid races), short-stemmed, or yellow-fleshed. The wild taxa of *Citrullus* in Africa are a useful source of genetic variation for watermelon breeders.

3. Cucumis L.
(Kirkbride, 1993)

Plants monoecious, annual (perennial elsewhere) vines with taproots. Stems 1–3 m or more long, relatively stout (2–4 mm in diameter, at least toward the base), coarsely roughened with stout, multicellular, pustular-based hairs, the tendrils unbranched. Leaves mostly long-petiolate, the petioles 1–9 cm long, lacking glands at the tip, coarsely roughened with stout, multicellular, pustular-based hairs. Leaf blades ovate to ovate-triangular or nearly circular to somewhat kidney-shaped in outline, palmately shallowly to moderately (3)5-lobed, often with 3 major lobes and 2 minor lobes, the lobes broadly triangular to more or less oblong to semicircular, rounded or sharply pointed at the tip and with broadly rounded (mostly more than 90°) sinuses, the margins otherwise finely toothed, the surfaces moderately to densely roughened with a mixture of minute and longer, pustular-based hairs. Flowers solitary or in small clusters in the leaf axils, the clusters with the main stalk absent or to 25 mm long, the individual flowers with stalks 1–8 mm long. Calyx lobes 1–3 mm long. Corollas 0.5–4.0 cm wide, bell-shaped or those of the staminate flowers sometimes saucer-shaped, 5-lobed, yellow to orangish yellow. Staminate flowers with the stamens distinct. Pistillate flowers with 3 staminodes, the ovary with numerous ovules per placenta, the stigma 3(–6)-lobed. Fruits solitary or less commonly (in *C. sativus*) in clusters of 2 or 3, juicy berries more than 5 cm long, the rind thin or thick, often somewhat leathery (but not hardened), indehiscent, variously shaped, with a stalk 3–25 mm long, the surface often hairy and sometimes with blunt, soft prickles when young, usually glabrous at maturity, smooth or sparsely warty (not spiny, but the bases of shed prickles sometimes still apparent) or with a dense network of ridges, green, pale grayish green, greenish yellow, or tan, sometimes with irregular stripes, glossy or dull. Seeds numerous (more than 20), 7–10 mm long, oblong-elliptic to oblong-obovate in outline, flattened, rounded to sharply pointed at the tip, the surface otherwise smooth, white or light yellow. About 32 species, native to Africa, Asia, south to Australia, introduced nearly worldwide.

The genus *Cucumis* is economically important for its edible fruits. The two most important species are those treated below as escapes from cultivation, the cantaloupes/muskmelons and the cucumbers/pickles. However, a third species is becoming popular in this country. *Cucumis metuliferus* E. Mey. ex Naudin was developed from plants native to Africa and southwestern Asia. It produces an oblong-ellipsoidal fruit 6–15 cm long whose yellowish orange to reddish orange surface has scattered, coarse, conical prickles and green pulp. Sold under the names horned melon and horned cucumber, the flavor of the cultivated strains is relatively bland (the wild relatives of most cultivated melons have a very bitter flavor), but the fruit has a long shelf life in supermarkets and its unusual appearance has presumed ornamental value.

1. Leaf lobes rounded; ovary and young fruit hairy but smooth; mature fruit with the rind smooth, somewhat flaky, or with a dense network of blunt ridges ... 1. C. MELO
1. Leaf lobes mostly sharply pointed; ovary and young fruit hairy and with blunt, soft, pustular-based prickles; mature fruit with the rind usually appearing irregularly bumpy or with scattered low warts (the persistent pustular bases of the prickles) .... 2. C. SATIVUS

1. Cucumis melo L. ssp. melo (muskmelon, cantaloupe)

Pl. 371 f, g; Map 1624

Leaf blades 3–10 cm long, 3–13 cm wide, broadly ovate to nearly circular or somewhat kidney-shaped in outline, the 5 shallow lobes more or less oblong or semicircular, with rounded tips. Staminate flowers solitary or in small axillary clusters, the individual flowers with slender stalks 1–8 mm long, the corolla lobes 2–12 mm long. Pistillate flowers solitary, with a stout stalk 2–15 mm long, the hypanthium and calyx densely hairy but without prickles, the corolla lobes 3.5–10 mm long. Fruits 8–12 cm long (much larger in some cultivated plants), spherical to ellipsoid or oblong-ovoid, the flesh and pulp yellow to orange or green, the rind usually relatively thick, smooth, somewhat flaky, or with a dense network of ridges, green, pale grayish green, greenish yellow, or tan, sometimes with irregular longitudinal stripes and/or grooves. $2n=24$. July–October.

Introduced, uncommon and sporadic (cultigen of Asian and African origin, introduced nearly worldwide, including the eastern and southern U.S.). Banks of streams and rivers and margins of ponds and lakes; also pastures, railroads, roadsides, and open, disturbed areas.

Wild populations of *C. melo* are found in the desert regions of Africa and southwestern Asia, where melons were domesticated by 2000 B.C. They were known to the Romans and Arabs and introduced into the New World by Columbus on his first voyage (Sauer, 1993). Like the watermelon, muskmelons were very popular among Native Americans and spread rapidly from tribe to tribe. Many cultivars have been developed, including disease-resistant sweet varieties with a netted rind (cantaloupe) or a smooth rind (honeydew, casaba). *Cucumis melo* is sometimes divided into subspecies or varieties (Whitaker and Davis, 1962; Kirkbride, 1993) based on fruit types or pubescence patterns of the pistillate flowers. In Missouri we have *C. melo* ssp. *melo,* which includes cantaloupes, muskmelons, honeydews, and casaba.

2. Cucumis sativus L. (cucumber)

Pl. 3371 h, i; Map 1625

Leaf blades 3–15 cm long, 3–13 cm wide, ovate to broadly ovate-triangular or sometimes nearly circular in outline, the 5 shallow to moderate lobes broadly triangular, with mostly sharply pointed tips. Staminate flowers solitary or in small clusters, the individual flowers with slender stalks 3–8 mm long, the corolla lobes 7–16 mm long. Pistillate flowers solitary or in small clusters, the individual flowers with stout stalks 3–8 mm long, the hypanthium and calyx moderately to densely hairy and with blunt, soft, pustular-based prickles, the corolla lobes 10–15 mm long. Fruits 5–20 cm long (much longer in some cultivated plants), oblong-cylindrical, sometimes slightly arched, the flesh and pulp pale green, the rind relatively thin, usually appearing irregularly bumpy or with scattered, low warts (the persistent pustular bases of the prickles), green to dark green, rarely with irregular longitudinal stripes. $2n=14$. July–September.

Introduced, uncommon, known thus far only from the city of St. Louis (cultigen of apparently Asian origin, introduced sporadically nearly worldwide, in the U.S. mostly east of the Mississippi River). Railroads.

Cucumbers most likely originated in subtropical areas bordering the Himalayan Mountains, where locally developed races are still found today. *Cucumis sativus* has been cultivated in India for at least 3,000 years and was introduced very early into Europe and China (Sauer, 1993). Numerous cultivars have been developed, including several for pickling and slicing. This species was first reported for Missouri by Mühlenbach (1979) based on his botanical inventories of the St. Louis railyards.

4. Cucurbita L.

Plants monoecious, annual or perennial vines. Stems 1–5 m or more long, relatively stout, 2–5 mm in diameter, coarsely roughened with stout, multicellular, pustular-based hairs, the tendrils branched. Leaves mostly long-petiolate, the petioles 3–12 cm long, lacking glands at the tip, coarsely roughened with stout, multicellular, pustular-based hairs. Leaf blades ovate to ovate-triangular or nearly circular in outline, unlobed or palmately shallowly to moderately or rarely deeply 3- or 5(7)-lobed, the lobes broadly triangular to more or less oblong to semicircular, rounded or bluntly to sharply pointed at the tip and with narrowly to more commonly broadly rounded or angled (mostly more than 90°) sinuses, the margins otherwise finely toothed, the surfaces moderately to densely roughened with mostly pustular-based hairs of varying length and thickness. Flowers solitary in the leaf axils or (in *C. pepo*) the staminate flowers occasionally in small clusters, the staminate flowers or flower clusters with longer stalks than those of the pistillate flowers. Calyx lobes 9–20(–30) mm long. Corollas 5–10 cm wide, deeply bell-shaped, 5-lobed, yellow to yellowish orange. Staminate flowers with the filaments fused into a tube except sometimes at the very base (the anthers fused into a headlike mass). Pistillate flowers with 3 staminodes, the hypanthium and calyx moderately to densely hairy, the ovary with numerous ovules per placenta, the stigma 3–5-lobed or more or less 6–10-

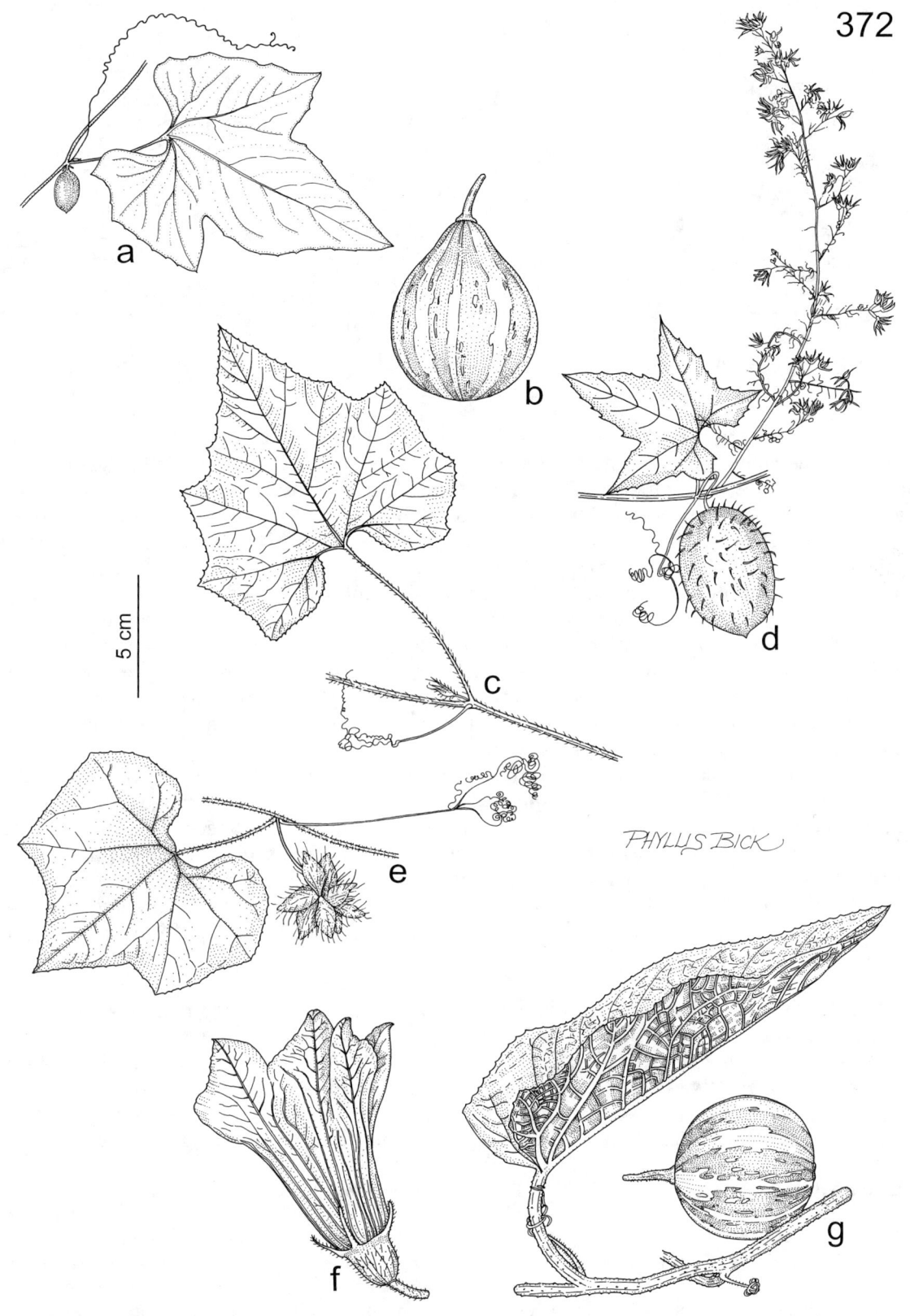

Plate 372. Cucurbitaceae. *Cayaponia quinqueloba*, **a**) fruiting branch. *Cucurbita pepo*, **b**) fruit, **c**) node with leaf and tendril. *Echinocystis lobata*, **d**) node with staminate inflorescence and fruit. *Sicyos angulatus*, **e**) fruiting branch. *Cucurbita foetidissima*, **f**) flower, **g**) fruit and node with leaf and tendril.

1626. Cucurbita foetidissima   1627. Cucurbita pepo   1628. Echinocystis lobata

lobed. Fruits solitary, modified berries 5–10 cm long (larger in some cultivated plants), with a pulpy, fibrous central portion, at least when young (drying out as the fruit matures in wild plants), and a relatively thin, hardened shell (the rind thicker and leathery in some cultivated plants), indehiscent, more or less spherical to ovoid or pear-shaped (variously shaped in cultivated plants), with a stalk 15–40 mm long, the surface glabrous at maturity, smooth (lacking prickles or warty outgrowths), green (variously colored in some cultivated plants), sometimes with irregular, longitudinal, light green or white stripes, becoming bleached to a yellow, tan, or ivory color, glossy or dull. Seeds numerous (more than 20), 6–10 mm long, elliptic to obovate in outline, flattened, sometimes with a pronounced, thickened rim, mostly rounded at the tip, the surface otherwise smooth, white, cream-colored, or tan, less commonly brown or black. About 13 species, native from North America to South America, introduced nearly worldwide.

The genus *Cucurbita* is of economic importance primarily for its edible and ornamental fruits but has also been used in crafts, as containers, and for soap (because some species produce appreciable amounts of saponins), among other things. Members of the genus have an extremely long history in the New World archaeological record (Nee, 1990) and were traded widely and transported throughout the Americas prior to European contact. Heiser (1979) has written eloquently and in great depth on the ethnobotany of the genus. Five different species are cultivated, including a great variety of types of pumpkins, acorn squashes, crooknecks, zucchinis, vegetable marrows, fordhooks, scallop squashes, and ornamental gourds. Common names such as squash and pumpkins are somewhat ambiguous as they have been applied to plants of more than one species.

1. Leaf blades triangular to ovate-triangular in outline, distinctly longer than wide, not lobed or with a pair of very shallow lobes, the surfaces appearing gray or strongly grayish-tinged. . . . . 1. C. FOETIDISSIMA
1. Leaf blades broadly ovate to nearly circular or somewhat kidney-shaped in outline, about as long as wide, usually distinctly lobed, at least the upper surface green, the undersurface sometimes appearing somewhat grayish-tinged . . . 2. C. PEPO

1. Cucurbita foetidissima Kunth (buffalo gourd, Missouri gourd)

C. perennis A. Gray

Pepo foetidissimus (Kunth) Britton

Pl. 372 f, g; Map 1626

Plants perennial vines with deep, stout, branched, somewhat tuberous roots, the trailing stems also sometimes rooting at the nodes, usually producing a strong aroma similar to that of garlic, especially when crushed or bruised. Leaf blades 10–40 cm long, 6–30 cm wide, triangular to ovate-triangular in outline, distinctly longer than wide, not lobed or with a pair of very shallow lobes toward the base, these broadly triangular, mostly sharply pointed at the tip, the surfaces densely roughened with short, stiff, pustular-based hairs (usually scattered, longer, stouter hairs also present), appearing gray or strongly grayish-tinged. Flowers all solitary. Fruits 6–10 cm long, spherical or nearly so, green with usually fairly strong, light green to white, irregular longitudinal stripes, bleaching to yellow, tan, or ivory-colored with age. $2n=40$, 42. June–August.

Possibly introduced, uncommon and sporadic (Nebraska to Texas west to California; Mexico; introduced sporadically east to Virginia and Florida). Banks of rivers; upland prairies; also edges of crop fields, railroads, roadsides, and open, disturbed areas.

Steyermark (1963) and Yatskievych and Turner (1990) treated this species as native in Missouri. Although native populations might plausibly occur in the northwestern portion of the state or in counties along the Missouri River, there is no strong evidence to support this hypothesis, in part because the older specimens have incomplete collection data on their labels. The oldest collections from Missouri were not made until the 1890s in the Springfield (Greene County), Sheffield (Jackson County), and Pacific (St. Louis County) areas, which is unusual for a species that is so robust and conspicuous where it occurs. Rare collections made along the Missouri River in later decades are more problematic, as the hard-shelled fruits are capable of long-distance dispersal by water. However, *C. foetidissima* has been encountered most commonly along railroads in the state. As of this writing, the species has not been collected in Missouri since 1956.

The fruits and seeds of buffalo gourd were used by Native Americans for food, as a soap substitute, and for ceremonial rattles, as well as medicinally for treatment of sores and various other pains (Moerman, 1998). The oil content of seeds is about 25–43 percent, and the protein content is 22–35 percent (Bemis et al., 1978). The roots of older individuals can be several meters long and weigh more than 80 kg (175 pounds), with up to 56 percent of the dry weight composed of starch. Because of its high protein, oil, and starch content and its very rapid potential growth rate, buffalo gourd has been identified as a potential new crop for arid regions of the Southwest and the tropics. Studies are ongoing to develop races suitable for livestock feeds, seed oils, and other food products, as well as to overcome potential problems, such as the somewhat unpleasant flavor of the flour made from the roots and the susceptibility of plants in cultivation to various diseases and insect pests (Gathman and Bemis, 1990).

2. Cucurbita pepo L. **var. ozarkana** D.S.
Decker-Walters (wild pear gourd, yellow-flowered gourd)
Pl. 372 b, c; Map 1627
Plants annual vines with taproots, the trailing or more commonly climbing stems not rooting at the nodes, not or only faintly producing a disagreeable aroma. Leaf blades 4–30 cm long, 5–30 cm wide, broadly ovate to nearly circular or somewhat kidney-shaped in outline, about as long as wide, shallowly to moderately 3- or 5(7)-lobed, the lobes broadly triangular or more or less oblong to semicircular, rounded or bluntly to sharply pointed at the tip, the surfaces moderately to densely roughened with short, stiff, pustular-based hairs, but at least the upper surface green, the undersurface sometimes appearing somewhat grayish-tinged. Flowers all solitary or the staminate flowers occasionally in small clusters. Fruits 5–8 cm long, spherical, ovoid, or somewhat pear-shaped, green with sometimes fairly faint, lighter green, irregular longitudinal stripes, bleaching to ivory-colored at full maturity. $2n=40$. August–October.

Scattered in the Ozark and Ozark Border Divisions (Illinois, Missouri, Arkansas, Oklahoma, and Louisiana). Banks of streams and rivers, sloughs, and bottomland forests; also old fields, fencerows, railroads, and open, disturbed areas.

Cucurbita pepo has a long history of cultivation. Remains of *C. pepo* have been found in archaeological sites dated to about 10,000 years old. Selection initially was for larger seeds and fruits, and later for nonbitter, less fibrous fruits with thinner rinds. The species exhibits an exceedingly wide range of variation in fruit types, including summer squash, zucchini, acorn squash, pumpkins, and ornamental gourds. Among the ornamental gourds, there exists a bewildering array of fruit sizes, shapes, colors, and surface ornamentation.

Traditionally, pumpkins, vegetable marrows, and a few ornamental gourds, all of whose ancestry can be traced back to Mexico, have been classified into ssp. *pepo*. Plants with small, egg-shaped gourds in northern Mexico and the United States were recognized taxonomically very early and given the name *C. ovifera* L. (later reclassified as *C. pepo* var. *ovifera* (L.) Alef. or ssp. *ovifera* (L.) D.S. Decker). The circumscription of this variant eventually came to include nearly all of the inedible ornamental gourds, as well as the crooknecks, scallop squashes, and acorn squashes (Bailey, 1929; Decker, 1988; Decker-Walters et al., 2002). Ornamental gourds are encountered rarely in disturbed habitats in Missouri as individual plants that do not persist for more than a single growing season. They are not treated formally in the present work.

At one time it was believed that domestication of *C. pepo* took place in central Mexico, and that cultivated strains were transported and traded by Native Americans north across Texas and into the eastern United States during pre-Columbian times (Smith et al., 1992). Free-living apparently wild gourds from Texas to Florida were considered to have been derived from escapes of these early introductions. More recently, an alternative hypothesis of the relationships in the complex has gained support from both taxonomic and archaeological research (Decker and Wilson, 1986, 1987; Decker, 1988; Decker and Newsom, 1988; Smith et al., 1992; Cowan and Smith, 1993; Newsom et al., 1993; Decker-Walters et al., 1993, 2002). These studies have shown that wild gourds apparently existed as native populations in a region from northeastern Mexico through the southeastern United States and into portions of the Ozark Highlands prior to the introduction of pumpkins and other central Mexican domesticates in prehistoric times. Thus, *C. pepo* was domesticated independently twice, once in central Mexico and again farther north.

The Texas wild gourd is found along rivers and streams in eastern Texas and has small, bitter, oval or pear-shaped fruits with green and white striping on the rind at maturity. These gourds were collected very early in Texas by Lindheimer and described as *Tristemon texanum* Scheele and were later transferred to *Cucurbita*. Bailey (1943) recognized *C. texana* (Scheele) A. Gray as a valid species and considered the possibility that it was the ancestor of ornamental gourds and some varieties of squash. However, specimens of wild small-fruited gourds also were collected outside of Texas (Cowan and Smith, 1993). For example, specimens from Missouri were identified by Steyermark (1963) as *C. pepo* var. *ovifera*. Evidence for multiple origins of domesticated squashes and gourds and the existence of a wild indigenous progenitor species in the southeastern and midwestern United States was provided by analysis of allozyme variation (Decker and Wilson, 1987; Decker-Walters et al., 1993). Wild populations of *C. pepo* north of Texas displayed a genetic profile differing from those in Texas. Plants in Arkansas and Missouri had more genetic similarities with the cultivated ornamental gourds and probably gave rise to them. The enzymatic data support the theory that genetic divergence in wild populations of *C. pepo* in the United States took place prior to domestication. This resulted in a taxonomic reclassification of the complex within ssp. *ovifera*. Ornamental, nonbitter cultivated gourds with yellow, orange, or other bright coloration and various shapes are known as var. *ovifera*. The wild gourds of Texas with bitter fruits that are green-and-white-striped at maturity are var. *texana* (Scheele) D.S. Decker. The populations elsewhere in the southeastern United States and northward into Missouri and Illinois with bitter fruits that are less prominently striped and turn ivory-colored at maturity are referred to as *C. pepo* var. *ozarkana*.

The allozyme data also indicated that introgression with cultivated *C. pepo* has occurred frequently, manifested by nonbitter fruits, larger seeds, and thick flower stalks. In addition to the wild gourd, plants that appear to reflect past hybridization or escapes from cultivation have been found sporadically in waste areas along railroads and stream banks in and around St. Louis and other urban areas in the state. These are recognized by their larger leaves and longer flower stalks. The commercial release of transgenic varieties of cultivated *C. pepo* varieties has raised the question of possible transfer of genes for disease resistance to wild, potentially weedy relatives (Decker-Walters et al., 2002). The var. *ozarkana* is said to be an aggressive weed in soybean and cotton fields in Arkansas, Louisiana, and Mississippi.

5. Echinocystis Torr. & A. Gray

One species, U.S., Canada.

1. Echinocystis lobata (Michx.) Torr. & A. Gray
(wild cucumber)
Sicyos lobatus Michx.
Micrampelis lobata (Michx.) Greene
Pl. 372 d; Map 1628

Plants monoecious annual vines with slender taproots. Stems to 5 m or more long, slender (1–2 mm in diameter), glabrous, not roughened, the tendrils branched. Leaves mostly long-petiolate, the petiole 1–4 cm long, lacking glands at the tip, glabrous or nearly so. Leaf blades 2–8 cm long, 3–12 cm wide, broadly ovate to nearly circular in outline, palmately moderately 5-lobed, the lobes triangular to oblong-triangular, with sharply pointed tips and mostly narrowly rounded (less than 90°) sinuses, cordate at the base, the margins otherwise sparsely toothed, the surfaces glabrous or slightly to moderately roughened with minute, nonsticky, pustular-based hairs (often only the small, hardened bases apparent). Flowers solitary or in small, few-flowered clusters (pistillate) or in well-developed racemes or racemelike panicles (staminate) 8–14 cm long in the leaf axils, the main stalk of the pistillate inflorescence 2–5 mm long, the clustered flowers with individual stalks 1–5 mm long. Calyx lobes 1.0–1.5 mm long. Corollas 8–12 mm wide, saucer-shaped to broadly bell-shaped, the usually 6 lobes 3–6 mm long, white to cream-colored. Staminate flowers with the filaments fused into a tube (the anthers usually free). Pistillate flowers usually lacking staminodes, the ovary with usually 2 ovules per placenta, the stigma bluntly 2- or 3-lobed. Fruits solitary, thin-walled berries, more or less inflated and juicy at first but becoming dry and fibrous inside a papery wall at maturity, dehiscing irregularly at the tip with age, 2.0–3.5 cm long, ovoid or ellipsoid, with a stalk 9–35 mm long, the surface green, covered with slender (often bristly), relatively soft, straw-colored to pale yellow prickles 3–6 mm long, otherwise glabrous. Seeds usually 4, 12–20 mm long, 6–9 mm wide, elliptic-obovate to more or less elliptic in outline above a short, stalklike base, flattened, usually pointed at the tip, the surface otherwise shallowly and irregularly pitted, dark brown, often finely mottled. $2n=32$. June–October.

Scattered mostly in the western half of the Unglaciated Plains Division sporadically south to Jasper and Greene Counties and east to the city of St. Louis (eastern U.S. west to Montana and Texas; Canada; introduced west to Washington and Arizona). Banks of streams and rivers and bottomland forests; rarely roadsides and disturbed areas.

Wild cucumber occasionally is grown as an ornamental on arbors for its showy staminate inflorescences and attractive foliage. It can escape from cultivation. Native Americans used an infusion of the species to treat fevers, rheumatism, headaches, and general pain, and to induce abortion (Moerman, 1998). With age, the large bristly fruits burst irregularly at the tip and the seeds are dispersed explosively under hydrostatic pressure.

6. Lagenaria Ser.

Six species, native to Africa and Madagascar, introduced widely.

1. Lagenaria siceraria (Molina) Standl. (bottle gourd, calabash, white-flowered gourd)

Pl. 371 c; Map 1629

Plants monoecious annual vines with taproots, usually producing a musky aroma when crushed or bruised. Stems to 5 m or more long, relatively stout (2–6 mm in diameter), densely pubescent with short and/or long, sometimes somewhat matted, spreading hairs, not roughened, the hairs minutely gland-tipped and sticky, the tendrils branched. Leaves mostly long-petiolate, the petiole 2–15 cm long, with a pair of distinctive, large, disc-shaped glands at the tip, with dense, spreading, sticky hairs. Leaf blades 4–20 cm long, 3–18 cm wide, ovate to more or less heart-shaped or kidney-shaped in outline, unlobed or more commonly shallowly 3- or 5-lobed, the lobes mostly broadly triangular, with bluntly or more commonly sharply pointed tips (occasionally tapered into a narrowly triangular, elongate point) and usually broadly rounded (much more than 90°) sinuses, cordate at the base, the margins otherwise irregularly and finely toothed and often somewhat wavy, the surfaces moderately to more commonly densely pubescent with short, tapered, soft hairs, 1 or both surfaces sometimes appearing somewhat grayish-tinged. Flowers solitary or paired in the leaf axils, the stalks 2–30 mm long. Calyx lobes 2–9 mm long (longer elsewhere). Corollas 40–100 mm wide, broadly bell-shaped to nearly saucer-shaped, the 5 lobes 15–30 mm long, white, sometimes with darker veins or cream-colored toward the base. Staminate flowers with the stamens free or the filaments sometimes partially fused (the anthers often appearing fused into a headlike mass). Pistillate flowers usually with 3 small staminodes, the ovary with numerous ovules per placenta, the stigma 3-lobed. Fruits solitary, modified berries mostly 10–40 cm long (but sometimes longer in cultivated plants), with a pulpy, fibrous central portion, at least when young (drying out as the fruit matures), and a relatively thick, hardened shell, indehiscent, highly variable in shape, the fruit body often abruptly tapered to an elongate, cylindrical basal portion, with a stalk 15–40 mm long, the surface glabrous at maturity, usually smooth (lacking prickles or warty outgrowths), green, often mottled or with irregular, longitudinal, light green or white stripes, becoming bleached to a yellow or tan color, more or less dull. Seeds numerous, 12–22 mm long, 6–15 mm wide, somewhat rectangular to broadly wedge-shaped in outline, the main body separated from a pair of thick, corky wings by longitudinal grooves or ridges on each face, flattened, mostly appearing notched or somewhat 3-lobed at the tip by extensions of the wings, the surface smooth to somewhat wrinkled

1629. Lagenaria siceraria   1630. Melothria pendula   1631. Sicyos angulatus

(especially on the wings), tan to dark brown, dull or occasionally somewhat shiny. $2n=22$. August–October.

Introduced, uncommon and sporadic (cultigen of African origin, introduced nearly worldwide, including the southern and eastern U.S.). Banks of streams and rivers; also railroads.

Fruits of *Lagenaria* have harder and more durable walls than those of the *Cucurbita* gourds. They have an extremely long history of use. The native range of the species apparently was in southern or eastern Africa, but the bottle gourd also was one of the earliest plants domesticated in the New World, with archaeological remains from Peru, Mexico, and Florida documenting its use up to 13,000 years ago (Heiser, 1979; Robinson and Decker-Walters, 1990), which predates the cultivation of maize and the manufacture of clay pottery. *Lagenaria* gourds can float in seawater, and the seeds remain viable for many months, thus the fruits may have drifted to the New World on ocean currents. Alternatively, the first Americans may have brought the species with them when they migrated to the New World across the Bering land bridge toward the end of the last Ice Age. *Lagenaria* produces gourds in a bewildering array of shapes and sizes, which have been used to make bottles, drinking cups, bowls, plates, ladles, pails, calabash pipes, snuffboxes, floats for fishing nets, musical instruments, birdhouses, ceremonial masks, and other handicrafts, sometimes with intricate engraving. Young fruits can be eaten like zucchini. Nonbitter edible races are known, and some are used for their oily edible seeds. The plants also have been used in various ways medicinally for a variety of ailments, as a purgative, and in the repair of skull fractures (Heiser, 1979). The white flowers open at night.

Several attempts have been made to divide *L. siceraria* into subspecies and varieties based on morphological differences in various parts of the plants, including leaf shape, flower size, fruit size and shape, and seed morphology. In part because of recent artificial hybridizations made between strains by plant breeders, formal taxonomic recognition seems unwarranted. However, recognition of a range of cultivars seems appropriate. Heiser (1973, 1979) grew plants of numerous accessions from nearly throughout the cultivated range of the species and followed Kobiakova (1930) in accepting two major lineages, those from Africa and the Americas, and those of Asian origin. Unfortunately, his use of the epithet "ssp. *asiatica*" for plants of the latter group appears to have been illegitimate, and this combination has not been validly published since then.

7. Melothria L.

About 12 species, North America to South America.

1. Melothria pendula L. (creeping cucumber)
Pl. 371 a, b; Map 1630

Plants monoecious perennial vines with fibrous roots and often a short, woody taproot. Stems to 4 m or more long, slender (1–2 mm in diameter), glabrous or sparsely (more densely on young growth) pubescent with short, nonsticky, spreading hairs, not strongly roughened, the tendrils usually unbranched. Leaves long-petiolate, the petiole 2–3 cm long, lacking glands at the tip, with sparse to moderate spreading hairs. Leaf blades 2–6 cm long, 2–9 cm wide, broadly ovate to nearly circular in outline, palmately shallowly to moderately (3)5-lobed with 3 major lobes and usually 2 minor lobes, the lobes broadly triangular to less commonly oblong-triangular or ovate, with sharply or occasionally bluntly pointed or rounded tips and usually broadly (more than 90°) or rarely narrowly rounded sinuses, cordate at the base, the margins otherwise sparsely toothed, the surfaces sparsely to moderately roughened with inconspicuous, short, nonsticky, pustular-based hairs. Flowers solitary (pistillate) or in small clusters (staminate) in the leaf axils, the main stalk 15–35 mm long, the clustered flowers with individual stalks 1–3 mm long. Calyx lobes 0.2–0.4 mm long. Corollas 3–5 mm wide, saucer-shaped to broadly bell-shaped, the usually 5 lobes 2–3 mm long, yellow to occasionally yellowish green. Staminate flowers with the stamens distinct. Pistillate flowers sometimes with 3 staminodes, the ovary with 3–6 ovules per placenta, the stigma 3-lobed. Fruits solitary, juicy berries, 0.7–1.2 cm long, ovoid to nearly spherical, the rind relatively thin and leathery, indehiscent, with a stalk 30–50 mm long, the surface mottled with darker and lighter green, sometimes yellowish green, becoming black with age, smooth (not spiny), glabrous, glossy. Seeds 10–14, 3–5 mm long, 2.0–3.5 mm wide, elliptic-obovate to obovate in outline, flattened, sometimes bluntly pointed at the tip, the surface smooth or finely wrinkled, white to off-white, usually shiny. $2n=24$. July–October.

Uncommon in the Mississippi Lowlands Division and the southern portion of the Ozarks (southeastern U.S. west to Kansas and Texas; Mexico). Bottomland forests, mesic upland forests in ravines, banks of streams and rivers, acid seeps, and bases and ledges of bluffs; also ditches, roadsides, and disturbed areas.

Steyermark (1963) noted that the seeds have a strong purgative property. In this species, functionally perfect flowers are produced occasionally at a few nodes in place of the more typical pistillate ones.

8. Sicyos L.

Forty to 50 species, North America to South America, Pacific Islands, Australia, New Zealand.

1. Sicyos angulatus L. (bur cucumber)

Pl. 372 e; Map 1631

Plants monoecious annual vines with slender taproots. Stems to 5 m or more long, slender (1–2 mm in diameter), lacking glands at the tip, sparsely to moderately pubescent with slender, spreading, multicellular hairs, not roughened, at least some of the hairs usually minutely gland-tipped but the stems not strongly sticky, the tendrils branched. Leaves mostly long-petiolate, the petiole 1–5 cm long, densely hairy. Leaf blades 2–20 cm long, 3–22 cm wide, broadly ovate to nearly circular or somewhat kidney-shaped in outline, palmately shallowly to moderately 5-lobed with usually 3 major lobes and 2 minor lobes, the lobes broadly triangular, with sharply pointed tips (those of the basalmost lobes occasionally rounded) and broadly rounded (more than 90°) sinuses, cordate at the base, the margins otherwise sparsely to moderately and finely toothed, the surfaces sparsely to densely pubescent with minute, nonsticky hairs, especially those of the upper surface often minutely pustular-based and thus somewhat roughened to the touch. Flowers in dense clusters (pistillate) or in short but long-stalked racemes or clusters (staminate) 4–15 cm long in the leaf axils, the main stalk of the pistillate inflorescence 10–40 mm long, the clustered flowers sessile or with individual stalks to 0.5 mm long. Calyx lobes 1.0–2.5 mm long. Corollas 8–12 mm wide, saucer-shaped to broadly bell-shaped, the 5 lobes 3–5 mm long, white to cream-colored. Staminate flowers with the filaments fused into a tube (the anthers fused into a headlike mass). Pistillate flowers usually lacking staminodes, the ovary with 1 ovule, the stigma 3(4)-lobed. Fruits in dense, headlike clusters, thin-walled modified berries, dry and with the papery wall relatively closely enveloping the seed, indehiscent (the outer wall sometimes tearing irregularly with age), 1.2–1.8 cm long, more or less ovoid or ellipsoid, somewhat flattened, with a stalk 15–60 mm long, the surface yellowish green to olive-colored, covered with slender (often bristly), relatively stiff, straw-colored to pale yellow prickles 3–7 mm long, otherwise usually moderately pubescent with relatively long, fine, spreading, multicellular hairs. Seed 1, 7–10 mm long, 4–7 mm wide, elliptic-obovate to more or less elliptic in outline with a small pair of thickened basal areas, somewhat flattened, rounded to bluntly pointed at the tip, the surface otherwise relatively smooth, greenish brown to olive-colored. $2n=24$. June–October.

Scattered nearly throughout the state but absent from most of the eastern half of the Glaciated Plains Division (eastern U.S. west to North Dakota and Texas; Canada; introduced in Europe, Asia). Banks of streams and rivers, bottomland forests, and bases, ledges, and tops of bluffs; also farmyards, railroads, roadsides, and disturbed areas.

This species is occasionally grown as an ornamental on arbors, but it is less attractive than *Echinocystis*. Although it is native to the eastern half of the country, it is listed as a noxious weed in Indiana and Delaware. Embryological studies have documented that *S. angulatus* is able to produce multiple parthenogenetic embryos in a single embryo sac (Dathan and Singh, 1990), but whether the species reproduces mostly sexually or apomictically is not known.

DIPSACACEAE (Teasel Family)

About 10 genera, 300–325 species, Europe, Asia, Africa.

1. Dipsacus L. (teasel)

Plants biennial. Stems erect, several-angled or ridged, prickly, often somewhat shiny. Leaves basal and opposite, those of the basal rosette sessile to short-petiolate, those of the stem sessile, those of a pair sometimes fused at the base. Stipules absent. Leaf blades simple, entire to bluntly toothed or irregularly pinnately lobed along the margins, sometimes also prickly, the undersurface prickly along the midvein, those of the basal leaves to 50 cm or more long, often withered or absent by flowering time, those of the stem leaves progressively reduced toward the stem tip. Inflorescences ovoid to cylindrical heads of numerous dense flowers, subtended by 1 or 2 whorls of involucral bracts of unequal lengths, these curved upward, prickly along the midrib and margins, and tapered to a spinelike tip. Flowers perfect, epigynous, somewhat zygomorphic, each subtended by 1 receptacular bract (this with the body lanceolate to oblong-ovate, tapered abruptly at the tip to a long [to 25 mm], slender, loosely ascending, spinelike awn, persistent after the fruits have been shed) and enclosed in a 4-ribbed involucel of fused bractlets (this about as long as the ovary, truncate or obscurely 4-toothed at the tip, shed with the fruit). Calyces cup-shaped, sometimes somewhat 4-lobed, somewhat 4-angled, densely silky-hairy, persistent at fruiting. Corollas 4-lobed, densely minutely hairy on the outer surface, with a long, slender tube, the lobes relatively short, oblong, rounded at the tip. Stamens 4, alternating with the corolla lobes, exserted, the filaments slender, attached at the corolla base, the anthers attached toward their midpoints. Staminodes absent. Pistil 1 per flower, of apparently 1 carpel (a second carpel aborting early in development). Ovary inferior, with 1 locule, the placentation terminal. Style 1 per flower, the stigma 1, positioned laterally at the style tip. Ovule 1. Fruits achenes, somewhat flattened, mostly hidden by the persistent involucel. Ten to 12 species, Europe, Asia, Africa, introduced in North America.

In Europe, the dried heads of fuller's teasel, *D. sativus* (L.) Honck., formerly were used (after the fruits had been shed) to raise the nap in fulling cloth and were cultivated on a limited basis. However, plastic substitutes now are generally used for this purpose. This species, which escapes only rarely in the northeastern United States, differs from *D. fullonum* in its receptacular bracts with strongly recurved awns. Both of the *Dipsacus* species naturalized in the state were officially designated as noxious weeds by the Missouri state legislature in 2000. Although these plants are occasionally cultivated in gardens, this practice is considered illegal and is to be discouraged. Similarly, the use of the fruiting heads in dried flower arrangements should be discontinued.

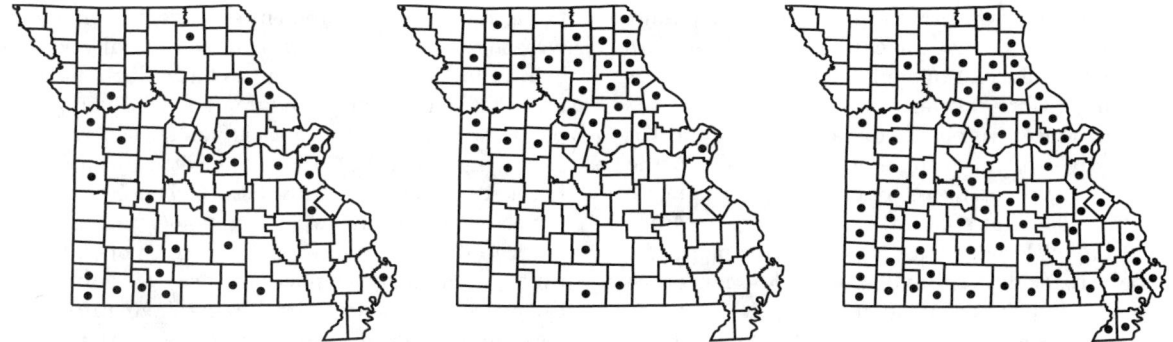

1632. Dipsacus fullonum   1633. Dipsacus laciniatus   1634. Diospyros virginiana

1. Stem leaves sessile at each node or sometimes inconspicuously fused at the very base, the margins scalloped to bluntly toothed; corolla lobes and anthers usually lavender, rarely white . . . . 1. D. FULLONUM
1. Stem leaves at each node prominently fused toward the base, forming a cuplike structure around the stem, irregularly pinnately lobed; corolla lobes and anthers white . . . . 2. D. LACINIATUS

1. Dipsacus fullonum L. (common teasel, wild teasel)

D. sylvestris Huds.

Pl. 373 b–d; Map 1632

Stems 0.4–2.5 m long. Basal leaves simple, oblong-oblanceolate, the margins scalloped to bluntly toothed and often also prickly. Stem leaves 3–35 cm long, sessile at each node or sometimes inconspicuously fused only at the very base, usually somewhat folded longitudinally, the lowermost oblanceolate to lanceolate, the margins scalloped to bluntly toothed and often also prickly, the uppermost linear, the margins usually entire. Involucral bracts at flowering 2–15 cm long, the longest usually longer than the head, narrowly linear, appearing angled or somewhat flattened in cross-section. Heads 3–10 cm long, ovoid to cylindrical-ovoid. Calyces 0.9–1.2 mm long. Corollas 10–15 mm long, the tube white, the lobes lavender, rarely white. Stamens with the anthers lavender, rarely white. Fruits (including the involucel but not the calyx) 3–5 mm long, linear, the involucel 4-angled, 8-ribbed, brown to dark brown. $2n=16, 18$. June–October.

Introduced, scattered nearly throughout the state, but apparently absent from the Mississippi Lowlands Division and most of the Glaciated Plains (native of Europe; widely naturalized in the U.S. and Canada). Banks of streams; more commonly roadsides, railroads, pastures, old fields, fallow fields, fencerows, and open or brushy disturbed areas.

Dipsacus fullonum has been established in Missouri for a relatively long time and was collected in widespread counties such as Greene, Cole, and St. Louis prior to 1880. In recent years, it appears to have declined in the state, perhaps because of competition from the more robust and aggressive *D. laciniatus*. Before Ferguson and Brizicky (1965) clarified the nomenclature of the *D. fullonum* complex, there was considerable confusion in the botanical literature over the proper application of names to the weedy wild plant now known as *D. fullonum* and its cultivated relative, *D. sativus*. Rare white-flowered plants have been called *D. sylvestris* f. *albidus* Steyerm. and were first reported for Missouri by Dunn (1982).

2. Dipsacus laciniatus L. (cut-leaved teasel)

Pl. 373 a; Map 1633

Stems 0.5–2.5 m long. Basal leaves irregularly pinnately lobed, oblong-lanceolate to narrowly ovate in outline, the lobes irregularly oblong, bluntly toothed or more commonly lobed again, sometimes also prickly. Stem leaves 3–50 cm long, prominently fused toward the base, forming a cuplike structure around the stem, usually somewhat folded longitudinally, irregularly pinnately lobed (the uppermost leaves occasionally entire or nearly so), oblong-lanceolate to narrowly ovate in outline, the lobes irregularly oblong to linear, bluntly toothed or more commonly lobed again, sometimes also prickly. Involucral bracts at flowering 1.5–10.0 cm long, the longest often longer than the head, narrowly lanceolate-triangular to linear, appearing flattened in cross-section. Heads 3–10 cm long, ovoid to cylindrical-ovoid. Calyces 1.2–1.6 mm long. Corollas 12–15 mm long, the tube and lobes white.

Stamens with the anthers white. Fruits (including the involucel but not the calyx) 4–5 mm long, linear, the involucel 4-angled, 8-ribbed, light brown. $2n=16, 18$. June–October.

Introduced, common but local north of the Missouri River, as yet uncommon and sporadic farther south (native of Europe; naturalized in the northeastern U.S. and adjacent Canada west to Minnesota and Colorado). Banks of streams; more commonly roadsides, railroads, old fields, fencerows, ditches, and disturbed, open places.

In Missouri, *D. laciniatus* tends to flower slightly later than does *D. fullonum*. It appears to be a relatively recent introduction in the state. The oldest specimen was collected in 1968 in Harrison County (Henderson, 1980), where the collector thought that it had spread by seed from a cemetery wreath to an adjacent fencerow. The actual mode(s) of introduction and the number of introductions into the state remain somewhat unresolved. During the past few decades, the species has spread rapidly along highways but appears mostly to have remained restricted to highly disturbed habitats. Only in recent years have Missouri biologists begun to observe it away from roads in various open habitats, particularly in northeastern Missouri and Boone County. In northern Illinois, *D. laciniatus* is becoming invasive in some wetlands, particularly fens. Thus it may cause problems in the future for land managers and conservationists in Missouri as it continues to spread into southern Missouri.

EBENACEAE (Ebony Family)

Two to 5 genera, 450–485 species, North America to South America, Caribbean Islands, Asia, Africa, Madagascar, Comoro Islands.

1. Diospyros L. (persimmon)

Four hundred to 475 species, North America to South America, Caribbean Islands, Asia, Africa, Madagascar.

A number of *Diospyros* species are important commercially for their edible fruits. The Japanese persimmon (*D. kaki* L.f.), a cultigen originating in eastern Asia, is frequently available in Missouri markets but tends to have a short storage life. It has a larger fruit than that of the native persimmon, but reputedly has an inferior flavor. Some species of *Diospyros* also are important timber trees, including ebony (principally *D. ebenum* J. König ex Retz., native to southern Asia).

1. Diospyros virginiana L. (persimmon, possumwood)
D. virginiana f. *atra* Sarg.
D. virginiana f. *pumila* E.J. Palmer & Steyerm.
D. virginiana var. *platycarpa* Sarg.
D. virginiana var. *pubescens* (Pursh) Dippel
Pl. 373 e–g; Map 1634

Plants dioecious, sometimes incompletely so, occasionally shrubs only 1–4 m tall, more commonly trees to 15 m tall, often suckering from the roots to form colonies, the bark deeply furrowed and often also separating into small squares with age, dark brown to dark gray or nearly black. Twigs dark gray to brown, glabrous or hairy, the winter buds ovate, bluntly pointed at the tip, with 2 main scales (sometimes 1 or 2 additional scales evident at the tip). Leaves alternate, usually short-petiolate. Stipules absent. Leaf blades simple, 4–15 cm long, ovate to oblong-ovate, tapered to a sharply pointed tip, narrowed, rounded, or cordate at the base, the margins entire, the surfaces glabrous or more commonly sparsely to densely pubescent with short hairs. Inflorescences axillary, solitary or (with staminate flowers only) in small clusters of 2 or 3. Flowers mostly imperfect (occasionally a few perfect flowers produced), actinomorphic, hypogynous, the staminate flowers appearing somewhat perigynous, short-stalked. Calyces 3.0–3.5 mm (staminate) or 7.0–7.5 mm (pistillate) long, deeply 4-lobed, the lobes ovate to broadly ovate, hairy (at least on the inner surface), those of the pistillate flowers

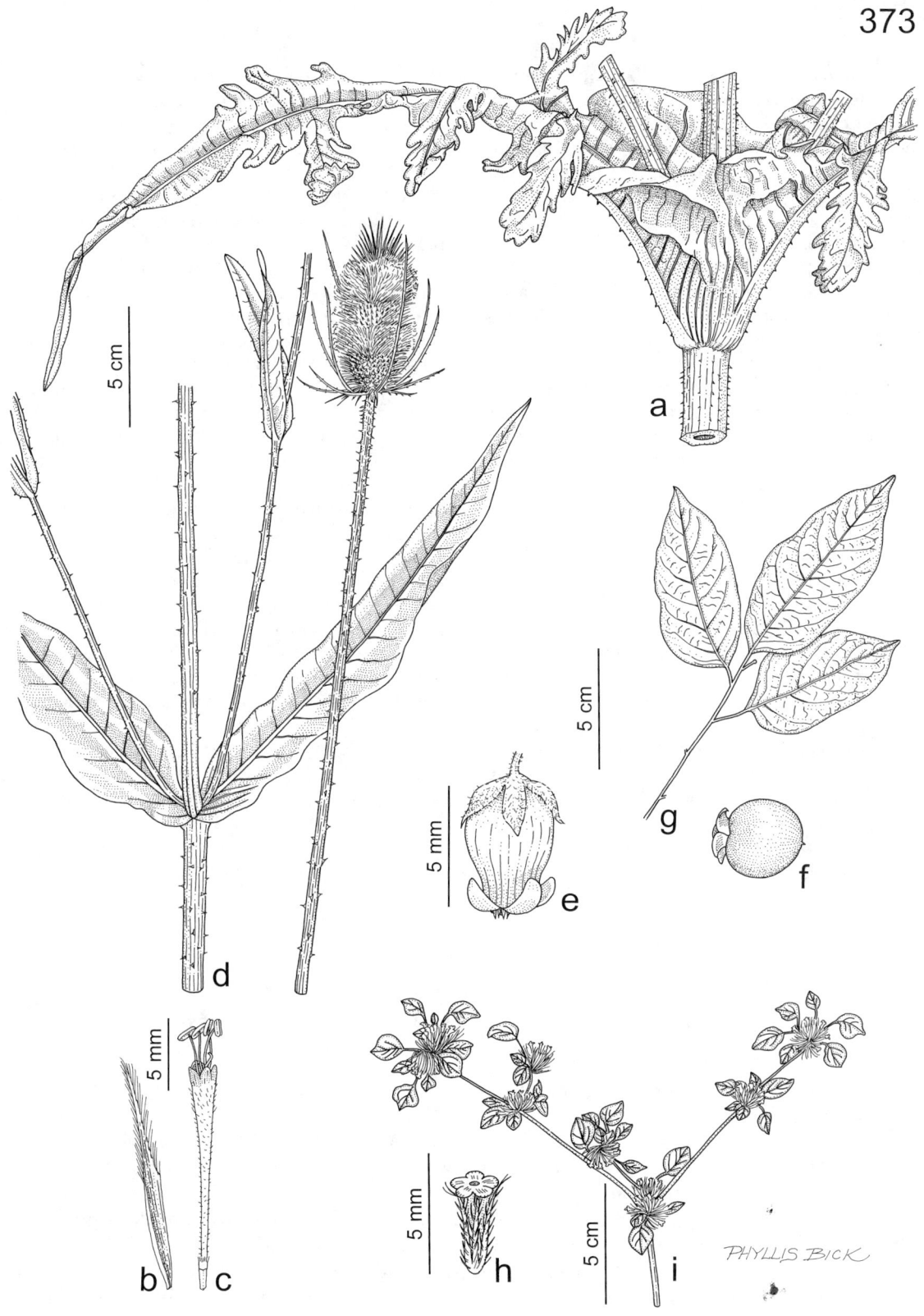

Plate 373. Dipsacaceae, Ebenaceae, Ehretiaceae. *Dipsacus laciniatus*, **a)** node with leaf. *Dipsacus fullonum*, **b)** receptacular bract, **c)** flower, **d)** node with leaves and inflorescence. *Diospyros virginiana*, **e)** flower, **f)** fruit, **g)** branch tip with leaves. *Tiquilia nuttallii*, **h)** flower, **i)** habit.

broader, persistent, and usually becoming papery and reflexed at fruiting. Corollas 5.5–9.0 mm (staminate) or 10–16 mm (pistillate) long, broadly urn-shaped, with short, spreading lobes, the tube pale yellow, the lobes with the inner surface darker yellow. Staminate flowers with a rudimentary nonfunctional pistil and usually 16 stamens, these about as long as the corolla tube, the filaments very short, hairy, attached in 2 rows toward the base of the corolla tube, the anthers attached at their bases, linear, dehiscing longitudinally. Pistillate flowers with 6–10 short, stamenlike staminodes and a single pistil, this with (4)8 locules. Ovary superior, of 4 fused carpels. Style 1, 4-lobed to below the midpoint, the stigmas capitate. Ovules 1 per locule. Fruits berries, 2–5(–7) cm in diameter, globose to depressed-globose, the surface glabrous, smooth or somewhat wrinkled with age, ripening to yellowish orange or yellowish brown, sometimes pinkish- or reddish-tinged, less commonly dark purple to bluish purple, usually somewhat glaucous. Seeds 3–6(–8) per fruit, 18–20 mm long, oval in outline, flattened, the surfaces finely granular, reddish brown. $2n=60, 90$. May–June.

Scattered to common nearly throughout the state, although uncommon or apparently absent from portions of the Glaciated Plains Division (eastern U.S. west to Nebraska and Texas, also Utah). Mesic to dry upland forests, savannas, glades, and less commonly banks of streams and rivers and bottomland forests; also old fields, fallow fields, fencerows, roadsides, and disturbed areas.

Diospyros virginiana exhibits a complex pattern of morphological variation across its range. Because of the variability of characters, such as the overall size of the plant, shape of the leaf blade base, pubescence of twigs and leaves, and size, shape, and color of the fruit, some authors have adopted a complex infraspecific classification of the species (Steyermark, 1963). However, these authors also admit that most of the pertinent characters show a more or less continuous spectrum of variation between populations, and there is little apparent correlation between characters. Many of the specimens are difficult to determine below the species level in the herbarium or field, thus no attempt is made here to formally recognize varieties.

The following synopsis is adapted from Steyermark's (1963) treatment. The fruits of persimmon are relished by wildlife, livestock, and humans. There is great variation in flavor between trees, and fruits eaten before they are entirely ripe are very astringent and unpalatable. Many connoisseurs refuse to pick the fruits before they have been exposed to a hard frost in late autumn, but Steyermark (1963) stated that this was erroneous, because many early-ripening forms exist, including some that ripen as early as July. The fruits are eaten raw or prepared into puddings, pies, syrup, jellies, and ice cream. They can also be used in a vinegar. Native Americans ground dried fruits into a type of flour. The leaves, which turn light yellow in the autumn, can be dried and used in teas. The seeds have been used in coffee substitutes. The seasoned wood of persimmon is among the hardest of North American woods. It has been used in making cabinets, parquet flooring, golf clubs, shuttles, billiard cues, and other items.

EHRETIACEAE (Ehretia Family)
Contributed by David J. Bogler and George Yatskievych

About 11 genera, 170 species, North America to South America, Caribbean Islands, Africa, Asia to Australia.

The Ehretiaceae have been treated as one of five subfamilies of the Boraginaceae by many botanists (Al-Shehbaz, 1991). Recent molecular work has suggested that the traditionally recognized family Hydrophyllaceae is nested within this group (Ferguson, 1998; Gottschling et al., 2001). One way of dealing with these data is to expand the definition of Boraginaceae to include the Hydrophyllaceae, as was advocated by the Angiosperm Phylogeny Group (1998, 2003), Judd et al. (2002), and Craven (2005). Arguing against this interpretation is that it creates a morphologically even more variable family that becomes accordingly more difficult to circumscribe in floristic manuals. Also, although the relationships among the subfamilies of Boraginaceae are still not fully understood, it is clear that the individual subgroups mostly are amply distinct based on both morphological and molecular data. Accordingly, Gottschling et al. (2001) and Diane et al. (2002) have taken the opposite approach and suggested the recognition

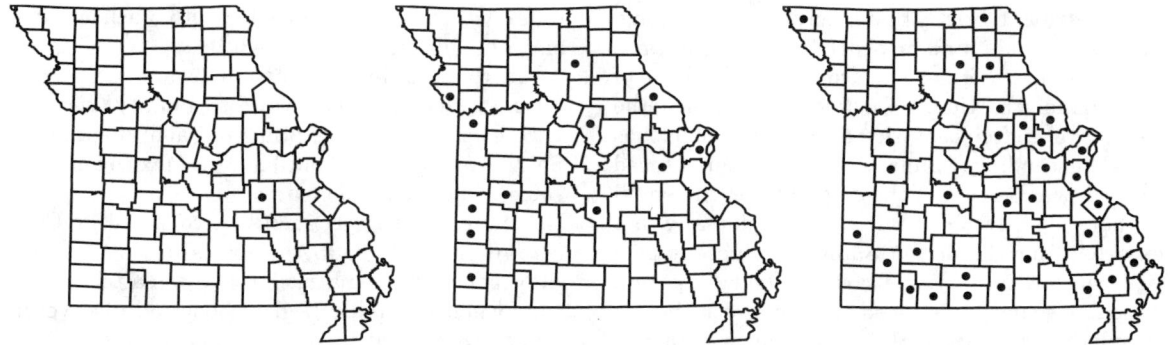

1635. Tiquilia nuttallii   1636. Elaeagnus angustifolia   1637. Elaeagnus umbellata

of segregate families, with the Boraginaceae restricted to the mostly temperate herbaceous genera formerly included in the subfamily Boraginoideae. Although future research undoubtedly will result in further refinements of classification within the Boraginaceae complex, it seems most expedient in the present work to deviate from the system of Cronquist (1981, 1991) and to segregate the Ehretiaceae and Heliotropiaceae from the Boraginaceae and Hydrophyllaceae.

The Ehretiaceae are primarily a tropical and subtropical family of mostly shrubs and small trees. The family generally differs from the Boraginaceae in its less strongly lobed ovary with a more terminal, 2-branched style, fruits that often are drupelike, and seeds with abundant endosperm, as well as several more esoteric features. The sole Missouri representative is in a genus that is atypical for some of these features.

1. Tiquilia Pers.
(Richardson, 1977)

Twenty-seven species, North America to South America.

Species of *Tiquilia* formerly were included in *Coldenia* L. That genus has become restricted to the single species, *C. procumbens* L., a weedy Old World annual that grows in mesic habitats, has asymmetrical leaf blades, and has flowers with the perianth 4-parted (Richardson, 1976, 1977).

1. Tiquilia nuttallii (Hook.) A.T. Richardson
(rosette crinklemat)
Coldenia nuttallii Hook.
Pl. 373 h, i; Map 1635

Plants perennial herbs, with a sometimes somewhat woody taproot. Stems 5–30 cm long, slender, more or less prostrate, forming mats or low mounds, with numerous more or less dichotomous branches (sometimes in whorls), brittle, pale brown to reddish or yellowish brown, moderately to densely pubescent with fine, stiff, mostly downward-curved hairs. Leaves clustered at the nodes and stem tips, short- to more commonly long-petiolate. Stipules absent. Leaf blades 3–10 mm long, 2–8 mm wide, ovate to broadly elliptic or nearly circular, angled or short-tapered at the base, angled or tapered to a bluntly or sharply pointed tip, the margins entire, relatively thick, sometimes somewhat curled under, and with scattered bristly hairs, usually appearing grayish, the midvein and 2 or 3 pairs of lateral veins strongly impressed on the upper surface and strongly ridged on the undersurface, the surfaces densely pubescent with short, curved to more or less appressed, minutely pustular-based hairs, the upper surface also with scattered to moderate longer, straight, bristly hairs. Inflorescences small clusters of flowers at the nodes and branch tips, the flowers occasionally solitary. Flowers actinomorphic, perfect, hypogynous, subtended by narrowly lanceolate to linear bracts. Cleistogamous flowers absent. Calyces 2.5–4.0 mm long, 5-lobed nearly to the base, the lobes linear to very narrowly lanceolate, moderately to densely pubescent with minute, curved or appressed hairs and scattered, longer, straight, bristly hairs. Corolla 3–4 mm long, lavender to pale pink or nearly white, sometimes yellowish on the inner surface toward the base of the expanded portion, more or less funnelform, the slender tube often constricted near top, the 5 lobes loosely ascending. Stamens 5, the filaments fused to the lower portion of the corolla tube, the small anthers slightly incurved, attached dorsally, not exserted from the flower. Staminodes absent. Pistil 1 per flower, of 2 fused carpels. Ovary superior, 4-locular, somewhat 4-lobed and concave at the tip, the placentation axile. Style 1 per flower, attached just below the tip of the ovary (in the shallow notch), 2-lobed to below the midpoint, each branch with a small, capitate stigma. Ovules 1 per locule. Fruits schizocarps splitting into 4 nutlets, these 1.0–1.5 mm long, 1-seeded, indehiscent, with a hardened outer wall, this brown, mottled with black or olive, smooth, shiny. $2n=16$. May–July.

Introduced, known thus far from a single historical collection from Crawford County (western U.S. east to Wyoming and Colorado; disjunct in South America; introduced in New Mexico and Missouri). Open, disturbed areas.

The Missouri specimen was collected by Cora Shoop in May 1935, before she and Julian Steyermark were married. Cora was a teacher at the former Steelville High School and discovered plants growing at the base of an open rocky slope on her school grounds near some train tracks. Steyermark (1963) noted that the site was later destroyed by a building project of the Works Progress Administration (WPA).

ELAEAGNACEAE (Oleaster Family)
(Graham, 1964)
Contributed by Alan E. Brant

Three genera, about 45 species, mostly temperate and subtropical regions of the Northern Hemisphere.

1. Elaeagnus L. (oleaster)

Plants shrubs or small trees, with a dense covering of silvery or rusty, peltate or stellate, scales or hairs on leaves, twigs, and buds. Leaves alternate, short-petiolate. Stipules absent. Leaf blades simple, the margins entire. Inflorescences small clusters of flowers or solitary flowers in the leaf axils. Flowers perfect but sometimes functionally unisexual, deeply perigynous, the receptacle elongated into a tubular or funnelform hypanthium with a pronounced constriction just above the ovary (which thus appears inferior), the nongreen portion above the ovary shed before fruiting. Sepals (3)4(–6), petaloid, spreading, similar in texture to the hypanthium, with stellate hairs at the tips. Stamens 4, attached near the tip of the hypanthium alternating with the sepals, the filaments very short. Petals absent. Ovary superior, 1-locular, with 1 ovule. Style 1, linear, with the stigma a linear area along the side toward the tip. Fruits achenes, but appearing drupelike or berrylike, the achene enclosed by (but not fused with) the expanded fleshy receptacle. About 40 species, North America, Europe, Asia.

The flowers of *Elaeagnus* have a strong, sweet fragrance and are unusual in that the hypanthium changes color and texture above the ovary, appearing confusingly similar to a corolla tube with four lobes at the tip. The fruits are similarly confusing, as the base of the hypanthium turns fleshy and berrylike. Roots of most species of *Elaeagnus* can form symbiotic relationships with soilborne bacteria of the genus *Rhizobium,* which have the ability to fix atmospheric nitrogen into nitrates that are absorbed as nutrients by plants, and they thus act to enrich the soil.

1. Leaves and branchlets with only silvery scales; hypanthium (beyond the constricted area above the ovary) slightly longer than to about 1.5 times as long as the sepals . . . . 1. E. ANGUSTIFOLIA
1. Leaves and branchlets with scattered to dense rusty brown scales; hypanthium (beyond the constricted area above the ovary) at least 2 times as long as the sepals . . . . 2. E. UMBELLATA

1. Elaeagnus angustifolia L. (Russian olive)
Pl. 374 h, i; Map 1636

Plants usually small trees, to about 8 m tall, the branches sometimes thorny. Bark of trunk and older branches thinly furrowed and sometimes somewhat peeling in thin strips, reddish brown to dark purplish gray or nearly black, the younger branches and especially twigs and branchlets silvery white with a dense covering of scales and stellate hairs. Leaf blades 30–70 mm long, 6–18 mm wide, lanceolate to narrowly elliptic, when young both surfaces silvery with a dense covering of scales, with age becoming green on the upper surface as the scales are shed and/or worn to stellate hairs (best seen with magnification), the undersurface retaining the silvery covering even in age. Flowers short-stalked (to 5 mm), 7–10 mm long at flowering. Hypanthium (beyond the constricted area above the ovary) slightly longer than to about 1.5 times as long as the sepals, silvery with scales on the outer surface, the sepals yellow on the inner surface. Mature fruits short-stalked (to 5 mm), 6–10 mm long, broadly oval in outline, yellow but appearing silvery with a more or less dense covering of scales. $2n=28$. April–June.

Introduced, uncommon and widely scattered in central Missouri (native of Europe, Asia; widely naturalized in the U.S. and Canada). Old fields, roadsides, railroads, and disturbed, open areas.

Although collections of Russian olive in Missouri date back to the early twentieth century, it apparently rarely escapes from cultivation in our area. Elsewhere in the United States, particularly in some parts of the northern Great Plains and in some southwestern states, the species has become invasive in natural habitats in similar fashion to the problems with *E. umbellata* in Missouri and other midwestern states.

2. Elaeagnus umbellata Thunb. (autumn olive)
Pl. 374 j, k; Map 1637

Plants shrubs to 3(–5) m, usually multiple-stemmed, rarely with a few thorns. Bark gray to reddish brown, the younger branches densely covered with both rusty and silvery scales. Leaf blades 30–80 mm long, 10–40 mm wide, narrowly elliptic to elliptic, silvery beneath, green above even when young, the silvery undersurface with at least a few scattered, rusty scales. Flowers short-stalked (to 5 mm), 6–8 mm long. Hypanthium (beyond the constricted area above the ovary) 2 or more times as long as the sepals, silvery with scales on the outer surface, the sepals cream-colored to yellow on the inner surface. Mature fruits short-stalked (to 5 mm), 5–8 mm long, circular in outline, more or less translucent, red, with scattered, rusty scales appearing as tiny dots to the naked eye. $2n=28$. April–June.

Introduced, scattered, mostly south of the Missouri River, but probably throughout the state (native of Asia; widely naturalized in the eastern U.S. and adjacent Canada west to Nebraska and Louisiana). Mesic upland forests, upland prairies, sand prairies, glades, savannas, and banks of streams; also old fields, pastures, fencerows, roadsides, and open, disturbed areas. April–June.

Autumn olive is popular in horticulture as an ornamental shrub. Although its introduction into North America dates back to the 1830s, beginning in about the 1940s, the USDA Soil Conservation Service (since renamed the Natural Resources Conservation Service) and various state conservation agencies began promoting its use as a valuable wildlife food plant and also as a windscreen and for use in erosion control. However, birds and small mammals have spread its seeds into various natural habitats, where it can take over the understory, and the species is now considered an invasive exotic in a number of states. In Missouri, its introduction is relatively recent; the earliest definitely determined collections of plants growing outside of cultivation date back to the 1980s (Yatskievych and Figg, 1989). Smith (1993) indicated that because of widespread planting, *E. umbellata* is probably now naturalized in most Missouri counties.

Steyermark (1963) tentatively included the superficially similar *E. multiflora* Thunb. (cherry silverberry) in his treatment, based on a single historical collection from St. Louis County. He stated, however, that this vegetative specimen "cannot be satisfactorily or definitely determined as this species or *E. umbellata*." As no further plants have been discovered to verify the presence of this species in Missouri, it is here excluded from the flora (the lone voucher is too incomplete to allow precise determination). *Elaeagnus multiflora,* which has escaped from cultivation sporadically in several eastern states, including Illinois, differs from *E. umbellata* in its longer flower and fruit stalks (10–15 mm) and in its shorter hypanthium (about as long as the sepals).

ELATINACEAE (Waterwort Family)

Plants annual. Stems loosely ascending, often rooting at the lower nodes. Leaves opposite, sessile or nearly so, the blades simple, the margins entire or finely toothed. Stipules scalelike. Inflorescences axillary, the flowers solitary or in small clusters of 2 or 3. Flowers inconspicuous, actinomorphic, perfect, hypogynous, sessile or short-stalked. Sepals 2–5(6), free or fused toward the base. Petals 2–5(6), shorter than to about as long as the sepals, free. Stamens as many or twice as many as the petals, in 1 or 2 whorls, the filaments free, the anthers ovoid, attached at the midpoint, yellow. Staminodes absent. Pistil 1 per flower, of 2–5 fused carpels. Ovary superior, with 2–5 locules, the placentation axile or more or less basal. Styles 2–5, short, the stigmas capitate. Ovules numerous. Fruits capsules, 2–5-valved, thin-walled, the valves separating from the partitions at dehiscence. Seeds oblong to elliptic in outline, somewhat asymmetric at the ends, the surface with a network of fine ridges and/or pits, sometimes obscurely so. Two genera, 40–50 species, nearly worldwide.

1. Sepals and petals 5(6); plants moderately to densely and minutely glandular-hairy; leaf margins with minute, gland-tipped teeth .... 1. BERGIA
1. Sepals and petals (2)3; plants glabrous; leaf margins entire .... 2. ELATINE

1. Bergia L.

Twenty to 25 species, North America, South America, Africa, Europe, Asia, south to Australia.

1. Bergia texana (Hook.) Seub. (bergia)

Pl. 374 f, g; Map 1638

Plants moderately to densely and minutely glandular-hairy, the hairs sometimes pinkish- to reddish purple–tinged. Stems 8–40 cm long, loosely ascending to erect. Leaves 7–40 mm long, the blades elliptic to oblanceolate, long-tapered to a short petiole, narrowed to a bluntly to sharply pointed tip, the margins with minute, gland-tipped teeth. Stipules 1.2–2.0 mm long, narrowly lanceolate to narrowly triangular, the margins with glandular teeth. Flowers 1–3 per leaf axil. Sepals 5(6), 2–3 mm long, lanceolate to ovate, with a thicker green central stripe and thinner white margins. Petals 5(6), 1.5–2.5 mm long, white. Stamens 5 or less commonly 10. Fruits globose or nearly so, with 5 locules. Seeds 0.3–0.5 mm long, more or less cylindrical, slightly curved, shiny, yellowish brown to brown, the surface obscurely pitted. June–November.

Uncommon in counties along the Missouri River floodplain (western U.S. east to Illinois and Louisiana). Banks of rivers; also crop fields, fallow fields, and open, disturbed areas, usually in sandy soil.

This inconspicuous species was known to Steyermark (1963) only from a few historical collections from Jackson County. More recent fieldwork has resulted in the discovery of a number of new populations across the state, and it apparently has benefited from disturbances along the Missouri River floodplain, especially the flood of 1993.

2. Elatine (waterwort)

Twenty to 25 species, nearly worldwide.

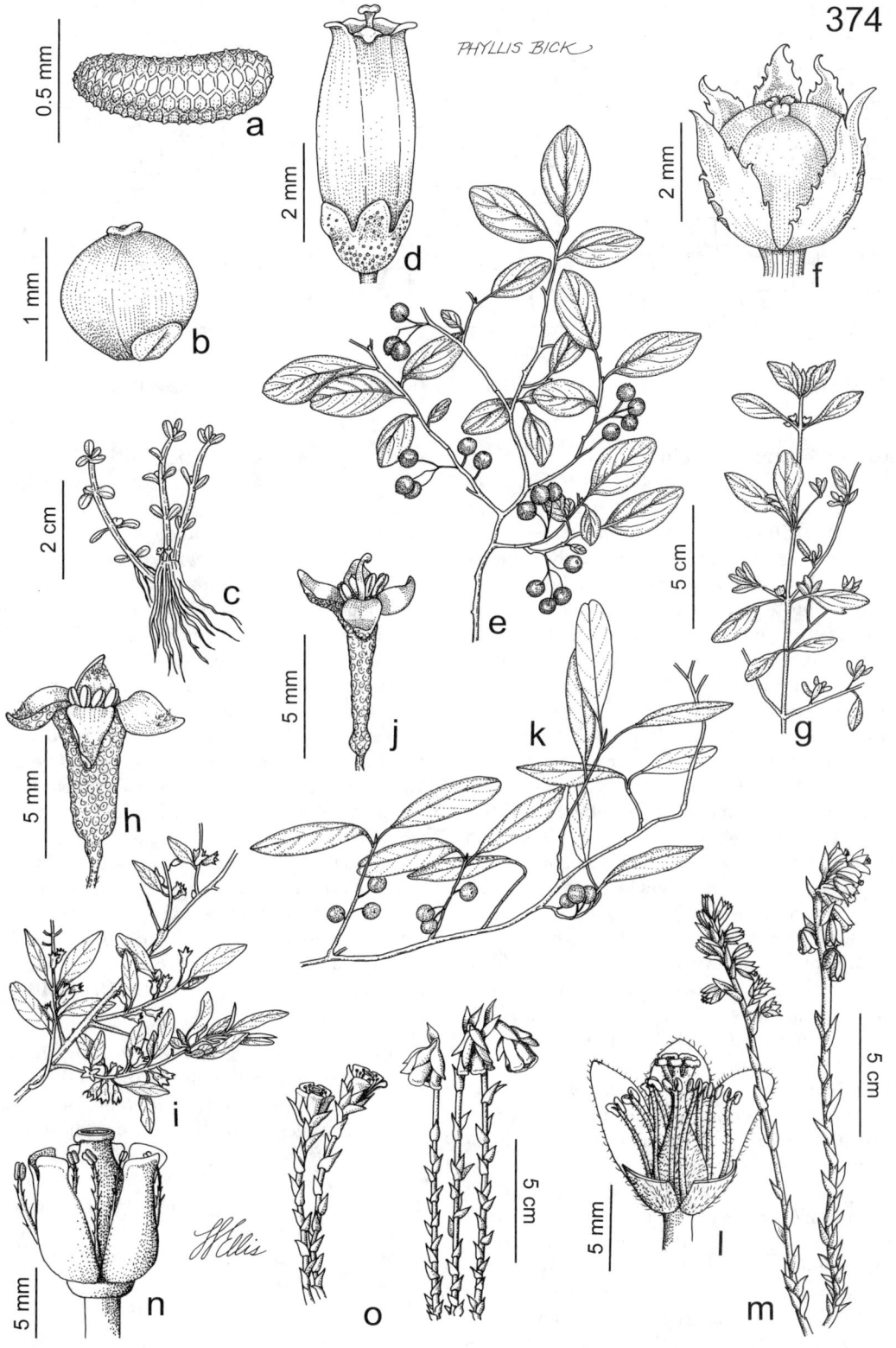

Plate 374. Elaeagnaceae, Elatinaceae, Ericacaeae. *Elatine triandra* var. *triandra*, **a)** seed. *Elatine triandra* var. *americana*, **b)** fruit, **c)** habit. *Gaylussacia baccata*, **d)** flower, **e)** fruiting branch. *Bergia texana*, **f)** fruit, **g)** habit. *Elaeagnus angustifolia*, **h)** flower, **i)** flowering branch. *Elaeagnus umbellata*, **j)** fruit, with persistent perianth and stamens, **k)** fruiting branch. *Monotropa hypopitys*, **l)** flower, **m)** habit. *Monotropa uniflora*, **n)** flower, with perianth partially removed to reveal the stamens and pistil, **o)** habit in fruit (left) and flower (right).

1638. Bergia texana   1639. Elatine triandra   1640. Gaylussacia baccata

1. Elatine triandra Schkuhr

Pl. 374 a–c; Map 1639

Plants glabrous, the vegetative parts sometimes reddish-tinged. Stems 1–6 cm long, prostrate to ascending. Leaves 3–8(–10) mm long, the blades slightly succulent, linear to spatulate, oblanceolate, or obovate, long-tapered to a sessile or short-petiolate base, rounded to bluntly pointed at the tip, sometimes with a minute, apical notch, the margins entire. Stipules 0.2–0.4 mm long, narrowly lanceolate to ovate-triangular, the margins entire. Flowers 1 per leaf axil. Sepals (2)3, 0.8–1.2 mm long, ovate, uniformly green or with narrow, white margins, often somewhat pinkish- or reddish-tinged. Petals (2)3, 0.8–1.2 mm long, greenish white, often pinkish-tinged. Stamens (2)3. Fruits globose to slightly depressed-globose, with (2)3 locules. Seeds 0.3–0.7 mm long, more or less cylindrical, straight or slightly curved, the surface usually somewhat iridescent, yellowish brown to brown, with a fine but usually noticeable network of pits, these mostly 6-angled (visible only under high magnification), becoming smaller toward the ends, in about 5–7 rows around the seed and 16–25 longitudinal rows. $2n=40$. June–October.

Uncommon, known only from a few widely scattered historical collections (discontinuously nearly throughout the U.S.; Canada, Europe, Asia). Submerged aquatic or terrestrial in or around ponds, sinkhole ponds, and rivers.

Plants of *Elatine* growing in aquatic habitats can be quite different in appearance than those growing in moist soil. Submerged plants are a darker green color and have stems with longer internodes and leaves that are longer and narrower, whereas terrestrial individuals tend to be lighter green (but frequently reddish-tinged) and have shorter internodes and shorter, broader leaves. Because of this morphological plasticity in response to habitat differences, most botanists have depended upon subtle differences in seed morphology to distinguish entities within the complex. Given that the plants apparently engage in a lot of self-pollination (Tucker, 1986), which can facilitate the fixation of morphological variations within populations, the significance of these characters in delimiting species requires further study. There is no modern monograph of *Elatine,* and the taxonomy of the species is poorly understood. The *E. triandra* complex as it exists in the eastern United States has been treated variously as a single species without infraspecific taxa (Duncan, 1964), one species consisting of two native and one introduced varieties (Fassett, 1939; Steyermark, 1963; Gleason and Cronquist, 1991), or three species (Fernald, 1941a; Tucker, 1986; Crow and Hellquist, 2000). The present treatment follows the "one species with multiple varieties" classification. In addition to the two varieties treated below, the other eastern North American variety is var. *brachysperma* (A. Gray) Fassett, which occurs sporadically from Ohio to Georgia and westward to the West Coast and differs in having seeds with only 9–15 (vs. 16–25) pits per longitudinal row.

1. Leaves mostly oblanceolate to obovate, usually rounded to bluntly pointed at the tip; seeds with the pits more or less rounded along the margins (best seen with high magnification) .... 1A. VAR. AMERICANA
1. Leaves mostly linear to spatulate, often minutely notched at the tip; seeds with the pits more or less 6-angled along the margins (best seen with high magnification) .... 1B. VAR. TRIANDRA

1a. var. americana (Pursh) Fassett

E. americana (Pursh) Arn.

Pl. 374 b, c

Plants often relatively densely colonial. Leaves mostly oblanceolate to obovate, usually rounded to bluntly pointed at the tip, rarely minutely notched.

Seeds with the pits more or less rounded along the margins (visible only under high magnification), the pits only slightly smaller toward the ends. June–October.

Uncommon, known only from historical collections from Jackson and Laclede Counties (eastern U.S. discontinuously west to Kansas and Louisiana; Canada). River floodplains and sinkhole ponds.

1a. var. triandra

Pl. 374 a

Plants usually relatively loosely colonial. Leaves mostly linear to spatulate, usually rounded at the tip, often minutely notched. Seeds with a network of mostly 6-angled pits (visible only under high magnification), the pits becoming smaller toward the ends, in about 5–7 rows around the seed and 16–25 longitudinal rows. $2n=40$. July–October.

Introduced, known only from a single historical collection from the city of St. Louis (western U.S. east to Maine and Alabama; Canada, Europe, Asia). Muddy bottoms of artificial ponds.

The North American distribution of this taxon is not fully understood. Some botanists separate American plants as *E. rubella* Rydb. and restrict true *E. triandra* to Eurasian plants (with introduced populations in Maine, New York, and Missouri).

ERICACEAE (Heath or Blueberry Family)
Contributed by David J. Bogler

Plants shrubs or small trees, sometimes evergreen, less commonly mycotrophic (receiving nutrients and water from associations with soilborne fungi) herbs lacking chlorophyll. Leaves alternate, simple, often somewhat thickened and leathery, sometimes reduced to scales, sessile or short-petiolate. Leaf blades simple, the margins entire or finely toothed. Stipules absent. Inflorescences terminal or axillary, mostly racemes, the flowers sometimes solitary, subtended by small bracts. Flowers actinomorphic to slightly zygomorphic, perfect, hypogynous or epigynous. Calyces deeply 5-lobed or of 4 or 5 free sepals (sometimes absent in *Monotropa*), usually persistent at fruiting. Corollas usually 5-lobed or of 3–6(–8) free petals, trumpet-shaped or tubular to urn-shaped. Stamens mostly 5 or 10, the filaments free or attached to the corolla base, the anthers attached more or less medially, becoming inverted during development such that the base becomes the apparent tip, sometimes with scalelike spurs near the filament-anther junction and/or awnlike extensions at the apparent anther tip, dehiscing mostly by pores near the apparent tip, these sometimes elongated and appearing slitlike. Pollen usually released in tetrads (groups of four) (except in *Monotropa*) and with viscin strands (sticky, cobwebby strands connecting the tetrads so the pollen tends to be shed in clumps). Pistils of usually 5 fused carpels. Ovary superior or inferior, with 1–10 locules, hollow and fluted, with axile or deeply intruding parietal placentation. Style 1 per flower, the stigma capitate to disk-shaped, sometimes (4)5-lobed. Ovules 1 to numerous. Fruits capsules or berries. Seeds 1 to numerous, usually small, sometimes winged. About 100 genera, about 3,000 species, nearly worldwide, often on acidic soils.

In the broad sense, the Ericaceae are a large and morphologically diverse family that includes trees, shrubs, epiphytes, and herbs. The group is so diverse that some botanists have broken it up into smaller families or subfamilies (Steyermark, 1963; Cronquist, 1981, 1991). *Vaccinium* and *Gaylussacia* sometimes have been segregated by a few workers as Vacciniaceae, along with other non-Missouri genera having an inferior ovary. *Monotropa* and related genera lacking chlorophyll have been placed in the Monotropaceae or have been included along with *Pyrola* (wintergreen) and other mycorrhizal but green, non-Missouri genera in the Pyrolaceae. The consensus of recent morphological and molecular studies is that these genera are indeed related and are best included in a single family (Stevens, 1971; Wallace, 1975; Judd and Kron, 1993; Kron, 1996), and the present treatment therefore deviates from the general practice in this manual of following Cronquist's (1981, 1991) familial classification system.

The Ericaceae are an economically important family. The genus *Vaccinium* is the source of blueberries and cranberries. A number of genera are cultivated as ornamentals, including *Arctostaphylos* (bearberry), *Epigaea* (trailing arbutus), *Erica* (heath), *Kalmia* (mountain laurel), *Pieris* (pieris, fetterbush), *Rhododendron* (azalea, rhododendron), and *Vaccinium* (there are also several additional genera not hardy in Missouri). All of the cultivated taxa require acidic soils, which makes them difficult to grow at many locations, given the widespread calcareous substrates in the state.

1. Plants herbs, lacking chlorophyll, variously white, yellow, or red .... 3. MONOTROPA
1. Plants shrubs or small trees, with chlorophyll; at least the leaves green
    2. Flowers relatively large; corollas more than 10 mm long, slightly zygomorphic, trumpet-shaped, widely spreading toward the tip; leaves tending to be clustered near the branch tips .... 4. RHODODENDRON
    2. Flowers relatively small; corollas less than 10 mm long, actinomorphic, tubular to more or less urn-shaped, usually somewhat constricted at the tip; leaves usually well spaced along the branches
        3. Stamens with the filaments elongate and S-shaped, the anthers unawned, dehiscing by short, terminal pores; ovary superior; fruit a dry, urn-shaped capsule with 5 strong ribs .... 2. LYONIA
        3. Stamens with the filaments short and straight; anthers tapered to tubular awns, dehiscing by more or less elongate pores on the tubes; ovary inferior; fruits dry drupes or fleshy berries
            4. Leaf blades with numerous minute, yellow resin glands, at least on the undersurface; ovary with 10 locules and 10 ovules; fruits drupes with 10 nutlets .... 1. GAYLUSSACIA
            4. Leaf blades without resin glands (rarely a few glands in *V. corymbosum*); ovary with 4 or 5 locules and numerous ovules; fruits berries with numerous seeds .... 5. VACCINIUM

1. Gaylussacia Kunth

About 50 species, North America, South America, most diverse in Brazil.

1. Gaylussacia baccata (Wangenh.) K. Koch
(black huckleberry)

Pl. 374 d, e; Map 1640

Plants shrubs, usually colonial by stolons and/or rhizomes. Stems 0.2–1.5 m tall, the bark more or less smooth, dark gray. Twigs usually sparsely to moderately and finely hairy, sometimes also glandular, green to reddish brown, becoming glabrous and gray to dark gray, the winter buds ovoid, with usually 3 outer scales, these dotted with minute, yellow resin glands (at least when young). Leaves well spaced along the branches, short-petiolate. Leaf blades 2.5–6.0 cm long, 1.0–2.5 cm wide, herbaceous to more or less leathery, elliptic to oblong-ovate, narrowed to a bluntly or less commonly sharply pointed tip, narrowed at the base, the margins entire or nearly so, the upper surface usually inconspicuously hairy along the midvein at maturity, usually with a few inconspicuous yellow resin glands, yellowish green, the undersurface sometimes inconspicuously hairy along the midvein, dotted with minute, yellow resin glands, otherwise pale green. Inflorescences short axillary racemes 1–3 cm long, with 2–7 flowers, each subtended by an inconspicuous bract 1.5–2.0 mm long, these usually shed as the flowers open. Flowers actinomorphic, epigynous. Hypanthium and calyx 1.4–2.0 mm long at flowering, fused to the ovary, dotted with minute, yellow resin glands. Calyces of 5 free sepals, 0.8–1.2 mm long, these triangular to broadly triangular, persistent at fruiting. Corollas 4–6 mm long, tubular to slightly urn-shaped, with 5 short, spreading to recurved lobes, white or greenish white, often pinkish-tinged. Stamens 10, not exserted, the filaments short, straight, attached at the corolla base, lacking spurs near the anther-filament junction, the anther sacs tapered into short tubes toward the apparent tip, dehiscing by a short,

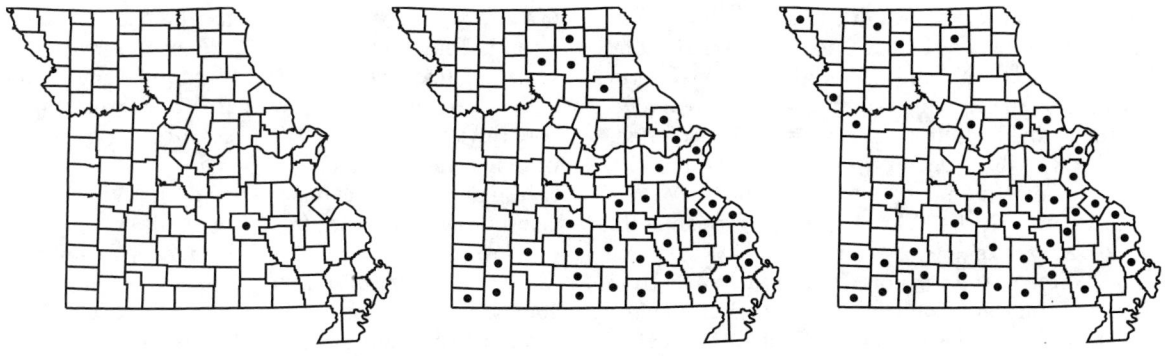

1641. Lyonia mariana   1642. Monotropa hypopitys   1643. Monotropa uniflora

slitlike, terminal pore. Ovary inferior, with 10 locules, the placentation axile. Styles 5–6 mm long, straight, exserted, the stigma capitate, often slightly lobed. Ovules 1 per locule. Fruits somewhat fleshy drupes, 6–10 mm in diameter, black, with 10 nutlets. Nutlets 1.5–1.9 mm long, 1-seeded, asymmetrically ovate in outline, flattened (wedge-shaped), rounded dorsally, the surface with dense, minute papillae, appearing pebbled, tan to yellowish tan. $2n=24$. April–May.

Uncommon, known thus far only from Montgomery and Perry Counties, with an unconfirmed report from Ste. Genevieve County (eastern U.S. west to Minnesota and Arkansas; Canada). Mesic to dry upland forests, usually on ridges above bluffs on acidic substrates.

In general appearance, this species is very similar to some *Vaccinium* species, especially *V. pallidum*. Thus, plants of *Gaylussacia* easily may be overlooked in the field. Huckleberries are readily distinguished from blueberries by their gland-dotted leaves and leathery fruit with 10 seedlike nutlets. The fruits are edible, but not very palatable or juicy. *Gaylussacia baccata* is pollinated by andrenid bees and honeybees. Nectar is secreted by nectaries located near the base of the style and is held in place by the filaments. In order to reach the nectar, the bee must stick its probiscis between the style and anthers, thereby getting pollen on its face. Bagging studies indicate that *G. baccata* is self-incompatible and is dependent on insects to cross-pollinate its flowers (Lovell, 1940).

2. Lyonia Nutt.

About 35 species, Northern Hemisphere.

1. Lyonia mariana (L.) D. Don (staggerbush)
Neopieris mariana (L.) Britton
Pl. 375 k–m; Map 1641

Plants shrubs, colonial by rhizomes. Stems 0.5–2.0 m tall, the bark gray, shallowly furrowed, and usually shredding in thin, narrow strips. Twigs glabrous to sparsely short-hairy, reddish brown to brown, becoming gray with age, often developing small black spots, lines, or streaks, the winter buds conic to ovoid, with 2–6 overlapping scales. Leaves well spaced along the branches, short-petiolate. Leaf blades 3–10 cm long, 1–5 cm wide, narrowly to broadly elliptic, narrowed or tapered to a sharply pointed tip, narrowed or less commonly rounded at the base, the margins entire, slightly thickened, the upper surface glabrous to sparsely short-hairy along the main veins, the undersurface with scattered minute, brown dots (actually minute papillae), also sparsely to moderately short-hairy along the main veins. Inflorescences of loose umbellate clusters along and at the tips of leafless branches, the flower stalks 5–20 mm long, with minute bracts at the very base. Flowers actinomorphic, hypogynous, fragrant. Calyces deeply 5-lobed, 4–7 mm long, the lobes narrowly oblong-triangular to nearly linear, sharply pointed at the tip, minutely glandular and sometimes also sparsely hairy. Corollas cylindrical to urn-shaped, 7–14 mm long, shallowly 5-lobed, the lobes spreading to recurved at flowering, white or less commonly pinkish-tinged, the outer surface sparsely glandular. Stamens 10, not exserted, the filaments with a prominent S-shaped curve, broadened at the base, densely hairy, with 2 short spurs near the anther-filament junction, the anthers lacking tubes or awns at the tip, but with a white, powdery deposit on the dorsal surface, dehiscing by 2 terminal pores. Ovary superior, glabrous, concave at the tip, with 5 locules, the placentation axile.

Style 5–6 mm long, straight, usually slightly exserted at flowering, the basal portion persistent at fruiting, the stigma capitate, slightly lobed. Fruits capsules 5–6 mm long, 4–5 mm wide, urn-shaped to pear-shaped, glabrous, brown with 5 thickened, pale brown ribs. Seeds 0.7–1.5 mm long, oblong-ovoid, somewhat angled, truncate at the tip, with a minute tail at the base, brown, the surface smooth. $2n=24$. May–June.

Uncommon and perhaps extirpated, known thus far only from Dent County (eastern U.S. from New York to Florida, also Missouri to Louisiana west to Oklahoma and Texas). Mesic upland forests in ravines.

This species was first reported for Missouri by Kucera (1953). The Missouri population, which has not been seen for many years, is somewhat disjunct from the closest populations in north-central Arkansas. The central United States sites in turn are disjunct from the species distribution in the eastern states, where it grows primarily in the Atlantic Coastal Plain. The leaves of *L. mariana* contain a series of toxic diterpenoid compounds, including grayanotoxin (andromedotoxin) and lioniatoxin (Burrows and Tyrl, 2001), which are poisonous to humans and livestock. These can have an intoxicating narcotic effect on sheep and cattle, hence the common name staggerbush (Judd, 1981).

The pollination mechanism in *Lyonia* is unusual among Missouri plants, but it is characteristic of some other non-Missouri genera of Ericaceae. The S-shaped filaments act as springs, initially positioning the anthers with the apical pores oriented inward. When a bee or other insect visits a flower and inserts its proboscis to sample nectar that collects at the corolla base, the anthers are first pushed outward, but then they spring back, dusting the insect's mouthparts with pollen. The stigma does not become receptive until after the pollen is shed (Judd, 1981).

3. Monotropa L. (Indian pipe, pinesap)
(Wallace, 1995)

Plants perennial mycotrophic herbs, lacking chlorophyll, the entire plant variously white to yellow or red, turning dark brown or black upon drying, glabrous to finely hairy or glandular-hairy. Stems unbranched. Leaves reduced to lanceolate bractlike scales, these shorter and mostly ovate toward the stem base, longer and lanceolate to oblong-elliptic higher on the stem, mostly sharply pointed at the tip. Inflorescences terminal racemes of 3–10 flowers or a solitary terminal flower, nodding during development and at flowering, becoming erect at fruiting. Flowers actinomorphic, perfect, hypogynous, bracteate, stalked. Calyx absent or of 4 or 5 scalelike, free sepals, usually shed as the flower opens. Corollas of 3–6(–8) free petals, broadly tubular or slightly urn-shaped, the petals somewhat pouched at the base, rounded to truncate at the tip. Stamens 6–10, in 2 whorls, not exserted, the outer whorl shorter than the inner whorl, the filaments attached between the pairs of lobes of a prominent nectary disk, usually hairy, the anthers without spurs or appendages, dehiscing by sometimes irregular slits across the tip. Ovary with (4)5(6) locules, superior, the placentation axile. Style 2–3 mm long, stout, straight, enlarging slightly and persistent at fruiting, the stigma obconic or disk-shaped, (4)5-lobed. Fruits capsules, globose to ovoid, the surface with shallow, longitudinal grooves, brown at maturity, dehiscing longitudinally. Seeds numerous, 0.8–1.2 mm long, 0.1–0.2 mm wide, very narrowly ellipsoid, tapered to a tail-like tip at each end, the surface smooth, light brown. Two species, North America to South America, Europe, Asia.

Monotropa and related genera are one of several groups of angiosperms known variously as saprophytes, mycotrophic plants, and epiparasites. In Missouri, the other main group of such organisms occurs in the Orchidaceae. These strange-looking plants have lost their chlorophyll and capacity for photosynthesis, deriving their water and nutrients via soilborne fungi that have established a mycorrhizal relationship with their roots (hence the term *mycotrophic*). The roots are individually poorly developed and are often clustered into dense, irregularly spherical masses of short roots. The term *saprophyte* is somewhat misleading, as plants of *Monotropa* do not, of themselves, derive sustenance directly from decaying organic matter. Instead, the fungal intermediates provide water and nutrients from decay processes and from tree species growing in the vicinity. In experiments involving the injection of radioisotopically labeled substances into the phloem of trees, Björkman (1960) demonstrated the passage of sugars and phosphates from the trees to *Monotropa*. This has given rise to the designation *epiparasite,* in recognition of the secondary parasitism of tree species via fungal connections, which is different from the true parasitism of plants such as mistletoes (*Phoradendron,* Viscaceae) that form direct connections with the vascular system of their hosts.

Species of *Monotropa* have been used as an eye tonic, sedative, analgesic, and general tonic, among other medicinal uses.

1. Inflorescences racemes of 3–10 flowers; stems usually yellow to red at flowering, usually minutely hairy .... 1. M. HYPOPITYS
1. Inflorescences of solitary flowers; stems usually white at flowering, less commonly reddish-tinged, glabrous .... 2. M. UNIFLORA

1. Monotropa hypopitys L. (pinesap, false beechdrops)

Hypopitys americana (DC.) Small

Pl. 374 l, m; Map 1642

Stems 8–25 cm long, often clustered, sparsely to densely and minutely hairy, variously yellow or yellowish brown to red, turning dark brown upon drying. Leaves 4–10 mm long, 2–5 mm wide, sparsely to densely and minutely hairy, the margins thin and somewhat irregular. Inflorescences racemes of 3–10 flowers, the flower stalks 4–8 mm long at flowering, elongating to 15 mm at fruiting. Sepals 4 or 5, rarely absent, 7–12 mm long, 1–3 mm wide, lanceolate, erect, minutely hairy on both surfaces and margins. Petals 8–17 mm long, 3–5 mm wide, elliptic to spatulate, minutely hairy on both surfaces and margins. Stamens with the anthers horseshoe-shaped, dehiscing by a single slit across the tip. Ovary 4- or 5-lobed, minutely hairy. Stigma more or less disk-shaped, with a small, circular depression ringed with a fringe of hairs. Fruits with the body 5–6 mm long, 4–5 mm wide, capped by the persistent style, the capsule wall segments relatively thin, usually shed as the fruit dehisces. $2n=16, 32, 48$. June–October.

Scattered nearly throughout the state, but absent or nearly so from the Unglaciated Plains Division (U.S., including Alaska; Canada, Mexico, Central America, Europe, Asia). Bottomland forests, mesic to less commonly dry upland forests, ledges of bluffs, and occasionally banks of streams and rivers.

The species epithet is sometimes spelled *hypopithys* in the botanical literature, based on this unintentional misspelling in Linnaeus's original description. *Monotropa hypopitys* is quite variable in such characters as size, color, degree of pubescence, leaf and bract margins, and proportions of ovary and style (Wood, 1961; Wallace, 1974, 1995). Seasonal variation also occurs in the appearance of this species in that plants blooming in the summer tend to be yellowish, whereas those blooming in the autumn are more pinkish or reddish. Rarely, individuals exhibit odd color patterns, such as candy-cane-like striping of red and white or red stems with yellow flowers. More than 80 segregates have been named, but none of these have been considered to warrant taxonomic recognition in the most recent monographs (Wallace, 1974, 1995). However, in a molecular study of Monotropaceae and their fungal associates, Bidartondo and Bruns (2001) found that North American populations both associated with different fungal groups than did Eurasian plants and were widely separated from them in a phylogenetic analysis of plastid gene sequence data. Clearly, further research is required to unravel the complicated patterns of variation in this complex. Future studies may show that two or more cryptic species have been masquerading under the name *M. hypopitys*.

Seed set in these plants is normally quite high. Bumblebees visit the flowers, but the plants may be mostly self-pollinated, as the anthers shed pollen directly onto the underside of the stigma.

2. Monotropa uniflora L. (Indian pipe, ghost flower)

Pl. 374 n, o; Map 1643

Stems 5–30 cm tall, solitary or in small clusters, glabrous, usually white, less commonly reddish-tinged, drying black. Leaves 5–14 mm long, 3–6 mm wide, glabrous, the margins entire or slightly irregular. Inflorescences of a solitary flower subtended by a single leaflike bract. Sepals apparently absent in Missouri material (4 or 5 sepals, 7–10 mm long elsewhere). Petals 13–18 mm long, 5–9 mm wide, broadly oblong, usually minutely hairy on the inner surface. Stamens with the anthers elliptic, dehiscing by 2 slits across the tip. Stigma obconic, with a relatively deep, irregular depression at the tip lacking a fringe of hairs. Fruits with the body 10–12 mm long, 8–10 mm wide, capped by the persistent style, the capsule wall segments thickened, persisting after dehiscence, often into the following year. $2n=32, 48$. August–October.

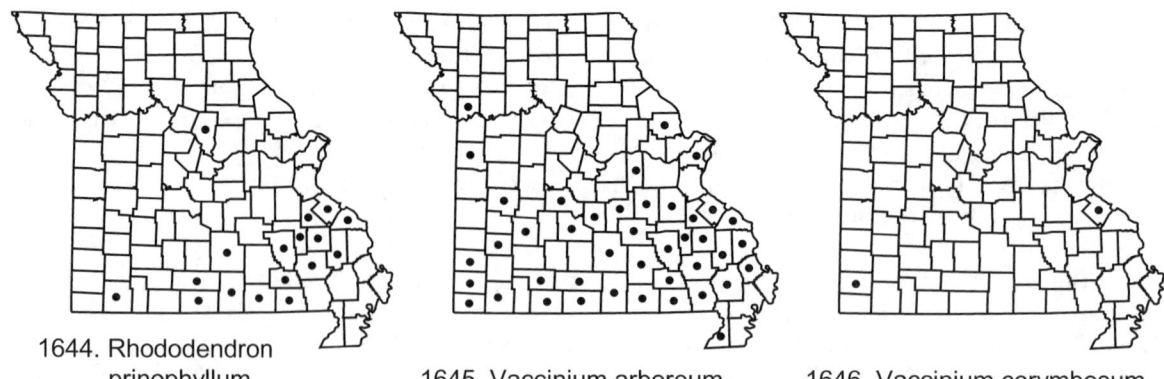

1644. Rhododendron prinophyllum   1645. Vaccinium arboreum   1646. Vaccinium corymbosum

Scattered nearly throughout the state but nowhere common (northern U.S. south to Florida, Texas, Montana, and California, also Alaska; Canada, Mexico, Central America, South America, Asia). Bottomland forests and mesic to dry upland forests.

4. Rhododendron L. (rhododendron, azalea)
(Kron, 1993)

About 850 species, North America, Europe, Asia, south to New Guinea, Australia.

Numerous species of *Rhododendron* are cultivated as ornamentals and many cultivars and hybrids have been developed. The group now informally called azaleas (once treated as the separate genus *Azalea* L.) tend to be species that have deciduous foliage, have mostly 5 stamens, and have funnelform corollas, whereas rhododendrons tend to have evergreen leaves, 10 stamens, and more bell-shaped flowers. Our native Missouri species, which is not often cultivated as it apparently is finicky about soils, is a typical azalea. As a group, rhododendrons are known to produce poisonous diterpenes and therefore should not be ingested.

1. Rhododendron prinophyllum (Small)
Millais (roseshell azalea, mountain azalea, wild honeysuckle, election pink)
Azalea prinophylla Small
R. roseum (Loisel.) Rehder
R. roseum f. *albidum* Steyerm.
R. canescens Porter (1889), not (Michx.) Sweet (1830)

Pl. 375 h–j; Map 1644

Plants shrubs or small trees, not colonial. Stems 0.8–3.0 m tall, the bark dark gray, often becoming scaly with small plates. Twigs densely and often minutely hairy, reddish brown, becoming brown to dark gray, the winter buds ovoid, with several overlapping scales, these sparsely and minutely hairy on the outer surface. Leaves tending to be clustered near the branch tips, short-petiolate. Leaf blades 4–9 cm long, 2–4 cm wide, ovate to obovate, the margins entire and short-hairy, sparsely hairy on both surfaces, the main veins of the undersurface densely hairy. Inflorescences short, umbellate racemes of 4–13 flowers, appearing before or with the leaves. Flowers slightly zygomorphic, hypogynous, fragrant. Calyces deeply 5-lobed, 1.0–1.5 mm long, the lobes rounded to bluntly pointed at the tip, densely glandular-hairy. Corollas trumpet-shaped, the tube 1.2–1.8 cm long, flaring to spreading lobes 1.1–1.6 cm long, rosy pink or rarely white, the outer surface moderately to densely glandular-hairy. Stamens 5, exserted, the filaments 4.0–4.5 cm long, arched upward, hairy toward the base, lacking spurs near the anther-filament junction, the anthers tiny, ovate, lacking tubes or awns at the tip, dehiscing by 2 terminal pores. Ovary superior, densely glandular-hairy, with 5 locules, the placentation axile. Style 4–6 cm long, arched upward or slightly S-shaped, exserted to slightly beyond the stamens, mostly not persistent at fruiting, the stigma obconic to disk-shaped, shallowly 5-lobed. Fruits capsules, 10–14 mm long, 3–5 mm wide, narrowly oblong-ovoid to cylindrical, sparsely to densely glandular-hairy, dehiscing longitudinally from the tip along the sutures. Seeds 2.0–3.5 mm long, 0.6–1.0 mm wide, irregularly ovate to ellip-

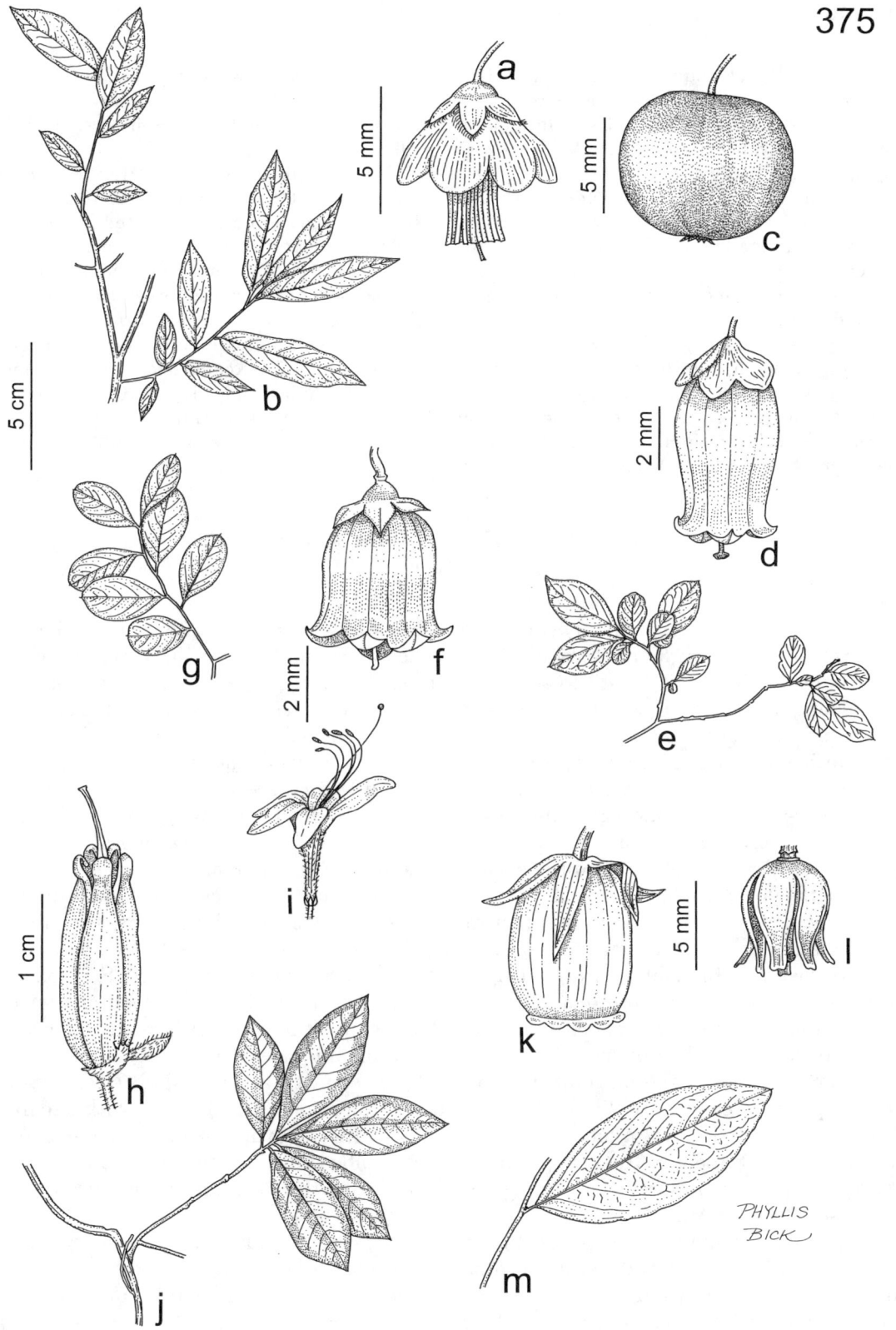

Plate 375. Ericaceae. *Vaccinium stamineum*, **a)** flower, **b)** branch with leaves. *Vaccinium pallidum*, **c)** fruit, **d)** flower, **e)** branch with leaves. *Vaccinium arboreum*, **f)** flower, **g)** branch with leaves. *Rhododendron prinophyllum*, **h)** fruit, **i)** flower, **j)** branch with leaves. *Lyonia mariana*, **k)** flower, **l)** fruit, **m)** node with leaf.

tic, strongly flattened and irregularly winged, the wing more or less entire at the tip and more or less fringed or dissected at the base, the surface more or less smooth, yellowish brown to light brown, shiny. $2n=26$. April–May.

Scattered and local in southeastern Missouri and the southern portion of the Ozark Division (eastern U.S. and adjacent Canada west to Missouri, Oklahoma, and Texas). Mesic upland forests and ledges of bluffs, mostly on north-facing exposures and acidic substrates.

This species has a disjunct distribution, occurring mainly in the Appalachians and the Ozarks. Overall, plants in the western part of the range tend to have longer corolla tubes than those in the eastern portion, but a number of exceptions are known from each region (Kron, 1993). This species was known as *R. roseum* in much of the older literature (Steyermark, 1963); however, that name was not validly published (Shinners, 1962; Kron, 1989).

The morphologically similar *R. canescens* (Michx.) Sweet (Piedmont azalea) is relatively widespread in the southeastern United States and has been documented from Union County, Illinois (adjacent to Cape Girardeau County, Missouri), as well as from several counties in western Kentucky and Tennessee. It differs from *R. prinophyllum* in its shorter and usually glandless flower stalks, narrower corolla tube expanded more abruptly into the lobes, and hairy fruits (Kron, 1993). This species can easily be mistaken for *R. prinophyllum* and may eventually be located in Missouri.

5. Vaccinium L. (blueberry)
(Vander Kloet, 1988)

Plants shrubs or less commonly small trees, sometimes evergreen, often colonial by stolons and/or rhizomes. Twigs glabrous or hairy, sometimes with small, orangish, warty pustules, the winter buds small, ovoid, with 3 to several outer scales, these usually glabrous. Leaves well spaced along the branches, sessile or short-petiolate. Leaf blades herbaceous to thick and leathery, the margins entire or finely and inconspicuously toothed, often also minutely hairy, the upper surface glabrous at maturity, the undersurface glabrous or inconspicuously hairy, lacking yellow resin glands (a few glands rarely present in *V. corymbosum*). Inflorescences mostly axillary, loose clusters or short racemes, less commonly reduced to solitary flowers, the flowers sometimes subtended by small, leaflike bracts at the base of the stalk, sometimes also with minute, scalelike bractlets toward the stalk midpoint. Flowers actinomorphic, epigynous. Hypanthium fused to the ovary, glabrous, usually somewhat glaucous, lacking yellow resin glands. Calyces deeply (4)5-lobed, glabrous, sometimes glaucous, the lobes triangular to broadly triangular, usually persistent at fruiting. Corollas tubular to urn-shaped or bell-shaped, (4)5-lobed, white or greenish white, often pinkish-tinged, the lobes ascending to recurved. Stamens (8)10, not exserted (except in *V. stamineum*), the filaments short, straight, attached at the corolla base, with or without spurs near the anther-filament junction, the anther sacs tapered into slender tubes toward the apparent tip, dehiscing by an oblique, terminal pore. Ovary inferior, with usually 5 locules, the placentation axile. Styles slender, at least slightly exserted, the stigma capitate, occasionally slightly lobed. Ovules numerous. Fruits berries, with 5 to numerous seeds. Seeds 0.8–2.0 mm long, asymmetrically ovate in outline, somewhat flattened and usually somewhat angular, the surface finely honeycombed with a network of minute, angular pits, orangish brown. About 400 species, North America to South America, Europe, Africa, Madagascar, Asia to Malesia.

Vaccinium is found in many parts of the world, usually on acidic, sandy, or peaty soils. The berries are an important food source for many species of birds, squirrels, and bears. Deer and rabbits browse on the young twigs and leaves. The genus is commercially important for its berries, including blueberries, cranberries, cowberries, lingonberries, and bilberries, which are eaten raw or cooked and are variously prepared into jams, pies, and juices. In North America, several species of blueberries have been improved by plant breeders and are cultivated commercially, mostly in the northern and northeastern United States and Canada. These include

V. corymbosum (highbush blueberry), *V. macrocarpon* Aiton (cranberry), and *V. angustifolium* Aiton (sweet lowbush blueberry). Many other species of blueberry are harvested from the wild, including *V. pallidum* in Missouri. Some species, such as *V. stamineum,* are generally considered unpalatable, but individual plants may yield good fruit.

The flowers of *Vaccinium* species are visited by large numbers of bees, primarily species of bumblebees and andrenids, but also honeybees. Pollinators receive both nectar (from a disk inside the base of the corolla) and pollen as their reward. The flowers are usually pendant, with the nectar held in place by the filaments. Flowers are usually protandrous; that is, the stamens mature first, with the styles maturing later, a mechanism to promote cross-pollination. Pollen accumulated in the long, narrow tubes falls out or is "buzzed out" when these are disturbed by an insect visitor (Crane et al., 1985). The stigmas become receptive after the pollen is dispersed.

1. Leaf blades with the tips mostly rounded (sometimes with a minute, abrupt point) or obtusely narrowed to a blunt point, thick and leathery, the upper surface shiny, glabrous, the margins often with widely spaced, minute, glandular teeth; stamens with spurs at the filament-anther junction, these slightly shorter than to about as long as the anther tubules; plants large shrubs or small trees
 . . . . 1. V. ARBOREUM
1. Leaf blades with the tips mostly sharply pointed or acutely narrowed to a blunt point, herbaceous to slightly thickened, the upper surface dull, glabrous or inconspicuously hairy, the margins entire or with small, nonglandular teeth; stamens lacking spurs at the filament-anther junction, or if spurs present then these much shorter than the anther tubules; plants low to tall shrubs
    2. Corollas bell-shaped, lobed to about the midpoint; stamens strongly exserted, with spurs at the filament-anther junction, these much shorter than the anther tubules; small, leaflike bracts present at the base of each flower stalk
 . . . . 4. V. STAMINEUM
    2. Corollas tubular to narrowly urn-shaped, shallowly lobed; stamens not exserted, lacking spurs at the filament-anther junction; leaflike bracts absent at the base of each flower stalk
        3. Plants medium to tall shrubs 1–3 m tall; corollas 8–10 mm long
 . . . . 2. V. CORYMBOSUM
        3. Plants low shrubs 0.2–1.0 m tall; corollas 4–7 mm long . . . . . . 3. V. PALLIDUM

1. Vaccinium arboreum Marshall (farkleberry, sparkleberry)

V. arboreum var. *glaucescens* (Greene) Sarg.
Batodendron arboreum (Marshall) Nutt.

Pl. 375 f, g; Map 1645

Plants medium to tall shrubs or small trees 2–5 m tall. Bark brown, longitudinally furrowed, peeling in thin plates with age. Twigs sparsely to densely hairy, green or reddish-tinged, becoming gray with age. Leaf blades 20–55 mm long, 12–30 mm wide, thick and leathery, oblanceolate to broadly elliptic-obovate, the tip usually rounded (sometimes with a minute, abrupt point) or obtusely narrowed to a blunt point, the margins often with widely spaced, minute, glandular teeth; less commonly entire, otherwise glabrous or hairy, the upper surface glabrous, waxy and shiny, the undersurface dull, glabrous to sparsely hairy along the midvein. Inflorescences short racemes or of solitary flowers. Flower stalks 4–10 mm long, with a conspicuous collarlike joint at the junction with the flower, often at least the lowermost stalks with small, leaflike bracts at the base. Calyx lobes 0.7–1.0 mm long, the margins sparsely hairy toward the tip. Corollas 2.5–3.0 mm long, 3–4 mm in diameter, broadly urn-shaped to nearly bell-shaped, white, shallowly lobed, the lobes reflexed. Stamens not exserted, with 2 yellow spurs at the filament-anther junction, these slightly shorter than to about as long as the anther tubules, the filaments flattened, hairy along the margins, the anthers tapered to tubules 1.5–2.0 mm long. Styles 4–5 mm, slightly exserted. Fruits 6–10 mm in diameter, black, shiny, not glaucous. $2n=24$. May–July.

Scattered to common in the Mississippi Lowlands, Ozark, and Ozark Border Divisions, extend-

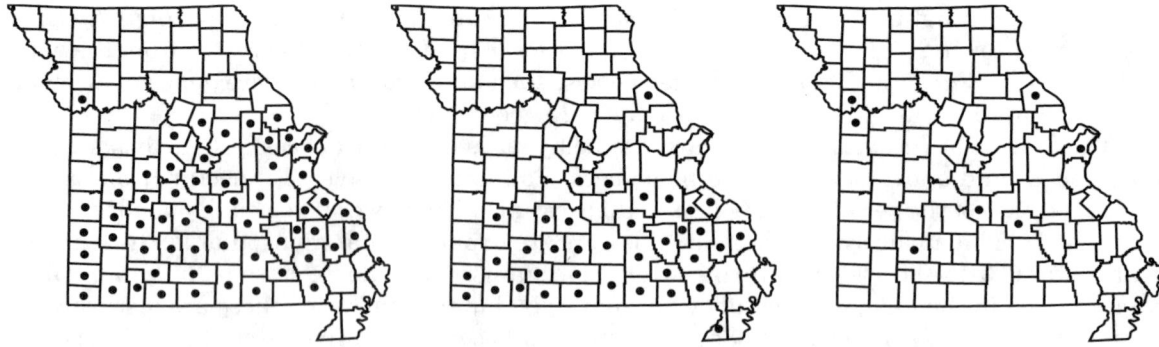

1647. Vaccinium pallidum   1648. Vaccinium stamineum   1649. Acalypha deamii

ing into the Lincoln Hills Section of the Glaciated Plains (eastern [mostly southeastern] U.S. west to Kansas and Texas). Mesic to dry upland forests, margins of glades, and tops of bluffs, occasionally banks of streams or margins of swamps, on acidic substrates.

The berries of *V. arboreum* have a mealy flesh and are considered unpalatable.

2. Vaccinium corymbosum L. (highbush blueberry, swamp blueberry)

Cyanococcus corymbosus (L.) Rydb.

Map 1646

Plants medium to tall shrubs 1–3 m tall. Bark reddish brown to grayish brown, longitudinally furrowed, peeling in thin plates with age. Twigs glabrous to densely hairy, sometimes glaucous, green to yellowish green or reddish green, becoming brown with age. Leaf blades 15–50(–80) mm long, 8–25 mm wide, relatively thin, not leathery, narrowly elliptic-oblanceolate to ovate, the tip usually sharply pointed (sometimes acutely narrowed to a blunt point), the margins entire or with small, nonglandular teeth toward the tip, sometimes curled under, otherwise usually glabrous, the upper surface glabrous, not waxy, dull, the undersurface dull, usually glabrous, pale or glaucous, rarely with a few tiny, yellow resinous glands. Inflorescences condensed, umbellate racemes, lacking leaflike bracts at the base of each flower stalk, but with small, scalelike, reddish bracts. Flower stalks 6–8 mm long, with a conspicuous collarlike joint at the junction with the flower. Calyx lobes 1.5–2.0 mm long, glabrous, often glaucous and/or waxy, rarely with a few tiny, yellow resinous glands. Corollas 8–10 mm long, 4–6 mm in diameter, tubular to narrowly urn-shaped, white to greenish pink, shallowly lobed, the lobes reflexed. Stamens not exserted, lacking spurs at the filament-anther junction, the filaments flattened, hairy along the margins, the anthers tapered to tubules 3–4 mm long. Styles 7–8 mm long, slightly exserted. Fruits 5–12 mm in diameter, blue, not shiny, glaucous. $2n=24, 48, 72$. April–May.

Possibly introduced, uncommon, known thus far from a historical, potentially native population in Newton County and a more recent, introduced occurrence in Ste. Genevieve County (eastern U.S. west to Wisconsin, Illinois, Oklahoma, and Texas). Dry upland forests on ridgetops, thus far only on sandstone substrate.

This species is an important fruit crop in the eastern and midwestern United States. More than 50 cultivars have been released, some with very large and juicy berries. At the site in Ste. Genevieve County, plants apparently became naturalized from adjacent plantings at homesites. The berries are eaten raw, cooked in pies, jams, and jellies, and dried for future use. Three distinct polyploid groups are known, and hybridization with *V. pallidum* is said to occur (Vander Kloet, 1988).

3. Vaccinium pallidum Aiton (lowbush blueberry, hillside blueberry)

V. vacillans Torr.
V. vacillans var. *crinitum* Fernald
V. vacillans var. *missouriense* Ashe
Cyanococcus vacillans Rydb.

Pl. 375 c–e; Map 1647

Plants low shrubs 0.2–1.0 m tall. Bark reddish brown to greenish brown, smooth, not peeling, sometimes becoming finely longitudinally ridged with age. Twigs glabrous to moderately hairy, sometimes glaucous, green to yellowish green, becoming brown with age. Leaf blades 20–50 mm long, 10–25 mm wide, relatively thin, sometimes somewhat stiff but not leathery, elliptic to ovate or obovate, the tip sharply pointed or acutely narrowed to a blunt point, the margins entire or with small, nonglandular teeth toward the tip, sometimes slightly thickened or curled under, otherwise glabrous, the upper surface glabrous or sparsely to moderately hairy along the midvein, not or only slightly waxy, usually at least somewhat shiny, the undersurface dull or slightly shiny, glabrous or sparsely hairy along the main veins, occasionally with reddish glands along the midvein, pale green. Inflorescences short racemes, sometimes reduced to solitary flowers, lacking leaflike bracts at the base of the flower stalks but with small, scalelike, reddish bracts on the stalks, these often shed as the flowers develop. Flower stalks 3–8 mm long, with a conspicuous swollen joint at the junction with the flower. Calyx lobes 1.2–1.6 mm long, glabrous, often glaucous. Corollas 4–7 mm long, 2–3 mm in diameter, tubular to narrowly urn-shaped, white to greenish white, often pinkish-tinged, shallowly lobed, the lobes reflexed. Stamens not exserted, lacking spurs at the filament-anther junction, the filaments flattened, hairy along the margins, the anthers tapered to tubules 1.5–2.0 mm long. Styles 5–6 mm long, slightly exserted. Fruits 4–6 mm in diameter, blue and glaucous, less commonly white or black, not shiny. $2n=24, 48$. April–May, occasionally reflowering in October.

Scattered to common, mostly south of the Missouri River (eastern U.S. and adjacent Canada west to Minnesota and Oklahoma). Mesic to dry upland forests and ledges and tops of bluffs; occasionally also old fields, on acidic substrates.

There is considerable variation in fruit color and pubescence in this species, but formal varieties based upon these characters are not recognized here. Although the yield is often low, the berries are delicious, and this species is harvested commercially in some southern states, including Arkansas. Specimens with white or black fruits are found only occasionally.

Steyermark (1963) discussed specimens determined by earlier botanists as *V. tenellum* Aiton and correctly dismissed these as misdeterminations of *V. pallidum* (as *V. vacillans*). *Vaccinium tenellum* is mostly restricted to the Coastal Plain and, except for its low stature, does not share many features with *V. pallidum*. Steyermark also excluded *V. pallidum* from the Missouri flora, which he considered distinct from *V. vacillans*. Vander Kloet (1978) studied this complex morphologically and through controlled crosses, concluding that *V. pallidum* and *V. vacillans* should be treated as a single taxon under the former name.

4. Vaccinium stamineum L. (deerberry, highbush huckleberry, southern gooseberry, squaw huckleberry)

V. stamineum var. *interius* (Ashe) E.J. Palmer & Steyer.
V. stamineum var. *melanocarpum* C. Mohr
V. stamineum var. *neglectum* (Small) Dean
Pl. 375 a, b; Map 1648

Plants shrubs 0.5–2.0 m tall. Bark gray to brown, relatively smooth, peeling in thin, papery strips with age. Twigs glabrous or less commonly sparsely to moderately hairy, green to yellowish green, becoming brown with age. Leaf blades 10–95 mm long, 8–30 mm wide, elliptic to oblanceolate, the tip usually sharply pointed, the margins entire or finely and indistinctly toothed, sometimes slightly curled under, otherwise hairy, the upper surface glabrous or sparsely to moderately hairy along the midvein, occasionally slightly glaucous, not or only slightly shiny, the undersurface dull, glabrous or sparsely hairy, light green, occasionally somewhat glaucous. Inflorescences racemes, with leaflike bracts present at the base of each flower stalk, these 6–18 mm long, 2–5 mm wide. Flower stalks 4–12 mm long, appearing continuous but actually with an inconspicuous joint (often visible only as an abrupt color change) at the junction with the flower. Calyx lobes 1.0–2.5 mm long, hairy along the margins, usually not glaucous. Corollas 4–6 mm long, 3–5 mm in diameter, bell-shaped, white, often purplish-tinged, lobed to about the midpoint, the lobes loosely ascending to more or less spreading. Stamens strongly exserted, with 2 yellow spurs at the filament-anther junction, these much shorter than the anther tubules, the filaments flattened, sometimes hairy, the anthers tapered to tubules 3–5 mm long. Styles 5–8 mm long, strongly exserted. Fruits 6–14 mm in diameter, blue and often glaucous, not shiny. $2n=24$. April–May, occasionally reblooming October–November.

Scattered almost entirely in the Ozark Natural Division and eastern portion of the Ozark Border (eastern U.S. and adjacent Canada west to Kansas and Texas; Mexico). Mesic to dry upland forests, glades, and tops of bluffs, on acidic substrates.

This is one of the most distinctive species of *Vaccinium* in North America, readily identified by its deeply lobed corolla and exserted stamens. It is also a highly polymorphic species, with variation in the pubescence of twigs and calyx, fruit color, and degree of waxiness. A number of species and varieties have been named based on these characters, but these do not breed true (Vander Kloet, 1988) and thus are not recognized here. *Vaccinium stamineum* is often associated with recently burned sites and may require fire conditions to promote seed germination (Ford, 1995).

EUPHORBIACEAE (Spurge Family)
Contributed by George Yatskievych and Mark H. Mayfield

Plants annual or perennial herbs or less commonly shrubs (trees, or lianas elsewhere), rarely twining, monoecious or dioecious, often with milky sap. Leaves alternate or less commonly opposite, simple (compound elsewhere), occasionally lobed, pinnately or less commonly palmately veined, the margins otherwise entire or toothed. Stipules usually present, but small, scalelike, hairlike, or glandular, and sometimes shed early. Inflorescences axillary or terminal, of solitary flowers, clusters, small panicles, spikes, or racemes, sometimes associated with prominent bracts, the basic unit in *Euphorbia* a highly modified cluster of small flowers grouped into a small, cuplike involucre called a cyathium (for more details, see the treatment of the genus). Flowers imperfect (appearing perfect in *Euphorbia;* see discussion under that genus), actinomorphic, hypogynous. Calyces of 3–12 sepals (absent in *Euphorbia*), these sometimes fused toward the base. Petals absent or 4–7, separate or rarely fused at the base, usually not showy. Staminate flowers with 1 to numerous stamens, the filaments distinct or less commonly fused into a tube toward the base, the anthers attached at the base or appearing attached between the somewhat spreading anther sacs. Nectar disc sometimes present in staminate and/or pistillate flowers, usually conspicuous, sometimes lobed or divided into segments. Pistil of usually 3 fused carpels, the ovary superior, usually 3-locular, the ovules 1 or 2 per locule, the placentation axile or more or less apical. Styles usually 3 (occasionally fused toward the base), each style often forked (unbranched in *Tragia,* with several irregular branches in *Acalypha*), the stigmas 1 per style branch (or branch fork), variously shaped but often consisting of an apical band along the inner surface of each style branch. Fruits appearing capsular but actually schizocarps (drupes, samaras, or berries elsewhere), usually lobed, dehiscent (indehiscent in some species of *Croton*) by the carpels splitting open usually from the tip along the inner margin elastically and longitudinally to expose the seeds, leaving a usually persistent central column. Seeds 1 or 2 per locule, often with a small, hardened, aril-like outgrowth of tissue (known as a caruncle) adjacent to the attachment point. In the broad sense, 300–400 genera, 7,000–8,000 species, worldwide, but most diverse tropical and subtropical regions.

The higher-order taxonomy of the Euphorbiaceae is still somewhat controversial. Some authorities have suggested splitting the family into five or more smaller families (see discussion in Wurdack et al. [2004]). Several authors have suggested, on the basis of molecular phylogenetic studies, that the group of about 55 genera and more than 1,500 species related to *Phyllanthus* (including *Andrachne* and *Phyllanthus* in Missouri) are not very closely related to the remainder of the Euphorbiaceae and should be treated as a separate family, Phyllanthaceae (Savolainen et al., 2000; Chase et al., 2002). However, the relationship of that group to other members of the order Malpighiales has not been settled (Wurdack et al., 2004). The phyllanthoid group is unusual within the Euphorbiaceae in having clear sap (lacking a milky latex) and in having two ovules per locule (Judd et al., 2002) and has been considered a separate subfamily within Euphorbiaceae by other authors (Webster, 1994; Govaerts et al., 2000; Radcliffe-Smith, 2001). The traditional, more inclusive circumscription of the Euphorbiaceae (Cronquist 1981, 1990; Webster, 1994; Govaerts et al., 2000; Radcliffe-Smith, 2001) is used in the present work with some reservation, as future studies likely will result in a better understanding of the phylogenetic relationships within the order Malpighiales. At the generic level, circumscriptions of *Andrachne, Euphorbia,* and *Phyllanthus,* and their segregates continue to be less than fully resolved (for further discussion, see the treatments of those genera).

The Euphorbiaceae are a large and morphologically diverse family. Although in the United States most species are herbaceous, the majority of the family consists of woody plants, including a number of succulents. The family is economically important for a variety of uses. *Hevea brasiliensis* (Willd.) Müll. Arg. (Para rubber; rubber tree) is the principal commercial source of latex for the production of rubber. *Manihot esculenta* Crantz (cassava) is cultivated widely in tropical regions for its starchy, tuberous roots, which are prepared as a dietary staple. The seeds of some genera contain commercially important oils, such as castor oil (*Ricinus communis* L.) and tung oil (mainly *Aleurites fordii* Hemsl.). Waxes have been extracted from some genera for candles and industrial uses. *Euphorbia pulcherrima* Willd. (poinsettia) is grown for its showy bracteal leaves, especially in conjunction with Christmas, but many other species are cultivated as houseplants, garden ornamentals, and specimen plants (especially the succulent taxa). The family is biochemically quite diverse, and various species have been used medicinally, especially for their alkaloids. By the same token, some species are poisonous, and a few genera have stinging hairs. The Neotropical genus *Hura* L. (cannonball tree) has woody fruits that are explosively dehiscent, flinging the shrapnel of 5–20 sharp-edged 1-seeded segments up to 15 m (these with sufficient force to shatter a window and cause a large gash in a metal file cabinet when an unwary botanist allows a ripe fruit to dry in his office); the immature fruits are used as go-cart wheels in parts of the Caribbean, and the toxic latex is used as a fish poison.

1. Leaves peltate, the blade large (10–90 cm long and wide), palmately lobed and veined, the petiole attached well away from the blade margin; plants robust herbs 1–5 m tall . . . . 6. RICINUS
1. Leaves not peltate, the blade smaller (to 10 cm long, but always much narrower than 10 cm), unlobed or pinnately few-lobed, the venation pinnate or only a solitary midvein visible, the petiole (if present) attached at the blade base; plants low shrubs (in *Andrachne*) or more commonly slender to robust herbs to about 1 m tall (usually much shorter)
    2. Plants with milky sap; flowers appearing perfect but the apparent flower actually an aggregation of several staminate flowers (each reduced to a solitary stamen) and a single central, usually short-stalked pistillate flower (this also lacking a perianth), these in a cup-shaped involucre (cyathium) with 1 or more marginal glands sometimes bearing small petaloid appendages . . . . 4. EUPHORBIA
    2. Plants with clear sap; flowers not appearing perfect, the staminate and pistillate flowers each with a calyx and sometimes also corolla, not grouped into a cuplike involucre
        3. Plants shrubs, the stems definitely woody above the base and not dying back to the ground each winter . . . . 2. ANDRACHNE
        3. Plants annual or perennial herbs, if perennial then sometimes from a somewhat woody rootstock but the stems not woody and dying back to the ground each winter
            4. Plants pubescent with branched or stellate hairs, these sometimes with branches more or less fused and then appearing as minute, scurfy, peltate scales; staminate and/or pistillate flowers with petals (except in *C. texensis*) . . . . 3. CROTON
            4. Plants glabrous or, if pubescent, then with only unbranched hairs; flowers all lacking petals
                5. Leaf blades with the margins entire
                    6. Stems and leaves hairy . . . . 1. ACALYPHA
                    6. Stems and leaves glabrous . . . . 5. PHYLLANTHUS
                5. Leaf blades with the margins toothed
                    7. The 3 styles free or fused only at the very base, each with several irregular branches; stems and leaves lacking stinging hairs; staminate flowers with 4–8 stamens . . . . 1. ACALYPHA
                    7. The 3 styles fused toward the base, each unbranched; stems and/or leaves with stinging hairs; staminate flowers with 2–4 stamens . . . . 7. TRAGIA

1. Acalypha L. (three-seeded mercury)

Plants annual (perennial herbs, shrubs, or trees elsewhere), monoecious (dioecious elsewhere), taprooted, with clear sap, pubescent with unbranched, nonglandular and sometimes also gland-tipped hairs; stinging hairs absent. Stems erect, branched or unbranched. Leaves alternate, short- to long-petiolate, the petiole attached at the base of the nonpeltate blade. Leaf blades variously shaped, angled or rounded at the base, angled or tapered to a usually sharply pointed tip, the margins entire or more commonly toothed, often more or less with 3 main veins from the base. Stipules scalelike, 0.5–1.5 mm long, tan to purple or sometimes green, usually shed early, linear to narrowly triangular, often with few to several bristly hairs at the tip. Inflorescences axillary and sometimes also terminal, usually associated with longitudinally folded or concave, persistent, lobed, leaflike bracts, the basic units small clusters of staminate or pistillate flowers, these arranged into spikes or racemes, the pistillate clusters either basal to the staminate clusters in the same spike or in separate terminal spikes. Flowers lacking a corolla and nectar disc. Staminate flowers sessile or nearly so, minute (less than 0.5 mm long), with 4 linear to narrowly triangular sepals (these hairy) and 4–8 minute stamens having short filaments (these free or fused at the very base). Pistillate flowers with 3(–5) minute, linear to ovate sepals, the ovary with 1, 2, or more commonly 3 locules and 1 ovule per locule, the 3 styles separate or fused only at the very base, each irregularly pinnately divided into several slender lobes. Fruits 2- or 3-lobed (except in *A. monococca*). Seeds ovoid, with a flattened, oblong to narrowly elliptic, small, white caruncle (or this apparently absent), the surface otherwise nearly smooth to shallowly pitted or with small tubercles, dark brown to light gray or tan, sometimes mottled. About 450 species, North America to South America, Caribbean Islands, Africa, Asia to Australia, Pacific Islands; introduced in Europe.

The large genus *Acalypha* is most diverse in the American tropics. Several species are cultivated as houseplants (and outdoors in warmer regions) for their foliage or inflorescences, especially the paleotropical shrub *A. hispida* Burm. f. (chenille plant, red-hot cattail), which has elongate pistillate spikes whose fuzzy texture and red coloration is caused by the feathery-branched, red styles.

1. Pistillate flowers in separate elongate terminal spikes, the staminate spikes axillary; at least some of the leaf blades shallowly cordate at the base, the margins with numerous (18–36 on each side) closely spaced teeth .... 4. A. OSTRYIFOLIA
1. Spikes all axillary, all or mostly with 1–3 basal pistillate flowers below the short to elongate portion with several to numerous staminate flowers; none of the leaf blades cordate, all tapered to broadly angled or slightly rounded at the base, the margins entire or more commonly with few to several (2–15 on each side) often broadly spaced teeth
    2. Bracts 5–9(–11)-lobed; leaves long-petiolate, the petiole slightly shorter than to slightly longer than the blade
        3. Fruits 2-locular, 2-seeded; seeds 2.2–3.2 mm long .... 1. A. DEAMII
        3. Fruits 3-locular, usually 3-seeded (rarely 1 of the ovules aborting); seeds 1.3–2.0 mm long .... 5. A. RHOMBOIDEA
    2. Bracts (9–)10–17-lobed; leaves relatively short-petiolate, the petiole 1/16–1/2 as long as the blade
        4. Leaves with the petiole 1/4–1/2 as long as the blade, mostly longer than the inflorescence bract(s) at the same node; bracts with the lobes triangular-ovate to broadly oblong, usually lacking stalked and sessile glands (rarely sparse glands present) .... 6. A. VIRGINICA

4. Leaves with the petiole ¹/₁₆–¹/₄ the length of the blade, mostly shorter than to occasionally about as long as the inflorescence bract(s) at the same node; bracts with the lobes linear to lanceolate, narrowly oblong, or triangular-ovate, usually with at least some of the pubescence of white-stalked and/or red sessile glands
    5. Fruits 3-locular, usually 3-seeded (rarely 1 of the ovules aborting); leaf blades narrowly lanceolate to more commonly oblong-lanceolate, oblong, or narrowly ovate; inflorescence bracts with the lobes mostly linear to lanceolate or narrowly oblong .... 2. A. GRACILENS
    5. Fruits 1-locular, 1-seeded; leaf blades linear to lanceolate; inflorescence bracts with the lobes mostly lanceolate to triangular-ovate ... 3. A. MONOCOCCA

1. Acalypha deamii (Weath.) H.E. Ahles (two-seeded mercury, large-seeded mercury)
  A. virginica L. var. *deamii* Weath.
  A. rhomboidea Raf. var. *deamii* (Weath.) Weath.

Pl. 376 i, j; Map 1649

Stems 20–60 cm long, glabrous or moderately to densely pubescent (sometimes in vertical lines) with short, strongly curved hairs. Leaves long-petiolate, the petiole slightly shorter than to slightly longer than the blade, much longer than the inflorescence bracts. Leaf blades 1–12 cm long, ovate to broadly rhombic, mostly broadly angled at the base, tapered to a sharply pointed tip, the margins with several (mostly 10–15 on each side) relatively closely spaced, usually blunt teeth, relatively thin-textured, the surfaces sparsely to moderately pubescent with relatively straight, more or less appressed hairs. Inflorescences entirely axillary spikes, 1–3 per node, each with 1–3 basal pistillate flowers below few to several nodes of staminate flower clusters, the tip of the spike not or only slightly extending beyond the bract. Inflorescence bracts 8–25 mm long, appearing more or less folded longitudinally around the inflorescence, with (5–)7–9 linear to lanceolate lobes, the margins sparsely to moderately hairy, the outer surface glabrous or sparsely hairy, sometimes some of the hairs gland-tipped, lacking minute, sessile, reddish glands. Fruits 2.5–3.4 mm long, 2-locular, 2-seeded, the surface moderately to densely hairy, lacking tubercles or slender projections at maturity. Seeds 2.2–3.2 mm long. $2n=40$. July–October.

Uncommon and sporadic (Iowa, Kansas, and Arkansas east to Pennsylvania and Tennessee). Edges and openings of bottomland forests and banks of streams and rivers.

This taxon was not treated by Steyermark (1963), who merely mentioned its existence (as *A. rhomboidea* var. *deamii*) in adjacent states. Miller (1964) mapped the species from three Missouri counties but provided no specimen citations to document her finds. Turner and Yatskievych (1992) first cited specimens from St. Louis and Pulaski Counties. The species remains relatively poorly known and may be overlooked in the field due to its vegetative similarity with *A. rhomboidea* (Becus, 2003). At present it appears to be rare throughout its range. Cooperrider (1984) suggested that there were problems in distinguishing *A. deamii* from *A. virginica* and treated it as a variety of that species. However, the consistently 2-seeded fruits with seeds more than 2 mm long are diagnostic for the species (Levin, 1999a). Gleason and Cronquist (1991) and some other authors have emphasized the abruptly drooping leaf blades as a unique feature of *A. deamii*, but this appears to be under seasonal or environmental influence (Becus, 2003). Plants observed during the present research in Pike County had spreading leaves similar to those of *A. rhomboidea* plants at nearby locations.

2. Acalypha gracilens A. Gray (slender three-seeded mercury)
  A. gracilens var. *delzii* Lill. W. Mill.
  A. gracilens var. *fraseri* (Müll. Arg.) Weath.

Map 1650

Stems 10–60 cm long, moderately to densely pubescent with short, strongly curved hairs. Leaves short-petiolate, the petiole ¹/₁₆–¹/₄ as long as the blade, shorter than to occasionally about as long as the inflorescence bracts. Leaf blades 1–7 cm long, narrowly lanceolate to more commonly oblong-lanceolate, oblong, or narrowly ovate, angled or slightly rounded at the base, angled or tapered to a usually sharply pointed tip, the margins nearly entire or more commonly with few to several (mostly 3–12 on each side) usually broadly spaced, blunt, minute teeth, often appearing shallowly scalloped, relatively thin-textured or somewhat thicker and stiffer, the surfaces sparsely to densely pubescent with short, straight to curved, loosely appressed hairs. Inflorescences entirely axillary spikes, 1–3 per node, each with 1–3 basal pistillate nodes (each with a separate folded bract) below several nodes of staminate flower clusters, the tip of the spike extending well beyond the bracts. Inflorescence bracts 1 per pistillate node, 4–25 mm long, appearing more

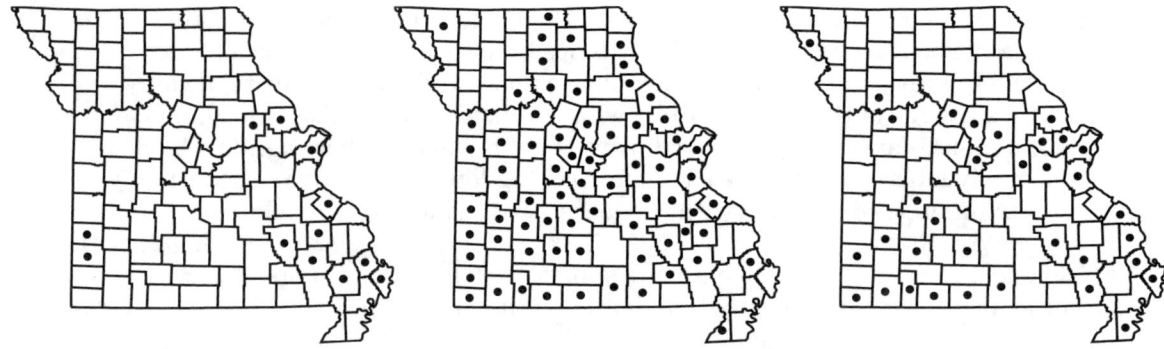

1650. Acalypha gracilens   1651. Acalypha monococca   1652. Acalypha ostryifolia

or less folded longitudinally around the inflorescence, with (9–)10–17 linear to lanceolate or narrowly oblong lobes, the margins sparsely to moderately bristly-hairy, at least some of the hairs usually gland-tipped, the outer surface sparsely to densely hairy, usually at least some of the hairs gland-tipped, usually also with sparse to moderate minute, reddish, sessile glands. Fruits 1.5–2.3 mm long, 3-locular, usually 3-seeded (rarely 1 of the ovules aborting), the surface moderately hairy and sometimes also with minute, sessile glands, lacking tubercles or slender projections at maturity. Seeds 1.2–2.0 mm long. June–October.

Uncommon in eastern Missouri and disjunctly in a few southwestern counties (eastern U.S. west to Iowa and Texas). Banks of streams and rivers, margins of ponds and lakes, savannas, glades, and sand prairies; also ditches, gardens, roadsides, and disturbed areas.

Miller (1964) speculated that midwestern populations of A. gracilens might be relatively recent introductions to the region. However, Levin (1999b) concluded that these populations had been in the area long enough for minor morphological differences to accumulate between them and other populations from the main portion of the species range farther to the south and east. In Missouri, although the species currently is found mostly at disturbed sites, the oldest specimens date back to George Engelmann's collections from the 1840s in seemingly natural habitats along the Mississippi River near St. Louis.

Levin (1999b) concluded, based on a detailed quantitative analysis of morphological variation in the A. gracilens/monococca complex, that two species should be recognized, but that recognition of varieties within A. gracilens was unwarranted. In Missouri, A. gracilens tends to be more robust than A. monococca, with bigger and broader leaves, as well as with longer and more strongly exserted spikes (with more pistillate nodes and a more elongate staminate portion with more nodes). Intergradation and possible hybridization between A. gracilens and A. virginica are discussed further in the treatment of that species.

3. Acalypha monococca (Engelm. ex A. Gray) Lill. W. Mill. & Ghandi (one-seeded mercury)

A. gracilens var. monococca Engelm. ex A. Gray

A. gracilens A. Gray ssp. monococca (Engelm. ex A. Gray) G.L. Webster

Pl. 376 g; Map 1651

Stems 10–45 cm long, sparsely to densely pubescent with short, strongly curved hairs. Leaves short-petiolate, the petiole $1/16$–$1/4$ as long as the blade, shorter than to occasionally about as long as the inflorescence bracts. Leaf blades 1–7 cm long, linear to lanceolate, angled or somewhat tapered at the base, angled or tapered to a usually sharply pointed tip, the margins entire or with relatively few (mostly 3–5 on each side) broadly spaced, blunt, minute teeth, sometimes appearing minutely scalloped, relatively thin-textured to somewhat thicker and stiffer, the surfaces sparsely to moderately pubescent with short, straight to curved, loosely appressed hairs. Inflorescences entirely axillary spikes, 1–3 per node, each with 1(–3) basal pistillate node(s) (each with a separate folded bract) below few to several nodes of staminate flower clusters, the tip of the spike usually extending well beyond the bracts. Inflorescence bracts 1 per pistillate node, 4–16 mm long, appearing more or less folded longitudinally around the inflorescence, with (9–)10–17 mostly lanceolate to triangular-ovate lobes, the margins sparsely to moderately bristly-hairy, at least some of the hairs usually gland-tipped, the outer surface sparsely to densely hairy, usually at least some of the hairs gland-tipped, also with sparse to moderate minute, reddish, sessile glands. Fruits 1.8–2.6 mm long, 1-locular, 1-seeded,

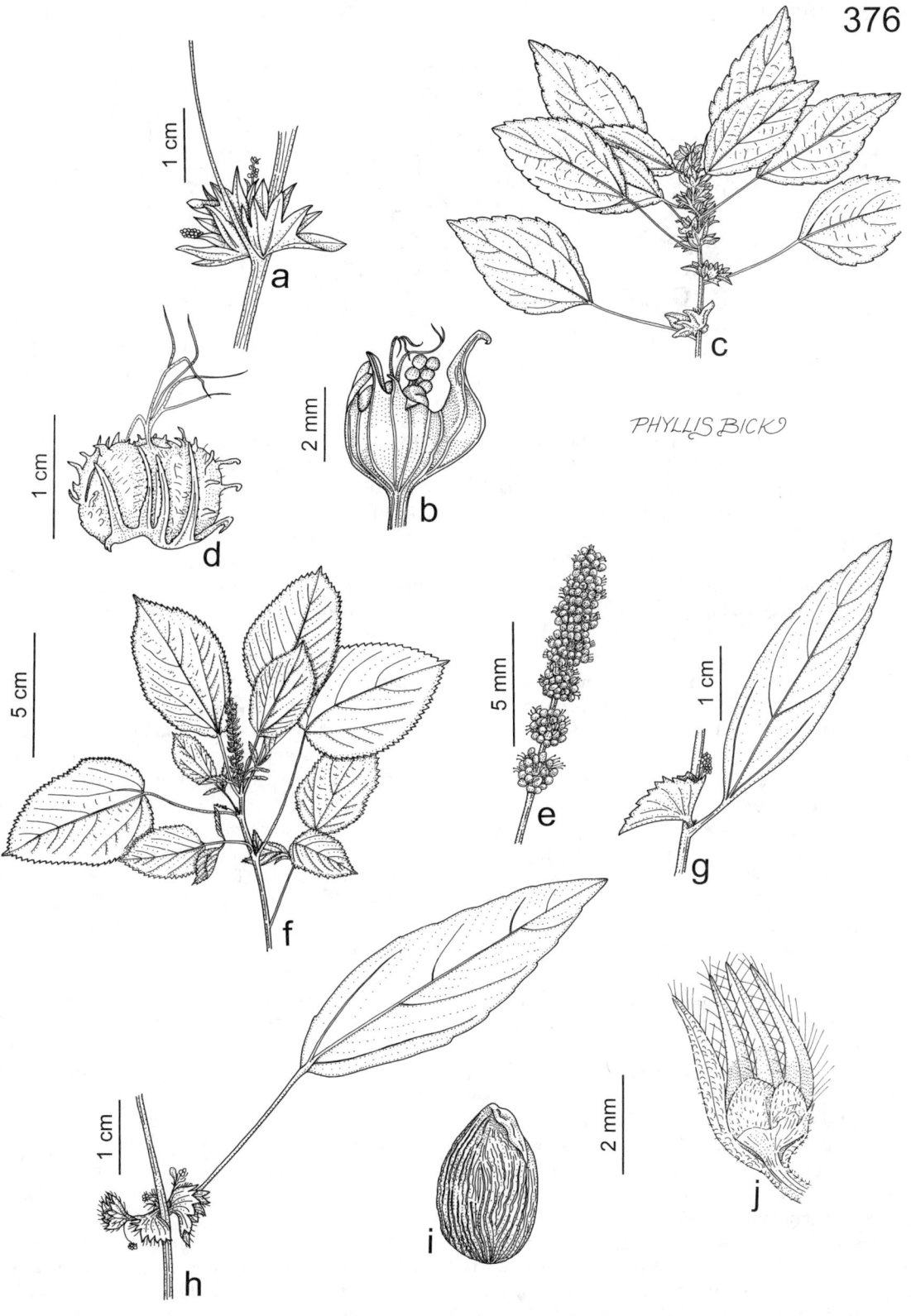

Plate 376. Euphorbiaceae. *Acalypha rhomboidea*, **a)** inflorescence at node, **b)** inflorescence, **c)** fertile branch. *Acalypha ostryifolia*, **d)** fruit with bracts, **e)** staminate spike, **f)** flowering branch. *Acalypha monococca*, **g)** node with inflorescence and leaf. *Acalypha virginica*, **h)** node with inflorescences and leaf. *Acalypha deamii*, **i)** seed, **j)** fruit with bracts (partially removed).

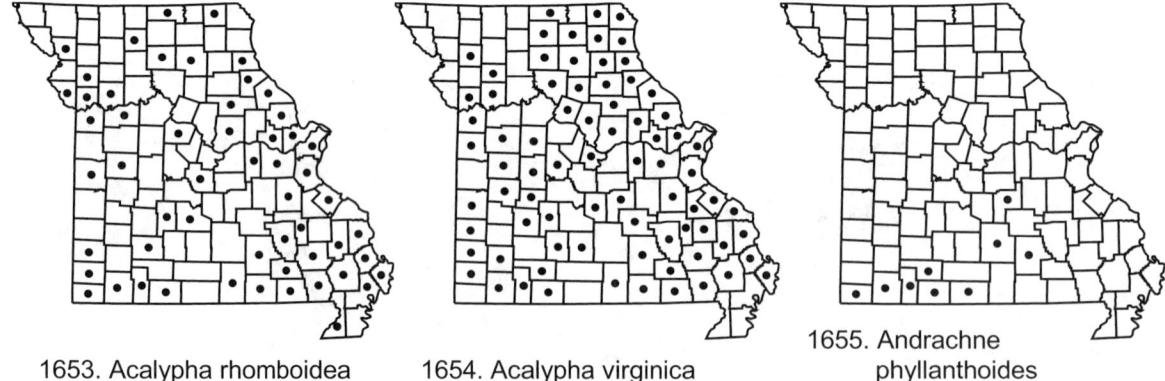

1653. Acalypha rhomboidea   1654. Acalypha virginica   1655. Andrachne phyllanthoides

the surface moderately hairy and occasionally also with sparse, minute, sessile glands, lacking tubercles or slender projections at maturity. Seeds 1.6–2.4 mm long. $2n=40$. May–October.

Scattered nearly throughout Missouri but uncommon or absent from most of the northwestern quarter of the state (Illinois to Iowa and Kansas south to Kentucky, Arkansas, and Texas). Glades, upland prairies, sand prairies, ledges and tops of bluffs, savannas, and openings of mesic to dry upland forest; also pastures, railroads, and roadsides.

For a discussion of the taxonomic problems involving this species and *A. gracilens,* see the treatment of that species.

4. **Acalypha ostryifolia** Riddell (roughpod copperleaf)

Pl. 376 d–f; Map 1652

Stems 10–70 cm long, often slightly zigzag above the midpoint, moderately to densely pubescent with short, strongly curved hairs, also with longer, straight hairs toward the tip, often many of these gland-tipped. Leaves short- to more commonly long-petiolate, the petiole ($^1/_{10}$–)$^1/_5$ as long as to slightly longer than the blade, much longer than the inflorescence bracts. Leaf blades 1–8 cm long, narrowly to broadly ovate or heart-shaped, those of at least some of the leaves shallowly cordate at the base (those of the others merely rounded), angled or tapered to a bluntly or sharply pointed tip, the margins with relatively numerous (18–36 on each side) closely spaced, usually sharp teeth, relatively thin-textured or somewhat thicker and stiffer, the surfaces (especially the undersurface) sparsely pubescent with straight to curved, more or less appressed hairs. Inflorescences of two types, the terminal spike with 5–20 nodes of pistillate flowers (1 or less commonly 2 bracts per node, 1 or rarely 2 pistillate flowers per bract), the axillary spikes 1 or rarely 2 per node, with few to several clusters of staminate flowers (rarely a few short pistillate spikes also in the upper leaf axils), the staminate spikes either lacking an inflorescence bract or with a minute, irregularly ovate bract. Pistillate bracts 3–12 mm long, appearing longitudinally concave around the pistillate flowers, with 13–17 linear or occasionally narrowly lanceolate lobes, the margins and outer surface of the lobes with moderate to dense, minute, sessile, yellowish glands and sometimes also sparse, stalked glands, the surface of the unlobed basal portion moderately to densely hairy with short, straight, spreading hairs, many of these gland-tipped, often also with sparse, sessile glands. Fruits 1.6–2.5 mm long, 3-locular, usually 3-seeded (rarely 1 of the ovules aborting), the surface with slender, elongate tubercles or projections, at least above the midpoint, otherwise glabrous or finely short-hairy, some of the hairs sometimes gland-tipped. Seeds 1.5–2.3 mm long. June–October.

Scattered mostly from the Missouri River floodplain southward (Pennsylvania to Florida west to Nebraska and Arizona). Banks of streams and rivers; also crop fields, fallow fields, gardens, railroads, roadsides, and open, disturbed areas.

This weedy species is found most commonly in disturbed habitats. It apparently has become much more abundant in the Missouri River floodplain in the years since Steyermark (1963) studied its distribution.

5. **Acalypha rhomboidea** Raf. (three-seeded mercury, rhombic copperleaf)

A. virginica L. var. *rhomboidea* (Raf.) Cooperr.

Pl. 376 a–c; Map 1653

Stems 15–60 cm long, glabrous or moderately to densely pubescent (sometimes in vertical lines) with short, strongly curved hairs, occasionally with a few longer, straight hairs toward the base. Leaves long-petiolate, the petiole slightly shorter than to slightly longer than the blade, much longer than the inflorescence bracts. Leaf blades 1–10 cm long, broadly lanceolate to ovate or rhombic, angled to broadly angled at the base, tapered to a sharply pointed tip, the margins with several (mostly 8–12 on each side) often relatively closely spaced, usually blunt teeth, relatively thin-textured, the surfaces sparsely pubescent with relatively straight, more or less appressed hairs. Inflorescences entirely axillary spikes, 1–3 per node, each with 1–3 basal pistillate flowers below few to several nodes of staminate flower clusters, the tip of the spike not or only slightly extending beyond the bract. Inflorescence bracts 4.5–25.0 mm long, appearing more or less folded longitudinally around the inflorescence, with (5–)7–9(–11) narrowly lanceolate to oblong-lanceolate lobes, the margins sparsely to moderately hairy, often some of the hairs gland-tipped, the outer surface glabrous or sparsely hairy, sometimes some of the hairs gland-tipped, lacking minute, sessile, reddish glands. Fruits 1.5–2.3 mm long, 3-locular, usually 3-seeded (rarely 1 of the ovules aborting), the surface moderately to densely hairy, lacking tubercles or slender projections at maturity. Seeds 1.3–2.0 mm long. July–October.

Scattered throughout the state (eastern U.S. west to North Dakota and Texas; Canada). Banks of streams and rivers, margins of ponds and lakes, bottomland forests, mesic upland forests, and moist depressions of upland prairies; also crop fields, fallow fields, old fields, railroads, roadsides, and disturbed areas.

Cooperrider (1984) pointed out problems in distinguishing A. rhomboidea from A. virginica and treated it as a variety of that species. However, based on his quantitative morphological studies, Levin (1999a) disagreed, noting that not all of the traditionally used qualitative features contained consistent differences between the species but that others served to uniquely differentiate them.

6. Acalypha virginica L. (Virginia copperleaf)
Pl. 376 h; Map 1654

Stems 15–60 cm long, sparsely to densely pubescent (sometimes in vertical lines) with short, strongly curved hairs, usually also with sparse to dense, longer, straight hairs. Leaves relatively short-petiolate, the petiole $1/4$–$1/2$ as long as the blade, usually longer than the inflorescence bracts. Leaf blades 1–12 cm long, lanceolate to narrowly ovate or narrowly rhombic, angled or slightly rounded at the base, angled or tapered to a sharply pointed tip, the margins with few to several (mostly 3–8 on each side) usually broadly spaced, blunt, shallow teeth, sometimes appearing shallowly scalloped or slightly undulate, relatively thin-textured, the surfaces sparsely pubescent mostly along the veins with short, straight to curved, more or less appressed hairs. Inflorescences entirely axillary spikes, 1–3 per node, each with 1–3 basal pistillate flowers below few to several nodes of staminate flower clusters, the tip of the spike often extending somewhat beyond the bract. Inflorescence bracts 4.5–20.0 mm long, appearing more or less folded longitudinally around the inflorescence, with (9–)10–15 triangular-ovate to broadly oblong lobes, the margins sparsely to densely bristly-hairy, usually lacking gland-tipped hairs, the outer surface sparsely hairy, usually lacking gland-tipped hairs, rarely with sparse, minute, reddish, sessile glands. Fruits 1.5–2.3 mm long, 3-locular, usually 3-seeded (rarely 1 of the ovules aborting), the surface moderately hairy and sometimes also with minute, sessile glands, occasionally with a few minute, low, warty projections when young but smooth at maturity. Seeds 1.3–2.0 mm long. $2n=40$. July–October.

Scattered nearly throughout the state (eastern U.S. west to South Dakota and Texas; introduced in Europe). Bottomland forests, mesic to dry upland forests, savannas, upland prairies, margins of ponds, lakes, and sinkhole ponds, and ledges and tops of bluffs; also fallow fields, old fields, pastures, cemeteries, ditches, gardens, railroads, roadsides, and open, disturbed areas.

Cooperrider (1984) pointed out problems in distinguishing A. rhomboidea, and A. virginica, treating these taxa as varieties of A. virginica. However, Levin (1999a) had no trouble in distinguishing them in his quantitative morphological studies of the group. Steyermark (1963), who suggested that A. virginica apparently was expanding its range northward during the decades of his research on the flora, also suggested occasional hybridization between A. virginica and A. gracilens. However, there has been no subsequent experimental confirmation of such hybrids. Steyermark also noted the existence of occasional specimens difficult to place in either A. rhomboidea or A. virginica. Such specimens—for example, occasional plants with exceptionally short petioles—will continue to vex Missouri botanists but fortunately are relatively rarely encountered. Although a great deal has been written about the morphology of the complex, it would benefit from additional biosystematic and population-genetic investigations. Even the chromosome numbers in the complex are not known with certainty, as Miller's (1964) counts based on $x=10$ contradict earlier reports based on $x=7$ (Webster, 1967).

Burrows and Tyrl (2001) noted that, of all the temperate North American species of Acalypha, A. virginica was the one most likely to cause problems with livestock. Plants in the genus contain diterpene esters that can act as irritants for soft tissues, mucous membranes, and the digestive tract.

2. Andrachne L.

About 15 species, North America, South America, Caribbean Islands, Europe, Africa, Asia. The generic limits of *Andrachne* have been broadened in recent years by some authors (Govaerts et al., 2000) to include the 10–25 species of *Leptopus* Decne., which occur from Asia south to Australia. *Andrachne* in the strict sense is mostly Mediterranean with a few species in North America, one of which also occurs disjunctly in portions of western South America. The two groups have been said to differ in a few relatively subtle characters such as details of ovule placentation. Molecular studies (Wurdack et al., 2004) have reinforced the relationship between the two groups but have not been able to resolve the controversy over generic limits thus far.

1. Andrachne phyllanthoides (Nutt.) Müll. Arg. (maidenbush, buckbrush)
Lepidanthus phyllanthoides Nutt.
Savia phyllanthoides (Nutt.) Pax & K. Hoffm.
Leptopus phyllanthoides (Nutt.) G.L. Webster
Pl. 377 a–d; Map 1655

Plants shrubs, dioecious (monoecious elsewhere), with clear sap, lacking stinging hairs. Stems 30–80 cm long, ascending, branched, glabrous or sparsely pubescent with inconspicuous, slender hairs toward the tip. Leaves alternate, sessile or more commonly very short-petiolate, the petiole attached at the base of the nonpeltate blade. Leaf blades (6–)10–18(–25) mm long, oblong-elliptic to broadly elliptic, broadly oval, or oblong-obovate, more or less rounded at the sometimes slightly asymmetrical base, rounded (occasionally shallowly notched) or broadly angled to a bluntly pointed tip, the margins entire or rarely slightly undulate, relatively thick-textured but not leathery, pinnately veined, the surfaces glabrous or inconspicuously short-hairy along the midvein. Stipules scalelike, 1.0–1.5 mm long, brown or purplish-tinged, mostly persistent, lanceolate to narrowly triangular, glabrous or with a few inconspicuous hairs. Inflorescences axillary, the staminate ones of solitary flowers or small, sessile clusters of 2–4 flowers, the pistillate ones of solitary flowers, usually with a small, stipulelike, fringed bract at the base, the flowers individually short- or long-stalked. Calyces deeply 5(6)-lobed, 1–2 mm long (in pistillate flowers becoming enlarged to 3.5 mm as the fruits mature), oblong to oblong-obovate, rounded at the tip, glabrous or the outer surface sparsely hairy, persistent at fruiting. Petals 5(6), 0.5–1.0 mm long (occasionally appearing absent in pistillate flowers), oblong, cream-colored to pale greenish yellow, rounded at the tip, the margins with a minute fringe of hairs. Nectar disc usually more or less divided into 5 lobes opposite the petals, these shallowly and broadly notched at the tip. Staminate flowers with 5 free stamens (the filaments arching outward) and an inconspicuous, nonfunctional ovary consisting of 3 peglike segments shorter than the filaments. Pistillate flowers with the ovary 3-locular and 2 ovules per locule, the 3 styles separate or nearly so, each 2-lobed toward the tip, each lobe broadened into a flattened, knoblike stigma. Fruits 4–5 mm long, 6.0–7.5 mm in diameter, not or only slightly lobed (slightly and very bluntly 3-angled in cross-section), dark purplish brown at maturity, glabrous or nearly so. Seeds up to 6 per fruit, 2.8–3.5 mm long, wedge-shaped, lacking a caruncle, the surface appearing finely pebbled, mottled with darker and lighter brown. $2n=26$. May–October.

Scattered in the Ozark Division east to Shannon County (Texas, Oklahoma, Arkansas, Missouri, and Alabama). Banks of streams and rivers (especially seepy rock shelves), ledges of bluffs, and glades, usually on limestone and dolomite substrates.

Andrachne phyllanthoides and the closely related *A. arida* (Warnock & M.C. Johnst.) G.L. Webster of Texas are morphologically isolated within the genus (Webster, 1967), and various authors have classified them into other genera in the phyllanthoid group of Euphorbiaceae, as evidenced by the synonymy cited above. Further studies are needed to confirm the placement of these taxa.

Steyermark (1963) speculated that the presence of *A. phyllanthoides* in Missouri represented an example of a species surviving in an Ozarkian refugium for warm-climate plants during Pleistocene glaciation. However, this view is at odds with modern theories of past vegetational patterns, which suggest that the present flora of Missouri resulted from northward migration of plant species as the glaciers retreated (see the introductory section on origins of the flora in volume 1 of the present work [Yatskievych, 1999]).

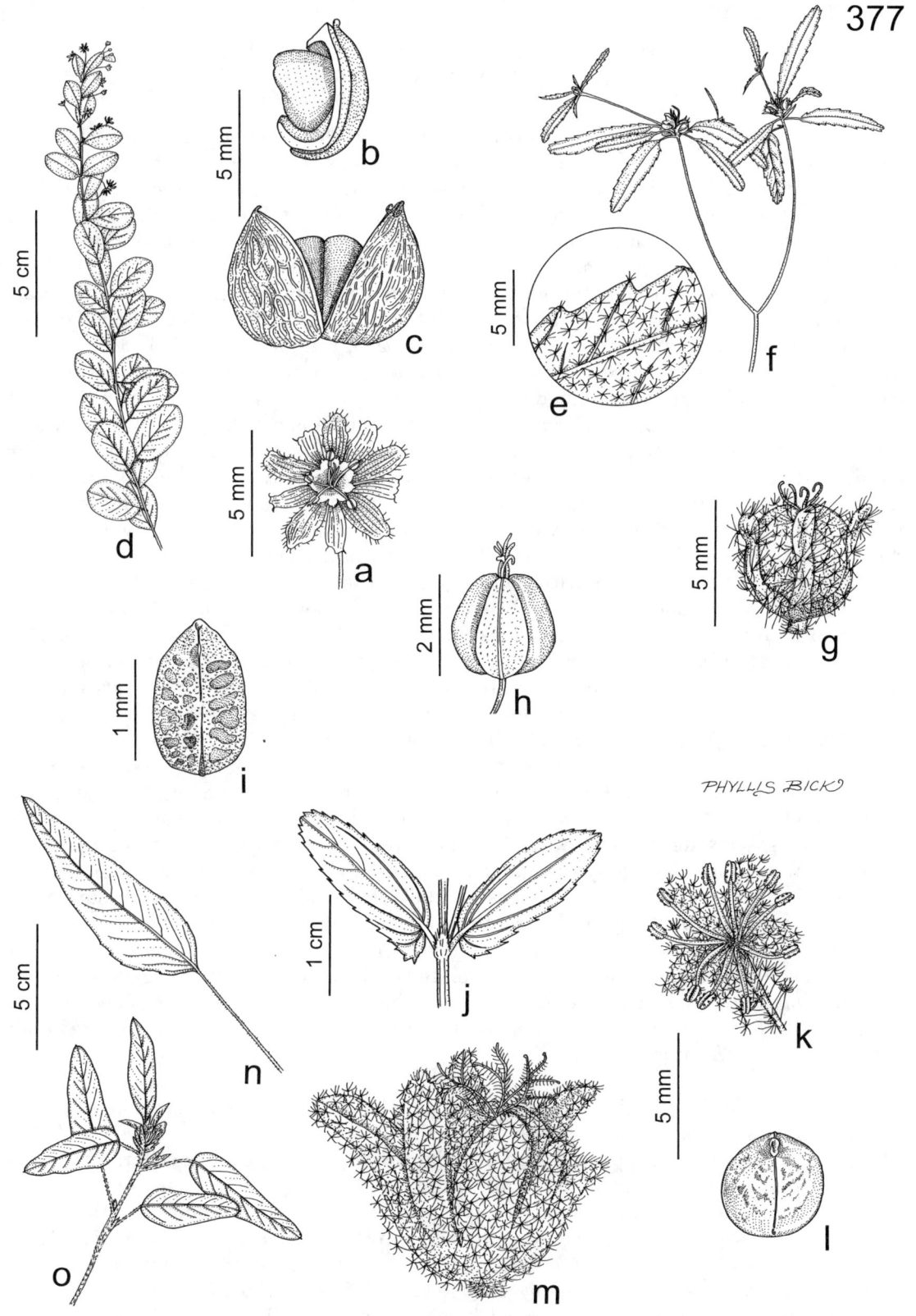

Plate 377. Euphorbiaceae. *Andrachne phyllanthoides*, **a)** flower, **b)** seed with capsule wall, **c)** dehiscing fruit, **d)** fertile branch. *Croton glandulosus*, **e)** leaf detail, **f)** fertile branch, **g)** fruit. *Euphorbia nutans*, **h)** fruit, **i)** seed, **j)** node with leaves. *Croton capitatus*, **k)** staminate flower, **l)** seed, **m)** fruit with calyx, **n)** leaf, **o)** fertile branch.

3. Croton L. (croton)

Plants annual (perennial herbs, shrubs, or trees elsewhere), monoecious or dioecious (in *C. texensis*), taprooted, with the sap clear (somewhat colored but not milky elsewhere), pubescent with branched or stellate hairs, these sometimes with branches more or less fused and then appearing as minute, scurfy, peltate scales; stinging hairs absent. Stems usually solitary, erect or ascending, unbranched or more commonly branched, sometimes from near the base. Leaves alternate but sometimes appearing opposite or whorled because of stem portions with very short internodes, sessile to long-petiolate, the petiole sometimes with 1 or 2 large, saucer-shaped glands at the tip, attached at the base of the nonpeltate blade. Leaf blades linear to nearly circular, tapered, angled, or rounded at the base, rounded or angled to tapered to a bluntly or sharply pointed tip, the margins entire or nearly so (regularly toothed in *C. glandulosus*), usually pinnately veined (sometimes only the midvein apparent). Stipules absent (in *C. michauxii* and *C. willdenowii*) or usually not apparent at flowering, then either minute (less than 1 mm long), lanceolate to ovate, tan scales that are shed early or small glandular dots that are obscured by the pubescence. Inflorescences terminal, axillary, and/or from the stem branch points, appearing as short, dense, spikelike racemes or dense clusters, the staminate and pistillate flowers variously positioned, each flower usually with an inconspicuous, short, slender bract (often threadlike, brown, and shed before flowering). Flowers with a nectar disc, this entire or with lobes equal to the number of calyx lobes. Staminate flowers with the calyces usually deeply 5-lobed, corollas of 5 petals (except in *C. texensis*) about as long as the calyx, the 5–20 small stamens (more elsewhere) with the filaments free. Pistillate flowers with the calyces usually deeply 5–12-lobed, corollas highly reduced or absent, the ovary with 1–3 locule(s) and 1 ovule per locule, the 1–3 style(s) free (or in some 3-styled species fused at the very base), each deeply 2-lobed, the lobes sometimes lobed again or branched. Fruits unlobed or slightly 3-lobed (circular in cross-section or nearly so), dehiscent except in *C. michauxii* and *C. willdenowii*. Seeds nearly spherical to oblong-ovoid or ovoid, the caruncle absent or a small, light-colored knob at the end adjacent to the attachment point, the surface smooth and often shiny, yellowish brown or reddish brown to dark brown, sometimes mottled. More than 800 species, nearly worldwide, but most diverse in tropical and subtropical regions.

The taxonomy and phylogeny of this large and morphologically variable genus still requires much further study. As noted by Webster (1967), for its size *Croton* contains surprisingly few species of economic importance. The common houseplants with brightly colored variegated leaves referred to as croton actually are *Codiaeum variegatum* (L.) A. Juss., a paleotropical member of the family not particularly closely related to the true crotons. The seeds of the Asian *Croton tiglium* L. are the source of croton oil, which has uses similar to those of castor oil (see the treatment of *Ricinus*). The stellate hairs sometimes are shed easily when plants are handled and can cause eye irritation.

1. Leaf blades with the margins finely toothed; petioles with 1 or 2 large, saucer-shaped glands at the tip . . . . 2. C. GLANDULOSUS
1. Leaf blades with the margins occasionally slightly wavy toward the base but otherwise entire; petioles lacking glands at the tip
    2. Plants dioecious; staminate flowers lacking a corolla; pistillate flowers with the 3 styles each divided nearly to the base into 4–6 lobes . . . . 6. C. TEXENSIS
    2. Plants monoecious; staminate flowers with a corolla; pistillate flowers with the 1–3 styles each 1 time deeply divided (2 or 3 times divided in *C. capitatus* but with the second and third order divisions usually well above the base)
        3. Ovary 3-locular; styles 3, each deeply 2-lobed, the lobes sometimes lobed again; fruits 3-seeded (rarely 2-seeded by abortion of 1 ovule)

4. Pistillate flowers with the calyx 6–9-lobed; styles each lobed 2 or 3 times (the total number of stigmatic branches thus theoretically 12–24 per flower, but in practice mostly 12–16); fruits 6–9 mm in diameter . . . . 1. C. CAPITATUS

4. Pistillate flowers with the calyx 5-lobed; styles each deeply lobed only 1 time (the total number of stigmatic branches thus 6 per flower); fruits 4–5 mm in diameter . . . . 3. C. LINDHEIMERIANUS

3. Ovary 1- or 2-locular; styles 2 or 3, each shallowly or deeply 2-lobed; fruits usually 1-seeded

5. Ovary 2-locular (1 of the locules usually aborting and the dehiscent fruit thus usually 1-seeded); styles 2, each deeply 2-lobed; seeds with a knoblike caruncle; leaf blades with the undersurface pubescent with stellate hairs (the calyx and ovary also with stellate hairs) . . . . 5. C. MONANTHOGYNUS

5. Ovary 1-locular (the indehiscent fruit thus 1-seeded); styles 3, each shallowly 2-lobed toward the tip; seeds lacking a caruncle; leaf blades with the undersurface pubescent with peltate, scalelike trichomes (these also on the stem, calyx, and ovary)

6. Leaf blades with the upper surface pubescent with minute, nonoverlapping, stellate hairs, the more or less equal branches 0.2–0.3 mm long; stems, undersurface of the leaf blades, and especially fruits with peltate scales consisting of a relatively narrow, fused central disc with a relatively long fringe of slender, stellate extensions beyond the disc . . . . 4. C. MICHAUXII

6. Leaf blades with the upper surface pubescent with short, overlapping, stellate hairs, the often unequal branches 0.6–1.0 mm long; stems, undersurface of the leaf blades, and especially fruits with peltate scales consisting of a relatively broad, fused central disc with a relatively short fringe of slender, stellate extensions beyond the disc . . . . 7. C. WILLDENOWII

1. Croton capitatus Michx. (woolly croton, hogwort)

Pl. 377 k–o; Map 1656

Plants monoecious, densely pubescent with short, stellate hairs, the branches 0.4–1.0 mm long, often slightly unequal. Stems 20–90 cm long, often sparsely to moderately alternately branched, but sometimes with a pronounced whorl of branches above the midpoint. Leaves all or mostly alternate, at least the lower and median leaves long-petiolate, the petiole without large, saucer-shaped glands at the tip. Leaf blades 2–12 cm long, lanceolate to oblong-lanceolate, triangular-lanceolate, or triangular-ovate, rounded or less commonly shallowly cordate at the base, rounded to angled or tapered to a bluntly or sharply pointed tip, the margins entire or slightly wavy below the midpoint, the undersurface usually paler than the upper surface. Inflorescences terminal at the branch tips (the uppermost branches sometimes short and these inflorescences then appearing axillary), short, dense, spikelike racemes (often appearing headlike or as dense clusters) with pistillate flowers toward the base and staminate flowers toward the tip. Staminate flowers with the calyx deeply 5-lobed, 0.8–1.2 mm long; the petals 5, 0.8–1.2 mm long, white to pale cream-colored; the stamens (7–)10–14. Pistillate flowers with the calyx 2–4 mm long at flowering, becoming enlarged to 6–9 mm long at fruiting, 6–9-lobed; the petals absent; the ovary 3-locular, the 3 styles each dichotomously lobed 2 or 3 times (the total number of stigmatic branches thus theoretically 12–24 per flower, but in practice mostly 12–16; the second and third order divisions usually well above the style base). Fruits 6–9 mm in length and diameter, nearly spherical, persistently densely hairy at maturity, 3-seeded (rarely 2-seeded by abortion of 1 ovule), dehiscent. Seeds 3.5–5.0 mm long, circular to oblong in outline, sometimes somewhat flattened, the caruncle present. $2n=20$. June–October.

Scattered nearly throughout the state but apparently absent from the far northwestern portion (eastern U.S. west to South Dakota and Texas; Mexico). Glades, upland prairies, and sand prairies; also pastures, dry ditches, old fields, fallow

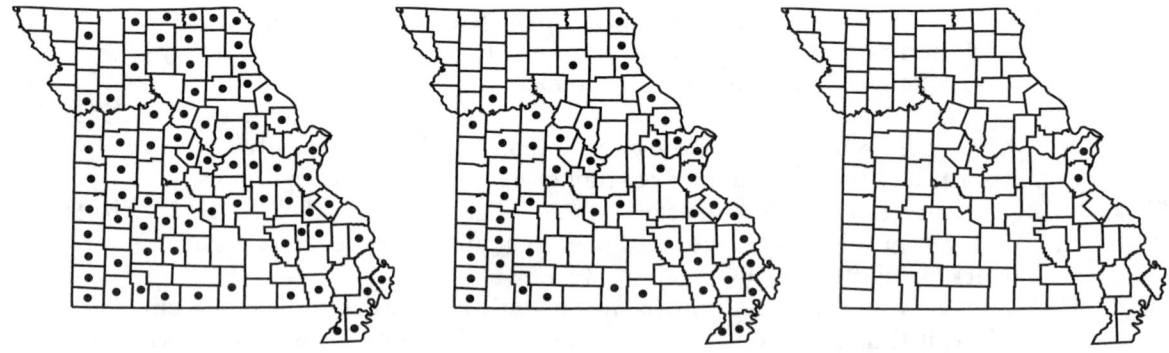

1656. Croton capitatus   1657. Croton glandulosus   1658. Croton lindheimerianus

fields, farmyards, railroads, roadsides, and open, sandy, disturbed areas.

Steyermark (1963) noted that cattle can become poisoned if they eat hay contaminated with this species, but that they tend to avoid living plants in pastures because of the bitter flavor that renders them relatively unpalatable. Burrows and Tyrl (2001) noted that the diterpenoid esters (croton oils) characteristic of the genus cause inflammation of the digestive tract.

Steyermark (1963) treated *C. capitatus* as comprising two varieties in Missouri. A third variety recognized by some authors (Johnston, 1958), var. *albinoides* (A.M. Ferguson) Shinners, grows in Texas and Mexico and differs only slightly from var. *lindheimeri* in its smaller (less than 4 mm long), more strongly obovate seeds. Johnston (1958) and Webster (1967) suggested, however, that the two taxa in Missouri might better be treated as distinct species. In spite of this, a number of seemingly intermediate specimens exist that appear to justify the use of a lower taxonomic rank. The characters used by Steyermark are also different than those used by Johnston, which may account for some of the ambiguously placed specimens, as Steyermark's use of leaf size and hair color does not appear to separate Missouri materials adequately.

1. Leaf blades with the tip rounded to bluntly or sharply pointed, if pointed then mostly tapered from about the midpoint (occasionally from the lower third in the largest leaves), occasionally rounded but with an abrupt, minute, sharp point; leaves all relatively long-petiolate, even those of the upper leaves; seeds about as long as wide (nearly circular in outline but somewhat flattened in profile), the surface smooth and usually not mottled
. . . . 1A. VAR. CAPITATUS

1. Leaf blades with the tip sharply pointed, tapered from near the base; lower leaves relatively long-petiolate, the petioles progressively shorter above the stem midpoint; seeds longer than wide (oblong in outline and sometimes somewhat flattened in profile), the surface finely but faintly roughened and usually mottled . . 1B. VAR. LINDHEIMERI

1a. var. capitatus

Leaves all relatively long-petiolate, those of the upper leaves not or only slightly shorter than those of the midstem. Leaf blades 2–8 cm long, lanceolate to oblong-lanceolate, mostly rounded at the base, the tip rounded to bluntly or sharply pointed, if pointed then mostly tapered from about the midpoint (occasionally from the lower third in the largest leaves), occasionally rounded but with an abrupt, minute, sharp point. Seeds about as long as wide (nearly circular in outline but somewhat flattened in profile), the surface smooth and usually not mottled. June–October.

Scattered nearly throughout the state but apparently absent from the far northwestern portion (eastern U.S. west to South Dakota and Texas). Glades and upland prairies; also pastures, dry ditches, old fields, fallow fields, railroads, roadsides, and open, sandy, disturbed areas.

1b. var. lindheimeri (Engelm. & A. Gray) Müll. Arg.
C. lindheimeri (Engelm. & A. Gray) A.W. Wood

Lower and median leaves relatively long-petiolate, the petioles progressively shorter above the stem midpoint. Leaf blades 3–12 cm long, lanceolate to triangular-lanceolate or triangular-ovate, rounded or shallowly cordate at the base, the tip sharply pointed, mostly tapered from near the base. Seeds longer than wide (oblong in outline and sometimes somewhat flattened in profile), the surface finely but faintly roughened and usually mottled. 2*n*=20. July–October.

Uncommon in the Mississippi Lowlands Division, west locally to Howell County; introduced in Moniteau County and the city of St. Louis (Texas to Missouri east to Tennessee and Florida; introduced in Kansas, Kentucky, and Indiana). Sand prairies; also pastures, dry ditches, old fields, fallow fields, farmyards, railroads, roadsides, and open, sandy, disturbed areas.

2. Croton glandulosus L. var. septentrionalis Müll. Arg. (sand croton, tropic croton)

Pl. 377 e–g; Map 1657

Plants monoecious, densely pubescent with short, mostly stellate hairs (the upper surface of the leaves sometimes with some of the hairs appearing unbranched or nearly so and spreading), the branches mostly 0.3–0.7 mm long and loosely appressed, but with 1 branch much longer (to 2 mm) and spreading. Stems 15–60 cm long, alternately branched but with some or most of the branches in irregular whorls. Leaves alternate, opposite, or whorled, short-petiolate, the petiole with 1 or 2 large, saucer-shaped (or 2-lipped), white to cream-colored glands at the tip. Leaf blades 1–7 cm long, narrowly oblong to oblong-lanceolate, oblong-oblanceolate, or less commonly oblong-ovate, mostly angled or tapered at the base at the base, mostly angled to a bluntly pointed tip (occasionally rounded or sharply pointed) tip, the margins finely toothed, the undersurface sometimes slightly paler than the upper surface. Inflorescences appearing terminal and between the stem branches, short, dense, spikelike racemes (often appearing as dense clusters) with pistillate flowers toward the base and staminate flowers toward the tip. Staminate flowers with the calyx deeply (4)5-lobed, 0.7–1.5 mm long; the petals (4)5, 1–2 mm long, white; the stamens 7–13. Pistillate flowers with the calyx 1.2–1.8 mm long at flowering, becoming enlarged to 3.5–4.5 mm long at fruiting, 5-lobed; the petals absent; the ovary 3-locular, the 3 styles each deeply 2-lobed (the total number of stigmatic branches thus 6 per flower). Fruits 3.5–5.5 mm in length, 4–5 mm in diameter, nearly spherical, 3-seeded (rarely 2-seeded by abortion of 1 ovule), dehiscent. Seeds 3–4 mm long, oblong-ovate to oblong-elliptic in outline, somewhat flattened and often slightly wedge-shaped, the caruncle present as a small knob. July–October.

Scattered nearly throughout the state but absent or uncommon in the western half of the Glaciated Plains Division (eastern U.S. west to Minnesota, Nebraska, and Texas). Glades, upland prairies, sand prairies, and openings of bottomland forests; also pastures, crop fields, fallow fields, ditches, levees, railroads, roadsides, and open, disturbed areas.

Infraspecific variation in *C. glandulosus* requires much more detailed study. Numerous varieties have been named in this widespread and variable species, which occurs from North America to South America and in the Caribbean Islands, but a comprehensive summary of variation in the species or even a key to determination of all of the varieties does not appear to exist. Johnston (1958) treated three varieties in Texas. Webster (1967) further noted the existence of several taxa described as species in Florida that are very closely related to *C. glandulosus,* some of which might better be classified as varieties. Missouri plants apparently correspond to var. *septentrionalis,* the northernmost phase, which is robust but with relatively small, narrower leaves having relatively sharply toothed margins, small seeds, and moderately coarse, spreading pubescence. Its relationship to the more tropical var. *glandulosus,* which apparently occurs as far north as Florida, remains to be elucidated. Interestingly, although the variety is considered of conservation concern in a few northern states, it is considered a crop weed in some southeastern states.

3. Croton lindheimerianus Scheele var. lindheimerianus (round-leaved woolly croton)

Pl. 378 h, i; Map 1658

Plants monoecious, densely pubescent with minute, stellate hairs, the branches 0.1–0.4 mm long, somewhat unequal and with 1 branch often raised from the surface. Stems 20–50 cm long, usually with 1 or few whorls of branches at or below the midpoint and repeatedly dichotomously branched above these. Leaves alternate, short- to long-petiolate, the petiole without large, saucer-shaped glands at the tip. Leaf blades 1–3(–6) cm long, nearly circular to ovate or oblong-ovate, rounded at the base, mostly rounded at the tip (sometimes bluntly pointed or with an abrupt, minute, sharp point), the margins entire or slightly wavy below the midpoint, the undersurface sometimes slightly paler than the upper surface. Inflorescences appearing terminal, axillary, and between the dichotomous upper stem branches, mostly short, dense, spikelike racemes (often appearing as dense clusters) with pistillate flowers toward the base and staminate flowers toward the tip. Staminate flowers with the calyx deeply (4)5-lobed, 1.5–2.5 mm long; the petals (4)5, 1.5–2.5 mm long, white; the stamens 7–9. Pistillate flowers with the calyx 2.5–3.0 mm long at flowering, becoming enlarged to 3.5–4.5 mm long at fruiting, 5-lobed; the petals absent;

1659. Croton michauxii   1660. Croton monanthogynus   1661. Croton texensis

the ovary 3-locular, the 3 styles each deeply 2-lobed (the total number of stigmatic branches thus 6 per flower). Fruits 4.0–5.2 mm in length, 4–5 mm in diameter, nearly spherical, 3-seeded (rarely 2-seeded by abortion of 1 ovule), dehiscent. Seeds 3.0–3.5 mm long, oblong-ovate to ovate in outline, slightly flattened, the caruncle present as a small knob. July–September.

Introduced, known thus far from Jefferson County and the city of St. Louis (Arizona east to Kansas and Arkansas; Mexico; introduced sporadically eastward to at least Indiana). Railroads and open, disturbed areas.

Although this species was not reported for Missouri until Viktor Mühlenbach (1979) documented it from the St. Louis railyards, a collection by Frederick Wislizenus from Kimmswick (Jefferson County) dating to July 1885 was discovered during the present research. In St. Louis, the species has resisted weed management efforts by the railroad companies and has become well established along some of the local tracks.

Most authors divide *C. lindheimerianus* into two varieties. The widespread (including Missouri) plants with relatively broad leaves, appressed hairs, and pendant fruits are var. *capitatus*. Johnston (1958) separated plants from Texas, Arizona, and Mexico with narrower leaves, somewhat shaggy pubescence, and ascending fruits as var. *tharpii* M.C. Johnst. Intermediates between the two varieties occur within the range of var. *tharpii* and even plants of var. *lindheimerianus* with so-called appressed pubescence frequently have one branch of each stellate hair raised from the surface.

4. Croton michauxii G.L. Webster (slender rushfoil)

Crotonopsis linearis Michx.

Pl. 378 c, d; Map 1659

Plants monoecious, moderately to densely pubescent with minute, peltate, scalelike trichomes (except on the upper surface of the leaf blades), these with a minute, raised, brown attachment point and a relatively slender, thin, white body, the slender, stellate extensions forming a noticeable fringe around the margins, this often slightly raised and thus providing a fuzzy appearance (especially on the fruits); the upper surface of the leaves with moderate minute, nonoverlapping, stellate hairs, the more or less equal branches 0.2–0.3 mm long. Stems (10–)15–40 cm long, usually sparsely alternately branched. Leaves alternate, sessile or very short-petiolate, the petiole without large, saucer-shaped glands at the tip. Leaf blades 1.0–2.5(–4.0) cm long, linear, angled or short-tapered at the base, angled or short-tapered to a sharply pointed tip, the margins entire, the undersurface paler than the upper surface. Inflorescences axillary, mostly elongate, loose spikes with 2–4(–6) pistillate flowers scattered toward the base and several staminate flowers toward the tip. Staminate flowers with the calyx deeply (4)5-lobed, 0.8–1.1 mm long; the petals (4)5, 0.6–1.0 mm long, white; the stamens 4–6. Pistillate flowers with the calyx 0.8–1.1 mm long at flowering, becoming very slightly enlarged at fruiting, (4)5-lobed; the petals absent; the ovary 1-locular, the 3 styles shallowly 2-lobed toward the tip. Fruits 2.5–3.0 mm in length and diameter, elliptic to oblong-ovate in outline, slightly flattened, 1-seeded, indehiscent, thin-walled. Seeds 2.5–3.0 mm long, elliptic to oblong-ovate in outline, slightly flattened, the caruncle absent. July–September.

Uncommon, known only from the Mississippi Lowlands Division and a single historical collection from St. Louis County (eastern [mostly southeastern] U.S. west to Iowa and Texas). Sand prairies, sand savannas; also roadsides and open, sandy, disturbed areas.

See the treatment of *C. willdenowii* for discussion of the transfer of *Crotonopsis* species into *Croton*.

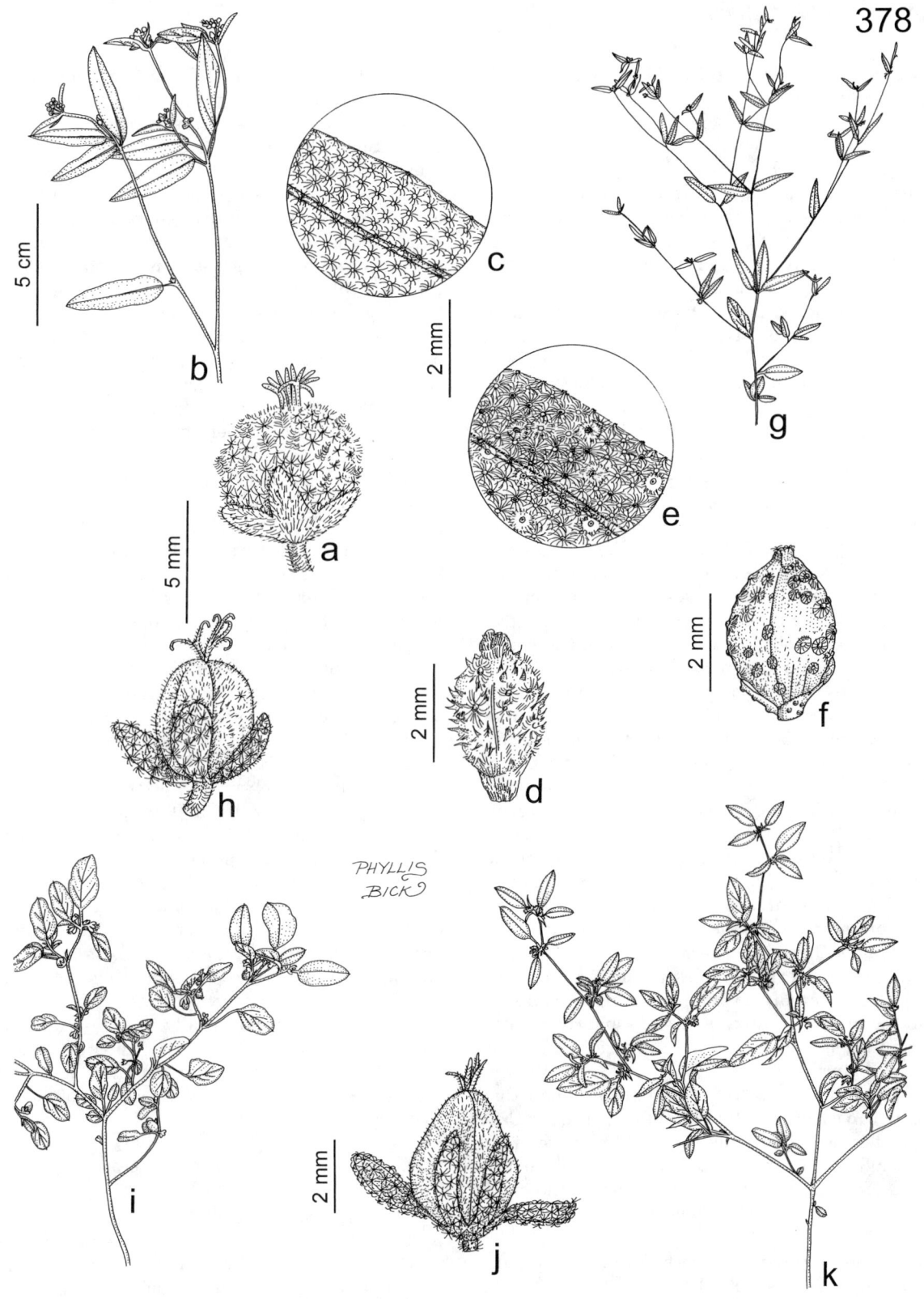

Plate 378. Euphorbiaceae. *Croton texensis*, **a)** fruit with calyx, **b)** fertile branch. *Croton michauxii*, **c)** leaf detail, **d)** fruit. *Croton willdenowii*, **e)** leaf detail, **f)** fruit, **g)** habit. *Croton lindheimerianus*, **h)** fruit with calyx, **i)** fertile branch. *Croton monanthogynus*, **j)** fruit with calyx, **k)** fertile branch.

5. **Croton monanthogynus** Michx. (one-seeded croton, prairie tea)

Pl. 378 j, k; Map 1660

Plants monoecious, densely pubescent with minute, stellate hairs (less densely on the upper surface of the leaf blades), the branches 0.2–0.3 mm long, sometimes slightly unequal and with 1 longer branch raised from the surface. Stems 20–40(–60) cm long, usually with 1 or few whorls of branches at or below the midpoint and repeatedly dichotomously branched above these. Leaves alternate, short- to long-petiolate, the petiole without large, saucer-shaped glands at the tip. Leaf blades 1–4 cm long, ovate to oblong-ovate or oblong-elliptic, rounded, angled or occasionally truncate at the base, mostly rounded at the tip (sometimes bluntly pointed or with an abrupt, minute, sharp point), the margins entire, the undersurface paler than the upper surface. Inflorescences appearing terminal, axillary, and between the dichotomous upper stem branches, mostly short, dense, spikelike racemes (often appearing as dense clusters) with pistillate flowers toward the base and staminate flowers toward the tip. Staminate flowers with the calyx deeply (3)4(5)-lobed, 1.5–2.5 mm long; the petals (4)5, 1.5–2.5 mm long, white; the stamens 3–8. Pistillate flowers with the calyx 1.5–2.0 mm long at flowering, becoming slightly enlarged to 2–3 mm long at fruiting, 5-lobed; the petals absent; the ovary 2-locular, the 2 styles each deeply 2-lobed. Fruits 3.5–4.0 mm in length, 2.5–3.5 mm in diameter, ovate in outline, not flattened, 1-seeded, dehiscent. Seeds 2.8–3.2 mm long, broadly elliptic to nearly circular in outline, slightly flattened, the caruncle present as a small knob. $2n=16$. May–September.

Scattered nearly throughout the state but absent or uncommon in the eastern half of the Glaciated Plains Division (eastern U.S. west to Nebraska and Arizona; Mexico). Glades, upland prairies, tops of bluffs, savannas, openings of dry upland forests, and occasionally banks of streams and rivers; also pastures, old fields, railroads, roadsides, and dry, open, disturbed areas.

Steyermark (1963) noted that, although this plant can poison livestock like other species of *Croton*, it is usually avoided by cattle because of its bitter flavor. He also noted that turkeys eat the seeds and that deer browse the foliage without apparent digestive tract problems.

6. **Croton texensis** (Klotzsch) Müll. Arg. (skunk weed, Texas croton)

Pl. 378 a, b; Map 1661

Plants dioecious, densely pubescent (sometimes only moderately so on the upper surface of the leaf blades) with more or less sessile, minute, stellate hairs, the branches 0.1–0.4 mm long, sometimes slightly unequal, the branches rarely fused toward the base and the hairs then appearing as peltate scales. Stems 20–90 cm long, often sparsely to moderately alternately branched, but sometimes with a whorl of branches above the midpoint. Leaves all or mostly alternate, mostly short-petiolate (the longest petioles less than ½ as long as the blade), the petiole without large, saucer-shaped glands at the tip. Leaf blades 1–5(–8) cm long, narrowly oblong to oblong-lanceolate or narrowly oblong-ovate, rounded or less commonly broadly angled at the base, rounded to angled or tapered to a bluntly or sharply pointed tip, the margins entire or slightly wavy below the midpoint, the undersurface usually paler than the upper surface. Inflorescences terminal at the branch tips (the uppermost branches sometimes short and these inflorescences then appearing axillary), short, dense, spikelike racemes (often appearing headlike or as dense clusters). Staminate flowers with the calyx 5-lobed, 1–2 mm long; the petals absent; the stamens 8–12. Pistillate flowers with the calyx 2.5–4.0 mm long at flowering, becoming slightly enlarged to 3–5 mm long at fruiting, 5-lobed; the petals absent; the ovary 3-locular, the 3 styles each divided nearly to the base into 4–6 lobes (the total number of stigmatic branches thus theoretically 12–18 per flower). Fruits 4–6 mm in length and diameter, nearly spherical, persistently densely hairy at maturity and sometimes finely warty, 3-seeded (rarely 2-seeded by abortion of 1 ovule). Seeds 3.5–4.0 mm long, oblong-ovate in outline, sometimes somewhat flattened, the caruncle present. May–October.

Introduced, uncommon, known only from historical collections from Jackson County (Wyoming to Arizona east to South Dakota and Texas; Mexico; introduced sporadically farther east). Banks of rivers; also railroads and open, sandy, disturbed areas.

This species is superficially similar in appearance to *C. capitatus*. According to Steyermark (1963), farther west it is a problem range plant and has caused cattle poisonings in portions of its native range.

7. **Croton willdenowii** G.L. Webster (common rushfoil)

Crotonopsis elliptica Willd.

Pl. 378 e–g; Map 1662

Plants monoecious, moderately to densely pubescent with small, peltate, scalelike trichomes (except on the upper surface of the leaf blades), these with a small, raised, brown attachment point and a relatively broad, thin, white body, the slender, stellate extensions forming a minute fringe around

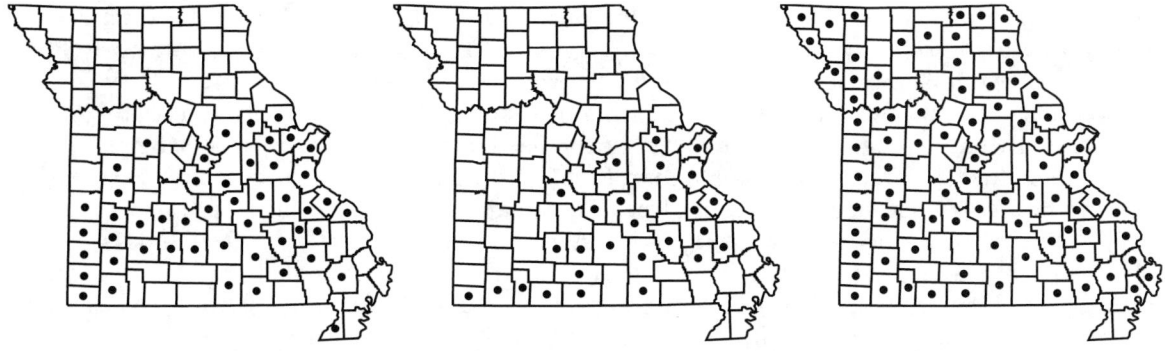

1662. Croton willdenowii 1663. Euphorbia commutata 1664. Euphorbia corollata

the margins, this appressed and not appearing fuzzy; the upper surface of the leaves with relatively dense, small, overlapping, stellate hairs, the often unequal branches 0.6–1.0 mm long. Stems 8–40 cm long, usually sparsely alternately branched. Leaves alternate, sessile or short-petiolate, the petiole without large, saucer-shaped glands at the tip. Leaf blades 0.7–2.0(–3.0) cm long, linear to narrowly ovate, angled or short-tapered at the base, rounded to angled or short-tapered to a usually sharply pointed tip, the margins entire, the undersurface paler than the upper surface. Inflorescences axillary, mostly short, loose spikes with 1 or 2 pistillate flowers at the base and several staminate flowers toward the tip. Staminate flowers with the calyx deeply (4)5-lobed, 0.8–1.1 mm long; the petals (4)5, 0.6–1.0 mm long, white; the stamens 4–6. Pistillate flowers with the calyx 0.8–1.1 mm long at flowering, becoming very slightly enlarged at fruiting, (4)5-lobed; the petals absent; the ovary 1-locular, the 3 styles shallowly 2-lobed toward the tip. Fruits 2.5–3.0 mm in length and diameter, elliptic to oblong-ovate in outline, slightly flattened, 1-seeded, indehiscent, thin-walled. Seeds 2.5–3.0 mm long, elliptic to oblong-ovate in outline, slightly flattened, the caruncle absent. June–September.

Scattered mostly south of the Missouri River (eastern U.S. west to Iowa, Kansas, and Texas). Glades, ledges and tops of bluffs, and openings of dry upland forests; also old fields; usually in nutrient-poor acidic soils.

This is the more widespread of the two rushfoils in Missouri. Previously, it and *C. michauxii* were classified into a separate genus of two species, *Crotonopsis,* based on their 1-seeded, indehiscent fruits. However, Webster (1992) argued that *Crotonopsis* represents merely a specialized subgroup within *Croton* related to the group of species that includes *C. monthogynous,* which has two-seeded fruits. In combining the genera, he found it necessary to coin replacement names for both species, as the original epithets from *Crotonopsis* already were in use for other species of *Croton* (*C. ellipticus* Geiseler and *C. linearis* Jacq.). Webster's view has been supported by preliminary molecular phylogenetic analyses (Berry et al., 2005), in which the two species of former *Crotonopsis* are nested within a more derived lineage within *Croton.*

4. Euphorbia L. (spurge, euphorbia)

Plants annual or perennial herbs (shrubs or trees elsewhere), monoecious but the flowers densely clustered into headlike units and appearing perfect, with milky sap, glabrous or pubescent with unbranched, nonglandular hairs; stinging hairs absent. Stems solitary to several, prostrate to erect, unbranched or more commonly branched, the branches alternate, opposite, or whorled. Leaves alternate, opposite, or whorled, sessile or short-petiolate, the petiole lacking glands, attached at the base of the nonpeltate (but sometimes cordate-clasping) blade. Leaf blades variously shaped, the margins entire, finely toothed, or pinnately few-lobed, appearing pinnately veined or with only the midvein apparent. Stipules absent or, if present, then either minute (less than 1 mm long), triangular scales (these usually persistent at flowering) or small glandular dots. Inflorescences with the basic unit a cyathium consisting of 1 central pistillate flower and several to numerous staminate flowers in a small, cup-shaped involucre with 1–5 conspicuous glands along the rim, the cyathia terminal, axillary, and/or from the stem branch points, appearing solitary or more commonly varying from small clusters to relatively large panicles. Flowers lacking a nectar disc (but the cyathium glands producing nectar), without a calyx and corolla. Staminate flowers consisting of 1 small stamen, this with a minute, jointed stalk, usually subtended by a minute bract at the base. Pistillate flowers consisting of 1 pistil, this with a short, jointed stalk (longer than those of the staminate flowers); the ovary with 3 locules and 1 ovule per locule, the styles free or fused at the very base, each shallowly to deeply 2(3)-lobed. Fruits somewhat 3-lobed (bluntly triangular in cross-section), often depressed-globose (slightly wider than long), dehiscent. Seeds nearly circular to oblong or oblong-ovate in outline, usually somewhat flattened in profile, sometimes slightly wedge-shaped, the caruncle absent or a small, light-colored knob or somewhat flattened bump at the end adjacent to the attachment point, the surface variously smooth, wrinkled, pebbled, or with minute warts or tubercles, reddish brown to dark brown, often all or partially covered with a thin, white to light gray, glaucous coating. In the broad sense, 1,500–2,000 species, worldwide.

The inflorescence unit known as the cyathium is unique in the Missouri flora. In a family noted for its small, inconspicuous flowers, the members of the tribe Euphorbieae have achieved the most extreme floral reduction imaginable. Staminate and pistillate flowers consist of a tiny solitary stamen and pistil respectively, without any calyx or corolla. However, as in other families in which individual flowers have become reduced to the point where they are ineffective at attracting insect pollinators, the flowers of Euphorbieae have become aggregated into dense clusters, each of which is associated with an involucre and mimics the function of a larger, showier flower. In the cyathium, the associated bracts are fused into a shallowly to deeply cup-shaped involucre containing a single central pistillate flower that appears stalked and several series of minutely stalked staminate flowers (each subtended by a very minute bract) radiating outward from the center. The rim of the involucre has 1–5 conspicuous glands of varying color and shape, and in some groups each of the glands has a white or pinkish-tinged petaloid appendage. The entire unit thus mimics a single flower. Because most of the Missouri species have small cyathia, care must be taken to recognize the nature of the floral assemblage to avoid miskeying the group.

Generic limits in the tribe Euphorbieae have remained controversial. Traditionally, the large and morphologically variable genus *Euphorbia* was regarded as comprising several well-marked subgenera but with a large number of residual unclassified species groups (Steyermark, 1963; Gleason and Cronquist, 1991; see also discussion in Webster [1967]). Over time, the trend became to recognize some of the most easily recognized infrageneric groups as segregate genera, especially *Chamaesyce* Gray (species 8–11, 13, 14, 16–18, 20 below; Webster, 1967, 1994; Yatskievych and Turner, 1990). Recent phylogenetic studies involving morphological characters (Park and Elisens, 2000) and molecular sequence data (Steinmann and Porter, 2002; Wurdack et al., 2005) have resulted in the recognition that, on the one hand, some of the segregate genera (like *Poinsettia* Graham) are unnatural assemblages of species related to different groups within *Euphorbia* in the broad sense, and, on the other hand, the removal of various segregate genera results in an unnatural classification of those species remaining in *Euphorbia*, with various species of *Euphorbia* in the strictest sense more closely related to some of the segregates than to other species of *Euphorbia*. The practical consequence of using a broad circumscription of *Euphorbia* to include nearly all of the members of the family that produce cyathia is that the genus becomes one of the largest genera of flowering plants in the world and includes incredible morphological diversity. In the future, we may expect further refinement of the classification within the tribe that will result in the breakup of *Euphorbia* into smaller generic units, but these genera will in most cases be circumscribed differently than the traditional segregates.

In the broad sense, *Euphorbia* includes plants ranging from tiny annuals to large trees. In dry portions of the Old World, succulent members of the genus have independently evolved growth forms convergent with those found in cacti (a nearly entirely New World group), differing conspicuously in their milky latex, paired (stipular) spines, and minute, imperfect flowers in cyathia. A large number of succulent euphorbias are grown as specimen plants by succulent plant enthusiasts, and some of the hardier woody and succulent species are cultivated outdoors in the southern United States. The genus also includes the cultivated poinsettia (*E. pulcherrima* Willd. ex Klotzsch), which is an economically important ornamental in the United States during the Christmas season. Other commercial uses include waxes removed by boiling from the stem surfaces of some species, including candelilla (*Euphorbia antisyphilitica* Zucc.), a native of the Chihuahuan Desert region. These waxes are used primarily in cosmetics and skin care products, varnishes, and polishes for shining certain kinds of leather. A number of species have been used variously for medicinal purposes around the world. Among the negative economic impacts, some species of *Euphorbia* are considered noxious weeds. In the northern and western United States, the invasive exotic leafy spurge (the Eurasian *E. esula* L.) aggressively outcompetes native grassland species, impacting the native biodiversity of more than 2.5 million acres and also rendering these areas less productive for grazing by cattle (Dunn, 1979, 1985). The spurges are notable for containing abundant and diverse diterpenoid esters that result in the plants potentially causing irritation to the skin, mucus membranes, and digestive tract. However, as noted by Burrows and Tyrl (2001), reports of severe poisoning by members of the genus are to some extent exaggerated.

Gleason and Cronquist (1991) and Kartesz and Meacham (1999) included Missouri in the range of *E. hexagona* Nutt. ex Spreng. (six-angled spurge). However, to date, no voucher specimens to substantiate this claim have been discovered. The source of confusion may be a report by Henry and Scott (1983) of an introduced population of the species from Mercer County (but Illinois, not Missouri). Although these authors included a statement in their discussion that *E. hexagona* had not been reported from Missouri, later authors unfamiliar with the local geography may have misinterpreted the report. *Euphorbia hexagona* occurs from Montana to New Mexico east to South Dakota, Iowa, Kansas, and Arkansas, and thus may eventually be discovered in western Missouri. The species is not particularly closely related to any of the Missouri spurges but resembles *E. corollata* superficially. It differs from that species in its relatively delicate annual habit, opposite leaves that are sharply pointed at the tip, and seeds with the surface pebbled, roughened, or finely tuberculate. Plants of *E. hexagona* would not key well to any of the Missouri species in the following key.

1. At least the uppermost leaves with conspicuous, broad, white margins, these occasionally slightly pinkish-tinged toward the outer edge .... 12. E. MARGINATA
1. None of the leaves with broad, white margins (the inflorescence bracts often bright red to reddish purple, at least toward the base, in *E. cyathophora*)
    2. Stem leaves alternate above the lowest node and below the inflorescence branches
        3. Inflorescence not an umbellate panicle with a whorl of leaves at the base, instead consisting of small clusters (these frequently paired) at the stem or branch tips; uppermost leaves sometimes pinnately few-lobed and bright red to reddish purple, at least toward the base; cyathia with the involucre having only 1(2) more or less 2-lipped marginal gland(s) .... 3. E. CYATHOPHORA
        3. Terminal portion of the inflorescence an umbellate panicle with a whorl of leaves at the base (additional smaller inflorescences often produced on branches below the main umbellate inflorescence); leaves all unlobed, the margins entire or finely toothed; cyathia with the involucre having 4 or 5 oval to crescent (viewed from the top) marginal glands

4. Leaves with the margins finely toothed (the teeth sometimes minute and visible only with magnification); cyathia with the involucre having oblong-oval to elliptic or nearly circular (viewed from the top) marginal glands
    5. Seeds 1.7–2.3 mm long, the surface smooth; cyathia with the involucre 1.2–1.5 mm long . . . . 15. E. OBTUSATA
    5. Seeds 1.3–1.7 mm long, the surface with a fine network of low ridges; cyathia with the involucre 0.6–0.9 mm long
       . . . . 19. E. SPATHULATA
4. Leaves with the margins entire; cyathia with the involucre having kidney-shaped to crescent or linear to narrowly oblong (viewed from the top) marginal glands, these sometimes with the central portion appearing tapered into a pair of horns
    6. Leaves closely spaced (crowded, especially toward the stem tip), those below the inflorescence linear to narrowly oblanceolate
       . . . . 4. E. CYPARISSIAS
    6. Leaves more widely spaced (not appearing crowded), lanceolate, oblong, obovate, or nearly circular, rarely a few of the leaves narrowly lanceolate in *E. corollata*
        7. Leaves along the inflorescence branches narrowly elliptic to elliptic or narrowly ovate (much longer than wide), not cupped around the cyathia; cyathia with the involucre having narrowly elliptic to nearly linear (viewed from the top) marginal glands, these lacking horns, but with showy, white petaloid appendages . . . . 2. E. COROLLATA
        7. Leaves along the inflorescence branches kidney-shaped to broadly triangular-ovate or broadly ovate (slightly longer than wide to somewhat wider than long), somewhat cupped around the cyathia; cyathia with the involucre having kidney-shaped to crescent (viewed from the top) marginal glands, these usually with the central portion appearing tapered into a pair of horns, but lacking petaloid appendages
            8. Stems 10–40 cm long; leaves 5–30 mm long; surface of the fruits smooth; seeds with the surface strongly pitted
               . . . . 1. E. COMMUTATA
            8. Stems 30–90 cm long; leaves (excluding the inflorescence bracts) 30–100 mm long; surface of the fruits finely warty or appearing roughened, especially on the angles; seeds with the surface smooth . . . . 7. E. ESULA
2. Stem leaves all or mostly opposite
    9. Cyathia with the involucre having a more or less fringed margin and only 1 or 2 more or less 2-lipped marginal gland(s), these yellow to greenish yellow, lacking petaloid appendages; stems pubescent with a mixture of dense, minute, downward-curved hairs and scattered, relatively long, spreading to downward-angled hairs; leaves short- to long-petiolate, the blade symmetrically tapered at the base, the margins coarsely toothed to nearly entire, mostly 15–60 mm long; seeds 2–3 mm long
        10. Leaf blades with the undersurface moderately pubescent with relatively stout hairs, these often with a minute, persistent pustular base; seeds angled in cross-section (both the oblique apical portion surrounding the caruncle and the longitudinal inner faces appearing angular), the surface

appearing relatively coarsely wrinkled or with poorly differentiated, low, broad warts (appearing lumpy or irregularly swollen) .... 5. E. DAVIDII
10. Leaf blades with the undersurface sparsely to moderately pubescent with relatively slender hairs, these not expanded at the base; seeds rounded in cross-section (the oblique apical portion surrounding the caruncle angled but the longitudinal inner faces appearing rounded), the surface appearing relatively finely warty or with fine tubercles............. 6. E. DENTATA
9. Cyathia with the involucre having a shallowly lobed or toothed rim and 4 or less commonly 5 unlobed marginal glands, these sometimes reddish purple to dark purple (sometimes green to yellowish green), with white or pinkish- to reddish-tinged petaloid appendages; stems glabrous or, if pubescent, then variously with either dense, minute, appressed or moderate to dense, upward-curved hairs; leaves sessile or very short-petiolate, the blade truncate to more or less rounded or angled at the base, often noticeably asymmetrically so (except in *E. geyeri* and *E. missurica*), the margins entire or finely toothed, 2–30 mm long; seeds 0.8–2.0 mm long
11. Stems and leaves glabrous
12. Leaf blades with the margins minutely toothed (best viewed with magnification), usually only above the midpoint
13. Seeds with 3 or 4(–6) coarse transverse ridges; involucres 0.6–0.9 mm long, each with 1–5 staminate flowers surrounding the solitary pistillate flower; stems not flattened or winged toward the tip
.... 9. E. GLYPTOSPERMA
13. Seeds smooth or with 1–4 indistinct, low cross-wrinkles, rarely appearing faintly roughened or pitted; involucres 0.8–1.2 mm long, each with 5–18 staminate flowers surrounding the solitary pistillate flower; stems often appearing somewhat flattened or narrowly winged toward the tip .... 18. E. SERPYLLIFOLIA
12. Leaf blades entire
14. Stems erect or ascending; leaf blades linear to narrowly oblong
.... 13. E. MISSURICA
14. Stems prostrate; leaf blades oblong or broadly oblong to broadly elliptic or nearly circular
15. Stipules from the adjacent leaf in each pair not fused into a single, small, scalelike structure on each side of the stem (or rarely fused on only 1 side at a few nodes), thus a pair of minute, free stipules usually positioned on each side of the stem between the leaf bases, the stipules often appearing irregularly fringed or lobed .... 8. E. GEYERI
15. Stipules from the adjacent leaf in each pair fused into a single, small, scalelike structure on each side of the stem positioned between the leaf bases, this often appearing irregularly fringed or lobed .... 17. E. SERPENS
11. Stems hairy, at least toward the tip, the pubescence sometimes in longitudinal bands on opposite sides of the stem; leaves usually hairy, at least when young (sometimes becoming glabrous at maturity), sometimes only near the base
16. Ovaries and fruits glabrous; stems erect or more commonly ascending, often arched at the branch tips .... 14. E. NUTANS
16. Ovaries and fruits hairy; stems mostly prostrate (sometimes loosely ascending near the tips), usually mat-forming

17. Styles entire or inconspicuously notched at the very tip; seeds with the surface usually mottled, finely pitted, some of the pits rarely forming shallow, irregular troughs and the seeds then appearing partially and irregularly few-ridged
 .... 20. E. STICTOSPORA
17. Styles 2-lobed or divided at least 1/4 of the way from the tip; seeds with the surface not mottled (sometimes all or partially with a thin, white coating in *E. prostrata*), smooth, roughened, or with cross-ridges
   18. Styles about 0.1 mm long, each deeply lobed nearly to the base; fruits moderately to densely pubescent with more or less spreading hairs toward the angles, less densely hairy to nearly glabrous between the angles; seeds with 4–7 relatively sharp, slender cross-ridges .... 16. E. PROSTRATA
   18. Styles 0.3–0.7 mm long, divided 1/4–1/2 of the way from the tip; fruits sparsely to moderately and relatively evenly pubescent with appressed or strongly incurved hairs; seeds with the surface smooth, finely roughened, or with 3 or 4 low, broadly rounded cross-wrinkles (species difficult to distinguish)
      19. Styles 0.5–0.8 mm long, divided about 1/2 of the way to the base; seeds with the surface smooth or appearing finely roughened, lacking cross-ridges... 10. E. HUMISTRATA
      19. Styles about 0.3–0.4 mm long, divided 1/4–1/3 of the way to the base; seeds with 3 or 4 low, broadly rounded cross-ridges .... 11. E. MACULATA

1. Euphorbia commutata Engelm. ex A. Gray
 (wood spurge)
Tithymalus commutatus (Engelm. ex A. Gray) Klotzsch & Garcke
E. commutata var. *erecta* Norton

Pl. 380 c–e; Map 1663

Plants perennial herbs, with fibrous roots. Stems 10–40 cm long, ascending from a frequently spreading base, often branched from near the base (spreading to loosely ascending overwintering shoots often produced in the late summer and autumn, these producing new growth the following spring), otherwise often unbranched below the inflorescence, the branches not flattened toward the tip, green to yellowish green, usually strongly reddish- or purplish-tinged especially toward the base, glabrous. Leaves alternate above the lowermost node and below the inflorescence branches (those of the inflorescence branches opposite) not appearing crowded, sessile or rarely minutely petiolate (those of the overwintering shoots mostly short-petiolate). Stipules absent. Leaf blades 5–30 mm long, unlobed, the margins entire, the surfaces glabrous, green to yellowish green, especially those of the lower leaves often strongly reddish- or purplish-tinged; those below the inflorescence oblanceolate to obovate, tapered at the base, rounded or abruptly short-tapered to a minute, sharply pointed tip; those along the inflorescence branches kidney-shaped to broadly triangular-ovate or broadly ovate (slightly longer than wide to somewhat wider than long), broadly rounded to cordate and clasping the stem or occasionally perfoliate, the terminal ones somewhat cupped around the cyathia, rounded or very broadly angled to a bluntly pointed tip. Inflorescences terminal, often umbellate panicles with opposite or whorled leaves at the base and each of the (2)3(4) primary branches usually branched 1–3 additional times, the cyathia solitary at the branch points and solitary or more commonly in small clusters at the branch tips. Involucre 1.7–2.5 mm long, glabrous, the rim shallowly 4-lobed, the margin usually minutely and inconspicuously hairy, the marginal glands 4, 0.8–1.3 mm long, more or less crescent, the oblong body tapered into a pair of slender, spreading horns, yellow, lacking petaloid appendages. Staminate flowers 9–15 per cyathium. Ovaries glabrous, the styles 0.9–1.3 mm long, each divided 1/2–2/3 of the way from the tip into 2 relatively slender lobes. Fruits 2.7–3.5 mm long, glabrous, smooth. Seeds 1.5–2.0 mm long, broadly oblong-elliptic to broadly ovate or nearly circular in outline, nearly circular in cross-section, rounded or slightly angled at the base, the surface strongly pitted, dull gray to olive gray, with a pale, irregularly winglike caruncle (shaped similar to a tiny fortune cookie). $2n=28$. April–June.

Scattered in the Ozark and Ozark Border Divisions (eastern U.S. to Iowa and Texas; Canada). Bottomland forests, mesic to dry upland forests, bases and ledges of bluffs, banks of streams and rivers, edges of glades, and rarely edges of fens.

This dainty species is relatively uniform in its morphology across its range. Some plants from the southeastern portion of the range (northward into Missouri) tend to have more persistent leaves on the overwintering shoots. They have been called var. *erecta,* but this variant is not sufficiently dis-

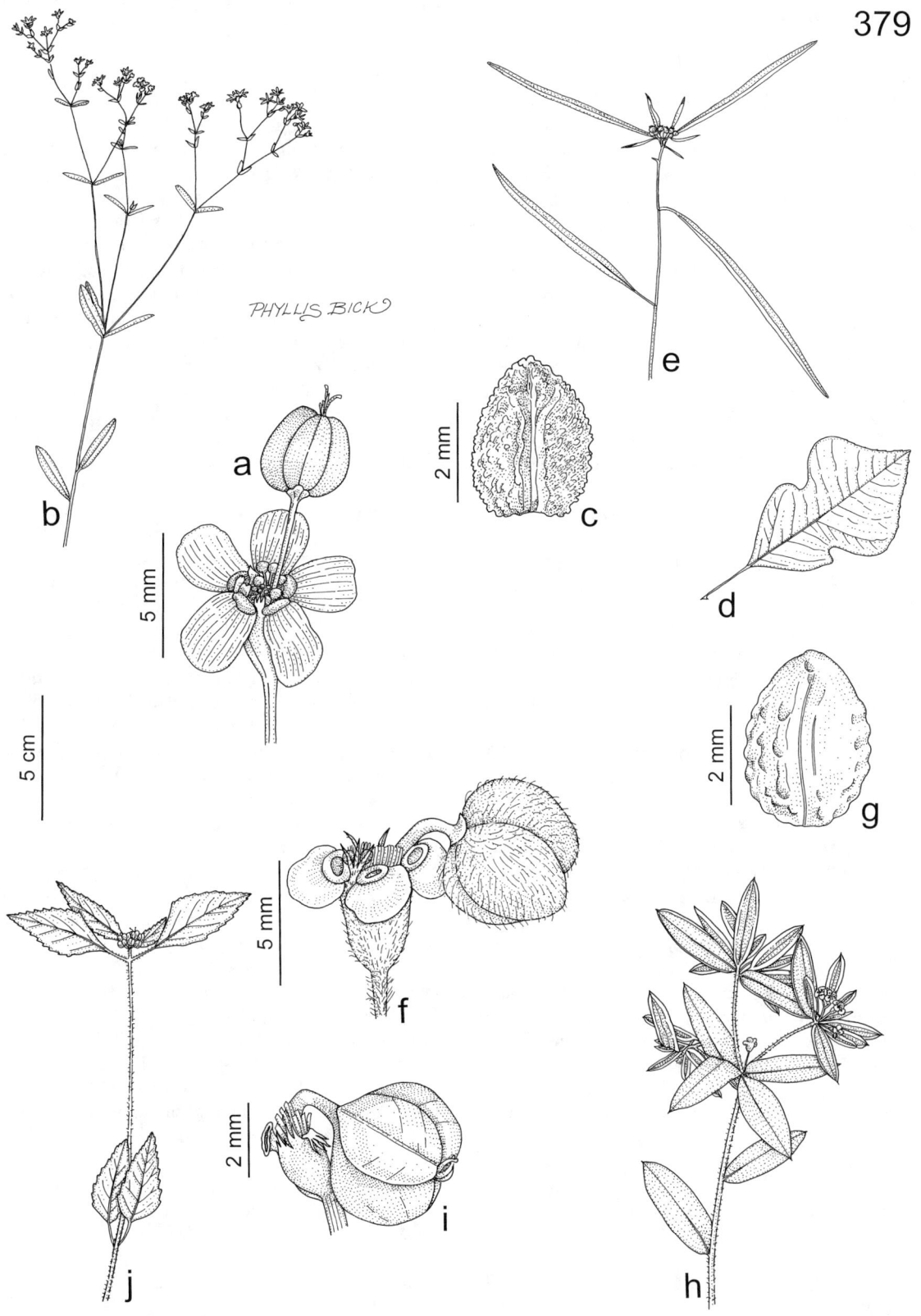

Plate 379. Euphorbiaceae. *Euphorbia corollata*, **a)** cyathium with fruit, **b)** habit. *Euphorbia cyathophora*, **c)** seed, **d)** lobed leaf, **e)** fertile branch with unlobed leaves. *Euphorbia marginata*, **f)** cyathium with fruit, **g)** seed, **h)** fertile branch. *Euphorbia dentata*, **i)** cyathium with fruit, **j)** fertile branch.

tinct to warrant recognition. However, some populations occurring from the Ouachita Mountains of eastern Oklahoma and adjacent west-central Arkansas northward to Barry County, Missouri, may represent an undescribed cryptic taxon. These plants, which currently are under study by Mark Mayfield of Kansas State University, differ from typical *E. commutata* in their more delicate habit and smaller, more deeply pitted, more oblong seeds. True *E. commutata* also tends to produce overwintering vegetative growth with distinctly petiolate oblanceolate leaves that commonly wither in the middle of spring, leaving a set of closely spaced leaf scars at the base (this phenomenon apparently is absent in the other taxon).

2. Euphorbia corollata L. (flowering spurge)

Agaloma corollata (L.) Raf.
Tithymalopsis corollata (L.) Klotzsch & Garcke
E. corollata var. *mollis* Millsp., misapplied
E. corollata var. *paniculata* (Elliott) Boiss., misapplied

Pl. 379 a, b; Map 1664

Plants perennial herbs, with a deep, spreading, usually branched rootstock. Stems 20–100 cm long, erect or ascending, unbranched below the inflorescence or occasionally few-branched, the branches not flattened toward the tip, green to yellowish green or tan, rarely reddish- or purplish-tinged, glabrous or less commonly sparsely to moderately pubescent with slender, mostly spreading hairs. Leaves alternate below the inflorescence branches (those of the inflorescence branches alternate, opposite, or whorled) not appearing crowded, sessile or rarely minutely petiolate. Stipules absent or a pair of minute, light brown, convex to conical, sessile glands. Leaf blades 10–60 mm long, unlobed, the margins entire, the surfaces glabrous or sparsely to moderately pubescent with slender, mostly spreading hairs, green to bright green; those below the inflorescence narrowly lanceolate to more commonly oblong, oblong-oblanceolate or elliptic, mostly tapered at the base, rounded or short-tapered to a sharply pointed tip; those along the inflorescence branches narrowly elliptic to elliptic or narrowly ovate (much longer than wide), rounded or angled at the base, angled to a bluntly or sharply pointed tip. Inflorescences terminal umbellate panicles with a whorl of leaves at the base and each of the 3–6 (–10) primary branches usually branched 3–5 additional times, the cyathia solitary at the branch tips and branch points. Involucre 1.2–1.8 mm long, glabrous or the outer surface minutely and inconspicuously hairy toward the rim, the rim shallowly 5-lobed, the margin minutely fringed or hairy, the marginal glands 5, 0.5–1.0 mm long, narrowly elliptic to nearly linear, yellowish green to green, with a relatively conspicuous petaloid appendage 1.5–4.5 mm long, this white. Staminate flowers 10–21 per cyathium. Ovaries glabrous, the styles 0.7–1.0 mm long, each divided $1/3$–$1/2$ of the way from the tip into 2 relatively stout lobes, these sometimes somewhat club-shaped. Fruits 2.5–4.5 mm long, glabrous. Seeds 2.5–2.8 mm long, oblong-elliptic to broadly ovate in outline, nearly circular in cross-section, rounded or somewhat angled at the base, the surface smooth or with shallow, faint pits, purplish brown but usually with a thin, white coating, lacking a caruncle. $2n$=26, 28, 30. May–October.

Scattered to common throughout the state (eastern U.S. west to South Dakota and Texas; Canada). Upland prairies, glades, ledges and tops of bluffs, savannas, openings of mesic to dry upland forests, and banks of streams and rivers; also pastures, old fields, fallow fields, fencerows, railroads, roadsides, and rarely open, disturbed areas.

More so than in any of the other species of *Euphorbia* in Missouri, in this attractive species the cyathia with their showy appendages resemble individual normal flowers, and the true nature of the cyathium is apparent only with close observation. Steyermark (1963) treated *E. corollata* as comprising three varieties differing in stem and leaf pubescence and cyathium size and density. However, the names that he used, var. *mollis* Millsp. and var. *paniculata* (Elliott) Boiss., technically are associated with a different species, the southeastern *E. pubentissima* Michx., which occurs no closer to Missouri than central Tennessee and northern Louisiana. That species differs from *E. corollata* in its shorter, more slender stems, usually downward-angled, mostly short-petiolate leaves, and involucral glands with somewhat smaller, oval to nearly circular petaloid appendages. Both Huft (1979) and Park (1998) included the epithets var. *mollis* and var. *paniculata* as synonyms of *E. pubentissima*, and both of these monographers of the group treated *E. corollata* and *E. pubentissima* as inherently variable species in which formal taxonomic recognition of varieties was not advisable.

3. Euphorbia cyathophora Murray (painted leaf, fire-on-the-mountain)

Poinsettia cyathophora (Murray) Klotzsch & Garcke
P. cyathophora var. *graminifolia* (Michx.) Mohl.
E. heterophylla L. var. *graminifolia* (Michx.) Engelm.

Pl. 379 c–e; Map 1665

Plants annual, with taproots. Stems 15–100 cm

1665. Euphorbia cyathophora   1666. Euphorbia cyparissias   1667. Euphorbia davidii

long, erect or ascending, unbranched or few- to several-branched, the branches not flattened toward the tip, usually green to yellowish green, occasionally reddish- to purplish-tinged, glabrous or with sparse multicellular hairs around the nodes. Leaves alternate above the lowest node and below the inflorescence branches (those of the lowermost node and the inflorescence branches usually opposite), mostly short-petiolate. Stipules absent or a pair of minute, light brown, convex, sessile glands. Leaf blades 15–150 mm long, highly variable in shape, linear to lanceolate, elliptic, ovate, or broadly elliptic, those of the upper leaves sometimes pinnately few-lobed (more or less fiddle-shaped), more or less symmetrically angled or tapered at the base, rounded or angled to tapered to a sharply pointed tip, the margins entire or toothed, the upper surface glabrous, bright green and (on the uppermost leaves) sometimes with a bright red to reddish purple (rarely pink, yellow, or white) region toward the base, the undersurface glabrous or sparsely pubescent with relatively stout, multicellular hairs, light green to pale green. Inflorescences terminal at the branch tips (not an umbellate panicle with a whorl of leaves at the base), of solitary or more commonly paired cyathia, sometimes appearing as small clusters. Involucre 2.0–2.5 mm long, glabrous, the rim irregularly lobed and fringed, the marginal glands 1 or less commonly 2, 0.7–1.5 mm long, appearing strongly concave and more or less 2-lipped, yellowish green to yellowish brown, lacking a petaloid appendage. Staminate flowers 30–50 per cyathium. Ovaries glabrous, the styles 0.8–1.1 mm long, each divided ½–⅔ of the way from the tip into 2 slightly club-shaped lobes. Fruits 3–4 mm long (nearly twice as broad), glabrous. Seeds 2.5–3.0 mm long, ovate to oblong-ovate in outline, more or less circular in cross-section, more or less flattened to slightly concave at the base, the surface with a network of low, sharp ridges or wrinkles and low, pointed tubercles, dark brown with the tips of the ridges and tubercles lighter brown, usually lacking a caruncle, a minute, discolored, slightly raised area occasionally present. $2n=28, 56$. July–October.

Scattered mostly south of the Missouri River (California to Florida north to Utah, South Dakota, Ohio, and Maryland [introduced in much of the western and northern portions of the range]; Mexico, Central America, South America, Caribbean Islands; introduced in Hawaii, Asia). Banks of streams and rivers, bases of bluffs, and bottomland forests; also fallow fields, old fields, gardens, roadsides, and open, disturbed areas.

This species was long known to midwestern botanists by the name *E. heterophylla* (*Poinsettia heterophylla* (L.) Klotzsch & Garcke). Dressler (1961) noted that true *E. heterophylla* is a different (Neotropical) species and that temperate North American plants are properly known as *E. cyathophora*. Botanists who have studied *E. cyathophora* have all remarked upon the extreme morphological plasticity within individual plants for characters such as leaf shape and coloration. Steyermark (1963) did not know the species from the Mississippi Lowlands Division, but it has become increasingly common in southern Missouri over the last few decades.

4. Euphorbia cyparissias L. (cypress spurge, graveyard spurge)

Pl. 380 a, b; Map 1666

Plants perennial herbs, with a fleshy rootstock and usually rhizomes. Stems 15–40 cm long, erect or ascending, unbranched below the inflorescence or often with several short, densely leafy vegetative branches in the median and upper leaf axils, the branches not flattened toward the tip, usually green to yellowish green, sometimes reddish- or purplish-tinged toward the base, glabrous. Leaves alternate below the inflorescence branches (those of the inflorescence branches usually opposite) but closely spaced (crowded, especially toward the stem tip), sessile. Stipules absent. Leaf blades 10–30 mm long, unlobed, the margins entire, the surfaces glabrous, yellowish green to green; those below the inflorescence linear to narrowly oblanceolate, rounded or angled at the base, mostly angled or short-tapered to a sharply pointed tip; those along the inflorescence branches broadly ovate to nearly heart-shaped or somewhat kidney-shaped, rounded to cordate at the base, mostly broadly angled to a sharply pointed tip. Inflorescences terminal umbellate panicles with a whorl of leaves at the base and with each of the up to 10 primary branches sometimes branched 1 or 2 additional times, the cyathia solitary at the branch tips. Involucre 2–3 mm long, glabrous, the rim shallowly 4-lobed to nearly entire, the marginal glands 4, 0.8–1.3 mm long, crescent with each end appearing as a short, outward-curved horn, greenish yellow to yellow, lacking a petaloid appendage. Staminate flowers 15–25 per cyathium. Ovaries glabrous, but the surface pebbled or minutely warty, the styles 0.5–1.0 mm long, each divided $1/5$–$1/4$ of the way from the tip into 2 somewhat club-shaped lobes. Fruits produced in relatively few cyathia, 2.5–3.0 mm long, glabrous but the surface pebbled or finely warty, especially toward the sutures. Seeds 1.5–2.0 mm long, oblong-elliptic in outline, nearly circular in cross-section, rounded or somewhat angled at the base, the surface smooth, grayish brown to silvery gray, with a small, nipplelike caruncle. $2n=20, 40$. April–August.

Introduced, scattered to uncommon, mostly in counties adjacent to the Missouri and Mississippi River floodplains (native of Europe, Asia, introduced widely but sporadically in the U.S., Canada). Glades, ledges and tops of bluffs, bottomland forests, and edges of mesic upland forests; also cemeteries, gardens, railroads, and roadsides.

5. Euphorbia davidii Subils

E. dentata Michx. var. gracillima Millsp.
E. dentata var. lancifolia Farw.

Map 1667

Plants annual, with taproots. Stems 20–70 cm long, erect or ascending, unbranched or few- to several-branched, the branches not flattened toward the tip, usually green to yellowish green, occasionally reddish- to purplish-tinged, densely pubescent with minute, downward-curved or downward-angled hairs, also with sparse to moderate longer, multicellular hairs. Leaves opposite (occasionally alternate at 1 or 2 of the uppermost nodes), short- to less commonly long-petiolate. Stipules absent or a pair of minute, light brown, convex, sessile glands. Leaf blades 10–100 mm long, highly variable in shape, linear to elliptic or sometimes lanceolate or ovate, not lobed, more or less symmetrically angled or tapered at the base, angled to tapered to a usually bluntly pointed tip, the margins irregularly and often relatively coarsely toothed or scalloped, the upper surface sparsely to moderately roughened with short, stiff, conical hairs having minutely pustular bases, green to dull grayish green and sometimes reddish- to purplish-tinged toward the margins or base, the undersurface moderately pubescent with relatively stout hairs, these often with a minute, persistent pustular base, and paler green than the upper surface. Inflorescences terminal, often a small, umbellate panicle with a whorl of leaves at the base, but this frequently reduced to 1–3 small clusters of cyathia. Involucre 2.5–3.0 mm long, glabrous, the rim irregularly lobed and fringed, the marginal glands 1 or less commonly 2, 0.7–1.2 mm long, appearing strongly concave and more or less 2-lipped, yellowish green to yellowish brown, lacking a petaloid appendage. Staminate flowers 25–40 per cyathium. Ovaries glabrous or with sparse, appressed hairs, the styles 1.0–1.5 mm long, each divided $1/2$–$3/4$ of the way from the tip into 2 slightly slender or club-shaped lobes. Fruits 3–5 mm long (somewhat broader), usually glabrous at maturity. Seeds 2.5–3.0 mm long, ovate to triangular-ovate in outline, bluntly angular in cross-section (both the oblique apical portion surrounding the caruncle and the longitudinal inner faces appearing angular), more or less flattened at the base, the surface appearing relatively coarsely wrinkled or with poorly differentiated low, broad warts (appearing lumpy or irregularly swollen), these sometimes denser toward the angles, mostly dark brown or nearly black, often appearing somewhat mottled, usually with a small but well-developed, pale caruncle. $2n=56$. July–October.

Scattered mostly in the western and northern portions of the state (Arizona to Texas north to Wyoming, Wisconsin, and Ohio; Mexico, Canada; introduced elsewhere in the U.S., South America, Australia). Banks of streams and rivers, bottomland forests, bottomland prairies, upland prairies, and sand prairies; also ditches, railroads, roadsides, and open, disturbed areas.

The name E. davidii was first applied to North American plants by Mayfield (1997), who elevated some of the several varieties described within E. dentata to species status and refined the set of morphological characters to distinguish them. Mayfield also first reported the taxon for Missouri. Although the apparently tetraploid ($2n=56$) E. davidii and apparently diploid ($2n=28$) E. dentata sometimes grow in mixed populations, they apparently do not hybridize readily. Crosses between plants of different ploidies would yield sterile off-

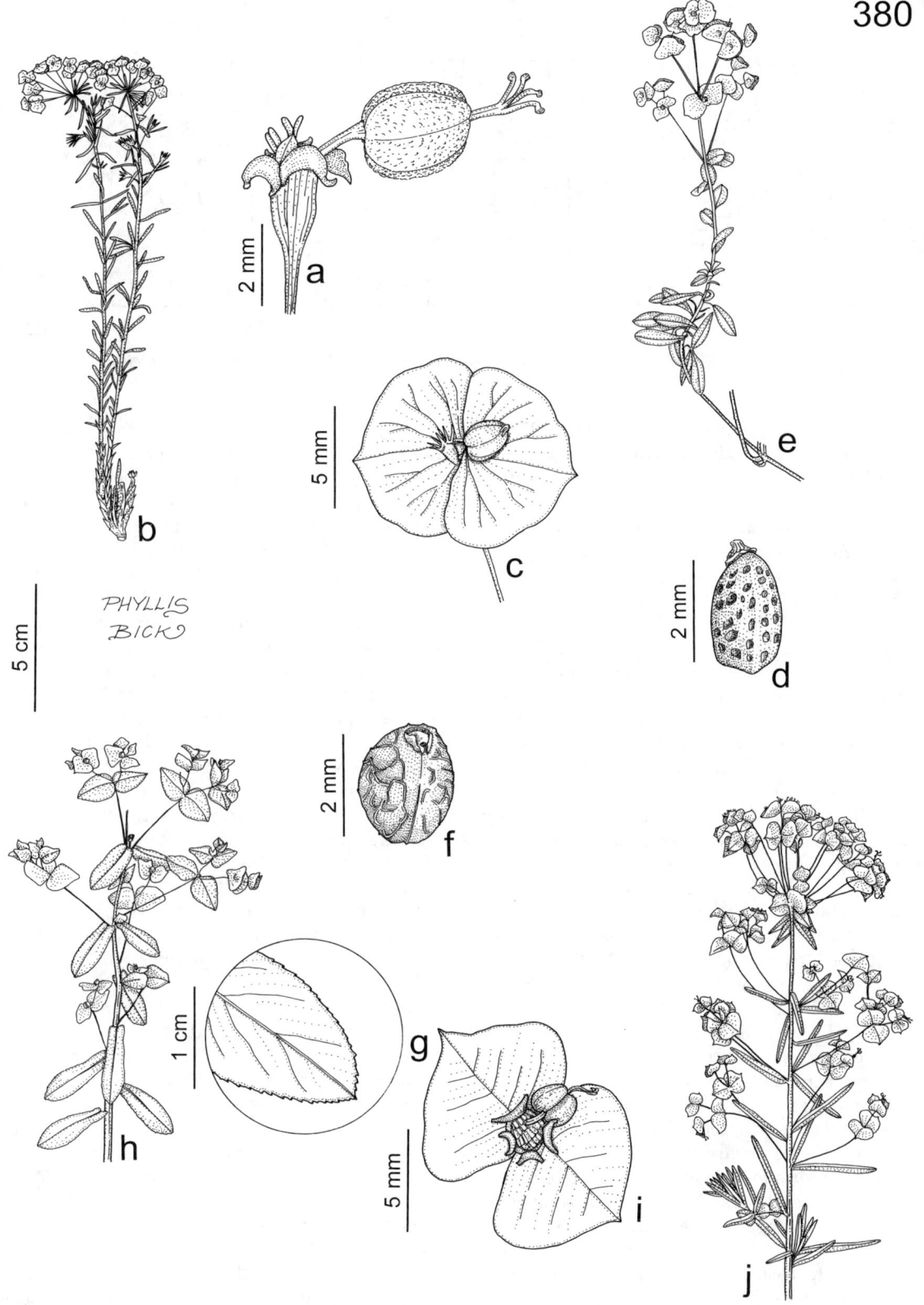

Plate 380. Euphorbiaceae. *Euphorbia cyparissias*, **a)** cyathium with fruit, **b)** habit. *Euphorbia commutata*, **c)** leaves and cyathium with fruit, **d)** seed, **e)** habit. *Euphorbia obtusata*, **f)** seed, **g)** leaf detail, **h)** habit. *Euphorbia esula*, **i)** leaves and cyathium with fruit, **j)** habit.

1668. Euphorbia dentata   1669. Euphorbia esula   1670. Euphorbia geyeri

spring. It should be noted that Mayfield (1997) regarded a single early mitotic chromosome count of $2n=14$ for *E. dentata* as having doubtful validity.

6. **Euphorbia dentata** Michx. (toothed spurge)
Poinsettia dentata (Michx.) Klotzsch & Garcke
E. dentata var. *linearis* Engelm. ex Boiss.
E. cruentata Graham

Pl. 379 i, j; Map 1668

Plants annual, with taproots. Stems 15–60 cm long, erect or ascending, unbranched or few- to several-branched, the branches not flattened toward the tip, usually green to yellowish green, occasionally reddish- to purplish-tinged, densely pubescent with minute, downward-curved or downward-angled hairs, usually also with scattered, longer, multicellular hairs. Leaves opposite (occasionally alternate at 1 or 2 of the uppermost nodes), short- to long-petiolate. Stipules absent or a pair of minute, light brown, convex, sessile glands. Leaf blades 10–60 mm long, highly variable in shape, linear to lanceolate, elliptic, ovate, or nearly circular, not lobed, more or less symmetrically rounded to angled or tapered at the base, rounded or angled to tapered to a usually bluntly pointed tip, the margins relatively coarsely and often irregularly toothed or less commonly finely toothed to scalloped, wavy, or nearly entire, the surfaces sparsely to densely pubescent or occasionally nearly glabrous, green to dull grayish green and sometimes reddish- to purplish-tinged toward the margins or base, the undersurface with somewhat longer, relatively slender hairs (these not expanded at the base) and paler green than the upper surface. Inflorescences terminal, often a small, umbellate panicle with a whorl of leaves at the base, but this frequently reduced to 1–3 small clusters of cyathia. Involucre 2.5–3.5 mm long, glabrous, the rim irregularly lobed and fringed, the marginal glands 1 or more commonly 2, 0.7–1.2 mm long, appearing strongly concave and more or less 2-lipped, yellowish green to yellowish brown, lacking a petaloid appendage. Staminate flowers 25–40 per cyathium. Ovaries glabrous, the styles 1.0–1.5 mm long, each divided $1/2$–$2/3$ of the way from the tip into 2 slightly club-shaped lobes. Fruits 3–5 mm long (somewhat broader), glabrous. Seeds 2.5–3.0 mm long, ovate to broadly ovate in outline, more or less rounded in cross-section (the oblique apical portion surrounding the caruncle angled but the longitudinal inner faces appearing rounded), more or less flattened to slightly concave at the base, the surface appearing relatively finely warty or with fine, relatively evenly spaced tubercles, light gray to dark brown or nearly black, often appearing somewhat mottled, often with a small but well-developed, pale caruncle. $2n=28$. July–October.

Scattered nearly throughout the state (Texas to Georgia north to Nebraska and Pennsylvania; Mexico; introduced elsewhere in the U.S.). Banks of streams and rivers, ledges and tops of bluffs, bottomland forests, mesic to dry upland forests, glades, and upland prairies; also crop fields, fallow fields, old fields, gardens, ditches, railroads, roadsides, and open, disturbed areas.

Steyermark (1963) referred two historical collections from Barton and Jackson Counties to *E. cuphosperma* (Engelm.) Boiss. (as *E. dentata* f. *cuphosperma* (Engelm.) Fernald; also known as *E. dentata* var. *cuphosperma* Engelm., *Poinsettia dentata* var. *cuphosperma* (Engelm.) Mohl.), based on their linear to narrowly lanceolate leaves. That taxon, which occurs from Arizona and New Mexico south through Mexico to Central America, is differentiated from all of the Missouri materials in the complex by its hairy (vs. glabrous) ovaries and fruits, and by its cyathia with relatively elongate involucral glands tapered to a stalklike base. Steyermark misapplied this name to Missouri specimens and the taxon is here excluded from the Missouri flora. The two specimens in question correspond better to var. *linearis,* a narrow-leaved form

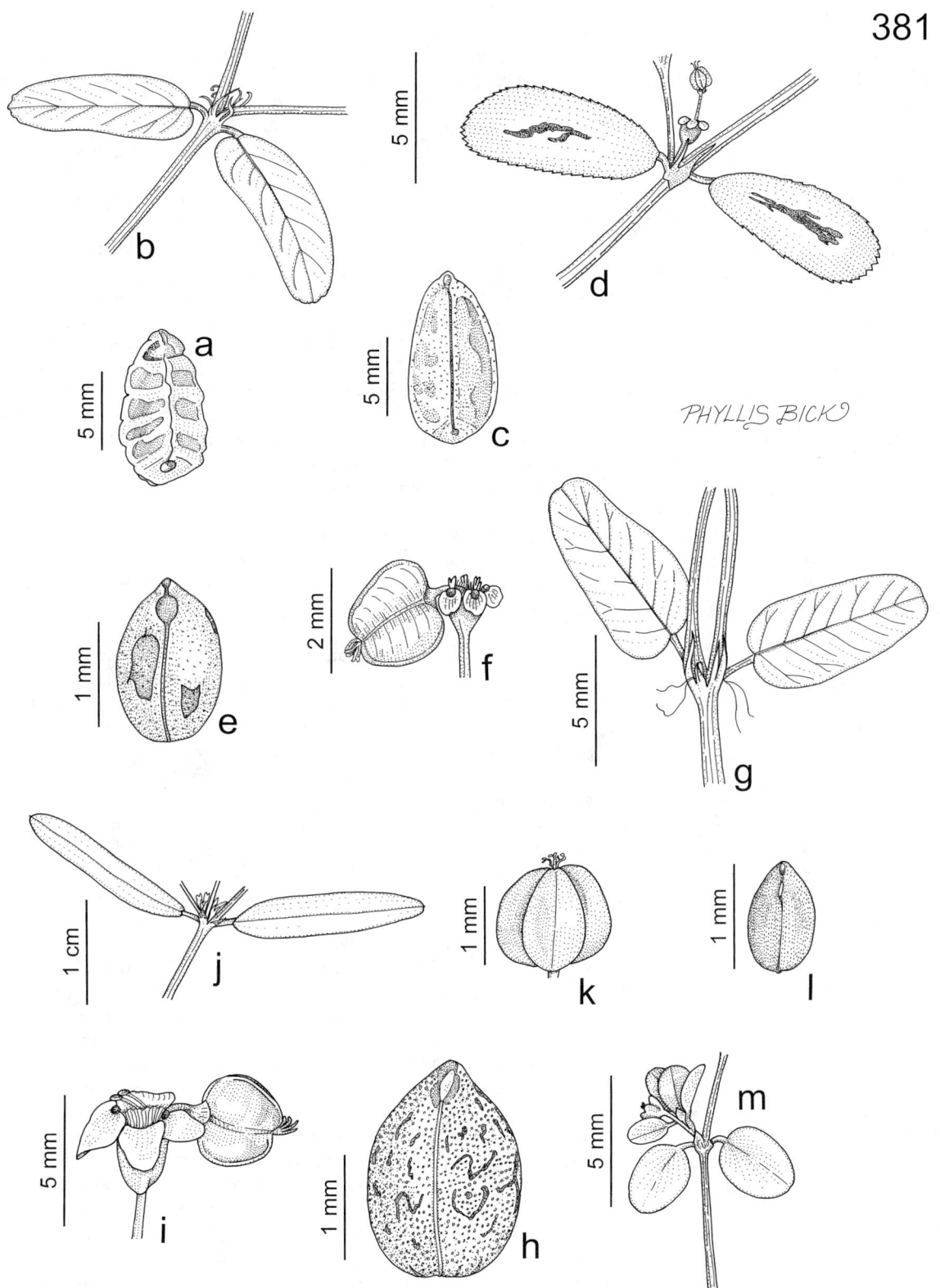

Plate 381. Euphorbiaceae. *Euphorbia glyptosperma*, **a)** seed, **b)** fertile node with leaves. *Euphorbia serpyllifolia*, **c)** seed, **d)** fertile node with leaves. *Euphorbia geyeri*, **e)** seed, **f)** cyathium with fruit, **g)** node with leaves. *Euphorbia missurica*, **h)** seed, **i)** cyathium with fruit, **j)** node with leaves and flowers. *Euphorbia serpens*, **k)** fruit, **l)** seed, **m)** fertile node with branchlet and leaves.

of *E. dentata* unworthy of formal taxonomic recognition that was first described from near St. Louis in adjacent western Illinois.

7. Euphorbia esula L. (leafy spurge, wolf's milk)

Pl. 380 i, j; Map 1669

Plants perennial herbs, with a deep, spreading, branched rootstock, usually also with rhizomes. Stems 30–90 cm long, erect or ascending, sometimes from a short-spreading base, often branched above the midpoint, the branches not flattened toward the tip, green to yellowish green, sometimes reddish- or purplish-tinged toward the base, glabrous. Leaves alternate above the lowermost node and below the inflorescence branches (those of the inflorescence branches opposite) not appearing crowded, sessile or very short-petiolate. Stipules absent. Leaf blades 30–100 mm long (those of the inflorescence branches 8–15 mm), unlobed, the margins entire, the surfaces glabrous, green to yellowish green; those below the inflorescence linear to narrowly oblong-lanceolate or narrowly oblanceolate, tapered at the base, rounded or abruptly short-tapered to a minute, sharply pointed tip; those along the inflorescence branches kidney-shaped to broadly ovate (slightly longer than wide to more commonly somewhat wider than long), broadly rounded or broadly angled to shallowly cordate and somewhat clasping the stem, the terminal ones somewhat cupped around the cyathia and strongly yellowish-tinged, rounded or more commonly very broadly angled to a minute, sharply pointed tip. Inflorescences terminal umbellate panicles with opposite or whorled leaves at the base and each of the 5–17 primary branches usually branched 1–4 additional times, the cyathia solitary or in small clusters at the branch tips. Involucre 1.5–3.0 mm long, glabrous, the rim shallowly 4-lobed, the margin usually densely and minutely cobwebby-hairy (the hairs attached on the inner side), the marginal glands 4, 1.2–1.9 mm long, more or less kidney-shaped to crescent, the oblong body often tapered into a pair of spreading horns, yellow to greenish yellow, lacking petaloid appendages. Staminate flowers 12–25 per cyathium. Ovaries glabrous, the styles 1.5–3.0 mm long, each divided $1/4$–$1/3$ of the way from the tip into 2 slightly club-shaped lobes. Fruits 2.5–3.5 mm long, glabrous, appearing finely warty or roughened especially along the angles. Seeds 2.2–3.0 mm long, oblong-elliptic to oblong-ovate in outline, slightly flattened to nearly circular in cross-section, rounded or slightly flattened at the base, the surface smooth, brown to silvery gray, usually finely mottled, with a pale, somewhat flattened to slightly winglike caruncle. $2n=20$, 60, 64. May–September.

Introduced, uncommon and sporadic (native of Europe, Asia; introduced in the northern U.S. south to California, New Mexico, and Virginia, also Canada). Pastures, railroads, roadsides, and open, disturbed areas.

Leafy spurge is a problem invasive exotic species throughout much of the northern half of the United States and adjacent Canada, particularly in the northern Great Plains. Dunn (1979, 1985) reviewed the history of its introduction into North America in the early 1800s and noted that more than 2.5 million acres are infested with this species. The plants have extensive rootstocks reported to reach more than 5 m deep and also produce long-lived seeds (Burrows and Tyrl, 2001). *Euphorbia esula* aggressively outcompetes most native grassland species, forming dense, often nearly pure stands. It has a large negative economic impact because the presence of this unpalatable species renders rangelands unfit for livestock grazing. Interestingly, young plants are palatable to sheep and goats, which are one means of controlling infestations.

The taxonomy of *E. esula* and its relatives (which include *E. cyparissias*), especially the nonnative North American populations, is still somewhat controversial. Radcliffe-Smith (1985) reviewed the taxonomy of the more than 75 taxa within the *E. esula/virgata* complex worldwide, concluding that about 20 species and hybrids either have been documented as growing somewhere in North America or are likely future introductions. He suggested that the most aggressively invasive member of the group was not true *E. esula*, but instead a hybrid between that species and another Old World species, *E. virgata* Waldst. & Kit. This supposed hybrid has been called *E.* ×*pseudovirgata* (Schur) Soó and was said to differ from *E. esula* in its narrow leaves generally widest below the midpoint. Steyermark's (1963) original Chariton County collection agrees with this morphology. Subsequently, Stahevich et al. (1988) and Crompton et al. (1990) conducted large-scale detailed cytological and morphometric analyses of the complex using both North American samples and additional plants from native Old World populations. They could find no chromosomal or morphological distinctions to separate *E. esula* and *E. virgata* (at least in North America) or any evidence that plants ascribed to *E.* ×*pseudovirgata* represented interspecific hybrids. These authors concluded that there are only five species and one hybrid present in North America and that the most widespread of these is *E. esula* in the broad sense (to include plants attributed to *E. virgata* and *E.* ×*pseudovirgata*). This conservative treatment is followed in the present work. The one surviving hybrid combina-

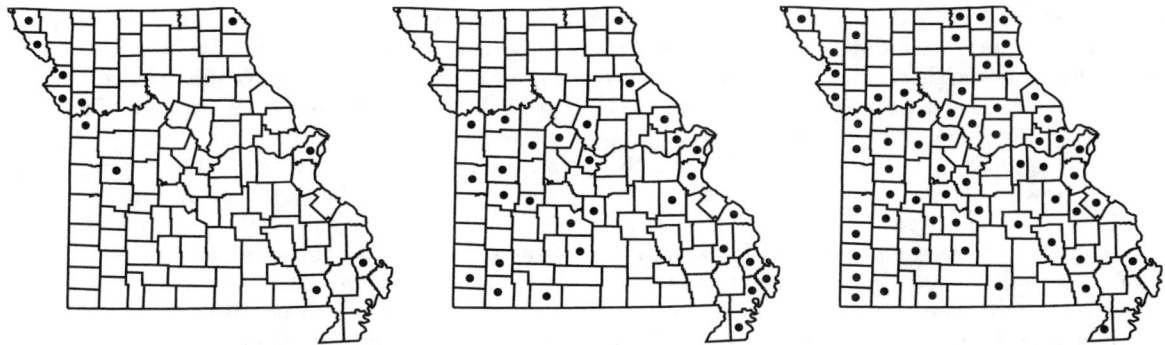

1671. Euphorbia glyptosperma   1672. Euphorbia humistrata   1673. Euphorbia maculata

tion, *E. esula* × *E. cyparissias* (*E.* ×*pseudoesula* Schur), has not yet been reported from Missouri but plausibly might be found in the state someday.

8. Euphorbia geyeri Engelm. var. geyeri
(Geyer's spurge)

Chamaesyce geyeri (Engelm.) Small

Pl. 381 e–g; Map 1670

Plants annual, with taproots. Stems 4–25(–45) cm long, prostrate, several- to many-branched, the branches often overlapping (plants loosely mat-forming), not flattened toward the tip, tan to yellowish green, sometimes slightly pinkish-tinged, glabrous. Leaves opposite, sessile or very short-petiolate. Stipules small scales 0.7–1.5 mm long, these not fused or, on 1 side of the stem, those from the adjacent leaf in each pair fused toward the base into a single small structure positioned between the leaf bases, mostly deeply and irregularly fringed or lobed. Leaf blades 4–12 mm long, oblong to oblong-obovate or oblong-elliptic, slightly asymmetrically angled or rounded at the base with the side toward the branch base usually slightly expanded (sometimes into a small, rounded auricle), broadly rounded or truncate and sometimes minutely notched at the tip, the margins entire, both surfaces glabrous, the upper surface usually light green to bright green, the undersurface usually pale grayish green. Inflorescences axillary, of solitary cyathia or appearing as small clusters on short axillary branches. Involucre 1.0–1.5 mm long, glabrous, the rim shallowly 4-lobed or 4-toothed, the marginal glands 4, 0.2–0.6 mm long and more or less equal in size, the body broadly oblong to nearly circular, yellowish green to green, with a relatively inconspicuous petaloid appendage 0.2–0.8 mm long, this white to somewhat pinkish- or reddish-tinged. Staminate flowers 5–17 per cyathium. Ovaries glabrous, the styles 0.2–0.5 mm long, each divided $1/3$–$1/2$ of the way from the tip into 2 slender lobes. Fruits 1.5–2.0 mm long, glabrous. Seeds 1.3–1.6 mm long, oblong-ovate to ovate in outline, bluntly angular in cross-section, slightly convex at the base, the surface smooth, white to reddish brown, lacking a caruncle. July–October.

Uncommon in the northern portion of the Mississippi Lowlands Division and in the northeastern corner of the state (Montana to New Mexico east to Michigan and Mississippi; Canada). Sand prairies and sand savannas; also sandy, open, disturbed areas; restricted to sandy soils.

This rare species is usually found growing in relatively deep sands where few other plants can survive to compete for space and nutrients. Steyermark (1963) reported it only from a single Clark County collection that he made in 1949, but Yatskievych discovered the species in 1992 in Scott County. Uncommon plants from New Mexico and Texas with the involucral glands lacking petaloid appendages have been called var. *wheeleriana* Warnock & M.C. Johnst. [*Chamaesyce geyeri* var. *wheeleriana* (Warnock & M.C. Johnst.) Mayfield].

9. Euphorbia glyptosperma Engelm. (sand mat, rib-seeded sand mat)

Chamaesyce glyptosperma (Engelm.) Small

Pl. 381 a, b; Map 1671

Plants annual, with taproots. Stems 5–40 cm long, prostrate, several- to many-branched, the branches often overlapping (plants mat-forming), not flattened toward the tip, yellowish green to tan, often pinkish- to purplish-tinged, glabrous, sometimes slightly glaucous. Leaves opposite, sessile or very short-petiolate. Stipules small scales 0.3–1.0 mm long, these not fused, mostly irregularly fringed or lobed. Leaf blades 3–15 mm long, oblong to oblong-obovate or oblong-ovate, asymmetrical at the base with the side toward the stem tip usually angled and the other side rounded to shallowly cordate and expanded into a small, rounded auricle, more or less rounded at the tip, the margins minutely few- to several-toothed (best observed with magnification), the surfaces glabrous, sometimes reddish-tinged or with an irregular reddish spot, the undersurface usually pale grayish green. Inflorescences axillary, of solitary cyathia or appearing as small clusters on short axillary branches. Involucre 0.6–0.9 mm long, glabrous, the rim shallowly 4-lobed or 4-toothed, the marginal glands 4, 0.1–0.3 mm long and more or less equal in size, the body narrowly oblong, greenish yellow, with a relatively inconspicuous petaloid appendage 0.1–0.4 mm long, this white or pinkish-tinged, the margin usually irregularly notched. Staminate flowers 1–5 per cyathium. Ovaries glabrous, the styles 0.1–0.4 mm long, each divided $1/3$–$1/2$ of the way from the tip into 2 somewhat club-shaped lobes. Fruits 1.2–1.8 mm long, glabrous. Seeds 1.0–1.4 mm long, oblong-ovate in outline, angular in cross-section, flat to slightly convex at the base, the surface with 3 or 4(–6) coarse transverse ridges (these mostly extending the full width of the face and through the angles), white to light tan, lacking a caruncle. June–October.

Uncommon, mostly in counties along the Missouri and Mississippi River floodplains (U.S. [except most southeastern states]; Canada, Mexico). Banks of streams and rivers, sand prairies, and loess hill prairies; also roadsides and open, disturbed areas.

10. Euphorbia humistrata Engelm.
Chamaesyce humistrata (Engelm.) Small

Pl. 382 g–i; Map 1672

Plants annual, with taproots (frequently also with adventitious roots at some of the nodes). Stems 5–45 cm long, usually prostrate, occasionally with ascending tips, several- to many-branched, the branches often overlapping (plants mat-forming), not flattened toward the tip, usually pale green to yellowish green, moderately to densely pubescent with short, appressed or incurved hairs, especially toward the branch tips, occasionally nearly glabrous toward the stem base, the hairs often in 2 bands along opposite sides of the stem. Leaves opposite, sessile or very short-petiolate. Stipules small scales 1.0–1.5 mm long, those from the adjacent leaf in each pair often fused into a single, small, scalelike structure on each side of the stem positioned between the leaf bases, this irregularly fringed or lobed. Leaf blades 5–17 mm long, oblong-obovate to obovate or oblong-ovate, occasionally some of the leaves narrowly oblong, asymmetrical at the base with the side toward the stem tip usually angled and the other side more or less truncate and expanded into a small, rounded auricle, rounded or less commonly broadly and bluntly pointed at the tip, the margins minutely few-toothed (best observed with magnification), the upper surface glabrous or nearly so and occasionally with a faint reddish spot, the undersurface glabrous or sparsely pubescent with loosely appressed to somewhat spreading, slender hairs and usually silvery to pale grayish green. Inflorescences axillary, of solitary cyathia or appearing as small clusters on short axillary branches. Involucre 0.6–1.0 mm long, sparsely hairy on the outer surface, the rim shallowly 4-lobed or 4-toothed, the marginal glands 4, 0.1–0.5 mm long and usually more or less equal in size, the body narrowly oblong to nearly linear, reddish purple to dark purple, with a relatively inconspicuous petaloid appendage 0.1–1.0 mm long, this white to strongly pinkish- or reddish-tinged. Staminate flowers 3–8 per cyathium. Ovaries hairy, the styles 0.5–0.8 mm long, each divided about $1/2$ of the way from the tip into 2 slender lobes. Fruits 1.0–1.5 mm long, sparsely to moderately and relatively evenly pubescent with appressed or strongly incurved hairs. Seeds 0.8–1.2 mm long, more or less oblong-ovate in outline, bluntly angular in cross-section, flat to slightly convex at the base, the surface smooth or appearing finely roughened, lacking cross-ridges, white to reddish brown, lacking a caruncle. June–October.

Scattered in the state but absent from portions of the Glaciated Plains and Ozark Divisions (Kansas to Texas east to Ohio and Florida; introduced sporadically east to New Jersey). Banks of streams and rivers, sloughs, saline seeps, and margins of ponds and lakes; also levees, railroads, roadsides, and open, disturbed areas.

Some plants of *E. humistrata* can be very difficult to distinguish from *E. maculata*. Steyermark (1963) discussed some of the differences between them, and Richardson (1968) tabulated a number of characters to separate the two taxa. These include the greater tendency of *E. humistrata* stems to root at the nodes and the deeper lobing of its involucres. *E. humistrata* also tends to have paler green stems and broader leaves that are less densely hairy on the undersurface and tend to lack a reddish spot on the upper surface. It also often has the involucre split about halfway down one side (in *E. maculata*, the involucre is entire or with a shallow split), with the stalk of the pistillate flower arched or recurved through the gap at maturity.

11. Euphorbia maculata L. (milk purslane, prostrate spurge)
Chamaesyce maculata (L.) Small
E. supina Raf.
C. supina (Raf.) Moldenke

Pl. 382 j–m; Map 1673

Plants annual, with taproots. Stems 5–45 cm long, usually prostrate, occasionally with ascending tips, several- to many-branched, the branches

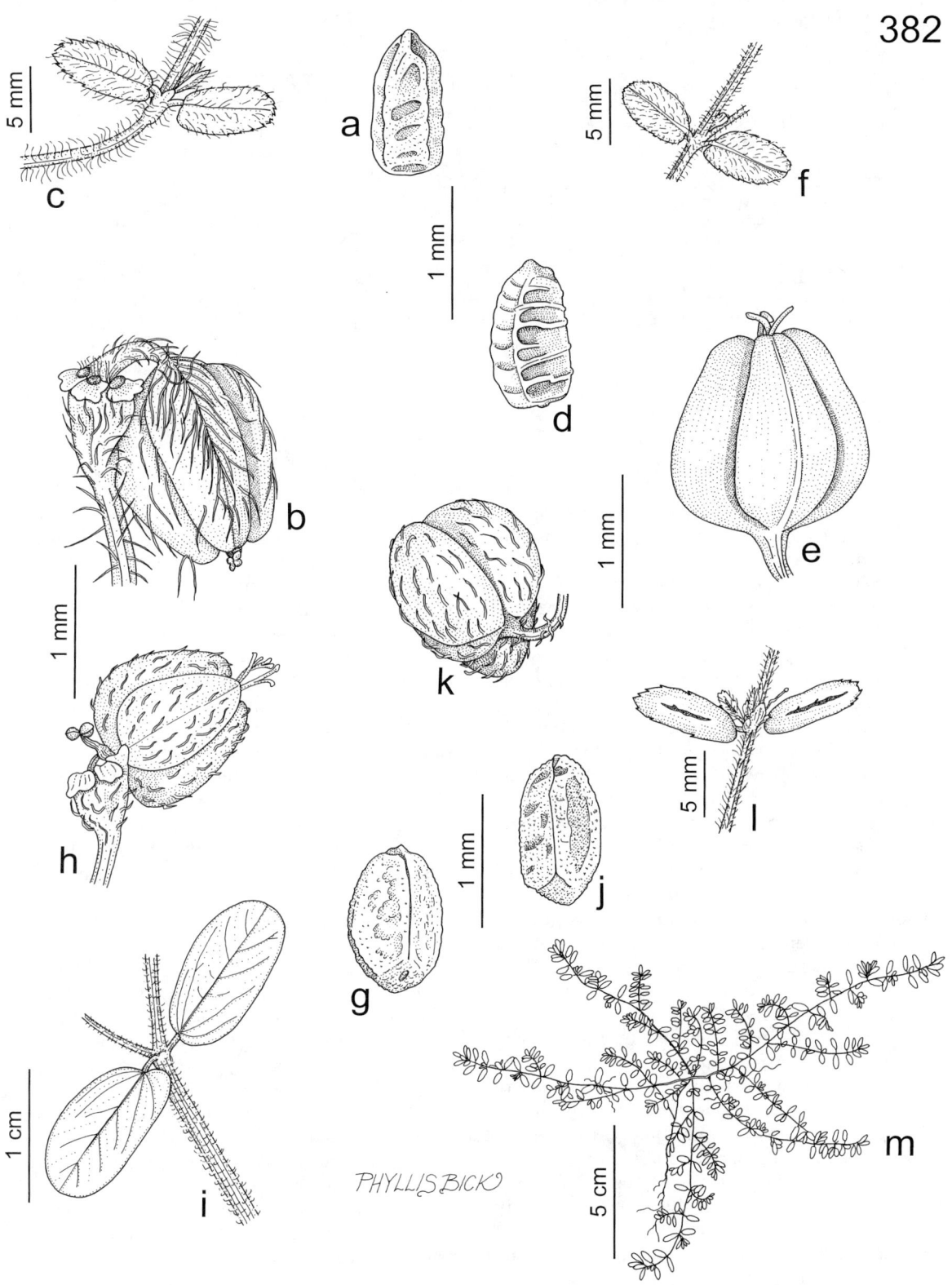

Plate 382. Euphorbiaceae. *Euphorbia stictospora*, **a)** seed, **b)** cyathium with fruit, **c)** node with leaves. *Euphorbia prostrata*, **d)** seed, **e)** fruit, **f)** node with leaves. *Euphorbia humistrata*, **g)** seed, **h)** cyathium with fruit, **i)** node with leaves. *Euphorbia maculata*, **j)** seed, **k)** fruit, **l)** node with leaves, **m)** habit.

1674. Euphorbia marginata 1675. Euphorbia missurica 1676. Euphorbia nutans

often overlapping (plants mat-forming), not flattened toward the tip, usually reddish brown, moderately to more commonly densely and evenly pubescent with short, appressed or incurved hairs. Leaves opposite, sessile or very short-petiolate. Stipules small scales 1.0–1.5 mm long, these not fused, often irregularly 2- or 3-lobed. Leaf blades 4–17 mm long, oblong-ovate to ovate-elliptic or oblong-elliptic, occasionally some of the leaves narrowly oblong, asymmetrical at the base with the side toward the stem tip usually angled and the other side more or less truncate and expanded into a small, rounded auricle, rounded or less commonly broadly and bluntly pointed at the tip, the margins minutely few- to several-toothed (best observed with magnification), the upper surface nearly glabrous to sparsely pubescent with relatively long, slender hairs and often also with an irregular reddish spot, the undersurface moderately to densely pubescent with somewhat appressed, sometimes somewhat woolly hairs and usually pale grayish green. Inflorescences axillary, of solitary cyathia or appearing as small clusters on short axillary branches. Involucre 0.8–1.0 mm long, sparsely hairy on the outer surface, the rim shallowly 4-lobed or 4-toothed, the marginal glands 4, 0.2–0.6 mm long and usually somewhat unequal in size, the body narrowly oblong to nearly linear, reddish purple to dark purple, with a relatively inconspicuous petaloid appendage 0.2–1.5 mm long, this white to strongly reddish-tinged. Staminate flowers 2–5 per cyathium. Ovaries hairy, the styles 0.3–0.4 mm long, each divided ¼–⅓ of the way from the tip into 2 slightly club-shaped lobes. Fruits 1.3–1.5 mm long, sparsely to moderately and relatively evenly pubescent with appressed or strongly incurved hairs. Seeds 1.0–1.2 mm long, more or less oblong-ovate in outline, angular in cross-section, flat to slightly convex at the base, the surface with 3 or 4 low, broadly rounded cross-wrinkles, white to light brown, becoming sticky when wet, lacking a caruncle. $2n=28$. May–October.

Scattered nearly throughout the state (eastern U.S. west to North Dakota and Texas; Canada; introduced farther west to Washington and California, also Hawaii, Europe, Asia). Glades, sand prairies, openings of mesic to dry upland forests, banks of streams and rivers, and receding margins of ponds, lakes, and sinkhole ponds; also crop fields, fallow fields, sidewalks, gardens, railroads, roadsides, and open, disturbed areas.

The epithet *E. maculata* was misapplied by most earlier botanists to plants with ascending stems that are called *E. nutans* in the present work. Burch (1966) reviewed the sources of data used by Linnaeus in his original descriptions of *Euphorbia* species and concluded that the name *E. maculata* was intended to apply to the prostrate-stemmed taxon. For further discussion of the distinctions between *E. maculata* and the closely related *E. humistrata*, see the treatment of that species.

12. Euphorbia marginata Pursh (snow-on-the-mountain)

Pl. 379 f–h; Map 1674

Plants annual, with a taproot. Stems 30–80 (–150) cm long, erect or ascending, usually unbranched below the inflorescence, the branches not flattened toward the tip, green to yellowish green or tan, often somewhat reddish- or purplish-tinged toward the base, moderately to densely pubescent with slender, spreading, somewhat tangled hairs, at least toward the tip, often glabrous or nearly so below the midpoint. Leaves alternate below the inflorescence branches (those of the inflorescence branches opposite or whorled), not appearing crowded, sessile or minutely petiolate. Stipules usually a pair of minute, narrowly conical, light brown glands but these shed as the leaves develop. Leaf blades 10–100 mm long, unlobed, the margins entire or slightly wavy, the surfaces glabrous or sparsely to moderately pubescent toward the base and along the midvein with slender, spreading, somewhat tangled hairs, light green to bright green;

those below the inflorescence broadly ovate to elliptic or oblong-obovate, rounded to angled or short-tapered at the base, angled or tapered to a bluntly or more commonly sharply pointed tip; those along the inflorescence branches narrowly elliptic to elliptic, oblanceolate, or narrowly obovate (much longer than wide), angled or tapered at the base, angled to a sharply pointed tip, at least the uppermost leaf blades with conspicuous, broad, white margins, these occasionally slightly pinkish-tinged toward the outer edge. Inflorescences terminal umbellate panicles with a whorl of leaves at the base and each of the 3 or 4(5) primary branches usually branched 2–5 additional times, the cyathia solitary or less commonly in small clusters at the branch tips and solitary at the branch points. Involucre 3–4 mm long, the outer (and inner) surface sparsely to densely hairy, the rim 5-lobed, the lobes deeply divided into a fringe of slender, pale lobes, the marginal glands 5, 1.0–1.8 mm long, oblong-elliptic to nearly circular and often somewhat concave, green to light green or yellowish green, with a relatively conspicuous petaloid appendage 2–4 mm long, this white. Staminate flowers 35–60 per cyathium. Ovaries densely pubescent with ascending hairs, the styles 1.0–2.5 mm long, each divided $^{2}/_{3}$–$^{3}/_{4}$ of the way from the tip into 2 relatively slender lobes. Fruits 4–6 mm long, moderately to densely hairy. Seeds 3–4 mm long, ovate to nearly circular in outline, nearly circular in cross-section, rounded at the base, the surface with a network of irregular wrinkles or blunt ridges and sometimes also low, rounded tubercles, orangish tan to olive gray, sometimes with a thin, white coating, lacking a caruncle. $2n=56$. June–October.

Scattered, mostly in the northern half of the state (Minnesota to Montana south to Texas and Arizona; Canada; introduced east to New Hampshire and Florida). Bottomland prairies and loess hill prairies; also pastures, gardens, railroads, and roadsides.

This species is cultivated as an ornamental for its showy, white-margined upper leaves. Steyermark (1963) noted that it grows well in hot, dry, clayey soils. Wherever it is grown, the species tends to escape sporadically.

13. Euphorbia missurica Raf. (prairie spurge, Missouri spurge)

Chamaesyce missurica (Raf.) Shinners

E. missurica var. *intermedia* (Engelm.) L.C. Wheeler

C. missurica var. *calcicola* Shinners

Pl. 381 h–j; Map 1675

Plants annual, with taproots. Stems 10–60 cm long, erect or ascending, several- to many-branched, the branching sometimes appearing dichotomous, not flattened toward the tip, tan to yellowish green, usually strongly pinkish- to purplish-tinged, glabrous, sometimes somewhat glaucous. Leaves opposite, sessile or very short-petiolate. Stipules small scales 0.7–1.5 mm long, those from the adjacent leaf in each pair sometimes fused toward the base on 1 or both sides of the stem into a single small structure positioned between the leaf bases, variously entire to deeply and irregularly fringed or lobed. Leaf blades (4–)8–30 mm long, linear to narrowly oblong or narrowly lanceolate-oblong, slightly asymmetrically angled or short-tapered at the base with the side toward the branch base usually slightly larger than the other side (not expanded into a distinct auricle), rounded to truncate at the tip, sometimes minutely notched or with an abrupt, minute, sharp point, the margins entire and sometimes somewhat curled under, both surfaces glabrous, the upper surface light green to bright green, the undersurface usually pale green. Inflorescences terminal at the branch tips and often also from between the branch points, of solitary cyathia or appearing as small clusters. Involucre 1.2–1.8 mm long, glabrous, the rim shallowly 4-lobed or 4-toothed, the marginal glands 4, 0.2–0.5 mm long and more or less equal in size, the body broadly oblong to nearly circular, yellowish green to green, with a relatively conspicuous petaloid appendage 0.4–2.5 mm long, this white or occasionally somewhat pinkish-tinged. Staminate flowers 24–50 per cyathium. Ovaries glabrous, the styles 0.5–1.5 mm long, each divided $^{1}/_{2}$–$^{3}/_{4}$ of the way from the tip into 2 slender lobes. Fruits 2.0–2.5 mm long, glabrous. Seeds 1.5–2.0 mm long, more or less ovate in outline, bluntly angular in cross-section, slightly convex to nearly flat at the base, the surface smooth, inconspicuously roughened or occasionally appearing slightly wrinkled, white to brown, often somewhat mottled, lacking a caruncle. May–September.

Scattered to uncommon in the western half of the state and south of the Missouri River (Montana to New Mexico east to Minnesota and Arkansas). Glades, ledges and tops of bluffs, and margins of dry upland forest; also rarely open, disturbed areas; usually on calcareous substrates.

This attractive slender-stemmed and slender-leaved species has involucral glands with relatively conspicuous petaloid appendages. The leaves sometimes are somewhat folded longitudinally.

14. Euphorbia nutans Lag. (nodding spurge, eyebane)

Chamaesyce nutans (Lag.) Small

E. preslii Guss.

E. maculata L., misapplied

Pl. 377 h–j; Map 1676

Plants annual, with taproots. Stems 20–80 cm

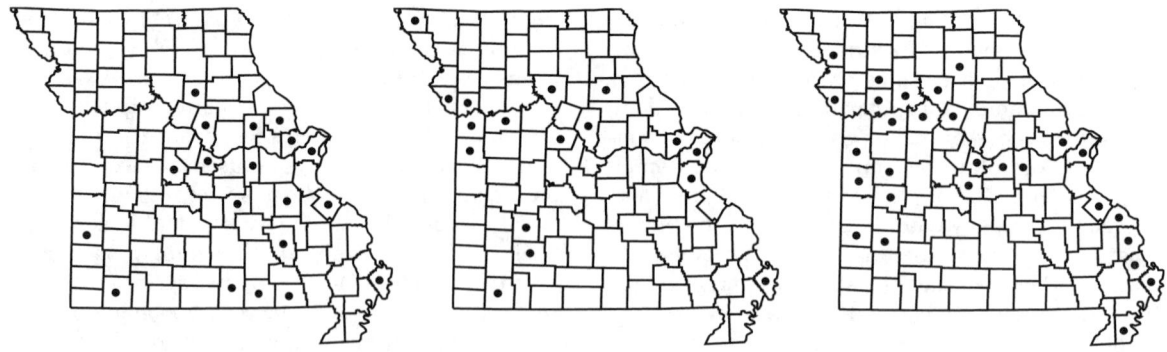

1677. Euphorbia obtusata 1678. Euphorbia prostrata 1679. Euphorbia serpens

long, erect or more commonly ascending, often arched at the branch tips, several- to many-branched, not flattened toward the tip, tan to reddish brown, sparsely to moderately pubescent with woolly hairs, sometimes mainly toward the stem tip and around the nodes, the hairs sometimes in 2 bands along opposite sides of the stem. Leaves opposite, sessile or very short-petiolate. Stipules small scales 1.0–1.5 mm long, these not fused or, on 1 side of the stem, those from the adjacent leaf in each pair fused toward the base into a single small structure positioned between the leaf bases, variously irregularly toothed, fringed, or divided. Leaf blades 8–40 mm long, oblong to oblong-lanceolate, asymmetrical at the base, with the side toward the stem tip usually angled or rounded and the other side more or less truncate to cordate and expanded into a small, rounded auricle, mostly angled to a bluntly pointed tip, the margins finely several-toothed, both surfaces glabrous or more commonly sparsely to moderately pubescent with curved to more or less spreading, slender hairs mostly toward the base, the upper surface usually reddish-mottled or with a conspicuous reddish spot, the undersurface variously pale green or faintly to strongly reddish-tinged. Inflorescences axillary, of solitary cyathia or appearing as small clusters on short axillary branches. Involucre 0.5–1.0 mm long, glabrous on the outer surface (often hairy along the inner margin), the rim shallowly 4-lobed, the marginal glands 4, 0.3–0.5 mm long and more or less equal in size, the body oblong to nearly circular, green or occasionally reddish purple, with a relatively inconspicuous petaloid appendage 0.2–1.5 mm long, this white or pinkish-tinged. Staminate flowers 5–28 per cyathium. Ovaries glabrous, the styles 0.6–2.5 mm long, each divided ⅓–½ of the way from the tip into 2 slender lobes. Fruits 1.6–2.3 mm long, glabrous. Seeds 1.0–1.6 mm long, elliptic-ovate to ovate in outline, angular in cross-section, slightly convex at the base, the surface finely and irregularly wrinkled or with indistinct shallow, rounded cross-ridges, sometimes faintly so, dark brown, sometimes with a thin, white coating, this often more persistent along the angles than the faces, lacking a caruncle. $2n=12, 14, 22$. May–October.

Scattered to common throughout the state (eastern U.S. west to North Dakota and New Mexico; Canada; introduced west to California, Europe). Banks of streams and rivers, exposed margins of ponds and lakes, edges of swamps, disturbed portions of upland prairies, and openings of mesic to dry upland forests; also pastures, fallow fields, margins of crop fields, railroads, roadsides, and open, disturbed areas.

Misapplication of the name *E. maculata* to the taxon here called *E. nutans* is discussed in the treatment of that species. Burch (1966) discussed the problems of assigning names to the four main entities in the nomenclatural complex and determined that the oldest valid name for the relatively robust temperate North American plants with ascending stems is *E. nutans*. Some earlier authors had used the name *E. hypericifolia* for this taxon, but Burch made a strong case for restricting that epithet to a different, mostly tropical species.

15. **Euphorbia obtusata** Pursh (blunt-leaved spurge)

Pl. 380 f–h; Map 1677

Plants annual, with taproots. Stems 20–70 cm long, erect or ascending, unbranched below the inflorescence or occasionally few-branched, the branches not flattened toward the tip, usually green to yellowish green, sometimes faintly purplish-tinged, glabrous. Leaves alternate above the lowest node and below the inflorescence branches (those of the inflorescence branches usually opposite, those of the basal node opposite or whorled but absent at flowering), sessile. Stipules absent. Leaf blades 10–45 mm long, oblanceolate to oblong-oblanceolate (those of the leaves of the inflorescence branches broadly ovate to broadly ovate-triangular), unlobed, rounded to truncate or shallowly cordate at the base and often somewhat clasping the stem, rounded or broadly angled to a bluntly pointed tip, the margins finely toothed mostly above the midpoint (the teeth sometimes minute and visible only with magnification), the surfaces glabrous, yellowish green to green. Inflorescences terminal umbellate panicles with a whorl of leaves at the base and each of the usually 3 primary branches often branched 1–3 additional times, the cyathia solitary at the branch tips and at the branch points. Involucre 1.2–1.5 mm long, glabrous, the rim shallowly 4- or 5-lobed to nearly entire, the marginal glands 4 or 5, 0.3–0.6 mm long, oblong-oval to elliptic or slightly kidney-shaped, usually red or reddish-tinged (occasionally greenish yellow), lacking a petaloid appendage. Staminate flowers 5–10 per cyathium. Ovaries glabrous, but the surface densely warty, the styles 0.7–1.2 mm long, each divided 1/3–1/2 of the way from the tip into 2 slightly club-shaped lobes. Fruits 3.0–3.5 mm long, glabrous but the surface finely warty. Seeds 1.7–2.3 mm long, broadly elliptic-ovate to nearly circular in outline, slightly biconvex in cross-section, rounded at the base, the surface smooth, reddish brown to dark purplish brown but often appearing slightly glaucous, with a pale, irregularly winglike caruncle, this often breaking off as the seeds are dispersed. May–July.

Scattered mostly in the Ozark and Ozark Border Divisions (Iowa to Texas east to Pennsylvania and South Carolina). Banks of streams and rivers and bottomland forests; also rarely fallow fields, ditches, railroads, roadsides, and open, disturbed areas.

The taxonomic status of this species is not entirely understood. Some authors include it in a broadly circumscribed *E. spathulata*.

Steyermark (1963) included *E. platyphyllos* L. (under the variant spelling *E. platyphylla*) in the Missouri flora with some reservations. He noted that a collection by B. F. Bush from Barry County was originally misdetermined as this species, but it instead represented material of *E. spathulata*, which left only a collection made by Earl E. Sherff in the city of St. Louis, and Steyermark admitted this was possibly from a cultivated plant. In his introduction to the Missouri flora, Steyermark (1963) discussed the series of specimens collected by Sherff in St. Louis that lack any further label data, noting that most of these originated from cultivated plants, and thus he excluded from the flora most of the species documented from Missouri only by these collections. As *E. platyphylla* has not been collected in the state since that time, the species is here excluded from the flora.

Euphorbia platyphyllos (Pl. 383 c, d) is a European species that has been introduced sporadically into the eastern (mostly northeastern) and central United States. It is morphologically most similar to *E. obtusata* but differs in its basally fused styles, each shallowly 2-lobed toward the tip. It differs from *E. spathulata* in its smooth seeds (vs. the surface with a network of raised ridges). It further differs from both of these species in its more sharply pointed leaf blades that are somewhat hairy on the undersurface and in the cyathia with a usually hairy involucre.

16. Euphorbia prostrata Aiton (groundfig spurge, milk spurge)
Chamaesyce prostrata (Aiton) Small

Pl. 382 d–f; Map 1678

Plants annual, with taproots. Stems 5–40 cm long, usually prostrate, occasionally with ascending tips, several- to many-branched, the branches often overlapping (plants mat-forming), not flattened toward the tip, usually reddish brown, moderately to densely and more or less evenly pubescent with short, incurved hairs toward the branch tips, often nearly glabrous toward the stem base. Leaves opposite, sessile or very short-petiolate. Stipules small scales 0.3–1.0 mm long, these not fused or, on 1 side of the stem, those from the adjacent leaf in each pair fused toward the base into a single small structure positioned between the leaf bases, this irregularly and deeply fringed or lobed. Leaf blades 3–10(–15) mm long, oblong to oblong-ovate, occasionally some of the leaves narrowly oblong, asymmetrical at the base with the side toward the stem tip usually angled or rounded and the other side more or less truncate and expanded into a small, rounded auricle, broadly rounded to occasionally broadly and bluntly pointed at the tip, the margins minutely several-toothed especially toward the tip (best observed with magnification) and usually reddish-tinged, the upper surface glabrous or nearly so and lacking a reddish spot, the undersurface sparsely to moderately pubescent with somewhat tangled, slender hairs and usually pale or light green. Inflorescences axillary, of solitary cyathia or appearing as small clusters on short axillary branches. Involucre 0.4–0.8 mm long, sparsely to moderately hairy on the outer surface, the rim shallowly 4-lobed, the marginal glands 4, 0.1–0.3 mm long and usually more or less equal in size, the body oblong to nearly circular, reddish purple to dark purple, with a relatively inconspicuous petaloid appendage 0.1–0.3 mm long, this white to strongly pinkish- or reddish-tinged. Staminate flowers 2–5 per cyathium. Ovaries hairy, the styles about 0.1 mm long, each divided nearly to the base into 2 slender lobes. Fruits 1.0–1.5 mm long, moderately to densely pubescent with more or less spreading hairs toward the angles, less densely hairy to nearly glabrous between the angles. Seeds 0.8–1.2 mm long, more or less oblong-ovate in outline, angular in cross-section, flat to slightly convex at the base, the surface with 4–7 relatively sharp, slender cross-ridges, light to dark brown, usually with a thin, white to pinkish white coating, this sometimes wearing away irregularly, lacking a caruncle. $2n=18, 20$. June–October.

Scattered, mostly in the floodplains of the Missouri and Mississippi Rivers (southern U.S. north to Wyoming, Illinois, and Massachusetts, introduced in most of the northern portion of this range; also native to Mexico, Central America, South America, Caribbean Islands; also introduced in Hawaii and other Pacific Islands, Europe, Africa, Asia). Banks of streams and rivers and sloughs; also levees, cracks and edges of pavement, railroads, roadsides, and open, disturbed areas.

Steyermark (1963) knew this species from a single historical specimen from Jackson County. Since that time, a number of collections have come to light, both previously misdetermined older specimens and newer ones. The native distribution of this widespread weedy species is not well understood.

17. Euphorbia serpens Kunth (round-leaved spurge)

Chamaesyce serpens (Kunth) Small

Pl. 381 k–m; Map 1679

Plants annual, with taproots. Stems (2–)5–30 (–40) cm long, prostrate, several- to many-branched, the branches often overlapping (plants loosely mat-forming), not flattened toward the tip, tan to yellowish green, sometimes slightly pinkish-tinged, glabrous. Leaves opposite, sessile or very short-petiolate. Stipules small scales 0.7–1.3 mm long, those from the adjacent leaf in each pair fused into a single, small, scalelike structure on each side of the stem positioned between the leaf bases, this often appearing irregularly fringed or lobed. Leaf blades 2–8 mm long, oblong-ovate to broadly ovate or nearly circular, asymmetrically rounded to shallowly cordate at the base with the side toward the branch base usually slightly expanded into a small, rounded auricle, rounded at the tip, the margins entire, both surfaces glabrous, the upper surface usually light green to yellowish green, the undersurface usually pale grayish green. Inflorescences axillary, of solitary cyathia or appearing as small clusters on short axillary branches. Involucre 0.7–1.0 mm long, glabrous, the rim shallowly 4-lobed or 4-toothed, the marginal glands 4, 0.3–0.6 mm long and more or less equal in size, the body oblong to narrowly oblong, yellowish green to green, with a relatively inconspicuous petaloid appendage 0.3–0.7 mm long, this white to somewhat pinkish-tinged. Staminate flowers 3–8(–12) per cyathium. Ovaries glabrous, the styles 0.2–0.4 mm long, each divided $^{1}/_{8}$–$^{1}/_{2}$ of the way from the tip into 2 slightly club-shaped lobes. Fruits 1.0–1.5 mm long, glabrous. Seeds 0.9–1.2 mm long, oblong-ovate to ovate in outline, bluntly angular in cross-section, slightly convex at the base, the surface smooth, white to brown, lacking a caruncle. $2n=22$. May–October.

Scattered, most commonly in counties adjacent to the Missouri and Mississippi River floodplains, but also sporadically in the Glaciated Plains and Unglaciated Plains Divisions (U.S. [introduced in some eastern and far-western states], Canada, Mexico, Central America, South America, Caribbean Islands). Banks of streams and rivers, exposed margins of ponds and lakes, sloughs, oxbows, and rarely moist depressions of glades; also ditches, margins of crop fields, roadsides and open, disturbed areas.

18. Euphorbia serpyllifolia Pers. (thyme-leaved spurge)

Chamaesyce serpyllifolia (Pers.) Small

Pl. 381 c, d; Map 1680

Plants annual, with taproots. Stems 5–40 cm long, prostrate to loosely ascending, several- to many-branched, the branches often overlapping (plants sometimes mat-forming), often appearing somewhat flattened or narrowly winged toward the tip, usually yellowish green, glabrous, sometimes slightly glaucous. Leaves opposite, sessile or very short-petiolate. Stipules small scales 0.5–1.5 mm long, these not fused, entire or irregularly fringed or lobed. Leaf blades 3–15 mm long, oblong to oblong-obovate, asymmetrical at the base with the side toward the stem tip usually angled and the other side rounded to shallowly cordate and expanded into a small, rounded auricle, more or less rounded at the tip, the margins minutely few- to several-toothed toward the tip (best observed with magnification), the surfaces glabrous, sometimes with an irregular reddish spot, the undersurface usually pale grayish green. Inflorescences axillary, of solitary cyathia or appearing as small clusters on short axillary branches. Involucre 0.8–1.2 mm long, glabrous, the rim shallowly 4-lobed or 4-toothed, the marginal glands 4, 0.1–0.3 mm long and more or less equal in size, the body narrowly oblong, greenish yellow, with a relatively inconspicuous petaloid appendage 0.1–0.3 mm long, this white or pinkish-tinged. Staminate flowers 5–18 per cyathium. Ovaries glabrous, the styles 0.2–0.5 mm long, each shallowly notched or divided less than $^{1}/_{3}$ of the way from the tip into 2 somewhat club-shaped lobes. Fruits 1.5–2.0 mm long, gla-

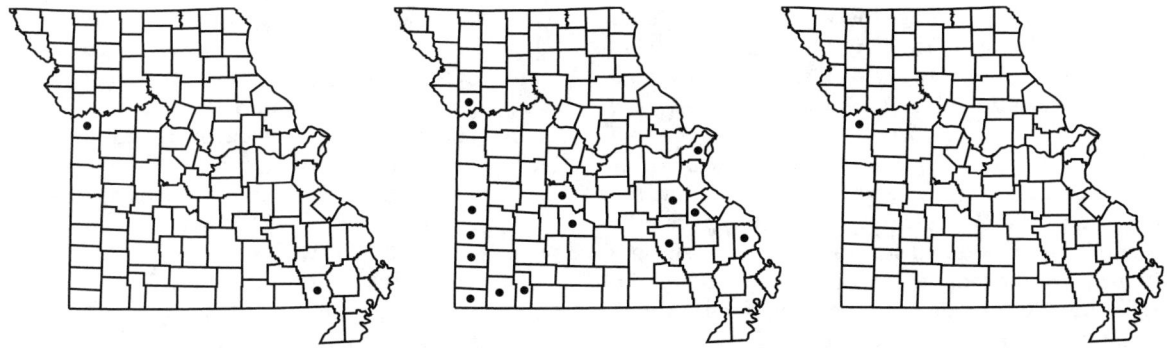

1680. Euphorbia serpyllifolia   1681. Euphorbia spathulata   1682. Euphorbia stictospora

brous. Seeds 1.0–1.6 mm long, ovate to oblong-ovate in outline, angular in cross-section, flat to slightly convex at the base, the surface smooth or with 1–4 low, indistinct cross-wrinkles (these not extending the full width of the face), rarely appearing faintly roughened or pitted, white to pinkish white, light tan, or pale purplish-tinged, lacking a caruncle. July–October.

Uncommon, known thus far from a presumably native historical collection from Jackson County and an introduced occurrence in Butler County (western U.S. east to Michigan and Texas; Canada, Mexico; introduced eastward to New Hampshire and Florida). Open, disturbed areas.

Steyermark (1963) was unsure whether the historical occurrence in Jackson County was native or not. The Butler County plants were growing in disturbed ground beneath a bird feeder.

19. Euphorbia spathulata Lam.

Pl. 383 g–i; Map 1681

Plants annual, with taproots. Stems 10–60 cm long, erect or ascending, unbranched below the inflorescence or occasionally few-branched, the branches not flattened toward the tip, usually green to yellowish green, sometimes faintly purplish-tinged, glabrous. Leaves alternate above the lowest node and below the inflorescence branches (those of the inflorescence branches usually opposite, those of the basal node opposite or whorled but usually absent at flowering), sessile. Stipules absent. Leaf blades 10–45 mm long, oblanceolate to oblong-oblanceolate (those of the leaves of the inflorescence branches broadly ovate to broadly ovate-triangular), unlobed, rounded to truncate or shallowly cordate at the base and sometimes somewhat clasping the stem, rounded or broadly angled to a bluntly pointed tip, the margins finely toothed mostly above the midpoint (the teeth sometimes minute and visible only with magnification), the surfaces glabrous, yellowish green to green. Inflorescences terminal umbellate panicles with a whorl of leaves at the base and each of the usually 3 primary branches often branched 1–3 additional times, the cyathia solitary at the branch tips and at the branch points. Involucre 0.6–0.9 mm long, glabrous, the rim shallowly 4- or 5-lobed to nearly entire, the marginal glands 4 or 5, 0.1–0.3 mm long, oblong-oval to elliptic or slightly kidney-shaped, greenish yellow to yellow, less commonly red or reddish-tinged, lacking a petaloid appendage. Staminate flowers 3–8 per cyathium. Ovaries glabrous, but the surface densely warty, the styles 0.3–0.8 mm long, each divided ¼–½ of the way from the tip into 2(3) slightly club-shaped lobes. Fruits 2–3 mm long, glabrous but the surface finely warty. Seeds 1.3–1.7 mm long, broadly elliptic-ovate to nearly circular in outline, slightly biconvex in cross-section, rounded at the base, the surface with a fine network of low ridges, reddish brown to dark purplish brown but sometimes appearing slightly glaucous, with a pale, irregularly winglike caruncle, this often breaking off as the seeds are dispersed. May–July.

Scattered mostly in the western half of the state and uncommon in the eastern portion of the Ozark Division (Minnesota to Texas west to Washington and California). Glades, upland prairies, ledges and tops of bluffs, and less commonly banks of streams and rivers and bottomland forests; also ditches and railroads.

20. Euphorbia stictospora Engelm. (mat spurge)

Chamaesyce stictospora (Engelm.) Small

Pl. 382 a–c; Map 1682

Plants annual, with taproots. Stems 5–45 cm long, usually prostrate, occasionally with ascending tips, several- to many-branched, the branches often overlapping (plants mat-forming), not flattened toward the tip, usually yellowish brown (sometimes slightly reddish-tinged), densely and evenly pubescent with spreading hairs. Leaves opposite, sessile or very short-petiolate. Stipules small scales 0.5–1.2 mm long, these not fused or, on 1 side of the stem, those from the adjacent leaf in each pair fused toward the base into a single small structure positioned between the leaf bases, variously entire to irregularly toothed or fringed. Leaf blades 3–10 mm long, oblong to oblong-obovate or occasionally nearly circular, asymmetrical at the base with the side toward the stem tip usually angled or rounded and the other side more or less truncate and expanded into a small, rounded auricle, broadly rounded to bluntly pointed or occasionally shallowly notched at the tip, the margins minutely or more deeply several-toothed (best observed with magnification), the upper surface sparsely to moderately pubescent with more or less spreading hairs and lacking a reddish spot, the undersurface moderately to densely pubescent with more or less spreading, slender hairs, often somewhat lighter green than the upper surface. Inflorescences axillary, of solitary cyathia or appearing as small clusters on short axillary branches. Involucre 0.7–1.0 mm long, moderately to densely hairy on the outer surface, the rim shallowly 5-lobed, the marginal glands 5, 0.1–0.3 mm long and often somewhat unequal in size, the body oblong, green to reddish purple, with a relatively inconspicuous petaloid appendage 0.1–0.3 mm long, this white to strongly pinkish- or reddish-tinged. Staminate flowers 3–9 per cyathium. Ovaries hairy, the styles 0.2–0.5 mm long, each entire or inconspicuously notched at the very tip. Fruits 1.5–2.3 mm long, moderately to densely pubescent with more or less spreading hairs, especially toward the angles. Seeds 1.0–1.5 mm long, oblong-ovate to ovate in outline, angular in cross-section, concave at the base, the surface finely pitted, some of the pits rarely forming shallow, irregular troughs and the seeds then appearing partially and irregularly few-ridged, light to dark brown, usually mottled, sometimes with a thin, white coating, this often wearing away irregularly, lacking a caruncle. July–October.

Introduced, uncommon, known only from historical collections from Jackson County (Arizona to Texas north to Wyoming and Iowa; Mexico; introduced sporadically east to New York). Dry, open, disturbed areas.

5. Phyllanthus L. (leaf-flower)
(Webster, 1970)

Plants annual or perennial herbs (shrubs or trees elsewhere), monoecious or occasionally dioecious (in *P. polygonoides*), with clear sap, glabrous (stinging and nonstinging hairs absent). Stems erect or ascending to arched, usually branched. Leaves alternate, sessile or more commonly very short-petiolate, the petiole attached at the base of the nonpeltate blade. Leaf blades narrowly oblong or oblanceolate to oval or obovate, angled or tapered at the base, rounded or broadly angled to a usually bluntly pointed tip, the margins entire, relatively thin-textured, inconspicuously pinnately veined (often only the midvein apparent). Stipules scalelike, 1–2 mm long, usually brown, sometimes shed early, narrowly lanceolate to ovate-triangular, the base sometimes with 1 or 2 minute, rounded auricles at the base. Inflorescences axillary, of solitary flowers or small, sessile clusters of 2–4 flowers (the staminate and pistillate flowers variously positioned on different plants but in our species the staminate ones often more common toward the stem tip), usually with a small, stipulelike bract at the base, the flowers individually mostly short-stalked. Calyces deeply (5)6-lobed, 0.5–1.5 mm long (in pistillate flowers sometimes becoming enlarged to 2.5 mm as the fruits mature), oblong to obovate, rounded at the tip, persistent at fruiting. Petals absent. Nectar disc entire or more or less divided into (5)6 lobes, these broadly rounded at the tip. Staminate flowers with (2)3 free stamens (the filaments more or less arching outward). Pistillate flowers with the ovary 3-locular and 2 ovules per locule, the 3 styles separate or nearly so, each deeply 2-lobed, each lobe slightly broadened into an inconspicuous terminal stigma. Fruits 1–3 mm long, 1.5–3.5 mm in diameter, not or only slightly lobed (circular or slightly and very bluntly 3-angled in cross-section), sometimes shallowly concave at the tip, tan to yellowish brown at maturity. Seeds up to 6 per fruit, wedge-shaped, lacking a caruncle, the surface with finely warty or with minute tubercles, gray to dark brown, sometimes slightly mottled. Perhaps 750–800 species, North America to South America, Caribbean Islands, Africa, Asia, Malesia, Australia.

The generic limits of *Phyllanthus* are still a controversial topic (Webster, 1994). Recent molecular studies (Wurdack et al., 2004; Kathriarachchi et al., 2005; Samuel et al., 2005) have

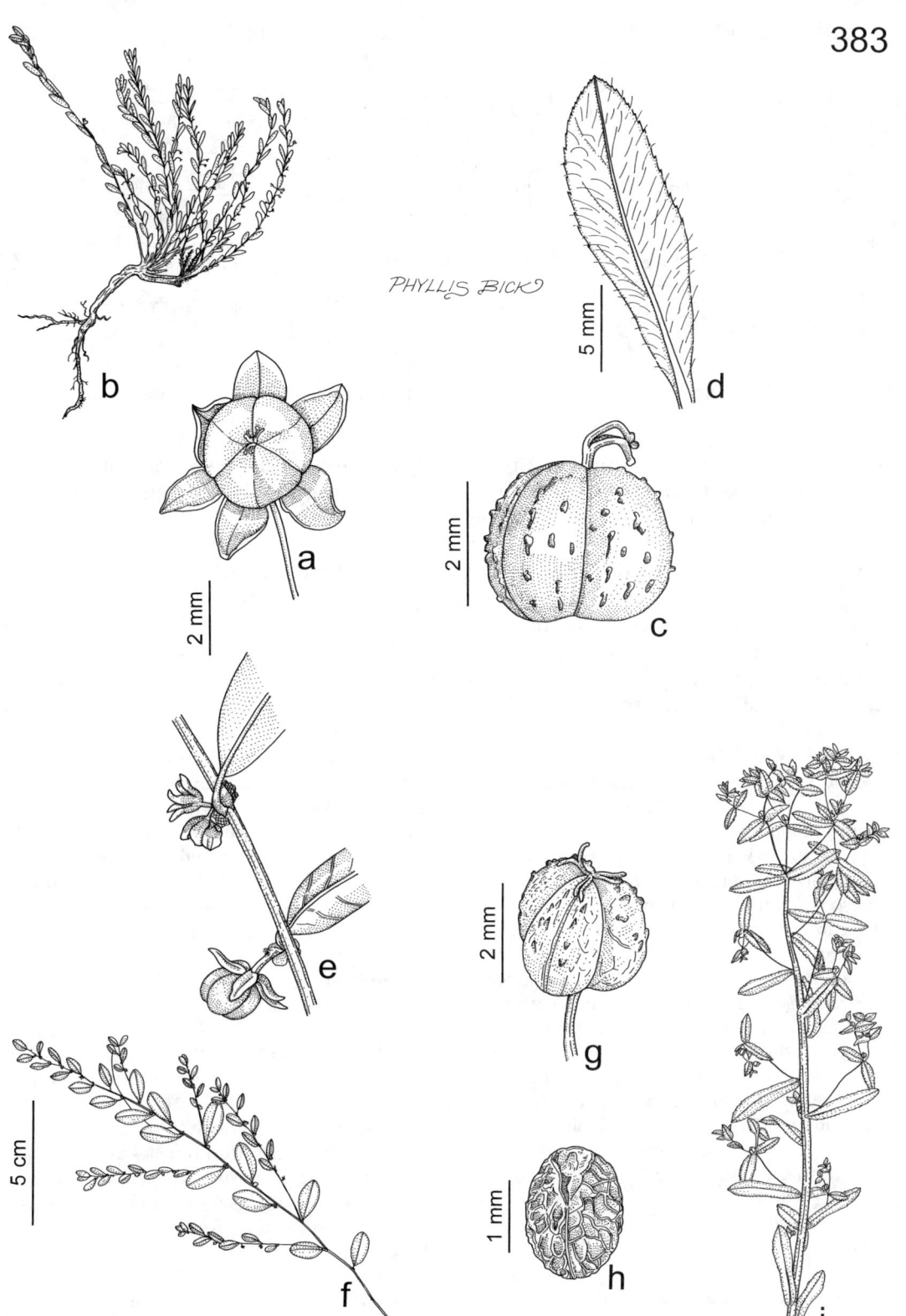

Plate 383. Euphorbiaceae. *Phyllanthus polygonoides*, **a)** fruit with calyx (top view), **b)** habit. *Euphorbia platyphyllos*, **c)** fruit, **d)** leaf. *Phyllanthus caroliniensis*, **e)** nodes with flowers and fruit, **f)** habit. *Euphorbia spathulata*, **g)** fruit, **h)** seed, **i)** habit.

1683. Phyllanthus caroliniensis   1684. Phyllanthus polygonoides   1685. Ricinus communis

suggested that either the group needs to be split into several smaller genera or the generic concept needs to be broadened to include three or four other groups currently recognized as separate genera. Further studies are needed.

1. Plants annuals with slender, fibrous roots; stems solitary, the branches and leaves arranged in 2 ranks (the plant thus appearing relatively flat); leaf blades oval to obovate, rounded at the tip .... 1. P. CAROLINIENSIS
1. Plants perennials with stout, usually woody roots; stems usually several, the branches and leaves arranged spirally; leaf blades narrowly oblong to oblanceolate, broadly angled to a usually bluntly pointed tip .... 2. P. POLYGONOIDES

1. Phyllanthus caroliniensis Walter ssp. caroliniensis (Carolina leaf-flower)

Pl. 383 e, f; Map 1683

Plants annual, with slender, fibrous roots. Stems solitary, 10–40 cm long, ascending or arched, with few to several spreading to loosely ascending branches, the branches arranged in 2 ranks. Leaves arranged in 2 ranks, giving the plant a flattened appearance. Leaf blades 6–25 mm long, oval to obovate, rounded at the tip, both surfaces green. Stipules usually uniformly tan to reddish brown at maturity. Axillary clusters with 1 staminate and often 1–3 pistillate flowers. Calyces of the staminate flowers 0.5–0.7 mm long, those of the pistillate flowers 0.8–1.5 mm long at flowering, the lobes narrowly oblong to narrowly oblong-oblanceolate, usually with a yellowish green central stripe and white margins. Fruits 1.5–2.0 mm in diameter. Seeds 0.8–1.1 mm long, gray to reddish brown. $2n=36$ [in ssp. *guianensis* (Klotzsch) G.L. Webster]. June–October.

Scattered mostly south of the Missouri River (Pennsylvania to Florida west to Kansas and Texas; Mexico, Central America, South America). Banks of streams and rivers, margins of ponds, lakes, and sinkhole ponds, bottomland forests, mesic upland forests in ravines, and moist depressions of upland prairies; also fallow fields, ditches, roadsides, and moist, disturbed areas.

Phyllanthus caroliniensis has an extremely broad latitudinal range, and several subspecies and varieties have been named. The other North American subspecies, ssp. *saxicola* (Small) G.L. Webster, which occurs in Florida and the Caribbean Islands, differs in its less-branched, roughened stems, somewhat thicker leaves, somewhat lobed (vs. entire) nectar disc, and slightly smaller flowers and fruits.

2. Phyllanthus polygonoides Nutt. ex Spreng. (buckbrush)

Pl. 383 a, b; Map 1684

Plants perennial, with a stout, often horizontal, usually woody rootstock. Stems several to numerous in a dense cluster, 10–40 cm long, erect or ascending, with relatively few strongly to loosely ascending branches, the branches arranged spirally. Leaves arranged spirally, the plants not appearing flattened. Leaf blades 5–10 mm long, narrowly oblong to oblanceolate, broadly angled to a usually bluntly pointed tip, the upper surface grayish green, the undersurface lighter-colored and somewhat silvery. Stipules with a brown or pinkish-tinged central stripe and relatively broad, white margins. Axillary clusters with usually 1–4 staminate and often 1 pistillate flowers (the plants occasionally appearing fully dioecious). Calyces of the staminate flowers 0.7–1.3 mm long, those of the pistillate flowers 0.8–1.5 mm long at flowering, the lobes oblong

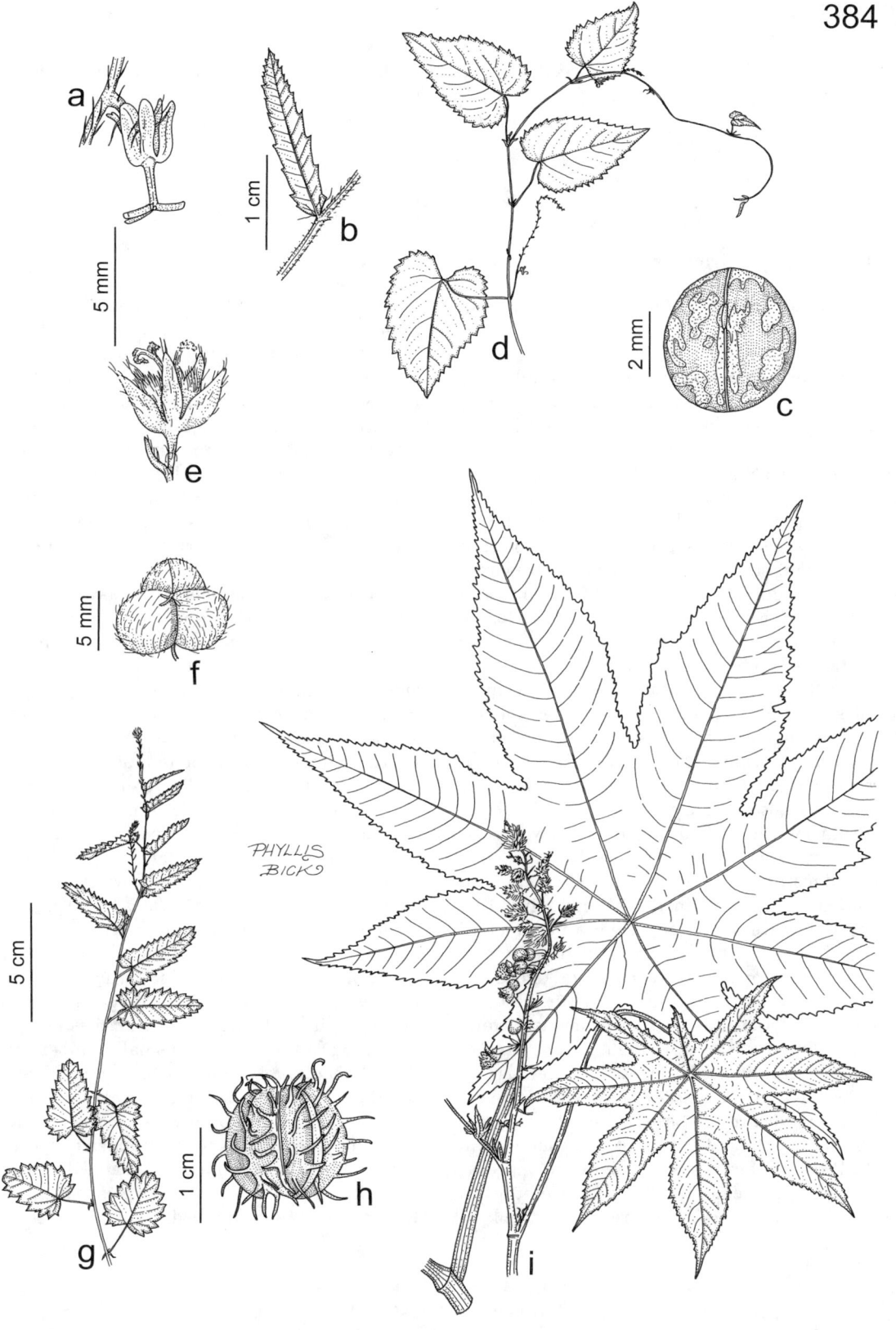

Plate 384. Euphorbiaceae. *Tragia ramosa*, **a)** pistillate flower, **b)** leaf. *Tragia cordata*, **c)** seed, **d)** habit. *Tragia betonicifolia*, **e)** flower, **f)** fruit, **g)** habit. *Ricinus communis*, **h)** fruit, **i)** flowering branch and leaves.

to obovate, entirely pale yellowish green or more often with narrow, white margins. Fruits 2.5–3.3 mm in diameter. Seeds 1.2–1.5 mm long, usually dark brown. 2n=16. June–October.

Uncommon, known thus far only from Stone County (Arizona to Missouri, Arkansas, and Louisiana). Dolomite glades.

6. Ricinus L.

One species, Africa, Asia, introduced nearly worldwide.

1. Ricinus communis L. (castor oil plant, castor bean)

Pl. 384 h, i; Map 1685

Plants annuals (becoming shrubs or small trees in tropical regions), monoecious, with clear sap, glabrous (lacking stinging and nonstinging hairs). Stems 100–500 cm long, erect or ascending, branched, sometimes somewhat glaucous, usually hollow between the nodes. Leaves alternate, long-petiolate, the petiole with usually several large, saucer-shaped glands positioned mostly at the base and tip, peltate, the petiole attached well away from the blade margin. Leaf blades 10–90 cm long and wide, more or less circular, palmately 6–11-lobed, the lobes triangular to more commonly elliptic or ovate, tapered to a sharply pointed tip, the margins coarsely and sharply toothed, pinnately veined. Stipules fused into a membranous 2-lobed sheath, 12–20 mm long, tan or purplish-tinged, shed early and leaving a circular scar around the stem. Inflorescences terminal but often appearing axillary or lateral and opposite the leaves, racemes or less commonly narrow, racemelike panicles, each with clusters of staminate flowers toward the base and clusters of all pistillate or mixed staminate and pistillate flowers toward the tip, the latter rarely with a few perfect flowers, subtended by a papery lanceolate to ovate bract at the base, the flowers individually sessile or relatively short-stalked. Calyces deeply 3–5-lobed, 2–5 mm long (in pistillate flowers usually shed as the flower opens), ovate, sharply pointed at the tip. Petals absent. Nectar disc absent. Staminate flowers with numerous free stamens, the filaments more or less fused into several clusters toward the base and irregularly several-branched, the anthers sometimes more than 1,000. Pistillate flowers with the ovary 3-locular and 1 ovule per locule, the 3 styles fused toward the base, bright red to pinkish red, each deeply 2-lobed, the lobes with densely papillose stigmatic regions for most of their length. Fruits 10–16 mm long, 12–16 mm in diameter, not or only slightly lobed (circular or slightly and very bluntly 3-angled in cross-section), red to purplish or pinkish red, the surface with dense, long, soft prickles. Seeds usually 3 per fruit, 8–12 mm long, more or less ellipsoid, slightly flattened, with a knoblike caruncle at the end adjacent to the attachment point, the surface appearing smooth, shiny, mottled with dark brown, light brown and white. 2n=20. June–October.

Introduced, uncommon and sporadic (native of Africa, India; introduced nearly worldwide, in the U.S. in the southwestern, southeastern, and northeastern quarters of the country). Railroads, roadsides, and open, disturbed areas.

Castor oil is extracted from the seeds of this species and is used in medicine as a purgative, but it is also important in the manufacture of paints, inks, plastics, soaps, and linoleum (Webster, 1967). Cultivation of *R. communis* as a crop in the United States has declined, but the species is widely grown as an ornamental, especially in the warmer portions of the country. The seeds are extremely poisonous because of concentrations of the phytotoxin ricin (a complex glycoprotein) in the seed coat. Because the hard seed coat sometimes can pass through the digestive tract intact, accidental ingestion of seeds sometimes does not lead to poisoning. However, if seeds are chewed or crushed, severe poisoning can result if even small numbers are ingested, leading to strong abdominal pain, reduced blood pressure, hypothermia, excessive salivation, diarrhea (sometimes bloody), weakness, trembling, anorexia, sweating, vomiting, sudden collapse, and sometimes seizures, coma, and death (Burrows and Tyrl, 2001). The leaves are also poisonous, but to a lesser degree than the seeds.

7. Tragia L. (noseburn, tragia)
(Miller and Webster, 1967)

Plants perennial herbs, monoecious, usually with a woody vertical rootstock, with clear sap, variously pubescent with shorter, unbranched, nonglandular hairs and at least some longer, needlelike, stinging hairs. Stems solitary or few to several, prostrate to ascending, sometimes twining, branched or unbranched. Leaves alternate, short- to long-petiolate, the petiole attached at the base of the nonpeltate blade. Leaf blades linear to lanceolate-triangular, ovate-triangular, or heart-shaped, rounded, truncate, or cordate at the base, rounded or more commonly angled or tapered to a usually sharply pointed tip, the margins toothed, usually pinnately veined (sometimes appearing somewhat 3- or 5-veined in *T. cordata*). Stipules scalelike or somewhat leaflike, 1–8 mm long, green or tan to brown, usually persistent, narrowly lanceolate to narrowly ovate-triangular, the margins usually with sparse to moderate spreading hairs. Inflorescences lateral and opposite the leaves (also terminal elsewhere), slender racemes with 1(2) pistillate flower(s) at the base and usually several to many nodes with solitary staminate flowers toward the tip, the nodes usually not crowded at flowering, each flower with a short, slender bract (this not folded longitudinally but sometimes slightly concave). Flowers lacking a corolla and nectar disc. Staminate flowers with the stalk jointed toward the base (the portion below the joint persistent, the upper portion shed with the flower), with 2 or 3 small stamens (more elsewhere) often having somewhat thickened filaments (these usually fused at the base). Pistillate flowers with the ovary having usually 3 locules and 1 ovule per locule, the 3 styles fused toward the base, not further lobed or branched. Fruits 3-lobed (rarely 1 of the carpels aborting and the capsule then with 2 larger lobes and 1 much smaller lobe), more or less explosively dehiscent, the seed usually dispersed with the associated portion of fruit wall. Seeds nearly spherical, the caruncle absent, the surface smooth (appearing roughened elsewhere), yellowish brown to dark brown, sometimes mottled with light yellow. About 150 species, North America to South America, Caribbean Islands, Africa, Asia, Australia.

Steyermark (1963) included only two species for Missouri, *T. cordata* and *T. urticifolia* Michx. The plants that he placed into the latter species are here treated as *T. betonicifolia* and *T. ramosa,* following Miller and Webster (1967). True *T. urticifolia* is similar to *T. betonicifolia* in its overall appearance but differs in its usually more conspicuously hairy stems and in having the persistent basal portion of the staminate flower stalks relatively long, longer than the subtending bract. It occurs from North Carolina and apparently Kentucky to Florida west to Arkansas and Texas.

The morphologically complex, multicellular stinging hairs of *Tragia* species are sometimes relatively sparse, and the burning reaction may be delayed for a few minutes after initial exposure. The burning sensation is caused by injection of a mixture of mostly complex proteins into the skin, although the specific compound(s) responsible remain undetermined (Burrows and Tyrl, 2001).

1. Stems twining, often climbing into adjacent shrubs and saplings; leaf blades broadly ovate to heart-shaped, cordate at the base, tapered to a usually sharply pointed tip, those of the largest leaves more than 5 cm long; fruits 5–7 mm long, 11–13 mm in diameter . . . . 2. T. CORDATA
1. Stems not or rarely only slightly twining, prostrate to ascending but not climbing into adjacent trees or shrubs; leaf blades linear to ovate-triangular, rounded, truncate, or cordate at the base, rounded or more commonly angled to a sharply or occasionally bluntly pointed tip, all 5 cm long or shorter; fruits 3–5 mm long, 6–9 mm in diameter
    2. Calyx lobes of pistillate flowers longer than the pistil; styles fused in the lower $1/3$–$1/2$, the stigmas papillose; leaf blades lanceolate to ovate-triangular . . . . 1. T. BETONICIFOLIA
    2. Calyx lobes of pistillate flowers shorter than the pistil; styles fused less than $1/3$ of length, the stigmas nearly smooth; leaf blades linear to narrowly oblong or lanceolate-triangular (those of the lowermost leaves sometimes ovate but these rarely persistent by flowering) . . . . 3. T. RAMOSA

1686. Tragia betonicifolia   1687. Tragia cordata   1688. Tragia ramosa

1. Tragia betonicifolia Nutt.
 T. urticifolia Michx. var. *texana* Shinners
Pl. 384 e–g; Map 1686

Stems 10–40(–60) cm long, prostrate or ascending, not twining or rarely slightly twining (but not regularly climbing in adjacent vegetation), pubescent with moderate short, straight to curved, softer, nonstinging hairs and sparse to moderate longer, stiff, spreading, stinging hairs. Leaves short- to long-petiolate, the stipules 2–5 mm long. Leaf blades 1–5 cm long, lanceolate to ovate-triangular, rounded, truncate, or cordate (often broadly so) at the base, rounded or more commonly angled to a sharply or occasionally bluntly pointed tip, the surfaces pubescent with usually sparse nonstinging hairs and sparse to moderate stinging hairs. Inflorescences with 1(2) pistillate flower(s) at the base below 14–75 staminate nodes. Staminate flowers with the bract 1–2 mm long; the flower stalk 0.7–1.0 mm long, the persistent lower portion 0.3–0.6 mm long; the calyx 3–5-lobed, 1.2–2.3 mm long; the stamens(2)3(4). Pistillate flowers with the bract 1.5–2.0 mm long; the flower stalk 0.7–1.0 mm long at flowering, elongating to 3–4 mm long at fruiting; the calyx (5)6-lobed, 1.5–3.0 mm long at flowering (longer than the pistil), enlarging to 3–5 mm long at fruiting; the styles fused in the lower $1/3$–$1/2$, the stigmas papillose. Fruits 4–5 mm long, 7–9 mm in diameter, moderately pubescent with mostly stinging hairs. Seeds 3–4 mm long. June–September.

Scattered south of the Missouri River, mostly in the Ozark and Unglaciated Plains Divisions (Kansas to Texas east to Missouri and Louisiana, possibly also Alabama). Glades, upland prairies, ledges and tops of bluffs, savannas, and openings of dry upland forests; often on limestone or dolomite substrates; also pastures, old fields, railroads, and roadsides.

As noted above, this species was first documented from Missouri by Miller and Webster (1967), who redetermined materials that Steyermark (1963) had referred to the closely related *T. urticifolia*. *Tragia betonicifolia* is the most commonly encountered member of the genus in Missouri. Broader-leaved individuals are sometimes misdetermined as *T. cordata*, but that species has a twining habit, larger, more heart-shaped leaves, and larger fruits and seeds, among other differences.

2. Tragia cordata Michx. (heart-leaved tragia)
Pl. 384 c, d; Map 1687

Stems 40–150 cm long, twining, often climbing into adjacent shrubs and saplings, pubescent with moderate short, mostly curved, softer, nonstinging hairs and sparse to moderate longer, stiff, spreading, stinging hairs. Leaves long-petiolate, the stipules 2–8 mm long. Leaf blades 2–12 cm long (those of the largest leaves more than 5 cm long), lanceolate to ovate-triangular, broadly ovate to heart-shaped, cordate at the base, tapered to a usually sharply pointed tip, the surfaces pubescent with usually sparse nonstinging hairs and sparse to moderate stinging hairs. Inflorescences with 1 pistillate flower at the base below (9–)18–40(–60) staminate nodes. Staminate flowers with the bract 1.5–2.0 mm long; the flower stalk 1.5–2.2 mm long, the persistent lower portion 0.7–1.0 mm long; the calyx 3(4)-lobed, 0.7–1.5 mm long; the stamens 3(4). Pistillate flowers with the bract 1.5–2.0 mm long; the flower stalk 1.0–1.5 mm long at flowering, elongating to 2.5–3.0 mm long at fruiting; the calyx 6-lobed, 1.5–2.0 mm long at flowering (about as long as or longer than the pistil), enlarging to 2–3 mm long at fruiting; the styles fused in the lower $1/4$–$1/3$, the stigmas papillose. Fruits 5–7 mm long, 11–13 mm in diameter, moderately pubescent with mostly stinging hairs. Seeds 4.3–5.5 mm long. July–September.

Uncommon, sporadic in the Ozark Division (Texas to Florida north to Missouri and Indiana). Bottomland forests and mesic upland forests in ravines.

Steyermark (1963) mapped this species from several additional counties in the Ozarks, but the voucher specimens to support these reports either could not be relocated during the present research or subsequently were redetermined as *T. betonicifolia*.

3. Tragia ramosa Torr.

Map 1688, Pl. 384 a, b

Stems 10–30(–50) cm long, prostrate or loosely ascending, not twining or rarely slightly twining (but not regularly climbing in adjacent vegetation), pubescent with moderate to dense, short, mostly curved, softer, nonstinging hairs and moderate longer, stiff, spreading, stinging hairs. Leaves short- to less commonly long-petiolate, the stipules 1–4 mm long. Leaf blades 1–4 cm long, linear to narrowly oblong or lanceolate-triangular (those of the lowermost leaves sometimes ovate with a cordate base but these withered or absent by flowering), rounded to truncate at the base, rounded or more commonly angled to a sharply or occasionally bluntly pointed tip, the surfaces pubescent with sparse to moderate nonstinging hairs and usually moderate stinging hairs. Inflorescences with 1(2) pistillate flower(s) at the base below 2–10(–20) staminate nodes. Staminate flowers with the bract 1–2 mm long; the flower stalk 0.7–2.0 mm long, the persistent lower portion 0.4–1.5 mm long; the calyx 3- or 4-lobed, 1.0–2.2 mm long; the stamens(2)3 or 4. Pistillate flowers with the bract 1.5–2.0 mm long; the flower stalk 1.0–1.5 mm long at flowering, elongating to 2.0–2.5 mm long at fruiting; the calyx (5)6-lobed, 0.8–2.5 mm long at flowering (shorter than the pistil), enlarging to 1.5–3.0 mm long at fruiting; the styles fused toward the base less than ⅓ of their length, the stigmas nearly smooth. Fruits 3–4 mm long, 6–8 mm in diameter, moderately pubescent with mostly stinging hairs. Seeds 2.5–3.5 mm long. $2n=44$. June–September.

Scattered in the western half of the Ozark Division, uncommon farther east (southwestern U.S. east to Missouri, Arkansas, and Texas; Mexico). Limestone and dolomite glades; also rarely dry pastures, ditches, and roadsides.

As noted above, this species was first documented from Missouri by Miller and Webster (1967), who redetermined materials that Steyermark (1963) had referred to *T. urticifolia*. *Tragia ramosa* tends to occur in very droughty exposed habitats and tends to appear smaller and grayer than does *T. betonicifolia*. However, both species are variable in overall appearance, and it is better to depend on the flower characters to separate them than the overlapping trends in vegetative differences.

FABACEAE (LEGUMINOSAE) (Bean Family)
Contributed by David J. Bogler and George Yatskievych

Plants annual or perennial, herbs, vines, shrubs, or trees, sometimes armed with spines, thorns, or prickles. Leaves alternate (occasionally appearing whorled from the tips of short spur shoots), sessile or petiolate, the petioles sometimes with 1 or more conspicuous glands. Stipules present, mostly herbaceous or scalelike, occasionally spiny, persistent or shed during leaf development. Leaf blades rarely simple, more commonly trifoliate or pinnately compound, less commonly palmately compound, the leaf blade or leaflet margins entire or less commonly toothed or lobed, the leaflets sometimes with small, scalelike or stipulelike structures (stipels) at the base, sessile or short-stalked, the stalks often with a thickened basal portion (pulvinus). Inflorescences axillary and/or terminal, variable, ranging from solitary flowers to clusters, racemes, spikes, heads, or panicles. Flowers usually perfect, actinomorphic, zygomorphic, or asymmetrical, hypogynous or perigynous, usually with a short, saucer- to bell-shaped hypanthium and/or a nectar disc. Sepals (4)5, free or fused, often persistent at fruiting. Petals (1–)5, free or less commonly fused toward the base, rarely attached to the stamens. Stamens 5, 10, or numerous, free or some or all of the filaments fused, the anthers with 2 locules, dehiscing by longitudinal slits, attached at the base or more commonly toward the midpoint. Pistil 1 per flower, with 1 carpel. Ovary superior or nearly so, with 1 locule containing 2 to many ovules usually in 2 longitudinal rows, the placentation lateral along 1 of the 2 sutures. Style 1, with 1 stigma, this terminal, minute. Fruits legumes, dehiscent longitudinally along both sutures, less commonly loments or indehiscent. Seeds 1 to numerous, often with a thick, hard seed coat. About 725 genera, about 19,000 species, worldwide.

Fabaceae (in the broad sense) are one of the three largest families of flowering plants in terms of species numbers (after the composites and orchids) and are the second most important family economically (after the grasses). Numerous species have diverse uses. A number of legumes are extremely important food crops, including soybeans (*Glycine*), beans (*Phaseolus*), peas (*Pisum*), peanuts (*Arachis*), chickpeas (*Cicer*), and lentils (*Lens*). Although in most cases it is the fruits or seeds that are eaten, a few genera have edible tubers, some of which were harvested for food or ceremonial purposes by Native Americans, including *Apios* and *Orbexilum*. Others are important for hay and livestock forage, including clovers (*Trifolium*), sweet clovers (*Melilotus*), and alfalfa (*Medicago*). Because many legume species have symbiotic relationships with bacteria of the genus *Rhizobium* in specialized root nodules and consequently have the capacity to fix atmospheric nitrogen into nitrates, numerous species are grown and subsequently plowed into the substrate as so-called green manures for soil improvement. The family also contains numerous genera grown as ornamentals for their attractive foliage, flowers, or fruits. The wood of most woody legumes is quite hard and dense, and the family contains a number of timber trees and species used for handicrafts and other wood products. The family also contains a number of species that have become invasive exotics in the United States following deliberate introduction as ornamentals, for soil stabilization, or pasture plants. The most notorious of these is kudzu (*Pueraria*), but several other genera contain species that cause problems in various regions.

Some authors (Cronquist 1981, 1991; Yatskievych and Turner, 1990) have treated the legumes as comprising three families (Caesalpiniaceae, Fabaceae in the strict sense, and Mimosaceae), but most botanists treat these groups as subfamilies of a single family. All three lineages have a long and complex fossil history. However, recent morphological and molecular phylogenetic analyses have indicated that the subfamily Caesalpinioideae, as traditionally circumscribed, is unnatural and that the mimosoid and papilionoid legumes represent specialized groups derived from some of the more primitive caesalpinioid genera (Hufford, 1992; Doyle et al., 1997; Bruneau et al., 2001; Kajita et al., 2001; Wojciechowski et al., 2004). Because a new classification has yet to be fully resolved and because at least in Missouri the three traditional groups have functional value for identification of genera, the present treatment continues to recognize three subfamilies within the Fabaceae. Note that the key to subfamilies below will not always work well for plants from surrounding states, where, for example, several additional simple-leaved species occur.

1. All or some of the leaf blades 2 times pinnately compound; flowers actinomorphic
    2. Inflorescences panicles or racemes; flowers mostly functionally staminate or pistillate; perianth forming the conspicuous part of the flower, the stamens shorter than to about twice as long as the calyx and corolla
    .... 1. CAESALPINIOIDEAE, P. 1059
    2. Inflorescences globose heads or dense spikes; flowers perfect; perianth inconspicuous; stamens forming the conspicuous part of the flower, many times longer than the calyx and corolla .... 2. MIMOSOIDEAE, P. 1075
1. Leaf blades trifoliate or 1 time pinnately compound, less commonly simple; flowers slightly to strongly zygomorphic (actinomorphic in *Gleditsia*, reduced to 1 petal in *Amorpha*)
    3. Leaf blades simple
        4. Plants shrubs or trees; leaf blades broadly heart-shaped, truncate or notched at the base .... 1. CAESALPINIOIDEAE, P. 1059
        4. Plants annual or perennial herbs; leaf blades variously shaped (but not heart-shaped), narrowed or rounded at the base .... 3. FABOIDEAE (Vol. 3)
    3. Leaf blades trifoliate or 1 time pinnately compound

5. Corollas only slightly zygomorphic (actinomorphic in *Gleditsia*), the uppermost petal positioned inside the lateral petals in buds and often somewhat smaller than the other petals, the petals all free; stamens not hidden by the petals, easily visible at flowering, the filaments not fused . . . .   1. CAESALPINIOIDEAE, P. 1059
5. Corollas usually strongly zygomorphic (except in *Amorpha* and *Dalea*), the uppermost petal positioned outside the lateral petals in buds and usually somewhat larger than the other petals (corolla reduced to 1 petal in *Amorpha*), the 2 lowermost petals often fused along their adjacent margins; stamens more or less hidden by the fused pair of lowermost petals (except in *Amorpha* and *Dalea*), some or all of the filaments fused . . . . 3. FABOIDEAE (Vol. 3)

1. Subfamily Caesalpinioideae Kunth

Plants trees, shrubs, or annual or perennial herbs, sometimes armed with thorns (spines or prickles elsewhere). Root nodules rarely present. Leaves alternate, sometimes appearing whorled in clusters on spur shoots, the petiole commonly with a swollen base (pulvinus) and often with glands. Stipules leaflike and persistent or small, scalelike, and shed early. Leaf blades pinnately compound or less often 2 times pinnately compound, rarely simple. Inflorescences various, axillary or terminal, small and inconspicuous or more commonly showy racemes or panicles, the flowers rarely borne directly on the trunk (cauliflory), usually with bracts. Flowers usually perfect or rarely incompletely monoecious or dioecious, often perigynous. Sepals 5, usually free, often unequal. Corollas well developed, zygomorphic or asymmetrical to nearly actinomorphic, usually of 5 free petals, these often large and showy, not differentiated into banner, wings, and keel (except in *Cercis*), the uppermost petal internal to the 2 lateral petals, the lower pair of petals separate. Hypanthium often present, elongate or short. Stamens 10 or fewer, all fertile or sometimes some of them reduced to staminodes, the filaments usually not fused, the anthers sometimes heteromorphic (some longer than others), attached at the base or toward the midpoint, dehiscent by slits or pores. Pistil often stalked, the ovules 2 to many. Fruits legumes (in Missouri taxa), dehiscent or indehiscent, variously shaped, but most commonly oblong to linear, flat, often dry. Seeds flattened or not, the seed coat hard, sometimes with an elliptic area defined by a fine, lighter-colored line or groove (pleurogram). One hundred fifty to 180 genera, 1,900–2,250 species, nearly worldwide, but most diverse in tropical regions.

The Caesalpinioideae produce legumes in a variety of sizes and shapes. They usually are elongate and with many seeds in our genera but are occasionally reduced in size and with a single seed. Most Caesalpinioideae have 1 or 2 times pinnately compound leaves. Very rarely are the leaves simple, as in *Cercis* or the mostly tropical orchid tree genus *Bauhinia* L. The leaves often are equipped with glands that act as extrafloral nectaries, which attract various insects, often ants. Floral morphology in Caesalpinioideae is diverse, and the subfamily is defined more by what it does not have than by characters shared by its members. Caesalpinioideae do not have the distinctive papilionaceous, butterfly-like (2-winged) corolla characteristic of the Faboideae, nor does the subfamily have the small, actinomorphic corollas and long, exserted stamens characteristic of the Mimosoideae. The petals of Caesalpinioideae are variously subequal or unequal in size, and the corollas vary from nearly actinomorphic to strongly zygomorphic or asymmetrical. The stamens exhibit a lot of variation in morphology, with a tendency to be heteromorphic, with some of the anthers noticeably larger than the rest. Less than half of the genera of Caesalpinioideae are nodulated with nitrogen-fixing bacteria, compared to the majority of genera in Mimosoideae and Faboideae (Allen and Allen, 1981). A few genera have seeds with an elliptical depressed area in the seed coat delineated by a fine, lighter-colored line or groove and known as a pleurogram. Of the genera in Missouri, a pleurogram is found only in *Senna*. A similar structure is found in many genera of Mimosoideae, but the pleurogram is absent in the Faboideae (Cowan, 1981; Gunn, 1991).

Caesalpinioideae traditionally have been divided into a number of tribes, but these have generally proven to be unsatisfactory, and molecular data are now demonstrating they are highly unnatural (Bruneau et al., 2001). Molecular studies also have elucidated general relationships of the genera in Missouri. *Cercis* appears to belong to an ancient lineage that, together with *Bauhinia,* is basal to the rest of Fabaceae (in the broad sense). *Gymnocladus* and *Gleditsia* are closely related. These two genera were once thought to be ancient taxa with primitively simple flowers (Polhill et al., 1981), but it now appears they are not as old as was previously thought and that some of their presumed primitive features may instead be derived (Schnabel and Wendel, 1998; Bruneau et al., 2001). *Senna* and *Chamaecrista* were included in *Cassia* by Steyermark (1963) and others, but Irwin and Barneby (1982) provided ample characters for treating them as discrete genera, an arrangement supported by molecular data (Bruneau et al., 2001). The taxonomic situation of *Pomaria* is complex, having been included in *Hoffmannseggia* and *Caesalpinia,* but molecular data support its recognition at the generic level (Simpson and Miao, 1997).

1. Plants small to large trees
    2. Leaves simple, unlobed; flowers appearing before the leaves; perianth whorls dissimilar, the sepals very short; corollas strongly zygomorphic, purplish pink .... 1. CERCIS
    2. Leaves compound; flowers appearing after the leaves; perianth whorls similar, the sepals well developed; corollas nearly actinomorphic, greenish white
        3. Trunk usually armed with thorns; twigs slender, zigzag; 2 times pinnately compound leaves produced only on long shoots (not on the short spur shoots, which have clusters of 1 time pinnately compound leaves), leaflets less than 1 cm wide, narrowly ovate; flowers in dense, spikelike racemes, hypanthium 2–3 mm long; fruits strongly flattened, papery to leathery, indehiscent; seeds strongly flattened .... 3. GLEDITSIA
        3. Trunks unarmed, lacking thorns; twigs thick, coarse, not noticeably zigzag; leaves all 2 times pinnately compound, leaflets more than 1 cm wide, broadly ovate; flowers in open racemes or panicles, hypanthium 6–11 mm long; fruits thick and only somewhat flattened, woody, dehiscent along the ventral margin with age; seeds somewhat flattened but turgid .... 4. GYMNOCLADUS
1. Plants annual or perennial herbs or small shrubs
    4. Lower leaf surface, stems, and flowers with conspicuous orange or black glandular dots; leaves 2 times pinnately compound; anthers attached near their midpoints, opening longitudinally; fruits crescent in outline, the surfaces with stellate protuberances .... 5. POMARIA
    4. Lower leaf surface, stems, and flowers lacking conspicuous glandular dots (but sometimes with inconspicuous, minute glandular hairs); leaves all 1 time pinnately compound; anthers attached at their bases, opening by terminal pores or slits; fruits relatively straight, the surfaces smooth
        5. Root nodules present; stipules conspicuous, persistent; flower stalks with 2 small bracts; stamens 10, all fertile, in 2 size classes; legumes elastically (explosively) dehiscent, the valves twisting abruptly; seeds lacking a pleurogram .... 2. CHAMAECRISTA
        5. Root nodules absent; stipules inconspicuous, shed early; flower stalks bractless; stamens 10, but the fertile stamens only 7, the upper 3 reduced to staminodes, overall in 3 size classes; fruits gradually dehiscent or indehiscent; seeds with a pleurogram .... 6. SENNA

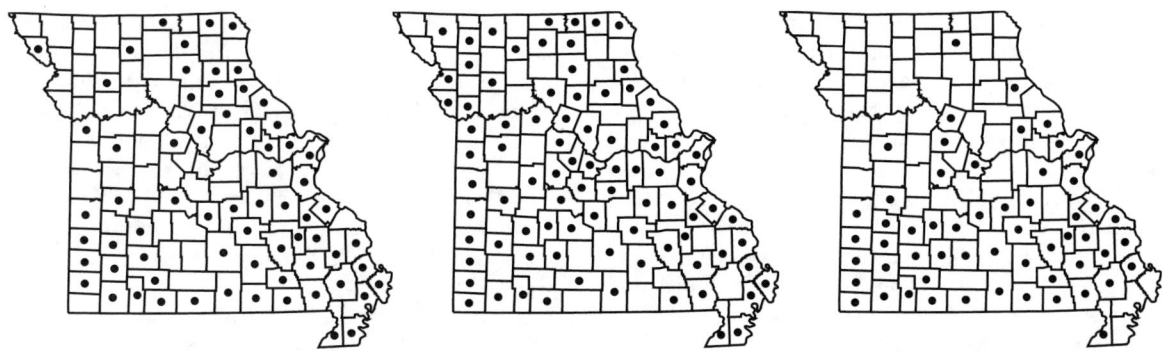

1689. Cercis canadensis   1690. Chamaecrista fasciculata   1691. Chamaecrista nictitans

1. Cercis L. (redbud)

Four to 14 species, temperate regions of North America, Europe, Asia.

Molecular studies have indicated that *Cercis* and the large tropical genus *Bauhinia* (orchid trees) are closely related and together occupy a basal position within the Fabaceae (Bruneau et al., 2001). The simple leaves of *Cercis* are thought to be derived from a pair of leaflets or lobes like those found in most *Bauhinia* species. Two species of *Cercis* have been described from North America: the wide-ranging *C. canadensis* in the eastern United States and Mexico, with more pointed heart-shaped leaves, and *C. occidentalis* Torr. & A. Gray in California, with more rounded, kidney-shaped leaves. Bruneau et al. (2001) concluded that *C. canadensis* is more closely related to the Eurasian *C. siliquastrum* L. than to *C. occidentalis*. The relationship probably dates back to the early Tertiary period when the two continents were much closer to another than in modern times (Davies et al., 2002).

1. Cercis canadensis L. var. canadensis
(eastern redbud, Judas tree)

C. canadensis f. *glabrifolia* Fernald

Pl. 385 c–e; Map 1689

Plants trees 3–12 m tall, unarmed, the crown widely spreading, rounded or flat-topped, the trunks single or multiple, often leaning with age, the bark smooth, dark reddish brown, becoming scaly and peeling with age, the twigs slender, zigzag, the winter buds small, rounded, with several scales; root nodules absent. Leaves appearing after the flowers, the petioles 3–6 cm long, with a swollen region at both ends. Stipules minute, scalelike, shed early. Leaf blades simple, 3–15 cm long, 3–15 cm wide, usually heart-shaped, the base cordate, angled or short-tapered to a bluntly to sharply pointed tip, the margins entire, the upper surface glabrous and often glossy, the undersurface glabrous to moderately but inconspicuously hairy, the venation palmate. Inflorescences umbellate clusters produced on short, spurlike shoots from second-year or older twigs and branches, sometimes directly from the trunk, the flower stalks 5–11 mm long. Flowers perfect, perigynous, strongly zygomorphic. Hypanthium 2–3 mm long, cup-shaped, pinkish-tinged. Calyces of 5 fused sepals, 0.5–1.0 mm long, with short, triangular teeth, persistent at fruiting. Corolla of 5 free petals, these 6–9 mm long, 3–4 mm wide, obovate, tapered to a stalklike base, pinkish purple or rarely white, the banner petal internal to the lateral petals in bud. Stamens 10, all fertile, the filaments not fused, 4–6 mm long, curved, hairy at the base, the anthers about 0.3 mm long, 0.2 mm wide, attached toward the midpoint, dehiscing by lateral slits. Style curved. Fruits legumes, 6–9 cm long, 1.0–1.5 cm wide, strongly flattened, short-stalked, winged along one margin, tapered at each end, papery to leathery, glaucous, few-seeded, without cross-partitions between the seeds, tardily dehiscent, persistent into the winter. Seeds 4.0–4.5 mm long, 3.0–3.2 mm wide, ovoid-elliptic to almost circular, flattened, hard, the surface smooth, reddish brown, shiny; pleurogram absent. $2n=14$. March–May.

Common nearly throughout the state (eastern U.S. west to South Dakota and New Mexico; Canada, Mexico). Mesic upland forests, tops and ledges of bluffs, margins of glades, and banks of streams and rivers; also margins of pastures, railroads, and roadsides.

Cercis canadensis is a well-known tree, much loved for its beautiful display of pinkish purple flowers in the spring. The flowers are pseudo-papilionaceous, which means that they resemble the papilionaceous flowers of members of the subfamily Faboideae in having a banner, lateral wings, and a keel. However, in *Cercis* the banner petal is smaller than the wings and positioned internal to them, and the pair of keel petals are not fused. The flowers produce nectar and are pollinated by long- and short-tongued bees. In nature, the redbud commonly is found as a small to medium-sized understory tree in late stages of forest succession. It often is associated with flowering dogwood (*Cornus florida* L., Cornaceae), which blooms at about the same time but begins slightly later.

Redbud is commonly cultivated. Its small size makes it suitable for suburban yards, but splitting of trunks and branches on mature trees is a problem. Redbud grows best in alkaline soils and partial shade, but it is tolerant of dry conditions and can be planted in the open. Plants grown in the open produce more flowers than those grown in the shade. The leaves turn yellow in autumn. A striking form with white corollas (f. *alba* Rehder) is rare in nature but popular in horticulture. Steyermark (1963) noted that trees growing wild in Franklin County were brought into cultivation and formed the source materials for horticultural white-flowered plants.

Some authors recognize three or more infraspecific taxa within *C. canadensis*. The most commonly segregated ones are the southwesternmost populations in Texas and Mexico, which have thicker leaves that tend to be slightly rounder and shinier. Isely (1998) recognized these as var. *mexicana* (Rose) M. Hopkins and var. *texensis* (S. Watson) M. Hopkins, which differ in details of leaf size and pubescence.

2. Chamaecrista (L.) Moench

Plants annual (perennial herbs or woody elsewhere), unarmed, with 1 to few stems; root nodules present. Leaves short- to long-petiolate, the petiole with a relatively large, sessile or stalked, more or less cup-shaped gland. Stipules well developed, with several prominent, more or less parallel, raised veins, persistent. Leaf blades evenly 1 time pinnately compound. Leaflets numerous, opposite, oblong to narrowly oblong or oblong-elliptic, asymmetrical at the base, the midvein usually somewhat off-center, the margins entire. Inflorescences axillary or lateral from just above the leaf axil, appearing as solitary flowers or small, loose clusters (actually short racemes), the flower stalks with 2 small bracts near or above middle, often twisted 180° so that the flower is resupinate. Flowers perfect, perigynous, somewhat asymmetrical. Hypanthium short, saucer-shaped. Calyces of 5 free sepals. Corollas of 5 free petals, these dissimilar in size and shape, the largest one appearing lowermost in the flower, yellow (rarely white), sometimes reddish-tinged toward the base. Stamens 5 or 10, usually all fertile, the filaments short, not fused, the anthers of different lengths, attached at the base, dehiscing by apical slits or pores. Styles curved. Fruits legumes, narrowly oblong, straight to curved, not twisted, strongly flattened, with 4–20 seeds, elastically dehiscent with coiling valves. Seeds nearly square to trapezoidal in outline, flattened, light to dark brown; pleurogram absent. About 265 species, widespread, primarily in the American tropics and temperate regions.

Although long recognized as a natural group, the generic status of *Chamaecrista* has been debated for over 200 years. *Chamaecrista* and *Senna* often have been included in the large genus *Cassia*, which in its broadest sense contains more than 600 species and is the largest genus in Caesalpinioideae (Isely, 1975; Irwin and Barneby, 1976). Bentham (1871) recognized 3 subgenera that correspond more or less to *Cassia* (in the narrow sense), *Senna*, and *Chamaecrista*. Irwin and Barneby (1981, 1982) provided a discrete set of characters for recognizing these at the generic level and made the required nomenclatural changes. In this narrow sense, *Cassia* consists of only 30 species of trees and shrubs that are confined to the tropics, lack extrafloral nectaries, have very long, sigmoidally curved filaments, and have anthers that dehisce by slits or basal pores. In contrast, both *Chamaecrista* and *Senna* are large genera, each with over 250 species of trees, shrubs, and herbs. Their stamens have very short filaments and dimorphic or trimorphic anthers that dehisce by terminal slits or pores. *Chamaecrista* is

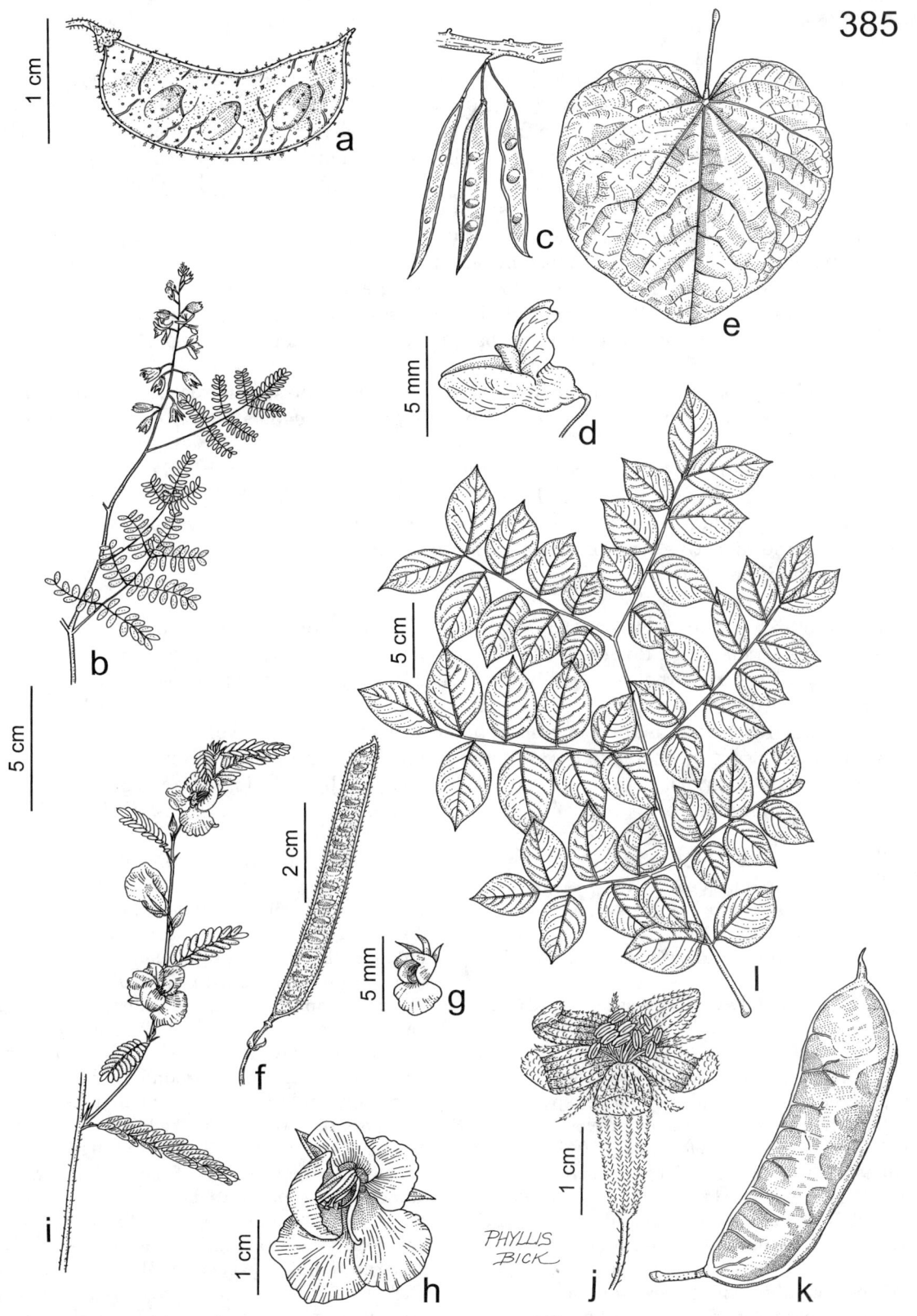

Plate 385. Fabaceae. *Pomaria jamesii*, **a)** fruit, **b)** flowering branch. *Cercis canadensis*, **c)** fruits, **d)** flower, **e)** leaf. *Chamaecrista nictitans*, **f)** fruit, **g)** flower. *Chamaecrista fasciculata*, **h)** flower, **i)** flowering branch. *Gymnocladus dioica*, **j)** staminate flower, **k)** fruit, **l)** leaf.

distinguished from *Senna* by the large, persistent stipules, 2 small bracts on the flower stalk, elastically dehiscent legumes, stamens with usually 2 sizes of anthers, and seeds lacking a pleurogram. *Senna* lacks bracts on the flower stalk, has tardily dehiscent, nonelastic legumes, a graded series of anthers, and seeds with a distinctive pleurogram appearing as a depressed area. Furthermore, root nodules with nitrogen-fixing bacteria are found in *Chamaecrista,* but not in *Senna* or the majority of other Caesalpinioideae.

The leaflets of *Chamaecrista* are somewhat sensitive to the touch, although not as sensitive as those of certain taxa of Mimosoideae. Uprooting the plant also causes the leaflets to close, and herbarium specimens almost always have the leaves in the closed position. On their own, the leaves close and pull upward at night into a so-called sleeping position, a phenomenon known as nyctinasty, common to many Fabaceae. This movement is thought to control water loss or afford protection from herbivores.

The seed dispersal mechanism is also interesting. When the seeds are fully mature and the legume is dry, the two valves separate suddenly, flinging the seeds a meter or more away (Lee, 1984).

The two Missouri species of *Chamaecrista,* particularly the more abundant *C. fasciculata,* are important wildlife food plants and sometimes are planted for this purpose. The foliage is nutritious for deer and livestock, although it contains anthraquinone compounds that can cause irritation of the digestive tract if eaten in large quantities (Burrows and Tyrl, 2001). Birds (especially quail and turkey) are fond of the seeds.

1. Stems 60–90 cm tall, unbranched or few-branched near or above middle; flower stalks 10–15 mm long; flowers showy, 25–30 mm in diameter, the petals subequal, the largest petal only slightly larger than the others; stamens 10, the anthers 6–10 mm long . . . . 1. C. FASCICULATA
1. Stems 30–40 cm tall, usually few-branched toward the base; flower stalks 2–3 mm long; flowers inconspicuous, 8–10 mm in diameter, the petals strongly dimorphic, the largest petal almost twice the size of other petals; stamens 5, the anthers 2–3 mm long . . . . 2. C. NICTITANS

1. Chamaecrista fasciculata (Michx.) Greene
 (showy partridge pea)
Cassia chamaecrista L.
Cassia fasciculata Michx.
Cassia fasciculata var. *robusta* (Pollard) J.F. Macbr.
Cassia fasciculata var. *depressa* (Pollard) J.F. Macbr.
Pl. 385 h, i; Map 1690

Stems 60–90 cm long, unbranched or few-branched from near or above middle, sparsely to moderately pubescent with short, curved or spreading hairs. Leaves 5–9 cm long, with 10–15 pairs of leaflets, the petiole 5–8 cm long, the petiolar gland 1.0–1.8 mm in diameter, sessile or occasionally short-stalked, located near the middle of the petiole. Stipules 7–16 mm long, 1–2 mm wide at the base, narrowly and usually asymmetrically lanceolate-triangular, long-tapered at the tip, with several prominent, parallel veins. Leaflets 10–20 mm long, 2–5 mm wide, narrowly oblong to oblong or occasionally oblong-elliptic, oblique at the base, abruptly tapered to a bluntly or sharply pointed tip, this sometimes with a minute, hairlike extension of the midvein, the margins with short, spreading hairs, the surfaces glabrous or hairy. Inflorescences axillary clusters of 1–3 flowers, with usually only 1 flower open at a time, the stalks 10–15 mm long, the bracts 3–6 mm long. Flowers somewhat asymmetrical, 25–30 mm in diameter. Sepals 10–12 mm long, 2–3 mm wide, lanceolate, sharply pointed, often hairy. Petals 15–20 mm long, 8–18 mm wide, broadly obovate, clawed, subequal, with one lateral petal curved around the stamens, the lower petal only slightly larger than the others, yellow or rarely white, some reddish-tinged toward the base. Stamens 10, unequal, with 9 smaller stamens grouped on one side of the pistil and 1 large stamen on the opposite side, the anthers 6–10 mm long. Ovary 4–5 mm long, hairy, the style 6–7 mm long. Fruits 3–6 cm long, 5–6 mm wide, sparsely to moderately hairy, especially along the margins, black when fully mature. Seeds 3.2–4.0 mm long, 2.0–2.4 mm wide, nearly square, the surfaces finely pitted, dark brown to black, not shiny. $2n=16$. July–October.

Scattered to common nearly throughout the state (eastern U.S. west to South Dakota and New

Mexico; Mexico). Glades, upland prairies, openings of mesic to dry upland forests, savannas, ledges and tops of bluffs, and banks of streams and rivers; also pastures, old fields, fallow fields, and roadsides.

Chamaecrista fasciculata is a widespread species, with populations that vary in growth habit and size, pubescence, leaflet number, gland characters, legume size, and anther color. These characters have been used singly and in combination to describe a number of varieties and forms, particularly in Florida, the Gulf Coast, and Texas (Turner, 1959; Correll and Johnston, 1970; Isely, 1975). However, Irwin and Barneby (1982) argued that most of this variation is not taxonomically useful, and they did not recognize any infraspecific taxa. Rare white-flowered plants from eastern Missouri have been called *Cassia fasciculata* f. *jensenii* E.J. Palmer & Steyerm.

Partridge pea has an interesting floral biology and ecology, reviewed by Gardner and Robertson (2000). The flowers open before dawn and close in the evening and have an unusual arrangement of floral parts. The petals are unequal in size, with the lowest petal the largest. There are 2 groups of stamens. One group of nine stamens arises from one side of the pistil. One of the lateral petals (called the cucullus) is rigid and curves around this group. The other stamen group consists of a single stamen that arises from the other side of the pistil, with a large, deflexed anther bent away from the other group. The style is thin and curved upward, to the left in some flowers and to the right in others. Styles and stigmas curved toward the left pick up pollen from the right side of the bee, and those curved toward the right receive pollen from the bee's other side. The flowers produce no nectar, and the major pollinators are bumblebees. The bees land directly on the anthers, curve their abdomens over the terminal pores, squeeze the anthers, and vibrate their bodies to release the pollen, an activity referred to as buzz pollination. It has been suggested that the nine anthers enclosed in the cucullus serve as feeding anthers to attract the bees, whereas the single curved anther is involved in fertilization, although pollen in each group is equally viable. Experimental removal of the cucullus does inhibit fruit set, suggesting that the cucullus guides the bees (Wolfe and Estes, 1992).

The function of the prominent extrafloral nectary on the leaf petiole also has been the subject of a number of investigations (Gardner and Robertson, 2000). The glands have direct connections to the phloem and secrete a nectar rich in sucrose. Many kinds of insects have been observed visiting the nectaries, but the majority of such visitors are ants. The ants remove the eggs and larvae of other insects that feed on the plants. When ants were experimentally excluded, the plants suffered greater damage from pests and produced fewer seeds than plants with the ants (Durkee et al., 1999).

2. **Chamaecrista nictitans** (L.) Moench **var. nictitans** (small-flowered partridge pea, sensitive pea)
Cassia nictitans L.
Chamaecrista procumbens (L.) Greene
Pl. 385 f, g; Map 1691

Stems 30–40 cm long, usually branched near the base, sparsely to moderately pubescent with short, curved or spreading hairs. Leaves 2–6 cm long, with 6–18 pairs of leaflets, the petiole 4–7 mm long, the petiolar gland 0.5–1.0 mm long, usually stalked, located slightly above the midpoint of the petiole. Stipules 3–6 mm long, 1.2–1.4 mm wide at the base, narrowly and usually asymmetrically lanceolate-triangular, long-tapered at the tip, with several prominent, parallel veins. Leaflets 6–10 mm long, 1–2 mm wide, narrowly oblong to oblong, oblique at the base, abruptly tapered to a minute, sharply pointed tip, the margins glabrous or with minute, ascending hairs, the surfaces glabrous or less commonly inconspicuously hairy. Inflorescences axillary, of solitary or paired flowers, the stalks 2–3 mm long, the bracts 1.5–2.0 mm long. Flowers relatively strongly asymmetrical, 8–10 mm in diameter. Sepals 3–4 mm long, lanceolate, sharply pointed, hairy. Petals dimorphic, with 1 larger petal 4–6 mm long, 3–4 mm wide, obovate, tapered to a stalklike base; and 4 smaller petals 2.6–3.0 mm long, 1.5–2.0 mm wide, yellow. Stamens 5 (rarely with 1 or 2 additional smaller staminodes), slightly unequal, oriented toward the lower side of the flower, the anthers 1.4–2.0 mm long. Ovary 1.5–2.0 mm long, hairy, the style 1.4–1.6 mm long. Fruits 2.0–3.5 cm long, 4–5 mm wide, finely hairy (glabrous elsewhere). Seeds 3.0–3.2 mm long, 1.5–2.0 mm wide, nearly square, the surfaces finely pitted, dark brown, shiny. $2n=16$. July–September.

Scattered, mostly south of the Missouri River (eastern U.S. west to Wisconsin, Kansas, and Texas; introduced in Hawaii). Glades, openings of mesic to dry upland forests, savannas, ledges and tops of bluffs, margins of sinkhole ponds, and banks of streams and rivers; also pastures, old fields, fallow fields, and roadsides.

Several varieties of *C. nictitans*, based on small differences in pubescence, were accepted in the older literature (Steyermark, 1963), but Irwin and Barneby (1982) did not recognize these. Instead they recognized a complex series of subspecies and varieties representing variation of different characters in various portions of the broad range of the species, which extends from North America to South America and various Caribbean Islands. Plants with fewer than ten stamens, relatively long-stalked petiolar glands, and short-stalked flowers with relatively small corollas and short styles correspond to their ssp. *nictitans*, which is widespread but the only subspecies whose range extends into the United States. Within ssp. *nictitans,* Irwin and Barneby separated five varieties based on differences in style length and shape, corolla morphology, and pubescence patterns, with the plants found in the eastern and midwestern United States corresponding to var. *nictitans.*

3. Gleditsia L.

Plants small to more commonly large trees, usually incompletely dioecious, the trunks and branches usually armed with simple or branched thorns, these often in clusters, the branches differentiated into short shoots with clustered leaves and elongate shoots with alternate leaves, the twigs of long shoots often somewhat zigzag, the winter buds inconspicuous and partially sunken into the twig; root nodules absent. Leaves appearing before the flowers, the petiole 1–3 cm long, the blade 1 or 2 times pinnately compound, the first leaves of the year evenly 1 time pinnately compound on the short shoots, the later-produced leaves 2 times pinnately compound on the long shoots. Stipules inconspicuous and scalelike, shed early. Leaflets alternate on the rachis, the margins sometimes minutely scalloped. Inflorescences spikelike racemes, solitary or clustered on the short shoots, arched or drooping; some trees with all staminate or all pistillate inflorescences, but often otherwise pistillate trees with some inflorescences having mixed imperfect and perfect flowers. Flowers perigynous, small, more or less actinomorphic. Hypanthium 2–3 mm long, cup-shaped to bell-shaped. Calyces of 3–5 sepals, these usually slightly unequal, similar in color and slightly shorter than the petals, not closing the flower in buds. Petals 3–5, slightly unequal, greenish white or occasionally slightly yellowish-tinged. Stamens 5–8, usually all fertile, the filaments 2–4 mm long, not fused, hairy toward the base, the anthers about 1.2 mm long, 0.5 mm wide, attached toward the midpoint. Style short, the stigma 2-lobed. Fruits legumes, relatively short or elongate, strongly flattened, short-stalked, straight or curved, sometimes twisted, 1- to many-seeded, indehiscent or dehiscing with age. Seeds elliptic to more or less circular, strongly flattened, brown to greenish brown; pleurogram absent. Twelve to 14 species, North America, South America, Asia.

Gleditsia is closely related to *Gymnocladus* (Gordon, 1966; Lee, 1976; Bruneau et al., 2001). Both are dioecious or incompletely dioecious trees with distributions disjunct between eastern Asia and eastern North America. The flowers of both genera are weakly differentiated into sepals and petals, and they share the unusual condition of sepals that do not cover the petals in the bud. These two genera were once considered primitive, ancient taxa dating back to the late Cretaceous (Polhill et al., 1981), but that idea is now changing with the advent of molecular data (Bruneau et al., 2001; Schnabel and Wendel, 1998) and reevaluation of the fossil record. It now appears that the simple, imperfect flower trait may be derived, and the disjunction may be of more recent origin, with the oldest reliable fossils from the Oligocene, approximately 25–35 million years ago (Schnabel and Wendel, 1998; Herendeen et al., 1992).

Gordon (1966) suggested that the two North American species of *Gleditsia* were each more closely related to different species occurring in ecologically similar regions of Asia than to each other. However, molecular studies have indicated that *G. triacanthos* and *G. aquatica* are sister species, and that they are in turn related to *G. japonica* Lodd. ex W. Baxter, a widespread Asian species (Schnabel and Wendel, 1998).

1. Petiole and rachis more or less glabrous; fruits 3–5 cm long, asymmetrically elliptic to ovate, usually with 1 nearly circular seed, lacking pulp .... 1. G. AQUATICA
1. Petioles and rachis distinctly hairy; fruits 18–35 cm long, elongate, with many elliptic seeds, these embedded in a jellylike pulp .... 2. G. TRIACANTHOS

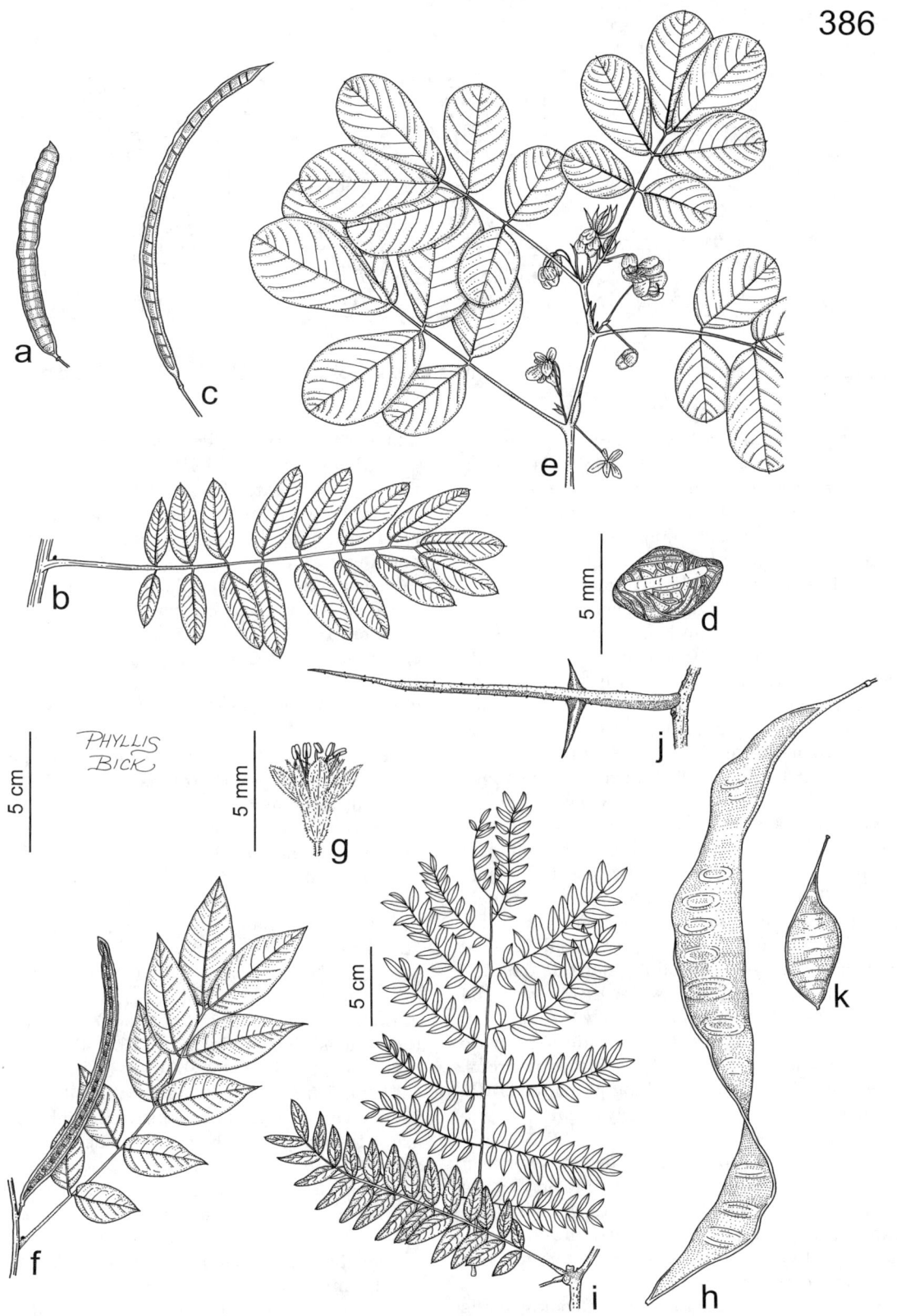

Plate 386. Fabaceae. *Senna marilandica*, **a)** fruit, **b)** leaf. *Senna occidentalis*, **c)** fruit, **d)** seed, **e)** flowering branch. *Senna obtusifolia*, **f)** fruiting branch. *Gleditsia triacanthos*, **g)** staminate flower, **h)** fruit, **i)** leaves. *Gleditsia aquatica*, **j)** branched thorn, **k)** fruit.

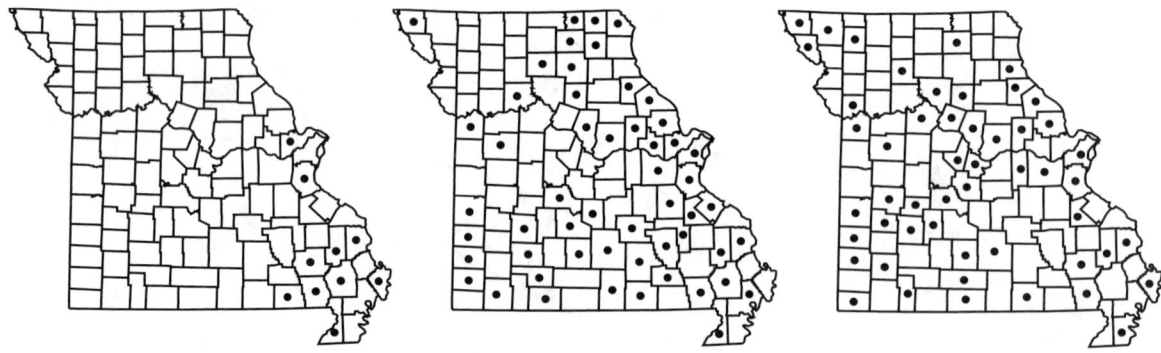

1692. Gleditsia aquatica   1693. Gleditsia triacanthos   1694. Gymnocladus dioica

1. Gleditsia aquatica Marshall (water locust)

Pl. 386 j, k; Map 1692

Plants trees 10–20 m tall, usually armed with conspicuous thorns, these 6–15 cm long, 2–6 mm in diameter at the base, simple or few-branched, the bark smooth, becoming finely grooved and occasionally somewhat scaly with age, dull gray to reddish brown. Leaves with the petiole and rachis glabrous or nearly so, the 1 time compound leaves 12–15 cm long, with 7–10 pairs of leaflets, these 2–5 cm long, 0.5–1.5 cm wide, narrowly ovate, rounded to bluntly pointed at the tip, the surfaces glabrous, somewhat shiny; the 2 times compound leaves with 4–7 pairs of pinnae each with 7–9 pairs of leaflets, these 1.0–2.5 cm long, 0.5–1.2 cm wide, narrowly ovate to ovate, rounded to bluntly pointed at the tip, the surfaces glabrous, green to yellowish green, the upper surface shiny. Inflorescences spikelike racemes 5–15 cm long with many flowers, the pistillate inflorescence with fewer and more widely spaced flowers than the relatively dense staminate ones. Flower stalks 1–3 mm long. Sepals 2.5–3.0 mm long, 0.8–1.5 mm wide, finely hairy. Petals 2.5–3.0 mm long, 1.5–2.0 mm wide. Fruits with the stalklike base 1–2 cm long, the body 3–5 cm long, 1.5–3.0 cm wide, asymmetrically elliptic or ovate (sometimes nearly circular), the surfaces glabrous, orange to reddish brown, lacking pulp. Seeds 10–12 mm long, 7–10 mm wide, nearly circular, brown. $2n=28$. May–June.

Scattered in the Mississippi Lowlands Division and north along the Mississippi River to St. Charles County (southeastern U.S. west to Missouri and Texas; introduced sporadically north to New York). Swamps, sloughs, and bottomland forests, frequently emergent aquatics.

Gleditsia aquatica is tolerant of flooding and is usually found in or near water. The fruits float and are probably dispersed by water. The trees can also grow on dry land and are sometimes planted in yards and parks. Forms lacking spines are occasionally found. The wood is durable, resists rotting, and has been used for fence posts and in cabinetry.

Although it is ecologically and geographically more or less isolated from *G. triacanthos,* putative hybrids between the two species have been reported from Arkansas, Indiana, Louisiana, Mississippi, and Texas. The hybrids, known as *G.* ×*texana,* are intermediate in most morphological characters and are always found with the parental species nearby (Gordon, 1966). So far, such hybrids have not been observed in Missouri.

2. Gleditsia triacanthos L. (honey locust)

Pl. 386 g–i; Map 1693

Plants trees 20–30 m tall or taller, usually armed with conspicuous thorns, these 2–7 cm long, 2–4 mm in diameter at the base, simple or often several-branched, the bark smooth and reddish brown on young trunks and branches, becoming deeply fissured and dark gray to nearly black on older trunks. Leaves with the petiole and rachis distinctly short-hairy, the 1 time compound leaves 12–14 cm long, with 7–16 pairs of leaflets, these 1–4 cm long, 7–14 mm wide, narrowly ovate, rounded at the tip, the upper surface glabrous and green to dark green, the undersurface short-hairy and pale; the 2 times compound leaves with 3–6(–8) pinnae, each with (2–)5–10 pairs of leaflets, these 1.3–2.5 cm long, 7–12 mm wide, narrowly ovate to ovate, rounded at the tip, the upper surface glabrous and green to dark green, the undersurface short-hairy and pale. Inflorescences spikelike racemes 5–10 cm long, with many fragrant flowers, the pistillate inflorescence with fewer and more widely spaced flowers than the relatively dense staminate ones. Flower stalks 0.5–3.0 mm long. Sepals 2–3 mm long, 1.0–1.4 mm wide, finely hairy. Petals 2.0–2.2 mm long, 1.4–1.8 mm wide. Fruits with the stalklike base 1–3 cm long, the body 18–35 cm long, 2.5–3.0 cm wide, elongate, flattened, curved or spirally twisted, the surfaces glabrous or hairy, purplish brown, the seeds embedded in a jellylike pulp, this eventually becoming dry and inconspicuous. Seeds

8–10 mm long, 6–7 mm wide, ovoid, olive green to brown. 2n=28. May–June.

Scattered to common nearly throughout the state (nearly throughout the U.S.; Canada; introduced in Australia). Bottomland forests, mesic upland forests, banks of streams and rivers, margins of sinkhole ponds, edges of glades, and edges of and drainages in upland prairies; also pastures, old fields, fencerows, and roadsides.

Honey locust is a very useful plant. It is very tolerant of urban and suburban conditions and makes an excellent shade tree, especially the thornless selections. The wood is hard and durable and is used for a variety of wood products, including fence posts and furniture. The legume pods are an important food for wildlife, and trees sometimes are planted in pastures for livestock. The pods have a sugar content of 13–30 percent, and the seeds have a protein content of 16–28 percent (Gold and Hanover, 1993). The pods can be gathered and ground to provide food for livestock with a nutritional value about that of alfalfa. Experimental plantings have produced over 6 tons of dry fruits per acre. Steyermark (1963) enjoyed the sweet pulp surrounding the seeds as a treat while hiking. The tree is fast-growing and very drought tolerant and is planted for windbreaks and erosion control. Unfortunately, although honey locust probably was mostly a bottomland tree originally, it is invasive in a variety of upland habitats, especially prairies in northern Missouri. It also tends to proliferate in disturbed bottomland forests.

The thorns of *G. triacanthos* represent a reduced branch system. They have been used in handicrafts, as large needles, as weapons, and for carding wool. The most commonly planted trees are thornless forms, which are found sporadically in nature throughout the range of the species; these have been called f. *inermis* Zabel. Cuttings taken from the upper, thornless, portions of otherwise thorny trees will also produce thornless individuals. However, the thornless condition appears to be genetically unstable. Offspring of thornless individuals sometimes produce thorns, and Michener (1986) has reported mature thornless specimens reverting to a thorny state.

The large spines and big fruits of *G. triacanthos* may be adaptations for large Pleistocene mammals such as horses, giant sloths, mastodons, and mammoths (Janzen and Martin, 1982; Barlow, 2000). These creatures were present in North America for millions of years and only became extinct in the last 10,000–20,000 years, perhaps at the hand of man. The spines may have protected the trees from these large animals. The pods of *G. triacanthos* have all the characteristic features of the so-called megafaunal dispersal syndrome discussed by Janzen and Martin (1981): they are large, more or less indehiscent, contain a nutritious pulp, and have hard seeds that would survive mastication. Passage of the seeds through the gut of large animals would serve to scarify them and speed germination. Present-day mammals such as horses and cattle eat them readily. No other large native mammals exist to disperse the seeds at present, which may be a factor in the often spotty distribution of the species. Other plants in the Missouri flora that similarly may have lost their original seed dispersers include Kentucky coffee tree (*Gymnocladus dioica*, Fabaceae), Osage orange (*Maclura pomifera*, Moraceae), pawpaw (*Asimina triloba*, Annonaceae), and persimmon (*Diospyros virginica*; Ebenaceae).

4. Gymnocladus Lam.

Three or 4 species, North America, Asia.

1. Gymnocladus dioica (L.) K. Koch (Kentucky coffee tree)
  G. canadensis Lam.

Pl. 385 j–l; Map 1694

Plants trees 10–20 m tall, usually incompletely dioecious, sometimes colonial from root suckers, unarmed, the bark shallowly grooved, silvery gray, often reddish-tinged, developing small, scaly ridges and becoming dark gray on older trunks, the branches not producing short shoots but the leaves sometimes tending to be clustered toward the branch tips, the twigs stout, the winter buds inconspicuous and strongly sunken into the twig; root nodules absent. Leaves appearing before the flowers, the petiole 10–20 cm long, the blade 30–90 cm long, 2 times pinnately compound, with 4–7 pairs of pinnae, each with 4–6 pairs of alternate leaflets, the lowermost pair of pinnae sometimes replaced with a pair of large leaflets. Stipules inconspicuous and scalelike, shed early. Leaflets 2–9 cm long, 1.5–5.0 cm wide, ovate to broadly elliptic, rounded to angled at the base, short-tapered or tapered to a sharply pointed tip, the margins entire and inconspicuously short-hairy, the upper surface glabrous, the undersurface finely hairy. Inflorescences racemes or more commonly narrow, racemose panicles, mostly appearing terminal, 14–20 cm long, ascending to spreading; some trees with all staminate or all pistillate inflorescences, but often otherwise with some inflorescences having mixed imperfect and perfect flowers. Flower stalks 10–35 mm long. Flowers perigynous, more or less actinomorphic, fragrant. Hypanthium 8–12 mm long, tubular to narrowly funnelform, densely hairy. Calyces of 5 sepals, these subequal, 4–7 mm long, narrowly oblong-elliptic, sharply pointed at the tip, moderately to densely hairy, not closing the flower in bud. Petals (3–)5, 4–10 mm long, 1.5–2.5 mm wide, greenish white, densely woolly. Stamens 10, unequal in 2 alternating long and short series, the filaments not fused, hairy at the base, the anthers 1.2–1.6 mm long, attached toward the midpoint. Style short, relatively straight, the stigma oblique. Fruits legumes, 9–15 cm long, 2–5 cm wide, 1.0–1.5 cm thick, more or less oblong, straight or slightly curved, rounded or short-angled at the base and sometimes with a short stalk to 0.5 cm long, abruptly short-tapered to a usually sharply pointed tip, 1–4-seeded, the valves velvety when young, becoming woody and glabrous, persistent into the winter, the valves dehiscing with age along the ventral suture, seeds with a stout attachment and embedded in a green, jellylike pulp. Seeds 15–20 mm in diameter, ovate to circular, somewhat flattened but turgid, dark reddish brown, hard, shiny; pleurogram absent. $2n=28$. May–June.

Scattered, but nowhere common, nearly throughout the state (eastern U.S. west to North Dakota and Texas; Canada). Bottomland forests, mesic upland forests, banks of streams and rivers, and bases of bluffs.

Although widely distributed, *G. dioica* is relatively infrequently encountered and apparently absent from many areas. It frequently occurs as a single tree or small colony in especially favorable sites, some of these possibly prehistoric settlements or campsites (Lee, 1976). For a discussion of dispersal problems for this and other species, see the treatment of *Gleditsia triacanthos* above.

The seeds are said to have been roasted and used as a substitute for coffee in the Revolutionary War and by the early settlers, but Thomas Nuttall (1821) reported that they were a poor substitute compared to chicory. The pulp, fresh seeds, and foliage are considered somewhat poisonous, apparently because of a variety of saponins, terpenoids, alkaloids, and unusual amino acids present in various parts of the plants (Burrows and Tyrl, 2001). *Gymnocladus dioica* is sometimes planted as a specimen tree, growing best in deep, rich soils with good drainage. It is one of the last trees to develop new leaves in the spring and is among the first to shed them in the autumn. The leaves turn a dusty yellow color in the fall. Some of the fruits are retained on the branches through the winter.

This generic name traditionally has been treated as feminine. However, according to the International Code of Botanical Nomenclature (Greuter et al., 1999) generic names formed from two words take the gender of the second word, in this case *cladus,* which is masculine (Robertson and Lee, 1976; Lee, 1976). Harriman (1998) submitted a proposal to conserve the generic name as masculine, but this became bogged down during discussion in the Committee for Spermatophyta and has yet to be resolved. Thus the species is spelled *G. dioicus* and *G. dioica* in various recent floristic and taxonomic treatments (Gleason and Cronquist, 1991; Isely, 1998).

5. Pomaria Cav.
(Simpson, 1998)

Twelve to 18 species, North America, South America, Africa.

1. Pomaria jamesii (Torr. & A. Gray) Walp.
(rush-pea)
Hoffmannseggia jamesii Torr. & A. Gray
Caesalpinia jamesii (Torr. & A. Gray) Fisher
Pl. 385 a, b; Map 1695

Plants perennial herbs, unarmed, with a thickened, sometimes somewhat woody taproot. Stems 5–40 cm long, erect to loosely ascending, usually branched, moderately to densely pubescent with short, downward-curved hairs and conspicuously dotted with small glands, these orange when fresh, blackening upon drying; root nodules lacking. Leaves with the petiole 1–2 cm long, the petiole and rachis short-hairy, the blade 3–5 cm long, 2 times pinnately compound, with 1–7 pairs of pinnae, each with 5–10 pairs of leaflets. Stipules conspicuous, 3–6 mm long, narrowly lanceolate to narrowly triangular, hairy, especially along the margins, the surfaces also gland-dotted. Leaflets 2–5 mm long, 1–2 mm wide, ovate to oblong-elliptic, angled at the base, rounded or less commonly bluntly pointed at the tip, the margins entire, the upper surface moderately hairy to nearly glabrous, the undersurface moderately to densely hairy and gland-dotted. Inflorescences terminal and lateral racemes, the lateral ones positioned opposite the

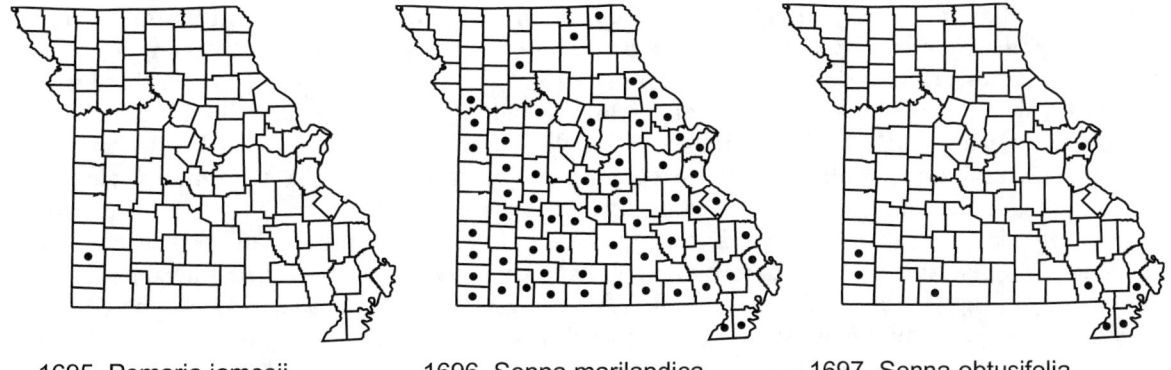

1695. Pomaria jamesii   1696. Senna marilandica   1697. Senna obtusifolia

leaves, 5–10 cm long. Flower stalks 2–5 mm long. Flowers perigynous, zygomorphic. Hypanthium 1–2 mm long, more or less saucer-shaped, hairy. Calyces of 5 sepals, 6–10 mm long, these appearing fused basally, unequal in size, the lowermost sepal the largest, 8–10 mm long, projecting between the two lower petals and cupping the stamens and pistil within, sharply pointed at the tip, hairy and gland-dotted. Petals 4–6 mm long, 2–3 mm wide, the innermost petal slightly larger than others and tapered to a short, stalklike base, yellow with a reddish-tinged base, drying orangish red, gland-dotted. Stamens 10, shorter than the petals, unequal in size, the filaments 4–5 mm long, not fused, densely hairy at the base, the anthers about 1 mm long, attached toward the midpoint. Ovary 2–3 mm long, gland-dotted, the style 2–3 mm long, curved, stigma lateral. Fruits legumes, 18–20 mm long, 6–10 mm wide, crescent in outline, strongly flattened, the surfaces gland-dotted and especially toward the margins with multicellular papillae covered with stellate hairs, 1- or 2(3)-seeded. Seeds 5–7 mm long, 4–5 mm wide, obovate in outline, strongly flattened, brown, glossy. Flowering times uncertain, but June–September according to Steyermark (1963).

Introduced, uncommon, known only from a single historical collection from Jasper County (Nebraska to Texas west to Arizona; Mexico). Habitat unknown with certainty, but "dumps and waste ground" according to Palmer and Steyermark (1935) and Steyermark (1963).

This species was classified in *Hoffmannseggia* by Steyermark (1963) and in *Caesalpinia* by Isely (1975, 1998). In the broad traditional sense, *Caesalpinia* is a heterogeneous assemblage of trees, shrubs, and perennial herbs with 2 times pinnately compound leaves, conspicuous yellow flowers in racemes, exserted stamens, and broad legumes (Isely, 1975, 1998). *Pomaria* and *Hoffmannseggia* were segregated long ago because of their herbaceous habit, but there has been no consensus on how to treat these taxa. Simpson and Miao (1997) presented data based on molecular studies of the chloroplast genome that caused them to reevaluate generic limits in the group, with the result that *Pomaria* and *Hoffmannseggia* were once again segregated as genera. The molecular data are correlated with a number of morphological features shared by species of *Pomaria*, including the presence of glandular dots on leaves and other organs, persistent stipules, sepals that are shed as the fruit matures leaving a ring at the fruit base, multicellular, elaborately branched trichomes on the fruit, and lateral stigma position (Simpson and Miao, 1997; Simpson, 1998).

6. Senna Mill. (senna)

Plants annual or perennial herbs (trees and shrubs elsewhere), unarmed, with 1 to several stems, the roots often blackish; root nodules lacking. Leaves short- to more commonly relatively long-petiolate, the petiole or rachis base with a large, variously shaped gland. Stipules small and scalelike to nearly hairlike, often shed early. Leaf blades evenly 1 time pinnately compound. Leaflets few to numerous, opposite, variously shaped, often asymmetrical at the base (with one side angled and the other rounded or cordate), the margins entire. Inflorescences axillary racemes, sometimes appearing aggregated in terminal clusters, the flower stalks lacking bracts. Flowers perfect, slightly perigynous, slightly to moderately asymmetrical, the buds usually nodding. Hypanthium very short and more or less disc-shaped. Calyces of 5 free sepals. Corollas of 5 free petals; these dissimilar in shape and position and usually slightly dissimilar in size, abruptly tapered at the base, rounded at the tip, yellow to orangish yellow, drying white, often with dark veins. Stamens 10 but the upper 3 reduced to staminodes, the filaments short, the anthers attached at the base, graded in size, the upper 3 strongly reduced and infertile, the middle 4 intermediate and fertile, the lower 3 large and fertile, all somewhat curved, the fertile ones dehiscing by an apical pore. Styles curved. Fruits legumes, elongate, not twisted, flattened to circular or rectangular in cross-section, narrowed to a short, stalklike base, indehiscent or, if dehiscent, then the valves not separating elastically or coiling. Seeds variously shaped and colored, usually somewhat flattened, with an elliptic pleurogram, this sometimes conspicuously different in color from the remainder of the seed. About 240–260 species, nearly worldwide, most diverse in the New World tropics.

Senna is a large, widespread, and morphologically diverse genus that traditionally was included in a broadly defined *Cassia* L. along with *Chamaecrista*. The differences between these taxa were discussed by Irwin and Barneby (1981, 1982), whose overall classification presently is followed by most botanists. In contrast to *Chamaecrista,* the stipules of *Senna* are small and weakly developed, and the flowers lack small bracts on the stalks; the stamens of *Senna* have relatively large, curved anthers of various sizes in a more or less graded series.

1. Leaflets obovate, mostly rounded at the tip; petiolar gland usually located between the lowermost pair of leaflets; fruits more or less circular to bluntly rectangular in cross-section, only slightly impressed between the seeds
 . . . . 2. S. OBTUSIFOLIA
1. Leaflets ovate to oblong-elliptic, angled or abruptly tapered to a sharply pointed tip; petiolar gland located near the petiole base; fruits strongly flattened, conspicuously impressed between the seeds
    2. Leaves with 3–5 pairs of leaflets, these angled or slightly tapered to a sharply pointed tip; anthers yellow; fruits arched upward, light brown, with a lighter-colored margin . . . . 3. S. OCCIDENTALIS
    2. Leaves with 8–10 pairs of leaflets, these abruptly short-tapered to a minute, sharply pointed tip; anthers purplish brown; fruits arched downward, dark brown to black, without a light margin . . . . 1. S. MARILANDICA

1. Senna marilandica (L.) Link (southern wild senna)

Cassia marilandica L.

C. medsgeri Shafer

Pl. 386 a, b; Map 1696

Plants perennial herbs, with a somewhat woody, often horizontal, branched rootstock, not or only slightly fragrant when bruised or crushed. Stems 1 to several, 100–200 cm long, erect or ascending, unbranched, glabrous or sparsely pubescent with spreading hairs toward the tip, sometimes somewhat glaucous. Leaves with the petiole 3–6 cm long, the petiolar gland positioned near the base, 1–2 mm long, ovoid to hemispherical or more or less short-cylindrical, appearing sessile and broadest at or below the midpoint. Leaf blades 14–20 cm long, with 8–10 pairs of leaflets. Leaflets 3.0–6.5 cm long, 10–25 mm wide, oblong to oblong-elliptic, oblique at the base, abruptly short-tapered to a minute, sharply pointed tip, the margins with a pale, narrow band and short, ascending hairs, the surfaces glabrous or the undersurface with scattered microscopic glandular hairs (at least when young), the undersurface also pale and somewhat glaucous. Inflorescences with 6–9 flowers, the flower stalks 10–15 mm long. Sepals somewhat unequal in size, variously 4–8 mm long, 3–4 mm wide, ovate, bluntly pointed at the tip, the margins short-hairy. Petals 7–12 mm long, 4–5 mm wide, oblanceolate to obovate. Stamens with the anthers purplish brown. Ovary 4–6 mm long, with short, appressed hairs, the style 2–3 mm long. Fruits 6–9 cm long, 7–10 mm wide, arched downward at maturity, strongly flattened, sparsely to moderately hairy when young, becoming glabrous at maturity, relatively conspicuously impressed between the seeds, dark brown to black at maturity. Seeds 4–5 mm long, 2.2–3.0 mm wide, oblong-obovate to obovate, slightly flattened, the surface often developing a fine network of cracks toward the margins at maturity, olive green to brown, more or less dull, the pleurogram usually slightly grayer than the remainder of the seed. $2n=28$. July–August.

Scattered south of the Missouri River, less commonly farther north (eastern U.S. west to Nebraska and Texas). Banks of streams and rivers, sloughs, bottomland and upland prairies, bottomland forests, mesic upland forests, bases, ledges, and tops of bluffs, glades, and savannas; also pastures, old fields, fallow fields, roadsides, and open, disturbed areas.

Senna marilandica occasionally is cultivated as an ornamental in gardens for its attractive foliage and flowers. It is closely related to *S. hebecarpa* (Fernald) H.S. Irwin & Barneby (northern wild senna), which is widespread in the northeastern United States and adjacent Canada west to Wisconsin, Illinois, and eastern Tennessee, possibly also sporadically farther south. The two species are northern and southern analogs that have sometimes been confused (Isely, 1998) but have a broad region of geographic overlap in which they appear to maintain themselves consistently without apparent intermediates or hybridization. *Senna hebecarpa* differs in its ovary with dense, somewhat tangled, spreading hairs (vs. appressed-ascending hairs) and its petiolar glands, which are more or less club-shaped and widest above the middle (vs. hemispherical to short-cylindrical). A single historical Missouri specimen of *S. hebecarpa* exists in the herbarium of the Missouri Botanical Garden. It was collected by Mrs. James R. Bettis from the parsonage grounds of a church in Webster Groves (St. Louis County) in 1925 and was originally determined as *Cassia marilandica*. This specimen was deaccessioned by Robert Woodson and sent to the University of Minnesota herbarium during the early 1950s as part of his infamous purge of so-called superfluous sheets from the Missouri Botanical Garden (Solomon, 1998). It was thus potentially unavailable for Julian Steyermark to examine during his research on the Missouri flora. The specimen was returned to St. Louis in 1993 as part of a generous effort on the part of the staff in Minnesota to repatriate some 75,000 specimens they had received from Woodson. Upon its return, the plant was redetermined correctly as *S. hebecarpa* by Ron Liesner of the Missouri Botanical Garden's herbarium staff. The circumstances surrounding the original collection cannot be determined from the sheet, but it seems likely that the plant from which the flowering branch tip was pressed was under deliberate cultivation, rather than a spontaneous weed. Thus, at least for now, this species remains excluded from the Missouri flora. However, botanists working particularly in northeastern Missouri eventually may discover it growing in a natural habitat in the state.

2. Senna obtusifolia (L.) H.S. Irwin & Barneby
(sicklepod, coffee weed)
Cassia obtusifolia L.
C. tora L.

Pl. 386 f; Map 1697

Plants annual (sometimes short-lived perennials farther south), producing a disagreeable odor when bruised or crushed. Stems 1 to several, (5–)30–100 cm long, erect or ascending, usually unbranched, with scattered, minute glandular hairs and sometimes also sparse, short, appressed or incurved hairs toward the tip. Leaves with the petiole 2–4 cm long, the petiolar gland positioned between the lowermost leaflets (rarely immediately below or somewhat above the lowermost pair), 1.5–2.0 mm long, narrowly columnar or slightly tapered from near the base, appearing sessile or more commonly short-stalked and angled toward the leaf tip. Leaf blades 5–8 cm long, with (2)3 pairs of opposite leaflets. Leaflets 2–6 cm long, 2–3 cm wide, broadly obovate to broadly obovate, oblique at the base, rounded or occasionally broadly angled to a very bluntly pointed tip, the margins with a pale, narrow band and short, ascending hairs, the surfaces glabrous or the undersurface with a few minute glandular or longer, nonglandular hairs toward the base. Inflorescences of solitary or paired flowers, sometimes also appearing as small clusters at the stem tip, the flower stalks 10–25 mm long, becoming elongated to 40 mm at fruiting. Calyces zygomorphic, the sepals variously 5–10 mm long, 2–3 mm wide, oblong-obovate to broadly elliptic, rounded to bluntly pointed at the tip, the margins with short, spreading hairs. Petals 7–14 mm long, 4–7 mm wide, oblong-obovate to obovate. Stamens with the anthers purplish brown. Ovary 4–6 mm long, with appressed or ascending hairs, the style 3–4 mm long. Fruits 9–16 cm long, 2–4 mm wide, arched downward at maturity, more or less circular to more commonly bluntly rectangular in cross-section, sparsely to moderately hairy when young, becoming glabrous at maturity, only slightly impressed between the seeds but with a pair of longitudinal ridges near the margins on each surface, greenish brown to brown at maturity. Seeds 3–5 mm long, 2.0–2.5 mm wide, somewhat rhomboidal to trapezoidal in outline, slightly flattened, the surface often developing a fine network of cracks toward the margins at maturity, reddish brown to dark brown, shiny, the pleurogram usually dull and slightly lighter than the rest of the seed. $2n=24$, 26, 28. July–September.

Uncommon in southeastern and southwestern Missouri, north sporadically to the city of St. Louis (native range poorly understood, probably originally native to tropical and warm-temperate regions of the New World, now widely distributed in both

1698. Senna occidentalis   1699. Acaciella angustissima   1700. Albizia julibrissin

hemispheres; in the U.S., present in the eastern states west to Nebraska and Texas, also California, Hawaii). Banks of streams and rivers, sloughs, and upland prairies; also railroads and open, disturbed areas.

The natural range of S. obtusiflora prior to the European colonization of North America is not fully understood. Steyermark (1963) considered at least some of the Missouri populations native and noted that Nicholas Riehl had collected the species as early as 1838 in a prairie in St. Louis. The species has a long history of association with humans, and farther south it can become a rank weed of pastures, farms, and waste places. As with many widespread, weedy species, there exists considerable genetic and morphological variation. The leaves have been used for food, as an adulterant of coffee, as a laxative and purgative, in poultices, and for dying cloth (Irwin and Barneby, 1982). The commercially available laxatives produced from anthraquinone extracts of Senna involve other tropical species. Burrows and Tyrl (2001) reviewed the toxicity of the genus, noting that both S. obtusifolia and S. occidentalis (but usually not S. marilandica) have been implicated in livestock poisoning, primarily when cattle have been fed fresh chopped forage containing relatively high concentrations of Senna, when cattle ingest wilted plants after first frost in the autumn, or when pigs and other livestock ingest grain contaminated by Senna seed.

3. Senna occidentalis (L.) Link (coffee senna, coffee weed)
Cassia occidentalis L.

Pl. 386 c–e; Map 1698

Plants annual, producing a disagreeable odor when bruised or crushed. Stems 1 to several, (5–)30–120 cm long, erect or ascending, usually unbranched, glabrous or with scattered, minute glandular hairs. Leaves with the petiole 2–3 cm long, the petiolar gland positioned near the base, about 1 mm long, more or less hemispherical, appearing sessile and broadest toward the base. Leaf blades 8–10 cm long, with 3–5 pairs of leaflets. Leaflets 2–7 cm long, 1.5–3.0 cm wide, ovate to broadly lanceolate, oblique at the base, angled or slightly tapered to a sharply pointed tip, the margins with a pale, narrow band and short, ascending hairs, the upper surface glabrous, the undersurface with scattered, minute glandular hairs. Inflorescences short racemes but often appearing as small clusters of 2–5 flowers or even solitary flowers, the flower stalks 3–6 mm long, becoming elongated to 10 mm at fruiting. Sepals unequal in size, variously 5–10 mm long, 2–3 mm wide, oblong to oblong-obovate, rounded to bluntly pointed at the tip, the margins often somewhat uneven, glabrous. Petals 9–14 mm long, 5–7 mm wide, oblanceolate to obovate. Stamens with the anthers yellow. Ovary 4–6 mm long, with loosely appressed or ascending hairs, the style 3–4 mm long. Fruits 7–14 cm long, 8–10 mm wide, arched upward at maturity, strongly flattened, sparsely to moderately hairy when young, becoming glabrous at maturity, relatively conspicuously impressed between the seeds, light brown with a lighter-colored margin at maturity. Seeds 4–5 mm long, 3–4 mm wide, oblong-obovate to obovate, slightly flattened, the surface often developing a fine network of cracks toward the margins at maturity, olive green to brown, more or less dull, the pleurogram usually slightly grayer than the remainder of the seed. $2n=28$. August–September.

Introduced, uncommon, mostly in the Mississippi Lowlands Division (native range poorly understood, probably originally native to tropical and warm-temperate regions of the New World; in the U.S., perhaps native as far north as North Carolina, Arkansas, and Oklahoma, introduced as far north as Kansas, Iowa, Illinois, Indiana, New York, and Massachusetts, also Hawaii). Banks of rivers; also fallow fields, railroads, roadsides, and sandy, open, disturbed areas.

Senna occidentalis is a widespread weedy species similar to *S. obtusifolia*. See the treatment of that species for a discussion of uses and toxicity of the two taxa.

2. Subfamily Mimosoideae (R. Br.) DC.

Plants trees, shrubs, or rarely perennial herbs, unarmed or armed with spines, thorns, or prickles. Root nodules usually present. Leaves alternate, occasionally appearing whorled in clusters on spur shoots, the petiole usually with a thickened basal portion (pulvinus), the petiole and/or rachis often with 1 or more glands. Stipules conspicuous or inconspicuous, sometimes modified into spines. Leaf blades usually 2 times pinnately compound, often with numerous leaflets. Inflorescences dense, headlike clusters, spikes, or spikelike racemes, these stalked, axillary or arranged in terminal racemose clusters, the bracts absent or minute and inconspicuous. Flowers perfect (rarely imperfect elsewhere), hypogynous. Calyces of 5 sepals, these sometimes minute, often fused most or nearly all of the way into a narrowly to broadly conical tube, this often with 5 small, triangular lobes, the lobes (when present) all similar in size and shape. Corollas relatively inconspicuous, actinomorphic, of 5 petals, these fused into a tube only toward the base or for most of the length. Stamens 5, 10, or numerous, none of them reduced to staminodes (or, in *Desmanthus,* occasional flowers having all nonfunctional stamens), free or the filaments fused toward the base, commonly white, yellow, or pink, long-exserted and relatively conspicuous, the anthers small, all similar in size and shape, dehiscing by longitudinal slits, usually attached toward the midpoint. Pistil sessile or short-stalked, the ovules 2 to many. Fruits legumes (in Missouri taxa), dehiscent or indehiscent, variously shaped, but usually narrowly oblong to linear, occasionally appearing coiled, often more or less flattened, dry to fleshy. Seeds flattened, variously shaped, the seed coat hard, usually with a U-shaped groove (pleurogram) on each side. Fifty to 65 genera, about 3,270 species, nearly worldwide, but most diverse in tropical and warm-temperate regions.

Members of the subfamily have flowers that are actinomorphic, with a relatively inconspicuous calyx and corolla. The greatly exserted stamens are the showy and colorful parts of the flower. The leaves in species of Mimosoideae almost always are 2 times pinnately compound, and many species have glands that act as extrafloral nectaries. Many species have leaves that are sensitive to touch or changes in internal turgor pressure, causing the pinnae and/or secondary pinnae to fold longitudinally and the leaf to bend downward abruptly. Because this movement occurs relatively quickly, herbarium specimens of mimosoid legumes often have leaves that are closed or appear wilted. The so-called sensitive plant that sometimes is cultivated as a houseplant and was the basis of some of the physiological studies on the causes of the leaf-closing phenomenon is *Mimosa pudica* L., a native of the New World tropics. A similar mechanism also operates in some genera causing the leaves to close at the end of each day and open again the following morning.

The molecular data from recent studies also indicate that, with the exception of a few problematic basal taxa, the genera of Mimosoideae form a natural group (Luckow et al., 2000; Bruneau et al., 2001). The subfamily has been divided into five or six tribes based on stamen and sepal characters (Elias, 1981), but molecular studies have suggested that these groupings are mostly artificial (Luckow et al., 2000). Delimitation of large genera such as *Acacia* and *Mimosa* also has been problematic (Elias, 1974).

1. Plants perennial herbs, usually from a woody rootstock
   2. Stems and petioles armed with prickles; leaf blades lacking glands on the rachis; flowers appearing pink to lavender pink; stamens 8–12; fruits often somewhat flattened but appearing more or less 4-sided, the surfaces with dense prickles . . . . 10. MIMOSA
   2. Stems and petioles unarmed; leaf blades with or without glands on the rachis; flowers appearing white or pale cream-colored; stamens 5 or numerous; fruits strongly flattened, not appearing 4-sided, the surfaces without prickles

3. Leaf blades lacking glands on the rachis; stamens numerous . . . 7. ACACIELLA
3. Leaf blades with a gland on the rachis between lowest pair of pinnae; stamens 5 .... 9. DESMANTHUS
1. Plants trees or shrubs
    4. Plants shrubs less than 1 m tall; leafstalk lacking glands between pinnae; leaflets 3–5 mm long; inflorescences headlike, the flowers appearing white or pale cream-colored .... 7. ACACIELLA
    4. Plants trees 3–7 m tall; leaf blades with 1 or more glands on the rachis between the pairs of pinnae; leaflets 8–50 mm long; inflorescences spikes or spikelike racemes or, if headlike, then the flowers appearing pink
        5. Leaves with 5–12 pairs of pinnae; inflorescences headlike, appearing pink; stamens 20–30, filaments united at the base; fruits 10–20 mm wide .... 8. ALBIZIA
        5. Leaves with 2 pairs of pinnae; inflorescences spikes or spikelike racemes, greenish yellow to yellow; stamens 10, free; fruits 8–10 mm wide .... 11. PROSOPIS

7. Acaciella Britton & Rose (acacia)

About 15 species, North America to South America, most diverse in Mexico.

As traditionally circumscribed, *Acacia* Mill. is the largest genus in the subfamily Mimosoideae, comprising about 1,350 species widely distributed in the tropics and warm-temperate regions of the world, and divisible into three large subgenera. Beginning in the 1960s, systematists studying variation in overall morphology, pollen morphology, and phytochemistry began making a case for separation of three or more groups from within *Acacia* as separate genera. These proposals were not accepted widely by the botanical community. With the advent of molecular studies on the mimosoid legumes, it became clear that these three large groups were more closely related to other genera of Mimosoideae than to each other. However, because of the complexity of the relationships involved and the inadequate number of species in this massive group sampled for molecular work to date, details of how these lineages should be classified remain unresolved (Miller and Bayer, 2003). Luckow et al. (2003) and Maslin et al. (2003) reviewed the studies published to that time that provided evidence for the breakup of *Acacia,* and they advocated the recognition of five to seven total genera, with four of these containing species native to the New World.

With the growing concern that three or more genera would be recognized in place of *Acacia* in the broad sense, botanists began to discuss which of the major lineages should retain the name *Acacia* and which should receive other names. This is of more than taxonomic importance, as numerous species in the overall group are cultivated around the world as ornamentals, for timber, and for various other uses. The thought that, regardless of which group remained under the well-known name *Acacia,* a large number of species would have to be renamed under other, more obscure generic epithets caused lengthy and heated nomenclatural discussions internationally. In the strict sense, the name *Acacia* was first typified to represent a group of about 160 species of nearly worldwide occurrence. However, Orchard and Maslin (2003) formally proposed that the name *Acacia* be conserved with the type species of *A. penninervis* Sieber ex DC., which belongs to a mostly Australian group traditionally known as *Acacia* subgenus *Phyllodineae* (DC.) Ser. They argued that because this represents by far the largest discrete assemblage (about 960 species) within the overall group, fixing the name *Acacia* on it would result in the fewest required transfers of species to other genera when *Acacia* is dismembered into more natural, smaller genera. After committees under the auspices of the International Association of Plant Taxonomy reviewed the proposal and gave it their tentative

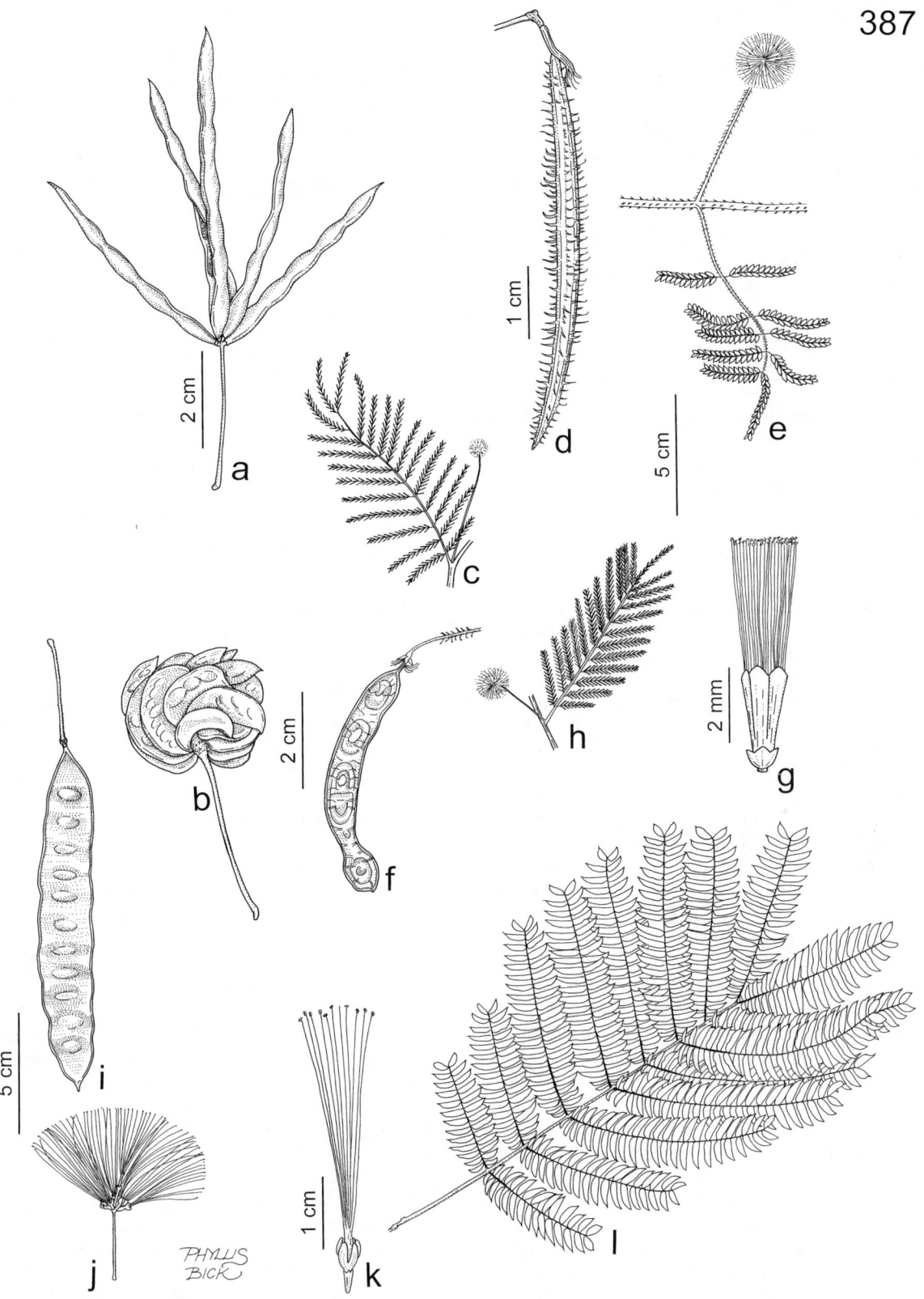

Plate 387. Fabaceae. *Desmanthus leptolobus*, **a)** fruits. *Desmanthus illinoensis*, **b)** fruits, **c)** flowering branch. *Mimosa quadrivalvis*, **d)** fruit, **e)** flowering branch. *Acaciella angustissima*, **f)** fruit, **g)** flower, **h)** flowering branch. *Albizia julibrissin*, **i)** fruit, **j)** inflorescence, **k)** flower, **l)** leaf.

blessing, it received lengthy discussion during the nomenclatural session at the most recent International Botanical Congress, held in 2005 in Vienna, Austria. There the delegates voted to formally approve the proposal. The result of this nomenclatural conservation is that the generic name *Acacia* becomes associated with a group of about 960 mostly Australian species formerly called subgenus *Phyllodineae* (DC.) Ser., including most of the species widely cultivated in warmer parts of the world.

The studies of Luckow et al. (2003), Maslin et al. (2003), and Miller and Bayer (2003) supported the transfer of the *A. angustissima* complex, traditionally called *Acacia* subgenus *Aculeiferum* Vassal section *Filicinae* Vassal, to the genus *Acaciella*. However, aside from confirming that this group is distinct from other American acacias, none of these workers was able to provide strong evidence for a close relationship with any particular group within the mimosoid legumes. Further research will be necessary to resolve this issue. *Acaciella* is notable for the absence of spines, thorns, prickles, and rachis glands, the presence of dehiscent fruits, seeds lacking an endosperm and an aril, and details of pollen ultrastructure and morphology.

1. Acaciella angustissima (Mill.) Britton & Rose (prairie acacia)
Acacia angustissima (Mill.) Kuntze
Acacia hirta Nutt.
Acacia angustissima var. *hirta* (Nutt.) B.L. Rob.
Acaciella hirta (Nutt.) Britton & Rose
Pl. 387 f–h; Map 1699

Plants shrubs, but in Missouri the stems mostly herbaceous and dying back to the ground in the winter (or nearly so), often clumped or colonial from a woody, rhizomatous rootstock, unarmed. Stems 30–70 cm long, arched or loosely ascending to less commonly nearly erect, the bark smooth or finely ridged toward the base, glabrous or sparsely pubescent with spreading hairs, green to reddish brown. Leaves with the petiole 1–3 cm long, the petiole and rachis sparsely to moderately pubescent with spreading and/or ascending hairs, lacking glands. Stipules inconspicuous, 3–4 mm long, linear to narrowly lanceolate. Leaf blades 5–12 cm long, with 9–15 pairs of pinnae, each with 18–30 pairs of leaflets. Leaflets 3–5 mm long, 0.8–1.2 mm wide, linear to oblong, oblique at the base, mostly angled to a bluntly pointed tip, sparsely to moderately hairy, especially along the margins. Inflorescences more or less spherical, headlike clusters (occasionally slightly elongate and then ovoid), axillary or sometimes appearing arranged in terminal racemose clusters, each about 1 cm in diameter (including the stamens), the stalk 2–5 cm long, with (4–)12–20 short-stalked flowers. Flowers appearing white to pale cream-colored. Calyces 0.4–0.6 mm long, broadly conical (appearing somewhat saucer-shaped), fused nearly to the tip, with inconspicuous, broadly rounded lobes, glabrous. Corollas 2.0–2.5 mm long, lobed to about the midpoint or nearly to the base, green to pale green, the lobes oblong to oblong-triangular, mostly bluntly pointed at the tip. Stamens numerous, the filaments 3–5 mm long, not fused at the base, white to pale cream-colored.

Fruits 4–7 cm long, 6–9 mm wide, narrowly oblong, tapered to a short, stalklike base, strongly flattened, straight to slightly curved, the margins undulating between the seeds, the surfaces sometimes somewhat constricted between the seeds, lacking prickles, glabrous or minutely hairy, reddish brown at maturity, dehiscent by two valves. Seeds 3–6, 3–4 mm in diameter, broadly ovate to nearly circular in outline, flattened, the surfaces smooth, brown, usually slightly mottled, somewhat shiny. $2n=26$. June–October.

Uncommon in the southwestern portion of the Ozark Division east to Oregon County (southwestern U.S. east to Missouri, Louisiana, and disjunctly to Florida; Mexico, Central America, South America). Glades and ledges and tops of bluffs, on limestone and dolomite substrates.

The description above applies only to Missouri plants, but the range listed is for the entire species. Variation within *A. angustissima* is complex, and several overlapping varieties have been described as varieties of *Acacia angustissima* (Isely, 1973; Turner, 1996). However, validly published combinations presently do not exist yet for most of these in the genus *Acaciella*. A number of varieties occur from Texas southward, where the species often grows as a large shrub or small tree. Plants in the northeastern portion of the species range, from Kansas to Texas east to Missouri, Louisiana, and Florida, have been described as *Acacia angustissima* var. *hirta,* which is unique in its nearly herbaceous habit and leaves with the rachises 6–10 cm long and 9–15 pinnae pairs, each with 18–30 pairs of leaflets. In Texas, this variety intergrades with var. *texensis* (Torr. & A. Gray) Isely, which occurs from Texas to Arizona and differs mainly in its shorter leaves with fewer pinnae and leaflets. Turner (1996) combined these two varieties and suggested that most of the differences between the idealized phases of the two were due to environmental differences.

8. Albizia Durazz.

Seventy to 150 species, Asia, possibly also Central America, South America, and Africa.

Generic limits of *Albizia* are controversial, with some authors restricting the group to 70–110 species occurring mostly in Asia and other authors broadening the circumscription to about 150 species in tropical and warm-temperate regions of both hemispheres (Elias, 1974; Nielsen, 1981). The solitary species introduced in Missouri is part of the core complex of the genus.

1. Albizia julibrissin Durazz. (silk tree, mimosa)

Pl. 387 i–l; Map 1700

Plants small trees or large shrubs to 10 m tall, often several-trunked and more or less flat-topped, unarmed, the wood soft, the bark smooth, gray and sometimes somewhat mottled. Twigs often somewhat zigzag, usually several ridged or finely fluted, green to brown or gray with scattered, small, lighter spots or bumps, glabrous, the winter buds small, more or less circular and slightly flattened, with several overlapping scales, the outer pair more or less obscuring the inner scales. Leaves with the petiole 3–6 cm long, the petiole and rachis moderately pubescent with short, curved hairs on the upper surface, with a large, lozenge-shaped gland near the petiole base and sometimes also between the uppermost pair of pinnae. Stipules shed during leaf development, small, linear. Leaf blades 10–22 cm long, with 5–12 pairs of pinnae, each with 13–30 pairs of leaflets. Leaflets 8–13 mm long, 1.5–4.0 mm wide, asymmetrically oblong to slightly crescent (the midvein near the upper margin), angled or short-tapered to a sharply pointed tip, the margins with minute, ascending hairs, the surfaces glabrous or sparsely hairy. Inflorescences conical to hemispheric, headlike clusters, these arranged in terminal racemes, each 2.5–4.0 cm in longest dimension (including the stamens), the stalk 2–5 cm long, with 15–30 or more sessile flowers. Calyces 2–4 mm long, narrowly conical to nearly tubular, fused most of the way to the tip, with bluntly to sharply pointed, broadly triangular lobes, glabrous or sparsely short-hairy. Corollas 7–8 mm long, shallowly lobed, white to pale pink. Stamens numerous (more than 20), the filaments 20–25 mm long, fused toward the base, pink to reddish pink. Fruits 8–20 cm long, 10–20 mm wide, narrowly oblong, tapered to a short, stalklike base, strongly flattened, straight, the margins sometimes slightly undulate and occasionally more strongly narrowed between a few of the seeds, the surfaces slightly constricted between the seeds, lacking prickles, finely hairy when young, usually glabrous or nearly so at maturity, light tan at maturity, indehiscent or slowly dehiscent with age. Seeds 12–18, 7–8 mm long, 4 mm wide, oblong-ovate, flattened, light brown. $2n=26$. June–August.

Scattered, mostly south of the Missouri River (native of Asia, widely introduced in the southern U.S. north to New York, Illinois, Utah, and California). Bottomland forests and mesic upland forests; also old fields, pastures, railroads, roadsides, and disturbed areas.

This species is grown as an ornamental for its gracefully spreading branches, delicate leaves, long flowering period, and attractive flower clusters. It is easily propagated from seeds or cuttings and grows rapidly. Unfortunately, these same features have led to its widespread establishment as an invasive exotic in the southern half of the state, particularly in southeastern Missouri. Steyermark (1963) merely mentioned *A. julibrissin* as a cultivated relative of *Acacia* but did not know it outside of cultivation. In his survey of legumes of the United States, Isely (1973) mapped it from southwestern and central Missouri as an escape from cultivation without any further details. Thompson (1980) documented the plant from an old pasture in Wayne County in the Mingo National Wildlife Refuge. Today, silk tree is present in most counties in the Ozark, Ozark Border, and Mississippi Lowlands Divisions, although it is still quite undercollected (as attested by the map for the species in the present work).

9. Desmanthus Willd.
(Luckow, 1993)

Plants perennial herbs (trees and shrubs elsewhere), with a thickened, woody taproot, unarmed. Stems few to numerous, erect to loosely ascending or occasionally spreading, often angled or ridged, at least toward the tip, glabrous or sparsely to moderately pubescent with short, curved or spreading hairs, sometimes sparsely and minutely roughened along the angles or ridges, green to olive green, often reddish brown toward the base. Leaves with the petiole short, 0.2–0.5(–1.0) cm long, the petiole and rachis glabrous or hairy, with a conspicuous or inconspicuous lozenge-shaped to somewhat barrel-shaped gland between the lowermost pair of leaflets, occasionally also between a few of the other leaflet pairs. Stipules inconspicuous, linear to hairlike, sometimes somewhat expanded at the base. Leaf blades 2–10 cm long, with 2–12 pairs of pinnae, each with 10–30 pairs of leaflets. Leaflets small, narrowly oblong to linear, oblique at the base, angled or short-tapered to a sharply pointed tip, glabrous or with sparse ascending hairs along the margins. Inflorescences more or less spherical, headlike clusters, axillary, up to 1 cm in diameter (including the stamens), the stalk 2–6 cm long, with 4–70 sessile flowers (sometimes a few of the outer flowers sterile or staminate in *D. illinoensis*), the bracts subtending each flower small but relatively conspicuous when the flowers are in bud. Flowers appearing white to pale cream-colored. Calyces conical to broadly conical, shallowly lobed, with triangular to broadly triangular lobes, glabrous. Corollas fused at or toward the base when young, but the petals appearing free at flowering, the petals or lobes linear to narrowly oblanceolate, pale green to greenish white. Stamens 5 (10 elsewhere), nonfunctional in sterile flowers (the filaments similar in length but the anthers shriveled or absent), the filaments 1.5–8.0 mm long, not fused at the base, white to pale cream-colored. Fruits 1–6 cm long, 2–6 mm wide, variously shaped, rounded or angled to a sometimes asymmetric sessile base, strongly flattened or turgid, the surfaces lacking prickles, glabrous, dehiscent by 2 valves. Seeds 2–5, 1.8–2.5 mm long, obovate or somewhat rhombic to narrowly obovate or narrowly oblong in outline, somewhat flattened, the surfaces smooth, reddish brown, dull or slightly shiny. Twenty-four species, North America to South America, Caribbean Islands.

1. Stems erect or ascending; leaves (3.5–)5.0–8.0 cm long; inflorescences with 20–70 flowers, the stalk 3–6 cm long at fruiting; fruits 1–2(–3) cm long, 5–7 mm wide, curved or twisted, the surfaces not constricted between the seeds; seeds 3–4 mm long... 1. D. ILLINOENSIS
1. Stems spreading or loosely ascending from a spreading base; leaves 2–4(–6) cm long; inflorescences with 4–10 flowers, the stalk 0.8–2.0 cm long at fruiting; fruits 2–6 cm long, 2–3 mm wide, straight, the surfaces constricted between the seeds; seeds 4–6 mm long... 2. D. LEPTOLOBUS

1. Desmanthus illinoensis (Michx.) MacMill. ex B.L. Rob. & Fernald (Illinois bundle flower, prairie mimosa)

Acuan illinoense (Michx.) Kuntze

Mimosa glandulosa Michx.

Pl. 387 b, c; Map 1701

Stems 20–120 cm tall, erect or ascending. Leaves mostly (3.5–)5.0–8.0 cm long, with 6–12 pairs of pinnae, each with 15–30 pairs of leaflets. Stipules 4–9 mm long, linear to hairlike, with a small, winglike, expanded base. Leaflets 1.5–4.0 mm long, 0.5–0.8 mm wide. Inflorescences 1.0–1.2 cm in diameter (including the stamens), with 20–70 flowers, the stalk 1–4 cm long at flowering, becoming elongated to 3–6 cm at fruiting. Flowers mostly perfect, but a few staminate and/or sterile flowers often produced, these usually smaller than the perfect ones. Buds obovate, rounded or very bluntly pointed at the tip. Calyces 1.4–2.0 mm long (as short as 0.6 mm in sterile flowers). Corollas (1.5–)2.0–3.3 mm long. Stamens 5, the filaments 4–6 mm long. Fruits 1–2(–3) cm long, 5–7 mm wide, asymmetrically oblong, slightly curved to strongly twisted, obliquely rounded at the base, strongly flattened, tapered to a sharply pointed tip, the margins not or only occasionally indented between the seeds, the surfaces not constricted between the seeds, reddish brown to black at maturity, dehiscent along both sutures but more slowly on one side than the other, with 2–5 seeds. Seeds 3–4 mm long, 2.0–2.5 mm wide, obovate to somewhat rhombic in outline. $2n=28$. June–August.

Scattered nearly throughout the state (eastern U.S. west to North Dakota, Nevada, and New Mexico). Glades, upland prairies, dry upland forests, savannas, banks of streams and rivers, and margins of ponds, lakes, and oxbows; also fencerows, pastures, edges of crop fields, railroads, roadsides, and open, disturbed areas.

This species is distinguished from all other species of *Desmanthus* by the dense cluster of curved and twisted fruits (Luckow, 1993). The degree of

1701. Desmanthus illinoensis   1702. Desmanthus leptolobus   1703. Mimosa quadrivalvis

curvature in the legume is variable, even on the same plant. *Desmanthus illinoensis* has a high protein content and is an important browse plant for livestock and deer. It is easily grown from seeds and is available commercially.

2. Desmanthus leptolobus Torr. & A. Gray
(prairie mimosa)
Acuan leptolobum (Torr. & A. Gray) Kuntze
Pl. 387 a; Map 1702

Stems 30–100 cm long, spreading or loosely ascending from a spreading base. Leaves 2–4(–6) cm long, with 4–9 pairs of pinnae, each with 15–22 pairs of leaflets. Stipules 3–7 mm long, linear to hairlike, not expanded at the base. Leaflets 1.5–4.0 mm long, 0.5–1.0 mm wide. Inflorescences 4–8 mm in diameter, with 4–10 flowers, the stalk 0.6–1.0 cm long at flowering, becoming elongated to 0.8–2.0 cm at fruiting. Flowers all perfect. Buds narrowly elliptic, sharply pointed at the tip. Calyces 0.5–1.0 mm long. Corollas 1.0–1.5 mm long. Stamens 5, the filaments 2–3 mm long. Fruits 2–6 cm long, 2–3 mm wide, linear, straight or nearly so, angled at the base, tapered to a sharply pointed tip, turgid, the margins indented between the seeds, the surfaces constricted between the seeds, dehiscent equally along both sutures, with 4–10 seeds. Seeds 4–6 mm long, 1.8–2.0 mm wide, narrowly obovate to narrowly oblong in outline. $2n=28$. May–August.

Introduced, uncommon, known thus far from Boone, Cape Girardeau, and St. Louis Counties (native of Kansas, Oklahoma, and Texas; introduced in Missouri). Railroads.

10. Mimosa L. (mimosa)

About 500 species, North America to South America, Caribbean Islands, Asia, Africa, Madagascar, most diverse in the New World tropics.

Traditionally, most authors have treated the Missouri species of *Mimosa* as part of the segregate genus *Schrankia* Willd. *Schrankia* was circumscribed to include about 15 or more species native from the central and southeastern United States to South America (Isely, 1971, 1973) and introduced sporadically in the Old World tropics. These were categorized as mostly prickly herbaceous plants with 4-angled legumes that appear to dehisce by 4 sutures. Since the time that *Schrankia* was described, many additional species of *Mimosa* have been described, and all of the characters that define this segregate also are found in other groups of *Mimosa* species. In his worldwide monograph of the group, Barneby (1991) chose to treat *Schrankia* as *Mimosa* section *Batocaulon* DC. series *Quadrivalves* Barneby. He further chose to reduce all of the species of *Schrankia* to a complex series of 16 varieties of a single species. Isely, who had been a strong proponent of splitting of *Schrankia* at both genus and species levels (Isely, 1971, 1973), later came to accept Barneby's conclusions in all regards (Isely, 1998). Nevertheless, the conservative species approach is accepted here with some reservations, and the complex is in need of more detailed taxonomic study.

1. Mimosa quadrivalvis L. **var. nuttallii** (DC. ex Britton & Rose) Beard ex Barneby (sensitive brier, catclaw sensitive brier, bashful brier)

Schrankia nuttallii (DC. ex Britton & Rose) Standl.

Leptoglottis nuttallii DC. ex Britton & Rose

S. uncinata Willd., misapplied and illegitimate

Pl. 387 d, e; Map 1703

Plants perennial herbs, with a thickened, woody taproot, armed with prickles. Stems 100–300 cm long, spreading or clambering into and over other vegetation, finely ridged toward the base, strongly ribbed above the midpoint, glabrous, yellow to yellowish brown, armed with numerous short, downward-curved prickles, these sparser toward the base, yellow to yellowish brown. Leaves with the petiole 2–5 cm long, the petiole and rachis armed with scattered to dense, short prickles similar to those of the stem, lacking glands. Stipules inconspicuous, 2–7 mm long, linear to hairlike. Leaf blades (3–)6–12 cm long, with 3–8 pairs of pinnae, each with 8–16 pairs of leaflets. Leaflets 3–9 mm long, 2–4 mm wide, oblong, oblique at the base, short-tapered to a bluntly pointed or minute, abrupt, sharply pointed tip, with sparse, loosely ascending hairs along the margins. Inflorescences more or less spherical, headlike clusters, axillary, each 2.0–2.5 cm in diameter (including the stamens), the stalk 1.5–4.0 cm long at flowering, becoming elongated to 3–10 cm long at fruiting, armed with scattered to dense, downward-curved prickles similar to those of the stem, with numerous (usually more than 75) short-stalked flowers (sometimes a few of the inner flowers staminate). Flowers appearing pink to lavender pink. Calyces minute, to 0.2 mm long, cup-shaped to conical, fused nearly to the tip, with microscopic irregular lobes, glabrous. Corollas 2.5–3.5 mm long, lobed less than 1/2 the way to the base, usually pink to purplish pink (rarely pale cream-colored with pinkish-tinged lobes), the lobes rounded to sharply pointed at the tip. Stamens 8–12, the filaments 7–10 mm long, not fused at the base, pink or lavender pink. Fruits 3–9(–12) cm long, 3–6 mm wide, linear to narrowly oblong, tapered to a sessile or less commonly very short, stalklike base, tapered to a slender beak at the tip, straight or slightly curved, often somewhat flattened but appearing more or less 4-sided, the margins more or less straight and somewhat thickened, the surfaces not constricted between the seeds, densely covered with curved prickles, otherwise glabrous, straw-colored to yellowish brown or light brown at maturity, dehiscent by 4 valvelike strips. Seeds 8–30, 3–4 mm long, more or less oblong in outline, slightly flattened, dark brown, smooth, somewhat shiny, sometimes covered with a thin, lighter outer layer that becomes cracked and peeling at maturity. $2n=26$. May–October.

Scattered nearly throughout the state, but apparently absent from the Mississippi Lowlands Division and the northwestern portion of the Glaciated Plains (Illinois to Louisiana west to North Dakota and New Mexico; introduced in Wisconsin, Michigan, and New York). Glades, upland prairies, savannas, and openings of dry upland forests; also old fields, ditches, railroads, roadsides, and rarely open, disturbed areas.

The pollination biology of this species has been described by Bernhardt (1990). Each prostate stem produces an inflorescence with an average of 26 heads of flowers. Only a few heads open on any one day over the course of the flowering period, and each head releases pollen only for a single day. The flowers open synchronously near dawn. The flowers produce a fragrance but no nectar, and a variety of different bees is attracted by the copious pollen that is released in tetrads. The bees scrape the pollen from the whole head as if it were a single flower.

The foliage (especially before the prickles harden) and seeds are eaten by a variety of wildlife. Some wildflower nurseries also sell sensitive brier as a garden ornamental and for ecological restorations. The dense balls of pink filaments tipped with tiny, cream-colored anthers are reminiscent of exploding fireworks. However, Steyermark (1963) cautioned that he experienced difficulty in cultivating plants either from transplanted individuals or from seeds.

Steyermark (1963) and some earlier authors applied the name *Schrankia uncinata* to this taxon. However, the original publication of that name listed the older epithet *S. horridula* Michx. in its synonymy and was thus illegitimate (Barneby, 1991; Isely, 1998). Further, the name *S. horridula* is now considered a synonym of *M. quadrivalvis* var. *angustata* (Torr. & A. Gray) Barneby, a closely related taxon that is endemic to the southeastern United States.

11. Prosopis L. (mesquite)

About 45 species, North America to South America, Asia, Africa.

The mesquites are an economically and ecologically important group of plants (Simpson, 1977). The pods and seeds are sweet and nutritious and were perhaps the most important food for native peoples in the Southwest (Felger, 1977). The pods also are eaten by cattle and wildlife. Mesquite wood is used for furniture, fence posts, handicrafts, charcoal, and many other things (Rogers, 2000). The mesquites also are an important member of a variety of desert, grassland, and riverbank plant communities in western North America (and elsewhere in the New World). In historical times, these plants have become much more abundant in the arid Southwest within their original geographic range because of suppression of wildfires, overgrazing of the range, and dissemination of the seeds by cattle (Fisher, 1977). Although they provide shade for livestock and a food source once grasses and other herbaceous plants have dried up or been consumed by cattle, many ranchers have attempted to eradicate populations on their lands. The species historically present in Missouri is the most abundant and widespread mesquite in the southwestern United States and adjacent Mexico.

1. Prosopis glandulosa Torr. var. glandulosa
(honey mesquite)

P. chilensis (Molina) Stuntz var. glandulosa (Torr.) Standl.

P. juliflora (Sw.) DC., misapplied

Map 1704

Plants shrubs or small trees 2–6 m tall, with a very deep rootstock, the trunks solitary or more commonly multiple, armed with mostly solitary, straight thorns 0.5–3.0 cm long at the branch nodes, the wood hard, the bark deeply longitudinally furrowed, the ridges tending to become shredded and peeling with age, dark gray and/or brown; root nodules present. Twigs and younger branches somewhat zigzag, reddish to grayish brown, glabrous, the older twigs and branches producing spur shoots (short shoots) at the nodes, the winter buds not apparent (more or less obscured on the spur shoots). Leaves alternate along young branches but becoming replaced by clusters of leaves on the spur shoots when these develop, the petiole 1.5–10.0 cm long, the petiole and rachis glabrous or with scattered, short hairs in 2 longitudinal lines along the ridges of the upper surface with a small, lozenge-shaped to oblong-spherical gland at the petiole tip between the bases of the pinnae. Stipules small and inconspicuous, linear to narrowly triangular. Leaf blades with 1 pair of loosely ascending pinnae, each pinna 4–18 cm long, with 6–18 pairs of loosely spaced leaflets. Leaflets 1–5 cm long, 2–5 mm wide, narrowly oblong to oblong, straight to slightly curved, somewhat asymmetric at the base, abruptly tapered to a minute, usually sharply pointed tip, the margins and surfaces glabrous. Inflorescences elongate spikes or spikelike racemes, these solitary or few from the spur shoots, each 6–10 cm long, the stalk 1–3 cm long, with numerous somewhat fragrant flowers. Calyces 1.2–1.5 mm long, conical to somewhat bell-shaped, fused most of the way to the tip, with triangular to broadly triangular lobes, these finely hairy toward the tip. Corollas 2.5–3.0 mm long, the petals free, densely hairy on the upper surface, greenish yellow. Stamens 10, the filaments 5–6 mm long, not fused toward the base, yellow. Fruits 8–20 cm long, 8–10 mm wide, narrowly oblong to more or less linear, tapered to a short, stalklike base, somewhat flattened, straight or occasionally slightly curved or twisted, the margins usually strongly narrowed between at least some of the seeds, the surfaces relatively thick, with a spongy inner layer and a hard outer layer, constricted and occasionally appearing somewhat jointed between the seeds, lacking prickles, finely hairy when young, glabrous at maturity, pale yellow to cream-colored, sometimes becoming bleached with age, indehiscent. Seeds 5–18, 6–7 mm long, 5–6 mm wide, oblong-elliptic to oblong-ovate, flattened, light brown to reddish brown. $2n=56$. Blooming date unknown for Missouri, but May–July in adjacent states.

Introduced, known only from a single historical collection from Jackson County (southwestern U.S. east to Kansas and Texas; Mexico; introduced in the Caribbean Islands, Asia, Australia). Disturbed areas.

The only known Missouri specimen of P. glandulosa was collected in 1918 by B. F. Bush from waste ground at or near Sheffield (Jackson County). Perhaps it was brought in from Texas or Oklahoma during cattle drives of the late 1800s. This specimen was overlooked by Steyermark (1963) and first reported by Yatskievych and Summers (1993), who mistakenly called it var. torreyana. The specimen originally was determined by the collector as P. juliflora, a name that traditionally was used in a broad sense to apply to a complex of species ranging from North America through the Neotropics to temperate South America. These species can be difficult to distinguish from herbarium material, and the cultivation of several members in both the New and Old Worlds with subsequent escapes outside their native ranges complicates the separation of taxa. Burkart (1976) restricted the name *P. juliflora* to one of the taxa in this complex that occurs in the New World tropics, from Mexico to northern South America and the Caribbean Islands. For plants of the United States and northern Mexico, he resurrected the name *P. glandulosa,* in which he recognized three varieties. The Missouri specimen is referable to var. *glandulosa,* with an ascending habit, relatively large, numerous leaflets, and mostly solitary thorns. This variety occupies the eastern portion of the species range and has a broad geographic overlap with the mostly western var. *torreyana* (L.D. Benson) M.C. Johnst. (with smaller, less numerous leaflets and mainly paired thorns). The rare var. *prostrata* Burkart was characterized by a prostrate rather than ascending habit and has been collected sporadically in Texas and Mexico.

1704. Prosopis glandulosa

GLOSSARY

a-. Prefix indicating lack of, as in asymmetrical, indicating lack of symmetry.

achene. Indehiscent, dry, single-seeded fruit with the seed attached to the fruit wall in only one place.

actinomorphic. Having radial symmetry, such as flowers with parts arranged so that opposite sides of a line drawn through the middle in any plane are mirror images.

acuminate. Gradually tapered to a sharp point with the sides somewhat concave near the tip.

acute. Narrowed to a pointed tip with the sides relatively straight.

adventitious. Developing in an irregular pattern, as in a root system lacking a single taproot or with roots produced along a stem.

allele. Variant of a gene locus differing in its expression or its attributes (molecular weight and/or charge) as resolved in an electrophoretic study.

allelopathy. The process whereby exudate chemical compounds in the soil from the roots or shed parts of a given plant species inhibit the germination and/or growth of other plant species.

allozyme. Genetically based variant of an enzyme produced by an organism that differs in size and/or charge of the molecule and can be separated from other forms of the same kind of enzyme (produced by the same gene) using the technique of gel electrophoresis. Allozymes are studied to document levels of genetic variation within and between populations and between species. See also isozyme.

alternate. With parts or structures attached singly at at each point or node, as in stems with a single leaf at each node. Contrasted with opposite and whorled.

anastomosing. Forming a network. In plants, usually applied to veins or nerves.

anatomy. Study of the internal organization and structures of plants. See also morphology.

androecium. Collective term for the stamens in a flower.

annual. Plant that germinates, grows, flowers, produces mature fruits, and dies during a single year. Also used generally for processes that occur during a single year. A variant known as biennial refers to plants that germinate during the autumn, overwinter in a rosette stage, and flower and fruit the following spring.

anther. Saclike, pollen-containing portion of the stamen, usually two- or four-lobed, the pollen-containing lobes known as anther sacs.

apical. Located at the tip of a structure.

apogamy. Type of asexual reproduction in which seeds are formed directly from maternal cells rather than following fertilization of an egg by pollen. Also referred to as apomixis (technically apogamy means without gametes and apomixis means without fertilization). Apomictic plants frequently are difficult or impossible to distinguish morphologically from sexually reproducing ones and apomixis must be confirmed through cytological and/or anatomical studies. One clue is that often in apogamous plants the proportion of flowers developing into fruits is very high.

apomixis. Synonymous with apogamy.

areole. In leaves with a network of anastomosing veins, an area of the surface surrounded by veins. In the Cactaceae, the differentiated patches along the plant surfaces (actually extremely short, highly modified branches) from which the spines are produced.

aril. Fleshy, often brightly colored appendage of the seed in some plants (for example, Celastraceae), sometimes appearing restricted at or near the hilum, sometimes covering all or nearly all of the seed. See also caruncle.

asexual reproduction. Any form of reproduction by which offspring are formed directly from maternal tissue without fertilization Includes apogamy, gemmae, bulblets, and stem and root divisions.

asymmetrical. Uneven, often applied to leaf bases in which the tissue is narrowed differently on opposite sides of the attachment point and to flowers in which the corolla is contorted or twisted such that the flower has neither bilateral nor radial symmetry.

-ate. Suffix indicating a two-dimensional shape, such as ovate, meaning a structure that is egg-shaped in outline.

attenuate. Gradually tapered to a slender, narrow tip or base.

auricle. Small, earlike lobe of tissue. Usually applied to the base of a leaf blade or leaflet.

awn. Slender, bristle-like structure, usually slightly broadened at the base, often attached to the tip of a leaf, fruit, or flower part. In the Asteraceae, a type of pappus segment.

axile. Kind of placentation in an ovary or fruit with two or more locules in which the ovules and seeds are attached centrally where the walls of adjacent locules intersect.

axillary. Located in the angle (axil) between a structure and its axis, such as the angle between a leaf and the stem to which it is attached.

axis. Stemlike longitudinal support structure, such as the rachis of a compound leaf or the stemlike central portion of an inflorescence.

backcross. Offspring of a mating between a hybrid and one of its parental taxa.

banner. Uppermost usually showy petal in many species of Fabaceae.

basionym. In botanical nomenclature, the initial publication of an epithet for a species or other level in a classification, which may be changed later as the classification becomes modified by transfer of the epithet into a new combination. For example, the wreath aster, *Symphyotrichum ericoides* (L.) G.L. Nesom, originally was described under the basionym *Aster ericoides* by Linnaeus before Guy Nesom upset traditionalists by reclassifying the species (and many others) into the segregate genus *Symphyotrichum*.

beak. Hard or firm, narrow point or projection, usually applied to appendages of fruits or seeds, less commonly to petals or other flower parts.

beard. Tuft or line of dense, long hairs.

berry. Fleshy fruit developing from a single pistil with several to many seeds (and lacking a stone or hard core).

bicolorous. Having two colors. Usually applied to petals, leaves, or scales with marginal or central bands different in color than the rest of the tissue. See also concolorous.

biconvex. Applied to structures, such as seeds and fruits, that are elliptic or oval in cross-section and thus appear convex on both sides. Structures that are unequally convex are rounded on one side and nearly flat on the other.

biennial. Plant that germinates and forms a rosette during the autumn, overwinters in this state, and produces flowers and fruits the following spring, eventually dying during the second growing season. See also annual.

bilabiate. Two-lipped. Usually applied to zygomorphic corollas or calyces of some plants that are lobed into an upper and lower main component (lip), one or both or which may be further toothed or more finely lobed

biosystematics. Branch of systematics involving the study of living plants and the use of comparative data from cytology, reproductive behavior, biogeography, and other sources to evaluate the evolutionary relationships within a taxonomic group.

bract. Reduced or otherwise modified leaf subtending a flower or part of an inflorescence. In the Asteraceae, each head potentially produces two kinds of bracts: 1) involucral bracts that are associated with the outer surface of the expanded receptacle outside or below the attachment points of the florets; 2) receptacular or chaffy bracts that subtend the individual florets on the the disc surface or upper portion of the receptacle.

bristle. Hairlike structure, often firm or stiff. In the Asteraceae, a type of pappus segment.
bulb. Underground modified leaf bud, usually covered with fleshy scales.
bulblet. Asexual reproductive structure appearing more or less bulblike, but produced on a leaf, stem, inflorescence, or flower.
calcareous. Limey, referring to substrates derived from limestone or dolomite bedrock.
calyx. Outermost perianth whorl of a flower, consisting of sepals. When only one perianth whorl is present, it is called a calyx by default.
capitate. Headlike, as in a dense headlike cluster of flowers.
capsule. Dry, dehiscent fruit composed of more than one carpel and usually containing two or more seeds.
carpel. In angiosperms, the modified fertile leaf (sporophyll) bearing the ovules. The pistil is composed of one or more carpels.
carpophore. In the fruit type known as a schizocarp, the usually persistent, sometimes Y-shaped stalk located between the mericarps (seed-containing structures) to which these remain attached as the fruit begins to break apart at maturity.
cartilaginous. With the texture of cartilage; tough or firm, but somewhat flexible.
caruncle. Outgrowth or appendage near the hilum of a seed, usually hard and not not fleshy. See also aril.
caudex. Short perennial stem at or just below the ground surface producing new aerial stems each year that die back at the end of the growing season.
cauline. Of the stem. Usually referring to leaves that are attached along an aerial stem rather than at the base of the stem.
chasmogamous. Open-flowering, referring to flowers whose buds open at flowering time to allow for dispersal of pollen and potential cross-pollination. Contrasted with cleistogamous.
chlorophyll. Green pigment in plants associated with photosynthesis.
chromosome. DNA-containing element of a cell nucleus, becoming condensed and visible prior to and during nuclear division.
circumboreal. With a continuous northern distribution in both North America and Eurasia.
clade. Natural (monophyletic) group in a phylogenetic study; that is, a group perceived to have a single genealogical origin, represented as a discrete branch of an evolutionary tree.
clasping. Enlarged at the base and partially surrounding the structure to which an object is attached, as in the bases of some leaves clasping the stem.
class. Rank in a taxonomic classification consisting of a collection of orders (groups of families) with a shared genealogy. For example, the dicots are class Magnoliopsida of the division Magnoliophyta (angiosperms).
clathrate. Lattice-like. Applied to scales composed of cells with the endwalls dark-colored and the centers transparent or nearly so.
cleistogamous. Closed flowered, referring to modified flowers of some species that remain closed and and set seed without exposing the stamens and stigmas, thus are obligately self-pollinated. Contrasted with chasmogamous (open-flowering).
cline. Variation of one or more characters along an environmental or geographic gradient, lacking a perceived compartmentalization of the variation into discrete or discontinuous character states and thus not useful for the recognition of taxonomic segregates within the complex of populations under study.
column. In the Asclepiadaceae and Orchidaceae, the structure composed of the staminal filaments and style fused together. In the Malvaceae and a few other families, the hollow tube formed from the fused filaments.
commissures. In the Apiaceae and other families producing fruits that are schizocarps, the sides or faces of two adjacent carpels that are joined prior to maturation of the fruit and between which the mericarps break apart at maturity.

compound. Divided into 2 or more discrete parts, as in a compound leaf divided into two or more distinct leaflets. Also applied to a pistil composed of more than one carpel.

concave. Referring to a surface that is hollowed out or sunken.

concolorous. Having a single color. See also bicolorous.

contorted. Asymmetrically and abruptly bent, curved, or looped, as in the corollas of some species.

conserve, conservation. Preservation of species and their habitats. In botanical nomenclature, the official retention of some name for a taxon (in order to stabilize the nomenclature) that is not the oldest validly published name for that taxon. This requires a formal printed proposal, a review of the situation by an official nomenclatural committee under the auspices of the International Association for Plant Taxonomy, and a vote on the proposal at an International Botanical Congress.

convex. Referring to a surface that is rounded, bulged outward, or dome-shaped.

cordate. Heart-shaped, with the point of attachment at the notched end. Usually applied to the base of a leaf (note that in some other texts this term refers to a heart-shaped leaf or structure, but in the present volume it is restricted to the base of a leaf or other structure).

corm. Short, solid, erect, underground stem with short internodes, usually thickened and with leaves closely overlapping, these sometimes modified and papery.

corolla. Innermost perianth whorl of a flower, consisting of petals.

corona. In some flowers, an extra whorl of petal-like structures located between the corolla and stamens.

corpusculum. In the Asclepiadaceae, the minute, central, longitudinally grooved, disklike structure of the wishbone-shaped apparatus that connects the two adjacent pollinium sacs and is involved in the specialized mechanism of pollen-mass dispersal by insects. in the family

corrugated. Folded or wrinkled into a regular series of ridges and valleys, as in some leaves that are zigzag in cross-section or some fruits with the surface having a regular series of parallel ridges and furrows.

cotyledon. Embryonic leaf in the seed, often developing into the first leaf of a seedling.

cultigen. Type of plant that arose from a cultivated progenitor. Such plants often have specialized morphological features that enhance their yield or harvestability, but inhibit their ability to survive in the wild.

cyathium. In some species of the Euphorbiaceae, a specialized structure consisting of one, central, stalked pistillate flower surrounded by few to numerous, small, staminate flowers (each with only one stamen) positioned on the concave upper surface of a small cup-shaped involucre with 1–5 conspicuous glands along the rim, the whole structure mimicking a single flower.

cyme. Branched determinate inflorescence, that is, one in which the terminal flowers bloom first. In the present Volume, cymes are included in the general definition of panicles.

cystoliths. Calcium carbonate inclusions in the cells of some plants. Frequently associated with the hardened, swollen bases of hairs in some families (Boraginaceae, Cannabaceae, etc.). Sometimes also used to refer to the swollen or pustular basal portion of such hairs.

cytogenetics. Study of the cytological aspects of inheritance and other genetic phenomena.

cytology. Study of cell function and structure. As applied to taxonomy (cytotaxonomy), mostly referring to the comparative study of chromosome number, morphology, and behavior at meiosis.

cytotype. Variant of a taxon differing in chromosome number from other members of the taxon.

deciduous. Referring to a structure that is shed at the end of the same growing season in which it is produced. Usually refers to leaves that are shed in autumn.

decurrent. Extended downward from the point of attachment, as in a leaf base that extends downward as two wings of tissue fused to the stem below the main attachment point of the leaf.

dehiscence. Method or pattern of breaking open, as in a fruit releasing its seeds.

depauperate. Stunted or poorly developed, as in a plant that is noticeably smaller than normal for the species because of adverse growing conditions.

dichotomous. Divided into two equal parts, as in a vein that forks evenly into two veins of equal size.

dicot. Member of the division Magnoliophyta class Magnoliopsida; a group delineated by a syndrome of features, among them the production of a pair of cotyledons by the embryo. Note that molecular phylogenetic studies have shown this to be an unnatural group because the monocots evolved from within it.

dikaryotic. Stage in the life history of some fungi in which cells are characterized by containing two discreet nuclei. Contrasted with homokaryotic.

dimorphic. With two morphologically different forms, as when leaves on various parts of a plant appear differently shaped.

dioecious. Plants that produces exclusively staminate or pistillate flowers, these located only on different individuals, not on the same plant.

diploid. Containing two sets of chromosomes in each cell; the basic level of ploidy in the sporophyte generation of plants. Usually contrasted with polyploid.

disc. Refers to the general shape of an object that is relatively flat and more or less circular in outline, as in the stigmas of some flowers. In the Asteraceae, disc has two specialized meanings: a) the expanded central portion of a head to which the florets and bracts are attached; b) the actinomorphic florets located centrally in the heads of most species (vs. the zygomorphic ray florets).

discoid. In the Asteraceae, heads containing only disc florets and no ray florets.

disjunct. Widely separated geographically from the rest of a taxon's distribution.

disturbophile. Organism that is adapted to living in highly disturbed habitats and often found only in such places.

division. Highest level in the taxonomic hierarchy of plant classification. Equivalent to the term phylum in animals.

dorsal. Referring to the back or outer (under) surface of a structure in relation to the axis. Also applied to the side of a leaf sheath where the leaf blade is attached to the sheath.

drupe. Usually fleshy, indehiscent fruit containing a stone. In drupes, the fruit wall is differentiated into three layers: the outer skin; a middle, usually fleshy layer; and a hardened inner layer surrounding the usually solitary seed.

elaiosome. Oily, usually white appendage on some seeds that provides food for ants that act as dispersal agents for the seed.

elastically dehiscent. The springlike dehiscence pattern of some fruits in which the fruit wall creates physical stress on portions of the fruit as it matures, resulting in sudden breaking open and often violent dispersal of the seeds.

ellipsoid. Three-dimensional shape about two to three times as long as wide, broadest at the middle, and tapering equally to each end.

elliptic. Two-dimensional shape about two to three times as long as wide, broadest at the middle, and tapering equally to each end.

embryology. Study of the morphology and development of embryos as well as ovules and the cells and nuclei of germinating pollen grains.

endemic. Restricted to a particular geographic area, as in Ozark endemics, which occur only on the Ozark Plateau of Arkansas, Missouri, and/or Oklahoma.

endosperm. Nutritive tissue surrounding the embryo in some seeds.

epigynous. Flowers having the perianth and stamens appearing attached at the tip of the ovary, the ovary appearing entirely inferior to the other floral parts.
epicalyx. Whorl of bracts closely subtending a flower and appearing similar to the calyx.
epiparasite. Organism that is parasitic on a host that is itself parasitic on some other organism.
epiphyte. Plant that is rooted on the surface of another plant, rather than in soil, but derives no water or nutrient from its host's tissues.
evergreen. Referring to a structure that persists for more than one growing season. Usually applied to leaves that remain green and function for two or more years.
exserted. Extending beyond the surrounding parts. Usually applied to stamens or styles that extend beyond the surrounding perianth. Contrasted with included.
extinct. No longer occurring anywhere; with no remaining living individuals.
extirpated. No longer occurring in a portion of its former range; eliminated from an area.
extrafloral nectary. Nectar producing gland located on some portion of the plant other than the flower.
family. Rank in a taxonomic classification consisting of a collection of genera with a shared genealogy and one or more cytological, morphological, or other shared distinguishing features.
farina. Glandular exudate with a powdery consistency.
fasciated. Unusual growth of plant tissues producing abnormally broadened and flattened structures; for example, the aberrant broadening and flattening of the receptacle to produce abnormal heads in some members of the Asteraceae.
fascicle. Tight bundle or cluster, often applied to clusters of leaves or stamens.
fertile. Capable of bearing spores, ovules, seeds, or stamens (contrasted with sterile).
filament. Stalklike or threadlike portion of a stamen to which the anther is attached.
filiform. Very slender and threadlike.
floret. One of the modified flowers aggregated into a head in the Asteraceae and some other families.
floriferous. Bearing flowers, as in the portion of an inflorescence or inflorescence branch on which flowers are produced.
flower. Reproductive structure of angiosperms, consisting of one or more pistils and/or stamens usually surrounded by one or two whorls of perianth (corolla and/or calyx).
follicle. Dry, dehiscent fruit composed of one carpel and usually containing two or more seeds that dehisces longitudinally along only one side.
form. Trivial variant within a species, often produced by a minor mutation, such as white-flowered individuals in an otherwise red-flowered species. Technically called forma.
fruit. Ripened ovary and associated structures; organ containing the seed(s) of angiosperms.
fusiform. A three-dimensional shape, spindle-shaped, broadest at the middle, and tapering equally to each end.
gall. Abnormal growth of plant tissue caused by insects, microorgansims, or physical injury, often appearing as a swollen area or an anomalous structure (sometimes superficially resembling a fruit).
gemma. Budlike asexual reproductive structure that becomes detached and dispersed, germinating to form a new plant.
genealogy. The study of the descent over time of an individual or taxon from its ancestors. In the context of plant evolution and phylogeny, the pattern of presumed ancestral relationships that gave rise to various taxa in a natural group.
genus. Rank in a taxonomic classification consisting of a collection of species with a shared genealogy.
glabrescent. Initially hairy, but becoming glabrous at maturity or with age.

glabrous. Lacking vestiture such as hairs or glands. Sometimes ambiguously referred to as smooth.
gland. Secretory structure.
glaucous. With a white, gray, or light blue waxy coating on the surface.
globose. A three-dimensional shape, globe-shaped or spherical.
glochids. Minute, bristlelike, barbed spines present in some genera of Cactaceae.
grain. Generally applied to small rounded structures, such as a unit of pollen or the seedlike structure on the surface of fruits in *Rumex*. Also sometimes applied to the fruit type in Poaceae (usually called a caryopsis); sometimes applied generally to plants producing seeds or fruits used in cereals and/or flour, including some species of Amaranthaceae and Poaceae.
gynoecium. Collective term for the carpels or pistils in a flower.
gynophore. In some flowers, a stalklike extension of the receptacle that elevates the pistil(s) above the attachment points of the the perianth and stamens.
gynostegium. In species of Asclepiadaceae, the complex structure in the center of each flower resulting from the aggregation of the carpels, stamens, and corona (hoods and horns).
habit. Growth form of a plant.
hastate. Halberd-shaped, as in leaf blades that are triangular with the basal lobes spreading rather than pointed downward.
haustorium. In parasitic plants, the organ that forms the physical connection between the parasite and the tissue of the host plant.
head. Dense cluster of flowers that are sessile or nearly so, sometimes surrounded by bracts. Basic inflorescence type of the Asteraceae.
herbaceous. Referring to plants that are not woody and/or die back to the ground every year. Also sometimes applied to tissues that are leaflike or relatively soft in texture (but not papery or membranous).
heterostyly. In some species, the production of flowers with more than one kind of reproductive morphology: some flowers with relatively elongate styles and short stamens such that the stigmas are positioned above the anthers; other flowers with shorter styles and longer stamens such that the stigmas are positioned below the level of the anthers. Rarely a third floral morphology is produced involving stamens of varying lengths in the same flower and short to intermediate length styles. Heterostyly is an adaptation to promote cross-pollination between flowers by limiting the ability of stamens to shed pollen directly onto the stigma(s) within the same flower.
heterozygous. Individual that, for a given gene has inherited different alleles from each of its parents. Also refers to the genetic condition of having different alleles present for a given gene locus in a particular individual.
hexaploid. Containing six sets of chromosomes in each cell. Usually contrasted with diploid.
historical. Event that occurred long ago; for purposes of the present volume, historical refers to events or collections having occurred more than 50 years ago.
homokaryotic. Stage in the life history of some fungi in which cells are characterized by containing only one nucleus. Contrasted with dikaryotic.
homozygous. Individual that, for a given gene has inherited identical alleles from each of its parents. Also refers to the genetic condition of having all of the same allele present for a given gene locus in a particular individual.
hood. A concave, hollow or arched covering, as in a modified petal in the corollas of some flowers. In the Asclepiadaceae, part of the gynostegium formed from a segment of a corona and sometimes partially enclosing a horn.
horn. In the Asclepiadaceae, an often arched or curved, tapered part of the gynostegium resembling the horn of a cow, formed from a segment of a corona and partially surrounded by a hood.

hyaline. Thin and usually somewhat transparent.

hybrid. Offspring of a mating between two parents belonging to different taxa. Interspecific hybrids often are sterile or have reduced fertility.

hypanthium. Ring or cup of tissue completely or partially surrounding (but not fused to) the ovary, formed either by the marginal expansion of the receptacle or the basal fusion of stamens and perianth

hypogynous. Flowers having the perianth and stamens appearing attached at the base of the ovary, the ovary appearing entirely superior to the other floral parts.

imbricate. Overlapping, as in a compound leaf with overlapping leaflets.

imperfect. Applied to an individual flower containing functional stamens or pistils, but not both (often termed unisexual). Plants with imperfect flowers are either monoecious or dioecious.

included. Not extending beyond the surrounding parts. Usually applied to stamens or styles that do not extend beyond the surrounding perianth and therefore can only be observed by dissecting the flower. Contrasted with exserted.

incurved. Referring to a structure with the tip curved inward, toward the middle, or toward the tip of the axis.

indehiscent. Not breaking open or apart at maturity to release its contents, as in some kinds of fruits.

indicator. In the present volume, referring to a taxon with strong fidelity to a particular habitat, such that the finding of the taxon at a given site is indicative of the presence of a particular community there.

inferior ovary. Ovary with the other flower parts (stamens, petals, and/or sepals) attached at its tip. Flowers with inferior ovaries are referred to as epigynous.

inflorescence. Flowering portion of a plant. The arrangement pattern of the flowers on a plant determine the inflorescence type.

infraspecific. In botanical classification the taxonomic categories within a given species, including subspecies, varieties, and forms.

infrastaminal scales. In some flowers with petals fused into a tube toward the base, small membranous outgrowths of tissue positioned along the inner surface of the corolla tube between and usually below the attachment points of the filaments of the stamens.

infructescence. Fruiting portion of a plant; the inflorescence at fruiting time. The arrangement pattern of the fruits on a plant determine the infructescence type.

integument. Covering of the ovule that develops into the seed coat.

internode. Portion of a stem, root, or inflorescence axis between the attachment points of structures such as a leaves and branches.

introgressive hybridization. Widespread hybridization over several generations between two at least partially interfertile taxa resulting in a mixed population of individuals with a range of morphologies variously intermediate between the parents.

invasive. Species that aggressively colonizes plant communities in which it did not occur naturally prior to human-mediated disturbances in recent times. Invasive exotics are non-native plant species that spread aggressively into native plant communities in a region.

involucel. Two or more bractlets subtending a secondary or ultimate umbel in a compound umbellate inflorescence.

involucre. Two or more bracts subtending an inflorescence. In the Asteraceae, referring to the few to numerous bracts in one to several series associated with the outer surface of the expanded receptacle of each head.

iridescent. Shiny with many colors; reflecting a rainbow pattern of colors.

isozyme. Genetically based variant of an enzyme produced by an organism that differs in size and/or charge of the molecule and can be separated from other forms of the enzyme (produced by different genes) using the technique of gel electrophoresis.

Isozymes are studied to document levels of genetic variation within and between populations and between species. See also allozyme.

keel. Prominent longitudinal ridge, like the keel of a boat. Also applied to the two, usually fused, lowermost petals of flowers in many species of Fabaceae.

lacerate. Irregularly cut or divided, appearing torn.

lanceolate. Two-dimensional shape, lance-shaped, about three to six times as long as wide, broadest below the middle, and tapering to the tip.

lateral. Attached to the side of a structure, as opposed to the tip. Often applied to inflorescences originating from the axils of leaves along the stem.

leaf axil. Angle between a leaf and the stem to which it is attached.

leaf blade. The expanded usually flattened portion of a leaf.

leaf sheath. Basal portion of some leaves that is wrapped around the stem.

legume. Dry fruit derived from a single carpel and usually dehiscent longitudinally along a pair of sutures along opposite margins of the fruit. The main fruit type in the Fabaceae. Also, a vernacular name applied generally to members of the Fabaceae.

liana. Woody climbing plant, for example a grape vine.

ligule. A floret with a perfect, zygomorphic, strap-shaped corolla in the heads of species of Asteraceae tribe Cichorieae. Also sometimes used to refer to the elongate strap-shaped portion of the zygomorphic corolla in ray florets of many Asteraceae.

ligulate. In the tribe Cichorieae of the Asteraceae, heads not containing a mixture of disc and ray florets, but instead, with only perfect, strap-shaped (ligulate) florets.

linear. Two-dimensional shape that is long and narrow (more than ten times as long as wide) with more or less parallel sides.

lip. Variously shaped projecting segment of a zygomorphic corolla or calyx, sometimes positioned lowermost in the flower. Corollas having both upper and lower lips are termed bilabiate.

lobe. Often rounded segment of a structure that is partially divided (not all the way to the base or center). Usually applied to a part of a leaf blade that is noticeably divided but not compound.

locule. Chamber or compartment of an ovary, fruit, anther, or other structure.

loess. Type of substrate composed of loosely compacted silty material thought to have been created by the grinding action of Pleistocene glaciers and later dispersed by wind across the landscape, creating deposits of uneven thickness.

loment. Dry fruit derived from a single carpel that breaks apart transversely at maturity into one-seeded segments enclosed in the fruit walls. Most commonly found in some genera of Fabaceae.

meiosis. Form of nuclear division normally resulting in four daughter cells, each with half as many sets of chromosomes as the parental cell. One of the two essential stages of sexual reproduction (along with fertilization).

membranous. Thin, soft, and more or less transparent.

mericarp. In the fruit type known as a schizocarp, the seed-containing structures that break apart from one another into separate units and eventually are dispersed intact as the fruit reaches maturity.

meristem. Tissue consisting of undifferentiated cells that divide actively to produce cells that become differentiated into various structures or tissues. Often located at the tip or margins of structures such as stems or leaves.

microspecies. In some plant groups, the formal taxonomic description at the species level of various populations that differ only slightly in morphology or other attributes from other populations of what many botanists consider more broadly to represent a single species. Microspecies often are described in genera in which various apomictic populations have become fixed for subtle morphological differences over time.

midnerve. Central nerve or vein of a structure other than a leaf, such as a sepal, petal, or fruit.
midrib. Central vein of a leaf, bract, or scale that is raised above the surface.
midvein. Central vein of a leaf with pinnate venation, often heavier or more conspicuous than the other veins.
mitosis. Form of nuclear division normally resulting in two daughter cells, each with the same number of sets of chromosomes as the parental cell; the cellular process that results in plant growth.
molecular. Dealing with molecules. In botany, usually referring to nucleic acids.
mono-. Prefix meaning one or single, as in a monotypic genus containing only a single species.
monocot. Member of the division Magnoliophyta class Liliopsida; a group delineated by a syndrome of features, among them the production of a single cotyledon by the embryo.
monoculture. Habitat or site containing only a single species.
monoecious. Plant that produces exclusively staminate and pistillate flowers, these both produced on the same plant, but sometimes in different positions.
monomorphic. Having only one type of shape or form. May refer to various structures, including leaves, fruits, and flower parts. Contrasted with dimorphic.
monophyletic. Having a shared genealogy; part of the same clade (branch in a phylogenetic tree).
montane. Of or pertaining to mountains.
morphology. Study of the external organization and structures of plants. See also anatomy.
morphometrics. Comparative study of taxa involving the statistical analysis of quantitative (and to some extent qualitative) morphological data.
morphotype. Morphological variant within a taxon.
mucilaginous. Slimy and moist; snotty.
mycorrhizal. Symbiotic relationship between plant roots and soilborne fungi. The plant receives nutrients and water from the fungus and the fungus presumably receives an enclosed, protected environment in which to grow.
mycotrophic. Organism that receives nutrients and water from soilborne fungi through a mycorrhizal association.
naturalized. Non-native taxon that has become established at a site and has formed a viable self-reproducing tpopulation.
nectary. Nectar-producing secretory structure, often but not always associated with the flower. Sometimes applied to an organ containing a nectar-producing gland.
nerve. Usually unbranched vein or rib of a structure other than a leaf, such as a corolla segment or fruit.
node. Place on a stem, root, or inflorescence axis at which structures such as a leaves and branches are attached.
nodulation. In many members of the Fabaceae (and some other plant families), a symbiotic relationship between the plant and bacteria of the genus of *Rhizobium* resulting in the development of specialized root nodules in which the bacteria fix atmospheric nitrogen into nitrates, the nitrogen-rich compounds used by plants as fertilizers.
nomenclature. System of names for a process or classification. As applied to botany, the system of names of taxa at all levels in the taxonomic classification.
nutlet(s). Appearing similar to small nuts. In some plant families, such as Boraginaceae and Lamiaceae, the one-seeded, hard-walled segments or lobes (mericarps) of the indehiscent dry fruits (schizocarps) that break apart at maturity and are dispersed intact.
ob-. Prefix indicating inversion, such as oblanceolate, meaning lanceolate, but with the attachment point at the narrow end and the widest point above the middle.

oblong. Two-dimensional shape about two to four times as long as wide, broadest at the middle, and rounded or tapering at each end, with the sides more or less parallel.

octoploid. Containing eight sets of chromosomes in each cell. Usually contrasted with diploid.

-oid. Suffix indicating similarity, such as petaloid sepals that are colored or otherwise resemble petals. Frequently applied to three-dimensional shapes, such as ovoid, meaning a structure that is egg-shaped in three dimensions.

opposite. With a pair of parts or structures attached on opposite sides of a single point or node, as in stems with two leaves at each node. Contrasted with alternate and whorled.

orbicular. Two-dimensional shape more or less circular in outline.

ovary. Expanded basal portion of the pistil containing the ovule(s).

ovate. Two-dimensional shape, egg-shaped, about two to three times as long as wide, broadest below the middle, and tapering to the tip.

ovoid. Three-dimensional shape, egg-shaped, about two to three times as long as wide, broadest below the middle, and tapering to the tip.

ovule. In seed plants, the structure containing the egg nucleus, associated tissue, and surrounding layer(s) of integument; after fertilization developing into the seed.

palmate. With all of the parts or divisions originating from a single point. Referring to leaves and other lobed or compound structures with the parts all attached to the tip of the subtending petiole or stem, or to a venation pattern with the main veins all arising from a single point at or near the base of the leaf blade.

panicle. Branched inflorescence, technically restricted to inflorescences with the ultimate branches having flowers that mature from the base of each branch toward the tip, but in the present volume referring generally to any inflorescence with the flowers attached to branches rather than to a single central axis.

papilionaceous. Referring to the flowers of species in the Fabaceae subfamily Faboideae, whose zygomorphic corollas are differentiated into a banner, two wings, and a keel.

papillose. Referring to a structure with tiny nipple-like projections (papillae) on the surface.

pappus. In the Asteraceae, the apparent outermost whorl of perianth in some or all flowers of most species, consisting variously of awns, bristles, and/or scales, sometimes reduced to a low crown or rim.

paraphyletic. In a phylogenetic study, a lineage that is considered unnatural because some taxonomically recognized portion of the group is revealed to have been derived as a specialized subgroup from within a portion of that lineage. For example, the dicots, as traditionally circumscribed, are considered paraphyletic because the monocots apparently originated as a specialized offshoot from within the dicots (as opposed to the two groups having evolved separately from some common ancestor.

parasitic. Organism that derives all or part of its food and/or water directly from the tissues of another organism.

parietal. Referring to the inner wall of a hollow structure. In botany, a kind of placentation in which ovules or seeds are attached to placentae along the outer wall of a locule (vs. axile placentation, in which the placentae are located centrally where the walls of adjacent locules intersect).

pebbled. With an uneven, sometimes roughened surface having a series of minute, low bumps or papillae.

pedicel (pedicillate). Stalk of an individual flower. Contrasted with peduncle.

peduncle. Stalk of an inflorescence or portion of an inflorescence. Contrasted with pedicel.

peltate. Umbrella-shaped, with a relatively flat structure having a stalk attached on the undersurface inward from the base.

pentaploid. Containing five sets of chromosomes in each cell. Because they contain an odd number of chromosome sets, pentaploid plants usually are either sterile or reproduce apomictically.

pepo. Fleshy indehiscent druit with a hardened or leathery rind; modified kind of berry.

perennial. Plant that lives for three or more years. Also sometimes applied to stems that do not die back in the winter.

perfect. Applied to an individual flower containing one or more functional stamens and pistils (often termed bisexual).

perfoliate. Referring to a leaf or leaves with the margins entirely surrounding the stem, such that the stem appears to pass through the leaf tissue.

perianth. General term referring collectively to any and all calyx (sepals) and corolla (petals) present in a flower.

pericarp. Wall of the fruit, derived from the wall of the ovary, sometimes differentiated into three layers.

periderm. Corky tissue produced by plant stems and roots in response to injury or aging.

perigynous. Flowers having the perianth and stamens appearing attached at the tip of a hypanthium, this saucer-shaped to deeply urn-shaped, sometimes more or less enclosing but not actually fused to the superior ovary.

persistent. Not shed with age, as in the calyx of some plant species that remains attached to the mature fruit.

petal. Structure or unit of the innermost perianth whorl of a flower, often white or brightly colored.

petaliferous. Bearing petals. In species producing both chasmogamous and cleistogamous flowers, the open-flowering chasmogamous flowers are sometimes called petaliferous to contrast them with the cleistogamous flowers, which do not produce apparent petals.

petaloid. Similar in appearance to a petal; for example the colored sepals of flowers in some species and the white or colored appendages of the cyathium glands in some species of *Euphorbia*.

petiole. Stalklike base of a leaf to which the blade is attached.

phellogen. Type of meristem (or the cells of this meristem) of plant stems and roots that results in the production of corky tissue.

photosynthesis. Physiological process by which plants convert water and carbon dioxide into sugars using light for energy.

phylogenetic. Referring to the genealogical history of a group of organisms as represented by ancestor/descendant relationships. Phylogenetics is the branch of systematics concerned with reconstructing phylogenies by hypothesizing and analyzing the primitive and derived states in a data set of various characters for a study group and its presumed closest relatives.

phylum. Highest level in the taxonomic hierarchy of animal classification; equivalent to the term division in plants.

physiology. Study of the processes and mechanisms by which organisms live and function.

phytochemistry. Study of chemical compounds produced by plants.

pinna. Primary division of a compound leaf, which may be further subdivided into pinnules.

pinnate. With the divisions originating at various points along opposite sides of an axis. Usually referring to compound leaves with the leaflets attached singly or in pairs along the central axis (rachis), or to a venation pattern with the main veins arising at different points along a longitudinal midvein.

pinnule. Ultimate division (leaflet) of a compound leaf. Usually applied to leaves that are two or more times compound.

pistil. Ovule-producing organ of a flower, consisting of an ovary (which contains one or more ovules) and usually a style and stigma.
pistillate. Referring to a flower bearing only one or more pistils and no functional stamens, or to a portion of a plant or inflorescence where the pistil-bearing flowers are situated. Sometimes also applied to scales or other structures associated with pistil-bearing flowers.
pith. Soft or spongy tissue at the center of some stems and roots.
placenta. Portion of the ovary to which the ovules are attached. The organization and position of this tissue in the ovary is called placentation.
pleurogram. In the seeds of some species of Fabaceae, a region of the seed coat differing in color and/or texture that is demarcated by a fine line or groove.
pollen. In seed plants, the minute structure containing and dispersing the usually immature, highly reduced male gametophyte; formed in an anther. Often referred to as a pollen grain.
pollinium. In species of Asclepiadaceae and Orchidaceae, a specialized saclike mass of waxy or coherent pollen grains that is dispersed as a unit (by insect pollinators, to which the pollinia adhere). In the Asclepiadaceae, the structure that is dispersed consists of a pair of pollinia from adjacent anthers connected by a wishbone-shaped apparatus, and this compound structure technically is called a pollinarium.
polyembryony. In some plant species, the production of more than one embryo within a single seed. In flowering plants, this unusual phenomenon often is associated with apomixis.
polymorphic. With several different forms, referring to plants or structures of plants that have different morphologies in different individuals or populations.
polyploid. Containing more than two sets of chromosomes in each cell. Usually contrasted with diploid.
prickle. Sharply pointed outgrowth of the epidermis or bark. Because prickles are surface structures, they often are relatively easily dislodged from the plant. See also, spine, thorn.
prostrate. Positioned flat on the ground.
protandrous. Condition in some flowers in which the the stamens mature and pollen is dispersed prior to the maturation and receptivity of the stigmas. This mechanism promotes cross-pollination.
pubescent. General term for a surface covered with any type of hair.
pulvinus. Thickened or swollen portion at the base of a petiole or leaflet stalk. Common in the Fabaceae.
punctate. Referring to a structure dotted with colored spots of pigment or with small glands (punctations) that are sessile or often sunken into shallow pits.
pustular-based. Referring to a hair with the base expanded into a small blisterlike mass (pustule), this usually persistent after the threadlike portion of the hair has worn away.
raceme. Unbranched inflorescence, technically restricted to inflorescences with stalked flowers maturing from the base toward the tip of an unbranched axis, but in the present volume referring generally to any inflorescence with stalked flowers attached to an unbranched central axis.
rachis. In a compound leaf, the stemlike axis to which the pinnae are attached. Sometimes also applied to the central axis of an inflorescence.
radiate. In the Asteraceae, heads containing both disc and ray florets.
radicle. Pertaining to the root of a plant. Often applied to the developing root of an embryo or seedling.
ray. Branch of an umbel or similar inflorescence. Also applied to the sterile or pistillate, zygomorphic florets located marginally in the heads of many species of Asteraceae (vs. actinomorphic disc florets).

receptacle. Expanded tip of a stem (pedicel) to which the flower parts are attached. Also applied to the expanded tip of a stem (peduncle) to which the flowers in a head are attached.

recurved. Referring to a structure with the tip curved outward, downward, or away from the tip of the axis.

reflexed. Referring to a structure abruptly bent backward or downward.

replum. In the modified capsular fruits of most species of Brassicaceae, the thin, placental band of tissue around the margins of the septum (the partition between the two locules).

resin-dotted. Referring to a structure dotted with small glands that are sessile or often sunken into shallow pits and produce a resinous exudate.

resupinate. Referring to flowers that are twisted at the base during development so that the top of the flower is oriented toward the bottom at maturity.

reticulate. Forming a network, as in anastomosing veins of a leaf. Also applied to a type of phylogeny in which new species evolve as the result of of interspecific hybridization followed by a stabilization of the hybrid and regaining of fertility, such that the offspring becomes a self-reproducing taxon reproductively isolated from both parental taxa.

rhizome. Modified stem occurring all or mostly underground, often creeping horizontally.

rhombic. Two-dimensional shape, more or less diamond-shaped.

rib. Prominent longitudinal vein or line raised above the surface of a structure such as a leaf, bract, scale, or fruit.

rootstock. General term for the underground structures of a plant including rhizomes and other rootlike stems and usually also the roots.

rosette. Group or cluster of leaves positioned in a radial pattern at or near gound level.

saccate. With a hollow pouch or spur, as in the modified spurred sepal in the genus *Impatiens*.

sagittate. Arrowhead-shaped, as in leaf blades that are triangular with the basal lobes pointed more or less downward, rather than spreading.

samara. Dry, indehiscent, usually single-seeded, winged fruit; for example, the fruits of *Acer* and *Fraxinus*.

saprophyte. Plant deriving water and food from an obligate mycorrhizal association, usually lacking chlorophyll. Because the soil-borne fungi are mostly species responsible for the decay of organic matter, saprophytes often are imprecisely defined as plants deriving nutrition from decaying organic matter. Also sometimes referred to as secondary parasites because the soil-borne fungi also have mycorrhizal associations with tree species, and nutrients may pass from the trees through the fungus to the saprophyte.

scale. Small, flattened, usually thin structure resembling the scale of a fish or reptile. Applied to structures covering some rhizomes, stems, leaves, or other plant parts that may be related to hairs or highly modified leaves. In the Asteraceae, a type of pappus segment.

scarious. Referring to structures or portions of structures that are dry, thin, and not green (usually white or somewhat transparent), as in the margins of some bracts.

schizocarp. A dry fruit formed from a compound ovary that breaks apart at maturity into two or more (depending on the number of carpels) indehiscent structures known as mericarps, which eventually are dispersed intact.

scorpioid. Coiled, like the tail of a scorpion. Usually applied to inflorescences with a central axis that uncurls from the tip as it elongates and develops, as in many species of Boraginaceae. For purposes of the present Volume, this category includes both the truly coiled inflorescences more precisely called helicoid and those derived from a more or less spirally zigzag axis to which the term scorpioid is restricted in some other works.

seed. Ripened ovule, consisting of an embryo and surrounding structures and/or layers.

sepal. Structure or unit of the outermost or only perianth whorl of a flower, usually green, but less commonly white or brightly colored.

sepaloid. Similar in appearance to a sepal, as in petals that are green rather than colored, or a bract that closely subtends a calyx.

septum. Partition between the chambers (locules) in an ovary or fruit. Also applied to one of the cross-partitions occurring in the stems or inflated leaves of some plant species.

sessile. Referring to structures that are not stalked, as in leaves lacking petioles.

sexual reproduction. Type of reproduction involving fusion of gametes, such as the fertilization of the egg and polar nuclei by sperm nuclei in angiosperms.

shrub. Woody plant with usually several main stems that are shorter and thinner (generally less than four inches in dbh [diameter at breast height]) at maturity than those of trees.

silicle. In the Brassicaceae, a modified kind of capsular fruit that is less than three times as long as wide. Contrasted with silique.

silique. In the Brassicaceae, a modified kind of capsular fruit that is more than three times as long as wide. Contrasted with silicle.

simple. In the context of the present volume, the opposite of compound. Usually applied to leaves with the blade entire or lobed but not fully divided into leaflets.

sinus. In plants, usually applied to the concave area between two adjacent lobes of a leaf or other divided structure.

somatic. Referring to all parts of the plant except the gametes (egg and sperm). In cytology, applied to chromosomes during mitosis.

spathe. One or a pair of large bracts surrounding an inflorescence and often partially concealing it.

spatulate. Spatula-shaped, widest near the broad rounded tip and tapering gradually to the base. Sometimes spelled spathulate in the older literature.

species. Basic unit of taxonomic classification; variously defined, but generally consisting of a group of usually interfertile populations of individual organisms with shared distinctive morphological or other features and a shared genealogy that maintain their discrete identity over time.

spherical. three-dimensional shape, globe-shaped or ball-shaped.

spike. Unbranched inflorescence, technically restricted to inflorescences with sessile flowers maturing from the base toward the tip of an unbranched axis, but in the present volume referring generally to any inflorescence with sessile flowers attached to an unbranched central axis.

spine. Sharply pointed structure derived from a modified leaf or stipule and arising from beneath the epidermis or bark. Often imprecisely applied to any sharply pointed structure on a plant. See also prickle, thorn.

spinescent. Modified into a sharply pointed, stiff or hard, spinelike tip, for example, the involucral bracts of many thistles (tribe Cardueae) and some other species of Asteraceae. Note that the term spinescent usually is not applied to true spines, but rather spinelike structures.

sporophyte. Spore-producing generation of the plant reproductive cycle, in vascular plants usually much larger than the gametophyte generation and usually with twice the number of chromosome sets; the conspicuous phase of the life cycle in vascular plants.

spur. In some flowers, a hollow saclike or tubular appendage of a calyx (or individual sepal) or corolla (or individual petal), often containing nectar.

spur shoot. Short branch with very congested nodes, such that the leaves produced appear whorled or clustered.

stamen. Pollen-producing organ of a flower, consisting of an anther (containing the pollen in usually two or four sacs) and often a stalklike or threadlike filament.

staminate. Referring to a flower bearing only one or more stamens and no functional pistils, or to a portion of a plant or inflorescence where the stamen-bearing flowers are situated. Sometimes also applied to scales or other structures associated with stamen-bearing flowers.

staminode. Sterile stamen (not producing pollen), usually modified in appearance from that of the fertile stamen(s).

stellate. Star-shaped, usually applied to hairs with several branches in a radial pattern from the base.

sterile. Unable to reproduce sexually, usually because of chromosomal problems during meiosis and/or failure of spores or fruits to mature.

stigma. Portion of the pistil that is receptive to pollen grains.

stipel. Small stipule-like structures at the base of a leaflets in a compund leaf. Common in the Fabaceae.

stipule. Appendage at a leaf base. May be leaflike, glandular, spiny, or shaped otherwise. Occurring in a pair on opposite sides of the leaf base and free, united with each other, or partially fused to the petiole or stem.

stolon. Modified stem elongated and arching or creeping above the ground, rooting at nodes or at the tip and giving rise to new plants asexually.

stomate. Pore in the surface of a plant structure involved in gas exchange between the plant and the atmosphere. The opening is surrounded by a pair of guard cells.

style(s). In some pistils, one or more structures connecting the ovary and stigma(s).

sub-. Prefix indicating below or less than. See subfamily or subtending in this glossary for examples.

subfamily. Infrafamilial rank in a taxonomic classification; a group of genera in a particular family with a shared genealogy and shared cytological, morphological, or other distinguishing features differing from other genera of the same family; sometimes divided into tribes.

subspecies. Infraspecific rank in a taxonomic classification; a group of plants or populations with characteristic cytology, morphology, or other distinguishing features differing from other members of the same species; often used interchangeably with variety, but sometimes used to denote an infraspecific taxon occupying a discrete portion of the range of a species; occasionally divided into varieties.

subtending. Attached or positioned immediately below some structure, such as a bract subtending a flower.

succulent. Referring to plants or structures that are thickened, fleshy, and juicy, as in the stems of most species of Cactaceae.

superior ovary. Ovary with the other flower parts (stamens, petals, and/or sepals) attached at its base or along the rim of a hypanthium that is not fused to the sides of the ovary. Flowers with superior ovaries are referred to as hypogynous (flower parts attached at the ovary base) or perigynous (flower parts attached to a hypanthium).

systematics. Study of the kinds and diversity of organisms and their interrelationships, including identification, classification, nomenclature, and taxonomy.

tapered. Gradually narrowed or drawn out to a usually sharp point, usually with noticeably concave sides. Contrasted with narrowed (wedge-shaped, with more or less straight sides) and rounded (not pointed, but with strongly convex sides).

taproot. Solitary vertical primary root, often with fine spreading branch roots.

tautonym. In botanical nomenclature, an illegitimate combination in which the specific epithet is identical to the generic epithet. For example, when the poison sumac, *Rhus toxicodendron*, was reclassified into the genus *Toxicodendron*, it was renamed *Toxicodendron pubescens* to avoid creating the tautonym *Toxicodendron toxicodendron*.

taxon. General term for an entity of any rank in a system of biological classification, such as a variety, species, genus, or family.

taxonomy. Technically, the study of the classification of objects into systems, including sets of principles, procedures, and rules, but in practice used interchangeably with the more general term systematics.

tendril. Slender twining organ derived from a modified stem or leaf and used to attach some climbing plants to supporting vegetation or other structures.

tepal. Segment of a perianth that is not differentiated into a calyx and corolla. May be applied to a perianth with numerous parts that grade gradually from green sepal-like segments into white or colored petal-like segments (as in many species of Cactaceae) or to a perianth in which the sepals and petals are identical in morphology and coloration.

terminal. Attached at the tip of a structure, as opposed to the sides, as in an inflorescence attached at the tip of a stem, a flower at the tip of an inflorescence or inflorescence branch, or a leaflet at the tip of a compound leaf.

ternate. Divided into three parts, as in some lobed compound leaves that are divided into lobes or leaflets in groups of threes.

thorn. Modified branch sharply pointed at the tip. Determinate thorns tend to cease growth after reaching a particular length, whereas indeterminate thorns potentially elongate with age into a normal branch with leaves and axillary buds. See also, prickle, spine.

tetraploid. Containing four sets of chromosomes in each cell. Usually contrasted with diploid.

translators. In the Asclepiadaceae, the two slender arms of the wishbone-shaped apparatus that connects the two adjacent pollinium sacs and is involved in the specialized mechanism of pollen-mass dispersal by insects in the family.

tree. Woody plant with usually one or few main stems that are taller and thicker (generally more than four inches in dbh [diameter at breast height]) at maturity than those of shrubs.

tribe. Infrafamilial rank in a taxonomic classification; a group of genera with a shared genealogy in a particular family that share cytological, morphological, or other distinguishing features differing from other genera of the same family.

trichome. General term for any kind of hair or hairlike structure on a plant surface.

trifoliate. Compound leaf with three leaflets. Trifoliate leaves may be pinnately or palmately derived, which usually can be determined by whether or not the terminal leaflet has a stalk that is different from the condition of the lateral leaflets.

trigonous. Referring to structures appearing more or less triangular in cross-section. Usually applied to stems or fruits.

triploid. Containing three sets of chromosomes in each cell. Because they contain an odd number of chromosome sets, triploid plants are either sterile or reproduce apomictically.

truncate. Appearing squared off or cut straight-across at the tip or base.

tuber. Underground stem, thickened and with several to numerous buds on the surface, modified for food storage. Also applied to a thickened, modified portion of a rhizome.

tubercle. Small rounded protuberance on a surface or structure.

turion. Small modified budlike branch that acts as an overwintering structure, germinating to regenerate a plant asexually the following spring, as in some species of Lemnaceae.

umbel. Unbranched, flat-topped or convex inflorescence with the stalked flowers attached in a more or less radial pattern at the tip of the main stalk. Also applied to branched inflorescences with a radial pattern of branching and the individual flowers in umbels at the tips of the ultimate branches. In compound umbels, the ultimate clusters of stalked flowers are referred to as umbellets. Also, a vernacular name applied to a member of the plant family Apiaceae.

undulate. Referring to structures with a regular wavy pattern, as in the margins of some leaves.

unisexual. Referring to imperfect flowers or plants producing imperfect flowers. Commonly used but technically incorrect, as angiosperm plants are sporophytes and sexuality applies to gametophytes.

utricle. Indehiscent, dry, thin-walled, more or less inflated or bladder-like, single-seeded fruit.

valve. In a capsule, one of the outer parts between the lines of dehiscence.

variety. Basic infraspecific rank in a taxonomic classification; a group of plants or populations with characteristic cytology, morphology, or other distinguishing features differing from other members of the same species.

vascular plant. Plant possessing vascular tissue (xylem and phloem for conducting water and food, respectively). Although a few mosses and algae have small amounts of vascular tissue, the term is usually reserved for pteridophytes, gymnosperms, and angiosperms.

vascular bundle. Group or cluster of vascular tissue (xylem and phloem) in some stems, leaves, and roots.

vegetative. Referring to structures or growth not associated with flowers or fruits.

ventral. Referring to the inner (upper) surface of a structure in relation to the axis. Also applied to the side of a leaf sheath opposite to the attachment point of the leaf blade.

versatile. Attached toward the middle, as in a stamen in which the filament is attached toward the middle of the anther.

whorl(ed). With three or more parts or structures attached at a single point or node, as in stems with a whorl of four leaves at each node. Contrasted with alternate and opposite.

wing. Thin or membranous, flattened projection or flap of tissue along a surface or the margin of a structure, such as some stems and fruits. Also applied to a lateral petal in the strongly zygomorphic flowers of many species of Fabaceae.

woody. Producing wood (quantities of hardened secondary vascular tissue). As applied in the present volume, indicating a tree, shrub, or perennial with a woody caudex.

zygomorphic. Having bilateral symmetry, such as flowers with parts arranged so that opposite sides of a line drawn through the middle are mirror images in only one plane.

LITERATURE CITED IN VOLUME 2

Abel, E. L. 1980. Marihuana: The First Twelve Thousand Years. Plenum Press, New York.
Abel, W. E., and D. F. Austin. 1981. Introgressive hybridization between *Ipomoea trichocarpa* and *Ipomoea lacunosa* (Convolvulaceae). Bull. Torrey Bot. Club 108:231–239.
Abrahamson, W. G., and K. D. McCrea. 1977. Ultraviolet light reflection and absorption patterns in populations of *Rudbeckia* (Compositae). Rhodora 79:269–277.
Abrams, M. D. 1998. The red maple paradox. BioScience 48:355–364.
Abuhadra, M. N. 2000. Taxonomic studies on the *Arenaria serpyllifolia* group (Caryophyllaceae). Fl. Medit. 10:185–190.
Adams, R. P., A. S. Tomb, and S. C. Price. 1987. Investigation of hybridization between *Asclepias speciosa* and *A. syriaca* using alkanes, fatty acids, and triterpenoids. Biochem. Syst. Ecol. 15:395–399.
Adams, W. P. 1957. A revision of the genus *Ascyrum* (Hypericaceae). Rhodora 59:73–95.
———. 1959. The status of *Hypericum prolificum*. Rhodora 61:249–251.
———. 1962. Studies in the Guttiferae. I. A synopsis of *Hypericum* section *Myriandra*. Contr. Gray Herb. 189:1–51.
———. 1972. Studies in the Guttiferae. III. An evaluation of some putative spontaneous garden hybrids in *Hypericum* sect. *Myriandra*. Rhodora 74:276–282.
Adams, W. P., and N. K. B. Robson. 1961. A re-evaluation of the generic status of *Ascyrum* and *Crookea* (Guttiferae). Rhodora 63:10–16.
Aellen, P. 1959. Familie Amaranthaceae. Pp. 461–532 in K. H. Rechinger, ed., [G. Hegi] Illustrierte Flora von Mitteleuropa. 2d ed., vol. 3(2), fascicle 1. Verlag Paul Parey, Berlin, Germany.
Alexander, E. J. 1955. *Thelesperma* Less. Pp. 65–69 in E. E. Sherff and E. J. Alexander, Compositae—Heliantheae—Coreopsidinae. N. Amer. Fl., ser. 2, 2:1–190.
Alexopoulos, C. J., C. W. Mims, and M. Blackwell. 1996. Introductory Mycology. 4th ed. John Wiley and Sons, New York.
Allen, O. N., and E. K. Allen. 1981. The Leguminosae: A Sourcebook of Characteristics, Uses, and Nodulation. University of Wisconsin Press, Madison.
Allison, J. R., and T. E. Stevens. 2001. Vascular flora of Ketona dolomite outcrops in Bibb County, Alabama. Castanea 66:154–205.
Al-Shehbaz, I. A. 1985. The genera of Brassiceae (Cruciferae; Brassicaceae) in the southeastern United States. J. Arnold Arbor. 66:279–351.
———. 1986. The genera of Lepideae (Cruciferae; Brassicaceae) in the southeastern United States. J. Arnold Arbor. 67:265–311.
———. 1988a. *Cardamine dissecta*, a new combination replacing *Dentaria multifida* (Cruciferae). J. Arnold Arbor. 69:81–84.
———. 1988b. The genera of Arabideae (Cruciferae; Brassicaceae) in the southeastern United States. J. Arnold Arbor. 69:85–166.
———. 1991. The genera of Boraginaceae in the southeastern United States. J. Arnold Arbor., suppl. ser. 1:1–169.
———. 2003. Transfer of most North American species of *Arabis* to *Boechera* (Brassicaceae). Novon 13:381–391.
Al-Shehbaz, I. A., and S. L. O'Kane Jr. 2002. Taxonomy and phylogeny of *Arabidopsis* (Brassicaceae) (August 22, 2002). Doi/10.1199/tab. 0001 (22 pp.) in C. R. Somerville and E. M. Meyerowitz, eds., The Arabidopsis Book. [Electronic publication.] American Society of Plant Biologists, Rockville, MD. http://www.aspb.org/publications/arabidopsis/.
Al-Shehbaz, I. A., and R. A. Price. 1998. Delimitation of the genus *Nasturtium* (Brassicaceae). Novon 8:124–126.
Al-Shehbaz, I. A., K. Mummenhof, and O. Appel. 2002. *Cardaria*, *Coronopus*, and *Stroganowia* are united with *Lepidium* (Brassicaceae). Novon 12:5–11.
Alston, A. H. G., and R. E. Schultes. 1951. Studies of early specimens and reports of *Ilex vomitoria*. Rhodora 53:272–279, pl. 1181.
Amato, J. A. 1993. The Great Jerusalem Artichoke Circus: The Buying and Selling of the Rural American Dream. University of Minnesota Press, Minneapolis.

LITERATURE CITED

Anderberg, A. A. 1991. Taxonomy and phylogeny of the tribe Gnaphalieae (Asteraceae). Opera Bot. 104:1–195.

———. 1994. Tribe Gnaphalieae. Pp. 304–364 *in* K. Bremer, Asteraceae: Cladistics and Classification. Timber Press, Portland, OR.

Anderberg, A. A., P. Eldenäs, R. J. Bayer, and M. Englund. 2005. Evolutionary relationships in the Asteraceae tribe Inuleae (incl. Plucheeae) evidenced by DNA sequences of *ndhF;* with notes on the systematic positions of some aberrant genera. Organisms Diversity Evol. 5:135–146.

Anderson, E. F. 2001. The Cactus Family. Timber Press, Portland, OR.

Anderson, E. [S.] 1936. An experimental study of hybridization in the genus *Apocynum*. Ann. Missouri Bot. Gard. 23:159–168.

Anderson, L. C. 1994. A revision of *Hasteola* (Asteraceae) in the New World. Syst. Bot. 19:211–219.

Anderson, L. C., and J. B. Creech. 1975. Comparative leaf anatomy of *Solidago* and related Asteraceae. Amer. J. Bot. 62:486–493.

Anderson, R. C., J. S. Fralish, J. E. Armstrong, and P. E. Benjamin. 1993. The ecology and biology of *Panax quinquefolium* L. (Araliaceae) in Illinois. Amer. Midl. Naturalist 129:357–372.

Angiosperm Phylogeny Group. 1998. An ordinal classification for the families of flowering plants. Ann. Missouri Bot. Gard. 85:531–553.

———. 2003. An update of the Angiosperm Phylogeny Group classification for the orders and families of flowering plants: APG II. Bot. J. Linn. Soc. 141:299–436.

Anonymous. 1998. White snakeroot, a potential poison. University of Missouri Extension Service Integrated Pest and Crop Management Newsletter [electronic publication], vol. 8, no. 16. http://ipm.missouri.edu/ipcm/archives/v8n16/ipmltr3.htm.

Antonio, T. M., and S. Masi. 2001. The Sunflower Family in the Upper Midwest. Indiana Academy of Science, Indianapolis.

Appel, O., and I. A. Al-Shehbaz. 2002. Cruciferae. Pp. 75–174 *in* K. Kubitzki, ed., Families and Genera of Vascular Plants. Vol. 5. Malvales, Capparales, and Non-betalain Caryophyllales (K. Kubitzki and C. Bayer, vol. eds.). Springer-Verlag, Berlin.

Applequist, W. L. 2002. A reassessment of the nomenclature of *Matricaria* L. and *Tripleurospermum* Sch. Bip. (Asteraceae). Taxon 51:757–761.

Applequist, W. L., and R. S. Wallace. 2001. Phylogeny of the portulacaceous cohort based on *ndhF* sequence data. Syst. Bot. 26:406–419.

Arnason, T., R. J. Hebda, and T. Johns. 1981. Use of plants for food and medicine by native peoples of eastern Canada. Canad. J. Bot. 59:2189–2325.

Arriagada, J. E. 1998. The genera of Inuleae (Compositae; Asteraceae) in the southeastern United States. Harvard Pap. Bot. 3:1–48.

Ashley, Mrs. H. V. 1963. Report for save the holly contributors. Garden Forum 25(6): 14-15.

Associated Press. 2002. Ragweed's re-emergence bad for farmers. Belleville News-Democrat, 26 August 2002, 3B.

Austin, D. F. 1978. The *Ipomoea batatas* complex—I. Taxonomy. Bull. Torrey Bot. Club 105:114–129.

———. 1990. Comments on southwestern United States *Evolvulus* and *Ipomoea* (Convolvulaceae). Madroño 37:124–132.

———. 1997. *Calystegia* (Convolvulaceae) in Texas. Sida 17:837–840.

———. 2000. Bindweed (*Convolvulus arvensis,* Convolvulaceae) in North America: From medicine to menace. J. Torrey Bot. Soc. 127:172–177.

Baagøe, J. 1974. The genus *Guizotia* (Compositae): A taxonomic revision. Bot. Tidsskr. 69:1–39.

Babcock, E. B. 1947a. The genus *Crepis:* Part 1, the taxonomy, phylogeny, distribution, and evolution of *Crepis*. Univ. Calif. Publ. Bot. 21:i–xii, 1–198, 1 pl.

———. 1947b. The genus *Crepis:* Part 2, systematic treatment. Univ. Calif. Publ. Bot. 22:i–x, 199–1030.

Bailey, L. H. 1929. The domesticated cucurbitas. Gentes Herb. 2:62–115.

———. 1943. Species of *Cucurbita*. Gentes Herb. 6:267–322.

Baldwin, B. G., B. L. Wessa, and J. L. Panero. 2002. Nuclear rDNA evidence for major lineages of helenioid Heliantheae (Compositae). Syst. Bot. 27:161–198.

Baldwin, J. T., Jr. 1945. Chromosomes of Cruciferae—II. Cytogeography of *Leavenworthia*. Bull. Torrey Bot. Club 72:367–378.

Ballard, R. 1986. *Bidens pilosa* complex (Asteraceae) in North and Central America. Amer. J. Bot. 73:1452–1465.

Barker, W. T. 1986. Brassicaceae Burnett. Pp. 293–333 *in* Great Plains Flora Association, eds., Flora of the Great Plains. University Press of Kansas, Lawrence.

Barker, W. T., and R. E. Brooks. 1986. Cannabaceae Endl. Pp. 123–125 *in* Great Plains Flora Association, eds., Flora of the Great Plains. University Press of Kansas, Lawrence.

Barkley, F. A. 1937. A monographic study of *Rhus* and its immediate allies in North and Central America, including the West Indies. Ann. Missouri Bot. Gard. 24:265–498.

Barkley, T. M. 1962. A revision of *Senecio aureus* Linn. and related species. Trans. Kansas Acad. Sci. 65:318-408.

——. 1963. The intergradation of *Senecio plattensis* and *Senecio pauperculus* in Wisconsin. Rhodora 65:65–67.

——. 1978. *Senecio*. N. Amer. Fl., ser. 2, 10:50–139.

——. 1980. Taxonomic notes on *Senecio tomentosus* and its allies (Asteraceae). Brittonia 32:291-308.

——. 1986. Asteraceae. Pp. 838–1021 *in* Great Plains Flora Association, eds., Flora of the Great Plains. University Press of Kansas, Lawrence.

——. 1999. The segregates of *Senecio* s.l., and *Cacalia* s.l., in the flora of North America, north of Mexico. Sida 18:661–672.

Barlow, C. 2000. The Ghosts of Evolution: Nonsensical Fruit, Missing Partners, and Other Ecological Anachronisms. Basic Books, New York.

Barneby, R. C. 1991. Sensitivae censitae: A description of the genus *Mimosa* Linnaeus (Mimosaceae) in the New World. Mem. New York Bot. Gard. 65:1–835, frontispiece.

Barth, H. J., C. Klinke, and C. Schmidt. 1994. The Hop Atlas: The History and Geography of the Cultivated Plant. Barth & Sohn, Nuremberg, Germany.

Baskin, J. M., C. C. Baskin, and M. E. Medley. 1983. The historical geographical distribution of *Onosmodium molle* Michx. subsp. *molle* (Boraginaceae). Bull. Torrey Bot. Club 110:73–76.

Baskin, J. M., K. M. Snyder, and C. C. Baskin. 1993. Nomenclatural history and taxonomic status of *Echinacea angustifolia, E. pallida,* and *E. tennesseensis* (Asteraceae). Sida 15:597–604.

Bassett, I. J., and C. W. Crompton. 1982. The genus *Chenopodium* in Canada. Canad. J. Bot. 60:586–610.

Bauer, R., and H. Wagner. 1990. *Echinacea,* Hanbuch für Ärzte, Apotheker und Andere Naturwissenschaftler. Wissenschaftliche Verlagsgesellschaft, Stuttgart, Germany.

Bayer, R. J. 1985a. Investigations into the evolutionary history of the polyploid complex *Antennaria neodioica* (Asteraceae: Inuleae). Pl. Syst. Evol. 150:143–163.

——. 1985b. Investigations into the evolutionary history of the polyploid complexes in *Antennaria* (Asteraceae: Inuleae). II. The *A. parlinii* complex. Rhodora 87:321–339.

——. 1993. A synopsis with keys for the genus *Antennaria* (Asteraceae: Inuleae: Gnaphaliinae). Canad. J. Bot. 71:1589–1604.

Bayer, R. J., and J. R. Starr. 1998. Tribal phylogeny of the Asteraceae based on two noncoding chloroplast sequences, the *trnL* intron and *trnL–trnF* intergenic spacer. Ann. Missouri Bot. Gard. 85:242–256.

Bayer, R. J., and G. L. Stebbins. 1982. A revised classification of *Antennaria* (Asteraceae: Inuleae) of the eastern United States. Syst. Bot. 7:300–313.

Beck, J. B., G. L. Nesom, P. J. Calie, G. I. Baird, R. L. Small, and E. E. Schilling. 2004. Is subtribe Solidagininae (Asteraceae) monophyletic? Taxon 53:691–698.

Becus, M. S. 2003. Observations on *Acalypha deamii* (Euphorbiaceae) in Ohio. Castanea 68:175–178.

Belcher, R. O. 1956. A revision of the genus *Erechtites* (Compositae) with inquiries into *Senecio* and *Arrhenechthites.* Ann. Missouri Bot. Gard. 43:1–85.

Bell, C. D., E. J. Edwards, S.-T. Kim, and M. J. Donoghue. 2001. Dipsacales phylogeny based on chloroplast DNA sequences. Harvard Pap. Bot. 6:481–499.

Belzer, N. F., and M. Ownbey. 1971. Chromatographic comparison of *Tragopogon* species and hybrids. Amer. J. Bot. 58:791–802.

Bemis, W. P., D. L. Curtis, C. W. Weber, and J. Berry. 1978. The feral buffalo gourd, *Cucurbita foetidissima.* Econ. Bot. 32:87–95.

Benson, L. 1982. The Cacti of the United States and Canada. Stanford University Press, Stanford, CA.

Bentham, G. 1871. Revision of the genus *Cassia.* Trans. Linn. Soc. London 27:503–591.

Berggren, G. 1981. Atlas of Seeds and Small Fruits of Northwest-European Plant Species with Morphological Descriptions, Part 3. Swedish Museum of Natural History, Stockholm.

Bernhardt, P. 1990. Anthecology of *Schrankia nuttallii* (Mimosaceae) on the tallgrass prairie. Pl. Syst. Evol. 170:247–255.

Berry, P. E., A. L. Hipp, K. J. Wurdack, B. Van Ee, and R. Riina. 2005. Molecular phylogenetics of the giant genus *Croton* and tribe Crotoneae (Euphorbiaceae sensu stricto) using ITS and *trnL–trnF* DNA sequence data. Amer. J. Bot. 92:1520–1534.

Bertin, R. I. 1982. Floral biology, hummingbird pollination, and fruit production of trumpet creeper (*Campsis radicans,* Bignoniaceae). Amer. J. Bot. 69:122–134.

Betancourt, J. L., A. Long, D. J. Donahue, A. J. T. Jull, and T. H. Zabel. 1984. Pre-Columbian age for North American *Corispermum* L. (Chenopodiaceae) confirmed by accelerator radiocarbon dating. Nature 311:653–655.

Bidartondo, M. I., and T. D. Bruns. 2001. Extreme specificity in epiparasitic Monotropoideae (Ericaceae): Widespread phylogenetic and geographical structure. Molec. Ecol. 10:2285–2295.

Biddulph, S. F. 1944. A revision of the genus *Gaillardia*. Res. Stud. Washington St. Coll. 12:195–256.

Bierner, M. W. 1989. Taxonomy of *Helenium* sect. *Amarum* (Asteraceae). Sida 13:453–459.

Binns, S. E., B. R. Baum, and J. T. Arnason. 2001a. Typification of *Echinacea purpurea* (L.) Moench (Heliantheae: Asteraceae) and its implication for the correct naming of two *Echinacea* taxa. Taxon 50:1169–1175.

———. 2001b. (1508) Proposal to conserve the name *Rudbeckia purpurea* (Asteraceae) with a conserved type. Taxon 50:1199–1200.

———. 2002. A taxonomic revision of *Echinacea* (Asteraceae: Heliantheae). Syst. Bot. 27:610–632.

Björkman, E. 1960. *Monotropa hypopitys* L.—An epiparasite on tree roots. Physiol. Pl. 13:308–327.

Blackwell, W. H., M. D. Baechle, and G. Williamson. 1978. Synopsis of *Kochia* (Chenopodiaceae) in North America. Sida 7:248–254.Blake, L. W. 1981. Early acceptance of watermelon by Indians of the United States. J. Ethnobiol. 1:193–199.

Blake, S. F. 1939. A new variety of *Iva ciliata* from Indian rock shelters in the south-central United States. Rhodora 41:81–86.

Blanca, G., and C. Díaz de la Guardia. 1997. Fruit morphology in *Tragopogon* L. (Compositae: Lactuceae) from the Iberian Peninsula. Bot. J. Linn. Soc. 125:319–329.

Bleeker, W., H. Huthmann, and H. Hurka. 1999. Evolution of hybrid taxa in *Nasturtium* R. Br. (Brassicaceae). Folia Geobot. Phytotax. 34:421–433.

Bleeker, W., C. Weber-Sparenberg, and H. Hurka. 2002. Chloroplast DNA variation and biogeography in the genus *Rorippa* Scop. (Brassicaceae). Pl. Biol. (Stuttgart) 4:104–111.

Bogle, A. L. 1970. The genera of Molluginaceae and Aizoaceae in the southeastern United States. J. Arnold Arbor. 51:431–462.

Boivin, B. 1971–1972. Flora of the Prairie Provinces, Part 3. Phytologia 22:315–398; 23:1–140.

Bolick, M. R. 1983. A cladistic analysis of the Ambrosiinae Less. and Engelmanniinae Stuessy. Pp. 125–141 *in* N. A. Platnick and V. A. Funk, eds., Advances in Cladistics. Vol. 2. Proceedings of the Second Meeting of the Willi Hennig Society. Columbia University Press, New York.

Bolli, R. 1994. Revision of the genus *Sambucus*. Diss. Bot. 223:1–227, pl. 1–29.

Boufford, D. E., and Q. Xiang. 1992. *Pachysandra* (Buxaceae) revisited. Bot. Bull. Acad. Sin. 33:201–207.Boulos, L. 1972. Révision systématique du genre *Sonchus* L. s.l., I: Introduction et classification. Bot. Notis. 125:287–305.

———. 1973. Révision systématique du genre *Sonchus* L. s.l., IV: Sous-genre 1, *Sonchus*. Bot. Notis. 126:155–196.

———. 1974a. Révision systématique du genre *Sonchus* L. s.l., V: Sous-genre 2, *Dendrosonchus*. Bot. Notis. 127:7–37.

———. 1974b. Révision systématique du genre *Sonchus* L. s.l., VI: Sous-genre 3, *Origosonchus:* Genres *Embergeria, Babcockia* et *Taeckholmia*. Species exclusae et dubiae. Index. Bot. Notis. 127:402–451.

Bowden, W. M. 1959. Phylogenetic relationships of twenty-one species of *Lobelia* L. section *Lobelia*. Bull. Torrey Bot. Club 86:94–108.

———. 1982. The taxonomy of *Lobelia* ×*speciosa* s.l. and its parental species *L. siphilitica* and *L. cardinalis* s.l. (Lobeliaceae). Canad. J. Bot. 60:2054–2070.

Bowen, W. R. 1962. Origin and development of winged cork in *Euonymus alatus*. Bot. Gaz. 124:256–261.

Bowles, M. L., J. L. McBride, and R. F. Betz. 1998. Management and restoration ecology of the federal threatened Mead's milkweed, *Asclepias meadii* (Asclepiadaceae). Ann. Missouri Bot. Gard. 85:110–125.

Bowles, M. [L.], G. Wilhelm, and S. Packard. 1988. The Illinois status of *Liatris scariosa* (L.) Willd. var. *nieuwlandii* Lunell, a new threatened species for Illinois. Erigenia 10:1–26.

Bradley, K. 2005. Glyphosate-resistant weeds. Integrated Pest Management Newsletter (University of Missouri–Columbia) 15 (1): article 1. http://www.missouri.edu/ipcm/archives/v15n1/ipmltr1.htm.

Bradley, T. R. 1968. Hybridization and variation in the species of *Triodanis* from North America. PhD diss., University of North Carolina, Chapel Hill.

———. 1975. Hybridization between *Triodanis perfoliata* and *Triodanis biflora* (Campanulaceae). Brittonia 27:110–114.

Brammal, R. A., and J. C. Semple. 1990. The cytotaxonomy of *Solidago nemoralis* (Compositae: Astereae). Canad. J. Bot. 68:2065–2069.

Bremer, K. 1987. Tribal interrelationships of the Asteraceae. Cladistics 3:210–253.

———. 1994. Asteraceae: Cladistics and Classification. Timber Press, Portland, OR.

Bremer, K., and C. J. Humphries. 1993. Generic monograph of the Asteraceae–Anthemideae. Bull. Nat. Hist. Mus. London, Bot. 23:71–177.

Britton, N. L. 1901. Manual of the Flora of the Northern States and Canada. Henry Holt, New York.

Brizicky, G. K. 1962. The genera of Anacardiaceae in the southeastern United States. J. Arnold Arbor. 43:359–375.

———. 1964a. The genera of Celastrales in the southeastern United States. J. Arnold Arbor. 45:206–234.

———. 1964b. Polyembryony in *Euonymus* (Celastraceae). J. Arnold Arbor. 45:251–259.

Brooks, R. E. 1986. *Corispermum* L. Pp. 173–174 *in* Great Plains Flora Association, eds., Flora of the Great Plains. University Press of Kansas, Lawrence.

Brouillet, L., and J. C. Semple. 1981. A propos du status taxonomique de *Solidago ptarmicoides*. Canad. J. Bot. 59:17-21.

Brown, D. K., and R. B. Kaul. 1981. Floral structure and mechanism in Loasaceae. Amer. J. Bot. 68:361–372.

Brown, D. T. 1998. Cannabis: The Genus *Cannabis*. Harwood Academic Publishers, Amsterdam.

Broyles, S. B., and R. Wyatt. 1993. Allozyme diversity and genetic structure in southern Appalachian populations of poke milkweed, *Asclepias exaltata*. Syst. Bot. 18:18–30.

Brummitt, R. K. 1965. New combinations in North American *Calystegia*. Ann. Missouri Bot. Bard. 52:214–216.

———. 1971. Relationship of *Heracleum lanatum* Michx. of North America to *H. sphondylium* of Europe. Rhodora 73:578–584.

———. 1980. Further new names in the genus *Calystegia* (Convolvulaceae). Kew Bull. 35:327–334.

———. 1988. Report of the Committee for Spermatophyta: 34. Synopsis of decisions Sydney 1981–Berlin 1987. Taxon 37:139–140.

———. 2000. Report of the Committee for Spermatophyta: 49. Taxon 49:261–278.

———. 2002. How to chop up a tree. Taxon 51:31–41.

Bruneau, A., F. Forest, P. S. Herendeen, B. B. Klitgaard, and G. P. Lewis. 2001. Phylogenetic relationships in the Caesalpinioideae (Leguminosae) as inferred from chloroplast *trnL* intron sequences. Syst. Bot. 26:487–514.

Burch, D. 1966. The application of the Linnaean names of some New World species of *Euphorbia* subgenus *Chamaesyce*. Rhodora 68:155–166.

Burckhalter, R. E. 1992. The genus *Nyssa* (Cornaceae) in North America: A revision. Sida 15:323–342.

Burk, C. J. 1961. Environmental variation in *Heterotheca subaxillaris*. Rhodora 63:243–246.

Burkart, A. 1976. A monograph of the genus *Prosopis* (Leguminosae subfam. Mimosoideae): Catalogue of the recognised species of *Prosopis*. J. Arnold Arbor. 57:219–249, 450–525.

Burrows, G. E., and R. J. Tyrl. 2001. Toxic Plants of North America. Iowa State University Press, Ames.

Cabrera, A. L. 1961. Observaciones sobre las Inuleae-Gnaphalinae (Compositae) de América del Sur. Revista Soc. Argent. Bot. 9:359–386.

Campbell, T. A. 1984. Agronomic and chemical evaluation of smooth sumac, *Rhus glabra*. Econ. Bot. 38:218–223.

Canne, J. M. 1977. A revision of the genus *Galinsoga* (Compositae: Heliantheae). Rhodora 79:319–389.
Carlson, A. W. 1986. Ginseng: America's botanical drug connection to the Orient. Econ. Bot. 40:233–249.
Castaner, D. 1984. An addition to the flora of Missouri found at the Missouri state fair grounds. Missouriensis 5:130–131.
Chambers, K. L. 1957. Taxonomic notes on some Compositae of the western United States. Contr. Dudley Herb. 5:57–68.
———. 1973. *Krigia cespitosa*. Pp. 52–53 *in* B. S. Vuilleumier, The genera of Lactuceae (Compositae) in the southeastern United States. J. Arnold Arbor. 54:42–93.
———. 2004. Taxonomic notes on *Krigia* (Asteraceae). Sida 21:225–236.
Channell, R. B. 1957. A revisional study of the genus *Marshallia* (Asteraceae). Contr. Gray Herb. 181:40–132.
Chase, M. W., S. Zmarzty, M. D. Lledó, K. J. Wurdack, S. M. Swensen, and M. F. Fay. 2002. When in doubt, put it in Flacourtiaceae: A molecular phylogenetic analysis based on plastid *rbcL* DNA sequences. Kew Bull. 57:141–181.
Chater, A. O. 1976. *Picris*. Pp. 316–317 *in* T. G. Tutin, V. H. Heywood, N. A. Burges, D. M. Moore, D. H. Valentine, S. M. Walters, and D. A. Webb, eds., Flora Europaea. Vol. 4. Plantaginaceae to Compositae (and Rubiaceae). Cambridge University Press, Cambridge.
Chater, A. O., and G. Halliday. 1993. *Arenaria*. Pp. 140–148 *in* T. G. Tutin, N. A. Burges, A. O. Chater, J. R. Edmondson, V. H. Heywood, D. M. Moore, D. H. Valentine, S. M. Walters, and D. A. Webb, eds., Flora Europaea. Vol. 1. Psilotaceae to Platanaceae. 2d ed. Cambridge University Press, Cambridge.
Christ, A. 1984. Missouri's interesting flora: *Myosotis stricta*. Missouriensis 5:96–97.
Chuang, T.-I., and L. Constance. 1969. A systematic study of *Perideridia* (Umbelliferae–Apioideae). Univ. Calif. Publ. Bot. 55:1–74.
Clausen, R. T. 1975. *Sedum* of North America North of the Mexican Plateau. Cornell University Press, Ithaca.
Clemants, S. E. 1992. Chenopodiaceae and Amaranthaceae of New York state. New York State Mus. Bull. 485:i–vi, 1–100.
Clevenger, S., and C. B. Heiser Jr. 1963. *Helianthus laetiflorus* and *Helianthus rigidus*: Hybrids or species? Rhodora 65:121–133.
Clevinger, J. A. 2004. New combinations in *Silphium* (Asteraceae: Heliantheae). Novon 14:275–277.
Clewell, A. F., and J. W. Wooten. 1971. A revision of *Ageratina* (Compositae: Eupatorieae) from eastern North America. Brittonia 23:123–143.
Clonts, J. A. 1972. A revision of the genus *Elephantopus*, including *Orthopappus* and *Pseudelephantopus* (Compositae). PhD diss., Mississippi State University, State College.
Clonts, J. A., and S. McDaniel. 1978. *Elephantopus*. N. Amer. Fl., ser. 2, 10:196–202.
Clute, O., and O. Palmer. 1893. Spurry *(Spergula arvensis)*. Bull. Michigan Agric. Exp. Sta. 91:3–8.
Cochrane, T. S. 1976. Taxonomic status of the *Onosmodium molle* complex (Boraginaceae) in Wisconsin. Michigan Bot. 15:103–110.
Coleman, J. R. 1965. Natural and artificial hybrids of *Cacalia atriplicifolia* and *C. muhlenbergii*. Rhodora 67:55–58.
———. 1966. A taxonomic revision of sect. *Ximenesia* of the genus *Verbesina* L. (Compositae). Amer. Midl. Naturalist 76:475–481.
Conrad, M. L. 1989. *Stylisma pickeringii* (Torr. ex M. A. Curtis) A. Gray var. *pattersonii* (Fern. & Schub.) Myint (Convolvulaceae): New to the flora of Missouri. Missouriensis 10:1–3.
Cook, L. M., P. S. Soltis, S. J. Brunsfeld, and D. E. Soltis. 1998. Multiple independent formations of *Tragopogon* tetraploids (Asteraceae): Evidence from RAPD markers. Molec. Ecol. 7:1293–1302.
Cook, R. E., and J. C. Semple. 2004. A new name and a new combination in *Solidago* subsect. *Glomeruliflorae* (Asteraceae: Astereae). Sida 21:221–224.
Cooperrider, T. R. 1984. Some species mergers and new combinations in the Ohio flora. Michigan Bot. 23:165–168.
———. 1989. The Clusiaceae (or Guttiferae) of Ohio. Castanea 54:1–11.
Core, E. L. 1941. The North American species of *Paronychia*. Amer. Midl. Naturalist 26:369–397.
Cornish, K., and D. J. Siler. 1996. Hypoallergenic guayule latex: Research to commercialization. Pp. 327-335 *in* J. Janick, ed., Progress in New Crops. American Society for Horticultural Science Press, Alexandria, VA.

Correll, D. B., and M. C. Johnston. 1970. Manual of the vascular plants of Texas. Contr. Texas Res. Found. 6:i–xii, 1–1881, 1 map.

Cosner, M. E., R. K. Jansen, and T. G. Lammers. 1994. Phylogenetic relationships in the Campanulales based on *rbcL* sequences. Pl. Syst. Evol. 190:79–95.Costea, M., and F. J. Tardif. 2003. Conspectus and notes on the genus *Amaranthus* in Canada. Rhodora 105:260–281.

Costea, M., A. Sanders, and G. Waines. 2001a. Preliminary results toward a revision of the *Amaranthus hybridus* species complex (Amaranthaceae). Sida 19:931–974.

———. 2001b. Notes on some little known *Amaranthus* taxa (Amaranthaceae) in the United States. Sida 19:975–992.

Cowan, C. W., and B. D. Smith. 1993. New perspectives on a wild gourd in eastern North America. J. Ethnobiol. 13:17–54.

Cowan, R. S. 1981. Caesalpinioideae. Pp. 57–64 *in* R. M. Polhill and P. H. Raven, eds., Advances in Legume Systematics, Part 1. Royal Botanic Gardens, Kew.

Cox, B. J. 1973–1974. A systematic comparison of *Catalpa* by polyacrylamide gel electrophoresis of seed proteins. Trans. Missouri Acad. Sci. 7–8:145–153.

Cox, B. J., and D. B. Dunn. 1973–1974. Catalpas occurring in the vicinity of Columbia, Missouri. Trans. Missouri Acad. Sci. 7–8:137–145.

Cox, P. B., and L. E. Urbatsch. 1990. A phylogenetic analysis of the coneflower genera (Asteraceae: Heliantheae). Syst. Bot. 15:394–402.

———. 1994. A taxonomic revision of *Rudbeckia* subg. *Macrocline* (Asteraceae: Heliantheae: Rudbeckiinae). Castanea 59:300–318.

Crane, J. H., G. C. Eickwort, F. R. Wesley, and J. Spielholz. 1985. Pollination ecology of *Vaccinium stamineum* (Ericaceae: Vaccinioideae). Amer. J. Bot. 72:135–142.

Craven, L. A. 2005. Malesian and Australian *Tournefortia* transferred to *Heliotropium* and notes on delimitation of Boraginaceae. Blumea 50:375–381.

Crawford, D. J. 1975. Systematic relationships in the narrow-leaved species of *Chenopodium* of the western United States. Brittonia 27:279–288.

———. 1977. A study of morphological variability in *Chenopodium incanum* (Chenopodiaceae) and the recognition of two new varieties. Brittonia 29:291–296.

Crawford, D. J., and H. D. Wilson. 1986. *Chenopodium* L. Pp. 186–173 *in* Great Plains Flora Association, eds., Flora of the Great Plains. University Press of Kansas, Lawrence.

Croat, T. B. 1970. Studies in *Solidago*. I. The *Solidago graminifolia–S. gymnospermoides* complex. Ann. Missouri Bot. Gard. 57:250–251.

———. 1972. *Solidago canadensis* complex in the Great Plains. Brittonia 24:317–326.

Crompton, C. W., A. E. Stahevitch, and W. A. Wojtas. 1990. Morphometric studies of the *Euphorbia esula* group (Euphorbiaceae) in North America. Canad. J. Bot. 68:1978–1988.

Cronin, J. T., and W. G. Abrahamson. 2001. Goldenrod stem galler preference and performance: Effects of multiple herbivores and plant genotypes. Oecologia 127:87–96.

Cronquist, A. 1943. The separation of *Erigeron* from *Conyza*. Bull. Torrey Bot. Club 70:629–632.

———. 1945. Notes on the Compositae of the northeastern United States. II. Heliantheae and Helenieae. Rhodora 47:396–403.

———. 1946. Notes on the Compositae of the northeastern United States. III. Inuleae and Senecioneae. Rhodora 48:116–125.

———. 1947a. Notes on the Compositae of the northeastern United States. IV. Solidago. Rhodora 49:69–79.

———. 1947b. Notes on the Compositae of the northeastern United States—V. Astereae. Bull. Torrey Bot. Club 74:142–150.

———. 1947c. Revision of the North American species of *Erigeron,* north of Mexico. Brittonia 6:121–300.

———. 1948. Notes on the Compositae of the northeastern United States. VI. Cichorieae, Eupatorieae, and Astereae. Rhodora 50:28–35.

———. 1952. Compositae. Pp. 323–545 *in* H. A. Gleason, The New Britton and Brown Illustrated Flora of the Northeastern United States and Adjacent Canada. Vol. 3. New York Botanical Garden, Bronx.

———. 1955. Compositae. Pp. 1–343 *in* C. L. Hitchcock, A. Cronquist, M. Ownbey, and J. W. Thompson, eds., Vascular plants of the Pacific Northwest. Univ. Washington Publ. Biol. 17 (5): 1–343.

———. 1959. *Boraginaceae.* Pp. 175–244 *in* C. L. Hitchcock, A. Cronquist, M. Ownbey, and J. W. Thompson, eds., Vascular plants of the Pacific Northwest, Part 4. University of Washington Press, Seattle.

———. 1980. Vascular Flora of the Southeastern United States. Vol. 1. Asteraceae. University of North Carolina Press, Chapel Hill.

———. 1981. An Integrated System of Classification of Flowering Plants. Columbia University Press, New York.

———. 1991. Synoptical arrangement of the subclasses, orders, and families of Liliopsida, as represented in our flora. Pp. lxxiii–lxxv *in* H. A. Gleason and A. Cronquist, Manual of Vascular Plants of Northeastern United States and Adjacent Canada. 2d ed. New York Botanical Garden, Bronx.

Crow, G. E. 1978. A taxonomic revision of *Sagina* (Caryophyllaceae) in North America. Rhodora 80:1–91.

Crow, G. E., and C. B. Hellquist. 2000. Aquatic and Wetland Plants of Northeastern United States. Vol. 1. Pteridophytes, Gymnosperms, and Angiosperms: Dicotyledons. University of Wisconsin Press, Madison.

Crowe, D. R., and W. H. Parker. 1981. Hybridization and agamospermy of *Bidens* in northwestern Ontario. Taxon 30:749–760.

Cruden, R. W. 1962. New combinations in *Silphium perfoliatum* L. Castanea 27:90–91.

Culwell, D. E. 1970. A taxonomic study of the section *Hypericum* in the eastern United States. PhD diss., University of North Carolina, Chapel Hill.

Dabydeen, S. 1997. Natural hybridization in the genus *Cirsium: C. altissimum* × *C. discolor*—Cytological and morphological evidence. Rhodora 99:152–160.

Daoud, H. S., and R. L. Wilbur. 1965. A revision of the North American species of *Helianthemum* (Cistaceae). Rhodora 67:63–82, 201–216, 255–312.

Dathan, A. S. R., and D. Singh. 1990. Embrology of the Cucurbitaceae. Pp. 185–199 *in* D. M. Bates, R. W. Robinson, and C. Jeffrey, eds., The Biology and Utilization of the Cucurbitaceae. Cornell University Press, Ithaca.

Daughtrey, M. 1993. Dogwood anthracnose disease: Native fungus or exotic invader? Pp. 23–33 *in* B. N. McKnight, ed., Biological Pollution: The Control and Impact of Invasive Exotic Species. Indiana Academy of Science, Indianapolis.

Davis, C. C., P. W. Fritsch, J. H. Li, and M. J. Donoghue. 2002. Phylogeny and biogeography of *Cercis* (Fabaceae): Evidence from nuclear ribosomal ITS and chloroplast *ndhF* sequence data. Syst. Bot. 27:289–302.

de Vries, I. M., and C. E. Jarvis. 1987. Typification of seven Linnaean names in the genus *Lactuca* L. (Compositae: Lactuceae). Taxon 36:142–154.

Dean, H. L. 1963. Further variation in style number and other gynoecial structures of *Lychnis alba* Mill. Phytomorphology 13:1–13.

Deardorff, D. C. 1977. Biosystematic studies in *Hieracium* (Asteraceae) section *Thyrsoidea*. PhD diss., University of Washington, Seattle.

Decker, D. 1988. Origin(s), evolution, and systematics of *Cucurbita pepo* (Cucurbitaceae). Econ. Bot. 42:4–15.

Decker, D. S., and L. A. Newsom. 1988. Numerical analysis of archaeological *Cucurbita* seeds from Hontoon Island, Florida. J. Ethnobiol. 8:35–44.

Decker, D. S., and H. D. Wilson. 1986. Numerical analysis of seed morphology in *Cucurbita pepo*. Syst. Bot. 11:595–607.

———. 1987. Allozyme variation in the *Cucurbita pepo* complex: *C. pepo* var. *ovifera* vs. *C. texana*. Syst. Bot. 12:263–273.

Decker-Walters, D. S., T. W. Walters, C. W. Cowan, and B. D. Smith. 1993. Isozymic characterization of wild populations of *Cucurbita pepo*. J. Ethnobiol. 13:55–72.

Decker-Walters, D. S., J. E. Staub, S.-M. Chung, E. Nakata, and H. D. Quemada. 2002. Diversity in free-living populations of *Cucurbita pepo* (Cucurbitaceae) as assessed by random amplified polymorphic DNA. Syst. Bot. 27:19–28.

DeJong, D. C. D. 1965. A systematic study of the genus *Astranthium*. Publ. Mus. Michigan State Univ., Biol. Ser. 2:429–528.

Delahaut, K. A. 1997. Aster Leafhopper. University of Wisconsin Extension Circular A3679. University of Wisconsin Division of Cooperative Extension, Madison.

Denison, E. 1977. The thistle that isn't. Nature Notes (Webster Groves) 49:7–9.

Desmarais, Y. 1952. Dynamics of leaf variation in the sugar maples. Brittonia 7:347–387.

Dewey, L. H. 1914. Hemp. Yearb. U.S. Dept. Agric. 1913: 283–346.

Diane, N., H. Förther, and H. H. Hilger. 2002. A systematic analysis of *Heliotropium, Tournefortia,* and allied taxa of the Heliotropiaceae (Boraginales) based on ITS1 sequences and morphological data. Amer. J. Bot. 89:287–295.

Díaz de la Guardia, C., and G. Blanca. 1988. Una especie poco conocida de *Tragopogon* L. (Compositae): *T. lamottei* Rouy. Lagascalia 15 (suppl.): 355–359.

———. 1992. Lectotypification of five Linnaean species of *Tragopogon* L. (Compositae). Taxon 41:548–551.

Dierker, B. 1992. A new *Eryngium* for Missouri. Missouriensis 13:25–26.

Diggs, G. M., Jr., B. L. Lipscomb, and R. O'Kennon. 1999. Shinners and Mahler's illustrated flora of north central Texas. Sida Bot. Misc. 16:i–xii, 1–1626.

Dilcher, D. L. 1989. The occurrence of fruits with affinities to Ceratophyllaceae in lower and mid-Cretaceous sediments. Amer. J. Bot. 76 (6, suppl.): 162.

Dille, D. P. 1976. A revision of the genus *Lactuca* (Compositae: Lactuceae) in the Great Plains. Master's thesis, Kansas State University, Manhattan.

Dillingham, F. T. 1907. The staff tree, *Celastrus scandens,* as a former food supply of starving Indians. Amer. Naturalist 41:391–393.

Donoghue, M. J., R. G. Olmstead, J. F. Smith, and J. D. Palmer. 1992. Phylogenetic relationships of Dipsacales based on *rbcL* sequences. Ann. Missouri Bot. Gard. 79:333–345.

Donoghue, M. J., T. Eriksson, P. A. Reeves, and R. G. Olmstead. 2001. Phylogeny and phylogenetic taxonomy of Dipsacales, with special reference to *Sinadoxa* and *Tatradoxa* (Adoxaceae). Harvard. Pap. Bot. 6:459–479.

Doolen, W. S. O. 1984. The vascular flora of Big Oak Tree State Park, Mississippi County, Missouri. Master's thesis, Southern Illinois University, Carbondale.

Downie, S. R., D. S. Katz-Downie, and K.-J. Cho. 1997. Relationships in the Caryophyllales as suggested by phylogenetic analysis of partial chloroplast DNA *ORF2280* homolog sequences. Amer. J. Bot. 84:253–273.

Downie, S. R., D. S. Katz-Downie, and M. F. Watson. 2000. A phylogeny of the flowering plant family Apiaceae based on chloroplast DNA *rpl16* and *rpoC1* intron sequences: Towards a suprageneric classification of subfamily Apioideae. Amer. J. Bot. 87:273–292.

Downie, S. R., F.-J. Sun, D. S. Katz-Downie, and G. J. Colletti. 2004. A phylogenetic study of *Perideridia* (Apiaceae) based on nuclear ribosomal DNA ITS sequences. Syst. Bot. 29:737–751.

Doyle, J. J., J. L. Doyle, J. A. Ballenger, E. A. Dickson, T. Kajita, and H. Ohashi. 1997. A phylogeny of the chloroplast gene *rbcL* in the Leguminosae: Taxonomic correlations and insights into the evolution of nodulation. Amer. J. Bot. 84:541–554.

Drapalik, D. J. 1970. A biosystematic study of the genus *Matelea* in the southeastern United States. PhD diss., University of North Carolina, Chapel Hill.

Dress, W. J. 1954. A revision of the genus *Chrysopsis* in eastern North America. PhD diss., Cornell University, Ithaca.

———. 1959. Notes on the cultivated Compositae 3. *Liatris.* Baileya 7:23–32.

———. 1961. Notes on the cultivated Compositae 6. The coneflowers: *Dracopis, Echinacea, Ratibida,* and *Rudbeckia.* Baileya 9:67–83.

Drew, W. B. 1944. Dew drop "grass" as a lawn plant in central Missouri. Ecology 25:246–247.

Dreyer, G. D., L. M. Baird, and C. Fickler. 1987. *Celastrus scandens* and *Celastrus orbiculatus:* Comparisons of reproductive potential between a native and an introduced woody vine. Bull. Torrey Bot. Club 114:260–264.

Drumke, J. S. 1964. A systematic survey of *Corylus* in North America. PhD diss., University of Tennessee, Knoxville.

Duke, S. O., K. C. Vaughn, E. M. Croom Jr., and H. N. Elsohly. 1987. Artemisinin, a constituent of annual wormwood *(Artemisia annua),* is a selective phytotoxin. Weed Sci. 35:499–505.

Duncan, W. H. 1964. New *Elatine* (Elatinaceae) populations in the southeastern United States. Rhodora 66:47-53.

Dunn, D. B. 1982. Problems in "keeping-up" with the flora of Missouri. Trans. Missouri Acad. Sci. 16:95–98.

Dunn, P. H. 1979. The distribution of leafy spurge *(Euphorbia esula)* and other weedy *Euphorbia* spp. in the United States. Weed Sci. 27:509–516.

———. 1985. Origins of leafy spurge in North America. Pp. 7–13 *in* A. K. Watson, ed., Leafy spurge. Monogr. Weed Sci. Soc. America 3:i–iv, 1–104.

Durkee, L. T., M. H. Harber, L. Dorn, and A. Remington. 1999. Morphology, ultrastructure, and function of extrafloral nectaries in three species of Caesalpiniaceae. J. Iowa Acad. Sci. 106:82–88.

Dvořák, F. 1991. Study on *Chenopodium borbasii* J. Murr. Feddes Repert. 102:351–373.

Easterly, N. W. 1957. A morphological study of *Ptilimnium*. Brittonia 9:136–145.

Ebinger, J. E., and W. E. McClain. 1991. Naturalized Amur maple *(Acer ginnala* Maxim.) in Illinois. Nat. Areas J. 11:170–171.

Edgington, Mrs. O. H. 1960. Save the holly. Garden Forum 21 (7): 14-16.

Edwards, A. L., R. Wyatt, and R. B. Sharitz. 1994. Seed buoyancy and viability of the wetland milkweed *Asclepias perennis* and an upland milkweed, *Asclepias exaltata*. Bull. Torrey Bot. Club 121:160–169.

Egolf, D. R. 1962. A cytological study of the genus *Viburnum*. J. Arnold Arbor. 43:132–172.

Elias, T. S. 1974. The genera of Mimosoideae (Leguminosae) in the southeastern United States. J. Arnold Arbor. 55:67–118.

———. 1981. Mimosoideae. Pp. 143–152 *in* R. M. Polhill and P. H. Raven, eds., Advances in Legume Systematics, Part 1. Royal Botanic Gardens, Kew.

Elmore, C. D. 1986. Mode of reproduction and inheritance of leaf shape in *Ipomoea hederacea*. Weed Sci. 34:391–395.

Emboden, W. A. 1974. *Cannabis:* A polytypic genus. Econ. Bot. 28:304–310.

Endress, M. E., and P. V. Bruyns. 2000. A revised classification of the Apocynaceae s.l. Bot. Rev. (Lancaster) 66:1–56.

Estilai, A., and J. G. Waines. 1990. Improved guayule germplasm for domestic production of natural rubber. Pp. 242–244 *in* J. Janick and J. E. Simon, eds., Advances in New Crops. Timber Press, Portland, OR.

Eyde, R. H. 1963. Morphological and paleobotanical studies of the *Nyssaceae* 1. Survey of the modern species and their fruits. J. Arnold Arbor. 44:1–59.

———. 1966. The Nyssaceae in the southeastern United States. J. Arnold Arbor. 47:117–125.

———. 1987. The case for keeping *Cornus* in the broad Linnaean sense. Syst. Bot. 12:505–518.

———. 1988. Comprehending *Cornus:* Puzzles and progress in the systematics of the dogwoods. Bot. Rev. (Lancaster) 54:233–351.

Fan, C. Z., and Q. Y. Xiang. 2001. Phylogenetic relationships within *Cornus* (Cornaceae) based on 26S rDNA sequences. Amer. J. Bot. 88:1131–1138.

Fang, R.-C., and R. K. Brummitt. 1995. *Calystegia*. Pp. 286–289 *in* Flora of China Editorial Committee, eds., Flora of China. Vol. 16. Gentianaceae through Boraginaceae. Science Press, Beijing.

Fassett, N. C. 1939. Notes from the herbarium of the University of Wisconsin—XVII. *Elatine* and other aquatics. Rhodora 41:367–377.

Faust, W. Z. 1972. A biosystematic study of the *Interiores* species group of the genus *Vernonia* (Compositae). Brittonia 24:363–378.

———. 1977. *Vernonia illinoensis* (Compositae): Species or hybrid? Castanea 42:204–212.

Felger, R. S. 1977. Mesquite in Indian cultures of southwestern North America. Pp. 150–176 *in* B. B. Simpson, ed., Mesquite: Its Biology in Two Desert Scrub Ecosystems. US-IBP Synthesis Series 4. Dowden, Hutchinson, and Ross, Stroudsburg, PA.

Feráková, V. 1977. The Genus *Lactuca* in Europe. Univerzita Komenshéko, Bratislava, Czechoslovakia.

Ferguson, D. M. 1998. Phylogenetic analysis and relationships in Hydrophyllaceae based on *ndhF* sequence data. Syst. Bot. 23:253–268.

Ferguson, I. K. 1966a. The genera of Caprifoliaceae in the southeastern United States. J. Arnold Arbor. 47:33–59.

———. 1966b. The Cornaceae in the southeastern United States. J. Arnold Arb. 47:106–116.

Ferguson, I. K., and G. K. Brizicky. 1965. Nomenclatural notes on *Dipsacus fullonum* and *Dipsacus sativus*. J. Arnold Arbor. 46:362–365.

Fernald, M. L. 1935. Midsummer vascular plants of southeastern Virginia. Rhodora 37:423–454.
———. 1941a. *Elatine americana* and *E. triandra*. Rhodora 43:208–211.
———. 1941b. Another century of additions to the flora of Virginia. Rhodora 43:485–657.
———. 1945. *Ruellia* in the eastern United States. Rhodora 47:1–38, 47–63, 69–90.
———. 1950. Gray's Manual of Botany. 8th ed. American Book, New York.
Fisher, C. E. 1977. Mesquite and modern man in southwestern North America. Pp. 177–188 *in* B. B. Simpson, ed., Mesquite: Its Biology in Two Desert Scrub Ecosystems. US-IBP Synthesis Series 4. Dowden, Hutchinson, and Ross, Stroudsburg, PA.
Fisher, T. R. 1957. Taxonomy of the genus *Heliopsis* (Compositae). Ohio J. Sci. 57:171–191.
———. 1958. Variation in *Heliopsis helianthoides* (L.) Sweet. Ohio J. Sci. 58:97–107.
———. 1959. Natural hybridization between *S. laciniatum* and *Silphium terebinthinaceum*. Brittonia 11:250–254.
———. 1966. The genus *Silphium* in Ohio. Ohio J. Sci. 66:259–263.
Fisher, T. R., and J. M. Speer. 1978. Systematic studies in the genus *Silphium:* Possible origin of *S. pinnatifidum* Ell. (Compositae). Pp. 451–463 *in* D. N. Sen, ed., Environmental Physiology and Ecology of Plants. Bishen Singh Mahendra Pal Singh, Dehra Dun, India.
Flora of North America Editorial Committee. 2006. Flora of North America, North of Mexico. Vols. 19–21. Magnoliophyta: Asteridae (in Part): Asteraceae, Parts 1–3. Oxford University Press, New York.
Fontes, E. M. G., D. H. Habeck, and F. Slansky Jr. 1994. Phytophagous insects associated with goldenrods (*Solidago* spp.) in Gainesville, Florida. Florida Entomol. 77:209–221.
Ford, B. A. 1995. Status of deerberry, *Vaccinium stamineum* L. (Ericaceae), in Canada. Rhodora 97:255–263.
Foster, S. 1991. *Echinacea,* the Purple Coneflowers. Botanical Series—301. American Botanical Council, Austin, TX.
Frankton, C., and I. J. Bassett. 1970. The genus *Atriplex* (Chenopodiaceae) in Canada. II. Four native western annuals: *A. argentea, A. truncata, A. powellii,* and *A. dioica*. Canad. J. Bot. 48:981–989.
Franzke, A., K. Pollmann, W. Bleeker, R. Kohrt, and H. Hurka. 1998. Molecular systematics of *Cardamine* and allied genera (Brassicaceae): ITS and non-coding chloroplast DNA. Folia Geobot. 33:225–240.
Frey, D., M. Baltisberger, and P. J. Edwards. 2003. Cytology of *Erigeron annuus* s.l. and its consequences in Europe. Bot. Helv. 113:1–14.
Fulling, E. H. 1943. Plant life and the law of man. IV. Barberry, currant and gooseberry, and cedar control. Bot. Rev. (Lancaster) 9:483–592.
Furlow, J. J. 1979. The systematics of the American species of *Alnus* (Betulaceae). Rhodora 81:1–121, 151–248.
———. 1987. The *Carpinus caroliniana* complex in North America. II. Systematics. Syst. Bot. 12:416–434.
———. 1990. The genera of Betulaceae in the southeastern United States. J. Arnold Arbor. 71:1–67.
Gadella, T. W. J. 1984. Notes on *Symphytum* (Boraginaceae) in North America. Ann. Missouri Bot. Gard. 71:1061–1067.
Gaiser, L. O. 1946. The genus *Liatris*. Rhodora 48:165–183, 216–263, 273–326, 331–382, 393–412.
Gandhi, K. N., and R. D. Thomas. 1991. Additional notes on the Asteraceae of Louisiana. Sida 14:514–517.
Garcia-Jacas, N., A. Susanna, V. Mozaffarian, and R. Ilarslan. 2000. The natural delimitation of *Centaurea* (Asteraceae: Cardueae): ITS sequences analysis of the *Jacea* group. Pl. Syst. Evol. 223:185–199.
Garcia-Jacas, N., A. Susanna, T. Garnatge, and R. Vilatersana. 2001. Generic delimitation and phylogeny of the subtribe Centaureinae (Asteraceae): A combined nuclear and chloroplast DNA analysis. Ann Bot. (London) 87:503–515.
Gardner, D. C., and K. R. Robertson. 2000. Effects of annual burning on populations of *Cassia fasciculata* (Fabaceae: Caesalpinioideae) with a review of its systematics and biology. Erigenia 18:22–29.
Gates, F. C., and S. F. Prince. 1938. *Taraxacum laevigatum* f. *scapifolium,* a new form of dandelion. Trans. Kansas Acad. Sci. 41:119–120.

Gathman, A. C., and W. P. Bemis. 1990. Domestication of buffalo gourd, *Cucurbita foetidissima.* Pp. 335–348 *in* D. M. Bates, R. W. Robinson, and C. Jeffrey, eds., The Biology and Utilization of the Cucurbitaceae. Cornell University Press, Ithaca.

Gentry, J. L., and R. L. Carr. 1976. A revision of the genus *Hackelia* (Boraginaceae) in North America, north of Mexico. Mem. New York Bot. Gard. 26:121–225.

Gervais, C. 1977. Cytological investigations of the *Achillea millefolium* (Compositae) complex in Quebec. Canad. J. Bot. 55:796–808.

Gillett, J. M., and N. K. B. Robson. 1981. The St. John's-worts of Canada (Guttiferae). Natl. Mus. Canada Publ. Bot. 11:i–iv, 1–40.

Gillis, W. T. 1971. The systematics and ecology of poison-ivy and the poison-oaks (*Toxicodendron,* Anacardiaceae). Rhodora 73:72–159, 161–237, 370–443, 465–540.

Gleason, H. A. 1947. The preservation of well known binomials. Phytologia 2:201–212.

Gleason, H. A., and A. Cronquist. 1963. Manual of Vascular Plants of Northeastern United States and Adjacent Canada. D. Van Nostrand, Princeton, NJ.

———. 1991. Manual of Vascular Plants of Northeastern United States and Adjacent Canada. 2d ed. New York Botanical Garden, Bronx.

Godfrey, R. K. 1952. *Pluchea,* section *Stylimnus,* in North America. J. Elisha Mitchell Sci. Soc. 68:238–271, pl. 20–23.

Gold, M. A., and J. Hanover. 1993. Honeylocust *(Gleditsia triacanthos),* a multipurpose tree for the temperate zone. Int. Tree Crops J. 7:189–207.

Gordon, A. 1999. The Salt Cave slippers: Fiber identification and content. Master's thesis, Washington University, St. Louis.

Gordon, D. 1966. A revision of the genus *Gleditsia* (Leguminosae). PhD diss., Indiana University, Bloomington.

Gottschling, M., H. H. Hilger, M. Wolf, and N. Diane. 2001. Secondary structure of the ITS1 transcript and its application in a reconstruction of the phylogeny of Boraginales. Pl. Biol. (Stuttgart) 3:629–636.

Gould, K. R., and M. J. Donoghue. 2000. Phylogeny and biogeography of *Triosteum* (Caprifoliaceae). Harvard Pap. Bot. 5:157–166.

Govaerts, R., D. Frodin, and A. Radcliffe-Smith. 2000. World Checklist and Bibliography of Euphorbiaceae (World Checklists and Bibliographies 4). 4 vols. Royal Botanic Gardens, Kew.

Graham, S. A. 1964. The Elaeagnaceae in the southeastern United States. J. Arnold Arbor. 45:274–278.

———. 1966. The genera of Araliaceae in the southeastern United States. J. Arnold Arbor. 47:126–136.

Green, P. S. 1966. Identification of species and hybrids in the *Lonicera tatarica* complex. J. Arnold Arbor. 47:75–88.

Greenway, J. 1793. An account of a poisonous plant growing spontaneously in the southern part of Virginia. Trans. Amer. Philos. Soc. 3:234–239.

Greer, L. F. 1997. Origins of polyploidy in *Thelesperma* (Asteraceae). Master's thesis, Sul Ross State University, Alpine, TX.

Gremaud, G. 1988. Three species new to Missouri. Missouriensis 9:15–17.

Greuter, W. 1995. *Silene* (Caryophyllaceae) in Greece: A subgeneric and sectional classification. Taxon 44:543–581.

———. 2003. The Euro+Med treatment of Cichorieae (Compositae): Generic concepts and required new names. Willdenowia 33:229–238.

Greuter, W., and H. M. Burdet. 1982. Caryophyllaceae. Pp. 185–191 *in* W. Greuter and T. Raus, eds., Med-checklist notulae, 6. Willdenowia 12:183–199.

Greuter, W., J. McNeill, F. R. Barrie, H. M. Burdet, V. Demoulin, T. S. Filgueras, D. H. Nicolson, P. C. Silva, J. E. Skog, P. Trehane, and D. L. Hawksworth, eds., 2000. International code of botanical nomenclature (St. Louis code). Regnum Veg. 138:i–xviii, 1–474.

Greuter, W., G. Wagenitz, M. Agababian, and F. H. Hellwig. 2001. Proposal to conserve the name *Centaurea* (Compositae) with a conserved type. Taxon 50:1201–1205.

Grierson, A. J. C. 1974. *Matricaria.* Pp. 252–264 *in* P. H. Davis, ed., Notes for a flora of Turkey XXX. Notes Roy. Bot. Gard. Edinb. 33:207–264.

Grotenhermen, F., and E. Russo, eds. 2002. Cannabis and Cannainoids: Pharmacology, Toxicology, and Therapeutic Potential. Haworth Integrative Healing Press, New York.

Gunn, C. R. 1991. Fruits and seeds of genera in the subfamily Caesalpinioideae (Fabaceae). Techn. Bull. U.S.D.A. 1755:i–v, 1–408.

Gustaffson, M. H. G., and K. Bremer. 1995. Morphology and phylogenetic interrelationships of the Asteraceae, Calyceraceae, Campanulaceae, Goodeniaceae, and related families (Asterales). Amer. J. Bot. 82:250–265.

Gustaffson, M. H. G., A. Backlund, and B. Bremer. 1996. Phylogeny of the Asterales sensu lato based on *rbcL* sequences with particular reference to the Goodeniaceae. Pl. Syst. Evol. 199:217–242.

Guthrie, R. L. 1969. A biosystematic study of *Taenidia* and *Pseudotaenidia* (Umbelliferae). PhD diss., West Virginia University, Morgantown.

Hall, G. W. 1967. A biosystematic study of the North American complex of the genus *Bidens* (Compositae). PhD diss., Indiana University, Bloomington.

Hall, H. M., and F. E. Clements. 1923. The North American species of *Artemisia*. Pp. 31–156 *in* H. M. Hall and F. E. Clements, The phylogenetic method in taxonomy. Publ. Carnegie Inst. Wash. 326:1–355, pl. 1–58.

Hall, J. C., K. J. Sytsma, and H. H. Iltis. 2002. Phylogeny of Capparaceae and Brassicaceae based on chloroplast sequence data. Amer. J. Bot. 89:1826–1842.

Hammer, K., P. Hanelt, and H. Knupffer. 1982. Vorarbeiten zur monographischen Darstellung von Wildpflanzensortimenten: *Agrostemma* L. Kulturpflanze 30:45–96.

Hampton, R., E. Small, and A. Haunold. 2001. Habitat and variability of *Humulus lupulus* var. *lupuloides* in upper midwestern North America: A critical source of American hop germplasm. J. Torrey Bot. Soc. 128:35–46.

Hancock, J. F., and R. E. Wilson. 1976. Biotype selection in *Erigeron annuus* during old field selection. Bull. Torrey Bot. Club 103:122–125.

Hansen, A. A. 1918. Canada thistle and methods of eradication. U.S.D.A. Farmer's Bull. 1002:1–15.

Hansen, E. J. P. 1966. The natural occurrence of a hybrid honeysuckle *(Lonicera ×bella)* in Ohio and Michigan. Michigan Bot. 5:211–217.

Harms, V. L. 1963. Variation in the *Heterotheca (Chrysopsis) villosa* complex east of the Rocky Mountains. PhD diss., University of Kansas, Lawrence.

———. 1965a. Cytogenetic evidence supporting the merger of *Heterotheca* and *Chrysopsis* (Compositae). Brittonia 17:11–16.

———. 1965b. Biosystematic studies in the *Heterotheca subaxillaris* complex (Compositae: Astereae). Trans. Kansas Acad. Sci. 68:244–257.

———. 1968. Nomenclatural changes and taxonomic notes on *Heterotheca,* including *Chrysopsis,* in Texas and adjacent states. Wrightia 4:8–20.

———. 1969. A tri-species hybrid population of *Vernonia* (Compositae). Amer. Midl. Naturalist 82:258–271.

———. 1974. A preliminary conspectus of *Heterotheca* section *Chrysopsis* (Compositae). Castanea 39:155–165.

Harms, V. L., J. H. Hudson, and G. F. Ledingham. 1986. *Rorippa truncata,* the blunt-fruited yellow cress, new for Canada, and *R. tenerrima,* the slender yellow cress, in southern Saskatchewan and Alberta. Canad. Field-Naturalist 100:45–51.

Harriman, N. A. 1965. The genus *Dentaria* (Cruciferae) in eastern North America. PhD diss., Vanderbilt University, Nashville.

———. 1980. *Leontodon* and *Hypochaeris* (Compositae) in Wisconsin. Michigan Bot. 19:93–95.

———. 1998. (1376) Proposal to conserve the name *Gymnocladus* (Leguminosae: Caesalpinioideae) with a conserved gender and type. Taxon 47:875–876.

Hart, T. W., and W. H. Eschbaugh. 1976. The biosystematics of *Cardamine bulbosa* (Muhl.) B.S.P. and *C. douglassii* Britt. Rhodora 78:329–419.

Hartman, R. L. 1986a. Apocynaceae Juss. Pp. 610–613 *in* Great Plains Flora Association, eds., Flora of the Great Plains. University Press of Kansas, Lawrence.

———. 1986b. Asclepiadaceae R. Br. Pp. 614–637 *in* Great Plains Flora Association, eds., Flora of the Great Plains. University Press of Kansas, Lawrence.

Hartman, R. L., R. K. Rabeler, and F. H. Utech. 2005. *Arenaria*. Pp. 51–56 *in* Flora of North America Editorial Committee, eds., Flora of North America, North of Mexico. Vol. 5. Magnoliophyta: Caryophyllidae, Part 2. Oxford University Press, New York.

Heard, S. B., and J. C. Semple. 1988. The *Solidago rigida* complex (Compositae: Astereae): A multivariate morphometric analysis and chromosome numbers. Canad. J. Bot. 66:1800–1807.

Heiser, C. B., Jr. 1947. Hybridization between the sunflower species *Helianthus annuus* and *H. petiolaris*. Evolution 1:249–262.

———. 1951. The sunflower among the American Indians. Proc. Amer. Phil. Soc. 95:432–448.
———. 1954. Variation and subspeciation in the common sunflower, *Helianthus annuus*. Amer. Midl. Naturalist 51:287–305.
———. 1973. Variation in the bottle gourd. Pp. 121–128 *in* B. J. Meggars, E. S. Ayensu, and W. D. Duckworth, eds., Tropical Forest Ecosystems in Africa and South America: A Comparative Review. Smithsonian Institution Press, Washington, DC.
———. 1976. The Sunflower. University of Oklahoma Press, Norman.
———. 1978. Taxonomy of *Helianthus* and origin of domesticated sunflower. Chap. 2, pp. 31–53 *in* J. F. Carter, ed., Sunflower Science and Technology. Agronomy 19:1–505.
———. 1979. The Gourd Book. University of Oklahoma Press, Norman.
———. 1985. Of Plants and People. University of Oklahoma Press, Norman.
Heiser, C. B., Jr., D. M. Smith, S. B. Clevenger, and W. B. Martin. 1969. The North American sunflowers *(Helianthus)*. Mem. Torrey Bot. Club. 22 (3): 1–218.
Henderson, N. C. 1980. Additions to the Flora of Missouri. Natural History Note. Missouri Department of Conservation, Jefferson City.
Henry, R. D., and A. R. Scott. 1983. New state records and other noteworthy collections for the Illinois vascular flora. Phytologia 52:331–335.
Herendeen, P. S., D. H. Les, and D. L. Dilcher. 1990. Fossil *Ceratophyllum* (Ceratophyllaceae) from the Tertiary of North America. Amer. J. Bot. 77:7–16.
Herendeen, P. S., W. L. Crepet, and D. L. Dilcher. 1992. The fossil history of the Leguminosae: Phylogenetic and biogeographic implications. Pp. 303–326 *in* P. S. Herendeen and D. L. Dilcher, eds., Advances in Legume Systematics, Part 4. The Fossil Record. Royal Botanic Gardens, Kew.
Hershkovitz, M. A., and E. A. Zimmer. 1997. On the evolutionary origins of the cacti. Taxon 46:217–232.
Higgins, L. C. 1979. Boraginaceae of the southwestern United States. Great Basin Naturalist 39:293–350.
Hilliard, O. M., and B. L. Burtt. 1981. Some generic concepts in Compositae—Gnaphaliinae. Bot. J. Linn. Soc. 82:181–232.
Hobbs, C. 1989. The *Echinacea* Handbook. Eclectic Medical Publications, Portland, OR.
Hodgdon, A. H. 1938. A taxonomic study of *Lechea*. Rhodora 40:29–69, 87–134, 3 pl.
Hoffmann, J. J., and S. P. McLaughlin. 1986. *Grindelia camporum:* Potential cash crop for the arid Southwest. Econ. Bot. 40:162–169.
Hoffmann, J. J., B. E. Kingsolver, S. P. McLaughlin, and B. N. Timmermann. 1984. Production of resins by arid-adapted Astereae. Pp. 251–271 *in* B. N. Timmermann, C. Steelink, and F. A. Loewus, eds., Phytochemical Adaptations to Stress. Plenum Press, New York.
Holmes, S. L. 1995. Final Report on the Missouri Natural Features Inventory of Cape Girardeau, Dunklin, Mississippi, New Madrid, Pemiscot, Perry, Scott, and Stoddard Counties. Missouri Department of Conservation, Jefferson City.
Holmes, W. C. 1981. *Mikania* (Compositae) of the United States. Sida 9:147–158.
Holzapfel, S. 1994. A revision of the genus *Picris* (Asteraceae, Lactuceae) s.l. in Australia. Willdenowia 24:97–218.
Hopkins, C. O., and W. H. Blackwell Jr. 1977. Synopsis of *Suaeda* (Chenopodiaceae) in North America. Sida 7:147–173.
Hou, D. 1955. A revision of the genus *Celastrus*. Ann. Missouri Bot. Gard. 42:215–302.
Hudson, S. 1997. Another new introduction for Missouri. Missouriensis 18:19–20.
Hufford, L. 1992. Rosidae and their relationships to other nonmagnoliid dicotyledons: A phylogenetic analysis using morphological and chemical data. Ann. Missouri Bot. Gard. 79:218–248.
Huft, M. J. 1979. A monograph of *Euphorbia* section *Tithymalopsis*. PhD diss., University of Michigan, Ann Arbor.
Hunt, D., ed. 1999. CITES Cactaceae Checklist. 2d ed. Royal Botanic Gardens, Kew.
Hunt, D., and N. Taylor, eds. 2002. Studies in the Opuntioideae (Cactaceae). Succulent Plant Research. Vol. 6. David Hunt, The Manse, Sherborne, Great Britain.
Hyatt, P. 2001. European chickweed invades Arkansas. Claytonia (Conway) 22(4): 6–7.
Iltis, H. H. 1952. A revision of the New World species of *Cleome*. PhD diss., Washington University, St. Louis.
———. 1958. Studies in the Capparidaceae—IV. *Polanisia* Raf. Brittonia 10:33–58.
———. 1966. Studies in the Capparidaceae—VIII. *Polanisia dodecandra* (L.) DC. Rhodora 68:41–47.

IRPCM Phytoplasma/Spiroplasma Working Team–Phytoplasma Taxonomy Group. 2004. '*Candidatus* Phytoplasma', a taxon for the wall-less non-helical prokaryotes that colonize plant phloem and insects. Int. J. Syst. Evol. Microbiol. 54:1243–1255.

Irwin, H. S., and R. C. Barneby. 1976. Notes on the generic status of *Chamaecrista* Moench (Leguminosae: Caesalpinioideae). Brittonia 28:28–36.

———. 1981. Cassieae. Pp. 97–106 *in* R. M. Polhill and P. H. Raven, eds., Advances in Legume Systematics, Part 1. Royal Botanic Gardens, Kew.

———. 1982. The American Cassiinae. Mem. New York Bot. Gard. 35:1–918 (2 parts).

Isely, D. 1971. Legumes of the United States: 3. *Schrankia*. Sida 4:232–245.

———. 1973. Leguminosae of the United States: I. Subfamily Mimosoideae. Mem. New York Bot. Gard. 25(1): 1–152.

———. 1975. Leguminosae of the United States: II. Subfamily Caesalpinioideae. Mem. New York Bot. Gard. 25(2): 1–228.

———. 1998. Native and Naturalized Leguminosae (Fabaceae) of the United States (Exclusive of Alaska and Hawaii). Monte L. Bean Life Science Museum, Brigham Young University, Provo, UT.

Ito, M., K. Watanabe, Y. Kita, T. Kawahara, D. J. Crawford, and T. Yahara. 2000a. Phylogeny and phytogeography of *Eupatorium* (Eupatorieae, Asteraceae): Insights from sequence data of the nrDNA ITS regions and cpDNA RFLP. J. Pl. Res. 113:79–89.

Ito, M., T. Yahara, R. M. King, K. Watanabe, S. Oshita, J. Yokoyama, and D. J. Crawford. 2000b. Molecular phylogeny of Eupatorieae (Asteraceae) estimated from cpDNA RFLP and its implication for the polyploid origin hypothesis of the tribe. J. Pl. Res. 113:91–96.

Jackson, R. C. 1960. A revision of the genus *Iva* L. Univ. Kansas Sci. Bull. 41:793–876.

Jacobson, H. A., J. B. Petersen, and D. E. Putnam. 1988. Evidence of pre-Columbian *Brassica* in the northeastern United States. Rhodora 90:355–362.

James, E., ed. 1823. Account of an Expedition from Pittsburgh to the Rocky Mountains, Performed in the Years 1819, 1820, by Order of the Hon. J. C. Calhoun, Secretary of War: Under the Command of Major Stephen H. Long, of the U.S. Top. Engineers. 2 vols. H. C. Carey and I. Lea, Philadelphia [also in 3 vols., Longman, Hurst, Rees, Orme, and Brown, London].

Jansen, R. K. 1981. The systematics of *Spilanthes* (Asteraceae: Heliantheae). Syst. Bot. 6:231–257.

———. 1985. The systematics of *Acmella* (Asteraceae: Heliantheae). Syst. Bot. Monogr. 8:1–115.

Jansen, R. K., and K.-J. Kim. 1996. Implications of chloroplast DNA data for the classification and phylogeny of the Asteraceae. Pp. 317–339 *in* D. J. N. Hind and H. Beentje, eds., Compositae: Systematics. Proceedings of the International Compositae Conference, Kew, 1994. Vol. 1. Royal Botanic Gardens, Kew.

Jansen, R. K., and J. D. Palmer. 1987. Chloroplast DNA from lettuce and *Barnadesia* (Asteraceae): Structure, gene localization, and characterization of a large inversion. Curr. Genet. 11:553–564.

Jansen, R. K., R. S. Wallace, K.-J. Kim, and K. L. Chambers. 1991. Systematic implications of chloroplast DNA variation in the subtribe Microseridinae (Asteraceae: Lactuceae). Amer. J. Bot. 78:1015–1027.

Janzen, D. H., and P. S. Martin. 1982. Neotropical anachronisms: The fruits the gomphotheres ate. Science 215:19–27.

Jeffrey, C. 1979. Note on the lectotypification of the names *Cacalia* L., *Matricaria* L., and *Gnaphalium* L. Taxon 28:349–351.

Johnson, M. F. 1971. A monograph of the genus *Ageratum* L. (Compositae—Eupatorieae). Ann. Missouri Bot. Gard. 58:6–88.

Johnson, S. A., L. P. Bruederle, and D. F. Tomback. 1998. A mating system conundrum: Hybridization in Apocynum (Apocynaceae). Amer. J. Bot. 85:1316–1323.

Johnston, I. M. 1952. Studies in the Boraginaceae, XXIII. A survey of the genus *Lithospermum*. J. Arnold Arbor. 33:299–366.

———. 1954. Studies in the Boraginaceae, XXVI. Further revaluations of the genera of the Lithospermeae. J. Arnold Arbor. 35:1–81.

Johnston, M. C. 1958. The Texas species of *Croton* (Euphorbiaceae). SouthW. Naturalist 3:175–203.

Jones, A. G. 1978a. The taxonomy of *Aster* section *Multiflorae* (Asteraceae), I. Nomenclatural review and formal presentation of the taxa. Rhodora 80:319–357.

———. 1978b. The taxonomy of *Aster* section *Multiflorae* (Asteraceae), II. Biosystematic investigations. Rhodora 80:453–490.

———. 1980. Data on chromosome numbers in *Aster* (Asteraceae), with comments on the status and relationships of certain North American species. Brittonia 32:240–261.

———. 1983. Nomenclatural changes in *Aster* (Asteraceae). Bull. Torrey Bot. Club 110:39–42.

———. 1989. *Aster* and *Brachyactis* in Illinois. Bull. Illinois Nat. Hist. Surv. 34:139–194, 1 map.

Jones, A. G., and P. Hiepko. 1981. The genus *Aster* s.l. (Asteraceae) in the Willdenow Herbarium at Berlin. Willdenowia 11:343–360.

Jones, G. N. 1957. On the distinctness of *Rudbeckia laciniata* and *R. ampla*. Madroño 14:131–133.

Jones, R. L. 1980. A revision of *Aster* section *Patentes* (Compositae). PhD diss., Vanderbilt University, Nashville.

———. 1983. A systematic study of *Aster* section *Patentes* (Asteraceae). Sida 10:41–81.

———. 1992. Additional studies of *Aster georgianus, A. patens,* and *A. phlogifolius* (Asteraceae). Sida 15:305–315.

Jones, S. B. 1972. A systematic study of the *Fasciculatae* group of *Vernonia* (Compositae). Brittonia 24:28–45.

Jones, S. B., and W. Z. Faust. 1978. *Vernonia*. N. Amer. Fl., ser. 2, 10:180–195.

Jones, S. B., W. Z. Faust, and L. E. Urbatsch. 1970. Natural hybridization between *Vernonia crinita* and *V. baldwinii* (Compositae). Castanea 35:61–67.

Karis, P. O. 1995. Cladistics of the subtribe Ambrosiinae (Asteraceae: Heliantheae). Syst. Bot. 20:40–54.

Jonsell, B. 1968. Studies in the North-west European species of *Rorippa* s.str. Symb. Bot. Upsal. 19(2): 1–221.

Judd, W. S. 1981. A monograph of *Lyonia* (Ericaceae). J. Arnold Arbor. 62:63–128, 129–209, 315–436.Judd, W. S., and I. K. Ferguson. 1999. The genera of Chenopodiaceae in the southeastern United States. Harvard Pap. Bot. 4:365–416.

Judd, W. S., and K. A. Kron. 1993. Circumscription of Ericaceae (Ericales) as determined by preliminary cladistic analyses based on morphological, anatomical, and embryological features. Brittonia 45:99–114.

Judd, W. S., Sanders, R. W., and M. J. Donoghue. 1994. Angiosperm family pairs: Preliminary cladistic analyses. Harvard Pap. Bot. No. 5:1–51.

Judd, W. S., C. S. Campbell, E. A. Kellogg, P. F. Stevens, and M. J. Donoghue. 2002. Plant Systematics: A Phylogenetic Approach. 2d ed. Sinauer Associates, Sunderland, MA.

Kajita, T., H. Ohashi, Y. Tateishi, C. D. Bailey, and J. J. Doyle. 2001. *rbcL* and legume phylogeny with particular reference to Phaseoleae, Millettieae, and allies. Syst. Bot. 26:515–536.

Karis, P. O. 1995. Cladistics of the subtribe Ambrosiinae (Asteraceae: Heliantheae). Syst. Bot. 20:40–54.

Karlsson, T. 2001. *Cerastium,* species 9–15. Pp. 149–158 *in* B. Jonsell and T. Karlsson, eds., Flora Nordica. Vol. 2. Chenopodiaceae–Fumariaceae. Royal Swedish Academy of Science, Stockholm.

Karpiscak, M. M., and O. M. Grosz. 1979. Dissemination trails of Russian thistle (*Salsola kali* L.) in recently fallowed fields. J. Arizona-Nevada Acad. Sci. 14:50–52.

Kartesz, J. T., and K. N. Gandhi. 1990. Nomenclatural notes for the North American flora. II. Phytologia 68:421–427.

———. 1991. Nomenclatural notes on North American flora. VII. Phytologia 71:87–100.

Kartesz, J. T., and C. A. Meacham. 1999. Synthesis of the North American Flora, ver. 1.0. North Carolina Botanical Garden, Chapel Hill, NC. CD-ROM.

Kathriarachchi, H., P. Hoffmann, R. Samuel, K. J. Wurdack, and M. W. Chase. 2005. Molecular phylogenetics of Phyllanthaceae inferred from five genes (plastid *atpB, matK, 3'ndhF, rbcL,* and nuclear *PHYC*). Molec. Phylogen. Evol. 36:112–134.

Kaul, R. B. 1986. Clusiaceae Lindl. Pp. 236–239 *in* Great Plains Flora Association, eds., Flora of the Great Plains. University Press of Kansas, Lawrence.

Keck, D. D. 1946. A revision of the *Artemisia vulgaris* complex in North America. Proc. Calif. Acad. Sci. 25:421–468.

Keith, E. L., J. R. Singhurst, and S. Cook. 2004. *Geocarpon minimum* (Caryophyllaceae), new to Texas. Sida 21:1165–1169.

Kelly, L. M. 2001. Taxonomy of *Asarum* section *Asarum* (Aristolochiaceae). Syst. Bot. 26:17–53.

Kephart, S. R., and C. B. Heiser Jr. 1980. Reproductive isolation in *Asclepias:* Lock and key hypothesis reconsidered. Evolution 34:738–746.

Kiers-van der Steen, A. M. 2000. Endive, chicory, and their wild relatives: A systematic and phylogenetic study of *Cichorium* (Asteraceae). Gorteria Suppl. 5:1–78.

Kim, J.-J., and B. L. Turner 1992. Systematic overview of *Krigia* (Asteraceae—Lactuceae). Brittonia 44:173–198.

Kim, S.-C., D. J. Crawford, and R. K. Jansen. 1996. Phylogenetic relationships among the genera of the subtribe Sonchinae (Asteraceae): Evidence from ITS sequences. Syst. Bot. 21:417–432.

Kimball, R. T., and D. J. Crawford. 2004. Phylogeny of Coreopsideae (Asteraceae) using ITS sequences suggests lability in reproductive characters. Molec. Phylogenet. Evol. 33:127–139.

King, C. C. 1981. Distribution of royal catchfly *(Silene regia)* with special reference to Ohio populations. Pp. 131–141 *in* R. L. Stuckey and K. J. Reese, eds., The Prairie Peninsula: In the "Shadow" of Transeau: Proceedings of the Sixth North American Prairie Conference. Biol. Notes Ohio Biol. Surv. 15:i–x, 1–279.

King, L. M. 1993. Origins of genotypic variation in North American dandelions inferred from ribosomal DNA and chloroplast DNA restriction enzyme analysis. Evolution 47:136–151.

King, L. M., and B. A. Schaal. 1990. Genotypic variation within asexual lineages of *Taraxacum officinale.* Proc. Natl. Acad. Sci. U.S.A. 87:998–1002.

King, R. M., and H. Robinson. 1970. Studies in the Eupatorieae (Compositae). XXV. A new genus *Eupatoriadelphus.* Phytologia 19:431–432.

———. 1987. The genera of Eupatorieae (Asteraceae). Missouri Bot. Gard. Monogr. Syst. Bot. 22:i–ix, 1–581.

Kingsbury, J. M. 1964. Poisonous Plants of the United States and Canada. Prentice-Hall, Englewood Cliffs, NJ.

Kirkbride, J. H., Jr. 1993. Biosystematic Monograph of the Genus *Cucumis* (Cucurbitaceae). Parkway Publishers, Boone, NC.

Kirschner, J., and J. Štěpánek. 1987. Again on the sections in *Taraxacum* (Cichoriaceae) (studies in *Taraxacum* 6). Taxon 36:608–617.

Klayman, D. L. 1985. Qinghaosu (artemisinin): An antimalarial drug from China. Science 228:1049–1055.

———. 1993 *Artemisia annua:* From weed to respectable antimalarial plant. Pp. 242–255 *in* A. D. Kinghorn and M. F. Balandrin, eds., Human Medicinal Agents from Plants. Amer. Chem. Soc. Symp. Ser. 534. American Chemical Society, Washington, DC.

Klips, R. A., and T. M. Culley. 2004. Natural hybridization between prairie milkweeds, *Asclepias sullivantii* and *Asclepias syriaca:* Morphological, isozyme, and hand-pollination evidence. Int. J. Pl. Sci. 165:1027–1037.

Knox, J. S. 1987. An experimental garden test of characters used to distinguish *Helenium virginicum* Blake from *H. autumnale* L. Castanea 52:52–58.

———. 1997. A nine year demographic study of *Helenium virginicum* (Asteraceae): A narrow endemic seasonal wetland plant. J. Torrey Bot. Club 124:236–245.

Knox, J. S., M. J. Gutowski, D. C. Marshall, and O. G. Rand. 1995. Tests of genetic bases of character differences between *Helenium virginicum* and *H. autumnale* (Asteraceae) using common gardens and transplant studies. Syst. Bot. 20:120–131.

Kobiakova, J. A. 1930. The bottle gourd. Trudy Prikl. Bot. 23 (3): 475–520.

Koch, M., and I. A. Al-Shehbaz. 2004. Taxonomic and phylogenetic evaluation of the American "*Thlaspi*" species: Identity and relationship to the Eurasian genus *Noccaea* (Brassicaceae). Syst. Bot. 29:375–384.

Koopman, W. J. M., E. Guetta, C. C. M. Van de Wiel, B. Vosman, and R. G. Van den Berg. 1998. Phylogenetic relationships among *Lactuca* (Asteraceae) species and related genera based on ITS-1 DNA sequences. Amer. J. Bot. 85:1517–1530.

Koopman, W. J. M., M. J. Zevenbergen, and R. G. Van den Berg. 2001. Species relationships in *Lactuca* s.l. (Lactuceae, Asteraceae) inferred from AFLP fingerprints. Amer. J. Bot. 88:1881–1887.Koyama, H., and D. E. Boufford. 1981. Proposal to change one of the examples in Article 57. Taxon 30:504–505.

Kral, R. 1960. A revision of *Asimina* and *Deeringothamnus* (Annonaceae). Brittonia 12:233–278.

Krings, A., and (J.) Q.-Y. Xiang. 2004. The *Gonolobus* complex (Apocynaceae: Asclepiadoideae) in the southeastern United States. Sida 21:103–116.

———. 2005. Taxonomy of the *Gonolobus* complex (Apocynaceae, Asclepiadoideae) in the southeastern United States: ISSR evidence and parsimony analysis. Harvard Pap. Bot. 10:147–159.

Krombein, K. V., P. D. Hurd, D. R. Smith, and B. D. Burks. 1979. Catalog of the Hymenoptera in America North of Mexico. Smithsonian Institution Press, Washington, DC.

Kron, K. A. 1989. *Azalea rosea* Loiseleur is a superfluous name. Sida 13:331–333.
———. 1993. A revision of *Rhododendron* section *Pentathera*. Edinburgh J. Bot. 50:49–364.
———. 1996. Phylogenetic relationships of Empetraceae, Epacridaceae, Ericaceae, Monotropaceae, and Pyrolaceae: Evidence from nucleotide ribosomal 18S sequence data. Ann. Bot. (London), new ser. 77:293–303.
Kucera, C. L. 1953. The genus *Lyonia* in Missouri. Rhodora 55:155.
Kurz, D. 1997. Shrubs and Woody Vines of Missouri. Missouri Department of Conservation, Jefferson City.
La Valva, V., and S. Sabato. 1983. Nomenclature and typification of *Ipomoea imperati* (Convolvulaceae). Taxon 32:110–114.
Lack, H.-W. 1974. Die Gattung *Picris* L., sensu lato, im ostmediterran-westasiatischen Raum. Diss. Univ. Wien 116:1–184, figs. 1–111, maps 1–21.
Ladd, D. 1990. Noteworthy collections: Missouri. Castanea 55:293–294.
———. 1994. Three plants new to Missouri. Missouriensis 15:28–30.
Lamarck, [J. B. P. A.] M. 1785. Encyclopédie Méthodique de Botanique. Vol. 1, part 2. Chez Panckouke, Paris.
Lamont, E. E. 1995. Taxonomy of *Eupatorium* section *Verticillata* (Asteraceae). Mem. New York Bot. Gard. 72:1–66.
———. 2004. New combinations in *Eutrochium* (Asteraceae: Eupatorieae), an earlier name than *Eupatoriadelphus*. Sida 21:901–902.
Lane, F. C. 1954. The genus *Triosteum* (Caprifoliaceae). PhD diss., University of Illinois, Urbana.
Lane, M. A. 1979. Taxonomy of the genus *Amphiachyris* (Asteraceae: Astereae). Syst. Bot. 4:178–189.
———. 1982. Generic limits of *Xanthocephalum, Gutierrezia, Amphiachyris, Gymnosperma, Greenella*, and *Thurovia* (Compositae: Astereae). Syst. Bot. 7:405–416.
———. 1985. Taxonomy of *Gutierrezia* (Compositae: Astereae) in North America. Syst. Bot. 10:7–28.
Lane, M. A., and R. L. Hartman. 1996. Reclassification of North American *Haplopappus* (Compositae: Astereae) completed: *Rayjacksonia* gen. nov. Amer. J. Bot. 83:356–370.
Lawrence, G. H. M. 1956. The cultivated ivies. Morris Arbor. Bull. 7:19–31.
Lee, I.-M., D. E. Gundersen-Rindal, R. E. Davis, K. D. Bottner, C. Marcone, and E. Seemüller. 2004. '*Candidatus* Phytoplasma asteris', a novel phytoplasma taxon associated with aster yellows and related diseases. Int. J. Syst. Evol. Microbiol. 54:1037–1048.
Lee, T. D. 1984. Effects of seed number per fruit on seed dispersal in *Cassia fasciculata* (Caesalpiniaceae). Bot. Gaz. 145:136–139.
Lee, Y. T. 1976. The genus *Gymnocladus* and its tropical affinity. J. Arnold Arbor. 57:91–112.
Leeuwenberg, A. J. M., and P. W. Leenhouts. 1980. Taxonomy. Chap. 2, pp. 8–96 *in* A. Engler, H. Harms, J. Mattfield, E. Werdermann, and T. Eckardt, eds., Die Natürlichen Pflanzenfamilien, Band 28BI, Angiospermae: Ordnung Gentianales Fam. Loganiaceae. 2d ed. Duncker and Humblot, Berlin.
LeJeune, K. D., and T. R. Seastedt. 2001. *Centaurea* species: The forb that won the west. Conservation Biol. 15:1568–1574.
Les, D. H. 1986. Systematics and evolution of *Ceratophyllum* L. (Ceratophyllaceae): A monograph. PhD diss., Ohio State University, Columbus.
———. 1993. Ceratophyllaceae. Pp. 246–250 *in* K. Kubitzki, ed., The Families and Genera of Vascular Plants. Vol. 2. Flowering Plants—Dicotyledons. Magnoliid, Hamamelid, and Caryophyllid Families. (K. Kubitzki, J. G. Rohwer, and V. Bittrich, vol. eds.) Springer-Verlag, Berlin.
———. 1994. Molecular systematics and taxonomy of lake cress (*Neobeckia aquatica*, Brassicaceae), an imperiled aquatic mustard. Aquatic Bot. 49:149–165.
Les, D. H., J. A. Reinartz, and E. J. Esselman, 1991. Genetic consequences of rarity in *Aster furcatus* (Asteraceae), a threatened self-incompatible plant. Evolution 45:1641–1650.
Levin, D. A. 1968. The structure of a polyspecies hybrid swarm in *Liatris*. Evolution 22:352–372.
Levin, G. A. 1999a. Notes on *Acalypha* (Euphorbiaceae) in North America. Rhodora 101:217–233.
———. 1999b. Evolution in the *Acalypha gracilens/monococca* complex (Euphorbiaceae): Morphological analysis. Syst. Bot. 23:269–287.
Lewis, W. H. 1988. Regrowth of a decimated population of *Panax quinquefolium* in a Missouri climax forest. Rhodora 90:1–5.
———. 1989. American Ginseng: A Forest Crop. 2d ed. Missouri Department of Conservation, Jefferson City.

Lewis, W. H., and R. L. Oliver. 1965. Realignment of *Calystegia* and *Convolvulus* (Convolvulaceae). Ann. Missouri Bot. Gard. 52:217–222.

Liede, S. 1996. *Cynanchum—Rhodostegiella—Vincetoxicum—Tylophora* (Asclepiadaceae): New considerations on an old problem. Taxon 45:193–211.

———. 1997. American *Cynanchum* (Asclepiadaceae): A preliminary infrageneric classification. Novon 7:172–181.

Lindsey, A. H. 1975. Classification versus characterization in *Thaspium* and *Zizia* (Apiaceae). Master's thesis, University of North Carolina, Chapel Hill.

———. 1982. Floral phenology patterns and breeding systems in *Thaspium* and *Zizia* (Apiaceae). Syst. Bot. 7:1–12.

Ling, Y.-R. 1995. 13. The New World *Artemisia* L. Pp. 255–281 *in* D. J. N. Hind, C. Jeffrey, and G. V. Pope, eds., Advances in Compositae Systematics. Royal Botanic Gardens, Kew.

Linnaeus, C. 1737. Genera Plantarum. 1st ed. Sumtibus Michaelis Antonii David, Paris.

———. 1753. Species Plantarum. 1st ed., 2 vols. Impensis Laurentii Salvii, Stockholm.

———. 1755. Flora Suecica. 2d ed. L. Salvius, Stockholm.

Lipow, S. R., and R. Wyatt. 1998. Reproductive biology and breeding system of *Gonolobus suberosus* (Asclepiadaceae). J. Torrey Bot. Soc. 125:183–193.

———. 1999. Floral morphology and late-acting self-incompatibility in *Apocynum cannabinum* (Apocynaceae). Pl. Syst. Evol. 219:99–109.

Lipscomb, B. L., and E. B. Smith. 1977. Morphological intergradation of varieties of *Bidens aristosa* (Compositae) in northern Arkansas. Rhodora 79:203–213.

Long, R. W. 1970. The genera of Acanthaceae in the southeastern United States. J. Arnold Arbor. 51:257–308.

Louda, S. M. 2000. *Rhinocyllus conicus*—Insights to improve predictability and minimize risk of biological control of weeds. Pp. 187–193 *in* N. R. Spencer, ed., Proceedings of the X International Symposium on Biological Control of Weeds. Montana State University, Bozeman.

Löve, D., and P. Dansereau. 1959. Biosystematic studies on *Xanthium:* Taxonomic appraisal and ecological status. Canad. J. Bot. 37:173–208.

Lovell, H. B. 1940. Pollination of the Ericaceae: V. *Gaylussacia baccata.* Rhodora 42:352–354.Lowden, R. M. 1978. Studies on the submerged genus *Ceratophyllum* L. in the neotropics. Aquatic Bot. 4:127–142.

Lowry, P. P., II, and A. G. Jones. 1984. Systematics of *Osmorhiza* Raf. (Apiaceae: Apioideae). Ann. Missouri Bot. Gard. 71:1128–1171.

Luckow, M. 1993. Monograph of *Desmanthus* (Leguminosae: Mimosoideae). Syst. Bot. Monogr. 38:1–166.

Luckow, M., P. J. White, and A. Bruneau. 2000. A phylogenetic analysis of the Mimosoideae (Leguminosae) based on chloroplast DNA sequence data. Pp. 197–220 *in* B. B. Klitgaard and A. Bruneau, eds., Advances in Legume Science, Part 10. Higher Level Systematics. Royal Botanic Gardens, Kew.

Luckow, M., J. T. Miller, D. J. Murphy, and T. Livshultz. 2003. Relationships among the basal genera of mimosoid legumes. Pp. 165–181 *in* P. S. Herendeen and A. Bruneau, eds., Advances in Legume Science, Part 9. Higher Level Systematics. Royal Botanic Gardens, Kew.

Luken, J. O., and J. W. Thieret. 1995. Amur honeysuckle (*Lonicera maackii;* Caprifoliaceae): Its ascent, decline, and fall. Sida 16:479–503.

———. 1996. Amur honeysuckle: Its fall from grace. BioScience 46:18–24.

Luken, J. O., J. W. Thieret, and J. R. Kartesz. 1993. *Erucastrum gallicum* (Brassicaceae): Invasion and spread in North America. Sida 14:569–582.

Lutz, H. J. 1943. Injuries to trees caused by *Celastrus* and *Vitis.* Bull. Torrey Bot. Club 70:436–439.

Ma, J. S. 2001. A revision of *Euonymus* (Celastraceae). Thaiszia 11:1–264.

Mabberley, D. J. 1997. The Plant Book. 2d ed. Cambridge University Press, Cambridge.

Mack, R. N. 1991. The commercial seed trade: An early disperser of weeds in the United States. Econ. Bot. 45:257–273.

Mackenzie, K. K. 1902. Manual of the Flora of Jackson County, Missouri. New Era Printing, Lancaster, PA.

———. 1906. *Onosmodium.* Bull. Torrey Bot. Club 32:495–506.

———. 1914. A new genus from Missouri. Torreya 14:67–68.

Mackenzie, K. K., and B. F. Bush. 1902. New plants from Missouri. Trans. Acad. Sci. St. Louis 12:79–89, pl. xii–xvii.

Mahler, W. F. 1975. Typification and distribution of the varieties of *Gnaphalium helleri* Britton (Compositae—Inuleae). Sida 6:30–32.

Maihle, N. J., and W. H. Blackwell Jr. 1978. A synopsis of North American *Corispermum* (Chenopodiaceae). Sida 7:382–391.

Manhart, J. R., and J. H. Rettig. 1993. Gene sequence data. Pp. 235–246 *in* H.-D. Behnke and T. J. Mabry, eds., Caryophyllales: Evolution and Systematics. Springer-Verlag, Berlin.

Maslin, B. R., J. T. Miller, and D. S. Seigler. 2003. Overview of the generic status of *Acacia* (Leguminosae: Mimosoideae). Austral. Syst. Bot. 16:1–18.

Mathias, M. E. 1938. A revision of the genus *Lomatium*. Ann. Missouri Bot. Gard. 25:225–297.

Mathias, M. E., and L. Constance. 1941a. *Limnosciadium,* a new genus of Umbelliferae. Amer. J. Bot. 28:162–163.

———. 1941b. A synopsis of the North American species of *Eryngium*. Amer. Midl. Naturalist 25:361–387.

———. 1944–1945. Umbelliferae. Fl. N. Amer. 28B:43–295.

———. 1951. Umbelliferae. Pp. 263–329, pl. 35–54 *in* C. L. Lundell and collaborators, Flora of Texas. Vol. 3. Southern Methodist University Press, Dallas, TX.

Mavrodiev, E. V., M. Tancig, A. M. Sherwood, M. A. Gitzendanner, J. Rocca, P. S. Soltis, and D. E. Soltis. 2005. Phylogeny of *Tragopogon* L. (Asteraceae) based on internal and external transcribed spacer sequence data. Int. J. Pl. Sci. 166:117–133.

Mayfield, M. H. 1997. A systematic treatment of *Euphorbia* subgenus *Poinsettia* (Euphorbiaceae). PhD diss., University of Texas, Austin.

McAllister, H. A., and A. Rutherford. 1990. *Hedera helix* L. and *H. hibernica* (Kirchner) Bean (Araliaceae) in the British Isles. Watsonia 18:7–15.

McGregor, R. L. 1968. The taxonomy of the genus *Echinacea* (Compositae). Univ. Kansas Sci. Bull. 48:113–142.

———. 1984. Studies in the genus *Apocynum* (Apocynaceae). Contr. Univ. Kansas. Herb. 9:1–12.

———. 1985a. Current status of the genus *Camelina* (Brassicaceae) in the prairies and plains of central North America. Contr. Univ. Kansas. Herb. 13:1–13.

———. 1985b. Studies on the validity of *Lobelia spicata* infraspecific taxa in the prairies and plains of central North America with notes on *Lobelia appendiculata*. Contr. Univ. Kansas Herb. 16:1–10.

———. 1986a. Chenopodiaceae Vent. Pp. 160–179 *in* Great Plains Flora Association, eds., Flora of the Great Plains. University Press of Kansas, Lawrence.

———. 1986b. Aceraceae Juss. Pp. 569–570 *in* Great Plains Flora Association, eds., Flora of the Great Plains. University Press of Kansas, Lawrence.

———. 1986c. Anacardiaceae Lindl. Pp. 571–575 *in* Great Plains Flora Association, eds., Flora of the Great Plains. University Press of Kansas, Lawrence.

———. 1986d. Apiaceae, Lindl. Pp. 584–604 *in* Great Plains Flora Association, eds., Flora of the Great Plains. University Press of Kansas, Lawrence.

———. 1986e. *Carduus*. Pp. 895–897 *in* Great Plains Flora Association, eds., Flora of the Great Plains. University of Kansas Press, Lawrence.

———. 1986f. *Evax*. P. 937 *in* Great Plains Flora Association, eds., Flora of the Great Plains. University of Kansas Press, Lawrence.

McInnis, N. C., L. M. Smith, and A. B. Pittman. 1993. *Geocarpon minimum* (Caryophyllaceae), new to Louisiana. Phytologia 75:159–162.

McKenzie, P. M. 2003. 50 CFR Part 17. Endangered and threatened wildlife and plants: Reclassification of *Lesquerella filiformis* (Missouri bladderpod) from endangered to threatened. Federal Register 68(199): 59337–59345.

McLaughlin, S. P. 1986. Mass selection for increased resin yield in *Grindelia camporum* (Asteraceae). Econ. Bot. 40:155–161.

McNeill, J. 1962. Taxonomic studies in the Alsinoideae: I. Generic and infra-generic groups. Notes Roy. Bot. Gard. Edinburgh 24:79–155.

———. 1978. *Silene alba* and *S. dioica* in North America and the generic delimitation of *Lychnis, Melandrium,* and *Silene* (Caryophyllaceae). Canad. J. Bot. 56:297–308.

———. 1980. The delimitation of *Arenaria* (Caryophyllaceae) and related genera in North America, with 11 new combinations in *Minuartia*.

McNeill, J., and H. C. Prentice. 1981. *Silene pratensis* (Rafn) Godron & Gren., the correct name for white campion or white cockle (*Silene alba* (Miller) E.H.L. Krause, nom. illeg.). Taxon 30:27–32.

McNeill, J., I. J. Bassett, and C. W. Crompton. 1977. *Suaeda calceoliformis*, the correct name for *Suaeda depressa* auct. Rhodora 79:133–137.

McVaugh, R. 1936. Studies in the taxonomy and distribution of eastern North American species of *Lobelia*. Rhodora 38:241–298.

———. 1945. The genus *Triodanis* Rafinesque, and its relationships to *Specularia* and *Campanula*. Wrightia 1:13–52.

———. 1948. Generic status of *Triodanis* and *Specularia*. Rhodora 50:38–49.

Mears, J. A. 1975. The taxonomy of *Parthenium* section *Partheniastrum* DC. (Asteraceae—Ambrosiinae). Phytologia 31:463–482.

Meehan, T. 1898. The plants of Lewis and Clark's expedition across the continent, 1804–1806. Proc. Acad. Nat. Sci. Phila. 1898:12–49.

Merrill, E. D., and S.-Y. Hu. 1949. Work and publications of Henry Muhlenberg, with special attention to unrecorded or incorrectly recorded binomials. Bartonia 25:1–66, 1 pl.

Messmore, N. A., and J. S. Knox. 1997. The breeding system of the narrow endemic, *Helenium virginicum* (Asteraceae). J. Torrey Bot. Club 124:318–321.

Meyer, F. K. 1973. Conspectus der "*Thlaspi*"-Arten Europas, Afrikas und Vorderasiens. Feddes Repert. 84:449–470.

———. 1979. Kritische Revision der "*Thlaspi*"-Arten Europas, Afrikas und Vorderasiens. Feddes Repert. 90:129–154.

———. 2003. Kritische Revision der "*Thlaspi*"-Arten Europas, Afrikas und Vorderasiens. Spezieller Teil. III. *Microthlaspi* F. K. Mey. Haussknechtia 9:3–59.

Miao, B., B. L. Turner, and T. Mabry. 1995a. Chloroplast DNA variations in sect. *Cyclachaena* of *Iva* (Asteraceae). Amer. J. Bot. 82:919–923.

———. 1995b. Systematic implications of chloroplast DNA variation in subtribe Ambrosiinae (Asteraceae: Heliantheae). Amer. J. Bot. 82:924–932.

Miao, B., B. Turner, B. Simpson, and T. Mabry. 1995c. Chloroplast DNA study of the genera *Ambrosia* s.l. and *Hymenoclea* (Asteraceae): Systematic implications. Pl. Syst. Evol. 194:241–255.

Michener, D. C. 1986. Phenotypic instability in *Gleditsia triacanthos* (Fabaceae). Brittonia 38:360–361.

Miller, J. T., and R. J. Bayer. 2003. Molecular phylogenetics of *Acacia* subgenera *Acacia* and *Acueiferum* (Fabaceae: Mimosoideae), based on the chloroplast *matK* coding sequence and flanking *trnK* intron spacer regions. Austral. Syst. Bot. 16:27–33.

Miller, K. I., and G. L. Webster. 1967. A preliminary revision of *Tragia* (Euphorbiaceae) in the United States. Rhodora 69:241–305.

Miller, L. W. 1964. A taxonomic study of the species of *Acalypha* in the United States. PhD diss., Purdue University, Lafayette, IN.

Miller, N. G. 1970. The genera of Cannabaceae in the southeastern United States. J. Arnold Arbor. 51:185–203.

———. 2001. The Callitrichaceae in the southeastern United States. Harvard Pap. Bot. 5:277–301.

Milstead, W. L. 1964. A revision of the North American species of *Prenanthes*. PhD diss., Purdue University, West Lafayette, IN.

Mitchell, R. S., and L. J. Uttal. 1969. Natural hybridization in Virginia *Silene* (Caryophyllaceae). Bull. Torrey Bot. Club 96:544–549.

Moerman, D. E. 1998. Native American Ethnobotany. Timber Press, Portland, OR.

Mohlenbrock, R. H. 1979. Botanizing Missouri in Illinois. Missouriensis 1(2): 8–9.

———. 2001. Flowering Plants: Pokeweeds, Four-o'clocks, Carpetweeds, Cacti, Purslanes, Goosefoots, Pigweeds, and Pinks. The Illustrated Flora of Illinois series. Southern Illinois University Press, Carbondale.

Moore, D. 1958. New records for the Arkansas flora. IV. Proc. Arkansas Acad. Sci. 12:12.

Moore, R. J., and C. Frankton. 1966. An evaluation of the status of *Cirsium pumilum* and *Cirsium hillii*. Canad. J. Bot. 44:581–595.

———. 1974. The thistles of Canada. Canad. Dept. Agric. Res. Branch Monogr. 10:1–111.

Morefield, J. D. 1992. Evolution and systematics of *Stylocline* (Asteraceae: Inuleae). PhD diss., Claremont Graduate School, Clarement, CA.

———. 2004. New taxa and names in North American *Ancistrocarphus, Diaperia,* and *Logfia* (Asteraceae: Gnaphalieae: Filagininae) and related taxa. Novon 13:463–475.

Morgan, J. T. 1967. A taxonomic study of the genus *Boltonia* (Asteraceae). PhD diss., University of North Carolina, Chapel Hill.

Mort, M. E., D. E. Soltis, P. S. Soltis, J. Francisco-Ortega, and A. Santos-Guerra. 2001. Phylogenetic relationships and evolution of Crassulaceae inferred from *matK* sequence data. Amer. J. Bot. 88:76–91.

Morton, G. H. 1973. The taxonomy of the *Solidago arguta—boottii* complex. PhD diss., University of Tennessee, Knoxville.

———. 1974. A new subspecies and other nomenclatural changes in the *Solidago arguta* complex. Phytologia 28:1–4.

———. 1984. A practical treatment of the *Solidago gigantea* complex. Canad. J. Bot. 62:1279–1282.

Morton, J. K. 1972. On the occurrence of *Stellaria pallida* in North America. Bull. Torrey Bot. Club. 99:95–97.

———. 1988. Variation in *Erigeron philadelphicus* (Compositae). Canad. J. Bot. 66:298–302.

———. 2005a. *Cerastium*. Pp. 74–92 *in* Flora of North America Editorial Committee, eds., Flora of North America, North of Mexico. Vol. 5. Magnoliophyta: Caryophyllidae, Part 2. Oxford University Press, New York.

———. 2005b. *Stellaria*. Pp. 96–116 *in* Flora of North America Editorial Committee, eds., Flora of North America, North of Mexico. Vol. 5. Magnoliophyta: Caryophyllidae, Part 2. Oxford University Press, New York.

———. 2005c. *Silene*. Pp. 168–216 *in* Flora of North America Editorial Committee, eds., Flora of North America, North of Mexico. Vol. 5. Magnoliophyta: Caryophyllidae, Part 2. Oxford University Press, New York.

Mosyakin, S. L. 1995. New taxa of *Corispermum* L. (Chenopodiaceae), with preliminary comments on the taxonomy of the genus in North America. Novon 5:340–353.

———. 1996. A taxonomic synopsis of the genus *Salsola* (Chenopodiaceae) in North America. Ann. Missouri Bot. Gard. 83:387–395.

Mosyakin, S. L., and S. E. Clemants. 2002. New nomenclatural combinations in *Dysphania* R. Br. (Chenopodiaceae): Taxa occurring in North America. Ukrayins'k Bot. Zhurn., n.s. 59:380–385.

Mühlenbach, V. 1979. Contributions to the synanthropic (adventive) flora of the railroads in St. Louis, Missouri, U.S.A. Ann. Missouri Bot. Gard. 66:1–108.

———. 1983. Supplement to the contributions to the synanthropic (adventive) flora of the railroads in St. Louis, Missouri, U.S.A. Ann. Missouri Bot. Gard. 70:170–178.

Mulligan, G. A. 1980. The genus *Cicuta* in North America. Canad. J. Bot. 58:1755–1767.

Mulligan, G. A., and I. J. Bassett. 1959. *Achillea millefolium* complex in Canada and portions of the United States. Canad. J. Bot. 37:73–79.

Mulligan, G. A., and D. B. Munro. 1984. Chromosome numbers and sexual compatibility in North America of *Rorippa sylvestris* (Cruciferae). Canad. J. Bot. 62:575–580.

Mummenhoff, K., H. Brüggemann, and J. L. Bowman. 2001. Chloroplast DNA phylogeny and biogeography of *Lepidium* (Brassicaceae). Amer. J. Bot. 88:2051–2063.

Murray, E. 1975. North American maples. Kalmia 7:1–20.

Murrell, Z. E. 1992. Systematics of the genus *Cornus*. PhD diss., Duke University, Durham, NC.

———. 1993. Phylogenetic relationships in Cornus (Cornaceae). Syst. Bot. 18:469–495.

Myint, T. 1966. Revision of the genus *Stylisma* (Convolvulaceae). Brittonia 18:97–117.

Nath, J., and S. N. Clay. 1972. Cytogenetic studies on some species of *Euonymus*. Caryologia 25:417–427.

Nee, M. 1990. The domestication of *Cucurbita* (Cucurbitaceae). Econ. Bot. 44 (3, suppl.): 56–68.

Neher, R. 1966. Monograph of the genus *Tagetes* (Compositae). PhD diss., Indiana University, Bloomington.

Nelson, P. W. 1979. A new halophyte for Missouri. Castanea 44:246–247.

———. 1985. The Terrestrial Natural Communities of Missouri. Missouri Natural Areas Committee, Jefferson City.

———. 2005. The Terrestrial Natural Communities of Missouri. 2d ed. Missouri Natural Areas Committee, Jefferson City.

Nesom, G. L. 1988. Synopsis of *Chaetopappa* (Compositae—Astereae) with a new species and the inclusion of *Leucelene*. Phytologia 64:448–456.

———. 1989. New species, new sections, and a taxonomic overview of American *Pluchea* (Compositae: Inuleae). Phytologia 67:158–167.

———. 1990a. Taxonomic status of *Gamochaeta* (Asteraceae: Inuleae) and the species of the United States. Phytologia 68:186–198.

———. 1990b. Further definition of *Conyza* (Asteraceae: Astereae). Phytologia 68:229–233.

———. 1990c. Studies in the systematics of Mexican and Texan *Grindelia* (Asteraceae: Astereae). Phytologia 68:303–332.

———. 1990d. Taxonomy of *Heterotheca* sect. *Heterotheca* (Asteraceae: Astereae) in México, with comments on the taxa of the United States. Phytologia 69:282–294.

———. 1990e. Taxonomy of *Solidago petiolaris* (Asteraceae: Astereae) and related Mexican species. Phytologia 69:445–456.

———. 1991. Union of *Bradburia* with *Chrysopsis* (Asteraceae: Astereae), with a phylogenetic hypothesis for *Chrysopsis*. Phytologia 70:109–121.

———. 1993a. Taxonomic infrastructure of *Solidago* and *Oligoneuron* (Asteraceae: Astereae) and observations on their phylogenetic position. Phytologia 75:1–44.

———. 1993b. Taxonomy of *Doellingeria* (Asteraceae: Astereae). Phytologia 75:452–462.

———. 1994. Review of the taxonomy of *Aster* sensu lato (Asteraceae; Astereae), emphasizing the New World species. Phytologia 77:141–297.

———. 1997. Review: "A revision of *Heterotheca* sect. *Phyllotheca* (Nutt.) Harms (Compositae: Astereae)" by J. C. Semple. Phytologia 83:7–21.

———. 2000. Generic conspectus of the tribe Astereae (Asteraceae) in North America and Central America, the Antilles, and Hawaii. Sida Misc. 20:1–100, 1 pl.

———. 2001. Notes on variation in *Pseudognaphalium obtusifolium* (Asteraceae: Gnaphalieae). Sida 19:615–619.

———. 2004a. New species of *Gamochaeta* (Asteraceae: Gnaphalieae) from the eastern United States and comments on similar species. Sida 21:717–741.

———. 2004b. New distribution records for *Gamochaeta* (Asteraceae: Gnaphalieae) in the United States. Sida 21:1175–1185.

———. 2005a. Taxonomic review of *Astranthium integrifolium* (Asteraceae: Astereae). Sida 21:2015–2021.

———. 2005b. Taxonomy of the *Symphyotrichum (Aster) subulatum* group and *Symphyotrichum tenuifolium* (Asteraceae: Astereae). Sida 21:2125–2140.

Nesom, G. L., and T. J. Leary. 1992. A new species of *Ionactis* (Asteraceae: Astereae) and an overview of the genus. Brittonia 44:247–252.

Nesom, G. L., and R. D. Noyes. 2000. *Batopilasia* (Asteraceae: Astereae), a new genus from Chihuahua, Mexico. Sida 19:79–84.

Nesom, G. L., and R. J. O'Kennon. 2001. Two new species of *Liatris* series *Punctatae* (Asteraceae: Eupatorieae) centered in north central Texas. Sida 19:767–787.

Nesom, G. L., Y. Suh, and B. B. Simpson. 1993. *Prionopsis* (Asteraceae: Astereae) united with *Grindelia*. Phytologia 75:341–346.

Newsom, L. A., S. D. Webb, and J. S. Dunbar. 1993. History and geographic distribution of *Cucurbita pepo* gourds in Florida. J. Ethnobiol. 13:75–97.

Nicely, K. A. 1965. A monographic study of the Calycanthaceae. Castanea 30:38–81.

Nielsen, I. 1981. Ingeae. Pp. 173–190 *in* R. M. Polhill and P. H. Raven, eds., Advances in Legume Systematics, Part 1. Royal Botanic Gardens, Kew.

Nienaber, M. A. 2005. *Geocarpon*. Pp. 150–151 *in* Flora of North America Editorial Committee, eds., Flora of North America, North of Mexico. Vol. 5. Magnoliophyta: Caryophyllidae, Part 2. Oxford University Press, New York.

Niesenbaum, R. A., M. G. Patselas, and S. D. Weiner. 1999. Does flower color change in *Aster vimineus* cue pollinators? Amer. Midl. Naturalist 141:59–68.

Nigh, T. A., and W. A. Schroeder. 2002. Atlas of Missouri Ecoregions. Missouri Department of Conservation, Jefferson City.

Nigh, T. A., S. G. Pallardy, and H. E. Garrett. 1985. Sugar maple–environmental relationships in the River Hills and Central Ozark Mountains of Missouri. Amer. Midl. Naturalist 114:235–251.

Nightingale, A. 1980. More *Alliaria* in our area. Missouriensis 1 (4): 26.

Nilsson, Ö. 2001. *Stellaria,* species 2–5. Pp. 117–121 *in* B. Jonsell and T. Karlsson, eds., Flora Nordica. Vol. 2. Chenopodiaceae–Fumariaceae. Royal Swedish Academy of Science, Stockholm.

Northington, D. K., III. 1971. Taxonomy of *Pyrrhopappus,* a cytotaxonomic and chemosystematic study. PhD diss., University of Texas, Austin.

———. 1974. Systematic studies of the genus *Pyrrhopappus* (Compositae, Cichorieae). Special Publ. Mus. Texas Tech Univ. 6:1–38.

Novak, S. J., D. E. Soltis, and P. S. Soltis. 1991. Ownbey's tragopogons: 40 years later. Amer. J. Bot. 78:1586–1600.

Noyes, R. D. 2000a. Biogeographical and evolutionary insights on *Erigeron* and allies (Asteraceae) from ITS sequence data. Pl. Syst. Evol. 220:93–114.

———. 2000b. Diplospory and parthenogenesis in sexual × agamospermous (apomictic) *Erigeron* (Asteraceae) hybrids. Int. J. Pl. Sci. 161:1–12.

Noyes, R. D., and L. H. Rieseberg. 1999. ITS sequence data support a single origin for North American Astereae (Asteraceae) and reflect deep geographic divisions in *Aster* s.l. Amer. J. Bot. 86:398–412.

Nuttall, T. 1821. A Journal of Travels into the Arkansas Territory, during the Year 1819. T. H. Palmer, Philadelphia, PA. Reprint, 1980, University of Oklahoma Press, Norman.

Nuzzo, V. A. 1991. Experimental control of garlic mustard (Alliaria petiolata (Bieb.) Cavara and Grande) in northern Illinois using fire, herbicide, and cutting. Nat. Areas J. 11:158–167.

———. 1993. Distribution and spread of the invasive biennial garlic mustard *(Alliaria petiolata)* in North America. Pp. 137-146 *in* B. N. McKnight, ed., Biological Pollution: The Control and Impact of Invasive Exotic Species. Indiana Academy of Science, Indianapolis.

Oberprieler, C. 2001. Phylogenetic relationships in *Anthemis* L. (Compositae, Anthemideae) based on nrDNA ITS sequence variation. Taxon 50:745–762.

Ochsmann, J. 1997. Ein Bestand von *Centaurea* ×*psammigera* Gáyer (*Centaurea diffusa* Lam. × *Centaurea stoebe* L.) am NSG Sonnenstein (Thüringen). Flor. Rundbr. 31:118–125.

———. 2001a. On the taxonomy of spotted knapweed (*Centaurea stoebe* L.) Pp. 33–41 *in* L. Smith, ed., The First International Knapweed Symposium of the Twenty-first Century. U.S.D.A. Agricultural Research Service, Albany, CA.

———. 2001b. An overlooked knapweed hybrid in North America: *Centaurea* ×*psammigera* Gáyer (diffuse knapweed × spotted knapweed). P. 76 *in* L. Smith, ed., The First International Knapweed Symposium of the Twenty-first Century. U.S.D.A. Agricultural Research Service, Albany, CA.

O'Donell, C. A. 1959. Las especies Americanas de "*Ipomoea*" L. sect. "*Quamoclit*" (Moench) Griseb. Lilloa 29:19–86.

O'Kane, S. L., Jr., and I. A. Al-Shehbaz. 1997. A synopsis of *Arabidopsis* (Brassicaceae). Novon 7:323–327.

———. 2002a. *Lesquerella* is united with *Physaria* (Brassicaceae). Novon 12:319–329.

———. 2002b. *Paysonia*, a new genus segregated from *Lesquerella* (Brassicaceae). Novon 12:379–381.

———. 2002c. Taxonomy and phylogeny of *Arabidopsis* (Brassicaceae). 22 pp. *in* American Society of Plant Biologists, eds., The *Arabidopsis* Book. doi: 10.1199/tab.0001. http://www.bioone.org/bioone/?request=get-toc&issn=1543-8120.

———. 2003. Phylogenetic position and generic limits of *Arabidopsis* (Brassicaceae) based on sequences of nuclear ribosomal DNA. Ann. Missouri Bot. Gard. 90:603–612.

O'Kane, S. L., Jr., I. A. Al-Shehbaz, and N. J. Turland. 1999. (1393) Proposal to conserve the name *Lesquerella* against *Physaria* (Cruciferae). Taxon 48:163–164.

Olsen, J. 1979. Taxonomy of the *Verbesina virginica* complex (Asteraceae). Sida 8:128–134.

Orchard, A. E., and B. R. Maslin. 2003. (1584) Proposal to conserve the name *Acacia* (Leguminosae: Mimosoideae) with a conserved type. Taxon 52:362–363.

Ørgaard, M. 1991. The genus *Cabomba* (Cabombaceae): A taxonomic study. Nordic J. Bot. 11:179–203.

Osborn, J. M., and E. L. Schneider. 1988. Morphological studies of the Nymphaeaceae sensu lato. XVI. The floral biology of *Brasenia schreberi*. Ann. Missouri Bot. Gard. 75:778–794.

Oskins, W. 1981. Who loves a swamp? Missouriensis 3 (3): 21–22.

Ownbey, G. B. 1951. Natural hybridization in the genus Cirsium—I. *C. discolor* (Muhl. ex Willd.) Spreng. × *C. muticum* Michx. Bull. Torrey Bot. Club 37:541–547.

———. 1952. Nuttall's Great Plains species of *Cirsium: C. undulatum* and *C. canescens*. Rhodora 54:29–35, pl. 1182–1183.

Ownbey, M. 1950. Natural hybridization and amphiploidy in the genus *Tragopogon*. Amer. J. Bot. 37:487–499.

Ownbey, M., and G. D. McCollum. 1953. Cytoplasmic inheritance and reciprocal amphiploidy in *Tragopogon*. Amer. J. Bot. 40:788–796.

———. 1954. The chromosomes of *Tragopogon*. Rhodora 56:7–21.
Oxelman, B., and M. Lidén. 1995. Genetic boundaries in the tribe Sileneae (Caryophyllaceae) as inferred from nuclear rDNA sequences. Taxon 44:525–542.
Oxelman, B., M. Lidén, and D. Berglund. 1997. Chloroplast *rps* 16 intron phylogeny of the tribe Sileneae (Caryophyllaceae). Pl. Syst. Evol. 206:393–410.
Oxelman, B., M. Backlund, and B. Bremer. 1999. Relationships of the Buddlejaceae s.l. investigated using parsimony jackknife and branch support analysis of chloroplast *ndhF* and *rbcL* sequence data. Syst. Bot. 24:164–182.
Oxelman, B., M. Lidén, R. K. Rabeler, and M. Popp. 2001. A revised generic classification of the tribe Sileneae (Caryophyllaceae). Nordic J. Bot. 20:743–748.
Paclt, J. 1998a. (1351) Proposal to amend the gender of *Euonymus,* nom. cons. (Celastraceae) to feminine. Taxon 47:473–474.
———. 1998b. (1378) Proposal to conserve the name *Celastrus* (Celastraceae) as being of feminine gender. Taxon 47:879–880.
Pak, J.-H., and K. Bremer. 1995. Phylogeny and reclassification of the genus *Lapsana* (Asteraceae: Lactuceae). Taxon 44:13–21.
Pallardy, S. G., T. A. Nigh, and H. E. Garrett. 1988. Changes in forest composition in central Missouri, 1968–1982. Amer. Midl. Naturalist 120:380–390.
———. 1991. Sugar maple invasion in oak forests of Missouri. Pp. 21–30 *in* G. V. Burger, J. E. Ebinger, and G. S. Wilhelm, eds., Proceedings of the Oak Woods Management Workshop. Eastern Illinois University, Charleston, IL.
Palmer, E. J. 1961. *Mentzelia albescens* and *Lonicera xylosteum* in Missouri. Rhodora 63:118–119.
Palmer, E. J., and J. A. Steyermark. 1935. An annotated catalogue of the flowering plants of Missouri. Ann. Missouri Bot. Gard. 22:375–758, 21 pl.
———. 1950. Notes on *Geocarpon minimum* Mackenzie. Bull. Torrey Bot. Club 77:268–273.
———. 1955. Plants new to Missouri. Rhodora 57:317.
Panero, J. L., and V. A. Funk. 2002. Toward a phylogenetic subfamilial classification for the Compositae (Asteraceae). Proc. Biol. Soc. Washington 115:909–922.
Park, K. 1998. Monograph of *Euphorbia* sect. *Tithymalopsis* (Euphorbiaceae). Edinburgh J. Bot. 55:161–208.
Park, K.-R., and W. J. Elisens. 2000. A phylogenetic study of tribe Euphorbieae (Euphorbiaceae). Int. J. Pl. Sci. 161:425–434.
Payne, W. W. 1964. A re-evaluation of the genus *Ambrosia* (Compositae). J. Arnold Arbor. 45:401–438.
———. 1970. Preliminary reports on the flora of Wisconsin no. 62. Compositae VI. Composite family VI. The genus *Ambrosia*—the ragweeds. Trans. Wisconsin Acad. Sci. 58:353–371.
Payne, W. W., and V. H. Jones. 1962. The taxonomic status and archaeological significance of a giant ragweed from prehistoric bluff shelters in the Ozark Plateau. Pap. Michigan Acad. Sci., Part 1 47:147–163.
Perdue, R. E., Jr. 1957. Synopsis of *Rudbeckia* subgenus *Rudbeckia*. Rhodora 59:293–299.
Perino, C. H. 1978. A revision of the genus *Lonicera* subgenus *Periclymenum* (Caprifoliaceae) in North America. PhD diss., North Carolina State University, Raleigh.
Perry, L. M. 1937. Notes on *Silphium*. Rhodora 39:281–297.
Pfeifer, H. W. 1966. Revision of the North and Central American hexandrous species of *Aristolochia* (Aristolochiaceae). Ann. Missouri Bot. Gard. 53:115–196.
Philbrick, C. T. 1984. Pollen tube growth within vegetative tissues of *Callitriche* (Callitrichaceae). Amer. J. Bot. 71:882–886.
———. 1989. Systematic studies of North American *Callitriche* (Callitrichaceae). PhD diss., University of Connecticut, Storrs.
Phillippe, L. R. 1978. A biosystematic study of *Sanicula* section *Sanicula*. PhD diss., University of Tennessee, Knoxville.
Pinkava, D. J. 1967. Biosystematic study of *Berlandiera* (Compositae). Brittonia 19:285–298.
———. 2003. Cactaceae Jussieu subfam. Opuntioideae Burnett. Pp. 102–150 *in* Flora of North America Editorial Committee, eds., Flora of North America, North of Mexico. Vol. 4. Magnoliophyta: Caryophyllidae, Part 1. Oxford University Press, New York.
Pireh, W., and R. J. Tyrl. 1980. Cytogeography of *Achillea millefolium* in Oklahoma and adjacent states. Rhodora 82:361–367.

Plunkett, G. M., D. E. Soltis, and P. S. Soltis. 1997. Clarification of the relationship between Apiaceae and Araliaceae based on *matK* and *rbcL* sequence data. Amer. J. Bot. 84:565–580.

Polhill, R. M., P. H. Raven, and C. H. Stirton. 1981. Evolution and systematics of the Leguminosae. Pp. 1–26 *in* R. M. Polhill and P. H. Raven, eds., Advances in Legume Systematics, Part 1. Royal Botanic Gardens, Kew.

Pons, A., and L. Boulos. 1972. Révision systématique du genre *Sonchus* L. s.l., III: Étude palynologique. Bot. Notis. 125:310–319.

Popov, M. G. 1974. Boraginaceae G. Don. Pp. 73–508 *in* B. K. Shiskin, ed., Flora of the U.S.S.R. Vol. 19. Tubiflorae. English translation by Israel Program for Scientific Translation, Jerusalem [originally published in Russian: Botanical Institute of the Academy of Sciences of the U.S.S.R. (Moscow) 19 (1953):98–691].

Powell, A. M. 1978. Systematics of *Flaveria* (Flaveriinae–Asteraceae). Ann. Missouri Bot. Gard. 65:590–636.

Pratt, D. B., and L. C. Clark. 2001. *Amaranthus rudis* and *A. tuberculatus*—one species or two? J. Torrey Bot. Soc. 128:282–296.

Prince, S. D., and R. N. Carter. 1977. Prickly lettuce (*Lactuca serriola* L.) in Britain. Watsonia 11:331–338.

Pringle, J. S. 2004. Nomenclature of the Virginia-bluebell, *Mertensia virginica* (Boraginaceae). Sida 21:771–775.

Pruski, J. F. 2005. Nomenclatural notes on *Cyclachaena* (Compositae: Heliantheae), resurrection and lectotypification of *Iva* sect. *Picrotus,* and report of *Iva* (syn. *Cyclachaena*) as new to Armenia. Compositae Newslett. 42:32–42.

Pryer, K. M., and L. R. Phillippe. 1989. A synopsis of the genus *Sanicula* (Apiaceae) in eastern Canada. Canad. J. Bot. 67:694–707.

Purseglove, J. W. 1969. Tropical Crops: Dicotyledons. Vol. 1. John Wiley and Sons, New York.

Pusateri, W. P., and W. H. Blackwell Jr. 1979. The *Echium vulgare* complex in eastern North America. Castanea 44:223–229.

Puttler, B., and W. C. Bailey. 2001. Biological and Integrated Control of Musk Thistle in Missouri. MU Guide IPM 1010. University of Missouri–Columbia.

Rabeler, R. K. 1985. *Petrorhagia* (Caryophyllaceae) in North America. Sida 11:6–44.

———. 1988. Eurasian introductions to the Michigan flora. IV. Two additional species of Caryophyllaceae in Michigan. Michigan Bot. 27:85–88.

———. 1992. A new combination in *Minuartia* (Caryophyllaceae). Sida 15:95–96.

———. 1993. Noteworthy collections: Arkansas. *Cerastium dubium* (Bast.) Guépin. Castanea 58:156.

———. 1996. *Sagina japonica* (Caryophyllaceae) in the Great Lakes region. Michigan Bot. 35:43–44.

Rabeler, R. K., and R. L. Hartman. 2005. Caryophyllaceae. Pp. 3–217 *in* Flora of North America Editorial Committee, eds., Flora of North America, North of Mexico. Vol. 5 Magnoliophyta: Caryophyllidae, Part 2. Oxford University Press, New York.

Rabeler, R. K., and J. W. Thieret. 1988. Comments on the Caryophyllaceae of the southeastern United States. Sida 13:149–156.

Radcliffe-Smith, A. 1985. Taxonomy of North American leafy spurge. Pp. 14–25 *in* A. K. Watson, ed., Leafy spurge. Monogr. Weed Sci. Soc. America 3:i–iv, 1–104.

———. 2001. Genera Euphorbiacearum. Royal Botanic Gardens, Kew.

Rauschert, S. 1974. Nomenklatorische Probleme in der Gattung *Matricaria* L. Folia Phytotax. Geobot. 9:249–260.

Ray, P. M., and H. F. Chisaki. 1957. Studies of *Amsinckia*. 1. A synopsis of the genus, with a study of heterostyly in it. Amer. J. Bot. 44:529–536.

Rebman, J. P. 1989. The vascular flora of Piney Creek Wilderness, Barry/Stone Counties, Missouri. Master's thesis, Southwest Missouri State University, Springfield.

Redfearn, P. L., Jr. 1980. *Silphium* hybrids. Missouriensis 2(2): 15.

Reveal, J. L. 1991a. Typification of six Philip Miller names of temperate North American *Toxicodendron* (Anacardiaceae) with proposals (999–1000) to reject *T. crenatum* and *T. volubile*. Taxon 40:333–335.

———. 1991b. *Rhus hirta* (L.) Sudworth, a newly revived correct name for *Rhus typhina* L. Taxon 40:489–492.

Reveal, J. L., and F. R. Barrie. 1992. *Matelea suberosa* (L.) Shinners (Asclepiadaceae)—Once again. Bartonia 57:36–38.

Richards, A. J. 1985. Sectional nomenclature in *Taraxacum* (Asteraceae). Taxon 34:633–644.

Richards, E. L. 1968. A monograph of the genus *Ratibida*. Rhodora 70:348–393.

Richardson, A. 1976. Reinstatement of the genus *Tiquilia* (Boraginaceae: Ehretioideae) and descriptions of four new species. Sida 6:235–240.

———. 1977. Monograph of the genus *Tiquilia* (*Coldenia,* sensu lato), Boraginaceae: Ehretioideae. Rhodora 79:467–572.

Richardson, J. W. 1968. The genus *Euphorbia* on the High Plains and Prairie Plains of Kansas, Nebraska, South and North Dakota. Univ. Kansas Sci. Bull. 48:45–112.

Rilke, S. 1999. Revision der Sektion *Salsola* s.l. der Gattung *Salsola* (Chenopodiaceae). Bibl. Bot. 149:1–190.

Robbins, H. C. 1968. The genus *Pachysandra* (Buxaceae). Sida 3:211–248.

Robbins, W. J. 1957. Physiological aspects of aging in plants. Amer. J. Bot. 44:289–294.

Roberts, M. L. 1985. The cytology, biology, and systematics of *Megalodonta beckii* (Compositae). Aquatic Bot. 21:99–110.

Robertson, C. 1889. Flowers and insects. III. Bot. Gaz. (Crawfordsville) 14:297–404.

Robertson, K. R. 1981. The genera of Amaranthaceae in the southeastern United States. J. Arnold Arbor. 62:267–314.

Robertson, K. R., and Y. T. Lee. 1976. The genera of Caesalpinioideae (Leguminosae) in the southeastern United States. J. Arnold Arbor. 57:1–53.

Robinson, H. 1978. Studies in the Heliantheae (Asteraceae). XII. Re-establishment of the genus *Smallanthus*. Phytologia 39:47–53.

———. 1999. Generic and subtribal classification of American Vernonieae. Smithsonian Contr. Bot. 89:i–iii, 1–116.

Robinson, R. W., and D. S. Decker-Walters. 1997. Cucurbits. Crop Production Science in Horticulture. Vol. 6. CAB International, Wallingford, Great Britain.

Robson, N. K. B. 1980. The Linnaean species of *Ascyrum* (Guttiferae). Taxon 29:267–274.

———. 1990. Studies in the genus *Hypericum* L. (Guttiferae) 8. Sections 29. *Brathys* (part 2) and 30. *Trigonobathys*. Bull. Brit. Mus. (Nat. Hist.), Bot. 20:1–151.

———. 1996. Studies in the genus *Hypericum* L. (Guttiferae) 6. Sections 20. *Myriandra* to 28. *Elodes*. Bull. Nat. Hist. Mus. London (Bot.) 26:75–217.

———. 2001. Studies in the genus *Hypericum* L. (Guttiferae) 4(1). Sections 7. *Roscyna* to 9. *Hypericum* sensu lato (part 1). Bull. Nat. Hist. Mus. London (Bot.) 31:37–88.

Rodman, J. E. 1990. Centrospermae revisited, part 1. Taxon 39:383–393.

———. 1993. Cladistic and phenetic studies. Pp. 279–301 in H.-D. Behnke and T. J. Mabry, eds., Caryophyllales: Evolution and Systematics. Springer-Verlag, Berlin.

Rodman, J. E., R. A. Price, K. Karol, E. Conti, K. J. Sytsma, and J. D. Palmer. 1993. Nucleotide sequences of the *rbcL* gene indicate monophyly of mustard oil plants. Ann. Missouri Bot. Gard. 80:686–699.

Roelfs, A. P. 1982. Effects of barberry eradication on stem rust in the United States. Pl. Dis. 66:177–181.

Rogers, C. B. 2003. Glyphosate resistance in horseweed *(Conyza canadensis)* from a western Kentucky farm. Proc. S. Weed Sci. Soc. 56:360 [abstract].

Rogers, K. 2000. The Magnificent Mesquite. University of Texas Press, Austin.

Rollins, R. C. 1942. A systematic study of *Iodanthus*. Contr. Dudley Herb. 3:209–215.

———. 1950. The guayule rubber plant and its relatives. Contr. Gray Herb. 172:1–72, 1 pl.

———. 1963. The evolution and systematics of *Leavenworthia* (Cruciferae). Contr. Gray Herb. 192:3–98.

———. 1993. The Cruciferae of Continental North America. Stanford University Press, Stanford, CA.

Rollins, R. C., and E. A. Shaw. 1973. The genus *Lesquerella* (Cruciferae) in North America. Harvard University Press, Cambridge.

Ronquist, F., and J. Liljeblad. 2001. Evolution of the gall wasp–host plant association. Evolution 55:2503–2522.

Rosatti, T. J. 1986. The genera of Sphenocleaceae and Campanulaceae in the southeastern United States. J. Arnold Arbor. 67:1–64.

———. 1989. The genera of suborder Apocynineae (Apocynaceae and Asclepiadaceae) in the southeastern United States. J. Arnold Arbor. 70:307–401, 443–514.

Roth, L., M. Daunderer, and K. Kormann, eds. 1994. Giftpflanzen, Pflanzengifte. 4th ed. Ecomed, Landsburg, Germany.

Roux, J., and L. Boulos. 1972. Révision systématique du genre *Sonchus* L. s.l., II: Étude caryologique. Bot. Notis. 125:306–309.

Ryder, E. J. 2002. The new salad crop revolution. Pp. 408–412 *in* J. Janick and A. Whipkey, eds., Trends in New Crops and New Uses. American Society for Horticultural Science Press, Alexandria, VA.

Samuel, R., H. Kathriarachchi, P. Hoffmann, M. H. J. Barfuss, K. J. Wurdack, C. C. Davis, and M. W. Chase. 2005. Molecular phylogenetics of Phyllanthaceae: Evidence from plastid *matK* and nuclear *PHYC* sequences. Amer. J. Bot. 92:132–141.

Sauer, J. D. 1955. Revision of the dioecious amaranths. Madroño 13:5–46.

———. 1957. Recent migration and evolution of the dioecious amaranths. Evolution 11:11–31.

———. 1967. The grain amaranths and their relatives: A revised taxonomic and geographic survey. Ann. Missouri Bot. Gard. 54:103–137.

———. 1972. The dioecious amaranths: A new species name and major range extensions. Madroño 21:426–434.

———. 1993. Historical Geography of Crop Plants: A Select Roster. CRC Press, Boca Raton, FL.

Sauer, J. D., and R. Davidson. 1961. Preliminary reports on the flora of Wisconsin. No. 45. Amaranthaceae—Amaranth family. Trans. Wisconsin Acad. Sci. 50:75–87.

Savolainen, V., M. F. Fay, D. C. Albach, A. Backlund, M. van der Bank, K. M. Cameron, S. A. Johnson, M. D. Lledó, J. C. Pintaud, M. Powell, M. C. Sheahan, D. E. Soltis, P. S. Soltis, P. Weston, W. M. Whitten, K. J. Wurdack, and M. W. Chase. 2000. Phylogeny of the eudicots: A nearly complete familial analysis based on *rbcL* gene sequences. Kew Bull. 55:257–309.

Schilling, E. E., J. L. Panero, and P. B. Cox. 1999. Chloroplast DNA restriction site data support a narrowed interpretation of *Eupatorium* (Asteraceae). Pl. Syst. Evol. 219:209–223.

Schmidt, G. J., and E. E. Schilling. 2000. Phylogeny and biogeography of *Eupatorium* (Asteraceae: Eupatorieae) based on nuclear ITS sequence data. Amer. J. Bot. 87:716–726.

Schmidt, K. A., and C. J. Whelan. 1999. Effects of exotic *Lonicera* and *Rhamnus* on songbird nest predation. Conservation Biol. 13:1502–1506.

Schnabel, A., and J. F. Wendel. 1998. Cladistic biogeography of *Gleditsia* (Leguminosae) based on *ndhF* and *rpL16* chloroplast gene sequences. Amer. J. Bot. 85:1753–1765.

Schneider, E. L., and J. M. Jeter. 1982. Morphological studies of the Nymphaeaceae. XII. The floral biology of *Cabomba caroliniana*. Amer. J. Bot. 69:1410–1419.

Schultes, R. E., and A. Hofmann. 1980. The Botany and Chemistry of Hallucinogens. 2d ed. Charles C. Thomas, Springfield, IL.

Schwegman, J. E., and R. W. Nyboer. 1985. The taxonomic and population status of *Boltonia decurrens* (Torrey and Gray) Wood. Castanea 50:112–115.

Schwegman, J. E., W. E. McClain, T. L. Esker, and J. E. Ebinger. 1998. Anthracnose-caused mortality of flowering dogwood (*Cornus florida* L.) at the Dean Hills Nature Preserve, Fayette County, Illinois, USA. Nat. Areas J. 18:204–207.

Scott, A. J. 1978. A revision of the Camphorosmioideae (Chenopodiaceae). Feddes Repert. 89:101–119.

Scott, R. W. 1990. The genera of Cardueae (Compositae; Asteraceae) in the southeastern United States. J. Arnold Arbor. 71:391–451.

Sell, P. D. 1993. *Scleranthus*. Pp. 178–179 *in* T. G. Tutin, N. A. Burges, A. O. Chater, J. R. Edmondson, V. H. Heywood, D. M. Moore, D. H. Valentine, S. M. Walters, and D. A. Webb, eds., Flora Europaea. Vol. 1. Psilotaceae to Platanaceae. 2d ed. Cambridge University Press, Cambridge.

Sell, P. D., and F. H. Whitehead. 1993. *Cerastium* (annual species). Pp. 164–175 *in* T. G. Tutin, N. A. Burges, A. O. Chater, J. R. Edmondson, V. H. Heywood, D. M. Moore, D. H. Valentine, S. M. Walters, and D. A. Webb, eds., Flora Europaea. Vol. 1. Psilotaceae to Platanaceae. 2d ed. Cambridge University Press, Cambridge.

Semple, J. C. 1977. Cytotaxonomy of *Chrysopsis* and *Heterotheca* (Compositae–Astereae): A new interpretation of phylogeny. Canad. J. Bot. 55:2503–2513.

———. 1978. The cytogeography of *Aster pilosus* (Compositae): Ontario and the adjacent United States. Canad. J. Bot. 56:1274–1279.

———. 1983. Range expansion of *Heterotheca camporum* (Compositae: Astereae) in the southeastern United States. Brittonia 35:140–146.

———. 1996. A revision of *Heterotheca* sect. *Phyllotheca* (Nutt.) Harms (Compositae: Astereae): The prairie and montane goldenasters of North America. Univ. Waterloo Biol. Ser. 37:i–iv, 1–164.

———. 2004. Miscellaneous nomenclatural changes in Astereae (Asteraceae). Sida 21:759–764.

Semple, J. C., and R. A. Brammall. 1982. Wild *Aster* ×*lateriflorus* hybrids in Ontario and comments on the origin of *A. ontarionis* (Compositae—Astereae). Canad. J. Bot. 60:1895–1906.

Semple, J. C., and J. G. Chmielewski. 1985. The cytogeography of *Aster pilosus* (Compositae—Astereae), II. Survey of the range, with notes on *A. depauperatus, A. parviceps,* and *A. porteri*. Rhodora 87:367–379.

———. 1987. Revision of the *Aster lanceolatus* complex, including *A. simplex* and *A. hesperius* (Compositae: Astereae): A multivariate morphometric study. Canad. J. Bot. 65:1047–1062.

Semple, J. C., and J. Ford. 1981. The phytogeography of leaf morphology in two species of North American asters: *Lasallea novae-angliae* and *L. oblongifolia* (Compositae). Brittonia 33:517–522.

Semple, J. C., V. Blok, and P. Heiman. 1980. Morphological, anatomical, habit, and habitat differences among the goldenaster genera *Chrysopsis, Heterotheca,* and *Pityopsis* (Compositae–Astereae). Canad. J. Bot. 58:147–163.

Semple, J. C., G. S. Ringius, C. Leeder, and G. Morton. 1984. Chromosome numbers of goldenrods, *Euthamia* and *Solidago* (Compositae: Astereae). II. Additional counts with comments on cytogeography. Brittonia 36:280–292.

Semple, J. C., J. G. Chmielewski, and R. A. Brammall. 1990. A multivariate morphometric study of *Solidago nemoralis* (Compositae: Astereae) and comparison with *S. californica* and *S. sparsiflora*. Canad. J. Bot. 68:2070–2082.

Semple, J. C., J. G. Chmielewski, and C. Leeder. 1991. A multivariate morphometric study and revision of *Aster* subg. *Doellingeria* sect. *Triplopappus* (Compositae: Astereae): The *Aster umbellatus* complex. Canad. J. Bot. 69:256–276.

Semple, J. G., J. Zhang, and C. Xiang. 1993. Chromosome number determinations in fam. Compositae, tribe Astereae. V. Eastern North American taxa. Rhodora 95:234–253.

Semple, J. C., G. S. Ringius, and J. J. Zhang. 1999. The goldenrods of Ontario: *Solidago* L. and *Euthamia* Nutt. 3d ed. Univ. Waterloo Biol. Ser. 39:i–vi, 1–90.

Semple, J. C., S. B. Heard, and L. Brouillet. 2002. Cultivated and native asters of Ontario (Compositae: Astereae): *Aster* L. (including *Asteromoea* Blume, *Diplactis* Raf. and *Kalimeris* (Cass.) Cass.), *Callistephus* Cass., *Galatella* Cass., *Doelleringia* Nees, *Oclemena* E.L. Greene, *Eurybia* (Cass.) S.F. Gray, *Canadanthus* Nesom, and *Symphyotrichum* Nees (including *Virgulus* Raf.). Univ. Waterloo Biol. Ser. 41:i–viii, 1–135.

Settergren, C., and R. E. McDermott. 1962. Trees of Missouri. University of Missouri–Columbia Agricultural Experiment Station, Columbia. [Reprinted several times without changes.]

Settle, W. J., and T. R. Fisher. 1970. The varieties of *Silphium integrifolium*. Rhodora 72:536–543.

Shan, R. H., and L. Constance. 1951. The genus *Sanicula* in the Old World and the New. Univ. Calif. Publ. Bot. 25:1–78.

Sheeley, S. E., and D. J. Raynal. 1996. The distribution and status of species of *Vincetoxicum* in eastern North America. Bull. Torrey Bot. Club 123:148–156.

Sherff, E. E. 1955. Compositae—Heliantheae—Coreopsidinae. N. Amer. Fl., ser. 2, 2:1–190.

Shetler, S. G. 1962. Notes on the life history of *Campanula americana,* the tall bellflower. Michigan Bot. 1:9–14.

Shildneck, P., and A. G. Jones. 1986. *Cerastium dubium* (Caryophyllaceae) new for the eastern half of North America (a comparison with sympatric *Cerastium* species, including cytological data). Castanea 51:49–55.

Shinners, L. H. 1946a. Revision of the genus *Chaetopappa* DC. Wrightia 1:63–81.

———. 1946b. Revision of the genus *Aphanostephus* DC. Wrightia 1:95–121.

———. 1946c. Revision of the genus *Kuhnia* L. Wrightia 1:122–144.

———. 1947. Revision of the genus *Krigia* Schreber. Wrightia 1:187–206.

———. 1949. Nomenclature of species of dandelion and goatsbeard (*Taraxacum* and *Tragopogon*) introduced into Texas. Field & Lab. 17:13–19.

———. 1950. The Texas species of *Thelesperma* (Compositae). Field & Lab. 18:17–24.

———. 1951. Revision of the North Texas species of *Heterotheca* including *Chrysopsis* (Compositae). Field & Lab. 19:66–71.

———. 1957. *Cayaponia quinqueloba* (Rafinesque) Shinners, comb. nov. Field & Lab. 25:32.

———. 1962. *Rhododendron nudiflorum* and *R. roseum* (Ericaceae): Illegitimate names. Castanea 27:94–95.

———. 1965a. *Ipomoea amnicola* (Convolvulaceae), a South American waif in Missouri. Rhodora 67:200.
———. 1965b. *Holosteum umbellatum* (Caryophyllaceae) in the United States: Population explosion and fractionated suicide. Sida 2:119–128.
———. 1971. *Kuhnia* L. transferred to *Brickellia* Ell. (Compositae). Sida 4:274.
Sieren, D. J. 1970. A taxonomic revision of the genus *Euthamia* (Compositae). PhD diss., University of Illinois, Urbana.
———. 1981. The taxonomy of the genus *Euthamia*. Rhodora 83:551–579.
Simpson, B. B., ed. 1977. Mesquite: Its Biology in Two Desert Scrub Ecosystems. US-IBP Synthesis Series 4. Dowden, Hutchinson, and Ross, Stroudsburg, PA.
Simpson, B. B. 1998. A revision of *Pomaria* (Fabaceae) in North America. Lundellia 1:46–71.
Simpson, B. B., and B. M. Miao. 1997. The circumscription of *Hoffmannseggia* (Fabaceae, Caesalpinioideae, Caesalpineae) and its allies using morphological and cpDNA restriction site data. Pl. Syst. Evol. 205:157–178.
Simurda, M. C., and J. S. Knox. 2000. ITS sequence evidence for the disjunct distribution between Virginia and Missouri of the narrow endemic *Helenium virginicum* (Asteraceae). J. Torrey Bot. Soc. 127:316–323.
Simurda, M. C., D. C. Marshall, and J. S. Knox. 2005. Phylogeography of the narrow endemic, *Helenium virginicum* (Asteraceae), based upon ITS sequence comparisons. Syst. Bot. 30:887–898.
Small, E. 1978. A numerical and nomenclatural analysis of morpho-geographic taxa of *Humulus*. Syst. Bot. 3:37–76.
———. 1979. The Species Problem in *Cannabis:* Science and Semantics. 2 vols. Corpus Publications, Toronto.
———. 1997. Cannabaceae Endlicher. Pp. 381–387 *in* Flora of North America Editorial Committee, eds., Flora of North America, North of Mexico. Vol. 3. Magnoliophyta: Magnoliidae and Hamamelidae. Oxford University Press, New York.
Small, E., and A. Cronquist. 1976. A practical and natural taxonomy for *Cannabis*. Taxon 25:405–435.
Small, J. K. 1903. Flora of the Southeastern United States. Published by the author, New York.
Smith, B. D., C. W. Cowan, and M. P. Hoffman. 1992. Is it indigene or a foreigner? Pp. 67–100 *in* B. D. Smith, Rivers of Change: Essays on Early Agriculture in Eastern North America. Smithsonian Institution Press, Washington, DC.
Smith, E. B. 1976. A biosystematic survey of *Coreopsis* in eastern United States and Canada. Sida 6:123–215.
———. 1988. An Atlas and Annotated List of Vascular Plants of Arkansas. 2d ed. Published by the author, Fayetteville.
———. 1994. Keys to the Flora of Arkansas. University of Arkansas Press, Fayetteville.
Smith, E. B., and H. M. Parker. 1971. A biosystematic study of *Coreopsis tinctoria* and *C. cardaminefolia* (Compositae). Brittonia 23:161–170.
Smith, M., and T. M. Keevin. 1998. Achene morphology, production and germination, and potential for water dispersal in *Boltonia decurrens* (decurrent false aster), a threatened floodplain species. Rhodora 100:69–81.
Smith, T., ed. 1993. Missouri Vegetation Management Manual. Missouri Department of Conservation, Jefferson City.
Solecki, M. K. 1983. Vascular plant communities and noteworthy taxa of Hawn State Park, Ste. Genevieve County, Missouri. Castanea 48:50–55.
Solomon, J. C. 1998. Specimen deaccessions from the Missouri Botanical Garden Herbarium during the tenure of Robert E. Woodson (1948–1963). Taxon 47:663–680.
Soltis, P. S., and D. E. Soltis. 1991. Multiple origins of the allotetraploid *Tragopogon mirus* (Compositae): rDNA evidence. Syst. Bot. 16:407–413.
Soltis, P. S., G. M. Plunkett, S. J. Novak, and D. E. Soltis. 1995. Genetic variation in *Tragopogon* species: Additional origins of the allotetraploids *T. mirus* and *T. miscellus* (Compositae). Amer. J. Bot. 82:1329–1341.
Soreng, R. J., and E. A. Cope. 1991. On the taxonomy of cultivated species of the *Chrysanthemum* genus-complex (Anthemidae; Compositae). Baileya 23:145–165.
Spellman, D. L., and D. B. Dunn. 1965. *Schmaltzia serotina* as a part of *Rhus trilobata*. Brittonia 17:286–288.

Spring, O., and E. E. Schilling. 1990. The origin of *Helianthus* ×*multiflorus* and *H.* ×*laetiflorus* (Asteraceae). Biochem. Syst. Ecol. 18:19–23.

Stahevich, A. E., C. W. Crompton, and W. A. Wojtas. 1988. Cytogenetic studies of leafy spurge, *Euphorbia esula,* and its allies (Euphorbiaceae). Canad. J. Bot. 66:2247–2257.

Stanford, N., and B. L. Turner. 1988. The natural distribution and biological status of *Helenium amarum* and *H. badium* (Asteraceae: Heliantheae). Phytologia 65:141–146.

Stearn, W. T. 1973. A synopsis of the genus *Vinca* including its taxonomic and nomenclatural history. Pp. 19–94 *in* W. I. Taylor and N. R. Farnsworth, eds., The *Vinca* Alkaloids: Botany, Chemistry, and Pharmacology. M. Dekker, New York.

Stebbins, G. L., Jr. 1939. Notes on *Lactuca* in western North America. Madroño 5:123–126.

———. 1993. *Leontodon.* Pp. 303–304 *in* J. C. Hickman, ed., The Jepson Manual: Higher Plants of California. University of California Press, Berkeley.

Stefanović, S., L. Krueger, and R. G. Olmstead. 2002. Monophyly of the Convolvulaceae and circumscription of their major lineages based on DNA sequences of multiple chloroplast loci. Amer. J. Bot. 89:1510–1522.

Steinmann, V. W., and J. M. Porter. 2002. Phylogenetic relationships in Euphorbieae (Euphorbiaceae) based on ITS and *ndhF* sequence data. Ann. Missouri Bot. Gard. 89:453–490.

Stevens, J. A. 1990. Response of *Campsis radicans* (Bignoniaceae) to simulated herbivory and ant visitation. Master's thesis, University of Missouri–St. Louis.

Stevens, P. F. 1971. A classification of the Ericaceae: Subfamilies and tribes. Bot. J. Linn. Soc. 64:1–53.

Steyermark, C. S. 1939. Distribution and hybridization of *Vernonia* in Missouri. Bot. Gaz. (Crawfordsville) 100:558–562.

Steyermark, J. A. 1934. Studies in *Grindelia.* II. A monograph of the North American species of the genus *Grindelia.* Ann. Missouri Bot. Gard 21:433–608.

———. 1941. A study of *Arenaria patula.* Rhodora 43:325–333.

———. 1950. Another station for *Geocarpon minimum.* Bull. Torrey Bot. Club 85:124–127.

———. 1952. New Missouri plant records (1949–51). Rhodora 54:250–260.

———. 1953. Another Coastal Plain relict in the Missouri Ozark region. Rhodora 55:15–17.

———. 1958. Another station for *Geocarpon minimum.* Bull. Torrey Bot. Club 85:124–127.

———. 1959. The taxonomic status of *Saxifraga palmeri.* Brittonia 11:71–77.

———. 1960. An unusual hybrid *Helenium.* Rhodora 62:343–346.

———. 1963. Flora of Missouri. Iowa State University Press, Ames. [With errata, printings 2–7, 1969–1996.]

Steyermark, J. A., J. W. Voigt, and R. H. Mohlenbrock. 1959. Present biological status of *Geocarpon minimum* Mackenzie. Bull. Torrey Bot. Club 86:228–235.

Stinner, B. R., and W. G. Abrahamson. 1979. Energetics of the *Solidago canadensis*–stem gall insect-parasitoid guild interaction. Ecology 60:918–926.

St. Pierre, M. D., R. J. Bayer, and I. M. Weis. 1990. An isozyme-based assessment of the genetic variability within the *Daucus carota* complex (Apiaceae: Caucalideae). Canad. J. Bot. 68:2449–2457.

Stratton, D. A. 1991. Life history variation within populations of an asexual plant, *Erigeron annuus* (Asteraceae). Amer. J. Bot. 78:723–728.

Stroh, G. 1935. *Myosotis micrantha* Pallas. Ein Beitrag zur Nomenklaturfrage. Notizbl. Bot. Gard. Berlin-Dahlem 12 (114): 471–473.

Strother, J. L. 1969. Systematics of *Dyssodia* Cavanilles (Compositae: Tageteae). Univ. Calif. Publ. Bot. 48:1–88.

———. 1986. Renovation of *Dyssodia* (Compositae: Tageteae). Sida 11:371–378.

———. 2000. *Hedosyne* (Compositae, Ambrosiinae), a new genus for *Iva ambrosiifolia.* Madroño 47:204–205.

Stuckey, R. L. 1972. Taxonomy and distribution of the genus *Rorippa* (Cruciferae) in North America. Sida 4:279–430.

Stuckey, R. L., and J. L. Forsyth. 1971. Distribution of naturalized *Carduus nutans* (Compositae) mapped in relation to geology in northwestern Ohio. Ohio J. Sci. 71:1–15.

Stuessy, T. F. 1972. Revision of the genus *Melampodium* (Compositae: Heliantheae). Rhodora 74:1–70, 161–219.

Suh, Y., and B. B. Simpson. 1990. Phylogenetic analysis of chloroplast DNA in North American *Gutierreza* and related genera (Asteraceae: Astereae). Syst. Bot. 15:660–670.
Suksdorf, W. 1931. Untersuchungen in der Gattung *Amsinckia*. Werenda 1:47–113.
Sullivan, J. 1991. Another alien has landed: *Evax prolifera* in Missouri. Missouriensis 12:1–3.
Sullivan, V. 1976. Diploidy, polyploidy, and agamospermy among species of *Eupatorium* (Compositae). Canad. J. Bot. 54:2907–2917.
———. 1978. Putative hybridization in the genus *Eupatorium* (Compositae). Rhodora 80:513–527.
Summers, B., and G. Yatskievych. 1990. *Aster macrophyllus* L. (Asteraceae), a new record for Missouri. Missouriensis 11:31–35.
Sundberg, S. D. 1986. The systematics of *Aster* subgenus *Tripolium* (Compositae) and historically allied species. PhD diss., University of Texas, Austin.
———. 2004. New combinations in North American *Symphyotrichum* subgenus *Astropolium* (Asteraceae: Astereae). Sida 21:903–910.
Sundell, E. 1981. The New World species of *Cynanchum* subgenus *Mellichampia* (Asclepiadaceae). Evol. Monogr. 5:1–62.
Susanna, A., N. Garcia-Jacas, D. E. Soltis, and P. S. Soltis. 1995. Phylogenetic relationships in tribe Cardueae (Asteraceae) based on ITS sequences. Amer. J. Bot. 82:1056–1068.
Sweeney, P. W., and R. A. Price. 2000. Polyphyly of the genus *Dentaria* (Brassicaceae): Evidence from *trnL* intron and *ndhF* sequence data. Syst. Bot. 25:468–478.
Swihart, B. 2000. Frost flowers. Missouri Conservationist 61(10): 10–13.
Swink, F., and G. Wilhelm. 1994. Plants of the Chicago Region. 4th ed. Indiana Academy of Science, Indianapolis.
Taylor, R. J. 1987. Populational variation and biosystematic interpretations in weedy dandelions. Bull. Torrey Bot. Club 114:109–120.
Tecic, D. L., J. L. McBride, M. L. Bowles, and D. L. Nickrent. 1998. Genetic variability in the federal threatened Mead's milkweed, *Asclepias meadii* Torrey (Asclepiadaceae), as determined by allozyme electrophoresis. Ann. Missouri Bot. Gard. 85:97–109.
Tenaglia, D., and G. Yatskievych. 2002. *Thlaspi alliaceum* (Brassicaceae), another non-native species new to Missouri. Missouriensis 23:39–41.
Theobald, W. L. 1966. The *Lomatium dasycarpum-mohavense-foeniculaceum* complex (Umbelliferae). Brittonia 18:1–18.
Thom, R. H., and J. H. Wilson. 1980. The Natural Divisions of Missouri. Trans. Missouri Acad. Sci. 14:9–23.
———. 1983. The natural divisions of Missouri. Natural Areas J. 3 (2): 44–51.
Thompson, R. L. 1980. Woody vegetation and floristic affinities of Mingo Wilderness Area, a northern terminus of southern floodplain forest, Missouri. Castanea 45:194–212.
Thompson, S. W., and T. G. Lammers. 1997. Phenetic analysis of morphological variation in the *Lobelia cardinalis* complex (Campanulaceae: Lobelioideae). Syst. Bot. 22:315–331.
Thorne, R. F. 1973. Inclusion of the Apiaceae (Umbelliferae) in the Araliaceae. Notes Roy. Bot. Gard. Edinburgh 32:161–165.
Thurman, C. J., and E. E. Hickey. 1989. Final Report: A Missouri Survey of Six Species of Federal Concern. Missouri Department of Conservation, Jefferson City.
Tomb, A. S. 1980. Taxonomy of *Lygodesmia* (Asteraceae). Syst. Bot. Monogr. 1:1–51.
Tooker, J. F., and L. M. Hanks. 2004. Endophytic insect communities of two prairie perennials (Asteraceae: *Silphium* spp.). Biodivers. & Conservation 13:2551–2566.
Trent, J. A. 1942. Studies pertaining to the life history of *Specularia perfoliata* (L.) A. DC., with special reference to cleistogamy. Trans. Kansas Acad. Sci. 45:152–164.
Tryon, R. M., Jr. 1939. The varieties of *Convolvulus spithamaeus* and of *C. sepium*. Rhodora 41:415–423, 2 pl.
Tucker, A. O., and N. H. Dill. 1989. Nomenclature and distribution of *Eupatorium* ×*truncatum*, with comments on the status of *E. resinosum* var. *kentuckiense* (Asteraceae). Castanea 54:43–48.
Tucker, G. C. 1986. The genera of Elatinaceae in the southeastern United States. J. Arnold Arbor. 67:471–483.
———. 2000. *Sagina* (Caryophyllaceae) in Illinois: An update. Rhodora 102:214–216.
Turland, N. 1996. Proposal to reject the name *Atriplex hastata* L. (Chenopodiaceae). Taxon 45:325–326.
Turland, N. J., and M. Wyse Jackson. 1997. Proposals to reject the names *Cerastium viscosum* and *C. vulgatum* (Caryophyllaceae). Taxon 46:775–778.

Turner, B. L. 1956. A cytotaxonomic study of the genus *Hymenopappus* (Compositae). Rhodora 58:163–186, 208–242, 250–269, 295–308.
———. 1959. The Legumes of Texas. University of Texas Press, Austin.
———. 1979. *Gaillardia aestivalis* var. *winkleri* (Asteraceae), a white-flowered tetraploid taxon endemic to southeastern Texas. SouthW. Naturalist 24:621–624.
———. 1984. Taxonomy of the genus *Aphanostephus* (Asteraceae—Astereae). Phytologia 56:81–101.
———. 1988a. Comments upon, and new combinations in, *Heliopsis* (Asteraceae, Heliantheae). Phytologia 64:337–339.
———. 1988b. A new species of, and observations on, the genus *Smallanthus* (Asteraceae: Heliantheae). Phytologia 64:405-409.
———. 1989. An overview of the *Brickellia (Kuhnia) eupatorioides* (Asteraceae, Eupatorieae) complex. Phytologia 67:121–131.
———. 1991. Texas species of *Ruellia* (Acanthaceae). Phytologia 71:281–289.
———. 1995. Synopsis of the genus *Onosmodium* (Boraginaceae). Phytologia 78:39–60.
———. 1996. Synoptical study of the *Acacia angustissima* (Mimosaceae) complex. Phytologia 81:10–15.
Turner, B. L., and K.-J. Kim. 1990. An overview of the genus *Pyrrhopappus* (Asteraceae: Lactuceae) with emphasis on chloroplast DNA restriction site data. Amer. J. Bot. 77:845–850.
Turner, B. L., and M. I. Morris. 1976. Systematics of *Palafoxia* (Asteraceae: Helenieae). Rhodora 78:567–628.
Turner, B. L., and M. Whalen. 1975. Taxonomic study of *Gaillardia pulchella* (Asteraceae–Heliantheae). Wrightia 5:189–195.
Turner, J., and G. Yatskievych. 1992. County record vouchers for vascular plant species newly recorded for Missouri since 1963. Missouriensis 13:1–24.
Tutin, T. G., and S. M. Walters. 1993. *Dianthus* Pp. 227–246 *in* T. G. Tutin, N. A. Burges, A. O. Chater, J. R. Edmondson, V. H. Heywood, D. M. Moore, D. H. Valentine, S. M. Walters, and D. A. Webb, eds., Flora Europaea. Vol. 1. Psilotaceae to Platanaceae. 2d ed. Cambridge University Press, Cambridge.
Tyrl, R. J. 1975. Origin and distribution of polyploid *Achillea* (Compositae) in western North America. Brittonia 27:187–196.
Unger, I. A. 1971. *Atriplex patula* var. *hastata* seed dimorphism. Rhodora 73:548–551.
Urbatsch, L. E. 1972. Systematic study of the *Altissimae* and *Giganteae* species groups of the genus *Vernonia* (Compositae). Brittonia 24:229–238.
Urbatsch, L. E., and R. K. Jansen. 1995. Phylogenetic affinities among and within the coneflower genera (Asteraceae, Heliantheae). Syst. Bot. 20:28–39.
Urbatsch, L. E., B. G. Baldwin, and M. J. Donoghue. 2000. Phylogeny of the coneflowers and relatives (Heliantheae: Asteraceae) based on nuclear rDNA internal transcribed spacer (ITS) sequences and chloroplast DNA restriction site data. Syst. Bot. 25:539–565.
Valentine, D., and A. J. Richards. 1967. Sexuality and apomixis in *Taraxacum*. Nature 214:114.
Van Alstine, N. E. 2000. Virginia Sneezeweed *(Helenium virginicum)* Recovery Plan. Technical/Agency Draft. U.S. Fish and Wildlife Service, Hadley, MA.
Van Bruggen, T. 1986. Crassulaceae DC. Pp. 356–358 *in* Great Plains Flora Association, eds., Flora of the Great Plains. University Press of Kansas, Lawrence.
Van Gessel, M. J. 2001. Glyphosate-resistant horseweed from Delaware. Weed Sci. 49:703–705.
Van Ham, R. C. H. J., and H. 't Hart. 1998. Phylogenetic relationships in the Crassulaceae inferred from chloroplast DNA restriction-site variation. Amer. J. Bot. 85:123–134.
Vander Kloet, S. P. 1978. The taxonomic status of *Vaccinium pallidum,* the hillside blueberries including *Vaccinium vacillans*. Canad. J. Bot. 56:1559–1574.
———. 1988. The Genus *Vaccinium* in North America. Publication 1828. Research Branch, Agriculture Canada, Ottawa, Ontario.
Voss, E. G. 1985. Michigan Flora. Part 2. Dicots (Saururaceae–Cornaceae). Bull. Cranbrook Inst. Sci. 59. Cranbrook Institute of Science, Bloomfield Hills, MI.
———. 1986. General committee report 1982–1985. Taxon 35:551–552.
———. 1996. Michigan Flora. Part 3. Dicots (Pyrolaceae-Compositae). Bull. Cranbrook Inst. Sci. 61. Cranbrook Institute of Science, Bloomfield Hills, MI.
Vuilleumier, B. S. 1973. The genera of Lactuceae (Compositae) in the southeastern United States. J. Arnold Arbor. 54:42–93.

Waddington, K. D. 1976. Pollination of *Apocynum sibiricum* (Apocynaceae) by Lepidoptera. SouthW. Naturalist 21:31–35.

Wagenitz, G., and F. H. Hellwig. 1996. Evolution of characters and phylogeny of the Centaureinae. Pp. 491–510 *in* D. J. N. Hind and H. J. Beentje, eds., Compositae: Systematics. Vol. 1. Royal Botanic Gardens, Gardens, Kew.

Wagenknecht, B. L. 1960. Revision of *Heterotheca* sect. *Heterotheca* (Compositae). Rhodora 62:61–76, 97–107.

Wagner, W. H., Jr. 1975. Notes on the floral biology of box-elder *(Acer negundo)*. Michigan Bot. 14:73–82.

Wagner, W. H., Jr., and T. F. Beals. 1958. Perennial ragweeds *(Ambrosia)* in Michigan, with the description of a new, intermediate taxon. Rhodora 60:177–204.

Wagner, W. L. 1979. Hitch-hiker from the West. Missouriensis 1 (3): 9–10.

Wahl, H. A. 1952–1953 [1954]. A preliminary study of the genus *Chenopodium* in North America. Bartonia 27:1–46.

Wallace, G. D. 1974. Studies of the Monotropoideae (Ericaceae): Taxonomy and distribution. PhD diss., Claremont Graduate School, Claremont, CA.

———. 1975. Interrelationships of the subfamilies of the Ericaceae and the derivation of the Monotropoideae. Bot. Notis. 128:286–298.

———. 1995. I. Ericaceae subfamily Monotropoideae. Pp. 13–27 *in* J. L. Luteyn, ed., Ericaceae. Part 2. The superior-ovaried genera. Fl. Neotrop. 66:1–560.

Ward, D. B. 1978. Keys to the flora of Florida—7, Campanulaceae. Phytologia 39:1–12.

Warners, D. P., and D. C. Laughlin. 1999. Evidence for a species-level distinction of two co-occurring asters: *Aster puniceus* and *Aster firmus*. Michigan Bot. 38:19–31.

Wasshausen, D. C. 1966. Acanthaceae J. St. Hil., Expos. Fam. 1:236. 1805. Pp. 223–282 *in* C. L. Lundell and collaborators, Flora of Texas. Vol. 1, part 3. Texas Research Foundation, Renner, TX.

———. 1998. Acanthaceae of the southeastern United States. Castanea 63:99–116.

Watson, L. E., and J. R. Estes. 1990. Biosystematic and phenetic analysis of *Marshallia* (Asteraceae). Syst. Bot. 15:403–414.

Webb, D. H. 1980. A biosystematic study of *Hypericum* section *Spachium* in eastern North America. PhD diss., University of Tennessee, Knoxville.

Webster, G. L. 1967. The genera of Euphorbiaceae in the south-eastern United States. J. Arnold Arbor. 48:303–361, 363–430.

———. 1970. A revision of the genus *Phyllanthus* (Euphorbiaceae) in the continental United States. Brittonia 22:44–76.

———. 1992. Realignments in American *Croton* (Euphorbiaceae). Novon 2:269–273.

———. 1994. Synopsis of the genera and suprageneric taxa of Euphorbiaceae. Ann. Missouri Bot. Gard. 81:33–144.

Weckman, T. J. 2002. Reinstatement of *Viburnum ozarkense* (Caprifoliaceae): An endemic taxon of the Interior Highlands of Arkansas, Missouri, and Oklahoma. Sida 20:849–860.

Wells, J. R. 1965. A taxonomic study of *Polymnia*. Brittonia 17:144–159.

———. 1969. A review of the varieties of *Polymnia uvedalia*. Rhodora 71:204–211.

Wen, J., and T. F. Stuessy. 1993. The phylogeny and biogeography of *Nyssa* (Cornaceae). Syst. Bot. 18:68–79.

Whitaker, T. W. 1944. The inheritance of certain characters in a cross of two American species of *Lactuca*. Bull. Torrey Bot. Club 71:347–355.

Whitaker, T. W., and G. N. Davis. 1962. Cucurbits: Botany, Cultivation, and Utilization. World Crops Books, Leonard Hill Limited, London.

White, O. E., and W. M. Bowden. 1947. Oriental and American bittersweet hybrids. J. Heredity 38:125–127.

Whitton, J., R. S. Wallace, and R. K. Jansen. 1995. Phylogenetic relationships and patterns of character change in the tribe Lactuceae (Asteraceae) based on chloroplast DNA restriction site variation. Canad. J. Bot. 73:1058–1073.

Wiegand, K. M. 1926. *Xanthium* P. 414 *in* K. M. Wiegand and A. J. Eames, The flora of the Cayuga Lake Basin, New York: Vascular plants. Mem. Cornell Univ. Agric. Exp. Sta. 92:1–414.

Wiese, M. 1979. From another birder. Missouriensis 1 (3): 13.

Wilbur, R. L. 1966. Notes on Rafinesque's species of *Lechea* (Cistaceae). Rhodora 68:192–208.

———. 1969. Cistaceae Juss. Pp. 1–17 *in* C. L. Lundell and collaborators, Flora of Texas. Vol. 1, part 1. Texas Research Foundation, Renner, TX.

———. 1970. Taxonomic and nomenclatural observations on the eastern North American genus *Asimina* (Annonaceae). J. Elisha Mitchell Sci. Soc. 86:88–96.

Wilbur, R. L., and H. S. Daoud. 1961. The genus *Lechea* (Cistaceae) in the southeastern United States. Rhodora 63:103–118.

Wilm, B. W., and J. B. Taft. 1997. *Trepocarpus aethusae* Nutt. (Apiaceae) in Illinois. Trans. Illinois State Acad. Sci. 91:53–56.

Wilmot-Dear, M. 1984. *Ceratophyllum* revised—A study in fruit and leaf variation. Kew Bull. 40:243–271.

Wilson, J. P. 1965. Variation in three complexes of the genus *Cornus* in eastern United States. Trans. Kansas Acad. Science 67:747–817.

Wilson, K. A. 1960. The genera of Convolvulaceae in the southeastern United States. J. Arnold Arbor. 41:298–317.

Witherspoon, J. T. 1974. Rediscovery of a hybrid *Lobelia* (Campanulaceae) in Missouri. SouthW. Naturalist 19:329.

Wofford, B. E. 1981. External seed morphology of *Arenaria* (Caryophyllaceae) of the Southeastern United States. Syst. Bot. 6:126–135.

Wojciechowski, M. E., M. Lavin, and M. J. Sanderson. 2004. A phylogeny of legumes (Leguminosae) based on analysis of the plastid *matK* gene resolves many well-supported subclades within the family. Amer. J. Bot. 91:1846–1862.

Wolfe, A. D., and J. R. Estes. 1992. Pollination and the function of floral parts in *Chamaecrista fasciculata* (Fabaceae). Amer. J. Bot. 79:314–317.

Wood, C. E., Jr. 1961. The genera of Ericaceae in the southeastern United States. J. Arnold Arbor. 42:10–80.

———. 1975. The Balsaminaceae in the southeastern U.S. J. Arnold Arbor. 56:413–426.

Wood, C. E., Jr., and P. Adams. 1976. The genera of Guttiferae (Clusiaceae) in the southeastern United States. J. Arnold Arbor. 57:74–90.

Woodson, R. E., Jr. 1928. Studies in the Apocynaceae. III. A monograph of the genus *Amsonia*. Ann. Missouri Bot. Gard. 15:379–434.

———. 1929. A new species of *Amsonia* from the south-central states. Ann. Missouri Bot. Gard. 16:407–410.

———. 1930. Studies in the Apocynaceae. I: A critical study of the Apocynoideae (with special reference to *Apocynum*). Ann. Missouri Bot. Gard. 17:1–212.

———. 1935. Studies in Apocynaceae. IV. The American genera of Echitoideae. Ann. Missouri Bot. Gard. 22:153–306.

———. 1941. The North American Asclepiadaceae. I. Perspective of the genera. Ann. Missouri Bot. Gard. 28:193–244.

———. 1943. A new *Amsonia* from the Ozarks of Arkansas. Rhodora 45:328–329.

———. 1954. The North American species of *Asclepias* L. Ann. Missouri Bot. Gard. 41:1–211, 3 pl.

———. 1962. Butterflyweed revisited. Evolution 16:168–185,

———. 1964. The geography of flower color in butterflyweed. Evolution 18:143–163, 1 pl.

Wuenscher, J. E., and A. J. Valiunas. 1967. Presettlement forest composition of the River Hills region of Missouri. Amer. Midl. Naturalist 78:487–95.

Wunderlin, R. P. 1972. New combinations in Compositae. Ann. Missouri Bot. Gard. 59:471–473.

Wurdack, K. J., P. Hoffmann, R. Samuel, A. de Bruijn, M. van der Bank, and M. W. Chase. 2004. Molecular phylogenetic analysis of Phyllanthaceae (Phyllanthoideae pro parte, Euphorbiaceae sensu lato) using plastid *rbcL* DNA sequences. Amer. J. Bot. 91:1882–1900.

Wurdack, K. J., P. Hoffmann, and M. W. Chase. 2005. Molecular phylogenetic analysis of uniovulate Euphorbiaceae (Euphorbiaceae sensu stricto) using plastid *rbcL* and *trnL-F* DNA sequences. Amer. J. Bot. 92:1397–1420.

Xiang, Q. Y., D. E. Soltis, and P. S. Soltis. 1998. Phylogenetic relationships of Cornaceae and close relatives inferred from *matK* and *rbcL* sequences. Amer. J. Bot. 85:285–297.

Yatskievych, G. 1990. Studies in the flora of Missouri, II. Missouriensis 11:2–6.

———. 1999. Steyermark's Flora of Missouri. Rev. ed., vol. 1. Missouri Department of Conservation, Jefferson City.

Yatskievych, G., and D. Figg. 1989. Studies in the flora of Missouri, I. New records for introduced taxa. Missouriensis 10:16–19.

Yatskievych, G., and R. Spellenberg. 1993. Plant conservation. Pp. 207–226 *in* Flora of North America Editorial Committee, eds., Flora of North America, North of Mexico. Vol. 1. Oxford University Press, New York.

Yatskievych, G., and B. Summers. 1991. Studies in the flora of Missouri, III. Missouriensis 12:4–11.

———. 1993. Studies in the flora of Missouri, IV. Missouriensis 14:27–42.

Yatskievych, G., and J. Turner. 1990. Catalogue of the flora of Missouri. Monogr. Syst. Bot. Missouri Bot. Gard. 37:i–xii, 1–345.

Yuncker, T. G. 1943. Convolvulaceae: *Cuscuta*. Pp. 123–150 *in* C. L. Lundell and collaborators, Flora of Texas. Vol. 3. Texas Research Foundation, Renner.

Zhang, J. J. 1996. A molecular biosystematic study on North American *Solidago* and related genera (Asteraceae: Astereae) based on chloroplast DNA RFLP analysis. PhD diss., University of Waterloo, ON, Canada.

Zijlstra, G., and J. Tolsma. 1991. (997) Proposal to conserve the spelling of (4618) *Euonymus* (Celastraceae). Taxon 40:137–139.

INDEX TO VOLUME 2

Principal entries for accepted scientific names have the page numbers in **boldface**. For scientific names treated as synonyms in the principal entries, both the names and page numbers are in *italics*. Vernacular names and all other occurrences of scientific names are in plain type.

A

Abelia, 765
Absinthe, 179, 180
Acacia, 1075, 1076, 1078
  angustissima, 1078
    var. hirta, 1078
    var. texensis, 1078
  hirta, 1078
  penninervis, 1076
  section Filicinae, 1078
  subgenus Aculeiferum, 1078
  subgenus Phyllodinae, 1076, 1078
Acacia, prairie, 1078
Acaciella, **1076,** 1078
  angustissima (Map 1699, Pl. 387 f–h), **1078**
  hirta, 1078
Acalypha, 943, **1012**
  deamii (Map 1649, Pl. 376 i, j), **1013**
  gracilens (Map 1650), **1013,** 1014, 1016, 1017
    ssp. monococca, 1014
    var. delzii, 1013
    var. fraseri, 1013
    var. monococca, 1014
  hispida, 1012
  monococca (Map 1651, Pl. 376 g), 1012, **1014**
  ostryifolia (Map 1652, Pl. 376 d–f), **1016**
  rhomboidea (Map 1653, Pl. 376 a–c), 1013, **1016,** 1017
    var.deamii, 1013
  virginica (Map 1654, Pl. 376 h), 1013, 1014, **1017**
    var. deamii, 1013
    var. rhomboidea, 1016
Acanthaceae, **2**
Acanthus family, 2
Acer, **7,** 10
  barbatum, 15
  floridanum, 15
  ginnala (Map 804), **9**
  negundo (Map 805, Pl. 196 a, b), 8, **9,** 48
    var.interius, 10, 12
    var. negundo, **10,** 12
    var. texanum, **10,** 12
    var. violaceum, 10
  nigrum, 15
    f. pubescen, 15

Acer nigrum (cont.)
    var. floridanum, 15
    var. glaucum, 16
    var. palmeri, 15
    var. schneckii, 16
  palmatum, 8
  pseudoplatanus, 8
  rubrum (Map 806, Pl. 196 c–f), 8, 9, **12**
    f. rotundata, 12
    f. tomentosum, 13
    var. drummondii, **13**
    var. rubrum, **13**
    var. tridens, 13
    var. trilobum, 12
  saccharinum (Map 807, Pl. 196 g–l), 8, **13**
  saccharum (Map 808, Pl. 196 m–o), 7, 8, 12, **13,** 14
    f. glaucum, 16
    f. rugelii, 16
    f. schneckii, 16
    ssp. floridanum, **15,** 16
    ssp. nigrum (Pl. 196 o), **15,** 16
    ssp. saccharum (Pl. 196 m, n), 15, **16**
    ssp. schneckii, 15, **16**
    var. floridanum, 15
    var. nigrum, 15
    var. schneckii, 16
    var. viride, 15
  tataricum, 9
    ssp. ginnala, 9
Aceraceae, **7,** 8
Acerates
  angustifolia, 152
  hirtella, 145
  viridiflora, 157
Achillea, **173,** 943
  millefolium (Map 937, Pl. 224 a, b), **173,** 174
    f. roseum, 174
    ssp. lanulosa, 173
    var. lanulosa, 173
  ptarmica (Map 938), **174,** 176
Acmella, **441**
  oppositifolia (Map 1141. Pl. 272 a, b), **441,** 442
    var. oppositifolia, 442
    var. repens, **441,** 442
  repens, 441
Acnida
  tamariscina
    var. prostrata, 32

Acnida tamariscina (cont.)
    var. tuberculata, 32
Acorn squash, 982, 983
Acroptilon, 322
  repens, 330
Actinomeris alternifolia, 562
Acuan
  illinoense, 1080
  leptolobum, 1081
Adoxa, 765
Adoxaceae, 765
Aegopodium
  podagraria (Pl. 202 c, d), 55
    var. variegatum, 55
African marigold, 556
Agaloma corollata, 1034
Ageratina, **394,** 401, 403
  altissima (Map 1104, Pl. 267 f, g), **395,** 410
    var. altissima, **395**
    var. angustatumn, 395
    var. roanensis, 395
Ageratum, **396,** 400, 401, 412
  conyzoides (Map 1105, Pl. 264 g, h), **396,** 412
    ssp. conyzoides, **396**
    ssp. latifolium, 396
Agoseris, 374
  cuspidata, 374
Agrimonia, 943
Agrostemma, 795, **797,** 829
  githago (Map 1447, Pl. 339 a, b), **797**
    var. githago, **797**
Aizoaceae, **17,** 795
Al Shehbaz, Ihsan A., 651
Albizia, **1079**
  julibrissin (Map 1700, Pl. 387 i–l), **1079**
Alder, 608, 609
  Black, 129, 609
  Common, 609
  European, 609
  Smooth, 609
  Tag, 609
Aleurites fordii, 1011
Alexanders
  Golden, 112
  Common golden, 114
  Heart-leaved golden, 114
Alfalfa, 1058
Alfalfa dodder, Large, 943
Alkali aster, Rayless, 284
Alkali goldentops, 482

Alkanet, 626
  Bastard, 628
Allamanda, 115
Allegheny spurge, 731
Alliaria, **658**
  officinalis, 658
  petiolata (Map 1309, Pl. 311 i, j), **658**
Allspice, Carolina, 740
Alnus, **608,** 940
  glutinosa (Map 1274), **609**
  serrulata (Map 1275, Pl. 303 e–g), **609**
    f. noveboracensis, 610
Alsinopsis patula, 818
Alternate-leaved dogwood, 959
Alyssum, **658,** 659
  alyssoides (Map 1310, Pl. 311 g, h), **659**
  desertorum (Map 1311, Pl. 311 c), **659**
  maritimum, 708
Alyssum, 658
  Hoary, 667
  Pale, 659
  Purple, 666
  Sweet, 708
Amanita, 343
Amaranth, 18, 19
  Cereal, 20
  Globe, 18
  Grain, 20
  Green, 28
  Prostrate, 24
  Purple, 25
  Rough green, 30
  Sandhills, 23
  Weedy, 20
Amaranth family, 18
Amaranthaceae, **18,** 795, 860
Amaranthus, **19,** 20, 24, 245, 871, 943, 944
  albus (Map 810, Pl. 198 a, b), **23**
    var. pubescens, 23
  arenicola (Map 811, Pl. 197 e, f), **23**
  blitoides (Map 812, Pl. 198 f), **24**
  blitum (Map 813) , **24,** 34
    ssp. blitum, 25
    ssp. emarginatus, **24**
    ssp. oleraceus, 25
    var. emarginatus, 25
    var. pseudogracilis, 25
  caudatus (Map 814, Pl. 198 e), 19, **25,** 26, 29
  cruentus (Map 815), 19, **26,** 29
  gracilis, 34
  graecizans, 24
  hybridus (Map 816, Pl. 198 g, h), 20, 26, **28,** 34
  hybridus × palmeri, 20
  hybridus × retroflexus, 20
  hybridus × rudis, 20

Amaranthus (cont.)
  hybridus × tuberculatus, 20
  hybridus var. hypochondriacus, 26, 28
  hypochondriacus (Map 817), 19, 26, **28,** 29
  lividus, 25
  palmeri (Map 818, Pl. 197 c, d), **29**
  powellii (Map 819, Pl. 197 h), **29**
    ssp. powellii, **29**
  retroflexus (Map 820, Pl. 197 g), **30**
  retroflexus × rudis, 20
  rudis, 20, *32*
  spinosus (Map 821, Pl. 198 c, d), 18, **31**
  section Blitopsis, 20
  subgenus Acnida, 20
  subgenus Albersia, 20
  subgenus Amaranthus, 20
  tamariscinus, 34
  torreyi, 24
  tricolor (Map 822), 20, **31**
    ssp. tricolor, 32
  tuberculatus (Map 823, Pl. 197 a, b, i–m), 20, **32,** 34
    var. prostrata, 34
    var. rudis, 32, 34
    var. tuberculatus, 34
  viridis (Map 824), 25, **34**
Ambrina
  ambrosioides, 875
  anthelmintica, 876
Ambrosia, 165, 245, 436, **442,** 443, 519, 568, 942, 943
  acanthicarpa (Map 1142, Pl. 271 f–h), **443**
  artemisiifolia (Map 1143, Pl. 271 d, e), **444**
    f. villosa, 444
    var. elatior, 444
  bidentata (Map 1144, Pl. 271 a–c), **444,** 446, 448
  coronopifolia, 446
  ×helenae, 444
  ×intergradiens, 444
  psilostachya (Map 1145, Pl. 272 f, g), **444,** 446
  tomentosa (Map 1146, Pl. 272 d, e), **446**
  trifida (Map 1147, Pl. 271 i, j), 444, 446, **448**
    f. integrifolia, 448
    var. texana, 448
American Energy Farming Systems, 516
American barberry, 604
American basket flower, 323
American bittersweet, 851
American bugseed, 894
American cress, 666
American feverfew, 526

American ginseng, 135, 136
American gromwell, 640
American hazelnut, 612
American holly, 128, 129
American smoke tree, 41
American spikenard, 132
Ammi, **62**
  majus (Map 838, Pl. 202 a, b), **62,** 63
Ammoselinum, **63**
  butleri (Map 839), **63**
  popei, 63
Ampelamus, 159
  albidus, 159
  laevis, 159
Amphiachyris, **202,** 237
  dracunculoides (Map 959, Pl. 228 a, b), **202**
Amphicarpaea, 943
Amsinckia, **624**
  barbata, 624
  hispida, 624, *625*
  idahoensis, 624
  lycopsoides (Map 1285, Pl. 305 a–c), **624**
  menziesii (Map 1286, Pl. 305 d–f), **625**
  micrantha, 625
  parviflora, 624
  retrorsa, 625
  tessellata (Map 1287, Pl. 305 j–l), **625**
Amsonia, **116**
  angustifolia, 117
  ciliata (Map 895, Pl. 215 f), **117**
    var. ciliata, 117
    var. filifolia, **117**
    var. tenuifolia, 117
    var. texana, 117
  illustris (Map 896, Pl. 215 c–e), **117,** 118
  salicifolia, 118
  tabernaemontana (Map 897, Pl. 215 h, i), **117**
    var. gattingeri, 118
    var. salicifolia, **118**
    var. tabernaemontana, **118**
Amsonia, Willow, 117, 118
Amur honeysuckle, 772
Amur maple, 9
Anacardiaceae, **40**
Anacardium, 40
  occidentale, 40
Anaphalis, **426**
  margaritacea (Map 1131, Pl. 293 a–d), **426**
Anchusa, **626**
  azurea (Map 1288), **626**
  canescens, 638
  officinalis (Map 1289), **626**
Andrachne, 1010, **1018**
  arida, 1018

Andrachne (cont.)
 phyllanthoides (Map 1655, Pl. 377 a–d), **1018**
Andrenid bee, 1001, 1007
Anethum, **63**
 graveolens (Map 840, Pl. 203 h, i), 55, **64**
Angelica, **64**
 venenosa (Map 841, Pl. 203 a, b), **64,** 66
Angelica
 Hairy, 64
 Wood, 64
Angelico, 88
Angiosperm, 1
Angle-pod, 160
Anise, 64, 84, 94, 180
 Sweet, 94
Anise root, 94
Anisillo, 556
Anisostichus capreolata, 615
Annonaceae, 1, **51,** 52, 1069
Annual bursage, 443
Annual fleabane, 220
Annual knawel, 828
Annual wormwood, 180
Ant, 616, 1065
Antennaria, **426,** 427
 howellii, 428
 longifolia, 428
 neglecta (Map 1132, Pl. 293 e–h), **428**
  var. campestris, 428
 neodioica, 428
 parlinii (Map 1133, Pl. 293 i–m), 427, **428,** 429, 433
  ssp. fallax, **429**
  ssp. parlinii, **429**
 plantaginifolia (Pl. 293 n, o), 428, 429
  var. ambigens, 429
  var. arnoglossa, 429
 racemosa, 428
 solitaria, 427
Anthemis, **176**
 americana, 442
 arvensis (Map 939, Pl. 224 f), **177**
  var. agrestis, 177
  var. arvenstis, 177
 cotula (Map 940, Pl. 224 c–e), **177**
 tinctoria, 176, 188
Anthriscus caucalis, **66**
Anthriscus
 caucalis (Map 842, Pl. 203 f, g), **66**
 scandicina, 66
Antistrophus, 374, 548
 pisum, 374
Anychia
 canadensis, 821
 fastigiata, 821

Aphanostephanus, **202,** 219
 skirrhobasis (Map 960), **202**
  var. kidderi, 204
  var. skirrhobasis, **203,** 204
  var. thalassius, 203
Apiaceae, **54,** 55, 64, 218
 Subfamily Hydrocotyloideae, 55
Apios, 940, 1058
Apium, **66**
 graveolens (Map 843, Pl. 203 c, d), 55, **66**
  var. dulce, 67
  var. graveolens, 67
Apocynaceae, **115,** 116, 141
Apocynum, **120,** 121, 943
 ambigens, 121
 androsaemifolium (Map 898, Pl. 216 j, k), **121,** 122
  var. androsaemifolium, 121
 androsaemifolium ×
  cannabinum, 121
 cannabinum (Map 899, Pl. 215 a, b, 216 f–h), 120, 121, **122,** 124
  var. glaberimmum, 122
  var. pubescens, 122
 cordigerum, 122
 ×floribundum, 120, 121, 122
 hypericifolium, 122
  var. cordigerum, 122
 leuconeuron, 122
 medium, 121
 ×medium, 122
 missouriense, 122
 sibiricum, 120, 122
  var. cordigerum, 122
Appalachian whitlow wort, 822
Apple, May, 607
Aquifoliaceae, **126**
Arabidopsis, **660,** 662
 lyrata (Map 1312, Pl. 311 d–f), **660**
  ssp. kamchatika, 660
  ssp. lyrata, **660**
 thaliana (Map 1313, Pl. 311 a, b), **660,** 662
Arabis, **662,** 731
 canadensis, 668
 dentata, 669
 glabra, 731
 hirsuta (Map 1314, Pl. 312 a, b), **662,** 664
  var. adpressipilis, **664**
  var. eschscholtziana, 664
  var. glabrata, 664
  var. pycnocarpa, **664**
 laevigata, 668
  var. missouriensis, 669
 lyrata, 660
  f. parvisiliqua, 660
 missouriensis, 669
  var. deamii, 669
 shortii, 669

Arabis shortii (cont.)
 var. phalacrocarpa, 669, 670
 virginica, 715
Arachis, 1058
Aralia, **131,** 940
 nudicaulis (Map 907, Pl. 218 b), **132**
 racemosa (Map 908, Pl. 218 a), **132**
 spinosa (Map 909, Pl. 218 c–f), **132,** 134
Araliaceae, 54, 55, **131**
Arbutus Trailing, 1000
Archilochilus colubris, 600
Arctium, **317**
 minus (Map 1037, Pl. 250 a–c), **317**
  f. pallidum, 318
  f. purpureum, 318
 tomentosum, 318
Arctostaphylos, 1000
Arenaria, 795, **797,** 817, 819
 lateriflora, 819
 leptoclados, 798
 muriculata, 818
 patula, 818
  f. media, 818
  f. pitcheri, 818
  f. robusta, 818
  var. robusta, 818
 serpyllifolia (Map 1448, Pl. 339 c, d), **798**
  ssp. leptoclados, 798
  var. serpyllifolia, **798**
  var. tenuior, **798**
 stricta, 817
  ssp. texana, 817, 818
  var. texana, 817, 818
Aristolochia, **138**
 reticulata, 139
 serpentaria (Map 912, Pl. 219 h, i), **138,** 139
  var. hastata, 138, 139
 tomentosa (Map 913, Pl. 219 a, b), **139**
Aristolochiaceae, 1, **138**
Arkansas lazydaisy
Arkansas leastdaisy, 211
Arkezostis quinqueloba, 976
Armoracia, **664,** 720
 aquatica, 719
 lacustris, 719
 rusticana (Map 1315, Pl. 313 c–e), **664,** 679
Arnoglossum, 576, **577**
 atriplicifolium (Map 1246, Pl. 296 a, b), **577,** 578, 590
 plantagineum (Map 1247, Pl. 296 h, i), **578**
 reniforme (Map 1248, Pl. 296 c–e), 577, **578**
Aromatic aster, 297
Aromatic sumac, 42

Arrowwood, 784
 Downy, 792
 Missouri, 789
 Northern, 792
 Ozark, 791
 Southern, 786
Artemisia, 172, **177,** 178, 180, 186, 187
 absinthium (Map 941, Pl. 226 a, b), **179**
 annua (Map 942, Pl. 225 a, b), , 178, **180**
 biennis (Map 943, Pl. 225 c, d), **180**
 campestris (Map 944), **182**
  ssp. campestris, 182
  ssp. caudata, **182**
  var. caudata, 182
 carruthii (Map 945, Pl. 226 e), **182**
 caudata, 182
 cernua, 184, 185
 douglasiana, 187
 dracunculoides, 185
 dracunculus (Map 946, Pl. 225 e, f), **184,** 185
  ssp. dracunculina, 184, 185
  ssp. glauca, 184, 185
 frigida (Map 947, Pl. 226 h, i), **185**
 glauca, 184, 185
  var. dracunculina, 184
 indica, 188
 ludoviciana (Map 948, Pl. 226 f, g), **185,** 187
  ssp. mexicana, 187
  ssp. redolens, 187
  var. gnaphalodes, 186
  var. ludoviciana, **186,** 187
  var. mexicana, **187**
  var. redolens, 187
 redolens, 187
 serrata, 186
 stellariana (Map 949), **187**
 vulgaris (Map 950, Pl. 226 c, d), **187,** 188
  var. glabra, 187
  var.kamtschatica, 188
  var. latiloba, 187
Artichoke, 316
 Globe, 316
 Jerusalem, 436, 515, 516
Arugula, 692
Asarum, **140**
 canadense (Map 914, Pl. 219 f, g), **140**
  var. acuminatum, 140
  var. ambiguum, 140
  var. reflexum, 140
Asclepiadaceae, 115, **140,** 141, 142, 159
Asclepias, **142,** 143, 939, 942

Asclepias (cont.)
 amplexicaulis (Map 915, Pl. 221 a, b), 143, **145,** 154
 exaltata, 143
 hirtella (Map 916, Pl. 221 c-e), **145**
 incarnata (Map 917, Pl. 220 g, h), 143, **146**
  f. albiflora, 146
  ssp. incarnata, **146**
  ssp. pulchra, 146
 longifolia
  ssp. hirtella, 145
 meadii (Map 918, Pl. 221 j, k), **146**
 perennis (Map 919, Pl. 221 h, i), 142, **148,** 149
 purpurascens (Map 920, Pl. 221 f, g), **149,** 154
 quadrifolia (Map 921, Pl. 220 a, b), **149**
 speciosa (Map 922), **150,** 152, 154
 stenophylla (Map 923, Pl. 222 c, d), **152**
 subverticillata (Map 924, Pl. 222 a), **152**
 sullivantii (Map 925, Pl. 220 i), **153,** 154
 syriaca (Map 926, Pl. 220 j, k), 149, 150, 152, **153**
  f. leucantha, 154
  f. inermis, 154
  var. kansana, 153
 tuberosa (Map 927, Pl. 220 c-f), 140, 142, **154,** 156
  f. lutea, 154
  ssp. interior, **154**
  ssp. rolfsii, 156
  ssp. terminalis, 156
  ssp. tuberosa, 156
  var. interior, 154
 variegata (Map 928, Pl. 222 g, h), **156**
 verticillata (Map 929, Pl. 222 b), 153, 154, **157**
  var. subverticillata, 152
 viridiflora (Map 930, Pl. 222 e, f), **157,** 158
  var. lanceolata, 157
  var. linearis, 157
 viridis (Map 931, Pl. 222 i, j), 142, **158**
Asclepiodora viridis, 158
Ascyrum, 920
 hypericoides, 918
  var. multicaule, 920
Ash-leaved maple, 9
Ashy sunflower, 506
Asimina triloba (Map 837, Pl. 202 g-i), **51,** 52, 1069
Aster, 198, **204,** 216, 226, 243, 278, 286, 315

Aster (cont.)
 alpinus, 203
 ×amethystinus, 290
 anomalus, 282
  f. albidus, 284
 azureus, 292, *300,* 301
  f. laevicaulis, 302
  var. poaceus, 302
 brachyactis, 284
 commutatus, 290
 cordifolius, 285, 286, 315
  ssp. sagittifolius, 285
  var. moratus, 286
  var. polycephalus, 285, 286
 drummondii, 286
 dumosus, 315
  var. dodgei, 287
  var. strictior, 287
 ericoides, 288, 290
  f. caeruleus, 290
  f. prostraus, 290
  var. prostratus, 290
 exilis, 311
 falcatus 290
  var. commutatus, 290
 firmus, 308, 309
 fragilis, 288, 296, *310,* 311
  var. subdumosus, 310
 furcatus, 226
 hemisphericus, 227
 hesperius, 294
 laevis, 291
  f. latifolius, 292
  var. geyeri, 292
 lanceolatus, 292, 294
  ssp. hesperius, 294
  var. interior, 295
  var. latifolius, 295
 lateriflorus, 295
 linariifolius, 243
  f. leucactis, 243
 lucidulus, 308
 macrophyllus, 227
 novae-angliae, 296
 oblongifolius, 297
  f. roseoligulatus, 298
  var. angustatus, 297
 ontarionis, 298
 oolantangiensis, 300, 301
  var. laevicaulis, 301, 302
  var. poaceus, 302
 paludosus, 227
  ssp. hemisphericus, 227
  var. hemisphericus, 227
 parviceps, 302
 patens, 303
  var. gracilis, 304
  var. patens, 304
  var. patentissimus, 304
 pilosus, 305
  f. pulchellus, 306
  ssp. parviceps, 302
  var. demotus, 305, 306

Aster pilosus (cont.)
　var. *platyphyllus, 305,* 306
　var. pilosus, 306
　var. *pringlei, 305*
praealtus, 306, 308
　var. *angustior, 306,* 308
　var. *subasper, 306*
ptarmicoides, 204, *269*
pubentior, 218
puniceus, 308, 309
　f. *lucidulus, 308*
　ssp. *firmus, 308*
　var. *firmus, 308*
sagittifolius, 285, 286, 292, *314,* 315
　f. *hirtellus, 314*
　var. *drummondii, 286*
sericeus, 311
　f. *albiligulatus,* 311
simplex, 292
　var. *interior, 295*
　var. *ramosissimus, 295*
subulatus
　var. *ligulatus,* 311
subasper, 308
tataricus (Map 961), **204**
turbinellus, 312
umbellatus, 218
　var. *pubens,* 218
urophyllus, 314, 315
vimineus, 278, 306, *310,* 311
　var. *subdumosus, 310*
Aster, 204, 219, 270, 278, 279
　Aromatic, 297
　Azure, 300
　Blue Wood, 285
　Decurrent false, 208
　Drummond, 286
　False, 206, 208
　Flat-topped white, 218
　Flax-leaved, 243
　Forked, 226, 227
　Freeway, 311
　Glossy-leaved, 308
　Grass-leaved golden, 239
　Inland saltmarsh, 311
　Large-leaved, 227
　Mexican, 469
　New England, 296
　Oblong-leaved, 297
　Ontario, 298
　Panicled, 292
　Prairie, 312, 314
　Rayless alkali, 284
　Silky, 311
　Single-stemmed bog, 227
　Small white, 302
　Smooth, 291
　Sneezewort, 269
　Southern prairie, 227
　Spreading, 303
　Stiff, 243
　Stiff-leaved, 243

Aster (cont.)
　Stokes, 591
　Tall white, 292
　Tatarian, 204, 205
　White heath, 305
　White prairie, 290
　White upland, 269
　White woodland, 295
　Willow-leaved, 306
　Wreath, 288
Aster yellows, 478
Aster yellows phytoplasma, 478
Asteraceae, **164,** 165, 166, 198, 245, 344, 374, 390, 942
　Subfamily Asteroideae, 166
　Subfamily Barnadesioideae, 166
　Subfamily Cichorioideae, 166
　Subfamily Liguliflorae, 166
　Subfamily Lactucoideae, 166
　Subfamily Tubuliflorae, 166
　Subtribe Ambrosiinae, 519
　Subtribe Centaureinae, 232
　Subtribe Sonchinae, 377
　Tribe Ambrosieae, 443
　Tribe Anthemideae, **171,** 172, 425
　Tribe Astereae, **198,** 425
　Tribe Cardueae, 166, **315,** 316, 488
　Tribe Cichorieae, 165, 166, **343,** 344, 516
　Tribe Cynareae, 166
　Tribe Eupatorieae, 166, 218, **393,** 400, 435, 436
　Tribe Gnaphalieae, 166, **425,** 572, 573
　Tribe Helenieae, 166, 435, 436
　Tribe Heliantheae, 166, 172, **435,** 436, 519
　Tribe Inuleae, 166, 425, **572,** 573
　Tribe Lactuceae, 166, **343**
　Tribe Plucheeae, 166, 572, **573**
　Tribe Senecioneae, 344, **576**
　Tribe Vernonieae, 166, 573, **590**
Asterologist, 164
Astranthium, 203, **205**
　ciliatum (Map 962, Pl. 229 h, i), **205**
　integrifolium, 205
　　ssp. *ciliatum,* 205
　　ssp. *integrifolium,* 205
　　var. *ciliatum,* 205
Atlantic ivy, 135
Atlantic mock bishop's weed, 98
Atocion, 795, **798,** 829
　armeria (Map 1449, Pl. 345 f), **800**
Atriplex, **862,** 866, 871, 876
　argentea (Map 1511), **863**
　　var. argentea, **863**
　hastata, 866
　heteropserma, 864, 865

Atriplex (cont.)
　hortensis (Map 1512, Pl. 352 b, c), **864**
　micrantha (Map 1513, Pl. 352 d), **864,** 865
　patula (Map 1514, Pl. 352 a), **865,** 866
　　var. *hastata,* 865
　　var. *triangularis,* 865
　prostrata (Map 1515, Pl. 352 e), **865,** 866
　rosea (Map 1516, Pl. 352 g), **866**
　subspicata, 866
　triangularis, 866
　truncata (Map 1517, Pl. 352 f), 863, **868**
　wrightii (Map 1518), **868**
Atriplex, Russian, 864
Aubrieta, **664,** 666
　deltoidea (Map 1316, Pl. 314 f, g), **666**
Austin, Daniel F., 932
Autumn olive, 995
Autumn sneezeweed, 490
Axyris, 869
　amaranthoides (Map 1519, Pl. 357 e, f), **869**
Azalea prinophylla, 1004
Azalea, 1000, 1004
　Mountain, 1004
　Piedmont, 1006
　Roseshell, 1004
Azure aster, 300

B

Bachelor's button, 324
Bald cypress, 125, 457
Ball cactus, 737
Ball mustard, 712
Balloon flower, 742
Balsam, 600
　Garden, 600
　Old-field, 434
Balsaminaceae, **599**
Balsamita
　major, 194
　　var. *tanacetoides,* 194
Banana
　Indiana, 52
　Missouri, 52
Barbara's buttons, 522
Barbarea, **666**
　verna (Map 1317), **666**
　vulgaris (Map 1318, Pl. 314 h–j), **667**
　　var. *arcuata,* 667
Barberry, 603, 604
　American, 604
　Common, 604
　Japanese, 604
Barberry family, 602

Barnaby's thistle, 330
Bashful brier, 1082
Basinger, Mark, 774
Basket flower, American, 323
Bassia scoparia, 896
Bastard alkanet, 628
Batodendron aboreum, 1007
Batopilasia, 206
Batschia
    canescens, 638
    caroliniense, 638
Battus philenor philenor, 138
Bauer, Bill, 836
Bauhinia, 1059, 1060
Baumgarten, Johann, 834
Bayer, Randall, 427, 428
Beach wormwood, 187
Beaked hawkweed, 354
Beaked hazel, 613
Bean, 1058
    Castor, 1054
    Indian, 619
Bean family, 1057
Bearberry, 1000
Beard, Goat's, 390
Bearsfoot, 555
Beauty, Rutland, 934
Bee
    Andrenid, 1001, 1007
    Carpenter, 618
Bee plant, Rocky Mountain, 910
Beech, Blue, 610
Beechdrops, False, 1003
Beehive cactus, 737
Beekeeper, 909
Beet, 860, 869
    Fodder, 870
    Garden, 869
    Sea, 870
Beetle, 548
    Chrysolinid, 924
Beggar-ticks, 449, 458–460
    Swamp, 459
Beggar's lice, 634
Begoniaceae, 974
Bellflower, 742
    Creeping, 744, 746
    Marsh, 744
    Rover, 744
    Tall, 743
Bellflower family, 741
Bellis, 198
Berberidaceae, **602**
Berberis, 603, 604
    aquifolium, 603
    canadensis (Map 1269, Pl. 302 b, c), **604**
    thunbergii (Map 1270, Pl. 302 a), **604**
    thunbergii × vulgaris, 604
    vulgaris (Map 1271, Pl. 302 d, e), **604**

Bergia, **996**
    texana (Map 1638, Pl. 374 f, g), **996**
Berlandiera, **449**
    texana (Map 1148, Pl. 272 c), **449**
Berteroa, **667**
    incana (Map 1319, Pl. 314 d, e), **667**
Berula, **67**
    erecta (Map 844), **67**, 218
    var. erecta, 68
    var. incisum, **67**, 68
Beta, 860, **869**
    vulgaris (Map 1520), **869**
    ssp. maritima, 870
    ssp. vulgaris, **869**, 870
Bettis, Mrs. James R., 1073
Betula, **610**, 943
    nigra (Map 1276, Pl. 303 a–d), **610**
Betulaceae, **607**, 608
Bicknese, Nina, 822
Bidens, 436, **449**, 450, 456, 460, 462, 939, 943
    alba (Map 1149, Pl. 274 e, f), 450, **451**
    var. radiata, **451**
    aristosa (Map 1150, Pl. 273 g–i), **452**, 454
    f. aristosa, 454
    f. fritcheyi, 454
    f. involucrata, 452
    f. mutica, 452, 454
    var. fritcheyi, 452, 454
    var. mutica, 452
    var. polylepis, 454
    var. retrorsa, 452, 454
    beckii (Map 1151, Pl. 284 g, h), 450, **454**, 456
    bipinnata (Map 1152, Pl. 274 i, j), **456**
    cernua (Map 1153, Pl. 273 a, b), **456**, 459, 460
    f. discoidea, 457
    var. elliptica, 456
    var. integra, 456
    comosa, 459, 460
    connata, 459, 460
    var. petiolata, 459
    coronata, 459
    discoidea (Map 1154, Pl. 274 g, h), **457**
    frondosa (Map 1155, Pl. 274 a, b), **458**, 460, 461
    laevis (Map 1156, Pl. 273 c, d), **458**, 459
    odorata, 452
    pilosa, 452
    var. radiata, 451
    polylepis, 452, 454
    var. retrorsa, 452
    trichosperma (Map 1157, Pl. 273 e, f), **459**

Bidens (cont.)
    tripartita (Map 1158, Pl. 273 j, k), 450, **459**, 460
    vulgata (Map 1159, Pl. 274 c, d), 458, **460**
    f. puberula, 460, 461
    var. schizantha, 461
Biennial wormwood, 180
Big blue lobelia, 748
Big marsh elder, 471
Big mock bishop's weed, 99
Bigflower coreopsis, 463
Bignonia, **615**
    capreolata (Map 1280, Pl. 304 f–h), **615**
Bignoniaceae, **614**
Bigroot morning glory, 952
Big-seeded scorpiongrass, 643
Bilberry, 1006
Bindweed, 936
    Field, 936
    Hedge, 932, 934
    Japanese, 933
    Low, 935
    Small, 936
Birch, 610
    Red, 610
    River, 610
Birch family, 607
Bird's rape, 674
Birthwort, 138
Birthwort family, 138
Bishop's weed, 62
    Atlantic mock, 98
    Big mock, 99
    Mock, 98
    Ozark mock, 99
Bitter cress, 678, 683
    Hoary, 682
    Northern, 682
    Pennsylvania, 683
    Small-flowered, 683
Bitter sneezeweed, 490
Bittersweet
    American, 851
    Bittersweet, Oriental, 851
Bitterweed, 376, 444, 490
Black alder, 129, 609
Black drink, 130
Black gum, 966, 967
    Swamp, 966
Black huckleberry, 1000
Black knapweed, 328
Black maple, 14
Black mustard, 674
Black snakereoot, 99, 102
Black stem rust, 603, 606
Black swallowwort, 159
Black tupelo, 966
Black-eyed Susan, 642
Blackfoot, hoary, 523
Blackhaw, 784, 791
Bladder campion, 841

Bladder catchfly, 841
Bladderpod, 712
    Missouri, 714
    Spreading, 714
Blanket, Indian, 484
Blanketflower, 483
    Prairie, 483
Blazing star, 413
Blessed milk thistle, 342
Blessed thistle, 322
Blite
    Coast, 888
    Sea, 902
    Strawberry, 880
Blitum capitatum, 880
Bloodleaf, 38
Bloodwort, 174
Blue beech, 610
Blue boneset, 400
Blue bottle, 324
Blue cardinal flower, 748, 750
Blue cohosh, 606
Blue devil, 300, 633
Blue lettuce, 370
Blue lobelia, 750
    Big, 748
Blue morning glory, 950
Blue mustard, 684
Blue sailors, 347
Blue scorpiongrass, 644
Blue star, 116, 117
    Ciliate, 117
    Shining, 117
Blue thistle, 633
Blue vine, 159
Blue wood aster, 285
Bluebell, 641, 746
Blueberry, 1000, 1006, 1007
    Highbush, 1007, 1008
    Hillside, 1008
    Lowbush, 1008
    Sweet lowbush, 1007
Blueberry family, 999
Blue-stemmed goldenrod, 254
Blueweed, 633
Bluntleaf milkweed, 145
Blunt-leaved spurge, 1046
Blunt-leaved yellow cress, 720
Boechera, 662, **668**
    canadensis (Map 1320, Pl. 312 c, d), **668**
    laevigata (Map 1321, Pl. 313 f, g), **668**
    missouriensis (Map 1322, Pl. 312 e, f), **669**
    shortii (Map 1323, Pl. 313 a, b), **669**
Boehmeria, 940, 943
Bog aster, Single-stemmed, 227
Bog yellow cress, 720
Bogler, David J., 40, 115, 348, 357, 371, 372, 377, 619, 741, 758, 850, 908, 957, 974,

Bogler, David J. (cont.)
    992, 999, 1057
Bollmann, Lia, 126
Boltonia, 198, **206,** 219, 279
    asteroides (Map 963, Pl. 231 f–i), **208,** 210
        f. rosea, 208
        var. asteroides, 208
        var. decurrens, 208
        var. latisquama), **208,** 210
        var. recognita, **208**
    caroliniana, 210
    decurrens (Map 964, Pl. 231 c–e), **208,** 209
    diffusa (Map 965, Pl. 231 a, b), **209,** 210
        var. caroliniana, 210
        var. diffusa, 210
        var. interior, 209, 210
    latisquama, 208
        var. decurrens, 208
        var. microcephala, 208
        var. occidentalis, 208
        var. recognita, 208
Bombus, 602
Bonamia, 957
Boneset, 406
    Blue, 400
    False, 397
    Late, 410
    Round-leaved, 407
    Upland, 410
Borage family, 619
Boraginaceae, **619,** 620, 621, 633, 992, 993
    Subfamily Boraginoideae, 993
Borago, 621
    officinalis, 621
Bottle, Blue, 324
Bottle gourd, 985, 986
Bouncing-bet, 827
Bowles, Marlin, 333
Box, 731
Box elder, 9, 10, 48
Box elder bug, 10
Boxwood, 731
Brachyactis
    ciliata, 284
        ssp. angusta, 284
Bradburia, **210,** 239
    pilosa (Map 966, Pl. 232 h–j), **210**
Bradbury, John, 210
Brant, Alan E., 599, 732, 858, 994
Brasenia, 732, **733**
    schreberi (Map 1398, Pl. 329 a, b), **733**
Brassica, **670,** 694, 724
    alba, 724
    arvensis, 725
    campestris, 675
    eruca, 692
    hirta, 724

Brassica (cont.)
    juncea (Map 1324, Pl. 314 a–c, 315 a–c), **672,** 674, 724
    kaber, 725
        var. pinnatifida, 725
    napus (Map 1325, Pl. 315 g–i), **672,** 674, 675
        var. napobrassica, 674
        var. napus, 674
    nigra (Map 1326, Pl. 315 d–f), 672, **674,** 724
    oleracea, 670, 672, 674
    rapa (Map 1327, Pl. 315 j, k), 672, **674**
        var. campestris, 675
        var. oleifera, 675
        var. rapa, 675
    tournefortii, 672
Brassicaceae, 390, **651,** 662, 689, 908, 909
Brauneria
    angustifolia, 475
    pallida, 475
    paradoxa, 476
Breweria, 957
    pickeringii
        var. pattersonii, 956
Brickellia, **396,** 397
    eupatorioides (Map 1106, Pl. 264 a–f), **397**
        var. corymbulosa (Pl. 264 e, f), **397**
        var. eupatorioides (Pl. 264 c, d), 397, **398**
        var. texana (Pl. 264 a, b), **398**
    grandiflora (Map 1107, Pl. 264 i–k), **398**
Brier
    Bashful, 1082
    Catclaw sensitive, 1082
    Sensitive, 1082
Bristly hawksbeard, 350
Bristly ox-tongue, 352
Bristly sunflower, 503
British yellowhead, 573
Broadhead, G. C., 500
Broadleaf goldenrod, 260
Broad-leaved pepper grass, 707
Broad-leaved toothwort, 680
Broccoli, 670
Bromus, 943, 944
Brook euonymus, 854
Broom snakeroot, 202
Broomweed, 202
Brown knapweed, 328
Brown mustard, 672
Brown-eyed Susan, 546
Brummitt, Richard K., 932
Brunnera, **627**
    macrophylla (Map 1290, Pl. 305 g–i), **627**
Brussels sprouts, 670
Bryonia boykinii, 976

Buckbrush, 780, 1018, 1052
Buffalo gourd, 982, 983
Buffalo weed, 448
Bug, Box elder, 10
Bugloss, 626
    Common, 627
    Italian, 626
    Showy, 626
    Siberian, 627
Bugloss fiddleneck, 624
Buglossoides, **628**
    arvense (Map 1291, Pl. 307 g–i), **628**
Bugseed, 893
    American, 894
    Common, 894
    Hairy, 894
Bull thistle, 341
Bumblebee, 602, 616, 1003, 1007, 1065
Bunch galler, Goldenrod, 252
Bundle flower, Illinois, 1080
Bupleurum, **68**
    rotundifolium (Map 845, Pl. 204 i, j), **68**
Bur chervil, 66
Bur cucumber, 987
Bur marigold
    Few-bracted, 457
    Nodding, 456
    Showy, 458
Burdock, 317
    Common, 317
    Cotton, 318
Burning bush, 854
Burra, 621
Bursage
    Annual, 443
    Perennial, 446
Bush
    Burning, 854
    Fire, 896
    Running strawberry, 856
    Snowball, 791
    Strawberry, 854
Bush, Benjamin F., 242, 336, 482, 484, 503, 508, 589, 603, 898, 926, 1047, 1083
Bush honeysuckle, 765
Bushy goldenrod Viscid, 230
Bushy wallflower, 696
Butterbur, 576
Buttercup pennywort, 86
Butterfly
    Eastern pipevine swallowtail, 138
    Milkweed, 142
    Monarch, 142
    Zebra swallowtail, 52
Butterfly weed, 154, 156
Butterweed, 586
Button
    Bachelor's, 324

Button (cont.)
    Spanish, 328
Button snakeroot, 82, 418, 422
Buttonbush, 457
Buttonbush dodder, 939
Buttons
    Barbara's, 522
    Golden, 195
Buxaceae, **731**
Buxus, 731
    sempervirens, 731

C

Cabbage, 670
Cabbage coneflower, 544
Cabomba, 732, **734**
    caroliniana (Map 1399, Pl. 329 c, d), 733, **734**
        var. caroliniana, **734**
        var. flavida, 734
        var. pulcherrima, 734
Cabombaceae, 733
Cacalia, 576
    atriplicifolia, 577, 578
    muhlenbergii, 578
    plantaginea, 578
    suaveolens, 582
    tuberosa, 578
Cacchione, Robert, 404
Cactaceae, **736,** 795
Cactoblastus, 736
Cactus, 1029
    Ball, 737
    Beehive, 737
    Cream, 737
    Missouri pincushion, 737
    Starvation, 740
Cactus family, 736
Caesalpinia, 1060, 1071
    jamesii, 1070
Caesalpiniaceae, 1058
Caesalpinioid legume, 1058
Calabash, 985
Calamintha, 943
Calliopsis, 466
Caltha, 218
    palustris, 218
Calycanthaceae, 1, **740**
Calycanthus, **740**
    floridus (Map 1403), **740,** 741
        var. floridus, **740**
        var. glaucus, 741
Calystegia, **932,** 936
    dahurica, 933
    fraterniflora, 935
    hederacea, 933
    macounii (Map 1577, Pl. 363 f), **933,** 936
    malacophylla, 933
    pellita, 933
    pubescens (Map 1578), **933**

Calystegia (cont.)
    sepium (Map 1579, Pl. 363 b, c), 933, **934,** 935
        ssp. americana, 934
        ssp. angulata, 934
        ssp. erratica, 934
        ssp. sepium, 934
    silvatica (Map 1580, Pl. 363 d), **935**
        ssp. fraterniflora, **935**
        ssp. orientalis, 935
        ssp. silvatica, 935
    spithamaea (Map 1581, Pl. 363 a), **935,** 936
        ssp. purshiana, 935
        ssp. spithamaea, **935**
        ssp. stans, 935
Camelina, **675**
    alyssum, 676
    microcarpa (Map 1328, Pl. 316 c, d), **675**
    sativa (Map 1329, Pl. 316 e, f), **676**
        ssp. microcarpa, 675
Campanula, **742,** 753
    americana (Map 1404, Pl. 330 f–h), **743,** 744
    aparinoides (Map 1405, Pl. 330 c), **744**
        var. aparinoides, **744**
        var. grandiflora, 744
    biflora, 754
    perfoliata, 757
    rapunculoides (Map 1406, Pl. 330 d, e), **744**
        var. rapunculoides, **744**
    rotundifolia (Map 1407, Pl. 330 a, b), **746**
Campanulaceae, **741,** 742
    Subfamily Campanuloideae, 742
    Subfamily Lobelioideae, 742
Campanulastrum americanum, 743
Camphorweed, 238, 242, 574
Campion, 828
    Bladder, 841
    Evening, 837
    Glaucous, 834
    Red, 836
    Snowy, 838
    Starry, 840
    White, 837
Campsis, **615,** 940, 943
    radicans (Map 1281, Pl. 304 i, j), **615,** 616
Camptotheca, 965
Canada fleabane, 214
Canada snakeroot, 100
Canada thistle, 335
Candelilla, 1029
Candidatus Phytoplasma, 478
Cannabaceae, 758
Cannabis, 758, **759,** 760
    indica 760

Cannabis (cont.)
 ruderalis, 759, 760
 sativa (Map 1418, Pl. 333 a, b), **759,** 760
  ssp. indica, 760
  ssp. sativa, 760
  var. spontanea, 759, 760
Cannonball tree, 1011
Canola oil, 674
Cantaloupe, 979, 980
Caper, 909
Capparaceae, 907, 909
Capparis spinosa, 909
Caprifoliaceae, **765,** 766
Capsella, **676**
 bursa-pastoris (Map 1330, Pl. 316 i–k), **676**
Carara didyma, 704
Caraway, 55, 68, 69
Cardamine, **678,** 711, 715
 arenicola, 683
 bulbosa (Map 1331, Pl. 316 g, h), **679,** 682
  f. fontinalis, 679
  var. purpurea, 682
 concatenata (Map 1332, Pl. 317 i, j), **680**
 diphylla (Map 1333, Pl. 317 g, h), **680**
 douglassii (Map 1334, Pl. 316 a, b), 679, **682**
 flexuosa (Map 1335), **682,** 684
 hirsuta (Map 1336, Pl. 317 f), **682,** 684, 715
 parviflora (Map 1337, Pl. 317 a–c), **683,** 684, 715
  var. arenicola, **683**
  var. parviflora, 683
 pensylvanica (Map 1338, Pl. 317 d, e), **683,** 684, 715
 rhomboidea, 679
 virginica, 715
Cardaminopsis lyrata, 660
Cardaria, 702
 chalepensis, 703
 draba, 706
  ssp. *chalepensis, 703*
  var. *elongata, 703*
 pubescens, 703
Cardinal flower, 747
 Blue, 748, 750
Cardoon, 316
Carduus, **318,** 320
 crispus (Map 1038, Pl. 250 d–f), **318**
 nutans (Map 1039, Pl. 250 g–i), **320,** 321
  ssp. leiophyllus, 321
  ssp. macrocephalus, 321
  ssp. macrolepis, 321
Carduelis, 335
Careless weed, 31, 471
Carissa, 115

Carnation, 809
Carolina allspice, 740
Carolina elephant's foot, 591
Carolina leaf-flower, 1052
Carolina thistle, 336
Carolina water shield, 734
Carpenter bee, 618
Carpet, Golden, 970
Carpinus, **610**
 caroliniana (Map 1277, Pl. 303 h, i), **610,** 612
  ssp. *virginiana, 612*
  var. caroliniana, **612**
  var. virginiana, **612**
Carrot, 55, 78
 Small wild, 79
 Wild, 78
Carrot family, 54
Carthamus, 316, 322
 tinctorius, 316
Carum, **68**
 carvi (Map 846, Pl. 204 f–h), 55, **68,** 69
Caryophyllaceae, 17, **794,** 813
Caryophyllales, 794
Casaba, 980
Cashew, 40
Cashew family, 40
Cassava, 1010
Cassia, 1060, 1062, 1072
 chamaecrista, 1064
 fasciculata, 1064
  f. jensenii, 1065
  var. depressa, 1064
  var. robusta, 1064
 marilandica, 1072, 1073
 medsgeri, 1072
 nictitans, 1065
 obtusifolia, 1073
 occidentalis, 1074
 tora, 1073
Castor bean, 1054
Castor oil, 1011, 1020, 1054
Castor oil plant, 1054
Catalpa, 614, **616**
 bignonioides (Map 1282, Pl. 304 e), **618,** 619
 bignonioides × ovata, 618
 bignonioides × speciosa, 618
 ×erubescens, 618
 ×hybrida, 618
 ovata (Map 1283, Pl. 304 d), 618, **619**
 ovata × speciosa, 618
 speciosa (Map 1284, Pl. 304 a–c), **619**
Catalpa, 616, 618, 619
 Chinese, 619
 Common, 618
 Hardy, 619
 Northern, 619
 Southern, 618
Catalpa sphinx moth, 618

Catawba tree, 619
Catchfly, 828
 Bladder, 841
 Forked, 834
 Large sand, 832
 Night-flowering, 838
 Royal, 840
 Sleepy, 831
 Small-flowered, 836
 Smooth, 834
 Sweet William, 800
Catclaw sensitive brier, 1082
Catfoot, 434
Catha edulis, 850
Catharanthus, 115, 116
 roseus, 115
Cat's ear, 357, 376
 Smooth, 358
 Spotted, 357
Cattail Red-hot, 1012
Cauliflower, 670
Caulophyllum, **606**
 thalictroides (Map 1272, Pl. 301 h–j), **606**
Cayaponia, **976**
 boykinii, 976
 grandifolia, 976
 quinqueloba (Map 1622, Pl. 372 a), **976**
Ceanothus, 943
Cecidomyid fly, 252
Celastraceae, 126, **850**
Celastrus, **850,** 851, 852
 alatus, 854
 articulatus, 851
 orbiculatus (Map 1502, Pl. 350 d, e), **851**
 scandens (Map 1503, Pl. 350 f, g), **851**
Celery, 55, 66
 Water, 92
Celosia, 18, **34**
 argentea (Map 825, Pl. 199 e), **35**
  f. cristata, 36
  var. argentea (Map 825), **35**
  var. cristata (Map 825, Pl. 199 e), 35, **36**
 cristata, 36
Centaurea, 187, **321,** 322
 americana (Map 1040, Pl. 251 a–c), 321, 322, **323**
 biebersteinii, 331
 centaurium, 322
 cyanus (Map 1041, Pl. 252 d, e), **324,** 326
 diffusa (Map 1042, Pl. 251 g), **326,** 327, 332
 diluta (Map 1043, Pl. 251 h), **326,** 332
 dubia, 330
  ssp. *vochinensis, 328*
 jacea, 328

Centaurea dubia (cont.)
    ssp. *nigra, 328*
  maculosa, 321
  melitensis (Map 1044, Pl. 251 i), **327**
  nigra (Map 1045, Pl. 252 f, g), **328**
  nigrescens (Map 1046, Pl. 252 j, k), **328**, 330
  ×moncktonii, 328
  ×pratensis, 328
  ×psammogena, 326, 332
  repens (Map 1047, Pl. 252 h, i), 322, **330**, 331
  solstitialis (Map 1048, Pl. 251 d–f), **330**
  stoebe (Map 1049), 326, 330, **331**
    ssp. micranthos, **331, 332**
    ssp. stoebe, 332
  vochinensis, 328, 330
Cephalanthus, 940, 943, 944
  occidentalis, 457
Cerastium, **800**, 802, 804
  arvense (Map 1450), **803**, 809
    f. oblongifolium, 809
    ssp. strictum, **803**
    ssp. velutinum, 809
    var. oblongifolium, 809
    var. villosum, 809
  brachypetalum (Map 1451, Pl. 340 g), **803**
    f. glandulosum, 803
    ssp. tauricum, 803
  brachypodum (Map 1452, Pl. 340 a, b), 802, **804**
  diffusum (Map 1453, Pl. 340 c, d), **804**, 806, 808
  dubium, 802
  fontanum (Map 1454, Pl. 340 h, i), **806**, 808
    ssp. fontanum, 806
    ssp. vulgare, **806**
  glomeratum (Map 1455, Pl. 341 b, c), **806**
  holosteoides
    f. glandulosum, 806
  nutans (Map 1456, Pl. 341 f–h), 802, 804, **807**
    ssp. nutans, **807**
    ssp. obtectum, 807
    var. brachypodum, 804
  pumilum (Map 1457, Pl. 341 d, e), 806, **807**, 808
    f. medium, 807
  semidecandrum (Map 1458, Pl. 341 a), 806, **808**, 809
  tetrandrum, 804
  velutinum (Map 1459, Pl. 340 e, f), 803, **809**
    ssp. velutinum, **809**
    ssp. villosissimum, 809
  viscosum

Cerastium velutinum (cont.)
    f. apetalum, 806
  vulgatum
    f. glandulosum, 806
Ceratomia catalpae, 618
Ceratophyllaceae, 454, **858**
Ceratophyllum, 454, 456, 734, **858**, 859
  demersum (Map 1509, Pl. 351 e–g), **859**
  echinatum (Map 1510, Pl. 351 c, d), **859**
Cercis, 1059, 1060, **1061**, 1062
  canadensis, **1061**, 1062
    f. alba 1062
    f. glabrifolia, 1061
    var. canadensis, **1061**
    var. mexicana, 1062
    var. texensis, 1062
  occidentalis, 1061
  siliquastrum, 1061
Cereal amaranth, 20
Chaerophyllum, 63, **69**, 943
  procumbens (Map 847, Pl. 204 a, b), **70**
    var. procumbens, 70
    var. shortii, 70
  tainturieri (Map 848, Pl. 204 c), **70**
    var. dasycarpum, 70
    var. floridanum, 70
  texanum, 70
Chaetopappa, **211**
  asteroides (Map 967, Pl. 232 f), **211**
    var. asteroides, **211**
    var. grandis, 212
Chamaecrista, 1060, **1062**, 1064, 1072
  fasciculata (Map 1690, Pl. 385 h, i), **1064**, 1065
  nictitans (Map 1691, Pl. 385 f, g), **1065**
    ssp. nictitans, 1066
    var. nictitans, **1065**, 1066
    procumbens, 1065
Chamaemelum, 192
  nobile, 192
Chamaesyce, 1028
  geyeri, 1041
    var. wheeleriana, 1041
  glyptosperma, 1041
  humistrata, 1042
  maculata, 1042
  missurica, 1045
    var. calcicola, 1045
  nutans, 1045
  prostrata, 1047
  serpens, 1048
  serpyllifolia, 1048
  stictospora, 1049
  supina, 1042

Chamomile, 172
  Corn, 177
  English, 192
  False, 190, 191
  German, 190, 191
  Roman, 192
  Stinking, 177
  Scentless, 192
  Scentless false, 196
  True, 192
  Wild, 190, 191
  Yellow, 188, 189
Chamomilla, 191
  recutita, 191
  suaveolens, 192
Chard, Swiss, 860, 870
Charlock, 724, 725
  Jointed, 716
Chenille plant, 1012
Chenopodiaceae, 18, 390, 795, **860**, 894
Chenopodium, 860, **870**, 871, 876, 887, 890, 943
  album (Map 1521, Pl. 353 f–h), 871, **874**, 875, 878, 884
    var. album 875
    var. berlandieri, 876
    var. lanceolatum 875
    var. missouriense, 884
  ambrosioides (Map 1522, Pl. 353 i–l), **875**, 876
    var. ambrosioides, **876**
    var. anthelminticum, **876**
  anthelminticum, 876
  atrovirens, 871
  berlandieri (Map 1523, Pl. 354 g, h), 875, **876**, 878, 879
    f. angustius, 878
    f. latifolium, 878
    f. neglectum, 878
    f. pedunculare, 878
    ssp. zschackei, 878
    var. berlandieri, 878
    var. boscianum, 875, **878**
    var. bushianum, 879
    var. farinosum, 878
    var. foetens, 878
    var. zschackei, 875, **878**
  bonus-henricus, 871
  ×borbasii, 887
  boscianum, 878
  botrys (Map 1524, Pl. 353 o, p), 876, **879**
  bushianum (Map 1525, Pl. 353 d, e), 875, 878, **879**, 880
  capitatum (Map 1526, Pl. 354 e, f), 870, **880**
  dakoticum, 878
  desiccatum (Map 1527, Pl. 354 a, b), **880**, 887
    var. leptophylloides, 887
  ficifolium (Map 1528, Pl. 355 a, b), **882**

Chenopodium (cont.)
  fremontii
    var. *incanum, 883*
  gigantospermum, 889
  glaucum (Map 1529, Pl. 353 a–c), **882,** 883
    var. glaucum, **883**
    var. salinum, **883**
  hybridum, 889
  incanum (Map 1530, Pl. 355 h, i), **883**
    var. elatum, 884
    var. incanum, **883,** 884
    var. occidentale, 884
  lanceolatum 875
  missouriense (Map 1531, Pl. 355 f, g), 875, **884**
  murale (Map 1532, Pl. 355 c–e), **884**
  opulifolium (Map 1533, Pl. 356 c–e), 875, **886**
  paganum, 879
  pallescens (Map 1534, Pl. 356 f, g), **887**
  pratericola (Map 1535, Pl. 354 c, d), **887**
  pumilio (Map 1536, Pl. 353 m, n), 870, 876, **888**
  rubrum (Map 1537, Pl. 356 l, m), **888**
    var. *humile, 888*
  salinum, 883
  simplex (Map 1538, Pl. 355 l, m), **889**
  standleyanum (Map 1539, Pl. 355 j, k), **889**
  strictum (Map 1540, Pl. 356 a, b), **890**
    var. glaucophyllum, **890**
    var. strictum, 890
  urbicum (Map 1541, Pl. 356 j, k), **890**
  ×variabile
    var. murrii, 878, 879
  watsonii (Map 1542, Pl. 356 h, i), **892**
Cherry, Cornelian, 958
Cherry silverberry, 995
Chervil, 69
  Bur, 66
  Wild, 70, 75
Chickpea, 1058
Chickweed, 844
  Clammy, 806
  Common, 846
  Common mouse-ear, 806
  Curtis's mouse-ear, 807
  Dark green mouse-ear, 804
  Field, 803, 809
  Five-stamen mouse-ear, 809
  Forked, 821
  Giant, 819, 820
  Gray mouse-ear, 803

Chickweed (cont.)
  Greater, 846
  Jagged, 814
  Lesser, 848
  Mouse-ear, 800
  Nodding, 806
  Prairie, 809
  Short-stalked mouse-ear, 804
  Small mouse-ear, 809
Chicory, 344, 347, 348, 390, 392
  Common, 347
Chigger, 599
Chigger flower, 154
Childing pink, 822
Chinchilla, 556
Chinese catalpa, 619
Chinese mustard, 672
Chinese spinach, 31
Chippewa, 276
Chittam-wood, 41
Cholla, Tree, 736
Chondrilla, 344
  juncea, 344
Chloracantha, 206
Chorispora, **684**
  tenella (Map 1339, Pl. 318 j–l), **684**
Christ, Art, 228
Chrysanthemum, 171, 172, 189, 194
  balsamita, 194
    var. *tanacetoides, 194*
  carinatum, 189
  coronarium, 189
  leucanthemum, 190
    var. *pinnatifidum, 190*
  parthenium, 195
  segetum, 189
Chrysanthemum
  Daisy, 190
  Garland, 189
  Portuguese, 190
  Tricolor, 189
Chrysolinid beetle, 924
Chrysopsis, 210, 238, 239
  camporum, 239
    var. *glandulissimum, 240*
  canescens, 240
  nuttallii, 210
  pilosa, 210
  villosa
    var. angustifolia, 242
  var. *camporum, 239*
  var. *canescens, 240*
Chung, Kuo-Fang, 363
Cicely
  Smooth sweet, 94
  Sweet, 93
  Woolly sweet, 93
Cicer, 1058
Cichorium, 343, 344, **347,** 390, 516
  endivia, 347

Cichorium (cont.)
  intybus (Map 1061, Pl. 255 h), **347,** 348
    f. album, 348
    f. roseum, 348
Cicuta, 55, **70,** 72, 79, 92, 103
  maculata (Map 849, Pl. 205 a, b), **70,** 72
    var. angustifolia, 72
    var. bolanderi, 72, **74**
    var. *curtissii, 74*
    var. maculata, 72, **74**
    var. victorinii, 72
Cigar tree, 619
Ciliate blue star, 117
Cineraria, 576
Cirsium, 321, **332,** 336
  altissimum (Map 1050, Pl. 253 a–c), **334,** 338
    f. moorei, 334
  arvense (Map 1051, Pl. 254 c, d), 315, 316, 332, **334**
    var. *horridum, 335*
  canescens (Map 1052), **336,** 341
  carolinianum (Map 1053, Pl. 253 f, g), 333, **336,** 340
  discolor (Map 1054, Pl. 253 d, e), 334, **338**
    f. albiflorum, 338
  hillii, 332, 333
  muticum (Map 1055, Pl. 254 a, b), 321, 332, **338**
  pumilum, 332
    ssp. hillii, 332
  texanum (Map 1056, Pl. 254 e, f), **340**
    var. texanum, **340**
  undulatum (Map 1057, Pl. 253 h), 336, **340,** 341
    var. *megacephalum, 340,* 341
  vulgare (Map 1058, Pl. 254 i–k), **341**
Cistaceae, **902**
Cistus, 903
Citrullus, **976,** 978
  lanatus (Map 1623, Pl. 371 d, e), **978**
    var. citroides, 978
    var. lanatus, **978**
  vulgaris, 978
City goosefoot, 890
Clammy chickweed, 806
Clammy goosefoot, 888
Clammy weed, 911
Clasping coneflower, 537
Clasping Venus' looking glass, 757
Clasping-leaved St. John's-wort, 917
Cleomaceae, **908,** 909
Cleome, **909**
  dodecandra, 911
  hassleriana (Map 1558, Pl. 333 j, k), **910**

Cleome (cont.)
   houtteana, 910
   serrulata (Map 1559, Pl. 333 h, i), **910**
   spinosa, 910
Cleome family, 908
Climbing dogbane, 124
Climbing hempweed424
Climbing milkweed, 159, 162
Clover, 1058
   Stinking, 910
   Sweet, 1058
Club, Hercules', 132
Clusiaceae, **912**, 913
Cluster-sanicle, 102
Cnicus, 322
Coast blite, 888
Coccinia grandis, 975
Cochineal, 736
Cockle
   Corn, 797
   Cow, 849
   Sticky, 838
   White, 837
Cocklebur, 568
   Common, 569
   Spiny, 569
Cockscomb, 35
Codiaeum variegatum, 1020
Coffee, Wild, 782
Coffee senna, 1074
Coffee tree, Kentucky, 1069
Coffee weed, 1073, 1074
Cohosh, Blue, 606
Coldenia nuttallii, 993
Collards, 670
Colorado greenthread, 560
Colt's foot, 576
Comfrey, 648
   Common, 650
   Wild, 632
Commelinaceae, 391
Common alder, 609
Common barberry, 604
Common bugloss, 627
Common bugseed, 894
Common burdock, 317
Common catalpa, 618
Common chicory, 347
Common chickweed, 846
Common cocklebur, 569
Common comfrey, 650
Common dandelion, 388, 390
Common dodder, 942
Common dwarf dandelion, 360
Common elderberry, 776
Common false flax, 676
Common flat-topped goldenrod, 230
Common golden Alexanders, 114
Common goldenrod, 256
Common groundsel, 590
Common hops, 762

Common horse gentian, 782
Common hound's tongue, 632
Common marigold, 556
Common meadow parsnip, 114
Common milfoil, 173
Common milkweed, 153
Common morning glory, 952
Common mouse-eared chickweed, 806
Common mugwort, 187
Common periwinkle, 126
Common ragweed, 444
Common rushfoil, 1026
Common sneezeweed, 490
Common sow thistle, 386
Common St. John's-wort, 924
Common stitchwort, 845
Common sunflower, 436, 497, 498
Common tansy, 195, 196
Common teasel, 989
Common water dropwort, 94
Common water hemlock, 70
Common winter cress, 667
Common wormwood, 179
Common yarrow, 173
Compact dodder, 940
Compass plant, 552
Compositae, **164**
Composite, 1058
Coneflower, 473, 534
   Cabbage, 544
   Clasping, 537
   Cutleaf, 543
   Drooping, 532
   Glade, 478
   Gray-headed, 532
   Great, 544
   Green prairie, 534
   Longhead prairie, 532
   Missouri, 545
   Narrow-leaved, 475
   Orange, 538, 540
   Pale purple, 475, 478
   Prairie, 531
   Purple, 478
   Rough, 540
   Shortray prairie, 534
   Showy, 540
   Sweet, 545
   Yellow, 476
Conium, 55, 66, 72, **74**
   maculatum (Map 850, Pl. 205 c, d), **74**
Conoclinium, **400**, 401, 412
   coelestinum (Map 1108, Pl. 267 a-c), **400**, 412
   f. album, 400
Conringia, **686**
   orientalis (Map 1340, Pl. 318 f, g), **686**
Conservation Commission, Missouri, 129

Convention on International Trade in Endangered Species of Wild Flora and Fauna, 136, 736
Convolvulaceae, **930.** 931, 934
Convolvulus, 932, **936**, 943
   arvensis (Map 1582, Pl. 363 e), 932, **936,** 938
      f. cordifolius, 936
   japonicus, 933
   pellitus
      f. anestius, 933
   sepium, 934
      f. coloratus, 933, 934
      f. malacophyllus, 934
      var. fraterniflorus, 935
      var. repens, 934
   spithamaeus, 933, 935
   wallichianus, 933
Conyza, **212**, 214, 219, 943
   canadensis (Map 968, Pl. 232 a, b), **214**
      var glabrata, 214
      var. canadensis (Map 968), **214**
      var. pusilla (Map 968), **216**
   ramosissima (Map 969, Pl. 232 c-e), **216**
Cook, Rachel, 256
Coontail, 858
Copperleaf
   Rhombic, 1016
   Roughpod, 1016
   Virginia, 1017
Coral honeysuckle, 774
Coralberry, 780
   White, 778
Cordiaceae, 620
Coreopsis, 436, 450, 456, **461**, 462, 559
   cardaminifolia, 466, 468
   grandiflora (Map 1160, Pl. 275 d), **463**
      var. grandiflora, 463
      var. harveyana, 463
   lanceolata (Map 1161, Pl. 275 a, b), **463**
      var. villosa, 463
   palmata (Map 1162, Pl. 275 e, f), **464**
   pubescens (Map 1163, Pl. 275 c), **464,** 466
      var. debilis, 466
      var. pubescens, **464,** 466
      var. robusta, 466
   tinctoria (Map 1164, Pl. 275 g), 461, **466,** 468
      f. atropurpurea, 468
      f. tinctoria, 468
      var. similis, 466
   trichosperma, 459
   tripteris (Map 1165, Pl. 276 a), 461, **468**

Coreopsis tripteris (cont.)
    var. deamii, 468
Coreopsis, 461
    Bigflower, 463
    Finger, 464
    Plains, 466
    Tall, 468
    Tickseed, 463
Coriander, 55, 75
Coriandrum, **75**
    sativum (Map 851, Pl. 205 e, f), 55, **75**
Corispermum, **893**
    americanum (Map 1543, Pl. 357 a–c), **894**
        var. rydbergii, 894
    hyssopifolium, 894
    nitidum, 894
    orientale
        var. emarginatum, 894
    pallasii (Map 1544, Pl. 357 g–i), **894**
    villosum (Map 1545, Pl. 357 d), 893, **894**
Corn chamomile, 177
Corn cockle, 797
Corn marigold, 189
Corn gromwell, 628
Corn spurrey, 842
Cornaceae, 457, **957**, 965, 1062
Cornelian cherry, 958
Cornflower, 324
Cornus, 957, 962
    alternifolia (Map 1606, Pl. 369 d, e), 958, **959**
    amomum (Map 1607), **959,** 962, 963
        ssp. amomum, 960
        ssp. obliqua, **959,** 960
        var. schuetziana, 959
    ×arnoldiana, 964
    asperifolia, 960, 963
    drummondii (Map 1608, Pl. 369 a–c), **960,** 962, 963
    florida (Map 1609, Pl. 369 f–h), 958, **962**, 963, 1062
        f. rubra, 963
    foemina (Map 1610, Pl. 369 j), 958, 960, **963**
        ssp. foemina, 963, **964**
        ssp. microcarpa, 963
        ssp. racemosa, 960, 962, 963, **964**
    kousa, 963
    mas, 958
    nuttallii, 963
    obliqua, 959, 964
    racemosa, 963, *964*
    stricta, 964
Coronopus, 702
    didymus, 704
Corylus, 607, **612,** 940

Corylus (cont.)
    americana (Map 1278, Pl. 303 l), **612,** 613
        f. missouriensis, 613
        var. indehiscens, 612, 613
Corylus
    avellana, 613
    cornuta, 613
Coryphantha missouriensis, 737
Cosmos, 436, **468,** 469, 484
    bipinnatus (Map 1166, Pl. 276 e), **469**
    parviflorus (Map 1167, Pl. 276 g, h), **470**
    sulphureus (Map 1168, Pl. 276 f), 469, **470**
        var. sulphureus, 471
Cosmos, 468
    Garden, 469
    Sulphur, 470
Costmary, 194
Cota, 176, **188**
    tinctoria (Map 951, Pl. 224 g, h), **188**
        ssp. tinctoria, **188,** 189
Cotinus, **40**
    americanus, 41
    coccigera, 41
    cotinoides, 41
    obovatus (Map 830, Pl. 200 e, f), **41**
Cottage pink, 812
Cotton burdock, 318
Cotton gum, 965
Cotton thistle, 341
Cottonweed, 36
Cottonweed
    Field, 38
    Slender, 38
    Swamp, 457
Cow cockle, 849
Cow cress, 703
Cow herb, 849
Cow parsnip, 85
    Giant, 85
Cow soapwort, 849
Cowbane, 94
    Spotted, 70
Cowberry, 1006
Cowslip, Virginia, 641
Craig, Moses, 546
Cranberry, 1000, 1006, 1007
    Highbush, 791
Crassulaceae, 967
Crawford, Daniel J., 871
Cream cactus, 737
Creeper
    Trumpet, 615, 616
    Virginia, 48
Creeping bellflower, 744, 745
Creeping cucumber, 986
Creeping St. John's wort, 915
Creeping yellow cress, 722

Crepis, **348**
    capillaris (Map 1062, Pl. 255 c–e), **349**
    pulchra (Map 1063, Pl. 255 a, b), **349**
    setosa (Map 1064, Pl. 255 f, g), **350**
    tectorum (Map 1065, Pl. 255 i, j), 349, **350**
Cress, 718
    American, 666
    Bitter, 678, 683
    Blunt-leaved yellow, 720
    Bog yellow, 720
    Common winter, 667
    Cow, 703
    Creeping yellow, 722
    Early winter, 666
    Field, 703
    Field penny, 730
    Globe-podded hoary, 703
    Heart-podded hoary, 706
    Hairy rock, 662
    Hoary, 703
    Hoary bitter, 682
    Lake, 719
    Lens-podded hoary, 703
    Marsh, 721
    Marsh yellow, 720
    Missouri rock, 669
    Mouse-eared, 660
    Northern bitter, 682
    Pennsylvania bitter, 683
    Penny, 710
    Perfoliate penny, 710
    Purple, 682
    Roadside penny, 730
    Rock, 662, 669
    Sand, 660
    Sessile-flowered, 721
    Shield, 707
    Small-flowered bitter, 683
    Smooth rock, 668
    Spreading yellow, 722
    Spring, 679
    Swine, 704
    Thale, 660
    Virginia rock, 715
    Wart, 704
    Winter, 666
    Yellow, 721
Crinklemat, Rosette, 993
Crinkleroot, 680
Cronquist, Arthur, 315
Crookneck, 982, 983
Cross
    St. Andrew's, 918, 920
    Widow's, 971
Cross vine, 615
Croton, 1010, **1020,** 1026
    capitatus (Map 1656, Pl. 377 k–o), **1021,** 1022, 1026, 1027

Croton capitatus (cont.)
    var. albinoides, 1022
    var. capitatus, **1022**
    var. lindheimeri, **1022**
  ellipticus, 1027
  glandulosus (Map 1657, Pl. 377 e–g), 1020, **1023**
    var. glandulosus, 1023
    var. septentrionalis, **1023**
  lindheimeri, 1021
  lindheimerianus (Map 1658, Pl. 378 h, i), **1023**
    var. capitatus, 1024
    var. lindheimerianus, **1023**
    var. tharpii, 1024
  linearis, 1027
  michauxii (Map 1659, Pl. 378 c, d), 1020, **1024,** 1027
  monanthogynus (Map 1660, Pl. 378 j, k), **1026,** 1027
  texensis (Map 1661, Pl. 378 a, b), **1026**
  tiglium, 1020
  willdenowii (Map 1662, Pl. 378 e–g), 1020, 1024, **1026**
Croton, 1020
  One-seeded, 1026
  Round-leaved woolly, 1023
  Sand, 1023
  Texas, 1026
  Tropic, 1023
  Woolly, 1021
Croton oil, 1020
Crotonopsis, 1027
  elliptica, 1026
  linearis, 1024
Crown daisy, 189
Crownbeard, 560
Crownbeard
  Golden, 562
  White, 566
  Yellow, 564
Crucifereae, **651**
Crunchweed, 725
Cryptotaenia, **75**
  canadensis (Map 852, Pl. 206 e, f), **75**
Cserei, Wolfgang von, 834
Cucumber, 979, 980
  Bur, 987
  Creeping, 986
  Horned, 979
  Wild, 984, 985
Cucumis, 975, **979**
  melo (Map 1624, Pl. 371 f, g), **979,** 980
    ssp. melo, **979,** 980
  metuliferus, 979
  sativus (Map 1625, Pl. 371 h, i), **980**
Cucurbita, 975, **980,** 982, 984
  foetidissima (Map 1626, Pl. 372 f, g), **982,** 983

Cucurbita (cont.)
  ovifera, 983
  pepo (Map 1627, Pl. 372 b, c), 980, **983,** 984
    ssp. ovifera, 983, 984
    ssp. pepo, 983
    var. ozarkana, **983,** 984
    var. ovifera, 983, 984
    var. texana, 984
  perennis, 982
  texana, 984
Cucurbitaceae, 974
Cudweed, 165, 430, 433
  Early, 432
  Fragrant, 434
  Purple, 432
Cultivated sunflower, 488
Cultivated Venus' looking-glass, 753
Cup
  Leaf, 529
  Pale-flowered leaf, 530
  Yellow-flowered leaf, 555
Cup plant, 552
Cup rosinweed, 552
Cupressaceae, 457
Curly thistle, 318
Curlytop gumweed, 236
Currant Indian, 780
Curtis' mouse-ear chickweed, 807
Cuscuta, 930, 931, **938**
  campestris (Map 1583, Pl. 365 h–k), **939**
  cephalanthi (Map 1584, Pl. 365 d–f), **939**
  compacta (Map 1585, Pl. 364 a–c), **940**
  coryli (Map 1586, Pl. 364 d), **940**
    var. stylosa, 942
  cuspidata (Map 1587, Pl. 364 e), **940**
  glomerata (Map 1588, Pl. 364 f, g), **942**
  gronovii (Map 1589, Pl. 365 a, b), **942**
    var. latiflora, 942
  indecora (Map 1590, Pl. 365 c), **943**
    var. neuropetala, 943
  pentagona (Map 1591, Pl. 365 g), **943**
  polygonorum (Map 1592, Pl. 364 h), **944**
  vulgivaga, 942
Cuscutaceae, 931
Cusp dodder, 940
Custard apple family, 51
Cutleaf coneflower, 543
Cut-leaved teasel, 989
Cut-leaved toothwort, 680
Cyanococcus
  corymbosus, 1008
  vacillans, 1008

Cyclachaena, **471,** 519
  xanthifolia, 472
  xanthiifolia (Map 1169, Pl. 284 a, b), **471,** 472
Cycloloma, **895**
  atriplicifolium (Map 1546, Pl. 357 j, k), **895**
Cylindropuntia, 738
  imbricata, 736
Cynanchum, 140, **158,** 159, 940
  laeve (Map 932, Pl. 223 a, b), **159**
  louiseae (Map 933), **159,** 160
  nigrum, 159
  rossicum, 160
  vincetoxicum, 160
Cynara, 316
  cardunculus, 316
  scolynmus, 316
Cynipid wasp, 374, 548
Cynipidae, 548
Cynoglossum, **630**
  officinale (Map 1292, Pl. 306 h, i), **632**
  virginianum (Map 1293, Pl. 306 e–g), **632**
Cynosciadium, **76,** 89
  digitatum (Map 853, Pl. 206 a, b), **76**
  pinnatum, 89
Cynoxylon floridum, 962
Cynthia, Two-flowered, 359
Cypress
  Bald, 125, 457
  Summer, 896
Cypress spurge, 1035
Cypress vine, 952

D

Dactylopius, 736
Daisy
  Crown, 189
  Daisy Doll's, 209
  Engelmann's, 480
  English, 198
  Ox-eye, 189
  Purple, 303
  Western, 205
  White, 189
Daisy fleabane, 220, 223
Daisy-chrysanthemum, 190
Dalea, 943
Dame's rocket, 696
Danaus plexippus, 142
Dandelion, 387, 388
  Common, 388, 390
  Common dwarf, 360
  Dwarf, 358
  False, 382
  Orange dwarf, 359
  Potato

Dandelion (cont.)
 Prairie, 374
 Red-seeded, 388
 Virginia dwarf, 362
 Western dwarf, 362
Darigo, Carl, 562
Dark green mouse-eared
  chickweed, 804
Datisca hirta, 46
Daucus, **78**
 carota (Map 854), 55, **78**, 79
  f. epurpuratus, 79
  f. roseus, 79
  ssp. carota, **78,** 79
  ssp. sativus, 79
 pusillus (Map 855, Pl. 206 g, h),
  79
Davidia, 958, 965
Davis, John, 562
Death angel mushroom, 343
Deciduous holly, 128
Decodon, 940
Decurrent false aster, 208
Deerberry, 1009
Dendranthema ×grandiflorum, 190
Dentaria, 678
 diphylla, 680
 laciniata, 680
Department of Conservation,
  Missouri, 126, 129, 136,
  822
Department of Transportation,
  Missouri, 469, 484
Deptford pink, 810
Descurainia, **686**
 pinnata (Map 1341, Pl. 319 a–c),
  686
  ssp. brachycarpa, **688**
  ssp. pinnata, **688**
  var. brachycarpa, 688
 sophia (Map 1342, Pl. 319 f, g),
  688
Desert goosefoot, 887
Desmanthus, 1075, **1079**, 1080
 illinoensis (Map 1701, Pl. 387 b,
  c), **1080**, 1081
 leptolobus (Map 1702, Pl. 387
  a), **1081**
Desmodium, 940
Devil, Blue, 300, 633
Devil's lettuce, 625
Devil's shoelaces, 615
Devil's shoestrings, 615
Devil's walking stick, 132
Dewdrop, Purple, 748
Dianthus, 795, **809**, 810, 822
 armeria (Map 1460, Pl. 342 c,
  d), **810**
  ssp. armeria, **810**
  ssp. armeriastrum, 810
 barbatus (Map 1461, Pl. 342 a,
  b), 809, **810**
  ssp. barbatus, **810**

Dianthus barbatus (cont.)
  ssp. compactus, 812
 deltoides (Map 1462, Pl. 342 e,
  f), **812**
  ssp. deltoides, **812**
  ssp. degenii, 812
 plumarius (Map 1463, Pl. 342 g,
  h), **812,** 813
  ssp. plumarius, **812,** 813
 prolifer, 822
Diaperia, **429**
 prolifera (Map 1134, Pl. 294 i, j),
  430
  var. barnebyi, 430
  var. prolifera, **430**
Dibothrospermum agreste, 198
Dichondra, 930, 931, **944**
 carolinensis (Map 1593, Pl. 366
  e, f), **944,** 946
 repens, 946
Dichondraceae, 931
Dicliptera, **2**
 brachiata (Map 798, Pl. 195
  d–g), **4**
Dicots, 1
Diffuse knapweed, 326
Dill, 55, 64
Dioscorides, 343
Diospyros, 940, **990**
 ebenum, 990
 kaki, 990
 virginiana (Map 1634, Pl. 373
  e–g), **990**, 992, 1069
  f. atra, 990
  f. pumila, 990
  var. platycarpa, 990
  var. pubescens, 990
Diplotaxis, **688**
 muralis (Map 1343), **689**
Dipsacaceae, 765, **988**
Dipsacales, 765
Dipsacus, **988**
 fullonum (Map 1632, Pl. 373
  b–d), 988, **989,** 990
 laciniatus (Map 1633, Pl. 373 a),
  989, 990
 sativus, 988
 sylvestris, 989
  f. albidus, 989
Discula destructiva, 963
Dixie stitchwort, 818
Dock, Prairie, 554
Dodder, 938
 Buttonbush, 939
 Common, 942
 Compact, 940
 Cusp, 940
 Field, 939, 943
 Hazel, 940
 Large alfalfa, 943
 Pretty, 943
 Rope, 942
 Smartweed, 944

Doellingeria, **216,** 219
 umbellata (Map 970, Pl. 230
  e–g), **218**
  var. pubens, **218**
  var. umbellata, 218
Dog fennel, 177, 402
 Yellow, 490
Dog mustard, 692
Dogbane, 120
 Climbing, 124
 Pink-flowered, 121
 Prairie, 122
 Spreading, 121
Dogbane family, 115
Dogwood family, 957
Dogwood, 958, 963
 Alternate-leaved, 959
 Flowering, 962, 963
 Gray, 963, 964
 Pagoda, 959
 Pale, 959
 Rough-leaved, 960
 Stiff, 963, 964
 Swamp, 959
Dogwood anthracnose, 963
Dogwood family, 957
Doll's daisy, 209
Dollar, Silver, 710
Dore, William G., 916
Dove tree, 958
Downy arrowwood, 792
Downy goldenrod, 268
Downy lettuce, 365
Draba, **689**
 aprica (Map 1344, Pl. 320 a, b),
  690
 brachycarpa (Map 1345, Pl. 320
  g, h), **690**
  var. fastigiata, 690
 cuneifolia (Map 1346, Pl. 320 i,
  j), **690**
  var. cuneifolia, **690**
 reptans (Map 1347, Pl. 320 e, f),
  691
  var. micrantha, 691
  var. stellifera, 691
 verna (Map 1348, Pl. 320 c, d),
  691
  var. boerhaavii, 691
Draba
 Shortpod, 690
 Wedgeleaf, 690
Dracopis, 537, 538
 amplexicaulis, 537
Dress, William, 242
Drink, Black, 130
Drooping coneflower, 532
Dropwort
 Common water, 94
 Water, 92
Drummond, Thomas, 308
Drummond aster, 286
Dunn, David, 848

Dusty miller, 187
Dutch flax, 675
Dutchman's pipe, 139
Dwarf dandelion, 358
    Common, 360
    Orange, 359
    Virginia, 362
    Western, 362
Dwarf fleabane, 216
Dwarf morning glory, 935
Dwarf St. John's-wort, 922
Dwarf sumac, 45
Dyssodia, **472**
    papposa (Map 1170, Pl. 276 c, d), **472**
Dysphania, 876
    ambrosioides, 875
    anthelmintica, 876
    botrys, 879
    pumilio, 888

E

Ear
    Cat's, 357, 376
    Hare's, 68
    Smooth cat's, 358
    Spotted cat's, 357
Early cudweed, 432
Early goldenrod, 263
Early scorpiongrass, 646
Early winter cress, 666
Eastern hop hornbeam, 613
Eastern pipevine swallowtail, 138
Eastern poison oak, 49
Eastern prickly pear, 738
Eastern redbud, 1061
Ebenaceae, **990,** 1069
Ebony, 990
Ebony family, 990
Echinacea, 436, **473,** 474, 476
    angustifolia (Map 1171, Pl. 277 a, b), 474, **475**
        var. strigosa, 475
    atrorubens, 478
        var. paradoxa, 476
    laevigata, 478
    pallida (Map 1172, Pl. 277 c, d), 474, **475,** 476, 479
        f. albida, 476
        var. angustifolia, 475
        var. simulata, 478
    paradoxa (Map 1173, Pl. 277 g), **476**
        var. neglecta, 478
        var. paradoxa, **476**
    paradoxa × simulata, 476
    purpurea (Map 1174, Pl. 277 e, f), 474, **478**
        f. liggetii, 478
    sanguinea, 476

Echinacea (cont.)
    simulata (Map 1175), 476, **478,** 479
    speciosa, 478
Echinocystis, **984,** 988
    lobata (Map 1628, Pl. 372 d), **984**
Echites difformis, 124
Echium, 620, **632,** 633
    menziesii, 625
    vulgare (Map 1294, Pl. 306 a, b), **633**
Eclipta, **479**
    alba, 479, 480
    prostrata (Map 1176, Pl. 278 a–c), **479,** 480
Eggert, Henry, 827
Ehretia family, 992
Ehretiaceae, 620, **992,** 993
Elaeagnaceae, **994**
Elaeagnus, **994**
    angustifolia (Map 1636, Pl. 374 h, i), **995**
    multiflora, 995
    umbellata (Map 1637, Pl. 374 j, k), **995,** 996
Elatinaceae, **996**
Elatine, **996,** 998
    americana, 998
    rubella, 999
    triandra (Map 1639, Pl. 374 a–c), **998**
        var. americana (Pl. 374 b, c), **998**
        var. brachysperma, 998
        var. triandra (Pl. 374 a), **999**
Elder
    Big marsh, 471
    Box, 9, 10, 48
    Marsh, 519, 520
Elderberry, 775, 776
    Common, 776
    Red-berried, 777
    Stinking, 777
Elecampane, 572
Election pink, 1004
Elephantopus, **591**
    carolinianus (Map 1260, Pl. 299 a, b), **591**
    nudatus, 591
    tomentosus, 591
Elephant's foot, 591
    Carolina, 591
Elm-leaved goldenrod, 276
Endangered Species Act, 148, 209, 226, 493, 714, 814
Endive, 344, 347
Engelmann, George, 192, 454, 1014
Engelmannia, **480**
    peristenia (Map 1177, Pl. 278 d, e), **480,** 482
    pinnatifida, 480
Engelmann's daisy, 480

English chamomile, 192
English daisy, 198
English ivy, 134, 135
Epigaea, 1000
Epilobium, 218
    leptophyllum, 218
Equisetum, 940
Erechtites, **580**
    hieracifolia, 580
    hieracifolius (Map 1249, Pl. 297 c–e), **580**
        var. hieracifolius, **580**
        var. intermedius, 580
        var. megalocarpus, 582
Ericaceae, **999,** 1000
Erigenia, **79**
    bulbosa (Map 856, Pl. 206 c, d), **79**
Erigeron, 203, 206, 214, **218,** 219, 942, 943
    annuus (Map 971, Pl. 233 g, h), **220,** 222, 224, 225
        ssp. *strigosus,* 223
        var. discoideus, 222
    canadensis, 214
    divaricatus, 216
    philadelphicus (Map 972, Pl. 233 a–c), **222**
        var. acaulescens, 223
        var. glaber, 223
        var. provancheri, 223
        var. scaturicola, 223
    pulchellus (Map 973, Pl. 233 k–n), 218, 219, **223**
        var. brauniae, 223
        var. pulchellus, **223**
        var. tolsteadtii, 223
    pusillus, 216
    ramosus, 223
    strigosus (Map 974, Pl. 233 I, j), 219, 220, 222, **223,** 224, 225
        var. beyrichii, 223, 224
        var. calcicola, 224
        var. discoideus, 223, 224
        var. dolomiticola, 224
        var. septentrionalis, 220, 224
        var. strigosus, **223**
    tenuis (Map 975, Pl. 233 d–f), **224,** 225
Erophila
    verna, 691
    ssp. *praecox, 691*
Eruca, **692**
    sativa, 692
    vesicaria (Map 1349, Pl. 319 h, i), **692**
        ssp. sativa, **692**
Erucastrum, **692,** 694
    gallicum (Map 1350, Pl. 319 d, e), **692,** 694
Eryngium, 54, **80**
    leavenworthii (Map 857), **80**

Eryngium (cont.)
  prostratum (Map 858, Pl. 207 d, e), **81**
  yuccifolium (Map 859, Pl. 207 f, g), **82**
Eryngo, 80
  Leavenworth, 80
  Spreading, 81
Erysimum, **694**
  asperum, 695
  capitatum (Map 1351, Pl. 321 c, d), **695**
    var. capitatum, **695**
  cheiranthoides (Map 1352, Pl. 321 e, f), **695**
    ssp. altum, 695
    ssp. cheiranthoides, 695
  inconspicuum (Map 1353, Pl. 321 a, b), **695**
    var. coarcticum, 696
    var. inconspicuum, 696
  repandum (Map 1354, Pl. 321 i-k), **696**
Escobaria missouriensis, 737
Euonymus, 850, **852,** 854
  alatus (Map 1504, Pl. 350 h, i), **854**
  americanus (Map 1505, Pl. 350 a-c), **854,** 855, 858
    var. prostratus, 856
  atropurpureus (Map 1506, Pl. 351 h-j), **855**
  europaeus, 855
  fortunei, 126, 856
  hederaceus (Map 1507, Pl. 351 a, b), 126, **856**
  japonicus, 856
  kiautschovicus, 856
  obovatus (Map 1508, Pl. 351 k, l), 855, **856,** 858
Euonymus
  Brook, 854
  Winged, 854
Eupatoriadelphus, 401
  fistulosus, 403
  maculatus, 404
    var. bruneri, 404
  purpureus, 407
Eupatorium, 393, **400,** 401, 412, 940, 943
  altissimum (Map 1109, Pl. 266 a, b), **402,** 404
  capillifolium (Map 1110, Pl. 266 e, f), 400, **402,** 403
  coelestinum, 400
  cuneifolium
    var. semiserratum, 408, 410
  fistulosum (Map 1111, Pl. 265 e-g), **403,** 407
  hyssopifolium (Map 1112, Pl. 266 g, h), 402, **403,** 408, 410
    var. calcaratum, **403,** 404

Eupatorium hyssopifolium (cont.)
    var. hyssopifolium, 404
    var. laciniatum, 404
  incarnatum, 412
  maculatum (Map 1113, Pl. 265 c, d), 218, 403, **404**
    ssp. bruneri, 404
    var. bruneri, **404,** 406
    var.foliosum, 406
    var. maculatum, 406
  perfoliatum (Map 1114, Pl. 268 a, b), 400, **406,** 407
    var. colpophilum, 407
    var. cuneatum, 407
    var. perfoliatum, **406**
  ×polyneuron, 407
  purpureum (Map 1115, Pl. 265 a, b), 403, **407**
    var. holzingeri, **407**
    var. purpureum, **407**
  resinosum, 407
    var. truncatum, 407
  rotundifolium (Map 1116), 404, **407,** 408, 412
    var. cordigerum, 407
    var. scabridum (Map 1116), **407**
  rugosum, 395
    f. villicaule, 395
    var. tomentellum, 395
  semiserratum (Map 1117, Pl. 266 c, d), 404, **408,** 410
  serotinum (Map 1118, Pl. 268 c, d), 407, **410**
  sessilifolium (Map 1119, Pl. 267 h, i), 407, **410**
    var. brittonianum, 410, 412
    var. sessilifolium, 412
  ×truncatum, 407
Eupelmidae, 548
Euphorbia, 939, 943, 1010, 1018, **1027,** 1028, 1029, 1034, 1044
  antisyphilitica, 1029
  commutata (Map 1663, Pl. 380 c-e), **1032,** 1034
    var. erecta, 1032
  corollata (Map 1664, Pl. 379 a, b), 1029, **1034**
    var. mollis, 1034
    var. paniculata, 1034
  cruentata, 1038
  cuphosperma, 1038
  cyathophora (Map 1665, Pl. 379 c-e), **1034,** 1035
  cyparissias (Map 1666, Pl. 380 a, b), **1035,** 1040
  davidii (Map 1667), **1036**
  dentata (Map 1668, Pl. 379 i, j), 1036, **1038,** 1040
    f. cuphosperma, 1038
    var. cuphosperma, 1038
    var. gracillima, 1036

Euphorbia dentata (cont.)
    var. lancifolia, 1036
    var. linearis, 1038
  esula (Map 1669, Pl. 380 i, j), 1029, **1040**
  esula × cyparissias, 1041
  geyeri (Map 1670, Pl. 381 e-g), **1041**
    var. geyeri, **1041**
    var. wheeleriana, 1041
  glyptosperma (Map 1671, Pl. 381 a, b), **1041**
  heterophylla, 1035
    var. graminifolia, 1034
  hexagona, 1029
  humistrata (Map 1672, Pl. 382 g-i), **1042,** 1044
  hypericifolia, 1046
  maculata (Map 1673, Pl. 382 j-m), **1042,** 1044, *1045,* 1046
  marginata (Map 1674, Pl. 379 f-h), **1044**
  missurica (Map 1675, Pl. 381 h-j), **1045**
    var. intermedia, 1045
  nutans (Map 1676, Pl. 377 h-j), 1044, **1045,** 1046
  obtusata (Map 1677, Pl. 380 f-h), **1046,** 1047
  platyphylla, 1047
  platyphyllos (Pl. 383 c, d), 1047
  preslii, 1045
  prostrata (Map 1678, Pl. 382 d-f), **1047**
  ×pseudoesula, 1041
  ×pseudovirgata, 1040
  pubentissima, 1034
  pulcherrima, 1011, 1029
  serpens (Map 1679, Pl. 381 k-m), **1048**
  serpyllifolia (Map 1680, Pl. 381 c, d), **1048**
  spathulata (Map 1681, Pl. 383 g-i), 1047, **1049**
  stictospora (Map 1682, Pl. 382 a-c), **1049**
  supina, 1042
  virgata, 1040
Euphorbiaceae, 525, **1010,** 1028
  Tribe Euphorbieae, 1028
Eurasian smoke tree, 41
Euritides marcellus, 52
European alder, 609
European fly honeysuckle, 768
Eurota solidaginis, 250
Eurybia, 219, **225,** 278
  furcata (Map 976, Pl. 230 c, d), **226**
  hemispherica (Map 977, Pl. 229 c-e), 225, **227**
  macrophylla (Map 978, Pl. 230 a, b), **227**

Eurybia (cont.)
　　paludosa, 227
Eurytomidae, 548
Euthamia, 198, **228,** 229
　　graminifolia (Map 979, Pl. 234
　　　g–i), **230,** 232
　　　var. graminifolia, 230
　　　var. major, 230
　　　var. nuttallii, 230
　　gymnospermoides (Map 980, Pl.
　　　234 d–f), **230,** 232
　　leptocephala (Map 981, Pl. 234
　　　a–c), **232**
Eutrochium, 401
　　fistulosum, 403
　　maculatum, 404
　　　var. bruneri, 404
　　　var. holzingeri, 404
　　purpureum, 407
Evax, 429
　　prolifera, 430
Evening campion, 837
Everlasting, 426, 430, 433
　　Pearly, 426
　　Sweet, 434
Evolvulus, **946**
　　alsinoides (Map 1594, Pl. 366 a,
　　　b), **946**
　　nuttallianus (Map 1595, Pl. 366
　　　g–i), **946**
　　pilosus, 946
Evonymus, 852
Eyebane, 1045
Eyes
　　Green, 449
　　Texas green, 449

F

Fabaceae, 962, **1057,** 1058, 1064,
　　1069
　　Subfamily Caesalpinioideae,
　　　1058, **1059,** 1060, 1064
　　Subfamily Faboideae, 1059,
　　　1062
　　Subfamily Mimosoideae, 1059,
　　　1064, **1075,** 1076
Fagus, 129
Falcaria, **82**
　　sioides, 82
　　vulgaris (Map 860, Pl. 207 a–c),
　　　82
False aster, 206, 208
　　Decurrent, 208, 209
False beechdrops, 1003
False boneset, 397
False chamomile, 191
False chamomile, Scentless, 196
False dandelion, 382
False flax, 675, 676
　　Common, 676
　　Littlepod, 675

False gromwell, 646
　　Western, 648
False Indian plantain, 582
False rampion, 744
False starwort, 208
False sunflower, 517
Family
　　Acanthus, 2
　　Amarath, 18
　　Barberry, 602
　　Bean, 1057
　　Bellflower, 741
　　Birch, 607
　　Blueberry, 999
　　Borage, 619
　　Cactus, 736
　　Carrot, 54
　　Cashew, 40
　　Cleome, 908
　　Custard apple, 51
　　Dogbane, 115
　　Dogwood, 957
　　Ebony, 990
　　Ehretia, 992
　　Fig-marigold, 17
　　Ginseng, 131
　　Goosefoot, 860
　　Gourd, 974
　　Heath, 999
　　Hemp, 758
　　Holly, 126
　　Honeysuckle, 765
　　Hornwort, 858
　　Maple, 7
　　Milkweed, 115, 140
　　Morning glory, 930
　　Mustard, 651
　　Oleaster, 994
　　Pink, 794
　　Rockrose, 902
　　Spurge, 1010
　　Staff-tree, 850
　　Stonecrop, 967
　　St. John's wort
　　Strawberry shrub, 740
　　Sunflower, 164
　　Touch-me-not, 599
　　Water shield, 732
　　Waterwort, 996
Fanwort, 734
Farkleberry, 1007
Fat hen saltbush, 865
Feather, Prince's, 28
Feather geranium, 879
Federated Garden Clubs of
　　Missouri, 129
Fennel, 84, 180
　　Dog, 177, 402
　　Hog, 94
Fetid marigold, 472
Fetterbush, 1000
Feverfew, 195
　　American, 526

Feverwort, 782
Few-bracted bur marigold, 457
Fiddleneck
　　Bugloss, 624
　　Small-flowered, 625
　　Tarweed, 624
Field bindweed, 936
Field chickweed, 803, 809
Field cottonweed, 38
Field cress, 703
Field dodder, 939, 943
Field hedge-parsley, 111
Field mustard, 674
Field penny cress, 730
Field pepper grass, 703
Field pussytoes, 428
Field snake-cotton, 38
Field sow thistle, 384
Field thistle, 335, 338
Fig-leaved goosefoot, 882
Fig-marigold family, 17
Filago, 429
Filbert, 613
Finches, 513
Finger coreopsis, 464
Fire bush, 896
Fire pink, 841
Fire-on-the-mountain, 1034
Fireweed, 580
Fire-wheels, 484
Fish and Wildlife Service, U.S.,
　　136, 148, 209, 493, 714
Five-stamen mouse-ear chickweed,
　　808
Flat-topped goldenrod, 228
　　Common, 230
　　Great Plains, 230
　　Mississippi Valley, 232
Flat-topped white aster, 218
Flaveria, **482**
　　campestris (Map 1178, Pl. 278
　　　f–h), **482**
Flax, 725
　　Common false, 676
　　Dutch, 675
　　False, 675, 676
　　Littlepod false, 675
Flax-leaved aster, 243
Flaxweed, 675
Fleabane, 218
　　Annual, 220
　　Canada, 214
　　Daisy, 220, 223
　　Dwarf, 216
　　Inland marsh, 574
　　Marsh, 573, 574
　　Philadelphia, 222
　　Spreading, 216
　　Stinking, 574
　　Whitetop, 220, 223
Fleischmannia, 400, 401, **412**
　　incarnata (Map 1120, Pl. 267 d,
　　　e), **412**

Flixweed, 688
Flores Olvera, Hilda, 863, 869
Florida lettuce, 365
Florida maple, 15
Flower
    American basket, 323
    Balloon, 742
    Blue Cardinal, 748, 750
    Cardinal, 747
    Chigger, 154
    Frost, 561
    Ghost, 1003
    Illinois bundle, 1080
    Spider, 910
    Tassel, 398
Flower head weevil, 320
Flowering dogwood, 962, 963
Flowering spurge, 1034
Fly, tephrid, 250
Fly honeysuckle, European, 768
Fodder beet, 870
Foeniculum, **84**
    vulgare (Map 861, Pl. 207 h, j), **84**
Foot
    Elephant's, 591
    Carolina Elephant's, 591
Fordhook, 982
Forget-me-not, 642, 644
    Giant, 632
    Small-flowered, 644
Forked aster, 226, 227
Forked catchfly, 834
Forked chickweed, 821
Forked scale-seed, 104
Fourleaf milkweed, 149
Fragrant cudweed, 434
Fragrant goldenrod, 267
Fragrant sumac, 42, 48
Frangipani, 115
Franseria, 443
    acanthicarpa, 443
    discolor, 446
Freeman, Craig, 475
Freeway aster, 311
French marigold, 558
Fringed quickweed, 486
Fringeleaf ruellia, 6
Froelichia, **36**, 943
    floridana (Map 826, Pl. 199 f), **38**
    var. campestris, **38**
    gracilis (Map 827, Pl. 199 g, h), **38**
Frost flower, 561
Frostweed, 566, 903
    Hoary, 904
Fuller's teasel, 988

G

Gaillardia, 436, **483,** 485

Gaillardia (cont.)
    aestivalis (Map 1179, Pl. 278 k, l), **483,** 484
    var. aestivalis, 484
    var. flavovirens, **483,** 484
    var. winkleri, 484
    lutea, 483, 484
    pulchella (Map 1180, Pl. 278 I, j), 469, **484**
    var. australis, 485
    var. picta, 485
    var. pulchella, **484,** 485
Gaillardia, 483
    Prairie, 483
Galinsoga, **485**
    ciliata, 485
    parviflora (Map 1181, Pl. 279 a, b), **486**
    var. semicalva, 486
    quadriradiata (Map 1182, Pl. 279 c, d), **486**
Galium boreale, 746
Gall moth, goldenrod, 252
Gallant soldier, 485
Gamochaeta, 165, **430,** 431
    argyrinea (Map 1135, Pl. 295 d), **431,** 432
    pensylvanica (Map 1136, Pl. 295 e), **432**
    purpurea (Map 1137, Pl. 295 a-c), 431, **432**
Garden balsam, 600
Garden beet, 869
Garden cosmos, 459
Garden goldenglow, 544
Garden lettuce, 370
Garden orach, 864
Garden pink, 812
Garden radish, 716
Garden rocket, 692
Garland chrysanthemum, 189
Garlic mustard, 658
Gayfeather, 418
    Narrow-leaved, 416
    Rough, 415
Gaylussacia, 999, **1000,** 1001
    baccata (Map 1640, Pl. 374 d, e), **1000,** 1001
Gentian, Horse, 780
Gentianaceae, 380
Geocarpon, 17, **813,** 842
    minimum (Map 1464, Pl. 343 a, b), **813,** 814
Geranium
    Feather, 879
    Mint, 194
German chamomile, 191
Geyer's spurge, 1041
Ghost flower, 1003
Ghost tree, 958
Giant chickweed, 819, 820
Giant cow parsnip, 85
Giant forget-me-not, 632

Giant ragweed, 448
Giant sloth, 1069
Giant St. John's wort, 915
Giardia, 712
Ginger, Wild, 140
Ginseng, 135
    American, 135, 136
Ginseng family, 131
Glade coneflower, 478
Gladecress
    Michaux's, 701
    Necklace, 700
Glasswort, 898
Glatfelter, Noah, 789
Glaucous campion, 834
Glaucous white lettuce, 380
Gleditsia, 1060, **1066**
    aquatica (Map 1692, Pl. 386 j, k), 1066, **1068**
    japonica, 1066
    ×texana, 1066
    triacanthos (Map 1693, Pl. 386 g-i), 1066, **1068,** 1069, 1070
    f. inermis, 1069
Glinus, 17
Glycine, 943
Globe amaranth, 18
Globe artichoke, 316
Globe-podded hoary cress, 703
Glory
    Morning, 948
    Wild morning, 934
Glossy-leaved aster, 308
Gnaphalium, 431
    helleri, 434
        var. micradenium, 434
    obtusifolium, 434
        var. micradenium, 434
    pensylvanicum, 432
    polycephalum, 434
    purpureum, 432
        var. spathulatum, 432
Gnorimoschema gallaesolidaginis, 252
Goat's beard, 390
Golden Alexanders, 112
    Common, 114
    Heart-leaved, 114
Golden aster, Grass-leaved, 239
Golden buttons, 195
Golden carpet, 970
Golden crownbeard, 562
Golden marguerite, 188
Golden ragwort, 584
Golden ragwort, Western, 588
Golden selenia, 723
Golden star, 238
Goldenaster, Soft, 210
Goldenglow
    Garden, 544
    Wild, 543
Goldenrod, 244, 263, 270

1158   INDEX

Goldenrod (cont.)
  Blue-stemmed, 254
  Broadleaf, 260
  Downy, 268
  Early, 263
  Flat-topped, 228
  Common, 256
  Common flat-topped, 230
  Elm-leaved, 276
  Fragrant, 267
  Gray, 264
  Great Plains flat-topped, 230
  Late, 261
  Mississippi Valley flat-topped, 232
  Missouri, 263
  Old-field, 264
  Ozark, 258
  Prairie, 275
  Rigid, 271
  Rough, 270
  Rough-leaved, 267, 274
  Rough-stemmed, 274
  Showy, 275
  Stiff, 271, 272
  Sweet, 267
  Tall, 248, 261
  Viscid bushy, 230
  Wreath, 254
Goldenrod bunch galler, 252
Goldenrod gall moth, 252
Goldenrod stem galler, 252
Goldentops, Alkali 482
Goldenweed, 233, 234
Goldfinch, 335
Gold-of-pleasure, 675
Gomphrena, 18
Gonolobus, **160**, 163
  gonocarpos, 160, 162
  suberosus (Map 934, Pl. 223 j-l), **160**, 162
    var. *granulatus,* 160
Good King Henry, 871
Gooseberry, Southern, 1009
Goosefoot, 870
  City, 890
  Clammy, 888
  Desert, 887
  Fig-leaved, 882
  Maple-leaved, 889
  Nettle-leaved, 884
  Oak-leaved, 882
  Pitseed, 876
  Red, 888
  Slim-leaved, 887
  Woodland, 889
Goosefoot family, 860
Gourd, 984
  Bottle, 985, 986
  Buffalo, 982, 983
  Ivy, 975
  Missouri, 982
  Ornamental, 982, 983

Gourd (cont.)
  Scarlet, 975
  Texas wild, 984
  White-flowered, 985
  Wild, 983, 984
  Wild pear, 983
  Yellow-flowered, 983
Gourd family, 974
Goutweed, 55
Grain amaranth, 20
Grape, Oregon, 603
Grape honeysuckle, 774
Grass, 1058
  Broad-leaved pepper, 707
  Field pepper, 703
  Green-flowered pepper, 704
  Orange, 917
  Pepper, 701, 704, 708
  Perennial pepper, 707
  Perfoliate pepper, 707
  Poor Man's pepper, 708
  Vernal whitlow, 691
  Virginia pepper, 708
  Whitlow, 689, 690
Grass-leaved golden aster, 239
Graveyard spurge, 1035
Gray dogwood, 063, 964
Gray goldenrod, 264
Gray mouse-ear chickweed, 803
Gray-headed coneflower, 532
Great coneflower, 544
Great flood, 209
Great Indian plantain, 578
Great lobelia, 750
Great Plains flat-topped goldenrod
Great ragweed, 448
Great St. John's-wort, 915
Great white lettuce, 380
Greater chickweed, 846
Greater periwinkle, 126
Greater St. John's-wort, 921
Green amaranth, 28
Green eyes, Texas, 449
Green milkweed, 157
  Tall, 145
Green osier, 959
Green prairie coneflower, 532
Green-flowered milkweed, 158
Green-flowered pepper grass, 704
Green-stemmed Joe-pye weed, 407
Greenthread
  Colorado, 560
  Stiff, 559
Grindelia, 165, **232**, 233, 234, 943
  camporum, 233
  ciliata (Map 982, Pl. 237 a, b), **233**
  lanceolata (Map 983, Pl. 235 f-h), **234**
    f. *latifolia,* 234
  nuda, 236
  papposa, 233, 234
  perennis, 236, 237

Grindelia (cont.)
  squarrosa (Map 984, Pl. 235 a-e), **236**
    var. nuda (Pl. 235 d, e), **236**
    var. quasiperennis, 236, **237**
    var. serrulata, 236, *237*
    var. squarrosa (Pl. 235 a-c), **237**
Gromwell
  American, 640
  Corn, 628
  False, 646
  Western false, 648
Groundfig spurge, 1047
Groundsel
  Common, 590
  Roundleaf, 586
  Texas, 590
Grove sandwort, 819
Guayule, 525
Guelder rose, 790, 791
Guizotia, 436, **488**
  abyssinica (Map 1183), **488**
Gum
  Black, 966, 967
  Cotton, 965
  Sour, 964, 966
  Swamp black, 966
  Tupelo, 965, 966
Gum plant, 232
Gumweed, 232
  Curlytop, 236
  Spiny-toothed, 234
Gutierrezia, 202, 229, **237**
  dracunculoides, 202
  texana (Map 985, Pl. 235 i-k), **237**
  var. *glutinosa,* 237
Guttiferae, **912**
Gymnocladus, 1060, 1066, **1069**
  canadensis, 1069
  dioica (Map 1694, Pl. 385 j-l), **1069**, 1070
  dioicus, 1070
Gypsophila, 795

H

Hackelia, **633**, 636
  virginiana (Map 1295, Pl. 306 c, d), **633**
Hairy angelica, 64
Hairy bugseed, 894
Hairy goldenrod, 262
Hairy lettuce, 365
Hairy nailwort, 821
Hairy parsley, 90
Hairy pinweed, 906
Hairy rock cress, 662
Hairy sunflower, 503
Haloragaceae, 454
Haplopappus, 226, 234
  ciliatus, 233, 234

Haplophyton, 116
Harbinger of spring, 79, 80
Hard maple, 13
Hardy catalpa, 619
Harebell, 746
Hare's ear, 68
Hare's-ear mustard, 686
Hartman, Ronald L., 794
Hashish, 760
Hasteola, 576, **582**
    suaveolens (Map 1250, Pl. 296 f, g), **582**
Hat, Mexican, 532
Haw, Possum, 128
Hawkbit, 371
    Lesser, 372
Hawksbeard, 348
    Bristly, 350
    Narrow-leaved, 349
    Small-flowered, 349
    Smooth, 349
Hawkweed, 353
    Beaked, 354
    Long-haired, 356
    Sticky, 357
Hawkweed ox-tongue, 376
Hazel, 612, 613
    Beaked, 613
    European, 613
Hazel dodder, 940
Hazelnut, 612
    American, 612
Heart-leaved golden Alexanders, 114
Heart-leaved meadow parsnip, 114
Heart-leaved tragia, 1056
Heart-podded hoary cress, 706
Heart-podded white top, 706
Heath, 1000
Heath aster, White, 305
Heath family, 999
Hedera, **134**
    helix (Map 910, Pl. 219 c), **134,** 135
        var. helix, 135
        var. hibernica, 135
Hedge bindweed, 932, 934
Hedge mustard, 726
    Tall, 726
Hedge-parsley, 110
    Field, 111
    Japanese, 111
    Knotted, 111
Hedyotis, 943
Heleastrum hemisphericum, 227
Helenium, 436, **489,** 490
    amarum (Map 1184, Pl. 279 g), 489, **490**
        var. amarum, **490**
        var. badium, 490
    autumnale (Map 1185, Pl. 279 e, f), **490,** 492, 493
        var. autumnale, 491

Helenium autumnale (cont.)
        var. canaliculatum, 490, 491
        var. parviflorum, 490, 491
    flexuosum (Map 1186, Pl. 279 h, i), **491,** 492, 493
    tenuifolium, 490
    virginicum (Map 1187), **492,** 493
Helianthemum, **903**
    bicknellii (Map 1553, Pl. 359 e, f), **904**
    canadense (Map 1554, Pl. 359 g–i), **904,** 905
Helianthus, 436, **493,** 494, 517, 548, 940, 942, 943
    angustifolius (Map 1188, Pl. 280 a, b), **497**
    annuus (Map 1189, Pl. 281 g, h), 165, 436, 488, **497,** 498, 500, 512, 516
        cv. 'Nanus Florepleno', 498, 500
        var. annuus, 500
        var. lenticularis, 497, 500
        var. macropocarpus, 497, 500
    ×cinereus
        var. sullivantii, 508
    decapetalus (Map 1190, Pl. 282 e, f), 493, **500,** 501
    divaricatus (Map 1191, Pl. 282 c, d), 494, **501,** 502, 504, 506, 508, 515
    dowellianus, 510
    formosus, 515
    ×glaucus, 506
    grosseserratus (Map 1192, Pl. 280 e, f), **502,** 508
        f. pleniflorus, 503
        ssp. maximus, 502
    hirsutus (Map 1193, Pl. 282 a, b), 494, 502, **503,** 504, 514, 515
        var. stenophyllus, 503
        var. trachyphyllus, 503
    ×intermedius, 503
    ×laetiflorus, 494, 511, 516
        var. rigidus, 510
    maximilianii (Map 1194, Pl. 280 g, h), 503, **504,** 506, 508
    microcephalus (Map 1195, Pl. 281 a, b), **506,** 508
    mollis (Map 1196, Pl. 282 g, h), 502, **506,** 508
        f. flavidus, 506
    occidentalis (Map 1197, Pl. 281 e, f), **508**
        ssp. occidentalis, **508,** 510
        ssp. plantagineus, 510
    occidentalis × grosseserratus, 510
    pauciflorus (Map 1198, Pl. 282 i, j), **510,** 511
        ssp. pauciflorus, **511**
        ssp. subrhomboideus, **511**

Helianthus pauciflorus (cont.)
        var. subrhomboideus, 511
    petiolaris (Map 1199, Pl. 281 c, d), 500, **512**
        ssp. petiolaris, **512**
    rigidus, 510, 511, 516
        ssp. subrhomboideus, 511
    salicifolius (Map 1200, Pl. 280 c, d), 497, **512**
    silphioides (Map 1201, Pl. 283 c, d), **513**
    strumosus (Map 1202, Pl. 283 a, b), 494, 501, 502, 504, **514,** 515, 516
    subrhomboideus, 511
    tuberosus (Map 1203, Pl. 283 e, f), 436, 494, 502, 504, 511, **515,** 516
    var. subcanescens, 515, 516
Heleastrum hemisphericum, 227
Heliopsis, 436, 493, **516,** 517
    helianthoides (Map 1204, Pl. 283 g–i), **517,** 518
        ssp. occidentalis, 517, 518
        ssp. scabra, 517, 518
        var. helianthoides, **518**
        var. occidentalis, 518
        var. scabra, **518**
Heliotrope, Winter, 576
Heliotropiaceae, 620, 993
Heliotropium, 620
Hell vine, 615
Helminthotheca , **352**
    echioides (Map 1066, Pl. 260 j, k), **352**
Helwingia, 957
Hemlock, 55, 74
    Poison, 74
    Spotted Water, 70
Hemlock chervil, 111
Hemp, 759, 760
    Indian, 120, 122
    Water, 32
Hemp family, 758
Hempweed, Climbing, 424
Hen saltbush, Fat, 865
Henderson, Norlan, 512, 554
Henry, Good King
Heracleum, 63, 66, **84,** 95
    lanatum, 85
    mantegazzianum, 85
    maximum, 85
    sphondylium (Map 862, Pl. 208 e, f), **85**
        ssp. montanum, **85**
        ssp. sphondylium, 85
Herb, Cow, 849
Hercules' club, 132
Hesperis, **696**
    matronalis (Map 1355, Pl. 322 f, g), **696**
Heterotheca, 210, **238,** 239

Heterotheca (cont.)
  camporum (Map 986, Pl. 236 a, b), **239**
    var. camporum (Map 986), **240**
    var. glandulissimum (Map 986), **240**
  canescens (Map 987, Pl. 236 c–e), **240**, 242
    latifolia, 242, 243
      var. arkansana, 242
    pilosa, 210
    psammophila, 243
    stenophylla (Pl. 236 i–k), 242
      var. angustifolia, 242
    subaxillaris (Map 988, Pl. 236 f–h), **242**, 243
      ssp. latifolia, 242
    villosa, 239
      var. angustifolia, 242
Hevea, 525
  brasiliensis, 525, 1010
Hibiscus, 943
Hieracium, 344, **353**, 377, 387
  caespitosum (Map 1067), 353, **354**
  floridanum, 354
  gronovii (Map 1068, Pl. 256 a, b), **354**
    var. foliosum, 354
  longipilum (Map 1069, Pl. 256 f, g), **356**
    f. eglandulosum, 357
  scabrum (Map 1070, Pl. 256 e), 356, **357**
    var. intonsum, 357
  venosum, 356
Highbush blueberry, 1007, 1008
Highbush cranberry, 791
Highbush huckleberry, 1009
Hillside blueberry, 1008
Hippocastanaceae, 8
Hoary alyssum, 667
Hoary bitter cress, 682
Hoary blackfoot, 523
Hoary cress, 703
  Globe-podded, 703
  Heart-podded, 706
  Lens-podded, 703
Hoary frostweed, 904
Hoary puccoon, 638
Hodgdon, Albion R., 906
Hoffmanseggia, 1060, 1071
  jamesii, 1070
Hog fennel, 94
Hogbite, 344
Hogweed, 214, 444
Hogwort, 1021
Hollow-stemmed Joe-pye weed, 403
Holly, 126
  American, 128, 129
  Deciduous, 128
Holly family, 126

Holosteum, 795, **814**
  umbellatum (Map 1465, Pl. 343 g, h), **814**
    ssp. umbellatum, **814**
Holzinger's Venus' looking glass, 754
Honesty, 710
Honewort, 75
Honey locust, 1068, 1069
Honey mesquite, 1083
Honeybee, 616, 1001, 1007
Honeydew, 980
Honeysuckle, 766, 782
  Amur, 772, 773
  Bush, 765, 772
  Coral, 774
  European fly, 768
  Grape, 774
  Japanese, 772
  Limber, 770
  Morrow, 768
  Red, 770
  Tatarian, 768
  Trumpet, 774
  Wild, 770, 1004
  Yellow, 770
Honeysuckle family, 765
Honeysweet, 39
Hooked scale-seed, 104
Hop hornbeam, 613, 614
  Eastern, 613
Hopi sunflower, 500
Hopi tea, 560
Hops, 760, 762
  Common, 762
  Japanese, 761
Hornbeam, 610
  Eastern hop, 613
  Hop, 613, 614
Horned cucumber, 979
Horned melon, 979
Hornwort, 858
Hornwort family, 858
Horse, 1069
Horse gentian, 780
  Common, 782
  Red-fruited, 781
  Yellow-flowered, 781
Horse purslane, 17
Horseradish, 664, 679, 680
Horsetail milkweed, 157
Horseweed, 214, 448
Hound's tongue, 630
  Common, 632
Huckleberry
  Black, 1000
  Highbush, 1009
  Squaw, 1009
Hudson, Stan, 86, 304
Hummingbird, 616, 748, 775
  Ruby-throated, 600
Humulus, 758, **760,** 762, 764
  americanus, 764

Humulus (cont.)
  japonicus (Map 1419, Pl. 333 f, g), **761**
  lupulus (Map 1420, Pl. 333 e), **762,** 764, 765
    var. lupuloides, 765
    var. lupulus, **764,** 765
    var. neomexicanus, **764**
    var. pubescens, **764,** 765
Hura, 1011
Hydrangea, 940
Hydrocotyle, 55, **85**
  ranunculoides (Map 863), **86**
  umbellata, 86, 88
  verticillata (Map 864, Pl. 208 c, d), **86,** 88
    var. verticillatum, 88
Hydrophyllaceae, 620, 992, 993
Hylotelephium, 970
  erythrostictum, 970
  telephioides, 971
Hymenoclea, 443
Hymenopappus, 172, **519**
  scabiosaeus (Map 1205, Pl. 227 a–c), **519**
    var. scabiosaeus, **519**
    var. corymbosus, 519
Hymenophysa pubescens, 703
Hypericaceae, 913
Hypericum, 912, **913,** 920, 924, 929, 943
  adpressum (Map 1561), **915**
  ascyron (Map 1562, Pl. 361 c, d), **915,** 916
    ssp. ascyron, 916
    ssp. gebleri, 916
    ssp. pyramidatum, **915**
  boreale, 924
  canadense, 922
    var. majus, 921
  cistifolium, 928
  ×dawsonianum, 921
  densiflorum, 921
    var. lobocarpum, 920
  dolabriforme, 928
  drummondii (Map 1563, Pl. 360 d–f), **916**
  gentianoides (Map 1564, Pl. 360 a–c), **917**
  gymnanthum (Map 1565, Pl. 360 g–i), **917,** 918, 924
  hypericoides (Map 1566, Pl. 361 i, j), 913, **918,** 920
    ssp. hypericoides, **920**
    ssp. multicaule, **920**
    ssp. prostratum, 920
    var muticaule, 920
  lobocarpum (Map 1567, Pl. 362 g, h), **920,** 921
  majus (Map 1568, Pl. 360 l, m), **921,** 922
  mutilum (Map 1569, Pl. 360 j, k), 918, **922,** 924

Hypericum mutilum (cont.)
    ssp. boreale, 924
    ssp. latisepalum, 924
    ssp. mutilum, **922,** 924
    var. parviflorum, 922, 924
    perforatum (Map 1570, Pl. 361 e, f), **924**
    prolificum (Map 1571, Pl. 362 c, d), 921, **925**
    pseudomaculatum (Map 1572, Pl. 361 g, h), **925,** 926
    punctatum (Map 1573, Pl. 361 a, b), **926**
        f. flavidum, 925
        f. subpetiolatum, 926
        var. pseudomaculatum, 925
    pyramidatum, 915, 916
    spathulatum, 925
    sphaerocarpum (Map 1574, Pl. 362 e, f), 913, **928**
        var. turgidum, 928
    stragulum, 920
    tubulosum, 929
        var. walteri, 930
    walteri, 930
Hypochaeris, 344, **357**
    glabra, 358
    radicata (Map 1071, Pl. 256 c, d), **357,** 358
Hypochoeris, 357
Hypopitys americana, 1003
Hyssop-leaved thoroughwort, 403

I

Ilex, **126,** 128
    decidua (Map 903, Pl. 217 j-n), **128**
    opaca (Map 904, Pl. 217 d-f), **128,** 129
        f. subintegra, 129
        f. xanthocarpa, 129
        var. arenicola, 129
        var. opaca, **128**
    verticillata (Map 905, Pl. 217 g-i), **129**
        var. padifolia, **129**
    vomitoria (Map 906), **130**
        var. chiapensis, 130
        var. vomitoria, **130**
Illinois bundle flower, 1080
Illinois pinweed, 906
Iltis, Hugh, 910
Impatiens, **599,** 942, 943
    balsamina (Map 1266, Pl. 301 a), **600**
    biflora, 600
    capensis (Map 1267, Pl. 301 b-d), **600,** 602
        f. albiflora, 602
        f. citrina, 602
        f. immaculata, 602

Impatiens capensis (cont.)
        f. peasei, 602
    fulva, 600
    pallida (Map 1268, Pl. 301 e-g), 600, **602**
        f. dichroma, 602
        f. speciosa, 602
Indian bean, 619
Indian blanket, 484
Indian currant, 780
Indian hemp, 120, 122
Indian mustard
Indian pipe, 1002, 1003
Indian plantain, 577, 578
    False, 582
    Great, 578
    Pale, 577
Indian tobacco, 428, 748
Indiana banana, 52
Inflated lobelia, 748
Inland marsh fleabane, 574
Inland saltmarsh aster, 311
International Association of Plant Taxonomy, 851, 852, 1076
International Botanical Congress, 851, 852, 854, 866, 1076
International Cactus Systematics Group, 736
International Code of Botanical Nomenclature, 1070
Inula, **572**
    brittanica, 573
    helenium (Map 1243, Pl. 294 a-c), **572,** 573
Iodanthus, **698**
    pinnatifidus (Map 1356, Pl. 322 c-e), **698**
Ionactis, **243**
    linariifolius (Map 989, Pl. 229 a, b), **243**
Ipomoea, **948**
    amnicola (Map 1596, Pl. 367 i-k), **949**
    batatus, 948
    cairica (Pl. 367 h), 948
    coccinea (Map 1597, Pl. 366 j, k), **949,** 950, 954
    cordato-triloba, 951
    hederacea (Map 1598, Pl. 368 h-j), **950,** 951, 952
        var. integriuscula, 950
    hederifolia, 950
    hirsutula, 952
    imperati, 934
    lacunosa (Map 1599, Pl. 367 a-d), **951**
        f. purpurata, 951
    ×leucantha, 951
    ×multifida, 954
    pandurata (Map 1600, Pl. 367 e-g), **952**
        f. leviuscula, 952

Ipomoea (cont.)
    purpurea (Map 1601, Pl. 368 e-g), 951, **952**
        var. diversifolia, 952
    quamoclit (Map 1602, Pl. 366 c, d), 930, 950, **952**
    stolonifera, 934
    trichocarpa, 951
    tricolor (Map 1603), **954**
        cv. 'Heavenly Blue', 954
Iresine, 18, **38,** 940
    rhizomatosa (Map 828, Pl. 199 a, b), **38**
Irish ivy, 135
Ironweed, 592
    Missouri, 598
    Prairie, 597
    Tall, 598
    Western, 596
    Yellow, 562
Ironwood, 613
Iroquois, 136
Isatis, **700**
    tinctoria (Map 1357), **700**
Italian bugloss, 626
Iva, 165, 436, 471, **519,** 520, 942
    annua (Map 1206, Pl. 284 c, d), **520**
        var. caudata, 520, 522
        var. macrocarpa, 522
    ciliata, 520
        var. macrocarpa, 522
    xanthiifolia, 471, 519
Ivy, 134
    Atlantic, 135
    English, 134
    Irish, 135
    Poison, 46, 48, 49, 599
Ivy gourd, 975
Ivy-leaved morning glory, 950

J

Jack-go-to-bed-at-noon, 392
Jacquemontia, **954**
    tamnifolia (Map 1604, Pl. 368 a, b), **954**
Jagged chickweed, 814
Japanese barberry, 604
Japanese bindweed, 933
Japanese hedge-parsley, 111
Japanese honeysuckle
Japanese hops, 761
Japanese maple, 8
Japanese persimmon, 990
Jeffersonia diphylla, 603
Jenny's stonecrop, 972
Jerusalem artichoke, 436, 515, 516
Jerusalem oak, 879
Jewelweed, 599, 600, 602
Jim Hill mustard, 726
Joe-pye weed, 401

Joe-pye weed (cont.)
    Green-stemmed, 407
    Hollow-stemmed, 403
    Spotted, 404
Jointed charlock, 716
Jones, Almut, 315
Judas tree, 1061
Justicia, **4,** 940, 943, 944
    americana (Map 799, Pl. 195 a–c), **5**
        var. subcoriacea, 5
    ovata (Map 800), **5**
        var. lanceolata, **5**
        var. ovata, 5

K

Kale, 670
Kalmia, 1000
Kansas sunflower, 512
Kartesz, John, 915
Keil, David J., 336
Kellogg, John, 242, 844
Kenguel seed, 343
Kentucky coffee tree, 1069
Kentucky viburnum, 789
Khakiweed, Tall, 556
Khat, 850
King, Robert Merrill, 393
King Henry, Good, 871
King-devil, Yellow, 354
Kinnikinnick, 958
Klamath weed, 924
Knapweed, 321, 332
    Black, 328
    Brown, 328
    Diffuse, 326, 332
    North African, 326
    Russian, 330
    Short-fringed, 328
    Spotted, 331, 332
    Tumble, 326
    Tyrol, 328, 330
Knawel, 827, 828
    Annual, 828
Knotted hedge-parsley, 111
Knox, John, 492, 493
Kochia, **895,** 895
    iranica, 896
    scoparia (Map 1547, Pl. 358 d–f), **896**
        var. culta, 896
        var. pubescens, 896
    sieversiana, 896
Kohl crop, 670
Kohlrabi, 670
Kowal, Robert, 587
Krigia, 344, **358**
    biflora (Map 1072, Pl. 257 j, k), 358, **359**
        f. glandulifera, 359
        var. biflora, **359**

Krigia biflora (cont.)
        var. viridis, 359
    cespitosa (Map 1073, Pl. 257 f), **360**
        f. gracilis, 360
        ssp. cespitosa, **360**
        ssp. gracilis, 360
    dandelion (Map 1074, Pl. 257 h, i), 358, **360**
        gracilis, 360
    occidentalis (Map 1075, Pl. 257 c–e), **362**
    oppositifolia, 360
    virginica (Map 1076, Pl. 257 a, b), **362**
Kudzu, 1058
Kuhnia
    eupatorioides, 397
        var. angustifolia, 398
        var. corymbulosa, 397, 398
        var. ozarkana, 398
        var. texana, 398

L

Lacinaria scariosa var. nieuwlandii, 420
Lactuca, 344, **363,** 368, 371
    altaica, 370, 871
    canadensis (Map 1077, Pl. 258 a, b), 363, **364,** 365, 366
        f. angustata, 365
        f. angustipes, 364
        f. exauriculata, 365
        f. stenopoda, 365
        f. villicaulis, 365
        var. latifolia, 365
        var. longifolia, 365
        var. obovata, 364, 365
    dregeana, 370
    floridana (Map 1078, Pl. 258 c, d), 363–365
        f. leucantha, 365
        f. villosa, 365
    graminifolia, 365
    hirsuta (Map 1079, Pl. 258 e, f), 363, **365**
        var. hirsuta, 366
        var. sanguinea, 366
    ludoviciana (Map 1080, Pl. 258 g, h), 363, **366**
    oblongifolia, 370
    pulchella, 370
    saligna (Map 1081, Pl. 259 a, b), **368**
        f. ruppiana, 368
        var. runcinata, 368
        var. saligna, 368
    sativa (259 e–g), 370
    scariola, 368
    serriola (Map 1082, Pl. 259 c, d), **368**

Lactuca serriola (cont.)
        f. integrifolia, 370
    tatarica (Map 1083, Pl. 259 h–j), 363, **370,** 371
        ssp. pulchella, **370,** 371
        ssp. tatarica, 371
Ladies' tobacco, 428
Lagenaria, **985,** 986
    siceraria (Map 1629, Pl. 371 c), **985,** 986
        ssp. asiatica, 986
Lake cress, 719
Lamb's quarters, 874
Lamiaceae, 620
Lanceleaf ragweed, 444
Lance-leaved water willow, 5
Laportea, 943
Lappula, **634,** 636
    echinata, 636
    lappula, 636
    occidentalis, 636
    redowskii (Map 1296, Pl. 307 f), **636**
        var. occidentalis, 636
    squarrosa (Map 1297, Pl. 307 a–c), **636**
    texana, 636
        var. cupulata, 636
    virginiana, 633
Lapsana, **371**
    communis (Map 1084, Pl. 260 a–c), **371**
        ssp. communis, 371
Large alfalfa dodder, 943
Large sand catchfly, 832
Large-leaved aster, 227
Large-seeded mercury, 1013
Lasallea, 298
Late boneset, 410
Late goldenrod, 261
Lauraceae, 1
Laurel, Mountain, 1000
Lazydaisy, 202
    Tiny, 211
Leaf, Painted, 1034
Leaf cup, 529
    Pale-flowered, 530
Leaf hopper, 478
Leaf mustard, 672, 724
Leaf-flower, 1050
    Carolina, 1052
Leafy spurge, 1029, 1040
Leavenworth eryngo, 80
Leavenworthia, 651, **700,** 701
    torulosa (Map 1358), **700,** 701
    uniflora (Map 1359, Pl. 321 g, h), **701**
Lechea, 905, 906
    intermedia, 906
    mucronata (Map 1555, Pl. 359 b–d), **906**
    racemulosa (Map 1556, Pl. 359 a), **906**

Lechea (cont.)
    tenuifolia (Map 1557, Pl. 359
        j–l), **906**
        var. occidentalis, 908
    villosa, 906
Legousia, 753
    speculum, 753
Legume, 1058
    Caesalpinioid, 1058
    Mimosoid, 1058
    Papilionoid, 1058
Leguminosae, **1057**
Lehmann, John G., 882
Lennoaceae, 620
Lens, 1058
Lens-podded hoary cress, 703
Lentil, 1058
Leontodon, 344, **371**
    taraxacoides (Map 1085), **372**
        ssp. longirostris, 372
        ssp. taraxacoides, **372**
Leopard plant, 576
Lepidanthus phyllanthoides, 1018
Lepidium, 651, **701,** 702
    appelianum (Map 1360, Pl. 318
        d, e), **703,** 704, 706
    campestre (Map 1361, Pl. 323 c,
        d), **703,** 728
    chalepense (Map 1362, Pl. 318
        c), **703,** 704, 706
    densiflorum (Map 1363, Pl. 323
        e, f), **704,** 708
    didymum (Map 1364, Pl. 318 h,
        i), **704**
    draba (Map 1365, Pl. 318 a, b),
        703, **706**
    latifolium (Map 1366, Pl. 323
        g–i), **707**
    perfoliatum (Map 1367, Pl. 323
        a, b), **707**
    ruderale (Map 1368), **707**
    virginicum (Map 1369, Pl. 323
        j–l), 704, **708**
Leptocoris trivittatus, 10
Leptoglottis nuttallii, 1082
Leptopus phyllanthoides, 1018
Lespedeza, 943
Lesquerella, 714
    filiformis, 714
    gracilis, 714
Lesser chickweed, 848
Lesser hawkbit, 372
Lesser star thistle, 326
Letterman, George, 701
Lettuce, 344, 363
    Blue, 370
    Devil's, 625
    Downy, 365
    Florida, 365
    Garden, 370
    Glaucous white, 380
    Great white, 380
    Hairy, 365

Lettuce (cont.)
    Prickly 368
    Rough white, 379
    Tall white, 378
    Western wild, 366
    White, 377, 378
    Wild, 364
    Willow-leaved, 368
    Woodland, 365
Leucanthemum, 172, **189,**
    190
    lacustre, 190
    maximum, 190
    ×superbum, 190
    vulgare (Map 952, Pl. 224 i, j),
        190
        var. pinnatifidum, 190
        var. vulgare, 190
Levisticum, **88**
    officinale (Map 865, Pl. 208 a,b),
        88
    paludapifolium, 88
Lewis and Clark, 719
Liatris, **413**
    aestivalis, 416
    aspera (Map 1121, Pl. 269 f–h),
        413, **415,** 420, 422, 424
        f. benkei, 415
    ×bebbiana (Pl. 270 d), 422
    cylindracea (Map 1122, Pl. 268
        g, h), 413, **415,** 423
        f. bartelii, 416
    ×frostii, 415
    hirsuta, 423
    ligulistylis, 413, 420
    mucronata (Map 1123, Pl. 269 i,
        j), **416,** 418
    ×*nieuwlandii, 420*
    novae-angliae
        var. nieuwlandii, 420
    punctata (Map 1124, Pl. 270 f,
        g), **416,** 418
        var. mexicana, 418
        var. mucronata, 416
        var. nebraskana, 418
        var. punctata, **416,** 418
    pycnostachya (Map 1125, Pl.
        270 a–c), 415, **418,** 420,
        422
        f. hubrichtii, 420
        var. lasiophylla, 420
        var. pycnostachya, **418**
    scabra, 420, 423
    scariosa (Map 1126, Pl. 269 a,
        b), 413, **420,** 422, 424
        var. nieuwlandii, **420**
        var. novae-angliae, 420
        var. scariosa, 420
    sphaeroidea, 415
    spicata (Map 1127, Pl. 270 e),
        413, 420, **422**
        var. resinosa, 422
        var. spicata, 422

Liatris (cont.)
    squarrosa (Map 1128, Pl. 268 e,
        f), 415, 416, **422,** 423
        var. glabrata, **423**
        var. hirsuta, 416, **423**
        var. intermedia, 415
        var. squarrosa, **423**
    squarrulosa (Map 1129, Pl. 269
        c–e), 415, 420, 422, **423,**
        424
Lice
    Beggar's, 634
    Nits and, 916
Liesner, Ron, 1073
Ligularia, 576
Ligusticum, **88**
    canadense (Map 866, Pl. 209 i,
        j), **88**
Limber honeysuckle, 770
Limestone ruellia, 7
Limnosciadium, **89**
    pinnatum (Map 867, Pl. 209 g,
        h), **89**
Linaceae, 725
Lincoln, Abraham, 396
Lincoln, Nancy Hanks, 396
Lindera, 940
Ling Yeou-ruenn, 186, 187
Lingonberry, 1006
Linnaeus, 851
Linum usitatissimum, 725
Liquidambar, 129
Lithospermum, 619, 628, **637**
    angustifolium, 640
    arvense, 628
    canescens (Map 1298, Pl. 308
        a–c), **638**
        f. pallida, 638
    caroliniense (Map 1299, Pl. 308
        d–f), **638**
        var. carolinense, 638
        var. croceum, 638
    croceum, 638
    incisum (Map 1300, Pl. 308 h, i),
        640
    latifolium (Map 1301, Pl. 308 g),
        640
    linearifolium, 640
    officinale
        var. latifolium, 640
Littlepod false flax, 675
Live-forever, 970, 972
Lobadium, 42
Lobelia, **746**
    appendiculata, 751
    cardinalis (Map 1408, Pl. 331 a,
        b), **747,** 748, 751
        f. alba, 748
        f. rosea, 748
        ssp. graminea, 747
    inflata (Map 1409, Pl. 331 c, d),
        748
    leptostachys, 752

Lobelia (cont.)
 puberula (Map 1410, Pl. 331 g, h), **748,** 750
  var. mineolana, 750
 siphilitica (Map 1411, Pl. 331 i, j), 748, **750,** 751
  f. albiflora, 751
  var. ludoviciana, 750
  var. siphilitica, 750
 ×siphilitica
  var. hybrida, 751
 ×speciosa, 751
 spicata (Map 1412, Pl. 331 e, f), **751,** 752
  f. campanulata, 752
  var. campanulata, 752
  var. hirtella, 752
  var. leptostachys, 751, **752**
  var. parviflora, 752
  var. spicata, **752**
 splendens, 747
Lobelia, 746, 747
 Big blue, 748
 Blue, 750
 Great, 750
 Inflated, 748
 Palespike, 751
 Spiked, 751
Lobularia, **708**
 maritima (Map 1370, Pl. 324 a, b), **708**
Locust
 Honey, 1068, 1069
 Water, 1068
Lomatium, **90**
 foeniculaceum (Map 868, Pl. 209 e, f), **90**
  ssp. daucifolium (Pl. 209 e, f), **92**
  ssp. foeniculaceum, **92**
  var. daucifolium, 92
Long-haired hawkweed, 356
Longhead prairie coneflower, 532
Long-leaved stitchwort, 845
Lonicera, 765, **766**
 ×bella (Map 1421, Pl. 335 c, d), **767,** 768, 770
 dioica (Map 1422, Pl. 334 g, h), **770**
  var. glaucescens, 770
 flava (Map 1423, Pl. 334 a, b), **770**
  var. flavescens, 770
 fragrantissima, 774
 japonica (Map 1424, Pl. 334 c), 766, **772**
 maackii (Map 1425, Pl. 335 e, f), 766, **772,** 773
 ×minutiflora, 768
 morrowii (Pl. 335 a, b), 768
 prolifera, 774
 ×purpusii (Map 1426), 766, **768**

Lonicera (cont.)
 reticulata (Map 1427, Pl. 334 f), **774**
 rhytidophylla, 774
 sempervirens (Map 1428, Pl. 334 d, e), **774**
  var. hirsutula, 775
 standishii, 784
 tatarica, 768
 xylosteum, 768, 770
Looking-glass
 Clasping Venus', 757
 Cultivated Venus', 753
 Holzinger's Venus', 754
 Prairie Venus', 757
 Slender-leaved Venus', 756
 Slimpod Venus', 756
 Small Venus', 754
Lophophora williamsii, 736
Lovage, 88
Love vine, 938
Love-lies-bleeding, 25
Low bindweed, 935
Lowbush blueberry, 1008
 Sweet, 1007
Luffa, 974
Lunaria, **710**
 annua (Map 1371), **710**
Lychnis, 795, **816,** 828
 alba, 837
 coronaria (Map 1466, Pl. 343 e, f), **816**
 dioica, 836
Lycopus, 940, 942, 944
Lygodesmia, 344, **372**
 juncea (Map 1086, Pl. 260 g–i), **372,** 374
Lyonia, **1001,** 1002
 mariana (Map 1641, Pl. 375 k–m), **1001,** 1002
Lysimachia, 940

M

Machaeranthera, 219, 226, 278
Maclura pomifera, 1069
Macrosteles, 478
Madagascar palm, 115
Madagascar periwinkle, 115
Magnoliaceae, 1
Magnoliid complex, 1
Magnoliopsida, **1**
Mahonia aquifolium, 603
Mahoney, Allison, 587
Maiden pink, 812
Maidenbush, 1018
Malabar spinach, 31
Malpighiales, 1010
Maltese star thistle, 327
Mammillaria missouriensis, 737
Mammoth, 1069

Mammoth Russian, 498
Mandrake, 607
Mangel-wurzel, 870
Mangifera, 40
 indica, 40
Mango, 40
Manihot esculenta, 1010
Man-of-the-earth, 952
Maple, 7
 Amur, 9
 Ash-leaved, 9
 Black, 14
 Florida, 15
 Hard, 13
 Japanese, 8
 Norway, 8
 Siberian, 9
 Silver, 13
 Soft, 13
 Southern sugar, 15
 Sugar, 12, 14
 Sycamore, 8
Maple family, 7
Maple-leaved goosefoot, 889
Marguerite, Golden, 188
Maria, Santa, 526
Marigold, 556
 African, 556
 Common, 556
 Corn, 189
 Fetid, 472
 Few-bracted bur, 457
 French, 558
 Mexican, 556
 Nodding bur, 456
 Showy bur, 458
 Water, 454
Marijuana, 759, 760
Marrow, Vegetable, 982, 983
Marsh bellflower, 744
Marsh cress, 721
Marsh elder, 519, 520
 Big, 471
Marsh fleabane, 573, 574
 Inland, 574
Marsh St. John's-wort, 928
Marsh yellow cress, 720
Marshallia, **522**
 caespitosa (Map 1207, Pl. 284 e, f), **522**
  var. caespitosa, **522**
  var. signata, 522, **523**
 obovata, 523
Master, Rattlesnake, 82
Masterwort, 85
Mastodon, 1069
Mat
 Rib-seeded sand, 1041
 Sand, 1041
Mat spurge, 1049
Matelea, 160, **162,** 163
 baldwyniana (Map 935, Pl. 223 h, i), **163**

Matelea (cont.)
  decipiens (Map 936, Pl. 223 c–g), **163,** 164
  gonocarpa, 160
  obliqua, 163, 164
  suberosa, 160
Matricaria, **190,** 191, 192
  chamomilla (Map 953, Pl. 227 h–j), **191,** 192
    var. coronata, 191, 192
  courrantiana, 191
  discoidea (Map 954, Pl. 227 f, g), **192**
  inodora, 196
    var. agrestis, 196, 198
    var. inodora, 196
  matricarioides, 192
  perforata, 196
  recutita, 191
  suaveolens, 191
Matted pearlwort, 826
Maximilian sunflower, 504, 506
May apple, 607
Mayfield, Mark H., 1010, 1034
Maytenus, 850
Mayweed
  Scentless, 196
McReynolds, Scott, 813
Meadow parsnip, 106
  Common, 114
  Heart-leaved, 114
Meadow salsify, 392
Mead's milkweed, 146, 148
Megalodonta beckii, 454, 456
Medicago, 943, 1058
Melampodium, **523**
  cinereum (Map 1208), **523**
    var. cinereum, 524
    var. hirtellum, 523
    var. ramosissimum, **523,** 524
Melilotus, 1058
Melon, Horned, 979
Melon-leaf, 976
Melothria, **986**
  grandifolia, 976
  pendula (Map 1630, Pl. 371 a, b), **986**
Menispermaceae, 603
  Large-seeded, 1013
  One-seeded, 1014
  Slender three-seeded, 1013
  Three-seeded, 1012, 1016
  Two-seeded, 1013
Mertensia, **641**
  pulmonariodes, 641, 642
  virginica (Map 1302, Pl. 309 f, g), **641,** 642
    f. berdii, 642
    f. rosea, 642
Mesquite, 1082, 1083
  Honey, 1083
Mexican aster, 469
Mexican hat, 532

Mexican marigold, 556
Mexican tea, 875
Michaux's gladecress, 701
Michaux's stitchwort, 817
Micrampelis lobata, 984
Microseris, 374
  cuspidata, 374
Microthlaspi, **710,** 711, 728
  perfoliatum (Map 1372, Pl. 327 g, h), **710,** 730
Mikania, **424,** 943
  pubescens, 424
  scandens (Map 1130, Pl. 270 h, i), **424**
Milfoil, 173
  Common, 173
Milk, Wolf's, 1040
Milk purslane, 1042
Milk spurge, 1047
Milk sickness, 395, 396
Milk thistle, 342, 343
  Blessed, 342
Milkweed, 142
  Bluntleaf, 145
  Climbing, 159, 162
  Common, 153
  Fourleaf, 149
  Green, 157
  Green-flowered, 158
  Horsetail, 157
  Mead's, 146, 148
  Narrow-leaved, 152
  Ozark, 158
  Poison, 152
  Poke, 115, 143
  Prairie, 145, 153
  Purple, 149
  Sand, 145
  Showy, 150
  Smooth, 153
  Smoothseed, 148
  Spider, 158
  Swamp, 146
  Tall green, 145
  Variegated, 156
  White, 148
  Whorled, 149, 157
Milkweed butterfly, 142
Milkweed family, 115, 140
Miller, Dusty, 187
Mimosa, 1075, **1081**
  glandulosa, 1080
  pudica, 1075
  quadrivalvis (Map 1703, Pl. 387 d, e), **1082**
    var. nuttallii, **1082**
  Section Batocaulon, 1081
  Series Quadrivalves, 1081
Mimosa, 1079, 1081
  Prairie, 1080
Mimosaceae, 1058
Mimosoid legume, 1058
Mint geranium, 194

Minuartia, 795, 797, **816,** 817
  michauxii (Map 1467, Pl. 339 i, j), **817**
  muriculata, 818
  muscorum (Map 1468), **818**
  patula (Map 1469, Pl. 339 e, f), **818**
    var. robusta, 818
Mississippi Valley flat-topped goldenrod, 232
Missouri, Federated Garden Clubs of, 129
Missouri, Women's League of, 129
Missouri arrowwood, 789
Missouri banana, 52
Missouri bladderpod, 714
Missouri coneflower, 545
Missouri Conservation Commission, 129
Missouri Department of Conservation, 126, 129, 136, 822
Missouri Department of Transportation, 469, 484
Missouri goldenrod, 263
Missouri gourd, 982
Missouri ironweed, 598
Missouri Native Plant Society, 975
Missouri Natural Heritage Database, 375
Missouri Natural Heritage Program, 148
Missouri pigweed
Missouri pincushion cactus, 737
Missouri rock cress, 669
Missouri spurge, 1045
Missouri viburnum
Missouri Wildlife Code, 136
Mist-flower, 400
Mistletoe, 1003
Mock bishop's weed, 98
  Atlantic, 98
  Big, 99
  Ozark, 99
Moehringia, 797, **819**
  lateriflora (Map 1470, Pl. 339 g, h), **819**
Molluginaceae, 17, 794
Mollugo, 17
Momordica lanata, 978
Monarch, 142
Money-plant, 710
Monocots, 1, 391
Monolepis, **896**
  nuttalliana (Map 1548, Pl. 358 i, j), **896**
Monotropa, 999, **1002,** 1003
  hypopithys, 1003
  hypopitys (Map 1642, Pl. 374 l, m), **1003**
  uniflora (Map 1643, Pl. 374 n, o), **1003**
Moraceae, 758, 1069

Mordellidae, 548
Mordellistema aethiops, 548
Morning glory, 948, 954
    Bigroot, 952
    Blue, 950
    Common, 952
    Dwarf, 935
    Ivy-leaved, 950
    Red, 949
    Red-centered, 949
    Scarlet, 949
    Small white, 951
    Wild, 934
Morning glory family, 930
Morrow honeysuckle, 768
Morton, John K., 837
Mosquito, 599
Mossy stonecrop, 970
Mosyakin, Sergei, 886, 894
Moth
    Catalpa sphinx, 618
    Gelechiid, 252
    Goldenrod gall, 252
    Parasitic, 616
Mountain azalea, 1004
Mountain laurel, 1000
Mouse-ear chickweed, 800
    Common, 806
    Curtis's, 807
    Dark green, 803
    Gray, 803
    Short-stalked, 804
Mouse-eared cress, 660
Mugwort, 177
    Common, 187
    Western, 185
Mühlenbach, Viktor, 203, 237, 238, 326, 328, 352, 392, 403, 468, 475, 560, 832, 868, 869, 948, 1024
Mulgedium, 363
Mullein-pink, 816
Mum
    Florist's, 189
    Garden, 189
Musclewood, 610, 612
Mushroom, Death angel, 343
Musk thistle, 320, 321
Muskmelon, 979, 980
Mustard, 670, 686
    Ball, 712
    Black, 674
    Blue, 684
    Brown, 672
    Chinese, 672
    Dog, 692
    Field, 674
    Garlic, 658
    Hare's-ear, 686
    Hedge, 726
    Indian, 672
    Jim Hill, 726
    Leaf, 672

Mustard (cont.)
    Sahara, 672
    Tall hedge, 726
    Tansy, 686, 688
    Tower, 730
    Treacle, 696
    Tumble, 726
    White, 674, 724
    Wild, 725
    Wormseed, 695
    Yellow, 724
Mustard family, 651
Myosotis, 620, **642**
    laxa, 644
    macrosperma (Map 1303, Pl. 309 a–c), **643,** 646
    micrantha, 644
    palustris, 644
    scorpioides (Map 1304, Pl. 307 d, e), **644**
        var. palustris, 644
    squarrosa, 636
    stricta (Map 1305, Pl. 309 h, i), **644,** 646
    verna (Map 1306, Pl. 309 d, e), 643, **646**
        var. macrosperma, 643
    virginiana, 646
        var. macrosperma, 643
Myosoton, 795, **819**
    aquaticum (Map 1471, Pl. 343 c, d), **819**
Myriophyllum, 454, 720
Myrtle, 126

N

Nabalus, 377
    albus, 378
    altissimus, 378
    asper, 379
    crepidineus, 380
    racemosus, 380
Nailwort, 820
    Hairy, 821
    Smooth, 821
    Yellow, 822
Naked-stemmed sunflower, 508
Nannyberry, 789
Narrow-leaved coneflower, 475
Narrow-leaved gayfeather, 416
Narrow-leaved hawksbeard, 350
Narrow-leaved milkweed, 152
Narrow-leaved pinweed, 906
Narrow-leaved puccoon, 640
Narrow-leaved sunflower, 497
Nasturtium, **711,** 715
    limosum, 721
    microphyllum, 711
    officinale (Map 1373, Pl. 325 i–k), **711**
    var. siifolium, 711

Natal plum, 115
Natural Heritage Database, Missouri, 375
Natural Heritage Program, Missouri, 148
Natural Resources Conservation Service, 995
Nature Conservancy, The, 129
Navajo tea, 560
Necklace gladecress, 700
Needles, Spanish, 456, 524
Neobeckia, 720
    aquatica, 719
Neobesseya missouriensis, 737
Neomammillaria missouriensis, 737
Neopieris mariana, 1001
Nerium, 115
Neslia, **712**
    paniculata (Map 1374, Pl. 324 j, k), **712**
Nesom, Guy, 203, 431, 434
Nettle-leaved goosefoot, 884
New England aster, 296
Niger seed, 488
Niger thistle, 436, 488
Night-flowering catchfly, 838
Nipplewort, 371
Nits and lice, 916
Nodding bur marigold, 456
Nodding chickweed, 807
Nodding spurge, 1045
Nodding thistle, 320
Nondo, 88
North African knapweed, 326
Northern arrowwood, 792
Northern bitter cress, 682
Northern catalpa, 619
Northern ragwort, 587
Northern wild senna, 1073
Norway maple, 8
Nosebleed, 174
Noseburn, 1054
Nothocalais, 344, **374**
    cuspidata (Map 1087, Pl. 260 d–f), **374**
Nuttall, Thomas, 186
Nuttall's sedum, 971
Nyctaginaceae, 795
Nyjer seed, 488
Nymphaeaceae, 1, 732
Nyssa, 940, 957, **964,** 965
    aquatica (Map 1611, Pl. 369 m, n), 457, **965**
    biflora (Map 1612), **966,** 967
    sylvatica (Map 1613, Pl. 369 k, l), **966,** 967
        var. biflora, 966
        var. caroliniana, 966
        var. dilatata, 066
        var. ursina, 967
    uniflora, 965
    ursina, 967

Nyssaceae, 957, 965

O

Oak
  Jerusalem, 879
  Poison, 46
  Eastern poison, 49
Oak-leaved goosefoot, 882
Oblong-leaved aster, 297
Oenanthe, **92**
  aquatica, 92
  javanica (Map 869), **92**
  pimpinelloides, 92
  sarmentosa, 93
Oenothera, 939, 943
Oil
  Canola, 674
  Castor, 1011, 1020, 1054
  Croton, 1020
  Tung, 1011
Oil plant Castor, 1054
Old Plainsman, 518
Old-field balsam, 434
Old-field goldenrod, 264
Oleander, 115
  Yellow, 115
Oleaster, 994
Oleaster family, 994
Oligoneuron, 244, 245, 270
  album, 269
  riddellii, 271
  rigidum, 271
    var. glabratum, 272
    var. humile, 272
Olive
  Autumn, 995
  Russian, 995
Onagraceae, 218
One-seeded croton, 1026
One-seeded mercury, 1014
Onopordum, 316, **341**
  acanthium (Map 1059, Pl. 254 g, h), **341,** 342
    ssp. acanthium, **341**
Onosmodium, **646**
  bejariense, 647
    var. hispidissimum, 648
    var. occidentale, 648
    var. subsetosum, 648
  hispidissimum, 648
  molle (Map 1307, Pl. 310 c–j), **647**
    ssp. bejariense, 647
    ssp. hispidissimum (Pl. 310 c–e), **648**
    ssp. molle, 647
    ssp. occidentale (Pl. 310 i, j), 641, 647, **648**
    ssp. subsetosum (Pl. 310 f–h), **648**
    var. hispidissimum, 648

Onosmodium molle (cont.)
    var. occidentale, 648
    var. subsetosum, 648
  occidentale, 648
  subsetosum, 648
Ontario aster, 298
Opuntia, 736, **737,** 738, 740
  arborescens, 736
  compressa, 738
    var. macrorhiza, 739
  humifusa (Map 1400, Pl. 329 f–i), **738,** 739
    var. ammophila 739
    var. austrina 739
    var. humifusa 739
  imbricata, 736
  macrorhiza (Map 1401, Pl. 329 j, k), **739,** 740
    var. macrorhiza, 740
    var. pottsii, 740
  polyacantha (Map 1402, Pl. 329 e), **740**
    var. polyacantha, **740**
  pottsii, 740
Orach, 862
  Garden, 864
  Tumbling, 866
  Two-seeded, 864
  Wedge-leaved, 868
Orange, Osage, 1069
Orange coneflower, 538, 540
Orange dwarf dandelion, 359
Orange grass, 917
Orange puccoon, 638
Orbexilum, 1058
Orchid, 1058
Orchid tree, 1061
Orchidaceae, 164, 1003
Orchidologist, 164
Oregon grape, 603
Oriental bittersweet, 851
Ormyridae, 548
Ornamental gourd, 982, 983
Orpine, 967
Osage orange, 1069
Osier, Green, 959
Osmorhiza, 93
  claytonii (Map 870, Pl. 209 a, b), **93**
  longistylis (Map 871, Pl. 209 c, d), **94**
    var. brachycoma, 94
    var. villicaulis, 94
Ostrya, **613**
  virginiana (Map 1279, Pl. 303 j, k), 612, **613**
    f. glandulosa, 614
    var. lasia, 613, 614
Owens, T., 736
Ownbey, Gerald B., 336
Ownbey, Marion, 391
Ox-eye, 516, 517
  Rough, 518

Ox-eye daisy, 190
Ox-tongue, 352, 375
  Bristly, 352
  Hawkweed, 376
Oxypolis, **94**
  rigidior (Map 872, Pl. 210 e, f), **94**
    var. ambigua, 94
Oyster, Vegetable, 391
Oyster plant, 391
Ozark arrowwood, 791
Ozark goldenrod, 258
Ozark milkweed, 158
Ozark mock bishop's weed, 99

P

Pachypodium, 115
Pachysandra, **731**
  procumbens (Map 1397), **731**
  terminalis, 731
Packera, 576, **582,** 583, 590
  aurea (Map 1251, Pl. 297 f, g), 583, **584**
  glabella (Map 1252, Pl. 299 i, j), **586**
  obovata (Map 1253, Pl. 298 g, h), 583, **586**
  paupercula (Map 1254, Pl. 298 i, j), 583, **587**
  paupercula × plattensis, 583
  plattensis (Map 1255, Pl. 298 d–f), 583, 587, **588**
  pseudaurea (Map 1256, Pl. 297 h, i), 583, 586, **588**
    var. pseudaurea, 588
    var. flavula, 588
    var. semicordata, **588**
  tomentosa (Map 1257, Pl. 298 a–c), 583, **589**
Pagoda dogwood, 959
Painted leaf, 1034
Palafoxia, **524**
  callosa (Map 1209, Pl. 284 i–k), **524**
Pale alyssum, 659
Pale dogwood, 959
Pale Indian plantain, 577
Pale purple coneflower, 475, 478
Pale sunflower, 500
Pale touch-me-not, 602
Pale-flowered leaf cup, 530
Pale-leaved sunflower, 514
Paleoherb, 1
Palespike lobelia, 751
Palm, Madagascar, 115
Palmer, Ernest J., 524, 537, 740, 751
Panax, **135,** 136
  quinquefolium, 136
  quinquefolius (Map 911, Pl. 219 d, e), **135,** 136

Panicled aster, 292
Papaveraceae, 603
Papilionoid legume, 1058
Para rubber, 1010
Parasenecio, 582
Paronychia, **820**
    canadensis (Map 1472, Pl. 344 j, k), **821**
    fastigiata (Map 1473, Pl. 344 g–i), **821**
        var. fastigiata, **821,** 822
        var. paleacea, 821, 822
    virginica (Map 1474, Pl. 344 a, b), **822**
    var. scoparia, 822
Parsley, 55
    Field hedge, 111
    Hairy, 90
    Hedge, 110
    Japanese hedge, 111
    Knotted hedge, 111
    Sand, 63
    Water, 103
    Wild, 90
Parsnip, 55, 95
    Common meadow, 114
    Cow, 85
    Giant cow, 85
    Heart-leaved meadow, 114
    Meadow, 106
    Water, 67, 103
    Wild, 95
Parthenium, 165, 474, **525,** 526
    argentatum, 525
    hispidum, 526, *528*
    hysterophorus (Map 1210, Pl. 285 e, f), **526**
    integrifolium (Map 1211, Pl. 285 a–d), **526,** 528
        var. hispidum (Pl. 285 c, d), **528**
        var. integrifolium (Pl. 285 a, b), **528**
    repens, 528
Parthenocissus, 48, 49
    quinquefolia, 48
Partridge pea
    Showy, 1064
    Small-flowered, 1065
Passifloraceae, 974
Pastinaca, **95**
    sativa (Map 873, Pl. 210 g, h), 55, **95**
Paulownia, 614
Paulowniaceae, 614
Pawpaw, 51, 52, 1069
Pea, 1058
    Sensitive, 1065
    Showy partridge, 1064
    Small-flowered partridge, 1065
Peanut, 1058
Pear
    Eastern prickly, 738

Pear (cont.)
    Plains prickly, 739
    Prickly, 737
Pear gourd Wild, 983
Pearlwort, 824
    Matted, 826
    Trailing, 826
Pearly everlasting, 426
Pennsylvania bitter cress, 683
Penny cress, 710
    Field, 730
    Perfoliate, 710
    Roadside, 730
Pennywort
    Buttercup, 86
    Water, 85, 86
    Whorled, 86
Penstemon, 943
Penthoraceae, 967
Penthorum, 943, 944, 967
Pepo foetidissimus, 982
Pepper and salt, 79, 80
Pepper grass, 701, 704, 708
    Broad-leaved, 707
    Field, 703
    Green-flowered, 704
    Perennial, 707
    Perfoliate, 707
    Poor man's, 708
    Virginia, 708
Pepperweed, Stinking, 707
Perennial bursage, 446
Perennial pepper grass, 707
Perennial sow thistle, 384
Pereskia, 736
Perfoliate penny cress, 710
Perfoliate pepper grass, 707
Pericallis, 576
Perideridia, **95**
    americana (Map 874, Pl. 210 a, b), **96**
    neurophylla, 95
Perilla, 939, 943
Peritoma
    integrifolia, 910
    serrulatum, 910
Periwinkle, 115, 125, 126
    Common, 126
    Greater, 126
    Madagascar, 115
    Rosy, 115
Persimmon, 990, 992, 1069
    Japanese, 990
Petasites, 576
Petrorhagia, 795, **822**
    prolifera (Map 1475, Pl. 344 e, f), **822**
Petroselinum, 55
    crispum, 55
Peyote, 736
Phaethusa
    helianthoides, 564
    virginica, 566

Philadelphia fleabane, 222
Phleum, 943
Phoradendron, 1003
Phyllanthaceae, 1010
Phyllanthoid group, 1010, 1018
Phyllanthus, 1010, **1050**
    caroliniensis (Map 1683, Pl. 383 e, f), **1052**
        ssp. caroliniensis, **1052**
        ssp. saxicola, 1052
    polygonoides (Map 1684, Pl. 383 a, b), **1052**
Physaria, **712,** 714
    filiformis (Map 1375, Pl. 322 h–k), **714,** 715
    gracilis (Map 1376, Pl. 322 a, b), **714**
        ssp. gracilis, **714,** 715
        ssp. nuttallii, 715
Phytolaccaceae, 795
Pickle, 979
Picris, 343, 344, 352, **375**
    altissima, 376
    echioides, **352**
    hieracioides (Map 1088, Pl. 260 l, m), **376**
        ssp. hieracioides, **376**
    integrifolia, 377
    rhagadioloides (Map 1089, Pl. 260 n, o), **376,** 377
    sprengerana, 377
Piedmont azalea, 1006
Pieris, 1000
Pigeonberry, 959
Pigweed, 19, 860, 870, 874
    Rough, 30
    Russian, 869
    Sandhills, 23
    Slender, 28
    Spreading, 24
    Winged, 895
Pilea, 939, 943
Pimpernel, Yellow, 106
Pimpinella anisum, 64, 84
Pincushion cactus, Missouri, 737
Pineapple weed, 192
Pinesap, 1002, 1003
Pineweed, 917
Pink, 809
    Childing, 822
    Cottage, 812
    Deptford, 810
    Election, 1004
    Fire, 841
    Garden, 812
    Maiden, 812
    Wherry's, 831
    Wild, 831
Pink family, 794
Pink queen, 910
Pink thoroughwort, 412
Pinkava, Donald, 740
Pinweed, 905

Pinweed (cont.)
　Hairy, 906
　Illinois, 906
　Narrow-leaved, 906
Pink-flowered dogbane, 121
Pipe
　Dutchman's, 139
　Indian, 1002, 1003
Pipevine, Woolly, 139
Pipevine swallowtail, Eastern, 138
Pistacia, 40
　vera, 40
Pistachio, 40
Pisum, 1058
Pitcher's stitchwort, 818
Pitseed goosefoot, 876
Pityopsis, 239
Plainleaf pussytoes, 428
Plains coreopsis, 466
Plains prickly pear, 739
Plains puccoon, 638
Plains stoneseed, 640
Plains sunflower, 512
Plainsman, Old, 518
Planodes, **715**
　virginica (Map 1377, Pl. 326
　　i–k), 683, 684, **715**
Plant
　Castor oil, 1054
　Chenille, 1012
　Compass, 552
　Cup, 552
　Gum, 232
　Leopard, 576
　Oyster, 391
　Rocky Mountain bee, 910
　Skeleton, 372, 374
　Snot, 733
　Telegraph, 242
Plantago, 943
Plantain
　False Indian, 582
　Great Indian, 578
　Indian, 577, 578
　Pale Indian, 577
　Robin's, 223
Plasmodium, 180
Platte thistle, 336
Platycodon, 742
　grandiflorum, 742
Plectocephalus, 322
Plectocephalus americanus, 323
Pleurisy root, 154
Pluchea, **573,** 943
　camphorata (Map 1244, Pl. 295
　　g–i), 573, **574**
　foetida (Map 1245, Pl. 295 f),
　　574
　　　var. foetida, **574**
　　　var. imbricata, 574
Plum, Natal, 115
Plumeria, 115
Plumeless thistle, 318

Poacynum, 120
Podophyllum, **607**
　peltatum (Map 1273, Pl. 302 f,
　　g), **607**
　f. deamii, 607
Poinsettia, 1011, 1028, 1029
　cyathophora, 1034
　　var. graminifolia, 1034
　dentata, 1038
　　var. cuphosperma, 1038
　heterophylla, 1035
Poison hemlock, 74
Poison ivy, 46, 48, 49, 599
Poison milkweed, 152
Poison oak, 46
　Eastern, 49
Poison sumac, 48
Poke milkweed, 143
Polanisia, **911**
　dodecandra (Map 1560, Pl. 333
　　c, d), **911**
　　ssp. dodecandra, **912**
　　ssp. trachysperma **912**
　　var. trachysperma, 912
　graveolens, 911
　trachysperma, 912
Polygonum, 939, 940, 943, 944
Polymnia, **529**
　canadensis (Map 1212, Pl. 285
　　g–i), **530**
　　f. radiata, 530
　cossatotensis, 529
　laevigata (Map 1213, Pl. 286 g,
　　h), **530**
　uvedalia, 555
　　var. densipilis, 555
Polyotus angustifolius, 152
Polytaenia, **96**
　nuttallii (Map 875, Pl. 210 c, d),
　　96
Pomaria, 1060, **1070,** 1071
　jamesii (Map 1695, Pl. 385 a, b),
　　1070
Pony-foot, 944
Poor man's pepper grass, 708
Populus heterophylla, 457
Portuguese chrysanthemum, 190
Portulacaceae, 736, 795
Possum haw, 128
Possumwood, 990
Potato, Sweet, 948
Potato dandelion, 360
Potato vine, Wild, 952
Poverty weed, 896
Powderhorn, 807
Prairie acacia, 1078
Prairie aster, 312, 314
　Southern, 227
　White, 290
Prairie blanketflower, 483
Prairie chickweed, 809
Prairie coneflower, 531
　Green, 534

Prairie coneflower (cont.)
　Longhead, 532
　Shortray, 534
Prairie dandelion, 374
Prairie dock, 554
Prairie dogbane, 122
Prairie gaillardia, 483
Prairie goldenrod, 275
Prairie ironweed, 597
Prairie milkweed, 145, 153
Prairie mimosa, 1080
Prairie parsley, 96
Prairie ragwort, 588
Prairie sagewort, 185
Prairie snakeroot, 416
Prairie spurge, 1045
Prairie sunflower, 512
Prairie tea, 1026
Prairie Venus' looking glass, 756
Prenanthes, **377**
　alba (Map 1090, Pl. 261 a, b),
　　378
　altissima (Map 1091, Pl. 261
　　e–g), **378,** 379
　　var. hispidula, 379
　　var. cinnamomea, 378, 379
　aspera (Map 1092, Pl. 261 h–j),
　　379, 380, 382
　crepidinea (Map 1093, Pl. 261 c,
　　d), **380**
　pendula, 377
　purpurea, 377
　racemosa (Map 1094, Pl. 261 k,
　　l), **380,** 382
　　ssp. multiflora, 382
Pretty dodder, 943
Prickly lettuce, 368
Prickly pear, 737, 738
　Eastern, 738
　Plains, 739
Prickly sow thistle, 386
Prince's feather, 28
Prionopsis, 234
　ciliatus, 233, 234
Prosopis, **1082**
　chilensis
　　var. glandulosa, 1083
　glandulosa (Map 1704), **1083,**
　　1084
　　var. glandulosa, **1083**
　　var. prostrata, 1083, 1084
　　var. torreyana, 1083, 1084
　juliflora, 1083, 1084
Prostrate amaranth, 24
Prostrate spurge, 1042
Protozoan, 712
Pruski, John H., 190
Pseudognaphalium, **433**
　helleri (Map 1138, Pl. 294 g, h),
　　434
　　var. micradenium, 434
　macounii, 433
　micradenium (Map 1139), **434**

Pseudognaphalium (cont.)
    obtusifolium (Map 1140, Pl. 294
        d–f), 433, **434**
Pterophyton helianthoides, 564
Pteromalidae, 548
Pterygopleurum, 95
Ptilimnium, **98**
    capillaceum (Map 876, Pl. 211
        h), **98,** 99
    costatum (Map 877, Pl. 211 a–c),
        99
    nuttallii (Map 878, Pl. 211 d, e),
        99
Puccinia graminis, 603, 604
Puccoon, 637
    Hoary, 638
    Narrow-leaved, 640
    Orange, 638
    Plains, 638
    Yellow, 640
Pueraria, 1058
Pulicaria, 572
Pulmonaria virginica, 641
Pumpkin, 982, 983
Purple alyssum, 666
Purple amaranth, 25
Purple coneflower, 478
    Pale, 475, 478
Purple cress, 682
Purple cudweed, 432
Purple daisy, 303
Purple dewdrop, 748
Purple milkweed, 149
Purple rocket, 698
Purple-headed sneezeweed, 491
Purse, Shepherd's, 676
Purslane
    Horse, 17
    Milk, 1042
    Sea, 17
Pussytoes, 426
    Field, 428
    Plainleaf, 428
Pycnanthemum, 943
Pyrethrum, 194
Pyrola, 999
Pyrolaceae, 999
Pyrrhopappus, **382**
    carolinianus (Map 1095, Pl. 262
        g, h), **382,** 383
    grandiflorus, 383

Q

Quamoclit
    coccinea, 949
    vulgaris, 952
Quarter vine, 615
Quarters, Lamb's, 874
Queen, Pink, 910
Queen Anne's lace, 78, 79
Qinghaosu, 180

Quickweed, 485
Quinine, Wild, 526

R

Rabeler, Richard K., 794
Rabbit tobacco, 430, 434
Radish, 716
    Garden, 716
    Wild, 716
Rafinesque, Constantine, 906
Ragweed, 442, 443
    Common, 444
    Giant, 448
    Great, 448
    Lanceleaf, 444
    Southern, 444
    Western, 444
Ragwort, 582
    Golden, 584
    Northern, 587
    Prairie, 588
    Western golden, 588
Raisin, Wild, 789
Rampion, False, 744
Ranunculaceae, 218, 603
Ranunculus aquatilis, 734
Rape, 672, 674
    Bird's, 674
Rapeseed, 672, 674
Raphanus, **716**
    raphanistrum (Map 1378, Pl.
        324 f), **716**
    sativus (Map 1379, Pl. 324 g–i),
        716
Rapistrum, **718**
    rugosum (Map 1380, Pl. 324
        c–e), **718**
    var. venosum, 718
Ratibida, **531,** 538
    columnaris, 532
    columnifera (Map 1214, Pl. 286
        a, b), **532**
        f. pulcherrima, 532
    pinnata (Map 1215, Pl. 286 e, f),
        531, **532**
    tagetes (Map 1216, Pl. 286 c, d),
        534
Rattlesnake master, 82
Rattlesnake root, 377–380
Rauvolfia, 115
Raveill, Jay, 820
Rayless alkali aster, 284
Red birch, 610
Red campion, 836
Red goosefoot, 888
Red honeysuckle, 770
Red maple, 12
Red morning glory, 949
Red scale, 866, 868
Red-berried elderberry, 777
Redbud, 1061, 1062

Redbud (cont.)
    Eastern, 1061
Red-centered morning glory, 949
Red-fruited horse gentian, 781
Red-hot cattail, 1012
Red-seeded dandelion, 388
Rhinocyllus conicus, 320, 321
Rhizobium, 994, 1058
Rhododendron, 1000, **1004**
    canescens, *1004,* 1006
    prinophyllum (Map 1644, Pl.
        375 h–j), **1004,** 1006
    roseum, 1004
        f. albidum, 1004
Rhombic copperleaf, 1016
Rhopalomyia solidaginis, 252
Rhus, **41,** 42
    americanus, 41
    aromatica (Map 831, Pl. 200 a,
        b), **42,** 48
        var. aromatica, **44**
        var. illinoensis, 44
        var. serotina (Map 831), **44**
    ×borealis, 46
    canadensis, 44
        var. serotina, 44
    copallinum (Map 832, Pl. 200 c,
        d), **45**
    var. latifolia, 45
    cotinoides, 41
    glabra (Map 833, Pl. 201 a–c),
        45
        f. laciniata, 46
        var. borealis, 46
    hirta, 46
    radicans, 49
        f. hypomalaca, 50
        f. negundo, 50
        var. radicans, 50
        var. vulgaris, 50
    toxicodendron, 49
        var. pubens, 51
        var. quercifolium, 49
    trilobata, 42
        var. serotina, 44
    typhina (Map 834, Pl. 201 d), **46**
        var. laciniata, 46
Rib-seeded sand mat, 1041
Ricinus, **1054**
    communis (Map 1685, Pl. 384 h,
        i), 1011, 1020, **1054**
Riehl, Nicholas, 1074
Rigid goldenrod, 271
Rimer, Rhonda, 493
Ringwing, Tumble, 895
River birch, 610
Roadside pennycress, 730
Roadside thistle, 334
Robin's plantain, 223
Robinson, Harold, 393
Rock cress, 662, 669
    Hairy, 662
    Missouri, 669

Rock cress (cont.)
    Smooth, 668
    Virginia, 715
Rock sandwort, 817
Rocket
    Dame's, 696
    Garden, 692
    Purple, 698
    Sand, 689
    Violet, 698
    Yellow, 667
Rocketweed, 692
Rockrose, 903
Rockrose family, 902
Rocky Mountain bee plant, 910
Roman chamomile, 192
Roman wormwood, 444
Root
    Anise, 94
    Pleurisy, 154
    Rattlesnake, 377–380
Rope dodder, 942
Rorippa, 711, 715, **718,** 719–722
    aquatica (Map 1381, Pl. 313 h, i), 718, **719**
    curvipes (Map 1382, Pl. 326 c, d), **720,** 723
        var. curvipes, 720
        var. truncata, 720
    islandica, 721
        var. fernaldiana, 720
    nasturtium-aquaticum, 711
    obtusa, 720
    palustris (Map 1383, Pl. 325 d-f), **720,** 722, 723
        ssp. glabra, 721
        var. fernaldiana, **720,** 721
        var. hispida, 721
    sessiliflora (Map 1384, Pl. 325 a-c), **721,** 722
    sinuata (Map 1385, Pl. 325 g, h), **722**
    sylvestris (Map 1386, Pl. 326 e, f), **722**
    tenerrima (Map 1387, Pl. 326 a, b), 720, **723**
    teres, 720, 723
    truncata, 720
Rosa, 940
Rose, Guelder, 790, 791
Rose-campion, 816
Roseshell azalea, 1004
Rosette crinklemat, 993
Rosette weevil, 320
Rosinweed, 547, 550
    Cup, 552
    Rough-leaved, 553
    Starry, 548
Rosinweed sunflower, 513
Rosy periwinkle, 115
Rough coneflower, 540
Rough gayfeather, 415
Rough goldenrod, 270

Rough green amaranth, 30
Rough ox-eye, 518
Rough pigweed, 30
Rough white lettuce, 379
Rough-leaved dogwood, 960
Rough-leaved goldenrod, 267, 274
Rough-leaved rosinweed, 553
Roughpod copperleaf, 1016
Rough-stemmed goldenrod, 274
Roundleaf groundsel, 586
Round-fruited St. John's-wort, 928
Round-leaved boneset, 407
Round-leaved spurge, 1048
Round-leaved woolly croton, 1023
Rover bellflower, 744
Royal catchfly, 840
Rubber, Para, 1010
Rubber tree, 525, 1010
Rubiaceae, 457
Rubus, 943
Ruby-throated hummingbird, 600
Rudbeckia, 436, **534,** 535, 538, 547, 940
    amplexicaulis (Map 1217, Pl. 276 b), 535, **537**
    bicolor, 542
    fulgida (Map 1218, Pl. 288 f, g), **538**
        var. fulgida, 539
        var. missouriensis, 545
        var. palustris, 540
        var. speciosa, 540
        var. sullivantii, **540**
        var. umbrosa, **540**
    grandiflora (Map 1219, Pl. 287 d, e), **540,** 542
        var. alismifolia, 542
        var. grandiflora, **540,** 542
    hirta (Map 1220, Pl. 287 b, c), **542**
        f. flavescens, 542
        f. homochroma, 542
        var. hirta, 543
        var. pulcherrima, **542,** 543
    laciniata (Map 1221, Pl. 288 e), 535, **543,** 544
        cv. 'Hortensis', 544
        var. ampla, 544
        var. heterophylla, 544
        var. hortensis, 544
        var. laciniata, **543,** 544
    maxima (Map 1222, Pl. 287 a), **544**
    missouriensis (Map 1223, Pl. 287 f, g), 534, 535, **545**
    nitida, 544
    purpurea, 478
    serotina, 542
    speciosa
        var. sullivantii, 540
    subtomentosa (Map 1224, Pl. 288 c, d), **545**
        f. craigii, 546

Rudbeckia (cont.)
    triloba (Map 1225, Pl. 288 a, b), 535, **546,** 547
        var. pinnatiloba, 547
        var. rupestris, 547
        var. triloba, **546,** 547
Ruellia, **5,** 6, 943
    humilis (Map 801, Pl. 195 m, n), **6**
        f. alba, 6
        f. grisea, 6
        var. expansa, 6
        var. frondosa, 6
        var. longiflora, 6
    pedunculata (Map 802, Pl. 195 j, k), **7**
        f. baueri, 7
    pinetorum, 7
    strepens (Map 803, Pl. 195 l), **7**
        f. alba, 7
        f. cleistantha, 7
Ruellia
    Fringeleaf, 6
    Limestone, 7
    Smooth, 7
Rumex, 871
Running strawberry bush, 856
Rush skeletonweed, 344
Rushfoil, 1027
    Common, 1026
    Slender, 1024
Rush-pea, 1070
Russian, Mammoth, 498
Russian atriplex, 864
Russian knapweed, 330
Russian olive, 995
Russian pigweed, 869
Russian thistle, 860, 899, 900
    Slender, 900
Rust, Black stem, 603, 606
Rutabaga, 672, 674
Rutland beauty, 934

S

Sabatia angularis, 380
Sabulina stricta, 817
Safflower, 316
Sage, 177
    White, 185
Sagebrush, 178
Sagewort
    Prairie, 185
    Sweet, 180
    Western, 182
Sagina, **824**
    decumbens (Map 1476, Pl. 345 a, b), 824, **826,** 827
        ssp. decumbens, **826**
        ssp. occidentalis, 826
        var. smithii, 826
    japonica, 824

Sagina (cont.)
   procumbens (Map 1477, Pl. 345 c), **826,** 827
Sailors, Blue, 347
Salicaceae, 457
Salicornia, **898**
   europaea (Map 1549), **898,** 899
   virginica, 899
Salix, 939, 940, 943, 944
Salsify, 344, 391, 392
   Meadow 392
   Western, 391
   Yellow, 391
Salsola, 860, **899,** 900, 944
   collina (Map 1550, Pl. 358 b, c), , 899, **900**
   iberica, 900
   kali, 902
      var. pontica, 902
      var. tenuifolia, 900
   pestifer, 900
   ruthenica, 900
   tragus (Map 1551, Pl. 358 a), 899, **900**
      ssp. pontica, 902
Saltbush, 862
   Fat hen, 865
   Wright's, 868
Saltmarsh aster, Inland, 311
Saltmarsh sand spurrey, 843
Saltwort, 900
Sambucus, 765, **775**
   canadensis (Map 1429, Pl. 335 g, h), **776**
      f. rubra, 776
      var. laciniata, 776
      var. submollis, 776
   nigra, 776
      ssp. canadensis, 776
   pubens (Map 1430, Pl. 336 h), **777**
   racemosa, 777
      ssp. pubens, 777
Samphire, 898
Sand catchfly, Large, 832
Sand cress, 660
Sand croton, 1023
Sand mat, 1041
   Rib-seeded, 1041
Sand milkweed, 145
Sand parsley, 63
Sand rocket, 689
Sand spurrey, 843
   Saltmarsh, 843
Sand vine, 159
Sandhills amaranth, 23
Sandhills pigweed, 23
Sandia, 978
Sandwort, 816, 818
   Grove, 819
   Rock, 817
   Slender, 798
   Thyme-leaved, 798

Sanicle, 99
   Cluster, 102
   Southern, 102
Sanicula, **99**
   canadensis (Map 879, Pl. 211 i, j), **100,** 102
      var. grandis, 102
   canadensis × marilandica, 102
   gregaria, 102
   marilandica, 102
   odorata (Map 880, Pl. 211 f), **102**
   smallii (Map 881, Pl. 211 g), 102
Santa Maria, 626
Sapindaceae, 8
Saponaria, 795, **827**
   officinalis (Map 1478, Pl. 345 d, e), **827**
   vaccaria, 849
Sarothra
   drummondii, 916
   gentianoides, 917
Sarsparilla, Wild, 132
Saururaceae, 1
Saururus, 939, 940, 943, 944
Savia phyllanthoides, 1018
Sawtooth sunflower, 502
Saxifragaceae, 967
Scale
   Red, 866, 868
   Silver, 863, 868
Scale-seed, 104
   Forked, 104
   Hooked, 104
   Western, 104
Scallop squash, 982, 983
Scarlet gourd, 975
Scarlet starglory, 949
Scentless chamomile, 192
Scentless false chamomile, 196
Scentless mayweed, 196
Schmaltzia, 42
   copallina, 45
   crenata, 44
   glabra, 45
   hirta, 46
   serotina, 44
Schrankia, 1081
   horridula, 1082
   nuttallii, 1082
   uncinata, 1082
Scleranthus, 795, 813, **827**
   annuus (Map 1479, Pl. 344 c, d), **828**
      ssp. annuus, **828**
Scorpiongrass, 642
   Big-seeded, 643
   Blue, 644
   Early, 646
   Water, 644
Scotch marigold
Scotch thistle, 341, 342
Scrophulariaceae, 2, 614

Sea beet, 870
Sea blite, 902
Sea purslane, 17
Secondatia, 125
Sedum, **967,** 968
   acre (Map 1614, Pl. 370 f), **970**
      ssp. acre, **970**
   alboroseum, 970
   erythrostictum (Map 1615, Pl. 370 h, i), **970,** 971
   fabaria, 972
   nuttallianum (Map 1616, Pl. 370 c, d), **971**
   pulchellum (Map 1617, Pl. 370 e), 971
   purpureum (Map 1618), **972**
   reflexum (Map 1619), **972**
   sarmentosum (Map 1620, Pl. 370 g), **973**
   telephioides, 971
   telephium, 970
      ssp. alboroseum, 970
      ssp.fabaria, 972
      ssp. purpureum, 972
   ternatum (Map 1621, Pl. 370 a, b), **973,** 974
Sedum, Nuttall's, 971
Seed
   Kenguel, 343
   Niger, 488
   Nyjer, 488
Seepweed, 902
Selenia ,**723**
   aurea (Map 1388, Pl. 326 g, h), **723**
Selenia, 723
   Golden, 723
Semple, John, 245
Senecio, 187, 576, 587, **589,** 590
   ampullaceus (Map 1258, Pl. 299 c, d), **590**
   aureus, 584
      var. gracilis, 584
      var. intercursus, 584
      var. semicordatus, 588
   cineraria, 187
   glabellus, 586
   obovatus, 586
      var. umbratilis, 586
   pauperculus, 587
      var. balsamitae, 587
   plattensis, 588
   *pseudotomentosus,*587, 588
   semicordatus, 588
   tomentosus, 589
   vulgaris (Map 1259, Pl. 297 a, b), **590**
Senna, 1059, 1060, 1062, 1064, **1071,** 1072, 1074
   hebecarpa, 1073
   marilandica (Map 1696, Pl. 386 a, b), **1072,** 1073, 1074

INDEX 1173

Senna (cont.)
    obtusifolia (Map 1697, Pl. 386 f), **1073**
    occidentalis (Map 1698, Pl. 386 c–e), **1074**
Senna, 1071
    Coffee, 1074
    Northern wild, 1073
    Southern wild, 1072
Sensitive brier, 1082
    Catclaw, 1082
Sensitive pea, 1065
Serinia, 360
    cespitosa, 360
    oppositifolia, 360
Seriphidium, 172
Serpentary, 138
Sessile-flowered cress, 721
Sesuvium, 17
    verrucosum, 17
Sheepberry, 789
Shepherd's purse, 676
Sherff, Earl E., 1047
Shield, Water, 733, 734
Shield cress, 707
Shining blue star, 117
Shining sumac, 45
Shoelaces, Devil's, 615
Shoestrings, Devil's, 615
Shoop, Cora, 994
Short-fringed knapweed, 328
Shortpod draba, 690
Shortray prairie coneflower, 534
Short-stalked mouse-ear chickweed, 804
Showy bugloss, 626
Showy bur marigold, 458
Showy coneflower, 540
Showy goldenrod, 275
Showy milkweed, 150
Showy partridge-pea, 1065
Shrubby St. John's-wort, 925
Sibaria, 715
    virginica, 715
Siberian bugloss, 627
Siberian maple, 9
Sicklepod, 668, 1073
Sickleweed, 82
Sicyos, **987**
    angulatus (Map 1631, Pl. 372 e), **987,** 988
    lobatus, 984
Silene, 795, 798, 816, **828,** 829
    anglica, 836
    antirrhina (Map 1480, Pl. 345 g, h), **831**
        f. apetala, 831
        f. bicolor, 831
        f. deaneana, 831
    armeria, 800
    caroliniana (Map 1481, Pl. 346 a, b), **831,** 832, 841
        ssp. wherry, 831

Silene caroliniana (cont.)
    var. caroliniana, 832
    var. pensylvanica, 832
    var. wherryi, **831,** 832
    conoidea (Map 1482, Pl. 348 d), **832**
    csereii (Map 1483, Pl. 347 e, f), 829, **834**
    cucubalus, 841
    dichotoma (Map 1484, Pl. 347 a, b), **834**
        ssp. dichotoma, **834,** 837
        ssp. racemosa, 834
    dioica (Map 1485, Pl. 348 c), 828, 829, **836,** 837
    gallica (Map 1486, Pl. 347 c, d), **836**
    latifolia (Map 1487, Pl. 348 g, h), 828, 829, 836, **837,** 838, 840, *841*
        ssp. alba, 837, 838
        ssp. latifolia, 838
    nivea (Map 1488, Pl. 346 c, d), **838**
    noctiflora (Map 1489, Pl. 348 e, f), 837, **838**
    pratensis, 837
    regia (Map 1490, Pl. 346 g–i), **840**
    Section Compactae, 798
    Section Silene, 798
    stellata (Map 1491, Pl. 347 g, h), 828, **840**
        var. scabrella, 840
    virginica (Map 1492, Pl. 346 e, f), 832, **841**
    vulgaris (Map 1493, Pl. 348 a, b), 834, **841,** 842
        ssp. vulgaris, **841,** 842
    wherryi, 831
Silk tree, 1079
Silky aster, 311
Silky wormwood, 184
Silphium, 493, **547,** 548, 550
    asperrimum, 548, 554
    asteriscus (Map 1226, Pl. 289 a, b), **548,** 550, 554
        var. asteriscus, **548,** 550
        var. scabrum, 548, 550
    gatesii, 548
    integrifolium (Map 1227, Pl. 289 d), **550,** 551, 554
        f. deamii, 551
        var. deamii, 551
        var. integrifolium (Map 1227), **551,** 552
        var. laeve (Map 1227), **551**
    laciniatum (Map 1228, Pl. 289 i), **552**
    perfoliatum (Map 1229, Pl. 289 e–h), 547, 551, **552**
        f. hornemannii, 553
        f. petiolatum, 553

Silphium perfoliatum (cont.)
    var. connatum, 553
    var. perfoliatum, 553
    radula (Map 1230), **553,** 554
    var. gracile, 554
    var. radula, **553,** 554
    speciosum, 551
    terebinthinaceum (Map 1231, Pl. 289 c), 547, 553, **554**
    var. pinnatifidum, 554
    var. terebinthinaceum, 554
Silver dollar, 710
Silver maple, 13
Silver scale, 863, 868
Silverberry, Cherry, 995
Silverrod, 262
Silybum, 316, **342**
    marianum (Map 1060), **342,** 343
Sinadoxa, 765
Sinapis, **724**
    alba (Map 1389, Pl. 327 e, f), 674, **724,** 725
        ssp. alba, **724**
        ssp. dissecta, 725
    arvensis (Map 1390, Pl. 327 a–d), **725**
Single-stemmed bog aster, 227
Sisters, Twin, 768
Sisymbrium, **725**
    altissimum (Map 1391, Pl. 328 a–c), **726**
    gallicum, 692
    loeselii (Map 1392, Pl. 328 h–j), **726**
    nasturtium-aquaticum, 711
    officinale (Map 1393, Pl. 328 f, g), **726,** 728
        var. leiocarpum, 726, 728
Sium, **103**
    suave (Map 882, Pl. 212 h–j), **103**
Six-angled spurge, 1029
Skeleton plant, 372, 374
Skeletonweed, Rush, 344
Skinner, Michael, 475
Skunk weed, 1026
Sleepy catchfly, 831
Slender cottonweed, 38
Slender pigweed, 28
Slender rushfoil, 1024
Slender Russian thistle, 900
Slender sandwort, 798
Slender snake-cotton, 38
Slender three-seeded mercury, 1013
Slender-leaved Venus' looking glass, 756
Slim-leaved goosefoot, 887
Slimpod Venus' looking glass, 756
Sloth, Giant, 1069
Small bindweed, 936
Small mouse-ear chickweed, 808
Small St. John's-wort, 917
Small Venus' looking-glass, 754

Small white aster, 302
Small white morning glory, 951
Small wild carrot, 798
Small woodland sunflower, 506
Smallanthus, 529, **555**
    uvedalius (Map 1232, Pl. 290 g, h), **555**
Smallflower morning glory
Smallflower wallflower, 695
Small-flowered bitter cress, 683
Small-flowered catchfly, 836
Small-flowered fiddleneck, 625
Small-flowered forget-me-not, 644
Small-flowered hawksbeard, 349
Small-flowered partridge-pea, 1064
Small-flowered St. John's-wort
Small-headed thistle, 336
Smartweed dodder, 944
Smentowski, Joe, 52
Smith, Tim, 751
Smoke tree
    American, 41
    Eurasian, 41
Smooth alder, 609
Smooth aster, 291
Smooth catchfly, 834
Smooth cat's ear, 358
Smooth hawksbeard, 349
Smooth milkweed, 153
Smooth nailwort, 821
Smooth rock cress, 668
Smooth ruellia, 7
Smooth sumac, 45
Smooth sweet cicely, 94
Smoothseed milkweed, 148
Snake-cotton, 36
    Field, 38
    Slender, 38
Snakeroot, 394
    Black, 99, 102
    Broom, 202
    Button, 82, 418, 422
    Canada, 100
    Prairie, 416
    Virginia, 138
    White, 378, 395
Snakeweed, 237
Sneezeweed, 174
    Autumn, 490
    Bitter, 490
    Common, 490
    Purple-headed, 491
    Southern, 491
    Virginia, 492
Sneezewort, 174
Sneezewort aster, 269
Snot plant, 733
Snowball bush, 791
Snowberry, 777
    Western, 778
Snow-on-the-mountain, 1044
Snowy campion, 838
Soapwort, 827

Soapwort (cont.)
    Cow, 849
Socrates, 74
Soft goldenaster, 210
Soft maple, 13
Soil Conservation Service, U.S.D.A., 773, 995
Solanum, 943
Soldier, Gallant, 485
Soldier's woundwort, 174
Solidago, 198, 204, 229, 230, **244,** 245, 270, 940, 942, 943
    altissima (Map 990, Pl. 241 a–c, g–j), 245, **248,** 250, 252, 258
        var. altissima (Pl. 241 h–j), **252**
        var. gilvocanescens (241 a–c, g), **252**
    arguta (Map 991, Pl. 239 a, b), **252,** 253
        var. caroliniana, 254
        var. arguta, **253**
        var. boottii, **253,** 254
        var. caroliniana, 253, **254**
        var. neurolepis, 253
        var. strigosa, 253
    asteroides, 269
    bicolor, 198, 244, 262
        var. concolor, 262
        var. hispida, 262
    boottii, 253
    buckleyi (Map 992, Pl. 238 a–c), , **254,** 269
    caesia (Map 993, Pl. 239 c–e), **254**
        var. caesia, 256
        var. zedia, 256
    canadensis (Map 994, Pl. 241 d–f), 245, 250, 252, **256,** 258, 262, 270
        var. canadensis, 258
        var. gilvocanescens, 252
        var. hargeri, **256**
        var. scabra, 252
    drummondii (Map 995, Pl. 241 k–m), **258**
    flexicaulis (Map 996, Pl. 239 f–h), **260**
    gattingeri (Map 997, Pl. 242 c, d), **260,** 264
    gigantea (Map 998, Pl. 242 a, b), **261,** 262
        var. leiophylla, 261
        var. serotina, 261
    graminifolia, 230
        var. graminifolia, 230
        var. media, 230
        var. nuttallii, 230
    gymnospermoides, 230
    hispida (Map 999, Pl. 238 d), 198, **262**
        var. hispida, **262**

Solidago (cont.)
    juncea (Map 1000, Pl. 243 c–e), **263**
        f. scabrella, 263
    leptocephala, 232
    ludoviciana, 253
    missouriensis (Map 1001, Pl. 242 e, f), **263**
        var. fasciculata, **263,** 264
        var. missouriensis, 264
    nemoralis (Map 1002, Pl. 243 a, b), **264,** 266
        ssp. decemflora, **266**
        ssp. nemoralis, **266**
        var. decemflora, 266
        var. haleana, 266
        var. longipetiolata, 266
    ×neurolepis, 253
    odora (Map 1003, Pl. 243 f–i), **267**
        var. chapmannii, 267
        var. odora, **267**
    patula (Map 1004, Pl. 240 e, f), **267**
        var. patula, **267**
        var. strictula, 268
    petiolaris (Map 1005, Pl. 238 h, i), 244, 254, **268,** 269
        var. angusta, 268, 269
        var. petiolaris, 269
        var. wardii, 268, 269
    ptarmicoides (Map 1006, Pl. 237 c, d), 244, 245, **269**
    radula (Map 1007, Pl. 241 n–p), **270**
        var. laeta, 270
        var. radula 270
        var. stenolepis, 270
    riddellii (Map 1008, Pl. 237 e, f), 245, 270, **271**
    rigida (Map 1009, Pl. 237 g, h), 245, 270, **271,** 272
        ssp. glabrata, **272**
        ssp. humilis, **272**
        ssp. rigida, 272, **274**
        ssp. sphagnifolia, 275
        ssp. villosa, 275
        var. glabrata, 272
        var. humilis, 272
        var. rigida, 272
    rugosa (Map 1010, Pl. 240 a, b), **274,** 276
        ssp. aspera, 260, **275**
        ssp. rugosa, **275**
        var. aspera, 275
        var. celtidifolia, 275
    speciosa (Map 1011, Pl. 238 f, g), **275**
        var. angustata, 276
        var. jejunifolia, 276
        var. pallida, 276
        var. rigidiuscula, **276**
        var. speciosa, **276**

Solidago (cont.)
strigosa, 253
ulmifolia (Map 1012, Pl. 240 c, d), 245, 253, 260, **277**
var. palmeri, **277**
var. ulmifolia, **278**
Solomon, James, 205
Soltis, Pamela and Douglas, 391
Sonchus, 344, **383,** 871, 944
arvensis (Map 1096, Pl. 262 e, f), **384**
ssp. arvensis, **384**
ssp. uliginosus, **384**
var. glabrescens, 384
asper (Map 1097, Pl. 262 a, b), 383, **386**
f. glandulosus, 386
ssp. asper, **386**
ssp. glaucescens, 386
oleraceus (Map 1098, Pl. 262 c, d), **386**
Sour gum, 964, 966
Southern arrowwood, 786
Southern blackhaw, 793
Southern catalpa, 618
Southern gooseberry, 1009
Southern prairie aster
Southern ragweed, 444
Southern sanicle, 102
Southern sneezeweed, 491
Southern sugar maple, 15
Southern wild senna, 1072
Sow thistle, 383
Common, 386
Field, 384
Perennial, 384
Prickly, 386
Spiny-leaved, 386
Soybean, 1058
Spanish buttons, 328
Spanish needles, 456, 524
Sparkleberry, 1007
Spear-scale, 865
Specularia, 753
biflora, 754
holzingeri, 754
lamprosperma, 756
leptocarpa, 756
perfoliata, 757
f. alba, 758
speculum, 753
Spergula, 795, **842**
arvensis (Map 1494, Pl. 349 f, g), **842,** 843
Spergularia, 795, **843**
marina, 843
salina (Map 1495), **843,** 844
Spermolepis, **104**
divaricata (Map 883, Pl. 212 d, e), **104**
echinata (Map 884, Pl. 212 f, g), **104**

Spermolepis (cont.)
inermis (Map 885, Pl. 212 a–c), **104**
Sphenoclea, 742
Sphenocleaceae, 742
Sphinx moth, Catalpa, 618
Spider flower, 910
Spider milkweed, 158
Spiked lobelia, 751
Spikenard, 131
American, 132
Spilanthes, 441
americana, 441, 442
var. repens, 441
Spinach, 860, 864, 871
Chinese, 31
Malabar, 31
Strawberry, 880
Spinacia, 18, 860
Spindle tree, 855
European, 855
Spiny cocklebur, 569
Spiny pigweed, 31
Spiny-leaved sow thistle, 386
Spiny-toothed gumweed, 234
Spotted cat's ear, 357
Spotted cowbane, 70
Spotted Joe-pye weed, 404
Spotted knapweed, 331
Spotted St. John's-wort, 925, 926
Spotted touch-me-not, 600
Spreading aster, 303
Spreading bladderpod, 614
Spreading dogbane, 121
Spreading eryngo, 81
Spreading fleabane, 216
Spreading pigweed, 24
Spreading yellow cress, 722
Spring, Harbinger of, 79, 80
Spring cress, 679
Spurge, 1027
Allegheny, 731
Blunt-leaved, 1046
Cypress, 1035
Flowering, 1034
Geyer's, 1042
Graveyard, 1035
Groundfig, 1047
Leafy, 1029, 1040
Mat, 1049
Milk, 1047
Missouri, 1045
Nodding, 1045
Prairie, 1045
Prostrate, 1042
Round-leaved, 1048
Six-angled, 1029
Thyme-leaved, 1048
Toothed, 1038
Wood, 1032
Spurge family, 1010
Spurrey, 842
Corn, 842

Spurrey (cont.)
Saltmarsh sand, 843
Sand, 843
Squash, 982
Acorn, 982, 983
Scallop, 982, 983
Summer, 983
Squaw huckleberry, 1009
Squaw weed, 584, 586, 588
St. Andrew's cross, 918, 920
St. John's wort, 913
Clasping-leaved, 917
Common, 924
Creeping, 915
Dwarf, 922
Giant, 913
Great, 915
Greater, 921
Marsh, 928
Round-fruited, 928
Shrubby, 925
Small, 917
Spotted, 925, 926
St. John's wort family, 912
Stachys, 944
Staff-tree family, 850
Staggerbush, 1001
Staghorn sumac, 46
Star
Blazing, 413
Blue, 116, 117
Ciliate blue, 117
Shining blue, 117
Star thistle, 321
Lesser, 326
Maltese, 327
Yellow, 330, 331
Star tickseed, 464
Starry campion. 840
Starry rosinweed, 548
Starvation cactus, 740
Starwort, False, 208
Staunchweed, 174
Stellaria, 795, 819, **844**
aquatica, 820
graminea (Map 1496, Pl. 349 a, b), **845**
var. latifolia, 845
longifolia (Map 1497, Pl. 349 c), **845**
media (Map 1498, Pl. 349 d, e), 844, **846,** 849
ssp. *neglecta,* 846
ssp. *pallida,* 848
neglecta (Map 1499), 844, **846**
pallida (Map 1500), 846, **848,** 849
pubera, 844
Stem galler, Goldenrod, 252
Stem rust, Black, 603, 606
Stevens, Jane, 354
Steyermark, Julian, 237, 312, 824, 994, 1041, 1073

Stick, Devil's walking, 132
Stickseed, 633, 634
  Two-row, 636
  Western, 636
Sticktight, 456, 459, 460
Sticky cockle, 838
Sticky hawkweed, 357
Stiff aster, 243
Stiff dogwood, 963, 964
Stiff goldenrod, 271
Stiff greenthread, 559
Stiff sunflower, 510
Stiff-leaved aster, 243
Stinking chamomile, 177
Stinking clover, 910
Stinking elderberry, 777
Stinking fleabane, 574
Stinking pepperweed, 707
Stinking wall-rocket, 689
Stinkweed, 556, 574, 730
Stitchwort, 844
  Common, 845
  Dixie, 818
  Long-leaved
  Michaux's, 817
  Pitcher's
Stokes aster, 591
Stokesia, 591
  laevis, 591
Stonecrop, 967
  Jenny's, 972
  Mossy, 970
  Three-leaved, 973
  Yellow, 973
Stonecrop family, 967
Stoneseed, Plains, 640
Strawberry blite, 880
Strawberry bush, 854
Strawberry bush, Running, 856
Strawberry shrub family, 740
Strawberry spinach, 880
Strophanthus, 115
Strophostyles, 944
Stylisma, **956**
  pattersonii, 956
  pickeringii (Map 1605, Pl. 368 c, d), **956,** 957
    pickeringii var. pattersonii, **956,** 957
    pickeringii var. pickeringii, 957
Stylophorum diphyllum, 603
Suaeda, **902**
  calceoliformis (Map 1552, Pl. 358 g, h), **902**
  depressa, 902
Sugar maple, 12, 13, 14
  Southern, 15
Sullivan, James, 562, 708
Sulphur cosmos, 469
Sumac, 41
  Aromatic, 42
  Dwarf, 45

Sumac (cont.)
  Fragrant, 42, 48
  Poison, 48
  Smooth, 45
  Staghorn, 46
  Winged, 45
Summer cypress, 896
Summer squash, 983
Summers, Bill, 404, 492, 493, 542, 714
Sump weed, 520
Sunchoke, 516
Sunflower, 165, 493
  Ashy, 506
  Bristly, 503
  Common, 436, 497, 498
  Cultivated, 488
  False, 517
  Hairy, 503
  Hopi, 500
  Kansas, 512
  Maximilian, 504, 506
  Naked-stemmed, 508
  Narrow-leaved, 497
  Pale, 500
  Pale-leaved, 514
  Plains, 512
  Prairie, 512
  Rosinweed, 513
  Sawtooth, 502
  Small woodland, 506
  Stiff, 510
  Thin-leaved, 500
  Tickseed, 452
  Willow-leaved, 512
  Woodland, 501
Sunflower family, 164
Sunflower heliopsis, 517
Susan
  Black-eyed, 542
  Brown-eyed, 546
Svida
  alternifolia, 959
  asperifolia, 960
  foemina, 964
Swallowtail, Zebra, 52
Swallowwort, Black, 159
Swamp beggar ticks, 459
Swamp black gum 966
Swamp blueberry, 1008
Swamp cottonwood, 457
Swamp dogwood, 959
Swamp milkweed, 146
Swamp thistle, 338
Swamp tupelo, 965, 966
Swedish turnip, 672
Sweet alyssum, 708
Sweet anise, 94
Sweet cicely, 93
  Smooth, 94
  Woolly, 93
Sweet clover, 1058
Sweet coneflower, 545

Sweet everlasting, 434
Sweet goldenrod, 267
Sweet lowbush blueberry, 1007
Sweet potato, 948
Sweet sagewort, 180
Sweet William, 810
Sweet William catchfly, 800
Sweet wormwood, 180
Sweetshrub, 740
Swine cress, 704
Swiss chard, 860, 870
Sycamore maple, 8
Symphoricarpos, 765, **777,** 944
  albus (Map 1431, Pl. 336 g), **778**
    var. laevigatus, 778
  occidentalis (Map 1432, Pl. 336 c, d), **778,** 780
  orbiculatus (Map 1433, Pl. 336 e, f), **780**
  f. leucocarpus, 780
Symphyotrichum, 206, 219, 226, **278,** 279, 290, 301, 315, 940, 942–944
  ×amethystinum, 290, 297
  anomalum (Map 1013, Pl. 244 a, b), **282,** 284
  ×batesii, 298
  ciliatum (Map 1014, Pl. 249 d, e), 278, **284**
  cordifolium (Map 1015, Pl. 244 c, d), **285,** 286, 287, 292
    var. polycephalum, 285
  divaricatum, 311
  drummondii (Map 1016, Pl. 244 e, f), **286,** 287
    ssp. drummondii, **286**
  dumosum (Map 1017, Pl. 246 a, b), **287,** 296, 311, 315
    var. dodgei, 287, 288
    var. dumosum, 288
    var. strictior, **287,** 288
  ericoides (Map 1018, Pl. 245 a, b), **288,** 290, 297, 298, 303
    var. ericoides, **290**
    var. pansum, 290
    var. prostratum, **290**
    var. stricticaule, 290
  falcatum (Map 1019, Pl. 245 c, d), **290**
    ssp. commutatum, **290,** 291
    ssp. falcatum, 291
    var. commutatum, 290
  firmum, 308, 310
  laeve (Map 1020, Pl. 248 e, f), **291,** 292, 301
    var. geyeri, 292
  lanceolatum (Map 1021, Pl. 246 e, f), **292,** 294, 296, 300, 308, 309
    ssp. hesperium, 294
    ssp. lanceolatum, 294
    var. hisuticaule, 294
    var. interior, 294, **295,** 311

Symphyotrichum lanceolatum (cont.)
    var. lanceolatum, 294, **295**
    var. latifolium, **295**
  lateriflorum (Map 1022, Pl. 246 g-i), 294, **295**, 296, 300, 308, 309, 311
    var. horizontale, 296
    var. lanceolatum, 296
  novae-angliae (Map 1023, Pl. 245 e, f), 290, **296**, 297, 298
  oblongifolium (Map 1024, Pl. 249 f, g), **297**
  ontarionis (Map 1025, Pl. 247 d, e), 294, 296, **298**, 300
    var. glabratum, 300
    var. ontarionis, **298**
  oolentangiense (Map 1026, Pl. 244 g, h), 292, **300**, 301
    var. oolentangiense, **302**
    var. poaceum, **302**
  parviceps (Map 1027, Pl. 247 f, g), **302**, 303, 306
  patens (Map 1028, Pl. 245 g, h), **303**, 304, 305
    var. gracile, **304**
    var. patens, **304**
    var. patentissimum, **304**
  pilosum (Map 1029, Pl. 247 h, i), 290, 300, 303, **305**, 306, 308
    var. pilosum, 306
    var. pringlei, 305, 306
  praealtum (Map 1030, Pl. 247 a-c), 294, **306**, 308, 309
    var. angustior, 306, 308
    var. subasperum, 306, 308
    var. praealtum, 308
  priceae, 306
  puniceum (Map 1031, Pl. 248 a, b), **308**, 309, 310
  racemosum (Map 1032), 278, 288, 290, 294, 300, 306, **310**, 311
    var. racemosum, 311
    var. subdumosum, **310**
  sericeum (Map 1033, Pl. 249 a-c), **311**
  subulatum (Map 1034, Pl. 248 c, d), **311**
    var. ligulatum, **311**, 312
    var. parviflorum, 312
    var. squamatum, 312
    var. subulatum, 312
  turbinellum (Map 1035, Pl. 229 f, g), 278, **312**
  urophyllum (Map 1036, Pl. 249 h, i), 286, 287, **314**, 315
Symphytum, **648**, 650
  officinale (Map 1308, Pl. 310 a, b), 632, **650**

Synosma suaveolens, 582

T

Tabernaemontana angustifolia, 117
Taenidia, **106**
  integerrima (Map 886, Pl. 213 a-c), **106**
Tag alder, 609
Tagetes, 436, **556**
  erecta (Map 1233, Pl. 290 e, f), **556**, 558
  minuta, 558
  patula (Map 1234, Pl. 290 c, d), **558**
  tenuifolia, 558
Tajo, Yerba de, 479
Tall bellflower, 743
Tall coreopsis, 468
Tall goldenrod, 248, 261
Tall green milkweed, 145
Tall hedge mustard, 726
Tall ironweed, 598
Tall khakiweed, 556
Tall thistle, 334
Tall thoroughwort, 402
Tall tickseed, 468
Tall white aster, 292
Tall white lettuce, 378
Tanacetum, 165, 172, 187, **194**
  balsamita (Map 955, Pl. 228 h, i), **194**
  cineariifolium, 194
  huronense, 192
  parthenium (Map 956, Pl. 228 c, d), **195**
  vulgare (Map 957, Pl. 228 e-g), **195**
    f. crispum, 196
Tansy, 194
  Common, 195, 196
Tansy mustard, 686, 688
Taraxacum, 344, **387**
  erythrospermum (Map 1099, Pl. 263 a), **388**
  laevigatum, 388
    f. scapifolium, 388
  laevigatum, 388
  officinale (Map 1100, Pl. 263 b, c), **388**
  Section Erythrosperma, 387
  Section Ruderalia, 387, 388
  vulgare, 388
Tarragon, 184
Tarweed fiddleneck, 624
Tassel flower, 398
Tatarian aster, 204, 205
Tatarian honeysuckle, 768
Taxodium distichum, 125, 457
Tea
  Hopi, 560
  Mexican, 875

Tea (cont.)
  Navajo, 560
  Prairie, 1026
Tear-blanket, 132
Teasel, 988
  Common, 989
  Cut-leaved, 989
  Fuller's, 988
  Wild, 989
Telegraph plant, 242
Tephrid fly, 250
Tetradoxa, 765
Texas croton, 1026
Texas green eyes, 449
Texas groundsel, 590
Texas thistle, 340
Texas wild gourd, 984
Thale cress, 660
Thaspium, **106**, 107, 114
  barbinode (Map 887, Pl. 213 f-h), **107**, 108
  trifoliatum (Map 888, Pl. 213 i-k), 107, **108**
    var. flavum, **110**
    var. trifoliatum, **110**
The Nature Conservancy, 129
Thelesperma, **558**, 559, 944
  ambiguum (Pl. 290 a, b), 560
  filifolium (Map 1235, Pl. 292 a, b), **559**, 560
    var. filifolium, 560
    var. intermedium, 560
  megapotamicum (Map 1236, Pl. 292 c, d), 559, **560**
    var. ambiguum, 560
  trifidum, 560
Thevetia, 115
Thin-leaved sunflower, 500
Thistle, 332
  Barneby's, 330
  Blessed, 322
  Blessed Milk, 342
  Blue, 633
  Bull, 341
  Canada, 335
  Carolina, 336
  Common sow, 386
  Cotton, 341
  Curly, 318
  Field, 335, 338
  Field sow, 384
  Lesser star, 326
  Maltese star, 327
  Milk, 342, 343
  Musk, 320, 321
  Niger, 436, 488
  Nodding, 320
  Perennial sow, 384
  Platte, 336
  Plumeless, 318
  Prickly sow, 386
  Russian, 860, 899, 900
  Scotch, 341, 342

Thistle (cont.)
    Slender Russian, 900
    Small-headed, 336
    Sow, 383
    Spiny-leaved sow, 386
    Star, 321
    Swamp, 338
    Tall, 334
    Texas, 340
    Wavy-leaved, 340
    Welted, 318
    Yellow star, 330, 331
Thistle tribe, 488
Thlaspi, 703, 711, **728**
    alliaceum (Map 1394), **730**
    arvense (Map 1395, Pl. 327 i–k), **730**
    campestre, 703
    perfoliatum, 710, 728
Thorny amaranth, 31
Thoroughwax, 68
Thoroughwort, 400, 406
    Hyssop-leaved, 403
    Pink, 412
    Tall, 402
Three-leaved stonecrop, 973
Three-seeded mercury, 1012, 1016
    Slender, 1013
Thyme-leaved sandwort, 798
Thyme-leaved spurge, 1048
Tickseed, 461
    Star, 464
    Tall, 468
Tickseed coreopsis, 463
Tickseed sunflower, 452
Tickweed, 566
Tidestromia, **39**
    lanuginosa (Map 829, Pl. 199 c, d), **39**
        ssp. eliassoniana, 40
        ssp. lanuginosa, 40
Tie vine, 954
Tim, Tiny, 813
Timme, Stephen, 764
Tinker's weed, 782
Tiny lazydaisy, 211
Tiny Tim, 813
Tiquilia, 620, **993**
    nuttallii (Map 1635, Pl. 373 h, i), **993**
Tithymalopsis
    corollata, 1034
    commutatus, 1032
Tobacco, 958
    Indian, 428, 748
    Ladies', 428
    Rabbit, 430, 434
Tofari, Sarah M. 206
Top, White, 703
Tongue
    Common hound's, 632
    Hound's, 630
Tooker, Midge, 732

Toothed spurge, 1038
Toothwort, 680
    Broad-leaved, 680
    Cut-leaved, 680
Top
    Heart-podded white, 706
    White, 703
Torilis, **110**
    arvensis (Map 889, Pl. 214 c–e), **111**
    japonica (Map 890), **111**
    nodosa (Map 891, Pl. 214 a, b), 110, **111**
Touch-me-not, 599
    Pale, 602
    Spotted, 600
Touch-me-not family, 599
Tower mustard, 730, 731
Toxicodendron, 42, **46**, 48, 940
    negundo, 50
    pubescens (Map 835, Pl. 201 e), 48, **49**
    quercifolium, 49
    radicans (Map 836, Pl. 201 f–h), 48, **49**, 50
        ssp. negundo, **50**, 51
        ssp. pubens, **51**
        ssp. radicans, **51**
        var. negundo, 50
        var. pubens, 51
        var. radicans, 50
    toxicarium, 49
    toxicodendron, 49
    vernix, 48
Trachelospermum, **124**
    difforme (Map 900, Pl. 216 a–e), **124,** 125
Trachomitum, 120
Tradescantia, 391
Tragia, 944, **1054,** 1055
    betonicifolia (Map 1686, Pl. 384 e–g), 1055, **1056,** 1057
    cordata (Map 1687, Pl. 384 c, d), 1055**1056**
    ramosa (Map 1688, Pl. 384 a, b), 1055, **1057**
    urticifolia, 1055–1057
        var. texana, 1056
Tragia, 1054
    Heart-leaved, 1056
Tragopogon, 343, 344, **390,** 391
    dubius (Map 1101, Pl. 263 d, e), **391,** 392
    lamottei, 392, 393
    mirus, 391
    miscellus, 391
    porrifolius (Map 1102, Pl. 263 f, g), 390, **391**
    pratensis (Map 1103, Pl. 263 h, i), 391, **392,** 393
Trailing arbutus, 1000
Trailing pearlwort, 826
Trautvetteria caroliniensis, 746

Treacle mustard, 696
Trécul, Auguste, 910
Tree
    Cannonball, 1011
    Catawba, 619
    Cigar, 619
    Dove, 958
    European spindle, 855
    Ghost, 958
    Judas, 1061
    Kentucky coffee, 1069
    Orchid, 1061
    Rubber, 525, 1010
    Silk, 1079
    Spindle, 855
    Wayfaring, 788, 789
Tree cholla, 736
Trepocarpus, **112**
    aethusae (Map 892, Pl. 213 d, e), **112**
Triadenum, 912, 913, **928,** 929, 940
    tubulosum (Map 1575, Pl. 362 i, j), **929,** 930
        var. walteri, 930
    virginicum, 930
    walteri (Map 1576, Pl. 362 a, b), 929, **930**
Trianthema, **17**
    portulacastrum (Map 809, Pl. 202 e, f), **17**
Tribe, Thistle, 488
Trichosirocalus horridus, 320
Tricolor chrysanthemum, 189
Trifolium, 944, 1058
Triodanis, **752,** 753
    biflora (Map 1413, Pl. 332 g), **754,** 758
    falcata, 753
    holzingeri (Map 1414, Pl. 332 f), **754,** 756
    lamprosperma (Map 1415, Pl. 332 h), **756**
    leptocarpa (Map 1416, Pl. 332 a, b), **756,** 758
    perfoliata (Map 1417, Pl. 332 c–e), 753, 754, 756, **757,** 758
        var. biflora, 754
Triosteum, 765, **780,** 781
    angustifolium (Map 1434, Pl. 337 a, b), **781**
        f. rubrum, 781
        var. eamesii, 781
    aurantiacum (Map 1435, Pl. 337 f, g), **781,** 782, 784
        var. aurantiacum, 782
        var. glaucescens, 782
        var. illinoense, **781,** 782
    illinoense, 781, 782
    perfoliatum (Map 1436, Pl. 336 a, b), **782,** 784
        var. aurantiacum, 782
Tripleurospermum, 191, 192, **196**

Tripleurospermum (cont.)
  inodorum (Map 958, Pl. 227 d, e), **196**
  maritimum, 192, 198
    ssp. inodorum, 196
Tristemon texanum, 984
Tropic croton, 1023
True chamomile, 192
Trumpet creeper, 615, 616
Trumpet honeysuckle, 774
Trumpet vine, 615
Tsuga canadensis, 732
Tunica prolifera, 822
Tumble knapweed, 326
Tumble mustard, 726
Tumble ringwing, 895
Tumbleweed, 23, 24, 899, 900
Tumbling orach, 866
Tung oil, 1011
Tupelo, 964, 965
  Black, 966
  Swamp, 965, 966
  Water, 457, 965
Tupelo gum, 965, 966
Turfgrass, 946
Turkey, 279
Turnip, 674
  Swedish, 672
  Wild, 718
Turnip weed, 718
Turritis, 662, **730**
  glabra (Map 1396, Pl. 328 d, e), **731**
    var. furcatipilis, 731
    var. glabra, **731**
Tussilago, 576
Twin sisters, 768
Twinleaf, 603
Twistwood, 788
Two-flowered Cynthia, 359
Two-row stickseed, 636
Two-seeded mercury, 1013
Two-seeded orach, 864
Tyrol knapweed, 328, 330

U

Umbelliferae, **54**
Upland aster, White, 269
Upland boneset, 410
Uropappus, 374
Urticaceae, 758
U.S.D.A. Soil Conservation Service, 773, 995
U.S. Fish and Wildlife Service, 136, 148, 209, 493, 714

V

Vaccaria, 795, **849**

Vaccaria (cont.)
  hispanica (Map 1501, Pl. 349 h, i), **849**
    ssp. hispanica, 849
    ssp. pyramidata, 849
    pyramidata, 849
Vacciniaceae, 999
Vaccinium, 999–1001, **1006,** 1007, 1009
  angustifolium, 1007
  arboreum (Map 1645, Pl. 375 f, g), **1007,** 1008
    var. glaucescens, 1007
  corymbosum (Map 1646), 1006, 1007, **1008**
  macrocarpon, 1007
  pallidum (Map 1647, Pl. 375 c–e), 1001, 1007, **1008,** 1009
  stamineum (Map 1648, Pl. 375 a, b), 1006, 1007, **1009**
    var. interius, 1009
    var. melanocarpum, 1009
    var. neglectum, 1009
  tenellum, 1009
  vacillans, 1008, 1009
    var. crinitum, 1008
    var. missouriense, 1008
Valerianaceae, 765, 766
Variegated milkweed, 156
Vegetable marrow, 982, 983
Vegetable oyster, 392
Venus' looking-glass
  Clasping, 757
  Cultivated, 753
  Holzinger's, 754
  Prairie, 756
  Slender-leaved, 756
  Slimpod, 756
  Small, 754
Verbena, 943, 944
Verbenaceae, 620
Verbesina, **560,** 561, 940, 943
  alternifolia (Map 1237, Pl. 291 a–c), 561, **562**
  encelioides (Map 1238, Pl. 291 d, e), 561, **562,** 564
    ssp. exauriculata, 562, 564
    var. exauriculata, 562, 564
  helianthoides (Map 1239, Pl. 291 f, g), **564**
  occidentalis, 562
  virginica (Map 1240, Pl. 291 h–j), 561, **566**
    var. laciniata, 566
    var. virginica, 561, **566**
Vernal whitlow grass, 691
Vernonia, **592,** 940, 942, 943
  altissima, 598
    f. alba, 598
    var. taeniotricha, 598
  arkansana (Map 1261, Pl. 300 g–j), **594,** 596, 597

Vernonia (cont.)
  baldwinii (Map 1262, Pl. 299 e–h), **596,** 598
    f. alba, 597
    f. albiflora, 597
    ssp. baldwinii, **596**
    ssp. interior, 596, **597**
    var. interior, 597
  crinita, 594
  drummondii, 598
  duggeriana, 596
  fasciculata (Map 1263, Pl. 300 a–c), **597**
    ssp. corymbosa, 597
    ssp. fasciculata, **597**
    var. corymbosa, 597
    var. nebraskensis, 597
  flavipapposa, 596
  gigantea (Map 1264, Pl. 300 d–f), 597, **598**
    ssp. gigantea, **598**
    ssp. ovalifolia, 598
  hymenolepis, 594
  ×illinoensis, 598
  interior, 597
    var. baldwinii, 596
  missurica (Map 1265, Pl. 300 k–m), 596, **598**
    f. swinkii, 598
  parthenioides, 596
  ×peralta, 596
  pseudobaldwinii, 596
  pseudodrummondii, 597
  ×sphaeroidea, 598
  reedii, 598
Viburnum, 765, **784,** 785, 789
  bushii, 791
  dentatum (Map 1437, Pl. 337 c–e), **786,** 788, 793
    var. deamii, 786, 788
    var. dentatum, 788
    var. indianense, 788
    var. lucidum, 792
  lantana (Map 1438), 784, **788,** 789
  lentago (Map 1439, Pl. 338 a, b), **789**
  molle (Map 1440, Pl. 338 g), **789,** 791
    f. leiophyllum, 790
  opulus (Map 1441), 784, **790,** 791
    var. americanum, **790**
    var. opulus, **790**
  ozarkense (Map 1442), **791**
  prunifolium (Map 1443, Pl. 338 c, d), 789, **791,** 792
    var. bushii, 791
  rafinesquianum (Map 1444, Pl. 338 h), 791, **792**
    var. affine, 792
  recognitum (Map 1445, Pl. 337 h, i), 788, **792**

Viburnum (cont.)
   rufidulum (Map 1446, Pl. 338 e,
      f), 789, 792, **793**
   Section Lentago, 789
   Section Odontotinus, 786
   Section Opulus, 791
   Section Viburnum, 789
   trilobum, 791
Viburnum, 784
   Kentucky, 789
Vinca, 115, **125**
   major (Map 901), **126**
   minor (Map 902, Pl. 217 a–c),
      126
Vincetoxicum, 159
   hirundinaria, 160
   nigrum, *159*, **160**
   rossicum, 160
Vine
   Cross, 615
   Cypress, 952
   Hell, 615
   Love, 938
   Quarter, 615
   Sand, 159
   Tie, 954
   Trumpet, 615
   Wild potato, 952
Violet rocket, 698
Virgin Mary, 343
Virginia copperleaf, 1017
Virginia cowslip, 641
Virginia creeper, 48
Virginia dwarf dandelion, 362
Virginia pepper grass, 708
Virginia rock cress, 715
Virginia snakeroot, 138, 139
Virginia sneezeweed, 492
Virginia whitlow wort, 822
Viscaceae, 1003
Viscid bushy goldenrod, 230
Vitis, 940, 943

W

Wahoo, 855
Walker, Randy, 160
Walking stick, Devil's, 132
Wallflower, 694
   Bushy, 696
   Smallflower, 695
   Western, 695
   Wormseed, 695
Wall-rocket, Stinking, 689
Wart cress, 704
Wasp Cynipid, 374, 548
Water celery, 92
Water dropwort, 92
   Common, 94
Water hemlock, 70
   Common, 70
Water hemp, 32

Water locust, 1068
Water marigold, 454
Water parsley, 103
Water parsnip, 67, 103
Water pennywort, 85, 86
Water scorpion grass, 644
Water shield, 733, 734
   Carolina, 734
Water shield family, 732
Water tupelo, 457, 965
Water willow, 5
   Lance-leaved, 5
Watercress, 711, 712
   Yellow, 720
Watermelon, 978, 980
Waterwort, 996
Waterwort family, 996
Wavy-leaved thistle, 340
Wayfaring tree, 788, 789
Webster Groves Nature Study
      Society, 562, 603, 732, 741
Wedgeleaf draba, 690
Wedge-leaved orache, 868
Weed
   Atlantic mock bishop's, 98
   Big mock bishop's, 99
   Bishop's, 62
   Buffalo, 448
   Butterfly, 154, 156
   Careless, 31
   Clammy, 911
   Coffee, 1073, 1074
   Green-stemmed Joe-pye, 407
   Hollow-stemmed Joe-pye, 403
   Horse, 448
   Joe-pye, 401
   Klamath, 924
   Mock bishop's, 98
   Ozark mock bishop's, 99
   Pineapple, 192
   Poverty, 896
   Skunk, 1026
   Spotted Joe-pye, 403
   Squaw, 584, 586, 588
   Sump, 520
   Tinker's, 782
   Turnip, 718
Weedy amaranth, 20
Weevil
   Flower head, 320
   Rosette, 320
Weigelia, 765
Wellstedia, 620
Wellstediaceae, 620
Welted thistle, 318
Western daisy, 205
Western dwarf dandelion, 362
Western false gromwell, 648
Western golden ragwort, 588
Western ironweed, 596
Western mugwort, 185
Western ragweed, 444
Western sagewort, 182

Western salsify, 391
Western scale-seed, 104
Western snowberry, 778
Western stickseed, 636
Western wallflower, 695
Western wild lettuce, 366
Wetter, Mark Allen, 237
Wheat, 603
Wherry's pink, 831
White aster
   Flat-topped, 218
   Small, 302
   Tall, 292
White campion, 837
White cockle, 837
White coralberry, 778
White crownbeard, 566
White daisy, 190
White heath aster, 305
White lettuce, 377, 378
   Glaucous, 380
   Great, 380
   Rough, 379
   Tall, 378
White milkweed, 148
White morning glory, Small, 951
White mustard, 674, 724
White prairie aster, 290
White sage, 185
White snakeroot, 378, 395
White top, 703
   Heart-podded, 706
White upland aster, 269
White whitlow wort, 691
White woodland aster, 295
White-flowered gourd, 985
Whitetop fleabane, 220, 223
Whitlow grass, 689, 690
   Vernal, 691
Whitlow wort, 689, 820
   Appalachian, 822
   Virginia, 822
   White, 691
Whittemore, Alan, 602, 607, 765
Whorled milkweed, 149, 157
Whorled pennywort, 86
Widow's cross, 971
Wild ageratum, 400
Wild Bird Feeding Industry, 488
Wild carrot, 78
   Small, 79
Wild chamomile, 190
Wild chervil, 70, 75
Wild coffee, 782
Wild comfrey, 632
Wild cucumber, 984, 985
Wild ginger, 140
Wild goldenglow, 543
Wild gourd, 983, 984
   Texas, 984
Wild honeysuckle, 770, 1004
Wild lettuce, 364
   Western, 366

INDEX

Wild morning glory, 934
Wild mustard, 725
Wild parsley, 90
Wild parsnip, 95
Wild pear gourd, 983
Wild petunia, 6, 7
Wild pink, 831
Wild potato vine, 952
Wild quinine, 526
Wild radish, 716
Wild raisin, 789
Wild sarsaparilla, 132
Wild senna
    Northern, 1073
    Southern, 1072
Wild teasel, 989
Wild turnip, 718
Wild wormwood, 182
Wildlife Code, Missouri, 136
Willam
    Catchfly sweet, 800
    Sweet, 810
Willow amsonia, 117, 118
Willow-leaved aster, 306
Willow-leaved lettuce, 368
Willow-leaved sunflower, 512
Winged euonymus, 854
Winged pigweed, 895
Winged sumac, 45
Wingstem, 560
Winter cress, 666
    Common, 667
    Early, 666
Winter heliotrope, 576
Winterberry, 129
Wintercreeper, 126, 856
Wintergreen, 999
Wislizenus, Frederick, 1024
Woad, 700
Wolfberry, 778
Wolf's milk, 1040
Women's League of Missouri, 129
Wood angelica, 64
Wood aster, Blue, 285
Wood spurge, 1032
Woodland aster, White, 295
Woodland goosefoot, 889
Woodland lettuce, 365
Woodland sunflower, 501
Woodson, Robert, 1073
Woolly croton, 1021
    Round-leaved, 1023
Woolly pipevine, 139
Woolly sweet cicely, 93
Works Progress Administration, 994
World War II, 525

Wormseed, 875
Wormseed mustard, 695
Wormseed wallflower, 695
Wormwood, 177, 178, 180
    Annual, 180
    Beach, 187
    Biennial, 180
    Common, 179
    Roman, 444
    Silky, 184
    Sweet, 180
    Wild, 182
Wort
    Appalachian whitlow, 822
    Clasping-leaved St. John's, 917
    Common St. John's, 924
    Creeping St. John's, 915
    Dwarf St. John's, 922
    Giant St. John's, 915
    Great St. John's, 915
    Greater St. John's, 921
    Marsh St. John's, 928
    Round-fruited St. John's, 928
    Shrubby St. John's, 925
    Small St. John's, 917
    Spotted St. John's, 925, 926
    Virginia whitlow
    Whitlow, 689, 820
    White Whitlow, 691
Wreath aster, 288
Wreath goldenrod, 254
Wright's saltbush, 868
Xanthium, 165, 436, 472, **568,** 569, 939, 944
    ambrosioides, 568
    chinense, 569
    inflexum, 569
    italicum, 569
    orientale, 570
    pensylvanicum, 569
    speciosum, 569
    spinosum (Map 1241, Pl. 292 e, f), 568, **569**
    strumarium (Map 1242, Pl. 292 g, h), 568, **569,** 570
        var. canadense, 569, 570
        var. glabratum, 569, 570
        var. strumarium, 570
    varians, 569
    wootonii, 569

X

Xanthocephalum dracunculoides, 202

Ximenesia encelioides, 562
Xylocopa, 618

Y

Yarrow, 173, 174
    Common, 173
Yatskievych, George, 40, 348, 357, 371, 372, 377, 619, 651, 758, 848, 992, 1010, 1041, 1057
Yaupon, 130
Yellow chamomile, 188, 189
Yellow coneflower, 476
Yellow cress, 721
    Blunt-leaved, 720
    Bog, 720
    Creeping, 722
    Marsh, 720
    Spreading, 722
Yellow crownbeard, 564
Yellow dog-fennel, 490
Yellow honeysuckle, 770
Yellow ironweed, 562
Yellow king-devil, 354
Yellow mustard, 724
Yellow nailwort, 822
Yellow oleander, 115
Yellow pimpernel, 106
Yellow puccoon, 640
Yellow rocket, 667
Yellow salsify, 391
Yellow star thistle, 330, 331
Yellow stonecrop, 973
Yellow watercress, 720
Yellow-flowered gourd, 983
Yellow-flowered horse gentian, 781
Yellow-flowered leaf cup, 555
Yellowhead, British, 573
Yerba de tajo, 479
Yucca, 82

Z

Zebra swallowtail butterfly, 52
Zigadenus elegans, 746
Zinnia, 436
Zizia, 107, 108, **112,** 114
    aptera (Map 893, Pl. 214 g, h), 107, **114**
    aurea (Map 894, Pl. 214 f, i), 107, **114**
    f. obtusifolia, 114
Zucchini, 982, 983, 986

INDEX TO FAMILIES TREATED IN VOLUME 2

Acanthaceae 2
Aceraceae 7
Aizoaceae 17
Amaranthaceae 18
Anacardiaceae 40
Annonaceae 51
Apiaceae 54
Apocynaceae 115
Aquifoliaceae 126
Araliaceae 131
Aristolochiaceae 138
Asclepiadaceae 140
Asteraceae 164
    Tribe Anthemideae 171
    Tribe Astereae 198
    Tribe Cardueae 315
    Tribe Cichorieae 343
    Tribe Eupatorieae 393
    Tribe Gnaphalieae 425
    Tribe Heliantheae 435
    Tribe Inuleae 572
    Tribe Plucheeae 573
    Tribe Senecioneae 576
    Tribe Vernonieae 590
Balsaminaceae 599
Berberidaceae 602
Betulaceae 607
Bignoniaceae 614
Boraginaceae 619
Brassicaceae 651
Buxaceae 731

Cabombaceae 732
Cactaceae 736
Calycanthaceae 740
Campanulaceae 741
Cannabaceae 758
Caprifoliaceae 765
Caryophyllaceae 794
Celastraceae 850
Ceratophyllaceae 858
Chenopodiaceae 860
Cistaceae 902
Cleomaceae 908
Clusiaceae 912
Convolvulaceae 930
Cornaceae 957
Crassulaceae 967
Cucurbitaceae 974
Dipsacaceae 988
Ebenaceae 990
Ehretiaceae 992
Elaeagnaceae 994
Elatinaceae 996
Ericaceae 999
Euphorbiaceae 1010
Fabaceae (excluding Faboideae)
  1057
    Subfamily Caesalpinioideae
    1059
    Subfamily Mimosoideae
    1075